In the Matter of J. Robert Oppenheimer:
Transcript of Hearing before Personnel Security Board and Texts of Principal Documents and Letters

United States Atomic Energy Commission

Foreword by Philip M. Stern

The MIT Press
Cambridge, Massachusetts, and London, England

Text reproduced from the original editions
In the Matter of J. Robert Oppenheimer: Transcript of Hearing before Personnel Security Board, Washington, D.C., April 12, 1954, through May 6, 1954,
and
In the Matter of J. Robert Oppenheimer: Texts of Principal Documents and Letters of Personnel Security Board, General Manager, Commissioners, Washington, D.C., May 27, 1954, through June 29, 1954, published in 1954 by the Government Printing Office, Washington, D.C.
Foreword and new material copyright © 1970 by
The Massachusetts Institute of Technology

First M.I.T. Press Paperback Edition, April 1971

Printed and bound in the United States of America by
Halliday Lithograph Corp.

ISBN 0 262 21003 7 (hardcover)
ISBN 0 262 71002 1 (paperback)

Library of Congress catalog card number: 79-138845

CONTENTS

Foreword by Philip M. Stern	v
Transcript of Hearing before Personnel Security Board	
April 12	1
April 13	51
April 14	99
April 15	161
April 16	221
April 19	283
April 20	349
April 21	427
April 22	475
April 23	535
April 26	569
April 27	631
April 28	675
April 29	729
April 30	791
May 3	841
May 4	907
May 5	947
May 6	965
Texts of Principal Documents and Letters	
Findings and Recommendations of the Personnel Security Board in the Matter of Dr. J. Robert Oppenheimer	997
Letter from General Manager K. D. Nichols, United States Atomic Energy Commission, to Dr. J. Robert Oppenheimer—Forwarding Findings and Recommendation of the Personnel Security Board	1023
Letter from Dr. Oppenheimer's Attorneys to General Manager K. D. Nichols, United States Atomic Energy Commission, Replying to the Latter's Letter to Dr. Oppenheimer Transmitting the Findings and Recommendation of the Personnel Security Board	1027
Letter from General Manager K. D. Nichols, United States Atomic Energy Commission, to Dr. Oppenheimer's Attorneys, Concerning Procedures in the Matter of Dr. J. Robert Oppenheimer	1035
Recommendations of the General Manager to the United States Atomic Energy Commission in the Matter of Dr. J. Robert Oppenheimer	1039
Decision and Opinions of the United States Atomic Energy Commission in the Matter of Dr. J. Robert Oppenheimer	1047

> NOTE.—Letter dated December 23, 1953, from General Manager K. D. Nichols, United States Atomic Energy Commission, to Dr. J. Robert Oppenheimer appears on pages 3 through 7.
> Letter dated March 4, 1954, from Dr. J. Robert Oppenheimer to General Manager K. D. Nichols, United States Atomic Energy Commission, appears on pages 7 through 20.

Name Index	1067
Subject Index	1075

FOREWORD

Having spent six eye-straining weeks in a line-by-line reading of this massive volume, I am not sure whether to pity or envy those who now have the opportunity, thanks to the initiative of The M.I.T. Press, and the fortitude to wade into it. But, on reflection, the envy far outweighs the pity, for lodged in those 750,000 words are history, drama, conflict, personalities, villains and heroes, injustice, brutality, courage, and eloquence — all of it grist for either historian or novelist.

For many who lived in the era of J. Robert Oppenheimer — and the shorter-lived notoriety of Senator Joseph McCarthy of Wisconsin — Oppenheimer and the ordeal he went through stir vivid recollections. More recently, however, Oppenheimer has become a rapidly receding figure in the public mind and his security trial is virtually unknown, so it may be helpful to set the factual stage for the remarkable proceeding you are about to read.

In 1943, J. Robert Oppenheimer, a shiningly gifted physics professor at Berkeley and then just 38, was chosen to head the Los Alamos laboratory which would lead a scientific race against the Germans to develop the world's first atomic bomb. Oppenheimer's prewar associations with the Left, including some Communists, aroused the suspicions of Army security officers, but he enjoyed the confidence of the Manhattan Project's director, General Leslie Groves, and survived several encounters with security investigators. After the war, revered as "the father of the atomic bomb," he was a ubiquitous advisor to the government on atomic and security matters, serving as a member of several advisory committees, chairing some of them — most notably the General Advisory Committee of the Atomic Energy Commission. In these posts, his outlook and his often abrasive personality collided with those in the Pentagon and elsewhere who were consumed by a fear of Soviet nuclear superiority and who passionately pressed for maximum-speed development of American military strength, especially nuclear. This conflict came most sharply into focus in 1949, just after the Soviets exploded their first atomic device (far sooner than most had expected), and the United States was forced to decide how vigorously to pursue the development of the "super" — the vastly powerful hydrogen bomb. Oppenheimer and the General Advisory Committee (GAC) wanted, at the least, to delay any major U.S. H-bomb program until there could be one last effort to reach an agreement with the Soviets on a mutual forswearing of the "super." President Truman decided otherwise.

Oppenheimer was not reappointed to the GAC in 1952 and, indeed, when Dwight Eisenhower and the Republicans took office in January 1953, Oppenheimer had all but disappeared from the Washington scene. His only official connection with the government was a consultant contract with the

AEC that was rarely invoked and that was scheduled to expire in June 1954, only a few months from the time his security case came before the AEC. Given that Oppenheimer's official influence had already all but vanished and that the AEC had an easy and graceful "out" (to let his consultant contract expire), the zeal with which the government pursued his official banishment through this security proceeding is one of the remarkable, mysterious, and disturbing elements of this whole affair.

The security case itself appears to have been triggered by a letter written to FBI Director J. Edgar Hoover on November 7, 1953, by William Liscum Borden, formerly the Executive Director of Congress's Joint Committee on Atomic Energy. Borden expressed the view that "more probably than not," J. Robert Oppenheimer not only had been, but continued to be, an "agent of the Soviet Union." Thirty days later, the FBI sent Borden's letter to the White House, where President Eisenhower ordered a "blank wall" erected between Oppenheimer and atomic secrets (many of which originated in work he had participated in) and directed the Atomic Energy Commission to study and resolve the matter. The AEC, in turn, named a special three-man board to hear the case, headed by Gordon Gray, president of the University of North Carolina and formerly Secretary of the Army. The two other panel members were retired industrialist Thomas A. Morgan, who had been president and chairman of the Sperry Corporation (CK), and Ward Evans, a professor of chemistry at Northwestern University.

That is the bare-bones factual setting in which this proceeding began on April 13, 1954.

Initially, the transcript of the hearing was never intended for publication. Indeed, from the opening moments of the case, Chairman Gray both admonished and assured those present that the proceedings were to be kept strictly confidential. But, as you will see, the existence of the trial became public knowledge on its second day, and the publication of both the AEC's charges and, later, of the Gray Board's adverse verdict caused wide-spread public controversy. The Gray Board opinion, in particular, came under particular fire for its apparent inconsistency in finding Oppenheimer both loyal and unusually discrete with secrets, but nonetheless a "security risk." Shortly after that opinion became public, the AEC decided to publish the transcript, notwithstanding Gray's assurances that this would not be done. This was rationalized in part by a purported fear that some extensive excerpts from the transcript, lost by one of the AEC Commissioners, might suddenly become public. But, since the missing document was recovered intact in a matter of hours, many believe that the AEC's real motive was to soften the public criticism of its prosecution of Oppenheimer and, particularly, of the Gray Board opinion. That view is reinforced by the fact that copies of the transcript sent out by the AEC Chairman's office were accompanied by a mimeographed sheet calling attention to certain passages that portrayed Oppenheimer in an unfavorable light (see pages 386–392, *The Oppenheimer Case**).

Although the Government Printing Office (which performed a printing

---

* Philip M. Stern, *The Oppenheimer Case: Security on Trial* (New York : Harper & Row, 1969), in collaboration with Harold P. Green.

miracle by typesetting, printing, and binding this massive book in about 48 hours) printed several thousand copies of the transcript and sold all of them, as the years went by the document became almost mysteriously fugitive. Although it was available in the public library, copies in the hands of private persons interested in the case almost always seemed to have disappeared, been thrown out or, often, lent to some forgotten (and forgetful) person and never returned. I am told that the Editors of The M.I.T. Press even had difficulty locating a copy from which to print this volume.

But they persevered, and the public has reason to be grateful to them for resurrecting it and for providing the useful index (I wish I had had it when spending countless hours trying to trace through this event or that). For what unfolded in the four weeks of this remarkable hearing has elements of interest for a variety of readers.

It is of interest to the historian, for in these pages the reader will find remarkable insights into, if not miniature portraits of, three distinct and important periods of twentieth century American history: first, the depression-ridden thirties, with the attendant disillusionment about the American system and the attraction toward the Left; second, the immediate postwar period, when our "gallant wartime ally," the Soviets, became our cold war adversaries and the political mood shifted sharply rightward; and third, the era of "McCarthyism," at the height of which the Oppenheimer proceeding took place. It is difficult, in retrospect, even for those who lived through it, fully to recapture the fear this powerful demagogue, Joseph McCarthy, aroused, and not merely among the unsophisticated. Men of strength, good sense, and self-respect suddenly found themselves accepting McCarthy's frame of reference and his "guilt by association" line of argument. As an example of the lengths to which this went, Democratic senators, searching for a lawyer to serve as counsel for their investigation of the so-called "Army-McCarthy" controversy, turned down a distinguished Pittsburgh attorney on the ground that his *firm* (although not he personally) represented Alcoa, which in turn had sponsored a television program critical of Senator McCarthy! And so if in these pages the reader feels that witnesses (especially Oppenheimer) succumbed in unseemly fashion to the browbeating of the prosecutor, Roger Robb,* he should bear in mind the fear that polluted the political atmosphere of 1954.

This volume also holds special interest for lawyers, for, as the only complete public transcript of a security hearing, it lays bare the procedural injustices inherent in the so-called loyalty-security system. That system can punish a man far more severely than most criminal laws permit; yet, it denies the "accused" many if not most of the rights he would enjoy in an ordinary court of law. Some of these unjust procedures have been mitigated since the Oppenheimer case, but the system retains, in principle, many features that do violence to the protections of personal liberty that the Founding Fathers sought to write into the Constitution.

In particular, this hearing revealed in unique detail the shocking extent to which the United States government in the name of "security" can and does pry into the most intimate details of a citizen's life. For eleven years

---

* Now a member of the U.S. Court of Appeals in Washington, D.C.

Oppenheimer's mail was opened; his telephone calls were monitored; his office and his home were "bugged"; his every movement was followed, even extending to a night spent with a former fiancée in Berkeley during the war. All of this official snooping is justified as *protecting* our "freedom" from totalitarian control.

Some lawyers may find the posture of Oppenheimer's attorneys unduly supine, especially in the face of Chairman Gray's passive acceptance of prosecutor Robb's assertive, aggressive, sometimes brutal tactics. In retrospect, knowing the outcome, one can speculate that the scientist's chief counsel, Lloyd Garrison, would better have served his client by protesting more vigorously — or even by withdrawing from the proceeding entirely, as he and Oppenheimer were urged to do after the first week of the hearing (see pp. 299–300, *The Oppenheimer Case*). One should bear in mind, though, the one-sided nature of the proceeding as well as Garrison's pre-hearing decision that Oppenheimer would not be able successfully to appeal an adverse Gray Board finding to the courts. Rather, Garrison hoped that a posture of reasonableness, rather than pugnaciousness, might be his client's only hope of winning a majority of the Gray Board.

Students of military policy will find this volume a rich vein to mine, for as Thomas Wilson has set forth in his excellent book, *The Great Weapons Heresy*, the postwar armaments debate — the "bigger bang" protagonists versus the disarmers — is dramatically enacted in these pages. Indeed, the suspicions and hatreds of Oppenheimer by the "cold warriors" were what triggered the case in the first instance and what provided the zeal and energy behind the government's often savage prosecution of Oppenheimer.

You will find, especially in the testimony of Vannevar Bush, an intense controversy over the inclusion in the AEC's "security-risk" charges of Oppenheimer's opposition to an all-out H-bomb program. The H-bomb charge was added, almost as an afterthought, at the initiative of Harold P. Green, the AEC lawyer who drafted the charges (see pp. 222–228, *The Oppenheimer Case*). Although Green says he intended only to test the physicist's veracity, the inclusion of the H-bomb charge, as Vannevar Bush heatedly pointed out, made the determination of Oppenheimer's "security-risk" status revolve in large measure around his *policy* views, creating a dangerous precedent that would inevitably inhibit the government from receiving honest policy advice. One wonders whether, if Green had not inserted the H-bomb matter into the charges, the Gray Board would ever have delved into the military policy questions that occupy such a large portion of the testimony and that seem of the most tenuous relevance to Oppenheimer's loyalty or his discretion with secrets — the only two meaningful criteria if the government is to persist in having a loyalty-security screening program.

Historians, who constantly struggle to portray past events in their proper setting and to avoid the distortions of looking backward at a different and unfamiliar era, may be intrigued by the manner in which this problem presented itself to (and was often poorly resolved by) the Gray Board. For here were three men, pulled abruptly from the milieu and perspective of a prosperous, Communist-hating, fear-ridden America of the mid-fifties and asked to sit in judgment on Robert Oppenheimer's radical associations and activities of the thirties, when the American economy was in total collapse, when Fascism was expanding unimpeded in Europe, and when American Communists, far from being treated as political lepers, as they were in 1954, were openly allied in common cause with the non-Communist American Left. It was almost as if the Gray Board members were looking back at the thirties through the wrong end of a telescope (one of the inherent inequities of the whole loyalty-security system). That would have been difficult enough even if they had shared Oppenheimer's radical views and experiences of the thirties; as it was, a man such as Gordon Gray, born to great wealth and almost wholly untouched by the Depression, seemed to have great difficulty seeing events against the backdrop of the thirties (see, for example, his questioning of Kitty Oppenheimer about her relations with the Communist party on page 917).

Novelists (and, perhaps, psychologists too) may be drawn to the human elements of this trial — especially the portrait of that most remarkable and enigmatic man, J. Robert Oppenheimer. His intellectual brilliance is almost legendary; so, to a lesser extent, is his reputation for arrogance. Why, then, did he permit himself to be bested, cornered, browbeaten by Roger Robb, and finally driven, bent over, wringing his hands, to blurt out that he had been "an idiot" (page 137)? Oppenheimer's self-deprecatory posture is the more remarkable when compared with his appearance, just five years earlier, before the House Un-American Activities Committee, when his self-assured answers to many of the same questions wholly persuaded the hard-bitten members of that Committee, including the then-Congressman Richard Nixon (see pp. 118–122, *The Oppenheimer Case*).

What prompted him to inform on close friends and former protégés, such as Haakon Chevalier and Bernard Peters — thus, ironically, making himself the instrument of the very loyalty-security system that was to bring his own downfall?

And what were his true feelings — or, more precisely, the whole of his feelings — about the "super" bomb? On the one hand, he is said to have suffered great remorse, along with other nuclear scientists, at having brought the atomic bomb into the world; and he recounts in this hearing his profound doubts about creating a vastly more destructive weapon. Yet, not just once, but twice in the course of the hearing, he said that this fearsome device (or, rather, Edward Teller's new notion for making it) was too "technically sweet" not to proceed with. His fellow scientists explain this as the voice of the purely intellectual, purely scientific part of the man. Well, if scientists as sensitive as Oppenheimer can indeed wall off their moral sensibilities so completely and successfully, then technology is an even more fearsome monster than most of us realize.

With all that *is* contained in these pages, there is much that is *not*. There is no clear picture, for example, of the two poles of Oppenheimer's remarkable personality — both of which, I believe, contributed to his downfall — his mesmeric charm on the one hand and his withering arrogance on the other. One senses the latter in the manner in which he offended Air Force scientist David Griggs; but there is little or no suggestion of the way he similarly infuriated Edward Teller and Lewis Strauss and AEC General Manager Kenneth Nichols — although the consequences of Oppenheimer's actions are manifest in Teller's unhappy and damaging testimony against his erstwhile friend, and in Strauss's* and Nichols's savage verdicts against Oppenheimer (wisely included in this new volume, although they were, initially, published separately by the Atomic Energy Commission).

Nor does the transcript fully portray the *political* context of the Gray Board proceeding — the extent to which in winning the White House in 1952 the Republicans generally, and General Eisenhower specifically, had exploited McCarthy's "throw-the-Communists-out-of-government" theme (see p. 197, *The Oppenheimer Case*). That in itself made it imperative that the new Republican Administration be "tough" in handling Borden's accusation that an "agent of the Soviet Union" was among the official consultants of the United States government. By coincidence, however, the political imperative was even more specific: the day before Borden sent his letter to the FBI, Herbert Brownell, the new Republican Attorney General of the United States, had accused Democratic President Truman of "knowingly promoting" a Soviet spy, despite warnings from the FBI (see pp. 218-219, *The Oppenheimer Case*). How would Brownell and his Department of Justice have looked had they failed to take vigorous action on so grave a warning as Borden's?

Borden's own background and personality, not even hinted at in his brief appearance before the Gray Board, form one of the most puzzling, intriguing, and novelesque aspects of this case. Model prep school student, academic leader in a distinguished Yale Law School class, self-styled liberal Democrat, he seems the least likely kind of person to level so grave and loosely documented an accusation. Yet he did, and in so doing, loosed the power of the United States government against J. Robert Oppenheimer.

Other integral elements of this fascinating story are not recounted in these pages: the sad aftermath for Oppenheimer, whose life and influence slowly wound down until his illness in 1966 and his death in February 1967 (pp. 450-458, *The Oppenheimer Case*); the even more tragic effects on the lesser-known scientists who play far lesser roles in this hearing (Frank Oppenheimer, Rossi Lomanitz, Bernard Peters, David Bohm) but who later were exiled, either from their professions or from the United States itself (pp. 430-440, *The Oppenheimer Case*); and the sort of "fairy-story justice" that Oppenheimer protagonists may find in the way in which those most responsible for Oppenheimer's downfall (Borden, Strauss, Teller) themselves later came to suffer grieviously because of their role in the Oppenheimer affair (pp. 441-449, *The Oppenheimer Case*).

What I find implicit in these thousand pages — and, especially, in the perversely illogical opinions rendered by the Gray Board and by the AEC

---

* Strauss was the author of the "majority opinion" (pages 1049 to 1052 of the "Principal Documents") although it was also signed by Commissioners Campbell and Zuckert.

majority — is not only the injustice but the fundamental senselessness of the entire loyalty-security screening system.

Some of the facts supporting that proposition are not contained in this specific hearing: the fact, for example, that of the nearly five million people screened between 1947 and 1952, 99.5% had *nothing* suspect in their backgrounds (which suggests that the half to three-fourths of a *billion* dollars the program has cost since the war was hardly an example of prudent spending); the fact that at least eleven persons later indicted for espionage had been *cleared* by the screening system (these facts and other criticisms of the loyalty-security system are more fully developed on pp. 470–493 of *The Oppenheimer Case*).

But the basic inanity of the system *is* wholly proven within these thousand pages — and it is this:

> The purpose of the screening program is to try to predict, to the extent possible, whether a person will *in the future* give away a secret or will otherwise betray his country. If one could devise the perfect test for this, it would probably consist of giving a person a secret, tempting him to give it away, and seeing what he does. By happenstance, that is precisely the test to which Oppenheimer was subjected: he was entrusted, not just with *a* secret, but with his country's most jealously guarded and important secret (the A-bomb project); his close friend, Haakon Chevalier, spoke to him about breaching that secret; his instantaneous and unequivocal reply was in the negative. Moreover, although even the most intimate details of his life were closely watched by the government over an 11-year period — not even his severest detractor, General Nichols, could find any evidence that he had *ever* given away a secret.
>
> And yet he was adjudged a "security risk." By that fact alone, the system must be said to topple itself, by the sheer weight of its monstrous, monumental illogic.

But that does not end the indictment of the system. Oppenheimer was accused, tried and convicted in the name of "security." But was the security of the United States really enhanced by depriving the United States government of his prodigious talents? What if the security officers had been listened to during the war when they recommended his dismissal from the government? By general acknowledgment, Oppenheimer, with his charismatic gift of leadership, was the man most singly responsible for our developing the atomic bomb which, rightly or wrongly, hastened the end of the war in Asia. Would it have increased national "security" to have fired him then?

Finally, these pages contain convincing evidence that Oppenheimer was brought down largely because of his *policy* views — certainly not because of any looseness with secrets. Therein lies the fundamental and greatest danger to the United States from the loyalty-security system, for until and unless the Federal government is forbidden (as it was for fifty years prior to World War II) to ask citizens about their political and religious beliefs and affiliations, the system can at any time be used to punish dissident views that do not happen to be in vogue at the time. Whether or not it is *actually* used for that purpose, the very existence of the system cannot encourage but can only inhibit the frank expression of criticisms of government policy. And in the absence of criticism lies the surest seed of erroneous policy.

Robert Oppenheimer died on February 18, 1967. On that day — one American citizen who during World War II worked his heart out for America's mortal enemy, the Nazis, enjoyed the full trust and confidence of the United States Government.* Yet, Robert Oppenheimer who, as these pages recount, worked his heart out for *his* government during World War II, died among those few Americans untrusted by his country.

The grotesqueness of that injustice must stand as an unerasable blot in American history. I can only hope for the sake of Oppenheimer and his memory that not only the injustice but the senselessness of the entire loyalty-security system come through loud and clear in this volume. For if Oppenheimer's ordeal serves to encourage the abolition of the system, then the trial recounted in these pages will not have been an entirely futile event.

October 1970   Philip M. Stern

---

* This is in no way to suggest that that man, Wernher von Braun, should not be trusted.

UNITED STATES ATOMIC ENERGY COMMISSION

In the Matter of

J. ROBERT OPPENHEIMER

TRANSCRIPT OF HEARING
BEFORE
PERSONNEL SECURITY BOARD

Washington, D. C.
April 12, 1954, through May 6, 1954

UNITED STATES ATOMIC ENERGY COMMISSION

PERSONNEL SECURITY BOARD

In the Matter of J. Robert Oppenheimer

Atomic Energy Commission,
Building T-3, Room 2022,
Washington, D. C., April 12, 1954.

The above-entitled matter came on for hearing before the board, pursuant to notice, at 10 a. m.

Personnel Security Board: Dr. Gordon Gray, chairman; Dr. Ward V. Evans, member; and Mr. Thomas A. Morgan, member.

Present: Roger Robb and C. A. Rolander, Jr., counsel for the board; J. Robert Oppenheimer, Lloyd K. Garrison, Samuel J. Silverman, and Allen B. Ecker, counsel for J. Robert Oppenheimer.

PROCEEDINGS

Dr. GRAY. I should like at this time to have the reporters sworn. For the information of Dr. Oppenheimer and his counsel, the reporter is Anton Papich, Jr., the transcriber Kenneth V. Bowers.

(The reporter and transcriber were thereupon duly sworn by Dr. Gray.)

Dr. GRAY. The hearing will come to order.

This board, appointed by Mr. K. D. Nichols, General Manager of the Atomic Energy Commission, at the request of Dr. J. Robert Oppenheimer, is composed of the following members: Gordon Gray, chairman, Ward V. Evans and Thomas A. Morgan. All members of the board are present, and board counsels Roger Robb and C. A. Rolander. Dr. and Mrs. Oppenheimer are present. Present also are Mr. Lloyd K. Garrison, counsel for Dr. Oppenheimer. Would you identify your associates?

Mr. GARRISON. Samuel J. Silverman, my partner, and Allen B. Ecker, associate of my firm.

Dr. GRAY. An investigation of Dr. J. Robert Oppenheimer conducted under the provisions of section 10 (b) (5) (B) (i–iii) of the Atomic Energy Act of 1946 has revealed certain information which casts doubt upon the eligibility of Dr. Oppenheimer for clearance for access to restricted data as provided by the Atomic Energy Act of 1946. This information is as follows:

This is a letter addressed to Dr. J. R. Oppenheimer, the Institute for Advanced Study, Princeton, N. J., dated December 23, 1953, reading as follows:

"DEAR DR. OPPENHEIMER: Section 10 of the Atomic Energy Act of 1946 places upon the Atomic Energy Commission the responsibility for assuring that individuals are employed by the Commission only when such employment will not endanger the common defense and security. In addition, Executive Order 10450 of April 27, 1953, requires the suspension of employment of any individual where there exists information indicating that his employment may not be clearly consistent with the interests of the national security.

"As a result of additional investigation as to your character, associations, and loyalty, and review of your personnel security file in the light of the requirements of the Atomic Energy Act and the requirements of Executive Order 10450, there has developed considerable question whether your continued employment on Atomic Energy Commission work will endanger the common defense and security and whether such continued employment is clearly consistent with the interests of the national security. This letter is to advise you of the steps which you may take to assist in the resolution of this question.

"The substance of the information which raises the question concerning your eligibility for employment on Atomic Energy Commission work is as follows:"

Let the record show at this point that Mr. Garrison asked to be excused for a few minutes.

"It was reported that in 1940 you were listed as a sponsor of the Friends of the Chinese People, an organization which was characterized in 1944 by the House Committee on Un-American Activities as a Communist-front organization. It was further reported that in 1940 your name was included on a letterhead of the American Committee for Democratic and Intellectual Freedom as a member of its national executive committee. The American Committee for Democracy and Intellectual Freedom was characterized in 1942 by the House Committee on Un-American Activities as a Communist front which defended Communist teachers, and in 1943 it was characterized as subversive and un-American by a special subcommittee of the House Committee on Appropriations. It was further reported that in 1938 you were a member of the Western Council of the Consumers Union. The Consumers Union was cited in 1944 by the House Committee on Un-American Activities as a Communist-front headed by the Communist Arthur Kallet. It was further reported that you stated in 1943 that you were not a Communist, but had probably belonged to every Communist front organization on the west coast and had signed many petitions in which Communists were interested.

"It was reported that in 1943 and previously you were intimately associated with Dr. Jean Tatlock, a member of the Communist Party in San Francisco, and that Dr. Tatlock was partially responsible for your association with Communist-front groups.

"It was reported that your wife, Katherine Puening Oppenheimer, was formerly the wife of Joseph Dallet, a member of the Communist Party, who was killed in Spain in 1937 fighting for the Spanish Republican Army. It was further reported that during the period of her association with Joseph Dallet, your wife became a member of the Communist Party. The Communist Party has been designated by the Attorney General as a subversive organization which seeks to alter the form of Government of the United States by unconstitutional means, within the purview of Executive Order 9835 and Executive Order 10450.

"It was reported that your brother, Frank Friedman Oppenheimer, became a member of the Communist Party in 1936 and has served as a party organizer and as educational director of the professional section of the Communist Party in Los Angeles County. It was further reported that your brother's wife, Jackie Oppenheimer, was a member of the Communist Party in 1938; and that in August 1944, Jackie Oppenheimer assisted in the organization of the East Bay branch of the California Labor School. It was further reported that in 1945 Frank and Jackie Oppenheimer were invited to an informal reception at the Russian consulate, that this invitation was extended by the American-Russian Institute of San Francisco and was for the purpose of introducing famous American scientists to Russian scientists who were delegates to the United Nations Conference on International Organization being held at San Francisco at that time, and that Frank Oppenheimer accepted this invitation. It was further reported that Frank Oppenheimer agreed to give a 6 weeks course on The Social Implications of Modern Scientific Development at the California Labor School, beginning May 9, 1946. The American-Russian Institute of San Francisco and the California Labor School have been cited by the Attorney General as Communist organizations within the purview of Executive Order 9835 and Executive Order 10450.

"It was reported that you have associated with members and officials of the Communist Party including Isaac Folkoff, Steve Nelson, Rudy Lambert, Kenneth May, Jack Manley, and Thomas Addis.

"It was reported that you were a subscriber to the Daily People's World, a west coast Communist newspaper, in 1941 and 1942.

"It was reported in 1950 that you stated to an agent of the Federal Bureau of Investigation that you had in the past made contributions to Communist-front organizations, although at the time you did not know of Communist Party control or extent of infiltration of these groups. You further stated to an agent of the Federal Bureau of Investigation that some of these contributions were made through Isaac Folkoff, whom you knew to be a leading Communist Party functionary, because you had been told that this was the most effective and direct way of helping these groups.

"It was reported that you attended a housewarming party at the home of Kenneth and Ruth May on September 20, 1941, for which there was an admission charge for the benefit of The People's World, and that at this party you were in the company of Joseph W. Weinberg and Clarence Hiskey, who were alleged to be members of the Communist Party and to have engaged in espionage on behalf of the Soviet Union. It was further reported that you informed officials of the United States Department of Justice in 1952 that you had no recollection that you had attended such a party, but that since it would have been in character for you to have attended such a party, you would not deny that you were there.

"It was reported that you attended a closed meeting of the professional section of the Communist Party of Alameda County, Calif., which was held in the latter part of July or early August 1941, at your residence, 10 Kenilworth Court, Berkeley, Calif., for the purpose of hearing an explanation of a change in Communist Party policy. It was reported that you denied that you attended such a meeting and that such a meeting was held in your home.

"It was reported that you stated to an agent of the Federal Bureau of Investigation in 1950, that you attended a meeting in 1940 or 1941, which may have taken place at the home of Haakon Chevalier, which was addressed by William Schneiderman, whom you knew to be a leading functionary of the Communist Party. In testimony in 1950 before the California State Senate Committee on Un-American Activities, Haakon Chevalier was identified as a member of the Communist Party in the San Francisco area in the early 1940's."

Let the record show that Mr. Garrison has returned to the hearing room.

"It was reported that you have consistently denied that you have ever been a member of the Communist Party. It was further reported that you stated to a representative of the Federal Bureau of Investigation in 1946 that you had a change of mind regarding the policies and politics of the Soviet Union about the time of the signing of the Soviet-German Pact in 1939. It was further reported that during 1950 you stated to a representative of the Federal Bureau of Investigation that you had never attended a closed meeting of the Communist Party; and that at the time of the Russo-Finnish War and the subsequent break between Germany and Russia in 1941, you realized the Communist Party infiltration tactics into the alleged anti-Fascist groups and became fed up with the whole thing and lost what little interest you had. It was further reported, however, that:

"(a) Prior to April 1942, you had contributed $150 per month to the Communist Party in the San Francisco area, and that the last such payment was apparently made in April 1942, immediately before your entry into the atomic-bomb project.

"(b) During the period 1942–45 various officials of the Communist Party, including Dr. Hannah Peters, organizer of the professional section of the Communist Party, Alameda County, Calif., Bernadette Doyle, secretary of the Alameda County Communist Party Steve Nelson, David Adelson, Paul Pinsky, Jack Manley, and Katrina Sandov, are reported to have made statements indicating that you were then a member of the Communist Party; that you could not be active in the party at that time; that your name should be removed from the party mailing list and not mentioned in any way; that you had talked the atomic-bomb question over with party members during this period; and that several years prior to 1945 you had told Steve Nelson that the Army was working on an atomic bomb.

"(c) You stated in August of 1943 that you did not want anybody working for you on the project who was a member of the Communist Party, since 'one always had a question of divided loyalty" and the discipline of the Communist Party was very severe and not compatible with complete loyalty to the project. You further stated at that time that you were referring only to present membership in the Communist Party and not to people who had been members of the party. You stated further that you knew several individuals then at Los Alamos who had been members of the Communist Party. You did not, however, identify such former members of the Communist Party to the appropriate authorities. It was also reported that during the period 1942–45 you were responsible for the employment on the atom-bomb project of individuals who were members of the Communist Party or closely associated with activities of the Communist Party, including Giovanni Rossi Lomanitz, Joseph W. Weinberg, David Bohm, Max Bernard Friedman, and David Hawkins. In the case of Giovanni Rossi Lomanitz, you urged him to work on the project, although you stated that you knew he had been very much of a Red when he first came to the University of California and that you emphasized to him that he must forego all political activity if he came to the project. In August 1943, you protested against the termination of his deferment and requested that he be returned to the project after his entry into the military service.

"It was reported that you stated to representatives of the Federal Bureau of Investigation on September 5, 1946, that you had attended a meeting in the East Bay and a meeting in San Francisco at which there were present persons definitely identified with the Communist Party. When asked the purpose of the East Bay meeting and the identity of those in attendance, you declined to answer on the ground that this had no bearing on the matter of interest being discussed.

"It was reported that you attended a meeting at the home of Frank Oppenheimer on January 1, 1946, with David Adelson and Paul Pinsky, both of whom were members of the Communist Party. It was further reported that you analyzed some material which Pinsky hoped to take up with the legislative convention in Sacramento, Calif.

"It was reported in 1946 that you were listed as vice chairman on the letterhead of the Independent Citizens Committee of the Arts, Sciences, and Professions, Inc., which has been cited as a Communist front by the House Committee on Un-American Activities.

"It was reported that prior to March 1, 1943, possibly 3 months prior, Peter Ivanov, secretary of the Soviet consulate, San Francisco, approached George Charles Eltenton for the purpose of obtaining information regarding work being done at the Radiation Laboratory for the use of Soviet scientists; that George Charles Eltenton subsequently requested Haakon Chevalier to approach you concerning this matter; that Haakon Chevalier thereupon approached you, either directly or through your brother, Frank Friedman Oppenheimer, in connection with this matter; and that Haakon Chevalier finally advised George Charles

Eltenton that there was no chance whatsoever of obtaining the information. It was further reported that you did not report this episode to the appropriate authorities until several months after its occurrence; that when you initially discussed this matter with the appropriate authorities on August 26, 1943, you did not identify yourself as the person who had been approached, and you refused to identify Haakon Chevalier as the individual who made the approach on behalf of George Charles Eltenton; and that it was not until several months later, when you were ordered by a superior to do so, that you so identified Haakon Chevalier. It was further reported that upon your return to Berkeley following your separation from the Los Alamos project, you were visited by the Chevaliers on several occasions; and that your wife was in contact with Haakon and Barbara Chevalier in 1946 and 1947.

"It was reported that in 1945 you expressed the view that 'there is a reasonable possibility that it (the hydrogen bomb) can be made,' but that the feasibility of the hydrogen bomb did not appear, on theoretical grounds, as certain as the fission bomb appeared certain, on theoretical grounds, when the Los Alamos Laboratory was started; and that in the autumn of 1949 the General Advisory Committee expressed the view that 'an imaginative and concerted attack on the problem has a better than even chance of producing the weapon within 5 years.' It was further reported that in the autumn of 1949, and subsequently, you strongly opposed the development of the hydrogen bomb; (1) on moral grounds, (2) by claiming that it was not feasible, (3) by claiming that there were insufficient facilities and scientific personnel to carry on the development, and (4) that it was not politically desirable. It was further reported that even after it was determined, as a matter of national policy, to proceed with development of a hydrogen bomb, you continued to oppose the project and declined to cooperate fully in the project. It was further reported that you departed from your proper role as an adviser to the Commission by causing the distribution separately and in private, to top personnel at Los Alamos of the majority and minority reports of the General Advisory Committee on development of the hydrogen bomb for the purpose of trying to turn such top personnel against the development of the hydrogen bomb. It was further reported that you were instrumental in persuading other outstanding scientists not to work on the hydrogen-bomb project, and that the opposition to the hydrogen bomb, of which you are the most experienced, most powerful, and most effective member, has definitely slowed down its development.

"In view of your access to highly sensitive classified information, and in view of these allegations which, until disproved, raise questions as to your veracity, conduct and even your loyalty, the Commission has no other recourse, in discharge of its obligations to protect the common defense and security, but to suspend your clearance until the matter has been resolved. Accordingly, your employment on Atomic Energy Energy Commission work and your eligibility for access to restricted data are hereby suspended, effective immediately, pending final determination of this matter.

"To assist in the resolution of this matter, you have the privilege of appearing before an Atomic Energy Commission personnel security board. To avail yourself of the privileges afforded you under the Atomic Energy Commission hearing procedures, you must, within 30 days following receipt of this letter, submit to me, in writing, your reply to the information outlined above and request the opportunity of appearing before the personnel security board. Should you signify your desire to appear before the board, you will be notified of the composition of the board and may challenge any member of it for cause. Such challenge should be submitted within 72 hours of the receipt of notice of composition of the board.

"If no challenge is raised as to the members of the board, you will be notified of the date and place of hearing at least 48 hours in advance of the date set for hearing. You may be present for the duration of the hearing, may be represented by counsel of your own choosing, and present evidence in your own behalf through witnesses, or by documents, or by both.

"Should you elect to have a hearing of your case by the personnel security board, the findings of the board, together with its recommendations regarding your eligibility for employment on Atomic Energy Commission work, in the light of Criteria for Determining Eligibility for Atomic Energy Commission Security Clearance and the requirements of Executive Order 10450, will be submitted to me.

"In the event of an adverse decision in your case by the personnel security board, you will have an opportunity to review the record made during your appearance before the board and to request a review of your case by the Commission's personnel security review board.

"If a written response is not received from you within 30 days it will be assumed that you do not wish to submit any explanation for further consideration. In that event, or should you not advise me in writing of your desire to appear before the personnel security board, a determination in your case will be made by me on the basis of the existing record.

"I am enclosing herewith, for your information and guidance, copies of the Criteria and Procedures for Determining Eligibility for Atomic Energy Commission Security Clearance and Executive Order 10450.

"This letter has been marked 'Confidential' to maintain the privacy of this matter between you and the Atomic Energy Commission. You are not precluded from making use of this letter as you may consider appropriate.

"I have instructed Mr. William Mitchell, whose address is 1901 Constitution Avenue NW., Washington, D. C., and whose telephone number is Sterling 3-8000, Extension 277, to give you whatever further detailed information you may desire with respect to the procedures to be followed in this matter.

"Very truly yours,

"K. D. NICHOLS, *General Manager.*

"2 Enclosures. 1. Criteria and Procedures. 2. Executive Order 10450."

I think at this time, then, it would be appropriate for the record to reflect Dr. Oppenheimer's reply of March 4, 1954. I shall now read Dr. Oppenheimer's reply.

This is a letter addressed to Maj. Gen. K. D. Nichols, General Manager, United States Atomic Energy Commission, Washington 25, D. C.

"DEAR GENERAL NICHOLS: This is in answer to your letter of December 23, 1953, in which the question is raised whether my continued employment as a consultant on Atomic Energy Commission work 'will endanger the common defense and security and whether such continued employment is clearly consistent with the interests of the national security.'

"Though of course I would have no desire to retain an advisory position if my advice were not needed, I cannot ignore the question you have raised, nor accept the suggestion that I am unfit for public service.

"The items of so-called derogatory information set forth in your letter cannot be fairly understood except in the context of my life and my work. This answer is in the form of a summary account of relevant aspects of my life in more or less chronological order, in the course of which I shall comment on the specific items in your letter. Through this answer, and through the hearings before the personnel security board, which I hereby request, I hope to provide a fair basis upon which the questions posed by your letter may be resolved.

"THE PREWAR PERIOD

"I was born in New York in 1904. My father had come to this country at the age of 17 from Germany. He was a successful businessman and quite active in community affairs. My mother was born in Baltimore and before her marriage was an artist and teacher of art. I attended Ethical Culture School and Harvard College, which I entered in 1922. I completed the work for my degree in the spring of 1925. I then left Harvard to study at Cambridge University and in Goettingen, where in the spring of 1927 I took my doctor's degree. The following year I was national research fellow at Harvard and at the California Institute of Technology. In the following year I was fellow of the international education board at the University of Leiden and at the Technical High School in Zurich.

"In the spring of 1929, I returned to the United States. I was homesick for this country, and in fact I did not leave it again for 19 years. I had learned a great deal in my student days about the new physics; I wanted to pursue this myself, to explain it and to foster its cultivation. I had had many invitations to university positions, 1 or 2 in Europe, and perhaps 10 in the United States. I accepted concurrent appointments as assistant professor at the California Institute of Technology in Pasadena and at the University of California in Berkeley. For the coming 12 years, I was to devote my time to these 2 faculties.

"Starting with a single graduate student in my first year in Berkely, we gradually began to build up what was to become the largest school in the country of graduate and postdoctoral study in theoretical physics, so that as time went on, we came to have between a dozen and 20 people learning and adding to quantum theory, nuclear physics, relativity and other modern physics. As the number of students increased, so in general did their quality; the men who worked with me during those years hold chairs in many of the great centers of physics in this country; they have made important contributions to science, and in many cases to the atomic-energy project. Many of my students would accompany me to Pasadena in the spring after the Berkeley term was over, so that we might continue to work together.

"My friends, both in Pasadena and in Berkeley, were mostly faculty people, scientists, classicists, and artists. I studied and read Sanskrit with Arthur Rider. I read very widely, must mostly classics, novels, plays, and poetry; and I read something of other parts of science. I was not interested in and did not read about economics or politics. I was almost wholly divorced from the contemporary scene in this country. I never read a newspaper or a current magazine like Time or Harper's; I had no radio, no telephone; I learned of the stock-market crack in the fall of 1929 only long after the event; the first time I ever voted was in the presidential election of 1936. To many of my friends, my indifference to contemporary affairs seemed bizarre, and they often chided me with being too much of a highbrow. I was interested in man and his experience; I was deeply interested in my science; but I had no understanding of the relations of man to his society.

"I spent some weeks each summer with my brother Frank at our ranch in New Mexico. There was a strong bond of affection between us. After my mother's death, my father came often, mostly in Berkeley, to visit me; and we had an intimate and close association until his death.

"Beginning in late 1936, my interests began to change. These changes did not alter my earlier friendships, my relations to my students, or my devotion to physics; but they added something new. I can discern in retrospect more than one reason for these changes. I had had a continuing, smoldering fury about the treatment of Jews in Germany. I had relatives there, and was later to help in extricating them and bringing them to this country. I saw what the depression was doing to my students. Often they could get no jobs, or jobs which were wholly inadequate. And through them, I began to understand how deeply political and economic events could affect men's lives. I began to feel the need to participate more fully in the life of the community. But I had no framework of political conviction or experience to give me perspective in these matters.

"In the spring of 1936, I had been introduced by friends to Jean Tatlock, the daughter of a noted professor of English at the university; and in the autumn, I began to court her, and we grew close to each other. We were at least twice close enough to marriage to think of ourselves as engaged. Between 1939 and her death in 1944 I saw her very rarely. She told me about her Communist Party memberships; they were on again, off again affairs, and never seemed to provide for her what she was seeking. I do not believe that her interests were really political. She loved this country and its people and its life. She was, as it turned out, a friend of many fellow travelers and Communists, with a number of whom I was later to become acquainted.

"I should not give the impression that it was wholly because of Jean Tatlock that I made leftwing friends, or felt sympathy for causes which hitherto would have seemed so remote from me, like the Loyalist cause in Spain, and the organization of migratory workers. I have mentioned some of the other contributing causes. I liked the new sense of companionship, and at the time felt that I was coming to be part of the life of my time and country.

"In 1937, my father died; a little later, when I came into an inheritance, I made a will leaving this to the University of California for fellowships to graduate students.

"This was the era of what the Communists then called the United Front, in which they joined with many non-Communist groups in support of humanitarian objectives. Many of these objectives engaged my interest. I contributed to the strike fund of one of the major strikes of Bridges' union; I subscribed to the People's World; I contributed to the various committees and organizations which were intended to help the Spanish Loyalist cause. I was invited to help establish the teacher's union, which included faculty and teaching assistants at the university, and school teachers of the East Bay. I was elected recording secretary. My connection with the teacher's union continued until some time in 1941, when we disbanded our chapter.

"During these same years, I also began to take part in the management of the physics department, the selection of courses, and the awarding of fellowships, and in the general affairs of the graduate school of the university, mostly through the graduate council, of which I was a member for some years.

"I also became involved in other organizations. For perhaps a year, I was a member of the western council of the Consumer's Union which was concerned with evaluating information on products of interest on the west coast. I do not recall Arthur Kallet, the national head of the Consumer's Union; at most I could have met him if he made a visit to the west coast. I joined the American Committee for Democracy and Intellectual Freedom. I think it then stood as a protest against what had happened to intellectuals and professionals in Germany. I listed, in the personal security questionnaire that I filled out in 1942 for employment with the Manhattan District, the very few political organizations of which I had ever been a member. I say on that questionnaire that I did not include sponsorships. I have no recollection of the Friends of the Chinese People, or of what, if any, my connection with this organization was.

"The statement is attributed to me that, while I was not a Communist, I 'had probably belonged to every Communist-front organization on the west coast and had signed many petitions in which Communists were interested.' I do not recall this statement, nor to whom I might have made it, nor the circumstances. The quotation is not true. It seems clear to me that if I said anything along the lines quoted, it was a half-jocular overstatement.

"The matter which most engaged my sympathies and interests was the war in Spain. This was not a matter of understanding and informed convictions. I had never been to Spain; I knew a little of its literature; I knew nothing of its history or politics or contemporary problems. But like a great many other Americans I was emotionally committed to the Loyalist cause. I contributed to various organizations for Spanish relief. I went to, and helped with, many parties, bazaars, and the like. Even when the war in Spain was manifestly lost, these activities continued. The end of the war and the defeat of the Loyalists caused me great sorrow.

"It was probably through Spanish relief efforts that I met Dr. Thomas Addis, and Rudy Lambert. As to the latter, our association never became close. As to the former, he was a distinguished medical scientist who became a friend. Addis asked me, perhaps in the winter of 1937-38, to contribute through him to the Spanish cause. He made it clear that this money, unlike that which went to the relief organizations, would go straight to the fighting effort, and that it would go through Communist channels. I did so contribute; usually when he communicated with me, explaining the nature of the need, I gave him sums in cash, probably never much less than a hundred dollars, and occasionally perhaps somewhat more than that, several times during the winter. I made no such contributions during the spring terms when I was in Pasadena or during the summers in New Mexico. Later—but I do not remember the date—Addis introduced me to Isaac Folkoff, who was, as Addis indicated, in some way connected with the Communist Party, and told me that Folkoff would from then on get in touch with me when there was need for money. This he did, in much the same way that Addis had done before. As before, these contributions were for specific purposes, principally the Spanish War and Spanish relief. Sometimes I was asked for money for other purposes, the organization of migratory labor in the California valleys, for instance. I doubt that it occurred to me that the contributions might be directed to other purposes than those I had intended, or that such other purposes might be evil. I did not then regard Communists as dangerous; and some of their declared objectives seemed to me desirable.

"In time these contributions came to an end. I went to a big Spanish relief party the night before Pearl Harbor; and the next day, as we heard the news of the outbreak of war, I decided that I had had about enough of the Spanish cause, and that there were other and more pressing crises in the world. My contributions would not have continued much longer.

"My brother Frank married in 1936. Our relations thereafter were inevitably less intimate than before. He told me at the time—probably in 1937—that he and his wife Jackie had joined the Communist Party. Over the years we saw one another as occasions arose. We still spent summer holidays together. In 1939 or 1940 Frank and Jackie moved to Stanford; in the autumn of 1941 they came to Berkeley, and Frank worked for the Radiation Laboratory. At that time he made it clear to me that he was no longer a member of the Communist Party.

"As to the alleged activities of Jackie and Frank in 1944, 1945, and 1946: I was not in Berkeley in 1944 and 1945; I was away most of the first half of 1946; I do not know whether these activities occurred or not, and if I had any knowledge of them at the time it would have been very sketchy. After Christmas of 1945 my family and I visited my brother's family for a few days during the holidays, and I remember that we were there New Year's Eve and New Year's Day in 1946. On New Year's Day people were constantly dropping in. Pinsky and Adelson, who were at most casual acquaintances of mine, may have been among them, but I cannot remember their being there, nor indeed do I remember any of the others who dropped in that day or what was discussed.

"It was in the summer of 1939 in Pasadena that I first met my wife. She was married to Dr. Harrison, who was a friend and associate of the Tolmans, Lauritsens, and others of the California Institute of Technology faculty. I learned of her earlier marriage to Joe Dallet, and of his death fighting in Spain. He had been a Communist Party official, and for a year or two during their brief marriage my wife was a Communist Party member. When I met her I found in her a deep loyalty to her former husband, a complete disengagement from any political activity, and a certain disappointment and contempt that the Communist Party was not in fact what she had once thought it was.

"My own views were also evolving. Although Sidney and Beatrice Webb's book on Russia, which I had read in 1936, and the talk that I heard at that time had predisposed me to make much of the economic progress and general level of welfare in Russia, and little of its political tyranny, my views on this were to change. I read about the purge trials, though not in full detail, and could never find a view of them which was not damning to the Soviet system. In 1938 I met three physicists who had actually lived in Russia in the thirties. All were eminent scientists, Placzek, Weisskopf, and Schein; and the first two have become close friends. What they reported seemed to me so solid, so unfanatical, so true, that it made a great impression; and it presented Russia, even when seen from their limited experience, as a land of purge and terror, of ludicrously bad management and of a long-suffering people. I need to make clear that this changing opinion of Russia, which was to be reinforced by the Nazi-Soviet Pact, and the behavior of the Soviet Union in Poland and in Finland, did not mean a sharp break for me with those who held to different views. At that time I did not fully understand—as in time I came to understand—how completely the Communist Party in this country was under the control of Russia. During and after the battle of France, however, and during the battle of England the next autumn, I found myself increasingly out of sympathy with the policy of disengagement and neutrality that the Communist press advocated.

"After our marriage in 1940, my wife and I for about 2 years had much the same circle of friends as I had had before—mostly physicists and university people. Among them the Chevaliers, in particular, showed us many acts of kindness. We were occasionally invited to more or less obviously leftwing affairs, Spanish relief parties that still continued; and on two occasions, once in San Francisco and once in Berkeley, we attended social gatherings of apparently well to do people, at which Schneiderman, an official of the Communist Party in California, attempted, not with success as far as we were concerned, to explain what the Communist line was all about. I was asked about the Berkeley meeting in an interview in 1946 with agents of the FBI. I did not then recall this meeting, and in particular did not in any way connect it with Chevalier, about whom the agents were questioning me; hence it seemed wholly irrelevant to the matter under discussion. Later my wife reminded me that the Berkeley meeting had occurred at the house of the Chevaliers; and when I was asked about it by the FBI in 1950, I told them so.

"We saw a little of Kenneth May; we both liked him. It would have been not unnatural for us to go to a housewarming for May and his wife; neither my wife nor I remember such a party. Weinberg was known to me as a graduate student; Hiskey I did not know. Steve Nelson came a few times with his family to visit; he had befriended my wife in Paris, at the time of her husband's death in Spain in 1937. Neither of us has seen him since 1941 or 1942.

"Because of these associations that I have described, and the contributions mentioned earlier, I might well have appeared at the time as quite close to the Communist Party—perhaps even to some people as belonging to it. As I have said, some of its declared objectives seemed to me desirable. But I never was a member of the Communist Party. I never accepted Communist dogma or theory; in fact, it never made sense to me. I had no clearly formulated political views. I hated tyranny and repression and every form of dictatorial control of thought. In most cases I did not in those days know who was and who was not a member of the Communist Party. No one ever asked me to join the Communist Party.

"Your letters sets forth statements made in 1942–45 by persons said to be Communist Party officials to the effect that I was a concealed member of the Communist Party. I have no knowledge as to what these people might have said.

What I do know is that I was never a member of the party, concealed or open. Even the names of some of the people mentioned are strange to me, such as Jack Manley and Katrina Sandow. I doubt that I met Bernadette Doyle, although I recognize her name. Pinsky and Adelson I met at most casually, as previously mentioned.

"By the time that we moved to Los Alamos in early 1943, both as a result of my changed views and of the great pressure of war work, my participation in leftwing organizations and my associations with leftwing circles had ceased and were never to be reestablished.

"In August 1941, I bought Eagle Hill at Berkeley for my wife, which was the first home we had of our own. We settled down to live in it with our new baby. We had a good many friends, but little leisure. My wife was working in biology at the university. Many of the men I had known went off to work on radar and other aspects of military research. I was not without envy of them; but it was not until my first connection with the rudimentary atomic-energy enterprise that I began to see any way in which I could be of direct use."

Let the record show that Mr. Oppenheimer has asked to be excused briefly.

"THE WAR YEARS

"Ever since the discovery of nuclear fission, the possibility of powerful explosives based on it had been very much in my mind, as it had in that of many other physicists. We had some understanding of what this might do for us in the war, and how much it might change the course of history. In the autumn of 1941, a special committee was set up by the National Academy of Sciences under the chairmanship of Arthur Compton to review the prospects and feasibility of the different uses of atomic energy for military purposes. I attended a meeting of this committee; this was my first official connection with the atomic-energy program.

"After the academy meeting, I spent some time in preliminary calculations about the construction and performance of atomic bombs, and became increasingly excited at the prospects. At the same time I still had a quite heavy burden of academic work with courses and graduate students. I also began to consult, more or less regularly, with the staff of the Radiation Laboratory in Berkeley on their program for the electromagnetic separation of uranium isotopes. I was never a member or employee of the laboratory; but I attended many of its staff and policy meetings. With the help of two of my graduate students, I developed an invention which was embodied in the production plants at Oak Ridge. I attended the conference in Chicago at which the Metallurgical Laboratory (to produce plutonium) was established and its initial program projected.

"In the spring of 1942, Compton called me to Chicago to discuss the state of work on the bomb itself. During this meeting Compton asked me to take the responsibility for this work, which at that time consisted of numerous scattered experimental projects. Although I had no administrative experience and was not an experimental physicist, I felt sufficiently informed and challenged by the problem to be glad to accept. At this time I became an employee of the Metallurgical Laboratory.

"After this conference I called together a theoretical study group in Berkeley, in which Bethe, Konopinski, Serber, Teller, Van Fleck, and I participated. We had an adventurous time. We spent much of the summer of 1942 in Berkeley in a joint study that for the first time really came to grips with the physical problems of atomic bombs, atomic explosions, and the possibility of using fission explosions to initiate thermonuclear reactions. I called this possibility to the attention of Dr. Bush during the late summer; the technical views on this subject were to develop and change from them until the present day.

"After these studies there was little doubt that a potentially world-shattering undertaking lay ahead. We began to see the great explosion at Alamogordo and the greater explosions at Eniwetok with a surer foreknowledge. We also began to see how rough, difficult, challenging, and unpredictable this job might turn out to be.

"When I entered the employ of the Metallurgical Laboratory I filled out my first personnel security questionnaire."

Let the record show that Dr. Oppenheimer has returned to the hearing room.

"Later in the summer, I had work from Compton that there was a question of my clearance on the ground that I had belonged to leftwing groups; but it was indicated that this would not prove a bar to my further work on the program.

"In later summer, after a review of the experimental work, I became convinced, as did others, that a major change was called for in the work on the bomb itself. We needed a central laboratory devoted wholly to this purpose, where people could talk freely with each other, where theoretical ideas and experimental findings could affect each other, where the waste and frustration and error of the many compartmentalized experimental studies could be eliminated, where we could begin to come to grips with chemical, metallurgical, engineering, and ordnance problems that had so far received no consideration. We therefore sought to establish this laboratory for a direct attack on all the problems inherent in the most rapid possible development and production of atomic bombs.

"In the autumn of 1942 General Groves assumed charge of the Manhattan Engineer District. I discussed with him the need for an atomic-bomb laboratory. There had been some thought of making this laboratory a part of Oak Ridge. For a time there was support for making it a Military Establishment in which key personnel would be commissioned as officers; and in preparation for this course I once went to the Presidio to take the initial steps toward obtaining a commission. After a good deal of discussion with the personnel who would be needed at Los Alamos and with General Groves and his advisers, it was decided that the laboratory should, at least initially, be a civilian establishment in a military post. While this consideration was going on, I had showed General Groves Los Alamos; and he almost immediately took steps to acquire the site.

"In early 1943, I received a letter signed by General Groves and Dr. Conant, appointing me director of the laboratory, and outlining their conception of how it was to be organized and administered. The necessary construction and assembling of the needed facilities were begun. All of us worked in close colloboration with the engineers of the Manhattan District.

"The site of Los Alamos was selected, in part at least, because it enabled those responsible to balance the obvious need for security with the equally important need of free communication among those engaged in the work. Security, it was hoped, would be achieved by removing the laboratory to a remote area, fenced and patrolled, where communication with the outside was extremely limited. Telephone calls were monitored, mail was censored, and personnel who left the area—something permitted only for the clearest of causes—knew that their movements might be under surveillance. On the other hand, for those within the community, fullest exposition and discussion among those competent to use the information was encouraged.

"The last months of 1942 and early 1943 had hardly hours enough to get Los Alamos established. The real problem had to do with getting to Los Alamos the men who would make a success of the undertaking. For this we needed to understand as clearly as we then could what our technical program would be, what men we would need, what facilities, what organization, what plan.

"The program of recruitment was massive. Even though we then underestimated the ultimate size of the laboratory, which was to have almost 4,000 members by the spring of 1945, and even though we did not at that time see clearly some of the difficulties which were to bedevil and threaten the enterprise, we knew that it was a big, complex and diverse job. Even the initial plan of the laboratory called for a start with more than 100 highly qualified and trained scientists, to say nothing of the technicians, staff, and mechanics who would be required for their support, and of the equipment that we would have to beg and borrow since there would be no time to build it from scratch. We had to recruit at a time when the country was fully engaged in war and almost every competent scientist was already involved in the military effort.

"The primary burden of this fell on me. To recruit staff I traveled all over the country talking with people who had been working on one or another aspect of the atomic-energy enterprise, and people in radar work, for example, and underwater sound, telling them about the job, the place that we were going to, and enlisting their enthusiasm.

"In order to bring responsible scientists to Los Alamos, I had to rely on their sense of the interest, urgency, and feasibility of the Los Alamos mission. I had to tell them enough of what the job was, and give strong enough assurance that it might be successfully accomplished in time to affect the outcome of the war, to make it clear that they were justified in their leaving other work to come to this job.

"The prospect of coming to Los Alamos aroused great misgivings. It was to be a military post; men were asked to sign up more or less for the duration; restrictions on travel and on the freedom of families to move about to be severe; and no one could be sure of the extent to which the necessary technical freedom of action could actually be maintained by the laboratory. The notion of disappearing into the New Mexico desert for an indeterminate period and under quasi military auspices disturbed a good many scientists, and the families of many more. But there was another side to it. Almost everyone realized that this was a great undertaking. Almost everyone knew that if it were completed successfully and rapidly enough, it might determine the outcome of the war. Almost everyone knew that it was an unparalleled opportunity to bring to bear the basic knowledge and art of science for the benefit of his country. Almost everyone knew that this job, if it were achieved, would be a part of history. This sense of excitement, of devotion and of patriotism in the end prevailed. Most of those with whom I talked came to Los Alamos. Once they came, confidence in the enterprise grew as men learned more of the technical status of the work; and though the laboratory was to double and redouble its size many times before the end, once it had started it was on the road to success.

"We had information in those days of German activity in the field of nuclear fission. We were aware of what it might mean if they beat us to the draw in the development of atomic bombs. The consensus of all our opinions, and every directive that I had, stressed the extreme urgency of our work, as well as the need for guarding all knowledge of it from our enemies. Past Communist connections or sympathies did not necessarily disqualify a man from employment, if we had confidence in his integrity and dependability as a man.

"There are two items of derogatory information on which I need to comment at this point. The first is that it was reported that I had talked the atomic-bomb question over with Communist Party members during this period (1942-45). The second is that I was responsible for the employment of the atomic-bomb project of individuals who were members of the Communist Party or closely associated with activities of the Communist Party.

"As to the first, my only discussions of matters connected with the atomic bomb were for official work or for recruiting the staff of the enterprise. So far as I knew none of these discussions were with Communist Party members. I never discussed anything of my secret work or anything about the atomic bomb with Steve Nelson.

"As to the statement that I secured the employment of doubtful persons on the project: Of those mentioned, Lomanitz, Friedman, and Weinberg were never employed at Los Alamos. I believe that I had nothing to do with the employment of Friedman and Weinberg by the Radiation Laboratory; I had no responsibility for the hiring of anyone there. During the time that I continued to serve as a consultant with the Radiation Laboratory and to advise and direct the work of some of the graduate students, I assigned David Bohm and Chaim Richman to a problem of basic science which might prove useful in analyzing experiments in connection with fast neutrons. That work has long been published. Another graduate student was Rossi Lomanitz. I remember vaguely a conversation with him in which he expressed reluctance to take part in defense research, and I encouraged him to do what other scientists were doing for their country. Thereafter he did work at the Radiation Laboratory. I remember no details of our talk. If I asked him to work on the project, I would have assumed that he would be checked by the security officers as a matter of course. Later, in 1943, when Lomanitz was inducted into the Army, he wrote me asking me to help his return to the project. I forwarded a copy of this letter to the Manhattan District security officers, and let the matter rest there. Still latter, at Lomanitz' request, I wrote to his commanding officer that he was qualified for advanced technical work in the Army.

"I asked for the transfer of David Bohm to Los Alamos; but this request, like all others, was subject to the assumption that the usual security requirements would apply; and when I was told that there was objection on security grounds to this transfer, I was much surprised, but of course agreed. David Hawkins was known to the personnel director at the laboratory, and I had met and liked him and found him intelligent; I supported the suggestion of the personnel director that he come to Los Alamos. I understand that he had had leftwing associations; but it was not until in March of 1951, at the time of his testimony, that I knew about his membership in the Communist Party.

"In 1943 when I was alleged to have stated that 'I knew several individuals then at Los Alamos who had been members of the Communist Party,' I knew of only one; she was my wife, of whose disassociation from the party, and of whose integrity and loyalty to the United States I had no question. Later, in 1944 or 1945, my brother Frank, who had been cleared for work in Berkeley and at Oak Ridge, came to Los Alamos from Oak Ridge with official approval.

"I knew of no attempt to obtain secret information at Los Alamos. Prior to my going there my friend Haakon Chevalier with his wife visited us on Eagle Hill, probably in early 1943. During the visit, he came into the kitchen and told me that George Eltenton had spoken to him of the possibility of transmitting technical information to Soviet scientists. I made some strong remark to the effect that this sounded terribly wrong to me. The discussion ended there. Nothing in our long standing friendship would have led me to believe that Chevalier was actually seeking information; and I was certain that he had no idea of the work on which I was engaged.

"It has long been clear to me that I should have reported the incident at once. The events that led me to report it— which I doubt ever would have become known without my report—were unconnected with it. During the summer of 1943, Colonel Lansdale, the intelligence officer of the Manhattan District, came to Los Alamos and told me that he was worried about the security situation in Berkeley because of the activities of the Federation of Architects, Engineers, Chemists, and Technicians. This recalled to my mind that Eltenton was a member and probably a promoter of the FAECT. Shortly thereafter, I was in Berkeley and I told the security officer that Eltenton would bear watching. When asked why, I said that Eltenton had attempted, through intermediaries, to approach people on the project, though I mentioned neither myself nor Chevalier. Later, when General Groves urged me to give the details, I told him of my conversation with Chevalier. I still think of Chevalier as a friend.

"The story of Los Alamos is long and complex. Part of it is public history. For me it was a time so filled with work, with the need for decision and action and consultation, that there was room for little else. I lived with my family in the community which was Los Alamos. It was a remarkable community, inspired by a high sense of mission, of duty and of destiny, coherent, dedicated, and remarkably selfless. There was plenty in the life of Los Alamos to cause irritation; the security restrictions, many of my own devising, the inadequacies and inevitable fumblings of a military post unlike any that had ever existed before, shortages, inequities, and in the laboratory itself the shifting emphasis on different aspects of the technical work as the program moved forward; but I have never known a group more understanding and more devoted to a common purpose, more willing to lay aside personal convenience and prestige, more understanding of the role that they were playing in their country's history. Time and again we had in the technical work almost paralyzing crises. Time and again the laboratory drew itself together and faced the new problems and got on with the work. We worked by night and by day; and in the end the many jobs were done.

"These years of hard and loyal work of the scientists culminated in the test on July 16, 1945. It was a success. I believe that in the eyes of the War Department, and other knowledgeable people, it was as early a success as they had thought possible, given all the circumstances, and rather a greater one. There were many indications from the Secretary of War and General Groves, and many others, that official opinion was one of satisfaction with what had been accomplished. At the tme, it was hard for us in Los Alamos not to share that satisfaction, and hard for me not to accept the conclusion that I had managed the enterprise well and played a key part in its success. But it needs to be stated that many others contributed the decisive ideas and carried out the work which led to this success and that my role was that of understanding, encouraging, suggesting and deciding. It was the very opposite of a one-man show.

"Even before the July 16 test and the use of the bombs in Japan, the members of the laboratory began to have a new sense of the possible import of what was going on. In the early days, when success was less certain and timing unsure, and the war with Germany and Japan in a desperate phase, it was enough for us to think that we had a job to do. Now, with Germany defeated, the war in the Pacific approaching a crisis, and the success of our undertaking almost assured, there was a sense both of hope and of anxiety as to what this spectacular development might portend for the future. This came to us a little earlier than to the public generally because we saw the technical development at close range and in secret; but its quality was very much the same as the public response after Hiroshima and Nagasaki.

"Thus it was natural that in the spring of 1945 I welcomed the opportunity when I was asked by Secretary Stimson to serve, along with Compton, Lawrence, and Fermi, on an advisory panel to his Interim Committee on Atomic Energy. We met with that committee on the 1st of June 1945; and even during the week when Hiroshima and Nagasaki were being bombed, we met at Los Alamos to sketch out a prospectus of what the technical future in atomic energy might look like: atomic war heads for guided missiles, improvements in bomb designs, the thermonuclear program, power, propulsion, and the new tools available from atomic technology for research in science, medicine, and technology. This work absorbed much of my time, during September and October; and in connection with it I was asked to consult with the War and State Departments on atomic-energy legislation, and in a preliminary way on the international control of atomic energy.

"I resigned as director of Los Alamos on October 16, 1945, after having secured the consent of Commander Bradbury and General Groves that Bradbury should act as my successor.

"There were then on the books at the laboratory, embodied in memoranda and reports and summarized by me in letters to General Groves, developments in atomic weapons, which could well have occupied years for their fulfillment, and which have in fact provided some, though by no means all, of the themes for Los Alamos work since that time. It was not entirely clear whether the future of atomic weapons work in this country should be continued at or confined to Los Alamos or started elsewhere at a more accessible and more practical site, or indeed what effect international agreements might have on the program. But in the meantime Los Alamos had to be kept going until there was created an authority competent to decide the question of its future. This was to take almost a year.

"The Post War Period"

"In November 1945, I resumed my teaching at the California Institute of Technology, with an intention and hope, never realized, that this should be a full-time undertaking. The consultation about postwar matter which had already begun continued, and I was asked over and over both by the Executive and the Congress for advice on atomic energy. I had a feeling of deep responsibility, interest, and concern for many of the problems with which the development of atomic energy confronted our country.

"This development was to be a major factor in the history of the evolving and mounting conflict between the free world and the Soviet Union. When I and other scientists were called on for advice, our principal duty was to make our technical experience and judgment available. We were called to do this in a context and against a background of the official views of the Government on the military and political situation of our country. Immediately after the war, I was deeply involved in the effort to devise effective means for the international control of atomic weapons, means which might, in the words of those days, tend toward the elimination of war itself. As the prospects of success receded, and as evidence of Soviet hostility and growing military power accumulated, we had more and more to devote ourselves to finding ways of adapting our atomic potential to offset the Soviet threat. In the period marked by the first Soviet atomic explosion, the war in Korea and the Chinese Communist intervention there, we were principally preoccupied, though we never forgot long-term problems, with immediate measures which could rapidly build up the strength of the United States under the threat of an imminent general war. As our own atomic potential increased and developed, we were aware of the dangers inherent in comparable developments by the enemy; and preventive and defensive measures were very much on our minds. Throughout this time the role of atomic weapons was to be central.

"From the close of the war, when I returned to the west coast until finally in the spring of 1947 when I went to Princeton as the director of the Institute for Advanced Study, I was able to spend very little time at home and in teaching in California. In October 1945, at the request of Secretary of War Patterson, I had testified before the House Committee on Military Affairs in support of the May-Johnson bill, which I endorsed as an interim means of bringing about without delay the much needed transition from the wartime administration of the Manhattan District to postwar management of the atomic-energy enterprise. In December 1945, and later, I appeared at Senator McMahon's request in sessions of his Special Committee on Atomic Energy, which was considering legislation on the same subject. Under the chairmanship of Dr. Richard Tolman, I served on a committee set up by General Groves to consider classification policy on matters of atcmic energy. For 2 months, early in 1946, I worked steadily as a member of a panel, the Board of Consultants to the Secretary of

State's Committee on Atomic Energy, which, with the Secretary of State's Committee, prepared the so-called Acheson-Lilienthal report. After the publication of this report, I spoke publicly in support of it. A little later, when Mr. Baruch was appointed to represent the United States in the United Nations Atomic Energy Committee, I became one of the scientific consultants to Mr. Baruch, and his staff in preparation for and in the conduct of our efforts to gain support for the United States' plan. I continued as consultant to General Osborn when he took over the effort.

"At the end of 1946 I was appointed by the President as a member of the General Advisory Committee to the Atomic Energy Commission. At its first meeting I was elected Chairman, and was reelected until the expiration of my term in 1952. This was my principal assignment during these years as far as the atomic-energy program was concerned, and my principal preoccupation apart from academic work.

"A little later I was appointed to the Committee on Atomic Energy of the Research and Development Board, which was to advise the Military Establishment about the technical aspects of the atomic-energy program; I served on it for 7 years; and twice was designated Chairman of special panels set up by the Committee.

"Meanwhile I had become widely regarded as a principal author or inventor of the atomic bomb, more widely, I well knew, than the facts warranted. In a modest way I had become a kind of public personage. I was deluged as I have been ever since with requests to lecture, and to take part in numerous scientific activities and public affairs. Most of these I did not accept. Some, important for the promotion of science or learning or of public policies that corresponded to my convictions, I did accept: the Council of the National Academy of Sciences, the Committee on the Present Danger; the board of overseers of Harvard College, and a good number of others.

"A quite different and I believe unique occurrence is cited as an item of derogatory information—that in 1946 I was 'listed as vice chairman on the letterhead of the Independent Citizens Committee of the Arts, Sciences, and Professions, Inc. * * * cited as a Communist front by the House Committee on Un-American Activities.' The fact is that in 1946, when I was at work on the international control of atomic energy, I was notified that I had been nominated and then elected as vice chairman of this organization. When I began to see that its literature included slogans such as 'Withdraw United States troops from China' and that it was endorsing the criticism enunciated by the then Secretary Wallace of the United States policy on atomic energy, I advised the organization in a letter of October 11, 1946, that I was not in accord with its policy, that I regarded the recommendations of Mr. Wallace as not likely to advance the cause of finding a satisfactory solution for the control of atomic energy, and that I wished to resign. When an effort was made to dissuade me from this course I again wrote on December 2, 1946, insisting upon resignation.

"Later in the postwar period an incident occurred which seems to be the basis of one of the items of derogatory information. In May 1950, Paul Crouch, a former Communist official, and Mrs. Crouch, testified before the California State Committee on Un-American Activities that in July 1941 they had attended a Communist Party meeting at a house in Berkeley, of which I was then the tenant. On the basis of pictures and movies of me which they saw some 8 years later, they said they recognized me as having been present. When the FBI first talked to me about this alleged incident, I was quite certain that no such meeting as Crouch described had occurred. So was my wife, when I discussed it with her. Later, when I saw the testimony, I became even more certain. Crouch had described the gathering as a closed meeting of the Communist Party. I was never a member of the party. Crouch said that no introductions had been made. I could not recall ever having had a group of people at my home that had not been introduced. In May of 1952, I again discussed this alleged meeting with the United States attorney in the Weinberg case (an indictment against Joseph Weinberg for perjury for having among other things denied membership in the Communist Party). I again said that I could not have been present at a closed meeting of the Communist Party because I was not a member of the party; that I had searched my memory and that the only thing that conceivably could be relevant was the vaguest impressions that someone on the campus might at some time have asked permission to use our home for a gathering of young people; that, however, I could recall no such gathering, nor any meeting even remotely resembling the one described by Crouch; that I thought it probable that at the time of the meeting, which by then had been fixed by Crouch as approximately July 23, my wife and I were away from Berkeley. Shortly thereafter, with the aid of counsel, we were able to establish that my wife and I left Berkeley within a few days after July 4, 1941, and did not return until toward the end of the first week in August.

"I need to turn now to an account of some of the measures which, as Chairman of the General Advisory Committee, and in other capacities, I advocated in the years since the war to increase the power of the United States and its allies to resist and defeat aggression.

"The initial members of the General Advisory Committee were Conant, then president of Harvard, DuBridge, president of the California Institute of Technology, Fermi of the University of Chicago, Rabi of Columbia University, Rowe, vice president of the United Fruit Co., Seaborg of the University of California, Cyril Smith of the University of Chicago, and Worthington of the duPont Co. In 1948 Buckley, president of the Bell Telephone Laboratories, replaced Worthington; in the summer of 1950, Fermi, Rowe, and Seaborg were replaced by Libby of the University of Chicago, Murphree, president of Standard Oil Development Co., and Whitman of the Massachusetts Institute of Technology. Later Smith resigned and was succeeded by von Neumann of the Institute for Advanced Study.

"In these years from early 1947 to mid-1952 the Committee met some 30 times and transmitted perhaps as many reports to the Commission. Formulation of policy and the management of the vast atomic-energy enterprises were responsibilities vested in the Commission itself. The General Advisory Committee had the role, which was fixed for it by statute, to advise the Commission. In that capacity we gave the Commission our views on questions which the Commission put before us, brought to the Commission's attention on our initiative technical matters of importance, and encouraged and supported the work of the several major installations of the Commission.

"At one of our first meetings in 1947 we settled down to the job of forming our own views of the priorities. And while we agreed that the development of atomic power and the support and maintenance of a strong basic scientific activity in the fields relevant to it were important, we assigned top priority to the problem of atomic weapons. At that time we advised the Commission that one of its first jobs would be to convert Los Alamos into an active center for the development and improvement of atomic weapons. In 1945–46 during the period immediately following the war, the purposes of Los Alamos were multiple. It was the only laboratory in the United States that worked on atomic weapons. Los Alamos also had wide interests in scientific matters only indirectly related to the weapons program. We suggested that the Commission recognize as the laboratory's central and primary program the improvement and diversification of atomic weapons, and that this undertaking have a priority second to none. We suggested further that the Commission adopt administrative measures to make work at Los Alamos attractive, to assist the laboratory in recruiting, to help build up a strong theoretical division for guidance in atomic-weapons design, and to take advantage of the availability of the talented and brilliant consultants who had been members of the laboratory during the war. In close consultation with the director of the Los Alamos Laboratory, we encouraged and supported courses of development which would markedly increase the value of our stockpile in terms of the destructive power of our weapons, which would make the best use of existing stockpiles and those anticipated, which would provide weapons suitable for modern combat conditions and for varied forms of delivery and which in their cumulative effect would provide us with the great arsenal we now have.

"We encouraged and supported the building up of the laboratory at Sandia whose principal purpose is the integration of the atomic warhead with the weapons system in which it is to be used. In agreement with the Los Alamos staff we took from the very first the view that no radical improvement in weapons development would be feasible without a program of weapons testing. We strongly supported such a program, helped Los Alamos to obtain authorization for conducting the tests it wished, and encouraged the establishment of a permanent weapons testing station and the adoption of a continental test station to facilitate this work. As time went on and the development of atomic weapons progressed, we stressed the importance of integrating out atomic warheads and the development of the carriers, aircraft, missiles, etc., which could make them of maximum effectiveness.

"We observed that there were opportunities which needed to be explored for significantly increasing our arsenal of weapons both in numbers and in capabilities by means of production plant expansion and by ambitious programs to enlarge the sources of raw materials. It was not our function to formulate military requirements. We did regard it as our function to indicate that neither the magnitude of existing plant nor the mode of operation of existing plant which the Commission inherited, nor the limitation of raw materials to relatively well known and high-grade sources of ore, need limit the atomic-weapons program

"The four major expansion programs which were authorized during the 6 years 1946 to 1952 reflect the decision of the Commission, the Military Establishment, the Joint Congressional Committee and other agencies of the Government to go far beyond the production program that was inherited in 1946. And the powerful arsenal of atomic weapons and the variety of their forms adaptable to a diversity of military uses which is today a major source of our military strength in turn reflect the results of these decisions. The record of minutes, reports and other activities of the General Advisory Committee will show that that body within the limits of its role as an advisory group played a significant, consistent, and unanimous part in encouraging and supporting and sometimes initiating the measures which are responsible for these results.

"As a committee and individually, our advice was sought on other matters as well. As early as October 1945 I had testified before a Senate committee on the Kilgore-Magnuson bill—the initial measure for a National Science Foundation; like most scientists I was concerned that steps be taken for recreating in the United States a healthy scientific community after the disruption of the war years. In the General Advisory Committee we encouraged the Commission to do everything that it properly could to support atomic science, both in its own laboratories and in the university centers to which we felt we must look for the training of scientists for advances of a basic character. Throughout the postwar period my colleagues and I stressed the importance of continuing support and promotion of basic science so that there might be a healthy balance between the effort invested in military research and applied science, and that invested in pure scientific training and research which is indispensable to all else. We supported the Commission's decision to make available for distribution in appropriate form and with appropriate safeguards the tracer materials, isotopes, and radioactive substances which have played so constructive a part in medicine, in biological research, in technology, in pure science, and in agriculture.

"We took an affirmative view on the development of reactors for submarines and naval propulsion not only for their direct military value but also because this seemed a favorable and forward-looking step in the important program of reactor development. We were, for the most part, skeptical about the initially very ambitious plans for the propulsion of aircraft, though we advocated the studies which in time brought this program to a more feasible course. We frequently pointed out to the Commission the technical benefits which would accrue to the United States by closer collaboration with the atomic energy enterprise in Canada and the United Kingdom.

"During all the years that I served on the General Advisory Committee, however, its major preoccupation was with the production and perfection of atomic weapons. On the various recommendations which I have described, there were never, so far as I can remember, any significant divergences of opinion among the members of the committee. These recommendations, of course, constitute a very small sample of the committee's work, but a typical one.

"In view of the controversies that have developed I have left the subject of the super and thermonuclear weapons for separate discussion—although our committee regarded this as a phase of the entire problem of weapons.

"The super itself had a long history of consideration, beginning, as I have said, with our initial studies in 1942 before Los Alamos was established. It continued to be the subject of study and research at Los Alamos throughout the war. After the war, Los Alamos itself was inevitably handicapped pending the enactment of necessary legislation for the atomic energy enterprise. With the McMahon Act, the appointment of the Atomic Energy Commission and the General Advisory Committee, we in the committee had occasion at our early meetings in 1947 as well as in 1948 to discuss the subject. In that period the General Advisory Committee pointed out the still extremely unclear status of the problem from the technical standpoint, and urged encouragement of Los Alamos' efforts which were then directed toward modest exploration of the super and of thermonuclear systems. No serious controversy arose about the super until the Soviet explosion of an atomic bomb in the autumn of 1949.

"Shortly after that event, in October 1949, the Atomic Energy Commission called a special session of the General Advisory Committee and asked us to consider and advise on two related questions: First, whether in view of the Soviet success the Commission's program was adequate, and if not, in what way it should be altered or increased; second, whether a crash program for the development of the super should be a part of any new program. The committee considered both questions, consulting various officials from the civil and military branches of the executive departments who would have been concerned, and reached conclusions which were communicated in a report to the Atomic Energy Commission in October 1949.

"This report, in response to the first question that had been put to us, recommended a great number of measures that the Commission should take the increase in many ways our overall potential in weapons.

"As to the super itself, the General Advisory Committee stated its unanimous opposition to the initiation by the United States of a crash program of the kind we had been asked to advise on. The report of that meeting, and the Secretary's notes, reflect the reasons which moved us to this conclusion. The annexes, in particular, which dealt more with political and policy considerations—the report proper was essentially technical in character—indicated differences in the views of members of the committee. There were two annexes, one signed by Rabi and Fermi, the other by Conant, DuBridge, Smith, Rowe, Buckley and myself. (The ninth member of the committee, Seaborg, was abroad at the time.)

"It would have been surprising if eight men considering a problem of extreme difficulty had each had precisely the same reasons for the conclusion in which we joined. But I think I am correct in asserting that the unanimous opposition we expressed to the crash program was based on the conviction, to which technical considerations as well as others contributed, that because of our overall situation at that time such a program might weaken rather than strengthen the position of the United States.

"After the report was submitted to the Commission, it fell to me as chairman of the committee to explain our position on several occasions, once at a meeting of the Joint Congressional Committee on Atomic Energy. All this, however, took place prior to the decision by the President to proceed with the thermonuclear program.

"This is the full story of my 'opposition to the hydrogren bomb.' It can be read in the records of the general transcript of my testimony before the joint congressional committee. It is a story which ended once and for all when in January 1950 the President announced his decision to proceed with the program. I never urged anyone not to work on the hydrogren bomb project. I never made or caused any distribution of the GAC reports except to the Commission itself. As always, it was the Commission's responsibility to determine further distribution.

"In summary, in October 1949, I and the other members of the General Advisory Committee were asked questions by the Commission to which we had a duty to respond, and to which we did respond with our best judgment in the light of evidence then available to us.

"When the President's decision was announced in January 1950, our committee was again in session and we immediately turned to the technical problems facing the Commission in carrying out the President's directive. We sought to give our advice then and in ensuing meetings as to the most promising means of solving these problems. We never again raised the question of the wisdom of the policy which had now been settled, but concerned ourselves rather with trying to implement it. During this period our recommendations for increasing production facilities included one for a dual-purpose plant which could be adapted to make materials either for fission bombs or materials useful in a thermonuclear program. In its performance characteristics, the Savannah River project, subsequently adopted by the Commission, was foreshadowed by this recommendation.

"While the history of the GAC opposition to a crash program for the super ended with the announcement of the President's decision, the need for evaluation and advice continued. There were immense technical complications both before and after the President's decision. It was of course a primary duty of the committee, as well as other review committees on which I served, to report new developments which we judged promising, and to report when a given weapon or family of weapons appeared impractical, unfeasible or impossible. It would have been my duty so to report had I been alone in my views. As a matter of fact, our views on such matters were almost always unanimous. It was furthermore a proper function for me to speak my best judgment in discussion with those responsibly engaged in the undertaking.

"Throughout the whole development of thermonuclear weapons, many occasions occurred where it was necessary for us to form and to express judgments of feasibility. This was true before the President's decision, and it was true after the President's decision. In our report of October 1949, we expressed the view, as your letter states, that 'an imaginative and concerted attack on the problem has a better than even chance of producing the weapon within 5 years.' Later calculations and measurements made at Los Alamos led us to a far more pessimistic view. Still later brilliant inventions led to the possibility of lines of development of very great promise. At each stage the General Advisory Committee, and I as its Chairman and as a member of other bodies, reported as faithfully as we could our evaluation of what was likely to fail and what was likely to work.

"In the spring of 1951 work had reached a stage at which far-reaching decisions were called for with regard to the Commission's whole thermonuclear program. In consultation with the Commission, I called a meeting in Princeton in the late spring of that year, which was attended by all members of the Commission and several members of its staff, by members of the General Advisory Committee, by Dr. Bradbury and staff of the Los Alamos Laboratory, by Bethe, Teller, Bacher, Fermi, von Neumann, Wheeler, and others responsibly connected with the program. The outcome of the meeting, which lasted for 2 or 3 days, was an agreed program and a fixing of priorities and effort both for Los Alamos and for other aspects of the Commission's work. This program has been an outstanding success.

"In addition to my continuing work on the General Advisory Committee there were other assignments that I was asked to undertake. Late in 1950 or early in 1951 the President appointed me to the Science Advisory Committee to advise the Office of Defense Mobilization and the President in 1952 the Secretary of State appointed me to a panel to advise on armaments and their regulation; and I served as consultant on continental defense, civil defense, and the use of atomic weapons in support of ground combat. Many of these duties led to reports in the drafting of which I participated, or for which I took responsibility. These supplement the record of the General Advisory Committee as an account of the counsel that I have given our government during the last eight years.

"In this letter, I have written only of those limited parts of my history which appear relevant to the issue now before the Atomic Energy Commission. In order to preserve as much as possible the perspective of the story, I have dealt very briefly with many matters. I have had to deal briefly or not at all with instances in which my actions or views were adverse to Soviet or Communist interest, and of actions that testify to my devotion to freedom, or that have contributed to the vitality, influence and power of the United States.

"In preparing this letter, I have reviewed two decades of my life. I have recalled instances where I acted unwisely. What I have hoped was, not that I could wholly avoid error, but that I might learn from it. What I have learned has, I think, made me more fit to serve my country.

"Very truly yours,

J. ROBERT OPPENHEIMER.

PRINCETON, N. J., MARCH 4, 1954."

Dr. GRAY. This board is convened to enable Dr. Oppenheimer to present any information he considers appropriate having a bearing on the documents just read and the information contained in them, this information being, of course, the same as that disclosed to Dr. Oppenheimer in Mr. K. D. Nichols' letter of December 23, 1953 to Dr. Oppenheimer and Dr. Oppenheimer's reply of March 4, 1954, and to provide a record as a basis for a recommendation to the General Manager of the Atomic Energy Commission as to Dr. Oppenheimer's eligibility for access to restricted data.

At this point, I should like to remind everyone concerned that this proceeding is an inquiry and not in the nature of a trial. We shall approach our duties in that atmosphere and in that spirit.

Dr. Oppenheimer, have you been given an opportunity to exercise the right to challenge any or all of the members of this Board?

Dr. OPPENHEIMER. I have, indeed.

Dr. GRAY. I should point out to you, sir, that if at any time during the course of this hearing it appears that grounds for challenge for cause arise, you will exercise your right to challenge for cause and the validity of the challenge will be determined in closed session by the members of the Board.

The proceedings and stenographic record of this board are regarded as strictly confidential between Atomic Energy Commission officials participating in this matter and Dr. Oppenheimer, his representatives and witnesses. The Atomic Energy Commission will not take the initiative in public release of any information relating to the proceeding before this board.

Now, at this time, Dr. Oppenheimer, you will be given the opportunity to present any material relevant to the issues before the board. At this point I think we shall find it necessary to exclude all witnesses except the one whose testimony is being given to the board under the provisions of the procedures which we must follow in this inquiry.

I shall read from the Security Clearance Procedures of the United States Atomic Energy Commission, dated September 12, 1950, under section 4.15, subsection (b):

"The proceedings shall be open only to duly authorized representatives of the staff of the Atomic Energy Commission, the individual, his counsel, and such persons as may be officially authorized by the board."

The Chairman would make the observation that counsel for the board has suggested that in the spirit of these regulations we should have present only the witness who is testifying or who is appearing.

Mr. GARRISON. Mr. Chairman, may I just say that I have a few preliminary remarks as counsel to make before Dr. Oppenheimer testifies, and it may well be that they will perhaps bring us to a suitable point of adjourning for this morning, so that Dr. Oppenheimer's testimony might begin this afternoon.

However, if you would prefer that Mrs. Oppenheimer not be present while I make these preliminary remarks which have to do largely with procedural aspects of what we propose to do, it would be quite satisfactory, of course, to me.

Dr. GRAY. Let us then proceed on that basis. So, Mrs. Oppenheimer, you are not at this moment excused.

I should like to ask Dr. Oppenheimer whether he wishes to testify under oath in this proceeding?

Dr. OPPENHEIMER. Surely.

Dr. GRAY. You are not required to do so.

Dr. OPPENHEIMER. I think it best.

Dr. GRAY. I should remind you, then, of the provisions of section 1621 of title 18 of the United States Code, known as the perjury statute, which makes it a crime punishable by a fine of up to $2,000 and/or imprisonment of up to 5 years for any person stating under oath any material matter which he does not believe to be true.

It is also an offense under section 1001 of title 18 of the United States Code, punishable by a fine of not more than $10,000 or imprisonment for not more than 5 years, or both, for any person to make any false, fictitious, or fraudulent statement or representation in any matter within the jurisdiction of any agency of the United States.

I think that before you proceed, Mr. Garrison, that it would be well to administer the oath to Dr. Oppenheimer.

J. Robert Oppenheimer, do you swear that the testimony you are to give the board shall be the truth, the whole truth and nothing but the truth, so help you God?

Dr. OPPENHEIMER. I do.

Dr. GRAY. May I also point out that in the event that it is necessary for anyone to disclose restricted data during his statements before this board shall advise the Chairman before such disclosure in order that persons unauthorized to have access to restricted data may be excused from the hearing.

Now, Dr. Oppenheimer, you may proceed, and I gather from what Mr. Garrison said, that he will at this point make a statement to the Board.

Mr. GARRISON. Mr. Chairman, members of the board, I would like to say at the outset that far from having thought of challenging any member of the board, we appreciate very much the willingness of men of your standing and responsibilities to accept this exacting and onerous job in the interests of the country. I express my appreciation to you.

We cannot help but be conscious of the fact that for the past week the members of the board have been examining a file containing various items about Dr. Oppenheimer to which we have had, and to which we shall have no access at all. I have been told that this is a large file, and I suppose a great deal of time has been spent on it. I am sure that it goes without saying that we are confident that the minds of the members of the board are open to receive the testimony that we shall submit.

If, as a result of going through the file, there are troublesome questions which have arisen, any items of derogatory information not mentioned in the Commission's letter of December 23, I know we can count on you to bring those to our attention so that we may have an adequate opportunity to reply to them.

I would take note at this point of section 4.15 (j) of the Rules and Regulations of the Commission, and of the second sentence, which reads, "If prior to or during the proceeding, in the opinion of the Board, the allegations in the notification letter are not sufficient to cover all matters into which inquiry should be directed, the board shall suggest to the manager concerned that in order to give full notice to the individual, notification letter should be amended."

If there are questions that you have in mind about these possible other items in the file that you would like to have cleared up, and shall go through the formality of the amendment of the letter, we will not press. But it would be helpful to us if we could at least be notified of any such items in a manner that would give us adequate time to study them and to prepare appropriate response.

Dr. GRAY. I think you need have no concern on that score, Mr. Garrison.

Mr. GARRISON. I am sure not. I would like at this point to read into the record a letter from Dr. Oppenheimer to Chairman Strauss of the Atomic Energy Commission, dated December 22, 1953. I would be glad to give copies to the members of the board.

I shall explain the purpose in a moment of reading this letter to you.

This letter is addressed to Adm. Lewis L. Strauss, Chairman of the Atomic Energy Commission, Washington, D. C., and is dated December 22, 1953, and reads as follows:

"DEAR LEWIS: Yesterday, when you asked to see me, you told me for the first time that my clearance by the Atomic Energy Commission was about to be suspended. You put to me as a possibly desirable alternative that I request termination of my contract as a consultant to the Commission, and thereby avoid an explicit consideration of the charges on which the Commission's action would otherwise be based. I was told that if I did not do this within a day, I would receive a letter notifying me of the suspension of my clearance and of the charges against me, and I was shown a draft of that letter.

"I have thought most earnestly of the alternative suggested. Under the circumstances this course of action would mean that I accept and concur in the view that I am not fit to serve this Government, that I have now served for some 12 years. This I cannot do. If I were thus unworthy I could hardly have served our country as I have tried, or been the Director of our Institute in Princeton, or have spoken, as on more than one occasion I have found myself speaking, in the name of our science and our country.

"Since our meeting yesterday, you and General Nichols told me that the charges in the letter were familiar charges, and since the time was short, I paged through the letter quite briefly. I shall now read it in detail and make appropriate response.

"Faithfully yours,

ROBERT OPPENHEIMER."

I have presented that, Mr. Chairman, simply to show that there has been no disposition on Mr. Oppenheimer's part to hold onto a job for the sake of a job. It goes without saying that if the Commission did not wish to use his services as a consultant that was all right with him. The point of this letter is that he felt that he could not in honor and integrity of his person simply resign and leave these questions unadjudicated. Fully realizing the terrible burden of going forward with this matter, and the natural risks in any proceeding of this character, including what may go on outside of these walls, nevertheless went forward.

He speaks in this letter of charges. I am glad that the chairman pointed out that word is not the appropriate word to be used here. We recognize that fact and have noted, indeed, earlier from a letter from General Nichols to me, dated January 27, 1954, in which I in a letter to him inadvertently used the word "charges," he said, "Please be advised that we do not consider that letter"—the one of December 23, 1953, the principal letter which you read into the record—"as being a statement of charges, but rather a statement of substantial derogatory information bearing upon his eligibility for AEC security clearance."

Gentlemen, for the last several months I have been immersed in talking with all of the people I could find who had worked with Dr. Oppenheimer over the years about their recollections of his activities and their impressions of him as a man and as a citizen, and I have immersed myself in his writings and in all of the details of the case.

I would just like to say that I have been struck by the instantaneous and warm and universal support which everybody that I talked with who has worked with him has given. It will be reflected in the testimony which we will bring here before you. I shall speak a little later about the scope of the testimony and the number of witnesses, but it has really quite impressed me.

I have also found among these gentlemen a great sense of anxiety about this case of what it may portend to the science program of the country if clearance in the end could be denied to a man who has tried to serve his country as Dr. Oppenheimer has served it; not so much a sense of what might happen to the scientists now in the Government service themselves, although this certainly has come to them as a great shock, but rather what it may do to the young scientists to whom the Government must turn in the future for aid and assistance in seeking to recruit personnel to the Government.

I mention this not because it has any precise bearing on the action and the findings in this case, but it is a part of the warp and woof then of the feelings with which the witnesses here will address you.

The case as it has looked to me stands out in sharp feature rather simply this way, that these derogatory items in the file mostly have to do with activities of Dr. Oppenheimer that go back to 12 to 15 years ago. A few have to do with 9 to 12 years. Since the war—since 1945—apart from the Crouch incident, which itself has to do with an alleged occurrence in 1941, there is in this letter of December 23—I think I am correct in saying—not a single item of derogatory information except the Independent Citizens Committee of the Arts, Sciences, from which he in fact resigned, the proffer of vice chairmanship, when he saw what it was up to and except for the hydrogen matter, which stands all by itself.

For from being to his discredit, far from casting doubt on his desire to serve his country as best he sees how to do it, I think our witnesses will persuade you beyond any doubt that his conduct in the hydrogen bomb matter was beyond any reproach; that it was an exercise of the most honest judgment done in the best interests of the country, and that his whole record since the war is rather astonishingly filled with a continuous series of efforts to strengthen the defenses of the United States in a world threatened by totalitarian aggression.

I was surprised to find that about half of his working time since 1945 has been devoted to service on Government boards and committees, from 1945 on, as a volunteer citizen, placing his talents at the service of the country. The richness and the variety of the services that he rendered in those capacities will be vividly brought out in the testimony.

I would like to say that everything he has done since the war, the hydrogen bomb and all the rest, has been done in a blaze of light. There has been not one thing that has not been done in the full daylight of the work of the Government and subjected to the most searching criticism of the ablest men in science and government, all doing each in their own way what they could do to serve the country.

I believe this record will be one which will persuade this board that to exclude Dr. Oppenheimer from the capacity that he continue to serve the Government as he has in the past would be contrary to the best interests of all of us.

Now a word about the procedure. We hope to present this case to you in terms of unrestricted data. It would be an unclassified case. We would like to present it in that direct lay fashion. I am not a scientist and except for Dr. Evans, the members of the board are not. We thought it would be best if we could avoid having to get involved in technical evidence of a very complex and difficult nature which would involve a great deal of time, and which would perhaps tend to lead us into the wrong path of exposing that the issue here is whether at a particular juncture Dr. Oppenheimer's scientific judgment was right or wrong. I am sure we all agree that the question here is not whether the advice that he gave at a particular time was from a scientific point of view, one with which this board might differ in the light of history. The real question is was his judgment an honest judgment; did he do the best he could for his Government.

I was a little fearful if we got into the whole realm of science that we would perhaps lost sight of that simple fact.

We want to on the other hand tell you as completely as we can—and I think it can be done within the limits of classification, the proper limits we can talk about here—exactly how the things were done which he did, and the procedures that were adopted and the way the tasks were gone about, the atmosphere in which they were conducted.

I would like to start, when we get into Dr. Oppenheimer's testimony, with a somewhat fuller account from him of his record of public service, beginning with the war years, and coming down to date.

I would like to start with that because the answer which he gave is less complete with respect to that portion of his life. With respect to the derogatory items of the early years, we have said about all that we can say except as you of course may wish to question him further, as I have no doubt you will, with regard to them. But I would like to have Dr. Oppenheimer tell you more than he has been able to do in the encompass of the answer about the way in which he has sought to serve the country since the war.

Our witnesses will mostly be bringing testimony about that service.

When we get through with that, there are a few supplementary things to be said about these earlier derogatory items, and some documentary evidence that we want to introduce.

The witnesses that we would like to call, after you yourself have finished questioning Dr. Oppenheimer, and when he is through—and I should say we will welcome questions as we go along, whatever you may wish to ask, as Dr. Oppenheimer testifies, and I hope you will, because I think it will make it easier for all of us if you would do that instead of leaving it all for the end—whenever we are through and the board is through with questioning Dr. Oppenheimer, then we would be prepared to invite a considerable number of witnesses to testify. There are as of this moment 27 witnesses whom we expect to call. There may be several more. There will also be 3 or 4 or 5—I don't know exactly how many—written documents from some witnesses who are simply unable to get here at all.

If the board would like, I should be glad to give you a list of the proposed witnesses, so that you may have it before you, and also a skeleton of the proposed times.

Dr. GRAY. I would think that would be very helpful, Mr. Garrison, if you would. It just possibly might have some bearing on the questions that might be put to Dr. Oppenheimer.

Mr. GARRISON. We will, I think, bring that in after lunch. It is not quite ready for presentation.

Dr. GRAY. Very well.

Mr. GARRISON. I would like at this time to give you—and I hope you might perhaps keep this handy—an exhibit called biographical data on Dr. J. Robert Oppenheimer.

Dr. GRAY. Are you offering that now, Mr. Garrison?

Mr. GARRISON. Yes.

Dr. GRAY. Would you mark the Oppenheimer exhibit No. 1, and we will receive it for the record.

(The document was marked "Oppenheimer Exhibit No. 1" and received for the record.)

Mr. GARRISON. I would say to the board that if you will turn to the first page, I, this is a concise summary of the major steps in Dr. Oppenheimer's career. It will be a handy guide for use when witnesses are testifying to particular committees or what not to see just at what stage they call.

Turning to the next page, II, you will see listed the various Government committees on which Dr. Oppenheimer has served, with the dates of service and the people who served with him on these various undertakings. This is since Los Alamos.

Dr. GRAY. May I interrupt? Just in the interest of keeping the record precisely clear, I wonder if that last statement is quite correct, because I believe you terminated your association with Los Alamos in the fall of 1945, and some of these committees overlap.

Mr. GARRISON. You are correct.

Dr. GRAY. I am not making it any more a particular point other than——

Mr. GARRISON. I appreciate the correction. I should make a further correction, that this is a partial list of the committees on which he served. They are the principal ones, the ones about which we shall mainly be talking here at the hearings.

Beginning with III and running all the way over is a detailed biography in which, to the best of our ability, we have put down year by year every association of which we have any record of his having joined or been a member of, every publication of his, every position that he has held on committees, either private or public, lectures that he has given, addresses. This is the outward and visible account of his activities, in short, as best we could compile them. If there are inaccuracies, they are entirely inadvertent.

One of the things that struck me as I went over this biography, which I asked to have prepared, was the quite evident fact that, during the prewar years in which most of these derogatory items arise, his energies were quite strongly devoted to almost entirely, really, his scientific work and scientific undertakings. They reveal really very little in the way of political interest or associations on his part just on the face of the record.

In the postwar period this becomes again apparent, and I would direct your attention to the fact that in this postwar period I do not think there is a single association of his that can possibly be questioned as derogatory by the Commission or by this board or anything, indeed, other than a rich record of association and devotion to his science and his service to the Government and his membership in various scientific and civic organizations of the highest standing.

There is, of course, also that notation about the Independent Citizens Committee of Arts and Sciences; but that, as he stated in his answer, he withdrew from because of its policy in replying to their proffer of an officer's position in the association and indicated his complete lack of sympathy with the kind of policies which it stood for.

I am going to, in the course of the testimony, introduce in evidence at pertinent places extracts from some of Dr. Oppenheimer's writings and addresses from the period 1945 to date. To attempt to introduce them all here would be beyond the obvious scope of this inquiry. But I want to assure the board that you will find a very consistent and very striking thread of continuous thought on Dr. Oppenheimer's part with respect to the strengthening of the defenses of this country, with respect to what has to be done to counter the Russian threat from abroad, with respect to building the strongest and soundest democratic America that man can do, consistent, unvarying, and very impressive.

This whole postwar period, everything in it, is really utterly inconsistent with any notion that this man could have been anything but a devoted supporter of the American system that we love.

I think that is all that I have to say of a preliminary character, Mr. Chairman. I appreciate your letting me say it. I think perhaps this would be an appropriate point to adjourn, and shall we come back at 2:30? Is that your schedule?

Dr. GRAY. Perhaps it should be a recess; I do not know, Mr. Garrison; and not an adjournment. I am sure we want to try to meet the convenience of everyone concerned and at the same time not to waste hours or minutes which could be useful in getting ahead with the inquiry.

Speaking for the board, I am sure we could be ready at 2 o'clock, but I do not want to press you and Dr. Oppenheimer.

Mr. SILVERMAN. I was suggesting that we could use a little extra time.

Dr. GRAY. Would you suggest 2:30?

Mr. GARRISON. Let us say 2:15.

Dr. GRAY. All right.

Mr. GARRISON. Before the recess, I want to read this into the record. This is a letter to me from Mr. William Mitchell, general counsel, dated January 15, 1954, that the Commission will be prepared to stipulate as follows for purposes of the hearing:

"On August 6, 1947, the Commission recorded clearance of Dr. J. Robert Oppenheimer, which it noted had been authorized in February 1947."

What that has reference to, as we will show more fully in the testimony, was the fact that in 1947 Dr. Oppenheimer's personnel file was sent to the Commission by Mr. J. Edgar Hoover, with the request that it be reviewed. This was at the time of the early days of the establishment of the Commission, and Dr. Oppenheimer had been appointed to the general advisory committee and had been elected its chairman. The Commission considered this entire file, which I believe the evidence will show contained substantially all that you have before you in the letter of December 23 except the Crouch incident, which occurred afterward, though it related to something in 1941 and except for the hydrogen bomb matters, and the Commission unanimously after discussing Dr. Oppenheimer's qualifications with many of the leading people who had had to do with him in the past and with officials of the Government reached the view that there was no question as to his clearance. That we will show later by testimony. I merely mention now that will be before you. I don't mean to import what I said into the stipulation which goes in front of what I have just said.

Dr. GRAY. We will now recess until 2:15.

(Thereupon, at 12:20 p. m., a recess was taken until 2:15 p. m. the same day.)

AFTERNOON SESSION

Dr. GRAY. We will begin the proceedings at this point again. Let the record show that Mrs. Oppenheimer is not present this afternoon.

Whereupon J. Robert Oppenheimer was called as a witness and, having been previously sworn, was examined and testified as follows:

DIRECT EXAMINATION

By Mr. GARRISON:

Q. Will you tell the board what your present position is, Dr. Oppenheimer, at Princeton?

A. My job is director of the institute of advanced study. For the most part this is not relevant to the hearing, but I will outline briefly some of the circumstances.

The institute is not part of Princeton University. It is a separate institute, very highbrow. It has about 130 members who are postdoctoral. Some of them are youngsters just out of graduate school; some are men of 50 and 60.

We try, though only in part, to patronize scholarship and science—science in the old sense of the word, meaning both the natural science and the humanities. I think the parts that are relevant to the welfare of the United States are fairly limited. We have a very good training ground for various students in pure mathematics, applied mathematics, and theoretical physics. Many of them who are at the institute are supported by the United States Government; many more go into work for a limited time or for a longer time on behalf of our Government.

We have a number of governmental undertakings. I think one of the more interesting is that we seem for the first time to be able to predict cyclogenesis— the generation of storms. This is of practical value, and the Government has instituted a program based on the research started at the institute.

We have one other function which I believe to be important at this time. We are as much as we can be, with our limited resources, an open house to scholars throughout the free world, from Europe, from Japan, from India— mostly, of course, from Europe. I think more than half of our people are from outside the United States. I think we go a long way toward persuading a very small fraction of the people abroad that the United States is a humane and civilized place, and programs about the institute have been carried by the Voice of America and in State Department bulletins, and I believe that the ill-fated glossy magazine that we put out in the Soviet Union, called America, published an account of our work.

In any case, there are many people in the Government who are proud of what we are doing, and I am proud of it.

Q. Do you have occasion to use classified material at the institute?

A. The institute has never accepted—I don't know how the board of trustees would respond—a classified contract. It has never been asked to accept one. There is work going on at the institute which is very close to classified work; but, by the time it is fed in to us, it is unintelligible and therefore declassified.

Individual members of the institute, of course, have done classified work. I am an obvious example. George F. Kennan is an example. Von Neumann is an example. I won't reel off the list of names. This is an affair between the individual and the Government. The institute interposes no objection.

Every year I get a letter from Los Alamos: "Do you object if we renew the contracts of these people," and I say that it is up to them and up to you.

Q. What security methods have you used at the institute in connection with your own classified materials in the past?

A. They have been very elaborate. When I came to Princeton the Atomic Energy Commission established a top-secret facility. I need not describe the rigmarole that goes into this, the warning systems and all the rest. There is a vault there. It has been moved recently, but it still is at the institute. I have never known the combination. The combination has been rigidly guarded. I believe our record is that we have never even displaced a document. I hope this continues to be true. That facility is still in existence for the benefit of other people who wish to use it.

Q. When did you come to the Institute at Princeton?

A. I came in the late summer, I think, of 1947. I had been a professor at California Institute of Technology and at the University of California at Berkeley. In late 1946 perhaps or early 1947, the present Chairman of the Atomic Energy Commission was chairman of the nominating committee to seek a new director to succeed Dr. Aydelott at the Institute and he offered me the job stating that the trustees and the faculty desired this.

I did not accept at once. I like California very much, and my job there, but I had, as will appear, not spent very much time in California. Also, the opportunity to be in a small center of scholarship across the board was very attractive to me. Before I accepted the job, and a number of conversations took place, I told Mr. Strauss there was derogatory information about me. In the course of the confirmation hearings on Mr. Lilienthal especially, and the rest of the Commissioners, I believe Mr. Hoover sent my file to the Commission, and Mr. Strauss told me that he had examined it rather carefully. I asked him whether this seemed to him in any way an argument against my accepting this job, and he said no, on the contrary—anyway, no—in April I heard over the radio I had accepted, and decided that was a good idea. I have been there since.

Q. You said you had not spent much time in California. That I take it was because of your engagement in public service in a rather continuous way?

A. Yes.

Q. That leads, I think, naturally into a discussion of your record of public service, and I would like to begin, Dr. Oppenheimer, with the war years, and have you tell the Board how you happened to get involved in atomic bomb work.

A. In the autumn of 1941 I was asked by Arthur Compton to attend a session of the special committee of the National Academy of Sciences, which had been set up to study the military uses of fission, the uranium project. I think that committee had other meetings. I attended a 2-day meeting. At that time—I need not go into details—I took an active part in the discussion.

Q. What was your position?

A. I was professor of physics at the University of California. I took an active part in the discussion primarily, I think, to be sure that the open questions were recognized as open and some sketch of a program understood. I believe everyone there was quite clear that we had to go ahead with this.

The next step was double. On the one hand, Earnest Lawrence, who was Director of the Radiation Laboratory at Berkeley, had on the trip to this meeting become more and more enthusiastic about the prospects for an electromagnetic separation of uranium isotopes, and we talked about that the whole way. When he got back, he started getting other people thinking about it, and I became a sort of adviser or consultant without appointment to that undertaking. I don't remember just when, but some time in the course of the next few months I had an idea which turned out to have been useful. It was not decisive, but it perhaps doubled or tripled the capacity, or halved or thirded the price of the plant they were building.

I met with them quite often at their steering committee and coordinating committee meetings, but never as an employee. I was still teaching and in fact teaching more than usual, because other people had gone off to work on radar and we were very badly understaffed.

Other things that I began to think more intensively and on my own about how to make atomic bombs and made some calculations on efficiency, design, probable amounts of material and so on, so that I got into it, and knew something about it. The result was that when I was called probably in the first days of 1942—anyway after Pearl Harbor—to Chicago, I was able to give a little information about this aspect of the problem. The main thing the Chicago people were up to was building reactors to make plutonium, or trying to see if they could build reactors to make plutonium. But in the original assignment of the responsibility, work on fast fission, which is what they used as a shorthand for the bomb, was also part of their job.

The man in charge of it was Gregory Breit. He had the wonderful code name of "Coordinator of Rapid Rupture." He wrote me some time in the spring, suggesting that we might have a conference in Minneapolis, that he was interested in work I had been doing, and perhaps might even want to come longer to Minneapolis. This never matured. I think Breit quit in June, I believe it was.

I went on to Chicago at Arthur Compton's request. I made arrangements to have Bethe and Teller and a few other people meet and also the heads of the subundertakings. that were trying to make measurements relevant to the design of bombs and specifications of bombs, and we had conferences for some days.

Fairly early in the game, Arthur Compton said would I take charge of this part of the work and I agreed to do so.

We also agreed that at that moment the job fell into two parts. One was the job of analysis and thinking, of theory, and we would set that up as a summer study in Berkeley. The other was to try to get some sense into the distorted and fragmentary work that was going on in a number of laboratories. There was a third part which was to get some new projects started.

Q. You spoke about the fragments and the scattering. I take it that was one of the factors that led you to think in terms of what ultimately was to become Los Alamos?

A. It did not go quite so fast. We spent this summer in study, and I traveled around and saw most of the laboratories. I had very good help from John Manley, who is not Jack Manley. We had a very heavy study, and began to see what was involved, not all of it, I would not say all of it until much later, and also spent a fantastically large fraction of our time on the thermonuclear program. That is the first time we really got into it.

What we then saw of the thermonuclear program was not very relevant to what you are reading in the papers today. But it excited us, and it seemed to make even more necessary that we understand what this was all about.

After our conferences were over, I went and reported to Compton who was off on a summer holiday about this aspect of it, as well as others. I then came on, I think, at his request, and saw Dr. Bush and told him about it. We also at about this time prepared a report on our views for transmission to the British.

There was a fairly complete interchange. We did not write about the thermonuclear program, but we wrote about some of the subleties of the atomic bomb program.

Then we began to notice how very much needed doing and how much the little laboratories were suffering from their isolation.

There was supposed to be security; anyway, there was good compartmentalization and the result was that people would not know what was going on anywhere else. Work was duplicated, and there was almost no sense of hope or direction in it.

By the fall of 1942, not only the theoretical people but anyone who knew the experimental situation realized that this had to be pulled together. It was not the first job. The first job was to make the stuff. But in hope that would come out all right, we had to have a place where we could learn what to do with it. This was not trivial. We therefore started chattering about should we have a laboratory in Chicago, should we have one at Oak Ridge. The prevailing notion was that there would be more or less a conventional laboratory until such time as we were really ready to get into almost ordnance experiments, and then we would go out and get a proving ground somewhere, which would be rather remote and a very few people involved.

This did not seem sound. It seemed to me and knowledgeable people, it was one package, ordnance, chemistry, physics theory, effects, all had to be understood together or the job would not get done. These were the considerations that led me to say to General Groves, who had by then been brought into the project as its head, that I thought a bomb laboratory was a good idea. That I thought it needed to have two characteristics; one, that it be free internally to talk about problems from one part of the job to the other, and that its external security be very, very good indeed, that it be isolated, if necessary guarded, and all the rest of it.

General Groves was very much interested from the beginning. I think I had a message from him to come on down to discuss the matter, and I remember that he and Colonel Nichols, and Colonel Marshall, and I got into some very limited place on the Twentieth Century Limited and talked about plans for such a post.

The original plans were much too small. They had in mind that it might be a useful thing if the key personnel of the laboratory were commissioned. I at that time, very foolishly I think, had no objection to it. I would have been glad to be an officer. I thought maybe the others would. But it was not very long before I talked to people who had to come to Los Alamos, especially those who had experience in radar and in military research, and they explained that it was hopeless to superimpose on a natural technical organization of a laboratory the irrelevant and sort of additional organization of the Military Establishment.

We had a long hassle about that. I think everyone agreed. In a letter which reached me early in 1943, signed by Groves and Conant, it was agreed that initially the workers in the laboratory would be civilians. It was contemplated that later at the more critical phases the key people would be commissioned. That plan was dropped, I think essentially because the numbers got so big and there was no need for it, and it became impractical.

About this time, in the autumn of 1943, Groves sent an engineer around to look for a place. He was around in the Southwest where I knew the country and in New Mexico, and I showed him and showed General Groves the city of Los Alamos. This appealed to General Groves very much, and he moved with unbelievable dispatch to acquire it, and we started construction.

It may be of interest to you that one of the first buildings built, and one of the first projects that we started was a measurement of the properties of tritium.

which is a conceivably important part of the thermonuclear program, and one of the first buildings built at Los Alamos had as its purpose the handling of materials that we thought might be of interest in thermonuclear work.

We put up a laboratory and a lot of houses which were hopelessly inadequate to our future needs, but at least did get us started. The real problem, of course, had nothing to do with that. It had to do with persuading people to come there. I think it true that at that time among scientists engaged in immediate military problems, radar, underwater sound, ordnance, and so on, the name of the uranium project was not good, because work had been going on for a number of years without very much sense of direction. There was great fear that this was a boondoggle, which would in fact have nothing to do with the war we were fighting Very, very few people turned us down coming to Los Alamos, but this was work for everyone. I think it was perhaps most work for me. I got a large group from Princeton, many people from the chemistry group in Berkeley, where we recruited the key chemical personnel. A group from Stanford. I won't bore you with the details of this. But it took from perhaps October or November 1942 until March of 1943 to get the rudiments of a laboratory. We stole a cyclotron from Harvard, some accelerators from Wisconsin. Everybody arrived with trunkloads of junk and equipment, and in this way we were able to be doing experiments—well, I got to Los Alamos toward the end of March, the equipment started coming a few days later, and by June we were finding out things that nobody knew before. That we thought was a fairly good record of speed.

We had a general notion at that time that all the work of the laboratory would be open to all the scientific members of the laboratory. This is a matter which General Groves, I think, concurred in, but which he never entirely liked. In other words, within the laboratory the competent people were supposed to know what the story was. It turned out over and over again this was a wise policy. Good ideas came from places that you would not have expected. Enthusiasm and understanding could be generated because people knew what it was all about.

On the other hand, we communicated very sparingly and through quite restricted external channels with other parts of the Manhattan District, the places that were making the materials, and the other laboratories, and I should say not at all or almost not at all with any other military research establishments, except those from whom we needed gear. We had some really fantastic security provisions. They were not in the end effective as we know. Families were supposed to come with their husbands if they wanted to, but they were not allowed to leave. We did have to let a couple of people leave the project, but the onus of doing this was very great and the pressure against it very great. We had all our phone calls monitored. It was illegal to mail a letter except in the authorized drops and ingoing and outgoing mail was censored.

Our names were not known and our drivers licenses were all made out under fictitious or artificial names. The laboratory was guarded within the post and the post was guarded. We went to precautions which did not do the trick, but which looked very formidable at the time.

I had partly the job of devising these idiotic things and partly the job of making them welcome. I engaged in several speeches why these precautions were necessary and desirable. I think I took most people along pretty well so that there was not too much kicking about the security regulations. I think we may have a letter which President Roosevelt wrote to me for the laboratory, I should think, and which gives some people two aspects of it. It was a sort of official statement that the security provisions, however irksome, were justified, and the other was that we better get on the job. We had enemies who might be up to it, and we better beat them to the draw. Shall we simply submit this?

Mr. GARRISON. I would like to read this letter into the record at this point, if I may. This is a letter from President Roosevelt to Dr. Oppenheimer, under date of June 29, 1943:

"MY DEAR DR. OPPENHEIMER: I have recently reviewed with Dr. Bush the highly important and secret program of research, development and manufacture with which you are familiar. I was very glad to hear of the excellent work which is being done in a number of places in this country under the immediate supervision of Gen. L. R. Groves and the general direction of the Committee of which Dr. Bush is Chairman. The successful solution of the problem is of the utmost importance to the national safety, and I am confident that the work will be completed in as short a time as possible as the result of the wholehearted cooperation of all concerned.

"I am writing to you as the leader of one group which is to play a vital role in the months ahead. I know that you and your colleagues are working on a hazardous matter under unusual circumstances. The fact that the outcome of your labors is of such great significance to the Nation requires that this program be even more drastically guarded than other highly secret war developments. I have therefore given directions that every precaution be taken to insure the security of your project and feel sure that those in charge will see that these orders are carried out. You are fully aware of the reasons why your own endeavors and those of your associates must be circumscribed by very special restrictions. Nevertheless, I wish you would express to the scientists assembled with you my deep appreciation of their willingness to undertake the tasks which lie before them in spite of the dangers and the personal sacrifices. I am sure we can rely on their continued wholehearted and unselfish labors. Whatever the enemy may be planning, American science will be equal to the challenge. With this thought in mind, I send this note of confidence and appreciation.

"Though there are other important groups at work, I am writing only to you as the leader of the one which is operating under very special conditions, and to General Groves. While this letter is secret, the contents of it may be disclosed to your associates under a pledge of secrecy.

"Very sincerely yours,

"FRANKLIN D. ROOSEVELT."

By Mr. GARRISON:

Q. This, I gathered, was in connection with your own efforts to impress upon the group the paramount needs of security and the importance of the work they were doing.

A. The importance I think there was very little doubt about. Everybody who was there who was a scientist knew it was important. We had a great deal of trouble with people who were not given information, with technicians, machinists, and so on, who found the conditions of life very disagreeable and no countervailing advantage of being associated with something they understood. But the scientists knew it was important.

Q. You were under a great deal of time pressure, were you? Was there a sense of urgency in the air?

A. My directive, I haven't got it, it is probably at Los Alamos as part of the record, was to lose no day in preparing an atomic bomb. The definition of an atomic bomb was that it should be at least equal to 1,000 tons of TNT in explosive force. This sense of pressure started at the beginning and never let up. I will come in a moment to how it was at the end.

If you want to ask anything about it, please do not hesitate to interrupt me.

Dr. GRAY. Yes. I think, Dr. Oppenheimer, we would prefer for you to go ahead. I do not want to say that no member of the board or counsel will not interrupt, and I think we are free to do so, but I think we would like you to proceed and if there is anything, we will inquire.

The WITNESS. Fine. We started out the job there with two sets of meetings. One was a large meeting——

Dr. GRAY. When was this?

The WITNESS. This would have been April 1943. A large meeting that I called all the people there in and a number of others whom I hoped to lure there, and many of whom were in fact later to come, to discuss the technical program. The other was a review committee that Groves appointed, more or less to find out what we were up to and to see that we were doing and what we were not doing. One of the things that the review committee recommended was that we immediately get into the ordnance problems.

This is something we felt very strongly. We immediately got into large scale chemical and metallurgical problems.

Another thing they recommended was that I not try to do everything myself, but get a personnel director, and some aides so that the place would run a little bit better.

We were building a town at the same time that we were building the laboratory. The program of the laboratory emerged from the technical meetings, and so did many of the people who were later to come there and play prominent parts. Some of them are probably going to appear before you. Bethe, an enormous, robust, and talented theoretical physicist; Admiral Parsons, who was the head of the Ordnance Division and is now gone. Fermi, who came in rather late and became an associate director and who among other things was in charge of those activities of the laboratory which were directly not relevant to the atomic bomb but looked further ahead. Bacher, who was in charge of one aspect of the physics of the bomb, and who will appear later. Cyril Smith, Zacharias, Hartley Rowe, who after he got back from General Eisenhower in Normandy landings advised us on engineering problems and helped set up the Sandia laboratory which has played such a large part since that time.

By Mr. GARRISON:
Q. That is Mr. Rowe of the United Fruit Co.?
A. Yes. Norman Ramsey, who was Parson's deputy. I mentioned some of the names of people who will appear here probably.

My job, I don't think too much should be made of it. It was the job of being sure that people understood and that the decisions were properly made, and there were many not easy decisions. We did this through a system of groups, divisions and coordinating councils and a steering committee which finally made the determination of laboratory policy. Sometimes on trivial things like on did we need another housing development, sometimes on very serious things which if made wrong would in fact have prevented our doing the job. We had quite complex relations in which Admiral Parsons was very helpful with the military services who in the end had to deliver this thing, and had to train for delivering it, and had to be sure that they knew all about it. We had to agree with them about the hardware, and be sure that the hardware we were developing would be usable by airmen when they were actually involved in it.

We had the problem of relations with the British. Bacher and I were called on to discuss with Chadwick and Peierls, the state of the British program and where a British mission was established at Los Alamos under the leadership of Chadwick, who is very, very famous and very forthright British scientist, a mission of some 20 people, extremely good. We had the problem of relations with the laboratories and plants that were providing us with military, the question of the specifications of the material and who was to do what. We had the normal administrative problems of a job that was quite unfamiliar, not as dangerous as President Roosevelt's letter indicated, but still capable of great danger as accidents occurring shortly after the war showed. It was very new and terribly exciting.

We had the job of keeping this rapid expansion and with the very end large group of brilliant individualistic and talented people in harmony and pulling on the same team. We had people there who were refugees from Germany and Italy. We had Englishmen, who had lots of Americans. It was in a funny way an international effort.

I need to say that it was not an international effort including Iron Curtain countries. I guess in those days there was only one iron curtain country. In a visit during the summer of 1943, Colonel Lansdale, head of Manhattan District security, in a talk, I think, to the key personnel of the laboratory, made it very clear how great weight the Government attached to maintaining this operation secure against Russian espionage or Russian intelligence.

Q. As the work progressed, you began to get goals and deadlines, I suppose, against which to produce the bomb, if you could?
A. The deadline never changed. It was as soon as possible. This depends on when we were ready, when the stuff was ready, and how much stuff we needed.
Q. Wasn't there a particular effort to get it done before the Potsdam Conference?
A. Yes, that was of course quite late. After the collapse of Germany, we understood that it was important to get this ready for the war in Japan. We were told that it would be very important—I was told I guess by Mr. Stimson—that it would be very important to know the state of affairs before the meeting at Potsdam at which the future conduct of the war in the Far East would be discussed.
Q. Discussed with the Russians?
A. I don't want to overstate that. It was my understanding, and on the morning of July 16, I think Dr. Bush told me, that it was the intention of the United States statesmen who went to Potsdam to say something about this to the Russians. I never knew how much. Mr. Stimson explained later that he had planned to say a good deal more than what was said, but when they saw what the Russians looked like and how it felt, he didn't know whether it was a good idea. The historical record as it is published indicates that the President said no more than we had a new weapon which we planned to use in Japan, and it was very powerful. I believe we were under incredible pressure to get it done before the Potsdam meeting and Groves and I bickered for a couple of days.

But in actual time it has been done enough times. There have been enough lurid news stories about that first test so that I need not repeat what it was like. In other context it should be said that it was as successful as we had any reason to hope, and I believe we got the job done as fast as we could. That is what we were told to do.

Mr. GARRISON. At this point I would like to read into the record a letter from General Groves——

Mr. ROBB. May I inquire, Mr. Garrison, these are copies, but you have the originals available?

Mr. GARRISON. We have the originals available and we would be very glad to show them to you.

Mr. ROBB. Thank you.

Mr. GARRISON. This is the letter of July 19, 1945, from General Groves and Dr. Oppenheimer reading as follows: [Reading:]

"Since I returned to Washington I have done little else but think about and talk about the truly magnificent results of the test conducted at Trinity last Monday morning."

Trinity was the code name for the place.

The WITNESS. Yes.

Mr. GARRISON (reading): "As time goes on and the test begins to take on its true perspective, I appreciate more and more the outstanding performance of you and your people in making the test so successful.

"General Farrell and I have discussed the project in all its many phases and have reviewed it from every possible angle. We both feel that the job is a high-water mark of scientific and engineering performance. Your leadership and skill and the loyal and able performance of all your subordinates made it possible.

"An immediate report was cabled to the Secretary of War on Monday on the great performance."

That would be to Potsdam, I take it?

The WITNESS. Yes.

Mr. GARRISON (reading): "He promptly cabled back heartiest congratulations to all concerned. This morning a fuller written report was sent to him by special courier and he should have our impressions of the test by the time you get this letter. I know that the President, the Secretaries of State and War and General Marshall who are so importantly engaged at Potsdam now will be as tremendously impressed as we were by the results of the test.

"I hope you will show or read the suitable parts of this letter to the men who did so much to make the job go so well and that you will extend to them my grateful thanks for a job well done.

"Again, with deepest thanks and every good wish for the continued success of our great project from both General Farrell and myself, I am,

Sincerely yours,

L. R. GROVES,
Major General, USA."

The WITNESS. Now, there are a few points I might make about this period. After the test but before the use of the bombs in Japan, I had a meeting with General Groves in Chicago to get some last minute arrangements fixed for the combat use of the weapon. I asked him at that time, how do you feel about this super—the super was our code name for what we then thought of the hydrogen bomb, and we don't know any more than we did when he came up, there was a little work but very inconclusive. As a matter of fact, the decisive measurements on the behavior of tritium were on my desk when I got home——

Dr. EVANS. What, sir?

The WITNESS. The decisive measurements on the tritium—these are declassified now, as you know—were on my desk when I got back from Trinity, General Groves was unclear whether his mandate and therefore mine extended to fiddling with this next project. I so reported to the people in the laboratory, who were thinking about it.

The second point I would not think to mention except that Mr. Garrison has asked me and that is whether there was any change in tempo after the war against Germany ended. There was, but it was upward. It was upward simply because we were still more frantic to have the job done and wanted to have it done so that if needed, it would be available.

In any case, we wanted to have it done before the war was over, and nothing much could be done. I don't think there was any time where we worked harder at the speedup than in the period after the German surrender and the actual combat use of the bomb.

The third thing is that I did suggest to General Groves some changes in bomb design which would make more efficient use of the material; and they have long since been done, of course. He turned them down as jeopardizing the promptness of the availability of bombs. He and I may not entirely agree about how long a delay would have been involved, but the very fact that any delay was involved was unacceptable.

Finally, there was, of course, a great deal of discussion—and I will return to the formal aspects of that—about the desirability of using the bombs in Japan. I think the hotbed of this discussion was in Chicago rather than in Los Alamos. At Los Alamos I heard very little talk about it. We always assumed, if they were needed, they would be used. But there were places where people said for the future of the world it would be better not to use them.

This problem was referred to me in a capacity different than director of Los Alamos. We did everything we could to get them out there and as fast and smooth as possible.

There was, however, at Los Alamos a change in the feel of people. I am talking vaguely because this is a community now of seven or eight thousand people, of whom maybe 1,000 or more are scientists and very close to each other, talking all the time. This was partly a war measure, but it was also something that was here to stay. There was a great sense of uncertainty and anxiety about what should be done about it.

The generation of that kind of public—of a concern very similar to the public concern—that followed Hiroshima and one natural outgrowth of which was our abortive effort to establish quite a new relation among nations in the control of atomic energy; that was not something that had its roots very far back; it started toward the end when the war was about over.

Hiroshima was, of course, very successful, partly for reasons unanticipated by us. We had been over the targets with a committee that was sent out to consult us and to consider them, and the targets that were bombed were among the list that seemed bright to us.

The Secretary of War deleted one target, and I have always been glad he did. That was the unbombed and culture capital of Japan, Kyoto. He struck that off. The two that were hit were among the targets selected. We sent a mission on out from Los Alamos to assemble, test the bombs on Tinian, and to fly with the B-29's that went out over the targets, and also to go in as soon as they could get clearance from General MacArthur.

That mission was under General Farrell, who might appear—I am not sure he can—to see what mess we made of those two towns.

When the war was over we came east, Dr. Bacher, Dr. Rabi, and I together. There was a rumor of some wonderful method of getting energy for nothing that the General Electric research people had discovered. Groves thought I ought to have a look at it. It turned out to be nonsense. In the course of this visit I talked with General Groves. There were at least two points that I ought to report.

One was that I told him, that as I had earlier suggested in outlining what the future work of the laboratory would be, I thought I should not continue as director. I was the director of an emergency. This was going to be something different, and I would not be the right person to preside over the change or the new effort.

In addition, there was not much left in me at the moment. We talked about my successor. This was not a trivial problem. It took a while. I talked to Commander Bradbury; I talked to General Groves. Everyone was pleased with that, and I think it was a very fine selection. I was therefore free to resign and did mid-October, October 16, or something like that.

The other thing is that General Groves told me very briefly that he had been told by Governor Byrnes—Justice Byrnes, I guess—who was then I think representing the President on the Secretary of War's interim committee, that with things as they were the work at Los Alamos ought to continue, but this did not apply to the super or didn't think this applied to the super.

I don't know whether I left out some things that would be illuminating. This is not a very vital part of our story from the point of view of the case, and I would like to get on.

Mr. GARRISON. I happen to have here, Mr. Chairman, the original of the United States of America Medal for Merit awarded to Dr. Oppenheimer, and I would just like to read it. It would only take a second. The citation is signed by President Truman to Dr. Oppenheimer "for exceptionally meritorious conduct in the performance of outstanding service to the War Department, in brilliant accomplishments involving great responsibility and scientific distinction in connection with the development of the greatest military weapon of all time, the atomic bomb. As director of the atomic bomb project laboratory in New Mexico, his initiative and resourcefulness and his unswerving devotion to duty have contributed immeasurably to the successful attainment of the objective. Dr. Oppenheimer's accomplishments reflect great credit upon himself and upon the military service." Signed, Harry Truman.

I am sorry I didn't have a copy for you to follow.

Mr. ROBB. That is already in the file.

Mr. GARRISON. This is January 12, 1946.

Dr. GRAY. You wish to read that in the record?

Mr. GARRISON. Yes. I think that is enough for the war period. I think we will now swing into the postwar problems that arose immediately out of the war and the way in which they involved Dr. Oppenheimer in the service of the country.

By Mr. GARRISON:

Q. You went back to Berkeley, of course, or you went back to Pasadena after you left Los Alamos.

A. We are not quite so far.

Q. What did you want to say previous?

A. In May I was asked to serve on the interim committee which Mr. Stimson set up.

Q. This prevented your leaving.

A. Yes; this was before I left Los Alamos. Lawrence, Fermi, and Arthur Compton were the other members of this panel. We met with the interim committee I think on the 1st of June—I am not certain—of 1945 for a very prolonged discussion which was attended by all members of the committee, all members of the panel, and for most of the time General Marshall.

Apart from trying to make as vivid as we could the novelty, the variety, and the dynamic quality of this field, which we thought very important to get across, that this was not a finished job and there was a heck of a lot we didn't know, much of the discussion resolved around the question raised by Secretary Stimson as to whether there was any hope at all of using this development to get less barbarous relations with the Russians.

The other two assignments which the panel had—one was quite slight. We were asked to comment on whether the bomb should be used. I think the reason we were asked for that comment was because a petition had been sent in from a very distinguished and thoughtful group of scientists: "No, it should not be used." It would be better for everything that they should not. We didn't know beans about the military situation in Japan. We didn't know whether they could be caused to surrender by other means or whether the invasion was really inevitable. But in back of our minds was the notion that the invasion was inevitable because we had been told that. I have not been able to review this document, but what it said I think is characteristic of how technical people should answer questions.

We said that we didn't think that being scientists especially qualified us as to how to answer this question of how the bombs should be used or not; opinion was divided among us as it would be among other people if they knew about it. We thought the two overriding considerations were the saving of lives in the war and the effect of our actions on the stability, on our strength and the stability of the postwar world. We did say that we did not think exploding one of these things as a firecracker over a desert was likely to be very impressive. This was before we had actually done that. The destruction on the desert is zero, as I think Mr. Gray may be able to remember. He had seen all these tests.

The other assignment brought me and the other members of the panel to Washington. They asked us to produce a prospectus about what needed to be done in atomic energy. We wrote a great big book. We called in all sorts of people—Allison—well, there is a list somewhere about; I won't try to remember—Allison, Rabi, Lawrence, Thomas—and tried to give as good an account of where the problem stood as we could.

This included the military applications. There was a special chapter on the thermonuclear problem written by Fermi, on the delivery problem, making weapons that were less clumsy than the ones we had, on the use of atomic energy for power and its use for propulsion, its use for instruments of scientific investigation, neutrons and radioactive tracers. Anyway, it was a fairly big fat book.

I suppose it is from that that the remark is quoted on the feasibility of the super that is ascribed to me in 1945. In any case that would have been my summary view of it at that time.

In connection with writing this report, I became involved in other activities here in Washington. The War Department was anxious to get legislation passed so that the atomic energy enterprise was not part of its budget and responsibility. General Marshall talked to me about it; and Mr. Harrison—who was Mr. Stimson's aide—talked to me about it; and others, as well. The matter seemed to be a bit stuck because, on the one hand, it was difficult to present legislation on the domestic control of atomic energy without saying whether you were going to do anything toward seeking an international control of some kind.

On the other hand, the State Department was not quite clear what it wanted to say about this for very understandable reasons. Therefore, I was asked to consult with Mr. Acheson and eventually with Mr. Byrnes, and the purposes of my visits were double. One was to explain how important it was for the survival of any atomic energy enterprise at all that there be some legislation and soon. That the people who were working on the job had some assurance of where they were going. And the second was to urge that, insofar as it could be with safety done, we explore the possibility of international control.

I did that, as I say, with Mr. Acheson and Mr. Byrnes. Then I went back to Los Alamos. We turned in our report from the interim committee. I was called back to testify on a matter not directly connected with the atom, and that was a pair of bills to set up a National Science Foundation by the joint committee called the Kilgore-Magnuson committee. I did so testify, and they asked me what the relation between the atomic energy undertaking and the National Science Foundation should be, and I think this is the first time I had public occasion to talk about the importance of unplanned and unprogramed scientific work, the enormous importance of training scientists, the importance of freedom in scientific world as opposed to the need for programatic and concentrated work on practical problems.

The next day I went up before Representative May's committee which was considering the May-Johnson bill. The May-Johnson bill was the outgrowth of the effort to get legislation adopted. The President had stated that he would seek international control, first talk with our allies, the British and Canada, and other nations, and he was considering a measure which would at least put our domestic house in order. This bill had been introduced in the House and Senate simultaneously. Hearings were being held on it in the House. Most scientists, and I think all the liberal press, was very mad at this bill. It sounded repressive. It had severe penalties for revealing information. It gave the Commission that was to handle the atom rather wide and rather undefined powers. I had a lot of confidence in the people who had drafted it and the people who would be administering it, and I testified in favor of it as an interim measure because I thought the sooner this got into organized hands the better chance that places like Oak Ridge and Los Alamos would be taken good care of, and after a year there would be plenty of chance to amend the legislation with whatever one had learned in between.

The newspaper PM had on the basis of my testimony the day before made one of their cartoons, Hats Off, on the basis of my testimony on this bill put in another cartoon Hats On. They didn't like it.

After that I went with Patterson—I think before this Stimson had left Washington. I saw him on the last day he was in office here, and he had indicated to me on that day that he thought it right and necessary to see if we could work out an international agreement on the regulation of the atom—I went with Patterson to talk to President Truman about it. He told me that he had invited King and Attlee to come and they would shortly be getting into it. By this time I moved to Pasadena.

I took up a job there as professor of physics. I did actually give a course, but it is obscure to me how I gave it now. The intention was to make that quite a full-time job and settle in Pasadena at least for that year. I still had the appointments at the University of California at Berkeley and the California Institute of Technology at Pasadena. I was called away from Pasadena to come back to Washington and testify before McMahon's committee. I was sort of reluctant to do it on the ground that I hoped to stay put. But I came back. He kept me over for several days to give both public testimony and secret testimony.

While that was going on, I was brought into conferences in the State Department——

By Mr. GARRISON:
Q. That committee of McMahon's was for wnat purpose?
A. The special Senate committee was trying to study the atom and draft legislation which was better than the May-Johnson bill—the committee that led to the McMahon Act under which we are operating even today. I was called into the State Department in the preliminary discussions of what the mission that was going over to Moscow might talk to the Russians about. The United States, England and Canada had issued a very resounding declaration about the need for international control of atomic energy consistent with safeguards, and the question was, What do we do next?

We discussed this at some length. I got the impression that we didn't have a very well thought through notion of what international control was or what we could say to the Russians, and I think it ended by our simply asking them to subscribe to the three-power declaration.

It is, I think, partly because of that that my interest in and to some extent my knowledge about the problem became known to people in the Department, and the result was that I was called back shortly after the opening of the next year for very serious work on the problem of international control.

I ought to mention one thing that occurred in Pasadena at that time. General Groves had this immense mass of technical information developed during the war. All of it was secret. Some was about lubricants, some about valves, and some about bombs. He wanted to get started on the job of sorting it out. What should be made public, what should by all means not be made public, and what should be worried about.

He appointed Dr. Richard Tolman in Pasadena as the chairman of the committee, and I was a member of it. You have a list of the other members. I think Lawrence and Urey were on it, to begin this process of sorting it out. We divided things into three classes: Those which were manifestly useful for science and the arts, and seemed to have no security value of any kind; those that were obviously connected with the military aspects of atomic energy and which should not be declassified unless there were international safeguards, an intermediate class of tough problems where we thought it would be dependent on the political assessment of the state of the enemy—it was not enemy in those days—of Soviet efforts and the prospects of conflict in a short time.

Our general philosophy was that if we are going to have a long, long period when we are not going to use these things and don't need these things, the more that is open, the better American technology and science will prosper. If the time is kind of short, then the advantages of our secretly developed information will be considerable.

Dr. GRAY. You say Dr. Tolman was chairman of this committee?
The WITNESS. That is right.
Dr. GRAY. What was this committee called?
The WITNESS. I have it down as declassification committee, but I am not sure. May I at this point interpolate that the biographical material that you were given late this morning was compiled by a very intelligent secretary. I did check with her on 1 or 2 things I remember. The records are good only since we came to the Institute. I wouldn't have you think that they are admirable records of the years during the war, because there just are no such things. It is the best we could do for your convenience.

By Mr. GARRISON:
Q. Then this takes us into the beginning of the plans for international control of atomic energy.
A. Yes.
Q. And the preliminary discussions within the Government about that?
A. I have talked about some of the preliminary discussions. I believe the background for the Lilienthal panel was the following. The Russians didn't want to talk about the atom at Moscow, but they did agree to this three-power declaration, and they threw the thing into the United Nations. There there was another resounding declaration and two Senators, Vandenberg and Connally, were disturbed that this might leak secrets, that we might not be adequately protected.

The Secretary of State said "No, there will be safeguards." When he got home he set up a committee under the chairmanship of Mr. Acheson, with General Groves, Dr. Bush, Dr. Conant, and Jack McCloy on it, and they were supposed to devise the safeguards. They started thinking about the safeguards and in Mr. Acheson's words, they soon found they were trying to devise a cowcatcher without ever having seen a locomotive, because nobody knew what was meant by international control. What sort of things would be, who would do what and what would the rules be. They appointed a panel of which Lilienthal was chairman, the membership you have in full there, Mr. Barnard was on it, and Mr. Winne was on it, and we were supposed to make a sketch of international control which would be sureful in coping with the atom and which would, if possible, be a step in carrying out that avowed intent of our action, namely, so to alter the relations between nations that war itself would be a lot less likely.

This was a pretty ambitious thing with all that in mind. It did not work, but people were talking that way in those days, and I must say that I was one of those who talked that way very freely.

Q. Did you about this time prepare a memorandum to Mr. Lilienthal containing your ideas?

A. The way it worked is that we met and in the first few weeks, a week or two, my job was that of teacher. I would get back at the blackboard and say you can make energy this way in a periodic table, and that way and that way. This is the way bombs are made and reactors are made. I gave in other words a course. I gave parts of this course also to Mr. Acheson and Mr. McCloy at night informally. Then we listened to parts of it that I didn't know anything about, where the raw materials were, and what kind of headache that was. Then everybody was kind of depressed the way people are about the atom, and we decided to take a recess.

Mr. Lilienthal asked everybody to write him a note if they had any ideas as to what might work and asked me in particular to write a primer on the subject so that people could have the facts at their disposal. I stayed in Washington and did both of these. I think the note is the thing to which you refer.

Q. Yes. I show you this document entitled, "Memorandum of February 2, 1946." It should be entitled, "Extract From Memorandum of February 2, 1946, From Dr. J. Robert Oppenheimer to David E. Lilienthal, Chairman of the Board of Consultants to the Secretary of State's Committee on Atomic Energy." This extract has been copied, has it not, from a carbon copy in your files from a memorandum which you gave to Mr. Lilienthal at the time?

A. So you tell me. There is no reason why the whole memorandum should not be available, but it is rather long.

Mr. GARRISON. I might say to the board that we will from time to time as we go along be offering you extracts from writings and articles and addresses of Dr. Oppenheimer. The full text of each of those will be available to the board, and the only reason for taking excerpts from them is to save time, and because they have a certain relevance to Dr. Oppenheimer's views at the time with respect to our foreign relations. This is an example of what we shall be doing. I would just like to read this, because it is quite a significant document. [Reading:]

"It is probable that the main desire of our Government is the achievement of safety and protection against the threat of atomic warfare. Even if it were possible to achieve this without considering such positive features as the extension of knowledge and its application to constructive purposes, it might be argued that such a course should not be followed. It is my belief that quite apart from its desirability, the provision for constructive development of the field of atomic energy will turn out to be essential for the operation of any system of safeguards. * * * In particular, it has become clear to us that not only politically but scientifically and technically as well, the field of atomic energy has witnessed very rapid change and very rapid progress. I believe that this will be the case in the future, too, and that no organization and no proposal can be effective which does not have a flexibility adequate to these changes. I further believe that any proposed organization must itself reflect the changing character of the problem and the constructive purposes which are a complement to control. * * *

"Almost everyone has, at one stage or another in his acquaintance with this problem, considered prohibiting further work on atomic energy, and devising a system of inspection adequate to insure that this prohibition is carried out. It is not only that this proposal would make impossible the application of existing knowledge to constructive ends; it would be so contrary to the human patterns of exploration and exploitation that no agreement entered into by heads of state could command the interest or the cooperation of the people of the world. An apparently less radical solution would be the separation of the functions and development and of control according to which the only responsibility of an international authority would be the inspection of work carried out under a purely national or private initiative, and the possible prohibition of some of this work. The negative approach to the problem of control would leave the inspecting agency with inadequate insight, both into the technical state of the subject, and into its motivation and the organic characteristics of its growth. * * *

"Against this background of the difficulties of control as an isolated and negative function, I have thought it essential at least to consider combining the functions of development and of control in a single agency. It is fairly certain that there are now, and will increasingly be, activities having to do with atomic energy which are not vital to control and which, for human, or organizational, or political reasons should not be included among the functions of the controlling authority; but there are certainly several such functions which, as matters now appear, should be so included among them: The development of raw materials, the exploration of atomic weapons, and the application, in its more dangerous forms, of atomic energy to power and technology. * * *"

Mr. ROBB. Do you have the original of that, Mr. Garrison, so that we can see the end of these sentences?

The WITNESS. We have only my own carbon of it, but we have it complete.

Mr. ROBB. That is what I mean.

The WITNESS. I am not ashamed of any aspect of the memorandum.

Mr. ROBB. I was not suggesting that you are, Doctor.

The WITNESS. I didn't want to burden you with it.

Dr. GRAY. May I ask a question there. Is your request there for the purposes of making the entire memorandum part of the record?

Mr. ROBB. Oh, no.

Mr. GARRISON. Quite probably we should have had it ready, and we will have it ready in a moment.

The WITNESS. Shall we save time by going on and we will have it as soon as it is available.

Mr. ROBB. Yes.

By Mr. GARRISON:

Q. Would you care to make any comment between the relationship of the ideas you expressed in this memorandum and the central philosophy of the Acheson-Lilienthal report as it finally emerged?

A. The comment seems to come inappropriately from me. I think they are identical. I think this is the heart of United States policy. I will say more. I think that any attempt at that time to establish control along these lines would, if accepted by the Soviets, have so altered their whole system and so altered their whole relations with the Western World that the threat which has been building up year after year since could not have existed. I think that no one at that time could with much confidence believe that they would accept these proposals. I think it was important to put them forward, and it was also important not to express too much doubt that they might be accepted.

In the U. N. we hammered away at this line, but there are some intervening complications.

Q. The central idea of this scheme, I take it, was that there should be not merely inspection of atomic-energy production and atomic-energy armaments, but actual ownership and control of that whole process by an international agency, so that purely national development of these atomic-energy programs would be ruled out, and that would have entailed in Russia as in other countries the actual ownership of productive facilities in that land, as in others, by an international agency, is that correctly stated?

A. That is correctly stated. I think it is part of the story. It would have meant that the Russian Government gave up control over things going on involving their citizens on their territory. It would have permitted free intercourse between Russian nationals and people of the rest of the world. It would have meant that there could be no Iron Curtain. How radical it was I may indicate by a comment that came much later. General Ridgway was on the Military Staff Committee at the U. N. at the time when I was on Mr. Brook's staff, and our people had looked at this proposal and said if it were to go through, they would recommend that all secret military establishments be abolished. This was quite a slug.

Q. Then work went forward on the report?

A. We worked very hard on it. I think I should say this, I have been on many committees. The last thing I want to persuade you is that I was the "big cheese" on these committees. I did have this idea. It does derive from me. But in other ways, the other members of the committee had similar ideas.

For instance, Dr. Winne and Dr. Thomas said when they heard about the raw material situation, we ought to get rid of the scramble for uranium. If we don't work together on this we will never catch up with the control problem. So each relying on his experience came to somewhat similar conclusions.

I think the implication that I am responsible and alone responsible for the report is wrong. I am responsible for writing a great deal of it; not all of it, but perhaps a half of it. It was, I think, persuasive document which both here and abroad spoke well of the generosity and prudence and sense of America.

Mr. GARRISON. I have here, Mr. Chairman, a copy of the Acheson-Lilienthal report, entitled "Report on the International Control of Atomic Energy" in case any members of the board would like to look at it now or later. I would like at this time to just read into the record three very short extracts from it.

Dr. GRAY. What is the date of that report?

Mr. GARRISON. March 16, 1946. It was prepared for the Secretary of State's Committee on Atomic Energy by a board of consultants, Chester I. Barnard, Dr. J. R. Oppenheimer, Dr. Charles A. Thomas, Harry A. Winne, David Lilienthal, chairman. I can put the page references into these excerpts.

Dr. GRAY. I don't think that is necessary.

Mr. GARRISON (reading): "International control implies an acceptance from the outset of the fact that our monopoly cannot last." (p. 53).

"It is essential that a workable system of safeguards remove from individual nations or their citizens the legal right to engage in certain well-defined activities in respect to atomic energy which we believe will be generally agreed to be intrinsically dangerous because they are or could be made steps in the production of atomic bombs." (p. 22).

"It therefore becomes absolutely essential that any international agency seeking to safeguard the security of the world against warlike uses of atomic energy should be in the very forefront of technical competence in this field. If the international agency is simply a police activity for only negative and repressive functions, inevitably and within a very short period of time the enforcement agency will not know enough to be able to recognize new elements of danger, new possibilities of evasion, or the beginning of a course of development having dangerous and warlike ends in view * * *" (p. 23.)

I think those three paragraphs are significant of the central thought of the report. I am sure if the board will at its leisure re-read again the memorandum to Mr. Lilienthal that Dr. Oppenheimer wrote on February 2, 1946, you will see that the same thought appears in that memorandum as appears in the final report.

Dr. GRAY. For the purposes of the record, these are not paragraphs which appear consecutively in this document. I don't know. I am asking for information. Are they separated? Is my question clear?

Mr. GARRISON. Yes, it is, indeed.

Mr. ECKER. I believe they do not appear consecutively where the quotes are closed.

Mr. GARRISON. Suppose we at the end of the hour put the page references in. They should be in.

Dr. GRAY. I think that is satisfactory.

By Mr. Garrison:

Q. Do you want to go now to your testimony before the McMahon committee?
A. I will go quickly. When the report was done, we had several conferences with Acheson's committee. In fact, the last and rather delicate chapter of the report which I largely wrote we did not originally have in. But the committee thought that some description of how you might get from where we were then to where we thought we would like to be was called for. This had the disadvantage that it tended to disclose some aspects of our negotiating position and made the publication of the report perhaps less wise than it would otherwise have been.

I went home and I was very soon called back for two reasons. The report was out and the newspapers greatly distorted and exaggerated the virtues of denaturing. We had said you could fix up fissionable material so it was not immediately useable in bombs. This was the headline. Probably when we wrote it we invited that distortion. In any case it occurred.

I came back partly to attend the meeting to get an agree statement out of a lot of echnical people as to what the truth was and partly to testify before McMahon's committee. I remember Senator Vandenberg saying "I like this." I think it was largely in that spirit that we went on with it. Baruch had been appointed to represent the United States in these negotiations and this was announced, I think, just about the time the report was done. I went back to California again, but before long I came back to talk with Mr. Baruch and Hancock and Eberstadt and tell them a little bit about how we had gone about it.

I then gave some lectures at Cornell on a rather broad subject, but one of the lectures was about the international control of atomic energy. It was reprinted rather widely, and was an advocacy of the position that we had adopted. I gave another talk the next day in Pittsburgh which was another job of advocacy of this set of proposals. It was reprinted in the New York Times. Mr. Baruch told me that I had scooped his speech that he was going to make at the opening of the U. N. That was not true. But it did have in it one element which was missing from the Lilienthal report and that was the remark that this business we were talking about was incompatible with a veto. You could not run a job like this and have Yugoslavia or Crete decide that they didn't like what was going on and stop it. This was the veto on operations; it was not the veto on sanctions, because nothing we discussed had to do with sanctions. That was the second of Mr. Baruch's points.

We met in Blair Lee House the next day and had a long discussion with Mr. Baruch and his staff. He asked me what we had done wrong in the report. I remember mentioning a few points, among them the failure to make clear the relation of what we proposed to the veto, and the invitation that we gave to the press and the public to exaggerate the value of the denaturing.

Very shortly thereafter I agreed to serve as one of the consultants to Mr. Baruch in preparation for and in the conduct of the U. N. negotiations. The senior consultant was Dr. Richard Tolman, whom I mentioned before. I think Dr. Robert Bacher and I were the most active next to Dr. Tolman, but Compton and Thomas and one or two other people were also involved.

We spent through the summer with him and with his staff, and tried to help. The main job we did was to get an agreed paper out of the International Commission that international control was technically feasible. This was something you could do. The Russian delegate, I think it was Gromyko, balked at signing this, but finally the Russians agreed that international control was technically feasible. I think it is the last time we have agreed with them on anything in the U. N., and certainly anything having to with the atom.

Q. They agreed that it was technically feasible, but the report did not say it was politically feasible.

A. They attacked the proposal. They attacked both the aspects which were prominent in the Acheson and Lilienthal thing, and that which Mr. Baruch added having to do with sanctions. I think they mostly attacked the main point, that this would have been a terrible invasion of their privacy, and they were not going to have it. This attack continued for years.

Dr. GRAY. May I interrupt you there, Dr. Oppenheimer. I want to know whether you want a break. You have been talking rather constantly.

Mr. GARRISON. I think he will be getting a break because I will be reading a few documents into the record, but I think the Board would like a break.

Dr. GRAY. I would like to see the point at which we will stop the hearing this afternoon.

(Discussion off the record.)

Dr. GRAY. Suppose we take a recess for 5 minutes.

(Brief recess.)

Dr. GRAY. I think we might as well proceed, Dr. Oppenheimer.

The WITNESS. After the summer of work with Mr. Baruch, it became difficult even for a dedicated optimist to think that anything would come of the negotiations in the sense of a real agreement. It was hard to believe that before it started, and the nature of the Soviet conduct, not only the kind of objections they made, but the nature of their dealings was extremely revealing to anyone who saw it for the first time.

In fact, it is worth recollecting that the Acheson-Lilienthal board was working in early 1946 at precisely the time when Stalin made the speech about their encirclement and their need to keep their guard up and to rearm.

I revert to the fact that it was healthy for us to attempt this, but that it should not be read into that time that we were going around in a mood of high optimism. I have seldom been as gloomy in my life; that even includes today.

Nevertheless there was a job to do and I continued to do it. The job was establishing to our friends in the U. N., to the governments and so far as possible to the officials and to the people of our friendly nations, that what we had put up made sense and was not a bluff and was not propaganda and that it had merit.

I don't know how important that job was but I stayed with the Baruch enterprise until he resigned, and then I was asked to serve as adviser to General Osborn, who took over in the spring of 1947. Osborn asked me to come up and spend some time with him talking it over. On the way I stopped at the State Department and Mr. Acheson showed me the President's speech on the Truman Doctrine. He wanted me to be quite clear that we were entering an adversary relationship with the Soviet, and whatever we did in the atomic talk we should bear that in mind.

I worked with Mr. Osborn intensively at first. I testified before the U. N. AEC, or one of its committees, on how you would go about on the international cooperative beneficial uses of atomic energy.

I continued to consult Mr. Osborn in company with Dr. Conant and General Farrell and General Groves, and maybe General Nichols, as long as the problem of atomic control was still a matter of debate in the United Nations until it was engulfed in the wider but also hopeless job of disarmament.

I would like at this time to say only two things. One is that the negative view of the possibility of any agreed solution with the Russians which came on us all then, as it has not gotten any different but gotten deeper, and I would like to refer to that again in connection with the work we did in 1952 for the State Department on the regulation of armaments, where the context was somewhat different.

The second is to say that incidental good did come of this effort. I think that, insofar as people paid attention to it, the United States proposals were recognized as indeed sensible, and we got lots of credit for them.

I ran into the representatives of the French and English and some other countries, too—however, primarily the French and English—and, though always keeping my own Government informed, as usual, I was, I think, able to do some useful jobs on the side. I talked to the French officials as well as the French scientists about the desirability of their building up a real scientific life in France and about the undesirability of their getting into any rivalry with us on the atomic business.

I said I thought we would be able to help and have more fellowships and laboratories, and we would get into lots of trouble if they were getting into sensitive areas from the point of view of security. I think I always reported and checked with the officials of AEC or the State Department when any such conversations occurred.

With the United Kingdom it was quite a different thing. There we had had an intimate partnership, as you read in the newspapers and know anyway, in the last few years and during the war. There were some excluded areas, but all the things I was concerned with the British knew about and contributed to.

I visited Europe in the summer of 1948. In the winter of 1949 we undertook to see what could be done to restore this partnership. You will hear testimony about this from other people. The problem kept arising because of raw-materials allocation, because of the dissatisfaction of the British, and because of the double problem that it was nonsense to have their best people duplicating what we were doing and that there was thought to be and perhaps was a security problem in working with them.

We had a meeting in Princeton for 2 or 3 days that I think was chaired by Mr. William Webster. The Commission was represented by the General Manager and General Counsel. The Military Establishment was represented by General Nichols and General Norstad; the State Department, by Mr. Kennan and Mr. Butler; and the interest at laity, by Dr. Conant and myself.

This was the beginning of an attempt which was abortive but which got quite far along to reuniting the relations between United States, England, and Canada in the atomic-energy business. It was abortive—I had better not say why because I was not in the politics of its abortion. But I have always regretted that failure, and I am not sorry for the efforts I made.

Mr. MORGAN. When was that?

The WITNESS. The meeting was in 1949. I read when I was out west in 1949 of the evening when the President called in the Senators to Blair House when he was leaving, and when they came out of the door the reporters talked to them and were told that the Senators heard something so dreadful that they could not speak about it. What they heard was about the wartime collaboration and that the British knew a lot about atomic bombs and could probably make them if they tried and that they were on the point of trying on their own. This is hearsay testimony, or testimony as to what I read in the papers.

As I say, our relations with the scientists of other countries and some effort to improve what we have learned to call the basis—the cordiality and strength of our alliances—these things did come out of these U. N. meetings. But it was pretty thin fruits compared to the vision of world government and permanent peace which some people had at the time.

I think now there is stuff to read.

By Mr. GARRISON:

Q. Dr. Oppenheimer, I have here a document, called atomic energy as a contemporary problem, by Dr. J. Robert Oppenheimer, presented at the National War College in Washington, September 17, 1947. This is a stenographic transcript of the remarks made by you on that occasion. This came from your files, I take it?

A. That is right.

Mr. GARRISON. I would be glad to hand it to counsel as I read an excerpt from it.

The WITNESS. This may not be published without the permission of the War College. It has no restricted data, but it cannot be published without the permission of the War College.

(Discussion off the record.)

Dr. GRAY. Will you proceed?

Mr. GARRISON. These excerpts are from pages 6 to 8 of that transcript.

Mr. ROBB. I have it.

Mr. GARRISON (reading): "At the same time, I think no one can take with any seriousness the hope or expectation that the Soviet Union will accede, or that it will come closer to acceding, to what is now the majority plan.

That is the United States plan. [Reading:]

"That is not too hard to understand. The cornerstone of our proposal is an institution which requires candidness and great openness in regard to technical realities and policy. It involves the working cooperation between peoples, irrespective of nationality. It involves a maximum effort to abolish national rivalries in the field of atomic energy, and in all dangerous areas of atomic energy it involves a total and genuine international action. It is clear that, even for the United States, proposals of this kind involve a very real renunciation * * *"

Mr. ROBB. Wasn't there an omission at that point?

Mr. GARRISON. There are three dots which I have indicated here, and if there is anything significant in the omission——

Mr. ROBB. No; I have not said there is.

Mr. GARRISON. I have indicated the omissions by dots.

Mr. ROBB. I think for the record it should be indicated.

Mr. GARRISON. Yes; the reporter will so indicate. [Reading:]

"But if, for the United States and the Western European powers, some sacrifices are required by these proposals, the sacrifices, the renunciation, required of Russia are of another order of magnitude; and that is because the proposed pattern of control stands in a very gross conflict to the present patterns of state power in Russia and because the ideological underpinning of that power—namely, the belief in the inevitability of conflict between Russia and the capitalist world, or the allegedly capitalist world—this underpinning, which is most difficult, I suppose, for a government to renounce, would be repudiated by a cooperation as intense or as intimate as is required by our proposals for the control of atomic energy. Thus what we have asked of the Russians is a very far-reaching renunciation and reversal of the basis of their state power and of their state power itself. It does not seem to me likely that we have found inducements or cajolery or threats which together are adequate to make them take this great plunge. That does not mean, I suppose, that this will never happen, but it will almost certainly not happen as a result of the discussions in the United Nations.

"The whole notion of international control presupposes a certain confidence, a confidence which may not be inconsistent with carrying a gun when you sit down to play poker but at least is consistent with sitting down to play poker. In the year and a half since the effort on these problems started, we have found ourselves forced by the Soviet moves and by the changing political situation throughout the world over and over again to take steps which were in essence a repudiation of that confidence; and the Soviet has taken ever more grave steps in repudiation of that confidence. * * * I therefore think that to believe seriously today [1947] that in 6 months, a year, or a year and a half we will have something resembling an ADA [Atomic Development Authority]. The cooperative development of atomic energy involves a kind of schizophrenia which can only lead to very bad political confusion. I even think the worry that one often hears discussed in unofficial, and sometimes official, circles— What would happen if the Russians suddenly reversed their stand, embraced our proposals, and started to work to put them in effect?'—that is an empty worry because it is in the nature of the proposals we have made—a protection afforded by our plans for the United States—that they cannot be implemented in very bad faith, that they presuppose a very large measure of peaceful intention, of cooperation, of confidence and candor before they can get started. I am therefore not very much alarmed that Mr. Gromyko will someday say to Mr. Osborn, 'We finally have understood your proposals and we think they are wonderful. We accept them in full.' I do not think this will happen."

The next excerpt is from an article in Foreign Affairs for January 1948, entitled "International Control of Atomic Energy," by J. Robert Oppenheimer. These are pages 12, 13, and 14 in that article.

Mr. Robb, do you have page 12 there?

Mr. ROBB. Yes.

Mr. GARRISON. This, you will see, is several months after the War College speech which we have just been through. [Reading:]

"Two aspects of this development need to be specially mentioned. One has to do with what may be called the aim of the United States policy—the sketch of our picture of the world as we would like to see it insofar as atomic energy was concerned. Here the principles of internationalization, openness, candor, and the complete absence of secrecy, and the emphasis on cooperative, constructive development, the absence of international rivalry, the absence of legal right for national governments to intervene—these are the pillars on which our policy was built * * * The second aspect of our policy which needs to be mentioned is that, while these proposals were being developed and their soundness explored and understood, the very bases for international cooperation between the United States and the Soviet Union were being eradicated by a revelation of their deep conflicts of interest, the deep and apparently mutual repugnance of their ways of life, and the apparent conviction on the part of the Soviet Union of the inevitability of conflict—and not in ideas alone, but in force. For these reasons the United States has coupled its far-reaching proposals for the future of atomic energy with rather guarded reference to the safeguards required, lest in our transition to the happy state of intenational control we find ourselves at a marked relative disadvantage. Natural and inevitable as these desires are, they nevertheless stand in bleak contradiction to our central proposals for the renunciation of sovereignty, secrecy, and rivalry in the field of atomic energy. Here again it is no doubt idle to ask how this country would have responded had the Soviet Union approached the problem of atomic-energy control in a true spirit of cooperation. Such a situation presupposes those profound changes in all of Soviet policy, which in their reactions upon us would have altered the nature of our political purposes and opened new avenues for establishing international control. * * *

"Questions will naturally arise as to whether limited but nevertheless worthy objectives cannot be achieved in this field. Thus, there is the question of whether agreements to outlaw atomic weapons more like the conventional agreements, supplemented by a more modest apparatus for inspection, may not give us some degree of security. Possibly, when the lines of political hostility were not as sharply drawn as they are now between the Soviet Union and the United States, we might have tried to find an affirmative answer to this question. Were we not dealing with a rival whose normal practices, even in matters having nothing to do with atomic energy, involve secrecy and police control which is the very opposite of the openness that we have advocated—and under suitable assurances offered to adopt—we might believe that less radical steps of internationalization could be adequate. * * * My own view is that only a profound change in the whole orientation of Soviet policy, and a corresponding reorientation of our own, even in matters far from atomic energy, would give substance to the initial high hopes."

By Mr. GARRISON:

Q. Dr. Oppenheimer, here is a letter to you from Mr. Chester Wood, the Secretary of the New York State Bar Association, enclosing a transcript of the remarks that you addressed to a meeting of the judicial section of that association, this being February 1948—the precise day is not clear. This was taken from your files, was it not?

A. It was certainly taken from my files. That is all I can say.

Q. Then you identify the document, I assume, do you not?

A. If I am to make a serious identification, I should see it.

Q. Yes [handing].

Mr. GARRISON. The excerpts which I have taken from that are at pages 7 to 10, inclusive.

The WITNESS. I do identify it.

Mr. GARRISON. Now I would like to read from this address. [Reading:]

"The proposals which the United States made and which are manifestly not going to be accepted were perhaps somewhat more radical even than the people of this country believed, perhaps even than some of the officers of this Government believed. The idea was not that one would fasten a scheme of control onto an otherwise unaltered pattern of the relations between sovereign states. The relation was rather that here appeared to be an opportunity, very pressing in its urgency and very rich in its technical patterns, for getting started, for making a very profound alteration in the relations between states, and one which might conceivably be sufficiently attractive to the Government of the Soviet Union to cause them to reverse what has been their long-standing policy of extreme secrecy, considerable terror, and very great latent hostility to the non-Soviet world.

"The changes that were implied or that would have been implied by the acceptance of our proposals, by the elaboration and implementation of our proposals, would have altered the face of the world. They would have done so in ways that no one is wise enough to predict but that surely would have led to a much greater openness, to a much greater candidness, to much more working cooperation between the peoples of various nations. * * * When you think, for instance, that so obvious a notion as the economic cooperation of the countries of Western Europe is still very far from a reality, you begin to realize that the formal agreement of the delegates was only the beginning of the problem. But one point overshadows this, and that is, however great the enunciation of what is for us a powerful action, however great the enunciation might appear to the British, who are concerned, as rapidly as possible, to reach the exploitation of atomic energy as a form of power, the sacrifices which the acceptance of these proposals would have meant to the Government of the Soviet Union went very much further than that, because it implied a repudiation of the philosophy by which that Government has come into being, has been living * * *"

Dr. GRAY. Do you suppose that word "enunciation" was improperly transcribed from your remarks?

The WITNESS. Yes. It was certainly "renunciation."

Mr. ECKER. It is a verbatim copy of the stenographer's transcript.

Mr. GARRISON. I am sure you are right, Mr. Chairman.

By Mr. GARRISON:

Q. Dr. Oppenheimer, I show you a manuscript entitled, "Address by J. Robert Oppenheimer before the Rochester Institute of International Affairs, December 11, 1948," at Rochester, N. Y., devoted to the prospects for world peace, and ask you if counsel selected that from your files?

A. He did.

Q. Will you hand it to counsel?

A. Yes [handing].

Mr. GARRISON. Mr. Chairman, I have a very short extract from that at page 3 [Reading:]

"Certainly there was little to inspire, and nothing to justify, a troubled conscience in the proposals that our Government made to the United Nations, as to the form which the international control of atomic energy should take. These proposals, and some detailed means for implementing them, were explored and criticized, elaborated, and recommended for adoption by 14 of the 17 member nations who served on the United Nations Atomic Energy Commission. They were rejected as wholly unacceptable, even as a basis for further discussion, by the three Soviet states, whose contributions to policy and to debate have throughout constituted for us a debasingly low standard of comparison."

Mr. GARRISON. I have here a reprint from the record of the Association of the Bar of the City of New York, volume 6, No. 3, for March 1951, containing an address by Dr. Oppenheimer, entitled "Contemporary Problems of Atomic Energy."

The excerpts which I am about to read to the Board appear at page 109 of this reprint from the record. [Reading:]

"Our proposals for the International Control of Atomic Energy, which were largely based on the technical realities of the field, were presented on our behalf to the United Nations by Mr. Baruch, and were widely accepted by the non-Communist nations. The implementation of these proposals would have required a profound alteration in some, at least, of those features of the Soviet system which are responsible for the great troubles we are in today. The failure to persuade the Soviet Government to alter its practices was anticipated by many. Yet we should not forget that this is an objective not only of the past but of the future as well.

"Let me mention 1 or 2 points. One, to my mind, the principal one, was that it was clear that no secure system could be developed for protecting people against the abuse of atomic weapons, unless the world were open to access, unless it was possible to find out the relevant facts everywhere in the world which had to do with the security of the rest of the world. This notion of openness, of an open world, is, of course, relevant to other aspects of the Soviet system. It is doubtful whether, without the newly terrible, yet archaic, apparatus of the Iron Curtain, a government like the Soviet Government could exist. It is doubtful whether the abuses of that Government could persist."

Mr. GARRISON. I have just one more short excerpt to read. This is from another article in Foreign Affairs of which we have a copy here for July 1953. This is quite recent. The excerpts are from pages 525 to 526 of that article. [Reading:]

"Earlier, shortly after the war's end, the Government of the United States had put forward some modest suggestions, responsive to these views, for dealing with the atom in a friendly, open, cooperative way. We need not argue as to whether these proposals were stillborn. They have been very dead a long, long time, to the surprise of only a few. Openness, friendliness, and cooperation did not seem to be what the Soviet Government most prized on this earth.

"It should not be beyond human ingenuity for us to devise less friendly proposals. We need not here detail the many reasons why they have not been put forward, why it has appeared irrelevant and grotesque to do so. These reasons range from the special difficulties of all negotiation with the Soviet Union, through the peculiar obstacles presented by the programmatic hostility and the institutionalized secretiveness of Communist countries, to what may be regarded as the more normal and familiar difficulties of devising instruments for the regulation of armaments in a world without prospect of political settlement.

"Instead we came to grips, or began to come to grips, with the massive evidences of Soviet hostility and the growing evidences of Soviet power, and with the many almost inevitable, yet often tragic, elements of weakness, disharmony and disunity in what we have learned to call the free world."

The WITNESS. I think we are through with this. I will leave it to counsel to say what it means, but I think that in every case I tried to explain that we could not take this path to people who insisted on thinking that we might, and yet not to talk publicly of the fact that we were giving up a position until the Government of the United States had in fact given it up.

There was a bit of discrepancy between our official position and reality and the opinion, let us say, of my colleagues in science. I tried to explain to them that the jig was up, because that was relevant to getting back to work. At the same time I could not come out and say, "This is a hopeless thing," because I had some official connection with the Government until the Government had itself said so. I think these dates will bear that out more or less.

Now we are through with this phase and entering on a new one. In late 1946, I was appointed by the President as a member of the General Advisory Committee to the Atomic Energy Commission. That is a long big job and I will talk about it. Shortly thereafter I was given a concurrent appointment which I held perhaps even a little longer. That was as a member of the Committee on Atomic Energy of the Joint Research and Development Board in the Military Establishment. This later became the Research and Development Board and the Chairmen varied. The initial arrangements were made by Dr. Bush who was head of this outfit.

Dr. Bush appointed Conant as Chairman, the members of the statutory military liaison committee as members, and as civilian members me and Crawford Greenewalt. There was some overlapping of membership between the Advisory Committee and this committee, and total overlapping of membership between the military liaison committee and this committee.

What we did on this committee I don't propose to go into in such detail, and I will try to finish with that this afternoon.

The initial job was to try to give direct technical information to the military on the military liaison committee. General Groves knew quite a lot about the atom and so did Admiral Parsons. The other members of the committee in those days were not very fresh to it. There was at that time not very much machinery for gathering information.

I think, as Dr. Bush explained it, it seemed like a good idea if the same technical considerations which were being made available to the Commission were being made available directly to the military. It was a liaison function. We had very little, if any, power, but we had the ability to talk about common problems.

The importance of this function declined very much because the military developed admirable ways of getting their own intelligence and their own knowledge and became as expert as anyone. But it did provide a continuing channel of discussion. Every once in a while we would stir something up in this committee which was useful.

I have in mind two examples. * * * I won't spell out the details but the question of getting from the hardware which the Commission provided and the hardware which the military services had to the point where you could really make effective use right away.

This was the time, I may remind you, when the feeling that war might break out, however erroneous—widespread war—was very, very general, and there was a war going on in Korea and it was not going too well.

Another example—our role was certainly not major in it—comes to mind, * * * There were two panels on this Board of which I acted as chairman. One was in the summer of 1948, and I think the members of it are listed on your paper, which was a general sorting operation. By then an enormous number of potentially useful applications of atomic energy to military things came up, some of them crazy, some of them sensible, some of them immediate and some of them very remote.

We sat down, the three generals, the admiral and I, and called in other people whose help would be useful and wrote our best opinion as to the relative time scales and absolute time scales of submarine propulsion and nuclear aircraft propulsion; how it was going with the deliverability of tactical weapons, what needed to be done here, what needed to be done there.

The description of the report, the contents of which I cannot tell you, is not going to be very interesting. I think it was a decent job.

The report that we wrote in late 1950 and early 1951—and I may remind you of who was on that committee. I was again the chairman.

Q. You are reading from what?

A. The third page of your notes. Bacher, Alvarez, Lawrence, Kelly, Parsons, Wilson, McCormack. There we took a somewhat deeper bite, because this was the time of the Chinese intervention and a time when as you may remember of daily alerts about the possibility of attack on the continental United States, a time of very great anxiety. We addressed ourselves to the question with what we have and can have soon, how rapidly we can get a really effective use of the atomic capability that we have developed. What can we do fast about this. You will hear testimony about this possibly from the other witnesses.

It is also a time at which technical prospects on the thermonuclear program were quite bleak. We so reported. I think it is interesting that there was no difference of opinion among us as to what we had solved.

This committee has continued until the Research and Development Board was abolished. I think these are the few points that I wanted to cover.

Now we have the GAC appointment and I suppose there it would be best to start up fresh in the morning. There is something to read. It is something that I came upon in the files during the period of getting them straight. It is a letter I wrote to Admiral McMorris of the General Board of the Navy, and it represents the view of our military problem which, at that time, and I believe before and after, was the view that I took into the General Advisory Committee and kept through it. It is not a committee statement. It is not a report of the GAC. It is my own thoughts. It may give some background for what we started out to do and what we did do in the descriptions we gave on the General Advisory Committee.

Q. These excerpts there come from this carbon from your files, is that correct?

A. That is correct.

Q. They begin on page 1.

Dr. GRAY. What is the date of this?

Mr. GARRISON. April 14, 1948. [Reading:]

"Whatever our hopes for the future, we must surely be prepared, both in planning and in the development of weapons, and insofar as possible in our 'force in being', for more than one kind of conflict. That is, we must be prepared to meet the enemy in certain crucial, strategic areas in which conflict is likely, and to defeat him in those areas. We must also be prepared, if need be, to engage in total war, to carry the war to the enemy and attempt to destroy him. One reason why we must keep both of these objectives in mind (and they call for quite definite plans and quite different emphasis as to equipment, troops and weapons) is that it may not be in our hands to decide. With this reservation, it seems appropriate to suggest that there may be two phases to the problem.

"At the present time (1948), to the best of my knowledge, the Soviet Union is not in a position to effectively attack the United States itself. Opinions differ and evidence is scanty as to how long such a state of affairs may last. One important factor may be the time necessary for the Soviet Union to carry out the program of atomic energy to obtain a significant atomic armament. With all recognition of the need for caution in such predictions, I tend to believe that for a long time to come the Soviet Union will not have achieved this objective, nor even the more minor, but also dangerous possibility of conducting radiological warfare."

The WITNESS. This was a bad guess.

Mr. GARRISON [reading]: "Insofar as the United States need not for some time to come fear a serious and direct attack on this country, it would seem to me likely that our primary objective would be to prevent the success of Soviet arms and Soviet policies, to carry out a policy of attrition, and not to engage in a total war aimed at destroying entirely the sources of Soviet power. There are many arguments for this and I have little to add to the obvious ones. Yet, the general political consideration that the consequences, even in victory, of a total war carried out against the Soviet Union would be inimical to the preservation of our way of life, is most persuasive to me.

"On the other hand, as time approaches, if it ever should, where as a result of political or military success in Europe or Asia, as a result of advancing technological development and improved industrial output, the Soviet Union becomes a direct threat to the United States, we shall no longer have this option. We should no longer have this option if the maintenance of a strategic area such as Western Europe or Japan could not be achieved without a direct attack on the sources of Soviet power.

"From this it seems to me that two conclusions would seem to follow: (1) That we must be prepared, in planning, in logistics, and in development, for more than one kind of war; and (2) that the very greatest attention must be given to obtaining reliable information about the state of affairs within the Soviet Union bearing on its military potential.

"One final comment: There is to my mind little doubt that were we today, with the kind of provocation which the Soviet Union almost daily affords, to attack the centers of Soviet population and industry with atomic weapons, we should be forfeiting the sympathy of many potential allies on whose cooperation the success of our arms and the fundamental creation of a stable peace may very well depend. These same people would no doubt be almost equally disturbed were we to renounce, irrespective of the development of Soviet power, recourse to such armament."

Are there any comments you would like to make on the views expressed there?

The WITNESS. I need to say two things. First, that this was apparently an answer to some inquiry. I don't know what the inquiry was. Second, that I was completely wrong in thinking that we could be relaxed about the Soviet atomic threat. I think I was in very general company. I think we all very soon rectified these views as the evidence came in. But this was a year and a half before the first Soviet explosion and the time when my views was, I think, quite the same as the general intelligence views.

By Mr. GARRISON:

Q. This opening paragraph, if I may go back to it for a moment, sounds to me rather like what Admiral Radford said the other day about the New Look. "We must be prepared to meet the enemy in certain crucial, strategic areas in which conflict is likely, and to defeat him in those areas. We must also be prepared, if need be, to engage in total war, to carry the war to the enemy and to attempt to destroy him." This has emphasis on flexibility, which I think is also apparent in that testimony by Admiral Radford.

Dr. GRAY. May I ask, did you read the beginning of this letter?

The WITNESS. No. I would like to have the beginning read, because the beginning states that I don't know anything about this subject.

Mr. ROBB. It occurred to me, Mr. Chairman, that the beginning and the end should be read to give the entire picture.

The WITNESS. I don't know what the beginning says.

Mr. ROBB. You are quite right, it says you don't know anything.

The WITNESS. Shall I do that. [Reading.]

"Thank you for your letter of March 31. In this you enclose the agenda for the study of the General Board, serial 315. You request specifically such comments as I can make on items 110, 118, and 120.

"Though I am aware of the great importance which attaches to this study, and the need for serious thought and effort on the part of many if the study is to be successful, I nevertheless must protest my almost total lack of qualification for speaking to the question which you have put. Such comments as I can make should be given no great weight, they rest on little experience and little knowledge.

"All three of the items referred to me have to do with the plans of the United States for waging war, and with the kind of war we should fight. Implicit in some questions and explicit in others, is the issue of weapons of mass destruction; should we use these, should we plan to use these, should we postpone the use of these. Implicit in the question is also the issue of a limited versus a total conflict; should the objective be destruction of the enemy, or his defeat in a specific area. Let me attempt to give my views on these matters."

Then it goes into what Mr. Garrison read.

The end is: "In conclusion, let me again remind you that these are in the nature of personal views, and that I can attach little weight to them, if, in matters which fall more closely within my field of competence, I can be of use to you, I shall of course be glad to do so."

Dr. GRAY. That is addressed to whom?

The WITNESS. Admiral McMorris, head of the General Board of the Navy. I am in a complete fog as to what it was all about, except insofar as this answer——

Dr. GRAY. Was this signed as Chairman of some panel?

The WITNESS. No, this was an individual opinion.

Dr. GRAY. Thank you.

Mr. GARRISON. It is simply introduced at this time to show his general approach to the whole policy of armament of this country.

The WITNESS. There is one small item before we get into the General Advisory Committee, and that is the following: There was set up under the contract with all three services, Army, Navy, and Air Forces, I think the operating contractor was the Army, a study at California Institute of Technology, Dr. DuBridge was in charge of it, under the name of project Vista, and its function was generally speaking to talk about ground combat and the support of ground combat. What that finally came down to was the study of the defense of Europe and what it came down to was the study of what you do to defend Europe at any time, as soon as possible, if necessary.

The men involved in this project worked very hard on it, and they kept asking me to come out and talk about the use of atomic weapons in this picture. I thought they knew as much as I did. Dr. Bacher was there. Dr. Lawrence was there, Dr. Christie was, and Dr. DuBridge was there. But they finally prevailed upon me, and I went out in the autumn of 1951, and we worked together on this problem.

Dr. Lauritsen and Dr. DuBridge went over with Mr. Whitman from the Office of the Secretary of Defense to visit General Eisenhower, Gruenther, Norstadt, and Hanley in Europe.

"What we attempted to do was to be sure it was clear to them how varied and useful atomic weapons could be in ways that are probably now quite obvious to you and ways which were not completely obvious then. General Eisenhower made 1 or 2 suggestions about things that he though it would be handy to have. The principal messages that we brought back to this country were a plea for more information as well as more hardware and to make atomic weapons available and for restriction of the limitations on discussions of military problems, with Allied Commanders. These were the things that made it hard to get on with these. I don't want to go into the technical aspects of it, though the antiair use of atomic weapons, their use to put out enemy airfields, both those that are near enough for combat planes and the deep lines strategic ones is an obvious example. This was the complement to the panel report I spoke of earlier on getting the atom to work on the battlefield as well as in the heartland. I think this may be a place to stop.

Dr. GRAY. Before we stop, I wonder if you can, Mr. Garrison, give an indication of the witnesses.

Mr. GARRISON. I thought we might discuss that informally off the record. I can bring this chart and show you about how it looks now.

Dr. GRAY. We will go off the record for a moment.

(Discussion off the record.)

Dr. GRAY. Are we prepared to say we will meet again tomorrow morning at 9:30?

Mr. GARRISON. We will undertake to be prompt.

Mr. ROBB. May I say, Mr. Chairman, as far as I am concerned, and Mr. Rolander, I cannot speak for the board, if it will accelerate matters and assist counsel to get some witnesses here, I would be very happy to come here earlier in the morning. I do not want to make that proposition too firm.

Dr. GRAY. Let the chairman speak for himself only and not for the other members of the board. If by meeting at 9 o'clock we could move along without inconvenience and so forth, I believe the board would be willing to meet at that time.

Mr. EVANS. You can say it for me, because time is important to me.

Mr. MORGAN. Yes.

Dr. GRAY. So would you bear that in mind, Mr. Garrison. Any telescoping we can do without inconvenience or harm we would be interested in doing.

(Thereupon at 5:13 p. m., a recess was taken until Tuesday, April 13, 1954, at 9:30 a. m.)

UNITED STATES ATOMIC ENERGY COMMISSION
PERSONNEL SECURITY BOARD
IN THE MATTER OF J. ROBERT OPPENHEIMER

ATOMIC ENERGY COMMISSION,
BUILDING T-3, ROOM 2022,
Washington, D. C., April 13, 1954.

The above-entitled matter came on for hearing, pursuant to recess, before the board, at 9:30 a. m.

Personnel Security Board: Mr. Gordon Gray, chairman; Dr. Ward V. Evans, member; and Mr. Thomas A. Morgan, member.

Present: Roger Robb, and C. A. Rolander, Jr., counsel for the board; J. Robert Oppenheimer, Lloyd K. Garrison, Samuel J. Silverman, and Allen B. Ecker, counsel for J. Robert Oppenheimer.

PROCEEDINGS

Mr. GRAY. I would like to call the proceeding to order.

The chairman of the board has a few observations to make, and I have a few questions to ask on behalf of the board.

I should like to read again for the record a statement which I made yesterday, that the proceedings and stenographic record of this board are regarded as strictly confidential between Atomic Energy Commission officials participating in this matter, and Dr. Oppenheimer, his representatives, and witnesses. The Atomic Energy Commission will not take the initiative in public release of any information relating to proceedings before this board.

The board views with very deep concern stories in the press which have been brought to the attention of members of the board. I personally have not had time to read the New York Times article, but I am told that both the Nichols letter to Dr. Oppenheimer, of December 23, and his reply of March 4, are reprinted in full. Without having any information whatsoever, I have to assume that this was given to the New York Times.

Dr. OPPENHEIMER. It says so in the paper.

Mr. GRAY. I do not suggest that represents a violation of security. I have a serious question about the spirit in keeping with the statement we made for the record yesterday about these proceedings being a matter of confidential relationship between the Commission and the board representing the Commission, and Dr. Oppenheimer and his representatives and witnesses.

We were told yesterday before this hearing began that you were doing all you could to keep this out of the press. You said you were late yesterday because you had "fingers in the dike," I believe was your expression, which I found somewhat confusing against subsequent events in the day when you say that you gave everything that you had to the press. We agreed yesterday that it would be very unfortunate to have this proceeding conducted in the press. There was no dissent from that view which was expressed, I believe, by all of us.

I think that it should be perfectly apparent, particularly to the attorneys involved, that this board faces real difficulties if each day matters about this proceeding appear, not on the basis of rumors or gossip, but on the basis of information handed directly to the press. I think it only fair to say for the record that the board is very much concerned.

I should like to ask some questions for the record about the authorized spokesman for Dr. Oppenheimer. I assume in addition to Dr. Oppenheimer that Mr. Garrison, Mr. Silverman, and Mr. Ecker are actively and officially associated in this proceeding.

I should like to ask who else is working on this who may be talking to the press?

Mr. GARRISON. Mr. Chairman, perhaps you could let me answer that question by a little history. The letter from the Commission was given on December 23. I came into the case early in January. Almost immediately, or perhaps the middle of January, it became quite apparent from inquiries that Mr. Reston addressed both to the Atomic Energy Commission and to Dr. Oppenheimer, that he already had information that clearance had been suspended, and that proceedings were going forward against Dr. Oppenheimer. He was most anxious to obtain background information from us.

We explained to him the nature of the proceedings and our earnest desire that this not be the subject——

Dr. OPPENHEIMER. May I correct that. Was this your conversation with Reston, because I believe the initial conversations were with me. He called and he was very persistent in calling. I tried to evade it. I knew what it would be about. After about 5 or 6 days of persistent telephoning, he talked to my wife, and said that he had this story and he wished I would talk to him.

I talked to him on the phone. I said I thought it contrary to the national interest that the story should be published, that I did not propose to discuss it with him, but if the time came when it was a public story, I would be glad to discuss it with him.

That was mid-January. I don't remember the date. I am depending on counsel's memory. I believe that was the substance of our talk. He told me two things. First, that my clearance had been revoked. That was the story he had heard. That this had been cabled, telegraphed, and broadcast to submarine commanders throughout the fleet and Army posts throughout the world, and second, that Senator McCarthy was fully aware of this and thought I ought to know that. That was the end of that discussion.

I was given to understand by proffers of kindness but not other sign that the Alsops knew the situation. Later this was confirmed by one of the prospective witnesses.

Mr. GRAY. You did not talk with either one of the Alsops?

Dr. OPPENHEIMER. I have not talked to either one of the Alsops until very recently, and I will describe those conversations. This was long ago, and it was my affair, and I thought my memory would be more vivid than yours.

Mr. GARRISON. Why don't you tell of your conversation with the Alsops?

Dr. OPPENHEIMER. That is not until very recently. Stewart Alsop called co-counsel, that is Herbert Marks, whose name should be in these proceedings—when would that have been, Saturday, Friday—quite recently, saying that they had the story and were frantic to publish, and that I should call Joe Alsop, who is up in Connecticut at a rest home.

Mr. GARRISON. In Garrison, N. Y.

Dr. OPPENHEIMER. I did call him there. I put on my spiel, the thing that I have said to everyone, that I thought this story coming out before the matter was resolved could do the country no good. Either I was a traitor and very, very important secrets had been in jeopardy over the last 12 years, or the Government was acting in a most peculiar way to take proceedings against me at this moment. This is the impression that I feared would be made. Neither impression could be good. Having both of them could be only doubly bad.

Therefore, not as far as I was concerned, but as far as what I thought was right, I urged Joe Alsop to hold his story, not to publish it. We did not discuss any substantive things except that Alsop told me how apprehensive he was that Senator McCarthy would come out with it. I believe that was all I said to Joe Alsop. He said he thought I was making a great mistake, but I said it was my mistake.

I recognized of course that he could publish any moment that he wanted to.

Mr. GRAY. May I ask, as of this time or 10 o'clock yesterday morning, had you given the New York Times these documents?

Dr. OPPENHEIMER. These documents were given to Reston by my counsel Friday night, I believe, without any instruction as to what he was to do with them, as background material.

Mr. GRAY. So that you knew when you made the statement here yesterday morning that you were keeping the finger in the dike that these documents, dated December 23, and March 4, were already in the possession of the New York Times.

Dr. OPPENHEIMER. Indeed we did.

Mr. GARRISON. Mr. Chairman, they were given to Mr. Reston with instructions not to be used unless it became essential for the Times to release the story because others were going to do likewise. We hoped even as of yesterday—the last word we had with Mr. Reston was after lunch—we hoped even as of yesterday that this could be held off, although I told you at the start that it might be only a matter of hours.

Mr. GRAY. You didn't indicate to me in any way—if you attempted to do so, it is a matter of my misinterpretation—that you had given documents which relate to these confidential proceedings and are part of these proceedings.

You mentioned Mr. Marks. Who else is authorized to speak for you, Dr. Oppenheimer?

Mr. GARRISON. No one else. Mr. Marks is not counsel of record in this proceeding. He has been associated with us from the start because of his knowledge of past history. I am still seeking his guidance and help.

Mr. GRAY. He is assisting, I take it, in preparing these documents which you present?

Mr. GARRISON. No; we did all that work ourselves.

Mr. GRAY. May I ask specifically for the record who prepared the excerpts about which I asked the question yesterday?

Mr. GARRISON. We did in our own office. I did. Mr. Ecker worked on them.

Mr. GRAY. I should like to know, Mr. Garrison, why it was yesterday that not one of the three of you could answer the question as to whether these paragraphs were consecutive or came from consecutive pages. It is apparent that someone else had prepared them.

Mr. GARRISON. No, Mr. Chairman.
Mr. GRAY. I have drawn a conclusion. If I am wrong——
Mr. GARRISON. I am very sorry that such thoughts should even occur to you. What happened was that some weeks ago I went through Dr. Oppenheimer's writings and I marked particular sections and passages from a lot of them that seemed to me to be worthy of presentation to the board, and I asked that they be extracted and copied out. I have not been over them for some time. To be frank with you, I have had so much else to do.
Mr. GRAY. My point in raising all this is that if there are a good number of people who are not appearing here who are going to be talking to the press, I would like to know what control or lack of control there may be in this situation. That is why I am raising this thing.
Mr. GARRISON. Yes.
Mr. GRAY. I think these stories are very prejudicial to the spirit of inquiry that I tried to establish as an atmosphere for this hearing as we started yesterday. I would very much regret that what would appear to be to the board possible lack of cooperation in conducting these proceedings in the press if that were prejudicial to what are the basic fundamental issues involved.
Mr. ROBB. Might I ask a question, Mr. Chairman?
Mr. GRAY. Yes.
Mr. ROBB. I don't think we have identified Mr. Marks.
Mr. GARRISON. Mr. Herbert S. Marks, former General Counsel of the Atomic Energy Commission, and a lawyer in Washington.
Mr. GRAY. He is an attorney and member of the District of Columbia Bar?
Mr. GARRISON. Yes.
Mr. GRAY. And do I understand he is of counsel to Dr. Oppenheimer?
Mr. GARRISON. He is associated wtih us as counsel.
Mr. GRAY. In the relationship of lawyer and client, is that correct?
Mr. GARRISON. Yes.
Mr. Chairman, may I just say another thing about the problem that we faced.
Mr. Reston from the middle of January has had the Alsops, and I don't know who else busy gathering information from anybody they could find and had developed so much of the story when Mr. Reston talked with us on Friday that it seemed to us that if the story had to break that rather than half a story or two-thirds of it or a quarter of it in fragments with constant demands afterwards from the press for the rest of it, that it was better that the basic documents be there for all to see.
This was not a happy decision or a pleasant one for Dr. Oppenheimer, believe me, to have the letter of charges displayed for the American public. It was something no man would ever wish to do. It was not until Mr. Reston told us yesterday afternoon that the thing absolutely could not hold, the stories were going to be published, Alsop said the same thing, that we said all right, go ahead then and print the documents.
Now, it is not our purpose to make any press comments upon this case. It is not our purpose to release any transcripts. If you will observe the Reston story, I am sure you will see that we have tried to avoid any kind of special pleading. Dr. Oppenheimer has made no statement. We are not trying to try this case ourselves in the press. I assure you with all earnestness that is true. I feel absolutely certain that it is better in the long run for the Government, for this board, and for us, that there be no suspicion about what is the scope of this case, whether the H-bomb is in it, and all those kinds of questions that would arise if the actual facts had not been disclosed.
Mr. SILVERMAN. May I point out, if I may interrupt, there was an item in the Reston story, however, it is understood that he, Dr. Oppenheimer, also put in evidence another secret document in the form of a memorandum. We haven't the faintest idea what they are talking about, nor did we give them any such information.
Mr. GRAY. Who is "we"? Who actually handed the documents to Mr. Reston?
Mr. GARRISON. I did myself, Mr. Chairman, personally.
Mr. ROBB. Did he also get a copy of this autobiography?
Mr. GARRISON. No.
Mr. ROBB. Mr. Garrison, may I ask another question? Didn't I understand you to say yesterday morning that explaining your tardiness at the hearing that you had been engaged in a press conference?
Mr. GARRISON. No, I had been engaged in threshing this problem out among ourselves, because the calls were coming in and putting us under the greatest pressure. In fact, right along we have been under pressure to make statements,

to initiate statements of our own and come forward with information. It has been a very, very difficult undertaking, Mr. Chairman.

Mr. GRAY. I am quite aware of that. On the other hand, you are quite aware also that the members of this board have been under pressure, and that we have I believe without fail said we will not discuss it. That will continue to be our position.

Mr. GARRISON. I should also like to say that we did not disclose to anybody—when I say "we", I mean every one of the counsel to my knowledge, and Dr. Oppenheimer—the names of this board or where the hearings were being held or anything else.

Mr. EVANS. Where did they get it?

Mr. GARRISON. I don't know. I have no idea.

Dr. EVANS. They called me up about 1:30.

Mr. GRAY. They called me, too, but I didn't answer the phone.

I would like to move to another point, if I may. I am sorry we are keeping Dr. Kelly waiting. This has to do with the schedule of hearings. You left a suggested typewritten schedule with us yesterday which was not made a part of the record. I think I should say that the Board cannot accept this as a schedule. I repeat, indeed, if it is necessary to repeat, that this is to be a fair inquiry, that Dr. Oppenheimer will be given full and adequate opportunity to make any presentation he has, and to present such witnesses as he desires, but as far as the schedule is concerned, the board feels that it is up to Dr. Oppenheimer and counsel to furnish the witnesses and information for the board.

We propose to sit from 9, if it is desired by Dr. Oppenheimer and his counsel, or from 9:30 to 12:30 and from 2 until approximately 4:30, give and take a little because of circumstances. Frankly, I think the board is unwilling to commit itself to a schedule which I am sure means that we will have some witnesses on a certain day who will be through and then there is nothing more for the board to do or for a part of the day. I should like to suggest, Mr. Garrison, that we inform you again that we will meet and we will hear the witnesses and and some approach be made to this problem from the point of view of the convenience of this board and not the convenience of the witnesses as would be true in most proceedings in the American tradition. If it seems to be necessary to hear a witness at a particular time in accordance with some prearranged schedule, some days in advance, I think you should be warned that the witness will probably be asked under oath whether this is the only time that he could appear, if we run into a situation where we must recess or delay proceedings because of a witness who has said, "I can come on a certain date."

We understand fully that Dr. Kelly can only be here this morning. We are very glad to hear him and we will hear him. Then I would very much prefer, and the members of the board would, if we could receive the remainder of Dr. Oppenheimer's presentation, and proceed with whatever period it seems desirable of questioning Dr. Oppenheimer, and then try to move forward with receiving testimony from the witnesses.

So I don't think that we wish to commit ourselves to a schedule which draws it out precisely as this is drawn. I am hopeful you will find that we will be reasonable and fair in hearing the witnesses.

Mr. GARRISON. Mr. Chairman, pursuant to your wishes that you expressed informally to us yesterday, I arranged for Dr. Bush to appear instead of this morning on Monday afternoon, the 19th, and I have arranged with Mr. Gordon Dean to appear Monday morning the 19th, in lieu of Wednesday afternoon.

Mr. GRAY. I would say, Mr. Garrison, that is quite all right with the board. This is a part of your responsibility of keeping witnesses and whatever else is to be presented to the board moving along as we sit and are available to hear them.

Mr. GARRISON. I have no doubt that we shall fill the afternoon session on the 19th, so that there will be no waste time of the board, because there are still several witnesses whom we have contemplated calling and we have not had a chance yet to talk with them.

Mr. GRAY. All right, sir.

Mr. GARRISON. For example, Mr. Conant, Mr. Bradbury, and several others. If you will indulge me, I would like to say one other word about counsel, because I think there has been some mystery, perhaps, cerated by Mr. Marks' relationship to the case. Mr. Marks is an old, very dear and very personal friend of Dr. Oppenheimer. They both came to see me when I was asked to serve as counsel. I am serving without fee in this case as a public service. To the best of my knowledge, Mr. Marks is serving without fee in this case as a gesture of very deep friendship and admiration for Dr. Oppenheimer. We have been working together, he and I, as one would work together in a matter of this sort without any really formal relationship except that it was understood that I would in effect try the case, conduct the proceedings and have the final decision and responsibility. He is now simply going about his law practice, and as I feel that I use his advice and need him, Dr. Oppenheimer leans very heavily on his opinions, we meet together and talk things over. It is that kind of a relationship.

It never occurred to me that it would be necessary or that I would be not frank with the Board in not entering his appearance here today, because actually we are the counsel conducting this proceeding, and I have the final decision. But I want you to be quite sure that Mr. Marks is not authorized by me to talk with the press or to exercise himself in any fashion on this matter. He is a friend and advisor and associate in that sense.

Dr. OPPENHEIMER. He is sometimes authorized to talk to the press in specific ways and with a specific message.

Mr. GARRISON. Both he and I have had conversations with Mr. Reston and Mr. Alsop and other newspapermen have called him up, but what I am trying to say is that Mr. Marks is not sitting in his office at my request conducting press conferences to spread information about this case. You can be just as sure as that——

Mr. GRAY. But he is authorized to speak to the press, at least those were Dr. Oppenheimer's words.

Mr. GARRISON. He is not authorized to conduct press conferences. He cannot avoid inquiries when they come to him. As far as I know, Mr. Chairman, we are all going to be battered—I was called at quarter to seven this morning.

Mr. GRAY. You can't avoid the call. But I can say to you on the basis of personal experience that it is possible not to talk.

Mr. GARRISON. That is what all of us have pledged each other to do, that is, not to talk.

Mr. GRAY. As of what time did you take that pledge?

Mr. GARRISON. We decided when the documents were made public that ends this matter as far as we are concerned.

Mr. GRAY. Fine. I am sorry we kept Dr. Kelly waiting. Would you get him in, if you are ready now to present Dr. Kelly.

Whereupon, Mervin J. Kelly, was called as a witness, and having been first duly sworn, was examined and testified as follows:

Mr. GRAY. Dr. Kelly, do you wish to testify under oath. You are not required to do so.

Dr. KELLY. I would be glad to testify under oath.

Mr. GRAY. Would you stand, then, please and raise your right hand.

Mervin J. Kelly, do you swear that the testimony you are to give to the board shall be the truth, the whole truth, and nothing but the truth, so help you God?

Dr. KELLY. I do.

DIRECT EXAMINATION

By Mr. GARRISON:

Q. Dr. Kelly, you are the president of the Bell Telephone Laboratory in New York City?

A. I am.

Q. And in 1950 to 1951, you served on a Research and Development Board panel under Dr. Oppenheimer's chairmanship?

A. That is correct.

Q. You had met Dr. Oppenheimer before that time?

A. Oh, yes.

Q. Could you say when you first met him?

A. It was at either a National Academy meeting—what is this thing in Philadelphia we belong to—the American Philosophical Society meeting in Philadelphia shortly after the war, late 1945, or early 1946. Oppie was addressing a meeting there at that time.

Q. Would you tell the board very briefly about your work with Dr. Oppenheimer on the Research and Development Board panel?

A. The Research and Development Board has had an Atomic Energy Standing Committee. At that time Robert LeBaron, Mr. William Webster was the head of the Research and Development Board. At Mr. Webster's request or suggestion Mr. LeBaron formed a panel in the late fall of 1949, as I remember. I had a letter from Mr. LeBaron in early November concerning serving on the panel, in which he told me that Dr. Oppenheimer was to be the chairman.

I accepted membership and then had relations with Dr. Oppenheimer from then on about it.

We had our first meeting early in December. The committee had 9 members, 3 military, 3 of the more academic scientists and 3 of the less academic. Gen. J. McCormick, who was then the military officer in the AEC, reporting to the General Manager, in charge of military programs, was ex officio and at all meetings.

The group was made up of Dr. Oppenheimer as chairman, Dr. Bacher, then of Cal Tech. He had been on the Commission. Dr. Louis Alvarez of the University of California. Prof. Charles Lauritsen of Cal Tech. Prof. Walter Whitman of MIT, and myself were the civilians. The three military members were Gen. K. D. Nichols of the Army, Adm. W. S. Parsons of the Navy, and Gen. R. C. Wilson of the Air Force.

The general charge to the committee was for it to view the status of atomic research in the Commission and its progress, the state of the stockpile, with the knowledge of the weaponry to come up with recommendations for the scope and emphasis in the military applications of the research and development program.

Mr. GRAY. Dr. Kelly, may I interrupt for a moment. I am afraid I failed to tell you that in the event that it is necessary for you to discuss any restricted data, I would appreciate your letting me know that you propose to do so.

The WITNESS. I don't propose to say anything here that in a closed hearing is not perfectly all right, whether the people are cleared or not.

Mr. GRAY. All right, sir.

The WITNESS. I was stating the scope of the examination as requested by Mr. LeBaron. I think I had completed by saying that we were going to look at what the military applications of the research and development program should be in the light of advancing knowledge in the atomic area, and the stockpile and the military situation. We had about 6 days in December of meetings and went over this whole matter. It was the first time that I had seen Dr. Oppenheimer in action in an operating sense in a responsibility of this kind.

He was an unusually able chairman. I have been on lots of committees and chairman of some, and I would put him right at the top in his patience in developing views and getting the views of everyone, and promoting full discussion, and yet giving the minimum of waste time for busy people that goes with committees of that size.

We came up, after much discussion, with very common views because it was in an area where, excepting for the enemy situation, there was generally a background of factual knowledge to work on.

After we had gotten to where we had a commonness of view as to what we should say the program should be in scope and emphasis, Dr. Oppenheimer undertook the job of preparing our report, which was an aid to all of us. I remember his staying on in Washington between meetings and beyond meetings for drafting the report. He drafted a report which with very minor modifications, I would say, all of us could sign as representing fully our own views as to what the military emphasis in research and development should be.

This was just at the threshold of the time where atomic basic knowledge had reached the point that it was possible to consider versatility. By that I mean extending the range of weapons well beyond that of the large free falling bombs. So this was rather a critical time.

That opportunity for extending the scope of weapons, that is, the range of versatility in military action, was a thing that needed very careful weighing and was weighed and our report encompassed the views on how that should be broadened. As a matter of fact, I know from my participation in the program that what happened in the succeeding years was very much along the line or substantially identical to the charter that we suggested as the research and development programing plan.

Mr. LeBaron wrote me, and no doubt other members of the committee, afterward, expressing appreciation and stating the way that it had been accepted favorably in both the Commission and the military. Throughout this, Dr. Oppenheimer was one of us in views, that is, had common views with us, as to the best military use of the fissionable materials and the kind of weapons that should be put into development, and in discussion there was every evidence of his dedication to the best use of this kind of power in the national interest possible. Any divergence in views as they developed were detailed and no greater difference in his views on that from one of us to the other than there would be between any two of us.

By Mr. GARRISON:

Q. Did you ever deduce that Dr. Oppenheimer ever overstated, in your opinion, the need of continental defense as distinguished from the production of offensive weapons and plans?

A. Quite the contrary. Dr. Oppenheimer's views on continental defense are so close to those that I have held from any close contact with it that I could not distinguish a difference.

In the late fall of 1952, Secretary Lovett asked me to head a civilian committee made up principally of top business leaders, such as Bob Wilson of Standard Oil, and top educational people, to survey the continental defense problem and to put it in proper perspective with the rest of our deterrent efforts. General McCormick, who had then come over into the Air Force, I succeeded in getting as a secretary to my committee.

During the progress of the committee's work which was in the first several months of 1953—the committee was then operating under Secretary Wilson, but Mr. Lovett had cleared with him when he appointed us in November that he wanted us to continue because it was going into the new administration of Mr. Wilson—and a number of times General McCormick for me, as I had a lot of other responsibilities, saw Dr. Oppenheimer. I know particularly of two visits. I remember two visits to Princeton where he discussed with Dr. Oppenheimer the evolving report and views. Of course, this could be said to be hearsay, but he recounted to me Dr. Oppenheimer's comments which were wholly favorable and differed only in insignificant detail. Dr. Oppenheimer felt it was a constructive judgment, which was in general, that while the country had not given proper emphasis to continental defense relatively, yet that our chief deterrent was strike, and that nothing should be done in bringing up to a proper level a continental defense effort that would weaken our strike. That was the general philosophy.

We recommended certain organizational and planning and procedural things to unify the program, but placed it second to strike in the general program of our best defense, and best deterrent aspect.

With the discussions that General McCormick had with him I could distinguish no difference. In fact, he spoke very complimentary, so General McCormick related to me, of the direction our thinking was taking.

I do not find the time to do a lot of talking about these things that are directly concerned, but in the Lincoln summer study, two of my members were on that study, and I know from them that the views of Dr. Oppenheimer, who was there occasionally, and others of the academic side, were very strong for looking into the Arctic line and the kind of implementation that was then in breadboard state, but in proper perspective.

I have since heard Dr. Oppenheimer discuss the defense aspect at closed meetings in the Council of Foreign Affairs—and this is in relatively recent months—and found his views there in general accord with the ones I have held and pushed for a stronger continental defense, better organized, unified, but done not at the expense of our strike power.

Q. What would you say as to Dr. Oppenheimer's reputation for straightforwardness, directness, veracity?

A. Among his peers, he is, first, known and recognized for his accuracy of thought and cleanness of expression. His words are considered generally well-weighed and meaningful because of their accuracy and temperate. I would know of no one that knew him as well as I that would feel that he overstated his position.

As to his veracity and dedication, I know of no one in the program, with the high clearances that he has had, and that I have, Q and top secret, everything he has done and said gives a full appearance to a great dedication, as full an appearance as any of us that are in and still cleared.

Q. Would you say that as chairman of this panel he made a contribution to the national welfare?

A. I am sure that he did. In the form that he writes all of his things, getting the views of the full committee that he shared, as to what the forward looking program should be, getting it clean, orderly, and well placed was a great contribution, as anyone working in the atmosphere of the Pentagon knows the great need for, that is, of getting direction and aim and purpose well spelled out. It was in this report of the panel which was his fine, clean writing, but which was the views of all of us which he shared.

Q. What have you to say as to his reputation for integrity and patriotism and your own personal feeling about that?

A. Among his peers, those who know him and know his work, I would say his reputation is the highest. As to my own personal belief, I know of no one in the program that I would have any more confidence in their integrity and dedication than I would of Dr. Oppenheimer.

Q. What would say as to the competence of the setup at Los Alamos and Sandia to handle the whole program during the years while Dr. Oppenheimer served on the General Advisory Committee, roughly 1947 to 1952.

A. I have known the situation there intimately since January 1949. That was my first entrance broadly into the atomic-weapon area. During the war we had quite a good-sized job at the laboratory in an area that did not concern Los Alamos directly, or Dr. Oppenheimer, and that was the research and early development of the membrane used at Oak Ridge for diffusion, a very difficult physical chemical job. In early 1949, the Commission asked me to make a study of the Los Alamos-Sandia combined operation and make recommendations as to any organizational changes. They had in mind not a complete satisfaction of the applied end of the weaponry, that is, after the nuclear job was completely done, the clothing of that with all the aerodynamic, electronic, and radar gear to make the completed weapon. That, as well as the nuclear, had been up at Los Alamos up until maybe a year or two, I was in in 1949, and then that part of it that had to do with the weaponry, exclusive of the explosive unit, was moved to Sandia to be close to the military people. But both operations were under Dr. Bradbury, and that was a contract with the University of California.

There was some question within the Commission, and Dr. Bradbury himself, as to the operations in Sandia. So I spent the greater part of 3 months looking searchingly at Los Alamos and at Sandia, and reported orally—I made the stipulation to the Commission that I must do it orally, as I could not take the time for a polished, finished report—giving my judgment of the very high competence of the Los Alamos operation, and the quality of the people in the program, the way they were attacking them, and while the buildings were temporary in the facilities for doing it.

The applications end of clothing the unit that has the explosive with the required aerodynamic and electronics, I found was not up to the capacities of the country in that kind of applied science and technology. So I recommended that part of the job be given to an industrial contractor, as there were components of engineering judgment and background at high levels that just were not in the program, and also knowing how to recruit the kind of people to build such a staff.

That recommendation was acted upon and Mr. Truman requested the A. T. & T. that we accept that Sandia operation, and a subsidiary corporation of the Bell System had been formed to do that.

The technical, the whole research and engineering side of it is my direct responsibility. I spend 1 week in 5—in fact, I am going out there tomorrow—so I have known the program intimately since 1949. I would say that the overall integrated program is the finest expression of American scientific and technical ability, and that we are where we are in the weapons program because of that plan for doing it, its competence and its relative freedom to operate as scientists and technologists do in our society, relieved from a lot of restrictions that come in from civil service, and other kinds of handling.

As I say, the only blemish on that program in 1949 was the inadequacy of the applied technology having to do with the aerodynamics, electronics, and so on.

Q. Based on your knowledge of Dr. Oppenheimer, your experiences with him, and his reputation as you know it, do you believe that his clearance would be clearly consistent with the interest of national security?

A. To the very best of my knowledge, I sincerely believe that, and I think that his absence from the programs and from the councils would be a distinct loss.

There is one observation, as I told you, that I would like to make, if this is an appropriate time, that I think is pertinent to the aspects of the problem that I can't testify directly on.

When scientists and applied scientists look into the crystal ball in the early stages when there is not enough known about the facts of nature, you can find quite wide and honest diversity of views which clear up and views become substantially common when enough knowledge of nature's laws and behaviorisms in the area come to light.

Taking an example, I was thinking last night from my earliest entrance into science at the graduate level in 1914 and 1918, I was Milliken's research assistant in Chicago. As I did, I did at great deal of the oil drop experimentation that he was doing, first to establish that there was an electron with a unique charge, and only 1 electron. During the early year of that there was quite a school of thought that there was not, that there were electrons of various sizes. I remember a distinguished professor at Vienna, whose name has slipped my mind, who published greatly on the subelectron. By 1917 there was enough accumulation of the facts that agreed there was only one electron, which is our primer today.

In this atomic area, as you know, the Atomic Energy Commission has not been blind through the years to the civilian application for power and, of course, have been looking at power applications for military with more vigor in the earlier stages of it than they were at the direct civilian-economy applications. But until the last year or so there were competent applied scientists who knew all of the facts that had evolved certainly up to a year and a half ago; and some of those that were right in the middle of it were of the views that the civilian applications, while certainly important to humanity, had a distant date because of economic considerations that you measure in decades.

One of the ones who was right in the program and so had all of the knowledge from that side that I frequently talked with about it in the last year and a half has changed his views completely and says that he has and he now feels confident that economic power will be with us in a decade. Yet, until there was more information that came from his programs, showing what economic factors could be, he was of the belief that it was a few decades at least away.

Dr. EVANS. You say you did work with Bob Milliken?

The WITNESS. Yes; I did all my graduate work with Milliken from 1914 to 1918 and then came to the Bell System, and have been there ever since.

Mr. GARRISON. That is all of Dr. Kelly unless the board would like to ask questions.

CROSS-EXAMINATION

By MR. ROBB:

Q. Dr. Kelly, may I ask what is your field?

A. I got my doctorate with a major in physics and minor in mathematics and came to the Western Electric laboratories in New York—and which later became Bell laboratories in 1925—as a research physicist and did my productive work as an applied scientist in the field of electronics. Since about 1936 I have been one with increasing scope of the technology that have looked at what others have done rather than doing it myself. So, over the whole field of telecommunications and science and technology, I would say that I am expert.

Q. Are you what is described as a nuclear physicist?

A. No; I am not a nuclear physicist. I have kept very conversant with it as an interested scientist, but there was in my student days and my active days, there was nuclear physics; and, as it evolved, I followed it closely. I have a number of nuclear physicists in my staff, among them Dr. Fisk, who was the first research director of the Atomic Energy Commission but knows as a participant the nuclear-fission field quite well. I have never practiced it, though.

Q. You would not offer yourself as an authority on nuclear physics?

A. No; just as one with an understanding of what others have done but not as an authority, because I have not practiced it—because again I limit myself in the amount that I look at.

Q. And, by the same token, I assume you would not offer yourself as an authority on the superbomb or the thermonuclear weapon?

A. No; that is right.

Q. Who are the leading authorities in the country on the thermonuclear weapon?

A. I would say that the outstanding nuclear physicists that are in the program, such as Bradbury and his immediate staff and Edwin Teller and Johnny von Neumann, would be names that would first come into my mind.

Q. Dr. Lawrence?

A. Yes. Again Dr. Lawrence is not a participant in the sense these men are but has a great understanding and came up through nuclear.

Q. I was not limiting myself to those who are not participating.

A. He would be one of great standing and the head of the laboratory doing a great deal in that field.

Q. Dr. Alvarez?

A. Dr. Alvarez, who was on this committee, is another; yes.

Q. Of course, Dr. Oppenheimer.

A. Dr. Oppenheimer, Teller, Bradbury, and von Neumann. Those are the first names that would come to my mind, but these that you add are in the same ball park.

Q. Probably Dr. Oppenheimer would be preeminent; would he not?
A. He would certainly be in the first four.
Q. Whether he would bat first or fourth you would not want to say, but he would be in the first four.
A. That is right. I would not be able to judge. I don't know that anyone could, because there are different qualities to it.
Q. Dr. Kelly, in this report that you spoke of that your panel made in 1950, would that have been the report dated December 29, 1950?
A. I would expect, without referring to the notes, that would be right. We finished our deliberations about the 22d or 23d, as I remember, and my letter from Mr. LeBarron is dated January 30. He talks of the report having been received and studied. That is January 30, 1951. So certainly it was issued some time after December 22 and before January 30.
Mr. GRAY. What was the date you mentioned?
Mr. ROBB. December 29, 1950.

By Mr. ROBB:
Q. Do you have any way of establishing that?
A. I could easily get it from the Department of Defense.
Q. Perhaps I can be of assistance. In your discussions in that panel, Doctor, did you and your colleagues discuss the so-called superweapon, the thermonuclear bomb?
A. No; we did not. It was not in the area of our cognizance. It was a research thing where it had not even been proved that it would be, and it was not in a stage where military application could be considered. So there was no discussion in committee at all about it.
Q. Would you say that again?
A. It was not in a stage of development where, as corresponded to the fission weapons, you could be talking about military applications knowledgeably and the different ways that you would use it. All the discussions, the formal discussions of the committee—if there were any others, it was individual and separate from the meetings I attended—was about fission and not fusion.
Q. In other words, you felt that the fusion weapon was something in the future; is that correct?
A. That is correct. We were working for the Department of Defense and not the AEC, and it was not ready to be considered at that stage.
Q. Did you make any comment in your report on the matter of thermonuclear warheads or fusion weapons?
A. I have not seen the report since it was issued. I would feel confident it was not there because it was not a matter of discussion. If it was, that is 4 years ago. I can't remember. It is three and a quarter years ago.
Mr. ROBB. Mr. Chairman, I would like to read the witness something from the report, which is classified.
The WITNESS. I have Q clearance; I can look at it.
Mr. GRAY. In that event, those who are not cleared in this hearing room will necessarily be excused.
Dr. OPENHEIMER. Since this is a report I wrote, is this one I may listen to?
Mr. ROBB. Absolutely, Doctor.
Mr. GARRISON. Mr. Chairman, we hoped that this might not arise, but if it is the feeling of the board that it is important to its own understanding of the case to put this kind of question, of course it is entirely acceptable to us, and we shall withdraw.
Mr. GRAY. I believe that would be best, Mr. Garrison.
(Counsel for Dr. Oppenheimer withdrew.)
(Classified transcript deleted.[1])
Mr. GRAY. Would you excuse me——
Mr. ROBB. I think counsel can come back now.
Mr. GRAY. That is what I was thinking. I don't want them excluded any more than necessary.
(Counsel for Dr. Oppenheimer returned to the hearing room.)
The WITNESS. It appears there is a reference to the thermonuclear job as being more than just in the future and my comments, Mr. Garrison, were that is is a complete blank in my memory, and I have not attempted to get a copy of that and read it before coming here. What I said was that the thermonuclear had not reached Sandia at all. While I knew the general situation and had not tried to follow it, so if it was discussed in the committee—I first said I had no memory

---
[1] Counsel not cleared for classified information.

of it, and I still haven't—but it must have been discussed, but I don't retain it. But at any rate, the thing it says there about the time of its development would have been a thing that I in signing it would have had to count on Dr. Oppenheimer, Dr. Alvarez, and Dr. Bacher as the nuclear physicists who would know and whose judgment I would have respected. But I can't recount because I don't remember any of the discussions between the three.

By Mr. ROBB:
Q. Dr. Kelly, were Dr. Alvarez and Dr. Bacher at that time, that is to say, 1950, close to the program of the Atomic Energy Commission?
A. Dr. Bacher had only recently resigned—I think it must have been within the year—from the Commission and gone out to Cal Tech. So he was pretty well up to date.
Q. How about Alvarez?
A. Alvarez was in the Radiation Laboratory and was very knowledgeable on nuclear phenomena generally, but what he would have known about this particular thing, having that knowledge, I would not know. He could well not be all current, but still capable of being so if he was given information. But Bacher certainly would have known, because he would have been a part of the deliberations. Alvarez may have known, but I don't remember what part he had in the program at the time, other than being at the Radiation Laboratory at Berkeley.
Q. Doctor, would you search your memory, please, and, sir, tell us was there any discussion in your meetings at that time as to whether or not the Atomic Energy Commission had the capabilities, the personnel, and so forth, to develop the thermonuclear weapon?
A. Any discussion of the thermonuclear problem is out of my mind. I have to say frankly that it was such a small part of the whole, and was so distant from the things that the committee itself could get hold of—I mean that the military could get hold of in the time immediately ahead—that it has not stuck with me as one of the more than minor things there. I just can't say.
Q. In other words, Doctor, is it fair to say that the thermonuclear problem, if we can call it such, was not a major part of your discussions and was not considered at that time to be important? Is that correct?
A. It was not considered at that time to be ready with enough knowledge about it to consider the emphasis in the military application area.
Q. I see.
A. It had not reached that state of development. I knew from visits from time to time up to Los Alamos and I had heard some discussions from Teller and others of the pros and cons about the development as people will discuss in that stage when there is insufficient data. Whatever discussion there was in this committee, I will have to say, not having refreshed my memory without reading it, I can't remember and would have said there was not discussion.
Q. Was there any discussion that you can recall of a second laboratory?
A. No, not in this committee at all.
Q. Doctor, when did you say you first met Dr. Oppenheimer?
A. It was at a meeting after the war in Philadelphia where he addressed either of those two societies that we belonged to. I can't remember which it was. It was very close after the war, because it had to do with these atomic problems, as I remember.
Q. I am not pressing for the exact date.
A. I would guess 1945 or 1946. It might have been even early 1947. I cannot remember without refreshing my mind. Do you remember when you made that talk in Philadelphia?
Dr. OPPENHEIMER. May I answer?
Mr. GRAY. Yes.
Dr. OPPENHEIMER. This was a joint meeting of the Philosophical Society and the Joint Academy of Sciences in mid-1945.
Mr. ROBB. We will give you the award for memory.
Dr. OPPENHEIMER. I made the speech.
The WITNESS. He made the speech. That is the first time I met him. I knew him by name.

By Mr. ROBB:
Q. How frequently have you seen him since?
A. It would average 4 or 5 or 6 times a year. Since I am only testifying directly as to one occurrence, this is the one occurrence where I had business relations, common obligations with Oppie, but I would see him at scientific

meetings or at universities 4 to 6 times a year, I would say would be a proper average.

Q. But the occasion about which you testified was your intensive experience with him.

A. That is right. This was one where I saw him in detailed action and taking a leadership as a good chairman should take.

Mr. ROBB. I think that is all I care to ask.

Mr. GRAY. Dr. Kelly, I am sorry, I don't think I can ask this question, because it involves the quotation.

May I ask this question: If there appeared in a report which you signed material which was not reflected in the discussions, would you have raised the question at the time?

THE WITNESS. Yes, I would be very meticulous about signing a thing if I didn't have views of my own from my own knowledge to sustantiate it. I would have asked afterward, or I would have had assurance from discussions that I do not now remember, that is, I would not have signed with that in there at the time I signed the report without a feeling that it reflected the judgments of exvperts in that area that I respected.

Mr. GRAY. I understand that, and I think that is quite appropriate, as you have said earlier, that you would have relied upon the three members of this committee who were particularly qualified in certain areas. I am afraid I perhaps did not phrase my question adequately.

I have no question about the reliability or your sense of dependence and confidence in the individuals concerned. My question really is, is it possible that this report could have reflected discussions which the committee did not actually engage in?

The WITNESS. I can't imagine that, because again knowing myself, I am confident that as of the time I signed it, I would not have signed it with something in that I either had not heard discussed and felt satisfied with or raised questions about. But my mind is just blank on that, because it was such a minor thing of the things to get hold of with the military. You must remember in a thing like this you had the combinations of expertness. There were questions talked about in there about tossed bombing. Lauritsen would know a lot about it. But Alvarez or Bacher would not know anything about it. So it was a combination of expertness in different areas adding up to the total. It just happens that my memory over the years had just dropped out completely whatever their discussions there were, even to the point of a comment as to the fusion weapon. Insofar as the military could do or the programing could do at that time it is somewhat gratuitous because it just was not ready for the military to get hold of.

Mr. GRAY. You felt as a committee member for one reason or another the military was not asking you to consider thermonuclear weapons.

The WITNESS. That is right. In the scope of the things that the military themselves would be concerned with, which really was the things at hand in the next year or so—there had been a meeting 2 years before, or a study of this kind 2 years before—it just was not in that ball park.

Mr. GRAY. Were you engaged in the earlier study?

The WITNESS. No; I was not in the earlier study. It was referred to. I don't remember what was in it but we had before us in the committee the study of the 2 years before. I remember having read it then, but I don't remember a thing that was in it now.

Mr. GRAY. Thank you, sir.

Dr. EVANS. Dr. Kelly, were you surprised how quickly they did develop the thermonuclear weapon after they started on it, or were you not?

The WITNESS. Sir, I was very much surprised. As a peripheral person on that and hearing the discussions about it before there were data up at Los Alamos and—they were not discussions like this war business, because I would not have been in them—but these were discussions preceding cocktail parties on the hill where Teller and others were engaged in speculations. The general views I had of the discussions there was that it was a long hard row.

Mr. GARRISON. What year was this?

The WITNESS. This was along in the 1950–51 time. I can't place it closer than that. I was up on the hill——

Mr. ROBB. May I interpose that you are in Washington. You are talking about the hill. You mean on the hill in Berkeley, Calif.?

The WITNESS. Down in Sandia we always speak of Los Alamos as on the hill. I would go up to Los Alamos about every other or every third trip to Sandia. At one of those in the early days of the nuclear physicists considering the structure and the problems involved, I remember a lot of cryogenic questions, just hearing those as a peripheral person cleared to hear it—the judgments I got and I well remember it was a thing we would not have to worry about for quite a while. "We" meaning the Sandia Corp.

Dr. EVANS. If you had to venture an opinion on it, your opinion would have been that it would have taken 2 or 3 years or longer than that?

The WITNESS. That is right. Frankly I was and am greatly surprised at the tempo of advance, and I believe that all in the program are somewhat surprised at some of the simplifications that are coming to light after you get hold of the things physically and can see them.

Dr. EVANS. Would you put the Englishman, Chadwick, in that list of people that know about it?

The WITNESS. Of course, Chadwick was out of the program. This is not the kind of thing that we can discuss with Englishmen after the Atomic Energy Act. I was not directly in the program during the war. But Chadwick, John Cockroft are among the names I would first mention in England of nuclear physicists who are very knowledgeable. But what they know about bombs, I don't know. While I see them at least once a year, we don't talk about bombs, because it is illegal.

Mr. GRAY. Do you have any further questions?

Mr. GARRISON. No.

Mr. GRAY. Thank you very much, Mr. Kelly. We appreciate your being here.

Mr. ROBB. Mr. Chairman, would it be in order for counsel to suggest a 5-minute recess.

Mr. GRAY. Yes, we will now take a short recess.

(Brief recess.)

Mr. GRAY. The proceeding will begin again.

Whereupon, J. Robert Oppenheimer resumed the stand as a witness, and having been previously sworn, was examined and testified further as follows:

DIRECT EXAMINATION—Continued

By Mr. GARRISON:

Q. Dr. Oppenheimer, would you care to make a comment about some of the matters touched on by Dr. Kelly in his testimony?

A. If the board would permit it, I would like very much to comment on it. This panel meeting about which Dr. Kelly has told you I referred to yesterday.

Mr. GARRISON. Could I interrupt a minute, please?

The board will find the reference to this panel on the second page of roman II, Membership on Government Committees, No. 5 (b).

The WITNESS. It was next to the last item in my testimony yesterday just before I told about Vista. I told you the personnel and the critical atmosphere of the war. I would like to stick as much as I can to nonclassified things.

I believe I told you yesterday two things about the period of this report. One was that it was the period after Chinese intervention in Korea when general war was very much in everybody's mind, not as a remote but as an immediate thing.

The second was that it was a low point in the prospects of the super. What you have heard read reflects that opinion. Dr. Kelly would certainly not have been more than a bystander in the formulation of this opinion. As he said, this was not his job. But the impression created in his testimony seems to me to need amplification.

Bacher was a member of the Atomic Energy Commission until sometime before. He was a continued consultant to Los Alamos and spent a good deal of time there.

General McCormick was the Director of the Division of Military Applications to the Commission, and was responsible for Los Alamos, received regular reports from the laboratory, talked with everyone involved that he wished to talk with and was well informed.

He is not a nuclear physicist, but he knew the views of nuclear physicists.

Lauritsen is a nuclear physicist. His whole life has been spent in nuclear physics except that part spent in atomic development. He was a consultant during the war and has been very close to the program of all forms of atomic development.

Alvarez is a nuclear physicist of distinction and was, I believe, one of the initial promoters of the crash program for the super, and has always had a great interest in the work.

General Parsons was a member of the evaluations group at that time. He had been at Los Alamos. His job was to keep in touch with current developments.

General Nichols—his status at that time I have forgotten, but I think he was in research and development in the Army.

All of these men had access to every document and report that existed and were knowledgeable not as to deep problems of contemporary physics, but as to the practical problems and evaluations which were current in the various places where work was going on or evaluation considered. Berkeley was one of them and Alvarez was there. I, therefore, think that there was a very substantial group of people, McCormick, Parsons, Bacher, Lauritsen, Alvarez, and myself, who knew what was believed at that moment and who had a chance to evaluate it critically.

Any judgment that was expressed about the thermonuclear program could have been expressed only with the consensus, the complete agreement of all members of that committee who knew about it and the undertaking on the part of those who didn't.

One other thing: Walter Whitman was a member of the General Advisory Committee and had complete access to all reports and so on, and he was, I think, a member of the committee.

The only thing I wish to protest is the suggestion that I was the only person competent to judge and that I sneaked a conclusion into the report that had not been thoroughly hashed out. I also concur with Dr. Kelly's statement, of course, that his primary interest was in other aspects of it.

Do you wish to question me about that at all?

Mr. GRAY. Mr. Robb, do you have any questions?

Mr. ROBB. No, not at this time.

Dr. GRAY. I think not. Dr. Oppenheimer. Would you proceed?

By Mr. GARRISON:

Q. Would you tell the board, now, Dr. Oppenheimer, about your appointment to the General Advisory Committee in 1946 and then something about its personnel and its purposes?

A. I thing I did describe my appointment which was in late 1946. Our first meeting was in early 1947. I was held up by bad weather. I thing Dr. DuBridge and I were both held up by bad weather and arrived late for the meeting.

Mr. GARRISON. This is on the first page of roman II, item 4.

The WITNESS. When I arrived I found the other members of the committee had held a meeting and elected me chairman. After consultation with the Commission itself, I accepted that position. We agreed that we would elect the chairman at every subsequent meeting, that is, the first meeting of each year. I was reelected at first without any concern on my part, but later with great concern. I will come to that when we come to that time in the history.

I think you have the names of the members of the committee.

Dr. EVANS. Yes.

The WITNESS. It is in my letter. It would only bore you to repeat the names.

Mr. GARRISON. They are right before the committee.

Mr. SILVERMAN. These were not all members at the same time.

The WITNESS. No. But I think that is spelled out in my answer. It is obviously an eminent committee and a varied committee. I can assure you that it was not a committee that regarded itself as subject to manipulation, or that it was subject to manipulation.

By Mr. GARRISON:

Q. What was the statutory function of the committee?

A. The law spells out that it is to advise the Commission on the scientific and technical aspects of research, development, production, materials, something along those lines, a rather clear mandate.

We, of course, from the very beginning recognized with relief that the job of decision making, the job of negotiation with other parts of the Government, the job of management, the final job of determination, rested elsewhere. It rested with the Commission, with the Department of Defense that was to establish military requirements, or rather, with the President who on the advice of the Department of Defense was to establish military requirements; with the Congress that carried out the appropriations. Our job was limited to advice.

A scientific adviser has, I think, one overriding obligation. It is his principal one in which he is delinquent if he fails, and that is to give the best fruits of his knowledge, his experience, and his judgment to those who have to make decisions.

He must attempt to study the problems that are put before him, to analyze them, to relate them to his own experience and to say what he thinks will happen and what he thinks won't happen; what he thinks experiments mean; what he thinks will happen if a program is developed along certain lines.

It is not possible to give this advice except against a background. That background is the kind of questions you ask; very often the things that are assumed in the questions you ask rather than state. If you are on your toes sometimes you can say that the question is not asked in the right way, that a different question should be asked. But by and large you will find yourself advising on what concerns the people to whom we are feeding advice. This through the years changed a great deal.

I have already testified that as of early 1947, the prospects of any meaningful international action in the field of atomic energy were largely gone. The problem that we faced then was to devise a program which would regain some of the wartime impetus and vigor, and above all to make available the existing know-how, the existing plant, the existing scientific talent, to make this available in the form of actual military strength.

It was not so available as of the first of January 1947. I need not go into the classified details. They are certainly available to you if you want them.

In the period characterized by the Russian bomb and the war in Korea and the Chinese intervention, the background of many questions was immediate readiness for general conflict, or the best we could do with regard to that.

In the last days of my service on the general advisory committee, one of the obvious questions was this: Since things are going quite well for us, what can we do, what should we do, to be prepared against enemy action? No doubt the enemy will have some time or other similar success.

These changes in the nature of the background were always there and I don't want to pretend that scientific advice in practical matters is like doing an experiment just for the purpose of satisfying your curiosity.

The GAC did not, strictly speaking, abide by its terms of reference. I would say in 2 or 3 ways it did not. In the first place in the early days we knew more collectively about the past of the atomic energy undertaking and its present state, technically and to some extent even organizationally or some parts of it, than the Commission did.

The Commission was new; its staff needed to be recruited. We knew about Los Alamos; we knew about Sandia, we knew about the Argonne Laboratory at Oak Ridge, and it was very natural for us not merely to respond to questions that the Commission put, but to suggest to the Commission programs that it ought to undertake; to suggest to the Commission things that needed doing of a technical sort.

Very frequently we would be asked, What will be the best way of organizing this; what will be the best conditions for recruiting scientists and for making their work productive? We never regarded that as a serious violation of our terms of reference.

As time went on and the Commission through its staff and actually in its membership knew more and more about the program, we tended to let the questions come from them. We would be confronted by great piles of documents and sometimes a set of questions about them at the beginning of every meeting. We would try to answer their questions rather than digging up from our own experience things that we knew.

This transition took place as the members of the committee became more remote from direct active participation in the program and as the Commission's understanding of its problems improved.

Sometimes the Commission would address to us questions which were not obviously related to scientific and technical advice. I would mention at the least three.

The Commission reviewed with us its security procedures, the procedures, I think, under which we are now sitting. I believe their interest in doing that was to find out whether these would seem fair and reasonable to scientists. I don't believe we responded in writing to that, but we probably said that this looked like a very fair setup.

The Commission reviewed with us very often the hassle about the custody of atomic weapons. The act provides that the President shall arrange their transfer from the Commission to the military services. This involved, I guess, both technical and political problems. We in this case confined ourselves to talking about the technical problems and pointing out that there were much more important political ones which it was not our job to pass on.

The very broad terms—and this, of course, I am coming to in a good deal more detail—in which the Commission addressed to us the question of the super bomb was another example, I think, where it did not consult us purely on the technical problem, but asked advice in which supposed technical competence and general good sense were supposed to be blended.

I haven't got all the examples, and I know many times we bowed out and did not answer the questions which were not technical and scientific. Often we were seduced into answering them.

The committee, during my chairmanship, met about 30 times in regular stated meetings. I think the most impressive thing—maybe we did some good—but the most impressive procedural thing is that the committee had 9 members; that means 270 attendances, and I believe there were not more than 5, or something close to that number of absences. That is, almost always everybody would be there and it was a rare meeting where two people, if there was such a meeting, would be absent. There were occasions where a member was abroad, as in the case of Dr. Seaborg in our meeting in October 1949. But they were not frequent.

This active interest and participation, I think, shows that the members of the committee, whatever the truth was, felt that what they were being asked to do was important to the Nation and they had a contribution to make.

We had several subcommittees appointed very early in the game, that is, into the natural divisions of the problem: A subcommittee on weapons, with Dr. Conant of Harvard as chairman; a subcommittee on reactors, of which, I think, Dr. Cyril Smith was chairman; and a subcommittee on research, of which Dr. DuBridge was chairman.

We also had an ad hoc subcommittee which lasted only a limited time to consider the problems of the best possible way in which existing or shortly to be available plant and existing raw material could be used to increase the quality and usefulness of the product, here, I think, only from the point of view of weapons; that is, how did you operate this plant? Did you operate them in parallel; were they independent units, and so on. That was under the chairmanship of Fermi, who was from the University of Chicago.

The committee as such had some foreign relations.

By Mr. GARRISON:

Q. By "foreign relations", you mean with other agencies of Government?
A. Thank you; with other agencies of Government.

We met quite frequently, especially in the early days, with the military liaison committee. It was usually present during our final report to the Commission.

The committee, at least once or more than once, appeared before the joint congressional committee. Its members appeared in open sessions during the spring of 1949 and in secret sessions.

We once, I think, called upon the President and wrote him an unclassified progress report. At the end of my service we wrote him a top secret progress report which I sent over and talked over with him when I visited him.

But by and large our relations were only those established by law to advise the Commission and we stuck pretty closely to that.

There is an important qualification to this. Many members of the committee were consultants to one or another of the laboratories. Rabi, for instance, was a founder of Brookhaven and very much interested in it. Fermi was a consultant to Los Alamos. So was Von Neumann, who came on later.

Many of the members of the committee had connections with Oak Ridge and the Argonne Laboratory. In addition to that we were, of course, a part of the general traffic of scientists. We knew each other. Therefore, we had another function besides advising the Commission on technical matters, and that was to represent to the Commission when it was a clear and obvious thing, the views of our colleagues and to represent to our colleagues the views of the Commission.

I mean by this, those who were engaged in the work, if the matters were classified; those who were not engaged in the work if it were such a thing as the support of basic science or a fellowship program or anything like that.

We got our information initially because we had it in our heads and had some reports left over from earlier times, overwhelmingly from Commission sources, but to some extent also by direct visits to the laboratories and by calling in directors of the laboratories, by calling in staff from the laboratories, so we tried to keep up to date.

I think we had Bradbury on very many times to tell us about the weapons work in the early days. Our secretary was John Manley, and he was Associate Director of Los Alamos, so he would bring a report to us, sometimes semiofficial and sometimes informal, of what was going on.

We consulted with the directors of all the laboratories at one time or another, and where relevant, with the people in charge of production plants.

We did one other thing which perhaps was not quite within the terms of the statute. Occasionally we would propose for the Commission, or rather, prepare for the Commission a statement of views which we would authorize them to make public. These were nonclassified statements in hearings before the Congress or in any way that they wanted.

I remember one such occasion when we thought a public statement would be desirable to set the atomic power problem in some kind of perspective so that people would not expect that coal and oil would be obsolete the day after tomorrow. We drafted a statement of this kind. First it was secret and then we got all the secret stuff out of it and handed it to the Commission. It used it in some way—I think not a terribly effective way—in a report to Congress. I think it was in regard to the use of isotopes, the fellowship programs, the promotion of basic research. We wrote several documents for the Commission to use if it would do them any good.

By Mr. GARRISON:

Q. When you say, Dr. Oppenheimer, that the committee acted beyond the statutory frame of reference, what you really mean, I take it, is that you did not act in violation of the statute?
A. Oh, no.
Q. But that it simply came about that the Atomic Energy Commission looked to your committee for help and guidance in ways that perhaps had not been forseen?
A. That is exactly right. The Commission relied on us very heavily, especially at the beginning, and relied on us for lots of things that were not provided for in the Act; where we felt we could help them we did. Our concern was to give them every possible encouragement and support.
Q. And then, as you testified a little earlier, as the Commission became more and more expert in its own field there was correspondingly less dependence for this kind of assistance from the committee?
A. That is right.
Q. Now, would you tell the board something about what the committee actually did and begin with the first meeting?
A. My recollection is not clear as to what happened at the first and what happened at the second meeting, but I think this is perhaps not too important.

Very early in the game we thought it important to see whether we agreed or had any views at all about what the job of the Commission was. That, of course, was the Commission's business to determine, but the nature of the advice we gave would be dependent on that.

Without debate—I suppose with some melancholy—we concluded that the principal job of the Commission was to provide atomic weapons and good atomic weapons and many atomic weapons. This referred to atomic explosives. There are other things, like the atomic submarine that you can call an atomic weapon, but that is not what we had in mind.

We thought it had three other undertakings. We thought from the first that however remote civil power might be, the Commission had an absolute mandate to do everything it could economically and fruitfully to get on with the exploration of it. We thought that the Commission needed to respond to requests from the military and needed to alert the military establishment as to other applications of atomic energy of military use, of which propulsion, radiological warfare may be two examples. I won't attempt to evaluate them at this moment.

The third thing that we felt—and it was not really third in our feelings, but simply in a budgetary and practical way—was that the Commission had a mandate to stimulate basic science in this country: The training of scientists; I guess just the acquisition of knowledge is what the law states.

At that time there existed in the Office of Naval Research one very good Government agency which was promoting basic science in many different fields with great forethought, wisdom and skill. Some of the things the Office of Naval Research did touched on the field that the Commission was in on atomic science. We never had any feeling that it was bad for the ONR to be in that. But this was to come up over and over again and I will return to it a little bit later.

These were the principal themes that occurred to us at the first meeting and the one that separated itself by urgency and importance in our own minds was the weapons field.

Q. That required attention first of all to the state of affairs at Los Alamos?

A. Yes. I think perhaps I should say that we did at one early meeting consider whether Los Alamos was the right place for weapons development.
Q. This is now 1947?
A. This would be early 1947. It was set up during the war for reasons which I went over yesterday. It is remote. It is expensive. It does not have very free access to a university or laboratories not under its control. There could have been arguments that a fresh start with something of the vigor that Los Alamos had when we began it might have been desirable.

We concluded at the first meeting that this was impractical; that Los Alamos had proved itself and its survival value by being there, by having a good staff, it was working on atomic bombs. It was not only working on atomic bombs but doing a lot of miscellaneous physics and chemistry. But it existed and the notion of starting up something else or tearing this down seemed to us full of dangerous delay.

So our first set of recommendations to the Commission was addressed—I think there were a lot at one time—but at any rate first among the recommendations were the recommendations to get Los Alamos going as a really first rate place.

The Commission had asked us either at our first or second meeting to review the report I described yesterday on the job in atomic energy which we had written for Mr. Stimson's panel. They asked us the question: Have any of these objectives been attained? They had not been. The time was rather short. The objectives were not easy. I think we said strictly speaking none has been attained. There are some now that ought to be added that have come up in the meantime. That report was not entirely complete.

We suggested that every inducement be made available to make work at Los Alamos attractive in the way of salaries and housing, but above all in the morale sphere in the sense of giving the men who were there the feeling that they were doing something vital for their country and in getting abroad in the country the sense that Los Alamos was not something left over from the last war, that work on the atomic bombs was somehow not an entirely creditable occupation, but quite the contrary feeling that there was nothing the nation needed more.

This did result in vast building programs at Los Alamos, in the expansion of the laboratory, in the availability to the laboratory of a great many people who were not trafficking there at earlier times. People go out now for the summer months and have been for the last 5 or 6 years and they come as consultants.

There is hardly a clear and qualified scientist in the country who is not available to Los Alamos for consultation or for such things as he is good for.

They have established a scheme of subcontracting which enables them to draw in even further resources than they can put on this relatively limited mesa.

I am not going to take all the recommendations of our early meetings. In the first place I have not looked them up and I don't have them in mind. I will rather follow the weapons themselves.

There had been, I think, some thought about weapons development after I left Los Alamos. There was one meeting which I could not attend on the thermonuclear program, and there were lots of things left over from the wartime to get people interested in making better weapons, better here meaning a whole lot of things. It means obviously getting more bang for a buck. It means more economy in the use of fissionable material. It means getting weapons which give you the maximum versatility in the kind of delivery system we have, so you don't have to use very big bombers and so on.

It means versatility in the size of weapons and their explosive effects. It means the ability to use the fissionable materials that are produced in some reasonable proportion to how they are produced and in some reasonable recognition of overall economy of neutrons and production facilities.

Very early in the game it became clear to us that nobody was going to pay attention to improving weapons. All that happened is that there were lots of blueprints and lots of models lying around and the only way to get this business really moving was through a testing program. The payoff with atomic weapons is to see if they really work as we think they do.

Sometimes you do this test to prove out a model which is essentially what you think it right. Sometimes you do it in order to see, as well as you can by experiment, how things are working in the explosion and guide you in future design. Good tests usually combine these features.

I believe we were extremely strong in urging that a test facility be established. I know that we worked quite hard to get accepted the initial Los Alamos program for the Eniwetok test which were a little more ambitious than was generally approved and where we felt they were really very much needed.

We were worried about the test site out in the Pacific as the only test site because of the cumbersomeness and the long advanced planning that was required. But the problem of getting a continental test site was one to which we could not contribute much except to say that it was very much needed and that we hoped it would be available.

Mr. GRAY. May I ask, when you say "we", you are always referring to GAC?

The WITNESS. For this field I am talking about the GAC. There were points on which we had differences of opinion. They were not very frequent. I believe in the weapons field they were not very major.

There were differences of opinion about the proper way to get reactor development going and perhaps some difference of opinion about the value of various forms of military propulsion. What I am reciting now I believe to be unanimous.

By Mr. GARRISON:

Q. Dr. Oppenheimer, in all of the recommendations that were made throughout these years from 1947 to 1952, during which you were Chairman, did you concur in those recommendations yourself personally? I mean to say that if there were differences of opinion, were there any instances in which recommendations were made in which you did not concur?

A. I think there may be that there were, but I don't remember them. They were not on points that seemed of great importance.

Mr. GRAY. May I ask as a matter of practice if the committee made a report and then if members had some difference of view they were reflected in a separate memorandum?

The WITNESS. The way it worked is the following: Maybe I had better go back to procedures. The meeting was generally opened by a meeting with the Commission, sometimes with the military liaison committee, at which the Commission would discuss with us what was on its mind, what advice it wanted.

There would be a period of briefings in which documents were brought in and the staff came and very often members of the various laboratories came and told their story. Usually there was more to consider than could be adequately considered in a 2 -or 3-day meeting.

We then would go into executive session, go over the program aloud and begin to talk about questions. Sometimes it was clear that the answer was obvious. Sometimes it was very tough. Sometimes we felt that the right answer would be very difficult for the Commission to carry out and we had the problem of giving our advice to the Commission in a way which was both honest and useful.

When we were about clear as to what we had to say we would meet again with the Commission, and occasionally with the military liaison committee, and at that point I would usually summarize out loud what our thoughts were and a record would be made of that. If I knew of divergences of opinion, I would call on those who had any divergent opinion to express their differences; If I didn't know about any, in any case I would go around the table asking for comments. There almost always were some comments because I had forgotten something, or I had given an emphasis which was not right, or some one wanted to strengthen what I had said.

This oral report I then made the basis of a letter to the Commission which was our immediate report to them. This was circulated to the members of the committee who could approve it and it was brought up for approval and amendment at the subsequent meeting as to whether it was an adequate expression of the Commission's views.

I remember one instance in which there was a dissent—one and only one instance—from my representation of the view of another member who said I had not gotten it straight and who wrote a letter amplifying.

We also, not always, but normally kept minutes. I say not always because I have the impression that the most controversial meeting in the light of history, that of October 1949, minutes were not kept. The meeting was too hectic, or something. The secretary never wrote them up, but wrote notes afterward. You know that better than I do.

The reports of the Commission, of course, though they usually were top secret or often top secret, were the Commission's property, and if it wanted to send them over to the Joint Congressional Committee, or the military liaison committee or anyone else, that was fine with us.

The minutes of the meeting, which often told what kind of hassles we had, what kind of arguments or considerations, we made available to the Commission to throw whatever light they could on what we knew and what we thought, but we asked them not to distribute the minutes since they identified individuals as saying this or that.

I think this is how the record was kept.

By Mr. GARRISON:

Q. I wanted the board to be sure, Dr. Oppenheimer, that when you recount, as you are about to do—and, indeed, as you have already begun doing—some of the important things that the committee recommended to the Commission and urged upon it in the national interest, they were all actions in which you yourself wholeheartedly approved.

A. If I had dissented, I would certainly have said so.

Q. So that the Board can understand that, you were really talking as much about your own views and contributions as you are about other people.

A. Yes, although I need to make one point clear. It is very important for a chairman to get everybody into the act and not to dominate a meeting. I think my normal practice was to bring up a question and then see what other members of the committee would say. I would not wish to testify, and I can't testify, that the views which I came out of the meeting with were always the same as the views I went into the meeting with. This was a matter of discussion. Sometimes new facts were brought to light, sometimes we learned things we had not known before; sometimes people talked me out of what I originally thought. But I certainly never incorporated in a report anything different than I thought was the best advice that I would give at that point.

Q. You have spoken now about the stress which the committee laid on the importance of tests for the development of atomic weapons. Do you want to say something about some of the other aspects of weapon improvement which you pressed for in those days.

Dr. EVANS. Pardon me, but may I ask one question about these tests before you leave that?

Dr. Oppenheimer, were there what we might call bad tests that did not come up to your mathematical calculations?

The WITNESS. I am not sure whether the answer to this is classified or not?

Dr. EVANS. Maybe I should not ask it.

The WITNESS. The security officer has left, but I will take the chance.

Dr. EVANS. I will hold the question.

The WITNESS. All right. The answer is of some interest, but not, I think, in connection with whether I am fit to serve the country.

Mr. GARRISON. If the chairman would like, we would be glad to step out.

The WITNESS. Let us not have any more classified stuff than we have to.

I ought to say that, at our first meeting or two—I don't remember which—we brooded to a very considerable length about the thermonuclear program. I think the state of affairs was that not much was known about it; it had not been pursued very vigorously, and the unknowns overwhelmed the knowns.

By Mr. GARRISON:

Q. Just to recapitulate, the work in the thermonuclear field began when at Los Alamos?

A. The theoretical work began in Berkeley in the summer of 1942. The thermonuclear work was pursued merely as a theoretical job and not a development job. I think it would naturally have been somewhat intensified after the war with the view of making better measurements and better calculations because it was one of the interesting things to do.

The question we tried to ask ourselves was, Is there enough in this so that it ought to be pushed, or is it something that will be a distraction from the very immediate job of getting some weapons into the places where they are needed? Our answer was, I think, the following: That it was a very interesting problem or set of problems; that if work were going on at Los Alamos it would attract first-rate theoretical physicists and that the probability was that if people studied the thermonuclear problems at Los Alamos this would help the other program rather than hurt it because it would have the effect of increasing the brains and resources of the laboratory.

I will have to give you a complete review of the thermonuclear thing, but this was our initial recommendation.

We made a number of other observations relevant to the weapons program. I think one of the important ones—I am not sure we were the first to do it—was to keep asking the Commission not how many bombs should they make, because that was not our job—that was the job of the Military Establishment—

but what were the real limits on how many they could make. How much material could be made available? Because, even though very great strides were made between 1947 and 1949 in the effectiveness with which material was used, there was still the question, Is the plant we have being used in the best possible way? Is there any inherent limitation on the plant? Is there enough raw material to sustain more plant? Is there any way in which you can relieve the limitation on raw material? Does this come back to a dollar limitation?

We addressed to the Commission from time to time questions intended to make clear to the Military Establishment that the requirements they were placing for atomic weapons were perhaps all that could be done right then with existing plant, raw material, operation, and bomb design, but by no means all that you could do if you really set to work on it.

The very large expansion programs which, of course, were not approved or formulated by us were certainly in part stimulated by the set of questions. There have been several expansion programs, and the whole atomic-weapons capacity has risen enormously. It took quite a while for this to take hold, but I think we started on it fairly early.

We were very concerned—I think probably this concern reached its maximum during the Korean war but started earlier and continued later—to adapting atomic warheads so that they could be used by a variety of carriers. This sometimes meant developing designs which were not, from the point of view of nuclear physics, the most perfect design, because you had to make a compromise in order to get the thing light or small or thin or whatever else it was that the carrier required. But experience showed that almost every improvement that you made in trying to make, let us say, a physically smaller atomic bomb was reflected in an improvement in the performance of the larger ones. So, as this thing began to unroll, you could not really tell whether an effort aimed at making an atomic bomb that you could shoot out of a machinegun— to take an obviously unclassified example—would not also help the very large bombs which are the most efficient.

This had something to do with trying to bring together the enormous program, of which our chairman surely knows a good deal, of missiles and the adaptation of weapons plans and missile plans. In this connection we welcomed the building up of Sandia that Dr. Kelly has described to you and tried generally to get as much coordination between the hardware side, the military application side, and the development of the atomic explosives themselves. I believe we were rather early in this preoccupation, which later became quite general.

We were concerned with flexibility and made a number of recommendations to the Commission which I need not spell out, the purpose of which was to be sure that if, during a war, you found out bombs you had were not exactly the one you wanted, you could do something about it. We felt that no amount of crystal-balling would make it certain that your stockpile corresponded to what you really needed in combat.

We suggested a variety of devices by which you could take advantage of what you learned in combat and come up quickly with what you needed.

I have listed these as some of the things about weapons. I have obviously left the hydrogen bomb for a separate item. I might run rather briefly through the other aspects of the Commission's work that I have mentioned.

The war almost stopped the training of scientists in this country and this started up again at an accelerated pace under the GI bill and the rest of it. But it was very clear that there were not enough people in the country to do the things that were needed. The couple of billion dollars which we now spend on research and development is not all spent on the salary of scientists, but it is very often bottlenecked by scientists.

It seemed to us that the source of all this was universities and university training. It seemed to us that the source of all this was the research in universities, in other words. It seemed to us that the source of the good work that had been done in the war was not in applied science but in the pure scientists who had learned their stuff in the hardest of all fields, the exploration of something that is really not known and really new.

We encouraged the Commission to take a number of steps which we thought would help this. They have, first of all, their regional laboratories, of which Brookhaven is a good example, Argonne is a good example, Oak Ridge and Berkeley. There we tried to get the Commission to do something which was only partially successful but has been quite successful in Brookhaven, and that is to separate as sharply as possible the secret and sensitive things which ought to be guarded and restricted and the things that are just published all the time in the journals and, therefore, to make it possible for these facilities to serve as wide a group of people as possible without involving delays and clearance procedures and in order to maintain really secure the things that were secret.

We tried very hard to get the Commission to support work which was not directly obviously related to the practical applications of atomic energy. There were arguments in those days that the Commission was so short handed, so in need of physicists, that the best thing they could do was to make it hard for physicists to get jobs so that they would come and work in the various laboratories. We thought that was quite wrong—that the best thing they could do was to support physics in the universities, that this would provide the young men—and it has, of course—who would be able to man their various laboratories in the years to come and they should do at least as well as the Office of Naval Research in those fields of science which by statute they were supposed to be responsible for—atomic science and chemistry, physics, geology. They have done this, and anyone who picks up a contemporary physics journal will see in it innumerable examples where it says that this work was supported by the Atomic Energy Commission.

The level of activity in physics, especially, but also in chemistry, has been very much raised by their efforts, and the number of people practicing has been enormously raised. What is more than that, if you now go to a contemporary Atomic Energy Commission Laboratory, a lot of the bright ideas and a lot of the best work is done by men whose names were not known 7 or 8 years ago and who have precisely come up through university training in the meantime. This is true of Los Alamos, and it is true of all the others.

I think on this we probably pushed the Commission and they regarded us as people who were, after all, largely professors and university presidents and we were pleading a special interest. We did plead a special interest, but we believed it to be in the national interest, too.

Where possible in basic science, we urged the Commission to make its unclassified facilities available on a worldwide basis. A good many scientists from friendly nations have come here to do experiments, to learn techniques, and also to teach us what they knew; and there are magnificent examples of international collaboration that have taken place in the Commission's laboratories. I think the most striking is probably known to you.

In 1947, I guess, the big accelerator at Berkeley started operation. Maybe it was 1946. People immediately looked to see whether the new high energies that were being provided were creating mesons which we knew were created in the cosmic rays but which were not artificially created before. They looked for months and months, and the reports were negative. This seemed very puzzling from the point of view of the theory.

A young Brazilian who had been studying in England arrived at the radiation laboratory, knew the technique used there, exposed a few photographic plates, and there were the mesons. This is a small illustration of the need from the scientific point of view of the international collaboration.

I think I need not point out that it is also a very limited but a very healthy element in the general structure of our alliances and in the good feelings that exist between people in other countries and here at home.

The Commission has, I think—and we so represented it—an obligation to make available to industry and to technology and medicine those facilities which by statute it and only it can operate. It has fulfilled this very well. The distribution of isotopes had been begun by the Manhattan District. It has been enormously expanded and speeded up and improved by the Commission. This is one example.

The use of reactors for both secret and nonsecret work is another example. I don't know how much you have found it profitable to leaf through the general advisory committee reports. I am sure you will find in them just countless occasions where either in general terms or in specific terms we tried to steer the Commission on a course which would enable it to do the maximum for American science.

I am not so proud of our record in the reactor program. This we never managed to give as effective advice about as I wished. We worried a lot about it, and you will find that if the advice was not good it was at least copious.

I think one reason for the difficulty is that progress in reactor development, whether for civil or military purposes, is a very expensive thing. It is the kind of thing you don't do in a small university laboratory. It is a big industrial enterprise. It may cost $10 million; it may cost $50 million. It is not something you can just try out for size.

We found it very hard to compose the conflict between the need for an orderly and comprehensive and intelligible program of reactor development and the inevitable enthusiasm which groups would get to have for their own pet baby and which maybe was a reactor which was not especially illuminating from the point of view of the program as a whole. We thought at one time that this could be helped by centralizing the reactor development work and so recommended to the Commission. This was one of the recommendations which was opposed. Fermi thought this was bad advice. In any case, it never happened. So we don't know whether it would have been good or not. We tried very hard to get some kind of policy committee of the people who knew about reactors, and that was formed, a committee of Oak Ridge and Argonne and General Electric scientists, so that they would get some agreement and not all push their own babies.

We strongly urged the Commission to get somebody in Washington who was an expert in reactors, and it turned out to be the Director of Reactor Development, Dr. Hafstad, who held that job from the beginning. I am not clear that he will be on any of your lists.

What in the end happened was that we began to sort out better and the Commission began to sort out better what the reactors were for, and therefore have more rational criteria of which ones to build. They were for production, the production of materials for bombs. They were for military propulsion. They were for learning about reactors so that you would know how to build the next ones better. These 3 purposes I think we recognized in 1947 or 1948.

After that I think the Commission's program began to take extremely good shape, and we have moved very far. We always liked the submarine reactor, not only because it would be a usable thing in warfare, but it looked close enough, to civil power, relevant enough to civil power, to be of interest from that point of view, too.

I believe we dragged our feet very much on the initial plans for flying aircraft with nuclear power. It seemed to us a very long-range thing and one that ought to be approached in the spirit of research rather than have a definite development and commitment. When I last heard about it, this was the state of affairs.

By Mr. GARRISON:

Q. This brings us logically to the report on the H-bomb in the fall of 1949. I don't know whether the board would think this was an appropriate point to adjourn or whether we should go ahead and start on it.

Mr. GRAY. I think we should start on it, Mr. Garrison, if you don't mind.

By Mr. GARRISON:

Q. The story begins, I take it, with the Russian explosion of an atomic bomb on September 23, 1949?

A. I don't think the story begins there. I will go back a little bit. We can begin in the middle and go both backward and forward.

In September of 1949, I had a call from either General Nelson or Mr. Northrop. * * *

A little later I came down to Washington and met with a panel. I see it says in my summary that this was advisory to General Vandenberg. I never was entirely clear as to who the panel was supposed to advise.

Mr. GRAY. This appears in the exhibit?

The WITNESS. That is right. This was Admiral Parsons, Dr. Bacher, Dr. Bush.

Dr. EVANS. Where is that?

Mr. ECKER. It is item 6, II.

Dr. EVANS. Yes; I have it.

The WITNESS. I think I had seen a good deal of the evidence before the panel was convened. In any case, we went over it very carefully and it was very clear to us that this was the real thing, and there was not any doubt about it. We so reported to whomever we were reporting. I think it was General Vandenberg. This was an atomic bomb, * * *

Yesterday you read evidence that in 1948 I was not thinking it would come so soon. * * *

I went over to the State Department where the question was being discussed— I was asked to go over by the Under Secretary—should this be publicly announced by the President and I gave some arguments in favor of that.

I don't know who finally resolved the matter, but the President did make a public statement. I was taken up to hearings before the Joint Congressional Committee. General Vandenberg certainly appeared and probably Admiral Hillenkoetter and other people whom I have forgotten. The committee was quite skeptical as to whether this was the real thing.

Mr. GRAY. Is this the GAC?

The WITNESS. No, the Joint Congressional Committee. They were quite skeptical and I was not allowed to tell them the evidence. It was understood that this was to be kept secret. All I could do was just sound as serious and convinced and certain about it as I knew how. I think by the time we left the Joint Congressional Committee understood that this event had been real. I do remember Senator Vandenberg's asking me, and it was the last time I met with him—he became ill not long thereafter—"Doctor, what do we do now?" I should have said I don't know. I did say we should stay strong and healthy, and we make sure of our friends. This was immediately before the General Advisory Committee meeting.

The Committee had a whole lot of stuff on its docket. I have forgotten the details. There was a docket for us. We disposed of that business, and we talked about this event. At that point Dr. Rabi returned. He had been in Europe on the UNESCO Mission. He read about this in the newspapers. The President had announced it. He said very naturally, "I think we ought to decide what to do. I think we ought to advise the Commission." I opposed that. I think most all other members of the Committee did on the ground that it might take a little while to think what to do and also on the ground that many of the things to do would be done against a framework of governmental decision as to which at that point we could only speculate.

During October or late September, I think October, a good many people came to see me or called me or wrote me letters about the super program. I remember three things. Dr. Teller arrived. He told me that he thought this was the moment to go all out on the hydrogen-bomb program.

Mr. GRAY. May I interrupt? I am sorry. This is following——

The WITNESS. Following the GAC meeting of September and prior to the meeting in October.

Mr. GRAY. Yes.

The WITNESS. Dr. Bethe arrived. I think they were there together or their visits partly overlapped, although I am not sure. He was very worried about it. He will testify.

By Mr. GARRISON:

Q. About what?

A. About the thermonuclear program, whether it was right or wrong; what his relations to it should be. I assume he will testify to that better than I can. It was not clear to me what the right thing to do was.

Mr. ROBB. You say to you or to him?

The WITNESS. To me. I had a communication. I can't find it as a letter, and I don't know whether it was a letter or phone call. It was from Dr. Conant. He said that this would be a very great mistake.

By Mr. GARRISON:

Q. What would be a great mistake?

A. To go all out with the super. Presumably he also will testify to this. He did not go into detail, but said if it ever came before the General Advisory Committee, he would certainly oppose it as folly.

The General Advisory Committee was called to meet in Washington, and met on two questions which were obviously related. The first was, was the Commission doing what it ought to be doing. Were there other things which it should now be undertaking in the light of the Soviet explosion.

The second was the special case of this; was it crash development, the most rapid possible development and construction of a super among the things that the Commission ought to be doing.

Now I have reviewed for you in other connections some of the earlier hydrogen-bomb tale. The work on it in the summer of 1942, when we were quite enthusiastic about the possibility, my report on this work to Bush, the wartime work in which there were 2 discoveries. 1 was very much casting doubt on the feasibility, and 1 which had a more encouraging quality with regard to the feasibility. Of the talks with General Groves in which he had indicated that this was not something to rush into after the war. Of the early postwar work, prior to the establishment of the Commission. Of our encouragement to the Commission and thus to Los Alamos and also directly to Los Alamos to study the problem and get on with it in 1947 and 1948.

The GAC record shows I think that there were some thermonuclear devices that we felt were feasible and sensible and encouraged. I believe this was in 1948. But that we made a technically disparaging remark about the super in 1948. This was the judgment we then had. I remember that before 1949 and the bomb, Dr. Teller had discussed with me the desirability of his going to Los Alamos and devoting himself to this problem. I encouraged him to do this. In fact, he later reminded me of that, that I encouraged him in strong terms to do it.

Now, the meetings on——

By Mr. GARRISON:

Q. The meeting of October 19?
A. The meeting of October 19, 1949. Have we the date right?
Mr. ROBB. October 29.
The WITNESS. October 29. I think what we did was the following. We had a first meeting with the Commission at which they explained to us the double problem: What should they do and should they do this? We then consulted a number of people. * * *

We had consultations not with the Secretary of State, but with the head of the policy planning staff, who represented him, George Kennan, as to what he thought the Russians might be up to, and where our principal problems lay from the point of view of assessment of Russian behavior and Russian motives. We had consultations with the Military Establishment, General Bradley was there, Admiral Parsons, I think General Hull or General Kyes, head of the Weapons Systems Evaluation Committee, General Nichols, probably. I won't try to recall all. Also Mr. LeBarron.

Prior to this meeting there had been no great expression of interest on the part of the military in more powerful weapons. The atomic bomb had of course been stepped up some, but we had not been pressed to push that development as fast as possible. There had been no suggestion that very large weapons would be very useful. The pressure was all the other way; get as many as you can.

We discussed General Bradley's analysis of the effects of the Russian explosion, and what problems he faced and with the staff, of course.

Then we went into executive session. I believe I opened the session by asking Fermi to give an account of the technical state of affairs. He has always been interested in this possibility. I think it occurred to him very early that the high temperatures of a fission bomb might be usable in igniting lighter materials. He has also an extremely critical and clear head. I asked others to add to this. Then we went around the table and everybody said what he thought the issues were that were involved. There was a surprising unanimity—to me very surprising—that the United States ought not to take the initiative at that time in an all out program for the development of thermonuclear weapons.

Different people spoke in different ways. I don't know how available to you the actual record of this conversation is or even whether it fully exists. But there was not any difference of opinion in the final finding. I don't know whether this is the first thing we considered or whether we considered the Commission's other question first. I imagine we went back and forth between the two of them.

To the Commission's other question, were they doing enough, we answered no. Have you read this report, because if you have, my testimony about it will add nothing.

Mr. GRAY. I believe that the report with two——
The WITNESS. Annexes.
Mr. GRAY. I don't know whether they are actually annexes, but two supplementary statements, I don't know whether that is in one page signed by two people or two separate sheets.
The WITNESS. The report itself you have.
Mr. GRAY. The report is available.

By Mr. GARRISON:

Q. I think you better say what you recollect of it.
A. I recollect of it that the first part of the report contained a series of affirmative recommendations about what the Commission should do. I believe all of them were directed toward weapons expansion, weapons improvement and weapons diversification. Some of them involved the building of new types of

plant which would give a freedom of choice with regard to weapons. Some of them involved just a stepping up of the amount. I don't think that this expressed satisfaction with the current level of the Commission effort.

On the super program itself, I attempted to give a description of what this weapon was, of what would have to go into it, and what we thought the design would be. I explained that the uncertainties in this game were very great, that one would not know whether one had it or not unless one had built it and tested it, and that realistically one would have to expect not one test, but perhaps more than one test. That this would have to be a program of design and testing.

We had in mind, but I don't think we had clearly enough in mind, that we were talking about a single design which was in its essence frozen, and that the possibility did not occur to us very strongly that there might be quite other ways of going about it. Our report had a single structure in mind—or almost a single structure—whose characteristics in terms of blast, of damage, of explosive force, of course, and certainly we tried in the report to describe as faithfully as we knew how. I think in the report itself we were unanimous in hoping that the United States would not have to take the initiative in the development of this weapon.

There were two annexes, neither of which I drafted. There is nothing of restricted data in those I believe, but perhaps we can't read them into the record anyway. Are there any restricted data?

Mr. ROLANDER. I think the question raised is whether other security information might be divulged.

The WITNESS. How many bombs we have and so on?

Mr. ROLANDER. Yes.

Mr. ROBB. Perhaps Dr. Oppenheimer could give us his summary.

The WITNESS. It is a long time since I read them. This ought to be in the record, ought it not? Could you let me read them?

Mr. ROBB. They have been available to Dr. Oppenheimer ever since the letter was sent to him. I think that was clearly understood, was it not, Doctor?

The WITNESS. I was told by counsel that I would be allowed——

Mr. ROBB. Any reports that you had prepared?

The WITNESS. That is right.

Mr. ROBB. So far you have not come down to avail yourself of it.

The WITNESS. I see. They are not here?

Mr. ROBB. We have extracts of them, yes, sir.

The WITNESS. I would think I might read the two annexes and paraphrase them.

Mr. GRAY. I think I am going to ask that we recess now, because there is not another matter to bring up not related to the testimony. I think in the meantime, Mr. Robb, the chairman would like to be advised about this.

Mr. ROBB. The security aspect?

Mr. GRAY. Yes. So we will recess now until two o'clock.

(Thereupon, at 12:25 p. m., a recess was taken until 2 p. m., the same day.)

AFTERNOON SESSION

Mr. GRAY. Gentlemen, shall we proceed.

(Thereupon, Albert J. Gasdor, the reporter, was duly sworn by the chairman.)

Whereupon, J. Robert Oppenheimer, the witness on the stand at the time of taking the recess, resumed the stand, and testified further as follows:

DIRECT EXAMINATION (Continued)

By Mr. GARRISON:

Q. You were in the course of commenting on the 1949 Report when we recessed.

A. Yes.

I find that the report has a letter of transmittal, that it has a section on affirmative actions to be taken, that it has a section on super bombs and that it has these two annexes of which you have heard.

As far as length is concerned, the section on affirmative actions and the section on super bombs are about equal, and I guess I can't tell you what is in the one on affirmative actions except in the very general terms I used before.

The first page of the page-and-a-half of the report on the super bomb is an account of what it is supposed to be, what has to be done in order to bring it about, and some semiquantitative notions of what it would take, what kind of damage it would do, and what kind of a program would be required. The essential point there is that as we then saw it, it was a weapon that you could not be sure of until you tried it out, and it is a problem of calculation and study, and then you went out in the proper place in the Pacific and found out whether it went bang and found out to what extent your ideas had been right and to what extent they had been wrong.

It is on the second page that we start talking about the extent of damage and the first paragraph is just a factual account of the kind of damage, the kind of carrier, and I believe I should not give it—I believe it is classified, even if it is not possibly entirely accurate.

I would like to state one conclusion which is that for anything but very large targets, this was not economical in terms of damage per dollar, and then even for large targets it was uncertain whether it would be economical in terms of damage per dollar. I am not claiming that this was good foresight, but I am just telling you what it says in here.

I am going to read two sentences:

"We all hope that by one means or another, the development of these weapons can be avoided. We are all reluctant to see the United States take the initiative in precipitating this development. We are all agreed that it would be wrong at the present moment to commit ourselves to an all-out effort towards its development."

This is the crux of it and it is a strong negative statement. We added to this some comments as to what might be declassified and what ought not to be declassified and held secret if any sort of a public statement were contemplated. If the President were going to say anything about it, there were some things we thought obvious and there would be no harm in mentioning them. Actually, the secret ones were out in the press before very long.

The phrase that you heard this morning, "We believe that the imaginative and concerted attack on the problem has a better than even chance of producing the weapon * * *"—I find that in this report, and in this report there is, therefore, no statement that it is unfeasible. There is a statement of uncertainty which I believed at the time was a good assessment. You would have found people who would have said this was too conservative, it could be done faster and more certainly, and you would find other people who would say that it could not be done at all; but the statement as read here, no member of the General Advisory Committee objected to, and I have heard very little objection to that as an assessment of the feasibility at that time.

This is the report itself, and there are parts of it which I think you should read but, for the record, there are parts that I cannot get into here.

Mr. ROBB. Mr. Chairman, I think it might be well for the record to show at this point that the board has read the entire report.

The WITNESS. I see. Then, what am I doing that for?

Mr. ROBB. Doctor, that is up to you.

Mr. GARRISON. I thought, Mr. Chairman, there was expressed a little doubt on the part of the board this morning as to just how completely it was recalled at this time, and I think also for that reason it is quite appropriate for Dr. Oppenheimer to perhaps tell the board in his own way what was in it.

Mr. GRAY. That is what I understood was the purpose of addressing his remarks as he is doing.

Mr. GARRISON. I am sure counsel was not mentioning that in the form of an objection.

Mr. ROBB. No, not at all. I was not offering that as an objection, and I do not object to anything. In fact, I might say that later on we might want to come back to this report.

The WITNESS. One important point to make is that lack of feasibility is not the ground on which we made our recommendations.

Another point I ought to make is that lack of economy, although alleged is not the primary or only ground, the competition with fission weapons is obviously in our minds. The real reason, the weight, behind the report is, in my opinion, a failing of the existence of these weapons would be a disadvantageous thing. It says this over and over again.

I may read, which I am sure has no security value, from the so-called minority report, Fermi and Rabi.

"The fact that no limits exist to the destructiveness of this weapon makes it very existence and the knowledge of its construction a danger to humanity as a whole. It is necessarily an evil thing considered in any light. For these reasons, we believe it important for the President of the United States to tell the American public and the world that we think is wrong on fundamental ethical principles to initiate the development of such a weapon."

In the report which got to be known as the majority report, which Conant wrote, DuBridge, Buckley and I signed, things are not quite so ethical and fundamental, but it says in the final paragraph: "In determining not to proceed to develop the super bomb, we see a unique opportunity of providing by example some limitations on the totality of war and thus of eliminating the fear and arousing the hope of mankind."

I think it is very clear that the objection was that we did not like the weapon, not that it couldn't be made.

Now, it is a matter of speculation whether, if we had before us at that time, if we had had the technical knowledge and inventiveness which we did have somewhat later, we would have taken a view of this kind. These are total views where you try to take into account how good the thing is, what the enemy is likely to do, what you can do with it, what the competition is, and the extent to which this is an inevitable step anyway.

My feeling about the delay in the hydrogen bomb, and I imagine you want to question me about it, is that if we had had good ideas in 1945, and had we wanted to, this object might have been in existence in 1947 or 1948, perhaps 1948. If we had had all of the good ideas in 1949, I suppose some little time might have been shaved off the development as it actually occurred. If we had not had good ideas in 1951, I do not think we would have it today. In other words, the question of delay is keyed in this case to the question of invention, and I think the record should show—it is known to you—that the principal inventor in all of this business was Teller, with many important contributions * * * other people, * * * It has not been quite a one-man show, but he has had some very, very good ideas, and they have kept coming. It is probably true that an idea of mine is embodied in all of these things. It is not very ingenious but it turned out to be very useful, and it was not enough to establish feasibility or have a decisive bearing on their feasibility.

The notion that the thermonuclear arms race was something that was in the interests of this country to avoid if it could was very clear to us in 1949. We may have been wrong. We thought it was something to avoid even if we could jump the gun by a couple of years, or even if we could outproduce the enemy, because we were infinitely more vulnerable and infinitely less likely to initiate the use of these weapons, and because the world in which great destruction has been done in all civilized parts of the world is a harder world for America to live with than it is for the Communists to live with. This is an idea which I believe is still right, but I think what was not clear to us then and what is clearer to me now is that it probably lay wholly beyond our power to prevent the Russians somehow from getting ahead with it. I think if we could have taken any action at that time which would have precluded their development of this weapon, it would have been a very good bet to take that, I am sure. I do not know enough about contemporary intelligence to say whether or not our actions have had any effect on theirs but you have ways of finding out about that.

I believe that their atomic effort was quite imitative and that made it quite natural for us to think that their thermonuclear work would be quite imitative and that we should not set the pace in this development. I am trying to explain what I thought and what I believe my friends thought. I am not arguing that this is right, but I am clear about one thing: if this affair could have been averted on the part of the Russians, I am quite clear that we would be in a safer world today by far.

Mr. GRAY. Would you repeat that last sentence. I didn't quite get it.

The WITNESS. If the development by the enemy as well as by us of thermonuclear weapons could have been averted, I think we would be in a somewhat safer world today than we are. God knows, not entirely safe because atomic bombs are not jolly either.

I remember a few comments at that meeting that I believe it best that people who are coming here to testify speak for themselves about; I am not sure my memory is right—comments of Fermi, of Conant, of Rabi, and of DuBridge as to how they felt about it.

Mr. GRAY. How many members of the GAC are being called by you—the members of the GAC at that time?

The WITNESS. Four or five, I think.

Mr. GARRISON. Mr. Conant, Dr. DuBridge, Dr. Fermi, Dr. Rabi, Mr. Rowe, Mr. Whitman, Professor Von Neumann——

The WITNESS. He was not there.

Mr. GRAY. It is a substantial membership.

Mr. GARRISON. We have a statement from Mr. Manley that we will probably introduce in written form to avoid the necessity of calling him from the State of Washington.

The WITNESS. I do not think we called Dr. Cyril Smith, but I will testify that he was an ardent signer of these documents.

Mr. GARRISON. Mr. Seaborg was away.

There were meetings after this.

The WITNESS. Yes. I think we have to keep strictly away from the technical questions. I do not think we want to argue technical questions here, and I do not think it is very meaningful for me to speculate as to how we would have responded had the technical picture at that time been more as it was later.

However, it is my judgment in these things that when you see something that is technically sweet, you go ahead and do it and you argue about what to do about it only after you have had your technical success. That is the way it was with the atomic bomb. I do not think anybody opposed making it; there were some debates about what to do with it after it was made. I cannot very well imagine if we had known in late 1949 what we got to know by early 1951 that the tone of our report would have been the same. You may ask other people how they feel about that. I am not at all sure they will concur; some will and some will not.

In any case, after this report, we had a series of further consultations. I remember that almost immediately afterward, I consulted with the Secretary of State—I think I consulted with him twice, perhaps alone and once with the head of the policy-planning staff—and we talked about this problem.

I remember that the Commission called us down sometime after our meetings, October 29 meeting, called only those members of the committee that were nearby, those on the east coast—Conant, Buckley, Rabi and me, four of us—and we went into it in a more informal session and that is the first time that I became aware of a division of opinion in the Commission and presumably we explained what we had in mind. There is no record of that meeting, or at least I have no record of it, and I have forgotten the details. I know they had another GAC meeting before the President's decision was made, and the Commission asked us to amplify those points. Presumably that was done and presumably you have access to those records, and I have no vivid recollection as to what was said.

In addition to that, toward the end of the period during which the President which making up his mind, I was called by the Joint Committee to come and explain what we had in mind. I was out in California at the time, but when I got back, I did appear before the joint committee. This was immediately before the President's decision was made, and I know how a decision was coming out, but I tried to explain what we had in mind as well as I could. That testimony is presumably also available to you. It is a fairly long statement, questions and answers from the Senators and Congressmen, and I think it stresses the same points as our first report; that is the impression I have. It is not accessible to me.

In any case, the GAC which had a habit of always being around when something was happening was in Washington when the President issued his announcement saying that we were going ahead with it.

Mr. GARRISON. The date of that was when?

The WITNESS. January 29, 1950. I remember two things: One is that in the relatively short interval between October 29 and January 29, the technical prospects for doing what we were planning to do had deteriorated. This was to continue for a long time, and the essential points had not yet come up. By that time, were also quite worried how to carry out the Presidential directive. I believe that our report of that meeting, January 29, 1950, said something like this: we are not going to go into the question of the wisdom of the decision. We now have to look at how to carry it out, and we pointed out that there were several things that the Commission needed to get very busy on if the program was to match. It had to make certain materials available in order to support the Los Alamos efforts, and it had to rearrange its programs in certain ways in order to get on with the job, and I think it was probably at that time that we got into the details of the Savannah River plant. The dual purpose of this seemed just right in view of the great technical uncertainties which were both qualitative and quantitative which then existed.

I believe that in every subsequent GAC report where we gave advice on the thermonuclear program, on the super part of it or the other parts of it, that the problem before us was what to do and how to get on with it, what made sense and what did not make sense, and that the morale and ethical and political issues which are touched on in these two annexes were never again mentioned, and that we never again questioned the basic decisions under which we were operating.

We tried, I think, throughout to point out where the really critical questions were. There was a tendency in this job, as in many others, to try to solve the easy problems and try to leave the really tough ones unworried about, and I think we kept rubbing on the toughest one, that this had to be looked into. That was done not completely; perhaps it is not absolutely done completely today, but the situation developed in a most odd way because, by the spring and summer of 1951, things were not stuck in the sense that there was nothing to do, but they were stuck in the sense that there was no program of which you could see the end.

Now, different people responded differently to that. Teller also pointed out quite rightly that there were other possibilities that might turn up and other people took a very categorical view that the whole business was nonsense.

Mr. GARRISON. Scientifically nonsense.

The WITNESS. Scientifically nonsense. I believe my own record was one that it looked sour but we have had lots of surprises and let's keep open-minded.

I was under very considerable pressure to report in bleak terms through the General Advisory Committee to the Commission and to the military on the prospects. I remember General McCormick saying that we had a duty to do this. At a later time, I remember Admiral Parsons saying that we had a duty to do this to the military rather than to the Commission. We were in somewhat of an uncomfortable position. We recommended against this; it was not going well, and we didn't quite think that it was right for us to say how badly it was going on the ground that this might not be credible, might not be convincing.

What we did do was hold a meeting—perhaps this was the weapons subcommittee of the GAC—out at Los Alamos at which we had talks by the people working on the job—Wheeler, Teller, Bradbury—I will not try to list them all—but, anyway, the people who were really doing the work, and we kept a transcript of these talks. We showed the transcript to the people whose views were represented and we asked them to edit the transcript and transmitted this transcript to the Commission, not as a report of ours but as a firsthand report of how things looked. I think this would have been in the summer of 1950 or it may have been somewhat later.

At the same time we went over the program with Los Alamos, there were weapons testing programs, their calculation programs, and I believe you will hear evidence that at least some people out there thought we were just the opposite of harmful but quite helpful in connection with this job.

We also kept in touch with and tried to help the production activities of the Commission, some of the engineering activities that went along with the basic research and development. It was partly, I think, in response to the sense that a report on this matter also needed to be available in military circles that the hydrogen super bomb was included in this report of the panel that we heard of this morning; it was toward the end of 1950, but it was all a part, that part of the advice or which seems to me is most central and basic and inescapable responsibility which is to tell what he knows of what is going on and what he knows of the truth. I feel that in this we did our duty rather well.

There are things that you probably want to question me about in some detail in the General Manager's letter. They have to do with unauthorized distribution of reports. We have an affidavit which we will introduce later which throws some light on it. To me, it was an utterly mysterious document. I did, of course—I won't say of course—in fact did show various GAC reports from time to time to a very few people who were actively engaged and responsibly engaged in the program. The purpose of this was certainly not to persuade them to come over to my views but to elicit their views and have a discussion.

I showed some of the reports on the super to Von Neumann at the Institute who is a very close friend and a very responsible man and whom I knew to be a great enthusiast for this program. I had no notion at all that this was going to change his mind but I thought it right to show them and he certainly was pleased that I did.

I showed nothing at Los Alamos. I wasn't there, and you will have a record of what happened, which I think will satisfy you as to why some of these documents were made available and how little that involved me—at least this is the story that I think will emerge.

It is also alleged that I kept people from working on the hydrogen bomb. If by that it is meant that a knowledge of our views which got to be rather widespread had an effect, I cannot deny it because I don't know, but I think I can deny that I ever talked anybody out of working on the hydrogen bomb or desired to talk anybody out of working on the hydrogen bomb. You will have some testimony on this, but since I don't know who the people are who are referred to in the General Manager's letter, what I say might not be entirely responsive. I know that in one case there was a very brilliant young physicist called Conrad Longmire. I think he was at the University of Rochester. In any case, he had applied to come to the Institute, and we granted him a membership there, and he said that he would like to go to Los Alamos for a year and I said, "Fine, go do that, and you can have your membership here at any time you want it," in an attempt to make the decision easy for him, because he didn't want to give up his Institute membership. I don't know but that there are other cases. Longmire is still there.

There are times when they communicated with me saying that it would be nice for him to spend a year at the Institute, but he has not come yet. I think we will have to get into the details if there is anything about my slowing down the work on the super, because, as a general allegation, I find nothing to take hold of there.

Mr. GARRISON. May I ask the board if it would suit your convenience to ask Dr. Oppenheimer questions that you have in mind about any of these portions, or would you rather do it at the end?

Mr. GRAY. I think we would rather do it at the end. I have not consulted with Mr. Robb about it.

Mr. ROBB. I think it would be preferable to ask the questions at the end.

Mr. GRAY. I think that would be preferable to get the continuity of Dr. Oppenheimer's testimony.

The WITNESS. I think it would be fair to say that between the first of 1950 and early 1951, my attitude toward this object was that we didn't know how to make it, and it was going to be very hard to make, but we had been told to dot it and we must try.

In the spring of 1951, there were some inventions made. * * * and from then on it became clear that this was a program which was bound to succeed. * * * Why none of us had them earlier, I cannot explain, except that invention is a somewhat erratic thing.

Teller had been working on this from 1942 on, his heart was in it, but it wasn't until 1951 that he thought about how to do it right.

Now, I have a few matters here which came in between. During the doldrums of the H-bomb, the war in Korea broke out, and a large part of GAC's and other committee's attention was, as I say, devoted to the very immediate and the very obvious, and, I would say, to using an atomic explosive not merely in a strategic campaign but also in a defensive or tactical campaign, and I think the record will bear out that that is what we were spending most of our time worrying about. That is the origin of the panel Kelly talked about this morning, the origin of the exercises which led to the development of a tactical capability in Europe, the origin of one at least of the threads, one at least of the reasons for the very great expansion in the atomic energy enterprise to support a much more diversified use of weapons, even leading some people to suggest—I think this was Gordon Dean—that maybe the atomic weapons on the battlefield would be so effective that it would not be necessary to use them strategically. I have never really believed that that was possible or believed that a sharp distinction between the two could be maintained or made intelligible.

In the late summer and autumn of 1950, I had an obvious personal worry. I had made as chairman, and had participated in, the recommendation against the development of the super. The super was a big item on the program. It wasn't going very well, and I wondered whether another man might not make a better chairman for the General Advisory Committee. This was particularly true since there were three new members added to the committee—Whitman, Murphree and Libby—and I felt a little uncomfortable about continuing in that office. I discussed it with several physicists. I remember discussing it with Teller and Bacher. Teller says that he does not remember discussing it with me. The general advice was: Let's all stick together as well as we can and don't resign and don't change your position.

Mr. ROBB. What was that date?

The WITNESS. In the summer of 1950.

When I got back in the autumn of 1950, the first meeting, I went to see Mr. Dean, who was Chairman of the Commission, and Commissioner Smyth and told them about my problem, and they said that obviously the chairman should be someone who would be comfortable with them—what would be their suggestions? They protested in very forceful terms that I should not quit as chairman, and that they would be very unhappy if I did, that I ought to carry on.

I also took the thing up with our committee, but our committee was not a very responsive group when it came to electing other chairmen, and I got no place. I did not feel that I ought to resign as chairman or refuse to serve. I thought

I ought to do what was comfortable for the Commission and the committee, and I tried to ascertain what that was.

Mr. GARRISON. How about your Princeton meeting?

The WITNESS. We are still on the subject of the H-bomb and its consequences. In the spring of 1951, I called—I am not sure whether I suggested it or whether Commissioner Smyth suggested it but we consulted about it—a rather large gathering for a couple of days at the institute in Princeton, and we had there, I think, all five Commissioners, the general manager and his deputy, the head of the Division of Military Applications, Bradbury and his assistants, Teller, Von Neumann, Bethe, Bacher, Fermi who was no longer a member of the committee, and Wheeler and one of his assistants, the people who were working on the program, and we had a couple of days of exposition and debate. I chaired the meeting, and I suppose I did the summarizing. It was not the full General Advisory Committee—the Weapons Subcommittee, essentially; the secretary of the committee was there and he took some notes but he did not write up an official report. At that time, I think we did three things. We agreed that the new ideas took top place and that although the old ones should be kept on the back burner, the new ones should be pushed. I believe there was no dissent from this; there was no articulated dissent. But later Commisisoner Murray asked if this wasn't a violation of the Presidential directive, and I could only respond that I didn't know as to what, but I thought it was a good course and, if it was, maybe the President would modify his directive.

At that meeting, I remember no dissent from that but there was a great deal of surprise at how things were changed. Fermi knew nothing of these developments and was quite amazed, and I think for the Commission it was quite an education to see what had happened in the meantime. At least that was the purpose, to get everybody together so that there was a common understanding.

The second thing was to recognize that some materials * * * might be handy to have, and the Commission was urged to get started on producing some of these materials. This was something that there was a little bit of objection to on the ground that everything changed so often in the past and maybe change in the future, and why get committed to a cumbersome operation on the basis of the then-existing state of knowledge, but I believe the prevailing opinion, and I know mine, was that the prevailing state of opinion was that it was a lot solider than anything that had occurred before and that they ought to go ahead and even at the risk of wasting a small amount of money.

The third thing we did was to talk about the construction and test schedules for these things, and there there were differences of opinions, having to do with whether the schedule should be aimed at a completed, large-scale explosion, or whether one should be aimed at componentry testing which presumably was supposed to have happened earlier and therefore might be illuminating with regard to the large-scale explosion.

As I say, there was not agreement, but the consensus was that unless the studies of the summer passed out on the feasibility of it, one should aim directly at the large-scale explosion, and the time scale of that operation from mid-1951 to late 1952 was, I think, a miracle of speed. I know there may be people who disagree, and I think it might have been done faster, but I can only reminisce and say that in the first days of Los Alamos, and in the fall of 1943, Bethe and Teller, two of the most brilliant theorists in this game and in their way most responsible men, said to me: "If we had the material now, we could have a bomb in 3 weeks." Actually, we were ready for the material just about when it arrived, which was not quite 2 years later, and the laboratory had doubled every 9 months in the interval and everybody was busy; and I think that the estimate of the theorists on how quickly you could do things that involve engineering and involve new chemistry and involve new metallurgy was likely to be a little optimistic.

I am continually impressed by the speed, sureness, certainty, skill and quality of the work that went into the preparation of this first large explosion and the subsequent work to exploit the development there established.

The next thing on which I had notes is that in the autumn of 1951——

Mr. GARRISON. That was at Princeton?

The WITNESS. This was the Princeton meeting that I have described. I think it was a very useful meeting. It might have been useful to me if we had made a record of it.

It was largely that it was not a formal type of GAC meeting and our secretary did not want to keep a record, but I believe a fairly good account of the substantive findings exist, and I believe Commissioner Smyth knows where to get hold of it. I don't know how to get hold of it.

In the autumn of 1951, there was an international conference in Chicago, and I attended it even though I was called away to testify for money for the National Science Foundation.

While there, I talked at some length with Teller and the summer's work had only made things look tied together. Teller expressed dissatisfaction with the arrangements made at Los Alamos. He didn't think the man whom Bradbury had put in charge of this development was the right man for the job, and he expressed to me the view that Fermi or Bethe or I would be the only people that he would be happy to work with. I don't know whether he meant me, but I said, "Well, that is fine," and he said that Bethe and Fermi wouldn't; "Would you be willing to?" I won't quote myself verbatim, but I remarked that that would depend on whether I would be welcomed by Bradbury. I had not planned to go back to Los Alamos. It seemed to me a bad thing for an ex-director to return. I was content with my job and work at Princeton, but I would communicate with Bradbury, and I called him and told him of the conversation and he gave no signs of wanting to have the ex-director back, and said that he had full confidence in the present man, and that was the end of that.

I don't believe that it would have been practical. I think you can't make an anomolous rise twice. I think I could create and guide Los Alamos during the war, but I think if I had returned there the situation would have been so different; I would have been ancient and not on my toes anymore, and I doubt if I would have felt appropriate, but, in any case, the success of this would have decisively depended on its being something that was actively in the desires and interests of the director, and it was not so.

The hydrogen bomb was not done, and during the winter of 1951–52 Los Alamos was working on it, and we kept in quite close touch. Bradbury came in quite frequently. He sent Froman and other people in to report to us, and I want to make it clear that I was not actually calculating out and working on it. I was merely trying to understand where the difficulties lay, if any, what the alternatives were, and to form a reasonable judgment so that I might give reasonable advice.

At that time, Teller's unhappiness with the arrangements became quite generally known, and we were frequently asked by the Commission, "Should there be a second laboratory?" We were asked, "Should this work be split off in some way from Los Alamos?" I don't know how many times that came up during the winter of 1951–52 as an item before the General Advisory Committee.

I think, on this point, we were not unanimous. I think Dr. Libby thought it would be a good idea to have a second laboratory at any time. The laboratory, the purpose of which would be to house Teller and bring you people into the program who were not working on it, even though this might take some people away from Los Alamos, even though it might interfere with the work then going on. The rest of us, I think, were fairly clear that the things were really going along marvelously well, and that if it was too difficult for Los Alamos to do the whole job, then steps should be taken to get some of their more routine operations moved to Sandia. We talked at great length about the rearrangement of the workload between the two places. Some of the suggestions we made were adopted.

We also talked to Bradbury about making within the framework of Los Alamos an advanced development section in which really radical ideas and wild ideas could be thought up and tried out. The Director thought it was feasible if he could get the right man. He tried very hard to get one man for it and, after some delay, this man turned him down, and I don't believe such a reform was undertaken then.

I believe that with the Commission's reluctance to establish a second weapons laboratory, there was some thought that the Air Force might directly establish one, and I think the Commission protested that but this is hearsay.

In any case, during the winter, our recommendations were to fix up Los Alamos so that it could do the job rather than start a separate establishment. Later, in the spring, perhaps in April, we learned that there had been some preliminary talks toward the converting of the laboratory at Livermore which had been engaged in an enterprise related to atomic energy, of which we the members of the GAC took a rather sour view of converting this, in part, so that it could get more weapons testing work with a special eye to the thermonuclear program. This we liked and this we endorsed.

The laboratory at Berkeley had often been involved in the instrumentation of weapons tests, and it seemed that this was a healthy growth which wouldn't weaken Los Alamos, which would bring new people into it where there was an existing managerial framework and where the thing could occur gradually, and, therefore, constructively, the notation of setting out into the desert and building a second site like Los Alamos and building a laboratory around Teller had always seemed to us to be something that was not going to work, given the conditions and given the enormous availability to Los Alamos of the talent that was needed for this problem.

In any case, the Livermore Laboratory was established sometimes perhaps in the summer of 1952, and has played its part in the subsequent work at the time when my clearance was suspended, the major and the practical, and the real parts of the program were still pretty much Los Alamos doing, but it was my hope, all our hope, that both institutions would begin pulling great weight. There had also been no serious friction between them.

Mr. GARRISON. Did you tell the board that Dr. Teller was in charge of the Livermore Laboratory?

The WITNESS. My understanding is that the director is Herbert York, but that this part of the laboratory's work was under the scientific direction of Teller. I think the board probably knows that better than I do at this point.

The super also—well, it was no longer the super—I forgot one thing, and it may be of some slight importance. This goes back—and I am sorry to have a bad chronology here——

Mr. GRAY. I think the record should show that Dr. Evans has just stepped out of the room.

Mr. ROBB. Dr. Evans has just stepped back into the room.

The WITNESS. At the time that the H-bomb problem first came up—I forgot to say two things.

I spoke of my later feeling that I should perhaps not be the chairman of the General Advisory Committee myself—but two things happened much earlier. I had some talks with the Secretary of State, too, I think, and so had Dr. Conant. Dr. Conant brought back, and so did Mr. Lilienthal, from the Secretary of State 2 messages; 1 was a message to Conant and me, for heck's sake not to resign or make any public statements to upset the applecart but accept this decision as the best to be made and not to make any kind of conflict about it. That was not hard for us to do because we hardly would have seen any way of making a public conflict, and the second part of the message was to be sure to stay on the General Advisory Committee; and that is what both of us did.

There was another item. He recognized, as has Mr. Lilienthal and as would any other sane man, whether or not a hydrogen bomb could be made, how soon we made it, the Russian possession of an atomic bomb raised a lot of other problems, military and political and upset a great many things.

The Government had been saying we had been expecting it, but now here it was—with regard to the defense of Europe; with regard to the usefulness of atomic retaliation in special conflicts, and I was called in to help in the preparation of the Security Council paper which was prepared that spring on the subject of which essentially was rearmament and the subject of which was how to solidify our alliances and increase the overall military power of the United States.

Mr. GRAY. This was the spring of 1950?

The WITNESS. This was the spring of 1950, in NSC 68 or 69, and you probably remember the number better than I do * * *

In any case, it needs to be testified by me that I was very aware of the fact that you couldn't, within the atomic energy field alone, find a complete or even a very adequate answer to the Russian breaking of our monopoly. I don't think I had a major part in this paper. It took months of staff work to do it. I wouldn't be surprised if—I don't know whether I had any part—but, in any case, I approved and helped with some parts of that and its purpose was the buildup which started some months later after Korea.

Mr. GRAY. Is that a good breaking point? Shall we take a 5-minute recess?

The WITNESS. It is fine since that is out of order and I apologize for putting it that way.

(Brief recess.)

Mr. GRAY. Shall we proceed?

The WITNESS. I have a few more words on the hydrogen bomb which are not very major. The hydrogen bomb once it looked like it got in Dr. Kelly's province, of course, came out in the Research and Development Board committee on which I served. * * *

I would like to summarize a little bit this long story I think you will hear from people who believed at the time, and believe now that the advice we gave n 1949 was wrong. You will hear from people who believed at the time and who even believe now that the advice we gave in 1949 was right. I myself would not take either of these extreme views.

I think we were right in believing that any method available consistent with honor and security for keeping these objects out of the arsenals of the enemy would have been a good course to follow. I don't believe we were very clear and I don't believe we were ever very agreed as to what such course might be, or whether such a course existed. I think that if we had had at that time the technical insight that I now have, we would have concluded that it was almost hopeless to keep this resource out of the enemy hands and maybe we would have given up even suggesting that it be tried. I think if we had had that technical knowledge, then we should have recommended that we go ahead full steam, and then or in 1948 or 1946 or 1945.

I don't want to conceal from you, and I have said it in public speeches so it would not make much sense to conceal from you the dual nature of the hopes which we entertained about the development of bigger and bigger weapons, first the atomic bomb, and then its amplified version, and then these new things.

On the one hand, as we said at the time, and as I now firmly believe, this stuff is going to put an end to major total wars. I don't know whether it will do so in our lifetime. On the other hand, the notion that this will have to come about by the employment of these weapons on a massive scale against civilizations and cities has always bothered me. I suppose that bother is part of the freight I took into the General Advisory Committee, and into the meetings that discussed the hydrogen bomb. No other person may share that view, but I do.

I believe that comes almost to the end except for one thing. I know of no case where I misrepresented or distorted the technical situation in reporting it to my superiors or those to whom I was bound to give advice and counsel. The nearest thing to it that I know is that in the public version of the Acheson-Lilienthal report, we somewhat overstated what could be accomplished by denaturing. I believe this was not anything else than in translating from a technical and therefore secret statement into a public and therefore codified statement, we lost some of the precision which should have gone into it, and some of the caution which should have gone into it.

I am now through with this.

By Mr. GARRISON:

Q. Dr. Oppenheimer, you said a little while back that you had shown GAC reports to several people. You mentioned von Neumann. I would like to clear up two things. One, to whom specifically do you recall having shown reports, and secondly, what was the character of these people in relation of the Government?

A. I will tell you what I remember. I showed our discussion of the reactor-development program to Wigner, who was the great expert in the field. I wanted to know what he thought. This may have been in 1947 or 1948. Wigner was, of course, an active participant in the reactor development work of the Commission, fully cleared and with very strong views of his own.

Q. He was not at Los Alamos?

A. No, his work was at Argonne and Oak Ridge. He was director of Oak Ridge, and he lives in Princeton. I did not go to any trouble to show it to him. I showed the one report that I was reading, the October 29 report, to von Neumann at the institute. He was one of the experts on the thermonuclear problem. He had talked with me, talked my ear off about it before, and also after. I may have shown it to Bethe but I am not sure.

Q. Bethe and he were again both cleared for top-secret information?

A. Yes. I doubt whether I showed it to Bethe, but I am not clear. I don't recollect. I would not have regarded it as improper. I would have regarded it as consistent with my job of attempting properly to advise the Commission and represent the scientific elite to the Commission—experts, not elite—and back and forth. I would have regarded it as proper on occasion and with discretion to show and discuss some of these problems with a cleared person. I am quite clear that a great deal of other showing was done in other ways, but that is something I had nothing to do with.

Q. With regard to the item of information in the Commission's letter that you caused to be distributed to key personnel at Los Alamos copies of the October 29, 1949, report with a view to influencing them against the H-bomb program, what have you to say about that specifically?

A. Specifically I deny it. I never did anything about having extra copies of reports made or sending them out or anything like that. I had no desire to influence Los Alamos. I certainly did not succeed in influencing Los Alamos

Mr. GARRISON. May I say to the board that I would like at this point to read into the record an affidavit from Dr. John Manley. I shall hand the original to the Chairman and then to counsel and copies to the members of the board, and then I will explain what it is about. I introduce this, Mr. Chairman, at this point because in the latter portion of this affidavit there is an account from Dr. Manley's records of what distribution was made at Los Alamos of the report in question. It will show, I think, conclusively that Dr. Oppenheimer had nothing to do with this.

By Mr. GARRISON:

Q. Dr. Oppenheimer, could you just tell the board in a few words who Dr. John Manley is?

A. Before the war he was professor of physics at the University of Illinois at Urbana. I knew him slightly. When I was asked by Arthur Compton to take charge of the bomb work, I didn't know much about experimental things and he asked Manley to be my deputy with regard to that. He was, an we worked very closely together. This would have been 1942–43. He helped build the Los Alamos Laboratory. He was in charge of the group at Los Alamos in the physics division of the laboratory.

He left Los Alamos after the war, returned to Los Alamos a year or so later, and became, I don't know how immediately, associate director. First he was in charge of the physics division. At that time, after our first meeting, the General Advisory Committee asked me to invite him to become our secretary. He was our secretary until what would have been 1950 or 1951—I have forgotten the date—at that time he left atomic-energy work and left Los Alamos and is chairman of the department of the University of Washington at Seattle. He is not Jack Manley.

Mr. GARRISON. Mr. Chairman, I am introducing this in affidavit form for a couple of reasons. One, Dr. Manley is in the State of Washington which is quite a little distance from us. Secondly, the part I want most to draw to the board's attention when I reach it in the affidavit has to do with an account of records of his. It is a little more precise to introduce it in written form, but needless to say, if the board would like to have us call Dr. Manley, we would be glad to do so. The program is rather crowded, and so there will be perhaps half a dozen written statements which perforce we will put in the record.

I would like to read this rapidly to the board now.

"STATEMENT OF DR. JOHN MANLEY:

"I live at 4528 W. Laurel Drive, Seattle 5, Wash. I am a professor of physics and executive officer of the department of physics of the University of Washington.

"I joined the Metallurgical Laboratory in January of 1942. This was before Dr. J. Robert Oppenheimer had anything to do with it. It was under the direction of Dr. A. H. Compton. In July of that year, Dr. Oppenheimer was selected to head the bomb phase of the project. I recall, for example, the expression of pleasure by Dr. Compton that he was able to get Dr. Oppenheimer to head this portion of the activities. At the same time, I was given responsibility for the experimental phase of the bomb project, Dr. Oppenheimer devoting his time to the overall problems and especially the theoretical aspects. (The first time I ever met Dr. Oppenheimer was in connection with this work in about July 1942. I had nothing to do with the selection of Dr. Oppenheimer for his post.)

"During the period from July 1942 to April 1943 I was responsible for the supervision of the experimental work under the direction of Dr. Oppenheimer with headquarters in Chicago. Although he was in residence in Berkeley at that time, he came east frequently for consultation on the detail work under numerous contracts. I was impressed at that time by his ready grasp of even minor details relating to the program.

"In the latter part of 1942, a decision was made to concentrate this phase of the program at Los Alamos, New Mexico. In this connection, I acted directly as an agent for Dr. Oppenheimer, who was to assume direction of the laboratory. Among other things I undertook the recruitment of personnel, to go to Los Alamos, from those groups who had already been engaged in experimental work.

"In April of 1943 I joined Dr. Oppenheimer at Los Alamos and assumed responsibility for one phase of the experimental program. During the period from 1942 to 1945 in which I continued to be closely associated with Dr. Oppenheimer, the clarity of the wisdom of the choice of him to lead this project increased. I am convinced that no one of my acquaintance possesses either the necessary broad technical knowledge and quick grasp of details or the sympathetic understanding of people which were so necessary to accomplish the project objective in a remote, isolated and self-contained community. I consider it a remarkable achievement, due in very great part to Dr. Oppenheimer's leadership, that this work was completed in time it was.

"During this period at Los Alamos, though I have no specific knowledge of the detailed matters of security procedures, personnel clearances, etc., I can recall no instance or situation which impressed me as suggesting laxity or slighting of security measures. There were, for example, specific instructions from Dr. Oppenheimer in 1943, when I was recruiting personnel concerning the secret nature of the project, and during the whole Los Alamos period, very evident support by him of restrictions imposed on civilian personnel, especially with respect to travel, correspondence, etc. As director of the laboratory, Dr. Oppenheimer was normally the recipient of most of the complaints from civilian personnel about security restrictions—restrictions on travel, etc., and I was impressed with the effectiveness of the job he did in persuading us of the necessity of these restrictions while in no wise relaxing the restrictions.

"I did not know anything about Dr. Oppenheimer's attitude on the question of employment of Communists, or ex-Communists, or pro-Communists; nor did I know whether any of the people employed were or had been Communists or pro-Communists. In my recruitment work I didn't have occasion to go into this question because (a) security was not my job, and (b) the recruitment that I had to do with was largely confined to individuals who were already working on various phases of the project and so had been cleared. I have not to this day heard any suggestions or even rumor of any security leakage with respect to the atomic-weapons program for which Dr. Oppenheimer could be charged with personal responsibility, or for which anyone ever suggested that Dr. Oppenheimer was even remotely responsible—unless the letter of the Commission dated December 23, 1953, suspending Dr. Oppenheimer's security clearance may be deemed to be such a suggestion.

"Although Dr. Oppenheimer left Los Alamos at the close of 1945, I continued there, and in 1946 was asked by him to spend part of my time as secretary to the General Advisory Committee of the Atomic Energy Commission, of which he was Chairman (as such secretary I was not a member of the General Advisory Committee). I accepted this duty and from that time until January 1951 I spent about one-fifth of my time in connection with the committee work, being at Los Alamos the remainder of the time, first as a division leader, and subsequently as technical associate director of the laboratory.

"In this period I know of no circumstances in which Dr. Oppenheimer attempted to influence in a direct personal way the course of events at Los Alamos (as distinct from the effect that the recommendations of the GAC might, in normal course, have on the work of the laboratory). In fact, I recall that on occasions when I would discuss laboratory problems with him he would frequently say 'But that's a problem for you and Norris.' (Norris Bradbury, the director of the Los Alamos Laboratory). Although Dr. Oppenheimer kept informed on the technical features of all phases of the weapons program and was often most helpful to the laboratory through the GAC or in personal contacts, I believe that he did not feel sufficiently familiar with the details of the laboratory operation to be able to advise appropriately on internal questions of use of personnel and facilities. It should be understood that many of the wartime senior personel of Los Alamos left at the close of the war, and those of us who stayed on felt a very direct challenge to assume all responsibility for the continuing program relying, of course, on occasion, on the technical advice of those individuals who had participated in the wartime program—individuals such as Dr. Oppenheimer, Dr. Fermi, Dr. Bethe, Dr. Bacher, and sot on. It should also be understood that the laboratory prepared its own program of activities and submitted those to the AEC for approval. In my own dual capacity as secretary to the GAC and one of the senior members of the Los Alamos Laboratory, I felt a special responsibility for liaison between that committee, so largely composed of former Los Alamos personnel, and the laboratory. It is my belief that this dual function of mine was considered valuable both by the committee and the laboratory.

"Shortly after the end of the war, there was considerable discussion among the people at Los Alamos as to whether it would be wise to continue the Los Alamos Laboratory, or whether it would be better to abandon the Los Alamos Laboratory because of its remoteness and the resultant complexity of the operation. It is my impression that Dr. Oppenheimer was not clear in his own mind as to what he thought would be wiser in the national interest. But it was my impression that there was no doubt in Dr. Oppenheimer's mind that the atomic-weapons program had to be continued, whether at Los Alamos or elsewhere, unless the international situation clearly indicated, by agreement, the abandonment of such activities.

"I should like to comment on the operation of the GAC as guided by its Chairman, Dr. Oppenheimer. A less conscientious committee could have considered only such matters as were presented to it by the Commission. The GAC, however, with many individuals senior to the Commission itself in atomic matters, considered it an obligation to supply such guidance to the Commission as its experience suggested might be in the national interest. Each meeting would be devoted to items specifically requested by the Commission and other items which the GAC deemed worthy of discussion. I recall several instances in which the GAC on its own initiative made recommendations for new programs long before the AEC found it possible to start such programs. The GAC was generally understood to be advisory, not simply in a formal sense to the Commission, but to its division and laboratories as well. This was accomplished by discussion with appropriate people in and out of GAC meetings and by visits to various laboratories. It was the method by which the GAC kept in close touch with key people and programs of the AEC.

"I should mention also that there was a very close similarity in the thinking of the members of the GAC and the top people at Los Alamos on most matters relating to weapons programs, so that if there were a division of opinion or doubt on any particular matter within the GAC, there would normally be the same division of opinion or doubt among the top people at Los Alamos. On the other hand, if there was unanimity of opinion and no doubt as to the proper course with respect to any particular question among the people at the GAC, there would normally be the same unanimity of opinion and lack of doubt as to those matters among the top people at Los Alamos. This was not primarily because either the people at Los Alamos took their lead from the GAC or the other way round (although of course each group normally would be, to some extent, influenced by the thinking of the other group); but the essential reason for the similarity was just that both groups had a common recognition of the national need and the limitations of facilities and personnel.

"This was true with respect to the debate concerning thermonuclear programs which became a subject of vigorous discussion at Los Alamos following the Russian explosion of an atomic bomb in September 1949. This debate continued until resolved by the President's announcement in January 1950. In this period there was, as in the past, informal exchange of views between members of the GAC and the senior personnel of the laboratory."

Now comes the part, Mr. Chairman, that is particularly pertinent to the question I put to Dr. Oppenheimer.

"I have been informed that it has been charged that Dr. Oppenheimer caused to be distributed separately and in private to personnel at Los Alamos certain majority and minority reports of the GAC having to do with the thermonuclear program. With reference to this matter, the following statements of my own knowledge are made:

"A. On November 10, 1949, while en route from Washington, D. C. to Los Alamos, I received a phone call from Carroll Wilson, AEC General Manager. The substance of this call was that Senator McMahon had requested copies of the GAC papers from the AEC and these had been sent to him. In view of the forthcoming visit of the Senator to Los Alamos, Mr. Wilson wished me to show the documents to Bradbury and C. L. Tyler (AEC manager at Los Alamos) and discuss their contents. He wished me also to show them to Wally Zinn (director, Argonne Laboratory), but as I was not carrying the documents, this was impossible. Mr. Wilson also asked if I would go on to Berkeley and talk to Earnest Lawrence (director. University of California Radiation Laboratory). I replied that since Bradbury would be away from Los Alamos all the following week, and I would be in charge, I could not comply with this request.

"B. Neither Bradbury nor Tyler were available when I arrived at Los Alamos on November 11, 1949, so the session with them was held the afternoon of November 12. At this session, I showed them the papers which had arrived by courier and tried to supply them with the background discussion which led to the papers.

"C. In view of the fact that Senator McMahon would be in Los Alamos the following week for discussion with senior laboratory personnel (Tech Board, except Dr. Bradbury, who left, I think, on November 13) I showed and discussed these papers with the following: J. M. B. Kellogg, evening November 12. Carson

Mark, morning November 13. Edward Teller, morning November 13. Robert Kimball, evening November 13. Alvin Graves, morning November 14. Darol Froman, morning November 14.

"I would add that I feel quite certain that the papers were shown to other members of the Tech Board who were to be present in the meeting with McMahon though my appointment list does not show this. In each case it was emphasized that the policy question was under consideration in highest governmental quarters and discussion of such matters should be strictly limited to senior personnel.

"D. The reports to which reference is made in this statement were the majority and minority reports prepared in the GAC meeting which ended October 30, 1949, and the report of the Chairman, GAC addressed to the Chairman, AEC on this meeting. In addition there was a report prepared by myself as secretary and directed to the Chairman, GAC. This report was prepared in lieu of minutes for the purpose of setting forth the secretary's impressions of the discussion of the GAC which led to the committee's documents, in order to provide additional background for interpretation of these documents. Since Dr. Oppenheimer, Dr. Fermi, and Dr. Smith were in Washington on November 7, they were consulted on the draft of my report and minor changes were made to represent their views with more correct emphasis. This report was completed and given to the Chairman, GAC on November 9.

"E. The meeting with Senator McMahon for which the 'distribution' of reports as described above was made, took place as Los Alamos November 15, 1949. The purpose of the meeting was to review the Los Alamos program including work on thermonuclear weapons. It was not for policy discussion concerning the thermonuclear program.

"From these items of fact it is clear that—(a) Revelation of these particular reports was authorized by the AEC in the person of the General Manager that the laboratories at Argonne, Berkeley, and Los Alamos be made aware of the GAC recommendations, (b) that the showing of the reports to members of the Tech Board was on the responsibility of Dr. Bradbury and myself in preparation for discussion with Senator McMahon who had seen them, (c) that the handling of the documents was in accord with established procedures, and (d) that Dr. Oppenheimer had nothing to do whatever with this matter.

"The discussion as to relative concentration on fission weapons and thermonuclear weapons had been a continuing one since 1942. It was recognized that the fission weapon would have to be made before the thermonuclear weapon would be possible. But even at the beginning it made an obvious difference in the program whether one were pointing toward a fission weapon, which should itself be used as the primary atomic weapon, or whether one were planning to make a thermonuclear weapon. There was also the question of whether it was better, as a military matter to improve and make larger numbers of fission weapons or to devote major time and effort to establish the possibility and practicality of some thermonuclear weapon. Wholly apart from the question of whether it would be technically possible to make a thermonuclear weapon, it was clear that the making of thermonuclear weapons would require the use of the same materials and personnel and money that might otherwise be devoted to making of improved fission weapons. In short, it would be a task comparable with the wartime development of the fission weapon. It was a matter of judgment as to the best way to utilize the materials, personnel, and money as between the fission-weapons program and the thermonuclear-weapons program.

"One of the difficulties that all concerned felt keenly in the effort to make up their minds on this question was that they did not have any really adequate appraisals of the military usefulness of the different weapons, nor were such appraisals supplied by the military.

"It is my impression that the GAC labored under the same difficulties as others on this problem, but that the GAC was certainly as active as any other group with respect to this problem. The GAC, and particularly Dr. Fermi, made an effort to evaluate the relative costs in terms of production facilities of the two types of weapons. It was not a military evaluation of worth.

"I normally attended meetings of the GAC, and it was my observation that Dr. Oppenheimer as Chairman took pains on all questions to sound out the views of the other members of the Committe before expressing his own. It was my impression that he did this because he was keenly conscious of the restraints of the chairmanship. It is my recollection that this was the way he conducted the October 1949 meetings that discussed the thermonuclear-weapons program. The matter of annexing both a majority and minority report to the report of the October 1949 meeting was, as I recall it, at Dr. Oppenheimer's suggestion and instruction because he wanted to be sure that the report fully reflected the views of all members of the committee. It was in the same spirit that he requested me to prepare a report on the meeting as a supplement to his report and those of the majority and minority.

"I find the suggestion that Dr. Oppenheimer attempted to or did retard the work of the Los Alamos Laboratory in any field, and specifically in the field of thermonuclear weapons, preposterous and without foundation. I had no feeling whatever that anybody at Los Alamos was holding back in effort on the thermonuclear weapon because of Dr. Oppenheimer's suggestion or example. (Indeed, I had no feeling that anyone was holding back on the work on the thermonuclear weapons once the President had decided the question by his announcement in January 1950. The work proceeded with willingness and cooperation from all concerned.) I know of my own knowledge that Dr. Oppenheimer never suggested to me that I should refrain from working on the thermonuclear-weapons program, or that I should go slow on it or anything like that.

"I never observed anything to suggest that Dr. Oppenheimer opposed the thermonuclear-weapons project after it was determined as a matter of national policy to proceed with the development of thermonuclear weapons, or that he failed to cooperate fully in the project to the extent that someone who is not actively working could cooperate. I do not recall anything in his subsequent conduct of the GAC meetings that suggested to me in the slightest that he was doing anything less than wholeheartedly cooperating. Neither have I ever heard from any scientists that Dr. Oppenheimer was instrumental in persuading that scientist not to work on the thermonuclear-weapons project.

"I have known Dr. Oppenheimer now since 1942. Until 1951 I worked very intimately and closely with him. I feel that I know him very well indeed. I consider that the work that he has done has been of the greatest possible value to the country; that if comparisons must be made, his contribution has probably been of more importance in the development of the atomic energy program than that of any other scientist in the country and perhaps than that of any other person in the country. I make this statement not only in recognition of the great contribution he made while he was director of the Los Alamos Laboratory, but also from by familiarity with his activities as Chairman of the GAC. He took an active part in the many complex problems of the whole atomic energy program. Its achievements are, I think, due in no small part to his activities. He has at all times had the national interest at heart and has never done anything that he thought or suspected might be contrary to the national interest.

"I am absolutely clear that he is in no sense whatever a security risk. I saw this both on the basis (a) of the fact that for over 10 years he was entrusted with the most secret information pertaining to the Nation's atomic developments and there was never the slightest leakage of secret information from or through him, or in any way related to him, and (b) on the basis of my intimate personal knowledge of him, his character and his views.

"My attention has been called to the fact that the letter of December 23, 1953, from the Atomic Energy Commission suspending Dr. Oppenheimer's clearance mentioned his having known someone named Jack Manley. I suppose I should record the fact that I assume that I am not the Jack Manley referred to because the letter refers to Jack Manley as a member or official of the Communist Party, and I have never been associated with the Communist Party. I do not recall that I have ever been known as Jack Manley. I do not know who Jack Manley is, nor do I know anyone of that name.

"JOHN H. MANLEY.

"Sworn to before me this 16th day of February 1954.

"MARY E. MOSSMAN, *Notary Public.*"

By Mr. GARRISON:

Q. Do you wish to make any comment on that affidavit, or does the Board wish to ask any questions of Dr. Oppenheimer relating to it?

Mr. GRAY. I am sure my question would be one which Dr. Oppenheimer could not answer, because it relates to the statement of Dr. Manley. I don't know what the significance of this is, but I would read this statement in parentheses on page 10. I don't take it that this refers to Dr. Oppenheimer, but in general it says: "Indeed, I had no feeling anyone was holding back on the work on thermonuclear weapons once the President decided the question." I get from that, it seems to me, the inference that there were those who were holding back. I repeat that does not refer to Dr. Oppenheimer in his language, but it seems to me that is a carefully worded observation. This is a reaction to it, however.

The WITNESS. Do you want to put a question to me about it? I will hazard an interpretation.

Mr. GRAY. Yes.

The WITNESS. The research calculations and experiments that were in course at Los Alamos would not be held back; they would be accelerated because there was a chance of going all out. Some arrangements of an engineering kind, of a production kind, of an administrative kind, you would make if you knew you were trying to make this thing as fast as possible but you would plan for but not make if you were uncertain as to that.

An example may be the Savannah River plants. Thinking began on them—should have begun earlier—but certainly began on them once the question was raised. The actual letting of the contract for design drawings and so on would presumably have waited the Presidential decision. I suppose it is this kind of thing. There was not any retardation compared to what went before. It was a failure to accelerate in those things which involved the commitment of funds.

Mr. GARRISON. Mr. Chairman, suppose we get in touch with Dr. Manley and either have a supplementary affidavit or ask him to come on. I think that is going to be a little awkward.

Mr. GRAY. May I not at this time, but later, consult with the counsel for the board on this point and perhaps we could pass on. I don't think it is fair to ask Dr. Oppenheimer to interpret what Dr. Manley had in his mind.

Mr. GARRISON. I agree with you.

Mr. SILVERMAN. May we go off the record for one moment?

Mr. GRAY. Yes.

(Discussion off the record.)

Mr. GRAY. Suppose we proceed, and if we wish anything further I will let you know, Mr. Garrison.

The WITNESS. I have three other items of national service. As far as I know, they are not controversial. I will outline them briefly.

In late 1950 and early 1951, Mr. William Golden was asked by the President's office to explore the question, Is the mobilization of scientists adequate? There was much talk during the Korean crisis of recruiting an emergency office like the Office of Research and Development. He talked with a lot of people, including me. I recommended that there be an advisory group to the National Security Council, if the National Security Council and the President wanted it, on technical matters, and there be standby plans for all-out mobilization. But, in view of the immense expansion of research and development in the Department of Defense, an emergency organization like Dr. Bush's in the last war would just not fit into anything.

After reflection, Golden persuaded his superiors that there should be an advisory committee. It was attached in a rather peculiar way to the Office of Defense Mobilization, and the invitations to join it suggested that this commitee would be advisory to the Director of Defense Mobilization, the then Mr. Charles Wilson, and the President. The chairman of the committee was Oliver Buckley. You have a list of its members.

Mr. GARRISON. It is item 7 on the second page of II of the biographical sheet.

The WITNESS. During approximately that first year, the committee met from time to time. It was seldom asked for advice. Dr. Buckley did a great many useful liaison jobs. We proffered very little advice. I think that our only function, perhaps, was to keep some balance between the needs for basic and universal research and training, on the one hand, and defense research and development, on the other.

Dr. Buckley resigned because of ill health and was replaced by Dr. DuBridge, who became chairman in 1952. I don't remember the date.

In the autumn of 1952 we had a 2- or 3-day meeting—probably 2 days—at Princeton of this full committee to see whether we had any suggestions to pass on to the new administration as to the mobilization of science. I think we concluded that we had been of no great use and that as constituted and conceived we should be dissolved.

We suggested some changes in research and development in the Defense Department, and they are pretty close, I think, to what has taken place in the reorganization of the summer of 1953. We also said that somehow or other the Security Council might need and should certainly have available to it technical advice of the highest order and must have access to the whole community of scientists so that, if anything they wanted to know that was relevant to their deliberaions, it might be available.

We said in that framework it is conceivable that another committee might be useful. We scribbled these things down on a piece of paper, and DuBridge was supposed to see that they somehow got to President-elect Eisenhower. The President-elect had a lot of other things to do; and we went together, DuBridge and I, to Nelson Rockefeller, who had been put in charge of a committee to suggest the reorganization of the executive branch of the Government. We talked a good bit about our good-for-nothing committee, handed him this memorandum, and he reported to me and DuBridge that they discussed it in the committee and gave it to the President and thought it made sense. We thought we were dead. We were, but not quite.

In the spring of 1953, I think at the request of Mr. Flemming and Mr. Cutler, we were reactivated and asked to convene. We met several times. The principal problems put before us were the proper use of scientific manpower, the very controversial and tough problem of continental defense, where there were several technical things that we were asked to look into and advise on and report on, and I think some other problems, but since I don't have the records of the committee I can't detail them.

The last meeting I attended was just before I left for Europe and not very long before my clearance was suspended, and our principal job there was to make sure that the Council and its staff knew of technical advances which were useful in early warning and in radar generally and that they understood that some of the arguments against the feasibility of early warning were obsolete because of discoveries that had been made in the meantime.

I have no further testimony on this committee.

By Mr. GARRISON:

Q. Then we come to certain studies of defense that you made or engaged in—defense against atomic warfare—perhaps you can say a word to the board about them.

A. Yes. This can be fairly brief.

The Department of Defense adopted during the Korean crisis a practice of letting our large segments of the defense problem as study projects to a university. The university would then call in competent people from the rest of the country. I have referred to project Vista as one such. There was one under contract, I think, only with the Air Force at MIT. Its code name was Charles. Its purpose was to have a look at air defense. I had the faintest connection with this. I believe I was present at some of the briefings. It led to the establishment of the Lincoln laboratory, which is a very large radar and air defense laboratory operated by MIT for the Air Force.

Another such study which I had suggested was set up through the Army and the NSRB, I guess, and that was to have a look at civil defense—a very tough and unstudied problem, really. I was not very active. I was on the advisory council or the policy council, but I met rather rarely. I did give one or two briefings, and I talked with General Nelson about the problems of writing an effective report. There were a great many recommendations; many of them have been made public. I think those which attracted the greatest attention were that, if civil defense was to be manageable at all, early warning and improved military interception, improved over what we than had or were planning, were an essential part of making civil defense manageable. With these conclusions I concurred.

The third item here is that, largely growing out of the work of some people on East River, and in particular Dr. Berkner and Dr. Rabi, there came a conviction not only that one had to have a better continental defense but quite a lot could be done about it.

I was consulted about the wisdom of it, and I agreed to hold a study during the summer of 1952, 2 months of intensive study, at the Lincoln laboratory, which would concern itself with both an evaluation of the prospects of continental defense and recommendations of how to get on with the job.

The Lincoln Laboratory was working very hard and very effectively on some aspect of this problem. The notion of the summer study was to look at parts that had not been adequately dealt with.

I attended the first week and I think the last week of meetings there. Radar is not the subject of my expertness. * * * There was a good deal of argument about interception. * * * There was certainly a great deal of discussion about the gravity of the problem and a great deal of discussion about the two-way relations between the Strategic Air Command and the continental defense, on the one hand the early warning, giving the Strategic Air Command a chance, and on the other hand the Strategic Air Command playing an essential part in reducing the severity of the attack.

The only part of the work that seemed to me undoubtedly successful were the proposals for early warning, the technical proposals about the equipment, and the general schemes about the location of the line and their extension. I regarded and don't know too much about the problems of interception and kill as fairly much unresolved at the end of the study.

These things came back, as I have said, to the Science Advisory Committee, and we picked up the recommendations there and did our best to explain them.

These almost all have to do with early warning. I believe that I have read in the papers that many steps have been taken to improve the situation. I think it is a very important contribution not to the security but to the deterrent value of our own offensive striking power and a deterrent to attack, at least during the period of limited enemy capability.

Those are the three projects.

The final assignment—and I assure you it is final—was of a somewhat different kind. In the spring of 1952 I had a letter from the Secretary of State appointing me or asking me whether I would serve as a member of a panel. The other members of the panel were Allen Dulles, John Dickey, Vannevar Bush, and Joe Johnson. The letter appointing us said that it seemed to be time that the delegate, who was then Benjamin Cohen, who was representing us in the disarmament conference, would like to advise and even more the people in the State Department who were responsible for our policy with regard to the regulation of armaments. We all went to a meeting with the Secretary of State, people of Defense—it was a great big meeting—somewhat puzzled as to whether there was any reality to the job we had been asked to assume but willing at least to listen.

At the meeting it was made clear by the Secretary that he would like any report, any study of the regulation of armaments—was it a feasible goal, was there any way to go about it, were there any tricks to it—similar to the Acheson-Lilienthal report of many years before, could armaments be regulated, and he would like us to help the people who were working diplomatically in this field. But he thought in addition that we ought to see whether we did not have something to say and get it written down.

Mr. GARRISON. This is item 8 of the memorandum.

Mr. GRAY. Yes.

The WITNESS. As to the consultations, they took place. I saw something of Mr. Cohen and maybe helped in some minor ways, and I think others did. We also talked with people in the Department of State. But there was clearly not much reality to the discussions of disarmament in the United Nations, and the most we could do was make a few helpful suggestions which would encourage our friends as to our good faith and interest.

It took a long while for the members of the panel to get cleared. But that happened sometime during the summer. We got George Brundy to be our secretary, who is now dean of Harvard College but was then professor of political science there. We had a look at what we had been asked to look at. We went over the studies of past efforts of disarmament. Mr. Dulles remembered them very vividly. It was very clear that you could not negotiate with the Russians much about anything and that nothing was harder to negotiate about than disarmament, and if you put these two things together it just was the bleakest picture in the world of getting anything effective down that line.

We took a look at the armament situation, getting some estimates of the growth of Russian capability and some estimates of our own as a measure for where they might be some time in the future. I think as always we thought we were being careful, but we were a little too conservative in estimating the speed and success of the Soviet program. We became very vividly and painfully aware of what an unregulated arms race would lead to in the course of years. We tended to think in the course of 5 or 10 years, but probably the time was shorter.

Our report was of course classified. We filed it in January of 1953. It had 5 recommendations, of which 2, I think I should not talk about because they had to do with the conduct of our diplomatic affairs and should be regarded as secret. They are not very ingenious.

The other three I embodied in an article that I published in Foreign Affairs. Before publishing it, I took it to the President. He showed it to Mr. Cutler. Mr. Cutler had no objection to my publication. He thought my publication would be helpful and encouraged me to go ahead with it.

These three were that the people of this country be given a better understanding of the dangers of the atomic arms race, that we attempt either through administrative practice or through revised legislation to work more closely with our allies on problems having to do with the offensive and defensive aspects of large weapons, and three, that we take further measures for continental defense as a supplement to our striking capability.

I was asked to report on these three things before the Jackson committee, I think it was on psychological strategy and so did rather briefly, and I was asked to report on these more or less as an advocate before the National Security Council, asked by the President, and I went to do that. At that time Dr. Bush and Commissioner Dean went with me. I presented the arguments, which I think are in Foreign Affairs, and which are still persuasive to me, in favor of these three steps.

I did mention the diplomatic points at the Security Council, because that was of course not a public meeting.

That brings me to the end of this fairly long spiel I have given you about my connection with the United States Government.

Mr. GRAY. Just one question. What was the date of that Foreign Affairs article?

The WITNESS. It was published in the July issue of 1953. It actually came out a little earlier, in June or something like that.

Mr. ECKER. That was submitted to you.

Mr. GRAY. I was not sure that was the same one.

The WITNESS. There are two.

Mr. ECKER. Yes.

(Discussion off the record.)

The WITNESS. Might I put one more statement into the record on my conduct as a part time public servant during these years.

Of course, these things were secret. They were not subject to the scrutiny of the press, and they were not generally open, but they were not secret in the sense that the people did not know what we were up to. We were constantly testifying before congressional committees, we were writing reports which were very widely circulated. We were under, I would say, a very intensive searchlight of scrutiny. We were always in a position where our advice could be countered, could be overruled or could be accepted. There was no opportunity for conspiracy in these things because the light of criticism was constantly shining on them.

Mr. GARRISON. Mr. Chairman, the first letter I should like to introduce into the record is from Gordon Dean, Chairman of the Atomic Energy Commission, to Dr. Oppenheimer dated June 14, 1952.

"Mr. J. ROBERT OPPENHEIMER,
  "*Institute for Advanced Study,*
    "*Princeton, N. J.*

"DEAR MR. OPPENHEIMER: I want to express my personal thanks to you for our talk of yesterday concerning the General Advisory Committee and its role as an advisory group to the Commission. It was most helpful.

"I want you to know that I fully appreciate the reasons behind your unwillingness to have your name considered for reappointment to the GAC. I would not have been quite so prepared for this had you not so long ago advised me of your intention to pass the baton on to another.

"It is impossible for me to magnify the contribution which, as Chairman of this distinguished group, you have made to the Commission and the country. It has been a magnificent one and we of the Commission will be forever grateful to you. The period covered by your chairmanship has been one in which this new agency needed very much the wisest possible guidance. This we have received and no one knows this better than myself.

"I am quite aware that there is no one who can adequately take your place, but your willingness to remain as a consultant to the Commission somewhat softens the blow of your departure from the GAC councils.

  "With every good wish,
    "Sincerely,
        "GORDON DEAN, *Chairman.*"

The second letter is signed Harry Truman, the White House, Washington, D. C., September 27, 1952:

"Dr. J. R. OPPENHEIMER,
  "*Director, the Institute for Advanced Study,*
    "*Princeton, N. J.*

"DEAR DR. OPPENHEIMER: Having in mind your strong desire, which you expressed to me last month, to complete your service on the General Advisory Committee to the Atomic Energy Commission with the expiration of your present term, I note with a deep sense of personal regret that this time is now upon us.

"As Chairman of this important committee since its inception, you may take great pride in the fact that you have made a lasting and immensely valuable contribution to the national security and to atomic energy progress in this Nation. It is a source of real regret to me that the full story of the remarkable progress that has been made in atomic energy during these past 6 years, and in which you have played so large a role, cannot be publicly disclosed, for it would serve as the finest possible tribute to the contribution you have made.

"I shall always be personally grateful for the time and energy you have so unselfishly devoted to the work of the General Advisory Committee, for the conscientious and rewarding way in which you have brought your great talents to bear upon the scientific problems of atomic energy development, and for the notable part you have played in securing for the atomic energy program the understanding cooperation of the scientific community.

"As director of the Los Alamos Scientific Laboratory during World War II, and as chairman of the General Advisory Committee for the past 6 years, you have served your country long and well, and I am gratified by the knowledge that your wise counsel will continue to be available to the Atomic Energy Commission on a consultant basis.

"I wish you every future success in your important scientific endeavors.

"Very sincerely yours,

"HARRY TRUMAN."

And the final letter is another one from Gordon Dean dated October 15, 1952.

"Dr. J. ROBERT OPPENHEIMER,
 "*Institute for Advanced Study,*
  "*Princeton, N. J.*

"DEAR OPPY: I cannot let your departure from the General Advisory Committee go by without expressing again my deep appreciation for the time and talent which you have so generously devoted to the work of the committee, and for the immensely valuable contribution you have made to the atomic energy program during the period I have been associated with it and before.

"I know that you are as fully aware as I am of the assistance the General Advisory Committee has given to the Commission during these past 6 formative years, and of the great scientific and technical strides that have been made in that time. I sincerely hope that some day, when the ills of the world are sufficiently diminished, the complete story of this progress can be told, so that the contribution of you and your colleagues may find its rightful place in the chronicle of our times.

"May I say that I shall always be grateful for your past work on behalf of the program, and for your willingness to continue to advise the Commission on a consultative basis.

"With every good wish,
  "Sincerely,

"GORDON DEAN, *Chairman.*"

There are, Mr. Chairman, several exhibits that I would like to introduce at this time having to do with Dr. Oppenheimer's views on the freedom of the mind and the human spirit. I introduce them to show a position which I think could not be tolerated for one moment behind the Iron Curtain.

Mr. GRAY. These are to be exhibits?

Mr. GARRISON. These will be extracts from original documents which I will hand the board. One is taken from a lecture, which 1 of the 3 was it, Dr. Oppenheimer, you gave?

The WITNESS. No; there were six. This is the last one.

Mr. GARRISON. Do you want to tell the board in 1 minute what those lectures were?

The WITNESS. Gladly. I was invited a year ago and then again this year to give lectures in England. They are named in honor of Lord Reith. They are on the home program and there is really a large audience, 15 million or something. They are meant to be quite serious. I think the first lectures were given by Russell, called Authority and the Individual. I called mine Science and a Common Understanding. I talked about it—I won't summarize them. That is irrelevant. The principal point was to indicate in what ways contemporary science left room for an integrated human community. Why it was not necessary specialized knowledge led to fragmentation in society. That was about it. The last lecture has something about that in it.

Mr. GRAY. My question is whether these are offered as exhibits. We have a couple of earlier documents.

Mr. GARRISON. I would like to treat these as the others, to have them available for the inspection of the board, so you may look at them in the whole.

Mr. ROBB. Are those the lectures published in a publication called The Listener?

The WITNESS. Yes.

Mr. ROBB. We have those.

Mr. GARRISON. The one I shall read into the record is a very short excerpt from a speech given to the University of Denver by Dr. Oppenheimer February 6, 1947. It is page 8 of the small reprint which I just handed to you. It reads as follows:

"And above all, I think, there stands the great conflict with Soviet communism. There may be people who believe that this (system)"—the insertion is our own for clarity—"originated in a desire to provide for the well-being of the people of Russia. * * * But whatever its origin, it has given rise to political forms which are deeply abhorrent to us and which we not only would repudiate for ourselves but which we are reluctant to see spread into the many areas of the world where there is great lability. * * *"

That word is "lability" and I understand it means flexibility.

Mr. GRAY. Thank you very much, Mr. Garrison.

Mr. GARRISON. Of course, Mr. Chairman, it is quite obvious—there is no mystery about these excerpts—I have quite plainly selected those which seemed to me relevant and that bore upon Dr. Oppenheimer's attitude toward the problem of our relation with Russia. They don't attempt therefore to be comprehensive excerpts of the whole speech but simply of those items which seem to me are utterly inconsistent with the notion that Dr. Oppenheimer could be, as depicted in the Commission's letter.

The next excerpt from the Reith lectures in The Listener, pages 1076 and 1077:

"It is true that none of us will know very much; and most of us will see the end of our days without understanding in all its details and beauty the wonders uncovered even in a single branch of a single science. Most of us will not even know, as a member of any intimate circle, anyone who has such knowledge; but it is also true that, although we are sure not to know everything and rather likely not to know very much, we can know anything that is known to man, and may, with luck and sweat, even find out some things that have not before been known to him. This possibility, which, as a universal condition of man's life is new, represents today a high and determined hope, not yet a reality; it is for us in England and in the United States not wholly remote or unfamiliar. It is one of the manifestations of our belief in equality, that belief which could perhaps better be described as a commitment to unparalleled diversity and unevenness in the distribution of attainments, knowledge, talent, and power.

"This open access to knowledge, these unlocked doors and signs of welcome, are a mark of a freedom as fundamental as any. They give a freedom to resolve difference by converse, and, where converse does not unite, to let tolerance compose diversity. This would appear to be a freedom barely compatible with modern political tyranny. The multitude of communities, the free association for converse or for common purpose, are acts of creation. It is not merely that without them the individual is the poorer; without them a part of human life, not more nor less fundamental than the individual, is foreclosed. It is a cruel and humorless sort of pun that so powerful a present form of modern tyranny should call itself by the very name of a belief in community, by a word 'communism' which in other times evoked memories of villages and village inns and of artisans concerting their skills, and of men of learning content with anonymity. But perhaps only a malignant end can follow the systematic belief that all communities are one community; that all truth is one truth; that all experience is compatible with all other; that total knowledge is possible; that all that is potential can exist as actual. This is not man's fate; this is not his path; to force him on it makes him resemble not that divine image of the all-knowing and all-powerful but the helpless, iron-bound prisoner of a dying world. The open society, the unrestricted access to knowledge, the unplanned and uninhibited association of men for its furtherance—these are what may make a vast, complex, ever-growing, ever-changing, ever more specialized and expert technological world nevertheless a world of human community."

Mr. GRAY. It is now I think 4:20. I wonder if there are any other exhibits. If not, this would seem to be a good breaking point.

Mr. GARRISON. Yes; I think so.

Mr. GRAY. Unless counsel for the board has something to say, we will then recess and meet again at 9:30 tomorrow morning.

(Thereupon at 4:20 p. m., a recess was taken until Wednesday, April 14, 1954, at 9:30 a. m.)

UNITED STATES ATOMIC ENERGY COMMISSION

PERSONNEL SECURITY BOARD

IN THE MATTER OF J. ROBERT OPPENHEIMER

ATOMIC ENERGY COMMISSION,
BUILDING T-3, ROOM 2022,
Washington, D. C., April 14, 1954.

The above entitled matter came on for hearing, pursuant to recess, before the board, at 9:30 a. m.

Personnel Security Board: Mr. Gordon Gray, chairman; Dr. Ward V. Evans, member; Mr. Thomas A. Morgan, member.

Present: Roger Robb, and C. A. Rolander, Jr., counsel for the board.

J. Robert Oppenheimer, Lloyd K. Garrison, Samuel J. Silverman, and Allen A. Ecker, counsel for J. Robert Oppenheimer.

Herbert S. Marks, cocounsel for J. Robert Oppenheimer. (Present for p. m. session only.)

PROCEEDINGS

Mr. GRAY. The presentation will begin.
Whereupon, J. Robert Oppenheimer, the witness on the stand at the time of taking the recess resumed the stand and testified further as follows:

DIRECT EXAMINATION (CONTINUED)

By Mr. GARRISON:
Q. Dr. Oppenheimer, will you tell the board something about your brother Frank, your relations with him?
A. He was 8 years my junior.
Q. It was just you and Frank in the family?
A. We were the only children. I think I was both an older brother and in some ways perhaps part of a father to him because of that age difference. We were close during our childhood, although the age gap made our interests different. We sailed together. We bicycled together. In 1929 we rented a little ranch up in the high mountains in New Mexico which we have had ever since, and we used to spend as much time there as we could in the summer. For my part that was partly for reasons of health, but it was also a very nice place.

My brother had learned to be a very expert flutist. I think he could have been a professional. He decided to study physics. Since I was a physicist this produced a kind of rivalry. He went abroad to study. He studied at Cambridge and at Florence. He went to college before that at Johns Hopkins.

When he came back to this country; he did take his doctor's degree at the California Institute of Technology.

We were quite close, very fond of one another. He was not a very disciplined young man. I guess I was not either. He loved painting. He loved music. He was an expert horseman. We spent most of our time during the summer fiddling around with horses and fixing up the ranch.

In the very first year he had two young friends with him who were about his age, and I was the old man of the party. He read quite widely, but I am afraid very much as I did, belles lettres, poetry.

Dr. EVANS. Was your father there at that time?
The WITNESS. My father was alive. He did occasionally visit at the ranch. His heart was not very good. This is almost 10,000 feet high, so he did not spend much time there. We could not put him up. It was a very primitive sort of establishment. There was of course the tension which a very intimate family relation of this kind always involves, but there was great affection between us.

He worked fairly well at physics but he was slow. It took him a long time to get his doctor's degree. He was very much distracted by his other interests.

In 1936, I guess it was, he met his present wife and married. I am not completely sure of the date, but I could check it. After that, a good deal of the warmth of our relations remained, but they were less intimate and occasionally perhaps somewhat more strained. His wife had, I think, some friends and connections with the radical circles in Berkeley. She was a student there. She had a very different background than Frank. She certainly interested him for the first time in politics and leftwing things. It was a great bond between them.

As I wrote in my answer, not very long after their marriage they both joined the Communist Party. This was in Pasadena. I don't know how long thereafter, but not very long thereafter, Frank came to Berkeley and told me of this. We continued to be close as brothers are, but not as it had been before his marriage.

He once asked me and another fellow to come visit one of the meetings that he had in his house, which was a Communist Party meeting. It is, I think, the only thing recognizeable to me as a Communist Party meeting that I have ever attended.

Mr. ROBB. I am sorry. Could we go back to where the doctor said he once asked me. I did not get the rest of the words.

The WITNESS. And another fellow. I would be glad to identify him, but he is not alive and not involved in the case.

By Mr. GARRISON:

Q. This was a professor?

Mr. ROBB. Was that Dr. Addis?

The WITNESS. No. This was Calvin Bridges, a geneticist at Cal Tech, and a very distinguished man, not a Communist as far as I know.

Dr. EVANS. This was not a closed meeting of the Communist Party?

The WITNESS. It was not closed because it had visitors. I understood the rest of the people were Communists. This was on the occasion of one of my visits to Berkeley and Pasadena. The meeting made no detailed impression on me, but I do remember there was a lot of fuss about getting the literature distributed, and I do remember that the principal item under discussion was segregation in the municipal pool in Pasadena. This unit was concerned about that and they talked about it. It made a rather pathetic impression on me. It was a mixed unit of some colored people and some who were not colored.

I remember vividly walking away from the meeting with Bridges and his saying, "What a sad spectacle" or "What a pathetic sight," or something like that.

Mr. GRAY. Did you give the approximate date of this, Doctor?

The WITNESS. I can give it roughly.

Mr. GRAY. I mean within a year.

The WITNESS. It would have been not before 1937 or after 1939. I think I ought to stress that although my brother was a party member, he did a lot of other things. As I say, he was passionately fond of music. He had many wholly non-Communist friends, some of them the same as my friends on the faculty at Cal Tech. He was working for a doctor's degree.

He spent summers at the ranch. He couldn't have been a very hard working Communist during those years.

I am very foggy as to what I knew about the situation at Stanford but my recollection is that I did not then know my brother was still in the party. He has testified that he was, and that he withdrew in the spring of 1941. He lost his job at Stanford. I never clearly understood the reasons for that, but I thought it might be connected with his communism.

We spent part of the summer of 1941 together at the ranch, about a month. That was after my marriage. He and his wife stayed on a while. Then they were out of a job. Ernest Lawrence asked him to come to Berkeley in the fall, 1 don't remember the date, but I think it is of record, and work in the radiation laboratory. That was certainly at the time not for secret work. He and I saw very little of each other that year.

My brother felt that he wanted to establish an independent existence in Berkeley where I had lived a long time, and didn't want in any sense to be my satellite. He did become involved in secret work, I suppose, shortly after Pearl Harbor. I don't know the precise date.

He continued with it and worked terribly hard during the war. I have heard a great many people tell me what a vigorous and helpful guy he was, how many hours he spent at work, how he got everybody to put their best to the job that was his. He worked in Berkeley. He worked in Oak Ridge. He came for a relatively brief time to New Mexico, where his job was as an assistant to Bainbridge in making the preparations for the test of July 16.

This was a job that combined practical experience, technical experience, a feeling for the country, and I think he did very well. He left very early—left long before I did—and went back to Berkeley. We did not see him again until the New Year's holidays in 1945 and 1946. After that, when we came back to Berkeley, we saw something of them, quite a little of them, until they moved to Minnesota.

As you probably know, he resigned from the University of Minnesota—his assistant professorship there—in the spring of 1949 at the time he was testifying before the House committee that he had been a member of the Communist Party. The university accepted his resignation. He has not been able to get a job since, or at least not one that made sense.

He had in the summer of 1948. maybe, or the winter of 1948–49, acquired a piece of property in southwest Colorado. It is also fairly high. It is in the Blanco Basin. I think he got it because it was very beautiful, and thought it would be nice to spend summers there. In any case, he and his wife and children moved up there, and have been trying to build it up as a cattle ranch ever since

They have been there, I think, with no important exceptions, from 1949 until today. This life is not what he was cut out for and I don't know how it will go.

I try to see him when I can. It does not come out to being much more than once a year. I think the last time I saw him was in late September or October of last year. Usually he would come down to Santa Fe, and we would have an evening together or something like that. I had the feeling the last time that I saw him that he was thoroughly and wholly and absolutely away from this nightmare which has been going on for many, many years.

These are at least some of the things that I wanted to say. I would like to say one more thing.

In the Commission's letter——

By Mr. GARRISON:

Q. Perhaps I could ask you about that.

On page 6 of the Commission's letter, which talks about Haakon Chevalier, there is a statement, I am quoting, "that Haakon Chevalier thereupon approached you either directly or through your brother, Frank Friedman Oppenheimer, in connection with this matter."

Was your brother connected with this approach by Chevalier to you?

A. I am very clear on this. I have a vivid and I think certainly not fallible memory. He had nothing whatever to do with it. It would not have made any sense, I may say, since Chevalier was my friend. I don't mean that my brother did not know him, but this would have been a peculiarly roundabout and unnatural thing.

Q. You spoke about attending at your brother's invitation that little Communist Party meeting in Pasadena somewhere in the late thirties, and that reminds me to ask you about another portion of the Commission's letter.

On page 3, I will just read a paragraph:

"It was reported that you attended a closed meeting of the professional section of the Communist Party of Alameda County, Calif., which was held in the latter part of July or early August, 1941, at your residence, 10 Kenilworth Court, Berkeley, Calif., for the purpose of hearing an explanation of the change in Communist party policy. It was further reported that you denied that you attended such a meeting and that such a meeting was held in your home."

Dr. Oppenheimer, did you attend a closed meeting of the professional section of the Communist Party of Alameda County which is said to have been held in your house in the latter part of July or early August, 1941?

A. No.

Q. Did you ever attend at any time or place a closed meeting of the professional section of the Communist Party of Alameda County?

A. No.

Q. Were you ever asked to lend your house for such a meeting?

A. No.

Q. Did you every belong to the professional section of the Communist Party of Alameda County?

A. I did not. I would be fairly certain that I never knew of its existence.

Q. Did you ever belong to any other section or unit of the Communist Party or to the Communist Party?

A. No.

Q. Apart from the meeting in Pasadena, to which we have just referred, have you ever attended a meeting which you understood to be open only to Communist Party members, other than yourself?

A. No.

Q. Have you ever had in your house at any time any meeting at which a lecture about the Communist Party has been given?

A. No.

Q. Do you recall any meeting in your house at any time at which a lecture about political affairs of any sort was given?

A. No.

Q. To sum up, Dr. Oppenheimer, do you deny the report set forth on page 3 of the Commission's letter which I read to you?

A. All but the denial; I deny the rest.

Mr. GARRISON. Just so the board understands, I read the statement to Dr. Oppenheimer, "It was further reported that you attended such a meeting and that such a meeting was held in your home.

The WITNESS. That I don't deny.

By Mr. GARRISON.

Q. The first sentence of the report you do deny.

A. Yes.

Mr. GARRISON. I would like to introduce, Mr. Chairman, at this point, copies of correspondence relating to the Independent Citizens Committee of the Arts, Sciences and Professions, which is mentioned in the Commission's letter on page 6, which reads that "it was reported in 1946 that you (that is, Dr. Oppenheimer) were listed as vice chairman on the letterhead of the Independent Citizens Committee of the Arts, Sciences and Professions, Inc., which has been cited as a Communist front by the House Committee on Un-American Activities.

I think in my earlier discussion with the board, I pointed out that in all the postwar period, this is the only association cited by the House Committee or in any other way challenged by any group in the Government as un-American with which Dr. Oppenheimer had any connection at all.

I now would like to introduce the correspondence which will show his resignation and his relationship to that committee which I think the board will agree was to his credit. I would like to read these into the record.

By Mr. GARRISON:

Q. Dr. Oppenheimer, I have here carbon copies of letter from you to the Independent Citizens Committee dated October 11, 1945, October 11, 1946, November 22, 1946, is an original letter from the committee to you, followed by a carbon of December 2, 1946, from you to them, and an original from the secretary to you of December 10, 1946. Do you identify these as having been in your files?

A. Yes, these were in my file, and I made them available to you.

Mr. GRAY. Mr. Garrison, I think perhaps for the record, at least what we have been handed, reflects nothing dated 1945. In your characterization of these documents, you said a letter of October something 1945.

Mr. ECKER. Excuse me. That is because it is a fuzzy date on the carbon.

Mr. GARRISON. It is my fuzziness, Mr. Chairman. The carbon shows its 1946.

Mr. GRAY. I am just trying to get the record straight.

Mr. GARRISON. I regret my eyesight was not equal to the carbon.

This first letter reads as follows, the letter of October 11, 1946, to the committee.

"INDEPENDENT CITIZENS COMMITTEE OF THE ARTS,
SCIENCES AND PROFESSIONS,
New York 17, N. Y.

"GENTLEMEN: Some months ago I was elected a vice chairman of the ICCASP. This has not been a very arduous responsibility, since I have had virtually no contact with the organization. I have, however, noted with a growing uneasiness over the past months ICCASP's statements on foreign policy.

"As examples, I may quote two programatic statements of the ICCASP policy: 'Maintain the Big Power veto in the Security Council', and 'Withdraw United States troops from China.'

"I do not wish to challenge the merits of the arguments that may be advanced for these two theses. They do not seem to me, at least in this bald form, to correspond to the extension of President Roosevelt's foreign policy; nor am I in accord with them.

"Most recently I have noted in the papers an item which disturbs me more, because it concerns the problem of atomic energy, with the outlines of which I am not unfamiliar, and for which I may even have a certain responsibility. I am, of course, aware that newspaper comments may often be misleading. As I understand it, the ICCASP at a recent convention in Chicago agreed to endorse the criticism of United States policy and procedure enunciated by Secretary Wallace in his letter to the President of July 23. Here again, I should not wish to argue that there was nothing sound in Mr. Wallace's comments, nor for a moment to cast doubt on the validity of his great sense of concern that a satisfactory solution for the control of atomic energy be achieved; but I cannot convince myself that, in the large, the suggestions made by Mr. Wallace would, if adopted, advance this great cause; and above all, I feel that the evidence which is now available, and which goes beyond that which was available on July 23 indicates the illusory nature of his recommendations.

"It is clear that I should not prejudge the position which the ICCASP is taking on these many important questions; but unless I am badly misinformed on what that position is, it seems to me I can no longer remain a vice chairman of that organization.

"Will you, therefore, accept this letter as a letter of resignation, unless it is clear to you, and you can make it clear to me, that it is based on a misunderstanding of the facts.
"Sincerely yours,

J. R. OPPENHEIMER."

Then comes the reply from the executive director, signed by Hannah Dorner, the executive director:

"INDEPENDENT CITIZENS COMMITTEE OF THE ARTS,
SCIENCES AND PROFESSIONS, INC.
"New York 19, N. Y., November 22, 1946.

"Dr. J. R. OPPENHEIMER,
"University of California, Berkeley 4, Calif.

"DEAR DR. OPPENHEIMER: Please forgive this delay in answering your letter, but I have been out of town a good deal and this is the first opportunity I have had.

"It would come as a great surprise to the members of the board of ICCASP that the organization can be found guilty of any contradiction of President Roosevelt's foreign policy. We have stated repeatedly that the organization was formed initially to re-elect Mr. Roosevelt and then reformed in order to provide a medium through which the members of the arts, sciences and professions could help to implement and carry out his program.

"In connection with the two programatic statements you refer to in your letter, unless I am very much mistaken the veto power is the core of the postwar foreign policy which Mr. Roosevelt outlined in conjunction with Churchill and Stalin. I don't know what Mr. Roosevelt would have said were he alive today about maintenance of United States troops in China. I do know that for years during the war he refused to send materiel into China because Chiang Kai-Shek was not using it against Japan but instead, saving it for the conflict he is currently engaged in. It is fairly common knowledge that the presence of United States troops and American materiel are being used to aid one side against another in a civil war. Without discussing the merits of either side, certainly it would seem that the American position should be one in which a real effort is made to create a democratic China instead of bolstering the position of military feudalism which Chiang Kai-shek and his supporters represent. I think Madame Sun Yat Sen's position is one which Americans might fairly support and the presence of our troops in China and our present policy are giving no encouragement to her views and to those Chinese who wish as she does for a truly democratic China.

"In connection with Mr Wallace's comments on atomic energy, let me make it clear that the statement on atomic energy at the Chicago conference was made by some 300 delegates representing many organizations, of which ICCASP was just one.

"You will have seen, I am sure, a further statement made since that conference on atomic energy by a coordinating committee of the Chicago conference, after Mr. Baruch clarified the points raised at the Chicago conference. As you unquestionably know, our science division has been working for some time both in New York and Chicago on an analysis of the atomic energy control program and as yet the ICCASP has not adopted a position since we are waiting on the final report of the science division. I assume that as a member of the division you will receive that report for your comment and criticism.

"In this letter I am attempting to answer the issues raised, with the hopes that they will clarify our position and that you will find yourself in substantial agreement with us. I realize that it is difficult for someone with as many demands upon his time as you to attend meetings of the ICCASP. It is unfortunate that this is so because you should participate with the rest of us in forming the policy, instead of getting it without the benefit of all of the full discussion that goes into arriving at these decisions.

"I hear frequently about how often you are in New York. If you would only let me know about these visits you could, I am certain, find a few hours to attend some of these meetings. I am sure it is quite unnecessary to make the point to you that the fate of a generation or two is being shaped today. The ICCASP is conscientiously trying to do what it can to make it a kinder fate. I am certain that all of us individually will disagree with the organization's position on one or two issues from time to time. The importance of the committee as a whole, what it has accomplished, and the need for keeping it alive and strong should transcend occasional differences.

"All of us value your continued association with the organization.

"Sincerely yours,

"HANNAH DORNER."

The reply by Dr. Oppenheimer, dated December 2, 1946, is as follows:

"Miss HANNAH DORNER,
"Independent Citizens' Committee of the Arts, Sciences and Professions, Inc.
"Hotel Astor, New York 19, N. Y."

I see the copy which we have handed the members of the Board, Dr. Oppenheimer's signature does not appear, nor does it appear on the carbon, but his initials are on the lower left and that of the typist.

"DEAR MISS DORNER: Thank you very much for your letter of November 22, in which you tried to explain to me how poor are the reasons I gave for resignation from the vice chairmanship of ICCASP in my letter of October 11. I wish that I might have been convinced by what you wrote for I share with you an appreciation for the many constructive and decisive things which the ICCASP is doing, and I am quite sure that I should not be moved to resign were it not for two circumstances. One is that I have a somewhat unreal position as vice chairman and might thus be thought to be far more influential and effective in shaping ICCASP policy than I have been or than I am likely to be in the near future. The second is that the matter of atomic energy is one of the very few on which I have more than the vaguest kind of views, is perhaps the only political issue on which I have a limited competence and have in the past borne some responsibility.

"I find nothing in the record to comfort me in the matter of atomic energy. The press release of the Chicago conference and its subsequent announcement are both very far from my views and were endorsed by ICCASP without qualifications. The last communication that I have received is dated Monday, September 23, and reached me after my letter of resignation. In it a resolution of the division of science and technology closely parallel to that adopted in Chicago was submitted to the executive committee of the ICCASP and approved. I have had no further communication since that time either with regard to atomic energy or to the functioning of the science division of the ICCASP, except for the proposed statement on the control of atomic energy which is undated and which likewise does not represent my views. I, therefore, feel that it is likely that there is a genuine difference of opinion on this matter between me and the executive committee of the ICCASP.

"For the reasons stated above I think it is not proper to continue to serve as vice chairman under these circumstances. I recognize that it is largely my own doing that I have not had a greater part in the formulation of ICCASP policy, but that should be a genuine reason of all of us not to accept a position of apparent responsibility without being willing to make the responsibility real.

"I should like to take this course of resignation since the alternative, to make public my dissident views, is repugnant to me and can help neither the ICCASP nor the cause of world peace which is surely our greatest common aim. I am, therefore asking you to accept my letter of resignation.

"Sincerely yours,

"JRO: cl."

Then the reply from Hannah Dorner.

Mr. ROBB. It is the same heading you had before.

Mr. GARRISON. Yes, it is the same heading as before. The date of this is December 10, 1946. It was on the original and should be on these copies. This is in reply to Dr. Oppenheimer's second letter insisting on resignation which I have just read to you.

"INDEPENDENT CITIZENS COMMITTEE OF THE ARTS,
SCIENCES AND PROFESSIONS, INC.,
"New York 19, N. Y.

"Dr. J. R. OPPENHEIMER,
"University of California, Berkeley 4, Calif.

"DEAR DR. OPPENHEIMER: "We accept with regret your resignation from the organization.

"We hope that some time again in the future you may want to rejoin us.

"Sincerely yours,

"HANNAH DORNER."

By Mr. GARRISON:

Q. Did you ever rejoin the organization, Dr. Oppenheimer?
A. No.
Mr. GRAY. Just as a matter of curiosity, did they ever take your name off the letterhead, do you know?
The WITNESS. They stopped sending me communications. I don't know.
Mr. GRAY. Your name apparently did not appear on these letterheads.
Mr. SILVERMAN. We did on the back. There are a lot of names on the back of the original.
Mr. GARRISON. We will hand this to the chairman in just a moment. I am just looking over these names. It shows Joseph E. Davies as the honorary vice chairman.
Mr. ROBB. Don't you think he ought to read them all?
Mr. GRAY. I think it would be well to read the whole.
Mr. GARRISON. This is on the back of the letterhead of the Independent Citizens Committee of the Arts, Sciences and Professions, Inc. This is the letter of December 10, 1946, accepting with regret Dr. Oppenheimer's resignation from the organization, and hoping some time again in the future he may want to rejoin them.
Mr. ROBB. Is that the same as the original letter of November 23, 1946? Is that the same list?
Mr. GARRISON. It appears on superficial observation the same. Mr. Robb, you can examine it at your leisure. I can see no difference.
Mr. ROBB. Why don't you let me take one of them and I will follow as you read, and we will know whether they are the same or not.
Mr. GARRISON. I am reading from the back of the letterhead, Independent Citizens' Committee of the Arts, Sciences, and Professions, Inc., Hotel Astor, New York 19, N. Y. Circle 6–5412.
Vice chairmen: Joseph E. Davies, honorary; Brig. Gen. Evans F. Carlson; Norman Corwin; Reuben G. Gustavson; Fiorello H. LaGuardia; J. Robert Oppenheimer; Paul Robeson; Harlow Shapley; Frank Sinatra.
Board of directors. Do you wish the board of directors?
Mr. GRAY. I think you better read it all.
Mr. GARRISON. Samuel L. M. Barlow, William Rose Benet, Leonard Bernstein, Walter Bernstein, Henry Billings, Charles Boyer, Henrietta Buckmaster, Eddie Cantor, Morris Llewellyn Cooke, Samuel A. Corson, John Cromwell, Bosley Crowther, Duke Ellington, Howard Fast, Jose Ferrer, Joan Fontaine, Allan R. Freelon, Dr. Channing Frothingham—a very dear friend of mine from Boston, Massachusetts, a distinguished physician—Dr. Rudolph Ganz, Ben Grauer, Marion Hargrove, Louis Harris, Moss Hart, Lillian Hellman, John Hersey, Melville J. Herskovits, J. Allen Hickerson, Thorfin R. Hogness, Walter Huston, Crockett Johnson, Gene Kelly, Isaac M. Kolthoff, Richard Lauterbach, Eugene List, Peter Lyon, John T. McManus, Florence Eldridge March, Dorothy Maynor, Stanley Moss, Ernest Pascal, Robert Patterson—I take it that was not the Secretary of War, but I guess we don't know.
The WITNESS. I know nothing about it.
Mr. GARRISON. I assume it was not. Linus Pauling, Virginia Payne, Dr. John P. Peters, Walter Rautenstrauch, Quentin Reynolds, Hazel Scott, A. C. Spectorsky, Carl Van Doren, Orson Wells and Carl Zingrosser.
Then follow a list of regional chapters. Shall I read those, Mr. Chairman?
Mr. GRAY. Is this just names of cities?
Mr. GARRISON. Yes, and addresses.
Mr. GRAY. I see no point in that. This is not related to the proceeding. But here is an organization accepting the resignation of one of its vice chairmen and apparently did not bother to strike his name off the letterhead on his letter of resignation. I really think this has no point, but from what I heard, it is very difficult to resign from some of these organizations once one seems to be a member.
Mr. GARRISON. I think you can take judicial notice of the fact that organizations reprint their letterheads at intervals, sometimes at considerable intervals.
Mr. ROBB. Mr. Chairman, I might say that the lists were identical so we have that in the record, too.

By Mr. GARRISON.

Q. Dr. Oppenheimer, do you adopt your answer consisting of your letter to Maj. Gen. K. D. Nichols, dated March 4, 1954, as your testimony in this proceeding?
A. Yes.

Mr. GARRISON. Mr. Chairman, that will be all the questions I wish to ask Dr. Oppenheimer. I may a little later as we proceed come back with some occasional questions, perhaps. That will be all at this point.

Mr. GRAY. They will be related to questions and discussions which will take place from now on. This is not going to circumscribe you in any way, but I take it Dr. Oppenheimer's presentation as you see it, and as he sees it, is complete now?

Mr. GARRISON. Yes. Mr. Chairman, there may be some detail that I have overlooked in the great press of preparing this which I might at a later stage ask to be inserted in the record, but so far as I am now aware, this completes the direct case. I assume we are not quite so rigid but what if I have overlooked something it may be later introduced?

Mr. GRAY. Yes.

Mr. GARRISON. There is no design to do so.

Mr. GRAY. I understand.

At this point, I think, then, we will suggest that counsel for the Board put to Dr. Oppenheimer the questions which he may have in mind.

CROSS EXAMINATION

By Mr. ROBB:

Q. Dr. Oppenheimer, did you prepare your letter of March 4, 1954, to General Nichols?

A. You want a circumstantial account of it?

Q. I assume you prepared it with the assistance of counsel, is that correct?

A. Yes.

Q. In all events, you were thoroughly familiar with the contents of it?

A. I am.

Q. And have read it over very carefully, I assume?

A. Yes.

Q. Are all the statements which you make in that letter the truth, the whole truth and nothing but the truth?

A. Yes.

Q. Those things which you state in there as of your personal knowledge are true to your personal knowledge?

A. That is right.

Q. And those things which you state of necessity on your information and belief, you do believe to be true?

A. That is right.

Q. Did you also prepare your exhibit 1, I believe it is, the biographical data?

A. The whole of it?

Q. Yes.

A. No, I did not.

Q. Who did prepare that, sir?

A. The long biographical account, the third part of it was prepared by Mrs. Katharine Russell, my secretary. I went over it and pointed out some things that were missing and that I knew were not in order. But I did not prepare it. I think I suggested most of the dates in the chronology, but some of them I don't know whether they came from, from counsel, presumably. As to the second, that was also prepared by Mrs. Russell.

Q. But you have, I assume, read it over pretty carefully?

A. No.

Q. You have not?

A. No. This was meant to be a helpful document containing what we could find in the files.

Q. Are you or are you not prepared to vouch for the accuracy?

A. No, I am not. It is everything we could find in the files or that I recollected in going over it.

Q. You have looked it over, have you not?

A. Sure.

Q. Is there anything in there that is not accurate to your knowledge?

A. No.

Q. Doctor, I am going to ask you to remember that you are under oath, and that therefore your oath must overweigh your modesty in answering the next few questions I am going to ask you. Will you do that, sir?

A. I will remember that I am under oath.

Q. Doctor, is it true that from 1943 until recently, at least, you were the most influential scientist in the atomic energy field in this country?

A. I think this is a question you will have to ask the people influenced.

Q. What is your answer?
A. With some people I was very influential. With others not at all. I was an influential physicist and put it anywhere you want.
Q. You were certainly——
A. I think Lawrence probably had in many ways more influence.
Q. Can you think of anyone else that you might say was more influential than you?
A. I should think the Commissioners, the physicists who were on the Commission, had more effect. Whether they had more influence or not, I don't know.
Q. You were certainly one of the most influential, were you not?
A. Of course.
Q. You might be described as one of the leading physicists in that field.
A. I have been so described.
Q. And you would concede in all modesty that is true. That is an accurate description, is it not?
A. Let me distinguish two things. One is the weight which was attached to my views, and that was considerable. The second is whether I was really very good at the subject and that I will have to leave to others to testify.
Q. Doctor, from 1943 until 1945, as director of the Los Alamos Laboratory, you were in direct charge of the atomic weapons program, were you not?
A. Of the program at Los Alamos, and some related things; yes.
Q. From 1943 until recently, sir, you had access to all classified information concerning the atomic-weapons program; is that true?
A. Yes. Probably not some aspects of atomic intelligence, but concerning our own program; yes.
Q. And from 1946 until 1952, while you were chairman of the General Advisory Committee, you had access to all classified information concerning the entire atomic-energy program, did you not?
A. I did.
Q. Doctor, in one way or another from 1943 until comparatively recently, you participated in all the important decisions respecting the atomic-weapons program, did you not?
A. I am not sure, but I will say yes, to be simple.
Q. Substantially all?
A. I won't embroider this. I don't know the deliberations of the interim committee, for instance. You may say I participated because we did give them some expressions of our opinion.
Q. That is why I said, Doctor, in one way or another.
A. Yes, I think that is probably fair.
Q. Is it a fair statement, Doctor, that until recently you knew more than anybody else about the atomic-weapons program?
A. I should think not. I should think Bradbury, who was in direct charge of it within the nature of things would have known a lot more about it.
Q. Prior to the time when you left Los Alamos in 1945 that was true, was it not?
A. Yes.
Q. Subsequent to 1945, Bradbury would probably be the only possible exception, would he not?
A. My feeling is that the people who do the job more than the kibitzers, and therefore some of Bradbury's top assistants—I may mention Froman, Holloway would have been more intimately versed. They would have certainly known more details and probably had as good a general picture.
Q. In all events, Doctor, you knew a great deal about it.
A. Yes.
Q. There is no question about that?
A. No, no.
Q. While you were chairman of the General Advisory Committee, were you frequently consulted by Mr. Lilienthal on a more or less personal basis for advice?
A. Not frequently, no.
Q. Sometimes.
A. Rarely, I think. I remember one occasion. I think the relations were committee to committee. I don't mean that we didn't discuss things. But I don't believe he put to me a problem, like shall we do this, or what shall we do about such and such a laboratory, as an individual. He occasionally talked to me about what to say in speeches.
Q. Did he used to call you on the telephone rather frequently?
A. I would say no, if you mean by rather frequently several times a month. I remember occasional telephone calls.

Q. Doctor, in your opinion, is association with the Communist movement compatible with a job on a secret war project?
A. Are we talking of the present; the past?
Q. Let us talk about the present and then we will go to the past.
A. Obviously not.
Q. Has that always been your opinion?
A. No. I was associated with the Communist movement, as I have spelled out in my letter, and I did not regard it as inappropriate to take the job at Los Alamos.
Q. When did that become your opinion?
A. As the nature of the enemy and the nature of the conflict and the nature of the party all became clearer. I would say after the war and probably by 1947.
Q. Was it your opinion in 1943?
A. No.
Q. You are sure about that?
A. That association——
Q. With the Communist movement.
A. The current association?
Q. Yes.
A. I always thought current association——
Q. You always thought that?
A. That is right.
Q. There had never been any question in your mind that a man who is closely associated with the Communist movement or is a member of the Communist Party has no business on a secret war project; is that right?
A. That is right.
Q. Why did you have that opinion? What was your reason for it?
A. It just made no sense to me.
Q. Why not.
A. That a man who is working on secret things should have any kind of loyalty to another outfit.
Q. Why did you think that the two loyalties were inconsistent?
A. They might be.
Q. Why?
A. Because the Communist Party had its own affairs, and its own program which obviously I now know were inconsistent with the best interests of the United States, but which could at any time have diverged from those of the United States.
Q. You would not think that loyalty to a church would be inconsistent with work on a secret war project, would you?
A. No.
Q. And of course that was not your view in 1943, was it?
A. No.
Q. Doctor, what I am trying to get at is, What specificallly was your reason for thinking that membership or close association with the Communist Party and the loyalties necessarily involved were inconsistent with work on a secret war project?
A. The connection of the Communist Party with a foreign power.
Q. To wit, Russia.
A. Sure.
Q. Would you say that connection with a foreign power, to wit, England, would necessarily be inconsistent?
A. Commitment would be.
Q. No; I said connection.
A. Not necessarily. You could be a member of the English speaking union.
Q. What I am getting at, Doctor, is what particular feature of the Communist Party did you feel was inconsistent with work on a secret war project?
A. After the Chevalier incident I could not be unaware of the danger of espionage. After the conversations with the Manhattan District security officers, I could not be but acutely aware of it.
Q. But you have told me, Doctor, that you always felt that membership or close association in the Communist Party was inconsistent with work on a secret war project. What I am asking you, sir, is why you felt that. Surely you had a reason for feeling that, didn't you?
A. I am not sure. I think it was an obviously correct judgment.
Q. Yes, sir. But what I am asking you is to explain to me why it was obvious to you.

A. Because to some extent, an extent which I did not fully realize, the Communist Party was connected with the Soviet Union, the Soviet Union was a potentially hostile power, it was at that time an ally, and because I had been told that when you were a member of the party, you assumed some fairly solemn oath or obligation to do what the party told you.

Q. Espionage, if necessary, isn't that right?
A. I was never told that.
Q. Who told you, Doctor?
A. My wife.
Q. When?
A. I don't remember.
Q. Prior to 1943?
A. Oh, yes.
Q. Doctor, let me ask you a blunt question. Don't you know and didn't you know certainly by 1943 that the Communist Party was an instrument or a vehicle of espionage in this country?
A. I was not clear about it.
Q. Didn't you suspect is?
A. No.
Q. Wasn't that the reason why you felt that membership in the party was inconsistent with the work on a secret war project?
A. I think I have stated the reason about right.
Q. I am asking you now if your fear of espionage wasn't one of the reasons why you felt that association with the Communist Party was inconsistent with work on a secret war project?
A. Yes.
Q. Your answer is that it was?
A. Yes.
Q. What about former members of the party; do you think that where a man has formerly been a member of the party he is an appropriate person to work on a secret war project?
A. Are we talking about now or about then?
Q. Let us ask you now, and then will go back to then.
A. I think that depends on the character and the totality of the disengagement and what kind of a man he is, whether he is an honest man.
Q. Was that your view in 1941, 1942, and 1943?
A. Essentially.
Q. What test do you apply and did you apply in 1941, 1942, and 1943 to satisfy yourself that a former member of the party is no longer dangerous?
A. As I said, I knew very little about who was a former member of the party. In my wife's case, it was completely clear that she was no longer dangerous. In my brother's case, I had confidence in his decency and straightforwardness and in his loyalty to me.
Q. Let us take your brother as an example. Tell us the test that you applied to acquire the confidence that you have spoken of?
A. In the case of a brother you don't make tests, at least I didn't.
Q. Well——
A. I knew my brother.
Q. When did you decide that your brother was no longer a member of the party and no longer dangerous?
A. I never regarded my brother as dangerous. I never regarded him—the fact that a member of the Communist Party might commit espionage did not mean to me that every member of the Communist Party would commit espionage.
Q. I see. In other words, you felt that your brother was an exception to the doctrine which you have just announced?
A. No; I felt that though there was danger of espionage that this was not a general danger.
Q. In other words, you felt—I am talking now about 1943—that members of the Communist Party might work on a secret war project without danger to this country; is that right?
A. Yes. What I have said was that there was danger that a member of the Communist Party would not be a good security risk. This does not mean that every member would be, but that it would be good policy to make that rule.
Q. Do you still feel that way?
A. Today I feel it is absolute.

Q. You feel that no member of the Communist Party should work on a secret war project in this country, without exception?
A. With no exception.
Q. When did you reach that conclusion?
A. I would think the same timing that I spoke of before as the obvious war between Russia and the United States began to shape up.
Q. Could you give us the dates on that?
A. Sure. I would have thought that it was completely clear to me by 1948, maybe 1947.
Q. 1946?
A. I am not sure.
Q. Doctor, let me return a bit to the test that you might apply to determine whether a member of the Communist Party in 1943 was dangerous. What test would you apply, or would you have applied in 1943?
A. Only the knowledge of the man and his character.
Q. Just what you yourself knew about him?
A. I didn't regard myself as the man to settle these questions. I am stating opinions.
Q. That is what I getting at. You have testified that your brother, to your knowledge, became a member of the Communist Party about 1936; is that right?
A. Yes, 1937, I don't know.
Q. When is it your testimony that your brother left the party?
A. His testimony, which I believe, is that he left the party in the spring of 1941.
Q. When did you first hear that he left the party?
A. I think in the autumn of 1941.
Q. In the autumn?
A. Yes.
Q. Is that when he went to Berkeley to work in the Radiation Laboratory?
A. Yes, on unclassified work.
Q. But he shortly began to work on classified work, is that right?
A. The time interval, I think, was longer.
Q. Shortly after that. Shortly after Pearl Harbor?
A. I am not clear about that. It was within a year certainly, probably about 6 months.
Q. You were satisfied at that time that your brother was not a member of the party any more?
A. Yes.
Q. How did you reach that conclusion?
A. He told me.
Q. That was enough for you?
A. Sure.
Q. Did you know that your brother at that time and for quite a while after that denied both publicly and officially that he had ever been a member of the Communist Party?
A. I remember one such denial in 1947.
Q. Did you know that your brother's personnal security questionnaire, which he executed when he went to work at Berkeley, failed to disclose his membership in the Communist Party?
A. No, I knew nothing about that.
Q. Did you ask him about that?
A. No.
Q. You knew, didn't you, sir, that it was a matter of great interest and importance to the security officers to determine whether or not anyone working on the project had been a member of the Communist Party?
A. I found that out somewhat later.
Q. Didn't you know it at that time?
A. It would have made sense.
Q. In 1941?
A. It would have made sense.
Q. Yes. Did you tell anybody, any security officer or anybody else, that your brother had been a member of the Communist Party? Did you tell them that in 1941?
A. I told Lawrence that my brother—I don't know the terms I used—but I certainly indicated that his trouble at Stanford came from his Red connections.
Q. Doctor, I didn't ask you quite that question. Did you tell Lawrence or anybody else that your brother, Frank, had actually been a member of the Communist Party?

A. I doubt it.
Q. Why not?
A. I thought this was the sort of thing that would be found out by normal security check.
Q. You were not helping the security check, were you, sir?
A. I would had if I had been asked.
Q. Otherwise not?
A. I didn't volunteer this information.
Q. You think your brother today would be a good security risk?
A. I rather think so.
Q. Beg pardon?
A. I think so.
Q. Doctor, will you agree with me that when a man has been a member of the Communist Party, the mere fact that he says that he is no longer a member, and that he apparently has no present interest or connections in the party, does not show that he is no longer dangerous as a security risk?
A. I agree with that.
Q. Beg pardon?
A. I agree with that.
Q. You agree with that.
A. I would add the fact that he was in the party in 1942 or 1938, did not prove that he was dangerous. It merely created a presumption of danger. This is my view, and I am not advocating it.
Q. In other words, what you are saying is that a man's denial that he is a member and his apparent lack of interest or connections is not conclusive by any means; is it?
A. No.
Q. Did you feel that way in 1943?
A. I would think so.
Q. Or 1942?
A. I would think so. I need to state that I didn't think very much about the questions you are putting and very little in the terms in which you are putting them. Therefore, my attempt to tell you what I thought is an attempt at reconstruction.
Q. Yes, but you couldn't conceive that you would have had a different opinion in 1943 on a question such as that, would you, Doctor?
A. No.
Q. Have you ever been told, Doctor, that it was the policy of the Communist Party, certainly as early as 1943, or say certainly as early as 1941, that when a man entered confidential war work, he was not supposed to remain a member of the party?
A. No.
Q. No one has ever told you that?
A. No.
Q. Can you be sure about that, sir? Does that statement come as a surprise to you?
A. I never heard any statement about the policy of the party.
Q. Doctor, I notice in your answer on page 5 you use the expression "fellow travelers." What is your definition of a fellow traveler, sir?
A. It is a repugnant word which I used about myself once in an interview with the FBI. I understood it to mean someone who accepted part of the public program of the Communist Party, who was willing to work with and associate with Communists, but who was not a member of the party.
Q. Do you think though a fellow traveler should be employed on a secret war project?
A. Today?
Q. Yes, sir.
A. No.
Q. Did you feel that way in 1942 and 1943?
A. My feeling then and my feeling about most of these things is that the judgment is an integral judgment of what kind of a man you are dealing with. Today I think association with the Communist Party or fellow traveling with the Communist Party manifestly means sympathy for the enemy. In the period of the war, I would have thought that it was a question of what the man was like, what he would and wouldn't do. Certainly fellow traveling and party membership raised a question and a serious question.
Q. Were you ever a fellow traveler?
A. I was a fellow traveler.

Q. When?
A. From late 1936 or early 1937, and then it tapered off, and I would say I traveled much less fellow after 1939 and very much less after 1942.
Q. How long after 1942 did you continue as a fellow traveler?
A. After 1942 I would say not at all.
Q. But you did continue as a fellow traveler until 1942?
A. Well, now, let us be careful.
Q. I want you to be, Doctor.
A. I had no sympathy with the Communist line about the war between the spring of 1940 and when they changed. I did not admire the fashion of their change.
Q. Did you cease to be a fellow traveler at the time of the Nazi-Russian Pact in 1939?
A. I think I did, yes.
Q. Now, are you changing ——
A. Though there were some things that the Communists were doing which I still had an interest in.
Q. Are you now amending your previous answer that you were more or less a fellow traveler until 1942?
A. Yes, I think I am.
Mr. GARRISON. Mr. Chairman, I think he testified that he tapered off; did he not?
Mr. ROBB. I said more or less a fellow traveler. I was trying to paraphrase.

By Mr. ROBB:
Q. Do you want to say something more, Doctor?
A. Yes.
Q. Doctor, I don't intend to cut you off at any time. If I ask a question and if you have not completed your answer, I wish you would stop me and finish your answer.
A. Let me give you a couple of examples.
Q. Yes, sir.
A. The Communists took an interest in organizing the valley workers. I think this was long after the Nazi-Soviet Pact. That seemed fine to me at the time. They took an interest in extricating and replanting the refugee loyalists fighters from Spain. That seemed fine to me at the time. I am not defending the wisdom of these views. I think they were idiotic. In this sense I approved of some Communist objectives. Beating the drums about keeping out of war, especially after the battle of France, did not seem fine to me.
Q. You continued your contributions to Communist causes through Communist channels until approximately 1942?
A. I don't remember the date. I have no reason to challenge the date in the Commission's letter.
Q. When did you fill out and file your first personnel security questionnaire?
A. It was in June or July, I guess, of 1942.
Q. Was that about the time when you ceased to be a fellow traveler?
A. No.
Q. How much before that?
A. I have tried to tell you that this was a gradual and not a sharp affair. Any attempt by me to make it sharp would be wrong. I tried in my answer to spell out some of the steps in my understanding, first, of what it was like in Russia. Second, the apparent pliability of American Communist positions to Russian interests, and my final boredom with the thing. It was not something that I can put a date on. I did not write a letter to the papers.
Q. Is it possible, Doctor, for you to set a date when you were sure you were no longer fellow traveling?
A. In that I had no sympathy for any cause the Communists promoted?
Q. Yes, sir.
A. I think I can put it this way. After the war and about the time of this letter——
Q. Which letter?
A. My letter to the Independent Citizens Committee, I was clear that I would not collaborate with Communists no matter how much I sympathized with what they pretended to be after. This was absolute. I believe I have not done so since.
Q. So that would be the Ultima Thule of your fellow traveling, that date?
A. Yes, but I think to call me a fellow traveler in 1944 or 1946 would be to distort the meaning of the word as I explained it.
Q. I think you have explained it pretty well.

A. That is right.
Q. Doctor, as a result of your experiences and your knowledge of Communists and communism, derived from your brother or wherever, were you able in 1942 and 1943 to recognize the Communist attitude and the Communist philosophy in a man?
A. In some cases, sure.
Q. Would you explain that a little bit?
A. My brother never talked Communist philosophy to me. I don't think it meant anything to him. I don't know. Some people did. They were interested in dialectical materialism and believed in the more or less determinate course of history and in the importance of the class war. I would have recognized that.
Q. You knew, of course, in 1943, and the years prior to that year, that Communists stood for certain doctrines and certain philosophies and took certain positions, did you not?
A. I don't know how much this is what I knew then, but it seems clear to me that there were tactical positions on current issues, which might be very sensible looking or popular or might coincide with the views of a lot of people who were not Communists. There was also the conviction as to the nature of history, the role of the classes and the changing society, the nature of the Soviet Union, which I would assume was the core of Communist doctrine, and I am not quite clear which of these you are talking about.
Q. What I am getting at, Doctor, and I will put it very plainly, do you think in 1942 and 1943 you were able to tell a Communist when you saw one?
A. Sometimes.
Q. What time do you think you would not have been able to?
A. In the case of a man who did not talk like one.
Q. What I am getting at is, how could you tell when a man was talking like one? What would a man who was talking like a Communist say?
A. In 1942 and 1943, I should think that an excessive pride and interest and commitment in the Soviet Union, a misstatement of their role, a view that they had always been right in everything they had done, these would have been some of the earmarks.
Q. Can you give us an example of such a man that you knew in those years?
A. I remember Isaac Folkoff talking about the wisdom of the Nazi-Soviet Pact, the strength of the Red Army, the certainty of Soviet victory at a time when I was very skeptical of the possibility of Soviet victory.
Q. And those were indicia to you that Folkoff was a Communist, is that right?
A. I knew it also, but they would have been.
Q. When was that, Doctor?
A. Obviously after the war started in Russia, probably in the winter of 1941 and 1942.
Q. Do you recall where you heard him make those statements?
A. I think it was at Berkeley.
Q. Where in Berkeley?
A. I don't remember. Not a public meeting.
Q. At someone's house?
A. Yes.
Q. Your house?
A. Conceivably.
Q. He was at your house?
A. I think so. My wife is sure not. I don't know.
Q. It would not have been unusual for him to be there; would it?
A. I don't believe he came more than once if he came at all. It would have been unusual.
Mr. GRAY. Excuse me. I would like to get that last. Did you say it would have been unusual?
The WITNESS. Yes.
Mr. GRAY. It would have been unusual?
The WITNESS. Yes.

By Mr. ROBB.
Q. Is there some particular occasion that you had in mind when he was at your house?
A. I remember this conversation I just repeated to you.
Q. Wasn't that at your house?
A. I think so. I am not sure.
Q. You think so?
A. Yes.

Q. What was the occasion that he was at your house, to the best of your recollection?
A. I have no recollection of what brought him. He had a son, I believe, living in Berkeley.
Q. Were there other people present?
A. Oh, surely, but I don't know who. There was no meeting of any kind, no conference, no conclave.
Q. Can you think of any other person that you recall now during those years of 1942 and 1943, maybe 1944, that talked and acted like a Communist so that you knew him to be one?
A. Obviously I knew Steve Nelson was, and I think he talked about the Red army sometimes. This wasn't a time at which Communist talk was very easily recognizable.
Q. Would you search your memory for any other example you might give us?
A. Possibly, though I don't think he was a member of the party, Bernard Peters would have talked along those lines.
Q. Did Peters ever tell you that he had been a member of the party at one time in Germany?
A. That was my impression, but he told me that I had misunderstood him. This was before the Nazis——
Q. Yes. Anybody else that you can think of that you can identify as a Communist by his talk and actions?
A. In a quite different way and not indicating Communist connections, Hawkins—this is David Hawkins—talked about philosophy in a way that indicated an interest and understanding and limited approval, anyway, of Engels, and so on.
Q. Of who?
A. Engels, who was a Communist doctrinaire, whom I have not read.
Q. Was that before Hawkins came to Los Alamos?
A. I don't remember when it was, but we have had several discussions.
Q. It was either before he came to Los Alamos or while he was at Los Alamos?
A. Yes.
Q. Anybody else?
A. That talked like a Communist?
Q. Somebody that you were able to identify by these tests that you have given us, these objective indicia of Communist sympathy or Communist connections?
A. Nothing is coming to my mind. If you have a specific person in mind, why don't you suggest it?
Mr. ROBB. Let us pass to something else.
Mr. Chairman, it seems to be 11 o'clock. If it meets with the board approval, we might take a brief recess.
Mr. GRAY. I think it would be well.
Dr. EVANS. I think it would be very wise.
(Brief recess.)
Mr. GRAY. The proceeding will resume.

By Mr. ROBB.

Q. Doctor, do you think that social contacts between a person employed in secret war work and Communists or Communist adherents is dangerous?
A. Are we talking about today?
Q. Yes.
A. Certainly not necessarily so. They could conceivably be.
Q. Was that your view in 1943 and during the war years?
A. Yes; I think it would have been. My awareness of the danger would be greater today.
Q. But it is fair to say that during the war years you felt that social contacts between a person employed in secret war work and Communists or Communist adherents were potentially dangerous; is that correct?
A. Were conceivably dangerous. I visited Jean Tatlock in the spring of 1943. I almost had to. She was not much of a Communist, but she was certainly a member of the party. There was nothing dangerous about that. There was nothing potentially dangerous about that.
Q. But you would have felt then, I assume, that a rather continued or constant association between a person employed on the atomic-bomb project and Communists or Communist adherents was dangerous?
A. Potentially dangerous; conceivably dangerous. Look: I have had a lot of secrets in my head a long time. It does not matter who I associate with. I

don't talk about those secrets. Only a very skillful guy might pick up a trace of information as to where I had been or what I was up to. Passing the time of day with a Communist—I don't think it is wise, but I don't see that it is necessarily dangerous if the man is discreet and knows what he is up to.

Q. Why did you think that social contacts during the war years between persons on the project—by the project, I mean the atomic-bomb project—and Communists or Communist adherents involved a possibility of danger?

A. We were really fantastic in what we were trying to keep secret there. The people who were there, the life, all of us were supposed to be secret. Even a normal account of a man's friends was something that we didn't want to get out. "I saw the Fermis last night"—that was not the kind of thing to say.

This was a rather unusual kind of blanket of secrecy. I don't think, if a Communist knows that I am going to Washington to visit the AEC, that is going to give him any information. But it was desired that there be no knowledge of who was at Los Alamos, or at least no massive knowledge of it.

Q. Did you have any talk with your brother, Frank, about his social contacts at the time he come on the project?

A. When he came to work for Earnest Lawrence, before there was any classified work, before I know about it and before he was involved in it, I warned him that Earnest would fire him if he was not a good boy. That is about all I remember.

Q. You didn't discuss with him his social contacts?

A. No.

Q, Either at that time or subsequently?

A. If you mean did he ever tell me that he had seen So-and-So, I don't know.

Q. No.

A. I don't believe we had a systematic discussion.

Q. Did you ever urge him to give up any social contacts who might have been Communists or Communist adherents?

A. I don't know the answer to that. It doesn't ring a bell.

Q. If you did, it made no impression on you?

A. Not enough to last these years.

Q. Doctor, referring to your answer—by the way, do you have a copy of your answer?

A. I have a copy of it.

Q. I think it would be well if you kept that before you because I might refer to it from time to time.

At pages 20 and 21 you speak of the statement in the letter to General Nichols that you secured the employment of doubtful persons on the project; and you mentioned Lomanitz, Friedman, and Weinberg. You say on page 21: "When Lomanitz was inducted into the Army, he wrote me asking me to help his return to the project. I forwarded a copy of his letter to the Manhattan District security officers and let the matter rest there."

I will show you the original of the letter signed by you, dated October 19, 1943, enclosing a copy of a letter apparently signed by Lomanitz of October 15, 1943, and I will ask you——

Mr. GARRISON. Mr. Robb, do you have a copy?

Mr. ROBB. Yes; we have those.

By Mr. ROBB:

Q. I will ask you if your letter is the one that you spoke of in your answer.

A. Yes.

Q. And the enclosure was the one you had received from Lomanitz?

A. I have not looked at the enclosure, but I have no reason to doubt it. Yes.

Q. Your original letter is on the stationary of "Post Office Box 1663, Santa Fe, N. Mex." That was the Los Alamos address, was it not?

A. That was the only address we had.

Q. The letter is dated October 19, 1943, and reads as follows:

"Lt. Col. JOHN LANSDALE,
    "*War Department, Washington, D. C.*

"DEAR COLONEL LANSDALE: I am enclosing a copy of a letter which I just received from Rossi Lomanitz. You will note that he states that Dr. Lawrence is interested in having him return to the project for work and suggests that I make a similar request.

"Since I am not in possession of the facts which lead to Mr. Lomanitz' induction, I am, of course, not able to endorse this request in any absolute way. I can, however, say that Mr. Lomanitz' competence and his past experience on the work in Berkeley should make him a man of real value whose technical service we should make every effort to secure for the project. In particular, Lomanitz has been working on a part of Dr. Lawrence's project in which historically I have a close interest and which I know is in need of added personnel.

"Sincerely yours,
"J. R. OPPENHEIMER."

This is Lomanitz' letter:

PRESIDIO OF MONTEREY,
October 15, 1943.

Prof. J. R. OPPENHEIMER,
 Los Alamos, Santa Fe, N. Mex.

DEAR OPJE: For 4 days now I've been a private in the Army, and to date it's not half bad.

We have taken examinations and had interviews in order to determine where we might best be assigned and are waiting for the assignment orders to come through from IX Corps area headquarters in Fort Douglas, Utah.

Before I left Berkeley, I spoke to Lawrence, and it was his idea for himself to put in a request that I be assigned back to work with him. He thought it might be quite effective if at the same time you were to ask for me, either to work with Lawrence or elsewhere.

I do not know whether or not you are in sympathy with this idea; it appeals to me, however; and, if you are interested, it might be wise to put in a request before assignment has been made by IX Corps area headquarters, which will certainly occur within a few days.

In any case, so far I'm rather enjoying the life here. Monterey is a beautiful place. Although they work us hard, they do it efficiently and with a purpose. The barracks, the messhall, the grounds are kept scrupulously clean. The food is excellent and abundant. There is a small library, a theater, and beer at the PX. And the men are easy to get along with.

I have not heard from Max since he got to Salt Lake City. I certainly hope he is getting along all right.

If I am shipped to another camp for basic training, I'll let you hear from me from there.

Respectfully yours,
Pvt. G. R. LOMANITZ,
A. S. N. 39, 140, 466, Company D; SCU 1930, Group 46.

By Mr. ROBB:

Q. Doctor, referring to your letter, you state, "I am, of course, not able to endorse this request in an absolute way."

What did you mean by that, sir?

A. The meaning to me, reading it now, is that I didn't know what the security problems were with Lomanitz. I had just been given a vague account that there were some. The phrase was that he had been indiscreet. I therefore could not judge whether there was a security hazard in his working on the project. If there was not, it seemed like a good idea.

Q. I see.

A. The thing that he was working on had been robbed of personnel because they came to Los Alamos. One of the men at Los Alamos was under great pressure to return to Berkeley, and we needed him at Los Alamos. This is what this recalls to me.

Q. Is this a fair statement? This meant that, so far as you knew, he was all right, but there was something else about him that you didn't know.

A. No. What it meant was that, as far as the technical side of things went, it would be a good idea to have him back. I would leave it to the security officer to decide whether there were overriding considerations.

Q. Did you know anything about him at that time that lead you to believe, except, as you have said, "vague stuff," that he was a security risk?

A. It was very vague. I knew one thing, and I reported it. That is that this whole business about Lomanitz had caused a big flap—his being inducted. I think more than one person wrote to me about it. Lansdale didn't tell me more than that he had been quite indiscreet.

In Berkeley I talked with the security officer, and either he suggested or he concurred in the suggestion that I talk with Lomanitz and see if I could not get him to come in and talk frankly about what the trouble was. He said there wasn't anything; there was nothing to talk about. This didn't reassure me.

Q. Of course, you would not have written that letter if you had known Lomanitz was a Communist, would you?

A. An active Communist?
Q. Yes.
A. No.
Q. Would you if you had known that he had previously been a Communist?
A. That would have depended on lots of things—what kind of a man he was, how long ago it was.
Q. In all events, you didn't know then, did you?
A. No.
Q. Would you have written that letter if you had known that Lomanitz had actually disclosed information about the project to some unauthorized person?
A. Of course not.
Q. All you knew was that Lansdale had said that in some way or another this Lomanitz had been indiscreet?
A. I knew that he was a relative of some one in Oklahoma, I think, who had been involved in a famous sedition case of some kind. As I said in my answer, I knew that he had been reluctant to take any part in the warwork.
Q. But certainly would not have wanted to have him around or suggested that he be around if you had known that he was a Communist or if you had known that he had revealed or disclosed information to some unauthorized person?
A. That is right.
Q. Beg pardon?
A. That is right.
Q. Your answer at page 21, you say that "in 1943 when I was alleged to have stated that 'I knew several individuals then at Los Alamos who had been members of the Communist Party,' I knew of only one; she was my wife," and so forth.
Are you sure that you knew only one person at Los Alamos that at that time who had been a member of the Communist Party?
A. I would not have written it if it had not been my best recollection.
Q. I thought so. How about Charlotte Serber?
A. I don't believe she ever was a member of the Communist Party.
Q. Was she at that time at Los Alamos?
A. Yes, and in a responsible position.
Q. You did not know?
A. No; I don't know today. In fact, I don't today believe.
Q. Pardon?
A. I don't today believe unless there is evidence that I have never heard of.
Q. It would be a great surprise to you to find that she had ever been a member of the party?
A. It would.
Q. Now, speaking of surprise, your answer at page 21, you state, "I asked for the transfer of David Bohm to Los Alamos, but this request, like all others, was subject to the assumption that the usual security requirements would apply. When I was told that there was objection on security grounds to this transfer, I was much surprised but, of course, agreed."
By that do you mean that, when you asked for the transfer of Bohm to Los Alamos, so far as you knew there was nothing wrong with him?
A. Absolutely.
Q. Otherwise you would not have asked; is that right?
A. I asked for the transfer of my brother, or at least concurred in it later, and there had been something wrong with him. But if I had known if there was anything wrong, I would certainly——
Q. I believe it was Colonel DeSilva that told you that, was it not?
A. No.
Q. About Bohm?
A. No; it was a coded telephone message from General Groves. When I asked what was wrong, I was told that he had relatives in Nazi Germany.
Q. So he might be subject to pressure from the Nazis?
A. I won't pretend that I fully believed this story. I didn't know what to think.
Q. That was the only thing that indicated that Bohm was not a fit man to come to Los Alamos?
A. What happened, this was a fairly dramatic thing and unique, so I remember it. I was in Santa Fe. General Groves and I had a little quadratic letter code. He called me up and told me in the code that Bohm could not come. That was that. I asked maybe a couple of people later what was wrong and they told me this story.
Q. About Nazi Germany.

A. Yes.
Q. Would DeSilva be one of those people?
A. I don't remember.
Q. He was your security man there, was he not?
A. Yes. I don't remember when he came. There was a first security man.
Q. Did you ever talk to DeSilva about Bohm?
A. I remember talking about Weinberg, Peters. Bohm may have been one of them. I think only in terms of a very general question on DeSilva's part, which of these is the most dangerous man in your opinion.
Q. Can you fix the approximate time when you got that information from General Groves about Bohm?
A. You mean that Bohm could not come?
Q. Yes.
A. That would have been late March.
Q. Of 1943?
A. That is right.
Q. Was there a man named Bernard Peters at the Berkeley radiation laboratory in 1943?
A. Yes.
Q. Did you know him?
A. Yes.
Q. How well did you know him?
A. Really fairly well.
Q. How had you come to know him?
A. He was a graduate student in physics and was interested in theoretical physics, so he was a student of mine. I knew both him and his wife personally.
Q. Was you relationship with Peters more than just the normal relationship of a professor and a student?
A. Yes.
Q. Social as well?
A. Yes.
Q. Was he a guest at your house from time to time?
A. Yes; he was.
Q. And his wife as well?
A. Yes.
Q. And were you and your wife guests at their house?
A. I am sure we were.
Q. How frequently did you see Peters outside of the normal contact that you had with him as a professor? I am talking now about the years 1942 and 1943, and so on.
A. I think after early 1943, not frequently.
Q. Because you were down at Los Alamos?
A. No, even before that. After it was clear that Peters was not going to Los Alamos, I had raised with him the question of whether he would.
Q. Raised with Peters?
A. Yes, of whether he would come. The fact that he was the right kind of physicist and that she was a doctor and we were short of doctors made this an attractive deal. They decided not to come. I think in 1941 we saw quite a lot of them.
Q. When did you first meet Peters?
A. I don't remember the date. It would have been in the late thirties, either at the time or shortly before the time that he came to study in the graduate school.
Q. When did he come to study there?
A. I can do a little dead reckoning.
Q. Approximately.
A. Approximately 1948 or something like that.
Q. I believe you said that you suggested to Peters he would be a good man to come down to Los Alamos.
A. I did.
Q. And Mrs. Peters, being a doctor, you thought she could be of help down there, too.
A. I certainly did.
Q. When was that, Doctor?
A. It would have been late 1942.
Q. Late 1942?
A. That is right.
Q. Mrs. Peters, you say, was a doctor. Did she ever act as your physician?

A. Yes; she did. I think only once in the spring of 1941. It may have been more frequent. I remember that time.
Q. But your relations with her were both professional and social, I take it.
A. Oh, yes.
Q. As of 1943 or 1942, what did you know about the background of Dr. Peters?
A. I knew that he had been caught as a student—his father was a professional man of some kind whom I met, they lived in Berkeley—that he had been caught, I believe, in Munich at the time of Hitler's rise to power; that he had taken part in that struggle. I would then have said—I have subsequently said—as a Communist. He had told me that this is an exaggeration. He was put in Dachau, that he managed to get out, that his wife and he escaped the country, that they came to this country, that they made some sort of a deal or agreement that he would work and she would go to medical school, and then she would work and he would go to college or to the university. These are in broad outlines the background.
Q. Did you regard Peters as in any way a dangerous man to be on a secret war project?
A. I am alleged to have said so.
Q. Did you say so?
A. I think I did.
Q. When?
A. At Los Alamos.
Q. When?
A. I think in 1943.
Q. 1943?
A. But I am not sure. I think not that he was a dangerous man to have on a secret war project, no. I think what I was asked by DeSilva, "Here are four names, Bohm, Weinberg, and somebody else and Peters; which of these would you regard as the most likely to be dangerous, and I think I answered Peters?
Q. Was that after you had suggested to Peters that he come to Los Alamos?
A. It was.
Q. How long after?
A. A year and a quarter, something like that.
Q. When had you formed that view that Peters might be a dangerous man?
A. During the period that he decided not to come to Los Alamos.
Q. What caused you to form that opinion?
A. The way he talked about things.
Q. Had he ever told you that he was a member of the Communist Party in Germany?
A. I believe that he had, or that I had been told it by a friend. I believed that he had. He told me later that I had misunderstood him.
Q. When did you believe that he told you that?
A. Early.
Q. When?
A. Late thirties.
Q. Who was the friend that you thought might have told you?
A. Possibly Jean Tatlock.
Q. Did she know Peters, too?
A. Yes.
Q. Quite well?
A. She knew Hannah Peters quite well.
Q. Did you know anything about Mrs. Peters' background?
A. Much less.
Q. What did you know about her?
A. That she also escaped from Germany, that she went to Italy, that she had been in medical school in this country.
Q. What did you know about her association with the Communist Party?
A. Literally nothing.
Q. Wasn't it pretty well known that Peters had been a Communist, and when I say wasn't it, I mean in 1941, '42 and '43?
A. I am not sure.
Q. What is you best judgment?
A. I would say it was not well known.
Q. You would say it was not?
A. But I am not sure.
Q. Did anyone else besides Miss Tatlock tell you anything about Peters' Communist connections?

A. No. The way in which this story came to me was that he had been involved in the great battle between the Communists and the Nazis in Germany; not that he was a member of the Communist Party in this country or anything like that. I think it came from him, and I don't think it came from Miss Tatlock, but I am not sure.

Q. Doctor, you have told us that to the best of your recollection Peters told you maybe in 1938 that he had been a member of the Communist Party. You testified, I think you said in 1942 or 1943, you suggested to him that he come to Los Alamos, is that correct?

A. That is right.

Q. What test did you apply at the time you suggested that he come to Los Alamos to satisfy yourself that he had severed any connection with the Communist Party?

A. I didn't think, and I don't think he had a connection with the Communist Party for 5, 6, 7, or 8 years, since he left Germany. That was a different Communist Party.

Q. What I am asking you, sir, is how did you reach that conclusion? What test did you apply?

A. He spoke disparagingly of the party.

Q. When was that?

A. From time to time all during this period. He never indicated any connection with it, though we often saw each other. I was just sure that he had no connection with the Communist Party.

Q. Did there come a time when you changed that opinion?

A. No.

Q. Are you satisfied that he never had any connection with the Communist Party?

A. I really know nothing about it after 1942. Therefore my satisfaction doesn't mean much except with regard to that time.

Q. Doctor, this young man, Giovanni Rossi Lomanitz, I believe you called him Rossi, didn't you?

A. That is the name he went by.

Q. He was a student of yours?

A. Yes.

Q. When?

A. Well——

Q. I might assist you with that.

A. Why don't you tell me?

Q. The record shows that he graduated at Oklahoma with a B. A. in physics in 1940. Then I believe he came to Berkeley and became a student of yours. Is that in accord with your recollection?

A. It could be.

Q. He went to work at the radiation laboratory at Berkeley on June 1, 1942. Is that in accord with your memory?

A. I have no recollection.

Q. But you would accept that?

A. Sure.

Q. The record also shows he was born October 10, 1921. Of course, you don't know that, but he was quite a young man.

A. He was extraordinarily young.

Q. Which would make him not quite 21 when he went to work at the laboratory.

A. Yes.

Q. Did he take his doctorate under you?

A. No; I don't think he got through with it. He was studying for it when the war interrupted. I am not certain on this point.

Q. Did you ask Lomanitz to come to work on the project?

A. Not in those terms. What I remember of it, I put down in my answer, that I endeavored to persuade him that he ought to be willing to do work on behalf of his country.

Q. It might be helpful to the board if we had an answer to a statement made to you in a letter to you from General Nichols at page 5.

Mr. GRAY. Which letter is this?

Mr. ROBB. Letter of December 23, 1953, page 5: "In the case of Giovanni Rossi Lomanitz, you urged him to work on the project."

By Mr. ROBB:

Q. Is that true?

A. I don't know. I urged him to work on military problems.

Q. The particular problem you had in mind was the atomic bomb, wasn't it?
A. Yes, but there were lots of other military undertakings. I believe that this report stems from my own account. I don't know where else it comes from. If that is true, I go ahead and accept it, but I don't remember at this point.
Q. I will continue the reading from the letter of General Nichols, "In the case of Giovanni Rossi Lomanitz, you urged him to work on the project. although you stated that you knew that he had been very much of a Red when he first came to the University of California."
Did you so state?
A. I have no recollection of it. I have no reason to doubt it.
Q. "And that you emphasized to him that he must forego all political activity if he came onto the project."
Did you so emphasize?
A. I doubt that.
Q. You doubt it?
A. Yes, because I never knew of any political activity.
Q. "In August 1943, you protested against the termination of his deferment." Did you do that?
A. Do we have anything on that, Mr. Garrison?
Q. Don't you have any recollection one way or another without assistance from the counsel?
A. I don't—that is, I don't have any recollection of to whom or in what terms. Did I communicate with Lansdale about that?
Mr. GARRISON. We have in our file a copy, I assume Dr. Oppenheimer will recall it, to Col. James C. Marshall, Manhattan District, New York City, dated July 31, 1943, "Understand that the deferment of Rossi Lomanitz, left in charge of my end of work for Lawrence project by me, requested by Lawrence and Shane, turned down by your office. Believe understand reasons but feel that very serious mistake is being made. Lomanitz now only man at Berkeley who can take this responsibility. His work for Lawrence preeminently satisfactory. If he is drafted and not returned promptly to project, Lawrence will request that I release 1 or 2 of my men. I shall not be able to accede to this. Therefore, urge you support deferment of Lomanitz or insure by other means his continued availability to project. Have communicated with Fidler and am sending this to you in support of what I regard as urgent request. Lomanitz deferment expires August 2."
Do you recall that now?
The WITNESS. It is obviously right. I didn't recall it.

By Mr. ROBB:
Q. You sent that telegram?
A. Sure.
Q. And you didn't recall that when I asked you the question whether you protested the deferment of Lomanitz?
A. No; I didn't.
Q. You had not seen that until your counsel read it?
A. I saw it at the time. I have not been over this file.
Q. You have not been over that?
Mr. GARRISON. Mr. Chairman, would it be proper for me to say that this was a file given to me by Mr. Marks who had very much earlier discussed this with Dr. Oppenheimer. I don't know at what point. I have not been over it with Dr. Oppenheimer myself.
Mr. ROBB. Mr. Chairman, may I inquire what other official papers that Mr. Marks had that he turned over to counsel for Dr. Oppenheimer.
Mr. GARRISON. Is this an official paper?
Mr. ROBB. It certainly is.
Mr. GRAY. I believe this is an official paper. I think at least I have a copy of it here.
Mr. ROBB. I have the original here. It is stamped confidential. It came from the records of the Manhattan District. I am slightly curious to know what Mr. Marks, a lawyer in private practice, is doing with parts of the files of the Manhattan Engineering District.
Mr. GRAY. Can you throw any light on this?
Mr. GARRISON. I don't know.
Mr. GRAY. Could you say whether by looking at that file there seem to be locuments of a classified nature in it?
Mr. GARRISON. I really don't know. I honestly looked at this just now. I do think I went over with great speed over that a minute or two ago.

Mr. GRAY. Perhaps the Chair should say that this is not a fair inquiry to put you to since Mr. Marks is not available, at least at this point, to answer the question. I think the record should reflect that at least there seems to be some reason for concern and inquiry as to how, as counsel said, there seems to be in the possession of a civilian lawyer in the community at least a document which is an official document, and which so far as this record shows is still marked "classified" with the classification of "confidential." I think it is unfair to expect you to answer that question.

I think, however, I should say for the record that this board may find it desirable to pursue this point further.

Mr. GARRISON. Mr. Chairman, I shall make diligent inquiry during the noon hour and tell you all that I can.

Mr. GRAY. Thank you.

Mr. ROBB. Mr. Chairman, if I might add, I trust that Mr. Garrison will inquire of Mr. Marks whether or not as General Counsel when he left his employment with the Commission as General Counsel, he took any other records or papers from the files.

The WITNESS. I believe that Mr. Marks would have gotten this in a very different way. If I had a file on this subject of Lomanitz, or if there were things around in my file and my secretary assembled them, he would have gotten it that way. I believe this to be correct.

By Mr. ROBB:

Q. Doctor, do you have in your files now any other Government records or papers which you have not returned to the Commission?

A. I was supposed to return everything. I directed my secretary to return everything, and I doubt very much if I have anything.

Q. I know you were supposed to return everything. My question was, sir, did you?

A. I signed a statement saying that I had directed my secretary to return to the Commission all classified documents.

Q. Doctor, I am sorry. I don't want to fence with you. Would you please answer my question. Did you return all the Government records you had in your possession?

A. From the Commission?

Q. From the Commission or any other source.

A. From the Commission.

Q. From the Commission? You still have some Government records from other sources?

A. Yes, they are in a vault. I don't have them accessible.

Mr. GARRISON. Because of my ignorance, I just raise the question whether a copy of this thing was Commission or Government property? I just don't know.

Mr. ROBB. I don't know. I am just curious to know.

Mr. GRAY. Is there any indication of a classification on the copy you have?

Mr. GARRISON. No.

Mr. ROBB. I have the original here of that teletype. It is marked "confidential."

By Mr. ROBB:

Q. Doctor, would that have been sent in code?

A. I don't know, but everything that went out of Los Alamos was confidential because we were confidential.

Q. Is there any question that this telegram was sent over a Government wire?

A. None.

Q. It was; was it not?

A. Sure.

Q. You didn't consider that telegram to be a part of your personal records, did you, sir, as distinct from the record of the Manhattan Engineering project?

A. If I took a copy of it, I did.

Q. But you have told us it was sent over a Government wire and presumably at Government expense on a matter of official business; is that right?

A. That is right.

Q. Now, getting back to the question that we started with, it is true that in August 1943, you protested against the termination of the deferment of Lomanitz; is that correct?

A. That is right.

Q. And it is true that you requested that he be returned to the project after his entry into the military service?

A. That is right.

Mr. GRAY. Excuse me, Mr. Robb. In Nichols' letter this is all in one sentence. It says, "In August 1943 you protested the termination of his deferment and requested that he be returned to the project after his entry into the military service."

This latter suggested action did not take place in August 1943. I think the record should show. In fact, I don't think there has been any testimony here about the request that he be returned to the project after he entered the military.

By Mr. ROBB:

Q. That was your letter of October 19, 1943, was it not, doctor?
A. That is right. That is the one I have before me.
Dr. GRAY. I beg your pardon. This is the letter that was read into the record.

By Mr. ROBB:

Q. That requested that he be returned.
A. If there were no security objections.
Mr. GRAY. That was dated October 19, 1943.
Mr. ROBB. Yes.

By Mr. ROBB:

Q. Doctor, how well did you know Lomanitz when he went to work at the Radiation Laboratory on June 1, 1942?
A. Not very well.
Q. Did you come to know him better thereafter?
A. No. Certainly somewhat better, because we would see each other from time to time.
Q. Did you have any relationship with him other than the relationship of professor and student?
A. Obviously this talk that I had with him was somewhat abnormal for the relation of professor and student. Otherwise not, I should think.
Q. Did he call you by your first name?
A. Robert? No.
Q. Did he call you "Oppy"?
A. He did in this letter.
Q. Did he do that habitually?
A. I don't know.
Q. What did you call him?
A. Rossi, I think.
Q. What did you know about his background, his past, at the time he came onto the project on June 1, 1942?
A. I knew but I no longer recall the connection in Oklahoma.
Q. Would you tell us about that?
A. He had an uncle or a relative who was tried on a sedition charge. It was a very major affair and was reported in the press shortly before he came to Berkeley. He was recommended as an extremely brilliant student.
Q. Who recommended him?
A. The people at the University of Oklahoma.
Q. Do you recall who they were?
A. No. Background beyond that—background when he came, nothing.
Q. When did——
Mr. GARRISON. Were you going to finish?

By Mr. ROBB:

Q. Had you finished?
A. This was as to the time when he arrived in Berkeley.
Q. No, I am asking you at the time when he went to work on the secret project on June 1, 1942, what you knew about him as of that time.
A. After that I knew something about his work. I knew he talked in a fairly wild way.
Q. What do you mean by that?
A. For instance, the statement that he didn't care, not that he didn't care, but it seemed to him that the war was so terrible that it didn't matter which side won, which I tried to talk him out of. That didn't seem to me a very sensible statement.
Q. Anything else?
A. I don't think so.

Q. Did you know at the time he came on the project that he had been what you described as a Red?
A. That was the story which he arrived with in Berkeley. Other graduate students told me that.
Q. Who?
A. I don't remember.
Q. Weinberg?
A. No.
Q. Bohm?
A. No.
Q. You are quite sure it was not Weinberg or Bohm?
A. Positive.
Q. But you can't recall who it was?
A. That is right.
Q. What was the name of that case in Oklahoma; do you remember?
A. I think it was Lomanitz.
Q. Was it the Allen case?
A. I am sorry, I don't know.
Q. You say it was a criminal sedition or syndicalism case?
A. I have not looked this up. It was hearsay at the time, or newspaper stuff. I can't tell you beyond the fact that it was a sedition or syndicalism case of some kind.
Q. Did you discuss it with Lomanitz?
A. I believe not.
Q. Beg pardon?
A. I believe not.
Q. You have mentioned several times a conversation you had with Lomanitz just prior to the time when he came to work on the secret project at Berkeley. Would you search your recollection and tell us all you can tell us about that conversation?
A. I told you that he explained that he wanted to continue to study physics, that he was not eager to participate in the war effort. I argued with him about it. I don't know whether I convinced him at the time.
Q. Is that all you recall about it?
A. Yes.
Q. Where did that conversation take place?
A. I think it was up in our home on Eagle Hill.
Q. When you say "our," you mean your home?
A. Yes. I think I asked him to come up to talk to me. I am not certain of that.
Q. Did you in that conversation discuss his radical political activities?
A. My memory is not.
Q. Was there anything said about him going to work in the shipyards?
A. I don't remember it. I think not.
Q. Did you know anything about his radical or political activities at that time?
A. No.
Q. Did you lay down any conditions to Lomanitz which you thought he should abide by in the event he went to work on the secret project at Berkeley?
A. This has a much more sinister sound than anything I could have said. I might have said he should behave himself.
Q. What did you mean by that?
A. He should not do anything wild or foolish.
Q. Such as what?
A. Such as make speeches.
Q. About what?
A. About the injustice of the world, the folly of the war, or any of the things that he shot his face off about.
Q. What led you to think that he might?
A. Because I had listened to him talk for a year or so.
Q. Where had you heard his talk?
A. This is not public speeches. I mean his conversation.
Q. Where had you heard those?
A. In the physics department.
Q. You mean in the classrooms?
A. No, in the offices.
Q. So at least to that extent your relationship with him had not been strictly that of a professor and a student, had it?

A. The relations between me and my student were not that I stood at the head of a class and lectured.
Q. I understand that, Doctor. Was it customary for your student to talk to you about the injustices of the world and things of that sort?
A. It was not uncustomary to talk to each other and me about anything that was on their minds.
Q. But you are quite sure that you knew nothing about Lomanitz's past radical or political activities at the time——
A. Activities, no.
Q. Why do you emphasize activities?
A. Because though I don't remember well, I do remember talk and not what he said but the general color of it.
Q. Do you remember any political talk?
A. No.
Q. You are quite sure that you laid down no conditions for him to abide by in the event he went to work on the secret project?
A. Beyond what I have said.
Q. Was there any reason for you to lay down such conditions?
A. I have told you that I knew nothing of political activity.
Q. That is what I thought. Now, prior to the time when Lomanitz went on the secret project in June 1942, did you discuss with any security officer anything that you knew about Lomanitz's background?
A. No, because—well, no.
Q. You didn't tell any security officer that you knew his family had been mixed up in a criminal case in Oklahoma involving sedition?
A. No.
Q. You may have answered this, Doctor, but how did you hear about that case?
A. I am not clear. Either by reading about it—no, somebody in the department told me about it.
Mr. GRAY. May I ask, did this decision involve the Communist Party?

By Mr. ROBB:
Q. It was a criminal syndicalism case.
A. I am not clear. It was sedition or criminal syndicalism.
Q. Did you understand it involved Communist activities?
A. It was not clear to me, but revolutionary activity, or alleged revolutionary activity
Q. It might have been Communist; is that correct?
A. Yes.
Q. As we have seen, there came a time, did there not, when you learned that Lomanitz was about to be inducted into the Army?
A. That is right.
Q. How did you learn that?
A. I first heard it in a letter from Dr. Condon.
Q. Dr. Who?
A. Condon.
Q. Condon?
A. Yes.
Q. What is his first name?
A. Edward.
Q. Edward Condon?
A. That is right.
Q. How did he happen to write you about it?
A. He had been at Los Alamos as associate director and left after a relatively short time and he transferred to Berkeley where he was involved in getting a transition from the laboratory work to the construction work under Westinghouse. He was director of research or associate director of research for Westinghouse. He was working in Berkeley.
One of the things he was working on was this invention that I mentioned a day or so ago. Why he wrote me about it, I don't know. He wrote me about it in a great sense of outrage.
Q. About when was that?
A. I don't recall.
Q. Do you have a copy of that letter?
A. I don't have a copy of that.
Mr. GARRISON. I don't know. I have not seen it.
The WITNESS. I doubt it.

By Mr. ROBB:
Q. This would be about when?
A. It would have been at the time the matter came up.
Q. That was about July.
A. That is right. Somewhat earlier, I think.
Q. A little earlier?
A. I think I went to Berkeley in July. I may have my dates mixed up.
Q. You made quite a stir about the matter; didn't you?
A. Apparently I did.
Q. You sent the teletype that we have seen.
A. That is right.
Q. Whom did you talk to about it?
A. Lansdale, when he was in Los Alamos.
Q. That is Colonel Lansdale?
A. That is right.
Q. The security officer of the District?
A. That is right, a security officer whose name I no longer remember in Berkeley.
Q. Would that be Captain Johnson?
A. It is not that you can refresh my memory. I really don't know.
Q. Would it be Colonel Pash?
A. I remember him.
Q. Did you talk to him about it?
A. That I think is possible.
Q. Anybody else?
A. I don't think so.
Q. During that period of time when this matter was under discussion and consideration did you talk to Lomanitz about it?
A. With the approval or the suggestion, I don't remember, of the security officer, I endeavored to persuade Lomanitz to get the thing straight with the security people. He assured me that there was nothing to get straight.
Q. Did you talk to him on the telephone?
A. I don't remember. I thought I talked to him in person.
Q. I think you did, but did you also talk to him on the telephone on several occasions?
A. I have no recollection of that, but you apparently know that I did.
Q. By the way, did you talk to Dr. Weinberg about Lomanitz's induction?
A. At that time?
Q. At that time or at any time?
A. I would be virtually certain not.
Q. At the time you discussed this matter with Colonel Lansdale, what did he tell you about it?
A. That Lomanitz had been indiscreet.
Q. Did Lansdale tell you what the indiscretion was?
A. No.
Q. Did Lansdale tell you or suggest to you that a rather thorough investigation was being made in connection with Lomanitz?
A. A thorough investigation?
Q. Yes, sir.
A. I don't believe so. Maybe he said we have looked into the matter very completely, or something like that.
Q. Did you understand either from Lansdale or anybody else that there was an investigation revolving around Lomanitz at that time?
A. I understood that there was an investigation—I won't say an investigation—but that something had been found out, and that people were worried, and they were trying to get it straightened out.
Q. Worried about what?
A. The alleged indiscretion.
Q. Worried about security?
A. Yes.
Q. Security meant espionage, didn't it?
A. Not to me.
Q. It didn't?
A. I didn't known what this was all about.
Q. But you knew there was some investigation going on, didn't you?
A. Yes.

Q. I notice in your answer at page 21, you say that you assumed that Lomanitz would be checked by the security officers as a matter of course. Is that correct?
A. I say that.
Q. Having that assumption in mind at the time Lomanitz joined the secret project, did you tell the security * * *
A. I knew very little about his background and I told them nothing.
Q. However much you knew, you told them nothing.
A. That is right.
Q. You didn't think that would have been appropriate for you to do?
A. I do today.
Q. You do today?
A. Yes.
Q. Why?
A. I think it would have been appropriate for me to tell the security officers anything I knew, but I didn't at that time volunteer any information.
Q. Why do you today think it would be appropriate?
A. I understand it as the proper relation of an employee to his Government.
Q. Doctor, what I am asking you is why do you so understand. What is your reasoning?
A. That part of the obligation of a Government employee is to make information available.
Q. You knew that the security of this project was of vital importance to the United States, did you not?
A. I did.
Q. And you had information, however little you think it was, which had a bearing upon whether or not Lomanitz was a good security risk, didn't you?
A. That is right.
Q. And you now understand, do you not, that it was your duty to make that information available to the security officers? Is that correct?
A. That is right.
Q. Especially in view of the fact that you had urged Lomanitz to join the project; is that correct?
A. That is right.
Q. But you didn't do it.
A. That is right.
Q. You have said that Lomanitz was not a close friend of yours.
A. That is right.
Q. So that your failure to make that information available was not because of any ties of friendship; was it?
A. No.
Q. I notice in your telegram, which Mr. Garrison has read, to Colonel Marshall—by the way, who was Colonel Marshall?
A. He was before General Groves took charge the head of the Manhattan District. What his position at this moment was, I am not clear.
Q. I notice in your telegram, in which you state that this is an urgent request, you say that Lomanitz was the only man in Berkeley who could take this responsibility, and so forth. Lomanitz at that time was 21 years old; wasn't he?
A. Twenty-two, I guess, by the record.
Q. After he left and went in the Army, did the project suffer very seriously?
A. I think it was taken over by Peters who had been doing something different.
Q. Lomanitz's job was taken over by Peters?
A. I believe so, but I am not sure. At that time I was pretty busy with my own troubles.
Q. Did you suggest Peters as a possibility for that job?
A. No.
Q. What I am getting at is, the project did not collapse after Lomanitz left; did it?
A. No. The things were put into the Oak Ridge plants. I don't know what arrangements were made.
Q. Yes, sir. Doctor, on page 22 of your letter of March 4, 1954, you speak of what for convenience I will call the Eltenton-Chevalier incident.
A. That is right.
Q. You describe the occasion when Chevalier spoke to you about this matter. Would you please, sir, tell the board as accurately as you can and in as much detail as you can exactly what Chevalier said to you, and you said to Chevalier, on the occasion that you mention on page 22 of your answer?

A. This is one of those things that I had so many occasions to think about that I am not going to remember the actual words. I am going to remember the nature of the conversation.

Q. Where possible I wish you would give us the actual words.

A. I am not going to give them to you.

Q. Very well.

A. Chevalier said he had seen George Eltenton recently.

Mr. GRAY. May I interrupt just a moment? I believe it would be useful for Dr. Oppenheimer to describe the circumstances which led to the conversation, whether he called you or whether this was a casual meeting.

Mr. ROBB. Yes, sir.

The WITNESS. He and his wife——

By Mr. ROBB:

Q. May I interpose, Doctor? Would you begin at the beginning and tell us exactly what happened?

A. Yes. One day, and I believe you have the time fixed better than I do in the winter of 1942–43, Haakon Chevalier came to our home. It was, I believe, for dinner, but possibly for a drink. When I went out into the pantry, Chevalier followed me or came with me to help me. He said, "I saw George Eltenton recently." Maybe he asked me if I remembered him. That Eltenton had told him that he had a method, he had means of getting technical information to Soviet scientists. He didn't describe the means. I thought I said "But that is treason," but I am not sure. I said anyway something, "This is a terrible thing to do." Chevalier said or expressed complete agreement. That was the end of it. It was a very brief conversation.

Q. That is all that was said?

A. Maybe we talked about the drinks or something like that.

Q. I mean about this matter, Doctor, had Chevalier telephoned you or communicated with you prior to that occasion to ask if he might see you?

A. I don't think so. I don't remember. We saw each other from time to time. If we were having dinner together it would not have gone just this way. Maybe he called up and said he would like to come.

Q. It could have been that he called you and you said come over for dinner; is that correct?

A. Any of these things could have been.

Q. You said in the beginning of your recital of this matter that you have described that occasion on many, many occasions; is that right?

A. Yes.

Q. Am I to conclude from that that it has become pretty well fixed in your mind?

A. I am afraid so.

Q. Yes, sir. It is a twice told tale for you.

A. It certainly is.

Q. It is not something that happened and you forget it and then thought about it next, 10 years later, is that correct?

A. That is right.

Q. Did Chevalier in that conversation say anything to you about the use of microfilm as a means of transmitting this information?

A. No.

Q. You are sure of that?

A. Sure.

Q. Did he say anything about the possibility that the information would be transmitted through a man at the Soviet consulate?

A. No; he did not.

Q. You are sure about that?

A. I am sure about that.

Q. Did he tell you or indicate to you in any way that he had talked to anyone but you about this matter?

A. No.

Q. You are sure about that?

A. Yes.

Q. Did you learn from anybody else or hear that Chevalier had approached anybody but you about this matter?

A. No.

Q. You are sure about that?

A. That is right.

Q. You had no indication or no information suggesting to you that Chevalier had made any other approach than the one to you?
A. No.
Q. You state in your description of this incident in your answer that you made some strong remarks to Chevalier. Was that your remark, that this is treasonous?
A. It was a remark that either said—this is a path that has been walked over too aften, and I don't remember what terms I said this is terrible.
Q. Didn't you use the word "treason"?
A. I can tell you the story of the word "treason."
Q. Would you answer that and then explain?
A. I don't know.
Q. You don't know now?
A. No, I don't know.
Q. Did you think it was treasonous?
A. I though it was terrible.
Q. Did you think it was treasonous?
A. To take information from the United States and ship it abroad illicitly, sure.
Q. In other words, you though that the course of action suggested to Eltenton was treasonous.
A. Yes.
Q. Since Eltenton was not a citizen, if it was not treasonous, it was criminal; is that correct?
A. Of course.
Q. In other words, you thought that the course of conduct suggested to Eltenton was an attempt at espionage; didn't you?
A. Sure.
Q. There is no question about it. Let me ask you, sir: Did you know this man Eltenton?
A. Yes; not well.
Q. How had you come to know him?
A. Perhaps "know" is the wrong word. I had met him a couple of times.
Q. How?
A. I remember one occasion which was not when I met him, but when I remember seeing him. I don't remember the occasion of my meeting him. Do you want me to describe the occasion I saw him?
Q. Yes, sir.
A. I am virtually certain of this. Some time after we moved to Eagle Hill, possibly in the autumn of 1941, a group of people came to my house one afternoon to discuss whether or not it would be a good idea to set up a branch of the Association of Scientific Workers. We concluded negatively, and I know my own views were negative. I think Eltenton was present at that meeting.
Dr. EVANS. What was that?
The WITNESS. I think Eltenton was present at that time. That is not the first time I met him, but it is one of the few times I can put my finger on.

By Mr. ROBB:
Q. Do you recall who else was present at that meeting?
A. The list is not going to be comprehensive and it may be wrong. I rather think Joel Hildebrand of the chemistry department at Berkeley, Ernest Hilgard of the psychology department at Stanford. There were several people from Stanford, 6 or 7 people from Berkeley.
Q. Was your brother Frank there?
A. I don't think so.
Q. Was David Adelson there?
A. I am not sure. I doubt it, but it is possible.
Q. He might have been?
A. Yes.
Q. Was a man named Jerome Vinograd there?
A. I don't think I knew him.
Q. Was he there whether you knew him or not?
A. I don't know.
Mr. ROBB. Mr. Chairman, I see it is half past 12. Would you want to adjourn now. This is a good stopping place now.
Mr. GRAY. I think so.
We will reconvene at 2 o'clock.
(Thereupon at 12:30 p. m., a recess was taken until 2 p. m., the same day.)

AFTERNOON SESSION

Mr. GRAY. We will begin the proceeding now.
The record should show the presence of Mr. Herbert S. Marks.
Mr. GARRISON. Mr. Chairman, I would like to ask Mr. Herbert S. Marks, associated with me as counsel in this matter to make a brief statement about how the copy of the teletype message that I read into the record this morning from Dr. Oppenheimer to Colonel Marshall came into first his possession and then mine.

Mr. GRAY. All right, sir.

Mr. MARKS. Shortly after the general manager's letter to Dr. Oppenheimer notifying him of this matter of the proceedings—shortly after that but considerably before Mr. Garrison came into this case—I began working in Dr. Oppenheimer's behalf in preparation for it.

On one occasion—I think it was the latter part of December—I was in Princeton and asked for whatever material Dr. Oppenheimer had there which might bear on any of the allegations in the letter.

As I recall, Dr. Oppenheimer's secretary gave me this particular folder or this particular batch of letters. The top one, which is a letter to Dr. Oppenheimer from Colonel Lansdale, dated October 22, 1943—this is a copy that I have—was marked "Confidential," but the word "Cancelled" was written over "Confidential." There also appeared a notation "Classification Cancelled through the Atomic Energy Commission, H. H. Carroll /s/ for the Chief, Declassification Branch."

I notice that the date under that cancelation is "1-29-53." I think that must be in error because this trip that I have reference to would have been in December of 1953 and not January 1953. The explanation, which as I remember Dr. Oppenheimer's secretary gave me was, this was just at the time when the Commission's representatives were in Princeton transferring or taking away files that Dr. Oppenheimer had there which were classified.

Dr. Oppenheimer's secretary explained further to me that in the course of her releasing these classified files to the Atomic Energy Commission, as she had been instructed to do, she went over them and identified certain items of correspondence which seemed to be of an essentially nonclassified character and made arrangements with the security officer for their declassification.

Without checking with her I can't be sure that this explanation is the one that accounts for all of the papers in this particular batch, of which the one referred to this morning, the teletype, which I believe was the one with the date of July 31, 1943, of which that item was one.

Without checking with Dr. Oppenheimer's secretary I can't be sure that this is the explanation, but I think it is.

Mr. GRAY. Is it your impression that the security officers declassified that whole file as of whatever date in December it was?

Mr. MARKS. You see, all I have, Mr. Gray, is the top letter of the batch with "Confidential" marked on it, and then canceled out and noted "declassified."

These are apparently copies of material which Dr. Oppenheimer's secretary made and I assume she kept whatever she copied from. The only thing I can conjecture is that that declassification must have been intended to apply to the whole batch, but perhaps Mr. Carroll of the Commission could be checking on that and we will also do so with Dr. Oppenheimer's secretary, if you wish.

Mr. GRAY. I think I should say for the record that although the original of the teletype message that we have been discussing—I have forgotten the date of it—is in the possession of the board and is itself marked "Confidential," of course, I have had no information as to when this was classified "Confidential"— whether when sent or some later date.

Mr. ROBB. I do not know.

Mr. GRAY. We do not know.

Dr. OPPENHEIMER. All teletypes out of Los Alamos carried the "Security" designation whatever their content.

Mr. GRAY. I would guess that, but I was not informed on that point. So I assume this was originally a confidential message. Again I assume this is the original.

Mr. GARRISON. I would like, Mr. Chairman, to give you the whole file for your inspection and that of the board.

Mr. GRAY. Of course, some of this is correspondence between Dr. Oppenheimer and Lomanitz, and includes these communications.

I don't think there is any point in dwelling on this at the moment, Mr. Garrison. I think Mr. Marks has given us the best explanation he can give. Unless some member of the board or counsel, Mr. Robb, has any questions of Mr. Marks, perhaps we better proceed with the hearing.

Mr. MARKS. I understand, Mr. Gray, that there was a question this morning as to whether I had any other file. I think there was this file and one other that could have been—1 or 2 more, although I doubt it—in any case when we decided to concentrate the final preparation of the case in Mr. Garrison's office, I simply scribbled on them as on this file, "Dr. Oppenheimer's own files," and turned them over to Mr. Garrison.

The only other file I remember of that character was the one dealing with the Independent Citizens Committee of the Arts, Sciences, and Professions, but my office will have a record of precisely what they were and I will check that.

Mr. GARRISON. In any event, that file, too, had nothing to do with Dr. Oppenheimer's relations with the Government at all, or his period of service at Los Alamos.

I, Mr. Chairman, certainly have no recollection of any file containing any correspondence of a quasi-governmental character except this one. The Independent Citizens Committee file which Mr. Marks turned over to us we have read completely into the record in toto. There may be 1 or 2 other files of that character. Again I am not quite sure, but I am quite certain on the quasi-governmental character.

Mr. GRAY. Yes. It would appear, and this is entirely supposition, that Dr. Oppenheimer had retained a file containing all of his correspondence with and relating to Mr. Lomanitz, and that the security officer apparently took that file and allowed Dr. Openheimer's secretary to make copies for another complete file on this.

This would be the impression I get from what Mr. Marks said.

Mr. MARKS. That is my impression of what occurred but I would have to check with Dr. Oppenheimer's secretary.

Mr. Garrison also mentioned to me that there was a question as to whether I had taken any files from the Atomic Energy Commission. I don't know whether that question was on the record or off, but for your reassurance I must say, of course not.

I took away from the Commission when I left in 1947 a great many papers that were mine or that were Government Printing Office documents, but all of my files were reviewed page by page by a security officer who then stamped the bundles that were transferred to me personally and gave me a certificate to the effect that there was nothing in them that belonged to the Commission or of a classified nature.

Mr. GRAY. Thank you very much. I understood Mr. Marks came for the purpose of making this statement; is that right?

Mr. GARRISON. I would like to have him remain this afternoon, Mr. Chairman.

Mr. GRAY. The record will show that he remains in his capacity of—how do you describe him—cocounsel?

Mr. GARRISON. Yes.

Mr. GRAY. So the record will reflect.

Mr. MARKS. That is the capacity I made this statement, I take it.

Mr. GRAY. There is no reason that the record should not reflect that.

Mr. ROBB. May I proceed?

Mr. GRAY. Yes; if you will.

Whereupon, J. Robert Oppenheimer, the witness on the stand at the time of taking the recess, resumed the stand and testified further as follows:

DIRECT EXAMINATION—Continued

By Mr. ROBB:

Q. Dr. Oppenheimer, while we are on the matter of the telegram about Mr. Lomanitz, I notice in the file that Mr. Garrison handed to the chairman a copy of a wire you sent to Mr. Lomanitz, dated July 31, 1943:

"Mr. G. R. LOMANITZ,
 "Radiation Laboratory, University of California, Berkeley, Calif.:
 "Have requested in proper places reconsideration of support for your deferment. Cannot guarantee outcome but have made strong request. Suggest you ask Fidler for current developments. Good luck.
"OPJE."

Q. Did you send that wire?
A. Evidently.
Q. Why was it so important to you that Lomanitz be not drafted?

A. I am not sure that it was so important to me. I had this outraged communication from Condon——
Q. You had what?
A. An outraged communication from Condon about it. We were very short of people. I doubt whether there was any more to it than that.
Q. Dr. Condon's opinions had a great weight with you?
A. They had some weight with me.
Q. I beg your pardon?
A. They had some weight with me. I thought it reflected a sense of trouble in Berkeley.
B. Is it your recollection that that communication was by way of a letter?
A. Yes.
Q. Did you put that in your file?
A. I don't have it.
Q. I didn't ask you that, sir. Did you put it in your file?
A. I don't know.
Q. Did you get any other letters from Rossi Lomanitz which are not in your filed?
A. I got some later.
Q. When?
A. Toward the end of the war. All of these were open and read, and there may be a record of them. I don't have any in mind. I had no further communications about his situation in the Army after I wrote a letter to his commanding officer.
Q. What were those communications about that you got from him later?
A. I think about coming back to Berkeley and studying after the war, that kind of thing.
Q. Did he ask your assistance in getting him back to Berkeley?
A. I don't recall. I don't see why that would be necessary.
Q. What do you mean by that?
A. I would not have had to get him into the university.
Q. Did you do anything about getting him a job or getting him placed after he got back from the Army?
A. I don't know. I wasn't there at that time.
Q. Wherever you were, did you do anything about it?
A. I have no recollection whatever. He would have come back as a graduate student, and I have no recollection at all of how he got back as a graduate student.
Q. If he had asked you, I assume there is no reason why you would not have helped him?
A. No.
Q. Doctor, do you have a file of correspondence with all of your graduate students who were working on this project with you?
A. No.
Q. Is there any particular reason why you preserved the file on Lomanitz?
A. Yes; there is. He was in some kind of trouble. I thought that somebody I might be asked about how I behaved.
Q. So you wanted to keep a record of it?
A. That is right.
Q. I assume you likewise charged your mind with the matter; is that correct?
A. No; I think I forgot it.
Q. Beg pardon?
A. I forgot it.
Q. You knew it was a matter that had to be handled with some care, did you not, because of the fact that he was in trouble?
A. I was aware of the fact that he was in trouble and thought I should keep what record I had.
Q. Doctor, before the noon recess we were talking about your acquaintanceship or friendship, whichever it was, with Mr. Eltenton. You told us, I believe, that he came to your home on one occasion for a meeting; is that right?
A. Yes.
Q. That was in the evening?
A. I think it was in the afternoon.
Q. Who had called that meeting?
A. I am not clear about that. I have tried to remember and I can't.
Q. Do you remember who presided?
A. No. Maybe I did.

Q. I believe I was asking you to try to remember who was there.
A. I identified probably fumblingly one or two people. It is possible that Addis was there.
Q. Who?
A. Addis. It is quite certain that Hilgard was there. It is probable that Hildebrand was there. I am not certain or very sure beyond that.
Q. When you said Addis, you meant Thomas Addis?
A. I did.
Q. Was David Adelson there?
A. You asked me that.
Q. Yes; I did. I don't think you answered.
A. I can't. I doubt it, but I am not certain.
Q. The last one I asked you about was Jerome Vinograd. Was he there?
A. Yes; you did. I answered that, not being acquainted with him, I don't know.
Q. How many people were there?
A. Fifteen.
Q. You are quite positive that Eltenton was there?
A. No; but I think so.
Q. Had you met Eltenton on many other occasions?
A. Oh, yes; I had met him before that.
Q. Where?
A. I don't remember.
Q. A social occasion?
A. Yes.
Q. Can you recall any of them?
A. No.
Q. Do you recall who introduced you to him?
A. No.
Q. Did Eltenton come to your house on any other occasion?
A. I am quite sure not.
Q. Did he come to your house in 1942 on one occasion to discuss certain awards which the Soviet Government was going to make to certain scientists?
A. If so, it is news to me. I assume you know that this is true, but I certainly have no recollection of it.
Q. You have no recollection of it?
A. No.
Q. Let me see if I can refresh your recollection, Doctor. Do you recall him coming to your house to discuss awards to be made to certain scientists by the Soviet Government and you suggesting the names of Bush, Morgan, and perhaps one of the Comptons?
A. There is nothing unreasonable in the suggestions.
Q. But you don't recall?
A. But I really don't remember.
Q. What did you know about Eltenton's background in 1943 when this Eltenton-Chevalier episode occurred?
A. Two things, three things, four things: That he was an Englishman, that he was a chemical engineer, that he had spent some time in the Soviet Union, that he was a member of the Federation of Architects. Engineers, Chemists, and Technicians—five things—that he was employed, I think, at Shell Development Co.
Q. How did you know all those things?
A. Well, about the Shell Development Co. and the Federation of Architects, Engineers, Chemists, and Technicians, I suppose he told me or someone else employed there told me. As for the background in Russia, I don't remember. Maybe he told me; maybe a friend told me. That he was an Englishman was obvious.
Q. Why?
A. His accent.
Q. You were fairly well acquainted with him, were you not?
A. No. I think we probably saw each other no more than 4 or 5 times.
Q. Did you see Eltenton after this episode occurred?
A. No.
Q. Have you ever seen him since?
A. No.
Q. Could that have been on purpose on your part? Have you avoided him?
A. I have not had to, but I think I would have.

Q. You have mentioned your conversation with Colonel Lansdale which I believe you said took place at Los Alamos?
A. Yes.
Q. In which he told you he was worried about the security situation at Berkeley. I believe we agreed that worry would naturally include a fear of espionage?
A. That is right.
Q. Did he mention any names in connection with that worry?
A. Lomanitz was obviously in the picture, and I believe that is the only one.
Q. Weinberg?
A. I don't think he did.
Q. But Lomanitz obviously?
A. Lomanitz.
Q. When did you first mention your conversation with Chevalier to any security officer?
A. I didn't do it that way. I first mentioned Eltenton.
Q. Yes.
A. On a visit to Berkeley almost immediately after Lansdale's visit to Los Alamos.
Q. Was that to Lieutenant Johnson; do you remember?
A. I don't remember, but it was to a security officer there.
Q. At Berkeley?
A. That is right.
Q. If the record shows that it was to Lieutenant Johnson on August 25, 1943, you would accept that?
A. I would accept that.
Q. You mentioned the Eltenton incident in connection with Lomanitz, didn't you?
A. The context was this. I think Johnson told me that the source of the trouble was the unionization of the radiation laboratory by the Federation of Architects, Engineers, Chemists, and Technicians. Possibly I had heard that from Lansdale. The connection that I made was between Eltenton and this organization.
Q. In your answer at page 22 you say, referring to the Eltenton episode: "It has long been clear to me that I should have reported the incident at once."
A. It is.
Q. "The events that lead me to report it, which I doubt ever would have become known without my report, were unconnected with it."
You have told us that your discussion with Colonel Lansdale encompassed the subject of espionage. Of course, you have told us also that the Eltenton matter involved espionage; is that correct?
A. Let us be careful. The word "espionage" was not mentioned.
Q. No?
A. The word "indiscretion" was mentioned. That is all that Lansdale said. Indiscretion was talking to unauthorized people who in turn would talk to other people. This is all I was told. I got worried when I learned that this union was connected with their troubles.
Q. But, Doctor, you told us this morning, did you not, that you knew that Lansdale was worried about espionage at Berkeley; is that correct?
A. I knew he was worried about the leakage of information.
Q. Isn't that a polite name for espionage?
A. Not necessarily.
Q. I will ask you now, didn't you know that Lansdale was concerned about the possibility of espionage at Berkeley?
A. About the possibility; yes.
Q. Yes.
A. That is right.
Q. So, Doctor, it is not quite correct to say that the Eltenton incident was not connected with your talk with Lansdale, is it?
A. I didn't mean it in that sense. I meant that it had nothing to do with Chevalier or Eltenton with respect to the events that aroused this.
Q. But your talk with Lansdale did have to do with the subject which included Chevalier and Eltenton, didn't it?
A. I have described it as well as I can. Chevalier's name was not mentioned; Eltenton's name was not mentioned; and espionage was not mentioned.
Q. I didn't say that. But it had to do with the subject which involved Chevalier or at least Eltenton?
A. Sure; that is why I brought it up.

Q. What did you tell Lieutenant Johnson about this when you first mentioned Eltenton to him?
A. I had two interviews, and therefore I am not clear as to which was which.
Q. May I help you?
A. Please.
Q. I think your first interview with Johnson was quite brief, was it not?
A. That is right. I think I said little more than that Eltenton was somebody to worry about.
Q. Yes.
A. Then I was asked why did I say this. Then I invented a cock-and-bull story.
Q. Then you were interviewed the next day by Colonel Pash, were you not?
A. That is right.
Q. Who was he?
A. He was another security officer.
Q. That was quite a lengthy interview, was it not?
A. I didn't think it was that long.
Q. For your information, that was August 26, 1943.
A. Right.
Q. Then there came a time when you were interviewed by Colonel Lansdale.
A. I remember that very well.
Q. That was in Washington, wasn't it?
A. That is right.
Q. That was September 12, 1943.
A. Right.
Q. Would you accept that?
A. Surely.
Q. Then you were interviewed again by the FBI in 1946; is that right?
A. In between I think came Groves.
Q. Pardon?
A. In between came Groves.
Q. Yes. But you were interviewed in 1946; is that right?
A. That is right.
Q. Now let us go back to your interview with Colonel Pash. Did you tell Pash the truth about this thing?
A. No.
Q. You lied to him?
A. Yes.
Q. What did you tell Pash that was not true?
A. That Eltenton had attempted to approach members of the project—three members of the project—through intermediaries.
Q. What else did you tell him that wasn't true?
A. That is all I really remember.
Q. That is all? Did you tell Pash that Eltenton had attempted to approach three members of the project——
A. Through intermediaries.
Q. Intermediaries?
A. Through an intermediary.
Q. So that we may be clear, did you discuss with or disclose to Pash the identity of Chevalier?
A. No.
Q. Let us refer, then, for the time being, to Chevalier as X.
A. All right.
Q. Did you tell Pash that X had approached three persons on the project?
A. I am not clear whether I said there were 3 X's or that X approached 3 people.
Q. Didn't you say that X had approached 3 people?
A. Probably.
Q. Why did you do that, Doctor?
A. Because I was an idiot.
Q. Is that your only explanation, Doctor?
A. I was reluctant to mention Chevalier.
Q. Yes.
A. No doubt somewhat reluctant to mention myself.
Q. Yes. But why would you tell him that Chevalier had gone to 3 people?
A. I have no explanation for that except the one already offered.
Q. Didn't that make it all the worse for Chevalier?
A. I didn't mention Chevalier.

Q. No; but X.
A. It would have.
Q. Certainly. In other words, if X had gone to 3 people that would have shown, would it not——
A. That he was deeply involved.
Q. That he was deeply involved. That it was not just a casual conversation.
A. Right.
Q. And you knew that, didn't you?
A. Yes.
Q. Did you tell Colonel Pash that X had spoken to you about the use of microfilm?
A. It seems unlikely. You have a record, and I will abide by it.
Q. Did you?
A. I don't remember.
Q. If X had spoken to you about the use of microfilm, that would have shown definitely that he was not an innocent contact?
A. It certainly would.
Q. Did you tell Colonel Pash that X had told you that the information would be transmitted through someone at the Russian consulate?
(There was no response.)
Q. Did you?
A. I would have said not, but I clearly see that I must have.
Q. If X had said that, that would have shown conclusively that it was a criminal conspiracy, would it not?
A. That is right.
Q. Did Pash ask you for the name of X?
A. I imagine he did.
Q. Don't you know he did?
A. Sure.
Q. Did he tell you why he wanted it?
A. In order to stop the business.
Q. He told you that it was a very serious matter, didn't he?
A. I don't recollect that, but he certainly would have.
Q. You knew that he wanted to investigate it, did you not?
A. That is right.
Q. And didn't you know that your refusal to give the name of X was impeding the investigation?
A. In actual fact I think the only person that needed watching or should have been watched was Eltenton. But as I concocted the story that did not emerge.
Q. That was your judgment?
A. Yes.
Q. But you knew that Pash wanted to investigate this?
A. Yes.
Q. And didn't you know, Doctor, that by refusing to give the name of X you were impeding the investigation?
A. I must have known that.
Q. You know now, don't you?
A. Well, actually——
Q. You must have known it then?
A. Actually the only important thing to investigate was Eltenton.
Q. What did Pash want to investigate?
A. I suppose the 3 people on the project.
Q. You knew, didn't you, Doctor, that Colonel Pash and his organization would move heaven and earth to find out those 3 people, didn't you?
A. It makes sense.
Q. And you knew that they would move heaven and earth to find out the identity of X, didn't you?
A. Yes.
Q. And yet you wouldn't tell them?
A. That is true.
Q. So you knew you were impeding them, didn't you?
A. That is right.
Q. How long had you known this man Chevalier in 1943?
A. For many years.
Q. How many?
A. Perhaps 5; 5 or 6, probably.
Q. How had you known him?

A. As a quite close friend.
Q. Had you known him professionally or socially?
A. He was a member of the faculty, and I knew him socially.
Q. What was his specialty?
A. He was a professor of French.
Q. How did you meet him; do you remember?
A. Possibly at one of the first meetings of the teachers union, but I am not certain.
Q. Were you a frequent visitor at his house?
A. Yes.
Q. And your wives were also friendly?
A. Right.
Q. Had you seen him at the meeting of leftwing organizations?
A. Yes. I think the first time I saw him I didn't know him. He presided at a meeting for Spanish relief at which the French writer Malraux was the speaker.
Q. Where was that meeting held?
A. In San Francisco.
Q. At whose house?
A. It was a public meeting.
Q. What other meetings did you see him at?
A. I am not sure that I can catalog them all. Parties for Spanish relief. The meeting was held at his house at which Schneiderman talked. The teachers union meetings, if they are counted as leftwing.
Q. What was the teachers union meeting about?
A. They had them regularly.
Q. Were those teachers union meetings held at private homes?
A. No.
Q. Some of them?
A. I don't think the union could have met in a private home.
Q. I don't know.
A. No. These were held in halls or, I think, in the International House.
Q. Any other meetings that you remember?
A. I would be certain there were, but they are not coming up.
Q. This meeting that you mentioned at which Schneiderman spoke—that was December 1, 1940, was it not?
A. I don't know the date, but I will accept it.
Q. Who was Schneiderman?
A. He was the secretary of the party in California.
Q. The Communist Party?
A. Right.
Q. This was held at Chevalier's house?
A. Yes.
Q. How many people were present?
A. Twenty, as a guess.
Q. In the evening?
A. Yes.
Q. Do you recall who was there?
A. Not very accurately and not with certainty. I didn't even recall the meeting until my wife refreshed my memory.
Q. Was Isaac Folkoff there?
A. It is possible.
Q. Was Dr. Addis there?
A. I think so.
Q. Was Rudie Lambert there?
A. I don't remember that, but possibly.
Q. Do you remember anybody else who was there?
A. Mr. Jack Straus.
Q. Who? How do you spell that?
A. S-t-r-a-u-s. I don't know whether it is one or two s's.
Q. Who was he?
A. A San Francisco businessman.
Q. Was he a member of the Communist Party?
A. Not to my knowledge.
Q. By the way, was Lambert a member of the Communist Party?
A. Yes.
Q. What was his function?
A. I never knew.
Q. You knew he was a member?

A. I knew he was a member and, in fact, had an official job.
Q. How often did you see Lambert?
A. Half a dozen times.
Q. In what connection?
A. Different one. Affairs like this: I had lunch with him once or twice with Folkoff. I saw him at a Spanish party.
Q. What was the purpose of those luncheons?
A. This was one of the times when they were telling me about why I needed to give them money.
Q. Money to what?
A. To them for use in Spain.
Q. Folkoff was a Communist?
A. Yes.
Q. What was his job in the party?
A. I think he was treasurer of something, but I never knew of what.
Q. Can you describe Lambert to us?
A. A lean, rather handsome man, moderate height, rather an effective speaker in conversation.
Q. What was the purpose, again, of this meeting at which Schneiderman spoke?
A. I suppose it was to acquaint the interested gentry with the present line or the then line of the Communist Party.
Q. Who asked you to go?
A. The Chevaliers.
Q. It was his house; wasn't it?
A. Yes.
Q. Did you know Chevalier as a fellow traveler?
A. I so told the FBI in 1946 and I did know him as a fellow traveler.
Q. He followed the party line pretty closely, didn't he?
A. Yes, I imagine he did.
Q. Did you have any reason to suspect he was a member of the Communist Party?
A. At the time I knew him?
Q. Yes, sir.
A. No.
Q. Do you know?
A. No.
Q. You knew he was a quite a "red", didn't you?
A. Yes. I would say quite Pink.
Q. Not Red?
A. I won't quibble.
Q. You say in your answer that you still considered him a friend.
A. I do.
Q. When did you last see him?
A. On my last trip to Europe. He is living in Paris, divorced and has been remarried. We had dinner with them one evening. The origin of this, or at least part of the origin——
Q. May I interpose? That was in December 1953?
A. Yes, December.
Q. Go ahead.
A. He wrote me a note saying that he had been at UNESCO and had run into Professor Bohr who told him I was coming to Europe—we were coming to Europe.
Q. Professor who?
A. B-o-h-r. He asked us to look him up if we got to Paris. We planned to do so. My wife called. He was out of town on a job. He got back and we had dinner together, the four of us.
The next day he picked us up and drove us out to visit with Malraux, who has had rather major political changes since 1936. We had a conversation of about an hour and he drove us back to the hotel.
Dr. EVANS. How long was Bohr in this country?
The WITNESS. Bohr?
Dr. EVANS. Yes.
The WITNESS. He has been here many different times.
Dr. EVANS. Just about the time that you began the work.
The WITNESS. He arrived early in 1944 and left about mid-1945; so that would be a year and a half.
Dr. EVANS. Did he go under the name of Bohr here?
The WITNESS. He had the code name of Nicholas Baker.

By Mr. ROBB:
Q. What kind of a code was that?
A. It was meant to conceal from people who should know that he was in this country and working on the atomic project.
Q. I see. Getting back to your visit with Chevalier in December 1953, was Dr. Malraux the gentleman who first introduced you to Chevalier?
A. He did not introduce me. He was the speaker at a meeting at which Chevalier presided. Malraux became a violent supporter of De Gaulle and his great brainman and deserted politics and went into purely philosophic and literary work. Our talk was purely of that.
Q. What was your conversation with Chevalier that you said you had for about an hour?
A. With Malreaux that was.
Q. It was not with Chevalier?
A. Chevalier took us there. We had dinner with him and his new wife the night before. The talk was personal, diffuse, and about how they were living and how we were living.
Q. Did you talk about Chevalier's passport?
A. No.
Q. Did you thereafter go to the American Embassy to assist Dr. Chevalier in getting a passport to come back to this country?
A. No.
Q. Do you know a Dr. Jeoffrey Wyman?
A. Yes, I do.
Q. Who is he?
A. He is the science attaché of the State Department in Paris. He is a man I knew at Harvard when I was a student there and Cambridge. He resigned from Harvard to accept this job.
The first day or so my wife and I were in Paris we called at the Embassy and we called on the Chargé d'Affaires, the Ambassador was ill and away, and Wyman asked us to lunch and we had lunch with him. This was a propriety. We didn't see Wyman again.
Q. Did you discuss with Wyman or anybody else the matter of Chevalier's passport?
A. I did not.
Q. At any time?
A. At no time.
Q. Let us move along to your interview with Colonel Lansdale on September 12.
A. Right.
Q. Did you tell him substantially the same story you told Colonel Pash?
A. I don't know whether he repeated it to me or I repeated it to him.
Q. In all events, if he repeated it to you——
A. I did not modify it.
Q. You affirmed it as the truth?
A. Yes
Q. So you lied to him, too?
A. That is right.
Q. Did he plead with you to give him the name of X?
A. He did.
Q. Did he explain why he wanted that name?
A. I suppose he did. I don't remember.
Q. You knew why he did?
A. It didn't need explanation.
Q. Did he explain to you that either X or Eltenton might have continued to make other contacts?
A. This would have been a reasonable thing to say.
Q Did you give him the name of X?
A. No.
Mr. GRAY. Suppose we break now for a few minutes.
(Whereupon, a short recess was taken.)
Mr. GRAY. May we resume.

By Mr. ROBB:
Q. Doctor, just so the record will be complete, do you recall in 1950 getting a letter from Dr. Chevalier who was then in San Francisco asking you to assist him by telling him what you testified before the House committee about the Chevalier-Eltenton incident?
A. Yes, I remember.

Q. Do you recall answering that letter?
A. I did answer it. I think I did not tell him what I testified, because it was in executive session, but referred him to a press account of what I testified. I am not quite certain on this point.
Q. At that time he was attempting to get a passport to leave the United States, was he?
A. I thought that was later, but I am not sure.
Q. That may have been. You did hear about it when he was attempting to get a passport; did you?
A. Yes.
Q. We will come to that later.
I will read you and ask you if this is the letter that you wrote to him. I am sorry I haven't a copy of it. On the stationery of the Institute for Advanced Study, Princeton, N. J., office of the director, February 25, 1950:

"DR. HAAKON CHEVALIER
 "*3127 Washington Street*
  "*San Francisco, Calif.*

"DEAR HAAKON: Thank you for your good letter of February 21. I can understand that an account of my testimony before the House committee could be helpful to you in seeking a suitable academic position at this time. I cannot send it to you because I have never myself had a transcript, and because the committee ruled at the time that they desired to keep, and would keep, the hearings secret. But I can tell you what I said. I told them that I would like as far as possible to clear the record with regard to your alleged involvement in the atom business. I said that as far as I knew, you knew nothing of the atom bomb until it was announced after Hiroshima; and that most certainly you had never mentioned it or anything that could be connected with it to me. I said that you had never asked me to transmit any kind of information, nor suggested that I could do so, or that I consider doing so. I said that you had told me of a discussion of providing technical information to the U. S. S. R. which disturbed you considerably, and which you thought I ought to know about. There were surely many other points; but these were, I think, the highlights; and if this account can be of use to you, I hope that you will feel free to use it.

"As you know, I have been deeply disturbed by the threat to your career which these ugly stories could constitute. If I can help you in that, you may call on me.

"Sincerely yours,

"ROBERT OPPENHEIMER."

Did you write that letter?
A. Oh, sure. I didn't recollect it.
Q. Was the account of your testimony which you gave there an accurate one?
A. I think it is fairly accurate.
Q. Dr. Chevalier thereafter used that letter in connection with his passport application.
A. I didn't know that.
Q. Did you talk to him about his passport application?
A. I did. He came to Princeton at the time and I referred him to counsel to help him with it.
Q. To whom did you refer him?
A. Joe Fanelli.
Q. In Washington?
A. Right.
Q. Is that the same Joe Fanelli who represented Mr. Weinberg in his criminal trial?
A. I believe it is.
Q. Was he a friend of yours, Fanelli?
A. No. I had not met him at the time I referred Chevalier to him, but he represented my brother at the time of his appearance before the House Un-American Committee. Wait just a minute——
Mr. ROBB. Mr. Chairman, I am sorry. I don't think counsel should coach the witness.
The WITNESS. You are quite right.
Mr. MARKS. I am very sorry.
Mr. ROBB. Will you resume?
The WITNESS. I did hear the correction.
Mr. ROBB. I hope it won't happen in the future.

Mr. GRAY. I think we should be careful, counsel, if you do not mind. I should repeat I think at this time because Mr. Marks has not been present before, that we consider under the regulations, spirit and letter that this it not a trial but an inquiry. Very considerable latitude, as you have observed and we have all experienced, is certainly allowed, and is to continue, in not trying to conform to rigid court procedures. But as far as the testimony of a witness is concerned, it must be his own testimony.

The WITNESS. I am sorry I did hear it. I was mistaken.

Mr. GRAY. The purpose of the inquiry is not entrapment.

The WITNESS. I understand that. I met Fanelli at one time, but I believe it was after I referred Chevalier to him. I met him first on the train going from Washington to Princeton where I was introduced by a friend, and I met him later in the preparation for the Weinberg case. But he had been recommended to me very highly, and I suggested him to Chevalier.

By Mr. ROBB:

Q. Dr. Chevalier came to Princeton to see you about the matter?
A. He came and stayed a couple of days. I don't think it would be right to say he came to see me about the passport problem. He had just been divorced. He talked of nothing but his divorce. But he was worried about whether to use an American passport or his French passport.
Q. About when was that, Doctor?
A. Could it have been the spring of 1951?
Q. I don't know.
A. It was immediately at the time he left the country.
Q. You had previously met Mr. Fanelli?
A. I believe I did not meet him until after this.
Q. Who, Doctor, had so highly recommended Fanelli to you?
A. I had heard him warmly spoken of by Mr. Marks. I think that is what it was.
Q. Who was the friend that was on the train with you?
A. Two, Sumner Pike, and Archie Alexander.
Q. I believe you said that your account of your testimony which you gave to Dr. Chevalier in your letter of February 24, 1950, was substantially accurate to the best of your recollection?
A. It was intended not to be misleading and to be reassuring.
Q. And had your testimony to which this letter referred been true? Was it the truth?
A. My testimony was certainly true.
Q. Doctor, I would like to go back with you, if I may, to your interview with Colonel Pash on August 26, 1943. I will read to you certain extracts from the transcript of that interview.

Colonel Pash said to you:

"Mr. Johnson told me about the little incident or conversation taking place yesterday in which I am very much interested, and had me worried all day yesterday since he called me.

"OPPENHEIMER. I was rather uncertain as to whether I should or should not talk to him, Rossi, when I was here. I was unwilling to do it without authorization. What I wanted to tell this fellow was that he had been indiscreet. I know that is right that he had revealed information. I know that saying that much might in some cases embarrass him. It doesn't seem to have been capable of embarrassing him, to put it bluntly."

Do you recall saying that?
A. Let me say I recognize it.
Q. In substance did you say that?
A. I am sure I did.
Q. So there was no question, Doctor, that this matter of the Eltenton incident came up in connection with your conversation about Lomanitz.
A. That is right.
Q. There is no question, is there, either, that at that time, August 26, 1943, you knew that Lomanitz had revealed certain confidential information?
A. I was told by Lansdale, that he had been indiscreet about information. It was not made clear to me——
Q. This says, "I know that is right that he had revealed information." So wouldn't you agree that you knew he had revealed information?
A. Yes.

Q. Very well. Pash said:

"Well, that is not the particular interest I have. It is something a little more in my opinion that is more serious. Mr. Johnson said that there was a possibility that there may be some other groups interested.

"OPPENHEIMER. I think that is true, but I have no first hand knowledge that it would be for that reason useful. But I think it is true that a man whose name I never heard, who was attached to the Soviet consul, has indicated indirectly through intermediate people concerned with this project that he was in a position to transmit without any danger of a leak or scandal or anything of that kind information which they might supply."

Do you recall saying that in substance?
A. I certainly don't recall it.
Q. Would you deny you said it?
A. No.
Q. Is there any doubt now that you did mention to Pash, a man attached to the Soviet consul?
A. I had completely forgotten it. I can only rely on the transcript.
Q. Doctor, for your information, I might say we have a record of your voice.
A. Sure.
Q. Do you have any doubt you said that?
A. No.
Q. Was that true. Had there been a mention of a man connected with the Soviet consul?
A. I am fairly certain not.
Q. You were very certain before lunch that there had not; weren't you?
A. Yes.
Q. You continue in that same answer: "Since I know it to be a fact, I have been particularly concerned about any indiscretions which took place in circles close enough to be in contact with him. To put it quite frankly, I would feel friendly to the idea of the Commander in Chief of informing the Russians who are working on this problem. At least I can see there might be some arguments for doing that but I don't like the idea of having it moved out the back door. I think it might not hurt to be on the lookout for it."

Do you recall saying something like that?
A. I am afraid I am not recalling very well, but this is very much the way I would have talked.
Q. Did you feel friendly to the idea of the Commander in Chief informing the Russians who were working on the problem?
A. I felt very friendly to the attempt to get real cooperation with the Russians, a two-way cooperation, on an official governmental level. I knew of some of the obstacles to it.
Q. Is this an accurate statement of your sentiments as of August 26, 1943: "I would feel friendly to the idea of the Commander in Chief informing the Russians who are working on this problem"?
A. The Russians who are working on this problem?
Q. Yes, sir.
A. I think that is not an accurate sentence.
Q. That is not the way you felt then?
A. No. I think I can say that I felt that I hoped that during the war good collaboration all along the line could be established with the Russians through governmental channels but I had no idea that there were any Russians working on the problem.
Q. On the problem, not the project. On the problem.
A. What problem?
Q. "I would feel quite friendly to the idea of the Commander in Chief informing the Russians who are working on this problem."

If you said that to Colonel Pash; did that express your sentiments?
A. What does it mean?
Q. I am asking you.
A. I don't know.
Q. That language is not intelligible to you?
A. On this problem? No.
Q. The problem of the atom bomb. Did you in 1943 feel friendly to the idea of the Commander in Chief of informing the Russians who were working on the problem of the atomic bomb?
A. I don't think there were any Russians working on the problem of the atomic bomb.

Q. Did you feel friendly in 1943 to the idea of the Commander in Chief giving the Russians any information about the work that was being done on the atomic bomb under your supervision?
A. If it had been a completely reciprocal and open affair with their military technology and ours, I would have seen arguments for it; yes, sir.
Q. In other words, you did feel friendly.
A. With these qualifications.
Q. You said here, "At least I can see there might be some arguments for doing that, but I don't like the idea of having it moved out the back door."
A. Right.
Q. Pash then said: "Could you give me a little more specific information as to exactly what information you have? You can readily realize that phase would be to me as interesting pretty near as the whole project is to.
"OPPENHEIMER. Well, I might say the approaches were always made through other people who were troubled by them and sometimes came and discussed them with me and that the approaches were quite indirect. So I feel that to give more perhaps than one name would be to implicate people whose attitudes were one of bewilderment rather than one of cooperation."
Do you recall saying something like that?
A. I don't recall that conversation very well.
Q. But you did, you are sure, tell Colonel Pash there was more than one person involved.
A. Right.
Q. Continuing: "I know of no case, and I am fairly sure in all cases where I have heard of these contacts would not have yielded a single thing. That is as far as I can go on that. There is a man whose name was mentioned to me a couple of times. I don't know of my own knowledge that he was involved as an intermediary. It seems, however, not impossible. If you wanted to watch him it might be the appropriate thing to do. He spent a number of years in the Soviet Union. I think he is a chemical engineer. He was, he may not be here, at the time I was with him here employed by the Shell Development. His name is Eltenton. I would think that there was a small chance—well, let me put it this way. He has probably been asked to do what he can to provide information. Whether he is successful or not, I do not know. But he talked to a friend of his who is also an acquaintance of one of the men on the project and that was one of the channels by which this thing went. Now, I think that to go beyond that would be to put a lot of names down of people who are not only innocent but whose attitude was 100 percent cooperative."
Do you recall saying that to Colonel Pash?
A. This sounds right.
Q. How much of that was not true? Approaching more than one person?
A. More than one person was not true.
Q. He talked to a friend of his, who is also an acquaintance of one of the men on the project. Who was the friend of his that you had in mind?
A. I can only guess, but that would be Chevalier and I would be the man on the project.
Q. Pash said to you: "However, anything we may get which would eliminate a lot of research work on our part would necessarily bring to a closer conclusion anything that we are doing."
In other words, he told you, didn't he, that they were going to have to do a lot of work to investigate this?
You answered, "Well, I am giving you the one name that is or isn't—I mean I don't know the name of the man attached to the consulate. I think I may have been told and I may not have been told. I have at least not purposely, but actually, forgotten. He is and he may not be here now—these incidents occurred in the order of about 5, 6, or 7 months ago."
You did tell Colonel Pash that there was a man from the consulate involved, didn't you?
A. I did.
Q. Was that true?
A. That there was a man in the consulate involved?
Q. Yes.
A. That I read since the end of the war?
Q. No. Did you know then that there was?
A. I am fairly sure not.
Q. Chevalier had not said anything to you about a man from the consulate, had he?
A. I have told you my sharp recollection of it.

Q. Further along you said, "I would feel that the people that they tried to get information from were more or less an accident, and I would be making some harm by saying that."
So you were talking about more than one person always, weren't you?
A. Yes; at that time.
Q. When you said "Well, I will tell you one thing. I have known of 2 or 3 cases, and I think 2 of the men are with me at Los Alamos. They are men who are closely associated with me."
"PASH. Have they told you that either they thought they were contacted for that purpose or they were actually contacted for that purpose?
"OPENHEIMER. They told me they were contacted for that purpose."
"PASH. For that purpose?"
Do you recall saying that to Pash in substance?
A. Yes.
Q. So you told him specifically and circumstantially that there were several people that were contacted.
A. Right.
Q. And your testimony now is that was a lie?
A. Right.
Q. Then you continue: "That is, let me give you the background. The background was, well, you know how difficult it is with relations between these two allies and there are a lot of people that don't feel very friendly towards Russia. So the information, a lot of our secret information, our radar and so on, doesn't get to them, and they are battling for their lives, and they would like to have an idea of what is going on, and this is just to make up in other words for the defects of our official communication. That is the form in which it was presented."
Did you tell Colonel Pash that?
A. I evidently did. This is news to me.
Q. Had the matter been presented to you in that form?
A. No.
Q. Had anyone told you that it had been presented in that form?
A. No.
Q. In other words, this also was a lie?
A. Yes, sir.
Q. Then you continue: "Of course, the actual fact is that since it is not a communication that ought to be taking place, it is treasonable."
Did you say that?
A. Sure. I mean I am not remembering this conversation, but I am accepting it.
Q. You did think it was treasonable anyway, didn't you?
A. Sure.
Q. "But it was not presented in that method. It is a method of carrying out a policy which was more or less a policy of the Government. The form in which it came was that couldn't an interview be arranged with this man Eltenton who had very good contact with a man from the Embassy attached to the consulate who is a very reliable guy and who had a lot of experience in microfilm or whatever."
Did you tell Colonel Pash that microfilm had been mentioned to you?
A. Evidently.
Q. Was that true?
A. No.
Q. Then Pash said to you: "Well, now, I may be getting back to a little systematic picture. These people whom you mention, two are down with you now. Were they contacted by Eltenton direct?"
You answered, "No."
"PASH. Through another party?
"OPPENHEIMER. Yes."
In other words, you told Pash that X had made these other contacts, didn't you?
A. It seems so.
Q. That wasn't true?
A. That is right. This whole thing was a pure fabrication except for the one name Eltenton.
Q. Pash said to you, "This would not involve the people, but it would indicate to us Eltenton's channel. We would have to know that this is definite on Eltenton."
In other words, Pash wanted to find out the channel, didn't he?

A. Yes.
Q. Pash said again. "The fact is this second contact, the contact that Eltenton had to make with these other people, is that person also a member of the project?"
You said "No." That was correct, wasn't it?
A. Yes.
Q. Again you said to Pash, "As I say, if the guy that was here may by now be in some other town, and then all I would have in mind is this. I understand this man to whom I feel a sense of responsibility, Lomanitz, and I feel it for two reasons. One, he is doing work which he started and which he ought to continue, and second, since I more or less made a stir about it when the question of his induction came up. This man may have been indiscreet in circles which would lead to trouble."
Did you say that to Pash?
A. Yes.
Q. Did you feel some responsibility for Rossi Lomanitz?
A. Evidently.
Q. Why?
A. Well, partly because I had protested his induction. Partly because he was a student of mine. Partly because I tried to persuade him to go into secret work.
Q. And you continue, "That is the only thing I have to say because I don't have any doubt that people often approached him with whom he has contact—I mean whom he sees—might feel it their duty if they got word of something to let it go further and that is the reason I feel quite strongly that association with the Communist movement is not compatible with a job on a secret war project. It is just that the two loyalties cannot go."
Doctor, who were the people that you thought Lomanitz had contact with or whom he saw who might feel it their duty to let the word go further?
A. I had no idea.
Q. You had none then?
A. I don't believe so. I certainly have none now.
Q. You did say that you thought association with the Communist movement was incompatible with work on a secret war project.
A. Right.
Q. Pash said to you again, "Were these two people you mentioned contacted at the same time?"
You answered, "No, they were contacted within a week of each other.
"PASH. They were contacted at two different times?
"OPPENHEIMER. Yes, but not in each other's presence."
Was that part of what you call a cock and bull story, too?
A. It certainly was.
Q. Pash said, "And then from what you first hear, there was somebody else who probably still remains here who was contacted as well?
"OPPENHEIMER. I think that is true."
Do you recall saying something like that?
A. No, but it fits.
Q. "PASH. What I am driving at is that there was a plan at least for some length of time to make these contacts and you may not have known all the contacts?
"OPPENHEIMER. That is certainly true. That is why I mentioned it. If I knew all about it, then I would say forget it. I thought it would be appropriate to call to your attention the fact that these channels at one time existed."
Doctor, is it now your testimony that there was no plan that you knew of?
A. This whole thing, except for the single reference to Eltenton I believe to be pure fabrication.
Q. In other words, your testimony now is that there was no plan that you knew about?
A. Right. I am certain of that.
Mr. GRAY. Excepting the Chevalier incident.
The WITNESS. Yes, yes. The only thing I mentioned here that has any truth to it is Eltenton.
Mr. GARRISON. Mr. Chairman, could I just make a short request at this point?
Mr. GRAY. Yes.
Mr. GARRISON. I appreciate the existence of the rule under which we cannot ask for access to the file and I am not going to protest that rule. I wonder, however, if it would not be within the proprieties of this kind of proceeding

when counsel reads from a transcript for us to be furnished with a copy of the transcript as he reads from it. This, of course, is orthodox in a court of law. I don't pretend that this is a court of law, but I do make the request because I don't know what else is in the transcript, and if parts of it are read from, it would seem to me that it would be proper for us to see what parts are not read from and to look at it as a whole. I don't want to make an argument. I put the question to you.

Mr. ROBB. Mr. Chairman, I don't know of any rule in the court of law that you must furnish counsel with the copy of the transcript you are reading of at the time. I might say that my thought would be at the conclusion of this examination to make the entire transcript a part of the record and let Mr. Garrison read it and see it, and then if he wants to ask anything about it on redirect, he can do so.

Mr. GRAY. I think that would be appropriate. I would like to indicate a caution—I don't know about this particular transcript—but I am not sure that in any case you could be able to make the whole thing a part of the record.

Mr. ROBB. I don't know, sir; this is presently marked "Secret" so I could not make it available to Mr. Garrison at this time.

The WITNESS. But it is being read into the record.

Mr. ROBB. That is right.

Mr. GRAY. Let us clarify that point for a moment. There is a classification officer who may at some time be present with us—I don't think he has been in the room—but he will be presented if he does come in and sit in the hearing, who is reading the transcript from the point of view of the classification necessities. So that all of the testimony is being read by him with the view to its treatment as open or classified matter. So that all of the tesimony will be so considered. I don't think that announcement has been made, and I think Dr. Oppenheimer and his counsel are entitled to know that.

Mr. GARRISON. Then do I understand the response to be that subject to check with the classification officer you propose to put the whole transcript in the record?

Mr. ROBB. I said that was my disposition, yes, sir, but Mr. Garrison, as you know, I am not an expert on the matters of classification myself. That is my disposition. This is something that Dr. Oppenheimer participated in, which I presume he knows about. I see no reason why it should not be made available to counsel. But as you know, as an amateur in the matter of classification, I will have to talk to other people about it.

Mr. GRAY. The record will reflect Mr. Garrison's request. I think the record should also reflect that the chairman has nothing to add beyond the exchange of conversation that has taken place here, because I don't know the answer, frankly, Mr. Garrison. We will consider the request and meet it the best we can.

Mr. GARRISON. I would like to make one further request in the interest of expedition, Mr. Chairman, and that is, if Mr. Robb could conveniently do so, it would be helpful if he checked with the classification officer the text of any further transcripts that he proposes to use, so that, assuming they do not contain Government secrets that can't be revealed in the interest of justice in this proceeding, we might have copies of them as soon as you have finished, or I would prefer while you were reading from them, because there has been, and I assume will continue to be, some time lag in the furnishing of transcripts to us. We have not yet had even the first day's transcript, which it is hard for me to believe could have contained anything of a classified nature and could have been read over rather shortly. I am not being querulous, Mr. Chairman, or complaining, but I just want to point it out.

Mr. GRAY. I understand.
Mr. ROBB. May I proceed, sir?
Mr. GRAY. Yes.

By Mr. ROBB:

Q. Doctor, one further item from the Pash interview. You said to Colonel Pash, according to this transcript, or Colonel Pash said to you, "I can see that we are going to have to spend a lot of time and effort which we ordinarily would not in trying to——

"OPPENHEIMER: Well——

"PASH. In trying to run him down before we even go on this.

"OPENHEIMER. You better check up on the consulate because that is the only one that Eltenton contacted and without that contact, he would be inefficient and that would be my——

"PASH. You say this man is not employed in the consulate?

"OPPENHEIMER. Eltenton?
"PASH. No, this man.
"OPPENHEIMER. I have never been introduced to him.
"PASH. Have you ever heard his name mentioned?
"OPPENHEIMER. I have never heard his name mentioned but I have been given to understand that he is attached to the consulate. But isn't it common practice for a consulate or legation to have someone attached to them?
"PASH. Yes. Military attachés are really run efficiently."
Dr. Oppenheimer, assuming that, don't you think you told a story in great detail that was fabricated?
A. I certainly did.
Q. Why did you go into such great circumstantial detail about this thing if you were telling a cock and bull story?
A. I fear that this whole thing is a piece of idiocy. I am afraid I can't explain why there was a consul, why there was microfilm, why there were three people on the project, why two of them were at Los Alamos. All of them seems wholly false to me.
Q. You will agree, would you not, sir, that if the story you told to Colonel Pash was true, it made things look very bad for Mr. Chevalier?
A. For anyone involved in it, yes, sir.
Q. Including you?
A. Right.
Q. Isn't it a fair statement today, Dr. Oppenheimer, that according to your testimony now you told not one lie to Colonel Pash, but a whole fabrication and tissue of lies?
A. Right.
Q. In great circumstantial detail, is that correct?
A. Right.
Q. Doctor, I would like to refer you again to your answer on page 21, in which you referred to David Bohm, and said that you were much surprised that you heard there was much objection to his transfer on security grounds. I believe we had some talk about that this morning.
A. We did.
Q. I want to read to you from a memorandum written by then Major DeSilva on March 22, 1944 in which he started off—this is file A—March 21, 1944, "Dr. Oppenheimer asked through his office for the purpose of relating certain incidents which took place at Berkeley, Calif., during Dr. Oppenheimer's recent visit there." It goes on to various matters and finally it comes to this:
"4. Oppenheimer went on to say that just as he was preparing to leave his hotel at Berkeley on his return trip, David Joseph Bohm came to see him. Bohm inquired about the possibilities of his being transferred to project Y on a permanent basis, stating that he had a 'strange feeling of insecurity' in his present surroundings. Oppenheimer stated he did not commit himself to Bohm but told him that he would let Bohm know if an opportunity were open at this project, and that if Bohm did not hear from Oppenheimer he should assume that such an arrangement was not workable and to forget the matter. Oppenheimer asked the undersigned if he would have objections to Bohm coming to project Y. The undersigned answered yes. Oppenheimer agreed and said the matter was therefore closed."
Does that memorandum refresh your recollection about your conversation with DeSilva?
A. There were two incidents. One was in March 1943 that I described this morning.
Q. Yes.
A. This is in March 1944, a year later, I take it.
Q. Yes.
A. I gather this is no more than my having been asked by Bohm could he come, my checking to see whether the objections to him still obtained.
Q. I see.
A. I think that is all.
Q. Was there any surprise, as you recall looking back, when you were told by DeSilva that the objection still obtained?
A. No.
Q. And the objections were what, now?
A. What I was told was that Bohm had relatives in Nazi Germany.
Q. Do you recall the circumstances of Bohm's coming to you at your hotel?
A. I did not recall them.
Q. Do you now?

A. No, but——
Q. You don't know whether he came to you in your room or where?
A. I don't know whether it was in the room or the lobby.
Q. Project Y was Los Alamos?
A. That is right.
Q. Let me see if I can refresh your recollection about the circumstances of Bohm coming to you. I will read you from a report of a surveillance of J. R. Oppenheimer, March 16, 1944, in Berkeley, Calif.:
"6:05 p. m. Subject and Frank left hotel."
That would be your brother?
A. Right.
Q. "And walked up and down Telegraph Avenue in front of the hotel. Both engaged in earnest conversation with each other.
"6:15 p. m. David Bohm walked south on Telegraph Avenue and met the Oppenheimers in front of the hotel. J. R. Oppenheimer and Bohm engaged in conversation for 5 minutes but Frank stood about 10 feet away from them and did not participate in the conversation."
Does that help to refresh your recollection?
A. No. I don't remember the incident. I don't see any reason to doubt it.
Q. Were you waiting for Bohm on the sidewalk there?
A. Since I don't remember the thing, I could not remember that. I don't know whether this was an appointment, an accident, or what.
Q. I might read you the next item:
"6:20 p. m. Subject and Frank entered car, license 53692, with Oppenheimers' luggage and drove to Fisherman's Wharf, San Francisco."
Would that indicate to you that you had waited for Bohm on the sidewalk?
A. It suggests it but I don't want to remember more than I do remember.

By Mr. ROBB:

Q. I don't want you to, Doctor. Let me read to you from a memorandum from Captain DeSilva at that time—he must have been promoted—dated January 6, 1944:
"Subject, DSM conversation with J. R. Oppenheimer. Capt. H. K. Calvert, United States Engineers Office, Post Office Box 1111, Knoxville, Tenn.
"1. During a recent conversation with Dr. Oppenheimer he brought up the subject of a situation at Berkeley, Calif. A general discussion followed, touching on such subjects as of AEC which Oppenheimer deplored, the Eltenton incident which he thought was reprehensible, and the contacts made by the professor which contacts he believed to be innocent. During the course of the conversation which took place en route to Santa Fe, Oppenheimer touched on the subject of what persons at Berkeley were in his opinion truly dangerous. He named David Joseph Bohm and Bernard Peters as being so. Oppenheimer stated, however, that somehow he did not believe that Bohm's temperament and personality were those of a dangerous person and implied that his dangerousness lay in the possibility of his being influenced by others. Peters, on the other hand, he described as a 'crazy person' and one whose actions would be unpredictable. He described Peters as being 'quite a Red' and stated that his background was filled with incidents which indicated his tendency toward direct action."
Do you recall that conversation?
A. I recall the conversation, though I don't recall these as accurate words. I remember only being asked by DeSilva, among these people, and I think there were four, which do you thing is the most dangerous, and saying Peters.
Q. Did you mention Bohm as truly dangerous?
A. I am quite certain I didn't. I think DeSilva mentioned Bohm, Weinberg and somebody else and Peters.
Q. You say you are quite sure you did not mention Bohm as dangerous?
A. I think so.
Q. You think you did?
A. I did not. I certainly never thought of him that way.
Q. You did not think of him as dangerous. If you had, you would not have spoken to DeSilva in March about bringing him to Y, would you?
A. I should hope not. I think there is a garble in this and also the whole tone is not I believe accurate. The conversation was initiated by DeSilva. He presented me with a list of names. I don't believe this is something that I dredged up for him.
Q. What did you know of Bohm's background?
A. I don't think I know anything about it.
Q. Nothing?

A. I have even forgotten where he comes from. I think I did not that. Was it Pennsylvania?
Q. There was nothing in Bohm's background to cause you to say to DeSilva that Bohm was a dangerous person?
A. No. My strong recollection is that I couldn't have said that, and didn't think so.
Q. You could not be mistaken about that?
A. I could be mistaken about almost anything, but this does not fit.
Q. But you had asked General Groves to transfer Bohm to Los Alamos.
A. In March of 1943, yes, before that.
Q. How long had you known Bohm? When did you first meet him?
A. I met him when he came as a graduate student to the department. I have forgotten when that was. A couple of years before 1943, probably.
Q. Where is he now, do you know?
A. Yes. He is in Sao Paulo, Brazil.
Q. He taught for a while at Princeton?
A. At the university, yes.
Q. You helped him to get his job there, didn't you?
A. I think I did.
Q. When?
A. 1946 or 1947.
Q. What job did he have there?
A. He was assistant professor of physics.
Q. Did you see him frequently when he was there?
A. He came to seminars. I saw him infrequently otherwise.
Q. Did you see him socially?
A. Infrequently. I went to a farewell party that Professor Wigner gave for him.
Q. When was that?
A. Just before he left for Brazil, probably 1949 or 1950.
Q. Do you recall in May 1949 when Bohm testified before the House committee here?
A. Yes. I remember meeting him on the street with Weinberg and Whitman and a couple of other people.
Q. What street?
A. Main street of Princeton, Nassau Street.
Q. Weinberg was up there?
A. He was up there.
Q. Was that before he testified?
A. Oh, no, Weinberg was not there. I am sorry. Lomanitz and Bohm were there.
Q. Was that before they testified?
A. Yes, I think so.
Q. Did you discuss with them or either of them what their testimony might be?
A. I said they should tell the truth.
Q. What did they say?
A. They said "We won't lie."
Q. Did you discuss with them whether they would claim their constitutional privilege?
A. No.
Q. You know now they did claim their constitutional privilege?
A. Yes, but I didn't know that at the time. I didn't know whether they knew it. This was a 2 minute brush on the street.
Q. Did they ask you for any advice about testifying?
A. No.
Q. Did they ask you to recommend counsel to them?
A. I am sure not.
Q. Did you recommend counsel to them?
A. I would have if they asked me.
Q. Who would you have recommended?
A. I am foggy on this. I might have recommended Durr, but this is not a recollection; it is a conjecture.
Q. Mr. Durr did in fact represent them, didn't he?
A. Right. I don't know that.
Q. You first said "Right." How did you know that? Did you hear he did?
A. It was certainly in the record.
Q. Did you read the record?
A. Yes.

Q. When?
A. Sometime afterward; I don't know.
Q. Why?
A. I was involved in the same investigation.
Q. You knew that they refused to answer upon the grounds of possible self-incrimination when asked about their Communist Party membership and activities.
A. I did, that is right.
Q. And espionage activities.
A. Did they refuse to answer about espionage, too?
Q. Doctor, I don't have it before me so I won't make a categorical answer. You probably know it better than I do.
A. I am not sure.
Q. Did you see Bohm after he testified?
A. I would assume so, since he came back to Princeton.
Q. How long after he testified was the farewell party?
A. Quite a long while.
Q. How long?
A. I think he spent a whole year at Princeton.
Q. Do you recall who else was at the party?
A. No, I remember the host and I remember that most of the physicists in the physics department were invited.
Q. Who was the host?
A. Eugene Wigner.
Q. Was Bohn fired at Princeton?
A. No, his contract was lapsed. It was not renewed.
Q. Did you assist him to get his job in Brazil?
A. I don't believe I had anything to do with that.
Q. Did you write him a letter of recommendation?
A. I don't remember.
Q. Would you think about that a minute?
A. It won't do any good.
Q. Would you have written him one if he had asked?
A. I am quite sure I would have written a letter of recommendation about his physics.
Q. Do you know how he did get his job in Brazil?
A. No.
Q. Do you know anybody in Brazil who is a physicist?
A. Caesar Lattes.
Q. Doctor, let me go back a moment. I am sorry I overlooked something. Did there finally come a time whein you did disclose the identity of Professor X?
A. Yes.
Q. When was that?
A. I don't remember when. In late summer or fall of 1943, I should think, at Los Alamos.
Mr. GRAY. May I in the interest of having the record perfectly straight, there is a Professor X who has been in the newspapers and I think that ultimately turned out to be a name that does appear in this record.
Mr. ROBB. Yes, I am sorry.
Mr. GRAY. So let us make it clear when Dr. Oppenheimer is asked about disclosing the identity of Professor X, actually in this case we are talking about Dr. Chevalier.
Mr. ROBB. Yes, we agreed that we would refer to him as X. I am talking about Dr. Chevalier.
Mr. GRAY. I am sorry. I guess it was Scientist X, but in any event, let us make it clear what we are talking about.
Mr. ROBB. I think your point is well taken.

By Mr. ROBB.

Q. There came a time at last when you did disclose that Haakon Chevalier was the intermediary.
A. Right.
Q. I find in the file, Doctor, a telegram signed, "Nichols" and addressed to the area engineer, University of California, Berkeley, Calif., attention Lt. Lyle Johnson, reading as follows:
"Lansdale advises that according to Oppenheimer professor contact of Eltenton is Haakon Chevalier. REF, EIDMMI-34. Classified secret. Oppenheimer states

in his opinion Chevalier engaged in no further activity other than three original attempts."

That wire is dated December 13, 1943. Would it be about December 13, 1943, that you disclosed the identity of Dr. Chevalier?

A. I thought it was earlier. It could have been that late. I thought it was considerably earlier.

Q. To whom did you make that disclosure?

A. To General Groves.

Q. And under what circumstances?

A. We talked in his room in Los Alamos.

Q. All right.

A. He told me that he simply had to know, and I surely told him that the story I told Pash was a cock and bull story at that time. That there were no three people.

Q. In other words, you lied to Groves, too?

A. No, I told him that the story I told Pash was a cock and bull story.

Q. You told Groves that you had told Pash a cock and bull story?

A. I am quite certain about that.

Q. You are sure about that?

A. Yes.

Q. You notice in this wire from General Nichols——

A. There are still the three people.

Q. You are still talking about the three people. I notice in the file of the same day General Nichols wired the Commanding Officer, United States Engineer Office, Santa Fe, New Mex., attention, Captain DeSilva. "Haakon Chevalier to be reported by Oppenheimer to be professor at RadLab who made three contacts for Eltenton. Classified secret. Oppenheimer believed Chevalier engaged in no further activity other than three original attempts."

On December 12——

Mr. GARRISON. That last wire was from whom?

Mr. ROBB. Nichols. On December 12, 1943, a wire to Capt. H. K. Calvert, Clinton Engineer Work, Clinton, Tenn. What was that, Oak Ridge?

The WITNESS. Yes.

By Mr. ROBB.

Q. "According to Oppenheimer professor contact of Eltenton is Haakon Chevalier. Oppy states in his opinion beyond original three attempts Chevalier engaged in no further activity. From Lansdale. DeSilva and Johnson to be notified by you."

Does that indicate to you that you told General Groves that there weren't three contacts?

A. Certainly to the contrary. I am fairly clear.

Q. You think General Groves did tell Colonel Nichols and Colonel Lansdale your story was cock and bull?

A. I find that hard to believe.

Q. So do I. Doctor, may we again refer to your answer, please, sir. On page 4: "In the spring of 1936, I had been introduced by a friend to Jean Tatlock, the daughter of a noted professor of English at the university, and in the autumn I began to court her, and we grew close to each other. We were at least twice close enough to marriage to think of ourselves as engaged. Between 1939 and her death in 1944, I saw her very rarely. She told me about her Communist Party memberships. They were on-again, off-again affairs and never seemed to provide for her what she was seeking. I do not believe that her interests were really political. She was a person of deep religious feeling. She loved this country, its people, and its life. She was, as it turned out, a friend of many fellow travelers and Communists, a number of whom I later was to become acquainted with."

Doctor, between 1939 and 1944, as I understand it, your acquaintance with Miss Tatlock was fairly casual; is that right?

A. Our meetings were rare. I do not think it would be right to say that our acquaintance was casual. We had been very much involved with one another, and there was still very deep feeling when we saw each other.

Q. How many times would you say you saw her between 1939 and 1944?

A. That is 5 years. Would 10 times be a good guess?

Q. What were the occasions for your seeing her?

A. Of course, sometimes we saw each other socially with other people. remember visiting her around New Year's of 1941.

Q. Where?

A. I went to her house or to the hospital, I don't know which, and we went out for a drink at the Top of the Mark. I remember that she came more than once to visit our home in Berkeley.

Q. You and Mrs. Oppenheimer?

A. Right. Her father lived around the corner not far from us in Berkeley. I visited her there once. I visited her, as I think I said earlier, in June or July of 1943.

Q. I believe you said in connection with that that you had to see her.

A. Yes.

Q. Why did you have to see her?

A. She had indicated a great desire to see me before we left. At that time I couldn't go. For one thing, I wasn't supposed to say where we were going or anything. I felt that she had to see me. She was undergoing psychiatric treatment. She was extremely unhappy.

Q. Did you find out why she had to see you?

A. Because she was still in love with me.

Q. Where did you see her?

A. At her home.

Q. Where was that?

A. On Telegraph Hill.

Q. When did you see her after that?

A. She took me to the airport, and I never saw her again.

Q. That was 1943?

A. Yes.

Q. Was she a Communist at that time?

A. We didn't even talk about it. I doubt it.

Q. You have said in your answer that you knew she had been a Communist?

A. Yes. I knew that in the fall of 1937.

Q. Was there any reason for you to believe that she wasn't still a Communist in 1943?

A. No.

Q. Pardon?

A. There wasn't, except that I have stated in general terms what I thought and think of her relations with the Communist Party. I do not know what she was doing in 1943.

Q. You have no reason to believe she wasn't a Communist, do you?

A. No.

Q. You spent the night with her, didn't you?

A. Yes.

Q. That is when you were working on a secret war project?

A. Yes.

Q. Did you think that consistent with good security?

A. It was, as a matter of fact. Not a word—it was not good practice.

Q. Didn't you think that put you in a rather difficult position had she been the kind of Communist that you have described her or talk about this morning?

A. Oh, but she wasn't.

Q. How did you know?

A. I knew her.

Q. You have told us this morning that you thought that at times social contacts with Communists on the part of one working on a secret war project was dangerous.

A. Could conceivably be.

Q. You didn't think that spending a night with a dedicated Communist——

A. I don't believe she was a dedicated Communist.

Q. You don't?

A. No.

Q. Did she go over to Spain?

A. No.

Q. Ever?

A. Not during the time I knew her.

Q. What was the occasion of her telling you about her Communist Party membership?

A. She would talk about herself rather freely, and this was one aspect of her life. She would tell me that she had been with a medical unit—I am making it up—with some kind of a unit, and it had been frustrating.

Q. What do you mean, you are making it up?

A. I mean I don't remember what kind of a unit, but she had been with some sort of a Communist unit and had left it. It had been a waste of time, and so on.

Q. By a medical unit, you mean a medical cell?
A. That is what I would have meant.
Q. You say here she was as it turned out a friend of many fellow travelers and Communists. Who were they?
A. Well, Addis was a friend of hers. Lambert was a friend of hers.
Q. Doctor, would you break them down? Would you tell us who the Communists were and who the fellow travelers were?
A. Lambert was a Communist. Addis is reported to be a Communist in the Commission's letter. I did not know whether he was a member of the party or not.
Q. You knew he was very close, didn't you?
A. Yes. Among fellow travelers, Chevalier. Among Communists or probable Communists, a man and his wife who wrote for the People's World.
Q. Who were they?
A. John Pitman and his wife. A lawyer called Aubrey Grossman, his wife she had known.
Q. Was she a Communist?
A. I don't know in the sense of party membership.
Q. But very close.
A. Close. Is the list long enough?
Q. I want you to give the ones you remember, Doctor. I assume when you wrote this sentence that she was, as it turned out, a friend of many fellow travelers and Communists, that you had people in mind.
A. I have gone over some of those I had in mind.
Q. Have you any more in mind?
A. There was another couple; yes. A girl called Edith Arnstein.
Q. Was she a Communist?
A. I believe so; yes.
Q. Anybody else?
A. I am sure there were more people.
Q. When did you first meet this group of Communists and fellow travelers who were friends of Miss Tatlock?
A. That came on gradually during 1937, maybe late 1936, not all at once.
Q. But they continued to be your friends?
A. Some of them.
Q. Chevalier still is your friend?
A. Chevalier is my friend.
Q. Addis was your friend until he died?
A. No. We had essentially, I think, no relations after the war.
Q. When did he die?
A. In 1950 or 1951.
Q. Do you recall, when you were interviewed by the FBI in 1950, you were asked about Dr. Addis?
A. Yes.
Q. And you declined to discuss Dr. Addis?
A. Yes.
Q. You said he was dead and couldn't defend himself.
A. I did say that.
Q. What did you think he had to defend himself against?
A. Being close to the Communist Party.
Q. Didn't you continue to see Dr. Addis periodically until he died.
A. No.
Q. But you say in your answer he did become a very close friend of yours.
A. Close would be wrong, I am sure. He became a good friend, I think I said.
Q. A friend.
A. A friend; that is more like it, I imagine.
Q. At least he was enough of a friend so you wouldn't discuss him with the FBI; is that correct?
A. Yes.
Q. That was in 1950?
A. I asked if it were important, and they thought not.
Q. They asked you about him, though, didn't they?
A. They were asking about me.
Q. Didn't they also ask about your friends?
A. Not much.
Q. You say in your answer at page 5, in describing your friendship with Miss Tatlock and meeting people through her, "I liked the new sense of companion-

ship." Who were the people whose companionship you enjoyed that you met through Miss Tatlock, the people that you just mentioned?
A. Oh, no. People who were in the teachers union, people in Spanish causes, great masses of people, in addition to some of those I just mentioned.
Q. Was the teachers union a Communist organization?
A. I think that there were Communists in it. I know there were some.
Q. Who were they?
A. Kenneth May, and I believe his first wife.
Q. Who else?
A. I have no certain knowledge of anyone else.
Q. You say in your answer at page 6, "I was invited to help establish the teachers union, which included faculty and teaching assistants in the university and schoolteachers of the East Bay." Who invited you?
A. We invited ourselves, I guess. A group of people from the faculty talked about it and met, and we had a lunch at the Faculty Club or some place and decided to do it. I don't know at whose initiative this was caused.
Q. About when was that, Doctor?
A. 1937 would be a fair guess.
Q. How long did you stay in that union?
A. Until 1941 or, I think, early 1941.
Q. Did you make a formal resignation?
A. No. That chapter of the union dissolved, and with its dissolution——
Q. Was Kenneth May an officer in the union?
A. I don't believe so.
Q. Do you recall who the officers were during your tenure as recording secretary?
A. I will remember some of them.
Q. Who?
A. Chevalier was president at one time. Margaret Ellis was president at one time.
Q. Was she a Communist?
A. I don't know.
Q. Was she close to it?
A. I think so. The reason I think so is that I had a letter from her about the Rosenberg affair not long ago.
Q. You mean asking your support?
A. Something like that.
Q. For the Rosenbergs?
A. Yes.
Q. That indicates to you that she is a Communist sympathizer, at least?
A. That is right.
Q. Who else among the officers?
A. A man called Fontenrose; Joe is the first name
Q. Was he a Communist sympathizer?
A. I don't think so.
Q. Who else?
A. I don't remember.
Q. Kenneth May was a Communist functionary in Alameda County, was he not?
A. That was later.
Q. Later?
A. Yes.
Mr. GRAY. Are you going to ask anything more about the teachers organization?
Mr. ROBB. I didn't have any questions in mind.
Mr. GRAY. I would just ask whether the dissolution related to any international event.
The WITNESS. The miserable thing fell apart because it grew into a debating society between the anti-interventionists and the interventionists, which had even less to do with teachers' welfare than what we had been doing before. I was strongly in favor of letting it collapse. It is my recollection that was not the pro-Communist view at that time, that they wanted it to continue.

By Mr. ROBB.
Q. You say in your answer on page 5: "I contributed to the strike fund of one of the major strikes of Bridges' union."
Do you recall about when that was, Doctor?
A. Could it have been 1938?

Q. I don't know.
A. Well, it couldn't have been before 1936, because I just didn't know or do anything of that kind before late 1936. It was probably 1938, 1937 or 1938.
Q. Do you recall about how much you gave?
A. I can guess.
Q. How much?
A. About $100.
Q. In cash?
A. I think so.
Q. Do you recall through whom you made that contribution?
A. I went to the wicket, the union wicket.
Q. Did you understand that Bridges was a Communist?
A. No, I understood to the contrary. I may have been fooled.
Q. You subscribed to the People's World, you say. When did you do that?
A. I don't recollect. It was for several years.
Q. How long did that subscription continue?
A. I would say for several years.
Q. Can you tell us about when it expired?
A. I can't of my own knowledge; no.
Q. Was it after you joined the project?
A. Since I don't know when it was, I can't answer that question.
Q. That was the west coast Communist newspaper; wasn't it?
A. That is right.
Q. Did you have that paper sent to you at your house?
A. Yes. I don't know whether I had it sent but anyway it came.
Q. And you paid for it?
A. Again I don't know whether I paid for it or whether it was distributed. I think I paid for it.
Q. Do you recall whether you canceled your subscription or whether you just let it expire or what?
A. I don't recall. I don't believe I canceled the subscription.
Q. Why did you subscribe to the People's World?
A. Well, I guess I took an interest in this formulation of issues; perhaps somebody asked me to.
Q. You read it, I take it?
A. Not fervently. It taught me to read—well——
Mr. GRAY. Would you repeat that?
The WITNESS. It was an interjection that was unnecessary.
Mr. GRAY. Excuse me.

By Mr. ROBB:
Q. You say "I contributed to the various committees and organizations which were intended to help the Spanish loyalist cause." What were they?
A. Wasn't there a North American committee?
Q. I don't know, Doctor; I am asking you.
A. I think there was a North American committee. There was another one. I don't know its name.
Q. Were those contributions fairly substantial?
A. I would think they were.
Q. What amounts would you say?
A. In the hundred dollar range.
Q. In cash?
A. Pardon me?
Q. In cash?
A. I would think so.
Q. I will come back to that in a minute.
You say, "I also began to take part in the management of the physics department, the selection of courses and the awarding of fellowships."
What do you mean by taking part in the awarding of fellowships, Doctor?
A. I was named to the graduate council of the university. The graduate council had a committee on graduate fellowships, and I served on that. This has nothing to do with communism.
Q. Were any fellowships awarded to any of your students?
A. I would hope so.
Q. Do you know whether or not Lomanitz or Bohm or Weinberg or Fred Man had a fellowship?
A. My recollection is that they did not.

Q. In all events, if they did, you didn't have any thing to do with it?
A. No, I think it was off the graduate council at that later date.
Q. You say on page 6: "I also became involved in other organizations. For perhaps a year I was a member of the Western Council of the Consumers Union."
Who composed the Western Council of the Consumers Union?
A. Chairman and the man I knew best was Robert Bradley, a professor of economics at the university.
Q. Was he a Communist sympathizer?
A. No, I don't think so. His wife was Mildred Eddy, and the two of them were what made this. They had enthusiasm for this.
Q. Did they recruit to it?
A. Yes, they asked me to come. It was a very inappropriate thing for me to do. I know nothing about the business.
Q. Who else was in the council?
A. I remember only one other man, and that is a man named Folkoff, who was not Isaak Folkoff.
Q. That is Richard Folkoff.
A. That could be.
Q. Was he a Communist?
A. I thought not, but I could be wrong.
Q. Anybody else?
A. There were other people, and I have forgotten them.
Q. What year was that that you were a member of that?
A. It says in my PSQ; I am afraid I can't improve on that. Could it have been 1937?
Mr. GARRISON. I think the biography will show, the one we submitted to you.

By Mr. ROBB:

Q. "I joined the American Committee for Democracy and Intellectual Freedom in 1937."
Did you also serve on the national executive committee of that organization?
A. The letterhead says so. I didn't meet with them.
Q. Do you know how you happened to get on the letterhead?
A. I supposed I accepted membership. I have no records of this except my own record—except what I said about it in the personal security questionnaire.
Mr. GRAY. 1938 is shown here as the date of the Consumers Union, 1938 to 1939.
Mr. ROBB. Yes.

By Mr. ROBB:

Q. When did you serve on the executive committee of the American Committee for Democracy and Intellectual Freedom?
A. I would assume that my dates 1937 to whatever it was that I gave in the personnel security questionnaire refer to that. I have no other record.
Mr. GARRISON. Again I think the biography may show that.
Mr. GRAY. It shows 1937 in the biography.
Mr. ROBB. The copy of the biography does not show a date when you ceased to be a member of that organization. When was it?
The WITNESS. I have no recollection of ceasing to be. It played no part in my life.

By Mr. ROBB:

Q. You mean you might still be a member of it?
A. I haven't heard from them for an awful long time.
Q. Your PSQ lists the American Committee for Democracy and Intellectual Freedom, 1937–, with an asterisk being at the foot of the page where you say, "It includes all organizations to which I now belong." So you were still a member in 1942; were you not?
A. Right.
Q. You have no idea how long after that you continued to be a member?
A. My membership involved no attendance in meetings, no activities that I could recall, and I certainly was not very active at Los Alamos.
Q. I see. You say, talking about your PSQ on page 6, "I say on that questionnaire, that did not include sponsorships." What is a sponsorship?
A. I am charged with a sponsor of this Friends of the Chinese People. I don't know what it means, but I think it means that you lend your name to something. I am sure that I lent my name to 1 or 2 parties or bazaars for Spanish war or

Spanish relief. I had no record of these and no good memory of them when I filled out my PSQ.

Q. A sponsorship was just something you lent your name, but did not become a formal member?

A. Yes. Maybe it was something you couldn't be a member of.

Q. Were there any other things that you think of now that you sponsored as distinguished from joining?

A. No, I can't.

Q. Now, coming to your questionnaire again, page 7, "The statement is attributed to me that while I was not a Communist, I'd probably belonged to every Communist-front organization on the west coast and had signed many petitions in which Communists were interested.

"I do not recall this statement nor to whom I might have made it, nor the circumstances. The quotation is not true. It seems clear to me that if I said anything along the lines quoted, it was a half jocular overstatement."

Assuming that it was a jocular overstatement, Doctor, had you belonged to any Communist-front organizations that you can think of?

A. We have just been over the Committee on Democracy and Intellectual Freedom, which has been so designated; Consumers Union which has been so designated; the Teachers Union, of which it could be so designated, I think. I think we have been over the list.

Q. That is what you had in mind? Had you signed any petitions?

A. I don't remember signing petitions. I think I may have or I would have signed petitions in the early days with regard to lifting the embargo on arms to Spain, or such a matter, but this is conjecture and not memory.

Mr. ROBB. Mr. Chairman, I have just a couple of more questions.

Mr. GRAY. Very well.

By Mr. ROBB:

Q. Doctor, I would like to read to you from a memorandum dated September 14, 1943, memorandum for the file.

"Subject: Discussion by General Groves and Dr. Oppenheimer, signed John Lansdale, Jr., Lt. Col., Field Artillery, Chief Review Branch, CIGMIS, reading as follows:

"During a recent train ride between Cheyenne and Chicago, General Groves and Dr. Oppenheimer had a long discussion which covered in substance the following matters:

"(f) Oppenheimer categorically stated that he himself was not a Communist and never had been, but stated that he had probably belonged to every Communist-front organization on the west coast and signed many petitions concerning matters in which Communists were interested."

Did you make such a statement to General Groves as reflected in this memorandum from Colonel Lansdale?

A. I remember the trip from Cheyenne to Chicago. I do not remember making the statement. I see no reason to deny it.

Q. Do you think if you did make it, you were just joking with General Groves?

A. I am pretty sure I was.

Q. Do you think General Groves misunderstood you maybe?

A. Maybe he didn't. Maybe in transmission it got garbled. I have no way of knowing.

Q. In that same paragraph while I am reading:

"He (meaning you) stated while he did not know, he believed his brother Frank Oppenheimer had at one time been a member of the Communist Party, but that he did not believe that Frank had had any connections with the party for some time."

Do you recall that statement?

A. I don't recall it. I did believe at that time that my brother had been out of the party for some time.

Q. Did you tell General Groves that while you did not know, you believed that your brother had at one time been a member of the party?

A. I should not have told him that.

Q. Did you tell him that?

A. I don't know.

Q. But you might have?

A. I should not have.

Q. If you did say that to General Groves, it was not strictly true?

A. No, I did know.

Q. Because you knew he had been a member.
A. I did know it.
Q. Yes, sir. Would you now deny that you made that statement to General Groves?
A. Oh, I couldn't.
Q. In other words, you might have told General Groves something that was not true?
A. Well, I hope I didn't.
Q. You might have; is that correct?
A. I hope I didn't.
Q. But might have; might you not?
A. Obviously I might have.

Mr. ROBB. It is half past four, Mr. Chairman.
Mr. GRAY. All right. I should like to say before we recess that one thing I neglected to say with respect to the transcript which we discussed earlier. In view of the fact, and especially referring now to the transcript of the first day, since there are so many references to other agencies, particularly the Defense Department, I am informed that it has been necessary to check not only with the security officers of the Commission, but with other Departments which in part explains the delay. There is no design, Mr. Garrison.
Mr. GARRISON. I am sure of that.
Mr. GRAY. We will meet again at 9:30.
(Thereupon at 4:30 p. m., a recess was taken until Thursday, April 15, 1954, at 9:30 a. m.)

UNITED STATES ATOMIC ENERGY COMMISSION

PERSONNEL SECURITY BOARD

In the Matter of J. Robert Oppenheimer

Atomic Energy Commission,
Building T-3, Room 2022,
Washington, D. C., April 15, 1954.

The above entitled matter came on for hearing, pursuant to recess, before the board, at 9:30 a. m.

Personnel Security Board: Mr. Gordon Gray, chairman; Dr. Ward V. Evans, member; and Mr. Thomas A. Morgan, member.

Present: Roger Robb, and C. A. Rolander, Jr., counsel for the board; J. Robert Oppenheimer, Lloyd K. Garrison, Samuel J. Silverman, and Allen B. Ecker, counsel for J. Robert Oppenheimer, and Herbert S. Marks, cocounsel for J. Robert Oppenheimer.

PROCEEDINGS

Mr. GRAY. The presentation will begin. I believe that General Groves is waiting.

General Groves, I should like to ask you whether you would like to testify under oath. You are not required to do so.

General GROVES. Whichever you prefer. It makes no difference to me.

Mr. GRAY. It is my guess that most everyone who appears will be testifying under oath.

General GROVES. It makes no difference in my testimony, but I would be very glad to.

Mr. GRAY. What are your initials?

General GROVES. Leslie R.

Mr. GRAY. Will you raise you right hand. Do you, Leslie R. Groves, swear that the testimony you are to give the board shall be the truth, the whole truth and nothing but the truth, so help you God?

General GROVES. I do.

Whereupon, Leslie R. Groves, was called as a witness, and having been first duly sworn, was examined and testified as follows:

DIRECT EXAMINATION

By Mr. GARRISON:

Q. General Groves, you are now vice president in charge of advance scientific research at Remington Rand?

A. No, I am not longer in charge of research. I am a vice president and director of Remington Rand.

Q. During the war, you headed the Manhattan Project in complete charge and development planning for use of the atomic bomb?

A. That is correct.

Q. During the postwar period you were Commanding General of the Armed Forces Special Weapons Project, 1947 to 1948?

A. Yes. My charge of the atomic work ended on the 1st of January 1947. I think you also should add that during the period from about March of 1947 until my retirement on the 29th of February 1948, I was a member of the Military Liaison Committee to the Atomic Energy Commission.

Q. You appointed Dr. Oppenheimer to be the director of the work at Los Alamos?

A. Yes, sir.

Q. You devolved great responsibility upon him?

A. Yes.

Q. Would you just say a word about the nature of that responsibility?

A. Complete responsibility for the operation of Los Alamos Laboratory, the mission of which was to carry on the research necessary to develop the design of a bomb, to develop the probabilities of whether a bomb was possible, and if the design would be feasible, and to develop what the power of the bomb would be. That was so that we would know at what altitude the bomb should be exploded.

Mr. GRAY. General, may I interrupt? I am sorry. If it becomes necessary in the course of your testimony to refer to any restricted data, I would appreciate your letting me know in advance that you are about to do so.

The WITNESS. All right, sir.

Not only design and make these experimental tests, but to actually produce the bombs which we expected to use in the war. It should be understood that as early—certainly before Yalta, because at that time I so informed President Roosevelt, or just before Yalta—I had concluded that we only needed two bombs to end the war.

Of course, I also proceeded on the theory that I might be wrong. For that reason we decided, or I decided that we would construct the actual bombs at Los Alamos. That included as matters developed the final purification of plutonium at Los Alamos.

Possibly—I am not certain—any final purification of U-235 that might be necessary.

In addition to that, as time went on throughout the project, I consulted with Dr. Oppenheimer frequently as to other problems with which I was faced. I think one of those is of such importance that it might be well to explain it to give a picture of the responsibilities which you might say he carried.

There was a very serious problem as to the purification of U-235. While this is not secret in any way, I would rather not have it talked about by anyone here, because it reflects to some extent on the wisdom of another scientist.

Mr. GRAY. There are no security implications involved?

The WITNESS. No security whatsoever. I will watch out for that. I have been watching out for that for so many years I don't think I will slip.

Mr. GRAY. Thank you, sir.

The WITNESS. There was a great question as to the electromagnetic process—how pure did the U-235 have to be to have an explosion. We could get no advice on that matter from the people that were responsible because nobody knew. All that was known was that the natural state of 0.707 percent of U-235 in uranium that it did not explode.

I felt * * * that we would have to have a * * * high * * * percentage of purity in order to have an explosive. Dr. Oppenheimer was used by me as my adviser on that, not to tell me what to do, but to confirm my opinion. I think it is important for an understanding of the situation as it existed during the war to realize that when I made scientific decisions—in case there are any questions that come in on that—that outside of not knowing all the theories of nuclear physics, which I did not, nobody else knew anything either. They had lots of theories but they didn't know anything. We didn't know whether plutonium was a gas, solid, or electric. We didn't even know that plutonium existed, although Seaborg, I believe it was, claimed to have seen evidences of it in the cyclotron.

We didn't know what any of the constants that were so vital were. We didn't know whether it could be made to explode. We didn't know what the reproductive factor was for plutonium or uranium 235. We were groping entirely in the dark. That is the reason that General Nichols and myself were able, I think, to make intelligent scientific decisions, because we knew just as much as everybody else. We came up through the kindergarten with them. While they could put elaborate equations on the board, which we might not be able to follow in their entirety, when it came to what was so and what was probably so, we knew just about as much as they did. So when I say that we were responsible for the scientific decisions, I am not saying that we were extremely able nuclear physicists, because actually we were not. We were what might be termed "thoroughly practical nuclear physicists".

As a result of this experience, maybe because Dr. Oppenheimer agreed with me and particularly because of other questions that were raised, I came to depend upon him tremendously for scientific advice on the rest of the project, although I made no effort to break down my compartmentalization. As you know, compartmentalization of information was my chief guard against information passing. It was something that I insisted on to the limit of my capacity. It was something that everybody was trying to break down within the project. I did not bring Dr. Oppenheimer into the whole project, but that was not only because of security of information—not him in particular, but all the other scientific leaders, men like Lawrence and Compton were treated the same way—but it was also done because if I brought them into the whole project, they would never do their own job. There was just too much of scientific interest, and they would just be frittering from one thing to another.

So Dr. Oppenheimer was used in many ways as a chief scientific adviser on many problems that were properly within his bailiwick. That included his final advice which brought up the question of the thermal diffusion separation process, which was the case, as you know probably by now, that we made this last ditch effort to bring that into the project.

We were late in bringing it in, because—again this is something that is not confidential, but I would rather not have it talked about—there had not been the proper cooperation by certain scientific personnel at the Naval Research Laboratory. There had been suspicion on the part of certain scientists that the figures that were talked about at the Naval Research Laboratory were not sound, * * * and we could not depend on them. The reason they felt this way was that the results were not in accord with scientific theory. It just gave the wrong answer. They were too favorable. We did not get into using that, to my recollection—I am not absolutely certain—but I believe it was Dr.

Oppenheimer who suddenly told me that we had a terrible scientific blunder. I think he was right. It is one of the things that I regret the most in the whole course of the operation. We had failed to consider this as a portion of the process as a whole. In other words, we considered this process as a process that would take uranium 235 from 0.707 up to the final purity instead of saying we will take it from 0.707 up to, say, 2 percent, and then put that in.

What we had done, everybody in the project—this was brought to my attention by I believe Oppenheimer—had failed to think about, well, after all, if you started off with uranium at 2 percent instead of 0.7 in any of our other processes, we would be crippling our output.

I tell you that not in praise of Dr. Oppenheimer, but more to give you a picture of how he was used throughout the process. I think that more or less answers Mr. Garrison's question.

If I talk too long, Mr. Gray, if you will just tell me to stop, it is your time and not mine.

By Mr. GARRISON.

Q. How would you rate the quality of his achievement as you look back on it?

A. Naturally I am prejudiced, because I selected him for the job, but I think he did a magnificent job as far as the war effort was concerned. In other words, while he was under my control—and you must remember that he left my control shortly after the war was over.

Q. If you had to make the decision again, would you make it in the same way with respect to the selection of Dr. Oppenheimer and devolving the responsibilities on him which you did?

A. I know of no reason why not. Assuming all the conditions are the same, I think I would do it.

Q. You saw him very closely during those years?

A. I saw him on the average, I would say, of anywhere from once a week to once a month. I talked to him on the phone about anywhere from 4 to 5 times a day to once in 3 or 4 days. I talked on all possible subjects of all varieties. During the time I spent a number of days, for example, on trains traveling where we might be together for 6 or 8 or 12 hours at a time.

Q. You were aware of his leftwing associations at the time—his earlier leftwing associations?

A. Was I or am I?

Q. Were you at the time you appointed him?

A. At the time I appointed him to the project, I was aware that there were suspicions about him, nothing like what were contained—and I might say I read the New York Times, the letter of General Nichols and Dr. Oppenheimer's letter. I was not aware of all the things that were brought out in General Nichols' letter at the time of the appointment, but I was aware that he was or that he had, you might say, a very extreme liberal background.

I was also aware of another thing that I think must be mentioned, that he was already in the project, that he had been in charge of this particular type of work, that is, the bomb computations, and that he knew all that there was to know about that. In general, my policy was to consider the fact that the man was already in the project, and that made it very questionable whether I should separate him and also whether I should separate him under what might be termed unpleasant conditions, because then you never know what you are going to do to him. Are you going to drive him over to the other side or not? As far as what I knew at the time of his actual selection, I knew enough to tell me that I would have considered him an extreme liberal with a very liberal background. Just how many of the details I knew at the time I don't know. I did know them all later.

Q. Based on your total acquaintance with him and your experience with him and your knowledge of him, would you say that in your opinion he would ever consciously commit a disloyal act?

A. I would be amazed if he did.

Q. Was there any leakage of information from Los Alamos to improper sources for which Dr. Oppenheimer had in your opinion any responsibility?

A. That is a very difficult question, because it brings up the fact that the scientists—and I would like to say the academic scientists—were not in sympathy with compartmentalization. They were not in sympathy with the security requirements. They felt that they were unreasonable. I never held this against them, because I knew that their whole lives from the time they entered college almost had been based on the dissemination of knowledge. Here, to be put in a strange environment where the requirement was not dissemination, but not talking about it, was a terrible upset. They were constantly under pressure from their fellows in every direction to break down compartmentalization. While I was always on the other side of the fence, I was never surprised when one of them broke the rules.

For example, I got through talking to Neils Bohr on the train going to Los Alamos for the first time, I think I talked to him about 12 hours straight on what he was not to say. Certain things that he was not to talk about out there. He got out there and within 5 minutes after his arrival he was saying everything he promised he would not say.

The same thing happened on one occasion with Ernest Lawrence, after he was told that he was not to say something; he got up to the blackboard with this group—it was a group of smaller size than this of the key people—and said "I know General Groves doesn't want me to say this, but" and then he went on and discussed what I didn't want him to say.

You may say what kind of military organization was that. I can tell you I didn't operate a military organization. It was impossible to have one. While I may have dominated the situation in general, I didn't have my own way in a lot of things. So when I say that Dr. Oppenheimer did not always keep the faith with respect to the strict interpretation of the security rules, if I could say that he was no worse than any of my other leading scientists, I think that would be a fair statement. It would not be right to say that he observed my security rules to the letter, because while I have no evidence of his violating them—after all, I am not stupid—I know he did. I could not say of my own knowledge that I never knew him just on the spur of the moment and I can't recall a case where he deliberately violated my security instructions.

That is different from violating what he knew that I would want. That was done by everybody in my organization, including the military officers because my organization was a peculiar one. A great deal of responsibility devolved on everybody. They all knew the goal. I know I was put in positions where I had to approve things, things people knew I didn't want to approve, but they got me in that corner. That was not limited as I say to scientific personnel. It applied to engineering personnel, that applied to military officers. They were the kind of men I wanted, and they were the kind of men that made the project a success. If I had a group of yes men we never would have gotten anywhere.

Q. The absence of compartmentalization on the Los Alamos project, General Groves, would you say that represented on Dr. Oppenheimer's part an honest judgment as to what in his opinion would produce the best operating results among the scientists on the project?

A. I always felt—I can't quite answer that—that Dr. Oppenheimer was led to that breakdown of compartmentalization at Los Alamos by a number of conflicting factors. Here I am just giving my surmise as to what I thought.

First, that he personally felt that was right in view of his background of academic work.

Second, that he felt it was necessary in order to attract the kind of men that he felt he had to have at Los Alamos. I agreed that it was a very decided factor and always thought it was in getting such men. I also felt that he was very much influenced at that time by the influence of Dr. Condon, who was for a very brief time the associate director there, and, as you all know, a very complete disappointment to me in every respect.

I would like to emphasize now before any question is asked that I was not responsible for the exact selection of Dr. Condon, but I was responsible for his selection because I insisted when Dr. Oppenheimer took the directorship that he have as his No. 1 assistant an industrial scientists, and we just made a mistake when we selected Dr. Condon. Who gave his name the first time I don't know, but Dr. Condon turned out to be not an industrial scientist, but an academic scientist with all of the faults and none of the virtues. That was my opinion. He did a tremendous amount of damage at Los Alamos in the initial setup. How much influence he had on Dr. Oppenheimer I don't know. But he was given certain responsibilities with my full approval—in fact, you might say my very insistent suggestion—that Dr. Condon with the industrial background should be the one to establish the working rules and the administrative scientific rules in the establishment, while Dr. Oppenheimer was thinking about how was the actual scientific work to be done.

I coupld never make up my own mind as to whether Dr. Oppenheimer was the one who was primarily at fault in breaking up the compartmentalization or whether it was Dr. Condon. I don't to this day know whether it was wise. I think it was a serious mistake and felt so at the time to have the lack of compartmentalization go on down the line. In other words, it was all right to have the leaders, maybe 20 to 30, but not to have as many men as were permitted to break down compartmentalization. * * *

They all, of course, had given an oath that they would support the security regulations, but that was not controlling. They wavered here and there.

I think that answers your question in general.

Q. How long was Dr. Condon on the project?

A. I think a very short time. The record would show, but my impression would be only 6 weeks to 2 months. I don't recall. A very short time. His departure, of course, was at his own volition. I always thought it was because he thought the project would fail, and he was not going to be associated with it. His record showed since then he has never been satisfied anywhere he was. He was always moving. It was a mistake to get him out there. It is a mistake for which the responsibility was maybe 75 percent mine and 25 percent Oppenheimer's or maybe my share was even more than that. But mine was very heavy, because he would never have been there if I had not told Oppenheimer what kind of assistant he should have.

Q. Apart from the question of compartmentalization as an operating policy, you had no occasion to believe that any leakage of information from Los Alamos occurred as a result of any conscious act of Dr. Oppenheimer's?

A. Oh, no. I don't consider that his compartmentalization was a conscious act that would tend to encourage the leak of information.

Q. You had complete confidence in his integrity?

A. During the operation of Los Alamos, yes, which was where I really knew him.

Q. And you have that confidence today?

A. As far as that operation went, yes. As I say, as far as the rest of it goes, I am, you might say, not a witness. I am really ignorant on that, excepting what I read in the papers.

Q. As the war neared its end, there was an even greater urgency to produce the bomb in time to use it, was there not?

A. No, because no one in this country conceived of the Japanese war ending as soon as it did, no one in responsible positions today, no matter what they say today or said since. There is not a soul that thought that the war was going to end within a reasonable time.

Q. Did Dr. Oppenheimer work as hard as a man could to produce that bomb in accordance with the deadline dates that you had projected?

A. Oh, yes, yes. In fact, he worked harder at times than I wanted him to, because I was afraid he would break down under it. That was always a danger in our project. I think it is important to realize in the case of Dr. Oppenheimer because I had a physical taken of him when we were talking about making it a militarized affair, and I knew his past physical record, and I was always disturbed about his working too hard. But I never could slow him down in any way.

Q. Do you recall your conversation with him about the Chevalier incident?

A. Yes, but I have seen so many versions of it, I don't think I was confused before, but I am certainly starting to become confused today. I recall what I consider the essential history of that affair. As to whether this occurred this time, where I was at the moment, I can't say that I recall it exactly. I think I recall everything that is of vital interest, as far as would be necessary to draw a conclusion as to that affair.

Q. Would you say what your conclusion was?

A. My conclusion was that there was an approach made, that Dr. Oppenheimer knew of this approach, that at some point he was involved in that the approach was made to him—I don't mean involved in the sense that he gave anything—I mean he just knew about it personally from the fact that he was in the chain, and that he didn't report it in its entirety as he should have done. When I learned about it, and throughout, that he was always under the influence of what I termed the typical American schoolboy attitude that there is something wicked about telling on a friend. I was never certain as to just what he was telling me. I did know this: That he was doing what he thought was essential, which was to disclose to me the dangers of this particular attempt to enter the project, namely, it was concerned with the situation out there near Berkeley—I think it was the Shell Laboratory at which Eltenton was supposedly one of the key members—and that was a source of danger to the project and that was the worry. I always had the very definite impression that Dr. Oppenheimer wanted to protect his friends of long standing, possibly his brother. It was always my impression that he wanted to protect his brother, and that his brother might be involved in having been in this chain, and that his brother didn't behave quite as he should have, or if he did, he didn't even want to have the finger of suspicion pointed at his brother, because he always felt a natural loyalty to him, and had a protective attitude toward him.

I felt at the time that what Oppenheimer was trying to tell me and tell our project, once he disclosed this thing at all—as I recall I had the feeling that he didn't disclose it immediately. In other words, he didn't come around the next day or that night and say to our security people, "Listen, some things are going on." I think he thought it over for some time. I am saying what I thought now, and not what we could prove, because we could never prove anything definite on this thing, because it all depended on the testimony of a man who was concerned in it.

I felt that was wrong. If I had not felt it was important not to have any point to protect Chevalier or to protect somebody else who was a friend, whom he felt that the man had made a mistake and he had adequately taken care of that mistake and more or less warned this man off.

I felt tht was wrong. If I had not felt it was important not to have any point of issue on what after all was a minor point with respect to the success of the project, I might have had quite an issue with him right then and there. As he told me very early in my conversation with him, he said, "General, if you order me to tell you this, I will tell you." I said, "No, I am not going to order you."

About 2 months later or some time later, after much discussion in trying to lead him into it, and having then got the situation more or less adjusted, I told him if you don't tell me, I am going to have to order you to do it. Then I got what to me was the final story. I think he made a great mistake in that. I felt so at the time. I didn't think it was great from the standpoint of the project, because I felt that I was getting what I wanted to know which, after all, I did know already, that this group was a source of danger to us. I didn't know that this group had tried to make this direct approach and pinpoint it that way, but I knew they were thoroughly capable of it, and I knew we had sources of danger in the Berkeley project.

I think that really was my impression of it, that he didn't do what he should have done. The reasons why were desire to protect friends and possibly his brother, and that he felt that he had done what was necessary in pinpointing. As far as I was concerned, while I didn't like it, after all it was not my job to like everything my subordinates did, or anybody in the project did. I felt I had gotten what I needed to get out of that, and I was not going to make an issue of it, because I thought it might impair his usefulness on the project.

I think that gives you the general story.

Mr. GARRISON. I think that is all that I would like to ask.

Mr. GRAY. Mr. Robb.

CROSS-EXAMINATION

By Mr. ROBB:

Q. General, you said this group; what group did you have in mind, sir? The group at Berkeley?

A. Oh, no. The group at the Shell Oil Co. laboratories. We never knew how many people were in that group. I didn't bring it to the attention of the Shell Oil Co. at the time, because I didn't want to disclose anything. I would rather have it there where I knew it. Of course, after the war, I brought it to the attention of various friends in the Shell Oil Co., and I believe that group was cleaned out in 24 hours.

Q. General, I find in the files a letter signed by you, dated November 14, 1946. I will read it:

ARMY SERVICE FORCES,
UNITED STATES ENGINEER OFFICE, MANHATTAN DISTRICT,
WASHINGTON LIAISON OFFICE,
Washington, D. C., November 14, 1946.

Mr. DAVID E. LILIENTHAL,
Chairman, Atomic Energy Commission, Washington, D. C.

DEAR MR. LILIENTHAL: I desire to bring to your attention that in the past I have considered it in the best interests of the United States to clear cerain individuals for work on the Manhattan project despite evidence indicating considerable doubt as to their character, associations, and absolute loyalty.

Such individuals are generally persons whose particular scientific or technical knowledge was vital to the accomplishment of the Manhattan project mission. In some instances, lack of time prevented my completely investigating certain persons prior to their working for the Manhattan project, so that in some cases individuals, on whom it was subsequently determined that derogatory information existed, had access to project information.

With the appointment of the Commission and the legal provisions for investigation of personnel by the Federal Bureau of Investigation, I see no reason why those persons on whom derogatory information exists cannot be eliminated. I unhesitatingly recommend that you give the most careful consideration to this problem.

The FBI is cognizant of all individuals now employed on the Manhattan project on whom derogatory information exists.

Sincerely yours,

L. R. GROVES, *Major General, USA.*

I find an answer to that from Mr. Lilienthal, dated December 4, 1946, which I will read:

UNITED STATES ATOMIC ENERGY COMMISSION,
Washington, D. C.

Maj. Gen. LESLIE R. GROVES,
Commanding General, Manhattan Project,
Washington, D. C.

DEAR GENERAL GROVES: This will acknowledge your letter of November 14, 1946, concerning continued employment of project personnel whose character, associations, and loyalty have been questioned by the Manhattan project but who have been employed nevertheless because they were considered vital to the accomplishment of the Manhattan project mission. This matter will receive the most careful consideration by the Commission. It would appear that, since the persons referred to in your letter had been continued somewhat beyond the accomplishment of the Manhattan project mission, that you do not regard their presence a source of critical hazard. On the other hand, if in your opinion a decision in this connection is urgent, I would appreciate your further views.

Sincerely yours,

DAVID LILIENTHAL, *Chairman.*

I find, then, your response to that letter, dated December 19, 1946:

WAR DEPARTMENT,
Washington, D. C., December 19, 1946.

MR. DAVID E. LILIENTHAL,
Chairman, Atomic Energy Commission,
Washington, D. C.

DEAR MR. LILIENTHAL: Reference is made to your letter of December 4, 1946, concerning the presence of certain individuals in the Manhattan project whose character, associates, and loyalty may be open to question. They could not be discharged summarily but, as I explained, their removal is of necessity a rather slow process, and whenever possible such removals have been effected by us through administrative means when the individuals could be conveniently relieved of such assignments. Considerable progress in reducing the number of such individuals has been made to date.

It would seem to me that, with the reinvestigation of all Manhattan project personnel by the Federal Bureau of Investigation, you could find it appropriate to effect the removal of the remaining individuals of questionable character.

Sincerely yours,

L. R. GROVES, *Major General, USA.*

General, do you recall writing the two letters and getting the answer from Mr. Lilienthal?

A. I recall writing a letter. You did very well. I didn't recall the other two. I recall writing one. I think it is appropriate, if I may, to insert that these letters were only written because previous verbal discussions which were very limited had proven unavailing and because Mr. Lilienthal had made it very plain that he wanted no advice of any kind from me. He wanted nothing whatsoever to do with me. He thought that I was the lowest kind of human being, and he was not going to get anything from me. This was written because I felt that it was the only way that I could adequately bring to the attention of the Commission the seriousness of this problem. Knowing Government procedure, I knew that, as long as it was verbal, nothing would be done. If I put it in writing, that they would always be thinking about the record. That is the reason that the letter was written.

I have never made a practice of trying to protect myself on the record, but I thought this was one time that I could secure action, and it was not written really with the idea of clearing my skirts for something that might come up, such as this, many years hence. It was to make him do it whether he wanted to do it or not.

Q. General, was Dr. Oppenheimer one of the "certain individuals" to whom you referred in those letters.

A. I don't believe so, because Dr. Oppenheimer was really out of the project at the time. Of course, he was retained as a consultant, but just what my consultant arrangements with him were I am not certain. It was more of a personal affair. I would say that he was not one of those that I was thinking about. I recall who I was thinking about in particular, and he was not the man. I don't think I was thinking about him.

If I may answer that you may ask next, but which is necessary for my answer, if he had been a member of the Manhattan project at the time, he would have been one of those about whom I was thinking.

Q. General, would you have cleared Dr. Oppenheimer in 1943 if you had not believed him to be essential to the project and if you had not known that he was already steeped in the project?

A. I think that I would not have cleared him if I had not felt that he was essential and if he had not already been so thoroughly steeped in the project. If the two were separated, I don't know. I can't say, because I was never faced with that, and it is awfully hard to try to recast it.

Q. I will show you a photostat of a letter bearing your signature, dated July 20, 1943, and ask if that is the letter whereby you did give clearance to Dr. Oppenheimer.

A. It is certainly my signature, because nobody has been able to forge it yet, and they have tried many times. Nobody could ever do it. I don't remember the exact wording. I do know that a letter of this general tenor was written. There is no question but what it was my letter.

Q. I might read this into the record. It is stamped "top secret," but it has been declassified:

WAR DEPARTMENT, OFFICE OF THE CHIEF OF ENGINEERS,
Washington, July 20, 1943.

Subject: Julius Robert Oppenheimer
To: The District Engineer, United States Engineer Office, Manhattan District, Station F, New York, N. Y.

1. In accordance with my verbal directions of July 15, it is desired that clearance be issued for the employment of Julius Robert Oppenheimer without delay, irrespective of the information which you have concerning Mr. Oppenheimer. He is absolutely essential to the project.

L. R. GROVES, *Brigadier General, CE.*

General, did your security officers on the project advise against the clearance of Dr. Oppenheimer?

A. Oh, I am sure that they did. I don't recall exactly. They certainly were not in favor of his clearance. I think a truer picture is to say that they reported that they could not and would not clear him.

Q. General, you were in the Army actively for how many years?

A. I don't know. 1916 to 1948, and of course raised in it, also.

Q. And you rose to the rank of lieutenant general?

A. That is right.

Q. During your entire Army career, I assume you were dealing with matters of security?

A. Never before this thing started. We didn't deal with matters of security in the Army, really, until this time. The Army as a whole didn't deal with matters of security until after the atomic bomb burst on the world because it was the first time that the Army really knew that there was such a thing, if you want to be perfectly frank about it.

Q. Certainly with your work in the Manhattan project you dealt intensively with matters of security?

A. I would say I devoted about 5 percent of my time to security problems.

Q. You did become thoroughly familiar with security matters.

A. I think that I was very familiar with security matters.

Q. In fact, it could be said that you became something of an expert in it?

A. I am afraid that is correct.

Q. I believe you said that you became pretty familiar with the file of Dr. Oppenheimer?

A. I think I was thoroughly familiar with everything that was reported about Dr. Oppenheimer; and that included, as it did on every other matter of importance, personally reading the original evidence if there was any original evidence. In other words, I would read the reports of the interviews with people. In other words, I was not reading the conclusions of any security officer. The reason for that was that in this project there were so many things that the security officer would not know the significance of that I felt I had to do it myself. Of course, I have been criticized for doing all those things myself and not having a staff of any kind; but, after all, it did work, and I did live through it.

Q. General, in the light of your experience with security matters and in the light of your knowledge of the file pertaining to Dr. Oppenheimer, would you clear Dr. Oppenheimer today?

A. I think before answering that I would like to give my interpretation of what the Atomic Energy Act requires. I have it, but I never can find it as to just what it says. Maybe I can find it this time.

Q. Would you like me to show it?

A. I know it is very deeply concealed in the thing.

Q. Do you have the same copy?

A. I have the original act.

Q. It is on page 14, I think, where you will find it, General. You have the same pamphlet I have.

A. Thank you. That is it. The clause to which I am referring is this: It is the last of paragraph (b) (i) on page 14. It says:

"The Commission shall have determined that permitting such person to have access to restricted data will not endanger the common defense or security," and it mentions that the investigation should include the character, associations, and loyalty.

My interpretation of "endanger"—and I think it is important for me to make that if I am going to answer your question—is that it is a reasonable presumption that there might be a danger, not a remote possibility, a tortured interpretation of maybe there might be something, but that there is something that might do. Whether you say that is 5 percent or 10 percent or something of that order does not make any difference. It is not a case of proving that the man is a danger. It is a case of thinking, well, he might be a danger, and it is perfectly logical to presume that he would be, and that there is no consideration whatsoever to be given to any of his past performances or his general usefulness or, you might say, the imperative usefulness. I don't care how important the man is, if there is any possibility other than a tortured one that his associations or his loyalty or his character might endanger.

In this case I refer particularly to associations and not to the associations as they exist today but the past record of the associations. I would not clear Dr. Oppenheimer today if I were a member of the Commission on the basis of this interpretation.

If the interpretation is different, then I would have to stand on my interpretation of it.

Mr. ROBB. Thank you, General. That is all.

Mr. GRAY. I would like to ask a question, General Groves. This relates to a question Mr. Garrison asked about the urgencies, whether the urgencies had been stepped up with respect to having these weapons ready toward the end of the war.

My recollection is that you said that there was not any acceleration as far as you were concerned?

The WITNESS. No. My mission as given to me by Secretary Stimson was to produce this at the earliest possible date so as to bring the war to a conclusion. That was further emphasized by his statement that any time that a single day could be saved I should save that day. The instructions to the project were that any individual in that project who felt that the ultimate completion, insofar as he understood it, was going to be delayed by as much as a day by something that was happening, it was his duty to report it direct to me by telephone, skipping all channels of every kind. So that urgency was on us right from the start.

Mr. GRAY. And any instructions with respect to that which went to the laboratory at Los Alamos would have come then from you?

The WITNESS. That is correct. I think, for your information, while the laboratory officially was under General Nichols, because the whole district was under Nichols, by an understanding between Nichols and myself, because that left me doing nothing but telling Nichols what to do, and it was beyond his capacity to do everything, in general a division of direct responsibility was made, and Nichols took over essentially Oak Ridge and the general administration.

With respect to Los Alamos, it was directly my responsibility in every way, everything that happened. The orders were issued direct. We tried to keep

Nichols informed to such extent as was necessary. So from a practical standpoint, although not on paper, the chain of command was direct from me to Dr. Oppenheimer.

Dr. GRAY. One other question now. Do you recall any key personnel in the project who left the project because of unsatisfactory record or promise as security risks?

The WITNESS. Oh, yes. There were some that were gotten rid of. A man named Hiskey, who very unfortunately happened to be a Reserve officer and was called to active duty and thus gotten out of it.

A man named Lomanitz' deferment on the draft was taken away. He was eventually drafted, although that took the utmost pressure. His draft board refused to remove the deferment. It became a matter of issue in which General Hershey had to issue direct orders that this exemption be removed and that he be drafted. If he was not drafted, he was going to get rid of the entire State board as well as the local board, which apparently was controlled by an element that were not in accord with what you and I think they should be. The board insisted on this man's being deferred.

There were other people that we wished to get rid of that we were unable to get rid of because of the effect upon the organization as a whole. Those were men—I don't think their names need be mentioned—about whom I had suspicions. Also, I think bearing on this there was an early conversation with the Secretary of War's office at the time before I started dealing with the Secretary direct, in which I asked if it was possible to intern a particular foreign scientist, an alien, and I was asked what evidence I had, and my reply was that I had no evidence other than intuition. I just didn't trust him. I knew he was a detriment to the project. I didn't accuse him of disloyalty or treason, but simply that he was a disrupting force and the best way out of it was to intern him.

I was told that this man didn't want to take it up with the Secretary. I insisted on it. He came back and said "General, the Secretary said we can't do that. General Groves ought to know that. I told the Secretary, of course, General Groves knew that would be your answer. He just still wanted to make a try." I think that is essential to realize.

In other cases, one of them at Berkeley, where I asked Dr. Lawrence or told him that I wanted a man to be gotten rid of, he said, "If I get rid of him—don't misunderstand me, if you order it, I always accept your orders—I want to warn you that if he is gotten rid of, there will be no work done in this laboratory for at least a month, no matter what I try to do myself, and the effect may last for 6 months or a year because of the attitude of the scientific world which did not appreciate the need of security."

I think that attitude was prevalent in the country as a whole. It was very touchy, and you could not run this thing and say a man is either black or white. If he is black or has any tinge of it, out he goes, and there is no question about it.

Mr. GRAY. Does the name "Weinberg" mean anything to you?

The WITNESS. Oh, yes.

Mr. GRAY. Would you mind——

The WITNESS. Weinberg was one of—I think some of the people over there could maybe amplify it a little—he was as I recall one of four young scientists at Berkeley. The other names, if they are mentioned, I think I could remember them.

Mr. ROBB. Might I mention them to assist: Weinberg, Bohm, Lomanitz, and Friedman.

The WITNESS. That sounds very familiar, and I think that is approximately right. Essentially they were a group about whom there was a great deal of question. I never had any confidence in them at all from the time that we started to get reports. They were not essential to the project. They were young men, and they could be replaced. But remember at that time there were not very many men and even a young man it was difficult to replace. But even so, we could get along without them.

Mr. GRAY. You did indeed in some cases.

The WITNESS. Oh, yes.

Mr. GRAY. The project was successful, and some of these men left the project?

The WITNESS. Yes; we got rid of them. But each one it was a terrible task to get rid of because it was not a case of my deciding he should go. First, the suspicion of the man, then a development enough to convince me, and then manipulation and just how were we going to do this thing. It was just as difficult as to get rid of a Cabinet officer in Washington that the country is behind, because you had all of the political play in there. Men who would become violently excited about the most minor thing. If I went on to the laboratory or on to a plant and failed to speak to somebody who was there or didn't see him—even at Oak Ridge I even had to go back at the expense of about three hours one day to speak to a superintendent that I had failed to see when I went through the plant and when he spoke to me, I had not answered him. When Nichols told me about it, I said "What is the damage?" He said, "You just got to go back." So it took about 3 hours with our location down there, and I went back. That was true. Everybody with the exception of a few of us, like Nichols and myself, whose physical resistance maybe was better, everyone was worked to the point where they were tense and nervous and they had to be soothed all the time.

I say that so you get the picture of why certain people were not removed. You say why didn't you remove them? Sure I wanted to remove them, but it was not wise. I think it is also important to state—I think it is well known—that there was never from about 2 weeks from the time I took charge of this project any illusion on my part but that Russia was our enemy and that the project was conducted on that basis. I didn't go along with the attitude of the country as a whole that Russia was a gallant ally. I always had suspicions and the project was conducted on that basis. Of course, that was so reported to the President.

Mr. GRAY. One other question about individuals. You said that Dr. Condon had been unsatisifactory in every respect. Does that include security? Did you have anything in mind on security in that regard, or loyalty?

The WITNESS. I would say not in giving any information, but in setting up. He set up the rules at Los Alamos—at least I always felt he was the man responsible for the rules—that tended to break down compartmentalization. He was the man who was primarily responsible for Los Alamos for the friction which existed. There would have been friction anyway. But the intensity of the friction that existed between the military officers who were trying to do the administrative operations out there so as to enable the scientists to work at science, Condon was the one who built all of that up.

The fact that he left there as he did and left this mess behind him, he left because of the reasons that he did leave. The fact that he—of course later when he worked at Berkeley—he didn't do what I term an honest day's work, I might add for your clarification that the work he was engaged on at Berkeley was something that required a man of his capabilities. Dr. Condon was a first-rate physicist. Don't misunderstand me. Lawrence and myself did not feel that this particular phase of the work was at all-interesting to us. We thought it was just no hope at all. But we also felt that we could not allow this field to go unexplored just because of a curbstone opinion which is really what Lawrence and mine were because we didn't know anything about it—I don't remember what it was now—it involved mathematics to see if this was feasible.

We had Condon working on that with a small group of juniors. By doing that we definitely proved that we were right in saying that we should neglect it. He was kept on there at Berkeley on a sort of parttime basis, traveling back and forth. He was very unsatisfactory there. In other words, he just didn't do an honest day's work in our opinion.

He would also be going to Pittsburgh for his own family convenience. He would be leaving Pittsburgh because he wanted to get out to Berkeley for personal reasons. Then of course the situation came up with his attempts to go to Russia just before the bomb exploded to that scientific conference where a member of our State Department kept the Army from knowing about these invitations. I found out about it because our scientists told me that they had received invitations. So we checked our project to see that none of our people would go, and then at the last minute when the plane was about to leave, we suddenly discovered that some industrial scientists, namely Condon and Langmuir of General Electric were going, and I then raised the question as to whether they should go with their top company officials.

After discussion with GE, I withdrew any objection to Dr. Langmuir going. Of course, Dr. Langmuir has since represented that, but that is all right. I did not withdraw the objection to Condon going. I had the fullest support from the corporations concerned. Condon's passport was withdrawn and he made a terrific battle to go. That battle was so unrealistic and so completely lacking in appreciation of what was the best interest of the United States that you couldn't help but feel that either he was such an utter fool that he could not be trusted, or else that he put his own personal desires above those of the welfare of the country and therefore he was in effect disloyal, even if it was not a case of deliberately going out to aid the enemy.

Mr. GRAY. One other question about Dr. Condon.

When he left Los Alamos and assumed this other relationship at Berkeley, did he have any responsibility for personnel at either place?

The WITNESS. He didn't leave directly for Berkeley. He was relieved from the project, and went back to the Westinghouse Co. It was later that he was picked up to go to Berkeley because we wanted to take a man that would not hurt the project in any way. As to his responsibilities for personnel at Los Alamos, that was one of his big responsibilities, to assist in recruiting personnel. The idea was that Dr. Condon, in my concept, and I believe Dr. Oppenheimer carried out that concept completely insofar as he felt that it was possible to carry it out because we both found out pretty soon that Condon was not competent—Oppenheimer was to think the scientific problems and to establish the schedule of scientific and technical work. Condon was to run everything connected with the procurement of personnel, the operation of the personnel, their relations with the military, and all that. The military was to run the housekeeping. As I say, Condon failed in that. Oppenheimer started to move into the personnel thing. Of course, Oppenheimer still had at the beginning to get the senior personnel, but building up and getting all the arrangements was supposed to be Condon's responsibility.

Mr. GRAY. This is while he was identified with the project.

The WITNESS. Yes.

Mr. GRAY. When he left, he had no responsibility?

The WITNESS. That is right. He had no responsibility. He left with, I would say—both Dr. Oppenheimer and myself—we had the utmost distaste for Dr. Condon. There was the utmost cooperation in getting this thing on a plane where you might say we had Dr. Condon on the record in a way that he has never liked to have it disclosed since, that he had not done a good job out there.

Mr. GRAY. My next question involves a considerable change of pace, General.

The WITNESS. That is all right, sir.

Mr. GRAY. Do you think that the Russian effort to develop this kind of weapon has in any way, as you look back on history, been accelerated by any information they may have gotten one way or another from our own people?

The WITNESS. Oh, yes. There is no question. If I can go into that a little bit, first they got information as to our interest essentially through espionage at Berkeley. These are all conclusions. You can't prove them, of course.

Mr. GRAY. I understand.

The WITNESS. They got the thought that we were interested there. They certainly had gotten before he ever came to the country—they must have gotten information from Fuchs that Britain was interested in this affair and that we were, too, because up until the time I came into control, there was a complete interchange of scientific information between Britain and America on this. If the British didn't know everything we were doing, it is because they were stupid, and they were not on the job. I don't think they did, but they knew most of it.

The next disclosure outside of that particular thing was that whatever Fuchs passed during the war, and I don't think he passed too much until near the end, they undoubtedly knew certain things—they had good espionage—and they knew a lot of things that were going on.

For example, when we had trouble at Hanford and our piles suddenly quit—I think that is generally known, again that is not secret, but I wouldn't like to have it repeated—we had trouble with our piles. The trouble existed because this was a sudden disclosure of a scientific effect that nobody had anticipated. The reason we had not anticipated that was because we had never operated our pile at Chicago, our preliminary work there, continuously. We had not operated continuously because my orders to the Chicago laboratory were directly and deliberately disobeyed. I had said that they will be operated continuously. We don't know what will happen. Let us find out. Of course I didn't anticipate this scientific problem, but after all, any engineer knows you ought to operate something continuously.

The power worked so well at Chicago that they operated it only during nice convenient hours. So we never got this effect that was so disastrous at Hanford. My officer in charge at Chicago failed because he didn't report that they were not carrying out my orders, which he should have done if he could not get them to comply.

When this thing happened at Hanford, it was known by people that had no right to know it within—I can't recall the exact time now—I think it was 48 hours. It was known in New York by somebody who was not in the project. To get to New York, I had to trace out this thing. I think it went from Hanford to Chicago, which was legitimate. * * * We found out that this man had an inkling that something had happened, and that was enough to show the extent of this kind of espionage.

There was a great deal of loose talk about it by scientific people, as I say, breaking down my compartmentalization rules.

Of course, I always knew that if you have this many people on a project, that somebody is going to be faithless and somebody is going to betray you, and that is why we had compartmentalization.

Then after the war when the May case broke in Canada, that of course was pure luck, what May had done. Apparently May gave to the Russians a sample of U–233 and a sample of something else. I think it was plutonium. I don't recall now. But the U–233 was all-important because that indicated to the Russians that we were interested in thorium, which could only be produced that way. The result of that was most unfortunate.

Then the next thing that happened was—I didn't know this until later—apparently there was a diary kept up there with certain names in it. I have never been able to get the truth of that, because people who were involved have clammed up. They were not people who were friendly to me in the main, anyway. They were not people who would disclose matters to me. But I believe there was a diary. I believe Fuchs' name was in that diary, a list of acquaintances or addresses, that was in the hands of somebody in that Canadian ring. I have always thought it was Fuchs. It has been told it was somebody else. Fuchs' name was in that. That list was supposedly disclosed to people in the United States, not in the project, but outside of the project, and the list was never shown to me, the one man who should have had it shown to him by all means.

There were attempts on the part of our Government to keep me from knowing about this Canadian affair. * * *

As I say, it was repeated and they knew what the story was, and yet they brought Fuchs over. Unfortunately Fuchs was in the delegation of British who came and discussed with us the gaseous diffusion process which was the one process we had that we really took our hair down and told them all about because the feeling was that they had initiated that process and they could be helpful.

There was also a very strong element, I would say 98 to 99 percent of the scientific personnel on the project, who considered the gas diffusion process a mistake, including the people who were actually responsible for the development. Dr. Urey, who was the head, violently opposed it. He said it couldn't possibly work. So it was not unreasonable to let the British look at it.

Of course, as you know and is well known, I was not responsible for our close cooperation with the British. I did everything to hold back on it. I would say perfectly frankly I did the things that I have sort of maybe by implication blamed on my scientists for doing. I did not carry out the wishes of our Government with respect to cooperation with the British because I was leaning over backwards.

That information that Fuchs gave was all important. The mistake that was made at Los Alamos in breaking down compartmentalization was vital to Fuchs, because Fuchs later went to Los Alamos, it was vital to Fuchs, and the information he passed to the Russians.

But in doing that, I think it is important to realize this with respect to Fuchs. If we had limited it to a small group, say just the top people, Fuchs might still have been in that group. Fuchs would also have worked on the hydrogen bomb as one of the subordinates, and would have passed that information.

With the British not being completely under my control, I think it would have been passed on by the British group to Fuchs, whether we had the compartmentalization strictly observed there or not. But irrespective of that, I feel that was one of the disadvantages of the breakdown of compartmentalization * * *.

On the situation as a whole, our reliance, when we first talked after the war about what the time limits were on the Russians and it is quite possible I talked to you about it when you were Secretary of the Army—I don't recall, I certainly made no bones about it—our reliance on what the Russians could or could not do was based on primarily the supplies of material which I felt would be available to them, that is raw material, and on the basis that there would be no general relaxation of security rules beyond the Smyth report, and the declassification study which said what could be released.

In that the criterion—and that criterion was established by a committee of eminent scientists, but like all committees, it was under pretty rigid control by me because I had the chairman, Dr. Tolman, who was in complete sympathy with me as far as I know, I had the secretary, who was an officer and a distinguished chemist handling that end—and they were told in advance what should be the criterion and they got the board to agree to that criterion. Nothing was recommended for declassification where it was felt that would be of any assistance to the Russians in developing the bomb.

Later, that has been stretched and stretched, and there has been a tremendous amount of data published. As you know I fought the battle. I did not win. The American people and the Congress and everybody else was opposed to me. It has always been said, get the information out, and there has been a great laxness there.

I think the primary reason was that the Russians got into these materials in Saxony. We didn't know about the material in Saxony. Of course, we knew about the material in Goachimstal. We were not worried about that. We never conceived that the whole area would be turned over to the Russians as they pleased, and to be able to mine on the basis they were, I don't know whether it was paid for by the American labor. The raw ore would very well have cost us $100 a pound at that time to get out the uranium * * *

Mr. GRAY. Just one other question which relates to my general question whether information actually went to the Russians. Of course, it has been a matter of considerable discussion in some quarters that one of the scientists by the name of Weinberg passed information to a Communist Party functionary.

The WITNESS. There is no doubt about that in my mind.

Mr. GRAY. I am sorry. My question I am sure has not been expressed. But there may have been other instances of that sort that you know nothing about.

The WITNESS. Of course, and I think there were. On Weinberg, I would like to emphasize that the information he passed was probably with respect to the electromagnetic process, and with respect to the fact that we were engaged in a big effort because that is all that he knew legitimately. He may have known some things illegitimately, and I am sure he did. We were never too much concerned about that, because I personally felt that the electromagnetic process was a process, while it was of extreme importance to us during the war, and we saved at least a year's time by doing it, that it was not the process we would follow after the war. That is one reason why we put silver in those magnets, because we knew we would get it out.

Dr. EVANS. General Groves, I would like to ask one question that is not very important, and maybe you can't answer it. There are some things that appear in magazines that is almost classified information. That article in Life, do you remember seeing that?

The WITNESS. No, I didn't read that.

Dr. EVANS. I think it was Life. It contained a lot of material that I did not think was unclassified. Did any of you people read that article?

The WITNESS. I have not read that, but I can tell you that I am constantly being shocked by what I see. With respect to that, to clarify a little my previous answer to Mr. Gray, because I am reminded of this by your question, during the war there were two things that came out that annoyed me tremendously. The last one was kind of funny but it still annoyed me. I thought that is an awfully cheap thing to do.

As you know, we had the utmost cooperation from the press. That is very definite. Our relationships were generally good. But on one occasion a newspaper wanted to print news about Hanford and what a tremendous development was out there. They had their reporter out and they had their story written and it was a bangup story. We found out about it, and they were told no, they could not permit it. Of course, that was handled through press censorship. We didn't deal directly with them. They said there are thousands of people that know it, and they would not agree with our philosophy which was that thousands of people could know it but that is no sign the Russians did, or the enemy—we could not talk about the Russians too much then. So that they agreed not to publish it.

About a month afterward a Congressman from Oregon, I think his name was Angell, suddenly made a speech on the floor of the House appealing for more appropriations for the Interior Department for, I think, installation of electric generators in Grand Coulee, or something of that kind, and among other things he said that there was this tremendous plant with great electrical demand at Hanford, Wash.

The paper came out with this. It was a little squib on the interior page. It said the Congressional Record contained the following today and it just quoted that absolutely. As I say, I thought it was awfully poor. I knew it had not been top management. I think it was somebody who got smart. But there was one very serious break that disclosed during the war—to me, if I had been a Russian I think if the intelligence of Kapitsa and the background or the intelligence of anyone else who was working on this project—it would have indicated that the way to produce an atomic bomb was in some way to take care that it might be based on implosion. I don't know if anyone else in the room saw tuat article. I think I probably discussed it with Dr. Oppenheimer at the time.

Dr. EVANS. I saw it.

The WITNESS. It was a terrible article. There just was not anything we could do. I was just as certain as I could be that somebody was just trying to get that information out. I don't know who was responsible. We, of course, did almost nothing about it, because that is the kind of thing you don't do anything about. We prevented in this country the republication of articles appearing abroad, particularly in Scandanavian papers, that disclosed ideas. We made no mention, for example, in the press dispatches when the heavy water plant was finally destroyed in Norway. They might be described in detail in the Scandanavian press. We objected and were successful in having them not reprinted on the ground that would indicate to the Russians some interest.

I don't know how successful we were in keeping the Russians from realizing what a tremendous effort this was, and how hopeful we were, and what the effects would be, but judging from the Russian attitude, I would say that they did not appreciate the strength of this weapon until it dropped on Hiroshima, and they were told of the effects. They still did not appreciate it until after Binkini, because the attitude of the Russian delegation at the United Nations, which of course was very responsive to Moscow as you know, changed completely, not immediately after the explosion, but within about 24 hours of the time that the ships returned to San Francisco, and the Russian observers who were there against my wishes—as you know, I did not control Bikini—got ashore and went to the Russian consulate. Within 24 hours to 48 hours, the whole attitude of the Russian delegation at the United Nations changed, and this became a very serious matter, instead of just being something, "Oh, well, it doesn't amount to much." That would indicate to me that they had not been convinced by their espionage of just how important this all was.

Mr. GRAY. Mr. Garrison.

REDIRECT EXAMINATION

By Mr. GARRISON:

Q. General, Dr. Oppenheimer had no responsibility for the selection or the clearance of Fuchs, did he?

A. No, not at all. He had no responsibility whatsoever, as far as I can remember. He had no responsibility for it, and I don't recall his ever having asked me to get an Englishman at the laboratory in any way, nor did he suggest their need. He acquiesced when I said I thought we should get them there in view of things, and because we desperately needed certain assistance that those men can give. They were a scientific reservoir. There was not any use in trying to keep them out, as I saw the picture. In other words, I tried to be reasonable about it. I didn't try to oppose the administration when I knew I was going to get licked. After all, I had been in Washington for many, many years.

Q. All this talk about espionage, you didn't mean to suggest by anything that you said with respect to it that Dr. Oppenheimer had anything whatever to do with espionage activities with foreign agents?

A. Oh, by no means. Dr. Oppenheimer was responsible as the director of the laboratory for assisting in every possible way our security and defense against espionage at Los Alamos. If you look down the chart, he might be responsible to a certain degree for operation of the security officer. It was more in the way of assisting that officer and of advising me or this officer's superiors if he thought the officer was not doing a good job. But the officer from a practical standpoint did not report to Oppenheimer excepting as a matter of courtesy.

Q. So you would not want to leave with this board even by the remotest suggestion that you are here questioning Dr. Oppenheimer's basic loyalty to the United States in the operation of the Los Alamos plant.

A. By no means and nothing about the espionage. I think it is very important if there has been any misunderstanding that Dr. Oppenheimer was not in any way responsible for anything to do with the protection of the United States against espionage, excepting cooperation which was natural as the head of the scientific effort out there. By no means was there any intent to imply. I hope I did not lead anybody to think otherwise for an instant.

Q. After Dr. Oppenheimer resigned as the director of the project, did he remain as a consultant for the Manhattan District?

A. Apparently he did. I didn't realize that until somebody asked me about it, or something was said here earlier. I think he did. I don't think he was on the payroll in any way. But certainly I would not have hesitated to ask him any questions or to discuss anything that was of a secret nature during that period I remained in control. For one thing, there was nothing that came up with which he was not already thoroughly familiar. There was no possibility of anything in that. So the question never arose. I think also as I recall he was a member of this declassification board although I am not certain of that. That would be in the record and of course he would know. That was the one chairmanned by Dr. Tolman.

Q. You have given us your interpretation of the requirements of the Atomic Energy Act, General Groves. Leaving the act to one side or supposing that it provided that the test of the employment of a man in Dr. Oppenheimer's position should be what is in the public interest, would you say that the revocation of his employment would be in the public interest if that is the way the act read?

A. The revocation under such extreme publicity as has occurred I think would be most unfortunate, not because of the effect on Dr. Oppenheimer—that I leave to one side—but because of what might be a very disastrous effect upon the attitude of the academic scientists of this country toward doing Government research of any kind, and particularly when there was not any war on. I think you can refer back to history as to the attitude of the average academic man in 1945 when the war was over. They were exactly like the average private in the Army who said to himself, the war is over, how soon can I get back home to mom and get out of this uniform. That was the way the average academic scientists felt. He wanted out. He wanted to be where he could resume his old academic life, and where he could talk and not have to be under pressure of any kind.

What happened is what I expected, that after they had this extreme freedom for about 6 months, they all started to get itchy feet, and as you know almost every one of them has come back into Government research, because it was just too exciting, and I think still is exciting. Does that answer your question?

Q. Yes. I have, General, a copy of a letter which I am sure you recall from yourself to Dr. Oppenheimer, dated May 18, 1950. I would like to read it, if I may, into the record. I am sure you have no objection to that.

A. No. Anything I wrote I have no objection to whatever.

Q. This is on the letterhead of Remington Rand, Inc., Laboratory of Advanced Research, South Norwalk, Conn. May 18, 1950.

"Dr. J. ROBERT OPPENHEIMER,
  "*The Institute for Advanced Study, Princeton, N. J.*

"DEAR Dr. OPPENHEIMER: If at any time you should feel that it were wise, I would be pleased to have you make a statement of the general tenor of that which follows:

" 'General Groves has informed me that shortly after he took over the responsibility for the development of the atomic bomb, he reviewed personally the entire file and all known information concerning me and immediately ordered that I be cleared for all atomic information in order that I might participate in the development of the atomic bomb. General Groves has also informed me that he personally went over all information concerning me which came to light during the course of operations of the atomic project and that at no time did he regret his decision.'

"I don't believe that you will find any need to make use of any such statement, but you might. You might wish to show it to some individual for his use in handling unpleasant situations, if any arise.

"I have been very much pleased with the comments that have been made by various persons in whose judgment I have more than average faith, such as the reported statement of Representative Nixon that he had 'complete confidence in Dr. Oppenheimer's loyalty.' This was made in a speech at Oakdale, Calif.

"I am sure of one thing, and that is, that this type of attack, while it is unpleasant, does not in the end do real damage to one's reputation.

"I wonder if you saw the editorial in the Washington Post to the effect that the way to cripple the United States atomic energy program would be to single out a few of the foremost nuclear physicists and dispose of them by character assassination. When I remember how the Post has written about me, it makes me wonder just who wrote this particular editorial.

"I do hope that you are finding life enjoyable and not too hectic and that I will have the pleasure of seeing you again before too long.

"My very best to Mrs. Oppenheimer.

"Sincerely yours" signed "L. R. Groves, Lt. General U. S. Army (Retired)."
General, if Dr. Oppenheimer had had occasion to make this statement public, needless to say it would have been the quoted portion as set forth in your letter. But I think it appropriate in this executive session to put the whole letter in the record and ask you if the expressions of confidence in him contained in this letter you wrote hold?

A. I think the letter is something that was absolutely what I thought at the time that I wrote it. I think if you interpret it in that light and know what has happened since, that you can draw your own conclusions as to what I feel today.

Mr. GARRISON. That is all.
Mr. ROBB. May I ask another question?
Mr. GRAY. Yes.

RE-CROSS-EXAMINATION

By Mr. ROBB:

Q. General Groves, I show you the memorandum which you wrote to the Secretary of War under date of March 24, 1947, and ask you if you recall writing that?

A. No; I don't recall. Oh, yes, surely I recall writing this. I know I wrote it because again my signature is there, and nobody ever successfully forged it.

Mr. ROBB. I think it might be well, Mr. Chairman, so the record would be complete, if I read this in the record, too.

"Memorandum to the Secretary of War.
"Subject: Loyalty clearance of Dr. J. R. Oppenheimer.
"WAR DEPARTMENT,
"*Washington, March 24, 1947.*

"In accordance with our telephonic conversation, I express below my views relative to the loyalty of Dr. J. R. Oppenheimer.

"When I was first placed in charge of the Atomic Bomb development in September 1942, I found a number of persons working on the project who had not received proper security clearances. One of these was Dr. Oppenheimer who had been studying certain of the theoretical problems concerning the explosive force of the bomb. The security organization, then not under my control, did not wish to clear Dr. Oppenheimer because of certain of his associations, particularly those of the past. After consideration of the availability and caliber of suitable scientists, I decided that it would be in the best interests of the United States to use Dr. Oppenheimer's services. Prior to this, I reviewed Dr. Oppenheimer's complete record personally. It was apparent to me that he would not be cleared by any agency whose sole responsibility was military security. Nevertheless, my careful study made me feel that, in spite of that record, he was fundamentally a loyal American citizen and that, in view of his potential overall value to the project, he should be employed. I ordered accordingly that he be cleared for the Manhattan project. Since then, I have learned many things amplifying that record but nothing which, if known to me at that time, would have changed my decision.

"In connection with the above statement, it must be remembered that the provisions of the Atomic Energy Act of 1946 did not control my actions prior to the enactment of that law. My decisions in respect to clearances of personnel were based on what I believed to be the best overall interests of the United States under the then existing circumstances. As I have long since informed the Atomic Energy Commission, I do not consider that all persons cleared for employment by the Manhattan District, while under my command, should be automatically cleared by the Atomic Energy Commission, but that that Commission should exercise its own independent judgment based on present circumstances."

Signed "L. R. Groves, *Major General, USA.*"

The WITNESS. Might I ask the date?

By Mr. ROBB:

Q. March 24, 1947. I thought I read that.
A. Oh, you did.
Q. Do you care to comment on that?
A. Yes; I would like to comment on that.
Q. Yes, sir.
A. It is my recollection, and particularly reinforced by those letters that you read previously and something that appeared in some paper which I know was true, that it was about this time that the Atomic Energy Commission reviewed

this question of Dr. Oppenheimer's usefulness on the project. They apparently, I think at that time that they actually reviewed it—and the paper stated it was March 8 that Lilienthal got a telephone call or that it was taken up by the Commission in response to a letter or something of information from J. Edgar Hoover—I believe I was in Florida at that time, because I had gone down there about that time to try to get away from Washington, and particularly to get away so that I would not be in Washington during the confirmation fight on the Hill on Lilienthal and the other Commissioners. The War Department insisted on my coming back. They thought, I think, 10 days was enough leave for me. They exerted all kinds of pressure on the Surgeon General, and I was finally sort of forced to come back much sooner than I wanted to come back. It was not health; it was just a case I wanted to be out of Washington during that time. I thought it was wise from the standpoint of everybody, including the national interest.

The Commission apparently cleared Dr. Oppenheimer on the basis of a letter, two letters—either 1 or 2—by Bush and Conant, who said more or less to the effect, as I recall, that Oppenheimer should be cleared because during the war I had used him on the Manhattan project, and everybody knew how insistent I was on security, and therefore he should be cleared. But the Commission never asked me.

As I say, Mr. Lilienthal never asked me anything anyway. So apparently after I got back, somebody woke up and they finally asked Paterson, and Mr. Paterson asked me to give him my views. That is why that letter was written..

I would like also to add for your information that all these letters that have been written, and in fact, almost every letter that was ever signed by me during the whole project, was personally written by me. It was not a staff prepared letter. I didn't have a staff in the first place, and I didn't write any letters that weren't important. All the letters you have heard today were undoubtedly written by me originally without a draft from anyone—possibly with some advice from people as to what do you think of this draft or something like that—but they were my letters in their entirety in every way.

Mr. ROBB. Thank you, General.
Mr. GRAY. Thank you very much, General. We are glad to have had you as a witness.
The WITNESS. Thank you very much for letting me come in.
Mr. GRAY. We will take a recess now, gentlemen.
(Brief recess.)
Mr. GRAY. Mr. Robb, are you ready?
Mr. ROBB. Yes, sir.
Whereupon, J. Robert Oppenheimer, a witness having been previously duly sworn, was recalled to the stand and testified further as follows:

CROSS-EXAMINATION—Resumed

By Mr. ROBB:

Q. Dr. Oppenheimer, yesterday we discussed for a little bit David Joseph Bohm. Do you recall that?
A. I recall most of it, I think.
Q. You testified that in accord with your letter of answer to General Nichols that you asked for the transfer of Bohm to Los Alamos. Do you recall that?
A. Surely.
Q. What did you know about David Joseph Bohm's academic background? In other words, his record as a scholar?
A. He was a good student, a very good student.
Q. Where had he been a student?
A. At Berkeley.
Q. Do you recall that his grades were not very good at Berkeley?
A. No. I think the grades he got from me were probably good. He has made a very great name for himself as a scientist.
Q. You testified, as I recall, that you had seen Bohm and Lomanitz at Princeton before they appeared and testified before the House committee.
A. This was pure accident. I was walking from the barber.
Q. Thereafter you read the transcript of their testimony.
A. Yes. I don't recall how carefully I read it, but I read it.
Q. It was a matter of interest to you, though, was it not?
A. Naturally.
Q. Did you notice that both Bohm and Lomanitz declined to answer upon the ground of possible self-incrimination when asked whether or not they knew Steve Nelson?

A. I recognize that.
Q. Did that make any particular impression upon you?
A. I concluded that they did know him.
Q. You also concluded, did you not, that the fact that they knew him might cause them to be incriminated in some criminal proceeding?
A. Right.
Q. It was not an unreasonable conclusion on your part, was it, that the criminal matter might be espionage?
A. I had been told in that interview in the spring of 1946 with the FBI that the investigation concerned their joining the Communist Party.
Q. But didn't you conclude when you read their testimony refusing to admit or answer whether or not they knew Nelson that they might have been involved in espionage with Nelson?
A. I didn't conclude that they were. I didn't conclude anything, sir.
Q. Didn't you conclude that they might have been?
A. I didn't draw any conclusion.
Q. What did you think they might have been incriminated in by their answers?
A. Membership in the Communist Party?
Q. Is that all?
A. That is all I knew about.
Q. Did you see Bohm after he testified?
A. I am sure I did.
Q. Did you talk with him about his testimony?
A. No.
Q. You did not cross him off your list of friends after he testified, did you?
A. We were in Princeton not really friends. We were acquaintances. I didn't cut him. I didn't run away from him. I don't believe there was any real problem.
Q. Was there any change whatever in your relationship with and your attitude toward Bohm after he testified?
A. I was worried about his testimony. I didn't like it.
Q. Was there any change in your relationship with Bohm or your attitude toward him?
A. My attitude I have just described.
Q. Was there any change in your relationship?
A. I find it hard to answer that question because the relationship was not a very substantial one.
Q. You said you were worried about his testimony. What do you mean by that?
A. I don't like it when people that I know have to plead the fifth amendment.
Q. But you testified yesterday that you would, had he asked you, given him a letter of recommendation after that.
A. A letter of recommendation as a competent physicist.
Dr. EVANS. Bohm is publishing scientific articles now, is he?
The WITNESS. He is.
Dr. EVANS. What university is he at?
The WITNESS. University of Technical Institute or something at Sao Paulo, Brazil.

By Mr. ROBB:
Q. Did you know a man by the name of Mario Schoenberg?
A. I think that is right. I was there last summer, and I didn't see Schoenberg.
Q. Do you know him?
A. No.
Q. Do you know anything about him?
A. He is reputed to be an active Communist.
Q. You have been told he was?
A. Yes.
Q. Did you and certain other persons sign a letter in his behalf in 1952, I believe it was?
A. Schoenberg?
Q. Yes, sir.
A. I don't remember it. I was told he was a Communist last summer when I was in Brazil.
Mr. GRAY. I would like to ask if this was referred to in General Nichols' letter; do you recall?

Mr. ROBB. Not specifically; no; but it was covered in general terms. May we pass on to something else while we try to find it?

The WITNESS. Let me stipulate. I learned of Schoenberg as a rather great scandal among the physicists in Brazil last summer. I don't know what the incident involving him was or what the problem involving him was, but obviously, if there is a petition or letter of record, I don't want to put you to the trouble of digging it up.

Mr. GARRISON. Yes.

The WITNESS. You want to see it?

Mr. GRAY. I just want to call attention to the fact that this letter was not specifically referred to.

Mr. GARRISON. This is totally new to us. We have never heard of the man as far as counsel is concerned.

Mr. GRAY. I am calling your attention to the fact that it is probably something new.

Mr. ROBB. We do not have it here. I will come back to it.

Mr. GRAY. Will you return to this?

Mr. ROBB. Yes, sir.

The WITNESS. I should not stipulate anything.

Mr. SILVERMAN. No; not as to a letter you couldn't remember.

The WITNESS. I don't remember.

Mr. GARRISON. Mr. Chairman, I would like to request at this point that, subject to check by you with counsel, that this whole matter of Dr. Oppenheimer's relations, if any, with this man Schoenberg be not considered a part of the record until the item has been checked.

Mr. GRAY. This portion of the record beginning with the first question about Schoenberg at this point will be stricken until you are prepared to read the letter.

Mr. ROBB. I have it here now, sir.

Mr. GRAY. Was that your suggestion?

Mr. GARRISON. No; I would like to make sure it does have some relation to Bohm or Lomanitz or some one of the people mentioned here. Otherwise it is completely new, and I think we should have a little notice of it, if we may. That is what I meant by a check.

Mr. GRAY. I think it would be well for counsel to read the letter and see whether you wish to make any suggestions.

Mr. ROBB. I will show this photostat to the doctor and ask him if he did in fact sign this letter.

I am sorry about the date, Doctor; it was in 1948.

Mr. GARRISON. Would you show it to us?

Mr. ROBB. Yes, indeed.

Mr. MARKS. Why don't you let us take a look at it first, Mr. Robb?

The WITNESS. I will identify my signature and the company, but I will also shut up.

Mr. GARRISON. Mr. Chairman, I think, strictly speaking, it is not within the purview of the letter, but we have no objection at all to its being read.

Mr. ROBB. Very well.

By Mr. ROBB:

Q. Doctor, I will read you this letter, or rather a photostat of it. At the top it bears the typewritten legend: "Dispatch No. 743, June 1, 1948. To Department—E. P. Keeler/elig." Below that, in printing, "Palmer Physical Laboratory, Princeton University, Princeton, N. J., May 20, 1948." Stamped "American Embassy, June 1, 1948."

"The Honorable HERSCHEL V. JOHNSON,
   "American Ambassador, Rio de Janeiro, Brazil.

"MY DEAR MR. AMBASSADOR: Prof. Mario Schoenberg, who was a guest in our laboratories at Princeton for several months a number of years ago, we have heard, to the dismay of all of us, has been imprisoned at Sao Paulo since March 30 without any formal accusation or any legal process. Can you do something to have his case reviewed? Schoenberg has made significant contributions to mechanics, classical and quantum electrodynamics, astrophysics, and cosmic-ray physics. He is the leader of the school of theoretical physics at Sao Paulo. His imprisonment has stopped not only the work of one of the leading Brazilian scientists but also his training of new Brazilian scientists, which is possibly even more serious. We have been told that Schoenberg is a Communist. It would appear most unfortunate if the apparently illegal imprisonment of Schoenberg could be used by Communists and fellow travelers to make him into a martyr for civil liberties. Both on this account and for the sake of science we hope you can do something either to get him freed directly or to have him brought to a fair trial

"Respectfully yours,

"P. A. M. DIRAE,
"*Professor of Mathematical Physics, Institute of Advanced Study.*
"S. BEFSCHETZ,
"*Chairman, Mathematics Department, Princeton University.*
"J. R. OPPENHEIMER,
"*Director, Institute for Advanced Study.*
"JOHN A. WHEELER,
"*Professor of Physics, Princeton University.*
"EUGENE P. WIGNER,
"*Professor of Mathematical Physics, Princeton University.*"

Did you sign that letter, sir?
A. My signature is authentic.
Q. Had you known Schoenberg before this?
A. It is my impression that I had not. I don't have an image of what he looks like. I was not in Princeton some years prior to that letter.
Mr. GARRISON. Mr. Chairman, if Mr. Robb is going to pursue a line of questioning about this which is, so far as we are concerned, new matter—we make no technical objection to its being introduced—I think it would be fair if we might have a 5-minute recess to discuss with Dr. Oppenheimer what he knows about this man.
Mr. ROBB. Why don't I defer this matter until after the luncheon recess?
Mr. GARRISON. All right.

By Mr. ROBB:

Q. Doctor, do you have before you your letter of answer to General Nichols again?
A. I do.
Q. Will you turn to page 7, the middle paragraph, where you state, "I contributed to various organizations for Spanish relief"—can you tell us what they were?
A. I mentioned the North American Committee yesterday afternoon. That is the one whose name sticks in my mind, but there were others.
Q. Do you recall any others?
A. I have forgotten the name of the other or rival organization. There was something about medical aid, an organization devoted to that.
Q. I believe you said your contributions were mostly in cash?
A. I think so. I am not very clear about it.
Q. You told us something of Dr. Addis yesterday and also Rudy Lambert, who is mentioned in the next paragraph. Addis was either a Communist or very close to a Communist.
A. Yes.
Q. Lambert was a Communist to your knowledge?
A. Right.
Q. You told us that Addis died, I think, in 1950; is that right?
A. I am not sure of that date.
Q. Approximately, then.
A. Approximately.
Q. You say here, "Addis asked me perhaps in the winter of 1937–38 to contribute through him to the Spanish cause."
Do you recall the circumstances under which he made that request to you?
A. He invited me to come to his laboratory to talk to me about it.
Q. And you went?
A. I went.
Q. Did you talk to him privately?
A. Yes.
Q. What did he say to you?
A. He said, "You are giving all this money through these relief organizations. If you want to do good, let it go through Communist channels, through Communist Party channels, and it will really help."
Q. Is that all he said?
A. That is the substance of it.
Q. Was there anything said about the amount of your contributions?
A. He said what I could.
Q. Did you tell him what you thought you could?

A. I don't think I made up my mind at that time.
Q. Did there come a time when you did?
A. No; except as we went on.
Q. Then you say, "He made it clear that this money, unlike that which went to the relief organizations, would go straight to the fighting effort." What do you mean by "the fighting effort"?
A. I understood that it meant getting men into Spain in an international brigade and getting equipment for them. That is what I understood. This was, I believe, an illegal operation, but I am not sure.
Q. Were you so advised at the time?
A. I was not advised; no.
Q. Is that why you made your contributions in cash?
A. I think it would have been a good reason for it. I ought to say that I did a great deal of my business in cash.
Q. Was there any other reason for making your contributions in cash?
A. I think I have stated it.
Q. You have stated the specific reason. Wasn't the reason in general that you wanted to conceal them?
A. I didn't want to advertise them, certainly.
Q. Reading further from your answer at the top of page 8: "I did so contribute usually when he communicated with me explaining the nature of the need."
How often would he communicate with you to explain the nature of the need?
A. I would think maybe 5 or 6 times during the time I was in Berkeley. A year.
Q. Five or six times a year?
A. Yes.
Q. What would be the nature of the need that he would explain?
A. First, it was the war, and then later it was something else. He would tell me about the fighting; he would tell me that they were hard up. He would paint the picture of the desperate situation as it rapidly developed and what money could do for it.
Q. You said later on it was something else. What was that?
A. That was the problem of getting the Spanish Loyalists out of the camps in France and getting them resettled. Don't misunderstand me. I am not talking of this in contemporary terms but in the terms that I understood in those days.
Q. What do you think now the need was?
A. I think probably, if the money went through Communist channels, the money was to rescue Communists.
Q. You knew it was going through Communist channels.
A. I knew it.
Q. For how many years did that go on?
A. You have fixed the date in early 1942. I have the feeling that is about right.
Q. You mean you think your last contribution was probably in early 1942?
A. Yes; in early 1942.
Q. Starting in 1937 or earlier?
A. Yes.
Q. In other words, it continued for approximately 4 years?
A. Yes.
Q. What was the average yearly amount that you gave through those channels?
A. I never totaled it up.
Q. I know that.
A. I should think more than $500 and less than $1,000.
Q. Doctor, I don't mean to pry into irrelevant matters of your personal life or affairs, but your income during those years was probably between fifteen and twenty thousand dollars a year, wasn't it?
A. No; that is on the high side.
Q. Would it have been $15,000?
A. I think my salary was $5,000. I have not looked it up. I believe we got about $8,000 or so in dividends and interest.
Q. Doctor, I am not trying to trap you.
A. No, no. It was not under $12,000 and not over $18,000.
Q. I have looked at your income-tax return for, I think, 1942, and it seemed to me to be about $15,000.
A. Good.
Q. That was your State income-tax return. So that it would be perfectly possible for you to give him $1,000 a year or even more, wouldn't it?

A. Sure. I was not using the money I had for my personal needs.
Q. You might have given him as much as $150 a month on the average?
A. That is a leading question.
Q. Yes; I know.
A. I could have as far as the money I had available.
Q. And you have no definite recollection as to just how much you did give him?
A. I remember once giving $300.
Q. In cash?
A. In cash.
Q. What was the need that he explained to you for that money?
A. I believe that was just before the end in Spain, that is, of the wa
Q. What was the need?
A. The need was to prevent defeat.
Q. You mean more cartridges or something?
A. More people.
Q. Your testimony is that Addis started you off on this, or rather your answer states that Addis started you off, and your testimony is, too, and there is a time when he brought in Isaac Folkoff.
A. Right. He told me he had been giving the money to Folkoff and Folkoff could explain things just as well.
Q. Was any reason given to you why Folkoff executed for Addis?
A. None.
Q. By the way, where did you usually give him this money—in your house, or where?
A. Sometimes when he was coming to Berkeley. More often I went to San Francisco and very often went to visit him in his laboratory or in his home. It wasn't a regular meeting. Sometimes we met casually and he talked to me and we would fix a meeting.

Mr. GARRISON. May I ask the clarification whether the "he" refers to Folkoff or Addis?

By Mr. ROBB:

Q. I am talking about Addis. Did you follow the same system with Folkoff?
A. Yes.
Q. Was there any difference?
A. No, except that Folkoff came less frequently to Berkeley.
Q. Did you ever go to Folkoff's house or office to give him money?
A. I don't remember his office or his house, but I won't at this stage deny it.
Q. About when was that when Folkoff came into the picture?
A. I don't remember. I can make a guess. In 1940. But it is a guess.
Q. You testified that Addis told you Folkoff would take over, and he would explain things to you, is that correct?
A. Yes.
Q. What did Folkoff explain to you?
A. With one or two exceptions it was all the business about the refugees, the camps in France, the resettlement problems, and how much it cost and how much it cost to get to Mexico, and all the rest. This was the campaign.
Q. What were the exceptions?
A. I remember one. The one I remember was a campaign—this occurred more than once—to organize the migratory labor in the California valley. I understood that Communists were involved in that.
Q. I was about to ask you a campaign by whom, and the answer would be by the Communists.
A. Right.
Q. You say in your answer, "Sometimes I was asked for money for other purposes. The organization of migratory labor in the California valleys, for instance." That is what you have reference to.
A. Right.
Q. What were any of the other purposes besides that?
A. Besides these three I mentioned, I don't recollect.
Q. You do recall there were others?
A. I have the impression there were others.
Q. Was it your procedure to cash a check and then turn the cash over to either Addis or Folkoff?
A. I presume I got the money from the bank.
Q. You had a checking account.
A. I had a checking account.

Q. You say in your answer, "In time these contributions came to an end. I went to a big Spanish relief party the night before Pearl Harbor; and the next day, as we heard the news of the outbreak of war, I decided that I had had about enough of the Spanish cause, and that there were other and more pressing crises in the world."

Doctor, the Spanish cause was identified in your mind with the Communist Party, wasn't it?.

A. Not as clearly as it has been since. The International Brigade, I think in fact was not purely Communist. It was certainly Communist organized.

Q. In all events, your contributions were strictly made to the Communists.

A. Absolutely.

Q. You did not feel any revulsion against the Communists until after Pearl Harbor?

A. I don't believe this indicates revulsion.

Q. Did you at the time of Pearl Harbor feel any revulsion against the Communist Party?

A. That is much too strong a word.

Q. You did not?

A. Not anything as strong as revulsion, no.

Q. You were not quite as enthusiastic as you had been previously, is that right?

A. Yes; I could put it a little more strongly than that and a little less strongly than revulsion.

Q. Very well. What was the reason why Pearl Harbor had any bearing on your attitude towards the Communist Party?

A. I think I should add something to what it says here, that is, I didn't like to continue a clandestine operation of any kind at a time when I saw myself with the possibility or prospect of getting more deeply involved in the war.

Q. There was no question in your mind that this was a clandestine operation, was there?

A. I don't think I concealed it from friends, but I didn't advertise it.

Q. You didn't conceal it from your Communist friends, certainly.

A. Or my wife or so on.

Q. What effect did the Nazi-Russian Pact of 1939 have on your attitude towards the Communist Party?

A. I hated the sudden switch that they made. I hoped that they would realize that this was a mistake. I didn't understand that the Communists in this country were not free to think, that the line was completely dictated from abroad.

Q. You didn't cease your contributions at that time, did you?

A. Contributions to this affair?

Q. Yes.

A. I don't think it had any effect.

Q. Pardon?

A. I think it had no effect.

Q. Doctor, coming to page 9 of your answer, you refer to your brother Frank, he told you in 1937, probably in 1937, that he and his wife Jackie had joined the Communist Party. What was the occasion for telling you that?

A. My memory is sharp, but it could be wrong. I think he drove up to Berkeley, spent the night with me, and told me about it then.

Q. What was the reason for telling you, do you know? Did he explain why he was telling you?

A. I was his brother, I suppose, and something of the fraternal relations was involved.

Q. Did he ask your advice about it?

A. Oh, lord, no. He had taken the step.

Q. Was it shocking to you?

A. My recollection, which may not be the same as his, is that I was quite upset about it.

Q. You say in the autumn of 1941 they, meaning your brother and his wife, came to Berkeley.

A. They moved to Berkeley.

Q. I am reading your answer.

A. Yes.

Q. "* * * and Frank worked for the Radiation Laboratory. At that time he made it clear to me that he was no longer a member of the Communist Party." How did he make it clear to you?

A. By saying so, I think.

Q. Just that?

A. It was presumably in a context. I don't remember the context.

Q. You mean he just said, "I am no longer a member"?

A. He probably said that he had not been since he left Stanford, which was some time earlier. No, I don't think he did. I don't think he did, because the Stanford thing I was not clear about.

Q. Did you talk with him about his left wing friends either then or later?

A. I may have.

Q. Why do you say you may have?

A. I don't recollect it. I may be wrong about this conversation with Frank, and it may be that I asked him, did he have any party connections.

Q. Why would you have asked him?

A. Ernest Lawrence had told me he would like to take Frank on. This was not secret work, but it was in the Radiation Lab. Lawrence had a very strong objection to political activity and to left wing activity. When Lawrence had talked to me about it, he said provided your brother behaves himself, or some such, and keeps out of these things. It would have been natural for me to inquire.

Q. You knew that if it were known that your brother was a member of the Communist Party, he could not get the job, didn't you?

A. Yes. My honor was a little bit involved because of my having talked to Lawrence.

Q. Did you know or did you believe that if it were known that your brother was a very recent member of the Communist Party, he might not get the job?

A. I didn't know and I don't know now what effect that would have had.

Q. Did you inquire?

A. No.

Q. Did you tell Lawrence that your brother had been a member of the Party?

A. I think I told him he had a lot of left wing activity.

Q. Did you tell him he had been a member of the Communist Party?

A. I don't think so.

Q. Your honor didn't require you to do that?

A. I didn't think so.

Q. You should have, should you not?

A. These things were not that way in those days, at least not in the community that I knew. It wasn't regarded, perhaps foolishly, as a great state crime to be a member of the Communist Party or as a matter of dishonor or shame.

Q. Now, continuing with your answer on page 9:

"As to the alleged activities of Jackie and Frank in 1944, 1945 and 1946: I was not in Berkeley in 1944 and 1945; I was away most of the first half of 1946; I do not know whether these activities occurred or not, and if I had any knowledge of them at the time it would only have been very sketchy."

Doctor, may I ask you, sir, you say if you had any knowledge; did you have any knowledge of them?

A. If I had known whether I had knowledge, I would have said so in here. I can't remember.

Q. You don't know whether you did or not?

A. That is right. I can't remember whether Frank referred to these things or not. I had no knowledge in the sense of a detailed or clear discussion and I didn't think it right to say that he couldn't have mentioned these lectures or something like that.

Q. Referring to your New Year's Day visit to Frank at his house, you were at Frank's house on New Year's Day in 1946?

A. I was. I believe that later in the day we went out to a reception, but this is my brother's recollection.

Q. Do you recall seeing Pinsky and Adelson there that day?

A. I certainly don't. I have written it here as is true.

Q. Do you recall that Mrs. Oppenheimer, I mean your wife, was ill that day?

A. I remember something which is not very clear. No, I don't recall. I thought maybe the evening before we had to come home early from New Year's Eve because she was not feeling well.

Q. Where were you staying at that particular time?

A. The whole of our family was staying with the whole of my brother's family. We had not seen each other for a long time, and we stayed in Berkeley.

Q. But you were not staying in the same house as your brother was in, were you?

A. We were in sort of a barn.

Q. That is correct. Don't you recall that Mrs. Oppenheimer was not feeling good, and she stayed in the barn and you went over to your brother's house and talked to Adelson and Pinsky?

A. I don't recollect it, no. I have no recollection of my wife's illness.

Q. You say "Pinsky and Adelson, who were at most casual acquaintances of mine"—how had you made their acquaintance, casual or otherwise?

A. Adelson I met, I believe, for the first time in his house—no, in the house of a friend, or in his house, I am not clear. That was many, many years earlier. They were thinking of starting this union at Shell, and they asked me to talk about how the Teachers Union had been.

A. I believe he had to do with the Federation of Architects, Engineers, Chemists, and Technicians.

Q. What was Adelson's work as far as you know?

A. He was at the Shell Development Co. as a scientist of some kind.

Q. Both Pinsky and Adelson you knew to be Communist sympathizers if not members?

A. I didn't know them to be members and I had so little contact with them at the statement that they were Communist sympathizers goes beyond what I know.

Q. Do you know a man named Barney Young?

A. Young?

Q. Yes.

A. I don't recollect.

Q. What did you see of Pinsky and Adelson subsequent to New Year's Day in 1946.

A. I don't think I saw them.

Q. Did you hear from them?

A. I don't remember.

Q. Or either of them?

A. I can't deny this because it has been a rather full life, but I don't recollect it.

Q. Do you recall in March 1946 when Adelson and Pinsky or either of them suggested that you run for Congress?

A. In March 1946?

Q. Yes.

A. March 1946, that I run for Congress?

Q. Suggested to Mrs. Oppenheimer.

A. I think this suggestion I heard about.

Q. That is right.

A. But I believe it was addressed to my brother.

Q. You are sure it was not to you?

A. Quite sure.

Q. How did you hear about it?

A. My brother told me. Not Pinsky and Adelson, but that somebody had put it up to him that he should run for Congress. You have a long record of folly here, but not that I ran for Congress.

Q. I was not insinuating that you accepted the suggestion, Doctor.

Doctor, you speak on page 10 of your letter of answer of the fact that your wife "for a year or two during her brief marriage to Dallet" was a Communist Party member. How long was her marriage with Dallet?

A. She will testify and you will get from her a real biography. The impression I have is that it started in 1934 or 1935, that he was killed in 1937. Something like 2 or 3 years. They were separated a part of this time. It is quite a complex story, and I don't want to make it more complex by my own unfamiliarity with it.

Q. I merely wish to find out what you meant by "brief marriage."

A. Right; 2 or 3 years.

Q. At page 10 of your answer, "I need to make clear that this changing opinion of Russia, which was to be reinforced by the Nazi-Soviet Pact, and the behavior of the Soviet Union in Poland and in Finland, did not mean a sharp break for me with those who held to different views. At that time I did not fully understand—as in time I came to understand—how completely the Communist Party in this country was under the control of Russia."

At that time, I assume you mean 1938 or 1939?

A. No; at that time refers to this period of the Nazi-Soviet Pact.

Q. I see. When did you come to understand that the Communist Party in this country was completely under the control of Soviet Russia?

A. I would give more or less the same answer to that, that I gave to your question about fellow traveling, that it was a gradual process. The shift in Communist position after the German attack on Russia, coming after the Nazi-

Soviet Pact, made a big impression. I guess during the war thinking about it and talking to people, I got that conviction pretty deep in me.

Q. Maybe 1946?

A. I think it was earlier than that.

Q. 1945, 1944?

A. Something like that. 1944 would be a good——

Mr. GRAY. Excuse me, Mr. Robb. It is 12:30. If you are about to go to some other question, I think we should now recess for lunch.

Mr. ROBB. Yes. I did not realize that.

Mr. GRAY. Yes. We will meet again at 2 o'clock.

(Thereupon, at 12:30 p. m., a recess was taken until 2 p. m., the same day.)

AFTERNOON SESSION

Mr. GRAY. The proceeding will begin.

I would like to say with respect to the proceedings today and tomorrow, I think we wil go ahead with the questioning of Dr. Oppenheimer this afternoon as expeditiously as possible. We would like to finish, if we can, the questioning of Dr. Oppenheimer and then put on these three witnesses tomorrow that are going to be here.

I understand that will be Colonel Lansdale, Mr. Glennon, and Dr. Compton. At the conclusion of their testimony we will then begin what would be referred to as redirect examination.

In this general connection, also, I express the hope that we can start at 9 o'clock in the morning.

Mr. GARRISON. I am sure that is possible.

Mr. GRAY. I think I would also like to say, Mr. Garrison, that I assume in a court that the general procedure would be that a judge would direct that the redirect examination proceed immediately upon the conclusion of the questioning on cross. However, in an effort to make sure we are giving every consideration possible to Dr. Oppenheimer and his counsel will take these witnesses out of order.

Mr. GARRISON. Thank you, Mr. Chairman. I believe it to be in the discretion of even a trial judge to do that. I also understand that this is not a trial but an inquiry.

Mr. GRAY. That is right, sir.

Mr. GARRISON. I have not been able to reach Colonel Lansdale yet. His plane is supposed to be arriving at 1:30.

Mr. GRAY. Then he has not been upset by any communication.

Mr. ROBB. May I proceed, Mr. Chairman?

Mr. GRAY. Yes.

Whereupon, J. Robert Oppenheimer, the witness on the stand at the time of taking the recess, resumed the stand and testified further as follows:

CROSS-EXAMINATION—Continued

By Mr. ROBB:

Q. Dr. Oppenheimer, would you refer to your letter of answer on page 11, where you say: "After our marriage in 1940, my wife and I for about 2 years had much the same circle of friends as I had had before—mostly physicists and university people."

Could you tell us, Dr. Oppenheimer, what names occur to you as your circle of friends during that period?

A. Many. Ed McMillan; the first night we were back in Berkeley we had dinner with the Lawrences; I had relatives there called the Sterns whom I had brought over from Germany—the Hands, the Chevaliers, the Edward Tolmans, the Meiklejohns, Jenkins.

Q. Is that David Jenkins?

A. No; that is Francis Jenkins. I can go on and on.

Q. I just wondered whom you had in mind.

A. This is not a bad example. The Addis'.

Q. The Kenneth Mays?

A. No; they were not close friends. I am not trying to name all the people that we occassionally saw.

Q. Did your circle of friends include some Communists or Communist sympathizers?

A. Oh, yes.

Q. Who were they?

A. Let us see about friends. The Chevaliers I have mentioned; the Addis' I have mentioned.

Among Communists, I don't think it would be right to call the Steve Nelsons friends, but we saw something of them. They were acquaintances. We did see the Mays—at least Ken May; I don't know that we saw his wife very much. Almost everybody in the physics department. The Hildebrands, the Peters'.

Dr. EVANS. Latimer?

The WITNESS. We saw him but he was not a personal friend.

Mr. GARRISON. Just for clarity——

The WITNESS. The Stephen Peppers.

Mr. GARRISON. When he said almost everybody in the physics department, would you determine whether he was referring to Communists or Communist sympathizers?

The WITNESS. No; not Communists.

Mr. ROBB. I understand you to mean you saw almost everybody in the physics department.

The WITNESS. That is right.

Mr. GRAY. The record will show that the witness did not say that everybody in the physics department was a Communist.

The WITNESS. That is right. The Peters'.

By Mr. ROBB:

Q. They were Communists.
A. I told you yesterday that they had no connection with the party.
Q. They were pretty close?
A. I think they had no connection with the party at all.

Mr. GRAY. There was one name that I didn't get and I don't know whether the reporter did, either. Was it Hand?

The WITNESS. George Hand.

By Mr. ROBB:

Q. Doctor, have you ever crossed anybody off your list or ceased to see them because of their Communist Party connections?
A. I can't put it that way. Since the war there are people with whom there has been a sense of hostility which I identified with their remaining close to the party.
Q. Who were those people?
A. This happened with the Peters'? It happened with a boy who was a doctor and a close friend of my brother's and used to spend summers at the ranch long ago.
Q. What is his name?
A. If you need his name I will give it to you. It is Roger Lewis. This is in a sense an estrangement, but it is not that I know they are members of the party and I no longer have anything to do with them. After the war I did not wish to have anything to do with party people in California. You mentioned the different Jenkins. That is Miss Arnstein's present name and I did not wish to see them and I didn't.
Q. She is the Miss Arnstein you mentioned yesterday?
A. Yes.
Q. Is she married to David Jenkins?
A. Yes.
Q. How well do you know Jenkins?
A. Not very well.
Q. Did you know him in 1943 and 1944?
A. 1944 certainly not.
Q. Did you know him in 1943?
A. I met him and don't have any recollection of seeing him in 1943.
Q. But you knew Miss Arnstein at that time?
A. From way back, yes.
Q. In what connection did you know her?
A. I think I told you she was one of Jean Tatlock's best friends.
Q. Did you see David Jenkins and Miss Arnstein or Mrs. Jenkins after the war?
A. No.
Q. What caused you to be estranged from her?
A. This is an example of people in the party. I have been searching to answer your question.
Q. You have searched your memory carefully and those are the names that came up?
A. I am not sure if I searched longer I would find others.

Q. You say on the same page: "We were occasionally invited to more or less obviously leftwing affairs, Spanish relief parties that still continued;" Doctor, why were they obviously leftwing?

A. If Schneiderman talked they were obviously leftwing. The Spanish Relief parties I think by then were obviously leftwing.

Q. What was there about them that indicated so clearly that they were leftwing?

A. I suppose the presence of many of the people whose names I have told you.

Q. In other words, you felt that those people would not have been at a party unless it was pretty obviously leftwing?

A. No, no; not at all. I don't think anybody would refuse to go to a party because it wasn't leftwing; but many people might refuse to go to a party if it were leftwing.

Q. You say on two occasions, "once in San Francisco and once in Berkeley we attended social gatherings of apparently well-to-do people, at which Schneiderman, an official of the Communist Party in California, attempted, not with success as far as we were concerned, to explain what the Communist line was all about."

Where were those parties held?

A. One that I talked about yesterday was at the Chevaliers. One that I did not talk about yesterday was at Louise Bransten's.

Q. Who is she?

A. She lived in San Francisco. I think she was separated from her husband, had some money and was a friend of Addis. I know very little about her but I believe she was a Communist sympathizer.

Q. Wasn't she a member of the Communist Party?

A. If she was I didn't know that. I didn't know anything about that.

Q. Did you ever hear that she was a mistress of a man named Keifits who was in the Russian Consulate?

A. No; I never heard that.

Q. How did you happen to meet Miss Bransten?

A. I don't remember.

Q. This party was held at her house?

A. Yes.

Q. In the evening?

A. Yes.

Q. How many people were present?

A. It was similar to the one at the Chevaliers, 20 people. I don't have a clear distinction between the two in mind.

Q. Can you recall about when that was?

A. No. It was after our marriage because my wife was there.

Q. After 1940?

A. I would say after the end of 1940.

Q. Subsequent, of course, to the Nazi-Soviet Pact?

A. Yes. Possibly subsequent—well, I don't remember.

Q. Who was present beside you?

A. I told you a few names at the Chevalier party, and I have no further memory or no very different memory about this group.

Q. You think it likely the same group?

A. Not identical, but overlapping.

Q. Can you tell us anybody who was there at Louise Bransten's house who was not either a Communist or a Communist sympathizer?

A. If you use the word "sympathizer" in a very loose sense, I can't.

Q. Have you ever described that meeting at Louise Bransten's house before in any testimony or in any statement that you have made?

A. Either my wife or I did to the Federal Bureau of Investigation.

Q. When?

Mr. MARKS. Mr. Chairman, could we have Dr. Oppenheimer's last preceding answer read, and also the question? I am trying to be sure I understood exactly what he said.

Mr. ROBB. Will the reporter read the question and answer, please.

(The question and answer were read by the reporter as herein recorded.)

By Mr. ROBB:

Q. My last question was when did you tell the FBI about the Louise Bransten party?

Mr. GARRISON. Mr. Chairman, isn't this an item not in the the Commission's letter?

Mr. ROBB. It is in Dr. Oppenheimer's answer. I think I have a right to explore it.

Mr. GARRISON. Did he mention Louise Bransten?

Mr. ROBB. He mentioned 2 parties, and I think I have the right to find out which they were and where they were held.

The WITNESS. I am not sure when. Conceivably the last time was in 1942. But that is easier to check on for you than for me.

By Mr. ROBB:

Q. Going back to the answer that Mr. Marks asked to have reread, the answer as to whether you could tell us anybody who was at Louise Bransten's who was not either a Communist or a Communist sympathizer, I will rephrase the question as follows: Can you tell us anybody there who was not either a Communist or a fellow traveler as you define that word?

A. I need to say that I cannot really remember who was there. I had trouble yesterday with the Chevalier meeting. I have a similar trouble here. I cannot help you out.

Q. Of those who you do remember being there, they were either Communists or fellow travelers, were they not?

A. I am not sure of Jack Straus.

Q. Jack who?

A. Straus. I am not sure where he stood. I am not absolutely certain whether he was at both of these meetings. He was at one of them. I think Mrs. Chevalier was not much of a Communist sympathizer. She was certainly at the one at her home, possibly at the one at Louise Bransten's.

Q. When you talked to the FBI agents in 1946, as you mentioned in your answer, is it your testimony that you did not recall one of these meetings had taken place at Chevalier's house?

A. That is right.

Q. And they asked you about certain meetings and you said that you thought they were completely irrelevant?

A. That is my recollection.

Q. Doctor, if you didn't remember at that time where the meeting had taken place, how did you know it was completely irrelevant?

A. It was a sudden change in questioning which had been about Chevalier and then there was a question as perhaps in this form: Do you remember attending a meeting at East Bay at which Schneiderman talked, or something like that.

Q. And you at once said that that is irrelevant?

A. I don't recollect. You have the record.

Q. My question is, sir, how could you be sure that the meeting was irrelevant if you didn't recall where it took place?

A. I couldn't be sure that I thought if it were relevant it would be explained to me. Instead the agent said that "we just do this sort of thing to test your veracity."

Q. When did you recall it took place at the Chevaliers?

A. I told my wife about this interview and she reminded me of it.

Q. When?

A. Very shortly thereafter.

Q. A day?

A. I don't remember.

Q. Within a day or two?

A. Very shortly thereafter.

Q. Did you then telephone the FBI to tell them that you remembered that it took place at the Chevaliers?

A. No; because the FBI had indicated that this was not a substantive question.

Q. Not what?

A. Not a question of substantive interest.

Q. When you recalled it had taken place at the Chevaliers, did you then think it was relevant?

A. Not terribly because I defined as well as I could Chevalier's political views.

Q. Did you think it had any relevance at all after you recalled where it had taken place?

A. I don't believe I put that question to myself.

Q. You were asked about the meeting again in 1950 by the FBI, is that correct?

A. Right.

Q. At that time you told them about the meeting at Chevalier's house.

A. Right.
Q. So you thought then it was relevant?
A. I don't remember the line of questioning. It was certainly relevant to their then questioning and they asked me about it.
Q. You next mention on page 11, Kenneth May. You knew he was an active Communist, didn't you?
A. I certainly knew it when it was public knowledge. I don't believe I knew it before that.
Q. When did that become public knowledge?
A. That is a matter of record, but not in my mind.
Q. Didn't you know he was a Communist Party functionary at any time of your association with him?
A. Yes.
Q. You knew that?
A. Yes. It was public knowledge that he was a Communist Party functionary during part of my association.
Q. I see.
A. But I don't remember the date when this occurred.
Q. In other words, while you were associating with him socially and otherwise, you knew that he was a Communist Party functionary because it was public knowledge?
A. Socially is better than socially and otherwise.
Q. Socially? Very well.
A. Sure.
Q. How did you come to know Dr. Weinberg?
A. In the most normal way. I knew all the graduate students who studied theoretical physics in the department of physics in Berkeley. I believe I called them all by their first names.
Q. Did you have any relationship with Weinberg other than that of professor and student?
A. I think I need to say several things in answer to that. The first simple answer is "No," until after the war when he was not a student but an instructor and when he and his wife—we saw them once or twice as was proper for dinner or tea or something.
The second thing is that with most of my students it would not be an uncommon thing for me to have dinner with them or to have lunch with them while we were working. I think my relations to Weinberg were much less close than with most of my graduate students.
Q. What was the occasion for you meeting with him and his wife after the war?
A. He was an instructor in the physics department in Berkeley. I think we probably had dinner or tea or something with every member of the department.
Q. Did he and his wife come to your house for social occasions?
A. Not more than once or twice.
Q. They did from time to time?
A. No.
Q. Well, once or twice?
A. Once or twice. I am not certain about this. I am speculating. We did see them as we saw everybody.
Q. You mentioned yesterday recommending counsel to Dr. Weinberg at the time of his criminal trial.
A. No. That is a misunderstanding.
Q. I beg your pardon.
A. That is a misunderstanding. I mentioned recommending counsel to Chevalier for his passport problem.
Q. I see.
A. It turns out that it was the same man or one of the 2 people who represented Weinberg in the course of his trial. I had nothing to do with his selection.
Q. Did you see Weinberg about the time of his criminal trial?
A. No, I did not. I saw him once very briefly. I can fix the time. It was the winter of 1952 at the American Physical Society meetings. I was with another past president and the president-elect of the society and he walked by, noticed us, shook hands and we passed the time of day.
Q. Did you ever discuss with Weinberg the matter of his criminal trial either before or after it took place?
A. I was represented by counsel.
Q. I know that.
A. There were no discussions between me and Weinberg.

Q. Your counsel and Weinberg's counsel presumably did discuss it?
A. That is right.
Q. Did you in any way help to finance Weinberg's defense in that case?
A. I did not.
Q. When did you first hear that Weinberg had been a Communist?
A. At the time of the 1946 interview with the FBI, the agents told me—they questioned me about Weinberg, Lomanitz and so on—and I said, "What is wrong with them?" He said, "There is a question of their membership in the Communist Party."
Q. Were you surprised to hear that?
A. A little bit but not much in the case of Weinberg.
Q. You are quite sure that is the first time you ever heard or had been told he was a Communist?
A. No. I had heard an earlier rumor.
Q. When?
A. When he came to Berkeley that he had been a member of the YCL, the Young Communist League in Madison, but it was hearsay.
Q. Who told you that?
A. I don't remember.
Q. Did you hear anything more about him at that time?
A. No.
Q. Did Weinberg and Lomanitz come to you to talk about Lomanitz' draft deferment?
A. No.
Q. Are you sure?
A. Let's see. The only time this might have been would have been at the time I talked to Lomanitz at the same time we talked so much of yesterday in the summer of 1943. I have no recollection of Weinberg being involved in that.
Q. Do you recall an occasion in Dr. Lawrence's office when you talked to both Weinberg and Lomanitz?
A. No, I don't.
Q. In all events, doctor, you are sure that until 1946, except for the rumor that you mentioned, you had no information to the effect that Weinberg was or had been a Communist?
A. No. I think that is right.
Q. You could not be mistaken about that?
A. One can be mistaken about anything. This is my best recollection.
Q. You say in your answer, "Hiskey I did not know."
A. No.
Q. Did you ever meet Hiskey?
A. There is this allegation that I met him at this party. I have no recollection of it and I don't know whether I was at the party or not. I didn't know him before the party; I didn't know him after the party; I am not clear whether I was at the party or not.
Q. Were you ever at any party at which either Hiskey or Weinberg was present?
A. I never had any recollection of Hiskey whatever until this story was brought up.
Q. How about Weinberg?
A. I am sure I was at parties at which Weinberg was present.
Q. What kind of parties?
A. Physics department, graduate school parties. I don't know what else.
Q. Leftwing parties?
A. I would not be surprised, but I don't remember.
Q. You would expect him to be at and to find him at some such party, would you not?
A. I would not have found it strange.
Q. When did you first meet Steve Nelson?
A. I don't know whether it was before my marriage to my wife or not. I think it was. She thinks that it was after our marriage.
Q. When did you think you met him, and what were the circumstances under which you met him?
A. I think it may have been in connection with a big Spanish party in the fall of 1939.
Q. Where?
A. In San Francisco.
Q. Do you recall talking to him on that occasion?
A. No.

Q. What is there about the occasion that makes Steve Nelson stand out in your mind?
A. He was a hero and there was either talk of him or I saw him, I don't know.
Q. What was he a hero for?
A. For his alleged part in the Spanish War.
Q. You knew he was a Communist Party functionary?
A. I knew he was a Communist and an important Communist.
Q. Thereafter, Steve Nelson was at your home on various occasions, was he not?
A. That was much later.
Q. When was that?
A. The times I remember—and I think they are the only times—were in the winter of 1941–42.
Q. What is the last date that you recall him being at your home?
A. I don't recall the dates. It probably was in 1942.
Q. 1942?
A. Yes.
Q. Summer, fall, spring, or when?
A. I don't know.
Q. Were you at that time working on the secret war project?
A. I was thinking about it if it was in the winter, and I was employed on it if it was the summer.
Q. I beg your pardon?
A. If it was in the winter I was thinking about it, and consulting about it; if it was in the summer, I was actually employed on it.
Q. In all events whether it was in the winter or summer, at the time Steve Nelson was at your house you had some connection with this project, did you not?
A. Oh, yes.
Q. How many times did Steve Nelson come to your house?
A. I would say several, but I do not know precisely.
Q. Did you ever go to his house?
A. I am not clear. If so, it was only to call for him or something like that.
Q. Call for him?
A. Yes.
Q. Why would you have called for him?
A. To bring him up to our house.
Q. Who else was present at your house on the occasions when Nelson was there?
A. I have no memory of this. These were very often Sundays and people would drop in.
Q. The occasions when he was there were not occasions when there was a large group of people?
A. No. We would be out in the garden having a picnic or something like that. It is quite possible that my brother and sister-in-law would come, but I have no memory of this.
Q. Can you give us any idea how long these visits were with Nelson?
A. A few hours.
Q. Each time?
A. The ones I am thinking of, and I think they are the ones you are referring to, and the only ones that occurred, are when he and his wife and his baby would come up.
Q. What did you have in common with Steve Nelson?
A. Nothing, except an affection for my wife.
Q. Did you find his conversation interesting?
A. The parts about Spain, yes.
Q. Was he a man of any education?
A. No.
Q. What did you talk about?
A. We didn't talk about much. Kitty and he reminisced.
Q. Reminisced about what?
A. My wife's former husband, people they had known in the party.
Q. Communist Party activities?
A. Past Communist friendships.
Q. Did Nelson tell you what he was doing in California?
A. No. I knew he was connected with the Alameda County organization.
Q. Did Nelson ever ask what you were doing?
A. No.

Q. Are you sure?
A. Positive. He knew I was a scientist.
Q. He knew that?
A. Yes.
Q. How did he know that?
A. It was well known in the community and we talked about it.
Q. Did you call him Steve?
A. I think so.
Q. Did he call you Oppy?
A. I don't remember.
Q. Probably?
A. I don't remember. He and my wife—she will tell you about it. They had close affectionate relationships and I was a natural bystander.
Q. Doctor, you knew a man named David Hawkins, did you not?
A. Yes.
Q. You speak of him on page 21 of your answer.
A. Right.
Q. How did you meet him?
A. I know that I—well, I better be careful because I never am quite clear or very seldom clear how I first meet people.
I believe we met him and his wife at my brother's at Stanford. I think it likely that I was at least acquainted with him on the Berkeley campus before that time, though I doubt I met his wife.
Q. Was the occasion that you think you met him at your brother's house at Stanford the occasion of some leftwing gathering?
A. No. It was a few people on the porch, or something like that.
Q. You say that you understood that Hawkins had leftwing associations?
A. Yes.
Q. How did you understand that?
A. I understood it in part from the conversations we had and in part from my brother. I am not sure where I got this information.
Q. When did you have the understanding first?
A. I don't know.
Q. Prior to 1943?
A. Prior to his coming to Los Alamos.
Q. What were the leftwing associations that you understood that he had?
A. Well, my brother was a good enough example.
Q. What others?
A. He and the Morrisons were closely acquainted.
Q. Who are the Morrisons?
A. Phillip Morrison was a student of mine and was very far left.
Q. He was very far left?
A. Yes.
Q. Was he a Communist?
A. I think it probable.
Q. Did he go to work on the project?
A. He did.
Q. With your approval?
A. With no relation to me.
Q. Did you ever make known to anyone that you thought that Phillip Morrison was probably a Communist?
A. No.
Q. Why not?
A. Well, let me say he was on the project in another branch quite independent of me. When he came to Los Alamos, General Groves let me understand that he knew Morrison had what he called a background and I was satisfied that the truth was known about him.
Q. Morrison came to Los Alamos?
A. That is right. When he came to Los Alamos we had this discussion.
Q. He was so far leftwing that you thought that the mere fact that Hawkins was a friend of his stigmatized Hawkins, too, did you not?
A. Not stigmatized him; gave him a leftwing association.
Q. What did Morrison do at Los Alamos? I don't mean in detail but in general.
A. He came late and he worked in what was called the bomb physics division. He worked with the reactor we had there. Then after the war he built a quite ingenious new kind of reactor.
Q. Did Phillip Morrison go over to Hiroshima to witness the drop?

A. He was over there. I think he was in Japan. He certainly was not at Hiroshima.
Q. Did you designate him to go to Japan?
A. I don't know. I don't believe so.
Q. Was your advice asked about him going there?
A. I am afraid to say to that I don't know the answer. I don't believe I would have interposed an objection.
Q. You would not have?
A. But I don't believe I was asked.
Q. Had you read Phillip Morrison's testimony before the House Committee?
A. I have.
Q. Was it House or Senate?
A. Senate.
Q. Are you satisfied from that testimony that he was a Communist?
A. Yes.
Q. Were you surprised when you read that testimony?
A. No.
Q. It accorded with what you previously knew?
A. With what I believed.
Q. Yes. What else did you know about Hawkins' leftwing associations?
A. I don't think I knew much more about it than I told you.
Q. Did you know anything about his wife?
A. I think he had a brother-in-law of whom I heard it said he was a Communist.
Q. Did you know a man named Parkman?
A. Yes.
Q. Did you know that Parkman was discharged from the Air Force because of his Communist leanings?
A. No.
Q. Did you ever hear that?
A. No.
Q. Did you know that Hawkins was a friend of Louise Brantsen?
A. No.
Q. What was Hawkins' training?
A. He was trained as a mathematician and philosopher.
Q. What was his major?
A. I don't know. I suppose philosophy.
Q. Philosophy?
A. I think so.
Q. Don't you know that?
A. He was a professor of philosophy. I didn't know him as a student.
Q. He was not a physicist?
A. No.
Q. By the way, how old was he, do you know?
A. No.
Q. Comparatively young, wasn't he?
A. Yes. I thing he was an instructor teaching mathematics at that time.
Q. You said: "I supported the suggestion of the personnel director that he, Hawkins, come to Los Alamos."
A. Yes.
Q. How did you support that suggestion?
A. Let me give a word of background. A committee of which Richard Toman was a member, possibly he was chairman, had come to review the state of affairs at Los Alamos in the spring of 1943. One of their recommendations was that we get a personnel director. There were a great many that I will not here record. One of their recommendations was that we get an aide to help the personnel director and me in the relations between the military establishment and the laboratory. The personnel director was William Dennis, a professor of philosophy at Berkeley. He did not stay terribly long but he came to help out in an emergency. What I heard indicated that Dennis proposed that Hawkins come as his aide and I approved it.
Q. How did you approve it?
A. I said I thought it was a good idea. However, I have relied somewhat on Hawkins' own testimoney of how he got to Los Alamos because I have very little—I have almost no direct memory of it.
Q. At the time you approved that suggestion you knew what you have told us about Hawkins' background and connection, didn't you?
A. I did.

Q. I find in the minutes of the governing board at Los Alamos for May 3, 1943, this entry: "Dr. Oppenheimer said he was going to try to get Lt. Col. Neil Asbridge added to Harmon's staff. He said Mr. Smith was leaving. He proposed to get David Hawkins from Berkeley to handle our relations with the post." Do you recall that?
A. Obviously.
Q. So you rather heartily approved of the suggestion that Hawkins come?
A. Oh, sure.
Q. What did Hawkins do when he got there?
A. I don't have the records available, but his first jobs were two. One was to handle the draft deferments which got to—and this was a job for the personnel division. He was a New Mexican. He knew the local head of the draft board.

The second job was to take up the complicated negotiations between the military authority and the scientists on the acceptance of a building, the installation of equipment, the completion of housing. That was the way it started out.

I also asked him to serve, along with Manley and Kennedy, on the Laboratory Security Committee, which had largely to do with physical security. I asked him after discussing the thing with General Groves to write the technical history of the laboratory. That was much later. By that time I knew him quite well and had come to have a sense of confidence in him—of great confidence.
Q. Hawkins wrote thhe manual of security for Los Alamos?
A. I don't remember that, but it would have been likely. I discussed security with him many times. His views and mine were in agreement.
Q. Hawkins became more or less your administrative assistant, didn't he?
A. For a while. The only person who had that title was David Dow.
Q. Wasn't Hawkins in fact, whether he had the title or not, pretty much your administrative assistant?
A. On the matters I have discussed, yes.
Q. Did Hawkins have access to all the secret information on the project at Los Alamos?
A. Most of it, I should think, yes.
Q. When he wrote the history, he had access to all of it, didn't he?
A. Most of it. I still think that some things like production rates, and so on, would not have come his way.
Q. Did his wife come to Los Alamos with him?
A. Yes.
Q. You know she was extremely left wing, if not a Communist, didn't you?
A. I didn't have that impression, but I may be wrong.
Q. You knew that her brother was, anyway?
A. Yes; I heard that.
Q. Did you ever make known to any security officer what you knew about Hawkins and his wife?
A. What I knew was not very substantial. When the question of the report came up I asked General Groves whether he regarded Hawkins' background as a reason for not doing this. I also discussed it at one other time in connection with a protest Groves made about one of his actions.
Q. You asked General Groves?
A. Right.
Q. Did you tell him what you knew?
A. I knew nothing beyond what was obvious that he had a left wing background.
Q. Did you tell him what you knew?
A. I don't remember.
Q. You say you don't remember?
A. No; I imagine I didn't in the light of the record in the other cases but I don't remember. I know we talked about it.
Q. What was there in Hawkins' background which led you to believe that he was qualified by training or experience to be an administrative assistant to you at Los Alamos?
A. For the jobs that I had in mind he had impressed me as a reasonable, tactful, intelligent person, interested in science, familiar with it. As far as I know, he was in fact very good.
Q. He was teaching philosophy, wasn't he?
A. No; he was teaching mathematics at that time. He knew a great deal about science. His philosophical interests were in science. I may add that he was certainly not the only person in the country for this job.

Q. Doctor, we spoke yesterday of your interview with Colonel Lansdale. I want to read you some extracts from the transcript of that interview, sir. Colonel Lansdale said to you, according to this transcript——

Mr. GARRISON. May we have the date?

Mr. ROBB. September 12, 1943. This is the interview that took place at the Pentagon. Colonel Lansdale said to you:

"We know, for instance, that it is the policy of the Communist Party at this time that when a man goes into the Army his official connections with the Party are thereupon ipso facto severed."

You answered: "Well, I was told by a man who came from my—a very prominent man, who was a member of the Communist Party in the Middle West, that it was the policy of the party there that when a man entered confidential war work, he was not supposed to remain a member of the party."

Who told you that?

A. I have no recollection at all, I will think, if you wish.

Q. I wish you would, sir.

A. From the Middle West.

Mr. GRAY. Read that again.

Mr. ROBB. "I was told by a man who came from my—a very prominent man who was a member of the Communist Party in the Middle West that it was the policy of the party there that when a man entered confidential war work he was not supposed to remain a member of the party."

By Mr. ROBB:

Q. Who was that man?

A. I recollect nothing about it. I will be glad to think about it.

Q. Do you want to think now?

A. I would prefer not to. If I can think about it and tell you tomorrow. It simply rings no bell.

Q. You don't recall anybody ever told you that?

A. No, I said yesterday I didn't recollect.

Q. I know you did. Does that serve to refresh your recollection in any way?

A. Quite to the contrary. From the Middle West?

Q. You then spoke about your brother.

Mr. MARKS. May I inquire, Mr. Chairman, if these transcripts are taken from recordings, just so we can understand what is being read?

Mr. ROBB. Yes. I have every reason to believe it is accurate.

Mr. MARKS. I don't question that, I just wondered what the origin was.

Mr. ROBB. I don't think that is necessarily a question counsel should have to answer.

Mr. MARKS. I asked the Chairman, sir.

Mr. GRAY. My answer is "I don't know." If you wish to discusse it further I would be glad to.

Mr. MARKS. I thought it was a matter that could be answered simply.

By Mr. ROBB:

Q. You spoke of your brother and said, "It is not only that he is not a member, I think he has no contact." Do you recall that?

A. No; I don't recall it, but that I can imagine saying.

Q. Lansdale said: "Do you know about his wife, Jackie?"

You answered: "I know I overwhelmingly urged about 18 months ago when we started that she should drop social ones which I regarded as dangerous. Whether they have in fact done that, I don't know.

Lansdale said, "Well, I am quite confident that your brother Frank has no connection with the Communists. I am not so sure about his wife."

You answered, "I am not sure either, but I think it likely some of its importance has left here. Also, I believe it to be true that they do not have any—I don't know this for a fact—but if they had, I didn't know it, any well established contacts in Berkeley. You see they came from Palo Alto, and they had such contacts there. Then my brother was unemployed for three very, very salutory months, which changed his ideas quite a lot, and when they started in Berkeley it was for this war job. I do not know but think it quite probable that his wife Jackie had never had a unit or group to which she was attached in any way. The thing that worried me was that their friends were very left wing and I think it is not always necessary to call a unit meeting for it to be a pretty good contact."

Doctor, who were the friends and social contacts that you might have had in mind when making that statement?

A. My sister-in-law in Berkeley?
Q. And your brother.
A. I am not sure who I did have in mind. My sister-in-law had a very old friend called Winona Nedelsky.
A. Who was she?
A. She was the wife of a physicist who left here—quite Russian—who had once been my stduent. She was a good friend of Jackie's. She earned her living in some Federal Housing Agency or Social Security Agency.
Q. Was she a Communist?
A. I believe so.
Q. Was she a friend of your sister-in-law in 1943?
A. I would think so. She was a friend. I don't know how much they saw each other.
Q. But in all events, you thought it cause for worry.
A. I would not have thought that a special cause for worry. I am having trouble in remembering what I could have had in mind and what I did have in mind.
Q. Can you think of anyone else that you might have had in mind as dangerous social contacts of your sister-in-law and your brother?
A. I don't know much about the life in Berkeley. I am afraid I can't.
Q. Lansdale said again, "To refer again to this business concerning the party, to make it clear the fact a person says they have severed connections with the party, the fact that they have at present no apparent interest or contact in it does not show where they have unquestionably formerly been members that they are dangerous to us."
You said, "I agree with that."
You still agree with that, do you?
A. Yes.
Mr. GARRISON. Mr. Chairman, I repeat the same request I made with respect to the previous transcript, that we would like to see a copy of the full transcript.
Mr. GRAY. May I say with respect to that that Dr. Oppenheimer will be given an opportunity to see documents reflecting conversations. They cannot be taken from the building.
Mr. GARRISON. We appreciate that. When may we have that opportunity?
Mr. GRAY. When the board and counsel have finished with the questioning.
Mr. GARRISON. You mean this afternoon?
Mr. GRAY. Whenever this is concluded.

By Mr. ROBB:
Q. Lansdale said to you, according to this transcript, speaking of your reluctance to disclose the name of Professor X: "I don't see how you can have any hesitancy of disclosing the name of the man who has actually been engaged in an attempt of espionage in time of war. I mean my mind does not run along those channels."
You said, "I know it is a tough problem and I am worried about it a lot."
That was a correct statement of your attitude, wasn't it?
A. I would assume so.
Q. Lansdale, referring again to your reluctance to disclose the name, says, "Well, if you won't do it, you won't do it, but don't think I won't ask you again. Now I want to ask you this, And again, for the same reason which implies you're here, you may not answer. Who do you know on the project in Berkeley who are now, that's probably a hypothetical question, or have been members of the Communist Party?"
You answered, "I will try to answer that question. The answer will, however, be incomplete. I know for a fact, I know, I learned on my last visit to Berkely that both Lomanitz and Weinberg were members. I suspected that before, but was not sure. I never had any way of knowing. I will think a minute, but there were other people. There was a, I don't know whether she is still employed or was at one time a secretary who was a member."
"LANSDALE. Do you recall her name?"
"OPPENHEIMER. Yes, her name was Jane Muir. I am, of course, not sure she was a member, but I think she was. In the case of my brother it is obvious that I know. In the cases of the others, it's just things that pile up, that I look at that way. I'm not saying that I couldn't think of other people, it's a hell of a big project. You can raise some names."
Doctor, having heard me read those lines, will you now concede that you knew at that time that both Lomanitz and Weinberg had been members of the Communist Party?

A. Evidently. Was I told by the security officers?
Q. I don't know. I have just read what you said. So when you wrote that letter of October 19, 1943, forwarding Lomanitz's request to be transferred back to the project from military service, you knew that he had been a Communist Party member, didn't you?
A. So it appears.
Q. And you knew as early as 1943 that Weinberg had been, too.
A. So it appears.
Q. Yes, sir.

Mr. GARRISON. Mr. Chairman, what troubles me about this whole method of examination is that counsel is reading from a transcript bits and parts without the full course of the conversation which took place to a witness whose memory at best, as anyone of ours would be, is very, very hazy upon all these things, and picking here a sentence and there a sentence out of context, and then holding him to the answer. I do think that this is a method of questioning that seems to me to be very unfair.

Mr. ROBB. Mr. Chairman, I don't mean to make any argument about the matter, but I assume that this Board is following this transcript. If the Board feels I am being unfair at any point, I suppose the Board will interpose.

Mr. GARRISON. Why shouldn't counsel be allowed to follow as any court of law, and this is not even a trial?

Mr. ROBB. As you no doubt know, I have tried a good many cases, and I don't think it would be in the ordinary course of a trial.

Mr. GARRISON. I disagree with you.

Mr. ROBB. I resent counsel's statement that I am trying to be unfair with this witness, because I assure you that I have made every attempt to be fair with him. In fact, were I trying to be unfair, I would not ask this witness any of these questions, but would leave it in the file for the Board to read. I am giving this witness a chance to make whatever explanation he wishes to make.

Mr. GARRISON. I still think that the fair thing would be to read the whole conversation and ask him what parts you want, instead of to pick isolated questions.

Mr. GRAY. On the point of picking isolated questions, without trying to look at this whole question at this moment, I think it is clear that this interview concerned itself with matters which are involved in the questions Mr. Robb has been putting to the witness, and which are generally, I think, not new material. General Nichols' letter of December 23, and Mr. Oppenheimer's reply of March 4, I think both address themselves in one way or another to these individuals, Lomanitz, Weinberg, Bohm, which have been the subject of these questions.

I would say, Mr. Garrison, that I don't think it would be helpful to you at this point to have the transcript. I have said, however, that Dr. Oppenheimer and his counsel will be entitled to examine it and certainly after examination if you wish to reopen any of this testimony, you will be given every opportunity to do so. I think it is the feeling of the chairman of the board that things are not taken here out of context in a way which is prejudicial. I think also that the board has heard Dr. Oppenheimer say that with respect to some of these matters he has no recollection, which at least to me is perfectly understandable, many of these things having taken place many years ago. I do not think that it is the purpose of counsel to develop anything beyond what the facts are in this case. At least that is my interpretation.

Mr. ROBB. That is my endeavor, Mr. Chairman.

By Mr. ROBB:
Q. May I ask you whether or not you recall this Jane Muir?
A. I remember her, not well.
Q. How did you happen to know her?
A. I met her and her husband through the Chevaliers some time before the war.

Mr. GARRISON. Mr. Chairman, I don't want to be captious or legalistic, but this is the example of the kind of problem. Jane Muir is not mentioned in the Commission's letter. Are we to be given a chance to remember all there is to remember about particular individuals? Now, Dr. Oppenheimer is being read aloud out of things that it is said he said a great many years ago, and new names come out which are not in the letter, and which we have never heard, and now he is asked all about them. That seems to me I submit not in keeping with the spirit of the letter. If he had volunteered the name of Jane Muir in testimony, that would be another matter. But this is something that is a complete surprise.

Mr. ROBB. Mr. Chairman, is it Mr. Garrison's position that he wishes time to consult with his client about the Jane Muir matter before we go into it?

Mr. GARRISON. With respect to any new name that is brought into this without any warning at all, we should be given a chance to have Dr. Oppenheimer reflect on what he remembers about it, and for use to have a chance to talk about it.

Mr. ROBB. We will let the Jane Muir go and come back to it at some future date if counsel feels that would be fair.

Mr. GARRISON. I think that would be fair with respect to every new name.

Mr. ROBB. We will go on to something else, then.

By Mr. ROBB:

Q. Then you were asked by Colonel Lansdale:
"—— can you tell me the names of anyone at Los Alamos that have been or are now party members?"
You answered: "I can't tell you the numbers of any who now are"—I assume that means names—"but I know that at least Mrs. Serber was a member. She comes from the Loef family in Philadelphia."

A. To the best of my knowledge this is not true.

Mr. GARRISON. That is the same question.

Mr. ROBB. I think not, Mr. Chairman. I think this is certainly in the scope of the letter of notification which Dr. Oppenheimer has challenged. Dr. Oppenheimer has said in his answer that the knew of no former member of the party at Los Alamos except his wife. He said that with some emphasis and repeated it here. I think I have the right to ask him whether he did know that Mrs. Serber was a member. I asked about Mrs. Serber yesterday.

Mr. GRAY. Mrs. Serber's name has appeared in this proceeding.

By Mr. ROBB:

Q. Don't you know that you did know in 1943 that Mrs. Serber had been a member of the party?

A. I don't know that she was a member of the party. I don't think she was a member of the party.

Q. You testified yesterday you would be very much surprised to find if she ever had been.

A. That is right, I would still be today.

Q. Have you any idea how this statement got in this transcript?

A. No.

Q. Do you know that Mrs. Serber came from the Leof family in Philadelphia?

A. That I know.

Q. When did you know that?

A. Long ago, 15 years ago.

Q. Beg pardon?

A. 15 years ago.

Q. How did you find that out?

A. She told me. My wife also knew her.

Mr. GRAY. Are you at a breaking point?

Mr. ROBB. Yes, sir.

Mr. GRAY. Let us take a recess.

(Brief recess.)

Mr. GARRISON. Mr. Chairman, forgive me for coming back to the same point, but during the recess I discussed this problem with my partner, Mr. Silverman, who has spent his life trying cases in the State of New York—I am not a trial lawyer, sir—our practice I am informed up there universally is that when counsel is cross examining a witness on a transcript he has never seen, counsel for the other side, if he asks the court for a copy, so he may read along with it, that request is granted. So if nothing else—I would not think of impugning this to Mr. Robb, and I hope he won't misunderstand me—I think it is the basis of the rule. That is the only reason I mention it. In other words, to make sure that the questions are in fact being read accurately from the transcript, and there are no interlineations or marks or matters of that sort that might perhaps raise a question as to the accuracy of what is there quite apart from the method by which the transcript was arrived at, and also to understand what the thread and continuity of the matter is. I merely report that to the Chairman. I don't want to put this on the basis of rules of law, because God knows, it is the rule here that this not a trial,, but an inquiry, and I should suppose that a fortiori, what is proper in court of law would be accorded to us here in an inquiry. I do not labor the point. I present it to you and I will rest upon it.

The WITNESS. May I make a comment?

Mr. GRAY. Surely.

The WITNESS. This last quotation about Mrs. Serber strikes me as so bizarre that I am troubled about the accuracy of the document. I am not certain——

Mr. GARRISON. Do you know, Mr. Robb, whether this was taken down by a stenographer or was it from a tape?

Mr. ROBB. Colonel Lansdale will be here. I might ask him. He is the one who conducted the interview.

Mr. GRAY. I would like to be excused with counsel for the Board for a moment, if you please.

Mr. ROBB. Mr. Chairman, I don't agree at all with the statement of law which has been made by Mr. Garrison although I confess I am not a New York trial lawyer. It has always been my understanding that when a witness is questioned about inconsistent statements, he is read the statements and he is asked if he made them. However, it is entirely immaterial to me whether counsel follows this statement or not. If the Chairman wants to have counsel have a copy of it, it is all right with me.

Mr. GARRISON. We would appreciate that.

Mr. ROBB. Very well.

Mr. GRAY. I am about to make the ruling that Mr. Robb follow reading this transcript as Mr. Robb reads it. Have you got a copy of it, Mr. Rolander?

Mr. ROLANDER. I just went out and asked the secretary to try to locate a copy from the original files. I though that might be most helpful.

Mr. ROBB. May the record now show, Mr. Chairman, that we are handing to Mr. Garrison the photostat copy of the interview with Dr. Oppenheimer by Lt. Col. Lansdale, 12 September 1943, consisting of 26 pages.

Mr. GARRISON. Thank you, Mr. Chairman.

By Mr. ROBB:

Q. I am reading from page 10. The transcript shows, Doctor, that you were asked this question by Colonel Lansdale:
"Now, do you know; was Mr. Serber a member of the party?"
You answered: "I think it possible, but I don't know."
Do you recall that question or answer?

A. No.

Q. Did you think at that time, think it possible that Mr. Serber was a member of the party?

A. That he was then a member of the party?

Q. Yes.

A. No.

Q. Did you think it possible that he had been?

A. Possible but unlikely.

Q. Were both Mr. and Mrs. Serber then at Los Alamos?

A. Right.

Q. What did you know about their background?

A. I knew them quite well.

Q. Did you know that they were leftish?

A. Yes.

Q. Did you know that they were at least fellow travelers?

A. I would say at most fellow travelers.

A. At most fellow travelers?

A. Yes.

Q. How did you acquire that information?

A. They were good friends of mine.

Q. From things they said to you and from activities you observed?

A. That is right.

Q. What activities?

A. Mrs. Serber was extremely active in Spanish relief at the time they were in Berkeley.

Q. What else?

A. Talk.

Q. What talk?

A. Just talk about affairs and politics.

Q. Communism?

A. Not as such.

Q. Had you seen the Serbers at these left wing gatherings that you mentioned?

A. Oh, yes.

Q. Frequently?

A. At the Spanish things very frequently.

Q. Along with the other people that you mentioned?
A. Right.
Q. What was Berber doing at Los Alamos?
A. He was a very prominent and constructive member of the theoretical physics division.
Q. Did he have access to classified information?
A. Indeed he did. He created it.
Q. Now, skipping down, you were asked: "How about Dave Hawkins?" And you said, "I don't think he was, I would not say so." Meaning "I don't think he was a member of the Communist Party." Do you recall that?
A. No. I clearly am not recalling this conversation at all.
Q. Was that your view in 1943 that you didn't think he was a member or had been?
A. I wish I knew what my view on these things was.
Mr. MARKS. Mr. Chairman, I think we ought to give the witness the question and the answer as it appears here.
Mr. GARRISON. It reads: "I don't think he was, I would not say so."
Mr. ROBB. That is what I read.
Mr. MARKS. You interpreted it.
Mr. ROBB. Very well. If you don't want me to give the witness the benefit of an interpretation, I won't do it.

By Mr. ROBB:
Q. The next question and answer:
"Now, have you yourself ever been a member of the Communist Party?" You answered, "No."
"LANSDALE. You've probably belonged to every front organization on the coast.
"OPPENHEIMER. Just about."
Doctor, do you recall that question and answer?
A. No, I don't. I don't recall this interview.
Q. If you said that to Colonel Lansdale, were you jocular?
A. I don't think I could have been jocular during this interview.
Q. "LANSDALE. Would you in fact have considered yourself at one time a fellow traveler?
"OPPENHEIMER. I think so. My association with these things was very brief and very intense."
Do you recall that at all?
A. I am not recollecting anything. You may find a phrase that I do recollect.
Q. In all events, Doctor, your answer, "I think so. My association with these things was very brief and very intense," it is now your testimony that was a correct statement of fact?
A. It was very intense; brief is a relative word.
Q. Colonel Lansdale said: "I should imagine the latter anyway."
Now, on page 11, you said, "It was historically quite brief and quite intense, and I should say I was——"
"LANSDALE. Now I have reason to believe that you yourself were felt out, I don't say asked, but felt out to ascertain how you felt about it, passing a little information, to the party.
"OPPENHEIMER. You have reason?"
"LANSDALE. I say I have reason to believe, that's as near as I can come to stating it. Am I right or wrong?
"OPPENHEIMER. If it was, it was so gentle I did not know it.
"LANSDALE. You don't know. Do you have anyone who is close to you, no that's the wrong word, who is an acquantance of yours, who may have perhaps been a guest in your house, whom you perhaps knew through friends or relatives who is a member of the Communist Party. By that I mean——
"OPPENHEIMER. My brother, obviously."
"LANSDALE. Well, no, I don't mean him.
"OPPENHEIMER. I think probably you mean someone who just visited for a few hours.
"LANSDALE. Yes.
"OPPENHEIMER. Yes, certainly, the answer to that is certainly, yes.
"LANSDALE. Would you care to give me any of their names?
"OPPENHEIMER. There is a girl called Eldred Nelson."
Do you know a girl called Eldred Nelson?
A. No, I know a man by that name.
Q. Who was he?

A. He was a student of mine. At the risk of finding I said something else 10 years ago, I would say he was not a Communist.

Q. Having heard Colonel Lansdale's question about whether you had any acquaintance of yours who might have been a guest in your hourse, whom you knew perhaps through friends and relatives, is it now clear to you who Colonel Lansdale was talking about?

A. I don't know. It might well be Steve Nelson.

Q. Isn't that pretty plain?

A. Yes.

Q. Why didn't you mention Steve Nelson?

A. I seem to have mentioned a Nelson.

Q. Eldred.

A. Eldred Nelson is not a girl. He is not a Communist. I think this only adds to my confusion about it.

Q. Very well. I will continue:

"LANSDALE. Suppose I've got a bunch of names here, some of them are right and some of them are wrong, you don't mind treating it that way, do you?

"OPPENHEIMER. No.

"LANSDALE. Did you know William Schneiderman?

"OPPENHEIMER. I know who he is. He is the secretary of the Communist Party. I have met him at cocktail parties.

"LANSDALE. You have no real personal acquaintance with him?

"OPPENHEIMER. No."

Now, you had met Schneiderman at these meetings where he lectured, had you not?

A. Yes.

Q. Were those cocktail parties?

A. Drinks were served.

Q. Were they cocktail parties?

A. No, I think they were evening parties.

Q. Did you tell Lansdale about that?

A. I don't know what I told him.

Q. If I tell you that the transcript shows you didn't, would you accept that?

"LANSDALE. Do you know a fellow named Rudy Lambert?

"OPPENHEIMER. I'm not sure, do you know what he looks like?"

"LANSDALE. No, I've never seen him. He's a member of the party. Do you know a Dr. Hannah L. Peters?"

You knew what Rudy Lambert looked like, didn't you?

A. Sure.

Q. Why did you ask Lansdale what he looked like?

A. I don't know that I did.

Q. If you did, Doctor, would it mean that you were ducking the question?

A. I would think so.

Q. The end of the question by Lansdale: "Do you know a Dr. Hannah L. Peters?

"OPPENHEIMER. Yes, I know her quite well.

"LANSDALE. Do you know that she's a Communist?

"OPPENHEIMER. I certainly knew that she was very close. I did not know she was a member.

"LANSDALE. You don't know what her position in the party is?

"OPPENHEIMER. No. I didn't even know she was a member.

"LANSDALE. Do you have any more than just an acquaintance with her?

"OPPENHEIMER. Yes, I know her quite well. Her husband is on the project."

That was true, wasn't it?

A. Yes.

Q. "LANSDALE. How about a fellow by the name of Isaac Folkoff?

"OPPENHEIMER. I don't know. I knew a Richard Folkoff who was a member of considerable importance."

A. Of what?

Q. I am reading you what the transcript shows.

A. A member of considerable importance?

Q. You knew that Isaac Folkoff was a member of considerable importance, didn't you?

A. Yes. I think that is a garble in the transcript again. Richard Folkoff was a member of Consumers Union as I told you yesterday.

Q. "LANSDALE. How about a man by the name of Steve Nelson?

"OPPENHEIMER. He was a professional party member. He's an organizer.

"LANSDALE. Did you know him well at all—under what circumstances did you know him?
"OPPENHEIMER. He was a friend of my wife's former husband who was killed in Spain. I have a thoroughly unprofessional acquaintance with him."
Do you recall saying anything like that?
A. I recall telling Lansdale about my wife and Nelson.
Q. "LANSDALE. How about Haakon Chevalier?
"OPPENHEIMER. Is he a member of the party?
"LANSDALE. I don't know.
"OPPENHEIMER. He is a member of the faculty and I know him well. I wouldn't be surprised if he were a member, he is quite a Red."
That is true?
A. He was quite a Red.
Q. You would not have been surprised to find he was a member of the party, would you?
A. I think I would during the period I knew him.
Mr. ROBB. Now I will skip over to page 14. Is there anything else in connection with those particular questions that you would like to have me read, Mr. Garrison?
Mr. GARRISON. I have not read all in between.
Mr. ROBB. I did not leave out anything for a couple of pages.

By Mr. ROBB:

Q. At the top of page 14:
"LANSDALE. Now, I want to ask you to go back to Lomanitz. You told me when I was down there that when you broke the subject to, what do you call him, Rossi?
"OPPENHEIMER. Rossi."
What do you mean by "down there"?
A. Lomanitz was never at Los Alamos.
Q. No. "You told me when I was down there"
A. I guess he means Los Alamos. I don't know.
Q. "When you first broke the subject to him about going on the place you stated that he was uncertain, he came up to your house and did what you characterized as a good deal of soul searching. I would like to know whether that soul searching or discussion of his own feelings had any relation to his work in the party?
"OPPENHEIMER. None whatever, I did not know he was a member of the party.
"LANSDALE. Until just recently.
"OPPENHEIMER. Yes. I knew he was extremely Red, but frankly I thought he was a member of the Trotskyite faction.
"LANSDALE. Which would ipso facto prevent him from——
"OPPENHEIMER. Being a member of the party. That's what I thought at that time. What he said he wanted at that time was to be a soldier and be one of the American people in that way and help mold their feelings by being a soldier, and wasn't that more worthwhile than working on this project. I told him he obviously had a lot of talent, he had training that he was throwing right away and that if he could make up his mind and it was a clearcut decision to use himself as a scientist and nothing else, that then that was the right thing to do.
"LANSDALE. What led you to exact from him a promise, or to make the condition of giving up political activities?
"OPPENHEIMER. Because he had distributed leaflets and because it was just generally obvious that he was a member of the union and radical societies."
Does that refresh your recollection as to what you knew about Lomanitz's background and what you said to him before he went on the project?
A. The union I guess is the FAECT. The leaflets I don't remember.
Q. Do you recall having any such conversation with Lansdale?
A. I didn't remember our discussing Lomanitz.
Q. Do you now recall that you did have a discussion with Lomanitz about his activities before he went on the project?
A. I think the substance of the conversation is that if he could be a scientist he should and he should do just that.
Q. Do you now recall laying down the conditions to him about giving up his previous political activities?
A. I don't recall it. We discussed it yesterday. I am not likely to recall that matter today. I don't mean to deny it, sir.

Q. "LANSDALE. Now, you have stated to me and also I think to General Groves that in your opinion membership in the party was incompatible with work on the project from a loyalty standpoint.
"OPPENHEIMER. Yes."
That was your viewpoint, wasn't it?
A. Yes.
Q. "LANSDALE. Now, do you also go so far as to believe that persons who are not actually members but still retain their loyalty to the party or their adherence to the party line are in the same category?
"OPPENHEIMER. Let me put it this way. Loyalty to the party, yes, adherence to the party line, maybe no. In that it need not necessarily, although it often is, be the sign of subservience. At the present time I don't know what the party line is in too much detail, but I've heard from Mrs. Tolman, Tolman's wife, that the party line at present is not to discuss postwar affairs. I would be willing to say that anyone who, well let me put it this way, whose loyalty is above all else to the party or to Russia obviously is incompatible with loyalty to the United States. This is, I think, the heart of it. The party has its own discipline."
Do you recall saying that?
A. No, I don't recollect much about this. This, however sounds like what I thought.
Q. You have no doubt that was your view at that time?
A. Substantially that was my view.
Q. Is there any difference between what I have read and what your view was at that time?
A. I don't know. It is a long couple of paragraphs. It is a long time ago. .I think it is substantially what I then thought.
Q. Lansdale then continued: "Now, I was coming to that. I would like to hear from you your reasons as to why you believe, let's stick to membership in the party, is incompatible to complete loyalty to the project. When, to state something a little bit foolishly membership in the Democratic Party certainly wouldn't be.
"OPPENHEIMER. It's an entirely different party. For one thing * * * I think I'd put it this way. The Democratic Party is the framework of the social customs * * * of this country, and I do not think that is true of the Communist Party. At least, I think that there are certainly many Communists who are above all decent guys, but there are also some that are above all Communists. It's primarily that question of personal honor that I think is involved. I don't know whether that answers the question but my idea is that being a Democrat doesn't guarantee that you're not a fourflusher, and also it has no suggestion just by virtue of your being a Democrat that you would think it would be all right to cheat other people for a purpose, and I'm not too sure about this with respect to the Communist Party."
Do you recall saying anything like that?
A. I don't.
Q. Would you say that did represent your views at that time?
A. I find nothing incompatible between it and what I remember. This is for me not a very easy line of questioning because I don't recollect what I said and I remember what I thought only in general terms.
Mr. GRAY. Before you turn the page, I think that as a matter of record at the end of the fourth line, the party, as you read it, I think, "The party has its own discipline," at least this copy seems to say "The party has its own disciples."
Mr. ROBB. Disciples, that is right.
Mr. GRAY. I don't think it is material.
Mr. GARRISON. I think this indicates the problems of accuracy, because it would probably make more sense as Mr. Robb read it.
Mr. ROBB. There seems to be a pen and ink interlineation.
Mr. GARRISON. The word "fourflusher" is not quite correct.
Mr. ROBB. Now I will turn to page 17. Mr. Garrison, is there something you want read?
Mr. GARRISON. Afterward, after I read it over.

By Mr. ROBB:

Q. Lansdale starts on page 16 and makes quite a little talk which I won't read because it is not essential to the context. At the top of page 17, he says:
"Here we are, we know that information is streaming out from this place every day. We know about some of it, how much of it is there that we don't know about?

"OPPENHEIMER. Places other than the west coast?
"LANSDALE. Sure, we know that definite efforts are being made to find out, they wouldn't be going to these efforts unless they really wanted it. Now, what shall we do? Shall we sit back and say well, my God, maybe the guy recanted, maybe he isn't at all.
"OPPENHEIMER. Hard for me to say because of my own personal trends, and as I say I know that the Serbers afford a good illustration of this I would hesitate to say to a stranger * * * about another closeup * * * person whose history was the same as that of Mrs. Serber's, sure she's all right but I know the Serbers and I am confident of them. Now I have worked on rather a personal basis. I don't know the Woodwards are members, I did not know that until General Groves mentioned it the other day that there was some question of it. I feel that in the case of the Serbers I could understand that very well. But I just don't know in a general case, it's impossible to say. I don't know any of these people in Berkley, I don't know Weinberg and Lomanitz well enough to swear——
"LANSDALE. Why is he moving heaven and earth to keep out of the Army?
"OPPENHEIMER. He told me he thought he was being framed, and I said I think that's nonsense, why would you be framed, and he said, 'Well, part of the general scheme * * * maybe they're after bigger game than the party.'
"LANSDALE. Did you ask him what the bigger game was?
"OPPENHEIMER. He said he thought you were after the union.
"LANSDALE. We're not.
"OPPENHEIMER. Well, I suggest you keep your eyes open. * * * I presuaded him, I think, that he should not try to stay on the project there."
Do you recall that portion of your conversation with Lansdale?
A. No.
Q. Would you deny that took place?
A. No.
Q. Your answer is no?
A. No.
Mr. GARRISON. Do you know whether these dots represent words that the stenographer didn't catch?
Mr. ROBB. I don't.

By Mr. ROBB:
Q. Now, we go to page 22. Lansdale said:
"Can you tell me any more, did Weinberg, it was Weinberg and Bohm who came to you, wasn't it?
"OPPENHEIMER. Yes, they came to me in Lawrence's office.
"LANSDALE. Yes. Did Weinberg and Bohm say anything? What did they say about the party?
"OPPENHEIMER. They didn't say anything about the party.
"LANSDALE. They didn't? Did they talk about the union?
"OPPENHEIMER. They talked, well they didn't even talk about the union. They talked about, I think I've given you a fairly good, I don't know what they might have said if we had met in the woods some place, but we met after all where there were two secretaries in the room.
"LANSDALE. Oh, they were there?
"OPPENHEIMER. I don't know whether the door was closed or not, but it was extremely open interview. I saw Lomanitz more or less, well I saw him first at one of the offices of a man, and we walked out to telegraph * * * but his discussion was a little bit more uninhibited than the others. These two fellows were concerned with only one thing, they said they had worked closely with Rossi, they thought he was a good guy and that they thought he was being framed for his activities in the union and his political sympathies, and they thought that because of this they were also in danger of such a nature that they should get out of the project into some other useful work or were they likely to be treated the same way.
"LANSDALE. Now let me ask you this. From what you stated to them, if they were in fact not fulfilling the conditions which you mentioned to them, which you said to them would have been tantamount to telling, then if you are doing that you'd better get out.
"OPPENHEIMER. Yes.
"LANSDALE. That is correct, isn't it?
"OPPENHEIMER. Yes, that is if they were violating any of the three rules which meant active in union, maintaining any contacts with Reds, not maintaining discretion, they were useless to the project."

Doctor, does that refresh your recollection about laying down some rules for Lomanitz, Bohm and Weinberg?

A. It refreshes to the extent that these three things said to them in the summer of 1943 would have been natural for me to say. I don't believe these were rules established by me long in advance. I think this refers to this conversation.

Q. Don't you recall now that you had a conversation with at least Lomanitz in which you had told him prior to his coming on the project that he would have to abide by these rules?

A. I don't remember those three rules at all. I think it natural that I talked to them at this time in those terms.

Q. I see. Your mind is a blank of any talk you might have had with Lomanitz prior to his going on the project?

A. No, it is not a blank. I have told you a little about it here, and I testified to the extent I can recall it.

Q. Aside from that, you could not recall anything?
A. Right.
Q. And your memory is not refreshed by what I read you?
A. No, on the whole it is confused by it.
Q. Very well. Doctor, did Naakon Chevalier tell you he had been interviewed by the FBI about the Eltenton Chevalier incident?
A. He did.
Q. When did he tell you that?
A. June or July of 1946.
Q. Shortly after he was interviewed?
A. Fairly shortly after.
Q. Did he tell you how long before that he had been interviewed?
A. I don't remember.
Q. Did you get the impression that it had been very recent?
A. I don't recollect the time interval. It was not a year or a month or month or anything—or a half year.
Q. What did he tell you about the interview?
A. He told me that he and Eltenton had been interviewed simultaneously, that they had questioned him about his approach to me.
Q. What else?
A. That they had asked him if he had approached anyone else, and I think— well, that they picked him up at Stimson Beach and had taken him into headquarters.
Q. Anything else?
A. That they pressed him about whether he talked to anyone else.
Q. Did he tell you what he said?
A. Not in any detail.
Q. How did he give you that information—in person or by telephone or by letter, or what?
A. What I recollect is that he came to our home.
Q. In Berkeley?
A. Yes.
Q. Was that before or after you were interviewed by the FBI?
A. It was quite a while before.
Q. So when you were interviewed by the FBI you knew substantially what Chevalier had told them?
A. Not in great detail.
Q. I said substantially.
A. Yes, I think so.
Q. Did you also learn what Eltenton had said?
A. No.
Q. Did you attempt to find out?
A. No.
Q. Now, you recall that you were interviewed by the FBI again in May 1950?
A. Yes.
Q. That was at Princeton, wasn't it?
A. Yes.
Q. Did you tell the agents on that occasion that you didn't know Weinberg was a Communist until it became a matter of public knowledge?

A. I may have said I was not certain. My own recollection of it is contrary to this interview with Lansdale which is that the first time I was alerted to it was by the FBI in 1946. But it is clear that I learned something about it or it may be clear that I learned something about it during the war.

Q. Didn't you tell the FBI agent on that occasion that you did not ynow that Weinberg was a Communist until it became a matter of public knowledge?
A. I don't remember.
Q. You don't remember whether you told them that or not?
A. No.
Q. When did it become a matter of public knowledge?
A. It is still not.
Q. Long after 1943, wasn't it?
A. He still denied it and I don't quite know what this refers to.
Q. Doctor, did you know prior to the time I began to read them to you that there existed transcripts of your interviews with Colonel Pash and Colonel Lansdale?
A. I imagined that.
Q. You think so?
A. Oh, yes.
Q. How do you think you found that out?
A. I didn't know it. I said I imagined it.
Q. You imagined it?
A. I thought almost certain that there would be a transcript of security talks.
Q. Had you asked anybody about it?
A. I don't think so.
Q. Were you interested?
A. I should have been. I don't think I was.
Q. In that same interview with the FBI in May 1950, did you tell the agent that you had made a big mistake in not dropping your Communist Party friend long before you did?
A. I certainly don't recognize that.
Q. Would you say you didn't tell them that?
A. I need to know more of the context and when this was and what it was about.
Q. It was May——
A. I don't mean the date of the interview, but the context of the interview, and what time we were referring to. Taken in this bald form——
Q. Did you say anything to that effect?
A. I don't know.
Q. Doctor, there came a time in 1949 when you testified before the House Committee on Un-American Activities concerning Dr. Peters, didn't you?
A. Right.
Q. Your testimony was thereafter reported in the public press, wasn't it?
A. Yes.
Q. Do you recall what paper it was that it came out in?
A. It was the Rochester paper.
Q. Did you see that item in the Rochester paper?
A. It was sent to me.
Q. Who sent it to you?
A. Oh, many people. Peters, probably Condon, several other people.
Q. Was your testimony accurately reported in the press?
A. As I remember, it was fairly accurately reported in the press. It was supposed to be secret testimony.
Q. I have before me a photostat—I am sorry I have not a copy, Mr. Garrison, but I will read most of it—a story of the Rochester Times Union, Rochester, N. Y., June 15, 1949. The headline was: "Dr. Oppenheimer Once Termed Peters 'Quite Red.'"
The leadoff paragraph is: "Dr. J. Robert Oppenheimer, wartime director of the atom bomb laboratory at Los Alamos, N. M., recently testified that he once termed Dr. Bernard Peters of the University of Rochester 'a dangerous man and quite Red,' The Times-Union Washington bureau reported today."
The story then continues after some explanatory paragraphs.
Mr. GRAY. Let me interrupt to say, have you another photostat of this news story?
Mr. GARRISON. I would like to see it if we could.
Mr. ROBB. Surely.
Mr. GARRISON. Is this question to be about Bernard Peters or Hannah Peters?
Mr. ROBB. Bernard.
Mr. GARRISON. The letter mentions Hannah Peters, and not Bernard.
Mr. ROBB. Are you sticking to that technicality, Mr. Garrison?
Mr. GARRISON. I was asking you if this was an inquiry into Bernard Peters' background.

Mr. ROBB. In Dr. Oppenheimer's testimony we have been talking about Bernard Peters for a couple of days.

By Mr. ROBB:

Q. The newspaper story continues after some explanatory paragraphs, "In his testimony, Dr. Oppenheimer said he became 'acquainted' with the existence of a Communist cell at Berkeley 'by disclosure of the intelligence agencies of the Government." The quotes are not closed, but I assume they should be there.

"Concerning Dr. Peters, Dr. Oppenheimer said he had known the scientists as a graduate student in the physics department in the late 1930's.

"Said Dr. Oppenheimer:

" 'Dr. Peters was, I think, a German national. He was a member of the German National Communist Party. He was imprisoned by the Nazis, and escaped by a miracle. He came to this country. I know nothing of his early period in this country. He arrived in California, and violently denounced the Communist Party as being a "do-nothing party." '

"Dr. Oppenheimer said he told Major DeSylva he believed Dr. Peters' background was filled with incidents that would point toward 'direct action.'

"Asked to explain this point, Oppenheimer observed:

" 'Incidents in Germany where he had fought street battles against the National Socialists on account of Communists; being placed in a concentration camp; escaping by guile. It seemed to me those were past incidents not pointing to temperance.'

"Questioned specifically on his reference to 'direct action', Dr. Oppenheimer said of Dr. Peters:

" 'I think I suggested his attack on the Communist Party as being too constitutional and conciliatory an organization, not sufficiently dedicated to the overthrow of the Government by force and violence.'

"Asked the source of his information that Dr. Peters had been a member of the Communist Party in Germany, Dr. Oppenheimer replied:

" 'It was well known. Among other things, he told me.'

"Dr. Oppenheimer said he could 'affirm that there is no connection between his (Peters') work and any application of atomic energy that falls within the jurisdiction of the (Atomic Energy) Commission. * * * I would believe that if Dr. Peters could teach what he knows to a young man capable of learning it, the country would be better off, because if Dr. Peters cannot be employed by the War Department, at least the young man could be employed by the War Department.' "

Doctor, are those quotations from the news story I have read you an accurate summary of your testimony?

A. They are fairly accurate. I didn't have the transcript at the time. I believe that a collation was made by Mr. Volpe, who had the transcript, to see how accurate they were.

Q. In other words, you checked it at the time to see if it was an accurate statement?

A. Somewhat later. It is not a very inaccurate statement.

Q. It is substantially accurate, isn't it?

A. I think so.

Q. Is there anything in there that you could point to and say that was out of line or inaccurate or incorrect?

A. Not without the transcript and perhaps not with the transcript.

Q. Did you at that time find anything to complain about in the accuracy?

A. No, not in the accuracy. The fact of the publication.

Q. Yes, you objected to that. Following the publication of that story, did you hear from Dr. Peters?

A. I did.

Q. How did you hear from him?

A. He wrote me—I was on the way west, and I learned that he had called my office at Princeton and my secretary reported to me what was bothering him; when I got to Berkeley there was a letter from him.

Q. What did he say?

A. He said that he was appalled and how could I have done him such harm. I don't remember the words of the letter.

Q. How long after you appeared and testified was that?

A. About 2 weeks.

Q. Did you hear from anybody else about this story?

A. I did indeed.

Q. Who?

A. Prof. Hans Bethe, Condon, my brother, Weiskopf, perhaps other people.
Q. What did Condon have to say?
A. He said I should not have hurt an innocent and loyal American in that way, that I must take him on at the institute if he lost his job, that if he lost his job, it would be wholly my doing. That I must try to make restitution, and that he hated to believe that I could have said such a thing, and in an attempt to protect myself. I knew very well if my file were ever made public, it would be a much bigger flap.
Q. Was that by way of a letter?
A. Yes, sir.
Q. Where was Dr. Condon at that time?
A. In Idaho Springs.
Q. What business was it of Condon's what you said about Peters?

Mr. GARRISON. Mr. Chairman, isn't this a new item of derogatory information that is being produced here?

Mr. ROBB. Dr. Oppenheimer brought the name up; I didn't.

Mr. GARRISON. Not in this connection.

Mr. ROBB. Certainly he did.

Mr. GARRISON. Not as an item of derogatory information.

Mr. ROBB. Mr. Chairman, I don't think we have to sift this through such a fine sieve as that.

Mr. GRAY. What is the objection, Mr. Garrison?

Mr. GARRISON. This is a wholly new transaction it seems to me. I don't know what Mr. Robb is leading up to, but it seems to be embarking on a course of discussion of Dr. Oppenheimer's relations with Dr. Condon. If that is to be regarded as an item of derogatory information whatever may be said of it, which I don't know, I should think that we should be entitled to some notice of it.

Mr. GRAY. I had not gathered, at least up to this point, that the discussion of Dr. Condon was of a derogatory nature. I believe Dr. Oppenheimer mentioned Dr. Condon's name. Do you need to refer at this point to Dr. Condon?

The WITNESS. I don't mind answering the question unless my counsel tells me not to.

Mr. GARRISON. I withdraw it.

The WITNESS. I don't know what business it was of Dr. Condon's, except that he was outraged at any harm brought to a scientist.

By Mr. ROBB:

Q. Is that the same Dr. Condon that wrote you about Lomanitz?
A. Yes, sir.
Q. And protested his draft deferment?
A. Yes, the same Dr. Condon about whom General Groves spoke this morning.
Q. Was Condon still on the project when he wrote about Lomanitz?
A. I don't know. He was cleared for it. He was an employee of Westinghouse, and I don't know his exact status. He was not part of the Los Alamos.
Q. Is that the same one as quoted as voicing absolute confidence and loyalty and integrity of Dr. Oppenheimer in the Princeton paper?
A. I would not be the least bit astonished but I don't know.
Q. You received that letter at Berkeley.
A. Right.
Q. At the same time did you see Peters at Berkeley?
A. Peters came to see me.
Q. Where did Peters come from?
A. He came from Idaho Springs.
Q. He went out with Condon?
A. There was a conference of physicists. I was supposed to go. I could not make it. I went straight to the west coast.
Q. Did your brother go?
A. No.
Q. Did Peters come alone?
A. His parents live in Berkeley.
Q. What did he say?
A. The general substance was: Was there any way in which I could help him to keep his job at the University. He also said I had misunderstood him about his being a member of the Communist Party in Germany. He worked with the Communists, he was not ashamed of it, but he was not actually a member and nobody could prove he was. He said, "You don't know my views about the American Communist Party" and I should not have quoted him.
Q. You were sure he had told you he had been a member?

A. I felt quite sure.
Q. You were sure he had told you he had been a member?
A. But I am not infallible in these things as is being made very clear in these proceedings.
Q. What did you do after you received the letter from Dr. Condon and talked to Dr. Peters?
A. I did a lot of things. I consulted Mr. Volpe over the phone who accompanied me to these hearings.
Q. Who is Mr. Volpe?
A. General counsel of the Atomic Energy Commission. He had accompanied me to the hearings. I told him of the great disturbance and I then wrote a letter to the Rochester papers which you no doubt have, the purpose of which was to undo any injury that I should not have done to Peters. I think I also wrote or communicated with the officials of the university saying that I would be glad to talk to them when I got back.
Q. Did you see Dr. Peters at Princeton before you saw him at Berkeley?
A. I believe I did.
Q. Did you tell him on that occasion that you had testified but that God had guided the questions so you didn't hurt him?
A. I certainly didn't.
Q. Or anything of that sort?
A. No.
Q. Did he come to see you at Princeton?
A. Yes, he did.
Q. About what?
A. I don't remember. He had been down to testify before this same committee. I imagine it was in that connection.
Q. He came to see you about your testimony and his testimony, didn't he?
A. I don't—I am sure he came to see me in connection with the testimony.
Q. And you discussed it?
A. I don't believe I discussed mine. It was in executive session.
Q. Did you discuss his? You didn't tell him in substance that God guided the questions or that fortunately the questions were such that you didn't do him any damage?
A. That would certainly not have been an accurate statement, and I don't remember making it.
Q. You then wrote a letter after you saw Peters and received Dr. Condon's letter to the paper.
A. Yes, sir.
Q. And in that letter you retracted some of the testimony you had given, did you not?
A. Right. I had that letter shown to the committee.
Q. Just for the record, I will read the letter. Do you have it there, Mr. Garrison?
Mr. GARRISON. No, but that is all right.
Mr. ROBB. This is a photostat from the Rochester, New York Democratic Chronicle, July 6, 1949. The letter is dated June 30, 1949, headed "Dr. Oppenheimer Explains."

"EDITOR, *Democrat and Chronicle:*

"Recently the Democrat and Chronicle published an article based on reports of my testimony before an executive session of the House Committee on Un-American Activities, which it seems to me could be damaging to the good name of Dr. Bernard Peters, of the University of Rochester.

"I first knew Dr. Peters about 12 years ago during his student days in California. I knew him, not only as a brilliant student, but as a man of strong moral principles and of high ethical standards. During those years his political views were radical. He expressed them freely, and sometimes, I thought, without temperance. This seemed to me not unnatural in a man who had suffered as he had at Nazi hands. I have never known Dr. Peters to commit a dishonorable act, nor a disloyal one.

"Dr. Peters has recently informed me that I was right in believing that in the early days he had participated in the Communist movement in Germany, but that I was wrong in believing—as the article stated—that he had ever held a membership in the Communist Party. That he has today no regrets for his actions in Nazi Germany he himself made clear in his statement that accompanied the publication of the article.

"From the published article one might conclude that Dr. Peters had advocated the violent overthrow of the constitutional government of the United States. He has given an eloquent denial of this in his published statement. I believe his statement.

"As indicated in the article, the questions which were put to me by the House committee with regard to Dr. Peters arose in part because of reports of discussion between me and the intelligence officers at Los Alamos. These Los Alamos consultations took place in connection with confidential wartime assignments. I wish to make public my profound regret that anything said in the context should have been so misconstrued, and so abused, that it could damage Dr. Peters and threaten his distinguished future career as a scientist.

"Beyond this specific issue, there is ground for another, more general, and even greater concern. Political opinion, no matter how radical or how freely expressed, does not disqualify a scientist for a high career in science; it does not disqualify him as a teacher of science; it does not impugn his integrity nor his honor. We have seen in other countries criteria of political orthodoxy applied to ruin scientists, and to put an end to their work. This has brought with it the attrition of science. Even more, it has been part of the destruction of freedom of inquiry, and of political freedom itself. This is no path to follow for a people determined to stay free.

"ROBERT OPPENHEIMER.

"BERKELEY, CALIF., *June 30, 1949.*"

By Mr. ROBB:
Q. Doctor, is that the letter you sent?
A. It is.
Q. How has your remark or testimony been misconstrued or abused?
A. Well, for one thing they were abused by being made public. This was an executive session. I should not have talked in executive session without thinking what they might do publicly.
Q. How had your testimony been misconstrued?
A. It was being misconstrued to mean that he should not keep his job. I had explicitly said that I thought it was good he keep his job.
Q. The report of your testimony was accurate, wasn't it?
A. Yes.
Q. And you have just told us that this letter was intended at least in part to repudiate that testimony, is that correct?
A. To repudiate is a little strong.
Q. Is it now your testimony, Doctor, that your testimony before the House committee to which we have referred was not true?
A. No, it is certainly not my testimony that it was not true. As to Peters' membership in the Communist Party in Germany, I have only really his word to go on. I am fairly sure of my initial recollection. I am very clear of his later denial. I don't think——
Q. Doctor, when you testified before the House committee, you knew for you to say that Dr. Peters told you that he had been a member of the Communist Party was a matter of great seriousness, didn't you?
A. Right.
Q. You would not have said that, would you, had you not been absolutely sure it was true?
A. I was convinced it was true, or I would not have said it.
Q. Beg pardon?
A. I was convinced it was true.
Q. And yet when Peters came to see you and you received a letter from Dr. Condon, you in effect repudiated that testimony, didn't you?
A. Does it say that I don't believe he was a member of the party?
Q. I have just read it to you.
A. I have forgotten.
Mr. GARRISON. Do you mind if I show it to him?
Mr. ROBB. Not at all.
The WITNESS. I don't say I believe his denial. I just say he denied it.

By Mr. ROBB:
Q. Very well. Isn't the implication of your letter that you were wrong in believing that he had been a member of the party?
A. I think it leaves the matter open.
Q. Was it your intention to convey that impression when you wrote the letter?
A. I think the sum total of my intention was not to get this guy fired from the University of Rochester because of intemperate remarks I made before the House committee.
Q. You thought your remarks were intemperate?

A. I think somewhat.
Q. You thought the truth was intemperate?
A. I think the phrasing of it was intemperate.
Q. Was it intemperate for you to testify, believing it to be true, that Peters had told you he had been a member of the Communist Party?
A. No.
Q. Wasn't it your intention in writing this letter on June 30 to convey to the public the impression that you had been mistaken in saying that he told you he had been a member of the party?
A. I simply gave his own statement.
Q. I know you did. But wasn't it your intention to give the public through the press the impression that you were mistaken?
A. I had no specific intention.
Q. When Dr. Condon wrote you about your own file what do you think he referred to by that?
A. I should think the material before you.
Q. Do you think you were being placed under any pressure by either Peters or Condon to retract what you said?
A. No, the real pressure came from people who were not belligerent at all, but who were regretful.
Q. Who were they?
A. Bethe, Weiskopf, my brother. They wrote very, very nice letters saying, this guy was being put—was suffering for something because I had done it and he should stay on his job.
Q. And you were influenced by that pressure, were you not?
A. Of course, I was.
Q. Where is Peters now?
A. He is in Tata Institute in Bombay.
Q. When did you hear from him last?
A. I had a note from him about physics, just an offprint, about a year ago.
Q. Did you help him get that job?
A. No.
Q. Did he——
A. Let's see. The man he works for—I didn't help him get the job. I know the man he works for.
Q. Who is that?
A. Bhabha is the name.
Q. Has he any Communist connection?
A. No, I don't know. He is an Indian, he is a millionaire. I don't know what he is.
Q. Do you hear from Dr. Peters frequently?
A. Very infrequently. I think the communications have been scientific papers, and one question, because I said he made a mistake and he wanted to know what I meant. I didn't answer it.

Mr. ROBB. Mr. Chairman, I have a subject that will take a few minutes. Would the Chair want me to continue or would you want to adjourn?

Mr. GRAY. It is now 25 minutes past 4, I believe. We usually sit until 4:30. I should like to inquire of Dr. Oppenheimer, and counsel, what their wishes are? The board is prepared to sit further.

Mr. GARRISON. Could we take a 3-minute recess?

Mr. GRAY. We certainly may, and we are prepared in the interest of moving along to sit further if Dr. Oppenheimer feels up to it.

Mr. GARRISON. Dr. Oppenheimer is ready to go on for about half an hour.

Mr. Chairman, I would just like to direct the Board's attention to a clipping I have just been handed from the New York Daily News for Thursday, April 15, 1954. The headline, "Two Letters Hit Oppenheimer as Informer." This is a news article by Jerry Green of the News Bureau, dated Washington, D. C., April 14. That is yesterday.

"The AEC problem of Dr. J. Robert Oppenheimer as a security risk tonight focused on two mysterious letters accusing the atomic scientist of turning informer in 1949 to protect himself. The letters bore the name of Dr. Hugh Condon"—
and so forth.

I know nothing more than to bring this to your attention.

Mr. ROBB. What paper is that in?

Mr. GARRISON. New York Daily News. It is today's paper but written dateline of yesterday, saying that this problem last night focused on the Condon letters, one of which at least has been the subject of testimony this morning.

Dr. EVANS. I read that last night. I wondered about it.

Mr. ROBB. They must have been clairvoyant. Could we proceed, Mr. Chairman?

Mr. GRAY. Yes.

By Mr. ROBB:

Q. Doctor, I would like to talk with you a bit about for convenience what I should call the Kenilworth Court incident. Are you familiar with what I am talking about?

A. I do indeed.

Q. You are quite familiar, are you not, with the testimony of Paul Crouch and Mrs. Crouch concerning that episode?

A. I have read it. I have gone over part of it with counsel. I am certainly familiar with its general import.

Q. I am not trying to trap you or anything, but merely trying to save time. If you are familiar with it, I won't need to rehearse it.

A. There may be points that I will be unfamiliar with but let me raise those.

Q. You recall that they testified in substance that they had been present at a closed Communist Party meeting in late July 1941—was it?

A. Right.

Q. At a house which you were the lessor at 10 Kenilworth Court, Berkeley, Calif. Crouch had addressed the meeting, explained the Communist Party line, and I believe they said that Joseph Weinberg was also present. Is that about it?

A. That is at least part of it.

Mr. GARRISON. The word is "lessee."

Mr. ROBB. I said lessor; the lessee.

By Mr. ROBB:

Q. I believe, Doctor, that the substance of your response in your answer at page 30 is that you were the lessee of the house at that time, but that you were in New Mexico, and did not attend such a meeting, is that correct?

A. That is part of my answer. The rest of it is that I also didn't attend such a meeting at any time.

Q. Yes, sir. You state in your answer, I believe, that you attempted to establish your whereabouts and with the assistance of counsel had found you were in New Mexico, is that correct?

A. I stated that. My answer, I think took—we will find the words, as near to July 23 as the central date.

Q. That is page 30 and 31.

A. I thought probably that at that time of that meeting which by then had been fixed by Crouch as approximately July 23, my wife and I were away from Berkeley. Shortly after with the aid of counsel——

Q. Is it your testimony now, Doctor, that you did not attend that meeting or any similar meeting at which Crouch made a talk?

A. Yes. A closed meeting of the Communist Party in my house at which people were introduced at which Crouch talked, I did not attend.

Mr. SILVERMAN. At which people were not introduced.

The WITNESS. Were not introduced.

By Mr. ROBB:

Q. Had you ever attended a closed Communist Party meeting of any kind?

A. I told you of the meeting at my brother's, which was not closed, because we were guests, but where everybody else I understood to be a Communist. I know of no other.

Q. Except for the identity of the speaker, Doctor, was there any substantial difference between the meeting at your brother's which you described and the meeting which Crouch described.

A. There was everything different. Crouch described a lecture. No one was introduced. It was at my house. It was to discuss, according to his description, high Communist policy. The meeting at my brother's house was a meeting at which people were introduced, at least to some extent, very friendly and not a lecture. They had literature. There was no talk of literature at the Crouch description. Everybody at this other meeting knew everybody else, except the two visitors, who were introduced as visitors. No similarity that I can see.

Q. What about the meeting at Louise Bransten's house? Was there a speaker there?

A. There was.

Q. Who was that?

A. Schneiderman.

Q. He was also the speaker at Haakon Chevalier's house, wasn't he?
A. Right.
Q. What was the difference between those two meetings and the meeting that Crouch described, of course leaving out the fact that Crouch described a meeting at your house which we know.
A. I had no impression and I know that the meetings at the Chevaliers and the Branstens were not intended as Communist Party meetings.
Q. But at both meetings, both the Chevalier and Bransten, you had a talk from a high Communist Party functionary about the Communist Party line, didn't you?
A. Right, absolutely.
Q. So at least to that extent those meetings were similar to the one described by Crouch.
A. Right.
Q. And they were both at night in a private house, is that correct?
A. Yes.
Q. And the one described by Crouch was at night and in a private house?
A. Yes.
Q. I believe you told us that so far as you recall most everybody at the Bransten meeting and the Chevalier meeting was either a Communist or a fellow traveler.
A. I said taking sympathizer in a broad sense, a sympathizer.
Q. So that the Bransten meeting and the Chevalier meeting and the one described by Crouch were all meetings where a high Communist Party functionary was going to explain and expound the Communist Party line, is that correct?
A. In that respect, correct. I believe there is a difference, because Crouch's description indicates that he was telling the comrades what to say and do. There was none of that quality in these other meetings.
Q. You mean Schneiderman didn't tell the comrades anything?
A. He said the party stands for this, the party decides, and so on, as a sort of exposition.
Q. Do you have any doubt that any comrade there present would have gathered from what Schneiderman said what Schneiderman was, what he was supposed to believe and say and do?
A. I can testify that it had no such meaning for me, because I was not a comrade.
Q. That was not quite my question, Doctor. Would you read my question back to the Doctor?
(Question read by the reporter.)
The WITNESS. I have a little doubt. I had more the feeling that this was a public relations show on Schneiderman's part.

By Mr. ROBB:
Q. You thought it was necessary for a Communist Party functionary to engage in public relations with comrades?
A. No, they weren't all comrades.
Q. A substantial number of them were, weren't they?
A. I don't know.
Q. Doctor, when you first heard about Crouch's testimony before the California committee, did you immediately deny categorially that you had been present at such a meeting?
A. I first heard about it from the Federal Bureau of Investigation, not from the testimony.
Q. When you first heard about it from the FBI, did you immediately categorically say, "No, I was not present?"
A. I said it pretty strongly. It took a long time. The investigators produced more and more detail and the more detail they produced, the more convinced I was that it had not occurred.
Q. Whose investigators?
A. The FBI investigators. I have forgotten their names.
Q. You mean it was not until after an investigation had been made——
Q. No, the FBI came to see me about this matter before the California testimony.
Q. Yes, sir.
A. They started interrogating me. As their account of the details of the meeting, the suggested details of the meeting, developed, it became quite clear that it had not occurred. I promised to talk it over with my wife and see if she

had any recollection of it, and I saw them a few days later, and said by then I was sure it did not occur.

Q. It took you a little while to make up your mind whether you had been present or not, didn't it?

A. I didn't know what it was that they were talking about. It did not come out in terms of a closed Communist meeting. You probably have the record of the interview, and I don't remember the details. But I remember that it wasn't until the thing had some shape that I knew what they were talking about.

Q. As soon as you were told that the question was whether you had been at a closed Communist Party meeting and Paul Crouch had made a talk——

A. I was not told about Paul Crouch.

Q. As soon as you were told by anybody that the question was whether or not you had been at a closed Communist Party meeting where somebody had made a talk, did you immediately say "No, it couldn't have been true; I was not there"?

A. I don't remember, and I don't know that is the form in which the incident occurred in the FBI interview.

Q. Was that prior to the time when Crouch testified?

A. Yes, I am sure that as a result of these two interviews with the FBI I did deny it. But I do not know in what sequence the itemization of this meeting occurred.

Q. But, Doctor, isn't it a fair statement to say that it took you some little time before you finally denied that you had been at such a meeting?

A. I don't—it probably took me some time, but I don't know what the facts withheld from me were until I denied it. This is available to you, but it is not in my memory. I will say one thing. I believe it was late in the interview that I said this didn't happen. But I don't know in what order things occurred.

Q. Do you recall telling the agents that you did recall somebody asking you to "give us your house for a meeting of young people"?

A. I have that in my answer.

Q. Do you recall telling the agents that?

A. No, I don't.

Q. Do you recall telling that the person who requested such permission could have been Kenneth May, but you don't recall that it was?

A. Yes. Now I recollect. I said that to somebody. Whether it was to the United States attorney or the agents, I don't remember.

Q. What I am asking you about is an interview with two special agents of the FBI.

A. Right.

Q. I believe at San Francisco.

A. No.

Q. Pardon. Wasn't it? Was it Princeton?

A. If it is the one I am thinking of, it was at Princeton.

Q. Do you recall making that statement about possibly loaning your house to some young people, possibly Kenneth May being involved, making that statement in May 1952, to perhaps the United States Attorney's office here?

A. Yes. In fact, I say that in my answer.

Q. Did you categorically deny ever having been at such a meeting, Doctor, until after your representatives had made the investigation in New Mexico?

A. I categorically denied it to the FBI in these two interviews.

Q. Doctor, your position is that you could not have been there because you were in New Mexico, is that right?

A. No. My position is a little more complicated than that. It is first that I recollect nothing about it, and that the circumstances are such that I don't believe I could fail to recollect it. It is second, that if I were there, it could not have been a closed meeting of the Communist Party, because I wasn't. It is, third, that at the time it is alleged to have occurred, and for a considerable time before and after that, we were not there.

The first point is important. I forget a lot, but the notion that I would forget a meeting in my own home at which a lecture has been given, I think that has never happened, is a little hard. The notion that I would forget a meeting in my own home filled with people at which no one was introduced is a little hard.

Q. Doctor, you purchased your house at one Eagle Hill there from a Mrs. Damon?

A. I did.

Q. Those negotiations were going on during July of 1941, weren't they?

A. Yes.

Q. Do you recall meeting with Mrs. Damon at Berkeley to divide up some furniture that was in the house?
A. Yes.
Q. In the new house?
A. Yes.
Q. Do you recall when that was?
A. Not with precision, but I should think the 8th of August was a pretty good guess. 10th of August, maybe. I am not sure.
Q. Did you go back to New Mexico after that happened?
A. No.
Q. Did you stay in Berkeley?
A. We stayed in Berkeley.
Q. It would have been physically possible for you to fly back from New Mexico for a day or two to Berkeley and then return, would it not?
A. Of course it would.
Q. It would not have taken you very long to get back from New Mexico?
A. No, especially for an important Communist meeting.
Q. But your testimony is that you didn't do that.
A. That is right.
Mr. ROBB. I thing that brings that item to a close.
Mr. GRAY. Do you have one other line of questioning?
Mr. ROBB. It will take quite a while.
Mr. GRAY. It is not yet five. Maybe I can ask two or three miscellaneous questions, if I may.
The WITNESS. Is there just one other major line of questioning?
Mr. ROBB. I think so, Doctor, but don't hold me to that.
Mr. GRAY. I was interested a while ago, and I suppose this is more curiosity than anything else, when you referred to being accompanied by counsel to the House Un-American Activities Committee, you mentioned Mr. Volpe.
The WITNESS. That is right.
Mr. GRAY. And identified him as General Counsel.
The WITNESS. Right.
Mr. GRAY. Was he at that time General Counsel?
The WITNESS. He was. And I was Chairman of the General Advisory Committee.
Mr. GRAY. It was in that capacity?
The WITNESS. I think the Commission agreed or decided that this was a proper arrangement.
Mr. GRAY. I don't know whether you know the answer to this question, Dr. Oppenheimer, but in reading the files, there appear references to closed meetings of the Communist Party. There also appear many references to meetings of people who were Communists or fellow travelers, which were referred to as social gatherings. Without implying that you are an expert in these matters, but from conversations with your brother, perhaps, or Mrs. Oppenheimer and others, is there any real difference between a closed meeting and a social gathering if the same people are involved?
The WITNESS. Let me tell you what I mean by the words. The words "closed meeting" mean to me one to which only members of the Communist Party can come. I think that is a rather sharp distinction if you are trying to identify who is and who isn't a member of the Communist Party. I should suppose that the difference between a meeting and a social gathering was rather wide. In a meeting it was business and it was transacted and there was probably a chairman and there might be dues collected and there might be literature. Anyway, this happened at the little meeting I saw at my brother's. I should think that a social gathering would be a lot of talk which could indeed be very bad talk, but which would not be organized or programatic. This is the sense in which I would interpret the words.
Mr. GRAY. So these two meetings which have been the subject of some discussion at both of which I believe Mr. Schneiderman spoke, in the terms of the definitions which you have given, they would really have been social gatherings?
The WITNESS. I would say they were neither. They were social gatherings ornamented by a special feature, namely, this lecture or speech. An ordinary social gathering I don't think has a lecture even in Communist jargon.
Mr. GRAY. I just had the impression about these functions that many of those that we referred to were social gatherings may have been meetings. That doesn't concern your attendance at all.
There is one question I have which relates to the security of the project itself. Very early in your testimony in some discussion about procedures or security measures which were taken after very careful thought, you made the observa-

tion obviously they did not succeed. Again this is not a direct quote. Do you mind amplifying on that just a moment?

The WITNESS. Yes. I think of the known leakages of information, Fuchs is by far the most grave. It occurred out of Los Alamos. I won't attempt to assess responsibility for the surveillance of personnel who moved around there. Facilities for surveillance were available, and they could well have been used in following Fuchs rather than somebody else. That would not have prevented his prior espionage, but it would have prevented the espionage at that time. I can't imagine any more pinpointed leakage than if Fuchs had simply communicated what he was working on. I don't mean that this was the only secret, but I can't imagine any single little point that would be more helpful to an enemy than the job he had himself. While not wishing to debate with General Groves either the necessity, the desirability or the dangers of compartmentalization, I would like to record that if Fuchs had been infinitely compartmentalized, what was inside his compartment would have done the damage.

Mr. GRAY. Here is a rather sharp change of pace again. Do you know what was the source of the allegation that you caused to be circulated the GAC report at Los Alamos?

The WITNESS. No. I would very much like to know the source of it. I have a conjecture, but I would prefer to be told and not to make the conjecture.

Mr. ROBB. Might I ask this one question, Mr. Chairman, suggested by a question that you asked?

Mr. GRAY. Yes.

By Mr. ROBB:

Q. Dr. Oppenheimer, when did you tell the FBI about this meeting at Louise Bransten's house.

A. I spoke of that earlier. I said at the latest 1952 and possibly earlier. I don't remember.

Q. At one of the interviews that we have talked about, either in 1946, 1950, or 1952, is that right?

A. Yes, in 1952 I think it possible that it was an interview with my wife at which I was present.

Q. You think that is when you told them about it?

A. I know that it occurred not later than that interview.

Q. Have you been interviewed since 1952?

A. Only minor ones, not protracted ones.

Mr. GRAY. Again a shift of subject, and you may have answered a question about this, Dr. Oppenheimer. I am sure you testified that your brother Frank had told you that he joined the Communist Party.

The WITNESS. Yes.

Mr. GRAY. Were you questioned about your knowledge about his severing connection with the Communist Party? I don't want to plow over some old ground.

The WITNESS. Was I questioned here? I think I was. I think it is in the transcript. As to the facts, I felt assured by talking to him in the fall of 1941 he was no longer a member. Whether that is because I asked him or because he told me——

Mr. GRAY. Yes, I do remember that was covered. Were you also asked about Mrs. Frank Oppenheimer? Did she follow the same course?

The WITNESS. I understood this went for both of them, but perhaps not quite so sharply.

Mr. GRAY. One other unrelated question. We talked yesterday about your having dinner or at least a social visit with Dr. Chevalier in Paris in November or December. There was in the morning press a statement attributed to Dr. Chevalier which had a St. Louis date line, I believe.

The WITNESS. Yes.

Mr. GRAY. Which indicated to me that he is in this country. Were you aware of the fact that he was back in this country?

The WITNESS. No. In fact, I am skeptical of it.

Mr. GRAY. I gather you are saying that you have no reason to believe that he is not still in Paris?

The WITNESS. That is right.

Mr. GRAY. I don't want to clutter up the record with the quotes of Dr. Chevalier, but am I right in thinking I read that in today's press?

Mr. ROBB. You are correct.

Mr. GRAY. It is now a couple of minutes after five. We will meet at nine in the morning and you will proceed with the questioning.

(Thereupon, at 5:30 p. m., a recess was taken until Friday, April 16, 1954, at 9 a. m.)

UNITED STATES ATOMIC ENERGY COMMISSION

PERSONNEL SECURITY BOARD

IN THE MATTER OF J. ROBERT OPPENHEIMER

ATOMIC ENERGY COMMISSION,
BUILDING T-3, ROOM 2022,
Washington, D. C., Friday, April 16, 1954.

The above entitled matter came on for hearing, pursuant to recess, before the board, at 9 a. m.

Personnel Security Board: Mr. Gordon Gray, chairman; Dr. Ward T. Evans, member; Mr. Thomas A. Morgan, member.

Present: Roger Robb, and C. A. Rolander, Jr., counsel for the board; J. Robert Oppenheimer, Lloyd K. Garrison, Samuel J. Silverman, and Allen B. Ecker, counsel for J. Robert Oppenheimer. Herbert S. Marks, cocounsel for J. Robert Oppenheimer.

PROCEEDINGS

Mr. GRAY. The proceeding will begin.

Mr. GARRISON. Mr. Chairman, before we begin, I want to make one procedure question.

When we adjourn this afternoon at half past three, it would be very helpful to us if we could have copies of all the transcripts of the testimony to date, whether they have been cleared or not, to work on. I assume this can be arranged. In other words, we could work on them in the other room with the understanding that they will not be taken out of the building, so that we can do some work on them over the weekend. I think perhaps this afternoon, this evening and tomorrow would pretty well do it. Are they still going the rounds?

Mr. ROBB. Frankly, Mr. Garrison, I don't know. I have had so many other things on my mind, I don't know what has been happening. I know somebody is reading them with a view to seeing what should be classified and what should not. Who had to do it and how many times it has to be read, I don't know. I have not read it myself.

Mr. GARRISON. This is the end of the first week now. Next week is going to be a very concentrated string of witnesses. This is about the only time that we shall have to do any work on them. There was such a jumble of dates and names that it is pretty hard just from scribbled notes here to——

Mr. GRAY. In response to the point raised by Mr. Garrison, I would have to say that I don't know what the situation is with respect to transcript and I will have to find out and we will respond.

Mr. GARRISON. That is why I raised the point at this point of time. I assume that the only problem is they are going out of the building, because as far as we are concerned, we have heard it all.

Mr. ROBB. It seems reasonable to me that Mr. Garrison should have access to them if they are available. As I say, there are higher powers than I.

Mr. GRAY. We will respond to the request.

Mr. ROBB. You have heard all the testimony, so why can't you read it?

Mr. GARRISON. It would be far better if they were released and we could keep them.

Mr. ROBB. Yes, but I don't have any control of that.

Mr. GARRISON. The next request has to do with the transcripts of the interviews with Pash and Lansdale. I have personally not had time to go over them, but my associates have, and I would like very much to have an opportunity to go over them myself at the end of the afternoon session.

Mr. ROBB. Surely.

Mr. GARRISON. Also, I would like to hear, and I think I should be entitled to hear, the recordings, because it appears from these transcripts there are places were they just don't seem to make sense at all. There were quite a number of gaps and statements when one doesn't know which voice is what, just from the grammatical structure of the thing. I don't want to make too much about this at all, but I am worried about it as counsel.

Mr. GRAY. We will receive this request along with the other, and we will respond to it in the course of the day.

Mr. GARRISON. I am told that the Pash transcript says in a little box at the top of it, some indication that this does contain errors and is substantially correct, or words to that effect. I feel this particularly on my conscience because I think it may well be that if we had the sense of what that transcript was like at the time Dr. Oppenheimer was testifying, I am not at all sure his testimony at all points would have been quite as it was. I don't want to overdo that point, but I want you to feel that sense of urgency that I as counsel do about it.

Mr. GRAY. I would make the observation, Mr. Garrison, that it is entirely possible that Dr. Oppenheimer's might not have been the same, but this is his testimony and not counsel's testimony.

Mr. GARRISON. That is right.

Mr. GRAY. I don't know that your having had an opportunity to read these transcripts in advance and advise Mr. Oppenheimer, if it had changed his testimony essentially, it would not have been in the interest—I don't suppose you meant to imply that.

Mr. GARRISON. No, I didn't mean having them in advance and advising him before, but simply having them before me as they were read so I might see what these gaps and garbles were. I did have the sense of the testimony in connection with the Lansdale one had rather a different quality and the line of questioning perhaps. But I don't want to carry the argument any further or push it an inch beyond what it is entitled to. I just want to express my sense of urgency as counsel to do a good job.

Mr. GRAY. I understand, and I have received the two requests, and we will respond to them.

Whereupon, J. Robert Oppenheimer, the witness on the stand at the time of taking the recess, resumed the stand and testified further as follows:

CROSS EXAMINATION—Resumed

By Mr. ROBB:

Q. Doctor, I have 1 or 2 miscellaneous questions. You mentioned Mr. and Mrs. Serber yesterday. Did you know them very well?
A. I did.
Q. How long have you known them?
A. He came as national research fellow to Berkeley. I think he held the fellowship 2 years. This may have been 1934–35. He stayed on as my research assistant I think for another 2 years. I got to know them during the period of this fellowship. I have known them ever since.
Q. Did you know his wife, Charlotte?
A. Sure.
Q. You mentioned that she had a rather important position at Los Alamos. What was it?
A. She was librarian.
Q. Did that mean she had charge of all the technical publications and technical materials in the project there?
A. She was in overall charge. The actual documentary stuff was in the immediate charge of another woman.
Q. Who was the other woman?
A. I have forgotten her name.
Q. Was Mrs. Serber's position one which would be described as highly sensitive?
A. Yes.
Q. She had access to a great deal of important classified information?
A. Yes.
Q. What did you know about her background so far as Communist connections were concerned?
A. I knew that she came of a radical family, the Leof family. I saw and heard in the transcript of my interview with Lansdale that I said she had been a member of the Communist Party. I have no current belief that this is true. I told you that she was very active in Spanish relief and that she and her husband had strong leftwing views.
Q. You knew that when she came to Los Alamos?
A. Yes.
Q. Were her activities and her beliefs such as those that we have described, I believe, as indicia of communistic tendencies? Do I make myself plain?
A. Only in part. I recollect, for instance, her expressing concern and dissatisfaction with the purge affairs, which I think was not a pro-Communist position. On the Spanish thing she was certainly very, very much engaged.
Q. On the leftwing side?
A. On the Loyalist side, which was also the leftwing side.
Q. How did you know about her family in Philadelphia?
A. I once met them. When I was in Philadelphia I met them on another affair. But this is something that over the years she gossiped about quite a lot.
Q. You said you knew she was quite radical, I believe.
A. Yes.
Q. Would you explain what you meant by radical?
A. I will try. I believe Leof was an old-time Socialist. Probably he was a Socialist when the various factions had no split. I believe that they also were very much concerned with the Spanish cause. I believe they also had leftwing friends. But I do not know any details.
Q. What did you mean when you spoke of the factions splitting, Doctor?

A. The Socialist Party, the Communist Party, the Trotskyite Party, the Stalin Party, and so on.
Q. Which faction did you understand that Leof went with?
A. I didn't understand.
Q. You were more or less familiar with those details of the factional disputes and debates in the party?
A. No, I was familiar with their existence.
Q. Was Mr. Serber also at Las Alamos?
A. Yes, he certainly was.
Q. What was his position?
A. He was head of a group in the theoretical physics division.
Q. Likewise, I assume, in possession of a great deal of classified information?
A. Indeed.
Q. Did you have anything to do with bringing them there?
A. Oh, yes, I was responsible.
Q. What did you do to bring them there?
A. I believe that they came to Berkeley for the summer study in 1942 along with the others that I mentioned. I think that they were still in Berkeley at the time we went to Los Alamos. They followed us there shortly after that.
Q. At your suggestion?
A. Yes.
Q. Where are the Serbers now?
A. At Columbia University.
Q. Do you see them frequently?
A. Very infrequently, to my regret.
Q. You still consider them your friends.
A. Oh, yes. I think they are no longer in any way leftwing.
Q. When did you last hear from them?
A. It is quite some time. Not a year, but they had personal difficulties this autumn, and we were in communication with them about that. I had a note from him on recommending a candidate more recently.
Q. Candidate for what?
A. A membership in the institute.
Q. You mentioned a man named Philip Morrison, doctor.
A. Yes.
Q. How well did you know him in 1943?
A. In 1943? I had known him well when he was in Berkeley. He was away, I don't remember quite how many years after leaving Berkeley. But I had known him very well at Berkeley.
Q. In what connection?
A. As a student and as a friend.
Q. You saw him socially and shall we say officially?
A. Yes. He was a student and then I believe he could not get a job, and we made some kind of an arrangement for him to stay on. I think he was probably in Berkeley 4 or 5 years.
Q. Did you see Morrison at many of these leftwing functions that you attended?
A. Not so many, I should think. He was not a person who was going to give much money to the Spanish cause. He had no money.
Q. What did you know about his political beliefs and affiliations in 1943?
A. As of then, or as of an earlier time?
Q. Beg pardon?
A. As of then I knew nothing.
Q. As of an earlier time.
A. As of an earlier time I knew that he was very close to the party and would have presumed that he might have been in the YCL or in the party.
Q. I believe you told us that yesterday. I believe you said yesterday that you either knew, or assumed that he was a member of the Young Communist League; is that right?
A. No; I didn't say that yesterday.
Q. Did you read Dr. Morrison's testimony before the Senate Committee on the Judiciary?
A. I did not read it carefully. I think I was away when he testified. I am not sure. I know the substance of it.
Q You know that he testified that he had been a member of the Communist Party.
A. Right.

Q. That didn't surprise you?
A. No.
Q. It was in accord with what you previously had known about him in general; is that correct?
A. It was.
Q. Morrison was a man who I believe you said went over to Japan before the drop on Hiroshima?
A. Not before. I think after.
Q. For what purpose did he go there?
A. I think to inspect damage. There was a team under General Farrell, and he wanted to see what the mess was that we had made.
Q. In other words, they wanted to see how the thing you made had worked.
A. Yes; and whether there was radiation; to make a good observation of the consequences.
Q. Who else from Los Alamos went over at that time?
A. Serber was also in Japan because he brought me a bottle from Nagasaki. I don't remember who else. Alvarez, I think.
Q. Did you select Serber and Morrison for those missions?
A. I don't recall how the selection was made. I would certainly not have been without responsibility for it, no matter how it was made. There may be a record of that.
Q. They would not have gone had you not approved it?
A. They would not have gone if I disapproved, that is certain.
Q. How recently have you seen Philip Morrison?
A. I think it may be a year ago.
Q. What were the circumstances?
A. I gave a lecture at the Rumford Bicentennial in Boston. I am not completely certain of this. I have not been in Ithaca, and he has not been in—well, I have not been in Ithaca, and he has not visited me at Princeton for something like a year.
Q. Has he visited you at Princeton since the war?
A. I don't recollect. It would have been very natural that he should have
Q. Why do you say it would have been very natural?
A. Princeton is a place that almost all physicists visit. He and I are old friends. I mean no more than that.
Q. And what?
A. I mean no more than that. He has not spent the night at our house or anything like that.
Q. But I assume that you had the occasion arise when you would have been happy to have offered him your hospitality for the night?
Mr. GARRISON. Mr. Chairman.
The WITNESS. This is not a question I feel capable of answering.

By Mr. ROBB:
Q. You still consider him your friend?
A. Yes. I don't feel very close to him. I suspect that though he is no longer at all close to the Communists, his views and mine do differ, and perhaps on matters on which he feels rather strongly.
Q. You say he is no longer at all close to the Communists?
A. That is my understanding.
Q. Where did you get that understanding?
A. We have many common friends.
Q. Who told you that he was no longer close to the Communists?
A. I don't think it is any one man. He worked at MIT last year, and several of the professors there talked to me about him and several of the people from Ithaca have talked to me about him.
Q. Did you base that understanding in any part upon Morrison's testimony which he gave before the Senate committee in May 1953?
A. No. Perhaps I should have, but I didn't.
Q. You have gone over that testimony?
A. I have gone over it this way (gesturing).
Mr. GARRISON. What was the answer?
The WITNESS. I have gone over it not in great detail.

By Mr. ROBB:
Q. I believe you said, Doctor, that you didn't think Morrison had visited you at Princeton during the last year. Was that your testimony?
A. That is my recollection; yes, sir.
Q. Had he visited you at Princeton prior to a year ago?

A. You asked me the question and I said I supposed it was likely. I have no recollection of a visit.

Q. Have you visited him or lunched or dined with him either in New York or Princeton or Ithaca or wherever since the war?

A. Yes. I had one dinner with him which I remember vividly. I think Mr. Marks——

Q. Mr. who?

A. Mr. Herbert Marks, Mr. Bacher, he and I had dinner together at the Hotel Brevoort. I may be wrong about Mr. Marks. Anyway, Bacher, Morrison, and I had dinner together, and I think Mr. Marks was there. This was during the time when he was on a committee appointed by General Groves——

Q. Who was on the committee?

A. Morrison—to consider the international control of atomic energy, and I was on a committee appointed by Mr. Byrnes to consider the international control of atomic energy. We were with encouragement as well as approval doing a little cross talking to see what ideas there were in the technical group.

I have also seen him at another time—certainly more than once he lectured at Cornell in the spring of 1946—and I would presumably have seen him then, though I don't specifically recollect it. I lectured at Cornell later, and I am sure I saw him at the reception which was given for me at the time. We have attended conferences of physicists and I am sure I have seen him then. This is probably not a complete list, but that is what comes to mind.

Q. Now, Doctor, I would like to turn to the matter of the thermonuclear problem.

A. Right.

Q. I think it might be helpful to the board, sir, if you gave, if possible, some categorical answers to some of the statements made in General Nichols' letter. I don't find that your letter of answer sharpens those issues, and I wonder if you can't sharpen them a bit. Do you have General Nichols letter before you?

A. I will get out General Nichols' letter. But to questions that are badly phrased, categorical answers are not always possible.

Q. Let us try, Doctor.

Page 6 of General Nichols' letter at the bottom of the page. Do you have it before you, sir?

A. I have it before me.

Q. "It was reported in 1945 you expressed the view that 'There is a reasonable possibility that it (the hydrogen bomb) can be made,' but that the feasibility of the hydrogen bomb did not appear on theoretical grounds as certain as the fission bomb appeared certain on theoretical grounds when the Los Alamos Laboratory was started."

Is that a true statement, Doctor?

A. You mean is this a true statement about the thermonuclear bomb or about my assertions?

Q. Your assertions.

A. It is a precise statement of what I thought.

Q. In 1945.

A. In 1945.

Q. Did you express that view in 1945?

A. I wrote a report. You see, I don't know to what document this refers. Is this in the Interim Committee report? If you will tell me where this is alleged to have been written, I will confirm it. It is an exact quotation, or purports to be an exact quotation. I have no objection to saying that it is a reasonable quotation, but how can I confirm it without knowing whether this is testimony before the Joint Congressional Committee, or an interview with Colonel Lansdale or a report I wrote.

Mr. GRAY. Can you identify the source of that?

Mr. ROBB. I am looking for it right now.

The WITNESS. Please don't misunderstand me. This is a good statement of what I believed. But I am being asked to say did I actually say it.

Mr. GARRISON. Mr. Chairman, I think we are entering an area here where, if this is an inquiry and not a trial, great latitude should be allowed the witness to explain his answers. I am sure that nothing could be more misleading than to have a simply yes or no as in a trial to things that simply overflow the landscape and their surrounding factors.

Mr. GRAY. I just make the observation that I don't recall, Mr. Garrison, at any point in this proceeding when the witness was interrupted in any way. Do you?

Mr. GARRISON. No.

The WITNESS. I was asked to make categorical answers and to some extent it might not be possible.

Mr. ROBB. I said it would be helpful to the committee.

The WITNESS. I will do the best I can.

By Mr. ROBB:

Q. You will agree it would be helpful to the board?

A. I do not agree on that second point. I will gladly state that this first statement is a good expression of my overall view in 1945, that I had occasion to report to the Government both to the congressional committee, McMahon's committee, and to the War Department, and no doubt to other places and I would have expressed my view, and since this was it, I have no objection to taking this as an expression of my view.

Q. Very well. That answers the question.

Now, to continue: "* * * and that in the autumn of 1949 the General Advisory Committee expressed the view that 'an imaginative and concerted attack on the problem has a better than even chance of producing the weapon within 5 years.'"

A. I think that is a direct quotation from the report of the October 29 meeting of the General Advisory Committee. I believe I wrote it myself. I think the committee had agreed with this statement ahead of time. I believe we discussed the statement and it is an expression of the views of the committee and of me.

Q. So that statement is true.

A. It is true.

Q. "It was further reported that in the autumn of 1949 and subsequently you strongly opposed the development of the hydrogen bomb: (1) On moral grounds; (2) By claiming it was not feasible; (3) by claiming that there were insufficient facilities and scientific personnel to carry on the development, and (4) that it was not politically desirable."

Is that statement true either in whole or in part?

A. It is true in part. It is out of context and it gives a very misleading impression.

Q. Now, would you please explain your answer and tell us what part is not true, what part is true?

A. I would say that in the official 1949 report, which you have read, we evaluated the feasibility, as it is stated up above, namely, that there was a better than even chance that if you worked hard on it and had good ideas you would have something in 5 years. That was then our view.

In the same report, which you have read, we pointed up the moral and political arguments against making an all-out effort. This was primarily in the annexes that were attached to the report, rather than in the official report which I prepared.

I think it possible that similar arguments were repeated in the report of the next meeting of the General Advisory Committee.

Q. Which would be when, Doctor?

A. Between the end of October and the 1st of January. Probably early December, or something like that. We did not at that time claim that it was not feasible, and I believe that I have never claimed that the hydrogen bomb was not feasible. But I have indicated, starting with early 1950 * * * very strong doubts of the feasibility of anything that was then being worked on. These doubts were right.

Q. Did you indicate such doubts prior to the GAC meeting of 1949?

A. In 1948 we had a GAC meeting and in that we didn't say it was not feasible, but I think we said it didn't look good. Something——

Q. Doctor, pardon me. I am talking about you. Did you say it was not feasible or it didn't look good?

A. As a member and Chairman of the General Advisory Committee, I said it didn't look good until some time in 1948.

Q. 1948.

A. Yes. This was a specific model and all of this is about a specific model. We will try to do this without classified stuff.

Q. Was that still your view at the time of the GAC meeting of October 29, 1949?

A. That it didn't look good?

Q. Yes.

A. If it had not been, we would not have said it would take 5 years and an imaginative and concerted attack.

Q. Doctor, would you come back to the centers we are talking about?
A. Right.
Q. I think you have mentioned the moral grounds. May I ask a question about that before we proceed to something else?
Did you continue your attitude in respect to the moral grounds subsequent to the GAC meeting of October 29, 1949?
A. I think we need to distinguish sharply as to whether I expressed in official reports or in dealings with the Government any desire to reraise the decision.
Q. Doctor, you and I are getting along fine. That was going to be my next question, so will you answer that, too?
A. I am quite sure we did not ask to have the decision reconsidered.
Q. Did you subsequent to the President's decision in January 1950 ever express any opposition to the production of the hydrogen bomb on moral grounds?
A. I would think that I could very well have said this is a dreadful weapon, or something like that. I have no specific recollection and would prefer it, if you would ask me or remind me of the context or conversation that you have in mind.
Q. Why do you think you could very well have said that?
A. Because I have always thought it was a dreadful weapon. Even from a technical point of view it was a sweet and lovely and beautiful job, I have still thought it was a dreadful weapon.
Q. And have said so?
A. I would assume that I have said so, yes.
Q. You mean you had a moral revulsion against the production of such a dreadful weapon?
A. This is too strong.
Q. Beg pardon?
A. That is too strong.
Q. Which is too strong, the weapon or my expression?
A. Your expression. I had a grave concern and anxiety.
Q. You had moral qualms about it, is that accurate?
A. Let us leave the word "moral" out of it.
Q. You had qualms about it.
A. How could one not have qualms about it? I know no one who doesn't have qualms about it.
Q. Very well. Clause 3 of that sentence, "By claiming there were insufficient facilities and scientific personnel to carry on the development." Is that true?
A. That is true in a very limited and circumscribed way. There were some conflicts of scheduling between fission weapon development and thermonuclear development. Where the thermonuclear development was directed toward the essential problem of feasibility, or what appeared clearly to me to be the essential problem of feasibility, I never had or could have any doubt that this should take priority, because that was the order under which we were operating.
Q. That this—which should take priority?
A. That the thermonuclear development. Where it was a question of what appeared to me a fruitless byline, there I did question the relative priority of such bylines and rather of immediate fission weapon developments.
Q. Did you ever claim that there were insufficient facilities and scientific personnel to carry on the development of the fusion weapon?
A. Certainly not in that bald form, because it was not true. I never believed it and I therefore don't believe I could have claimed it.
Q. "(4) and that it was not politically desirable." Did you make such a claim?
A. That was certainly a better statement of the general import of the GAC report—of the annex to the GAC report—than moral grounds.
Q. Did you continue to express those views subsequent to the President's decision of January 1950?
A. After the President's decision, I appeared on a broadcast program with Mrs. Roosevelt and Lilienthal and Bethe, and what I said indicated I was not entirely happy, perhaps, with the procedures by which the decision was arrived at.
Q. Would you tell us what you said?
A. I can get hold of it.
Q. Give us your best recollection of it, Doctor.
A. I said that the decision is like the decision to seek international control of atomic energy or the decision to proceed with the hydrogen bomb had compli-

cated technical background, but they also had important moral and human consequence; that there was danger in the fact that such decisions had to be taken secretly, not because the people who took the decisions were not wise, but because the very need, the very absence of criticism and discussion tended to corrode the decision making process. That these were hard decisions, that they were dealt with fearful things, that sometimes the answer to fear could not lie in explaining away the reasons for fear. Sometimes the only answer for fear lay in courage.

This is probably not very accurate, but we can easily provide you with that.

Q. About when was that, Doctor, that you made those statements?

A. I would guess that it was within 2 months of the 1st of February 1950.

Q. Did you make any other public statements along those same lines?

A. Not quite. In addressing the Westinghouse talent search here in Washington—this is a group of young people ostensibly who get rewarded for doing well in high school and get sent on to college, attended by dignitaries—I talked about science and in the initial paragraph I said that I was not going to talk to them about the problem of the statutory requirements for AEC fellowships, or the problem of the hydrogen bomb. These were things that I hoped would not be in their minds very much when they grew up. I was going to talk to them immediately about pure science.

Q. Did you make any other public statements along those lines? Pardon me. About when was that that you made that statement?

A. I believe I said no more than this, but we also have a record of that.

Q. About when did you make that statement?

A. That would have been in the spring of 1950.

Q. Did you make any other public statements along those lines?

A. We have an almost complete record—I think a complete record—of everything public. I am not remembering anything else right now.

Q. Doctor, you know, do you not, that you are a physicist who is largely admired and whose words have great weight with other physicists, don't you?

A. With some.

Q. Beg pardon?

A. With some physicists.

Q. With many physicists; don't you?

A. Right.

Q. And that is especially true of younger physicists?

A. I know some old physicists.

Q. Some old physicists; too.

A. I don't think it is essentially true of younger physicists, because I am not longer in a very extensive—the people who study with me or even under my auspices are not as they were before the war, a large fraction or a substantial fraction of the theoretical physicists in the country. They are a very small fraction.

Q. But as of 1950, you were certainly——

A. No, this is still true.

Q. Pardon?

A. This was true then.

Q. But in 1950 you were pretty much a hero to a substantial group of physicists in this country; weren't you?

A. I should think that your knowledge of that was as complete as mine.

Q. Wouldn't you agree with that statement, Doctor, laying aside your modesty?

A. Well, you read to me yesterday—no, you told me yesterday—and could today have read in the papers a letter from one physicist who seems not to have regarded me as a hero by 1950.

Mr. GARRISON. If you don't mind my interrupting a second about procedure, I think this can be off the record.

Mr. GRAY. Yes.

(Discussion off the record.)

Mr. GRAY. Would you proceed.

By Mr. ROBB:

Q. Doctor, we were talking about your standing and influence with psysicists as of 1950. Would you not agree, sir, that you were a hero to a very substantial party of physicists as of 1950?

(Mr. Garrison left the room.)

The WITNESS. I don't know. I would think a judgment of what my position was in others' eyes should be left out of this.

By Mr. ROBB:
Q. What?
A. A judgment of how I appeared to people should be left to those to whom I appeared, rather than to me.
Q. Well, let us put it this way. Wouldn't you agree that anything said by you would have great weight with a great number of nuclear physicists?
A. Would have some weight with quite a few people, physicists and nonphysicists.
Q. Doctor, let me ask you, sir, do you think that public statements which you have told us about and which you have summarized, tended to encourage other physicists to work on the hydrogen bomb?
A. I should think that they were essentialy neutral. I coupled the hydrogen bomb and the decision to seek international control of atomic energy first, so that there was no substantive criticism of the decision. In the effect I merely referred to the fact that the hydrogen bomb had been a very controversial thing as had the National Science Foundation fellowships.
Q. You certainly didn't think those expressions by you were going to encourage physicists to work on the project?
A. They were not intended to affect what physicists did on the project at all.
Q. Doctor, I didn't ask you what you intended. I am asking you what you reasonably believe would be the result of those statements.
A. I reasonably believe that the result of those statements would be nil as far as the activities of professional physicists on the hydrogen bomb project or any other aspect of the Atomic Energy Commission work.
Q. Had a great many physicists at or about that time asked you your views on whether or not the hydrogen bomb should be produced?
A. Not a great many; no.
Q. Had some?
A. Before the President's decision?
Q. Yes.
A. Yes; some had.
Q. Who?
A. I told you about Bethe and Teller and their visit. Lawrence sent on Serber. That was about the same time. This was before the GAC meeting. Alvarez discussed it with me. Bacher discussed it with me. Lauritsen discussed it with me. Von Neumann discussed it with me.
Q. Rabi?
A. Rabi was a member of the general advisory committee.
Q. Did he discuss it with you before the meeting?
A. At least we referred to it. I don't know much of a discussion we had.
Q. DuBridge?
A. Before the meeting?
Q. Yes.
A. I have no recollection of that. It is possible; I think it unlikely.
Q. Conant? Of course, I know Conant is a chemist and not a physicist.
A. Conant told me he was strongly opposed to it.
Q. Did you express any views to Conant?
A. I believe not.
Q. In other words, he told you what his views were before you expressed yours to him?
A. He told me what his views were before mine were clearly formulated.
Q. I believe you testified the other day that at the time you heard from Conant, either by mail or orally, that you were in some doubt about the matter, that you had not made up your mind.
A. Yes; that is right.
Q. How long before the GAC meeting was that?
A. I don't remember. Certainly not more than a month. It could not have been more than a month, and it probably was of the order of a week.
Q. The next sentence of General Nichols' letter: "It was further reported that, even after it was determined as a matter of national policy to proceed with development of the hydrogen bomb, you continued to oppose the project and declined to cooperate fully in the project."
Are the statements made in that sentence true?
A. Let us take the first one.
Q. Yes, sir.
A. I did not oppose the project. Let us take the second one.
Q. You mean after——
A. After the decision was made, I did not oppose the project.

Q. Very well. Let us take the second one.

A. I would need to know what cooperate fully, who asked me to cooperate and what this meant, was before I could answer it. I did not go out to Los Alamos and roll up my sleeves, and maybe that is what cooperating fully means. I would like to know what this does mean.

Q. Did you ever tell Teller that you could not work on the project?

A. I told him I was not going out to Los Alamos to work on it.

Q. Did you ever tell him that you could not work on it at all?

A. That is far more sweeping than turned out to be true, and I doubt if I would have said it.

Q. What work did you do on the project?

A. I did my official job of learning about it and advising about it and thinking about it.

Q. You mean official job as chairman of the GAC?

A. Right, and of other committees.

Q. Of learning about it?

A. And of advising about it and of thinking about it.

Q. Whom did you advise?

A. The Atomic Energy Commission.

Q. You mean the members of the Commission?

A. The Commission as a body.

Q. Did you do any scientific work on the project? By that I mean calculations. The kind of scientific work you did on the atom bomb.

A. No; not with anything like that intensity. I checked some qualitative things so I would be fairly sure I understood them. I did very little scientific work on the atom bomb after I assumed the direction of the Los Alamos laboratory.

Q. You made the decisions there, didn't you, Doctor?

A. I did. In this case I won't say I made the decision—it was not my responsibility—but I certainly helped to make the decision which I believe got the thing started in the right direction. I didn't have the ideas. There were a great many ideas I didn't have about the atom either.

Q. The next sentence, I believe, you already commented on. That refers to the statement that you caused the distribution of the report at Los Alamos. You said that you did not do that; is that right?

A. Right.

Q. The next sentence refers or is the statement that you were instrumental in persuading other outstanding scientists not to work on the bomb. I believe you deny that; is that correct?

A. I think I would be glad to deny it. I would like to know what outstanding scientist I might have persuaded not to work on the bomb.

Mr. GRAY. I suppose the question could be answered. Did you attempt to persuade anyone not to work on the hydrogen bomb?

The WITNESS. No.

By Mr. ROBB:

Q. I will read you the last clause of that: "The opposition to the hydrogen bomb, of which you are the most experienced, most powerful, and effective member, has definitely slowed down its development."

Let us break that down. Would you agree that you are or were the most experienced, most powerful, and most effective member of the opposition to the hydrogen bomb?

A. What time are we talking about?

Q. At any time.

A. Well, I would say I was not the most powerful, I was not the most experienced, and I was not the most influential. But, if you take all three factors together, perhaps I combined a little more experience, a little more power, and a little more of influence than anyone else.

Q. At what time?

A. I am thinking of the period between the Russian test and the President's decision.

Q. How about after the President's decision?

A. There was not any opposition to the hydrogen bomb.

Q. Weren't you still opposed to the development of the hydrogen bomb?

A. No.

Q. Do you think your opposition and the opposition of the group of people who agreed with you prior to the President's decision slowed down the development of the hydrogen bomb?

A. I find it very hard to judge. I have testified—let me testify as follows: There are two parts to a development like this. One is to have sensible ideas. These are partly a matter of scientific analysis and partly a matter of invention. The other is to get plants built, material produced, equipment shoved around, and a host of technical and technological developments carried out.

With the atom bomb the pacing factor was the second. We could have had the atom bomb as far as ideas went considerably earlier than we could have it as far as hardware went.

(Mr. Garrison returned to the room.)

The WITNESS. With the hydrogen bomb I believe that the pacing factor was good ideas. If they had occurred earlier, the physical development of the weapon would not have been quite as rapid as it was in fact, coming at a time when a great many of the auxiliary things had already been done. If they had occurred later, the development of technology which had occurred would not have done us any good. I therefore do not believe that any substantial delay in the actual date of our first successful thermonuclear test, * * * derived from the 3 or 4 months of deliberations. Whether the GAC was responsible for these 3 or 4 months of deliberations or whether that would have occurred in any case, I do not know.

By Mr. ROBB:

* * * * * * *

Q. I believe you testified that you learned that Fuchs had told the Russians that we were working on the hydrogen bomb; is that right?
A. No. What I learned was that Fuchs had told them of some technical points.
Q. Having to do with the hydrogen bomb?
A. Having to do with the hydrogen bomb.
Q. I believe Fuchs was present and took part in a conference at Los Alamos in the spring of 1946; is that correct?
A. Right. I don't know the date. I couldn't go to it. I was invited, but I could not go.
Q. Did you see a report of it?
A. I believe I did, not a very detailed report.
Q. That conference reviewed——
A. What was then known.
Q. What was then known?
A. It was full of mistakes.
Q. In all events, presumably, what Fuchs knew, the Russians knew.
A. Right.
Q. Now, I have a note here, Doctor, that you testified that there was a surprising unanimity—I believe that was your expression—at the GAC meeting of October 29, 1949, that the United States ought not to take the initiative at that time in an all-out thermonuclear program. Am I correct in my understanding of your testimony?
A. Right.
Q. In other words, everybody on the committee felt that way about it?
A. Everybody on the committee expressed themselves that way.
Q. Beg pardon?
A. Everybody on the committee expressed themselves that way.
Q. How many people were on the committee?
A. There were 9 on the committee; 1 man was absent in Sweden.
Q. Who was that?
A. Seaborg.
Q. Where was he from, Doctor?
A. University of California. He worked during the war at the University of Chicago.
Q. He did not get to Washington at all?
A. Not at that meeting.
Q. So you didn't know how he felt about it?
A. We did not.
Q. You didn't know either how he felt about it. He just was not there.
A. He was in Sweden, and there was no communication with him.
Q. Beg pardon?
A. He was in Sweden, and there was no communication with him.
Q. You didn't poll him by mail or anything?
A. This was not a convenient thing to do.

Q. No, sir. I believe, Doctor, that you afterward testified along those same lines before the Joint Committee of the House and Senate on Atomic Energy, that there was unanimity but that Dr. Seaborg was not heard there; is that right?
A. It is true, and I suppose I was asked.
Q. I see.
A. I may add that at later meetings, which Seaborg did attend, he expressed himself with great reserve and indicated that he would prefer not to say anything one way or the other on the hydrogen-bomb issue.
Q. Now, Doctor, I believe you testified the other day that in 1942 you foresaw the possibility of developing a thermonuclear weapon; is that right?
A. Yes; we discussed it much of the summer of 1942.
Q. That was at Berkeley?
A. Yes.
Q. Did you also discuss it at a meeting at Chicago?
A. I don't recollect that, but it is quite likely.
Q. I believe you said that you were quite enthusiastic at that time about the possibilities; is that correct?
A. I think it would be better to say that we thought it would be much easier than it was.
Q. The thermonuclear weapon was worked on at Berkeley?
A. Thought about—just thought about.
Q. When you got down to Los Alamos the thermonuclear was one of the first things that you began to work on?
A. It never occupied a large part of the laboratory's effort. It could not. But it was kept on the back burner throughout the war.
Q. I believe you said you had one building, one of the first buildings constructed was—what do you call it, cryogenics building?
A. Cryogenics building, which we used for quite different purposes.
Q. But it was built for the purposes of working on the thermonuclear, wasn't it?
A. Yes.
Q. Work continued on the thermonuclear at Los Alamos under your direction throughout the war, didn't it?
A. Yes.
Q. Then in 1944, Doctor, you applied for a patent on the thermonuclear bomb, didn't you?
A. I have forgotten that.
Q. Did you?
A. We discussed it, and I do not know whether this actually went through. Was this with Teller and Bethe? If it was with Teller and Bethe, then I think it went through.
Q. The patent was granted in 1946, I believe.
A. Yes.
Q. Do you remember that now?
A. Yes. I was simply not sure whether we had gone through with it or not.
Q. And then, I believe, your testimony was that even after you left Los Alamos in 1945, the work on the super continued there?
A. Yes; it did.
Q. And of course that had your approval and support?
A. Yes; it did.
Q. I believe you testified at the first meeting of the GAC the matter of the thermonuclear was discussed, is that correct?
A. Right.
Q. And you encouraged the Commission to get on with the work, as you put it, is that right?
A. Yes. I think specifically what I testified was that we considered whether this long-range and very unsure undertaking—it is very difficult and which we thought of then as 5 years or more—whether thinking about that and working on it would hurt or harm the other jobs at Los Alamos. We decided that it would probably not hurt or harm, but on the contrary help.
Q. So they should get ahead with it.
A. So we encouraged them to do this.
Q. We use the expression "thermonuclear weapon." By that you meant a weapon of vastly more power than the atom bomb, did you not?
A. The original picture was that. Other pictures came in during the first year or so of the Commission and also looked very practical.
Q. When we say——

A. You would like to leave out the small thermonuclear weapons if there are such things.

Q. Yes. But the thing you were talking about in 1942, and working on at Los Alamos——

A. Would be a very big explosive.

Q. A tremendous explosive. I don't know whether it is classified or not but 10,000 times the power of the atom bomb, or something like that.

A. Anyway, very large.

Q. That would not be an exaggeration, would it, 10,000 times?

A. This I think is classified.

Q. Very well. Some weapon to use the technical expression in what we call the megaton range, is that right?

A. That is right.

Q. That is what you had in mind beginning in 1942?

A. That is right.

Q. Doctor, in your work and discussions in 1942, in your work on the thermonuclear weapon at Los Alamos in 1943 to 1945 and in your application for the patent of 1944, and in your advice which you as chairman of the GAC gave to the Commission to get on with the work on this thermonuclear, at all those times and on all of those occasions, were you suffering from or deterred by any moral scruples or qualms about the development of this weapon?

A. Of course.

Q. You were?

A. Of course.

Q. But you still got on with the work, didn't you?

A. Yes, because this was a work of exploration. It was not the preparation of a weapon.

Q. You mean it was just an academic excursion?

A. It was an attempt at finding out what things could be done.

Q. But you were going to spend millions of dollars of the taxpayers' money on it, weren't you?

A. It goes on all the time.

Q. Were you going to spend millions if not billions of dollars of the taxpayers' money just to find out for yourself satisfaction what was going on?

A. We spent no such sums.

Q. Did you propose to spend any such sums for a mere academic excursion?

A. No. It is not an academic thing whether you can make a hydrogen bomb. It is a matter of life and death.

Q. Beginning in 1942 and running through at least the first year or the first meeting of the GAC, you were actively and consciously pushing the development of the thermonuclear bomb, weren't you? Isn't that your testimony?

A. Pushing is not the right word. Supporting and working on it, yes.

Q. Yes. When did these moral qualms become so strong that you opposed the development of the thermonuclear bomb?

A. When it was suggested that it be the policy of the United States to make these things at all costs, without regard to the balance between these weapons and atomic weapons as a part of our arsenal.

Q. What did moral qualms have to do with that?

A. What did moral qualms have to do with it?

Q. Yes, sir.

A. We freely used the atomic bomb.

Q. In fact, Doctor, you testified, did you not, that you assisted in selecting the target for the drop of the bomb on Japan?

A. Right.

Q. You knew, did you not, that the dropping of that atomic bomb on the target you had selected will kill or injure thousands of civilians, is that correct?

A. Not as many as turned out.

Q. How many were killed or injured?

A. 70,000.

Q. Did you have moral scruples about that?

A. Terrible ones.

Q. But you testified the other day, did you not, sir that the bombing of Hiroshima was very successful?

A. Well, it was technically successful.

Q. Oh, technically.

A. It is also alleged to have helped end the war.

Q. Would you have supported the dropping of a thermonuclear bomb on Hiroshima?

A. It would make no sense at all.
Q. Why.
A. The target is too small.
Q. The target is too small. Supposing there had been a target in Japan big enough for a thermonuclear weapon, would you have opposed dropping it?
A. This was not a problem with which I was confronted.
O. I am confronting you with it now, sir.
A. You are not confronting me with an actual problem. I was very relieved when Mr. Stimson removed from the target list Kyoto, which was the largest city and the most vulnerable target. I think this is the nearest thing that was really to your hypothetical question.
Q. That is correct. Would you have opposed the dropping of a thermonuclear weapon on Japan because of moral scruples?
A. I believe I would, sir.
Q. Did you oppose the dropping of the atom bomb on Hiroshima because of moral scruples?
A. We set forth our——
Q. I am asking you about it, not "we."
A. I set forth my anxieties and the arguments on the other side.
Q. You mean you argued against dropping the bomb?
A. I set forth arguments against dropping it.
Q. Dropping the atom bomb?
A. Yes. But I did not endorse them.
Q. You mean having worked, as you put it, in your answer rather excellently, by night and by day for 3 or 4 years to develop the atom bomb, you then argued it should not be used?
A. No; I didn't argue that it should not be used. I was asked to say by the Secretary of War what the views of scientists were. I gave the views against and the views for.
Q. But you supported the dropping of the atom bomb on Japan, didn't you?
A. What do you mean support?
Q. You helped pick the target, didn't you?
A. I did my job which was the job I was supposed to do. I was not in a policymaking position at Los Alamos. I would have done anything that I was asked to do, including making the bombs in a different shape, if I had thought it was technically feasible.
Q. You would have made the thermonuclear weapon, too, wouldn't you?
A. I couldn't.
Q. I didn't ask you that, Doctor.
A. I would have worked on it.
Q. If you had discovered the thermonuclear weapon at Los Alamos, you would have done so. If you could have discovered it, you would have done so, wouldn't you?
A. Oh, yes.
Q. You were working toward that end, weren't you?
A. Yes. I think I need to point out that to run a laboratory is one thing. To advise the Government is another.
Q. I see.
A. I think I need to point out that a great deal that happened between '45 and '49—I am not supposed to say to what extent—but to a very, very massive extent, we had become armed atomically. The prevailing view was that what we had was too good—too big—for the best military use, rather than too small.
Q. Doctor, would you refer to your answer, please, sir? One further question before we get into that.
Am I to gather from your testimony, sir, that in your opinion your function as a member and chairman of the GAC included giving advice on political policies as well as technical advice?
A. I have testified as to that.
Q. Would you repeat it for me, sir?
A. I will repeat it. Our statutory function was to give technical advice.
Q. Yes, sir.
A. We were often asked questionss which went outside of this narrow frame. sometimes we responded, sometimes we didn't. The reason why the general advice, I would call it, editorializing rather than political advice, contained in our annexes was in the annexes and not in the report because it did not seem a proper function for the General Advisory Committee to respond in these terms to the question that had been put to them.

Q. Doctor, is it a fair summary of your answer—and I refer you to page 37, and the following pages of your answer—that what the GAC opposed in its October 29, 1949, meeting was merely a crash program for the development of the super?
A. Yes. I think it would be a better summary to say we opposed this crash program as the answer to the Soviet atomic bomb.
Q. What did you mean by a crash program?
A. On the basis of what was then known, plant be built, equipment be procured and a commitment be made to build this thing irrespective of further study and with a very high priority. A program in which alternatives would not have an opportunity to be weighed because one had to get on and because we were not going to sacrifice time.
Q. Doctor, isn't it true that the report of the GAC you wrote, didn't you ——
A. I wrote the main report; yes.
Q. Isn't it true that the report of the GAC and the annex to which you subscribed unqualifiedly opposed the development of the super at any time?
A. At that time.
Q. At any time?
A. No. At least, let us say we were questioned about that in a discussion with the Commission, and we made it quite clear that this could not be an unqualified and permanent opposition. I think that in the reading of the report without the later discussions and reports it could be read that way. But in the light of what was later said, it could not be read that way.
Q. Didn't the annex to which you subscribed say in so many words, "We believe a superbomb should never be produced"?
A. Yes; it did.
Q. It did say that?
A. Yes.
Q. Do you interpret that as opposing only a crash program?
A. No. It opposed the program. Obviously if we learned that the enemy was up to something, we could not prevent the production of a super bomb.
Q. What did you mean by "never"?
A. I didn't write those words.
Q. You signed it, though, didn't you?
A. I believe what we meant—what I meant was that it would be a better world if there were no hydrogen bombs in it. That is what the whole context says.
Q. Doctor, don't you think a fair interpretation of the record and the annex which you signed was an unqualified opposition to the production of super at any time or under any circumstances?
A. No;. I don't.
Q. That is your view?
A. Yes.
Q. In all events, Doctor, you did say in your report that no one could tell without an actual test whether the super would work or whether it wouldn't, is that right?
A. Yes.
Q. You testified that you had no intimation from Dr. Seaborg prior to the GAC meeting of October 29, 1949, as to what his views on the subject were. I am going to show you a letter taken from your files at Princeton, returned by you to the Commission, dated October 14, 1949, addressed to you, signed Glenn Seaborg, and ask you whether you received that letter prior to the meeting of October 29, 1949.
A. I am going to say before I see that that I had no recollection of it.
Q. I assumed that. May I interrupt your reading of it a moment?
A. Yes.
Mr. ROBB. Mr. Chairman, I have been told by the classification officer that there are two words here that I must not read. They are bracketed, and I am showing them to Dr. Oppenheimer, and when I read the letter I shall leave them out, but I want Dr. Oppenheimer to see them.
The WITNESS. I would be sure of one thing, and that is if that letter reached me before the meeting, I read it to the committee.

By Mr. ROBB:

Q. The letter was dated October 14, 1949.
A. So it almost certainly reached me.
Q. So presumably unless it came by wagon train, it reached you, didn't it?
A. Right.

Q. I will read this letter:

"UNIVERSITY OF CALIFORNIA,
"RADIATION LABORATORY,
"*Berkeley 4, Calif., October 14, 1949.*

"DR. J. ROBERT OPPENHEIMER,
"*The Institute for Advanced Study,*
"*Princeton, N. J.*

"DEAR ROBERT: I will try to give you my thoughts for what they may be worth regarding the next GAC meeting, but I am afraid that there may be more questions than answers. Mr. Lilienthal's assignment to us is very broad; and it seems to me that conclusions will be reached, if at all, only after a large amount of give and take discussion at the GAC meeting.

"A question which cannot be avoided, it seems to me, is that which was raised by Ernest Lawrence during his recent trip to Los Alamos and Washington. Are we in a race along this line and one in which we may already be somewhat behind so far as this particular new aspect is concerned?"

Q. He was talking about the thermonuclear, wasn't he?
A. It would be obvious to me he was.

Q. Continuing: "Apparently this possibility has begun to bother very seriously a number of people out here, several of whom came to this point of view independently. Although I deplore the prospects of our country putting a tremendous effort into this, I must confess that I have been unable to come to the conclusion that we should not.. Some people are thinking of a time scale of the order of 3 to 5 years which may, of course, be practically impossible and would surely involve an effort of greater magnitude than that of the Manhattan project. My present feeling would perhaps be best summarized by saying that I would have to hear some good arguments before I could take on sufficient courage to recommend not going toward such a program.

"If such a program were undertaken, a number of questions arise which would need early answers. How would the National Laboratories fit into the program? Wouldn't they have to reorient their present views considerably? The question as to who might build neutron producing reactors would arise. I am afraid that we could not realistically look to the present operators of Hanford to take this on. It would seem that a strong effort would have to be made to get the duPont Company back into the game. It would be imperative that the present views of the reactor safeguard committee be substantially changed.

"I just do not know how to comment, without further reflection, on the question of how the present 'reactor program' should be modified, if it should. Próbably, after much discussion, you will come to the same old conclusion that the present four reactors be carried on, but that an effort be made to speed up their actual construction. As you probably know, Ernest is willing to take on the responsibility for the construction near Berkeley of a"—and then I omit the two words—"heavy water natural uranium reactor primarily for a neutron source and on a short time scale. I don't know whether it is possible to do what is planned here, but I can say that a lot of effort by the best people here is going into it. If the GAC is asked to comment on this proposal, it seems to me clear that we should heartily endorse it. So far as I can see, this program will not interfere with any of the other reactor building programs and will be good even if it does not finally serve exactly the purpose for which it was conceived; I have recently been tending toward the conviction that the United States should be doing more with heavy water reactors (we are doing almost nothing). In this connection, it seems to me that there might be a discussion concerning the heavy water production facilities and their possible expansion.

"Another question, and one on which perhaps I have formulated more of a definite opinion, is that of secrecy. It seems to me that we can't afford to continue to hamper ourselves by keeping secret as many things as we now do. I think that not only basic science should be subject to less secrecy regulation but also some places outside of this area. For example, it seems entirely pointless now to hamper the construction of certain types of new piles by keeping secret certain lattice dimensions. In case anything so trivial as the conclusions reached at the recent international meeting on declassification with the British and Canadians at Chalk River is referred to the GAC I might just add that I participated in these discussions and thoroughly agree with the changes suggested, with the reservation that perhaps they should go further toward removing secrecy.

"I have great doubt that this letter will be of much help to you, but I am afraid that it is the best that I can do at this time.

"Sincerely yours, Glenn" and below that in typing, "Glenn T. Seaborg."

So, Doctor, isn't it clear to you now that Dr. Seaborg did express himself on this matter before the meeting?

A. Yes, it is clear now. Not in unequivocal terms, except on one point, and on that point the General Advisory Committee I think made the recommendation that he desired.

Q. But he did express himself, didn't he?

A. Absolutely.

Q. In a communication to which he apparently had given some thought, is that correct?

A. Right, and to which no doubt at the time I gave some thought.

Q. That is right. You have no doubt that you received this before the General Advisory Committee meeting, is that correct?

A. I don't see why I should not have.

Q. Why did you tell the Joint Congressional Committee on Atomic Energy when you testified on January 29, 1950, that Dr. Seaborg had not expressed himself on the subject prior to the meeting?

A. I am sure because it was my recollection.

Q. That testimony was given in January 1950, wasn't it?

A. That is right.

Q. And this letter had been received by——

A. Let me add one point. We had a second meeting on the hydrogen bomb which Seaborg attended and we asked him how he felt about it, and he said he would prefer not to express his views.

Q. But weren't you asked, Doctor, or didn't you tell the joint committee that Dr. Seaborg had not expressed himself on this subject prior to the meeting of October 29, 1949?

A. I would have to see the transcript. I don't remember that question and the answer.

Q. If you did make that statement, it was not true, was it?

A. It is clear that we had an expression, not unequivocal, from Seaborg, before the meeting of October 29.

Q. Doctor, did you hear my question?

A. I heard it, but I have heard that kind of question too often.

Q. I am sure of that, Doctor, but would you answer it, nevertheless?

Mr. MARKS. Isn't Dr. Oppenheimer entitled to see the testimony which is being referred to, instead of answering a hypothetical question?

Mr. ROBB. It is not a hypothetical question.

By Mr. ROBB:

Q. If you told the joint committee, sir, that Dr. Seaborg had not expressed himself prior to the meeting of October 29, 1950, that was not true, was it?

A. It would depend, entirely.

Q. Yes or no.

A. I will not say yes or no. It would depend entirely on the context of the question. The only two things in this letter that Seaborg is absolutely clear about is that we ought to build certain kinds of reactors and we ought to have less secrecy. On the question of the thermonuclear program he can't find good enough arguments against it, but he does have misgivings.

Q. All right, Doctor. You told this Board this morning that Dr. Seaborg did not express himself prior to the meeting of October 29, 1949.

A. That is right. That was my recollection.

Q. Was that true?

A. No, that was not true.

Q. You told the board this morning——

Mr. GRAY. Are you pursuing the Seaborg matter now?

Mr. ROBB. I thought I would come back to it, sir.

Mr. GARRISON. Mr. Chairman, I think it would be fair since the question was raised, because of the implications that may be left that the actual questions put to Dr. Oppenheimer by the joint committee about Dr. Seaborg should be read into the record with sufficient context to show what it was about. Otherwise, we are left with a possible misapprehension as to what really did take place. I don't know. I have never seen the transcript.

Mr. ROBB. Mr. Chairman, that is impossible unless we have a meeting of the joint committee and they authorize that to be done. But Dr. Oppenheimer this morning as the board no doubt heard, recalled that he had so testified before the joint committee.

The WITNESS. I had testified; I had not so testified.

Mr. ROBB. The record will show what the doctor testified.

The WITNESS. If I testified that I recall so testifying, I would like to correct the transcript.

Mr. ROBB. That was not correct, either?

Mr. SILVERMAN. He didn't say it.

Mr. ROBB. All right. The record will show what he testified to.

Mr. GARRISON. What is the procedural requirement for reading into the record the questions from that transcript?

Mr. ROBB. That transcript will not be released, as I understand it, without the vote of the committe to do so, Mr. Garrison, which is why I was not able to read Dr. Oppenheimer what he said.

The WITNESS. I think a lot depends on the nature of the question. Had Dr. Seaborg made up his mind, had he concurred with your view, or so on. It is clear from this letter he wanted to hear a discussion about it. That he saw it was a very tough question.

Mr. ROBB. May I ask the doctor one more question before we take a break on this Seaborg matter.

Mr. GRAY. Yes.

By Mr. ROBB:

Q. Doctor, are you sure that you read Dr. Seaborg's letter to your committee, the GAC committee, at the meeting of October 29, 1949?

A. Since I forgot the existence of the letter, obviously I cannot remember reading it. I always read communications on matters before us to the committee.

Q. Is there any reflection in the report of the committee that Dr. Seaborg had expressed himself in any way about this matter?

A. No, there certainly is not.

Q. I beg pardon?

A. There isn't.

Mr. ROBB. All right.

Mr. GARRISON. May I ask the chairman whether the board has before it the transcript of the joint committee testimony? I ask merely because of the fact that if it has been released to the board——

Mr. GRAY. Let me respond to your question this way, Mr. Garrison, and say that after recess, which I propose to call in a moment, I should like to respond to that.

We will now recess.

(Brief recess.)

Mr. GRAY. I would like to pursue the question which Mr. Garrison raised just before the recess.

The board does not have before it a complete transcript of the testimony which was under discussion.

(Mr. Marks not present in the room.)

Mr. GRAY. However, I can say to Dr. Oppenheimer and his counsel that the board does understand from a source it believes to be reliable that Dr. Oppenheimer was asked a question with respect to the extent of unanimity of the views of the members of the GAC with respect to what we have been describing as the crash program. I am not sure whether it was so referred to in the testimony, but there was this question.

In response to the question Dr. Oppenheimer stated that he thought it was pretty unanimous view, that one member of the committee, Dr. Seaborg, was away when the matter was discussed, and that he had not expressed himself on it, and further saying that the other members will agree with what he has said.

The WITNESS. That is a little different from what I was told I said. I was told I said explicitly that Seaborg had said nothing about the matter before the meeting. This was several months after the meeting and I was asked whether Seaborg had expressed his views in connection with this meeting. I would think that the proper answer to that was not so far from what you quoted me as saying.

Mr. GRAY. We are trying to develop what actually the facts were in the case, and I believe you did testify that you had no communication with respect to this matter from Dr. Seaborg or at least you said you did not recall a communication, I believe.

The WITNESS. Is that what it says in the transcript?

Mr. GRAY. No; I think that is what you said earlier this morning.

The WITNESS. I would like to make a general protest. I am told I have said certain things. I don't recall it. I am asked if I said these what would that be. This is an extremely difficult form for me to face a question. I don't know what I said. It is of record. I had it in my own vault for many years. It is not classified for reasons of national security, this conversation, and I have no sense that I could have wished to give any impression to the joint congressional committee other than an exposition because when I testified I knew for a fact that the decision had been taken, I testified in order to explain as well as I could to the committee the grounds for the advice, the color of the advice, the arguments that we had in mind. It was not an attempt to persuade them. It was not in any way an attempt to alter the outcome. It was an attempt to describe what we had in mind. A few minutes after I testified, I believe, or shortly after I testified, the Presidential announcement came out, and I knew what it was going to be. So this was not a piece of advocacy. It was a piece of exposition.

I would like to add one other thing. Having no recollection of the Seaborg letter, I cannot say that I did this. But it would have been normal practice for me at one of the meetings with the Commission not merely to read the letter to the committee, but to read the letter or parts of it relevant to our discussion to the Commission and the committee.

By Mr. ROBB:
Q. In other words, Doctor, if you didn't read this Seaborg letter to your committee, it would have been quite unusual?
A. Yes.
Q. Doctor, will you help me a little bit on physics. I notice Dr. Seaborg in this letter talks about the reactor program. Was that program a necessary step in the development of the thermonuclear weapon?
A. It was thought to be.
Q. What was done, or what did the General Advisory Committee advise or urge to be done in respect of a reactor program subsequent to the President's decision of January 1950.
A. Already in the October 29 report we urged that a reactor program to produce these neutrons, the number of which is classified, be expedited. We, however, said that this should be done not for the purpose of the super program, but for many other purposes. We urged that the thing be built.

I believe after the Presidential decision, we urged that the reactor program be flexible because it was already apparent at that time that the ideas as they existed in October 29, 1949, were undergoing very serious modification. If you wish me to refresh my memory on the precise points, I would be glad to. I have not done so.
Q. Doctor, am I correct in my memory of your earlier testimony that the reactor program was one thing that you are now and were at that time dissatisfied with and did not go very well?
A. That is quite a different thing. That is the development of reactors for power.
Q. That was something else?
A. That is something quite different. This is a production reactor. I would not say that we were satisfied with the production reactor picture.
Q. It is a heavy water reactor, is what you need for this program?
A. No, not neecssarily. It is a possible way of going about it.
Q. What progress was made in developing the reactors that were necessary for the hydrogen bomb?
A. That were then thought to be necessary?
Q. Yes.
A. Great progress.
Q. They were built, were they?
A. Yes.
Q. At Hanford?
A. No.
Q. Is that classified, Doctor?
A. It is in all the papers. They were built at Savannah River.
Q. I see.
A. They were built I think with the early development and study undertaken at the Argonne Laboratory and the duPont laboratory facing into the engineering and construction phases.
Q. Doctor, I want to show you a copy of a letter also taken from your files that you had at Princeton and turned back to the Commission. This is a copy

of a letter dated October 21, 1949, bearing the typewritten signature Robert Oppenheimer, addressed to Dr. James B. Conant, president, Harvard University: "Dear Uncle Jim:" I ask you if you wrote that letter.

A. October 21, 1949?

Q. Yes, sir.

A. I would like to look it over.

Q. Certainly. That is why I handed it to you, Doctor. I want you to look it over carefully. Take your time.

A. I wrote this letter.

Q. You wrote that letter.

A. Can we read it in full?

Q. I am going to. You sent this letter on or about October 21, 1949.

A. I have no reason to doubt it.

Q. Doctor, in this letter as in the other, the classification officer has expurgated a few words which are indicated by brackets. Will you look at them now so you will know what they are when I read it?

A. Yes. Could we paraphrase this by saying for a number of applications of military importance?

Q. I will tell you what, Doctor. When I get to that point, I will stop and you paraphrase it, because you can paraphrase that sort of stuff better than I can.

"Dear Uncle Jim:

"We are exploring the possibilities for our talk with the President on October 30th. All members of the advisory committee will come to the meeting Saturday except Seaborg, who must be in Sweden, and whose general views we have in written form. Many of us will do some preliminary palavering on the 28th.

"There is one bit of background which I would like you to have before we meet. When we last spoke, you thought perhaps the reactor program offered the most decisive example of the need for policy clarification. I was inclined to think that the super might also be relevant. On the technical side, as far as I can tell, the super is not very different from what it was when we first spoke of it more than 7 years ago: a weapon of unknown design, cost, deliberability and military value. But a very great change has taken place in the climate of opinion. On the one hand, two experienced promoters have been at work, i. e., Ernest Lawrence and Edward Teller. The project has long been dear to Teller's heart; and Ernest has convinced himself that we must learn from Operation Joe that the Russians will soon do the super, and that we had better beat them to it."

What was Operation Joe, the Russian explosion?

A. Right.

(Mr. Marks entered the room.)

By Mr. ROBB:

Q. Of September 1949?

A. Right.

Q. Continuing your letter: "On the technical side, he proposes to get some neutron producing heavy water reactors built; and to this, for a variety of reasons, I think we must say amen since"—now would you paraphrase?

A. There were three military applications other than the super which these reactors would serve.

Q. "* * * and many other things will all profit by the availability of neutrons.

"But the real development has not been of a technical nature. Ernest spoke to Knowland and McMahon, and to some at least of the joint chiefs. The joint congressional committee, having tried to find something tangible to chew on ever since September 23d, has at least found its answer. We must have a super, and we must have it fast. A subcommittee is heading west to investigate this problem at Los Alamos, and in Berkeley. The joint chiefs appear informally to have decided to give the development of the super overriding priority, though no formal request has come through. The climate of opinion among the competent physicists also shows signs of shifting. Bethe, for instance, is seriously considering return on a full time basis; and so surely are some others. I have had long talks with Bradbury and Manley, and with Von Neumann. Bethe, Teller, McCormack, and LeBaron are all scheduled to turn up within the next 36 hours. I have agreed that if there is a conference on the super program at Los Alamos, I will make it my business to attend.

"What concerns me is really not the technical problem. I am not sure the miserable thing will work, nor that it can be gotten to a target except by ox cart. It seems likely to me even further to worsen the unbalance of our present war plans. What does worry me is that this thing appears to have caught the imagination, both of the congressional and of military people, as the answer to the problem posed by the Russian advance. It would be folly to oppose the exploration of this weapon. We have always known it had to be done; and it does have to be done, though it appears to be singularly proof against any form of experimental approach. But that we become committed to it as the way to save the country and the peace appears to me full of dangers.

"We will be faced with all this at our meeting; and anything that we do or do not say to the President, will have to take it into consideration. I shall feel far more secure if you have had an opportunity to think about it.

"I still remember my visit with gratitude and affection.

"ROBERT OPPENHEIMER.

"Dr. JAMES B. CONANT,
"*President, Harvard University, Cambridge 38, Mass.*"

Doctor, would it appear to you from that letter that you were in error in your previous testimony that you had not expressed your views to Dr. Conant before the meeting of October 29, 1949?
A. Yes.
Q. Beg pardon?
A. Yes.
Q. Do you wish now to amend your previous answer that Dr. Conant reached the views he expressed to you without any suggestion on your part?
A. I don't know which preceded which.
Q. Is there any indication to you in this letter which I have just read that Conant had previously expressed any views to you?
A. I would say there is an indication that there had been discussion between us. I am not clear.
Q. Why were you writing to Dr. Conant before the GAC meeting on this thing?
A. I think the letter explains that.
Q. You were not trying to propagandize him, were you?
A. No.
Q. Do you agree with me that this letter is susceptible of that interpretation that you were trying to influence him?
A. Not properly; not properly so susceptible.
Q. You notice in this letter, Doctor, that you referred to Dr. Seaborg's letter, so you had it at that time, didn't you?
A. Right.
Q. And that must have been the letter we read this morning, is that correct?
A. I would assume so.
Q. Would you agree, Doctor, that your references to Dr. Lawrence and Dr. Teller and their enthusiasm for the super bomb, their work on the super bomb, that your references in this letter are a little bit belittling?
A. Dr. Lawrence came to Washington. He did not talk to the Commission. He went and talked to the joint congressional committee and to members of the Military Establishment. I think that deserves some belittling.
Q. So you would agree that your references to those men in this letter were belittling?
A. No. I pay my great respects to them as promoters. I don't think I did them justice.
Q. You used the word "promoters" in an invidious sense, didn't you?
A. I promoted lots of things in my time.
Q. Doctor, would you answer my question? When you use the word "promoters" you meant it to be in a slightly invidious sense, didn't you?
A. I have no idea.
Q. When you use the word now with reference to Lawrence and Teller, don't you intend it to be invidious?
A. No.
Q. You think that their work of promotion was admirable, is that right?
A. I think they did an admirable job of promotion.
Q. Do you think it was admirable that they were promoting this project?
A. I told you that I think that the methods—I don't believe Teller was involved, Lawrence promoted it—were not proper.
Q. You objected to them going to Knowland and McMahon?
A. I objected to their not going to the Commission.
Q. Knowland and McMahon, by that you meant Senator Knowland and Senator McMahon.
A. Of course.

Q. Did you go to any Senators about this?
A. I appeared before the Senate at their request in my statutory function.
Q. Did you go to any Senators privately about it?
A. Certainly not before discussing it with the Commission. I do not know whether I discussed it with Senator McMahon. If so, it was at his request.
Q. You said certainly not before discussing it with the Commission. Did you after discussing it with the Commission go to any Senators privately about it?
A. Privately?
Q. Yes, sir.
A. I don't remember whether I talked to McMahon or not.
Q. Did you go to the President about it?
A. No.
Q. You mention in this letter a meeting with the President. Did that take place?
A. No.
Q. Did you ever talk to the President about the matter?
A. No.
Q. Do you know whether or not Mr. Lilienthal did?
A. It is in the public press that he did and he told me that he did.
Q. Did you discuss the matter with him before he went to see the President?
A. The time that is in the public press is when he and Acheson and Johnson went over to call on the President.
Q. That was just prior to the President's decision?
A. Yes.
Q. Did you discuss the matter with Lilienthal before that meeting?
A. Before the meeting of October 29?
Q. Before he went to see the President.
A. We discussed it many times between October 29 and the President's decision.
Q. Did you brief Mr. Lilienthal on your views about the thermonuclear weapon before he went to see the President?
A. We talked over and over again—I don't believe it was ever a question of briefing—and I don't have—I am fairly sure that this description of any talk we had was wrong.
Q. Is there any doubt in your mind that when he saw the President, Mr. Lilienthal expressed to the President your views on this matter?
A. That he spoke my views to the President?
Q. Yes.
A. I have no idea.
Q. Did you talk with him after he had seen the President?
A. At this meeting of three people?
Q. Yes.
A. Yes. He came back and told us about it. I think this was actually the general advisory committee, rather than me.
Q. Didn't Mr. Lilienthal report to you in substance that the views he expressed to the President were the same ones you entertained?
A. I don't remember that way of saying it. If it was, it would have been the committee and would have referred to the mass of documents, reports and so on, between the 29th of October and that time.
Q. Was there any doubt in your mind that Mr. Lilienthal shared your views on this matter of the thermonuclear?
A. We knew that he was opposed to the crash program. I was never entirely clear as to the components of this opposition.
Q. Was there any question is your mind that in reaching that view Mr. Lilienthal gave great weight to your advice?
A. He gave some weight to it. I doubt if he gave inordinate weight to it.
Q. Aren't you sure, Doctor, that Mr. Lilienthal necessarily relied very heavily on you for advice in this matter?
A. The matters that engaged his interest were not primarily the technical ones. On technical things of course he relied on our advice.
Q. Doctor, you begin your letter to Mr. Conant, whom you address as "Dear Uncle Jim" with this sentence: "We are exploring the possibilities for our talk with the President on October 30."
Wouldn't that indicate to you that you were opening this subject with him for the first time, that is, with Dr. Conant for the first time?
A. That would indicate that we had discussed it earlier.
Q. It would?

A. Yes, sir. Otherwise, I would have said we are thinking of going to see the President, or what would you think of going to see the President. It refers toward the end to a visit.

Mr. GRAY. May I ask, is this visit to the President a visit of the GAC?

The WITNESS. Sure. We went to see him occasionally. This was a terrible flat. We had in mind that maybe we ought to go over to see him. We decided that this had better be handled through the responsible organs of the Government and not by a group of outside advisors, and we did so. Whether this was the Commission's view or our view, I don't remember.

By Mr. ROBB:

Q. Doctor, how did you know that Dr. Lawrence had talked to Senator Knowland and Senator McMahon, and some at least of the joint chiefs?

A. This was gossip and I have forgotten who gave it to me. Possibly Rabi, but I am not sure. I know that Lawrence talked to Rabi on his way home from Washington and I would assume that he told him something about it.

Q. You say here, "The climate of opinion among the competent physicists also shows signs of shifting." What did you mean the "the climate of opinion"?

A. What people were thinking.

Q. What were they thinking?

A. What they were thinking about the desirability of stepping up this program, I should think.

Q. You mean that up to then competent physicists had been opposed to it?

A. Had not been excited by it.

Q. Had not been enthusiastic.

A. Right.

Q. Now they were beginning to get more enthusiasm for it, is that correct?

A. Yes. I don't know whether enthusiasm or a feeling of necessity or so. I don't know the detail.

Q. Did that cause you alarm?

A. No.

Q. Wasn't that what you were expressing to Dr. Conant in this letter?

A. I was telling him in what form that I thought the problem would come before us, what the surrounding circumstances were.

Q. How did you know that Bethe was seriously considering return on a full-time basis?

A. He came to visit me at Princeton and talked to me.

Q. "And so surely are some others"; whom did you have in mind?

A. From the way that sounds I would say I had no one specific in mind.

Q. Doctor, how many reactors of any kind were built while you were chairman of GAC?

A. I don't know. I will start to think. A dozen and a half or something like that.

Q. How many physicists did you discuss this matter of the thermonuclear with prior to the meeting of October 29, 1949?

A. I clearly can't answer that question.

Q. A large number?

A. No; not a large number. I have tried to think of the ones that stuck in my memory. I have forgotten some things.

Q. Did you talk to Dr. Rabi?

A. Yes.

Q. When did you see him and where?

A. Either in Princeton or New York.

Q. Did he come to see you?

A. I don't remember. We saw a great deal of each other.

Q. What was his attitude on the thermonuclear at the time you talked to him prior to the meeting?

A. I believe, to put it as accurately as I can, it was one of somewhat quizzical enthusiasm.

Q. What did you say when you found that out?

A. I don't think I said much.

Q. Did you encourage him in his enthusiasm?

A. I don't see how I could have, but I don't remember the words I used.

Q. You said you talked to Dr. Serber.

A. Yes.

Q. He came to see you at Princeton, didn't he?

A. He was sent by Lawrence

Q. Sent by Lawrence and Alvarez?

A. Sent by Lawrence.
Q. Serber told you he was going to work on the thermonuclear, didn't he?
A. No.
Q. Did he come to ask you whether you would work on it or not?
A. I never fully understood the mission. He said he had come to discuss it.
Q. Do you know whether or not, prior to his seeing you, Serber had said that he would join the project and work on the thermonuclear?
A. I don't know. I had the impression that he had not made a commitment of such a kind and didn't intend to.
Q. Didn't he tell you he had come to see you to enlist your responsibility for the project?
A. To enlist my support for it.
Q. Yes, sir.
A. No; I don't think so.
Q. What had Lawrence sent him to see you for?
A. To discuss it with me.
Q. Just to discuss it with you?
A. Yes.
Q. That is all?
A. Yes.
Q. Did you encourage Serber to work on it?
A. No; I don't think I did.
Q. Did you discourage him?
A. No; I don't think I did.
Q. Did he work on it?
A. No; I don't believe he did. He may have a little.
Q. Did you talk with Dr. DuBridge about the matter before the meeting?
A. I think so, but I am not quite sure.
Q. Do you know what his view on it was before the meeting?
A. No.
Q. You didn't hear?
A. I don't remember.
Q. Did you talk with Bacher about the program before the meeting?
A. Is that one of the names that is in the list?
Q. What list?
A. The list in my letter to Conant. I have forgotten.
Q. No. You talked with Bradbury, Manley, and von Neumann, you say in this letter.
A. Right.
Q. Do you recall whether you talked to Bacher at all?
A. No; I don't. I did talk to him at a later stage, I remember very well.
Q. Were your long talks with Bradbury, Manley, von Neumann individual talks, or did you talk in a group?
A. With von Neumann, since he was right next door, it would be alone, and with Bradbury and Manley it would have been together.
Q. Can you tell us anything about what you said to them?
A. No; I can't. I would guess I mostly asked them.
Q. Would it not be reasonable, Doctor, to conclude that you expressed to them substantially the same views you expressed to Dr. Conant in this letter of October 21?
A. The situation was a little different. I would think that I would have got Bradbury to tell me as much as he could rather than to tell him what I thought.
Q. Doctor, you say here you have had long talks; presumably you talked too, didn't you?
A. I always do.
Q. Yes. So isn't it a fair conclusion, Doctor, that in your long talks with Bradbury, Manley, and von Neumann you expressed the same feelings and the same views which you set out in writing to Dr. Conant?
A. I very strongly doubt it. The relations were quite different. With Conant we had a problem of advice before us. The views that I expressed there are not the views the committee adopted. The background was something I thought he ought to know about. I would guess that with von Neumann, Bradbury, and Manley—anyway, with Bradbury and von Neumann—the talk would have been much more on technical things. I remember von Neumann saying at this time: "I believe there is no such thing as saturation. I don't think any weapon can be too large. I have always been a believer in this." He was in favor of going ahead with it.
Q. Did he afterward work on the project?

A. He did.

Q. Do you recall what views you expressed to Serber when he came to see you at Princeton?

A. I would think possibly not far from those I expressed here, that this was a thing that one had to get straight, but it was not the answer. I am conjecturing now. An honest statement would be to say I don't recall.

Q. Did you talk to Dr. Alvarez about the thermonuclear program about this time?

A. I think I did more than once.

Q. What views did you express to him about it?

A. I remember once when I expressed negative views, but I think in a rather indiscreet form of telling him what other people were saying.

Q. Would you tell us about that occasion and when it was?

A. The occasion I remember is during the GAC meeting. Alvarez and Serber and I had lunch together. The discussion was in midprogress, and we had not reached a conclusion. I said quite strongly negative things on moral grounds were being said.

Q. Did you specify what those negative things were?

A. I don't remember.

Q. Those were your views, too, weren't they?

A. They were getting to be in the course of our discussion.

Q. You felt strongly negative on moral grounds, didn't you?

A. I did as the meeting came to an end. I think the views that are expressed in the letter to Conant probably are as measured and honest as any record could be, and I think my attempt to reconstruct what I thought at one or another moment in this time of flux would be less revealing than what you have read out loud.

Q. Do you recall what Serber's attitude was at the time of this luncheon?

A. No.

Q. Do you recall whether or not Serber subsequently opposed the development of the thermonuclear?

A. I know of no such opposition.

Q. In all events he did work on it.

A. He worked on it very little but not very hard or effectively.

Q. But not what?

A. Not very hard or effectively.

Q. Doctor, you have testified, I believe the report of the GAC reflects, that it was impossible to tell without a test whether a thermonuclear device would work or not; is that correct?

A. Right.

Q. Did there come a time when some tests of a thermonuclear bomb were scheduled.

A. In October of 1952? That is the time?

Q. I think so; yes.

A. Right.

Q. Did you suggest that that test be postponed?

A. I would like to haul off.

Q. Like to what?

A. I would like to pull back a little back.

Q. Very well.

A. I was then a member of this panel of the State Department. Another member was Dr. Bush. He told me right before—well, very early in the meeting of the panel—that he had been to see the Secretary of State about his anxieties of the timing of this test. I did nothing whatever about it. When the panel was meeting during the summer and late autumn, we discussed this matter as relevant to our terms of reference in great detail. The panel insisted that we make our views known as to the advantages and disadvantages of the scheduled date to the Secretary. So we did.

I also inquired of Bradbury about what a postponement of a week or 2 weeks or so on would mean in a technical sense. I believe this is the summary of all that I had to do with it. The scheduled date was November 1, before the presidential election. It was at a time when it was clear that whatever administration was coming in was different from the outgoing administration.

Q. You did favor the postponement of the test; is that right?

A. No; I think that is not right. I think I saw strong advantages in not holding it then and many strong disadvantages. I reported both.

Q. You were at that time a member of the State Department panel on disarmament; is that right?

A. Yes.
Q. In fact, you were chairman of the panel, weren't you?
A. I was.
Q. Did your panel make a report on this matter of the postponement of the test?
A. It discussed it with the Secretary of State. It made no report.
Q. You made no written report?
A. Right.
Q. Didn't you favor the postponement of the test, Doctor?
A. I have explained to you that I saw strong arguments for it and strong arguments against it. I didn't think it was my decision or my job advocacy.
Q. I understand that, Doctor. I am asking for your opinion at the time. I think it is a rather simple, plain question. Did you or did you not favor postponement of the test?
A. My candid opinion was that it was utterly impractical to postpone the test but that we nevertheless owed it to the Secretary of State what we thought was involved in holding it at that time.
Q. Was one factor which you thought perhaps made a postponement advisable the reaction of the Soviet to the test?
A. We thought that they would get a lot of information out of it.
Q. How long was it suggested that the test be postponed, it if was postponed?
A. Until the new administration either before or after its assumption of office could conduct it or could be involved in the responsibility for it.
Q. Doctor, we are agreed, I take it, that in the absence of a test it was impossible ever to determine whether a thermonuclear would or would not work; is that right?
A. To be sure. At that stage, let me say we had quite different designs. I reported to the President that, although you could not be certain of the performance of any one design, it was virtually assured that this could be done. The situation was wholly different in 1949 where the doubts would have been of a very much more acute character with that model. However, you don't have a weapon until you proof-fire it.
Q. No. Even in 1949, Doctor, could anybody have said that the thermonuclear would not work in the absence of a test?
A. I could say a specific model would not work, and that has been said, wholly without a test.
Q. Could you in 1949 have said that no model of a thermonuclear could be made that would work?
A. Of course not. You can't say that nobody will ever think of anything. I have the memorandum of the panel on this subject. It has no restricted data in it. If the panel would like a copy of that memorandum, I can make it available.
Mr. GARRISON. You mean the board.
The WITNESS. If the board would like a copy of the memorandum, I can make it available. I don't have it with me because, although not free of restricted data, it obviously is a classified document.
Mr. GRAY. Yes.

By Mr. ROBB:
Q. One further matter, Doctor, so the record will be complete. It is a fact, is it not, that you opposed the establishment of a second laboratory?
A. The General Advisory Committee and I opposed the plans during the winter of 1951–52—the suggestion then made—but we approved the second laboratory as now conceived because there was an existing installation, and it could be done gradually and without harm to Los Alamos. There is a long record of our deliberations.
Q. I understand that. There was a proposal made in 1951 to establish a second laboratory for the purpose of working on the thermonuclear.
A. Right.
Q. And for various reasons which you have explained you and the committee opposed the establishment of that laboratory.
A. That is correct.
Q. Do you think now that the reasons that you advanced then were sound ones?
A. Yes. I think if we had thought that it was possible to take an existing Commission facility that was working on something that didn't amount to anything and convert it gradually into a weapons facility, the arguments we had then would not have applied. The proposal was to found something new in

some new desert, and this we thought could not be done without taking a big bite into Los Alamos.

Q. Who proposed establishing it in some new desert?

A. This is the way in which the Commission presented it to us—a second Los Alamos.

Q. The fact that it was established in some new desert would have made it much more difficult to get personnel, would it not?

A. That is right.

Q. Did you suggest an alternative that they might establish it in some place other than a desert?

A. No. We suggested lots of places that were open to the Commission to get work on various aspects of this problem, and that Los Alamos used some contracting and delegation to a very much greater extent than they had. This is different only in a minor way from the arrangement now made in California.

Q. Doctor, at the outset of your testimony, you took an oath to tell the truth, the whole truth, and nothing but the truth, so help you God.

A. Yes.

Q. Are you fully conscious of the solemn nature of that oath?

A. Yes.

Mr. GARRISON. Mr. Chairman, is this necessary?

Mr. GRAY. I think the chairman would have to say that the witness took the oath and had read to him the penalties prescribed. I see no reason for the record to reflect this question being asked again.

Mr. ROBB. Very well. That is all I have at the moment, Mr. Chairman.

Mr. GARRISON. Perhaps we could take a 5-minute recess.

Mr. GRAY. It will be perfectly all right, because I have a couple of questions that I would like to ask and maybe the board members do. But a recess is quite satisfactory.

Mr. GARRISON. You would like to continue questioning Mr. Oppenheimer.

Mr. GRAY. Yes.

The WITNESS. Let us get that over with.

Mr. GRAY. Part of this, Dr. Oppenheimer, to complete what seems to be a slight gap—at least my first question, this was in relation to the statutory function and mission of the GAC, and the question of whether there were departures from the technical and scientific advice.

I think twice you observed that the GAC on occasion failed to respond to questions.

The WITNESS. Yes.

Mr. GRAY. Or did not respond. There is no implication in my question.

The WITNESS. Did not respond to nontechnical questions.

Mr. GRAY. That is correct. Could you give an example of that kind of thing?

The WITNESS. Yes. We were asked whether the Armed Services or the Commission should have custody of atomic weapons. We didn't answer that question. We simply gave a few technical comments on it. We were asked sometimes questions about organization.

Mr. GRAY. I see. I think that is what I had in mind.

My next question is one which was not fully developed, I think, in the questioning of counsel. I don't think it is a new matter, and I think it is pertinent to the whole problem.

Is it your opinion, Doctor, that the Russians would not have sought to develop a hydrogen bomb unless they knew in one way or another, or from one source or another, that this country was proceeding with it?

The WITNESS. That was my opinion in 1949. As of the moment I have no opinion. I don't know enough about the history of what they have been doing.

Mr. GRAY. I don't think my question relates so much to historical events as to a view of the international situation and the problems with which this country was confronted. Would it not have been reasonable to expect at any time since the apparent intentions or the intentions of the U. S. S. R. were clear to us that they would do anything to increase their military strength?

The WITNESS Right.

Mr. GRAY. Whatever it might be.

The WITNESS. Oh, sure.

Mr. GRAY. So you don't intend to have this record suggest that you felt that if those who opposed the development of the hydrogen bomb prevailed that would mean that the world would not be confronted with the hydrogen bomb?

The WITNESS. It would not necessarily mean—we thought on the whole it would make it less likely. That the Russians would attempt and less likely that they would succeed in the undertaking.

Mr. GRAY. I would like to pursue that a little bit. That is two things. One, the likelihood of their success would we all hope still be related to their own capabilities and not to information they would receive from our efforts. So what you mean to say is that since they would not attempt it they would not succeed?

The WITNESS. No. I believe what we then thought was that the incentive to do it would be far greater if they knew we were doing it, and we had succeeded. Let me, for instance, take a conjecture. Suppose we had not done anything about the atom during the war. I don't think you could guarantee that the Russians would never have had an atomic bomb. But I believe they would not have one as nearly as soon as they have. I think both the fact of our success, the immense amount of publicity, the prestige of the weapon, the espionage they collect, all of this made it an absolutely higher priority thing, and we thought similar circumstances might apply to the hydrogen bomb. We were always clear that there might be a Russian effort whatever we did. We always understood that if we did not do this that an attempt would be made to get the Russians sewed up so that they would not either.

Mr. GRAY. Further with respect to the hydrogen bomb, did in the end this turn out to be a larger weapon than you felt it might be when it was under discussion and consideration in 1942 and 1943?

The WITNESS. We were much foggier in 1942 and 1943. I think your imaginations ranged to the present figures.

Mr. GRAY. I think I should disclose to you what I am after now. I am pursuing the matter of the moral scruples. Should they not have been as important in 1942 as they might have been in 1946 or 1948 or 1949?

The WITNESS. Yes.

Mr. GRAY. I am trying to get at at what time did your strong moral convictions develop with respect to the hydrogen bomb?

The WITNESS. When it became clear to me that we would tend to use any weapon we had.

Mr. GRAY. Then may I ask this: Do you make a sharp distinction between the development of a weapon and the commitment to use it?

The WITNESS. I think there is a sharp distinction but in fact we have not made it.

Mr. GRAY. I have gathered from what you have said, this was something that underlay your thinking. The record shows that you constantly, with greater intensity at varying times perhaps, encouraged the efforts toward some sort of development, but at the point when it seemed clear that we would use it if we developed it, then you said we should not go ahead with it. I don't want to be unfair, but is that it?

The WITNESS. That is only a small part of it. That is a part of it. The other part of it is, of course, the very great hope that these methods of warfare would never have to be used by anybody, a hope which became vivid in the fall of 1949. The hope that we would find a policy for bringing that about, and going on with bigger and bigger bombs would move in the opposite direction. I think that is apparent in the little majority annex to the GAC report.

Mr. GRAY. Was it your feeling when you were concerned officially and otherwise with a possible disarmament program that the United States and its allies would be in a better bargaining position with respect to the development of some sort of international machinery if it did not have the hydrogen bomb as a weapon in the arsenal, or is that relevant at all?

The WITNESS. The kind of thing we had in mind is what one would do in 1949 and 1950.

Mr. GRAY. This is quite a serious line of questioning as far as I am concerned, because it has been said—I am not sure about the language of the Nichols letter—at least in this proceeding and later on in the press, that you frustrated the development of the hydrogen bomb. That has been said. There have been some implications, I suppose, that there were reasons which were not related to feasible, to cost, et cetera.

The WITNESS. Right. I think I can answer your question.

Mr. GRAY. Very well.

The WITNESS. Clearly we could not do anything about the nonuse or the elimination of atomic weapons unless we had nonatomic military strength to meet whatever threats we were faced with. I think in 1949 when we came to this meeting and talked about it, we thought we were at a parting of the ways, a parting of the ways in which either the reliance upon atomic weapons would increase further and further or in which it would be reduced. We hoped it would be reduced because without that there was no chance of not having them in combat.

Mr. GRAY. Your deep concern about the use of the hydrogen bomb, if it were developed, and therefore your own views at the time as to whether we should proceed in a crash program to develop it—your concern about this—became greater, did it not, as the practicabilities became more clear? Is that an unfair statement?

The WITNESS. I think it is the opposite of true. Let us not say about use. But my feeling about development became quite different when the practicabilities became clear. When I saw how to do it, it was clear to me that one had to at least make the thing. Then the only problem was what would one do about them when one had them. The program we had in 1949 was a tortured thing that you could well argue did not make a great deal of technical sense. It was therefore possible to argue also that you did not want it even if you could have it. The program in 1951 was technically so sweet that you could not argue about that. It was purely the military, the political and the humane problem of what you were going to do about it once you had it.

Mr. GRAY. In further relation to the October 29 meeting of the GAC, I am asking now for information: From whom did the GAC receive the questions which the Commission wished the GAC to answer?

The WITNESS. The Commission met with us. I think there was probably a letter to me from Mr. Lilienthal. This is not certain, but probable. But the record will show that.

In supplement of the letter calling us to the meeting, we were addressed by the Commission at the outset.

Mr. GRAY. This communication signed by Mr. Pike, Acting Chairman, the date of the letter was the 21st.

The WITNESS. Right.

Mr. GRAY. So in part your instructions, if I may use that term, at least came from a letter. I am unable to read it. In this letter there were raised a lot of questions. In your reply I believe to General Nichols and certainly your testimony here, you say that the GAC was asked to consider two questions: One, are we doing all we should; two, what about the crash program.

My question is was it in a meeting with the Commission that the agenda or proposed agenda items were refined to these two?

The WITNESS. I would think that we would have been charged, so to speak, by the Commission with its formulation of what it wanted us to do.

Mr. GRAY. And it was your clear understanding as Chairman that what they wanted you to do in that meeting——

The WITNESS. Was to answer those two questions. I would be unhappy if many of the questions in Mr. Pike's letter remained unanswered in our answer, but I don't remember. It doesn't matter.

Mr. GRAY. I would like to ask about one of these questions. This is not surprise material for Dr. Oppenheimer.

Do you remember, Dr. Oppenheimer, whether, when you went into your meeting, you expected to consider cost of the super in terms of scientific personnel, physical facilities and dollars?

The WITNESS. We outlined in our answer—I don't know whether we expected to, I have seen our answer just two days ago—in our answer we have four items saying what it would require to carry out the program.

Mr. GRAY. I see.

The WITNESS. Perhaps not the dollars. We were not very good on dollars.

Mr. GRAY. May I ask you now to turn your mind to an entirely different kind of thing, the Chevalier incident, in which it would appear that at that time and under those circumstances within the framework of loyalty generally—loyalty to an individual, broader loyalty to a country, and I am not talking about espionage—in that case considerations of personal loyalty might have outweighed the broader loyalties.

The WITNESS. I understand that it would appear that way. It is obvious from my behavior that I was in a very great conflict. It is obvious that I decided that with regard to Eltenton the danger was conceivably substantial and that I had an obligation to my country to talk about it. In the case of Chevalier, I would not think that I regarded it as a conflict of loyalties, but that I put too much confidence—put an improper confidence in my own judgment that Chevalier was not a danger.

Mr. GRAY. Another instance which has been discussed in the proceeding, the testimony with respect to Dr. Peters and your subsequent letter to the Rochester newspaper. In writing that letter, which perhaps was motivated by a desire not to hurt the individual, or to make restitution——

The WITNESS. Not to get him fired, anyway.

Mr. GRAY. Not to get him fired—again was this the same kind of conflict that you had with respect to——

The WITNESS. No; I think this was almost wholly a question of public things. Personal things were not involved. He was a good scientist doing according to everyone's account no political work of any kind, doing no harm, whatever his views. It was overwhelming belief of the community in which I lived that a man like that ought not to be fired either for his past or for his views, unless the past is criminal or the views lead him to wicked action. I think my effort was to compose the flap that I had produced in order that he could stay on and that this was not a question of my anguish about what I was doing to him.

Mr. GRAY. As you know, this board is asked to consider present and future circumstances. Do you feel that today where there became a conflict between loyalty to an individual and a desire to protect him and keep his job or have him keep his job—whatever it might be—and a broader obligation, and I consider it to be broader is the reason I put it that way, that you would follow this same kind of pattern with respect to other individuals in the future?

The WITNESS. The Chevalier pattern; no, never. The Peters pattern I do not believe that I violated a broader obligation in writing the letter. It was for the public interest that I wrote it.

Mr. GRAY. You make a distinction between what is said about a man in executive session—we are talking in terms of loyalty—and what is said about a man for public consumption. Do you think on the basis of the same facts it is appropriate to say one thing in executive session, and another thing for public consumption?

The WITNESS. It is very undesirable. I wish I had said more temperate, measured and accurate words in executive session. The it would not have been necessary to say such very different words publicly.

Mr. GRAY. I suppose my final question on that is related to the view you held at one time that a cessation—correct me if I mistake this—of Communist activities, as distinguished from Communist sympathies, was important in considering a man for important classified work. Is that your view today?

The WITNESS. No; I have for a long time been clear that sympathy with the enemy is incompatible with responsible or secret work to the United States.

Mr. GRAY. So it would not be sufficient to say to a man, stop making speeches, stop going to meetings; that would not be enough?

The WITNESS. It was not in fact sufficient before. It was sufficient only if it was a man whose disengagement was dependable.

Mr. GRAY. Disengagement as far as activities are concerned.

The WITNESS. And to some extent conduct. Today it is a very simple thing, it seems to me, and has been for some years. We have a well-defined enemy. Sympathy for him may be tolerable, but it is not tolerable in working for the people or the Government of this country.

Mr. GRAY. One other question, which relates to the record, and your reply to General Nichols, and that is with respect to whose initiative it was which led to the employment of Dr. Hawkins as assistant personnel officer or whatever his title was. Do you now recall whether you simply endorsed the notion of his employment, or whether you——

The WITNESS. No; I said in my earlier testimony that I relied rather heavily—that I relied on Hawkins' testimony under oath—that he had been asked for by the personnel director. I don't recall how the discussion started.

Mr. GRAY. Finally, and this is much less important than some of these other questions, when in 1946 you resigned from the ICCASP, in your letter of resignation you referred to your disagreement with their current position with respect to the extension of President Roosevelt's foreign policy, despite the many constructive and decisive things that this organization was doing; I wondered what you had in mind.

The WITNESS. I wondered when I heard it. There is in my file a reference to a panel of the committee that was advocating and speaking for a National Science Foundation; though that is only one thing, it has always seemed a constructive one.

Mr. GRAY. Because you had testified that you did not know too much about what they were doing and had not been active.

The WITNESS. This seems to be the only record I have.

Dr. EVANS. Dr. Oppenheimer, did the Condon letter have much weight with you in changing your position on that security committee?

The WITNESS. The Peters thing?
Dr. EVANS. Yes.
The WITNESS. No. The letters that had weight with me were from Bethe and Weiskopf. They were written in very moderate and dignified——
Mr. EVANS. Condon did write a letter about it?
The WITNESS. He did, and it has been published in the papers. It made me angry.
Dr. EVANS. Another question: From a political point of view, did you consider the super a bad project even if it could be made?
The WITNESS. I think your record says that if we could have a world without supers it would be a better world.
Dr. EVANS. Did you consider the fact that there would not be many targets for a super?
The WITNESS. We did indeed. We discussed that. We said we had many more than the Russians. We said we were more vulnerable to it, and went into the questions of delivering it by ship and so on.
Dr. EVANS. There is one other question that I want to ask and perhaps you won't answer this and can't, and I wouldn't want you to in that case. Did you reach the conclusion that the super would work purely from a mathematical point of view. In other words, you had not tested it as yet?
The WITNESS. At what stage is this? When I did reach that conclusion?
Dr. EVANS. Yes.
The WITNESS. Yes. I didn't reach the conclusion that the precise designs and details embodied in our first thing would work as well as it might, but I reached the conclusion that something along these lines could be made to work.
Dr. EVANS. That is all.
Mr. GARRISON. Could we just have the last question read?
Dr. EVANS. I can restate it. Did you reach the conclusion that the super would work from a purely mathematical point of view because they had not made the test.
Mr. GARRISON. Excuse me.
The WITNESS. I believe in our report to the President we said though there is always in matters of this kind the possibility that a specific model will fail, we are confident that this program is going to be successful.
Dr. EVANS. There was a delicate boundary there that you could not be quite sure?
The WITNESS. You can never be quite sure of anything in the future.
Mr. GRAY. It is 12:15 and you asked for a recess.
Mr. GARRISON. I don't think a recess is necessary.
(Discussion off the record.)
(Witness excused temporarily.)
Mr. GRAY. Dr. Glennan, do you care to testify under oath? You are not required to do so.
Dr. GLENNAN. I don't understand you.
Mr. GRAY. Do you care to testify under oath?
Dr. GLENNAN. I would be glad to.
Mr. GRAY. All right, sir. Would you be good enough to stand and hold up your right hand? What is your full name?
Dr. GLENNAN. Thomas Keith Glennan.
Mr. GRAY. Thomas Keith Glennan, do you swear that the testimony you are to give the board shall be the truth, the whole truth, and nothing but the truth, so help you God?
Dr. GLENNAN. I do.
Whereupon, Thomas Keith Glennan was called as a witness, and having been first duly sworn, was examined and testified as follows:
Dr. GRAY. Now, you will forgive for an elementary lesson but I think I should remind you of the provisions of section 1621 of title 18 of the United States Code, known as the perjury statute, which makes it a crime punishable by a fine of up to $2,000 and/or imprisonment up to 5 years for any person stating under oath any material matter which he does not believe to be true. It is also an offense under section 1001 of title 18 of the United States Code, punishable by a fine of not more than $10,000 or imprisonment for not more than 5 years or both for any person to make any false, fictitious, or fraudulent statement or representation in any matter within the jurisdiction of an agency of the United States.

I should also like to make the request that in the event it is necessary for you to discuss any restricted data in your testimony, that you let the chairman know before any disclosure for reasons which probably are obvious.

I think those are the instructions I am to give you, so you may proceed.

DIRECT EXAMINATION

By Mr. GARRISON:

Q. Mr. Glennan, would you state your full name for the record?
A. Thomas Keith Glennan.
Q. You are president of Case Institute in Cleveland?
A. I am.
Q. Dr. Glennan, there was handed to me just now an affidavit by you which I will give you.

Mr. GARRISON. Mr. Chairman, I had not thought to suggest to Dr. Glennan that he would read the statement which he has prepared because I had preferred to go along in the ordinary way by question and answer, but in the recess I discussed the matter with Dr. Glennan, and I thought in the interest of time it might be well if he would read this and then respond to any questions that anybody might like to put to him.

Mr. GRAY. We should be glad to have him read it.

By Mr. GARRISON:

Q. Mr. Glennan, is this statement that you prepared your own in toto?
A. Without question.
Q. Did you receive any drafting assistance from anyone representing Dr. Oppenheimer?
A. No. The only drafting assistance I received was a question that I raised myself as to restricted data that might be in here, and with the help of Mr. Beckerley this morning I changed part of one sentence to remove that.
Q. I simply ask you to speak of the work you have done with Dr. Oppenheimer, your relations with him and your views about him.
A. That is right.
Q. Perhaps you would read this statement to the Board.
A. My name is Thomas Keith Glennan. I am 48 years old and I am president of Case Institute of Technology in Cleveland, Ohio. From October 1, 1950, until November 1, 1952, I was on leave of absence from Case, and served during that period as a member of the United States Atomic Energy Commission. I have read somewhat hastily the pertinent parts of a letter addressed recently by the General Manager of the Atomic Energy Commission to Dr. J. Robert Oppenheimer. That letter recites certain incidents reported by the FBI presumably which have caused serious questions to be raised by certain persons concerning the loyalty of Dr. Oppenheimer to the United States of America.

Shortly after taking office as a Commissioner, I met Dr. Oppenheimer for the first time. During the ensuing years our meetings were limited to those days when the General Advisory Committee was in session, to discussions at Princeton in mid-June 1951, which I shall mention later, and to such other group meetings which may have occurred at the offices of the Commission in Washington during the period noted. My contacts with Dr. Oppenheimer since November 1952 have been limited to correspondence at infrequent intervals.

My earliest recollection of a General Advisory Committee meeting had to do with a review in late 1950, as I recall it, of the first two important Atomic Energy Commission expansion programs.

If I might interpolate, I would say the first of the two important Atomic Energy Commission programs.

I was impressed as a new member of the Commission by the expressions of satisfaction on the part of Dr. Oppenheimer and other members of the General Advisory Committee, and I recall comments to the effect that the General Advisory Committee under Dr. Oppenheimer's chairmanship had been urging expansion in the fissionable materials and weapons field for some time. About this same time I first became aware of the problems posed for the Commission and in particular for the Los Alamos laboratory by the findings of the theoretical group there, that requirement for special materials appeared to be such that there would result a substantial reduction in the production of fissionable materials.

Q. Perhaps you could read it a little more slowly.
A. I learned, too, something of the disagreements that had taken place in late 1949 within the Commission itself, and within the General Advisory Committee, on the question of pursuing vigorously prosecution of the thermonuclear program. While it was apparent that certain moral questions had been raised in addition to questions of technical feasibility in these earlier debates, it seemed clear to me that the technical problems and the tremendous cost in terms of decreased

plutonium production had been of very great concern to the scientists involved. In the balance was the question of exploiting at all possible speed the very promising developments in the fission field, and the rapid buildup of a stockpile of great effectiveness against the diversion of effort and material to an as yet unproven thermonuclear device.

It is to be remembered that theoretical studies and calculations were proceeding during this period following on the President's decision to proceed with the diffusion program in early 1950.

In the late spring of 1951, certain studies made at Los Alamos by Teller, Nordheim, and others, began to show promise. A meeting was called, I believe jointly by the Commission and the General Advisory Committee, for the purpose of reviewing these new propulsions. The meeting was held at the Institute for Advanced Study at Princeton around the 19th and 20th of June 1951. The top level of scientific personnel available to the Commission were in attendance, as were all the Commissioners. It was this meeting that gave new hope to all for the thermonuclear program. It is my recollection that Dr. Oppenheimer participated with vigor and that there was never apparent to me at that time or subsequently anything in his actions or words that indicated anything other than a recognition of important new theoretical findings, and the necessity for pursuing vigorously these promising new leads.

It is true that Dr. Oppenheimer opposed the immediate establishment of a second weapons laboratory. So did I, and on the ground that Los Alamos was in the best possible position to push forward on the new propositions. To create a new laboratory would have been a crushing blow to the morale of the Los Alamos staff members and much valuable time would have been lost. Need for expansion of research effort was apparent, however, and studies were begun shortly thereafter to determine the best methods by which such expansion could be accomplished.

In the meantime Los Alamos pressed forward with great urgency to develop fusion devices for early tests of the new theories.

I cite these instances because it may be that accusations of disloyalty have been made against Dr. Oppenheimer——

Mr. ROBB. Mr. Chairman, I hate to interrupt the witness, but I feel it my duty to call to the attention of the Chairman, the provisions of the procedure that no witness will be permitted to make an argument from the witness stand. I apprehend that Dr. Glennan is about to make such an argument. I am not of course intending to suggest that Dr. Glennan is not doing anything he does not believe to be entirely proper, but the board procedures do provide under section 4.15, paragraph (f) "nor will the board permit any person to argue from the witness stand." I merely want to bring that to the board's attention, for whatever it might be worth.

Mr. GARRISON. Mr. Chairman, Dr. Glennan, I believe, is about to state his opinion. Surely the Chair will not consider this to be an argument.

Mr. GRAY. I would like to ask Dr. Glennan if all of his statement is directing himself specifically to the paragraph in the Nichols letter which you referred to at the outset, reporting certain positions, attitudes, and so on, of Dr. Oppenheimer, with respect to the development of the hydrogen bomb.

The WITNESS. Since, Mr. Chairman, my knowledge of these matters is limited largely or limited wholly, I should say, to the time I was on the Commission, I am dealing principally with that question.

Mr. GRAY. Do you have something farther?

Mr. GARRISON. I was going to make the general observation, Mr. Chairman, that in the case of many, if not most of the witnesses who will follow Dr. Glennan, I have asked them to recall the circumstances under which they had occasion to work with Dr. Oppenheimer, the extent to which they knew him, what they did together, what their views of him as a man and an American were as a result of their contacts with him, and all this item to be highly pertinent to the question, the ultimate question of judgment which this Commission has to make.

Mr. ROBB. Mr. Chairman, I thoroughly agree with that.

Mr. GARRISON. One of the basic questions in weighing a man's loyalty and citizenship is what sort of things has he done for his country in a time when the country is hard beset by foreign intentions.

Another test is what men of standing and eminence and character believed him to be on the basis not merely of reputation—community reputation—but on the basis of actual contacts with him.

I can't conceive that any question would arise in the mind of the Chairman as to the relevance of testimony of this character.

Mr. GRAY. I don't believe that counsel, Mr. Robb, has raised a question of relevance.

Mr. ROBB. Of course not.

Mr. GRAY. He has addressed himself to the procedure which is not generally too well defined. Did you want to say something?

Mr. ROBB. I thoroughly agree with all that Mr. Garrison has said. I have no intention of suggesting that those matters should not receive full discussion before this Board. I merely felt it my duty, Mr. Chairman, as I apprehended that Dr. Glennan was launching into what can be described as an argument, rather than a recital of facts and circumstances. Of course, I am afraid that this is something we get into when a witness does read a prepared statement. It is rather difficult for counsel to control what he says and it is very apt to become an argument or a speech rather than testimony.

Mr. GARRISON. Mr. Chairman, I really am amazed that this question should be raised.

Mr. GRAY. I think in this case, if only in the interest of economy of time, I am going to ask the witness to proceed with his prepared statement and we can argue these procedural questions later.

Mr. GARRISON. I myself often thought of that provision of the rules, Mr. Chairman, during some of the questioning that has taken place, but I have refrained from raising it.

The WITNESS. May I proceed, sir?

Mr. GRAY. If you will.

The WITNESS. I cite these instances because it may be that accusations of disloyalty have been made against Dr. Oppenheimer in part because of his disagreements with others because of the feasibility of one technical program compared with another, or one method of attack on a problem as compared with another. At no time did I then nor do I now know of any evidence that would indicate that Dr. Oppenheimer had been disloyal. Disagreements of this kind on technical and administrative matters are not sufficient ground for accusations such as have been made. Rather they are the normal phenomena in development matters of this nature.

Of the history of Dr. Oppenheimer prior to 1950 I have only limited knowledge and can make no comment. In light of his diligence in the prosecution of the Commission's program and insofar as my personal contacts with him have been revealing, I believe Dr. Oppenheimer to be a loyal citizen of the United States.

By Mr. GARRISON.

Q. And on the basis of these contacts, would you say that his continued employment as a consultant would be clearly consistent with the interests of national security?

A. I would.

Mr. GARRISON. That is all, Mr. Chairman.

Mr. ROBB. I have no questions.

Mr. GRAY. Does any member of the board have any questions?

Mr. MORGAN. No.

Dr. EVANS. No.

Mr. GRAY. Thank you very much.

The WITNESS. Thank you.

(Witness excused.)

Mr. GARRISON. If it is agreeable with the board, Dr. Compton will not take long. Would you like to hear him now?

Mr. GRAY. I think we might proceed with Dr. Compton.

(Discussion off the record.)

Mr. GRAY. Do you wish to testify under oath? You are not requested to do so.

Mr. COMPTON. I am perfectly willing to do so.

Mr. GRAY. Will you stand, please, and raise your right hand? What are your initials?

Dr. COMPTON. K. T.

Mr. GRAY. K. T. Compton, do you swear that the testimony you are to give the board shall be the truth, the whole truth, and nothing but the truth, so help you God?

Dr. COMPTON. I do.

Mr. GRAY. I must call your attention to the provisions of the perjury statutes which make it a crime punishable by fine up to $2,000 and/or of imprisonment up to 5 years for any person to state under oath any material matter which he does not believe to be true, and also call your attention to the fact that it is an offense under the statutes punishable by fine of not more than $10,000 or imprisonment for not more than 5 years or both for any person to make any false, fictitious or fraudulent statement or representation in any matter within the jurisdiction of an agency of the United States.

I should also, Dr. Compton, if it becomes necessary for you to make any reference to or to disclose restricted data in your discussion here, ask that you inform me in advance of the necessity to do so.

Finally, I should point out to you that we treat the proceedings of this board as a matter which is confidential as between the Atomic Energy Commission and its officials and agencies and Dr. Oppenheimer and his representatives, and we hope that witnesses will be guided accordingly, as far as the press and others are concerned. Mr. Garrison.

DIRECT EXAMINATION

By Mr. GARRISON.

Q. Dr. Compton, you were the president of the Massachusetts Institute of Technology from 1930 to 1948, I believe?
A. Yes, sir.
Q. Would you state briefly some of the positions which you have held in the Government having to do with the defense effort?
A. Probably most important of those was as a member of the National Defense Research Committee from 1940 to 1945 when I was in general charge of the developments in radar, fire control and instruments. Part of that time and only an early part of that time had to do with the atomic energy program.

I was later in 1945—in the first half of the year—a member of Secretary Stimson's Committee on Atomic Energy which was advising President Truman. That was the committee which George Harrison of New York Life was chairman.

Then in 1946, I was Chairman of the Joint Chiefs of Staff Evaluation Board on the first Bikini atom bomb test, and a member of the President's Evaluation Committee on that same test.

Then between a year and 2 years ago I was a member of the Committee under Lewis Strauss which was appointed by the late Senator McMahon to consider certain problems having to do with the capital facilities for atomic energy.

In that connection we made some appraisal of the work at Savannah River and at Paducah. The committee was disbanded, however, shortly after Senator McMahon's death.

I should also mention that I was in 1947 and 1948 chairman of the Research and Development Board in the Department of Defense immediately following Dr. Bush in that position.

I think those are the principal positions.
Q. Thank you. You first met Dr. Oppenheimer at Goettingen, I think you told me, in 1926.
A. That is right, November and December 1926. He was there as a post graduate student. I was there as a visitor working on a manuscript, and I saw quite a bit of Dr. Oppenheimer at that time.
Q. You yourself were trained as a physicist?
A. Right.
Q. Would you tell the board the nature of the undertakings in which Dr. Oppenheimer and you have worked together?
A. Starting with Goettingen, our first undertaking—we were a committee of some 20 American graduate students—to organize a Thanksgiving dinner to pay back the social debt to our German professors who had been very hospitable to us. That had its amusing incidents, but it has nothing to do with the atomic energy work.

I have met Dr. Oppenheimer at professional meetings frequently from time to time. The last meeting with him until this morning that I can recall was at Princeton in his office where I had been asked by Miss Shaver, the president of Lord and Taylor, to try to prevail on Einstein to accept one of the Lord and Taylor awards, and I called on Dr. Oppenheimer for advice on how best to approach Professor Einstein. My only contact that I can recall with Dr. Oppenheimer having to do with the atomic energy project was while on Secretary Stimson's committee in 1945.

One of the problems before us was to try to estimate the amount of time that it would take a foreign country, and particularly Russia, to produce an atomic bomb. At that time we called in 2 groups on 2 separate days. One group consisted of the presidents or chief engineers of the industrial companies that had been most engaged in the production of the atomic bomb plants, that is Eastman, duPont, Carbide and Carbon Chemicals, Westinghouse, as I recall.

The other meeting was with a group of scientists—Fermi, Oppenheimer, Ernest Lawrence, and my brother, Arthur. There may have been one other. I am not quite sure. It was at that meeting that as a result of those conferences that we came to the very rough estimate that it would require Russia a minimum of 5 years and a maximum of 20 and probably 10 to produce an atomic bomb.

In that connection, the predominant factor was not scientific information, because we realized that the Russians could get that as well as we could, but it had to do with industrial capacity—machine tools, to make tools, production of electronic control equipment, capacity to produce certain chemicals with the desired degree of purity, and things of that sort.

Q. I think Dr. Oppenheimer was a member of the Atomic Energy Committee of the Research and Development Board under William Webster when you were Chairman of the Research and Development Board.

A. That is correct.

Q. Based on your acquaintance with Dr. Oppenheimer, your knowledge that you have of him, what would you say as to his loyalty to the United States?

A. I have never had any question of it. I have no question of it now. He is completely loyal.

Q. Again based on your experience with him and your knowledge of him, would you say that his continued employment as a consultant to the Atomic Energy Commission would be clearly consistent with the interests of national security?

A. So far as I know the situation, I would say yes. I think I would have to qualify that by this fact. While my personal impression, my faith is sound, it would have to be subject to derogatory evidence that I don't know anything about, which I take it is the purpose of this committee to investigate.

Q. Of course, that goes without saying. I am asking you for your judgment simply based on your own personal feeling about him and knowledge of him.

A. Yes.

Q. As to that, you are clear in your mind.

A. Perfectly clear, yes.

Q. What in your judgment would be the effect, if any, on the scientific community if Dr. Oppenheimer's clearance were to be revoked?

A. I believe—and I feel very certain of this—that there would be a shock, there would be a discouragement, there would be confusion. I think the result would be very bad.

Q. Bad for the country?

A. For the country.

Mr. GARRISON. That is all.

Mr. ROBB. I have no questions, Mr. Chairman.

Mr. GRAY. Thank you very much, Doctor.

The WITNESS. I would like to say this. If anything should come up later in connection with things in which my past contact with Oppenheimer might raise questions for future evaluation, I would of course be glad to come down and appear if I can be of any help.

(Witness excused.)

Mr. GRAY. We are now in recess. I hope we can start at 2 o'clock.

(At 12:55 p. m., a recess was taken until 2 p. m.)

AFTERNOON SESSION

Mr. GRAY. The proceeding will begin. Do you wish to testify under oath, Mr. Lansdale? You are not required to do so.

Mr. LANSDALE. I have no wish at all in that respect. I leave that to counsel or to the Board.

Mr. GRAY. I might say to you the board imposes no requirement. All the witnesses to this point have testified under oath.

Mr. LANSDALE. Then let us keep it uniform.

Mr. GRAY. Would you stand and raise your right hand.

John Lansdale, Jr., do you swear that the testimony you are to give the Board shall be the truth, the whole truth and nothing but the truth, so help you God?

Mr. LANSDALE. I do.

Mr. GRAY. Now, Mr. Lansdale, I am required to call your attention to the provisions of the United States Code which make it a crime punishable by fine and imprisonment for any person to state under oath any material matter which he does not believe to be true and to remind you it is also an offense under the code punishable by a fine or imprisonment or both for any person to make any false, fictitious or fraudulent statement or representation in any matter within the jurisdiction of an agency of the United States.

I should like to ask that in the event it becomes necessary for you to disclose what you believe to be classified data during your testimony you should advise me before such disclosure in order that we may take certain steps.

Mr. LANSDALE. May I in that regard rely on Mr. Rolander because it has been since 1945 that I have had any acquaintance with what is classified or what is not.

I have heretofore adopted the practice that I considered everything I did was classified. I know that is not really true any more.

Mr. ROLANDER. Mr. Chairman, I think specific reference is being made with reference to restricted data, which is more in terms technical data. Mr. Lansdale, with respect to matters which were previously classified would probably be considering investigative data which was at that time classified. That would not at this time be considered as classified.

Mr. GRAY. Perhaps I was in error to raise the question here. But you will be on the alert, Mr. Rolander.

Mr. ROLANDER. Yes, sir.

Mr. GRAY. Finally, Mr. Lansdale, I should point out to you that this Board considers the proceedings strictly confidential between the Commission and its officials and Dr. Oppenheimer and his representatives and witnesses. This Board takes no initiative in the release of any information. Speaking for the Board I express the hope that witnesses will take the same view of the situation.

Mr. LANSDALE. This witness will.

Mr. GRAY. Mr. Garrison.

Whereupon, John Lansdale, Jr., was called as a witness, and having been duly sworn, was examined and testified as follows:

DIRECT EXAMINATION

By Mr. GARRISON:

Q. Mr. Lansdale, you are presently a member of the law firm of Squire, Sanders, and Dempsey in Cleveland, Ohio?

A. Yes, sir.

Q. You attended Virginia Military Institute and after that the Harvard University Law School?

A. Yes, sir.

Q. And during the war you were the security officer for the Manhattan District at Los Alamos?

A. The question is inaccurate. I was responsible to General Groves for the overall security and intelligence of the atomic bomb project, not technically the Manhattan District which was an administrative organization.

Q. But you were the top security officer for the atomic bomb project?

A. Yes, sir.

Q. Would you tell the board how you happened to get into the security work which you were charged with by General Groves?

A. I believe General Groves advised me that he requested me to take charge of that work because I had previously, before the Army had been given responsibility for the atomic bomb project, made a security investigation at Dr. Conant's request at Berkeley and thus by that accident I was one of the very few Army officers who had any knowledge of the existence and nature of the project.

Q. Dr. Conant asked you to undertake this study of the situation at Berkeley in 1941, as I recall.

A. It was either in December 1941 or January 1942. My recollection is a little fuzzy on the precise date but it was right in that time.

Q. And you were attached at that time to Gen. Robert Lee in G-2?

A. Yes.

Q. In the counterintelligence work?

A. I was in the so-called Counterintelligence Branch of the Office of Assistant Chief of Staff, G-2, War Department General Staff.

Q. Will you tell the board about your discussions with General Groves about Dr. Oppenheimer's background and about his clearance?

A. I cannot recall precisely when we first began to discuss Dr. Oppenheimer.

Q. May I interrupt you one minute?

A. Yes.

Q. I would just like to ask you if you have discussed the subject matter of your general scope of testimony here today with representatives of the Commission who are assisting the board in its deliberations as well as with us?

A. That is right. I think it fair to say——

Q. I don't mean every question I am going to put has been discussed.

A. I think it fair to say that I have not discussed with the Commission staff my testimony as such. I have very briefly last night and at greater length some days or weeks ago answered to the best of my ability every question that I could that they had about this background.

Mr. ROBB. Mr. Chairman, for the sake of continuity in the record, I wonder if I might put one question at this point?

Mr. GRAY. Yes.

Mr. ROBB. Did we not also permit you to refresh your recollection by looking at certain portions of the file with which you had been concerned?

The WITNESS. Oh, yes.

Mr. ROBB. That is all.

The WITNESS. There were several documents which you gave me to read to refresh my recollection and to mutually try to arrive at facts which were not apparent in the record.

In any event, Dr. Oppenheimer had been on the project prior to the time that the Army took over. When the Army took it over, the security was virtually nonexistent and the program of personnel clearance was practically nonexistent. I won't say it did not exist because it did, but it was very incomplete. One of the first things that we did was to attempt to get some investigation and set up some program for the clearance of the personnel that were received with the project, as it were.

I, myself, never was until fairly late in the game transferred to the Manhattan District. I remained with G-2 and performed my duties as a supervisory matter along with my other duties in G-2.

Then Lieutenant Calvert was assigned to the Manhattan District as the security officer and he conducted the clearance program.

In connection with that we received reports, primarily from the Federal Bureau of Investigation, as I remember, concerning Dr. Oppenheimer's associations and relatives, as well as himself. These caused us, needless to say, a great deal of concern. I may be inexact in my dates, but my recollection is that this took place about the time that Los Alamos was being established and my recollection is that they had not yet moved up on the Hill, but still had the office or laboratory down in Santa Fe while we were constructing a road up there.

I brought up these, because of Dr. Oppenheimer's prominent position as the head of the Los Alamos laboratory, to the attention of General Groves and we discussed them at some length.

General Groves' view was (a)—I wonder if I am permitted to say—I don't know what his view was, of course, as I only know what he told me.

Mr. GRAY. You certainly can say what he told you.

The WITNESS. I would like to correct that. Obviously I don't know what was in the man's mind. All I know is what he told me.

General Groves' view, as I recall expressed, was (a) that Dr. Oppenheimer was essential; (b) that in his judgment—and he had gotten to know Dr. Oppenheimer very well by that time—he was loyal; and (c) we would clear him for this work whatever the reports said.

I will confess that I myself at that time had considerable doubts about it. Because of our worry, or my worry, let us say, about Dr. Oppenheimer, we continued to the best of our ability to investigate him. We kept him under surveillance whenever he left the project. We opened his mail. We did all sorts of nasty things that we do or did on the project.

I interviewed him myself a number of times. As I recall, the recommendations of the security organization headed up by Captain Calvert were adverse to Dr. Oppenheimer. They recommended against clearance.

By Mr. GARRISON:

Q. Who was Captain Calvert?

A. I think his official title was District Security Officer. He was on General Nichols', then Colonel Nichols, staff. In any event, I fully concurred with General Groves as our investigation went on with the fact that Dr. Oppenheimer was properly cleared.

Now, you asked to relate our discussions. That is difficult. Our discussions spread over many, many months. They continued when the name——

Mr. GRAY. Excuse me, please. Did you say I asked to relate the discussions?

The WITNESS. No; Mr. Garrison did.

Mr. GRAY. Excuse me. A moment ago I thought when you asked whether you were privileged to say what General Groves said, I said that was all right.

The WITNESS. No. I think that was your question, wasn't it?

Mr. GARRISON. Yes, it was.

The WITNESS. I remember that I asked General Groves early in the game what would he do if it turned out that Dr. Oppenheimer was not loyal and that we could not trust him? His reply was that he would blow the whole thing wide open.

I do not mean to imply by that, that our conclusions as to clearance were necessarily dictated by indispensability. I wish to emphasize it for myself. I reached the conclusion that he was loyal and ought to be cleared.

By Mr. GARRISON:

Q. You did have certain employees, did you not, that the project had at Los Alamos who were kept on the basis of what might be called a calculated risk?

A. Yes; that is true. That is true of Los Alamos and other parts of the project.

Q. Certain people who were known or believed to be Communists?

A. Yes, sir.

Q. Why did the project employ some people of that character?

A. My only answer to that is that we continually had to exercise judgment as between obvious all out security and the necessities of the project. It must be remembered that the Germans were far ahead of us in the development of an atomic bomb. We believed that the nation which first obtained one would win the war. We were under, believe me, very terrible feeling of pressure. Every security decision we made with reference to important people was made in that background.

We had a number of persons who we believed were very likely to be Communists, who we were persuaded were doing such useful work and such important work, that good judgment required that we keep them and let them do their work and surround them and insulate them to the best extent of our ability. That is what we did in a number of cases.

I can't answer it any better than that.

Q. Dr. Oppenheimer was not in that category of calculated risk, I take it?

A. Not in my judgment, no.

Q. Did you ever know of any leakage of information from any of the persons of the sort you have mentioned to the outside?

A. We never discovered any leakage of information from those persons that we deliberately kept as a calculated risk. I don't mean to assert that there was none. We discovered none and we used every effort we could to make it difficult for them.

For example, with many of them we made it perfectly obvious that we were watching their every move so as to be sure that if they desired to pass information they would go to extraordinary lengths to do so and thus make it easier to detect.

Q. Did you know of any leakage from the Los Alamos project, apart from that which has become public property?

A. Apart from the inexcusable Greenglass case, I now recall none that we knew of. Oh, we had a mail censorship program set up and we were continually picking up the things in letters that we thought ought not to go out and which we intercepted. Those were the kind of things which my recollection is that we didn't regard as deliberate attempts at security violation.

Q. Los Alamos was operated on a noncompartmentalized basis, was it not?

A. Generally speaking that is true. Indeed that was the purpose of the establishment of Los Alamos.

Q. Would you explain that a little more fully?

A. My recollection is that we had originally planned to put Los Alamos at Oak Ridge. Indeed that was the original purpose of the acquisition of so much of the land down there before we understood how big the project was going to be.

Dr. Oppenheimer and various of his associates were quite strong in their feeling that the fastest and best progress could be made if we could find a place where the principle of compartmentalization which we had adopted generally for the whole project could be laid to one side, at least so far as the important people on the project were concerned.

It was believed that the establishment of the laboratory in an isolated place where means of egress and ingress could be easily controlled and means of communication monitored should be done, if feasible.

We did have certain compartmentalization there. As I remember, we had the so-called technical area where the actual laboratories were, and as I recall we had two kinds of badges, for example; those who could get in there and those who could not. To that extent we had compartmentalization.

Q. You visited the project frequently, did you not?

A. Many times, yes.

Q. Did you form any judgment as to the wisdom in an overall point of view of the establishment of Los Alamos as a community in which work could be carried forward in the relatively free and less cramped manner that you have described than would have been the case at Oak Ridge, for example?

A. Let me answer that this way. I do not conceive that I had then, nor do I have now sufficient technical knowledge to enable me to measure the difference between the speed of accomplishment and not. To my mind then, simple logic dictated that it must be so, and I saw no reason to change my mind.

I wish to add that I thought then and later events as the project went on proved that this theory of an insulated city in the middle of a desert is more easily postulated in theory than it is carried out in actual practice. But nevertheless we did a fairly good job in that regard.

Q. Was the job of administering this community a difficult one in your judgment as you observed it?

A. It certainly was. The commanding officers were changed very rapidly.

Q. What would you say as to the nature of the scientists and their human characteristics, as you saw them at work on the project in relations to the problem of administration?

A. The scientists en masse presented an extremely difficult problem. The reason for it, as near as I can judge, is that with certain outstanding exceptions they lacked what I called breadth. They were extremely competent in their field but their extreme competence in their chosen field lead them falsely to believe that they were as competent in any other field.

The result when you got them together was to make administration pretty difficult because each one thought that he could administer the administrative aspects of the Army post better than any Army officer, for example, and didn't hesitate to say so with respect to any detail of living or detail of security or anything else.

I hope my scientist friends will forgive me, but the very nature of them made things pretty difficult.

Q. They were slightly restive under the confinement of the isolated city.

A. Very. As time went on, more so. Toward the latter stages it became increasingly difficult to sit on the lid out there. During the early stages, no.

Q. What was Dr. Oppenheimer's policy as an administrator in relation to keeping the morale going and keeping the natural restiveness of these people within bounds? Was he helpful?

A. So far as I observed it, he was very helpful. The difficulty primarily arose from those that were one step below him, let us say, in the scientific side. Dr. Oppenheimer himself so far as security matters with which I was particularly concerned was extremely cooperative.

Q. Could you give the board a little picture of the actual security measures which were enforced there at Los Alamos?

A. Yes. In the first place, physically we had—I have forgotten how many—some troops, a guard company or two companies, wasn't it, and we maintained patrols around the perimeter. We established a system of monitoring telephone calls and mail. We established a post office, you might say, down in Santa Fe in an office. We censored all mail on a spot check basis, and the mail of the more important scientists and those upon whom we had derogatory information 100 percent. We maintained, at least in the early days—later it became a spot check basis, as I remember—a continual monitoring of all means of communications, telephone calls, and the like.

We attempted to be as careful as we could in the clearance of personnel who were sent there. It is quite true that there, as in other places, we stretched our clearance procedure when the pressure was on for personnel.

Those who have not been through it cannot conceive, again I say, the extreme pressure we were under—when the recruitment program was on, and when we were actually building the weapon, not to let people go, because the clearance procedure took a long time, or it seemed so to those who were responsible for getting ahead with the job.

I have forgotten precisely what our restrictions were on visitation, but people were not permitted to go on trips unless it was officially necessary. We had rather rigid restrictions even on visiting Santa Fe. Those, I remember, were among the restrictions that we simply had to relax as the project went on. We countered that by placing men of our organization in all of the hotels in town as desk clerks and the like and covering the city of Santa Fe as best we could.

We tried to make it the securest of our institutions. The inexcusable Greenglass case indicates that it was not so secure after all.

Q. What do you recall of your interview with Dr. Oppenheimer on what we call here the Chevalier incident, if you know what I have reference to?
A. Yes. That is one of the things which I have had the advantage of reading the transcript of some weeks ago and glancing at one page of it again last night. I should say that I talked to Dr. Oppenheimer many times. In that particular case the interview was when he was in Washington and I now believe that the interview took place in General Groves' office, although that is a reconstruction. I have no precise recollection of it except that it was in Washington. Do you wish me to relate the substance of it?
Q. Yes.
A. The substance of it was that Dr. Oppenheimer had advised our people on the west coast that an approach had been made to someone on the project to secure information concerning the project, and that the approach had been made by one Eltenton who was well known to us—from Eltenton to a third person and from the third person to the project.
From reading the transcript and having my attention called to memoranda by Mr. Robb and Mr. Rolander, the information was that the contact was with three persons. It is perfectly obvious that was the story. It is a curious trick of memory but my recollection was one and that the one person was Dr. Oppenheimer's brother, Frank Oppenheimer. I have no explanation as to how I translate it from three into one.
I called General Groves last night and discussed it with him in an attempt to fathom that and I can't figure it out. But the record shows clearly that there were three.
My effort was to get Dr. Oppenheimer to tell me the identity of the person that was later identified as Chevalier. In that I was unsuccessful. Perhaps I was not as resourceful a questioner as I might have been. In any event I could not get him to tell me. That is the sum and substance of it.
I came back and told the general that it was up to him, that he just had to get the information for us, which the general undertook to do and later reported back the information. That goes on for pages. I am quite sure that I interrogated him concerning other persons on the project. I am quite sure it is a long statement as I read it in the transcript. Our discussion covered a wide range. That is my present recollection.
Q. Was there any other instance in which Dr. Oppenheimer did not give you information that you asked for?
A. I don't recall any.
Q. Would you class this incident as an illustration of the characteristic of the scientific mind that you spoke of a while back as deciding in their own minds what properly they should do, what was required to be done in the public interest?
A. Yes, I think that is a fair statement. I think this whole incident is a good illustration of that. I will confess that I was pretty fed up with Dr. Oppenheimer at that moment because of the background against which we were working of the Weinberg case out on the west coast and the difficulties that we were having with this Federation of Architects, Engineers—what is the name of that thing—FAECT—who were well organized in one of the oil company laboratories out there and had been making efforts to organize the radiation laboratory at Berkeley.
I had previously in connection—let me say it this way—in connection with Dr. Oppenheimer's recruitment program, the names of one or two persons who figured prominently in the attempted or actual espionage incident on the west coast were among those that were slated for transfer to Los Alamos.
In order that there could be the least possible furor about it, I went out to Los Alamos to talk to Dr. Oppenheimer so that there would be no pressure upon the part of him or his people to bring these persons out there. At that time I told Dr. Oppenheimer something of our difficulties in Berkeley. How much I cannot now recall except that I would have told him as little as I thought I needed to.
The fact that I had to do that indicates the kind of people we were dealing with, because these persons, and Dr. Oppenheimer was no exception, believed that their judgment as to what people needed to know, as to what was security and the like was as good or probably better than others.
It was subsequent to that conversation that Dr. Oppenheimer then, I assume, realizing the seriousness of the situation, advised our people on the west coast of this attempt coming out of the FAECT, because Eltenton was well known to us as a Communist, active in the Communist apparatus on the west coast, and a member of this laboratory group, this FAECT.

Dr. Oppenheimer then told us that Eltenton had made this approach. It was perfectly plain that Dr. Oppenheimer believed that it was quite unnecessary to our security problem to know the names of the person or persons—the one who later turned out to be Chevalier—got this contact with.

To my mind it was a sad exhibition of judgment, and an exhibition of ego that is quite unwarranted, but nevertheless quite common. That is the way I regarded it then. It did not endear him to me at the time. That is the sort of incident that it appeared to me to be.

Q. He did regard it as important and in the national interest for him to impart information that had come to him about Eltenton?

A. I assume that he did, otherwise he would not have done it.

Q. He took the initiative in doing that?

A. That is my recollection. My recollection is that he went to Lyle Johnson who was then the security officer at the Radiation Laboratory. Am I correct about Lyle's position then? I believe he was the security officer there. We had a very large organization on the West Coast, the investigative organization headed by Boris Pash, and I think Lyle Johnson was the security officer of the laboratory. In any event he was in the security organization at that time.

Q. Dr. Oppenheimer has testified to a visit that he paid Jean Tatlock in 1943 on a trip away from the project. Dr. Oppenheimer knew that he was under surveillance like everybody else when he left the project, did he not?

A. I assume that he did. We never told him. But I assume he realized it.

Q. Was it common knowledge that these security regulations applied to travel outside the project?

A. That question I can't answer. It was certainly common knowledge that travel outside of the project was not permitted except upon official business and prior terms. There were certain persons that we made no effort to conceal the fact from that they were under surveillance for the reason I mentioned. Dr. Oppenheimer was not in that category. We never advertised to him that he was under surveillance.

Our people, as I recall, who were handling that problem believed that he was aware of it.

Q. Did he make any other visits outside of the one I mentioned to other people that you know of?

A. Of course, he made visits to many people.

Q. Let me take that back. Any people unconnected in some way with the governmental effort?

A. As I recall, his trips at that time were primarily for recruiting personnel. We were aware of his visit to Miss Tatlock, I guess it was, and I do not now recall any other visit to persons that might be on the suspect list, let me put it that way. The record may dispute me on that, but I certainly don't recall any.

Q. To go back a moment to the Chevalier conversation, it has been testified here that after Dr. Oppenheimer told General Groves about Chevalier that certain telegrams were sent by you and General Nichols, I think in December 1943, still referring to three contacts.

A. That is right. One of those was shown to me last night.

Q. Do you have any explanation of that? Is it possible that you yourself having had three in mind may have concluded that still obtained, or was there anything more precise about it that you can remember?

A. I have been dredging my memory yesterday and today particularly about that. Unquestionably Dr. Oppenheimer told us there were three. The record shows that beyond dispute. There is no question that at a later time—at least at a later time—we were informed that there was one only and that one was Frank Oppenheimer, because I remember distinctly going over to the FBI and visiting Mr. Tamm who was then, I believe assistant to J. Edgar Hoover, and Mr. Whitson, who was the FBI Communist expert, that it was Frank Oppenheimer and that we had got that information, or that General Groves had obtained that information on the express term that it would not be passed on.

General Groves told me that, but I found it necessary to violate General Groves' direction in that regard and to give to the Bureau the identity of Frank Oppenheimer.

Whether the General went back again at my request, or on his own and talked further with Dr. Oppenheimer, whether the General and I reached the conclusion that it must have been Frank, I don't know—we discussed it many times—yet I distinctly remember this condition of secrecy.

Dr. EVANS. Did you say General Groves told you that it was Frank Oppenheimer?

The WITNESS. Yes, sir; it is my recollection that General Groves told me it was Frank Oppenheimer. What mystifies me, gentlemen, is that the record shows three and there is a complete gap there. There is no record at all of Frank or anything else. Yet nothing could be clearer in my memory than of that incident of going over at night and talking to Tamm and Whison. Nothing could be clearer in my memory than General Groves' direction that I was not to pass it on to anybody, which I promptly violated in a very unmilitary manner.

That gap or jump I have no explanation for. My memory is a complete blank.

By Mr. GARRISON:

Q. Would it refresh your recollection or still further confuse you if I were to say to you that my recollection of the conversation with General Groves about this was that Dr. Oppenheimer named Chevalier to him as the man, but that he General Groves, suspected that it was?

Mr. ROBB. Mr. Chairman, may I interpose here? We had General Groves here yesterday called by Mr. Garrison and he was not asked about this. It is entirely all right with me if Mr. Garrison wants to put testimony in this way. But if this is to be done, I think General Groves should be brought back and asked about it personally.

Mr. GARRISON. I am not putting in this as testimony.

Mr. ROBB. That is the effect of it.

Mr. GARRISON. As I recall General Groves' testimony yesterday he said that the whole thing was so confused in his mind that he could not make head or tail out of it.

Mr. ROBB. If that is the case, then I don't think Mr. Garrison ought to attempt to refresh the recollection of a witness by quoting General Groves.

The WITNESS. Can I say this, or volunteer it? Last night it was around 11 o'clock when I left here and got back to the hotel room when I called General Groves on the telephone for the purpose of rehashing this very thing. As a result of my conversation with him, I am no further informed than I testified to. That is all I can say.

By Mr. GARRISON:

Q. Is it possible that General Groves told you that he thought it was Frank rather than that it was Frank?

A. Yes; it is possible. I say it is possible because it would have been characteristic of the General. The General had superb judgment in that regard. He was frighteningly right an immense number of times in making such judgments. It is possible. It would have been characteristic of him.

Q. Do you know of any other instance in which Dr. Oppenheimer was approached by anybody on the subject of obtaining information of improper character?

A. No.

Q. I don't mean to imply that in this particular interview about Chevalier about which he has testified that Chevalier asked him for information. He has testified to the contrary. I didn't mean to imply my my question any doubt as to that testimony.

What I merely wanted to ask you is whether in your surveillance of him outside of the project, did you have any occasion or did any approach to him come to your knowledge on the part of anybody with respect to the subject of obtaining information outside of what we are talking about?

A. Not to my knowledge.

Q. Do you know whether Dr. Oppenheimer requested the employment of his brother, Frank, on the project?

A. No; I don't know. My impression is that Frank was already on it when the Army took over, but I would not be sure about that.

Q. You had many interviews with both Dr. Oppenheimer and his wife during the course of the work on the project?

A. Yes.

Q. Did you endeavor in these interviews to form the most accurate and thorough going judgment possible as to his political orientation? I will come to Mrs. Oppenheimer later. Did you search to find out what you could about his attachment or lack of attachment to Communist ideology?

A. Yes, sir; that was the purpose of my talks with him. I was working on that all the time.

Q. What judgment did you form as to his political convictions at this time, that it, at the time of the project?

9. May I qualify your question? You asked me as to my judgment as to his political convictions. I formed the judgment that he was not a Communist.

Q. How did you form that judgment?

A. I would like to continue with that. My working definition of a Communist is a person who is more loyal to Russia than to the United States. That is the definition I formed very early during my work on the Communist problem in the War Department, and which I still think is a sound definition. You will note that has nothing to do with political ideas.

Unquestionably Dr. Oppenheimer was what we would characterize—and as hide bound a Republican as myself characterizes—as extremely liberal, not to say radical. Unfortunately, in this problem of determining who is and who is not a Communist, determining who is loyal and who is not, the signs which point the way to persons to be investigated or to check on are very frequently political liberalism of an extreme kind. The difficult judgment is to distinguish between the person whose views are political and the person who is a Communist, because communism is not a political thing at all.

Q. You had an extensive experience in that kind of interrogation throughout the war, did you?

A. Yes, sir; I certainly did.

Q. Did you have enough experience at it to feel as confident as men can be about their judgments?

A. I believe so. I was a lot younger then than I am now, and I am sure I had more confidence in my judgment then than I have now.

Q. About many things?

A. About many things. But my job in the War Department and up until the time I officially moved over to the atomic bomb project and severed all connections with the War Department in January 1944, was primarily concerned with the formation of judgment as to who were or were not Communists in the loyalty sense in the Army.

Q. You were satisfied on the basis of these interrogations and of all that you knew about Dr. Oppenheimer from surveillance and all other sources that he was not a Communist as you have defined one in the sense of being more loyal to Russia than to the United States?

A. Yes.

Q. You were satisfied that he was a loyal American citizen?

A. Yes.

Q. Putting the interests of his country first?

A. I believed that.

Q. Did you form the same judgment about Mrs. Oppenheimer?

A. Yes, in a different sense. Mrs. Oppenheimer, I believed then had unquestionably been either a member of the Communist Party or so close to it as to be substantially the same thing. Her first husband had been——

Q. You say "had been." When?

A. In the thirties, as I recall. As I recall, she had been an organizer out in Ohio somewhere during the depression. Her first husband had been—what is his name?

Q. Dallet.

A. Dallet. Had been in the Abraham Lincoln Brigade during the Spanish War. That was always, particularly those who went in early and stayed long, a pretty fair index of then current attitude of people. Her background was not good. For that reason I took as many occasions as I could to talk to Mrs. Oppenheimer.

As I recall Mrs. Oppenheimer's background and associations subsequent to the thirties, they had not been different from that of Dr. Oppenheimer—or materially different—from that of Dr. Oppenheimer.

Mrs. Oppenheimer impressed me as a strong woman with strong convictions. She impressed me as the type of person who could have been, and I could see she certainly was, a Communist. It requires a very strong person to be a real Communist.

I formed the conviction over many interviews with her and many discussions with her that she had formed the conviction that Dr. Oppenheimer was the most important thing in her life and that his future required that he stay away from Communist associations and associations with people of that ilk.

It was my belief that her strength of character—I think strength of character is the wrong word—her strength of will was a powerful influence in keeping Dr. Oppenheimer away from what we would regard as dangerous associations.

Q. Did you have any doubt as to her own disassociation from the Communist Party?

A. No; I don't think I did.
Q. And to her prior disassociation from the party before coming to the project?
A. That is right.
Q. You regarded Mrs. Oppenheimer on the basis of your interrogation of her and all that you knew about her as loyal to the United States of America?
A. Yes; I did. I want to qualify that by saying that I think—no, I won't qualify that at all. The answer is yes.
Q. If you had the decision to clear or not to clear Dr. Oppenheimer today, based upon your experience with him during the war years and up until the time when your association with him ended, would you do so?
A. I will answer that, yes, based upon the same criteria and standards that we used then. I am making no attempt to interpret the present law. Those criteria were loyalty and discretion.
Q. What would you have to say as to his discretion as you saw it?
A. I think it was very good. We always worried a little bit about how much he talked during his recruitment efforts. Certainly there were times when as a security officer I would have judged the amount of information that he felt he had to give to induce somebody to come on to the project to have been indiscreet. That is always a question of judgment and it was in the line of duty. so to speak.
Q. Apart from the problem of recruitment, what would you say?
A. Yes; I believed him to be discreet. I thought it was indiscreet of him to visit Miss Tatlock.
Mr. GARRISON. That is all at the moment, Mr. Chairman.
Mr. GRAY. Mr. Robb.

CROSS EXAMINATION

By Mr. ROBB:
Q. As I understand it, Mr. Lansdale, you are not offering any opinion as to whether or not you would clear Dr. Oppenheimer on the basis of presently existing criteria?
A. That is a standard that is strange to me. I don't know what it is. If somebody would interpret it for me—isn't it getting pretty hypothetical?
I believed on the basis of information I had then that Dr. Oppenheimer was loyal and discreet. I have not changed my mind, although I have no knowledge of events transpiring since sometime in 1945.
Q. You said that you thought Oppenheimer's discretion was very good, is that correct?
A. Yes, sir.
Q. You had no doubt, did you, that Jean Tatlock was a Communist?
A. She was certainly on our suspect list. I know now that she was a Communist. I cannot recall at the moment whether we were sure she was a Communist at that time.
Q. Did your definition of very good discretion include spending the night with a known Communist woman?
A. No; it didn't. Our impression was that that interest was more romantic than otherwise, and it is the sole instance that I know of.
Q. Were there some people called Barnett that you knew about on your suspect list?
A. That name doesn't ring a bell with me.
Q. Were the Serbers on your suspect list?
A. Yes; sir.
Q. High on it?
A. Fairly so; yes.
Q. Was Dr. Oppenheimer intimate with them?
A. They were on the project at Los Alamos. The social life of that project, isolated as it was, was very close. The Serbers were, as I remember, friends of the Oppenheimers.
Q. They were friends of his?
A. That is my recollection.
Q. Were there some people named Morrison on your suspect list?
A. Yes.
Q. High on it?
A. Phillip Morrison?
Q. Yes, sir.
A. I think so. I don't think he was out at Los Alamos. Was he?
Q. Yes; I believe he was. Were they also good friends of the Oppenheimers?
A. That I don't recall. May I stop? I am not supposed to interrogate the interrogator, am I?

Q. Was there a man named David Hawkins on your suspect list?
A. Yes; I believe so in a mild sort of way. I mean he was one of those persons we felt uneasy about without having anything definite.
Q. You have since learned that Hawkins had been a Communist?
A. From what I read in the newspapers.
Q. Was he an intimate of Dr. Oppenheimer?
A. I don't now recall him as being. I know he came out to the project for personnel work.
Q. Wasn't he sort of Dr. Oppenheimer's legman and assistant?
A. My recollection was that David Hawkins was regarded as extremely important to the recruitment program which was one of Dr. Oppenheimer's primary responsibilities and in that sense, yes.
Q. He was working right close to Dr. Oppenheimer.
A. So far as recruitment is concerned, that is my recollection.
Q. Were there some people named Woodward on your suspect list?
A. I believe so, although I remember nothing more about them.
Q. By the way, when you say suspect list, you mean people who were suspected of being Communists or close to Communists?
A. Or that we were uneasy about it. Perhaps suspect list should be security list.
Q. Yes.
A. People concerning whom we took more or less risk on, depending on the circumstances and the times.
Q. Were the Woodwards intimates of Dr. Oppenheimer?
A. As I told you, I remember the name. I remember nothing more about them now. Perhaps if you could recall some instance to me, I would remember.
Q. Was a man named Lomanitz on your suspect list?
A. Oh, very much.
Q. He was one of your top suspects, wasn't he?
A. Yes, sir.
Q. Was he close to Dr. Oppenheimer?
A. I don't recollect that he was. My recollection of Rossi Lomanitz is that he was a student of Dr. Oppenheimer. He was at the Radiation Laboratory until we had him inducted into the Army and thus got him off the project.
Q. Do you recall Dr. Oppenheimer protesting about his induction?
A. I recall Dr. Oppenheimer raising a question about it. Indeed if I recall that was the occasion of at least one of my talks with Dr. Oppenheimer, that is, to ask him "for goodness sake to lay off Lomanitz and stop raising questions about it."
Q. In other words, he had been raising questions about it?
A. My recollection was that he had. Lomanitz was regarded as a brilliant young man and the people like Earnest Lawrence and Dr. Oppenheimer did not want to lose him. I remember Earnest Lawrence yelled and screamed louder than anybody else about us taking Lomanitz away from him.
Q. Was a man named Bohm on your list?
A. Yes.
Q. Was he a friend of Dr. Oppenheimer?
A. I have no recollection about that. He also came from Barkeley. I assume Dr. Oppenheimer must have known him.
Q. Was a man named Weinberg on your suspect list?
A. Right at the top of the list.
Q. In fact, Weinberg gave information to Steve Nelson, didn't he?
A. That is our belief. We proved to our satisfaction that he gave information to Steve Nelson for money.
Q. What was the relationship between him and Dr. Oppenheimer?
A. My recollection is about the same as Lomanitz.
Q. Now, Mr. Lansdale, when did you come into the Army?
A. May 1941.
Q. What had been your expereince prior to going into the Army?
A. Lawyer.
Q. How old were you then, sir?
A. I was born in 1912. 29, wasn't it?
Q. Had you had any previous experience as a security officer or investigator?
A. No, sir, not other than in connection with trying law suits. I was a trial lawyer.
Q. In other words, you were not a professional?
A. I certainly was not.

Q. How long had you been in the Army before you went on to this security work?
A. About 3 minutes.
Q. What was your rank when you started out?
A. First Lieutenant. I was a reserve officer in the field artillery by reason of graduation from VMI.
Q. After your interview of September 12, 1943, with Dr. Oppenheimer did you submit a copy of that to General Groves?
A. The record so shows. I have no present recollection of it.
Q. You have no doubt that you did?
A. I have no doubt that I did.
Q. Did you have any doubt that prior to doing that you read it over to make sure it was an accurate reflection of what had been said in your interview?
A. I have no doubt that I read it over and I would have made any changes that I felt were erroneous in substance, but as I remember that was a recording. I would have made no attempt to correct English or reconstruct garbled portions.
Q. But had you found anything in there which was not in accord with what had been said? In other words, had the stenographer not correctly transcribed the recording, you would certainly have made the correction, would you not?
A. I can only say I am sure I would. We are reconstructing now. I have no present recollection.
Q. I don't expect you to recall now independently, Mr. Lansdale. But as your past memory recorded, you have no doubt that transcript was accurate, do you?
A. No; I really don't.
Q. Mr. Garrison asked you some questions about the scientific mind in relation to that interview that you had with Dr. Oppenheimer and you responded, I think, that Dr. Oppenheimer's attitude might well have been a manifestation of the workings of a scientific mind; is that correct?
A. Oh, yes; of which I came up against many examples.
Q. Dr. Oppenheimer has testified here before this board that he lied to you in that interview. You would not say that lying was one of the manifestations of a scientific mind, would you?
A. Not necessarily, no.
Q. It is not a characteristic——
A. It was certainly a characteristic -to decide that I didn't need to have certain information.
Q. No. But the question is, Mr. Lansdale, you would not say that scientists as a group are liars, would you?
A. No. I don't think persons as a group are liars.
Q. No.
A. I certainly can't over emphasize, however, the extremely frustrating, almost maddening, let me say, tendency of our more brilliant people to extend in their own mind their competence and independence of decision in fields in which they have no competence.
Q. You were undertaking at the time you interviewed Dr. Oppenheimer to investigate what you believed to be a very serious attempt at espionage, it that right?
A. Yes. Let me put it this way. No. "Yes" is a fair answer.
Q. And Dr. Oppenheimer's refusal to give you the information that you asked him for was frustrating to you?
A. Oh; certainly.
Q. You felt that it seriously impeded your investigation, didn't you?
A. Certainly. But he wasn't the first one that impeded my investigation, nor the last.
Q. Mr. Lansdale, do you have any predisposition or feeling that you want to defend Dr. Oppenheimer here?
A. I have been trying to analyze my own feelings on that.
Q. I notice you volunteered that last remark, and I wondered why.
A. I know, and it was probably a mistake. I have attempted as nearly as I can—as nearly as it is possible—to be objective.
Q. Yes, sir.
A. I do feel strongly that Dr. Oppenheimer at least to the extent of my knowledge is loyal. I am extremely disturbed by the current hysteria of the times of which this seems to be a manifestation.
Q. You think this inquiry is a manifestation of hysteria?
A. I think——
Q. Yes or no?

A. I won't answer that question "Yes" or "No." If you are tending to be that way—if you will let me continue, I will be glad to answer your question.
Q. All right.
A. I think that the hysteria of the times over communism is extremely dangerous. I can only illustrate it by another dangerous attitude which was going on at the same time we were worrying about Dr. Oppenheimer's loyalty.

At the same time over in the War Department I was being subjected to pressure from military superiors, from the White House and from every other place because I dared to stop the commissioning of a group of 15 or 20 undoubted Communists. I was being vilified, being reviewed and rereviewed by boards because of my efforts to get Communists out of the Army and being frustrated by the blind, naive attitude of Mrs. Roosevelt and those around her in the White House, which resulted in serious and extreme damage to this country.

We are going through today the other extreme of the pendulum, which is in my judgment equally dangerous. The idea of what we are now doing, what so many people are now doing, are looking at events that transpired in 1940 and prior in the light of present feeling rather than in the light of the feeling existing then.

Now, do I think this inquiry is a manifestation of hysteria? No. I think the fact that so much doubt and so much—let me put it this way. I think the fact that associations in 1940 are regarded with the same seriousness that similar associations would be regarded today is a manifestation of hysteria.
Q. Now, Mr. Lansdale, it is true, is it not—
A. By golly, I stood up in front of General McNary then Deputy Chief of Staff of the Army and had him tell me that I was ruining peoples' careers and doing damage to the Army because I had stopped the commissioning of the political commissar of the Abraham Lincoln Brigade, and the guy was later commissioned on direct orders from the White House.

That stuff that went on did incalculable damage to this country, and not the rehashing of this stuff in 1940. That is what I mean by hysteria.
Q. How do you know what this board is doing is rehashing old stuff?
A. I don't know. That is what I have been——
Q. That is what?
A. That is all that can be had from me because that is all I know.
Q. Mr. Lansdale, it is true, is it not, that the security officers down the line below you in the Army hierarchy were unanimous in their opposition to the clearance of Dr. Oppenheimer?
A. Virtually so, yes. I say virtually so because I cannot precisely now recall that it was unanimous. Certainly Captain Calvert—I believe he was then a captain, who was then the security officer—I am quite certain recommended against it. He was Colonel Nichols' security officer. I am quite certain Colonel Pash felt so. I should think that the answer was yes.

Let me add this: That had I been confined to the bare record, I might possibly have reached the same conclusion. In other words, if Dr. Oppenheimer had not been as important as he was, I would certainly have stopped with the record and used my every endeavor to persuade the General that Dr. Oppenheimer ought to be dispensed with.

However, in view of his importance to the project we made a tremendous effort to reach a settled conclusion in our own minds. At least I did, and I am sure the General did.
Q. You mean if he had not been an important figure you would just have discarded him as a nubbin and gone on to something else?
A. Oh, absolutely.
Q. Did you receive reports from the security officers at Los Alamos and Berkeley?
A. I undoubtedly received many reports from them. Let me say this. Our organization administratively was that all of those reports went to Oak Ridge which was the district engineer's office—first to New York and then when they moved to Oak Ridge, there. All of those reports did not come up to me. However, from Los Alamos they all came directly to me because we held that more or less outside of the ordinary course of administration.
Q. By the way, Mr. Lansdale, you said a little while ago that you now believed your interview of September 12 took place in the office of General Groves. How did you have your memory refreshed about that?
A. Well, by this process. My memory was that it was in Washington. My memory was clear on that. It was reported to me that Mrs. O'Leary, who was the general secretary, seeing that transcript, believed that it looked like her

typing. We had a concealed microphone in the General's office which we had set up for these purposes.

Dr. Oppenheimer tells me his recollection is that it was in the General's office. He recalled to my mind that I had met him at the train and gone up with him and that I clearly remember.

That is how I reconstruct it. It certainly happened either there or in my office at the Pentagon.

Q. When did Dr. Oppenheimer tell you that?
A. Last night.
Q. He remembered that?
A. He remembered that it had occurred in General Grove's office.
(Discussion off the record.)

Mr. ROBB. Mr. Chairman, I would be an hour or so more in examining Mr. Lansdale. I see that it is almost 3:30, the time that the board indicated it planned to recess. However, I am at the disposal of the board.

The WITNESS. It is pretty important that I get through today.

Mr. ROBB. I am sure you realize there are some things that are pretty important to go over.

The WITNESS. Certainly. I am at your disposal and prepared to go into the night or return at a later date, but I have some court commitments for the rest of next week.

Mr. GRAY. Off the record.
(Discussion off the record.)
Mr. GRAY. Let us break now for 5 minutes.
(Whereupon, a short recess was taken.)
Mr. GRAY. Will you proceed?

By Mr. ROBB:

Q. Mr. Lansdale, in referring to the scientific mind, were you basing your appraisal of a scientific mind upon your experience with that mind as represented by people like Lomanitz, Bohm, and Weinberg?
A. No. People like Ernest Lawrence and Fermi and Oppenheimer, and A. H. Compton, and the numerous people in the Metallurgical Laboratory.
Q. Karl Compton?
A. Karl Compton I had very little contact with.
Q. Mr. Lansdale, I want to show you a memorandum dated September 2, 1943, entitled, "Subject: J. R. Oppenheimer. Memorandum for Lt. Col. Pash, and a covering memorandum from Colonel Pash to you, signed P. de S., dated September 6, and ask you if that came to you in your official capacity?
A. Yes, my initials are on it, also General Groves' initials. I have no present recollection, you understand, of it.
Q. I understand.
A. But unquestionably it did.
Mr. GARRISON. May we see that?
Mr. ROBB. Yes.

By Mr. ROBB:

Q. You have no doubt that you gave consideration to that memorandum in your appraisal of Dr. Oppenheimer?
A. I didn't examine the content of it.
Mr. ROBB. Could the witness see the memorandum, please, Mr. Garrison. You are going to have plenty of time to look at it. I am trying to get along here in a hurry, Mr. Garrison.
Mr. GARRISON. I understand.
The WITNESS. Oh, yes, I am quite sure.

By Mr. ROBB:

Q. Who was "P. de S."?
A. That undoubtedly was Peer de Silva, who for some period of time, and I assume during this time, was security officer at Los Alamos.
Q. Was he a regular Army officer?
A. That is right. I believe he was a first lieutenant. He may have been a second lieutenant.
Q. He was afterward Colonel de Silva.
A. That I don't know.
Q. He was a professional; was he not?
A. Oh, yes. He was a professional soldier. He was not a professional security officer, if that is the implication, except that we were all professionals.

Q. He was certainly more of a professional than you were; wasn't he, Colonel?
A. In what field?
Q. The field he was working in, security.
A. No.
Q. No?
A. No.
Q. He was a graduate of West Point; wasn't he?
A. Certainly. I am a graduate of VMI, too. You want to fight about that?
Q. No, sir, I don't want to fight with you. I will show you a memorandum dated August 12, 1943, memorandum for General Groves, Subject: J. R. Oppenheimer; signed John Lansdale. Did you write that memorandum?
A. That is unquestionably my signature. Let me read it.
Mr. ROBB. May we go off the record.
(Discussion off the record.)
The WITNESS. Yes, I not only wrote this memorandum; I now recall the interview. As a matter of fact, this is the——

By Mr. ROBB:
Q. You are talking now about the memorandum of what?
A. Memorandum dated August 12, 1943, from me to General Groves, concerning J. R. Oppenheimer. This appears to be when I went out and made a trip to quiet people down about Lomanitz. We were having a great deal of trouble with Ernest Lawrence about taking Lomanitz away from him. Then Dr. Oppenheimer got in the picture, and I just went out to quiet things down.
Q. Colonel, I detect a slight tendency on your part to blame Lawrence for Lomanitz exclusively. Isn't it a fact that Dr. Oppenheimer was also very much exercised about Lomanitz?
A. I don't recall that he was exercised about Lomanitz—yes, he was exercised about Lomanitz. We got word through Peer de Silva as I recall that Oppenheimer was raising a question about us permitting Lomanitz to be inducted into the Army. I suspect he didn't know we were moving heaven and earth to get him inducted. Our main row with Lawrence, we had more trouble with Ernest Lawrence about personnel than any four other people put together.
Q. I will show you a photostat of a memorandum dated September 14, 1943, a memorandum for the file, subject "Discussion by General Groves and Dr. Oppenheimer," which bears the typewritten signature "John Lansdale, Jr., Lt. Col., Field Artillery." Did you write that memorandum?
A. I unquestionably did. Unquestionably I did write it.
Q. Does that memorandum or can you say with assurance that that memorandum accurately reflects that you had been told by General Groves about a discussion which he had had about Oppenheimer?
A. Of course not. All I can say is that I would have attempted as accurately as I could to record the substance of our conversation.
Q. That is what I mean.
Mr. GARRISON. Mr. Chairman, I really am getting disturbed about the problem we face of not knowing really what these questions are about. We haven't been supplied with copies of these.
Mr. ROBB. I will hand it to you right now.
Mr. GARRISON. If we can stop for a minute while we read them—I don't want to delay matters, but I have to protect my client as a lawyer.
Mr. ROBB. That is right. I have plenty of time.
Mr. GRAY. You will have an opportunity to read them. As I understand the questions which Mr. Robb has put they have been questions for identification, rather than substance.
Mr. GARRISON. He is beginning to ask him questions about them, and I haven't the slightest idea what is in them.
Mr. GRAY. You have them before you now.
Mr. ROBB. You have them all now; have you not?
Mr. GARRISON. There is an awful lot to read, Mr. Chairman.
Mr. GRAY. Do you propose to ask questions about the contents of the memoranda?
Mr. ROBB. No.
Mr. GARRISON. We may ourselves wish to ask questions now that they have been introduced. You have asked to put them all into the record instead of reading them in, off the record, with knowledge that Colonel Lansdale apparently can come back next week.
Mr. ROBB. I don't know whether he can or not. You called him here. I didn't call him.

Mr. GARRISON. Yes, it is your request to put them in the record without reading them.

Mr. ROBB. Do you want me to read them into the record and keep Colonel Lansdale here? I will do it.

Mr. GRAY. Just a minute. We will read them into the record.

Mr. GARRISON. I think we should.

Mr. ROBB. It is entirely all right with me. I am trying to accommodate Mr. Lansdale.

Mr. GRAY. The record will show at this point that Mr. Morgan, a member of the board, is forced to leave the proceedings.

Mr. GARRISON. Not permanently.

Mr. GRAY. No.

(Mr. Morgan left the room.)

Mr. GARRISON. Mr. Chairman, just for the sake of regularity even for 3 minutes, do you think it wise to proceed with a board member absent? There may be questions on this——

Mr. GRAY. I take it that he will have the record before him, or the record will be available to him. He is leaving town.

Mr. GARRISON. He is not coming back now.

Mr. GRAY. No, he is not coming back this afternoon. The board is trying to accommodate you and your witness. We can easily adjourn at this time and ask Mr. Lansdale to come back Monday.

Mr. GARRISON. We were told, Mr. Chairman, that you were going to adjourn at 3:30 this afternoon.

Mr. ROBB. And we sat overtime to accommodate Mr. Lansdale.

Mr. Chairman, it is quite obvious that it is going to take me I don't know how long to read these memoranda into the record, and it is now quarter of four, and I don't see any possibility of finishing with Colonel Lansdale this afternoon.

Mr. GRAY. Certainly as far as Mr. Morgan's absence is concerned, it cannot be affected by reading memoranda into the record which he could read. There is no point in his hearing the memoranda. I am sure of that. So would you proceed?

Mr. ROBB. Yes, sir. I will read the first memorandum that I showed Colonel Lansdale.

"HEADQUARTERS WESTERN DEFENSE COMMAND AND FOURTH ARMY,
"OFFICE OF THE ASSISTANT CHIEF OF STAFF G–2,
"Presidio of San Francisco, Calif.

"In reply refer to: (CIB) September 6, 1943, Subject: J. R. Oppenheimer.

"To: Lt. Col. John Lansdale, Jr., room 2C 654 Pentagon Building, Washington, D. C.

"1. Enclosed is a report on the evaluation of J. R. Oppenheimer, prepared in this office by Capt. Peter deSilva, now engaged in evaluation of the DSM project.

"2. This Office is still of the opinion that Oppenheimer is not to be fully trusted and that his loyalty to a Nation is divided. It is believed that the only undivided loyalty that he can give is to science and it is strongly felt that if in his position the Soviet Government could offer more for the advancement of his scientific cause he would select that Government as the one to which he would express his loyalty.

"3. This Office does not intend to evaluate the importance or worth of Dr. Oppenheimer as a scientist on the project. It is the responsibility of this Office to evaluate him from any possible subversive angle. Because of this the enclosed report is being submitted for your information.

"(Signed)   "Boris T. Pash.
              "BORIS T. PASH,
"Lt. Col. M. I., Chief, Counter Intelligence Branch.
              "For the A C of S. G–2."

"HEADQUARTERS WESTERN DEFENSE COMMAND AND FOURTH ARMY,
"OFFICE OF THE ASSISTANT CHIEF OF STAFF G-2,
"Presidio of San Francisco, Calif.

"In reply refer to: (CIB) September 2, 1943, Memorandum for: Lt. Col. B. T. Pash.

"Subject: J. R. Oppenheimer.

"1. With regard to recent developments in the espionage case centering about the DSM project, the part played by J. R. Oppenheimer is believed to take on a more vital significance than has heretofore been apparent. Briefly, it may be said that subject has just recently brought himself to the fore by volunteering scraps of information which are of vital interest to the investigation being conducted by this Office. In conversation with Lt. L. E. Johnson, he had said that he had good reason to suspect that the Soviet Union was attempting to secure information about the project. In a subsequent conversation with Lt. Col. Pash, subject elaborated on the matter and disclosed that about 4 months ago a Shell Development employee, one Eltenton, on behalf of a Soviet consular attaché, had contacted a U. C. professor who in turn had attempted, on at least three occasions, to secure sources of information within the project who would transmit the information to Eltenton, who in turn would supply it to the Soviet consular agent, all to be done informally in order to circumvent a State Department policy of not cooperating with the Soviet Union, which policy is influenced by certain unnamed State Department officials who were supposed to be anti-Soviet and who would not allow such action to be taken openly. Oppenheimer claims he does not condone such methods, and is satisfied that no information was passed by those channels. He did not disclose the name of the professor, as he thought that such an action would be unethical and would merely disturb some of his associates who were in no way guilty of any wrongdoing. Roughly, the above has been the extent of Oppenheimer's most recent activity.

"2. The writer wishes to go on record as saying that J. R. Oppenheimer is playing a key part in the attempts of the Soviet Union to secure, by espionage, highly secret information which is vital to the security of the United States. An attempt will be made to show the reasons for the above statement. It has been known, since March 29, 1943, that an overt act of espionage was committed by the Soviet Union. Subject's statements indicate that another attempt has been made, through Eltenton, Oppenheimer, himself having a rather lengthy record of Communist sympathy and activity. has actively engaged in the development of a secret project. Most of his friends and professional associates are Communists or Communist sympathizers. He himself has gone on record as saying on two occasions, to Lieutenant Colonel Lansdale and to Lieutenant Colonel Pash, that Communist activity on the part of a project employee is not compatible with the security necessary to the project. To quote him, "and that is the reason I feel quite strongly that association with the Communist movement is not compatible with the job on a secret war project, it is just that two loyalties cannot go." To Lieutenant Colonel Lansdale, he said that he knew that two Los Alamos employees had at one time been Communists, but that he was satisfied that they no longer were. Yet during the long period during which he has been in charge of the project, and in spite of the fact that he is perfectly competent to recognize the Communist attitude and philosophy, and further in spite of the fact that he, by choice as well as by professional necessity, is close to his key associates, and again in spite of the fact that he claims, in effect, not to feel confident of the loyalty of a Communist—in spite of all this, Oppenheimer has allowed a tight clique of known Communists or Communist sympathizers to grow up about him within the project, until they comprise a large proportion of the key personnel in whose hands the success and security of the project is entrusted. In the opinion of this officer, Oppenheimer either must be incredibly naive and almost childlike in his sense of reality, or he himself is extremely clever and disloyal. The former possibility is not borne out in the opinion of the officers who have spoken with him at length.

"3. To go further, the supposition will be raised that subject has acted reasonably, according to his own viewpoint, and has voluntarily come forward and proferred valuable information (re Eltenton, etc.). To examine the background for such an action we find several incidents which may have had an influence on his action. First, the news of Lomanitz's cancellation of deferment was made known to Oppenheimer, together with the surmise, on Lomanitz's part, that his (Lomanitz's) radical activities had been investigated. Shortly thereafter, an officer from the Military Intelligence Service, War Department, called on him at Los Alamos. Both of the above actions were necessary and desirable, but nevertheless they could not avoid indicating to Oppenheimer that, very probably, some sort of a general investigation, more extensive than a routine security check, was under way. If he is disloyal, as believed by the writer, the most obvious and natural move would have been exactly what he actually did do—on his next trip to Berkeley he let it drop to Lieutenant Johnson the piece of information indicating knowledge of an attempt at espionage, knowing that he would subsequently be contacted for further details by someone probably connected with the investigation. As it was absolutely necessary and such contact was made, whereupon subject elaborated on the incident, but in such a manner as to indicate that there was nothing seriously wrong, and never once indicating that espionage might have been involved.

"Although he had every opportunity to do so, he did not mention the fact that Steve Nelson visited him and solicited cooperation; instead, he revealed the channel of communication in which Eltenton played a part. He declined to name the professor involved, possibly intending to dole out that bit of information at a later date. He determined very definitely that Military Intelligence was conducting an investigation, and chose to cooperate to a certain extent, disclosing only what he desired to and relying on this apparent spirit of cooperation, together with his importance to the project, to protect himself. It is not inconceivable that he could, by intelligent manipulation, actually exercise a strong control over the extent and direction of the investigation. Add to the above proposition the fact that Oppenheimer, until alerted to the fact that an investigation was in progress, made absolutely no attempt to inform any responsible authority of the incidents which he definitely knew to have occurred and which he claims, he did not approve. To go further, he apparently made no attempt to resolve, for his own conscience and satisfaction, any doubts concerning the Communist affiliations of some of his employees whom he knew to have been so affiliated at one time. At no time, to the knowledge of this officer, has Oppenheimer attempted, in any way, to report any such affiliation, known or suspected, for the information of the Army, nor has he taken anyone into his confidence concerning his views on the subject. None of this was done until it became obvious to him that an investigation was being conducted, and that unless he made the first move, he would ultimately be questioned, and would not be in the favorable position of having offered the information.

"4. It is the opinion of this officer that Oppenheimer is deeply concerned with gaining a worldwide reputation as a scientist, and a place in history, as a result of the DSM project. It is also believed that the Army is in the position of being able to allow him to do so or to destroy his name, reputation, and career, if it should choose to do so. Such a possibility, if strongly presented to him, would possibly give him a different view of his position with respect to the Army, which has been, heretofore, one in which he has been dominant because of his supposed essentiality. If his attitude should be changed by such an action, a more wholesome and loyal attitude might, in turn, be injected into the lower echelons of employees. It is not impossible that a thorough review of the general opinion holding Oppenheimer irreplaceable might result in lending strength to the argument that he is a citizen working for the United States, in this case represented by the War Department, and not an individual who cannot be held or restricted, while continuing independent scientific endeavor, to the normal definition of loyalty to his country.

"(Signed) P. DE S."

I wonder if Mr. Rolander might spell me on this reading, Mr. Chairman.

Mr. GRAY. Very well. Would you identify it?

Mr. ROLANDER. This memorandum is dated August 12, 1943.

"WAR DEPARTMENT,
"MILITARY INTELLIGENCE SERVICE,
"*Washington.*

"Memorandum for General Groves:
"Subject: J. R. Oppenheimer.

"1. Upon a recent visit to Los Alamos this officer had an opportunity for some private discussion with J. R. Oppenheimer on matters of general interest. During the course of this discussion the subject of the withdrawal of the deferment of G. R. Lomanitz came up. Mr. Oppenheimer stated that his interest in Lomanitz was purely scientific. He stated that Lomanitz was about to be made a group leader, and that he was engaged upon a type of work with which only 2 other persons were thoroughly familiar, and these 2 persons are now working for Oppenheimer. Oppenheimer believed that if Lomanitz's services were lost, E. O. Lawrence would request Oppenheimer to release one of the latter's men for work at Berkeley. This Oppenheimer is unwilling to do, and wishes to avoid any issue in this respect with Lawrence.

"2. Oppenheimer stated that he knew very little about Lomanitz and had not, except upon one occasion, had any relationship with him other than that of professor and student, and, subsequently, employer and employee on the project. Oppenheimer stated that at the time he asked Lomanitz to come on the project, Lomanitz visited Oppenheimer at his home and did what Oppenheimer characterized as 'a good deal of soul searching.' Oppenheimer stated that he meant by this that Lomanitz was of the opinion that a very terrible weapon was being developed, and was fearful that there would not be adequate international control of this weapon. Furthermore, Lomanitz wondered whether his sense of duty did not require him to make a more direct contribution to the war effort by joining the Army or working in the shipping yards or some similar establishment. Oppenheimer stated that he gave Lomanitz 'a good talking to' and told him very definitely and strongly that the project was important to the war effort, and that it must have his complete loyalty. Oppenheimer further stated that he told Lomanitz that he must forego all political activity if he came on to the project. Oppenheimer stated that he put this very strongly. He had previously stated that he knew that Lomanitz had been very much of a Red as a boy when he first came to the University of California, but he professed to have no knowledge of his activities.

"3. Oppenheimer stated that 2 days later Lomanitz told him that he wanted to go onto the project, and accepted all of the conditions laid down by Oppenheimer. Oppenheimer was curious as to why we were taking the action that we did, and also wondering if after Lomanitz was inducted into the Army he could not be returned to the project either as a Reserve officer or as a soldier.

"4. This officer told Dr. Oppenheimer that it was believed to be necessary to avoid making any further requests for deferment for Lomanitz because he had been guilty of indiscretions which could not be overlooked or condoned. This officer stated that these had nothing to do with any political activity. Oppenheimer was further told, however, that since the occurrence of the indiscretion upon which action was based, steps have been taken to determine rather completely Lomanitz's activities, and that it could be said that in the course of this investigation it had been learned that Lomanitz had not ceased his political activities. Oppenheimer said 'that makes me mad.'

"5. There then ensued a general discussion of the Communist Party. Oppenheimer was told that from a military intelligence standpoint we were quite unconcerned with a man's political or social beliefs, and we were only concerned with preventing the transmission of classified information to unauthorized persons, wherever that person's loyalties might lie, or whatever his social, political, or religious beliefs might be. He we told that the underlying principles behind all of our security measures were that the United States so far as the development of any device or technique was concerned, was the sole party interested, although the benefit of the employment of any devices would, of course, redound to the benefit of all persons on the same side as this country.

"6. Oppenheimer concurred in the general principles stated, but stated that he did not agree with us with respect to the Communist Party. He stated that he did not want anybody working for him on the project that was a member of the Communist Party. He stated that the reason for that was that 'one always had a question of divided loyalty.' He stated that the discipline of the Communist Party was very severe and was not compatible with complete loyalty to the project. He made it clear he was not referring to people who had been members of the Communist Party, stating that he knew several now at Los Alamos who had been members. He was referring only to present membership in the Communist Party.

"7. A general discussion then ensued in which Oppenheimer deplored the manner in which the Russians had let their people 'down in France and in the United States.'

"7a. The opportunity to secure the names of the former members of the party known to Oppenheimer did not present itself, due to the entrance of a third party.

"Note: J. R. Oppenheimer gave every appearance of sincerity in this discussion. He was, however, extremely subtle in his allusions, and there was a good deal of delicacy evidenced both by this officer and by Dr. Oppenheimer in pursuing this discussion. Upon reviewing the discussion after leaving Dr. Oppenheimer, this officer came to the conclusion that what Dr. Oppenheimer was trying to convey was, in the case of Lomanitz, that Lomanitz had been worried about his obligations to the party, and that Oppenheimer had told him that he must give up the party if he came on the project. This officer also had the definite impression that Oppenheimer was trying to indicate that he had been a member of the party, and had definitely severed his connections upon engaging in this work. On the whole, it seemed that Oppenheimer, in a rather subtle way, was anxious to indicate to this officer his position in that regard.

"(Signed) JOHN LANSDALE, Jr.,
"*Lieutenant Colonel, Field Artillery, Chief, Review Branch, CIG, MIS,*
"(For the A. C. of S., G–2)."

The next communication is dated September 14, 1943.
"Memorandum for the file.
"Subject: Discussion by General Groves and Dr. Oppenheimer.

"1. During a recent train ride between Cheyenne and Chicago, General Groves and Dr. Oppenheimer had a long discussion which covered in substance the following matters:

"(*a*) Dr. Oppenheimer stated that because he felt responsible for the employment of Giovanni Rossi Lomanitz, and had secured a promise from him as a condition of employment to cease all outside activities and particularly those of a political nature, he wanted to have a talk with him. While Oppenheimer did not know the cause of objection by the Army to Lomanitz he did know that he had been indiscreet and that he was still engaged in political activities. Dr. Oppenheimer said that the interview with Lomanitz was very unsatisfactory, and that Lomanitz was defiant. Oppenheimer was sorry that he had ever had anything to do with him, and he did not desire any further connection with him.

"(*b*) Oppenheimer also had a talk with Joseph Weinberg and David Bohm. This interview was sought by the latter two persons. They stated to Oppenheimer that they were disturbed by the evident pressure being brought to bear to force the induction of Lomanitz into the Army, and that inasmuch as they were close to Lomanitz and interested in union activities they wanted Oppenheimer's advice as to whether they should resign their positions and seek employment elsewhere where their talents would be more appreciated.

"*c*. Oppenheimber told them that if they had continued to adhere to the promise made by them to him that they would cease all political activities, including Communist Party activities, then they had nothing to fear. Oppenheimer called E. O. Lawrence in at this point and secured from Lawrence confirmation of his previous statements. Dr. Oppenheimer stated at one point that Weinberg had expected to go to site Y but that it was never his (Oppenheimer's) intention to have him there.

"*d*. Some discussion was had about Dr. Oppenheimer's previous relations to Colonel Pash and Lieutenant Johnson about the Soviet attempt to secure information which had come to Oppenheimer's attention some time ago. Oppenheimer's attitude was that he would give the name of the intermediate contact at the University of California if pressed to do so, and told by General Groves that we had to have it, but that he did not want to do so because he did not believe that any further contacts had been made and was confident that the contacts that had been with the project had not produced any information. He intimated further that it was a question of getting friends of his into difficulties and causing unnecessary troubles when no useful purpose could be served. In this connection it should be noted that General Groves asked Oppenheimer generally about several people at the University of California, among whom might be the contact, which had been supplied to him by Colonel Pash. Among these names was A. Flannigan, who now appears from subsequent developments to be the contact. With respect to Flannigan, Oppenheimer stated that he did not know him except casually, but that he had the reputation of being a real 'Red.'

"Oppenheimer stated that Mrs. Charlotte Serber came from a Communist family in Philadelphia, and probably at one time had been a Communist herself. However, he did not think that she was at this time. It is thought that he said that he had no intimation that Professor Serber was or had ever been connected with the Communist Party. Oppenheimer reiterated his previous statements that membership in the Communist Party was incompatible with employment on the project because of the divided loyalty which it involved. He expressed the opinion that transmission of information to any outside person or party on the part of the people on the project would amount to treason.

"*f*. Oppenheimer categorically stated that he himself was not a Communist and never had been, but stated that he had probably belonged to every Communist-front organization on the west coast, and signed many petitions concerning matters in which Communists were interested. He stated that while he did not know, he believed that his brother, Frank Oppenheimer, had at one time been a member of the Communist Party, but that he did not believe that Frank had had any connections with the party for some time.

"*g*. He stated that his wife, Katherine, was born and raised in Germany, was a first cousin once removed of General Kietel of the German Army and that her mother had at one time been engaged to marry him and that her family were still on amicable terms with the Kietel family. He stated, also, that his wife's first husband had been killed in Spain while fighting for the Loyalist armies, and that he understood that he had been a 'good guy.' No opinion was expressed by Oppenheimer as to whether this first husband had been a Communist.

"JOHN LANSDALE, Jr.,
"*Lt. Colonel, Field Artillery, Chief, Review Branch, CIG, MIS.*"

Mr. ROBB. Mr. Chairman, I have perhaps a dozen more questions that I should like to ask Mr. Lansdale. If Mr. Garrison will agree to go ahead in the absence of Mr. Morgan, I will do so with the Chair's consent. Otherwise I will hold them up.

Mr. GARRISON. Quite agreeable.

Mr. GRAY. I want to make certain of this now, Mr. Garrison. You raised the question.

Mr. GARRISON. I did, and I now waive it, Mr. Chairman, in the interest of proceeding.

Mr. GRAY. All right.

By Mr. ROBB:

Q. Colonel, you spoke of your surveillance you instituted at Los Alamos and Berkeley; is that correct?

A. Yes.

Q. Is is not your testimony or your belief, is it, Colonel, that that surveillance would have prevented the passing of information?

A. No; it would have, we hoped, detected and provided us with the opportunity to prevent it.

Q. No surveillance could prevent a man from passing a note to another man at some time during the 24 hours of the day, could it?

A. No necessarily; no. Of course not.

Q. This man David Greenglass that you mentioned was an employee at Los Alamos?

A. He was in a military organization we called the SED. I cannot recall what those initials stand for.

Dr. OPPENHEIMER. May I coach the witness? Special Engineer Detection.

The WITNESS. That is correct.

By Mr. ROBB:

Q. How long was he there?

A. I don't know. I don't remember. He was in a group of technicians, as I remember, machinists and the like. We formed this organization due to the shortage of personnel in order to recruit from the Army people with special skills that were needed at Los Alamos. All that I recall about David Greenglass is what I read in the papers or what I heard from Rolander and others during the Rosenberg trial. He is certainly an example of one we missed.

Q. You certainly learned, didn't you, that Greenglass and Fuchs from Los Alamos had given to the Russians the entire story of our work at Los Alamos, or substantial portions of it?

A. I certainly learned that they passed information. The characterization as the whole story or not, I have no——

Q. You certainly learned that they passed vital information to the Russians?

A. I certainly learned that they passed a sketch, as I remember, of the implosion device—is that the right term? All I know is what Rolander told me in New York.

Q. You learned that subsequently from talking to Mr. Rolander and reading the newspapers?

A. That is right.

Q. You did not learn it while you were the chief security officer?

A. I sure didn't.

Mr. ROBB. That is all. Thank you.

Mr. ROLANDER. Mr. Chairman, may the record indicate that discussions relative to Greenglass and Rosenberg was during the preparation of the case that was presented in New York.

The WITNESS. Yes; the Rosenberg spy trial.

Mr. GRAY. I see. I have a couple of questions.

Do I understand that the security measure which were instituted, that trips away from Los Alamos, I understood you to say, had to be cleared in advance, and did you also say it had to be on official business?

The WITNESS. Yes. My recollection may not be exact on this. I know we attempted particularly at the very start to restrict any trips away from Los Alamos to official business or something like a death in a person's family where it appeared to be necessary to let them go. As time went on, that became more relaxed. I can't measure the precise time, of course.

Mr. GRAY. This is a change of subject now. In your discussion of the characteristics of scientists, I think I am correct in my recollection that you said you felt that Dr. Oppenheimer was making a decision which he felt he was competent to make with respect to the disclosure of the names of the persons who were approached by the unknown intermediary?

The WITNESS. Yes.

Mr. GRAY. I am simply asking this for the record. Weren't you seeking the name of the person who approached the person?

The WITNESS. I was seeking both; yes, sir.

Mr. GRAY. That answers my question. You were seeking both the name of the three, if there were three, and also the individual who subsequently turned out to be——

The WITNESS. Chevalier. Certainly, that we regarded obtaining that as more important than obtaining the ones that were approached, although I don't want to say we didn't regard that as important.

Mr. GRAY. I believe that clears the record. I believe when you read the transcript the emphasis was on the other.

The WITNESS. I see.

Mr. GRAY. This reverts to your observations about the swing of the pendulum.

The WITNESS. Yes.

Mr. GRAY. Certainly I think you are entitled to and should express your opinions about such matters. However, I wonder if you know the statute under which, or the regulations under which, this board is created?

The WITNESS. At one time I have read them, sir. I was familiar with them at the time they were enacted, but I have not looked at that in years. The other night one of these gentlemen here told me what the language was, but for the life of me I could not quote it now.

Mr. GRAY. I really wouldn't expect you, frankly, to be familiar with it. We are charged, as I understand it, to consider the problem put before us with respect to the character, loyalty, and associations of an individual. These are the criteria in the act.

The WITNESS. Character, loyalty, and association.

Mr. GRAY. My question of you is perhaps of a philosophical nature. I think you rather suggested that this board should not concern itself with associations perhaps in the thirties or forties?

The WITNESS. I did not intend to convey that. Certainly the board should concern itself with that. What I intended to convey was that the appraisal or evaluation of associations in the forties must be viewed in the light of the atmosphere existing then and not in the light of the atmosphere existing at the present time.

Mr. GRAY. You did not mean to suggest that it was your opinion that you could only consider current associations in determining problems of this kind?

The WITNESS. Of course not. Always our starting point, our leads to people who are disloyal, are such things as associations. For example, you can hardly put your finger on a scientist or a university professor or people who tend to get into civic affairs, you can hardly find one anywhere who is now in his fifties or so that has not been on at least one list of an association which was later determined to be subversive or to have leanings that way. Nevertheless, those associations are most frequently the starting point or the leads for investigation go to further. You always have the question of determining the significance of those: (a) the signficance at the time of them, (b) whether, assuming that there was a sinister significance it has continued.

I have never, strongly as I have felt and acted with reference to communism, never adopted the assumption, once a Communist sympathizer, always a Communist sympathizer. One of the finest things that Soviet Russia ever did for us was the quick switch of the on again off again with Germany. That did more than anything else to tell the men from the boys in the Communist Party. It would be a terrible mistake to assume that, once having had sinister associations, a man was forever thereafter damned. Yet, once you uncover those, you must always exercise judgment. That judgment is always made up of a large body of intangibles. It is seldom you get anything concrete.

I am being a little vague, I know, but the whole subject is vague.

Mr. GRAY. Because of your observation—I don't disagree with what you state as a philosophy at all—I am pointing out that you have come a long way to be a witness to testify with events with which you are familiar, all of which took place some years ago.

The WITNESS. Oh, yes.
Mr. GRAY. But, under the terms of the act and the regulations under which this board was constituted, they are all relevant. That was my point.
The WITNESS. I don't mean to convey they are not.
Dr. EVANS. Colonel, I think you overstretched the meter when you said all professors have something like that in their background.
The WITNESS. I said "almost."
Dr. EVANS. That is not true. Did you find men like Compton, Conant, Fermi, Bohr, and Hildebrandt, the peculiar type scientists?
The WITNESS. What I referred to as the scientific mind?
Dr. EVANS. Yes.
The WITNESS. I would except from that A. H. Compton. A. H. Compton in my opinion frankly is one of the finest men I ever knew. He has breadth and judgment.
Dr. EVANS. You are talking about Arthur and not Karl?
The WITNESS. Yes. I scarcely knew Karl Compton. I just met him.
Dr. EVANS. I would like to ask you another thing. Do you think loyalty to an individual is of more importance than loyalty to a country?
The WITNESS. No, sir; I don't. One of the characteristics of war and near war is the existence of that fact—loyalty to the country takes in my judgment, and ought to take, precedence above all. There are those that feel differently. We are all familiar, of course, with the device of placing a person in the position of choosing between loyalty to someone near and dear and loyalty to country, and different people react differently to it, depending upon their strength of character and feeling of patriotism and the like.
Dr. EVANS. I would like to ask you one more question.
The WITNESS. I have never been in that position, so I can only speak theoretically.
Dr. EVANS. Do you as a rule dislike the scientific mind? Is it a peculiar thing?
The WITNESS. I will say this, that during the war I came very strongly to dislike the characteristics which it exhibited.
Dr. EVANS. That is all.

REDIRECT EXAMINATION

By Mr. GARRISON:
Q. I have just one question, Mr. Lansdale.
Referring back to the confused incident of the Chevalier matter, what would you say, on the basis of your total experience with Dr. Oppenheimer, would be your general opinion as to his veracity?
A. There is no question that—I don't believe that he lied to us except about this one incident—my general impression is that his veracity is good. I don't know of any other incident.
Q. Just so there is no possible implication in the record, he had no responsibility for Mr. Greenglass in any way, shape, or form, did he?
A. I don't believe so. I will take full responsibility for that one. That was the outstanding blunder of the century.

CROSS-EXAMINATION

By Mr. ROBB:
Q. Colonel Lansdale, as a lawyer are you familiar with the legal maxim, "Falsus in uno, falsus in omnibus"?
A. Yes; I am. Like all legal maxims, it is a generalization, and not of particular significance when applied to specifics.
Q. When you are trying a jury case and the veracity of a witness is in question, do you request the court to give an instruction on that subject?
A. Oh, certainly; don't you?
Q. Certainly, I want to know what you do.
A. The instruction usually is that the jury may, but does not have to, take that as an indication, and the judgment is to be exercised in the particular case.
Q. And when you are trying a jury case and you examine a witness on the opposite side and you demonstrate that he has lied, don't you argue to the jury from that that they should disregard his evidence?
A. You are speaking now as to what I as an advocate do?
Q. Yes.
A. It depends on circumstances; usually I do.
Q. Sure. Any lawyer worth his salt would.
A. Particularly if it is my belief.

Q. Yes, sir.

Mr. ROBB. That is all.

Mr. GRAY. The testimony will be made available to you here in the building. That, I think, answers the one question we discussed.

Mr. GARRISON. This afternoon.

Mr. GRAY. Yes.

Mr. ROLANDER. I don't know whether it will be available this afternoon. I understood he wanted to review the material tomorrow. Will that be inconvenient?

Mr. GARRISON. I want to get on it this afternoon so we perhaps can get done with it by tomorrow.

Mr. GRAY. The second question was, you asked for permission to hear the recordings. As I understand, there is available to the board a recording of the Pash interview. So far as I know, the recording of the Lansdale interview is not available; but, if you desire, the board with Dr. Oppenheimer and counsel will listen to the record on Monday if this is important to you before you start redirect examination.

Mr. GARRISON. As to the Pash recording, how are we to hear that?

Mr. ROBB. Right here.

Mr. GRAY. I think we must hear it in the proceeding. I believe that disposes of the two questions you asked?

Mr. GARRISON. Yes. Thank you very much.

Mr. GRAY. We will meet again at 9:30 on Monday morning.

(Thereupon at 4:35 p. m., a recess was taken until Monday, April 19, 1954, at 9:30 a. m.)

UNITED STATES ATOMIC ENERGY COMMISSION

PERSONNEL SECURITY BOARD

In the Matter of J. Robert Oppenheimer

Atomic Energy Commission,
Building T-3, Room 2022,
Washington, D. C., Monday, April 19, 1954.

The above entitled matter came on for hearing, pursuant to recess before the board, at 9 : 30 a. m.

Personnel Security Board: Mr. Gordon Gray, chairman; Dr. Ward V. Evans, member; and Mr. Thomas A. Morgan, member.

Present: Roger Robb, and C. A. Rolander, Jr., counsel for the board; J. Robert Oppenheimer, Lloyd K. Garrison, Samuel J. Silverman, and Allen B. Ecker, counsel for J. Robert Oppenheimer; Herbert S. Marks, cocounsel for J. Robert Oppenheimer.

PROCEEDINGS

Mr. GRAY. We will start the proceedings.

I should like the record to reflect that Mr. Morgan, who, as you all recall, found it necessary to leave the proceedings before we completed our work on Friday, has read carefully the transcript made of the proceedings, especially that portion which took place in his absence, and is thoroughly familiar with what transpired. Is that correct?

Mr. MORGAN. That is correct.

Mr. GRAY. I think the record ought to show that.

Now, Mr. Robb.

Mr. ROBB. Yes, Mr. Chairman.

Mr. Rathman is here and will play these records for us. Counsel for Dr. Oppenheimer have been furnished with two copies of the transcript to follow. I would suggest that in the event that anyone at any time wishes any portion of the recording played again, so that we may check it, if you will just so indicate, we will do that. Of course, that includes the reporter. I understand it is most difficult for a reporter to take this down on a machine. So if the reporter wishes to stop and have something played over again, that will be done. Of course, Mr. Garrison, anything that he wishes to be played over if he will just indicate it will be done.

Mr. GARRISON. Suppose we find, Mr. Chairman, as we listen to this, what seems to us to be variances between sound and text. Should we make a note of those as we go along?

Mr. GRAY. I think that would be the proper procedure.

Mr. ROBB. I think that would be the only way to do it, Mr. Garrison.

Mr. Rathman, would you begin to play the records?

I might say, gentlemen, for your benefit, to assist you, at the beginning of this record you will hear some door slamming and seat creaking and so on, and some introductory gabble, which is not important here. I suppose people are coming into the room and sitting down. The transcript, which begins, "This is a pleasure," does not begin for perhaps 30 seconds.

Mr. GARRISON. Mr. Chairman, could we have read into the record the first paragraph of the transcript that will not appear in the sound?

Mr. ROBB. Do you wish me to do that, Mr. Chairman?

Mr. GRAY. If you would.

Mr. ROBB. "San Francisco, California, August 27, 1943.

"Memorandum for the Officer in Charge.

"Subject: D. S. M. Project.

"Re: Transcription of Conversation between Dr. J. R. Oppenheimer, Lt. Col. Boris T. Pash, and Lt. Lyall Johnson.

"Transmitted, herewith, is the transcript of conversation between Dr. J. R. Oppenheimer, Lt. Col. Boris T. Pash, and Lt. Lyall Johnson held in Lt. Johnson's office in the new class room building, University of California, Berkeley, Calif., on August 26, 1943. It is to be noted that in some places the conversation was very indistinct and that the running commentary may be indecisive in these places, but the substance of the material discussed is, herewith, presented:

"PASH. This is a pleasure, because I am interested to a certain extent in activities and I feel I have a certain responsibility in a child which I don't know anything about. General Grove has, more or less, I feel placed a certain responsibility in me and it's like having a child, that you can't see, by remote control. I don't mean to take much of your time——

"OPPENHEIMER. That's perfectly all right. Whatever time you choose.

"PASH. Mr. Johnson told me about the little incident, or conversation, taking place yesterday in which I am very much interested, and it had me worried all day yesterday since he called me up. I thought if he could——

"OPPENHEIMER. I was rather uncertain as to whether I should or should not talk to him. I am unwilling to do it without authorization. What I wanted to tell this fellow was that he had been indiscreet. I know that he had revealed information. I know that saying that much might in some cases embarrass him. It doesn't seem to have been capable of embarrassing him, to put it bluntly.

"PASH. That is not the particular interest I have. It is something a little more, in my opinion, more serious. Mr. Johnson said that there was a possibility that there may be some other groups interested.

"OPPENHEIMER. I think that is true, but I have no first-hand knowledge and that would not be, for that reason, very useful to me. I think it is true that a man whose name I never heard who was attached to the Soviet consul has indicated indirectly through intermediaries people concerned in this project, that he was in a position to transmit, without any danger of a leak or anything of that kind, or a scandal, information which they might supply."

Dr. EVANS. That is one correction that you passed over. That "intermediary" and not "intermediaries."

Mr. ROBB. And that is true, instead of that.

(Discussion off the record.)

Mr. GARRISON. The only comment I would make, Mr. Chairman, is that in quite a number of places, I think I marked 1, 2, 3, 4, 5, 6, there were scraps of talk that were not recorded here because of the speed. Also here a word and there a word was either dropped out in the speed of the transcription or the order was sometimes inverted a little bit. I am not saying that this alters the substance, but I do think that if there comes a passage——

Mr. ROBB. That is true. Mr. Chairman, I think this suggestion is a very excellent one. I am sure if there is any matter of substance which counsel finds of recording which he feels is different from the transcript, I trust he will indicate, that we may play the record again, and also that we will agree on it. Will you do that, Mr. Garrison?

Mr. GARRISON. I want to make it clear that we are not attempting, and we can't on one playing, to authenticate the entire record.

Mr. ROBB. Very good. Shall we go ahead, Mr. Garrison?

(Mr. Rathman resumed playing back the recording.)

"OPPENHEIMER. Since I know it to be a fact——"

Mr. GARRISON. There seem to be some words in the conversation which do not appear in the transcript immediately prior to the sentence reading, "Since I know it to be a fact." This conversation apparently dealing in some way with the Soviet consulate.

[Recording:]

"Since I know it to be a fact, I have been particularly concerned——"

Mr. GRAY. I think what was said there is that Dr. Oppenheimer is saying it might be assumed that a man attached to the Soviet consul might be doing this. "But since I know it to be a fact, I have been particularly concerned." That is my interpretation.

Mr. GARRISON. It is something like that, Mr. Chairman. I was not exactly clear. We might have it once more, if you don't mind.

[Recording:]

"I will take it assumed that a man attached to the Soviet consul might be doing this. But since I know it to be a fact I have been particularly concerned about any indiscretions which took place in circles close which might be in contact with it. To put it quite frankly, I would feel friendly to the idea of the Commander in Chief informing the Russians who are working on this problem."

Mr. MARKS. May we stop at this point?

Mr. GARRISON. Mr. Chairman, I think it quite clear from the recording that the sentence read, "I would feel friendly to the idea of the Commander in Chief informing the Russians that we are working on this problem."

Mr. GRAY. I would have to ask that it be played again.

[Recording:]

"I will take it to be assumed that a man attached to the Soviet consul might be doing this, but since I know it to be a fact, I have been particularly concerned about any indiscretions which took place in circles close to the consul or which might come in contact with it. To put it quite frankly, I would feel friendly to the idea of the Commander in Chief informing the Russians that we were working on this problem. At least I can see that there might be some arguments for doing that, but I do not feel friendly to the idea of having that—I think that it might not hurt to be on the lookout for it."

Mr. GARRISON. May we stop at that point. Is the chairman satisfied that the phrase was "informing the Russians that we were working on this problem"?

Mr. ROBB. That is the way I heard it.

Mr. GRAY. It is not clear to me, but I think it is clear that the word simply was not "who." Precisely what the word or words might have been, I am not sure, but my inclination is to feel that it is as you suggest.

Mr. GARRISON. Counsel would agree?

Mr. ROBB. That was my understanding of it, Mr. Garrison.

Mr. GARRISON. That it did read "informing the Russians that we were working."

Mr. ROBB. I think it is.

Mr. GARRISON. Either one, it doesn't matter, Mr. Chairman. I would point out that in the cross-examination of Dr. Oppenheimer, this particular phrase was picked out of the transcript about informing the Russians, as it reads here, "who are working on this problem," as if there were something sinister about it. It came as a great surprise to Dr. Oppenheimer, and I think the record now should explicitly show that this was an error in transcription and that any notion that the Russians were then working on this problem was simply not suggested in this conversation.

Mr. ROBB. I wouldn't go so far as that, sir. I think the record shows that the recording says. I think that is as far as we can go.

Mr. GRAY. I think there is agreement between counsel as to what seems to be the correct transcript now on this point. Certainly speaking for the Board, I don't think we can draw any conclusions into the record at this point, Mr. Garrison. I think the record ought to be clear as to what the language was.

Mr. GARRISON. I want to make clear that any inference drawn from the previous cross-examination is now to be wiped out.

Mr. GRAY. I should think that you would wish—on redirect, if I can use that term—to come back to this point. Certainly the record now will reflect what the concensus is as to this language. I am just hesitant to accept an interpretation of counsel as a part of a board conclusion at this time. What we are doing is correcting the record as I understand it. You are certainly free to come back to this.

Mr. ROBB. Will you start at the beginning?

[Recording:]

"OPPENHEIMER. I probably know this. I will take it is to be assumed that a man attached to the Soviet consul might be doing this, but since I know it to be a fact, I have been particularly concerned."

Mr. ROBB. Mr. Chairman, might I interpose at this point. I think it is pretty clear now, Mr. Garrison, that the beginning of that sentence is, "I would take it that it would be assumed that a man attached to the Soviet consulate might be doing this, but since I know it to be a fact"; isn't that the way you heard it?

Mr. GARRISON. That is about the way I heard it.

Mr. ROBB. Did you hear it any differently than that?

Mr. GARRISON. I think that is about correct.

Mr. GRAY. While we are in this interruption, my interpretation of the recording is that the word "aides" should have been "circles".

Mr. GARRISON. Yes, Mr. Chairman.

Mr. ROBB. I had already corrected that in my transcript.

Mr. GRAY. Would you proceed, Mr. Rathman.

[Recording:]

"I would take it that it is to be assumed that a man attached to the Soviet consulate might be doing it, but since I know it to be a fact, I have been particularly concerned about any indiscretion which took place in circles close enough to come in contact with it. To put it quite frankly, I would feel friendly to the idea of the Commander in Chief informing the Russians that we were working on this problem. At least I can see that there might be some arguments for doing that, but I do not feel friendly to the idea of having it moved out the back door. I think that it might not hurt to be on the lookout for it.

"PASH. Could you give me a little more specific information as to exactly what information you have? I mean, you can readily realize that phase would be, to me, probably of interest as pretty near the whole project is to you.

"OPPENHEIMER. Well, I might say that the approaches were always through other people, who were troubled by them, and sometimes came and discussed them with me; and that the approaches were always quite indirect so that I would feel that to give—well, to give more, perhaps, than one name, would be to implicate people whose attitude was one of bewilderment, rather than one of cooperation. I know of no case, and I am fairly sure that in all cases where I heard of it, these contacts would not have yielded a single thing. That is as far as I can go on that.

"Now, there is a man whose name was mentioned here a couple of times. I do not know of my own knowledge he is involved as an intermediary. It seems, however, not impossible, and if you wanted to watch him, it might be the appropriate thing to do. He spent a number of years in the Soviet Union. I think he is a chemical engineer. He was—he may not be here—he was at the time I was with him here employed at the Shell Development. His name is Eltenton. I would think that there was a small chance that—well, let me put it this way—I think he has probably been asked to do what he can to provide information. Whether he is successful or not, I don't know. But if he talked to a friend of his who was also an acquaintance of one of the men on the approach, that was one of the channels by which this thing went on.

Now, I think that * * * asked to do what he could to provide information. Whether he was successful or not I couldn't know. But he talked to a friend of his who was also an acquaintance of one of the men on the project, and that was one of the channels by which this thing went. Now, I think that to go beyond that would be to put a lot of names down of the people who are not only innocent, but whose attitude is 100 percent effective.

"PASH. Now, here's a point. You can readily realize that if we get information like that we have to work in an absolutely discreet manner. In other words, we can't afford to indicate——

"OPPENHEIMER. That you are concerned.

"PASH. That we are concerned or through whom we get information.

"OPPENHEIMER. Naturally.

"PASH. However, any——"

[End of recording.]

Mr. GRAY. I should like to record my observation about some of these words here.

Mr. ROBB. Yes, sir.

Mr. GRAY. First of all, it is pretty clear to me going back to this earlier paragraph that the language should be "informing the Russians that we were working on this project."

Mr. ROBB. That is correct.

Mr. GRAY. Then in the third paragraph, I believe in the first sentence, it should read, "Well, I might say that the approaches were always to other people" rather than "through other people."

Mr. ROBB. I would like to have that played back.

Mr. GRAY. Will you play the beginning of this again, please?

Mr. GARRISON. Mr. Chairman, so that we do this in the same order, I listened to thees words that were in the middle of the sentence beginning, "At least I can see that there might be some arguments for doing," I then heard these words, "I don't know whether it could or could not be done, but I don't like the idea of having them moved out the back door."

Mr. ROBB. I don't know. There are some words in there that I didn't get. Let us see, and we will play it again. I hope these records don't get worn out while we are playing them.

Mr. GARRISON. I hope we don't have to play the whole thing through just for this one thing.

Mr. ROBB. No; that is right at the beginning.

[Recording:]

"It must be assumed that a man attached to the Soviet consulate might be doing this, but since I know it to be a fact, I have been particularly concerned about any indiscretions which took place in circles close to the consul or which might come in contact with it, because to put it quite frankly, I would feel friendly to the idea of the Commander in Chief informing the Russians that we were working on this problem. At least, I can see that there might be some argument for doing that. I don't know whether it could or not have been done, but I don't like the idea of having it moved out the back door. I think that it might not hurt to be on the lookout for it."

"PASH. Could you give me a little more specific information as to exactly what information you have? I mean, you can readily realize that phase would be, to me, probably as interesting as pretty near the whole project is to you.

"OPPENHEIMER. Well, I might say that the approaches were always to other people, who were troubled by them, and sometimes came and discussed them with me."

Mr. ROBB. Mr. Chairman, it is quite plain that the sentence reads, "I might say that the approaches were always to other people"; is that correct, Mr. Garrison?

Mr. GARRISON. Yes.

Mr. GRAY. Then a few minor ones.

Mr. ROBB. May I say with Mr. Garrison's help I do find the phrase "I don't know whether it could or could not be done, but" comes in.

Mr. GRAY. "I am not friendly to the idea of having it move out the back door."

Mr. ROBB. That is right.

Mr. GRAY. Further in that third paragraph, I think that the third sentence would read, or portions of it, "and that the approaches were always quite indirect." The word "always."

Mr. GARRISON. Yes, Mr. Chairman.

Mr. GRAY. In the fifth line, very minor, the word "attitudes" should be "attitude," singular.

Mr. ROBB. Yes.

Mr. GRAY. Down about the middle of that paragraph, "He spent quite a number of years in the Soviet Union."

Mr. GARRISON. I also heard the words "He is an Englishman" in there.

Mr. ROBB. That is in here. Isn't it?

Mr. GARRISON. No, "He spent"—I have interlineated "He is an Englishman" or "He is English".

Mr. ROBB. I think that is in there some place, but that is not very important.

Mr. GARRISON. No.

Mr. GRAY. There are a couple of other places. The word "is" should have been "was" and "the project" should be "this project."

Mr. GARRISON. The sentence reading, "I think there is a small chance," I think the "is" there, that the word was "was." That is after the word "Eltenton." "His name is Eltenton." "I would think there was a small chance."

Mr. GRAY. That is correct.

Mr. ROBB. Mr. Chairman, may I ask Mr. Garrison, is there any question that the voice we hear in the paragraphs marked "O" is Dr. Oppenheimer?

Mr. GARRISON. Not so far.

Mr. GRAY. Are we ready to proceed? While he is fixing that record, a very minor one, I think Colonel Pash said "absolute discreet manner" rather than "absolutely".

Mr. GARRISON. I am not bothered with that type of correction, Mr. Chairman.

[Recording:]

"That we might get which would eliminate a lot of research work on our part would necessarily lead to the conclusion anything we are doing.

"OPPENHEIMER. I am giving you the one name I think—I mean I don't know the man attached to the consulate. I think I may have been told or I may not have been told. But I have actually forgotten. He is—and he may not be here now. These incidents occurred in the order of about 5, 6 or 7 months.

"JOHNSON. I was wondering, Dr. Oppenheimer, if there was a particular person, maybe a person on the project that you were trying to pump information from—that if we knew who those were, would at least know where to look for a lead, not from the standpoint of fellow hate, but looking at a certain picture.

"PASH. Here is the point that I would feel——

"OPPENHEIMER. I would feel that the people that tried to get information from were more or less an accident and I would be making some harm by saying that.

"PASH. Yes. Here's the thing. We of course assume that the people who bring this information to you are 100 percent with you, and therefore, there is no question about their intentions. However, if——

"OPPENHEIMER. Well, I will tell you one thing——"

[End of recording.]

Mr. EVANS. Was that word "lead" or "leak".

Mr. GARRISON. I thought it was "leak".

Dr. EVANS. It is "lead" here.

Mr. GARRISON. Yes. It sounded like "leak" to me.

Dr. EVANS. It sounded like "leak" to me. "I was wondering, Dr. Oppenheimer, if there is a particular person—maybe a person on the project that you were trying to pump information from—that if we knew who those were, would at least know where to look for a leak"——

Mr. ROBB. Play that again, please.

[Recording:]

"These events occurred of the order of five, six, or seven months ago.

"JOHNSON. I was wondering, Dr. Oppenheimer, if there is a particular person,—maybe a person on the project that you were trying to pump information from,—that if we knew who those were, would at least know where to look for a leak, not from the standpoint of fellow hate, but looking at a certain picture.

"PASH. Here's the point that I would feel——

"OPPENHEIMER. I would feel that the people that if they tried to get information were more or less an accident and I believe I would be making some harm by saying that.
"PASH. Yes. Here's the thing—we of course assume that the people who bring this information to you are 100 percent with you, and therefore, there is no question about their intentions. However, if——
"OPPENHEIMER. Well, I will tell you one thing. I have known two or three cases, and I think two of them are the men with me at Los Alamos. They are men who are very closely associated with me.
"PASH. Have they told you that either they thought they were contacted for that purpose or they actually were contacted for that purpose?
"OPPENHEIMER. They told me that they were contacted.
"PASH. For that purpose.
"OPPENHEIMER. That is, let me give you the background. The background was—well, you know how difficult it is with the relations between these two allies, and there are a lot of people who don't feel very friendly toward Russia, so that the information—a lot of our secret information, our radar and so on, doesn't get to them, and they are battling for their lives and they would like to have an idea of what is going on. This is just to make up in other words for the defects of our official communication. That is the form in which it was.
"PASH. Oh, I see.
"OPPENHEIMER. Of course, the actual fact is that it is not a communication that ought to be taking place. But it is matter of carrying out a policy which was more or less a policy of the government and the form in which it came as that could an interview be arranged with this man Eltenton who had very good contact with a man from the Embassy attached to the consulate who was a very reliable guy. That is his story. And who had a lot of experience in microfilm work.
"PASH. Well, now, I may be getting back to a systematic picture here. But do you mind? These people whom you mentioned, two are down with you now. Were they contacted by Eltenton direct?
"OPPENHEIMER. No.
"PASH. Through another party?
"OPPENHEIMER. Yes.
"PASH. Well, now, could we know through whom that contact was made?
"OPPENHEIMER. I think it would be a mistake——"
Mr. GARRISON. Mr. Chairman, could we stop?
Mr. ROBB. I have several corrections, Mr. Chairman.
Mr. GRAY. Very well.
Mr. ROBB. In the first paragraph on that page, Mr. Garrison, we pretty well agreed on, and the second.
Mr. GARRISON. Except I would just like to note the phrase "not from the standpoint of fellow hate," that there were quite indistinguishable words that accompanied that. I don't know what the words were.
Mr. GRAY. I would question myself that the words were "fellow hate."
Mr. ROBB. I don't know.
Mr. GARRISON. There were other words not in there.
Mr. ROBB. I might say that the "J" indicated there is Lieutenant Johnson who was also present. In the third line on that page of the transcript, as I heard it, it is, "These instances occurred of the order of about 5, 6, or 7 months ago," is that correct?
Mr. GARRISON. That is right.
Mr. ROBB. As I heard it in the paragraph marked "H", the word "lead" should be "leak". The words "fellow hate" I don't pick that up.
The next paragraph marked "O", as I heard it, read "I would feel that the people that they tried to get information from." Did you get that, Mr. Garrison?
Mr. GARRISON. Yes.
Mr. GRAY. I think that was clear.
Mr. ROBB. The next paragraph marked "O".
Mr. GARRISON. While we were on that paragraph, after the word "accident," there were some words interpolated by Mr. Pash that did not come through on the transcript.
Mr. ROBB. Yes; that is correct. The next paragraph marked "O", as I got it, reads, "Well, I will tell you one thing. I have known of two or three cases, and I think two of the men were with me at Los Alamos." Did you get that?
Mr. GARRISON. Yes.
Mr. ROBB. "They are men who are very closely associated with me."
Mr. GARRISON. Correct.

Mr. ROBB. Then the next large paragraph marked "O", reads as I got it in the third and fourth lines, "There are a lot of people that don't feel very friendly to Russia" instead of "toward the Russians". Did you get that?
Mr. GARRISON. I did.
Mr. GRAY. It is "a lot of people who don't feel very friendly."
Mr. ROBB. Yes, sir.
Mr. GARRISON. That is right.
Mr. ROBB. In the last paragraph on that page, as I got it, it reads, "Of course, the actual fact is that since it is not a communication which ought to be taking place, it is treasonable, but it was not presented in that method."
Mr. GARRISON. Right after that word "method" I had some words——
Mr. ROBB. That didn't come through. Yes, sir. "It is a method of carrying out a policy which was more or less a policy of the Government and the form in which it came was that an interview be arranged with this man Eltenton who had very good contacts with a man from the embassy attached to the consulate who was a very reliable guy, and who had a lot of experience in microfilm work or whatever."
Mr. GARRISON. That "or whatever," I would like to have it played again. I think there was another word or two after the word "whatever."
Mr. ROBB. I think so, but I didn't get it. "In microfilm work" and also after the word given there were 2 or 3 words that I didn't get.
[Recording:]
"* * * a policy which was more or less a policy of the Government, and the form in which it came was that could an interview be arranged with this man Eltenton, who had very good contacts with a man from the embassy, attached to the consulate, who was a very reliable guy, that is his story, and who had a lot of experience in microfilm work, or whatever.
"PASH. I may be getting back to a little systematic picture"——
Mr. ROBB. Will you stop there? Mr. Garrison, I don't know whether you got it the way I did, but I thought I heard "who was a very reliable guy," a kind of parenthetical story, "That is his story."
Mr. GARRISON. Yes.
Dr. OPPENHEIMER. After "whatever" it said "the hell."
Mr. ROBB. Thank you.
Mr. GARRISON (reading). "A lot of experience in microfilm work, or whatever the hell."
Mr. ROBB. Dr. Oppenheimer is certainly the best expert on his own voice.
Mr. GRAY. In the next paragraph when he plays that, I think the word "two" right in the middle, on the top of page 4. "two are down there" should be "who". Will you play that again?
[Recording:]
"PASH. Well, now I may be getting back to a little systematic picture, but do you mind. These people whom you mentioned, who were down there with you now, were they contacted by Eltenton?
"OPPENHEIMER. No.
"PASH. Through another party?
"OPPENHEIMER. Yes.
"PASH. Well, now, could we know through whom that contact was made?
"OPPENHEIMER. I think it would be a mistake"——
Mr. ROBB. I still got a "two".
Mr. GARRISON. I thought it was "who."
Dr. EVANS. I thought it was "who."
[Recording:]
"PASH. These people whom you mentioned, who were down with you now, were they contacted by Eltenton direct?
"OPPENHEIMER. No.
"PASH. Through another party?
"OPPENHEIMER. Yes.
"PASH. Well, now, could we know through whom that contact was made?
"OPPENHEIMER. I think it would be a mistake"——
Mr. ROBB. I don't know.
Dr. EVANS. I would like to know how many of us thought it was "who" and how many thought it was "two." I thought personally it was "who."
Mr. GRAY. Let us make this the last time.
[Recording:]
"I may be getting back to a little systematic picture here, but do you mind? These people who you mentioned, two are down there with you now, were they contacted by Eltenton direct?

"OPPENHEIMER. No.
"PASH. Through another party?
"OPPENHEIMER. Yes.
"PASH. Well, now, could we know through whom that contact was made?
"I think it would be a mistake"——
Mr. GRAY. Mr. Morgan thinks it is "two" and I could flip a coin.
Mr. ROBB. I don't know that it is terribly important.
Mr. GARRISON. I don't know, Mr. Chairman.
Mr. ROBB. Why don't we put "who?" and "two?" in the transcript. Is that all right, Mr. Garrison?
Mr. GARRISON. It is all right with me. I would note also there are some words after systematic picture indicated by the dots that don't appear.
Mr. ROBB. That is something like, "getting back to a little systematic picture, if you don't mind."
Mr. GARRISON. Something like that. I would observe that those are the first dots we have seen in this transcript although we have all agreed that there are some words and passages that don't appear in quite a number of places.
Mr. ROBB. All right.
[Recording:]
"I think I have told you where the initiative came from and that the other things are almost purely accidental, and it would involve people who ought not to be involved in this.
"PASH. Yes. Well, this would not involve the people but it indicates to us Eltenton's channel. We would have to know that this is definite on Eltenton, and we of course naturally——
"OPPENHEIMER. It is not definite in the sense that I have seen him do the thing.
"PASH. No.
"OPPENHEIMER. He may have been misquoted.
"PASH. That is right.
"OPPENHEIMER. I don't believe so. Now, Eltenton is a member of the FAECT. Whether or not——
"PASH. That is the union?
"OPPENHEIMER. That is the CIO. He is a man whose sympathies are certainly very far left, whatever his affiliations, and he may or may not have regular contacts with a political group. I doubt it. In any case, it is a safe thing to say that the channels that would be followed in this case are those involving people who have generally been sympathetic to the Soviet and somehow connected peripherally with the Communist movement in this country. That's obvious. I don't need to tell you that.
"PASH. Yes. The fact is this second contact—the contact that Eltenton had to make with these other people is that person also a member of the project?
"OPPENHEIMER. No.
"PASH. That also is an outsider?
"OPPENHEIMER. It's a member of the faculty, but not of the project.
"PASH. A member of the faculty here? Eltenton made it through a member of the faculty to the project.
"OPPENHEIMER. As far as I know, these approaches were—there may have been more than one person involved. I don't know.
"PASH. Here's how I feel about this leftist inclination. I think that whether a man has 'left' or 'right' inclinations, it is his character which is back of it—if he is willing to do this, it doesn't make any difference what his inclinations are. It is based on his character primarily and not——
"OPPENHEIMER. Yes. A thing like this going on, let us say, with the Nazis would have a somewhat different color. I don't mean to say it would be any more deserving of attention or any more dangerous, but it would involve probably different motives.
"PASH. Yes.
"OPPENHEIMER. I'm pretty sure that none of the guys here with the possible exception of the Russian, who is doing probably his duty by his country—but the other guys that were just—they didn't do anything, but they were considering the step which they would have regarded as thoroughly in line with the policy of this Government, and just making up for the fact that there were a couple of guys in the State Department who would block such communications. You may or may not know that in many projects we share information with the British and some we do not, and there was a great deal of feeling about that and I don't think that the issues involved here seem to people very different

except that of course the people on the project realize the importance and the whole procedure gets away from them.

"PASH. Now, do you feel"——

[End of the recording.]

Mr. ROBB. I noticed a few minor corrections, but none I think that is worth talking about, unless Mr. Garrison has some.

Mr. GRAY. I have one that may be minor, but perhaps it should be noted. In the paragraph that the CIO union, in the fourth line, I believe that the language was "a safe thing to say that the channels that would be followed in this case" instead of "to be followed." Did you get that?

Mr. ROBB. I didn't get that. Did Mr. Garrison get that?

Mr. GARRISON. No. Mr. Marks said he did. We accept that. Could I in the same paragraph note that after the words "I doubt it" by Mr. Oppenheimer, I heard an interjection by Mr. Pash, saying, "Here is the way I feel about this case," and then it carries on with Mr. Oppenheimer saying, "It is a safe thing to say."

Mr. ROBB. I think that is true.

Mr. GARRISON. I mention that because here is the word "case" which is put in Dr. Oppenheimer's mouth which in fact came from Mr. Pash. I don't think it alters the substance.

Mr. ROBB. I think Dr. Oppenheimer did use the word "case." It appeared that Colonel Pash, interrupting Dr. Oppenheimer and Dr. Oppenheimer keeping on talking, I heard Dr. Oppenheimer's voice saying, "In any case", although I don't know that it is important.

Mr. GARRISON. You heard the word "case" twice.

Mr. ROBB. Yes.

Mr. GARRISON. You heard the word "case" again?

Mr. ROBB. I thought I did, yes. Do you want to play it over again?

[Recording:]

"He may have been misquoted.

"PASH. That is right.

"OPPENHEIMER. I don't believe so. Now Eltenton is a member of the FAECT. Whether or not——

"PASH. That is the union——

"OPPENHEIMER. That is the CIO union. He is a man whose sympathies are certainly very far left, whatever his affiliation is, and he may or may not have regular contacts with a political group. I doubt it.

"PASH. Here is the way I feel.

"OPPENHEIMER. In any case, it is a safe thing to say that the channels that will be followed in this case are those involving people who have generally been sympathetic to the Soviet——"

Mr. ROBB. I don't know who said it, Mr. Garrison.

Mr. GRAY. It is my impression that there was an interruption by Colonel Pash, and Dr. Oppenheimer did say "In any case, it is a safe thing." I don't know that it is important.

Mr. ROBB. I don't think it is important.

Mr. GRAY. I do think there are 2 things I should point up in the fifth paragraph, about the middle of that paragraph, where I believe Dr. Oppenheimer said, "might block such communications" rather than "would."

Mr. ROBB. Yes, I heard that, too.

Mr. GARRISON. Yes, Mr. Chairman.

Mr. GRAY. In the last line of that paragraph, I don't think the words "gets away from them" are correct. The word "gets" is not correctly transcribed, but I can't tell what it was.

Mr. GARRISON. That whole last line to me is rather indistinct. There were some words that don't appear and I don't quite get the sense of it.

Mr. ROBB. I don't either, Mr. Garrison, but I don't think it is terribly important.

Mr. GARRISON. Mr. Chairman, I would just make this suggestion perhaps in view of the time pressure under which we are all laboring. Possibly the chairman in order to save the time of the board would think it appropriate that we might make an arrangement with counsel on the other side to continue this playing at some time that would not take up the time of the board, and bring to the board and read into the record any changes that we agree upon. I think we probably would have no difficulty in doing that. I don't press that.

Mr. ROBB. That might be possible, unless the board wishes to participate in this.

Mr. GRAY. I am sorry to engage in a time consuming procedure, Mr. Garrison, but I am inclined to think that if there are to be any changes in the record, the board regrettably must hear them. I am sorry about the delay involved.

[Recording:]

"PASH. Do you feel that would affect—and there could be continued attempts now to establish this type of contact?

"OPPENHEIMER. I haven't any idea.

"PASH. You haven't any idea?

"OPPENHEIMER. As I say, if the guy that was here may by now be in some other town and all that I would have in mind is this. I understood that this man to whom I feel a sense of responsibility, Lomanitz, and I feel it for 2 reasons. One, he is doing work which he started and which he ought to continue, and second, since I more or less made a stir about it when the question came up, that this man may have been indiscreet in circles which would lead to trouble. That is the only thing that I have to say. I don't have any doubt that people often approached him, with whom he has contacted, I mean whom he sees, might feel it their duty if they got word of something, to let it go further and that is the reason why I feel quite strongly that association with the Communist movement is not compatible with the job on a secret war project, it is just that the 2 loyalties cannot go.

"PASH. Yes. Well——

"OPPENHEIMER. That is not an expression of political opinion. I think that a lot of very brilliant and thoughtful people have seen something in the Communist movement, and that they maybe belong there, maybe it is a good thing for the country. They hope that it doesn't belong on the war project.

"PASH. I get your point. I don't want to seem to you insistent. I want to again I think explore the possibility of getting the name of the person on the faculty. I will tell you for what reason. Not for the purpose of taking him to task in any way whether it is unofficially, officially, or openly or what, but to try to see Eltenton's method of approach. You may not agree with me, but I can assure you that that is one of the more important steps.

"OPPENHEIMER. I have to take the following points of view: I think in mentioning Eltenton's name I subsequently said about the man that I think that he may be acting in a way which is dangerous to his country, and which should be watched. I am not going to mention the name of anyone in the same breath, even if you say that you will make a distinction. I just can't do that, because in the other cases, I am convinced from the way in which they handled the thing that they themselves thought it was a bad business.

"PASH. These other people, yes, I realize. But here is the point, doctor; if that man is trying to make other contacts for Eltenton.

"OPPENHEIMER. Yes.

"PASH. You see, it would take us some time to try to——

"OPPENHEIMER. My honest opinion is that he probably isn't, that he ran into him at a party and they saw each other or something and Eltenton said, "Do you suppose you could help me. That is a very serious thing, because we know that important work is going on here, and we think this ought to be available to our allies, and would you see if any of those guys are willing to help us with it, and then it wouldn't have to be so much." [Inaudible.]

Dr. EVANS. There was one place there, "not for the purpose of taking him to task in any way, whether it be unofficially, officially or openly."

Mr. GARRISON. Yes.

Mr. ROBB. I think one of the more important steps Colonel Pash said, one of the most important steps. Did you get that?

Mr. GARRISON. I didn't have it.

Mr. SILVERMAN. Yes.

Mr. ROBB. There was an overriding remark of Dr. Oppenheimer in which he said, "I understand that." Did you gentlemen catch that?

Mr. GARRISON. Yes.

Mr. MARKS. Yes. The word "subsequently" I understood as "essentially."

Mr. GARRISON. Yes. "Subsequently" should read "essentially" in the next paragraph. "I think in mentioning Eltenton's name I essentially said about the man."

Mr. ROBB. I didn't get that.

Mr. GARRISON. In the paragraph at the top where he said that is not an expression of political opinion, I think a lot of very brilliant and thoughtful people have seen something in the Communist movement, and that they maybe belong there, and that maybe it is a good thing for the country.

Mr. ROBB. I think so. I thought I heard instead of "they hope it doesn't belong," "I hope it doesn't belong on the war project." Did you get that?
Mr. GARRISON. I didn't.
Mr. SILVERMAN. It was very indistinct.
Mr. ROBB. I think it was "I."
Mr. GARRISON. I heard some words after "war project" that I couldn't get. Also, some of the words in the next Pash paragraph at the end after one of the more important steps.
Mr. ROBB. Most important steps.
Mr. GARRISON. Most.
Mr. ROBB. Dr. Oppenheimer said "I understand" after that.
Mr. GARRISON. Yes. Then instead of "I have to take," it is "I wish"—did you get that—I understand that, but I have to take the following point. That is already your correction.
Mr. ROBB. Yes.
Mr. GRAY. Are there any other suggestions about that portion? I have no more. Will you proceed, Mr. Rathman.
[Recording:]
"PASH. Were these two people you mentioned, were they contacted at the same time?
"OPPENHEIMER. They were contacted within a week of each other.
"PASH. They were contacted at 2 different times.
"OPPENHEIMER. Yes, but not in each other's presence.
"PASH. That is right. And then from what you first heard, there is someone else who probably still remains here who was contacted as well.
"OPPENHEIMET. I think that is true.
"PASH. What I am driving at is that there was a plan, at least for some length of time, to make these contacts—and we may not have known all the contacts.
"OPPENHEIMER. That is certainly true. That is why I mentioned it. If I knew all about it, then I would say forget it. I thought it would be appropriate to call to your attention the fact that these channels at one time existed.
"PASH. Yes.
"OPPENHEIMER. I really think that I am drawing a line in the right place.
"PASH. You see, you understand that I am sort of—you picture me as a bloodhound on the trail and that I am trying to get out of you everything I possibly can.
"OPPENHEIMER. That is your duty to a certain extent.
"PASH. You see what I mean.
"OPPENHEIMER. It is also my duty not to implicate these people, acquaintances, or colleagues of whose position I am absolutely certain—myself and my duty is to protect them.
"PASH. Oh, yes.
"OPPENHEIMER. If I thought that—I won't say it—it might be sligthly off.
"PASH. Well, then, here's another point, doctor, if we find that in making these various contacts that we get some information which would lead us to believe that certain of these men may have either considered it or still are considering it, mind you. I do not even know these men, so it can't be personal.
"OPPENHEIMER. No. Well, none of them that I had anything to do with even considered it. They were just upset about it. They have a feeling toward this country and have signed the Espionage Act; they feel this way about it for I think that the intermediary between Eltenton and the project, thought it was the wrong idea, but said that this was the situation. I don't think he supported it. In fact, I know it.
"PASH. He made about at least 3 contacts that we know of.
"OPPENHEIMER. Well, I think that's right, yes.
"PASH. And 2 of those contacts are down there. That means we can assume at least that there is one of these men contacted still on the project here.
"OPPENHEIMER. Yes, I believe that this man has gone or is scheduled to go to site X.
"PASH. This third man?
"OPPENHEIMER. I think so.
"PASH. Well, why can't you cross that line. I certainly appreciate this much.
"OPPENHEIMER. I think it is a thing you ought to know.
"PASH. Oh, no doubt.
"OPPENHEIMER. I think it is probably one of those sporadic things and I do not think—I may have no way of thinking it was systematic but I got from the way it was handled, which was rather loosely, and frankly if I were an agent I would not put much confidence in people who are loose-mouthed or casual."

Mr. GRAY. Are there any observations about that portion of the transcript?

Dr. EVANS. The word "Oppenheimer" was after "Doctor".

Mr. GARRISON. The sixth paragraph, "Dr. Oppenheimer: I really think I am drawing a line in the right place." That phrase "a line in the right place" I didn't get.

Mr. ROBB. Something about a line.

Mr. GARRISON. Something about it.

Mr. ROBB. If he plays it over enough, it will come out in the right place, but I don't know. I have not played it over enough. Do you want to play it again?

Mr. GARRISON. I don't think so, unless we find something more difficult. I just want to say I didn't even get it.

Mr. ROBB. I think in the paragraph below that where it says, "It is also my duty not to implicate these people, acquaintances, or colleagues" and so on—I think the and so on is correct.

Mr. GARRISON. That is correct. And after the people "and who are".

Mr. ROBB. I think so.

Mr. GARRISON. There are some indistinct words in Mr. Pash's previous 2 sentences at the end. Then coming down, "Dr. Oppenheimer: If I thought that—I won't say it—It might be slightly off," and some indistinct words.

Mr. ROBB. That is right.

Mr. GARRISON. Then the next paragraph, "They were upset about it".

Mr. ROBB. That is right.

Mr. GARRISON. Then some indistinct words followed that.

Mr. ROBB. I think so.

Mr. GRAY. On that paragraph——

Mr. GARRISON. All the rest of it seemed to me just fuzzy.

Mr. GRAY. The word "even", I think, was not in that paragraph in the first line. While none of them that I had anything to do with considered it, they were just upset about it, is the way I heard it.

Mr. GARRISON. Yes.

Mr. ROBB. Do you want that paragraph played again?

Mr. GARRISON. I am not sure it would do any good.

Mr. ROBB. Let us try it.

[Recording:]

"PASH. Certain of these men may have considered it or are still considering it (mind you, I don't even know these men, so it can't be personal).

"OPPENHEIMER. None of these that I had anything to do with even considered it.

"PASH. Yes.

"OPPENHEIMER. They just were upset about it. They have a feeling toward this country and have signed the Espionage Act; they feel this way about it for I think that the intermediary between Eltenton and the project, thought it was the wrong idea, but said that this was the situation. I don't think they supported it. In fact, I know it.

"PASH. He made about at least 3 contacts that we know of."

Mr. ROBB. Mr. Garrison, I got "have a feeling" "Espionage Act", "intermediary between Eltenton and the project thought is was wrong idea," "was the situation" and there are some words in between there that are indistinct. Is that the way you heard it?

Mr. GARRISON. More or less. I am frank to say I would not feel confident.

Mr. ROBB. I did hear "intermediary."

Mr. GARRISON. I hear that.

Mr. ROBB. "Project" and "wrong idea." "I don't think he supported it. In fact, I know it." I heard that.

Mr. GARRISON. Yes.

Mr. ROBB. Perhaps it is not too important.

Mr. GARRISON. Now, on the next page, the third and fourth paragraphs. "This third man?" "That is right." I am not quite sure of that.

Mr. ROBB. Shall we have it again?

Mr. GARRISON. Yes.

[Recording:]

"PASH. He made about at least three contacts that we know of.

"OPPENHEIMER. I think that's right, yes.

"PASH. And 2 of these contacts are down there. That means we can assume at least there is one of these men contacted still on the project.

"OPPENHEIMER. Yes. I believe that this man has gone or is scheduled to go to site X.

"PASH. This third man?

"OPPENHEIMER. I think so."
Mr. ROBB. All right. Mr. Garrison?
Mr. GARRISON. I heard the words "This third man". I heard some indistinct words at the end of the preceding sentence. "That is right," I didn't hear.
Mr. ROBB. That is unquestionably there. Will you play it again?
[Recording:]
"PASH. This third man?
"OPPENHEIMER. I think so."
Mr. ROBB. That is right.
Mr. GARRISON. I heard something like picture.
Mr. ROBB. I think that is the picture.
Mr. GARRISON. Something like that.
Mr. ROBB. It could be.
Mr. GARRISON. Why don't we pass it?
Mr. ROBB. One thing, Mr. Chairman. I noticed on the other page.
Mr. GARRISON. Could I have it once again.
[Recording:]
"Yes, I believe that this man has gone, or is scheduled to go to site X.
"PASH. This third man?
"OPPENHEIMER. I think so."
Mr. SILVERMAN. I thought he said, "I think so."
Mr. GARRISON. It sounded this time more like, "I think so." I really just don't know.
Mr. GRAY. It would appear, would it not, whether Dr. Oppenheimer said, "That is right", or "That is the pitcure", or "I think so", that he was not indicating disagreement with Colonel Pash at that point?
Mr. GARRISON. I wouldd take that to be so.
Mr. ROBB. Mr. Chairman, I did notice one thing in the record. It mentions on page 7 of the transcript on the fourth line from the top, as I heard it, it reads, "What I am driving at is that means that there was a plan."
Dr. EVANS. I thought it was "is". It doesn't matter at all.
Mr. ROBB. Did you get that, Mr. Garrison?
Mr. GARRISON. No.
Mr. ROBB. You don't want to hear that again?
Mr. GARRISON. No.
Mr. ROBB. You won't agree on that?
Mr. GARRISON. I don't think it is important enough, to play again.
Dr. EVANS. I don't either.
Mr. ROBB. All right.
Mr. GRAY. Are we ready to proceed with the next portion?
[Recording:]
"I would not think that this was a very highly organized or very well put together plan but I don't know and I was very much afraid when I heard of Lomanitz' indiscretion that it might very well be serious. I hope that isn't the case.
"PASH. You mentioned that this man may be a member of the FAECT. Do you think, as a representative of the organization, he would sort of represent their attitude or do you think he is doing that individually?
"OPPENHEIMER. Oh, the FAECT is quite a big union and has all sorts of people in it. I am pretty sure and I don't think it is conceivable that he could be representing the attitude of the union——
"PASH. Well, I don't know enough about it to——
"OPPENHEIMER. I think that—well, I don't know. I think at one time they had a strong branch up at the Shell Development Research Laboratories, the FAECT, and I believe it is the union which has got organized on the hill.
"JOHNSON. Yes, it has been around for some time.
"PASH. This man Eltenton is a scientist?
"OPPENHEIMER. I don't know. I would guess he is some sort of a chemical engineer.
"PASH. Would he be in a position to understand the information furnished him?
"OPPENHEIMER. I don't know that either. It would depend on how well it was furnished. I mean he has some scientific training and certainly if you sat down with him and took a little time. My view about this whole damn thing, of course, is that the information we are working on is probably known to all the governments that care to find out. The information about what we are doing is probably of no use because it is so damn complicated. I don't—I mean I don't agree that the security problem on this project is a bitter one, because if one means by the security problem preventing information of technical use to another country

from escaping. But I do think that the intensity of our effort and our concern of the international investment involved—that is information which might alter the course of the other governments, and I don't think it would have any effect on Russia [inaudible]. It might have a very big effect on Germany, and I am convinced about that and that is as everyone else is.

"PASH. Oh.

"OPPENHEIMER. To give it roughly what we're after and I think they don't need to know the technical details because if they were going to do it they would do it in a different way—they wouldn't take our methods—they couldn't because of certain geographical differences, so I think the kind of thing that would do the greatest damage if it got out would just be the magnitude of the problem and of the time schedules which we think we have of that kind.

"PASH. To answer your question—Eltenton if you were picking a man which would be an intermediary he wouldn't be a bad choice, I would mention he had some kind of chemical engineering job in Russia. He was trained in England, also in Russia 4 or 5 years and things like that. Does he speak Russian, do you know?

"OPPENHEIMER. I don't know. I don't know. He speaks with a slight English accent.

"PASH. If it is necessary would you mind and would it interfere with your work if I would have——

Mr. ROBB. Mr. Chairman, I know the paragraph marked "P" in this transcript on page 9 about a third of the way down is actually Dr. Oppenheimer speaking, "To answer your question" and so on. Colonel Pash made some interruption and then Dr. Oppenheimer continued. Did you get that?

Mr. GARRISON. No; I didn't.

Mr. ROBB. Page 9, "To answer your question—Eltenton if you were picking a man which would be an intermediary he wouldn't be a bad choice." That is obviously Dr. Oppenheimer.

Mr. GARRISON. Dr. Oppenheimer's voice does come in there.

Mr. ROBB. That is Dr. Oppenheimer speaking there and not Colonel Pash.

"Mr. GARRISON. I am not sure the words "To answer your question"——

Mr. ROBB. Could we play that?

Mr. GRAY. Before we play it back, let me make a couple of other observations.

In the first paragraph on this page, the fifth line from the end of the paragraph, "and our concern of the 'national' investment involved," rather than the "International" investment.

Mr. ROBB. Yes, sir.

Mr. GARRISON. Our concern with, I think it was also.

Mr. GRAY. Yes.

Mr. GARRISON. And some words after the word "escaping" that were indistinct, and before the word "but."

Mr. ROBB. I think so.

Mr. GARRISON. And the dots after the word "Russia" contained some words.

Mr. GRAY. Would you play that portion again?

Mr. ROLANDER. The last third.

Mr. GARRISON. Before we do that, perhaps we could make one or two observations so that we can be listening to it.

Mr. GRAY. Yes.

Mr. GARRISON. In the next Oppenheimer paragraph, there are some indistinct words to begin with, and "to give it roughly," I thought it read "To give the Russians" or "To give to Russia."

Mr. ROBB. It could be.

Mr. GARRISON. And I think they don't—that seemed to me fuzzy.

Mr. GRAY. Let us listen to that again.

[Recording:]

"[Inaudible.] It might have a very big effect on Germany, and I am convinced about that and that is as everyone else is.

"[Inaudible.] And I think they don't need to know the technical details, because if they were going to do it, they would do it in a different way. They wouldn't take our methods (inaudible) so I think the kind of thing that would do the greatest damage if it got out would just be the magnitude of the problem and of the time schedules which we think we have, that kind of thing.

To answer your question, Eltenton——

"PASH. Uh huh.

"OPPENHEIMER. To answer your question—Eltenton if you were picking a man to be an intermediary would not be a bad choice. He had some kind of chemical engineering job in Russia. He was trained in England, he was in Russia for 4 or 5 years (inaudible).

"PASH. Does he speak Russian, do you know?

"OPPENHEIMER. I don't know. (Inaudible) with a slight English accent."

Mr. ROBB. Mr. Garrison, did you catch that now? that the "P" paragraph should be really Dr. Oppenheimer?

Mr. GARRISON. Yes.

Dr. EVANS. And that is "roughly" and not "Russia"?

Mr. ROBB. I think it is.

Mr. GRAY. I think in that paragraph the language "He was trained in England, was in Russia 4 or 5 years," rather than "also in Russia."

Mr. ROBB. I got it "and in Russia."

Mr. GRAY. It does make a little difference to say he was trained in Russia or was in Russia.

Mr. SILVERMAN. I heard it the way the Chairman did.

Dr. EVANS. So did I.

Mr. ROBB. Was in Russia.

Mr. GRAY. He was trained in England, was in Russia 4 or 5 years.

In the preceding paragraph, in the interests of grammar, I think actually what Dr. Oppenheimer said at the end of that paragraph, "and of the time schedules which we have, that kind of thing," this is very unimportant.

Mr. Robb (reading) : "which we have—that kind of thing."

Mr. GARRISON. Mr. Chairman, I wonder if we might take a 5-minute recess. We have a very serious problem about our witnesses. Dr. Bethe is here in town ready to testify. So is Mr. Gordon Dean. Dr. Kennen is here from out of town, Dr. Buckley, you remember we talked about last week, is here. Dr. Fisk is here from New York, and General Osborne is also here. I just at this point don't know what to suggest. Obviously if we go through this at the rate we are, it will consume most of the rest of the morning and some of this testimony will be quite of considerable length and I think quite important to the Board. I know it would be informative to the Board.

Mr. GRAY. I would like to ask the board members a question about a ruling that you may recall I made earlier about the necessity for us to hear with counsel the remainder of this transcript. My reaction was that, as I stated, if there were to be any changes, we should hear the discussion, but it does occur to me after having thought about it, if counsel agree, there is no problem. In the event there is disagreement and it seems to be a material matter, then perhaps we should hear those portions about which there is disagreement. I would want to make sure that the Board would agree with that different kind of ruling on that question.

Mr. MORGAN. Yes.

Mr. GRAY. Is that all right, Dr. Evans?

Dr. EVANS. I was certainly in accord with you that we ought to go over this thing together, but if it is necessary, I shall agree to do it the other way.

Mr. GRAY. I am sure that counsel will be diligent. To the extent that counsel can agree, I think it would appear to be pretty clear and if you cannot, perhaps we shall have to hear the disputed portions. Is that satisfactory to you?

Mr. ROBB. Yes, sir. I might suggest that in view of the fact that we will be changing our methods of operation as it were, I think we ought to attempt to get a complete transcript on which we can agree, so it will be all set out at one part of the record because the record will be hard to understand.

Mr. GARRISON. I think it is important that the record indicate what has taken place.

Mr. ROBB. Yes.

Mr. GARRISON. I think we should agree and stipulate on the changes we should make and bring that back to the Board for its approval and incorporation in the record, and that the whole document in its original form should go in the record.

Mr. ROBB. Mr. Chairman, I might say I also think that the Lansdale transcript should also be set up in the record at the same time. I don't think there is any need to read that, because counsel has had it and has read it. Mr. Lansdale testified about it on Friday.

Mr. GARRISON. I would like when we have time to read it into the record, because there are some comments, Mr. Chairman, that I would like to make about some passages in it as we go along. I think the transcript as a whole gives a rather fresh impression, and rather a different one of the whole interview. There are some things in it that are really quite worth a moment of thought as we go along. Not for the purpose of correction, but for the purpose of illustrating what I think took place.

Mr. ROBB. Mr. Chairman, I have some questions as to whether counsel should read a transcript and at the same time make an argument about it. It seems to me that the transcript ought to be before the board for such use as the Board wants to make of it. I assume that there will be an appropriate time at the close of these proceedings when counsel can make his argument.

Mr. GARRISON. All right. I withdraw that, Mr. Chairman. But I would like to have it read, because I think it is important for the board to hear it.

Mr. GRAY. The board has read it, I assume. You want to read it aloud?

Mr. GARRISON. Yes.

Mr. GRAY. If a request is made for that procedure, I think we will follow it so that it will at the appropriate time be read. I do not think we ought to interrupt at this point to read it.

Mr. ROBB. No, sir.

Mr. GRAY. Let us take a recess in any event.

Mr. GARRISON. Mr. Chairman, at the end of the recess, I think we would be prepared to have Dr. Bethe.

Mr. GRAY. All right.

(Brief recess.)

Mr. GRAY. Do you wish to testify under oath? You are not required to do so.

Mr. DEAN. I would be happy to, if that is the custom.

Mr. GRAY. All the witnesses have.

Mr. DEAN. I shall be glad to.

Mr. GRAY. Would you stand and raise your right hand. Gordon Dean, do you swear that the testimony you are to give the board shall be the truth, the whole truth and nothing but the truth, so help you God?

Mr. DEAN. I do.

Whereupon Gordon Dean was called as a witness, and having been first duly sworn, was examined and testified as follows:

Mr. GRAY. It is my duty, Mr. Dean, to say to you that in the event that it becomes necessary for you to discuss restricted data in your testimony, you should advise the chairman of the board of any such disclosure. We would appreciate your cooperation in that respect.

A further observation I should like to make to you is that the proceedings and record of this board are regarded by us as strictly confidential between the Commission and its officials and Dr. Oppenheimer and his representatives and associates, and that the Commission will take no initiative in the public release of any information relating to these proceedings. I think on behalf of the board, I express the hope that witnesses may take the same attitude about it.

I think perhaps for the record also that it is my duty, Mr. Dean, to remind you of the penalties under the perjury statutes. I should be glad to read a summary of those provisions, but I assume you are thoroughly familiar with them.

The WITNESS. I am familiar with them.

Mr. GRAY. Mr. Garrison.

DIRECT EXAMINATION

By Mr. GARRISON:

Q. Mr. Dean, you are a member of Lehman Brothers in New York?
A. I am.
Q. And you served on the Atomic Energy Commission from May 1949 to June 1953?
A. That is correct.
Q. And you were appointed Chairman, when was that, August 1950?
A. I think it was the latter part of August—no, the early part of August or the last part of July. I have forgotten the exact date. It was the summer of 1950.
Q. When did you first become acquainted with Dr. Oppenheimer?
A. I had never met Dr. Oppenheimer until I came to the Commission. I met him for the first time when I as a member of the Commission met with the General Advisory Committee of which he was then the Chairman.
Q. Could you give the board a general picture of the positive work of the General Advisory Committee during Dr. Oppenheimer's chairmanship, as you saw it. By positive, I mean what the GAC did to build up and strengthen the military position of the country.
A. I assume that some of this may be repetitious. The General Advisory Committee was established by law. The members were appointed by the President. They selected their own chairman. This was the way in which Dr.

Oppenheimer, having once been appointed by the President, was made chairman of that committee.

They used to meet about every month and a half to 2 months. I think the minimum requirement was four times a year, but they met much more frequently than that. They sometimes have special called meetings so that they would get together on occasions as much as perhaps 3 weeks apart if the occasion justified it.

They also worked through subcommittees of the General Advisory Committee. There was one on weapons. The General Advisory Committee is essentially a committee of senior scientific people. There were a few exceptions. There were from time to time outstanding businessmen on it. But primarily it is a senior scientific advisory group to the Commission, and so specified in the law.

They have been very active. They were every moment from the time I went on the Commission. It was a very important committee and contributed very much in guidance to the Commission on very difficult problems that we had, particularly scientific problems.

Q. What was the attitude of the committee under Dr. Oppenheimer's chairmanship with respect to the expansion of our atomic facilities?

A. In every case—and I might say this to give you just a little bit of history—the Atomic Energy Commission underwent a series of expansions of its facilities. By expansions, I mean this: The design, the construction, and the putting into operation of large reactors, such as those out at Hanford, to produce plutonium or tritium or other products. The expansion of the large gaseous diffusion plants which gives you your uranium 235. In other words, when you are talking about facilities, you are talking about facilities which give you * * * the plutonium and U-235 fissionable material.

All of these expansions were blessed by the General Advisory Committee. I know of no instance where there was an expansion program beginning with the summer of 1949 when we went into building a new gaseous diffusion plant at Oak Ridge, up until the latest big expansion of 1953, which was a $3 billion expansion program, I know of no instance when the expansion program was not thoroughly backed by the General Advisory Committee and heartily backed.

Q. Did they help to suggest and initiate expansion programs?

A. This I would almost have to go back and refer to the minutes of meetings to tell you where an expansion program initiates. It is very hard to put your finger on it. A need arises, and there are many huddles. Probably the records would show that some had originated with the GAC but on this I am not sure. We certainly consulted with them each time when we were thinking of an expansion program. They always blessed it.

Q. You spoke of the weapons subcommittee. Was Dr. Oppenheimer a member of that?

A. I think he was a member of the weapons subcommittee the entire time I was on the Commission. He was certainly very active in it, it was the most active committee of the GAC. I should say this so far as the GAC and weapons are concerned: I would think that at least 50 percent, and perhaps much more of its time was spent in the weapons field. There was far more interest on the part of GAC on the weapons program at Los Alamos and the production of fissionable materials than in any other phase.

Q. Do you recall a conversation with Dr. Oppenheimer in the spring of 1950 about a bucket of neutrons?

A. I do.

Q. Can you say something about the significance of that and of Dr. Oppenheimer's view about what ought to be done?

A. The reference to neutrons was really a suggestion. He spoke of it in the slang term—a bucket of neutrons. What he really meant was that what the Commission needed more than anything else were some reactors in which neutrons could be put to their best use. This was in a sense the idea behind the Savannah River design and the Savannah River reactors, which were dual purpose. I am not sure whether that is classified or not. Let us end it there. That was the reason behind the Savannah River reactors.

It was in the spring of 1950 that we were considering an expansion program which could carry us either into a strong A-program or a strong H-program, depending on what our research and development program showed.

Mr. GRAY. Did you say the spring of 1950?

The WITNESS. The spring of 1950. That is when we were getting together and wrapping up the kind of expansion program in order to take care of a stronger A and H program. This is when we first began to think of how we could build the Savannah River reactors. It was an entirely new design. That was put through Congress, as I recall, in the matter of about 90 days in the late spring and early summer of 1950.

Mr. GRAY. This is before you became a member of the Commission.

The WITNESS. It began to be discussed while I was a member, and then I had to present the program to the Congress in either the late summer—it could have been early fall of 1950.

By Mr. GARRISON:

Q. You became a member of the Commission in May 1949?
A. Yes; in May 1949.

Mr. GRAY. I beg your pardon. I had the years confused. You were on the Commission when all of this developed.

The WITNESS. Yes; I am not testifying to anything I did not see or experience myself.

By Mr. GARRISON:

Q. Dr. Oppenheimer was helpful in connection with this strengthening of the program you have devised?

A. Always. There was one big problem that we had and that was precisely what kind of design for the Savannah River reactors. * * * That went back and forth many times, but it was a question simply of the economics of buying neutrons, so to speak.

Q. There was a meeting in June 1951 at Princeton in connection with the H-bomb program?

A. There was. If I could give you a little history before we get to that June meeting, I would like to go back to the fall of 1949. I think it is necessary to have in the back of your mind before you talk about this June meeting in Princeton, in the fall of 1949, the Russians, we learned, this was September, had exploded their first A bomb. Dr. Oppenheimer, along with 2 or 3 other persons, were brought in here * * * and came up with the conclusion that there was no question but that the Russians had exploded an A bomb.

Then the question became one of having lost our monopoly, if we ever had it, what should we do to intensify the atomic energy program of this country.

Many things were suggested, including bringing in certain corporations with certain know-how, such as the duPont Co., which was done, and they did eventually build the Savannah River reactors.

Work on the thermonuclear weapon, many other things, I can't list them all, they can be found in a classified statement which I made before the Joint Committee on Atomic Energy in a closed session. If you have occasion to refer to that, I remember being asked the question, "What do you do now" and I listed about 8 or 10 things.

Mr. GRAY. What was the approximate date of that?

The WITNESS. That would be in the fall of 1949. This started quite a discussion inside the AEC as to what priority should be given to a thermonuclear weapon.

The only thing that we knew about in this field at that time was one method of approach, which unfortunately if it is to remain classified, I cannot describe, but I will try to do it in unclassified language.

There was one way of approaching the problem. Nobody had ever built such a gadget. Nobody had ever accumulated enough materials to actually fire a gadget of this kind, as it was then thought of. Nevertheless, there was a feeling on the part of some, including myself, that an effort to go into the thermonuclear or fusion field was something that we could not overlook.

Here was a new field. Here was a potential source of great energy. While we didn't know what the gadget might look like when we got through, certainly it should have a high priority in the shop. There were others who felt differently. This was a matter of much discussion. There were discussions at that time between the General Advisory Committee and the Atomic Energy Commission. Most of the General Advisory Committee, all of them, decided that we should not go ahead under a high priority in the thermonuclear field at that time.

The reasons as I recall them were several. There was, I think, in the background on the part of some what I would call a visceral reaction——

Mr. ROBB. Pardon me?

The WITNESS. Visceral, tummy—of going into a field such as this at this point, when these people had developed an A bomb. They had seen it used successfully. Our A bombs were getting stronger every year. Our stockpile was growing.

By Mr. GARRISON:
Q. Excuse me.
A. I am trying to describe the events of 1949 and relate them later to the June meeting.
Q. I think since we started on this fall of 1949, we better postpone the discussion of the Princeton meeting. I asked you about that only to give the general picture of the work of the GAC.
A. All right.
Q. I think it is best we continue now. Since you started on this, I think perhaps it is more appropriate anyway chronologically to take it. Are you now beginning to describe the attitudes of the members of the GAC at their October 1949 meeting?
A. I am as best I recall them.
Q. Then suppose we have it understood that you are now telling the board the general nature of what the GAC reported to the Commission. I would like to go just for a minute into the question of the scope of the report of the GAC to the AEC, and ask you whether in your opinion the GAC exceeded its statutory functions or just how you looked upon the role of the GAC as an adviser to the Commission.
A. The GAC used to be concerned sometimes that it was perhaps exceeding its strict statutory functions. This was never too important to me. I always felt that if we could get the wisdom of the people who were on the General Advisory Committee, we should have it. So what their statutory function as a committee was was not important to me.
In this instance, in the fall of 1949, it was not a question of anybody exceeding authority. The then Chairman of the Commission, Mr. Lilienthal, had asked the General Advisory Committee very specifically to review this question of whether we should attach a high priority to a thermonuclear or fusion program. They were asked this question. They were asked to consider it at their meeting which took place in October 1949.
They did consider it. I think they considered little else, I think for about 3 days, than this issue. They came in with their report to the effect that they felt it was a mistake.
The reasons that they gave I suppose appear in the minutes of the General Advisory Committee, but we had many discussions and those don't appear in the minutes.
The reasons were many. I said there was one, a visceral reaction at first. If I am not departing from the role of witness, I would like to give you my understanding of that reaction.
These were men who had developed the A bomb. Oppenheimer had the big hand in it, as you know. He also had a hand in the measures for the international control of atomic energy, and served on the board, and was a coauthor——
Q. By the board, you mean the Lilienthal panel?
A. The Lilienthal panel which later substantially was turned into the Baruch Plan in the U. N. They were hopeful at that time that you would not have the world in the position where you had two great powers simply stockpiling weapons and no solution to the problem. Consequently, after 2 or 3 years of rather frustrating dealings with the Russians, when this proposal of building another bigger one hit them, as some said, as the answer to our national security, I think it rather floored them and disgusted them. They lived through the A bomb. They tried to get international control. If this was the only answer to the problem, namely, of building bigger H bombs, this was not a satisfactory answer for those people. I think it was a stomach reaction along those lines.
I did not agree with it, but I think I can understand it.
Q. You are referring to those members of the GAC who were atomic scientists.
A. That is right, and specifically I would say to Oppenheimer and also to Fermi and others who sat on the board and Conant, because they had all been in the program. There were other reasons, however, beyond the tummy reasons for opposing it at that time. You don't decide to manufacture something that has never been invented. Nothing had been invented. No one had any idea what the cost of this thing would be in terms of plutonium bombs. As the debate or discussions waged in the fall of 1949, we had so little information that it was very difficult to know whether this was the wise thing to do—to go after a bomb that might cost us * * * plutonium bombs, and then after 2 or 3 years effort find that it didn't work. That was the kind of problem. So there were some economics in this thing.

There was another reason. This was how much of a diversion of Los Alamos—energies, scientific energies, could you safely divert to a project which might or might not succeed when the ball was rolling so beautifully in your A bomb program, and we were getting more bang out of our fissionable material, more weapons for the same amount of fissionable material.

Those were all considerations. There may have been others in there that I have overlooked, but those are the principal ones.

The unknown quantity was very much there. You don't build bombs by memoranda. We could write and discuss and interchange papers all night long and still we were in the dark on this thing.

Mr. Strauss and I at that time felt quite strongly we nevertheless should embark on this.

Q. This is after the GAC report?

A. This is after the GAC report. The GAC had another meeting shortly after the October meeting. I think they came together in a matter of 3 or 4 weeks, and as I recall they reiterated their stand of the October meeting.

The Commission realized—if I can turn from that now for the chronology—this was a decision which could not be and should not be made alone by the Atomic Energy Commission. It was something that had to be resolved eventually by the President. He should make it only after consulting with the Secretary of Defense and the Secretary of State. So instead of taking a vote, a vote as such as I recall was never taken on this issue—we did get together and try to write a paper for the President's guidance—we, as the Commission, in that we attempted to find as many things as we could agree on, premises that we believed to be true, and we wrote those down first. Then we wrote down what might be called a majority report and a minority report. Then we all added individual opinions. So the President could have everything before him.

The paper was given to the Secretary of State, the Secretary of Defense. They had a meeting and the Commission was ordered to go on a high priority thermonuclear research development, and this was done.

Once the President made the decision, I know of no instance where it could be said that the members of the General Advisory Committee, or any individual, opposed that program. I know of many instances where they helped it and at great pains.

This leads me, I think, into the June meeting.

Q. Perhaps just before we get there, there was a problem of recruitment of physicists.

A. There was a problem.

Q. After the President's go ahead order, was the GAC, specifically Dr. Oppenheimer, helpful in that respect?

A. The story did come to me once through Dr. Teller that he was fearful that he would not get much help out of Dr. Oppenheimer in this recruitment program. I said to Dr. Teller, "I think what you should do is go up and see Dr. Oppenheimer, and see if he cannot give you some help." So he did. He went to Princeton. My recollection is—I can't give you the exact date on this—I do recall his going to Princeton and I do recall Dr. Oppenheimer giving him a list of some 10 or 12 names at least of people he thought would be helpful in this program. Teller later advised me that these people were all either at Princeton or the advanced institute, and that he was not able to get any of them to leave. That is the story on recruitment.

I did several times in appearing before the General Advisory Committee in the summer of 1950 and the spring of 1951, the winter of 1950, ask them for names of people that we could get into the program from universities, from private industry and so forth, and some names were given to me. Some we were successful in getting; others we were not. I know of no instance, however, where anyone was discouraged from working on the program by Dr. Oppenheimer.

During the spring and summer of 1950, some rather striking developments came along in the A-bomb program. Remember our stockpile at that time was not as big as we would like to have had it. These developments were very big. I think the GAC went out to Los Alamos in the summer of 1950, the weapons committee, and worked with Dr. Bacher, who was then on leave from Cal Tech, and spending some time as a consultant at Los Alamos. Out of this summer's work and it is hard to credit to any one person, came some very significant developments which as I say* increased * * * our stockpile of A bombs. This was happening at the same time that the H program looked very discouraging.

---

*Supplied in declassifying to clarify deletion.

Some studies had been made by Dr. Ulam at Los Alamos and he ran some samplings which made it look as though an H bomb built along the lines that were talked about in the fall of 1949 just could not be done, or if done it would be at such a great cost in A bombs that you couldn't pay the price.

These things were happening. The H-bomb program looked bad. Every result was discouraging. The A-bomb program was improving. However, in the spring of 1951, we started a series of tests. By that I mean test explosions. We opened in a jury rig fashion on the Nevada proving ground. As I recall in that year we shot something like 14, 15, maybe 16 bombs altogether. Four at Eniwetok in the spring of 1951, and quite a few in Nevada. Some of these bore some relationship to a possible H program, and notably one shot which was fired in May of 1951 at Eniwetok, which I can't describe without using classified information.

After that explosion I thought it was high time that we got together all the people who had any kind of a view on H weapons. Of course, there were many views among the scientists. By views, I don't mean views as to whether you could have one, but views of whether you could have one and how you would get it.

I talked, as I recall, to 2 or 3 of the Commissioners and said wouldn't it be good if we could get them all around a table and make them all face each other and get the blackboard out and agree on some priorities.

We did do that. We asked Dr. Oppenheimer, as chairman of the Weapons Committee of the GAC, to preside at the meeting. We had at that meeting in Princeton in June of 1951 every person, I think, that could conceivably have made a contribution. People like Norris Bradbury, head of the Los Alamos laboratory, and 1 or 2 of his assistants, Dr. Nordheim, I believe, was there from Los Alamos very active in the H program. Johnny von Neumann from Princeton, one of the best weapons men in the world, Dr. Teller, Dr. Bethe, Dr. Fermi, Johnny Wheeler, all the top men from every laboratory, sat around this table and we went at it for 2 days.

Out of the meeting came something which Edward Teller brought into the meeting with his own head, which was an entirely new way of approaching a thermonuclear weapon. * * *

I would like to be able to describe that but it is one of the most sensitive things we have left in the atomic energy program * * *. It was just a theory at this point. Pictures were drawn on the board. Calculations were made, Dr. Bethe, Dr. Teller, Dr. Fermi participating the most in this. Oppy very actively as well.

At the end of those 2 days we were all convinced, everyone in the room, that at least we had something for the first time that looked feasible in the way of an idea. * * *

I remember leaving that meeting impressed with this fact, that everyone around that table without exception, and this included Dr. Oppenheimer, was enthusiastic now that you had something foreseeable. I remember going out and in 4 days making a commitment for a new plant. * * * We had no money in the budget to do it with and getting this thing started on the tracks, there was enthusiasm right through the program for the first time. The bickering was gone. The discussions were pretty well ended, and we were able within a matter of just about 1 year to have that gadget ready.

It had to be shipped to Eniwetok. We had to lay it on the task force and it was fired in November 1952.

Since then there have been many others fired out in the Pacific in this field.

That is the significance of the June meeting. It was the first time that all competent people in this program that could contribute anything sat around the same table and finally came up with something they all agreed on. That is when it began to roll and it rolled very fast then.

That is the chronology of it.

Q. Mr. Oppenheimer was the chairman of the meeting and presided?

A. He presided at the meeting and participated actively in the meeting and left the meeting enthusiastic. I recall talking with him afterward, and he was, I could say, almost thrilled that we had something here that looked as though it might work. * * * I might say, that the gadget which we originally thought of in 1949 probably never would work and would have cost in terms of A bombs a price we could never have paid.

Q. You remember the Crouch incident with which the board here is familiar?

A. The first recollection I have of that, I guess the only one——

Q. I am not asking you to recite what it was, because the board knows all about it.

A. Yes, I remember the Crouch incident. If you mean by that his testimony in California.
Q. Yes.
A. Yes.
Q. After that was brought to the attention of the Commission, did the Chairman ask you to go through Dr. Oppenheimer's personnel file and inquire into the whole question of his clearance?
A. I wonder if you could refresh my recollection on the date. Was this about the summer of 1950?
Q. It was in May of 1950, in the spring of 1950.
A. As I recall it, it was before I became chairman. I may have been acting chairman that day in the absence of the chairman. The Crouch incident was brought to my attention. I thought it was something that we ought to talk to Dr. Oppenheimer about.

I asked our general counsel, Mr. Volpe, to talk to Dr. Oppenheimer about this Crouch incident. I wanted it delicately done in the first place. I had no idea whether Crouch was telling the truth or not. He did, and reported back to me that he had gone into this at great length with Dr. Oppenheimer, and that no such meeting as Crouch had described, which was as I recall a kind of a meeting of a Communist cell to recite the party line, that was supposed to have taken place somewhere in Berkeley back in 1940 or so, no such meeting had ever taken place.

He said, "I won't say that I didn't meet Crouch at some cocktail party or something like that, because we had plenty of people around the place, but no such meeting as this, you can be sure."

"I never sat in on any Communist meeting or Communist cell meeting." This picture as I recall is a small group of 4 or 5 people had gone off in a room in a house and talked over the Communist Party line.
Q. Did you go through Dr. Oppenheimer's personnel file?
A. I did. This is the first occasion I ever had to look at Dr. Oppenheimer's personnel file. Ordinarily Commissioners don't go through the files of people unless there is some real reason. Here, however, was a person who was chairman of the committee; he had been cleared in 1947 by the Commission, and I for the first time picked it up and went through it personally myself.

I then asked Dr. Oppenheimer if he could come in and see me about this, and I personally asked him about the Crouch incident. He said substantially what I have said he said in reply to Mr. Volpe, and I believed him.
Q. Did you continue to read matters that went into his personnel file after this?
A. I told the security officer, I believe, or perhaps my secretary, that anything coming from the FBI concerning Dr. Oppenheimer I wanted to see, and file in my own mind at least.

Two or three did come in. Because here was a file with a lot of early association evidence, I thought he was too important a man for me to overlook him, and it was my responsibility as Chairman, also. So I did see, I am sure, every memorandum from the FBI. But there were only 2 or 3, and there was nothing particularly new in them, as I recall, from that point on.
Q. What was your belief as to Dr. Oppenheimer's loyalty after you had been through the file and had talked with him?
A. There was no question in my mind—I must say when I first looked at the file, I had doubts, largely growing out of these early associations—but there was never any doubt in my mind after I examined the file and based partly on my knowledge of Dr. Oppenheimer, which was very close, there was never any doubt as to his loyalty in my opinion. None. That decision had to be made one way or the other. It could not be half way. There were some very unpleasant early associations when you look at them in retrospect, but as far as his loyalty I was convinced of it, not that the file convinced me so much, but the fact that here was a man, one of the few men who can demonstrate his loyalty to his country by his performance. Most people illustrate their loyalty in negative terms. They did not see somebody. Here is a man who had an unusual record of performance. It is much broader than I have indicated so far.
Q. Would you state to the board your general impression of his character as well as his loyalty, his integrity and sense of discretion? How would you rate those qualities?
A. I would say that he is a very human man, a sensitive man, a very well educated man, a man of complete integrity in my association with him. And a very devoted man to his country, and certainly to the Commission. No question of these things in my mind.

Q. Would you say a word about Dr. Oppenheimer's interest in military defense in late 1952 and early 1953 in connection with operation Lincoln, for example? I don't want you to go into great detail.

A. I will just say a word about that because I was not particularly identified with project Lincoln. Dr. Oppenheimer had many advisory posts to the Secretary of State, Secretary of Defense, and advisor, if not a member, of the Committee on Atomic Energy of the Research and Development Board, and others, and participated in many studies. When he left—when his term had expired—as chairman of the GAC in the summer of 1952, he particularly turned his attention to defense measures against A bombs and spent a very large share of his time on such questions as the necessity for an adequate radar net, early warning radar system, on certain civilian defense measures, and on the importance of interception, and as always the importance of our capacity to deliver our bombs.

From the very beginning I recall this is one of Dr. Oppenheimer's great worries, that our Air Force would not stay up at the level of our bomb production, that some day we might find ourselves short of delivery. So he was concerned with all four of those things.

Q. You have sat on the Security Council since President Eisenhower's election?
A. Several times on special things.
Mr. GRAY. Would you repeat that?
Mr. GARRISON. I asked him if he had sat on the Security Council under the present administration.
Mr. GRAY. The National Security Council?
Mr. GARRISON. Yes.

The WITNESS. The Chairman is not a member of it. But as questions came up touching on atomic energy the Chairman of the AEC, which I was at the time, was invited over to participate. I guess there were four or five occasions, perhaps more, in the spring of last year when I did sit in on the National Security Council on atomic matters.

By Mr. GARRISON:

Q. Did Dr. Bush and Dr. Oppenheimer come before the Council when you were sitting on it?
A. They appeared one day, yes. They made a presentation, the nature of which I am sorry I am a little hazy on. I think it had mostly to do with what at that time was perhaps loosely called operation Candor, and with civilian defense and other defensive devices.

Q. In all of your contacts with Dr. Oppenheimer, has he ever underestimated the Russian threat in your opinion?
A. Never. From the very earliest times Oppenheimer has been worried very much about, first of all, the lack of reliability of the Russians. He showed some frustration in our inability in the early days to work out a system and he never underestimated the Russians. A lot of our people have, but this is one man who never did.

Q. Do you remember a discussion with Dr. Oppenheimer in the fall of 1950 about his chairmanship of the GAC?
A. Yes. This was after I was Chairman. Dr. Oppenheimer came to me one day—his term had to run until August of 1952, I think.

Q. As a member?
A. As a member. He was then Chairman. He said he knew that we had had quite a disagreement on the H-bomb program back in 1949 and whether it should have a high priority. He told me that he thought that this had perhaps hurt his effectiveness on the General Advisory Committee, and that he was prepared to get off if for one moment I thought that this effectiveness had been so hurt that he could not serve.

I thought about it for a few moments—in fact, I had thought about it before—and I told him that I thought that the General Advisory Committee would definitely lose, and so would the Commission, if we lost him from it at that time, and that I felt as one who had disagreed with him on the thermonuclear program that his effectiveness perhaps had been hurt in some quarters and some people's opinions, but not in mine. I would miss him very much if he left.

When 1952 came around, he had served his time and he said, "I have been on too long. I think newer heads should be brought into the program," and he said, "I hope you would not urge the President to reappoint me." So I sent a letter to the President saying that these three members, Conant, DuBridge, and Oppenheimer were leaving. I prepared a draft of the letter for the President to sign for each one of them thanking them for their services, and that was the end of Dr. Oppenheimer's term.

Q. Summing up your convictions about Dr. Oppenheimer, you have testified to his loyalty and to his integrity and character with full knowledge of what you told us about your reading of his personnel file. I take it, also, that it goes without saying that you have read the Commission's letter which initiated this proceeding?
A. The charges? Yes, I have.
Q. The Commission refers to them as items of derogatory information, and not as charges.
A. That is right. I read that letter.
Q. On the basis of that knowledge and your experience with him, in your opinion is he or is he not a security risk?
A. He is not a security risk in my opinion. If I had so considered him a security risk, I would have initiated such a hearing long, long ago. I think his usefulness has been impaired by all this. I don't know how much he can contribute further to his country, but I would hope we would get the maximum out of him. I am certain that he is devoted to his country and if given an opportunity to serve, will serve and effectively as always.
Mr. GARRISON. That is all, Mr. Chairman.

CROSS-EXAMINATION

By Mr. ROBB:
Q. Mr. Dean, Dr. Oppenheimer has testified before this board in substance that in 1943 he became aware of an attempt at Russian espionage against the atomic bomb project. He has further testified that when interviewed about this matter by intelligence officers of the United States Army, he told these officers a fabrication and tissue of lies.
He has also testified——
A. May I ask, are you quoting from some testimony?
Mr. GRAY. Just a minute, please.
Mr. GARRISON. Mr. Chairman, I want to object in the strongest terms to the form of the question which counsel has put. I think it is impossible to present to this witness the questions about the Chevalier incident without really thoroughly going into the whole case and incident in all its ramifications. I think the question gives an utterly false summation of what actually happened in the total Chevalier incident which is the only way that it can be looked at.
Mr. ROBB. Mr. Garrison can go into it if he wishes, I think I have the right to put the question to the witness in the form of an assumption, if not otherwise.
Mr. GRAY. I take it you are objecting to the question, Mr. Garrison?
Mr. Garrison. I am objecting to any question to this witness that tries to put to him the Chevalier incident without going into it in the kind of shape that the matter has come to this board. It involves the whole question of his relations with Chevalier, of his initiating the information about Eltenton, of the views of General Groves and Colonel Lansdale. This whole thing has a very long and complicated story. To say here to this witness as a fact that Dr. Oppenheimer did this and that in respect to the Chevalier incident seems to me most unfair.
Mr. ROBB. Mr. Chairman, there is not the slightest doubt that Mr. Oppenheimer did testify that he lied to Colonel Pash and Colonel Lansdale, not once, but many times, and that his statements——
Mr. GARRISON. Mr. Chairman——
Mr. ROBB. May I finish—and his statements to those officers constituted a fabrication and tissue of lies, and he knew when he was lying, he was impeding the investigation in progress. There is no question in the world that the record shows that.
Mr. GARRISON. Mr. Chairman, this whole business of the so-called lies over and over again was in fact nothing but one story. He told this story to Colonel Pash. He told part of it, that we have reference to here, to Colonel Lansdale. By breaking up the component parts of that story into separate questions, counsel in his cross examination made this appear as if one lie after another had been told.
It lies heavy on my conscience that I did not at that time object to the impression that was trying to be conveyed to this board of a whole series of lies when in fact there was one story which was told.
Mr. GRAY. Let me ask Mr. Garrison this question. Is it clear that the record shows that there was a fabrication?
Mr. GARRISON. Yes.
Mr. GRAY. I wonder if Mr. Robb can proceed from that point on his question in a way that it would not be objected to?

Mr. ROBB. I can't keep Mr. Garrison from objecting, Mr. Chairman. Just so we have no doubt about it, I will read from the record at page 488:

"Isn't it a fair statement today, Dr. Oppenheimer, that according to your testimony now you told not one lie to Colonel Pash, but a whole fabrication and tissue of lies?

"A. Right.

"Q. In great circumstantial detail, is that correct?

"A. Right."

I submit my question on the basis of that is perfectly fair.

The WITNESS. I don't know what the question is at this point.

Mr. ROBB. Of course you don't.

Mr. GARRISON. Mr. Chairman, it really does not convey at all what this was about. The question of whether Chevalier told 3 men or 1, whether Eltenton had a contact at the consulate or didn't, whether the consulate had some microfilm or didn't, all that was of an irrelevant character of what the security officer wanted to find out, which was Chevalier's name. The substance of this whole thing is that Dr. Oppenheimer did not for a long time, and he has regretted and has said so explicitly, revealed the name of Chevalier, which was what the security officers wanted. These incidental details about whether there were 3 men or 1 had nothing to do with the problem that the security officers were faced with. I think that is the question that counsel has put to Dr. Oppenheimer in that form was an unfair one which distorted the record, and I should have objected to it at that time.

Mr. GRAY. I would like to say, Mr. Garrison, that frankly the Chairman of the Board does not know what the question is, and I have heard the witness observe that he does not. I don't know what the question is. The argument to the Chairman by counsel in the presence of the witness pretty well established a background perhaps to which you are objecting to in the first place. There has been a discussion of this incident. I should like to ask if Mr. Robb will put his question, and I will give Mr. Garrison an opportunity to object to the question.

By Mr. ROBB:

Q. Mr. Dean, I am going to ask you to assume that Dr. Oppenheimer testified before this board that in 1943 he became aware of an attempt at Russian espionage against the atomic energy project, and assume that he further testified that when interviewed about this matter by intelligence officers of the United States Army, he told these officers a fabrication and tissue of lies, and assume that he further testified that when he told these lies, he knew that by telling them, he was impeding the investigation of Russian espionage.

Now, if Dr. Oppenheimer so testified in substance, would that cause you to change your opinion about him?

A. As a security risk, then, or a security risk today?

Q. Now.

A. None. There must have been some reason for it.

Mr. GARRISON. Mr. Chairman, I think the assumptions in his question amount to the same thing as putting to the witness a question as to something which is only a fraction of Dr. Oppenheimer's testimony. One would have to add to that and assume that he initiated the whole matter by bringing to the attention of the security officers that there was a man called Eltenton who ought to be watched because he had a contact and a way of transmitting information.

One would have to assume also that the contact was a colleague at the University of Dr. Oppenheimer's in whom he had complete personal confidence, and ultimately told the name of that friend of his, notwithstanding his belief in his innocence, to General Groves. All of that has also to be assumed because all of that is part of this thing we are talking about.

Mr. GRAY. I should like to ask in view of the answer of the witness whether it doesn't make any difference now.

The WITNESS. I am sorry I answered before you had the opportunity to object.

Mr. GRAY. I don't think as far as this witness is concerned the additional fact which then would bring on certain others in fairness in the record, for example, the disclosure of the name was under orders, and things of that sort, but I think all of that, Mr. Garrison, in view of the answer of the witness——

Mr. GARRISON. Mr. Chairman, I will yield on this point. I didn't actually hear the witness' answer.

Mr. GRAY. I would gather the witness' answer was favorable to Dr. Oppenheimer. It was so intended, was it not?

The WITNESS. Yes. My answer was, do you mean a security risk then or now. The questioner said "A security risk now", and I said none.

By Mr. ROBB:

Q. I believe you added he must have had a reason for it, is that right?
A. I don't know all the circumstances. When I say he must have had, I would say I would think there would be some reason for it, is a better way to put it.

Mr. GRAY. I don't know whether we could get into the question for the reason for it without going into the whole record.

The WITNESS. I frankly don't know the reasons.

By Mr. ROBB:

Q. Would you have thought he was a security risk at that time?

Mr. GARRISON. Mr. Chairman, that is a highly hypothetical question based on a complete lack of understanding what this is about. How can he possibly testify what his judgment was on an incomplete fragment of the record.

Mr. ROBB. I though the witness had some distinction in his mind. I thought it fair to ask him what it was.

Mr. GRAY. The witness indicated a distinction about his testimony, and has said that he would find it difficult to address himself to that question without knowing the circumstances, if I understood his testimony.

The WITNESS. That is it.

By Mr. ROBB:

Q. Now, Mr. Dean, you spoke of a conversation you had with Dr. Edward Teller concerning Dr. Oppenheimer.
A. Yes. In connection with recruitment?
Q. Yes.
A. Yes.
Q. Could you fix the date of that conversation?
A. I had difficulty in fixing the date of it. It would be some time in 1950 or 1951. That I am sure. I can't quite place it, though, because Dr. Teller was in and out of Los Alamos so many times during this period, back at the University of Chicago, out to California, back at Los Alamos, that I don't recall the exact times when he was trying to recruit. It may have been in 1951 at a time when he was trying to get support for a second laboratory. It may have been that late.
Q. Who was trying, sir?
A. Teller.
Q. Would you tell us the substance of that conversation? You mentioned it, but I don't think you told us very much about it.
A. That particular conversation is only one little piece in a long story of the second laboratory, and I had many with Teller.
Q. Did you have many with Dr. Teller in which Dr. Oppenheimer was mentioned?
A. I would not say many. His name probably came up in two or three conversations.
Q. Would you give us the substance of those conversations?
A. I wouldn't want to quote on these. I can give you the tenor or the setting for these conversations. That is about all I can do. Teller undoubtedly felt that Oppenheimer was wrong in his original decision on the thermonuclear program in 1949.
Q. You mean to oppose it?
A. To have voted against giving it that priority at that time. Teller was an optimist in this field and thought that things could be done. He was very active in recruiting. He told me that he thought he would not get much help out of Oppenheimer. He may even have intimated that Oppenheimer would discourage people from coming.
Q. Did he so intimate?
A. Yes, I think that is a fair statement.
Q. What did he say about that?
A. He said he feared that he might. I said the way to resolve that is to go up and see him.
Q. Did he say why he feared that?
A. No. If he did, I can't recall precisely why.
Q. Did you ask him?

A. No, because I knew the two personalities so well. Two men that had little different views on things and how to do things. I was anxious to keep Teller and I was anxious to get the most out of Oppenheimer. So I said, "Go up and ask Oppenheimer if he will give you some names." Oppenheimer, as I recall it, gave him a list of 10 or 12 names. Then Teller came back and reported that they were all people at Princeton, which would be normal to have most of the names at least picked from the place where he was teaching, and that he was unable to get any of them to come.

Q. You said at Princeton; you mean they were all working under Dr. Oppenheimer at Princeton?

A. Not necessarily. They were either at Princeton University or the Institute of Advanced Studies.

Q. If they were at the institute, they were under Dr. Oppenheimer.

A. Yes, that is right.

Q. And Teller reported back he could not get any of them to come?

A. That is right.

Q. Did he say what reasons they had given him for not coming?

A. No.

Q. Did he attribute their not coming to Dr. Oppenheimer's influence?

A. No, he did not. He left an inference that Oppenheimer might have been responsible, but he did not say so in so many words.

Q. You gathered that from what Teller stated?

A. Yes. But I also knew the difficulty of getting anybody at that time to go to work with Dr. Teller at a laboratory which had not been created, and which was completely unplanned, site unselected, the organization for which had not been outlined, and so forth.

Q. You mentioned a second laboratory.

A. Yes.

Q. That question came up, I believe, in the fall of 1951, did it not?

A. That is about the time.

Q. Subsequent to the Princeton meeting.

A. Yes.

Q. I believe you told us at the Princeton meeting everybody agreed that you had the right gadget to make thermonuclear.

A. Well, we hoped we did. It looked promising.

Q. In all events, Dr. Oppenheimer thought so?

A. That is right. Everyone around the table did.

Q. In the fall of 1951, Mr. Dean, the GAC recommended against the establishment of a second laboratory, didn't it?

A. I would have to refer to the minutes. I would caution you on this, if I may. When you refer to second laboratory, I think it is well to define the terms, because the second laboratory, so-called, had been mentioned to many people. To some it meant a possible second Los Alamos at a new site in an isolated spot with some 2,000 to 3,000 scientific people in the laboratory, and equipment which would be necessary, which means a capital investment of $110 million. That is what Los Alamos is. To some people it meant that kind of a lab. To other people, it meant a very small laboratory specializing in nuclear fission, low temperature and metallurgy, and to be rather hastily put together, perhaps in Colorado, a place somewhere near Denver and Boulder. To other people it meant an Air Force laboratory at Chicago, which would be turned into a thermonuclear lab. This had some support from the Air Force people.

To me—and we debated this at some length—it had to be if it was ever going to work a place that was already established if you were going to save time. It had to be a place where you had to have a man in there who commanded respect, that Teller would work for and work with, and be comfortable working with. There was only one place that I could finally fasten on that fitted this, and this was to work under Ernest Lawrence at an established place, that is, you had a Radiation Lab. You had another site which we were using for other purposes, some 30 miles away at Livermore and that is eventually what was done.

So when people speak of second labs, and the controversy concerning second labs, I think it is important that in each case to make them define their terms. They meant entirely different things, some of which in my opinion were wise and some of which were not wise.

Q. I understand that Mr. Murray, one of the Commissioners, and Dr. Teller, did present to the General Advisory Committee a proposal for a second laboratory in December 1951. Do you recall that?

A. Not specifically, but it is quite possible.

Q. In their memorandum which I believe was prepared by Dr. Teller, it was stated, "The very rapidity of recent progress"——

Mr. GARRISON. Mr. Chairman, may we see this?

Mr. ROBB. No, sir.

Mr. GARRISON. I am asking the chairman if we may have a copy of this document to see what is being read from and what the nature of it is.

Mr. GRAY. I will have to inquire as to the security.

Mr. ROLANDER. The document itself is classified. I think the portion he is reading may be read without disclosing security information.

Mr. GARRISON. I submit that the document be shown to Mr. Dean who is cleared for security information.

Mr. ROBB. Much of this may be obviated if Mr. Garrison would wait until I complete my question before interrupting me.

Mr. GRAY. I would suggest that Mr. Robb read his question and see if you feel that there is any difficulty about it, Mr. Garrison.

Mr. ROBB. Mr. Chairman, I might say in general that I had understood that this was not a court proceeding, and I was going along on that assumption. I think it is clear I have refrained from making the slightest objection to any of Mr. Garrison's questions or testimony or anything. I don't care what form Mr. Garrison puts his questions. I assume this is not a court proceeding. But if Mr. Garrison is going to stick on technicalities and turn this into a proceeding according to the strict rules of evidence, I think we ought to have it understood here and now.

Mr. GARRISON. Mr. Chairman, it certainly was our understanding that this was to be an inquiry and not a trial.

Mr. GRAY. That is correct.

Mr. GARRISON. I do most earnestly say to the Board that the only objections that I have raised with respect to cross examination which at times in this room has taken on the atmosphere of a prosecution than anything else, which I have not though perhaps I should have objected, the only questions I have raised have seemed to me to be of a rather basic character, where a scrap of a document has been read from without either the witness or ourselves knowing what was the content of it. I think if you will examine the Pash and Lansdale transcripts, as I know you will in full, you will find that the things taken out of context have been given not really a fair impression. This to me is rather elementary and not a technical matter.

Mr. GRAY. With respect to those documents, of course they will be in the record. The Chair does not know from what document Mr. Robb is reading or whether it can be made available. I repeat, I suggest that he read the question and if the witness finds it confusing or alien to him, he can so indicate. Then if you object to the question being put, I should like to hear from you.

Mr. GARRISON. All right.

The WITNESS. Could you tell me again Mr. Robb what is the memo purported to be dated, to and from?

By Mr. ROBB:

Q. It is a memorandum from Commissioner Murray and Dr. Teller to the GAC. What I am going to do is ask you if you recall this was the position those two gentlemen took. If you don't, that is the end of it. I am advised that they stated in their memorandum: "The very rapidity of recent progress is evidence of potentialities which have been neglected for years, and which will not be fully exploited unless a new laboratory is established."

Do you recall any such argument as that being made for a new laboratory?

A. There were many arguments along this line, and it is quite possible that some such thing was said. I am sure that the matter came before the GAC in one form or another. It is consistent that Tom Murray and Teller should be for a second laboratory of some kind because they both felt very strongly about it, as I did, as a matter of fact, but it was a question of where, when and who.

Q. The second lab which Teller and Murray were for was the second lab to work on the thermonuclear, is that right?

A. This is not clear. Perhaps the document may clear it up as to what kind of lab they are talking about at that point.

Q. What is your best recollection about it, sir?

A. You see, there are two kinds of labs you could have to work on the thermonuclear. One is an across the board laboratory such as Los Alamos, with all of its departments: a test division, physics division, a chemistry division, a metallurgical division, and all the other divisions which make an integrated laboratory. This is one way, and perhaps this is the best way to have a thermonuclear laboratory if you had the time, because so many of the problems that touch on fission bear on fusion. * * *

So theoretically if you had time and money and everything, you would build another Los Alamos.

We didn't feel we had that. I don't know whether this particular proposal was that broad or whether it was the kind of proposal that Teller and I talked about several times, which was simply a very specialized laboratory emphasizing low temperature work and some metallurgy. That is why I think you have to define your terms on this. If there is anything in the document which describes what type of lab at that point, then I can do it. But just a second lab, with all the labs we had at that time, was——

Q. In any event, Mr. Dean, did there come a time when the General Advisory Committee did take a position on the establishment of a second laboratory for whatever purpose?

A. Yes, I believe they did. I cannot recall at this moment, and I wish my recollection could be refreshed by some document, what the issue was before the GAC at that time. If it was an across the board, another Los Alamos, I am sure they said no to it, because I do recall many discussions saying who would you get to run it? Where would you recruit the men who knew about weapons, who were all at Los Alamos at that time except a few people in specialties at some of the universities, such as Ohio State, which had a very strong low temperature group and so forth. But virtually all the people that would contribute to this would be people who were working for us in the weapns lab with a few outsiders. So I am sure that the GAC at that time, however the issue was presented to them, concluded that just didn't make sense.

Q. Do you remember when that was?

A. I don't really. It could have been either the fall of 1951 or all the way through 1952, because it seems to me it was a matter——

Q. Did there come a time——

Mr. GARRISON. Could we have the date of that memorandum?

Mr. ROBB. This memorandum is a compilation of memoranda, but the particular memorandum I was referring to was prepared December 19, 1951.

By Mr. ROBB:

Q. Mr. Dean, did there come a time when you yourself as Chairman of the AEC wrote to the Joint Congressional Committee on Atomic Energy respecting the establishment of a second laboratory?

A. I am sure there must have been, but I don't have the documents.

Q. Do you recall when that might have been?

A. If you will give me some hint as to what I said in it, I could perhaps time it.

Q. Frankly I don't have your entire letter, but I will try to give you a hint.

A. All right.

Q. I don't know whether this is classified or not. Is it, Mr. Rolander?

A. I did write a letter on this whole thing on the second lab at one time. Just what must be done about it, and how to do it, and so forth. But it seems to me that was internal.

Q. I am told I can read this.

"January 9, 1952." Does that accord with your memory?

A. No, it doesn't, but go ahead. We will get the substance.

Q. "The creation of a dual laboratory such as Los Alamos would dilute scientific talent and introduce difficult problems of coordination. Further, because of the disassociation of talent and effort between two laboratories, the rate of progress would be reduced.

"We further feel that the division of talent between Los Alamos and a competing laboratory would at this time retard rather than accelerate the development program. Scientists of the caliber necessary to man and administer another laboratory for the prosecution of the development programs similar in scope to Los Alamos are limited in number."

Do you recall something about that?

A. Yes, that would be my view today.

Q. That was a fair statement of your position at that time?

A. Yes. If you are talking about another Los Alamos, and I think I kept saying such as Los Alamos all the time.

Q. Where had you received your information as to the availability of scientists necessary to man another laboratory?

A. This I had to live with every day.

Q. Had you obtained some of it from Dr. Oppenheimer?

A. I don't recall. We may have had conversations, but you can be sure that I also talked to all of the top scientists about this topic. This would be Von Neumann and members of the Commission like Smyth. It would be Dr. Rabi, who was quite helpful in recruiting a few people for us. It certainly would not be attributed to one man. This was something you had to keep on top of all the time.

Q. Do you remember subsequent to that, I think in February 1952, when the General Advisory Committee again recommended against the establishment of a second laboratory?

A. I don't recall that specific date, but it would not be inconsistent with what I know, that they took a position against another Los Alamos.

Q. Did the General Advisory Committee ever recommend in favor of a second laboratory of whatever kind, Los Alamos or anything else?

A. I reported eventually—I can't say this was my solution, undoubtedly some of the other Commissioners helped with it—my idea was, finally after this had all been worked out, that if you were going to have a place where you could put Teller and some of the other people to the best use, and if they were not at Los Alamos, the best place was under Dr. Ernest Lawrence out on the west coast. I did not quite know what the shape of this laboratory was going to be when it first started out.

Q. Excuse me, Mr. Dean. I asked you whether or not the GAC made any recommendation, not what you said. What the GAC did.

A. About the second lab?

Q. Yes.

A. If by the second lab you mean the Berkeley ——

Q. Any kind of lab.

A. They certainly did not frown on putting Teller out under Lawrence, and that is what the second lab as it turned out to be was.

Q. When did that take place?

A. This took place, I would say, about a year after the spring of 1951 shot. I fix that date because the men who went out from Lawrence's lab—I am sorry I have forgotten this man's name—Dr. York went out and did some tests in the spring of 1951 on the shot that dealt somewhat with thermonuclear processes. They came back, quite an equipped group of able young men. Here was a nucleus, and it happened to be in Lawrence's lab. You could put Teller in there. I recall we put this to the GAC and everybody felt that it was fine. We had found a place where Teller was happy and could work. I think this was the reaction.

Q. About when was that, do you remember?

A. I say I think this must have been about a year after the shot in the spring of 1951, which would put it somewhere over perhaps May of 1952.

Q. Where was that place that Teller worked—Livermore?

A. That is where he was put to work, yes, sir.

Q. Was there an establishment set up there?

A. There had been an establishment there before. That is another attractive thing about it. It had some buildings. We had some barracks. We had a big armory that we could use immediately. We got some instruments in very fast. It recruited very rapidly under Lawrence, York, Teller.

Q. Did you expand those facilities substantially?

A. Yes, they were expanded very fast.

Q. Did you spend a great deal of money on it?

A. We spent, I think perhaps—I could not give you the dollar figure—I imagine something like $11 million—No, I don't dare risk a figure.

Q. In other words, it became a very substantial establishment?

A. Yes.

Q. I believe you said you had no difficulty in recruiting for it.

A. We were recruiting an entirely different group at this point. We were recruiting men for that laboratory, I would say practically all of whom came immediately out of school. They were young Ph. D.'s and some not Ph. D.'s. We did not get in that laboratory any of the people, as I recall, that we originally thought of as being available for use on a thermonuclear project, like Seitz—oh, the names slip me. None of those people went to Berkeley. What they did was under Lawrence's administration, with Teller as the idea man, with York as the man who would pick up the ideas and a whole raft of young imaginative fellows you had a laboratory working entirely—entirely—on thermonuclear work.

Q. That laboratory was devoted entirely to thermonuclear?
A. Yes, but one thing that must not be forgotten is that—throughout the whole second lab thing—may I elaborate here on a background?
Q. Go ahead. I am not sticking to the rules of evidence.
A. One of the big problems of setting up a second lab and what kind of a lab was this, was always this: The morale of Los Alamos. Los Alamos, let us not forget this, is the laboratory which has been responsible for all of the research in the development of our A bombs, and all of the research and development until recently on the thermonuclear weapons. They have all come out of Los Alamos. I always feared and many others feared that if you made any drastic move which struck at Los Alamos morale, or if you inferred for one moment that they were not working their hearts out, because believe me, they were proving it with their results, you had a real problem in this weapons development field. That was one of the touchy things about setting up a new laboratory. Who was going to man it. What kind of a lab would it be, and what would it do to Los Alamos? That was the big fear I had.
Q. Mr. Dean, when you did set up this laboratory at Livermore——
A. It worked out very well.
Q. It didn't impair morale at Los Alamos?
A. No. Los Alamos rolled very fast as it always has rolled. I think it worked largely because of Teller getting along very well with Dr. Lawrence. These things are a question of human relations. They got along extremely well.
Q. In fact, Mr. Dean, wasn't there a healthy spirit of competition between Los Alamos and Livermore after you set up Livermore?
A. I don't know it produced any more weapons, but there probably was a good sense of competition.
Q. And you have already said you had no trouble of getting personnel.
A. Not of the type I described. These were not the specialists that we wanted to get in the early days for the real rush. Very few of those ever came. Those people never went to California later on.
Q. How long would you say that the discussion went on before you finally established that lab at Livermore? How long did this discussion about establishing a second laboratory, whatever you want to call it, go on—a year?
A. It could have been a year, yes. In the meantime, however, Los Alamos was doing the work and that is what we are testing today in the Pacific.
Q. Is that laboratory at Livermore independent of Los Alamos, or was it?
A. Yes, it is independent except that you have to have a very close liaison for test schedules and everything else.
Q. I understand that.
A. Very close relations. But it is independent of Los Alamos. It should be pointed out that the University of California is the contractor for Los Alamos, and the University of California is also the contractor to the Commission for the Livermore lab, but only in that sense are they related.
Mr. GRAY. Let me interrupt. How much longer do you think you will take?
Mr. ROBB. Probably 10 or 15 minutes, maybe more, depending on Mr. Dean.
(Discussion off the record.)

By Mr. ROBB:

Q. Mr. Dean, I have in my notes that you testified that all expansions of the atomic program were blessed by the GAC beginning in 1949. You were not thinking about the second lab in that connection, were you?
A. No. When I speak of expansion programs, I am speaking of the erection of facilities with which to make bombs, and that is either plutonium or U-235 for the most part.
Q. You mentioned Dr. Oppenheimer's attitude on project Lincoln. That was the project to discuss the defense of the continental United States, was it?
A. Yes; I believe that is a fair description. This was not done under the auspices of the Commission, and I was not close to Lincoln. I just read the report when it was finished.
Q. That was done under the auspices of the Army?
A. Somewhere in the Department of Defense. Which agency did it, I don't know. Perhaps Air Force.
Q. Did you learn what theory Dr. Oppenheimer espoused in that connection?
A. I would hate to be examined on the Lincoln report. I really don't have a good recollection of it.
Q. Do you recall—if you don't, you can say so, of course—do you recall whether or not he espoused what might be described as a Maginot line type of defense?

A. I don't know what Maginot line means.
Q. I mean a fixed defense.
A. No. I mean does it mean in terms of radar defense?
Q. No; a fixed defense as distinguished from a strong offensive striking force.
A. I can't say.
Q. You don't recall?
A. I just don't recall.
Q. You mentioned Dr. Oppenheimer's connection with the * * * detection program. Were you familiar with that?
A. Oh, yes.

*   *   *   *   *   *   *

Q. May I interpose here, Mr. Dean, I am reminded that, if you get into classified material, would you be good enough to indicate so that we can take the appropriate steps?
A. I won't consciously do it. Sometimes I have a question because these things change from day to day. Why don't we just call it Air Force?
Mr. ROBB. This has been gone into on direct examination, and I would like to ask some questions about it.
Mr. GARRISON. I was not under the impression I had.
Mr. ROBB. Yes. Didn't you say something about Dr. Oppenheimer's connection with the * * * detection program?
The WITNESS. In this one instance. I stated he was called back in the fall of 1949 to make an appraisal of the Russian bomb.

By Mr. ROBB:
Q. And you told something about what his appraisal was, didn't you?
A. That they had shot one.
Q. I want to ask same questions about that.
Mr. ROBB. I am told that these questions will involve confidential classified material.
Mr. GRAY. In that event we will have to excuse counsel and anyone else who is not cleared for the disclosure of classified material.
Mr. GARRISON. Is the classification officer clear that this has to be answered only off the record?
(Discussion off the record.)

By Mr. ROBB:
Q. Mr. Dean, do you recall that there were * * * methods of * * * detection which were discussed at that time?
Mr. BECKERLEY. I don't see how one can get into this without disclosing information presently considered by the Department of Defense as classified.
Mr. GRAY. I am sorry, Mr. Garrison; you will have to be excused, although Dr. Oppenheimer will remain.
Mr. GARRISON. I assume that this has relevance.
Mr. ROBB. I would not ask the question if I did not think so, Mr. Garrison.
(Counsel left the room.)
(Classified transcript deleted.)
(Counsel for Dr. Oppenheimer returned to the room.)
Mr. GRAY. I think counsel for Dr. Oppenheimer should know that in the judgment of the chairman of the board—I would ask the other board members to listen to this—nothing transpired of consequence in absence of counsel. Do you agree?
Dr. EVANS. I do.
Mr. MORGAN. I do.
Mr. ROBB. The witness said he didn't know much about that subejct.
Mr. GRAY. That is correct.

By Mr. ROBB:
Q. Mr. Dean, you testified somewhat about conversations you had with Edward Teller about Dr. Oppenheimer, did you not?
A. As to recruitment; I think so.
Q. I have the impression that Dr. Teller was inclined to be critical of Dr. Oppenheimer in those conversations; is that right?
A. The best I can recall is the inference he left from what he said that he feared that Oppenheimer would not be helpful in recruiting men for him.
Q. That is what I thought you said. Didn't you have any conversations with other nuclear physicists who were also critical of Dr. Oppenheimer because of Dr. Oppenheimer's attitude on the thermonuclear?

A. I can recall three people, and I would urge you very much to have them in front of you as witnesses and for the Board to call them. One would be Dr. Ken Pitzer—I say three, and I can't recall them—I would certainly say Teller. Those were about the only two.

Q. Alvarez?

A. I never heard Luis Alvarez speak——

Q. Lawrence?

A. No, No. I don't recall any conversations with Ernest Lawrence about Oppenheimer. There may have been some.

Q. You mentioned Dr. Pitzer. What was his conversation about Dr. Oppenheimer, and when did it take place?

A. It was not a conversation. It was some letters and a speech he made in which he voiced criticism of the GAC. I think he may have named Dr. Oppenheimer and Conant by name, but it was quite clear he meant these two gentlemen.

Q. What was the substance of that criticism?

A. It was general—I think the speech was given to refer to one document—before some teachers in the Long Beach schools shortly after Pitzer left the Atomic Energy Commission. The general criticism, although he was somewhat specific in certain particulars, was that the General Advisory Committee was not imaginative enough. I remember 2 or 3 things that he criticized. One was that they were very conservative. He may have implied criticism of the Commission as well. I have forgotten this. That is, in setting the standards for the reactors, that we required too much isolation. I seem to recall that he advocated a reactor that would blow up so we would find out what would happen, so we would finally know.

There were other items in the speech that were critical. I don't recall any of them in the weapons field, however.

Q. Critical of whom?

A. GAC.

Q. That would include Dr. Oppenheimer.

A. Yes. He undoubtedly meant Oppenheimer, as I read it.

Q. You said that he suggested that the Commission was requiring too much isolation on your reactors.

A. That is right. I remember specifically he mentioned the Wahluke slope, which is a large area of sagebrush on the other side of the Columbia River from the Hanford works. He thought we should have completely opened up the Wahluke slope to irrigation and therefore farming and therefore to people. We were far too cautious in having a safe area around reactors.

Q. Was his suggestion that by locating reactors so far out in the wilderness you were making it difficult to get personnel?

A. No.

Q. Then why did he object to putting reactors out in wilderness?

A. He thought the day was coming very fast when you ought to be thinking about central-station powerplants. If you are going to get in that region, you ought to have them down near the middle of town. If we took undue safety measures, we were far too cautious.

Q. Did any scientist ever suggest to you that while you were on the Commission that you were putting your laboratories and reactors too far away, way out in the woods, so you couldn't get people?

A. No; not to get people, but that we were just unduly safe and we required too much land. We had some 200,000 acres in Idaho and the same at Hanford.

Q. Mr. Dean, what I was trying to get it, why did it make any difference to a scientist if you had 10 acres, 10,000, or 10 million acres?

A. I think Pitzer at that time was arguing that we were far too cautious in our safety standards, far too cautious. That was the general gist of his criticism. His speech is available. He wrote in some other periodical. I can't place it.

Q. Now, Mr. Dean, you testified that you read what you described the file on Dr. Oppenheimer.

A. The then file.

Q. That was the Atomic Energy Commission's clearance file?

A. Yes. It was a collection of all FBI reports and things that came in.

Q. How many volumes was it?

A. I don't recall.

Q. One?

A. I have no idea. I have no idea whether it was 1 or 2 or 3.

Q. How long did it take you to read it?

A. I have no idea.

Q. An hour or 2 hours?
A. I just don't recall. I have no idea. I may have taken it home. I have forgotten.
Q. Did you read any files of the Manhattan Engineering District?
A. Yes. I recognized when I saw this list of derogatory information many things that were in that file.
Q. In that file?
A. Yes.
Q. Where did you get that file?
A. From our security officers, as I recall.
Q. It was in your file?
A. In the shop.
Q. You are sure about that?
A. Yes; it was.
Q. If I told you that file had been over at the FBI since 1946, would that change your answer?
A. It would not, because that is not the file I read. I read the file, if you are imputing to me or suggesting that I did not read a file on Dr. Oppenheimer, including the early derogatory, you are wrong.
Q. Of course you had a file. I am trying to find out which one you read.
A. This I couldn't tell you.
Q. You don't know whether you read the Manhattan Engineering District files or not?
A. I can't recall whether it was so labeled. It had material in it dealing with the earlier days.
Y. Yes; those were reports.
A. As I recall, FBI reports.
Q. Did you see any transscripts of question-and-answer interviews with Dr. Oppenheimer?
A. I have seen interviews with the FBI in that file.
Q. But those were not questions and answers, were they?
A. As I recall, they were summaries of testimony.
Q. You said you, I believe, talked to Dr. Oppenheimer about this Crouch matter?
A. Yes.
Q. And he told you he had never sat in on any Communist meeting?
A. There was a specific meeting that was referred to by Crouch.
Q. I know. I have a note here in quotes: "I never sat in on any Communist meeting." Did he tell you that?
A. I wouldn't be sure of the exact words. What he did tell me was that he never sat in any meeting such as this that Crouch had talked about.
Q. Did you ask him whether he had ever been to any Communist meeting?
A. I don't recall.
Q. Would that have been a natural question to ask him?
A. It might have been.
Q. But you don't recall whether you did or not?
A. I don't. I was dealing entirely with new evidence which came up in the Crouch episode.
Q. Yes.
A. That is all I interrogated him on.
Q. But you were undertaking to evaluate him as a security risk, weren't you?
A. I had to do that every day.
Q. As a part of that evaluation, wasn't it important to you to know if he had not attended the meeting described by Crouch, whether he might have attended the other similar meetings?
A. It was not particularly important to me to know what kind of meetings he attended in 1941. I had known the man 11 years before.
Q. As far as you were concerned, he might have attended a dozen Communist Party meetings in 1941?
A. I had no evidence from the file.
Q. But you said it was not important to you what meetings he attended in 1941; is that right?
A. It certainly was not important to me at that time because the only question in my mind was, Is Crouch telling the truth about a specific meeting?
Q. That is what I am getting at. If he had not attended the Crouch meeting, you were not concerned with how many other similar meetings he attended.
A. That was not the issue before me at that time.

Q. Would you answer my question?
A. All right.
Q. If he had not attended the Crouch meeting, you were not concerned with how many other Communist meetings he might have attended?
A. At that time, no; I was not.
Q. Mr. Dean, I will read you the question and answer. I have before me, Mr. Dean, which I received from a reliable source, and having been Washington——
A. There are both kinds, reliable and unreliable.
Q. You will know that the testimony at the executive sessions of Congress cannot be released without a vote of the committee, and so forth.
A. Yes.
Q. I have reason to believe that the following took place at a meeting of the Joint Committee on Atomic Energy on February 21, 1952. I will read you this and ask you if you remember it and if you have any comment to make on it.
Mr. GARRISON. Mr. Chairman, I make the same objection to reading from documents that can't be shown and looked at. He says he has reason to believe they represent what took place.
Mr. GRAY. I would ask counsel whether he can summarize what he understands to be the situation.
Mr. ROBB. I will try it that way.

By Mr. ROBB:

*   *   *   *   *   *   *

Mr. GRAY. Let me interrupt again. It is now after 1 o'clock.
Mr. ROBB. That is my last question. It is entirely up to Mr. Dean how long he talks, unless he suggests something in his answer that makes me ask another question.
The WITNESS. It depends on what you mean by inertia. It was known in the Los Alamos lab and long before the Los Alamos lab, perhaps 20 years before Los Alamos was created, and maybe much before that, that one way of getting terrific amounts of energy was through the fusion process. But no one foresaw a way to do it because you could not get the heat to fuse. There is a story I remember running into when I first went to the Commission about a scientist going to the mayor of Moscow, I believe, and saying, "If you will give me all the electrical energy that lights the city of Moscow, in one night, I will somehow concentrate this and bring about a fusion reaction." It had been written about. I don't know how he was going to do this, but that was the idea. It has been written about in popular form. The principle was well known.

But at the end of the war, as you know from history, Los Alamos just like everything else in our defense effort slowed down. The boys came home and literally the scientists went home . It was at a low state of morale. We were working on a A-bomb. There was no incentive particularly at this time to develop an H-bomb.

The incentive came, it seemed to me, for the first time—although some theoretical work was being done—you just don't make an H-bomb. You go out and measure the cross sections of various elements and combinations of elements. There is a lot of fundamental work that goes on before you even think of developing a gadget.

The incentive came in 1949 to develop the H-bomb. This came almost entirely from the A-bomb explosion by the Russians. What do you mean by inertia at Los Alamos? The A-bomb program was going. Should you divert your people to an H-bomb program at that point? Nobody else did anywhere in our Defense Establishment. We cut down our Navy, we put it in mothballs and all these things happened because the war was over. You don't get incentives out of a peacetime situation such as you had at the close of the war.

The Russians gave us an incentive to work on something, and we went to it. The delays or inertia are not realistic. They are not good words to describe what transpired. We never saw anything that really had a chance until the Princeton meeting in June of 1951. We never had a chance.

You will always get some scientists to say, "I think I know a way to do," and you always give him a chance to draw it out and spell it out and interrogate him. But all the competent people in our program, even those enthusiastic for H-bomb program, say, in the spring of 1950, were pretty blue people because it didn't look like you could do it unless you came up with a new idea.

The new idea came up in the Princeton meeting in June of 1950. I don't know what words you apply to situations like that that are complicated. Inertia probably is not a good word. There was no incentive to do it. There was every incentive to make your A stockpile strong.

Mr. ROBB. One further question is suggested, Mr. Chairman.

By Mr. ROBB:

Q. Mr. Dean, do you know anything about an order in December 1952, to move Dr. Oppenheimer's papers back to Washington?

A. December 1952? Yes, it was done on my orders.

Q. Why was that?

A. Because his job had terminated as chairman of the GAC. I thought that it was only proper that all the papers which Dr. Oppenheimer had in his capacity as chairman should be brought into Washington. As I recall, I sent up Mr. LaPlante and Mr. Roy Snapp, Mr. Rolander may have gone up, 1 or 2 people to inventory those papers and remove from the files the GAC things. We were actually thinking of another facility for the then chairman, Dr. Rabi, and we were exploring whether it should go to Columbia.

Q. When did Dr. Oppenheimer's job as chairman expire?

A. August 1952.

Q. He was at that time consultant to the Commission, was he not, in 19 December 1952?

A. In December 1952, he was a consultant. I don't know to what extent he consulted, but he was a cleared consultant.

Q. Did you move those GAC papers out or did you change your mind about it?

A. No, we moved out all those that dealt with that. We did not take things that he had to have as a consultant and he was a consultant at four or five places. None of this was done for security matters, but as a matter of good administration. The papers necessary for the Chairman of GAC we removed because we had a new Chairman.

Mr. ROBB. That is all.

Mr. GRAY. I have a couple of questions, Mr. Dean, if I may. I am sorry to hold you here.

The WITNESS. It is all right.

Mr. GRAY. I am confused—not altogether from your testimony—but I am generally confused about the instructions to the GAC for the October 29, 1949, meeting.

The WITNESS. I did not formulate those so my recollection is a little hazy.

Mr. GRAY. Were you then Chairman?

The WITNESS. I was not chairman. I was a member. I went on in May as a member, Mr. Lilienthal was then chairman. Then the meeting took place in October 1949.

Mr. GRAY. From what you say you can't clear up my confusion very much, if you don't know.

The WITNESS. I might be able to if I saw a document. Is there a document?

Mr. GRAY. There is a letter.

The WITNESS. Usually we write the GAC, Mr. Chairman, in advance of a meeting and we say, "We would like to have your views on so many topics," and it is usually about a page or two page letter. That is customary. I would think that was done in connection with the October meeting.

Mr. GRAY. There was a letter signed by an Acting Chairman at one point. I don't recall whether he is to be a witness here or not, Mr. Pike.

Mr. GARRISON. I believe so, Mr. Chairman. I have not heard from him.

Mr. GRAY. Actually you were not in charge at that time.

The WITNESS. No, I was not.

Mr. GRAY. You said following the Princeton meeting, the members of the GAC, or some of them, if this is a correct quotation, went to great pains to help out in the H-bomb program. Do you remember saying that?

The WITNESS. Yes, they were certainly enthusiastic. It would be hard for me to explain precisely what they did.

Mr. GRAY. The great pains is adequate for me. I am trying to identify people rather than effort. Was Dr. Oppenheimer one of those who went to great pains in your judgment?

The WITNESS. He certainly expressed enthusiasm.

Mr. GRAY. I will put my question this way: Did you mean to include him in that earlier statement?

The WITNESS. Yes, definitely. I never saw a meeting and we had many with more unanimity.

Mr. GRAY. This is not a meeting. This is following the meeting. You said in your direct testimony that many of the GAC members present at that meeting helped at great pains with the program and you intended to include Dr Oppenheimer?

The WITNESS. I did, yes.

Mr. GRAY. This is a change of pace. Do you recall who the security officer of the Atomic Energy Commission was at the time you examined the files, whatever they were, with respect to Dr. Oppenheimer?

The WITNESS. We had a hiatus in thereabout that time which would be Admiral Bingrich. I am not sure when I had this conversation with Dr. Oppenheimer on the Crouch matter that Captain Waters had taken office or not.

Mr. GRAY. I had in mind the time that you examined the files.

The WITNESS. That is the one I have reference to.

Mr. GRAY. I suppose the record will reflect.

The WITNESS. I just don't recall. The record will reflect who was the security officer.

Mr. GRAY. You made the observation that Dr. Oppenheimer was cleared in 1947 by the Atomic Energy Commission.

The WITNESS. Yes.

Mr. GRAY. I am a little confused as to any direct action on that point, although the record may show something to the contrary.

The WITNESS. The only reason I can say anything about it is that I had occasion to look it up once to make sure about the clearance.

Mr. GRAY. It is your recollection——

The WITNESS. I was not there, but I remember something in the files showing that there had been Commission action in 1947.

Mr. GRAY. That is what I was confused on as to specific Commission action.

Mr. GARRISON. Mr. Chairman, I have a letter from the General Manager about that, stating that there was official action. I should like to introduce it in the record after lunch.

Mr. ROBB. I think you already did.

Mr. GARRISON. I guess I did. I think you are right.

Mr. SILVERMAN. There is a stipulation on the first day, Mr. Chairman.

Mr. GRAY. Mr. Dean, we had an earlier witness before the Commission who testified that never once in his long service in the Government did he have any drafts of communications prepared for him, and he never wrote a letter which he had not personally written and so forth. You said that you drafted a letter for the President to send to Dr. Oppenheimer. Did he use the draft?

The WITNESS. He did.

Mr. GRAY. You were asked the question whether you felt that Dr. Oppenheimer today is a security risk, and your answer was clearly in the negative. There seemed to be no question.

The WITNESS. That is correct.

Mr. GRAY. That is against the framework of the Atomic Energy Commission Act of 1946, the terms of which you are familiar with?

The WITNESS. Yes.

Mr. GRAY. One final question and I am through. At one point you said that there was criticism of the GAC by some scientists and you named one or two.

The WITNESS. Dr. Pitzer, I think.

Mr. GRAY. In that connection you said of course that meant Dr. Oppenheimer. Did you mean to say that meant Dr. Oppenheimer? Did you mean to say it included him?

The WITNESS. I meant it included Dr. Oppenheimer.

Mr. GRAY. I would like to be clear on this point.

The WITNESS. Yes; I am sorry I left that impression. I know it certainly included Dr. Oppenheimer. It included Dr. Conant and it may have included all of the GAO. It would certainly have included Dr. Oppenheimer. I don't have the document, but that is certainly the impression I had at the time it came out.

Dr. EVANS. Mr. Dean, I am somewhat confused, and I am asking for information to clear my own mind, as to what we are doing here. I have been on a number of these committees, and we had certain qualifications to go by, loyalty, association and character. Let us take loyalty. Just what does that mean?

The WITNESS. To me?

Mr. EVANS. Yes. Does that mean loyalty to your friend, loyalty to your country, or both?

The WITNESS. No. When I use the term "loyalty" and when I testified that I believed Dr. Oppenheimer without equivocation was loyal, I meant loyal to his country, that he has given very strongly of his time and energies, that he has so far as I have been aware always attempted to come up with the answer to any trouble, and there have been hundreds which were presented to him which were strengthening his country, rather than weakening his country. That is all I mean by loyalty.

Dr. EVANS. Associations; do we have to go by that?

The WITNESS. I think associations definitely must be weighed in any of these things. It is a question of the weight that is attached.

Dr. EVANS. I am just asking for information.

The WITNESS. Yes.

Dr. EVANS. It becomes so fogged up in my own mind I don't quite know what I am doing.

The WITNESS. If you wish my opinion on this, I think associations have to be weighed. I think they have to be weighed, however, very carefully in the light of the circumstances and the time in which the association was made. I am not sure had I first seen Dr. Oppenheimer as a young man in the late twenties and early thirties, and met him in the atmosphere of Berkeley in 1939 and 1940, whether I could clear Dr. Oppenheimer. I feel entirely different about him having watched him closely over a period of 4 years, and having evaluated quite carefully his service to his country, and certainly to the Atomic Energy Commission. I think the associations must be weighed in those circumstances.

Dr. EVANS. Of course, all scientific people know the value of a man like Dr. Oppenheimer. I am just trying to get these things cleared up so that I can act like an intelligent individual. If a man would be more loyal to his friends than he would be to his country, I don't know where I am.

The WITNESS. I think Dr. Oppenheimer, if I can volunteer this, the loyalty of which I speak is loyalty to his country. I think that is uppermost. I might even say he might be more loyal to his country than his friends. I am sure he would be because he ranks it higher.

Dr. EVANS. I have no more questions.

Mr. GARRISON. Mr. Chairman, I realize how late the hour is. Would you indulge me in two questions and that will be all?

Dr. GRAY. Yes.

REDIRECT EXAMINATION

By Mr. GARRISON:

Q. I think you said in interviewing Dr. Oppenheimer about the Crouch incident you had no concern—this was a question put to you by counsel about inquiring into other possible meetings of the past that he might have it. Did you mean by that that you didn't care about past associations?

A. No. I didn't mean to leave that impression. The impression I meant to leave was this, that I had no occasion at this time to evaluate anything particularly except this current piece of information. I did take occasion, however, to go back on the file which had already been in the Commission and on which he had been cleared in 1947, and went through it, and I saw this association evidence.

Q. You were asked whether you knew well the two personalities of Dr. Teller and Dr. Oppenheimer. This question was asked in connection with Dr. Teller's attempt to recruit men for the project. You have already talked about Dr. Oppenheimer's personality. Would you give us your impression of Dr. Teller's personality, particularly with reference to the problem of recruitment?

A. Dr. Teller is a very, very able man. He is a genius. There is no question about it. He has contributed much in the way of ideas to our weapons development. He is a very good friend of mine, and I admire him. He is a very difficult man to work with, as sometimes happens. Dr. Teller did not work well at Los Alamos, and left there on two occasions. I was responsible on both occasions for getting him to go back. I was finally responsible, I think in part, for finding a haven for Dr. Teller, because we needed him. But you can't break up a whole Los Alamos laboratory for one man, no matter how good he is, and that was a problem. I don't want to be too derogatory in my appraisal of Teller, because I could not sing his praises enough as to his contributions. He is not an administrator. I am sure when he went out to recruit there are people saying, "All right, Edward, we will work with you some place, but somebody has to run the show. Somebody has to pick up the papers and take the administrative load." Teller is not that type of person. So I can conceive that Edward would have great difficulty in recruting people. I think if you will call upon some of the people from Los Alamos, they will give you the same impression. It is not that they don't like Edward. It is not a question of likes or dislikes. It is a question of his personality. You have to find a peculiar environment in which he does his best work.

RE-CROSS-EXAMINATION

By Mr. ROBB:

Q. Was one reason why Dr. Teller left Los Alamos that he felt that not enough emphasis was being given to the thermonuclear?
A. I suppose that might be said to be true. He certainly differed with the head of the laboratory, Dr. Bradbury, as to the organization of a thermonuclear setup. I tended to think that Bradbury's judgment on the organization of which he was head was better than Teller's so far as organization went. Yet we wanted Teller in the program somewhere. That was a problem.
Q. Dr. Teller did think that not enough emphasis was being given to the thermonuclear, didn't he?
A. Yes. Dr. Teller has thought of some other things which were not true.
Q. Yes.
A. Such as the type of weapon which was thought of in 1949 was a good weapon. We know today it would never have been a good weapon. I admire him for his enthusiasm and optimism and pushing the frontiers of knowledge in order to get some kind of gadget, but I am glad we didn't go after that particular weapon.
Q. Now, Mr. Dean, the answer to my question was "yes," wasn't it?
A. State it again.
Q. Dr. Teller felt that not enough emphasis was being given to the thermonuclear?
A. Oh, I suppose that is true; not his kind of emphasis.
Mr. ROBB. That is all. Thank you.
Mr. GRAY. We will recess now, gentlemen.
I would like to thank Mr. Dean for coming before us.
Mr. DEAN. It's a pleasure.
Mr. GRAY. We will reconvene at 2:30.
(Thereupon, at 1:25 p. m., a recess was taken until 2:30 p. m., the same day.)

AFTERNOON SESSION

Mr. MARKS. Mr. Garrison has asked me to say that he had told this panel at one of the hearings last week, which I did not attend, what my part was in this case. Subsequently he consulted with me again, and told me that the conduct of the proceedings was turning out to be a very much more massive undertaking than he had planned for and asked me whether I would help more actively, and I agreed to. He has asked me this afternoon to carry on until he can rejoin the proceedings. He is now taking care of some other business connected with the case.
Mr. GRAY. The next witness is Hans Bethe. Do you wish to testify under oath, Dr. Bethe?
Dr. BETHE. Yes; I do.
Mr. GRAY. You are not required to, but all the witnesses have. Would you be good enough to stand and raise your right hand, please?
Hans Bethe, do you swear that the testimony you are to give the board shall be the truth, the whole truth and nothing but the truth, so help you God?
Dr. BETHE. I do.
Whereupon, Hans Bethe was called as a witness, and having been first duly sworn, was examined and testified as follows:
Mr. GRAY. Will you be seated, sir.
It is my duty to remind you of the penalties provided by the United States Code, by the statute known as the perjury statute. I should be glad to read those, if you think it necessary, but I gather you are familiar with them?
The WITNESS. I don't think it would be necessary.
Mr. GRAY. Second, I should like to request that if in the course of your testimony you should disclose any restricted data, I would ask you to notify me in advance, so that we might take appropriate steps if the circumstances require.
I was about to say we have a classification officer present, but we don't, so I would ask you therefore to be careful in that respect.
Finally, I should say to you that the proceedings and record of this board are regarded as confidential between the Commission and its officials, and Dr. Oppenheimer and his representatives and witnesses. The Atomic Energy Commission will not take the initiative in any public releases relating to these proceedings, and on behalf of the board, I express the hope that witnesses will take the same view of the matter.
Would you proceed now, please.

DIRECT EXAMINATION

By Mr. MARKS:

Q. Dr. Bether, will you please identify yourself and give a little account of your professional background?

A. I am a professor of physics. I have been a professor at Cornell University since 1935. I have been at Cornell all the time except during the war years when I was absent on war work, including a prolonged stay at Los Alamos Scientific Laboratory.

I am also this year the president of the American Physical Society.

Q. You are a member also of the National Academy of Science?

A. I am.

Q. Are you an American citizen?

A. Yes.

Q. By naturalization?

A. Yes.

Q. When did you come to this country?

A. In 1935.

Q. And where did you come from?

A. I came originally from Germany. I left Germany in 1933 because of the Nazi persecutions when I knew that I could not hold office under the Nazi regime. I first went to England and then came to this country.

Q. Since the war years have you had an connection with the atomic energy program?

A. Yes; I have. I have been a consultant to several laboratories of the Atomic Energy Commission. I have spent most of my consultations for the Los Alamos Laboratory. I have almost regularly spent summers at Los Alamos since 1949, I believe. I have taken off a whole semester in the spring of 1952 to help the Los Alamos work.

Q. How would you describe your role, as that of a consultant?

A. I am a consultant on matters of theoretical physics. I believe I am one of the chief consultants in theoretical physics to Los Alamos.

Q. What was your first acquaintance with Dr. Oppenheimer?

A. I first met Dr. Oppenheimer very briefly during a meeting of the German Physical Society at a regional section of it in 1929.

Q. When was your next connection with him?

A. The next that I remember was in 1940 on the occasion of a meeting of the American Physical Society at Seattle, Wash.

Q. What have been your associations or contacts with him since that time?

A. I have seen him quite frequently, especially we had a very deep association during the Los Alamos time, during the war, when he was the director of the laboratory, and I was the leader of the theoretical division of Los Alamos.

Q. How often have you seen him since the war?

A. I would say an average of perhaps three times a year, some years more, some years less.

Q. Have these contacts since the war had anything to do with your official connections with the atomic energy program?

A. Yes, many of the contacts had. Some were purely on matters of physics outside the atomic energy program, but many of our contacts have been connected with the atomic energy program, I in my capacity as consultant to Los Alamos, and he in his capacity as chairman of the General Advisory Committee, not that these contacts were also formal in a meeting of the Advisory Committee, but we often talked about these matters.

Q. Dr. Bethe, have you read the letter of General Nichols and Dr. Oppenheimer's reply?

A. Yes, I have.

Q. How far back does your own familiarity with Dr. Oppenheimer's political associations and activities go?

A. I——

Q. And what do you know about them?

A. I heard about his political inclination in 1938 from some good friends of ours, Dr. Weisskopf and Dr. Placzek, who is mentioned in Dr. Oppenheimer's answer letter, and I understood from them that he was inclined rather far to the left.

Q. Coming to the work on the atomic bomb, would you tell us briefly about the part that you and he played in the work on this subject before Los Alamos was formed and then subsequently during the Los Alamos days?

A. Our association began in 1942, on this matter. Dr. Oppenheimer called together a group of theoretical physicists, to discuss the way how an atomic bomb could be assembled. This was a small group of about seven people or so. We met in Berkeley for the summer of 1942. We first thought it would be a very simple thing to figure out this problem and we soon saw how wrong we were.

Q. What about Los Alamos? When did you join the Los Alamos group?

A. Between that time and Los Alamos, the first was the time when Los Alamos was being created. It was a very hard task to create this laboratory. Most scientists were already involved in war work very deeply and it required somebody of very great enthusiasm to persuade them to leave their jobs and to join the new enterprise of Los Alamos. I think nobody else could have done this than Dr. Oppenheimer. He was successful in getting together a group of really outstanding people.

At Los Alamos, as I mentioned before, we had very close relations because I was the leader of one of the divisions, one I believe of seven divisions. We met almost daily, certainly at least once a week.

In Los Alamos again I want to say how difficult a job it was and it seems to me that no enterprise quite as hard as this had ever been attempted before. I believe that Oppenheimer had absolutely unique qualifications for this job and that the success is due mostly to him and mostly to his leadership in the project.

Q. What were some of the factors that made it so difficult?

A. There were many. One was in the technical work itself.

Q. I simply wanted to indicate the nature of the difficulty.

A. It was that all the time new difficulties came up in different connections, new technical difficulties which had to be solved.

Q. Apart from technical difficulties.

A. Apart from that, one great difficulty was that scientists are great individualists, and many of the scientists there had very different ideas how to proceed. We needed a unifying force and this unification could only be done by a man who really understood everything and was recognized by everybody as superior in judgment and superior in knowledge to all of us. This was our director. It was also a matter of character, of devotion to the job, of the will to succeed. It was a matter of judgment of selecting the right one among many different approaches. It was a matter of keeping people satisfied that they had a part in the laboratory, and we all had the feeling that we had a part in the running of the laboratory, and that at the same time at the head of the lamoratory somebody who understood more than we did.

Q. Was there any notable exceptions to this?

A. There were a few notable exceptions. There were people who were dissatisfied. Among them was Dr. Teller.

Q. Why was he dissatisfied?

A. He had——

Q. By the way, am I right that he was on your staff?

A. He was on my staff. I relied—and I hoped to rely very heavily on him to help our work in theoretical physics. It turned out that he did not want to cooperate. He did not want to work on the agreed line of research that everybody else in the laboratory had agreed to as the fruitful line. He always suggested new things, new deviations. He did not do the work which he and his group was supposed to do in the framework of the theoretical division. So that in the end there was no choice but to relieve him of any work in the general line of the development of Los Alamos, and to permit him to pursue his own ideas entirely unrelated to the World War II work with his own group outside of the theoretical division.

This was quite a blow to us because there were very few qualified men who could carry on that work.

Q. Turning to another subject, Dr. Bethe, what was the attitude of Dr. Oppenheimer with respect to the requirements of security at Los Alamos?

A. He was very security minded compared to practically all the scientists. He occupied a position very much intermediate between the Army and the scientists. The scientists generally were used to free discussion and free discussion of course was allowed in the laboratory completely and this was one of the reasons for putting it at the remote place. However, many of us did not see sometimes the need for the strictness of the requirements and Dr. Oppenheimer was, I think, considerably more ready to see this need and to enforce security rules.

Q. Is that what you mean by occupying a position intermediate between the scientists and the Army.
A. That is what I mean.
Q. Let me ask you, Dr. Bethe, if you can speak of it, what views did the scientists have about the moral or humane problems that many people have discerned in the atomic bomb program at Los Alamos.
A. I am unhappy to admit that during the war—at least—I did not pay much attention to this. We had a job to do and a very hard one. The first thing we wanted to do was to get the job done. It seemed to us most important to contribute to victory in the way we could. Only when our labors were finally completed when the bomb dropped on Japan, only then or a little bit before then maybe, did we start thinking about the moral implications.
Q. What did you think about that or what did the scientists generally think about it?
A. There was a general belief that this was a tremendous weapon that we had brought 'into the world and that we might have been responsible for incredible destruction in the future. That we had to do whatever we could to tell people, especially the people of the United States, what an atomic bomb meant, and that we should try as much as possible to urge an international agreement on atomic weapons in order to eliminate them as weapons from war if this could be agreed to by all the major nations.
Q. I would like to come back to that subject, Dr. Bethe, but first let me ask you whether you were familiar at the time—that is, at the close of the war—with the problems that were posed by the so-called May-Johnson bill for domestic control of atomic energy?
A. I was, yes.
Q. Was that bill a subject of interest and discussion at Los Alamos, and if so, in what terms?
A. It was to a considerable extent, although not as much as some other laboratories of the Manhattan District. Most of the scientists at Los Alamos were opposed to the May-Johnson bill.
Q. Why?
A. It perpetuated Army control which we had felt was rather irksome and work was perhaps not conducive to the best results in research during peacetime. It included a lot of very severe and unprecedented stipulations as to punishments for almost any move a scientist might make. Finally, it seemed to us that it made it very much harder than necessary to achieve international control, which seemed to us the most important aim.
Q. Do you know what position Dr. Oppenheimer took on this subject?
A. Yes. Dr. Oppenheimer supported the May-Johnson bill, and he was very much attacked for this by some of his colleagues. I personally did not feel very strongly, by the way. He supported the May-Johnson bill because he thought that this was the only way to preserve the laboratories as running units to continue the work for the time being, rather than to have an interim during which the laboratories might disintegrate.
Q. Dr. Bethe, I would like to return now to this subject of international control of atomic energy which you mentioned. Did you observe as time went on, that is, from the close of the war during the next couple of years, any change in attitudes on the part of scientists and on the part of Dr. Oppenheimer on this subject?
A. Yes; definitely so.
Q. Would you speak of that?
A. Dr. Oppenheimer was one of the members of the Lilienthal board which worked out the American plan for international control.
Q. What date was that?
A. That was in the spring of 1946. I can't put it very much closer. In the early spring of 1946. Then he was an advisor to Mr. Baruch who was the American representative to the United Nations. At all these times he put a great effort into working out a plan which would give this country some measure of security from future atomic war.
However, the actual negotiations started in the United Nations Atomic Energy Commission and it was soon evident——
Q. That would still be in 1946?
A. That was still in 1946. It started in June 1946, I think. It was soon evident, at least to Dr. Oppenheimer, that the Russian attitude was very inflexible.

Q. How do you know that, Dr. Bethe? It was soon evident, you say.
A. I have known it as a fact only as of January 1947.
Q. What happened then?
A. In January 1947, I made a visit to Berkeley to give some lectures, and Dr. Oppenheimer and I had some conversations—quite long conversations—about the fate of the atomic energy control plan. He told me then that he had given up all hope that the Russians would agree to a plan which would give security and in particular——
Q. Security to whom?
A. To all of us. To us, I suppose, as well as to them. Particularly he pointed out how much the Russian plan was designed to serve the Russian interests and no other interests, namely, to deprive us immediately of the one weapon which would stop the Russians from going into Western Europe; if they so chose, and not give us any guarantee on the other hand that there would really be a control of atomic energy, not give us any guarantee that we would be safe from Russian atomic attack at some later time.
I have heard him talk about this subject quite often, the first time in January of 1947.
Q. What were your own views at that time?
A. I had not seen things very clearly. I still had considerable hope that international agreement could be achieved, and I know now that I was quite wrong. In fact, I saw right then that I was quite wrong. I was quite pessimistic at that time, but I thought this was such an important subject that the Russians would finally have to see that it was in their interest, as well as ours, to have a real control plan with some teeth in it.
Q. Did your own views change?
A. My own views changed, and I think perhaps partly influenced by the discussion I had with Dr. Oppenheimer. I certainly thought that there was not much hope and I certainly agreed that the Russian plan was all that Dr. Oppenheimer had represented it to be.
Q. Dr. Bethe, let me go back for a moment. I think you said that you had been told in the late thirties that Dr. Oppenheimer's, I think you used the phrase "extreme" left wing political views. That was between the time when you first met him in 1929 and your later closeness to him?
A. Yes.
Q. When you again met Dr. Oppenheimer, after this brief meeting that you described in 1929, what were your own observations about his political orientation?
A. They were very surprising to me.
Q. When would this have been?
A. That was in 1940. At the Physical Society meeting in Seattle, Wash., we had a long evening in which political matters were discussed. This was in late June, I believe, of 1940. It was just after the fall of France, and I felt very deeply that a great catastrophe had happened to the world. At this conversation, Dr. Oppenheimer talked for quite a long time in this same sense.
(Mr. Garrison entered the room.)
The WITNESS. He told all of us how much France meant to the western world, and how the fall of France meant an end of many things that he had considered precious and that now the western civilization was really in a critical situation, and that it was very necessary to do something to save the values of western civilization.

By Mr. MARKS:

Q. Coming back, now, to the postwar period, you told us that you were consultant at Los Alamos after you left Los Alamos. In that connection did you observe what, if any, influence the General Advisory Committee of the Atomic Energy Commission had on the course of events at Los Alamos Laboratory?
A. I could observe this to some extent, perhaps not enough because I was not at Los Alamos between January of 1946 and the summer of 1947.
Q. Just tell us about the period from 1947 on. What was the influence, if you know of it, of the General Advisory Committee on the course of events.
A. I know that the General Advisory Committee always was very helpful to Los Alamos, and that the Los Alamos people repeatedly told me that one could always get support for the best ideas in weapons development at the General Advisory Committee. It was that organization in the Government which had the greatest knowledge about these matters, and from what I know about the Los Alamos work, every important development in weapons which was done at Los Alamos was strongly supported by the General Advisory Committee.

Q. From the end of the war to the latter part of 1949, did you have any part in thermonuclear research at Los Alamos?

A. Not much. I mainly worked on fission weapons. However, there was some minor application of thermonuclear principles which was worked on at Los Alamos during the summer of 1949, and in which I participated. This turned out very useful later on.

Q. After the explosion of the Russian A bomb, was there any change in the character of your work?

A. Yes.

Q. Would you describe what happened?

A. Should I——

Q. As to yourself.

A. In October of 1949 I had a visit from Dr. Teller at Los Alamos.

Q. You were at Los Alamos?

A. No, he was at Los Alamos. I was in Ithaca. He came to visit me as he was also visiting several other scientists, and he tried to persuade me to come to Los Alamos full time, and to help evolve full scale thermonuclear weapons.

Q. Dr. Bethe, there has been some talk in these proceedings about the General Advisory Committee meeting towards the end of October of 1949.

A. May I go on?

Q. I beg your pardon. I am sorry.

A. At the time Dr. Teller visited me, I had very great internal conflicts what I should do. Dr. Teller was presenting to me some ideas of his technical ideas which seemed to make technically more feasible one phase of the thermonuclear program. I was quite impressed by his ideas.

On the other hand, it seemed to me that it was a very terrible undertaking to develop a still bigger bomb, and I was entirely undecided and had long discussions with my wife.

Q. When did this occur?

A. This was early in October, as far as I remember. It may have been the middle of October, but some time between early and middle of October. What I should do? I was deeply troubled what I should do. It seemed to me that the development of thermomnuclear weapons would not solve any of the difficulties that we found ourselves in, and yet I was not quite sure whether I should refuse.

Q. Did you consult Dr. Oppenheimer about what to do and if so, approximately when?

A. I did consult Dr. Oppenheimer. In fact, I had a meeting with him together with Dr. Teller. This was just a few days later, I think only 2 days later, or 3, than my first meeting with Dr. Teller. So this would again be around the middle of October, and perhaps a little earlier. I found Dr. Oppenheimer equally undecided and equally troubled in his mind about what should be done. I did not get from him the advice that I was hoping to get. That is, I did not get from him advice from either direction to decide me either way.

He mentioned that one of the members of the General Advisory Committee, namely Dr. Conant, was opposed to the development of the hydrogen bomb, and he mentioned some of the reasons which Dr. Conant had given. As far as I remember, he also showed me a letter that he had written to Dr. Conant. As far as I remember, neither in this letter nor in his conversation with us did he take any stand.

Q. What did you do about the invitation that Teller had extended you?

A. About 2 days after talking to Dr. Oppenheimer I refused this invitation. I was influenced in making up my mind after my complete indecision before by two friends of mine, Dr. Weisskopf and Dr. Placzek. I had a very long and earnest conversation with Dr. Weisskopf what a war with the hydrogen bombs would be. We both had to agree that after such a war even if we were to win it, the world would not be such, not be like the world we want to preserve. We would lose the things we were fighting for. This was a very long conversation and a very difficult one for both of us.

I first had a conversation with Dr. Wiesskopf alone and then with Weisskopf and Placzek together on the drive from Princeton to New York. In this conversation essentially the same things were confirmed once more. Then when I arrived in New York, I called up Dr. Teller and told him that I could not come to join his project.

Q. When would this have been, approximately?

A. I still can't give you any much better date than before. It was certainly quite some time before the General Advisory Committee meeting. I don't know whether it was 2 weeks before or 10 days before. It may have been 3 weeks before. I could establish the date if this is important.

Q. Since that time, however, you have done work on the thermonuclear program, on the H bomb?

A. I have indeed.

Q. When did that begin?

A. This began after the outbreak of the Korean war.

Q. What have you done since then, describing it just in general terms?

A. In June of 1950, when the Korean war broke out, I decided that I should put a full effort on Los Alamos work and in particular should work also on thermonuclear weapons. I offered to Los Alamos to do active work at times when I was at Los Alamos, but also when I was at Cornell. This offer was accepted. I have done work with an assistant who I supplied from among my own students. I believe this work has been recognized as contributing.

Q. Are you saying that continuously from the outbreak of the Korean trouble——

A. Essentially continuously. I worked of course only part time as long as I was at Cornell. Then I was at Los Alamos at more frequent intervals since then. I mentioned before that I spent a whole 8 months there from February 1952 to September, which was a critical period in the development of the first full scale thermonuclear test which took place in November of 1952, as you well know.

I also went there at other times during the summer. I went usually for a month in the winter, and I worked in between at Ithaca.

Q. When you did finally decide in the summer of 1950 to go to work on the thermonuclear program, what became of the inner troubles that you had previously that contributed to turning down Teller's original offer?

A. I am afraid my inner troubles stayed with me and are still with me, and I have not resolved this problem. I still feel that maybe I have done the wrong thing, but I have done it.

Q. You have done the wrong thing in what?

A. The wrong thing in helping to create a still more formidable weapon, because I don't think it solves any of our problems.

Q. During the early part of 1950, that is, after you turned down Teller's invitation, but before you went to work at Los Alamos, on the thermonuclear program, you made some public statements, I believe, in the press. You wrote an article which I believe was published in the Scientific American, and the Bulletin of Atomic Scientists, setting forth your views about the thermonuclear problem.

Would you describe briefly what you regarded as the alternative to going ahead with the thermonuclear program?

A. Yes, sir.

Q. I am speaking now of the period from the end of 1949 to the middle of 1950.

A. Yes. I thought that the alternative might be or should be to try once more for an agreement with the Russians, to try once more to shake them out of their indifference or hostility by something that was promising to be still bigger than anything that was previously known and to try once more to get an agreement that time that neither country would develop this weapon. This is enough of an undertaking to develop the thermonuclear weapon that if both countries had agreed not to do so, that it would be very unlikely that the world would have such a weapon.

Q. Can you explain, Dr. Bethe, how you reconciled that view just described of wanting to make another try at agreement with Russia, with the view that you described a little while ago in which you expressed the feeling that negotiations with Russia on the A bomb were hopeless?

A. Yes. I think maybe the suggestion to negotiate again was one of desperation. But for one thing, the difference was that it would be a negotiation about something that did not yet exist, and that one might find it easier to renounce making and using something that did not yet exist to renounce something that was actually already in the world. For this reason, I thought that maybe there was again some hope. It also seemed to me that it was so evident that a war fought with hydrogen bombs would be destruction of both sides that maybe even the Russians might come to reason.

Q. Didn't you feel that there was a risk involved in taking the time to negotiation which might have given the Russians the opportunity to get a head start on the H-bomb?

A. There had to be a time limit on the time that such negotiations would take, maybe a half year or maybe a year. I believe we could afford such a head start even if there were such a head start. I believed also that some ways could have been found that in the interim some rsearch would go on in this country. I believed that also our armament in atomic bombs as contrasted to hydrogen bombs was strong enough and promised to be still stronger by this time, that, is, by the time the hydrogen could possibly be completed, so that we would not be defenseless even if the Russians had the hydrogen bomb first.

Q. Do you have any opinion, Dr. Bethe, on the question of whether there has been in fact any delay in the development and the perfection of thermonuclear weapons by the United States?

A. I do not think that there has been any delay. I will try to keep this unclassified. I can't promise that I can make myself fully clear on this.

Q. Try to, will you?

A. I will try. When President Truman decided to go ahead with the hydrogen bomb in January 1950, there was really no clear technical program that could be followed. This became even more evident later on when new calculations were made at Los Alamos, and when these new calculations showed that the basis for technical optimism which had existed in the fall of 1949 was very shaky, indeed. The plan which then existed for the making of a hydrogen bomb turned out to be less and less promising as time went on.

Q. What interval are you now speaking of?

A. I am speaking of the interval of from January 1950 to early 1951. It was a time when it would not have been possible by adding more people to make any more progress. The more people would have to do would have to be work on the things which turned out to be fruitful.

Finally there was a very brilliant discovery made by Dr. Teller. * * * It was one of the discoveries for which you cannot plan, one of the discoveries like the discovery of the relativity theory, although I don't want to compare the two in importance. But something which is a stroke of genius, which does not occur in the normal development of ideas. But somebody has to suddenly have an inspiration. It was such an inspiration which Dr. Teller had * * * which put the program on a sound basis.

Only after there was such a sound basis could one really talk of a technical program. Before that, it was essentially only speculation, essentially only just trying to do something without having really a direction in which to go. Now things changed very much * * *. After this brilliant discovery there was a program.

Q. Dr. Bethe, if the board and Mr. Robb would permit me, I would like to ask you somewhat a hypothetical question. Would your attitude about work on the thermonuclear program in 1949 have differed if at that time there had been available this brilliant discovery or brilliant inspiration, whatever you call it, that didn't come to Teller until the spring of 1951?

A. It is very difficult to answer this.

Q. Don't answer it if you can't.

A. I believe it might have been different.

Q. Why?

A. I was hoping that it might be possible to prove that thermonuclear reactions were not feasible at all. I would have thought that the greatest security for the United States would have lain in the conclusive proof of the impossibility of a thermonuclear bomb. I must confess that this was the main motive which made me start work on thermonuclear reactions in the summer of 1950.

With the new * * * ([1] idea) I think the situation changed because it was then clear, or almost clear—at least very likely—that thermonuclear weapons were indeed possible. If theromnuclear weapons were possible, I felt that we should have that first and as soon as possible. So I think my attitude might have been different.

Q. One final question, Dr. Bethe. I should have asked you this. I have referred you to the press statements and the article that you published in the late winter and spring of 1950, expressing critical views of the H-bomb program. Did you ever discuss those moves, that is to make such statements and write such articles, with Dr. Oppenheimer?

A. I never did. In fact, after the President's decision, he would never discuss any matters of policy with me. There had been in fact a directive from Presi-

---

[1] Supplied for clarity.

dent Truman to the GAC not to discuss the reasons of the GAC or any of the procedures, and Dr. Oppenheimer held to this directive very strictly.

Q. Did you consult him about the article?

A. I don't think I consulted him at all about the article. I consulted him about the statement that we made. As far as I remember, he gave no opinion.

Q. On the basis of your association with him, your knowledge of him over these many years, would you care to express an opinion about Dr. Oppenheimer's loyalty to the United States, about his character, about his discretion in regard to matters of security?

A. I am certainly happy to do this. I have absolute faith in Dr. Oppenheimer's loyalty. I have always found that he had the best interests of the United States at heart. I have always found that if he differed from other people in his judgment, that it was because of a deeper thinking about the possible consequences of our action than the other people had. I believe that it is an expression of loyalty—of particular loyalty—if a person tries to go beyond the obvious and tries to make available his deeper insight, even in making unpopular suggestions, even in making suggestions which are not the obvious ones to make, are not those which a normal intellect might be led to make.

I have absolutely no question that he has served this country very long and very well. I think everybody agrees that his service in Los Alamos was one of the greatest services that were given to this country. I believe he has served equally well in the GAC in reestablishing the strength of our atomic weapons program in 1947. I have faith in him quite generally.

Q. You and he are good friends?

A. Yes.

Q. Would you expect him to place his loyalty to his country even above his loyalty to a friend?

A. I suppose so.

Mr. MARKS. That is all.

CROSS EXAMINATION

By Mr. ROBB:

Q. Doctor, when Dr. Teller came to see you in 1949, were you at Ithaca then, sir?

A. Yes.

Q. And then you and Dr. Teller went down to Princeton to see Dr. Oppenheimer?

A. We went down separately, but we met again in Princeton.

Q. May I ask, Doctor, why did you pick Dr. Oppenheimer to consult about this matter?

A. Because we had come to rely on his wisdom.

Q. Doctor, you spoke of Dr. Teller at Los Alamos as always suggesting new * * * (ideas).

A. Yes.

Q. It was a new * * * (idea) suggested by Dr. Teller which resulted in your success in producing the thermonuclear; wasn't it?

A. This may be true, and some of his suggestions certainly were extremely valuable.

Q. Yes, sir.

A. There were other suggestions which turned out to be very much to the contrary. Dr. Teller has a mind very different from mine. I think one needs both kinds of minds to make a successful project. I think Dr. Teller's mind runs particularly to making brilliant inventions, but what he needs is some control, some other person who is more able to find out just what it is the scientific fact about the matter. Some other person who weeds out the bad from the good ideas.

* * * as soon as I heard of Dr. Teller's new invention, I was immediately convinced that this was the way to do it, and so was Dr. Oppenheimer. I should mention a meeting which took place in 1951, in June, at which Dr. Oppenheimer was host. At this meeting the final program for the thermonuclear reactions was set up. At this meeting Dr. Oppenheimer entirely and wholeheartedly supported the program.

Q. Doctor, how many divisions were there at Los Alamos?

A. It changed somewhat in the course of time. As far as I could count the other day, there were 7, but there may have been 8 or 9 at some time.

Q. Which division was Klaus Fuchs in?

A. He was in my division which was the Theoretical Division.

Mr. ROBB. Thank you. That is all.

Mr. GRAY. I have some questions, Dr. Bethe. Early in your testimony in response to a question from Mr. Marks about cooperation and happy atmosphere—these were not your words or his—you said there were certain notable exceptions.

The WITNESS. Right.

Mr. GRAY. You named Dr. Teller.

The WITNESS. Right.

Mr. GRAY. Could you name certain other of the notable exceptions?

The WITNESS. I can recall only one person. That was Dr. Felix Bloch, who left the project after some time and went to a radar project instead. He was at Los Alamos only for a short time. Otherwise, I can't recall any exceptions.

Mr. GRAY. This is a matter of information, perhaps. I was interested, however, in one of your objections to the May-Johnson Act, on the ground, and I think I use your words, "that it provided punishments for almost any move a scientist might make."

The WITNESS. Right.

Mr. GRAY. What do you have in mind?

The WITNESS. When you read the document—I am afraid I didn't read it from beginning to end—the thing which was most conspicuous to us was that that listed a large number of things that were to be considered a security violation and set down very harsh penalties, unprecedented penalties, I believe, for these.

Mr. GRAY. What kind of things were the penalties imposed for? That is what I am trying to get at. I am not familiar, I am sorry to say, with the provisions.

The WITNESS. I don't know that this is terribly important, and I should not insist on it too much. It said if you betray some secret—if some secrets leak out by negligence, then you go to prison for 10 years. If you do it with the intent to hurt, the United States, the penalty is death, and so forth and so on. The things that were mentioned were definitely things that should be punished. It only seemed to us that the punishment was perhaps a little harsh and a little too much emphasized in the bill.

Mr. GRAY. I don't want to pursue this too far, but your characterizations of these actions as almost any move a scientist might make, you mean any treasonable move which by carelessness might be the equivalent, I suppose?

The WITNESS. No, it would require much less than that. It would require an act of slight negligence rather than any callousness.

Mr. GRAY. Yes.

The WITNESS. But I don't wish to insist on this.

Mr. GRAY. Very well. Also in response to a question from Mr. Marks you said that you were very much surpised in 1940, soon after the fall of France—I believe you said this—Dr. Oppenheimer's political reorientation—the phrase is mine, not yours—and you cited as an example and you cited as conviction that an extraordinary effort needed to be put forth to save western civilization.

The WITNESS. Yes.

Mr. GRAY. I believe you were asked that question by Mr. Marks in the context of Dr. Oppenheimer's earlier very leftwing views.

The WITNESS. Precisely.

Mr. GRAY. Other than the fall of France were there indications in his conversation—the long conference you had—in 1940 which would indicate a change in these extreme or very leftwing views?

The WITNESS. I am not sure that I recall any other motivation. I am sure that the fall of France was uppermost in all our minds, and that this was the dominant theme. I don't know what other motivation Oppenheimer went through to change his mind.

Mr. GRAY. I really was not concerned so much with motivations as whether you sensed a modification of the extremeness of his leftwing views.

The WITNESS. I certainly did not. It did not come up even as a part of the conversation that his views were leftwing at this time. That is, as you recall, this was in the time of the Russo-German Pact. I don't believe the pact was mentioned. Maybe it was. If so, it must have been mentioned in the same context, in the same spirit as the fall of France, namely, that it was a most deplorable thing. There was nothing in the conversation which indicated any leftwing orientation at that time. In fact, the opinions of experts were the exact opposite of the party line.

Mr. GRAY. May I again refer to your conversations with Dr. Teller and with Dr. Oppenheimer in October 1949, at which time you were deeply troubled as to whether you should go back or should again work—what was it—at Los Alamos?

The WITNESS. Right.

Mr. GRAY. Did you get far enough along in your thinking, Dr. Bethe, and in your discussions with Dr. Teller, to talk in terms of what the salary might be if you went back to Los Alamos?

The WITNESS. We did discuss this. Even though I was not at all decided whether I wanted to go, I wanted to discuss the things sufficiently so that at least external circumstances would be reasonable if I went.

Mr. GRAY. So that at the time your mind was at least open to the point that the shape and form and nature of the job was interesting at least.

The WITNESS. Right.

Mr. GRAY. May I ask, then, how long after this conversation with Dr. Teller in which salary and other conditions were discussed was it that you began making speeches and writing the bulletins opposing work on the hydrogen bomb, or is that a clear question?

The WITNESS. That is perfectly clear. This was 3 months later.

Mr. GRAY. Three months later?

The WITNESS. Three months and a little.

Mr. GRAY. Mr. Marks asked you a question about——

The WITNESS. May I make one more remark in this connection?

Mr. GRAY. Yes, indeed.

The WITNESS. During the time when the Government was in the process of deciding whether to go ahead with the program, I felt and I think all scientists felt that we must not make speeches. This does not mean that we held any different opinion. But during this time it was a secret deliberation of the Government and it was not in the public domain and we therefore restrained ourselves from expressing our opinion—meticuously—in any way.

Mr. GRAY. So that your speeches in opposition came after the President's decision?

The WITNESS. They came after the President's decision. They could not come before the President's decision. This does not mean that the President's decision changed my mind in any way.

Mr. GRAY. I think you made it clear in your testimony that you feel that following the Presidential decision there was no delay in the development of the hydrogen bomb.

The WITNESS. Yes.

Mr. GRAY. Can you say the same thing about the period from 1945 until January 1950?

The WITNESS. This is a very difficult question. I think one would have to take the periods apart. I believe, let me say in the beginning, first of all that there was in the end no delay.

Mr. GRAY. You mean taking the years from 1945 to 1950, or whenever it was?

The WITNESS. 1952.

Mr. GRAY. That there was no delay?

The WITNESS. Yes. One of the ingredients in my saying so is that in order to have a successful thermonuclear weapon you first need to have an extremely good fission weapon. You cannot make a success of a thermonuclear weapon without that. As you know, the fission weapon is used as a trigger to provide the heat in the thermonuclear weapon. This is public knowledge. Anything beyond that I cannot say. It is necessary to have extremely good fission weapons and what Los Alamos laboratory did in all the time until 1950, early in 1950 and indeed later, too, was a continuous and very spectacular improvement in fission weapons, so much so, as President Eisenhower announced in his United Nations speech, that the power of the fission weapon has increased 25 fold since Hiroshima and Nagasaki.

Therefore, this work was all in the direction that was necessary to bring success in the thermonuclear program.

Now, then, in the first period from the end of the war to the beginning of the AEC, that is, to January of 1947, Los Alamos was in a state of disintegration, and Los Alamos, just like our Armed Forces, was declining in strength. All of us wanted to go home just as all the boys from oversea wanted to go home, and as their mothers wanted them to come home. So everybody wanted to go home.

Also, we wanted to give a chance to the international organizations. This changed completely when the AEC took over in the beginning of 1947, and from then on really a strong program in weapons development was started.

I should say in all fairness that in all this program Dr. Teller played a very important role and did not show any deviations, as I criticized during the war time period. Even so, it needed some time to build up the strength of the laboratory.

It was impossible for the laboratory to do very many things at the same time in 1947 or 1948, let us say. All the same, some research was going on all the time on some phases of thermonuclear reactions.

I mentioned before that one particularly promising, although minor application of such reactions, was actively worked on in the summer of 1949 when I was there, and it had then been worked on for some time. It actually turned out that this was more useful in the end than would have been a concerted attack on what was then believed to be the main subject.

Mr. GRAY. You think that the demonstration of genius on the part of Dr. Teller * * *—I know I am asking a question that you can't answer, but I will ask it anyway—do you think if the GAC in 1947, when it was constituted, had concluded as the President concluded in January 1950, that it is possible that Dr. Teller's stroke of genius might have come sooner * * *. It had no relation to the atmosphere, facilities and those things. I know this is a very difficult question.

The WITNESS. Yes.

Mr. GRAY. If it is not clear to you, I am addressing myself to the point that it has been said in many places that the attitude of the GAC did in fact delay successful work. I believe this has been said. You are familiar with that.

The WITNESS. I am familiar with that.

Mr. GRAY. I am trying to address myself to that point.

The WITNESS. It is awfully hard to answer. It is true certainly that a stroke of genius does not come entirely unprepared and that you get ideas only on the subjects that you are working on. If you are working on other subects, let us say fission weapons, you probably won't have any inspiration about thermonuclear weapons. It is true on the other hand that two quite important suggesttions or discoveries were made on thermonuclear problems during the time when Los Alamos was not actively working on these. I cannot name them in an unclassified session.

One of them was the thing that I mentioned repeatedly, the minor application, as I call it, of thermonuclear principles. I think it is quite obvious that only when there is a concerted effort can there be the atmosphere in which you can have big ideas. Whether we would be farther ahead or less far ahead, I don't know.

Mr. GRAY. I was aware that was a difficult question. I have only two more, Doctor.

You testified that at one period you were hoping that it might be possible to prove that thermonuclear weapons were just simply not possible.

The WITNESS. Yes.

Mr. GRAY. I assume, then, that you were hoping that if they were not possible in this country they could not be possible in the U. S. S. R.?

The WITNESS. Precisely.

Mr. GRAY. Did you have any reason to hope that the Russians were not taking a contrary view to yours? You were hoping that it could not be possible.

The WITNESS. Yes.

Mr. GRAY. Would it be unreasonable to suppose that the Russians might have been taking the contrary view?

The WITNESS. That they were hoping that it was possible?

Mr. GRAY. Yes.

The WITNESS. I am quite prepared to assume that, but I don't know.

Mr. GRAY. So that there was a double hope that we couldn't do it and also that they couldn't, but we had no basis for believing that they would not make every effort, I assume?

The WITNESS. That is true. In the times when everybody was very pessimistic about the outcome of our own effort, that is, in the year 1950 essentially, I was often hoping that the Russians would spend their efforts on this problem and that they would waste their efforts on this problem.

Mr. GRAY. My final question, I think, relates to Mr. Marks' last question to you.

In the light of your intimate personal acquaintanceship with Dr. Oppenheimer and within the framework of the Atomic Energy Act of 1946, you have no doubts about him with respect to his loyalty, his character, his discretion, which were the three areas which Mr. Marks put the question to you.

The WITNESS. Yes.

Mr. GRAY. In order to complete the record, because there is another consideration which the act imposes and that is, associations, would you answer also affirmatively to the question including the test of associations?

The WITNESS. Those associations that I personally know about I certainly heartily approve. The associations which I mentioned——
Dr. EVANS. What was that?
The WITNESS. I said that——
Mr. GRAY. The associations he knows about he would heartily approve.
The WITNESS. The associations in the dim past of the late 1930's and maybe early 1940's I certainly cannot approve, but I think they are superseded by a long record of faithful service and that one has to judge a man according to his actions, recent actions, which are, as far as I know, all in the public domain and all perfectly known and open to scrutiny.
Mr. GRAY. Thank you.
Mr. ROBB. Mr. Chairman, I have a couple of questions but would you rather I save them until Dr. Evans finishes?
Mr. GRAY. Yes. Dr. Evans, do you have any questions?
Dr. EVANS. Yes. Dr. Bethe, for the record—we can look it up, but you can tell us—where did you do your university work?
The WITNESS. I studied at the Universities of Frankfort and Munich in Germany and got my Ph. D. in Munich in 1928.
Dr. EVANS. Have you taken out any patents on these types of weapons?
The WITNESS. I believe I have a patent or two on fission weapons. I don't believe I have any on the thermonuclear weapons.
Dr. EVANS. What are your political views? You are a citizen of the United States?
The WITNESS. Yes. This is perhaps——
Mr. GRAY. Excuse me.
Dr. EVANS. Do I have a right to ask that?
Mr. GRAY. Perhaps it may be that the witness would be entitled to have a little bit of understanding. I don't know that the question—excuse me. If the witness objects to answering, he can.
Dr. EVANS. Maybe I should not ask this question.
The WITNESS. I have no objection at all. I have never had any association with a leftwing organization whatsoever. My political views are best described by Adlai Stevenson's views.
Dr. EVANS. He is from Chicago.
The WITNESS. Right.
Dr. EVANS. I want to ask you one other question.
Being a normal man and a good man, I take it, do you still in the back of your head have these moral scruples about these things?
The WITNESS. I do.
Dr. EVANS. That is all.
Mr. ROBB. I think there are a couple of questions suggested by the chairman's questions.
Mr. GRAY. If you will, I want to see if Mr. Marks has any questions.
Mr. MARKS. I have a couple, but either way.
Mr. GRAY. Suppose you proceed.

REDIRECT EXAMINATION

By Mr. MARKS:

Q. I neglected to ask you one question, Dr. Bethe. When was the next time after your meeting with Dr. Oppenheimer in 1940 that you saw him?
A. I saw him for a day in 1941.
Q. When and where as that?
A. That was in New Mexico on his ranch on the 24th of July, as far as I know.
Q. You had some reason for fixing that date?
A. Yes. There was a previous case in which Dr. Oppenheimer was accused of having attended a meeting in Berkeley some time in July. He asked me to establish the date of my visit, and I tried to do that. I must confess that I came only within 2 or 3 days, and the exact date was supplied to me.
Q. Supplied to me?
A. Supplied to me by a friend of Dr. Oppenheimer. But I came within 2 or 3 days.
Q. You mean——
A. On my own investigation. I did not——
Q. I think you better tell us the whole thing, because I don't know it.
A. Well, I was asked to find out when I had visited Dr. Oppenheimer so that I could, if necessary, testify to that. I made some searching of my own memory,

I could establish a date of the 1st of August when I met Dr. Teller for a summer vacation, and I calculated back that some time in the early twenties of July I had been at Dr. Oppenheimer's ranch.

I furthermore knew the hotel in which I stayed the night after, and I made sure that I could find out from the hotel register what day we had stayed there, if need be, but they were reluctant to do this, because it was some 12 years back. Then before I had any chance to go further into this, I was told that it was the 24th of July.

Q. You mean that the hotel register was the 24th of July?

A. The hotel register was not searched, but an independent search—there were other events which took place during my visit—namely, Dr. Oppenheimer was kicked by a horse. It was possible to establish that date.

Q. And you remember that you were there?

A. I remembered that.

Q. So when you said a friend of Dr. Oppenheimer's supplied the date, what you meant was that a friend of Dr. Oppenheimer told you what date it was the horse kicked him?

A. Right.

Q. I think I may have phrased a question ineptly in relation to your speeches and your articles in early 1950. In answer to a question of mine did you say that those statements and articles opposed work on the H bomb program?

A. No; they did not. They only deplored that such a thing would be made, and they expressed the hope that we would never use it. The statement said that we were hoping that the United States would never use the H bomb until it was used against us first. I don't know whether that is a good scheme. I think it should be understood as a desperate attempt to reconcile an accomplished fact of the H bomb program, which we did not want to oppose with our deeply troubled conscience.

Mr. MARKS. May I identify for the record, in case the board should wish to refer to this article, so we are sure that we are all talking about the same thing. I have reference to a press release which was reported in the New York Times, Sunday, February 5, 1950, and I have reference to an article that was published in the April issue of 1950 of Scientific American, and a reprint of that article which appeared in the April issue of the same year in the Bulletin of the Atomic Scientists.

Mr. ROBB. Do you want to show them to the witness, Mr. Marks?

Mr. MARKS. Yes, I think that would be a good idea.

The WITNESS. Yes, I remember these.

Mr. ROBB. I might say, Mr. Chairman, I think those are articles in the file that you have before you.

Mr. ROLANDER. The Scientific American article is in the file.

Mr. MARKS. Mr. Silverman has suggested that he recalls some reference earlier in the proceeding to a patent that was mentioned relating to thermonuclear devices in which you and Dr. Teller and Dr. Oppenheimer were all involved. Do you have any recollection of that?

The WITNESS. I am sorry, I don't.

Mr. MARKS. That is all

RE-CROSS-EXAMINATION

By Mr. ROBB:

Q. Doctor, I suppose it is a truism that you don't have ideas about things like thermonuclear weapons on the atomic weapons without working on them?

A. That is certainly true.

Q. And you really can't tell whether they will work without experiment, can you?

A. You can tell pretty well by purely theoretical investigations.

Q. But in the last analysis you have to try them out?

A. In the last analysis you have to try them out. I think it is a matter of record that the General Advisory Committee has always been strongly recommending tests on atomic weapons.

Q. Yes, sir. Doctor, I am a little bit confused about the two periods we have been talking about. One was between the spring of 1946 and I think January 1950, is that right?

A. Yes, sir.

Q. I am not entirely clear as to just what was going on during that period at Los Alamos in respect to the thermonuclear. I don't mean the technical details, but who was working on it at Los Alamos, and how much work were they doing? Could you help us on that?

A. I will try. It was definitely a matter of very minor priority. It was carried on, that is, one line of work was carried on mainly by summer consultants, particularly by Dr. Nordheim. Another line of work was proposed by Dr. Teller. I don't remember exactly when, but it was probably about in 1948. I am not sure. That was worked out quite actively. I would say some 20 percent or so of the work of the theoretical division went into that from then on.

Q. I find in the file here some notation, which I can't vouch for, but perhaps you can tell me whether it is right or not, "That during that period from the spring of 1946 until January 1950, the work being done at Los Alamos on the thermonuclear was being done by Dr. Richtmyer, who worked for approximately 8 months on the problem.

A. Yes.

Q. Dr. Nordheim who worked approximately a month, and Dr. Teller, who worked approximately 2 months, and in addition there were perhaps 2 or 3 computers who worked for a full year. Would that sound about right to you?

A. This would sound about right for the one development that I spoke of, which I said was done by summer consultants. I would have thought from my recollection that Dr. Nordheim had worked on it more than that. But on the other hand, it is true that Dr. Richtmyer worked on it. I think this is a fairly good description of what went on on this one development. The other development——

Q. Would you wait just a moment? I am told I cannot ask you the question. Go ahead.

A. The other development—I am perfectly prepared to later on after we finish the unclassified part, to answer classified questions.

Q. We are trying to stay unclassified, and it is quite easy with me, because I don't know much about this technical part of it.

A. Yes, sir. The other development which I talked about which I called the minor application of thermonuclear principles was really one of the functions of the theoretical division. That is, of the division which generally was in charge of doing the thinking, the theoretical thinking before matters were put into the development stage. This, as far as I remember, was supported by the GAC. On this I gave the figure which I mentioned before of about 20 percent. I know of 2 people who worked on this, Dr. Langmire and Dr. Rosenbluth.

Q. Was that at Los Alamos?

A. Yes; at Los Alamos.

Q. How long did they work on it?

A. To the best of my recollection about a year before the fall of 1949. However, I may be wrong.

Q. That would be from the fall of 1948 to the fall of 1949.

A. Yes; this may be about right. I am afraid I don't remember it in all detail.

Q. So you would add those two gentlemen and their period of work to the names and the periods I read you.

A. Not only that, but there were several others of less standing involved in this, and I should mention that these two, apart from Dr. Richtmyer, and apart from the head of the theoretical division, are probably the most able members of the theoretical division.

Q. One further matter. Do you remember perhaps in May 1952 preparing a history of thermonuclear development?

A. I certainly do.

Q. For whom did you prepare that?

A. I prepared it for Mr. Dean, who was then the Chairman of the Atomic Energy Commission. However, not on his request, but rather to state the history as I saw it, and as most people at Los Alamos saw it.

Q. At whose request did you prepare it?

A. At my own.

Q. Did Dr. Oppenheimer discuss it with you?

A. No.

Q. He did not talk with you at all about it?

A. No.

Mr. ROBB. That is all. Thank you.

Mr. MARKS. May I ask a couple of more questions?

Mr. GRAY. Yes.

REDIRECT EXAMINATION

By Mr. MARKS:

Q. I would like to be sure, Dr. Bethe, that I understand the sense in which you made the statement about which the chairman has also questioned you, I believe, that the motivation you had in going back to work in the summer of 1950 on the thermonuclear problem was the hope that you could prove it would not work. Did you mean that you hoped you could prove by argument that it would not work, or that you could discover it as a law of nature in the sense of the theory of relativity or another scientific theory that it was impossible?

A. Hardly quite as conclusively as the theory of relativity, but rather that I could make an argument that the methods that we could conceive of for such development would all not work. That there were laws of nature which doomed such an attempt to failure.

Q. Would that process which you now describe of work on which you launched have been an indispensable part of discovering what would work?

A. I think so; yes. I don't know whether it was indispensable because Teller dispensed with it. Teller was able to make his invention without having had a conclusive discussion of all the possibilities.

Mr. MARKS. That is all.
Mr. GRAY. Thank you very much, Dr. Bethe.
(Witness excused.)
Mr. GRAY. We will take a recess.
(Brief recess.)
Mr. GRAY. Dr. Bethe, we have asked you to come back to clear up something for the record which I think needs to be done. Mr. Robb will do it very quickly.

Dr. Hans Bethe resumed the stand as a witness, having previously been duly sworn, was examined and testified further as follows:

FURTHER RE-CROSS-EXAMINATION

By Mr. ROBB:

Q. Doctor, I revert to your talk about the report you prepared in May 1952.
A. Yes.
Q. I asked you if you talked with Dr. Oppenheimer before you prepared it and you said that you didn't and I accepted that, of course.
A. Yes.
Q. I find after you left the room in the file a letter which is marked "Top Secret," but I want to show it to you. It is addressed by you to Dr. Oppenheimer on May 28, 1952, and apparently attached to a copy of your report to Mr. Dean, dated May 23, 1953.

Just so there be no misunderstanding in the record I want you to look at this and give any comment you may have.

A. This seems to say that we did talk about it. As far as I remember, it was merely that I reported to him that I was writing such a document. It was certainly not initiated by him and the contents that should be in it were not discussed with him.

Q. And you sent him a copy of the report?
A. I did.
Mr. ROBB. Mr. Classification Officer, if I get into something here, will you please stop me.

By Mr. ROBB:

Q. I notice you refer to a visit to Griggs. Who was Griggs?
A. Mr. Griggs was the chief scientist to the Air Forces.
Q. What was that visit about? You say "yesterday morning I visited Griggs." That was apparently May 27.
A. At the time there was much discussion of the past record of Los Alamos and much discussion of the question whether a second laboratory for weapons work should be opened. It seemed to me that some rather false information was current with some people, particularly in the Air Forces and one of the persons whom I knew to be an exponent of this section of Air Force opinion was Dr. Griggs. Therefore, I went to see him to clear up the past Los Alamos record, and also to discuss generally the function of thermonuclear weapons in warfare.
Q. What was his view on the second laboratory?
A. He was very much for it.
Q. You were against it?
A. Yes.

Q. You say in your letter to Dr. Oppenheimer: "As you know, I visited Griggs yesterday morning." Had you talked to Dr. Oppenheimer about visiting Griggs before you went to see him?
A. I obviously had.
Q. Do you remember what your conversation with Dr. Oppenheimer was?
A. I am afraid I don't but it probably ran somewhat similar to what I just told you: That I wanted to clear up these matters. By the way, Dr. Oppenheimer opposed the second laboratory much less than I did, if at all.
Q. I will show you the letter so you will get the context and maybe I can phrase a question that won't be overruled by our friend the security officer.
Doctor, I notice you speak of the atmosphere calming down considerably in this matter. What were you talking about there?
A. I will try to remember. I am not sure I can.
Q. If you don't, just tell us you don't and that is the end of it.
A. I may have referred to the general controversy of whether atomic weapons work was being pursued—no; I mean thermonuclear work—was being pursued sufficiently effectively at Los Alamos. This was our main concern at the time. It was believed by Dr. Oppenheimer and myself and by the members of the Atomic Energy Commission that Los Alamos was doing a very good job on thermonuclear weapons at that time, and this was borne out by the success of the test in November of 1952.
Dr. Teller, I think, was conducting a campaign to establish the contrary. I believe this was the matter I am talking about.
Q. I notice in here again referring to your visit to Griggs, which you say took from 10 a. m. to 12:30, you said you were surprised because your conversation with him was quite pleasant?
A. Right.
Q. What did that refer to, Doctor?
A. Dr. Griggs had been very much of an exponent of the view that Los Alamos was not doing its job right and very much an exponent of the view that thermonuclear weapons and only the biggest thermonuclear weapons should be the main part of the weapons arsenal of the United States. I had very much disagreed with this, with both of these points, and so I expected that we would have really a very unpleasant fight on this matter. We didn't.
Q. I see you mention in the first paragraph of your letter "very peaceable and enjoyable dinner with Dr. Oppenheimer" and the talk you had with him. Had you discussed with him at that dinner your forthcoming visit to Griggs. I don't know whether this was discussed at the dinner or otherwise. That was up at Princeton, was it?
A. I remember the dinner was here in Washington during the meeting of the American Physical Society. I may be wrong.
Mr. ROBB. That is all. Thank you, Doctor.

FURTHER REDIRECT EXAMINATION

By Mr. MARKS:
Q. Dr. Bethe, what kind of a second laboratory did Griggs favor?
A. Well, he favored a second laboratory to work on weapons and such a laboratory was then established at Livermore very shortly after all these conversations took place, namely, in July of 1952.
This laboratory has been getting all the credit for thermonuclear development, which is unjustified.
Q. What do you mean by that?
A. I mean by that that the majority of the weapons which have been developed and which are being tested now in the Pacific and the most powerful of them were developed exclusively by the Los Alamos Laboratory.
Q. Dr. Bethe, you said, as I understood your remarks, that you disagree with Griggs about the desirability of relying exclusively on thermonuclear weapons?
A. I did not say exclusively. Predominantly.
Q. Was this because of moral considerations?
A. Yes. It was my belief that if and when war ever comes that it is most important not to overdestroy the enemy country, but to fit the weapon in each case to the target and to attempt the best accuracy that one can on bombing so as to make a minimum of destruction compatible with gaining the objective. It was on this that we disagreed.
Q. I am afraid I don't understand you. Did you mean atomic weapons could do the job?

A. Yes, sir. Supposing you have, for instance, a city which contains two industrial plants which you want to bomb, each of which could be knocked out by a 100 kiloton atomic weapon correctly placed, you could also use a 5 million ton thermonuclear weapon to hit them both, which would reduce the problem for the Air Forces because they would have to fly only 1 plane instead of 2.

It seemed to me that both from moral considerations and for the consideration of the state of the enemy country after the war, which we traditionally take care of in some way, it was important to choose the former alternative and not the latter.

Mr. MARKS. That is all.
Mr. GRAY. Thank you very much, Dr. Bethe.
(Witness excused.)
Mr. GARRISON. Mr. Chairman, I will ask Dr. Fisk to come in. In the division of labor, I will ask my partner, Mr. Silverman, to put the questions to him.
Mr. GRAY. May I have your initials?
Dr. FISK. James B. Fisk, F-i-s-k.
Mr. GRAY. Do you wish to testify under oath?
Dr. FISK. Yes.
Mr. GRAY. Would you then please stand and raise your right hand? James B. Fisk, do you swear that the testimony you are to give the board shall be the truth, the whole truth and nothing but the truth, so help you God?
Dr. FISK. I do.
Mr. GRAY. Will you sit down, please, sir. I must remind you of the existence of the perjury statutes. I will be glad to describe to you the penalties imposed if you wish, but I assume you are familiar with them.
Dr. FISK. I think I am familiar with them.
Mr. GRAY. Second, I should ask that if it becomes necessary in the course of your testimony to make any reference to or disclose any restricted data, I would ask that you notify the board in advance so that we can take appropriate steps if that becomes necessary.

Finally, I should say that we consider the proceedings and record of this board as a confidential matter between the Commission and its officials and Dr. Oppenheimer and his representatives and witnesses, and that the Commission will not take the initiative in releasing anything about these proceedings to the press.

We express the hope that will also be the attitude of the witnesses who are appearing.

Mr. Silverman, would you proceed.
Whereupon, Dr. James B. Fisk was called as a witness, and having been duly sworn, was examined and testified as follows:

DIRECT EXAMINATION

By Mr. SILVERMAN:
Q. What is your present position in private employment?
A. I am vice president in charge of research at the Bell Telephone Laboratories.
Q. What is your present position with the Atomic Energy Commission?
A. Member of the General Advisory Committee.
Q. How long have you been a member of the General Advisory Committee?
A. I was appointed in the fall of 1952.
Q. Will you tell us what previous positions you have held with the Atomic Energy Commission?
A. From February 1, 1947 until September 1948, I was Director of the Division of Research of the Atomic Energy Commission. Subsequent to that I was for a year or perhaps a little longer a consultant to the General Manager.
Q. When did you first meet Dr. Oppenheimer?
A. The first time I met Dr. Oppenheimer in any other than a very casual way was in January 1947.
Q. That was just before you assumed your position as Director of the Research Division?
A. Yes.
Q. After that did you work with Dr. Oppenheimer with any degree of closeness?
A. During the time I was Director of the Division of Research I saw Dr. Oppenheimer on many occasions, usually in connection with the work of the Commission.

Q. At that time he was chairman of the GAC?
A. At that time he was chairman, yes.
Q. Did you also serve on committees with Dr. Oppenheimer?
A. Subsequently to that time I have been on the Science Advisory Committee in the Office of Defense Mobilization of which he was an initial member, and on various committees, such as visiting committees to Harvard University and things of that sort.
Q. What was the visiting committee to Harvard University?
A. Physics Department. In fact, those are the only two that I should cite.
Q. You mean the Science Advisory Committee and the visiting committee to Harvard?
A. Yes.
Q. What is the period of your joint service on the Science Advisory Committee as far as you can recall?
A. I should say the overlap was a matter of something under a year. Approximately a year, I would say.
Q. When was that?
A. Quite recently.
Q. Do you recall about the time that you assumed your position as Director of the Division of Research in the early part of 1947 there was discussion about what to do about Los Alamos?
A. This was, I believe, the most important problem that I came in contact with at that time in the Commission—the health and vigor of Los Alamos.
Q. Would you tell us something about what the problem was and what was done about it with particular reference to what you know about what Dr. Oppenheimer did?
A. I can say a few rather general things here which I think may be of significance. The Commission had just, of course, come into existence. Los Alamos at that stage, with all of the rehabilitation of science and scientists following the war, many people going back to their normal pursuits and normal homes, Los Alamos was in a state where there was a real question as to whether or not it could survive. There was a great deal of attention paid to its growth, regaining of strength not only on the part of the Commission itself and the staff, but on the part of the General Advisory Committee. This was a very principal question.

Many of us spent many, many days in Los Alamos with people, attempting to keep the staff together, formulating its program and doing the things that would give it real life and vigor. In all of these activities, the General Advisory Committee was extremely helpful. Dr. Oppenheimer in particular was extremely helpful and thoughtful about the cirrcumstances which could bring the laboratory back to life.

It seems to me, if I may add this, that the health and vigor of Los Alamos today is a very direct result of the activities of those times. I believe it is the strongest laboratory the country has.

Q. Did Dr. Oppenheimer play any part in connection with the research work that was done by the AEC?
A. The principal activity for which I had a direct responsibility in the Commission, although all of us were doing a great variety of things in those days, was the research program. This was something that was inherited in part from the Manhattan District but it was something in another sense that had to be started in some parts anew. There were new national laboratories being formed, such as Brookhaven, such as the new Argonne Laboratory, and the whole problem here was to generate a research program that would keep American science and particularly the science that was relevant to the Commission's activities strong and vigorous.

There were many problems that came up day by day and in many, many cases, as Director of the Research Division, I turned to the General Advisory Committee for assistance and always got very good advice and very strong support for those things that made sense in my judgment.

Q. Did Dr. Oppenheimer play a role in this advice and assistance that you got from the GAC?
A. A very prominent role, both in terms of the formal activities of the General Advisory Committee itself and in terms of many informal contacts where I felt free to call upon him and where I saw him in the Washington offices.

The examples that one could cite are almost too numerous to detail, but all of this added to a feeling for strengthening science in the United States and science in its relevance to the Commission's overall program.

Q. Have you formed an opinion as a result of your contact with Dr. Oppenheimer, and your knowledge of Dr. Oppenheimer with respect to his integrity, his loyalty and any other factors that might bear on his being a security risk?

A. Yes, I have. I have a very high opinion of all of these factors and I would go on to say that I know of no more devoted citizen in this country.

Mr. SILVERMAN. I have no further questions.

CROSS EXAMINATION

By Mr. ROBB:

Q. Doctor, is the Bell Laboratory the one that Dr. Kelly is the head of?
A. He is the president, yes.
Q. Is he the Dr. Kelly that appeared here the other day?
A. Yes. He is my senior.

Mr. ROBB. Thank you; that is all I care to ask.

Mr. GRAY. Thank you very much, Dr. Fisk. We very much appreciate your coming.

(Witness excused.)

Mr. GARRISON. Mr. Chairman, I would like to ask General Osborn to testify next because I think we can get through with him this afternoon.

Mr. GRAY. All right.

May I ask for your initials?

General OSBORN. Frederick H. Osborn. I usually don't use the initial. Frederick Osborn.

Mr. GRAY. Do you wish to testify under oath?

General OSBORN. Yes, I do, sir.

Mr. GRAY. Would you be good enough to stand and raise your right hand?

Frederick Osborn, do you swear that the testimony you are to give the Board will be the truth, the whole truth and nothing but the truth, so help you God?

General OSBORN. I do.

Mr. GRAY. Thank you. Will you be seated, sir?

It is my duty to remind you of the penalties provided by the statutes, that is, the so-called perjury statutes. Unless you wish, I will not recite these penalties. I assume you are familiar with them.

General OSBORN. Yes.

Mr. GRAY. I should also like to ask that if in the course of your testimony it becomes necessary for you to discuss restricted data you will inform the Chairman in advance.

Finally, I should say that we consider these proceedings as a confidential matter between the Atomic Energy Commission and its officials and representatives and Dr. Oppenheimer, his representatives and associates, and that the Commission will not take the initiative in releasing anything publicly about these proceedings. We express the hope that this will be the view of the witnesses as well.

Whereupon, Frederick Osborn was called as a witness, and having been duly sworn, was examined and testified as follows:

DIRECT EXAMINATION

By Mr. GARRISON:

Q. General Osborn, I just want to ask you a few questions about your wartime experience and service. You were a member and chairman of the President's Advisory Committee on Selective Service in 1940, were you not?
A. Yes.
Q. And chairman of the Joint Army and Navy Committee on Welfare and Recreation beginning in March 1941?
A. I was appointed by Mr. Stimson to that post.
Q. Then you were promoted by General Marshall to brigadier general?
A. I was commissioned by General Marshall.
Q. As brigadier general and later you were promoted to major general in 1943?
A. That is right.
Q. You were Director of the Information and Educational Division of the USA, the United States Army?
A. Yes.
Q. And resigned from the Army in 1945?
A. That is right.
Q. And then you were appointed Deputy Representative of the United States on the United Nations Atomic Energy Commission, 1947 to 1950?
A. That is right.

Q. And it was in that connection that you had a close acquaintance and working relationship with Dr. Oppenheimer?
A. That is right.
Q. I want to just ask you a few questions about that experience of yours and I would like to ask you rapidly a few questions that will bring us to the historical point about which you are to testify.
The Baruch plan had been presented to the United Nations Atomic Energy Commission in the fall of 1946, is that right?
A. That is right.
Q. The plan was approved in December 1946, by every one but the Russians and their satellites?
A. There may have been some small nations in abstention—I forgot—but it was approved in effect by all the nations except the Russians and their satellites.
Q. And after that, the Commission adjourned?[1]
A. That is right.
Q. And Mr. Baruch shortly afterwards resigned his position as the United States representative on the Commission?
A. Yes.
Q. And then during the winter the resignation was not filled?
A. For the first 2 months it was not filled.
Q. And you were asked by General Marshall to take on the job of Deputy United States Representative on the Commission, succeeding Mr. Baruch?
A. That is right.
Q. That was in February 1947?
A. He told me he had a job for me and would I take a job, was I free to take a job on February 22, and I said I always would take a job, whatever he asked me to do. Then he sent for me to come to Washington on the 28th of February and told me what the job was. I was sworn in on Friday——
Q. March 7?
A. Yes, March 7, Friday, a week later. All Fridays.
Q. And at about that time on that day or shortly thereafter, did Dr. Oppenheimer get in touch with you at Acting Secretary Acheson's request?
A. I had come down on Friday, March 7. I remember these dates because Washington's birthday was when we gave an honorary degree to General Marshall at Princeton, which was the 22d. I came down the following Friday. I came down to be sworn in on Friday, March 7.
While I was in Dean Acheson's office, or waiting outside—I was back and forth from Dean Acheson's office; he was then Assistant or something or other Secretary of State to General Marshall, tiding over [1]—Dr. Oppenheimer called me from San Francisco. I had not known Dr. Oppenheimer before. I had simply known his name and knew who he was. He said that he wanted to see me.
Shall I go on?
Q. Yes.
A. I said "I am new to this job; I know nothing about it. Would it be better if we waited a couple of weeks until I was acquainted with the job."
He said, "No, I want to see you right now. Will you be in Washington or New York tomorrow?"
I said, "Yes, of course, I will see you if you want to come on, but it is a long trip to take."
He said, "I would like to see you."
We made a date to meet at the offices of the Atomic Energy Commission in New York the next day, Saturday. I went back to New York. Dr. Oppenheimer flew in from the coast and arrived early Saturday morning and met me at half past 11. I had a car and drove him to my country place up at Garrison across from West Point because I was spending the weekend there. He spent Saturday and Sunday with me. We drove back at 3 o'clock on Sunday because I was going to meet with Baruch.
Q. Did he get in touch with you at Acheson's request?
A. No. He knew from Acheson I had been appointed. It was on his initiative, I am pretty clear, that he wanted to see me.
What he wanted to see me about is this. The general tenor I remember quite clearly. I don't remember the exact words. I remember the general tenor clearly because it was very interesting.

Inaccurate.

Q. What was the key question facing the United States representative at that time? In fact, what was the key question facing the Commission?

A. I know now what the question was. I am not sure that I knew then because I was just getting started. The key question was whether the negotiations should be continued.

Q. With the Russians?

A. With the Russians in the United Nations, the Russians having turned down the Baruch plan. They had not vetoed it; they had abstained but not agreed to it.

The Baruch plan was very general in its statement. It was not a detailed plan. They had turned down the general principles of the plan. The question was, should the United States agree to continue the negotiations in the United Nations?

Dr. Oppenheimer came on to tell me that there were two very serious dangers in continuing negotiations. The general background was that he was now certain, after watching the Russians for 3 or 4 months, that the Russians had no intention of accepting any plan for the control of atomic energy—international control of atomic energy—which would mean lifting the Iron Curtain. He had come to the conclusion that their behavior showed that they were not going to lift the Iron Curtain, * * * for to do so would mean the end of the regime.

Yet he felt certain that, if the Iron Curtain was not lifted, any plan of international control would be exceedingly dangerous to the United States. What he was afraid of was that if we continued these negotiations we would make some compromises which without our fully realizing it would put us in the position of having accepted an agreement for the control of atomic energy, possibly with prohibition of bombs, without in reality the Russians having lifted the Iron Curtain.

There would be some system where we would accept compromises which would put the United States in a very dangerous position of not really knowing what was going on in Russia, whereas the Russians would know all about what was going on here.

This was the first danger he foresaw, and he talked about this. This was the purpose of his trip. He also felt that this continuation of negotiations was something that the Russians would be glad to use the United Nations as a medium for propaganda, and this propaganda they could use against us, and it would be just as effective as any propaganda we would get by insisting on the Baruch plan.

So he was for discontinuing the negotiations.

Q. Then you consulted representatives of the French——

A. I went back to New York and I saw McNaughton, the Canadian representative, a very able man, and Cadogan, the British representative, and Parodi, the head of the French delegation.

Q. What was their attitude?

A. They all felt very strongly that the negotiations should continue. They said they really had not a good look at the Baruch plan, they had not taken much part in drawing it, they did not know what it would look like if it was put in more detailed form. They said they would be in an impossible position in their own countries if they agreed to calling off the negotiations.

Senator Austin told me that he had been called to Washington to attend a meeting of the President's Executive Committee on the Regulation of Armaments, which was commonly called RAC, composed of Patterson, Forrestal, Lilienthal, Acheson, possibly Lovett. I think Lovett was not present at that meeting. I know he was not present at the meeting, and I am not sure whether he was still Secretary for Air. I guess he wasn't. Anyway, it was composed of Patterson, Forrestal, Lilienthal and Acheson.

I asked Austin if I could go to this meeting with him in Washington. He said yes, he would take me along. Austin felt very strongly that we should continue negotiations. He came to this meeting and he said that we should continue negotiations * * *.

Forrestal said, "This is a lot of bunk," and so did Patterson.

Q. How about Acheson?

A. Acheson didn't take any part.

Mr. ROBB. This is a lot of what?

The WITNESS. A lot of bunk. Forrestal was perfectly outspoken, and so for that matter was Patterson * * * He said we should not go on with the negotiations.

I asked if I might speak. I said I agreed with Austin that we should continue the negotiations for quite different reasons. I felt the Russians had no intention seriously and they would not agree to any form of control that we could accept, but that I had talked to * * * the British, the French and Canadian representatives and these men were very insistent that we continue negotiations.

I thought if we were properly on our guard we need not make any bad mistakes or endanger the situation, and it would be very injurious to our international position to take a lone position, refusing to negotiate.

Forrestal said, "That makes sense to me; what do you thing, Bob?" Paterson said, "I think we should go ahead if this is the reason and if we do it without eyes open."

Acheson said he was opposed to our going ahead. Lilienthal said that he agreed. Acheson said, "If you feel this way, it is all right * * * to go ahead."

So the next time I saw Dr. Oppenheimer—I forget when it was, fairly soon— I told him I had a part in this decision to go ahead notwithstanding his advice. I told him the reasons. He said, "Well, I had information which he had not taken into account, that he had not talked to any of the representatives of the other countries, naturally."

He said, "I was the boss of this situation and, if this was the decision, this was the decision. He would go along with it and I could count on him for any help I felt he could give us."

I then asked Dean Acheson if I might appoint a committee of consultants. I think it was on quite an informal basis—simply consultants who would meet with me in New York when I felt I needed their advice. On that committee I asked to serve Dr. Oppenheimer, Jim Conant, General Groves, Bacher, and Dr. Tolman, who died a year later (and I think Lincoln Gordon).[1]

By Mr. GARRISON:
Q. Chester Barnard?
A. Yes.
Q. And General Farrell?
A. Yes. (And Lincoln Gordon was on it. He had been on Baruch's staff. He was a professor at Yale or Harvard, and I think he was on it for a while.)[1]

This committee was wonderful and also the attention and interest they gave it. Oppenheimer and Conant said that any time we needed them they would drop anything they were doing and would come on for consultation.

When we were coming close to a decision as to what detailed spelling out we should do of the Baruch plan, I always consulted this committee. While I don't remember particular things that were said at committee meetings, Dr. Oppenheimer's position consistently through the first year when we were spelling out the Baruch plan in detail was that we must be very careful not to give up anything. If we lost the proposal in the Baruch plan which had already been too much weakened—the original proposal of the Acheson-Lilien- that plan that there must be international ownership and management of these plants—if we lost this, we would begin to get in an increasingly weakened position and he would be very scared of it.

So I think we strengthened the position that had already been weakened. It had already gone to the question of whether there should be inspection being left a little indefinite. I think we strengthened it under Dr. Oppenheimer's urging and that of other members of the committee.

Q. By the summer of 1949, or in the summer of 1949, did Dr. Oppenheimer make any comments in your consultant's committee which you have just been describing about the state of affairs in relation to the Baruch plan and the Russians?

A. By the summer of 1948 we went to the General Assembly in Paris with quite a well completed detailed outline of the Baruch plan—still called the Baruch plan, if you want, but it was the United Nations plan by this time— and under instructions of General Marshall, very specific and written, that we would try to call off the negotiations and, if we could not call them off entirely, then further talks should be put in the hands of the six sponsoring powers so it would not any longer be done in public and these ridiculous meetings which the Russians were using wholly for propaganda by this time would be avoided.

[1] Inaccurate.

We were successful to have the negotiations transferred to the so-called six sponsoring powers who had originally sponsored the setting up of a commission.
During 1949 we held occasional meetings of the sponsoring powers. I had my committee and the individual members of it in from time to time.

* * * * * * *

Q. From these contacts with Dr. Oppenheimer during these 2 years, 3 years actually, did you form an impression of his character and his quality as an American citizen?
A. I formed the impression of a man most consistent and determined in his desire to protect the United States against what he considered a very dangerous situation, a great number of dangers in these negotiations, and willing to take infinite pains to see that we didn't fall into any of these traps.
Hence I considered him a man of real patriotism and very consistent character and great loyalty because, after all, the very first thing I did, knowing nothing about this situation—when 2 weeks after he had taken the trouble to fly out from California—I had gone against his advice without telling him what I was doing. This made no difference. He just stuck at what he considered his job of seeing that we didn't fall into any pitfalls on this thing.
Q. By going against his advice, you have reference to your testimony that, after consulting with the British, French, and Canadians, you favored continuing negotiations with the Russians?
A. Yes. He remained intensely loyal. It has always struck me. I have been in a good many jobs, and this is not always the case when you cross a man at the beginning.
Mr. GARRISON. That is all.

CROSS-EXAMINATION

By Mr. ROBB:
Q. General, that was before the Russians exploded their A-bomb, was it?
A. Yes; all of this was before. I think they didn't explode their A-bomb until 1950.
Q. Yes.
Mr. GARRISON. 1949.
Mr. ROBB. 1949; I beg your pardon.
The WITNESS. Was it December of 1949?
Mr. GARRISON. September.
Mr. ROBB. September.
The WITNESS. I don't remember any activity on the part of the consulting powers after that time. We had really stopped meeting. I was on part time then. I resigned in December or the 1st of January, effective January 31, 1950.

REDIRECT EXAMINATION

By Mr. GARRISON:

* * * * * * *

Q. Do you remember a talk which Dr. Oppenheimer gave to the United Nations Committee in 1947? You don't have any recollection of that?
A. I remember his appearing. I am trying to think when that was. I remember that we asked him—one of the things that McNaughton, of Canada, wanted to do was to get Dr. Oppenheimer to appear, but I forget just what period it was in our negotiations. I think we had several scientists speak to the Commission to inform them about the situation. I don't remember what he said; I am sorry.
Mr. GRAY. Dr. Evans, do you have any questions?
Dr. EVANS. No.
Mr. GRAY. Thank you very much, General Osborn. We appreciate your appearance.
The WITNESS. Thank you, sir.
(Witness excused.)
Mr. GARRISON. Mr. Chairman, might I take 1 minute on the record. I would like to renew my request that copies of the transcript be given to us daily. I made arrangements with the reporter for two copies from now on.
Mr. ROLANDER. That is correct.
Mr. GARRISON. The point I would like to make is this. It is very, very difficult for us to work on these transcripts in the anteroom outside. I spent the Easter weekend in there, and it is not easy for us to work outside of our offices on these things, as you can well understand. I really don't know what this

problem about classification is, but it does seem to me that we should be in the position every day to have transcripts and to have them so counsel might take them out of the building and work on them because it is very, very difficult for us here.

Mr. ROLANDER. May I say that I and the classification officer also worked this weekend to try to get these transcripts reviewed. I think we are in a position to give you volumes 1 and 3 tonight. Some of the other volumes are creating some problems. We find we may, in order to eliminate any need for a classification stamp, have to scissor or remove a sentence or two or a paragraph or two, of course with the knowledge of counsel. I think this might aid us in getting the review completed more quickly.

Mr. GARRISON. I would rather take something that had some scissor holes in it if I could take it out of the building, assuming I could know what the scissor holes consisted of.

Mr. ROLANDER. Our problem is that so many other agencies have been mentioned in this proceeding. Although everyone has attempted to refrain from discussing restricted data, information having interest to other agencies and a programmatic interest has come up in the record which we feel is necessary to examine quite carefully.

We will attempt to scissor these transcripts and see if we can't move them a little more quickly. But as of tonight I think we can only assure you volumes 1 and 3, and by working tonight perhaps tomorrow we can assure you other transcripts.

Mr. GARRISON. I am sure you have been working hard on it. Aren't these references to other agencies chiefly in Dr. Oppenheimer's direct testimony?

Mr. ROLANDER. Of course, General Groves' testimony, Mr. Dean's testimony this morning also had certain items. I don't have the transcript in front of me, Mr. Garrison. We have, of course, provided a man here so that you can work at any hour that you want to. You are aware of that, of course.

Mr. GARRISON. Yes. I appreciated Mr. Williams' being here all day yesterday. I brought him lunch in a bag.

Mr. ROBB. I might say that I have felt the same difficulty because I have to come down to the safe to look at anything. I can't take anything home with me. Frankly, I have not had time to read the transcript.

Mr. GARRISON. May we take out of the building the Pash and Lansdale interviews? They are marked unrestricted.

Mr. ROBB. It is all right with me.

Mr. ROLANDER. It is not a part of the record yet.

Mr. ROBB. I see no objection.

Mr. GARRISON. We were going to try to agree on that. It is kind of late now. Do you think we should do it this afternoon? That is, on the recording.

Mr. ROBB. Yes; I understand.

Mr. GARRISON. I feel kind of weary.

Mr. ROBB. I do, too.

Mr. ROLANDER. May I say one other thing about the transcript. We will place on top of the transcript a list of obvious errors. We, of course, have not tried to correct a misspelling unless it is an error of substance. If you see any errors in addition to the ones we have noted, you can tell us.

Mr. GARRISON. I assume, Mr. Chairman, if counsel can give the board a stipulation of correction of obvious errors in the record, it won't be necessary to take the time of the board during the proceedings.

Mr. GRAY. The Board would be glad to receive it in that manner.

Mr. ROBB. I might say that I think on the whole the reporter has been doing a splendid piece of work.

Mr. GARRISON. I join in that.

Mr. GRAY. If there is nothing further at this time, we will recess until 9:30 in the morning.

(Whereupon, at 5:15 p. m., a recess was taken, to reconvene at 9:30 a. m., Tuesday, April 20, 1954.)

UNITED STATES ATOMIC ENERGY COMMISSION

PERSONNEL SECURITY BOARD

In the Matter of J. Robert Oppenheimer

Atomic Energy Commission,
Building T-3, Room 2002,
Washington, D. C., Tuesday, April 20, 1954.

The above entitled matter came on for hearing pursuant to recess, before the board, at 9:30 a. m.

Personnel Security Board: Mr. Gordon Gray, chairman; Dr. Ward V. Evans, member; and Mr. Thomas A. Morgan, member.

Present: Roger Robb, and C. A. Rolander, Jr., counsel for the board; J. Robert Oppenheimer, Lloyd K. Garrison, Samuel J. Silverman, and Allen B. Ecker, counsel for J. Robert Oppenheimer; Herbert S. Marks, cocounsel for J. Robert Oppenheimer.

PROCEEDINGS

Mr. GRAY. First of all, I have a telegram from an individual named C. S. Kuntz, 4507 North Dover Street, Chicago 40, Illinois. His telegram to me said, "Please wire date of Oppenheimer hearing as wish to testify. Was employed under Dr. Oppenheimer 1943 through 1946, respectively."

I am informed that Mr. Kuntz has indicated that he wishes to testify for, if you will allow me to put in that way, Dr. Oppenheimer. I pass it along for whatever value it may be to you. If you do not propose to call him, perhaps I should communicate and tell him that we will not need him as a witness. Can you answer that question now?

Dr. OPPENHEIMER. Can I hear the spelling of the name?
Mr. GRAY. K-u-n-t-z.
Dr. OPPENHEIMER. I don't recollect him.
Mr. GRAY. Obviously he was someone who worked with you in the laboratory.
Mr. GARRISON. Thank you, Mr. Chairman, for bringing it to our attention.
Mr. GRAY. Perhaps you would want to talk about this.
Mr. GARRISON. As far as I now know, we certainly have no intention of calling him.
Mr. GRAY. Very well. The other thing I want to discuss is the question of the redirect examination of Dr. Oppenheimer. You will recall, I guess it was on Friday, we had a discussion of this and because of the presence of witnesses, specifically including Mr. Lansdale and Dr. Glennan, I guess, one of whom I think was already on his way, or who was in the city at the time, out of consideration for their problems, we said that we would hear them. You will recall, however, that I indicated that the board wished to proceed with the redirect examination of Dr. Oppenheimer at the conclusion of the cross examination with these interruptions for convenience.

The board feels very strongly that a good deal of confusion, at least in our minds, could be eliminated by getting back to Dr. Oppenheimer right away and seeking in so far as possible to have his testimony, redirect, and any recross examination, not fragmented and interrupted.

I understand that Mr. Kennan is here and at the moment is waiting to testify, and of course we will hear him. Perhaps I should ask if there are any other witnesses on hand this morning?

Mr. GARRISON. Yes, Mr. Chairman. Mr. Lilienthal is here from New York. Mr. Sumner Pike is here from Maine. Dr. Fermi is arriving at noontime from Chicago. Professor Zacharias is here from Boston. Dr. Conant has cabled over that at 2 o'clock this afternoon he had to make plans because of the complicated nature of his witness here. Professor Ramsey from Harvard is here.

What our problem very simply has been is this: When we got the transcripts Friday afternoon, it was just physically not time over the weekend—I didn't get through more than a portion of one volume of the cross examination myself—with the difficulties of trying to arrange for these witnesses and all the rest impinging, and also the problem of analyzing what had become quite a complicated record with the bringing in of the elaboration of various names and places and dates of recollections, and then the procedure that had been adopted of asking Dr. Oppenheimer questions on the basis of what he recalled in the past and then producing documents, some of which had been taken from his own files on a classified basis, and suddenly declassified, all this produced a very complicated and difficult record to analyze.

It has been purely a problem of time. I don't want to ask Dr. Oppenheimer to testify until we really are prepared, and he is thoroughly ready to do it. I want very much to cooperate with the board in this. I would think that today the calendar is really beyond our——

Mr. GRAY. It would sound so.

I would suggest, then, that we proceed with the witnesses who are here or on their way here today and start tomorrow morning with Dr. Oppenheimer again.

I want to ask about Dr. Conant. You started to mention him.

352

Dr. EVANS. Is he here in Washington?

Mr. GARRISON. Yes, he is. He will be here today at 2 o'clock. We can talk over our problem during the lunch hour.

Mr. GRAY. All right. May I ask at this point only for an indication as to what is involved in time, because I think the board will probably call some witnesses, and they have to be alerted, I suppose, or should be, I am sure.

Could you give an indication of who are yet to come before the board as witnesses called by Dr. Oppenheimer? This is not for any purpose other than time.

Mr. GARRISON. I understand. Yes, we have 10, I think, outside of those I mentioned today. General McCormack, Walter Whitman, Dr. Rabi, Dr. von Neumann, President DuBridge, Dr.. Bacher, Dr. Lauritsen, I think President Killian, Mr. Hartley Rowe, and Mr. Harry Winne, and Norris Bradbury.

Dr. EVANS. That makes how many in all?

Mr. GARRISON. That is 11 with Bradbury.

Dr. EVANS. In addition to the ones for today.

Mr. GRAY. That is after today.

Mr. GARRISON. I would rather doubt, if it is possible for us to get prepared for redirect and I suppose some more cross examination of Dr. Oppenheimer this week, I should think that it would not be possible to hear all these witnesses this week. We had hoped to be able to do so, because we knew how much the board desired not to have to come back nor do we ourselves wish to come back, either. But I do think as we are going it means that there will be probably one day or possibly a day and a half overflow of testimony beyond what we can do in this one week.

Mr. GRAY. I think it is impossible to say that with any certainty at this point.

Mr. GARRISON. Does the board have any present idea as to when it will reconvene for the sake of hearing witnesses that it wishes to call.

Mr. GRAY. I would expect at this point that we would proceed next week for that purpose. I don't think there will be too many witnesses. Perhaps we should proceed.

Mr. GARRISON. Could you inform us who they are going to be?

Mr. GRAY. Yes, we will give you an indication. The board has not come to any final conclusion. For one thing, up until this point I don't think that we have known all the witnesses for certain whom Dr. Oppenheimer wishes to call. I am sure some of these the board would have called if he had not called them.

Mr. GARRISON. I think this is the original list that I gave you away back, Mr. Chairman. I think there may have been one or two additions since the very first day.

Mr. GRAY. Yes.

Dr. EVANS. Did you mention Dr. Bush?

Mr. GARRISON. I forgot him.

Dr. EVANS. I thought you did.

Mr. GARRISON. That makes 12. He can come on 15 minutes notice, so I had him on the side.

Mr. Chairman, I think this is just about it as I gave it to you the first day.

Mr. GRAY. All right. Can we proceed with Mr. Kennan.

Mr. GARRISON. Mr. Marks will examine Mr. Kennan.

Mr. GRAY. All right.

What is your full name?

Mr. KENNAN. George Frost Kennan.

Mr. GRAY. Do you wish to testify under oath? You are not required to do so.

Mr. KENNAN. I would be quite prepared to testify under oath.

Mr. GRAY. All the witnesses have to this point. Would you then stand, please, and raise your right hand?

George Frost Kennan, do you swear that the testimony you are to give the board shall be the truth, the whole truth, and nothing but the truth, so help you God?

Mr. KENNAN. I do.

Whereupon, George Frost Kennan was called as a witness, and having been first duly sworn, was examined and testified as follows:

Mr. GRAY. Will you be seated, please.

It is necessary for me to point out to you the existence of the so-called perjury statutes. I shall be glad to give you an indication of the penalties if you wish.

The WITNESS. I don't think it is necessary, sir.

Mr. GRAY. I want to bring them to your attention.

I should also like to ask that in the event, probably unlikely in this case, that it is necessary for you to disclose any restricted data in your testimony, that you advise me before such disclosure, in order that we might excuse any unauthorized persons.

Finally, I should like to point out to you that the proceedings and record of this board are regarded as strictly confidential between the Atomic Energy Commission and its officials participating, and Dr. Oppenheimer, his representatives and witnesses. The Commission will take no initiative in the public release of any information relating to these proceedings, and we express the hope that will be the attitude of witnesses.

Mr. Marks.

Direct Examination

By Mr. Marks:

Q. Mr. Kennan, will you please identify yourself and give the board briefly your professional history? I am told that you should be addressed as Ambassador Kennan, but that does not come quite naturally to me. If I may, I would like to call you Mr. Kennan.

A. I am now a Foreign Service Officer retired. I grew up in Wisconsin. I had my early schooling there, went to Princeton University and went almost immediately from Princeton University into the Foreign Service of the United States, where I served for 27 years without interruption. I retired from the Foreign Service last July, and am now on a regular retired status.

Q. And what is your present position, if any?

A. I am at present a member of the Institute for Advanced Study in Princeton, and in residence at the institute, engaged in certain research and writing work—scholarly work.

Q. Will you please describe in a little more detail the highlights of your experience in the Foreign Service?

A. I served first for a year or two in Geneva and Hamburg as a vice consul, and then was selected—this was away back in 1928—as one of the first group of men to be trained for special work in the Soviet and Russian field. We were at that time given rather a thorough course of training, usually 3 or 4 years of it. I was sent to the University of Berlin where I spent 2 years and took the diploma of the Oriental Seminary in Berlin, and after that I was put as a reporting officer in Berlin first in the legation of Riga before we had any relations with the Soviet Union, and after 1933 in the Embassy in Moscow.

Since that time I have had the status in the service of a specialist on Soviet affairs. I have served on 4 different occasions in the Embassy in Moscow in various ranks, the last time as ambassador.

Q. When was that?

A. That was in 1952. I have had other service relating to the Soviet Union, such as the so-called Russian desk in the State Department in 1937 and 1948. So I have been pretty closely in contact with Soviet problems for most of my career.

Q. What other main Foreign Service have you had?

A. I have served for many years in Germany.

Q. When was that?

A. In addition to my studies there in the earlier years, I studied during the war—or rather I served during the war—from the outbreak of war until Pearl Harbor, after I was interned for a time, so that I had nearly 3 years of wartime service in Germany. Also service in Austria and Czechoslovakia and in Portugal.

Q. Prior to your retirement, what was your last position in the Department of State here?

A. My last position was Ambassador to the Soviet Union.

Q. Prior to that?

A. Prior to that I was from 1947 to 1950 Director of the Policy Planning Staff of the Department of State and from the beginning of 1950 until the middle of that year counselor of the Department of State. My last official position was that of counselor of the Department of State, a position that I had only for 6 months in 1950.

Q. Could you describe in just a very few sentences what your responsibilities were as head of the policy planning staff and as counselor of the State Department?

A. The Policy Planning Staff was established by General Marshall in the spring of 1947. I was asked to found it and to determine its composition and its procedures and to head it initially, and did so. We were an advisory staff to the Secretary of State. We were there to advise him on questions with regard to which he might seek our advice or on such major questions of foreign policy and especially long term policy as we ourselves thought were in an advisory opinion to the Secretary.

The staff is still in existence. Its methods of work have changed from time to time, but has remained as a permanent unit of the Department of State, and is the only body as far as I am aware in the framework of the staff which has a universal competence. Its competence is not restricted to any geographic area or functional area.

Q. When you were counselor, what does the term "counselor" mean? Is that like Assistant Secretary or Under Secretary?

A. Counselor again is the senior advisor to the Secretary of State who has no operational responsibility in the sense that he has no division or administrative apparatus under him. He is in purely an advisory capacity. The title has existed for many decades, and is usually a man who is kept there simply on the basis of his personal experience and qualities which it is felt might be useful to the Secretary, possibly to the President as an advisor.

Q. In the hierarchy, am I right in understanding that the post of counselor is equivalent to or just under that of Under Secretary?

A. Yes, sir. It has varied. I would say it has always been between the 2d and the 4th place in the Department of State, depending on the Secretary of State and the counselor and the arrangements made at the moment.

Q. I should have asked you at the beginning a matter which I suppose everybody knows about, but which we ought to inquire about for the record.

Are you the author, I think it is fair to say, of a rather famous article, called "Sources of Soviet Conduct"?

A. I am. The article was written privately for Mr. Forrestal in December 1946 and January 1947. He had asked me for a review of another paper that he had obtained from another source on this subject, and I told him I can't comment on that, but I would be glad to give him my own views, and did in this paper.

Later Mr. Armstrong, who was head of the Council on Foreign Relations and editor of the magazine Foreign Affairs, asked me if I had anything along this line that I could submit for publication, and I did on the condition that it would be published anonymously.

To my horror, the article actually appeared after I had taken over the policy planning staff under General Marshall, and the authorship of it leaked, and it caused quite a sensation when it did appear. As far as I know, it did no damage. It had been duly cleared by the Department of State so General Marshall never held it against me. I was the author of it.

Q. I don't think Mr. Robb will object to my asking you if it has not been generally regarded as a rather robust statement of the situation of the United States with respect to the peril that we faced vis a vis the Russians.

A. It was an attempt to analyze the reasons for a pattern of Soviet behavior which surprised many people in this country in the months immediately following the war, and to suggest——

Q. You mean an unfriendly policy?

A. Yes, and to suggest an approach to this problem on our part that would be hopeful and helpful.

Q. I would like to turn now to a more specific subject and ask you what you have had to do with the problem of Soviet espionage, Soviet infiltration of agents into the United States, problems of security?

A. In the early days before our recognition of the Soviet Government when a number of us worked on the Baltic States in reading the Russian press——

Q. Do you speak and read Russian?

A. I do, sir, yes. We were rather shocked to observe the names and statements of Americans or people who held themselves out as Americans, but who were giving statements for the press in Moscow of an extraordinary nature, and ones that indicated that their allegiance was to the Soviet Union and not to this Government.

Mr. ROBB. Could we have the date on this, Mr. Marks?

The WITNESS. I would say roughly in the years between 1929 and 1933. At that time we were concerned about it. I personally brought some of these names to the attention of the Government back here, and raised the question as to whether passports could not be denied to these people because it seemed to me evident that they had expatriated themselves in every sense of the word, subjectively.

We ran up against the snarls of legislative provisions and procedural provisions of the Government, and I don't believe anything was done about it at that time.

After the recognition of the Soviet Union during the thirties, this continued. I must say, to be a source of concern to practically all of us, I think, who were professional officers in this field, and serving in Moscow. We saw people about whose intentions and activities we had great doubt. There was not much that we could do about it then from our position, except to try to see to it that those people were not used in the Embassies and that they were handled with due discretion by Embassy people. In other words, our concern there was primarily with the security of our own mission.

I may say that I think the Moscow Embassy was the first mission of our governmental service to institute proper security precautions in time of peace. We were the first people so far as I am aware who always had our code books accompanied day and night by an American in the room and never left them in the safes alone, and things of that sort.

By Mr. MARKS:

Q. When did you start these security practices?

A. From the day the mission arrived in Moscow in March 1934. We were aware of the fact——

Q. You went in with the first mission?

A. I was there prior to it. I made the physical arrangements for its arrival. I was there in the fall of 1933. We were very much aware that we could not depend on the Moscow employees, that we had to assume that all employees were sent by the Soviet police, and we could depend on no custodial employees to be secure, and we had to rely on our own sources.

We brought 9 Marine sergeants with us and tried to set this thing up on our own hook as a sound show from the point of security.

Q. What experience, if any, have you had with intelligence work using that in the somewhat broader sense than you have been speaking?

A. What I am about to say is a matter which I think violates no classification or any document in the Government, but is not one which I have ever spoke about publicly, and I say it only for the information of the board.

* * * * * * *

Q. Were you also mindful in those experiences of the earlier insights and perceptions that you had had with the nature of and difficulty of relations with Soviet Russia?

A. I felt that the earlier experience with Soviet problems and especially the security problems concerned with work in the Soviet Union stood me in a very good stead in Portugal. Russian espionage then was not our problem. It was Germans we were facing during the war. But it was, I believe, partly because of the experience I had with Soviet matters that I was selected to do this job of wartime coordination.

Q. In what connections have you known Dr. Oppenheimer?

A. I first met Dr. Oppenheimer so far as I can recall when I was Deputy for Foreign Affairs. That is equivalent to Deputy Commandant for Foreign Affairs at the National War College here in Washington in 1946. Dr. Oppenheimer lectured there. I was in charge of political instruction generally. I heard the lecture and was very much impressed by the eminence, clarity and precision and scrupulousness of thought by which it was characterized.

I then took over this responsibility as head of the Policy Planning Staff in the Department of State, and in the ensuing years until the summer of 1950, when I left the Department of State, I met Dr. Oppenheimer on numbers of occasions in the course of my work. Those occasions were practically all ones or almost all ones on which we had to work on the formulation of foreign policy in fields that required the collaboration of other departments of Government and notably the Atomic Energy Commission and the Department of Defense.

The main fields with which I was concerned were those of the international control of atomic energy, and the straightening out of our relations with our own allies, particularly the British, and the Canadians, in matters affecting our ability to obtain raw materials for the conduct of our atomic energy program here.

In attempting to meet the problems of foreign policy that arose out of those questions, we found it necessary to sit down together with representatives of the Defense Department and the Atomic Energy Commission, and to work as a group in determining our governmental positions. In the course of those consultations, I sat several times at least in rooms here in Government offices with Dr. Oppenheimer and participated in consulations in which he also participated. Some of those meetings I chaired. I remember at least one which he chaired. It depended on where we met.

Q. Were these matters on which you sat of importance? How would you describe them?

A. I would described them as matters of the greatest delicacy and of, I think, quite vital importance to the conduct of our entire atomic energy program in this country. They were all matters which were given the highest possible security classification at the time, and I do not recall that we ever had any leaks about them. They were conducted in complete secrecy.

Q. Were these matters in which issues arose involving actual or potential conflicts between positions and alternatives that our Government was considering and those that you would have expected or knew that the Russians were taking?

A. Only with respect to the international control of atomic energy was that true. I must say the bitterest problems after the time that I came in, the ones that preoccupied me most, were ones involving our effort to straighten our relations with our own allies and to place them on a satisfactory basis.

Q. In connection with the latter type of problems, were the positions that you were working toward, positions that you expected or knew to be uncongenial to the Russians or hostile to the Russians?

A. The very reason we worked so hard on these matters and took them so seriously——

Q. When you say "these matters"?

A. The questions with relation to our allies at that time. The very reason we worked so hard on them and took them so seriously was that we were aware that if the questions involved were not solved in some satisfactory manner, the only people who could gain by that would have been the leaders of the Soviet Union. They would have derived the greatest possible satisfaction and profit to their own foreign political purposes had these negotiations not been successful and had real differences and ugly differences been permitted to develop between ourselves and the British and the Canadians. I think the reasons for that are obvious. * * *

Q. These problems that you are talking about, then, concerned the raw materials or at least in part concerned the raw materials problem?

A. That is correct.

Q. Raw materials for atomic energy.

A. That is correct. They did. A collaboration was required between the 2 governments, and at the time that I came into these matters in 1947, it seemed evident to me that that collaboration was very seriously threatened by the way that events had developed to date, and it was time that both our Government and the British Government gave them the most serious thought. We did that. I think it fair to say that we were successful in tiding these relationships over a very crucial and difficult period, primarily the period of the years of 1948 and 1949.

Q. You are confident that the Russians would have profited greatly if the result had been opposite?

A. Yes. I can assure you that the source of my own alarm and concern about these matters was the conviction that if we failed to solve the problems involved, the Russians would be the gainers.

Q. Did Dr. Oppenheimer have a role of any importance in these deliberations that you have described?

A. He was one of a number of officials, people in our governmental establishment, who were concerned with these matters. I say in our governmental establishment; I do not recall exactly what his position was at that time, but he was in councils of the Government about such matters, sat in on a number of these discussions, at least 2 or 3 that I recall specifically, I think.

Q. That is on raw materials?

A. On raw materials. It is my recollection and a very vivid recollection that his participation was extremely helpful to us, so must so that I am not sure really whether we would have been able to do what we did at all without his help.

Q. I would like to remind you, Mr. Kennan, that I think during the period of years that you are referring to Dr. Oppenheimer was for the most of the time Chairman of the General Advisory Committee of the Atomic Energy Commission.

A. Yes.

Q. As a result of your experience with Dr. Oppenheimer in the cases that you have reference to, what convictions, if any, did you form about him?

A. I formed the conviction that he was an immensely useful person in the councils of our Government, and I felt a great sense of gratitude that we had his help. I am able to say that in the course of all these contacts and deliberations within the Government I never observed anything in his conduct or his words that could possibly, it seemed to me, have indicated that he was animated by any other motives than a devotion to the interests of this country.

Q. Did you ever observe anything that would possibly have suggested to you that he was taking positions that the Russians would have liked?
A. No. I cannot say that I did in any way. After all, the whole purpose of these exercises was to do things which were in the interest of this country, not in the interests of the Soviet Union, at least not in the interests of the Soviet Union as their leaders saw it at that time. Anyone who collaborated sincerely and enthusiastically in the attempt to reach our objectives, which Dr. Oppenheimer did, obviously was not serving Soviet purposes in any way.
Q. Have you said that he contributed significantly to the results?
A. I have, sir.
Q. Mr. Kennan, is there any possibility in your mind that he was dissembling?
A. There is in my mind no possibility that Dr. Oppenheimer was dissembling,
Q. How do you know that? How can anybody know that?
A. I realize that is not an assertion that one could make with confidence about everyone. If I make it with regard to Dr. Oppenheimer it is because I feel and believe that after years of seeing him in various ways, not only there in Government, but later as an associate and a neighbor, and a friend at Princeton, I know his intellectual makeup and something of his personal makeup and I consider it really out of the question that any man could have participated as he did in these discussions, could have bared his thoughts to us time after time in the way that he did, could have thought those thoughts, so to speak, in our presence, and have been at the same time dissembling.

I realize that is still not wholly the answer. The reason I feel it is out of the question that could have happened is that I believed him to have an intellect of such a nature that it would be impossible for him to speak dishonestly about any subject to which he had given his deliberate and careful and professional attention.

That is the view I hold of him. I have the greatest respect for Dr. Oppenheimer's mind. I think it is one of the great minds of this generation of Americans. A mind like that is not without its implications.
Q. Without its what?
A. Implications for a man's general personality. I think it would be actually the one thing probably in life that Dr. Oppenheimer could never do, that is to speak dishonestly about a subject which had really engaged the responsible attention of his intellect. My whole impression of him is that he is a man who when he turns his mind to something in an orderly and responsible way, examines it with the most extraordinary scrupulousness and fastidiousness of intellectual process.

I must say that I cannot conceive that in these deliberations in Government he could have been speaking disingeniously to us about these matters. I would suppose that you might just as well have asked Leonardo da Vinci to distort an anatomical drawing as that you should ask Robert Oppenheimer to speak responsibly to the sort of questions we were talking about, and speak dishonestly.
Q. Mr. Kennan, in saying what you have just said, are you saying it with an awareness of the background that Dr. Oppenheimer has, the general nature of which is reflected in the letter which General Nichols addressed to him, which is the genesis of these proceedings, and his response?
A. I am, sir.
Q. How do you reconcile these two things?
A. I do not think that they are necessarily inconsistent one with the other. People advance in life for one thing. I saw Dr. Oppenheimer at a phase of his life in which most of these matters in General Nichols' letter did not apply. It seems to me also that I was concerned or associated with him in the examination of problems which both he and I had accepted as problems of governmental responsibility before us, and I do not suppose that was the case with all the things that were mentioned in General Nichols' letter about his early views about politics and his early activities and his early associations.

I also think it quite possible for a person to be himself profoundly honest and yet to have associates and friends who may be misguided and mislead and for who either at the time or in retrospect he may feel intensely sorry and concerned. I think most of us have had the experience of having known people at one time in our lives of whom we felt that way.
Q. I think one might interpret this correspondence that I have referred to as going even further than that. I won't go into what has been testified here or a characterization of that which has been said in this room, but in the correspondence itself, an incident is referred to—I assume you have read the correspondence?

A. I have in a cursory way as a newspaper reader reads it in the newspapers.

Q. An incident is referred to in 1943, in which it is said that an approach to Dr. Oppenheimer was made under circumstances suggesting that the approach was somehow connected with a possible effort by the Russians to secure information or to secure information in their behalf, and that for some months thereafter he failed to report this incident.

What effect does that failure on his part which he freely admits was wrong have on your present thinking about it?

A. Mr. Marks, I have testified about him here as I have known him. I can well understand that at earlier periods in his life conflicts of conscience might have arisen as I think they could with any sensitive person between his feelings about his friends—perhaps his pity for them—and his governmental duties. On the other hand, I would also be inclined to bear in mind the fact that in 1943 the Soviet Union was hardly regarded by our top people in our Government as an enemy. That great masses of American materials were being prepared for shipment to the Soviet Union, many of them I assume involving the transmission of official secrets. I could imagine that the implications of this may not at that time have appeared to be so sinister as they do today in retrospect, and I could also imagine if after all the information was not given in this particular instance, the man in question might have felt that no damage had been done to the Government interest, and that the question of the men who had initiated such a request might be better perhaps left to their own consciences and to the process of maturity in their own development.

I don't know. I can imagine those things. For that reason I would hesitate to make definite judgments on the basis simply of what I read in the letter of indictment.

Q. Would it change your opinion if I were to suggest to you that when Dr. Oppenheimer did report this incident to security officers on his own initiative, as it turned out, he didn't tell them everything about it. He still withheld the name of the friend and told them a story that was not the whole truth.

A. Mr. Marks, I do not think that that would alter anything on the statement that I just made prior to your question. I might only add to it that I could well conceive that Dr. Oppenheimer might have done things which he would think in retrospect were mistakes or which others would conclude in retrospect were mistakes, but that would not preclude in his own instance any more than it would in the case of any of the others the process of growth and the ability to recognize mistakes and to learn from them and to make fewer in the future. What I have said about his activities, his personality, the cast of his mind during the years when I knew him would I think not be affected.

Q. These convictions that you have expressed about him, the confidence that you have expressed in him, what part is played in that judgment by the experience that you had as a Soviet expert?

A. I think a considerable part. One of the convictions that I have carried away from such experience as I have had with these matters in the field of Soviet work concerning the Soviet Union is that these things cannot really be judged in a fully adequate way without looking at the man as an entirety. That is I am skeptical about any security processes that attempt to sample different portions of a man's nature separate from his whole being. I must say as one who has seen Robert Oppenheimer now over the course of several years, and more latterly outside of Government, that I have these feelings and entertain them on the basis of my estimate of his personality and his character as a whole.

Q. Are they feelings or are they convictions?

A. They are on my part convictions, sir.

Q. Mr. Kennan, let me turn now to a quite different subject. In your capacity as head of the policy planning staff in the State Department, were you ever consulted about the problem of the hydrogen bomb which came up, to refresh your recollection of the date, toward the end of 1949?

A. Yes, I was consulted by the Secretary of State in that connection, although I was not asked and could not really properly have been asked to give an opinion to him officially as to whether we should or should not proceed to the development of this weapon.

My recollection is that——

Q. Would you wait just a minute? I need to ask Mr. Garrison a question. May I have a 30-second interval here? I need to ask Mr. Garrison about a matter.

Mr. GRAY. Yes.

Mr. MARKS. Thank you very much.

By Mr. MARKS:

Q. I was about to ask you what were the circumstances under which you were consulted.

A. I can only give my recollection here, and I must say my recollection of all these official matters at that time are somewhat telescoped and entirely capable of being in error with regard to details. But the recollection is simply this. When it was first made known to the Secreaty of State that there was a technical possibility of going ahead with the development of this weapon, at least to the extent the Government now had before it a decision as to whether to develop the weapon or not——

Q. The question of making it.

A. The question of making a decision as to whether to attempt to develop the weapon or not. When that state of affairs was first brought to the attention of the Secretary of State, he at a very early stage there asked me into his office. My recollection is that Dr. Oppenheimer was there, and there may possibly have been one or two other people, but I do not remember who they were. We spoke about this and the only thing I can remember, I think, of that conversation is that we were all agreed that regardless of how the decision might fall, it was important that this Government should reexamine its position with respect to the international control of atomic energy to make sure that nothing had been left undone from our side to get international agreement about these weapons, before we proceeded with this program of the hydrogen bomb.

In other words, we wanted to make absolutely certain that before launching on this new phase of the atomic weapons race, our position in the United Nations on the international control of atomic energy was the best position that we could devise, and most hopeful one.

The Secretary of State asked me to reexamine this question, to have another look at our international negotiation position as we had exposed it in the United Nations bodies with regard to the international control of atomic energy, and to see whether that was still sound, whether anything had happened in the circumstances of the preceding 2 or 3 years since we had advanced it to change the assumptions on which it rested, whether there was anything more that we could now propose which might have a chance of putting an end to the atomic weapons race instead of facing us with the necessity of going ahead with this.

I did look at this problem in the course of the ensuing weeks and my recollection is that I gave my opinion to the Secretary of State in January 1950 on that subject.

Q. I take it that on at least one or perhaps more occasions in the course of carrying out this assignment or at least the initiation of it you heard Dr. Oppenheimer express his views.

A. I recall going to Princeton in the fall of 1949 on one occasion. I had several things to do there. I called on Dr. Oppenheimer at the Institute if my memory is correct, and we discussed it then. I was also once at some time in that period—I don't know exactly when—asked to appear before the General Advisory Committee of the Atomic Energy Commission, simply as a consultant. They wanted to hear my views. They asked me questions. The questions related primarily to the present state of our relations with the Soviet Union, the state of what we called the cold war. I replied as frankly as I could to them.

Q. What impression did you get, if your remember it, of Dr. Oppenheimer's views?

A. I would not be able to quote his views in memory or in any detail or in any great accuracy. I can only say that the general impression I carried with me was the impression of a man who was greatly troubled by what he felt to be the extremely solemn implications of this decision.

Q. That is the pending decision?

A. The pending decision. Who realized that it was one the implications of which might carry very far. That it was almost impossible to predict where we might end up if this sort of a race with weapons of mass destruction were to go on indefinitely, and therefore was greatly troubled and concerned to arrive at the most enlightened and sound decision that could be made.

Q. Did he try to sell you on any view?

A. It is not my recollection that he did. I fear that I talked more about my own views here than he did about his with regard to this subject. But I do not have the recollection that he endeavored to persuade me that any answer to this problem was the right one or the wrong one. To me, then, we were still at a preliminary stage in it. The entire effort really on the part of both of us then was to try to identify the considerations that were relevant to the problem to see what we had that we could really hang onto in approaching the decision.

Q. When it came time for you to give the Secretary of State your views or your analysis of the problem, what did you report to him, and when was it approximately?

A. I reported to him approximately in the month of January, I would think around the middle of the month or shortly after.

Mr. ROBB. 1950?

The WITNESS. 1950, yes. The gist of my own views was simply this: I felt that this Government was in no way in good position to make any great decisions with regard to either the international control of atomic energy or actually with regard to its own weapons program before it gained greater clarity in its own mind as to the purposes for which it was holding what were sometimes called the A, B, C, weapons in general. By that I am thinking of the weapons of mass destruction, the atomic, chemical, and so forth. It seemed to me that there was unclarity in the councils of our Government as to the reasons why we were cultivating and holding these weapons. The unclarity revolved around this question. Were we holding them only as a means of deterring other people from using them against us and retaliating against any such use of these weapons against us, or were we building them into our military establishment in such a way that we would indicate that we were going to be dependent upon them in any future war, and would have to use them, regardless of whether they were used against us first.

By Mr. MARKS:

Q. Have we not taken the position that we would only use them for purposes of retaliation?

A. It is not my impression that we have, and it was not my impression at that time that there was any such determination in the councils of the United States Government.

On the other hand, if I remember correctly, I was able to cite statements that had been made by some of our high military leaders—I think both in the councils of this Government and in the NATO councils of Europe—which indicated very strongly that we were getting ourselves into a position where we would have to use these weapons as forward military weapons, regardless of whether they were used against us.

The point that I tried to emphasize to the Secretary of State related, of course, directly to the question of international control about which I had been asked. I told him that I thought we ought first to face this problem. It was my belief that we should hold these weapons only for purposes of retaliation and as a deterrent to their use against us. That anything else would get us into a race with these mass destruction weapons to which I could see no end, which I was afraid would distort the thinking of the public mind about problems of foreign policy and military policy in this country if it were permitted to proceed. So as I say, I favored the holding of these weapons only for purposes of retaliation and as a deterrent.

Whether that came out clearly in my report to the Secretary of State, I do not know, because that was not actually the question that was asked me. But I am sure it was implicit in what I said to the Secretary, and by the same token I think it was implicit that we ought really to make this other decision before we made decisions about the hydrogen bomb.

Q. Mr. Kennan, you will have to explain a little more to me at least what you conceived to be the relevance of clarification of this question to the question of whether or not we ought to proceed with making hydrogen bombs.

A. Yes. As I saw it, the relevance was this. If you were asked, should we or should we not proceed to the development of a whole new range of more powerful atomic weapons which was involved in the hydrogen bomb decision, you had to ask yourself how much do we need the weapons of mass destruction in general. That is the first question that had to be faced, because if you already had enough, perhaps you didn't need the hydrogen bomb at all. I could not see how you could answer the question of how much do we need until you had answered the question of why are we holding these weapons anyway, and what do we expect to accomplish with them.

If you were holding them as deterrents and for purposes of retaliation, really for purposes in order that they might not be used against you, then what you needed was merely enough to make it an unprofitable and unpromising undertaking on the part of anyone else, the Russians in particular, to use these weapons against us.

If on the other hand you were going to regard them as an integral part of forward American military planning and something on which we would be dependent in a future war, regardless of the circumstances of the origin of that war, then you came up with a different answer or you might come up with a different one in regard to the hydrogen bomb.

Q. So the point you are making is not that you were opposed to the hydrogen bomb necessarily, but only it seemed to you that it was essential first that this other subsidiary question should be clarified?

A. That is correct. I must say that personally while I was not competent to form a finished opinion on this and was never called upon to do so, I had not at that time seen the evidence that what we already held in the old and regular atomic bomb, I may speak of it that way, was not enough to make it a fruitless undertaking from the standpoint of Soviet policy to launch a war on us with these weapons.

In other words, I considered the burden of proof to rest on that point. It seemed to me you would have to prove that we could not do the job with the weapons we already had, and to my knowledge that was never demonstrated to me at the time. Perhaps the answer might have been one thing or the other, but I had never seen the proof.

Mr. MARK. I think that is all, Mr. Robb.

CROSS-EXAMINATION

By Mr. ROBB:

Q. Mr. Kennan, that was a most interesting discussion. I certainly have enjoyed it.

A. Thank you, sir.

Q. Mr. Kennan, I was interested in your description of your security precautions which you took over in Russia. I believe you said you brought in six Marine sergeants to assist.

A. That is correct, sir.

Q. How did you happen to turn to the Marines, rather than the State Department?

A. The person who deserves the credit for that was Ambassador Bullitt, our first ambassador to the Soviet Union. Mr. Bullitt had very strong feelings about security and had, I believe, had something to do with the Navy. I asked to be excused here; at one time or another he was Assistant Secretary of the Navy, or in any case he knew people in the Naval Establishment, and he asked President Roosevelt to arrange it and get Marine sergeants.

Q. He was something of an expert on Russian espionage, wasn't he?

A. At least he was very security conscious, by that time, and was helpful, I must say, in that way.

Q. Did you give these Marines a pretty thorough checking over before you brought them into the Embassy?

A. I don't believe so. These things were rather primitive compared to our present standard today. We left that to the command staff of the Marine Corps.

I must say, though, I think they were very hearty and loyal Americans, the fellows we got. Our difficulties with them were not ones of security. They were other kinds.

Q. I can imagine that. Supposing you had learned that one of these Marines or anybody else who had to deal with your security matters said that he had recently been a member of the Communist Party, but had left the party just before coming to your Embassy; would you have had him around?

A. I think our tendency would certainly have been to urge that he not be in the Moscow Embassy at that time. He would presumably have had still some contacts with people in Moscow which would have been undesirable.

Q. Or if he had any close conections with the Communist Party, I assume you would not have been very enthusiastic about having him around them, would you?

A. That is correct, for our purposes there in the embassy.

Q. Have you had much experience, Mr. Kennan, with Communists—I just don't know how to express it—are you familiar with Communist dogma or technique?

A. I think I am, sir. I have had about 20 years of reading the Soviet press and some times other press organs with the view to determining whether they reflected that type of dogma or not. I feel I have a certain familiarity with it.

Q. Would you place much weight in a statement of a Communist that he just left the party or had disassociated himself with it before coming on some secret work for the Government?

A. I would certainly regard it as a factor very seriously relevant to fitness for office, but one to be examined individually. You asked a moment ago about the case of our Embassy out there. Mr. Bullitt for whom I had the greatest respect, and about whose security I never had the faintest doubt, had been married to the widow of John Reed, who was the first prominent American Communist, I suppose, in this country. We didn't find that a source of worry with regard to Mr. Bullitt.

Q. No, I am talking rather than matrimonial association, more active association with the Communist Party. Would you tend to view with considerable skepticism a statement of a man who admitted that he had been an active member of the Communist Party or had been active in Communist affairs, a statement of such a man that he had just left the Communist Party or left the Communist affairs on the eve of coming to work in the Embassy? Wouldn't you view that statement with some skepticism?

A. I think we would have regarded it as a factor which meant that there was a certain burden of proof to demonstrate that the man's value to us was very great, and that this could be satisfactorily explained away, and we had something that we could depend on in judging that he was now a person whose loyalty we didn't need to worry about.

Q. Just for the record, Mr. Kennan, I think it is plain, but was it 1946 that you had these discussions with Dr. Oppenheimer down at the War College?

A. I don't recall discussions down there except possibly after his lecture, but it was in 1946 to my recollection that he lectured there, and that I first met him.

Q. Was that the year when you were taking various positions which Mr. Marks said would not be accepted by the Russians with much favor? Was that the year 1946?

A. No; it was the following year.

Q. 1947?

A. 1947.

Q. Mr. Kennan, of course you don't know anything about what Dr. Oppenheimer testified before this board, do you, sir?

A. I know nothing whatsoever about it, sir. I have not discussed it with anyone.

Q. Coming to your discussion of the problem which confronted you gentlemen when you were deciding whether or not to go ahead with the hydrogen bomb, do I understand, Mr. Kennan, that your thought is that whether we wanted the hydrogen bomb merely for retaliation or whether we wanted it for affirmative action, if I may put it that way, in either event we wanted the bomb?

A. No. My feeling is that until you decided that first question, you didn't know whether you wanted the bomb or not.

Q. I see.

* * * * * * *

Q. Which bombs are you talking about?

A. Even the old-fashioned kind. You must remember that these men since the Revolution in these 38 years that have transpired since the Revolution have with great trouble and pain succeeded in building up a certain amount or a considerable amount of industry in Russia. That is their pride and joy politically. That is the thing that they claim they were going to do, to industrialize this country. Their aim has been to catch up with and overtake America, and their great boast is that in a primitive and partially underdeveloped country, they have succeeded pretty much with their own resources in producing now major industry.

* * * * * * *

Q. Mr. Kennan, did you have any view in 1950 as to whether or not the Russians would attempt to develop the hydrogen bomb whether we did or not?

A. I do not recall specifically. I think I may have doubted that they would proceed to the development of it, and I think I may have been in error on that point, as I look at it today.

Q. Do you have any doubt now that they would have whether we did or not?

A. I am still not sure that they would have because I am not sure—I don't know enough about the scientific and the economic aspects of this problem—to know how worthwhile they would have regarded it. It may perfectly well be that they would have said the hydrogen bomb will call for this and this amount of investment in scientific personnel and materials, and perhaps we would be better off to put that investment into the older type of atomic weapons.

Q. That was more of a scientific question that you were not qualified to deal with.

A. I was not qualified to deal with it.

Q. I would like to ask you a question as an expert on diplomacy, Mr. Kennan. Supposing the Russians had developed the hydrogen bomb, and had got it and we didn't have it; what would then be our position vis-a-vis the Russians in any negotiations?

A. That, of course, is a key question and a very penetrating one. It is one which I have had occasion to argue many times with my friends here in Washington. I do not think that the position would have been so much different from what it is today. The Russians have for reasons which I don't think include any altruism or any thing like that, or idealism, but they have been very, very careful not to use the weapons of mass destruction as a threat to other people. I don't recall any time that the Russians have ever threatened as a means of political pressure to use these weapons, to use these weapons against anybody else. On the contrary, their position has been consistently all along that they were holding them—whether this is true or not, it has been their public position—that they were holding them for purposes only of retaliation and deterrents and would not use them unless they were used against them.

It would be a change of Soviet policy if they were to attempt to use any of these weapons as a means of pressure. I have also always held doubts—I realize this is a very difficult thing to express—as to whether the fact that perhaps one party had weapons of this sort a little more destructive or greatly more destructive than the other would nevertheless change this situation so vitally. We did, after all, have the old type of bomb. We had some means of delivery. I think the world would have gone along pretty much the same. I have in mind in making that judgment the fact that atomic weapons are not the only weapons of mass destruction that exist. There are also extremely ugly and terrible biological and chemical weapons, at least we have been allowed to think there are, and if the Russians want to create destruction in this country solely for the sake of destruction, I think there are other means by which they can do it than the hydrogen bomb.

Q. You don't feel, then, that we would have been at any disadvantage as against the Russians if they had the hydrogen bomb and we had not?

A. I am not absolutely certain. I cannot give you a flat negative answer to that. Perhaps we would have been. Perhaps I have been wrong about this. But I think that our position with regard to them has depended much less on the mathematical equation of who has this and who has that in the way of weapons of mass destruction than we think it has. After all our problems with them as I have seen them on the political side were very much the same in the days when we had the monopoly of the atomic weapon as they are today to my way of thinking. They are pretty much the same old problems. I really do not suspect these people, Mr. Robb, of a desire to drop this thing on us just out of some native contrariness or desire to wreak destruction for destructions' sake in this country. I think they are people who fight wars for very specific political purposes, and usually to get control over some area or territory contiguous to what they already have.

I have often had occasion to say that there is only one real question that interests these people, I mean the Soviet leaders, and that is the question of who has the ability to haul people out of bed at three in the morning and cause them to disappear without giving any accounting for them, and where. In other words, who can exercise totalitarian police power over a given territory, and where can you do it. That is what they are interested in knowing. They think that everybody else rules the way they rule. They are always interested in the territorial problem. For that reason I don't think that these weapons play such a part in their thinking as they play in ours. They want to know not only how to destroy territory, but how to get control of it, and dominate it and run people.

Q. Of course, you will agree that if you were mistaken in that evaluation, it would be a very serious mistake.

A. I agree and for that reason I have, I believe, always had a certain caution with regard to my own views.

Q. Yes, sir. Mr. Kennan, you spoke of the Russian policy as manifested to you. Do you believe the Russians were sincere in their manifestations to you of their policy?

A. Oh, no. We have never drawn our judgments of their policy from a literal interpretation of their words. There is no reason why these people should ever have been sincere in anything that they said to a capitalist government. They may have been on occasions, but there is no real reason for it.

Q. Putting it in the language of the ordinary man, you just can't trust them, isn't that right?

A. That is correct. They do not really expect to be trusted.

Mr. ROBB. Thank you very much.

Mr. GRAY. May I ask you some questions, Mr. Kennan?

First of all, may I assume that you are familiar in general terms with the Atomic Energy Act of 1946, and therefore some of the framework within which this Board is operating? I would be glad to go into it, if you wish.

Against that background, and with all the facts which are coming before us in these proceedings, you are aware, of course, that this Board faces very difficult decisions. I don't want to make statements for you but would you think that we face very difficult decisions in this proceeding?

The WITNESS. I do. There is no doubt about it.

Mr. GRAY. I am sure you are here to be helpful in this inquiry, I trust, therefore, that you will not misapprehend any questions I ask which are quite serious and relate to some of the deeper issues involved.

You have testified, I think, without reservation as to your judgment of Dr. Oppenheimer's character and loyalty as you have known him and on the basis of your knowledge.

The WITNESS. That is right, sir.

Mr. GRAY. In your experience in Government, have you ever known well any persons whose loyalty and character you respected and admired about whom it developed that you perhaps were later mistaken on account of issues we are talking about in this inquiry?

The WITNESS. I am wracking my memory here. I can recall people I have respected and admired who later turned out to be even in my own opinion unfit for Government service by virtue of personal weaknesses. I do not recall anyone who was ever a friend of mine and with whom I had any degree of association in the discussion of political matters relating to the Soviet Union who later turned out to be a person unfit for Government service by virtue of any disloyalty or of any ideological weakness. I cannot recall any such person.

There have been 1 or 2 times, Mr. Chairman, when I have been obliged to draw to the attention of the Government circumstances with regard to Government employees which seemed to me to point to a likelihood that they were not loyal American citizens. I have done that on occasions. I was not competent to make a final decision as to whether they were or were not. But I have had to report circumstances which looked to me to be suspicious and I believed were. But those were people with whom I was not closely associated. They were minor employees. What I happened to know about them were things I was able to observe in the course of official work.

Mr. GRAY. If you were today director of the policy planning staff and there came to you from a staff member or from some other source, perhaps even the Secretary of State, that a certain individual had been made a member of the policy planning staff who had had close Communist associations as late as the late thirties or perhaps early forties, would you seriously consider adding such a person to your staff today?

The WITNESS. It would depend, Mr. Chairman, on what I would think were his possibilities for contribution to the staff and to what extent the negative points on his record had been balanced out by a record of constructive achievement and loyalty. I might say by way of example that when I first set up the staff I rejected one man who had been recommended to me actually by higher authority in the Government because he had appeared as a character witness for a man who was convicted as being a Communist, and I thought at best his judgment was bad. But I rejected in that instance this man who had no previous record of experience in the Government; I was not under the impression that his contribution would be a major one, or that it would be worthwhile doing it in that case.

I must say if it were a person of outstanding capabilities and especially a person who had in addition to the negative factors rendered distinguished service to the Government, then I would want to look at it very hard.

Mr. GRAY. I assume that if it were a secretary, for example, or clerical assistant, that it would be easier for you to decide that the person should not be employed.

The WITNESS. I would think that would be correct.

Mr. GRAY. So I gather that you feel that perhaps the application of individual judgment increases with the stature and importance of the individual concerned. That is perhaps not a clear question.

The WITNESS. I do feel this, that the really gifted and able people in Government are perhaps less apt than the others to have had a fully conventional life and a fully conventional entry, let us say, into their governmental responsibilities. For that reason I think that while their cases have to be examined with particular care, obviously for the reasons of the great responsibilities they bear and the capabilities for damage in case one makes a mistake, nevertheless it is necessary to bear in mind in many cases, especially people who have great intellectual attainments—because those attainments often it seems to me do not always come by the most regular sort of experience in life, they are often the result of a certain amount of buffeting, and a certain amount of trial and error and a certain amount of painful experience—I think that has to be borne in mind when one uses people of that sort.

I agree it presents a special problem, not an easy one for the Government. I have the greatest sympathy for the people who have to face it.

Mr. GRAY. You in your testimony referred to the possible conflicts of conscience a man might have and you used the expression, I think, pity for friends who perhaps have been misguided. I am not sure those were the words, but the general import.

You perhaps are aware that under the act, one of the criteria imposed by the language of the act seems to be the associations of an individual. I know you feel that past associations must be weighed in the light of more recent conduct and other factors you have stated.

Would you feel continued association with individuals falling in this category for whom one would have pity and with respect to whom one might have had conflicts of conscience, was important at all in the situation?

The WITNESS. I would think, Mr. Chairman, that it is a thing which would have to be explained, but I find great difficulty in accepting the belief that a man must rule out all those associations, whether or not they engage in any way his official responsibilities. I think there are certainly times when they are to be avoided. I suppose most of us have had friends or associates whom we have come to regard as misguided with the course of time, and I don't like to think that people in senior capacity in Government should not be permitted or conceded maturity of judgment to know when they can see such a person or when they can't. If they come to you sometimes, I think it is impossible for you to turn them away abruptly or in a cruel way, simply because you are afraid of association with them, so long as what they are asking of you is nothing that affects your governmental work.

I myself say it is a personal view on the part of Christian charity to try to be at least as decent as you can to them.

I realize that it is not advisable for a man in a position of high security to be seen steadily with people about whose loyalty there is a great doubt, unless they happen to be intimates in his family or something like that.

Mr. GRAY. But when you say intimates of his family, you mean blood relationships?

The WITNESS. Something of that sort.

Mr. GRAY. Or marital relationships and things of that sort.

The WITNESS. Yes.

Mr. GRAY. You said an individual should not decline to see such a person if the approach were made by such other person. Would you think it would be questionable if a person in a high position took the initiative himself in seeing one of his former associates about whom there might be some question?

The WITNESS. It is difficult for me to judge in the absence of the knowledge of the circumstances.

Mr. GRAY. I understand.

The WITNESS. I am aware of this as a very difficult problem of professional ethics. It seems to me once or twice I have had conflicts of this sort myself, but I know that in these cases I would always like to have felt that my superiors in Government had enough confidence in me to let me handle that problem according to my own best conscience. I do worry about the sort of schoolboy relationship to one's friends and acquaintances which gets involved if you apply too rigid standards of security in that respect.

Mr. GRAY. But you would always feel that in any conflict between loyalty to a friend and obligation to government, it would not be a conflict difficult to resolve?

The WITNESS. No, sir; it would not. There is only one way in which it can be resolved, and that is in favor of the Government. If that is impossible, then I would say a man should resign. He should not permit himself to remain in the Government with any conflict of loyalties of that sort.

Mr. GRAY. One of the hard facts of our times of course is the inevitable conflict of the requirements of what we generally refer to as security and what we like to think of unlimited freedoms of man's mind and conscience. This is maybe a major dilemma of our times, at least in this country.

The WITNESS. May I add one thought to what I said before in reply to your question?

Mr. GRAY. Yes.

The WITNESS. I see as one of the most difficult aspects of this problem the trouble that the individual Government official has in arriving at an assessment of the reliability of his friends. I have continued to accept as friends some people who have been criticized publicly and on whose reliability some suspicion has been thrown publicly in this country, because I myself have never seen yet the proof that those charges were correct, and have not considered myself in a position to arrive at a negative judgment about this. I have felt that until it is demonstrated to me that people who are friends of mine really have been guilty of some genuine dereliction of their duty to the Government or their loyalty to the Government, it is not for me to jump to conclusions about it, and out of a timidity lest my name be affected with theirs to cut off social relations with them.

I must say when it is demonstrated to me that anyone has been so derelict, then I have no desire to continue the friendship or the association, and especially if I were in Government service I would consider it quite out of the question. But there have been many instances in which one has been torn between the fact that doubts have been raised, but proof has not been given. There I feel that the burden of proof so far as one's relations with one's friends is concerned is on the accuser. Unless it is demonstrated to me that my friend in some way offended against the law or against his governmental duty, I am slow to drop my friend myself.

Mr. GRAY. I would like to move back to the question of your attitudes toward the development of the hydrogen bomb in the period before the President's decision to proceed in January of 1950. Had you been told, Mr. Kennan, in 1949, for example, by a scientist whose judgment and capability you respected that it was probable that a thermonuclear weapon could be developed which would be more economical in terms of the use of material and cost and the rest of it than the equivalent number of atom bombs, would you have then been in favor of developing the hydrogen bomb?

The WITNESS. I would not have favored developing it at least until a real decision had been made in this Government about the role which atomic weapons were to play generally in its arsenal of weapons. I would have had great doubts then about the soundness of doing it. That comes from philosophic considerations partly which I exposed to the Secretary of State, which did not I might say meet with his agreement or with that of most of my colleagues and the future will have to tell, but it seemed to me at the end of this atomic weapons race, if you pursued it to the end, we building all we can build, they building all they can build, stands the dilemma which is the mutually destructive quality of these weapons, and it was very dangerous for us to get our public before the dilemma, that the public mind will not entertain the dilemma, and people will take refuge in irrational and unsuitable ideas as to what to do.

For that reason I have always had the greatest misgivings about the attempt to insure the security of this country by an unlimited race in the cultivation of these weapons of mass destruction and have felt that the best we could do in a world where no total security is possible is to hold just enough of these things to make it a very foolish thing for the Russians or anybody else to try to use them against us.

Mr. GRAY. So you would have been in favor of stopping production of the A bomb after we had reached a certain point with respect to the stockpile?

The WITNESS. That is correct.

Mr. GRAY. Whatever that might have been?

The WITNESS. No; and I didn't consider myself competent to determine exactly what that point was. I have never known the number of our bombs nor the real facts of their destructiveness or any of those things.

Mr. GRAY. Knowing the Russians as you do—perhaps as well as any American—would you have expected them to continue to improve whatever weapons they may have within limitations of economy, scientific availability and so forth?

The WITNESS. My estimate is that they would have cultivated these weapons themselves primarily for the purpose of seeing that they were not used, and would have continued to lay their greatest hopes for the expansion of their power on the police weapons, the capacity to absorb contiguous areas, and on the conventional armaments as a means of intimidating other people and perhaps fighting if they have to fight.

Mr. GRAY. I have one final question. Were you opposed to the use of the atom bomb?

The WITNESS. I knew nothing about it sir, until I read it actually in the Soviet papers in Moscow, that it had been used.

Mr. GRAY. You were in Moscow?

The WITNESS. I was in Moscow at the time and therefore could not look at it—I could look at it only retrospectively. I must say that personally I am not at all sure that we were well advised to use it. I have great fears of these things.

Mr. GRAY. Do you think we perhaps were ill advised to develop it?

The WITNESS. No; that I don't think.

Mr. GRAY. I said I had just one question and I am sorry I am going to ask you another. The atom bomb was many times as powerful as any explosive we had prior to its development. The same is true, I suppose, of the H bomb. I don't know what the geometric progressive relationship would be, but that is unimportant. You had a serious question about proceeding with the hydrogen bomb. No question that we should have done what we did with respect to the development of the atom bomb.

Is the different attitude on this due to the fact that perhaps an atom bomb properly placed could take care of a target and that a larger bomb would be unnecessarily large. Is it size? Is that the distinction you make? Is it because the civilian population may be involved more deeply?

The WITNESS. It is because of the wonder on my part as to whether we did not already have enough of this sort of terrible ability to commit destruction. At least I had not seen it proven to me that we needed more perhaps. Perhaps there again with some of us civilians it becomes hard for us to absorb the mathematics of destruction involved in these things. To my mind the regular old bomb made a big enough bang, as big as anybody could want. I found it difficult—you see what has worried me, Mr. Chairman, about going ahead with this is that we would come to think of our security as embraced solely in the mathematics of whatever power of destruction we could evolve, and we would forget our security lies still very largely in our ability to address ourselves to the positive and constructive problems of world affairs, to create confidence in other people.

I am convinced that the best way to keep our allies around us is not to pay outwardly too much attention to the atomic weapons and to the prospect of war, but to come forward ourselves with plans that envisage the constructive and peaceful progress of humanity. I realize that while we do that we have to preserve an extremely alert and powerful defense posture at all times. But I believe in preserving that posture to the maximum, and talking about it to the minimum, and then limiting ourselves in our foreign policy primarily to the constructive rather than negative objectives.

I have feared that if we get launched on a program that says the only thing we are concerned to do in the development of atomic weapons is to get as much as possible as rapidly as possible, that the attentions of the public and the Government will become riveted to that task at the expense of our ability to conduct ourselves profitably in positive aspects of foreign policy. That has been the nature of my worry.

I have never felt a great degree of certainty about this and I have always realized it was a very difficult problem. But it did seem to me at that time, and it seems to me still in retrospect, that one could doubt the desirability of going ahead with this weapon then from motives which were very serious and respectable motives. In other words, one could doubt it out of a devotion to the interests of our country. At least I feel that I did. Very often today when I read the papers, it seems to me that some of the things I feared at that time are beginning to develop in some degree.

Dr. EVANS. Mr. Kennan, there are a couple of questions I want to ask you. You will admit, I suppose, that at one time in his career, Dr. Oppenheimer displayed that he was a rather naive individual. You will admit that, won't you?

The WITNESS. That I think is apparent from the exchange of correspondence that I read in the papers.

Dr. EVANS. Now, another question. Because a man has had some communistic connections, he might be placed sometimes in an entirely different position in regard to security from a man that had not had those connections would be placed, is that true?

The WITNESS. I think that is corect. It appears in a different light.

Dr. EVANS. You understand the position that this board is in, don't you?

The WITNESS. I believe I do, sir.

Dr. EVANS. We have to decide on these things in regard to character, associations and loyalty. This is not a job that any of us sought. You understand that.

The WITNESS. I do.

Dr. EVANS. We didn't want it.

The WITNESS. I do.

Dr. EVANS. I don't want it today. We all know Dr. Oppenheimer's ability. Nobody knows better than I do. This act mentions certain things—character, associations, and loyalty. It doesn't say in there anything about the outstanding ability which is mentioned here so much. You understand that point, don't you?

The WITNESS. Yes.

Dr. EVANS. Perhaps the act ought to be rewritten. I don't know. I just want you to understand the position we are in. It is not a pleasant position.

The WITNESS. I do, sir.

Dr. EVANS. Now, just one other question. You opposed this hydrogen bomb on two grounds—on moral grounds and on the fact it was so big it would be like using a sledge hammer to kill a mosquito. Is that true?

The WITNESS. I have never conceived them really as just the moral ground because I didn't consider that. After all, we are dealing with weapons here, and when you are dealing with weapons you are dealing with things to kill people, and I don't think the considerations of morality are relevant. I had real worries, sir, about the effects of this on our future policy and suitability of our future policy.

Dr. EVANS. That is all.

REDIRECT EXAMINATION

By Mr. MARKS:

Q. Mr. Kennan, I would like to follow up briefly the question that you were asked by Dr. Evans about the problem which this board faces, and the test it has to apply in discharging its rather awesome responsibility, is one in which it has to assess, as I read the act, character, associations and loyalty of the individual, advise the Commission whether the Commission should determine that permitting the individual to have access to restricted data—a term which I believe you understand—will not endanger the common defense and security.

In answer to a question—I think it was addressed to you by the Chairman—about the relationship between a case involving a stenographer—Mr. Robb asked you about a case involving a Marine—the natural question also arises whether different standards should apply to an extraordinary individual.

I would not suggest to you any question which implied that different standards should apply, but I would like to explore your own views about what standards you had in mind when you said that in relation to gifted individuals, it was common to find that they had unconventional backgrounds, and that therefore, as I understood it, a different type of inquiry was required for evaluation. Could you explain a little bit more fully what you had in mind?

A. It is simply that I sometimes think that the higher types of knowledge and wisdom do not often come without very considerable anguish and often a very considerable road of error. I think the church has known that. Had the church applied to St. Francis the criteria relating solely to his youth, it would not have been able for him to be what he was later. In other words, I think very often it is in the life of the spirit; it is only the great sinners who become the great saints and in the life of the Government, there can be applied the analogy.

I have often said it is the people who have come to their views through the questioning of other things who have the highest and firmest type of understanding in the interests of the Government. At any rate, it seems to me that the exception people are often apt not to fit into any categories of requirements that it is easy to write into an act or a series of loyalty regulations.

I feel that one ought to bear that in mind. I realize the problem for the Government as to how it is to do it, and technically it is not always easy. It is a dangerous thing to talk exceptions because nobody can define again by category who is an exceptionally gifted person and who is not. The attempt is often invidious and involves the creation of an invidious distinction.

I am not sure it can be formalized, but I have always felt that the United States Government has to realize that it has a real problem here, particularly with the people who have the greater capacities. There is need here for considerable flexibility, and as I say at the outset, I think for a looking at the man as a whole and viewing his entire personality and not judging portions of it.

I am afraid that may not be a very clear answer to what you asked.

Q. Many people would say, Mr. Kennan, that you are a gifted individual. I know of nothing to suggest that you came to the Government and remained in it for so many long years of great service as the result of any unconventional background. How do you reconcile those things?

A. I consider myself to be a fortunate man. At the age of 23, at a time when many American young people of good education were drifting into what I think was an unsound approach to life, I was sent out to the Baltic States. I saw the square where the Bolshevik commissars had only recently been shooting their hostages. I saw the building on Elizabeth Street in the cellar of which they had done their torturing. I was affected from the beginning by a sense of the grotesque injustice of taking a whole class as they did, the bourgeoisie of these countries, and punishing them just because they were classifiable as bourgeoisie. I must say I was so affected by what I saw of the cruelty of Soviet power that I never could receive any of its boasts about social improvement with anything other than skepticism. I think that experience helped me a great deal at an early date, and helped me to avoid mistakes that I might otherwise have made.

Later it fell to me very deeply in Russian literature and German literature, and I have had to go through all that. It has developed in me as I think in long foreign residence it does—I was abroad 18 years, and a deep acquaintanceship with the thinking of other people—it has involved me sometimes in conflict when I come home. I find myself tending to be critical sometimes of conditions in our country more than other people are, and it is a thing which I have had to fight within myself. Probably what you can say in reply to your question is that I have been lucky in the first place, and secondly, I have been able to conceal the difficulties on the intellectual road I have gone more than other people have been able to, to keep them within myself and fight them out myself.

Q. Let us leave you out of it.

A. Yes.

Q. Do I understand what you have been saying is that in your experience more frequently than not the extraordinarily gifted individual realizes the fulfillment of his potential as a result of background that has involved many unconventional elements?

A. I think it is often that you get that. I must say that when people are really gifted, those who have what you might call genius of some sort, intellectual or artistic, it is hard for them to arrange their relationships to live in minor matters and in a manner which is wholly conventional. I think we have seen that all through time. Again, I would like to emphasize I do not underrate the seriousness of the problem that it poses for the Government when these people are used for Government work. But I think it is a problem that should be regarded as such.

Q. Mr. Kennan, you have been asked questions in a framework that implies at least that they are addressed to you by the board in the light of the rigorous requirements of the Atomic Energy Act. I hope I am not out of order in saying that as a lawyer I cannot believe that the Atomic Energy Act intended to deny to the Atomic Energy Commission the services of gifted people.

I ask you to consider in the light of that statement this question: In your opinion, and based on all of the experience which you have described here this morning, are the character, associations and loyalty of Dr. Oppenheimer such as to bring you to a determination that permitting him to have access to restricted data will not endanger the common defense and security?

A. Mr. Marks, I cannot anticipate, of course, the judgment of this board, and the same information is not available to me as is available to the board. I would consider my own opinion one not founded as well as will be the opinion of the board. I can only judge on the basis of what I have seen, which is a portion of the evidence.

Q. Of course.

A. On that basis, I may say that I myself have no doubt whatsoever about this, and on the basis of what I know I would be entirely in favor. I think it flows from what I have said here earlier. I have forgotten how your question was worded.

Mr. ROBB. Could we have it read back?
(Question read by the reporter.)
The WITNESS. May I then simply rephrase my answer from the beginning here, and ask that it be regarded as the answer to this question.

On the basis of what is known to me of Dr. Oppenheimer's qualities, his personality and his activities during the period that I have known him, I would know of no reason why he should not be permitted to have access to restricted data in the Government.

RE-CROSS-EXAMINATION

By Mr. ROBB:

Q. Mr. Kennan. I gather that you say—and I think quite properly—that of course you don't know what information may be available to the board.
A. Quite so.
Q. Of course, you would agree there might be things known to the board which if known to you would change your answer to the question.
A. Certainly.
Q. Mr. Kennan, we have discussed somewhat the criteria and so on of these security procedures and tests. In a case where the question of individual security clearance was involved, assume that the evidence was more or less in equipoise, who do you think ought to have the benefit of the doubt—the individual or the Government?
A. I think unquestionably, sir, the Government should have the benefit of the doubt. In saying that, if I may just say so, I am animated by the reflection that the Government's interest might also be torn, that the Government might have need of the man, and that interest should also be recognized.
Q. I am not saying to you, sir, that the evidence here is in equipoise; I am just assuming that.
A. I understand.
Q. Mr. Kennan, I would like to ask you another question in your role as an expert on diplomacy which I perhaps should have asked you before.

What in your opinion would be the effect and would have been the effect in 1950 on our allies if the Russians had had the thermonuclear weapon and we had not. Do I make myself clear, sir?
A. Yes, sir. I imagine that it might to some extent have been an unfortunate one. I do not think decisively unfortunate. I think it would have depended on what we might have been able to say to them about the adequacies of our existing stockpile of atomic weapons.
Q. Would you tell us what you mean by "unfortunate"?
A. Unless we were able to demonstrate to them that what we already held in the way of atomic weapons was sufficient to make it most unlikely that even the Russian hydrogen bomb would be used against ourselves or our allies, then I would consider that the effect on our allies might have been unfortunate. But I would remember that the allies have never been, it seems to me, as conscious of the importance of atomic weapons as we have.
Q. Putting it again in the language of the well known man of the street, if the Russians had had the thermonuclear weapon and we had not, the result might have been that some at least of our allies would have been scared off from us, is that right?
A. Yes, sir. That is certainly one of the considerations that would have had to be taken into account in deciding whether to go ahead with the weapon or not.
Q. Mr. Kennan, you mentioned—I don't recall the exact language you used but I think the substance of it was—that there were some friends of yours that you suspended until their guilt was proven, or something of that sort?
A. Yes.
Q. Would you mind telling us who you had in mind?
A. A number of my colleagues in the Foreign Service have had the experience of seeing charges or insinuations advanced against them in the public print here, and of having to face congressional charges or congressional investigations of one sort or another. That is the only point I wish to make. I have not done anything to terminate my associations with those men just on the basis of the fact that the charges were raised against them. I have waited to see whether anything would be proven. I prefer to give my friend the benefit of the doubt until something was——
Q. Have there been some in respect of whom the charges have been proven?
A. There have been two who have left the Department of State—two or three—but I am not sure that charges were proven. I really would have to

ransack my memory to recall exactly the way these cases went. I believe they all left in an honorable way. Doubts were raised and their names were mentioned publicly.

Q. You had faith in them?

A. In every case that I have in mind here I have had—at least I have never seen the evidence that these men were not loyal Government servants, and in the absence of that evidence I tried not to jump to any conclusion.

Q. Now, would you mind telling us who they were?

Mr. GARRISON. Mr. Chairman, I think it is an unfair question to ask this witness to discuss other people in the Government, and I don't see what possible relevance it can have to the inquiry of this board.

Mr. ROBB. Mr. Chairman, I asked the witness if he would mind. That is why I asked him that way.

Mr. GRAY. The witness certainly would be given the privilege of declining to answer this question if he wishes without any significance being attached to it.

The WITNESS. Mr. Chairman, if at any time the board feels the need for the names of these people, I would be very happy to give it. But otherwise, I think at the present time I would prefer not to mention them. The names are fairly well known ones.

By Mr. ROBB:

Q. May I ask you this, sir, and certainly this gentleman has been much mentioned in the public press.

Mr. GARRISON. Mr. Chairman, I really object to this. I see this proceeding into a line of questioning which by some form of suggestion as to names of people who have been adversely discussed in the press being brought in here with some suggestion that this somehow is connected with Dr. Oppenheimer.

Mr. GRAY. I make this observation, Mr. Garrison, that the testimony of a witness which has been given with clarity and conviction and I think rather eloquently is based, as I understand it, on his own subjective judgment with respect to the character, loyalty and associations of Dr. Oppenheimer. Is that a fair statement?

The WITNESS. Yes.

Mr. GRAY. I think that if there are cases or similar situations in which the judgment of the witness has proven not to be borne out by the facts, that it is pertinent to this inquiry.

Mr. GARRISON. I withdraw the objection, Mr. Chairman.

Mr. GRAY. I want to make it clear that I am sure that the board does not wish a lot of names brought in here by the heels.

Mr. ROBB. Oh, no.

Mr. GRAY. I don't interpret that to be the point.

Mr. ROBB. No, that is exactly the theory I put the question to.

By Mr. ROBB:

Q. Was Owen Lattimore one of your associates or friends?

A. No, he was not. I never had any personal acquaintance with him.

Q. He would not be one that you included?

A. No, he would not be included. The men I had in mind were associates of mine in the Foreign Service, and one in particular who has been in a number of congressional and loyalty board hearings. I have testified in those hearings as I have in this one. So far to my knowledge he has never been found guilty by any board or formal branch of an agency of the Government of anything reprehensible to him. I have continued to see him and know him as a friend.

Q. Were you called as a witness by him, sir?

A. Yes, but if I may say so, initially over my own objections because I was then an official of the Department of State, and I felt that the loyalty board should ask me as an official for my opinion, feeling that I owed my loyalty entirely to the interest of the Government, and not to the man as a party in a dispute.

Q. Have you testified in any other so-called loyalty hearing?

A. Yes. I testified in one. Again it was the case of a Foreign Service officer who asked me to testify in his behalf and to read 1,200 pages of his reports, and to tell the board that they did not contain evidence of Communist loyalty.

I told him that I would prefer, as an official of the Department of State, not to do that at his request, but would be happy to do it at the board's request. He did get a letter from the board asking me to do that. The result is that I had to go through 1,200 pages of material and gave the board an opinion.

Q. I am sure the board here understands the difficulty you had in reading 1,200 pages. That is all I care to ask.

Mr. GRAY. Dr. Evans I believe has 1 or 2 questions.

Dr. EVANS. Mr. Kennan, in answer to one of the questions that was asked you, I think you stated in effect, or at least you implied that all gifted individuals were more or less screwballs.

The WITNESS. Let me say that they apt to be, if I may.

Dr. EVANS. Would you say that a large percentage of them are?

The WITNESS. No, sir; I would not say that they are screwball, but I would say that when gifted individuals come to a maturity of judgment which makes them valuable public servants, you are apt to find that the road by which they have approached that has not been as regular as the road by which other people have approached it. It may have had zigzags in it of various sorts.

Dr. EVANS. I think it would be borne out in the literature. I believe it was Addison, and someone correct me if I am wrong, that said, "Great wits are near to madness, close allied and thin partitions do their bounds divide."

Dr. Oppenheimer is smiling. He knows whether I am right or wrong on that. That is all.

Mr. GRAY. Mr. Kennan, you certainly would not be prepared to testify that all professors are screwballs, would you?

Dr. EVANS. I am worried about that, because it has been brought up 2 or 3 times. I am getting a little sore about it.

Mr. GRAY. One further serious question. These gifted people about whom there has been a very considerable discussion here, as you say, in many cases arrived at judgments, attitudes, convictions after all sorts of experience. You feel, however, that the unusual person or gifted person who has traveled perhaps a different road than most other people can at one point reach a stability on the basis of which there can be absolute predictability as to no further excursions?

The WITNESS. Let me say at a point where there can be sufficient predictability to warrant his being accepted by the Government for public service.

Mr. GRAY. Thank you very much, Mr. Kennan. We appreciate you being here.

(Witness excused.)

Mr. GRAY. We will take a recess, gentlemen.

(Brief recess.)

Mr. GRAY. Do you wish to testify under oath? You are not required to do so. Most all witnesses have.

Mr. LILIENTHAL. I prefer to.

Mr. GRAY. Would you stand, please, and raise your right hand.

David E. Lilienthal, do you swear that the testimony you are to give the Board shall be the truth, the whole truth and nothing but the truth, so help you God?

Mr. LILIENTHAL. I do.

Whereupon, David E. Lilienthal was called as a witness, and having been first duly sworn, was examined and testified as follows:

Mr. GRAY. You are no doubt familiar with the so-called perjury statutes. I should be glad to read the penalties, if you wish.

The WITNESS. I am familiar with them.

Mr. GRAY. I should like to say to you, Mr. Lilienthal, that if during the course of your testimony, it should develop that you are about to discuss restricted data, I would appreciate your letting me know so that the necessary security precautions might be taken.

The further observation I would have to you is that we treat these proceedings as confidential between the Atomic Energy Commission officials and Dr. Oppenheimer and his representatives and witnesses. The Commission will initiate no public releases with respect to these proceedings and we express the hope that witnesses will follow the same course.

The WITNESS. Yes, I certainly shall not initiate any public statement.

DIRECT EXAMINATION

By Mr. SILVERMAN:

Q. Mr. Lilienthal, what is your present occupation?

A. I am in private business in New York City as adviser on industrial matters to financial and industrial enterprises. I am also a corporate officer as chairman of the board of the Minerals and Chemical Co.

Q. Do you have any Government employment or position at this time?

A. I do not.

Q. You were formerly Chairman of the Atomic Energy Commission?

A. I was between late October 1946 and the 15th of February 1950.
Q. I think you said at 5 p. m.
A. Yes; at 5 p. m.
Q. When did you first meet Dr. Oppenheimer?
A. On the occasion of the bringing together of a board or panel—a board of consultants or panel—by the Department of State in January 1946. That panel was organized under a committee called the Secretary of State's Committee.
The purpose of the panel and the directions of the panel were to seek to find some basis for a plan or program for the international control of atomic weapons. There were five members of this panel designated, I think, by the Secretary of State, or perhaps by the Under Secretary of State, Mr. Acheson, and Dr. Oppenheimer was one of those panel members.
Q. And you were the chairman of the panel?
A. I was the chairman of the panel. The other members were Mr. Harry Winne, vice president of General Electric Co. at that time, Dr. Charles A. Thomas, who was then executive vice president of the Monsanto Chemical Co. and now its president. Mr. Chester Barard, then president of the New Jersey Telephone Co., and Dr. Oppenheimer.
Q. Will you tell us something of how much contact you had with Dr. Oppenheimer during the work on this panel?
A. The panel was convened and met briefly with the Secretary of State's Committee. Perhaps I should indicate the personnel of that committee. This was the first meeting with Dr. Oppenheimer. That committee consisted of Under Secretary Acheson—perhaps Assistant Secretary at that time—John McCloy, Gen. Leslie R. Groves, President Conant of Harvard, and Dr. Vannevar Bush, Chairman of the Joint Research Board of the Defense Establishment.
That meeting with this top committee was briefed and then this board of consultants virtually lived together for 6 or 7 weeks until we finally presented our report to the committee which we reported.
Q. When you say you virtually lived together, you mean you spent substantially all the time together?
A. Yes, with the exception of a few breaks, we had committed ourselves to devote all of our time to this problem until we either said we couldn't think of anything useful or came up with a report.
This we did, so we worked together here as a group. We traveled about the country, seeing the various atomic energy installations, for some of us the first time—Oak Ridge, Los Alamos, Hanford, and so on.
In this process of course we came to know each other quite well. We then came back to Washington and spent a good many days in the process of jointly drafting our report.
Q. Was Dr. Oppenheimer active in this work?
A. Yes, he was indeed.
Q. Would you tell us something about the positions that Dr. Oppenheimer took and the work he did in the drafting of that report, particularly as it bears on his attitude toward the problem presented by our relations with Russia?
A. I think the theme of this group in which Dr. Oppenheimer's views contributed substantially was that we should try to absorb the facts about atomic energy and see if we could not come up with some practical, we hoped, and workable and acceptable system of control and protection for the United States and for the world. So Dr. Oppenheimer's approach as the rest of us was first to ascertain the facts as a matter of technology and so on. Of course, in that respect he and Dr. Thomas were really teachers for the rest of us. Then as to policy, I can recall perhaps a few illustrative instances.
Dr. Oppenheimer—and there was unanimity on this but he certainly probably initiated the idea, and certainly pressed it and elaborated it—which relates to the attitude of Russia and Soviet communism, the first idea we discussed was that of international inspection of countries in the United Nations, to see whether they were carrying on atomic weapon enterprises.
This we rejected and an important part of our reasoning for rejecting it was that it was not a foolproof method. Something more than inspection would be necessary, that without international ownership and control of the raw materials and the operations in the atomic energy field, the United States could not trust the Russians merely by inspection to comply with the requirements of this scheme.
The actual development of this idea that inspection was inadequate to protect ourselves from the Russians or was an inadequate idea to go before the world—the protection of the world—was largely formulated by Dr. Oppenheimer

and technical associates of his like Dr. Bacher, who had studied the physical problem of the ease with which inspection could be avoided by an operating organization in Russia as distinguished from having a United Nations operating and management team running the plant, that periodic inspection was not a foolproof system.

Q. In your view was the report of that panel one that was reasonably soft or what have you in respect to the hope of cooperation or with respect to what one could expect from the Russians?

A. We tried to make it as nearly foolproof as we could. There was early discussion that any proposal that a United Nations operating organization should operate a gaseous diffusion plant within Russia would obviously conflict with the Russian views about the Iron Curtain and access of foreigners and so on.

The question was raised first by Mr. Winne, as to whether it made any sense to make a proposal which we were pretty sure the Russians would reject. We concluded, and I took responsibility for this idea initially, that we should present an idea we could stand for, leaving the question of whether it should be submitted to the Russians with a rather strong likelihood of it being rejected, to others.

It was our job to develop a workable foolproof system. Therefore, to answer your question about denominating this, I think we did devise what would be called a tough program. This was reviewed later by Mr. Baruch and his associates. They accepted these essentials and they too were insistent on what Mr. Baruch called a foolproof system, a tough system.

Q. And Dr. Oppenheimer was in accord with this tough system?
A. Yes, and contributed a great deal to it.
Q. When did you say you became Chairman of the AEC?
A. I think it was the 28th of October 1946.
Q. Some time after you became Chairman was the question of Dr. Oppenheimer's past associations and his left wing activities and so on called to your attention?
A. Yes, it was.
Q. Will you tell us the circumstances of that, please?
A. The board will recall that there is a kind of grandfather clause in the Atomic Energy Act, by which those who had been cleared under the Manhattan District continued to hold their clearances—I have not looked at this provision for some time—but the effect is to hold their clearances until a reexamination by the FBI was made, and the question is reexamined on the basis of new additional information, or something to that effect. So we had a number of such reexaminations coming to us.

I have located the date of March 8 as being the date on which I appeared—give or take a day or so—a call from Mr. Hoover saying he was sending over by special messenger an important file involved in this reexamination.

I received this file. It related to Dr. Oppenheimer. It contained in it a great deal of information from the Manhattan District, and perhaps some subsequent investigation. I called the commissioners together on the 10th. The day of Mr. Hoover's call appears to be Saturday. In any event, I called the commissioners together on a Monday, March the 10th, in the morning, I believe.

The existence of this sort of information I did not know up until that time, and I don't think any of us did, unless perhaps Dr. Bacher did.

Q. You say you called the Commission together. Who was present at the meeting?
A. My recollection is that all the commissioners were present. This would be Dr. Robert F. Bacher, who was at the Los Alamos project during the war, Sumner T. Pike, Louis L. Strauss, and Wesley W. Waymack.

Q. Will you tell us what happened at that Commission meeting?
A. Commission conference would be the best description because it continued for some time. It was very informal. We had this file which I requested all the commissioners to read. It was not necessary to request them to because it was obviously a matter of great interest and importance. Instead of delegating this to someone else, it seemed clear that we should do the evaluating, since the responsibility of deciding what should be done, if anything, was ours. So we did begin a reading of this file around the table in my office in the New State Building, and then later as time went on, members would take all or parts of their file to their offices and so on.

One of the first things that was observed was that although this file did contain derogatory information going back a number of years, it did not contain any reference, as far as I recall, or at least any significant reference, to the work that Dr. Oppenheimer had done as a public servant.

Q. Let me interrupt you for a moment. You have seen the Commission's letter of December 23, 1953, which suspended Dr. Oppenheimer's clearance.

A. I have.

Q. So far as you can recall what is the relationship between the derogatory information contained in that letter and the material that was before you sent to you by Mr. Hoover in 1947?

A. From my careful reading of the Commission's letter and my best recollection of the material in that file, and the charges cover substantially the same body of information——

Q. Except for the hydrogen bomb stuff, of course.

A. Yes, up to the point of 1947, I suppose.

Q. You were saying that you found that the file contained derogatory information, but did not contain affirmative matter, shall we say?

A. It did not contain any information about those who worked with Dr. Oppenheimer in the Manhattan District. So we asked Dr. Vannevar Bush, who we knew had been active in the pre-Manhattan District enterprise, as well as since that time, and Dr. James Conant, both who happened to be in town, to come in and visit us about this file. They expressed themselves about Dr. Oppenheimer and his loyalty and character and associations and particularly the degree to which he had contributed to the military strength of the United States.

I called Secretary Patterson, or someone did, to ask him to request General Groves, under whom Dr. Oppenheimer had served, be asked to supply a statement about his opinion about Dr. Oppenheimer and the circumstances under which he was selected and kept as director of the laboratory.

We discussed this with Dr. Bush and Dr. Conant during that day and I think into the next day.

Q. Did you ask Dr. Bush or Dr. Conant for anything in writing?

A. I don't know whether they volunteered or whether we asked, but certainly they did provide written statements more or less following the line of their oral statements.

Yesterday I had an opportunity to read these and refresh my recollection on them. I take it they are in the files.

Mr. SILVERMAN. Does the board have Dr. Bush's letter and Dr. Conant's letter?

Mr. GRAY. I am sure we do.

Mr. ROLANDER. They are a part of the files.

Mr. GRAY. Was there also a written statement by General Groves?

Mr. SILVERMAN. That is already in the record at page 582, or something like that. Unless there is some other written statement I don't know about.

Mr. GRAY. I am asking for information.

Mr. ROLANDER. General Groves' statement was read into the record the other day.

The WITNESS. Then there was a letter from Secretary Patterson to us on the same subject.

Mr. ROLANDER. That is a part of the file.

Mr. SILVERMAN. I wonder if this might not be an appropriate time to read those into the record, sir.

Mr. ROBB. Go ahead.

Mr. SILVERMAN. I don't have them.

Mr. ROBB. The file contains a letter of General Groves which has been read into the record. A letter from Mr. Conant and a letter from Mr. Patterson. I am sure somewhere in here there is a letter from Mr. Bush. The chairman has it now. It also contains for your information—when that was received, I am not sure—the citation which accompanied the medal for merit which was awarded to Dr. Oppenheimer.

Mr. SILVERMAN. That I think has already been read into the record.

Mr. ROBB. I think so. But that is in the file.

The WITNESS. As I recall, this was on the recommendation of General Groves. I probably had seen it at that time.

Mr. ROBB. The medal for merit citation apparently was sent to Mr. Lilienthal and a letter from George M. Elsey dated March 14, 1947.

Mr. SILVERMAN. Did it say who Mr. Elsey was?

Mr. ROBB. I can't read it here, "Commander, USNR." I guess he was secretary or something of the board.

The WITNESS. If I may, I can identify him. He was in the White House staff, assistant to Clark Clifford.

Mr. ROBB. I will read it if you want.

"THE WHITE HOUSE.
"Washington, March 14, 1947.

"Memorandum for: Mr. Lilienthal.

"The members of the Medal for Merit Board who recommended to the President in January, 1946, that Dr. J. R. Oppenheimer be awarded the Medal for Merit, were: Owen J. Roberts, Chairman, William Knudsen, and Stephen Early.

"Richmond B. Keech, Administrative Assistant to the President, was secretary to the board.

"General Knudsen has since resigned as a member of the board and has been succeeded by Chief Justice D. Lawrence Groner of the United States Court of Appeals for the District of Columbia. Mr. Keech, although no longer Administrative Assistant to the President, remains secretary to the board.

"A copy of the Executive order and a copy of the regulations governing the Medal for Merit are enclosed.

"Respectfully,

"GEORGE M. ELSEY,
"Commander, USNR."

Mr. SILVERMAN. I would like at this point to read the letters from Dr. Bush and Dr. Conant and Secretary Patterson.

Mr. ROBB. I have here the original of the letter from Dr. Bush dated March 11, 1947. The original of the letter from General Groves dated March 27, 1947, which I believe is already in the record.

Mr. SILVERMAN. That is already in the record.

Mr. ROBB. The original of the letter from Mr. Patterson dated March 25, 1947. The original of the letter from Mr. Conant dated March 29, 1947. I will hand these all to you, sir.

Mr. SILVERMAN. Thank you very much. With the permission of the board——

Mr. GRAY. Excuse me for the interruption. I think we will proceed with the presentation of these letters and then break for lunch. I am afraid if our experience with other witnesses is any indication, we will probably have to ask you to come back after lunch, Mr. Lilienthal.

The WITNESS. I would like to express the hope that I would be able to finish today so I can get back to work, but I, of course, will be back after lunch.

(Discussion off the record.)

Mr. SILVERMAN. It is agreed that these letters will go into the record.

Mr. Robb has already identified the letters. They will be read into the record. They will be transcribed into the record at this point, the letter of March 11, 1947, on the letterhead of the Joint Research and Development Board to Mr. David E. Lilienthal, Chairman, signed "V. Bush, Chairman." The letter of March 24, 1947, on the War Department letterhead already read into the record once, and I see no reason to read that in again.

The letter of Robert P. Patterson, Secretary of War, on War Department letterhead, stamped secret, I may say, to Hon. David E. Lilienthal. The letter of March 29, 1947, from Dr. Conant to Mr. Lilienthal.

(The letters are as follows:)

"THE JOINT RESEARCH AND DEVELOPMENT BOARD,
"Washington 25, D. C., March 11, 1947.

"Mr. DAVID E. LILIENTHAL,
"Chairman, Atomic Energy Commission,
"New War Department Building, Washington 25, D. C.

"DEAR MR. LILIENTHAL: At our conference yesterday you asked me to comment concerning Dr. J. Robert Oppenheimer, and I am very glad to do so.

"Dr. Oppenheimer is one of the great physicists of this country, or of the world for that matter. Prior to the war he was on the staff of the University of California, and was regarded as leader in the theoretical aspects of atomistics and similar subjects of physics. Shortly after the Army entered into the development of atomic energy, he was given a very important appointment by General Groves. This appointment made him director of the laboratory at Los Alamos, which was in all probability the most important post held by any civilian scientist in connection with the entire effort. General Groves undoubtedly made this appointment after a very careful study of the entire affair from all angles, as this was his custom on important appointments.

"Subsequent developments made it very clear that no error had been made in this connection, for Dr. Oppenheimer proved himself to be not only a great physicist, but also a man of excellent judgment and a real leader in the entire effort. In fact, it was due to the extraordinary accomplishments of Oppenheimer and his associates that the job was completed on time. Subsequent to the end of the war Dr. Oppenheimer has had a number of important appointments. He was invited by Secretary Stimson as one of the scientists consulted by the secretaries of War and Navy in connection with the work of the Interim Committee. He was appointed by the State Department as a member of the board which drew up the plan on which Mr. Baruch based his program. He has recently been appointed by the President as a member of the General Advisory Committee of your organization. I have appointed him a member of the Committee on Atomic Energy of the Joint Research and Development Board. All of this has followed from his extraordinary war record in which he made a unique and exceedingly important contribution to the success of the war effort of this country.

"I know him very well indeed and I have personally great confidence in his judgment and integrity.

"Very truly yours,

"(Signed) V. BUSH, *Chairman.*"

"WAR DEPARTMENT,
"*Washington, March 25, 1947.*

"Hon. DAVID E. LILIENTHAL,
"*Chairman, Atomic Energy Commission,*
"*Public Health Service Building, Washington, D. C.*

"DEAR MR. LILIENTHAL: In connection with your inquiry about Dr. J. Robert Oppenheimer, a member of the General Advisory Committee to the Atomic Energy Commission, I am glad to furnish the following information:

"It is my understanding that Dr. Oppenheimer is a leading physicist of the world. During the war he held the key post of Director of the Los Alamos Laboratory under the Manhattan district project, which as you know was the enterprise under the War Department responsible for development of the atomic bomb. His performance in that post, under direction of General Groves, was a brilliant success.

"For his exceptionally meritorious service, he was recommended by General Groves to receive the Medal for Merit in August 1945. This recommendation was approved by Secretary of War Stimson, and the award was made by the Medal of Merit Board appointed by the President.

"Dr. Oppenheimer was also appointed by the War Department to be a member of the Advisory Panel of Scientists, to assist the Interim Committee designated by Secretary Stimson in May 1945, to recommend policies in regard to the atomic bomb and to suggestion legislation concerning atomic energy. I met Dr. Oppenheimer several times in the course of this work and received a most favorable impression of his ability, judgment, character, and devotion to duty.

"Dr. Oppenheimer was recently appointed by Dr. Vannevar Bush, Chairman of the Joint Research and Development Board of the War Department and Navy Department, to be a member of the Committee on Atomic Energy under that board.

"I am enclosing with this letter a memorandum submitted to me by General Groves relative to the loyalty of Dr. Oppenheimer.

"In conclusion, I should say that from my knowledge of the work that he has done toward making the atomic bomb a success and in other matters related to atomic energy, I have confidence in his character and loyalty to the United States.

"Sincerely yours,

"(Signed) ROBERT P. PATTERSON, *Secretary of War.*"

"HAVARD UNIVERSITY,
"*Cambridge, Massachusetts, March 27, 1947.*

"MR. DAVID E. LILIENTHAL,
"*Chairman, Atomic Energy Commission,*
"*Washington 25, D. C.*

"DEAR MR. LILIENTHAL: I am writing you this letter because l understand certain inquiries have been made in regard to the loyalty of Dr. J. Robert Oppenheimer.

"It is quite unnecessary for me to recite in this letter the tremendous contribution to the war effort made by Dr. Oppenheimer. As director of the Los Alamos Laboratory he carried a heavy responsibility both of a scientific and technical nature and as an administrator faced with an extremely difficult problem. He fulfilled his duties in an admirable manner. I think it can be said that he is 1 of the 3 or 4 men whose combination of professional knowledge, hard work, and loyal devotion made possible the development of the bomb in time to end the Japanese war.

"My first personal acquaintanceship with Dr. Oppenheimer started in the summer of 1941. From then until the present day I have seen him intimately and discussed with him all manner of questions. During the war I visited Los Alamos frequently and in so doing came to know him very well. Since the war, I have discussed not only atomic energy for industrial and military purposes, but all phases of the international problem of control. Likewise, our conversation has ranged over the whole field of American politics and foreign policy. Therefore, I feel sure that the statements that I make about him are based on an intimate knowledge of the man, his views, and his emotional reactions.

"I can say without hesitation that there can be absolutely no question of Dr. Oppenheimer's loyalty. Furthermore, I can state categorically that, in my opinion, his attitude about the future course of the United States Government in matters of high policy is in accordance with the soundest American tradition. He is not sympathetic with the totalitarian regime in Russia and his attitude towards that nation is, from my point of view, thoroughly sound and hard headed. Therefore, any rumor that Dr. Oppenheimer is sympathetically inclined toward the Communists or toward Russia is an absurdity. As I wrote above, I base this statement on what I consider intimate knowledge of the workings of his mind.

"At the time of Dr. Oppenheimer's entering the work on atomic energy, I heard that there was some question of his clearance by the security agencies. I understand that was based on his associations prior to 1939 and his 'left wing' sympathies at that time. I have no knowledge of Dr. Oppenheimer previous to the summer of 1941, but I say unhesitatingly that whatever the record might show as to his political sympathies at that time or his associations, I would not deviate from my present opinion, namely, that a more loyal and sound American citizen cannot be found in the whole United States.

"Very sincerely yours,

"(Signed) JAMES B. CONANT, *President.*"

Mr. GRAY. The record will show that the members of the board have read these letters. Each member of the board has read all of these letters.

Can you give me an indication, Mr. Silverman, of what length of time your questioning may consume? I am not going to try to hold you to it.

Mr. SILVERMAN. I understand that, sir. All lawyers know that there is no more unreliable answer than that of a lawyer as to how long he is going to take. I would guess that our direct examination will probably consume approximately an hour. I used the word "guess" in its sharpest meaning.

Mr. GRAY. Let us proceed now for 15 minutes and then take a break for lunch.

By Mr. SILVERMAN.

Q. Who was present at this conference of the Commissioners on March 10th or thereabouts?

A. My recollection is that all the Commissioners, and that from time to time we may have—I am not clear on this—but we probably called in the acting security officer and other staff people. I am rather vague on that.

Q. Who were the members of the Commission that were present?

(Discussion off the record.)

The WITNESS. All the members of the Commission.

By Mr. SILVERMAN:

Q. Did you give the names of them?

A. Yes.

Q. Thank you. Did you then take the matter up with anyone in the office of the President?

A. Yes. This would be March 11, on the Tuesday following the Monday I have referred to, Dr. Bush and I made an appointment with Clark Clifford, the President's counsel, and asked him to call the President's attention to this file.

The reason for doing this was that we were a little uncertain about our role here. The members of the GAC under the law were appointed by the President and not by the Commission. They were not subject to Senate confirmation, but they were Presidential appointees. It seemed important to call this matter to his attention to make sure that the President was made aware of this file. This was the purpose of our call.

It was left that Mr. Clifford would advise the President and would send word to us if there were further questions that the President had.

The reading of the memorandum from Commander Elsey refreshes my recollection on one of the things that was said either over the phone or in conversation, that his Medal for Merit Commission had knowledge about Dr. Oppenheimer. I didn't know there was this citation that was sent at that time. I think it was later that week that Mr. Clifford phoned me, or Dr. Bush, and I may have gone back—I am not too clear on this—but in any case, the President was advised and the President didn't express any views about what should be done. He did not express the view that the clearance should be canceled or that he should remove Dr. Oppenheimer or anything of that kind.

Q. Did you do anything further in an effort to decide what you should do about this problem?

A. We discussed the matter together. We interrogated and conversed with our associate, Dr. Bacher, because he was an intimate associate—he was actually deputy to Dr. Oppenheimer at Los Alamos during most of the period of the Los Alamos Laboratory, and had therefore a day to day working knowledge of him—and he expressed his view about Dr. Oppenheimer. I think I called on the chairman of the Military Liaison Committee, who at that time was General Brereton, and advised that the question had been raised. General Groves was a member of the Liaison Committee at that time and could inform the committee to the extent that the committee wanted further information.

I think that covers what was done at that time, except that we reached a conclusion. The conclusion was that on the whole set of circumstances, there did not seem to be any occasion for cancelling or withdrawing the clearance or taking any other action.

Q. This of course was after the Atomic Energy Act was in effect?

A. Yes. We were organized under the Atomic Energy Act. There was one further thing that we decided we should do, and that was to communicate with Mr. Hoover, the director of the Federal Bureau of Investigation, for the purpose of seeing whether there was anything that had come in this file since he called me or whether we were properly construing the facts in the file. I did call on Mr. Hoover. I have refreshed my recollection on this obviously or I wouldn't remember it without it. It was on the 25th of March. My office diary shows that I called on him on the 25th of March, and discussed this file with him.

Q. Will you tell us of that discussion?

A. Whether Mr. Hoover had one of his associates there or not, I am not sure, but from the Commission it was Mr. Joseph Volpe, Jr., Deputy General Counsel at that time. I am not too clear, but I think the acting security officer was with us, whose name was Tom Jones. My recollection is not too clear here. My recollection of that conversation is as follows:

First there seemed to be general agreement, or I expressed the view that here was a man who had certainly contributed a great deal to the military strength of the United States under circumstances of great difficulty and so on. Everyone we had consulted who had worked with him and naming them, Dr. Bush, General Groves and so on, were clear that this was true, that he had done a good job. Mr. Hoover said there could not be any question about that.

Then the question was discussed as to the relevance, as to the weight to be given this long series of associations with left wing and crackpot and communistic sorts of organizations or people of which the record contained a great deal of information. On this I reported to Mr. Hoover that we would like to know whether there was something in this that we had missed but that our evaluation of it was that on the whole record in view of what had happened since that time that Dr. Oppenheimer had proved by his work, by his activities, by the things he had done for this country, that he was not only loyal, but that he had character that made him suitable as an employee of the Atomic Energy Commission.

Then Mr. Hoover said—this is my impression—of course, Mr. Hoover makes it a point not to evaluate these reports. I have dealt with him on a number of these things. He very likely did not evaluate it. But when I asked him if there was anything that we had missed or any implication that we had not seen but perhaps he, with his closer knowledge of the file might see, he said, well, the only reservation he had was that he didn't like that episode about—what is his name, a French name.

Dr. EVANS. Chevalier.

The WITNESS. Yes, Chevalier. That Oppenheimer did report it finally, but he waited an awful long time, and he critized that. He was quite critical of it. Of course, I completely agreed with that.

Beyond that there was no further comment about the file. So we left with no suggestion from Mr. Hoover that further investigation ought to be carried on or that the file was incomplete, that there were things we didn't know about.

I think that is the last——

By Mr. SILVERMAN:

Q. Was there any suggestion by Mr. Hoover that the explanation Dr. Oppenheimer had given of that incident was not correct, or don't you remember?

A. My recollection is that his criticism was that he should have reported this to the authorities at once, instead of waiting. I have forgotten how long it was, but it was an intolerable period. It was weeks, I think. That was the point of his comment.

Q. After that, did you report the result of this interview to your fellow Commissioners?

A. Yes. Either I wrote a memorandum about it, or Mr. Volpe did. I inquired at the Commission yesterday and find that they were not able to locate such a memorandum, but did locate a memorandum to the files which I had not seen, from Mr. Jones. That is the only one that they have been able to dig up. I think there is a report by Volpe as well, but it has not been located yet.

Mr. SILVERMAN. May I inquire, do you have Mr. Volpe's report?

Mr. ROBB. No, sir, I have one by Mr. Jones. Do you want to read it in the record?

Mr. SILVERMAN. No, I think not at this point.

The WITNESS. I did read that yesterday and I am familiar with its contents.

By Mr. SILVERMAN:

Q. As a result of Mr. Volpe's report—was Mr. Jones present at the conference of the Commissioners?

A. He didn't say so in his memorandum and I am not clear on this. I am rather assuming that he must have been but he reports what he understood went on. He very likely was. I am a little fuzzy about that.

Mr. SILVERMAN. In the interest of continuity, we might just as well put Mr. Jones' memorandum in.

Mr. ROBB. I might say, Mr. Chairman, by way of explanation that of course ordinarily the Atomic Energy Commission treats as confidential any discussions between it and its representatives and Mr. Hoover. However, I think since the witness has gone into this matter, that it is entirely appropriate for me to read this memorandum:

"Office Memorandum United States Government.

"Date: March 27, 1947.

"To: File.

"From: T. O. Jones (ink initials TOJ).

"Subject: J. Robert and Frank Oppenheimer.

"At a meeting held on Tuesday, March 25, 1947, between representatives of the Atomic Energy Commission and the Federal Bureau of Investigation and attended both by Mr. Lilienthal and Mr. Hoover, there was some discussion on the case of the Oppenheimer brothers. Certain comments made by Mr. Hoover appear of particular interest.

"Concerning Frank, Mr. Hoover consistently expressed himself as feeling that there was no question of his—" the word as typed did not have the "un" in front of it and the "un" is written in longhand, and beside the word is written in again in longhand "TOJ"—"undesirability. Although Mr. Hoover would doubtless dislike to be put in the position of evaluating the information on Frank, nevertheless it is felt that the impression he left at this meeting should be carefully considered if at any future time it is proposed to reinstate Frank's clearance for Restricted data.

"In the case of J. Robert, those present all seemed keenly alive to the unique contributions he has made and may be expected to continue to make. Further, there seemed general agreement on his subversive record * * * that while he may at one time have bordered upon the communistic, indications are that for some time he has steadily moved away from such a position. Mr. Hoover himself appeared to agree on this stand with the one reservation, which he stated with some emphasis, that he could not feel completely satisfied in view of J. Robert's failure to report promptly and accurately what must have seemed to him an attempt at espionage in Berkeley.

"Mr. Lilienthal mentioned that the general question of J. Robert Oppenheimer's clearance had been discussed with Secretary Patterson, General Groves and Drs. Bush and Conant, and that all four were writing letters to him endorsing J. Robert Oppenheimer. Mr. Hoover said he would be glad to have such letters for the completion of his files and was told that he would be provided with copies.

"(NOTE.—Original copy placed in J. Robert's File. Cc placed in Frank's File.)"

By Mr. SILVERMAN:
Q. Does that accord with your recollection?
A. Not quite, but it certainly is not very far off. Where my recollection varies with this is not that Mr. Hoover was not critical of the Chevalier incident and the laxity of reporting, but I don't recall his saying that he was not satisfied with the man. My recollection was that he was not satisfied with the way Oppenheimer had behaved in not reporting this promptly. Except for that, I think roughly—I certainly left with the impression that Mr. Hoover would have said about J. Robert what he said about Frank if he felt that this qualification was a very strong one. He certainly did not say that.

Q. You reported back to your fellow Commissioners this conversation with Mr. Hoover?
A. Yes. I am a little puzzled why this memorandum was not sent to the Commission or me and why it went to the files and that is why I think there is also a memodanum from Volpe to the files because that was our regular practice.

Q. As a result of this review and discussion with Mr. Hoover and so on, that you have described, what did the Commission do?
A. The Commission concluded not to cancel or whatever the term is, not to cancel the clearance—I suppose that is the way to say it. I believe the form this took was a decision to continue the clearance. I think that is the way it took. Actually it was not until August that I find in the minutes of the Commission any reference to that action. The reference in August relates to February, and I am sure the Secretary was wrong. It was actually in March.

Mr. ROBB. How is that again?
The WITNESS. In August——
Mr. SILVERMAN. Perhaps we could have this a little more precise. I will call the attention of the board and of Mr. Robb to pages 80 and 81 of the record in this case at which point Mr. Garrison referred to a letter he had from Mr. William Mitchell, general counsel, dated January 15, 1954, that the Commission will be prepared to stipulate as follows for purposes of the hearing:

"On August 6, 1947, the Commission recorded clearance of Dr. J. Robert Oppenheimer, which it noted had been authorized in February 1947."

Mr. ROBB. That is correct. I thought Mr. Lilienthal thought that was not right.
Mr. SILVERMAN. Mr. Lilienthal will now state what he was saying.
The WITNESS. I find that the minute entry of this clearance which should have appeared in the minutes of February or March, at the time that it took place, actually appears as far as I can tell in the minutes of August as of March.
Mr. ROBB. As of February.
The WITNESS. As of February.
Mr. ROBB. It is February, is it not?
Mr. SILVERMAN. I got it.
Mr. ROBB. I am agreeing with you, but apparently Mr. Lilienthal is not.
Mr. SILVERMAN. Mr. Lilienthal is agreeing with all of us. The minutes show the thing and his recollection of the date is different.
The WITNESS. I think there is a very easy explanation.

By Mr. SILVERMAN:
Q. Would you give it, please?
A. The Secretary of the Commission in August was Mr. Bellsley, who had the chore of bringing the minutes up to date during the period when the Commission was on the Hill most of the time on confirmation hearings. He himself had not attended these early and informal meetings, and I think this probably accounts for the fact he thought it was February. It could not have been February, because the file did not reach us until March. But the minutes, of course, are right and state it was in February.
Q. Was the Joint Committee on Atomic Energy familiar with Dr. Oppenheimer's report?
A. Yes; you mean the commitee with this file?
Q. Not this file, but this record.
A. They were certainly familiar with Dr. Oppenheimer. He was an advisor to the committee.
Q. Did they know about his left wing activities, or don't you know?
A. I don't know about them as of this time. Later on this file was transmitted to the Joint Committee and examined by them, along with a lot of other files.
Mr. GRAY. If you are not on the file any further, I will ask that we recess for lunch.
(Thereupon at 12:50 p. m., a recess was taken until 2:00 p. m., the same day.)

AFTERNOON SESSION

Mr. GRAY. The proceeding will begin.
Mr. GARRISON. Mr. Chairman, could I talk to you for a moment about procedure?
Mr. GRAY. Yes.
Mr. GARRISON. I have had literally 3 minutes to talk to Dr. Oppenheimer about this problem and no time to consult with my associates. This is what I would like to suggest to the board about Dr. Oppenheimer's direct examination. I would like to state the request first and then give you my reasons.
I would like to ask the privilege of the board to adjourn at lunch time tomorrow and carry through with the witnesses in the morning, some of whom are going to have to spend the night here even though they were scheduled for today, and give us an afternoon off so we can do some work and then put Dr. Oppenheimer back on the stand, if that is the right term, on Thursday morning and as much of the day as you wish to devote to him.
I think our redirect examination will not be very extensive. I know that there will be cross-examination after that.
Now I would like to state the reason for this. It is not that I have any need of elaborate preparation, but I and my associates are too physically and emotionally worn down by having to do all the work outside of these hearings at night that I am just not really in a position to do what counsel should do, which is to sit down quietly with Dr. Oppenheimer and go over the transcript and make a preparation.
I don't want to expand on that plea unless the board wants me to go into it further. I am just telling you that I am nearly at the end because of the pressures that have come upon us all.
I know of the board's most earnest desire, and I know the time problem that Mr. Robb faces, cut off from his practice—I am cut off from mine, too—and I don't want any special favors. I just want a chance to have a little time. I would be most loathe to make that request if I felt that thereby I would be forcing the hearings into another week. But as I look at the calendar and note what progress we have been making I think it is now clearly not going to be possible to have both redirect and the balance of the witnesses concluded by Friday night.
I might suggest a possible alternative which would not lose even half a day, and that is to put Dr. Oppenheimer back on the stand Monday morning. Personally I would prefer to have it on Thursday and then be done with it. But I just can't do this thing tonight for tomorrow. I just can't, Mr. Chairman.
Mr. GRAY. You made the request of the board and I think before I respond to it, I would feel that I should consult the other members of the board about the request. I don't want the record to reflect my consulation with the members of the board.

Mr. ROBB. May I say before the Board withdraws that Mr. Garrison courteously presented this to me in private conversation and asked me my views upon it, and I told him that although I could recognize his difficulties that I nevertheless could not endorse his request because I felt that such a procedure was somewhat out of the ordinary and unusual and I was extremely anxious that this hearing should both in substance and in form take a normal course.

Is that about what I said to you?

Mr. GARRISON. Yes, indeed, Mr. Robb, and it is a fair comment.

I think I said in return that this was an inquiry and not a trial, and Mr. Robb agreed with that. I understand that in a trial the reasons for continuing the redirect of a witness after the cross is the natural thing to do before a jury or a judge. Most trials at law involve fairly simple issues of fact. Here we are dealing with a man's whole life and it is quite a different thing, Mr. Chairman.

Mr. GRAY. The board will withdraw for a moment.

(Whereupon, the board withdrew from the hearing room.)

Mr. GRAY. Mr. Garrison, the members of the board have discussed your request, and as I have expressed to you several times, we are concerned about the fragmentation which has been involved in the proceeding and continue to be concerned about it.

Our personal preference, I think, as individuals and collectively would be to proceed and not to take an afternoon off because every day we take off now, it seems to us, adds another one on the end, and that involves problems for everybody concerned.

On the other hand, I believe the record of these proceedings reflects clearly to this point our desire that every courtesy and consideration and every possible effort as fairness be demonstrated to Dr. Oppenheimer. I am authorized by my colleagues on the board to say for them and for myself that whereas we regret very much this development and this kind of interruption, failure to be able to keep witnesses on as the whole story unfolds as they can tell it, nevertheless we, pursuant to your request and out of consideration for Dr. Oppenheimer will recess at the lunch hour tomorrow or as soon thereafter as time may be required to finish the witnesses who are here and will proceed on Thursday morning with the redirect examination of Dr. Oppenheimer.

Mr. GARRISON. Thank you, Mr. Chairman. I appreciate that very much. Part of the problem that we have faced with the witnesses was our desire to follow your own request at the start of the proceedings so we have no gap at all, and so we have tended to overload the witness schedule.

Mr. GRAY. I understand.

Mr. GARRISON. Since they do come from out of town, it makes it unusually difficult. For example, Dr. Fermi is here as scheduled and he has a Chicago plane to make. Mr. Lilienthal has to be in Camden tonight. Dr. Conant has appointments with the Secretary, and so forth.

If it is possible to do so, we would like to put Dr. Fermi on after Dr. Conant because he is going to be very short and has only one thing to talk about, but if you would very much prefer to resume with Mr. Lilienthal we will do it that way.

Mr. ROBB. Mr. Chairman, I am doing the best I can to accommodate Mr. Garrison, but I am supposed to examine these witnesses to develop facts. Although I appreciate Mr. Garrison's problem, it is extremely difficult for me to listen to 1 witness for 15 minutes and then have him leave the stand and hear some other witness and ask him questions and then come back to the first witness and maybe have him go off and finally get around to examining the first witness. It is almost impossible to do that properly.

I do think that once we get Mr. Lilienthal back on the stand he ought to stay there until he is through.

Mr. GARRISON. I didn't mean to break him up in fragments again. I meant to sandwich Dr. Fermi in between the two fragments of Mr. Lilienthal. I don't press it at all, Mr. Chairman.

Mr. GRAY. It seems to me that once Mr. Lilienthal is off the stand, Dr. Conant and Dr. Fermi come in. I am sorry to use the expression on the stand. As a witness in the proceeding. We with respect to any other witness we just won't interrupt them any more, Mr. Garrison.

Mr. GARRISON. Thank you, Mr. Chairman.

(Discussion off the record.)

Mr. GRAY. Do you wish to testify under oath?

Dr. CONANT. Yes, I would be glad to.

Mr. GRAY. You are not required to, but all witnesses to this point have.

Would you then please stand and raise your right hand?

James B. Conant, do you swear that the testimony you are to give the board shall be the truth, the whole truth and nothing but the truth, so help you God?

Dr. CONANT. I do.

Mr. GRAY. Would you be seated, please, sir.

It is my duty to remind you of the existence of the so-called perjury statutes. I should be glad to give you more detailed information about them if that is necessary. I assume that it is not.

Dr. CONANT. Quite so.

Mr. GRAY. I should like to ask you, Dr. Conant, if it becomes necessary in your testimony to refer to restricted data, that you let me know in advance so that we may take necessary security precautions.

My final observation to you at this point is that we treat these proceedings as confidential between the Commission and its officials and Dr. Oppenheimer and his representatives and witnesses. The Commission will take no initiative in any public release with respect to these proceedings. We are expressing the hope that each witness will take the same view.

Dr. CONANT. Good.

Whereupon, Dr. James B. Conant was called as a witness, having been duly sworn, was examined and testified as follows:

DIRECT EXAMINATION

By Mr. GARRISON:

Q. Dr. Conant, just for the record, you are the United States High Commissioner to Germany?

A. That is correct.

Q. And formerly president of Harvard University?

A. Quite so.

You are appearing here at our request?

A. Yes, sir, at your request.

Q. Would you state very briefly the course of your acquaintance with Dr. Oppenheimer?

A. As I recall it, I must have met Dr. Oppenheimer for the first time in the early discussions of the atomic bomb affair and then followed, of course, his work at Los Alamos and my capacity as scientific advisor to General Groves in which I was at Los Alamos quite often.

After the end of the war I saw him again in connection with the so-called Acheson-Lilienthal report. I consulted with him occasionally in that connection and then again when the General Advisory Committee to the Atomic Energy Commission was established we met and he was elected chairman.

From then on I saw him quite often in connection with those meetings. Then later when he was elected an overseer of Harvard I saw him in that connection. I should say a few years ago he was on an informal committee of which I was chairman, the committee on the present danger.

Q. You have read the Commission's letter of December 23, 1953, which initiated these proceedings containing the derogatory information about Dr. Oppenheimer?

A. Yes, I have read it.

Q. Have you a comment to make on it?

A. Yes, I have. I would like to comment on it. I would like to comment on one section particularly. Somewhere in the letter it says that the substance of the information which raises the question concerning your eligibility for employment, referring to Dr. Oppenheimer, on atomic energy work, is as follows, and then later it says that it was further reported that in the autumn of 1949 and subsequently you strongly opposed the development of the hydrogen bomb; one, on moral grounds; two, by claiming it was not possible; three, by claiming that there were insufficient facilities and scientific personnel to carry on the development; and four, that it was not politically desirable.

Well, it seems to me that letter must have been very carelessly drafted, if I may say so, because if you take those two statements together, of course, it would indicate that anybody who opposed the development of the hydrogen bomb was not eligible for employment on atomic energy work later.

I am sure that no one who drew that letter could have intended that, because such a position would be an impossible position to hold in this country; namely, that a person who expressed views about an important matter before him, as a member of the General Advisory Committee, could then be ineligible because of a security risk for subsequent work in connection with the Government. I am sure that argument would not have been intended. If it did, it would apply to me because I opposed it strongly, as strongly as anybody else on that committee, that is, the development of the hydrogen bomb. Not for the reasons that are given there.

If I might say so they are a rather caricature of the type of argument which was used in the committee in which I participated. I should say I opposed it as strongly as anybody on a combination of political and strategic and highly technical considerations. I will go into that later to some degree although I don't think this is the place to justify the conclusions of the General Advisory Committee. It would be a long story.

It seems to me that clearly the question before you here is the question rather, is the implied indictment, I submit, namely, because of the information in the first part of this letter—Dr. Oppenheimer's association with alleged Communist sympathizers in the early days in his youth—that that somehow created a state of mind in Dr. Oppenheimer so that he opposed the development of the hydrogen bomb for what might be said reasons which were detrimental to the best interests of the United States, because they were interests of the Soviet Union which he in one way or another had at heart.

That, I take it, is the issue which I take it is before you in part in considering this letter. It is to that that I would like to speak for, I think, I have some evidence that convinces me that any such charge is completely ill founded.

If it were true that Dr. Oppenheimer's opposition to the development of the hydrogen bomb were in any way connected with a sympathy which he might have had with the Soviet Union, or communism, then surely many other actions and decisions which he was involved in over the period of years in which I was associated with him would have likewise been influenced by any such point of view.

The record is quite the contrary. I just call your attention to a few facts probably already before you—actions of Dr. Oppenheimer, participation in decisions, all of which were strongly detrimental to the interests of the Soviet Union after the close of the war.

We can start with the time shortly after the Acheson-Lilienthal report when an attempt was made through the United Nations to get an agreement with Russia on the control of atomic bombs.

As I recall it, Dr. Oppenheimer was early associated with Mr. Baruch and then later with Mr. Osborn in that series of negotiations. I was only tangentially associated, I was called in from time to time by Mr. Osborn. I remember sitting in one or two meetings. I can't give you the dates because I haven't had time to look any of this up, and I don't keep records.

At that time we had a number of discussions which were early, you see, in the development of the postwar period, with Dr. Oppenheimer and with others. At that time it seemed to me that Dr. Oppenheimer's appraisal of the Russian menace, of the Soviet situation, was hard headed, realistic, and thoroughly anti-Soviet, designs which even then were quite clear with their expansion into the free world.

That would be my first basis for believing that his attitude at that time was thoroughly loyal to the United States and thoroughly opposed to the Soviet Union and communism in every way.

Then coming to the period when he became chairman of the General Advisory Committee. Again this is probably well known to you. There is no restricted information here. I am going to speak in general terms.

Yet, as Winston Churchill later said, it was the possession of the atomic bombs in our hands that prevented, so he believes, Russia being at the channel ports during that period of history. There was a great deal to be done. Dr. Oppenheimer was a vigorous proponent as chairman of the committee of getting ahead and putting that shop in order.

Los Alamos was revivified. From then on all the decisions of the committee, with possibly the exception of this controversial thing about the hydrogen bomb would, I think, be shown entirely on the side of arming the United States. There was only one possible enemy against whom it was being done—it was the Soviet Union.

There are many other matters if I had a chance to go over the records of the General Advisory Committee.

As seems implied in this indictment that Dr. Oppenheimer was influenced by pro-Soviet and anti-United States views, he would not have taken the views he did. I named just two that come to me.

One is a matter on which I think I can take some credit of calling to the attention of the Advisory Committee of getting ahead rapidly on methods of detecting any explosion that might occur in the atomic field by the Russians. I remember Dr. Oppenheimer may have picked that up before I did; he may have had the suggestion before I did, although I don't think so, and taking steps in the committee to see that something would be done in that regard.

Clearly anybody that was influenced by any point of view in favor of the Soviet Union could hardly have done that.

Another matter—the development of smaller atomic bombs which could be used for tactical purposes; support of the ground troops which in my judgment of military strategy seemed to me of great importance. That was a matter which I know he pushed vigorously in the committee. He made strong statements about it. I think he was very active.

There again it seems to me is an illustration of a definite action taken by this man which contradicts what seems to me the implied thesis in this part of the indictment.

There is a final matter which is not connected with the General Advisory Committee but which is of rather a personal nature. I spoke to the committee on the present danger. That was a group of men that came together informally to make a public committee, started in the fall of 1950. The Korean War was going in a bad way. We believed that the United States Government was not taking proper steps to put itself in a strong military position, particularly with respect to the defense of Europe on the ground.

Late that year or early in 1951 we put out some statements urging Universal Military Service and urging that we send more troops to Europe, generally the policy which has become the policy of the United States. Dr. Oppenheimer was asked to join that committee. He joined it. He subscribed to all those doctrines which were most vigorously anti-Communist. He spoke to at least one, I think, informal gathering where we were trying to raise some money to get ahead with a little of our propaganda work. Perhaps it is unnecessary to put on the record that I must admit that we had no success with our doctrine of Universal Military Service, but that is another story.

As far as the defense of Europe on the ground is concerned, things have followed the way we at least advocated.

Q. There was put in evidence here, Dr. Conant, a letter which Dr. Oppenheimer identified as one written to you shortly before the meeting of the General Advisory Committee in October 1949, in which he addressed you as "Uncle Jim" and talked about the question of the hydrogen bomb and the forthcoming meeting.

When I showed you that letter, as I did ——
A. Yes, you showed me that last night.
Q. A copy of it, I mean. Did you have any recollection of having received it?
A. No, I had not. I did not remember it. I couldn't say that when I saw it. I suppose it was delivered. It must have been a classified document. I was very fussy about not taking classified documents when they came to the office. If I received it, I must have taken it right down to Washington. I don't say I didn't see it, but I have no remembrance of it. I would not have known about it if you had not called it to my attention.
Q. Do you remember any discussion with Dr. Oppenheimer one way or another before the October meeting?
A. No, I am afraid my detailed recollection of that period is very hazy. I think there were two meetings at least of the General Advisory Committee.
Q. One in October and one in December?
A. Yes. There was certainly plenty of discussion in those meetings. Those I remember pretty well, but when and where I first discussed this matter, where I first heard of it, is not clear in my mind. Whether I walked into it, or whether, as implied by that letter, it was before, or whether it was some other source of information, I am sorry I just don't remember.
Q. How did Dr. Oppenheimer as chairman of the General Advisory Committee conduct the meetings?
A. He was an excellent chairman, but I hope he won't take it amiss if I say he ran them like a faculty meeting. There was a great deal of discussion and a great deal of talk. They were the most lengthy meetings I ever sat in on in my life. They consumed an un-Godly amount of time, but they covered the ground from A to Z.
Q. Coming now to the meeting of October 29, 1949, when you first discussed the hydrogen bomb, upon whose technical advice did you rely?

A. I can't be sure of that meeting because, as I say, my memory of that period is not accurate enough to spot the meeting and the discussions, and so on. As I said, in my comment here a minute ago, I was moved in my opposition to this in signing the statement of the General Advisory Committee, which I have not seen since, by a mixture of political, strategic, and technical considerations. Those technical considerations are extremely detailed, but judging from some things I have read in general in the press, completely misunderstood.

Of course they concerned the question of what kind of large weapon to make and what was the cost and what were the opportunities of doing it, and what were the probabilities.

When it came to a question of the nuclear physics in which I am by no means an expert, I always counted on Dr. Fermi's judgment. With all due respect to all the other members of the committee, I felt he was both experimentally and theoretically the man whose judgment was to be relied on. Indeed his record during the development of the atomic bomb I consider one of the most extraordinary pieces of scientific correct calculations I can image. The story is a perfectly amazing one.

Q. Would you state very briefly for the board the reasons which lead you to make the recommendation which you did make on the subject of the hydrogen bomb?

A. It is a very complicated thing. I think it would take a long time to do a detailed inquiry into that. Some day if the Government wants to set that up, I should be glad to take the time, but I would have to go back into the record.

Therefore, what I shall do is only a general sketch. With all due respect to Dr. Gray and his colleagues, this would take a board, which included a nuclear physics expert, to assess the questions of whether the technical part of this decision was right or wrong.

On the general strategic and political grounds there were some of the same reasons which we subsequently brought to a head on the committee on the present danger, namely, this was supposed to be an answer to the fact that the Russians had exploded an atomic bomb.

Some of us felt then, and I felt more strongly as time went on, that the real answer was to do a job and revamp our whole defense establishment, put in something like Universal Military Service, get Europe strong on the ground, so that Churchill's view about the atomic bomb would not be canceled out.

One of the considerations was that this was sort of a Maginot Line psychology being pushed on us. On the technical ground the question was the investment in preparing certain materials which I am not going into, which are restricted, which seemed at that time necessary; the use of materials which I don't want to mention, which would be used up.

The question was when you expended a certain amount of manpower and energy and material, would you actually from the point of view of delivering blows against a potential enemy be very much better off even if this line worked?

Of course, to do an assessment on whether we were right on the technical ground you would have to then go into the subsequent developments which I don't know about because I ceased being on the committee in August 1952. But judging from what I read in the papers, some things have worked and presumably along different lines from what we were then thinking.

Q. In March 1947 did Mr. Lilienthal as chairman of the Commission ask you for your opinion with respect to Dr. Oppenheimer's loyalty?

A. Yes. I recall that this was at the time when Mr. Wilson who was general manager, Mr. Lilienthal and the other members were up for confirmation in the Senate. I think that is the right time. I remember Mr. Wilson and I think Mr. Lilienthal coming to me, saying that we have been apprised that there are some things in the record of Dr. Oppenheimer which indicate association with alleged communists, some things of that sort, and we want to know whether you are prepared to make a statement in regard to his loyalty.

I am pretty sure I didn't examine the file. I am sure I didn't. I said that "I don't know about the past, but I am glad to put on record what I now believe, based on my knowledge of him since the early days of the war," and there is such a letter in existence. I have not seen it.

Q. I have it here, Dr. Conant. It was brought into evidence this morning.

May I, with the board's permission, just read you the last two paragraphs. The first four of the letter have to do with a recital of your acquaintance with Dr. Oppenheimer and the circumstances of your writing the letter. Then you went on to say: "I can say without hesitation that there can be absolutely no question of Dr. Oppenheimer's loyalty. Furthermore, I can state categorically that, in my opinion, his attitude about the future course of the United States

Government in matters of high policy is in accordance with the soundest American tradition. He is not sympathetic with the totalitarian regime in Russia and his attitude towards that nation is, from my point of view, thoroughly sound and hard headed. Therefore, any rumor that Dr. Oppenheimer is sympathetically inclined towards the Communists or towards Russia is an absurdity. As I wrote above, I base this statement on what I consider intimate knowledge of the workings of his mind. [Reading:]

"At the time of Dr. Oppenheimer's entering the work on atomic energy, I heard that there was some question of his clearance by the security agencies. I understand that was based on his associations prior to 1939 and his 'left-wing' sympathies at that time. I have no knowledge of Dr. Oppenheimer previous to the summer of 1941, but I say unhesitatingly that whatever the record might show as to his political sympathies at that time or his associations, I would not deviate from my present opinion, namely, that a more loyal and sound American citizen cannot be found in the whole United States."

You wrote that?

A. Yes, I wrote that. I have every reason to believe I wrote it.

Q. Dr. Conant, you formed your judgment at that time on your appraisal of Dr. Oppenheimer as a total man?

A. Yes. That was based clearly on my acquaintance with him during the Los Alamos project and this other period which I mentioned in which we discussed the whole question of the control of the bomb, which gave me a chance to explore many political problems which we would not have explored at Los Alamos.

Q. Having in mind the Commission's letter of December 23, 1953, to which we have referred on the one hand, and what Dr. Oppenheimer has done since March 1947 when this letter was written, do you have reason to modify or alter the view which you expressed about him in March 1947?

A. No. I would think on the contrary the actions and decisions which I put on the record here seem to me to make quite clear that he was party to many actions on the part of the General Advisory Committee which were strongly opposed to any Soviet policy. It makes more certain the statements I then made based on what was after all a shorter acquaintance with him.

Mr. GARRISON. That is all, Mr. Chairman.

Mr. GRAY. Mr. Robb.

CROSS-EXAMINATION

By Mr. ROBB:

Q. Dr. Conant, at the outset of your statement you quoted from the letter from Mr. Nichols to Dr. Oppenheimer, did you not?

A. Yes. Did I quote correctly?

Q. Did you quote that from memory?

A. No. I wrote it.

Q. Did you copy it?

A. I copied it from the New York Times.

Q. Would you read it to me again?

A. As I wrote it, there is a place somewhere about a third of the way down which says that the substance of the information which raises the question concerning your eligibility for employment on atomic energy work is as follows— is that correct?

Q. Go ahead.

A. Then there are a lot of other things and then comes: "It was further reported that in the autumn of 1949 and subsequently you strongly opposed the development * * *"

Q. That is fine. That word "further" indicates, does it not, that that sentence is tied in with other sentences in the same paragraph?

A. Yes, and to that extent it is the simplification that I spoke of. If you don't emphasize the "further," it would appear—that would be an impossible thing, and I am sure nobody intended it to mean so. Therefore, it was the implication I was speaking to.

Q. Certainly you would agree that sentence must be taken in its context with the rest of the paragraph?

A. Quite so. It was to that that I was speaking when I attempted to put in evidence that which made me think such an implication was wrong.

Q. That is a rather long paragraph, taking almost a page of single spacing in the letter, is it not?

(No response.)

Q. Doctor, referring to your letter to the Commission or to Mr. Lilienthal on March 29, 1947, which Mr. Garrison read to you, the last sentence especially?
A. May I look at it?
Q. Yes. Have you a copy of it?
A. It is the first time I have seen it since I wrote it.
Q. "I have no knowledge of Dr. Oppenheimer previous to the summer of 1941, but I say unhesitatingly that whatever the record might show as to his political sympathies at that time or his associations * * *" and so forth. By that did you mean, sir, that even though the record might have shown that he had been an actual member of the Communist Party that would not alter your opinion?
A. If he had been an actual member I would have been willing to bet that he would have renounced the membership and be one of these people who had changed his point of view, as some people have. I had no reason to believe that any such charges were in there.
Q. No. I am trying to explore if I might just how far you were going.
A. Political sympathies is not the same word as political associations.
Q. I understand that.
A. It was political sympathies that seemed to be charged at that time.
Q. I am trying to find out how you defined the term.
A. You are asking me now rather than when I wrote that letter, because it is pretty hard for me to say what I thought when I wrote this.
Q. I assume that you still stand back of that sentence now?
A. That is right.
Q. Would it be your testimony now that even though Dr. Oppenheimer might have belonged to the Communist Party in 1941 you still would make the same statement about him?
A. Yes; provided there was not anything in the record to show that he continued to be a member and he was an agent and so on. If you brought out a lot of those facts which to my mind would be impossible considering the actions he had taken, of course, anybody can be mistaken on those things.
Q. Of course, Doctor, you don't know what the testimony before this board has been?
A. No, I don't.
Q. Nor do you know what the record or file before the board discloses?
A. No. I only know what is in the letter of General Nichols.
Q. You spoke of the meeting in March 1947, which for your information I will tell you, I think we agreed, was March 10, 1947. You met with the Commission. Do you remember that? I believe you said Mr. Wilson was there.
A. Did I? Where was this? I am sorry.
Mr. GRAY. I thought Dr. Conant said Mr. Wilson and somebody came to see him.
Mr. ROBB. Was that it?
The WITNESS. All I remember is that Carroll Wilson who was then the general manager and I think Mr. Lilienthal came to see me.

By Mr. ROBB:

Q. I see.
A. That is all I have a clear remembrance on.
Q. And asked your opinion?
A. They said here we are told something in the record is doubtful about Dr. Oppenheimer, what do you think of him, and will you write a letter, and I wrote it.
Q. Did they at that time show you the record they were talking about?
A. To my memory they didn't.
Q. Doctor, you spoke in your testimony of three factors which you consider in connection with the atomic bomb: Political, strategic, and technical; is that right?
A. Yes.
Q. When you referred to the strategic factor, did you refer to the military strategy?
A. Yes. Military and political strategy run together pretty heavily, but I had in mind the sort of thing I spoke of a moment ago. The strategic considerations which are implied in that statement of Winston Churchill.
Q. Of course, Doctor, you don't pretend to be an expert on military strategy, do you?
A. No, but I think I am entitled to an opinion on it.
Q. Certainly. Most of us do. I am just trying to find out what your political field is.

A. Remember at that time all of us talked a great deal about the military strategy involving atomic bombs and so on. I certainly am no expert on it.

Q. Were there men who were qualified military strategists trained in that particular field at that time who disagreed with you on your views?

A. I don't know that we had them on the committee. After all this thing was a recommendation of a general advisory committee. It was going to be bounced over many other hurdles. We were not setting policy. We were giving our views. I am not prepared to say. I don't know that I discussed it at that time. Subsequently I was quite aware of people who had different views.

Q. Subsequently you found out they did?

A. Certainly.

Q. Doctor, I am a little bit curious, if you don't mind, as to why you gentlemen on the committee undertook to give advice as to military strategy?

A. I don't know. The General Advisory Committee is concerned with a great many things. Nobody has to take the advice if they don't want to. It turned out they didn't.

As I recall the report, it was largely centered on the technical. I have not seen the report since the day it was written. Certainly the things run together terrifically on the question of what you expend money and manpower for.

We were a general advisory committee, not a technical advisory committee. We ventured even to suggesting how they should reorganize the Commission.

Q. Yes. Your comments, it seems to me, though, Doctor, had to do not with technical matters or matters concerning the development of the bomb, but rather with the use use which the military might make of the bomb. Would you agree with that?

A. No. A great deal about the question of the use of manpower and money and fissionable material—I guess that is not restricted—in the best use to make weapons. The question of delivery of the weapons has always been a concern of the General Advisory Committee.

Q. By delivery you mean on the enemy?

A. Yes. One of the things the General Advisory Committee most concerned itself with from the start was the whole question of getting a position where bombs could be delivered satisfactorily. It was also true at Los Alamos.

Q. In all events you did feel and you do now feel that considerations of military strategy properly came within the function of your committee?

A. As a question of advice surely. They didn't have to take it if they didn't want to.

Mr. ROBB. I think that is all I care to ask, Mr. Chairman.

Mr. GRAY. May I ask a question or two.

Is your recollection that you didn't see any files?

The WITNESS. My recollection is that I didn't see any file on Dr. Oppenheimer at all.

Mr. GRAY. I don't know what the facts are, but I wouldn't want to leave the record fuzzy with respect to this conference, whenever and wherever it took place.

We had before us briefly this morning Mr. Lilienthal who testified, I believe, that he called a meeting in his office with you and Dr. Bush to discuss this matter. Do you remember discussing it in the presence of Dr. Bush?

The WITNESS. I certainly discussed the subject with Dr. Bush.

Mr. GRAY. Your recollection is not good?

The WITNESS. I have nothing to indicate that it is not correct. I remember Bush at the time wrote a letter, too.

Mr. GRAY. Yes, he did.

The WITNESS. Bush and Groves both wrote letters. I don't remember whether I discussed it with Bush in the presence of the Commission. That is not clear in my mind. I remember writing the letter very clearly

Mr. GRAY. Therefore, since you didn't see the file and you don't remember any real discussion of the file, your judgments with respect to Dr. Oppenheimer are based entirely on your association with him?

The WITNESS. Entirely on my associations with him from the period of 1941 on and based on my discussions with him, particularly after the war, on what I would consider matters much more than technical but ones which would bring out very definitely a man's sympathies or latent sympathies with the Soviet Union, which you will recall at that time was a subject of considerable debate with many people.

I found it refreshingly, from my point of view, hard headed and anti-Soviet, which was my view at the time and always has been.

Mr. GRAY. At this time you were president——
The WITNESS. Yes.
Mr. GRAY. Did you have experience of having put faith in a member of your faculty, having supported him and defended him, not simply on the issue of academic freedom but a little beyond that—your own faith in him—and then find yourself experiencing a situation such as at Minnesota with respect to a man referred to as Scientist X, who appears in this proceeding? I am not trying to relate it in that way.
Mr. ROBB. Weinberg.
The WITNESS. No. Obviously it could happen to anyone.
Mr. GRAY. I know of no instance myself.
The WITNESS. No, I don't recall.
Mr. GRAY. That could happen?
The WITNESS. It could happen, of course.
Mr. GRAY. Are you familiar, Dr. Conant, with the provisions of the Atomic Energy Act of 1946 within the framework of which we must carry on the proceedings of this board and the fact that under the act the criteria apparently are character, associations and loyalty of an individual. That a determination must be made with respect to those with a finding that permitting an individual to have access to restricted data will not endanger the common defense or security.
I have expressed that badly, but these criteria are established and then they must be met for clearance for access to classified material.
I am not asking you at this time to comment on the wisdom of these criteria, but to ask you if you are aware that these are the criteria?
The WITNESS. I was not aware of the exact phraseology but some such things.
Mr. GRAY. I would ask you, then, whether you feel that any board or any Government official in trying to make an evaluation, as we believe we are called upon to do, must take into account associations over a period of years in order to make a finding with respect to this criterion of associations.
I ask this because I believe that your stated view, which was clear and convinced, was that early associations were unimportant in the light of later conduct.
The WITNESS. And by later associations, surely. I would have said quite clearly that since the period I have known him that the associations from all the evidence I had, yes.
What you are saying is that associations beyond a certain period might lead you to inquire into later ones. But certainly it does not say in the act how distant those associations. I assume it means the present character and the present associations. Therefore, you are going back into the past in order to extrapolate it into the present.
If I had known he had any associations—or a suspicion—I don't believe he would—that would not have affected my statement; it is not only his views, but talking with his then associations.
Mr. GRAY. You would not have too much information about associations.
The WITNESS. Not undercover in nature.
Mr. GRAY. Or generally speaking, would you? You met Dr. Oppenheimer in GAC meetings and panels.
The WITNESS. You are talking about the letter, or are you talking about my present knowledge of him?
Mr. GRAY. At the time that you knew him.
The WITNESS. As of now, which is much greater than at the time in 1947 when after all I had known him only 6 years. Since then I have seen a great deal more of him.
Mr. GRAY. We are called to make a recommendation as of now and not as of 1947.
The WITNESS. Quite so.
Mr. GRAY. So that we, you understand, have to take into account all the material which seems to be substantiated which is before us, perhaps some of which you are not at all familiar with.
The WITNESS. Quite so. I am presenting to you, to sum up, the evidence which seems to me makes extremely improbable the hypothesis called for by that word "further", Mr. Robb, which you called to my attention, of the hydrogen bomb with the consequence of the early associations set forth in the letter. That is what I was speaking to.

Mr. GRAY. A summary of your testimony might be that so far as you have any knowledge about anything and on the basis of your best judgment you consider that Dr. Oppenheimer's character, loyalty and associations are such that he should have access to restricted data.

The WITNESS. Quite so. And I would give the specific items in which his judgment was such that if he had been influenced by pro-Communist views, or pro-Soviet views, he would not have taken those actions or decisions, and they were quite serious. In other words, this is not a general expression of belief based on casual conversations, but participating in a great many, I would say, fairly powerful anti-Soviet actions.

Mr. GRAY. Do you have any questions, Dr. Evans?

Dr. EVANS. Yes, a few.

Dr. Conant, you understand the position this board is in on this matter.

The WITNESS. I beg your pardon.

Dr. EVANS. You understand our position. We didn't seek this job.

The WITNESS. I can readily understand that.

Dr. EVANS. We are trying to do the best we can.

The WITNESS. Quite so.

Dr. EVANS. Perhaps this advice to us should be rewritten now and say something about the present. What do you think about that? Do you think we should go by this thing at all?

The WITNESS. I am really not here to advise you on what you should do. I pray that is beyond my competence.

Dr. EVANS. In regard to character, associations and loyalty.

The WITNESS. You are probably a lawyer——

Dr. EVANS. That is what it says here.

The WITNESS. I should imagine lawyers would argue what present, past and so on, meant. Far be it from me to enter into that argument.

Dr. EVANS. I have nothing more.

REDIRECT EXAMINATION

By Mr. GARRISON:

Q. Just one question, Dr. Conant. Supposing that you were told that early in 1943 during the wartime project on which Dr. Oppenheimer served he had been approached by a friend—I think you have heard of the Chevalier incident?

A. It is in the letter.

Q. That this friend had told him of Eltenton's channel for transmitting information to Russians, that Dr. Oppenheimer rejected emphatically any suggestion that activity of this sort should be engaged in and spoke of it as treasonous; supposing that some months later, after a delay of some months, Dr. Oppenheimer volunteered the information about Eltenton to security officers but refused to disclose at their request and their urging the name of his friend who was the intermediary and indeed suggested that the intermediary might have been some unnamed other people; that later when he, having persisted in this refusal to name this friend, knowing that the security officers were very anxious to ascertain who it was, General Groves asked him to tell him, that he declined to tell General Groves, that unless General Groves ordered it and General Groves said he didn't want to order it, but to think it over and later General Groves did tell him that he would order him unless he told him, and that Dr. Oppenheimer then revealed the name of Chevalier; would the judgment which you have expressed here about Dr. Oppenheimer's loyalty, about his character, be altered?

A. It seems to me if I followed this hypothetical—I assume it is hypothetical, the way you are stating it—incident, if I sum it up, in that case the question would have been that he had been negligent in taking steps necessary to bring into prosecution somebody who had attempted to get information? Is that roughly what the charge would have been?

This is a fairly complicated story you are telling me with a good many yeses, ands, and buts in it.

Q. There was the element of delay in reporting it; there was the delay of not frankly stating it and the circumstances when he did report it; there was the element of declining to name the friend after he had been pressed to do so; but there was the element finally of his revealing the name and also of his having initiated the whole business of revealing Eltenton's name.

A. Of course, any such thing like that would depend on the number of instances. You are assuming this is the one instance.

Q. For the purpose of the question, yes.
A. I would suppose that the question that would be presented then with that is, What were the motives at that time, and what did that show about his subsequent attitude in regard toward the Soviet Union? Did he do that at that time for reasons of trying to protect the Soviet Union agent who was trying to get information and did that indicate that he would continue to have an attitude from then on about various matters connected with atomic energy which would be not in the interest of the United States?

In view of all the things I mentioned, I would say that it didn't change it for that reason. It stood by itself and had nothing else but conversation with the man. You have to take the summation of evidence as you see it. If I were merely testifying here that I had known Dr. Oppenheimer in talks over these years, and so on, and I thought he was a loyal citizen, I don't think my evidence would be of the sort that I hope it is. By having participated with him in what I believe to have been effective actions against the Soviet Union.

Mr. ROBB. May I ask one more question?

RE-CROSS-EXAMINATION

By Mr. ROBB:

Q. Dr. Conant, as a distinguished scientist and scholar——
A. I am not a distinguished scientist, but I am willing to be considered a scholar; thank you.
Q. As a scholar, you would agree, would you not, sir, that any conclusion, that any opinion about a given problem, to be reliable, must be based on all the relevant facts and all the relevant evidence?
A. Surely.
Q. And any opinion or conclusion which is not based on all the relevant facts and all the relevant evidence might be fallible?
A. Yes; but as a scholar I know perfectly well there is no such thing as all the relevant; all the human beings can do is give their evidence and statement on what seem to them the relevant things at the time.
Q. Precisely.
A. Therefore, I don't quite like the word "all" there, because that implies an omniscience.
Q. All the available evidence?
A. All the available evidence.
Q. With that amendment, you would answer yes to both my questions?
A. Yes.
Mr. ROBB. Thank you.
Mr. GRAY. May I pursue this hypothetical question of Mr. Garrison's for a moment, Dr. Conant? You suggested what issue that hypothetical situation might raise, namely, that this might be an indication of an interest in protecting the Soviet Union. I am not sure these were your remarks.
The WITNESS. Or an act of the Soviet Union, if I got the quick summary of it correctly.
Mr. GRAY. Or it might be interpreted as simply a desire to protect a friend.
The WITNESS. Yes. I would say a mistaken idea that you had to protect a friend in those circumstances.
Mr. GRAY. If in this hypothetical situation as I think Mr. Garrison indicated, the security officer was pressing for this information, very important perhaps to the security officer who was charged with the security and who would not have any reason to believe that perhaps friendship was involved, the question again—and I am relating this to the present and to the act—or I suppose a question is: In any situation involving a divided loyalty or a conflicting loyalty, the protection of a friend, and to the obligation one owes to one's government, is there any question as to which should be——
The WITNESS. Not in my mind. That is why as you recall, I said I wanted to answer that question in the context that this was one incident and not many. I think we all recognize in reviewing a long history of a person, people can make errors. If they are single, they are one thing; if they are multiplied, they are quite a different picture.
Dr. EVANS. Dr. Conant, if you had been approached by someone for security information, wouldn't you have reported it just as quickly as you could?
The WITNESS. I think I would have, yes. I hope I would have; let us put it that way.
Dr. EVANS. That is all.
Mr. ROBB. May I ask one more question?
Mr. GRAY. Yes.

By Mr. ROBB:
Q. When you did report it, Doctor, you would have told the whole truth about it?
A. I hope so.
Q. I am sure you would. Thank you.

REDIRECT EXAMINATION

By Mr. GARRISON:
Q. Dr. Conant, suppose that in the hypothetical question Dr. Oppenheimer had sincerely believed that his friend was incapable of lending himself to activity of this character, and that loyalty to his Government was not in fact involved, so that the fault was one of asserting his own judgment and deciding for himself whether the interests of the country were involved, rather than following the assurance of the security agent that it was, would you feel that the culpability or the fault was of a different order than protecting a friend about whose loyalty he was in doubt?
A. I take it that even this hypothetical question I am not asked to pass a moral judgment on. I would be concerned with what does that action indicate in regard to a question which I take it is here, which is the security risk of the man in question. It seems to me that is what you have to put it in context with. I am not going into the fine moral things as to whether people do things this way or that way. Conflicting loyalties were involved. You asked me the question how I would have resolved myself. I am quite frank to say I would have resolved these the way I answered. If the question is having somebody else resolve them, what does that show in view of a total record in regard to a security question.
Mr. GRAY. Thank you very much.
The WITNESS. Thank you. I appreciate you for allowing me to come in at this moment, because I am on a tight schedule, as you say.
(Witness excused.)
Mr. GRAY. Let us proceed with Dr. Fermi, if he is here.
Dr. Fermi, do you wish to testify under oath?
Dr. FERMI. I would be glad to.
Mr. GRAY. The other witnesses have. You are not required. May I have your full name?
Dr. FERMI. Enrico Fermi.
Mr. GRAY. Would you be good enough to stand and raise your right hand?
Enrico Fermi, do you swear that the testimony you are to give the board shall be the truth, the whole truth and nothing but the truth, so help you God?
Dr. FERMI. I do.
Whereupon, Enrico Fermi was called as a witness, and having been first duly sworn, was examined and testified as follows:
Dr. GRAY. Would you be seated, please, sir.
I must point out to you the existence of the perjury statutes. I assume you are familiar generally with those?
The WITNESS. More or less, yes.
Mr. GRAY. I should be glad to disclose the penalties if you wish.
The WITNESS. I will try not to be involved with them.
Mr. GRAY. May I ask if in the course of your discussion here it becomes necessary for you to disclose restricted data, will you advise me before the disclosure, because there are certain steps we would find it necessary to take in that event.
Also I say to each witness that we consider that these proceedings are a confidential matter between the Atomic Energy Commission and its officials, and Dr. Oppenheimer, his witnesses and representatives. The Commission will take no initiative in release to the press anything about these proceedings and the testimony, and we express the hope each witness will take the same view of the situation. Mr. Garrison.
Mr. MARKS. Mr. Chairman, in the interest of getting back to the interrupted witness as quickly as possible, I will ask just a very few questions of Dr. Fermi.

DIRECT EXAMINATION

By Mr. MARKS:
Q. Dr. Fermi, would you be good enough to identify yourself for the record?
A. My name is Enrico Fermi. I am at present professor of physics at the University of Chicago.
Q. Were you a member of the General Advisory Committee of the Atomic Energy Commission?

A. I was a member of the General Advisory Committee for a period of a little bit short of 4 years, until December of 1950.

Q. You participated then in the deliberations of that committee concerning the advice to the Commission on the thermonuclear program in the fall of 1949?

A. I did.

Q. Would you tell the board briefly what you can in an unclassified way about those deliberations, the positions taken, the reasons for them?

A. Yes. I should perhaps mention the matter goes back to about 5 years, and my recollection is partly vivid, partly a little bit uncertain, but I think I remember the essentials, which are about this way: That the committee was confronted with forming an opinion whether it was the right time to start an all out program for developing the hydrogen bomb.

Q. This would have been the meeting of October 29, 1949?

A. That I understand is the date, although I don't remember it on my own. So we were confronted with this decision. I can testify naturally to my feelings in this matter better than I can to those of other people. As far as I could see the situation, I had the concern that the pressure for this development was extremely inordinate, or at least so it seemed to me. I was concerned that it might weaken the development of conventional atomic weapons which was then picking it up and essentially set it back for what seemed to me at the time a not quite decided advantage on the other side. For that reason, and I believe that these views must have been shared more or less by everybody in our group, because a decision that it was not the right time to go in an absolutely overriding way in that direction was, as far as I remember, unanimous.

There was a subsequent point on which some difference of opinion arose, and I found myself in this connection in the minority together with Rabi. Again I have no absolutely clear recollection. I have no doubt that the board has available the records of those meetings presumably where things are spelled out in full detail. My recollection is that this divergence of opinion was on whether to essentially declare or establish the policy not to go ahead with the program or whether some circumstances could make us go ahead.

My opinion at that time was that one should try to outlaw the thing before it was born. I sort of had the view at that time that perhaps it would be easier to outlaw by some kind of international agreement something that did not exist. My opinion was that one should try to do that, and failing that, one should with considerable regret go ahead.

Q. Do you remember, Dr. Fermi, whether or not there was opportunity at those meetings late in October 1949 with the freest and fullest discussion among you—consistent with the rather brief time, few days?

A. Yes, I think so. I think everybody had a right to his own opinion and to defend his own opinion.

Q. Was there a great deal of discussion and debate?

A. No doubt there was. I think we had some trouble and some soul searching, all of us.

Q. There has been introduced in the record here a letter which was written by Dr. Seaborg, around the middle of October 1949 to Dr. Oppenheimer which dealt with the subject of the thermonuclear problem among other things. The letter has been variously interpreted as to what it means. Do you have any recollection at all of that letter?

A. No, not from that time. In fact, as far as I am aware, the first time I learned it from you was this afternoon.

Q. Seaborg was absent from that meeting?

A. Seaborg was absent, yes.

Q. Shortly after this meeting in October 1949, am I right that there was another meeting of the GAC?

A. Yes.

Q. Within a month or so?

A. I don't remember, but within a relatively short time.

Q. And was Seaborg present at that next meeting?

A. I think so, yes. In fact, I remember, or I have an impression or he gave me the impression to be somewhat happy not to have been confronted with the difficulties of contributing to what was a difficult decision. That was the impression that he gave me at least.

Q. Shortly after this time—that would have been the end of 1949—it was not long after that you left GAC?

A. In the following summer. I suppose the last meeting must have been in the late spring.

Q. Do you have any memory of actions which the GAC took in that rather brief interval?

A. My general impression is that we all had the concern that the conventional weapons program should not be weakened and we tried to see that the various provisions that were taken for furthering the hydrogen program would not be of such a nature of interfering seriously with the conventional weapons program. Actually I believe that this could be done and I am not aware that there has been such a weakening.

Q. Do you have any impression that these actions that you took had the effect of interfering with the program for the thermonuclear development?

A. No.

Q. Going back to the earlier period when you were a member of the GAC, prior to the meeting on the thermonuclear device, would you describe very briefly the position that Dr. Oppenheimer took with respect to the development, perfection and refinement of atomic weapons?

A. Yes. I think I can say very definitely that I always saw him push for all the measures that could improve our positions in conventional atomic weapons, and this includes seeing to it that exploration of ores would go ahead vigorously, that production of primary materials would be expanded, that all the various gadgets that go into this weapon would be streamlined as much as possible, that varieties of weapons that could conceivably improve our military position would be investigated and developed. I don't in fact in this respect remember any instance in which I disagreed on essential points. We always found ourselves very much together pushing in that direction together with the help of our colleagues. But perhaps Oppenheimer first and I, in somewhat second line, knew perhaps more about the technical details of weapons than most other people of the board knew, so that this task naturally fell more precisely in our province.

Q. Would you say that these measures with respect to which you and Oppenheimer had a primary concern and role have had any significant effect on the military power of the United States?

A. I would think so.

Q. Could you amplify that at all?

A. It is very hard to know what would have happened if something had not happened. Still I feel that this action certainly has contributed, I think, in focusing the attention of the Commission on the importance of certain actions, in breaking certain bottlenecks that were retarding or limiting the production. Advice I don't suppose is comparable to action in importance, but as far as advice is of importance, I think it was in that direction definitely.

Q. One final question. In his role as chairman of the General Advisory Committee and conducting the meetings and the affairs of that committee, what opportunity did Dr. Oppenheimer afford to the other members of the committee to express fully their views and to exert their influence?

A. I think perfect opportunity. Of course, he is a person who knows a great deal about these things and knows how to express what he knows with extreme efficacy, so naturally many questions just because of this preeminence and not because so much of his sitting in the chair, he would naturally take a leading role. But certainly everybody had a perfect freedom to act with his own mind and according to his conscience on any issue.

Mr. MARKS. That is all, Mr. Robb.

CROSS-EXAMINATION

By Mr. ROBB:

Q. Doctor, how long were you on the General Advisory Committee?

A. About 4 years.

Q. Did you write the reports of the committee?

A. Did I do what?

Q. Did you write any of the committee's reports?

A. No; I don't remember that I did.

Q. Who did?

A. Mostly the chairman, and he was helped by the secretary of the committee, who was at that time Dr. Manley.

Q. Dr. Oppenheimer and Dr. Manley were the ones who took care of that?

A. I think in most cases, as far as I know, the reports were written by them.

Q. And the report of the October 29, 1949, meeting, did Dr. Oppenheimer write that?

A. Yes, I presume so. I imagine probably Rabi and I jointly wrote——

Q. You wrote a separate report?

A. Wrote our brief minority opinion on a very partial issue of that meeting.
Q. When I said separate report, you wrote a minority opinion.
A. Yes; something of that kind.
(Discussion off the record.)

Mr. GRAY. Dr. Fermi, Dr. Conant has appeared before this board in the proceeding, and he was, I believe, at the same time a member of the General Advisory Committee.

The WITNESS. That is correct.

Mr. GRAY. He testified that being primarily a scholar and secondarily a scientist, he relied upon you for technical advice in these matters.
Can you recall, did he talk with you prior to that October 29th meeting about the subject matter which was to be taken up at the meeting? Did he come to you or seek your views on this principal issue which was to be before that meeting?

The WITNESS. I don't remember that he did. My recollection would be that we came into the meeting and some sort of general discussion started right away in the open meeting. That is my impression. At least I don't remember of any private conversations.

Mr. GRAY. You don't recall any conversation?

The WITNESS. No.

Mr. GRAY. Would you guess now on the basis of recollection that most of the people who came to that meeting had their minds pretty well made up about this issue, or do you think that they arrived at the conclusions which were reflected in the various reports they signed as a result of the meeting?

The WITNESS. I would not know. I had and I imagine that many other people had sort of grave doubts. It was a difficult decision. Even now with the benefit of 5 years of hindsight, I still have doubts as to what really would have been wise. So I remember that I had in my own mind definite doubts, and I presume my ideas and I imagine those of other people, too, must have gradually been crystallizing as the discussion went on. However, I have no way of judging.

Mr. GRAY. I know it is difficult to answer that question. The fact is that in this particular case, Dr. Conant did not take your advice.

The WITNESS. I don't remember that we had any particular discussion outside the meeting.

Mr. GRAY. He didn't take the same position you did in this meeting.

The WITNESS. In that particular we were on different sides, that is correct.

Mr. GRAY. I would like to have asked Dr. Conant this question. This is being discussed in his absence.

Mr. GARRISON. May I ask a question for clarification relating to the chairman's question? It is my recollection that Dr. Conant said he looked to you for guidance on matters of nuclear physics, and for your judgment in those matters, that is, primarily to you. When you say you took different sides in this meeting, I want to make quite clear whether you mean with respect to what ought to be done internationally and so forth, by the country on the one hand, and what the technical situation was on the other.

The WITNESS. I see. I don't remember of any essential disagreement on the technical situations. I suppose I think we expressed our opinion in terms, if I remember correctly, of a somewhat better than even probability. I think it was a fair opinion at that time. I don't think one could have said or could have guessed better than in those terms. In other words, it was not a foregone conclusion by any means, and we knew and we said that it was not a foregone conclusion.

On the other hand, it was to be expected that perhaps just with development and with some amount of technical luck the thing might be pushed through. That was about the situation at the time; that, as far as I can recollect, we all agreed with the situation. I don't believe there was any difference of opinion on this line.

Dr. EVANS. For the benefit of the record, for some people that may not know you as well as I have known you, would you state where you were educated?

The WITNESS. Where I was educated?

Dr. EVANS. Yes.

The WITNESS. I was educated in the University of Pisa in Italy.

Dr. EVANS. And you taught over there?

The WITNESS. I taught not in Pisa; I taught first in Florence, and then in Rome for many years, until I came to this country, and I taught in this country for 2 years in 1939—for more than 2 years, 4 years or so at Columbia University, since 1939, and then after the war interlude, I have been teaching at the University of Chicago.

Dr. EVANS. You were at Columbia University when the first knowledge came out about the fission of uranium.

The WITNESS. Yes, that is right.

Dr. EVANS. Do you believe, Dr. Fermi, that scientific men should be sort of circumscribed in regard to scientific information that they may discover?

The WITNESS. I am sorry, I am not sure I got the question.

Dr. EVANS. Do you believe in circumscribing the scientific men in regard to scientific information that they discover, that is, not permitting them to publish it?

The WITNESS. I see. The matter was this. In ordinary times, I would say that scientific discoveries should be made public. At that particular time with the war impending and critical political situations and so on, I joined with a group of others, the leader of the group or the most active member of that group was Leo Szilard, in a voluntary censorship to keep certain results that could lead in the direction of the atomic bomb.

Dr. EVANS. Do you believe it is actually possible to conceal this kind of information?

The WITNESS. Well, for a very limited time, yes. Forever, no.

Dr. EVANS. That is, you could have guessed a lot of this stuff if you had been over in Rome?

The WITNESS. I think I might possibly have guessed some things, at least.

Dr. EVANS. That is all.

Mr. GRAY. Thank you very much, Dr. Fermi.

(Witness excused.)

Mr. GRAY. We will recess for a few minutes.

(Brief recess.)

Mr. GRAY. Will you proceed, Mr. Silverman.

Whereupon, David E. Lilienthal, a witness having been previously duly sworn, resumed the stand and testified further as follows:

DIRECT EXAMINATION—Resumed

By Mr. SILVERMAN:

Q. Mr. Lilienthal, would you care to describe briefly what situation you found in general in the atomic energy establishment when you became chairman in 1946?

A. Perhaps some chronology will help. The war was concluded in early August of 1945, and at that time the Congress began considering what should be done with the atomic energy enterprise. It was a big concern without any guidance given it by legislation or otherwise until over a year later, when the McMahon Act was passed. So that in that period, there was the period of the Manhattan district acting in a sense as a caretaker and the uncertainty resulted in things that we found when we came into the enterprise.

When I first saw it was when the board of consultants visited the projects in February of 1946. Deterioration had set in as one might expect. Scientists had left the project in large numbers. Contractors had declined to go forward, such as duPont. DuPont turned in its contract at Hanford. There was great uncertainty. Morale was badly shot. At Los Alamos we found the most serious situation because although some very able men remained, the top management of that project had left for the universities. We found a great many health hazards and fire hazards that were very damaging to morale. * * *

From a management point of view, it was extremely difficult because the Army had insisted that their officers should move back into their military posts. This meant we had to try to find people to take their place. There was no inventory of the properties. There was no accounting. This whole thing had been done so hastily that it had not been possible to do that. These things made it very difficult for the men who were operating to make head or tail of what they were doing. The net effect of that was a very depressed state of mind.

As I say, this can be annotated at some length. This is what we found at the time we began the enterprise in January 1946.

Q. Did you consider one of your first tasks and the most important task was the rehabilitation of the atomic energy program?

A. Yes; that was our duty. Beginning with personnel and trying to get people back who had left and get additional people in, both management and technical.

Q. Was the GAC helpful on that?

A. Yes; they were. By reason of the fact that the GAC included men of real distinction in the scientific world and that the Chairman of the GAC had been the former head of the Los Alamos project, they spent a good deal of time as individuals and as a group trying to induce people to return to Los Alamos or other undertakings in the Commission. We did make use of them in that way.

Q. What was the function of the GAC, as you understood it to be?

A. The law defines it as an advisory body on technical and scientific matters. That was the role that by and large was followed. It was independent of the Commission, set up as a statutory advisory body as distinguished from perhaps the score of advisory bodies that we set up by administrative action. It had its own secretariat. The secretariat acted between meetings.

The dealings with the Commission were rather formalized. But by and large the roles were of two kinds. One, to review technical and scientific matters, and second, to initiate scientific and technical matters.

Q. Did you feel that the GAC under Dr. Oppenheimer's chairmanship performed that function during your incumbency in office?

A. Yes; I thought as an advisory group it worked very well, I don't mean to say that we always agreed with the advice and this of course we didn't. The GAC was very diligent in meeting frequently and in documenting their recommendations and in keeping contact with the division heads and operating people in the Commission between their meetings.

Q. Do you care to state the role and attitude of the GAC with respect to some of the problems that faced you during your incumbency?

Mr. ROBB. Could I have that question read back?

(Question read by the reporter.)

The WITNESS. One can only select a few examples to respond to tnat.

In the weapons field they were most active. This was because the weapons problems were the primary problems of the Commission in part and partly because these men had special qualifications in that direction. They either initiated or reviewed such things as efforts to revise the design of weapons in order to get more weapons for the same amount of material, to increase the destructive power of weapons, to boost their destructive power, to improve their combat effectiveness in the direction of lightness and field manageability, matters of that kind.

I thing the board will find problems of this kind treated in some detail all the way through the GAC letters to us, and reports to us and our request to them and the operations between the secretariat and Division of Military Applications of the Commission, the Military Liaison Committee and others.

These are examples of the sort of thing they did.

By Mr. SILVERMAN:

Q. We have gone into that in the record with other witnesses.

I want to turn now to the situation as it existed after the Soviet atomic explosion, I think, of September 23, 1949. Would you tell us very briefly what our defense posture was as far as you can in unclassified terms with respect to the AEC's function and responsibility?

A. I will try to summarize this. The details of course are available to the board.

Mr. ROBB. Mr. Chairman, may I interpose just so the record may be clear. I am not making any objection, of course. May I ask if the witness is about to read a statement?

The WITNESS. No. I have some notes that would hasten the presentation.

Mr. ROBB. That is entirely all right. I just wanted the record to reflect if you were reading a statement.

The WITNESS. I am not reading a statement, but from notes.

Mr. ROBB. Which I assume you made.

The WITNESS. Yes; notes in my handwriting.

The situation on September 23, which I believe is the date which President Truman announced the atomic explosion in Russia, as far as the AEC's program for weapons was concerned was something like this:

A program for the expansion of weapon production had been under study by the Military Establishment and the AEC over a period of months, probably beginning in February, and continuing through October 19, when President Truman formally approved this expansion program. This was encouraged by the GAC, and it was certainly a program that included additions to Oak Ridge and elsewhere, additions to Los Alamos and so on.

As to the improvement of weapons, here too there was a program which had been recommended by our Division of Military Application, had been approved and amended in some ways by the General Advisory Committee, by Los Alamos

Laboratory, and it had a number of parts. These are rather important. These are found in these records, but I think it might serve to spell it out a little in lay terms.

I have consulted with Mr. Beckerley privately about classification problems, and he assures me that the way I will put it will not involve any classification problem.

Mr. ROBB. If it does, Mr. Chairman, I assume Mr. Beckerly will raise his hand or something?

The WITNESS. Yes. I have rehearsed this with him.

Mr. ROBB. I have no doubt that you will be all right.

The WITNESS. I want to be very careful about it, and that is why I have asked him in advance.

This weapons improvement program which was in effect—that is, the program had been approved or was actually in operation at Los Alamos and Sandia—was of several parts. Among these parts were a program for an increase in the numbers of atomic weapons through new design, an increase in the numbers of weapons through greater material production, an increase in the numbers of weapons through programs relating to raw materials, a program for increasing the destructive power of the weapons over those at Hiroshima and Nagasaki by a substantial factor, an improvement in the combat usefulness of the weapons by reengineering these weapons.

This led to the establishment of the Sandia operation and my soliciting the aid of the Bell Laboratories and the Western Electric on behalf of the Commission and the President to take over that operation in order that we might have weapons that had field usefulness, as distinguished from weapons that it almost took a Ph. D. in physics to handle, instead of a sergeant.

This is an important story and I only refer to it. The details, I am sure, are in the file.

An improvement in problems associated with delivery. This concerns size and weight and other matters of that kind of great importance. And finally, plans for greatly stepped up power of weapons by a very large factor, by certain innovations of design that had been worked on for some time, but were at the point where a program for building such weapons was just around the corner.

The product of this stepped up program for this greatly heightened destructive power of weapons would produce a weapon which was so much larger than the original weapons that we were advised that one such bomb would take out almost any target in the world, and two would take out any target.

I have consulted with Mr. Beckerley and I make this statement after that consultation with him that President Eisenhower in his United Nations speech on December 8 spoke of an attainment of a fission bomb—an A-bomb type—of 25 times the power of the original bombs with an energy release of the order of 500,000 tons of TNT equivalent. Whether that bomb is the bomb that was recommended by the GAC and the Division of Military Applications, and was part of the program at the time of the Russian A-bomb, I don't know. I state these facts, and I am assuming that this must be the fission bomb that was planned at that time.

That was the program roughly that we had at the time of September 23. I ought also to say that to the best of my knowledge the Commission had not received, nor had any of the divisions any request from the Defense Establishment for a weapon of unlimited size or destructive power, nor any request for a weapon of greater destructive power than the stepped up fission bomb to which I have just referred. That the Commission did not have a military evaluation at that time of the military value of a hydrogen bomb or a bomb of size without definite limit. That it had not before it no diplomatic or political evaluation of the effect of such a weapon pro or con, on such matters of the cold war, or the effect on our alliances and other diplomatic and international relations.

The board is familiar with the fact, and the records are here that the Commission asked the GAC to assemble, especially to consider certain questions affecting the Commission's duties grew out of the announcement about the Russians' success with an A-bomb. Those questions roughly seemed to me something like this:

Is this program that we now have and have under way adequate to fulfill our duties? If not, what modification or what alternative course or alternative courses should be pursued? Among those alternative courses, should an all-out H-bomb program be instituted in order that we should adequately and properly fulfill our duty?

By Mr. SILVERMAN:

Q. Mr. Lilienthal, the Chairman of the Board, I think it was, perhaps it was Mr. Robb, called our attention in the course of these hearings to a letter signed by Mr. Pike as Acting Chairman of the AEC to the GAC, giving them their instructions. Were those the only instructions that the Commission sent or gave to the GAC with respect to this meeting?

A. My recollection is that prior to Mr. Pike's letter I wrote a letter, a rather brief letter, setting out or asking them to assemble for consultation on the consequences as far as the Commission's duties were concerned on this Russian A bomb. When the GAC did meet on October 29, the Commissioners or some of us met with them initially, and I suppose to them orally indicating—not attempting to limit their considerations to technical matters alone, although it was assumed that technical matters would be the basis for other recommendations. There are two letters, therefore. One letter by Mr. Pike is more in the nature of the usual letter we sent prior to every meeting in which certain specific things are asked. The letter that I wrote is of more general character.

Q. Did you also speak to Dr. Oppenheimer orally or don't you recall?

A. I think I called him by phone to ask him to sound out the committee members, what was the earliest date when all the members could be present. This, I think, was about the 8th or 9th of October.

Q. Have you recently seen that letter you wrote Dr. Oppenheimer?

A. Yes, I saw it yesterday.

Mr. SILVERMAN. Does the board have that?

Mr. GRAY. I don't think I have seen that letter.

Mr. ROLANDER. What is the date of that, do you know?

The WITNESS. No, but it would perhaps be the 15th of October, or something like that. I did see it yesterday in the big file.

By Mr. SILVERMAN:

Q. Proceed.

Mr. ROBB. If you wait just a moment, perhaps we can get that letter. I don't know.

The WITNESS. It is a fairly short letter.

Mr. ROBB. This seems to be it. October 11. I am told by Mr. Beckerley you can read that into the record if you want to.

By Mr. SILVERMAN:

Q. Would you do so, please?

A. This is dated October 11.

Mr. ROBB. 1949.

The WITNESS. 1949. [Reading:]

"Dear Robert:

"We quite understand the General Advisory Committee's wish at its last meeting to postpone making any specific recommendations to the Commission, but rather to express its readiness to be called upon whenever it might appear that it could help. We are very appreciative of that offer and we want and need to avail ourselves of your counsel and guidance.

"The Commission is, of course, asking itself afresh in the light of operation Vermont if the present, and presently planned, program constitutes doing everything that it is reasonably possible for us to do for the common defense and security.

"This is, I realize, a very large question, but it is the essential measure of the Commission's responsibility and the question to which we are trying to make certain there is a clear and affirmative answer. To that answer the Committee has important contributions to make, and we would welcome your advice and assistance on as broad a basis as possible. Do you think it would be possible to assemble the Committee in the very near future to meet with the Commission?"

Mr. ROBB. Mr. Pike's letter was subsequent, but I am told by Dr. Beckerley that involves classified material, Mr. Lilienthal.

The WITNESS. Then there appears to be a memorandum from my secretary indicating that Dr. Oppenheimer had phoned concerning this letter and suggesting dates.

Mr. ROBB. Mr. Pike's letter was October 21, 1949.

The WITNESS. Yes.

Mr. ROLANDER. Mr. Chairman, I am informed that operation Vermont refers to Joe I, which was the first Russian explosion, for the clarity of the record.

Mr. GRAY. I hope it clears the record.
Mr. ROBB. Do you want Dr. Oppenheimer's answer to that letter in the record?
Mr. SILVERMAN. I haven't seen it. Do you think it bears on it, Mr. Robb.
Mr. ROBB. It might.
Mr. SILVERMAN. Let us have it in the record if there is any question about it.
Mr. ROBB. I will show it to Dr. Beckerley.
Mr. GRAY. While they are looking at the record, why was the Pike letter written as Acting Chairman? This is a thing I am just curious about.

The WITNESS. I think I was probably away at the time the letter was prepared. It perhaps was before each of these GAC meetings, our staff and the GAC secretariat would get together and prepare a kind of agenda in the form of a letter, questions that either they wanted to raise with us or either that we wanted to raise to them. We sent this kind of staff letter and the Commissioners signed it. If I were there, I would have signed it.

Mr. GRAY. I see.
Mr. ROBB. Dr. Beckerley says it is all right.
The WITNESS. This is October 14, 1949.

"Dear Mr. Lilienthal:

"Thank you for your good letter of October 11. I can well understand the desire of the Commission to have us consider the overall program at this time. We shall do our best to do so.

"It has proven possible to call the meeting on the 29th and 30th of October; that is the first day on which both President Conant, who is quite busy, and Professor Fermi, who is in Rome, can possibly attend. It is not possible to schedule a meeting date on which Dr. Seaborg can be with us since he has long planned a trip to Sweden. I have, however, made arrangements to obtain from him in writing, and, if necessary, consultation, his views on the subject of the meeting. With the exception of Dr. Seaborg, I expect that all members of the advisory committee will be able to come. Some of us will plan to be in Washington on the 28th for preliminary consultation. I think it best, however, that the formal meeting not be called until the morning of the 29th. I regret that this is a weekend; that seems to be inherent in the makeup of the GAC.

"May I suggest that if there are any materials that it would be wise for us to examine before meeting with the Commission on Saturday morning you arrange to have them transmitted as early as possible; but in any event in time to permit study before we actually come together. The secretary of the committee, Dr. Manley, will be in Washington next week, and will, I am sure, be glad to consult with the staff of the Commission on the preparations for the forthcoming meeting.

"With every warm good wish, Robert Oppenheimer.
"Copy to Dr. John Manley."

By Mr. SILVERMAN:

Q. Will you tell us what happened thereafter?
A. There was one other thing. It is known to the board, but I want to make that in my remarks I take full cognizance of it, that the occasion for the precise occasion for considering the H bomb either as a part of the program or a supplement to the existing program was a memorandum from our fellow Commissioner, Mr. Strauss, dated about October 5 or 6, which is in the record. All of these documents added together represented the frame of the Commission's thinking at the time of the meeting on October 29th and 30th.

Q. Now, what happened at that meeting, as far as you can recall, or whatever impressed you about it.
A. Some of the Commissioners, perhaps all, but certainly I, attended the opening meeting or part of the opening meeting of the GAC. It was their meeting. Their practice was to ask us in as observers or to ask us questions. If we wanted to meet with them as our meeting, we would ask them to come to the Commission's room. In this way it preserved the identity of the meeting being as either a GAC meeting or AEC meeting. This was a GAC meeting.

I opened the conference by repeating as well as I can recall the substance of the paragraph in the letter that has been read into the record indicating that we wanted their advice on whether our program as it had been approved, the present program, the program in planning to which I referred, met the requirements of our duty, and if not, how it should be supplemented and in particular should it be supplemented by an all-out program on the H-bomb as proposed by Commissioner Strauss.

The GAC's report is in your record. The points that most impressed me were two. One, the technical considerations that were discussed in the time while I was in their meeting which did not by any means include the whole meeting. Most of their meeting was in executive session, but there were considerations of diversions of materials to another program, the H-bomb program, which was problematical, discussion of whether such a weapon as the hydrogen, deuterium, tritium, et cetera, weapon that was then under consideration would improve our retaliatory strength sufficiently to justify the risks involved in diversion of materials and other related points.

There was discussion of whether a weapon larger than the 500,000 tons fission weapon that was in the works, half a million tons of TNT equivalent, whether a weapon larger than that didn't go beyond the point of diminishing returns in terms of the destruction it would effect.

There was a consideration of whether our program then was not the best way to use the materials and the manpower that we had. These technical considerations impressed me very much.

The second point that impressed me a good deal was one I had thought about myself and others, of course, and that was a consensus among a number of GAC members that launching of a weapon larger than the stepped up weapon would not give us a false impression of security and illusion of security that we had gained a decisive or absolute weapon, an illusion of security which a number of the GAC members attributed to our possession of the A bomb, an overvaluation of the security that could be secured from large bombs alone as distinguished from a balanced military establishment.

In any case the GAC's views and the AEC's views were submitted to the President in writing on November 9. They are of course in this record.

By Mr. SILVERMAN:

Q. They may be in the files and not in this record.
A. Yes, they are in the files. In this report we tried to make the President's job as easy as possible by agreeing on as many things as we could about the facts. This was largely a staff paper prepared which we approved. There is agreement in this report which you will find that went to the President on a number of things——

Mr. ROBB. This is the report that went to the President from the Commission?
The WITNESS. Yes.
Mr. ROBB. Not the GAC report.
The WITNESS. No. The GAC report was included in it. The Commission's report began with an agreement, "Mr. President, we are in agreement completely on a number of the basic facts about this situation."
Mr. ROBB. Excuse me for interrupting.
The WITNESS. I am sure this is a document if it is relevant is not so long that the board may read it. It is a classified report, of course.

Then we recognize, that is, the AEC, that this is not a question which the AEC could decide. This is a question for the President. But we do indicate what our views are. Mr. Strauss indicated, as indicated earlier, for an all out program. Three of us, Commissioners Pike, Smyth, and myself, said in one sentence we are not for this program—we are not at this time, I think are the words that are used—and Mr. Dean had a position which I think might be described as not quite at this time.

There was a preliminary thing that ought to be gone through. This is spelled out in his own words in the report, and I won't take the time to review it if you wish me to.

Just as an individual, if I may say so, I don't conceive that the question to which I am to address myself is the wisdom or unwisdom of either of these courses. At that time this represented the best judgment that each of us could summon to this question prior to the consultations which took after this at which time I had another chance to look at the problem in the light of the State and Defense Department views.

By Mr. SILVERMAN:

Q. I think it might be of some interest to know to what extent the Commissioners and the Commission were relying on the GAC report. Also I am going to ask you about the National Security Council, or perhaps you will come to that in your testimony directly, to what extent that relied on the GAC report.
A. In this case I can only speak for myself. The other Commissioners either have or will indicate the extent to which they relied on the GAC. It was my

view that technical considerations advanced by the GAC in the first part of the report which deals with technical matters was very persuasive. I recognized I was a layman but these were men of great competence, and the things that they said were most persuasive to me. They included in their report statements about matters that were not technical but which they asserted were related to technical considerations, strongly planted, or expressions of that kind.

Some of these impressed me, one of them particularly, that there was a point of diminishing returns, that to announce publicly as apparently it was necessary, the building of a weapon of almost unlimited size would be in conflict—would put us in the eyes of our friends and potential friends in an unfavorable light without compensating advantages to us, and similar considerations of that kind.

Some of the members expressed themselves in various ways and which seemed to me to have some validity. In my first report of views to the President I laid considerable stress on that. Also on the concern I had then which was increased a great deal after I served on the committee with the State Department and Defense Department to which we were relying almost entirely upon atomic weapons, upon large weapons.

That brings me then to the final stage in my own participation in this.

On November 19, that is 10 days after this report of the AEC and the views of its individual Commissioners, the GAC report, and the views of its members, went to the President, the President created a subcommittee of the National Security Council to advise him further on this matter. That committee consisted of the Secretary of State, Mr. Acheson, the Secretary of Defense, Mr. Johnson, and myself. I would say that I had resigned and my resignation had been approved by the President early in November to be effective, I think, the first of December, but he asked me to stay on until this particular chore was finished.

May I interupt to say that the report of November 9 and the record will show or the file will show did not contain as of that date I think the views of Mr. Smyth and Mr. Strauss, except as to their conclusions. They sent their memoranda a few days later or some time later, in any case. I consider that the November 9 report supplemented by these subsequently filed statements as the views of the AEC.

Returning then to the National Security Council subcommittee, this subcommittee was set up by a letter from the President to the members of the subcommittee, which is in the file, that I examined yesterday, and therefore is available to the members of the board. It set up the considerations the President wanted weighed. It began a series of staff studies and consultations, recognizing that the issue was not really an AEC issue but a broad issue, as broad as the powers and the functions of the Chief Magistrate himself.

We had meetings of this kind. I met along with Commissioner Smyth, whom I asked to accompany me, because he was a scientist, and a technical man, as well as a member of the Commission, and we met with General Bradley and others of the Military Establishment. I should say that what impressed me most in this consultation was later set out in the argument I sought to make to the National Security Council and that was that General Bradley stated rather flatly that they had no reserve except the A bomb in the event of aggression against us any place in the world. Later General Bradley stated this publicly in a speech in Chicago in November before the Executives Club, I believe. It was harrowing experience to me to be told this, and it made a great impression on me in this respect. Right or wrong, this was the reaction I had. We had, it seemed to me, falsely relied upon the security of simply a stockpile of A bombs, that we had impoverished our Military Establishment—this was the period of an economy drive—we were closing military establishments. Instead of drafting boys, we were reversing the process. We were bringing our national budget away down. This seemed to me really quite harassing in the light of the fact that trouble might break out anywhere and as indeed it did break out in June in Korea at which time, of course, our reliance on the atomic bomb was certainly not a sufficient one.

From that time on a consideration was immediately given to a broadening of our Military Establishment, instead of relying entirely on weapons of this kind, and we moved in the other direction.

I mention this because I would like if it meets with the approval of the board, if they were to read—not that they won't have enough to read—but there is in the file a memorandum of expression of my views to the National Security

Council on this point. It is not the wisest expression in the world, but it is certainly a reflection of the effect upon me of these various discussions within the Government.

The thing that especially impressed me was that our earlier discussion of what kind of a program we should have did not have the advantage of knowing the limitations of the Military Establishment at that time. This has been photostated and is in the file. It was originally classified by me as top secret. It has been recently declassified but then reclassified as security information. I am not just sure what that means. But it is not classified under the Atomic Energy Act. If it is consistent with the procedures of the board, if portions of that which represent only expression of my views rather than quotations from State or Defense Department documents, if that could be read by the board, or included in the record, I think it would complete the whole picture, and my own reaction is to this as a consequence of the considerations begun September 23.

Mr. ROBB. May I make a statement about that?

Mr. GRAY. Yes.

Mr. ROBB. I believe you now have the original of this?

The WITNESS. Yes, it was transmitted to me by the secretary with Mr. Dean's approval.

Mr. ROBB. Yes, sir. I have in my hand what I think is a photostat of that; would you look at it and see if it is?

The WITNESS. Yes, that is a photostat of that document.

Mr. ROBB. I am advised, Mr. Chairman, that this memorandum which as the witness has stated was originally classified top secret was thereafter changed in classification on the side of the photostat which I have where there appears the notation "Classification changed to confidential security information by authority of Office of Classification by William E. Riley, Chief, Documents Control Branch, Division of Security, April 1, 1953". Below that are some words I can't make out.

On the bottom of that appears the notation in longhand signed by R. B. Snapp, April 20, 1954, "The control records indicate this memo was retained per D. E. Lilienthal's request by R. B. Snapp under unbroken seal until September 22, 1952, when with D. E. L.'s permission it was transferred to the general files per Commission direction at meeting April 4, 1953."

I am informed that the photostat which I hold was made at the time the original was turned over to Mr. Lilientfial for his personal file. Is that right?

The WITNESS. I didn't know that it was.

Mr. ROBB. Mr. Chairman, I am informed that Mr. Nichols, the general manager, states that this memorandum contains so many references to matters concerning other agencies than the AEC, that it is impossible to declassify it so that it can be read in the open record of these proceedings. In other words, it does contain restricted information. However, I think that Mr. Lilienthal's suggestion that the board should consider it is sound, and I suggest that it might be included in a separate classified record. It occurs to me that since the board might want to inquire of Mr. Lilienthal about it, that it would not be amiss to read it in such a record so that Mr. Lilenthal might be asked any questions which might help the board in connection with this memorandum.

The WITNESS. May I interrupt. You used the term "restricted data." I believe that is in error. I have a note from Mr. Beckerley which states there is no restricted data within the meaning of the Atomic Energy Act in the memorandum.

Mr. ROBB. Mr. Lilienthal, I am just repeating what I was told by the General Manager, by Mr. Mitchell, the General Counsel who took it up with the General Manager. Would Mr. Mitchell care to correct me on that?

Mr. MITCHELL. You are quite right.

(Discussion off the record.)

Dr. BECKERLEY. The document is classified by virtue of its containing security information other than restricted data.

Mr. ROBB. I think, Mr. Chairman, that this does contain information that the board might well wish to have. I think that since it was prepared by Mr. Lilienthal it would be appropriate that he could be here when the board is considering it so they might ask any questions that might appear to be relevant.

Mr. SILVERMAN. Mr. Chairman, I don't know what is in this thing because I have never seen it. I am really concerned primarily just with the question of the extent to which the decision that was ultimately made was one that was based on GAC advice, and to what extent it was based on other considerations. What you decide to do about this memorandum, since I know nothing about it, I really have no views about it. I would prefer to finish my direct examination. If it then seems desirable to the board to read this into the record, there is nothing I can do about it, because I know nothing about it.

Mr. GRAY. Suppose we proceed with direct examination. I am sure there will be some questions that you will be asking and perhaps the board members, of Mr. Lilienthal, and perhaps before we start that, we might take a look at this and see if we wish to ask him any questions.

Mr. ROBB. That is right. It occurred to me that it would be well to have it read, so Mr. Lilienthal could hear it, and have it fresh in mind so we might ask any question against the background of Mr. Lilienthal hearing the memorandum and against the background of having ourselves heard it.

Mr. GRAY. Will you proceed.

By Mr. SILVERMAN:

Q. Would it be fair to say that in the decisions that were ultimately made reliance was placed on the GAC at least by yourself as to technical matters——

Mr. ROBB. I hate to interrupt you, but may I interpose one further remark that I myself saw this memorandum for the first time I think probably during the midmorning and I have not yet myself had a chance to read it very carefully.

Mr. SILVERMAN. You are 8 hours ahead of me.

Mr. ROBB. I have seen enough of it to know that the board ought to have it before it.

By Mr. SILVERMAN:

Q. Would it be fair to say that your reliance on the GAC was great as to technical matters and the further away it got from technical matters, the more your reliance was on other agencies, and, on your own judgment and on other departments of the Government?

A. During the first phase of my participation in this matter before we had any important contact with the military or any contact with the State Department—obviously that didn't contribute to any views I had—I did have great respect for the views of the GAC on technical matters. I took very much to heart their statement that their conclusions were planted in technical considerations. I had such respect for the wisdom of men like Conant and Oppenheimer and Fermi and other men that I certainly paid close attention to what they said on matters that were not technical. I think the best evidence I came out with were the things I wrote at the time, some of which they would not endorse and were not included in their views. It is hard to divide on these things. I am sure of the importance I assessed to the technical view, and the rest is another matter that is hard to define.

Q. This memorandum was dated January 31?

A. Yes.

Q. And you resigned February 15?

A. It took——

Q. At least your resignation took effect February 15?

A. It was the third stage, that is right.

Q. You did not ask to have your clearance continued?

A. No.

Q. So I take it you do not know whether the hydrogen bomb that we hear about in the newspapers has any relation if any to the things talked about in 1949?

A. No, I have had no access to restricted data since that time, and no occasion to use it.

Q. As a result of your experience with Dr. Oppenheimer and your knowledge of him, have you formed an opinion as to his loyalty, his integrity, his character, all the other factors that go into forming a judgment as to his loyalty, security?

A. Yes, I have.

Q. What is your opinion?

A. I have no shadow of a doubt in my mind that here is a man of good character, integrity and of loyalty to his country.

Q. How would you assess him as a security risk?

A. I did not regard him up until the time my knowledge of the program ceased, and had no occasion to regard him as a security risk.

Q. I think you already indicated that in March 1947 you consciously assayed the situation and came to the conclusion that he was not a security risk?

A. Yes. At that time we had this file before us and that was my conclusion, that in the light of the overall picture, taking everything into account, the minus signs were very few indeed, and the plus signs very great indeed, and I thought he was a contribution to the security of the country. I have had no occasion since that time to change that view.

Q. Has your experience with him confirmed that view?

A. My experience from that time did confirm that view. I am sure that it is clear that he has made great contributions to the security of the country.

Mr. SILVERMAN. I have no further questions.

Mr. ROBB. Mr. Chairman, it is about a quarter to five. May I ask the pleasure of the board about proceeding?

(Discussion off the record.)

Mr. GRAY. I think we will take a recess for a few minutes and then proceed with the examination of the witness.

Mr. ROBB. Yes, sir.

(Brief recess.)

(Classified transcript deleted.)

By Mr. ROBB:

Q. Mr. Lilienthal, Dr. Oppenheimer, just before his counsel came back in, suggested a question to be asked of you and I believe it was, do you recall a discussion or a statement by General Bradley before the General Advisory Committee at the October 29, 1949, meeting, is that right?

The WITNESS. The only comment that I recall was in response to a question about the military value of a bomb of virtually 1,000 times Hiroshima and his response was, as I recall, that it would be principally psychological. I don't recall how he defined that. That is the only recollection that comes to my mind of that discussion.

Dr. OPPENHEIMER. May I ask one more question?

Do you recollect his account of our military position as of October 29, 1949?

The WITNESS. No, I recollect that description in a later phase of my activities in the National Security Council subcommittee, but not at the meeting of October 29.

By Mr. ROBB:

Q. Mr. Lilienthal, were the views expressed by you in this memorandum of January 31, 1950, so far as you knew, in accord with the views of Dr. Oppenheimer at that time?

A. No, I don't know to what extent they were in accord. Consultations I had with Dr. Oppenheimer in the GAC were more or less terminated after the report. But there are things in this position that do relate to the views of the GAC, such as the overreliance on large bombs.

Q. Did you discuss your appearance at this meeting with Dr. Oppenheimer before you went there?

A. No; I don't recall I did at all.

Q. Did you report to him afterward about it?

A. I will give you the rest of the events in answer to that. After this meeting referred to we did go to the President. The President made his decision. I then went back to the Atomic Energy Building where the GAC was in session and reported the decision. That is the last I have had to do with the subject.

Q. Did you talk to Dr. Oppenheimer personally about this conference that you had?

A. I don't believe so. I think the GAC or most of the members were in session. It could be that I did. I do have the recollection of reporting to the group as a whole. Whether I saw him separately, I am not clear.

Q. Do you have any reason to believe that the views expressed by you in this memorandum differ from the views held by Dr. Oppenheimer at that time?

A. I haven't any way of really knowing. I can identify some of the views that grew out of GAC recommendations in which Dr. Oppenheimer either led or took part. But taking it as a whole I have no way of identifying it in that way.

Q. Did you believe on January 31, 1950, when you addressed your remarks to this meeting, that the views you were expressing were in accord with the views previously expressed by the GAC in their report?

A. It seems to me the GAC report, except as to its conclusion, and the views I expressed in this memorandum and to the National Security Council subcommittee do not coincide. They are not in conflict in some places, but they certainly cannot be said to be identical. An examination of the GAC's report I think will make that clear.

Q. Wherein do they differ?

A. I can't answer that without having the report before me, which you can do as well as I. There are many points in here—for example, the powerless state of our defense at this time was not included in the GAC report to the best of my recollection.

Q. I will reframe my question, then.

Did you believe at the time you addressed these remarks to this meeting that the views you expressed with respect to the thermonuclear program were in accord with the views of the GAC?

A. You see, I didn't think the issue was the thermonuclear program.

Q. I am asking you now.

A. I don't quite see how one can answer the question put that way. I didn't think that was the issue. I hope I have made it plain in this memorandum. I didn't think that was the central issue. I thought the central issue was getting busy strengthening the security of this country which was in bad shape.

Q. You mean you were not talking about the thermonuclear program at this meeting?

A. Of course I was, but I didn't think that was the central question.

Q. Whether it was the central question or not, you talked about it, didn't you?

A. Yes, of course.

Q. Do you have any doubt that what you had to say about the thermonuclear program was in accord with the views of the GAC?

A. It certainly was in accord with the views as to the result that a crash should not be instituted. But the reasons for that and the conditions that I had suggested grew out of my discussions with the Military Establishment and with Dr. Smyth.

Q. I understand that. The GAC made a report to your Commission setting forth their views about what should be done with respect to the thermonuclear, didn't they?

A. Yes.

Q. And you talked about that when you appeared at this meeting on January 31, 1950, didn't you?

A. Yes.

Q. Now, was what you said at the meeting in any respect different from what you understood to be the views of the GAC on the thermonuclear program?

A. I have tried to answer that by saying that as to the result——

Q. The difficulties——

A. It is the reasoning that I adduced was not the reasoning in substantial part the reasons that are stated in the GAC's report and that is evident by reading it.

Q. Were your conclusions the same?

A. The net result was very close to being the same, namely, that we should not proceed. But the alternative that I proposed was not the alternative that the GAC proposed. I mean that is a very important distinction. I want to be sure it is said that I benefited a good deal in my view from the discussions and the GAC report, but the net result is quite a different argument.

Q. You have told us you were not in favor of a test program.

A. That is right, until we got ourselves in shape.

Q. Was the GAC?

A. No.

Q. So you were in accord on that.

A. That is right. The thing GAC didn't say, at least as I recall its report, was to make this point that before we decide this question and commit ourselves further to overreliance on weapons of this kind, we should make a stern reexamination of our position. That they did not say. That is what I have been trying to say. I think that is an important qualification.

Q. Wasn't the GAC pretty unqualifiedly against developing a thermonuclear at any time?

A. The best record of that is what they said, and I think the answer to that is that 6 of them were flatly against it.

Q. Did you take any advice or get any information from the experts of the GAC as to the feasibility of the thermonuclear?

A. Yes. They did supply us with their conclusion about whether it was feasible or not.

Q. What was it?

A. That conclusion is written in this report to the President of November 9. I would not undertake—it is before you. As I recall it says that the chances of its being feasible are 50–50, or something of that sort.

Q. Maybe a little bit better than that, doesn't it say that?

A. I have forgotten but it is there in the report.
Q. Did you get any opinion as to the possible thermonuclear bomb capability of the Russians?
A. I didn't understand you. Did I get from where?
Q. Did you get any opinion from the GAC or anybody else as to the possible thermonuclear bomb capability of the Soviet Union?
A. I don't recall, except that the assumption was, without any discussion, that the Russians were capable. It was only safe to assume that the Russians were capable of producing a hydrogen bomb.
Q. It was just a question of time, isn't that right?
A. Yes, something of that sort. The only safe assumption was to believe that in time they could do it.
Q. From whom did you get that information?
A. It was not a matter of information. It was an assumption that was adopted.
Q. Didn't you check with the scientists? You did not know yourself, did you?
A. No, there were no intelligence reports that I can recall.
Q. Did you talk to any of the experts on the GAC about whether or not the Russians might produce a thermonuclear?
A. I don't recall anything except that we started from that premise that in time they could do it.
Q. Who is "we"?
A. Everyone who was discussing the matter, GAC, the AEC and so on. I think that is what we advised the President. We were all agreed that was probably the case.
Q. In other words, that was the opinion of the GAC, was it not?
A. Yes, that is right. Opinion is not quite the word because we didn't have any facts. We just said we have to assume that they are capable of doing it.
Q. You were not an expert on such matters.
A. No, I think the term "possible capabilities" was one way of expressing it. I think we went further than that, and thought it was better to assume that it was not only possible, but that they could do it.
Q. Mr. Lilienthal, the question of whether or not the Russians could make a thermonuclear is a pretty important factor.
A. Yes. You are using thermonuclear and we were talking about a hydrogen bomb.
Q. You used the expression thermonuclear in your memorandum of January 31, 1950.
A. Yes, but I think the GAC referred to it as the "Super", which was the hydrogen bomb.
Q. The question of whether or not the Russians could make the super was a pretty important factor.
A. It was.
Q. So I assume you get the best opinion you could.
A. It was not a question of fact.
Q. It was a question of opinion.
A. That is right.
Q. Whose opinion did you take?
A. I don't recall but I assume it would be scientists or intelligence officers. Probably the scientists, probably the GAC.
Q. Probably Dr. Oppenheimer?
A. I rather you would not push me after I said I don't remember.
Q. I am sorry I have to push you because I want to get responsive answers.
A. I don't remember, but I am saying that this was the assumption on which we proceeded.
Q. It would be reasonable that you did consult the men who knew most about such matters, wouldn't it?
A. You can say that. I have not said it.
Q. Wouldn't it?
Mr. SILVERMAN. Aren't we in argument now?
Mr. ROBB. I think the record is clear.
The WITNESS. Look, we told the President that is the basis on which we were proceeding.
Mr. ROBB. What I am trying to find out is where you got your information. I assume you did not get it from me, or Mr. Garrison.
Mr. GARRISON. He said he had no information.

By Mr. ROBB:

Q. I am trying to find out why you made the assumption.
A. I agreed with you that probably the opinion came from the GAC, but we didn't have any information.
Q. Mr. Lilienthal, just so the record will be clear, this memorandum of January 31, 1950, you wrote and put in the AEC files, is that right?
A. That is right.
Q. And at that time it was classified as top secret.
A. I classified it, yes.
Q. And then there came a time on April 1, 1953, apparently when that was classified to confidential security information.
A. Yes.
Q. Did there come a time when you were given the original of that memorandum from the AEC files?
A. Yes. I called on the Chairman of the Commission, Gordon Dean, and consulted with him about this. He asked the classification division and the Secretary of the Commission to look into the matter, and some weeks later it was sent to me by the Secretary of the Commission with a letter and a note from Mr. Beckerley, the head of the classification division.
Q. Do you want to read that in the record? It is up to you if you want to read it.
Mr. SILVERMAN. Does it advance our inquiry?
Mr. ROBB. I don't think so.
The WITNESS. The only thing that bothers me is whether we have to ask counsel to leave. In any case, it expresses Mr. Beckerley's view about the reclassification.

By Mr. ROBB:

Q. When did you receive this from the Commission?
A. I don't have Mr. Snapp's letter.
Q. Would it appear on Mr. Berkerley's note?
A. No, this seems to be undated.
Q. About when?
A. I am sorry, it is dated; October 6, 1953 is his note, and it was probably mailed to me some time after that. That would be last November.
Q. Why did you want to get this from the Commission?
A. It was a statement of my views and I was quite anxious for my own protection to have access to a statement that I had written about my own views. It seemed to me very important, and it is even more important now.
Q. Protection from what?
A. Protection of my record as to what my views were at that time. The reason being that my views have been extemperized on in the press and elsewhere, and I felt much easier having a record of just exactly what it was I said.
Q. I assume you have kept this confidential?
A. Yes, and I have kept it in a safe and so on. I plan to return it to the Commission now that I know you have a photostat.
Q. Beg pardon?
A. I think now that it is in the record and you have a photostat of it, it is probably just as well for me to return it to the Commission, or put it in a lock box.
Q. I see. You thought when you got it back that this was the only copy?
A. No. I knew it had been photostated.
Mr. GRAY. Excuse me. I feel it my duty to point out to the witness that he has made conflicting statements on the question of photostating. I don't care what your answer is, but earlier you said you had not known it was photostated. This is in your interest.
The WITNESS. Yes. The facts are these: that this was put in a sealed envelope and filed. Then I inquired of the Secretary of the Commission, what about that sealed envelope and he told me that it later had been opened and had been put into the files of the Commission and had been photostated.
Mr. GRAY. I may have misunderstood you. I am sorry. But I believe the earlier transcript perhaps when we were in executive session will reflect an observation, maybe casual, that you had not known it had been photostated.
The WITNESS. I thank the Chairman. These are the facts.

By Mr. ROBB:

Q. Again, in the interest of clarifying the record, didn't you just say that now that you have learned that the Commission had photostated this document, you might as well return the original?

Mr. SILVERMAN. Mr. Chairman, does this relevance the inquiry relating to Dr. Oppenheimer?

Mr. GRAY. The Chairman will make this observation. He is trying to do his best to conduct a fair hearing, and when it appears to the Chairman that a witness through inadvertence or somewhere else is in a position of perjuring himself, I am going to call it to his attention.

Mr. GARRISON. You are right, Mr. Chairman.

Mr. GRAY. I am sorry to use the word "perjury," but if at one point of the testimony a witness says one thing and at another point he says directly contrary, at one point the testimony is in error. I don't think it advances anything, the protection of Mr. Lilienthal's appearance as a witness in this case.

Mr. SILVERMAN. I was not referring to your inquiry. I was referring to Mr. Robb's question, Mr. Robb's question which was not related to that.

Mr. ROBB. Will you read the question back, Mr. Reporter?

(Question read by the reporter.)

The WITNESS. Yes.

Mr. ROBB. Just a minute. Did the Chairman hear the question?

Mr. GRAY. The Chair is a little confused. Does the witness object to answering this question?

Mr. SILVERMAN. Whatever is the easiest and the quickest way to do it.

Mr. GRAY. Let us clear the record.

The WITNESS. I had been told in a conversation with the secretary of the Commission that he opened the envelope and put this in the file and had photostated it. It was only yesterday that I saw that this was true, that I saw the photostat in the file that was supplied to me yesterday afternoon in Mr. Snapp's office. I was then reassured that there was a photostat. I didn't want to leave it simply on Mr. Snapp's general assertion. I now know that there is such a photostat, because I have seen it.

Mr. ROBB. May I repeat the question. In the interest of clarifying the record, didn't you say a little while ago that now that you know that there has been a photostat made, you may as well return the original.

The WITNESS. That is right. I now know because I have seen the photostat.

By Mr. ROBB:

Q. You mean you learned for the first time yesterday that there had been a photostat made?

A. For the first time I saw it, and was sure the information supplied me was correct.

Q. Now, may I, sir, go back to the beginning of your testimony in which you gave an account of the events which took place in March 1947? I believe you said that the file was delivered to you on a Saturday, March 8, is that right?

A. That is my recollection, yes.

Q. By messenger?

A. I don't recall. It was delivered to the Commission. No, I do recall now. My recollection is that I had a call from Tom Jones.

Q. Who was he for the record?

A. Tom Jones, who was the acting security officer at that time. My recollection is that he phoned me—I think this was a Saturday, a Saturday afternoon—the file Mr. Hoover referred to in his telephone conversation to me had been received.

Q. And then it was delivered to you?

A. It was delivered to me Monday morning.

Q. Monday morning?

A. That is my recollection. Monday the 10th. It was delivered to the Commission, that is to Mr. Jones or some one on the 8th.

Q. And there was with that, I assume, a covering letter from Mr. Hoover, is that correct?

A. I assume so.

Q. What you have referred to here as the file was the material you got from Mr. Hoover, is that right?

A. Yes, that is right.

Q. On either March the 8th or March 10th, whichever day you received it?

A. Yes.

Q. That is what you refer to as the file?

A. That is correct.

Q. What did that consist of?

A. I can't recall except that was a very substantial file, that it contained the kind of—a great deal of material from the Manhattan District, Intelligence Division, or whatever it was called, counterintelligence. It was a typical FBI file. A typical FBI personnel file.

Q. I have before me what you received, Mr. Lilienthal. It appears to be a 12-page summary memorandum on J. Robert Oppenheimer, and a 15-page summary memorandum on Frank Oppenheimer. Is that in accord with your recollection of what you received?

A. No, it is not. I am sure you are obviously correct. My recollection was that we had a big file. I didn't recall that there was a summary from the FBI.

Q. Is it now your testimony that you had received something in addition to this summary memorandum from the FBI?

A. My recollection is that we did get—this is quite a while ago and I don't recall the exact form in which it came.

Q. Would you describe these two reports as a file?

A. You mean as distinguished from a report?

Q. Yes, sir.

A. In view of what you have told me, a file or report I should think would be equally descriptive. What you are suggesting is that this was a summary of the content of the file, rather than the raw material of the file, and that apparently is what is the case if that is what you say.

Q. The letter from Mr. Hoover, Mr. Lilienthal, see if this refreshes your recollection, dated March 8, 1947, addressed to you:

"MY DEAR MR. LILIENTHAL: In view of developments to date I thought it best to call to your attention the attached copies of summaries of information contained in our files relative to Julius Robert Oppenheimer, who has been appointed as a member of the General Advisory Committee, and his brother, Frank Friedman Oppenheimer, who was employed in the Radiation Laboratory at Berkeley, Calif., until recently. It will be observed that much of the material here contained in the attached memoranda was obtained from confidential sources."

Having heard that, do you agree that what you got was the two summaries?

Mr. GARRISON. Is that the whole letter?

The WITNESS. I don't know the distinction between the summary and the report. But whatever you have there, if you have it, I received. In order to refresh my recollection of this hearing, I asked for this file yesterday and was told it was an FBI file and I could not see it. If I had seen it, my recollection would have been refreshed.

By Mr. ROBB:

Q. You know, don't you, Mr. Lilienthal, that the rules for security hearings, which I believe were adopted while you were chairman, provide that the contents of FBI reports may not be disclosed?

A. Yes, but the rules of the Commission, as I understand, permit Commissioners to have access to anything they had access to during the period of their commissionership.

Q. I don't want to debate that with you.

A. I apparently am wrong if that is the regulation now, but that is what I asked for.

Mr. GARRISON. Mr. Chairman, since this is now the subject of discussion of this record, I would like to request that we be furnished a copy of this summary.

Mr. ROBB. No, sir, I am sorry, Mr. Chairman, I would have to object to that. I think we are in agreement with what was furnished, Mr. Lilienthal.

The WITNESS. You have it there.

Mr. GARRISON. Mr. Lilienthal has not received it, and you have told him he received it, but he doesn't know what it is.

Mr. GRAY. I can't make a ruling about the availability of FBI documents. I can't rule affirmatively in response to your request. As of this minute I will have to be guided by the security officer and the attorneys in this, Mr. Garrison.

Mr. ROBB. Mr. Chairman, the rules under which these hearings are conducted provide that reports of the Federal Bureau of Investigation shall not be disclosed to the individual or to his representatives.

Mr. ROLANDER. Mr. Chairman, we have a new reporter to spell the other reporter. Could he be sworn?

(The reporter, Harold B. Alderson, was thereupon duly sworn by the Chairman.)

By Mr. ROBB:

Q. Now, after you received this material from Mr. Hoover, on Monday morning, do I understand your testimony that you presented it to the Commission, is that right?

A. That is my recollection.
Q. And each of them read the material, is that correct?
A. During the course of succeeding hours, or a couple of days, each of them did read it.
Q. Didn't they read it right then?
A. That was my recollection.
Q. That they did?
A. They sat down and began passing it around, and took it to their offices, and so on.
A. I think some of them stayed, and some of them took it to their offices for further reading, and so on.
Q. Who was present at that meeting?
A. My recollection is that all of the members of the Commission were there, and I have something of a recollection that Mr. Jones was there, Tom Jones.
Q. Was Mr. Wilson, the general manager, there?
A. I don't recall.
Q. Was anybody else?
A. I really don't recall, and I know the Commissioners were there. I am quite sure they were.
Q. After you had digested this material that Mr. Hoover had sent you, did you form any opinion as to whether or not the information contained in Mr. Hoover's material was true or false?
A. Well, I don't know how to answer that. The information was like other information and we had no way of determining whether it was true or false and we did not see the people and the informants were anonymous and so on, and so I don't know how to answer that question.
Q. Well, from that point on, did you proceed on the assumption it was true, or did you proceed on the assumption it was false?
A. Well, I proceeded on the assumption, we proceeded to try to evaluate it, some of it having a ring of veracity and some of it—for example as I recall one of the reports, and I think it is in this report, the informant turned out to be a nine-year-old boy. If that is true in this case, it may not be, then obviously you would say, "Well, this probably is not anything to rely on." But in other cases the report would say that the informant "X" is someone the bureau has great confidence in, and you would assume that that was true.
Q. Was the nine-year-old boy referred to in the material Mr. Hoover sent you on March 8?
A. I had an impression, but this may have been some other file and as I remember that as an illustration of how you have to evaluate these things.
Q. Well, now, having this material before you, I assume that contained certain allegations against Dr. Oppenheimer, didn't it?
A. It constituted derogatory information about Dr. Oppenheimer, that is right.
Q. And you say you proceeded to evaluate it?
A. We did our best to evaluate it.
Q. What did you do to evaluate it?
A. Well, in general, speaking for myself, I followed this kind of a rule, that assuming that part of this material that has the ring of veracity to it is to be true, and discarding that that looks rather unimportant, or perhaps not true, does this derogatory information balanced against all of the other things one knows about the man indicate that he is a security risk or he is a man who would endanger the security of the United States. That is on the whole case.
Q. When did you go through that process?
A. As we were reading the file.
Q. You mean that morning?
A. Well, in the process of considering it, yes.
Q. When did you reach your conclusion on it?
A. I don't recall exactly. It was I think, probably, during the course of that week, after we talked to Dr. Conant and these other people that knew Dr. Oppenheimer well. There was a consensus that there was no occasion for us to cancel this clearance by anything that we had seen. I don't think that there was any question raised by anyone to the contrary, but in any case that was the feeling that I had.
Q. Didn't you reach that conclusion the same afternoon?
A. That isn't my recollection because we did go to the President or Dr. Bush and I went to the President the next day, but it could be that.
Q. In the process of reaching that conclusion, sir, did you go back to Mr. Hoover to ask him for further details about this matter?

A. We didn't immediately, no. We recognized the responsibility, and Mr. Hoover had transmitted the most recent information he had and the responsibility for evaluating and the conclusion was ours, and we did later think that it would be wise to go and see whether we were misinterpreting some of this, and that was the purpose of the visit later in March.

Q. But did you communicate with Mr. Hoover and say, "Mr. Hoover, here is an item here that we wonder about. What is your evidence to back this up?"

A. No, I don't think we did.

Q. You didn't do that in respect to any of these items, did you?

A. I don't think that was the practice.

Q. Did you do it?

A. No, I don't think we did.

Q. And I believe you have testified there were some items that you accepted as true, and some you had doubt about?

A. Yes. I can't remember which was which, but I have the recollection that some of these things were stronger and more clear than others, but the whole picture was that of derogatory information about the man's post associations, and one episode that was worse than that.

Q. Which was that?

A. Involving Chevalier.

Q. What do you mean, "worse than that," Mr. Lilienthal?

A. Well, this struck me as being the only thing, the thing in the whole record, that would give the gravest concern, and for that, and the thing that dismissed that concern from my mind was the fact that General Groves and Mr. Lansdale, the security officer, at the time this happened examined this man on the question, and were apparently satisfied that this was not or did not endanger the national security, and the evidence to that was they kept him on. I can't add anything to that. That seemed to me a very conclusive kind of a judgment about whether he was dangerous or not.

Q. Now, on that same day, this is March 10 again, in the afternoon, you met and talked to Dr. Bush, didn't you?

A. About what?

Q. Dr. Bush?

A. What is that?

Q. Didn't you meet and talk to Dr. Bush about Dr. Oppenheimer?

A. Yes, and Dr. Bush was invited to meet with the Commission, and I don't know whether it was that day or not, but it was about that time.

Q. And you wanted to get his opinion?

A. Yes, sir.

Q. Did you show him this material from Mr. Hoover?

A. I can't recall.

Q. Then I believe you called in Dr. Conant, didn't you?

A. That is my recollection, yes, sir.

Q. Did you show the material to him?

A. I don't recall, I certainly discussed the context of it, but I doubt whether he was asked to read the file.

Q. You mean you made Dr. Conant familiar with the material?

A. We tried to communicate to him what the nature of the derogatory information was, and I am now, my recollection is not precise about it, but that is my best recollection. We certainly conveyed to him the problem this report or file represented.

Q. Isn't it true, Mr. Lilienthal, that that very day, March 10, 1947, after talking with Dr. Bush and Dr. Conant, that you concluded that there was no doubt as to Dr. Oppenheimer's loyalty?

A. I don't recall whether it was that day, I am satisfied as to what the ultimate conclusion was, but we did not entertain any doubts for any length of time, and I for one entertained no doubt, speaking for myself, entertained no doubts at all.

Q. Now, thereafter, I believe you testified you talked to Mr. Clifford at the White House about it?

A. Yes.

Q. And what was the purpose of your conference with him?

A. Well, we had in mind that Dr. Oppenheimer was an appointee of the President, and unlike employees of the Commission he was an appointee of the President as a member of the General Advisory Committee, and we ought to make sure the President knew of the existence of this derogatory information, and so as I recall Dr. Bush and I conveyed this information to him, and I believe it was on the following day.

Q. By the way, Dr. Oppenheimer was appointed by the President in February, wasn't he?
A. I don't recall, I thought it was earlier than that.
Q. At all events, it was prior to the time you received this information from Mr. Hoover, wasn't it?
A. That is my recollection.
Q. Did you suggest to Mr. Clifford that a special board be convened to review this material?
A. No, we did not.
Q. Was that ever discussed with Mr. Clifford?
A. No, I believe not.
Q. Are you sure about that?
A. I am not sure, but I have no recollection of it.
Q. Was there any reason that you knew of for the appointment of a board of any kind to review this material?
A. No. It didn't seem to me and I don't recall it seemed to anyone that there was that much question about it. The reason for that of course is that this man subsequent to the time of these events and these associations had done a great deal for his country and to prove by his conduct that he was a loyal citizen of the country. He wasn't just an ordinary unknown individual whose achievements were not well known to us and to the people we consulted.
Q. As to the creation of a board of any sort to evaluate this material, it was never discussed between you and Mr. Clifford?
A. I don't recall, it could be, but I don't recall that. Mr. Clifford, my impression is Mr. Clifford said he would advise the President, but Mr. Clifford did not seem to take this seriously, and to the extent of requiring procedure of that kind, but I could be quite wrong about that.
Q. Now, you were asked by—I forget which one of counsel was pitching at that time, was it Mr. Silverman, I guess—but you were asked what the Joint Committee knew about this material, and you said, as I recall, you didn't know whether they did or not, is that right?
A. I said that at a later date, I am sure they did.
Q. Well, did you advise the Joint Committee of this development?
A. I don't recall, and I just don't have any recollection of that.
Q. Isn't it a fact that you did not?
A. Well, it may be, I just don't recall.
Q. Was there any reason why you shouldn't have?
A. Any reason why we should or should not have?
Q. Should not have?
A. Well, if we had had doubts about our responsibility in the matter, I am sure we would have done so, but we didn't.
Q. Did you discuss the question of whether or not you should advise the Joint Committee?
A. I don't recall, and we couldn't have submitted a file to them in any case, because at that time the President's regulations forbade it, and later on when the President's regulations were amended this file was available to them.
Q. Now, I believe you were asked whether or not the FBI statement that you received from Mr. Hoover contained all of the information about Dr. Oppenheimer, is that correct?
A. I am sorry, I didn't understand you.
Q. I will strike that question, it isn't very clear. I believe you were asked whether or not on March 8 or 10, whichever it was, you had the complete story or file from Mr. Hoover, and you said you did, is that right?
Mr. SILVERMAN. Is that a question or an answer?

By Mr. ROBB:

Q. Did you receive any further information from Mr. Hoover after you received the first information on March 8 or 10?
A. My recollection is that we didn't, but I wouldn't be positive about it. My recollection is that this was the whole of the information, whatever it was, the file or report, it was delivered, and it is my recollection that that was the sum total of what was delivered to us.
Q. Did you discuss this matter with counsel at the time, to get their opinion on it?
A. I don't recall. It sounds reasonable one should, but at a later date we certainly discussed it with Mr. Volpe, because Mr. Volpe accompanied me on the visit to Mr. Hoover.

Q. Who was Mr. Volpe?
A. At that time he was deputy general counsel.
Q. Who was the general counsel?
A. He was Mr. Herbert Marks.
Q. Who is here?
A. Yes, sir.
Q. Did you go over it with Mr. Marks?
A. Well, I don't recall. I am sure we went over it with Mr. Jones, he was present as the security officer and whether we went over it with counsel, except this occasion of this visit to Mr. Hoover, I just don't remember. I think that I might say at this point, the Senate Committee on Atomic Energy was holding hearings on the confirmation of the Commission, and we were spending for 13 straight weeks most of my time up there, so that my recollection of the operations of the Commission are not as clear as they might be.

Mr. GRAY. Let me ask a question on this point, Mr. Robb. Mr. Lilienthal, this was important enough to go and talk with Clark Clifford at the White House about, and was important enough for you to go back and talk with Mr. Hoover about it, and are you sure you didn't discuss it with the deputy counsel of the Commission?

The WITNESS. I would think that, I assumed I did.

Mr. GRAY. Wouldn't it be unreasonable to think that you had not discussed it with counsel if you went to the White House, and to the Department of Justice with it?

The WITNESS. I really——

Mr. GRAY. I am not asking you to recall something you can't recall. Well, I am sorry, if you can't recall——

The WITNESS. It depends, Mr. Chairman, on the functions of the general counsel's office at that time, in relation to security matters. If they had functions in that field——

Mr. GRAY. Do you recall whether they did or not?

The WITNESS. My recollection is that those functions were confined to legal questions rather than questions of evaluating the file. The security offce had the responsibility for assisting the Commission in a staff sense on evaluation of files. I think that that was the practice from then continuously, actually.

By Mr. ROBB:
Q. Now, the Atomic Energy Act required an FBI investigation of all personnel, does it not?
A. Yes, sir.
Q. Did you have the FBI investigation which had been made pursuant to the act at the time you made this decision in March?
A. I can only tell you what my impression is. The Atomic Energy Act required a reexamination by the FBI and the bringing up to date of those people who had had clearance under the Manhattan District. Dr. Oppenheimer of course did have such, and I have been assuming what I have been saying here in my recollection is that this was that reexamination, his clearance up to this point having been a Manhattan District clearance, and I could be wrong about that.
Q. I believe you testified in response to a question by Mr. Silverman, that you had read the letter of General Nichols to Dr. Oppenheimer?
A. To General Groves.

Mr. ROBB. This is the difficulty of switching witnesses back and forth, and you get confused.

Mr. SILVERMAN. The witness has only been here during his testimony.

Mr. ROBB. I understand, but I have been here through them all, and I am getting confused.

Mr. SILVERMAN. You are not confused. General Nichols is the one.

By Mr. ROBB:
Q. Have you read the letter from Mr. Nichols to Dr. Oppenheimer?
A. Yes, I have.
Q. Do you recall in there the statement that Dr. Oppenheimer had contributed $150 a month to the Communist Party up to about April of 1942?
A. No, I don't recall that.
Q. Well, do you recall whether or not you had any such allegation as that before you in March of 1947?
A. Oh, no, I couldn't remember as fine a point as that, no, I don't recall it.
Q. Do you recall whether or not in March of 1947 you had at the AEC the old Manhattan District files?

A. I know that we were supposed to have them, because the President's order, I believe, directed their transfer, and the President's order of December 31, 1946. It transferred the properties and so on of the Manhattan District to the AEC, and that presumably included the files of the Manhattan District.

Q. Just so we can be clear about that, I have before me a memorandum dated March 12, 1947, Memorandum to the File, signed by Bernard W. Menke, staff security officer. Do you remember him?

A. I don't.

Q. It refers to the FBI files, and I think under the rules I am not permitted to read it, but I will read the part pertinent to this particular point.

"The complete Manhattan Engineering District files concerning J. R. Oppenheimer were sent to the FBI about July or August of 1946, at the time he left employment controlled by the Manhattan Engineering District. This action was apparently in accordance with some agreement, the parties to which are said to have been General Groves, J. E. Hoover, and the Attorney General; under which agreement the FBI, upon assuming exclusive investigative jurisdiction of a person who departed from project work, received the full Manhattan Engineering District investigative file pertaining to that person. For this reason the pertinent files are not available for reference in analyzing the instant summaries."

From that you conclude, wouldn't you, that the old Manhattan Engineering District files were not in your shop, so to speak?

A. No, apparently they had been transferred to the FBI.

Mr. GRAY. Do you think you ever saw the Manhattan District files, Mr. Lilienthal?

The WITNESS. I am beginning to doubt it, and if I had looked at this file before I came to testify, I would be a little clearer.

Mr. GRAY. I don't think that there is anything here that indicates whether you did or not see the Manhattan District file.

Mr. ROBB. I don't want to trap the witness, I am sure he didn't, because they were not there.

Mr. GRAY. I believe in your testimony in response to questions from Mr. Silverman this morning, you did refer to the Manhattan District files. Is it possible that you could have seen them at some other time, or some other channel?

The WITNESS. I don't think so. I was referring to this report as counsel denominated it, as containing a summary as he points out of what all the Manhattan District files contain.

By Mr. ROBB:

Q. Now, was Mr. Carroll Wilson present at the meetings which were held concerning this matter?

A. I don't recall. My recollection is that these were executive meetings and those Mr. Wilson would not attend, but he might have attended. I don't really recall.

Q. I have before me, taken from the files, the original of the memorandum from Carroll L. Wilson, general manager, to the file, and I will read it to you.

"United States Atomic Energy Commission, Washington, D. C."——

Mr. GARRISON. Shouldn't the reporter note the withdrawal of Dr. Evans?

Mr. ROBB. He is back here, so it is academic.

"United States Atomic Energy Commission, Washington, 25, D. C. Memorandum to the File: Carroll L. Wilson, general manager, Subject: J. Robert Oppenheimer." There is a longhand note, "March 10, 1947, copy 1 and 2, Series 'B', typed Security Office, D. Dean."

"March 10, 1947: Summaries of information received on March 8 from the FBI regarding J. Robert Oppenheimer and his brother Frank F. Oppenheimer, were considered by the Commissioners in closed session this morning. This file was accompanied by a letter dated March 8 from Mr. J. Edgar Hoover and the file was delivered to Mr. Jones by the FBI on Saturday morning, March 8.

"The letter from Mr. Hoover transmitted a copy of what was described as a summary of the FBI files concerning J. Robert Oppenheimer and his brother, Frank F. Oppenheimer. The summary consisted of material usually referred to as derogatory.

"The Commission met in closed session and each of the Commissioners read the rather voluminous summary and noted from the incompleteness of the account as contained in the summary that either it did not reflect the results of a full investigation or did not contain all information bearing on the matter. The Commission also noted that the evidence summarized which, as stated in Mr. Hoover's letter, came from confidential sources, could seriously impeach Dr. Oppenheimer and that as a consequence this matter was one in which not only the Commission but also Dr. Bush as chairman of the Joint Research and Development Board and Dr. Conant as chairman of the Atomic Energy Committee of the JRDB were also concerned. Furthermore, in view of the role of both Dr. Bush and Dr. Conant in connection with the Manhattan project during the war, and their association with Dr. Oppenheimer while he was director of the Los Alamos Laboratory of the Manhattan project, it was felt that they should be consulted promptly. Dr. Bush was reached by telephone and it was arranged that he meet the Commission at 3:15 p. m.

At 3:15 p. m. the Commissioners met and were joined by Dr. Bush. The delivery of this file and the fact that it contained derogatory information were reported to Dr. Bush although he was not shown the file. Dr. Bush stated that he was not familiar with Dr. Oppenheimer's background prior to his joining the Manhattan project in 1942, but that he had concurred in the choice of Dr. Oppenheimer by General Groves for the important post of the Los Alamos Bomb Laboratory and that he felt that Dr. Oppenheimer's exceptional performance as director of that laboratory and subsequently in other roles advising the Government on the subject of atomic energy had clearly demonstrated his loyalty as a citizen of the United States and his integrity.

"Inasmuch as Dr. Conant had been closely associated with Dr. Oppenheimer in connection with the Manhattan project, he was invited to sit with the Commission and Dr. Bush for discussion of this matter and he joined the meeting at about 3:45. Dr. Conant stated that his association with Dr. Oppenheimer dated from the beginning of Dr. Oppenheimer's connection with the Manhattan project and that he was not familiar with the contents of any investigative files concerning Dr. Oppenheimer's background. He stated that General Groves had taken full responsibility for selection of Dr. Oppenheimer to head the Los Alamos Laboratory and that it was certainly a matter of public knowledge that this laboratory under Dr. Oppenheimer's brilliant and driving leadership had made an enormous contribution to the war effort.

"Inasmuch as General Groves had made the original selection of Dr. Oppenheimer, the chairman attempted to reach him by telephone but was advised that he was en route by automobile from Florida to Washington and could not be reached.

"Drs. Bush and Conant shared the views of the Commission that the record of Dr. Oppenheimer's contributions to the country in this field during the last 4 or 5 years have been so outstanding that it could leave no doubt as to his loyalty. It was further stated that, in view of Dr. Oppenheimer's unique position as an authority in this field, if anything were permitted to occur which might cause him to be lost to the Government in connection with atomic energy, it would be a very serious blow to our progress in this field and would have very serious consequences in the attitude of his fellow scientists towards this project. Even if no precipitous action were taken which would affect Dr. Oppenheimer's continuance in his present efforts with respect to this project, any public disclosure, either of the information contained in files or of the fact that such information exists which is the subject of serious concern as to Dr. Oppenheimer's qualifications, the consequences upon the leading scientists engaged in the project would still be serious.

"In view of the interest of the War Department and the role of the War Department in bringing Dr. Oppenheimer originally into this project, Drs. Conant and Bush arranged to see Secretary Patterson later in the afternoon. They did see him and he promptly agreed to contact General Groves.

"March 11, 1947: The Commission met this morning for further consideration of the matters discussed yesterday in connection with Dr. Oppenheimer. The Commission concluded tentatively (1) that on the basis of the——

Mr. GARRISON. What is that?

Mr. ROBB (reading), "The Commission concluded tentatively, (1) that on the basis of the information supplied by Dr. Bush and Dr. Conant concerning Dr. Oppenheimer's outstanding contributions in this project and his consistent concern for the security of this country in connection with his services as a member of the JRDB Committee on Atomic Energy and as an adviser to the Department of State, Dr. Oppenheimer's loyalty was prima facie clear despite material contained in the FBI summary; (2) that as a result of his work for the Government during the last 4 years he is now 1 of the best, if not the best-informed scientist in regard to 'restricted data' concerning atomic energy; (3) that while under these circumstances the questions raised by the summary did not create an issue or any immediate hazard, it was essential to undertake promptly a full and reliable evaluation of the case so that it could be promptly disposed of in one way or another.

"As a first step, it was decided to secure as promptly as possible written expression of views from Dr. Bush, Dr. Conant, and General Groves as to Dr. Oppenheimer's loyalty. As a second step, it was decided that the chairman should confer with Dr. Bush and Mr. Clifford of the White House concerning the establishment of an evaluation board of distinguished jurists to make a thorough review and evaluation of the case. Inasmuch as Dr. Oppenheimer is a Presidential appointee to the General Advisory Committee to the Commission, the case is one in which the White House has a definite interest. In addition, the matter is of interest to the Department of State inasmuch as Dr. Oppenheimer has served as an adviser to the Department of State on many phases of atomic energy, including serving as a member the Board of Consultants to the Department of State in the preparation of a plan for the international control of atomic energy, and subsequently as an adviser to Mr. Baruch and more recently as adviser to Mr. Frederick Osborne.

"At 3 p. m. today Dr. Bush and the chairman"—that was you, wasn't it?

The WITNESS. Yes, sir.

Mr. ROBB (reading). "Dr. Bush and the chairman met with Mr. Clifford and advised him of the circumstances in connection with this case and discussed with him the desirability of having a review of this case by a board of distinguished jurists or other citizens. The chairman proposed that there be considered for membership on this board judges of the Supreme Court. Mr. Clifford stated that he was decidedly opposed to any move which would draw members of the court into outside activities and felt that this case did not warrant an exception to that policy. This policy would not preclude selection of other jurists for temporary service on such an evaluation board if it were deemed desirable that such a board be established. Mr. Clifford stated that he would discuss the matter with the President and communicate with the chairman and Dr. Bush on Wednesday.

"The results of the discussion with Mr. Clifford were reported to the Commission at a meeting at 5 p. m. this afternoon. At that meeting the general manager reported that a detailed analysis of the FBI summary was in process of preparation by the Commission's security staff as an aid to evaluation."

Have you any comment on that, Mr. Lilienthal?

The WITNESS. No. I haven't. It is quite evident that Mr. Clifford in the end did not favor the idea of such a board, or perhaps we changed our minds, but I had forgotten that recommendation.

By Mr. ROBB:

Q. You had forgotten that?

A. Yes. I think the thing that this does confirm is that the initial reaction of the Commissioners was as stated, on the whole case, in view of the record of service to his country, this did not raise questions in our minds but was a case or matter that should be very carefully dealt with, and dealt with very carefully in the evaluation process.

Q. But you would agree, would you not, sir, that in 1947 you and the Commission seriously considered, and in fact were of the view that a board should be impaneled to consider this matter?

A. It is quite evident from this memorandum that this was considered.

Q. And you thought enough of it to go to Mr. Clifford at the White House and so recommend?

A. That is right.

Q. In other words, you recommended in 1947 that the exact step which is now being taken, be taken then?

A. We suggested it, and I think perhaps that is the import of the memorandum as I recall, we suggested this to the White House.

Q. That step did not strike you as fantastic or unreasonable, did it?

A. No.

Q. Now, did you talk with Mr. Clifford again about that matter of the board?

A. I don't recall, and I really don't.

Q. I will show you the original of a memorandum, on March 12, 1947, 11:25 a. m., report of telephone conversation, at 11:20 with Clark M. Clifford, Special Counsel for the President. That is dated March 12, and it has "DEL" on the bottom. Did you write that, Mr. Lilienthal?

(Whereupon, the document was handed to the witness.)

Mr. GARRISON. Did you say, "Did he write it?"

Mr. ROBB. Did he dictate it?

Mr. GARRISON. This is a record that he purportedly made?

By Mr. ROBB:

Q. Your answer is that you did dictate it?
A. Yes, it would appear that I didi, and may I read it?
Q. I am about to read it to you.
A. All right.

Mr. ROBB. I will ask Mr. Rolander to read it.

Mr. ROLANDER (reading). "March 12, 1947, 11:25 a. m. Report of telephone conversation at 11:20 with Clark M. Clifford, Special Counsel to the President: I put in the call to Clifford. I told him that following the conference yesterday afternoon with him, Dr. Bush and myself concerning an FBI report on a member of the General Advisory Committee, it was Dr. Bush's and my understanding that the status of the matter was as follows:

"That having presented the matter to the President by the method of presenting it to Mr. Clifford, Mr. Clifford would lay the matter before the President and advise us whether we should proceed to submit this matter to a board of review of judges or other outstanding citizens outside the Government, or what course should be followed.

"I said that until we heard from him it was Dr. Bush's and my understanding that the record would be supplemented by statements from Dr. Bush, Dr. Conant, etc., and an analysis of the report made within the Commission, but that no steps would be taken with respect to a board of review in this case. He said that Dr. Bush's and my understanding in this respect was correct. He said that after our conference he had briefly discussed the matter with the President; that it was a matter the President would want to think over; that it was presented at a time when the President (here I am relying on my shorthand notes, taken during the conversation) was exceedingly busy on an all-important matter, * * * Clifford said he, the President, will want to think it over some; that the next few days the President will be away from Washington. He will have time to think it over and determine if the board of review idea is the proper way to go about it.

"I told Mr. Clifford that we had not reported the receipt of this report on a Presidential nominee to the Joint Committee or to its chairman; I asked if he had any comment on that. 'You have put it up to the President through me and are awaiting his reaction after he has had time to give it some thought. You have done the 2 things that are right to do: (1) The Commission has made an immediate check with the 3 individuals who know most about the situation—Dr. Bush, Dr. Conant, and General Groves, and (2) you have presented the matter over here. So far as I know that is all that you are under any reasonable obligation to do." He said that we should therefore let the matter stand until we hear from him. He said that if I had not heard from him by the time the President returns to Washington. I should call and remind him about it. He said that if absolutely essential he could interrupt the President and get some decision about the board of review at any time but that he didn't want to do so unless it was absolutely urgent.

"I said that the man in question had been awarded a Medal of Merit, the highest civilian award, for his war work; it was my impression that these awards were made by the President on the basis of recommendation by a distinguished reviewing board. He said that the board initiated the recommendations and reviewed them and then the President acted upon them. He said further that he would supply a copy of this recommendation which could be made a part of the record in this matter."

By Mr. ROBB:

Q. Now, Mr. Lilienthal, this was a matter of grave import to you, wasn't it?
A. Yes it was an important matter, one of many important matters, that is right.
Q. It was of sufficient importance, and important to you, that you took shorthand notes on this conversation, and then dictated a memorandum about it, is that right?
A. That is right.
Q. But it is now your testimony that you had completely forgotten any discussion with Mr. Clifford about a board of review?
A. It is.
Q. And you had completely forgotten that you even considered such a board?
A. It is. I must say it just entirely escaped my mind.

Mr. GARRISON. Mr. Chairman, I would like to point out that it seems to me that the practice that is adopted here of asking, this was the same case with Dr. Oppenheimer on his cross-examination, the Government in possession of documents taken from here and there, including from their own files, in Dr. Oppenheimer's case in Princeton, and knowing that they had, first asking them to testify about something quite a while ago, without warning and without reading the documents, and presenting them and saying, "Tell us what happened," and it seems to me that this is designed to try to make the witness look to the board in as unfavorable a light as possible, and to make what is a lapse of memory seem like a deliberate falsification. I regret that this kind of procedure which is quite suitable in criminal prosecution and a court of law, when that attempt is being made before a jury, I am sorry that it has to be made here.

Mr. ROBB. Mr. Chairman, may I reply to that, as I take it to be some reflection upon my professional integrity and my professional methods. Let me say——

Mr. GARRISON. I have not questioned your integrity.

Mr. ROBB. I have no apology to make for the methods I am pursuing in the cross-examination of these witnesses. It is an axiom that the greatest invention known to man for the discovery of truth is cross examination, and I am pursuing what Mr. Garrison should know are orthodox, entirely proper and entirely legitimate methods of cross-examination. I make no apology to Mr. Garrison or anyone else for the method I am pursuing, and I submit that I have been entirely fair.

I asked the witness and I have taken him over these matters which I submit are matters which, well, I won't make an argument on that point, and he has said he did not remember them, and now I have read him these papers, and he says that he forgot them.

The WITNESS. Mr. Chairman, may I make this comment, that in the great multiplicity of things that went on at that time, it is not at all impossible that I should not remember even as important a matter as this, but a simple way to secure the truth and accuracy would have been to have given me these files yesterday, when I asked for them, so that when I came here, I could be the best possible witness and disclose as accurately as possible what went on at that time. I am a little confused about the technique. The board wants the facts, and the facts are in the file, and I asked for the file so I could be a better witness, and if was denied me. So I just have to rely on memory during a very troubled and difficult time on matters that are obviously important, but they are not as important as many other things we were concerned with at that time. It would help me a good deal, and I could be a much better witness if I saw the files that I helped to contribute to make.

Mr. ROBB. Mr. Chairman, I think Mr. Garrison would agree that it is an entirely fair comment to make that it is demonstrated that the memory of the witness was not infallible.

The WITNESS. I would be the first to insist on that.

Mr. ROBB. Since we are depending largely on memory, I think it is a fair test.

Mr. SILVERMAN. Why, when we have documents.

Mr. GARRISON. I thought the notion of an inquiry and not trial was to get at the truth by the shortest possible route, and it seems to me the attempt to make a witness seem to be not telling the truth, or his memory is not to be relied on by this board, by the surprise production of documents, is not the shortest way to arrive at the truth. It seems to me more like a criminal trial than it does like an inquiry and I just regret it has to be done here.

Mr. GRAY. Well, the board certainly will take cognizance of the comments of counsel in respect to this matter, and I think that if counsel is not permitted to engage in cross examination and simply relies on notes the witnesses may take from documents in a file, there may be some difficulty in arriving at some evaluations, and now on this particular point, it seems to me pertinent at least against general and public discussions, with which counsel cannot be unaware, including the New York Times story, the information for which was furnished by counsel, it is repeatedly and publicly stated that the Commission and others cleared Dr. Oppenheimer at the time that these were old charges rehashed, and completely considered and evaluated at the time. It does seem important to me, at least as chairman of this board, to find out exactly what did take place at that time.

Mr. GARRISON. I agree with you, Mr. Chairman, in full. I want nothing but the truth brought out here. And all of the truth about all of the things, and I want complete cross examination, and I raise only the question of surprising the witnesses with documents they themselves prepared which are in the file and which the Government has, and it seems to me a shorter way of arriving at the truth and a fairer way where a witness has prepared a document which the Government has in its possession is to ask him if he prepared that document, and to read it into the record, rather than confuse him first by asking him about things that he doesn't remember. That is the only point I make, and that limited point, and I wish in no way to confine this inquiry. But it is an important point though limited.

Mr. ROBB. May I proceed? I have two more questions.

Mr. GRAY. You are not going to confront the witness with any more documents?

By Mr. ROBB:

Q. Was any board ever convened?
A. No, I am sure of that.
Q. Did you hear any more from Mr. Clifford about it?
A. I don't recall. We certainly didn't have a recommendation from him that a board be convened or such a board would have been convened.
Q. Now, you testified, I believe, that I think in 1949 you were working on an A bomb of vastly increased power, is that right?
A. Yes, sir.
Q. About 500,000 tons of TNT, is that right?
A. My recolection which I think is correct is that this was in the order of 20 to 25 times the Hiroshima which would work out to four or five hundred thousand.
Q. One of those bombs—pardon me.
A. You used the words "working on" and what I think I said was a program approved and being accelerated to that end, and I had assumed that the fission bomb referred to by the President on December 8 was presumably that bomb.
Q. And one of those bombs would take out a small city and two would take out a big one, was that right?
A. My recollection of the estimates that were made at that time to us by technical people, Dr. Bradbury, and so on, was that one such bomb would take out all targets in the United States except perhaps a two to five—most of the large cities of the United States, and two would take out any large city.
Q. Was there any reluctance or any hanging back on the part of scientists to work on that bomb because of what we call moral grounds?
A. No.

Mr. ROBB. That is all.

The WITNESS. I wanted to ask a question about this document.

By Mr. ROBB:

Q. Which one, sir?
A. The document that I wrote, not that one, but the one that was referred to or read into the record in the closed session. In that regard there is the disposition. I suggested, and I don't know whether this is in the record or not, but I would like to get it clear that I suggested that it would be easier for me and more convenient for me if this stayed in the Commission's files, and so I won't have the responsibility of its protection, which is a fairly complicated business. Although I have very great question in view of what Mr. Beckerley said about the confidential nature of it, I don't want to take any chances on it. I haven't shown this to anyone, but I have relied on its contents and an article appeared in the October 4 issue of the New York Times, and I want to be sure that this is not a surprise to the board. In the course of that article, I did not quote from this, I recited the kind of arguments that were made at the time, but I did not disclose any of the confidential information, I am confident. But I would prefer to have the document here, and I would like to renew my request, the request I made to Mr. Snapp, that an effort be made to separate out these things which are clearly not confidential at all and simple expressions of my views, and those things which they regard as possibly confidential, because they had not consulted the State Department or the Department of Defense.

Then I would just like to leave this here, and not have the responsibility of it.

Mr. GRAY. I have no objection to your recital. Actually I would suggest that this is a matter between you and the Atomic Energy Commission, or at least the security people in the Government, and not with the board as to what disposition is made of the original document. So I think whatever requests you make, don't rely on this board to see that they are carried out with respect to the treatment of the document.

I should also say, Mr. Lilienthal, that I for one did not know of the existence of this document until we started our discussion of it, whenever it was today.

Mr. GARRISON. May I say the same for counsel on this side.

Mr. ROBB. I didn't either. The first I heard of the document was this morning.

The WITNESS. There is one point, and the reason I mentioned it, is because I did not want any question about the fact that I have relied upon the substance of the statement of my view in this piece and relied upon the expression of those views as reflected in this document. For an effort to state clearly in a public article in the New York Times, what my position was at that time, and the reason being that that position was, I thought, being unfairly presented, and I just want to be clear as far as the board is concerned that that is no failure to disclose that at the time I turned this back.

Mr. GRAY. I am sorry to have to address another couple of questions to you. I apologize to everybody including my colleagues on the board and counsel. I am still confused about the instructions to the General Advisory Committee for the October 29, 1949, meeting. This, I think, is pertinent to the inquiry because Dr. Oppenheimer, in his reply, says that the Atomic Energy Commission called a special session of GAC, and asked to consider and advise on two related questions.

First, it was whether, in view of the Soviet success, the Commission's progress was adequate. Now, that is covered, I believe, in the letter which was read into the record which you wrote the General Advisory Committee. Am I correct in that?

Mr. ROBB. I think so.

Mr. GRAY. I am not trying to trap you.

The WITNESS. I must say that I am getting a little—that is my recollection of it; it was a short letter.

Mr. GRAY. And, if not, in which way it should be altered or increased, and I think that that was correct; and, second, and now I am reading from Dr. Oppenheimer's reply, whether a "crash" program for the development of the super should be a part of any new program.

Now, in your letter which was read into the record, and in my recollection of the letter signed by Mr. Pike, as Acting Chairman, I haven't yet found any reference to this specific question as to whether a crash program in relation to the super was put to the Commission.

Now, it is entirely possible.

The WITNESS. Was put to the GAC?

Mr. GRAY. I am sorry, was put to the GAC, and it may be fruitless to pursue this at this point, and I would like somehow to be informed as to how that second question actually was put to the GAC. It is Dr. Oppenheimer's recollection clearly that they were asked, or I believe he so testified, and he put it in his letter, and I am not suggesting that they were not asked, but I am trying to find out how they were asked. If you do not have any recollection, I do not want to pursue it further with you now.

The WITNESS. I am sure it was presented to the GAC, but I must say I cannot say exactly in what form.

Mr. GRAY. In security problems, generally, Mr. Lilienthal, was Mr. Volpe a person whom you frequently consulted? He accompanied you, I believe, to Mr. Hoover's office in connection with this matter. Did you frequently consult him generally and was he your security sort of person?

The WITNESS. He was consulted because legal questions frequently arose, and he probably, and I can't recall precisely, he may well have been consulted on the general questions of policy. This is just too vague in my recollection to know just how that division of responsibility was made.

Mr. GRAY. I have one further question.

The WITNESS. I think perhaps if it is important I could dig into the files and try to illuminate that, but I haven't any recollection.

Mr. GRAY. I have one further question, which relates to your feeling in early 1950 that it would be unwise to proceed with a program which would lead to a test of the super; is that stated correctly?

The WITNESS. Yes; I stated it more extensively than that, but I thought something ought to be done beforehand.

Mr. GRAY. I am about to come to that; until there had been a rigorous reexamination of military plans and policies, were these things in your judgment mutually exclusive. In other words, could not the reexamination have gone forward simultaneously with steps which might determine the feasibility of the super?

The WITNESS. In this memorandum and in my statement to the National Security Council, I tried to indicate why I felt that they could not.

Mr. GRAY. That they could not?

The WITNESS. That going ahead with this program would prejudice that reexamination, and I could well have been wrong about it, but that was the view I had and that is what I said. In fact, no reexamination was made, but in any case my concern was that once that decision was made the reexamination wouldn't take place. Whether I was right or not, it was the view I had.

Mr. GRAY. I didn't understand that. Did it occur to you that, as it did to some people who were active and informed in this program, proceeding with further development might prove that the super was infeasible, or was not feasible, or did you assume that if we really went ahead with it we could do something about it?

The WITNESS. Well, I was as much concerned as anything with the effect of an announcement that we were going in to an all-out program of that kind, that that would prejudice the reexamination, and whether it came out that we could make it or couldn't, that that would confirm the course we then pursued or reliance, not upon really taxing ourselves, and really going to town with an important military program, but going off on this same course again.

Mr. GRAY. That suggests that if you had to make a guess as to the feasibility you would have guessed it was feasible.

The WITNESS. I think that I can't improve on the way in which we prevented our conclusion on this to the President, that we were assuming that it could be done.

Mr. GRAY. That answers my question.

Mr. SILVERMAN. I have no questions, except for one I would like to ask Mr. Robb.

Do we now have all of the documents on this clearance thing in 1947, or are there later documents?

Mr. ROBB. There is one thing in the file, and do you want me to read it now, if I can find it?

Mr. SILVERMAN. We might just as well have it complete.

Mr. GRAY. Is this something that needs to be read at this time?

Mr. ROBB. I can read it the first thing in the morning.

Mr. GRAY. Is it something that must be read in Mr. Lilienthal's presence?

Mr. ROBB. It may be. It is dated July 18, and I don't know whether it had to do with this or not. I will read it if you want me to right now.

"Confidential.
"Office Memorandum, United States Government.
"To: G. Lyle Billsley.
"From: T. L. Jones.
"Subject: J. Robert Oppenheimer.
"Date: July 18, 1947.

"Herewith a complete investigative file on J. Robert Oppenheimer, upon whom it is believed the Commission may not have formalized their decision. If the Commission meeting minutes contain indication of Commission action, would you kindly so advise? If they do not, I presume that you will wish to docket this case for early consideration. Each Commissioner and the General Manager have seen every report in this file with the exception of a summary of July 17, and my memorandum for the file dated July 14, 1947."

In longhand there is "Joe Volpe: Time flies. Will you please go to work on this?"

(Signed) "G. L. B."

Then also in longhand, "August 2d. Ret to Mr. Billsley by hand." Underscored twice. "Lyle: I looked over this file after you left it with me last night. My impression is that the Commission saw no need for formal action following the meeting they had with Mr. Hoover referred to in Lilienthal's letter of April 3 to the FBI Director. I assume that the information which has come in since that time has been circulated among the (over) Commissioners for their information. If Tom thinks the summary of July 17 and his file memorandum of July 14 should be circulated, that should be done. In addition, I think that you should check my impression of the status of this case with the Commission itself. J. V. Jr."

And the next thing is on August 11, which I believe we had, but just so that it will be all complete, I will read it:

"Office Memorandum.
"Date: August 11, 1947.
"From: T. L. Jones" Initials "TLJ."
"To: William Unna.
"Subject: J. Robert Oppenheimer.

"Authorization for granting final Q type security clearance to the subject is contained in minutes of the meeting of the Atomic Energy Commission at 10:30 a. m., Wednesday, August 6, 1947. It is reflected in the minutes that this clearance was granted during February 1947, but was reaffirmed because previous minutes failed to reflect the action. In addition, as you know, Dr. Oppenheimer was previously cleared by the Manhattan District. Would you please make the appropriate entry in your records."

Now, is that all there is?

Mr. SILVERMAN. What about the memorandum of July 14?

Mr. ROBB. That is the summary of an FBI file which I don't think I can read. That is July 17.

Mr. SILVERMAN. And July 14, also.

Mr. ROBB. July 14:

"To File
"From: T. L. Jones." (Signed) 'TLJ'.
"Date: July 14, 1947.
"Subject: J. Robert Oppenheimer and Philip Morrison.

"July 10, 1947, in the course of a conversation with John Lansdale, Jr., former chief of intelligence and security for Maj. Gen. Ralph R. Groves, you mentioned the two subject cases with both of which he had contact during the war. In both cases, in fact, Lansdale himself interviewed the men at some length. I did not ask Mr. Lansdale for an official opinion on either case, and no doubt before giving one should this ever be considered desirable, he would wish an opportunity to review the cases and apprise himself of recent developments. However, his rather casual comments seemed of interest and worth preserving in the files.

"These were that he was absolutely certain of the present loyalty of J. Robert Oppenheimer, despite the fact that he doubtless was at one time at least an avid fellow traveler, but that he felt that Morrison was a Communist. Lansdale has not of course had occasion to review the recent reports on either man, as his remarks should probably be interpreted as reflecting his judgment at the time of his most recent review of each case."

Mr. ROLANDER. That is all.

Mr. SILVERMAN. I want to apologize to the Commission for piling my straw on top of it.

Mr. GRAY. We will recess until 9:30 tomorrow morning. Thank you very much, Mr. Lilienthal.

The WITNESS. Thank you for your consideration in seeing me through today.

(Witness excused.)

Mr. GRAY. We will adjourn.

(Thereupon at 7:45 p. m., a recess was taken until Wednesday, April 21, 1954, at 9:30 a. m.)

UNITED STATES ATOMIC ENERGY COMMISSION

PERSONNEL SECURITY BOARD

In the Matter of J. Robert Oppenheimer

Atomic Energy Commission,
Building T-3, Room 2022,
Washington, D. C., Wednesday, April 21, 1954.

The above-entitled matter came on for hearing, pursuant to recess, before the board, at 9:30 a. m.

Personnel Security Board: Mr. Gordon Gray, chairman; Dr. Ward T. Evans, member; and Mr. Thomas A. Morgan, member.

Present: Roger Robb and C. A. Rolander, Jr., counsel for the board; J. Robert Oppenheimer; Lloyd K. Garrison; Samuel J. Silverman; Allen B. Ecker, counsel for J. Robert Oppenheimer; and Herbert S. Marks, counsel for J. Robert Oppenheimer.

PROCEEDINGS

Mr. GRAY. The proceeding will resume.
Mr. Pike, do you wish to testify under oath? You are not required to do so.
Mr. PIKE. I would rather testify under oath.
Mr. GRAY. What is your full name, sir?
Mr. PIKE. Sumner T. Pike.
Mr. GRAY. Sumner T. Pike, do you swear that the testimony you are to give the board shall be the truth, the whole truth and nothing but the truth, so help you God?
Mr. PIKE. I do.
Mr. GRAY. Will you be seated, please, sir.
May I, perhaps unnecessarily, call your attention to the existence of the perjury statutes. I am sure you are familiar with them.
I should like to request that if in the course of your testimony it becomes necessary for you to discuss or disclose restricted data you will notify the chairman in advance so we can take necessary steps under those circumstances.
Mr. PIKE. I may have to ask Mr. Rolander whether things are restricted or not because I have been away from this thing for 2½ years and I don't know what has been released.
Mr. GRAY. Please be free to make any inquiry about it.
Mr. ROLANDER. We have Dr. Beckerley with us.
Mr. GRAY. The other thing I should like to say to you, sir, is that we treat these proceedings as a confidential matter between the Commission and its officials and Dr. Oppenheimer and his representatives and witnesses. The Commission will initiate no releases with respect to this proceeding. We are expressing the hope that each witness will take the same attitude.
Mr. PIKE. It bothers me a little bit in case there should be leaks as to what attitude shall I take, but as far as it seems reasonable and possible, I will go along with your feeling on it. I will be the source of no leaks.
Mr. GRAY. I have simply stated the position of this board.
Mr. Garrison, would you proceed.
Whereupon, Sumner T. Pike was called as a witness, having been duly sworn, was examined and testified as follows:

DIRECT EXAMINATION

By Mr. GARRISON:
Q. Mr. Pike, what is your present position?
A. I am chairman of the Public Utilities Commission in the State of Maine.
Q. Appointed to that by the Governor?
A. Yes.
Q. By the Governor and council?
A. Yes. The council follows very much the same confirmation procedure as the Senate.
Q. Are you engaged in business in Maine, also?
A. Yes. I am a part owner of a couple of businesses.
Q. Sardines?
A. Sardines.
Q. You served on the Atomic Energy Commission from 1946 to the end of 1951, did you not?
A. Exactly to December 15, 1951.
Q. You were Acting Chairman the last 4 months?
A. No. It was between the time Mr. Lilienthal left which I think was in February 1950, until Gordon Dean was appointed, I believe, in July of that same year, for a few months.
Q. During this period you were well acquainted with Dr. Oppenheimer?
A. Yes.
Q. Would you say something about the nature and extent of that acquaintanceship?

A. I first met Dr. Oppenheimer, I think, at the first meeting of the General Advisory Committee—I don't remember the date of that—during that period, which must have been late 1946 or early 1947 until the day I left. It happens to be the day that they were meeting. I also saw him when they met in Washington, which was, I think, oftener than the statutory minimum of four times a year. I sometimes saw him outside of the meetings and I sometimes saw him when he was in Washington not at a meeting of the committee.

He was in town at times as a member of other boards and committees and sometimes perhaps as just an individual. Outside of the office I saw him—well, let me see, there was a period I think during the summer of 1947 when we boarded at the Bohemian Grove Forest out in California. We were there 3 or 4 days.

Q. Us being whom?

A. The Commission, its laboratory heads, some of the General Advisory Committee and several scientists like Dr. Wigner. I don't remember whether he was a laboratory head or not at that time. At that period we were put around at the various cottages and Dr. Oppenheimer, Mr. Lilienthal, and I were put in the same cottage.

There were other times, once perhaps, or oftener, when the committee was here I had them up to dinner between their meetings.

Q. The GAC?

A. The GAC, yes; I saw Dr. Oppenheimer, I think, at Dr. Symth's house. I believe that day I left I was going to see him again but there was a bad snowstorm and nobody could get to Dr. Smyth's. I have not seen him from then until yesterday.

Q. The relations between the Atomic Energy Commission and the GAC were of a fairly close character, would you say?

A. I should think so.

Q. Did you attend meetings of the GAC?

A. Yes. Their custom was to ask us in, usually once or twice during their meetings, and then almost invariably at the end of their meetings, at which time Dr. Oppenheimer would give us an oral review of the things that they had been taking up and the results they arrived at. Later, as I remember it, he would send a written summary.

I don't think there was any meeting they had here that I did not attend in part, except possibly when I was away on vacation or on Commission business.

Q. Did you attend a meeting in Princeton in the early summer of 1951 over which Dr. Oppenheimer presided?

A. Yes.

Q. The purpose of that meeting was to push forward with——

A. It was to pull together, as I remember it, various ideas that had developed about hydrogen or fusion weapons. It was quite a substantial meeting. As I remember it lasted the better part of 2 days.

Q. Mr. Pike, there has been a good deal of testimony here about the work of the GAC, and I am going to try to avoid duplicating the record, so I will just ask you a general question.

Based on your observations and of the knowledge of the work of the GAC and of Dr. Oppenheimer's chairmanship of it, did you form any impression as to his own contribution to strengthening the country in the 6 years that you have been talking about?

A. Yes; I think the GAC under his chairmanship made a major contribution to the work of the Commission and the Commission, I take it, was trying to work for the good of the country.

Q. You have read the Commission's letter of December 23, 1953, which initiated these proceedings, containing the derogatory information about Dr. Oppenheimer?

A. Yes, I read the New York Times which I take it gave the full letter.

Q. On the basis of your knowledge of Dr. Oppenheimer and your experiences with him, what is your opinion as to his loyalty?

A. I never had any question about his loyalty. I think he is a man of essential integrity. I think he has been a fool several times, but there was nothing in there that shook my feeling. As a matter of fact, it was a pretty good summary, it seemed to me, of the material that was turned over to us early in 1947 by the FBI, all except the last thing about the hydrogen bomb. Of course, that was not in then.

Q. The letter and, I assume, the file contained data about past associations of his.

A. Yes.

Q. In your judgment is his character and the associations of the past and his loyalty such that if he were to continue to have access to restricted data, he would not endanger the common defense or security?

A. No, I don't think he would endanger the common defense or security the least bit.

Q. You read about the Chevalier incident in the Commission's letter and Dr. Oppenheimer's answer?

A. Yes.

Q. It is not clear as to how much of that story was in the file that you went over in 1947. I assume you went over whatever the file was?

A. Yes.

Q. Personally?

A. Personally.

Q. And participated in the discussions with the other Commissioners?

A. Oh, yes.

Q. Do you or do you not have any clear recollection of the Chevalier incident as of that time? If you don't, don't try to——

A. I don't think I have much beyond the summary of the letter of last week, which was published last week. There was a lot more there. It was a pretty thick file, but I don't remember exactly what was in the file.

Q. May I put to you a hypothetical question which I put to you, I think, last night in order that you might have an opportunity to reflect on it. Supposing that it were established in addition to the description of the incident as it appears in the Commission's letter that after the conversation between Chevalier and Dr. Oppenheimer in which Chevalier had informed him that Eltenton was in a position to transmit secret data to Russia, that for several months Dr. Oppenheimer failed to report the matter to the security officers; that thereafter he did on his own initiative report to the security officers, but revealed only the name of Eltenton, and declined when pressed to do so to reveal the name of Chevalier, was not frank in describing the exact circumstances of what had taken place, added to the story about Chevalier without mentioning many certain facts which were not in the picture; that later when again pressed to reveal the name of Chevalier he again declined; that General Groves asked him to reveal the name and he said he would not do so unless ordered; that General Groves said he didn't want to order him to do it, asked him to think it over and met with him again and said he would have to order him unless Dr. Oppenheimer would tell him the name and Dr. Oppenheimer finally revealed the name of Chevalier.

Assuming that this were established would this alter the opinion that you have expressed here to the board about your present views of Dr. Oppenheimer's loyalty and the propriety of his having continued access to restricted data?

A. No.

Q. Do you want to say why it wouldn't alter your opinion?

A. I think it was a bad incident. Taken alone it would have bothered me very much. I suspect I have been party to incidents in my life that I rather not have certainly taken out of context. This, woven into the context, however, of performance under closer observation for him, many years and achievements of such size as to warrant the gratitude of this country, I don't think it should be given much weight at all.

Q. Turning to another topic of the H bomb for a moment, without going into the details about which there is a great deal in the record, as I understand it in reporting to the President the views of the Atomic Energy Commission about whether to go forward with an all-out H-bomb program or not, following the Russian explosion in the fall of 1949, there were several separate reports, were there not?

A. There were. I think there were four.

Q. Would you just say what they were? I mean who made these four reports?

A. Strauss made one definitely for going ahead; Dean made another in which he recommended some prior——

Q. He already has testified.

A. Smyth and Lilienthal made another.

Q. Mr. Lilienthal has testified about that.

A. I agreed with them that this was not the time to go on an all-out effort but put in a supplementary memorandum which, as I remember, I had to put in somewhat later on account of being on the coast. I had to take a trip at that time.

Q. You went to the coast after the discussion?

A. After the discussion. I don't think I put in my separate memorandum until I got back. That must have been about the middle of November.
Q. That was about 10 days after the meeting or something like that?
A. I think so.
Q. That went to the White House?
A. Yes. Whether it went to the Security Council or the President, I don't know. I have to perhaps say here that I had not realized that I had any access to records so I have not looked at any records since I left the Commission in 1951 and, of course, took none with me. I am relying completely on my memory as to the time and dates.
Q. Do you remember the substance of the points that you made in that memorandum?
A. I think so.
Q. Would you state them?
A. One of them was that we had no knowledge that the military needed such a weapon. Another one was that the cost of producing tritium in terms of plutonium that might otherwise be produced looked fantastically high—80 to 100 times, probably, gram for gram.

The third one, and this sort of tied into the first, was, as we all know, that the damage power of the bomb does not increase with the size of the explosion, and it seemed that it might possibly be a wasted effort to make a great big one where some smaller ones would get more efficiency.

I think I put in another one: That as between the fission work we were doing and the fusion thing in question here, there were some good things about the fission things. Up to that time and up to the present nobody has brought up anything useful for mankind out of the fusion.
Q. Out of the fusion?
A. The fusion. In other words, I have never yet heard of any possibility of anything beneficial coming from the hydrogen end of it.
Q. In terms of useful energy?
A. Other than as a weapon. Again I am going entirely from memory, but I think that is what I put in my memorandum.
Q. These were your own independent views?
A. They were my own. They could not be completely independent because the Lord knows we had been talking and discussing and, let us say, arguing for well over a month and at that time, possibly nearer 2 months. So the views were the result of a great deal of discussion. I think they were my own. I came with a slightly different set of reasons than the others, although I did come out with the same recommendation as Lilienthal and Smyth did.

Mr. GARRISON. I think that is all, Mr. Chairman.
Mr. ROBB. I have no questions.
Mr. GRAY. I have some questions, Mr. Pike.

I have been in the course of these proceedings pursuing something that has been illusory and evasive as far as I am concerned, and it may be just because I don't comprehend what has been said. This perhaps involves a matter of recollection on your part, so, of course, you can testify only what you recall about it.

In Dr. Oppenheimer's reply, dated March 4, to General Nichols' letter, he referred to the October 29, 1949 meeting of the General Advisory Committee and indicated that this meeting was called to consider two questions. One was the general questions in the light of the news about the Soviet success, was the Commission doing all it should do, and if not, in what way should it alter its course.

The second was to pursue the question of whether there should be a "crash" program with respect to the Super.

The record shows that the then Chairman, Mr. Lilienthal, wrote a letter— I am sorry I don't remember the date—to the General Advisory Committee, which raised this first question.

Then the record shows that later in the same month, that is, October, there was a letter—I can refer to that?

Mr. ROLANDER. Yes, certainly. You can show it to him.
Mr. GRAY. A letter dated October 21, 1949, signed by you as acting chairman of the Commission, to Dr. Oppenheimer with respect to this October meeting and asking certain questions, I believe, that the committee should address itself to.

Mr. ROLANDER (Handing letter to witness).
The WITNESS. I would not have remembered this in detail, but questions of this sort were certainly running through our minds at the time.

Mr. GRAY. Yes. Of course, there are a lot of questions raised in this letter.
The WITNESS. That is right.
Mr. GRAY. I have not looked at it very carefully recently, but I don't believe this second question which I referred to and which appears in Dr. Oppenheimer's reply of March 4 certainly was asked in that form in this letter.
The WITNESS. Would you repeat that second question for me, Mr. Gray? In reading I forgot what the second point was.
Mr. GRAY. Yes. Let me give you the exact language of that.
Dr. Oppenheimer's reply indicates that the Commission asked the General Advisory Committee to consider and advise on two related questions, the second of which is, and I am now reading from his letter, "whether a 'crash' program for the development of the Super should be a part of any new program?"
What I have been trying to identify for my own information in that accord is how this second question got asked in that form. I don't believe it is raised in that form by your letter.
The WITNESS. I don't see it there.
Mr. GRAY. I might say to you that I believe that Mr. Lilienthal testified that his recollection was not good on this point. Am I correct on that? If I didn't ask him this question it is because it was late and I was tired, because I have really been trying to find out about it.
Mr. GARRISON. I think I remember, Mr. Chairman, that he testified he had written the letter that raised the first of these two questions, and I myself don't remember very clearly.
Mr. GRAY. My question of you is: Do you recall whether you met with the committee and asked this second question about the "crash" program?
The WITNESS. I remember very distinctly the phrases "crash program" and "all out program" being used almost interchangeably for some months. If I had to rely on my unaided memory, and I guess I do, I would think that phrase arose with Mr. Strauss. At least in my mind it ties in with what he wanted to do.
In the meetings of the General Advisory Committee— of course, I am sure you are aware from previous testimony—they were not held to the things which the Commission asked them to do. I think there were several times when they got here and either took up things not on the previously prepared agenda of their own motion or something had happened between the time of the calling of the meeting and the time that they got there that would be discussed.
As I remember it, they were reasonably formal and kept pretty full notes, but I don't think there was any reason why a thing should not be discussed and considered even though it had not been put on any agenda, like the calling of a meeting of a board of directors or stockholders, you tell them what you know should be discussed and then you leave room for anything new that may come up.
It seems to me that knowing there was a very strong recommendation for a heavy program on what we now call the Super, I guess—but that is an old name—this would inevitably have come up in the discussion called for by this letter. I don't know whether I am helping you out or not.
The "crash program" or the "all out program," let us say, was an extreme of one position. It seems to me, let us say, that was the position that Mr. Strauss took.
Mr. GRAY. It was the position that the Government of the United States ultimately took.
The WITNESS. I am saying at this time. The program as laid down by the President in 1950, 2 or 3 months later, * * * was very shortly embodied in a budget that was set up, an emergency deficiency bill, a very large size, in addition to the one which we had already sent up that year which had already strained the imagination of the Appropriations Committee pretty strongly.
That was a heavy program, yes. I am trying to answer your question. I am afraid I haven't very well.
Mr. GRAY. I am afraid You haven't and I won't take any more of your time in pursuing it. I don't think you can answer it. I think you have indicated your memory is not clear as to the letter or instructions.
The WITNESS. My impression is that this crept into the discussion and probably got the name crash some where along the line because it was a convenient handle, just as the name of Super came along—I don't know where it came from—but it became a convenient handle.
Mr. GRAY. I would like to turn to something else, if I may, Mr. Pike, and that is the consideration given by the Commission to the clearance of Dr. Oppenheimer in, I believe, March 1947.
The WITNESS. I think that is right.

Mr. GRAY. Is it your recollection that the Commission took formal action to clear Dr. Oppenheimer? I might say that there is some confusion about this.

The WITNESS. I don't have any clear recollection that we took formal action to clear him then. I think you are all aware that was a period of extreme confusion.

Mr. GRAY. Yes.

The WITNESS. When the confirmation hearings were going on on the Hill, when the Commission was going through the initial throes of organization and really had not started to organize. My memory is that even the minutes themselves had to be rewritten some months later, that is, the minutes of the meetings. I may be wrong about that. But if you told me that something was not on the record as of that time, I would say I would not be at all surprised.

Mr. GRAY. I think the fact is that in August something was written which purported to reflect action taken in March.

Mr. ROBB. February.

Mr. GARRISON. It said February.

Mr. GRAY. It said February when indeed whatever took place actually took place in March. So there is a good deal of confusion. I don't think the record is clear that there was formal action which cleared Dr. Oppenheimer in 1947.

I am just asking you whether you are surpised to hear me say that the record is not clear on that point?

The WITNESS. No, sir; I am not. I think both Mr. Beckerley and Mr. Rolander were here during that period. This is off the particular subject of Dr. Oppenheimer. But as I remember it, Lyle Bellesly was succeeded by Roy Snapp as secretary and Bellesly's records were in unsatisfactory shape and unsatisfactory to everybody. I think Snapp went right back and took what he had, what he could find, and wrote up things. There were a lot of ex post facto things in the record.

I think you will find if you go through it there were a lot of things picked up and a lot of things missed that should have been picked up.

Mr. GRAY. The fact is that the Chairman of the Commission discussed this matter with people in the White House at about the time that the Commission read these files.

The WITNESS. I am quite sure about that.

Mr. GRAY. Is it possible that this kind of thing could have happened: That the Commission knew that the Chairman had consulted the White House; that the Chairman was perhaps expecting some further word from the White House; that no further word ever came from the White House and that in fact nothing was ever done about the action on the clearance?

The WITNESS. I suppose that is possible. Of course, that "as of" date was before the delivery of this dossier; the February date, if I am not mistaken.

Mr. GRAY. Yes; the February date could not, I think, be correct.

The WITNESS. I am not sure that it couldn't.

Mr. GRAY. You mean it is possible that the clearance might actually have been considered in February?

The WITNESS. I am not sure. For instance, the clearance of all the members of the General Advisory Committee might have been made and considered in February. I am not sure that it might not have happened that this was the only case where a question was raised. This may have been kept in abeyance to see whether that should have been confirmed until August.

I am no clearer on the thing than our records are, but I think that is all in the realm of possibility.

Mr. GRAY. Did you consider, however, this a serious thing at the time?

The WITNESS. Oh, yes, I did. I am sure we all did. There were five of us on the Commission. As I remember it, this was a unanimous action.

Mr. GRAY. I am going to change my course now a little bit, Mr. Pike.

You testified that one of your reasons for not being enthusiastic about the all-out program was the fact that there had been expressed no military need for this kind of weapon.

The WITNESS. Yes.

Mr. GRAY. Do you think it possible that a military need had not been expressed to the Commission at that time because the military did not have any reason to believe that it was feasible? The reason I ask that is that once it became feasible there seems to be no question that the military people think there is a need. I believe that is correct.

The WITNESS. I think you will find, or there should be in the documentation of the Commission and probably in that of the military liaison committee, the first expression of the military that such a thing was desirable. I don't remember the date of it, of course. I remember distinctly seeing such a paper. Whether it was in a meeting of the military liaison committee meeting or a Commission meeting, I don't remember. * * *

I remember frankly in the back of my head thinking that I would like to get these boys on the line. I think later they came on the line. You are perfectly familiar with that. You were in that rat race at one time.

Mr. GRAY. My recollection is about the same as yours.

I suppose people in the military liaison committee at that time perhaps can answer the question I put to you better than you could.

I want to ask you one other serious question. You say that as of 1949 and indeed as of today so far as you know, there seems to be no use other than a military which might come out of these processes?

The WITNESS. Yes, sir; I believe that is correct.

Mr. GRAY. I am asking for information. I don't believe we had any testimony on that.

The WITNESS. I am sure that there had been none suggested then. If there have been any suggested since, I am unaware of it.

Mr. GRAY. In your official position you would be very much interested in that.

The WITNESS. I would. All I am saying is that a good many things have happened since December 15, 1951, and, of course, I would not be aware of those. I have had no security clearance. I think I have been in the Commission Office once at their request and that was when the question came up of power plant for the Paducah operation.

Mr. GRAY. Are there any questions, Dr. Evans?

Dr. EVANS. Yes. Mr. Pike, I understand that you did say that Dr. Oppenheimer made a number of mistakes?

The WITNESS. I think so, yes.

Dr. EVANS. I want to ask you another question. If you had been in Dr. Oppenheimer's position when he was approached in this matter about giving information to our enemies, you would have reported that immediately, would you not?

The WITNESS. In 1943 I think I would have. I fortunately was not in the position of having that question put up to me. But I think I would have.

Dr. EVANS. I wish you would explain. Do you think there is any military need today for a Super?

The WITNESS. Yes, I believe there is, Doctor. I think if you go back and get the document I think exists, you will see one or two reasons that I didn't have in mind.

One of them, for instance, is that you get a much larger margin of error for a miss. Something, for example, that will take a radius of 10 miles rather than 1 mile.

Another one is that if you can get through you only put at risk 1 or 2 or 3 planes as against a flock of them to destroy a big target.

I can rationalize uses for the Super. I felt that the military desirability of the Super ought to be estimated by military people rather than a bunch of civilians like ourselves. I still think their views would be quite authoritative with me.

Dr. EVANS. You wished in your own account here to go rather slow on this Super, didn't you?

The WITNESS. I wished to get, as I testified later before the Joint Committee, to get more facts before going out on a crash program.

I would like to bring in here one thing that was not very well considered in the period we were talking about but had to come up some months later. I think it was after Mr. Lilienthal left. I remember I was on that committee of the National Security Council.

The order had been given and the question was not whether to go ahead but how to go ahead. I brought up at that meeting my point of view which was that this country could be in no more miserable position than to have a successful development on our hands and then to have to spend 3 or 4 years in building factories to produce the thing.

Therefore, in going ahead with the development we had to at the same time go ahead with our factories or plants just as though we were sure we were going to have a successful development. That seemed to me always to be an inherent part of the development question.

You see why we would be in a miserable position. We had proven that it could be done, and somebody else could have easily proven the same thing at the same time and 3 or 4 years to build plants would be a pretty tough period.

So it involved a major expenditure of time and money, effort and manpower and it was not a thing to be gone into lightly. I wanted to get some important facts into the picture, all the facts that could be gotten, and I was not willing to recommend a drive program until we had some of those facts.

Some of them came in if I am not mistaken. I think we got that military appraisal or at least a military appraisal before the January 1950 decision from the White House. I am not completely sure of that, but I think that was in.

I don't know whether I have answered your question or not.

Dr. EVANS. Yes, I think you have. The thing that I was trying to get your opinion on was as to whether A bombs as big as this and as costly as this would mean that we ought to have a lot of targets on which to use them, whereas if we only had a couple it would be like killing a mosquito with a sledge hammer.

The WITNESS. This was in my mind. I am afraid to give numbers would be to get into a security point.

*　　　*　　　*　　　*　　　*　　　*　　　*

Dr. EVANS. Thank you.

Mr. GRAY. I have just one question suggested by Dr. Evans' question.

Your view was that we ought to know more about it. You were not just unalterably opposed?

The WITNESS. No. I think I put it in my memorandum which you should have the qualification "at this time."

Mr. GRAY. Is it your recollection that most of the members of the General Advisory Committee were opposed at any time?

The WITNESS. No, that is not my recollection, although I would, of course, have to refresh my memory. That is not my recollection. I think they brought in, as perhaps they properly should, some, let us say, political and strategic and moral questions which frankly did not weigh very big with me. As far as I am concerned there was not then and there is not now a great deal of difference in morality between one kind of warfare and another. This stuff never affected me very much. But I think the GAC did give it perhaps more consideration than I did.

Mr. GRAY. Do you have any questions?

Mr. GARRISON. Just a few, Mr. Chairman.

FURTHER DIRECT EXAMINATION

By Mr. GARRISON:

Q. Speaking of what was before the GAC at their meeting on October 29, 1949, in response to a question by the Chairman you said something to the effect that the question of the crash program crept into the discussion, as I recall the phrase.

I wonder if when you were talking about discussion you had reference to the preliminary meeting between the members of the Commission and the members of the GAC which started off the meeting, as I understand it, in accordance with the regular practice?

A. No. I think what I was referring to was the various meetings of the Commission during, let us say, the month or a little more than a month between the announcement of the Russian bang and this GAC meeting.

Q. In other words, in the Commission's discussions before the GAC meeting the question of a crash program for the H-bomb was to the fore?

A. I think so. Let me bring another group in on that. Don't forget that we had a large and a very able staff. We had the heads of the various divisions in Washington and we had at our various outposts people who came in on short notice. I am sure the Commission minutes will show who was at various meetings and when, but I am completely clear in my memory that there had been a lot of discussions. I am not completely clear in my memory exactly when they took place and who was present at each one. That is a matter of record and can be verly clearly and easily got at.

Q. When the question was put to the GAC in Mr. Lillienthal's letter asking that consideration be given to whether in view of the Soviet success the Commission's program was adequate and if not, in what way it should be altered or increased, would it or would it not have been a natural outgrowth of that question, considering the times and the discussions that you had, to consider the question of the hydrogen crash program?

A. I think it would have been a natural thing. If you will remember, the hydrogen question had never been dropped. It had been in charge of a small group headed by Ed Teller. Dr. Teller was never one to keep his candles hidden under bushels. He was kind of a missionary. I might say that perhaps John, the Baptist, is a little overexaggeration. He always felt that this program had not had enough consideration. Teller in my view was a pretty single-minded and devoted person. I would guess that it would have suited him completely if we had taken all the resources we had and devoted it to fusion bombs.

He is a very useful and a very fine man, but I always thought he was kind of lopsided, as a good man specialists are. This was one of the things that would naturally have come into any involved discussion of what we ought to be doing. I don't know whether I have answered your question or not.

Q. Just two more questions.

After President Truman gave the go ahead on the H-bomb program, did the GAC, as you recall, cooperate with the Government and accept that decision and move forward?

A. Yes. When you say move forward, one has to remember that some of the developments in the early months were quite disappointing. The thing was attacked, I think, wholeheartedly and we were not happy, not about cooperation, but not happy about the results for some time.

Q. Did Dr. Oppenheimer, so far as you yourself knew, do anything to delay or obstruct the program?

A. Oh, no; rather the reverse.

Q. One final question.

When the Chairman was talking with you about the question of the 1947 clearance, you used the phrase "unanimous action." I would like to ask you, leaving aside the question of dates and minutes, what you recollect of what the commissioners actually did do. Did they sit around the table together and consider the matter of Dr. Oppenheimer's clearance and come to some view about it, or how was it done?

A. They did what you suggest. I want to go back to a fundamental question of Commission organization which came up very early when we met. I had something to do with the result of it. There was a question as to whether we should not organize, let us say, something like the Interstate Commerce Commission—so and so be in charge of this, and so and so be in charge of that, and sort of departmentalize ourselves. That question was answered in the negative and I was instrumental—I do not say I was the dominant factor, but I had this experience on the Securities and Exchange Commission just after they had abandoned that sort of division or labor system and the very unsatisfactory results of that were in front of my mind—so that while naturally each one perhaps would give a little more attention to the thing he knew best—Bacher, let us say, was the physicist, I knew something about mining and raw materials, and so on. Yet, our actions were taken together and our responsibility was both joint and separate and complete. In other words, while we asked for advice and asked for help in a great many areas, the final responsibility was always ours, and it was always joint and if anybody had a dissent, it was recorded in those meetings. So, if there was no dissent recorded, each one of us was in on the decision and each agreed on it.

Do I answer your question?

Mr. GARRISON. That is all, Mr. Chairman.

Mr. GRAY. Was any member of the Commission interested particularly in security problems? You were interested in mining, for example. Do you remember whether any Commissioner at that time was?

The WITNESS. I would say that Commissioner Strauss had some background in security problems when he was over at the Navy and perhaps took a more direct interest than the rest of us. This security problem, I say say, was the most nagging problem of all in a good many months of the Commission's existence.

If you remember the law, it not only required an FBI investigation of new employees but also required going over everybody who had been cleared by the Manhattan project who was still working. This dragnet brought up quite a few customers. I probably am exaggerating but it seemed to me as though we took over half our time for the first 7 or 8 months on these distinct personnel security problems.

Of course, there were physical security problems, such as a barbed wire fence had rusted, or the grass had grown so that a fellow could slither through it near one of the plants. This could not all be corrected at once. This was part of the general neglect into which the project had fallen during the year or so

Congress had been trying to make up its mind as to what law to pass and the further 3 months Mr. Truman was trying to draft five people willing to serve on this Commission. The war was over, let us say, in August 1945, the Commission was appointed in late October—I would guess the 28th—of 1946, and there was a period of slowdown which looked at that time when we came on as though it might culminate disastrously. There were a lot of problems that had come up. * * *

Mr. GRAY. Mr. Lilienthal testified that the Deputy General Counsel of the Commission, Mr. Volpe, was active with him in considering Dr. Oppenheimer's clearance.

Do you recall whether counsel of the Commission participated in this, Mr. Pike?

The WITNESS. No, I don't.

Mr. GRAY. Would he have normally sat with the Commission when they considered these security cases?

The WITNESS. No. The counsel of the Commission like every other officer of the Commission was called in when the Commission felt it needed him. Of course, Volpe was a natural for this thing because he had done some security work for General Groves before and had a general acquaintance I think with the security problems in the Manhattan District.

During this period, as I say, we had no security officer, or if we had one, I don't remember who it was. You picked on the fellow who might be of some help and Joe Volpe had some background in this sort of business.

Mr. GRAY. It was not because he was assistant counsel or deputy counsel, but more because he had a background.

The WITNESS. That would be my belief; yes, sir.

Dr. EVANS. Mr. Pike, you spoke about the trouble you had with investigating the security. Did it seem to you that there was really more screwy people in here than you would have expected to find ordinarily?

The WITNESS. No; I don't think so, Doctor. As I remember it, a great many of the star customers had already gone. My best recollection is that of about 60,000 people on the job at that time, we had around 60 or 65—it sticks in my memory as one-tenth of 1 percent—of people about whom there were questions coming from a vague doubt to a fairly substantial doubt. Those figures may not be exact but that is the range, I am sure.

Dr. EVANS. Thank you; that is all.

Mr. GRAY. Thank you very much, Mr. Pike.

The WITNESS. Thank you, sir.

(Witness excused.)

(Discussion off the record.)

Mr. GRAY. Let us get back on the record.

Mr. ROLANDER. As I said, Mr. Chairman, we have not had an opportunity to review all of the transcripts of what we had hoped to be unclassified portions of the hearing, so we have permitted Dr. Oppenheimer and his counsel to review the transcripts here in the AEC building. As we complete our review we turn the transcripts over to them by receipt. I have also permitted them to use secretaries for the purpose, as I understood, to assist them in preparing questions and what material they needed o continue their presentation. I am somewhat concerned, however, that if they bring stenographers in here that they not make copies of the transcripts until thehy have been approved from a classification standpoint.

I wanted to go on record as noting that some information may have to be classified from a national defense standpoint. This information should be protected from that standpoint as well as the confidential relationship between Dr. Oppenheimer and the Commission.

Mr. GARRISON. Mr. Chairman, Mr. Ecker and Mr. Topkis from my office have been at my request making summaries of various portions of the transcript and have the transcript as a whole in the room assigned to us and with a stenographer at intervals to whom they have been dictating. Up to this point I had assumed that there was no problem about this at all.

I suppose in the nature of things there are bound to be where something in the record needs to be cleared up a quotation here and there directly from the transcript dictated to the stenographer to write up so that we can study it. It is awfully hard for us to work here ourselves in that room.

Now, if there is a security question about the contents or about quotations from the transcript, I would like to know what it is so we could have an understanding about it.

Do I understand that these transcripts that we have been working on are still in some way being reviewed?

Mr. ROLANDER. Yes, they are. They are being reviewed not only by our own classification officer, but by representatives of other agencies.

Dr. BECKERLEY. May I make a comment for the record.

We have made arrangements with the Department of Defense for review of certain portions of the transcript. Two or three people are coming over at 1 o'clock today. I hope we will be able to clean up all of the Defense Department questions with respect to the transcripts to date at that time.

There is some intelligence data that has crept in in a few spots. I have taken steps to have that reviewed. In addition there is some material which may have sensitivity in the Department of State. This is also being reviewed at the present time.

Mr. GARRISON. Have you any suggestion to offer about it?

Dr. BECKERLEY. I would be happy to define the areas which I am quite sure there are some questions about. Whether there is any classified information in these particular sections, I don't know.

Mr. GARRISON. Could you mark the portions of the transcript that are being reviewed for security purposes and then have it understood that we would not make any quotations from those portions of the transcript?

Dr. BECKERLEY. I certainly could; yes. I can identify the areas where there is some possible sensitivity but in view of the fact that these are matters outside of the purview of the Commission I have no way of knowing whether these are or are not sensitive.

Mr. GARRISON. Could that be done with some expedition?

Dr. BECKERLEY. Yes, I could do that right now, as a matter of fact from my notes.

Mr. GARRISON. So that at lunch, let us say, we would know what those passages or portions are?

Dr. BECKERLEY. Yes. Could you indicate which parts of the transcript or are you doing them in sequence?

Mr. GARRISON. We are doing them in sequence. I asked Mr. Topkis to begin at the beginning, page 1, and give us a summary.

Mr. ROLANDER. Mr. Garrison has received volumes one and three. So it would only pertain to volumes 2, 4, 5, 6, and 7.

Dr. BECKERLEY. Two is one of our more troublesome ones since it concerned the witness' activities with the Defense Department.

Mr. GARRISON. One and three are completely clear and can be taken out of the building.

Mr. ROBB. Those you have, Mr. Garrison?

Mr. GARRISON. Yes.

Mr. GRAY. Let me suggest that I believe we are discussing matters which really should be between counsel and the Atomic Energy Commission and its officials on which I think this board can't make any ruling. I don't mind hearing the discussion, but I think we are taking the time of the board to cover material with which you ought to deal with Mr. Rolander.

Mr. GARRISON. Yes. There are volumes 5 and 6 of the transcript. When will we get those?

Dr. BECKERLEY. Five has some material which I have asked State to look at. I can define the areas.

Mr. GARRISON. We had them to work on last night but not this morning.

Mr. GRAY. If this conversation is going to be pursued I am going to have the board excused and let Mr. Rolander and Mr. Garrison discuss it.

Mr. GARRISON. It is relevant to the board because it is a part of the whole procedural problem we do face, which we have to bring to the board's attention, Mr. Chairman.

Mr. GRAY. Anything that is under the jurisdiction of this board should be brought to the board's attention, but I cannot make a ruling on security matters.

Mr. GARRISON. Mr. Chairman, just one statement for the record. If there is anything—we will make our copies available to you of everything that we have dictated or written up to this point—that we have extracted from the minutes that has a security question, we want to make it perfectly clear that we will return that to you.

Mr. ROLANDER. Fine.

Mr. GARRISON. That, then, can be worked out.

Mr. GRAY. We will take a short recess.

(Whereupon, a short recess was taken.)

Mr. GRAY. Would you stand and raise your right hand, please?
Mr. RAMSEY. Yes, sir.
Mr. GRAY. Do you wish to testify under oath?
Mr. RAMSEY. Yes, sir.
Mr. GRAY. You are not required to do so, but all of the witnesses have.
Mr. RAMSEY. I am perfectly willing.
Mr. GRAY. Norman Foster Ramsey, Jr., do you swear that the testimony you are to give the board shall be the truth, the whole truth and nothing but the truth, so help you God?
Mr. RAMSEY. I do.

Whereupon, Norman Foster Ramsey, Jr. was called as a witness, and having been first duly sworn, was examined and testified as follows:

Mr. GRAY. Will you be seated, please, sir.
It is my duty to call your attention to the existence of the perjury statutes. I assume you are familiar with them.
The WITNESS. Yes, sir.
Mr. GRAY. In the event, Professor Ramsey, it becomes necessary for you to refer to restricted data in your testimony, I would ask you to let me know in advance. so that we may take certain appropriate and necessary steps.

I should also observe to you that we consider this proceeding a confidential matter between Atomic Energy Commission representatives and Dr. Oppenheimer, his witnesses and representatives, and the Commission will make no public releases. It is our custom to express the hope to the witnesses that they will take the same view.

The WITNESS. I might add one thing sir, that the chairman of my department called in great concern that a newspaper reporter called him yesterday and asked him if by chance I were to be a witness, and he said he wasn't sure, or something like this, and this got reported in the paper that Professor Bainbridge said I was to be a witness here. This is certainly not my fault and certainly not his.

Mr. GRAY. Yes.

DIRECT EXAMINATION

By Mr. GARRISON:
Q. You are a professor of physics at Harvard University?
A. Yes, sir.
Q. You come from a military background?
A. Through my father. My father enlisted at the age of 16 in the Spanish-American War. He then went to West Point. He served in World War I and World War II, and is now retired a brigadier general.
Q What were your wartime positions? Would you just run over those briefly?
A. I was consultant to the National Defense Research Committee. I was doing radar research at the Massachusetts Institute of Technology, at MIT Radiation Laboratory. I was an expert consultant to the Secretary of War in the Pentagon Building with the Air Force during about 1942–43, and I was at Los Alamos from 1943 to the end of the war, during which time I actually was officially employed as an expert consultant to the Secretary of War, though I worked completely within the Los Alamos location.
Q. What positions in the Government do you now hold?
A. No full-time position. I am a consultant to a number of the services, that is, I am a member of the Air Force Scientific Advisory Board. I am a member of the newly established Defense Department Panel on Atomic Energy.
Q. Excuse me. Is that panel in substance the successor to the atomic energy responsibilities of the Research and Development Board?
A. Not in a certain sense a strict successor, but with the reorganization this is what has been substituted for it. * * *
Q. When did you first meet Dr. Oppenheimer?
A. I first met Dr. Oppenheimer in the summer of 1940.
Q. This was at a meeting of the American Physical Society?
A. That is correct, the Seattle meeting of the American Physical Society, which was also on my honeymoon and Professor Zacharaias, who had a car, we had been riding with him, and Dr. Oppenheimer rode with us from Seattle to Berkeley, and we stayed at Dr. Oppenheimer's house for approximately 2 days in the early summer of 1940.
Q. This was at the time of the collapse of France in World War II?
A. Yes, sir.
Q. Did you have any conversations with Dr. Oppenheimer about that?

A. We had a number of conversations, and it is certainly difficult to reconstruct all of them in any detail.

Q. I wouldn't ask you to.

A. On the other hand, I do remember some. In particular there were some on that at which Dr. Oppenheimer expressed a very grave concern for the French and the British and particularly a rather fondness for Paris, and the trouble which it was very actively in at that time, though this was at the time of the Russian-Nazi pact.

Q. At Los Alamos, when you were there from 1943 to 1945, what was your particular job?

A. I was head of the so-called delivery group, which meant that this was the group that was concerned with making sure that the Los Alamos weapon was a real weapon, that is, something that could be carried in an airplane and dropped from same.

Also, this meant I had charge of the relationship with the Army Air Forces, and the 509th Bombardment Group, both in the testing of same and then ultimately actually I was chief scientist at Tinian, where we assembled the two atomic bombs used during the war. Actually the late Admiral Parsons was head of the group at Tinian, and I was chief scientist under Admiral Parsons.

Q. To what extent was there compartmentalization at Los Alamos and what would your observation be as to the general policy which was adopted there about the division of labor among the groups?

A. I would say for the basic scientific developments, there was very little compartmentalization for very good reasons. This was also true at the MIT Radiation Laboratory. It had been discovered quite early in the war in a number of laboratories that inefficiency went up very rapidly with excessive compartmentalization. Actually at Los Alamos my own group, being somewhat more over the direct scientific developments and also being considered one of the most top secret things—particularly the fact that we were so far along that anyone had any interest in relationships with the Air Force—for this reason we were to a considerable degree compartmentalized. That is, we were never invited to give reports at the staff seminars on what we were doing.

Likewise, when we were away from the place, we were in fact required by security regulations to some degree to our embarrassment to be untruthful as saying where we came from. We were not allowed to say we came from Los Alamos. In fact, we had to say we came from other places.

Q. Would you care to make any comment upon the quality of Dr. Oppenheimer's leadership at Los Alamos? I don't want a great deal of detail, but just your impression.

A. Yes, sir. I saw it very obviously through the work and was most impressed in every way. I think he did a superb technical job, and one which also made all of us acquire the greatest of respect and admiration for his abilities and in view of this hearing I might also add his loyalty and his integrity.

Q. At the end of the war was there a problem of holding Los Alamos together?

A. Yes, a very great problem in that most of the key people in the laboratory, like myself, were men fundamentally interested in pure science. For patriotic motives we had by then been devoting 4 or 5 years of our lives since we had really started in 1940 before the work working on things. We were indeed very eager to get back to our research laboratories where we would do the fundamental research that we were here to do.

As a result everyone was very eager to get away. It was chiefly some rather elloquent pleas on the part of Dr. Oppenheimer that kept many there together. Actually I know of this in two ways. One, by the fact that for the initial pleas in this direction I was not at Los Alamos since I was in charge of the group at Tinian. Most of us there thought all of our friends would be rushing away from Los Alamos with terrific rapidity. We arranged by cablegram for moving vans, asked our wives to arrange the moving. As a matter of fact, when we got back, we were in some degree of disgrace with the rest of our friends who had the benefit of Oppenheimer's lecture of the importance of staying on.

I was actually one of the first people getting away from Los Alamos, and I have been somewhat embarrased about this ever since. I was also told off about this.

Q. During the controversy about whether to go ahead full steam on the H bomb program or not, that is to say, roughly in the fall of 1949, and continuing on until President Truman's announcement in January of 1950, you were a member of the Air Force Science Advisory Board?

A. That is correct.

Q. You did not take any official part in the formulation of policy about the H bomb?
A. No, sir.
Q. I just want to ask you one question——
A. We were, however, informed to a considerable degree of the technical status of it. That is, we were given review meetings at Sandia.
Q. I want to ask you one question as a matter of interest. How did your own mind at that time run on the question?
A. I found it a very difficult problem that I worried about a great deal, even though I did not contribute to it. I would say roughly I was in the state of schizophrenia, which was best described by saying I was actually 55 percent in favor of going ahead, that is, I felt it was a development even with a crash program was appropriate to, and 45 percent in my own mind against it. Again this I also record as 100 percent loyalty. It was not a matter of loyalty versus disloyalty, certainly from what I had been presented; it was not a very useful looking weapon that was being described * * * I better not go much further.
Q. During the past 4 or 5 years, Dr. Oppenheimer, I think, has been chairman of the committee of the Harvard Overseers to visit the Harvard physics department?
A. Yes, sir.
Q. Have you had some association with him in that connection?
A. Yes, I have had quite a few, chiefly on two different problems. The first one was immediately following the outbreak of hostilities in Korea. Our department was very much concerned and worried with what was the best way for our department to contribute to the country when the country was in a state of emergency, at the same time doing its very important work also for the country of training students. We had a number of discussions among ourselves, and a particularly enlightening discussion with our visiting committee under the chairmanship of Dr. Oppenheimer—the visiting committee includes chiefly the various industrial physicists—and I think the help we got from them was very great.

During these conversations, Dr. Oppenheimer particularly eloquently expressed the problem that the United States was faced with, the threat that was there from Russia and emphasized the importance of our doing work, particularly by taking leave from Harvard for consultation and also urged with the President and provost, at least I am told of it later, the importance of allowing members of our staff to take such leave. Indeed, they have been taking it.

I think on the whole we have averaged one or two men, usually about two men, at any one time from our department on leave on one or another defense project. Some, for example, on the H bomb. There is one at Livermore at the present time.

Q. Did you have any discussions with Dr. Oppenheimer in his capacity as chairman of the visiting committee about the question of Professor Wendell Furry?
A. Yes, sir. We had numerous discussions. For background I should add that our department had the misfortune of having one of the more famous of the cases in one of the congressional investigations, namely, a member of the physics department at Harvard, Professor Wendell Furry, in some early hearings of the congressional committees, using the Fifth Amendment. He is no longer using the Fifth Amendment. He did in the early hearings. His first use was without consultation with anyone. In fact, his lawyer said don't discuss this case with anyone. They don't have immunity privileges. He is on his own, I am afraid, on this kind of a matter, not too bright a fellow. He thought he should use the Fifth Amendment which I personally greatly regret.

After this was done we had extensive conversations with several members of our visiting committee, particularly Oppenheimer as chairman. Oppenheimer very vigorously deplored to both some of us in the department and also to Furry himself the unwisdom of Furry's choice, and even the wrongness of Furry's choice in using the Fifth Amendment.

He also during the course of this expressed rather strong feelings about the fact that Furry had been for really a fantastically long time a member of the Communist Party.

I must admit that during these discussions which were quite extensive, the kind in which we each shared views, to the best of my knowledge Oppenheimer's views and my views, completely independently arrived at, we each had those views at the time we first got together, were essentially identical.

Dr. EVANS. Did you suspect Furry of being a Communist before that time?

The WITNESS. I actually did not know Furry during the period he was a Communist. He was out of the Communist Party when I first met him. I certainly was not too surprised he was. Even in the first 2 years I knew him—he has changed quite markedly—even those views were a little bit wild in my opinion. I did not know and neither did other members of the department know that he had actually been a member of the Communist Party.

By Mr. GARRISON:

Q. I think he had been a graduate student at Berkeley?
A. He had, but I believe I am correct in saying he had not been a member of the Communist Party at that time. I believe he joined only after he came to Harvard.
Q. You were a consultant on Project Lincoln?
A. Yes, sir.
Q. Did you have occasion in that capacity——
A. There were several meetings. Actually I was a consultant in a sense that did not work very hard on the project. I was chiefly called in on various policy discussions. * * * I was consultant of this and chiefly sat in an various meetings at intervals discussing policy.
Dr. Oppenheimer had been on the summer study group there which group I was not a member of but which came out with I think some very important suggestions for the defense of the United States, * * *.
In the same policy discussions we certainly discussed these to a fair extent. Throughout these again I had reaffirmed what I had known all along, the deep feeling of loyalty and of concern which Dr. Oppenheimer felt for the United States and very clearly that the thing of which he was afraid, the country of which he was afraid, was Russia.
It was just as much as in the Pentagon Building. It was a case a Russian bomber can take off from here and get through. It was not any sort of saying, "Well, now, we better not consider the Russians to be our potential enemy."
Mr. GARRISON. That is all.

CROSS-EXAMINATION

By Mr. ROBB:

Q. Doctor, when did you first learn that you were going to be a witness here?
A. I first learned that I was to be a witness, I would say—it is hard to say—roughly 3 weeks ago. I had heard of the charges—not of the charges—I had heard that Dr. Oppenheimer's clearance had been suspended prior to that time. I heard about that officially through the Air Forces in conjunction with my work in the Scientific Advisory Board.
Q. How did you learn you were going to be a witness?
A. I learned by phone call from Mr. Garrison asking for an appointment, which I admit I had no idea and we had the appointment. I can look up the exact date in my calendar if it is important.
Q. It is not important. Did you discuss the matter of testifying with your superiors?
A. No, sir. Universities operate in funny ways. I don't think we have particular superiors in this kind of matter.
Q. Did you tell anybody in the department?
A. I only told the chairman of my department as I was leaving to come here.
Q. Who is that?
A. Professor Kenneth Bainbridge, who incidentally was the scientist in charge of the first atom bomb tests in New Meximo.
Q. You mentioned Dr. Furry, is it?
A. Yes; that is right.
Q. He was at Harvard for some time?
A. I think he came to Harvard—the two dates I will get mixed—I would say he came to Harvard in 1936, and joined the Communist Party in 1938. No, he would not have joined in 1940. He came in about 1936.
Q. When did you know him?
A. I may have met him, it is one of these things you can't be sure when you meet a person, I met him during the war at a Physical Society meeting but my first knowledge of meeting him to attach a name to him and know the man was when I arrived at Harvard in the fall of 1947.
Q. And you knew him from then as an associate?
A. I knew him as an associate and very well.

Q. As a colleague?
A. A colleague, that is right.
Q. Did you suspect that he either was or had been a Communist?
A. If there had been any member of our department who would have been, he would have certainly been the one. I must admit that it seemed to me somewhat in some of our political arguments in my opinion he is not terribly sound on them. I would like to get in the record I am a very strong opponent of the Communists and have been.
Q. I gathered that.
A. On the other hand, Furry is being confronted with a real tough problem. He has completely changed. Of this I know. He is also now an opponent of the Communists.
Q. I see in Dr. Oppenheimer's list of publications on his PSQ a lot of publications.
Mr. GRAY. Perhaps you better identify for the record what a PSQ is.
Mr. ROBB. Personnel security questionnaire. There are a lot of articles and things.

By Mr. ROBB:
Q. I see one here on the Theory of Electron and Positive, W. W. Furry, Phys. Rev. 45, 245–262, February 15, 1934. Also Phys. 45, 34–43, 34–44, March 1, 1934. Would that W. Furry be Wendell?
A. This is the same Furry. I should add one thing on the basis of sworn testimony on several committees from Furry, he was not a member of the Communist Party at that time, and was not a member until 4 years subsequent to that time. He joined in 1938. This is in the testimony of the McCarthy hearing in Boston.
It is also in Furry's testimony to the Harvard Corporation which was investigating his case.
Q. Do you recall whether he said where he was when he joined the party?
A. He said he was at Harvard. I know which came first, but I don't know the time sequence. I am sure it was in 1938 he joined actually.
Dr. EVANS. Did you have a communistic cell at Harvard?
The WITNESS. According to the testimony of practically everyone who was in it there was a group of, I guess, about ten or so people in the period of around 1938, chiefly, who were indeed members of the Communist Party. There has been quite a lot of testimony about that group, sir, and by people all of whom were away out of the Communist Party at the present time, and it indeed emphasizes the point there are all sorts of ways of being Communist. This was a high and idealistic group of people, completely foolish in my opinion, naive and stupid, to have gotten into it, but nevertheless, they were a very highminded group which by the sworn testimony of all concerned, if anyone had ever approached them and asked them to do anything even remotely treasonable, they would not only have refused to do it, but they would have after a certain degree of soul search, would have felt obligated to reporti t at that time. There are just many ways of being foolish.

By Mr. ROBB:
Q. You were at Los Alamos from 1943 to 1945?
A. Yes, sir.
Q. Was that a pretty closely knit group down there?
A. Yes, sir.
Q. I suppose among the physicists——
A. A very fine group, too, I should say.
Q. And among the physicists everybody knew everybody else pretty well?
A. Fairly well, although as the lab got bigger, there were a number whom you certainly did not know. I will name one offhand I did not know although you subsequently get the impression that this is the most important scientist we had, and this is Fuchs. To the best of my knowledge he was never at the lab. I had never seen him.
Q. Did you know a man by the name of Philip Morrison?
A. Yes; I did.
Q. How well did you know him?
A. I would say only moderately. He was not in my group. On the other hand, he worked quite closely with us at times. Incidentally, to the best of my knowledge, he did a very good job there. Incidentally, he is at the present moment a professor at Cornell University.
Q. Did you see any indication of Communist leanings on his part?

A. Yes; I would say not necessarily at that time. There were many subjects which we would argue and I would disagree. But they were friendly disagreements. He thought I was a little naive and I thought he was a little naive.

Q. When did you discern indications of Communist tendencies on his part?

A. I don't know. I think I probably always considered him leftist and I certainly never knew he was more than that. I might add by reputation even before I met Dr. Oppenheimer, he had the reputation of being leftish. I certainly never heard anyone say he had been a member of the Communist Party. I think the same is true of Morrison.

Q. Did you know Charlotte Serber at Los Alamos?

A. Yes; I know her.

Q. What can you tell us about her Communist tendencies or otherwise?

A. I must admit on that I did not know that she had them. There is a certain mannerism. Sometimes she had a characteristic of, or, maybe a little intellectual snobbery at intervals, which I think some people have had, which incidentally she has gotten completely over subsequently. I think there is nothing in the political discussions that would have implied it. Actually I got to know her better since the war than I did at Los Alamos so we lived more closely together then. I have seen her as recently as a month ago.

Q. Where are they now?

A. Professor Serber is a professor of theoretical physics at Columbia University. He is also a consultant at the Brookhaven Laboratory for the Atomic Energy and presumably thereby cleared.

Q. What about David and Frances Hawkins, did you know them at Los Alamos?

A. I knew them. Again they were not among my intimate friends, but we knew them. They seemed to be doing a good job, or he did. Actually she I can place, and this is about all and I certainly—actually I had—I would not have suspected—I was quite surprised when I learned Dave had been a member of the Communist Party. In the case of Morrison, I had more political discussions. I knew we disagreed more on things than with Dave. Actually it quite startled me in his case. I don't think Dave and I ever had a political argument.

Q. What was his job down there?

A. He was in an administrative position. Here I better make sure I am truthful on which my memory is a little vague. I think partly vague because of this peculiar arrangement I had. I was there as a consultant to the Secretary of War, and I did not go through the personnel channels. It is my impression he had dominantly to do with personnel problems and sort of administrative help and this kind of thing. He may have had to do with housing, though I don't know.

Q. What about his wife. Did you know her?

A. Very little only. I would recognize her if I see her. That is about all.

Q. Did you know a man there named Robert Davis?

A. Yes; I did.

Q. What was his job?

A. Again he was in more the administrative. Later I knew his job best near the end of the war, when he was indeed writing up something of the history or something of this kind of the project.

Q. Was that Davis or Hawkins?

A. I would have said Davis had something to do with that.

Q. Maybe he did.

A. Maybe I should appeal to higher authority. I am a little vague on that. I might add on this it was felt that our end of the project was too secret and it never got written up. I think I do know what Davis is doing now. Hawkins was probably on the history. I would say that Davis was concerned, subject to correction later, with editing a series of books on the technical projects developed in the lab, the kind of thing that was to be published openly subsequent to the war. It was perfectly clear that my end of the work was never going to be published which it never has been and I had very little to do with it.

Q. Did you come in contact with Davis very much down there?

A. I would say a reasonable amount at the end. We were not particularly compatible people, not particularly incompatible.

Q. Did you see any indication on his part of Communist tendencies?

A. Not of Communist tendencies, of a slight glumness at intervals.

Q. Slight what?

A. Glumness. Perhaps an undue reserve. I don't know if this has to be a Communist tendency. I didn't see anything. That is true of all concerned.

Q. You never suspected him of being a Communist?
A. I would never suspect. This is true of Morrison. He was more left in his political views than I, but I would not suspect him to be a member of the Communist Party.
Q. You would not have suspected that Hawkins was either, would you?
A. No; that is right.
Q. Did you know a woman down there named Shirley Barnett?
A. Yes.
Q. Who is she?
A. She was the wife of the medical doctor. He was our pediatrician.
Q. Did she have a job there?
A. She may have. There was a period of time when it was felt for economy of housing the wives were urged vigorously to take jobs within the technical area. It was later realized in part that this was not as good economy as we thought because the husband then at intervals had to wash the dishes, so the wife could do less important work. I think for a period of time she probably was employed.
Q. I don't expect you to remember all these things.
A. I will do my best.
Q. Do you recall at one time she was one of Dr. Oppenheimer's secretaries?
A. That may be. Pricilla Duffield was the principal of Dr. Oppenheimer's secretaries. She was the one to whom we always went. It may very well be she was.
Q. Did you know Shirley Barnett well?
A. Moderately; the best good summary is that we probably spent a total of 4 hours or 5 hours in conversation. You get to know a person fairly well, but you don't get to know everything.
Q. Did you ever see any indication of Communist tendencies on her part?
A. No; there was no chance for a conversation to get that far. She is not one who—some people you get to know well enough you can do it—in Oppenheimer's case, I would know it much better. None of these people did I know as nearly as well as I knew Dr. Oppenheimer or Wendell Furry.
Q. Did you know Dr. Oppenheimer's brother Frank at all?
A. Yes; I did.
Q. How did you know him?
A. He was an employee at Los Alamos and an assistant to Dr. Bainbridge.
Q. Did you know that Frank had ever been a Communist?
A. Only after I read it in the newspapers.
Q. Were you surprised when you heard that?
A. Yes, although—yes, I was certainly surprised by this. There were probably other people at the lab I might have been more surprised about, including myself.
Q. Did you know Mrs. J. Robert Oppenheimer?
A. Mrs. J. Robert Oppenheimer? Yes, though not too well.
Q. Did you know she had ever been a Communist?
A. No, sir. Well, I did not know at Los Alamos. I was indeed told by Oppenheimer himself, in fact in conjunction with the discussions pertaining to Furry a year or so ago, that she had been a member of the Communist Party.
Q. Were you surprised when you heard that?
A. Well, I mean there is a surprise in each direction. It is quite conceivable; on the other hand I had no reason to anticipate it, and since the number is small, I would say yes, I was generally surprised.
Q. Did you know Mrs. Frank Oppenheimer, whose name was Jackie?
A. I know her chiefly by name. I did not know her well; no, sir.
Q. Did you know some people down there named Woodward?
A. Woodward?
Q. Yes.
A. Not at Los Alamos or not well enough to be sure.
Q. Doctor, I wonder if you can help us a little bit. You said that you were a consultant or advisor to the Air Force in connection with an atomic matter.
A. I am a member of the Air Force Scientific Advisory Board. I am on the Armament Panel. * * *
Q. How long have you been doing that for the Air Force?
A. I have been doing that for the Air Force I would say since about 1946, practically since the end of the war.
Q. Doctor, could you tell us in 1949 there was a lot of discussion about whether we would try the thermonuclear or whether we would not; what was the position of the Air Force on this?

A. Our panel was consulted on it officially. On the other hand, this was one on which we were given more information because of the relationship to ourselves, the official advising group for the Air Force, technical people within the Air Force doing it. In general, certainly as the briefings were presented to us of what was then available from the Air Force point of view, the delivery point of view and what kind of Air Force could be useful, it was a pretty dismal proposition. * * * This looked like a long time proposition.

Q. Did the Air Force want the thermonuclear weapon?

A. There were different people within it, and we saw the men who briefed us, and they were of both opinions. It is my impression that the Air Force official policy was yes, * * *.

Q. I am just asking for information because it had not been clear to me.

A. Particularly within the working groups of the Air Force with which we operated, * * *.

I would not be surprised the same way I divide it within myself, sort of 55 percent probably for and 45 percent against.

Q. Did the Air Force finally take an official position as to whether they wanted the weapon or whether they didn't?

A. This I cannot comment on. It was never referred to us. If I knew, I don't remember. Eventually they have. They have a position now very strongly. They very much want it now. This has been in our discussion. At what year and at what time they decided they wanted it, I am completely unclear.

Q. Was there some debate, Doctor, about a strategic Air Force against a so-called Maginot Line defense that you had anything to do with?

A. I had problems to do with the Air Force since about, since I went to the Radiation Lab in 1941. Ever since that time there has been a very vigorous debate about strategic bombing versus tactical versus air defense. This is a very real problem the Air Force has to face. How does it distribute its funds. Within the Air Force there is at all times a considerable amount of dissention on the matter, ranging from the Strategic Air Command—each group essentially saying it has the important thing. * * *

Q. Just so the record will be clear, Doctor, when you speak of a strategic air force——

A. We all agreed you need a strategic air force. Then it is essentially a matter of how you cut a pie. Do you put practically everything in the strategic air force with only a token air defense? Do you put an equal distribution or how do you do it? I think most people will agree you need to have a large and strong strategic air force. On the other hand, there are tactical problems.

Mr. GARRISON. Mr. Chairman, I don't want to shut off discussion, and this is all very interesting, but is it relevant to the problem before the board? I ask this question only in the interest of time, because we have two more witnesses waiting.

Mr. ROBB. I thought it was, Mr. Chairman, or I would not have gone into it. I think there has been something said in the record by Dr. Oppenheimer, something about project Vista. Didn't it have to do with that?

The WITNESS. Project Vista had to do with essentially the ground forces, not the Air Force. Essentially the problem of project Vista was given at Korea. How do you do something about it. This was very closely related also to the Air Force. It was a joint project supported by the Air Force, as well as the ground forces. * * *

By Mr. ROBB:

Q. Project Lincoln.

A. That was another aspect of the same thing. It is an air defense problem, continental defense. * * *

Q. Doctor, just so the record may be clear, may I ask you this question: When you speak of a strategic air force, what is meant is not a striking force as distinguished from a defensive force?

A. Well, no. It is a striking force in general to strike rather deep. A tactical air force is the one that strikes near the front lines of combat. The strategic one is the one that bombs the cities and bombs the industrial sources. They get confused. In the heat of battle they throw everything wherever it is most needed.

Q. Doctor, to pull this in briefly, do you know what Dr. Oppenheimer's position was on these questions?

A. I believe I know. I have had a number of discussions with him on it. I think I know fairly closely. This was the belief as mine that you need all, you need a balanced force, not exclusively or too overwhelmingly one. You need a very strong strategic air command. I believe, however, he felt that too large a fraction of the Air Force's moneys were going to that compared to the very small amount that was going to the problem of air defense. I must admit I agree with him. I am not sure that we would necessarily agree as to how much connection needs to be made. He may want to do it more or less than I.

On this I am in complete agreement and so are many members of our advisory committee board.

Q. In other words, the scientists tend to favor rather the continental defense theory, is that it?

A. No; I would say they favor the balanced force theory which many people in the military also favor, * * * I don't know of any scientist concerned with military things who thinks that we should drop the strategic air force. Almost all I know and it is my impression that Dr. Oppenheimer would also argue that it should be the biggest part of the air force, but not the whole thing.

Q. I am cautioned that I should avoid getting into classified material on that matter.

A. I think what we have said so far is all right, but we are getting close, I agree.

Q. Did Dr. Oppenheimer have any part in that?

A. Yes, sir. * * *

Q. Can you without getting into classified material give us Dr. Oppenheimer's position?

A. I think approximately. I think his position was that the defense of the country as well as its ability to retaliate was a very important thing, which was being underdeveloped. * * *

In order to strengthen our country, we needed to put more support behind this. I might add that this is now to the best of my knowledge part of the official policy of the United States.

Q. Was there more than one technique without getting into classified material?

A. There are a number of intermixed techniques in this. You use all. I would say that the most important of the new ideas is the one you referred to and I will avoid having to refer to it myself.

Q. Were there three fundamental techniques, Doctor?

A. The usual thing when you categorize things—if you name them, I will agree with them maybe.

Q. I will ask a question that maybe will kind it up. Was there any technique that Dr. Oppenheimer opposed?

A. I don't know. It is on the record that at least one time he opposed development of an H-bomb.

Q. I am talking about this long range detection?

A. I don't know of any, no, sir. There may be, but I certainly do not know it.

Mr. OPPENHEIMER. I know this is not a classroom, but the counsel and the witness are talking about two quite distinct things and therefore they are not understanding each other.

Mr. ROBB. I realized that, too, on the last question. I don't think the witness understood my question.

By Mr. ROBB:

Q. I was talking about this long range detection matter, Doctor. I asked you whether there was more than one technique for long range detection, and I believe you said there was.

A. Sure.

Q. The question I asked you was there any technique that Dr. Oppenheimer opposed?

A. Not to my knowledge. I thought you meant a nondetection technique.

Q. One further question. Was there a man down at Los Alamos while you were there named David Greenglass?

A. I never met him, but I obviously read about him in the paper. I believe he was a machinist.

Q. You didn't know him?

A. Never saw him.

Mr. ROBB. Thank you very much.

Mr. GRAY. Dr. Ramsey, with respect to the compartmentalization versus noncompartmentalization, I believe you indicated that this was a technique which had been used in some other laboratories, and was found to be useful as far as the expedition of work was concerned at Los Alamos?

The WITNESS. Correct.

Mr. GRAY. Am I right, however, in recalling that you said that you were in a compartmentalized area?
The WITNESS. I would say semicompartmentalized.
Mr. GRAY. Because of the extreme secrecy?
The WITNESS. And also from the lack of necessity of knowledge of technical development. The point of view that certainly most of us adopted was in the best interests of the country, what will speed things versus what will risk security. In my own group there wasn't much advantage to have the interchange that was so necessary to the development in the rest of the group, and there was also this particular secret aspect that my group indicated how far we were coming along.
Mr. GRAY. So in the absence of the desirability on the ground of expedition of the work, compartmentalization was a security measure which was adhered to?
The WITNESS. Yes; I incidentally believe that what was done on the compartmentalization there was very good indeed, and the noncompartmentalization. I think it would have been vastly later had it not been for that.
Mr. GRAY. One other question about Los Alamos. You were not allowed to leave the premises without permission, is that correct?
The WITNESS. This varied a little from time to time. We always had to show passes at the gate.
Mr. GRAY. No.
The WITNESS. For any extensive visit you had. I think you could go to Santa Fe to do shopping without higher authority.
Mr. GRAY. Who was in charge of that?
The WITNESS. We showed our passes to the guard at the gate. I would say probably Colonel de Silva.
Mr. GRAY. It would be the security people.
The WITNESS. Yes; it would be the security people.
Mr. GRAY. On your formula 5545, had you served on a committee or in some other capacity at that time and in such capacity been required to vote on the crash program, I assume that the 55 percent——
The WITNESS. That is correct.
Mr. GRAY. There comes a time when a man——
The WITNESS. Has to make a decision, that is correct. One important argument might have reversed the 55 the other way. I would have to face that. That is correct I would have voted in that time in favor of it.
Mr. GRAY. You pretty well knew the various arguments?
The WITNESS. I think I knew most of them. I did not know all of them. I certainly respected those people. There were many who disagreed with me.
Mr. GRAY. Yes; I understand that. Just in the interest of my understanding the record, in talking about Dr. Furry, you said he could not have joined the Communist Party in 1940. What did you mean by that?
The WITNESS. I can tell you what I mean. I realized this when I said 1940. This was the time of the Nazi-Soviet agreement, and I do know also from the testimony that he almost got out at that time. Actually he didn't get out at that time. But he almost did. Essentially by that argument I am saying that I think it would have been very unlikely that would have been the moment at which he initiated the move of getting in. It is because also I remember he had been in before that period.
Mr. GRAY. Yes. I was trying to get that clear. Whether you are saying that it could not have been 1940 had to do with your recollection or had to do with an international situation.
The WITNESS. I would say it actually had to do with both. I think it was dominantly recollection. As I started to say this, I remember the 1938 date. But what I know of him I think this would not have been the date he would have chosen. It is the period of the collapse of France and the Nazi-Soviet Pact. I am sure he would not have chosen that as joining. He was very upset about it, and in fact dropped going from all meetings.
Mr. GRAY. You said he almost resigned.
The WITNESS. Yes. As a matter of fact, if it were not that he moved so slowly—it took him about a year to make up his mind to drop out by which time Russia was an ally.
Mr. GRAY. There have been a lot of allegations about the fact that people at Harvard and other institutions have been involved—I don't mean to single out Harvard—but they have been.
The WITNESS. That is correct.
Mr. GRAY. Of course, Dr. Furry's name has appeared publicly along with three others at the same time.

The WITNESS. There have been a total of three. Actually one of them is no longer teaching at Harvard. He was on a temporary appointment. One has an appointment terminating this year. Furry is the only permanent member of the tenure appointment in the Harvard faculty for which this is true.

Mr. GRAY. Were these others known to you?

The WITNESS. No; I never met any of them. Incidentally, Kaneman, our other most conspicuous case, Furry has never met him. I am sorry they saw each other at a hearing.

Mr. ROBB. Is that Martin Kaneman?

The WITNESS. It is a good question. I think it is Leon. I am quite sure.

Dr. EVANS. Dr. Ramsey, would you tell us about your undergraduate and graduate education and where you had them.

The WITNESS. Yes, sir. I received my bachelor's degree from Columbia University. I was given a traveling fellowship by Columbia University to go Cambridge University where I did the peculiar thing—the universities are different—I received another bachelor's degree from Cambridge University, subsequently a master's degree. I came back and got my Ph. D. degree from Columbia.

Dr. EVANS. Did you meet a Bernie Peters down there at Los Alamos?

The WITNESS. I certainly didn't meet him at Los Alamos. I met him at Rochester subsequently, and I didn't realize he had been at Los Alamos.

Dr. EVANS. Did you meet Lomanitz down there?

The WITNESS. No, sir.

Dr. EVANS. Rossi?

The WITNESS. No, sir.

Dr. EVANS. Did you meet Weinberg down there?

The WITNESS. At Los Alamos?

Dr. EVANS. He was at Berkeley.

The WITNESS. I think he was at Berkeley. As a matter of fact, I never met Weinberg.

Dr. EVANS. Did you meet Mr. Flanders down there?

The WITNESS. Yes; he was a mathematician.

Dr. EVANS. Yes; he was an electronic mathematician.

The WITNESS. He was in the computing. It was mathematics at first. It gradually developed into electronics.

Dr. EVANS. Did he have his beard?

The WITNESS. He had his beard, and it startled the security guards no end.

Dr. EVANS. You say you knew Fuchs?

The WITNESS. Fuchs, under sworn testimony I would have to say to the best of my knowledge I have never seen the man, and I couldn't even prove he was ever at Los Alamos.

Mr. GRAY. Forgive me for reminding you, that you are giving sworn testimony.

The WITNESS. That is correct. I was about to say if I were, and realized that I am.

Dr. EVANS. Some of these people that you knew down there in this cell at Harvard, a number turned out later to be Communists.

The WITNESS. Yes; actually the only member of the group at Harvard that I ever met was Furry. This was subsequent to his membership.

Dr. EVANS. You knew Hawkins, you said.

The WITNESS. Yes. I am sorry. At Los Alamos I knew the people I have enumerated, including Hawkins.

Dr. EVANS. From what you know now, and thinking back, would you think you are a very good judge as to whether a man is a Communist or not?

The WITNESS. I would say yes; I think on the following, I mean since you were not trying to judge, you can guess some people might be and some were not. I don't think you can explicitly with someone you don't know terribly well as with all the ones I have enumerated, my conversation runs to maybe a total of 4 or 5 hours, I certainly would have had no claim with anyones enumerated would I ever have felt in a position of saying they weren't. I would not have been in a position to claim they were or were not. Simply I didn't know them well enough. I don't think ability to judge enters there. A person whom I never met I can't say anything. A person whom I met only casually, chiefly to talk about the physics problems, is no way to judge.

Dr. EVANS. That is all.

Mr. GRAY. Mr. Garrison?

Mr. GARRISON. No.

Mr. GRAY. Thank you very much.

The WITNESS. Thank you. Sorry to have taken so much of your time.
(Witness excused.)
(Brief recess.)
Mr. GRAY. Dr. Rabi, do you wish to testify under oath?
Dr. RABI. Certainly.
Mr. GRAY. Would you be good enough to raise your right hand. I must ask for your full name.
Dr. RABI. Isadore Isaac Rabi.
Mr. GRAY. Isadore Isaac Rabi, do you swear that the testimony you are to give the board shall be the truth, the whole truth, and nothing but the truth, so help you God?
Dr. RABI. I do.
Whereupon, Isadore Isaac Rabi was called as a witness, and having been first duly sworn, was examined and testified as follows:
Mr. GRAY. Would you be seated, please, sir?
I must remind you of the existence of the perjury statutes. I am prepared to give you a description of the penalties if you wish, but may I assume you are generally familiar with the perjury statutes?
The WITNESS. I know that they are dire.
Mr. GRAY. I would also ask, Dr. Rabi, that you notify me in advance about the possible discussion or disclosure of any restricted data which you may get into or find necessary to get into your testimony.
The WITNESS. I hope to have the help of Dr. Beckerley on that.
Mr. GRAY. He is here and I am sure will be alert.
The WITNESS. I am confused about what has been declassified that I want technical professional help.
Mr. GRAY. Finally, I should point out to you that we regard the proceedings of this board as a matter confidential in nature between the Commission and its officials and Dr. Oppenheimer, his representatives and witnesses. The Commission will make no public release of matters pertaining to these proceedings, and on behalf of the board, I make it a custom to express to the witnesses the hope that they may take the same attitude.
The WITNESS. Yes, sir.

DIRECT EXAMINATION

By Mr. MARKS:
Q. Dr. Rabi, what is your present occupation?
A. I am the Higgins professor of physics at Columbia University.
Q. What official positions do you have with the Government?
A. Let me see if I can add them all up.
Q. Just the most important.
A. At present as chairman of the General Advisory Committee, as successor to Dr. Oppenheimer. I am a member of the Scientific Advisory Committee to ODM, which also is supposed to in some way advise the President of the United States.
I am a member of the Scientific Advisory Committee to the Ballistics Research Laboratory at Aberdeen Proving Ground. I am a member of the board of trustees of Associated Universities, Inc., which is responsible for the running of Brookhaven Laboratory. I am a consultant to the Brookhaven National Laboratory.
I was a member of the project East River, but that is over. I was at one time the chairman of the Scientific Advisory Committee to the policy board of the joint research and development board, and a consultant there for a number of years. I am a consultant to project Lincoln.
That is about all I can remember at the moment.
Q. That is enough. Speaking roughly, how much of your time do you devote to this official work?
A. I added up what it amounted to last year, and it amounted to something like 120 working days. So you might ask what time do you spend at Columbia.
Q. How long have you been a member of the General Advisory Committee?
A. Since its inception. I don't remember the exact date of my appointment but I have been to every meeting. I may have missed one since the first.
Q. When did you become chairman?
A. I became acting chairman when Dr. Oppenheimer's term was out. By our own custom the chairman is elected at the first meeting of the calendar year, and I was elected chairman by the committee at the first meeting which I think was in January of last year. I am not sure of the date of the meeting.

Q. Dr. Rabi, to what extent has your work as consultant in various capacities in the Government overlapped or coincided with work that Dr. Oppenheimer was performing at the same time and in the same general field?

A. Chiefly of course the General Advisory Committee and also to a degree in project Lincoln, and particularly the summer study of, I believe, 1952.

Q. Summer study where?

A. This was a summer study at Cambridge on the question of continental defense of the United States.

Q. How long have you known Dr. Oppenheimer?

A. I think we first met in the end of 1928 and we got to know one another well in the winter and spring of 1929. I have known him on and off since. We got together very frequently during the war years and since.

Q. Do you know him intimately?

A. I think so, whatever the term may mean. I think I know him quite well.

Q. Dr. Rabi, if you will indulge me I would like to skip around somewhat because as nearly as possible I would like to avoid too much repetition of things that have already been gone into by others.

Will you describe the extent that you can what took place in the fall of 1949 insofar as the GAC was concerned or you are concerned in respect of the question of thermonuclear program for the Atomic Energy Commission?

A. I can only give my own view and my own recollection. I have not prepared myself for this by studying the minutes. I intended to, but I am on in the morning rather than the afternoon. So I can give you just my own recollection.

The thermonuclear reaction or as it was called the super was under intense study from my very first contact with Los Alamos.

Q. When was that?

A. About April 15, 1943. At the establishment of the laboratory, Dr. Oppenheimer called together a group of people to discuss the policy and technical direction of the laboratory, and I was one of those who was invited to that discussion. All through the war years and following that, that was a subject of discussion and consideration by some of the very best minds in physics.

The problem proved to be an extremely difficult, very recalcitrant problem, because of the many factors which were involved where the theory, the understanding of the thing, was inadequate. It was just a borderline. The more one looked at it, the tougher it looked.

Following announcement of the Russian explosion of the A bomb, I felt that somehow or other some answer must be made in some form to this to regain the lead which we had. There were two directions in which one could look; either the realization of the super or an intensification of the effort on fission weapons to make very large ones, small ones, and so on, to get a large variety and very great military flexibility.

Furthermore, a large number, a large increase in the production of the necessary raw materials, the fissionable materials and so on, or one could consider both. There was a real question there where the weight of the effort should lie.

Q. When would you say that this question that you are now describing began o become acute in your thinking?

A. Right away.

Q. You mean with the Russian explosion?

A. As soon as I heard of the Russion explosion. I discussed it with some colleagues. I know I discussed it with Dr. Ernest Lawrence, with Luis Alvarez, and of course with the chairman of our committee, Dr. Oppenheimer. In fact, I discussed it with anybody who was cleared to discuss such matters, because it was a very, very serious problem.

That question then came up at the meeting of the General Advisory Committee.

Q. That would have been the meeting that began on October 29, 1949?

A. Yes. I do not recollect now whether this was the first meeting after the announcement of the Russian explosion or whether there was an intervening meeting.

Q. To refresh your recollection, Dr. Rabi, I think it has been in the record here that there was a regular meeting of the General Advisory Committee just after or just at the time when the Russian explosion was being evaluated.

A. Yes. I recollect now. In fact, I was coming up on the airplane and there was Dr. Cockroft, the director of Harwell—he didn't tell me what it was—but he said you will read something very interesting in the newspaper.

Q. You were coming on the airplane from where?

A. From New York to Washington on the airplane. I ran into Dr. Cockroft, and he told me I would read something very interesting in the noon paper. When I stepped off the plane there was the *Star* with this announcement.

Q. This meeting which you identified was more or less contemporaneous by the official announcement of this Government that there had been a Russian explosion, was there any discussion at that time of the thermonuclear?

A. I would have to refresh my memory on that. I can not say. I would be astonished if there were not. I cannot say. I could go back and look. In fact, we talked about it at every meeting.

Q. In all events, the interval between that meeting and the one on the 29th, was very much on your mind?

Yes, sir.

Q. Do you have any recollection or impression as to the form in which the question of what to do about the thermonuclear problem came up in your meeting that began on October 29?

A. The way I recollect it now, without perusal of the minutes—in fact, I think we kept no minutes of that meeting which is somewhat unfortunate under the present circumstances—the way I recollect ——

Q. Do you know why no minutes were kept?

A. Because the discussion ranged so very widely. We were concerned during that period, as I remember and we consulted with the Joint Chiefs of Staff, we consulted with representatives of the State Department and a whole lot of stuff was there which we didn't feel should be distributed around. We decided not to keep accurate minutes of the meeting.

What was the question again.

Q. I asked you whether you had any recollection or any impression as to the form in which the question concerning the thermonuclear problem came before you, that is, the GAC, at the meeting which began on October 29, 1949.

A. As I recollect it now—it is 5 years ago—the chairman, Dr. Oppenheimer, started very solemnly and as I recall we had to consider this question. The question came not whether we should make a thermonuclear weapon, but whether there should be a crash program. There were some people, and I myself was of that opinion for a time, who thought that the concentration on the crash program to go ahead with this was the answer to the Russian thermonuclear weapon. The question was, should it be a crash program and a technical question: What possibilities lay in that? What would be the cost of initiating a crash program in terms of the strength of the United States because of the weakening of the effort on which something which we had in hand, namely, the fission weapons, and the uncompleted designs of different varieties, to have a really flexible weapon, the question of interchangeability of parts, all sorts of things which could be used in different military circumstances.

Then there was the question of the military value of this weapon. One of the things which we talked about a great deal was that this weapon as promised which didn't exist and which we didn't know how to make, what sort of military weapon was it anyway? What sort of target it was good for. And what would be the general political effect.

In other words, we felt—and I am talking chiefly about myself—that this was not just a weapon. But by its very nature, if you attacked a target, it took in very much more. We felt it was really essential and we discussed a great deal what were you buying if you got this thing. That was the general nature of the discussion.

Technical, military, and the combination of military political.

Q. Dr. Rabi, if in the state of mind that you have described the question among others had been put to you by the Commission or its chairman to consider an appraisal of the then program of the Atomic Energy Commission of whether it was adequate and if not, what to do about it, what you would have considered a question in those general terms embraced.

A. Are you referring specifically to the thermonuclear weapon or to the whole program?

Q. I am referring to anything that you think of. Would that have embraced the thermonuclear?

A. The thermonuclear weapon at Los Alamos went through ups and downs. We spent a lot of time talking of how we could get some very good theoretical physicists to go to Los Alamos and strengthen that effort. We thought at times of the effort as being such a distant thing that working on that kind of research because it was a distant thing and new ideas would evolve and would really act as a ferment and sort of spark the laboratory. It was one of those things where you really didn't know how to find a way. Where experiments were really difficult to make and tremendously expensive.

With the ideas in hand it was very hard to know how to go at this thing, even how to set up a crash program. But what we were concerned about on the other hand, we felt that there was a very great inadequacy in the Commission's program with respect to the production process, the amounts of fissionable material, and the amounts of raw material which were being produced, that we were not spending enough money on that.

We felt almost from the very beginning of an increase in Hanford. We made a technical recommendation at the time of how more could be gotten out of Hanford. About hastening the construction of certain chemical plants for the purification of the material. It was our feeling that the resultant controversy when the President ordered Savannah River that the whole controversy was worth the thing.

Q. You are getting ahead of me.
A. You asked such a broad question.
Q. I am losing track of this. Just once more, to search your memory, and if you haven't got any, all you have to do is say so—search your memory as to the form in which, the nature of the circumstances in which there was before the General Advisory Committee in the capacity as such at the October 29——
Mr. ROBB. 1949 meeting.
Mr. MARKS. I am sorry.

By Mr. MARKS:

Q. At the October 29, 1949, meeting. The sense that you were appropriately considering the question of a crash program for the super. If you haven't got any memory, say so.
A. The sense of whether we were considering a crash program for the super?
Q. Do you have any memory as to how that question was before you? Among lawyers we say how did the question come up in the case.
A. You mean in detail how it came up? You mean who said what to whom, when? That I don't remember. I am sure it was before us.
Q. You don't know who presented it?
A. How it was presented, whether it was first presented with our preliminary meeting with the Commissioners, whether it was first suggested by Dr. Oppenheimer, and then confirmed in the preliminary meeting with the Commissioners, and so on. I really don't remember. At other meetings we have minutes and all this would have been spelled out.
Q. To the extent that you can tell it without getting into any classified material, what was the outcome of the GAC meeting of October 1949?
A. I will try to give it as best I can.
Q. Let me break it down. First, is it fair to say that the committee was in agreement with respect or essentially in agreement with respect to the technical factors involved in the thermonuclear situation?
A. It was hard to say whether there was an agreement or not because what we are talking about was such a vague thing, this object, that I think different people had different thoughts about it. You could just give a sort of horseback thing and say, maybe something would come out in 5 years. It is that sort of thing. I know in my own case I think I took the dimmest technical view of this, and there are others who were more optimistic.
Q. I think it has been indicated here that there was some statement in the report of the GAC at that time to the effect that it was the opinion that a concerted imaginative effort might produce—that there was a 50–50 chance of success in 5 years.
Mr. ROBB. In the interest of accuracy, I think the report says a better than even chance. Let me check it to make sure.
Mr. GARRISON. That is correct.

By Mr. MARKS:

Q. Was that supposed to be a consensus of the views?
A. More or less. When you are talking about something as vague as this particular thing, you say a 50–50 chance in 5 years, where you don't know the kind of physical factors and theory that goes into the problem. I just want to give my own impression that it was a field where we really did not know what we were talking about, except on the basis of general experience. We didn't even know whether this thing contradicted the laws of physics.
Q. You didn't know what?
A. Whether it contradicted the laws of physics.

Q. In other words, it could have been altogether impossible.
A. It could have been altogether impossible. The thing we were talking about. I want to be specific.
Q. I understand.
A. We were talking within a certain definite framework of ideas.
Q. To the extent that you can describe them now and confining yourself to that meeting, to the extent that you can describe them without trespassing on classified material, what were the recommendations of the GAC?
A. They were complicated. We divided into two groups. No, there were some recommendations to which I think we all agreed, which were specific technical recommendations.
Q. Can you say what they had to do with in general terms?
A. Certain improvements in weapons, the production of certain material which would be of great utility in weapons and which we felt at the time might be fundamental if a super were to be made. We recommended sharply a go-ahead on that. We recommended certain directions of weapons and there was a third important recommendation which I don't recollect now of a technical nature.
Q. You have spoken of a division. What had you reference to there?
A. In addition to that there were supplementary reports on which Dr. Fermi and I formed a minority, and the other six members present the majority. That had to do with this sphere where the political and the military impinge. One group felt—I don't like to speak for them because the record is there, but my impression was—that this projected weapon was just no good as a weapon.
Q. You mean the particular weapon?
A. I am not talking from the technical but the military opinion. That it was not of great military utility. The possible targets were very few in number, and so on. I could elaborate on that if I should be asked, but I am speaking for somebody else, and there is a record.
Q. That was the group with which you did not join?
A. Yes. Of this specific design, Dr. Fermi and I as I recollect it now felt that in the first place as far as we could see from the question of having a deliverable weapon one did not gain a tremendous amount. Secondly, we felt that the whole discussion raised an opportunity for the President of the United States to make some political gesture which would be such that it would strengthen our moral position, should we decide to go ahead with it. That our position should be such that depending on the reaction, we would go ahead or not, whatever going ahead were to mean.
Q. What made you think that it was appropriate for you to speak about these rather nontechnical but more political, diplomatic and military considerations?
A. That is a good question. However, somehow or other we didn't feel it was inappropriate. In our whole dealing with the Commission, we very often, or most often, raised the questions to be discussed. In other words, we would say we want to discuss this and this thing. Would you please provide us with documents, would you bring individuals to talk to us on this, and we would address the Commission on questions.
On the other hand, we didn't feel badly if they didn't act on our suggestions. Sometimes they did and sometimes they didn't. So we did not feel that this was inappropriate. It would be very hard for me to tell you now why we thought it was appropriate, but we thought so.
Q. After this meeting of the GAC, the outcome of which you described——
A. I might add, to add to your feeling on this, the Joint Chiefs consented to come and talk to us, and gentlemen from the State Department came and talked to us. So we did not have the feeling all along that we were going far beyond our terms of reference otherwise these people would not have showed up.
Q. If you can properly so so, Dr. Rabi, to what extent and in what way did the appearance of the Joint Chiefs or their representatives affect the course of your thinking and your expression of view?
A. Oh, dear; that is very hard to remember. I can only talk for myself. I, myself; I don't want to talk for anybody. * * *

By Mr. MARKS:
Q. Did the GAC have any responsibility for seeing to it that the Joint Chiefs were briefed?
A. No. We did meet fairly frequently with the Military Liaison Committee.
Q. Is it fair to say that the GAC tried to keep the Military Liaison Committee fully informed?

A. Our job was not to inform the Military Liaison Committee. Our job was definitely to talk to the AEC and as we interpreted it on the suggestion of the chairman of the AEC at one time, to the President on some very special occasion. We have tried then and since not to be the servant of the MOC or to work directly through them or the joint congressional committee. Our job is to work with the AEC as specified in the law and possibly with the President.

Q. After the President announced the decision to go ahead with the hydrogen bomb in January of 1950, what attitude and what steps, if any, did the GAC take with respect to the subject from then on?

A. I think we started talking about the best ways and means to do it. It was a very difficult question, because here is a statement from the President to do something that nobody knew how to do. This was just a ball of wax. So we were really quite puzzled except insofar as to try to get people to go and look at the problem.

Q. In that connection, did the GAC itself try to look into the problem?

A. Insofar as we could; yes. We had people who were quite expert and actually worked on it, chiefly of course Dr. Fermi, who went back to Los Alamos, summers and so on, and took a lot of time with it. So we had a very important expert right on the committee. Of course, Dr. Oppenheimer knew very well the theoretical questions involved.

Q. Do you think the GAC had any usefulness in helping the work on this particular subject?

A. I think it did; I think it had a great usefulness some way indirect and some way direct, ways of trying to bring out the solid facts. It is awfully hard to get at those facts. I recall particularly one meeting, I think it was in the summer of 1950 at Los Alamos, I am sure of the dates, where we actually got together all the knowledgeable people we could find, I think Dr. Bethe was there and Fermi, to try to produce some kind of record which would tell us where we stood. This was before the Greenhouse test.

Q. You mean what the state of the art was at that time?

A. What the state of the art was, and where do we go from here.

Q. How many of he laws of nature on the subject were available?

A. What ideas and what technical information was available. We got this report and it was circulated by the Commission in various places because there was some kind of feeling that here the President is given the directive and somehow something is going to appear at the other end and it was not appearing.

Q. If you can tell, Dr. Rabi, what was the connection or relation between the meeting you have just described at Los Alamos and another meeting that has been testified here which took place, I believe, in 1951, in the late spring at Princeton?

A. That was an entirely different meeting. At that meeting we really got on the beam, because a new invention had occurred. There we had a situation where you really could talk about it. You knew what to calculate and so on, and you were in the realm where you could apply scientific ideas which were not some extrapolation very far beyond the known. This is something which could be calculated, which could be studied, and was an entirely different thing.

Q. Why did it take that long?

A. Just the human mind.

Q. There was the President's directive in January 1950.

A. Why it took this long? One had to get rid of the ideas that were and are probably no good. In other words, there has been all this newspaper stuff about delay. The subject which we discussed in the 1949 meeting, that particular thing has never been made and probably never will be made, and we still don't know to this day whether something like that will function.

This other thing was something quite different, a much more modest and more definite idea on which one could go.

Q. I interrupted you a while back when you displayed some enthusiasm with the Savannah River project. Would you try to fix in point of time when you intended that expression of enthusiasm?

A. Just as soon as we got some more money to make more plants which would make fissionable material and really here was a policy of containment. * * *

I am quite sure that would have been unanimous in the committee. Also, there were certain technical devices to increase the production and we pressed on that. There was a very long delay just because of conservatism, and a new contractor, and so on, in doing some of those things. But the pressure of the GAC all along——

Q. When you say all along, what do you mean, 1947, 1948, 1949?

A. At almost every meeting.
Q. Through all of those years?
A. That is right. Increased production of both fissionable material and of raw material, and particularly we kept on recommending a facility for the production of neutrons which we knew would be very useful in some way or other without particularly specifying where the use would come.
Q. Was Savannah River regarded by you as one of the great answers to that need which you have just described?
A. Oh, yes; I regard Savannah River as the way we answered the Russian success.
Q. I don't know whether you said earlier what Dr. Oppenheimer's view was about that.
A. I am quite sure that he was never in disagreement with that.
Q. Was never in disagreement?
A. Yes.
Q. Did he evidence your enthusiasm?
A. I think so. He is not the same enthusiastic fellow as I am, but I was quite sure he believed that it was a correct step.
Q. Dr. Rabi, there has been some questioning and some talk by other witnesses about a subject which is somewhat obscure to me, but perhaps if I just identify it, you may be able to say something about it, namely, the question of a so-called second laboratory. Is there anything that you can say properly on that subject?
A. I will try and let Dr. Beckerley watch me on it.
Q. Maybe you better consult with Dr. Beckerley first.
A. That question came up again and again. Los Alamos is an awkward place and so on, and various people kept on saying——
Q. May I interrupt you, Dr. Rabi. When the term "second laboratory" is used, is it fair for me to assume that what is being talked about is the second laboratory which will have something to do primarily with weapons?
A. That is what I am talking about. I am just giving you my recollection of a whole series of discussions which came up from time to time. That competition is good. Los Alamos has been criticized for being too conservative and stodgy. The suggestion that some other group utilizing talent which for some reason or another was unobtainable at Los Alamos would be a good thing.

I, myself, I may say was not in favor of that, and my own reason was—and I think Dr. Oppenheimer shared this reason, at least in part—that Los Alamos was a miracle of a laboratory. If you had looked at the dope sheet of the people that were there, you would not have expected in 1945 that it would be just a tremendously successful laboratory and of such a very high morale. It was really a terrific laboratory, just a miracle of a place.

As a result of establishing another laboratory, I was afraid that it would be taken at Los Alamos as a criticism and taking chances of spoiling morale. Those laboratories, as I think Mr. Morgan will know, largely depend upon the few key people. If you are to lose them, you have lost the lab. So my own feeling was, they are doing remarkably well and why upset the applecart. There was a possibility also that they would lose some personnel in a sort of general division.

Finally it turned out in the expansion of the activities of Los Alamos, these various tests and so on, that they used a lot of the contractors all over the place. They do a tremendous amount of subcontracting all over the place
Q. All over the place?
A. All over the United States. One very good group in instrumentation was developed at Berkeley by Dr. York.

Then there was an additional circumstance that some important contract on a subject which I won't even enter was canceled there, and personnel became available, and I think it was a suggestion of the GAC that that group should be combined and another laboratory made whose chief terms of reference would be in the realm of instrumentation for the study of explosions.

Subsequently, and I think not on the direct recommendation, although I am not sure about the record, but this is my recollection, the terms of reference of that laboratory were expanded, so that it became an actual second weapons laboratory. I think in popular opinion such as Time magazine, and so on, it is that laboratory which produced the thermonuclear weapon. That is a lie.
Mr. GRAY. That is what, Doctor?
The WITNESS. That is a lie.

By Mr. MARKS:

Q. Do you mean by that to say that what has been produced came out of Los Alamos?

A. Yes, sir.

Q. There has been a good deal, I think, of official information about the present strength of the United States in relation to nuclear weapons, fission and fusion. Is that in your opinion the result of work at Los Alamos?

A. Yes, it is my unqualified opinion.

Q. And not the second laboratory?

A. Not the second laboratory. The second laboratory has done very good work on instrumentation.

Q. There has also been some talk as a result of questioning in these proceedings about the question of continental defense. Is there anything that you can say properly about that subject, about your attitude on it, and about Dr. Oppenheimer?

A. I can suggest the motivation and I think Dr. Oppenheimer and I agreed. It is threefold. One, we think that to protect the lives of Americans is worth anybody's while. Two, that one is in a stronger position in a war if one is fighting from a protected citadel, rather than just being open and just a slugging match with no defense guard put up. Thirdly, and it is more political, that the existence of such a defense would make us less liable to intimidation and blackmail.

Behind this were some brand new ideas, at least new to me, which came from some individuals in Cambridge, particularly Dr. Zacharias, which made such a defense line possible at a reasonable cost.

Q. Who is Dr. Zacharias?

A. Dr. Zacharias is a professor of physics at MIT. He is the head of their division of nuclear science. During the war he was at the Radiation Laboratory at MIT on radar. He spent a certain amount of time at Los Alamos. He was the head of the * * * summer study for the Navy. * * *

Q. Are you sure you are not mistaken? * * *

A. Zacharias was the head of the Hartwell study. Then also he ran the summer study.

Q. When you speak of the summer study, you means the one that is popularly called project Lincoln?

A. No, project Lincoln is a big project and laboratory which exists. The summer study was a special group brought together for a limited period of time of experts in different fields to look into the technical military question of the possibilities of the defense of the United States.

Q. Were you and Dr. Oppenheimer concerned at all with that?

A. I think we each spent a week or so at the beginning and a week or so at the end of this. We were not actually members of the working party.

Q. You were consultants?

A. Consultants.

Q. Does the attitude that you have described on the subject of continental defense mean that you are opposed to a powerful strategic air policy?

A. As far as I am concerned, I certainly am not.

Q. Am not what?

A. Opposed to it. I am very much in favor of it. I would like to see it more effective than it is. * * *

Q. Are the two things compatible, the continental defense you are talking about, and the strategic?

A. Absolutely. These are the 2 arms. One is the punching arm and the other the guard. You have to have both, in my opinion.

Q. Do you know whether Dr. Oppenheimer's views are materially different from yours on the subject?

A. I don't think they are. I think his emphasis might be somewhat different. I don't think the views are different. I think the emphasis might be different.

Q. In what way?

A. Now we are getting into things which I would prefer not to answer.

Q. Why?

A. Because it comes into questions of actual strategy and tactics of which we have special knowledge and I don't want to go into any details of that sort.

Q. All right. Just so that I will understand what you are saying, I take it that you strongly favor, and to your knowledge Oppenheimer strongly favors, a powerful strategic air policy.

A. Yes.

Q. And that you also favor an effective continental defense.
A. That is right.
Q. And that you regard the two things as not incompatible?
A. No, no. I think they are just absolutely complementary. They both have to be there. To put it in a word, a strategic air arm unless you are going to prevent a war is a psychological weapon, a deterrent. But the other fellow may not be the same and you have to have some kind of defense before he does you irreparable damage, and furthermore, your plans may not go as you expect. They may miscarry. Unless you have a defense, you are not getting another chance.

Dr. GRAY. Let me interrupt for a moment to ask you how much longer do you think your direct will take?

Mr. MARKS. Just two more questions. If you would rather——

Mr. GRAY. No, proceed.

By Mr. MARKS.

Q. Doctor, it can be gathered from the nature of these proceedings that this board has the function of advising the Commission with respect to a determination that the Commission must make on whether permitting Dr. Oppenheimer to have access to restricted data will not endanger the common defense and security.

In formulating this advice, the considerations suggested by the Atomic Energy Act to be taken into account are the character, associations, and loyalty of the individual concerned.

Do you feel that you know Dr. Oppenheimer well enough to comment on the bearing of his character, loyalty and associations on this issue?

A. I think Dr. Oppenheimer is a man of upstanding character, that he is a loyal individual, not only to the United States, which of course goes without saying in my mind, but also to his friends and his organizations to which he is attached, let us say, to the institutions, and work very hard for his loyalties; an upright character, very upright character, very thoughtful, sensitive feeling in that respect.

With regard to the question of association, I might say that I have seen the brief form of what would you call it, the report of Dr. Oppenheimer?

Q. What is that?
A. It is some document about 40 pages which is a summary.
Q. When did you see it?
A. Some time in January.
Q. How did you happen to see it?
A. The Chairman of the Commission asked me to take a look at it.
Mr. GARRISON. What year?

The WITNESS. This year. I would say that in spite of the associations in there, I do not believe that Dr. Oppenheimer is a security risk, and that these associations in the past should bar him from access to security information for the Atomic Energy Commission.

By Mr. MARKS.

Q. The report you speak of, is that in amplification of the letter of allegations or derogatory information which you have read of General Nichols to Dr. Oppenheimer?

A. I don't know whether it was made as an amplification.
Q. I am just trying to get some sense of what it is.
A. I don't know. I understood it to be a digest of a very big file.
Q. I didn't understand clearly, Dr. Rabi. You used the phrase "bar him." Would you mind repeating what you had in mind?

A. I will put it this way. If I had to make the determination, after having read this and knowing Dr. Oppenheimer for all the years I would know him, I would have continued him in his position as consultant to the Atomic Energy Commission, which he was before.

Mr. MARKS. That is all.
Dr. GRAY. Are you ready to proceed with the examination?
Mr. ROBB. Mr. Chairman, it is now about 1:15. I am going to take 45 minutes anyway, and of course we have no lunch. I would much prefer to take a brief break to get a cup of coffee and a sandwich before proceeding.

(Discussion off the record.)

Dr. GRAY. We will now recess until 2 o'clock.

Mr. GARRISON. Mr. Chairman, is there any more news about the schedule for next week? You said the board might be calling witnesses, and would let us know what you have decided.

Mr. GRAY. I am afraid we will have to talk about that some at lunch, because I don't have anything new at the moment.

(Thereupon at 1 : 05 p. m., a recess was taken until 2 : 00 p. m., the same day.)

AFTERNOON SESSION

Mr. GRAY. Shall we resume?

Mr. MARKS. It is agreeable to Dr. Oppenheimer that the proceedings continue this afternoon without his presence.

Mr. GRAY. I just want to make it clear that it is a matter of his own choosing, and of Mr. Garrison, that they are not present this afternoon for the remainder of these proceedings.

Mr. MARKS. That is correct. He may be back before we finish, but this is a matter of his own choosing.

Mr. GRAY. Would you proceed, Mr. Robb.

CROSS-EXAMINATION

By Mr. ROBB:

Q. Dr. Rabi, you testified that in the fall of 1949, the problem of the super program had your attention quite considerably.
A. Yes.
Q. And I believe you said that you talked with Dr. Lawrence and Dr. Alvarez about it.
A. Yes.
Q. Could that have been in October, just before the meeting of the GAC.
(Dr. Oppenheimer entered the room.)
Mr. GRAY. You are back now, Dr. Oppenheimer.
Dr. OPPENHEIMER. This is one of the few things I am really sure of.
The WITNESS. I can't remember the exact date. I think it was in the fall. It was before the GAC meeting.

By Mr. ROBB.

Q. It was before the GAC meeting?
A. I am quite sure.
Q. Did Dr. Alvarez and Dr. Lawrence come to see you in New York?
A. That is right.
Q. Together or did they come separately?
A. Together.
Q. What was the purpose of their visit to you, sir?
A. Well, we are old friends. I don't remember what the purpose was that they wanted to come up which I didn't find extraordinary. Physicists visit one another. Both are people I have known for a long time. But we did talk on this thing which was in our mind.
Q. Yes. To save time, didn't they come to see you with special reference to the thermonuclear question or the super question?
A. That may have been in their minds. It may have been in their minds. We got going on it right away.
Q. In all events, you talked about it?
A. That is right. What was in their minds, I don't know.
Q. Do you recall what their views were on it as they expressed them to you then?
A. Their views were that they were extremely optimistic. They are both very optimistic gentlemen. They were extremely optimistic about it. They had been to Los Alamos and talked to Dr. Teller, who gave them a very optimistic estimate about the thing and about the kind of special materials which would be required. So they were all keyed up to go bang into it.
Q. They thought we ought to go ahead with it?
A. I think if they had known then what we knew a year later, I don't think they would have been so eager. But at that time they had a very optimistic estimate.
Q. To help you fix the time, was that after the Russian explosion?
A. After the Russian explosion.
Q. Was that the main reason why they thought we ought to get along with the thermonuclear program?
A. I don't know.

Q. Beg pardon?
A. I would suppose so. As I testified before, what I testified was that we felt we had to do something to recover our lead.
Q. Did you express your view to them on that subject?
A. Yes, that we had to do something, and I think that I may have inclined— this is something which I kept no notes and so on.
Q. I understand, doctor.
A. I think I may have inclined toward their view on the basis of the information they said they had from Dr. Teller.
Q. Did you find yourself in any substantial disagreement with their views as they expressed them then?
A. It wasn't the case of agreement or disagreement. I generally find myself when I talk with these two gentlemen in a very uncomfortable position. I like to be an enthusiast. I love it. But those fellows are so enthusiastic that I have to be a conservative. So it always puts me in an odd position to say, "Now, no. There, there," and that sort of thing. So I was not in agreement in the sense that I felt they were as usual, which is to their credit—they have accomplished very great things—overly optimistic.
Q. Except for that you agreed with their thought that we ought to do something, as you put it, to regain our position?
A. That is right. I felt very strongly. I spoke to everybody I could properly speak to, as I said earlier, talking about what we could do to get back this enormous lead which we had at that time. This of course was one of the possibilities.
Q. Was it before that or after that you talked to Dr. Oppenheimer?
A. I really don't remember the sequence of events at that time and when I saw Dr. Oppenheimer, whether he was away for the summer or I was, or what. I wish I could testify. I don't keep a diary.
Q. I understand. All I want is your best recollection, doctor. Whenever you talked to Dr. Oppenheimer, did he express his views on this matter?
A. It is very hard to answer. I just don't recollect to tell you a specific time at a specific place where I spoke to Oppenheimer.
Q. May I help you a little bit? It is difficult to separate what he might have told you before the meeting with what he said at the meeting.
A. To which meeting are you talking?
Q. The meeting of October 29.
A. I don't really remember that we met before the meeting or immediately before the meeting, or that he told me something of that sort. I just don't remember. My actual recollection is that I learned the purpose of the meeting at the meeting, but I am not certain. I just can't tell.
Q. At all events, the views expressed by Dr. Oppenheimer at the meeting were not in accord with those expressed to you by Alvarez and Lawrence, were they?
A. No, the meeting was a very interesting one. It was a rather solemn meeting. I must say that Dr. Oppenheimer as chairman of the meeting always conducted himself in such a way as to elicit the opinions of the members and to stimulate the discussion. He is not one of these chairmen who sort of takes it their privilege to hold the floor; the very opposite. Generally he might express his own view last and very rarely in a strong fashion, but generally with considerable reservations. When he reported to the Commission, it was always a miracle to the other members on the committee how he could summarize three days of discussions and give the proper weight to the opinion of every member, the proper shade, and it rarely happened that some member would speak up and say, "This isn't exactly what I meant." It was a rather miraculous performance.
Q. Doctor, as chairman of the GAC, do you have custody of the minutes of the GAC?
A. In what sense do you mean, sir? Do I possess them in my office in New York?
Q. Yes, sir.
A. No, sir.
Q. Where would those be?
A. In the AEC building in our office.
Q. In all events there were no minutes of this October 29th meeting?
A. I don't think there were minutes. There was a report.
Q. Yes.
A. When we got down to a sort of settled procedure, we had the minutes. But at the end of the meeting there was a verbal report from the chairman GAC to the Chairman AEC and then a written report summarizing certain

conclusions and recommendations, and if there were differences of opinion, trying to give the proper shade and tone, telling the date of the next meeting, and if we know, the kind of questions we would like to take up at the next meeting.

Q. Do you recall any mention at that meeting of October 29, 1949, of a communication from Dr. Seaborg about the problem under discussion?

A. I can't recollect. I don't know. I might add it would not have been very significant, because my feeling is now that we came into the meeting without any clear ideas, that in the course of an extremely exhausting discussion to and fro, examining all the possibilities we each became clearer as to what this thing meant. So anybody who didn't participate in the discussion wouldn't have gotten what we conceived at that time to be that kind of clarity.

Q. You said somebody from the Joint Chiefs came to talk to you. Do you remember who that was?

A. As I remember it, I think it was General Bradley. * * *

Q. Doctor, whose business was it to brief General Bradley, anyway?

A. I suppose the Military Liaison Committee.

Q. I see. You mean between the AEC and the Joint Chiefs?

A. Yes, that is the way of communication, I presume.

Q. Who was on that Committee?

A. That is a matter of record. I am sorry, I can't remember who happened to be the chairman. The military personnel changed all the time. The chairman changed all the time. For the life of me, I can't remember at present who it was then. * * *

Mr. GRAY. Just at that point, you mean with respect to A bombs, if I can refer to it that way?

The WITNESS. The materials are similar.

Mr. GRAY. So you had in mind also the thermonuclear?

The WITNESS. Yes, sir, everything. * * *

Mr. ROLANDER. For clarity, you said Iowa; did you mean Ohio?

The WITNESS. I meant Ohio. Thank you. Portsmouth.

By Mr. ROBB:

Q. You spoke of a meeting at Princeton in 1951, is that right?

A. Yes.

Q. That was after Dr. Teller's discovery, if we may call it such, wasn't it?

A. At that point I wouldn't call it Dr. Teller's discovery. I think Dr. Teller had a very important part in it, but I would not make a personal attribution.

Q. I was not trying to decide that, but merely to identify it. It was after some discovery was made which was extremely promising.

* * * Was there any discussion at that meeting as to whether or not the President's directive to proceed with the thermonuclear permitted you to go ahead with the development of that invention? Do I make myself clear?

A. No.

Q. Was there any discussion about whether or not you could go ahead with the work on that invention, with the exploitation and development of it in the terms of the President's orders or directive?

A. The only discussion, as I recall, sir, were the ways and means of going ahead, and how to get certain questions settled. There were certain technical questions of what would happen under certain circumstances in this design. It was amenable to theoretical calculations by some very good man, I think Dr. Bethe went and did it.

Q. But there was no discussion about whether or not the terms of the President's directive permitted you to go to work on that invention?

A. No, I don't recall any. It would be hard for me to see why there should have been.

Q. Doctor, I notice this sentence in the report of the GAC of the October 29 1949, meeting, which I am told I may read aloud:

"It is the opinion of the majority that the super program itself should not be undertaken and that the Commission and its contractors understand that construction of neutron producing reactors is not intended as a step in the super program."

Doctor, were the neutron producing reactors to which you had reference there the same type that were constructed at Savannah?

A. Yes, sir. They were constructed with that in mind. They were dual purpose. The design could be optimized in one direction or another direction and a balance was made, as I remember.

Q. Is it appropriate to ask the doctor when they were constructed?
Dr. BECKERLEY. I think that is a matter of public record.
The WITNESS. It is a matter of record, and I would not try to test my memory on that.

By Mr. ROBB:
Q. In all events, when they were constructed, they were constructed with a view that they would be a step in the super program?
A. That they could be a step in the super program. We were in a wonderful position, we could go one way or the other.
Q. Doctor, you said that the chairman of the Atomic Energy Commission, Mr. Strauss, in January of this year had asked you to take a look at the FBI report which he had on Dr. Oppenheimer.
A. Yes.
Q. Did you mean to say by that that he asked you to come to his office for that purpose?
A. We talked about the case, of course. He informed me of the thing.
Q. Yes.
A. He thought as chairman of the General Advisory Committee I ought to know the contents of that report. I think if I had asked for the full report, I would have gotten it. I may say that that record is not something I wanted to see.
Q. No, I understand that.
A. In fact, I disliked the idea extremely of delving into the private affairs in this way of a friend of mine, but I was finally convinced that it was my duty to do so.
Q. Certainly. What I had in mind, doctor, was that you did not mean to suggest that Mr. Strauss sent for you and said to you in effect, "Look what I have now."
A. Oh, no.
Q. I was sure of that.
A. No.
Q. Did you go to see him on that occasion on your own volition or did he send for you?
A. I go see him every time I am in Washington and spend an hour or two with him discussing all sorts of problems which refer to the GAC, AEC relations. I am going to see him this afternoon if I get away from here in time.
Q. Certainly. Doctor, don't answer this question unless you want to, but did you go to see Mr. Strauss on one occasion more or less in behalf of Dr. Oppenheimer?
A. Just specially for that purpose?
Q. Well, among other purposes. You may have had other purposes.
A. We have talked about this every time I met him.
Q. Yes, I can quite understand that.
A. Yes. I have talked to Mr. Strauss on this certainly in behalf of Dr. Oppenheimer, but even more in behalf of the security of the United States. To tell you frankly, I have very grave misgivings as to the nature of this charge, still have, and the general public discussion which it has aroused, and the fear that as a result of such a discussion important security information absolutely vital to the United States may bit by bit inadvertently leak out. I am very much worried about that.
Q. Doctor, do you approve of Dr. Oppenheimer's course of giving the letter from General Nichols and his reply to the newspapers?
A. I don't know his motives on that. In his position, I think I would have done the same thing.
Q. I just wanted to get your views on it.
A. Yes.
Q. You said, sir, that you would rather not answer with respect to the matter of continental defense?
A. No, I did not.
Q. May I finish my question? As to the difference in emphasis between you and Dr. Oppenheimer?
A. No, sir, I don't recall I said that.
Q. I misunderstood you.
A. It was a possible difference in emphasis of the method of employment of a strategic air force.
Q. I see. That is what I was trying to say.

A. In the method of employment. In other words, this is a kind of military question and runs into problems of target selection, things of that sort. For that reason, since this is not just an AEC question for which I understand the members of this panel are cleared, but refers to DOD questions, I would rather not talk about it.

Q. In other words, you feel that would be classified information which you should not disclose even to the members of this board?

A. That is right. I don't want to skirt around and maybe fall into something.

Mr. ROBB. I see. I think that is all I would like to ask, Mr. Chairman.

Mr. GRAY. Dr. Rabi, you mentioned this morning that at the October 1949 meeting of the GAC, General Bradley came, to the best of your recollection, and you said also there was a State Department man. Do you remember who that was?

The WITNESS. I think it was Mr. Kennan.

Mr. GRAY. You mentioned a meeting at Los Alamos in the summer, I believe, of 1950?

The WITNESS. Yes.

Dr. GRAY. That was before the Princeton meeting, of course, to which you referred?

The WITNESS. Yes.

Dr. GRAY. Was Dr. Oppenheimer at the meeting in the summer of 1950?

The WITNESS. Yes, indeed. I don't remember exactly. The meeting, I think, was a meeting of the Subcommittee on Weapons. I think there were three subcommittees of the General Advisory Committee which were sort of specialized, one weapons, one on reactor and one on research. I think that was the Weapons Subcommittee. I don't recall the full attendance at that meeting, but Dr. Oppenheimer was there.

Dr. GRAY. With respect to the development of the H bomb—I don't know how to refer to it exactly, but you know what I am talking about—and the issue of who was for and who was against, was it your impression that Dr. Oppenheimer was unalterably opposed to the development?

The WITNESS. No, I would not say so, because after we had those two statements, which were written by different groups which were put in, I distinctly remember Dr. Oppenheimer saying he would be willing to sign both.

Mr. GRAY. My question was bad, because "unalterably" is a pretty strong word, and you have already testified that subsequent to the President's decision he encouraged the program and assisted in it.

The WITNESS. Yes, sir.

Mr. GRAY. So I think this was a bad question.

The WITNESS. I was really testifying as to that time, that there were two statements of attitudes which differed, and he said he would be ready to sign either or both.

Mr. GRAY. He would have been willing to sign the one which you signed?

The WITNESS. That Fermi and I did, yes.

Mr. GRAY. Would you have considered those two reports absolutely consistent?

The WITNESS. No.

Mr. GRAY. Yourself?

The WITNESS. No. I just answered your question about being unalterably opposed.

Mr. GRAY. There was a real difference?

The WITNESS. Yes, sir, there was a real difference. There was no difference as far as a crash program was concerned. That they thought was not in order.

Mr. GRAY. I have one other question. You testified very clearly, I think, as to your judgment of Dr. Oppenheimer as a man, referring to his character, his loyalty to the United States, and to his friends and to institutions with which he might be identified, and made an observation about associations.

As of today would you expect Dr. Oppenheimer's loyalty to the country to take precedence over loyalty to an individual or to some other institution?

The WITNESS. I just don't think that anything is higher in his mind or heart than loyalty to his country. This sort of desire to see it grow and develop. I might amplify my other statement in this respect, and that is something we talked of through the years. When we first met in 1929, American physics was not really very much, certainly not consonant with the great size and wealth of the country. We were very much concerned with raising the level of American physics. We were sick and tired of going to Europe as learners. We wanted to be independent. I must say I think that our generation, Dr. Oppenheimer's and my other friend that I can mention, did that job, and that 10 years later we were at the top of the heap, and it wasn't just because certain refugees came out of Germany, but because of what we did here. This was a conscious motivation. Oppenheimer set up this school of theoretical physics which was a tremendous contribution. In fact, I don't know how we could have carried out the scientific part of the war without the contributions of the people who worked with Oppenheimer. They made their contributions very willingly and very enthusiastically and singlemindedly.

Mr. GRAY. Perhaps I could get at my question this way. You are familiar, if you have read the Nichols letter and read the summary of a file which Chairman Strauss handed you, with the Chevalier episode to some extent, I take it.

The WITNESS. I know of the episode, yes.

Mr. GRAY. Would you expect Dr. Oppenheimer today to follow the course of action he followed at that time in 1943?

The WITNESS. You mean refuse to give information? Is that what you mean?

Mr. GRAY. Yes.

The WITNESS. I certainly do. At the present time I think he would clamp him into jail if he asked such a question.

Mr. GRAY. I am sorry.

The WITNESS. At the present time if a man came to him with a proposal like that, he would see that he goes to jail. At least that is my opinion of what he would do in answer to this hypothetical question.

Mr. GRAY. Do you feel that security is relative, that something that was all right in 1943, would not be all right in 1954?

The WITNESS. If a man in 1954 came with such a proposal, my God—it would be horrifying.

Mr. GRAY. Supposing a man came to you in 1943.

The WITNESS. I would have thrown him out.

Mr. GRAY. Would you have done anything more about it?

The WITNESS. I don't think so. Unless I thought he was just a poor jackass and didn't know what he was doing. But I would try to find out what motivated him and what was behind it, and get after that at any time. If somebody asked me to violate a law and an oath——

Mr. GRAY. I hope you are not taking offense at my asking this question, but this is a perfectly serious question because you have testified without equivocation, I think, and in the highest possible terms of Dr. Oppenheimer's character, his loyalty, and with certain reservations about his early associations. As Mr. Marks pointed out in the question leading to this testimony, these are things which the Atomic Energy Act says must be taken into account in this matter of clearance. I trust you understand this is a very solemn duty that this Board has been given.

The WITNESS. I cerainly do, sir.

Mr. GRAY. There have been those who have testified that men of character and standing and loyalty that this episode should simply be disregarded. I don't think that is an unfair summary of what some of the witnesses have said. Do you feel that this is just a matter that is of no consequence?

The WITNESS. I do not think any of it is of no consequence. I think you have to take the matter in its whole context. For example, there are men of unquestioned loyalty who do not know enough of the subject—I am talking now of the atomic energy field—so that in their ordinary speech they don't know what they are saying. They might give away very important things.

Mr. GRAY. That would be true of me, I am sure.

The WITNESS. It certainly has been true of a lot of military stuff that you see published. It makes your hair stand on end to see high officers say, and people in Congress say some of the things they say. But with a man of Dr. Oppenheimer's knowledge, who knows the thing completely, and its implications and its importance, and the different phases, believing as I do in his fundamental loyalty, I think to whomever he talked he would know how to stay completely clear of sensitive information.

Mr. GRAY. In any event, I suppose——

The WITNESS. I think there is a very large distinction there.

Mr. GRAY. In any event, I believe you did testify that you would be quite convinced—I am not sure you did—are you quite convinced that as of today Dr. Oppenheimer's course of action would be in accord with what you would do, rather than what he did in respect to the matter of this sort. I can't say what a man will do, but we only can apply subjective tests in these matters as far as your testimony as to character, loyalty and so forth, are concerned. So this is all subjective, but would you expect without any real question in your mind that today Dr. Oppenheimer would follow the kind of course that you would approve of today with respect to this matter?

The Witness. I think I can say that with certainty. I think there is no question in my mind of his loyalty in that way. You know there always is a problem of that sort. I mean the world has been divided into sheep and goats. I mean the country has been divided into sheep and goats. There are the people who are cleared and those who are not cleared. The people against whom there has been some derogatory information and whatnot. What it may mean and so on is difficult. It is really a question in one's personal life, should you refuse to enter a room in which a person is present against whom there is derogatory information. Of course, if you are extremely prudent and want your life circumscribed that way, no question would ever arise. If you feel that you want to live a more normal life and have confidence in your own integrity and in your record for integrity, then you might act more freely, but which could be criticized, either for being foolhardy or even worse.

In one's normal course at a university, one does come across people who have been denied clearance. Should you never sit down and discuss scientific matters with them, although they have very interesting scientific things to say?

Mr. Gray. No, I would not think so.

The Witness. That is the sort of question you are putting, Dr. Gray, and I am answering to the best of my ability.

Mr. Gray. I am wondering whether it is, Dr. Rabi. Let me say this. I think there is not anybody who is prepared to testify that he can spot a Communist with complete infallibility. I know that there have been people who surprised me that I had an acquaintanceship with who turned out to be Communists. I don't think it is unfair to say that witnesses including Dr. Oppenheimer himself have testified that there were people who later turned out to be Communists, to their surprise, who they identified.

I am asking against the background of the security of this country which must be paramount, it seems to me, perhaps unhappily, to any other consideration or personal institution, can we afford to make it a matter of individual judgment as to whether a person is dangerous, in this case Mr Chevalier. I don't know that he has ever appeared before any committee or anything else. I don't know whether he is a member of the Communist Party or not. It is conceivable that he might have been. I am afraid I am making an argument now, but it is all a part of this question. Against what I believe to be the commitments involved in joining the Communist Party, can it be a matter of individual judgment whether it does no harm to either fail to report what seems to be an espionage attempt or to discuss in however clear terms information which is of a classified nature. That is the most confused question you ever had put to you, and I think I should eliminate the last part in any event, because the Chevalier incident did not, as I understand it, involve disclosure of information. There was none of that involved. I don't want the record to make it appear that I am implying that. This was simply, a question of not taking immediate security precautions either in respect to reporting the incident, a later matter of declining to disclose the name of the man who made the approach and certain other less than frank aspects. 1 believe you said you did not think that was a proper course to follow, and you would expect Dr. Oppenheimer to follow a different course today.

The Witness. Yes.

Mr. Gray. Which implies, certainly, I think that you think he should follow a different course today.

The Witness. I can't say anything but yes. We have all learned a whole lot since that time. A lot of things which were quite different at one time but different in another. You have to become accustomed to life in this kind of life when you are involved in this kind of information.

Mr. Gray. You are saying that in your judgment Dr. Oppenheimer has changed?

The Witness. He has learned.

Mr. Gray. All right.

The Witness. I think he was always a loyal American. There was no doubt in my mind as to that. But he has learned more the way you have to live in the world as it is now. We hope at some future time that the carefree prewar days will return.

Dr. Evans. Dr. Rabi, would you tell us something about your early education?

The Witness. I am a graduate of Manual Training High School in Brooklyn, a graduate of Cornell University with a degree of bachelor of chemistry—we are fellow chemists.

Dr. EVANS. I am glad you had some chemistry.

The WITNESS. I had an awful lot of chemistry. Then I worked after that for a year in analytical laboratories, the Pease Laboratory, which were an affiliate of the Lederle Laboratories in New York, and then for various things for a few years. I went back to Cornell, I think it was in 1923, for graduate work in chemistry, but during the course of setting up my program, I decided to change to physics. I spent a year at Cornell in graduate work and then went to Columbia where I transferred, where I took my doctor's degree in 1927. I am older than Dr. Oppenheimer, but his degree, I think, is older than mine, or about the same vintage.

During that period I supported myself by instructing in physics at the College of the City of New York. Then I got a fellowship from Columbia, and went to Europe to study theoretical physics, first at Munich and then to Copenhagen, and then to Hamburg.

While there I had an idea for an experimental problem and changed back to doing experimental physics. After my experiment was done, I went to Leipzig with Professor Heisenberg back to theoretical physics, where I first met Dr. Oppenheimer briefly on his visit, and after Dr. Heisenberg went to the United States for a lecture tour, I went to Zurich, where Dr. Oppenheimer was working on Stellar, and we found ourselves sympathetic.

At the end of that summer I went to Columbia as a lecturer in physics. I have been at Columbia ever since, except for a 5-year period during the war. I enlisted—enlisted is the wrong word—I left Columbia in November 1940 to join the radiation lab at MIT, which was concerned with the production of microwave radar, the research and development of microwave radar, and stayed there throughout the war.

My connection with Los Alamos, I was never on their payroll, but went there as a radiation lab man.

Dr. EVANS. Let me ask you another question that has nothing particularly pertinent to this proceeding. Is George Pegram still active?

The WITNESS. Wonderfully. He is doing two men's work. He is 78, you know. Recently he has had a heart attack. He is chairman of a committee which handles all the research contracts which amount to many, many millions for the university.

Mr. GRAY. I think the record will have to show that he is a native North Carolinian.

The WITNESS. Yes, sir, a graduate of Trinity College. His father was professor of chemistry.

Dr. EVANS. I wish you would tell him that Dr. Evans asked about him.

The WITNESS. I would be delighted to.

Dr. EVANS. Now, another question. Were you as a scientific man particularly surprised when you heard that the Russians had fired a bomb, or would you have expected it?

The WITNESS. I was astonished that it came that soon. I will tell you this was a peculiar kind of psychology. If you had asked anybody in 1944 or 1945 when would the Russians have it, it would have been 5 years. But every year that went by you kept on saying 5 years. So although I was certain they would get it——

Dr. EVANS. You were certain they would get it?

The WITNESS. I was certain that they would get it, but it was a stunning shock.

Dr. EVANS. You would be pretty certain right now that they will get the thermonuclear?

The WITNESS. In time. What I am afraid of is this controversy over this case may hasten the day because of the sort of attrition of the security of technical information, all sorts of stuff appearing in the newspapers and magazines and so on that sort of skirts around it. You know you have a filter system for information. You put bits and pieces together. They already know something. * * *

Dr. EVANS. You understand, of course, our position on this board, do you not?

The WITNESS. Yes; it is not your problem, but I think it is the problem of the Government of the United States.

Dr. EVANS. Did you know that some of the people that were educated with Dr. Oppenheimer, listened to his lectures, and turned out to be Communists?

The WITNESS. Educated with him?

Dr. EVANS. It was in that school that he conducted——

The WITNESS. You mean who studied with him?

Dr. EVANS. Yes.
The WITNESS. I have heard that, but I can't—this is not direct information.
Dr. EVANS. You have met some Communists, have you, Dr. Rabi?
The WITNESS. I have met people who later said they were Communists. At Los Alamos I met Mr. Hawkins, who said he had been a Communist, and this other chap, what is the name, I can't remember at this moment. I certainly knew Frank Oppenheimer from the time he was a kid in high school.
Dr. EVANS. You didn't meet any of those at the Radiation Laboratory like Bernie Peters?
The WITNESS. I met Peters just fleetingly once or twice. I don't recall any actual conversations with Peters.
Dr. EVANS. Dr. Rabi, if you were approached by someone attempting to secure from you security information, would you report it immediately, or would you consider it for quite a long time?
The WITNESS. Are you talking about April 21, 1954?
Dr. EVANS. Oh, no.
The WITNESS. What date are you talking about?
Dr. EVANS. I am talking about the Chevalier incident. What date was that?
Mr. ROBB. Late 1942 or early 1943.
The WITNESS. I would like to have the question, since this is a crucial question, put more fully so that I can answer the point rather than make up the question, so to speak.
Dr. EVANS. You are giving me a big job, aren't you?
The WITNESS. This is not child's play here.
Dr. EVANS. If you had been working on security material, material that had a high priority, and someone came to you and told you that they had a way of getting that material to the Russians, what would you have done immediately?
The WITNESS. You mean if it was just someone that I didn't know?
Dr. EVANS. No; someone that you knew. Suppose I was a friend of yours and I came and told you.
The WITNESS. And I thought that you were a completely innocent party or not? I think that is the nub of the question, what I would have done at that time. I can't say what would have done at that time. I kind of think I would have gone after it and found out just what this was about.
Dr. EVANS. That is all.

REDIRECT EXAMINATION

By Mr. MARKS:

Q. Dr. Rabi, what do you mean you would have gone after it and found out what this was about?
A. I would have tried to see that the proper authorities found out what these people meant to do, what the thing was. I know a number of times during the war I heard funny noises in my telephone and got the security officers after it.
Q. Dr. Rabi, Mr. Robb asked you whether you had spoken to Chairman Strauss in behalf of Dr. Oppenheimer. Did you mean to suggest in your reply—in your reply to him you said you did among other things—did you mean to suggest that you had done that at Dr. Oppenheimer's instigation?
A. No; I had no communication from Dr. Oppenheimer before these charges were filed, or since, except that I called him once to just say that I believed in him, with no further discussion.
Another time I called on him and his attorney at the suggestion of Mr. Strauss. I never hid my opinion from Mr. Strauss that I thought that this whole proceeding was a most unfortunate one.
Dr. EVANS. What was that?
The WITNESS. That the suspension of the clearance of Dr. Oppenheimer was a very unfortunate thing and should not have been done. In other words, there he was; he is a consultant, and if you don't want to consult the guy, you don't consult him, period. Why you have to then proceed to suspend clearance and go through all this sort of thing, he is only there when called, and that is all there was to it. So it didn't seem to me the sort of thing that called for this kind of proceeding at all against a man who had accomplished what Dr. Oppenheimer has accomplished. There is a real positive record, the way I expressed it to a friend of mine. We have an A-bomb and a whole series of it, * * * and what more do you want, mermaids? This is just a tremendous achievement. If the end of that road is this kind of hearing, which can't help but be humiliating, I thought it was a pretty bad show. I still think so.

By Mr. MARKS:

Q. Dr. Rabi, in response to a question of the Chairman, the substance of which I believe was, was Dr. Oppenheimer unalterably opposed to the H-bomb development at the time of the October 1949 GAC meeting, I think you said in substance no. and then you added by way of explanation immediately thereafter the two annexes or whatever they were——

A. During the discussion.

Q. During the discussion he said he would be willing to sign either or both. Can you explain what you meant by that rather paradoxical statement?

A. No, I was just reporting a recollection.

Q. What impression did you have?

A. What it means to me is that he was not unalterably opposed, but on sum, adding up everything, he thought it would have been a mistake at that time to proceed with a crash program with all that entailed with this object that we didn't understand, when we had an awfully good program on hand in the fission field, which we did not wish to jeopardize. At least we did not feel it should be jeopardized. It turned out in the events that both could be done. Los Alamos just simply rose to the occasion and worked miracles, absolute miracles.

Mr. MARKS. That is all.

RE-CROSS-EXAMINATION

By Mr. ROBB:

Q. Doctor, on the occasion when you were in Mr. Strauss' office, and he showed you the report that you testified about, how long would you say that meeting lasted?

A. I can't remember.

Q. A few minutes?

A. I don't know whether it was a few minutes or half an hour. If you were Mr. Strauss, there are calls coming in all the time from all over, from the White House, and what not.

Q. Did you look at the report in Mr. Strauss' office?

A. No. I put it in an envelope and went to our GAC office. I read it there, and then brought it back.

Q. Dr. Rabi, getting back to the hypothetical questions that have been put to you by the Chairman and Dr. Evans about the Chevalier incident, if you had been put in that hypothetical position and had reported the matter to an intelligence officer, you of course would have been told the whole truth about it, wouldn't you?

A. I am naturally a truthful person.

Q. You would not have lied about it?

A. I am telling you what I think now. The Lord alone knows what I would have done at that time. This is what I think now.

Q. Of course, Doctor, as you say, only God knows what is in a man's mind and heart, but give us your best judgment of what you would do.

A. This is what I think now I hope that is what I would have done then. In other words, I do not—I take a serious view of that—I think it is crucial.

Q. You say what?

A. I take a serious view of that incident, but I don't think it is crucial.

Q. Of course, Doctor, you don't know what Dr. Oppenheimer's testimony before this board about that incident may have been, do you?

A. No.

Q. So perhaps in respect of passing judgment on that incident, the board may be in a better position to judge than you?

A. I have the highest respect for the board. I am not going to make any comment about the board. They are working very hard, as I have seen.

Q. Of course, I realize you have complete confidence in the board. But my point is that perhaps the board may be in possession of information which is not now available to you about the incident.

A. It may be. On the other hand, I am in possession of a long experience with this man, going back to 1929, which is 25 years, and there is a kind of seat of the pants feeling which I myself lay great weight. In other words, I might even venture to differ from the judgment of the board without impugning their integrity at all.

Q. I am confining my question to that one incident, Doctor. I think we have agreed that the board may be in possession of information from Dr. Oppenheimer's own lips about that incident which is not now available to you, is that correct?

A. This is a statement?
Q. Yes.
A. I accept your statement.
Q. And therefore it may well be that the board is now in a better position than you, so far as that incident is concerned, to evaluate it?
A. An incident of that sort they may be. I can't say they are not. But on the other hand, I think that any incident in a man's line of something of that sort you have to take it in sum.
Q. Of course.
A. You have to take the whole story.
Q. Of course.
A. That is what novels are about. There is a dramatic moment and the history of the man, what made him act, what he did, and what sort of person he was. That is what you are really doing here. You are writing a man's life.
Q. Of course, but as a scientist, Doctor, and evaluating, we will say, an explosion you perhaps would be in a better position to evaluate an explosion having witnessed it and having first-hand knowledge about it than somebody who had not, is that right?
A. If you put it in that way, I don't know the trend of your question. I am not fencing with you. I really want to know what you are getting at.
Q. I am not fencing with you either.
A. If you are saying that an eyewitness to something can give a better account of it than a historian, that I don't know. Historians would deny it. It is a semantic question, but if you want to be specific about it——
Q. I will put it this way. As a scientist, you would say that one having all the facts about a particular physical manifestation or reaction would be in a better position to evaluate that than somebody who did not have all of the facts or might not know one of the facts?
A. A lot of the things about this are not the sort of things which you term just facts. We have Mr. Morgan here, for example, who has been the head of a big business which he built up. He gets as many facts as possible, but I am sure beyond that there is a lot of experience and color which make his judgment. In a court of law it might be something else. Ultimately you go to a jury who have facts, and then they add a whole lot of things which your heart identifies as facts and their experience in life to a situation. I was afraid your question was tending to put me in the position of a so-called fiction scientist who looks at certain facts and measurements, and we are not talking about such a situation.
Q. Let me get back again to the concrete. Would you agree, Doctor, that in evaluating the Chevalier incident one should consider what Dr. Oppenheimer says happened in that incident, together with the testimony of persons such as yourself?
A. Wait a minute. I didn't testify to that incident because I have only heard about it.
Q. Together with testimony of persons such as yourself about Dr. Oppenheimer.
A. Yes, that is right.
Q. Very well; therefore, one who had heard Dr. Oppenheimer describe the incident and had heard your testimony would be in a better position to evaluate it than one who had not heard Dr. Oppenheimer describe it, is that correct?
A. I will put it this way. I think this committee is going into this and they will be in as good a position as it is humanly possible to be for people who have never met this man before to make a judgment about it. I certainly reserve the right to my own opinion on this, because I am in the possession of a long period of association, with all sorts of minute reactions. I have seen his mind work. I have seen his sentiments develop. For example, I have seen in the last few years something which surprised me, a certain tendency of Dr. Oppenheimer to be inclined toward a preventive war. Nothing went all the way. But talking and thinking about it quite seriously. I have to add everything of that sort. All sorts of color and form my own opinion. But I am not on this board, and I think this board is trying to do what it can in this business of getting testimony, the kind of people to come talk to them, the evaluation of the people and the kind of insight, whether they are just loyal people or whether they have thought about the problem, and so on. It is a tough job. Bpt nevertheless, I say I will still stick to my right to have my own opinion.
Q. Certainly, Doctor. To sum up, I suggest to you what I did to Dr. Conant, and he agreed, that in deciding about a matter such as the Chevalier incident, one must consider all the available relevant evidence, is that right?

A. Certainly.
Q. And that would include what actually happened and what people such as yourself, who know Dr. Oppenheimer, say about Dr. Oppenheimer.
A. You are talking about the job of the committee; yes.
Q. Yes.
Mr. ROBB. Thank you, Doctor.
Mr. GRAY. Do you have any more questions?
Mr. MARKS. I think I better ask one more question, if the board will indulge me.

REDIRECT EXAMINATION

By Mr. MARKS:
Q. Dr. Rabi, in view of the quite serious questions which quite properly have been asked you in regard to this so-called Chevalier episode, I would like to try to summarize for you what I understand the testimony to be, and ask you how that would affect the opinions you have expressed.
As I understand the testimony, it is that Chevalier who was an old friend of Dr. Oppenheimer, a member of the faculty in romance languages at the University of California, was at his house on an occasion in the early part of 1943, and at that time Dr. Oppenheimer found himself at one point in the visit alone with Dr. Chevalier, who said that he understood from Eltenton that Eltenton had a way of getting information to the Russians. I thing it is fair to say that the testimony is that Oppenheimer reacted emphatically in rejecting as wrong any consideration of such a matter, and used very strong language to Chevalier, and that Oppenheimer was thereafter convinced that Chevalier had entirely dropped the matter.
Some months later after Los Alamos had been set up and Oppenheimer was there as director, the security officer, Lansdale, mentioned to Oppenheimer that there was trouble of some kind at Berkeley. The indication was that some of the young physicists had committed indiscretions.
On the occasion of Dr. Oppenheimer's next visit to Berkeley he sought out the security officers there, told them that he understood that there was trouble of some kind, said that he thought that a man Eltenton would bear watching.
The next day the security officers asked Oppenheimer to talk to them further about the incident. At that time they asked him to explain the circumstances which had moved him to suggest the name Eltenton. Dr. Oppenheimer said that there had been an intermediary.
The security officers asked him to name the intermediary. He declined to do so. The security officers asked him whom the intermediary had approached. Oppenheimer said people on the project, and in the course of a long interview it appears that they suggested there were two or three such people. He did not name himself or Chevalier as the people concerned.
In the course of a long conversation at that time with the security officers, he mentioned also that a man at the Soviet consulate was involved, and there was some reference to microfilm, although the transcript of the conference between Oppenheimer and the security officers is not clear as to the context in which microfilms were mentioned.
Later Colonel Lansdale, a few weeks later, again interviewed Oppenheimer and asked him to name the intermediary. Oppenheimer again declined, and on all of these occasions he gave as his explanation that he didn't want people to get in trouble who had acted properly and innocently, that he thought he was revealing the name of the only person who could possibly be guilty of real wrongdoing.
Some time after he refused to give the true story to Lansdale or give the names to Lansdale, General Groves talked to him and asked him to name the intermediary. On that occasion Oppenheimer said, "I won't give you the names unless you order me to." Groves said, "I don't want to order you. Think about it."
Shortly after that, Groves again came to Oppenheimer and said, "I need to have the name. If you don't give it to me, I will have to order you to," and at that time Oppenheimer gave the name of Chevalier as the intermediary.
In the course of questioning Dr. Oppenheimer about these circumstances, counsel for the board put the question to him whether the story that he had told the security officers on the occasion of the interview that I have described at Berkeley wasn't a fabrication and a tissue of lies, and to this, I think, Oppenheimer responded, "Right."
A. Right it was.
Q. He accepted counsel's characterization. I may say that this occurred in the course of a very thorough cross-examination.

Mr. ROBB. Have you finished, Mr. Marks?
Mr. MARKS. Yes.
Mr. ROBB. Mr. Chairman, that was a rather long question.
Mr. MARKS. I was about to ask a question.
Mr. ROBB. I thought you were finished. It was a rather long statement, and I don't want the record to show that I am accepting as a completely accurate statement the entire circumstances but of course I am not going to object to it. I have not objected to any question, and I don't intend to.
Mr. GRAY. Certainly it will be obvious in the record that this was stated as Mr. Marks' summary.
Mr. ROBB. Yes, certainly.
Mr. GRAY. On that basis he will now ask the question.
Mr. ROBB. Certainly. I am sure Mr. Marks understands.
Mr. MARKS. I understand perfectly.
Mr. GRAY. That was not the question.
Mr. MARKS. No, it wasn't. I thought Mr. Robb wished to make a correction. I understand exactly your point, Mr. Robb.

I ask you, Dr. Rabi, whether this account of my impression of the essentials of what has been brought out here leads you to wish to express any further comment?

The WITNESS. The only comment I can make on this right off is that it is part and parcel of the kind of foolish behavior that occurred in the early part of the record, that there were very strong personal loyalties there, and I take it in mentioning Eltenton he felt he had discharged his full obligation. My comment is that it was a very foolish action, but I would not put a sinister implication to it. The record is full of actions before Oppenheimer became the sort of statesman he is now of that sort of thing.

By Mr. MARKS:

Q. Are you confident or are you not confident, Dr. Rabi, whichever it is—let me put it this way. Are you confident that Dr. Oppenheimer would not make the kind of mistake again?
A. I certainly am. He is a man who learns with extraordinary rapidity.
Q. Would you agree that incident involved a conflict in loyalties?
A. The question is whether to my mind, whether it involved a conflict of loyalties within his own heart. I don't think it did in his own heart, at least from what you tell me, and taking the sum total. Apparently Chevalier was a man of whom he was very fond personally. They shared a mutual interest, I presume, of French literature. I don't think I have met the gentleman. By pointing the finger at Eltenton I think he felt that he had done the necessary thing for the protection of security. I think if he thought about it more profoundly at the time, and were not so tremendously occupied and burdened by the Los Alamos problems, he might have seen that and this was certainly something that he could not hope to keep quiet. It was a great mistake in judgment and everything else. He should have swallowed that bitter pill at once. But I read no sinister implication in it.
Q. Would you be confident or would you not be confident that today he would resolve the question of his responsibility on the one hand to the country or the public in a way that you would?
A. I think he would be very conscious of his position, not to impair his usefulness to the United States. Even though he might not have shared certain fears, he would not have taken that particular responsibility of withholding that information and have run that particular personal danger of doing it. I think he is just a much more mature person than he was then.
Mr. MARKS. That is all.
Mr. ROBB. May I ask one more question?

RE-CROSS-EXAMINATION

By Mr. ROBB:

Q. This is a purely hypothetical question, Doctor. I just want to get your reaction to it.
Suppose on all the evidence this Board should not be satisfied that Dr. Oppenheimer in his testimony here has told this board the whole truth; what would you say then about whether or not he ought to be cleared?
A. It depends on the nature of the sort of thing he withheld. There may be elements of one's private life that do not concern this board or anybody else.
Q. Suppose the board should not be satisfied that he had told the truth or the whole truth, about some material matter; what would you say then?

A. It would depend again on the nature of the material matter. If I agreed that the matter was material and germane to this, then I would be very sorry.
Q. What?
A. I would be very sorry.
Q. You mean you would feel that they could not clear him?
A. I feel it would be a very tough question.
Q. Wouldn't you feel that they couldn't clear him, or would you rather not answer that?
A. It is the sort of hypothetical question which to me goes under the terms of a rather meaningless question, with all due respect, in the sense that I want to know the material fact, and I would want—the reason we don't have an individual but a board is that I would want to discuss it with others to help bring out our own feelings, and so forth.
Q. Certainly.
A. So therefore I feel that to answer a hypothetical question in this way without putting myself into the position as a member of the board, and what would be the outcome of my discussions and weighing of this thing with the other members of the board, I think an answer to that sort of thing is something I could not give, because I haven't got the circumstances under which to answer it.
Q. But the circumstances might be such——
A. If you want to set me up on the board, then I would come out with an answer.
Q. No, let me ask you one more question. The circumstances might be such that you would feel that the board should not clear him if that happened?
A. There certainly are circumstances which I can picture where the board could not clear him. You know the sort of evidence that Thoreau refers to of finding a trout in the milk; I am pretty sure it is adultery. I am not saying there is no evidence where I would be doubtful. I would rather be more specific about it.

Mr. GRAY. I am sure that Dr. Rabi understands that this board has reached no conclusion. The board has no review or position, and will reach none until the hearings are concluded. I am not suggesting that counsel's question was improper. I wish, however, to say for the record that it clearly is a hypothetical question.

Mr. ROBB. That is why I prefaced it by saying it was hypothetical.

Mr. GRAY. I know you did. I know you didn't intend to lead Dr. Rabi to the conclusion that the board had reached a conclusion on anything. I don't mind counsel giving their view of the testimony on either side. I do object to anything that suggests that this board has reached any kind of conclusion.

Mr. ROBB. Of course I had no such intention. That is why I prefaced my question by saying this is indeed a hypothetical question.

I think that is all, Doctor. Thank you.

Mr. GRAY. We can now thank you very much, Dr. Rabi.

The WITNESS. Thank you, sir.

(Witness excused.)

Mr. GRAY. Does that conclude your witnesses for today?

Mr. GARRISON. Yes.

Mr. GRAY. We will recess until 9:30.

Mr. GARRISON. Could we make it 10?

Mr. GRAY. I would be glad to talk to the board about it. My inclination is against it.

I am sorry. I would like to accommodate you, but the board feels we should start at 9:30.

(Thereupon at 3:25 p. m., a recess was taken until Thursday, April 22, 1954, at 9:30 a. m.)

UNITED STATES ATOMIC ENERGY COMMISSION

PERSONNEL SECURITY BOARD

In the Matter of J. Robert Oppenheimer

Atomic Energy Commission,
Building T-3, Room 2022,
Washington, D. C., Thursday, April 22, 1954.

The above-entitled matter came on for hearing, pursuant to recess, before the Board, at 9:30 a. m.

Personnel Security Board: Mr. Gordon Gray, chairman; Mr. Ward T. Evans, member; and Mr. Thomas A. Morgan, member.

Present: Roger Robb, and C. A. Rolander, Jr., counsel for the board; J. Robert Oppenheimer, Lloyd K. Garrison, Samuel J. Silverman; and Allan B. Ecker, counsel for J. Robert Oppenheimer; Herbert S. Marks, cocounsel for J. Robert Oppenheimer.

PROCEEDINGS

Dr. GRAY. Gentlemen, we will start. I am sure this is unnecessary, but I would like to remind the witness that he is still testifying under oath in the proceeding.

Mr. GARRISON. Mr. Chairman, after a superficial examination of the record, which was not really quite completed, we reached the conclusion last night, rather late last night, that the questions we had thought had not perhaps been sufficiently covered, and that might need amplification or some further explanation had been covered at one point or another in the record, and wishing to avoid any unnecessary duplication or repetition of what has gone past, we decided not to have any formal redirect examination, but to ask Dr. Oppenheimer to sit where he is sitting this morning and to respond to all questions which you might wish to put to him upon any of the subjects of the inquiry.

Of course, he will be available for your questioning at any other time, also.

Mr. GRAY. The board accepts your decision as to procedure, of course, in this matter. Do I understand that you have no questions to ask?

Mr. GARRISON. That is right, but we would welcome questions from the board at this time or any time.

Mr. GRAY. I see. Mr. Robb, do you have any questions?

Mr. ROBB. I have nothing further to ask Dr. Oppenheimer.

Mr. GRAY. Dr. Evans?

Dr. EVANS. No.

Mr. GRAY. Mr. Morgan?

Mr. MORGAN. No.

Mr. GRAY. I don't believe the board has any questions at this time, Mr. Garrison. I wonder if we are ready to proceed with other witnesses?

Mr. GARRISON. I think after a very short recess, we shall be able, sir. I am sorry to waste any time of the board, but I think you will understand.

Mr. GRAY. Absolutely, yes.

Mr. GARRISON. Professor Whitman will be shortly here, I believe, and I think Dr. Bradbury will also be shortly here. We will see what else we can do so as not to needlessly waste time.

Mr. GRAY. Let us consider ourselves in recess until your next witnesses appear.

(Brief recess.)

Mr. GRAY. I think we may as well proceed at the moment, even in Mr. Morgan's absence, because I am sure he will return by the time we get to any substantive testimony.

Do you wish to testify under oath, Dr. Bradbury?

Dr. BRADBURY. Yes.

Mr. GRAY. What is your full name?

Dr. BRADBURY. Norris Edwin Bradbury.

Mr. GRAY. Would you stand and raise your right hand.

NORRIS EDWIN BRADBURY, do you swear that the testimony you are to give the board shall be the truth, the whole truth and nothing but the truth, so help you God?

Dr. BRADBURY. I do.

Whereupon, Norris Edwin Bradbury was called as a witness, and having been first duly sworn, was examined and testified as follows:

Mr. GRAY. Would you be seated.

I shall briefly call your attention to the existence of the perjury statutes. May we assume that you are familiar that there are such statutes with penalties?

The WITNESS. Yes.

Mr. GRAY. I should like to request that in the course of your testimony if it becomes necessary for you to disclose or advert to restricted data, you let me know in advance so we may take necessary and appropriate steps.

(Mr. Morgan entered the room.)

Mr. GRAY. Finally, I should say to you that we consider these proceedings as a confidential matter between the Atomic Energy Commission and its officials and Dr. Oppenheimer, his representatives and witnesses. The Commission will initiate no releases about these proceedings. In each instance on behalf of the board I express the hope that witnesses will take the same view.

The WITNESS. It is understood.

DIRECT EXAMINATION

By Mr. SILVERMAN:

Q. Dr. Bradbury, what is your present position?
A. I am Director of the Los Alamos Scientific Laboratory, Los Alamos, N. Mex.
Q. Do you also hold any academic position?
A. I am professor of physics at the University of California.
Q. How long have you been Director of the Los Alamos Scientific Laboratory?
A. Since October 1945.
Q. Dr. Bradbury, you have read the Commission's letter of December 23, 1953, which suspended Dr. Oppenheimer's clearance.
A. Yes.
Q. Have you read his answer, too?
A. Yes, at least as I have seen it in the press.
Q. I want to draw your attention to that portion of the letter or direct your attention to the matter relating to development of a thermonuclear device, the hydrogen bomb as it has been called.

First, would you tell us, or would you describe for us something of the nature of the thermonuclear research that went on at Los Alamos. I don't mean for you to tell us what was done, but whether it was a matter that proceeded by jumps, whether there were long periods when there was no thermonuclear research, or whether it was continuous, and so on.
A. The possibility of using cheap fuels to make effective military explosion——
Q. Excuse me. Could we have dates on this where possible so it would be clearer to the board?
A. I will try to put dates in this.
Q. Yes, sir, so the board will follow you.

Mr. GRAY. Since you are interrupted, I am sorry, the security officer is always properly quite nervous.

The WITNESS. I will be equally careful about this. In fact, I suspect I am as conscious of these things as anyone.

Mr. GRAY. I am sure you are.

Mr. ROLANDER. I did not mean to suggest that.

The WITNESS. The possibility of using cheap fuels of which the so-called hydrogen bomb is an example was of interest at Los Alamos from its inception. There was active research, investigation and exploration in this field during the war years.

This interest continued after the war in a very active way; not only was basic fundamental nuclear fission done, in the relevant nuclear field, but experimental groups having to do with techniques that might be applicable were carried on and carried on actively. There were a number of conferences held during the years immediately following the war. There was actually a system, essentially thermonuclear in nature, devised shortly after the war in 1946–47 for which techniques were then not possible or appropriate to bring to fruition.

A number of people in our theoretical division kept an active interest in this field. The basic difficulty which confronted everybody at that time was the calculation difficulty, and indeed, no calculating machines existed that would permit some of the particular problems to be explored.

This interest in the field was continuous and lasted up to the present time. There were no gaps in it. I will say that following the Russian explosion in 1949, the laboratory on its own initiative, of course, actively explored all its areas of development, areas of research, to see if there were any that should be given still further attention or more active attention in an attempt to reestablish the lead which we thought we had enjoyed in the years following the close of the war.

Certainly the thermonuclear field in general at that time offered the only outstanding promise, of reestablishing the technical lead if indeed it were a possible field to bring to fruition.

At that time there were, let us say, grave technical concerns, not only with the actual nature of the systems which had been thought of, that is to say, whether or not they would indeed work in an effective fashion, but whether they would be useful in terms of vehicles that might be expected to employ such devices.

As is the case with any technical development, further knowledge sometimes brought increased pessimism or sometimes it brought optimism. The thermonuclear field went through cycles of this sort.

The one thing that was clear at all times was that unless there was active thought in this field, active exploration of it, that potentially useful ways to make such a device would not be found.

Is that enough to answer your question?

Q. I think it does, Dr. Bradbury.

I think it does. Would you say that there was active thought and active exploration of this field continuously at Los Alamos both before and after the fall of 1949?

A. Yes.

Q. Was the fall of 1949 some sort of a crossroads in that?

A. The fall of 1949 was really a crossroads in the atomic energy business. As I said in my earlier remarks, at that time it became clear that a step had been accomplished by Russia. Naturally we explored our own activities to make sure that our own technical progress was devoted as well as we could see it to maintaining the lead which we had thought we had.

Q. What would you say as to the cooperation or lack of cooperation that was evidenced by specifically Dr. Oppenheimer and generally by the General Advisory Committee with respect to the thermonuclear program?

A. Both the General Advisory Committee and Dr. Oppenheimer, I always found from my personal knowledge extremely helpful and cooperative—I am seeking an appropriate word—actively cooperative with the Los Alamos Laboratory in this field. This was, of course, not a unique thing in the thermonuclear field. The GAC and Dr. Oppenheimer had always to my knowledge been an active friend and been active friends of the laboratory, and had been helpful and had worked closely with us in all our discussions relevant to Los Alamos, or many discussions relative to Los Alamos. They invited the staff of the laboratory to meet with them. I met with them myself on many occasions.

Their comments were always helpful. Their advice was always helpful. I never knew them or Dr. Oppenheimer to take a stand or a position or to give advice which was other than useful and helpful to the laboratory.

Q. By the way, in general did you and the people at Los Alamos, perhaps, if you can speak for them, agree or disagree with the position taken by the GAC in October 1949?

A. I think that if we disagreed, we disagreed perhaps in flavor rather than in a substantive way. We felt extremely strongly that the thermonuclear field had to be explored, had to continue to be explored, that indeed it had grave obstacles in its way at that time, but that no decisions as to the wisdom or morality of making or stockpiling H bombs could be possibly undertaken by this country unless there was a complete knowledge of all the facts.

It was equally important that this country know what the potentialities were in this field from, let us say, a defensive point of view. In other words, we must know, we had to know, what the Russians might be able to accomplish in this field.

Accordingly, the philosophy of the laboratory was that we did not wish to enter into the debate as to whether or not this course was wise or moral or politically sound. We regarded ours the technical responsibility to know as much as it was possible to know and as rapidly as it was possible to know it, about what was broadly called the H-bomb.

This is not a very satisfactory terminology, but if it is read as relevant to the thermonuclear field, I think this will correctly describe our position.

There was, as I have said, active interest in this field and had been. It seemed to us unfortunate that the way the issue came out in the public was that here was a crossroads, and that the country or the laboratory went this way or that way. Frankly it would have been impossible to have stopped the active consideration and exploration of this field by any fiat. You cannot stop people from thinking. It was an exciting field. It apparently violated no laws of nature and inventive and ingenious scientists are bound to think about and do the work which is relevant to this activity.

We, of course, agreed with the publicly announced decision that this work should indeed go ahead and go ahead vigorously. Whether or not this was at variance with the general flavor of the GAC's thinking at that time, I would not want to say.

Q. Do you recall a meeting at Princeton in the spring or summer of 1951?

A. Yes, I do.

Q. You were present at that meeting?
A. I was present.
Q. Would you care to say something about the role played by Dr. Oppenheimer there, particularly in connection with what it may indicate to the board as to his cooperation in the thermonuclear program?
A. The meeting of the General Advisory Committee in June, I believe it was, of 1951, was called following an Eniwetok operation. It was called following, let me say, the discovery at Los Alamos of some extremely promising ideas in this field, and at that time the exploitation of these ideas seemed to us at Los Alamos and to others of our consultants and associated with us in the field warrant some attention by the Commission to certain decisions, let me say, of production, which were extremely important, and could well be quite expensive.

We as the laboratory made this proposal. We found the General Advisory Committee and Dr. Oppenheimer extremely enthusiastic both about this idea and about the general proposals which were needed to implement this idea, particularly insofar as they required Commission action. Indeed, I think it fair to say that the General Advisory Committee and Dr. Oppenheimer were willing to go further than the laboratory in support of this, let us say, new approach to the problem, and that their recommendations to the Commission were at least as enthusiastic as ours, and actually went somewhat beyond, in terms of support, what we had originally drafted.

I would regard this myself as very positive evidence of the interest and enthusiasm which the GAC was showing and showed in this field.
Q. You have read the portion of the Commission's letter of December 23, 1953, which referred to the circulation and distribution of the General Advisory Committee report?
A. Yes.
Q. What was the practice at the laboratory with respect to information as to the work recommendation and reports of the General Advisory Committee?
A. If I may go back to 1946 or 1947, I guess, when the General Advisory Committee was first set up, I believe it was widely recognized that the atomic weapons field was that field in which the Commission had its greatest immediate concern at that time. They were extremely anxious to support the Los Alamos Laboratory and to make sure its work was in the most fruitful directions, and had the maximum amount of assistance from the Commission.

To this end they asked the GAC to pay particular attention to Los Alamos and they requested of me that I loan to GAC as its recording secretary Dr. John Manley, who was then my associate director for research. Manley was an outstanding physicist and had long experience with many phases, in fact almost all phases, of the atomic energy program since its inception in the early 1940's. His selection was motivated both by his qualities as an individual and by the fact that he was intimately aware of the activities of the laboratory and this intimate awareness was regarded as extremely useful to the GAC in their deliberations.

In consequence of Manley's relation both to me and to the GAC, it was customary as I have indicated earlier both for me and members of my staff to meet with GAC when problems of Los Alamos were being discussed.

It was also customary for me at least to see in draft form those portions of the GAC minutes which were relevant to Los Alamos. I probably would have been unable to find any specific piece of paper which said this is indeed the request of either the Commission or the GAC. However, I am quite personally certain that it had the knowledge and at least the tacit consent of all concerned. As I say, it was frequently the occasion when we met with the GAC and to see the results of our remarks or deliberations in the draft form which were not surprising.
Q. Did you also see them in final form?
A. Probably so, because Mr. Manley's drafts were generally as good as his final form.
Q. Now, with respect to the GAC report of the meeting of October 1949, do you recall whether you saw that specifically, and if so, whether there was anything unusual about it, whether it was the normal practice, or what happened?
A. I presume I did. I cannot give any precise date that I remember seeing this precise document. But I would regard it as most likely that I did see it. Certainly we had met with the GAC in discussing some of these matters either at that time or in the general vicinity of that time, and I was well aware of the general concern of the GAC in these matters. It would have been quite natural for me to have seen these and discuss them with Manley and for members of my senior staff to have seen them.

Q. So far as you observed was there anything that Dr. Oppenheimer did to cause, as far as you know or ever heard, any unusual distribution of this GAC report?
A. Not to my knowledge.
Q. Did he play any role in the distribution of the report?
A. Not to my knowledge.
Q. Perhaps distribution is a word of art. I understand it means giving people copies and so on. Did the circulation, the showing, the knowledge of the GAC report, shall we say, so far as you know, cause anybody to change his opinion at the Los Alamos Laboratory about working on the thermonuclear program?
A. Not to my knowledge. The laboratory scientists in general, and those who contributed conspicuously to this field are strong-minded individuals and generally reach their own conclusions about matters of this nature. While I think that we regretted what seemed to be in some degree—I won't say opposition, but some degree of divergence from what might have been the flavor, let us say, of the GAC approach to it, I know of no senior person directly concerned with the weapons program at Los Alamos who left the laboratory. Indeed Dr. Manley did leave the laboratory some time in, I think, the latter part of 1950. This could be found from the record, of course—to accept the position of chairman of the department at the University of Washington, Chairman of the Department of Physics. Manley had not been directly connected with the weapon program, and the weapon development program in the laboratory. I think his title was associate director for research.

Another senior individual did leave the laboratory in 1951, that was Dr. Edward Teller, but in view of Teller's connection with this whole matter, I think you may guess it was not because of any feeling he may have had about the position of the GAC.
Q. He certainly didn't leave because he didn't want to work on thermonuclear.
A. This, I believe, is correctly said.
Q. You have seen the portion of the Commission's letter in which the statement is made, "It was further reported that you, Dr. Oppenheimer, were instrumental in persuading other outstanding scientists not to work on the hydrogen project, and the opposition of the hydrogen bomb of which you are the most experienced, most powerful and most effective member has definitely slowed down its development."

What would you say about the statement that the program was slowed down because of Dr. Oppenheimer's opinion or activities?
A. It is not my opinion that the program was slowed down, as I have said. Of course, if he himself had been in a position or wished to work on it directly and personally, this would undoubtedly have been a great help. However, it is my opinion that the program went and has gone with amazing speed, particularly in view of the predictions made regarding the difficulty of this program throughout the years 1945 to 1949. I know of no case, if you wish me to pursue these remarks, where Dr. Oppenheimer persuaded anyone not to work in this field

As I have remarked, scientists of this caliber generally make up their own minds about wishing to work or not to work in this field. A number of outstanding people whom we would like to have brought into this program felt that their best contribution to the country was to remain in university circles and contribute to the training of graduate students.

With this point of view, one can hardly differ. Of course, Los Alamos Laboratory had a selfish approach to it.
Q. Would you say that Dr. Oppenheimer's attitude, opinions, activities with respect to the development of thermonuclear weapons in any way indicated that there were some malevolent or sinister motives about it?
A. Absolutely not. As I have remarked, from 1946 on, I have never known him to act in a way other than was a help to the laboratory. In one specific instance—and doubtless others if I could recall them—outstanding young men, this was in 1949, incidentally, an outstanding young theoretical physicist by the name of Conrad Longmire had been offered an appointment by Dr. Oppenheimer at the institute. This of course is evidence itself of the outstanding caliber of this individual. It turned out he was always willing to consider coming to Los Alamos, and we were extremely anxious to have him. Dr. Oppenheimer very graciously extended or postponed his appointment to the institute indefinitely to permit him to come to Los Alamos. Indeed, Dr. Longmire never did return to the institute, and even in the last year we have explored with Dr. Oppenheimer the possibility of Longmire taking a sabbatical at the institute, and Dr. Oppenheimer has been willing to consider this.

He has given us frequently prospects, outstanding young individuals, whom we might be able to approach particularly in the field of theoretical physics to join the laboratory.

With me personally he has never been other. From October 1945 on and during the war years, other than encouraging, helpful, congratulatory, and generally both a personal friend and a friend of the laboratory.

Q. How long have you known Dr. Oppenheimer?

A. I knew him as an instructor when I was a graduate student at Berkeley in 1932–31, probably, somewhere through there. I knew him as director of Los Alamos Scientific Laboratory from June of 1944 until October 1945. I knew him thereafter as chairman of the General Advisory Committee and saw him regularly, I would say, several times a year, in that capacity. He visited Los Alamos, I would again say, at least once a year or perhaps twice, in connecton with his responsibilities as chairman of the General Advisory Committee.

Q. How well do you think you know him as a man, his character, and so on, the kind of person he is?

A. I would think I would know him as well as one knows any individual with whom one has had friendly and professional contact over quite a long number of years, and perhaps better than the average having seen him in his capacity as director of the laboratory, in which I then had an assisting subordinate position.

Q. Do you have an opinion as to Dr. Oppenheimer's loyalty to the country, and as to whether he would be a security risk?

A. I do have such an opinion and it is a very strong one.

Q. Would you state it, please?

A. I would regard him from my observation as completely loyal to this country. In fact, I would make a statement of this sort, I think, that while loyalty is a very difficult thing to demonstrate in an objective fashion, if a man could demonstrate loyalty in an objective way, that Dr. Oppenheimer in his direction of Los Alamos Laboratory during the war years did demonstrate such loyalty. I myself feel that his devotion to that task, the nature of the decisions which he was called upon to make, the manner in which he made them, were as objective a demontration of personal loyalty to this country as I myself can imagine.

Q. As to this business of a security risk, which I take it is perhaps a little different from loyalty, do you have an opinion on that?

A. I do not regard him as a security risk.

Mr. SILVERMAN. I have no further questions.

Mr. GRAY. Mr. Robb.

CROSS-EXAMINATION

By Mr. ROBB:

Q. Doctor, Dr. Oppenheimer in his answer at page 25, "I resigned as director of Los Alamos on October 16, 1945, after having secured the consent of Commander Bradbury and of General Groves that Bradbury should act as my successor."

Would you tell us about what happened in that connection? I assume that is true, is not not?

A. This statement is true. I had been assigned to the Los Alamos Laboratory as a commander in the United States Naval Reserve in June of 1944. I had been on active duty since 1941 on leave of absence as professor of physics at Stanford University. Frankly to my great surprise and equally frankly still to my surprise, some time in September—I don't remember the precise date—Dr. Oppenheimer called me in and asked if I would be willing to undertake the direction of the Los Alamos Laboratory, that he himself intended to leave and return to academic work and asked me, as I have said, to undertake this task.

The only specific reason for this, as far as I can see, was that in the course of my duties there from 1944 to 1945, I had had contact with a number of activities in the laboratory. My background was in physics, at least, and partly in nuclear physics. I did not agree to do this at that particular moment when he asked me. I asked time to think about it. I wanted to speak further with General Groves. I wished to consult with some of the senior members of the laboratory, Fermi, Bethe, and others, and ask them their opinion of my competence of this task, and what they foresaw of the problem.

I was personally extremely concerned—this is purely a personal opinion—that the laboratory continue its task. Its task in the war years had been outstandingly accomplished, but there were a number of avenues that remained to be explored. There was certainly my personal conviction that in the exploration of these avenues still further avenues would be found that it would be necessary to go into. I regarded it is inevitable that with the disclosure of the world that such bombs could be made, that other countries would undertake this activity, and that the United States would have to be the leader in this field in so far as it could make itself sure of this.

So I had a deep personal conviction that the laboratory should continue. I ultimately agreed to undertake the task for a period of 6 months or until some more logical successor could be found. Apparently no more logical successor could be induced to take the task, and I also became then convinced that it was impossible for a short-time man, a man on a short-time basis, with the announced intention of leaving, to build a permanent and enthusiastic laboratory. Whereupon I agreed to remain on an essentially indefinite basis.

Q. Doctor, you will forgive me. I am not a physicist so I don't know too much about such matters, but we have heard a number of times here reference to work on a thermonuclear device or work on a fission device. I wonder if you can tell us without getting into classified detail just what does a physicist, when he works on such a device? Does he just lock himself up in a darkroom and think, or what does he do?

A. No. I am afraid to answer your question directly would require a detailed discussion of how a laboratory works.

Q. I don't want that. I am wondering what you do when you work on these things.

A. No one man, I think it is fair to say, works on a fission bomb. Let me give you just a broad example here. One group of people, theoreticians, mathematicians, computers, will be exploring the behavior of a number of, let us say, possible systems.

Mr. GARRISON. Just for clarity, you ask about thermonuclear. He used the word fission just now.

Mr. ROBB. I said thermonuclear or fission.

Mr. GRAY. Did you intend to say fission?

Mr. ROBB. Yes.

The WITNESS. My words will be essentially applicable to both. Let us use fission and fusion indistinguishably here, because I think my remarks would be applicable to both. Working on designs for possible systems and computing, as far as the techniques of the time permit their behavior.

Another group of people, experimentalists, technicians, mechanics, shop people, will be making relevant experiments on quantities which have to go into these calculations.

Still another group of people will be working on the techniques of making the actual parts which will be required and obtaining them in the proper physical form or the proper purity, or whatever is required. All these activities follow along and periodically come to pyramids of accomplishment.

Another group of people will be doing actual, let me say, nuclear weapon engineering. That is, making out of a theoretician's schematic drawing a practical operable system. So when you speak of a person working on an atom bomb, whether it be fission or fusion, you can hardly be speaking of a person doing this. It is a group of persons whose activties have to be correlated, some at the broad base of research looking toward problems in the future; others which are involved in activities leading to a specific weapon accomplishment.

By Mr. ROBB:

Q. That helps me very much, Doctor. In other words, the development of a fission device or a fusion device requires a lot more than just thinking about it.

A. This is absolutely true.

Q. Doctor, between 1946 and 1950, how many people at Los Alamos were working on the thermonuclear as distinguished from just thinking about it?

Mr. SILVERMAN. I am not sure that the witness indicated that thinking was not a part of working.

Mr. ROBB. I think we can define our terms here.

Mr. GRAY. This is a very intelligent witness, and I am sure he is not easily confused.

By Mr. ROBB:

Q. I am not trying to confuse you, Doctor.

A. I am sure you are not.

Q. I am trying to find out, because it has always been foggy to me.

A. I understand the import of your question, but it will be necessary to answer it in a somewhat ambiguous fashion for this reason.

Let me take an example which will certainly be obvious, and certainly unclassified. The hydrogen bomb is widely known to potentially utilize one or more isotopes of hydrogen. The nuclear cross sections of these isotopes have to be known in the various energy spectrums with great accuracy for the computations. Accordingly, during the war and even after we had active groups, actively engaged in exploring the nuclear properties of the light elements, the elements which might possibly be effective or utilizable in the fusion of thermonuclear field. Those people were doing physics. They were also engaged in research which was relevant to the thermonuclear weapon.

Another example which will be difficult for me again to give because of security reasons, but I will try to guard my words—certain aspects of the so-called fission field are directly relevant, intimately related to the fusion field. If you wish to have an unclassified example of this, again it is widely known in the comic strips, that apparently some sort of primary bomb, trigger mechanism as it is called, is apparently required. How then does one distinguish developing very unique and specialized skills in primary bombs as an example?

Is this directly related to the fission field where it is immediately applicable or directly related to the thermonuclear field where it becomes applicable as soon as the techniques become sufficiently skilled?

I cannot answer your question as to what group was engaged in thermonuclear work and what people were engaged in fission work. The fields intermingle to such an extent that while we have been asked this question for a period of years by a variety of bodies, no definite answer is possible without going into detail; this man was doing this and it had that applicability and it had that applicability.

Q. Had you finished?
A. Yes.
Q. Was there any particular group at Los Alamos during that period from 1946 until 1950, or team that was working on the thermonuclear particularly?
A. There were a number of people in our theoretical division supported by computers and computing machinery that were particularly concerned with the exploration of various phenomena that would be relevant to the behavior of thermonuclear systems.
Q. Am I right in your explanation that the fission bomb is one step toward the thermonuclear; is that right?
A. I am quoting commonly accepted——
Q. Yes, sir. Were Dr. Richtmyer and Dr. Nordheim and Dr. Teller on that team that was working definitely on the thermonuclear at Los Alamos?
A. Dr. Richtmyer devoted a good portion of his time to this matter, but also a good portion of his time to the fission field. At one time Dr. Richtmyer served as alternate division leader, so he had other interests. One of his major interests was a certain type of system which may be described properly as thermonuclear, although this should not be construed to be a specific definition of it.

*　　　*　　　*　　　*　　　*　　　*　　　*

Q. And Dr. Teller?
A. Dr. Teller the same thing. Dr. Teller had been interested in this field very much, and probably a major portion of his time during the war was devoted to the exploration of this type of system. It was not uniquely so, and was not during his contact with Los Alamos after the war. But it was always one of his enthusiasms.
Q. Was anybody else, if I may use the expression, during the period of 1946 to 1950 at Los Alamos specializing on the thermonuclear?
A. How should I describe the position of people who were measuring the cross sections of deuterium?
Q. I don't know, Doctor.
A. I don't know either. You ask me were they specializing in thermonuclear.
Q. Yes, sir.
A. There were those people, if I wish to do so, that could be described as particularly interested in the thermonuclear field. I would not so describe them. They were doing fundamental research in physics, which was relevant to the thermonuclear field. Another group of experimentalists I prefer not to describe in detail who were doing work which might have been undertaken by the laboratory as general research, but was undertaken undoubtedly by the laboratory because of its probable relevance at that time to the technology of thermonuclear devices. Were they doing work in the thermonuclear field specifically, or were they not, and I cannot answer your question directly.

I am trying to make it clear that the thermonuclear field had active support both in the theoretical side, and in the relevant experimental and technological side during the war and thereafter.

Q. Could you give us any idea of how long Dr. Richtmyer devoted to the thermonuclear as distinguished from his other work?

A. You mean the percentage of his personal time?
Q. Yes.
A. I suppose roughly 50 percent so distributed.
Q. How long was he down there, sir?
A. He has been there since the war up until last year. He is still on our payroll. He is currently assigned by us to the computing center at New York University. He is shortly going to assume the directorship of that group, it is my information. He has become extremely interested in the techniques of computation.
Q. Is Dr. Nordheim still there?
A. Dr. Nordheim was and is a consultant to us. He spent roughly a year with us on leave of absence from Duke University.
Q. I believe he was down there one summer.
A. He has spent summers with us. He has spent 1 year and a good part of another on leave of absence with us.
Q. Was it during the summer that he was actively interested in the thermonuclear?
A. Certainly during the summers and during the year he spent with us. He was engaged in the computations, let us say, and trying to formulate a design for a specific type of thermonuclear system.
Q. Did you have some computers who were working on the thermonuclear problem?
A. Computers are an essential part of any thermonuclear computation. They have a very great task to play because the computations in this field are not things you make with a slide rule or a small pad of paper. As I believe I remarked earlier, one of the stumbling blocks in the years 1943 or 1948 or 1949 was the absence of computing machinery, the so-called electronic brains of sufficient capacity and magnitude to handle the type of computations which were involved. Only recently, with the development of machines such as the Maniac, the computer at Princeton, IBM computers, have we had machines which even begin to attack the problem which was confronting us during the 1944–49 era.

Dr. EVANS. They are differential equations that have no integral?
The WITNESS. They are only attackable by essentially calculation methods, by approximation methods.

By Mr. ROBB:
Q. Doctor, you mentioned in 1949 Dr. Longmire had an appointment at Princeton, but came down to Los Alamos and stayed. Did that take place before or after the Russian explosion?
A. His arrival at Los Alamos was in August or September of 1949. This is clearly almost coincidental with the Russian explosion. So his decision to come there I think must have preceded the actual knowledge of the Russian explosion.
Q. Doctor, what was your position after the Russian explosion on the question of whether or not we should develop the thermonuclear bomb? Were you for it or against it?
A. I was under the impression I had made some remarks on that subject. When you say develop the thermonuclear bomb, may I qualify my remarks to this extent. I felt, as I believe I said earlier, extremely strongly that the laboratory must undertake all possible attacks upon the thermonuclear system to see what there was of utility in this field. Now, it seems easy now to say thermonuclear bomb has been developed by public announcement; it seems obvious that there must always have been such a device in the obvious cards. This was not the case. The state of knowledge of thermonuclear systems during the war, and thereafter, and really up until the spring of 1951, was such as to make the practical utility or even the workability in any useful sense of what was then imagined as a thermonuclear weapon extremely questionable. This does not mean that——in fact, it meant very much to us that one must find out what is there in this field. Only by work in it will one find out. It is possible that we would have explored the field and out it was not, that we could not find a useful military system in it. But without this exploration, it is clear you wouldn't know.
We felt very strongly that we had to know the fact. In 1949–50 the state of knowledge at that time would certainly permit one to be very pessimistic about the practical utility of what was called a hydrogen bomb.
Q. Did you think that the Russians would certainly try to find out?

A. I was personally certain that no group of people knowing the energy which was available in the so-called fusion type of reaction would fail to explore this field.

Q. Therefore you thought we ought to also?

A. I certainly feel this way, yes, felt and feel.

Mr. ROBB. Thank you, Doctor. That is all I care to ask the Doctor.

Mr. GRAY. Dr. Bradbury, you referred to regaining a lead which we had had. I believe this was your expression with respect to this kind of thing we are talking about today.

The WITNESS. Yes.

Mr. GRAY. I suppose if in that context one refers to the thermonuclear weapons, it is a question of size. Is that a fair statement?

The WITNESS. You mean size of bang?

Mr. GRAY. That is right, yes.

The WITNESS. I am afraid it is more complicated than that.

Mr. GRAY. What I am trying to get at, Doctor, did your approach to this problem involve any kind of moral consideration or was this purely technical on the ground of practicability and usability?

The WITNESS. You are inquiring as to my personal opinions in this matter?

Mr. GRAY. That is correct. During this period that we are talking about from 1946 when you became director of the laboratory up until the present time. I may be making an effort at distinction which can't be as clearly made as I am trying to do it. But let us take a very simple matter. I suppose any ordinary conventional weapon, with respect to that, the question of making it more efficient is not a moral question at all. If you assume the weapon you have already swallowed the moral implications, I suppose.

What I am trying to get at is what you meant by regaining the lead.

The WITNESS. I meant by this only the fact that in, I think, the general guesses that people made that the Russians in the development of both the actual fact of atomic weapons and the related production enterprises had been expected to be something of the order of 5 or more years behind us. The appearance of a Russian atomic explosion in September 1949 was generally regarded, I believe, as a year or 2 or 3 earlier than one might have reasonably expected the Russians to reach this accomplishment. They were clearly therefore working at a high rate of speed, even granting what I think became evident later, the treachery of Fuchs.

At the time of course, we were not aware, as I recall, that Fuchs had indeed passed information on. Perhaps this made it seem a little more plausible that they had made such rapid progress. But at any rate it was clear at that time that—I am now only quoting my own thinking and opinion in this matter—it seemed to me that we were in the position of 2 runners in the race, where it was quite clear that your opponent was running and running quite fast. It was probably you were ahead of him in actual distance. It was not obvious that he was not running faster than you were. Our own objectives at that time had to be as far as we could make them to be sure we were running as fast as he was.

Mr. GRAY. And successful work on thermonuclear weapons might have been considered one of our legs.

The WITNESS. This I would definitely so consider. As you are aware, the thermonuclear field has two obvious military characteristics. One, apparently that in a single strike the destructive effort to deliver would be presumably very great; two, that if the materials that went into this system were indeed cheap and available, that the cost of such systems and therefore their number would not be subject to the same sort of restrictions that so-called fissions are subject to. Both these characteristics are of obvious military interest.

There are other characteristics of thermonuclear systems or any weapons systems for that matter which have to do with essentially deliverability. In other words, a weapon is no good if it is of such a character that it can't be delivered. Hence any weapons system must be looked at in terms of its net operational worth, in terms of its cost, its effects and its relation to the vehicle system appropriate to it.

All of these questions with respect to fusion systems had to be explored. They were not known at any time in 1949, certainly, and it was possible, I will not guarantee, that effort in this field would lead to something which would have military utility. However, I would like to emphasize that this was at that time a technological question. It was not guaranteed by Los Alamos or anyone else that indeed there would be a feasible or effective useful thermonuclear system.

Mr. GRAY. But on this matter of lead, thermonuclear weapons certainly were a part of that picture.

The WITNESS. Very definitely so. There were also leads that had to be established in the fission field or were being established in the fission field. This was another part of the military strength of the country.

Mr. GRAY. As a matter of hindsight, suppose there had been a Presidential directive in 1945 or at some later date, perhaps, but earlier than January 1950; is it possible that we might have had the invention or discoveries earlier?

The WITNESS. My personal opinion in answer to that question is in the negative. I would like to say as much as I can within the bounds of security as to why.

Could I consult just a moment on the question, Mr. Rolander?

(Consultation.)

Mr. GARRISON. Mr. Chairman, I will leave it to the board if the board would like to, after hearing what Dr. Bradbury has to say, explore it in classified terms. We would withdraw.

Mr. GRAY. Thank you. I hope that won't be necessary.

The WITNESS. I believe I can make my remarks in a fashion which will be acceptable. The only line of attack which had occurred to us on this problem throughout the years 1942 onwards seemed to be a line of attack during 1945–49 which would be fraught with enormous technical difficulties, that is, practical technological difficulties.

There was also a grave question as to whether or not the systems then thought of would have any behavior that would be at all, let us say, effective in terms of their probable complexity, probable size and probable cost. Had we endeavored to explore those fields in that state of knowledge, we would have had in my opinion two extremely undesirable courses, one of which would have been, I believe, almost fatal. We would have spent time lashing about in a field in which we were not equipped to do adequate computational work. We would have spent time exploring with inadequate methods a system which was far from certain to be successful, * * * I am getting here on thin ice, but if you will let me stick by my earlier remarks that skill and ingenuity in the fission field is an essential prerequisite to the success in the thermonuclear field, the progress of the laboratory during the years following the war in the understanding and development, and indeed, some systems of very close relevance to the thermonuclear system as we know them today, were an essential part of the ultimate actual ability to make an effective thermonuclear weapon.

Hindsight is a difficult thing. Perhaps the statement I am making is self-serving. But my own personal opinion is that the course of action pursued by the laboratory is right. I regret to make this statement in this fashion, perhaps because it was partly I presume my decision. But in retrospect I cannot see how we could have reached our present objectives in a more rapid fashion by any other mechanism except the mechanism by which we went.

Mr. GRAY. You think there has not been delay in any event. You reject the notion that there has been delay in the development of this weapon?

The WITNESS. I reject this notion. I also think that it is perhaps correct to say that at any time, particularly in 1945, 1946, and 1947, there were certain fundamental objectives at the laboratory that simply had to be met. If we had, let us say, retained our 1945 technology in weapons through the next 3 or 4 years, with or without thermonuclear systems, this country would have been enormously deficient in strength compared to what it was actually at that time because of the efforts of the laboratory in the fission field. These efforts also made possible subsequent developments in the thermonuclear field.

Mr. GRAY. In your conversations with Dr. Oppenheimer in 1945 with respect to the possibility of your becoming director, did you discuss what policy of the laboratory might be with respect to this matter we have been talking about, do you recall?

The WITNESS. With respect to the development of thermonuclear systems?

Mr. GRAY. Yes.

The WITNESS. No, we did not discuss this. Let me say I have no recollection of discussing this. I would like to make one additional comment in that connection. Shortly after I assumed the directorship of the laboratory, I had a meeting of all the staff members then present and one to which I was essentially talking—let me say the senior staff members, the corodinating council of the laboratory, at that time I discussed my own philosophy of the laboratory and included in that philosophy was the continuation of the exploration which we had been doing in the thermonuclear field.

Mr. GRAY. Do you recall any change of attitude on Dr. Oppenheimer's part towards the development of thermonuclear systems at any time during your association with him?

The WITNESS. I mentioned earlier the developments, the ideas in this field which occurred during the spring of 1951, prior to the meeting at Princeton in June of 1951. I think I would be correct in saying that these ideas seemed technically sound to Dr. Oppenheimer and that he upon hearing of them, regarded the prospect of success in the field as extraordinarily more likely. I think his opinions expressed at the meeting in Princeton reflected this opinion, if you wish, that here was a technique or an idea which cast a new light on the practicality of such systems.

Mr. GRAY. But you don't recall anything at the time, for example, of the use of the atomic weapon in the late months of the war that reflected any changed attitude toward thermonuclear weapons?

The WITNESS. I don't believe I ever discussed the use of any atomic weapon in war with Dr. Oppenheimer. Certainly not at that time. It would not have been my position in the laboratory to do so. We probably had discussed the GAC meetings later on of how such weapons might be employed, what vehicles might be used for them, the problems of vehicles, questions of that sort.

Mr. GRAY. Dr. Bradbury, you mentioned Dr. Teller's departure from the laboratory. I am not familiar with the circumstances of that. Could you very briefly indicate what the circumstances were?

The WITNESS. If I could do so in what I might regard as administrative confidence. This is not restricted data, but on the other hand, it has to do with personal relationships between Teller and myself.

Mr. GRAY. I don't know important it is to have that.

The WITNESS. Perhaps I can answer this without any serious difficulty, but again I would like to say that this is essentially—could I make it off the record, if you wish?

Dr. EVANS. I think Dr. Bradbury doesn't have to answer.

Mr. GRAY. Let us go off the record.

The WITNESS. I don't care whether it is in the record or not. All I would like to say is that Teller and I disagreed as to the most effective method of the administration of the thermonuclear program at Los Alamos for its most rapid accomplishment, and ultimately we disagreed on essentially a matter of triviality, that is to say, the projection in point of time in advance, a date for a definitive test operation. I think for some time prior to that, Dr. Teller and I had had some differences of personal opinion not regarding the importance of the program or the general way in which it should be going, but we had differences of opinion regarding the best way to administer it. These were differences of a rather fundamental nature in the administration of a laboratory, and since the administration of the laboratory was essentially my responsibility, I had to do it in a way that seemed best to me.

Ultimately Teller left. Our relations are personally friendly. He was a consultant to the laboratory thereafter. He still spends occasional time with us, although his primary interests are now with another group.

Mr. GRAY. At the time of the close of the war, there were varying views as to what should be done with the laboratory, I believe. There were some who wished to close it up, some who wished to continue full speed, some who favored its removal to some other place. Is that a correct statement of the varying views among the staff?

The WITNESS. I am afraid that I probably would not have a complete cross section of all the views. My own opinion was obviously strong, and my own, that the laboratory should not be closed up. It is unlikely that very many people came to argue with me that it should be closed up.

Mr. GRAY. Did you ever hear Dr. Oppenheimer express the view that it should be closed?

The WITNESS. I never did. In fact, I would probably be the last person to have heard him make such a statement inasmuch as he was instrumental in me taking it over. It would be unlikely that he would say at the same time to close it up. I was aware, and this was the proper question at the time, was Los Alamos, New Mexico, the best place to operate this laboratory. This question was actively explored by the Manhattan District in the year following the war, and the ultimate decision was that it was probably the best place to operate it.

Mr. GRAY. Dr. Bradbury, I don't want you or anyone else to misunderstand the next question I am going to ask. It points to no conclusion certainly in my mind about anything at all. It has to do with perhaps the most serious underlying implication involved in these proceedings. That has to do with loyalty to country.

I think your statement in response to a question from counsel was that you had no question about Dr. Oppenheimer's loyalty, and you based it at least in part on his very remarkable accomplishments during the war years as Director of the Laboratory. I think there are those perhaps who questioned Dr. Oppenheimer's loyalty and who might argue that an individual who was sympathetic to the USSR could very consistently have gone far beyond the call of normal duty in his war work, which was beneficial to the interests of the United States, and still have felt that sympathetic interests for the Soviet Union were also being served. That is at least an argument can be made, and I am sure you are familiar with it.

The WITNESS. Yes.

Mr. GRAY. In your testimony about Dr. Oppenheimer's loyalty, are you prepared to give your judgment to the war years? In other words, do you think that his actions since the war are of the same character and nature as to lead you to a conclusion about his loyalty?

The WITNESS. I do, and I have the same opinion. I think it can be supported by the same sort—perhaps not quite the same sort of objective evidence. I am well aware that it is possible to attribute ulterior motives to almost any human action. It is possible to argue these questions in perpetuity along those lines. Referring to my statement about his behavior as Director of Los Alamos Laboratory, in my own opinion, this to me constitutes as strong objective evidence as one can hope for, of loyalty. I have to base this not only upon the technical accomplishments of the laboratory, but upon the way in which these accomplishments were done, upon the manner in which he sought and made use of advice from his senior staff, essentially upon a sort of subjective impression which you can only get by seeing a man look worried, that indeed the success of this laboratory and its role in the war that was then going on were objectives which were uppermost and suprpassed all others in his mind I was not looking in his mind, and I cannot say this of course from definite knowledge. You can never say anything about a man's loyalty by looking at him except what you feel. I would feel from everything that I could see of his operation at Los Alamos during the war years that here is a man who is completely and unequivocally loyal to the best interests of this country.

I would make the same remark about the associations I had with him after the war years. I suppose it is true, although he can say this better than I, that he had deep personal concerns about the actual role of atomic weapons in the national security. I think anyone is entitled and should have this same sort of concern. What personal decisions one makes in the long run is of course a personal matter. But certainly his chairmanship of the GAC after the war years never questioned the fact or never questioned the assertion that the Los Alamos Laboratory should continue, should be strengthened, should proceed along lines of endeavor which were of military effectiveness. Every decision that I can recall that the GAC made with respect to the laboratory, with the possible exception of what may have been their opinion regarding thermonuclear development, seemed to me to be the right decision. In other words, there was never to my knowledge any degree of difference of opinion between myself, my senior staff, and the positions taken by the GAC.

This was particularly the case that the laboratory felt extremely strongly that actual test of nuclear weapons were a fundamental part of the progress in this field. We still feel that way extremely strongly. The GAC supported us in this. Had they not done so, our progress would have been enormously slower or almost zero. This could have been a point where one might have taken a contrary position perhaps. The GAC did not do so.

I believe the question which I tend to believe was exaggerated at the time in the public press and got into erroneous importance at the time through the efforts of a number of people—it assumed an erroneous stature in public debate—was on a case where we might have found ourselves in a difference of opinion with the GAC. Whether this difference was real or not, I am not prepared to say. But I have stated what the opinion of the laboratory was as strongly as I can.

I do not personally believe that if there was this difference of opinion, and I presume there was some difference of opinion here, that it was based on malevolent motives.

I believe and still believe that the apparent position of the GAC was based upon a defendable argument although one with which I might not personally agree. I might not have personally agreed with one of the conclusions of the question of policy that some members of the GAC arrived at. Nevertheless, I do not regard them as opinions which are either malevolent or subversive. I positively regard them as opinions which can be held and which were held as matters relating to the safety of the United States.

The safety of the United States I am convinced was uppermost in the minds of all members, including the chairman, of the GAC. We may have differed as to the best methods of obtaining the safety. I think such differences are an essential part of any democratic system. I never had then nor do I now have the slightest feeling that these differences were motivated by any other than a direct deep and sincere concern for the welfare of the country.

That was only substantiated by the actions of the GAC after the President's decision, which again were in strong support of this whole field which we characterize as thermonuclear. Basically the GAC supported the laboratory as a weapons laboratory in all fields. If there was a difference of opinion in 1949–50, it had to do with perhaps the technical question of emphasis on one or another line of attack in the weapons field in general.

Does that answer your question?

Mr. GRAY. I think probably it does. I think your answer is in the affirmative. I think my question was that you feel that the character and nature and intensity of Dr. Oppenheimer' loyalty has been as great in postwar years as you saw it in the war years.

The WITNESS. That is my feeling.

Mr. GRAY. Are there any questions?

Dr. EVANS. Yes. Dr. Bradbury, where did you have your undergraduate and graduate education?

The WITNESS. I received the bachelor of arts degree from Pomona College in Clairmont, Calif., in 1929. I received the Ph. D. from the University of California in 1932. Then for 2 years I was research fellow at MIT. Thereafter I was on the academic staff at Stanford University, first as assistant professor, associate, and then full professor.

Dr. EVANS. Are you a Communist?

The WITNESS. No, sir.

Dr. EVANS. Have you ever been?

The WITNESS. No, sir.

Dr. EVANS. Have you ever been a fellow traveler?

The WITNESS. No, sir.

Dr. EVANS. There were a lot of organizations that the Attorney General listed as under communistic control, Doctor; do you know that list?

The WITNESS. I have seen that list.

Dr. EVANS. Are you a member of any of those organizations?

The WITNESS. I am not. I think it would be an awful time to find out if I were.

Dr. EVANS. Were you surprised when the Russians fired the bomb?

The WITNESS. In 1949?

Dr. EVANS. Yes.

The WITNESS. Yes, sir.

Dr. EVANS. You were surprised?

The WITNESS. I was surprised.

Dr. EVANS. Do you think the knowledge that Fuchs might have given them helped them in that?

The WITNESS. I now think so. I was surprised at the time that it came so early. It is now my personal impression although I have no evidence to support this, of course, that probably they were assisted along these lines by the information Fuchs appears to have given them.

Dr. EVANS. You do think that scientific men should be required to keep their discoveries secret when they might affect the country and not publish them?

The WITNESS. That is a very difficult question to answer, sir. It is very difficult for a scientist doing basic research to be sure that in the course of time this particular technical report, paper, invention or discovery may affect the security.

May I give you an example of this? It would have been a perfectly normal thing for a scientist to do, although somewhat difficult, to measure certain neutron cross sections of deuterium in 1932, 1934, 1936, and 1938, and so on. It would have been a nice task and perfectly good nuclear science at that time. At the present time, such cross section measurements are, of course, carefully guarded secrets because they are relevant to a thermonuclear problem. How in 1934 or 1936 would one have known that these cross sections are going to be something that would affect national security? I can't give you an answer to your question.

I think if an individual knows or believes that his discovery is immediately relevant to national security, he has definite responsibility to the country in that connection.

Dr. EVANS. Do you think that scientific men as a rule are rather peculiar individuals?

The WITNESS. When did I stop beating my wife?

Mr. GRAY. Especially chemistry professors?

Dr. EVANS. No, physics professors.

The WITNESS. Scientists are human beings. I think as a class, because their basic task is concerned with the exploration of the facts of nature, understanding, this is a quality of mind philosophy—a scientist wants to know. He wants to know correctly and truthfully and precisely. By this token it seems to me he is more likely than not to be interested in a number of fields, but to be interested in them from the point of view of exploration. What is in them? What do they have to offer. What is their truth. I think this degree of flexibility of approach, of interest, of curiosity about facts, about systems, about life, is an essential ingredient to a man who is going to be a successful research scientist. If he does not have this underlying curiosity, willingness to look into things, wish and desire to look into things, I do not think he will be either a good or not certainly a great scientist.

Therefore, I think you are likely to find among people who have imaginative minds in the scientific field, individuals who are also willing, eager to look at a number of other fields with the same type of interest, willingness to examine, to be convinced and without a priori convictions as to rightness or wrongness, that this constant or this or that curve or this or that function is fatal.

I think the same sort of willingness to explore other areas of human activity is probably characteristic. If this makes them peculiar, I think it is probably a desirable peculiarity.

Dr. EVANS. You didn't do that, did you?

The WITNESS. Well ——

Dr. EVANS. You didn't investigate these subversive organizations, did you?

The WITNESS. No. Perhaps my interest lay along other lines. I don't think one has to investigate all these political systems.

Dr. EVANS. Do you go fishing and things like that?

The WITNESS. Yes, I have done a number of things. Some people, and perhaps myself among them, I was an experimental physicist during those days, and I was very much preoccupied by the results of my own investigations.

Dr. EVANS. But that didn't make you peculiar, did it?

The WITNESS. This I would have to leave to others to say.

Dr. EVANS. Younger people sometimes make mistakes, don't they?

The WITNESS. I think this is part of people's growing up.

Dr. EVANS. We all do.

The WITNESS. That is, take actions which turn out to be wrong later on. Whether they were mistakes at the time may be a debatable question.

Dr. EVANS. Do you think Dr. Oppenheimer made any mistakes?

The WITNESS. My personal feeling here with regard to the situation specifically to the question of organizations is that these are actions which in the light of history, in the light of subsequent developments, turn out to have been undesirable. I would not like to say that I regard them as either right or wrong. I say that simply they turn out to have been bad for him to have done at this time. At the time they were done, I regard them as potentially at least without significance. They reflected a certain area of interest, an interest which as you recall was held by a number of people at that time. The Spanish war was of concern to a number of people.

Dr. EVANS. That is potentially they should have been of no interest to this Board?

The WITNESS. No, I cannot say that. I don't wish to make a speech. It is unfortunate that the number of objective examples which one has of, let us say, people who are disloyal is extremely small. You can count them on the fingers of one hand. In every case these people seem to have been drawn from a certain type of background in which at least some degree of interest in liberal, leftwing or Communist activities was a part. Therefore, I have to agree that where this background of interest in these affairs occurs, that a query at least is indicated.

It is a fact of life, but I think it perhaps regrettable that because a few people out of thousands have been discovered in this particular area, that thousands or tens of thousands are automatically thereby put potentially in the same category. I think the question has to be raised because of the things which Fuchs, Alan Nunn May, Greenglass have done. Perhaps it is one of the most serious things they have done, to cast a shadow of suspicion on those who were interested in these activities for completely humanitarian or intellectual motives.

I think therefore this question has to be raised. I myself do not regard the matter of membership in such societies or interest in them as particularly significant in the light of the times—let me say necessarily significant in the light of the times. I think it is a question which must be raised, must be explored. It may turn out to have meaning. It might be in this case it does not have meaning.

Dr. EVANS. You spoke of loyalty. Would you put loyalty to your country above loyalty to your friends?

The WITNESS. I would.

Dr. EVANS. That is all I have.

REDIRECT EXAMINATION

By Mr. SILVERMAN:

Q. Dr. Bradbury, from your knowledge of Dr. Oppenheimer, today, do you think he would put loyalty to his country above loyalty to a friend?

A. I believe he would.

Mr. SILVERMAN. That is all.

RE-CROSS-EXAMINATION

By Mr. ROBB:

Q. Doctor, I have one question suggested by your discussion with the Chairman about what might be the result had there been a Presidential directive in 1945 or 1946 to undertake all out work on the H-bomb.

It has been testified here, Doctor, that something happened in the spring of 1951, and that accelerated the successful development of the thermonuclear so that work came to a successful conclusion maybe 18 months thereafter.

My question is, Supposing that something had happened in 1945 or 1946, what would have been the result? How soon do you think you would have had the thermonuclear weapon perfected?

A. We had this idea——

Q. Is that an intelligent question?

A. This is a question that I would answer this way. Had this idea occurred in 1945, 1946, 1947 or 1948 or almost any time before it did occur, we would not have known how to use it in an effective military fashion. We were already pursuing in the years following the war those techniques, specifically in the fission field, which made the implementation of this idea a practical thing. We had already conducted experiments. I can't describe them for security reasons. They were in the fission field, and bore directly upon this field. Frankly, if I may go back to one of your other potential questions, had there been a Presidential directive to proceed along thermonuclear lines in 1945, I would almost doubt in retrospect that we would have done or could have done anything much different than we did. In other words, the active exploration of the fission field was a necessary and essential prerequisite known all along to the fusion field. Had there been such a hypothetical decision, it is impossible to answer. Had there been, we would have done exactly as we did. We might have been persuaded otherwise, and I think if we had we would have found ourselves farther behind in 1954 than we are.

Q. I am not sure your answer—and that is my fault and not yours.

A. It is my fault.

Q. Your answer about not knowing how to use this discovery in 1946 or 1947, could you explain that a bit further?

A. I would have great difficulty in doing so without going into restricted data. Let me think for a moment to see if I can find some way around this.

There would be two possibilities. We would not have been able to make the relevant calculations for mechanical reasons. We would not have been able to make them for let us say technological reasons, because only in the course of those years did we begin to get some understanding of how to compute atomic or fission bombs. * * *

Q. Doctor, in the years between 1946 and 1950, did you have the staff and the equipment then to do what you did subsequent to this discovery in 1951?

A. Between when did you say, 1945 and 1950?

Q. Yes, sir. In other words, assuming this discovery in 1945, 1946, or 1947, did you then have the staff to do what you did with the discovery in 1951 and 1952?

A. As you are doubtless aware, in 1945 the laboratory of course was partly civilian and partly military. We had a couple of thousand SED, special engineering detachment of the military personnel. We had a number of officers. In 1945 and early 1946, a great part of our civilian personnel left to return to school, to their industrial and academic jobs. The size of the laboratory reached its minimum roughly in September of 1946, at which time its size was roughly half, perhaps a little less than half of its size at the present time. From that time on it has grown steadily up to about the present time.

There were admittedly difficulties in taking the laboratory through the transition period prior to the Atomic Energy Act, while personnel straightened themselves out in their own desires. In 1946, throughout the entire year, or at least until the adoption of the Atomic Energy Act, perhaps we were lucky to keep ourselves alive. We had the Crossroads Operation to carry out, and life was far from easy. I don't say it has ever been easy, but in those days certainly our task was not simple. We were devoting, as I have said earlier, our major directed effort, the efforts which come to the peaks of these pyramids of development, two things which would make the production capacity of the United States as effective in a military way as it possibly could be right then and there. We were also devoting our efforts to making atomic weapons as they then existed more effective as part of a weapons system for the country; in other words, an effort to maximize the immediate potential of the country.

As I have said earlier this was not to the exclusion of thermonuclear work but it was the focus of achievement which was in the fission field. We would have had a hard time and unprofitable time and I think in the light of subsequent events, and it would have been an error and mistake to try to hash about in a field for which none of the basic technologies then existed, and at a time when there were very clear things to be done in the fission field.

Q. Beginning with the Presidential directive in January 1950, did you thereafter receive additional personnel and additional funds and additional assistance in your work?

A. The laboratory has never lacked for funds. The actual request for funds has always been supported by the Commission and the Congress. The growth of the laboratory has been as rapid as we could make it subject to housing and the ability to draw personnel into our isolated area, and into the classified field. There was no immediate change in either dollars or personnel before or after the President's recommendation. It was a matter of growth. We did at that time carry out an active campaign to enlist the services of a number of the senior scientists of the country who had been with the project during the war, to see if they could come back on a year's leave of absence, and we were successful in a number of these cases, and in a number we were not because they felt their task was more urgent in the instruction of graduate students.

Q. Whether it was immediate or not, as a result of the Presidential directive, was there an expansion in your facilities and personnel and funds?

A. As a result of the Presidential directive, I can't say there was. I would say there has been an expansion and an increase of our funds continuously in the years from 1945 on onward. I would have to look at a graph of the actual dollars per year spent. I don't have it with me. I would doubt if such a graph of dollars spent would show any significant fluctuation in the period we were talking about, except as a result of a test activity occurring in this year or not in this year. By this I do not mean that we lack support. We have always received from the Commission and the Commission from the Congress as much support as we could see our way clear to use in a justifiable fashion.

Mr. ROBB. Thank you, sir.

Mr. GRAY. I am sorry, Dr. Bradbury, that I am not through with my questions. When did you go to Los Alamos?

The WITNESS. I arrived July 4 or just about July 4 of 1944. I first visited there some time in June 1944 when I was about to be transferred there. Prior to that I was at the United States Proving Ground at Dahlgren, Va.

Mr. GRAY. I have forgotten at what time some of these people whom we discussed in earlier proceedings, such as Lomanitz, left. I guess he left before you arrived?

Mr. ROBB. He was not at Los Alamos.

Mr. BECKERLEY. He was at Berkeley. Did you know that man?

The WITNESS. No.

Mr. GRAY. Did you know David Hawkins?

The WITNESS. Yes.
Dr. EVANS. Weinberg?
The WITNESS. Weinberg, no.
Mr. GRAY. What were some of the other names?
Mr. SILVERMAN. I don't believe Weinberg was at Los Alamos.
Dr. EVANS. No, he wasn't.
Mr. GRAY. You knew Hawkins?
The WITNESS. I knew David Hawkins, yes.
Mr. GRAY. Did you know anything about his sympathies?
The WITNESS. At that time, no. I was unaware of his background until it was about to appear in the public notice.
Dr. EVANS. That is, it is perfectly possible to be about a man quite a long time and not know anything about his background?
The WITNESS. It is perfectly possible. I knew David Hawkins in a friendly fashion. I presume I have had cocktails with him. I presume I have been to dinner with him. I never discussed politics with him and found him a very loyal supporter of our activities there.
Mr. GRAY. Were you surprised when you read or heard that he had been a member of the Communist Party?
The WITNESS. I would say I was surprised, yes. I don't wish to have this interpreted that I was shocked. I have no idea of this. I had no reason to have any idea.
Mr. GRAY. Did you know Philip Morrison?
The WITNESS. Yes.
Mr. GRAY. Do you know anything about his sympathies?
The WITNESS. I would say my personal contact with him was the same as with David Hawkins. I had more technical contact with him because he was very active in the design of one of our research tools, the so-called fast reactor. We valued his professional advice extremely highly. I never recall discussing with him political problems. I was, I think, indirectly aware that he was not entirely sympathetic to the development of the atomic bombs. But I don't think he was unique in this feeling among people who were about to leave Los Alamos.
Mr. GRAY. This would indicate that you could know an individual and see him frequently, as Dr. Evans said, in complete ignorance of membership in the Communist Party?
The WITNESS. I am sure this is certainly true. I knew Fuchs well.
Dr. EVANS. You did know him?
The WITNESS. I wouldn't say well. I am sure Fuchs has been a guest at my house, and has had cocktails at my house or perhaps even eaten dinner at my house.
Mr. GARRISON. In Los Alamos?
The WITNESS. Yes. I must say in that case I was deeply shocked by what appeared to have been Fuch's activities at the time. This was a great shock to all of us at Los Alamos.
Dr. EVANS. It was a great shock to everybody.
Mr. GRAY. There seems to be no question that he had a commitment to a foreign power, does there?
The WITNESS. I perhaps might have a slightly different interpretation of it. I think it must be said in fairness to Fuchs that he worked extremely hard and effectively for Los Alamos and this country. He appears to have a divided or double loyalty. I think his accomplishments at Los Alamos it must be said were very effective.
Mr. GRAY. This was the point I was trying to make in the question I asked you earlier, and when I asked you not to misunderstand the import of the question, that here is an example, Fuchs, himself, who at the same time could want Los Alamos to be a marvelously successful laboratory, and still have loyalty to another country.
The WITNESS. I never saw in Fuchs anything other than to indicate a hardworking, effective, skilled physicist. I think it is agreed that his accomplishment at Los Alamos did assist the laboratory in the attainment of his objectives.
Dr. EVANS. He was a Dr. Jekyl and Mr. Hyde.
The WITNESS. I have to admit a complete failure to understand Mr. Fuchs.
Mr. GRAY. Thank you very much, Dr. Bradbury.
(Witness excused.)
Mr. GRAY. We will take a little recess.
(Brief recess.)

Mr. GRAY. Let us resume. Dr. Evans is out for a moment but will be back. Dr. Whitman, do you wish to testify unde oath? You are not required to do so.
Dr. WHITMAN. I am perfectly willing to.
Mr. GRAY. All the witnesses have so testified.
(Dr. Evans entered the hearing room.)
Dr. WHITMAN. Yes, I will be glad to.
Mr. GRAY. Would you be good enough to stand and raise your right hand, please. What is your full name?
Dr. WHITMAN. Walter G. Whitman; Walter Gordon Whitman.
Mr. GRAY. Walter Gordon Whitman, do you swear that the testimony you are to give the board shall be the truth, the whole truth and nothing but the truth, so help you God?
Dr. WHITMAN. I do.
Mr. GRAY. Will you be seated, please, sir.
It is my duty to remind you of the existence of the so-called perjury statutes, Dr. Whitman. May we assume that you are familiar with their existence and penalties?
Dr. WHITMAN. Yes.
Mr. GRAY. I should like to ask that in the course of your testimony if it becomes necessary for you to disclose or refer to restricted data that you notify me in advance so we may take necessary and appropriate steps.
Finally, Dr. Whitman, we treat these proceedings as a confidential matter between the Atomic Energy Commission and its officials and Dr. Oppenheimer, his witnesses, and representatives. The Commission will initiate no public releases with respect to these proceedings. It is my custom to express on behalf of the board a hope that witnesses will have the same view.
Mr. Silverman, will you proceed.
Whereupon, Walter Gordon Whitman was called as a witness, and having been duly sworn, was examined and testified as follows:

DIRECT EXAMINATION

By Mr. SILVERMAN:
Q. Dr. Whitman, will you state what your profession is, please?
A. I am a chemical engineer and the head of the chemical department of the Massachusetts Institute of Technology.
Q. Do you hold any governmental position?
A. I am a member of the General Advisory Committee of the Atomic Energy Commission.
Q. How long have you been such a member?
A. Since the summer of 1950.
Q. I understand that you were formerly chairman of the Research and Development Board of the Department of Defense?
A. Yes, sir. I came down under General Marshall in the summer of 1951, served under him, Mr. Lovett and Mr. Wilson for 2 years.
Q. Will you tell us something about your association with Dr. Oppenheimer.
A. My first meeting with Dr. Oppenheimer came in 1948 at a time when I was the director of the so-called Lexington project which MIT ran for the Atomic Energy Commission to determine or pass upon the feasibility of nuclear powered flight.
In connection with that project I met Dr. Oppenheimer in June of 1948 at the time we were getting background information. The contact was not important. My real contact began in September 1950 at the first meeting of the General Advisory Committee after my appointment.
I knew him in General Advisory Committee work quite intimately for the next 2 years until the termination of his 6-year term on the General Advisory Committee. He was, of course, the chairman of the committee, as you know.
I had very close association with him also when I accepted the position as chairman of the Research and Development Board because he was then a consultant to me and a member of my committee on atomic energy, a committee composed of high ranking military officers from the three services concerned with atomic energy and certain civilians, Dr. Oppenheimer, Dr. Bacher, and a few others.
Q. Dr. Bacher was chairman?
A. Dr. Bacher was chairman of that committee. That association was very close from August 1, 1951, for the next 2 years while I was in the Pentagon. I also served on a special panel headed by Dr. Oppenheimer in December 1950. This was in the Pentagon under the Research and Development Board before I

became chairman and the purpose of this special committee was to review the status of atomic energy and military applications and try to point out the lines of research and development which should be followed in a wider exploitation of atomic energy for military purposes. It was a look into the future.

I also had one special connection with Dr. Oppenheimer in December 1951 on a trip to visit SHAPE headquarters and General Eisenhower to discuss with him the findings of the so-called VISTA report. The VISTA report carried out at the California Institute of Technology for the military was headed by Dr. Lée DuBridge. Dr. DuBridge, Prof. Charles Lauritsen and Dr. Oppenheimer went over to discuss this report with General Eisenhower and others—General Gruenther, General Norstad, under the general sponsorship of the research and development board, of which I was chairman. So I accompanied them on this 1-week trip with the approval of Mr. Lovett, the Secretary of Defense.

I would say that my other contact official connection which is of less importance was as a fellow member of the Science Advisory Committee from about the fall of 1951 until December of 1953.

Mr. GRAY. Science Advisory Committee of what?

The WITNESS. Of the Office of Defense Management.

Of these various contacts my close association on the General Advisory Committee, the trip to Europe in connection with the VISTA report and the close association as my consultant in the research and development board and a committee member are the significant ones.

By Mr. SILVERMAN:

Q. Would you tell us something about how Dr. Oppenheimer ran, if that is the correct word to use, the meetings of the GAC so far as bringing out or permitting expressions of views of the members and so on is concerned?

A. In the first place Dr. Oppenheimer worked very hard in advance of the meeting in order to prepare a most worth while agenda for consideration by the committee. Some of the items were suggested by the Commission itself and others were brought up by study by other members of the committee, particularly by Dr. Oppenheimer. He was very careful to outline the problem and to see to it that we had authoritative presentations of the situation on which we were to give advice. I may say that he made it quite a point to assure the participation and the expression of views by all members of the committee, not to initially state his own views and try to coerce others to those views.

I think we were all, at least I was, remarkably impressed by his ability to summarize the conclusions and the thinking of the committee in the presentation before the Commissioners themselves at the end of the 3-day meeting.

Perhaps I should say that initially we would meet with the Commissioners and discuss the subjects that would be brought up. They would point out particular things on which they would like our views and advice.

Q. This was an oral discussion?

A. This was an oral discussion. The last item of the 3-day meeting was a meeting with the Commissioners themselves at which was presented the conclusions and thinking of the committee.

During the progress of the meeting very frequently individual Commissioners would come in to participate in the discussions which we were holding.

In his final summarization of the committee advice, Dr. Oppenheimer had a remarkable ability to pull it together and he would also make quite a point of asking individual committee members to explain more at length their views, which might be entirely in accord with his summary, or might represent a different position. So I always had a feeling that as the chairman of the meeting he was most anxious that the Commission get the benefit not only of the summary which the chairman of the committee could give, but also the views which might represent differing shades of opinion or even disagreement.

Q. You, of course, were not a member of the General Advisory Committee at the famous October 1949 meeting on the hydrogen bomb?

A. No, I had nothing to do with that, knew nothing of it and didn't enter the scene until a year later when the President's decision had been announced and many months had elapsed.

Q. During the period from the time you became a member of the General Advisory Committee in September 1950 until Dr. Oppenheimer's term expired in the summer of 1952, would you care to say anything about Dr. Oppenheimer's attitude and contributions, if any, toward the work of the GAC in connection with the hydrogen bomb?

A. This subject came up again and again at our meetings. Frankly, I was shocked to read any comment that there was an attempt to obstruct progress after the decision was made, because all the way through I had the feeling that he not only was not obstructing but that he was working hard toward helping toward the early success of the hydrogen program.

Q. Do you recall a meeting at Princeton in the late spring or early summer of 1951 on the hydrogen bomb?
A. I do.
Q. Can you tell us anything about that and particularly Dr. Oppenheimer's roll there?
A. Dr. Oppenheimer was the moderator of that meeting, which consisted of him, if not all of us on the General Advisory Committee, some of the Commissioners, people like Dr. Teller, Dr. Bradbury, and at that time there was a very thorough consideration of what the status was today, what the hopes and prospects were and at the conclusion of it, a program was discussed with which the meeting was in pretty general agreement on pushing ahead the lines that should be pushed hardest.

I should say frankly that I, not being a nuclear physicist, found that when Dr. Teller, Dr. Oppenheimer, Dr. Bethe, and Dr. Fermi got talking about some of the technical problems, it was a bit over my head. I, however, was in a position, I believe, to sense the significance of what was being discussed and to concur wholeheartedly in the conclusions which were reached.

Q. Was Dr. Oppenheimer's position at that meeting one of actively being in favor of going ahead with whatever line of development was there agreed upon?
A. Yes. He very much took the position of being the moderator of the meeting to be sure that all of the facts were brought out, that the discussion was active between some of these very brightest minds of the country, and to see to it that the thing was pulled together in the way of a conclusion as to future action.

Q. Have you from time to time discussed with Dr. Oppenheimer and worked with Dr. Oppenheimer on the matters involving the proper use of atomic weapons?
A. Oh, yes. This was a very important part of his function as advisor to me in the Department of Defense.
Q. Would you care to say something about Dr. Oppenheimer's work or contributions in developing the concept of tactical use of atomic weapons?
A. Yes. Dr. Oppenheimer fully realized that atomic materials—the raw materials for nuclear explosions—would become increasingly abundant and increasingly cheaper. There had been in the early days of scarcity a very strongly held belief that the bomb was useful in strategic bombing and there had been very little thought given to the expansion of the use of the bomb for other military purposes.

I should say that always Dr. Oppenheimer was trying to point out the wide variety of military uses for the bomb, the small bomb as well as the large bomb. He was doing it in a climate where many folks felt that only strategic bombing was a field for the atomic weapon.

Q. Strategic bombing is a large bomb somewhere where the Army is not?
A. * * * I should say that he more than any other man served to educate the military to the potentialities of the atomic weapon for other than strategic bombing purposes; its use possibly in tactical situations or in bombing 500 miles back. He was constantly emphasizing that the bomb would be more available and that one of the greatest problems was going to be its deliverability, meaning that the smaller you could make your bomb in size perhaps you would not have to have a great big strategic bomber to carry it, you could carry it in a medium bomber or you could carry it even in a fighter plane.

In my judgment his advice and his arguments for a gamut of atomic weapons, extending even over to the use of the atomic weapon in air defense of the United States has been more productive than any other one individual. You see, he had the opportunity to not only advise in the Atomic Energy Commission, but advise in the military services in the Department of Defense.

The idea of a range of weapons suitable for a multiplicity of military purposes was a key to the campaign which he felt should be pressed and with which I agreed.

I think it rather significant to realize that in the days of scarcity there was such a strong——
Q. Scarcity of what, sir?
A. Scarcity of fissionable material. In the early days there was such a strong feeling that the bomb was the peculiar and sole property of the Strategic Air Command. It was very necessary to open up to the minds of the military

the other potential uses of this material which was going to become more available and cheaper all the time, and that deliverability was going to be a vital factor.

Q. On what occasion did Dr. Oppenheimer express and urge these views?

A. The first time I ran into them was on the special panel over in RDB in September of 1950 on the forward look to the atomic weapon in the Department of Defense. At that time I didn't have enough background, frankly, to contribute very much to it. Subsequently when I became chairman of RDB this was rather a key point in my own determination of emphasis in research and development.

Q. Was Dr. Oppenheimer opposed to the use of atomic weapons for strategic purposes?

A. That is a hard thing to say. He was certainly not opposed to the development of atomic weapons useful for strategic purposes. This is what I would like to say specifically. I saw no evidence of obstruction in the development. I think many of us felt that if and when the atomic weapon is really loosed in a strategic campaign, which would be on both sides, it is the end of civilization as we know it, and that the efforts must be predominantly to prevent any such thing from happening. But the necessity for being strongly armed for strategic air I have never questioned Dr. Oppenheimer's realization.

Q. Perhaps I have not expressed it too clearly, but what I would like is for you to comment on Dr. Oppenheimer's views as to emphasis on one branch or another of the use of atomic weapons, or as to a feeling that it is a matter of balance or what have you?

A. Yes. I think very definitely he felt that great emphasis should be put on having a spectrum in the arsenal of atomic weapons; that there were so many potentialities to this new material. He recognized as practically everybody has that the strategic use was being pushed with utmost speed.

He felt it quite incumbent—I am interpreting, this is my feeling of how he felt—to emphasize the many other potentialities of the atomic weapon, and since that was not being talked about by others he was peculiarly conscious of his responsibility.

Q. Did that cause some trouble for him in the Department of Defense?

A. The Strategic Air Command had thought of the atomic weapon as solely restricted to its own use. I think that there was some definite resentment at the implication that this was not just the Strategic Air Command's weapon.

Q. Did Dr. Oppenheimer urge this view of balanced defense and the gamut of atomic weapons on this trip to SHAPE that you mentioned also?

A. Yes. In the talks which were held with them, General Eisenhower, General Gruenther, General Norstad. General Eisenhower, of course, at that time with the defense of Europe was particularly interested in the views as to what the developments might be and how they could be employed in his mission.

Q. How well do you feel you know Dr. Oppenheimer as a man, with respect to his loyalty and character and so on?

A. I feel I know him quite well.

Q. Do you have an opinion as to Dr. Oppenheimer's loyalty to the United States and as to whether he is a security risk?

A. I have a very strong opinion.

Q. Would you state that opinion, please?

A. I have an opinion that he is completely loyal and that he is not any more of a security risk than I am. Perhaps I should explain.

I feel that anyone who has secret information is to a degree a security risk, which would be illustrated by the fact that if I were unfortunately in Communist hands and they elected to torture me, I have no confidence in my ability to refrain from disclosure. Under those circumstances I think almost any of us would be security risks and the more information we have the greater the risk. But with the exception of this, which is common to all of us, I do not regard Dr. Oppenheimer as any more of a security risk than I regard myself.

Q. And even that is not an exception, I take it. I will withdraw that.

A. At least I have some confidence in myself.

Q. Have you read the letter of the commission dated December 23, 1953?

A. I have.

Q. Referring to the one suspending Dr. Oppenheimer's clearance and your answer is that you have?

A. I have.

Q. That contains certain items of derogatory information.

A. Yes; it does.
Q. Does that letter change your views as to Dr. Oppenheimer's loyalty or his being a security risk?
A. It does not.
Q. Were you familiar with those items of derogatory information, except for the hydrogen bomb as to which you said you were rather shocked, prior to the Commission's letter?
A. I was.
Q. Will you tell us the circumstances under which you became familiar with that?
A. In my position in the Pentagon, Dr. Oppenheimer's case was brought to my personal attention through the security officers. This was close to the completion of my term in the Pentagon. I said that I would personally review the whole case and leave for my successor my recommendation in terms of whether or not Dr. Oppenheimer should be reappointed for another year as a consultant in the Department of Defense.
Mr. ROBB. Could we have the date on this?
The WITNESS. That was early July 1953.

By Mr. SILVERMAN:
Q. Dr. Whitman, that was pursuant to the President's Executive order requiring a review of all such cases?
A. That was in line with the President's order which required a review of cases which had significant derogatory information.
Q. What was your position at that time?
A. I had been Chairman of the Research and Development Board until the reorganization plan went into effect on the 20th of June 1953. My successor, who was to be appointed as Assistant Secretary of Defense, Research and Development, was not going to take office until the latter part of the summer.
Mr. GRAY. What was his name, for the record?
The WITNESS. Donald Quarles. He subsequently took office on the first of September. In the meantime I continued operating with the same functions which I had, but under the official designation of Special Assistant to the Secretary of Defense for Research and Development. I took a Saturday when no one else was around to study the file very thoroughly. As I understand it, it was a summary by the FBI of the material in Robert's folder. It was a file that may have had 50 or 60 pages in it.

By Mr. SILVERMAN:
Q. How long did it take you to read it?
A. It took me at least 2 hours, and I think more, because I was reading it very carefully and re-reading to feel that I had the significance of the file. At the conclusion I wrote longhand a memorandum pointing out that I had been——
Q. Do you have a copy of that memorandum?
A. I have a copy of the memorandum.
Q. Perhaps it would be simpler to read the memorandum than for you to tell what it said.
A. Regarding Dr. J. Robert Oppenheimer. I have known for some time of the general nature and salient features of the information contained in this file. It discloses nothing which would cause me to modify my previous confidence in his loyalty.
"Based on extensive associations with Dr. Oppenheimer over the past 3 years in the General Advisory Committee of the AEC and in the Office of Defense Management Science Advisory Committee, and in the Research and Development Board, I am convinced that he can be of great service as a consultant to the research and development work of the Department of Defense.
"I unqualifiedly recommend his reappointment as a consultant."
Q. I take it nothing has happened between the date of that memorandum and today that would cause you to change your opinion as to Dr. Oppenheimer's loyalty or being a security risk?
A. No, sir; I would make the same recommendation today.
Mr. GRAY. What was the date?
The WITNESS. The date of that was July 10, 1953.

By Mr. SILVERMAN:

Q. So far as you can now recall, are there any items of derogatory information in the Commission's letter of December 23, 1953, other than the hydrogen bomb, that was not included in the file that you then examined?

A. To the best of my recollection everything except the references to the hydrogen bomb was in the file which I examined.

Mr. SILVERMAN. I think I have no further questions to ask Dr. Whitman.

Mr. GRAY. All right.

CROSS-EXAMINATION

By Mr. ROBB:

Q. Doctor, do you know whether he was reappointed?

A. I do not know for certain. I left at the end of July. This is hearsay. I think that the case was really brought up to the attention of Mr. Wilson some some time in the fall after the new Assistant Secretary, Mr. Quarles, had taken office on the first of September.

Q. Who would have made the appointment—Mr. Wilson?

A. It had been previous practice for me to make the reappointments. The practice was in process, I think, of change during the summer of 1953, following the President's Executive order; and I frankly do not know what the present procedure is, whether Mr. Quarles makes the appointment or whether Mr. Wilson does.

Q. Or maybe Mr. Quarles recommends and Mr. Wilson makes the appointment.

A. I just don't know.

Q. I seem to recall seeing a statement in the press the other day from Mr. Wilson to the effect that he will not have Dr. Oppenheimer over there. Did you see that?

A. I saw Mr. Wilson's press statement. In fact, I have a copy of the whole thing.

Q. If that were accurately reported, it would indicate that he was not reappointed.

Mr. GARRISON. Mr. Chairman, I didn't think Dr. Oppenheimer's name was mentioned.

Mr. ROBB. Apparently the witness understood it as I did.

Mr. GRAY. I think the chairman would make this observation. Perhaps Mr. Garrison is technically correct, but I believe there seems to be no question in the minds of any of us that Mr. Wilson in every likelihood was referring to Dr. Oppenheimer.

Mr. ROBB. I have forgotten what the pending question was.

Mr. EVANS. Do we have a copy of that?

Mr. ROBB. Dr. Whitman says he has a copy of it. Do you have a copy?

The WITNESS. I have a copy of his statement which was sent to me, or at least of the press conference. I think I have. This is entitled, "Excerpts From Department of Defense, Office of Public Information, Minutes of Press Conference Held by the Honorable Charles E. Wilson, Secretary of Defense, Wednesday, April 14, 1954."

By Mr. ROBB:

Q. Do you want to read the pertinent portion to us, or do you want me to read it?

A. It is rather extensive. I would just as soon give it to you for the committee if you care to have it.

Q. Thank you.

A. It is not significantly different from the report that came out in the New York Times.

Mr. ROBB. It is quite long, as the witness says. It is five pages. So I will not attempt to read it now.

Mr. GRAY. The state of the record now would indicate that Mr. Wilson would not have accepted your recommendation in all probability, at least that is the impression. If counsel want to straighten it out——

Mr. SILVERMAN. I have no information on the subject. The only comment I wish to make is that it is perfectly possible that Mr. Wilson reviewed the file. I have no idea what Mr. Wilson did. I do think there is a difference in the weight to be given to a determination and a recommendation made by a man who reads through a file with the duty of trying to make a recommendation, and with all due respect to Cabinet officers and even ex-Cabinet officers, the statements that they make in a press conference.

Mr. GARRISON. Mr. Chairman, I think since the matter has been brought up, I would request that the press conference be read into the record.

Dr. EVANS. I think that is very wise.

Mr. GRAY. As Dr. Whitman indicated, this is entitled, "Excerpts From Department of Defense, Office of Public Information, Minutes of Press Conference Held by the Honorable Charles E. Wilson, Secretary of Defense, Wednesday, April 14, 1954, 3 p. m., Room 3E–869, the Pentagon, Washington, D. C."

There are some dots. I am not clear what that indicates, but following the dots:

"The PRESS. Mr. Wilson, can you discuss the Dr. Oppenheimer situation at all?

"Secretary WILSON. No. I'd class this in the same category. That is apparently going to be reviewed by a board. I shouldn't comment on that either.

"I would like to comment, without referring to people or any particular incidents. On this question of security risks and loyalty, they are distinctly different things. If a man is accused by being disloyal or subversive, that is some kind of an act against the country. The security risk business is simply trying to eliminate the people that are more than average security risks, so that you don't get them in the wrong place where they might do some damage. In other words, we are trying to prevent the trouble instead of getting into trouble and then accusing somebody of disloyalty or subversive activities and trying them or court martialing them like we would in the Army. That is a distinct difference and it should be understood.

"I might explain it. It is a little bit like selecting a teller in a bank. The president of a bank selects a teller. If the man frequents gambling joints and has contacts with the underworld, you ordinarily don't hire him. Or if you found out after you did hire him that at one time he had been convicted of theft or something like that, maybe he is reformed and all, but you still don't expose him again. You don't wait until he has stolen money from the bank and then try to do something about it. You try to get people that are qualified and are not financial risks in that sense.

"Now, the American people, I am sure, would like to get the people that are security risks out of their armed services. It is too important a matter. So, if you men could clarify this business for the benefit of the public, the difference between accusing a man of being disloyal to his country and of subversion, in which case he could go to jail or have all kinds of things done to him for the crimes that he had committed, the other thing is that just on account of his association and his train of thought and his previous activities he is a bad risk, so you don't expose him to a place where he might do the wrong thing."

Then there are some more dots.

"The PRESS. This hypothetical question concerns, say, some specialist in a field that the military services might require. He is one of the 3 or 4 men in the country who is qualified to handle a certain problem that concerns weapons that the Defense Department is interested in, and the project is a very important one, a top priority project. This man as a young man may have had some Communist connections or sympathies and at the present time he indicates that he no longer has them. His services are important to the Defense Department. What would you do about bringing him in to work on that project?

"Secretary WILSON. I'd look at the other 2 or 3 if he is 1 out of 3 or 4. [Laughter.]

"The PRESS. Let's add another point. Suppose that he is the key man in that situation and without him you could not get any success in the project.

"Secretary WILSON. This is an awfully big country and I doubt if there are any such people.

"The PRESS. Mr. Secretary, I'll ask you a specific question on the same lines. I believe it is correct that the Army and possibly the Air Force brought to this country a great number of German scientists to work on guided missiles development, men with a record of recent past association with the Nazis. How does that square with what you are saying, or do you think that was a mistake?

"Secretary WILSON. There is no way I can pass on it broadly. You'd have to look at each case on its own.

"The PRESS. Would you say, sir, that we have reached the stage in our atomic weapons development so that we no longer need the services of important theoretical physicists and mathematicians, that it is now largely an engineering or applications engineering problem?

"Secretary WILSON. No; I wouldn't say that.

"The PRESS. In other words, we still need the type of scientist that I was referring to earlier?

"Secretary WILSON. That's right."

More dots.

"The PRESS. Mr. Secretary, have you expressed yourself about the various reports that the H-bomb development might have been unduly delayed?

"Secretary WILSON. No. I have never made any comment on it.

"The PRESS. Do you have one?

"Secretary WILSON. No.

"The PRESS. Do you know of any such delays?

"Secretary WILSON. See, I wasn't even here in my present position, and that one also comes under this category of something that is being reviewed. So, I shouldn't try to get into the play from the sidelines.

"The PRESS. Sir, has the Defense Department brought down a blank wall between any other scientists and its atomic weapons research besides Dr. Oppenheimer?

"Secretary WILSON. Well, we are carefully going over everything in connection with our present security regulations for civilians and military people as well. The directive I put out last Thursday clarified the thing somewhat in the military establishment and was an effort to have the uniform procedures and step them up and handle the thing more promptly than we had.

"The PRESS. But nothing has been done in the case of any individual?

"Secretary WILSON. Well, of course they are being worked on all the time.

"The PRESS. Has there been any attention——

"The PRESS. Any more top attention, someone, say, as of great prominence as Dr. Oppenheimer? Do you know of anyone else?

"Secretary WILSON. No; I don't.

"The PRESS. Mr. Wilson, there has been a suggestion——

"Secretary WILSON. See, actually we are not trying to hurt anybody or smear anybody. We are just trying to do a good job for the country as quietly as we can and quite frankly, I have great sympathy for people that have made a mistake and have reformed, but we don't think we ought to reform them in the military establishment. They ought to have a chance somewhere else.

"The PRESS. Does that mean that Dr. Oppenheimer will no longer be admitted to miiltary bases——

"Secretary WILSON. Well——

"The PRESS. —or military secrets?

"Secretary WILSON. His case is being reviewed by a proper board that has been appointed for the purpose, I understand."

More dots.

"The PRESS. Mr. Secretary, is Dr. Oppenheimer on any advisory boards or committees in connection with special weapons or research and development in the armed forces?

"Secretary WILSON. No, he was a consultant to the Research and Development Board until that was abolished last July after we got the Reorganization Plan No. 6 in effect for the Department of Defense.

"The PRESS. Why was he dropped then?

"Secretary WILSON. We dropped the whole board. That was a real smooth way of doing that one as far as the Defense Department was concerned. [Laughter.]"

More dots.

"The PRESS. Mr. Secretary, if the Defense Department needed a scientist—this is a hypothetical question—who had questionable association in his past and where the Defense Department thought that the services they could get from that scientist would outweigh the harm he might do because of possible bad associations, would you take him on?

"Secretary WILSON. Well, I suppose the answer there would depend on how critical the thing was and what the degree of past record was and so forth. That is one I might put up to Moses. [Laughter.] Any of you remember reading how Moses' father-in-law told him how to organize the children of Israel for effective operation?

"The PRESS. Well, how about Saint Paul——

[Laughter.]

"Secretary WILSON. I don't know whether you would refer that one to Moses or not."

More dots.

"The PRESS. Mr. Secretary, another Moses question. During the time that this has been up, this current problem we have with the AEC and so on, has anybody figured out how to keep secrets from men who probably put the secrets in in the first place?

"Secretary WILSON. Well, maybe I should tell you a story on that one."
That is the end of the document which I have.

By Mr. ROBB:
Q. Doctor, do you agree with Mr. Wilson's philosophy or theory respecting security risks as expressed in that press conference?
A. I would find it quite difficult to say what Mr. Wilson's philosophy is from this press conference.
Q. May I ask you another question along those same lines? You said that you reviewed this file. From that am I to take it that some question had arisen which you were asked to answer?
A. Yes. The President's Executive order had come out. This file was referred to my attention because it obviously fell under the President's security order. It was obvious to the security officers of ODM. They felt that this was a case to be reviewed.
Q. That is what I am getting at. You did not read the President's order and automatically get the file. Somebody brought it to you because of the President's order?
A. That is correct; yes.
Q. Am I to gather that whoever it was that brought it to you expressed the view that this file on its face raised some question about Dr. Oppenheimer?
A. Yes.
Q. Doctor, you spoke of the Vista project and your trip to see General Eisenhower. Had you participated in the writing of that report?
A. I had not.
Q. Was that the report that was prepared in Pasadena in the fall of 1951?
A. Yes.
Q. Had Dr. Oppenheimer taken any part in that as far as you know?
A. I am quite sure that Dr. Oppenheimer had worked with the Vista Project to some degree, particularly in the section dealing with atomic energy.
Q. Do you know what part he had played in connection with that section?
A. I am not too clear on that, but I believe he had quite a significant part in helping in the drafting of that chapter.
Q. Did you ever discuss it with him?
A. Yes.
Q. Could you tell us from your discussion with him what his views were on that subject?
A. I know that he felt that the atomic weapon had a potentially very important part in the problem of ground operations, particularly in the defense of Western Europe. He felt that there were many opportunities to exploit the atomic weapon which should be aggressively developed.
Q. I assume that these questions relate to the fall of 1951. Did he give you his views at that time in connection with this report, about how he thought the available stockpile of atomic weapons should be divided?
A. Yes. With the growing stockpile he very definitely felt that a range of the smaller weapons which would be useful for tactical purposes should be increased in numbers * * *.
Q. Did Dr. Oppenheimer express the opinion that the proportion of atomic weapons to be assigned to the Strategic Air Command should be kept the same, increased, or decreased?
A. Frankly I don't recall.
Q. Did he express any opinion to you as to whether there should be any announcement by the United States with respect to the possibility of a strategic atomic attack on Russia?
A. I am going to try to answer this as carefully as I can.
Q. Yes, sir.
A. In the course of our trip over to SHAPE—We flew over and we had discussions and we met with General Eisenhower as I say, and we had other discussions—many facets of the atomic weapon utilization were discussed among the four of us who were there. As is customary in such discussions, almost every shade of opinion was expressed in exploring the future of the atomic weapon.

For example, I would probably present the arguments one way and then turn around and try to present them the other way. Dr. Oppenheimer certainly expressed many views about the most effective utilization of the atomic weapon in the problems of our military strength. * * *

Q. You and Dr. Oppenheimer?

A. Dr. DuBridge and Dr. Lauritsen.

Q. You used the word "retaliate," Doctor. Was there any discussion about whether or not the United States should announce that it would not initiate a strategic bombing of Russia?

A. Frankly I don't remember. It could have been discussed. I say probably it was because we were exploring all of the facets of it.

Q. Can you tell us what Dr. Oppenheimer's view was on that question?

A. No.

* * * * * * *

Q. Yes, sir. What I am attempting to direct my question to now, sir, is a question not of retaliation, but of using the atomic weapon first.

A. I don't believe that any of us really discussed that. To me in my own view it doesn't seem like the right way to go at it, and I don't believe we discussed that.

Q. Did you have any discussion about the value of the thermonuclear weapon?

A. No. We were concerned at this stage with the Vista Report dealing with the ground forces and the defense of Western Europe and the concept of the thermonuclear weapon being involved in the immediate defense of Western Europe didn't seem pertinent. We knew at that time, of course, that thermonuclear weapons of great magnitude—well, we felt they would find their usefulness in the strategic campaign, rather than the tactical.

Q. Did you have a copy of this Vista Report with you when you went over there?

A. Yes; a draft of it.

Q. A draft of it?

A. Not the final Vista Report. In fact, might I interject one of the main reasons for going on this trip was so that General Eisenhower and others over there could be apprised of the Vista findings and tentative conclusions and could express their judgment before the report was quite finalized.

Q. Did the draft that you had with you include the section to which you referred on atomic weapons?

A. Yes.

Q. Was that section later changed?

A. I think it was. I think practically everything in that draft—I mean many of the salient features of that draft—were changed. That was the purpose of the visit.

Q. Can you tell us anything about what led up to the change in the section of that report having to do with atomic weapons?

A. I think that the discussions at that time were an important part of the process of bringing the report into final form. May I emphasize the main purpose of this was to go over with a rough draft and see what the final report should say.

Q. What was the date when you went over? I don't mean the exact date.

A. It was early December of 1951.

Q. Before you went over, do you recall talking to Mr. William Burden and Mr. Garrison Norton about the report?

A. Yes.

Q. They came to see you in your office, did they?

A. They did.

Q. And they discussed the section of the report having to do with atomic weapons, didn't they?

A. Yes.

Q. Did they tell you that they were disturbed about it?

A. Yes.

Q. Did they tell you why?

A. Yes.

Q. What did they say?

A. They were very much concerned——

Q. May I interrupt before you start that? Will you tell who those gentlemen were?

A. Mr. Burden was the special assistant to Tom Finletter who was the Secretary of the Air Force. Mr. Garrison Norton, I believe, was assistant to Mr. Burden.

* * * * * *

Q. Did you have a copy of the draft before you when you talked with them?
A. No, I think not.
Q. Did they tell you who had prepared the particular section to which they took exception?
A. They said that chapter had been written primarily by Dr. Oppenheimer.
Q. Did you tell them you were disturbed too about it?
A. I said I was disturbed because they were disturbed and that I would have an opportunity to discuss this with Dr. Oppenheimer and Dr. DuBridge.
Q. Did you express the view that efforts should be made to have this section modified?
A. I certainly said that if it contained the implications which they were worried about, there should probably be some modification. You must realize that I was not familiar at that time with what the chapter said.

Mr. ROBB. That is all I care to ask, Mr. Chairman.
Mr. GRAY. I have a couple of questions.
I would like to continue now, because I think we are so nearly through we won't have to call you back after lunch.
For the record under whose auspices was the Vista contract made?
The WITNESS. The Vista contract was administered under one of the branches of the Army. It may have been the Signal Corps. I am not sure.
Mr. GRAY. But not under the Research and Development Board?
The WITNESS. No.
Mr. GRAY. So you had no responsibility for the Vista report?
The WITNESS. I had only this responsibility, that the general problem of coordinating the research and development was a responsibility of my office and this was a project which, administered by the Army, nevertheless had great Air Force and a little Navy interest in it. It was so full of suggestions on research and development that there was a distinct interest and responsibility on my part in terms of the nature of the report and the subsequent implementation of the research and development features.
Mr. GRAY. I didn't mean to imply by that that you were dealing with something which was not your concern. But it was not your direct responsibility.
The WITNESS. That is correct, although Mr. Lovett and I, talking over the question of the visit to SHAPE agreed that this Vista report was of such significance in research and development that the particular visit should be arranged as a Research and Development Board visit with me in attendance as the chairman of the RDB. So we really went over under the sponsorship of the RDB rather than of the Army.
Mr. GRAY. In your testimony, Dr. Whitman, you said that Dr. Oppenheimer more than any other man had educated the military as to the true potentiality of atomic weapons or something to that effect.
The WITNESS. That is my belief. From my observation I would so say.
Mr. GRAY. I don't question it. I am interested to know how was this educational process carried out? What were the mechanics? Who were the people? Who was it that needed to be educated?
The WITNESS. Practically all of the officers. After all, this was really a very new field. Dr. Oppenheimer was able to carry out that education considerably by virtue of his connection with the Research and Development Board as a member of the Committee on Atomic Energy, which contained such people as Admiral Parsons, who subsequently has died. As Captain Parsons he dropped the bomb over Hiroshima. General Nichols, now the manager of the AEC, General Bunker of the Air Force, men of that ilk.
Mr. GRAY. Military people.
The WITNESS. I might say also General McCormack who at that time was in the AEC in charge of the Military Division, but who subsequently went back into the Air Force. Men of that ilk who were leaders in the field and lots of others who were coming along. There has been a tremendous problem of education in this entirely new weapon.
Mr. GRAY. But it was in Dr. Oppenheimer's relationship to the Research and Development Board that these educational processes took place?
The WITNESS. I would say that was an important part of it. He, of course, has had many contacts with the military in other ways. This is the one I had the best opportunity to observe.
Mr. GRAY. Dr. Whitman, I don't suggest anything sinister about this, but I think you are the third witness who has said that he felt that the use of hydrogen weapons in an all-out war would mean "the end of civilization as we know it." This is I think the precise language. This language appears in a report some

place in which you participated. I don't want to pursue this too far, but I was just struck by the fact——

The WITNESS. I don't recall it in any report. But in my conversations when Mr. Lovett was Secretary of Defense, in our circle, I reiterated this point and brought it up again and again as indicating the relative emphasis which we must follow in the Defense Department, particularly in research and development, but in other ways. In other words, what things come first. I have had occasion to appraise this and * * * and lots of other things, and these are rather testing appraisals over a period of 2 years when I was responsible there, * * * I do feel that the future of civilization——

Mr. GRAY. I don't question your feeling. I don't want to pursue it.

I have two questions now, and I am through.

In your testimony earlier you said that the reading of the Nichols letter of December 23 does not change your mind at all or would not change your position which you took in July of 1953, with respect to clearance of Dr. Oppenheimer, for classified information. I would just like to have it clear, is that on the assumption that the derogatory information contained therein is true, or that it is not true, or do you make any assumption about that?

The WITNESS. Might I explain why I say this?

Mr. GRAY. Yes, I would like for you to.

The WITNESS. General Nichols' letter contains for the most part material which I had already reviewed and had rather prayerfully reached my own conclusion. It contains in addition what I regard as a very serious charge, that Robert Oppenheimer obstructed and tried to delay progress on the hydrogen bomb. Because my own association with him started in 1950, and had been quite intimate since that, when he would have put in the obstructions after the President's decision if he were obstructing it, my own personal experience with him convinces me that is false. So the only additional information above the file is something on which I have a right to a strong personal opinion by association.

Mr. GRAY. I think that is a clear statement.

My question now is, did you come to your conclusion with respect to the other derogatory information, other than the hydrogen bomb obstruction, on the assumption that all of that might have been true, and nevertheless you felt there was no security problem?

The WITNESS. Yes. I realized of course that it could not all be true, because some of it is contradictory. I was willing to assume that the damaging statements in there could have been true and still reached the conclusion.

Mr. GRAY. Or today you would say assuming it is true, you would still reach this conclusion?

The WITNESS. Yes.

Mr. GRAY. Did the security officers in the Military Establishment make any recommendations to you with respect to your position, which is reflected in the memorandum you wrote?

The WITNESS. I think they made the recommendation that this is a case which I must review under the President's order. I don't know. In fact I don't recall ever having had them say that "We think"—I mean express the judgment—that he should not be reappointed. They may well, but I don't recall it. I wouldn't be advised if they had, because security officers are notably careful as policemen to take the negative point of view.

Mr. GRAY. Aren't Government officials generally careful?

The WITNESS. I am afraid they are too much. This is why I said I rather prayerfully thought this whole thing over before I came out with the unqualified recommendation that he be reappointed.

Mr. GRAY. Yes. Your recommendation is very clear.

Do you have any questions?

Dr. EVANS. Are you a Communist?

The WITNESS. No, sir.

Dr. EVANS. You have never been, have you?

The WITNESS. No, sir.

Dr. EVANS. Are you a fellow traveler?

The WITNESS. No, sir.

Dr. EVANS. You never have been?

The WITNESS. No, sir.

Dr. EVANS. Have you belonged to those subversive organizations mentioned by the Attorney General?

The WITNESS. No, sir.

Dr. EVANS. Have you met any Communists?

The WITNESS. I have met Russians during the war when I was with the War Production Board where I had to deal with them on issues of supplies for Russia.
Dr. EVANS. Have you met any Americans that turned out to be Communists?
The WITNESS. I don't recall that I ever have, Dr. Evans.
Dr. EVANS. I have no more questions.
Mr. SILVERMAN. I have, I think, two questions.

REDIRECT EXAMINATION

By Mr. SILVERMAN:

*   *   *   *   *   *   *

Q. Did you feel that Dr. Oppenheimer's views as to relative division of fissionable materials between strategic bombing uses and other uses were motivated by anything other than considerations for the security and defense of the United States?
A. Not at all.
Mr. SILVERMAN. I have no further questions.
Mr. ROBB. No further questions.
Mr. GRAY. Thank you very much.
The WITNESS. Thank you.
Mr. SILVERMAN. Mr. Chairman, there is one question I overlooked. May I ask it?
Mr. GRAY. Counsel has another question for you.
Mr. SILVERMAN. I am sorry.

By Mr. SILVERMAN:

Q. Dr. Whitman, did you have an informal or formal security board that looked into the question or looked at your recommendation afterwards with respect to Dr. Oppenheimer?
A. I had a security board setup under me to give me advice. This particular board did not look at Dr. Oppenheimer's case prior to my receiving it. Now, by hearsay I understand that that board was continued by my successor, and did review the case and my recommendation, but that is purely hearsay.
Q. Do you know whether they agreed with your recommendation?
A. Hearsay, they did.
Q. Who were the members of the board?
A. Dr. Robert W. Cairns, who at the time was my vice chairman. Dr. L. T. E. Thompson, who at the time was my vice chairman, and General John Hines, who was my senior Army officer.
Mr. SILVERMAN. That is all.

RE-CROSS-EXAMINATION

By Mr. ROBB:

Q. You mean you already decided the case before they reviewed it?
A. No, I think I explained that because my term was going to be over at the end of the month, and I realized that this case would not be finally decided until the new Assistant Secretary came in, what I did was reviewed the case and gave my recommendation which by hearsay subsequently Mr. Quarles referred to this same informal committee that I had appointed.
Mr. ROBB. That is all. Thank you.
Mr. GRAY. Thank you very much.
(Witness excused.)
Mr. GRAY. We will reconvene at 2: 15.
Mr. GARRISON. Mr. Chairman, could we have a little bit longer, because we have a problem with witnesses. Could we make it 2: 30?
Mr. GRAY. We will make it 2: 30.
(Thereupon at 1: 15 p. m., a recess was taken until 2: 30 p. m., the same day.)

AFTERNOON SESSION

Mr. GRAY. Mr. Rowe, do you wish to testify under oath? You are not required to do so. I should tell you that all the witnesses to this point have.
Mr. ROWE. I would prefer to.
Mr. GRAY. Would you be good enough to stand and raise your right hand? What is your full name?
Mr. ROWE. Hartley Rowe.

Mr. GRAY. Hartley Rowe, do you swear that the testimony you are to give the board shall be the truth, the whole truth and nothing but the truth, so help you God?

Mr. ROWE. I do.

Whereupon, Hartley Rowe was called as a witness, and having been first duly sworn, was examined and testified as follows:

Mr. GRAY. Would you be seated, please, sir.

It is my duty to remind you of the existence of the perjury statutes. I trust we need not discuss those here. You are familiar with them?

The WITNESS. I have read them several times, yes, sir.

Mr. GRAY. In the event, sir, that in the course of your testimony it becomes necessary to disclose restricted data, I should like to ask that you notify me in advance, so that we might take appropriate steps.

Finally, I point out to you that we consider these proceedings a confidential matter between the Atomic Energy Commission, its officials on the one hand, and Dr. Oppenheimer, his representatives and witnesses on the other hand. The Commission will take no initiative in releasing material to the press about these proceedings, and on behalf of the board, I express the hope to each witness that he will take the same view.

DIRECT EXAMINATION

By Mr. MARKS:

Q. Mr. Rowe, will you please identify yourself for the record?
A. In just what manner?
Q. Your present position.
A. I am vice president and director of the United Fruit Co.
Q. What is your profession?
A. I am an engineer.
Q. Will you describe very briefly your professional career in just a few sentences?
A. I started after graduation from college as an engineer with the Isthmian Canal Commission, which was later termed the Panama Canal Commission, and served there 15 years.

I came back to the United States at the end of that time and entered in consulting service with a firm by the name of Lockwood, Green & Co., first in Detroit and then in Boston. I was with them about 7 years, and then went to the United Fruit Co. as their chief engineer, and I have been with them ever since.

Q. When did you become a vice president of United Fruit?
A. 1928.
Q. Will you also describe briefly your original connection with war work, that is, World War II, and what it consisted of?
A. In 1940 I was connected with the National Defense Research Committee, headed up by Dr. Vannevar Bush, Dr. Karl Compton, and Dr. Conant. That was later made into the Office of Research and Development.

I was chief of Division 12, which handled mobile equipment and naval architecture from 1940 until the conclusion of the war, and the conclusion of our reports in 1946. I was also a consultant to the Secretary of War. I was a consultant on the Rubber Division of the War Production Board and several other short time jobs that I don't recall at the moment.

Q. What developments did you have a share in while you were with the NDRC and its successor?
A. The one that gained the most notoriety was the Duck, from that the Weasel, which was a very slight snow vehicle traveling over snow and over marshy ground.

Q. In your capacity as a consultant for the Secretary of War, did you have any overseas assignments?
A. Yes. In May 1944 I was assigned to General Eisenhower's staff as a technical adviser primarily for the purpose of bringing to the attention of the field commanders and the troops there the military things that had been developed by OSRD up to that time. I served with him for about 7 months.

Q. Served with him?
A. With SHAEF for about 7 months.

Q. What were the conditions under which you took that assignment?
A. There were two. One ordinary condition is that I requested I be introduced to General Eisenhower and his staff by a general officer, and second, that I thought I could be most effective operating out of channels and directly by a pipeline to Washington.

Q. Why were you interested in that latter?
A. Principally because I don't know how to operate through military channels. Secondly, that I felt I could be more effective and save a great deal of time—time was of the essence—and be much more effective to the field commanders.
Q. What were some of the things with respect to which you had any influence in that assignment?
A. Radar and radar controlled guns, the proximity fuse, and its introduction to combat the buzz bomb, the infrared instruments that were used by the paratroopers to collect together after a drop.
Q. You have any difficulties persuading them to adopt these measures?
A. None whatever.
Q. After your assignment with SHAEF, what was your next connection with war work?
A. As soon as I returned to the United States from that work, I was notified that they wanted me to go to the Pacific and do the same kind of work for General MacArhtur. It had all been arranged with his consent under the same conditions. Before I could get away, Dr. Conant and Major General Leslie Groves came to me and said they had a job they wanted done and I told them I was afraid I couldn't do it, because I had already signed up, and they said this takes priority over everything you have been assigned to, so you better do what we want you to do.
The only question I asked was whether or not the assignment would be in the continental United States or whether it would still be abroad.
Q. What was that assignment?
A. I was assigned as a consultant to General Groves and Dr. Oppenheimer in the procurement of materials in the development of the A-bomb, trying to be of what assistance I could to bring it to a conclusion on a predetermined date.
Q. Where did you do that work?
A. In Los Alamos.
Q. How much time did you spend on it?
A. I spent a greater portion of my time commuting between Los Alamos and my office in Boston. I usually spent the weekends in Boston and spent from Monday to Friday in Los Alamos, or in some other city in connection with the work.
Q. During that period how well did you come to know Dr. Oppenheimer?
A. I was reporting more to him than I was to anyone else. I became very well acquainted with him.
Q. I take it during all of this period you continued your connection with the United Fruit Co.?
A. Yes, sir. The only time I had a leave of absence was when I was in Europe.
Q. After the war what connections did you have in any role with the Government?
A. I was made a member of the first General Advisory Committee in 1946, I believe, and served for the 4-year term to which I was appointed, from 1946 to 1950. I think the initial date was August or September and it ended in August or September.
Q. That is the initial date of the term was August 1946?
A. Yes.
Q. But you actually began your service early in 1947?
A. No. As soon as I was appointed, I think we met within the next month. I am quoting entirely from memory, because I kept no papers of any kind covering any of this confidential or secret work that I did.
Q. In connection with your work on the General Advisory Committee in those first 4 years of its existence, did you again work closely with Dr. Oppenheimer?
A. After the conclusion?
Q. No, in that 4-year period.
A. We met once a month for 2 or 3 days and 2 or 3 nights.
Q. Do you recall the meeting of the General Advisory Committee at the end of October 1949?
A. Yes, sir.
Q. That would have been not long after the announcement of the Russian explosion of the atomic weapon?
A. I don't know whether they had their first atomic explosion or not, but your records must show.
Q. To refresh your recollection, the announcement of the Russian explosion was at the end of September 1949. In all events, do you recall the session of

the GAC at which the subject of a crash program for the hydrogen bomb was the subject of debate?
A. Quite vividly, yes.
Q. Do you recall how the question came to you, how the question came to the General Advisory Committee?
A. My recollection is that it was brought up by the then chairman of the committee, and asked for——
Q. The then chairman of the committee?
A. Of the Commission, asking for the advice of the General Advisory Committee on whether or not we should enter into a crash program looking toward the development of the H-bomb.
Q. Do you have any recollection whether that would have been an oral or a written request from the chairman of the Commission?
A. I couldn't say. I never saw the written request that I know of.
Q. Would you give an account, as far as you can on the basis of your memory, and without getting into classified materials, of that meeting of the GAC, of its discussions and of your own views on the subject of the crash program for an H-bomb?
A. My recollection is that it was a pretty soul searching time, and I had rather definite views of my own that the general public had considered the A-bomb as the end of all wars, or that we had something that would discourage wars, that would be a deterrent to wars. I was rather loath to enter into a crash program on the H-bomb until we had more nearly perfected the military potentialities of the A-bomb, thinking that it would divert too large a portion of the scientific world and too large a portion of the money that would be involved to something that might be good and it might be bad.
Q. As far as you yourself were concerned, did you have any qualms about the development of an H-bomb or the use of it if it could be developed?
A. My position was always against the development of the H-bomb.
Q. Could you explain that a little?
A. There are several reasons. I may be an idealist but I can't see why any people can go from one engine of destruction to another, each of them a thousand times greater in potential destruction, and still retain any normal perspective in regard to their relationships with other countries and also in relationship with peace. I had always felt that if a commensurate effort had been made to come to some understanding with the nations of the world, we might have avoided the development of the H-bomb.
Q. Did you oppose the actions that the Atomic Energy Commission was taking and with respect to which the General Advisory Committee was advising during the period between 1947 and 1950 to realize the full potential of the A-bomb?
A. Will you state the question again?
Q. Did you oppose the efforts that were made to realize the full potential of the A-bomb during the period 1947 onward?
A. Not knowingly, no. We were in that, and my earnest opinion was that we should make the best of it.
Q. If you can, would you explain why on the one hand you supported the development of A-bombs to their full potential, but at the same time held views that were in opposition to the H-bomb?
A. I thought the A-bomb might be used somewhat as a military weapon in the same order as a cannon or a new device of that sort, and that we perhaps could use it as a deterrent to war, and if war came, if we had all the potentialities of it developed, we would be in a stronger position than if we only had the bomb itself without any of the other characteristic military weapons that were developed later.
Q. Why did you distinguish between that and the H-bomb?
A. Purely as a matter of the order of destruction. The H-bomb, according to the papers, this is not classified, is a thousand times more destructive than the A bomb, and you haven't yet reached the potentiality of it.
Q. I am not clear whether you are saying that you felt that the H-bomb was big enough for our needs.
A. I think the A-bomb was exploited to its full capacity, yes. I don't like to step up destructiveness in the order of 1,000 times.
Q. There has been talk that the H-bomb had unlimited capacity for stepping up destructiveness. * * *
Q. Could you describe, if you have any recollection, what influence other members of the GAC had on your thinking about the H-bomb?
A. Very little, if any.
Q. Did any of them have any particular influence?

A. I think I arrived at my conclusions even before the discussion came before the committee.

Q. After the President announced his decision in January 1950, to proceed with an all-out program to develop an H-bomb, you served on the General Advisory Committee for some months?

A. Yes.

Q. During that period can you state what your attitude was and what the GAC's attitude was about cooperating in this program which the President had announced?

A. I can only state definitely what my attitude was, and that was that we had received a directive and we had to go ahead. From my observations of the other members of the committee, I don't think there was any lag anywhere in either thought or deed. There were great scientific discussions which must necessarily take place before you can organize a procedure and ask for funds for the development of something that was as obscure at the moment as that was.

Q. Did you ever notice anything that Dr. Oppenheimer did that was contrary to the course you have just described?

A. No, sir.

Q. I would like to turn now, Mr. Rowe, to a quite different subject. Have you had any experience with communism?

A. You may be getting me into trouble, because I don't think so. I have had for many years, and recently renewed, was my Q clearance. One of the questions I was asked at that time was whether I ever knew or associated with Communists. My answer was that I knew Communists in Central America, but I had not associated with them. I didn't either know or associate knowingly with any Communists in the United States. Knowing that, I can answer your question.

Q. Let us confine the question to Central America. What experience have you had in Central America? How often have you been down there?

A. I went to Central America first in 1904 and served 15 years in those countries, and then came back and later went with the United Fruit Co. in 1926, and I have made an annual trip to the tropics, with the exception of 2 war years—1 of them was 1944 and the other was 1946.

Q. When you make this anual trip, how much time do you spend in the various Central American countries?

A. I have to cover 7 or 8 countries, and it is usually 2 or 3 weeks in each country.

Q. Don't answer this question if there is any reason from your own standpoint why you should not. Let me ask you: Is it a matter of business interest to you to know what is going on in these Central American countries politically?

A. No, absolutely.

Q. Would you say that you are familiar with the situation in Guatemala?

A. I am familiar with all of the principal things that have taken place there. I don't know of the every day detail in the country. I do know their pattern and that is, it follows a very distinct pattern. In my experience in other countries it always follows the same pattern. They start out by wanting to do something for the common people, and they usually pass what they call an agrarian law, which allows the Government to take up any lands that are not being used for other purposes for distribution among the population. * * *

By Mr. MARKS:

Q. Mr. Rowe, I think it fair to say that the problem before this board is one of formulating advice to the Atomic Energy Commission on the question of whether it would endanger the common defense and security if Dr. Oppenheimer were permitted to continue to have access to restricted data. In formulating that advice, the board has to take account of the provisions of the Atomic Energy Act, which stated that the determination should be made on the basis of a man's character, loyalty and associations. Do you have an opinion on this subject?

A. Yes; I do.

Q. Would you state what your opinion of Dr. Oppenheimer is in the background of the question I have asked?

A. I can only speak from my acquaintance with Dr. Oppenheimer during these years that I have outlined to you. So far as I am personally concerned, and so far as my own observations go, Dr. Oppenheimer is no greater risk than any other American citizen except for one thing, and that is he has a greater knowledge of atomic fission than anyone else that I know of in the country.

If you are put in a position of knowing secret and top secret information, the more you know, the greater risk you become, if you are ever in circumstances where you, as our boys have been in Korea—I don't know how I would react, and I don't know how Dr. Oppenheimer would react to brutal treatment. But in the course of his associations in the United States, I would have no reservation whatever.

Q. Are you saying that you have no question as to the loyalty, character or associations?

A. None whatever, based on my association with him.

Q. Have you taken into account in expressing this personal opinion the fact that at least up to some time in the early forties there is what is described technically as derogatory information, which means that there is an extensive record of associations with leftwing and with Communist personalties and affairs?

A. I haven't reviewed that testimony thoroughly. I have only read what is in the papers. I have never discussed it with Dr. Oppenheimer at all. Until I knew some more of the surrounding facts and reasons and the climate of public opinion at those times, I would not modify my statement.

Q. Would it surprise you if he had such associations and engaged in such activities as I have indicated in that period that the man you know, Dr. Oppenheimer, is a changed man?

Mr. ROBB. How is that again?

(Question read.)

The WITNESS. There are really two questions there.

Mr. MARKS. I think it is not a good question. Would you strike it out.

Mr. ROBB. Mr. Chairman, I am not objecting and don't intend to, but a thought does occur to me that sometimes the questions are a little bit leading.

Mr. MARKS. I think I have asked enough questions, Mr. Robb.

Mr. GRAY. You are not making any objection?

Mr. ROBB. I am not making any objection. I am just calling attention to that fact for whatever it may be worth.

I have no questions, Mr. Chairman.

Mr. GRAY. Mr. Rowe, I was very much interested in your description of your feelings in late 1949 about the development of the H-bomb. I think you made it very clear how you felt about it.

I would like to ask you whether you ever, in thinking about our problem and what we should do in this country, whether it was a source of concern to you that the Soviet Union might be working and perhaps successfully, towards the development of this kind of weapon. Perhaps my question is does that make any difference to you at all?

The WITNESS. It makes some difference, yes; but I would place more reliance on the proper use of the A-bomb without the H-bomb unless it developed as it did later that we had to go into it as a deterrent. I don't think it will ever be used against our enemies. I am quite concerned as to whether we would ever use the A-bomb or the A-bomb artillery or other military weapons.

Mr. GRAY. Some witnesses who have come before this board have testified that the news of the Soviet success in early fall, whenever it was, September, announced in September——

The WITNESS. You mean last year?

Mr. GRAY. No; I mean in 1949, the A-bomb of the Soviet.

The WITNESS. Yes.

Mr. GRAY. Some witnesses have testified that at that point they felt that we should do something to regain our lead, is the way it has been expressed, I believe; that we had a margin of advantage we thought over a possible enemy, and the one with whom we would most likely be engaged in conflict if we became so engaged, that with the announcement of the Soviet explosion it appeared that the lead we had might dwindle and perhaps not continue to be a lead, and therefore something should be done to regain it. Do I understand your testimony correctly in thinking that you felt that proper exploitation of the weapon we already had and the knowledge we already had would have enabled us to maintain the lead, or was that important?

The WITNESS. I wasn't thinking so much of the lead, but I thought it would be more effective, and we would have a better balanced military arm, the Army, the Navy, and the Air Force. Whatever you take away from any one of those three is going to unbalance them. A trade of the effort being put on the H bomb would detract from the things that needed to be done to get new weapons so that in the next world war we would not be fighting the war with the weapons of the previous war, as we have in the last two. It seemed to me we had a much better chance militarywise in perfecting our A bomb weapons. You understand what I mean by the different kind of weapons?

Mr. GRAY. Yes, sir.

The WITNESS. Than it would be to devote that effort to producing something that was a thousand times worse in explosive power at least, and can only be used in my opinion in retaliation. I don't think it has any place in a military campaign at all. Then if you used it in retaliation, you are using it against civilization, and not against the military.

I have that distinction very clearly in my mind. I don't like to see women and children killed wholesale because the male element of the human race are so stupid that they can't get out of war and keep out of war.

Mr. GRAY. I would like to turn to something else for a moment. You have read General Nichols' letter and Dr. Oppenheimer's reply?

The WITNESS. Yes, sir.

Mr. GRAY. Do you feel that your present conviction about Dr. Oppenheimer's character, loyalty and associations, would be the same if you knew that the information contained in the Nichols letter by early associations was true? Would your reply still be the same?

Let me repeat, Mr. Rowe, I am not saying that it is or is not true. Can you assume that derogatory information and still arrive at the answer you gave to Mr. Marks' question?

The WITNESS. I think my answer to that would be I would make it just that much stronger because people make mistakes and people in the climate of public opinion in those days which was quite different than it is now—we know a great deal more than we did then—I think a man of Dr. Oppenheimer's character is not going to make the same mistake twice. I would say he was all the more trustworthy for the mistakes he made.

Mr. GRAY. Let us not use Dr. Oppenheimer's name in the next question or in reply to it. Do you feel that a man might have been in the late thirties or early forties a member of the Communist Party and in 1954 not be a security risk with respect to the most highly classified information?

The WITNESS. That is rather hard to answer categorically, but a great many men would be a better risk. I would not say that they would all be a better risk.

Mr. GRAY. What you are saying is that it is possible for a man to have been a Communist and to have so completely renounced that that he would not be a security risk in later years?

The WITNESS. Yes, sir; that is what I am trying to say. Remember we all had an opinion during the depression days that our Government was lacking in some respects. It was discussed in almost every meeting of men that got together. We did not seem to know how to cope and cope quickly with a condition that was facing us. There were all sorts of opinions, that we should follow the British Constitution, that we should do this, that we should do that, we should do the other. One characteristic solution that I heard was that you should arm every other man with a pistol and let him go out and shoot one man, and that would cure the unemployment in every short order. Those points don't come from the heart or from the mind. They are just discussion.

Mr. GRAY. I believe you indicated that you felt that a man who had had no Communist associations might logically be expected generally speaking to be a better security risk than one who might have had such connections. I don't want to make a statement that does not represent your view at all.

The WITNESS. I can't answer that for everybody.

Mr. GRAY. I think you were careful to say that it would be important to know who the individual was.

The WITNESS. And how he reacted to a mistake.

Mr. GRAY. I think I can ask my next question which will cover what I am driving at. You would urge that the Government would take whatever chance there was in a situation with an individual who might have had these associations and who apparently had renounced them. You would say if there is any chance the Government ought to take it?

Excuse me, Mr. Rowe; I am really trying to get what your view is. This obviously is the kind of question that this board must ask itself.

The WITNESS. I understand your predicament.

Mr. GRAY. I am doing a very poor job of putting my questions. I am not experienced in this kind of procedure.

The WITNESS. In a great many instances the man would be a better risk knowing more about the Communistic Party. I think if I had known more about it in 1930 and 1940, I would have acted quite differently in my business in connection with my company and in treatment of Government officials * * *. 1 would have a better understanding of what the thing was all about.

Mr. GRAY. Again, without asking you to consider that this refers to Dr. Oppenheimer, would your reaction as a citizen of the United States be necessarily unfavorable if you knew that the United States Government had given access to classified material to a former Communist if you were satisfied with the individual?

The WITNESS. No, sir; that wouldn't worry me a bit.

Mr. GRAY. I think you have answered the question which I have had quite a time putting to you. Dr. Evans.

Dr. EVANS. I have just one question. You understand the position that this committee is in, don't you?

The WITNESS. Yes, sir; I believe I do.

Dr. EVANS. I hope you do. You are a man that has had experience, and you know what you are talking about. I have just one question to ask you. It is not quite the same as the Chairman was asking you.

If you had a lot of secret information in your mind, and you had some friends that were Communists, would you be in a more dangerous position than if you didn't have those Communist friends?

The WITNESS. You probably would; yes, sir.

Dr. EVANS. That is all I have to ask.

Mr. GRAY. Mr. Marks, do you have any other questions?

By Mr. MARKS:

Q. You used the expression, Mr. Rowe, in answer to some questions that were asked by the Chairman "better security risk."

I am not sure I understood what you meant by the term "better security risk." Let me put it this way. What is the difference between a man who is not a security risk in your opinion and a man who is a better security risk?

A. His character.

Q. Which of those two men would you trust most?

A. The man I thought had the best character.

Q. What I am trying to get at is—it is just that I don't quite understand the sense in which you are using the term—would you trust most the man that you regard as a better security risk or the man whom you simply regarded as not a security risk?

A. What I was trying to bring out is that there are different degrees of security risks. The more secret information a man has, the more likely he is to get in difficulties if then it came to a point where he was subject to torture. That is what I was trying to distinguish between a small amount of secret information and a large amount of secret information.

Mr. GRAY. A man with the greater amount would involve a greater security risk, that is what you said?

The WITNESS. That would be his personal risk.

By Mr. MARKS:

Q. Do you think based on your experience with Dr. Oppenheimer he would have any difficulty, as you know him today, in exercising discretion not to reveal secret information or information he ought not to reveal to unauthorized individuals?

A. I certainly do. I trust him implicitly.

Mr. ROBB. I have no questions.

Mr. GRAY. Thank you very much, Mr. Rowe.

The WITNESS. Yes, sir.

(Witness excused.)

Mr. GARRISON. May we have a short recess?

Mr. GRAY. Yes.

(Brief recess.)

Mr. GRAY. Do you wish to testify under oath?

Dr. DUBRIDGE. As you wish: whichever you prefer.

Mr. GRAY. You are not required to, but every witness who has come has done so.

Dr. DUBRIDGE. Yes, I will be glad to.

Mr. GRAY. What is your full name?

Dr. DUBRIDGE. Lee Alvin DuBridge.

Mr. GRAY. Lee Alvin DuBridge, do you swear that the testimony you are to give the board shall be the truth, the whole truth and nothing but the truth, so help you God?

Dr. DUBRIDGE. I do.

Whereupon, Lee Alvin DuBridge was called as a witness, and having been first duly sworn, was examined and testified as follows:

Mr. GRAY. Would you sit down, please, sir.

I must mention to you the existence of the perjury statutes. I assume you are familiar with them and it is not necessary to review them.

The WITNESS. Yes.

Mr. GRAY. I should like to ask that if at any time during your testimony it becomes necessary to refer to or disclose restricted data that you will notify me in advance so that we might take certain appropriate and necessary steps.

The WITNESS. Yes, sir. You wish the answer even if it does include restricted data.

Mr. GRAY. Yes, that is correct. If you cannot accept a question without referring to something of that sort, let us know and then we will find out whether to put the question or not to put it. I should point out to you that we consider this proceeding a confidential matter between the Atomic Energy Commission and its officials on the one hand, and Dr. Oppenheimer and his representatives and witnesses on the other. The Commission will undertake no initiative in release of information about these proceedings. On behalf of the board, I express the hope to each witness that he will follow the same course.

Mr. Garrison.

DIRECT EXAMINATION

By Mr. GARRISON:

Q. Dr. DuBridge, will you state your present position?

A. I am the president of the California Institute of Technology in Pasadena, Calif.

Q. Would you tell the board what Government positions you have held and now hold?

A. The list that I have held is somewhat long.

Q. Just the main ones.

A. I don't have the complete list before me, but among them, the ones I would consider pertinent are the following: I was appointed by the President in 1946 as a member of the General Advisory Committee of the Atomic Energy Commission for a 6-year term which expired in 1952. This term was coincidental with the term of Dr. Oppenheimer and Dr. Conant.

I am now Chairman of the Science Advisory Committee of the Office of Defense Mobilization, a committee which was established under the chairmanship of Dr. Oliver Buckley, some 2 or 3 years ago, and I succeeded Dr. Buckley as chairman a little over a year ago. Dr. Oppenheimer has been a member of this committee also.

I was for a term a member of the Naval Research Advisory Committee of the Department of the Navy and a member of the Advisory Panel of the United States Army.

For a term I was also a member of the Science Advisory Board to the United States Air Force.

Those I think are the principal advisory positions I have held since the war in the Government.

Q. What has been the general nature of your acquaintance with Dr. Oppenheimer? About when did you first meet him?

A. I met him first some time in the thirties as a physicist at Physical Society meetings and seminars. My first clear recollection is hearing him talk at a seminar at the University of Minnesota. I saw him occasionally during the thirties at Physical Society meetings, but was not intimately acquainted with him.

In 1939 I spent the summer doing research at the Radiation Laboratory at the University of California, just as a summer period of relaxation and refreshment, and work and became a little bit better acquainted with him personally at that time. At least on one occasion I was invited to his home.

During the war I was at MIT in the Radiation Laboratory there which had nothing to do with the Radiation Laboratory of the University of California. We were working on radar. I did not see Dr. Oppenheimer during that period very much, since he was at Berkeley and later Los Alamos.

The beginning of what I would call our close friendship, however, occurred in May 1945, when he requested that I come to Los Alamos with one of the members of our Radiation Laboratory to consult with the Los Alamos staff on some of the electronic and production problems which were being faced by the Los Alamos group, and particularly to discuss which members of the electronics group at MIT might be transferred to Los Alamos to assist in their work. I spent a week at Los Alamos at that time.

Following the war when we both became members of the General Advisory Committee, we also became what I consider to be good friends, and our friendship has continued since that time.

During the last years since 1946, I have frequently been a guest in his home and have seen him in Washington, of course, at many meetings where we have spent long hours together in the meeting room and outside. He has visited Pasadena. He was incidentally a member of the faculty of the California Institute of Technology when I arrived there as president in the summer of 1946. However, shortly thereafter he left to assume his present position at the Institute for Advance Study. So for a short time we were associated in Pasadena. Does that cover the situation?

Q. Yes. Of course, he has been with you on the Science Advisory Committee. I think you said?

A. That is correct, yes.

Q. I want to ask you a little about the work of the General Advisory Committee from its inception up to the October 1949 meeting. I want to ask you a few questions about that meeting and then a few questions about what happened in the GAC after President Truman gave the go-ahead on the all-out program for the H-bomb.

We have a good deal of testimony already on these subjects. I don't expect an exhaustive discussion from you, but I would like you to tell the board a few of the things that stand out in your memory during the period from the beginning of the GAC up to October 1949 in the way of recommendations made by the GAC to the Commission and what part Dr. Oppenheimer played in that effort.

A. As you are aware, this is a very large subject, and I can only repeat a few things that come to mind that would seem to me to be pertinent. If I may say so, Mr. Chairman, it is my understanding that the object of this hearing is to secure information that casts light on Dr. Oppenheimer as a loyal citizen of the United States, and as a good security risk.

Some of the things that might have happened in GAC arguments back and forth, I think are irrelevant to that question.

Mr. GRAY. Did you say irrelevant?

The WITNESS. Irrelevant to the question of security risk and loyalty. But I will start back with the beginning and hit a few points that occur to me.

When the General Advisory Committee first was assembled, at its first meeting early in 1947, it was apparent to us largely from the reports which Dr. Oppenheimer presented to the General Advisory Committee, but also reports we received directly from the Director of Los Alamos, that the Los Alamos Laboratory was in a state of very considerable disruption. The end of the war had brought about the desire on the part of the scientists there, a large number of them, to return to their universities or their industrial positions, and to resume their normal scientific careers and a very large number of course did that.

This left the top level positions of Los Alamos, many of them vacant. They were quickly filled by bringing up younger men, but these were men with lesser experience and less maturity. The departure of many key scientists of course left the laboratory in a state of demoralization.

There had been a year's lag between the end of the war and the passage of the Atomic Energy Act, a year in which uncertainty about the future of Los Alamos and the atomic energy project was current. The members of the Los Alamos Laboratory did not know what their future was to be as individuals or their function in atomic energy work. This was true of other laboratories, too. Therefore, the General Advisory Committee considered this as an important function in getting started and this came in a question asked by the Atomic Energy Commission: How can we restore, reestablish, strengthen the Los Alamos Laboratory as an effective weapon development laboratory.

It was evident at that time the most important thing that the Atomic Energy Commission faced was how to bring the atomic weapons work back to full strength. It was evident to us that peacetime applications of atomic energy were somewhat remote, would be somewhat difficult to proceed with at that time and that in view of the shortage of raw materials, the shortage of scientists, it was clear that the weapons program was the most important program to push forward, and the major job was how to strengthen Los Alamos, get better men there, and give the men who were there the maximum amount of scientific help.

Repeatedly this question came before the General Advisory Committee in session after session during those 2 years. It was always evident that the Chairman of the General Advisory Committee was among the most insistent, that this was our job, to help Los Alamos and strengthen the weapons program at Los Alamos.

A special weapons committee was appointed, a subcommittee of GAC, which I was not a member of, which paid visits to Los Alamos following the weapon program. Dr. Oppenheimer and Dr. Rabi and Dr. Conant were on the committee, and have or will tell you more about the work of that committee.

The objective of all members of the General Advisory Committee, especially under the leadership of our Chairman, was the strengthening of the United States military position in the field of atomic weapons, and doing this by using our scientific experience and technological work in process in Commission laboratories bearing on the weapons program especially at Los Alamos.

It was also evident to us that a critical bottleneck in the production of more and better atomic weapons was the availability of raw materials, plutonium particularly. So we discussed and made recommendations to the Commission at various times at various meetings for the expansion and improvement of the production facilities at Hanford. We felt it was quite important to increase the rate of production of plutonium and to expand the neutron yield of the Hanford reactors, and to increase the plutonium production there.

At various times we made recommendations, some of which eventually were adopted; others were not.

These matters of improving our weapons position and our fissionable materials position engaged a very large section of the attention of the General Advisory Committee during those days. We discussed also how the general scientific picture of the country would be strengthened especially in the nuclear physics and nuclear science areas through the Atomic Energy Commission support of scientific activities, through a fellowship program and so on. But never far beneath the surface of our discussions was the question of military strength of the United States in the atomic weapons field.

I may say that throughout the discussion on the General Advisory Committee we had many long and earnest discussions. We usually met for 3 days at a time and often went through the evenings, always informally in the evenings if not formally, and it was a very hardworking Committee. Always was the feeling of urgency and of concern that we should advise the Commission properly in ways that would strengthen the United States.

There were disagreements at times, of course, among members of the Committee. That is the reason you have a Committee rather than one person, so that different points of view can be represented. These points of view were brought forward frankly and given full discussion in all cases. But in the end almost invariably the recommendations of our Committee were unanimous.

There were occasional minority reports. These were never suppressed. But they were also written up when they seemed important and wished by the minority members and sent to the Commission along with the majority report of the members of the Committee.

This is the general tone and tenor of the discussions of our Committee.

Q. Do you have any comment on Dr. Oppenheimer's part in all this?

A. Even if Dr. Oppenheimer had not been officially elected Chairman each year, and if I may say so, he resigned or attempted to resign each year, feeling that a new Chairman should be elected, the Committee unanimously rejected his recommendation every year, and asked him to continue to serve as Chairman. He was so naturally a leader of our group that it was impossible to imagine that he should not be in the chair. He was the leader of our group first because his knowledge of the atomic energy work was far more intimate than that of any other member of the Committee. He had obviously been more intimately involved in the actual scientific work of the Manhattan project than any other person on our Committee. He was a natural leader because we respected his intelligence, his judgment, his personal attitude toward the work of the Commission, and the Committee. Of course, without saying we had not the faintest doubt of his loyalty. More than that, we felt, and I feel that there is no one who has exhibited his loyalty to this country more spectacularly than Dr. Oppenheimer. He was a natural and respected and at all times a loved leader of that group.

At the same time I should emphasize that at no time did he dominate the group or did he suppress opinions that did not agree with his own. In fact, he encouraged a full and free and frank exchange of ideas throughout the full history of the Committee. That is the reason we liked him as a leader, because though he did lead and stimulate and inform us and help us in our decisions, he never dominated nor suppressed contrary or different opinions. There was a free, full, frank exchange, and it was one of the finest Committees that I ever had the privilege to serve on for that reason.

Q. Coming now to the October 29, 1949 meeting at which the question of the crash program for the H-bomb was discussed at great length, do you recall how the topic of the so-called crash program for the H-bomb came up to the GAC?

A. This is a matter of recollection of a particular thing that happened. I will have to tell it in rather general terms though I am sure the records of the Committee must be available to you.

It is my recollection that as the Committee assembled for this meeting, we were informed by the Chairman that a question which was before us for consideration was whether a large undertaking should be initiated by the United States.

Q. You say the Chairman?

A. The Chairman stated to the Committee.

Q. The Chairman of——

A. The Chairman of our Committee, Dr. Oppenheimer, stated to the Advisory Committee that a matter we should consider was the question of whether the United States should embark upon a large production program aimed at the production of hydrogen weapons, and the particular version of the hydrogen weapon which was then called the Super. This production program involved first——

Q. May I go back a minute to ask you whether the members of the AEC met with the GAC before you went into your meeting? Let me ask you the question, are you talking now about the meeting of the GAC members themselves, or are you talking about the beginning of the session which, as I understand it, the practice was that the members of the AEC met with the GAC.

A. I am sorry I don't recall that particular meeting. Sometimes we met with the members of the Atomic Energy Commission at the beginning of our session, sometimes in the middle or at the end, or sometimes several times. I just simply do not recall whether in this particular session we met with the Commissioners first. I am sorry I do not recollect that. I do have a vivid recollection of Dr. Oppenheimer presenting to us the question, when the Commissioners were not present, only the Committee was assembled. Dr. Oppenheimer presented to the Committee this question: Shall we advise the Commission to embark upon this program? This proposal involved the construction of large reactors designed for the production of tritium.

At this point, Mr. Chairman, I am not sure whether what I want to say contains restricted material or not.

Q. I think I could perhaps just ask you a few questions that will avoid that, because we have had quite a little testimony about what happened, and I want to bring out just a few points.

I would like to ask you a few questions about the report itself. I understand about the report, but what I want to ask you about is the 2 annexes, one signed by yourself and Dr. Conant and Dr. Oppenheimer and Mr. Rowe, as I recall, and the other by Dr. Rabi and Dr. Fermi. Perhaps I have left out somebody of the majority. Do you recall who drafted the so-called majority annex?

A. I think it went something like this. May I go back just a moment? After this question was posed by Dr. Oppenheimer to the Committee for its consideration—and I will not attempt to state the full technical content of that question at the moment—Dr. Oppenheimer asked the members of the Committee if they would in turn around the table express their views on this question. The way in which the Committee happened to be seated at the table, I was either the last or the next to last to express my views.

The Chairman, Dr. Oppenheimer, did not express his point of view on this question until after all of the rest of the members of the Committee had expressed themselves. It was clear, however, as the individual members did express their opinions as we went around the table, that while there were differing points of view, different reasons, different methods of thinking, different methods of approach to the problem, that each member came essentially to the same conclusion, namely, there were better things the United States could do at that time than to embark upon this super program.

Q. These discussions I take it ranged over several days.

A. This particular phase was in one session in one half day. Later after we had gone around the table and expressed our opinion, we then elaborated and explored, wrote up drafts, argued about them, redrafted and so on, for at least 2 days. But to get the problem before us, the Chairman simply asked each member of the Committee to make a brief statement, and I suppose each person took 5 to 10 minutes or thereabouts to express his views.

After they were all on the table, the Chairman said he also shared the views of the Committee. We then discussed the question of how to state our views and our recommendations most effectively to the Commission.

It was on this subject of how our general conclusions could be most effectively and clearly stated that a very substantial discussion went forward for the next day or two.

It is my recollection that Dr. Conant and myself and possibly at least one other were on one Committee to make a draft, and that Dr. Rabi and Dr. Fermi were asked to make another draft. These 2 groups retired and prepared their respective drafts, and came back to the Committee meeting and read them.

We criticized each other's draft, made suggested changes and discussed the question at greater length and eventually came out with these 2 versions.

Q. There has been testimony here as to the views of different members of the GAC. I don't want to ask you to attempt to reconstruct in detail the majority annex which is not in the record, but I would like to have you state to the board as simply as you can your recollection of the position which you held at the time on this subject, how you felt about it, and why.

A. Recalling as nearly as I can, projecting my thoughts back 5 years or four and a half, it went something like this: First, though I was not intimately familiar with the technique of the atomic and hydrogen bomb design, it was my impression that the super design, which was then being considered, was * * * in too early a stage to embark on a large and expensive program. In other words, there were technical reasons why a crash program at that time seemed unwise.

Secondly, it was clear in my mind that the fission weapon program was progressing quite well, that better designs of fission weapons had been developed over the 2 or 3 years immediately preceding that time, that both larger in point of view of energy and smaller fission weapons had been evolved, and were designed and still further progress was rapidly being made. That we were, in other words, rapidly attaining a position of great strength in the fission weapon field. That some of these fission weapons were very much larger in their energy release than the original fission weapons exploded over Japan. That very much more efficient ways of using our fissionable material had been found so that our stockpile with a given number of pounds of fissionable material had greatly multiplied, and was in the process of being further multiplied.

Therefore, it was to the best interest of the United States to proceed as rapidly as possible to continue this development and improvement of our fission weapons so that our stockpile would be more effectively used, and our weapon strength would be further increased for a variety of military purposes. Small weapons for tactical purposes, and very large weapons for strategic purposes.

Mr. ROBB. Is this the majority report?

The WITNESS. This is my view as I recall it at the time.

Mr. ROBB. This is the separate opinion of Dr. DuBridge, and the other gentlemen who joined with him?

Mr. GARRISON. I am not asking him to recollect in detail the precise order of language and so forth in the majority report.

Mr. ROBB. I understand. I want to have it clear in the record if we can which particular report he was talking about.

The WITNESS. As I understood the question, it was to give my own views as to the hydrogen weapon at the time. To some extent these were reflected in the report, to some extent they were not.

Mr. ROBB. I see.

The WITNESS. If we made any mistake in our reports, the mistake was in not amplifying and giving our views. I think we made our reports too brief, and therefore they were not understood. Therefore, much of what I am saying is opinion I held as I recall it, and I am not sure just how much was written down. Only a small part of that actually. Therefore, there were technical reasons for not thinking that the super was ready for production. There were important reasons for thinking that there were more fruitful things at Los Alamos, and the other laboratories could proceed on the fission program.

The fission weapon program was that such that a very large destructive power was in our hands, and it was not clear to me that the thermonuclear weapons would add in significant ways to that destructive power.

Finally, there was a question of whether the United States could not find a better way of strengthening, rather than deteriorating its moral position with the rest of the world. It seemed to me and to some other members—I think all of the members of the Committee—that if the United States, instead of making a unilateral announcement that it was proceeding with this new and terribly destructive weapon, should instead say to the world that such a weapon may be possible, but we would like to discuss methods of reaching agreements where no nation would proceed with the design and construction of such a weapon.

It seemed to me at the time that the moral position of the United States in the face of the rest of the world would be better if we took that kind of a stand rather than making a unilateral announcement that we were proceeding with this new weapon of mass destruction. That as I recollect it was the background of my thinking at that time.

I must say that I cannot claim credit for originality in these thoughts. These thoughts evolved from my discussions with the other members of the Committee. But as nearly as I can reconstruct my thoughts at that time, that is it.

By Mr. GARRISON:

Q. After this October meeting you had another meeting in the first week in December and resumed the discussions, did you not?
A. Yes.
Q. After President Truman's direction to proceed with the program, did the GAC under Dr. Oppenheimer's chairmanship cooperate and try to carry out the President's directive, and if so, in general what did it do?
A. During the October meeting and during all the time immediately following that before our next meeting, I should make it clear that the only objective of the committee—and I am confident, of its chairman—was to increase the strength of the United States. All of the arguments and recommendations were aimed at that end. There was not the slightest question about that in any of our minds. If there were differences of opinion, these were honest differences of opinion as scientists and had nothing to do with our objectives in improving the position of the United States morally and physically. Though our recommendations as transmitted to the Commission were not accepted by the President of the United States, when we next met the announcement of the President of the United States was made, as I recall, during our meeting, and it was then clear to us that the decision of the United States had been made, that it was our job then to collaborate and cooperate fully in carrying forward this decision.

From that time forward I recall of no argument within the committee but that we had only one duty, and that was to implement the decision of the Government in proceeding with this project.

Q. Did Dr. Oppenheimer agree with that?
A. Fully and completely.
Q. Do you want to say anything more about what the committee itself actually did to help implement the program?
A. This was a matter mostly of technical assistance to the Los Alamos Laboratory in which I personally was not competent to participate. By discussing the program with the members of the Los Alamos Laboratory and others like Dr. Hans Bethe, I think substantial assistance was rendered by members of our committee individually and collectively to the program. I think a conspicuous piece of assistance to the thermonuclear program was a conference which Dr. Oppenheimer called at Princeton, I believe in June of 1951, at which time the purpose of this conference was to review the entire technical status of the thermonuclear program.

The members of the General Advisory Committee were all invited to this conference, and the members of the Commission. In addition, a number of the key staff members of Los Alamos including Dr. Bradbury, consultants of Los Alamos, including Dr. Bethe and Dr. Teller. This conference lasted 2 or 3 days, I have forgotten which, and was a long and extensive and intensive examination of the technical problem of the thermonuclear program.

There were many technical ideas which had been considered which were then being considered and being examined, and these were all laid out, and discussed in great detail, with an attempt to find out where is the best and most promising line of procedure with what was known at that time.

I believe that this conference was held at a critical time and was a critical and important assistance to clarifying ideas of the technical problems involved, and illustrating the next steps in the theoretical and experimental program of the laboratory. At various times during the months and years that followed, we were asked to give technical opinions on various aspects of the thermonuclear program and we did this as earnestly and carefully as we could.

Our objective was always to help the Commission in its work and since its job was to carry forward this program we considered it our job to help. In this, as in every other matter, the Chairman was our leader in this effort.

Q. Mr. Walter Whitman testified this morning about visiting SHAEF in connection with the Vista report. I believe you were the head of the Vista project?

A. Yes, sir.

Q. And you accompanied Mr. Whitman and Dr. Oppenheimer and Dr. Lauritsen on this trip to Europe?

A. That is correct.

Q. I don't want to go into the details because there was a good deal of testimony about it. I would just ask you in a general way whether Dr. Oppenheimer contributed in any respect to the usefulness of this project?

A. I think if I may, I would like to say a word about the Vista project. This was a project which the Air Force, the Army, and the Navy asked the California Institute of Technology to undertake, to examine some of the problems being faced by the Air Force and the Army, * * *

A substantial group was assembled at the California Institute of Technology during the summer of 1951 to examine these problems. We made extensive trips to Army, Air Force, and Navy installations, had a very large number of Army, Navy, and Air Force officers visit the institute to discuss and give us information and background on these problems. As the late summer came along, the group which had been assigned under the chairmanship of Dr. Robert Bacher, then a member of the California Institute staff, * * *.

Dr. Bacher and Dr. Christie and the others on this group suggested to me that it would be useful to them if Dr. Oppenheimer could be invited to come out and spend a little time with the Vista group to consult further on this subject. At our invitation Dr. Oppenheimer did come to Pasadena, and we discussed this subject at great length. He was of assistance in taking the draft of the chapter which had already been prepared and discussing the best method of presenting it, and threshing out further ideas and assisting the group in clarifying this idea and preparing a final draft.

During the cause of the Vista discussions, many problems came up in regard to the battle of Western Europe where we did not have the information about organization, forces, the NATO structure and the NATO problems, and we thought it would be helpful, after assembling our own ideas, if we could go over to Europe and consult the leaders, General Eisenhower and the other leaders, of the United States forces in Europe, to get the information which they had available and to discuss with them their thoughts about the battle of Western Europe, if it should occur.

I think it was during a discussion at which Dr. Oppenheimer was present, at which we were exploring ideas with John McCone, who either at that time still was or had just retired as Assistant Secretary of Air Force under Mr. Finletter, John McCone urged this trip and offered to assist us in arranging it, and it was finally arranged through the Secretary of Defense, Mr. Lovett, that a group of the Vista project people headed by myself and after some discussion the other members of the group to include Dr. Oppenheimer and Dr. Lauritsen, to go to Europe, and Dr. Lovett offered the facilities of the Department of Defense to make this trip possible, appointed Mr. Whitman, who was then Chairman of the Research and Development Board, to make all the administrative arrangements and to accompany us on behalf of the Secretary of Defense.

The four of us then went to Paris in the fall of 1951, I think November. We went to Paris. We saw General Eisenhower on two occasions and we went up to Weissbaden and met with General Norstadt and Air Force officials. We went to Heidelberg and met with United States Army commanders, returned— I am sorry, General Norstadt has headquarters not at Weissbaden, but at Fontainebleau, south of Paris, where General Norstadt was located, and we discussed things with him there.

Through all these discussions with the Army as to their problems * * * the problems which the Air Force faced in having enough airplanes, the right kind of airplanes, cooperating with the Army and so on, in all these discussions all four of us took an active part. I felt these discussions were very illuminating. They helped us form our own ideas that went into the final Vista report.

General Eisenhower's thoughts were particularly helpful. We had lunch with him and a long discussion with him on the general problem of the defense of Western Europe. It was obvious that the group was well picked, I felt. Dr. Lauritsen and Dr. Oppenheimer and Mr. Whitman were all important contributors to the effectiveness of our discussions. * * *

Q. You have read General Nichols' letter of December 23, 1953. You have read the items of derogatory information in it. Assuming that those items of derogatory information were true and without saying whether they are or not, what would your opinion be as to the loyalty of Dr. Oppenheimer, except for the hydrogen bomb allegation which I left out for purposes of this question.

A. You prefer to leave them out.

Q. Yes. I think that is of a different character.

A. It has always been, ever since reading this letter of General Nichols, difficult for me to see how any of the allegations therein had any significant relevance to the question of the loyalty and integrity of Dr. Oppenheimer. Some of the statements made in that letter having to do with acquaintances and associations and friends Dr. Oppenheimer has said were, of course, true.

Q. May I just for a moment remind you that the Atomic Energy Act requires the board to consider character, associations, and loyalty. Having this frame of reference that the board here must consider, the character, associations and loyalty of Dr. Oppenheimer, in determining whether or not his continuance of his clearance would endanger the national safety, having in mind the past associations set forth in the letter, having in mind what you know about Dr. Oppenheimer's character, having in mind what you say that the continuance of his clearance would to any degree endanger the national safety?

A. In no degree whatsoever.

Q. On what do you base this judgment?

A. In the first place, these associations that are mentioned were those of many, many years ago. As I understand it, they have largely long since been terminated, in at least one case by death. In the second place, these were rather natural associations of a person who had strong human interests, interests in human rights and human liberties and human welfare, who had strong revulsions against the growth of dictatorship in Germany, Spain, and Italy, and who wanted to express his opposition to such violations of human liberty as he regarded these dictatorships. He therefore found himself among others of like minds, some of whom it turned out were possible members of the Communist Party. But this was only a natural exhibition of his deep interest in human beings and in human liberty and had nothing to do with his devotion to this country, or nothing adverse to do with this country.

In the second place, it seems to me that to question the integrity and loyalty of a person who has worked hard and devotedly for his country as Dr. Oppenheimer has on such trivial grounds is against all principles of human justice. It seems to me whatever his ideas and associations were in 1935, is quite irrelevant in view of the last years since 1941–42, during which he has shown such a devoted interest to the welfare, security and strength of the United States. Whatever mistakes, if they were mistakes, and I do not suggest that they were, that were made in the thirties have well been washed out and the value of a man like Dr. Oppenheimer to his country has been adequately and repeatedly proved.

It would be in my opinion against all principles of justice to now not recognize the way in which his loyalty has been proved in a positive way through positive contributions. Furthermore, this country needs men of that kind, and should not deprive itself of their services.

Q. I think I should put this question to you because it is something that I want you to bear in mind when I ask you to give me your final judgment.

You are familiar with the Chevalier incident as recited in the Commission's letter.

A. That is my only familiarity, what I read in the letter.

Q. Supposing that it had been shown here that after Dr. Oppenheimer had had the conversation with Chevalier that for several months he did not report the incident to security officers, that after he had heard from the security officers at Los Alamos that they were concerned about espionage at Berkeley that on his next trip to Berkeley he told the security officers about Eltenton, did not reveal the name of Chevalier and declined to do so. Supposing it was further established that he told the security officers that his friend whose name he would not reveal had contact with the Russian consulate and that there were microfilm facilities for transmitting information, and that the friend had approached

3 different persons, 2 or 3, 3 I think, and suppose that these were untrue statements about the consulate, the microfilm and the 3 persons, suppose that he was again urged after having been urged by the security officers at Berkeley to reveal the name of his friend, he was again urged by Colonel Lansdale and again declined, he was again urged by General Groves and said he would not do so unless ordered; General Groves said he didn't want to order him to do it, asked him to think it over; General Groves saw him again and said he would have to order him if he would not reveal the name, and at that point Dr. Oppenheimer revealed the name of Chevalier.

I am not trying to ask you now to do anything more than to assume that you had that set of facts before you. Would your conclusion still be the same as you have expressed it here to the board?

Mr. ROBB. Mr. Chairman, I don't object to the question, but I wish it to be recorded that my failure to object does not imply or import that I endorse the complete accuracy or fullness of the hypothesis stated by Mr. Garrison.

Mr. GARRISON. I quite understand that. To carry it further I would have to read the whole testimony.

Mr. ROBB. I understand. I don't want to debate it.

Mr. GARRISON. I want to give Dr. DuBridge the nature and character of the problem.

The WITNESS. May I ask one question on your assumption? In what year was this supposed to have taken place?

By Mr. GARRISON:

Q. 1943. You would regard that seriously, I take it?

A. I would want to examine this situation very seriously and what you said about the assumption obviously does not include all the facts. I assume therefore you wish me to answer this from the point of view of my knowledge of Dr. Oppenheimer's character and integrity, and my statement would be without hesitation that I would say that these acts which he is supposed to have committed in no case stem from any disloyalty to the United States, but possibly a mistaken but nevertheless a sincere and honest belief that this was the best thing to do at the time. I just know that Dr. Oppenheimer is loyal to his friend and loyal to his country, that he is honest, but has a humane feeling, that if he did these things it was with a sense that a loyalty to a friend was important but was not in conflict with any loyalty to the country at that time.

Q. Do you think that today if he were asked by security officers to reveal information which they believe to be important for the security of the country, that he would decline to do so even if a friend were involved?

A. I am sure that at any time if he had felt a loyalty to his country was involved, he would have done what seemed to be the proper thing to reinforce that loyalty.

Q. I am asking you today, leaving aside whether he thought that his friend was innocent or not, if he were told by security officers that in their judgment the interests of the country required knowledge which he had about a friend, would he put the interests of his country ahead of the friendship?

A. I am confident that he would. We have all learned a great deal about security problems in the last 10 years.

Mr. GARRISON: That is all.

CROSS-EXAMINATION

By Mr. ROBB:

Q. Doctor, do you think that loyalty to a friend justifies the giving of false information to a security officer?

A. I would not wish to do that myself.

Q. You would not do it, would you?

A. I don't think so.

Q. In fact, you can't conceive of any circumstances under which you would not?

A. I wouldn't say that.

Q. It is hard to think of any?

A. First, it is hard to project ourselves back 10 years as to what the situation was like then. None of us had any very keen appreciation of the problems of security and secrecy at that time or what was involved. I cannot say under no circumstances would I be reluctant to give away or give information about a friend if I were personally convinced that this information had nothing to do with the country's welfare. I would try to cooperate with security officers under all conditions but I cannot say that under no conditions would I be reluctant to give such information.

Q. That was not quite my question. My question was whether or not you would feel that loyalty to your friend justified you in lying to a security officer.

A. No, I would not feel so.

Q. The standards of honesty were the same in 1943 as they are now, weren't they?

A. Presumably.

Q. Doctor, I was interested in your discussion of the Vista matter. As I understand it, what was it called—a committee?

A. It was called a project.

Q. That project took place in the summer of 1951.

A. That is correct. Our report was completed in early 1952.

Q. You said Dr. Oppenheimer was not there when the project commenced, is that it?

A. That is right. He was a member of the staff of the project only for a relatively short period.

Q. I believe he came out in about November?

A. I believe it was before that, but I do not remember the dates.

Q. I don't know exactly either.

A. I think it was the latter part of the summer, September.

Q. Do you recall it was chapter 5 of the report that dealt with atomic bomb matters?

A. That is correct.

Q. Did Dr. Oppenheimer prepare an introduction to that chapter?

A. Dr. Oppenheimer collaborated with the other members of the committee that were responsible for chapter 5 in developing chapter 5. He did not write either the first or the last draft of that chapter. He assisted in the preparation of 1 or 2 intermediate drafts.

Q. Was there a time in November when the group was reviewing the report as a whole with you presiding?

A. Immediately after our return from Europe?

Q. No, sir, I am talking about before you went to Europe.

A. We had weekly meetings reviewing various chapters and various parts of the report. I don't know which one you are referring to.

Q. I realize it is hard to project yourself back.

A. We had many meetings and I was chairman of most of them.

Q. Perhaps I can refresh your memory. I am informed that on November 13, 1951, when the group was reviewing a draft of the report that you announced that Dr. Oppenheimer had prepared a portion of the introduction to the report. Do you remember that?

A. I don't recall the exact incident but it is quite possible I did, because he did prepare a draft of a part of chapter 5 at that time. It was not the final draft, but it was an intermediate one.

Q. I am informed that you stated that you considered that to be a great document, and you felt confident it would be accepted without amendment. Do you remember that?

A. No, I don't.

Q. I am not trying to lead you into something, but trying to find out whether that coincides with your memory.

A. I don't remember that meeting or the statement. At the time I certainly did have the opinion that the draft that Dr. Oppenheimer helped prepare of the introductory portion of the chapter was a fine contribution to the Vista work. I believed that and I still believe it.

Q. Was that draft which Dr. Oppenheimer helped to prepare incorporated in the draft which you took to Europe?

A. It certainly was incorporated in it, but I am sure there were probably changes in the wording between that time and the time we went to Europe. In other words, there were continuous changes in the wording of all parts of the report.

Q. By the way, at those meetings in November, was General Quesada present?

A. General Quesada participated in some meetings.

Q. Did General Quesada undertake to make available to your group his report on the so-called Greenhouse test?

A. I don't recall.

Q. In the draft Dr. Oppenheimer helped to prepare, the introduction to the report, was any reference to thermonuclear weapons made?

A. In the introduction to chapter 5?
Q. Yes, sir.
Q. This is a matter of record whether there was or was not. I don't recall. Certainly in some drafts, and I believe in the final report there was a reference to thermonuclear weapons.

Mr. ROBB. I might say, Mr. Chairman, I am undertaking to do this on an unclassified basis for the benefit of counsel. I suppose ultimately I will have to ask the Doctor some questions on a classified basis and read some extracts I have here, but I don't want to do it if I can help it, because I want Mr. Garrison to hear it.

The WITNESS. Do you have notes on it?
Mr. ROBB. Yes, I have.
Mr. GARRISON. I don't want anything withheld from the board.
Mr. ROBB. No, but I am trying to keep out of the classified area.
Mr. GRAY. Let us see if you can do it unclassified.
The WITNESS. Did I make clear in my answer to that question? I don't recall at what stage or what draft reference to the thermonuclear weapons came in, but there was a reference and only a passing one, as I recall.

By Mr. ROBB:
Q. Do you recall that subsequent to the November meeting, that draft of chapter 5 or the introduction to it was amended?
A. It was amended many times.
Q. Was it amended subsequent to that meeting in November?
A. Since I don't recall the particular meeting, I can't answer that specifically. I cannot even recall at the moment the date on which we departed on our trip to Europe. May I ask if that date is available? I don't have that date.
Mr. GRAY. Mr. Robb, do you have it?
Mr. ROBB. I am looking now to see if I can find it.
The WITNESS. These were matters of continuous study and drafting and redrafting and changing and finally we got a version which we took to Europe. We redrafted pieces of it on various chapters while we were in Europe as a result of our discussions. We came back and redrafted many parts again in the light of what we had learned, and finally got a report which we all agreed was the best we could do, which was submitted then to the Defense Department.

By Mr. ROBB:
Q. I have a note here, Doctor, which may assist you that you returned from your visit to Paris and reported to the Vista group on the 18th of December, 1951. That might help you fix the date when you went to Paris. At your meetings in California in the summer and fall of 1951, did you confer with General Quesada?
A. Yes, we asked General Quesada to come and discuss these various matters with us and at our invitation he did come.
Q. Did you have any report from General Quesada on the Greenhouse test?
A. As I say, I just don't remember. We certainly talked with General Quesada about atomic tests. Whether the Greenhouse test was specifically reported on as such, I don't recall.
Q. Was the Greenhouse test exclusively atomic or wasn't that thermonuclear in part?
A. I don't know.
Mr. BECKERLEY. The public record is that it included experiments in thermonuclear.
Mr. ROBB. The answer was that he didn't know.
(Record read by the reporter.)
The WITNESS. One reason for not recalling is that I never can remember the code words for these various tests.

By Mr. ROBB:
Q. I can't either.
A. Whether General Quesada reported or not, we certainly knew through various channels because I was still a member of the General Advisory Committee at the time about the Greenhouse test.
Q. Doctor, do you remember—I don't expect you to remember the date, but I will give it to you to assist you—on April 30, 1952, having lunch with Dr. Rabi, Mr. David Griggs, Mr. Garrison Norton, and Mr. William Burden at Mr. Burden's house here in Washington?
A. Yes. I can't confirm the date, but I remember approximately that time and I have only had lunch there once with that group.

Q. Do you recall that you and Dr. Rabi on that occasion expressed some opinions concerning H-bomb development?

A. We had a very vigorous discussion of this question, yes.

Q. Would you undertake, please, sir, to give us the opinions that you and Dr. Rabi expressed?

A. It is a little difficult to try to recall a conversation of 2 years ago. If I do recall, they were not substantially different from the ones I have already expressed here previously in regard to whether or not the thermonuclear weapons were important additions or were not to the military potential of the United States, and questions, if so, under what conditions they could be used. If you have any specific questions about the statements I made——

Q. I can understand how hard it is to remember. Do you recall you and Dr. Rabi saying in substance that you thought that there were two things that were more important than H-bomb development, the first being a concerted effort of the best minds in this country toward peace with Soviet Russia. Do you recall something like that?

A. That is quite consistent with what I might have said.

Q. Do you recall Dr. Rabi saying together with Dr. Oppenheimer and Dr. Lauritsen that he, Dr. Rabi, would press for action in accordance with plans that they were preparing, and that they were already in touch with the State Department, on the subject.

A. I don't recall that.

Q. Do you recall anything like that?

A. I have a faint recollection at this time that there was a committee at work in the State Department on exploring new approaches to an agreement with Russia. I had nothing to do with that committee. Though it is quite possible that Dr. Rabi said something about it, I am inclined to feel that I probably did not express any opinion about it since I did not have personal knowledge about it.

Q. I was not suggesting that you did. I was asking if you recall Dr. Rabi saying something about going to the State Department on the subject.

A. It is not impossible that he made such a remark.

Q. Do you recall that was Dr. Rabi's feeling at the time?

A. I think it probably was, namely, that because of the terrifying implications of A bombs and thermonuclear weapons, it was desirable to make another attempt to find a way to avoid using them.

Q. Do you recall either you or Dr. Rabi or both of you expressing an opinion that the second thing which was more important than H-bomb development was more emphasis on having a good air defense?

A. We certainly emphasized the importance of an air defense, yes.

Q. I believe at that luncheon meeting you said you had quite a go-around with these gentlemen.

A. We had a very vigorous discussion with Mr. Griggs.

Q. Yes, you put it more delicately than I did.

A. I didn't mean it that way. Our discussion was primarily with Dr. Griggs, who disagreed with Dr. Rabi and myself very violently on some points.

Q. Dr. Griggs contended that Dr. Oppenheimer had got the GAC to soft pedal the thermonuclear development, didn't he, and you said that was not so?

A. That is correct.

Mr. GARRISON. Mr. Chairman, could I just ask what is the general nature of the document that Mr. Robb is reading from?

Mr. ROBB. I am sorry, it has top secret stamped all over it, Mr. Garrison.

Mr. GRAY. Do you wish to make any point of this?

Mr. GARRISON. No.

By Mr. ROBB:

Q. That was the bone of contention between you in general, that Dr. Griggs said Oppenheimer and the GAC had not fully supported work on the thermonuclear and you and Dr. Rabi contended that the GAC had consistently supported it and emphasized it?

A. Essentially that is correct. Griggs made what we considered to be false statements, that the GAC had impeded thermonuclear development. We both emphasized strongly that neither Dr. Oppenheimer nor the GAC had impeded the development of thermonuclear weapons. On the contrary, from almost the opening day of the GAC's existence, its chairman and its members had recommended to the Commission that thermonuclear research proceed and the implemented and strengthened at Los Alamos. We did not feel at the time that the time 1950 was ripe for the production effort, but we always advocated the research and development effort. Our difference of point of view with Dr. Griggs, as I recall, was that he felt that the thermonuclear weapon development and production was No. 1 priority for the country. We felt that improving our fission weapon program and improving our defense were just as important, if not more important at that time.

Q. This was 1952?
A. Yes.
Q. Was that the view of Dr. Oppenheimer at that time, too?
A. It is a little hard to speak as to what his opinions were at any particular moment. I think in general we have agreed with each other. These were technical matters of priority and I must insist that at all times Dr. Oppenheimer, myself, Dr. Rabi and the others had only one objective in mind; that was strengthening the moral and physical and military position of this country. We had no other thought.

Mr. GRAY. Excuse me. At one point I am going to ask for a recess, but I don't want to cut you off in the middle of one thing you want to pursue.
Mr. ROBB. I have one question and then I think we might take a recess.

By Mr. ROBB:

Q. Doctor, you testified that the recommendations of the Vista Report were carried out and are still being carred out, is that right?
A. In so far as the use of atomic weapons is concerned. There are some other recommendations which were not. There are others that had nothing to do with atomic weapons which are being carried out.
Q. Were those recommendations to which you referred the same as the recommendations in the draft which Dr. Oppenheimer helped prepare in the fall of 1951 at Pasadena?
A. I believe so.
Mr. ROBB. This is a good time to stop.
Mr. GRAY. Let us take a few minutes recess.
(Brief recess.)
(The following portion of testimony, numbered pages 1747 through 1758, is classified, and appears in a separate volume.)

By Mr. ROBB:

Q. Doctor, I want to ask you a couple of questions and I want to assure you that when I ask you, I have not the slightest intention of being offensive or suggesting the slightest impropriety on your part. Did you volunteer to be a witness here?
A. I am trying to recall how it came about. I would have been glad to volunteer. I think I probably said to Dr. Oppenheimer or his counsel that if there is anything I could do to help, I would be glad to do so.
Q. Did you in that connection with helping undertake to raise a fund to assist Dr. Oppenheimer in this matter?
A. The newspaper reports in that connection are mistaken. As near as I can tell, the origin of that statement was that at the Cosmos Club here in Washington one day a few weeks ago, several friends said, "Would it not be nice if Oppenheimer's friends chipped in $100 each to raise a fund to assist him in the expenses of his hearing?" We agreed that this would be nice, and maybe somebody should see the best way of doing it. The matter dropped there, and that is the last I heard of it until I saw the statement in the paper. I do not know where they got that information that I was organizing a fund. I did not and was not and am not. After the thing appeared in the paper I received many letters, however, with checks from individuals who read it in the paper and sent in their contributions.
Q. I was sure you wanted to have the record clear on it.
A. I returned all these checks to the donors.
Q. Were the friends you were talking to any of the other witnesses who appeared here?
A. Some were and some were not.
Q. Who were the ones who were witnesses?
A. I do not know who else have been witnesses, as a matter of fact.
Q. Could you tell us who the friends were?
A. Dr. Rabi, I believe, was present at the time the discussion went under want, and Dr. Bacher.
Q. Dr. Fermi?
A. Dr. Fermi was not present. Mr. Trevor Gardner.
Q. Who is he, sir?

A. He is the Assistant to the Secretary of the Air Force for Research and Development. I believe that is his title. He is a civilian engineer who was formerly associated with the General Tire and Rubber Co.

Q. Was that the group?

A. Dr. J. R. Zacharias of MIT was another member. I think it was actually Dr. Zacharias who raised the question.

Q. Was that luncheon for the purpose of discussing this case, if we can call it such?

A. No. This was just an informal grouping at the Cosmos Club. The occasion was the last meeting of the Advisory Committee, ODM, of which I am chairman. These others that I have mentioned, except Mr. Gardner, are members of that committee and we happened to be in town together. Gardner had at our request appeared before the committee that day to discuss some matters so he joined some of us at the Cosmos Club for dinner, I believe. This was a friendly discussion, wouldn't it be nice if we could help our friend.

Q. Yes, certainly. About when was that, Doctor, in March?

A. May I refer to my diary?

Q. Yes, sir.

A. I think I can give you the exact date of that last meeting. I believe it was the 12th or 13th of March.

Q. Did you see or talk to Dr. Oppenheimer about that time?

A. Did I see or talk to him?

Q. Yes.

A. I believe I called him on the telephone just to ask how things are going and to wish him well.

Q. Was he in Washington?

A. He was in Princeton. I am sorry, no. I called him at Princeton, but they found him somewhere in Washington and I talked to him on the phone.

Q. Did you see him?

A. I did not see him.

Q. What was the substance of your conversation?

A. I just said "Robert, how are things going?" It was only a friendly conversation, attempting to express confidence in him and cheer him up if possible.

Q. Did Dr. Oppenheimer tell you how things were going?

A. He only said it was not a very pleasant experience that he was going through.

Q. Anything more?

A. Nothing more relating to the substance of this case.

Q. That is what I mean. Substance?

A. That is right.

Q. What was said about the case in addition?

A. Just what I said, as I recall. It was not a very pleasant experience for him to be going through.

Q. Would it be on that occasion that you suggested to him that you testify or had you previously?

A. I had already previously discussed testimony with his counsel before that time.

Q. Have you since discussed your testimony with counsel and with Dr. Oppenheimer?

A. I have discussed the testimony?

Q. Yes.

A. I have not seen Dr. Oppenheimer just before I came here today. I have discussed of course the testimony with his counsel.

Q. You understand I am not trying to pry into your affairs, but I think these are matters which the Board ought to have on the record.

A. Yes, sir.

Q. Did you discuss the case after that with Mr. Gardner?

A. Did I discuss the Oppenheimer case?

Q. Yes, sir.

A. After that time?

Q. Yes, sir.

A. I have not seen Mr. Gardner—I think he did come to Pasadena shortly afterwards—yes, he did, on another business trip, and I think we probably did discuss it. Mr. Gardner has been very much interested in it, very much disturbed that a man so fine and so loyal should be accused, and he has been very anxious to discuss the case. We did discuss it.

Q. Has he been active in assisting Dr. Oppenheimer, do you know?

A. Has Mr. Gardner been active in assisting?

Q. Yes, sir, in any way.
A. I do not know whether he has seen Dr. Oppenheimer or not, or his counsel. I just don't know.
Q. Has he ever told you that he was doing some work for Dr. Oppenheimer?
A. No, he never has.
Mr. ROBB. That is all I care to ask.
Mr. GRAY. Dr. DuBridge, I am going back now briefly to October 29, 1949. Would you consider the two annexes to the GAC report in conflict with one another?
The WITNESS. Certainly not. Their conclusions were the same. They were slightly different approaches to these conclusions. Dr. Rabi and Dr. Fermi emphasized one aspect of the argument, and the rest of us emphasized another aspect. It was my feeling that these were definitely not in conflict, but only bringing out different points of view, which led essentially to the same conclusion and recommendation.
Mr. GRAY. There is something in your testimony that led me to ask whether we could make this kind of distinction with respect to what we have been calling the crash program. You know what I mean when I say that.
The WITNESS. Yes.
Mr. GRAY. I think this kind of distinction has only perhaps just come clear to me. Could there be a distinction between a crash program for the development of a thermonuclear weapon as distinguished from a crash program for the production of same?
The WITNESS. Of course, yes.
Mr. GRAY. Your position was that there should be no crash program for the production?
The WITNESS. That is correct.
Mr. GRAY. Did you favor a crash program for development?
The WITNESS. We favored the continuation of the research and development program at Los Alamos. We felt that it was going along pretty well. We recommended against at that time a crash program for production. In this case research and development I use together because both aspects are involved. But the research and development programs were in progress at Los Alamos.
Mr. GRAY. That was going as fast as it possibly could?
The WITNESS. We thought it was going along reasonably.
Mr. GRAY. There was nothing that could be done to speed up that to get into a crash-production program? I am trying to get it clear in my mind because I am still a little confused by the different points of view that are expressed about this thing. I can understand it better if this is a valid distinction.
The WITNESS. In my opinion it is. The research program began away back—there was some talk about thermonuclear programs during the war, as you know. I am informed though I was not present at the first discussion of thermonuclear programs was at a session in 1942 of which Dr. Oppenheimer was in charge in which the ideas of thermonuclear reaction were discussed. When I was in Los Alamos in 1945, the idea of thermonuclear explosions was then described to me in the general nature of the kind of reaction one might have. At various times we received reports from Los Alamos in the General Advisory Committee meetings as to the progress on research on thermonuclear reactions. It was my impression that this research was going forward, that there were some very difficult technical obstacles, but that the research and development was moving forward. It was not my intention at least in making this recommendation and signing it that this research and development effort should in any way be slowed down, but should be continued——
Mr. GRAY. At the same pace?
The WITNESS. At the same pace, and if possible, expanded if additional people could be found. We did not at any time recommend stopping the effort at Los Alamos.
Mr. GRAY. That is clear to me that you didn't stop it.
The WITNESS. Or slowing it down.
Mr. GRAY. Or slowing it up. I am wondering whether it was a matter of discussion in the GAC as to whether something more might be done in research and development short of production than was being done.
The WITNESS. Again, it is a little difficult to project the opinions back to that time, but as I recollect my own views on it they were that the thermonuclear program was proceeding satisfactorily, that it was a difficult decision of priority as to whether additional effort—that means men—should be transferred into the thermonuclear program as compared to the fission program, which was also pro-

ceeding beautifully, and was resulting in substantial improvements in our stockpile position on fission weapons.

There was a delicate balance there as to whether more good people—it took very good people at that time to make any good contribution to the thermonuclear program—should be asked to transfer from the fission to the thermonuclear program. I think it should also be made clear that these two programs are by no means independent; * * *. The thermonuclear and fission programs were very closely related, and going forward hand in hand as they must necessarily do.

In our opinion it was not a matter of real conflict but there was a matter of balance. We felt that very important fission programs were under way that should not be slowed down.

By Mr. GRAY:

Q. And they might have been slowed down by more emphasis on research and development with respect to the other weapons?

A. They could have been.

Q. I want to discuss a little bit with you, if I may, your views with respect to loyalty. This follows some direct questions put to you.

It is my recollection that you stated at one time in the day that you felt that former associations were irrelevant. If that is not a fair summary, I wish you would correct me. In any event, you felt that in this particular situation they are not relevant.

A. I was confining my remarks to the particular associations mentioned in the allegation in this case and to the individual in this case.

Q. Dr. DuBridge, Cal. Tech. has a lot of Government-sponsored research.

A. Yes.

Q. Is some of it classified?

A. There are two parts to our research, if I may explain. One large project which is operating off the campus about 5 miles at the Government-owned installation. Cal. Tech. operates it. That is a classified project on rockets.

On the campus where our students are, we have essentially no classified work in progress. We avoid it on the campus. There are one or two pieces of equipment, wind tunnels, to which classified models are occasionally brought for test and so for a while a classification screen has to be set up around. But by and large, we do not have classified research going on on the campus.

Q. At the off-campus center, which does have classified work, you must have certain employment policies with respect to people there. I assume you don't knowingly employ a person who is currently a member of the Communist Party?

A. Obviously not.

Q. That would be pretty clear, I think. Are the prospective employees or personnel on that project asked if they have ever been members of the Communist Party?

A. I am not sure I can answer that. I don't know what questions the personnel officer asks. No one is employed on that project, however, until we have received from the Army a clearance saying that this man is cleared for confidential work. This is a project under the sponsorship largely of the Army Ordnance Corps. There is a local ordnance office in Pasadena. All prospective employees are referred to them for screening and clearance. I am sure that they would not clear anybody who was a member of the Communist Party.

Q. Currently.

A. Yes.

Q. Would they clear anybody who had been a member of the Communist Party?

A. We had one case a few years ago where they did clear a person who had been a member of the Communist Party. When they found it out, however, they withdrew his clearance.

Q. Would you make a distinction between the type of clearance needed for someone who is going to join the faculty on the campus where there is not classified information and someone who would join the other project where there is classified?

A. Yes.

Q. You would apply a more rigid test on the off-campus center?

A. Yes. Further, on the off-campus center, we say as a university we are not competent to judge the security risk of prospective employees. We therefore refer these questions to the Army.

Q. So, as president you don't take responsibility securitywise for the people employed on that project?

A. That is right. We naturally are careful in our employment policies to not get prospective employees referred to the Army that are obvious security risks even to us. We would not employ anyone until we were sure first he was an honest man, second he was an able scientist or engineer, and third, that his former employees and associates felt that he was a good man to work in such a group. We would give this kind of general screening of ability and integrity first. But we would not attempt an FBI investigation.

Q. I understand. You get applications for employment at that center, and if you think the individual is a good prospect for employment, you ask the Army to clear him?

A. That is right.

Q. If you knew that a man was a member of the Communist Party, would you even send his name over?

A. I would not consider it at all.

Q. If you knew he had been a member of the Communist Party, would you send his name over?

A. If he was an applicant for a job at the classified research laboratory, that is a little difficult, because it would depend a little on the circumstances as to what the man had done in the meantime. Whether he had told us honestly he had been a member and had resigned, or whether he had hidden it and we had found it out in some other way.

Q. In the latter case, there would not be much question?

A. Yes.

Q. But you are not sure about in a case——

A. If a man came to us and said, "I was a member of the Communist Party 20 years ago, I resigned for the following reasons," we would probably say, "Well, everything else being acceptable, we will not put you at work, but we will put your name in for clearance, and we will see what the Army thinks of your connection."

Q. In testifying about associations earlier today, you indicated an understanding that in a particular case the associations ceased. I believe at least that was true.

Let me say that this board has reached no conclusion, and I want to make clear that I am trying to establish your philosophy, and not to ask you to pass judgment on any set of facts.

Suppose some of these associations continued, would that change the answers you gave?

A. If they had continued in an active way, and if the associations, the individuals involved had continued themselves an active association with the Communist Party, I would think this was a proper matter to be further investigated.

Q. So in that case associations would be very relevant?

A. That is correct. If they were continuing, and if the individuals involved were continuing their association with the party.

Q. I have just one final question which relates to your discussion of the atmosphere and times in the late thirties and early forties when people were concerned with what was happening in Germany and Spain. You indicated that at least part of this deep concern was a reaction to dictatorship and therefore some people turned to the Communist Party in reaction to revulsion against dictatorship. Wasn't it pretty well understood in this country at that time that the Soviet Union was a dictatorship?

A. It is a rather curious situation that the most active verbally opposition to Hitler at that time came from members of the Communist Party. It is now obvious to all of us that this was a piece of hypocracy, since their own regime was a dictatorship all the time. I think, however, in the early 1930's it was not so clear as it is now that the Communist Party in the United States was really a part of the Soviet Government apparatus, nor was it so clear that the type of dictatorship was the same. I think those who thought that were wrong and mistaken, but it was nevertheless true. Wasn't it half a million people voted for the Communist candidate for President in the thirties, apparently under the illusion that the Communist Party had a solution to the depression problems, or something and we were not aware of the nature of the world conspiracy which was developing at that time. But it is certainly true that I believe many people joined the Communist Party, or became associated with those who were members because the members did express an active opposition to Hitlerism, to Nazism, to Fascism generally and a support of the Spanish Loyalists.

Q. I don't pose as an expert. You asked me a question. I think you will not find that we ever had a time in the political history of this country where a half million people voted for the Communist Party candidate. I believe that you would find that in the depression years, to use the words of the Democratic candidate last year, almost a million people voted against capitalism. Again just to make sure I don't accept that statement of the situation, the vast majority of those were votes for Norman Thomas, the Socialist candidate, and I am guessing—I don't know whether I am sworn here—I am guessing that very considerably less than half a million ever voted for the Communist Party. I think we are engaged in an excursion.

A. Yes, I think so. I hope my figures there will not be taken seriously. But there was a substantial vote for the Communist Party.

Q. Yes, certainly more than would be true today, I think.

A. Yes.

Mr. GRAY. Dr. Evans, do you have any questions.

Dr. EVANS. Dr. DuBridge, let us go back again to that Chevalier incident. You remember about it. I want to ask you this question. Was it Dr. Oppenheimer's job to decide whether the security of his country was involved, rather than to report the incident?

The WITNESS. Would you repeat that?

Dr. EVANS. Yes. Was it Dr. Oppenheimer's job to decide for himself whether the security of the country was involved rather than report the incident immediately?

The WITNESS. I think possibly Dr. Oppenheimer was mistaken in his judgment at that time. I am sure it is a mistake he will not repeat.

Dr. EVANS. You would not have done it the way Dr. Oppenheimer did?

The WITNESS. Knowing what I do now, today, I would not. What I would have done in 1940, I cannot say.

Dr. EVANS. That is all.

Mr. GRAY. Mr. Garrison.

REDIRECT EXAMINATION

By Mr. GARRISON:

Q. I have just one question to clear up what may or may not be a misunderstanding.

When you were being asked about the luncheon, I think at Mr. Burden's in Washington, and the discussion with Mr. Griggs, and so on, I think the question was put to you whether you said anything at that luncheon to the effect that you regarded the development of continental defense and of atomic weapons, fission weapons, as more important at that time than the H-bomb. I wanted to ask you whether you meant to convey to the board—if you did, you should say so—that you had in mind at that time or indeed at any time that there should be any lessening of the effort to produce the H-bomb, or any lessening of cooperation with the letter and spirit of President Truman's "go-ahead."

A. It was not my understanding then or now that President Truman's decision meant that no other military program should go forward other than the H-bomb program, or that even that the H-bomb program would have overriding priority over all others. It seems to me then that of more immediate concern to the strength of the country was the continued development of our fission stockpile and the methods for delivering it, plus the continued development of a method of defending this country against a fission bomb attack which then was as now certainly possible on the part of the Russians. It was not our thought that giving attention and effort to the fission program or especially to the continental defense program need in any way detract from the essential part of the effort on the H-bomb program.

I think what we were trying to get across at that time there were many people, it seemed to us, who were of the opinion that the only thing that could save this country was to get an H bomb right now, and that all other things would sink into insignificance by comparison. I felt that was not a fair evaluation of this country's military situation. That it was important that the fission program go ahead and the continental defense go ahead. The continental defense is now going ahead on a large scale, and it is recognized that it is an important enterprise, and indeed its importance has increased by virtue of the H-bomb effort on the part of the enemy.

In other words, we were trying to get a proper balance in the military program of the United States, and arguing for a proper balance.

Q. You said H-bomb development on the part of the enemy. You don't know personally that they are working on the H-bomb now?

A. I meant the H-bomb because it is my understanding that the Atomic Energy Commission has detected evidence of a thermonuclear explosion in Russia.

Dr. EVANS. Thank you.

By Mr. GARRISON:

Q. Is it unclassified to say when?

Mr. BECKERLEY. It was announced.

The WITNESS. It was announced.

Mr. GARRISON. When was it announced?

Mr. BECKERLEY. August 1953.

The WITNESS. That is, of course, the time this was being discussed. What I was referring to was also after. I was saying that the continental defense now that is going ahead was even more important because of the thermonuclear explosion by Russia in 1953.

By Mr. GARRISON:

Q. I think when I was asking you about your opinions regarding Dr. Oppenheimer's loyalty, when I put to you a very long question about the Chevalier incident, I also asked you to assume that all the derogatory information in the December 23 letter of the Commission was true, leaving aside the items about the H-bomb, and you answered the question leaving aside the items about the H-bomb.

I just wanted to make sure—and I think it is probably sure by now, but perhaps not—that with respect to the items of information about the H-bomb in the Commission's letter, do you have any opinion with regard to those particular items?

A. Yes. In the first place, I think ——

Q. Let me refer to it a little more explicitly. What I have reference to are the suggestions that Dr. Oppenheimer ——

A. May I refer to a copy of that letter?

Mr. ROBB. Surely.

By Mr. GARRISON:

Q. He caused to be distributed and so forth, copies of the report, that he discouraged people from working on the project, and that he delayed the production of the work on the bomb. I am paraphrasing it. You have the exact language there.

A. In the first part of this paragraph, which is on page 6 of the original letter, the paragraph starting, "It was reported that in 1945, you expressed the view" and so on, certain statements are made about Dr. Oppenheimer's opinion on the feasibility and desirability of an H-bomb program.

Q. What I have reference to are the reports at the top of page 7.

A. I would like to make a report about the first part.

First, it seems to me that those statements about his opinions, even insofar as they are true, could perfectly possibly and indeed I believe were the opinions of a perfectly loyal American seeking to increase and not decrease the military establishment of his country.

"Further reported that even after it was determined as a matter of national policy to proceed with the development of a hydrogen bomb, you continued to oppose the project and not cooperate fully in the project."

To the best of my knowledge that statement was false. "It was reported that you departed from your proper role in the distribution of the reports of the General Advisory Committee for the purpose of trying to turn such top personnel against the development of the hydrogen bomb." To the best of my knowledge that is false.

I think it is quite probable that copies of GAC reports did reach the top people of Los Alamos as all our reports did by normal channels, but that the chairman of the committee departed from his proper role or did this with the purpose of trying to turn personnel against the hydrogen bomb is in my opinion false.

"It was further reported that you were instrumental in persuading other outstanding scientists not to work on the hydrogen project, and your opposition to the hydrogen bomb of which you are the most experienced and most powerful has definitely slowed down its development," that is also false. Quite the contrary, I believe Dr. Oppenheimer's efforts and the efforts of the GAC were intended solely to improve the position of this country, with no other objective, purpose or result.

Mr. GARRISON. That is all.

Re-Cross-Examination

By Mr. Robb:

Q. Just to have the record clear, what you have done is to give your opinions without knowing definitely the facts?

A. I said to the best of my knowledge in each case.

Mr. Robb. Thank you.

Mr. Gray. Thank you very much, Dr. DuBridge.

(Witness excused.)

Mr. Gray. We are in recess until 9:30 tomorrow morning.

(Thereupon at 6:10 p. m., a recess was taken until Friday, April 23, 1954, at 9:30 a. m.)

UNITED STATES ATOMIC ENERGY COMMISSION

PERSONNEL SECURITY BOARD

IN THE MATTER OF J. ROBERT OPPENHEIMER

ATOMIC ENERGY COMMISSION,
BUILDING T-3, ROOM 2022,
Washington D. C., Friday, April 23, 1954.

The above-entitled matter came on for hearing, pursuant to recess, before the board, at 9:30 a. m.

Personnel Security Board: Mr. Gordon Gray, chairman; Dr. Ward V. Evans, member; and Mr. Thomas A. Morgan, member.

Present: Roger Robb and C. A. Rolander, Jr., counsel for the board; J. Robert Oppenheimer, Lloyd K. Garrison, Samuel J. Silverman, and Allan B. Ecker, counsel for J. Robert Oppenheimer; Herbert S. Marks, cocounsel for J. Robert Oppenheimer.

PROCEEDINGS

Mr. GRAY. The proceeding will begin.

I suggest we open the proceedings with your request or statement, Mr. Garrison.

Mr. GARRISON. Mr. Chairman, I was informed by you yesterday afternoon that some witnesses would be called this coming week by the board. I had assumed from prior discussions that we would be informed of the names of these witnesses, but whether or not that assumption was correct, I asked you at the close of the session yesterday for the names of the respective witnesses in order that we might have time to prepare for cross examination, if cross examination seemed to be indicated with respect to one or more of them.

I would like to state very briefly the reasons why it seemed to me this request is a proper one to make on behalf of Dr. Oppenheimer.

The purpose of this inquiry which is not a trial is to arrive at the truth as nearly as truth can be arrived at. I don't think it takes any argument to point out that cross examination is one of the ways of bringing out the truth. I appreciate fully that there is no question here of denying the right of cross examination, but there is, as I am sure the board knows, oftentimes a need of preparation in cases where there may be an element of surprise in the calling of a witness, or in cases where a witness who one might perhaps think it possible the board might call we would know in advance would require a great deal of preparation, and in the press of other work, we would not want to undertake that uselessly if the person were not to be called. But in the main it is to have an opportunity to consider who is going to be called and to inform ourselves as to what we need to do.

With respect to our own witnesses, we have I think from the very first day, and from time to time gladly supplied the board with a list of people whom we expected to call. There have been changes in the schedule. Some inevitable additions and some who could not make it because of conflict of things and so forth, but in general I have tried to keep the board as accurately informed as I could.

It is quite clear that in the case of at least some of these witnesses substantial preparation for cross examination was made ahead of time and in the case of several others opportunity was had for the representatives of the board to discuss matters with these witnesses themselves, a process to which we had not the slightest objection at all.

Now, it seems to me that the same kind of notice and the same opportunity for preparation both in fairness to Dr. Oppenheimer and in the interest of developing the true state of affairs be accorded to Dr. Oppenheimer.

Therefore, on his behalf I request that we be informed of the witnesses whom the board proposes to call.

Mr. ROBB. Mr. Chairman, unless ordered to do so by the board, we shall not disclose to Mr. Garrison in advance the names of the witnesses we contemplate calling.

I should like briefly to state the reasons which compel me to this conclusion in the very best of spirit, and I am sure Mr. Garrison will take it that way.

In the first place, I might say, Mr. Chairman, that from the very inception of this proceeding, I think Dr. Oppenheimer has had every possible consideration. Going back to December, subsequent to the receipt by him of the letter from General Nichols, the time for his answer to be sent in was extended several times at his request, and without any objection whatever, because it was thought that was a reasonable request.

At the proceedings before this board, I am sure the record will show that the board has extended every courtesy and consideration to Dr. Oppenheimer and his witnesses. The board has permitted the testimony of several witnesses to be interrupted in order that others might be called to suit their convenience. The board has sat long hours for that purpose. One evening, as I recall, we sat until 7:45, and I cross examined the witness for the last 2 hours of that session. On one occasion we adjourned early so that Mr. Garrison might confer with his client with a view to putting him on for redirect examination.

Counsel has made no objection to any questions, although I say frankly that some questions might have been objectionable, but witnesses have been permitted to argue from the witness stand without objection, and tell the board in rather forceful terms about what the board ought to do about the problem, without objection.

Mr. Rolander has worked late at night and on Saturday and Sunday in order to get the record in shape so that it might be taken by Mr. Garrison and his associates.

I mention all these things, Mr. Chairman, only to illustrate what I think the record abundantly shows, which is every effort has been made to make this a full and a fair hearing, and to accord Dr. Oppenheimer every right, and I am sure that has been done.

Mr. Chairman, the public has an interest in this proceeding also, and of course the public has rights which must be looked out for. In my opinion, and it is a very firm opinion, the public interest requires that these witnesses be not identified in advance. I will say frankly that I apprehend, and I think reasonably apprehend, that should that be done, the names of these witnesses would leak, and the result then would be the embarrassment and the pressure of publicity.

I think furthermore, and I will be frank about it, that in the event that any witnesses from the scientific world should be called, they would be subject to pressure. They would be told within 24 hours by some friends or colleagues what they should or should not say. I say specifically and emphatically I am not suggesting that would be done by Dr. Oppenheimer, his counsel or anybody representing him. But I think the record abundantly shows here the intense feeling which this matter has generated in the scientific world. I think it perfectly reasonable to believe that should there appear here today that Scientist Y was to testify, inside of 24 hours that man would be subject to all sorts of pressure.

Now, Mr. Garrison has said there would be no leak. Perhaps so, Mr. Chairman, but the New York Times of the day after this hearing began, and the column which appeared in the Washington Post this morning do not lead me to rely with any great assurance upon any such statements. I think it would be a serious danger that the orderly presentation of testimony, the truthful presentation of testimony would be impeded were these witnesses to be identified.

Mr. Garrison speaks of the preparation for cross examination. In the first place, I didn't ask Mr. Garrison for the names of his witnesses in advance. It was entirely immaterial to me whether he gave them to me or not. We talked, of course, to General Groves, Mr. Lansdale—I think that is all of the witnesses—because both of them wanted to look at the files to refresh their recollection. Most of the witnesses who were called here I never saw before in my life.

I will let Mr. Garrison in on a little trade secret. In the case of almost all of the witnesses, my only advance preparation for cross examination was a thorough knowledge of this case. I am sure that Mr. Garrison has an equally thorough knowledge of the case. He has been working on it, I am sure, as long as I have. He has the assistance of Dr. Oppenheimer. Dr. Oppenheimer is the one man in the world who knows the most about Dr. Oppenheimer, his life, and his works. He also knows as much, I think, as anybody else about the subject of nuclear physics, which has been under discussion.

Mr. Garrison also has the assistance of three able counsel in this room, and I believe one other lawyer who is reading the transcript and making a digest of it for him.

As for surprise, I am sure any witness who testifies here within the scope of the issues of this case will not be unfamiliar to Mr. Garrison, nor will the subject matter of his testimony be unfamiliar to Mr. Garrison.

I am sure Mr. Garrison can do just as well as I did, however well that may have been. Maybe he wants to do better, if he can, fine.

Mr. Chairman, to sum up, my position is simply dictated by the public interest which I think would not be served by a disclosure in advance of the names of these witnesses for the reasons I have stated. I think that fairness to Dr. Oppenheimer does not require such a disclosure.

Mr. GRAY. Do you care to respond to any of that?

Mr. GARRISON. Mr. Chairman, I don't want to make an argument. I just want to make one or two observations.

First, with regard to the procedure of the board, the only thing that I have objected to that I still regard with all due respect as not in keeping with the spirit of the regulations is the questioning of witnesses, particularly Dr. Oppenheimer, as to their recollection of things past when the Government had in its possession papers, some of them taken in Dr. Oppenheimer's case from his own file as classified, and then declassified and read to him after the questions had been put in a way that could be calculated to make the witness appear in as poor a light as possible. The sort of thing I can make no objection to on orthodox legal rules of trial behavior in a court room, but which seem to me not appropriate here. I simply have to say that lest by silence I seem to acquiesce.

I also might say that in a court room that state of affairs can scarcely arise because of the nature of the documents and the source from which they came in this case. So it is perhaps an altogether novel situation and all the more I think not in keeping in the spirit of inquiry as distinct from a trial.

Now, with respect to leaks, I think all of us have done what we can to prevent them. I know we have. I have not seen the column in the Washington Post this morning. I have not read it. I have heard of it. I understand it is something to do with General Osborne's testimony and stated in quite an erroneous fashion, in a way that certainly could not have been put out by anybody connected with Dr. Oppenheimer in any way.

It was also stated in that column that Dr. Oppenheimer's representatives are not available to the press, which is certainly the case as far as giving out of information is concerned. I think the only actual leak that is difficult to explain about these proceedings since we began was Jerry Green's column about the Condon letters published actually the night before they were produced in evidence here, a statement about which on information which only could come from somewhere within the Government.

If it be the conclusion of the chair that in the light of this discussion the names of witnesses should still be withheld, I would then—perhaps I should ask the chair to first rule on that, and then make another request if I need to.

Mr. ROBB. I have nothing more to say, Mr. Chairman.

Mr. GRAY. I can respond on behalf of the board, because we have had some discussion of it this morning. I am going to advert to several things that counsel said here, so my statement may be in the nature of random observations in part.

I think that since the column in the Washington Post it has become a matter of this record in fairness to the chief counsel for Dr. Oppenheimer, it should be said that he has been hard to get hold of, specifically by name, and I am sure that is correct.

With respect to a reference to the Condon letters, it was my recollection that we had a Condon letter in this record. I didn't know there was more than one letter that appeared in this record. I suppose, however, that is not too material because I am quite convinced in my mind that nobody connected with this proceeding released those communications to anybody.

I might say the reason I am confident is that if for example the counsel for the Government and the board were interested in releasing information to the press which would be detrimental to Dr. Oppenheimer, I would not guess that the Condon letters referred to would be perhaps the most significant material for that purpose.

Now, it is true, Mr. Garrison, that you have at all times attempted to keep the board and Mr. Robb informed as to your general course of action with respect to witnesses. It is a courtesy which has been appreciated. It was not something that was required by the board.

I would like to say a little bit about this matter of calling witnesses. In our earlier discussion, I think I have loosely used the phrase witnesses to be called by the board. Actually I don't think at this moment that the board intends to call any witnesses. I do not consider that we have called those who have testified to this point, and the witnesses whom Mr. Robb will examine in direct examination will be called by him. For that purpose, this board considers you the attorney for Dr. Oppenheimer, Mr. Robb the attorney for the Atomic Energy Commission. He was appointed by the Atomic Energy Commission, as I understand it.

The board would be very much concerned if Dr. Oppenheimer's interests were in any way adversely affected by anything in the nature of surprise. I would guess from what Mr. Robb has told me that there probably will not be an element of surprise in the sense that we have in mind in this discussion. If, however, there is, the board will wish to be informed by counsel for Dr. Oppenheimer, and can give you assurance on behalf of the board that we will so conduct the proceeding that any disadvantage to Dr. Oppenheimer by reason of surprise as may be related to cross examination may not continue.

The board is interested in developing the facts, and if you are unable under the circumstances to perform your functions—very important functions—as counsel for Dr. Oppenheimer, we want to hear about it, and take the necessary steps.

The proceedings under which we operate, which are familiar to you, I know, require that the board conduct the proceedings in a way which will protect the interests of the individual and of the Government. The representetive of the Government in this case feels with some conviction that the interests of the Government could possibly be prejudiced by furnishing a list of witnesses at this time.

My ruling after consultation with the board is that Mr. Robb will not be ordered by the board to furnish these names. I couple to that ruling ,however, a repeated assurance that we wish to hear you at any time that you think you are at a disadvantage by not having had the names of the witnesses.

I would make one further observation, and that is in preparation for any cross examination, no attorney—or it is a very rare thing if an attorney knows what the testimony on direct examination is going to be. I suspect we have had so much of a record in this case that there is hardly anything that might be in any way related to it that has not been in some way discussed in this hearing.

I have one other observation. You have expressed unhappiness with the cross examination of witnesses, particularly of Dr. Oppenheimer. I hope that it will be unnecessary to say to you, Mr. Garrison, that the membes of this board, with the exception of a very brief period one afternoon when Mr. Morgan was unavoidably absent, have heard all of the testimony, the circumstances under which it has been given, the board will have available to it therefore not only the transcript, but a very vivid recollection of the circumstances under which the testimony was given. Without in any way making any observation about the merits of this suggestion you have made about the manner of examination, certainly the board will consider what has been adduced here, and not be particularly impressed, for example, with the fact that a witness failed to recollect a meeting or writing a letter or something of that sort. I think we will try to consider these things in balance and perspective.

Mr. ROBB. Mr. Chairman, may I say one further thing?

Mr. GRAY. Yes, sir.

Mr. ROBB. Lest my silence be misinterpreted, I wish to say that nobody connected with the Commission, as far as I know, had the slightest thing to do with the release of the so-called Condon letter. I think it is quite apparent on the face of the news story that it came from some other department of the Government.

Mr. GRAY. Or perhaps some other branch.

Mr. ROBB. Some other branch of the Government is what I meant; yes, sir.

Mr. GARRISON. I think on that counsel on that occasion referred to Mr. Green as perhaps clairvoyant.

Mr. CHAIRMAN. May I make a final observation?

Mr. GRAY. Yes, you may.

Mr. GARRISON. I want to thank you for the courtesy with which this proceeding has been uniformly conductd. I know the spirit of fairness which animates the members of the board. What you have said about considering any request we might make for time to prepare for cross-examination if we were disadvantaged by the calling of some particular witness meets what I was going to say after the chairman had made his ruling.

I just feel I must make one comment, not in criticism of the board, but with respect to the procedure. The notion that counsel for the Commission is to call his own witnesses in a proceeding which therefore takes on the appearance of an adversary proceeding with the board sitting as judges, and counsel for the Government on the one hand, and counsel for the employee on the other, is not quite a true picture of the actual shape of affairs. Unlike in an ordinary adversary proceeding before a judge in a courtroom, counsel here is possessed of documents taken from Dr. Oppenheimer's files in some cases which we have no opportunity to see in advance of their reading, and all the rest of which we have no opportunity ever to see.

It differs further in that the board itself is in possession of all these documents which it has had a week's opportunity to examine before the hearing began. This, then is not like an ordinary adversary proceeding. This is what we have to bear, Mr. Chairman. I am sure the board is aware of the problem that this presents to a person whose whole career and in a way his whole life is at stake.

I think I have no more to say.

Mr. GRAY. Let me make one further comment. I am sure all members of the board are aware of the difficulties involved for Dr. Oppenheimer. The board is certainly aware of the agonized character of these proceedings as far as Dr. Oppenheimer is concerned. This is not for any of us involved a pleasant kind of task. We are sympathetic to the difficulties. Some of these are inherent difficulties. I am sure we would all agree as to that.

I should explain further the view, so far as I know now, that witnesses will be called by counsel. First of all, I think it would be unreasonable to suppose that you would call witnesses for Dr. Oppenheimer who would do other than support his position and him as an individual. There obviously is division of opinion with respect to this matter or it would not be before us. Certainly the board must hear from people who may be in disagreement, perhaps, or who can shed further light beyond that thrown on the matter by representatives of Dr. Oppenheimer.

I am very anxious that it not appear that this board has called any witness as a board witness who had come here in a sense on behalf of prosecution. This is why I am making this distinction.

I think I should further say that if you read the regulations, the board does have power to call witnesses. We interpret that this way. It is conceivable that a witness who might normally be expected to testify for Dr. Oppenheimer would not be called by you. I am sure this is not the situation but my illustration could well be Mrs. Oppenheimer. I take it under these proceedings the board would have the power to call Mrs. Oppenheimer.

On the other hand, it is conceivable that there might be someone identified with the Atomic Energy Commission in an official capacity who would not be called by Mr. Robb, or whom the Atomic Energy Commission might not wish to be called. In that event, I take it that this board has the power to say we must hear from that witness.

I know of no such situation and that is why I have said at this point that the board would not call any witnesses and that is why I distinguish the matter of the development of opposed views in these matters.

I invite any further comment from counsel.

Mr. GARRISON. Mr. Chairman, we welcome the calling of witnesses either by the board or Mr. Robb or both to the extent that they can throw light upon the problem before the board. We feel rather relieved in fact that this is to be done, because I think it will bring out what we are confident will be the true situation, which we believe to be one which would lead to a sound conclusion here regarding Dr. Oppenheimer's clearance.

With respect to Mrs. Oppenheimer, we, of course, expected to call her as a witness and are expecting to put her on Monday morning—put her on is not the phrase—invite her to testify on Monday. She came as the board will recall on the first day on crutches as a result of a broken ankle, and she subsequently has had what appears to have been a case of German measles. But she is now all right and will testify, barring accidents, on Monday.

Mr. GRAY. Of course, we should be glad to hear from her. I knew it had been your intention to bring Mrs. Oppenheimer before the board, and that is why I used this as an illustration, because I am sure it would not develop into the kind of situation I described.

Mr. GARRISON. I would like to put one question to Mr. Robb. In the New York Journal American of last week—I am sorry I don't have the clipping, and this is just by hearsay—I am informed in Howard Rushmore's column last week Mr. and Mrs. Crouch were quoted as saying that they had been told that they would be called here as witnesses. I wonder of counsel could give me any information pertaining to that.

Mr. ROBB. I didn't see the column and don't know anything about it, Mr. Garrison, so I don't think I should comment on it. I am not responsible for what somebody writes in New York. I don't know anything about it.

Mr. GARRISON. I understand that. Could you say within the keeping of the chairman's ruling whether or not you expect to call them, because there is a great labor of preparation there.

Mr. ROBB. It is rather difficult to say at this time, because I don't know what is going to develop here from here on, Mr. Garrison. I would just rather not comment at this time.

Mr. GARRISON There is not any notion that physicists would pressure on the Crouches?

Mr. ROBB. Not a bit, no, sir.

Mr. GARRISON. Is there any reason why we should not be informed if they are to be called?

Mr. ROBB. If they are or if they are not.

Mr. GARRISON. Either way. If they are not, it will relieve us of a considerable amount of unnecessary work. If they are, we should have time to prepare for it.

Mr. GRAY. I would like to make an observation about that particular request. The board felt that Mr. Robb's point about some of these witnesses was well taken and that is why we gave the ruling we did. I don't see, Mr. Robb, why in this case you can't.

Mr. ROBB. I don't either. I will say that is a reasonable request. No, I have no intention at this time of calling Mr. or Mrs. Crouch. I will tell you that frankly. But as you realize, I can't project myself into the middle of next week. I don't know what will develop.

Mr. GARRISON. I assume if you change your intention you will notify us?

Mr. ROBB. I will do so, yes, sir.

Mr. GRAY. Mr. Garrison, do you have a witness?

Mr. GARRISON. Yes.

(Discussion off the record.)

Mr. GRAY. Mr. Winne, do you care to testify under oath? You are not required to do so.

Mr. WINNE. I would be glad to testify under oath, Mr. Gray.

Mr. GRAY. Would you stand and raise your right hand, please?

Mr. WINNE. Harry Alonzo Winne.

Mr. GRAY. Harry Alonzo Winne, do you swear that the testimony you are to give the board shall be the truth, the whole truth, and nothing but the truth, so help you God?

Mr. WINNE. I do.

Whereupon, Harry Alonzo Winne was called as a witness, and having been first duly sworn, was examined and testified as follows:

Mr. GRAY. Would you be seated, please, sir, and indulge me while I remind you of the existence of the perjury statutes. I should be glad to discuss them with you, but may I assume you know about them?

The WITNESS. I know there are such things. I don't know the details, but it is not necessary.

Mr. GRAY. I should like to request, Mr. Winne, that if in the course of your testimony it becomes necessary to refer to or disclose restricted data, you notify me in advance so that we may take certain steps which are appropriate and necessary?

The WITNESS. Yes, sir.

Mr. GRAY. Finally, I should like to say to you that the board treats these proceedings as confidential matter between the Commission and its officials on the one hand, and Dr. Oppenheimer and his representatives and witnesses on the other. The Commission will make no release of matter with respect to these proceedings. On behalf of the Board it is my custom to express the hope to each witness that he or she will take the same view.

The WITNESS. I so understand and I agree, Mr. Gray.

Mr. GRAY. I might say we had some discussion before you came in on procedural matters, and somehow there crept into the record a conversation about a column which appeared in the Washington Post this morning which I read and which said that the board is demanding secrecy. The board desires not to have leaks, of course, but I remind you if you read that column, that I simply expressed a hope to you.

The WITNESS. Surely.

Mr. GRAY. Mr. Marks.

DIRECT EXAMINATION

By Mr. MARKS:

Q. Mr. Winne, what is your present position?

A. May I start back a little? I retired from employment with the General Electric Company at the end of 1953. Now I am retired but I have a number of activities which keep me pretty busy, one of which is as Chairman of the Technical Advisory Panel on Atomic Energy, in the Office of the Assistant Secretary of Defense for Research and Development.

I am also a member of two committees of the National Science Foundation, and then I have various community activities and so forth in my home area, as trustee of three different colleges, and things like that.

Q. What was your professional career with the General Electric Company?
A. I started with General Electric as soon as I left college in 1910, and was with General Electric until December 31, 1953, filling various positions on the way up to becoming in 1941 vice president in charge of apparatus engineer, and then in 1945 vice president in charge of engineering policy so-called, which was essentially a coordinating and policy directing position for the engineering effort of the company as a whole, which position I held under a slightly different title, vice president, engineering, until November 1, 1953, when I was assigned to a certain special problem, which I worked on until the end of the year.

I might mention also because I think it is pertinent here that during the war years, starting with either the end of 1942 or early 1943, I devoted a good deal of time to coordinating and directing in a general way the efforts of General Electric Company in connection with the atomic energy program. The General Electric Company produced a lot of equipment, particularly for the magnetic separation process at Oak Ridge, and also the gaseous diffusion process at Oak Ridge, with both of which I was quite familiar, spending a few days at different times at Berkeley and some time at Oak Ridge.

Then in 1946, when General Electric took over the operation of the Hanford Works, I was appointed chairman of the so-called nucleonics committee of the company, which from that time for several years directed the general policy and the operation of the company in the atomic energy field, that is, the operation of the Hanford Works, the construction and operation of what was called the Knolls Atomic Power Laboratory at Schenectady, and other activities in the atomic energy field.

Q. I recall that last fall, I believe it was, you received some industrial award. Can you remember what that was?
A. That was last summer. It was the so-called McGraw Award for men in the electrical manufacturing industry, as distinguished from a similar award for men in the utility industry, and so forth. I received the award for the manufacturing man in the electrical industry last summer.

Q. When did you first know Dr. Oppenheimer?
A. To the best of my knowledge, I first met him in Mr. Acheson's office, I think in late January or early February of 1946, when I was asked to serve as one member of the five-man board of consultants to the Assistant Secretary of State's Committee on Atomic Energy in endeavoring to propose some plan for international control of atomic energy.

Q. How well did you get to know him as a result of that, or other work?
A. I feel quite well, Mr. Marks, because during the period of discussion and final drawing up of this plan for international control of atomic energy, that board of consultants met almost continuously for about 8 weeks, I think it was, except for weekends and even sometimes on weekends.

Q. How many hours a day did you work together on that?
A. Very often it was a matter of all day and dinner and evening, starting at 8:30 or 9 o'clock in the morning. So I felt that during that experience I got to know him, I feel, very well.

Since that time I have had—I can't state definitely just how many contacts. He and I were both members of the Committee on Atomic Energy of the Research and Development Board, as I recall, starting in early with my membership in early 1952, or possibly late in 1951.

Q. Research and Development Board of what agency?
A. I think it was called the Department of Defense at that time. Even prior to that in connection with the activities of the MLC, the Military Liaison Committee—I was not a member of that committee—I was invited to make at least two trips to the West Coast visiting various installations with that committee. It started at the time that Donald Carpenter was chairman of the committee. My contacts continued with it while Bill Webster was also chairman.

I remember one of the trips Karl Compton was along. On those trips—I don't recall whether on every one—at least one I recall meeting Dr. Oppenheimer at Berkeley and serving on a subcommittee of which he was chairman, which I think was set up by Mr. Carpenter, although I am not absolutely sure of that, to consider the matter of radiological warfare.

I visited Princeton once at least since his taking over the direction of the institute there. It was a more or less social session of the members of the board of consultants at the institute I suppose I have seen him 15 or 20 times, possibly more, since the days of the board of consultants.

I have visited at his home in Berkeley, I think, twice as a part of one of these groups which were making these trips to the west coast, not privately, I mean, but as a group of several at a cocktail party or something of that nature at his home in Berkeley. So as I say, I feel I know him quite well.

Q. The 15 or 20 times that you are speaking of, are those including the work on the State Department board in 1946?

A. No, since that time.

Q. Have most of these occasions been social or have they been working relations?

A. No, most of them have been in connection with work of the Committee on Atomic Energy or as I say, the trips with the MLC, and so forth.

Q. Speaking in a very general way, with what subject has the work of this Committee of the Research and Development Board been concerned?

A. Primarily with the use of atomic energy in military preparedness of the country, both in the form of weapons and also of propulsion equipment of naval vessels and aircraft.

Q. During the war, when you were working on aspects of the atomic energy project in the Manhattan District, who were your contacts there?

A. During the war?

Q. Yes, at the time, who were your contacts with?

A From the Manhattan District General Groves, at that time Colonel Nichols, Colonel Walter Williams, a few contacts with General Groves' predecessor whose name I cannot recall at the moment, and then with the Kellex Corp. people, Dobie Keith, Al Baker and others in connection with the gaseous diffusion plant, and with Stone & Webster, A. C. Klein and others of that organization, and the Carbide & Carbon people operating Oak Ridge—too numerous to mention.

Q. If you happen to know, can you say who suggested your name for membership on the Board of Consultants to the State Department on international control of atomic energy in 1946?

A. I do not know. I have always suspected that General Groves is the one who suggested it, because I did not know Mr. Acheson or Mr. Byrnes, nor the other members of the State Department's Committee on Atomic Energy at that time. So I have always suspected General Groves did, but I do not know that

Q. In your work on that committee, concerned with the problem of international control of atomic energy, what was your major worry about or what country or what countries?

A. Our major consideration, of course, was the protection of the United States, that is, of devising a scheme of control of atomic energy which would ultimately, we hoped, prevent the use of atomic bombs and might lead to—this may have been wishful thinking—abolition of warfare entirely, but always without sacrificing the protection of the United States.

Q. In those deliberations and in that work, what was your attitude, and if you can say, what was the attitude of your colleagues about Russia?

A. I think I can say we looked upon Russia as the most probable enemy of the United States. We looked upon her as the country which would be working hardest on trying to produce atomic weapons. I think none of us foresaw that she would produce these as early as ultimately turned out to be the case. We had hopes—again this as it turned out was probably wishful thinking—that Russia might be willing to go along with the plan which we ultimately evolved and succeeded.

Q. What did you think of the efficacy of that plan as a measure of protection for the United States?

A. We thought it was the best we could devise. We recognized that the detection of possible operations in the production of atomic weapons would at best be difficult, but we thought that the plan which we finally evolved could successfully do that.

Q. What part did the respective members of that board play, you and your four colleagues, in the development of the plan that you ultimately recommended?

A. That is a difficult question to answer, because there was so much back and forth discussion and give and take. I think that the germ of the idea—the first suggestion of the idea of the international development authority came from Dr. Charles Thomas, who is now president of Monsanto Chemical Company. We were all searching for some method which would not forestall the peaceful development of atomic energy and of the use of atomic energy which I felt was so very important. You may remember that in the early stages of the discussion, someone suggested that perhaps the only thing to do was to stop all work entirely. That the only hope for preventing the use of atomic weapons in warfare

Q. Did Dr. Oppenheimer suggest that?
A. No, I think that was Mr. Lilienthal. I said if that was the aim of the board of consultants, this was no place for me, because I thought that the development had to go forward. We had to devise, if possible, some means for controlling the development in such a way as to prevent the use of atomic energy for weapons.

Mr. GRAY. I am sorry. May I ask you to repeat that suggestion that Mr. Lilienthal made? My attention wandered for a moment, Mr. Winne.

The WITNESS. As I recall it, this was in the first 1 or 2 days of our discussion, and we were all of us somewhat appalled by the immensity of the problem which we faced in trying to arrive at some solution to this question. Mr. Lilienthal suggested—I am not sure that it was 100 percent serious, but perhaps in partly a joking tone—maybe the only recommendation we could come up with would be to outlaw all development in atomic energy. The only way we could hope to prevent the use of it in warfare was that. I recall I spoke up and said if that was to be our objective this was no place for me, because I wanted to see atomic energy developed for peacetime industrial use, primarily.

By Mr. MARKS:

Q. What view ultimately prevailed in the formulation of the report after the 2 months or whatever it was of deliberations and discussion?
A. The view that peacetime development should go forward and that we should set up, as you will recall from the report, this atomic development authority, which could exercise enough supervision to prevent the use of atomic energy in weapons, or at least to give forewarning to all nations in case any nation undertook the development or the manufacture of atomic weapons.

Q. When you say forewarning, what do you mean by that?
A. I mean we felt that the conversion from peacetime development to the production of actual weapons would take a certain amount of time measured in months, at least, and that the authority could be aware of this reasonably soon after it was undertaken by any nation, and could thereby warn the other nations of the United Nations community that such and such a nation was in effect abrogating the pact, and going ahead with the development of weapons so that the other nations could, if they desired, do likewise.

Q. What gave you any hope that under the plan you devised, the international authority of which you speak, would have had early enough warning of sinister developments in Russia or other countries?
A. We felt that it was absolutely necessary that all countries be open to inspection by this international authority, inspection which would be broad enough to permit the detection of supposedly clandestine operations in the production of atomic weapons.

Q. In the later deliberations in which you participated with Dr. Oppenheimer on the Atomic Energy Subcommittee of RDB, to what, if any, extent did problems concerned with the potential menace of Russia enter into your considerations?
A. I would say to a great extent. Always in the backs of our minds and frequently in the discussion was the question as to what Russia was doing, what her atomic stockpile might amount to, and as to when she might start a war in which atomic weapons would probably be used. That was always one of the main considerations which guided our discussion, and thinking. It may be well to state that on this committee there were not only civilian members, such as Dr. Bacher, who was chairman, Dr. Oppenheimer and Dr. Bethe, Mr. William Hosford, formerly vice president of Western Electric—I don't remember if there were other civilians—but there were also representatives of each of the armed services. General Yates of the Air Force, Admiral Withington and later Admiral Wright of the Navy, and General—I can't think of his name, from the Army, but usually two representatives from each of the services. So military considerations were the prime matters which we were discussing of course.

Q. In the course of these working relations and other relations you had with Dr. Oppenheimer, did you form any opinion about his loyalty to the United States, and his character?
A. Yes; very definitely. I have no question at all as to his loyalty to the United States. I think he is a man of high character. I have great respect and admiration for him.

Q. What led you to this opinion?
A. I can't cite specific instances, but his discussion, his remarks during the deliberations of first the board of consultants in 1946, and at later meetings of the Committee on Atomic Energy. As I say, I can't specify remarks, specify

comments, but there just developed within me a conviction as to his great concern for our country and his loyalty to it, his great concern for the safety of our country.

Q. What, if any, attitude did you observe with respect to Russia?

A. The feeling that Russia is the country which we have to guard against, a country maybe certainly our enemy and maybe the one to start a war against us, and one against which we must be on our guard at all times.

Q. When did you first form this impression?

A. I can't cite any particular date or time. It gradually developed.

Q. 1946, 1947?

A. It developed in the days of our board of consultants meetings in 1946, Mr. Marks, and, has, if anything, been strengthening since that time.

Q. Mr. Winne, have you read the letter of December 23, 1953, from General Nichols to Dr. Oppenheimer, which is the genesis of these proceedings?

A. As it appeared in the New York Times, yes; and then I again glanced through it this morning, or rather the copy which you have, and which you left with me as you came in here.

Q. Placing to one side the statements in that letter relating to the subject of the so-called hydrogen bomb and assuming that the derogatory information otherwise—and I am asking you only to assume not to consider whether it has been established in this proceeding that it is true or not—assuming that it is essentially true, the derogatory information other than that concerning the hydrogen bomb, what effect does that have on the conviction you have expressed with respect to Dr. Oppenheimer's loyalty and character?

A. I am still convinced of his loyalty to the United States and of his character. I am glad you said placing to one side the statements with reference to the hydrogen bomb. I have no objection to the first part of the statement with reference to the hydrogen bomb, but if it should be true that he really worked against the development of the hydrogen bomb, which I do not believe, after the President had decided to go ahead with it, that I could not understand. If that proved to be true, it would bother me a great deal.

The statements to the effect that he was opposed to the development before the President decided to go ahead with it do not bother me particularly, and it may be well that I state here that in the early days in the talk about the hydrogen bomb I personally had grave misgivings as to whether it was wise at that time to go ahead with that development. Those misgivings were based on two factors. One, that the development of the hydrogen bomb at that time, it seemed to me, would detract from what we might term our atomic capability because the development of one important ingredient would reduce——

Q. What do you mean by ingredient, if you can describe it in unclassified terms?

A. I don't know whether this is classified or unclassified.

Q. Did you mean a material?

A. A material, yes; because the production of that one required material would decrease the production of plutonium for the atomic bombs. Of course, as I say, this was several years ago and presumably our stockpile of atomic bombs at that time was not nearly so great as it is now. I knew from our operations at Hanford that the production of this material would make serious inroads on the production of plutonium.

So that raised the question as to whether it was desirable to go ahead with it at that time. It would also require the time and attention of a great many physicists and engineers.

Then I also had this question as to its military usefulness as compared to the atomic bombs; that is, whether a sufficient number of targets which would justify the use of so powerful a weapon as the hydrogen bomb. Two, even if there were, it seemed to me that there was a good possibility that it might be better to attack with, say, 25 planes, each carrying—and I use 25 to pull a number out of the hat, it might be 50, 100, or 10—each carrying 1 or possibly 2 atomic bombs, or to attempt to attack it with 1 or 2 planes each carrying a hydrogen bomb.

It seemed to me that the chances that a considerable number of the atomic bomb carrying planes would get through were so much greater than the chance that 1 or 2 carrying hydrogen bombs would get through, that the effectiveness of the greater number of planes with atom bombs might be considerably greater than 1 or 2 planes with a hydrogen bomb.

So I had that question. Of course, not being a military man, I am not competent to really pass on that sort of thing.

I recall that in discussing this matter with Ernest Lawrence——

Q. When would that have been?

A. I don't know. It was probably somewhere around 1950 or 1951. I don't know the exact date, Mr. Marks, but in discussing it with Ernest Lawrence, I mentioned these misgivings. When I first said that I had some misgivings as to whether it was wise to go ahead with the hydrogen bomb development, he expressed surprise. Then when I explained why, he said, "Oh, you mean that." He said, "I thought perhaps you might have the ethical or moral misgivings that some people have." I said, no, I did not, that it was entirely on a practical basis.

As I say, I had those same misgivings. Developments have, I think, shown that those misgivings were pretty largely unfounded, because at Hanford we have been able, as has been told publicly, to so greatly increase the production of plutonium from the piles which when we took over were supposed to be about ready to quit, that the production of the material for the hydrogen bomb has not seriously interfered with the production of sufficient plutonium. The costs in the equivalent of atom bombs have proven to be much lower. So that the program on the hydrogen bomb is working out much better than I had expected it would. I think that is true of many people. Many people thought at that time that it was going to make serious inroads in the production of atomic bombs, and that the hydrogen bombs would be extremely expensive. Of course, they are expensive in any ordinary terms.

Q. At the time you speak of, whether it was in 1950 or 1951——
A. It could have been in 1949. I don't remember, Mr. Marks. But I recall distinctly the conversation. I have seen Ernest Lawrence many times, and I can't tell you which time it was.

Q. At the time you speak of, what if any responsibility did you personally have for the operation at Hanford?
A. I was at that time still chairman of the nucleonics committee of the General Electric Co., which was the policy setting committee for all of our operations in the atomic energy field. As such I held a very real responsibility for the Hanford Works. In fact, at the particular time that the hydrogen bomb or that we began to produce at Hanford material for the hydrogen bomb, our organization had been changed somewhat so that the Hanford Works operation reported through a vice president located there directly to me, whereas previously it had been a part of the chemical division of the General Electric Co.'s operation, simply guided by the nucleonics committee. So I was pretty well aware of what was going on at Hanford and what the changes in production might be.

Q. In describing your misgivings that you held and you expressed you say to Dr. Lawrence about proceeding with the hydrogen bomb program, misgivings relating to the possible inroads that such a program might make on production of materials needed for A bombs, I think you said you were thinking particularly about production at Hanford.
A. Production at Hanford and the military usefulness of the hydrogen bomb.

Q. I think you said it turned out that production at Hanford for atomic bombs did not in fact suffer?
A. To say that it did not suffer is probably a correct statement, because had we not produced some material for hydrogen bombs, we would have produced more plutonium. But even with producing the material for the hydrogen bombs, we had increased the production of plutonium to such a great extent that the atomic bomb production was maintained at a very high rate.

Q. Hod did you bring that about to the extent that you can say in unclassified terms?
A. Of course, there are a tremendous amount of technical details, most of which are classified, covering the changes in operations which we made there which enabled us to step up the production of the existing piles very materially, and also to reduce the cost of the operation.

Q. Why didn't you foresee that at the time you talked to Dr. Lawrence?
A. Those changes came along rather gradually, and it is not always possible to foresee just what can be done. As a matter of fact, as I say at the time when we took over in 1946, it was thought that the piles would be out of commission in a very few years, and have to be completely replaced, whereas today they are still running and producing at a very much higher rate.

Q. In general who had responsibility for bringing about the changes or improvements, whatever they were, at Hanford that enabled you to keep up your production for A bombs in a manner that you had thought impossible or improbable if the H bomb program were adopted?
A. It was the General Electric Co. organization at Hanford primarily.

Q. You just didn't foresee that would be possible?
A. That is right.

Q. Did the General Electric people who were responsible to you at Hanford foresee it?

A. They may have foreseen more of it than I did, because they were closer to the job, but they were certainly not willing to go out on a limb and say that the things which were accomplished would be accomplished. As I say, there were gradual developments in the operation and whole technology of the pile operation which permitted us to do that.

Q. I think you said that reading this letter from General Nichols and assuming that the derogatory information, except for that part of it which you specifically excluded relating to the hydrogen bomb, relating to part of the information, you said I think that would not alter the conviction you expressed with respect to Dr. Oppenheimer's character and loyalty to the country?

A. That is true.

Q. General Nichols' letter also speaks of a variety of associations which Dr. Oppenheimer is said to have had with Communists, with left wing organizations, with causes which have been identified with Communist objectives. How do you reconcile your expression of confidence in Dr. Oppenheimer with this array of associations?

A. I think Dr. Oppenheimer's reply explains those associations. It explains how they developed and how he ultimately cast them off as he became more acquainted with the aims and objectives of those associations, of the Communist Party, of Russia. I think his subsequent efforts on behalf of the country, his thinking and the discussions he participated in in the meetings of the board of consultants of the Committee on Atomic Energy of the Research and Development Board, indicate to me that he is completely free of perhaps what you might call illusions or lack of understanding which he had in those earlier days. I think they do not affect his basic loyalty to the country.

Q. Suppose it appeared in these proceedings that at least some of the associations referred in in the Nichols letter—or that some of the people referred to about whom questions have arisen—were people that he still on occasion saw. I think it appears in evidence here, or perhaps in the answer, I have forgotten which, that as recently as last November in Paris, when Dr. Oppenheimer was abroad, he saw at the request of his old friend Chevalier, he saw Chevalier. Does that worry you?

A. No.

Q. Why?

A. I know nothing about the association between Dr. Oppenheimer and Chevalier, except through what I read in these 2 letters, 1 from General Nichols and 1 from Dr. Oppenheimer. But it appears that Chevalier was a close friend of his in the early days at Berkeley, and even though Chevalier may have been proved to be a Communist, and to have had the wrong kind of ideas, shall we say, I would not hold it against Dr. Oppenheimer's loyalty to the country at all, if he should, on Chevalier's request, see him to discuss whatever Chevalier wished to discuss with him. I feel sure he would not have divulged to Chevalier anything which would be inimical to the interests of this country.

Q. Do you think in making that determination of what would or would not be inimical to the United States, Dr. Oppenheimer would make the decision on the basis of his judgment or on the basis of the rules of the Government?

A. I think he would make it on the basis of the rules of the Government insofar as the rules cover the situation. Beyond that he would use his own judgment in which I would have confidence.

Q. Doesn't it worry you that a man who has as much classified information as Dr. Oppenheimer would even see a person like Chevalier?

A. No, Mr. Marks, that does not worry me, because, as I say, I have confidence in the loyalty and in his judgment. His judgment in his younger days, it may be claimed, was faulty. Instead of judgment, it may have been a lack of understanding of these organizations and so forth. But from the period of my knowledge of Dr. Oppenheimer and my acquaintance with him, I have no cause whatsoever to doubt his loyalty or his good judgment in political as well as technical matters.

Q. Let us take another case. I have forgotten if there are more in either the letter or the proceedings here, but one I remember is Dr. Morrison. Do you know who he is?

A. Yes, I know who he is because in connection with the board of consultants in the early days of 1946, we visited several of the installations and I remember meeting Dr. Morrison. I can't remember where, whether it was at Los Alamos or where, but I remember meeting him at that time.

Q. I think it has come out in congressional hearings that Dr. Morrison was once a Communist. Would it bother you in the connection in which we have been speaking if Dr. Oppenheimer had seen Dr. Morrison in recent years?
A. No, it would not.
Q. I don't think his name has been mentioned in the Nichols letter, but I think another name that has cropped out in these proceedings is that of a Dr. Seber, at Columbia.
A. As far as I know, I don't know him at all.
Q. I would like to make sure that you have in mind the full import of some of what appears to be the more important derogatory information in the Nichols letter. I would like to read to you, in order to be sure that you have it vividly in mind, one paragraph of this letter, and then I would like to ask you to make a comment.

In the letter that General Nichols sent to Dr. Oppenheimer, the following appears:

"It was reported that prior to March 1, 1943, possibly 3 months prior, Peter Ivanov, secretary of the Soviet Consulate, San Francisco, approached George Charles Eltenton for the purpose of obtaining information regarding work being done at the radiation laboratory for the use of Soviet scientists; that George Charles Eltenton subsequently requested Haakon Chevalier to approach you concerning this matter; that Haakon Chevalier thereupon approached you, either directly or through your brother, Frank Friedman Oppenheimer, in connection with this matter; and tht Haakon Chevalier finally advised George Charles Eltenton that there was no chance whatsoever of obtaining the information. It was further reported that you did not report this episode to the appropriate authorities until several months after its occurrence; that when you initially discussed this matter with the appropriate authorities on August 26, 1943, you did not identify yourself as the person who had been approached, and you refused to identify Haakon Chevalier as the individual who had made the approach on behalf of George Charles Eltenton; and that it was not until several months later, when you were ordered by a superior to do so, that you so identified Haakon Chevalier. It was further reported that upon your return to Berkeley following your separation from the Los Alamos project, you were visited by the Chevaliers on several occasions; and that your wife was in contact with Haakon and Barbara Chevalier in 1946 and 1947."

I would also like to read Dr. Oppenheimer's reference to this episode in his answer on page 22 of the answer:

"I knew of no attempt to obtain secret information at Los Alamos. Prior to my going there my friend Haakon Chevalier with his wife visited us on Eagle Hill, probably in early 1943. During the visit, he came into the kitchen and told me that George Eltenton had spoken to him of the possibility of transmitting technical information to Soviet scientists. I made some strong remark to the effect that this sounded terribly wrong to me. The discussion ended there. Nothing in our long-standing friendship would have led me to believe that Chevalier was actually seeking information; and I was certain that he had no idea of the work on which I was engaged.

"It has long been clear to me that I should have reported the incident at once. The events that led me to report it—which I doubt ever would have become known without my report—were unconnected with it. During the summer of 1943, Colonel Lansdale, the intelligence officer of the Manhattan District, came to Los Alamos and told me that he was worried about the security situation in Berkeley because of the activities of the Federation of Architects, Engineers, Chemists, and Technicians. This recalled to my mind that Eltenton was a member and probably a promoter of the FAECT. Shortly thereafter, I was in Berkeley and I told the security officer that Eltenton would bear watching. When asked why, I said that Eltenton had attempted, through intermediaries, to approach people on the subject, though I mentioned neither myself nor Chevalier. Later, when General Groves urged me to give the details, I told him of my conversation with Chevalier. I still think of Chevalier as a friend."

Refreshing your mind about that incident, what effect does that have on your opinion about Dr. Oppenheimer?
A. It does not change my opinion as to his basic loyalty to the country. I think that had I been in his place, I would have reported the incident immediately with the name, although one cannot at this date put himself back in the frame of the situation as it existed in 1943, and say definitely what he would have done. It seems to me that I would have reported it at that time.

As I say, it still does not affect my belief and my conviction in Dr. Oppenheimer's strong loyalty to our country. I think it was an error on his part not to report it immediately with the full details, but all of us make mistakes at some times.

Q. How does it affect your opinion about his character?
A. It does not affect that either. I still think his character is very high.
Q. As I recall, you said, Mr. Winne, that you are at present chairman of a committee on atomic energy.
Q. It is a technical advisory panel on atomic energy in the Department of Defense, reporting to Assistant Secretary Quarles. It, together with a so-called coordinating committee made up—this is a civilian committee—there is also a coordinating committee on atomic energy which is made up entirely of military personnel which reports to Secretary Quarles. The panel of which I am chairman is purely advisory. We have no power whatsoever other than the power of facts as we may develop them. It, together with that military committee, in effect replaces the old Committee on Atomic Energy of the Research and Development Board, of which Dr. Oppenheimer was a member at the time I became a member. Incidentally, I would be very glad to have Dr. Oppenheimer as a member of the panel today if he is cleared by this Board. I have that faith in his loyalty to the country and his outstanding ability as a scientist, which needs no testimony. We need that kind of people on such a panel.
Q. As chairman of that committee, do you feel any personal and official responsibility?
A. Very, very definitely, responsibility to do everything we can to assist the military organization of the country in developing the most effective use of atomic energy for military purposes. Of course, incidentally, protecting the interests of this country very fully from the standpoint of classified information and so forth.
Q. Do you feel any responsibility in that capacity for the security of the sensitive information that flows to you?
A. Very, very definitely. I feel a very high sense of responsibility.
Mr. MARKS. That is all, Mr. Robb.
Mr. GRAY. I am going to ask that we recess very briefly.
(Brief recess.)
Mr. GRAY. Mr. Robb.

CROSS-EXAMINATION

By Mr. ROBB:

Q. Mr. Winne, Mr. Marks read you two paragraphs from Dr. Oppenheimer's answer and in particular one sentence which I will reread for clarity: "When asked why, I said that Eltenton had attempted through intermediaries to approach people on the project, though I mentioned neither myself nor Chevalier."
Dr. Oppenheimer has testified before this board, sir, that what he said on that occasion was in certain respects untrue; specifically, that he said that there were three people who were approached whereas in fact there was only one; that he reported that there had been conversation about microfilm with Chevalier, whereas in fact there had not; that he reported that Chevalier had spoken of making a contact through someone in the Russian consulate, although in fact that was not true.
Does that disturb you, sir?
Mr. MARKS. Mr. Robb, would you mind if I ask you to identify the time at which these statements that you described were made and to whom?
Mr. ROBB. I am talking about the occasion referred to in this letter when Dr. Oppenheimer reported to the security officer about this episode with Chevalier. In that interview he has testified before this board he made certain misstatements of fact knowingly.

By Mr. ROBB:

Q. Does that disturb you, sir.
A. It disturbs me to some extent that he should have done that at that time. As I say, as I look at that incident I would have reported the whole thing immediately and in the true aspects of it. I don't know why he did not. He has since in his letter admitted that he should have or thinks he should have. It is a rather disturbing incident, there is no question of that.
But on the other hand from my almost living with him and the other members of the panel for 8 weeks and quite a lot of contacts since in deliberations on

weapons and that sort of thing, I still have no question about his loyalty to the country.

Q. Yes, sir; you speak of loyalty. Would the fact that he deliberately lied to the security officer about this matter in certain respects in your opinion have a very material bearing upon his character?

A. Obviously if a man deliberately lies it does have some bearing on his character. Of course, in connection with that the full situation at the time should be known. It is impossible for me to look back 10 years and to visualize just what the situation was in his respect at that time, although I can see no reason why he should have lied about it if that is what he did at that time.

Q. Suppose Mr. Winne you had an employee at GE who undertook to report some such incident to you and you subsequently found out that he lied to you abut certain material parts of it, would you be disturbed about it?

A. Yes, I would be disturbed and I would endeavor to find out just why and what all the circumstances were. But it would by no means necessarily be reason for firing him and his subsequent conduct would have much greater bearing on my feelings toward him than would that particular incident.

Q. Is it a fair statement that unless he could give you a pretty satisfactory explanation of why he lied to you you would fire him?

A. It would depend on what the situation was, what he was lying about and that sort of thing.

Q. Assume it was a very important matter.

A. If it was a very important matter and he could not give a convincing reason as to why he felt it was necessary at that time, it is quite probable that disciplinary action would be taken.

Q. Assume that the matter arose that you were looking into—you wanted to find out all you could about it for the good of GE—and you talked to an employee about it and he lied to you about it, and those lies impeded you in finding out about it and made it more difficult for you to run the matter down, wouldn't that disturb you very greatly?

A. It would disturb me, yes.

Q. And it would be very likely that when you found out about it under those circumstances you would fire him, wouldn't you?

A. Again it would depend on what the matter was; it would depend on his value to the company, his ability and several factors like that. Certainly the act of lying about an important matter would be considered as a black mark, you might say, against him.

Q. It would be something that you would require some explanation for, wouldn't you?

A. Yes.

Q. How well do you know Dr. Morrison, Mr. Winne?

A. I just met him, as I say, either on a trip or maybe he appeared before the board of consultants in some capacity to explain. You see, many of us on that board of consultants——

Q. Pardon me. I don't mean to cut you off but perhaps I can save a little time by coming to the point.

Do you know anything about his background?

A. No; other than what has appeared in the newspapers.

Q. You mean about his Communist connections?

A. I understand he has at least been accused. I don't recall I have ever seen that it was proved that he had Communist leanings or was a member of the Communist Party.

Q. He has admitted that he was.

A. I didn't know that unless it was brought out in the questioning by Mr. Marks. I forget. I knew he was at least under suspicion. I didn't know it at the time that I met him.

Q. I understand that.

A. I can't say I know him well at all because I have seen him once or twice.

Q. Knowing what you do about Dr. Morrison, do you think you would employ him on a GE confidential project?

A. On a matter like that I would have to know more about him and more about his subsequent actions and more about—I would have to know him much better than I do now to say whether or not I would be willing to employ him.

Q. You would want to look into it?

A. Very definitely.

Q. Just the way this board is looking into Dr. Oppenheimer.

A. I suppose so.
Q. Thoroughly, in other words.
A. Yes.
Q. In other words, you think that his background would raise some question which ought to be resolved.
A. Yes, sir.
Q. One further question on this subject about Dr. Oppenheimer. Suppose it should appear that Dr. Oppenheimer in some respects has not told the whole truth to this board in his testimony or in his answer, would that disturb you greatly?
A. Yes, it would.
Q. That would have a very material bearing on your judgment of him, of course, would it not?
A. I think it would. Again when you say "told the truth", it is a matter of if he has given incorrect information through mistake.
Q. No.
A. You mean if he deliberately lied about some important matter.
Q. Yes.
A. That would have a very definite bearing in my opinion.
Q. Doctor, G. E. has had many confidential war projects which have come under your supervision.
A. It has had a great many war projects, some more or less directly under my supervision and many more about which I have known in general and have had advisory contact with and that sort of thing.
Q. If you found that the man in charge of one of those projects had a number of Communist friends or friends who were either Communists or fellow travelers, would that trouble you somewhat?
A. If I had any doubts about the man himself, yes, it would. On the other hand, there are many of our scientists and some of our top engineers who are of the turn of mind as so many of the scientists—a very inquiring type of mind, very curious about everything—and I would not be at all surprised to find that some of them may have attended Communist meetings, may have had discussions with Communists just to find out what line the Communists are using and what their aproach is to world conditions and so forth. That would not necessarily be disturbing.
Q. Has anyone suggested to you that is what Dr. Oppenheimer did in this case?
A. No.
Q. Taking our hypothetical superintendent again, suppose you found that that man had brought a number of his Communist or fellow traveler friends along to work with him on your project, how would you feel about that?
A. That would bother me, but I would have to give consideration to the question of whether or not he could get people of ability to do the project, whether he was making a judgment as between getting the project done at all or getting it done with some degree of risk by bringing in such people.
Q. Do I understand that the security officers on any project that you are familiar with would have permitted on the project people that they knew to be either Communists or fellow travelers?
A. No, I don't think they would have. I don't recall any case where that kind of a situation has arisen. But one does have to some times, if a job just has to be done, make some compromises in the way that he gets the job done.
Q. Of course, you would assume our hypothetical superintendent would have told the security officers all about these fellows?
A. Yes.
Q. That would be his duty, wouldn't it?
A. Yes.
Q. And that would not be an unreasonable duty to impose on him?
A. No, it would not.
Q. Even though they happened to be his friends?
A. No, that would not be an unreasonable duty to impose upon him even though they happen to be his friends.
Q. Mr. Winne, you mentioned the meeting in 1946 in Secretary Acheson's office. Do you recall who was present at that meeting when you met Dr. Oppenheimer?
A. As I recall it, all of the people who were to be ultimately members of this board of consultants were present, which included Dr. Oppenheimer, David Lilienthal, Chester Barnard, Charles Thomas and myself.
I don't remember whether at that meeting Mr. Marks and Mr. Carroll Wilson were present. I do not remember whether they were. They were then, or we

met them soon afterwards, I don't remember which, because they acted as secretaries and so forth for the board of consultants.

Q. Which Mr. Marks is that?
A. Herbert Marks.
Q. This Mr. Marks who is here?
A. Yes.
Q. What was his connection with the Committee?
A. He was in the Department of State at that time and he and Carroll Wilson were assigned to the Committee to help us with writing up the ultimate report and getting information as we might ask for it and that sort of thing.
Q. Is that the Mr. Carroll Wilson who was later secretary of the AEC?
A. He was later General Manager of the AEC.
Q. That is when you first got to know Mr. Marks?
A. Yes, that is right. I think I had met Mr. Marks once or twice, perhaps, prior to that when he was in, I think it was called, the power section of the War Production Board, or something of that order. I believe he visited Schenectady with a group and I met him at that time.

Mr. ROBB. That is all I care to ask, Mr. Chairman.
Mr. GRAY. Mr. Winne, your convictions are pretty deep about this matter. That is apparent.
The WITNESS. Yes.
Mr. GRAY. I know you are here to be helpful to this board in the discharge of a really very difficult task. There has been some discussion about the Nichols letter and Dr. Oppenheimer's reply which quite apart from the record of this proceeding establish certain facts. There are certain things reported and adverted to in General Nichols' letter and which are said to be true in Dr. Oppenheimer's reply.

Mr. Winne, against the background of the exchange of letters, I would like to read you certain pertinent excerpts from the personnel security clearance criteria for determining eligibility which was issued by the Commission and which we are required, as I understand it, to consider in the course of these deliberations.

I would be glad if counsel for Dr. Oppenheimer would watch me closely in this because I don't want to leave out anything that might be pertinent and therefore mislead Mr. Winne.

This is a very serious question I am addressing to you. This document establishes the fact, or rather, recites the fact that the Commission in September 1950, issued its procedure for administrative review—that is the reason for which we are convened—and points out also that this procedure places considerable responsibility on the managers of operations, and it is to provide uniform standards for their use that the Commission has adopted the criteria described herein.

I might interrupt to say that I am sure it is true that managers of operations here would be in this case the General Manager of the Commission, General Nichols.

Mr. ROBB. That is correct.
Mr. GRAY. Then reading from the document:
"Under the Atomic Energy Act of 1946, it is the responsibility of the Atomic Energy Commission to determine whether the common defense or security will be endangered by granting security clearance to individuals either employed by the Commission or permitted access to restricted data."

Then omitting some language: "Cases must be carefully weighed in the light of all the information and a determination must be reached which gives due recognition to the favorable as well as to the unfavorable information concerning the individual and which balances the cost of the program of not having his services against any possible risks involved."

I believe you, in your testimony, put some emphasis on the point of great services and values that Dr. Oppenheimer has been to the program.

The WITNESS. Yes.
Mr. GRAY. Then it says, "To assist in making these determinations on the basis of all the information in a particular case, there are set forth below a number of specific types of derogatory information. The list is not exhaustive, but it contains the principal types of derogatory information which indicate a security risk." Then it says that they are divided into two categories.

Category (A) includes certain things. I am going to read paragraph No. 1 and parts of paragraph No. 3.

"Category (A) includes those cases in which there are grounds sufficient to establish a reasonable belief that the individual or his spouse has:

"1. Committed or attempted to commit, or aided or abetted another who committed or attempted to commit, any act of sabotage, espionage, treason, or sedition.

* * * * * * *

"3. Held membership in or joined any organization which has been declared by the Attorney General to be totalitarian, Fascist, Communist, subversive * * * or, prior to the declaration by the Attorney General, participated in the activities of such an organization in a capacity where he should reasonably have had knowledge as to the subversive aims or purposes of the organization;".

* * * * * * *

"6. Violated or disregarded security regulations to a degree which would endanger the common defense or security;".

There are a lot of other types of derogatory information which I am not reading. I hope it does not distort it to take those out of context. Then I would go to the last two or three paragraphs of this document:

"The categories outlined hereinabove contain the criteria which will be applied in determining whether information disclosed in investigation reports shall be regarded as substantially derogatory. Determination that there is such information in the case of an individual establishes doubt as to his eligibility for security clearance.

"The criteria outlined hereinabove are intended to serve as aids to the Manager of Operations in discharging his responsibility in the determination of an individual's eligibility for security clearance. While there must necessarily be an adherence to such criteria, the Manager of Operations is not limited thereto, nor precluded in exercising his judgment that information or facts in a case under his cognizance are derogatory although at variance with, or outside the scope of the stated categories. The Manager of Operations upon whom the responsibility rests for the granting of security clearance, and for recommendation in cases referred to the Director of Security, should bear in mind at all times, that his action must be consistent with the common defense or security."

I suppose it is true that the Executive order of the President, which I think has somewhat more restrictive criteria, must also be taken into account in these proceedings. I will not take the time now to take you through all of those.

I have indicated this is a serious inquiry and I am asking for your help to this board.

The WITNESS. Yes.

Mr. GRAY. It seems to me pretty clear that some of these criteria have been met, if you will, by the exchange of letters that I read. Would you agree with that?

The WITNESS. It seems to me that the exchange of letters indicates that in the earlier years under consideration—I think it is 1942 and earlier—that Dr. Oppenheimer—I forget the exact wording there—did support to some extent some of the organizations which have since been declared subversive or perhaps were at that time. I do not know.

Mr. GRAY. This is quite a serious question. One of our difficulties is that it does not say "is a member."

The WITNESS. I recognize that.

Mr. GRAY. It says "The individual or his spouse," and then "done these things."

The WITNESS. Of course, Dr. Oppenheimer does admit that his wife had been a member of the Communist Party.

Mr. MARKS. That is correct.

The WITNESS. That is in the letter. So taking the strictly legal interpretation perhaps you have no alternative there.

Mr. MARKS. Mr. Chairman, I would like to interrupt.

Mr. GRAY. Surely.

Mr. MARKS. Because I feel that there is a really very important technical question of interpretation that is involved in the question.

Mr. GRAY. I would be glad if you would state it.

Mr. MARKS. I do not think that the criteria which you read mean or are intended to mean that the establishment——

Mr. ROBB. Mr. Chairman, might I interrupt? Would it not be well to have the witness step out while this is going on?

Mr. MARKS. We would be glad to have that done.

Mr. ROBB. I don't know whether the witness is going to be confused or not.

Mr. GRAY. I really think actually the argument should not be given in answer to a question by the witness. I will rephrase my question and see if I can take care of your difficulty.

We have had witnesses before the board, Mr. Winne—men of great stature and eminence—who have been inclined to treat very lightly these matters which we have been discussing here, I think with sincerity and conviction, on the ground of what they think they know of Dr. Oppenheimer all this washes out anything that happened in the past.

I will now ask this question: Has anything here said since your direct testimony made you wish to alter your direct testimony as a result of hearing what I read?

Is that a fair question?

The WITNESS. I feel it does not change my opinion, Dr. Gray. As I was about to say, and this is an entirely gratuitous remark and perhaps I should not make it, but it seems to me that it may be possible that you have no alternative but to make a certain finding here. But even if you make a finding adverse to Dr. Oppenheimer, my personal feeling still is that he is loyal to the country, that he would be an asset to the whole atomic and hydrogen weapons project for the country.

You may, because of the wording of the law, be forced to make a decision adverse. I hope you will not, but you may be forced to.

Mr. GRAY. Just for the sake of the record now, and perhaps to ease Mr. Marks'——

Mr. MARKS. No; this is perfectly all right.

Mr. GRAY. I am making no assumption of any kind.

The WITNESS. No; I recognize that.

Mr. GRAY. The board has reached no conclusions and I certainly would say that we cannot say that any alternatives or set of courses of action are necessarily inevitable in this thing. I don't want to have any misunderstanding on that point.

There is substantial and widespread ignorance about the procedures and the requirements of the law in these cases, I believe. I don't mind saying that I am deeply troubled by these things that are before us. However, I don't want to pursue it with you further because I think you have made it absolutely plain that you would go as far as the law would allow you to go to grant Dr. Oppenheimer security clearance. That is the sum of it, isn't it?

The WITNESS. Yes, that really is the sum of it, Dr. Gray. To express my own belief, I think it is not necessary to assume that because a man several years ago—I am not referring to Dr. Oppenheimer now, but anyone—was supporting the Communist Party, particularly if he was a youngster in college at the time, that should disqualify him for security clearance today. I hope most of us have changed our ideas about many subjects as we have gone along through life. I think in many cases it would be found that if the true facts could be gotten at, especially the youngsters in college who have supported the Communist Party to some extent or joined it or something like that, really did not realize that they were acting inimical to the interests of the country. I think all of those things should be taken into consideration.

I know it is an almost insuperable job for a board such as yours with the law as it exists. You, of course, have to abide by the law.

Mr. GRAY. I had one other question which is entirely unrelated to what we have just been discussing and I guess it is more for my information than anything else. It is an uninformed question.

Are there developments which are useful for the welfare of mankind as opposed to wars of destruction which may come out of the hydrogen bomb discoveries and inventions and development, in your judgment?

The WITNESS. I do not know that, Dr. Gray. Based on the long history of science I would bet that there will be rather than that there will not. But I do not know of any in the immediate future.

Mr. GRAY. Dr. Evans, have you any questions?

Dr. EVANS. Yes. Mr. Winne, you feel that in these atomic developments with the fission and the fusion bomb we are just scratching the surface of what we will know years from now.

The WITNESS. Certainly we will know a tremendous amount more than we know now if we keep on with our developments. Whether you mean by that that we will develop much more powerful bombs and weapons and so forth, I do not know that. But we will certainly know much more about them and be able to produce them at lower cost and much less effort and so forth.

Dr. EVANS. I merely mean this: Do you remember Faraday's experiment with the coil of wire before the Royal Society?

The WITNESS. I remember it rather vaguely.

Dr. EVANS. Let me refresh your memory. He put a coil of wire between two magnets and the coil of wire was carrying an electric current and the wire turned like this [indicating]. Gladstone said to him, "But of what possible use can it be?" Faraday said, "Mr. Gladstone, you may be able to tax it." Rather interesting, isn't it?

The WITNESS. Yes.

Dr. EVANS. Someday we very likely will be able to tax this. You also feel that we should be smart enough to have international agreement on these things rather than to allow them to destroy us, don't you?

The WITNESS. I feel we must exert every effort to prevent weapons of any kind from destroying us—every reasonable effort—without sacrificing anything as material as a Nation. Whether that should be by international control or whether simply the fear which I think is gradually being generated in all people, the fear of the use of these weapons, is going to prevent their actual use.

Dr. EVANS. You don't feel that threatening the use of these weapons is going to do the thing. It has to be done by some other way.

The WITNESS. No. I think it is possible that the mere threat of the results from the use of these weapons may prevent their actual use, Dr. Evans. I would feel still safer if we had some really workable system of providing for international disarmament, but it has to be a workable system and one which will really protect all the countries if it is really to work.

Dr. EVANS. You realize when we begin to deal with this sort of thing with these enormous temperatures and pressures, we are beginning to deal with the kind of things that make and destroy worlds, isn't that true?

The WITNESS. I am not enough of a scientist to say whether or not that is true, Dr. Evans, but it seems as though we may be approaching that point.

Dr. EVANS. Do you have any ethical or moral scruples when you think about these terrible things today?

The WITNESS. I would hope that we will not have to use the atomic and hydrogen bombs in war for the destruction of other peoples. On the other hand, unless and until some reasonable system of control for actual prevention of their use is in effect, I think our country has no course but to go ahead with their development and try to develop the very best weapons than can be made.

Dr. EVANS. I quite agree with you. You will admit, Mr. Winne, and I think you did, that Dr. Oppenheimer was indiscreet on occasion.

The WITNESS. Yes, on the basis of the information particularly with reference to his not disclosing this instance when he was approached.

Dr. EVANS. I want to ask, you are not a Communist?

The WITNESS. No, sir.

Dr. EVANS. Have you ever been a fellow traveler?

The WITNESS. No.

Dr. EVANS. Have you any Communist friends?

The WITNESS. No. Well, I don't know, but not that I know of.

Dr. EVANS. Would you, if you were on a security committee, go to see a Communist friend?

The WITNESS. If I were on a security committee?

Dr. EVANS. Yes.

The WITNESS. Would I go to see a Communist friend?

Dr. EVANS. Yes.

The WITNESS. That is a question that is very difficult to answer.

Dr. EVANS. You don't have to answer.

The WITNESS. Without having all the circumstances, that is.

Dr. EVANS. You don't have to answer it. Perhaps it is a bad question. Strike it.

The WITNESS. If I had a friend who had committed a serious crime and was in prison, I might go to see him if he was a close friend, to try to find out from him just why he did it and what the circumstances were and to be of some moral support to him in trying to rehabilitate himself and that sort of thing. One can't answer a general question like that yes or no.

Dr. EVANS. You would not have done this thing in regard to this Chevalier incident in just the way Dr. Oppenheimer did.

The WITNESS. I think not, Dr. Evans, to the best of my knowledge and belief. I think I would not have done it.

Dr. EVANS. That is all.

Mr. GRAY. Mr. Marks.

REDIRECT EXAMINATION

By Mr. MARKS:

Q. I think you have probably answered this, Mr. Winne, but just to be sure that we are clear as to your own thoughts I would like to go over some ground.

Mr. Robb was inquiring of you as to what you would do as one of the responsible chief officials of the General Electric Co. in various contingencies relating to conduct of an employee. I need to ask you whether, if it came to your attention that an important employee and a trusted employee had many years ago in different times and circumstances committed acts of the kind that Mr. Robb described in relation to the Chevalier incident or some other incident that you can imagine involving the truth and refusal to cooperate in an investigation, that it happened many years ago and there had been a long intervening period of faithful service to the General Electric Co., what consideration would you give or how would you seek to weigh considerations that you would have to judge in determining his future with the General Electric Co.? That is his future, if any.

A. That again is a question the answer to which would vary under different circumstances. First I would endeavor, as it seems to me this Board is doing, to find out all I could about the circumstances in the early years, to see just what caused the employee to do whatever he had done. Then I would investigate very carefully all of his actions with the company since that time, talk with him, and if this were an important employee, talk with the higher officers of the company and then come to a decision as to what we should do about it. I don't think one can say right offhand whether we would fire him or keep him. It would depend on a lot of circumstances.

Q. Just one other question, and I just have no idea whether or not you know the answer to it.

I ask you whether at the time which you have referred to that you had some contact with Dr. Morrison, while you were a member of the Board of Consultants of the State Department, did you know of the capacity in which he was then connected with the Manhattan District?

A. I think I probably did, Mr. Marks, but I can't recall definitely that I did, nor do I recall now just what capacity he was employed in the Manhattan District, if he was employed.

Mr. MARKS. Mr. Robb, will you permit me to ask a question that I am afraid is leading but is intended to refresh——

Mr. ROBB. I am afraid of most anything you ask, Mr. Marks, but go ahead.

Mr. GRAY. Mr. Robb is glad for you to ask a leading question.

Mr. ROBB. I don't think this witness will be led, Mr. Marks. I think the witness will answer the question in his own way.

Mr. MARKS. I am sure of that.

By Mr. MARKS:

Q. Do you know, Mr. Winne, whether or not at the time Dr. Morrison had his contacts with your board of consultants he was then serving under a designation or appointment from General Groves as a member of General Groves' Committee on International Control of Atomic Energy?

A. I cannot recall, Mr. Marks, whether that was the case or not. I cannot recall.

Mr. MARKS. That is all.

Mr. ROBB. I have nothing further, Mr. Chairman.

Mr. GRAY. Thank you very much, Mr. Winne. I am sorry we kept you so long.

The WITNESS. Thank you. That is perfectly all right.

(Witness excused.)

Mr. GARRISON. Mr. Chairman, I would like to make a couple of statements for the record. I have also one or two affidavits to read into the record.

I would suggest that since it is now quarter past twelve, or approximately that, and Dr. Bush has agreed to testify at 2 o'clock this afternoon, I don't think there is any use in starting with Dr. Bacher who is ready to testify because he can wait over until Monday. We will have to go into next week anyway.

My thought would be, Mr. Chairman, to adjourn very shortly so that we might have Dr. Bush promptly at 2 and then, I think, the arrangement we made yesterday, which would enable the board to adjourn at a reasonable hour.

Mr. GRAY. If you have some affidavits, can we read those into the record now and that will save a little time?

Mr. GARRISON. Yes. If I might just make a statement for the record on one or two things that have come up and I think are worth saying.

First just one word about this Drew Pearson column which I have seen now. It is entitled "Veil Over Oppenheimer Case", and the first paragraph describes how nobody can find out where Dr. Oppenheimer is living. I may say that was arranged deliberately, Mr. Chairman, by all of us for the very purpose of avoiding statements to the press.

"Lloyd Garrison, attorney for the atomic scientist is just as mysterious as his client."

If there was left any implication that I am the only one of the attorneys associated in this case who is as mysterious as the client, I want most emphatically to reject that implication. As a matter of fact, Mr. Marks, who has cut himself off completely from his office—he has not even received a telephone call since last week—and I have been living together except to separate to go to bed at night.

Mr. Ecker has been with us almost continuously except when he has been down here working on transcripts. Mr. Topkis is going back to New York after a couple of days of help. Dr. Oppenheimer has been almost continuously with us. I just say to you, sir, that there is not a one of us who has had contacts with the press in this time and since the early calls bombarded us, in which we said that we cannot give information and returned the calls as a matter of courtesy.

How this came to be is a mystery to all of us. I want to say this most emphatically for cocounsel and my associates in this matter as well as for myself.

Mr. GRAY. Thank you. May I address a question to you. Do you want to leave the record in such a state that all counsel for Dr. Oppenheimer are mysterious? That is a facetious observation.

Mr. MARKS. I would like to say that when calls from the press come to me and I am available, I take them. I try to be civil and courteous and I refuse to make any comments of any kind about this proceeding, even as to whether the proceedings are in progress.

Mr. MORGAN. The only question I had was whether you believe what Mr. Pearson writes or not? It may not be pertinent to this hearing.

Mr. GRAY. I doubt if it is worthwhile pursuing that. Would you proceed, Mr. Garrison.

Mr. GARRISON. Of course, we don't believe this stuff. I don't believe any of it.

Mr. Chairman, just a word about these criteria which I am so glad that you raised. It has been on my own mind to say something about it, but I didn't want to interrupt the flow of the testimony.

I would like to read into the record and just for a moment bring to the attention of the board rather forcibly the two paragraphs that follow the rescription of the general nature of the Atomic Energy Act. These are taken from the Atomic Energy Commission criteria for determining eligibility from which the chairman read particular excerpts from category (A).

"Under the act, the Federal Bureau of Investigation has the responsibility for making an investigation and report to the Commission on the character, associations, and loyalty of individuals who are to be permitted to have access to restricted data. In determining any individual's eligibility for security clearance other information available to the Commission should also be considered, such as whether the individual will have direct access to restricted data or work in proximity to exclusion areas, his past association with the atomic energy program, and the nature of the job he is expected to perform (certainly something we have here before us). The facts of each case must be carefully weighed and determination made in the light of all the information presented whether favorable or unfavorable. The judgment of responsible persons as to the integrity of the individuals should be considered. The decision as to security clearance is an overall, commonsense judgment, made after consideration of all the relevant information as to whether or not there is risk that the granting of security clearance would endanger the common defense or security. If it is determined that the common defense or security will not be endangered, security clearance will be granted; otherwise, security clearance will be denied.

"Cases must be carefully weighed in the light of all the information, and a determination must be reached which gives due recognition to the favorable as well as unfavorable information concerning the individual and which balances the cost to the program of not having his services against any possible risks involved. In making such practical determination, the mature viewpoint and responsible judgment of Commission staff members, and of the contractor concerned are available for consideration by the general manager."

I think that last sentence, of course, is particularly pertinent to the general manager's consideration, but I am sure that this board is expected to provide the general manager with all of this kind of information that is here set forth.

This would include, for example, responsible judgment of a man like Dr. Bradbury who is a Commission staff member.

I would like to stress in summary that it seems to me that quite pertinent to this proceeding is Dr. Oppenheimer's past association with the atomic energy program, the nature of his job as a consultant, the judgment of responsible persons who have appeared here and will appear here as to his integrity and the responsible mature viewpoint and responsible judgment of Commission staff members who have testified—only one of them actually—and that the case must be carefully weighed in the light of all of the information.

There is one other thing I would like to point out. That is, if category (A) is considered, as, of course, it must be, it is said to include those classes of derogatory information which establish a presumption of security risk.

I take it that it is quite clear from this that if the board should find a derogatory item which it felt had been established under category (A), which I hope the board will not and believe it should not on the evidence—but if it should— that would establish a presumption which, I take it under this overall judgment that is referred to here, would be rebuttable by other evidence such as what Dr. Oppenheimer has actually done for his country and the opinion of responsible people who know him and the like. In other words, it is not a final and conclusive matter but a rebuttable presumption.

Mr. GRAY. I assume, Mr. Garrison, that at the conclusion of the testimony you possibly may wish to address yourself to some of these matters. I would not at this time respond to any request for an interpretation of the criteria either in this document or in the President's order.

I frankly have received this statement of yours at this time in the record because I initiated all this by bringing it up with Mr. Winne. I think I would like to say why I did that.

I believe it is true and I say this now not in the presence of any witness that we have had some witnesses who have come before the board and in effect have said, "I know this man to be loyal; clear him." That is the sum of some of the testimony we have had.

There has been an inclination to be impatient with procedures and regulations and things of that sort. I just wanted to make clear that everybody understands that the board must take into account all rules, regulations, and procedures in the course of its proceedings and I would not wish you to draw any conclusion now from anything I might have said in talking to Mr. Winne.

Mr. GARRISON. Mr. Chairman, speaking for Dr. Oppenheimer, we agree that any light waving aside of what are serious matters or what may be requirements of the regulations we are not in sympathy with. We take this just as seriously as does the board. That goes for all of us.

I think apart from that, the mere testimony from a witness that having known Dr. Oppenheimer closely for many years he has a conviction about his loyalty, I would say that in itself is pertinent.

Mr. GRAY. I quite agree it is pertinent. Speaking at least for one member of the board, these deep convictions held by responsible people are important in these deliberations. They are important to me and I am sure to the other members of the board.

Mr. ROBB. Mr. Chairman, might I interpose since we are talking about these criteria. We might at this point refer to section 4.16 of the procedures, which also refers to them: "Recommendations of the board. The board shall carefully consider all material before it, including reports of the Federal Bureau of Investigation, the testimony of all witnesses, the evidence presented by the individual, and the standards set forth in AEC Personnel Security Clearance Criteria for Determining Eligibility. In considering the material before the board, the members of the board, as practical men of affairs, should be guided by the same consideration that would guide them in making a sound decision in the administration of their own lives. In reaching its determination, the Board shall consider the manner in which the witnesses have testified before the board, their demeanor on the witness stand, the probability or likelihood of their testimony, their credibility, the authenticity of documentary evidence, or the lack of evidence upon some material points at issue."

Mr. GARRISON. That is all I have to say.

Mr. GRAY. Do you have some affidavits at this time, Mr. Garrison?

Mr. GARRISON. Yes. I wonder, Mr. Chairman, if we might adjourn for lunch. It is almost 12: 30. I will proceed, however, if you wish.
Mr. GRAY. How long are they?
Mr. GARRISON. I would say it would probably take 10 or 15 minutes.
Mr. GRAY. I think we should recess for lunch, then, and be here at 2 o'clock.
(Whereupon, at 12: 25 p. m. a recess was taken, to reconvene at 2 o'clock this day.)

AFTERNOON SESSION

Mr. GRAY. Do you wish to testify under oath?
Dr. BUSH. Whatever is customary.
Mr. GRAY. All the witnesses have.
Would you stand and raise your right hand, please. What is your full name?
Dr. BUSH. Vannevar Bush.
Mr. GRAY. Vannevar Bush, do you swear that the testimony you are to give the board shall be the truth, the whole truth and nothing but the truth, so help you God?
Dr. BUSH. I do.
Whereupon, Vannevar Bush was called as a witness, and having been first duly sworn, was examined and testified as follows:
Mr. GRAY. It is my duty to remind you of the existence of the so-called perjury statutes. I assume we don't need to discuss those in any detail.
The WITNESS. No, I think I know about them.
Mr. GRAY. I should like to request that if in the course of your testimony it becomes necessary for you to refer to or disclose restricted data, let me know in advance so we may take certain necessary and appropriate steps.
The WITNESS. Yes.
Mr. GRAY. Finally, I would like to say to you that we consider these proceedings a confidential matter between the Atomic Energy Commission and its officials on the one hand, and Dr. Oppenheimer and his counsel and witnesses on the other.
The WITNESS. I have already said to the press several times that I would not discuss this subject while it was before this Board.
Mr. GRAY. We just express the hope that it will be your position.
The WITNESS. Yes, sir.

DIRECT EXAMINATION

My Mr. GARRISON:
Q. Dr. Bush, would you state for the record your present position, and after that, the principal Government offices which you have held and now hold?
A. I am president of the Carnegie Institution in Washington. At the present time I hold no Government post except membership on one or two committees. I don't think you need to have them.
I was chairman of the National Advisory Committee for Aeronautics for several years, about 1939.
I was chairman of the National Defense Research Committee when it was formed in June of 1940.
I was a Director of the Office of Scientific Research and Development when it was formed in June of 1941, through the war, and until after it was closed out after the war.
During the war I was chairman of the New Weapons Committee of the Joint Chiefs of Staff.
After the war I was chairman of the Joint Research and Development Board of the Army and Navy, and then when that board was made permanent by statute, I was chairman of the Research and Development Board until 1949.
I think those are the principal appointments, sir.
Q. About how long have you known Dr. Oppenheimer?
A. I have known him well since the early days of the war. I undoubtedly met him in gatherings of physicists before that time, but have no specific recollection of the first date that I met him.
Q. What was your connection with his appointment to the Manhattan District?
A. There were appointments before then. At that time General Groves, who was in charge of the Manhattan District, reported to a body of which I was chairman, and which I omitted to list. It is rather hard to get all of these in. It was the Military Policy Committee, of which I was chairman. Dr. Conant was my deputy. General Groves took up all of his programs and policies with that group.

At the time General Groves made the appointment of Dr. Oppenheimer at Los Alamos, he took that matter up with us. In my memory he took it up informally, not in a formal meeting, and discussed it with Dr. Conant and with me.

Q. What recommendation did you make?

A. General Groves said he had in mind appointing Dr. Oppenheimer. He review for us orally what he knew of Dr. Oppenheimer's prewar record. I don't remember that we looked at any file or any written records. He recited some of the previous history. Then he asked the opinion of me and Dr. Conant in regard to the appointment, and I told him I thought it was a good appointment.

Q. Did you have any discussion about any prior left-wing associations that he had?

A. Yes, we did. He recited previous associations.

Q. When you say "he," you mean whom?

A. General Groves.

Q. About when was this?

A. I noted down a few dates. I can't say, gentlemen, that my memory for dates and the like is good. In fact, it is a little bad. I have that date here somewhere. Oppenheimer was chosen in November of 1942.

Q. Did you have opportunity to observe his work at Los Alamos?

A. In a sense which I was responsible for it. The structure at that time, you remember, was this: OSID started this work and continued it for a considerable period. It continued parts of it in fact after that date. I originally carried the full responsibility for it, reporting to the President. On my recommendation when the matter came to the construction of large facilities, the matter was transferred to the War Department. Secretary Stimson and I conferred, and the Manhattan District was set up. Groves was made head of it.

After that the Military Policy Committee reviewed his recommendations on which I was chairman, and there was also a policy committee appointed by the President which consisted of the Vice President, Secretary Stimson, General Marshall, Dr. Conant and myself, I believe. That was appointed by Mr. Roosevelt at my request. When I was carrying the full responsibility, I told him I would prefer to have some group of that sort, and that committee was appointed. It never was formally dissolved.

Q. Would you say a word as to your view of his achievement at Los Alamos?

A. He did a magnificent piece of work. More than any other scientist that I know of he was responsible for our having an atomic bomb on time.

Q. When was your next governmental connection with him, do you recall?

A. There have been so many I am not sure which one.

Q. Let me go back a minute and ask you another question about the Los Alamos work.

What significance would you attach to the delivery of the A-bomb on time, or was it delivered on time?

A. That bomb was delivered on time, and that means it saved hundreds of thousands of casualties on the beaches of Japan. It was also delivered on time so that there was no necessity for any concessions to Russia at the end of the war. It was on time in the sense that after the war we had the principal deterrent that prevented Russia from sweeping over Europe after we demobilized. It is one of the most magnificent performances of history in any development to have that thing on time.

Q. You were connected with the effort of this country to control international atomic energy before the United Nations?

A. Yes. After the war, very soon after the war, you remember that there was a so-called Atlee Conference, when Mr. Atlee came over and the Prime Minister of Canada came down. At that conference was prepared a declaration. I managed that affair for Secretary Byrnes and John Anderson, and I wrote that declaration. That is where it was decided to take this matter to the United Nations.

The next step was the Secretary of State's committee of which I was a member. That committee appointed a panel of which Dr. Oppenheimer was a member. That panel prepared what later became known as the Baruch Plan. After it was prepared, it was approved by the Secretary of State's committee, and it was presented to the United Nations by Mr. Baruch at the President's request.

Q. Did you see something of Dr. Oppenheimer during that period?

A. Certainly. We had a number of discussions between the main committee that was drafting the agreement.

Q. Did you form any opinion as to his contribution at that time?

A. His contribution was substantial in the thinking that went into that very difficult matter.

Q. When you became chairman, I think, of the Joint Research and Development Board in 1947, did you set up an Atomic Energy Committee?

A. That is right. I appointed Dr. Oppenheimer as chairman of it, as I remember.

Q. What would you say as to his services in that connection?

A. I think I can save time by saying that I have worked with him on this general subject in many capacities. Two have been mentioned. He was also on the panel which reviewed the evidence before Mr. Truman made the announcement of the Russian atomic explosion. He and I were both members of a panel set up by the Secretary of State which worked a year ago last summer, I believe, on general disarmament matters. I think there were probably one or two other occasions. I worked with him on many occasions on this general subject.

Q. In connection with the Secretary of State's panel, did you have occasion to visit the Secretary of State in the summer of 1952?

A. I will not try to be exact on dates on that. But when the panel had gotten to a point where it was about to draft a report, we met with the full panel and the Secretary of State, and went over some of our conclusions orally, as I remember.

Q. Before that time did you have occasion to talk with the Secretary of State about the question of postponing the test of the H-bomb?

A. I did. That had nothing to do with that panel, however. That was a personal move that was made, as a matter of fact, before the panel was in operation. The clearances on the panel were delayed. In that interim I visited the Secretary of State and gave my personal opinion in regard to that test. Before so doing I talked with a number of my friends.

Q. Who did you talk to among others?

A. Mr. Elihu Root. I also talked with three or four members that were waiting to go to work on the panel. John Dickey, Joseph Johnson, Allan Dulles, Robert Oppenheimer. I undoubtedly discussed it with one or two others. In every case it was discussing the matter in generalities, without going into confidential matters. It was not necessary in order to do that.

I then visited the Secretary of State and gave him my personal opinion on that matter.

Q. Without revealing any matters that you consider confidential, could you state what your position at the time was with respect to that test?

A. Wait a minute. I gave the Secretary of State a memorandum which gave him my personal views. I made no copy of that memorandum. Nobody knows the exact content of that memorandum as far as I know except the Secretary of State and anyone he may have told about it. It has never been made public. It seems to me that it would be quite improper for me to give you the content. I will lean on the judgment of the chairman. My inclination is that I should not reveal this before this board.

Mr. GRAY. Dr. Bush, I think you should not discuss the contents of the memorandum, but I see no reason why if you expressed your views to a number of people at that time, why you can't——

The WITNESS. Quite right. I can readily say what moved me to go at all, and what the general tenor of my thinking was, much as I discussed it then.

There were two primary reasons why I took action at that time, and went directly to the Secretary of State. There was scheduled a test which was evidently going to occur early in November. I felt that it was utterly improper—and I still think so—for that test to be put off just before election, to confront an incoming President with an accomplished test for which he would carry the full responsibility thereafter. For that test marked our entry into a very disagreeable type of world.

In the second place, I felt strongly that that test ended the possibility of the only type of agreement that I thought was possible with Russia at that time, namely, an agreement to make no more tests. For that kind of an agreement would have been self-policing in the sense that if it was violated, the violation would be immediately known. I still think that we made a grave error in conducting that test at that time, and not attempting to make that type of simple agreement with Russia. I think history will show that was a turning point that when we entered into the grim world that we are entering right now, that those who pushed that thing through to a conclusion without making that attempt have a great deal to answer for.

That is what moved me, sir. I was very much moved at the time.

By Mr. GARRISON:
Q. Turning now to the matter of the controversy in the fall of 1949 over whether or not to proceed with an all-out program for the development of the H-bomb, did you have any official participation in the actions that were taken at that time?
A. No. I did not. I had no official connection with the matter. I would like to make one thing clear. There have been statements in the paper that at that time I expressed opinions on that matter. I did not do so. In fact, I very carefully refrained from doing so. There was some talk in the press of a review body on that matter. I was named as a possible chairman. I said to one or two men on Capitol Hill that I felt that would be a mistake, to establish such an affair. In the first place, the General Advisory Committee had been set up by law for the explicit purpose of reviewing such matters, and second, a review panel would constitute new men, and it would take months of work before it could understand the technical matters involved and pass reasonable judgment. Hence I declined to give any personal estimate of the matter at the time.
Q. Would you care to express a judgment about it now?
A. I have never reviewed in detail all of the considerations. No, I am not going to express an opinion on that today. Let me say with all due respect that I don't think this board could arrive at the question of whether reasonable judgment was shown at that time. There are some exceedingly difficult things that come into such a question. I can certainly recite things that would need to be considered.

For one thing I think it is fully evident that the hydrogen bomb was of great value to Russia—much greater value to Russia than to us. I think I can also be sure that a test by us of a hydrogen bomb would be of advantage to Russia in the prosecution of their program. There are two considerations that might weight very heavily indeed in such a consideration. The other one, of course, is feasibility.
Q. Turning to another topic, at the time of the establishment of the Atomic Energy Commission and the General Advisory Committee, or several months after the establishment of them both, did the Chairman of the Atomic Energy Commission consult you about Dr. Oppenheimer's clearance?
A. Yes, I remember that he did. Mr. Lilienthal consulted me, and I wrote him a letter about it.
Q. Do you have a copy of that with you?
A. What I have is this. I have no record in my files of these matters. All of my records in the Office of Scientific Research and Development were of course turned over to the Defense Department. All of my records in the Research and Development Board remain there. I have not gone back to those files.
From stenographic notebooks I have a transcript of the body of that letter.
Q. Isn't that the one we have already read in the record?
A. Quite likely. I could not find a copy, sir. Would you want to look at it to see if it is?
Mr. GARRISON. Would there be any objection to reading it again?
Mr. GRAY. No, there would be no objection.
Mr. ROBB. No, of course not.
The WITNESSS I could not find a copy anywhere, but my stenographer had his old notebooks and that is where I got it from. Isn't is quicker for me to read it?
Mr. GRAY. Why don't you read it?
The WITNESS. "At our conference yesterday you asked me to comment concerning Dr. J. Robert Oppenheimer, and I am very glad to do so. Dr. Oppenheimer is one of the great physicists of this country or of the world for that matter. Prior to the war he was on the staff of the University of California, and was regarded as the leader of theoretical aspects of atomistics and similar subjects of physics. Shortly after the Army entered into the development of atomic energy he was given a very important appointment by General Groves. This appointment made him director of the laboratory at Los Alamos, which was in all probability the most important post held by any civilian scientist in connection with the entire effort. General Groves undoubtedly made this appointment after a very careful study of the entire affair from all angles, as this was his custom on important appointments. Subsequent developments made it very clear that no error had been made in this connection, for Dr. Oppenheimer proved himself to be not only a great physicist, but also a man of excellent judgment and a

real leader in the entire effort. In fact, it was due to the extraordinary accomplishments of Oppenheimer and his asociates that the job was completed on time. Subsequent to the end of the war Dr. Oppenheimer has had a number of important appointments. He was invited by Secretary Stimson as one of the scientists consulted by the Secretaries of War and Navy in connection with the work of the Interim Committee. He was appointed by the State Department as a member of the board which drew up the plan on which Mr. Baruch based his program. He has recently been appointed by the President as a member of the General Advisory Committee of your organization. I have appointed him a member of the Committee on Atomic Energy of the Joint Research and Development Board. All of this has followed from his extraordinary war record in which he made a unique and exceedingly important contribution to the success of the war effort of this country.

"I know him very well indeed and I have personally great confidence in his judgment and integrity."

Mr. ROBB. I have the original now.

By Mr. GARRISON:

Q. At the time you wrote that letter, had you been through Dr. Oppenheimer's personnel file, the FBI reports?

A. I don't think I ever went through Dr. Oppenheimer's FBI file. If I did, I certainly do not remember.

Q. Did you understand at the time that you wrote that letter that he had left wing associations?

A. I understood that at the time of his first appointment was made at Los Alamos. I had an exposition of the entire affair from General Groves.

Q. You read the letter of General Nichols dated December 23, 1953, to Dr. Oppenheimer, containing the items of derogatory information?

A. Yes, I read that as it appeared in the press.

Q. Is there anything in that letter which would cause you to want to qualify the letter which you wrote to Mr. Lilienthal that you have just read?

A. Now, let me answer that in two parts. I had at the time of the Los Alamos appointment complete confidence in the loyalty, judgment, and integrity of Dr. Oppenheimer. I have certainly no reason to change that opinion in the meantime. I have had plenty of reason to confirm it, for I worked with him on many occasions on very difficult matters. I know that his motivation was exactly the same as mine, namely, first, to make this country strong, to resist attack, and second, if possible, to fend off from the world the kind of mess we are now getting into.

On the second part of that, would I on the basis of that document, if those allegations were proved, change my judgment. That is what I understand this board is to decide. I don't think I ought to try to prejudge what they might find out.

Q. I would not want to ask you to do that, and my question is not designed to do that.

A. My faith has not in the slightest degree been shaken by that letter or anything else.

Mr. GARRISON. I think that is all, Doctor.

Mr. GRAY. Mr. Robb?

Mr. ROBB. I have no questions, Mr. Chairman.

Mr. GRAY. I have one question which relates to the development of the hydrogen bomb in general, and it is prompted by something you said in answer to a question put to you by Mr. Garrison, I think.

I believe you said that you felt that that test in the fall of 1952 was of value to the Russians in their own program. Did I understand that correctly?

The WITNESS. I am sure it was.

Mr. GRAY. And this is for technical reasons?

The WITNESS. I am sure of it for one reason because when we reviewed the evidence of the first Russian atomic explosion, we didn't find out merely that they had made a bomb. We obtained a considerable amount of evidence as to the type of bomb, and the way in which it was made. If they had no other evidence than that from their own test and the like, they would have derived information. * * *

Mr. GRAY. Would it have been your guess that the Soviets would have attempted to develop this kind of weapon?

The WITNESS. Why, certainly, because it is very valuable indeed to them. To us, with 500 KT fission bombs we have very little need for a 10 megaton hydrogen bomb. The Russians, on the other hand, have the great targets of New York and Chicago, and what have you. It is of enormous advantage to them.

Mr. GRAY. So they probably would have sought to develop this in any event unless some international control machinery had been in effect.

The WITNESS. That is right.

Mr. GRAY. And our not proceeding, as some people thought we should not, probably didn't have any relation to what the Russians might do about it.

The WITNESS. I think it has relation to what the Russians might do about it because whether we proceeded or not determined to some extent the speed with which they could proceed. Let me interpose a word there, Mr. Chairman.

Mr. GRAY. Yes, sir.

The WITNESS. It was not a question, as I understand it, of whether we should proceed or not. It was a question of whether we should proceed in a certain manner and on a certain program. I have never expressed opinions on that. But certainly there was a great deal of opinion which seemed to me sound that the program as then presented was a somewhat fantastic one. So it was not a question of do we proceed or do we not. I think there was no disagreement of opinion as to whether we ought to be energetic in our research, whether we should be assiduously looking for ways in which such a thing could be done without unduly interfering with our regular program. The question of whether we proceeded along a certain path—may I say one more word on that, Mr. Chairman, quite frankly, and I hope you won't misunderstand me, because I have the greatest respect for this board. Yet I think it is only right that I should give you my opinion.

I feel that this board has made a mistake and that it is a serious one. I feel that the letter of General Nichols which I read, this bill of particulars, is quite capable of being interpreted as placing a man on trial because he held opinions, which is quite contrary to the American system, which is a terrible thing. And as I move about I find that discussed today very energetically, that here is a man who is being pilloried because he had strong opinions, and had the temerity to express them. If this country ever gets to the point where we come that near to the Russian system, we are certainly not in any condition to attempt to lead the free world toward the benefits of democracy.

Now, if I had been on this board, I most certainly would have refused to entertain a set of charges that could possibly be thus interpreted. As things now stand, I am just simply glad I am not in the position of the board.

Mr. GRAY. What is the mistake the board has made?

The WITNESS. I think you should have immediately said before we will enter into this matter, we want a bill of particulars which makes it very clear that this man is not being tried because he expressed opinions.

Mr. GRAY. Are you aware, Dr. Bush, how this got in the press and was spread throughout the world?

The WITNESS. Yes, I know how it was released.

Mr. GRAY. Do you know who released it?

The WITNESS. I believe this gentleman on my right released it.

Mr. GRAY. I don't think you can blame the board. We had quite a discussion about that.

The WITNESS. It was bound to be released sometime when you made your report.

Mr. GRAY. It might have leaked. I don't think it was bound to be released. I assure you, and I am sure that we are all sure that whatever the outcome, this board is going to be very severely criticized.

The WITNESS. I am sure of that, and I regret it sincerely, sir, because I fear that this thing, when your report is released, will be misinterpreted on that very basis whatever you may do.

Dr. EVANS. Dr. Bush, you don't think we sought this job, do you?

The WITNESS. I am sure you didn't, and you have my profound sympathy and respect. I think the fact that a group of men of this sort are willing to do as tough and as difficult a job as this augurs well for the country. It is in stark contrast with some of the things that we have seen going on about us in similar circumstances. Orderly procedure and all of that is good. I merely regret that the thing can be misinterpreted as it stands on the record, and misinterpreted in a way that can do great damage. I know, of course, that the executive branch of the United States Government had no intention whatever of pillorying a man for his opinions. But the situation has not been helped, gentlemen, recently by statements of the Secretary of Defense. I can assure you that the scientific community is deeply stirred today.

The National Academy of Science meets this next week, and the American Physical Society meets, and I hope sincerely that they will do nothing foolish. But they are deeply stirred. The reason they are stirred is because they feel that a professional man who rendered great service to his country, rendered service beyond almost any other man, is now being pilloried and put through an ordeal because he had the temerity to express his honest opinions.

Mr. GRAY. Dr. Bush, are you familiar with the Atomic Energy Act of 1946 at all?

The WITNESS. I have read it.

Mr. GRAY. Are you familiar with the fact that the Commission has a published set of procedures which for these purposes have the effect of law?

The WITNESS. Yes. I am not quarreling with the procedure, Mr. Chairman.

Mr. GRAY. As I understand it, and I can be corrected by counsel, the writing of a letter to Dr. Oppenheimer with specifications is required under these procedures.

The WITNESS. I have been a friend of General Nichols for many years. He wrote the letter. I quite frankly think it was a poorly written letter and should have been written in such a way that it made it absolutely clear that what was being examined here was not the question of whether a man held opinions and whether those were right or wrong, whether history has shown it to be good judgment or poor judgment. I think that should have been made very clear.

Mr. GRAY. I would also point out just in the interest of having a record here, and I don't consider myself in any argumentation with you, for whom I have a very high regard, personally and professionally, that there were items of so-called derogatory information—and that is a term of art—in this letter, setting aside the allegations about the hydrogen bomb. There were items in this letter which did not relate to the expression and holding of opinions.

The WITNESS. Quite right, and the case should have tried on those.

Mr. GRAY. This is not a trial.

The WITNESS. If it were a trial, I would not be saying these things to the judge, you can well imagine that. I feel a very serious situation has been created, and I think that in all fairness I ought to tell you my frank feeling that this has gotten into a very bad mess. I wish I could suggest a procedure that would resolve it.

Mr. GRAY. The proceeding, of course, is taking place in accordance with procedures, and I was glad to hear you say a few moments ago that you felt that this was a fair kind of proceeding. I am not sure I am quoting you correctly.

The WITNESS. You can quote me to that effect. I think some of the things we have seen have been scandalous affairs. I think in fact the Republic is in danger today because we have been slipping backward in our maintenance of the Bill of Rights.

Mr. GRAY. Dr. Evans.

Dr. EVANS. Dr. Bush, I wish you would make clear just what mistake you think the board made. I did not want this job when I was asked to take it. I thought I was performing a service to my country.

The WITNESS. I think the moment you were confronted with that letter, you should have returned the letter, and asked that it be redrafted so that you would have before you a clearcut issue which would not by implication put you in the position of trying a man for his opinions.

Dr. EVANS. I was not confronted with that letter, and I don't think it would have made any difference if I had been. I was simply asked if I would serve on the board. What mistake did I make when I did that?

Mr. GARRISON. Mr. Chairman, might I make a remark for myself here, speaking for Dr. Oppenheimer? I have the deepest respect for Dr. Bush's forthright character, for his lifelong habit of calling a spade a spade as he sees it. I simply want to leave no misunderstanding on the record here that we share the view that this board should not have served when asked to serve under the letter as written.

The WITNESS. I can assure you, Mr. Chairman, that the opinions being expressed are my own. They usually are.

Mr. GRAY. I have never heard it suggested that you didn't express your own opinion, Dr. Bush.

Dr. EVANS. Dr. Bush, then your idea is that suppose I was asked to serve on this board, and I didn't know anything about it—I had not seen any of this material—after I had agreed to serve, and saw this material, I should have **resigned?**

The WITNESS. No, I think you simply should have asked for a revision of the bill of particulars.
Dr. EVANS. I am just anxious to know what you think my procedure should have been.
The WITNESS. That is what I think. Now, I don't see how you can get out of this mess.
Mr. MORGAN. Doctor, on what ground would you ask for a bill of particulars if you didn't know the record?
The WITNESS. I think that bill of particulars was obviously poorly drawn on the face of it, because it was most certainly open to the interpretation that this man is being tried because he expressed strong opinions. The fact that he expressed strong opinions stands in a single paragraph by itself. It is not directly connected. It does have in that paragraph, through improper motivations he expressed these opinions. It merely says he stated opinions, and I think that is defective drafting and should have been corrected.
Mr. MORGAN. In other words, we want to prejudge the case before we know anything about it.
The WITNESS. Not at all. But I think this board or no board should ever sit on a question in this country of whether a man should serve his country or not because he expressed strong opinions. If you want to try that case, you can try me. I have expressed strong opinions many times, and I intend to do so. They have been unpopular opinions at times. When a man is pilloried for doing that, this country is in a severe state.
Mr. MORGAN. I have no more questions.
Mr. GARRISON. I should like to ask one more question.
The WITNESS. I hope it is a gentle one. Excuse me, gentlemen, if I become stirred, but I am.

By Mr. GARRISON:

Q. Dr. Bush, have you had some experience in handling security questions in the past?
A. Throughout the war, I was responsible for security in the Office of Scientific Research and Development. The formal situation was this. All the appointments I was responsible for clearance in the organization. On appointment on the staff of contractors, the contractor himself was responsible. Of course, you realize that to a contractor was given only the information within his field. No question was raised in connection with contractors unless either the Army or the Navy cautioned about them. On appointments to OSRD, I had advice from both the Army and the Navy, but the responsibility was mine.
I might say in passing that there were a good many appointments, and I know of no case in which an appointment on OSRD was made in which disloyalty has since been proved. I am proud of that record. I think our procedure in clearance at that time was a sane and reasonable one and effective one.
Mr. GARRISON. That is all.
Mr. ROBB. May I ask one question.

CROSS-EXAMINATION

By Mr. ROBB:

Q. I am going to ask you a question which I am sure you will describe as a gentle one. Let me tell you I never saw this letter in question until 2 months after it was written. I am not asking this for personal reasons.
A. I am sure you didn't write it.
Q. I am sure you didn't mean to imply that. Would you make a distinction between the question of whether a man's opinions were right and wrong, and the question of whether a man's opinions were expressed in good faith or bad faith?
A. Yes; a very great difference. If this paragraph that I referred to had said by improper motivation because this man had allegiance to another system than that of his own country, he expressed these opinions in an attempt to block the program, then I would not have objected.
Q. If the paragraph was interpreted to question the good faith of the opinion, then you would have no objection to it.
A. No, if it was done explicitly enough, certainly not.
Q. Thank you.
A. The trouble is of course that the public will not read and will not interpret gently or sympathetically. The public is going to read this in the worst possible interpretation.
Mr. GRAY. Thank you very much, Dr. Bush.

The WITNESS. Thank you, sir.
(Witness excused.)
Mr. ROBB. That is all we have to do today.
Mr. GRAY. Do you have some affidavits?
Mr. GARRISON. I think they could go over until Monday. It won't take very long.
Mr. GRAY. We will recess for the weekend and meet again Monday morning at 9:30.
(Thereupon at 2:50 p. m., a recess was taken until Monday, April 26, 1954, at 9:30 a. m.)

UNITED STATES ATOMIC ENERGY COMMISSION

PERSONNEL SECURITY BOARD

In the Matter of J. Robert Oppenheimer

Atomic Energy Commission,
Building T-3, Room 2022,
Washington, D. C., Monday, April 26, 1954.

The above entitled matter came on for hearing pursuant to recess before the board, at 9:30 a. m.

Personnel Security Board: Mr. Gordon Gray, chairman; Dr. Ward T. Evans, member; and Mr. Thomas A. Morgan, member.

Present: Roger Robb, and C. A. Rolander, Jr., counsel for the board; J. Robert Oppenheimer; Lloyd K. Garrison; Samuel J. Silverman; and Allan B. Ecker, counsel for J. Robert Oppenheimer; Herbert S. Marks, cocounsel for J. Robert Oppenheimer.

PROCEEDINGS

Mr. GRAY. Before we start, Mr. Garrison, Dr. Evans has a statement he would like to make for the record. With your consent, I should like this to appear in the record at this point.

Dr. EVANS. Mr. Chairman, for the record, I would like to state that I think Dr. Bush was in error when he stated that the members of the board made a mistake when they agreed to serve on this board unless the letter from General Nichols was rewritten. Personally I knew very little about this case when I agreed to serve on it at considerable inconvenience to myself, and I did so because I thought it was my duty to serve.

Mr. GRAY. Mrs. Oppenheimer do you wish to testify under oath?

Mrs. OPPENHEIMER. Yes.

Mr. GRAY. Would you be good enough to stand and raise your right hand. Your name is Katherine Oppenheimer?

Mrs. OPPENHEIMER. Yes.

Mr. GRAY. Katherine Oppenheimer, do you swear that the testimony you are to give the board shall be the truth, the whole truth, and nothing but the truth, so help you God?

Mrs. OPPENHEIMER. I do.

Whereupon Katherine Oppenheimer was called as a witness, and having been first duly sworn, was examined and testified as follows:

Mr. GRAY. Will you be seated, please.

Mrs. Oppenheimer, it is my duty to remind you of the existence of the perjury statutes. We will assume that you are familiar with them.

I should also like to say to you what I have said to the other witnesses, and that is that we consider these proceedings a confidential matter between the Atomic Energy Commission and its officials on the one hand, and Dr. Oppenheimer and his witnesses and representatives on the other. The Commission will issue no public releases, and we express the hope that witnesses will take the same view.

The WITNESS. Right.

Mr. GRAY. Mr. Garrison, will you proceed?

DIRECT EXAMINATION

By Mr. SILVERMAN:

Q. Mrs. Oppenheimer, you are the wife of Dr. J. Robert Oppenheimer?
A. I am.
Q. What were you doing in the autumn of 1933?
A. I was attending the University of Wisconsin.
Q. You were attending the University of Wisconsin?
A. That is right.
Q. As an undergraduate student?
A. Yes.
Q. What did you do during the Christmas holidays of 1933?
A. I went to stay with friends of my parents in Pittsburgh.
Q. Will you tell us the circumstances of your meeting Joe Dallet?
A. Yes. I have an old friend in Pittsburgh, a girl called Selma Baker. I saw quite a bit of her at that time. It was Selma who said she knew a Communist, and would we like to meet him. Everybody agreed that would be interesting. There was a New Year's party. Selma brought Joe Dallet.
Q. Did you and he fall in love during that holiday period?
A. We did.
Q. Did you decide you would be married?
A. We did.
Q. Did you fix a date for that?
A. Yes. I decided to go back and finish my semester at Wisconsin and then join Joe in Youngstown and get married there.
Q. Is that what you did?
A. Yes.

Q. The semester ended at the end of January, I suppose, of 1934, and you went to Youngstown?
A. Early February. I don't know.
Q. Joe Dallet was a member of the Communist Party?
A. He was.
Q. And you knew that he was?
A. Yes.
Q. During your life with him, did you join the party?
A. Yes, I did.
Q. Will you tell us why you joined the party?
A. Joe very much wanted me to, and I didn't mind. I don't know when I joined the party. I think it was in 1934, but I am not sure when.
Q. Did you do work for the party?
A. Yes.
Q. What kind of work?
A. I mimeographed leaflets and letters. I typed. I did generally office work, mostly for the steel union that was then in existence.
Q. What were most of your activities related to?
A. Mostly to the union at first, and later anything that came up, I was sort of general office boy.
Q. Did you pay dues to the party?
A. Yes.
Q. How much were the dues?
A. I believe mine were 10 cents a week.
Q. Would you describe the conditions under which you lived with Joe Dallet as those of poverty?
A. Yes.
Q. How much rent did you pay?
A. Five dollars a month.
Q. As time went on, did you find that you became devoted to the party or more devoted or less devoted or more attached or less attached?
A. I don't think I could ever describe it as a devotion or even attachment. What interest I had in it decreased.
Q. Did Joe's interest decrease?
A. No, not at all.
Q. Was that a cause of disagreement between Joe and yourself?
A. I am afraid so.
Q. Did you and Joe ultimately separate?
A. We did.
Q. When was that?
A. About June of 1936.
Q. Would you say that your disagreement with Joe about your lack of enthusiasm, shall we say, for the party, had something to do with the separation?
A. I think it was mostly the cause of the separation. I felt I didn't want to attend party meetings or do the kind of work that I was doing in the office. That made him unhappy. We agreed that we couldn't go on that way.
Q. Did you remain in love with him?
A. Yes.
Q. Where did you go when you separated?
A. I joined my parents in England.
Q. That was about June of 1936?
A. I think it was June.
Q. Did a time come when you wrote Joe that you were willing to rejoin him?
A. Yes. I wrote him probably very early in 1937, saying that I would like to rejoin him.
Q. Did he answer you?
A. He answered saying that would be good, but he was on his way to Spain to fight for the Republic cause, and would I please instead meet him in Paris.
Q. Where did you meet him?
A. I met him at Cherbourg aboard the Queen Mary as it docked.
Q. That was in 1937?
A. Yes. I think it was March. I am not sure.
Q. Did you go with him then to Paris?
A. We took the boat train and went to Paris.
Q. How long did you stay in Paris with him.
A. I would think about 10 days. It could have been a week, it could have been 2 weeks, but roughly——

Q. Do I understand that he had a furlough or some time off or something because of the reunion?
A. That is right.
Q. What did you do during that 10 days or so in Paris?
A. We walked around and looked at Paris, went to restaurants, the sort of thing one does in Paris. We went to the museums and picture galleries. We went to one large political meeting, a mass meeting, where they were advocating arms for Spain.
Q. Who was the speaker?
A. Thorez.
Q. He was a Communist?
A. Yes.
Q. Do you recall any other political activities if that might be called one during that period or that 10 days or so?
A. I think one should describe as a political activity that one place I saw where people who were going to Spain were being checked in and told how to do it. I went there once.
Q. As a spectator?
A. I had nothing to do.
Q. Then Joe went off to Spain.
A. Yes.
Q. During that period did you meet Steve Nelson?
A. Yes. I met him in Paris. I saw him several times. I think Joe and I had meals with him occasionally.
Q. What did you talk about with him?
A. I don't know; all kinds of things. I think among other things the only thing that interests this board is the fact that we talked of various ways of getting to Spain, which was not easy.
Q. Then Joe went to Spain at the end of that 10 days os so?
A. Yes.
Q. What did you do?
A. I went back to England.
Q. Did you try to do anything about joining Joe?
A. Yes, I wanted to very much.
Q. What was your plan as to how you would join Joe?
A. I was told that they would try to see if it were possible, and if it were, I would hear from someone in Paris and then go to Paris, and be told how to get there.
Q. Was there talk of your getting a job somewhere in Spain?
A. Yes. I don't know what, though.
Q. Were you ultimately told that it was possible?
A. I got a letter from Joe saying that he found me a job in Albacrete.
Q. Did you then go to Paris?
A. First I stayed in England and waited quite a while, until October.
Q. What year was this?
A. 1937. I then got a wire saying I should come to Paris, and I went. Do you want me to go on?
Q. What happened when you got to Paris?
A. When I got to Paris, I was shown a telegram saying that Joe had been killed in action.
Q. What did you do then?
A. I was also told that Steve Nelson was coming back from Spain in a day or two, and I might want to wait and see what Steve had to say. He had a lot to tell me about Joe.
Q. Did Steve come?
A. Yes.
Q. And met you in Paris?
A. Yes.
Q. Did you talk with Steve?
A. Yes, I spent at least a week there. I saw Steve most of the time.
Q. What did you talk about with him.
A. Joe, himself, myself.
Q. Would you say that Steve was kind to you and sort of took care of you during that period?
A. He certainly was, very.
Q. Did you discuss with Steve what you would do now?
A. I did.

Q. Will you tell us what that discussion was?
A. For a little while I had some notion of going on to Spain anyway.
Q. Why?
A. I was emotionally involved in the Spanish cause.
Q. Did Joe's death have something to do with your wanting to go on anyhow?
A. Yes, as well as if alive he would have.
Q. Did you discuss this with Steve?
A. I did, but Steve discouraged me. He thought I would be out of place and in the way. I then decided that probably I would go back to the United States and resume my university career.
Q. Is that what you did?
A. Yes.
Q. After you returned to the United States, did you continue to see any of the rriends that you had with the Communists?
A. When I first got back I saw some friends of Joe's in New York who wanted to know about him and to whom I wanted to talk. I saw some other members of the Communist Party in New York. I went to Florida with three girls. I know one was a Communist. I think another one was, and the third one I don't remember.
Q. Did that relationship with Communist friends continue?
A. No, it did not.
Q. What happened?
A. I visited a friend of mine in Philadelphia. I had planned to go to the University of Chicago, and got back to the United States to go back to their second trimester. I don't know whether they still have that system. I knew no one there. I met a lot of people in Philadelphia, and they said, "You know all of us, why don't you stay here?" I stayed in Philadelphia and entered the University of Pennsylvania, the spring semester of the year 1937–38.
Q. What kind of work did you do at the university?
A. Chemistry, math, biology.
Q. Was biology your major?
A. It became my major interest.
Q. Did you continue to do professional work as a biologist?
A. I did graduate work later and some research.
Q. Ultimately you had a research fellowship or assistantship?
A. Both.
Q. Where?
A. University of California.
Q. Did you remarry?
A. Yes.
Q. Would you give us the date of your remarriage and the man whom you married?
A. I married Richard Stewart Harrison, an English physician, in 1938, in December or November.
Q. Was he a Communist?
A. No.
Q. He was a practicing physician?
A. He had been, I think, in England. He had to take all his examinations in this country and do an internship and a residency before he could practice here.
Q. Did he go to California?
A. Yes.
Q. And you went with him?
A. No. He went to California much earlier than I to take up his internship.
Q. Did you go out there to join him?
A. Yes.
Q. After graduation in June of 1939? When did you meet Dr. Oppenheimer?
A. Somewhere in 1939.
Q. When were you divorced from Dr. Harrison?
A. In the first of November 1940.
Q. You then married Dr. Oppenheimer?
A. Yes.
Q. Did there come a time after you married Dr. Oppenheimer when you again saw Steven Nelson?
A. Yes.
Q. Will you tell us the circumstances of that?
A. I will as best I can remember. I remember being at a party and meeting a girl called Merriman. I knew of her. She was in Albacrete, and her husband

also got killed in action there. The reason I remembered her name is that I had been asked to bring her some sox when I came. When I met her at this party, she said did I know that Steve Nelson was in that part of the country. I said no, and then expressed some interest in his welfare. Some time thereafter Steve Nelson telephoned me, and I invited him and his wife and their small child up to our house.

Q. What did you talk about?
A. We had a picnic lunch. The Nelsons were very pleased that they finally had a child, because they tried for a long time to have one without success. We talked about the old days, family matters.
Q. Did you see him again?
A. I think that they came out to our house two times.
Q. Was it all just social?
A. Yes.
Mr. GRAY. What was the date of this period, approximately? If you have said, I have forgotten.
The WITNESS. I didn't say, Mr. Gray, because I am a bit vague.

By Mr. SILVERMAN:
Q. Can you give it as closely as you can?
A. Yes. I would guess it was late 1941 or perhaps in 1942. I don't know.
Q. Are you fairly clear it was not later than 1942?
A. Fairly clear.
Q. Have you seen Steve Nelson since 1942?
A. Since whenever it was?
Q. Yes.
A. No.
Q. You are no longer a member of the Communist Party?
A. No.
Q. When would you say that you ceased to be a member?
A. When I left Youngstown in June 1936.
Q. Have you ever paid any dues to the party since then?
A. No.
Q. Will you describe your views on communism as pro, anti, neutral?
A. You mean now?
Q. Now.
A. Very strongly against.
Q. And about how far back would you date that?
A. Quite a long time. I had nothing to do with communism since 1936. I have seen some people, the ones that I have already described.
Mr. SILVERMAN. That is all.
Mr. ROBB. No questions.
Mr. GRAY. Mrs. Oppenheimer, how did you leave the Communist Party?
The WITNESS. By walking away.
Mr. GRAY. Did you have a card?
The WITNESS. While I was in Youngstown; yes.
Mr. GRAY. Did you turn this in or did you tear it up?
The WITNESS. I have no idea.
Mr. GRAY. And the act of joining was making some sort of payment and receiving a card?
The WITNESS. I remember getting a card and signing my name.
Mr. GRAY. Generally speaking, as one who knows something about communism as it existed at that time in this country and the workings of the Communist Party, and therefore a probable understanding of this thing, what do you think is the kind of thing that is an act of renunciation? That is not a very good question. In your case you just ceased to have any relationships with the party?
The WITNESS. I believe that is quite a usual way of leaving the party.
Mr. GRAY. When you were in the party in Youngstown, or when you were in the party at any time, did you have a party name?
The WITNESS. No. I had my own name, Kitty Dallet.
Mr. GRAY. Was that the usual thing for people to use their own name?
The WITNESS. I knew of no one with an assumed name. I believe that there must have been such people, but I knew of none.
Mr. GRAY. I think the record shows that in some cases there were people who had some other name.

The WITNESS. I think there were people who lived under an assumed name and had that name in the party, but then that was the only name I would have known.

Mr. GRAY. When you saw Steve Nelson socially in whatever year this was, 1940, 1941 or 1942, did you discuss the Communist Party with him? Did he know that you were no longer a member of the Communist Party?

The WITNESS. Yes, that was perfectly clear to him.

Mr. GRAY. Did he chide you for this or in any way seek to reenlist your sympathy?

The WITNESS. No.

Mr. GRAY. He accepted the fact that you had rejected communism?

The WITNESS. Yes. I would like to make it clear that I always felt very friendly to Steve Nelson after he returned from Spain and spent a week with me in Paris. He helped me a great deal and the much later meeting with him was something that was still simply friendship and nothing else.

Mr. GRAY. The people you dealt with in Paris or that you saw there were members of the Communist Party. I have in mind any discussions you had about going to Spain, both before and after your husband's death?

The WITNESS. I wouldn't know who was or wasn't then. Many people were going to Spain who were not members of the Communist Party. I think, however, that probably most of the people I saw were Communists.

Mr. GRAY. But at that time you were not?

The WITNESS. No.

Mr. GRAY. This was following your leaving the party in Youngstown?

The WITNESS. That is right.

Mr. GRAY. Do you suppose they were aware of the fact that you had left the Communist Party?

The WITNESS. I am sure they were. I mean such as knew me.

Mr. GRAY. This is a question not directly related to your testimony, but we have had a witness before the board recently—I might say I am sorry I didn't ask him this question—and this witness referred to Soviet communism in a general discussion here before the board. In your mind as a former member of the Communist Party in this country, can a distinction be made between the Soviet communism and communism?

The WITNESS. There are two anwsers to that as far as I am concerned. In the days that I was a member of the Communist Party, I thought they were definitely two things. The Soviet Union had its Communist Party and our country had its Communist Party. I thought that the Communist Party of the United States was concerned with problems internal. I now no longer believe this. I believe the whole thing is linked together and spread all over the world.

Mr. GRAY. Would you think that any knowledgeable person should also have that view today?

The WITNESS. About communism today?

Mr. GRAY. Yes.

The WITNESS. Yes, I do.

Mr. GRAY. I was puzzled by this reference to Soviet communism in April 1954. But in any event, you would not make a distinction.

The WITNESS. Today, no; not for quite a while.

Mr. GRAY. But in those days you in your own mind made the distinction?

The WITNESS. Yes.

Mr. GRAY. At that time the American Communist Party was not known to you to be taking its instructions from Russia?

The WITNESS. No.

Mr. GRAY. You testified that today you are opposed to the Communist Party and what it stands for.

The WITNESS. Yes.

Mr. GRAY. I am getting back now to whatever action of renunciation is. Do you think these days that a person can make a satisfactory demonstration of renunciation simply by saying that there has been renunciation?

The WITNESS. I think that is too vague for me, Mr. Gray.

Mr. GRAY. All right. I am afraid it is a little vague for me, too. I won't pursue it.

Do you have any questions?

Dr. EVANS. Just one. Mrs. Oppenheimer, I have heard from people that there are two kinds of Communists, what we call an intellectual Communist and just a plain ordinary Commie. Is there such a distinction, do you know?

The WITNESS. I couldn't answer that one.

Dr. EVANS. I couldn't either. Thank you. I have no more questions.
Mr. GRAY. Thank you very much, Mrs. Oppenheimer.
(Witness excused.)
Mr. GARRISON. May we take a 5 minute recess?
Mr. GRAY. Yes.
(Brief recess.)
Mr. GRAY. Do you wish to testify under oath, Dr. Lauritsen?
Dr. LAURITSEN. I would like to.
Mr. GRAY. Will you raise your right hand, please, sir. What is your full name?
Dr. LAURITSEN. Charles Christian Lauritsen.
Mr. GRAY. Charles Christian Lauritsen, do you swear that the testimony you are to give the board shall be the truth, the whole truth, and nothing but the truth, so help you God?
Dr. LAURITSEN. I do.
Whereupon, Charles Christian Lauritsen was called as a witness, and having been first duly sworn, was examined and testified as follows:
Mr. GRAY. Will you be seated; please, sir.
Dr. Lauritsen, it is my duty to remind you of the existence of the so-called perjury statutes with respect to giving false information, and so forth. Is it necessary for me to review those provisions with you, or may we assume you are familiar with them?
The WITNESS. I would be very glad to hear the essentials.
Mr. GRAY. The provisions of section 1621 of title 18 of the United States Code, known as the perjury statute, make it a crime punishable by a fine of up to $2,000 and/or imprisonment of up to 5 years for any person to state under oath any material matter which he does not believe to be true.
It also is an offense under section 1001 of title 18 of the United States Code, punishable by a fine of not more than $10,000 or imprisonment for not more than 5 years, or both, for any person to make any false, fictitious, or fraudulent statement or misrepresentation in any manner within the jurisdiction of any agency of the United States.
If, Dr. Lauritsen, in the course of your testimony it should become necessary for you to refer to or disclose restricted data, I would ask you to notify me in advance so that we might take certain necessary and appropriate steps.
Finally, I should say to you that we consider these proceedings a confidential matter between the Atomic Energy Commission and its officials on the one hand, and Dr. Oppenheimer, his representatives and witnesses on the other. The Commission will make no release about this proceeding and this testimony, and we express the hope that witnesses will take the same view.
Mr. Marks.

DIRECT EXAMINATION

By Mr. MARKS:
Q. Dr. Lauritsen, what is your present position, and where?
A. I am professor of physics at the California Institute of Technology.
Q. How long have you held that post?
A. I believe as full professor since 1936. I have been at the California Institute since 1926, first as a graduate student, later as assistant professor, and subsequently associate professor, and full professor.
Q. Are you an experimental or theoretical physicist?
A. Experimental.
Q. Dr. Lauritsen, will you describe briefly the nature of the more important war work that you did during World War II? Let me suggest that you leave out the preliminaries and just describe as what you regard the most important.
A. All right. Starting in July 1940, I came to Washington and joined the National Research Defense Committee which had just been formed in June. The organization consisted of four divisions, and I was appointed by Dr. Bush as vice chairman of division A on armor and ordnance. More important things that we worked on in that division initially were proximity fuse sand rockets.
Q. Will you tell the board about your work on rockets during the war?
A. Yes, I will be glad to. In the early summer of 1940—I am sorry, 1951—Dr. Hafstad and I were sent to England.
Q. Who is Dr. Hafstad?
A. Dr. Hafstad at the present time is head of the Reactor Division of the AEC. Dr. Harry Hafstad. He and I were sent by the NDRC to England to discuss proximity fuses with them. We brought over the first samples of the proximity fuses we made in this country.

Mr. ROBB. I can't hear the witness. Will you speak louder?

The WITNESS. Shall I repeat? Dr. Hafstad and I were sent by the National Defense Research Committee to England on proximity fuses. I had also been intersted in the development of rockets in this country and the program was in my opinion not very satisfactory at that time, although I was responsible for it. I knew nothing about the subject at that time. So while we were in England in the early summer of 1941, I obtained all the information that I could on rockets in England and on the British rocket program. At the same time I also obtained all the information I could about the British atomic energy program.

When I came back I reported to Dr. Vannevar Bush on these two subjects. You wanted particularly to hear about the rocket program?

By Mr. MARKS:

Q. Yes.
A. As a result of my report to Vannevar Bush, he asked me to organize an expanded effort on producing of rockets for the armed services. This I tried to do first here in the East without very much success, and in the fall of 1941, I went back to Pasadena and started a program at the California Institute for the development of rockets.

A number of my colleagues had been here in Washington up to that time working on proximity fuses. They went back to Pasadena with me and started this rocket program.

The result of this was that ultimately we produced all the rockets that were used in World War II by the Navy and the Marines and the Air Force.

Q. When you say "we produced," who do you mean by "we"?
A. I mean this rocket project at the California Institute of Technology.
Q. Who was the head of that?
A. I was the technical director of that program and responsible for the technical program.
Q. You mean you produced at the project in Pasadena?
A. At the project we developed the first type of rockets that we thought were necessary and that we could get interest in that the military thought they needed, particularly the Navy. I worked personally very closely with the Bureau of Ordnance. We then developed these and tested them, and when they were approved, we produced them until such time as large companies could get into production.

A typical example was the 5-inch rocket, which you read so much about that was used in Korea. This one we developed and we manufactured in Pasadena something considerably over 100,000, which were used in the European theater, and later on in the Pacific war. * * *

Q. Did I understand you to say, however, that your project in Pasadena produced all the rockets that were used in World War II.
A. All the rocket types, not the individual rockets that were fired. We produced them only until large companies could take over production, which was usually something like a year. We made all the rockets used in the African landings, in the Sicilian landings, and in a number of the landings in the Pacific, like Iwo Jima, and many of the others. Altogether several hundred thousand rockets. Our total project added up to about $80 million spent at the project.

Q. How many people did you have under you in that production work?
A. I am not quite certain, but I believe the number was something like 3,500 at the maximum. This, of course, did not include contractors and subcontractors. These were the people employed by the California Institute for this purpose.

Q. Can you tell what importance you attached to the rocket program and why?
A. Personally I like to think that the most important thing was the landings in the Pacific which ultimately became a matter of walking ashore. There were very few casualties due to the heavy bombardment of the shore defenses just before landings were made. A number of landings were made in the Pacific with almost no losses. Of course, the same thing was true at Inchon. The coast line was——

Q. Inchon when?
A. During the Korean war. The coast line was heavily bombarded and there was no opposition when we landed. This is of course not entirely due to rockets, but until they started using the rockets in large numbers, the losses on landing operations were very heavy.

Q. I don't quite understand what part the rockets played.
A. The advantage of the rocket is that you can unload almost a whole ship's cargo of rockets in a very short time and no shore installations can withstand such bombardment. Rockets can be fired in huge numbers at one time.

Another application was the application of the 5-inch rocket that I just mentioned to airplanes. This made a very powerful weapon out of the carrier based airplanes as well as the small support aircraft used by the Air Force for supporting ground troops. They are for all practical purposes equivalent to a 5-inch naval gun.

Q. Did the use of rockets represent any change in the nature of firepower?

A. It is an enormous increase in firepower at the moment you need it, in a very short time. You can fire thousands of them in 1 minute. It would not be possible to provide enough guns to deliver that fire at one time in a short period. * * *

Q. Apart from the work that you did on development and production of these new weapons, did you have anything to do with the introduction of their use into military operations?

A. Yes.

Q. Would you describe that?

A. Whenever a weapon was accepted for service use, and we produced the ammunition, we usually sent a man along with the equipment to the various theaters to be sure that it was received with some understanding and used in a reasonable way and that the equipment was kept in operation, and that the crews were trained. It was usualy necessary to spend some time training crews. As an example, I might mention that I was together with one or two of my colleagues to Normandy in 1944 to introduce these rockets to the Air Force. We equipped some squadrons and trained them in their spare time, usually at night after they had been carrying out their daytime missions and operations. They were enthusiastic enough about it to work on learning how to use them during the evenings. They would go back from Normandy, sometimes, over to England to practice on a field that we had borrowed from the British. It was necessary to stay with an operation like this long enough until the weapon was properly used.

Q. How much experience have you had in this kind of field work in the introduction of the use of new weapons that you have been concerned with?

A. Usually I did not personally go out with all the equipment. I did personally go out with some of the first submarine weapons that we developed, and I took part in many of the training exercises on the shore bombardment rockets and on the aircraft rocket I was frequently involved. But often other members of our organization were the ones that went out in the field to help with these things.

Mr. GRAY. Do you mind if I interrupt for a minute, Mr. Marks?

Mr. MARKS. Surely.

Mr. GRAY. Are you going to relate this to the present inquiry?

Mr. MARKS. Yes.

The WITNESS. I think I have not said anything that is classified.

Mr. MARKS. If the chairman prefers, I would be glad to get directly to the issue on the present inquiry and then go back?

Mr. GRAY. I don't want to restrict you at all, but we are in a little different field than we have been discussing in these hearings.

Mr. MARKS. I don't want the testimony to be unintelligible through any point of mine.

Mr. GRAY. Let us proceed.

By Mr. MARKS:

Q. Let us leave this subject and let me ask you what later work did you do in World War II apart from the rocket work?

A. During 1944 it became apparent to us that the war was coming to an end, and that there would probably not be time to dream up any very important new rocket weapons that could be produced in quantity to have much effect on the war. The Navy agreed with us. At that time they decided that they wented to take over the operation of the facilities that had been developed for our purpose, namely, the large test and development station at Inyokern, Calif. We had been operating that station during the war. In 1944, the Navy decided that they would take that over gradually and also to take over the future development of rocket weapons.

Q. What did you do then?

A. At the request of Dr. Bush and Dr. Oppenheimer and General Groves, I went to Los Alamos to help with the final stages of the atomic bomb.

Q. What did you do in that capacity?

A. Most of what I did was talking, I am afraid. I attended numerous meetings of the various divisions when they had meetings trying to make decisions. I would usually attend these meetings. I attended meetings of the various steering committees and in general tried to assist Dr. Oppenheimer in any way that I could on making decisions, particularly on hardware.

Q. What do you mean by hardware?

A. Hardware is all the things that are required to produce a weapon and all the components that are necessary for the weapon itself. They may be electronic gadgets or castings or machine parts or production tools. We had a considerable part of the responsibility for producing the explosive components, that is, the conventional explosive components, and the various tools necessary and installations necessary for producing these.

Q. Did you have a title at Los Alamos? Where did you work?

A. I had no title. I worked directly in Dr. Oppenheimer's office.

Q. How long was that?

A. Just about 1 year. I agreed to stay 1 year, and at the end of 1 year the war was over.

Q. When you say you worked in his office, you mean in the office which he occupied?

A. Yes.

Q. You and he occupied an office together?

A. That is right.

Q. How long had you known Dr. Oppenheimer before that?

A. I have known Dr. Oppenheimer since he came back from Europe, from Goettingen, which I believe was in 1928 or 1929. I am not certain about the date. In 1928 or 1929, when he came to Pasadena.

Q. In the years since that date, how well have you known him?

A. I have known him as well as I have known any member of our faculty.

Q. Commencing when?

A. Very soon. In fact, I probably saw him the date he arrived because it happened accidentally that we had been interested in the same problem, he in the theory of it, and I in the experimental aspects of it. So he looked me up very soon after he came there, I believe.

Q. What date was that approximately?

A. Either in 1928 or 1929. I could certainly get that date, but I could not be very certain at the moment.

Q. Did you become close friends?

A. I would say so, yes.

Q. Has that friendship continued?

A. Yes, sir.

Q. Did you observe Dr. Oppenheimer during the thirties and the forties, and can you say anything about his political views and activities during that time?

A. I cannot say very much about it. I knew very little about it until, I think, about the time of the Spanish war. This was the first time that I knew that he had any political interest. Up to that time I have no recollection that we ever discussed political questions of any in interest or serious nature.

Q. What impression did you come to have of his political interests?

A. It is a little difficult to say because I think they changed a great deal with time. I would say that at one stage he was very deeply interested in the Spanish Loyalist cause, and took the attitude that was taken at that time by many liberals, the hope that they could do something about it, and that they would like to help the Spanish Loyalist cause.

You spoke of his changing views. What do you mean by that?

A. I think it was probably a gradual increase in interest in social causes, a compassion for the underdog, if you like. The attitude that many liberals took at that time.

Q. Did you observe in him an identification with views that were regarded as Communist views or with which the Communists were associated?

A. I think at that time very few of us and perhaps very few Americans had very little idea about what communism was. I think most of us that were concerned about political things and international things were considerably more concerned about fascism at that time than we were about communism. Fascism seemed the immediate threat, rather than communism. Also, I think perhaps my own views were colored by the fact that I was born and raised in Denmark, where Germany was the natural enemy, rather than Russia. I think for that reason we did not pay as much attention to the evils of communism as we should have done.

Q. Were you mixed up in any communistic activity?

A. No.

Mr. ROBB. Mr. Chairman, I don't think the witness quite answered the question Mr. Marks propounded to him. I wonder if we might have it read back so the witness could have it in mind.

(Question read by the reporter.)

The WITNESS. I frankly did not know, just what characterized the Communist view at that time. When they talked about improving the lot of the working people, I believe Oppenheimer and probably many other people thought this was a good beginning. But that this was not the whole story of the Communist ideology I think was not realized by very many people at that time. Does that answer the question?

Mr. ROBB. Yes.

The WITNESS. It did not occur to me at that time or at any other time that he was a Communist Party member. In fact, at the date we are talking about, namely, the early part of the Spanish-American War, I didn't know there was such a thing as——

By Mr. MARKS:

Early part of what war?

A. The Spanish War. Did I say Spanish-American?

Q. Were you in the Spanish-American War?

A. No. These are words that have just been associated so long. I was not aware that there was such a thing as a secret membership of the Communist Party. I don't know if other people were but I was not.

Q. I asked you, I think at the time Mr. Robb reminded me that you had not answered an earlier question, whether you were mixed up in Communist activities yourself.

A. I was not.

Q. Was there a difference between yourself in that respect and what you observed of Oppenheimer at that time?

A. I think I was more pessimistic about what liberals could accomplish, even if they were trying to accomplish good things. I was less optimistic about what you could do about these activities. Therefore, I took no part in them.

Q. As time went on, did you notice any change in Dr. Oppenheimer's attitude about these matters which you have indicated as being more optimistic than yours?

A. How far along are we now? Are we still in the thirties?

Q. Let us take the period, Dr. Lauritsen, from the late thirties to the early forties.

A. In the late thirties, the event that I remember best, the discussion that I remember best, is the discussion we had, I believe it was on the day that Russia signed the agreement with Germany. This was an event that shocked me very deeply and we discussed it at considerable lengths.

Q. When you say "we," who do you mean?

A. I was thinking about conversations Dr. Oppenheimer and I had on that day, I believe, or at least very shortly after. I was very convinced that this was the beginning of a war and during our conversation I am quite certain that Dr. Oppenheimer agreed with this point of view, and was as concerned about it as I was.

Q. Concerned in what way?

A. Afraid that this would lead to war, realizing what a bad situation, what a dangerous situation for the rest of the world, this combination of Russia and Germany could be.

Q. Did anything else happen as time went on of that nature?

A. This was 1933, was it not? Shortly after that the war started. The war was a reality. That is Germany went into Poland.

Q. What if anything did you observe about Dr. Oppenheimer's attitude as these events of the late thirties and early forties progressed?

A. You must realize that our most intimate contacts at this time during the late thirties were limited to the spring term, because Dr. Oppenheimer spent most of the year in Berkeley, and only the spring term, part of May and June, in Pasadena. So there were considerable intervals when we did not spend a great deal of time together.

The next thing that I recall was in 1940, and it was in the spring of 1940 that we at the California Institute realized that we would have to change our way of life, and that sooner or later we would have to get into war work. As I have already related, in June 1940, NDRC, the National Research Council, was organized and in July I joined. So there were long periods after that when I did not get back to California and when I did not see Dr. Oppenheimer.

Q. Do you have anything to say—if you don't, of course, just don't say it—about your observation of his views during the period 1940, 1941, 1942, 1943?

A. I can only say that at that time in 1940–41, I saw Dr. Oppenheimer only rarely, probably only 2 or 3 times. I do remember at one time—I think it was in 1941—he did not tell me what he was working on, I did not tell him what I was working on, but he did ask me if I thought that there would be an opportunity possibly later of his contributing to the work that we were working on. When I say "we," I meant in 1941, division A, Professor Tolman, who was a very good friend of both of us, who was the chairman, and I was the vice chairman. Dr. Oppenheimer expressed the desire perhaps to join us, because of our old associations.

Q. Could you date that time?

A. I could not be sure but I think it was 1941.

Q. Was it the early or latter part of 1941?

A. It was almost certainly either during the spring term, namely, June and July, early summer, or else Christmas, because those were the two times when we were most likely to be in Pasadena at the same time. As I say, the rest of the year he was as far as I know in Berkeley.

Q. So that the next intimate, if you can call it that, contact you had with Dr. Oppenheimer is when you came to Los Alamos into his office?

A. That is right.

Q. Did you observe anything about his political attitude then?

A. At that time politics didn't seem very important. The job was to win the war.

Q. What did you do after the war?

A. After the war I went back to teach school at California Institute of Technology.

Q. And how long did you do that without extensive outside interests?

A. This continued without too much interference from the outside until the start of the Korean war.

Q. What change occurred in your own work after the commencement of the Korean war?

A. Actually some of these activities started before the Korean war. I may have a little difficulty in getting the actual date, but I will at least get the sequence.

The first so-called study project that I was asked to join was called the Hartwell project. * * *

After this study, which was according to the Navy people that I know quite satisfactory and quite useful to them, these studies became a habit.

Q. Became a habit with whom?

A. With the military, and a number of such studies were originated by the military.

Q. How many did you engage in?

A. Hartwell, as I say, was the first one. The next one was called Charles. * * * I believe it was in the summer of 1951, which resulted in the setting up of the Lincoln laboratory. * * * Perhaps this was before. It was already in the summer of 1950 that we undertook at the California Institute of Technology a study that was called the Vista study. * * *

Q. From whom did you get that assignment?

A. From all three services. It was originally suggested in somewhat modified form by the Air Force, but before we undertook the program the Army and the Navy joined, and it was done jointly for all 3 services and under the direction the 3 services.

Q. What other connections have you had with military work since 1950?

A. In fact, ever since the war I have spent a little time in an advisory capacity at the naval ordnance station at Inyokern. In the beginning it was merely because of personal and friendly relations with the technical director up there. Frequently I visited at his request. Somewhere around 1949, he requested from the Chief of the Bureau of Ordnance an advisory board which was set up on a formal basis. It was setting up this thing for the same purpose, but on a formal basis. This board was organized, I believed, in 1949. I was the first chairman of that board. The board still meets about three times a year and I am a member of that board, but no longer chairman. We rotate the chairmanship.

This was the only direct connection I had with military affairs, as I say, until 1950.

The next thing I was requested to do was to go to Korea for the Secretary of Defense's office, the Weapons System Evaluation Group. I went there in October and November of 1950. The Korean war started, I believe, in June 1950. Assuming this is correct, then it was in October or November of 1950.

Q. What else have you done along these lines?

A. I am still on two panels of the Scientific Advisory Board of the United States Air Force, one panel on explosives and ordnance, the other panel on nuclear weapons. I am a member of an advisory board to the Research and Development Command of the Air Force in Baltimore. I am a member of a panel on armament.

Q. Since 1950, how much of your time has been devoted to this work connected with military affairs?

A. Including homework and travel, it is probably about half my time.

Q. Since 1950?

A. Yes.

Q. In these connections, did you have anything to do with much top secret material?

A. Yes.

Q. Do you have what is called a Q clearance in all of these matters?

A. Yes.

Q. In this work commencing with the Korean war, what associates have you had with Dr. Oppenheimer and to what extent has his work and yours overlapped or coincided?

A. I believe the first contact on these problems was in connection with an ad hoc committee in the Research and Development Board. It was an ad hoc committee of the Committee on Atomic Energy in the Research and Development Board. We had meetings around Christmastime or I guess January 1951—probably December 1951 and January 1952—the purpose of this was to make recommendations to the Research and Development Board and the Military Liaison Committee on long range planning and production of atomic weapons. I think this was the first contact.

I was, as far as I know, still am, a consultant to this permanent committee of the Research and Development Board.

Q. What other connections did you have with Dr. Oppenheimer in this work? By "this work" I mean the general activities of yourself in the military field since the Korean war.

A. I believe the next connection or perhaps this was even before was on the Vista project, where we asked Dr. Oppenheimer to help us on a particular chapter on which he was better informed than most of the rest of us.

Q. In connection with the work at MIT on continental defense, did you have any association with Dr. Oppenheimer?

A. Yes, I do not recall whether Dr. Oppenheimer was present or took part in the first study, the one I referred to as the Charles study, but after the Lincoln Laboratory got under way, there was a subsequent study the following summer at which Dr. Oppenheimer and I were both present part of the time. The main purpose of this study was to see if the Lincoln Laboratory could somehow be improved, whether they were doing the right things, and whether we were covering all the important aspects of continental defense.

Q. In your observations, do you care to make any comment about the nature of his contributions to these various endeavors that you have described?

A. I think they were very important. It is always hard in a large group like that to know who contributes most. It is a joint effort.

Q. What was your own purpose in all of the military work that you have been doing since Korea, speaking generally?

A. My own purpose is to contribute to avoiding a war if we possibly can. To be somewhat more specific, I think my general thinking was very much influenced by the detailed objectives of the Vista study. * * *

Q. Again speaking generally, how did you give expression to this purpose in the work you did and the policies you advocated?

A. We did that by getting a great deal of help from the military, especially from the people that had fought over there in World War II. * * *

Q. What part did consideration of atomic weapons play in this work?

A. We felt that if atomic weapons——

Mr. ROBB. Excuse me. The witness said "We." Could he identify who he is talking about for clarity?

The WITNESS. I am talking about the Vista project.

Mr. ROBB. I mean the individuals, in the interest of clarity.

By Mr. MARKS:
Q. Would you try to do that?
A. I will be very glad to try to do that. The way the Vista project was operated was that a group of us would be together usually for a couple of hours every morning, and discuss what we were trying to do—trying to formulate what our understanding of the problem was. This group consisted of Dr. DuBridge, the director of the project, Dr. Fowler, and a number of the senior members of the institute faculty, like Dr. Bacher and myself, and also the heads of the various subdivisions of the project. With us would usually be visiting people or people that we could somehow persuade to spend time with us. As an example, I might mention that Dr. Wedemyer spent several weeks with us, and General Quesada spent a good deal of time with us. We kept notes during these discussions and tried to write up what we thought was a sensible program as a result of these discussions.
Q. What about Dr. Oppenheimer?
A. I doubt if he was ever present at any of the daily strategy sessions. If he was, it was only one or two occasions.
Q. What part did he play in the ultimate formulation of the Vista report?
A. He played an important part in expressing our ideas on * * * one particular chapter, called chapter 5. On that chapter he was very helpful.
Q. Who were the other who worked on that chapter?
A. Most of the preliminary planning and writing was done—the discussion was usually between Dr. Bacher, Dr. Christie, and myself, and I think most of the preliminary writing was done partly by Dr. Bacher and part by Dr. Christie.
Q. Did Dr. Oppenheimer help on that?
A. Not in the preliminary stages, but later on in arranging the material and presenting it in the final form, he helped. He made a very important contribution.
Q. If you can put it briefly, what was the essential point of this chapter 5 on atomic weapons that you have been talking about?
A. The essential point was that we felt that without the use of atomic weapons to support ground operation, to destroy mass attacks like we have seen in World War II so often on the Russian front, and like we have seen in Korea. To hold it was necessary to have atomic weapons that could be delivered on short notice, and with high accuracy in all kinds of weather. We felt that with the growing stockpile, it was wise and besides, that it was necessary in order to solve our problem. But in any case, we felt that it was wise to use part of our stockpile or to divide part of the stockpile so that it could be used for this purpose if it was necessary. We believe it was necessary in order to resist aggression.
Q. Did this mean that you were opposing strategic air use of atomic weapons?
A. Not at all. It meant that we felt that * * * it would not be wise to devote all of our stockpile to strategic weapons.
Q. There has already been testimony in these proceedings about a trip to SHAPE which was made by yourself and a number of others in connection with the Vista report in the latter part of 1951. Could you describe the circumstances of that trip, why it was made and what you did?
A. This was at the time when the Vista report was nearing completion. It was in what we considered very close to its final form. Some of the people on the project, especially Dr. DuBridge, felt that it would be very important to discuss the proposals, especially the more radical proposals in tactical change in airport with the staff of the supreme headquarters. Some of the Secretaries, I believe especially Secretary Lovett, thought this was sound before the report went in officially. So the arrangements were made and Dr. DuBridge, Dr. Oppenheimer, and Dr. Whitman and I went over to discuss the Vista report with the planning staff at supreme headquarters.
Q. Why did you take Dr. Oppenheimer along?
A. Because we felt that people would be more likely to believe what he said about what we could do about atomic weapons than what any of the rest of us said. Also, he learns very fast, and we thought we might learn something while we were over there.
Q. Was there any difference between your views and his about the use of atomic weapons?
A. Not that I know of.
Q. As a result of your visit to supreme headquarters toward the end of 1951, did you make any essential changes in your report?
A. There were no essential changes. There were changes in wording, in expression, and the way the material was presented perhaps in some places, but there was no change in the essential idea. * * *

Q. When you speak of using atomic weapons in support of ground troops, are you speaking simply of a different kind of use of atomic weapons, or are you speaking of different kinds of atomic weapons than those which would be used in strategic?

A. They would in general be quite different.

Q. Had they been developed?

A. They had not been developed and tactics for delivering them had not been developed at that time.

Q. What were you advocating then?

A. We were advocating the development and use of weapons that would be suitable for precise delivery at close range from our troops and in all kinds of weather. * * * We felt that by increasing the accuracy you could economize on material. You see if accuracy is poor, you must have a very large explosive to destroy a target. If the accuracy is high you can get along with a much smaller weapon.

Q. I think there has been a suggestion in these proceedings that in the course of going through various drafting stages, the Vista report changed substantially from time to time in respect of chapter 5. Would you comment on that?

A. I hope it improved due to discussions with very many people. The purpose was still the same. There was nothing changed that made it less useful for our purpose. There was no significant change in the methods proposed, as far as I know. I can say I was still happy with the final version of the report. I think it would accomplish our purpose.

Q. How about some of the intervening drafts?

A. There were a number of those. There were not nearly as many as there would have been if I had been writing it. Most people need more than one draft. As a result of discussions some wording was changed, or perhaps some emphasis was changed, but the general purpose and important ideas in that chapter were not changed as far as I know.

Q. Just so that we can be clear, Dr. Lauritsen, will you say again what that essential purpose was?

A. That essential purpose was to try to develop weapons—in particular in chapter 5, of atomic weapons—for supporting ground operations. * * *

Q. In any of these drafts did the essential attitude that you have described with respect to strategic use of atomic weapons as contrasted with the new tactical uses change?

A. I think not. Certainly not to the extent that this was lost sight of because this is what made the Vista proposals as a whole possible.

Q. When you speak of this, I am not sure I know what you mean.

A. I mean the tactical use * * *.

Q. Did you at all times think that was consistent with the maintenance of a strong strategic air force?

A. Yes. I think this is even more true now * * *. But even at that time I think it was sound to start on this development.

Q. Again by this development, you mean tactical use?

A. Weapons that could be delivered with high accuracy in any kind of weather * * *.

Q. Will you turn your mind now, Dr. Lauritsen, to the studies that you referred to having to do with continental defense at MIT. What was the relation in your mind between the efforts that were being made in those studies, what was the relation in your mind between those studies and those efforts and the policies that you advocated with respect to tactical and strategic use of atomic weapons or any kind of weapons?

A. I am not sure I understand your question. I am not sure there is any relation except as far as our overall military effort is concerned. May I say why I was interested in Lincoln? Is that what you are trying to get at?

Q. That is correct.

A. We knew by that time, by the time of the first study on Lincoln, namely, the Charles study, that the Russians had, or very soon would have, a very considerable capability of striking us * * *. We knew they had an air force that was capable of coming over here and delivering those weapons. We felt it was important first of all to get as early warning as we could of a possible attack. Second, that it was important to be able to shoot down as many of these bombers before they reached our strategic airfields and our principal cities. * * *

Mr. ROBB. Could we have a date on this, Mr. Marks, approximately?

The WITNESS. The Charles study? I can certainly find that. It was either 1950 or 1951. * * *

By Mr. MARKS.
Q. Dr. Lauritsen, let me see if I can get at the question that I put to you rather badly a moment ago. Do you believe that we need a strong strategic air?
A. I do.
Q. Do you believe that we need strong developments and strong capabilities in respect of tactical use of atomic weapons?
A. I do.
Q. Do you believe that we need a strong continental defense?
A. Yes.
Q. Taking into account what you know of the relation between scientific development and military affairs, and taking into account what you know of our capabilities and potential, do you regard these three views that you have expressed as consistent or inconsistent?
A. I think they are consistent.
Q. To what extent can you say of your own knowledge that the views you have just described are similar to or different from the views that you know Dr. Oppenheimer to hold?
A. I believe he agrees with me. He has worked hard on all of these three things. I think his purpose has been the same as mine. He may have sometimes thought of it differently in different details. The aims have been the same, I am convinced, and we have agreed in general.
Q. Dr. Lauritsen, what opinion do you have about Dr. Oppenheimer's loyalty and character? By loyalty I mean loyalty to the United States.
A. I have never had any reason to doubt it.
Q. Do you think you could be mistaken about this?
A. I suppose one can always be mistaken, but I have less doubt than any other case I know of.
Q. Less doubt than in any other case?
A. Than in any other person that I know as well.
Q. Do you know many people better?
A. Not many. I suppose I know my own son better, but I don't trust him any more.
Q. To what extent would you trust Dr. Oppenheimer's discretion in the handling of classified information, restricted data?
A. You are referring now to recent years when he understood these problems, I hope. In that case I think I would trust his discretion completely. I think in the early thirties very few of us knew anything about discretion and were not very conscious of security. Whether he had been indiscreet at that time, I don't know. It is possible. It is possible I have been indiscreet. But I am sure after he understood what security meant, and what was involved, that he has been as discreet as he knew how.
Q. What do you mean by as discreet as he knew how?
A. As discreet as it is possible to be and try to get some work done.
Q. Do you have any idea about whether your views about the needs for and the possibilities of being discreet are any different than his?
A. I think they are no different now, certainly.
Q. Let us take the period commencing in 1944, when you went to Los Alamos. Is that the span of years you are talking about?
A. During that period this would apply. At that time he knew the importance of the information we had.
Mr. MARKS. That is all, Mr. Robb.
Mr. GRAY. I think it would be well to break for a few minutes at this point.
(Brief recess.)
Mr. GRAY. Mr. Robb, will you proceed.

CROSS-EXAMINATION

By Mr. ROBB:
Q. Doctor, do I understand that you have known Dr. Oppenheimer both professionally and socially?
A. That is correct.
Q. Have you visited him from time to time at his ranch in New Mexico?
A. I have visited him I think twice.
Q. When was that, sir?
A. About the middle thirties—1935 or 1936, I believe.
Q. Do you also know Dr. Oppenheimer's brother Frank?
A. I do.
Q. When did you meet him, sir?

A. I believe I met him for the first time at the ranch in 1935 or 1936. I may have seen him once before, but I am not quite sure.
Q. Was he on the faculty at Cal. Tech.?
A. He was a graduate student.
Q. Under you?
A. Yes.
Q. Did you get to know him pretty well, too?
A. I got to know him quite well in the laboratory.
Q. And you saw him on the ranch, also, I take it?
A. Yes.
Q. Did you know him at Los Alamos?
A. Yes, I did.
Q. Have you seen him since then very frequently?
A. Not frequently. I have seen him. Most recently last year at a meeting of the physical society in Albuquerque, N. Mex.
Q. Up until the end of it, did you have any reason to bel'eve that Frank was a Communist or had been a Communist?
A. No, I had no reason to believe that until he made that statement himself.
Q. What would you say about Frank's loyalty?
A. I have no reason to doubt his loyalty.
Q. And his character?
A. His character is very good.
Q. You would make about the same answer about him that you do about Dr. Oppenheimer?
A. Yes, I would think so. His judgment was perhaps not as good as Dr. Oppenheimer's.
Q. Yes, sir. I notice that you made some little distinction between Dr. Oppenheimer's present appreciation of security and his appreciation in the past of security.
A. I think that applies to all of us.
Q. Yes, sir. You suggested that there might have been some change in Dr. Oppenheimers' attitude on those matters.
A. On how important you think it is, how seriously you take it.
Q. Would you care to tell us, Doctor, when you think that change took place?
A. I think we all learned about it during the war.
Q. You think Dr. Oppenheimer learned about it during that period?
A. That would be my judgment. I think this was true of most of us that had had little to do with military things until that time.
Q. I see. Did you know many of Dr. Oppenheimer's friends?
A. I knew of his friends in Pasadena and some of his friends in Berkeley.
Q. That is up until the war years, is that right?
A. That is right.
Q. Did you know a man named Frank K. Malina at Pasadena?
A. Yes, I did.
Q. Who was he, Doctor?
A. He was first, I believe, a graduate student and later a research fellow in the aeronautics department on a special project that had to do with rocket developments.
Q. Was he working under you?
A. No, he was not. When I first went to Pasadena I knew of his work. I did not know of him personally. I hoped to get him to undertake the rocket work at Pasadena. However, we did not agree on what should be done, so I dropped the subject and went to Pasadena myself to do the work. We had no connection with their development, which incidentally resulted in the so-called assisted takeoff system which is not a weapon for a method for getting aircraft with overload or from too short trips.
Q. What they called JATO?
A. That is right.
Q. Did you suspect at any time that Frank Malina had any Communist connections?
A. I had no way of knowing. I did not know him personally.
Q. You did not suspect that?
A. I had no way of knowing. I did not know him socially. I never have associated with him. I have only talked with him a few times when I tried to get him interested in this project.
Q. At least you knew him well enough to suggest that he come help you on the project?

A. That is right, I knew of his work.
Q. Did you suspect that he had any Communist connections?
A. I had no idea, no.
Q. Did you know a man at Pasadena named Martin Summerfield?
A. I had a student at one time in a class, not in my laboratory, by that name. I believe he later worked at the jet propulsion laboratory but I have had no connection with him since he was a student.
Q. That jet propulsion laboratory was a part of Cal. Tech.?
A. It is a contract with the Army Ordnance Corps that is administered by Cal. Tech., but it has no other connection with Cal. Tech. It has the same relation as Los Alamos has with the University of California.
Q. You never had any suspicion, of course, that Martin Summerfield had any Communist connections?
A. I had no way of knowing.
Q. Did you know Dr. Thomas Addis at Berkeley?
A. No.
Q. Did you know a David Adelson at Berkeley?
A. No.
Q. Did you know a couple named Henry Barnett and Shirley Barnett?
A. At Los Alamos?
Q. Yes.
A. Yes.
Q. Who were they?
A. He was a doctor, I believe.
Q. Who was she?
A. I think she was a secretary.
Q. To whom?
A. To Dr. Oppenheimer, I believe, or assistant secretary.
Q. Did you ever have any reason to suspect that they had any Communist connections?
A. No.
Q. Did you ever suspect that?
A. No. I would have no way of knowing.
Q. Did you know a man named David Bohm at Berkeley?
A. No, sir.
Q. Did you ever meet a woman named Louise Bransten?
A. Not that I know of.
Q. Did you ever know a man called Haakon Chevalier?
A. No, sir.
Q. You never met him?
A. No, sir.
Q. Did you know a man named Robert Raymond Davis at Los Alamos?
A. I can't recall. I can't be sure. Davis—the name sounds familiar. Was he a physicist at Los Alamos?
Q. I believe so. In all events, you didn't know him well if you knew him?
A. I certainly did not know him well. I have no recollection of knowing him.
Q. Did you ever know a man up at San Francisco named Isaac Folkoff?
A. No, sir.
Q. Did you know a man at Berkeley named Max Friedman?
A. No, sir.
Q. Did you know David and Francis Hawkins at Los Alamos?
A. I knew David Hawkins, not Francis Hawkins.
Q. Who was David Hawkins?
A. David Hawkins was, I guess you would call him a historian. When I knew him he was writing the history of the project.
Q. Did you know him before he began to write the history of the project, Doctor?
A. I knew him probably from the time I joined Los Alamos, that is, from September 1944.
Q. You didn't join until September 1944?
A. That is right.
Q. Do you recall what Hawkins was doing then?
A. No, I don't. I think he was already then thinking about this history, but I am not quite sure what he was doing.
Q. Did you ever suspect that he ever had any Communist connections?
A. No, sir, I did not know him personally.
Q. Did you know a man at Berkeley named Alexander S. Kaun?
A. No, sir.

Q. Did you know a man at San Francisco named Rudy Lambert?
A. No, sir.
Q. Did you know a man at San Francisco named Lloyd Lehmann?
A. No, sir. If any of these people are physicists it is quite possible I have met them at one time or another, but I have no recollection of knowing them.
Q. When I say San Francisco, I mean the area of San Francisco to include Berkeley.
A. Yes.
Q. Did you know a man named Giovanni Rossi Lomanitz up there?
A. No, sir.
Q. Did you know a man in San Francisco named Kenneth May?
A. No, sir.
Q. Did you know a man named Philip Morrison at Los Alamos?
A. Yes.
Q. What was he?
A. He was a theoretical physicist.
Q. Did you come in contact with him frequently?
A. I saw him quite frequently. In fact, I knew him before I went to Los Alamos. He was a student of Dr. Oppenheimer who occasionally while he was a graduate student came during the spring term to Pasadena with Professor Oppenheimer.
Q. Did you ever suspect he had any Communist connections?
A. No, sir.
Q. Did you know a man at Los Alamos named Eldred Nelson?
A. I knew Nelson, but I do not recall him at Los Alamos. I recall him the year after. He was in Pasadena the year—in late 1945 and 1946.
Q. Did you know anything about his background and associations?
A. Not any.
Q. Did you know a man at Los Alamos named Bernard Peters, and his wife, Hannah Peters?
A. Is he the physicist who later was at Rochester?
Q. Yes, sir.
A. I did not know him at Rochester, but I met him since the war. I mean I didn't know him at Los Alamos.
Q. Did you know him at Berkeley?
A. No, sir.
Q. Have you ever suspected that he has ever had any Communist connections?
A. I heard that after the war.
Q. After the war?
A. Yes; I did not know him before.
Q. How did you happen to hear that?
A. The way I heard about it was that 2 years ago the physical society had a meeting in Mexico City, and I was president—no, I was elected—I was vice president of the physical society. Dr. Rabi was the president of the physical society. He was at that time in Italy at UNESCO meeting, so I was requested to represent the physical society in Mexico City. There was an invited paper on the program to be given by a physicist who had worked with Peters. It was joint work that was to be presented. This invited speaker died in the meantime or was killed in an accident, I believe, and some of Peters' colleagues requested that the physical society should appoint him the invited speaker to give this paper. They referred to me as the highest official in the country at the time.
Q. The senior officer present.
A. That is right. However, I referred it back to the secretary of the society, who habitually handled all of those things. So I avoided the decision. But this was the first time as far as I remember that I had met Dr. Peters.
Q. How did the Communist business come into it?
A. It came in because someone told me to be careful about this, because he might not be able to get permission to go into Mexico. So that is the reason I did not want to invite him.
Q. Did he appear?
A. He did, but be appeared without official invitation from the society.
Q. That was 2 years ago?
A. I think so, 2 or 3.
Q. Did you ever know a man up in the San Francisco area, named Paul Pinsky?
A. No, sir.
Q. Did you ever hear of him?

A. I think I have heard the name, but it doesn't mean anything to me, and I certainly do not known him personally.

Q. Did you ever know a man in the San Francisco area named William Schneiderman?

A. No, sir.

Q. Did you ever hear of him?

A. I have heard the name.

Q. It doesn't mean anything to you?

A. No, sir.

Q. Did you know Robert and Charlotte Serber at Los Alamos.

A. Yes, sir.

Q. Do you know anything about their political background?

A. No. There again I had known them long before the war. Dr. Serber was again one of the students that came down during the spring term with Dr. Oppenheimer. So I got to know them long before the war, and saw quite a bit of them at Los Alamos.

Q. Did you ever suspect that Mrs. Serber had any Communist connections?

A. No; I did not suspect that she had Communist connections, I would say that I thought she was again what I would call an optimistic liberal.

Q. That is as far as your suspicion, if you can call it such, went?

A. That is right.

Q. Did you ever know a man named Joseph Weinberg?

A. In Pasadena or in Berkeley?

Q. In Berkeley or Pasadena.

A. No. I know only of what I read in the papers. He is Scientist X, is that not correct? I do not know him. As far as I know, I have never met him.

Q. You don't know anything about him?

A. No.

Q. Do you know anything about the organizations to which Dr. Oppenheimer may have belonged in the late thirties and early forties?

A. No; I do not. I assume you do not refer to the physical society or the National Academy?

Q. No, sir. I mean the other organizations.

A. No.

Q. Doctor, you said that Dr. Oppenheimer played an important part in expressing the ideas of your group in chapter 5 of the Vista report. Would you tell us just what that part was that Dr. Oppenheimer played?

A. I think you know that Dr. Oppenheimer is very articulate.

Q. Yes, sir.

A. He is very good at expressing ideas clearly and understandably.

Q. Yes.

A. This is primarily what I had reference to.

Q. You mean he drafted that part of the report?

A. There were several drafts before he came out the first time, and then there were many discussions afterward. The wording was modified more or less continuously until the final version was accepted.

Q. Dr. Oppenheimer's part was in preparing that final draft?

A. The final draft and possibly some intermediate drafts where the wording was somewhat different, perhaps the emphasis somewhat different, but as far as I know, the main theme was the same.

Q. Was that final draft presented at a meeting out in Pasadena?

A. Yes. In fact, even earlier drafts were presented to the whole group that was working in the field.

Q. There certainly came a time when the finished product was presented to the meeting, is that right?

A. That is right.

Q. Do you recall who it was who presented it, Doctor?

A. I believe Dr. DuBridge presented it.

Q. Referring to that draft as it was prepared by Dr. DuBridge, do you remember if that said anything about thermonuclear weapons?

A. They may have been mentioned, but they were not part of our proposal for close support, for Army support.

Q. Would you explain that a bit to me, Doctor? Why weren't they?

A. In the first place, at that time the feasibility of hydrogen weapons had not yet been established, and we did not feel that this could be part of our proposal. Development work was still going on and the investigation of the technical feasibility of a hydrogen bomb was still going on.

Q. This was in May 1952?
A. No; this was in November 1950, I believe, was it not?
Q. We were both wrong. It was November 1951.
A. November 1951, that is correct. That is the time that we discussed the version that we took to Europe with us. That was November 18, 1951, I believe.

* * * * * * *

By Mr. ROBB:
Q. Doctor, do you remember in the spring of 1952, specifically in May, when there was discussion about so-called Ivy Shot?
A. I remember the Ivy Shot; yes.
Q. The Ivy Shot was supposed to be a test of some thermonuclear device.
A. That is right.
Q. Did you take any position on whether or not that should be postponed or canceled?
A. I thought it was an important time to see if some agreement could be reached for avoiding future tests, or if there was some way of reaching agreements on control of weapons of that character. I thought a study should be made, and consideration should be given to the possibility of making use of this important event to accomplish this purpose.
Q. Was it your position that the Ivy test——
Mr. MARKS. Mr. Robb, what was the date of the Ivy test?
Mr. ROBB. It was in the fall of 1952, wasn't it, Doctor?
The WITNESS. That is my belief and recollection.
Mr. ROBB. We are talking now about the spring of 1952.

By Mr. ROBB:
Q. Was it your position in the spring of 1952 that the Ivy test should not take place?
A. It was my position that some effort should be devoted that summer to studying the question of whether we could take advantage of this possibility of trying to reach some sort of agreement on the limitation of the use of thermonuclear weapons.
Q. Agreement with whom?
A. With the Russians.
Q. Did you oppose the Ivy test in the absence of such an effort to make an agreement?
A. I did not oppose it in any official capacity. I thought it was very unfortunate.
Q. Were you opposed to the development of the so-called H-bomb?
A. Yes.
Q. Had you finished?
A. I have finished. I think I have said all I can say unless we go into classified material.
Q. Were you opposed to the development of the H-bomb as of the spring of 1952?
A. You refer to a hydrogren bomb, is that correct?
Q. Yes, sir.
A. I thought it would be very unfortunate to devote an effort to that that would be so large that it would interfere with the weapons that we have discussed earlier, namely, the weapons that the Vista study indicated were needed for ground support and for resisting aggression in Western Europe.
Q. Doctor, I don't want to be unfair with you, but am I to conclude from your answer that you were opposed to the development of the H-bomb?
A. I was not opposed to a study of the technical feasibility of an H-bomb, That was the question that was being considered at that time, I believe.
Q. In May 1952?
A. I think so. I believe this was the President's directive, that a study of the technical feasibility should be made. This I was in favor of.
Q. Doctor, when was the President's directive?
A. I believe there were two directives, one on the 30th of January, and the second, an official newspaper announcement, on the 31st of January, 1950.
Q. That was 2 years before May 1952.
A. That is right.
Mr. ROBB. I think that is all I care to ask, Mr. Chairman.
Mr. GRAY. Dr. Lauritsen, do you feel as of today a member of the Communist Party, that is, a man who is currently a member of the Communist Party, is automatically a security risk?

The WITNESS. I think so.
Mr. GRAY. You don't have any question in your mind about that, do you?
The WITNESS. No; not if I can believe what I have been told about the Communist Party, and I do believe it.
Mr. GRAY. In testifying earlier, I think you said you considered Dr. Frank Oppenheimer loyal in every respect, and with no reservations about this character or trustworthiness?
The WITNESS. That is right.
Mr. GRAY. Are you aware that Dr. Frank Oppenheimer has stated at an earlier period in his life he was a member of the Communist Party?
The WITNESS. Yes, sir I am aware of that now.
Mr. GRAY. But still you say you have no reservations about his loyalty or character?
The WITNESS. No, I have not.
Mr. GRAY. Would you explain to the Board why you conclude that you would trust him with any secret, which I believe is the effect or import of what you say, today, although you believe that a member of the Communist Party is automatically a security risk? Would you explain that?
The WITNESS. I believe he has resigned from the Communist Party, and he is no longer under the discipline of the Communist Party. I believe he was cleared for work on war projects during the war and including nuclear weapons work.
Mr. GRAY. This was not after it was known he was a member of the Communist Party?
The WITNESS. This I have no way of knowing. I do not know what turned up in his investigation.
Mr. GRAY. Would you feel that if it had been known at the time that he was a member of the Communist Party he should have been cleared for war work?
The WITNESS. If he had not resigned previously, I would certainly not recommend his clearance. If he had resigned previously because he no longer wanted to be a member of the Communist Party because he had found out that the Communist Party was not what it appeared to be, then I would still be inclined to say that he would be reliable.
Mr. GRAY. Today on classified projects for which you have some responsibility, including a security responsibility, if a man comes to the project seeking employment, who is known to you to have been a former member of the Communist Party, would you employ him simply on his statement that he no longer was a member of the Communist Party?
The WITNESS. No; not without appropriate clearance through official channels.
Mr. GRAY. What would your recommendation be?
The WITNESS. If he had resigned from the Communist Party when he found out what the purpose of the Communist Party really was, and had been a member only as long as he had been under misconceptions about these things, then I would not hold that against him.
Mr. GRAY. You would accept as evidence of that his own statement?
The WITNESS. Not necessarily. I think some people you can trust, and others you can't trust. I think it depends on what other activities he has been involved in and what he has been doing. In Frank's case, I think he demonstrated that he wanted to work for this country. Other people perhaps have not demonstrated that. I think there is a great deal of difference between being a Communist in 1935 and being a Communist in 1954. I don't think very much of us knew, I certainly did not know what the Communist Party was up to and how it operated.
Mr. GRAY. Let me ask this question: Would it be a rather accurate summary of at least parts of your testimony to say that you never really understood very much about the Communist Party or its workings?
The WITNESS. That I did not?
Mr. GRAY. That is right.
The WITNESS. At that time.
Mr. GRAY. Because each of these people that Mr. Robb asked you about, who I think were later identified as having been in the party or close to it, you testified that this was something you had no knowledge or suspicion about.
The WITNESS. That is right.
Mr. GRAY. Have you ever known anybody that you thought was a Communist?
The WITNESS. Not personally, no.
Mr. GRAY. So membership in the Communist Party is something you really have not concerned yourself with in any way?

The WITNESS. That is right.
Mr. GRAY. Did you know Mr. Fuchs?
The WITNESS. I knew him at Los Alamos.
Mr. GRAY. You didn't suspect he was a Communist?
The WITNESS. No. I did not know him well. My contacts with him were limited to our having lunch together in the same dining room occasionally. Apart from that, I did not know him.
Mr. GRAY. But you had no more suspicion of him than you did of the others whose names have been mentioned here?
The WITNESS. No.
Mr. GRAY. If asked at the time would you have said that he was loyal to the war effort?
The WITNESS. I would not have said it. I did not know him well enough to have an opinion. I had nothing to do with his work.
Mr. GRAY. If he worked very hard at Los Alamos and contributed effectively, that is in a sense a demonstration of his loyalty?
The WITNESS. I would say it would be one in his favor, but perhaps not conclusive.
Mr. GRAY. In the light of developing facts.
The WITNESS. That is right. I could not have testified against him if I had been asked to because I did not have the information.
Mr. GRAY. You would not consider yourself an expert on communism in any sense of the word?
The WITNESS. No.
Mr. GRAY. Have you any questions, Dr. Evans?
Dr. EVANS. Yes. Doctor, you said you were born and raised in Denmark.
The WITNESS. That is correct.
Dr. EVANS. Would you tell us just where you were educated?
The WITNESS. I studied in a technical school called Odense. I graduated from there in 1911. Then I studied at the Royal Academy of Arts in Copenhagen subsequently.
Dr. EVANS. You got a degree from there?
The WITNESS. I got a degree from this technical school, what probably here would be called structural engineering. I think that would be the nearest approach to it. I was at that time planning to be an architect.
Dr. EVANS. You are not a Communist?
The WITNESS. I am not a Communist, no.
Dr. EVANS. Have you ever been what is called a fellow traveler?
The WITNESS. No, sir.
Dr. EVANS. Have you belonged to any of these subversive organizations that appear on the Attorney General's list?
The WITNESS. No, sir.
Dr. EVANS. Doctor, do you believe that a man can be perfectly loyal to his country and still be a security risk?
The WITNESS. I suppose so, yes.
Dr. EVANS. You have faith in Dr. Oppenheimer's discretion, you say?
The WITNESS. I do.
Dr. EVANS. Have you ever been approached for security information?
The WITNESS. Yes.
Dr. EVANS. Men have approached you?
The WITNESS. FBI, yes.
Mr. GRAY. I want to make sure the witness understands this question.
Dr. EVANS. I don't mean the FBI.
The WITNESS. I am sorry.
Dr. EVANS. I mean somebody that might be a Soviet agent.
The WITNESS. No.
Dr. EVANS. You have never been approached?
The WITNESS. Not to my knowledge, no.
Dr. EVANS. You are not always able, Doctor, to tell these Communists when you meet them, are you?
The WITNESS. That is right.
Dr. EVANS. It apparently is not easy to recognize them.
The WITNESS. That is right.
Dr. EVANS. It is particularly apparent for a professor not to know whether people are Communists, is that true?
The WITNESS. I think it is true of anybody. I don't think professors are any better or any worse than any other people.

Dr. EVANS. I don't know, Doctor, since I have been on this board. That is all.
Mr. GRAY. Mr. Marks.

REDIRECT EXAMINATION

By Mr. MARKS:

Q. Dr. Lauritsen, looking back over the span of the last 25 years, do you know Robert Oppenheimer or Frank Oppenheimer better?
A. Robert much better.
Q. Would you explain that?
A. We had more professional things in common and were if not of the same age, at least more nearly the same age. It was only reasonable both being members of the faculty that I should know him better. Also, I have known him a longer time and a greater fraction of the time.
Q. You said you don't consider yourself an expert on communism.
A. No; I don't.
Q. Do you consider yourself an expert on Dr. Oppenheimer's trustworthiness?
A. No, I don't know what an expert on that is, or how you get to be an expert on that. I only know what my own feelings and belief are, and it is very deep.
Q. There was a long list of names read to you. Some of them you said you didn't know.
A. As far as I know, as far as I remember.
Q. Some of them you described, such acquaintance as you had with them. Are there any people on that list that was read to you by Mr. Robb with respect to whom your knowledge was as great as that of Dr. Oppenheimer?
A. No; I think not.
Q. Considering the fact, Dr. Lauritsen, that you extensively engaged in military work of a top secret nature, would you consider it a departure from discretion if you were to visit with Dr. Morrison today?
A. No.
Q. Would you consider it a departure from discretion if you visited with the Serbers?
A. No.
Q. Would you visit with them?
A. I would like to very much.
Q. Would you say the same of Dr. Morrison?
A. I know him very little. I know the Serbers fairly well. I have no knowledge that they are Communists.
Mr. GRAY. I don't know whether you had completed your questioning.
Mr. MARKS. I think so.

RE-CROSS-EXAMINATION

Q. How about Peters? Would you visit him?
A. I don't know him personally, but I feel that it would be very wrong for the Physical Society to throw him out of the society. It is not a political society.
Q. No, but from what you have heard about Peters, would you feel that you were being discreet to associate with him?
A. I really don't know enough about him to be sure about that.
Q. You couldn't be sure either way?
A. I don't have enough information.
Q. One question that I overlooked, Mr. Chairman. Do you recall, Doctor, anything in the Vista report, either in the draft as it was read in final form at Pasadena or later, about an announcement by the United States that no strategic atomic attack would be made against Russia unless such an attack were first started by Russia, either against the Zone of Interior or against our European Allies, or something of that sort?

* * * * * * *

Q. Was there anything about any announcement to that effect being made by the United States? Was there any recommendation?
A. Was a recommendation in the Vista report?
Q. I am asking you to search your recollection for it.
A. I think it is possible that we pointed out that we felt that the tactical support should be available if such a statement was made, that we would not use the strategic capability except in retaliation.
Q. Was that your view?
A. Yes.
Q. Was it Dr. Oppenheimer's?

A. I believe so.
Q. Did your views and Dr. Oppenheimer's pretty generally coincide during this period?
A. I think so.
Q. Was that true in May of 1952, also?
A. In May of 1952?
Q. When you were talking about the Ivy test.
A. I think so, yes.
Q. Did you discuss that with him?
A. Yes.
Mr. ROBB. I think that is all.
Mr. GRAY. I have just one other question, Dr. Lauritsen. Would it be fair for me to assume that your view with respect to a Communist, former Communist, and so forth, is that you really prefer not to have to make these determinations, and you would rely on the security people for it?
The WITNESS. Yes.
Mr. GRAY. In fairness, isn't that your statement, that you would just prefer not to have to go into it?
The WITNESS. That is right.
Mr. GRAY. I don't want to put my statement in your month.
The WITNESS. No, I agree this is the point of view. We have machinery for handling these cases, and I think it would be quite wrong for me to make the decisions.

REDIRECT EXAMINATION

By Mr. MARKS:
Q. Dr. Lauritsen, accepting the view that you have just described, that we have machinery for deciding the kind of issues that the Chairman has mentioned about the Communists, I would like to have you distinguish between the operation of that machinery in the large, the operation of that machinery in general, and the opinion that you hold with respect to Dr. Oppenheimer, and I would like to ask you whether you have any hesitation in making the judgment on this matter with respect to Dr. Oppenheimer—personal judgment—that is, a personal judgment with respect to his character, loyalty, discretion.
A. Would you say what the question is?
Q. My question, is, bearing in mind your view that it is appropriate for the machinery of government to determine questions of who is and who is not a Communist, who is and who is not a security risk, I would like now to ask you whether in view of that opinion you have any hesitation in expressing what your own convictions are about Dr. Oppenheimer?
A. I think I have already done so. I take it we are in the middle of the operation of this machinery, and I have made statements that I would have no hesitation to recommend complete clearance.
Q. I was not asking you that.
A. I thought that is what you asked.
Q. I think you answered the question.
Mr. GRAY. I think his earlier testimony pretty adequately answered that. I don't want to keep him from saying it again, but I think it is perfectly clear.
Mr. MARKS. No, I just wanted to be sure that there was a distinction. Just one more question.

By Mr. MARKS:
Q. The testimony that has already been given by others here suggests to me that it is not inappropriate for me to ask the question that I am about to ask. If, however, the answer to it in any way involves classified information, you will have to say so.

* * * * * * *

A. May I state it a little differently? It is a little hard to answer directly.
Q. Answer it as best you can, if you can without getting into classified material.
A. The best I can say is that from what I know about the discussions that have appeared in the newspapers, the discussion has been on the basis of whether you are for or against a crash program on hydrogen bombs. This expression was not used as far as I know in any directive by the President. The President's directive did not mention crash program. It did not mention hydrogen bomb. I believe it mentioned the order of investigation of the technical feasibility of thermonuclear weapons.

* * * * * * *

Does that answer the question? I was never opposed to carrying out what I understood to be the President's directive but it has been discussed in very different terms, it seems to me.

Mr. MARKS. I think that is all.

Mr. ROBB. That is all.

Mr. GRAY. Thank you very much, Dr. Lauritsen.

The WITNESS. Thank you.

(Witness excused.)

Mr. GRAY. Who is your next witness?

Mr. GARRISON. I told Mr. Buckley that I arranged for him to testify at 2 o'clock.

Mr. GRAY. Do you have anybody here now?

Mr. GARRISON. We have Dr. Zacharias here.

Mr. GRAY. Is he likely to be a long witness?

Mr. MARKS. I hope not.

Mr. GRAY. Could we get started with Dr. Zacharias?

Mr. GARRISON. The problem you will recall about Mr. Buckley——

Mr. GRAY. Yes, I would say in this case because of Mr. Buckley's health and circumstances of his being here, if we don't finish with Dr. Zacharias, we will interrupt his testimony. But I would like to get ahead with it if we can unless you object to that.

Mr. ROBB. No, indeed.

Mr. GRAY. Dr. Zacharias, do you wish to testify under oath? You are not required to do so. However, I think I should point out to you that every witness who has appeared to this point has chosen to do so.

Dr. ZACHARIAS. Yes, I do.

Mr. GRAY. Would you be good enough to stand and raise your right hand. What is your full name?

Dr. ZACHARIAS. Jerrold R. Zacharias.

Mr. GRAY. Jerrold R. Zacharias, do you swear that the testimony you are to give the board shall be the truth, the whole truth, and nothing but the truth, so help you God?

Dr. ZACHARIAS. I do.

Whereupon, Jerrold R. Zacharias was called as a witness, and having been first duly sworn, was examined and testified as follows:

Mr. GRAY. Will you be seated, please, sir?

It is my duty to remind you of the existence of the perjury statutes, and the fact that there are penalties with respect to violation of those statutes. Do I need to review those with you, Doctor?

The WITNESS. No, sir.

Mr. GRAY. I should like to request that if in the course of your testimony it becomes necessary for you to refer to or disclose restricted data, you notify me in advance so that we may take certain appropriate steps in the interest of security.

Finally, I should say to you, as I say on behalf of the board to all witnesses, that we consider this proceeding a confidential matter between the Atomic Energy Commission and its officials on the one hand, and Dr. Oppenheimer, his representatives and witnesses on the other. The Commission is making no releases with respect to these proceedings, and we express the hope that the witnesses will take the same view.

Mr. MARKS. May we pause just a minute. I am not sure you expressed the hope to Dr. Lauritsen.

Mr. GRAY. Yes, I did.

Mr. MARKS. I am quite sure he understands that.

Mr. GRAY. No, I am not sure that I did.

Mr. MARKS. I think he understands in any case but I just wanted to be sure.

Mr. GRAY. Will you proceed.

DIRECT EXAMINATION

By Mr. MARKS:

Q. What is your present position?

A. I am professor of physics at the Massachusetts Institute of Technology, and director of the laboratory of nuclear science there.

Q. What connections have you had with military work commencing with World War II. State this very briefly, if you will.

A. I worked during World War II primarily in the radiation laboratory at MIT. For a shore period during that time at the Bell Telephone Radar Laboratory at Whippany, N. J. I spent about 4 months at the end of the war, just overlapping VJ-day, at Los Alamos. Then I spent a fair amount of time on a number of study projects for the military and for things associated with the military.

Q. What are those projects?

A. The first one was a study of nuclear powered flight sponsored by the Atomic Energy Commission, a project headed by Walter Whitman, and known as Project Lexington. I think it was probably the first of the things that we call summer studies.

The second one was Project Hartwell, which I directed, * * *.

Then Project Charles, which was a study at MIT, headed by F. W. Loomis. I was the associate director of that study.

* * * * * * *

Then out of that study there grew a laboratory at MIT called the Lincoln Laboratory, * * *.

I was for a time associate director of the Lincoln Laboratory in its first year or so.

Q. When was that?

A. The laboratory started in about June, July or so of 1951. I was involved for a short time—not very long—in Project Vista, which I am sure has entered into these discussions before. Then as a member of Project Lincoln, I was in charge of a study on defense of the North American Continent, a project that had no name. We were trying not to let it be a project, but it got to be known as the summer study of Project Lincoln. That was in the summer of 1952. I think that is about it.

Q. In connection with this last project that you have described, did you personally make any special contribution to it that you can describe without getting into classified material?

A. I was director of the project, and therefore involved in almost all phases of it. I think without getting into rather involved technical discussion which might turn out to be classified, that is, a frank discussion of which might go off into classified channels, I think it would be best not to be too specific about personal contributions.

I would be glad to if necessary.

Q. How long have you known Dr. Oppenheimer?

A. As I remember it, I met him when he was a student abroad. It was in the summer of 1926 or 1925. It was the summer of 1926 at a meeting in the University of Leyden in Holland, and talked to him a bit.

Q. What is your association with him since that time?

A. Since that time I would say it has been very scanty up until my working at Los Alamos. However, I did meet him again in 1940—the summer of 1940—Norman Ramsey and I met him at Seattle, and together we drove south to San Francisco and Dr. Oppenheimer joined us.

Q. Putting to one side such casual associations, what is the period during which you have had close associations with him?

A. The close association, I would say, substantially started at Los Alamos in July of 1945.

Q. Since that time have you had frequent occasion to work with him?

A. Yes, and mostly on things that involve the military. To some extent on general policies, regarding the support of science.

Q. Regarding the support of science where?

A. Support of science in this country generally. Let us call it financial support of science and the trends that physics takes.

Q. Just to be sure I understand you, you are speaking now, I take it, about two different aspects of your postwar association with Dr. Oppenheimer?

A. Yes.

Q. One is military.

A. One is you might say military in matters of national policy and that sort of thing, and the other has to do with support of—let us be specific—of nuclear physics. It being fairly expensive, there has been a fair amount of discussion about how such an expensive thing can be properly supported, and the directions on which it ough to go. On those subjects we have had considerable discussion.

Q. Did you have much contact with Dr. Oppenheimer in connection with Vista?

A. Not really. I saw him there. I was at that project only about 2 or 3 weeks. He was there at the time that we were working on substantially different things, and although I saw him there, I wasn't very close to the particular thing he was working on.

Q. How about Project Lincoln?

A. On Project Lincoln, I think the most important thing to mention would be the study during the summer of 1952. Could I go into a little detail on that?

Q. Yes, bearing in mind the Chairman's caution about classified information.

A. I think the story of that summer study is probably worth putting into the record, and I will try to do it as quickly as I can, because it has been to a certain extent a moderately controversial thing.

* * * *

Q. Let me interrupt you there to ask you if you can say, was there any policy with respect to continental defense before the summer study of which you are speaking?

A. Surely. The Air Force had then and has considerable interest in continental defense, and was going along certain technical lines, and with the buildup of a certain amount of counter force for the protection of the continent. In fact, the Lincoln Laboratory itself, which was by then a year and a half to 2 years old, is a laboratory that is under contract to the Air Force. It is a joint Army, Navy, Air Force laboratory, but the Air Force holds the contract and is the major contributor.

Q. I interrupted you when you were about to tell the story of what happened as a result of the summer study.

A. The Lincoln Laboratory set up to work on technological and technical aspects of continental defense. In fact, air defense of any sort. Just prior to the summer of 1952, Dr. Lauritsen and I had a long discussion about the trend in continental defense, whether the buildup was great enough, * * *.

Dr. Lauritsen and I decided that it might be a very good thing if we looked into these technical, military, and economic questions again during that summer. We decided that we should talk this over with certain others whom we knew very well. First of all, Dr. Hill, who was then the Director and is now the director of Lincoln Laboratory. We decided we would talk it over with Dr. Oppenheimer and Dr. Rabi.

Q. Why did you talk to Dr. Oppenheimer?

A. In my experience it is always profitable to talk to Dr. Oppenheimer. His head is so clear on questions of this sort that when you flounder for months to try to formulate your ideas, you get to him and he can listen and help state clearly what you and he and others have decided is the germ of what you are thinking. This is true in all of my contacts with Dr. Oppenheimer on this kind of question.

We decided, then, that it would be a good idea to start such a study, that Dr. Oppenheimer, Dr. Rabi, and Dr. Lauritsen agree to work on this study in part. The reason is that it is very difficult to recruit men of stature, men of ability into any kind of study. They are doing what they think is adequate and they have some sense of urgency but they also have the feeling, why don't we let somebody else do the work.

Dr. Hill, who is the director of the Lincoln Lab, and I felt that if Dr. Oppenheimer, Dr. Rabi, and Dr. Lauritsen agreed to work on this in part, that it would be easier for us to recruit a number of very brilliant people and some of the more experienced people to do the job. Indeed, that turned out to be true. So that directly within the Lincoln Laboratory and sponsored by the Air Force, as I say, we set up a study.

We came out with three recommendations, one of which I would like to say something about. And the other two I will just mention and not go into more deeply because of security classification.

* * * *

Q. Let us not go into those matters, Dr. Zacharias. You spoke of resistance to these ideas in some quarters in the military and civilian circles. Has this resistance persisted?

A. I am not sure that I said there was resistance. However, I am sure in the newspapers it is clear that continental defense is a subject that has a lot of emotion in it. * * * I don't want to bring in the confusion of post hoc ergo propter hoc, but it is true just before the time of the study and before the discussions that followed it, there was not a strong policy, and there now is a strong policy.

In other words, I don't want the summer to be credited with change of national policy no matter what I happen to think when I am by myself.

Q. Did you conceive the recommendations of this summer study that you have referred to as being inconsistent or to be in conflict with any national policy with respect to what is described as strategic air policy?

A. The only conflict is of a funny sort. Let me begin it this way. Certainly part of any defensive system in this country is what we call our offensive plan. One doesn't think of protecting the continent by conventional defensive means. That is, destruction of enemy bases is just as important and every bit as important as local defense. It was the feeling of a number of us who worked on this summer study that the amount of money and the amount of effort that the Government would have to put into overall defense was larger or is larger than was being put in then. Many people interpreted our strong recommendations for defense as an unfortunate method of cutting into appropriations for Strategic Air Command. This was not the case in our recommendations and we believed then and I still believe that the money is going to have to come from other sources, and not from cuts from the military except in the matter of pruning certain military things that are not terribly fruitful.

Objections to try to build up continental defense from the point of view of people who are trying to build up offensive power alone, simply that if you work with a limited number of dollars and a limited amount of effort, naturally if you build one thing up, you would have to build the other down. Whereas, I am firmly of the opinion that we are going to build the whole thing up, and our economy will have to stand it, and I am assured that it will. Does that answer your question?

Q. You mean that you had both strategic air and also continental defense?
A. Yes, sir; and other military things, too, as events of the present show.
Q. In this work that you have been describing——
Mr. GRAY. Are you still on the continental defense, or are you about to leave?
Mr. MARKS. I was about to get to a final couple of questions.
Mr. GRAY. Please proceed if you are that close to finishing. My question was related to whether we should stop now for lunch.
Mr. MARKS. I think I could finish in just a few minutes.
Mr. GRAY. Let us go ahead.

By Mr. MARKS:

Q. The work which you described in which Dr. Oppenheimer participated on continental defense and other military and scientific affairs, who did you conceive to be the enemy that we needed to be worried about?
A. There is no question in anybody's mind, and there was no question in the mind of anyone who participated or was closely associated with any of these discussions, Soviet Union, and the word "enemy," or "Russia" and the word "enemy" are sort of interchanged freely. It is that deeply imbedded in everybody's thinking, including that of Dr. Oppenheimer.
Q. What was your general purpose in devoting yourself to this work?
A. That is a simple question. This is the only country we have, and these are tough times, and we want to help it.
Q. As a result of your association with Dr. Oppenheimer have you formed an opinion or conviction as to his character and his loyalty to the United States?
A. I am completely convinced of his loyalty to the United States. Can I add a little way of saying it?

When you are gathered in a group of men who are discussing the details on how to combat the Russians, how to contain the Russians, how to keep them from overrunning the rest of the world, and so on, the loyalties come out very, very clearly. There just is not any question in my mind that Dr. Oppenheimer's loyalty is for this country and in no way or shape by anything other than hostility toward the U. S. S. R.

Q. What about his character?
A. His character? Ethical, moral is first rate.
Q. Do you have any views as to his capacity to exercise discretion in dealing with classified and restricted data and military secrets?
A. In my opinion, he is always discreet and careful and has regarded the handling of secret documents and secret ideas and so on with discretion and understanding. You might thnk it is not the easiest thing in the world to carry around a head full of secrets and go about in public, too, and talk about burning questions of the day. It is difficult. I believe that Dr. Oppenheimer has showed in every instance to my knowledge that he can do this kind of thing.
Mr. MARKS. That is all.
Mr. ROBB. I can finish in 2 minutes, I think.
Mr. GRAY. If we can, let us go ahead.

CROSS-EXAMINATION

By Mr. ROBB:
Q. Doctor, are you in the group that is called ZORC?
A. Yes, except let me say that this name was never heard of by the members of that group, by any one of those four until it appeared in the national magazine.
Q. I was going to ask you if you could tell us what you know about the origin of that nomenclature.
A. I have no knowledge of the origin of that nomenclature. I do know one friend of mine went around to a meeting of the Physical Society and hunted for people who had heard of it. Found one and I would rather not mention the name because it has nothing to do with this thing. He may have heard it or it may have been the invention of the man who wrote the article.
Q. I think for our purpose, the name is not popular. Was there a group consisting of yourself, Dr. Oppenheimer, Dr. Rabi, and Dr. Lauritsen?
A. No, no more than there would be a group of any four people who respect each other despite the fact that they hold slightly different ways of looking at things—a community of interests and a slight disparity of approach. These four people, I think, are very different.
Q. Were you four people the nucleus of that Lincoln summer study?
A. No, sir. The four were not. I would say the nucleus, as I tried to clarify before, were Dr. Hill and myself. That is, the director of the Lincoln Laboratory. The first discussions were with Dr. Lauritsen. Dr. Oppenheimer and Dr. Rabi agreed that it would be a good thing to go ahead with it and they were willing to lend their prestige to help pull in some people into it, but this is far from being the nucleus of the thing.
Q. That is what I am trying to find out because it has been rather fuzzy in my mind. Were you four people—Dr. Oppenheimer, Dr. Lauritsen, Dr. Rabi, and you—peculiarly active in that summer study? Were you the leaders of it?
A. Let me say this. I ran it. I was the director of it. So, I was in it. There are no two ways about that. Dr. Rabi, Dr. Oppenheimer and Dr. Lauritsen spent a small fraction of their time. However, let me say this. We had for the first week of that study a briefing for 4 days, as I remember it, that was packed with as much meat as you can get into any 4 days of technical briefing. I wanted a summary of that technical briefing, and there were about 65 people there, all very fully informed, and the only man I could turn to give a summary, who could pull the thing together, was Dr. Oppenheimer. He did a masterful job. It was perfectly clear to everybody in that group how Oppenheimer felt about all of the issues, so that if you questioned any one of those you could find a statement of what he believed.
Q. Was there any discussion, Dr. Zacharias, about the comparative morality of a so-called fortress concept, on the one hand, and a strategic air force to wage aggressive war on the other?
A. Not in that summer study. I am afraid that wars are evil. I do not think there is anyone in the room who would take exception to that. It is not a very meaningful statement. But the question of morality, one way or the other, you do not have time for when you are trying to think how you fight.
Q. Was there any conclusion reached as to the relative importance of a strategic air force on the one hand and an impregnable air defense on the other hand and, if so, what was it?
A. I know of no one who really knows the inside of the military who believes that it is possible to have either an impregnable and all overwhelming and completely decisive strategic air command, and I know of no one in the know who thinks you can have a completely impregnable defense. What the country needs is a little of both and one has to supplement the other. That was clearly stated in the conclusion of this report.
Mr. ROBB. That is all I care to ask.
Mr. GRAY. I have a couple of questions. I am going to reverse my procedure and call on Dr. Evans.
Dr. EVANS. I have no questions.
Mr. MORGAN. I have no questions.
Mr. GRAY. Dr. Zacharias, have your own associations been in question? Have you, for example, been identified with any groups which the Attorney General has listed in these various publications?
The WITNESS. No, sir. Let me make one statement about this, which I have written on all security questionnaires so you will know.
In the late thirties, sometime, there grew up something called the American League Against War and Fascism. I may have been a member. I would now,

thinking back on it, believe that I should have been. It is an organization which became Communist-dominated. What I have had to say in any security questionnaire is this: that if their rolls say I was, I was. If their rolls say I wasn't, I wasn't. It was not something that I had much time for or much traffic with. This is the only thing of any sort remotely associated with this kind of thing. Mind you, it was not a Communist-dominated organization when I was looking into it and thinking that it might be a good thing to back.

Mr. GRAY. I think that is a very fair statement.

May I just ask this one further question. At one time, did you begin to be conscious that association with the Communist Party had elements of danger? Is that a clear question?

The WITNESS. Yes, but like the question that is not completely clear, the answer will take a couple of minutes. Yes or no will not quite do.

Mr. GRAY. I understand.

The WITNESS. I went to college in New York, at Columbia, having come from the South. I learned about that, that there was such a thing as Communism, as a college student naturally. I lived in New York as a graduate student at Columbia and as a member of the teaching staff of one of the municipal colleges, Hunter College. There was Communist argument all around. I could never really understand any of the Communist arguments and always fought bitterly, intellectually with all of the people who tried to hand out the Communist line, so I would say that at no time since even my first discovery of Communism did I ever think there was anything very sensible about it.

I remember even what I thought as a freshman in college. At no time did I ever think there was anything sensible about it, so there was never any sudden becoming aware. However, the buildup of the Communist talk was something that a number of us in New York would always fight off and I can remember some bitter battles with the pinks of the 1930's.

Mr. GRAY. As of the time the fighting started in Europe, would it have been clear to you that communism might have involved some security problems? I am not sure that is a fair question. What I am trying to get at is whether you as a scientist were conscious at all of communism either in relationships or its threats or dangers, or whether it was something that really did not cross your path at all.

The WITNESS. I do not think one could have claimed that he was awake and live in New York City in the thirties and not know that there was communism. I think a lot of people did not regard is as the threat that it turned out to be. Russia was small, it was experimental, it was backward, and so on. I do not think any people who were backing it then knew that it would capture half of the globe by 1954. Does that answer your question?

Mr. GRAY. Yes. I think perhaps I will put one other to you.

Is there any question in your mind that employing a Communist today on matters involving security would be a mistake, one who is now a member of the Communist Party?

The WITNESS. Let me get this straight.

Mr. GRAY. Let me put the question this way: In your mind, would party membership be an automatic bar to a man who was being considered for work of a classified nature?

The WITNESS. Certainly.

Mr. GRAY. Would this have been true in your mind in the war years of World War II?

The WITNESS. A then member of the Communist Party, I would have thought the same, because I had such a low opinion of their attitudes. In the case of some whom you might call American Communists, there was a fanaticism that left little doubt about whether you would want to have them on a secret project. There are many who saw the light and when they did—the Russo-German Pact certainly cut a lot of those—and the less fanatical ones were probably hireable.

Mr. GRAY. It follows, I suppose, from what you have already said that you feel that today a man who might have been a member of the Communist Party can be in 1954 a perfectly safe person securitywise. That is possible?

The WITNESS. Yes, I think so. I think also that in giving a security clearance one should look at the depth of his involvement and what sort of involvement there was.

Mr. GRAY. Dr. Evans.

Dr. EVANS. There was quite a number of Communists around Hunter College at that time.

The WITNESS. I do not know how many. I knew that Bella Dodd, who was the head of the Teachers Union there was likely to be a Communist. Remember, it is hard to know who is a Communist if you are not in it, but I was never surprised when Bella Dodd confessed that she was a Communist.

Dr. EVANS. You can meet a lot of people and talk to them and know them in a certain way and not know they are Communists.

The WITNESS. It depends on how you define it. Some people want to be very specific and try to say a dues-paying member. You might not know whether a man was a dues-paying member unless you happened to have some mechanism for knowing it. A man is not likely to show you a red card and say, "Look, I am a member of the Communist Party." But you can certainly tell the flavor of a man's opinions by what he says. There are many people that I would call Russophilic American Communists—lovers of Russia. You could tell this by talking to them, I am sure.

Dr. EVANS. You have never been approached by anyone trying to get classified information from you, have you?

The WITNESS. No, sir.

Dr. EVANS. I have no further questions.

REDIRECT EXAMINATION

By Mr. MARKS:

Q. Just three questions.

Dr. Zacharias, in response to a question of the chairman, as to whether you would consider someone who once had been a Communist or perhaps he said close to Communists, but who no longer was, considering his present hirability for secret work, you said you would have to take account of the extent of his involvement in the Communist movement. Would you also take into account his record since then?

A. Certainly. Whenever you sign a petition saying, "I give this man clearance to work on such-and-such secret project," this is a positive statement, and I think should be backed up with good, full knowledge and appreciation, pro and con.

Q. In response to a question by Mr. Robb about continental defense and strategic offensive, I think you said that what you were advocating and what your group in the summer study was advocating was a little of both.

A. Maybe I should have said a lot of both.

Q. Just one other question. Do you have any connection with the Science Advisory Committee of the Office of Defense Mobilization?

A. Yes, sir. I am either a consultant or a member depending on whether the names have been changed in the last month or two. There are so many people who are members of the Science Advisory Committee and so many people called consultants and it was decided to switch the titles of the groups.

Q. Do you attend those meetings regularly?

A. Yes, sir.

Q. Could you make any comment on the value of Dr. Oppenheimer's contributions in that organization.

A. There are very few people who have Dr. Oppenheimer's ability to synthesize the additions of others along with the ideas of himself. He has that wonderful ability. Meetings that have gone on without Dr. Oppenheimer, in my opinion, have suffered somewhat from this lack. Mind you, there are people on that Committee who have a real gift for summary, but they are not the equal of Robert Oppenheimer. In particular, DuBridge and Killian, two college presidents. Maybe that is part of the equipment of a college president, but neither one of them will focus the ideas quite as well as Robert Oppenheimer.

Dr. EVANS. I did not get what you said about the equipment of college presidents.

The WITNESS. The ability to bring ideas into a clear focus. I am afraid it sounded——

Dr. EVANS. You say that is the ability or is not the ability?

The WITNESS. It is the ability.

Dr. GRAY. He said it may be.

Mr. MARKS. That is all.

Mr. ROBB. I have nothing further.

Mr. GRAY. Thank you very much, Dr. Zacharias.

(Witness excused.)

Mr. GRAY. We will recess until 2:15 p. m.

(Thereupon, at 1:10 p. m., a recess was taken to reconvene at 2:15 p. m. this day.)

AFTERNOON SESSION

Mr. GRAY. Mr. Buckley, do you care to testify under oath? You are not required to do so.

Mr. BUCKLEY. I am quite willing to do so.

Mr. GRAY. All the earlier witnesses have done so. If you do wish to, would you raise your right hand and stand please? May I have your full name?

Mr. BUCKLEY. Olliver E. Buckley. If you wish the middle name, it is Ellsworth—Olliver Ellsworth Buckley.

Mr. GRAY. Olliver Ellsworth Buckley, do you swear that the testimony you are to give the board shall be the truth, the whole truth, and nothing but the truth, so help you God?

Mr. BUCKLEY. I do.

Mr. GRAY. Will you be seated, please, sir.

I am required to remind you of the existence of the so-called perjury statutes. May I assume you are familiar with those? I am prepared to review with you the penalties for falsification or fabrication under oath.

Mr. BUCKLEY. I realize they are severe. I could not state them.

Mr. GRAY. I think that is adequate.

Mr. BUCKLEY. I should like to ask, sir, if the course of your testimony should indicate to you that it is necessary to advert to or disclose restricted data you let me know in advance so that we may take certain necessary and appropriate steps.

Finally I should say to you what I have been saying on behalf of the board to each of the witnesses, and that is, that we considered these proceedings a confidential matter between the Atomic Energy Commission and its officials on one hand and Dr. Oppenheimer, his representatives and witnesses on the other, and that the Commission is making no releases with respect to these proceedings and we express the hope that the witnesses will take the same view of the situation.

Would you proceed, Mr. Garrison.

Whereupon, Olliver E. Buckley was called as a witness, and having been duly sworn, was examined and testified as follows:

DIRECT EXAMINATION

By Mr. GARRISON:

Q. Mr. Buckley, would you state your present position?

A. I am retired. I was formerly chairman of the board of Bell Telephone Laboratories.

(Mr. Morgan left the hearing room.)

A. Before that, I was president of Bell Telephone Laboratories—president for a period of 10 years and chairman for a period of 1. I am still a member of the board of directors of Bell Telephone Laboratories.

Q. Your training has been that of what?

A. I hold a doctor's degree in physics, and after obtaining that at Cornell University, went to the Bell System—really with the Western Electric Co. engineering department which later was merged into Bell Telephone Laboratories—and spent my whole professional career in that organization in one way or another, except for a period of 1 year in the Signal Corps in the First World War.

Q. During the Second World War, did you hold a defense position?

A. I was a member of the Guided Missile Section or Division—I forget just how they labeled it—of the National Defense Research Committee and Chairman of the particular branch of that that had to do with applications of television to guided missiles. I was also for a time a member of the Communications Division of NDRC.

(Mr. Morgan reentered the hearing room.)

By Mr. GARRISON:

Q. Then after your service in World War II, would you state the governmental committees on which you served in connection with our defense work?

A. There was another committee—an ad hoc committee—that I served on for a short time during the war that perhaps deserves mention. That was the National Academy of Science Review Committee on Atomic Energy, which was, I think, for a short period in 1941. After the war, I served on the Industrial Advisory Committee of the Atomic Energy Commission from October 1947 to August 2, 1948, when I was appointed to the General Advisory Committee and dropped off the Industrial Advisory Committee.

Q. And you served on the General Advisory Committee for 6 years?
A. It will be 6 years the first of August. I am nearing the end of my statutory term.

In April of 1951 I was appointed Chairman of the Science Advisory Committee of the Office of Defense Mobilization, which office I held until May 15, 1952, when I resigned because of illness, though remained at the request of the President in my position as a member of that committee. I am still a member of that committee.

Q. When did you first meet Dr. Oppenheimer, in what year and what connection?
A. I am not certain. I recall Dr. Oppenheimer as a younger man in presenting papers to the American Physical Society which I attended. The first definite memory I have of meeting him was while I was on the Industrial Advisory Committee of the Atomic Energy Commission.

Q. In 1947?
A. That would be 1947, when the GAC met with the Industrial Advisory Committee on one occasion.

Q. Were you closely associated with him—I know you were on the GAC—in the work of the Science Advisory Committee of the Office of Defense Mobilization?
A. I was, quite, because I sought his advice at the time I was considering acceptance of that appointment. The Committee, as it was originally proposed by some people working in the Government, was not one I thought I could accept, but with some modifications I came to the conclusion that I ought to accept it if it could be cut to fit my ideas a bit better. I consulted Dr. Oppenheimer in this connection and he was very helpful in working out some of my problems in this connection.

Q. You remember, of course, the October 1949 meeting of the General Advisory Committee that had to do with the H-bomb program.
A. I have refreshed my memory on that occasion by looking up some notes in the AEC and recall some things about it.

Q. Did you join in the so-called majority report at the October meeting?
A. I did.

Q. Did you later at the next meeting in December or before then submit an additional statement of your own?
A. Yes. That was the meeting early in December—December 3. I wrote up a separate attachment that did not in my opinion reverse the position I had taken, but elaborated on it from my point of view. There was no attempt in that statement to express the views of other members of the committee, but rather my own interpretation of what the committee statement signified.

Q. Would you care to summarize as briefly as you can for the board what your position in the matter was?
A. I shall have to refer to my notes to do that. I haven't a transcript of that thing.

Mr. ROBB. Excuse me, Doctor, is that your letter of December 3, 1949?
The WITNESS. That is right.
Mr. ROBB. Would you like to see that?
The WITNESS. I have seen it. I saw it the other day over at the AEC. I don't know whether there is anything in there that is regarded as classified material at the present time.
Mr. ROLANDER. I will have to consult the classification officer.
Mr. GARRISON. I didn't intend to ask Mr. Buckley to go into much detail but just state the essence of his position without reading from the text.
Mr. ROBB. He could certainly have it before him if he wishes to have it while he is testifying. It is marked "Top Secret."
The WITNESS. This is the difficulty with its label. I felt at liberty to make a few cryptic notes about it.
Mr. ROBB. Yes, indeed.
Mr. GARRISON. Do you wish to have the text before you?
The WITTNESS. No, I don't have to have the text before me.
Mr. GARRISON. I didn't ask for anything very elaborate.
The WITNESS. Is there a security officer present?
Mr. ROLANDER. I am the security officer. I have asked for the classification officer. But I think if you talk in general terms you won't have any difficulty here.
The WITNESS. Will you check me if I do go beyond bounds?
Mr. ROLANDER. I will try to be of service.

The WITNESS. I see no danger in discussing it, but I don't wish to violate any security regulation.

Mr. GARRISON. Perhaps while we are waiting for him I could ask you one or two preliminary questions.

By Mr. GARRISON:

Q. How did you come to write a statement of your own?

A. As I recall it—my memory is not entirely clear on this point—I thought that our statement of October had been misinterpreted and I thought that what I meant at any rate in signing the statement needed more explanation than the mere statement itself gave.

Shall I proceed.

Mr. ROLANDER. The classification man is here now, Dr. Buckley, so if you would like to proceed you can check with him any question that might arise.

The WITNESS. This memorandum was based on the question of an immediate all out effort on what was called the super, which was a hypothetical kind of a weapon at that time, as I recall. I was at the time still opposed, as I had been something a month earlier to a crash program to produce something that we didn't understand and the consequences of which we did not understand. I based my opinion on certain assumptions which I enumerated: (1) our ignorance of how to build the super or whether it could in fact be built at all; (2) the great cost in money which it represented and the diversion of effort from the A-bomb program which it must mean; (3) the small, if any, adidtion to military effectiveness as I then viewed this hypothetical weapon; (4) if we can do it, the Russians can also do it, but they cannot do it so quickly.

I assumed those things were so. I noted that others might not agree with those assumptions. It was the way it looked to me. I endeavored to appraise what I would call the good versus the harm of this development. It was, I thought, a possible retaliatory weapon, one of doubtful value. It represented the diversion of effort from the area of practical military weapons to the end only of extensive genocide and ruthless destruction. It might have an adverse effect on the acceleration of Russian development. It might lead to a false sense of security and it represented some loss of moral and political value in limiting defense activity to instruments of military effectiveness. Those, as I recall, with the aid of my notes, were questions in my mind based on the assumptions which I had made.

Weighing the pros and cons as best I could, I favored very careful systems analysis of the "super" program, and an active program of research—doing everything that we could see needed to be done to establish whether this thing could be done and how—so that we could know what we were making policy about. This was one of the things that troubled me: That we were advising on policy about a thing that we didn't understand and see our way through on. I thought that we ought to see our way through and not be hysterical about an all-out development and production of a weapon of which we knew so little and without compromising our position and restricting production to weapons of predominantly military value. My notes are not too clear on this point. I am rather cryptic and I would refer you to the document itself.

I favored strongly building a large stock of A-bombs at the same time that we pursued this super idea further in the laboratory and by test shots of various sorts that would lay a sound engineering foundation for doing the job.

That is what I scratched in an obscure way out of my notes and the document may not be entirely consistent with those words, but the general idea that I had was that I thought we ought to proceed with research and development parts of these things rather than an all out production immediately of something we didn't understand either physically or with regard to its probable consequences.

By Mr. GARRISON:

Q. After President Truman directed in January, 1950, that work in connection—I am not trying to state exactly what his directive was, but I think you know what I mean—that work on the thermonuclear weapon should move forward actively, what would you say as to the cooperation or lack of cooperation of the members of the GAC, particularly Dr. Oppenheimer, with the national policy?

A. I think all members of the GAC accepted the President's decision as a definite determination of policy to which we were bound and all of us, along with Dr. Oppenheimer, conducted ourselves accordingly from thence on. There was no argument about it. That was the policy. However, we did persist in our opinion that the A-bomb stockpile should be enlarged and that development should proceed in that field as well, which I think was consistent with the President's order.

Q. Do you feel that your associations with Dr. Oppenheimer in the years that you served with him on GAC and your service with him on the Science Advisory Committee were sufficiently close to enable you to form a judgment as to his character and loyalty to the United States?

A. The question never arose in my mind as to whether he was loyal to the United States. I believed and believe that he was loyal to the United States. I just don't recall any event that even raised that issue in my mind.

Q. Would you have any comment as to the quality of his service in those years to the country?

A. This is in the postwar years you are speaking of?

Q. Yes.

A. I think it was extraordinary service to the country. The job of being chairman of the GAC is a very heavy and time consuming job. He was our unaminous chairman during the period that my service overlapped his and he was so outstandingly good in that position that if you give value to the services of the GAC you must also give great value to the service of its chairman who was an excellent chairman.

Q. What would you say as to his discretion or lack of discretion, particularly with reference to his knowledge of classified material of a very secret character?

A. I assumed and believed him to be discreet with reference to such material.

Q. You read the Commission's letter of December 23, 1953, to Dr. Oppenheimer which initiated these proceedings.

A. I read it in the newspaper.

Q. Do you have the same confidence in him today that you had when you served with him in the postwar years?

A. Yes.

Mr. GARRISON. I think that is all, Mr. Chairman.

Mr. GRAY. Mr. Robb.

CROSS-EXAMINATION

By Mr. ROBB:

Q. Doctor, are you a nuclear physicist?

A. I am not, sir.

Q. So in respect of the question of the feasibility of a superbomb, I suppose you had to rely on the opinions of others, didn't you?

A. That is right.

Q. Whose opinion did you rely on, Doctor?

A. I gave great weight to Dr. Oppenheimer's opinion. I subsequently to the letter of which I just spoke visited Los Alamos and heard a discussion of it by Dr. Teller and got a briefing on it, you might say. I could not analyze that or criticize it as a physicist, of course.

Let me say that so far as I could understand it, it was consistent with the opinion that I had formed after hearing from Dr. Oepnheimer and others, that it was one of these things that had a speculative chance. It was a hypothetical kind of thing and not the kind of a thing that was developed later.

Q. Doctor, you said that you felt that your subscription to the majority report of the GAC of the October 1949 meeting had been misinterpreted, I believe you said.

A. I think that is stated in the document that I wrote and, I think, misunderstood.

Q. Would you explain that to us a little bit, Doctor?

A. Yes. As I look back on it, that statement doesn't fully reflect our discussion at that meeting because I believe that it was the general opinion that research in the direction of thermonuclear weapons should be heavily pushed. I can't prove that but I think that was the position. I believe that I thought it was the position at the time I wrote this memorandum. But further than that, I can't recall. That was not brought out in the October statement, you see.

Q. I see.

A. As a matter of fact, there was work going on already and work planned ahead at the time of this thing being set up. It was down the thermonuclear alley: The question at issue was a crash program to build a hypothetical super, as I recall it. My memory may not be accurate but that is the best I can recall.

I think that memorandum which I endeavored to sum up is consistent with that point of view because in the memorandum I did not take exception to the prior statement. I was in my mind elaborating on it. I did not attempt, as I said, to reflect the opinion of all the others. But I believe on that point it was consistent with the position that the GAC took at that time and had taken previously.

Q. Doctor, do you recall in your later memorandum making some reference to a public commitment not to develop the thermonuclear weapon?

Mr. GARRISON. Would you make that a little more clear?

The WITNESS. I don't recall offhand.

By Mr. ROBB:

Q. Let me show it to you.

A. That is a statement, I think, of my opinion at that time.

Q. Having looked at this do you now remember that you did make some reference as to whether there should or should not be a public commitment not to develop the weapon?

Mr. GARRISON. Could you read the sentence?

Mr. ROBB. May I read this, Mr. Classification Officer?

Mr. MARSHALL. May I see it, please?

Mr. ROBB. Yes [handling].

By Mr. ROBB:

Q. I want you to explain this and what caused to put that in. The two sentences I have in mind are these: "Whatever course of action is adopted in the development of superbombs I do not wish at this time to recommend for or against a public commitment not to develop the weapon, nor have I any specific recommendation as to declassification. Some public announcement of policy may be necessary or desirable but I do not feel able to advise wisely."

Would you mind explaining what you had in mind?

A. It seems to me it hardly needs explaining. I think that is a clear statement.

Q. I just wondered if there had been some discussion in the GAC as to whether there should be a public commitment or not.

A. I don't recall any. There may have been, but I don't recall it.

Q. Had there been any discussion as to declassification?

A. I don't recall that there was any at all.

Mr. ROBB. That is all I care to ask, Mr. Chairman.

Mr. GRAY. Mr. Buckley, you have made a distinction, I think, in your testimony between research and development or partial development on the one hand, and an all-out production effort on the other. This is a distinction I believe you made and I believe you have stated that you were opposed to what has been called the crash or all-out effort on the super. At least this was your position and was the majority position of the GAC in the October 1949 meeting.

I think you also testified that you felt, however, that we should have an active program of research. I believe those were your words.

Did you later feel that the interpretation of the written report of the October 1949 meeting lead people to believe that you had been opposed as a committee to active research? Is that one of the reasons you felt that you wanted to make a clarifying statement later?

The WITNESS. I now believe, or, as I recall, that was my position on the thing. I wasn't aware that there was any great difference in the committee on this thing. I wanted to state it more explicitly. Perhaps in that committee I had been rather often making the point that we ought to do what I called systems analysis to see as far as we could where we are going before we embark on a heavy development program.

Mr. GRAY. I am a layman. Would systems development be the same thing as active research?

The WITNESS. No. Systems development would be a paper study, generally speaking—those supported by experiments—to determine systematically ends and possible means of achieving those ends in the nature of a technical survey and enlarging the technical grounds for planning a program with these ends in view.

I thought we ought to see our way through just as far as we could and build up as good a technical background for a program as we possibly could and that this would be the economical and speedy way to do the job, whatever job appeared to be good to do.

Mr. GRAY. Would you forgive me just a moment while I glance at your letter.

Your feeling is that your participation as a member of that October meeting did not in any way commit you against the development of this weapon although you did oppose all-out production?

The WITNESS. You could say an all-out development and production program. I through that a more careful study of the problem based on further experimenting than had been done and based on our military objectives might lead to some major modification of the program, but it was not to my mind a determination advice on our part not to pursue the study of thermonuclear weapons. Is that clear?

Mr. GRAY. Yes, I think you have answered the question.

The WITNESS. That is the way I now recall my position which I think is fairly set forth in that letter which I wrote.

Mr. GRAY. Mr. Garrison, do you have any further questions?

Mr. GARRISON. No.

Mr. ROBB. I have no further questions, Mr. Chairman.

Mr. GRAY. Thank you very much, Doctor; we appreciate you being here.

(Witness excused.)

Mr. GRAY. Who is the next witness, Mr. Garrison?

Mr. GARRISON. Dr. Bacher, Mr. Chairman.

Mr. GRAY. Dr. Bacher, do you wish to testify under oath?

Dr. BACHER. I would be very glad to, if you so wish.

Mr. GRAY. You are not required to, but all other witnesses have done so.

Dr. BACHER. I should be glad to do so.

Mr. GRAY. Would you stand and raise your right hand, please, and also give me your full name?

Dr. BACHER. Robert Fox Bacher.

Mr. GRAY. Robert Fox Bacher, do you swear that the testimony you are to give the board shall be the truth, the whole truth, and nothing but the truth, so help you God?

Dr. BACHER. I do.

Mr. GRAY. Would you be seated, please, sir?

I am required to call your attention to the existence of the so-called perjury statutes. May I assume you are familiar with them and their penalties and it is unnecessary to review them?

Dr. BACHER. I think I am.

Mr. GRAY. I should like to ask, Dr. Bacher, if in the course of your testimony you find it necessary to refer to or disclose restricted data that you notify me in advance so that we might take certain appropriate and necessary steps.

I should also make the same observation to you that I have tried to remember to make to all the witnesses, that we consider these proceedings a confidential matter between the Atomic Energy Commission and its officials on the one hand and Dr. Oppenheimer and his representatives and witnesses on the other.

The Commission is making no releases to the press and on behalf of the board I express the hope that the witnesses will take the same course of action.

Mr. Garrison, will you proceed.

Mr. GARRISON. Yes, Mr. Chairman.

DIRECT EXAMINATION

By Mr. GARRISON:

Q. Dr. Bacher, would you state your present position?

A. I am Chairman of the Division of Physics, Mathematics, and Astronomy and professor of physics at Cal Tech.

Q. Where did you receive your academic training?

A. I went as an undergraduate to the University of Michigan, took a bachelor's degree, and later a doctor's degree in physics in 1930.

Q. How long have you known Dr. Oppenheimer, approximately?

A. Approximately since 1929 or 1930 when he visited the University of Michigan during the summer to give some lectures there in the summer symposium in theoretical physics.

Q. When did you first get to know him very well?

A. That was somewhat later. I know him through the thirties. If I recall correctly, he lectured in Ann Arbor once or twice more in the early thirties and I think I was present at that time. During the fall of 1930 I was national research fellow at the California Institute of Technology and he was lecturing there during the fall term. I saw him quite frequently during that period. Later than that I saw him only occasionally at meetings or at other times. I remember at one time seeing him in the winter of 1934 in New York when I was an instructor at Columbia and he was visiting his father there. Between then and the war period I think I saw him only occasionally at scientific meetings. My close association with him began just prior to the establishment of the Los Alamos Laboratory.

Q. Suppose you just state what your Government service has been beginning with your work at Los Alamos.

A. I came to Los Alamos from the radiation laboratory at MIT where I had been for 2 years and a half and on the occasion of the starting of the laboratory at Los Alamos. There was a conference when that laboratory was started. I attended the conference. It was decided during the conference that I would join the laboratory and I did, in charge of the Division of Experimental Physics.

In the summer of 1944 the laboratory was reorganized and I became the head of the Bomb Physics Division, which was a position I held until the end of the war. This involved in both capacities very close contact with Dr. Oppenheimer and this contact was, I would say, daily and very close.

Q. What was your next Government service?

A. My next Government service, if I recall correctly, was on a committee having to do with declassification which was set up by the Manhattan District at the end of the war. I think I served on one other committee for the Manhattan District and I don't recall exactly what the title of that committee was. Then during the summer of 1946 I served as a scientific adviser to the United States delegation to the United Nations Atomic Energy Commission.

Q. In that connection you had an opportunity to see Dr. Oppenheimer some more?

A. Yes.

Q. What next after that?

A. After that in October of that year, or it was the first of November, I became a member of the Atomic Energy Commission and was a member of the Atomic Energy Commission until I left in mid-May 1949.

Q. Have you had Government service since then?

A. Since then I have been an adviser to the Atomic Energy Commission and am still an adviser to the Atomic Energy Commission.

I have been first a member of a panel on long-range objectives, I thing it was called—this may not be quite the right title for it—of the Committee on Atomic Energy of the Research and Development Board from spring 1951 until its dissolution in 1953. I was Chairman of the Committee on Atomic Energy of the Research and Development Board.

Q. Was Dr. Oppenheimer a member of that Committee?

A. Dr. Oppenheimer was a member of that committee. I am presently a member of the Technical Panel on Atomic Energy of the Office of Assistant Secretary of Defense for Research and Development. There may be some others which I have forgotten for the moment.

Q. Going back to the Los Alamos period, how much did you see of Dr. Oppenheimer in those years from April 1943 to the close?

A. A very great deal. Much of the work for which I was responsible was very close to the heart of our problem of making an atomic weapon. The demand was for much information from other parts of the laboratory and in particular needed a great deal of guidance from the theoretical people.

As a consequence of this, in particular, I saw a great deal of Dr. Oppenheimer. It would be hard for me to estimate how much I saw him but it seems to me looking back on it that there was scarcely a day going past that I did not spend an hour or more with him.

Q. When he went away did you from time to time act as acting director of the project?

A. I think not in any official capacity, but I believe sometimes when he left the laboratory he did leave me in charge.

Q. Did you yourself go on any official missions with him?

A. On a number of occasions I went on official missions with Dr. Oppenheimer, trips to the east and in some cases to the west coast, where we needed to get information for the project.

Q. Do you have any recollection of his political views in those years as he may have expressed them to you in talks that you may have had?

A. We were pretty busy trying to make an atomic bomb and we didn't talk about many other things. I was aware of the fact that Dr. Oppenheimer seemed to be a Democrat and views that one would associate with his being a Democrat. I was an upstate New York Republican, and we used to joke about this from time to time. But we didn't have much political discussion.

Q. Coming to the period of your service on the Atomic Energy Commission, I would like to ask you to recall what you can of the actions that were taken with respect to Dr. Oppenheimer's clearance in 1947.

A. I might say in this respect that I did refresh my memory on this point by consulting some of the minutes of the Commission, because when I started to think about it, I found I didn't have all of it so clear in my mind.

The consideration of the appointment of the General Advisory Committee to the Commission was taken up at one of the early meetings of the Commission. In fact, if my memory serves me now on this refreshing of this morning, it was at the second meeting at which this was discussed.

This had to do with who were to be the members of the General Advisory Committee.

Q. This is about what time?
A. This was about the 20th of November, I think.
Q. Of what year?
A. Of 1946. Then a little later——
Q. Before the appointment of the GAC?
A. Yes. Then a little later there was some discussion of the question of making some announcement about this, of the appointments which had been made by the President. I have forgotten exactly when that was, but I presume in the interim period recommendations had been made to the President, and he had approved these and actually appointed the members of the Committee.

Q. Let me just make sure I understand. The Atomic Energy Commission recommended some names to the President for appointment to the GAC?
A. That is right. It was a Presidential appointment.
Q. Were the people appointed by the President the same as those who had been recommended?
A. If I remember correctly, that is so.
Q. In any event was Dr. Oppenheimer among those recommended?
A. It was, yes.
Q. This was a recommendation of the Commission as a whole?
A. This was a recommendation of the Commission as a whole.
Q. Now, coming to the clearance and the actions that had to do with his clearance, would you say what you can remember of that?
A. If I recall correctly, clearance at the start of the Commission activities was for the most part just carrying over clearance that had been given under the Manhattan District. Also, if I recall correctly, all members of the General Advisory Committee had during the war some access to activities in the Manhattan District, and some of them had been employees for an extensive period and continued to hold Manhattan District clearance up to that time. If I remember correctly, this clearance was then just continued, because it took some time to get clearance procedures, and so on, under the Atomic Energy Act into full operation. So this was the first basis of clearance. For new employees, there had to be from the time the Atomic Energy Commission took over investigation under the act.

Q. What do you next remember about Dr. Oppenheimer's clearance.
A. I recall that during the spring of 1947 this questiion was discussed. I am not precisely sure in response to what, but I think in response to a query to the Commission. I remember that we looked at various times through that period, first a summary of information from the FBI, and later a quite voluminous file. Exactly when that is done, I am afraid I don't remember.

Q. Do you have a recollection of having examined then both the summary and some kind of a file?
A. Yes.
Q. Do you have any recollection at all as to the approximate dimensions of these documents?
A. I am afraid I don't, except that the file, I remember, was a fairly thick document. I don't know, something like this [indicating].

Mr. GARRISON. Mr. Chairman, in connection with the examination of Mr. Lilienthal, there was put into the record at page 1409 of the transcript a memorandum from Mr. Jones, the security officer, to Mr. Bellesly, which contained a reference of which I would just read one sentence. This is a note by Mr. Volpe in longhand on the file, and it says—this is dated July 18—"My impression is that the Commission saw no need for formal action following the meeting they had with Mr. Hoover, referred to in Lilienthal's letter of April 3, to the FBI Director."

We asked for the documents pertaining to this matter when we were in the course of examining Mr. Lilienthal.

Mr. ROBB. What was that?

Mr. GARRISON. This is a letter of Mr. Lilienthal of April 3 to Mr. Hoover, referred to in Mr. Volpe's longhand note on the Jones memorandum to Bellesly of July 18.

Mr. ROBB. I am sorry. I fell off on the first turn of that, Mr. Garrison. What was the question?

Mr. GARRISON. What I was going to ask the chairman was to have the letter of April 3 in the record so that we might see what it was that Mr. Lilienthal wrote to Mr. Hoover because I think it might help to clarify the matter under discussion.

Mr. ROBB. I have it before me. Shall I read it? This is a copy. I assume it is the one of April 3, 1947:

"TOJ/D," in the upper right-hand corner.

"Hon. J. EDGAR HOOVER,
    "Federal Bureau of Investigation,
        "United States Department of Justice, Washington, D. C.

"DEAR MR. HOOVER: As agreed at our recent meeting I am forwarding for your information copies of letters in the possession of the Atomic Energy Commission concerning Dr. J. Robert Oppenheimer, as well as papers relating to the award of the Medal of Merit to Dr. Oppenheimer.

"Sincerely yours,

"DAVID E. LILIENTHAL, *Chairman.*"

("Enclosures: cc Mr. Lilienthal. File 2.")

Then some longhand notes: "Enclosures, papers on Medal of Merit, letters from Conant, Patterson, Groves, Bush." That is in longhand.

"Distribution: One and two, to Mr. Hoover. Three and four, to Mr. Lilienthal. Five, reading file. Six, records section file."

Mr. GRAY. That is the longhand note?

Mr. ROBB. The one Mr. Garrison read, "My impression is that the Commission saw no need for formal action following the meeting that they had with Mr. Hoover, referred to in Lilienthal's letter of April 3, to the FBI Director."

That apparently was sending the Medal of Merit award we had here, and the letters from Patterson, Groves, Conant and the others.

Mr. GARRISON. This seems to refer to a meeting with Mr. Hoover.

Mr. ROBB. That was a meeting on which there was a memorandum written by Mr. Jones, which was read into evidence, on March 27, 1947. That is in the record some place.

Mr. GARRISON. Mr. Chairman, I have some more requests for information that I think the Commission can give us about the history of these events that I would like to submit to the board, but I don't want to take the time now while Dr. Bacher is on the stand. I thought possibly the particular letter might throw a little more light.

Mr. ROBB. Maybe I can throw some light on it, if I might.

Mr. GRAY. If you are going to pursue questioning of Dr. Bacher about those events, or if you are, Mr. Robb, I think it might be helpful to Dr. Bacher to have his recollection refreshed because people seem not to remember this period very clearly.

The WITNESS. Thank you, Mr. Chairman.

Mr. GARRISON. What is there, Mr. Robb?

Mr. ROBB. I don't know whether I am at liberty under the rules to tell you, but apparently a number of people were interviewed concerning Dr. Oppenheimer. I think Dr. Bacher was interviewed. I think that material was in the file before the board.

Mr. GRAY. One thing it seems to me that Mr. Garrison is perhaps groping for is the possibility that there may have been a meeting of the full Commission with Mr. Hoover. Mr. Lilienthal testified, did he not, about a conversation?

Mr. ROBB. That is right.

Mr. GRAY. That is, with Mr. Hoover at a time when he was accompanied by the deputy counsel of the Commission. It would be my guess on the basis of anything I have heard, Mr. Garrison, that there was not a full meeting of the Commission with Mr. Hoover, but this I am not sure about.

Mr. ROBB. If there was, I find no reflection of it in this file.

Mr. ROLANDER. The only record in the file of such a meeting was the one discussed and introduced in the record when Mr. Lilienthal testified.

Mr. GRAY. And this involved a visit to Mr. Hoover's office of Mr. Lilienthal and Mr. Volpe.

Mr. ROLANDER. That is right.

Mr. GRAY. I would guess the Commissioners would remember if they went in a body to Mr. Hoover.

Mr. ROLANDER. The memorandum in discussing the meeting, it refers to meeting between representatives of the Atomic Energy Commission and the Federal Bureau of Investigation. Whether that includes all members of the Commission, I just don't know.

Mr. GARRISON. Could we have read into the record the portion of the minutes of August 6, 1947, relating to the matter of Dr. Oppenheimer's clearance?

Mr. ROBB. I thought this thing that had Mr. Volpe's note on it was all there was on it.

Mr. GARRISON. Mr. Volpe's note was before that.

Mr. ROBB. Here is a paper here, August 11, 1947, from T. O. Jones to William Uanna, "Subject: J. Robert Oppenheimer."

Mr. GARRISON. This refers to the meeting and I think that was read into the record.

Mr. ROBB. "Authorization for granting final Q type clearance, August 6."

Mr. GARRISON. What I would like to have is the actual August 6 meeting.

Mr. ROLANDER. I think we had in the record a stipulation as to what the minutes reflected. Isn't that satisfactory?

Mr. GARRISON. It did not seem to me to be a quotation from the minutes, but rather a stipulation by the Commission that clearance be recorded, or something of that matter. At least it did not on its face appear to be a quotation from the minutes.

Mr. ROBB. I don't know. Frankly I did not concern myself with it in view of the stipulation. I have never looked at the minutes.

Mr. ROLANDER. I don't think we can state the actual Commission minutes. The Commission minutes as such, I don't believe it proper for us to quote them. Therefore, at that time the Commission did, early in the proceedings, agree to a stipulation as to what took place. That is what we had hoped to make a part of this record, and has already been made a part of the record.

Mr. GARRISON. Mr. Chirman, I am not asking for any portion of the minutes which might have to do with extraneous matters, but only that portion which relates to Dr. Oppenheimer's clearance, and it seems to me that is a piece of information very relevant to this proceeding, and certainly can evolve no matter of improper information to be read into a record like this.

Mr. GRAY. I am not informed about the minutes or about the procedures of the Commission not making its minutes available. I think in this case I will have to rely on the representative of the Commission, Mr. Rolander, who says that you do not think the pertinent portions of the minutes can be read into the record?

Mr. ROLANDER. That is my understanding, yes.

Mr. ROBB. I might say, Mr. Chairman, that I don't think either I or Mr. Rolander would have authority in the light of what I take it to be policy to make any commitment. I think it probably should be submitted to the Commission for its ruling.

Mr. GRAY. If you wish, Mr. Garrison, now to make a request of that sort, I certainly will transmit it. I don't think anybody here has authority to grant it.

Mr. GARRISON. I would like to make a formal request of that sort, Mr. Chairman. As I read the rules of these proceedings, I must say I see nothing in them that would stand in the way of that. On the contrary, it seems to me that the emphasis on obtaining all relevant information which is set forth explicitly in the rules should make this information available both to the board and to us.

Mr. ROBB. I am not debating that with Mr. Garrison, Mr. Chairman. We would be happy to transmit the request to the Commission, but I don't think I have the opportunity to say whether or not they will do it.

Mr. GARRISON. Then we have made the request, Mr. Chairman.

Mr. GRAY. Yes.

Mr. GARRISON. I would like to proceed with Dr. Bacher on this matter, and ask him to remember.

By Mr. GARRISON:

Q. You told us now you have the recollection of having gone over a summary. Do you recollect anywhere near at all how many pages that may have been?

A. No; if I had to make an estimate I would guess around 30 or 40 pages or something of that sort.

Q. In addition to that, a thicker file?

A. At a later date, if my memory serves me correctly, I believe we went over a very much thicker file, and I believe it was reviewed by the other Commissioners, too.

Q. Do you remember discussing this with other Commissioners?

A. Yes.

Q. What do you remember?

A. I don't remember very much about the discussion with the other Commissioners, except that I remember either before or during the Commission meeting referring to various parts of it which seemed to be relevant to happenings in the past that we thought we ought to know about. I can't remember very much at the moment just what was said about that. But we did review that and discuss it in the Commission meeting.

Q. Do you recollect any decision on the matter or any conclusion?

A. My memory is that when a query was addressed to the Commission, it seemed appropriate to us to consult with some of the people with whom Dr. Oppenheimer had worked during the war other than ourselves. I can't remember exactly who was consulted, but I am relatively sure that Dr. Bush and Dr. Conant were consulted. I don't remember who else was consulted. After consultation with these people and a review of the file, the question was discussed by the Commission and I think the conclusion was arrived at that the Commission saw no reason in view of the information which had been brought up to take any different action on the clearance of Dr. Oppenheimer than that which had already been taken.

Q. Do you know Mr. Serber?

A. I do.

Mr. ROBB. Did you say Mr. or Mrs.?

Mr. GARRISON. I will ask about both.

By Mr. GARRISON:

Q. Do you know Mr. and Mrs. Serber?

A. Yes.

Q. Where did you first know them?

A. I can't remember when I first met them. I presume that I knew them before the war, but if so, only very slightly. The first I knew them really at all well was at Los Alamos. Dr. Serber was a member of that laboratory and was there when I arrived.

Q. Did you know anything of their political background at the time?

A. I would say "No."

Q. Did the question of Dr. Serber's clearance come up when you were a member of the Atomic Energy Commission?

A. It did.

Q. What was done about it?

A. If I recall correctly, Mr. Serber's clearance came up as part of the reinvestigation of all contractors' employees. There was a certain amount of derogatory information in the file that appeared. I have forgotten exactly what happened in the local office out there, but it was concluded that there ought to be a hearing board set up on this.

Q. The local office where?

A. The local office on the Pacific coast. A hearing board was set up on the Pacific coast, I believe out of the San Francisco office, and I can't remember the members of that hearing board, but I if remember correctly, Admiral Nimitz was the chairman of it. The hearing board made a report which I believe was transmitted to the Commission, and the Commission acted favorably on clearance after the hearing.

Q. Did the panel recommend clearance?

A. If my memory serves me correctly, they did.

Mr. GARRISON. Mr. Chairman, our information is, I feel quite certain, that the Atomic Energy Commission records will bear this out. I would simply like to state for the record subject to verification, which I am sure can be made by Mr. Mitchell or Mr. Rolander, that the panel in addition to Admiral Nimitz as chairman, consisted of Mr. John Francis Neyland, regent of the University of California, and a lawyer, well known. I think he was counsel to the Hearst interests in San Francisco. And Major General Joyce, of the Marines. If I could just state that in the record and ask if that could be checked.

Mr. ROBB. I believe that is correct, Mr. Garrison.

By Mr. GARRISON:
Q. What was the date of that?
Mr. ROBB. I don't have it.

By Mr. GARRISON:
Q. Do you recall about when this was after the start of the Commission? Would you date it from there?
A. I would think this was 1947 or perhaps the beginning of 1948. I am not clear on the date.
Q. Do you have occasion to see Dr. Serber now from time to time?
A. Yes. He is professor of physics at Columbia University, and I see him from time to time when I go to New York.
Q. Do you see Mrs. Serber from time to time?
A. Occasionally.
Q. When you say when you go to New York, in connection with what would this normally be?
A. In connection with Physical Society meetings or other scientific meetings in New York. Professor Serber is now spending, I beileve, one day a week out at Brookhaven Laboratory, in particular in the interpretation of some of the work they are doing with their high energy accelerator out there, their cosmotron. This is related to work that I am closely interested in, so I see him from time to time because he has the most interesting information on what is going on there.
Q. Do you know whether a Q clearance is called for by that sort of work?
A. I don't know. I presume he must have some sort of clearance to be a regular consultant to the Brookhaven Laboratory, but what sort of clearance he has, I don't know. I never have any questions concerned with classified information to discuss with him.
Q. What was the character of the clearance which the AEC granted in 1947 or 1948, whenever it was?
A. I believe this was a Q clearance that he was granted at that time.
Q. Have you ever heard of any action changing that?
A. No.
Mr. ROBB. This is Dr. Serber, and not Mrs. Serber.
Mr. GARRISON. Yes. I don't believe she is a physicist or works on Government projects.
Mr. ROBB. No.

By Mr. GARRISON:
Q. Isn't that correct?
A. No, she is not physicist.
Q. As a member of the Atomic Energy Commission, did you have occasion to observe closely the work of the GAC?
A. Yes, I think that during the period I was in Washington I probably followed the work of the General Advisory Committee more closely than any other member of the Commission. This was natural because I was the only one with a scientific and technical background, and the work of the General Advisory Committee was mostly scientific and technical. I frequently attended much of their meeting and read their reports very carefully. They were very valuable to us in getting the atomic energy enterprise back on its feet and getting some of the work established that we thought ought to get established.
Q. Would you make a comment on Dr. Oppenheimer's work as chairman of that committee?
A. It was outstanding. He was appointed a member of the General Advisory Committee. The members of the General Advisory Committee themselves elected him chairman of that committee. Until he left, the committee, I believe, he continued to be chairman. He had had the closest connection with the weapons development work of any of the members of the General Advisory Committee.
In that period in early 1947 when the General Advisory Committee was set up, our greatest problem was to try to get the Los Alamos Laboratory in the development of weapons into a sound shape. The General Advisory Committee, I might add, was vigorous on this point, and very helpful in getting the laboratory into shape both by reason of the recommendations which they made, and also the direct help that they gave us in connection with personnel for the laboratory.
Q. What about Dr. Oppenheimer's individual contribution in this effort?
A. I would say in this effort Dr. Oppenheimer's individual contribution was the greatest of any member of the General Advisory Committee. He took his

work on the General Advisory Committee very seriously. He usually came to Washington before the meetings to get material ready for the agenda and usually stayed afterward to write a report of the meeting.

During the course of the meeting prolonging discussion at great length so everybody would express his views, nevertheless after the views had been expressed, he had a very great clarity in focusing these views of what would be a report of the committee.

Q. What was your normal routine when the General Advisory Committee would meet in Washington? When I say your routine, I mean the routine of the Atomic Energy Commission. Did you meet with the GAC or how did that work?

A. If I recall correctly, usually the members of the Commission came in at the start after the meeting at least for a little while and then usually before the end of a meeting there was a session of the General Advisory Committee with the Commission. Sometimes this might occur on a Sunday afternoon, but usually there was a session at the end of the General Advisory Committee so that there could be discussion of what appeared to be their recommendations. At such time it was usual that Dr. Oppenheimer would give a verbal summary in the presence of all the members of the General Advisory Committee, and of the Commission of their findings, and then these would be discussed.

Q. What was the character of the initial meeting between the members of the Atomic Energy Commission and the GAC? At the start of the meeting, in other words?

A. I think this initial meeting was apt to be somewhat less regular. Usually most of the members of the Commission went down; if I remember correctly, the Chairman, Mr. Lilienthal, would generally convey to the committee questions which had come up either within the Commission or from members of the staff to be proposed to the committee.

Q. There was verbal discussion?

A. There was verbal discussion.

Q. You left the Atomic Energy Commission in May of 1949?

A. Mid-May 1949.

Q. So you were not present at the October meeting.

A. No.

Q. Did you remain as a consultant after you left the Commission?

A. Yes. I have been an adviser to the Commission since I left in 1949 and still am.

Q. At the time of the Russian explosion, did you have to do with assessing the information about that?

A. Yes.

Who else had to do with that?

A. If I recall correctly, Dr. Bush was chairman of a group called together in mid-September 1949 to assess the information which was relevant to the determination of whether the Russians exploded an atomic bomb. The other members of the group, if I recall correctly, were Admiral Parsons, Dr. Oppenheimer, and myself, and I believe Dr. Arthur Compton was supposed to be there, as a member of the group, but could not come. If I recall correctly there were just four members of the panel that were set up to assess this information.

I can't give you the exact date on this, but it must have been about the 15th of September.

Q. After President Truman's declaration in January 1950 about the thermonuclear program, did you make a speech on the subject of the program?

A. I made a speech called, "The Hydrogen Bomb," in the end of March 1950. This is open and available for the record and I am sure that looking this over will be much better than any memory I have of what is in that speech.

Q. I just want to ask you two general questions about it. Were you, in that speech, critical of President Truman's declaration?

A. No.

Q. What was the principal point you made in that speech?

A. I would say there were two points, but here I would like any remarks that I make to be subject to referral to the speech itself for anyone to judge what the speech says. I would say there were two principal points. One, I had misgivings about over-reliance in a weapon which seemed to me to not add much beyond large fission weapons to our national arsenal, and second, I was very much concerned that there was not more information available to the public on which sensible opinions could be formed.

Q. You said, I think, that you served on the Committee on Atomic Energy of the Research and Development Board?
A. Yes.
Q. And that you became chairman of it and served as chairman from 1951 to 1953?
A. Yes.
Q. Did that committee convene a panel in late 1950 or early 1951 to consider our weapons program?
A. If I recall that is about the time that a panel was convened for that purpose.
Q. And you were a member of it?
A. I was a member of a panel that was convened about that time for studying our weapons program.
Q. And Dr. Oppenheimer was a member of it?
A. Yes.
Q. And members of the military?
A. Yes.
Q. Yes. If I recall correctly, Mr. Oppenheimer was chairman of that panel and other members were General Nichols and Admiral Parsons, and I think General Wilson from the Air Force, Dr. Alvarez, Dr. Lauritsen, and myself. Some of these may not be correct, but I think they are.
Q. Do you have any particular commitment on Dr. Oppenheimer's service on both the committee and on the panel.
A. If I recall correctly, the panel met for 2 or 3 days to discuss what might be the important areas for progress. We then divided up the various areas to study somewhat further to find out a bit more about it and came back at a subsequent day to write a report, and incorporate the views of the various days' smaller groups at that time. With his unusually great clarity Dr. Oppenheimer succeeded in turning out a report that stated very accurately what the panel thought in draft form. This was then discussed essentially word by word by the panel, and a report finally appeared which presumably is available somewhere.
Q. From your vantage point, if I may call it that, of the chairman of the Committee on Atomic Energy in the years 1951 to 1953, have you any judgment which you could express to the board regarding any alleged or possible delays in the production of thermonuclear weapons.
A. I am. My impression is that this went ahead pretty fast. At least as far as the research and development work went, all of the effort that could be put on this was put on it. After a job is done, it is always easier to look back and say if we had not done this, we would have saved some time. I believe that almost everything that was done either in fission weapons or in thermonuclear weapons was very relevant to the job of making a thermonuclear weapon.
Q. You are still a consultant to the Department of Defense?
A. Yes.
Q. You had to do with the Vista program?
A. Yes.
Q. Were you chairman of the Vista project?
A. No. Dr. DuBridge was chairman of the Vista project.
Q. What was your share of it?
A. I was responsible for one section of the project which had to do with atomic weapons.
Q. You were in charge of that section?
A. Yes.
Q. There has been a good deal of testimony about this project and I don't want to duplicate the record about Dr. Oppenheimer's participation in it, and so forth. I would just like to ask one or two questions about it. Was there a question of allocations as between the Strategic Air Command and tactical air group with respect to the materials that would go into tactical weapons?
A. Yes; but I believe it would take a little further discussion to make clear just what was meant by that. I am not exactly sure on this point whether one does not get into classified information. I think it could be answered without getting into classified information but if there is someone here whom I could consult on that point——
Q. I am not going to ask you any questions of that character. I would like to have your judgment as fairly as you can express it without going into classified materials as to whether the recommendations of this chapter on atomic energy would have affected the hydrogen bomb program then under way, whatever its nature may have been.

A. I know of no way in which it would have affected that.
Q. Was there any purpose to affect that program in any way?
A. I am not even sure I understand the question.
Q. I am not sure I do either. What I am trying to bring out is was this question of allocation related in any way to the thermonuclear work that was going forward?
A. Not that I know of.
Q. It was a question of the allocation of then existing fission materials?
A. Could I say a word about what the purpose of this section of the report was, because otherwise I think it is not even clear what you would like me to answer.
Q. I don't want you to answer anything except what you know.
A. I won't. * * *
Mr. ROLANDER. I think that is all right.

By Mr. GARRISON:
Q. Would you say a word about Dr. Oppenheimer's contribution to the results of this report?
A. The Vista project was started in April 1951, if I recall correctly. Is that correct?
Q. I think that is right.
A. Is that a correct date?
Mr. ROBB. I think so.
The WITNESS. I believe it is correct. It continued through the summer.
Mr. ROBB. That is right.
The WITNESS. It was started in April 1951. It continued through the summer, * * *. The other people who worked with this group were Dr. Lauritsen and Dr. Christie. Dr. Thorndyke from the Brookhaven Laboratory was there during most months of the summer. Dr. Hayworth from the Brookhaven Laboratory was there for a period of a week or so, and a few other people helped us from time to time during that period. During the summer we got a good many of our ideas in line and during the fall started to formulate these so that we could write a report.

I think that by fall much of the background information was beginning to be clear, and many of our ideas were beginning to be a little clearer. It was very difficult to formulate these ideas because all of the points we wished to recommend were interrelated and we found ourselves in difficulty.

I think it was about this time, I don't remember the date, October or November, that we were fortunate to get Dr. Oppenheimer to come and spend a week or 10 days with us. He was very helpful to us in formulating these ideas. I think that we had a first draft of the report actually written down at that time, but it was not in very good form. After 2 more days of discussion with him, he had some ideas of how these things could be better formulated, and helped very much in bringing them to a focus.

Subsequently this draft then went through several revisions. I don't even remember how many. It was finally revised in late December of that year and the final report, I think, appeared or was proposed shortly after Christmas.
Q. Dr. Bacher, you are familiar with the Commission's letter of December 23, 1953, to Dr. Oppenheimer initiating these proceedings?
A. I have read it.
Q. Apart from the allegation or the reports about the H-bomb, did the rest of it come to you as a surprise?
To put it another way, how much, if any, of the matters in this letter apart from the H-bomb would you say you had been over previously at the time of the 1947 clearance?
A. It is, of course, hard to give a categorical answer to a question like that, but I didn't find any parts of it that seemed surprising to me in view of the things I had read before.
Q. How well do you feel that you know Dr. Oppenheimer?
A. I feel I know him very well. I have worked very closely with him during the war, have seen him frequently since the war, and feel I know him really very well. I just don't think it would be possible to work with a man as closely as I worked with Dr. Oppenheimer during the war without knowing him very well.
Q. What is your opinion as to his loyalty to the United States?
A. I have no question at all of his loyalty.
Q. On what do you base that? Is that purely a subjective judgment?
A. I think opinions of that sort are always subjective judgments. In this case I put great credence in my own judgment, naturally, because I know him very

well. But this is essentially an assessment on my part based on knowing him for a great many years. I have the greatest confidence in his loyalty.

Q. What would you say as to his sense of discretion in the use that he would make of the knowledge that has come to him and will continue to come to him assuming that he continues in Government work?

A. I found Dr. Oppenheimer to very discreet. I can remember during the war once when we had to go out on a trip together and it was essential that he carry a memorandum, that even in note form was classified, and he was so careful and he pinned it in his hip pocket. I thought here is a man who really is very careful about these things. But to say more generally as to his discretion, I have always found Dr. Oppenheimer to be very discreet in his handling of classified information.

Q. Is there anything else you care to say to this board about his character as a man and as a citizen?

A. I have the highest confidence in Dr. Oppenheimer. I consider him to be a person of high character. I consider him to be a man of discretion, a good security risk and a person of full loyalty to the country.

Mr. GARRISON. That is all, Mr. Chairman.

CROSS-EXAMINATION

By Mr. ROBB:

Q. Dr. Bacher, you were asked by Mr. Garrison what you knew about Dr. Oppenheimer's political views at the time you were in Los Alamos, and you answered, I believe, that you knew him to be a Democrat.

Did you know anything about his interest in other political philosophies?

A. As I think I answered Mr. Garrison, too, we didn't have very much time to discuss politics at Los Alamos.

Q. Whether you discussed it or not, did you know?

A. Not much. I had been aware of the fact that he had leftish sympathies before the war, but I didn't really know very much about it, and I didn't discuss it with him.

Q. Did you ever state to anyone that you knew that between 1934 and 1942, Dr. Oppenheimer became interested in various political philosophies and was interested as many others were at the time in the experiment being conducted by the Soviet Government in Russia?

A. I don't know, but it sounds as if I might have.

Q. Did you know that?

A. That is a difficult question to answer, because I am not exactly sure what it would take to know that. I was aware that this was commonly discussed.

Mr. ROBB. Mr. Chairman, there is in the file before the board a memorandum to the files, dated March 14, 1947, the subject is stated to be a study of a report on J. Robert Oppenheimer, or an analysis of a report on J. Robert Oppenheimer. Much of this analysis has to do with FBI reports which I am not allowed to discuss or disclose here.

Mr. GARRISON. This is an analysis by whom?

Mr. ROBB. It is not signed, Mr. Garrison, strangely enough. But it is in the AEC files under that date.

Mr. GARRISON. Is that a document used in connection with the clearance discussions?

Mr. ROBB. I assume it was. I don't know. It is March 14, 1947.

Mr. GARRISON. An unsigned document?

Mr. ROBB. That is correct.

Mr. GRAY. It is on AEC stationery?

Mr. GARRISON. Are you going to read portions of that to Dr. Bacher?

Mr. ROBB. Yes, sir. Mr. Chairman, as I say, I am not permitted to read those portions which reflect FBI reports. I would like, however, to read a certain portion which does not necessarily involve such reports, and wherein some minor instances there are some references——

Mr. GARRISON. I am sorry. I did not hear that.

Mr. ROBB. I would like to read certain portions which do not involve reference to FBI reports. In some instances where there is reference to FBI reports, I would like to delete or paraphrase, so as not to get into FBI reports. I wish the board would follow me so I am not distorting.

Mr. GARRISON. Mr. Chairman, is there anything to show that this may not be simply a kind of memorandum exchanged between security officers?

Mr. ROBB. I don't know what it is. It is a memorandum to file.

Mr. GRAY. There is not anything to show the authorship of this report.

Mr. GARRISON. I am a little troubled about reading into the record matter from a document whose purpose, nature, origin, authenticity; we have no knowledge at all.

Mr. ROBB. Could you want the board to consider it without your hearing it?

Mr. GARRISON. I would like to hear everything that the board considers. I know that to be beyond the possibilities, greatly as I regret it.

Mr. ROBB. May I proceed, Mr. Chairman. I am reading from page 4 of this memorandum, starting at the bottom—"It is known"——

Mr. GARRISON. Mr. Chairman, could we have this read first off the record to see what we can make of it, and then see if it belongs in a part of the record which conceivably one day may become public? I am not saying that there is any plan to make it public, but this is a record of some historic character, and I think——

Mr. GRAY. I would like to ask Mr. Robb whether this is going to be the basis of a question to Dr. Bacher?

Mr. ROBB. I think it relates to Dr. Bacher's testimony, and I want to put some questions to him about this.

Mr. GARRISON. Does it relate to him personally?

Mr. ROBB. Not at all.

Mr. GARRISON. Why can't you put your question without reading it from an unknown document?

Mr. ROBB. Because I am conducting this questioning and I would like to do it in my own way.

Mr. GARRISON. I am conducting my question to the Chair.

Mr. ROBB. You asked me and I answered it.

Mr. GRAY. Where is that?

Mr. ROBB. Starting at page 4 of the report, at the bottom of the page, the next to the last paragraph.

Mr. GRAY. And how much?

Mr. ROBB. Reading from there through the first full paragraph on page 6.

Mr. GRAY. I am going to allow counsel to read these portions he has indicated.

Mr. ROBB. Mr. Chairman, may I suggest that there are certain minor references in here to FBI reports which we are not permitted to disclose which is why I was going to undertake to read it to give counsel the benefit of it with those references deleted.

This board, as I understand it, is to base its decision in this matter upon the whole file before it. If counsel does not want to hear this, and wants the board to go ahead and consider it without him hearing it, that is all right with me.

Mr. GARRISON. Mr. Chairman, what I object to is reading into the record what I take to be allegations about Dr. Oppenheimer's past which are unsupported by anything approaching a signature, without any knowledge of the use to which this was put, or the source of it, without any possible means of our knowing what it is going to say. It seems to me to read an anonymous allegation of that kind about Dr. Oppenheimer into the record——

Mr. GRAY. I don't believe that the portion that Mr. Robb proposes to read makes allegations with respect to Dr. Oppenheimer. Am I correct?

Mr. ROBB. It concerns certain individuals employed on the project. I apprehend that this report was before Dr. Bacher at one time or another.

Mr. GRAY. This report clearly came out of the Atomic Energy Commission files. As Mr. Robb said, I think it is safe to assume that even though Dr. Bacher may not remember seeing this particular document, that at one time he certainly had seen it in connection with the clearance procedures involved.

Mr. GARRISON. Mr. Chairman, if this was a part of the material which Dr. Bacher went over, why can't it be shown to him now, and then questions put to him about individuals, rather than reading this into the record. There certainly can be no objection to a former member of the Commission reading something from the Commission's files, as I understand it, particularly if he has already read them in the past.

Mr. ROBB. I certainly would not expect Dr. Bacher to remember this offhand.

Mr. GARRISON. Why can't you show it? Mr. Chairman, wouldn't that be the appropriate procedure to let Dr. Bacher look at this, and then if counsel wants to ask him questions about particular individuals, he can.

Mr. ROBB. Mr. Chairman, I want to ask Dr. Bacher questions about this memorandum. I think the record ought to reflect what it is before I start to ask him questions about it.

Mr. GRAY. I think I shall have to talk with my colleagues on the board. I understand you are objecting to the reading.

Mr. GARRISON. Yes, sir. I don't object if it is shown to Dr. Bacher so he may read it, and then questions put to him about particular individuals, whatever questions that counsel wants to ask. I just have this feeling that to read into the record these anonymous passages about particular people is not sound procedure.

Mr. ROBB. Of course, Mr. Chairman, I can't quite follow my friend because this report is before the board in its entirety. I can't see why putting a portion in the record seems to be such a horrible step to take. The only thing that will happen if I read this is that counsel will get to hear it.

Mr. GARRISON. It also will become a part of the transcipt, which may become a permanent record.

Mr. ROBB. I assume these files are a public record.

Mr. GARRISON. It may become public.

Mr. ROBB. It won't become public through us.

Mr. GRAY. I think it is not unreasonable to assume that some time this transcript may become a public record. I would hope not, but I think we can make no guarantees. I would like to have a consultation with my colleagues on the board. I think we will just move into the other room briefly so we won't have to send all of you out of the room.

(The board withdrew.)

(The board reentered the room.)

Mr. GRAY. After conferring with my colleagues on the board, I am going to suggest that Mr. Robb show this document to Dr. Bacher, and if he wishes to point out particularly the paragraphs which he is now concerned with and then to ask him to question Dr. Bacher on the basis of these paragraphs without reading them into the record.

Mr. ROBB. Mr. Rolander, is it all right for Dr. Bacher to make references to FBI?

Mr. ROLANDER. Yes, but Dr. Bacher should not refer to references in discussion.

Mr. ROBB. May we take time out while he reads it?

Mr. GRAY. Yes.

Mr. ROBB. May we proceed?

Mr. GRAY. Yes.

By Mr. ROBB:

Q. Dr. Bacher, you have read the paragraphs in that analysis to which I referred you?

A. Yes.

Q. Doctor, if the statements made in this analysis about Charlotte Serber are fact, would you have had her on the project at Los Alamos?

A. Could I see this thing again to refer to?

Q. Yes, sir. [Handing.]

Mr. GARRISON. Mr. Chairman, I would like to note for the record that Dr. Bacher's answer to that question, whether he answers yes or no, scarcely seems to me to be relevant to the subject of this inquiry for it has absolutely no bearing on the question of whether Dr. Oppenheimer knew those facts to be true or not, whatever these facts may be. This is a question in the cark about the witness' opinion about something not in the record about some member of the project. I fear that the inference which the question may wish to have drawn is that if the witness answers the question in the negative somehow that will be taken as directed to Dr. Oppenheimer. It just seems not to belong in the record, but I don't want to seem to be argumentative about this, but I do put it to the Chairman very seriously.

Mr. GRAY. Your observation about it is in the record, and I am certain the board will take into account all of the circumstances, including the nature of the memorandum under discussion, and the related matters you pointed out about it.

The WITNESS. Would you repeat the question?

(Question read by the reporter.)

The WITNESS. In order to answer that question, Mr. Robb, I think it is necessary to go back and make a bit of a statement about what the basis for security clearance was at Los Alamos.

We as technical people at Los Aalmos did not put ourselves in the position in any case of making a judgment as to whether scientific people should or should not be a member of the project. This was a question which was left up to the security officers. For example, to take the case of Philip Morrison—I happen to remember this, and it is referred to in the same document which you have just asked me to look at—in his case he was a member of the metallurgical laboratory at the University of Chicago. Some time in the summer of 1944 I was on a recruiting trip for the Los Alamos laboratory. We were desperately trying to get people from other sections of the project to help us in the work out there. I went to the metallurical laboratory, I went to the SAM laboratories in New York, and if I recall correctly, I went also to Oak Ridge. At each of these places I talked to people and approached them with reference to coming to Los Alamos. Finally after finding that some of the people whom I had initially approached were unavailable for security reasons at Los Alamos, I took the precaution of not talking to people until I cleared it with the security officer. In other words, it was clear from this that the responsibility for as to who came to Los Alamos was held with the security office and not with the scientific director or any member of the scientific staff.

In the case of Philip Morrison I interviewed him in Chicago. Subsequently, if I recall correctly, a question was raised as to whether it was advisable for him to come to Los Alamos. We pointed out that he was a very able man, would help us more in our work out there than most of the other people that we might get, and after review somewhere, it was decided that he would come to Los Alamos and he did, and made a number of valuable contributions to the project.

I think this is only to indicate that judgment as to what had to be taken for fact in these matters and the decision as to what ought to be done on that was something which was in the hands of the security officer at Los Alamos.

By Mr. ROBB:

Q. May I interpose, since we digressed a little bit, you have here, have you not, given a judgment on Dr. Oppenheimer as a security matter?

A. I have given my personal opinion.

Q. Yes, sir. Would you give that same personal opinion in respect of Charlotte Serber, assuming that the statements you have read about her in this memorandum are true?

A. I will say this. I don't think Dr. Oppenheimer would not have had her at Los Alamos if he did not think she was reliable.

Q. Would you please answer the question? I am asking for your opinion.

A. I believe I would have relied on the security officer to make a decision on this.

Q. Suppose the security officer told you the facts set out in this memorandum, and asked you for your opinion as to whether she should be there or whether she should not, what would you have done?

A. In any security case, there are lots of acts and these may only be a part of the facts. A security judgment, as I understand it, is as a matter of balancing one thing against another.

Q. In other words, you don't think you are qualified to give an opinion?

A. I do think I am qualified to give an opinion.

Q. Would you give one on Charlotte Serber?

A. In answer to that question, I think you need all the facts and not just what you have given me.

Q. Assuming that these facts were given to you, do you think that taking those facts as data that she had any business on that project?

A. It seems to me that these are not necessarily facts. They are stated in the form of it as an opinion.

Q. I am asking you to assume that they are facts.

A. Could I read them again, please?

Q. Yes.

Mr. GRAY. I would say that the witness does not have to assume they are facts, but for the purpose of a question only you may. This is not to get you on record.

Mr. ROBB. No, I am not asking you to say they are facts. I am merely trying to explore the witness' criteria of security standards.

The WITNESS. Mr. Chairman, as you can see from my answers, I am a little reluctant to answer hypothetical questions.

Mr. GARRISON. I think, Mr. Chairman, that when counsel put the question to Dr. Bacher, I thought he was making a comparison or parallel between that question to Dr. Bacher about Mrs. Serber, and the question I put to Dr. Bacher about his opinion of Dr. Oppenheimer. Quite clearly his opinion about Dr. Oppenheimer is based on many long years of intimate association in Government work, and I think to analogize that to an opinion about Mrs. Serber based on a hypothetical set of facts is quite misleading.

Mr. ROBB. I don't think the Doctor is misled. Have you now read that again, Doctor?

The WITNESS. I have now read it again.

I think, Mr. Robb, that there is a great difference between assuming that is a fact, and proceeding on the basis. I think the real question comes up as to whether that is a fact or not.

By Mr. ROBB:

Q. Assume that you knew that these statements were the truth about Mrs. Serber, would you then be of the opinion that she should be cleared for service on a secret war project such as Los Alamos?

A. In the case that all those facts are correct as stated, and were current at the time, I would say no.

Q. Yes, sir. What was Mrs. Serber's job down there?

Mr. GARRISON. Mr. Chairman, please believe me, I am not trying to delay or obstruct. I think since we now have had put to the witness questions about these facts, those facts now ought to go in the record. I hoped when counsel had shown this document to Mr. Bacher that the course of questioning would have allowed a different line. But the record as it now reads is absolutely blind and incapable of evaluation by us. While I had hoped to avoid this kind of reading of this raw undigested anonymous material into the record I now see no recourse but to have it done, because otherwise the transcript is left in a meaningless state of affairs. I think it better go in. I am sorry it has taken this turn. But I didn't suppose that the questions would bring about that result.

Mr. ROBB. I am perfectly satisfied with the record as it stands, Mr. Chairman. Mr. Garrison didn't want it read. I wanted to read it. I foresaw exactly what would happen. Now he wants it read.

Mr. GARRISON. Mr. Chairman, I really think it should go in. I have thought from the argument that the question of counsel would put would be of an entirely different category than to say assuming these facts to be true, what would your opinion have been. I think we now ought to have the facts in the record. I would like to have them read into the record so we know what we are talking about.

Mr. GRAY. The Chair proposes to suggest that these paragraphs be read into the record, but first I would like to know whether either of my colleagues feel that is not a proper procedure.

Dr. EVANS. It is all right. If Mr. Garrison wishes to have it read, it is all right with me.

Mr. GARRISON. I do think the end result is an objectionable one, but it is less objectionable now to have it in than to leave it blank.

Mr. ROBB. Mr. Chairman, I am a little bit confused. Am I to read just the section dealing with the Serbers, or all the paragraphs I have in mind?

Mr. GARRISON. I would just do the Serber one.

Mr. ROBB. All right. I will have to leave out certain portions.

Mr. GARRISON. Would you indicate where the portions are left out.

Mr. ROBB. "It is known"—I am leaving out something—"that subject was responsible for the employment on the project at Los Alamos of a number of persons"—I left out a word—"known to be either Communists or active Communist sympathizers"—omissions—"Robert and Charlotte Serber. With respect to the persons mentioned above, it is known that Charlotte Serber's family is prominent in Communist Party ranks in Philadelphia, Pa.; that she herself was probably a party member and possibly a member of the Comintern, and that she has always been active in radical activities and front organizations wherever she has lived. Her husband, Robert Serber, perhaps under her influence, has been active in the same circles since he married her, although there is no conclusive evidence that he is a party member. Robert Serber"—blank, blank—"were graduate students of the University of California under subject."—blank, omissions. "It is known that all of them"—referring to certain other persons and the Serbers—"perhaps influenced by subject were extremely active in Communist activities on the campus at Berkeley during this time. After finishing their studies all"—blank—"of the men went to the University of Illinois where they are also known to have associated with known Communists, and to have taken part in Communist activities. When the Manhattan Project came into being, the Serbers were employed at Los Alamos by subject"—omissions—"all of these people were very close personally to subject and there is little room to doubt that he was aware of their sympathies and activities. In evaluating this information, it must be kept in mind that both"—

blank—"and Serber were technically very well qualified for the work for which subject wanted them, despite their youth."

I think that is all on Serber.

Mr. GRAY. Just one other place. After a blank, "Serber, too, is highly regarded."

Mr. ROBB. Yes. "Serber, too, is highly regarded."

Mr. GRAY. I think the record should show that this, without omissions that are important to this discussion, represents excerpts from a memorandum in the Atomic Energy Commission files on Atomic Energy Commission stationery, entitled, "Memorandum to files. Subject: Analysis of Report on J. Robert Oppenheimer." Unsigned, and dated March 14, 1947, and with no identification as to its author.

Mr. GARRISON. Mr. Chairman, may I now point out what seems to be the vice in this matter of using as a hypothetical case to Dr. Bacher—I have no objection to putting hypothetical cases to him to see how his mind works on these things—but here are some people called Serber. All we know on the record is that Dr. Serber was cleared by a distinguished panel of which Admiral Nimitz was chairman, and cleared by the Atomic Energy Commission itself for top secret Q clearance. Presumably this material was taken into account. It is certainly clear from the Commission's criteria that in evaluating Professor Serber's qualifications, his wife's background must also have been taken into account. Here now are two people that I don't know from Adam, but it seems to me most unfair to use them as a framework for a hypothetical question. A document of this kind, anonymous and full of blanks, in the case of people who have been cleared by Admiral Nimitz and Mr. Neyland and General Joyce, and by the Commission itself. To me it serves no purpose in proceeding and is most unfair to all concerned. It leaves the inference in the record that in spite of the subsequent clearance of the Serbers that——

Mr. ROBB. Of the who?

Mr. GARRISON. If Dr. Oppenheimer ever sees them at all it is something very wrong. This is a backhanded accusation against the Serbers in this record—I am not defending them at all—but I am questioning the validity of this procedure. I would specifically request the chairman that hypothetical cases to Dr. Bacher be put in the form of X or what have you, and not names of people to be used for material of this character.

Mr. ROBB. Mr. Chairman, Mr. Garrison keeps arguing about the clearance of the Serbers. So far as I know, Mrs. Serber has never been cleared by the Atomic Energy Commission but she was employed at Los Alamos as a librarian. She had access to all the classified information that was there. My questions to Dr. Bacher were directed at his opinion of Mrs. Serber. I read the matter about Mr. Serber just because I felt sure if I didn't read it all, Mr. Garrison would say I should have read it all. I have not asked him anything about Mr. Serber yet. May I proceed, Mr. Chairman.

Mr. GRAY. I think the witness has already answered the question.

Mr. ROBB. Yes.

By Mr. ROBB:

Q. Doctor, do you know what evidence might have been presented to the board which cleared Dr. Serber?

A. No, I was not present.

Q. May I ask you, Doctor, do you recall whether or not in 1947 the Commission had its security officer prepare some analysis of the FBI reports in the file for you?

Mr. GARRISON. Which file is this, Dr. Oppenheimer's file?

Mr. ROBB. Yes.

The WITNESS. Mr. Robb, I remember reading a summary but I don't believe I remember anything that would allow me to answer your question either in the affirmative or negative.

By Mr. ROBB:

Q. I notice here in Mr. Jones' memorandum to the file which refers to entries which is March 10, 1947, the last page of that contains this notation, "The results of the discussion with Mr. Clifford were reported to the Commission at a meeting at 5 p. m. this afternoon." That would be March 11.

Mr. GARRISON. Is this the document read into the record before?

Mr. ROBB. Yes, sir. "At that meeting the general manager reported that a detailed analysis of the FBI summary was in process of preparation by the Commission's security staff, as an aid to evaluation."

By Mr. ROBB:

Q. Assuming such summary was made, no doubt you had it before you?
A. It sounds so, but I don't remember it, Mr. Robb.
Q. I was not there, but my thought is that probably this paper that I showed you which purports to be an analysis of the report on Dr. Oppenheimer was the analysis referred to in that note of March 11.
A. I am afraid I can't help you on that.
Mr. Chairman, could I make an observation on this last discussion?
Mr. GRAY. You certainly may.
The WITNESS. In view of the fact now that this has been read into the record, I tried in my answer to you about Mrs. Serber on the hypothetical question, to make it clear that if that information was (a) fact, and (b) current, that the answer I gave then applied. I think the question that I had in my mind, and the reason I found it so difficult to answer the hypothetical question which you posed, was that I would assume that the board and also the Commission in reviewing a case did not believe that was either (a) fact, or (b) current. I think these are the pertinent questions in making a decision.

By Mr. ROBB:

Q. Are you talking about the Commission or the board considering Dr. Oppenheimer's case?
A. No, I am talking about the Serber case, which is the question you asked me about.
Q. Of course, Mr. Serber's case was distinct from that of Mrs. Serber. My question related to Mrs. Serber, and perhaps to make it perfectly clear whether I am getting at it, I will ask you this: If you had that data before you in 1942 and 1943, and had to make a decision as to whether Mrs. Serber would come to Los Alamos, would you have decided that she should come or that she should not come?
A. Once again my answer to you would be that I would leave that to a full investigation by security officers under those circumstances, because this does not constitute a full record.
Q. But assume that the investigation disclosed that those statements were true, and you then had to make the decision, what would it have been?
A. I said if they were true facts and were current, that is, applied as of that day, which is not clear, I might add, from the record you have read, then I would say no.
Mr. ROBB. I think that answers my question. Thank you. That is all I have to ask.
Mr. GRAY. Dr. Evans.
Dr. EVANS. Dr. Bacher, did you have a graduate student at your school by the name of Sheehan in the last 2 years?
The WITNESS. It could be, but I don't recall him in physics.
Dr. EVANS. He was a chemistry student, but he took a lot of physics. He was one of my students, and I just wondered if you knew him.
The WITNESS. I did not know him.
Dr. EVANS. Dr. Bacher, you have never been a Communist?
The WITNESS. No.
Dr. EVANS. Never been a fellow traveler?
The WITNESS. No.
Dr. EVANS. Have you belonged to any of those subversive organizations that the Attorney General listed?
The WITNESS. As fas as I know I have never belonged to any organization that is on the Attorney General's list.
Dr. EVANS. Do you think that a man can be completely loyal to his country and still be a security risk?
The WITNESS. Yes. If he is a drunkard, he might be a security risk and be completely loyal.
Dr. EVANS. Just suppose because of his associates.
The WITNESS. It seems to me that on this question of association that is a different question. If you have full confidence in a man's character and his integrity and his discretion, I don't believe that one can rule him out as a security risk on the basis of his knowing people who have in the past had connection with the Communist Party, mostly because I don't believe there would be many people left in the United States that would satisfy that criterion.
Dr. EVANS. Then you are answering the question this way. You think a man can be completely loyal, and if he is completely loyal, he is not a security risk? Is that what you are saying?

The WITNESS. I believe I specified a little more than that, Dr. Evans. I said, if I recall correctly, that if he is a person of high character, a person of integrity, and a person who is discreet, and is at the same time a person who is clearly loyal, then he is not a security risk, assuming of course that other criteria such as he is not a drunk or things of that sort are included.

Dr. EVANS. You think Dr. Oppenheimer is always discreet?

The WITNESS. I do.

Dr. EVANS. Do you think he was discreet when he refused to give the name of somebody that talked to him? Do you remember that Chevalier incident?

The WITNESS. I don't remember the point you refer to, I am afraid.

Dr. EVANS. Someone approached Dr. Oppenheimer about getting security information, and Dr. Oppenheimer refused to give the name of the man that approached him.

The WITNESS. I thought he did give the name, Dr. Evans.

Dr. EVANS. He refused twice I think, and for quite a long time he didn't give it. Am I right on that?

Mr. ROBB. I believe that is correct.

Mr. GARRISON. That is right.

Dr. EVANS. Was that discreet?

The WITNESS. Could you ask the question again, Mr. Evans?

Dr. EVANS. Yes. If you were on a project, and you had access to a lot of secret information, and I came to you and told you that there was somebody that knew that I could give information to if you would give it to me, would you have gone and told somebody that I had approached you?

The WITNESS. I think that should have been reported.

Mr. GARRISON. Mr. Chairman.

Dr. EVANS. Maybe I put the question very badly.

Mr. GARRISON. All right. I accept it as a hypothetical question.

Dr. EVANS. You have never been approached by people?

The WITNESS. No, never.

Dr. EVANS. Do you believe a man should place loyalty to his country before loyalty to a friend?

The WITNESS. Yes.

Dr. EVANS. That is all I want to ask.

Mr. GRAY. Dr. Bacher, did you know—I am not sure whether this was covered in earlier testimony—David Hawkins?

The WITNESS. Yes.

Mr. GRAY. Did you know him well?

The WITNESS. I met him first at Los Alamos, Mr. Chairman, when he was a member of that laboratory. I cannot remember exactly when he came to Los Alamos. I would guess some time in the last part of 1943 or early 1944. I met him there, knew him fairly well at Los Alamos, and have known him a bit since the war. He lived in Washington for a time and did some work, I think, at the end of the war in finishing up a history that he had been preparing of the Los Alamos project. I knew him a bit while he was here in Washington. I have not seen him now for some time. I believe he is in Colorado.

Mr. GRAY. At the time you knew him at Los Alamos or later, did you have any information about his what I believe are sometimes referred to as political affiliations? Did you know anything about his connections?

The WITNESS. I did not discuss politics with him. I believe I read some testimony since that he has had and I must say I was very surprised at what came out in that testimony, because I believed Hawkins and believe him today to be a person of character, and I don't believe one who could today subject himself to the rigid control that would be required if he were to have the affiliations of which I believe he has testified since then.

Mr. GRAY. I don't believe he has testified to any current affiliation.

The WITNESS. No, I meant in the past.

Mr. GRAY. You testified that you interviewed Philip Morrison.

The WITNESS. Yes.

Mr. GRAY. With respect to his employment.

The WITNESS. Yes, that is right.

Mr. GRAY. Did you know anything about his political affiliations?

The WITNESS. I didn't at that time, no.

Mr. GRAY. Would it surprise you if he had had Communist associations or connections as a personal matter?

The WITNESS. Today?

Mr. GRAY. Perhaps I am not making my question clear. My question is whether it would surprise you today to know that he then at the time you interviewed him had political connections which you would feel would not make him a good security risk today?

The WITNESS. After all, Mr. Chairman, in the meantime I have read some of these things so I could not easily be surprised by it.

Mr. GRAY. Were you surprised when you read them?

The WITNESS. I was surprised when I found out in that particular case.

Mr. GRAY. When you interviewed people for the laboratory this kind of question was not asked?

The WITNESS. No, I had no relation to that. Any interview by a scientific person was concerned entirely with the question of whether that man would be an appropriate addition to the laboratory on scientific and technical ground. The question of whether he came to the laboratory or not was left to the security officer to pass on.

Mr. GRAY. That was the system you used; that probably is not the system today, is it? Everybody concerned with the project is expected to take some interest in security?

The WITNESS. Yes, I would say also at the time I interviewed Morrison, I didn't know anything at all about his background.

Mr. GRAY. On the question of identification of people and with no conclusions to be drawn from the question, did you know Fuchs well?

The WITNESS. I knew him reasonably well at Los Alamos, because he was a member of the Theoretical Division and did a certain amount of work for the Division for which I was responsible there. I didn't know him well outside work, but within the laboratory there I saw him fairly frequently. I probably knew 8 or 10 members of the Theoretical Division better than I knew Fuchs, and my knowledge of him was entirely through the work of the project.

Mr. GRAY. He was considered to be doing a good job?

The WITNESS. He did a good job, I believe.

Dr. EVANS. You were very surprised when that came out?

The WITNESS. I was certainly surprised.

Dr. EVANS. You might have lost a little faith in your own judgment of people?

The WITNESS. I didn't know him very well personally, that is, I didn't spend many hours with him. I saw him mostly in a scientific and technical capacity. So I didn't have an opportunity to form a personal judgment of Fuchs very much. He was a very quiet, very retiring person.

Mr. GRAY. Would you say, Dr. Bacher, that aside from the security aspect, you were responsible for the employment of Philip Morrison as a member of the project? I asked that badly. You have already testified that you didn't concern yourself with the security angle.

The WITNESS. Yes.

Mr. GRAY. Did Dr. Oppenheimer suggest Morrison as a prospect?

The WITNESS. I don't recall that he did. I think as a matter of fact that I interviewed him at the metallurgical laboratory and how I got the list of people that I interviewed at the metallurgical laboratory, I just don't remember. I think it was presented by the metallurgical laboratory of people on the project whom they thought would be helpful in the work at Los Alamos, and who in the emergency they could manage to get along without or were willing to get along without.

Mr. GRAY. In any event, you were exercising your own best judgment in interviewing Morrison for possible employment?

The WITNESS. Yes, I think a question was raised about Morrison. If I recall correctly, we from Los Alamos said he was one of the people that would be most useful to us from the scientific and technical end. The question was reviewed, I don't know whether by local security people or whether in Washington, and Morrison then came to Los Alamos. I think this was along about in the early fall of 1944.

Mr. GRAY. Do you have any more questions?

Mr. GARRISON. May I ask one more question about Morrison?

REDIRECT EXAMINATION

By Mr. GARRISON:

Q. Did you interview a group of young men at the metallurgical laboratory?
A. Yes.
Q. And he was one of a group?
A. Yes.

Q. And in interviewing them what did you seek to find out?
A. I sought most to find out what their work had been at the metallurgical lab, and whether they would fit into the work that we had to do at Los Alamos and in part to find out whether they would be willing to pick up their belongings and their families and move out to New Mexico to undertake work on that project.

Most of the people wanted to know quite a little bit about what the circumstances were, because they didn't have very good information on this point, and they were unwilling to make a decision in the matter until they learned a little more about the physical surroundings, and so on.

Q. And had all of these young men been cleared for work on the metallurgical project?
A. Yes.
Q. Have you ever been fooled in your judgment of the loyalty of anybody whom you have known as long and as intimately as Dr. Oppenheimer?
A. No.
Q. Do you think you could be?
A. I doubt it.
Mr. GRAY. Do you have any more questions?
Mr. GARRISON. Since Dr. Evans put a hypothetical question about the Chevalier case, I think I would like to read from the Commission's letter and put a question myself.

By Mr. GARRISON:

Q. I am reading, Dr. Bacher, from the Commission's letter of December 23, 1954, on page 6 which you testified you had read, but I want to refresh your memory of it.

"It was reported that prior to March 1, 1943, possibly 3 months prior, Peter Ivanov, Secretary at the Soviet Consulate, San Francisco, approached George Charles Eltenton for the purpose of obtaining information regarding work being done at the Radiation Laboratory for the use of Soviet scientists; that George Charles Eltenton subsequently requested Haakon Chevalier to approach you concerning this matter; that Haakon Chevalier thereupon approached you, either directly or through your brother, Frank Friedman Oppenheimer, in connection with this matter; and Haakon Chevalier finally advised George Charles Eltenton that there was no chance whatsoever of obtaining the information. It was further reported that you did not report this episode to the appropriate authorities until several months after its occurrence; that when you initially discussed this matter with the appropriate authorities on August 26, 1943, you did not identify yourself as the person who had been approached, and you refused to identify Haakon Chevalier as the individual who had made the approach on behalf of George Charles Eltenton; and that it was not until several months later, when you were ordered by a superior to do so, that you so identified Haakon Chevalier. It was further reported that upon your return to Berkeley following your separation from the Los Alamos project, you were visited by the Chevaliers on several occasions; and that your wife was in contact with Haakon and Barbara Chevalier in 1946 and 1947."

In Dr. Oppenheimer's answer at page 22, he said as follows:

"I knew of no attempt to obtain secret information at Los Alamos. Prior to my going there my friend Haakon Chevalier with his wife visited us on Eagle Hill, probably in early 1943. During the visit, he came into the kitchen and told me that George Eltenton had spoken to him of the possibility of transmitting technical information to Soviet scientists. I made some strong remark to the effect that this sounded terribly wrong to me. The discussion ended there. Nothing in our long-standing friendship would have led me to believe that Chevalier was actually seeking information; and I was certain that he had no idea of the work on which I was engaged.

"It has long been clear to me that I should have reported the incident at once. The events that led me to report it—which I doubt ever would have become known without my report—were unconnected with it. During the summer of 1943, Colonel Lansdale, the intelligence officer of the Manhattan District, came to Los Alamos and told me that he was worried about the security situation in Berkeley because of the activities of the Federation of Architects, Engineers, Chemists and Technicians. This recalled to my mind that Eltenton was a member and probably a promoter of the FAECT. Shortly thereafter, I was in Berkeley and I told the security officer that Eltenton would bear watching. When asked why, I said that Eltenton had attempted, through intermediaries, to approach people on the project, though I mentioned neither myself nor Chevalier. Later, when General Groves urged me to give the details, I told him of my conversation with Chevalier. I still think of Chevalier as a friend."

Supposing that the evidence here showed that Dr. Oppenheimer's statement about the approach by Chevalier included a statement by him to the security officers to whom he initiated the mention of the name of Eltenton the fact that Chevalier, whom he did not name, had approached three people; that actually Chevalier, according to Dr. Oppenheimer's testimony, approached him only; that he invented the fact that there were three people and not one; that in his discussions with the security officers he said that Eltenton had a contact with the Russian consulate and that there was somebody that had microfilm or some other method of getting secret information to Russia and that those details were also inventions.

Taking al that now into account, and taking further into account the fact that General Groves pressed Dr. Oppenheimer for the name of the intermediary, namely, Chevalier, that Dr. Oppenheimer said he would tell him if ordered and General Groves said that he did not want to order him and asked him to think it over and that later General Groves said he must have the name and that if it were not told to him he would have to order it, that Dr. Oppenheimer revealed the name of his friend Chevalier to General Groves. Taking all of that into account and assuming for the purpose of this question that this is the record before you, would your previous answer about your confidence in Dr. Oppenheimer's loyalty be altered in any way in your mind?

Mr. ROBB. May I just enter my usual caveat to the record as to the accuracies of the hypothesis, Mr. Chairman.

Mr. GARRISON. Quite right.

Mr. GRAY. That means, Dr. Bacher, that Mr. Robb does not necessarily accept——

The WITNESS. I fully understand that.

Mr. GRAY. It puzzled Dr. Bacher.

The WITNESS. Thank you.

Mr. GRAY. This is Mr. Robb's statement for the record and now you can proceed with the answer.

By Mr. GARRISON:

Q. I say that is my version of the hypothesis.
A. Would you restate the question, not the whole hypothesis.
Q. Having all of this before you now, you previously testified that on the basis of your experiences with Dr. Oppenheimer, you were confident of his loyalty to the United States and also that you considered him to be a good security risk.

I ask you now, accepting what I told you to be the case for the purpose of the discussion, would your conviction about the matters that you expressed about his loyalty and his security be the same.

A. No. I think he made a mistake in not reporting it immediately, but this does not change my judgement of Dr. Oppenheimer.

Q. When you say no, you mean by that——

Mr. ROBB. I think he meant yes, if there is any question.

The WITNESS. The question was, Did it change my opinion?

Mr. GARRISON. That is correct.

The WITNESS. The answer is "No." I believe Dr. Oppenheimer made a mistake in not reporting that incident immediately, but what you have told me and read into the record does not change my judgment given previously.

By Mr. GARRISON:

Q. Do you think that Dr. Oppenheimer would today do what he did in 1943 in this incident if the facts I have told you are the case?
A. I do not. I think he realizes he made a mistake on that by your statement there.
Q. I do not want you to accept my statement.
A. By the statement in the record and I believe the same thing of my own knowledge.

Dr. EVANS. That is, he was not particularly discreet at that time.

The WITNESS. I think this is more a question of judgment rather than discretion.

Dr. EVANS. He did not have good judgment at that time. How is that?

The WITNESS. It seems to me this is more a question of judgment than discretion.

Dr. EVANS. I do not know the difference.

By Mr. GARRISON:

Q. Was there involved in this case, Dr. Bacher, as I put it to you any leakage of information by Dr. Oppenheimer?

A. No, not that I know of. The word discretion is usually used in security matters with reference to someone saying something that might conceivably be classified where someone can hear it who is not authorized to receive the information. That is why I made that difference.

Q. Would you say it was the fact here that quite contrary to the leaking of information, Dr. Oppenheimer declined to have anything to do with even a notion of leaking information and after much delay revealed finally the names of the people above?

A. He seems to have reported the incident fully, judging from what you read me. The only question seems to be one of time.

Mr. GARRISON. That is all.

Mr. ROBB. That is all. I have no further questions.

Mr. GRAY. I have one question and this won't take long.

There are those in the scientific community today, Dr. Bacher, who think that the fact of this proceeding is an outrage. There are some, I say, would feel that way. You have heard the view expressed?

The WITNESS. I have heard that it has been expressed.

Mr. GRAY. I do not say that it a universally held view but there are those who hold it.

As a former member of the Commission, I would like to ask you whether you feel that this matter is of such serious consequences that this kind of hearing is a good thing. I am not talking about the publicity angles and the rest of it. I mean in the interest of the Government and of the individual himself.

I will put it this way: If such a hearing had been had in 1947, it would not have been an outrage, would it?

The WITNESS. I find it very difficult to answer that question, Mr. Chairman. In the first place, it is hard to know what one means by an outrage.

Mr. GRAY. That is my characterization and I agree that it is bad to have it in the record. I should not perharps express it this way, but to say that there are those in the scientific community who see absolutely no justification for this hearing, is that an exaggeration of a point of view which exists?

The WITNESS. It may exist. I have tried rather hard not to talk to too many people before testifying here and I do not have a good view of what people think, so I cannot answer your question really very well on that.

With respect to the procedures that AEC has for handling security cases, these, of course, were worked up rather carefully by the Commission over a long period of time. Our generatl counsel pointed out to us that the essence of a proper system for handling security cases was the procedure and, therefore, the Commission in setting up the present procedure tried hard to follow as nearly as possible those procedures which over the years have come to be recognized in courts of law. This can't be followed fully where questions touching on classified information and involving classified information must appear. This poses very grave difficulties.

I can think of no way, for example, in which hearings of the present sort could be held in public as some people have requested. I just do not know how a thing like that could be done. I am not sure that I get the flavor of your question.

Mr. GRAY. That wasn't directly responsive, but do you feel that having established the procedures, I suppose while you were a member of the Commission——

The WITNESS. Yes.

Mr. GRAY. The Commission having established them and I assume your having felt at the time that they were fair, do you as a former Commissioner and as a scientist and as a former associate and a friend of Dr. Oppenheimer feel that the Commission should not have instituted this proceeding?

The WITNESS. That, Mr. Chairman, would depend on my assessment of whether there has been substantial new derogatory information brought to bear about Dr. Oppenheimer. I have not seen such in reading the set of charges that have been brought up and listed by the general manager that were not known before. There may be information which I do not have. But on the list of charges that were there, I did not see any substantial amount of new derogatory information.

Mr. GRAY. Without in any way endorsing or rejecting the information about the hydrogen bomb, that certainly is new since 1947. I am speaking now of the material in the general manager's letter.

The WITNESS. Yes.

Mr. GRAY. Also, I think that it is true that there are files which are in existence which were not available to the Commission in 1947.

The WITNESS. You see, I am not aware of that, Mr. Chairman.

Mr. GRAY. Again, I am not suggesting that there is anything that should or should not be concluded from those files, but that is the fact.

Finally, I suppose the question of formal action of clearance of Dr. Oppenheimer in 1947 remains to be a matter surrounded by some mystery. Would counsel accept that in view of the fact that the reference to this action which apparently finally was written down in August referred to action which took place in February, although in fact any clearance which may have been passed upon by the Commission must have been done by it in March and there is some confusion. I do not cite this as having a bearing on the ultimate question of Dr. Oppenheimer's clearance as much as having a bearing on the propriety of these proceedings.

If I seem to be making an argument, it is not my intention, but I was interested actually in having your view because, in a sense, you have been on both sides of this kind of thing.

The WITNESS. Let me see if I can answer your question this way: If what I read in the papers has been correct and most of my information on this does come from reading newspapers, there seems to be two possible ways in which the case could be handled. Either the Commission could have, on the occasion of the case being raised, again recommended to the President that there be an administrative clearance, either by the Commission or directly by the President, or as the second alternative a hearing could be set up.

I presume from what I have read in the papers that the President made the decision that there should be a hearing. These, I think, are the only two alternatives as far as I know that exist. There may be others with which I am not familiar. With that decision, I think a hearing is being held under all of the regulations that have been set up and the procedures of the AEC.

I find it very difficult to answer hypothetical questions without all of the information that went into this decision.

Mr. GRAY. I think I should, as chairman, make an observation for the record, that an assumption about the participation of the President of the United States in this matter is the assumption of the witness.

The WITNESS. It was only what I read in the paper, Mr. Chairman.

Mr. GRAY. We will allow the witness certainly to report anything of his recollection of what he has read in the press, but I do not want to involve the President of the United States in this proceeding, because I have no information in that regard myself.

Mr. GARRISON. Mr. Chairman, I think a little while back you put a question to counsel on this side which remained unanswered, when you said wouldn't counsel agree that there was confusion as to whether Dr. Oppenheimer had been cleared——

Mr. GRAY. Whether it was formal action.

Mr. GARRISON. If I might be permitted to respond——

Mr. GRAY. You certainly may.

Mr. GARRISON. I would say at this point we simply do not know. I do know precisely what the course of action was that was taken. I made a request a little earlier today for a copy of the minutes of the August meeting relating to it which has been taken under advisement. I have some other questions having to do with the record which I would like to put to the board in the morning. I do not want to take your time this afternoon.

Mr. ROBB. Mr. Chairman, as I advised the chairman, I have to leave. I would like to leave as soon as I may. Does Mr. Garrison have any more questions?

Mr. GARRISON. No, sir.

(At this point, Mr. Robb departed from the hearing.)

Dr. EVANS. If you had been a free agent and not connected with these projects, just an ordinary of the country, and you had been asked to serve on this panel as we have been, would you have thought it your duty to do so?

The WITNESS. Yes.

Mr. GRAY. Thank you very much, Dr. Bacher. We appreciate your coming here.

We are recessed now until 9:30 in the morning.

(Thereupon, the hearing was recessed at 5:30 p. m., to reconvene at 9:30 a. m., Tuesday, April 27, 1954.)

UNITED STATES ATOMIC ENERGY COMMISSION

PERSONNEL SECURITY BOARD

IN THE MATTER OF J. ROBERT OPPENHEIMER

ATOMIC ENERGY COMMISSION,
BUILDING T-3, ROOM 2022,
Washington, D. C., Tuesday, April 27, 1954.

The above-entitled matter came on for hearing, pursuant to recess, before the board, at 9:30 a. m.

Personnel Security Board: Dr. Gordon Gray, chairman; Dr. Ward T. Evans, member; and Mr. Thomas A. Morgan, member.

Present: Roger Robb, and C. A. Rolander, Jr., counsel for the board; J. Robert Oppenheimer, Lloyd K. Garrison, Samuel J. Silverman and Allan B. Ecker, counsel for J. Robert Oppenheimer; Herbert S. Marks, cocounsel for J. Robert Oppenheimer.

PROCEEDINGS

Mr. GRAY. The chairman wishes the record to show that following Dr. Bacher's appearance as a witness, the chairman conferred with Mr. Garrison and Mr. Robb. The chairman suggested to counsel that the board was willing to strike that portion of Dr. Bacher's testimony which related to the memorandum in the AEC files, dated March 14, largely on the ground that the memorandum in question was unsigned and unidentified.

The chairman stated that his suggestion was also related to Mr. Garrison's objection that the memorandum in question introduced into the record statements about the Serbers which were unidentified in origin. The chairman made it clear to counsel that the board does not feel there is any question of impropriety, but wished to take into account fully every possible consideration of fairness as far as the record is concerned.

Mr. Robb indicated that he had no objection to this procedure. Mr. Garrison felt that it would be a mistake, once the record was formed, to strike this portion of the record.

Is that correct, or is any of that incorrect? I would like help on this, because I am simply trying to reflect what the facts are.

Mr. ROBB. It is entirely correct as far as I am concerned.

Mr. GARRISON. Mr. Chairman, I think I would just say that while I objected to the introduction of the document and the questions based on it, I still hold the views that I then expressed. The matter in fact having come before the board and testimony having been had before us, I think that it should stand in the record.

Mr. GRAY. Under the circumstances, the record will stand.

Mr. GARRISON. Mr. Chairman, before we begin with the witness, I would just like to ask if the minutes of that August 6, 1947, meeting that I asked for yesterday are available?

Mr. ROBB. Mr. Chairman, I am informed by Mr. Mitchell, the General Counsel, that he has taken the matter up with the Commission. Both he and I have recommended that they be made available, but they will not meet until this afternoon, at which time they will make the decision.

Mr. GRAY. I think the record ought to clearly show that only the Commission can make this decision.

Mr. ROBB. That is correct.

Mr. GARRISON. The board can, however, Mr. Chairman, I take it join in the request to the Commission and make it available.

Mr. GRAY. I think it is understood that the board did join in that request.

Mr. ROBB. I think there is no question about that.

Mr. GRAY. General, I would like to ask whether you wish to testify under oath. You are not required to do so. I think in fairness I should say that all witnesses have so testified.

General MCCORMICK. I am perfectly willing.

Mr. GRAY. Would you be good enough to raise your right hand, General? What is your full name?

General MCCORMACK. James McCormack, Jr.

Mr. GRAY. James McCormack, Jr., do you swear that the testimony you are to give the board shall be the truth, the whole truth, and nothing but the truth, so help you God?

General MCCORMACK. I do, sir.

Whereupon James McCormack, Jr., was called as a witness, and having been first duly sworn, was examined and testified as follows:

Mr. GRAY. Will you be seated, please.

I am required to remind you of the existence of the so-called perjury statutes. May we assume that you are familiar generally that there are perjury statutes?

The WITNESS. I am familiar; yes, sir.

Mr. GRAY. I am prepared to review the penalties, if you wish.

The WITNESS. It won't be necessary.

Mr. GRAY. May I ask, General McCormack, if in the course of your testimony it becomes necessary for you to refer to or disclose restricted data, you notify me in advance so that we might take the necessary and appropriate steps in the interest of security?

The WITNESS. All right, sir.

Mr. GRAY. Finally, I should say to you what I try to remember to say to all witnesses, that we consider these proceedings a confidential matter between the Commission and its officials on the one hand, and Dr. Oppenheimer and his representatives and witnesses on the other. The Commission is making no releases about these proceedings. On behalf of the board, I express the hope that witnesses will take the same view of the matter.

The WITNESS. If I may ask, this is as regards public statements.

Mr. GRAY. That is correct.

The WITNESS. Thank you.

Mr. GRAY. I should say further that in your case, there is no military requirement involved about participating in these proceedings and what you might say about them. I think I covered it as well as I could by saying that the board considers these proceedings a confidential matter between the Commission and Dr. Oppenheimer, and their various representatives.

The WITNESS. I had not meant to confuse, sir, but before coming, I told my immediate commander where I was going and the purpose. I wanted you to know that.

Mr. GRAY. That is essential, I think. You have to tell him when you return where you have been and what you have been doing, perhaps?

The WITNESS. Your experience would indicate that.

DIRECT EXAMINATION

By Mr. SILVERMAN:

Q. General McCormack, for the record will you state your rank and branch of service, and your present post, please?

A. I am a major general in the United States Air Force. My present position is vice commander of the Air Research and Development Command, stationed at Baltimore, Md.

Q. You are appearing as a witness at the request of the attorneys for Dr. Oppenheimer?

A. That is right.

Q. Could you tell us a little bit about your present command, what the Air Research and Development Command is?

A. The Air Force, unlike the Army or Navy, has consolidated all of its research and development creative engineering activities in a single command, and all of its procurement, production, supply, and service activities in another. The first is the Air Research and Development Command, and the second is the Air Materiel Command. These two functions are put together in separate packages in the Navy, in the Bureau of Ordnance, Bureau of Aeronautics, and so forth.

The Air Force as the field operating organization and not as Washington policy staff, we have purview over all research and development activities directly supported and sponsored by the Air Force, and are responsible for liaison with corresponding corollary complementary activities of interest to us in other services and indeed in science and industry and throughout the Government.

Q. Does this command include such portions of research and development as have to do with development and use of atomic weapons?

A. We carry the Air Force responsibility in that field, although the major responsibility of course rests with the Atomic Energy Commission.

Q. About how large is the personnel of this command?

A. Approximately 40,000 total on the Government rolls, roughly half military and half civilian, of whom some 25,000 could be said to be engaged in research, development, and testing activities. The rest are supporting groups.

Q. You formerly were Director of the Division of Military Application of the Atomic Energy Commission?

A. From February 1947 to August 1951.

Q. And was it in that connection that you had your contacts with Dr. Oppenheimer?

A. Yes; principally. I have seen him a few times since leaving the Commission, but not at all during the past year.

Q. During the time when you were Director of the Division of Military Application of the Atomic Energy Commission, did you have occasion to observe the work of the General Advisory Committee insofar as that affected matters with which you were familiar, and particularly the work of Dr. Oppenheimer?

A. I would say I got a rather good view of it. It was the usual practice—I don't know how many departures there were—to invite me or my staff in when the General Advisory Committee was discussing in preliminary fashion matters affecting my operating responsibility.

Q. Would you care to comment on the contribution of the General Advisory Committee, and particularly of Dr. Oppenheimer toward helping the atomic energy program, and in particular as far as you could observe it.

A. I have worked with a number of advisory committees in my business. I think the General Advisory Committee was the outstanding one of my experience in terms of its qualifications, its interest in the work, and its consistent effort to be helpful in broadening the base of weapons development, of pushing out into other areas of military interest, generally to the full extent.

I speak in terms of my own responsibility which was below the policy level as regards the Commission. Just generally I would say the committee was continuously interested in doing the very best they could by the weapons program. A committee, of course, is limited in the impact it can have as opposed to the administrative organization.

Q. Did you work fairly closely with the committee and Dr. Oppenheimer during this 4 years or so that you were Director of the Division of Military Application?

A. I saw the committee very frequently. The record would indicate how many meetings they held during that period but I have it in mind it must have been 4 to 6 a year. In addition to that, I saw members of the committee passing through Los Alamos, through the Commission building in Washington.

Q. And that included Dr. Oppenheimer?

A. Yes; I saw a great deal of Dr. Oppenheimer.

Q. Were you familiar with Dr. Oppenheimer's views on the relative division of atomic weapons between strategic air use and use for tactical purposes and continental defense, or is that a very mixed up question? If you understand, will you answer the question I should have put?

A. I take your question to relate to the division of weapons in the stockpile or the division of effort for developing new weapons.

Q. Perhaps you will answer both.

A. I don't think the General Advisory Committee or Dr. Oppenheimer were concerned with the division of actual weapons in stockpile. That is much the question of the design of the weapon for the purpose for which it was created which was one in accordance with military requirements and the program laid out on that basis.

With regard to contemplating future uses of fissionable material when weapons might be developed and fabricated, my recollection is that the General Advisory Committee and of course Dr. Oppenheimer as its leader and spokesman, were very strongly in favor of developing new types which would open new uses for tactical applications, particularly. My recollection may be faulty on this point but I think up to the time I left the Commission, the use of atomic weapons in air defense was not a clear enough picture for any strong views one way or another.

Q. Was it your impression that Dr. Oppenheimer was in favor of limiting the use of atomic weapons for strategic air purposes or strategic air bombing?

A. Setting up a limit which would be effective in a campaign? Not to my knowledge.

Q. Was he in favor of cutting down the proportion of fissionable materials that went into strategic air bombing?

A. As best I can remember this arose only once during the period of my association with Dr. Oppenheimer, and it had more to do with contemplated future uses, if I can make this clear. I recall Dr. Oppenheimer's being a proponent of the school that if you are to get the full military developmental and operational interest in atomic weapons for tactical use, you had to give them something realistic to put in their thinking, such as an understanding that as these uses are developed, material will be available.

This is my statement of the thesis, not Dr. Oppenheimer's. My recollection is that this was a line of his thinking as I understood it.

Q. Did that involve cutting down the amount of material available for strategic air bombing or did he think there would be enough for both?

A. I had not recalled the thesis as being an arbitrary reduction as against some future date, but rather as a factor for planning. War plans are different.

Q. I am not talking about war plans.

A. What you use weapons for when you actually start using them is what the situation requires. I don't recall Dr. Oppenheimer ever denying that.

Q. Did he indicate that this use of atomic weapons was an ever-expanding business, and you have enough materials both for tactical uses and strategic?
A. That I think is a fair statement.
Q. By the way, your present command covers both so-called continental defense and tactical and strategic use of atomic weapons insofar as the Air Force is concerned?
A. That is right.
Q. In the course of your meeting and acquaintanceship with Dr. Oppenheimer, did you feel you came to know him quite well?
A. Oh, yes.
Q. Do you have an opinion as to Dr. Oppenheimer's loyalty to the United States?
A. I never had a question as to it.
Q. Do you have an opinion as to whether he is a security risk, as to his discretion in the use of classified materials, whether it is safe to trust him with such?
A. Nothing in my associations with him would raise the question with me.
Mr. SILVERMAN. That is all.

CROSS-EXAMINATION

By Mr. ROBB:
Q. General, you spoke of the role of the GAC towards helping the atomic energy weapons program. I assume that you followed the debate in the GAC in October 1949 with respect to the development of the thermonuclear weapon?
A. Yes; although I was not specifically present at the time. Perhaps I was not entirely clear in my previous answer, although I think I was. That debate was a debate at the Commission's policy level. I was speaking of my relations with the General Advisory Committee in carrying out the policies that were decided upon.
Q. Were you supposed to represent the views of the military at that time in respect to the thermonuclear weapon?
A. No. The Military Liaison Committee is the normal organization under the law. I was an operating staff officer of the Commission. I did my best to carry military thinking into the Commission, yes, but the formal responsibility rested with the Military Liaison Committee.
Q. What was the military thinking in October 1949 with respect to the development of the thermonuclear weapon?
A. In my understanding the military interest was a very definite interest in going forward with it if indeed it proved to be technically feasible, although questions of scale and rate of effort and what you cut off your programs to encompass new efforts, these were questions. But on the broad question of going ahead, I think the military interest was solid.
Q. In other words, it was a weapon the military wanted?
A. If it could be made; yes.
Q. After the meeting of the GAC of October 29, 1949, and the report which they made on that meeting, did you read the report of the GAC?
A. I must have, although I don't recall any of its particulars. The only thing that is sharp in my memory is that there was a dissent, but even the details of the dissent, I would not be a very competent witness on.
Q. In all events, you were familiar in general with the decision of the GAC?
A. I was generally familiar with it, yes, although I should definitely stipulate that it was not entirely clear to me at the time, nor would it be now, because I have not been in the business for some time, precisely what the question was that the GAC had before it. Whether it was a yes or no decision, shall we or shall we not, or crash versus no increase in the program. I imagine it was a rather complex question.
Q. Was the position of the GAC on the thermonuclear pleasing to the military and to you as a member of the military?
A. I beg your pardon. I didn't hear it.
(Question read by the reporter.)
The WITNESS. I was in disagreement with it.

By Mr. ROBB:
Q. Would you mind telling us why you were in disagreement with it?
A. I think my thought was just about as simple as this. If the weapon is there, if it can be had, how can we afford not to try for it.
Q. Have you remarked, General, in substance that the position of the GAC in that matter was either silly or sinister?

A. I thought as a sort of a professional staff officer that the quick action on a problem which obviously loomed so large, if I had to choose between the words, I would say silly. I drew no sinister implications. Indeed, I could not have stayed with the Commission had I done so, because some of my bosses——

Q. I am not suggesting that you did draw such implications, but have you not remarked that either one of two alternatives was offered; either it was silly or it was sinister?

A. I think that is about it, yes, sir.

Q. So in respect of that action by the GAC, it could not be said that the GAC was in your opinion doing its best by the military weapons program, could it?

A. I had not thought that was necessarily a part of the package. I speak of the General Advisory Committee, and the help they tried to give me in the programs for which I was responsible as being consistent throughout. There was a very large policy question up for discussion. The General Advisory Committee talked it out among themselves, and with the Commission, and initially recommended against a full blast ahead program, anyway. Once the decision was over, I suppose those who had reservations continued to hold them, and certainly enthusiasm for the program fluctuated as the prospects of early technical success fluctuated. But I was not aware of any member of the General Advisory Committee trying to hold back the program.

So far as my efforts to push the program forward, I would always have liked to have had more help from everybody, budget and everything else. I was not aware of anyone trying my feet.

Q. I was talking to you about the decision. I think you have answered the question.

General, you were asked your opinion with respect to Dr. Oppenheimer's trustworthiness and whether you trust him, and you said you would, is that right?

A. From any facts known to me, I would, sir.

Q. Beg pardon?

A. From all the facts known to me, I would; yes, sir.

Q. Have you heard anything about the episode which occurred in 1943 when Dr. Oppenheimer had a conversation with a man named Chevalier in which the possibility of passing information to the Russians was mentioned?

A. I know what I have heard about it since this board was established; that is all.

Q. What have you heard, General?

A. I have heard that Chevalier, who was a friend of Dr. Oppenheimer's in some rather obscure way suggested that there was a channel through which information on the project which Dr. Oppenheimer was by then in charge of, I believe, at Los Alamos, although I think it had not grown up, there was a channel for passing information from this project to the Communist apparatus. I have heard that Dr. Oppenheimer told him that was a horrid idea, but that he waited until some time later before he reported it then to the security organization of the Manhattan project, and having reported it, then, tried for a while anyway to shield his friend, Chevalier, whom he thought was not really involved in it until General Groves asked him a direct question at which time he told the whole story. I am repeating my recollection of reading newspapers and hearing conversations on the matter.

Q. Of course, you are not familiar with what Dr. Oppenheimer may have testified about that incident here in these hearings, are you?

A. Not in specific detail; no.

Q. I would like to read you a portion of Dr. Oppenheimer's testimony and get your views on that. I might tell you so that this will be intelligible to you that Dr. Oppenheimer was interviewed by Colonel Pash of the security organization about this matter, and then by Colonel Lansdale.

Mr. GARRISON. Mr. Chairman, I think this raises the same question that I raised earlier. I have no objection—we all have been putting hypothetical questions to witnesses—but it seems to me to extract a piece of the testimony and only one piece, and then to ask opinions upon that without having the whole testimony. That is an unfair method of procedure. I think I made this objection at the outset, and it was after that that the questions began to be put in a different form. I do very much object to just a piece of the transcript being read from the evidence without the context of the whole.

Mr. ROBB. I am going to read a rather substantial piece. Of course, Mr. Chairman, Mr. Garrison framed his hypothetical questions, and that has been all right with me. I think I have a right to ask this witness on the basis of questions and answers right in this record whether his answer would be the same.

Mr. GARRISON. This is not a hypothetical question.

Mr. ROBB. No; this is a definite question.

Mr. GARRISON. This is a slice out of the transcript.

Mr. GRAY. I would like to ask Mr. Garrison whether his point is that the witness is not hearing everything that Dr. Oppenheimer testified before this board, or whether the witness is not hearing everything he said with respect to this particular incident?

Mr. GARRISON. Everything he said before the board with respect to this incident. It seems to me to lift a part of it out, and ask the witness' opinion about that is to present him only a fraction of the total in what could be a misleading light. I don't know what fraction it is. I think it is quite different from putting a question if it has been established here that such and such took place before the board. I think that is different. It is quite clearly put as not the evidence itself. I never attempted to say to a witness what the evidence here had been.

Mr. ROBB. I think my method is more accurate. I am going to read him the actual questions and answers.

Mr. GARRISON. In my questions I tried to summarize the best I could the way it looked to me. I appreciate that on each occasion Mr. Robb quite properly reserved his own feeling or position that the story as he might relate it would be a little different.

Mr. GRAY. I am going to ask Mr. Robb if he can put his question in hypothetical terms as he would see the question and not be confined to any hypothetical questions which counsel for Dr. Oppenheimer would.

Mr. ROBB. Very well. I will attempt to summarize the testimony which I have in mind.

By Mr. ROBB:

Q. General, I will ask you, sir, to assume that when questioned before this board about that episode and his interview with Colonel Pash, he was asked whether he told Colonel Pash the truth about the episode and he said no. He was asked if he lied to Colonel Pash and he said yes. When asked why he did that he said "Because I was an idiot." He said "I was also reluctant to mention Chevalier" and somewhat reluctant to mention himself.

Assume further that he was asked whether or not if the story he told to Colonel Pash had been true, it would have shown that both Chevalier and Dr. Oppenheimer were deeply involved in an espionage conspiracy. He agreed that was so.

A. May I ask you to repeat this last statement of yours?

Q. He was asked whether or not if the story which he told to Pash had been true, instead of as he said false, that story would have shown that both Dr. Oppenheimer and Chevalier were deeply involved in an espionage conspiracy.

Mr. GARRISON. Mr. Chairman, I certainly don't recall that.

Mr. ROBB. Since my friend objects——

Mr. GRAY. I would say to Mr. Garrison that he certainly has the privilege of making a statement that Mr. Robb has made in each case with respect to a hypothetical question.

Mr. GARRISON. Yes, but this is so obviously a paraphrase of the transcript. It is not an attempt at a summary. It seems to me it doesn't event attempt to give the witness a picture of what took place.

Mr. ROBB. I can see, Mr. Chairman, I should have interrupted Mr. Garrison's question and raised technical questions about it, too, but I didn't do it.

Mr. GRAY. Proceed, Mr. Robb.

By Mr. ROBB:

Q. Did you have the last in mind, General?

A. If I have heard you correctly in answer to a question whether had he told the truth it would have shown him, Dr. Oppenheimer, and Mr. Chevalier to be deeply in espionage.

Q. Yes.

A. And he answered yes, he would have.

Q. Yes.

Mr. SILVERMAN. No.

Mr. GARRISON. Mr. Chairman——

Mr. ROBB. Wait a minute, Mr. Garrison.

Mr. SILVERMAN. You misunderstood.

By Mr. ROBB:

Q. I am going to explain it. Assume that the story he actually told Colonel Pash was true, then would that not have shown that he was deeply involved in an espionage conspiracy? Do I make myself plain?

Mr. GARRISON. Mr. Chairman, I don't think it is plain, and I don't think it is in the record.

Mr. ROBB. Very well. I will read this to you: "In other words, if X [meaning Chevalier] had gone to three people, that would have shown, would it not—
"OPPENHEIMER. That he was deeply involved.
"That he was deeply involved. That is, was not just a casual conversation.
"OPPENHEIMER. Right."
Now, am I justified?

Mr. GARRISON. No, because you indicated that Dr. Oppenheimer would then be involved. That is what I very deeply object to.

Mr. ROBB. Page 488:
"Q. You will agree, would you not, sir, that if the story you told to Colonel Pash was true, it made things very bad for Mr. Chevalier?
"A. For everyone involved in it.
"Q. Including you.
"A. Right."
Now, may I proceed?

Mr. GARRISON. Mr. Chairman, he said that the story was an invention and the implication here to the witness is that he lied about something which would have implicated himself in espionage. I don't think that implication ought to be in this record at all.

Mr. ROBB. That is exactly what he said.

By Mr. ROBB:

Q. General, will you further assume——
Mr. GRAY. Well——
Mr. ROBB. Pardon me.
Mr. GRAY. Could you state the last assumption that you made?

By Mr. ROBB:

Q. Would you assume that the testimony was to that effect?
A. I am clear on what this point is now.
Q. Fine. Would you further assume, sir, that Dr. Oppenheimer knew that by refusing to name the man we referred to as "X", who afterwards turned out to be Chevalier, Dr. Oppenheimer knew by refusing to name him, he was impeding the investigation by the security officers into this espionage conspiracy?
Assuming those things, General, would you care to amend your answer with respect to the trustworthiness of Dr. Oppenheimer?
A. I spoke of my opinion in the period in which I was associated with him, and knowledge from my associations. From that time, 1943, I would have said this was a very foolish action. I could not have—I could not now believe that Dr. Oppenheimer would have acted that way at the later time when I was associated with him. I think probably he had learned a great deal about the mechanics of security in the intervening years.
Q. Does it come as a shock to you to hear that occurred?
A. When I first read it in the newspaper——
Q. No, sir, I am talking about what I have just told you about it. Does it come to you as a shock to hear that happened?
A. It is not a comfortable thought that one should have been, to use Dr. Oppenheimer's word, such an idiot at that time. It would certainly come as a shock to me if there were evidence that he still operated that way in 1947 and afterward when I knew him.
Q. As a military man, General, and a professional soldier, suppose you found out that someone in your command had conducted himself in that way in an interview with a security officer; what would you do?
A. As of now in the context of the past, I would want to get all the facts bearing on it before I spoke. Years have passed.
Q. Suppose you found out today that someone in your command had conducted himself in that way last week in an interview with one of your security officers; what would you do?
A. I would take immediate action.
Q. You would court-martial him, wouldn't you.

A. The formalities are that I would suspend him and turn his case over to the OSI.
Q. For an investigation?
A. Yes.
Q. Looking to a court-martial, would you not?
A. Depending on the facts.
Q. Because you would take a very serious view about it?
A. I would, indeed.
Q. To a military man, General, lying is never justified. I mean to one of your own security officers. You could not justify that, could you?
A. False official statements are not condoned; no.
Mr. ROBB. That is all. Thank you.
Mr. GRAY. General McCormack, your recent experience has been a very great deal of research and development, is that correct? That has been your primary concern in recent years?
The WITNESS. From the administrative side. I am not a technical person.
Mr. GRAY. I understand. This is one of your responsibilities insofar as you have ultimate responsibility. One of them is in the general field of research and development. I am going to ask you a question now which reflects some confusion on my part about the well known October 1949 meeting of the General Advisory Committee, and the circumstances surrounding it—the events leading up to it, and subsequent events.
It is clear, I believe, that the recommendation of the General Advisory Committee was not to proceed with an all out program for the production of this weapon. Is that a fair statement as you understood it?
The WITNESS. That was surely a part of the decision, yes, sir. The other things that went around, I would have to go and read the record.
Mr. GRAY. But that was clear. Another alternative, I suppose, which would have been at any time before the GAC was the alternative of not proceeding at all with research, development or production, leading to the weapon under discussion.
The WITNESS. In theory that was an alternative, yes, sir. In practical fact, science goes on, of course.
Mr. GRAY. Is there in your opinion anything that the General Advisory Committee might have done in October 1949 which would have represented a middle ground between these two extreme positions?
The WITNESS. Yes, sir.
Mr. GRAY. Do you mean by that, that perhaps they could have emphasized more strongly and recommended more enthusiastically research and development perhaps short of the all-out production program which was at least one issue with respect to which they took a position?
The WITNESS. Oh, yes, sir. There is a vast amount of middle ground between the two alternates as you stated them at the end of the spectrum.
Mr. GRAY. Did you feel at the time that perhaps the GAC might consistent with the technical uncertainties, which clearly existed, have recommended more of an effort that this action of October 1949 seemed to you to suggest?
The WITNESS. Yes, sir, I do. In fact, the program as it proceeded was a question of picking up steam as you could do it. Greater expenditures of effort as useful places to expend that effort appeared in the course of the research.
Mr. GRAY. In your judgment could the GAC have at that time recommended actions involving this greater effort without serious impairment or without impairment of the on-going fission program at the time?
The WITNESS. The question of scale and rate of effort, yes, sir. Anything that we did immediately that we had not been doing before required either new resources to do it with, or it had to displace something. So the phasing out of the old programs and the phasing in of expanded effort in the thermonuclear field was more or less a normal process, although it proceeded at very high priority, as high as we could put on it.
Mr. GRAY. I am now going to ask a question with respect to which you have not testified this morning, that is, do you feel that the military at that time was well informed about the possible and appropriate and sensible use of atomic weapons?
The WITNESS. Knowledge was far less complete than it is today, and probably less complete today than it will be at some time in the future.
Mr. GRAY. Do you feel that the lack of knowledge on the part of the military was a factor in whatever delay there might have been in the development of this weapon?

The WITNESS. Lack of knowledge on the part of the military services as to just what the technical prospects were, I would say, yes, sir. I would give the same reply, I think, with regard to the more advanced fission weapons that have come out since that time. So much of this was——

Dr. EVANS. You mean the fusion weapon or the fission?

The WITNESS. I said fission, then sir. It is all a part of it——

Dr. EVANS. Yes, I understand.

The WITNESS. Of a single problem. The atomic weapon field has gone forward very rapidly compared in contrast with our experience in the development of the other machines of war that the foreseeing uses, the techniques of their use, their application to given battle situations, had to be developed as the weapons developed. It was my constant experience as long as I was with the Commission that the invention had to precede in part a clear and detailed plan for its use.

Take the use of the weapon carried under a fighter aircraft, for example; you had to have some idea of what you had in the way of energy release in the weapon before you could develop the fighter tactics and before the fighter tactics are clear in mind, the Air Force is in a poor position to say to the infantry on what you can do in putting atomic explosions down on the battle line.

Mr. GRAY. What was the function of the Military Liaison Committee?

The WITNESS. Under the original law, it was appointed by the Secretary of War and Navy and in the revision of the law that happened after the unification of the Armed Forces, the Chairman of the Military Liaison Committee was made an appointment for confirmation by the Senate, and he represents the Secretary of Defense.

Mr. GRAY. Was it one of the functions of that Committee to keep the military advised in these respects with respect to the matters about which you said they might have known more than they did?

The WITNESS. Theirs was the formal responsibility. There were many of us working on it, of course. It was in large part a process of mutual education.

Mr. GRAY. In your opinion and recollection, General McCormack, is it possible that we would ever have found ourselves in a period or at a position in this Government in which the military might have been stating no military requirements because they believed there was no technical feasibility and the Commission and its agencies might have been not pressing for development on the ground that there was no military demand?

The WITNESS. I think in the practical working of the organization as it then existed, sir, that insofar as we were wise, insofar as we knew what to do, this gap could not have existed. I, for instance, could not have sat in my office in the Commission knowing that there was a prospect in any field that might be of some military interest without seeking out the military service, or that segment of one of the military services that might be most interested in and make sure they got as clear a look at it as they could have had at that time. This was on the informal basis. Our formal dealings through the Military Liaison Committee will reflect the big issues. They will not reflect the myriad of contacts and interchange, the stationing of military officers at Los Alamos, the loaning of Los Alamos personnel to the target planners in the Pentagon, the interchanges of visits and so on. This was a very broad thing. By these means we tried to grow up with the situation as fast as the situation was growing.

Mr. GRAY. As a practical matter, you think the answer to my question would be no?

The WITNESS. To the limit of our wisdom and ability to do it.

Mr. GRAY. I am making the assumption that those concerned with the program were of course doing their duty as they saw it best under the circumstances. I mean by that it is unreasonable to suppose that many in the military could understand some of the technical implications, especially those who were not themselves scientists. You do not feel that there was delay because of any possible misunderstandings by the military about scientific feasibility and at the same time misunderstanding by the scientific advisers as to military requirement?

The WITNESS. I think an honest answer in the light of history, sir, is that there must have been delays. I would not know how to put my finger on them. Had we known where they existed at the time, we would have cured them. But in fact, they must have existed in a sense not entirely different from the normal business where I am now, where there are delays getting a new aircraft in operation because its operating characteristics exceed the experience of the pilots until they have had a chance to work on it. Therefore, you go down to the production line with things that you have to re-do, and this introduces delays.

Mr. GRAY. Thank you, sir.

REDIRECT EXAMINATION

By Mr. SILVERMAN:

Q. With respect to the question Dr. Gray was asking you about delays, and your answer, would you say that the delays in the development of the thermonuclear weapon, so far as you knew about them, were greater than just the normal delays that one would expect, because one is venturing into a new field?

A. First, other than counting off the period of the debate as a delay, if you wish, I am not aware of any delays in the thermonuclear program that occurred for any reason other than just not knowing how to do the next step. I know the resources that were available to us to put in the program were freely available at all times. Los Alamos competence built up, and we drew in others to work on it. With that stipulation, I don't have any delay to put my finger on. I would have to say that the thermonuclear program went well indeed, even with shifts in the lines of technical attack. It still kept apace which I thought was admirable at the time and met the expectations that were at least apparent to us at the beginning.

Q. I think Dr. Gray asked you about whether you felt the military was well informed with respect to the development of thermonuclear weapons and the possibility of developing thermonuclear weapons—something of that order— and as I got your answer, it was that we are better informed today and we will be better informed at some future date.

A. I was merely trying to indicate that being informed and not being informed is a very relative term if you are going back to the beginning of a program of inventions which had not yet been invented.

Q. With respect to the period of October 1949, did you feel that the military was well informed as to the feasibility and the possibilities of use of atomic and thermonuclear weapons in the light of what was then known with respect to the feasibility of such weapons?

A. If there was anything known in the Commission organization or its laboratories of importance about the prospects of thermonuclear weapons feasibility that was not known to the military services, I was certainly unaware of it. But little was known as a fact. We were dealing with very large conjectures.

Q. With respect to the chairman's question of a possible middle ground between the two ends of the spectrum, was it your feeling that the GAC was in favor of a program of research on the feasibility of thermonuclear weapons?

A. There was a research program in thermonuclear weapons and had been since I first reported in to the Commission. It had not picked up much headway until the whole situation was catalyzed by the news of the Russian fission explosion. I have no specific memory at this time of the reaction of the General Advisory Committee, or any of its individuals, as to the degree to which this program might be expanded, yet falling short of the program which they recommended against.

Is this responsive? It is to me a very complicated question.

Q. I think it is probably as responsive as you can make it to me. With respect to the remark about the GAC report or recommendation being silly or sinister——

Mr. ROBB. Did you say "and" or "or"?

Mr. SILVERMAN. I said "or."

Mr. ROBB. That is what I said. That is what he said.

By Mr. SILVERMAN:

Q. What did you think it was?

A. Perfectly frankly, I thought the rush action was silly.

Q. Did you think it was sinister?

A. I did not. As I stated earlier with several of my immediate superiors in the Commission holding a view which I understood to be similar to the General Advisory Committee. I would have moved out immediately had I thought there was a sinister implication in the opposition.

Q. With respect to how well informed the military was on the prospects of the thermonuclear weapon, do you recall a panel report to the War Department prepared late in 1945 describing the prospects of the super?

A. I do not recall a report under that name. There were papers in the Commission which had been prepared some time back when I joined it at the beginning of 1947, and this was a new paper prepared at the beginning of 1947 for the use of the new Commission which rounded them up as they then appeared and all of these papers in my memory anyway read about the same as the state of knowledge, as far as I understood it certainly as far as I recall it, had not advanced substantially from 1945 to 1947. Nor indeed was there any big breakthrough from the research program between 1947 and the time after the program had been accelerated, although there were new ideas coming along.

Mr. SILVERMAN. I have no further questions.

RE-CROSS-EXAMINATION

By Mr. ROBB:

Q. General, when you spoke a minute ago of the rush action, did you refer to the action by the committee?
A. It all happened very quickly.
Q. Yes.
A. Yes, as I recall the committee and the Commission acted jointly, and went to the President with their combined opinion or separate opinions. They were not unanimous, of course.
Q. General, I take it you are not a nuclear physicist?
A. I am not, sir.
Q. You said I think in response to a question by Mr. Silverman that the thermonuclear program went very well indeed.
A. In terms of timing and eventually meeting our expectations. It had its ups and downs, of course.
Q. What time were you referring to when you said that?
A. From the beginning of 1950 until what I regarded as a successful milestone just before I left the Commission in 1951.
Q. Yes. That is what I thought.
Was there a considerable stepping up in the efforts to develop the thermonuclear subsequent to the President's statement in January 1950?
A. Indeed there was, sir. We stepped it up in all ways of which we were capable.
Q. Would you care to give us an opinion, recognizing of course that you are not a nuclear physicist, as to what might have been the result had that stepped-up program been started in 1945 or 1946? Might you have gotten the end result sooner?
A. Putting the same effort into it that we were able to put into it in 1950?
Q. Yes, sir.
A. Speaking nonexpertly from the scientific point of view in any event, I think it could not have helped speeding the time when there would have been a thermonuclear weapon, looking back on it. I can easily see why General Groves and the Commission later with all of the other urgent work to do in rebuilding Los Alamos and getting the fission weapon program straightened out, did not feel up to making a gamble certainly as early as 1945.

Mr. ROBB. I am not debating that. I merely want to get your opinion as to the time element.

Thank you very much, General.

Mr. GRAY. Thank you very much, General McCormack.

(Witness excused.)

Mr. GRAY. We will take a recess.

(Brief recess.)

Mr. GRAY. Dr. von Neumann, do you wish to testify under oath?

Dr. VON NEUMANN. Yes.

Mr. GRAY. You are not required to do so. The other witnesses have.

Dr. VON NEUMANN. I am quite prepared.

Mr. GRAY. Would you be good enough to stand and raise your right hand, and give me your full name?

Dr. VON NEUMANN. John von Neumann.

Mr. GRAY. John von Neumann, do you swear that the testimony you are about to give the board shall be the truth, the whole truth, and nothing but the truth, so help you God?

Dr. VON NEUMANN. I do.

Whereupon John von Neumann was called as a witness, and having been first duly sworn, was examined and testified as follows:

Mr. GRAY. Will you be seated, please.

I am required to remind you of the existence of the so-called perjury statutes. I shall be glad to review them with you if necessary.

The WITNESS. I think I am aware of them.

Mr. GRAY. May I ask if in the course of your testimony it becomes necessary for you to refer to or disclose restricted data, you notify me in advance, so we can take appropriate and necessary steps in the interest of security.

Finally, Doctor, I would say to you, as I say to each of the witnesses on behalf of the board, that we consider these proceedings a confidential matter between the Atomic Energy Commission and its officials on the one hand, and Dr. Oppenheimer, his representatives and witnesses on the other hand. The Commission is making no releases with respect to these proceedings, and we express the hope that the witnesses will take the same view.

The WITNESS. Yes, sir.

DIRECT EXAMINATION

By Mr. SILVERMAN:

Q. Dr. von Neumann, what is your present nongovernmental position?

A. I am professor of mathematics at the Institute for Advanced Study at Princeton.

Q. How long have you been that?

A. Since 1933.

Q. That was before Dr. Oppenheimer came there?

A. Yes.

Q. I understand you were for 2 years president of the American Mathematical Society.

A. That is correct.

Q. You have been a member of the National Academy of Science, I understand, ince 1937?

A. Yes, since 1937.

Q. Will you state your present governmental position?

A. I am a member of the General Advisory Committee of the Atomic Energy Commission. I have been that since 1952. I have been a consultant to the Los Alamos Laboratory since 1943. Outside the Commission, I am a member of the Scientific Advisory Board of the Air Forces. I have also a few other governmental advisory positions.

Q. Would you tell us the story of when you first knew Dr. Oppenheimer and what contacts you have had with him since?

A. I think that Dr. Oppenheimer and I first met in Germany in 1926. It was in Goettingen, to be precise. We were both I think immediately after our respective Ph. D.'s and we were both there. There was a great center of theoretical physics in Goettingen, and we were both there at the time.

Then between 1926 and 1940, we may or may not have met. I think we did not, although I knew about Dr. Oppenheimer and I knew about his work.

In 1940 we met in Los Angeles, and we had several conversations. We also met at that time in Seattle. We met again in early 1943, at which time Dr. Oppenheimer told me that he wanted me to join a project which he could not describe at that moment.

Then I went to England and came back in the fall, and then I was asked officially to go to Los Alamos. After that, our contacts have been practically continuous, with a slight interruption between 1945 and 1947, when we both had left Los Alamos and Dr. Oppenheimer had not yet come to Princeton.

Q. Since 1947 you have both been?

A. I would say our association has been practically continuous since 1943.

Q. You referred to meeting Dr. Oppenheimer in 1940 in Los Angeles, and did you say at Seattle, also?

A. Yes, it was outside of Seattle.

Q. Was that after the fall of France, or about the time of the fall of France?

A. This was in May or June of 1940. It was in the period during which France was collapsing, and the conversation I had mentioned we then had and which I assume is relevant in this context, it was one we had about the political situation then. What I do recall very clearly is that Dr. Oppenheimer was for intervention on the side of the western allies. This was of course a very acute question at the moment, and I asked practically everybody I met how he felt so this I remember quite clearly.

Q. There has been, I guess, a fair amount of testimony that would be an understatement—about the GAC report of October 1949, with respect to the hydrogen bomb and the thermonuclear program. Dr. von Neumann, did you agree with the GAC report and recommendations?

A. No. I was in favor of a very accelerated program. The GAC at that point recommended that the acceleration should not occur.

Q. Very accelerated hydrogen bomb or thermonuclear program?
A. Yes, it is all the same thing.
Q. Would it be fair to say one might say in the opposite camp on the question?
A. Yes, that is correct.
Q. Did you consider that the recommendations of the GAC and in particular Dr. Oppenheimer were made in good faith?
A. Yes, I had no doubt about that.
Q. Do you have any doubt now?
A. No.
Q. You knew, of course, that Dr. Oppenheimer was not the only person who was opposed to the program?
A. No, the whole group of scientists and military who were keenly in this matter—of course, there had been a lot of discussion and practically everyone of us knew very soon fairly precisely where everybody stood. So we know each other's opinions, and very many of us had discussed the matter with each other. Dr. Oppenheimer and I had discussed it with each other, and so we knew each other's views very precisely.

My impression of this matter was, like everybody else, I would have been happy if everybody had agreed with me. However, it was evidently a matter of great importance. It was evidently a matter which would have consequences for the rest of our lives and beyond. So there was a very animated controversy about it. It lasted for months.

That it lasted for months was not particularly surprising to my mind. I think it was perfectly normal that there should be a controversy about it. It was perfectly normal that emotions should run rather high.

Q. Have you yourself participated in the program of the development of thermonuclear weapons and the hydrogen bomb?
A. Yes.
Q. After the President's decision in January of 1950, is it your impression that the GAC and particularly Dr. Oppenheimer was holding back in the effort to develop the bomb?
A. My impression was that all the people I knew, and this includes Dr. Oppenheimer, first of all took this decision with very good grace and cooperated. The specific things I know were various actions which were necessary in 1951. At that time there were a number of technical decisions that had to be made about the tehcnical program. I know in considerable detail what Dr. Oppenheimer did then, and it was certainly very constructive.
Q. Can you tell us any of that in unclassified terms?
Mr. ROBB. Excuse me. Could I ask what date he is referring to?
The WITNESS. I am referring particularly ot a meeting in Princeton in June 1951.
Mr. ROBB. Thank you.

By Mr. SILVERMAN:

Q. I don't know whether you can expand on this in unclassified terms or not.
A. I think the details of why there was a need for technical decisions at that moment and exactly how far they went and so on, I assume is classified, unless I am otherwise instructed. But it is a fact. You must expect in any program of such proportions that there will be as you go along serious technical decisions that have to be made. This was one. There was a meeting at Princeton which was attended I think by part of the GAC. I think it was the weapons subcommittee of the GAC which is in fact about two-thirds of the group, plus several Commissioners, plus several experts which included Dr. Bethe, Dr. Teller, myself, Dr. Bradbury—I am not sure whether Dr. York was there—Dr. Nordheim and possibly others. This meeting was called by Dr. Oppenheimer and he certainly to the extent which anybody was directing it, he was directing it. This was certainly a very necessary and constructive operation.
Q. At that meeting did he express himself as being in favor of going ahead?
A. In all the discussions at that point there was no question of being or not being in favor. In other words, it was a decided technical policy. I didn't hear any discussions after 1950 whether it ought to be done. There certainly were no such discussions at this meeting. The question was whether one should make certain technical changes in the program or not.

All I am trying to say is that at that point there was a need for technical changes. If anybody wanted to misdirect the program by very subtle means, this would have been an occasion.

Q. Did Dr. Oppenheimer cooperate in making it easier for you and others to work at Los Alamos for Los Alamos on the hydrogen bomb program?

A. I certainly never had the slightest difficulty. One thing is that I think if Dr. Oppenheimer had wanted to create difficulties of this kind, as far as I am concerned, it would have been possible. Also, our relations would probably have deteriorated. There was absolutely nothing of that. Our personal relations stayed very good throughout. I never experienced any difficulty in going as much to Los Alamos as necessary.

Q. There was no suggestion by Dr. Oppenheimer that this was interfering with your work at the institute?

A. None whatsoever, absolutely none.

Q. And did you spend a good deal of time at Los Alamos?

Mr. ROBB. Could we have the times fixed on these? I am sorry to keep interrupting.

By Mr. SILVERMAN:

Q. After 1949?

A. Yes. It may have averaged 2 months a year. Not all in one, but say in two pieces of 3 weeks and various shorter visits. I must say this was uniform from 1945 to almost now. I have been somewhat less in Los Alamos lately because I have other commitments.

Q. And I take it there was no objection to your doing any work that might be helpful to Los Alamos at Princeton?

A. Absolutely none whatsoever.

Q. Did Dr. Oppenheimer attempt to dissuade you from working on the hydrogen-bomb program?

A. No. We had a discussion. Of course, he attempted to persuade me to accept his views. I equally attempted to persuade him to accept my views, and this was done by two people who met during this period. I would say apart from the absolutely normal discussion on a question on which you happen to disagree, there was absolutely nothing else. The idea that this might be pressure I must say did not occur to me ever.

Q. Do you now think that it was pressure?

A. No. I think it was the perfectly normal desire to convince somebody else.

Q. During what period was this discussion?

A. This was in 1949, December 1949. I remember quite clearly two discussions, one which was about half an hour at which time I saw the GAC opinion and we discussed it.

Q. You had a Q clearance at that time?

A. Yes. We discussed the same subject again about a week later, again for about 20 minutes or half an hour, I don't know. We probably also talked about the subject on other occasions, but I don't recall.

Q. Wasn't the discussion about whether you personally should work on the hydrogen-bomb program?

A. Absolutely not. The only question was whether it was or was not wise to undertake that program.

Q. You have known Dr. Oppenheimer, I think you said, substantially continuously since 1943 to the present date?

A. Yes.

Q. With the exception of the period from 1945, the end of the Los Alamos days, until 1947, when Dr. Oppenheimer came to the institute as director.

A. That is correct.

Q. During that period you have really lived in the same small town?

A. Yes.

Q. And been friends and known each other quite well during all that time?

A. Yes.

Q. Both professionally and socially?

A. Yes, that is correct.

Q. Do you have an opinion about Dr. Oppenheimer's loyalty to the United States, his integrity?

A. I have no doubts about it whatever.

Q. Your opinion I take it is quite clear and firm?

A. Yes, yes.

Q. Do you have an opinion as to Dr. Oppenheimer's discretion in the handling of classified materials and classified information?

A. Absolutely. I have personally every confidence. Furthermore I am not aware that anybody has questioned that.

Q. There seems to be some question among my associates whether I asked this. Do you have an opinion about Dr. Oppenheimer's loyalty?

A. Yes.

Q. What is that?
A. I would say he is loyal.
Q. Do you have any doubt on that subject at all?
A. No.
Mr. SILVERMAN. I have no further questions.

CROSS-EXAMINATION

By Mr. ROBB:
Q. Dr. von Neumann, you stated that Dr. Oppenheimer attempted to persuade you to accept his views, and you attempted to persuade him to accept your views in December 1949?
A. Yes.
Q. Would you tell us briefly what his views were as you understood them?
A. Well, that it would be a mistake to undertake an acceleration of the hydrogen bomb, the thermonuclear program for the following reasons: Because it would disorganize the program of the AEC because instead of developing fission weapons further, which one knew how to do and where one could predict good results fairly reliably, one getting back on a crash program which would supersede and damage everything else, and the results of the crash program would be dubious. That furthermore, from the military point of view, making bigger explosions was not necessarily an advantage in proportion to the size of the explosion. Furthermore, that we practically had the lead in whatever we did, and the Russians would follow, and that we were probably more vulnerable than they were for a variety of reasons, one of which is that we can probably saturate them right now—I meant right then—whereas they could not at that moment. Therefore, a large increment on both sides would merely mean that both sides can saturate the other. Also, that since there was now this possibility of a large increment in destructive power, this was now for the second time, and possibly for the last time an opportunity to try to negotiate control and disarmament. I think this was by and large the argument. There are a few other angles which are classified which I think are not very decisive.
Q. Doctor, was there anything in his argument about the immorality of developing the thermonuclear?
A. I took it for granted that it was his view. It did not appear very much in our arguments, but we knew each other quite well. My view on that is quite hard boiled, and that was known.
Q. What was Dr. Oppenheimer's view, soft boiled?
A. I assume, but look, now, I am going by hearsay. I have not discussed it with him.
Q. I understand.
A. I assume that one ought to consider it very carefully whether one develops anything of this order of destruction just per se.
Q. Yes, sir. Doctor, in response to a question from Mr. Silverman, you said you had no question about Dr. Oppenheimer's integrity, did you not?
A. Yes.
Q. By that you meant his honesty, did you not?
A. Yes.
Q. Doctor, do you recall having heard anything about an incident which occurred between Dr. Oppenheimer and a man named Chevalier?
A. Yes, but that was lately. I do not know for absolutely sure when I first heard it. I saw the letter of charges and there it occurs. When I read it, I had the vague impression that I had heard this before, but I think that this was in the last few years.
Q. You saw the letter of General Nichols and Dr. Oppenheimer's response?
A. Yes. I am not absolutely certain whether I saw the complete original or whether I saw somebody's excerpts of relevant parts.
Q. What is your present understanding about that incident that I referred to—the Chevalier incident? What do you have in mind about what happened?
A. What I understand happened was—and pleace correct me if my recollection is inexact—my impression is that Chevalier was a man who had been Dr. Oppenheimer's friend in earlier years, who in 1942, I think, or early 1943, when Dr. Oppenheimer was already associated with the atomic energy project which was not yet the Manhattan district, made an approach and suggested to him that somebody else, whose name I have forgotten, was working for Russia and would be able to transmit scientific and technical information to Russia.

I understood that Dr. Oppenheimer essentially told him to go to hell, but did not report this incident immediately, and that when he later reported it, he did not report it completely for some time, until, I think, ordered by General Groves to do so.

By Mr. ROBB:

Q. Your memory is pretty good, Doctor, Do you recall the name of the other person was Eltenton?

A. Yes, Eltenton.

Mr. ROBB. Mr. Chairman, I would like to ask the witness a hypothetical question. I assume, Mr. Garrison would file a caveat to it but I venture to suggest in the interest of entirety to assist the board and the witness, it would be most helpful if Mr. Garrison allowed me to state my question before he made his objections.

By Mr. ROBB:

O. I want you to assume now, Dr. von Neumann, that Dr. Oppenheimer reported and discussed this incident with two security officers, one named Colonel Pash and one named Colonel Lansdale, and will you please assume that Dr. Oppenheimer has testified before this board that the story of the Chevalier incident which he told to Colonel Pash on August 26, 1943, and affirmed to Colonel Lansdale on September 12, 1943, was false in certain material respects.

Assume that he has testified here that the story he told to Pash and Lansdale was a cock and bull story, that the whole thing was pure fabrication, except for the one name Eltenton; that he told a story in great detail that was fabricated, that he told not one lie but a whole fabrication and tissue of lies in great circumstantial detail.

Assume that he has further testified here that his only explanation for lying was that he was an idiot, and he was reluctant to name Dr. Chevalier and no doubt somewhat reluctant to name himself.

Assume he has further testified here that if the story he told to Colonel Pash had been true, that it showed that Dr. Chevalier was deeply involved in a conspiracy; that the conversation or the remarks of Dr. Chevalier were not just a casual conversation and it was not just an innocent contact, but that it was a criminal conspiracy on the part of Dr. Chevalier.

Assume that he testified further that if the story that Dr. Oppenheimer told to Colonel Pash was true—if it was true—then it made things look very bad for both Dr. Chevalier and Dr. Oppenheimer.

Mr. GARRISON. Mr. Chairman, I wish the record to show that I do not accept this assumed version of the testimony as being an accurate summary of the testimony.

Mr. GRAY. The record will show that counsel for Dr. Oppenheimer does not accept the question as put. The witness will consider this a hypothetical question.

The WITNESS. May I ask, Mr. Chairman, I have not quite understood the meaning of this exchange. Does this mean that the question ought to be answered?

Mr. GRAY. Let me state it this way, Dr. von Neumann. You must not assume that this board has reached any conclusions with respect to any matter before it. Therefore, in statements to you by counsel, either Mr. Garrison or Mr. Robb, and questions put to you by either Mr. Garrison or Mr. Robb which are said to you to be hypothetical, you are asked to reply to that question on an assumption that the facts are true for the purpose of this question, and not to assume that this is a conclusion of the board.

The WITNESS. Yes.

Mr. GARRISON. May I ask if the question might be reread at this point?

The WITNESS. I would also like to ask a few elucidations about the question. For one thing, Mr. Robb, you have described a hypothetical situation, but if I did not get mixed up, you did not ask the question.

Mr. ROBB. I have not asked the question. I wanted to give Mr. Garrison a chance to object. Would you like the question read back to you?

The WITNESS. No. I will ask you a few things about the hypothetical question, because it is pretty complicated.

Mr. GARRISON. Before we go further, I want to emphasize my point that I want it clearly understood that the question that was put involved asking the witness if the false story which he had told had been true, there would have been a criminal conspiracy and make it clear that even if the false story that was true there was no suggestion by Dr. Oppenheimer that he was involved in espionage.

Mr. ROBB. Mr. Garrison, I will ask the witness.
Mr. GRAY. I would suggest you proceed with your question.
Mr. ROBB. Thank you.

By Mr. ROBB.

Dr. von Neumann, my question is, assuming that Dr. Oppenheimer testified before this board as I have indicated to you, would that shake your confidence in his honesty?
A. May I ask you again, if I understood correctly——
Q. Yes.
A. If I understood correctly, the hypothetical representation to the board would have been something like this: That a false statement was made because Dr. Oppenheimer wanted to avoid naming Mr. Chevalier and himself. I understood your description first as saying that he said that he is supposed to have said that he made these statements to security officers because he did not want to mention Chevalier's name and did not want to mention his own name. Is this correct?
Mr. ROBB. I wonder if we might have the question read back to the witness?
The WITNESS. Please read it back.
(Question read by the reporter.)
The WITNESS. In other words, the hypothetical testimony is that his conduct was first of all due to a desire to make things easier for Chevalier and possibly for himself, but on the other hand, it actually made it much worse. Is this the idea?
Mr. ROBB. I hesitate to instruct the witness, Mr. Chairman, beyond the statement of the hypothesis.
Mr. GARRISON. I think that is right.
Mr. SILVERMAN. You asked the witness a hypothetical question. If the witness is not entirely clear as to the hypothetical question, if the witness' understanding of it is at all different from that of the hearers, it make his answer not very competent, and therefore it is important to have it clear.

By Mr. ROBB:

Q. I think it is clear to say that part of the assumption is that Dr. Oppenheimer testified that one of his explanations for this conduct was that he was reluctant to mention Dr. Chevalier and somewhat reluctant to mention himself.
A. But at the same time, he now realized that his statements if true would actually be much worse for Chevalier.
Q. I think that is a fair statement, yes, sir.
A. So this was an attempt to achieve something of which it actually achieved the opposite, is that the idea?
Q. That might be inferred, yes.
A. Look, you have to view the performance and the character of a man as a whole. This episode, if true, would make me think that the course of the year 1943 or in 1942 and 1943, he was not emotionally and intellectually prepared to handle this kind of a job; that he subsequently learned how to handle it, and handled it very well, I know. I would say that all of us in the war years, and by all of us, I mean all people in scientific technical occupations got suddenly in contact with a universe we had not known before. I mean this peculiar problem of security, the fact that people who looked all right might be conspirators and might be spies. They are all things which do not enter one's normal experience in ordinary times. While we are now most of us quite prepared to discover such things in our entourage, we were not prepared to discover these things in 1943. So I must say that this had on anyone a shock effect, and any one of us may have behaved foolishly and ineffectively and untruthfully, so this condition is something ten years later, I would not consider too serious. This would affect me the same way as if I would suddenly hear about somebody that he has had some extraordinary escapade in his adolescence. I know that neither of us were adolescents at that time, but of course we were all little children with respect to the situation which had developed, namely, that we suddenly were dealing with something with which one could blow up the world. Furthermore, we were involved in a triangular war with two of our enemies had done suddenly the nice thing of fighting each other. But after all, they were still enemies. This was a very peculiar situation. None of us had been educated or conditioned to exist in this situation, and we had to make our rationalization and our code of conduct as we went along.

For some people it took 2 months, for some 2 years, and for some 1 year. I am quite sure that all of us by now have developed the necessary code of ethics and the necessary resistance.

So if this story is true, that would just give me a piece of information on how long it took Dr. Oppenheimer to get adjusted to this Buck Rogers universe, but no more. I have no slightest doubt that he was not adjusted to it in 1944 or 1945.

Q. Had you completed your answer?
A. Yes.
Q. In 1943, Dr. Oppenheimer was the director of the Los Alamos Laboratory, wasn't he?
A. Yes.
Q. I believe at that time he was 39 years old?
A. Yes.
Q. You wouldn't say he was at that time an adolescent, would you?
A. No. I was trying to make this clearer. There are certain experiences which are new for an adolescent, and where an adolescent will behave in a silly way. I would say these experiences were new for a man of 39, if he happened to be 39 at that moment in history.
Q. Do you think, Doctor, that honesty, the ability and the desire to tell the truth, depends upon the international situation?
A. It depends on the strain under which you are.
Q. The strain?
A. Yes.
Q. You mean a man may lie under certain strains when he would not under ordinary circumstances?
A. Yes, practically everybody will lie under anesthesia.
Q. Do you think, Doctor, if you had been placed in the same situation that Dr. Oppenheimer was in 1943, in respect of this matter, that you would have lied to the security officers?
A. Sir, I don't know how to answer this question. Of course, I hope I wouldn't. But—you are telling me now to hypothesize that somebody else acted badly, and you ask me would I have acted the same way. Isn't this a question of when did you stop beating your wife?
Q. I don't think so, Doctor, since you asked me. You do feel that Dr. Oppenheimer as you put it acted badly in the matter?
A. The hypothetical action, I take it, is a bad action.
Q. Quite serious, isn't it?
A. That depends on the consequences, yes.
Mr. Robb. I think that is all I care to ask, Mr. Chairman.
Mr. Gray. Dr. von Neumann, you went to Los Alamos in the fall of 1943?
The Witness. Yes.
Mr. Gray. Did you stay there throughout the war years?
The Witness. Yes. I was not there continuously, but I spent there about 1 month out of 3, and this up to the end of the war.
Mr. Gray. In 1943, did you consider that people who were identified with the Communist Party had any kind of commitment to a foreign power, specifically to the Soviet Union?
The Witness. I think that if somebody was a party member and under party discipline, yes.
Mr. Gray. My question is not what you believe now, but what you would have believed then.
The Witness. I so then believed. If somebody was under party discipline, yes.
Mr. Gray. So you were aware in 1943 of the threats to the security of the country which might come from allowing members of the party to have access to classified information?
The Witness. It certainly was a security risk, yes. I certainly felt that as a security risk. May I say I had the feeling that this was definitely a three-way war. At that moment two of the enemies had to all advantage got into a fight of their own. It was perfectly proper to exploit this. That as far as developing the atomic bomb was concerned, what all of us had in mind in 1943 and 1944 was this. Of course, the German science and technology was enormous. We were all scared to death that the Germans might get the atomic bomb before we did. We found out later that they had somewhat neglected this area, and they didn't get as far as we thought they would get. I don't think anybody could foresee that. I think it would have been a great mistake to bank on it in 1943 and 1944. We all were actuated by a desire which was primarily one to

get, if it is possible, an atomic bomb before anybody else does. We certainly all had the feeling that this was paramount, and that it was quite proper to take calculated risks in this regard.

I must say that I considered Russia an enemy from the beginning to the end, and to now, and the alliance with Russia is a fortunate accident that two enemies had quarreled. However, I think it also was perfectly fair to take advantage of this, that the military commander could perfectly well decide that one should take calculated risks on this, and employing a Communist might at that moment accelerate getting an atomic weapon ahead of Germany.

Of course, it would later be a bad problem from the security point of view. But then the German danger was there, and the other thing was remote, and military information obsolesces rapidly anyhow. So I think it was not unreasonable to take such a step.

Mr. Gray. You might have applied a different test with respect to the calculated risk in 1943 than you would apply today?

The Witness. Entirely.

Mr. Gray. Were you acquainted during your service at Los Alamos with Dr. Hawkins?

The Witness. Yes, I knew him.

Mr. Gray. With your awareness of the existence of the Communist Party, did you in any way have reason to believe that he was a member of the Communist Party at that time?

The Witness. You see, it is a little difficult to be quite sure in 1945 whether you think you learned around 1944, you learned 6 months earlier or later. I am fairly sure I had no idea if his Communist affiliations when I came to Los Alamos and first met him. He was not a particularly well known man and not to me. I think I learned that he had some kind of Communist connection before I left Los Alamos. Exactly how he had that connection I did not learn at that time.

Mr. Gray. And if someone had asked you at that time, this would be one of the calculated risks?

The Witness. I would say this was a calculated risk, yes.

Mr. Gray. From what you knew of Dr. Hawkins at the time, was he pretty well an indispensible member of the team out there?

The Witness. If I am not mistaken he was a project historian.

Mr. Gray. I think that was in part——

The Witness. He was not indispensible in the sense in which a man who is primarily interested in a technical sense. He was not a physicist. He was not a chemist or an applied mathematician. I think he was a philosopher.

Mr. Gray. And a mathematician.

The Witness. And some experience in sciences. He was a perfectly suitable person for being a project historian. Exactly how hard or easy it was to get a man who is qualified to do this thing I did not know at that time. I would say it is a job which requires a special kind of talent, and is not quite easy to fill.

Mr. Gray. Did you know Philip Morrison?

The Witness. Yes, I know Philip Morrison.

Mr. Gray. Did you then know anything about his political affiliations?

The Witness. I am fairly sure that I learned the fact that he had close Communist ties later.

Mr. Gray. And not at the time?

The Witness. This must have been in mid-1945 that I learned this.

Mr. Gray. Were you acquainted with Fuchs?

The Witness. Yes, I knew Fuchs quite well.

Mr. Gray. Did you have any reason to suspect his integrity or dependability or whatever was involved in the subsequent disclosures?

The Witness. Not particularly. He was a rather queer person, but then under these conditions queer persons occur. I did not suspect him particularly. He was clearly not an ordinary person.

Mr. Gray. What I am getting at is whether you had reason to believe he was a Communist.

The Witness. I think I did not know about him, no. I did not know about him, that he was a Communist practically until the whole affair broke.

Dr. Evans. Practically what?

The Witness. Until it became known, until he confessed, or rather until he was shown.

Mr. Gray. At the time you learned about it, were you surprised?

The Witness. Look, I was not surprised in this sense, that he clearly was a peculiar person. So if it turns out about an ordinary run of the mill person that he is a conspirator and spy, you are shocked and surprised. He was a very peculiar person with respect to whom I didn't have much experience. Of course, I was surprised by the fact that there had been such a thing, that a spy had been so well placed.

Mr. GRAY. When you said a few moments ago that you didn't know about it until practically at the time the disclosures were made public, does that mean that there was information available to people at Los Alamos about him, about his Communist connections, before the story was known here in Washington?

The WITNESS. I don't think so.

Mr. GRAY. I didn't think you intended to say that.

The WITNESS. No.

Mr. GRAY. I want to make clear that the record did not reflect it until you intended to say it.

The WITNESS. No, no; absolutely not.

Mr. GRAY. You think in that case if people in charge had known that Fuchs was a member of the Communist Party or had a Communist commitment, that this is the kind of calculated risk that you felt was desirable to take in those days? Was the calculated risk worth it in the case of Fuchs?

The WITNESS. Clearly not. I don't quite get the question. In the light of hindsight, clearly not.

May I say this was of course a highly empirical subject. Fuchs made a contribution. Of course, the damage he made outweighs the contribution by far, probably. Exactly what concentration of spies one would find among the people with Communist backgrounds nobody knew ahead of time, and quite particularly the technical people didn't know. So I would say this was a decision for security and for whatever branch of the Government was involved, which deals with counterespionage to make.

Mr. GRAY. Today you would not recommend employment on a sensitive project of someone known now to be a member of the Communist Party.

The WITNESS. No.

Mr. GRAY. Suppose there was recommended to you an individual for employment who some years ago had what you believed to be close Communist affiliations; what would your response be today?

The WITNESS. I would certainly not employ him in a sensitive job.

Mr. GRAY. A person who had had close Communist affiliations in an earlier period of his life?

The WITNESS. How early? I thought you said a few years ago. I mean how early. I would say if somebody had close affiliations with the Communist Party after 1945 or later, then I would certainly not employ him in a sensitive job. If he had close affiliations with the Communist Party in the late 1930's, then I would say if he was never a party member, then I would view the entire situation and I think if there is prima facie evidense of a probability that he had changed his views, I certainly would. If he was an actual member of the party, I would say that the burden of proof that he is no longer a member is on him. In other words, on his general conduct since then. I think you must consider the total personality and the total life and the probable motivation and interests of the person after 1940.

Mr. GRAY. Do you pick 1940 as a particular year?

The WITNESS. No. It is a vague thing. It is somewhere between 1940 and 1944, I would say.

Mr. GRAY. That close affiliations as late as 1944——

The WITNESS. I would begin to get worried, in fact, seriously worried. The great watershed is evidently the Second World War. There are all sorts of things happening there. For instance, the possibility for error is greater in 1943 and 1944 when the Russians were allies, than in 1940, when they were cooperating with the enemy. So I think dating between 1940 and 1944 is very difficult. But I would say definitely that I would take a lenient view of things before 1940, and a very hard view of things after 1944.

Mr. GRAY. Suppose at Los Alamos someone had come to you—this is purely hypothetical—and said, although the British are our allies and the official policy of the United States Government is to share military information of the highest degree of secrecy with the British, this policy is being frustrated in Washington, now I have a way of getting to the British scientists information about what we are doing here in Los Alamos, and don't you think it is up to us to make sure that official policy is not frustrated, and you knew that this person was interested in the British, what would your position have been at that time, Dr. von Neumann?

The WITNESS. For one thing, I would certainly not have given him information, but I asume that the main question is would I have reported him right away.

Mr. GRAY. Yes; let me ask that question. The British were allies, it was official policy, this man frankly said that then if the information were made available, it could be transmitted through channels which were not official channels.

The WITNESS. I would probably have reported him. I realize, however, that this can lead to a bad conflict. If I am convinced that the man is honest in his own benighted way, that is an unpleasant conflict situation, I would probably have reported him anyway.

Mr. GRAY. The reason I asked the question is not to get an answer from you on the basis of a hypothetical question, but to really ask next whether you would have made a distinction at that time between an approach on behalf of the Russians and an approach on behalf of the British.

The WITNESS. Yes. I think the probability of being at war with Russia in the next 10 years was high, and the probability of being at war with England in the next 10 years was low.

Mr. GRAY. Thank you, Dr. Evans.

Dr. EVANS. Dr. von Neumann, where were you born?

The WITNESS. Budapest, Hungary.

Dr. EVANS. I think you did tell us, but I want to know again, just where were you educated?

The WITNESS. I studied chemistry in Berlin and Zurich and graduated as an engineer of chemistry in Zurich.

Dr. EVANS. Zurich?

The WITNESS. Yes; in Switzerland. After that I got a Ph. D. in mathematics in Budapest, Hungary. This was in 1926.

Dr. EVANS. When did you come to this country?

The WITNESS. 1930.

Dr. EVANS. Are you a citizen of the United States?

The WITNESS. Since 1937.

Dr. EVANS. And were you professor here at any time in any institute?

The WITNESS. Yes; I was professor of mathematical physics at Princeton University until 1933. At that time the Institute for Advanced Study began to operate in Princeton and I was then appointed to the Institute for Advanced Study.

Dr. EVANS. You first met Dr. Oppenheimer in Goettingen?

The WITNESS. It was either Zurich or Goettingen in 1926.

Dr. EVANS. Doctor, do you think a man can be loyal to his country and still, due to his associates, be a security risk?

The WITNESS. That is possible; yes.

Dr. EVANS. Do you think a scientific man—a man trained in mathematics, like yourself—after any country had exploded an atomic bomb, a scientist like yourself in Russia, could guess a good bit about it?

The WITNESS. That depends when. I think in 1943, hardly. Pardon me. Just from the fact of the explosion?

Dr. EVANS. You knew it was an atomic bomb explosion, and you knew the room to the atom had been unlocked, and we knew the structure in there, and the quantum mechanics connected with it, you would be able to guess a good bit?

The WITNESS. Surely. Knowing about nuclear fission and knowing that somebody else had been able to make a detonation, one could go ahead on that basis, but it takes a large organization.

Dr. EVANS. Yes; it does. Do you believe scientific men should be required not to publish this discovery?

The WITNESS. In which era?

Dr. EVANS. Any time.

The WITNESS. Forgive me, sir, I have not understood. You mean that no discovery should be published?

Dr. EVANS. Yes; a scientific man makes a discovery; should we keep it secret or should we publish?

The WITNESS. No; it ought to publish. There are military areas, there are areas of classification and I think apart from this, one ought to publish.

Dr. EVANS. Apart from that?

The WITNESS. Yes.

Dr. EVANS. You do think there are some that should be kept secret?

The WITNESS. Oh, yes.

Dr. EVANS. If someone had approached you and told you he had a way to transport secret information to Russia, would you have been very much surprised if that man approached you?
The WITNESS. It depends who the man is.
Dr. EVANS. Suppose he is a friend of yours.
The WITNESS. Well; yes.
Dr. EVANS. Would you be surprised?
The WITNESS. Yes.
Dr. EVANS. Would you have reported it immediately?
The WITNESS. This depends on the period. I mean before I got conditioned to security, possibly not. After I got conditioned to security, certainly yes.
Dr. EVANS. You would.
The WITNESS. I mean after quite an experience with security matters and realizing what was involved; yes.
Dr. EVANS. I am sure you would now, Dr. von Neumann.
The WITNESS. There is no doubt now.
Dr. EVANS. You don't know some years ago whether you would have or not?
The WITNESS. What I am trying to say is this, that before 1941, I didn't even know what the word "classified" meant. So God only knows how intelligently I would have behaved in situations involving this. I am quite sure that I learned it reasonably fast. But there was a period of learning during which I may have made mistakes or might have made mistakes. I think I didn't.
Dr. EVANS. Would you put loyalty to a friend above loyalty to your country at any time?
The WITNESS. No.
Dr. EVANS. Have you met any Communists?
The WITNESS. Oh; yes.
Dr. EVANS. That you knew were Communists?
The WITNESS. Oh; yes.
Dr. EVANS. Have you any friends that are Communists?
The WITNESS. At this moment; no.
Dr. EVANS. Do you always know a Communist when you meet him?
The WITNESS. No.
Dr. EVANS. I guess that is all.

REDIRECT EXAMINATION

By Mr. SILVERMAN:
Q. Perhaps particularly in view of Dr. Evans' question about whether you ever met any Communists, I hope you will forgive me if I ask you 1 or 2 personal questions.
Was your family in Hungary at or about the time of the Soviet state there?
A. Yes.
Q. And did they leave in part because they didn't like it?
A. We left Hungary very soon after the Communists seized power. The Communist regime in Hungary lasted 130 days. This was in 1919. We left essentially as soon as it was feasible, which was about 30 or 40 days later, and we returned about 2 months after the Communists had been put down. I left Hungary later than this, to be exact 2 years later in order to go to college.
I first intended to become a chemical engineer, and if I had become a chemical engineer I might have returned to Hungary. Since I decided to become a mathematician and then the academic outlook in Hungary was not at all promising whereas in Germany at that time it was very promising indeed, I then decided to go to Germany.
Q. As you grew up, did you and your family regard Russia as a sort of natural enemy of Hungary?
A. Russia was traditionally an enemy of Hungary. There was a seed of war betweeen Hungary and Russia in 1948 which according to the Hungarian version, which is what I know, the Hungarians put down the Russian army. After this they were not friendly. This trauma lasted after the First World War. After the First World War everybody had reason to worry about it. But I was a child of 9 when the First World War broke out. So Russia was traditionally the enemy. After the First World War and the second war, there is quite a pattern. I think you will find, generally speaking, among Hungarians an emotional fear and dislike of Russia.
Q. I want to go to another subject. Would you say that the development of computers was an important or essential part of the hydrogen-bomb program?

A. The way the thing went, it was very important. Whether one could have done without it is a different question. I have been a very strong proponent of computers and their use so I don't want to overevaluate it, but I think it made an important difference, let us say.

Q. Could you elaborate on that? Perhaps the view to indicating to what extent the development of computers at the particular time the hydrogen bomb was being developed contributed to it.

A. You mean what the role of very fast computers was or who developed them and why?

Q. Was it a fact that there were developments, important developments in computers during the period.

A. Very high speed computing came into reasonably general use just about during those years. I would say——

Q. When you say "those years," what do you mean?

A. When the hydrogen bomb was developed. I would say about two-thirds of the development took place under conditions like this, that the heavy use of computers was made, that they were not yet generally available, and that it was necessary to scrounge around and find a computer here and find a computer there which was running half the time and try to use it, and this was the operation I was considerably interested in. I would say the last third of the development, computers were freely available and industrially produced, and by now this is not a scarce commodity. It was very scarce during more than the first half of the hydrogen-bomb project.

Q. Was there also a question of some kind of computers not perhaps developed yet?

A. The art is better now than it was then. I would say by now what passes for a fast computer is 3 or 4 times as fast as 3 or 4 years ago. There were few of them and there were fewer people who knew what to do with them, and they were less reliable.

Dr. EVANS. Did you know my friend Mr. Flanders?
The WITNESS. Yes, I know him well.
Dr. EVANS. Did you know a chemical engineer named Adelaneau?
The WITNESS. No.
Dr. EVANS. He was connected with gas. Was there such a thing as the Roumanian-English Oil Co. over there, do you know?
The WITNESS. Probably. I know there was a lot of oil in Rumania, and I know the English companies were the ones exploiting it.
Dr. EVANS. I wondered if you knew him as I knew him personally very well.
The WITNESS. No.

By Mr. SILVERMAN:

Q. Would you say anything about the role done at the Institute with respect to the development of computers?

A. We did plan and develop and build and get in operation and subsequently operate a very fast computer which during the period of its development was in the very fast class.

Q. Did Dr. Oppenheimer have anything to do with that?

A. Yes. The decision to build it was made 1 year before Dr. Oppenheimer came, but the operation of building it and getting it into running took approximately 6 years. During 5 of these 6 years, Dr. Oppenheimer was the Director of the Institute.

Q. When was it finally built?

A. It was built between 1946 and 1952.

Q. When it was complete and ready for use?

A. It was complete in 1951, and it was in a condition where you could really get production out of it in 1952.

Q. And was it used in the hydrogen bomb program?

A. Yes. As far as the Institute is concerned, and the people who were there are concerned, this computer came into operation in 1952, after which the first large problem that was done on it, and which was quite large and took even under these conditions half a year, was for the thermonuclear program. Previous to that I had spent a lot of time on calculations on other computers for the thermonuclear program.

Q. You were asked if there were an incident that looked like an approach to espionage to you, you indicated you would report it, and now you indicated you certainly would and at other times you hoped so.

A. I would. It is possible to define a transitional period in everybody's life where he is not fully aware of the problem being present. How well anybody behaves in the period is in part a question of fortitude and in part a question of luck. There is always a relation of these things

Q. If such an approach were made to Dr. Oppenheimer today, what do you think his reaction would be?

A. I have no doubt that he would report it.

Q. Immediately?

A. I think so, yes. May I say I can summarize my views on this. I think after about a year's experience with military security and implications of security and the things which make it necessary, I think every one of us and I am convinced of Dr. Oppenheimer, and I, and everybody who I take seriously, would act the same way, namely, follow the rules which exist.

Q. Do you think that Dr. Oppenheimer would place loyalty to a friend above loyalty to his country?

A. I would not think so.

Q. Dr. Evans asked you about whether it is possible for a man to be loyal to his country, and yet be a security risk because of his associations.

A. Yes.

Q. I think you answered "Yes." Do you feel you know Dr. Oppenheimer's associations reasonably well?

A. I rather think so.

Q. Do you think that Dr. Oppenheimer is a security risk because of his present associations?

A. No, I don't think so.

Mr. SILVERMAN. That is all.

Mr. ROBB. One further question.

CROSS-EXAMINATION

By Mr. ROBB:

Q. Doctor, you have never had any training as a psychiatrist, have you?

A. No.

Mr. ROBB. That is all.

Mr. GRAY. Thank you very much, Dr. von Neumann.

(Witness excused.)

Mr. GRAY. We will recess until 2 o'clock.

(Thereupon at 12:35 p. m., a recess was taken until 2 p. m., the same day.)

AFTERNOON SESSION

Mr. GRAY. Do you wish to testify under oath?

Dr. LATIMER. I am willing.

Mr. GRAY. You are not required to do so, but all the witnesses have.

Dr. LATIMER. I am willing.

Mr. GRAY. Would you hold up your right hand, and give me your full name?

Dr. LATIMER. Wendell Mitchell Latimer.

Mr. GRAY. Wendell Mitchell Latimer, do you swear that the testimony you are to give the board shall be the truth, the whole truth, and nothing but the truth, so help you God?

Dr. LATIMER. I do.

Whereupon, Wendell Mitchell Latimer was called as a witness, and having been first duly sworn, was examined and testified as follows:

Mr. GRAY. Would you be seated, please, sir.

Dr. Latimer, it is my duty to remind you of the existence of the so-called perjury statutes. I should be glad to review them with you if necessary, but may we assume you are familiar with them.

The WITNESS. I think I am in general familiar.

Mr. GRAY. All right, sir. I should like to request that if in the course of your testimony it becomes necessary for you to refer to or disclose restricted data, you notify me in advance so we may take necessary and appropriate steps in the interest of security.

The WITNESS. I hope if I step over at any time that somebody would check me, because I am not always sure as to what is restricted, and what is not.

Mr. GRAY. We have, Dr. Latimer, a security officer of the Commission present, and I suppose available a classification officer, if we need to call him in. So if there is some question in your mind, we will try to answer the question.

Finally, I should like to say to you that we consider these proceedings a confidential matter between the Atomic Energy Commission, its officials and witnesses on the one hand, and Dr. Oppenheimer and his representatives on the other. The Commission is making no release with respect to these proceedings, and we express the hope to every witness that he will take the same view.

DIRECT EXAMINATION

By Mr. ROBB:

Q. Dr. Latimer, would you tell the board what your present position is, sir?
A. At present I am professor of chemistry at the University of California, and associate director of the Radiation Laboratory.
Q. Located where, sir?
A. At Berkeley, Calif.
Q. And you live in Berkeley, Calif.?
A. I live in Berkeley.
Q. Could you give the board some account of your education and background?
A. I have an A. B. from the University of Kansas. I have a Ph. D. from the University of California. I have been at the University of California on the staff since 1919. I was dean of the College of Chemistry for 8 years. Is there anything else that you want?
Q. What is your specialty in science, Doctor?
A. My specialty is thermodynamics and inorganic chemistry.
Q. Have you held any positions or offices in the National Academy of Science?
A. I am a member of the National Academy and I was chairman of the chemistry section for one term.
Q. Are you the author of any books?
A. Yes, I have several textbooks. I also edited a series of books for the Prentiss Hall Publishing Co.
Q. On what?
A. Chemistry in general.
Q. Do you know Dr. Oppenheimer?
A. Yes, I do.
Q. How long have you known him, sir?
A. Oh, a great many years; ever since he came to the University of California.
Q. Beg pardon?
A. Ever since he came to the University of California I think we have been acquainted.
Q. Did you know him when he was on the faculty there?
A. Yes, I did, both before and after the war.
Q. Has your acquaintance been both social and official?
A. Not very highly social. I believe I was at his house for cocktails at one time. Officially, early in the Los Alamos program my group made a few hundred milograms of plutonium for their project. I think it was the first plutonium that they had. During that period I saw him several times.
Q. Doctor, you somewhat anticipated my next question, which was whether or not there came a time when you and your group at Berkeley did some work on the A-bomb.

Mr. SILVERMAN. Would you mind, I don't quite understand this reference to Dr. Latimer's group.

Mr. ROBB. I was going to ask him to explain that, too.

The WITNESS. Plutonium was discovered in our laboratory by Professor Seaborg and his group, and after Seaborg went to Chicago to work in the metallurgical laboratory there, I continued to direct a group on the chemistry of plutonium, and in the early days our principal source of plutonium was from our cyclotrons. So we worked up as large samples as we could of plutonium in order to study its chemistry.

The group I was directing did a lot of the early work on the chemistry of plutonium.

By Mr. ROBB:

Q. When you say your group, Doctor, to what do you refer?
A. I guess we had about 25 men working on the chemistry of plutonium.
Q. You mean working under you in your department?
A. Yes.
Q. When you refer to the cyclotron, where was that located?
A. There were two cyclotrons at Berkeley. The one that was used largely was the 60-inch cyclotron on the campus.
Q. At Berkeley?

A. At Berkeley.
Q. That is what was called the radlab.
A. Yes; it is called the radlab.
Q. The Radiation laboratory?
A. Radiation laboratory.
Q. When did this work on plutonium go on, Doctor?
A. I started Dr. Libby working on radioactive problems about 1933. Between that and 1940, we had built up quite a group, Seaborg and Kennedy, and at the time the war broke out, we had probably the best group of young nuclear chemists all over the country, so it was just a gradual transition from our research program that we had underway to applications for the Manhattan District.
Q. Doctor, I would like to ask you a question for the record. What is the connection between plutonium and the atom bomb?
A. Plutonium was one of the elements which were fissioned with slow neutrons, and therefore it is a material which can be used to sustain chain reactions, and was one of the materials used in the B-bombs.
Q. In connection with your work on plutonium and your production of plutonium, did you come in contact with Dr. Oppenheimer during the war?
A. As I mentioned, we did make the first sample of plutonium for the Los Alamos Laboratory. I believe I am correct in that. We did other work for them. We made various ceramic materials for them in which to melt plutonium. We tried to be as helpful as we could although we were working closer with the Chicago laboratory. Still we did jobs for Los Alamos as best we could when they requested it.
Q. How frequently did you have occasion to see or meet Dr. Oppenheimer during the war period?
A. Not very frequently. As you know, after they went to Los Alamos, they were pretty well tied down there. We didn't see many of the men after that.
Q. Did you follow the work that was being done at Los Alamos?
A. Not very closely. We were interested in the production of plutonium, and they were fabricating it into bombs. We didn't follow that side of it.
Q. Doctor, did there come a time when you began thinking about a weapon which is called the H-bomb?
A. Yes.
Q. When was that?
A. I suspected I started worrying about the H-bomb before most people. Just as soon as it became evident to me that the Russians were not going to be cooperative and were distinctly unfriendly.
Q. Would you keep your voice up just a bit, Doctor?
A. I felt that it was only a question of time that the Russians got the A-bomb. I haven't much confidence in secrecy keeping these things under control very long. It seemed to me obvious that they would get the A-bomb. It also seemed to me obvious that the logical thing for them to do was to shoot immediately for the super weapon, that they knew they were behind us in the production of a bomb. It seemed to me that they must conclude shooting ahead immediately in making the super weapons. So I suspect it was around 1947 that I started worrying about the fact that we seemed to be twiddling our thumbs and doing nothing.
As time passed, I got more and more anxious over this situation that we were not prepared to meet, it seemed to me, a crash program of the Russians. I talked to a good many people about it, members of the General Advisory Committee.
Q. Do you recall who you talked to about it?
A. I talked to Glenn Seaborg for one. I didn't get much satisfaction out of the answers. They seemed to me most of them on the phony side.
Q. Doctor, may I interpose right here before we go on to ask you a couple of questions, first, why did it seem obvious to you that the Russians would proceed from the A-bomb to the H-bomb?
A. They knew they were behind us on the A-bomb, and if they could cut across and beat us to the H-bomb or the super weapons, they must do it. I could not escape from the conclusion that they must take that course of action. It was the course of action that we certainly would have taken if we were behind. I could not escape from that conclusion.
Q. The second question is, you said that we seemed to be twiddling our thumbs in the matter. What was the basis for that feeling on your part?
A. In the period between 1945 and 1949 we didn't get anywhere in our atomic energy program in any direction. We didn't expand our production of uranium

much. We didn't really get going on any reactor program. We didn't expand to an appreciable extent our production of fissionable material. We just seemed to be sitting by and doing nothing.

I felt so certain that the Russians would get the A-bomb and shoot for the H-bomb that all during that period I probably was overanxious, at least compared to most of the scientists in the country. But it seemed to me that such an obvious thing would happen.

Q. Reverting again to your narrative, you said you talked to Dr. Seaborg and others about going ahead with the H-bomb, and their answers, you said, seemed to be phoney. What did you mean by that?

A. I can't recall all the details during that period. When the Russians exploded their first A-bomb, then I really got concerned.

Q. What did you do?

A. In the first place, I got hold of Ernest Lawrence and I said, "Listen, we have to do something about it." I think it was after I saw Ernest Lawrence in the Faculty Club on the campus, the same afternoon he went up on the Hill and Dr. Alvarez got hold of him and told him the same thing. I guess the two of us working on him at once with different impulses got him excited, and the three of us went to Washington that weekend to attend another meeting, and we started talking the best we could, trying to present our point of view to various men in Washington.

On that first visit the reception was, I would say, on the whole favorable. Most people agreed with us, it seemed to us, that it should be done.

Q. Could you fix the approximate date of this?

A. I would say within 2 or 3 weeks after the explosion of the Russian bomb. I don't remember the date of that.

Q. That was in September 1949.

A. Shortly after that.

Q. And you said your reception seemed to be on the whole favorable. Do you recall whom you saw on that occasion?

A. Around the Commission I think Dean was the only Commissioner there. I talked largely to the chemistry group there, to Dr. Pitzer, and Dr. Lauritsen, and Dr. Lawrence and Dr. Alvarez talked to a good many other men. They talked to, as I recall, members of the joint congressional committee, and to various men in the Air Force and Army.

Q. Do you recall whether you talked to any other scientists who were not with the Commission?

A. Yes. I talked to Dr. Libby and Dr. Urey in Chicago. I talked to everybody I could, but I don't remember now. I tried to build up pressure for it. I definitely tried to build up pressure for it.

Q. What was the reception of your suggestions received at that period of time? I am speaking of the time 2 or 3 weeks after the Russian explosion.

A. It was favorable, I would say. We met practically no opposition as I recall.

Q. Will you tell us whether or not that situation changed?

A. It definitely changed.

Q. When?

A. Within a few weeks. There had been a lot of back pressure built up, I think primarily from the Advisory Committee.

Q. Would you explain that to us a bit?

A. I don't remember now all the sources of information I had on it, but we very quickly were aware of the fact that the General Advisory Committee was opposed.

Q. What was the effect of that opposition by the Committee upon fellow scientists, if you know?

A. There were not many scientists who knew the story. I frankly was very mystified at the opposition.

Q. Why?

A. Granted at that time the odds of making a super weapon were not known, they talked about 50–50, 10 to 1, 100 to 1, but when the very existence of the Nation was involved, I didn't care what the odds were. One hundred to one was too big an odd for this country to take, it seemed to me, even if it was unfavorable. The answers that we kept getting were that we should not do it on moral grounds. If we did it, the world would hate us. If we didn't do it, the Russians wouldn't do it. It was too expensive. We didn't have the manpower. These were the types of argument that we got and they disturbed me.

Q. Did you ascertain the source of any of this opposition?

A. I judge the source of it was Dr. Oppenheimer.

Q. Why?

A. You know, he is one of the most amazing men that the country has ever produced in his ability to influence people. It is just astounding the influence that he has upon a group. It is an amazing thing. His domination of the General Advisory Committee was so complete that he always carried the majority with him, and I don't think any views came out of that Committee that weren't essentially his views.

Q. Did you have any opinion in 1949 on the question of the feasibility of thermonuclear weapons?

A. Various calculations seemed to show that it might go if you could just get the right conditions or the right mechanical approach to it. The odds didn't look good, but as I say, I didn't care what the odds were, if there was a possibility of it going, I thought we must explore it, that we could not afford to take a chance not to. The stakes were too big. The very existence of the country was involved and you can't take odds on such things.

Q. Was there any way that you knew of to get the answer without experiment and tests?

A. No, I am sure all the calculations showed that the only way it could ever be settled was by trying it.

Q. Have you followed the progress of the thermonuclear program since 1949?

A. In a rough way, yes. In the past 2 years, we have been working on some of the problems at the Radiation Laboratory.

Q. At Berkeley?

A. At Berkeley.

Q. Dr. Latimer, this board is required within the framework of the statute to determine upon its recommendation to the general manager as to whether or not the security clearance of Dr. Oppenheimer should be continued and the standards set up by the statute for the board are the character, the associations and the loyalty of Dr. Oppenheimer. Would you care to give the board, sir, any comments you have upon the basis of your knowledge of Dr. Oppenheimer as to his character, his loyalty and his associations in that context?

A. That is a rather large order.

Q. I know it is, Doctor.

A. His associations at Berkeley were well known. The fact that he did have Communist friends. I never questioned his loyalty. There were elements of the mystic in his apparent philosophy of life that were very difficult to understand. He is a man of tremendous sincerity and his ability to convince people depends so much upon this sincerity. But what was back of his philosophy I found very difficult to understand.

A whole series of events involved the things that started happening immediately after he left Los Alamos. Many of our boys came back from it pacifists. I judged that was due very largely to his influence, this tremendous influence he had over those young men. Various other things started coming into the picture.

For example, his opposition to the security clause in the atomic energy contracts, opposition on the floor of the National Academy which was very intense and showed great feeling here. These various arguments which were used for not working on the H-bomb, the fact that he wanted to disband Los Alamos. The fact of the things that weren't done the 4 years that we twiddled our thumbs. All these things seemed to fit together to give a certain pattern to his philosophy. A man's motives are just something that you can't discuss, but all his reactions were such as to give me considerable worry about his judgment as a security risk.

Q. I will put it in very simple terms, Doctor. Having in mind all that you have said, and you know, would you trust him?

A. You mean in matters of security?

Q. Yes, sir.

A. I would find—trust, you know, involves a reasonable doubt, I would say.

Q. That is right.

A. On that basis I would find it difficult to do so.

Q. Doctor, it has been suggested here that Dr. Oppenheimer is so valuable to this country's weapons program that he should be continued in his present status. What can you say about that?

A. He could be of tremendous value to this country. His leadership of the scientists of the country has been extremely valuable. As far as his value in continuing the atomic energy program, I would say it is largely in the influence he has upon other scientists. One of the things that annoys a great many scientists more than anything else is this statement that he alone could have built the A-bomb, or that he alone could have carried on the program. One very prominent engineer said to me yesterday that statement just gets me down. Sure, I can pick out a half-dozen young men that could do the job.

Whenever you do anything new the first time it seems awfully hard, but later you discover that all you have done is taken a long roundabout road to get there. Actually there is a shortcut and you get there in a hurry. So one always tends to magnify the difficulties the first time you do a thing. If you have enough good men working on it, you are almost sure to find a shortcut.

I think the developments in the super weapon that have occurred recently show that this went along without very much—at least the key ideas were not supplied by him.

Q. What?
A. The key ideas were not supplied by him.
Mr. SILVERMAN. By Dr. Oppenheimer.
The WITNESS. That is right.

By Mr. ROBB:

Q. But you would not say that he was indispensible?
A. No, I couldn't say that.
Mr. ROBB. That is all I care to ask, Mr. Chairman.
Mr. SILVERMAN. May we take about 5 minutes recess to consult with my colleagues?
Mr. GRAY. Was there anything said you didn't hear, Mr. Silverman?
Mr. SILVERMAN. No, sir.
Mr. GRAY. I think we might as well proceed.

Let me say this. My commitment on behalf of the board with respect to cross-examination of witnesses whose direct examination has been conducted by Mr. Robb is that if there are instances in which Mr. Garrison felt that he was disadvantaged by surprise, we would consider any reasonable request. But it doesn't seem to me necessary to take a recess for purposes of cross-examination unless there is something that you——

Mr. SILVERMAN. Mr. Chairman, I don't press the point particularly. There are 1 or 2 places when I was talking to Dr. Oppenheimer when Mr. Marks heard something and I asked what was said, and he says he has it down. It is that sort of thing.

Mr. GRAY. If you feel at any point you cannot properly represent Dr. Oppenheimer's interest, I would want you to inform the board.

Mr. SILVERMAN. I will do my best to represent Dr. Oppenheimer's interest. We will just take a minute here if that is all right.

Mr. GRAY. Go ahead.

CROSS-EXAMINATION

By Mr. SILVERMAN:

Q. Dr. Latimer, Dr. Oppenheimer left Berkeley in 1947, didn't he, to go to the Institute for Advanced Study?
A. I don't remember the date.
Q. How often would you say you have seen Dr. Oppenheimer since 1947?
A. Not very frequently. I have seen him at the academy meetings. He has been back to Berkeley on visits, but it has been infrequent.
Q. Woud you say you have seen him 10 times, 5 times?
A. Let us say five times.
Q. Were those in fairly large groups?
A. I would certainly at least meet him and shake hands with him and maybe pass a few words.
Q. Just social?
A. These were casual meetings.
Q. You met him a few times casually since 1947?
A. That is right.
Q. And before that, did you meet him frequently?
A. We never had an intimate relationship. We saw each other on the campus.
Q. You were members of the same faculty.
A. We were members of the same facility and had the normal contacts as between faculty members.
Q. Did he ever visit your home?
A. No.

Q. And the only time you have a recollection of visiting his home is that one time you went to a cocktail party?
A. I believe that is all I recollect.
Q. You say you started worrying, I think was the phrase you used, about the hydrogen program and about the fact that we seemed to be twiddling our thumbs about 1947, when your worries began?
A. I can't date it, but at the end of the war I was not content for us to stop going ahead. I did not trust the Russians and I immediately started worrying about keeping ahead. I can't date it, but let us say I suggested it even before it became obvious to everybody that the Russians were not going to be friendly. I started worrying about it.
Q. Did you know whether there was work being done on thermonuclear research, and research on thermonuclear weapons at Los Alamos during the war?
A. Yes, I knew that the program, that a start had been made on it.
Mr. ROBB. Have you finished the answer?
The WITNESS. I knew a start had been made on it. I knew they had not gotten very far, but that calculation had been made and various possible approaches were being investigated.

By Mr. SILVERMAN:
Q. Did you know that research continued?
A. Yes, it continued without much pressure on it.
Q. How did you know what was being done?
A. I saw Teller occasionally. I don't suppose I had a very clear idea at that time except that it is not hard to form an impression of the magnitude of a program from many different sources.
Q. What I am concerned about is to what extent these sources were matters of which you had some fairly direct personal knowledge.
A. I don't know what you mean quite by direct personal knowledge. I was not down to Los Alamos during that period, and I didn't talk to the men working on the program during that period. But our general impressions around the radiation laboratory, the general impressions I got from talking to men in Washington, was that things were not moving ahead.
Q. Did you have some sort of responsibility for any part of the atomic weapons program?
A. During those years?
Q. Yes.
A. No.
Q. Did you have any official connection with it?
A. I was still associate director of the radiation laboratory, and the men together in this laboratory talked over between them many problems. There is a pretty general amount of information on these programs.
Q. What I am concerned about is, was what you knew pretty much what you picked up in a sort of general way, or was it something that it was your business to know something about, and that you made fairly direct efforts to find out?
A. It was not directly my business to know about it except as a citizen of this country who had a certain amount of information on that subject, and was greatly concerned about what was being done. I would ask questions as high up as I could to find out what was being done. Maybe the answers were often vague, but still anyone can form a pretty definite impression by such methods.
Q. Quite so. I would not for a moment question your right to form an opinion. Indeed a very natural interest would lead to it. What I am trying to arrive at was the opinion or impression you had formed the impression of an interested citizen without very direct access or responsibility to the problem, or was it that of a man whose job it was to be working on the problem?
A. It was not my job to be working on it, but I had a lot of information about the nuclear program. I had a lot of sources of classified information. I think I might say that my suspicions over that period had been verified by evidence that has come out later.
Q. What you had was suspicions?
A. It was obvious during those years we were not doing anything of any significance.
Q. Did the radiation laboratory do any substantial work on atomic weapons during the years 1945 to 1949?
A. No.
Q. Did you know what General Groves' views were as to whether it was desirable in the years 1947 on—in the early years there—as to whether it was desirable to concentrate on fission weapons rather than on thermonuclear?

A. I suppose I heard his views. They seemed to coincide with that of the General Advisory Committee pretty much. I suspect again under the influence of Dr. Oppenheimer.

Q. You don't of course question General Groves' patriotism or his good faith?

A. I don't question the patriotism of any of the members on that committee. Of course, he was not on the committee. Not only General Groves, but the other members on the committee, Conant and the other members, they were under the influence of Dr. Oppenheimer, and that is some influence, I assure you.

Q. Were you under Dr. Oppenheimer's influence?

A. No, I don't believe I was close enough contact to be. I might have been if I had been in closer contact.

Q. You think that General Groves was under Dr. Oppenheimer's influence?

A. Oh, very definitely.

Q. Have you ever spoken to General Groves?

A. About this problem?

Q. At all.

A. Oh, yes; I saw him frequently during the war.

Q. On what do you base your judgment that General Groves was under Dr. Oppenheimer's influence?

A. I wouldn't go too far in answering that question, because I don't know how much General Groves' opinions have changed in recent years. The statements that I have heard attributed to him seemed to follow the same—at least for a while, I have not seen his statements very recently—but during part of this period he seemed to be following the Oppenheimer line.

Q. What I am curious about is how do you know that Dr. Oppenheimer was not following the Groves line?

A. That is ridiculous.

Q. Pardon?

A. Knowing the two men, I would say that is ridiculous. Oppenheimer was the leader in science. Groves was simply an administrator. He was not doing the thinking for the program.

Q. I am trying to arrive upon what it is that you base your—I think you said it was a suspicion, but perhaps I am wrong, that General Groves was under Dr. Oppenheimer's influence. Is it simply the fact of your knowledge of Dr. Oppenheimer and the fact that he is a leading scientist and a man of great gifts.

A. I know these things were overwhelming to General Groves. He was so dependent upon his judgment that I think it is reasonable to conclude that most of his ideas were coming from Dr. Oppenheimer.

Q. How do you know he was so dependent?

A. I don't. I don't know, but I have seen the thing operate.

Q. There were other scientists at Los Alamos, weren't there?

A. Yes, there were.

Q. And General Groves has had contact with other scientists.

A. Yes, but there were no other scientists there with the influence that Dr. Robert Oppenheimer had and moreover this close association with Groves certainly one would normally conclude that he still had tremendous influence over him. It may be an unreasonable conclusion, but it doesn't seem so to me.

Q. Forgive me, but no man considers his own view unreasonable.

A. That is right. You must accept these as my personal opinions and nothing more than that.

Q. I am trying to arrive on what you base these personal opinions.

A. Various things that go into a man's judgment are sometimes difficult to analyze.

Q. I am trying to find out to what extent objective facts——

A. I had studied this influence that Dr. Oppenheimer had over men. It was a tremendous thing.

Q. When did you study this influence?

A. All during the war and after the war. He is such an amazing man that one couldn't help but try to put together some picture.

Q. Tell us about these studies that you made about Dr. Oppenheimer's influence. You said after the war.

A. He has been a most interesting study for years. Unconsciously, I think one tries to put together the elements in a man that make him tick. Where this influence comes from, what factors in his personality that give him this tremendous influence. I am not a psychoanalyst. I can't give you how my picture of this thing was developed, but to me it was an amazing study, just thinking about these factors.

Q. For a long time you have been thinking about Dr. Oppenheimer's influence on people.
A. Yes, particularly during this period when he was able to sway so many people, so many of his intimate——
Q. What is the period here?
Mr. ROBB. Wait a minute. He has not finished.
Mr. SILVERMAN. Sorry.
The WITNESS. During this period of discussion as to whether one should work on the H-bomb and the super weapons. I was amazed at the decision that the committee was making, and I kept turning over in my mind how they could possibly come to these conclusions, and what was in Oppenheimer that gave him such tremendous power over these men.

By Mr. SILVERMAN:
Q. Did you talk to any of these men over whom Dr. Oppenheimer had this tremendous power?
A. Occasionally, yes.
Q. Would you tell us whom you talked to, please?
A. The man on the Commission I was most intimately associated with was Dr. Seaborg, since he was a member of my department. I talked to him very frequently about the problem.
Q. Did Dr. Seaborg say he just couldn't stand up to Dr. Oppenheimer's influence.
A. He didn't stand up to him very well.
Q. What did he say?
A. That is years ago. I can't remember.
Q. I am trying to distinguish between your judgment and what you were told.
A. These were my judgments, I would say. I have seen him sway audiences. It was just marvelous, the phraseology and the influence is just tremendous. I can't analyze it for you, but I think all of you know the man and recognize what I am talking about.
Q. I think you said that you judged that the source of the opposition to the hydrogen bomb, the back pressure, I think you referred to it as, was Dr. Oppenheimer.
A. That is right.
Q. Would you tell us on what you based that judgment?
A. As chairman of the Committee he wrote all the Committee reports and the decisions became pretty apparent. I don't remember how the decisions leaked out but the fact that they recommended to the President that no work be done. Surely nobody could conclude it wasn't largely Dr. Oppenheimer's opinion which was being presented.
Q. Have you ever met Dr. Conant?
A. Yes, I know Dr. Conant.
Q. Would you say that he is a man of fairly firm character?
A. I have known him a long time. He is a man of force, but in matters pertaining to theoretical physics, I think he trusted Dr. Oppenheimer completely.
Q. And on what do you base that?
A. The fact that he followed along so consistently.
Q. Do you know whether Dr. Conant's judgment in connection with the hydrogen bomb was based on a technical evaluation—I don't mean, a technical evaluation—a judgment as to the nuclear aspects of the problem, the scientific nuclear aspect of the problem?
A. Those were the reasons which were given in the report. They were expressed in technical terms. I was by no means convinced that those were the real reasons behind the decision.
Q. Have you read that report?
A. I don't know as I ever have. I may have in recent years seen in the atomic energy office copies which would confirm my opinions, but certainly the essence of the report was known, that they were opposed to the thermonuclear weapons. We didn't have the manpower for it. It would detract from our A-bomb work—a number of reasons like that. I don't know. Technical reasons were given.
Q. You consider those technical reasons relating to nuclear physics?
A. They sounded pretty phony to me.
Q. That was not my question, precisely. My question was whether you considered those reasons related to nuclear physics, and on which therefore Dr. Conant might be relying on Dr. Oppenheimer?

A. Yes, those were the obvious reasons given, I believe.
Q. Did you consider that those were reasons related to nuclear physics on which Dr. Conant would therefore be relying on Dr. Oppenheimer?
A. Those would have been legitimate reasons if he had been exercising his free judgment and not overwhelmed by his great confidence in Dr. Oppenheimer's judgment. I doubt if it was a free judgment on his part.
Q. My question, sir, is not whether it was free judgment or whether it was legitimate reasons or anything like that. My question is whether you consider manpower a problem of nuclear physics.
A. It was in this case. In this case if it was true we didn't have the manpower to do it, it was a legitimate reason. But I believe we did have the manpower to do it as subsequent events showed.
Q. Is that the problem that Dr. Conant was relying on Dr. Oppenheimer, as to whether we had the manpower?
A. I judge he offered that as one of the reasons.
Q. You don't know now whether you have ever read the GAC 1949 report, or do you?
A. I don't recall. I have talked to a good many men who have seen it. I have talked to Dr. Pitzer and Dr. Seaborg and probably a half dozen others who have seen it. Whether I read it or not, I don't recall, but the essence of it was obvious.
Q. Do you know whether these reasons you have given were stated in the 1949 report of the GAC?
A. I can't at this moment say definitely, but they were, as I recall, approximately the arguments given.
Q. You say as you recall. As you recall it from what?
A. As I recall it from the discussion which was occurring at that time. That has been a number of years ago.
Q. Discussion with whom, sir?
A. With everybody concerned in the program and that was concerned in this decision. There was general discussion among the scientists on the atomic energy program whether the thing should go. These arguments were tossed back and forth very freely among hundreds of men on the program.
Q. What I am concerned about, sir, is the reasons given in the GAC report.
A. Yes, sir.
Q. Do you know what the reasons that were given in the GAC report were?
A. I can't at the moment quote the reasons given, but the intent of the report was obvious. Four or five years ago I could have given you many of the details, but today all I can recall in detail is the intent of the report.
Q. And you think that the report did contain this argument about diversion of manpower?
A. You see, there were so many arguments being given by members of the General Advisory Committee, many of them verbally, and what was actually written down in that report at this moment, confusing all these arguments that are given, I could not definitely state.
Q. You came to Washington in an effort, I think you put it, to build up pressure for the hydrogen bomb.
A. I came to Washington on another mission, but while I was here, I did everything I could to build up pressure for the work.
Q. Did you know that the General Advisory Committee would be consulted on this problem?
A. Why, surely.
Q. How many members of the General Advisory Committee do you know personally?
A. I forget now. Many of them I did not know intimately. Rabi, I knew fairly well. Fermi I had a speaking acquaintance with. Seaborg, I don't remember the exact composition of that Committee at that time.
Q. Did you attempt to communicate your views to any member of the General Advisory Commitee?
A. I certainly worked hard on Seaborg.
Q. Didn't Dr. Seaborg tell you that he was not going to be at the meeting?
A. He wrote a letter, I believe.
Q. Didn't he tell you he was not going to be at the meeting?
A. Yes, but he still had influence.
Q. Did you speak to anyone else who was going to be at the meeting?
A. I have not directly.
Q. I don't understand what you mean by not directly.

A. I worked on a good many of my friends around the Commission, such as Ken Pitzer. I told him my point of view.
Q. Dr. Pitzer did not have to be convinced of your point of view, did he?
A. It didn't take very long to.
Q. Did you try to speak to Dr. Oppenheimer about it?
A. I did not.
Q. Did you then hold the view that Dr. Oppenheimer was a very influential member of the GAC?
A. Oh, that was obvious.
Q. Did you then hold the view that whatever Dr. Oppenheimer's view was would ultimately be the GAC view?
A. The majority, I believe. I believe there was occasionally a dissent, but certainly the majority followed his opinion.
Q. Didn't it occur to you that it might be useful to call up Dr. Oppenheimer and try to present your point of view in the hope that GAC would be influenced?
A. I didn't think my opinion would have much influence upon him.
Q. In matters as important as this, did it really matter what the chances were of your being able to influence Dr. Oppenheimer?
A. It was merely a matter of procedure. I was trying to accomplish my objectives, but one makes judgment as to how is the best way to accomplish these objectives. I talked to Admiral Strauss and gave him detailed statements of what I thought he could use with the President to make the decision.
Q. Did you think that Admiral Strauss' influence was greater than that of Dr. Oppenheimer?
A. When he got the Army and Navy and others behind him it turned out it was.
Q. Did you then think that Admiral Strauss' influence would be greater?
A. I did.
Q. Didn't you think it would be a good idea if you could get the GAC to go along?
A. I hoped they would.
Q. Did you do any more?
Mr. ROBB. Wait a minute. He has not finished his answer.
Mr. SILVERMAN. I am sorry I keep interrupting.
The WITNESS. Let it go at that. I hoped he would. But I didn't feel with very many members of the GAC I didn't have much influence. After all, a chemist does not have much influence with theoretical physicists.
Mr. SILVERMAN. I believe there is one chemist in this room that has a certain amount of influence.
The WITNESS. Not directly.

By Mr. SILVERMAN:
Q. Wasn't there a chemist on the GAC?
A. Seaborg.
Q. How about a fellow named Conant?
A. He was a college president.
Q. You didn't think that speaking to Dr. Conant there would be any rapport between you and Dr. Conant?
A. No. In fact—well, I guess it doesn't matter.
Q. You said some of the boys came back from Los Alamos pacifists, and you judged that to be due to Dr. Oppenheimer's influence. On what did you base that judgment?
A. Their great devotion to him. They were capable of independent judgment, but it looked to me like a certain amount of indoctrination had taken place. That matter I would not put too much weight on, but it was just an observation that they had.
Q. Forgive me, Dr. Latimer. This is a terribly serious matter, this whole proceeding.
A. I realize it. I feel terrible about it.
Q. I understand that, sir. Is it your considered judgment that boys came back pacifists from Los Alamos due to Dr. Oppenheimer's influence?
A. That was the conclusion I came to. I may be wrong, but that was my conclusion.
Q. And you gave that conclusion in your direct testimony.
A. Yes.
Q. On what did yo base that conclusion?

A. It is difficult to analyze it. I talked to them. This was years ago, though. I can't recall all the details of it. That was the conclusion I came to. I don't remember now what went into my judgment at the time.

Q. Dr. Latimer, let me put it to you as frankly as I can, and I would like you honestly, and I know you will, to consider this point of view. Would you say that your judgment that these boys were influenced to become pacifists by Dr. Oppenheimer is based essentially on your judgment that Dr. Oppenheimer is a very persuasive person, and that very few people come in contact with Dr. Oppenheimer without being influenced by him?

A. That is certainly an important factor in my decision.

Q. And that therefore if someone comes back after having a contact with Dr. Oppenheimer with a view which to you appears to be Dr. Oppenheimer's view, it is in your judgment reasonable to suppose that Dr. Oppenheimer influenced them?

A. I would conclude from the devotion of these boys to him that would not be contrary to his own opinions and probably expressed.

Q. Did you know what his opinions were on the question of pacifism?

A. Let me phrase this a little differently. Let us not put the general pacifism, but an unwillingness to build weapons or to work on any research involving weapons. I believe that was a more careful statement of the opinions they voiced.

Q. Dr. Latimer, that is a very different thing from being pacifists, is it not?

A. It amounts to the same thing, I would say. We have to have weapons to fight. If we don't have weapons, we don't fight.

Q. Wasn't it true that many scientists after the explosion at Hiroshima and perhaps even before that—many scientists after the explosion at Hiroshima were terribly troubled by this weapon?

A. Oh, yes.

Q. Weren't you, sir?

A. I was more troubled by what the Russians might do along the same line.

Q. I would like to ask you whether you were troubled by this weapon.

A. No.

Q. Were you troubled by the fact that 70,000 people were killed at Hiroshima?

A. I felt that you might even have saved lives. I had been in the Pacific and I had seen something of the difficulty of getting the Japanese out of caves. I went over there on a special mission that involved that problem. I felt that if we had to land our boys on the coast of Japan, and knowing what I knew about the difficulty of getting Japanese out of underground positions, that the loss of life might be very much greater.

Q. I think we all understand that consideration, Dr. Latimer, and I think we all share it. What I would like to know is whether you were troubled by the fact that 70,000 people were killed at Hiroshima.

A. I suppose I was troubled to the same extent that I was troubled by the great loss of life which occurred in our fire bombs over Tokyo. The two things were comparable in my mind. I am troubled by war in general.

Q. Don't you think that perhaps boys who had worked on the atom bomb and who perhaps felt some responsibility for the bomb might have felt that trouble in perhaps even more acute form?

A. I grant that is correct; they might have.

Q. Now, I think you said that you referred to Dr. Oppenheimer's opposition to the security clause.

A. Beg pardon?

Q. I think you referred to Dr. Oppenheimer's opposition to the security clause.

A. This was just part of the pattern that seemed to be developing. There was quite a group in the academy who fought the security clause in the AEC contracts, and I think many of them were sincere in it. I just said this was a part of the picture. Dr. Oppenheimer being more eloquent and speaking more forcefully before the academy, seemed to be carrying the lead in the attack. This is not in itself important, because he was joined by many others, especially an eminent astronomer from Harvard.

Dr. EVANS. Harold Shapeley?

The WITNESS. Yes.

By Mr. SILVERMAN:

Q. Was also a member of the United States Senate one who joined him?

A. Yes.

Q. Senator McMahon?

A. I suppose so. I didn't approve of it. But I thought he was a little vociferous.

Q. You say Dr. Oppenheimer opposed a security clause. You mean he opposed a security clause for the building of works?

A. No; these were fellowships. I think he probably had the right position. But I said this simply fits into a pattern. He was in good company in making his objections. I think he was probably right in this particular case.

Mr. GRAY. What was the date of this, roughly?

The WITNESS. I can't give the date. In itself I would not attach any great significance to it. I simply said this fits into a pattern.

By Mr. SILVERMAN:

Q. This is something you mentioned on direct examination as a reason why I think you had some trouble about it.

A. I think this is part of the picture. I can expand the picture a little more if you would want me to.

Q. I would like to concentrate for a moment on this particular item. So the opposition to the security clause was an opposition to a security clause with respect to AEC fellowships?

A. The fact that they had to take a loyalty oath. There was a division in the academy. I just mention this as indicating the side that he was always on. In itself I would not attach any intention except as part of a general picture.

Q. I think in view of the fact that you mentioned it and referred to it as a security clause in an AEC contract, it is desirable that the record be clear now as to what it is he was opposed to.

A. He was opposed to an oath which all holders of AEC contracts must take. I believe that was a more direct statement.

Q. All holders of AEC contracts?

A. No; all holders of AEC fellowships. Let me get my phraseology correct.

Q. I think it is important. And these were fellowships in basic science?

A. They were.

Q. Were they fellowships in the building of weapons?

A. No; they were just part of the pattern which had been set up by Congress. The item is not highly significant in itself.

Q. You did consider that Dr. Oppenheimer's position was right on that, wasn't it?

A. I felt that the act of Congress was unfortunate but in view of the act, I didn't feel that one should offer this strenuous objection that he offered.

Q. So though you thought that he was right in his position, your objection was that he stood up too strongly for his position?

A. I would say this, that I didn't approve of it, either, but since the act of Congress set this up, I thought the strenuousness—it was the intensity of his objections, rather than whether it was right or wrong.

Q. Now I don't understand. I thought at one point you said that Dr. Oppenheimer was right in his opposition.

A. I think the loyalty clause in the contract was wrong.

Q. Do you think that Dr. Oppenheimer was right in his opposition to that clause?

A. I didn't oppose it on the floor of the academy. I think I voted against the resolution.

Q. Did you think that Dr. Oppenheimer was right or wrong?

A. I thought he was within his rights in offering the objections.

Q. I thought you said in answer to an earlier question that he was probably right in opposing it. That is not what you meant?

A. I thought I tried to make myself plain.

Q. Excuse me. It was not entirely clear to me, and I would like you to make it clear.

A. I felt that the thing basically was not good, but I was somewhat struck by the intensity of his opposition.

Q. What was basically not good? The thing that Dr. Oppenheimer opposed?

A. Correct.

Q. Surely you don't draw any unfavorable inferences from the fact a man intensely opposes that which he believes to be wrong?

Mr. ROBB. Mr. Chairman, I think the witness has explained 4 or 5 times what his view on that was.

Mr. SILVERMAN. Perhaps that is right.

Mr. GRAY. Proceed.

By Mr. SILVERMAN:
Q. Let me ask just one more thing. Was it a loyalty clause that Dr. Oppenheimer opposed, or was it an FBI investigation in this connection, and was it for classified fellowships or for unclassified?
A. For unclassified.
Q. For unclassified?
A. As I recall.
Q. Did he make a distinction between classified and unclassified?
A. I do not recall that.
Q. I think you said that Dr. Oppenheimer wanted to disband Los Alamos?
A. As I recall it it was essentially that. He wanted to move it to Chicago, I believe. At least it would have appeared to have been a serious interruption of the program.
Q. How do you know that he wanted to disband Los Alamos?
A. That impression was built quite a number of years ago, and I am not sure that I remember all the details that went into my knowledge, but it was correct, wasn't it?
Q. One of the advantages of being a lawyer is that I don't have to answer questions.
A. I may have been misinformed, but I believe I wasn't.
Q. Was one of the details that went into your knowledge of Dr. Oppenheimer's decision a conversation with Dr. Oppenheimer on this point?
A. No.
Q. Don't you think that might have been the most reliable source of information on that point?
A. I think my judgment was reliable.
Q. I think you referred to the fact that many scientists were annoyed at the notion that Dr. Oppenheimer alone could have built the atom bomb. I take it you were among those scientists, or weren't you?
A. I certainly appreciate his very great contributions. They were tremendous. But I certainly think it would be erroneous to assume that it could not be done by anybody else.
Q. My question, sir, was not that. My question is whether you were among the scientists who have been annoyed at that notion?
A. I am annoyed at that statement which has been appearing in the newspapers. Every time I pick up a newspaper and read that, I am definitely annoyed. A great many other scientists I know are equally annoyed.
Q. Do you know whether Dr. Oppenheimer has ever taken that position?
A. I do not. He is a very modest man. I assume he would not take that position.
Q. Have you read Dr. Oppenheimer's answer to the Commission's letter in this preceeding?
A. I have read it.
Q. Do you know whether he said anything on that point?
A. I don't recall that he did.
Q. I think you said that the key ideas with respect to the hydrogen bomb were not supplied by Dr. Oppenheimer. That is what you said, wasn't it.
A. I believe I did put it that way. Maybe it could be better phrased than that.
Q. Perhaps you would phrase it better then because I think it would be desirable to have your notion as clearly as possible on this record.
A. This gets on the verge of classified information, of coure, but I think one can say without going into classified information that the idea which made it work easily was not supplied by him.
Q. The idea that what?
A. That made it much easier to build was not supplied by him.
Q. If it makes it easier there has been testimony in this record that Dr. Teller and Dr. Ulam made very great contributions.
A. Yes.
Q. Did you understand that Dr. Oppenheimer claimed that he had supplied the key ideas?
A. No, I did not. I had not heard that he had.
Q. I was just sort of wondering why you found it necessary or desirable to refute a statement which apparently had not been made.
Mr. ROBB. Mr. Chairman, I don't think that is hardly a fair question inasmuch as I asked him to make his comments with regard to Dr. Oppenheimer, and it was in response to that question that he made that remark.
Mr. SILVERMAN. I see.

By Mr. SILVERMAN:
Q. I want to return for a moment to the GAC as constituted in 1949 to your fear of not being able to influence them. I think you gave as one of your reasons that chemists might not have much influence with nuclear physicists. Dr. Cyril Smyth was on the GAC?
A. That is right, he was. I had forgotten he was on there. If I had gotten hold of him, I would have certainly talked to him.
Q. This was a terribly important thing, wasn't it, the problem of influencing the country's national policy on the building of the hydrogen bomb?
A. We got the right answer, too.
Q. Didn't you think it was worth your while to call Dr. Smyth?
A. I worked through other methods.
Q. You did not work through the GAC?
A. If I had failed I would have said it was certainly unfortunate, but as long as we didn't lose the battle, I guess it was not so important.
Q. What I am merely asking, sir, is do you think it is fair to say that the GAC was influenced completely in its opposition to the hydrogen bomb by Dr. Oppenheimer's domination without having talked to some of the members of the GAC who participated in the discussion?
A. I think it is fair.
Q. I think you suggested that we made very little or no progress in atomic armament from 1947 to 1950. That comes as something of a surprise to me. I think there has been some testimony in the record that would seem to be the other way. But perhaps I am wrong. How do you know that nothing had happened of value?
A. You keep asking me to go back and analyze my judgments. The reactor program did not move forward, the development of our natural uranium supplies did not move forward rapidly, the expansion of Hanford was slow, the expansion of production of U-235 did not move much, this sort of thing.
Q. Which reactor program did not move forward? The reactor program for weapons?
A. No, the general reactor program which of course related to the program as a whole. Weapons are not entirely independent of the reactor program.
Q. Wasn't it true there were expansions, large and important expansions in the reactor program between 1947 and 1950 with respect to weapons?
A. It was delayed at least a year by busting up the Los Alamos group and arguing where it was going and a lot of scientists got discouraged and quit.
Q. Wondering what?
A. I forget the details of whether it was going to be moved out to Chicago or Idaho. You broke up a competent group at Los Alamos and delayed the whole program for a while.
Q. Don't you recall that there was a delay in over a year after the war before an act was passed by Congress?
A. Yes, but still there was plenty of delay after that.
Q. Wasn't it during that period until an act was passed by Congress that the great deterioration occurred at Los Alamos?
A. I don't remember the exact date. A lot of deterioration occurred during that period. But certainly the reactor program didn't move forward.
Q. Do you recall that Dr. Oppenheimer testified in favor of early legislation in order to prevent the deterioration of Los Alamos?
A. No, I do not.
Q. Do you know whether there was a sizable growth in the stockpile of fissionable material and of atomic weapons in the period of 1947 to 1950?
A. Under existing facilities there should have been a sizable growth.
Q. Do you know whether there was or wasn't?
A. Those figures are confidential and I don't have access to them, but knowing in general about what the production capacities were, one could conclude that the normal production went on, but there was no reasonable expansion of the program.
Q. And on what do you base your conclusion that there was no reasonable expansion of the program?
A. None of my friends disappeared to work on projects anywhere. If there were any such projects set up, they were kept awfully secret to me.
Q. Can you tell us to what extent work on the atom bomb done after the war was helpful or perhaps essential as a precondition to the physics in the development of the hydrogen bomb?

A. I think Dr. Teller could answer that question much better than I. It is his particular field. My impressions would be based very largely on what Dr. Teller has told me, and it would be second hand. I place considerable reliance on it.

Q. You did say that you thought there had been no progress in atomic weapons from 1947 to 1950.

A. I said very little progress. You had a program and you kept it going, but there was no——

Q. Would you tell us what Dr. Teller told you as to whether work on atomic-bomb development was helpful as a precondition to the physics of the hydrogen bomb?

A. I think he would say he got some encouragement, but he had a small group, 2 or 3 or 4 men working with him, something of the sort.

Q. I am afraid you are not answering my question.

A. I thought I was.

Q. What I was asking was whether what Dr. Teller told you about the extent to which postwar work on the atom bomb, not necessarily by him, was helpful as a precondition to the physics of the hydrogen bomb.

A. I can't give you more than the general feeling that he didn't get much encouragement during that period.

Mr. SILVERMAN. I have no further question.

Mr. GRAY. Dr. Evans.

Dr. EVANS. Dr. Latimer, I might say I relied on Latimer and Hildebrand for a great many years.

The WITNESS. It is very kind of you to say so.

Dr. EVANS. When the fission bomb was fired, is it correct in saying you were worried about the other end of the curve that Harkins wrote about many years ago?

The WITNESS. Yes. It, of course, became obvious to everyone that energetically such things were possible and being a student in thermodynamics, when something is possible, it is probable that somebody can make it work.

Dr. EVANS. Have you ever been approached for secret information?

The WITNESS. No.

Dr. EVANS. Have you known any Communists?

The WITNESS. Yes, I have known Communists. They planted a Communist secretary on me at one time during the war until the FBI discovered her. The Army sent her to me. That is the only intimate connection that I recall.

Dr. EVANS. Did you know Fuchs?

The WITNESS. No, I did not.

Dr. EVANS. Dr. Latimer, anyone that knows him and his work would not call Dr. Conant a nuclear physicist by any stretch of the imagination?

The WITNESS. No, he is an organic chemist.

Dr. EVANS. Thank you.

Mr. GRAY. I have a question just in the interest of finding out what happened to the debate. Was the security clause with respect to fellowships retained or rejected?

The WITNESS. Let's see. I forget the outcome of that. I think the academy refused to administer them, but I am not sure now as to the outcome of it.

Mr. ROBB. I have one question.

REDIRECT EXAMINATION

By Mr. ROBB:

Q. Doctor, was there a young man named Kennedy whom you knew who had been at Los Alamos?

A. Yes.

Q. Is he now in your department?

A. No, he is not.

Q. Did he return to you after he worked at Los Alamos?

A. No, he went to the University of Washington at St. Louis.

Q. Did you have any conversation with him after he worked at Los Alamos?

A. Yes, I had conversation with him.

Q. Will you state whether or not you observed Dr. Kennedy had any of these feelings that you mentioned with respect to working on weapons?

A. I believe to the best of my memory that he was one of the group that said he would no longer work on weapons.

Q. Did that strike you as unusual?

A. Not in itself. I would say I was a little surprised, a Texan taking that point of view.

Q. He is a Texan?
A. I believe so.
Mr. ROBB. That is all. Thank you.

RE-CROSS-EXAMINATION
By Mr. SILVERMAN:
Q. Did Dr. Kennedy say to you that he had talked to Dr. Oppenheimer about the question of working on weapons?
A. I cannot recall that he did.
Mr. SILVERMAN. That is all.
Mr. GRAY. Thank you very much, Dr. Latimer.
(Witness excused.)
Mr. GRAY. I would like to ask Mr. Garrison if he wants to offer those affidavits at this time?
Mr. GARRISON. Yes, I think it is a good time.
Mr. ROBB. Are you going to read them, Mr. Garrison?
Mr. GARRISON. I would like to. They are rather short. I would like the board to hear them.
I have a very short statement, Mr. Chairman, by Walter G. Whitman, dated April 23, 1954, entitled, "Corrections to Testimony of Walter G. Whitman given April 22, 1954." He sent this to me on his own initiative. I am sorry I don't have copies of this.
Mr. ROBB. May I see it before you read it into the record?
Mr. GARRISON. Yes. I also have one from Dr. Killian, Mr. Robb, of which I regret to say I don't have copies.
Mr. ROBB. I don't think it is a matter of much substance, but on Dr. Killian, he has not testified before.
Mr. GARRISON. No. Mr. Whitman says:
"Dr. Evans asked me a question as to whether I had personally known any Communists or persons who were subsequently shown to be Communists. My answer should be amended to include the following information.
"I have known Professor W. T. Martin, who was a member of a faculty committee at MIT which I chairmanned in 1949–51. Professor Martin testified in 1953 before a congressional committee that he was a member of the Communist Party about 1938 and that he left it in about 1946. My association with him did not involve any consideration of political philosophy, or any matters of security.
"I have known Professor I. I. Amdur very casually since about 1934. It is my understanding that Professor Amdur testified at the same congressional hearing that he had been a member of the Communist Party over somewhat the same period of time as Professor Martin had.
"I regret that I overlooked these two cases when I was testifying.
"(Signed) WALTER G. WHITMAN."
There is no objection, Mr. Chairman, to adding that to the record as a correction supplement?
Mr. GRAY. The chairman sees no objection.
Mr. ROBB. I have none whatever.
Mr. GARRISON. I have here an original affidavit signed by James R. Killian, Jr., which I would like to read. I am sorry I don't have copies, Mr. Chairman.
"Sworn to before me this 20th day of April 1954.
"RUTH L. DAWSON, *Notary Public.*"
It begins:
"COMMONWEALTH OF MASSACHUSETTS,
County of Middlesex, ss:
"James R. Killian, Jr., being duly sworn, deposes and says:
"I am president of the Massachusetts Institute of Technology in Cambridge, Mass. I am a member of the Science Advisory Committee of the Office of Defense Mobilization, a committee in which both J. Robert Oppenheimer and I have been members since it was appointed by President Truman in 1951. I have attended about 10 meetings of this committee at which Dr. Oppenheimer was present, including formal gatherings associated with these meetings. Once when the committee met in Princeton, the members of the committee dined at Dr Oppenheimer's home.

"In the course of these meetings I have observed no action or suggestion on the part of Dr. Oppenheimer that seemed to me to be against the interest of the United States, or to give any support to the charges against him in General Nichols' letter. On the contrary, he impressed me in these meetings as a man deeply devoted to strengthening the security of the Nation and fertile of ideas for promoting the national welfare. Every aspect of his work on this committee sustained my confidence in his loyalty and integrity.

"To my knowledge this committee never discussed the desirability of making hydrogen bombs. Certainly I never heard any statement by Dr. Oppenheimer that reflected opposition on his part to the decision that had been made by the administration to go ahead on this development.

"I recall being with Dr. Oppenheimer on 1 or 2 occasions other than the meetings described above, and these meetings were casual or social. He came to MIT in 1947 to deliver the Arthur D. Little Memorial Lecture. He gave this lecture before a large audience which seemed absorbed by his ideas and moved by his sincerity.

"Dr. Oppenheimer was a participant in the summer study project of the MIT Lincoln Laboratory in the summer of 1952. He was able to give the project only a very slight amount of time, as I recall, and I was not present at any of the meetings in which he participated. He did give a briefing to the group undertaking the study on the meaning of atomic warfare. * * *

"An earlier project, known as Project Lexington, carried out by the Atomic Energy Commission under contract with MIT sought information from Dr. Oppenheimer which has been described elsewhere by the director of this project, Prof. Walter Whitman. No information I have about Dr. Oppenheimer's relationship to either of these projects has given cause to question his integrity and loyalty.

"(Signed) JAMES R. KILLIAN, Jr."

Mr. GRAY. That affidavit becomes a part of the record.

Mr. GARRISON. I have copies of this supplemental affidavit of Dr. Manley. This was to clear up a question that arose in his testimony. I think the chairman put the question.

Mr. GRAY. Yes. I might say with respect to that, or at least one portion of that, which involved a round use of the words "instrumental in persuading" rather than "attempts to persuade." Later on in reading General Nichols' letter that was General Nichols' own language in the letter and if I had realized at the time I probably would not have raised the question. There is no reason why this amplification should not be made.

Mr. GARRISON. I think this relates to an additional question, Mr. Chairman. It is an additional one, because it also covers "instrumental." This is a supplemental statement signed by Dr. John H. Manley, "Sworn to before me this 16th day of April 1954. Mary E. Mossman, notary public.

"I have been requested to clarify portions of my statement of February 16, 1954. This request reached me on April 15, 1954 by personal visit of Mr. Walters and Mr. Chipman of the Seattle FBI office with a teletype inquiry originating with the AEC and by a letter informing of Mr. Lloyd K. Garrison's offer to Mr. Gordon Gray to ask me for clarification. All questions refer to statements on page 10 of the reference document. I was informed that the AEC inquiry was for clarification of the following excerpts:

"1. 'Indeed, I had no feeling that anyone was holding back on the work on thermonuclear weapons once the President had decided the question by his announcement in January 1950.'

"2. 'I never observed anything to suggest that Dr. Oppenheimer opposed the thermonuclear weapons project after it was determined as a matter of national policy to proceed with development of thermonuclear weapons, or that he failed to cooperate fully in the project to the extent that someone who is not actively working could cooperate'.

"3. 'Neither have I heard from any scientists that Dr. Oppenheimer was instrumental in persuading that scientist not to work on the thermonuclear weapons project'.

"Mr. Garrison's inquiry related to the first excerpt.

"I do not now have a copy of the charges against Dr. Oppenheimer, but I recall that one was the accusation of opposition to H-bomb development after the Presidential decision of January 1950. My statements (1) and (2) above were directed to this charge and therefore contain specific reference to the President's decision. It is completely incorrect to asume that the converse statement was true before January 1950.

"With respect to excerpt (1) I call attention to the two preceding sentences of my statement which have no time qualification and which, I hope, are unambiguous. To say that no one held back at any time would be ambiguous because, as I tried to show in preceding pages, the question was one of relative effort and anyone fully occupied with A-bomb problems was in effect being held back from H bomb work, not because of Dr. Oppenheimer but because of laboratory program and AEC direction before aJnuary 1950. I know of no case of an individual connected with the weapons program who could be accused of 'holding back' from improper, malicious or disloyal motives. This includes Dr. Oppenheimer.

"With respect to excerpt (2), it is a matter of AEC record that Dr. Oppenheimer and others opposed a top-priority program to develop thermonuclear weapons before January 1950. The reasons are also a matter of record. I add that the approved programs of the Los Alamos Laboratory for a considerable period prior to this date included such work, that these programs were normally reviewed by the General Advisory Committee, Dr. Oppenheimer, Chairman, and that I can recall no instance of his opposition, formal or informal, direct or indirect, to the thermonuclear investigations proposed in these programs and carried forward by the laboratory. On the contrary, I know of specific assistance on his part in certain examinations of theoretical questions.

'With respect to excerpt (3) I can state that I never heard from any scientist that Dr. Oppenheimer ever attempted to persuade or was instrumental in persuading that scientist not to work on the thermonuclear weapons project. Neither did I ever hear Dr. Oppenheimer make such an attempt nor did I at any time see any evidence that would lead me to believe that any scientist was so approached or influenced either by Dr. Oppenheimer or by anyone else. My position was such that I believe any such attempt would have come to my attention.

"(Signed) JOHN H. MANLEY."

Mr. GRAY. Thank you, Mr. Garrison.
Mr. GARRISON. That is all we have, sir.
Mr. GRAY. We will now recess until 9:30 tomorrow morning.
(Whereupon at 3:47 p. m., a recess was taken until Wednesday, April 28, 1954 at 9:30 a. m.)

UNITED STATES ATOMIC ENERGY COMMISSION

PERSONNEL SECURITY BOARD

IN THE MATTER OF J. ROBERT OPPENHEIMER

ATOMIC ENERGY COMMISSION,
BUILDING T-3, ROOM 2022,
Washington, D. C., Wednesday, April 28, 1954.

The above-entitled matter came on for hearing, pursuant to recess, before the board, at 9:45 a. m.

Personnel Security Board: Mr. Gordon Gray, chairman; Dr. Ward T. Evans, member; and Mr. Thomas A. Morgan, member.

Present: Roger Robb, and C. A. Rolander, Jr., counsel for the board; J. Robert Oppenheimer, Lloyd K. Garrison, Samuel J. Silverman, and Allan B. Ecker, counsel for J. Robert Oppenheimer; Herbert S. Marks, cocounsel for J. Robert Oppenheimer.

PROCEEDINGS

Mr. GRAY. I should like to read into the record a communication to me from the general manager with respect to the minutes of August 6, 1947, meeting of the Commission:

"UNITED STATES ATOMIC ENERGY COMMISSION,
"*Washington 25, D. C.*

"Memorandum for Mr. Gordon Gray, Chairman, Personnel Security Board.

"On February 19, 1954, Mr. Mitchell wrote Mr. Garrison referring to a meeting of the Commission on August 6, 1947, at which the question of the continuance of the clearance of Dr. Oppenheimer was considered. I understand Mr. Garrison, as counsel for Dr. Oppenheimer, has now requested the precise text of these minutes.

"The minutes show that at the meeting held on August 6, 1947, Commissioners David E. Lilienthal, Sumner T. Pike, Lewis L. Strauss, and W. W. Weymack were present. Following is the full text of that part of the minutes which reflect the action taken regarding Dr. Oppenheimer:

" 'Mr. Bellsley called the Commission's attention to the fact that the Commission's decision to authorize the clearance of J. R. Oppenheimer, chairman of the General Advisory Committee, made in February 1947, had not previously been recorded. The Commission directed the secretary to record the Commission's approval of security clearance in this case and to note that further reports concerning Dr. Oppenheimer since that date had contained no information which would warrant reconsideration of the Commission's decision.'

"(Signed)   K. D. NICHOLS,
"*General Manager.*"

Mr. GARRISON. Mr. Chairman, could I look at that again, or could you read the last?

Mr. GRAY. There is no reason why you should not look at it. Do you propose to discuss this?

Mr. GARRISON. I would like to make a brief comment about it.

Mr. GRAY. If it is in the nature of argument on the part of counsel, I don't think this is the appropriate time. This was read into the record pursuant to your request to be read in the record. It was not done earlier because the Commission had to make the decision with respect to the request. At a time when the board is considering testimony with respect to the matters involved in this memorandum, or at a time when you as counsel are addressing the board, it would be perfectly appropriate to discuss it, but I don't want the record now to involve a discussion of this particular meeting and the circumstances surrounding it.

Mr. GARRISON. Mr. Chairman, there is a statement in this memorandum about myself to which I would like to respond at this point of time, and I can scarcely respond to it without a little discussion. I believe this is the time to do it, and it won't take long.

Mr. GRAY. I take it you are not now wishing to discuss the matter involved in the memorandum, but some reference to you in the memorandum?

Mr. GARRISON. Yes, but they are intertwined. Perhaps I could tell you what I have in my mind, and you can stop me if you wish me to go no further.

Mr. GRAY. All right.

Mr. GARRISON. I will begin by saying this, that I am sure Mr. Mitchell will remember a conference which Mr. Marks and I had with Mr. Mitchell and General Nichols, I think around the 12th of February, or the 13th, at which we brought with us a list of documentary material—items of documentary material—which we believed would be relevant to the proceeding here for the Commission to make available to us and be helpful to the board.

Mr. Marks had a typewritten list which was left with counsel as I recall. I had a handwritten short list of which I have the original with me, but in it was explicitly contained a request for the minutes of the Atomic Energy Commission meetings relating to the clearance of Dr. Oppenheimer in 1947, and a request for all pertinent documents having to do with that whole matter.

Mr. Mitchell and General Nichols said that they would take these matters under advisement and would notify me how much of the documentary material that we asked for could be made available.

The next thing that I heard of that was the letter from General Nichols of February 19, which contained the stipulation—letter from Mr. Mitchell, addressed to me, and saying, "This will confirm our telephone conversation of today. The Commission will be prepared to stipulate as follows for purposes of the hearing:

"On August 6, 1947, the Commission recorded clearance of Dr. J. Robert Oppenheimer, which it noted had been authorized in February 1947.

"Furthermore, Dr. Oppenheimer will be given an opportunity to read the minutes of the GAC meeting of October 1949."

We had asked that they be made available in some summarized form.

"by coming to the Commission's offices for his convenience. Arrangements for this purpose may be made with Mr. Nichols.

"Sincerely yours."

I am sure also Mr. Mitchell will recall the telephone conversation referred to in that letter in which he explained to me that our requests for documentary materials had all been declined—every one of them—and the only information of a documentary character which we could have relating to documentary material relating to the actions of the Commission in 1947 was this one stipulation.

I expressed a natural disappointment, but we didn't have any argument about it. But that is the way it was left.

The situation then is that back in the middle of February, I did ask for these minutes. They were denied. I was given a stipulation which I think the record here will show was misleading because even yourself, Mr. Chairman, in these proceedings a little while back quite doubted whether there had actually been any clearance.

Mr. GRAY. I would say for the record that I still feel that there is very considerable mystery about it. I don't want to get into an argument about it now, but I don't want to leave the impression that what we have now read into the record clears up my mind on it.

Let me say on this matter that the situation now is, and the record of this proceeding will show, that in the course of the conduct of these proceedings, and in the context of matters before this board, you requested the inclusion of the full minutes in the record. Counsel for the board and the board then asked the Commission to consider whether they would depart from what I understand to be policy in the Commission with respect to minutes, and would furnish the actual transcript of the minutes of that meeting, and that has now been done. I don't want to interrupt whatever you are saying about it.

Mr. GARRISON. I appreciate the cooperation of the board in making this available. What disturbs me is that the very significant words "further reports concerning——"

Mr. GRAY. Now you are getting into discussion of a material nature which I don't want to appear in the record at this time. You will not be denied an opportunity to go into that.

Mr. GARRISON. May I make one other comment, then, Mr. Chairman. Seven different documents relating to——

Mr. GRAY. Is this related to the minutes we have read into the record?

Mr. GARRISON. It is related to a request I would like to make to the board.

Mr. GRAY. I would like to say I don't think that type of thing is before the board at this time. We really are responding to a request that the transcript reflect the minutes of this particular meeting which has been the subject of considerable discussion in these proceedings and about which there possibly will be further discussion. If you have any observations to make for the record or otherwise about other documents, about your relationship with the Commission, or anything that is pertinent to this hearing, you will be given an opportunity to do so, but I don't want to go into it at this time.

Mr. GARRISON. When may I go into that, Mr. Chairman?

Mr. GRAY. I wouldn't want to establish a precise time. I should think, Mr. Garrison, that it would be satisfactory for the Chairman to assure you that you will have the opportunity. We are in the middle of testimony from witnesses now, and I don't want to get into a long discussion of a matter that is not related to their testimony.

Mr. GARRISON. When may I make a request of you for further information relating to this clearance?

Mr. GRAY. I don't know that I have ever denied you the opportunity to make a request at any time in these proceedings. I am unable to answer that. I want to give you my assurance that you will be given an opportunity to discuss anything pertinent to this proceeding, and beyond that, I am unable or unwilling to do it at this moment.

Mr. GARRISON. I will make this request without argument, Mr. Chairman.

Mr. GRAY. Make your request. If you are going to make a request——

Mr. GARRISON. For further information.

Mr. GRAY. I would say this. I would suggest that any request for materials which are not in the record and which are in the hands of any Government agency to which you do not have access should be made to the agency itself, and I should be glad to discuss this with you, Mr. Garrison, but I do not want to get into an argument at this time in this proceeding about matters which are not pertinent to the testimony that is being given to this board.

Mr. GARRISON. Mr. Chairman, I will follow your instructions. I would like to make the request for the cooperation of this board and the assistance of this board in obtaining information. I understand that counsel for this board did, on behalf of the board, ask the Commission for the minutes of the August meeting. I think it appropriate indeed that this board should make a similar request in connection with the further information which I have in mind. I will state that request at any time you wish.

Mr. GRAY. I will now rule that we will not discuss this matter at this time, Mr. Garrison. You will forgive me for becoming impatient. I have made it abundantly clear that we are in the middle of testimony from witnesses, and I am not going to have this reflect at this time discussions about your relationships with Government agencies. I repeat my assurance that you will be given an opportunity to say anything that is pertinent to this proceeding, and I think the record will show abundantly that the board has given every possible cooperation.

I would like now to proceed with the witnesses who will be before us this morning.

Mr. ROBB. Mr. Chairman, may I make one brief comment? We will, of course, attempt to keep the first coming to the mill. I am advised, however, of one matter which I think I should tell the board about.

The Commission has been advised by Mr. Reynolds, who is the business manager for the radiation laboratory at the University of California, that because of illness and ill health, Dr. Ernest O. Lawrence, who is the director of the radiation laboratory, and who had been expected to appear here, who I believe has gotten as far as Oak Ridge, we have been advised, will not be able to appear, and he has had to return to the west coast. I mention that now to explain why we may perhaps have a gap. I hope we won't. I hope we will be able to keep the board running at full time. I expect we will.

Mr. GRAY. Thank you.

General Wilson, do you wish to testify under oath? You are not required to do so.

General WILSON. I would prefer to do so.

Mr. GRAY. Would you give me your full name?

General WILSON. Roscoe Charles Wilson, major general, United States Air Force.

Mr. GRAY. Would you raise your right hand? Roscoe Charles Wilson, do you swear that the testimony you are to give the board shall be the truth, the whole truth, and nothing but the truth, so help you God?

General WILSON. I do, sir.

Whereupon, Roscoe Charles Wilson was called as a witness, and having been first duly sworn, was examined and testified as follows:

Mr. GRAY. Would you be seated, General.

Allow me to remind you of the existence of the so-called perjury statutes. May we assume that you are familiar with them?

The WITNESS. Yes.

Mr. GRAY. I should like also to request, General Wilson, that if in the course of your testimony it becomes necessary for you to disclose or refer to restricted data, you notify me in advance so that we may take the necessary and appropriate steps in the interest of security.

The WITNESS. Yes, sir.

Mr. GRAY. Finally, I should say to you that we consider these proceedings a confidential matter between the Atomic Energy Commission and its officials and witnesses on the one hand, and Dr. Oppenheimer and his representatives on the other. The Commission is making no news releases. I express the hope on behalf of the board that witnesses will take the same view of the situation.

The WITNESS. May I make a statement, please, sir?

Mr. Chairman, I would like the record to show that I am appearing here by military orders, and not on my own volition.

DIRECT EXAMINATION

By Mr. ROBB:

Q. General, would you tell the board what your present assignment is, sir?

A. I am in the process of change of station. I have just been relieved as commandant of the Air War College, and am in transit to my new command, which is commander of the Third Air Force in England.

Q. Would you tell us what the Air War College is, sir?

A. The Air War College is an adult school to which the military sends selected colonels or Navy captains, members of the State Department and CIA, and certain foreign officers who have completed about 15 years of service. These people are schooled in international relations, in military matters, particularly air matters, and in grand strategy. The purpose is to prepare them for positions of high responsibility in the military.

Q. How long did you serve as the commandant or president of that college, sir?

A. About 2½ years, sir.

Q. Where is that located?

A. Montgomery, Ala.

Q. Would you tell the board, General, something of your previous military background and history?

A. I was appointed to the Military Academy as a result of competitive examination by President Hoover. I graduated in 1928, and from the flying schools in 1929. I was sent to postgraduate engineering school from which I graduated—a 1-year course—in 1933.

Q. Where was that, sir?

A. That was at Dayton, Ohio. I was an instructor in natural and experimental philosophy at the Military Academy at West Point in 1938 and 1939, and was assistant professor of natural and experimental philosophy there in 1940. I was chief of experimental aircraft design at Dayton and when the war started, was brought into Washington as chief, bombardment engineering, and later became chief of development engineering for the staff in Washington, that is the air staff.

Q. How long have you held your present rank of major general, sir?

A. I was made a major general in 1951.

Q. General, during the war, what, if any, connection did you have with the atomic-bomb program?

A. Sir, in 1943, I believe it was, I was directed by General Arnold to make certain that the support of the Army Air Forces was given Gen. Leslie Groves. I served General Groves as a liaison officer while still maintaining my position as chief of development engineering in the Air Force. My duty was to assist him in procuring materials, scarce items, especially electronic equipment, to make certain that if a bomb were developed that there would be an aircraft to carry it, and later on to make certain that an organization was assembled, trained, and equipped to deliver the weapon.

My association with General Groves was not directly under his command, but in his support.

Q. What did you do in that connection? Where did you go and what did you do after you got that assignment?

A. In Washington I principally with Captain Parsons of the Navy and with Dr. Norman Ramsey and with General Groves, my duty largely was assembling material and getting equipment together, and arranging later on for aircraft to be modified.

In the spring of 1944 I was sent by General Groves to Los Alamos, and there I talked again with Ramsey and Parsons and with Dr. Oppenheimer, and with others who were concerned with the external configuration of the weapons. The idea was to make certain that the aircraft had an equipment in which the bomb would fit, and also to make such minor modifications to the exterior of the weapon as might be necessary to make it fit.

Later on that year, General Groves sent me again to Los Alamos, this time to see if an airdrome could be built on a plateau, and also to recommend to him if I could an area in which some tests might be made. My impression was that he had several people doing both of these, but I did it also as an independent mission.

Q. Did you make such recommendations?

A. Yes, sir; I did.

Q. What site did you recommend?
A. I picked the bombing range at Alamagordo, N. Mex.
Q. In that connection did you have occasion to confer with Dr. Oppenheimer?
A. I am quite certain I met Dr. Oppenheimer at that time.
Q. Following that, what duty did you perform?
A. Sir, I monitored the Air Force portion of the program until December. By monitoring I mean I selected the commander of the organization, I made sure that he had personnel, I followed the modification of the aircraft, the supply of the aircraft, and helped where I could to supply the then Manhattan District with the equipments and the military assistance that they desired.

In December, I was relieved and sent to a bombardment wing, and in the summer of 1945 was sent overseas. I remained at Okinawa until both bombs were dropped on Japan, and then I was hurriedly brought home and sent out to Japan again where I joined the party to look at the wreckage.

Q. Then there came a time when you returned to the United States?
A. Yes, sir. I came back in August or September of 1945, and was assigned as the deputy to General LeMay, who was then Deputy Chief of Staff for Research and Development.
Q. What did you do in that connection?
A. He had been brought in to revitalize research and development in the Air Force, and I assisted him in programing where we could.
Q. How long did you carry on that work?
A. I was there, sir, as I recall until 1947, at which time I was assigned as the deputy to General Groves, who was then Chief of the Armed Forces Special Weapons Project.
Q. What were your duties in that connection?
A. They were to reflect in the activities of this joint agency Air Force thinking to the extent it was possible for me to do so. The Armed Forces Special Weapons Project was and is unusual in that it is a service which is common to all of the armed services, and the chief of it is the subordinate of each of the chiefs of service, but not the subordinate of the Joint Chiefs.
Q. How long did you stay on that duty, General?
A. I stayed there until 1950.
Q. Did General Groves stay that long?
A. No, sir. He retired. My notes and my mind are a little hazy on this, but he was succeeded by General Nichols in this period, and I served as General Nichols' deputy.
Q. Is that General K. D. Nichols, who is presently general manager of the AEC?
A. Yes, sir.
Q. Did there come a time when you served on the Research and Development Board and the Military Liaison Committee?
A. Yes, sir.
Q. When was that, General?
A. In the latter part of 1948 if my memory is firm. Certainly during 1949 and a part of 1950 I served as a member of the Committee on Atomic Energy of the Research and Development Board. Throughout all this period up until the middle of 1951 I was a member of the Military Liaison Committee.
Q. What was your duty in those two connections? What did you do in general? I don't mean a daily diary.
A. Yes, I understand. The Military Liaison Committee to the Atomic Energy Commission is an agency which is charged with making certain that the military interests of the Nation are properly reflected in the activities of the Commission. It served also as a group—I am oversimplifying this, sir—which kept the Defense Department advised of the potentials of the developments of the Atomic Energy Commission.

The Committee on Atomic Energy of the Research and Development Board was a coordinating group designed to establish programs to prevent overlap and unnecessary duplication in research and development. In the Committee on Atomic Energy, our duties were confined to the field of atomic energy.

Q. In connection with your work on those two groups, will you tell us whether or not you came in contact with Dr. Oppenheimer?
A. Yes, sir.
Q. What were your contacts with him?
A. Dr. Oppenheimer was a member of the Committee on Atomic Energy. I think I saw him at almost all of the meetings during 1949. He also served as the chairman of a long-range objective panel on which I had the honor to serve in 1948, and chairman again of a similar panel or the same panel reconvened

in 1950. Of course, he was a member of the General Advisory Committee of the Atomic Energy Commission, and occasionally we saw him in that capacity also.

Q. Were your contacts with him rather frequent?

A. I would not say frequent, but rather regular. Perhaps I saw Dr. Oppenheimer once every month or so. He was very kind to me, and when our panel met out in California he invited me to his home; this sort of an association.

Q. General, are you familiar with the history of the position of the military and, in particular, the Air Force with respect to the thermonuclear weapon?

A. Sir, I would like to refer to my notes, if I may.

Q. Certainly. Have you recently refreshed your recollection about that matter?

A. I did indeed. I struggled with this very problem yesterday.

Q. General, I think it would be helpful to the board if you could give us in your own way something of the history of the position of the military and the Air Force on this matter. You may of course refer to notes to refresh your recollection.

A. I find it a little difficult to pinpoint some of these things. For instance, I am aware of a meeting at Los Alamos which had been requested by the scientists to discuss matters of military interest. I remember at that meeting General LeMay was asked what size bomb do you want. There had been a great deal of discussion about smaller bombs.

* * * * * * *

I have a lot of this sort of information in my mind, and I am embarrassed that I can't put dates to it. But I do have a few dates.

I have a statement that I found in a document marked top secret, sir, but the statement itself is not top secret. This is a little confusing to me, but it does indicate—I think it is safe to say it—that in 1948 both the Research and Development Board, and the Joint Chiefs of Staff had expressed an interest in continuing research on the thermonuclear weapon. This is the first written statement I can find in my own records—in 1948.

On September 23, 1949, we had the announcement of the Russian A-bomb, and that I really think sparked off the military interest in this larger weapon.

In the early part of October, Drs. Bradbury and Lawrence visited the Armed Forces Special Weapons project, where they talked to General Nichols and at the same time Dr. Edward Teller visited the Air Force, where they talked to a group at which I was present on the possibilities of a thermonuclear weapon. They urged that the military express its interest in the development of this weapon.

Mr. SILVERMAN. Pardon the interruption. Would you mind giving the names of the people who were present again?

The WITNESS. Drs. Bradbury and Lawrence visited the Armed Forces Special Weapons project. This was early in October 1949. Perhaps I better clarify something. I am not sure whether Teller's visit to the Air Force was at the same time or shortly thereafter. This is a little hazy in my mind. But in the same general period of October 1949.

On October 13 of 1949—and I am sure as a result of the urging of Dr. Bradbury and Dr. Lawrence—General Nichols, who was of course the subordinate of General Vandenberg, went to General Vandenberg with General Everest of the Air Force, and urged General Vandenberg as the No. 1 bomber man to express again the military's interest in a large weapon.

General Vandenberg directed Nichols and Everest to express his point of view to the Joint Chiefs of Staff that afternoon, since Vandenberg was not going to be present at that meeting. This they did.

On October 14, 1949, the Joint Chiefs met with the joint congressional committee on Atomic Energy, where General Vandenberg, speaking for the Joint Chiefs, strongly urged the development of this thermonuclear weapon. I have a copy of the excerpts of the notes of the meeting covering General Vandenberg's statement if the committee wishes it to be read.

Mr. ROBB. I might say, Mr. Chairman, that has been released by formal action of the joint committee, confirmed to General Nichols by letter which we received this morning.

Mr. GRAY. You may read it.

The WITNESS. "Page 1792. One of the things which the military is preeminently concerned with as the result of the eary acquisition of the bomb by Russia is its great desire that the Commission reemphasize and even accentuate the development work on the so-called super bomb. General Vandenberg discussed this subject briefly and stated that it was the military point of view that the super bomb should be pushed to completion as soon as possible, and that the General Staff had so recommended. In fact, his words were, 'We have built a fire under the proper parties,' which immediately brought forth the comment, who are the right parties? General Vandenberg replied that it was being handled through the Military Liaison Committee. He further stated that having the super weapon would place the United States in the superior position that it had enjoyed up to the end of September 1949 by having exclusive possession of the weapon. There followed a series of questions, somewhat of a technical nature about the super weapon, which General Nichols answered for the Chiefs of Staff. He stated that it was the opinion of the scientists that the possibility of a successful super weapon is about the same as was the possibility of developing the first atomic weapon at the 1941–42 stage of development. He stated that the military fears that now the Russians have a regular atomic weapon, they may be pushing for the super weapon, and conceivably might succeed prior to success in this country of the same project. * * *

This was on October 14, sir. On October 17, the JCC wrote a letter to the Committee on Atomic Energy and this letter is on file in the Military Liaison Committee, in which they requested further information on the big weapon and expressed some concern that the committee had not asked for funds to prosecute the project.

Mr. GRAY. Which committee?

The WITNESS. I beg your pardon. The Atomic Energy Commission. This was a letter to the Commission and a copy of this letter came to the Military Liaison Committee.

At that same meeting, the chairman of the Military Liaison Committee informed that committee of his visit, together with General McCormack and Dr. Manley to Dr. Oppenheimer at Princeton where they had discussed the super and other problems to be taken up by the General Advisory Committee.

At that same meeting the Military Liaison Committee approved a directive to reconvene the long-range objectives panel. This was the second panel on which I had the honor of serving with Dr. Oppenheimer.

On October 27, there was a joint meeting of the Atomic Energy Commission and the Military Liaison Committee, at which the Commission announced that it had asked the General Advisory Committee to consider the super weapon in the light of recent developments. Then of course on the 28th and 29th of October was the meeting of the GAC.

On November 8, 1949, the MLC at its meeting heard a report from the Secretary that in accordance with the directive to reconvene the long-range objectives panel, he had been determining the availability of membership of the panel, and that he had discussed the panel with Dr. Oppenheimer on the 29th of October, and that Dr. Oppenheimer agreed that the panel should meet but "felt strongly that this should not be done until a great deal more information was available, probably not before February of 1950."

November 9, 1949, is the letter from the AEC to the President.

November 19 was the letter from the President to Admiral Sowers of the National Security Council, and during this period a military committee or subcommittee was set up to advise Admiral Sowers in determining the position on the thermonuclear development. This was a committee composed of General Nichols, Admiral Hill, and General Nordstad of the Air Force.

On the 13th of January 1950 there is a letter to the Secretary of Defense from General Bradley in which the military views are set out. I do not have that document. I have a hazy recollection of what might have been in it, sir. I do know that it expressed concern lest the Russians come up with this bomb before the United States did, and the feeling that this situation would be intolerable, since it would reverse the advantage we had had in this country prior to the Russian A-bomb explosion.

The rest of my notes are to the effect that in February the Air Force announced that it had undertaken the development of an aircraft to carry a weapon of this sort, and a program which it was coordinating with the AEC.

On February 18—and I would like to say that my memory of this date is not certain—I have noted February 18, 1950, to the best of my knowledge, the long-range objectives panel was completed and submitted to the Committee on Atomic Energy.

By Mr. ROBB:

Q. Can you tell us about that report, General?

A. This panel was composed of a group of military people, of which I was one, and the chairman was Dr. Oppenheimer. Another member was Dr. Bacher, and another Dr. Luis Alvarez. The panel contained some conservative statements on the possibility or the feasibility of an early production of a thermonuclear weapon. These reservations were made on technical grounds. They were simply not challengeable by the military. They did, however, cause some concern in the military.

It is hard for me to explain this, except to say that most of us have an almost extravagant admiration for Dr. Oppenheimer and Dr. Bacher as physicists, and we simply would not challenge any technical judgment that they might make. But I must confess, and I find this exceedingly embarrassing, sir, that as a result of this panel and other actions that had taken place in the Committee on Atomic Energy, that I felt compelled to go to the Director of Intelligence to express my concern over what I felt was a pattern of action that was simply not helpful to national defense.

Q. Action by whom?
A. By Dr. Oppenheimer.
Q. Would you explain what that pattern was?
A. I would like first to say that I am not talking about loyalty. I want this clearly understood. If I may, I would like to say that this is a matter of my judgment versus Dr. Oppenheimer's judgment. This is a little embarrassing to me, too. But Dr. Oppenheimer was dealing in technical fields and I was dealing in other fields, and I am talking about an overall result of these actions.

First, I would like to say, sir, that I am a dedicated airman. I believe in a concept which I am going to have to tell you or my testimony doesn't make sense.

The U. S. S. R. in the airman's view is a land power. It is practically independent of the rest of the world. I feel that it could exist for a long time without sea communications. Therefore, it is really not vulnerable to attack by sea. Furthermore, it has a tremendous store of manpower. If you can imagine such a force, it could probably put 300 to 500 divisions in the field, certainly far more than this country could put into the field. It is bordered by satellite countries upon whom would be expended the first fury of any land assault that would be launched against Russia, and it has its historical distance and climate. So my feeling is that it is relatively invulnerable to land attack.

Russia is the base of international communism. My feeling is that the masters in the Kremlin cannot risk the loss of their base. This base is vulnerable only to attack by air power. I don't propose for a moment to say that only air power should be employed in case of a war with Russia, but I say what strategy is established should be centered around air power.

I further believe that whereas air power might be effective with ordinary weapons, that the chances of success against Russia with atomic weapons or nuclear weapons are far, far greater.

It is against this thinking that I have to judge Dr. Oppenheimer's judgments. Once again, his judgments were based upon technical matters. It is the pattern I am talking about.

I have jotted down from my own memory some of these things that worried me.

First was my awareness of the fact that Dr. Oppenheimer was interested in what I call the internationalizing of atomic energy, this at a time when the United States had a monopoly, and in which many people, including myself, believed that the A-bomb in the hands of the United States with an Air Force capable of using it was probably the greatest deterrent to further Russian aggression. This was a concern.

*   *   *   *   *   *   *

To do this the Air Force felt that it required quite an elaborate system of devices. Some were relatively simple to produce, some of them were exceedingly difficult to produce, and some of them were very costly. Dr. Oppenheimer was not enthusiastic about 2 out of 3 of these devices or systems. I do not challenge his technical judgment in these matters, but the overall effect was to deny to the Air Force the mechanism which we felt was essential to determine when this bomb went off. In our judgment, this was one of the critical dates, or would be at that time, for developing our national-defense policy.

Dr. Oppenheimer also opposed the nuclear-powered aircraft. His opposition was based on technical judgment. I don't challenge his technical judgment, but at the same time he felt less strongly opposed to the nuclear-powered ships. The Air Force feeling was that at least the same energy should be devoted to both projects.

*   *   *   *   *   *   *

The approach to the thermonuclear weapons also caused some concern. Dr. Oppenheimer, as far as I know, had technical objections, or, let me say, approached this conservatism for technical reasons, more conservatism than the Air Force would have liked.

The sum total of this, to my mind, was adding up that we were not exploiting the full military potential in this field. Once again it was a matter of judgment. I would like to say that the fact that I admire Dr. Oppenheimer so much, the fact that he is such a brilliant man, the fact that he has such a command of the English language, has such national prestige, and such power of persuasion, only made me nervous, because I felt if this was so it would not be to the interest of the United States, in my judgment. It was for that reason that I went to the Director of Intelligence to say that I felt unhappy.

Mr. ROBB. That is all I care to ask. Thank you, General.

CROSS-EXAMINATION

By Mr. SILVERMAN:

Q. General, you said you are not raising a question of loyalty?
A. No, sir.
Q. You do not question Dr. Oppenheimer's loyalty?
A. I have no knowledge in this area at all, sir.
Q. Do you——
Mr. ROBB. Wait a minute. Let him finish his answer.
The WITNESS. I have no knowledge one way or another.

By Mr. SILVERMAN:

Q. Have you any information to indicate that Dr. Oppenheimer has been less than discreet in the handling of classified information?
A. No, sir; I haven't. Maybe I talk probably too much.
Q. Please.
A. I read an article on the way up to Washington in the U. S. News & World Report, and this was a considerable surprise to me——
Q. Excuse me. If you are going to tell us something that you know about, we are all interested to hear it.
A. I beg your pardon?
Mr. ROBB. Wait a minute.
Mr. SILVERMAN. Let me finish, Mr. Chairman. If Mr. Robb or the chairman thinks what I am saying is wrong——

By Mr. SILVERMAN:

Q. I would think if all you would do is tell us about an article in U. S. News & World Report, we would do better reading the article.
Mr. ROBB. I think what the general refers to is the letter of General Nichols and Dr. Oppenheimer's letter, which has been frequently referred to.
The WITNESS. Yes; this is what I am speaking of. This was news to me. I assume you are speaking of the period in which I served with Dr. Oppenheimer, and my answer to your question is "No; I do not."
Mr. GRAY. I did not understand that.
The WITNESS. I was not aware of any indiscretion on the part of Dr. Oppenheimer in the handling of classified material in the period in which I served with him.

By Mr. SILVERMAN:

Q. I have some notes on some of the things you said, and I think I would like to run through them and ask for elaboration where questions arose in my mind, sir.
A. Yes, sir.
Q. I think you said you are appearing on military orders and not on your own volition?
A. Yes, sir.
Q. I take it you didn't ask for these orders?
A. I certainly did not.
Q. What was the first intimation that you had, sir?
A. I was telephoned about 3 or 4 days ago by General McCormack, of personnel of the Air Force, saying that by verbal orders of the Chief of Staff of the Air Force I was to report to this committee.
Q. And you then reported to——
A. I then reported to Mr. Robb.
Mr. ROBB. That is the first time any major general ever reported to me.

By Mr. SILVERMAN:

Q. You stated, I believe, you went through your notes and gave various dates of expressions of military interest in the hydrogen bomb.

A. Yes.

Q. I think you said that the Russian explosion of September 23, 1949, really sparked off the military interest in the hydrogen bomb or some such phrase?

A. Some such phrase. The interest was there, but this certainly in my opinion, at least from where I saw it, the little piece in the Air Force, this certainly at least gave impetus to the interest.

Q. By that I take it you mean that the Air Force was much more actively and intensely interested after September 23, 1949, than before?

A. Yes, that is a fair statement.

Q. I think you said that the long range objectives panel was completed. I take it unless my notes are wrong the report of the panel was completed and submitted to the Committee on Atomic Energy on February 18, 1950, I thought you said.

A. That is the best of my recollection, February 18, 1950.

Q. Could it be 1951?

A. I am very sorry. This is the one date on which I am really worried. I regret that I had to do some rather hard research and I must say it could have been 1951. Wait. February 1951?

Q. Is there some way you could find out rather readily? There is no desire here by anybody to trap you on anything. I just want to get the facts.

A. I apologize, sir, this was a bad date. I could find out if I could make a phone call, sir.

Mr. ROBB. I may be able to help you on that. Is that the one Dr. Kelly was on?

The WITNESS. Yes, he was on that panel. This was a panel of the Committee on Atomic Energy.

By Mr. SILVERMAN:

Q. I have here some references to a report of the panel in the testimony, part of which was classified, and therefore I don't have it, but I think in the unclassified portion a date was given of December 29, 1950. I think that is the date you gave, Mr. Robb. Let me see if I can find the place? On page 196 of the record, Mr. Robb, you referred to a report which of course I have not seen, dated December 29, 1950, and I do not know whether that is the report the General is talking about. I just don't know.

Mr. ROBB. The report entitled, "Military Objectives on the Use of Atomic Energy, to the Atomic Energy Committee of the Research and Development Board of the Department of Defense."

The WITNESS. This sounds right. Is there a cover sheet with the list of members?

Mr. ROBB. I don't have that here. That was December 29, 1950. Then January 6, 1951, if I might give this, Mr. Silverman, to assist you, the General Advisory Committee considered that report and commented that it stated the military objectives with clarity and keen insight into the reality of the present situation. Mr. Whitman and Dr. Oppenheimer participated in the report, abstaining from taking action on the matter.

Mr. GRAY. I should like to ask counsel if he wants to establish this date, perhaps we could recess and let General Wilson make his telephone call.

Mr. SILVERMAN. I think that would be the sensible thing to do. I think that is the easiest way to do it.

Mr. GRAY. We will recess for a few moments.

(Brief recess.)

Mr. GRAY. Would you proceed.

By Mr. SILVERMAN:

Q. Have you now ascertained that date, General?

A. I have. The correct date is January 18, 1951, which is the date of the approval by the Committee on Atomic Energy.

Q. General, I would like you to cast your mind back now to that period as well as you can. Do you know whether that was the time at which the feasibility of the thermonuclear weapon technically appeared to be at almost its low period?

Mr. ROBB. What period are we talking about now?

Mr. SILVERMAN. January 1951.

The WITNESS. Of course, you realize I am guessing. It was pretty low in my opinion. It was similar to most projects of this sort. There is a certain optimism, then there is a period of pessimism, and then the optmism grows again. My feeling is that it became lower a little later, and it became lower because of some doubt as to the amount of a very scarce and costly material.

By Mr. SILVERMAN:
Q. Was it lower then, do you recall, as to the prospects of feasibility than it had appeared, say, a year earlier which was the time of the President's directive?
A. Sir, you are asking me to pass judgment on a technical matter.
Q. If you don't know, say so.
A. I don't know.
Q. And you don't recall discussions at that time?
A. Yes, sir; I can remember discussions among the scientists.
Q. What did the discussions among the scientists indicate to you?
A. You see, my oracle in this matter was Dr. Oppenheimer and they indicated that this was a difficult job. I speak of oracle as Chairman of this Board. He was the expert.
Q. Do you recall who the other members of that panel were?
A. I recall some of them. I didn't write down their names. Dr. Oppenheimer, Dr. Bacher, Dr. Alvarez, Dr. Kelly, I was a member, Gen. James McCormack was a member, General Nichols was a member, but I don't believe he attended the meetings as a member. He was in the process of transfer about this time.
Q. I have here a list which might be helpful to you, sir. Dr. Lauritsen was a member of the Committee?
A. Yes, sir.
Q. I think you mentioned Dr. Whitman, did you not?
A. Dr. Whitman was; yes.
Mr. SILVERMAN. That is the list, Mr. Chairman. It is item 5 on Dr. Oppenheimer's biography in section 2, "Membership on Government Committees." It is item 5 (b).

By Mr. SILVERMAN:
Q. Did you mention Admiral Parsons?
A. And Admiral Parsons. I beg your pardon.
Q. Instead of our doing this the hard way depending on my recollection of what you said, let me read the list as I have it, and see if that accords with your recollection.
A. All right.
Q. Dr. Oppenheimer was Chairman; Dr. Bacher, Dr. Alvarez, Dr. Lauritsen, Dr. Kelly, Dr. Whitman, General Nichols, Admiral Parsons, yourself, General McCormack, with David Beckler as secretary.
A. That is correct. I also recollect that Nichols did not act as a Committee member. I do think he appeared on that, but he was not a member.
Q. Those people in addition to Dr. Oppenheimer's scientific knowledge, Dr. Bacher is an eminent physicist, is he not, and a great man who had great knowledge in this field?
A. That is correct.
Q. And he joined in the report, did he not?
A. That is correct.
Q. Did he question the statement about the feasibility of the hydrogen bomb as it then appeared?
A. I am searching my memory pretty hard, but my recollection is that Dr. Bacher supported Dr. Oppenheimer in this view.
Q. They all signed the report?
A. This is something else I don't recall. I don't recall signing a report. I recall that the report was prepared and it contained a statement that there was no substantial difference in opinion or no important disagreement or something of that sort. It was then submitted to the Committee on Atomic Energy which voted to accept it or otherwise. I don't recall signing it.
Mr. SILVERMAN. Perhaps Mr. Robb, you could clarify that point, because I take it you have the report?
Mr. ROBB. I don't have it.
Mr. SILVERMAN. I am sorry. I thought when you questioned Dr. Kelly on the basis of having signed the report——
Mr. ROBB. No.
The WITNESS. It would be normal to sign the report, but I don't recall that this is an important point.

By Mr. SILVERMAN:
Q. Dr. Lauritsen was an eminent physicist, was he not?
A. Yes, sir.
Q. And a man very well informed on matters of nuclear weapons?
A. Each of these civilians really was in a similar class.
Q. Did he join in the technical judgment as to the feasibility of the hydrogen bomb?
A. I don't recall the discussion.
Q. Did he disagree?
A. The statement in the report was to the effect that there was no substantial disagreement in the report as finally drafted.
Q. Dr. Alvarez was an eminent physicist; was he not?
A. Yes.
Q. And a man who is very familiar with matters of nuclear weapons?
A. Yes.
Q. He was a man who rather favored the development of the hydrogen bomb. He took a different view from the members of the General Advisory Committee; did he not?
A. I am sorry, sir, I don't recall.
Q. In any event, he was very familiar with matters of nuclear weapons.
A. As far as I know; yes, sir.
Q. And you do not recall that he expressed a dissent on this point?
A. No one dissented. As I recall there was discussion in the meeting but when the report finally was drafted, it was submitted with the statement that there was no important difference of opinion in the report as submitted.
Q. You have no doubt that was a correct statement?
A. I think that was a correct statement. But I would like to make this reservation, sir.
Q. Yes.
A. Certainly I, as a military man, did not engage in the technical part of this discussion. I don't think the military people were in a position to debate the technical judgment.
Q. We, of course, all realize that while your knowledge of these matters is doubtless greater than you perhaps like to admit for reasons of modesty, your knowledge is certainly not that of these scientists by a long shot.
A. That is correct.
Q. We don't for a moment question that fact. What about Admiral Parsons, was he quite familiar with these matters?
A. I would say Admiral Parsons was probably as close to a scientist as we had in our group.
Q. And he had been at Los Alamos, too, had he not?
A. Yes, sir.
Q. Short of being one of these four nuclear physicists that I have mentioned, he really was very familiar with the problems of nuclear weapons and the scientific aspects of them?
A. I think among military men he was certainly as well informed as anyone.
Q. He, too, of course, joined in the report. They all joined?
A. There was no important disagreement.
Q. General Nichols—you said that he did not really actively participate.
A. That is my recollection, sir.
Q. Yes. It occurs to me that this matter of the date of that panel has perhaps another important bearing which I would like to suggest to you, and see whether I am right. January 1951—indeed I think December was the date of the report itself, December 29, or something like that.
A. Yes, sir.
Q. We were already in the Korean war; is that not correct, General?
A. Yes.
Q. And that started, I think, in June of 1950?
A. In June.
Q. When did the Chinese intervention come?
A. Oh, my goodness, sir, I regret I just don't remember.
Q. Wasn't it just about that time, or just a little before that?
A. I don't recall.
Q. Wasn't it in December, I think, of 1950?
A. I am sorry, I don't remember. I would have to refresh my memory.
Q. Let me suggest this to you. If this doesn't refresh your recollection, it doesn't. Had there not been alerts of possibility of enemy aircraft at about that time, shortly after the Chinese intervention? Do you recall that?

Mr. ROBB. Mr. Silverman, I am not quite clear what the question means. Maybe the general is. You mean alerts of enemy aircraft here or in Korea?
Mr. SILVERMAN. I think perhaps in the North American Continent.
The WITNESS. I was not aware of any such thing.

By Mr. SILVERMAN:
Q. Was not the panel concerned, the Chinese intervention—I am merely suggesting this to you.
A. I will accept this for lack of notes of my own.
Q. I regret to say I have not myself looked it up. That is my information. Was not the panel concerned at that point about the possibility of an eruption of a general war in the near future?
A. Yes; that is fair. This is almost a constant state of mind, sir.
Q. Well——
Mr. ROBB. Let him finish.
The WITNESS. We are always worried in the Pentagon about an accident which might start trouble. Surely this was a tense period.

By Mr. SILVERMAN:
Q. More so than an earlier year?
A. Yes.
Q. Was not the problem of the panel one of trying to make suggestions as to the use of atomic weapons in the event of an emergency which might arise in the very near future?
A. No, sir; that would be a military judgment, and this panel was a technical panel which was attempted to develop guidance for research and development projects.
Q. Were suggestions made at that panel as to the possible use of atomic weapons that might be feasible and usable in the quite near future, much nearer than it looked as if a hydrogen bomb could be developed?
A. I have no recollection, frankly, sir, but I would very much be surprised if this group of people at that time didn't discuss those things.
Q. Do you recall discussions of the use of the atom in some versatile way in an emergency which might occur very soon, at that panel discussion?
A. You mean as a radiological warfare type of operation? I am afraid I don't understand.
Q. I am afraid I don't know myself. I am thinking of recommendation for the use of smaller atomic weapons to be carried on a small airplane.
A. Yes.
Q. Was that discussed at that time?
A. Yes.
Q. Was that recommended?
A. Yes; this program was recommended. There has always been an interest in this field.
Q. And that was a matter which looked a good deal more feasible in the quite near future than the hydrogen bomb, did it not?
A. Yes; I think that is a fair statement.
Q. Do you recall you were rather enthusiastic about the prospect of that?
A. I am first of all a big-bomb man, but I do recognize the potential value of the so-called tactical weapons. Here was an opportunity to increase the stockpile of weapons. This, of course, was something of importance to all of us. This had more to do than simply developing weapons of smaller size. These were still very potent weapons.
Q. I gather that even the smaller atomic weapons are very potent weapons.
A. I am not expressing myself well. These are still large weapons to be carried by large aircraft. There was a technical development which promises to still increase the number of bombs. This was of great importance to us at that time.
Q. And that was discussed at that time?
A. Yes.
Q. Was that a development that looked as if it would be usable in the event of an emergency in the near future but more likely to be usable than, say, a hydrogen bomb?
A. Yes.
Q. I think you said that you are a big-bomb man, and at an earlier stage you referred to yourself as a dedicated airman. I assume that the two are not quite the same, but those are both parts of your views.

A. I mean that I believe in the theories of Douhet and Mitchell and Admiral Mahan as modified to fit the present war. This is a belief that the objective of war is not the defeat of the enemy's army, but the defeat of the enemy's will to wage war. That this comes about only after failure to win the real victory, which is the prevention of war.

Q. The views you have expressed I take it are your views as a dedicated airman and a believer in big bombs.

A. That is correct.

Q. I don't mean for a moment that you get any pleasure out of the dropping of big bombs. You understand that, of course.

A. That is correct.

Q. Are the views you expressed pretty much unanimous views among the informed people of the Air Force?

A. There are a great number of people who belong to this school of thought. They might not subscribe to my views precisely as I have expressed then to you. I don't want to be coy or overcautious here, but I would not speak for the whole Air Force. But there are members of this group.

Q. Are there people in the Air Force who don't agree with you?

A. Yes, there are.

Q. People of good faith?

A. That is correct.

Q. You refer to yourself as a dedicated airman. I take it that a dedicated naval officer might have somewhat different views?

A. I hope that we are all dedicated Americans. When I say this, I mean our dedication is to the preservation of the United States. I don't want to sound sentimental to you, but this is the idea. I have oversimplified my statement by saying I am a dedicated airman. I believe that proper defense lies along the line that the Air Force proposes, or that I suggest. I know that the other services have other views.

Q. And you are not surprised that the civilians have perhaps still other views.

A. No, sir.

Q. Do you recall that just about the time of the GAC meeting, just a couple of weeks before it, I think, there was some testimony before the Committee on Armed Services of the House of Representatives. I think the newspapers may have called it the Battle of the Admirals, or something. It was the discussion of the B–37. Do you remember testimony of Admiral Ofstie?

A. Yes. Let me say I recall that he did testify. I don't recall just what it was. I know I didn't like it at the time.

Q. Is this part of what you didn't like? I am reading from page 183 of the hearings before the Committee on Armed Services, House of Representatives, 81st Congress. The dates of the hearings run October 6 to 21, 1949. I don't have here unfortunately the number of the document as such. It is page 183— it is somewhere in October. I can't tell without running through it which precise day it was. Page 183, Admiral Ofstie was testifying:

"There is a widely held belief that the Navy is attempting to encroach on strategic air warfare, and that this was the principal consideration in planning the so-called supercarrier. This is a misconception which is quite at variance with the facts. We consider the strategic air warfare as practiced in the past and as proposed in the future is militarily unsound and of limited effect, morally wrong and decidedly harmful to the stability of a possible world war."

I take it that is part of the statement with which you disagree?

A. I don't agree with any part of it from start to finish.

Q. You don't question Admiral Ofstie's good faith in making the statement?

A. I most seriously question his good judgment in making such a statement.

Mr. GRAY. I would like to ask if your purpose is getting somebody else's views in this record, or whether you are questioning the general about something that he can testify about.

Mr. SILVERMAN. The general did not testify about this, sir. At least not that I know of. What I am attempting to do, sir, is to see if the general will agree with me that it is possible in good faith for people whose patriotism is unquestioned to hold these views.

Mr. GRAY. He has stated unequivocally for the record that he does agree with you. I want to make myself clear in my question to you. We have allowed, so far as I know, almost unlimited latitude in what has been brought before the board, hearsay, documents which at times seemed to the Chairman to be really unrelated to the inquiry, but if you feel that is important to further establish the fact that the general agrees with you, I would let you argue for your point,

but I believe he has stated clearly that it is possible for people of good faith to be in disagreement on these matters. There is no question in your mind about it?
The WITNESS. There is no question in my mind, no, sir.
Mr. GRAY. If that is your point, I think it has been well made.

By Mr. SILVERMAN:
Q. I think you questioned Dr. Oppenheimer's judgment on a number of matters. You said that Dr. Oppenheimer was interested in the internationalizing of atomic energy at a time the United States had a monopoly of it, and that was the greatest deterrent to Russian aggression. I take it you concede—excuse me, I am not meaning to be sarcastic at all—I am sure you do concede that Dr. Oppenheimer did play a great role in the development of the atomic bomb which did become this great deterrent to Russian aggression?
A. Yes, sir.
Q. Did you hear at the time of this discussion of internationalizing of atomic energy that it was the view of many scientists that Russia would have the atomic bomb in time anyhow?
A. Yes; I think I understood this to be the case.
Q. And therefore perhaps it might be better to internationalize it while there was a chance to do so?
A. I had never heard that argument.
Q. You did not?
A. No.
Q. You did know that many people of good faith did urge that point?
A. I am not aware. I don't believe I have ever heard that argument.
Q. I did not make myself clear: You stated you had not heard that argument, and I did not therefore make my next question clear. You did know that many people of good faith, many informed people, were in favor of what came to be known as the Acheson-Lilienthal and later the Baruch plan?
A. I don't think you are speaking of quite the same thing. The Baruch plan had certain safeguards in it which change it from what I believed to be Dr. Oppenheimer's earlier program. It was less general, let us say.
Q. Would it surprise you to learn that there are those who think that it was more general?
A. That is possible.
Q. Do you know that Dr. Oppenheimer supported the Baruch plan?
A. Yes.
Q. And, of course, the Russians opposed it?
A. Yes.
Q. Had you heard that it was Dr. Oppenheimer's view that inspection is not enough, that you could not be sure that the Russians would not evade inspection, and therefore it was necessary to have an international agency that would itself be the only one that could?
A. I didn't know this as a fact, I am sorry.
Q. I think you said on technical grounds, Dr. Oppenheimer did not support the full long-range detection program of the Air Force?
A. That is my recollection; yes, sir.
Q. That he was not enthusiastic about 2 out of 3 of these devices.
A. Yes.
Q. I think you also said you do not challenge Dr. Oppenheimer's judgment?
A. That is correct.
Q. As to the 2 out of 3 devices that Dr. Oppenheimer did not support, do you recall that he was always in favor of continued research on them?
A. Oh, definitely. My recollection is that in most of these matters Dr. Oppenheimer always favored research. I have never heard him at any time say that the field was closed and we needed no more study or research.
Q. Did you understand Dr. Oppenhemier's lack of enthusiasm for these two devices was based on the then state of technical development of those devices?
A. Yes; I believe that I understood that this was why he was not enthusiastic.
Q. Are these two devices that Dr. Oppenheimer was not enthusiastic about now in effect?
A. Yes, sir.
Q. Were they bettered by research?
A. Of course.
Q. You said 2 of 3 devices. I would like to turn to the third device, the one that he was enthusiastic for.
A. Well, yes; all right.

Q. I don't want to put words in your mouth.
A. It is hard for me to talk about these things. We are not naming names. They were three. They were of relative degrees of development or lack of development. The one that appeared to be most immediately promising, the one that perhaps we had the most information on was the one that Dr. Oppenheimer supported to the greater degree.
Q. Do you recall the circumstances of the development of that method?
A. Only vaguely. That it was during the war. I was conscious that it was being done, because I had been asked for aircraft to assist in some of the experiments. This is the limit of my knowledge.
Q. Do you know that it was under Dr. Oppenheimer's direction at Los Alamos that that first system for long-range detection of atomic explosions was initiated?
A. I don't know that as a fact, but I am not surprised, sir.
Q. And that it was done substantially at the same time as we were developing the atomic bomb?
A. I knew the activities were about the same time. Of this I was aware.
Q. Was it done at Los Alamos?
A. This I don't know, sir, because of the compartmentalization of that project. I don't know who was doing it.
Q. Do you know whether Dr. Oppenheimer directed the first trial of that method?
A. No, sir, I don't.
Q. I think you said Dr. Oppenheimer opposed nuclear powered aircraft and was less strongly opposed to nuclear powered ships.
A. On technical grounds. My statement was that he was opposed to these in this order. He had a time scale. As I recall it was the orderly development of these in series appealed to him. I am trying to say why one was ahead of the other. So far as I know it was only on technical grounds that he objected or opposed these.
Q. And you do not question his technical judgment?
A. No, sir.
Q. Was he alone in this technical judgment?
A. No, sir.
Q. Were there other well informed scientists who joined with him.
A. Yes; I am sure there must have been, because there was a great deal of controversy in this area.
Q. Was his opposition in committees or did he make public statements?
A. These were in committees. I don't recall any public statements on the matter.
Q. And these committees did have other scientists on them?
A. Yes.
Q. With respect to Dr. Oppenheimer's opposition to nuclear-powered flight and the apparent support of nuclear-powered ships——
A. Perhaps opposition is not the word. I wish we could find a better word?
Q. Lack of enthusiasm?
A. Lack of confidence in the timely success, or something of that sort. I don't think I have ever heard Dr. Oppenheimer doubt that this would be accomplished, but it was always 15 to 20 years, so far away that there were many other things that we could do more profitably now.
Q. Was there not a statement made, perhaps by the Chief of the Air Force, in any event by a very important official of the Air Force—I don't happen to know the name—within the last year or so in which he said that nuclear-powered flight looked like something we might have in about 20 years?
A. I don't know what, sir, I am sorry.
Q. Do you recall the Lexington study on nuclear-powered flight?
A. Yes; I do.
Q. What was their conclusion?
A. This was a study to make a statement, if possible, on the feasibility of achieving nuclear-powered flight. The report was rendered by Dr. Whitman, I believe, who was the chairman, and immediately there was a controversy as to what the report meant. The Air Force maintained that the Whitman report stated that nuclear-powered flight was feasible provided certain things were done. The opponents to the project said that these things that had to be done were of such a nature as to render the program infeasible. This is my recollection of it.

I personally think that the Whitman report or Lexington report stated that the project was feasible.

Q. Did the report say anything about the time scale in which one could hope to have nuclear-powered flight?
A. I am sure it did, but I don't recall what it was. It was not tomorrow. I don't want to give the impression that I feel that if we had poured all the money we had available into this project we could have had a nuclear airplane in a matter of months. We knew it was going to take time. But our argument was that the sooner we got to it, the better off we would be.
Q. Do you remember what Dr. Oppenheimer's participation was in the Lexington study?
A. I am sorry, I do not.
Q. Do you remember whether he did more than give a few briefings to the committee?
A. I really do not know.
Q. It has been the consistent position of the Air Force that nuclear-powered flight should be pushed?
A. Yes.
Q. In fact, however, have the Air Force come up with different programs for nuclear-powered flight from time to time?
A. We have to my knowledge come up with two. The first one failed to gain the scientific support essential. It was then reorganized on a different basis which promised greater support, especially from the Atomic Energy Commission. These are the two that I know of.
Q. What are the dates of those?
A. I am very sorry.
Q. Did the second program substantially revise the first one?
A. Yes; I think that it changed the time scale. I had left this business before really I could see it get under way so I am not too competent to discuss it, but it did revise the time scale, setting up a program somewhat longer than the Air Force would have liked.
Q. Do you know what the time scale was, that is, the revised time scale?
A. No, sir; I do not.
Q. Would it be fair to say that the revision of the program was to bring it more into accord with what appeared to be the technical realities of the situation?
A. I can only make an assumption here. I assume that it did.
Q. As to the difference between nuclear powered aircraft and nuclear powered ships, do you doubt that the possibility and the time scale of nuclear powered ships is very different indeed from that of nuclear powered flight?
A. This is an area of debate. You can find a lot of answers to this. As far as I am concerned, I recognize that the problem is more difficult in the airplane. There were at that time and still are a large number of aeronautical engineers who could have been put to work on this project. My own feeling is that it probably would have lagged behind the submarine but that if we had given it a real push, it might have not lagged too far behind.
At any rate, such an airplane would be of such importance to this Nation that my own feeling is that we should have prosecuted it vigorously from the start.
Q. Would you concede that it was possible for men of good faith, technically informed on the subject, to feel that it made sense to proceed with the nuclear powered ships at a somewhat faster pace than nuclear powered flight?
A. I have heard that discussion, yes, sir, and I will concede that.
Q. The fact is, is it not, that at least the scientists seem to feel that there are fewer technical difficulties with respect to nuclear powered ships than with nuclear powered flight?
A. Yes, sir.
Q. I don't suggest either of them is easy. I think the newspapers indicate that the submarine has been produced.
In any event, certainly Dr. Oppenheimer did press for continued research in both areas and particularly in the area of nuclear powered flight.
A. I can't answer that in the affirmative. I think that Dr. Oppenheimer pressed for continued research and experiment in reactors which in time might have contributed to nuclear powered flight. I won't say that Dr. Oppenheimer pressed for nuclear powered flight.
Q. I didn't mean that. I mean pressed for research.
A. Yes: that is correct.
Q. He did keep saying let us find out about as much of this as we can.
A. Yes

Mr. ROBB. Wait a minute.
The WITNESS. In reactors.

By Mr. SILVERMAN:
Q. Didn't he say let us find out what we can about nuclear powered flight, too?
A. My recollection is that he didn't. I am not even sure that he showed an interest in flight. This is my recollection.
Q. These reactors, of course, were essential for nuclear powered flight?
A. Yes.
Mr. SILVERMAN. I have no further questions. Thank you, General.
Mr. GRAY. General, I would like you not in any way to take offense at my first question of you.
You stated for the record that you were here under orders.
The WITNESS. Yes, sir.
Mr. GRAY. I think all of us understand what that means. But by that, do you mean that your presence here is a result of military orders, and am I correct in assuming that your testimony is your own, and not in any way involved with military instructions?
The WITNESS. My testimony is my own, sir. By this I meant, and I expressed myself very awkwardly, that I find this a very painful experience because of my admiration for Dr. Oppenheimer. I am exceedingly sorry that this is taking place, and I don't think I would have volunteered to come up here to make statements of this sort.
Mr. GRAY. I think that the board is aware of the painful nature of the matter.
General Wilson, approximately when did you feel impelled to go to the Chief of Air Force Intelligence?
The WITNESS. This was after this long-range objectives panel, the date of which I had confused. It was in January of 1951. I went to Intelligence and I remember going actually from one of these panel meetings, rather than to the Provost, because my feeling was not one of making charges, but I was uncomfortable. I was worried about something I could not put my hands on. I saw somebody to consult with.
Mr. GRAY. I am a little confused by that last answer and your reference to some officer other than the Chief of Air Force Intelligence.
The WITNESS. If I had thought that there had been an overt act or a deliberate move to obstruct the proper defense of the country, something of that sort, I would probably have appealed to the Provost Marshal. This would have been my duty to do so and make charges. But this is not a matter of charges. This was a matter of really worry that a general pattern of activity coming from a man of such stature seemed to me to be jeopardizing the national defense. Once again this is bluntly understated, but it was a worry, a concern. I wanted to discuss it with someone I thought was knowledgeable in this sort of an area.
Mr. GRAY. You felt that the security of the country might be somehow involved?
The WITNESS. Yes, sir.
Mr. GRAY. You stated, General Wilson, on the basis of your association— I believe you stated—with Dr. Oppenheimer, you did not doubt his loyalty to the United States?
The WITNESS. I have no knowledge of this at all, sir. I certainly have observed nothing nor have I heard him say anything that I personally would say was disloyal. In fact, sir, it seems to me that he has demonstrated his loyalty, once again in a private opinion, in the tremendous job he has done for this country. I have just no knowledge of this.
Mr. GRAY. I should like to ask you another question on this point. It may be a difficult one to answer. Is it possible, do you think, for an individual to be completely loyal to the United States, and yet engage in a course of conduct which would be detrimental to the security interests of the United States?
The WITNESS. Yes; I do.
Mr. GRAY. I would like to refer now to what you described as a pattern of conduct. You mentioned several things. The internationalization of atomic energy has not been accomplished. With respect now to the long-range detection system, have these other two that have been under discussion here been developed, and are they now in use?
The WITNESS. Yes, sir; they have been developed and are in use. It was a bitter wrangle to get them developed, but they are in use.

Mr. GRAY. With respect to nuclear powered aircraft—I don't know what the security problem is in this next question—may I ask you whether this is a promising field at the present time?

* * * * * * *

Mr. GRAY. I suppose I should state frankly the purpose of this series of questions. You have stated that you do not question Dr. Oppenheimer's technical judgment and competence.

The WITNESS. Yes.

Mr. GRAY. You made that very clear. I am trying to find out really whether in these several things that you referred to as constituting what might be a pattern of conduct, whether events have shown technical judgment in this case to have been faulty. Let me say for the record this board is not asked to pass upon the technical judgment of anybody, and is not competent to pass upon it. But it seems to me an answer to my question is pertinent to the part of the inquiry that we are engaged in. So I ask whether in these areas subsequent events have proved the validity or otherwise of these technical judgments which you accepted more or less without question, I believe you said, from Dr. Oppenheimer. We know that internationalization of atomic energy has not been accomplished. With respect to the others——

The WITNESS. Of course, the long-range detection program has been accomplished. I don't recall that Dr. Oppenheimer ever said that this couldn't be done. It was just perhaps that we ought to concentrate on the portions that could be done readily and quickly. I don't remember exactly the argument. It was essentially that—do what we can and perhaps that is the best we can do, this sort of thing, and for the rest let us experiment. The Air Force was frantic because it was charged with the job of detecting this first explosion and it felt all three methods had to be developed and put in place or it would fall down on its job.

Mr. GRAY. I think I won't press you on the answer to the question as I asked it, because it is not a good question.

The WITNESS. Yes, sir; I am sorry.

Mr. GRAY. General Wilson, with respect to what might be called the philosophy of strategy in a conflict with the Soviet Union, is it your view, as a dedicated airman today, knowing what you know about our capabilities in the field of nuclear weapons, that these weapons are important?

The WITNESS. Vastly, yes, sir.

Mr. GRAY. And as an airman, would you feel that even with improvements in the atomic weapons, which may have taken place in these years we have been discussing, these are still important weapons, that is, the thermonuclear?

The WITNESS. Yes, sir.

Mr. GRAY. You feel as an airman, knowing air capabilities, that they have direct useful application in the course of a conflict with the Soviet Union in particular?

The WITNESS. I think that they are vital, sir, to deterring a war, and I think that they are vital to winning a war should such a thing come. Further than this, it would seem intolerable to me that the Russians have such a weapon and the United States not. This is to get back to this area again. I would have reversed essentially our position when we had a monopoly on the atomic bomb—not entirely, but to a large degree. Involved as we are in a nonshooting war, this could have been a tremendous defeat for the United States.

Mr. GRAY. We have had testimony given to this board by scientists who were involved in some of these discussions to the effect that thermonuclear weapons are more useful to the enemy than they are to us. By that I believe they meant to say that we are more vulnerable, assuming that both powers have these weapons, than are the Russians. Do you share that view?

The WITNESS. Of course, it depends on the perimeters of our problem. Stated just as you have stated it, I would share that view. But think what would happen if we did not have the bomb and they did. The fact that we are troubled does not mean we should have this weapon in my view.

Mr. GRAY. I may get you into a classified difficulty so let me know if I do. Is a part of your conviction that these weapons are vital to our security based on considerations of numbers of aircraft that might be involved in any use of these weapons? Is that a clear question?

The WITNESS. Do you mean, sir, that by having these weapons fewer airplanes might be required?

Mr. GRAY. That is part of it, yes. Is that an important military consideration?

The WITNESS. It is to a degree. In order to be effective an enemy's defenses must be saturated. By this there must be a certain number of attacks made to confuse and confound his defense. This establishes really the minimum number of aircraft. This is sort of "get rich quick" air tactics. Added to that is the matter of flexibility to take care of local situations. This also could require a number of aircraft. What I am trying to say is that if you have a weapon that is 10 times as great as your old weapon, you cannot reduce your number of aircraft by 10 automatically. There are other considerations.

Mr. GRAY. I think I have only one more question. During the period with respect to which you have testified—perhaps I should be more specific—during the period 1947 to January 1950, did you have a serious question in your mind, based on what information you had, that the Air Force might have difficulty in developing a carrier which was capable of transporting and delivering the weapon which was under debate?

The WITNESS. This is the atomic bomb in that period and the thermonuclear bomb coming up?

Mr. GRAY. That is correct.

The WITNESS. Of course, there was no question about carrying the atomic bomb.

Mr. GRAY. Yes.

The WITNESS. There was no question among the combat bombardiers about their ability to deliver it. There was a great deal of impassioned debate on this subject, but I have never heard a bombardment commander say he could not deliver the weapon.

Mr. GRAY. This is the atomic weapon?

The WITNESS. This is the atomic weapon. We didn't know what the size and the weight and shape of this thermonuclear weapon would be, but as soon as the President directed that we determine the feasibility of it, the Air Force went immediately into a study of deliverability, and we were prepared with a series of devices to carry it. Some of them were not good, but they were a start. * * *

Mr. GRAY. In October 1949, based on what you know—how much or how little—about the technical difficulties in bringing about such a weapon which the Air Force might use, was there any doubt in your mind about your ability to design a plane, a carrier which would be effective?

The WITNESS. That a plane could be designed?

Mr. GRAY. Yes.

The WITNESS. No, sir; I don't think there was any such doubt. You can design as big an airplane as you want, I am sure.

Mr. GRAY. I am asking you this question because you are an airman.

The WITNESS. Yes, sir. My answer is, No, there was no doubt of the ability of the aircraft industry to design an airplane to carry almost anything. The important thing is that we get to work on it, and that we work together with the Atomic Energy Commission so that we can keep the size and shape together to come up with a good device in a timely manner.

Mr. GRAY. Dr. Evans.

Dr. EVANS. General Wilson, it has been mentioned a number of times in this meeting this morning that you were a dedicated airman. I wish to state for the record that this board does not think there is any approbrium, and I don't think anybody in this room thinks there is any approbrium connected with being a dedicated airman.

The WITNESS Thank you, sir. I invented the term.

Mr. SILVERMAN. If there was any suggestion that I meant any such thing, I certainly did not.

Mr. GRAY. I think Dr. Evans wishes everybody here to take judicial notice that there may have been people present who may have been interested in the Army at one time.

The WITNESS. I understand, sir.

Dr. EVANS. One of the possible reasons there may have been opposition to this thermonuclear weapon was possibly that Russia had fewer targets for that thing than we had. Was that ever mentioned? It would be like killing a mosquito with a sledge hammer.

The WITNESS. I have heard this sort of debate, but not seriously in official circles; no, sir.

Dr. EVANS. Do you have an idea now that the thermonuclear weapon was developed far more quickly than you would have had reason at one time to think it might be?

The WITNESS. Yes. I was agreeably surprised. Yes, sir.

Dr. EVANS. That is all I have.

REDIRECT EXAMINATION

By Mr. ROBB:

Q. General, there are a couple of questions suggested by the chairman's questions.

We have heard some discussion here by various witnesses about tactical bombing versus strategic bombing. I wonder if you could give us a little information about what the distinction is, what the two kinds of bombing are, so we have it from somebody who knows what he is talking about?

A. There is no real distinction. It is an over-simplification of terms. I think that what is meant by tactical bombing is bombing in immediate support of ground troops, somethng of this sort. Actually my view and the view of my school is that all bombing is directed toward a strategic goal, and that bombing done on the battlefield should be timed with bombing done against the enemy's will to resist, so that both will be mutually supporting. Short of a lecture, sir, I hope that will suffice.

Q. Is the thermonuclear weapon considered to be a tactical weapon or a strategic weapon, or both?

A. If you will accept my definition, which is not an accurate one, that a tactical weapon is in support of ground troops on the battlefield, then you would assume that a thermonuclear weapon would be a strategic weapon. We don't like to use these terms. We prefer not to, because they are all directed to a strategic end.

Q. Is the nuclear powered ship, using the term perhaps unprofessionally, a strategic or tactical weapon?

A. For the same reason you can't differentiate. It would be a highly flexible performing airplane.

Q. I am talking about a ship.

A. Oh, a ship. I beg your pardon. I don't think you can differentiate there either. It depends on how they are employed.

* * * * * * *

Mr. ROBB. That is all.

Mr. SILVERMAN. I think I have one question.

RE-CROSS-EXAMINATION

By Mr. SILVERMAN:

Q. I think the chairman asked you about whether you had any question in October 1949 about the possibility of determining an aircraft large enough to carry a thermonuclear weapon. I am not sure in my own thinking. We are talking about a big hydrogen bomb?

A. I understand, sir.

Q. I think you said you didn't have any doubt that it could be done?

A. It could be designed, yes.

Q. Will you give us some idea about how long it takes from design of a plane to production?

A. It varies of course. The cycle used to be about 3 years. When I left the business it had crept up to about 5 and I don't know how long it is, but it is a goodly period. That is from the drawing board to the production and rolling them off, and not a modification.

Q. If it were a much bigger plane than anything that had been had before it might be presumably longer?

A. It might be longer if it is from the original concept of production. If it is a modification, it is different.

Q. And one couldn't tell what you needed until you saw the size and shape of the thing you had to carry?

A. Yes, sir.

Mr. SILVERMAN. Thank you.

Mr. ROBB. Thank you, General.

Mr. GRAY. Thank you very much, General Wilson.

(Witness excused.)

Mr. GRAY. We will recess until 2 o'clock.

(Thereupon at 12:05 p. m., a recess was taken until 2 p. m., the same day.)

AFTERNOON SESSION

Mr. GRAY. Dr. Pitzer, do you wish to testify under oath? You are not required to do so.

Dr. PITZER. I would be very happy to do so if that is customary.

Mr. GRAY. All the other witnesses have.

Will you raise your right hand and give me your full name?

Dr. PITZER. Kenneth Sanborn Pitzer.

Mr. GRAY. Kenneth Sanborn Pitzer, do you swear that the testimony you are to give the board will be the truth, the whole truth, and nothing but the truth, so help you God?

Dr. PITZER. I do.

Whereupon Kenneth Sanborn Pitzer was called as a witness and, having been first duly sworn, was examined and testified as follows:

Mr. GRAY. Will you be seated, please.

It is my duty to remind you of the existence of the so-called perjury statutes. May we assume that you are familiar with them?

I should also like to request, Dr. Pitzer, if in the course of your testimony it becomes necessary for you to refer to or to disclose restricted data, you will notify me in advance, so that we may take the necessary steps in the interests of security.

Finally, I should like to say to you that we consider this proceeding a confidential matter between the Atomic Energy Commission, its officials, and witnesses on the one hand, and Dr. Oppenheimer and his representatives on the other. The Commission is making no releases to the press, and we express the hope that witnesses will take the same view.

The WITNESS. Surely.

Mr. GRAY. Mr. Robb, would you proceed?

DIRECT EXAMINATION

By Mr. ROBB:

Q. Doctor, would you tell us what your present post or position is?

A. My present post is professior of chemistry and dean of the college of chemistry, University of California, at Berkeley.

Q. Would you tell us something of your academic training and background, please, sir?

A. My undergraduate training was at the California Institute of Technology, with a bachelor's degree and a Ph. D. at the Univeristy of California in Berkeley.

Q. In what?

A. Physics and chemistry; officially chemistry. My general work has been what is sometimes described as a borderline area between physics and chemistry for the most part, although my professional affiliation has been with the Chemical Society pirmarily.

I am a member, indeed, a fellow, of the American Institute of Physics, as well as affiliated with the Chemical Society.

Q. Would you say when you took your Ph. D.?

A. 1937.

Q. Do you know Dr. Oppenheimer?

A. Certainly.

Q. How long have you known him, sir?

A. I at least knew of him when I was at Cal Tech in the period 1931 to 1935. More personal acquaintanceship developed gradually during the period from 1935 on at Berkeley and in the later years I was, of course, a professional colleague, and I was a member of the staff in chemistry and in physics.

Q. Have you ever been employed by the Atomic Energy Commission?

A. Yes. I was director of the Division of Research of the Atomic Energy Commission from approximately the beginning of 1949 to the middle of 1951.

Q. You left your academic duties and came on to take that position; is that right?

A. Yes, I was asked to do this. The only basis which seemed reasonable and agreeable to me was on a leave of absence basis, because I wished to maintain as a primary career actual direct scientific work and teaching at the university.

The Commission originally asked me to come for 2 years and leave was arranged on that basis. As a later step it was extended for another 6 months.

Q. When your leave was up, you went back to California?

A. When my leave was up I went back to California. The only difference was that they asked me to take over the deanship. At that time I had been just professor of chemistry previously.

Q. What connection have you now if any with the atomic energy program?

A. My principal connection now is as consultant and affiliate of the radiation laboratory at the University of California, including the program at Livermore, as well as the campus.

Q. Is the Livermore side Dr. Teller's laboratory?

A. It is commonly known as that. I have taken special pains to be sure that the chemistry and chemical engineering program at the Livermore laboratory was adequately staffed and in a healthy state, including the loaning of members of our departmental staff to that program.

Q. I should have asked you in sequence, but I will ask you now, what were your duties as director of research of the Atomic Energy Commission?

A. I am glad you came back to that. My line duties, as it were, concerned responsibility for basic or fundamental research in the physical sciences, including mathematics, chemistry, physics, metallurgy. In what might be described as a staff capacity, I was, shall we say, scientific adviser to other division directors, such as production, military applications, and in general wherever scientific—let me say advice in the physical sciences was useful to the Commission.

Q. And you undertook those duties, I believe you said, in 1950?

A. No, January 1949.

Q. I beg your pardon.. Doctor, coming to September 1949, will you state whether or not you had any knowledge of any questions arising or interest in a socalled thermonuclear weapon about that time?

A. Yes, I think it was about that time that my colleagues from Berkeley, Latimer, Lawrence, and Alvarez, came in in connection with some other meeting, and drew my attention particularly to the importance of a more vigorous program in this area.

Q. When you say came in, you mean came to Washington?

A. Yes. That is, they had come to Washington, two of them being members of another panel in some other field, and arriving the day before the meeting, came in to see me and talked about the potentialities in this area.

Q. And you said their thoughts were what about it?

A. Their thoughts were that this represented an important area in which the defense of the United States could be improved by a vigorous program of research and development leading to what has now become commonly termed the hydrogen bomb.

Q. Was that before or after the Russian explosion?

A. It was after the Russian explosion.

Q. Did you thereafter have occasion to see Dr. Oppenheimer?

A. The event that I recall was on a weekend, some time in October—the exact date could be developed if desired, but I do not remember it now—in which I had been up in that area, particularly giving an address to the Chemical Society meeting at Reading, Pa., and I dropped by and visited with Dr. Oppenheimer.

Q. Where?

A. At his home in Princeton, or his office, too, and we discussed this subject, and also the subject of the Atomic Energy Commission fellowship program which was having certain difficulties at that time. I would not say that either one or the other was necessarily the principal reason for the visit.

Q. What was said by Dr. Oppenheimer about the thermonuclear?

A. I was very much surprised to find that he seemed not in favor of a vigorous program in this area.

Q. Do you recall whether or not he gave a reason for that feeling?

A. I am a little vague in my memory as to the reasons and the details of the discussion then. As nearly as I can recall the reasons were substantially the same as are stated in the General Advisory Committee report of October 30, wasn't it?

Q. 29th, I believe it was.

A. And in particular in the appendix or substatement that was signed by Dr. Oppenheimer with others.

Q. Was this occasion on which you saw Dr. Oppenheimer before or after that meeting of the GAC?

A. This was before the GAC meeting. I am quite positive of that.

Q. Do you recall whether or not any mention was made by Dr. Oppenheimer of the views of any other scientists?

A. I am quite sure there was mention at that time of discussion or communication between Dr. Oppenheimer and Dr. Conant, and an indication that Dr. Conant was taking a view similar to that being expressed by Dr. Oppenheimer.

Q. Before we go further in point of time, were you familiar at that time in the fall of 1949 with the work which was going on, prior to the Russian explosion, at Los Alamos in respect to the thermonuclear?

A. I would not say I had a detailed acquaintanceship with that. I knew there was a small study program of some sort and that Dr. Teller was the figure that was regarded as the principal expert in the field. As I recall, he spent a portion of the time from year to year in Los Alamos. I don't recall the details. I did visit the Los Alamos laboratory in 1949 and reviewed its program in some detail, at least in the areas of which I had particular cognizance or competence, and it was apparant that there was no extensive program in the thermonuclear field.

Q. Would you say that the work that was going on was significant or otherwise in point of magnitude and intensity of effort?

A. It was certainly not what you would call a vigorous program. It was a sort of very subsidiary exploration of a few people—I don't know just how many.

Q. You saw, did you, the report of the GAC of the October 29–30 meeting?

A. Yes. I have forgotten just how long after it was issued.

Q. Were you here in Washington at the time of that meeting?

A. Yes.

Q. Will you tell us whether or not you had prepared any material or any presentation to make to the committee in respect of the thermonuclear problem?

A. No; I don't believe I had any particular presentation prepared at that time. I don't recall any such.

Q. Were your views on the matter solicited by the GAC?

A. I don't recall the detail, but I do not believe that they were, although I am not sure about that point. I do recall having come down at one period and then having had Carroll Wilson, then general manager, apologize and say that the attendance at the forthcoming session was being more highly restricted than he had anticipated. At least this particular session I did not attend. I am not very clear as to the exact detail.

Q. Had your views been solicited or received by the committee on other matters?

A. Oh, indeed.

Q. Prior to that time?

A. Yes.

Q. And were they solicited on other matters subsequent to that time?

A. Yes.

Q. You have stated or have told us about your conversation with Dr. Oppenheimer prior to the GAC meeting and you told us about seeing the report of the GAC meeting. Were you aware subsequent to the GAC meeting of any significant change in Dr. Oppenheimer's views as he had expressed them to you orally, and as they were expressed in the report of the GAC meeting?

A. Over what period of time do you mean?

Q. Any time subsequent to that?

A. I am sure there was some change in detailed view, but I don't recall any marked or major or sudden change.

Q. I am speaking particularly of his attitude with respect to the advisability of going ahead with the thermonuclear program. Were you aware of any significant change in that or any increase of enthusiasm?

A. Certainly not any very marked increase in enthusiasm. There was no major or sudden change that I was aware of.

Q. Doctor, would you say that you are pretty familiar with the nuclear scientists, physicists, and chemists in the country? Are you generally familiar with them?

A. I have reasonably wide acquaintanceship, more of course on the chemical side, but I am acquainted with many nuclear physicists.

Q. Given Dr. Oppenheimer's attitude and feelings as you have described them, what can you tell us about what would be the effect in your opinion upon the scientific world of such attitudes and feelings so far as either increasing or decreasing enthusiasm for the thermonuclear program? That is a long question. I hope it is clear. I am trying not to lead you.

A. I hope you will permit me to make a statement of my general impressions of that time. After the President made the decision and announced it to the papers, I was rather surprised to find that Dr. Oppenheimer did not in some manner or another disqualify himself from a position of, shall we say, technical leadership of the program. I had the feeling that if my advice on a major subject of this sort had been so—if the decision had been so much in reverse from my advice, let us put it that way—that I would not have wanted to be in a position of responsibility with respect to the subsequent pursuit of the program.

As to just what course of action would have been most appropriate, there are various alternatives. I think this would have led to a clearer and more vigorous program had some other arrangement of this sort been had.

Q. Why do you think that, Doctor?

A. It would have been clear that the Commission was by this time thoroughly behind the program and that the fullest support was going to be given to it because special arrangements had been made to be sure that the leadership would be vigorous.

Q. Do you think the fact that Dr. Oppenheimer stayed on entertaining the views which you have told us about discouraged other physicists from going ahead on the program with vigor?

A. I can only say to this that I am afraid it may have. I am not aware of detailed negotiations or influences on particular individuals, but I do know there was difficulty in that early period in obtaining the staff that would have seemed desirable to me and as I believe Dr. Teller felt was desirable at that time, particularly in the theoretical physics area. To have had other advisory leadership that was known to be enthusiastic for the program would, I think, have assisted.

Q. You suggested other advisory leadership. Did you have in mind a specific step that might have been taken either by Dr. Oppenheimer or by the Commission to get such leadership?

A. As I said before, it seemed to me that there were several alternatives there. If the most extreme change had seemed desirable, there was a possibility of full changes of membership in the Statutory Advisory Committee. Other possibilities could have been the appointment of some special panel in this field, and of course a marked and clearcut change in the viewpoints of certain individuals would have assisted the program.

Q. In your opinion did Dr. Oppenheimer do everything he might have to further the program after the President's decision?

A. Again in an inferential sense, I am afraid I must say that he did not.

Q. Would you explain that to us a little bit?

A. As I indicated earlier, it seemed to me that had he enthusiastically urged men in the theoretical physics field to go to Los Alamos or other points as indicated for this program that the difficulties in staffing it would have not arisen. I am sure he had great influence over individuals in that field.

On the other hand, as I say, this is simply an inference, and not something that I know from day to day and man to man.

Q. I understand. What was Dr. Oppenheimer's influence in the physics field during that period to your knowledge?

A. He was unquestionably a most influential individual in dealings with other physicists, particularly theoretical physicists, but also experimental men.

Q. Doctor, did there come a time when Dr. Libby was appointed to the General Advisory Committee?

A. Yes.

Q. Did you have anything to do with that appointment?

A. I don't know how much I had to do with the appointment, but at that time I discussed problems with the then Chairman, Gordon Dean.

Q. Could you give us the approximate date of that?

A. I am trying to think when those appointments were made. This must have been in the late spring or summer of 1950, I would infer.

Q. Would you go ahead? I am sorry I interrupted you.

A. At that time I pointed out to Mr. Dean, as I indeed had pointed out earlier, that there was a considerable body of scientific opinion of the very distinguished and able men that was more enthusiastic with respect to the thermonuclear weapons program and had undoubtedly different views in a number of respects than were represented on the Advisory Committee as of that time. I urged him to appoint to that Committee at least one individual who had been from the beginning enthusiastic for the thermonuclear program and who would assure him of advice based on that point of view.

Q. Whom did you suggest, if anyone?

A. I suggested a number of names, including Dr. W. F. Libby, of the University of Chicago, and eventually Dr. Libby was appointed to the Committee.

Q. Was there a weapons subcommittee of the General Advisory Committee?

A. I believe there was; yes.

Q. Who chose that weapons subcommittee?

A. I have never been a member of the Committee, and I cannot state as a matter of knowledge what the Committee procedure was. I presume that the selection was very likely on nomination by the Chairman and confirmation by the Committee, although it may have been by the full Committee action in some other mechanism.

Q. However, it was done, was Dr. Libby ever appointed to that weapons subcommittee to your knowledge?

A. I don't know whether he was ever appointed to the Committee, but I am substantially certain that he was not appointed to the Committee in the fall of 1950.

Q. There has been quite a bit of testimony here about a meeting held at Princeton, I believe, in the spring of 1951. Are you familiar with that meeting in general, and did you hear about it?

A. Yes, I heard about that meeting.

Q. Was Dr. Libby invited to that meeting as far as you know?

A. As far as I know, he was not.

Q. What can you tell us about the importance or the essentiality to the atomic weapons and the thermonuclear weapons program today of Dr. Oppenheimer, in your opinion?

A. Let me develop this in a number of facets.

Q. That is why I asked the broad question so you can answer it in your own way.

A. I would like to discuss these briefly from three points of view. One is in terms of immediate scientific work. That is the calculations, theoretical derivations and this sort of thing. This by and large is done by younger people, particularly in the field of theoretical physics. I haven't the slightest doubt that Dr. Oppenheimer would be valuable to such work but, by and large, from that tradition and experience in theoretical physics, this sort of thing is done by people in their twenties or thirties.

The second aspect is that of leadership among men in this field. I have no doubt that Dr. Oppenheimer's influence and importance in the sense of leadership among men is of the highest order. He would have a great deal of influence and could be of a great deal of assistance in persuading able people to work at certain places and at certain times and in selecting people for this.

The third phase that I would mention would be that on what might be called policy advice. This is the sort of thing that the Commission and other nontechnical management people need. Personally, I would not rate Dr. Oppenheimer's importance in this field very high for the rather personal reason, I suspect, that I have disagreed with a good many of his important positions and I personally would think that advisers in the policy field of greater wisdom and judgment could be readily obtained.

Q. You say very honestly that you personally disagree. Let me ask you whether or not events have proved that you were right or Dr. Oppenheimer was right.

A. That is a difficult question. I think personally that we were right in going into a vigorous thermonuclear program at the time we did. I would not want to question the possibility of a perfectly sincere and reasonable judgment to the contrary at that time. I want to make it perfectly clear that I am emphasizing here essentially need, or in the extreme, indispensability of the advice than some other feature. Possibly it would be just fair to say that in the policy area I certainly do not regard Dr. Oppenheimer as having any indispensability.

Q. One final question, Doctor. You are not here as a witness, are you, because you wanted to be?

A. Certainly not. Thank you for asking that. I am here only at the very specific and urgent request of the general manager and with a feeling that as one of the senior scientific personnel of the Commission at a critical time that it was only reasonable that I should accede to the general manager's request.

Q. Doctor, I am asking this next question so that the record will be plain and not intending to insinuate anything.

Although you are here at the specific request of the general manager, your views which you have expressed are your own independent views, aren't they?

A. Indeed. I am expressing only precisely my own views and I think anyone that knows me would be pretty certain that I would not express anybody else's views no matter how they were put.

Q. In other words, the general manager's request brought you physically here but did not give you the ideas which you expressed.

A. That is correct.

Mr. ROBB. That is all I care to ask, Mr. Chairman.

Mr. GRAY. Mr. Silverman.

Mr. SILVERMAN. Yes, sir.

CROSS-EXAMINATION

By Mr. SILVERMAN:

Q. I think you said that at your visit to Dr. Oppenheimer in Princeton in October of 1949, I thought you said you were rather vague as to Dr. Oppenheimer's statement of his views but that as far as you can recall, they were about the same as in what has come to be known as the majority appendix to the GAC report.

A. What I believe I said was that I was surprised that he was opposing a vigorous program and that as nearly as I can recall for it were substantially those in this majority appendix.

Q. Do you recall specifically that he then told you the reasons and what they were? I am not trying to trap you into anything. Or do you think it possible that you are now reading back the reasons stated in the GAC report, and they did not surprise you very much when you heard them as Dr. Oppenheimer's views?

A. I am sure we did discuss the problem, not at great length, but at appreciable length, and that the reasons must have been offered. I frankly can't be sure exactly which argument came into the picture at which time.

Q. You were asked about the extent of the thermonuclear program work that was being done in that field up to September of 1949. I think you said that you didn't think there was a very extensive program, or something of that kind?

A. Yes.

Q. If I am wrong, don't hesitate to correct me. It is all right. Would you say that Dr. Bradbury, who was the director of the laboratory at Los Alamos, was perhaps in a better position to give a statement of the extent of the thermonuclear work that was being done than you were?

A. Oh, indeed. Dr. Bradbury had more detailed information concerning the size of the program, as did Dr. Teller and others.

Q. Your position was director of research. Am I correct that weapons development or research was not a part of your responsibility?

A. The situation with particular respect to weapons was as follows. The line authority for the Los Alamos Laboratory and the remainder of the weapons development, as well as production program, was in the Division of Military Applications under the directorship then of General McCormack. My function in that area was strictly a staff function to be of whatever assistance and advice I could be since General McCormack was not himself a scientist.

Q. If and when you were asked for scientific advice, you would give it, and find out what you could, and so on?

A. Yes. In fact, I would go further. I am not particularly bashful. I would frequently make suggestions on my initiative, and I was invited to make suggestions on my initiative.

Q. I am not suggestion that you were not, nor that your suggestions were not entirely welcome. I am sure they were. I am just trying to establish the lines of responsibility.

A. That is correct.

Q. And that, in fact, the development of weapons would be more a matter that perhaps General McCormack would know more about, and perhaps Dr. Bradbury would.

A. In terms of the details or in General McCormack's case, the administrative side of the program, that statement would be appropriate.

Q. And in terms of what was actually done in the development of the weapons.

A. I wouldn't argue that.

Q. I am not trying to argue with you either. I think you said that you did not think that your views were solicited by the General Advisory Committee at the time of the October 1949 report. Do you recall whether there was a subsequent time, fairly shortly after the General Advisory Committee report, when they did solicit your views?

A. As I recall, there was a subsequent meeting, possibly in early December, in which this subject was reviewed again. If I remember correctly, General McCormack and I were both invited to that meeting and invited to essentially speak our peace, since we were by that time believed to be in substantial disagreement with the Committee. As I recall, General McCormack testified at greater length and I supported the view contrary to the Committee's report briefly.

Q. You said testified; spoke, I take it you mean. It was a discussion.

A. Yes.

Q. I think you said you were rather surprised that Dr. Oppenheimer did not disqualify himself from a position of technical leadership of a program with which he apparently disagreed. Do you know whether Dr. Oppenheimer did in fact offer to resign from the chairmanship of the General Advisory Committee at that time?

A. I have no information on that.

Q. You have not heard that he offered to the Chairman, Mr. Dean, to resign?
A. I don't believe I heard that; no.
Q. And you don't know what Mr. Dean's reaction was. You just never heard of it?
A. I never heard about it.
Q. I think there has been testimony here about it, so I think the record is clear enough on it.
A. At least, if I heard of it, I do not recall at this time.
Q. I take it you would be less critical of Dr. Oppenheimer's attitude if that were the fact, if he offered to resign and was urged to remain?
A. Certainly so. I think, however, that his position today would be better if he had insisted on at least some degree of disqualification in this field at that time.
Q. I wish you would elaborate on that.
A. Let me put it this way. I am extremely sorry to see this issue concerning advice which on hindsight proved not too good brought up in connection with a security clearance procedure. I feel very strongly that scientists should feel free to advise the Government and not be held to account if their advice proves not the best afterward. This should have no relevance to security clearance procedure. If Dr. Oppenheimer had seen fit to insist upon stepping out of the position of advising on the hydrogen program, this could not be introduced into this argument at this time. I am very sorry to see that it does have to come up at this time.
Q. I need hardly say that I entirely agree with you.
I think you said that you thought that Dr. Oppenheimer's attitude may have discouraged people from working on the thermonuclear program, and you were very frank in saying you didn't have details of that, and so on. I suppose your greatest familiarity would be with the situation at Berkeley, would it not?
A. I certainly had some degree of familiarity with the situation at Berkeley at the time I was in Washington, both because the radiation laboratory was more immediately under the Research Division, and because all of my personal contacts with the Berkeley staff.
On the other hand, I would assure you that I took a very definite interest in this thermonuclear program and visited Los Alamos on occasion, and visited with Professor Teller and others when he was in Washington in order to see how it was going, and in order to offer my assistance at any time.
Q. I think you were asked whether you thought Dr. Oppenheimer did everything he might have done to further the President's thermonuclear development program after the President's decision, and I think you said you thought he might not have. Everything that a man might have done is a relative thing.
Mr. ROBB. Mr. Chairman, I hate to interrupt but it seems to me that the witness ought to do the testifying and not Mr. Silverman.
Mr. SILVERMAN. That is an introduction to the question.
Mr. GRAY. I think it is true that you have been expressing your views quite frequently, Mr. Silverman, in this cross-examination, and I have not stopped you, again in the interest of not being too rigid in our procedures. But I think it well for me to make a request at this time that you confine your introductory statements to the necessities of the question, because the record should primarily reflect the views of witnesses, rather than counsel.
Mr. SILVERMAN. I have tried to do so, sir, and I will try to be more careful of that.
Mr. GRAY. Thank you.

By Mr. SILVERMAN:

Q. Would you say that doing everything that one might have done is a relative matter?
A. It is a relative matter, and in my earlier answer to the question I was not trying to slice close to the line. I felt that the events of that period were sufficiently wide of a narrow borderline to justify the critical statement.
Q. In one sense, and I am not criticizing you, sir, you did not do everything you might have to further the program.
A. No. There are things on hindsight one can always figure out one could have done more. I suppose one could have done many things differently, but I certainly carried it as a high priority among my duties, particularly considering that it was not a line, but rather a staff problem, and I regarded the program since as something that demanded my attention whenever anything substantial could be contributed to it.

Q. You didn't consider that it was necessary for you not to return to the university, for instance?
A. No. But I delayed the return for 6 months very substantially on that account.
Q. Believe me, I am not criticizing you, sir. I think you are entirely within your rights. You have taken the position as a consultant which I take it is a part time position.
A. Yes; I think since you are pursuing this matter, I would like to say a little further that I am not myself a nuclear physicist. The chief contributions which I can make to this program are to be sure that the chemical engineering components that need to go into the various units are made to the exact specifications that are required, and so on. My position is the administrative position in chemistry at the University of California at Berkeley, and I have thought my best contribution would be to see that the proper people were working on the proper jobs at the proper time, rather than I should necessarily go and do them with my own hands.
Q. Don't you think, sir, that the decision as to how much of one's own efforts and time one puts into some program is a matter for personal judgment of a man?
A. Yes; I was considering these judgments earlier in very appreciable degree with respect to the adequacy of staffing of a given program and the ability of a particular person to take steps to assure that the program was adequately staffed. In my own position the sort of thing I could do was to essentially say, "Look, Mr. So-and-so, we will get along without you in the department, half time or full time, next semester. This is an extremely urgent job." Of someone not associated with the university initially, but in my general field I can advise him of the importance of the program and urge him strongly to serve if offered an appropriate position. It is in this frame of reference that my earlier comments were made.
Q. Don't you think that service on the General Advisory Committee is itself quite an important contribution?
A. It is, indeed, an important position.
Q. Returning to your statement that you thought that you thought Dr. Oppenheimer's attitude may have discouraged people from working on the thermonuclear program, there, of course, have been other factors in the difficulty of getting staff, were there not?
A. There are always other factors. The question is the relative importance of this task as compared to others, and the sense of urgency which is imparted to a man who is considering either going to this program or not going to the program.
Q. I think you said that you saw no marked increase in Dr. Oppenheimer's enthusiasm as to going ahead with the hydrogen bomb. Was that during the period you were here?
A. Yes; that was during the period I was in Washington. I have seen Dr. Oppenheimer only most infrequently since I left Washington.
Q. When did you leave Washington?
A. This was the summer of 1951.
Q. Are you in a position to say as to whether his enthusiasm increased with the later improved outlooks for the feasibility of the hydrogen bomb?
A. I am not in a position to say anything about that.
Q. You referred to the appointment of Dr. Libby to the General Advisory Committee. I think you said that Dr. Libby was one of a number of names that you had suggested. Do you know that Dr. Libby was on a list that Dr. Oppenheimer submitted to Chairman Dean for membership on the General Advisory Committee?
A. I have no knowledge of that.
Q. Before you came to your position with the Atomic Energy Commission as director of research, did Dr. Oppenheimer have a conversation with you in which he urged you or asked you whether you would be willing to spend some time in Government work in Washington?
A. It is very likely that this was the case. I am not sure.
Q. In your testimony earlier about a meeting at Princeton—there have been so many meetings at Princeton—I am talking about the weekend you spent at Princeton when you spoke to Dr. Oppenheimer about the hydrogen-bomb program in the fall of 1949, and also the fellowship program.
Mr. ROBB. I don't think he spent a weekend there.
Mr. SILVERMAN. As long as we have the time. As to the length of time, it doesn't matter. I am making no point about it being a weekend at all.

By Mr. SILVERMAN:

Q. What was the fuss about the fellowship program?
A. This is a long story. The essence of it was that the Congress of the United States introduced a rider in the Appropriation Act which required investigation and a decision as to loyalty for all fellows under the program in the future.
Q. What was your view on that?
A. I was very sorry to see such a requirement introduced into the program.
Q. Was Dr. Oppenheimer's view in accord with yours on that?
A. In that general way, yes. I was sorry to see it introduced. I was equally sorry and disturbed by the events and situations which had come to the attention of the Congress and which led them to introduce it.
Q. Were you against this requirement?
A. As I say, I was opposed to the introduction of a requirement for full investigation. I was hoping that the situation could be handled by some loyalty oath or some other procedure which would not require a full field investigation, but which would still give a case of reasonably substantial certainty of loyalty to the United States.
I might add that this was the course taken with respect to the National Science Foundation later.
Q. Were you critical of the work of the Reactor Safeguard Committee?
A. Yes; I have been critical of that.
Q. Do you recall who the chairman of that committee was during the period when you were critical of it?
A. Surely. My good friend Edward Teller. I have argued with him in a friendly fashion on many times.
Q. And you don't for a moment question his good faith and what he did there?
A. Not at all.
Mr. SILVERMAN. Thank you.
Mr. GRAY. Dr. Pitzer, are you familiar with the exchange of letters between General Nichols and Dr. Oppenheimer? Have you read them?
The WITNESS. I have read that double-page spread in the New York Times, which contains I believe what you are referring to.
Mr. GRAY. I suppose that was accurate. I never checked it. I would like to read you a part of General Nichols' letter. This is in a paragraph which in its entirety related to the hydrogen bomb, starting about the middle of the paragraph:
"It was further reported that even after it was determined as a matter of national policy to proceed with development of a hydrogen bomb, you continued to oppose the project and declined to cooperate fully in the project."
That is a sentence in that paragraph. In order to get a clearer view of your opinion in my own mind, may I assume that it is an accurate reflection of your testimony that this suggestion is not borne out by your understanding of events, that is, you have not testified that Dr. Oppenheimer continued to oppose the project?
The WITNESS. I am forced to say that my impressions of that period were more consistent with the hypothesis that he was still personally opposing the project than with the hypothesis that he had made a major change in his views and was now strongly supporting the project.
Mr. GRAY. I suppose there is a difference of finding oneself in personal opposition and finding oneself opposing. I must say I had not thought of a distinction of this sort in this language until this moment. But I would like to know what you feel. Let us assume that this means actively opposed as distinguished from holding to personal views in opposition. Is that a clear distinction in your mind?
The WITNESS. I must admit that I am likewise trying to make a finer distinction than I thought about commonly before. What I mean to say is essentially this: I have no personal knowledge of Dr. Oppenheimer going to Mr. X and saying don't work at Los Alamos, or of his making a technical recommendation obviously and distinctly contrary to the demonstrable good of the program.
On the other hand, I have great difficulty believing that the program would have had certain difficulties that it did have at that time if he had enthusiastically urged individuals to participate in the program, because as I said before, he was a great personal influence among theoretical physicists at that time. I am afraid the distinction is primarily one of ignorance.

Mr. GRAY. It is clear that you have said that you feel that Dr. Oppenheimer failed to encourage people or did not encourage people—I don't mean to use a word that is loaded—did not encourage people to work on the project. You have said you didn't know of any instances in which he actively sought to discourage people from working on the project.

The WITNESS. At least not at this time. Part of my impressions may have carried over from instances known in greater detail at a date nearer the time of events.

Mr. GRAY. You could not name anyone that you thought had failed to work on the project because of Dr. Oppenheimer's persuasive powers?

The WITNESS. I know, for example, there was much discussion about Hans Bethe at that time. It is entirely plausible to me that had Dr. Oppenheimer encouraged Dr. Bethe he might have very likely entered the program actively at that time. This is supposition. I was certainly not present at the conversations between Dr. Bethe and Dr. Oppenheimer. I mention Dr. Bethe in part by way of example.

Mr. GRAY. Would you return for a moment to the second GAC meeting in late 1949—I have forgotten when that was. December, I think.

The WITNESS. I believe so.

Mr. GRAY. At which time you and General McCormack were invited to present your views to the General Advisory Committee. I believe you said that General McCormack spoke at some length and you supported his views. What was General McCormack's view and yours at the time? What was expressed to the GAC as well as you recall it?

The WITNESS. My recollection is rather vague of that particular time, and I am somewhat reluctant to try to put words in General McCormack's mouth after this lapse, but the view that I believe I would have been attempting to present at that time was essentially the one, that one could not improve the national defense by remaining in ignorance in an area where there are developments of potentially very great importance to the national defense. I was unable to see how a policy of intentionally not pursuing a vigorous program could possibly be consistent with optimum defense of the country.

Mr. GRAY. You referred to what you supported as a more vigorous program than was in effect at that time. It is clear that the General Advisory Committee recommended in October and again in December against an all out production effort of the so-called super. That was clearly one of the recommendations, as I understand it.

The WITNESS. Yes.

Mr. GRAY. I would like to put to you a question I have put to other witnesses with very little success, and it may be my ignorance or just my failure to ask a question properly. In your judgment was there something that the GAC could have recommended at this time which was short of an all-out production program but more than was recommended?

The WITNESS. Oh, indeed; obviously, to me.

Mr. GRAY. Was that your position at the time, or were you for the all-our production? You see I am a little confused when you say a more vigorous program.

The WITNESS. Let me put it this way. I was for a very vigorous program, one which would have the highest possible priority, subject to reasonable continuation of other important programs. In other words, I was not in favor of stopping a lot of other important activities, but I was thoroughly convinced that the necessary manpower could be recruited, the necessary facilities provided, for a very vigorous program of the general nature that was being discussed and advocated at that time by Dr. Lawrence and Dr. Teller and others. I believe I said at that time—I am sure I felt—that this business of a crash program was largely what we called a strawman. In other words, it seems to me that the General Advisory Committee was clearly in a position to have recommended a program of intermediate intensity if such had been their judgment.

The recommendations that were actually made, as you gentlemen have them, are almost entirely negative in character. They are in terms of not doing this and not doing that.

Mr. GRAY. The reason I started to smile is I think you answered my question, the question I have been trying to ask, at least you have given me your opinion about it, and you made it clear to me that perhaps there is a valid distinction, at least in your mind, between something that was all out and something that was more vigorous than was then in progress.

May I turn now to another thing about which you testified very briefly, Dr. Pitzer. You referred to your unhappiness with respects to events that led up to congressional action in attaching the rider to the appropriation bill. What are these events that you have in mind?

The WITNESS. The sequence began with a young man by the name of Friestad.

Mr. GRAY. I didn't mean to bring my university into this hearing.

The WITNESS. I am sorry; the facts are that way.

Mr. GRAY. I honestly did not know this is what you were talking about.

The WITNESS. He was first essentially exposed and discussed as essentially, I believe, an admitted Communist and holding a fellowship. Hearings were held and there was a great deal of discussion in the press, and as it were, one thing went on to another, until, the Senate in due time attached this rider to the bill and the House accepted it.

Mr. GRAY. Prior to this time when the Congress established the requirement which you found yourself unhappy about, did you participate in any kind of discussions with respect to what should be required of these fellows in the way of disclosure of political offiliations as we seem to refer to them in this hearing?

The WITNESS. Yes. There were discussions within the Commission at that time. I have forgotten exactly the details. I certainly participated in such discussions.

Mr. GRAY. Would the GAC have participated in this kind of discussion? You, of course, were not a member of the GAC.

The WITNESS. I don't recall the chronology. This fellowship business happened pretty fast, and I rather doubt if there happened to occur a GAC meeting in that period. I believe I recall that the then Chairman, Mr. Lilienthal, got in contact with Dr. Oppenheimer and possibly other members of the GAC by telephone—they may have to come to Washington specially—and it may have been that a meeting was held, but I don't recall such.

Mr. GRAY. Let us leave the GAC out of it at this point and let me approach it from another angle the thing that I am trying to get clear in my mind.

There were discussions, I assume, in which a suggestion was made that there should be no inquiry put to an Atomic Energy Commission fellow with respect to his political affiliations. This was the view of some people at that time, is that correct?

The WITNESS. I believe such views were held at that time.

Mr. GRAY. My question is this: Was this the view of the Commission at that time, or could the Commission be said to have had a view?

The WITNESS. I don't believe the Commission could be said to have had a view at that time. At least if as a Commission it reached any decision, I am not aware of it now.

Mr. GRAY. I don't think I will pursue that any further, Dr. Pitzer. Dr. Evans?

Dr. EVANS. Dr. Pitzer, you said you were not a nuclear physicist, is that right?

The WITNESS. That is correct.

Dr. EVANS. Would you call yourself a physical chemist or a physicst?

The WITNESS. I would call myself a physical chemist; yes, sir.

Dr. EVANS. I want to ask you if you met a man in recent years, a graduate of Cal. Tech., by the name of Sheehan? It is one of my students that I sent out there. I thought he was particularly brilliant. He got a Ph. D. degree.

The WITNESS. I have met, I believe, casually, a young Sheehan, but I don't know enough about his background to complete the identification with certainty.

Dr. EVANS. Have you met any Communists in the course of your career, that you knew were Communists?

The WITNESS. It may well have happened. They didn't have Communist labels pinned on them at the time.

Dr. EVANS. They don't often have, do they?

The WITNESS. No, they don't often have.

Dr. EVANS. Did you know David Hawkins?

The WITNESS. The name is familiar to me. If I ever met him, I do not recall it.

Dr. EVANS. Did you know Bernie Peters?

The WITNESS. Again if I ever met him personally, I do not recall it, although I recall very vividly the case of getting him a passport to India that took a definite Commission action, so that his name is definitely familiar to me.

Dr. EVANS. Did you know Fuchs?

The WITNESS. I don't believe I ever knew Fuchs, or ever met him. I knew of him from the scientific literature.

Dr. EVANS. I have no further questions.

REDIRECT EXAMINATION

By Mr. ROBB:

Q. Doctor, is it or is it not true in your opinion that in the case of a scientist as influential as Dr. Oppenheimer a failure to lend enthusiasm and vigorous support to a program might constitute hindrance to the program or opposition to the program?

A. There is a certain element of semantics in that question, but I would say yes.

Mr. ROBB. Thank you.

Re-cross-examination by Mr. SILVERMAN:

Q. I think I have just one more question. You testified about the difficulty of obtaining staff on the thermonuclear program. I think you indicated that Dr. Oppenheimer was not helpful. Is Dr. Karplus at Cal. Tech.?

A. I believe so.

Q. Do you know whether he is a man that Dr. Oppenheimer recommended to go there?

A. I don't know the details.

Q. He is or has been from time to time a temporary member of the Institute for Advanced Study, has he not?

A. As I say, I am not familiar with the details in that case. The staffing at Livermore in the physics area has been in the very able hands of Ernest Lawrence and other physicists, including Edward Teller. I simply have not felt it necessary or needful to pay attention to details in that area.

Mr. SILVERMAN. That is all.

Mr. ROBB. That is all.

Mr. GRAY. Thank you very much, Dr. Pitzer.

(Witness excused.)

Mr. GRAY. We will recess now, gentlemen, for a few minutes.

Mr. GRAY. Dr. Teller, do you wish to testify under oath?

Dr. TELLER. I do.

Mr. GRAY. Would you raise your right hand and give me your full name?

Dr. TELLER. Edward Teller.

Mr. GRAY. Edward Teller, do you swear that the testimony you are to give the board shall be the truth, the whole truth, and nothing but the truth, so help you God?

Dr. TELLER. I do.

Whereupon, Edward Teller was called as a witness, and having been first duly sworn, was examined and testified as follows:

Mr. GRAY. Will you sit down.

Dr. Teller, it is my duty to remind you of the existence of the so-called perjury statutes with respect to testifying in a Government proceeding and testifying under oath. May I assume that you are generally familiar with those statutes?

The WITNESS. I am.

Mr. GRAY. May I ask, sir, that if in the course of your testimony it becomes necessary for you to refer to or to disclose restricted data, you let me know in advance, so that we may take appropriate and necessary steps in the interests of security.

Finally, may I say to you that we consider this proceeding a confidential matter between the Atomic Energy Commission, its officials and witnesses on the one hand, and Dr. Oppenheimer and his representatives on the other. The Commission is not effecting news releases with respect to these proceedings, and we express the hope that witnesses will take the same view.

DIRECT EXAMINATION

By Mr. ROBB:

Q. Dr. Teller, may I ask you, sir, at the outset, are you appearing as a witness here today because you want to be here?

A. I appear because I have been asked to and because I consider it my duty upon request to say what I think in the matter. I would have preferred not to appear.

Q. I believe, sir, that you stated to me some time ago that anything you had to say, you wished to say in the presence of Dr. Oppenheimer?

A. That is correct.

Q. May I ask you, sir, to tell the board briefly of your academic background and training.

A. I started to study in Budapest where I was born, at the Institute of Technology there, chemical engineering for a very short time. I continued in Germany, first in chemical engineering and mathematics, then in Munich for a short time, and finally in Leipzig in physics, where I took my doctor's degree.

After that I worked as a research associate in Goettingen, I taught in London. I had a fellowship, a Rockefeller fellowship in Copenhagen.

In 1935 I came to this country and taught for 6 years at the George Washington University, that is, essentially until the beginning of the war.

At that time I went to Columbia on leave of absence, partly to teach and partly in the very beginnings of the war work in 1941–42, as I remember, and then I participated in the war work. After the war I returned to teach in Chicago at the University of Chicago, which also was interrupted with some work for the AEC, and now for the last year I am at the University of California in Berkeley.

Q. Dr. Teller, you know Dr. Oppenheimer well; do you not?

A. I have known Dr. Oppenheimer for a long time. I first got closely associated with him in the summer of 1942 in connection with atomic energy work. Later in Los Alamos and after Los Alamos I knew him. I met him frequently, but I was not particularly closely associated with him, and I did not discuss with him very frequently or in very great detail matters outside of business matters.

Q. To simplify the issues here, perhaps, let me ask you this question: Is it your intention in anything that you are about to testify to, to suggest that Dr. Oppenheimer is disloyal to the United States?

A. I do not want to suggest anything of the kind. I know Oppenheimer as an intellectually most alert and a very complicated person, and I think it would be presumptuous and wrong on my part if I would try in any way to analyze his motives. But I have always assumed, and I now assume that he is loyal to the United States. I believe this, and I shall believe it until I see very conclusive proof to the opposite.

Q. Now, a question which is the corollary of that. Do you or do you not believe that Dr. Oppenheimer is a security risk?

A. In a great number of cases I have seen Dr. Oppenheimer act—I understood that Dr. Oppenheimer acted—in a way which for me was exceedingly hard to understand. I thoroughly disagreed with him in numerous issues and his actions frankly appeared to me confused and complicated. To this extent I feel that I would like to see the vital interests of this country in hands which I understand better, and therefore trust more.

In this very limited sense I would like to express a feeling that I would feel personally more secure if public matters would rest in other hands.

Q. One question I should have asked you before, Dr. Teller. Are you an American citizen, sir?

A. I am.

Q. When were you naturalized?

A. In 1941.

Q. I believe you said that about 1941 you began to work on the atomic bomb program.

A. I don't think I said that. Certainly I did not intend to say it.

Q. I will rephrase the question. When did you begin to work on the atomic bomb program?

A. That again I am not sure I can answer simply. I became aware of the atomic-bomb program early in 1939. I have been close to it ever since, and I have at least part of the time worked on it and worried about it ever since.

Q. Did you work during the war at Los Alamos?

A. I did.

Q. When did you go there, sir?

A. In April 1943.

Q. What was the nature of your work there?

A. It was theoretical work connected with the atomic bomb. Generally speaking—I do not know whether I have to go into that in any detail—I was more interested by choice and also by directive in advanced development, so that at the beginning I think my work was perhaps more closely connected with the actual outcome or what happened in Alamagordo, but very soon my work shifted into fields which were not to bear fruition until a much later time.

Q. Will you tell the board whether or not while you were in Los Alamos in 1943 or 1944, you did any work or had any discussions about the so-called thermonuclear weapon?

A. Excuse me, if I may restate your question. I got to Los Alamos in early April 1943. To the best of my recollection, although I might be wrong—I mean my date might not be quite precise—I left at the beginning of February 1946. Throughout this period I had very frequent discussions about thermonuclear matters.

Q. Will you tell us whether you ever discussed the thermonuclear method with Dr. Oppenheimer?

A. I discussed it very frequently indeed with him. In fact my discussions date back to our first association in this matter, namely, to the summer of 1942.

Q. What was Dr. Oppenheimer's opinion in those discussions during those years about the feasibility of producing a thermonuclear weapon?

A. This is something which I wish you would allow me to answer slightly in detail, because it is not an easy question.

Q. Yes, sir.

A. I hope that I can keep my answer in an unclassified way. I hope I am not disclosing a secret when I say that to construct the thermonuclear bomb is not a very easy thing, and that in our discussions, all of us frequently believed it could be done, and again we frequently believed it could not be done. I think Dr. Oppenheimer's opinions shifted with the shifting evidence. To the best of my recollection before we got to Los Alamos we had all of us considerable hopes that the thermonuclear bomb can be constructed. It was my understanding that these hopes were fully shared by Dr. Oppenheimer.

Later some disappeared and perhaps to counterbalance some things that might have been said, I think I have made myself some contributions in discovering some of these difficulties.

I clearly remember that toward the end of the war Dr. Oppenheimer encouraged me to go ahead with the thermonuclear investigations. I further remember that in the summer of 1945, after the test at Alamogordo it was generally understood in the laboratory that we are going to develop thermonuclear bombs in a vigorous fashion and that quite a number of people, such as the most outstanding, like Fermi and Bethe, would participate in it.

I also know that very shortly after the dropping of bombs on Japan this plan was changed and to the best of my belief it was changed at least in good part because of the opinion of Dr. Oppenheimer that this is not the time to pursue this program any further.

I should like to add to this, however, that this also thoroughly responded to the temper of the people in the laboratory, most of whom at that time understandably and clearly and in consonance with the general tempo of the country, wanted to go home.

Q. Did you have any conversations with Dr. Oppenheimer at or about September 1945 about working on the thermonuclear?

A. We had around that period several conversations and in one of them, to the best of my recollection, Oppenheimer and Fermi and Allison and I were present. Oppenheimer argued that this is not the time at which to pursue the business further, that this is a very interesting program, that it would be a wonderful thing if we could pursue it in a really peaceful world under international cooperation, but that under the present setup this was not a good idea to go on with it.

I perhaps should also like to mention that to the best of my knowledge at that time there was a decision by a board composed of several prominent people, one of them Dr. Oppenheimer, which decided in effect that thermonuclear work either cannot or should not be pursued that it at any rate was a long-term undertaking requiring very considerable effort. To my mind this was in sharp contrast to the policy pursued a short time before.

But I also should say that this sharp contrast was at least in part motivated by the fact that in Los Alamos there was a crew of exceedingly able physicists who could do a lot and at the end of the war were trying to get back to their purely academic duties, and in this new atmosphere it might have appeared indeed hard to continue with such an ambitious program.

One member of the board which made this decision, Fermi, and who concurred in that decision, told me about that decision and told me that he knew that I am likely to disagree with it, and asked me to state my opinion in writing. This I did, and I gave my written statement to Oppenheimer, and therefore, both the opinion that the thermonuclear bomb at that time was not feasible, and my own opinion that one could have proceeded in this direction are documented.

Q. Did there come a time when you left Los Alamos after the war?

A. That is right. As I mentioned, I left in February 1946. May I perhaps add something here if we are proceeding in a chronological manner?

Q. Yes.

A. Perhaps if I might interject this not in response to one of your questions.
Q. That is perfectly all right, sir.
A. I would like to say that I consider Dr. Oppenheimer's direction of the Los Alamos Laboratory a very outstanding achievement due mainly to the fact that with his very quick mind he found out very promptly what was going on in every part of the laboratory, made right judgments about things, supported work when work had to be supported, and also I think with his very remarkable insight in psychological matters, made just a most wonderful and excellent director.
Q. In that statement were you speaking of Dr. Oppenheimer's ability as an administrator or his contribution as a scientist or both?
A. I would like to say that I would say in a way both. As an administrator he was so busy that his purely scientific contributions to my mind and in my judgment were not outstanding, that is, not insofar as I could see his original contributions. But nevertheless, his scientific contributions were great by exercising quick and sound judgment and giving the right kind of encouragement in very many different cases. I should think that scientific initiative came from a great number of other excellent people whom Oppenheimer not let alone but also to a very great extent by his able recruiting effort he collected a very considerable number of them, and I should say that purely scientific initiatives and contributions came from many people, such like, for instance, von Neumann, Bethe, Segre, to mention a few with whom I am very closely connected, and very many others, and I cannot begin to make a complete list of them.
Q. Coming back to a previous question, Doctor, you say you did leave the laboratory in January 1946?
A. I believe February 1946, but it might be the last days of January. I do not remember so accurately.
Q. Would you tell us whether or not before that happened you had any conversations with Dr. Bradbury and Dr. Oppenheimer about the question of whether you should leave or not?
A. I had several conversations.
Q. Would you tell us about those conversations?
A. Of this kind. I am not at all sure that I can mention them all to you. One was to the best of my recollection in August of 1946, at which time the laboratory was still apparently going at full tilt. Dr. Oppenheimer came to see me in my office.
Q. You said August 1946?
A. August 1945. Thank you very much for catching this mistake.
He had a long conversation with me from which it became clear to me that Dr. Oppenheimer thought that the laboratory would inevitably disintegrate, and that there was not much point in my staying there, at least that is how I understood him. I had been planning to go to Chicago where I was invited to go, and participate in teaching and research work, which I was looking forward to. Then somewhere during the fall of 1945, I believe, Bradbury asked me to take on the job of heading the Theoretical Division.
I was very much interested in seeing the continuation of Los Alamos in a vigorous manner, and in spite of my desire to go back to academic work, I considered this very seriously. I asked Bradbury about the program of the laboratory and in effect I told him—I certainly do not remember my words—that I would stay if 1 or 2 conditions would be met, not both, but one of them. Either if we could continue with the fission program vigorously and as a criterion whether we would do that or not, I said let us see if we could test something like 12 fission weapons per year, or, if instead we would go into a thorough investigation of the thermonuclear question.
Bradbury, I think realistically, said at that time that both of these programs were unfortunately out of the question. I still did not say no. Oppenheimer was going to come and visit the laboratory shortly after, and I wanted to discuss it with him.
I asked him or I told him that Bradbury had invited me, and asked him whether I should stay. Oppenheimer said that I should stay and he also mentioned that he knows that General Groves is quite anxious that I should. Then I mentioned to him the discussion with Bradbury. I said something to this effect. This has been your laboratory. This is your laboratory. It will not prosper unless you support it, and I don't want to stay here if the laboratory won't prosper.
Q. If what?
A. If the laboratory will not prosper. I think I said, I know that there can be no hard and fast program now, but I would like to know whether I can count on your help in getting a vigorous program somewhere along the lines I mentioned established here.

Again I am sorry I cannot quote any literal reply by Oppenheimer, but my recollection of his reply was that it meant that he is neither able nor willing to help in an undertaking of this kind. I thereupon said that under these conditions I think I better leave the laboratory.

Oppenheimer's statement was that he thought that this was really the right decision, and by leaving the laboratory at that time, I could be of greater service to the atomic energy enterprise at a later period.

I remember having seen Oppenheimer the same evening at some party. I forget in whose house it was. He asked me then whether having made up my mind, I don't feel better, and I still remember that I told him that I didn't feel better. But that was where the matter rested at that time.

I think this tied in more or less with my general impression that Oppenheimer felt at least for 1 year after the laboratory that Los Alamos cannot and probably should not continue, and it is just as wise and correct to abandon it.

I am exceedingly glad that due to the very determined action of Bradbury, who was not deterred by any prophecies of this kind, the laboratory was not abandoned, because I am sure had that been done, we would be now in a much worse position in our armament race than we happen to be.

Q. Do you recall any remark by anybody to the effect that the laboratory should be given back to the Indians?

A. I heard this statement attributed to Oppenheimer. I do not remember that he ever said so to my hearing.

Q. Thereafter, you did in fact leave Los Alamos, Doctor?

A. I left Los Alamos, but I did go back very frequently as a consultant.

Q. Where did you go from Los Alamos?

A. To the University of Chicago.

Q. When you went back as a consultant what was the particular problem you were working on?

A. Actually I have been working on quite a number of problems as required. I, of course, continue to be very much interested in the thermonuclear development, and I did continue to work on it, as it were, part time. This, however, at that time was a very minor portion of the enterprise of the laboratory. I would say that on the average between 1945 and 1949—I don't know—a very few people worked on it steadily. I would not be able to say whether this number was 3 or 4 or 5 or 6 out of a thousand or more than a thousand in the laboratory. But this was the order of magnitude, and therefore popularly expressing and crudely expressing the state of affairs, in spite of my working there and in spite of some reports being issued, I can say that the work was virtually at a standstill.

Those were also the years when after some initial hesitation, the testing program was resumed. I understand that this resumption of the testing program was encouraged by the General Advisory Committee on which Oppenheimer was the Chairman. I was also a little bit involved in planning the first extensive test after the war. I don't mean now the Bikini test, but the following one, which I think was called Sandstone. So I would like to say that even the fraction of the time which was considerably less than one-half, which was one-third, it perhaps was not even as much as one-third, I was spending at Los Alamos. Perhaps one-third of my time went into Atomic Energy Commission work, and this was divided between thermonuclear work and other supporting work for Los Alamos, and work on an appointment which I got on the recommendation, I believe, of the General Advisory Committee, on the safety of reactors.

So I would say that of my own time a really small fraction has gone into thermonuclear development during those years and that altogether the effort was very, very slow, indeed.

Q. You were familiar with the effort that was being put in at Los Alamos in respect of thermonuclear?

A. I was.

Q. Doctor, let me ask you for your opinion as an expert on this question. Suppose you had gone to work on thermonuclear in 1945 or 1946—really gone to work on it—can you give us any opinion as to when in your view you might have achieved that weapon and would you explain your opinion?

A. I actually did go to work on it with considerable determination after the Russian bomb was dropped. This was done in a laboratory which at that time was considerably behind Los Alamos at the end of the war. It is my belief that if at the end of the war some people like Dr. Oppenheimer would have lent moral support, not even their own work—just moral support—to work on the thermonuclear gadget, I think we could have kept at least as many people in Los Alamos as we then recruited in 1949 under very difficult conditions.

I therefore believe that if we had gone to work in 1945, we could have achieved the thermonuclear bomb just about 4 years earlier. This of course is very much a matter of opinion because what would have happened if things had been different is certainly not something that one can ever produce by any experiment.

Q. That is right.

A. I think that statements about the possible different course of the past are not more justified but only less hazardous than statements about the future.

Q. Doctor, it has been suggested here that the ultimate success on the thermonuclear was the result of a brilliant discovery or invention by you, and that might or might not have taken 5 or 10 years. What can you say about that?

A. I can say about it this. If I want to walk from here to that corner of the room, and you ask me how long it takes to get there, it depends all on what speed I am walking with and in what direction. If I start in that direction I will never get there, probably. It so happened that very few people gave any serious thought in this country to the development of the thermonuclear bomb. This was due to the fact that during the war we were much too busy with things that had to be done immediately in order that it should be effective during the war, and therefore not much time was left over.

After the war the people who stayed in Los Alamos, few and discouraged as they were, had their hands full in keeping the laboratory alive, keeping up even the knowledge of how to work on the simple fission weapons. The rest of the scientists were, I think, equally much too busy trying to be very sure not to get into an armament race, and arguing why to continue the direction in which we had been going due to the war would be completely wrong. I think that it was neither a great achievement nor a brilliant one. It just had to be done. I must say it was not completely easy. There were some pitfalls. But I do believe that if the original plan in Los Alamos, namely, that the laboratory with such excellent people like Fermi and Bethe and others, would have gone after the problem, probably some of these people would have had either the same brilliant idea or another one much sooner.

In that case I think we would have had the bomb in 1947. I do not believe that it was a particularly difficult thing as scientific discoveries go. I do not think that we should now feel that we have a safety as compared to the Russians, and think it was just necessary that somebody should be looking and looking with some intensity and some conviction that there is also something there.

Q. Is this a fair summary——

A. May I perhaps say that this again is an attempt at appreciating or evaluating a situation, and I may be of course quite wrong, because this is clearly not a matter of fact but a matter of opinion.

Q. Is this a fair summary of your opinion, Doctor, that if you don't seek, you don't find?

A. Certainly.

Q. Do you recall when the Russians exploded their first bomb in September 1949? Do you recall that event?

A. Certainly.

Q. Will you tell the board whether or not shortly thereafter you had a conversation with Dr. Oppenheimer about the thermonuclear or about what activity should be undertaken to meet the Russian advance?

A. I remember two such conversations. One was in the fall and necessarily superficial. That was just a very few hours after I heard, returning from a trip abroad, that the Russians had exploded an A-bomb. I called up Oppenheimer who happened to be in Washington, as I was at that time, and I asked him for advice, and this time I remember his advice literally. It was, "Keep your shirt on."

Perhaps I might mention that my mind did not immediately turn in the direction of working on the thermonuclear bomb. I had by that time quite thoroughly accepted the idea that with the reduced personnel it was much too difficult an undertaking. I perhaps should mention, and I think it will clear the picture, that a few months before the Russian explosion I agreed to rejoin Los Alamos for the period of 1 year on leave of absence from the University of Chicago.

I should also mention that prior to that Oppenheimer had talked to me and encouraged me to go back to Los Alamos, and help in the work there. I also went back to Los Alamos with the understanding and with the expectation that

I shall just help along in their normal program in which some very incipient phases of the thermonuclear work was included, but nothing on a very serious scale.

I was quite prepared to contribute mostly in the direction of the fission weapons. At the time when I returned from this short trip abroad, and was very much disturbed about the Russian bomb, I was looking around for ways in which we could more successfully speed up our work and only after several weeks of discussion did I come to the conclusion that no matter what.the odds seemed to be, we must at this time—I at least must at this time put my full attention to the thermonuclear program.

I also felt that this was much too big an undertaking and I was just very scared of it. I was looking around for some of the old crew to come out and participate in this work. Actually if anyone wanted to head this enterprise, one of the people whom I went to visit, in fact the only one where I had very strong hopes, was Hans Bethe.

Q. About when was this, Doctor?
A. To the best of my recollection it was the end of October.
Q. 1949?
A. Right. Again I am not absolutely certain of my dates, but that is the best of my memory. I can tie it down a little bit better with respect to other dates. It was a short time before the GAC meeting in which that committee made a decision against the thermonuclear program.

After a somewhat strenuous discussion, Bethe, to the best of my understanding, decided that he would come to Los Alamos and help us. During this discussion, Oppenheimer called up and invited Bethe and me to come and discuss this matter wih him in Princeton. This we did do, and visited Oppenheimer in his office.

When we arrived, I remember that Oppenheimer showed us a letter on his desk which he said he had just received. This letter was from Conant. I do not know whether he showed us the whole letter or whether he showed us a short section of it, or whether he only read to us a short section. Whichever it was, and I cannot say which it was, one phrase of Conant's sticks in my mind, and that phrase was "over my dead body," referring to a decision to go ahead with a crash program on the thermonuclear bomb.

Apart from showing us this letter, or reading it to us, whichever it was, Oppenheimer to the best of my recollection did not argue against any crash program. We did talk for quite awhile and could not possibly reproduce the whole argument but at least one important trend in this discussion—and I do not know how relevant this is—was that Oppenheimer argued that some phases of exaggerated secrecy in connection with the A-bomb was perhaps not to the best interests of the country, and that if he undertook the thermonuclear development, this should be done right from the first and should be done more openly.

I remember that Bethe reacted to that quite violently, because he thought that if we proceeded with thermonuclear development, then both—not only our methods of work—but even the fact that we were working and if possible the results of our work should be most definitely kept from any public knowledge or any public announcement.

To the best of my recollection, no agreement came out of this, but when Bethe and I left Oppenheimer's office, Bethe was still intending to come to Los Alamos. Actually, I had been under the impression that Oppenheimer is opposed to the thermonuclear bomb or to a development of the thermonuclear bomb, and I don't think there was terribly much direct evidence to base this impression on. I am pretty sure that I expressed to Bethe the worry, we are going to talk with Oppenheimer now, and after that you will not come. When we left the office, Bethe turned to me and smiled and he said, "You see, you can be quite satisfied. I am still coming."

I do not know whether Bethe has talked again with Oppenheimer about that or not. I have some sort of a general understanding that he did not, but I am not at all sure that this is true.

Two days later I called up Bethe in New York, and he was in New York at that time, and Bethe then said that he thought it over, and he had changed his mind, and he was not coming.

I regretted this very much, and Bethe actually did not join work on the thermonuclear development until quite late in the game, essentially to put on the finishing touches.

I do not know whether this sufficiently answers your question.

Q. Yes, sir. Then, Doctor, the record here shows that on October 29 and 30, 1949, the GAC held its meeting, and thereafter reported its views on the thermonuclear program. Did you later see a copy of the report of the GAC?
A. I did.
Q. Would you tell us the circumstances under which you saw that?
A. Immediately following the meeting, the decision of the General Advisory Committee was kept very strictly confidential. I have seen at least one member of the committee namely, Fermi, who in spite of our very close relationships and the general support of my work in Los Alamos and his knowledge of my almost desperate interest in the undertaking, said that for the time being he just could not even give me an indication of what is happening except from the general tenor of his remarks it was clear that whatever decisions were reached were not terribly favorable to a crash program.

I sort of understood that some kind of action or discussion was under way which can proceed properly only if it is kept in the very smallest circles. This also, of course, became known in Los Alamos, and caused quite a bit of worry there.

After passage of a little while—and I do not know how much time, but I would say roughly 2 weeks—the secretary of the General Advisory Committee, Dr. Manley, who also was associate director in Los Alamos, returned to Los Alamos. He called me into his office and showed me both the majority and minority report of the General Advisory Committee, and in showing me these reports, he used words which I at least at that time interpreted as meaning that Oppenheimer wanted me to see these reports, which I thought was kind. My general understanding was that these reports were also shown to something like half a dozen or dozen of the senior people in the laboratory.

At any rate, the contents of the report were known without my telling it to people. It was just public knowledge among the senior people practically then and there. Of course I was just most dreadfully disappointed about the contents of the majority and minority reports, which in my eyes did not differ a great deal.

I also should say that in my opinion the work in Los Alamos was going to be most seriously affected by the action of the General Advisory Committee, not only as an official body, but because of the very great prestige of the people who were sitting on it. Therefore, it seemed to me at that time, and it also seems to me now entirely proper that this document should have been made available in Los Alamos.

Q. Doctor, in what way did you think that the work would be affected by the report?
A. I would say that when I saw the report, I thought that this definitely was the end of any thermonuclear effort in Los Alamos. Actually I was completely mistaken. The report produced precisely the opposite effect.
Q. Why?
A. Immediately, of course, it stopped work because we were instructed not to work, but it gave people in Los Alamos much greater eagerness to proceed in this direction and from discussions I had in Los Alamos in the following days, I gathered the following psychological reaction:

First of all, people were interested in going on with the thermonuclear device because during the war it had been generally understood that this was one of the things that the laboratory was to find out at some time or other. It was a sort of promise in all of our minds.

Another thing was that the people there were a little bit tired—at least many, particularly of the younger ones—of going ahead with minor improvements and wanted to in sort of an adventurous spirit go into a new field. However, I think the strongest point and the one which was a reaction to this report was this: Not only to me, but to very many others who said this to me spontaneously, the report meant this. As long as you people go ahead and make minor improvements and work very hard and diligently at it, you are doing a fine job, but if you succeed in making a really great piece of progress, then you are doing something that is immoral. This kind of statement stated so bluntly was not of course made in the report. But this kind of an implication is something which I think a human being can support in an abstract sense. But if it refers to his own work, then I think almost anybody would become indignant, and this is what happened in Los Alamos, and the result was that I think the feelings of people in consequence of this report turned more toward the thermonuclear development than away from it.

Q. You means it made them mad.
A. Yes.

Q. Doctor, in the absence of the President's decision of January, would that anger have been effective?
A. No.
Q. Let us go back for a moment——
A. There is no doubt about it. The laboratory just could not put aside a major fraction of its effort on a program of this kind unless we were going to be instructed to do it. Actually, I am pretty sure the anger in a way would have been effective in that more people would have been willing to put aside a little part of their time and worry about it and think about it, and so perhaps it would have been a little effective. But I think that still would have been a very slow and painful progress and probably even now we would be just nowhere.
Q. Dr. Manley has submitted an affidavit here to the effect that he showed you those reports as a result of an impending visit to Los Alamos by Chairman McMahon, Chairman of the Joint Congressional Committee on Atomic Energy. Would you comment on that, and tell us just what it was that Dr. Manley said that gave you the impression that it was Dr. Oppenheimer who wanted you to see the report and tell us whether or not Dr. Manley's remarks were susceptible of the interpretation that it was Chairman McMahon who wanted you to see them?
A. I must say this is possible. To the best of my recollection, I was even struck at that time by these words—Manley said something of that kind, that our Chairman, or the Chairman, I don't know which, sends his regards and wants you to see this. Now, this is to the best of my recollection, and I don't remember that Oppenheimer's name was mentioned. At that time I interpreted this as meaning that it was the Chairman of the General Advisory Committee—that is Oppenheimer. I am quite sure that Manley did not say explicitly that it was McMahon, and to refer to him as simply Chairman would seem to me to be a little remarkable. However, Manley has been showing this document to quite a few people, and perhaps in repeating the phrase a few times parts of the phrase got dropped off. I interpreted it at that time as meaning that Oppenheimer wanted me to see the document. I think it is not excluded that it was Senator McMahon who wanted me to see the document; and if Manley says this, then it must be so.
Q. Did you know Senator McMahon?
A. Yes.
Q. Let me ask you whether or not in that conversation with Manley he mentioned Senator McMahon by name.
A. To the best of my memory, no. I do remember that Senator McMahon came out shortly afterward. I believe I heard about his visit only later, but I might be mistaken.
Q. On the subject of Senator McMahon, will you tell the board whether or not you had proposed to see Senator McMahon about the thermonuclear matter?
A. I did.
Q. When was that?
A. This was quite shortly after the meeting of the General Advisory Committee.
Q. Did you see him?
A. I did.
Q. Did you have any conversation with Dr. Manley before you saw him?
A. I did.
Q. Tell us about that.
A. I had two conversations with him; the one which I think is more relevant, and which certainly strikes more clearly in my mind, was a telephone conversation. This was after the meeting of the General Advisory Committee. I was on my way from Los Alamos to Washington. The main purpose of my visit was to see Senator McMahon. On the way I stopped in Chicago and saw Fermi in his office. It was at that time that I got the impression which I mentioned to you earlier. During my conversation with Fermi, Manley called and asked me not to see Senator McMahon. I asked why. He said that it would be a good idea if the scientists presented a united front—I don't know whether he used that word—I think what he really said was something of this kind, that it would be unfortunate if Senator McMahon would get the impression that there is a divided opinion among the scientists, or something of that kind. I said I had an appointment with Senator McMahon and I wanted to see him. Manley insisted that I should not. Thereupon I made the suggestion that I would be willing to call up Senator McMahon and tell him that I had been asked not to see him, and for that reason I would not see him.

At that point Manley—I don't knew whether I said to Manley that I had been asked by him or whether I would just say I had been asked—and thereupon Manley said, "All right; you better go and see him." That was essentially the contents of my discussion with Manley over the phone.

When I arrived in Washington, Manley met me at the station. I had already the feeling from the discussion with Fermi that at least Fermi's private feelings were not for a crash program. I knew what was in the wind, but I did not know what the decision was. Manley had originally in Los Alamos agreed that we should proceed with the thermonuclear weapon. At least, that was my clear understanding.

He received me on the station with these words, "I think you sold me a gold brick." I remember this particularly clearly, because my familiarity with the English language not being excellent, I did not know what he meant, and I had to ask him what a gold brick is, which he proceeded to explain.

Q. What did he explain, Doctor?

A. A brick covered with gold fill which is not as valuable as it looks.

Q. What did you understand him to refer to?

A. To the thermonuclear program, which, in my opinion, was what we should do, what would be the effective way for us to behave in that situation. Manley implied that in the discussions of the General Advisory Committee another proposal emerged, which was much better, much more hopeful, a better answer to the Russian proposals—excuse me, to the Russian developments—he, however, would not tell me what it was. I was a little mystified. I then went to see Senator McMahon. He did not tell me what was in the report of the General Advisory Committee, but he used some very strong words in connection with it, and did so before I had opened my mouth, words to the effect, "I got this report, and it just makes me sick," or something of that kind.

I did then say that I hoped very much that there would be some way of proceeding with the thermonuclear work, and Senator McMahon very definitely said that he will do everything in his power that it should become possible.

Q. What was your purpose in seeing Senator McMahon?

A. May I say very frankly I do not remember. One of my purposes, I am quite sure, was a point not connected with the thermonuclear development. It was this, that at some earlier time—I am not sure whether it was a year or earlier or when—Senator McMahon was in Los Alamos at the time when I was visiting there. I had an opportunity to talk to him. Senator McMahon asked me to talk with him, and he asked me what I thought would be the best method to increase effectiveness of Los Alamos. I made a few general remarks at that time, which I do not recall, but I remember very clearly that Senator McMahon asked me a question, which I answered, and the answer to which question I regretted later. It was whether the salary scale in Los Alamos was adequate.

Later, when I got a little bit closer back and talked with people, I felt that I had given the wrong answer and I wanted to correct this, and therefore I wanted to see Senator McMahon. However, by the time I actually went to see him, the thermonuclear discussion had gone, as I have indicated, to a point where it was perfectly clear to me that I wanted to talk with him about that question and certainly even by the time I left Los Alamos and before Manley's telephone conversation, I fully hoped to discuss this matter with him because by that time it was quite clear to me that this was one of the very important things that was going on in Los Alamos. This is to the best of my recollection. But I am not at all sure. It may even be possible that I had seen Senator McMahon about another matter at an earlier time. I believe, however, that all this took place in the same conversation.

Q. In January 1950, the President decided that we should go ahead with the thermonuclear program. Do you recall that?

A. I do.

Q. After that decision was announced, did you go to work on the thermonuclear?

A. I most certainly did.

Q. Was the program accelerated?

A. It was.

Q. What was done in general to accelerate it?

A. A committee was formed which for a strange and irrelevant reason was called a family committee.

Q. Who was on that committee?

A. I was the chairman and there were a number of people representing various divisions in the laboratory, and this committee was in charge of developing some thermonuclear program and within a very short time this committee made a number of proposals directed toward some tests which were to give us information about the behavior of some phenomena which were relevant.

At the same time I exerted all possible effort and influence to persuade people to come to Los Alamos to work on this, particularly serious because theoretical work was very badly needed.

Q. What was done in respect of the number of personnel working on the thermonuclear? Was it increased, and if so, how much?

A. It was greatly increased. As I say prior to that there was at most half a dozen people working on it. I am not able to tell you how many people worked on the thermonuclear program in that period. I would say that very few people worked on it really full time. I am sure I didn't work on it full time although in that time the major portion of my effort was directed toward the thermonuclear work. I believe that Los Alamos has prepared an official estimate in response to a question, and that would be, I think, the best source of how many people worked on the thermonuclear program at that time. I would guess, but as a very pure guess, and I should not be surprised if that document would disprove me, that the number of people working on the thermonuclear program increased then to something like two, three, or four hundred, which still was something like 10, 20, or perhaps a little more percent of the laboratory's effort. Perhaps it was closer to 20 percent. I might easily be mistaken.

Q. At all events it was a very large increase.

A. It was a very large increase. As compared to the previous one it was just between standing still and starting to go.

Q. Did you at or about that time, that is, shortly after the President's decision, have any discussion with Oppenheimer as to whether or not he would assist you?

A. I had two discussions with him, but one was shortly before. I would like to quote it a little. Actually the time when President Truman made the announcement I happened to be in Los Angeles and was planning to stay there, in fact had accepted an appointment at UCLA which I at that time had to postpone at any rate because I saw this in the paper. You see, I was not going to stay in Los Alamos much longer, and the fact that there came this announcement from President Truman just changed my mind. Prior to the announcement, preceding it perhaps by 2 or 3 days, I saw Dr. Oppenheimer at an atomic energy conference concerning another matter, and during this meeting it became clear to me that in Dr. Oppenheimer's opinion a decision was impending and this decision would be a go-ahead decision.

At that time I asked Oppenheimer if this is now the decision, would he then please really help us with this thing and help us to work, recalling the very effective work during the war. Oppenheimer's answer to this was in the negative. This was, however, very clearly before President Truman's decision. However, I also should say that this negative reply gave me the feeling that I should not look to Oppenheimer for help under any circumstances.

A few months later, during the spring, I nevertheless called up Oppenheimer and I asked him not for direct help, but for help in recruiting people, not for his own work but for his support in recruiting people. Dr. Oppenheimer said then, "You know in this matter I am neutral. I would be glad, however, to recommend to you some very good people who are working here at the Institute," and he mentioned a few. I wrote to all of these people and tried to persuade them to come to Los Alamos. None of them came.

Q. Where were those people located?

A. At the Institute of Advanced Study in Princeton.

Q. There has been some testimony here that a scientist named Longmire came down to Los Alamos to assist you with the cooperation of Dr. Oppenheimer. Do you recall whether he came down there before the H-bomb conference or afterward?

A. I should like to say first of all that Dr. Longmire did help in the H-bomb development and helped very effectively indeed. I should say helped in fission work and in the thermonuclear work, and is now one of the strongest members of Los Alamos. He came before all this happened. I remember that I tried to get him on the recommendation of Bethe some time early in 1949. I also remember that a little later in the spring or early in the summer I learned—I think it was in May—that Longmire had declined an invitation to Los Alamos, and I also learned that the salary offered him was some 20 percent less than the salary I had recommended. I thereupon talked with the appropriate people in Los Alamos and got them to make a second offer to Longmire at the original salary level, and after I secured agreement on that I called up Longmire and told him that we can offer him this salary and would he please come. Longmire said

"Yes." He would come. However, he had accepted an invitation in the meantime at the Institute of Advanced Study and he now no longer could change his mind. Thereupon I said, "Well, what about it if I try to get this chance? Come with us anyway for a year. After a year you can go back to the Institute. I will talk to Oppenheimer about this." Longmire said, "If Oppenheimer will agree to this, I will consider coming very seriously."

I thereupon called up Oppenheimer on the phone, and at least I believed I approached him directly, I am not sure, somebody approached him, but I think I did it directly, and I remember on that occasion Dr. Oppenheimer was exceedingly cooperative and did give whatever formal assurances he could give. It was not terribly formal. He gave assurances that after a year if Longmire wanted to come back to the Institute, he would be very welcome, and if he wants to go to Los Alamos, that is a very good idea, and so on, and after this was arranged, Longmire did come.

Q. This was when?
A. This was all, however, before anyone of us dreamed about the Russian explosion. That was in the early summer or late spring of 1949. I should also say that after Longmire got to Los Alamos, he not only worked effectively, but liked it so much that then on his own choice he really just stayed there, and is still there, although in the meantime he also taught for certain periods in Rochester, I believe, or in Cornell.

Q. Except for giving you this list of names that you have told us about of people all of whom refused to come, did Dr. Oppenheimer, after the President's decision in January 1950, assist you in any way in recruiting people on the thermonuclear project?
A. To the best of my knowledge not in the slightest.

Q. After the President's decision of January 1950, did Dr. Oppenheimer do anything so far as you know to assist you in the thermonuclear project?
A. The General Advisory Committee did meet, did consider this matter, and its recommendations were in support of the program. Perhaps I am prejudiced in this matter, but I did not feel that we got from the General Advisory Committee more than passive agreement on the program which we evolved. I should say passive agreement, and I felt the kind of criticism which tended to be perhaps more in the nature of a headache than in the nature of enlightening.

I would like to say that in a later phase there is at least one occurrence where I felt Dr. Oppenheimer's reaction to be different.

Q. Would you tell us about that?
A. I will be very glad to do that. In June of 1951, after our first experimental test, there was a meeting of the General Advisory Committee and Atomic Energy Commission personnel and some consultants in Princeton at the Institute for Advanced Study. The meeting was chaired by Dr. Oppenheimer. Frankly I went to that meeting with very considerable misgivings, because I expected that the General Advisory Committee, and particularly Dr. Oppenheimer, would further oppose the development. By that time we had evolved something which amounted to a new approach, and after listening to the evidence of both the test and the theoretical investigations on that new approach, Dr. Oppenheimer warmly supported this new approach, and I understand that he made a statement to the effect that if anything of this kind had been suggested right away he never would have opposed it.

Q. With that exception, did you have any indication from Dr. Oppenheimer after January 1950 that he was supporting and approving the work that was being done on the thermonuclear?
A. My general impression was precisely in the opposite direction. However, I should like to say that my contacts with Oppenheimer were infrequent, and he might have supported the thermonuclear effort without my knowing it.

Q. When was the feasibility of the thermonuclear demonstrated?
A. I believe that this can be stated accurately. On November 1, 1952. Although since it was on the other side of the date line, I am not quite sure whether it was November 1 our time or their time.

Q. What?
A. I don't know whether it was November 1 Eniwetok time or Berkeley time. I watched it in Berkeley.

Q. Did you have a conversation with Dr. Oppenheimer in the summer of 1950 about your work on the thermonuclear?
A. To the best of my recollection he visited Los Alamos in the summer of 1950 and then in the early fall the General Advisory Committee met in Los Alamos—I mean he visited in Los Alamos early in the summer, and then they

met in Los Alamos sometime, I believe, in September, and on both occasions we did talk.

Q. What did Dr. Oppenheimer have to say, if anything, about the thermonuclear?

A. To the best of my recollection he did not have any very definite or concrete advice. Whatever he had tended in the direction that we should proceed with the theoretical investigations, which at that time did not look terribly encouraging, before spending more money or effort on the experimental approach, which I think was at that time not the right advice, because only by pursuing the experimental approach, the test approach, as well as the theoretical one did we face the problem sufficiently concretely so as to find a more correct solution. But I also should like to say that the opinion of Dr. Oppenheimer given at that time to my hearing was not a very decisive or not a very strongly advocated opinion, and I considered it not helpful, but also not as anything that need worry us too much.

I must say this, that the influence of the General Advisory Committee at that time was to the best of my understanding in the direction of go slow, explore all, completely all the designs before looking into new designs, do not spend too much on test programs, all of which advice I consider as somewhat in the nature of serving as a brake rather than encouragement.

Q. Doctor, I would like to ask for your expert opinion again.

In your opinion, if Dr. Oppenheimer should go fishing for the rest of his life, what would be the effect upon the atomic energy and the thermonuclear programs?

A. You mean from now on?

Q. Yes, sir.

A. May I say this depends entirely on the question of whether his work would be similar to the one during the war or similar to the one after the war.

Q. Assume that it was similar to the work after the war.

A. In that case I should like to say two things. One is that after the war Dr. Oppenheimer served on committees rather than actually participating in the work. I am afraid this might not be a correct evaluation of the work of committees in general, but within the AEC, I should say that committees could go fishing without affecting the work of these who are actively engaged in the work.

In particular, however, the general recommendations that I know have come from Oppenheimer were more frequently, and I mean not only and not even particularly the thermonuclear case, but other cases, more frequently a hindrance than a help, and therefore, if I look into the continuation of this and assume that it will come in the same way, I think that further work of Dr. Oppenheimer on committees would not be helpful.

What were some of the other recommendations to which you referred?

A. You want me to give a reasonably complete list? I would be glad to.

Q. Yes.

A. And not distinguish between things I know of my own knowledge and things I know from hearsay evidence?

Q. Yes.

Mr. ROBB. May I go off the record just a moment?

(Discussion off the record.)

Mr. GRAY. We will take a short recess.

(The last question and answer preceding the recess were read by the reporter.)

By Mr. ROBB:

Q. Doctor, in giving your answer, I wish you would give the board both those items that you know of your own knowledge and the others, but I wish you would identify them as being either of your own knowledge or on hearsay.

A. Actually, most of them are on some sort of hearsay. I would like to include not only those things which have occurred in committee but also others.

I furthermore felt that I should like at least to make an attempt to give some impression of the cases in which Dr. Oppenheimer's advice was helpful. His first major action after the war was what I understand both from some part of personal experience and to some extent of hearsay, as I have described, his discussions which led at least to some discouragement in the continuation of Los Alamos. I think that it would have been much better if this had not happened.

Secondly, Oppenheimer published shortly after in connection with the Acheson-Lilienthal report a proposal or supported a proposal, I do not know which, which was based on his scientific authority to share denatured plutonium with others with whom we might agree on international control. I believed at that time and so did many others that denaturing plutonium is not an adequate safeguard.

One of the first actions of the General Advisory Committee—this is hearsay——

Q. Excuse me, doctor. Have you finished your discussion of the other matter?

A. I intended to have it finished but I will be glad to stop and answer questions.

Q. Let me ask a question in that connection as to whether or not Dr. Oppenheimer either at that time or subsequently recommended some inspection of the Russian atomic plants.

A. My understanding is that inspection was an integral part of the Acheson-Lilienthal report, and that, in turn, Dr. Oppenheimer had very actively participated in drafting this report.

I should like to say that in my personal opinion—perhaps I should have said that right away—the Acheson-Lilienthal proposal was a very good one, would have been wonderful had it been accepted, and the inspection to my mind was a very important portion of it. I did not follow these things very closely but I believe it was something with which Dr. Oppenheimer had also agreed or recommended. Which ever the case was, if I am not mistaken in this matter, I really should include that among the very valuable things he did after the war.

Q. Excuse me, and now go ahead.

A. Thanks for bringing up this matter.

One of the first actions of the General Advisory Committee was to advise that reactor work at Oak Ridge should be discontinued and the reactor work should be concentrated at the Argonne Laboratory in Chicago. That was recommended, as I understand, by a great majority.

I also understand that Fermi opposed this recommendation. All this is hearsay evidence but of the kind which I heard so often and so generally that I think it can be classed as general knowledge within AEC circles.

Now, I should like to say that it appeared to many of us at the time, and I think it has been proved by the sequel, that this recommendation was a most unfortunate one. It set our reactor work back by many years. Those exceedingly good workers who left—the great majority of those very good workers who left Oak Ridge—did not find their way into the Argonne Laboratory but discontinued to work on atomic-energy matters or else worked in a smaller group on the side very ineffectively. The very small and determined group which then stayed behind in Oak Ridge turned out in the long run as good work as the people at the Argonne Laboratory, and I feel that again being a little bit uncertain of what would have happened if this recommendation had not been and would not have been accepted, we would be now a couple of years ahead in reactor development. I would like to count this as one of the very great mistakes that have been made.

I understand, having finished with this one, that among the early actions of the General Advisory Committee was, after it was decided that Los Alamos should go on, to recommend strong support for Los Alamos and particularly for the theoretical group. I understand that Oppenheimer supported this and I again think that this was helpful. I have a little personal evidence of it, although it is perhaps somewhat presumptuous of me to say so, that Oppenheimer was active in this direction, for instance, by advising me unambiguously to go back at least for a limited period. I know similarly that in that period he helped us to get Longmire. I also have heard and have heard in a way that I have every reason to believe that in a number of minor but important details in the development of fission weapons, Oppenheimer gave his expert advice effectively, and this included the encouragement of further tests when these things came along.

Q. Tests on what?

A. Tests of atomic bombs, of fission bombs.

Now, the next item is very definitely in the hearsay category, and I might just be quite wrong on it, but I have heard that Dr. Oppenheimer opposed earlier surveillance, the kind of procedures——

Mr. SILVERMAN. I did not understand. Opposed what?

The WITNESS. Earlier surveillance, the sort of thing which was designed to find out whether or not the Russians have detonated an atomic bomb. If this should prove to be correct, I think it was thoroughly wrong advice. Then I think generally the actions of the General Advisory Committee were adverse to the thermonuclear development, but to what extent this is so and why I believe that it is so, we have discussed and I do not need to repeat any of that.

Finally, when, about 3 years ago, the question arose whether this would be a good time to start a new group of people working in a separate laboratory, along similar lines as Los Alamos and competing with Los Alamos, the General Advisory Committee, or the majority of the General Advisory Committee and in particular Dr. Oppenheimer, was opposed to this idea, using again the argument which was used in the case of Oak Ridge, that enough scientific personnel is not available. In this matter I am personally interested, of course, and I was on the opposite side of the argument and I believe that Dr. Oppenheimer's advice was wrong. Of course, it is quite possible that his advice was right and mine was wrong. In the meantime, however, we did succeed in recruiting quite a capable group of people in Livermore. .I think this is essentially the extent of my knowledge, direct or indirect, in the matter. I think it would be proper to restrict my statements to things in close connection with the Atomic Energy Commission and to disregard advice that I heard that Oppenheimer has given to other agencies like the Armed Forces or the State Department. This would be hearsay evidence of a more shaky kind than the rest.

By Mr. ROBB:

Q. Doctor, the second laboratory, is that the one in which you are now working at Livermore?

A. That is one at which I had been working for a year and at which I am now working part time. I am spending about half my time at the University of California in teaching and research and half my time in Livermore.

Q. Did you have any difficulty recruiting personnel for that laboratory?

A. Yes, but not terribly difficult.

Q. Did you get the personnel you needed?

A. This is a question I cannot really answer, because it is always possible to get better personnel. But I am very happy about the people whom we did get and we are still looking for very excellent people if we can get them, and I am going to spend the next 3 days in the Physical Society in trying to persuade additional young people to join us.

Q. Numerically at least, you have your staff; is that right?

A. I would say numerically we certainly have a staff but I do not think this answer to the question is relevant. It is always the question of whether we have the right sort of people and I do believe we have the right sort of people.

Q. Is that laboratory concerned primarily with thermonuclear weapons or is that classified?

A. To the extent that I can believe what I read in Time magazine, it is not classified, but I would like to say that my best authority on the subject is Time magazine.

Q. What does Time magazine say about it?

Mr. SILVERMAN. Well——

Mr. ROBB. I will skip that.

By Mr. ROBB:

Q. I will ask you this, Doctor: Will you tell us whether or not the purpose of establishing a second laboratory was to further work on the thermonuclear?

A. That was a very important part of the purpose.

Mr. ROBB. Mr. Chairman, that completes my direct examination, and it is now 5:30.

Mr. GRAY. I think we had better ask the witness to return tomorrow morning at 9:30.

Mr. GARRISON. Mr. Chairman, we only have one or two questions.

The WITNESS. I would be very glad to stay for a short time.

Mr. GRAY. I have some questions, but I do not think it will take too long, and if you only have a few——

Mr. SILVERMAN. We have so very few, I am almost tempted not to ask them.

CROSS-EXAMINATION

By Mr. SILVERMAN:

Q. You were just testifying about the Livermore Laboratory?

A. Right.

Q. Did Mr. Oppenheimer oppose the Livermore Laboratory as it was finally set up?

A. No. To the best of my knowledge, no.

Q. His opposition was to another Los Alamos?

A. It was to another Los Alamos, and when the Atomic Energy Commission, I think, on the advice from the military did proceed in the direction, the General Advisory Committee encouraged in particular setting up a laboratory at the site

where it was set up. But prior to that, I understand that the General Advisory Committee advised against it.
Q. That is when there was a question of another Los Alamos?
A. Right.
Q. Dr. Teller, when was Livermore set up in its present form?
A. This is something which is more difficult——
Q. You think that is classified?
A. No. It is more difficult to answer than the question of when a baby is born because it is not born all at once. I think the contracts were signed with the Atomic Energy Commission sometime in July 1952. There was a letter of intent sent out earlier and the work had started a little before that. Actually, we moved to Livermore on the 2d of September 1952 and work before that was done in Berkeley.
Q. Do you now have on your staff at Livermore some people who had been or who are members of the Institute for Advanced Study? I am thinking particularly of Dr. Karplus.
A. The answer is no. Dr. Karplus has been consulting with us for a period. He had accepted an invitation to the University of California and he is maintaining his consultant status to the Radiation Laboratory in general, of which Livermore is a part. I believe, but this is again a prediction about the future and my expectation, that Dr. Karplus in the future will help us in Livermore by consulting, but I also believe that for the next couple of years, if I can predict his general plans at all and I talked a bit with him, this is likely not to be terribly much because he will have to adjust himself to the new surroundings first.
Q. Do you know whether Dr. Oppenheimer recommended that Dr. Karplus go to work at Livermore?
A. I have no knowledge whatsoever about it. It is quite possible that he did.
Mr. SILVERMAN. I have no further questions.
Mr. GRAY. Dr. Teller, I think earlier in your testimony you stated that in August 1945, Dr. Oppenheimer talked with you and indicated his feeling that Los Alamos would inevitably disintegrate. I believe those were your words, and that there was no point in your staying on there. Is my recollection correct?
The WITNESS. Yes. I am not sure that my statement was very fortunate, but I am pretty sure that this is how I said it.
Mr. GRAY. Would you say that his attitude at that time was that it should disintegrate?
The WITNESS. I would like to elaborate on that for a moment. I think that I ought to say this: I do not like to say it. Oppenheimer and I did not always agree in Los Alamos, and I believe that it is quite possible, probably, that this was my fault. This particular discussion was connected with an impression I got that Oppenheimer wanted me particularly to leave, which at first I interpreted as his being dissatisfied with the attitude I was taking about certain questions as to how to proceed in detail. It became clear to me during the conversation—and, incidentally, it was something which was quite new to me because prior to that, while we did disagree quite frequently, Oppenheimer always urged no matter how much we disagreed in detail I should certainly stay and work. He urged me although on some occasions I was discouraged and I wanted to leave. On this occasion, he advised me to leave. I considered that at first as essentially personal matters. In the course of the conversaiton, it became clear to me that what he really meant at that time—I asked him—we disagreed on a similar thing and I forget the thing, but I do remember asking him in a similar discussion that, 3 months ago—"You told me by all means I should stay. Now you tell me I should leave." He said, "Yes," but in the meantime we had developed these bombs and the work looks different and I think all of us would have to go home—something to that effect. It was at that time that I had the first idea that Oppenheimer himself wanted to discontinue his work very rapidly and very promptly a Los Alamos. I knew that changes were due but it did not occur to me prior to that conversation that they were due quite that rapidly and would affect our immediate plans just right then and there. I do not know whether I have made myself sufficiently clear or not.

I failed to mention this personnel element before. I am sorry about that. I think it is perhaps relevant as a background.
Mr. GRAY. Do you think that Dr. Bradbury has been an effective director of the Las Alamos Laboratory?
The WITNESS. I am quite sure of that.
Mr. GRAY. It is my impression that he was selected by Dr. Oppenheimer. Do you know about that?

The WITNESS. I heard that statement. I also heard the statement that it was General Groves who recommended Bradbury. I have not the least information upon which to decide which of these statements or whether any of these statements are correct. Perhaps both of them are correct.

Mr. GRAY. It could be. Were you aware of the presence of any scientists on the project following the January 1950 decision who were there for the purpose of proving that this development was not possible rather than proving that it was possible?

The WITNESS. I certainly would not put it that way. There have been a few who believed that it was not possible, who argued strongly and occasionally passionately for it. I do not know of any case where I have reason to suspect intellectual dishonesty.

Mr. GRAY. Excuse me, Dr. Teller. I would like the record to show that it was not my intention to impute intellectual dishonesty to anybody, but you have no knowledge of this.

The WITNESS. I would like to say that on some visits when Bethe came there, he looked the program over someway critically and quite frankly he said he wished the thing would not work. But also he looked it over carefully and whatever he said we surely agreed. In fact, we always agreed.

Mr. GRAY. Yes, I think that clears it up perhaps.

You talked with Dr. Fermi soon after the October 1949 meeting of the GAC, and whereas he was not at liberty to tell you what the GAC decided, you got the impression that they were not favorable to a crash program, as you put it.

The WITNESS. Actually, Dr. Fermi gave me his own opinion, and this was an essential agreement with the GAC. This discouraged me, of course. He also gave me the impression that the GAC really decided something else, something essentially different.

Mr. GRAY. You subsequently saw the GAC report?

The WITNESS. I did.

Mr. GRAY. Is my impression correct that the tenor of the report was not altogether only a question of not moving into a crash program but was opposed to the development of the weapon altogether.

The WITNESS. This was my understanding. In fact, that is definitely my recollection.

Mr. GRAY. Now, Dr. Teller, you stated that the GAC report stopped work at Los Alamos. I assume you meant work on thermonuclear devices.

The WITNESS. I said that and may I correct it, please. What I really should have said was prevented the start of work because work really did not get started.

Mr. GRAY. I think that is important because I thought I heard you say that you instructed not to work. What you mean is that you were instructed not to start anything new.

The WITNESS. That is correct. I am sorry if I expressed erroneously.

Mr. GRAY. Was a result of the GAC report that the 6 or 8 or 10 or whatever it was people who were then working, did they stop their work?

The WITNESS. No, certainly not. In fact, there was an increase of people working right then and there, which was in the relatively free community. Not all of this work was directed in this relatively free atmosphere. It was evident that some work would continue. It was quite clear that in the period November–December–January, we did do some work and more than we had done earlier. However, we did not make a jump from, let us say, 6 people to 200, but we made a jump of from 6 people to 12 or 20. I could not tell you which.

Mr. GRAY. Dr. Teller, General Nichols' letter to Dr. Oppenheimer, which I assume you have some familiarity with——

The WITNESS. I read it. That is, I read the New York Times. If that is assumed to be a correct version——

Mr. GRAY. As far as I know, it is correct. There is one sentence which reads as follows:

"It was further reported that you departed from your proper role as an advisor to the Commission by causing the distribution, separately and in private, to top personnel at Los Alamos of the majority and minority reports of the General Advisory Committee on development of the hydrogen bomb for the purpose of trying to turn such top personnel against the development of the hydrogen bomb."

If this conversation you had with Dr. Manley about which you have testified and in which he referred to our chairman or the chairman was the source of this report, am I right in assuming that your testimony is that you are not prepared to say that Dr. Oppenheimer did cause the distribution of this?

The WITNESS. My testimony says that I cannot ascertain that Dr. Oppenheimer caused distribution. I have presented in this matter all that I can remember.

Mr. GRAY. Dr. Teller, you are familiar with the question which this board is called upon to answer, I assume.

The WITNESS. Yes, I believe so.

Mr. GRAY. Let me tell you what it is and invite counsel to help me out if I misstate it. We are asked to make a finding in the alternative, that it will or will not endanger the common defense and security to grant security clearance to Dr. Oppenheimer.

I believe you testified earlier when Rr. Robb was putting questions to you that because of your knowledge of the whole situation and by reason of many factors about which you have testified in very considerable detail, you would feel safer if the security of the country were in other hands.

The WITNESS. Right.

Mr. GRAY. That is substantially what you said?

The WITNESS. Yes.

Mr. GRAY. I think you have explained why you feel that way. I would then like to ask you this question: Do you feel that it would endanger the common defense and security to grant clearance to Dr. Oppenheimer?

The WITNESS. I believe, and that is merely a question of belief and there is no expertness, no real information behind it, that Dr. Oppenheimer's character is such that he would not knowingly and willingly do anything that is designed to endanger the safety of this country. To the extent, therefore, that your question is directed toward intent, I would say I do not see any reason to deny clearance.

If it is a question of wisdom and judgment, as demonstrated by actions since 1945, then I would say one would be wiser not to grant clearance. I must say that I am myself a little bit confused on this issue, particularly as it refers to a person of Oppenheimer's prestige and influence. May I limit myself to these comments?

Mr. GRAY. Yes.

The WITNESS. I will be glad to answer more questions about it to you or to counsel.

Mr. GRAY. No, I think that you have answered my question. I have, I think, only one more.

I believe there has been testimony given to this board to the effect—and again I would like the assistance of counsel if I misstate anything—that the important and significant developments in the thermonuclear program since January of 1950 have indeed taken place at Los Alamos and not at Livermore. Am I wrong in stating that?

Mr. ROBB. Somebody said that.

Mr. GRAY. Do you recall?

Mr. SILVERMAN. My recollection is that there was testimony that the important developments in the thermonuclear bomb which have thus far been tested out and which were the subject of the recent tests were developed at Las Alamos. I think that was the testimony.

Mr. GRAY. Will you assume that we have heard something of that sort? Do you have a comment?

The WITNESS. Is there a ruling that I may answer this question in a way without affecting security? I would like to assume that. I think I should.

Mr. ROLANDER. If you have any worry on that point, perhaps the board may wish you to give a classified answer on that.

The WITNESS. I mean I would like to give an unclassified answer to it and if you think it is wrong, strike it later. I understand that has been done before. I would like to make the statement that this testimony is substantially correct. Livermore is a very new laboratory and I think it is doing a very nice job, but published reports about its importance have been grossly and embarrassingly exaggerated.

Dr. EVANS. I have one question.

Dr. Teller, you understand——

The WITNESS. May I leave that in the record? I would like to.

Mr. ROLANDER. Yes.

Dr. EVANS. You understand, of course, that we did not seek the job on this board, do you not?

The WITNESS. You understand, sir, that I did not want to be at this end of the table either.

Dr. EVANS. I want to ask you one question. Do you think the action of a committee like this, no matter what it may be, will be the source of great discussion in the National Academy and among scientific men in general?

The WITNESS. It already is and it certainly will be.
Dr. EVANS. That is all I wanted to say.
Mr. ROBB. May I ask one further question?

RE-DIRECT EXAMINATION

By Mr. ROBB:
Q. Dr. Teller, you did a great deal of work on the thermonuclear at the old laboratory, too, at Los Alamos.
A. Certainly.
Mr. SILVERMAN. I have one question.

RE-CROSS-EXAMINATION

By Mr. SILVERMAN:
Q. I would like you, Dr. Teller, to distinguish between the desirability of this country's or the Government's accepting Dr. Oppenheimer's advice and the danger, if there be any, in Dr. Oppenheimer's having access to restricted data. As to this latter, as to the danger in Dr. Oppenheimer's having access to restricted data without regard to the wisdom of his advice, do you think there is any danger to the national security in his having access to restricted data?
A. In other words, I now am supposed to assume that Dr. Oppenheimer will have access to security information?
Q. Yes.
A. But will refrain from all advice in these matters which is to my mind a very hypothetical quetion indeed. May I answer such a hypothetical question by saying that the very limited knowledge which I have on these matters and which are based on feelings, emotions, and prejudices, I believe there is no danger.
Mr. GRAY. Thank you very much, Doctor.
(Witness excused.)
Mr. GRAY. We will recess until 9: 30 tomorrow.
(Thereupon, the hearing was recessed at 5: 50 p. m., to reconvene at 9: 30 a. m., Thursday, April 29, 1954.)

UNITED STATES ATOMIC ENERGY COMMISSION

PERSONNEL SECURITY BOARD

In the Matter of J. Robert Oppenheimer

Atomic Energy Commission,
Building T-3, Room 2022,
Washington, D. C., Thursday, April 29, 1954.

The above-entitled matter came on for hearing, pursuant to recess, before the board, at 9:30 a. m.

Personnel Security Board: Dr. Gordon Gray, chairman; Dr. Ward T. Evans, member; and Mr. Thomas A. Morgan, member.

Present: Roger Robb, and C. A. Rolander, Jr., counsel for the board; J. Robert Oppenheimer, Lloyd K. Garrison, Samuel J. Silverman, and Allan B. Ecker, counsel for J. Robert Oppenheimer; Herbert S. Marks, cocounsel for J. Robert Oppenheimer.

PROCEEDINGS

Mr. GRAY. Do you wish to testify under oath, Mr. McCloy. You are not required to do so. I think I should say to you the every witness appearing has so testified.

Mr. McCLOY. Yes.

Mr. GRAY. Would you stand and raise your right hand, and give me your full name.

Mr. McCLOY. John J. McCloy.

Mr. GRAY. John J. McCloy, do you swear that the testimony you are to give the Board shall be the truth, the whole truth, and nothing but the truth, so help you God?

Mr. McCLOY. I do.

Whereupon, John J. McCloy was called as a witness, and having been first duly sworn, was examined and testified as follows:

Mr. GRAY. It is my duty to remind you of the existence of the perjury statutes. May I assume you are familiar with them and their penalties?

The WITNESS. Never personal, but I am familiar with them.

Mr. GRAY. I would like to make one other statement to you in behalf of the board, that is, we treat these proceeding as a confidential matter between the Atomic Energy Commission and its officials on the one hand, and Dr. Oppenheimer and his representatives and witnesses on the other. The Commission is making no releases with respect to these proceedings, and on behalf of the board, I express the hope that witnesses will take the same view.

The WITNESS. I will be glad to do so.

DIRECT EXAMINATION

By Mr. GARRISON:

Q. Mr. McCloy, will you state for the record your present position?

A. I am presently the chairman of the board of the Chase National Bank.

Q. Would you also state for the record the positions that you held in the Defense Establishment during the war?

A. In the summer of 1940 I came to the War Department as a consultant to the Secretary of War, and remained in that position until I became Assistant Secretary of War some months later. I remained as the Assistant Secretary of War throughout the entire period of the war, and I left the War Department in the fall of 1945.

Then I have been on various special committees in connection with the defense. I was on the President's Committee—I forgot the name of it—it was the one upon which Mr. Acheson and General Groves served, dealing with the question of the control of atomic weapons.

I think that completes my defense experience.

Q. Will you tell the board your contacts with the atomic energy program during the war and your acquaintance with Dr. Oppenheimer?

A. During the war I was very closely associated with Mr. Stimson. In the early days of the war, I had many conversations with him in regard to the menace of a possible German development of an atomic weapon. He had been in conversation with the President and had deeply interested himself in this particular matter. Although I was not on any particular committee nor was I in direct charge of any element of the atomic development, as a result of my position with Mr. Stimson as a general consultant with him, he frequently talked to me about the state of the program, character of the threat, and what we should do about it. Generally these conversations took place at the house here in Washington which was called Woodley at the close of the day after the normal routine of the Department was over.

This contact lasted throughout the war and on into the conference at Potsdam, until finally I left the Department, as I did shortly after his departure.

I think I ought to say that I was also in contact with General Groves from time to time. I visited not all the establishments, but some of the establishments which had been erected, and from time to time helped in connection with the priorities and the allocations to insure that the atomic project was given the fullest of priorities and the greatest of cooperation and support so far as the War Department was concerned.

I think that sketches it.

Q. Did you have any occasion to talk with Mr. Stimson or General Groves about Dr. Oppenheimer?

A. Yes; not at the beginning of the war. I did not hear of Dr. Oppenheimer until well toward the end of the war. I can't exactly fix the dates in my mind, but I do recall that some substantial period before we left for the Potsdam conference we learned of the real progress that had been made at Los Alamos, and the name of Dr. Oppenheimer was mentioned in that connection. Somewhere I should say around 1944, or perhaps as early as 1943, I heard the name, but in 1944, and the beginning of 1945, it was a rather prominently mentioned name. Frequently Mr. Stimson referred to the work that Dr. Oppenheimer was doing, and the great possibility that at Los Alamos things were developing which would shortly and within the measurable future produce rather spectacular results.

I may volunteer the information that it was only in respect of it that Dr. Oppenheimer was making. There was no question of security in that regard, althought I do remember General Groves speaking to me from time to time as he sometimes did about his problems, saying that he did have some security preoccupations. I am trying to remember back as best I can. They were, as far as I can remember, confined to a concern that information that the English were getting from our atomic developments might be leaked to the French, where General Groves had real suspicions, particularly because of the association of Dr. Curie with the atomic development in France. He referred somewhat to his security precautions and indicated to me that he had dismissed one or two or a few people from Los Alamos, but never was the question raised in any regard to Dr. Oppenheimer, nor did Mr. Stimson, as I say, have anything but great admiration and praise for the achievements that Dr. Oppenheimer was accomplishing.

Q. Did you come into contact with Dr. Oppenheimer at the time of the Acheson-Lilienthal report?

A. Yes. I would say there were 3 phases of my experience with Dr. Oppenheimer. The first I have already described, which I would say was the Stimson-War Department contact, and that was a very slight personal contact, but I knew him, and I knew his name, and knew what was going on in general.

The second was the Lilienthal-Acheson Committee report, and the third, apart from some intermittent contacts of no consequence, was my association with Dr. Oppenheimer on the so-called Soviet study group, which is a group set up by the Council of Foreign Relations in New York City, which was erected in consultation with the State Department to see what we could do by gathering together a group of knowledgeable people—a rather small group, but well experienced and somewhat distinguished group—that would quietly study this whole problem of our relations with the Soviet, to see if we could do anything that would be of benefit to the Government or to general public opinion in that field.

Dr. Oppenheimer was a member of that group. He was selected primarily because of his outstanding reputation in the atomic field, and since the atomic element was important in the consideration of our relations with the Soviet we felt that we should have someone on the board who was well equipped to advise us in that connection.

Incidentally, in respect to that second phase, I think I probably should say that apart from Dr. Oppenheimer's membership on the panel, I think we called it—a panel which was composed, as well as I remember—you would have the records of it.

Q. I think that is in the record.

A. Winne, Thomas, Barnard, and so forth. Apart from his expositions to the Committee at that time of the technical aspects of the problem, I endeavored to learn a little something about the art so that I would be more familiar and more capable of understanding some of the technical expositions and better equipped to discuss the whole problem. He undertook to tutor me in the art, I don't think with any great success. But that was not his fault.

During the course of that experiment on his part, I got to know him fairly well, and that was just a side comment on the extent of my relationship with him.

Q. Do you know anything at first hand about his attitude toward Russia and the whole problem at that point of time?

A. Growing out of the concern we all had after the successful dropping of the bomb on Hiroshima and Nagasaki, and the great preoccupation that particularly Mr. Stimson, as well as many others, had as to what we were going to do from here out, there were many discussions in Washington and from time to time I was consulted by members of the Government as to what I thought about it.

After Mr. Stimson retired, we talked about it a good bit in his home at Long Island. As you know, this committee was set up and a report was made.

In connection with the committee's action, as I say, Dr. Oppenheimer was a member of the panel and we looked to him for the technical expositions. Our technical questions were mainly directed to him. There was then a very intriguing problem of the possibility of denaturing this material so that it would not have an explosive or at least a lethal weapon effect. It was thought that by a certain process you could denature it or delouse it in such a way so that it would not have the harmful effects that the weapon itself might have. That was gone into at some length and Dr. Oppenheimer explained the limitations and possibilities of that.

I remember at that time, or at least I gathered the impression at that time, that he was quite alert to the interests of the United States in connection with this. The proposals that were made for international control were to be hemmed about by certain provisions which we thought would secure the interests of the United States, as best we could consistent with the overall philosophy of having an international control in effect. I generally am of the impression that Dr. Oppenheimer at that time was as sensitive as I should say any one was in regard to the security interests of the United States.

There were, as I recall it, one or two points of difference in the committee on which I don't believe Dr. Oppenheimer, if he had any knowledge at all, certainly expressed no view. There was a question as to whether we would publish the report. Some members of the committee were in favor of publishing it, and others were opposed to it. I think a vote was taken and we decided not to oppose it, and then somehow or other it did see the light of day, but I never knew how it got out.

Q. Not to oppose, but not to publish.

A. I meant to say not to publish. There were some questions that developed in the Committee as to whether we ought to be a little more rigid than we were with regard to security provisions. In that I remember General Groves differed somewhat with some of the other members of the Commission. But as I recall it, General Groves' position, which was supported by me and others, prevailed. I can't recall any participation by Dr. Oppenheimer in that discussion. I do recall very definitely in responding to questions, it seemed to me that he was very objective in just what we could expect in the way of safety precautions and what we could not.

So I did gain the impression that he was alert to the necessity of protecting in so far as it was possible to protect the interests of the United States, as I say, consistent with the concept of international control.

There is one other contact with Dr. Oppenheimer that I am a little vague about, and I am not absolutely certain that he was present at a meeting that took place well before Potsdam in the War Department in Secretary Stimson's office, where we discussed with the Committee that Mr. Stimson had set up, and with some scientists. I have the impression—I know Van Bush was there—that Dr. Oppenheimer was there, and that was as to whether or not we should drop the bomb and generally where this whole thing was leading, where we were going with it.

I recall either as a result of my presence at that meeting, or Dr. Oppenheimer's presence at that meeting, or from what Mr. Stimson told me, that all of the scientists, I believe, but certainly Dr. Oppenheimer, were in favor, all things considered, of dropping the bomb.

Mr. ROBB. May I interpose and ask which bomb we are talking about?

The WITNESS. I am talking about whether we should drop it on the Japanese.

Mr. ROBB. Yes. We have had so many bombs.

The WITNESS. Yes. I am talking about the first one. At that time we had not even picked the target. There was a good bit of discussion about the target before we left abroad and some further discussion at Potsdam about it.

By Mr. GARRISON:

Q. Coming down to the Soviet study group which you mentioned in the Council of Foreign Relations, you were the presiding officer of that group?

A. Yes, I was the presiding officer.

Q. And Dr. Oppenheimer was a member of the group?
A. Yes.
Q. And who were some of the other members?
A. I don't know that I have a list of the members. I think I can remember them mainly from memory.

Ferdinand Eberstadt was a member. Averill Harriman was a member. Dr. Wriston, president of Brown, is a member. Devereux Josephs, president of the New York Life Insurance Co., was a member. Professor Fainsod of Harvard, who was the head of the Russian studies at Harvard, is a member.

We have observers from the Government there who were not strictly members, but who have asked to sit in and who do sit in. General Lemnitzer is one of them. Mr. Bowie, adviser to the Secretary of State, and professor at Harvard Law School, acts as observer, and Mr. Allen Dulles or his deputy from CIA.

There are other members of the group, but I suppose I better get you a complete list.

Q. Would you just say a word about Dr. Oppenheimer's participation in the group, and particularly the character of the views which he has expressed in his discussions with respect to our relations with Russia?

A. We have adopted a rule in that group not to give any publicity to the views expressed around the table there, and certainly not to attribute anything in respect of a particular individual. But I suppose if I have Dr. Oppenheimer's consent, I can go ahead.

Q. Yes.
A. I feel a certain responsibility as chairman of that group, and being so insistent upon the fact that there should not be attributions and no leaks from that group, I don't like to be the first one to violate it. We selected in the first place, as I have already indicated, Dr. Oppenheimer, because of his knowledge in this field, because of the pronounced importance of this whole subject in regard to our relations with the Soviet. He at one meeting expounded to us at considerable length.

Q. Would you say about what year this was?
A. I suppose that was last fall, I think.
Q. That is near enough.
A. Last fall, yes. He has been a member of the group from the beginning, but he was abroad.
Q. When did the group begin?
A. It began at the beginning of 1953. It has been going for a year, and it will probably go for another year. He was selected at the outset and attended one or two meetings and then he went to lecture abroad so we didn't have him present at a substantial number of meetings. Then he did give us a picture of where he thought we stood generally in relation to the Soviet in respect to atomic development.

Q. Without going into the details of what he said, what impression did his talk leave on you about his general attitude toward the situation?

A. The impression that I gathered from him was one of real concern that although we had a quantitative superiority, that that didn't mean a great deal. * * * We were coming to the point where we might be, he used the graphic expression like two scorpions in a bottle, that each could destroy the other, even though one may have been somewhat larger than the other, and he was very much concerned about the security position of the United States. He pressed vigorously for the continued activity in this field, and not letting down our guard, so to speak. Taking advantage of any opportunity that really presented itself that looked as if it was substantial, but if there was to be any negotiation, be certain that we were armed and well prepared before we went to such a conference. Indeed, I have the impression that he, with one or two others, was somewhat more, shall I say, militant than some of the other members of the group. I think I remember very well that he said, for example, that we would have to contemplate and keep our minds open for all sorts of eventualities in this thing * * *.

In the course of this, I think I should say that he was questioned by the members of the group from time to time. In a number of cases, he refused to reply, saying that he could not reply because in doing so that would involve some security information. His talk was generally in generalities, to some extent following the line that he took in an article which I saw later on published in Foreign Affairs.

I got the very strong impression of Dr. Oppenheimer's sensitivity to what he considered to be the interests of the United States and to the security of the United States.

Q. Based on your acquaintance with Dr. Oppenheimer, and your experiences with him, would you give the board your opinion as to his loyalty and as to his security risk or want of risk?

A. In the first place, just to get it out of the way, let me say that there is nothing that occurred during the entire period of my contact with Dr. Oppenheimer which gave me any reason to feel that he was in any sense disloyal to the United States. But I would want to put it more positively than that, and also add that throughout my contacts with him, I got the impression, as one who has had a good bit of contact and experience with defense matters, that he was very sensitive to all aspects of the security of the United States.

I gathered the impression that he was deeply concerned about the consequences of this awful force that we had released, anxious to do what he could toward seeing that it was not used or did not become a destroyer of civilization. He was somewhat puzzled as to what form that would take and still be consistent with the interests of the United States. That perhaps more than a number of others who were, so to speak, laymen in this field, who were members of that study group, was aware of the techniques of the defense of the United States. He was a little more aware than those who had not been really associated with the Defense Department of the military position of the United States somewhat apart from the atomic situation. So much for loyalty.

I can't be too emphatic as to my impression of Dr. Oppenheimer in this regard. I have the impression of his being a loyal, patriotic citizen, aware of his responsibilities and that I want to accent.

As to his security risk—to use the current phrase—I again can state that negatively certainly. I know of nothing myself which would make me feel that he was a security risk. I don't know just exactly what you mean by a security risk. I know that I am a security risk and I think every individual is a security risk. You can always talk in your sleep. You can always drop a paper that you should not drop, or you can speak to your wife about something, and to that extent no human being is an absolutely secure person. I don't suppose we are talking about that.

I never heard of any of Dr. Oppenheimer's early background until very recently, and so that has never been an element in my thinking. I have only thought of him as being a figure whom I feel I know, and I feel I am somewhat knowledgeable in this field, and one I feel I know is as much responsible as anybody else if perhaps not more than anybody else in this particular field of the weapon for our preeminence in that field. Too many reports came in to us as to the work that he was doing, the difficulties under which he was laboring, and they were difficulties because there had to be very great security precautions and a lot of barbed wire and what not which introduced serious human problems in connection with the plants where he was operating, and the reports all were that in spite of all this, and in spite of the little squabbles that took place among this confined group of scientists, there was a certain inspiration to their work and enthusiasm and a vigor and energy that many ascribed to Dr. Oppenheimer, and which I am quite clear played a major part in bringing about the achievement of the weapon at the critical point, and time that it was achieved.

There is another aspect to this question of security, if I may just go on, that troubles me and I have been thinking about it a good bit since I have read the charges and the reply of Dr. Oppenheimer, and have talked to a number of people who are somewhat familiar with this whole subject. It seems to me that there are two security aspects. One is the negative aspect. How do you gage an individual in terms of his likelihood of being careless with respect to the use of documents or expressions, if he is not animated by something more sinister? There is also for want of a better expression the positive security. I remember very vividly the early days when the warnings that Neils Bohr—I was not in Washington when Neils Bohr first came over, but I saw him from time to time after that—when he announced to us and to the President that the uranium atom had been split, and we might look forward with some concern to the possibility that the Germans would have an atomic weapon, and our eagerness at that time to take on, practically speaking, anyone who had this quality of mind that could reach in back of and beyond, from the layman's point of view, at least, and deal with this concept and reduce it to reality.

As I try to look back to that period, I think we would have taken pretty much anybody who had certainly the combination of those qualities, the theoretical ability, plus the practical sense, to advance our defense position in that field. In those days we were on guard against the Nazis and the Germans. I think we would have grabbed one of them if we thought he had that quality, and surrounded him with as much security precautions as we could. Indeed, I think we would have probably taken a convicted murderer if he had that capacity. There again is this question of the relative character of security. It depends somewhat on the day and age that you are in.

I want to emphasize particularly this affirmative side of it. The names we bandied about at that time included a number of refugees and a number of people that came from Europe. I have the impression—I may be wrong about it—but I have the impression that a very large element of this theoretical thinking did emanate from the minds of those who immigrated from this country, and had not been generated here as far as it had been in Europe. There were names like Fermi and Wigner and Teller, Rabi, another queer name, Szilard, or something like that—but I have the impression they came over here, and probably embued with a certain anti-Nazi fervor which tended to stimulate thinking, and it is that type of mind that we certainly needed then.

We could find, so to speak, practical atomic physicists, and today there are great quantities of them being trained, and whether we are getting this finely balanced imagination which can stretch beyond the practicalities of this thing is to my mind the important aspect of this problem. The art is still in its infancy and we still are in need of great imagination in this field.

In a very real sense, therefore, I think there is a security risk in reverse. If anything is done which would in any way repress or dampen that fervor, that verve, that enthusiasm, or the feeling generally that the place where you can get the greatest opportunity for the expansion of your mind and your experiments in this field is the United States, to that extent the security of the United States is impaired.

In other words, you can't be too conventional about it or you run into a security problem the other way. We are only secure if we have the best brains and the best reach of mind in this field. If the impression is prevalent that scientists as a whole have to work under such great restrictions and perhaps great suspicion, in the United States, we may lose the next step in this field, which I think would be very dangerous for us.

From my own experience in Germany, although they were very backward in this field, and in that respect there is a very interesting instance which I have seen referred to in print——

Mr. GRAY. Mr. McCloy, may I interrupt you for a minute? As a lawyer, you must observe we allow very considerable latitude in these hearings, and we have tried in no way to circumscribe anything that any witness wishes to say, and in fact, almost anything the lawyers wanted to say has gone into the record. You were asked a question, I believe, by Mr. Garrison, about Dr. Oppenheimer's—it has been a long time and I have forgotten.

Mr. GARRISON. Loyalty, and him as a security risk.

Mr. GRAY. Yes. Whereas I think your views are entitled to great weight on these matters generally, I would respectfully and in the most friendly spirit, suggest that we not wander too far afield from this question.

The WITNESS. I didn't mean to wander too far.

Mr. GRAY. Yes, sir.

The WITNESS. I did want to make one point. I have been asked this recently in New York frequently: Do you think that Dr. Oppenheimer is a security risk, and how would I answer that. This is long before I had any idea I was going to be called here. What do you mean by security, positive, negative, there is a security risk both ways in this thing. It is the affirmative security that I believe we must protect here. I would say that even if Dr. Oppenheimer had some connections that were somewhat suspicious or make one fairly uneasy, you have to balance his affirmative aspect against that, before you can finally conclude in your own mind that he is a reasonable security risk, because there is a balance of interest there; that he not only is himself, but that he represents in terms of scientific inquiry—I am very sorry if I rambled on about that and I didn't mean to.

Mr. GRAY. I don't want to cut you off at all, but you were getting back about something of the Nazis during the war.

The WITNESS. Yes. Let me tell you why I did that, if I may.

Mr. ROBB. Mr. Chairman, may I interpose one thought. I think the rules do provide that no witness will be allowed to argue from the witness stand. I think the witness should bear that in mind, if I might suggest it.

The WITNESS. Yes. I don't mean to argue. I am trying honestly to answer the question whether this man is a security risk in my judgment from what I know of him.

Mr. ROBB. I understand.

The WITNESS. Take the case—and perhaps I should not argue and maybe this ought to be off the record.

Mr. ROBB. The rule is quite specific, Mr. Chairman, that is the only reason I bring it up.

Mr. GRAY. Mr. Robb is correct that the regulations by which this proceeding is governed state that no witness shall be allowed to argue.

The WITNESS. I am trying to think out loud rather than argue.

Mr. GRAY. May I ask that you proceed.

The WITNESS. I will come to the point on it. I think I could give a rather vivid example of what I am trying to say, but I won't refer to that. I will say that as far as I have had any acquaintance with Dr. Oppenheimer, I have no doubt as to his loyalty, and I have absolutely no doubt about his value to the United States and I would say he is not a security risk to the United States.

Mr. GARRISON. Thank you.

Mr. GRAY. Do you have any questions, Mr. Robb?

CROSS-EXAMINATION

By Mr. ROBB:

Q. How long have you been president of the Chase National Bank?
A. A little over a year.
Q. Had you previously had experience in the banking business?
A. I was president of the so-called International Bank for Reconstruction and Development, which is known as the World Bank.
Q. Chase is the largest bank in the world?
A. No; it is the third. The Bank of America and National City are larger.
Q. Have you a great many branches?
A. Yes; 28.
Q. As far as you know, Mr. McCloy, do you have any employee of your bank who has been for any considerable period of time on terms of rather intimate and friendly association with thieves and safecrackers?
A. No; I don't know of anyone.
Q. I would like to ask you a few hypothetical questions, if I might, sir.

Suppose you had a branch bank manager, and a friend of his came to him one day and said, "I have some friends and contacts who are thinking about coming to your bank to rob it. I would like to talk to you about maybe leaving the vault open some night so they could do it," and your branch manager rejected the suggestion. Would you expect that branch manager to report the incident?
A. Yes.
Q. If he didn't report it, would you be disturbed about it?
A. Yes.
Q. Let us go a little bit further. Supposing the branch bank manager waited 6 or 8 months to report it, would you be rather concerned about why he had not done it before?
A. Yes.
Q. Suppose when he did report it, he said this friend of mine, a good friend of mine, I am sure he was innocent, and therefore I won't tell you who he is. Would you be concerned about that? Would you urge him to tell you?
A. I would certainly urge him to tell me for the security of the bank.
Q. Now, supposing your branch bank manager, in telling you the story of his conversations with his friend, said, "My friend told me that these people that he knows that want to rob the bank told me that they had a pretty good plan. They had some tear gas and guns and they had a car arranged for the getaway, and had everything all fixed up," would you conclude from that it was a pretty well-defined plot?
A. Yes.
Q. Now, supposing some years later this branch manager told you, "Mr. McCloy, I told you that my friend and his friends had a scheme all set up as I have told you, with tear gas and guns and getaway car, but that was a lot of bunk. It just wasn't true. I told you a false story about my friend." Would you be a bit puzzled as to why he would tell you such a false story about his friend?
A. Yes; I think I would be.

Mr. ROBB. That is all.

Mr. GRAY. Mr. McCloy, for the record, you were speaking about Mr. Stimson's report as to the position of the scientists with respect to the dropping of the first bomb?

The WITNESS. Yes.

Mr. GRAY. As I recall it there was some sort of interruption, and I don't believe the record reflects what you were about to say the position of the scientists was on that matter.

The WITNESS. That they were in favor of dropping the bomb, and that Dr. Oppenheimer was one of those who had been in favor.

Mr. GRAY. Perhaps the interruption was in my own mind.

Mr. ROBB. I think I asked him which bomb, and then he said it was the Japanese bomb, and Dr. Oppenheimer favored the dropping of it.

Mr. GRAY. Yes. Pardon my lapse.

Second, I think the record ought to reflect all the names of the members of this group you were discussing.

The WITNESS. I think I may have it in my brief case if I may look it up. My brief case is in the other room. This is the Council of Foreign Relations that you are referring to?

Mr. GRAY. Yes.

The WITNESS. I certainly can supply you with that.

Mr. GRAY. We will get that from you.

(The list is as follows:)

MEMBERSHIP OF THE STUDY GROUP ON SOVIET-AMERICAN RELATIONS

John J. McCloy, chairman, Chase National Bank.
Frank Altschul, General American Investors Corp.
Hamilton Fish Armstrong, Foreign Affairs, Council on Foreign Relations.
McGeorge Bundy, Harvard University, resigned from group in 1953.
Arthur Dean, Sullivan and Cromwell, joined group, Spring, 1954.
William Diebold, Council on Foreign Relations.
F. Eberstadt, F. Eberstadt & Co., Inc.
Merle Fainsod, Harvard University.
William T. R. Fox, Columbia University.
George S. Franklin, Jr., Council on Foreign Relations.
W. A. Harriman, Former Ambassador to the Soviet Union.
Howard G. Johnson, Ford Foundation.
Devereux C. Josephs, New York Life Insurance Co.
Milton Katz, Ford Foundation.
Mervin J. Kelly, Bell Laboratories.
William L. Langer, Harvard University.
Walter H. Mallory, Council on Foreign Relations.
Philip E. Mosely, Russian Institute, Columbia University.
J. Robert Oppenheimer, Institute for Advanced Study.
Gerold T. Robinson, Columbia University.
Dean Rusk, Rockefeller Foundation.
Charles M. Spofford, Davis, Polk, Wardwell, Sunderland & Kiendl.
Shepard Stone, Ford Foundation.
Jacob Viner, Princeton University, inactive because of ill health.
Henry M. Wriston, Brown University.

GOVERNMENT OBSERVERS

Robert Amory, Jr., Central Intelligence Agency.
Robert R. Bowie, Department of State.
Lyman L. Lemnitzer, General, Department of the Army.

RESEARCH STAFF FOR THE STUDY GROUP ON SOVIET-AMERICAN RELATIONS

Henry L. Roberts, research secretary, Council on Foreign Relations.
Gerhart Niemeyer, formerly with the Department of State.
Marina S. Finkelstein, formerly with the research program on the U. S. S. R.
Perry Laukhuff, formerly with the Department of State (with group for 5 months).
A. David Redding, formerly with the Rand Corp.
Donald Urquidi, former student at the Russian Institute, Columbia.
Paul E. Zinner, formerly at Harvard University.

Mr. GRAY. Have you read the letter of December 23 from General Nichols to Dr. Oppenheimer, and Dr. Oppenheimer's reply perhaps as they appeared in the press?

The WITNESS. Yes, I didn't read them critically, but I know pretty much what is in them, because I read them rather hastily.

Mr. GRAY. Is this the first knowledge you had of the reported associations of Dr. Oppenheimer?

The WITNESS. No. I think I heard somewhere about a year ago, and I can't place where I heard it, that there was some question about Dr. Oppenheimer's early associations, that his brother or wife had been a Communist. It was within a year that I heard it.

Mr. GRAY. Mr. McCloy, following Mr. Robb's hypothetical question, for the moment, let us go further than his assumption. Let us say that ultimately you did get from you branch manager the name of the individual who had approached him with respect to leaving the vault open, and suppose further that your branch manager was sent by you on an inspection trip of some of your foreign branches, and suppose further that you learned that while he was in London he looked up the man who had made the approach to him some years before, would this be a source of concern to you?

The WITNESS. Yes; I think it would. It is certainly something worthy of investigation, yes.

Mr. GRAY. Now, Mr. McCloy, you said in referring to Dr. Oppenheimer that he more than perhaps anybody else is responsible for our preeminence in the field of the weapon. You are referring now to the atomic bomb?

The WITNESS. Yes; the atomic bomb.

Mr. GRAY. Could you make the same statement with respect to the H-bomb?

The WITNESS. I don't know enough about the development of the H-bomb. That occurred after I left the Defense Establishment.

Mr. GRAY. So you are confining your testimony to the development of the atomic bomb.

The WITNESS. Yes; to the development of the atomic bomb.

Mr. GRAY. On the basis of what you know, which specifically includes of course your associations with Dr. Oppenheimer, and on the basis of what you read in the newspapers, would you feel that any further investigation in this matter was necessary at all? Would you be prepared to say that the Atomic Energy Commission should just forget all about it?

The WITNESS. I don't know what I read in the newspapers really. This thing that Mr. Robb questioned me about, I have imagined that relates to some incident in connection with Dr. Oppenheimer's past or has some bearing on it. I am not familiar with that. If that was in the answer and the reply I didn't read it critically. It was about some approach but it didn't stay in my mind. I just read it going downtown in the morning.

No, I would say that anyone in the position of Dr. Oppenheimer with his great knowledge on this subject, the very sensitive information that he has, most of which I guess is in his own brain, if association which was suspicious turned up in connection with him, I think it would be incumbent upon this group or some other group to investigate it. I don't suggest in any way that it should not be investigated or that it can be cast off casually. All I say is that I think you have go to look at the whole picture and the contributing factors of this man, and what he represents, before you determine the ultimate question of security.

Mr. GRAY. So that you would say as of today that it is appropriate and proper to have this kind of an inquiry?

The WITNESS. As far as I know, certainly if you have something there that trips your mind, you ought to make an inquiry about it.

Mr. GRAY. I meant this proceeding that we are involved in.

The WITNESS. Yes.

Mr. GRAY. Would you take a calculated risk with respect to the security of your bank?

The WITNESS. I take a calculated risk every day in my bank.

Mr. GRAY. Would you leave someone in charge of the vaults about whom you have any doubt in your mind?

The WITNESS. No, I probably wouldn't.

Mr. GRAY. My question I can put in a more straightforward way, and it is one of the basic issues before the country, and certainly one involved in this country. And that is, when the paramount concern is the security of the country, which I believe is substantially the language of the Atomic Energy Act, can you allow yourself to entertain reasonable doubts?

Before you answer, let me say if this leads you to think that I or the members of the board have any conclusions about this matter at this point, I wish you would disabuse yourself of that notion.

The WITNESS. Surely.

Mr. GRAY. What I am trying to get at is this relates yourself in your discussion about the other things you have to take into consideration.

The WITNESS. Surely. That brings me back again on this problem which I was checked a little because I was going a little far afield, and I don't think I can get the pat analogy to the bank vault man. But let me say, suppose that the man in charge of my vaults knew more about protection and knew more about the intricacies of time locks than anybody else in the world, I might think twice before I let him go, because I would balance the risks in this connection.

Take the case of the bank teller business, because I saw Mr. Wilson's remark, and I pricked up my ears when he said that, because I am a banker, and he was comparing my profession to this thought of reforming a bank teller. This was the incident I was about to speak of, if I may now inntroduce it with your consent.

Mr. GRAY. Yes.

Mr. ROBB. Mr. Chairman, may I make myself plain? I have no objection to Mr. McCloy giving a full explanation of any of his answers.

The WITNESS. One of my tasks in Germany was to pick up Nazi scientists and send them over to the United States. These Nazi scientists a few years before were doing their utmost to overthrow the United States Government by violence. They had a very suspicious background. They are being used now, I assume—whether they are still, I don't know, because I am not in contact with it—on very sensitive projects in spite of their background. The Defense Department has been certainly to some extent dependent upon German scientists in connection with guided missiles. I suppose other things being equal, you would like to have a perfectly pure, uncontaminated chap, with no background, to deal with these things, but it is not possible in this world. I think you do have to take risks in regard to the security of the country. As I said at the beginning, even if they put you—I won't be personal about it—but let us say put Mr. Stimson or anybody in charge of the innermost secrets of our defense system, there is a risk there. You can't avoid the necessity of balancing to some degree.

So I reemphasize from looking at it, I would think I would come to the conclusion if I were Secretary of War, let us balance all the considerations here and take the calculated risk. It is too bad you have to calculate sometimes. But in the last analysis, you have to calculate what is best for the United States, because there is no Maginot Line in terms—it is just as weak as the Maginot Line in terms of security.

Mr. GRAY. Do you understand that it is beyond the duty of this board to make the ultimate decision as to who shall be employed by the Government on the basis of his indispensability or otherwise?

The WITNESS. Surely.

Mr. GRAY. We are more narrowly concerned with the field of security as we understand the term.

The WITNESS. I understand that.

Mr. GRAY. I think I have no more questions. Dr. Evans.

Dr. EVANS. Mr. McCloy, you say you talked to Bohr?

The WITNESS. Yes; Neils Bohr.

Dr. EVANS. Where did you talk to Neils?

The WITNESS. I talked to him abroad and here. He visited Washington, you know.

Dr. EVANS. I know. Did he tell you who split the uranium atom over there?

The WITNESS. Wasn't it Hahn and Straussman?

Dr. EVANS. Yes. I am just giving you a little quiz to find out how much you associated.

The WITNESS. You terrify me.

Dr. EVANS. Did you read Smyth's book?

The WITNESS. Yes; I did. I was also tutored by Rabi; I may say that when Dr. Oppenheimer gave me up as a poor prospect.

Dr. EVANS. And you think we should take some chances for fear we might disqualify someone who might do us a lot of good?

The WITNESS. Yes; I do.

Dr. EVANS. You do?

The WITNESS. Yes.

Dr. EVANS. There is nothing in the regulations applying to this board that mentions that point.

The WITNESS. Yes.

Dr. EVANS. You understand this is not a job we tried to seek.

The WITNESS. Goodness knows, I know that.

Dr. EVANS. You think that there are very few scientists that could do Dr. Oppenheimer's work?

The WITNESS. That is my impression.
Dr. EVANS. That is, you think he knows perhaps more about this, as you mentioned in your vault business, than anybody else in the world?
The WITNESS. I wouldn't say that; no. But I would certainly put him in the forefront.
Dr. EVANS. And you would take a little chance on a man that has great value?
The WITNESS. Yes, I would; particularly in the light of his other record, at least insofar as I know it. I can't divorce myself from my own impression of Dr. Oppenheimer and what appeals to me as his frankness, integrity, and his scientific background. I would accept a considerable amount of political immaturity, let me put it that way, in return for this rather esoteric, this rather indefinite theoretical thinking that I believe we are going to be dependent on for the next generation.
Dr. EVANS. That is, you would look over the political immaturity and possible subversive connections and give the great stress to his scientific information?
The WITNESS. Provided I saw indications which were satisfactory to me, that he had reformed or matured.
Dr. EVANS. I have no more questions.
Mr. GRAY. Mr. Garrison?
Mr. GARRISON. I would like to put one question, if I may.
Mr. GRAY. Yes.

REDIRECT EXAMINATION

By Mr. GARRISON:

Q. Having in mind the question that Dr. Evans last put to you, I would just like to read you a paragraph from the Atomic Energy Commission's criteria for determining eligibility, which is a guide to the board here, as I understand it, and ask you if this is something of what you yourself had in mind when you talked about positive and negative security:
"Cases must be carefully weighed in the light of all the information and a determination must be reached which gives due recognition to the favorable as well as unfavorable information concerning the individual, and which balances the cost to the program of not having his services against any possible risks involved."
I also should read you the section from the Atomic Energy Act which provides that, "No individual shall have access to restricted data until the FBI shall have made an investigation and report to the Commission on the character, associations, and loyalty of such individual and the Commission shall have determined that permitting such person to have access to restricted data will not endanger the common defense or security."
Having read the portion of the Commission's criteria which I read to you and the section of the statute which I read to you, would you or would you not say that your observations about positive, as well as negative, security have a place within this framework?
A. Yes; I would say so.
Mr. ROBB. Mr. Chairman, may I just point out for the record—I don't wish to get into any debate about the matter—the section that Mr. Garrison read from the criteria, I believe, applies to the decision which is to be made by the general manager as an administrative matter in determining whether the subject is to be kept on.
Dr. EVANS. It is not the action of this board.
Mr. ROBB. It does not refer to this board.
Dr. EVANS. This board doesn't have to do that.
Mr. GRAY. I think it is sufficient in the presence of this witness to simply raise that question. I think otherwise there would appear as a part of Mr. McCloy's testimony very considerable argument about the meaning and provisions of this.
The WITNESS. May I say I was not familiar with that provision.
Mr. GRAY. That is one reason I don't want to debate it while you are in the witness chair, Mr. McCloy. I think I ought to say to you that there are a good many other provisions in this criteria document which was referred to by Mr. Garrison, establishing categories of derogatory information, et cetera, and I would just call your attention to the fact that these other things appear and the discussion you have is by no means conclusive as to the duties of this board.
Mr. ROBB. That is all I wanted to point out.

Dr. EVANS. Mr. McCloy, our business is simply to advise. We don't make the decision.

The WITNESS. I see. You make an advisory report to the general manager.

Mr. GRAY. We make a recommendation.

Dr. EVANS. And sometimes the recommendations of a board like this are not carried out at all.

By Mr. GARRISON:

Q. I would like to put one final question to you. Is it your opinion that in the light of the character, associations, and loyalty of Dr. Oppenheimer as you have known him, that his continued access to restricted data would not endanger the common defense and security?

A. That is my opinion.

Mr. GARRISON. That is all.

Mr. ROBB. That is all. Thank you, Mr. McCloy.

(Witness excused.)

Mr. GARRISON. May I read one sentence from the criteria into the record, not by way of argument, but simply because I would like to respond to it.

Mr. GRAY. I have no objection to your reading one sentence from the criteria, but I don't want to get into a discussion of the meaning of these regulations. You may read your sentence and if Mr. Robb wants to read a sentence, I will give him one crack.

Mr. GARRISON. This is section 4.16 of the United States Atomic Energy Commission Rules and Regulations. This is entitled: "Recommendations of the Board:

"(a) The Board shall carefully consider all material before it, including reports of the Federal Bureau of Investigation, the testimony of all witnesses, the evidence presented by the individual and the standards set forth in AEC personnel security clearance criteria for determining eligibility."

Mr. GRAY. We will recess for a short period.

(Short recess.)

Mr. GRAY. Mr. Griggs, do you wish to testify under oath? You are not required to do so, but all witnesses have.

Mr. GRIGGS. Yes.

Mr. GRAY. What is your full name?

Mr. GRIGGS. David Tressel Griggs.

Mr. GRAY. Would you raise your right hand, please. David Tressel Griggs, do you swear that the testimony you are to give the board shall be the truth, the whole truth, and nothing but the truth, so help you God?

Mr. GRIGGS. I do.

Whereupon, David Tressel Griggs was called as a witness, and having been first duly sworn, was examined and testified as follows:

Mr. GRAY. It is my duty to remind you of the existence of the so-called perjury statutes. I should be glad to review those with you if you feel the need of it, or may we assume you are generally familiar with them?

The WITNESS. I am not familiar with it.

Mr. GRAY. Forgive me if I briefly tell you that section 1001 of title 18 of the United States Code makes it a crime punishabl eby a fine of not more than $10,000 or imprisonment for not more than 5 years or both for any person to make any false, fictitious, or fraudulent statement or representation in any matter within the jurisdiction of any agency of the United States.

Section 1621 of title 18 of the United States Code makes it a crime punishable by a fine of up to $2,000 and/or imprisonment of up to 5 years for any person to state under oath any material matter which he does not believe to be true.

Those are in general the provisions of the statutes to which I had reference.

The WITNESS. Thank you.

Mr. GRAY. I should like to request, Mr. Griggs, that if in the course of your testimony it becomes necessary for you to refer to or to disclose restricted data, you let me know in advance so that we may take the necessary steps in the interest of security.

The WITNESS. May I ask, sir, does this apply to only restricted data or any classified matters?

Mr. GRAY. I think clearly it applies to restricted data. If you find yourself getting into matters with respect to which there is a serious classification, as contrasted with what I used to know as the restricted label not in the atomic energy sense, I don't think you need to bother about that. But if you get into secret matters, I think you better let me know you are entering into that field.

The WITNESS. I understood that I had a measure of protection in this in that there was a person here who would——
Mr. GRAY. If any question arises and no one here can give you the answer to it, a classification officer can be made available.
Mr. ROLANDER. That is right.
Mr. GRAY. Finally, I should say, Mr. Griggs, that we consider this proceeding a confidential matter between the Atomic Energy Commission, its officials and witnesses on the one hand, and Dr. Oppenheimer and his representatives on the other. The Commission is making no release with respect to this proceeding and on behalf of the board, I express the hope to all the witnesses that they will take the same view.
The WITNESS. Yes, sir.

DIRECT EXAMINATION

By Mr. ROBB:
Q. Mr. Griggs, where do you live at present, sir?
A. My home address is 190 Granville Avenue, Los Angeles, Calif.
Q. You are appearing here today in response to a subpena?
A. Yes, I am.
Q. You are not here, Mr. Griggs, because you want to be here?
A. No. I do feel it is my duty to testify as requested, however. The reason that I am glad that there is a subpena in the case is because some of the testimony that I may have to give may involve matters of Air Force concern.
Q. You said you felt it was your duty to testify as requested. Just to make it clear, you don't mean that you had been requested to testify in any particular way, do you?
A. No.
Q. Mr. Griggs, what is your present occupation or employment?
A. I am professor of geophysics at the University of California at Los Angeles.
Q. How long have you been in that position?
A. Since May of 1948.
Q. Would you tell us something of your academic training and background?
A. I graduated from Ohio State University in 1932, and stayed theer for a year taking a master's degree. I went to Harvard where for 7 years I was a member of the Society of Fellows. In approximately June of 1940, I left to be a member of the Radiation Laboratory at MIT. You have asked only about my academic training. That includes my academic training.
Q. Just for the benefit of those of us who are not experts, would you tell us what you mean by geophysics? What kind of physics is that? We have heard about nuclear physics and physical chemistry. What is a geophysicist? I don't means a complete explanation.
A. In general it is the application of physical methods to the problems of the earth.
Q. You mentioned that you began work on radar in 1940?
A. In 1940, yes.
Q. At MIT?
A. Excuse me. I beg your pardon. I made a mistake. This is in 1941. I hope the record can be corrected on that.
Q. How long did you stay there in that work?
A. I was there until August of 1942.
Q. What did you do after that? Would you go ahead now and in your own way tell us chronologically what you did after that?
A. Yes. During my time at the radiation laboratory I was concerned primarily with the development of airborne radar. In August of 1942, I was requested to come down to the War Department to serve as an expert consultant in the Office of the Secretary of War, and particularly within the office of the Secretary I was working in the office of Dr. Edward L. Bowles. My duties there were to do what I could to insure the integration of our new weapons, principally radar, since that was the subject with which I was familiar, into the operational units of the War Department, and since the Air Force was the principal customer of this, I worked primarily with the Air Force.
I went overseas for extensive periods and spent between 2½ and 3 years, I believe, overseas in the European theaters, and after V-E Day I was transferred to the Far Eastern Air Forces, where I was served as chief of the scientific advisory group to the Far Eastern Air Forces, still, however, on assignment from the Office of the Secretary of War.
Q. Who was the head of your group over there in the Far East?

A. I was the head of the scientific advisory group directly under General Kenney as the commanding general of the Far Eastern Air Forces.

Q. Was Dr. Compton over there?

A. After V-J Day, Dr. Compton headed a mission of which I was a part——

Mr. SILVERMAN. Which Dr. Compton?

The WITNESS. Dr. K. D. Compton.

This was called the scientific intelligence advisory section, I believe, of GHQ, General MacArthur's command based in Tokyo after the occupation.

I was there for 2 months and returned to the United States in November of 1945.

By Mr. ROBB:

Q. What did you do then?

A. I had looked for the end of the war hoping that I could immediately return to my academic pursuits. After having seen so much destruction of principally urban destruction, both in Germany and Japan, I had hoped that the world would have come to a realization that steps necessary to prevent war must be taken. I left the War Department and spent perhaps 6 weeks trying to get back into the swing of things. I had no position to return to at that time, so I was looking for an academic position.

Then I became convinced that as a result, I think, largely of the activities of the United Nations with regard to Persia, that we were in for a long-term military problem. Because of my nearly unique experience in integrating new weapons into the military, I felt that I should remain in that work for some time until a new group of people could be brought along. For that reason I responded in the affirmative when I was asked to join what later became the Rand project in the Rand Corp., and I did join them in February 1946.

Q. In what capacity?

A. I was the first full-time employee of the Rand project and as the project grew and divided into sections, I was head of the atomic energy section, I believe it was called, at that time. It is now called the nuclear energy division of the Rand Corp.

Q. Go ahead.

A. I remained there until May of 1948, when I left to go to the University of California. At that time the section had been built up to the point where I felt that if anything, it could carry on better after I left than it had been doing.

Q. You went back to the University of California where?

A. At Los Angeles.

Q. In what capacity?

A. As I have already said, I was professor of geophysics in the Institute of Geophysics at Los Angeles.

Q. Did you entirely terminate your relationship with Rand or not?

A. No. My agreement with President Spraulle at the time I joined the university, I felt free to and did act in consulting capacity on defense problems. I have been ever since consultant to the Rand Corp. with the exception of the 1 year I served here in the Air Force, and at various times I have been consultant to the Armed Forces Special Weapons project, to the radiation laboratory at the University of California, to the Air Force, and the Corps of Engineers.

Q. Will you tell us whether or not Rand was doing work for the United States Government in the field of nuclear weapons?

A. Oh, yes.

Q. You mentioned that you were with the Air Force. When did that start?

A. I left on leave of absence under a strong request from the Chief of Staff of the Air Force to serve as chief scientist of the Air Force, which I did for the period of September 1, 1951, through June 30 of 1952.

Q. In that capacity did you concern yourself with the thermonuclear problem?

A. Yes.

Q. May I interrupt the course of your narrative for a moment to ask you whether or not you met Dr. Oppenheimer?

A. Oh, yes.

Q. When?

A. I can't be sure of the first time that I met him, but I have seen him on a number of occasions since 1946.

Q. In other words, you know Dr. Oppenheimer?

A. Oh, yes.

Q. And you have known him since about 1946?

A. Yes. I think I did not know him before.

Q. Getting back to your work with the Air Force in respect of thermonuclear matters, what was your first connection with that when you were with the Air Force?

A. I should say that through my Rand connections largely I had been following as well as I could from afar the course of developments in this field at Los Alamos and about the time I came to Washington there was, as you have abundant testimony, intensification of this program and reason for much more optimism than had been generally present in the past.

Q. You mentioned that we had abundant testimony. Of course you have not been present. What did you mean by that?

A. I referred to the implications I got from conversations with you and Mr. Rolander.

Q. All right. Go ahead.

Mr. MARKS. What was the testimony about? I am very sorry.

Mr. ROBB. Mr. Griggs said "as you have abundant testimony there was optimism about the program in 1951." I merely wanted to draw from him what he meant by the testimony.

By Mr. ROBB:

Q. You mean in the course of interviewing you as a witness, we took it for granted that there was in 1951 increased optimism in respect of the thermonuclear program, is that it?

A. I would have assumed this whether you said anything to me or not. I presume you have been getting into this business pretty thoroughly and I certainly hope that the board has.

Q. All right, sir. Go ahead. I am sorry I interrupted your course of thought. You were about to tell us about what you had to do with the thermonuclear program, and I believe you were explaining why you were interested in it when you came to the Air Force.

A. Shortly after I started work in the Air Force at that time as chief scientist, it became apparent that it was possible to think of actual weapons of this family, and there were estimates as to performance of these weapons which made them appear to be extraordinarily effective as weapons for the Air Force. If these estimates could be met, it was perfectly clear to my colleagues in the Air Force that it was of the utmost importance that the United States achieve his capability before the Russians did.

In this regard the opinions of the Air Force coincided with the opinions expressed by General Bradley for the Joint Chiefs of Staff in his memorandum of October 1949.

Q. Go ahead.

A. This is a long story.

Mr. SILVERMAN. What is the question?

Mr. ROBB. I asked him to tell us about his connection with the thermonuclear program, and just what you first did when you came with the Air Force.

By Mr. ROBB:

Q. What was the first step you took in respect to the thermonuclear program?

A. I can hardly remember what the first step I took was. The first step I took was to get additional information as to the status.

Q. To whom did you go for that information?

A. To the Office of Atomic Energy of the Air Force and to the Atomic Energy Commission.

Q. What did you find out about the status of the program?

A. As I have already testified, everything I found at that time gave indication or gave promise of the fairly early achievement of an effective weapon.

Mr. GARRISON. Could I understand what time was this?

By Mr. ROBB:

Q. Was this in the fall of 1951?

A. Yes.

Q. Did there come a time when you had some discussion about the establishment of a second laboratory?

A. Yes; we were verly deeply concerned in this.

Q. Why?

A. In the President's directive of January 31, 1950, it was stated the rate and scale of effort on thermonuclear weapons should be jointly determined by the Department of Defense and the Atomic Energy Commission. It was therefore a part of our responsibility as a part of the military to make known our views on this matter. We felt at the time we are speaking of, namely, late 1951 and early 1952, the effort on this program was not as great as the circumstances required under the President's directive.

Q. So what did you do?

A. I personally first tried to find out from the AEC what action they were taking in this direction. The things that I found out led me to believe——

Q. Well, pardon me. Go ahead.

A. You were going to ask a question?

Q. I was going to save time. Did the Air Force commend the establishment of a second laboratory?

A. The Air Force did. So did the Department of Defense.

Q. Did you at that time ascertain what the position of Dr. Oppenheimer was on that?

A. I did not talk as near as I can recall to Dr. Oppenheimer about this question. By hearsay evidence, I formed a firm impression that he was opposed to it. I have since read the appropriate minutes of the General Advisory Committee, and believe that this is substantiated in those minutes.

Q. Did there come a time when a project known as Vista was carried out?

A. Yes.

Q. Were you familiar with that project?

A. Yes, surely.

Q. Would you tell us what you can of the origin of that and its history?

A. May I volunteer a statement?

Q. Yes, indeed, sir.

A. The testimony that I have to give here before this board, as I understand the line that your questions are following, is testimony which will be concerned at least in part with two very controversial issues on which I was a participant in the controversy in my clear understanding on the opposite side of this controversy from Dr. Oppenheimer.

Q. And you wish what you have to say to be taken in that context?

A. Yes. I want to make it clear that I was an active participant in the controversy, and may not be fully capable of objectivity.

Q. Because you were an active participant, we have asked you to come here because you know about it. Now, would you go ahead, sir, and tell us what you know about the origin of this Vista project, and in particular reference to any connection Dr. Oppenheimer had with it, and then what happened in the Vista project?

A. I am not hesitant to answer this question and I don't want that impression to be conveyed if I can avoid it. However, I do feel the need of some clarification of what is obviously going to follow from your present trend of questions, because a great many of my scientific colleagues are involved in this controversy and on both sides. In my mind there existed at the time and today a possible distinction between the position of my other scientific colleagues and that of Dr. Oppenheimer.

Q. When did you first become aware of the starting of the so-called Vista project?

A. The Vista project was started, as near as I can remember, in the spring or summer of 1951, largely through the activities of Dr. Ivan A. Getting and Dr. Louis N. Ridenour, who were at that time serving full time with the Air Force. Dr. Getting was serving as assistant for evaluation in the Office of the Deputy Chief of Staff for Development. Dr. Ridenour was serving as chief scientist. In other words, as my predecessor. They, after a very considerable persuasive effort, induced the California Institute of Technology to undertake the Vista project which can be briefly characterized as a project to study the tactical warfare * * *. This project was undertaken by Cal. Tech. as a joint project between the three services—the Army, Air Force, and the Navy.

Q. Were there various meetings of scientists in Pasadena in connection with this study?

A. Yes.

Q. When did those meetings come to a close, approximately?

A. As nearly as I can remember, the Vista report was submitted in January of 1952, and the Vista project was terminated essentially with the presentation of the Vista report.

Q. Was there a section of that report, section 5, I believe, which dealt with atomic and nuclear matters?

A. Chapter 5.

Q. Did you attend any of the sessions in California?

A. Yes; I did.

Q. Were you present at the sessions about the middle of November 1951?
A. I visited the Vista project about the middle of 1951; yes, sir.
Q. Will you tell us whether or not you recall an occasion when a draft of chapter 5 was presented to the assembly?
A. Yes; I do recall.
Q. Do you recall who it was who presented it?
A. Some of us from the Air Force were there to have a preview of the Vista report as it then existed in draft form—partially at least in draft form—and this included Mr. William Burden, who was assistant to the Secretary of the Air Force for Research and Development; Mr. Garrison Norton, who was deputy to Mr. Burden; Lt. Col. T. F. Walkowicz, and myself. We had a session which was officially presided over, I think, by Dr. Fowler, but in which Dr. DuBridge as senior member of Cal. Tech. took the leading role, essentially, and in which Dr. Lauritsen, Dr. Milliken, and Robert Bacher were active. There were doubtless others there. Your question, I believe, was who presented this draft?
Q. Yes, sir.
A. I don't remember in detail, but I think the proper answer to your question is that parts of it were presented by all of these people.
Q. Do you recall anyone making any statement as to who prepared the introduction to this draft?
A. There was a part of the Vista in draft form which we were told had been prepared by Dr. Oppenheimer, and we were told that what we were shown was a verbatim draft as he had prepared it.
Q. Who told you?
A. We were told that by DuBridge, Bacher, Lauritsen, and perhaps others.
Q. Did you examine that draft?
A. Yes.
Q. Was there anything about it which impressed itself on your mind?
A. Yes; indeed.
Q. What was it?
A. There were three things about this general area of the Vista report that I regarded as unfortunate from the standpoint of the Air Force. I can't be sure that all three of these things were in the draft that was written by Dr. Oppenheimer, but I think they were. However, the first and perhaps most controversial point as far as we in the Air Force were concerned, I am quite sure, was in the part that was said to have been prepared by Dr. Oppenheimer.

* * * - * * *

Q. Was there anything else in that draft that struck you?
A. As I said, there were two other points. I can't swear to it that these were in the draft written by Dr. Oppenheimer, but I am sure that he was aware of these points.
Q. Did you understand that Dr. Oppenheimer approved these points?
A. Yes; I did. I think there is no question about that.

* * * * * * *

Q. What is SAC?
A. The Strategic Air Command. And because of these facts, I considered this to be contrary to the national interest.
Q. What was the third point which impressed itself upon you?
(No response.)
Q. I might ask you this question. Was there anything in the draft at that time concerning the feasibility or the use of thermonuclear weapons?
A. May I say before I respond to your last two questions that coupled with this second point, * * * there was a recommendation as to the specific nature of the weapons which should form a stockpile. This recommendation was substantially different from the recommendation of the Joint Chiefs of Staff and the Department of Defense, and in my mind, coupled with the other recommendation of the tripartite allocation, had that second recommendation as to the specific nature of the weapons to be stockpiled been accepted, it would also have acted to restrict our military atomic capability.
Q. Yes, sir.
A. Now, as to the third point of the Vista report which troubled me, there was the statement to the effect that in the state of the art it was impossible to assess the capabilities of thermonuclear weapons adequately to evaluate their tactical significance. Bear in mind this was in the late fall of 1951. As near as I can recall, this particular piece was written by Dr. Oppenheimer, according to the testimony as I have already cited.
Mr. SILVERMAN. You mean according to what you heard?

The WITNESS. According to the testimony of DuBridge and Bacher. I am using testimony in too loose a word.

Mr. SILVERMAN. You don't mean their testimony.

The WITNESS. No.

Mr. GRAY. Let me suggest that you will have the opportunity to cross-examine.

Mr. SILVERMAN. This was not intended as cross-examination. It seemed to me that there was a slight error which I thought—if I am wrong, fòrgive me—that the witness would like to have corrected.

The WITNESS. I do appreciate clarification of that point. I meant what we had been told by DuBridge, Bacher, Lauritsen, and others at the time.

This statement seemed to me to be quite contrary to the technical expectations in the field of thermonuclear weapons at that time, with which Dr. Oppenheimer as Chairman of the General Advisory Committee should certainly have had complete familiarity. I have said that poorly, but I hope the sentence is clear.

I might say further on that, that Dr. Teller had previously spent a period of a few days, I believe, at the Vista project, * * * There have since been other analyses of this specific problem * * *.

Have I made clear what I am talking about?

By Mr. ROBB:

Q. I am told I may not ask you specifically what the final recommendations of the Vista report were—at least not in open session here—but I would like to ask you whether or not the statements which you have told us about the draft were substantially modified or changed?

A. Yes; they were. These statements that I have talked about.

Q. Yes.

A. These were ones which our party—the people I have named from the Air Force who were there—felt very strongly about and which Mr. Finletter felt strongly about and General Vandenberg, and I believe as a result of their action, in part directly with Dr. Oppenheimer, these statements were revised.

Q. May I ask you, sir, was there any particular reason at that time why you paid especial attention to any recommendations or views of Dr. Oppenheimer?

A. This is what you would call a leading question?

Q. I don't think so.

A. May I interrupt to say some other things about the Vista Report?

Q. Yes, sir.

A. With the exception of these three statements—perhaps a few other things—we found, the Air Force, and I as a part of the Air Force, that the Vista Report was a very fine job, and particularly in connection with the recommendations for the use of atomic weapons. This contrasted to thermonuclear weapons. The activities of the Air Force at that time were aided in this direction by the Vista Report, and specifically, I think, it is quite appropriate to say that Dr. Oppenheimer's contribution in this direction was helpful to the Air Force. This is a matter that I personally know to have extended over a period of several years.

Have I made what I am trying to say clear?

Mr. GRAY. Yes.

Mr. ROBB. Read the question, please.

(Question read by the reporter.)

The WITNESS. It seems to me this question can be answered only in broad context, if you will allow me.

Mr. GRAY. Yes; you may answer it any way that seems best to you, Mr. Griggs.

The WITNESS. It seems obvious to me that what you are asking as I understand it is one of the purposes of these hearings, namely, to investigate loyalty. I want to say, and I can't emphasize too strongly, that Dr. Oppenheimer is the only one of my scientific acquaintances about whom I have ever felt there was a serious question as to their loyalty. The basis for this is not any individual contact that I have had with Dr. Oppenheimer or any detailed knowledge that I have had of his actions. But the basis is other than that and perhaps it is appropriate that I say what it is.

I first warned about this when I joined the Rand project, and was told that Dr. Oppenheimer had been considered during the Los Alamos days as a calculated risk. I heard very little more about this until I came to Washington as chief scientist for the Air Force.

In that capacity I was charged with working directly with General Vandenberg, who was then Chief of Staff of the Air Force, on matters of research and development, and I was charged with giving advice as requested to the Secretary of the Air Force, who was then Mr. Finletter. I worked closely with General Doolittle, who was Special Assistant to the Chief of the Air Force.

Shortly after I came to Washington I was told in a way that showed me it was no loosely thought out—let me correct that statement. I was told in a serious way that Mr. Finletter—or rather, I was told by Mr. Finletter that he had serious question as to the loyalty of Dr. Oppenheimer. I don't know in detail the basis for his fears. I didn't ask. I do know that he had access to the FBI files on Dr. Oppenheimer, at least I think I am correct in making that statement. I had this understanding.

I subsequently was informed from various sources of substantially the information which appeared in General Nichols' letter to Dr. Oppenheimer, which has been published. I feel I have no adequate basis for judging Dr. Oppenheimer's loyalty or disloyalty. Of course, my life would have been much easier had this question not arisen.

However, it was clear to me that this was not an irresponsible charge on the part of Mr. Finletter or on the part of General Vandenberg, and accordingly I had to take it into consideration in all our discussions and actions which had to do with the activities of Dr. Oppenheimer during that year.

By Mr. ROBB:
Q. You mentioned General Vandenberg; did you have conversations with him about the matter?
A. Oh, yes.
Q. Tell us about that.
A. I had numerous conversations with General Vandenberg about this.
Q. To shorten it up, could you tell us whether or not the purport of what General Vandenberg said was similar to what was said by Mr. Finletter?
A. Yes.
Q. Mr. Griggs, did there come a time when a project known as the Lincoln Summer Study was undertaken?
A. Yes.
Q. Can you tell us briefly what that was and when it took place?
A. May I answer a broader question in my own way?
Q. Yes. I am merely trying to bring these matters up and let you tell us about them in your own words.
A. It became apparent to us—by that I mean to Mr. Finletter, Mr. Burden, and Mr. Norton, that there was a pattern of activities all of which involved Dr. Oppenheimer. Of these one was the Vista project—I mean was his activity in the Vista project, and the things I have already talked about. We were told that in the late fall, I believe, of 1951, Oppenheimer and two other colleagues formed an informal committee of three to work for world peace or some such purpose, as they saw it. We were also told that in this effort they considered that many things were more important than the development of the thermonuclear weapon, specifically the air defense of the continental United States, which was the subject of the Lincoln Summer Study. No one could agree more than I that air defense is a vital problem and was at that time and worthy of all the scientific ingenuity and effort that could be put on it. We were, however, disturbed at the way in which this project was started.

* * * * * * *

It was further told me by people who were approached to join the summer study that in order to achieve world peace—this is a loose account, but I think it preserves the sense—it was necessary not only to strengthen the Air Defense of the continental United States, but also to give up something, and the thing that was recommended that we give up was the Strategic Air Command, or more properly I should say the strategic part of our total air power, which includes more than the Strategic Air Command. The emphasis was toward the Strategic Air Command.

It was further said in these initial discussions with people who it was hoped would join the project that the Lincoln Summer Study would concern itself with antisubmarine warfare.

I hope it is clear to the board. If it is not, I should like to make clear why it is that I felt upset by the references to the relative importance of the Strategic Air Command and the Air Defense Command, and to the suggestion that we, the United States, give up the Strategic Air Command. Should I amplify that?

Mr. GRAY. Yes; if you will.

The WITNESS. The reason that I felt this was unfortunate as a part of the Lincoln Summer Study is similar to the reason that I felt that a similar suggestion which I have already referred to was unfortunate in the case of the Vista study, namely, that neither of these two studies had the background nor were charged with the responsibility of considering in any detail or considering at all the fact of the activities of the Strategic Air Command. I felt that for any group to make such recommendations it was necessary that they know as much about the Strategic Air Command and the general strategic picture as they knew about the Air Defense Command.

Also we have learned to be a little cautious about study projects which have in mind making budget allocations or recommending budget allocations for major components of the Military Establishment gratuitously, I might say. There are of course groups charged with this, but the Lincoln Group was not charged with this.

There was another aspect of the initial phases of the Lincoln Summer Study which upset me very greatly, and that is that the way in which it was first started gave considerable promise—considerable threat, I might say—of destroying the effectiveness of the Lincoln project. The Lincoln project was one which the Air Force relied on to a very great extent in developing the future air defense capability of the United States Air Force, and of the United States in large measure.

Sir, if I am getting too detailed about this——

Mr. GRAY. No, you proceed.

By Mr. ROBB:

Q. Had you completed your answer on that?
A. Yes, unless you desire amplification.

May I say one more thing in that connection? I probably have not made it very clear, but as near as we could tell the Lincoln Summer Study came about as one of the acts of this informal committee of three which I mentioned of which Dr. Oppenheimer was one.

Q. Who were they?
A. As I have said, Dr. Oppenheimer and two other scientists.
Q. Who were the other scientists?
A. Dr. Rabi and Dr. Lauritsen.
Q. There has been some mention of a group called ZORC. Was there any such group as that that you knew about?
A. ZORC are the letters applied by a member of this group to the four people, Z is for Zacharias, O for Oppenheimer, R for Rabi, and C for Charlie Lauritsen.
Q. Which member of the group applied it?
A. I heard it applied by Dr. Zacharias.
Q. When and under what circumstances?
A. It was in the fall of 1952 at a meeting of the Scientific Advisory Board in Boston—in Cambridge—at a time when Dr. Zacharias was presenting parts of a summary of the Lincoln Summer Study.
Q. In what way did he mention these letters? What were the mechanics of it?
A. The mechanics of it were that he wrote these three letters on the board——
Dr. EVANS. Did you say three letters?
The WITNESS. Four. You said three.

By Mr. ROBB:

Q. That was my mistake. Wrote them on what board, a blackboard?
A. Yes.
Q. And explained what?
A. And explained that Z was Zacharias, O was Oppenheimer, R was Rabi, and C was Charlie Lauritsen.
Q. How many people were present?
A. This was a session of the Scientific Advisory Board, and there must have been between 50 and 100 people in the room.
Q. To sum up, Mr. Griggs, in the Lincoln Study did they come up with a report of some sort?
A. I don't know.
Q. There has been some——
A. When I say I don't know, I mean I don't know whether there was a formal written report.
Q. Did you attend the sessions or any of the sessions?
A. I attended only the initial sessions, the first three days or so of the summer study. That was while I was still chief scientist of the Air Force, and after I left I had no further contact with it. That is, no further attendance at these meetings.

Q. There has been estimony here, I think, to the effect that the burden of thinking of the Lincoln study was that there should be a balance between an offensive or strategic air force and the continental defense of the Untied States. Would you care to comment on that?

A. I have already tried to give the board the impression that I may not be a thoroughly objective witness in controversial matters, and this was a controversial matter, but the impression I had was that there was a strong element in the Lincoln Summer Study activities and subsequent activities which can best be described as being similar to the article by Joseph Alsop, I believe, in the Saturday Evening Post, about the Lincoln Summer Study. As I recall it, this article recommended a Maginot Line type of concept in which we depend on air defense rather than our retaliatory capability. I think in this article the impression was given that through the technological breakthroughs, which had been exploited in the Lincoln Summer Study, it would be possible if their recommendations were followed to achieve a very high rate of attrition on attacking aircraft.

This, of course, can easily be checked by referring to the article. But as I recall it, rates of attrition approaching 100 percent were considered to be possible in that article.

This article reflected, as near as I could see, the spirit of a part of the Lincoln Summer Study. From what I knew then and from what I know now, I think that any such optimism is totally unjustified, and if we based a national policy on such optimism, we could be in terrible trouble.

Q. Now, Mr. Griggs, coming to May 1952, I will ask you whether you recall visiting Dr. Oppenheimer at Princeton?

A. Yes.

Q. In general what was your purpose in going to see him?

A. Do you mind if I answer this again fairly fully?

Q. No, sir.

A. During the meetings of the National Academy of Science in Washington in the spring of 1952, we had a luncheon meeting at Mr. Burden's house at which Dr. DuBridge and Dr. Rabi were present, as well as Mr. Burden, Mr. Norton of the Air Force, whose name I have mentioned before, and myself.

The purpose of this meeting was to allow Mr. Burden and Mr. Norton, who were charged with important recommendations with respect to our thermonuclear program, to talk to two eminent people who were familiar with aspects of the activities of the Atomic Energy Commission bearing on the thermonuclear problem—much more familiar with these—than I was and who were on the opposite side of this particular controversy which has already been mentioned, namely, the second laboratory controversy, who were on the opposite side of that than I was.

During that meeting, I made some statements to DuBridge and Rabi as to what I thought of the activities of the General Advisory Committee of the AEC with respect to the development of the thermonuclear weapons. These statements of mine were such as to imply that I didn't feel that the General Advisory Committee had been doing anywhere near as much as it could do to further the development of the thermonuclear weapon, nor anywhere near as much as it should, under the President's directive, and the subsequent directives which came out setting the rate and scale of effort on the thermonuclear program.

When I made these statements, based on as good information as I was able to obtain prior to that time, Dr. Rabi said that I was quite wrong, and that my sources of information had been inadequate. I responded, as near as I can recall, that I would be glad to get all the information I could so that I would have a proper view of the activities of the General Advisory Committee in this respect.

He then said that I couldn't get a clear picture of this without reading the minutes of the General Advisory Committee. I responded that I would be very happy to have the opportunity to read these minutes, and asked how I could get access to them, and whether I should request clearance for this by a member of the Atomic Energy Commission.

He responded, very much to my surprise, that the Atomic Energy Commission was unable to grant access to the minutes of the General Advisory Committee, that these were the personal property of the Chairman, Dr. Oppenheimer.

Mr. SILVERMAN. Who was it that this conversation was with?

The WITNESS. This was Dr. Rabi. I don't recall exactly the next thing in the conversation, but before we parted, Dr. Rabi suggested that he arrange a meeting at Princeton with Dr. Oppenheimer and myself and himself, Dr. Rabi, at which time I would have a chance to review the minutes of the General Advisory Committee so that I would be set straight on these matters.

That meeting turned out to be impossible, because Dr. Rabi had an illness at the time when we tentatively set up the date, and somewhat after that time I was in Princeton on other business, and called Dr. Oppenheimer, reminding him of this and suggesting that I would be happy to meet with him on this general subject if he so desired. Thereupon, we had this meeting.

By Mr. ROBB:
Q. What was the subject of your discussion when you did meet with him?
A. I, of course, brought up this background and the reason for my interest, as I recall it. I didn't really expect that I would be allowed to read the minutes of the General Advisory Committee, and it turned out that this was not offered by Dr. Oppenheimer.
Q. Did you ask?
A. Yes.
Q. What did he say?
A. I don't recall.
Q. In all events, you didn't get to read them?
A. No. I was shown by Dr. Oppenheimer at that time, two documents which have been referred to in Dr. Oppenheimer's letter in response to General Nichols. These were the documents with which I am sure the board is familiar, submitted, I believe, as annexes to the report of the General Advisory Committee in late October of 1949. These were the recommendations as to action in the thermonuclear weapon and the 2 documents were, 1 signed by—perhaps I need not go into this.
Q. I think it is pretty clear in the record already. This was in May 1952?
A. I would have to check my records on this. I can find out exactly when it was. I recall only that it was in the late spring of 1952.
Q. What, if anything, did Dr. Oppenheimer say in response to your suggestion that the GAC had not been doing everything possible in furtherance of the thermonuclear program?
A. We had, as near as I can recall, a fairly extensive or fairly lengthy discussion which I would estimate lasted something like an hour. This was, of course, one of the main topics of our discussion. So we both said quite a lot. So I can't answer your question simply.
Q. In general, did he accept your suggestion or did he say on the contrary that he thought they had been doing everything possible?
A. I am reasonably sure that I am accurate in saying that he attempted to convince me that they had, in fact, been doing everything possible. He mentioned specifically at that time the actions of the General Advisory Committee—I may not have this technically right when I say the actions of the General Advisory Committee—but the actions taken by people, including members of the General Advisory Committee, at a meeting in Princeton following the Greenhouse tests.
Q. In the course of that conversation that you told us about, will you tell us whether there was anything said by you about certain remarks which you attributed to Dr. Oppenheimer about Mr. Finletter?
A. I don't believe I attributed remarks to Dr. Oppenheimer during this dismussion. However, I did have a question as to the origin of a story which I had heard repeated from a number of sources, I believe, including Dr. Oppenheimer, about Finletter.
Q. Would you tell us what was said between you and Dr. Oppenheimer about that subject?
A. First I better repeat the story or the burden of the story.
Mr. SILVERMAN. Mr. Chairman, I assume Mr. Robb knows what is coming, and he things it has some bearing on this, because I am having a great deal of difficulty even in trying to guess.
Mr. ROBB. So far as anybody can know the workings of another man's mind, I think I know what the testimony will have to be. I spent until half past 1 o'clock this morning trying to find out.
Mr. SILVERMAN. It is hard for me to see, but all right.

By Mr. ROBB:
Q. Would you go ahead and answer the question?
A. During the spring of 1952, there had been a series of briefings within the Defense Department on the thermonuclear weapon possibilities and on their military effectiveness. The story to which I refer is said to have occurred or was said to have occurred during one of these briefings. As near as I could find

out the story was supposed to have reported a statement said to have been made by Mr. Finletter during one of these briefings.

The story was that Mr. Finletter had said in the course of the briefing, if we only had * * * of the bombs we could rule the world. This story had been told in my hearing in a context which suggested that we had irresponsible warmongers at the head of the Air Force at that time.

I was anxious to find out what part Dr. Oppenheimer had in spreading this story, and what basis there was for such a story. I asked specific questions——

Q. Of whom?
A. Of Dr. Oppenheimer.
Q. On this occasion?
A. Yes. I specifically asked Dr. Oppenheimer as I recall it if he had repeated this story. His answer as near as memory serves was that he had heard the story. I then tried to question him as to the person to whom these remarks which I had already quoted were attributed. While I don't think he said so by name, he left no doubt in my mind that these remarks were supposed to have been made by Mr. Finletter. I believe I assured Dr. Oppenheimer—excuse me. May I say one other thing first?

I tried to get enough information in this conversation with Dr. Oppenheimer to be sure in my own mind at which one of these several briefings these remarks were supposed to have been made. This remark was supposed to have been made. I became convinced that this was supposed to have been made at a briefing of Mr. Lovett by Dr. Teller and the Rand group at which I had been present, and which I still remember clearly the list of all those people who had been present. I believe I told Dr. Oppenheimer that Finletter made no such remark, and that insofar as I knew anything about Finletter's feelings on the matter, nothing could have been further from Mr. Finletter's thoughts. And I think I knew Mr. Finletter well enough to be sure of this. I was certain that no such remark had been made.

Dr. Oppenheimer said to me, I believe, that his source was one which he could not question. In other words, I clearly got the impression that he believed that Mr. Finletter said these remarks, and that my story of the occasion was not correct.

Q. Let me ask you whether you had ever heard Dr. Oppenheimer repeat this story?
A. I believe I have, although here my memory does not suffice, but according to my notes of the time which I looked at yesterday they say that I had heard him say that.
Q. Did you at that time make some memorandum of this matter?
A. Yes. No. Excuse me. I did not at that time make a memorandum, but on a later occasion I did.
Q. Either at that time or shortly thereafter?
A. Yes. I did as I recall a few weeks thereafter. The reason, as I recall it, for my making a memorandum at all, and I may point out that this memorandum I typed myself, and put on "Eyes only" classifications on it, because I thought it should be kept very close. The reason I made this memorandum was because Mr. Finletter was scheduled to have a meeting with Dr. Oppenheimer and because of what I had been told as to the possible nature of subject to be discussed. I thought he ought to have this information as accurately as I could describe it.
Q. In that conversation with Dr. Oppenheimer at Princeton was there any mention of a statement or announcement by the United States with respect to the development of the thermonuclear—any public announcement as to whether we would go ahead with it or not?
A. As I have already mentioned, Dr. Oppenheimer showed me these documents of the General Advisory Committee which were on this subject.
Q. In that context, did you follow up that matter with Dr. Oppenheimer in any way, and if so, what response did he make?
A. Let me make clear or let me emphasize that at this time I was on the opposite side of the controversy with respect to he second weapons laboratory, and Dr. Oppenheimer knew full well I was on the opposite side.
Q. I will put the question to you directly.
A. Excuse me, but let me say hence I was surprised that he would show me these documents. They were shown to me as near as I can recall in the context of the actions of the General Advisory Committee, and to me they seemed wholly had. In other words, I have not mentioned this before, but my view was and is that if the policy recommended by the General Advisory Committee had been adopted, it could be a national catastrophe.

Q. Do you recall whether or not you expressed some such view to Dr. Oppenheimer on that occasion?
A. I don't think I used words like that, but I made it quite clear I am sure that these documents semed to me unfortunate.
Q. What was his response to that?
(No response.)
Q. I will put the question to you directly.
Was there any discusion between you and Dr. Oppenheimer about your views on his loyalty?
A. Yes, there was.
Q. What was that?
A. I have forgotten the sequence of these things. I have of course forgotten the details of it, but I believe at one point Dr. Oppenheimer asked me if I thought he was pro-Russian, or some word of this sort, or whether he was just confused. As near as I can recall, I responded that I wished I knew. I might say that is my position today, and I hope that all of us who have question will be reassured by the proceedings of this board one way or the other. Does that answer your question?
Q. Did Dr. Oppenheimer say anything further in that context?
A. I believe it was after this that he asked me if I had impugned his loyalty to high officials of the Defense Department, and I believe I responded simply, yes, or something like that. If I were to answer that question—I think that before an answer should have been given, because as I understand the literal meaning of this word, I had not impugned his loyalty, but his loyalty had been impugned in my hearing, and we had discussed this—I had discussed this with high officals of the Defense Department, as I have already said, Mr. Finletter and General Vandenberg.
Q. Do you recall whether Dr. Oppenheimer had any comment to make on your mental process?
A. Yes, he said I was a paranoid.
Mr. ROBB. That is all I care to ask.
Mr. GRAY. I think we better recess now and meet again at 1:45.
(Thereupon at 12:25 p. m., a recess was taken until 1:45 p. m., the same day.)

AFTERNOON SESSION

Mr. GRAY. Are you ready to proceed, Mr. Silverman?
Whereupon, David Tressel Griggs, the witness on the stand at the time of taking the recess resumed the stand and testified further as follows:

CROSS-EXAMINATION

By Mr. SILVERMAN:
Q. Dr. Griggs——
A. Excuse me, Mr. Griggs.
Q. Mr. Griggs, I think you testified about a dispute about a second laboratory.
A. Yes, sir.
Q. Did you at first favor a separate Air Force laboratory?
A. I can only answer that question properly since we have not laid the foundation for it by a rather extensive answer. Is that all right?
Q. Let me ask you this first.
A. In other words, you don't want me to make an extensive answer.
Q. If you can fairly do so.
A. I would like to, because if I answer the specific question out of context, I think it might give the wrong impression.
Q. I assure you you will have your opportunity to answer quite in context and immediately. I just want to know whether there was a time when you favored a separate Air Force laboratory.
A. There was a time at which we suggested that the Air Force, if necessary, undertake a separate laboratory.
Q. Now, do you feel that you want to add something to that?
A. Yes. In late January or nearly that time——
Q. Which year, sir?
A. Excuse me, of 1952. I tried to find out what the status of the effort was within the AEC in terms of furthering the nuclear weapon development. I found that there had been a suggestion for the formation of a second laboratory that went under a variety of names at that time. If we need not qualify it further than that, I won't.

At one stage in the proceedings preliminary negotiations had been undertaken with the University of California, specifically with Dr. Ernest Lawrence, to this end.

In my discussions with Commissioner Murray on this subject, I confirmed my suspicion, speaking loosely, that roadblocks are being put in the way of this development. Unless I misinterpreted what he said, he confirmed my fear that the General Advisory Committee, and specifically Dr. Oppenheimer had been interfering with the development of the institution or the initiation of the second laboratory.

We in the Air Force waited a period to see what was going to happen and when progress was not positive in this direction, we then discussed with Dr. Teller the possibility of forming a second laboratory. One of the things that motivated us in this was that Dr. Teller was no longer working regularly at Los Alamos on the project. Knowing his ability and contributions in the past, I felt and it was felt by the Air Force that he should be encouraged to participate.

We felt further that the effort that was then being applied at Los Alamos was not commensurate or was not large enough to be commensurate with the need for effort in order properly to pursue the President's directive and the subsequent directives setting the rate and scale of effort.

The question had already been looked into within the Air Force as to whether it was appropriate—whether it was legal for the Air Force to establish such a weapons laboratory. Our legal advice from the Air Force counsel was that the provisions of the Atomic Energy Act placed a responsibility on the Air Force as a branch of the military services to insure that the weapon development was adequate.

It was further the legal opinion of our counsel that it was legally possible within the framework of the Atomic Energy Act for the Air Force to establish a second laboratory.

We knew as a practical matter that this would be a very difficult way in which to increase our effectiveness in the development of nuclear weapons. We further knew that although it might be legally possible to set up a second laboratory, it could not have any possible chance of success unless this activity received the real blessing and support of the Atomic Energy Commission. We did, however, look into the possibilities of setting up a second laboratory and had preliminary negotiations about this with the University of Chicago, who had an Air Force contract, at which university Dr. Teller was at that time.

Dr. Teller already had relations with this Air Force contract at the University of Chicago, and he had confidence of the ability of the people on this project to undertake the development of a second laboratory, and felt that he could get support—in fact, he had discussed with his colleagues, Fermi and others—who could be very helpful in such a laboratory, and there were preliminary discussions with the administration of the University of Chicago already preparatory in the forms of staff work to see if the Air Force could accept such a responsibility if the Atomic Energy Commission desired it.

Does that answer your question?

Q. I think your first answer answered my question. The rest of the explanation was what you wanted to make.

Mr. ROBB. I am sorry, I can't hear.

Mr. SILVERMAN. The witness asked me if I thought he answered my question and I said the first answer answered my question, and the next was the explanation he wished to give.

The WITNESS. May I ask the chairman, since I am not too familiar with your procedure, whether such an explanation on my part is desirable from your standpoint, or whether you would rather get on with the proceedings?

Mr. GRAY. Mr. Griggs, our procedures are very flexible, here, and we are not in any way adhering to ordinary rules which would apply in a court of law, and therefore within limits a witness can say anything he believes to be pertinent to the question asked him, except that he is not supposed to engage in argument.

In reply to your question as it related to that answer, it was perfectly appropriate for you to say that you would not want to answer that question without explanation.

The WITNESS. I want to follow your desires, sir. If you will stop me when I get too extensive, I would appreciate it.

Mr. SILVERMAN. It is the desire of all of us that the testimony given shall be as clear and as truthful and as full as possible. I think on that there is no doubt that we all join. If you have some doubts that something you are being asked may result in a misleading answer, try to answer the question, and if you think you want to add something, tell me so.

The WITNESS. Yes. I felt a little bad because this was the first question you asked me, and I had gone into this extent.

By Mr. SILVERMAN:

Q. Are you now satisfied that Livermore is a good solution of the second laboratory problem?

A. Livermore is the solution of the second laboratory problem adopted by the AEC. I have been, although not actively, a consultant to the Livermore project, and hence I am not without bias in this field. What I have heard and what I have experienced at the Livermore project convinces me that it is a very fine effort in that direction.

I might specifically say that one of the objections which was raised to the formation of a second laboratory was the impossibility or stated impossibility of recruiting personnel, that is, appropriately trained personnel. I think Livermore Laboratory has been spectacularly successful in this respect.

Q. I take it the purport of your answer is that you think Livermore is a good solution to the second laboratory problem?

A. Yes.

Q. Do you know whether Dr. Oppenheimer opposed the Livermore solution?

A. Of my direct knowledge, I do not.

Q. There has been testimony here that he did not oppose it. Does that surprise you?

A. You mean surprise me that there has been testimony to that effect?

Q. Yes.

A. No, but I certainly would not be surprised if there had been testimony to the effect that he had opposed it, either. I think it depends on who you ask.

Q. You have no personal knowledge on that subject?

A. No; not to my recollection.

Q. And I take it you would agree that the testimony of the people who did have personal knowledge would perhaps be the most reliable guide?

A. If all of the testimony that has been given before this board indicates that Dr. Oppenheimer did not oppose this laboratory then I would feel that you didn't have all the expert opinion in.

Q. Did Dr. Oppenheimer tell you at Princeton that he favored the Livermore solution?

A. I don't recall that he discussed this. I would be almost certain that he didn't tell me that he favored the Livermore solution.

Q. In that discussion at Princeton at which this story about Secretary Finletter came up, I think you said that you mentioned the story first?

A. That is right. In the discussion at Princeton.

Q. Yes, that is what I meant.

A. Yes.

Q. And Dr. Oppenheimer said he had heard some such story?

A. He said he had heard the story.

Q. Did he say that he had heard that story with respect to Mr. Finletter, or did he say that there was a story around the AEC that somebody in the Air Force had said something like that?

A. I think you will find my testimony on that is fairly explicit, and with the hope that I don't contradict that I said before——

Q. Just tell what your best recollection is.

A. My best recollection is that he did not mention the name of Mr. Finletter in connection with this story, but the things that he did say left no doubt in my mind that it was Finletter to whom the story was supposed to have been attributed.

Q. What did he say?

A. You see, I was anxious to find out who was supposed to have made these remarks and hence I asked a number of leading questions. I was first interested in discovering at which one of the several briefings this remark is supposed to have been said. From what Dr. Oppenheimer said, I became satisfied that it was the briefing of Mr. Lovett in Mr. Lovett's office at which this took place.

Q. Excuse me; if you can tell us what it was he said?

A. I can't tell you what he said. Do you expect me to be able to remember word for word what he said?

Q. Of course not. I am asking you to try to recall the substance of what he said. You said from what he said you got the impression that he was talking about Mr. Finletter.

The WITNESS. Mr. Chairman, since the question seems to be going beyond the ability of my memory—it seems that way to me—I do have notes on this subject which are in my files at the Pentagon. I was unable to bring them with me. If you wish amplification of this, the best record is what are in my files at the Pentagon.

By Mr. SILVERMAN:
Q. When did you make these notes?
A. They were made at a time shortly after our discussion.
Q. Can you give any idea of about how long after the discussion you made these notes?
A. Excuse me. The document I was referring to is the one that you have here.
Mr. ROBB. That we have a photostat of.
The WITNESS. I think so.
Mr. SILVERMAN. If it will refresh the witness' recollection. [Document handed to witness.]
Mr. ROLANDER. I don't think he can read this memorandum. I will have to check with the classification officer.
Mr. SILVERMAN. If the witness is going to testify from a document used to refresh his recollection, which I cannot see, I would rather skip the testimony.
Mr. ROBB. As far as I am concerned you can see it, Mr. Silverman. I would like to have it read into the record.
Mr. SILVERMAN. If you want to read it into the record, that is fine, but I do not wish to be in the position of examining a witness who is testifying from a document I cannot see.
Mr. GRAY. What is the security problem?
Mr. ROLANDER. May I check it with the classification officer?
Mr. GRAY. Yes.
Mr. ROLANDER. This memorandum is satisfactory from a security standpoint if one item, a number, is deleted, a numeral.
Mr. SILVERMAN. This numeral will have nothing to do with this.
Mr. ROLANDER. That is right.
Mr. SILVERMAN. It is all right with me. The witness will read this into the record, I assume, because otherwise I will not be able to know what is in it.
Mr. ROBB. If you will ask him, I am sure he will.
Mr. SILVERMAN. I don't know whether they will let me.
Mr. ROBB. Sure.
Mr. SILVERMAN. Put your finger over the number.
Mr. ROBB. Mr. Chairman, might the witness read it into the record since it has been discussed?
Mr. GRAY. It is my understanding that is why we delayed to let the security officer check it, to be read into the record. Do you object to it being read into the record?
Mr. SILVERMAN. I would as soon like to see it. I don't know what is in the document.
Mr. GRAY. There has been enough discussion about this conversation. I take it this document relates to the conversation you had. Is that correct?
The WITNESS. Yes.
Mr. GRAY. Does this document relate about this conversation about which you cannot recall precisely?
The WITNESS. Yes.
Mr. GRAY. I think the Chair will ask the witness to read it.
The WITNESS. You want me to read it verbatim including the title?
Mr. GRAY. Leave out the number.
The WITNESS. This is a memorandum to Mr. Finletter "Eyes Only" classification, June 21, 1952:

"1. In view of your possible meeting with Oppenheimer I want to record as accurately as I can my recollection of parts of my conversation with him on May 23, 1952.

"2. I said that I had heard from associates of his a story, as follows: 'At one of the briefings given by Teller on the implications of the H-bomb, a high official of the Department of Defense exclaimed, "If only we could have blank of those (H-bombs) we could rule the world." ' Oppenheimer said that he was familiar with the story, said that it had occurred at the briefing of Mr. Lovett."

Then there is an asterisk, and a list of the people as far as my recollection served who were present at that particular briefing. I was one of them.

"I told him that I was present at that briefing, and that nothing could be further from the actual reaction of those present. He then stated that he had confidence in the reliability of his information, and further, that it was 'my boss' who is supposed to have said it." The "my" of course refers to me. "On further questioning, he left no doubt in my mind that it was you to whom he was referring, although he did not use your name.

"3. I have heard this story used by him and others as an illustration of the dangerous warmongers who rule the Pentagon, and who are going to precipitate this Nation into a war unless a few scientists can save it.

"4. After he had showed me the GAC recommendation of December 1949 that the United States not intensify H-bomb development, but publicly renounce its development, and when I was pressing the point that such a course of action could well be disastrous to this country, Oppenheimer asked if I thought he were pro-Russian or just confused. After a moment I replied frankly that I wish I knew. He then asked if I had 'impugned his loyalty.' I replied I had." In my testimony this morning I expanded that. "He then said he thought I was paranoid. After a few more pleasantries our conversation came to an end."

Signed by me. Shall I read the footnote?

Mr. GRAY. Yes.

The WITNESS. This refers to the Lovett briefing: "This briefing took place in March 19, 1952. Those present, as far as memory serves, were: Lovett, Foster, Finletter, Pace, Whitehair, LeBaron, Nash, Burden, Norton, Griggs, Teller, Collbohm, Henderson, Blesset, Hitch, and Brodie."

At the bottom of the page it says, "This is the only copy of this memorandum," but since I am reading a certified true copy, that obviously is not so.

Does that answer your question?

By Mr. SILVERMAN:

Q. You were asked by the chairman to read the memo.

A. No; you asked the question to which I was trying to respond, and this is for the purpose of refreshing my memory. Does that answer your question?

Q. That is your best recollection?

A. Yes.

Q. Thank you.

Mr. ROBB. Mr. Chairman, I don't know whether the witness knows it or not, but this is on the stationery of the Department of the Air Force, Washington.

The WITNESS. Should I have read that into the record?

Mr. ROBB. I don't know.

The WITNESS. I really don't think that applies, because this is not the original.

Mr. ROBB. I get it.

By Mr. SILVERMAN.

Q. You testified to being present at a session or some sessions in California in, I think, November 1951 with respect to the Vista report.

A. Yes.

Q. How did you happen to go there?

A. Of course, since the Air Force had been instrumental in establishing the Vista project, we were very much interested in the results of their extensive studies, and we also, of course, were interested in seeing the shape of the report at this, which was the draft stage, for two reasons, of course, both obvious reasons. One, that we wanted to be able to act on any recommendations which were favorable before waiting for the formal report. We made frequent visits to the Vista project. This was not our first. It had been after some interval and things were happening at a substantial rate there.

And second, of course, as we always are, we were interested in reviewing the document to see if it contained any things to which we violently objected so that we could discuss these with the authors at that time.

Q. Had Mr. John McCone suggested to Secretary Finletter that somebody go out there to confer with the people who were working on Vista?

A. I should not be surprised if he had. You can get more accurate testimony from others on this.

Q. Mr. McCone was formerly the Under Secretary of the Air Force?

A. I believe that is correct.

Q. Was it your understanding that he had seen a draft of the Vista report, and called Mr. Finletter?

A. You are asking me about a matter of which I have no personal knowledge.

Q. There have been a certain number of things in your testimony on which you did not have personal knowledge.

A. No. What I mean is I don't think—at least my memory is not adequate to tell me whether I had heard that Mr. McCone had been over a draft of the Vista report.

Q. Let me complete my question, and then if you don't recall, you don't recall.
A. I do clearly that Mr. McCone had been in touch with Mr. Finletter, and I think that he had been in touch with him in connection with the Vista report, but my memory does not suffice—in fact, I am not sure I knew at the time the details that you are asking me.
Q. Did you know or did you understand that Mr. McCone had said that the Vista report had a lot of good things in it, and that the Air Force ought to be interested in it?
A. As I say, this is the same as the last question.
Q. If you don't recall——
A. I don't know this, but I would expect that he would if that is helpful. As I tried to say in my testimony, the Vista report had a lot of things in it, and as I also tried to say, I am reasonably sure that some of the things I regarded as favorable in the Vista report were in some measure at least the product of Mr. Oppenheimer's contribution.
Q. There was a draft of chapter 5 presented at this session in November 1951 which you testified to. I think you said that there were points which you found most controversial which I take it is your polite way of saying you disagreed with most strongly. The first point was a recommendation that the President of the United States announced that the United States would not use the strategic Air Force in an attack on cities or urban areas except in retaliation.
A. Those are not my exact words, but certainly this is the substance, except in response to an attack by the Russians on us, not in retaliation. This is quite a difference. On our cities.
Q. I thought you used the word "retaliatory" but it is all right.
A. I did use the word "retaliatory," but not in this connection.
Q. I just didn't want to mislead you as to what I though you had said. How sure are you that recommendation was in a draft of chapter 5?
A. I am sure as I can be of anything which I studied extensively 2 years ago, and which was of considerable concern to me.
Q. You actually saw this in a document?
A. Oh, yes.
Q. Would it surprise you to learn that Dr. Oppenheimer never advocated such an announcement, and was opposed to any such announcement?
A. Yes.
Q. Bearing in mind my last question, and the obvious implication of it, how confident are you that Dr. Oppenheimer was responsible for such a suggestion in the Vista report?
A. The basis for my belief that he was responsible for it I have already given in my testimony, namely, that we were told by DuBridge, Bacher, and Lauritsen, possibly others, that the document we were shown was a draft of an introduction prepared by Oppenheimer, and it was, word for word, his text.
Q. Did these gentlemen say that was Dr. Oppenheimer's suggestion?
A. No; they said this was his text. It follows it was his suggestion. I may have answered that last question wrong. I would rather think that they did say it was his suggestion. When I answered the question, I was thinking of what they said as they gave us this report. But we had a considerable discussion of this point with them afterward, and it is quite possible; in fact, I would certainly expect that they had said it was his suggestion in our discussion, but not in presenting the document to us.

* * * * * * *

Q. Do you recall whether the draft made the point that there might be circumstances in which it might be unwise to use our full strategic airpower, and yet it might still be important to use atomic bombs for tactical uses?
A. I believe it contained information to the effect.
Q. Did it contain a recommendation that we, therefore, be prepared with some degree of flexibility to be able to use either strategic airpower or tactical, whichever, or both, might be desirable in the light of the circumstances which might arise?
A. Yes: I am quite sure it contained strong emphasis on the desirability for flexibility in the use of atomic weapons.

* * * * * * *

Q. Was that not also Dr. Oppenheimer's recommendation?
A. I don't know that for a fact, but I certainly would expect that Dr. Oppenheimer would have made such a recommendation, in view of what I knew of his activities at the time, and his beliefs. If it is appropriate to mention it again, I saw Dr. Oppenheimer on a number of occasions in the general time period

advocating strongly the development of weapons for tactical use. On each one of these occasions when I saw him in this role I was impressed with his forcefulness, and I was also impressed with the fact that I agreed with the stand that he was taking on the use of tactical weapons.

I also should say, as I said this morning, I felt very strongly about this point, and I was urging within the Air Force, although my colleagues in Vista would not believe it, the development of the capability of delivering tactical weapons, and there are lots of stories that go with this.

Q. Mr. Griggs, the suggestion that we be prepared to use both strategic airpower and tactical would hardly be consistent with the suggestion to abolish, to give up our strategic airpower, would they?

A. No. One of the troubles I have is lack of consistency, as I mentioned before. However, there was no statement in this Vista document that I saw which suggested that we give up strategic airpower. There was this suggestion which I have said, which had it been adopted, would have restricted the use of the Strategic Air Force.

Q. You understood later from Dr. Oppenheimer—I don't want to put words in your mouth, sir—in connection with the Lincoln study, I think you said, that you had heard that some people were saying that it was necessary to give up strategic power of our airpower.

A. In order to get world peace. This was the way it was said. I should amplify that, I think. This statement was made, not by Dr. Oppenheimer, to my knowledge, but by Dr. Zacharias. It was made, however, after considerable discussions of this matter with Dr. Oppenheimer.

Q. Do you know whether Dr. Oppenheimer was ever in favor of giving up the strategic part of our airpower?

A. I have seen numerous indications that Dr. Oppenheimer felt that it is necessary for the United States to give up something in order to achieve world peace. Perhaps that is a little too loose; but if it is adequate for you, I won't expand. That is, the world-peace thing.

Q. Did you ever see——

A. Just a moment. I am sorry. This was merely an introduction to your question. It is clear that this was a position taken in the recommendation for the H-bomb.

Q. Which was the position?

A. That we must give up something. It was recommended in the case of the H-bomb that we give up the H-bomb, which to me, as I have indicated, could have been national calamity if the Russians got that first, as I was sure that they would if we didn't press. I don't think I have any reason—I can't recall any reason—other than this indication from the talk of Dr. Zacharias that Dr. Oppenheimer had advocated giving up the Strategic Air Force. That is one reason I was interested in the matter, because this was going a little further than he had according to my understanding of the past.

I believe it is recorded in the minutes of the meeting of the State Department panel of consultants that Dr. Oppenheimer suggested that since it was necessary for the United States to give up something in order to achieve world disarmament, that we consider giving up strategic missiles.

Q. Have you seen those minutes?

A. I have seen those minutes.

Q. And have you seen that statement of Dr. Oppenheimer?

A. According to my memory, I have seen that statement of Dr. Oppenheimer. This is subject to check by looking up the minutes of the first meeting of the panel.

Q. When did you see those minutes?

A. I saw them shortly after the meeting.

Q. You mean in 1946?

A. No. This was in the panel which was established in the spring of 1952, by the State Department, as announced by the Alsops' column.

Q. Whose column?

A. Joe Alsop.

Q. You saw this yourself?

A. I am just identifying the panel. I don't remember the exact title, but it was essentially on the subject of nonatomic disarmament, if I recall correctly. It was a panel of the State Department. It included Dr. Oppenheimer, Dr. DuBridge, Dr. Bush, and others.

Q. Where did you get your information as to the membership of this panel?

A. As I say, I have seen the minutes.

Q. Who were the members again?

A. My first information as to the membership of the panel came from the Alsop column.
Q. You saw the minutes?
Mr. ROBB. Let him finish the answer.
The WITNESS. I told you I saw the minutes. You asked me another question. I said my first information as to the membership of the panel I believe came from the Alsop column, which as near as my memory serves described this panel as having been brought into being as the result of activities of Drs. Oppenheimer, Rabi, and Lauritsen.

By Mr. SILVERMAN:
Q. You gave some of the members of the panel a minute or two ago. Would you mind telling us again?
A. Yes. I said I believe this panel included Dr. Oppenheimer, Dr. DuBridge, Dr. Conant, and others. I think the complete membership of the panel should be available.
Q. And where did you get the information as to the membership of the panel?
A. You have asked me three times.
Q. Yes; and you said the minutes, and then you went to the Alsop column.
Mr. ROBB. Then you cut him off.
The WITNESS. Would you mind repeating?
Mr. GRAY. What do you want repeated?
The WITNESS. He has asked this question three times. I have answered it in two different ways. I am not communicating very well. I don't know what your difficulty is. Since it takes time to read these minutes suppose I try again.

By Mr. SILVERMAN:
Q. You know what my question is, sir?
A. How I knew about the membership of the panel. My first knowledge of this, as I have said, I think came from the Alsops' column. It turned out to be substantially correct when I was able to check it both by contacts in the State Department and by reading the minutes, which recorded, of course, the membership.
Q. And the members were who?
A. I have testified so far as my memory serves me Dr. Oppenheimer, Dr. DuBridge, Dr. Bush were members, and others.
Q. I think you also mentioned Dr. Conant?
A. Did I mention Dr. Conant? I am not perfectly clear on this. I should like to refresh my memory. I think Dr. Conant was—no, I am sorry I just can't remember.
Q. You did mention Dr. Conant, didn't you?
A. Pardon?
Q. You did mention Dr. Conant as a member?
A. The people that I meant to mention were Oppenheimer, DuBridge, and Bush. If I mentioned Conant—as I say, right now I am not clear whether he was a member or was not a member. It would be real easy to find out.
Q. It is easy to find out. I have the list here. Would you be surprised to find that Dr. DuBridge was not a member?
A. It would certainly indicate that my memory is in error if Dr. DuBridge was not a member.
Q. Would it surprise you to find out that there are no minutes of that panel?
A. That would surprise me very much.
Q. Where did you see these minutes?
A. I asked for them and had them sent over to me, minutes of the first meeting.
Q. Whom did you ask for these minutes?
A. As near as I can recall I asked my executive officer at the Pentagon, Colonel Walcowicz.
Q. Where did he get them from?
A. We have a liaison contact with the State Department.
Q. Where are those minutes now?
A. I haven't got any personal knowledge.
Q. When did you see them?
A. In the spring of 1952.
Q. Can you obtain those minutes for this board?
A. I haven't any idea, but I can obtain them if they are in my own files.
Q. Will you please do so?

Mr. ROBB. Wait a minute, Mr. Chairman. I don't know how this witness can be asked to obtain minutes from the State Department. I don't think that is fair.

Mr. GRAY. I think the point is well taken. If the witness is referring to something in his own files, he can be asked. But the witness cannot be asked to obtain documents from the State Department.

The WITNESS. I am sorry, when I said my own files, I meant my old files from the Pentagon, and I was told yesterday that I cannot get anything out of there except from the Liaison Division of the Air Force. I am sure if this document is in my file or if it is in the Air Force or can be tracked down, those documents can be made available to this board. But I am not clear what the best way of doing it is.

By Mr. SILVERMAN:

Q. Was the document minutes?
A. That is my recollection.
Q. You are not talking about a report now?
A. No; I am not talking about a report.
Q. I want to return now to the third of the controversial points in the Vista report.
A. Yes.
Q. As I have it here it is that in the state of the art as it then existed, it was impossible to assess the capabilities of thermonuclear weapons with respect to their tactical use.
A. Yes.
Q. Do you know whether Dr. Oppenheimer put that statement into the report?
A. No, I don't know.
Q. Do you think that Dr. Oppenheimer's judgment——
A. May I amplify that. The whole of this chapter 5 on atomic weapons which we have referred to as it was then presented to us was comprised of two parts. It was comprised of a part, essentially the body of the chapter, which had been written by the people of Vista, I believe, prior to Dr. Oppenheimer's visit, or at least he was not the direct author of that part. Then there was a separate document which, as near as I can recall, bore the title only of introduction, which was composed of a few pages. That is the part that was said to have been written by Dr. Oppenheimer. Because of the similarity in the subject matter of these two reports, I can't be sure which thing I associate with Vista was in which one of these two documents. What I have just said indicates that my memory is that the third point was in the main body of the Vista report. The main body of chapter 5 was in the Vista report, rather than in the piece written by Dr. Oppenheimer. I think there was some confusion about this when I first testified, because there were two reports, and I would like to make that clearer.

Q. Do you recall what other nuclear physicists participated in the Vista project?
A. There were quite a few. Do you want me to name as many as I can?
Q. Name a few; yes.
A. Of course, you asked nuclear physicists; there were Dr. Bacher, Dr. Lauritsen——
Q. I should say I am referring specifically to those who participated with respect to chapter 5.
A. All right, Dr. Fowler.
Q. Dr. Lauritsen and Dr. Bacher participated in chapter 5?
A. Yes, I think so. Dr. William Fowler. Dr. DuBridge participated. I don't think he took an active writing part. I believe he could be classed as a nuclear physicist.
Q. Do you think that these people were in a pretty good position, or perhaps in a better position than you, to judge as to the technical capabilities of the thermonuclear weapon as they appeared in November of 1951?
A. Yes, I think—you mean these latter people?
Q. Yes.
A. With the exception of Dr. Bacher, no; and I am not sure what his state of knowledge was.
Q. Dr. Lauritsen.
A. It is, however, clear to me that Dr. Oppenheimer was better informed than I was.
Q. How about Dr. Lauritsen?
A. Lauritsen I would think no. As I mentioned before, Dr. Teller, who I think was better informed than any of these people, had visited the Vista project

not very long before this, and had attempted to persuade the Vista people that a thermonuclear weapon was in such a state that it should be included in studies of * * * atomic warfare. As I mentioned also before, there were other agencies who at nearly the same time came to roughly the same conclusion that Teller did.

Q. With respect to the Lincoln study, do you know what part Dr. Oppenheimer played in the actual study?

A. As I have said, my attendance at the Vista study was limited to, I believe, the first 3 days. At that time Dr. Oppenheimer was present and participated fairly actively.

Q. Who appointed the people who made the Lincoln study?

A. Who appointed them?

Q. Yes, did they appoint themselves, or what?

A. As in the history of all these things, there is a little complicated genesis. It was pretty clear in the lines of the group who were pressing for this action which I have already mentioned as to who were most useful and likely candidates. The appointment of the group itself I do not know in detail but I would certainly presume that the appointment of these was made by the Lincoln project. I believe I have seen letters of invitation—that is a form of a letter of invitation that was sent out to the participants in the Lincoln summer study. Does that answer your question?

Q. And who signed those letters?

A. I believe they were signed by Dr. Hill, who was then the director of the Lincoln project.

Q. I think you used the phrase about the Lincoln group being in favor of a Maginot line type of defense.

A. I believe I mentioned this in connection with the Alsop article.

Q. Do you know whether Dr. Oppenheimer favored such a thing?

A. I did not hear Dr. Oppenheimer use any such word.

Q. Do you know what Dr. Oppenheimer's views were about the possible effectiveness of continental air defense at that time?

A. My last direct knowledge of this came from the contacts during the first 3 days of the sessions and this is all as far as Dr. Oppenheimer's personal views are concerned. At that time it was too early in the study to say with any definiteness what the views would be after the study. It was certainly the hope of all of us that as a result of the summer study the effectiveness of our air defense would be materially improved. I should say what I don't believe I did say this morning, that I believe that as a result of the Lincoln summer study our air defense is materially improved.

Q. Was that the main object of the Lincoln summer study, to find ways to improve our air defense?

A. Yes, sir.

Q. And did the Lincoln study ever recommend the giving up of any part of our strategic airpower?

A. No, not to my knowledge.

Q. I think you have already said so far as your knowledge goes, Dr. Oppenheimer did not recommend that.

A. That is right. I would like to amplify my answer on that for the benefit of the board, since this is the first mention of the summer study in this much detail.

We were concerned by the thing I have already mentioned, that is, the fear that the summer study might get into these things which we regarded as inappropriate for Lincoln, and as of questionable value to the Air Force—I refer to the giving up of our strategic air arm, and the allocation of budget between the Strategic Air Command and the Air Defense Command—but we were also very much concerned in the early days of the formation of the Lincoln Summer Study, because it was being done in such a way that had it been allowed to go in the direction in which it was initially going, every indication was that it would have wrecked the effectiveness of the Lincoln laboratory. This was because of the way the thing was, the summer study was being handled administratively.

So far as I know, it was not because of any direct action on the part of Dr. Oppenheimer. On the other hand, I felt at the time that Dr. Oppenheimer should have been well enough informed and alert enough to see that this would be disastrous to the Lincoln Summer Study.

After having reported this to the Secretary of the Air Force, Mr. Finletter, who had been actively concerned with the summer study, and had been very much—excuse me, I made a mistake—I said Mr. Finletter had been actively concerned with the summer study. I meant to say he had been concerned with project Lincoln. He had been in touch with President Killian, and Provost Stratton of MIT on the prosecution of project Lincoln. So I reported this to Mr. Finletter, and he essentially charged me with trying to find out if the summer study was going to be conducted in such a way as to result in a net gain to the effectiveness of Lincoln or a net loss.

If it looked to me as though it were going to be a net loss, I was asked to inform him so that steps could be taken to correct this condition, or to cancel the summer study if that were necessary.

I got in touch with Provost Stratton at MIT. I found that he hardly knew about the existence of the plan for the summer study. He undertook to look into it. I told him the things that worried me and worried Mr. Finletter about it. He did look into it. Some corrective action was taken in terms of discussions with people most involved and in terms of changing the organizational structure by which the summer study was to be introduced into the Lincoln project, and at a slightly later date Mr. Killian of MIT called me and told me that he was satisfied partly as a result of the recent activities that he and Dr. Stratton had been engaged in, which I have already mentioned, and that the Lincoln Summer Study would operate to the benefit both of Lincoln and the interests of the Air Force.

He further said, since I had mentioned that one of the things we were afraid of was that the Lincoln Summer Study results might get out of hand, from our standpoint, in the sense that they might be reported directly to higher authoriy, such as the National Security Council, President Killian reassured me that he had taken steps so that he was sure that the summer study would be—I think his words were "kept in bounds."

On the basis of this assurance we had no further—that is, Mr. Finletter, myself, and General Yates, and the other Air Force people—had no further immediate worries about the summer study and we encouraged it.

Q. Will you tell us what part did Dr. Oppenheimer play in this?

A. Oppenheimer played the part in it that I have already mentioned, in that the summer study, as near as my information goes, was conceived at a meeting at which he was present, that he allowed his name, and I believe encouraged the use of his name, in recruiting for the Lincoln Summer Study. That he was closely associated with the people who were recruiting for the summer study and who were preparing its plans. I think that covers the question.

Q. Was the idea of the Lincoln Summer Study to be a study of continental air defense?

A. No; that is too narrow a definition.

Q. What was it?

A. There had already been a study of continental air defense * * * only 1 or 2 years before, so one of the things that we were concerned with in the Air Force was whether this was to be a going over the same ground, or what new ground it was intended that this study cover.

Q. Would you just tell us what was it you found that the Lincoln Summer Study was supposed to do?

A. I believe in the literature that was sent out—I should not say literature—in the letters of invitation that were sent out that the Lincoln Summer Study should consider the problems of air defense * * *.

Q. Didn't you agree that it was a good idea to consider that?

A. I am still referring to your earlier question, if I may.

Q. Which one?

A. Your last question.

Q. Which question?

A. The question you asked just before.

Q. Will you tell me what it is because I have forgotten.

A. You asked me as to the subject matter of the Lincoln Summer Study. I responded that this was the information that was contained in the letter of invitation that was sent out. However, I had other information which gave me concern about some aspects that were considered for the programing of the Lincoln Summer Study. Particularly I had been present at a preliminary meeting before the existence of the summer study project in which it seemed to me that there was perhaps too much emphasis assigned to the development of an early warning line across—is there any security problem involved here?

Mr. ROLANDER. I don't think so.

Mr. MARSHALL. That is all right.

The WITNESS. Across our northernmost approaches, and that this problem—I should say that one reason that this problem received such particular emphasis at that time was because of the rather exciting new developments, technological developments in this field, which had been brought forward to my knowledge principally by Dr. Lloy Berkner. However, I was worried because it seemed to me and to some of the responsible people in project Lincoln that I talked to that it was necessary to consider this in context of our whole air defense system, and this was not being done, to my mind, adequately in the early discussions which I heard on this subject.

By Mr. SILVERMAN:
Q. Did you hear Dr. Oppenheimer in these early discussions?
A. No; he was not in this particular early discussion to which I referred.
Q. You did agree, I assume, that it was a good idea to study the feasibility of an early warning line?
A. The feasibility of an early warning line had been studied before by more than one agency. It certainly seemed to me a good idea in the light of recent technological development which I mentioned.
Q. Isn't that exactly what the Lincoln study did do?
A. The Lincoln Summer Study?
Q. Yes.
A. It did do this. It did not restrict its activities to this, as far as I am aware. As I have testified, my detailed knowledge of the Lincoln Summer Study activities is very incomplete.
Q. What troubles me is that you were worried that the result might be disastrous, that the direction in which it was going might be disastrous. Which direction was it going?
A. I have tried to make clear, perhaps I have not adequately, that the things I was worried about were that first there would be a diversion of effort created in the Lincoln Laboratory, which could have an adverse effect of the total program of Lincoln Laboratory. This diversion of effort I have tried to illustrate by the suggested consideration of the relative role of the Strategic Air Command and the Air Defense Command, by the suggested introduction of antisubmarine warfare into the Lincoln project, which had no bearing on the Air Defense problem as I saw it, and more importantly by the possibility, at one time a probability, that if the Lincoln Summer Study proceeded as it was then planned, there was substantial indication that it would wreck the laboratory in terms of its adverse effect on the people who were then contributing to the effort. I can go into more detail on this, if you wish.
Q. You did not wish them to study the problem of antisubmarine defense?
A. As I have said, I considered this inappropriate to project Lincoln. I am certainly in favor of studying antisubmarine warfare. Bear in mind the Lincoln project was supported roughly 85 percent—although it was a three service contract—it was supported between 80 and 90 percent by Air Force funds.
Q. Did you ever hear that Dr. Oppenheimer was in favor of studying antisubmarine warfare in connection with the Lincoln study?
A. No. As I have told you, my information on that came from suggestions by Dr. Zacharias in approaching people to work at the Lincoln summer study.
Q. Do you know what Dr. Oppenheimer's views were at that time, or are now as to the effectiveness of continental air defense?
A. At which time, sir?
Q. I asked about both the time of the Lincoln study and now.
A. What do you mean by the time of the Lincoln study? You mean the beginning or the end?
Q. We will start with the beginning. Do you know what his views were at the beginning of the Lincoln summer study?
A. I think his views were the same as mine and I believe the same as all of us that we were hopeful that there would be really substantial improvement in the air defense capability of the United States.
Q. Did you ever talk to him about that?
A. Yes, I think so.
Q. Was it his view that you could not have a 100 percent defense?
A. I don't know. As I have said, this was at the beginning of the study. Whether he thought it was possible or not would not have had any effect on me.
Q. Do you know what his views were at the end of that study?
A. I do not.
Q. Do you know what his views are today?
A. I do not.
Q. Did you ever hear Dr. Oppenheimer say that it was possible to have a 100 percent continental air defense?

A. No; I have had no contact with Oppenheimer so far as memory serves, as far as I now recollect, since that first session at the beginning of the Lincoln summer study.

Q. And you did not stay through to the end of the Lincoln summer study because you left?

A. I came there as part of my duties in the Air Force and I left the Air Force on the 1st—I left Washington on the 1st of July 1952.

Q. Returning to this visit in Princeton in May of 1952, what was the purpose of that visit?

A. I was asked that question I believe by Mr. Robb, and I tried to answer it as clearly as I could. Did you not understand it, or do you wish me to amplify it, or do you wish me to answer it again?

Q. I would like you to answer my question, sir.

A. In my answer to this question, which as near as I can recall was almost an identical question this morning, I said as a part of the discussion that we had had at lunch at Mr. Burden's house between Dr. DuBridge and Dr. Rabi, Mr. Burden, Mr. Norton, and myself, it had been mentioned by Dr. Rabi that in order to correct impressions that I had I should read the minutes of the General Advisory Committee. He told me that these minutes were the personal property of the chairman, that I could see them only by Dr. Oppenheimer's permission. He undertook to see if a meeting could not be arranged at Princeton to provide me the opportunity to study these minutes for this purpose. As I testified this morning, this tentative plan was not possible because of the illness that Dr. Rabi contracted.

When I was in Princeton for other purpose, therefore, in May of 1952, I called Dr. Oppenheimer and reminded him of this with the object of seeing whether it would be possible for me to see the minutes in his office or—this was in my mind—if that was not possible, to discuss these matters on which there seemed to be very considerable divergence of opinion between himself and me.

Does that answer your question?

Mr. GRAY. Does that answer your question, Mr. Silverman, or did you hear his answer?

Mr. SILVERMAN. I heard his answer.

By Mr. SILVERMAN:

Q. Did Dr. Oppenheimer tell you that the minutes were his personal property rather than the property of the Commission?

A. No; I didn't say Mr. Oppenheimer. As I testified this morning, Dr. Rabi told me that.

Q. Aren't you certain that Dr. Rabi didn't tell you that the minutes were the property of the committee, as distinct from the property of the chairman?

A. No, sir; as far as recollection serves.

Q. You said Dr. Oppenheimer did show you the majority and minority annexes to the October 1949 report?

A. That is correct.

Q. Did Dr. Oppenheimer say to you that it was the practice of the committee not to show minutes of the committee to any person without the consent of the members of the committee in order that the discussion might be quite free at committee meetings?

A. I don't recall whether or not he said that to me. Since I didn't expect him to show the minutes to me anyway, it would not make much impression.

Q. Did you expect him to show the report to you?

A. No; frankly I didn't.

Q. Had you tried to see the report before?

A. No; not to my recollection.

Q. Did you know that there was a copy of the report in the Defense Department?

A. I don't think I did know that.

Q. I think that Mr. Robb asked you a question about whether in that conversation in May of 1952 with Dr. Oppenheimer there was any mention of a public announcement as to whether we would go ahead with the thermonuclear developments and my notes don't show the answer to that question.

A. My answer, as I recall it, was that this subject was mentioned in one of the two annexes, and that we might have discussed this in connection with that, but I don't recall with any degree of reliability that we did discuss this particular subject.

Q. There had in fact been a public announcement as to our going ahead with thermonuclear developments 2 years before?

A. What is your question?

Mr. SILVERMAN. Mr. Reporter, would you mind reading it?

(Question read by the reporter.)

Mr. ROBB. Mr. Chairman, I think in fairness to the witness I should say that my recollection is that my question had to do with whether there was any discussion of an announcement that we would renounce the H-bomb.

Mr. SILVERMAN. I don't want to get into a dispute with Mr. Robb about our respective recollections. We are all trying to get the record clear on it. My own notes are the other way.

The WITNESS. My recollection jibes with what has just been said.

Mr. SILVERMAN. Then perhaps in the interest of clarity would it not be desirable to read my last question and the answer, and if the witness misunderstood my question and gave an answer——

Mr. GRAY. I suggest that you ask the witness the question you want to put to the witness, Mr. Silverman, and I would suggest that you listen to his reply. You have been so busy taking notes that is one reason you have missed some of these questions. I don't mind your asking the witness any question if you are trying to develop any point, including anything concerned with the veracity of the witness, but I think it is wasting the time of the board to ask an identical question of the witness, and go through these long answers when the transcript already reflects the question and answer.

Mr. SILVERMAN. Mr. Gray, I do not wish to be in a position of differing with you sharply on a matter as perhaps as relatively unimportant as this. My own recollection is that the answer was not precisely given before and if I am mistaken and taking up the time of the board, I am sorry.

The WITNESS. May I ask, Mr. Silverman, if you were going to ask for my reply to Mr. Robb's question that we go back to his original question, since I think there is a difference of opinion as to what his original question was. Is that what you want to do?

Mr. SILVERMAN. It is fine by me.

Mr. GRAY. You ask any question you want, Mr. Silverman.

Mr. SILVERMAN. I have asked the question. I have been told in effect that I have misstated Mr. Robb's question. I am sorry that Mr. Robb should feel that. My note is rather clear as to what Mr. Robb's question was.

Mr. ROBB. Mr. Question, may I just say this: I don't want to take up too much time. It is perfectly obvious that my question was directed to the first sentence of the fourth paragraph of a memorandum which the witness has read into evidence, which reads as follows: "After he showed me the GAC recommendation of December 1949 that the United States not intensify H-bomb development, but publicly renounce its development, and when I was pressing the point that such a course of action could well be disastrous to this country, Oppenheimer asked if I thought he were pro-Russian or just confused."

It is perfectly obvious that my question was bringing out from the witness that portion of his discussion with Dr. Oppenheimer.

Mr. SILVERMAN. It is perfectly obvious, and it seems to me that portion you have just read is exactly what I was asking about, and not at all the question you had thought you had asked, Mr. Robb.

The WITNESS. Just a moment. You said in following this up that there was a public announcement, did you mean that there was any such public announcement as the one mentioned there.

Mr. SILVERMAN. Yes. There was a public announcement by the President that we would go ahead with thermonuclear development.

The WITNESS. That is not what it says there.

By Mr. SILVERMAN:

Q. There was the recommendation of a public announcement the other way.
A. That is right. That is quite different.
Q. Was there any discussion in 1952 at your meeting with Dr. Oppenheimer in Princeton that there should now be an announcement, in 1952, that we would not go ahead with thermonuclear development.
A. I don't think so.
Q. And the discussion that you had with Dr. Oppenheimer in 1952 was about the recommendation in the GAC committee report in 1949.
A. As near as memory serves insofar as our discussion had anything to do with public announcements, it was.

Q. And you knew of course that question had already been resolved and that the President had announced we were going ahead with the thermonuclear development?

A. If what I said this morning gave any impression to the board that in May of 1952 Dr. Oppenheimer was pursuing in his discussion with me a recommendation that we at that time in May of 1952 publicly renounce the H-bomb, I think that such an impression would be false, and I did not mean to give that. You were attempting to clarify this.

Q. I was attempting to clarify what that discussion was about, yes.

I think you said in your direct testimony, did you not, that such question as you have as to Dr. Oppenheimer's loyalty was not based on any individual contact or detailed knowledge by you of his acts?

A. That is correct.

Q. I think you went further and said you did not feel that you really had an adequate basis for judging his loyalty or disloyalty.

A. That is certainly correct, and I think it is correct that I said it and it is certainly correct that I feel it.

Q. I think you also said that based on hearsay you have been suspicious or troubled about it for some time.

A. Troubled, yes.

Q. Would it be fair to say you have been suspicious of it for some time?

A. The circumstances which I pieced together by hearsay evidence, as I think I testified, were substantially similar to those that were listed among the allegations in General Nichols' letter were sufficient to cause me grave concern.

Q. Weren't you suspicious back at the time when you were first warned about Dr. Oppenheimer's loyalty when you joined the Rand project?

A. This, as I testified this morning, it was said to me that Dr. Oppenheimer during Los Alamos days had been considered a calculated risk. This statement was made to me by a person that I respect and it was not made as an idle statement. I took it seriously.

Q. And thereafter in your contacts with Dr. Oppenheimer you could not help being a little bit on your guard?

A. That is correct.

Q. And perhaps trying a little bit to see what might be beneath the surface of what Dr. Oppenheimer was saying?

A. That is correct. May I amplify this point?

Q. Certainly.

A. As I testified, particularly during my term with the Air Force as chief scientist for the Air Force—I don't want to emphasize this chief scientist business, because it doesn't mean anything, but this is just to identify the time that I am referring to—as I testified, I was on the opposite side of a pretty violent controversy from Dr. Oppenheimer in at least two cases. I was also on the opposite side—I mean on his side about people as to whom I had no question as to loyalty or motives. I have been involved in a great many—not a great many, but a number of pretty strong controversies in the military, and I think it is a fair general observation that when you get involved in a hot enough controversy; it is awfully hard not to question the motives of people who oppose you. This, I am sure, could not but have colored my views on the subject.

The nagging uncertainty in this particular case was the fact that I had heard the loyalty question raised by responsible people in a serious way.

If it ever comes to the day when we can't disagree and disagree violently in public and on national policy, then of course I feel that it will be a calamity for our democracy. I think perhaps I have said enough.

Q. I think since you candidly told us much of the information you have given is based not on your personal knowledge, I would like to review with you the items relating to Dr. Oppenheimer that you have of your own knowledge and see if those are correct. I will just run through them and see if they are correct as to your personal knowledge.

.That you visited Vista and you heard a draft report.

A. Read.

Q. Read. With which you disagreed as to three points.

A. Which was said to have been written by Oppenheimer.

Q. That it was said to have been written by Oppenheimer. You realize that of course would be hearsay.

A. Yes.

Q. Your personal knowledge is——

A. My personal knowledge includes the fact that the three people in whom I have the utmost confidence said it was written by Dr. Oppenheimer, as my personal knowledge.

Q. Was Dr. Oppenheimer there?

A. No, he was not there.

Q. Dr. Oppenheimer contributed or made valuable contributions in the Vista report which were helpful to the Air Force. I think you said you personally know that.

A. I can't say I know this in detail, but I am reasonably sure that this is so. I extended that of course to include the other fields of activity, fields of activity other than Vista as well.

Q. Dr. Oppenheimer's views with respect to the Lincoln summer study, you know only by hearsay?

A. Except as they were expressed during the first 3 days of the study, yes.

Q. In those first 3 days, he didn't say anything about giving up strategic airpower?

A. No.

Q. And you know that Dr. Zacharias——

A. I might point out that after the first session—I think it was the first session—in which Dr. Oppenheimer had taken a fairly active part and he came up to me afterward and said, "Did I do all right?"

Q. And what did you say?

A. I said "Yes," or words to that effect.

Q. Were you just being polite?

A. No.

Q. And you were present when Dr. Zacharias wrote the initials "ZORC" on the blackroad?

A. Yes.

Q. And you went to see Dr. Oppenheimer and you have told us of the conversation with him in May of 1952?

A. Yes.

Q. And, of course, you were there and you heard that conversation and participated in it.

Mr. SILVERMAN. That is all. Thank you.

Mr. GRAY. Mr. Griggs, if I though you could make the 3:30 plane, I would not ask you a couple of questions, but you have missed that plane.

The WITNESS. I am at your service, sir.

Mr. GRAY. I don't have very much actually. On the ZORC thing, you saw Dr. Zacharias write the things on the board. Had you before heard these letters used together?

The WITNESS. Yes.

Mr. GRAY. You may have testified about this, but do you remember when you first heard them?

The WITNESS. I did not testify about it. As near as I can recall, I learned about this abbreviation first in a telephone conversation with George Valley, and I would guess that this was roughly halfway through the summer study. But I can't be sure about that.

Mr. GRAY. The summer study was in 1950?

The WITNESS. 1952.

Mr. GRAY. When did this meeting take place at which Dr. Zacharias wrote the letters on the board, if you remember?

The WITNESS. That was at the Scientific Advisory Board meeting in Cambridge in, I believe, September of 1952. It was after the completion of at least the formal phases of the summer study, and it was on the occasion at which Dr. Zacharias was presenting some of the conclusions of the Lincoln summer study to the Scientific Advisory Board of the Air Force.

Mr. GRAY. The magazine article you mentioned came out later than either of these events?

The WITNESS. I don't know, sir. I would have to look it up.

Mr. GRAY. Was this name in 1952 well known among physicists, that is, the summer and fall of 1952?

The WITNESS. Well known among the physicists, speaking of. the physical profession?

Mr. GRAY. That is right.

The WITNESS. No; I don't think it was well known.

Mr. GRAY. Do you know that it had appeared publicly in print at the time that you saw Dr. Zacharias use it? My question should be, Do you know whether it had? I don't know myself.

The WITNESS. I am afraid, sir, I would have to check dates on that. As near as I can recall, it did appear in print in the Fortune article and whether that was before or after the Scientific Advisory Board meeting, I would really have to check.

Mr. GRAY. Do you know the origin of the putting of those letters together?

The WITNESS. No more than I have told you and Zacharias on explaining of what the letters stood for, which coincided with what George Valley had told me over the telephone.

Mr. GRAY. A question now about the Vista report. You have been questioned a good deal about the meeting you attended in November 1951, I suppose it was.

The WITNESS. Yes, sir.

Mr. GRAY. And the first draft or the draft of the introduction to chapter 5, were there substantial changes in that introduction between the time you heard it read at this meeting and when the report finally appeared and was published?

The WITNESS. Yes, there were. There were some very substantial changes. The first time I referred to was deleted. If you are going to get into this question, however, I should point out that there were two versions of the printed Vista report, one of which was called back, I believe, for cecurity reasons. The first edition was called back for security reasons, I believe, and later reissued. The changes to which I refer, as near as I can recall, and I am reasonably sure with regard to this grst point, that was deleted in both of these published versions.

Mr. GRAY. So that the two version really are not important in trying to get at the question as to whether there were substantial changes.

The WITNESS. No.

Mr. GRAY. Could you agree with the description that the changes were only an emphasis and not in substance?

The WITNESS. One of the changes which I was most concerned was the deletion of this particular statement with respect to withholding the use of our Strategic Air Force until—the Strategic Air Force for attack on their cities until our cities were attacked. That was deleted. I would say this is a change in substance, if I understand your question.

Mr. GRAY. Do you have any questions?

Dr. EVANS. No.

Mr. MORGAN. No.

Mr. ROBB. No.

Mr. SILVERMAN. I am just wondering on this business of Dr. Zacharias writing on the blackboard the initials ZORC.

By Mr. SILVERMAN:

Q. Is it possible that the occasion of his doing that might have been after the magazine article?

A. As I say, I would have to check dates to find out.

Mr. SILVERMAN. Thank you.

The WITNESS. I am reasonably sure that—in fact, I am as sure as I can be of anything in my memory—that my first hearing of these initials, which as I said came in a telephone conversation to the best of my memory, that was prior to any publication of these initials in this connection that I saw.

Mr. SILVERMAN. I have no further questions.

Mr. GRAY. Thank you very much.

(Witness excused.)

Mr. GRAY. Dr. Alvarez, do you wish to testify under oath? You are not required to do so.

Dr. ALVAREZ. I would like to testify under oath, sir.

Mr. GRAY. Would you give me your full name.

Dr. ALVAREZ. Luis Walter Alvarez.

Mr. GRAY. Would you raise your right hand? Luis Walter Alvarez, do you swear that the testimony you are to give the board shall be the truth, the whole truth, and nothing but the truth, so help you God?

Dr. ALVAREZ. I do.

Mr. GRAY. Would you be seated, please.

Whereupon, Luis Walter Alvarez was called as a witness and, having been first duly sworn, was examined and testified as follows:

Mr. GRAY. It is my duty to remind you of the so-called perjury statutes. Are you familiar with them?

The WITNESS. In a broad way I am, yes.

Mr. GRAY. May I ask that if in the course of your testimony here it becomes necessary for you to disclose or refer to restricted data you notify me in advance so that we may take the necessary steps in the interest of security.

Finally, I should say to you that we treat these proceedings as a confidential matter between the Atomic Energy Commission and its officials and witnesses, on the one hand, and Dr. Oppenheimer and his representatives on the other. The Commission will make no releases about these proceedings. On behalf of the board, I express the hope that the witnesses will follow the same course.

The WITNESS. Yes.

Mr. GRAY. Mr. Robb, will you proceed.

Mr. ROBB. Thank you.

DIRECT EXAMINATION

By Mr. ROBB:

Q. Where do you live at present, Dr. Alvarez?
A. I live at Berkeley, Calif.
Q. What is your present occupation or position?
A. I am professor of physics at the University of California.
Q. How long have you been there?
A. I have been at the university for the past 18 years with time off for war-work.
Q. Would you tell us something about your academic training and background, please, sir.
A. I went to the University of Chicago both for my undergraduate training and also my graduate work in physics. In my graduate career, I was very fortunate in having as my research professor Dr. Arthur Compton who is perhaps best known to this board as the director of the wartime Metallurgical Laboratory. I worked with him in the field of cosmic rays. I took my doctor's degree in the field of optics.
Q. In the field of what?
A. Optics. After I left the University of Chicago with my Ph. D.
Q. Did you publish any papers?
A. I published 2 or 3 papers during that period, one of them as coauthor with Dr. Compton.
Q. Very well, go ahead.
A. After I received my Ph. D. degree, I had the opportunity to go to the radiation laboratory at the University of California at Berkeley. This was probably the most important thing that happened to me in my scientific career. I became associated with Professor Lawrence and got into the field of nuclear physics, which I had not been in before.

For the first 2 years there in Berkeley, I was a research assistant in the laboratory and then I was asked to join the faculty of the university, first as an instructor and then working up through the ranks to the position of professor of physics, which I was given in 1946 just after the war. I have been professor of physics ever since.

Q. You mentioned an interim period during the war. Did that begin in about 1940?
A. Yes; in November 1940. The National Research Defense Council set up a laboratory at MIT to work on microwave radar. This was a field which had been developed by the British. We in this country had nothing in that field and so this laboratory was set up. I was one of the charter members.
Q. With whom did you work there?
A. The director of the laboratory was Dr. Lee DuBridge and there were many other nuclear physicists, roughly of my age, who worked in the laboratory.
Q. How long did you stay there?
A. I stayed there until the summer of 1943 at which time the main radar projects in which I was concerned were well along toward production or in production, and since my primary usefulness is not in the field of production but rather in research and development, I felt this was a natural time to leave and join the Manhattan District.
Q. How did you happen to join the Manhattan District?
A. I had had several offers from men in the district. I had at least one from Dr. Oppenheimer, I had one from Arthur Compton, and I had conversations with Professor Lawrence about joining his staff.
Q. Do you recall any particular conversation you had with Dr. Oppenheimer at about that time with respect to whether or not you would join the Manhattan District?
A. Sometime, I believe, in 1942, Dr. Oppenheimer asked me to come down to New York from Boston to talk with him about problems in the field of the Manhattan District. He was anxious that I join him in his work, and I remember a most interesting afternoon we spent together, during which time he told me for the first time the possibility of building a thermonuclear weapon.

Q. What did he tell you about it?
A. He told me in some detail of the scientific design, as he then envisaged it, and pointed out how it would be triggered. * * *

By Mr. ROBB:
Q. Would that have been a weapon of great power, in the megaton range?
A. Yes. As Dr. Oppenheimer pointed out to me, there was no apparent limit to the magnitude of the explosion, whereas there appeared to be a limit to the magnitude of the explosion from what we now call an atomic bomb.
Q. Did Dr. Oppenheimer in that discussion raise any question with you either about the feasibility or the morality of constructing such a weapon?
A. He certainly raised no question about the morality of the thing. We had a technical discussion to which I contributed essentially nothing about the feasibility of it from the scientific point of view.
Q. By the way, how long have you known Dr. Oppenheimer?
A. I have known him for 18 years.
Q. Are you here as a witness today because you want to be here or because you were asked to come?
A. I certainly find it an unpleasant duty but I consider it to be a duty to be here. I was asked by General Nichols to come.
Q. Following that discussion, did you go to Los Alamos?
A. Not following that discussion; no.
Q. I mean subsequently to it.
A. Subsequently to it, I did go to Los Alamos, yes; but not as a result of that conversation, no.
Q. How long afterward was it?
A. In the spring of 1943 I went to California in connection with the radar work and stopped at Berkeley, which was the first time I had been in Berkeley since 1940, and I spent a week with Professor Lawrence looking at the work that was going on at Berkeley in the isotope separation and asked Professor Lawrence whether it would not be a good idea to join. I was homesick for the kind of work in physics which was going on there and it had great appeal to me. I told Professor Lawrence that my usefulness at the radiation laboratory at MIT was almost coming to an end, and I could make a break at this point. He said he would be very happy to have me come and we made a tentative arrangement that I would come as soon as I got back from a trip to England which I had to make in the summer. Shortly after that, Dr. Bacher and Dr. Bainbridge, who were both at the radiation laboratory at that time, talked with me and told me that they thought it would be better for me to go to Los Alamos where they were going. They were both leaving the radiation laboratory at about this time and said if I were shifting to the atomic program, it would be better to go to Los Alamos where the problems were more difficult rather than to Berkeley where the problems were essentially solved. So, I agreed with them and made arrangements with Dr. Oppenheimer to go to Los Alamos. When I was in England, I received a wire from Dr. Oppenheimer asking me if I would, for a while, work with Fermi at Chicago. Apparently Fermi had been trying to get his former student, Segre, who was then at Los Alamos to come to Chicago to help him, and the professor suggested I go instead of Segre because Segre was deep in business.
Q. So you paused at Chicago?
A. So I went to Chicago for 6 months and then proceeded to Los Alamos.
Q. You arrived at Los Alamos approximately when?
A. In the spring of 1944.
Q. When you got to Los Alamos, will you tell us whether or not you found there constructed a liquid hydrogen plant?
A. Yes, * * *.
Q. Was the liquid hydrogen plant a facility for making a fission weapon?
A. I can think of no importance that it had in that connection.
Q. How long did you stay at Los Alamos?
A. I stayed there until approximately November of 1945.
Q. What was your duty there?
A. When I first arrived, I was assigned as a sort of assistant to Dr. George Kistiakowsky who was in charge of the explosives work in connection with the implosion weapon.
My first technical job was to set up an experiment designed to test some important features of the implosion method. Then, shortly after that, some young men working with me and I got into the field of the detonating mechanism

for the high explosive, and I think that this was my most important contribution at Los Alamos in the system of setting off the bomb. I do not believe it could have been done without this contribution.

Q. Did there come a time when you made a rather long airplane flight?

A. Yes; in the spring of 1945 when our detonator system was through its development and was to proceed to production it was turned over to Dr. Bainbridge to put into final form, and I was essentially out of a job at that point. I went to Dr. Oppenheimer and asked him what I should do now that this first job of mine was complete, and I said that I hoped he could get me a job which would get me overseas. He said that the laboratory wanted to have some method of testing the effectiveness of the bomb over enemy territory.

You see, normally a military weapon is tested on a proving ground. Many rounds are shot and one knows all its characteristics. But, in this particular case, the weapon was so expensive and there were so few of them that it seemed more reasonable to take the proving ground over the enemy territory to measure the blast wave, the pressure shock waves and thereby to measure the efficiency of the bomb.

So, I took that job on in the spring of 1945.

Q. What did you do?

A. A small group working with me designed equipment which could be fitted into a parachute-borne pressure gage which could be dropped over the point where the bomb was released, and then these pressure gages had radio transmitters which would send signals back to an airplane where they could be recorded on cathode ray oscilloscopes by photography, and when the films were analyzed later, one could measure the peak pressure in the shock-wave and by scaling laws in aerodynamics one could then compute the blast of the bomb.

Q. Did you go to Japan?

A. I spent about 2½ months on Tinian Island and I rode in the observation plane during the raid on Hiroshima.

Q. How far behind the plane that dropped the bomb were you?

A. As I remember, we flew formation approximately a quarter-of-a-mile behind from the time we left Iwo Jima until we got back from the Japanese coast on the way out.

Q. And you measured the effect of this explosion?

A. Yes. I had to be adjusting the receiving apparatus for this instrumentation during our sharp turn after our bomb was dropped and our getaway run. We were essentially running away from the shock-wave with our airplane. So I was quite preoccupied during this time.

Q. And thereafter you returned to Los Alamos?

A. As soon as I got back from Tinian, I packed up my household goods as quickly as possible and moved my family back to Berkeley; yes. There was nothing essentially for me to do at Los Alamos. Both of my jobs were complete.

Q. And you resumed you academic career?

A. Yes; I did.

Q. Did you continue any work as a consultant for the Atomic Energy Commission or the radiation laboratory?

A. For the first 2 years after the war, I believe that most, if not all, of my salary was paid by the Atomic Energy Commission. Since then, one-third of it has been paid by the University of California for one-third teaching duties that I now exercise and the other two-thirds is paid by the Atomic Energy Commission through the University of California as a contractor.

Q. Doctor, directing your attention to September 1949 when the Russians exploded their first atomic bomb, did that cause some concern on your part?

A. Yes; it caused a great deal of concern on my part. I tried to make up my mind what was the right thing to do. I had been spending 4 years doing basic research again. I think of it as sort of being recharged after 5 years of military development work. I had to take awhile to get back into the frame of mind of a practicing physicist. I had been concentrating my attention on that phase of my career and now, suddenly, it appeared that a crisis had arrived and perhaps I should get back into the field of atomic energy.

Q. Why did you think a crisis had arrived?

A. The Russians had exploded an atomic bomb, and I thought that your own program had not been going terribly fast. It certainly had not been going at nearly the rate it had during the war, but this is quite natural.

Q. Did you discuss with any of your colleagues what ought to be done?

A. Yes; I did. I saw Professor Lawrence the next day, and I told him that I thought we should look seriously into the business of constructing the super weapon which had, as far as I knew, been neglected in this 4-year period. I had not followed the situation closely enough to be sure that it had been neglected but that was my impression.

Q. Did you make any inquiry to see whether or not your feeling was correct as to whether it had been neglected?

A. Yes. Professor Lawrence and I got on the phone that afternoon and called Edward Teller at Los Alamos and asked him if we could come down and talk to him in the near future, and, as I remember, within a day or two, we took a plane to Los Alamos where we did talk to Dr. Teller and found out the present rather inadequate status of the super program.

Q. Beginning at about that time and the next few weeks, Doctor, did you keep any notes in the form of a diary as to what your activities were in respect of a program for the development of the super bomb?

A. Yes; I did. I would like to explain how I came to do that. I am not by nature a particularly methodical person, and I have never kept a diary except for a few months when I was in high school and one other rather important occasion, and that was when I was in charge of coordinating the activities during the first few months of the radiation laboratory at MIT. Dr. DuBridge put me in charge of meeting schedules and during that period I kept a detailed diary of everything that was going on in the laboratory, the state of development, so that I knew where things were.

At the end of the war, Dr. DuBridfge told me that this turned out to be one of the most valuable documents they had because there was no other record of the early days of the laboratory. Later on, there were lots of notes, memoranda, nad reports, but in the first 3 months, the only record that was there was my diary of the laboratory. It turned out to be of great use in the patent field and it had a lot to do with clarifying the ideas of the person who wrote up the history. So, I was aware of the fact that I had done this once to good avail and it seemed now that a new program was about to be started and I might as well keep a diary again. That is my reason for doing it.

Q. Do you have it with you, Doctor, the original of that diary?

A. Yes, I have my typewritten sheets here. They cover the period of about 3 weeks from the time the Russian bomb was dropped.

Q. Typewritten or longhand?

A. They are in longhand.

Q. Doctor, the security officer using my jackknife has removed 2 or 3 words from the typewritten copy.

Mr. ROBB. Mr. Rolander, I wonder if you would hand that copy with those excisions which haveto do with technical matters to our friends across the table.

Mr. SILVERMAN. Can we take a minute to look at this? Are you going to question him about it?

Mr. ROBB. Yes, right now.

Mr. SILVERMAN. Let us take a minute or two to glance over it.

Mr. GRAY. All right .

Mr. SILVERMAN. Unless you are going tor ead it into the record——

Mr. ROBB. I am going to read it item by item and ask the witness to explain it.

By Mr. ROBB:

Q. I would like to run this through with you and ask you to amplify.

"October 5, 1949. Latimer and I independently thought that the Russians could be working hard on the super and might get there ahead of us. The only thing to do seems to get there first—but hope that it will turn out to be impossible."

Would you explain to us what you meant by that "hope that it will turn out to be impossible."

A. By that I meant that there might be some fundamental reason in the physics of the bomb that would prevent anyone from making it work just in the same sense that people have often said that you cannot make a thermonuclear weapon that will burn up the atmosphere and the ocean. I hoped that some such law would prevail and keep anyone from building it, because then our stockpile of atomic weapons gave us the lead on the Russians.

Q. You mean if it turned out that it would violate some law of nature the Russians could not make it either?

A. That is right, because if they did make it, that would give them a great jump ahead of us and essentially nullify our stockpile of atomic weapons.

Dr. EVANS. The laws of thermodynamics might tell you it could not be done?

The WITNESS. Yes, something of that sort.

By Mr. ROBB:
Q. You thought you ought to find out.
A. I said we can't trust this hope, but let us find out.
Q. Who is Latimer?
A. He is dean of chemistry at the University of California.
Q. Is there anything you have to add to that first item?
A. No, I can't think of anything.
Q. "October 6, 1954: Talked with E. O. L. about the project and he took it very seriously—in fact he had just come from a session with Latimer. We called up Teller at Los Alamos to find out how the theory had progressed in the last 4 years. Since E. O. L. and I were to leave tomorrow for Washington, we decided to go a day earlier and stop in for a day at Los Alamos to talk with Teller. Left San Francisco at 7:30 p. m."
Q. Who was E. O. L.?
A. E. O. L. is the director of the radiation laboratory at the University of California, Prof. Ernest O. Lawrence.
Q. Have you any recollection of what Dr. Teller told you in the call that you mentioned about how the theory had progressed in the last 4 years?
A. You mean during the visit, not during the telephone call? He obviously could not tell us on the telephone.
Q. I will get to October 7. Is there anything further to add to that item, October 6?
A. No.
Q. "October 7, 1949: Arrived Albuquerque 3 a. m., and spent rest of night in Hilton Hotel. Left by Carco plane for Los Alamos at 10 a. m., and spent rest of day talking to Teller, Gamov, Manley, and Ulam. They give project good chance if there is plenty of tritium available. There must be a lot of machine calculations done to check the hydrodynamics, and Princeton and L. A. are getting their machines ready. We went back to Albuquerque with Teller and talked until bedtime. We agreed that a conference should be called at L. A. next month to see what should be done. L. A. had been talking about one for early next year. We can't wait too long. Teller brought up D_2O pile as easy way to get excess neuts. E. O. L. and I said we would get going on that at once. Left Albuquerque at 3:30 a. m."
In your talk with Teller, Dr. Manley, Gamov, and Ulam, did you ascertain from them how much work had been done on thermonuclear?
A. Yes. As far as I can recall, Dr. Teller told us that he had been working on the program essentially since the end of the war. Dr. Gamov had been there for approximately a year on leave from George Washington University. Dr. Ulam had done some work on it and there had been a modest program of machine calculations to check hydrodynamics. But that is essentially all. The program had essentially not been of any magnitude worthy of the name.
Q. Beg pardon?
A. The program essentially did not exist except for Teller.
Q. You mention "must be a lot of machine calculations done to check the hydrodynamics, and Princeton and L. A. are getting their machines ready." What did you mean by that?
A. I referred there to the so-called Maniac, an electronic calculating machine invented by Dr. Von Neumann of the Institutte of Advanced Study which was being built at Princeton and a copy being built at Los Alamos to do these terribly involved calculations.
Q. Was that the machine at Princeton under Dr. Oppenheimer's auspices?
A. I don't know. I know Dr. Von Neumann is a member of the Institute and, therefore, is under Dr. Oppenheimer, but I do not know whether the machine was the property of the Institute or the property of the University of Princeton.
Q. Did you have any reason to believe at that time that Dr. Oppenheimer would not be ready to go ahead with this program?
A. Of course not. The most enthusiastic person I had ever met on the program of the super weapon was Dr. Oppenheimer.
Q. Is there anything further to add in connection with this October 7 entry? L. A., I assume, means Los Alamos.
A. That is right.
Q. Los Alamos?
A. It means Los Alamos. I would like to say something about this program, about the D_2O pile. This is a heavy water reactor and it has virtue because in a heavy-water reactor there are lots of free neutrons available that are not available in the graphite moderated reactors which the Commission then owned

almost entirely. As Teller pointed out, tritium was * * * material for the production of hydrogen bombs. To produce tritium, one needs excess neutrons and, therefore, Professor Lawrence and I, who were looking for something to do to help the program along, said we would start a program to build such piles for the Commission.

Q. "October 8, 1949: Arrived Washington after lunch. Went to AEC and talked with Pitzer, Gen. McCormack, Latimer, and Paul Fine. Told them what we planned to do and got good response.

"Had dinner with Alfred and Mannette Loomis at Carleton Hotel."

Pitzer, who he is?

A. He is Dean Kenneth Pitzer who was then Director of Research of the Atomic Energy Commission.

Q. General McCormack, who was he?

A. I believe he was the head of the Military Liaison Committee.

Q. Latimer is the same Latimer?

A. Yes.

Q. And Paul Fine, who was he?

A. Paul Fine was, I would guess, a sort of administrative assistant to Dr. Pitzer. I might say that I was somewhat surprised at Dr. Fine's reaction, because he was the first person that I had met since the Russian bomb went off who was not enthusiastic about the problem of building the Super weapon. I attributed this to the fact that he had all during the war and was still then sort of an administrative assistant and I put him down as a person with essentially no imagination and discounted this.

Dr. EVANS. He was not enthusiastic?

The WITNESS. No. He was not, but knowing his nature, I was not upset by this.

By Mr. ROBB:

Q. "Told them what we planned to do." What was that?

A. That we planned to go into a vigorous program of building heavy water moderated supplies to supply free neutrons to make tritium.

Q. The item about dinner does not refer to the thermonuclear program, I assume.

A. No.

Q. "October 9, 1949—Sunday: Had breakfast with Mr. LeBaron—Deputy Secretary of Defense for Atomic Affairs. Told him of our plans. Went to R. W. panel meeting for most of the day. Program approved but probably nothing will happen. 'Gram of neutrons' recommended—that ties in well with our program. (At noon, E. O. L. heard he was a father for the 6th time.) Spent afternoon and evening with Mr. and Mrs. LeBaron and talked with him about several phases of the situation."

"Told him of our plans"; are those the same plans you referred to?

A. The plans to build a heavy-water reactor.

Q. "Went to RW panel meeting." What does that mean?

A. That was an ad hoc panel on radiological warfare. This was a subject which was very close to Professor Lawrence's heart. He had made serious proposals in the Defense Department that warfare could be waged effectively by the use of radioactive products. I was not a member of the RW panel but Professor Lawrence asked me to come along since I was part way there after my trip to Los Alamos.

Q. "Program approved but probably nothing will happen." What did you mean by that?

A. People agreed that the idea of radiological warfare was attractive in many ways but again the country had no supply of free neutrons * * *.

Q. "Gram of neutrons recommended." Is there any comment to make about that?

A. No. The panel said that it believed the Atomic Energy Commission should initiate a program to provide this gram of neutrons; and when I say this fits in well with our program, our program to build heavy water piles would provide we hoped considerably more than a gram of neutrons. Therefore, we would have available either tritium or radioactive warfare agents.

Q. What was the attitude of Mr. LeBaron with respect to your proposals?

A. He was, of course, quite enthusiastic about it.

Q. I guess there is nothing further to add about that item, is there?

A. No.

Q. "October 10, 1949: Saw Ralph Johnson at AEC and made arrangements to go to Chalk River to see their pile. Talked with General McCormack about plans. Went to Capitol and had lunch with Senator McMahon and Representative Carl Hinshaw. Told them of our plans and got good reactions. Stressed need for cooperation between British, Canadians, and ourselves. They said they would be in Berkeley within 10 days. Also said to call them if anything held up our plans. Back to AEC—saw Lilienthal. He was only lukewarm to proposition. Saw all four other Commissioners, who seemed to like what we were setting out to do. They weren't too happy about our going to Chalk River but finally agreed to give us their blessing, and make it official. We had planned only a personal visit to Bernard Kinsey. On way to plane stopped in to see RCA color television demonstration."

Who was Ralph Johnson?

A. He was one of the administrative people at the AEC. I do not remember him in detail.

Q. What was the pile at Chalk River?

A. Chalk River is the Canadian atomic energy establishment where they had built the outstanding heavy water pile. There was only one in this country; it was a very low-power pile at the Argonne Laboratory. The Canadian one was the one which we planned to use as a prototype of the ones which we were contemplating building, and we thought as long as we were in the East we should have a look at this thing. We had only seen pictures and heard descriptions of it.

Q. "Talked with General McCormack about plans." He is the same one you mentioned before?

A. Yes.

Q. "Went to Capitol and had lunch with Senator McMahon and Representative Carl Hinshaw." Would you tell us about that?

A. Yes; I would like to do that because various members of the scientific fraternity at various times told me that Professor Lawrence and I used undue influence by going to see Senator McMahon and various Congressmen to try to influence them to get the hydrogen bomb program started. What actually happened was that about a month before this, and before the Russian explosion, Carl Hinshaw, who is the leading Member of Congress in the field of aviation and air navigation and things of that sort, called at the laboratory and he and I had a very long discussion on the present state of the air navigational art in this country. This is a field in which I got some competence during the war. Mr. Hinshaw found that my views on the subject were somewhat different than the official CAA views and asked me if I would write him a detailed letter explaining my views. I prepared a 35-page typewritten document with lots of diagrams expressly for his personal use, and I had this with me when I arrived in Washington. So, I called up Congressman Hinshaw and told him that I had the document and I would like to bring it to him at the Capitol. I mentioned that Professor Lawrence and I were there together. As soon as he heard that he said, "Please hold down and I will call you back in about 5 minutes." He called back and said, "I have just spoken with Senator McMahon, who would like you and Professor Lawrence to have lunch with him at his chambers in the Capitol today if you can do so." And that is why we had our conference with Senator McMahon and Congressman Hinshaw.

Q. The next sentence, "Told them of our plans and got good reactions." What can you tell us about that?

A. Both of these gentlemen told us that they thought we were doing the right thing. They were very happy to see some action in the field of thermonuclear weapons. They both expressed concern about the fact that so little was going on in the AEC in this field. They said, "We hope you can get something going."

Q. I guess the next sentence or two needs no explanation unless you think they do, "Stressed need for cooperation between British, Canadians, and ourselves."

A. By that I meant that the Canadians were far ahead of us in the heavy water pile technology and that if we were to be able to move rapidly, we would need cooperation of the Canadians.

Q. "They said they would be in Berkeley within 10 days. Also said to call them if anything held up our plans."

Was there any discusion about what might hold up your plans?

A. I can't remember anything of that nature.

Q. "Back to AEC—saw Lilienthal. He was only lukewarm to proposition." Have you any added comment to make about that?

A. I must confess that I was somewhat shocked about his behavior. He did not even seem to want to talk about the program. He turned his chair around and looked out the window and indicated that he did not want to even discuss the matter. He did not like the idea of thermonuclear weapons, and we could hardly get into conversation with him on the subject."

Q. "Saw all 4 other Commissioners, who seemed to like what we were setting out to do. They weren't too happy about our going to Chalk River, but finally agreed to give us their blessing and make it official."

Is there any comment on that?

A. I do not know the reasons for them not wanting us to go, but I assume it had something to do with the political situation, and I have nothing to add there.

Q. "We had planned only a personal visit to Bernard Kinsey." Who is he?

A. Dr. Bernard Kinsey is one of the chief physicists at the Chalk River laboratory, and he was a member of the radiation laboratory in 1953 and 1936 and, therefore, a personal friend of both Professor Lawrence and me.

Q. I guess the RCA color television demonstration is immaterial to this.

A. To this, yes, sir.

Q. "October 11, 1949: In New York, found we were unable to get seats to Ottawa. We went to see Rabi and found him very happy at our plans. He is worried, too. I took plane home and arrived in Berkeley at 11 p. m."

What can you tell us about your conversation with Dr. Rabi?

A. I think I can sum it up best by trying to paraphrase what Dr. Rabi said. It was somewhat complimentary and I hope you excuse it if I say it. What he said was essentially that "It is certainly good to see the first team back in." He said, "You fellows have been playing with your cyclotron and nuclei for 4 years and it is certainly time you got back to work, and I am awfully happy to see you back in the business."

Q. What was he worried about?

A. I can't remember that he was worried about anything.

Q. You said that he was worried, too.

A. He was worried about the Russian explosion and the fact that our lead in the field of atomic energy had apparently been cut. He agreed with us that the hydrogen bomb program was a very good program, and he was happy we were doing something to get it reactivated.

Q. "October 12, 1949: Told some of the men at the lab of our trip. Don Cooksey, Brobeck, McMillan, Serber, Seaborg, Thornton, Gordon, Fidler. All said they would join new project."

By the laboratory, you meant what laboratory?

A. I mean the top man at the radiation laboratory at the University of California.

Q. Who is Don Cooksey?

A. Associate director of the laboratory.

Q. Brobeck?

A. Assistant director and chief engineer.

Q. McMillan?

A. Professor of physics and Nobel prize winner in physics.

Q. Serber?

A. Professor of theoretical physics at the university.

Q. Seaborg?

A. Professor of chemistry, also a Nobel prize winner.

Q. Was he a member at that time of the GAC?

A. Yes; he was.

Q. Thornton?

A. Robert Thornton, professor of physics and in charge of the 184-inch cyclotron.

Q. Gordon?

A. He was Brobeck's assistant in the engineering department.

Q. Fidler?

A. He was, I believe, at that time AEC representative in the Bay Area.

Q. "All said they would join new project."

A. That means the project of building heavy water piles. I might point out that this meant quite a change for all of them. Mr. Brobeck was at that time busily engaged in designing the bevatron which recently ran for the first time and everyone else was busily engaged on a program that he would much rather do than build heavy water piles, but all agreed that it was the right thing to do at that time.

Q. Project for building heavy water piles was for the purpose of developing the thermonuclear; is that right?

A. It was for the purpose of supplying tritium for tests of the thermonuclear weapon; yes, sir.

Q. Is there anything else to add about that entry?
A. I can't think of any.
Q. "October 13: E. O. L. returned and we had long conference about plans. Discussed site and technical plans." E. O. L., I assume, is Dr. Lawrence?
A. That is right.
Q. Site for what?
A. That was the site for the heavy water piles. The main requirement there is lots of cooling water.
Q. "October 14: Larry Hafstad, Head of Reactor Division of AEC, was present—we had called him from Washington. Dave Griggs and Bob Christie were present also. Decided sea water cooling O. K. and decided put pile on ocean north of S. F. and south of Tomales Bay. Hafstad will be in Chicago on Monday and will send out some pile experts as soon as possible next week. Decided to build pile in units, to give chance for rapid change. Probably H_2O cooling O. K. as at Chalk River. Took Hafstad to airport and went to Woodside to see Mr. Neylon. Home at midnight."
How did you happen to call Mr. Hafstad, or Dr. Hafstad?
A. Dr. Hafstad was the Director of the Reactor Division of the AEC, and we were people who wanted to build piles but who had no technical qualification in that field. We had never been in the reactor business. We thought the one thing we could supply was the ability to build large-scale apparatus and build it fast. This is what Professor Lawrence's laboratory did during the war, and the instance of the Oak Ridge isotope separation plant.
Q. What was the status at that time of the reactor program so far as you knew?
A. I thought it was in the doldrums. I don't know precisely how many piles had been built since the war. These records are available, but essentially no new additional piles had been built for several years after the war as contrasted with the fact that during the war there was the original Chicago pile, the Oak Ridge pile of a different design, the Hanford piles, water-cooled gravity moderated piles, and the heavy water pile at Chicago, four different kinds of piles had been built in a very short space of time, and in several years after the war no pile had been built.
Q. "Dave Griggs and Bob Christie were present also." Who was Dave Griggs?
A. Dave Griggs was the gentleman who just came out of this room. He was professor of geophysics of the University of California at Los Angeles.
Q. Christie?
A. Bob Christie is professor of physics at California Institute of Technology, and is the man who did the theoretical design on the Nagasaki bomb.
Q. What was their function at this meeting?
A. Dave Griggs was there because we hoped that he would want to join us. He is an enthusiastic person who likes to get things done in a hurry. He was sympathetic to our point of view that such piles should be built. Bob Christie was there because he was an expert in the field of neutron diffusion and pile technology. He designed to so-called water boiler at Los Alamos.
Q. "Decided sea water cooling O. K. and decided to put pile on ocean, north of S. F." I guess that means San Francisco.
A. Yes.
Q. "And south of Tomales Bay." That is near San Francisco?
A. No.
Q. Have you anything to add to that?
A. No; it was not a good decision and we changed it in a couple of days.
Q. "Hafstad will be in Chicago on Monday and will send out some pile experts as soon as possible next week." Am I to gather with that that Dr. Hafstad was with you?
A. It certainly seemed that way to us. He came out himself and he said he would send people who were competent in the field of pile design to help us. One of his great difficulties, as I see it, was that piles were not getting built because apparently people wanted to design the perfect pile and build the perfect pile and not take it in easy steps. We on the other hand were a group who said we don't care about the niceties of the thing; we are not experts. We want to get some piles built, and we will build them fast. It was a different approach than had been used before.
Q. You mean you wanted to find out and didn't think you had the chronometer to do it?
A. That is precisely it.
Q. "Decided to build pile in units, to give chance for rapid change." What do you mean by that "give chance for rapid change"?

A. I believe Professor Lawrence thought we should build a very large concrete shield with a number of tanks in it to hold heavy water, and to provide facility for changing the geometrical arrangement. This philosophy is now incorporated in the so-called swimming pool reactors where one can make changes easily whereas the first piles were built so that no fundamental changes could be made. The geometry was set in the design.

Q. "Probably H_2O cooling O. K. as a Chalk River." I assume that speaks for itself.

A. Yes, that means you can cool the pile with ordinary water rather than with heavy water. The Chicago pile was cooled with heavy water. The Chalk River was moderated with heavy water and cooled with light water.

Q. "Took Hafstad to airport and went to Woodside to see Mr. Neylon."

A. Mr. Neylon is a member of the Board of Regents of the University of California, and at that time was the chairman of the Radiation Laboratory Committee of the Regents.

Q. What was your purpose in seeing him?

A. Professor Lawrence wanted to tell him that the radiation laboratory was thinking of embarking on a large-scale construction program, and he thought it right that Mr. Neylon should know that such a thing was in the wind.

Q. Did he approve?

A. Yes, he approved. We didn't tell him anything about tritium. I don't know whether he was cleared.

Professor Lawrence said this was an important thing from the national standpoint, and Mr. Neylon agreed it was the right thing to do.

Q. "October 15: Cal beat USC. Parties at Jenkins and Serber. Long talk with Dave Griggs at latter. He thinks we are doing the right thing, but isn't ready to join yet."

Who is Jenkins that you mention?

A. He was professor of physics at the University of California.

Q. And Serber?

A. I have already mentioned him.

Q. And Dave Griggs is the same Dave Griggs you mentioned?

A. That is right. I would like to point out here that the reason that we didn't try to get Dave Griggs to work with us is that he alone of all the people in the field of radar had stayed on in war work for 2 years after the war. He was through all in setting up the Rand project at Santa Monica which is doing such a fine job for the Air Force. I had once told Griggs privately that if there was another war he had 2 years of credit in my book, that he didn't have to come in for 2 years, because he had stuck out the last war for 2 years overtime. So we didn't try to ask him to join.

Q. "October 16: Sunday—Rest. Drew Pearson's first mention of 'H-bomb'." I guess there is no need for explanation of that.

"October 17. Monday: Talked with Hafstad, Zinn and Pitzer this afternoon on phone. Things are going as well as possible. Zinn will send out someone toward the end of this week. He hopes to be here after the Oak Ridge info. meeting, which starts in about a week. He says he has ideas about how to do the job, and is not sure we should just start off copying Chalk River. Talked to Teller at Los Alamos. Notes on all conversations in file."

Start with that last item "Notes on all conversations in file." Are those notes still available, or have you destroyed them?

A. I think I have destroyed them. I could not find them the last time I looked.

Q. Coming back to the beginning, you have already stated who Hafstad is. Who is Zinn?

A. Walter Zinn, director of the Argonne laboratory and probably the country's leading technical man in the design of reactors of all sorts.

Q. Pitzer?

A. Director of Research at AEC.

Q. What was the subject of that conversation?

A. I can only tell by refreshing my memory in looking at the notes.

Q. Yes, sir.

A. I gather that Zinn thought that we should build one of the more exotic types of piles which he had under construction. This is a natural reaction from a man in his position who is concerned with the proper design of piles. We on the other hand were not concerned with that at all. We wanted to build some piles, and we knew that the Chalk River design was sound, and we thought we would go ahead and build those.

Q. Was there any question that Dr. Hafstad and Dr. Zinn and Dr. Pitzer were behind you?

A. I didn't think there was, no.
Q. "Talked to Teller at Los Alamos." Do you recall anything about that?
A. No, I don't.
Q. The next item, "October 18: E. O. L. said I had been elected to carry out our program. He looked at sites on Sunday and Monday, and favors some land east of Benicia fronting on Suisun Bay. He says I will be director of the Suisun Laboratory. I am therefore going on almost full time as director of a nonexistent laboratory on an unauthorized program. Cleared out my desk in the linac building and had my file moved down to the director's office in the new building. Decided to talk with L. A. DuBridge and R. F. Bacher tomorrow in Pasadena."
Is there any comment to make on that item, Doctor?
A. This day was the day that I felt I stopped being a physicist after 4 years, and went back to war work. I moved my office out of my research building and became an office worker.
Q. Linac.
A. Linear accelerator. That is the abbreviation.
Q. October 19——
Mr. GRAY. Are you moving to another date? I just want to ask for clarification, you referred to moving into the director's office in the new building.
The WITNESS. Yes.
Mr. GRAY. Was there an existent place known as the Suisun Laboratory then?
The WITNESS. No. Suisun Bay is the north stem of San Francisco Bay, and we had picked out a tentative site on the north shore of that bay where it was far enough from inhabited buildings that we thought it would be safe to put our piles. We wanted to have them close enough to the lab so we could go there very often.

By Mr. ROBB:

Q. What was the new building?
A. The new building was the building which had been erected by the Atomic Energy Commission on the radiation laboratory land in the past few months and was just being occupied as a laboratory and administrative building at that time.
Q. You mean it was new as compared to other buildings which had been built previously.
A. Yes, people were just moving in, and I moved into the director's office.
Q. "October 19: Spent all day in Pasadena discussing project with L. A. D. and R. F. B. They had no objections and I felt they were impressed with the seriousness of the situation, and thought we were doing the right thing."
Who were L. A. D. and R. F. B?
A. L. A. D. is Dr. E. A. DuBridge, who was then and is now president of the California Institute of Technology. He was also a member of the General Advisory Commission, and had been my boss at the radiation laboratory at MIT for 3 years. I had talked with Professor Lawrence a great deal, and I wanted to check up with my other wartime boss to get his ideas and also to see whether he thought that this program we had in mind was something that would be attractive to the General Advisory Commission.
Q. How well did you know Dr. DuBridge?
A. I would say that there are very few people that I know better. One of the reasons for this is that Dr. DuBridge and I for 3 years during the war were members of a 3 man driving club to conserve gasoline. We drove to work every morning and drove back home again every night for 3 years, and I think one gets to know a person very well under those circumstances.
Besides this, of course, we had our association as director and member of the laboratory staff.
Q. R. F. B. who was he?
A. He is Prof. Robert Bacher, who at that time was professor of physics at Cal. Tech., and who had previously been a member of the Atomic Energy Commission, one of the original Commissioners.
Q. How well did you know him?
A. I knew him exceedingly well. We had worked together for 3 years at the radiation laboratory at MIT. We had worked together at Los Alamos. We were close personal friends. Our wives were goods friends. Whenever I went to Cal. Tech. I always stayed at the Bacher home, and whenever he came to Berkeley he stayed in our guest room. We were very close friends.
Q. Without going into great detail, did you explain to these gentlemen what your plans were?
A. Yes, I explained in considerable detail.

Q. Was there any doubt in your mind that they approved?

A. No, there was absolutely no doubt in my mind that they approved. I know them so well that we had a real meeting of the minds. They expressed their interest and approval in many ways and I am sure that they thought it was a fine idea.

Q. "October 20: George Weil and Henry Ott, from the AEC Reactor Division arrived. Spent most of the day with them. Inspected the Suisun sight for the first time—it looks very attractive. George had to leave tonight as he is due in London on Monday. Ott is staying for a few days to help out on pile design."
These two men came from where, Washington?

A. Washington.

Q. And were sent by whom?

A. Mr. Hafstad, I assume.

Q. In other words, at that time, October 20, you were really getting moving?

A. We were getting all the cooperation we could ask for.

Q. Is there anything further to add to that entry?

A. I don't think so.

Q. "October 21, Friday: Spent most of the day reading reports on piles, and relearning elementar pile theory."
"October 22, Saturday: More report reading."
Is there anything to add to those two items?

A. Just the fact that for 4 years or 5 years I had not thought anything about piles or reactors. I had worked with Fermi at Chicago in 1943, and had some acquaintance with piles, and their theory, but I had forgotten the essential points.

Q. "October 24, Monday: Made several telephone calls. Hafstad (at Oak Ridge Conference) says nothing has happened in the last week about our program. This is very disappointing in view of Hafstad's enthusiasm last week when he left. Talked to Pitzer—also at Oak Ridge—for the first time in a week. He had just come from afternoon meeting with Zinn-Weinberg, etc., to discuss our program. Apparently Zinn has thrown a lot of doubts into peoples' minds about the wisdom of our program. Have sensed this from conversations last week with Zinn and Hafstad. Pitzer wants us to present our plans at GAC meeting this weekend in Washington. Agrees with me that had better be done in person than by letter.

"Had lunch with E. O. L. and Mr. Neylon in S. F. Mr. N. said things were moving well, as witness unfreezing of AEC funds by Congress. Advised us essentially to keep our shirts on.

"Talked with Teller, who had just met Fermi at airport in Chicago. No reaction from Fermi, as he was tired from his long trip from Italy. Said he felt he could count on Bethe. Felt Oppie was lukewarm to our project and Conant was definitely opposed. Said Los Alamos was trying to set up conference for Nov. 7.

"E. O. L. talked to Senator Knowland—has date for Senator to come up the hill on Friday at 11 a. m."
Coming back to the first of that entry for October 24, would you explain to us a little bit the entry about Mr. Hafstad's apparent change in attitude? What did you mean by that?

A. I think it is clear that I concluded from what he said that he was no longer as enthusiastic as he had been. The fact that Zinn was thinking that perhaps we were doing the wrong thing, I think is a very natural reaction on his part. After all, he had been designing piles for 4 years since the end of the war, and he had seen none of these being reproduced in hardware. Now if a lot of money was to be made available to build piles, I can appreciate his point of view that he would like to see some of his ideas get into the piles, and not have his merely copy what he probably considered to be an outmoded design of the Canadians.

Q. You mentioned Weinberg here. Which Weinberg is this?

A. This is Alvin Weinberg, director of the Oak Ridge Laboratory.

Q. It is not Joe?

A. Definitely not Joe.

Q. Pitzer wants us to present our plans at GAC meeting this weekend in Washington. "Agrees with me that had better be done in person than by letter."
Who was the "us" that he spoke of?

A. I assume he meant Professor Lawrence, Mr. Brobeck and myself.

Q. Did you at or about that time start to get ready to go to Washington to present your plans?

A. Yes. Mr. Reynolds, who is our business manager, worked day and night preparing cost estimates for the project and Mr. Brobeck was busy on the design features of it, and we had a presentation to make, and we were getting prepared for it.

Q. "Had lunch with E. O. L. and Mr. Neylon in S. F. Mr. N. said things were moving well, as witness unfreezing of AEC funds by Congress."
Does that require any amplification?

A. Perhaps it does. As I recall, Professor Lawrence and I were both getting worried about the fact that there seemed to be a lack of enthusiasm suddenly pervading the scene and we were worried about this, whether it was a change in climate in Washington or what was happening, so we went to a man with some experience in the political field, and asked him whether he thought that this was bad enough that we should be worried about it, and he reassured us and said no, things are moving well. Congress is showing its enthusiasm for an expanded AEC program by unfreezing some funds. He said, "Keep your shirts on, boys, it is going to be all right."

Q. You talked with Teller and so forth. Where did you talk with him?
A. I can't recall.
Q. Was it by phone or in person?
A. I suppose it was by phone, but I really could not be sure. I gather from the entries on this Monday that I was in Berkeley, and I don't recall that Teller came to Berkeley in that period, so I assume it was by phone.

Q. Do you recall whether you knew why he thought he could count on Bethe?
A. I assume that he had had conversations with Bethe and Bethe agreed that the super program should be reactivated. I can't give any definite testimony because he just told me that.

Q. The next item: "Felt Oppie was lukewarm to our project and Conant was definitely opposed."
Does that require any amplification?

A. This is quoting Dr. Teller if I read my notes correctly. I had no conversation with Dr. Oppenheimer on this subject, and I had no reason to feel that he would not be enthusiastic about it. In fact, I assumed he was enthusiastic as were all the other people with whom I talked.

Q. "Said Los Alamos was trying to set up conferences for November 7." Conferences for what?

A. This was the conference that I believe was referred to in one of the first day's notes. Dr. Teller said he thought it would be an excellent idea to bring together all of the men who had thought about problems of the super during the war, together with new theoretical physicsts, young ones who had appeared on the scene since the war, and to discuss the present state of the art, to see what new things had come in, just a sort of reorientation conference, I think.

Q. Did that conference come off?
A. That conference as far as I know never did come off.
Q. "E. O. L. talked to Senator Knowland—has date for Senator to come up the hill on Friday at 11 a. m."

A. This is up the Berkeley hill to the radiation laboratory. Senator Knowland is an alumnus of the University of California and Professor Lawrence met him at the Faculty Club one day and invited him to come up the hill. He was there on other business.

Q. "October 25, 1949—Tuesday: Decided to go to Chicago—Argonne—with Brobeck and Gordon, leaving tomorrow. Should get to Argonne Thursday morning when Zinn returns from Oak Ridge. After 2 days there should go to Washington for GAC meeting. Talked to Serber about GAC meeting. He volunteered to see Oppie before the meeting. Called Oppie who said he had hoped to be able to talk to him. Therefore Serber is going with us tomorrow and will continue to Princeton and have a day with Oppie, before he leaves for meeting in Washington.

"Reynolds working on cost figures for presentation to GAC. My thinking about pile is along direction of fewer larger fuel rods. Called Gale Young at Nuclear Development Associates in New York City. He was out of town. We would like to get him as a consultant on our project."

Mr. SILVERMAN. I think you read "we."
Mr. ROBB. I think that is what it is. Will you look at the original and see whether it would be "we" or "he."
The WITNESS. In the case of "we would like to get him," it is "we."
Mr. SILVERMAN. It is evidently a typograhpical error.
Mr. ROBB. Yes.

By Mr. ROBB:

Q. "Chicago meeting—then on to Washington—talked with all GAC and most of AEC Commissioners. Particularly interesting talk with Oppie just after he briefed Bradbury and Norstad at GAC meeting. Pretty foggy thinking."
That is the last entry in your diary?

A. That is right, because after that the project was dead.

Q. Going back to the beginning of that entry, which apparently covered several days——

A. Yes. This I wrote up after I got back from the trip to Washington.

Q. What was your purpose in going to Chicago to the Argonne Laboratory?

A. As I said earlier, Dr. Zinn is the leading designer of piles in the country and they were most cooperative and said they would supply us with any information they had available that would help us in modernizing slightly the Chalk River pile.

Q. Brobeck, I believe you identified.

A. Brobeck is the chief engineer of the laboratory and Gordon his assistant went along with me to communicate and talk with the pile designers at the Argonne.

Q. In other words, you did go to Chicago.

A. Yes, sir.

Q. As you planned.

A. Yes, sir.

Q. Talked to Serber about GAC meeting. Where did that conversation take place?

A. That took place in Berkeley. Could I expand a bit on that?

Q. Would you do that, please, sir?

A. Yes, As I said earlier, Dr. Serber was one of the group that had expressed a willingness to work hard on the program of building heavy-water piles. He was to be our chief theoretical adviser, and we were counting on his help. There is one thing in here which is not written down, and I think I am correct in remembering it this way. I believe I called Dr. Oppenheimer from Berkeley and asked him if I could see him before the General Advisory Commission meeting to talk over our plans. You will note that in this whole diary there is no mention of any talks between me and Dr. Oppenheimer. I was anxious in view of the fact that I had heard that he was lukewarm to the program to have a chance to brief him on the program and if possible to get a little enthusiasm on his part.

As I remember it, Dr. Oppenheimer said he would be very glad to see me in Princeton, and in fact invited me to stay overnight in their guestroom.

Then it turned out that our time in Chicago was limited and I thought I had better stay and talk pile design because I had spoken with Dr. Serber about this meeting with Oppenheimer and Serber said he would be glad to present our case to Dr. Oppenheimer and try to convince him of its worthwhileness. So essentially I deputized Dr. Serber to transmit my point of view to Dr. Oppenheimer. In fact, I was glad to do so, because Dr. Serber and Dr. Oppenheimer are somewhat closer friends than Dr. Oppenheimer and I. They have been closer personally. Dr. Oppenheimer and I were certainly excellent friends at the time and Dr. Serber, I thought, could perhaps do a little better job than I could. I thought and felt strongly that he would present the point of view which was the laboratory point of view at that time, namely, that this was a very worthwhile program and we should get it going.

Q. You had no doubt at all about Dr. Serber's enthusiasm for your program?

A. Aboslutely none.

Q. Do you know whether Dr. Serber did go to Princeton to see Dr. Oppenheimer?

A. Yes, he did.

Q. We will come to that a little later.
"Reynolds working on cost figures for presentation to GAC." You have already told us of that.

A. Yes.

Q. "Called Gale Young at Nuclear Development Associates." Who was he?

A. Gale Young was a very competent theoretical physicist in the field of pile design. He and I had been classmates and he was one of the leading men at the Metallurgical Laboratory in Chicago during the war, on the design of the Hanford reactors. He had for awhile after the war worked for the Atomic Energy Commission, and then he and a group of his friends set up a company to do consulting work on pile design. Dr. Lawrence and I felt that if we were to make too much use of the Argonne Laboratory and the Oak Ridge Laboratory in the design of our piles that people could criticize us for taking effort away from those laboratories which were designing piles, and we thought it would be much better if we could get a company which was set up to advise people, and was interested in making money by doing this, and if we could get them as essentially auxiliary to our design department.

Q. The next item: The Chicago meeting you have already told us about that.

A. Yes. This was purely a technical meeting in which I was pretty much in the background. It was an engineering meeting to a large extent.

Q. And then on to Washington. "Talked with all of GAC and most of AEC Commissioners." What can you tell us about that?

Mr. GARRISON. Could we ask the date of that?

The WITNESS. The date of that meeting is in the record some place. I don't happen to have it down. I believe we spent 2 days in Chicago; if I were to hazard a guess it would be the 27th plus or minus a day.

By Mr. ROBB:

Q. Of October 1949?
A. Yes.

Q. Will you tell us about your talking with GAC and most AEC Commissioners?

A. Since I have no notes, I can't remember any details of those conversations.

Q. You did see them all and did present your program?

A. Yes; before the meeting. This normal procedure before you go into a meeting with a formal plan to talk it over formally to get peoples' views and to clarify any misunderstandings they might have about it.

Q. You mention here, "Particularly interesting talk with Oppie just after he briefed Bradbury and Norstad at GAC meeting." Were you at that GAC meeting?

A. No; I had no reason to be at that GAC meeting. That was a closed meeting, if I remember correctly, at which time the Commissioners met with the GAC, and the top military men in the country.

Q. Where were you?

A. I was standing inside the main entrance to the Atomic Energy Commission building and I watched my friends go upstairs, and I saw the famous military men whom I recognized from their pictures follow along. The meeting lasted for some while. I watched the people come back out again and in a few minutes Dr. Oppenheimer came along and invited Dr. Serber and I, who were standing together outside the building, to have lunch with him.

Q. Did you have lunch with him?

A. Yes. We went to a small restaurant in the immediate neighborhood of the Commission building, and that was the first occasion that Dr. Oppenheimer told me of his views on the building of the hydrogen bomb.

Q. What did he tell you?

A. He said that he did not think the United States should build the hydrogen bomb, and the main reason that he gave for this if my memory serves me correctly, and I think it does, was that if we built a hydrogen bomb, then the Russians would build a hydrogen bomb, whereas if we did not build a hydrogen bomb, then the Russians would not build a hydrogen bomb.

I found this such an odd point of view that I don't understand it to this day. I told Dr. Oppenheimer that he might find that a reassuring point of view, but I didn't think that very many people in the country would accept that point of view?

Q. Was Dr. Serber present?

A. Dr. Serber was present and agreed with Dr. Oppenheimer and this surprised me greatly in view of the fact that 2 or 3 days before he had gone to see Dr. Oppenheimer telling me that he would try to convert Dr. Oppenheimer's lukewarmness into some enthusiasm for our project.

Q. What was the impact of all this on you?

A. Well, for the first time I realized that the program that we were planning to start was not one that the top man in the scientific department of the AEC wanted to have done. We thought that we were doing this as a public service. We were interrupting our own work to do this job. We certainly were not going to try to force anybody to take these piles. We had thought all along that everyone would be enthusiastic about having a big source of free neutrons.

Q. Did you stay in Washington until the end of the GAC meeting?

A. I believe I left right away after my conversation with Dr. Oppenheimer. I have no way of refreshing my memory on that. I felt that the program was dead, and that is the reason the diary ends at this point.

Q. Until revived by the Presidential pronouncement in January 1950, was the program dead?

A. Dr. Teller was still working at Los Alamos and as far as I know that was all that was going on in the program.

Q. What did you do?

A. As I remember I went back to doing physics.

Q. Did you reflect on this development which you observed in your conversation with Dr. Oppenheimer?

A. Yes, I did. Of course, I later became aware of the contents of the GAC policy memorandum to the Atomic Energy Commission. I was not allowed to read it because there was no particular reason for me to do so, but I was told that the GAC had said that the United States should not build the hydrogen weapon. I have since heard a great deal of talk about the fact that the GAC was opposing a crash program, but after rereading some of the document last night that is not my impression of what it said.

Q. Which document do you refer to?

A. The GAC policy report.

Q. I will ask your opinion, Doctor. Suppose the thermonuclear program had gone ahead full steam beginning in 1946, how soon do you think we would have gotten the weapon?

A. That is a very difficult question to answer, but I would add to the date 1946 the number of years that it took after the Presidential directive was given and arrive at an answer which would probably not be off by more than a year.

Q. Which would be what?

A. Would you do the arithmetic?

Q. It has been suggested here that the achievement of the thermonuclear weapon was the result of a brilliant invention or discovery which might have taken many years or might have taken a very brief time, and therefore it is impossible to project the length of time that it might have taken had the program begun 2 or 3 or 4 years earlier than it did. What could you tell us about that suggestion?

A. I think brilliant inventions come from a concentrated effort on a program. The reason there were not any brilliant inventions in the thermonuclear program for 4 years after the war is that there was no climate to develop in. Lots of people were not thinking about the program. Essentially one man was, and it is very hard to generate ideas in a vacuum.

Q. Were there further inventions which speeded up and furthered development of the atomic weapon?

A. Yes. I would like to give one instance of that. When I arrived at Los Alamos, as I say, my job was to help Dr. Kistiakowsky in the development of the implosion weapon. Dr. Kistiakowsky was the country's leading expert in the field of high explosives. He had been director of the Bruceton Laboratory of Army Ordnance, and Dr. Oppenheimer exerted great effort to get him to Los Alamos, and fortunately was successful. I had a number of conversations with Dr. Kistiakowsky on the feasibility of the implosion weapon and on every occasion for quite some time Dr. Kistiakowsky said that he felt Dr. Oppenheimer was mad, almost, to think that such an absurd object could ever be made to work. Here was the leading explosive expert saying that Dr. Oppenheimer was just wrong, this thing could not be built, and yet it was built.

Dr. Oppenheimer was absolutely right, and he was right because he set up a group of people that put a concentrated effort on the program and 2 or 3 brilliant inventions did come out which made this thing possible. Dr. Oppenheimer always said that the implosion program would work and he was right and he had good reasons for saying it would work, even though at that time the technology did not permit it.

The technology was developed because of the climate at Los Alamos, enthusiastic people who said we don't care what the experts say, we will make it work. This was the thing that was missing in the hydrogen bomb program after the war, and the thing which came into it some while after the Presidential directive.

Q. Now, directing your attention to a time perhaps a couple of months after your return from Washington in 1949, I will ask you if you will recall a conversation with Dr. Vannevar Bush about Dr. Oppenheimer?

A. Yes.

Q. Could you tell us what that was and the circumstances?

Mr. GARRISON. When was this?

Mr. ROBB. Perhaps a couple of months after his return from Washington in October 1949.

The WITNESS. I can give you some information that will place this conversation to within a day, because Dr. Bush was in California to inspect one of the Carnegie Institution facilities at Stanford University. As you know, Dr. Bush is director of the institution. I remember that when I arrived home after our conversation with Dr. Bush, I found in the mailbox a copy of Life magazine which had a condensation of the book Modern Arms and Free Men. So that places the date within a day.

What Dr. Bush said to Professor Lawrence and me was that he had been appointed by the President to head an ad hoc committee to assess the evidence for the Russian explosion. The Atomic Energy Commission and the Armed Forces, particularly the Air Force, had collected a good deal of information, all of which tended to indicate that the Russians had exploded a bomb, but before announcing that to the public the President wanted to make sure that the evidence was conclusive. If I remember Dr. Bush correctly, he said that he was made chairman of that. If I can paraphrase Dr. Bush's statement and give them in the first person, they went something like this. He said, "You know, it is a funny thing that I should be made head of such a committee, because I really don't know the technical facts in this field. I am not an atomic physicist, and I am not the one to assess these matters." But, he said, "I think the reason the President chose me is that he does not trust Dr. Oppenheimer and he wants to have someone in whom he has trust as head of this committee."

Dr. Bush then said that the meetings of the committee were very interesting. In fact, he found them humorous in one respect, because he said, "I was ostensibly the chairman of the committee. I called it to order, and as soon as it was called to order, Dr. Oppenheimer took charge as chairman and did most of the questioning." I believe Dr. Bush said that Dr. Oppenheimer wrote the report. This was the first time that I had ever heard anyone in my life say that Dr. Oppenheimer was not to be trusted.

Dr. EVANS. Would you make that statement again?

The WITNESS. This was the first time that anyone had ever said in my presence that Dr. Oppenheimer was not to be trusted.

By Mr. ROBB:

Q. You and Dr. Lawrence and Dr. Bush, you say, were driving some place?

A. This was driving back from Stanford to Dr. Bush's hotel in San Francisco.

Mr. GRAY. We will have a recess for 2 minutes.

(Short recess.)

By Mr. ROBB:

Q. Dr. Alvarez, coming now to the winter of 1950, did you serve on a committee called the Long Range Planning Committee?

A. Yes; I did. I did that at the request of Dr. Oppenheimer who called me and said, "We are having a meeting of a committee to try to find out the future of the military applications of atomic energy." He said, "I would like to have you on this committee because I know you represent a point different from mine, and I think it would be healthy to have you on this committee." I felt very happy about this. I thought Dr. Oppenheimer was being very fair in inviting me to join this committee, and I accepted the appointment.

Q. Who else was on the committee?

A. The scientific members were Dr. C. C. Lauritsen, of Cal. Tech., Dr. Bacher, of Cal. Tech., I believe Dr. Whitman was on the committee, General Nichols attended one meeting of the committee, but he did not sign the report, Dr. M. J. Kelly was on the committee. I should say I have refreshed my memory on this by reading the report, and I would not have remembered all of these gentlemen without doing so.

Q. What was the purpose of that committee, again?

A. This committee was a committee of the RDB, the future of the atomic weapons program for periods ranging from 2 to 5 or 10 years.

Q. Where did you meet?

A. We met in Washington in the Pentagon.

Q. How long a period did you meet?

A. I believe it was 2 days.

Q. What can you tell us about the discussion that went on with respect to atomic weapons and the thermonuclear?

A. As I had expected from the makeup of the committee there was great enthusiasm for small-scale weapons for tactical use.

Q. Great enthusiasm on the part of whom?

A. Dr. Lauritsen particularly. I had been on a committee the summer before with Dr. Lauritsen which investigated antisubmarine warfare and I had talked at some length with him on the subject,.and I knew that he had a great enthusiasm for this program which was not then a part of the atomic-energy program which I had not thought very much about, and I had no strong views one way or another. I went on the theory that if Charlie Lauritsen thought it was a good idea, it was a good idea, because I had such great respect for his judgment in the field of scientific weapons.

Q. Now, would you go ahead and tell us what happened? I interrupted your recitation.

Mr. SILVERMAN. Could we have the date of that meeting? I think we had winter of 1950.

The WITNESS. Yes, I think it was December 1950.

Mr. ROBB. We have had a lot of testimony about it.

By Mr. ROBB:

Q. Go ahead, Doctor.

A. There was a good deal of discussion about tactical weapons, small weapons, using small amounts of fissionable materials. There was discussion of the tactical use of these weapons. General Nichols briefed us on the present status of the guided-missiles program, of which he was then Deputy Director, since there was much interest in the use of atomic warheads on guided missiles. This part of the program I thought was in competent hands so I didn't have much to say one way or the other. I thought Dr. Lauritsen and Oppenheimer handled this part of the program very well, and I had no disagreement with this.

I found, however, that I was in serious disagreement with them one one point and that was that they thought that the hydrogen program was going to interfere seriously with the small-weapons program by taking away manpower at Los Alamos which could otherwise be put on the hydogen bomb. My view was that the things were not mutually exclusive, if I can use the scientific phraseology. That is, there was no reason to say we have to have hydrogen bombs and not small weapons and vice versa. It seemed to me that there were great resources of scientific manpower in the country and that one could have both of these programs simultaneously. I did not object to the small-weapon program because it would interfere with the hydrogen bomb and I was surprised that they objected to the hydrogen-bomb program because it would interfere with the small-weapons program.

Q. Did Dr. Oppenheimer have anything to say specifically about the hydrogen-bomb program being carried on?

A. I remember one statement that Dr. Oppenheimer made because it shocked me so greatly and I repeated it to several people when I got home. I remember telling Professor Lawrence about it, and I believe I told Dr. Cooksey. Again if I can be excused for paraphrasing and using first person, Dr. Oppenheimer said essentially this: "We all agree that the hydrogen-bomb program should be stopped, but if we were to stop it or to suggest that it be stopped, this would cause so much disruption at Los Alamos and in other laboratories where they are doing instrumentation work that I feel that we should let it go on, and it will die a natural death with the coming tests"—which were the Greenhouse tests—"when those tests fail. At that time will be the natural time to chop the hydrogen bomb prorgam off."

I assumed I had been put on this committee to present views in favor of the hydrogen bomb because I had been always of that point of view. I didn't object to Dr. Oppenheimer's statement, because he said that he was not planning to stop the program. My feeling at the time was that if the Greenhouse test failed, and then Dr. Oppenheimer or the GAC did something to stop the hydrogen-bomb program, then would be a good time to fight. It seemed to me to be quite useless to express disapproval of this because nothing was being don to stop the program.

However, I found later much to my dismay that my own political naivete in matters of this kind led me astray and I found that the report which I signed, and I am sorry to say I signed, did do the program great harm.

Q. Why?

A. Dr. Teller saw me several months later, and he said, "Louis, how could you have ever signed that report, feeling the way you do about hydrogen bombs?" I said, "Well, I didn't see anything wrong with it. It said the hydrogen-bomb program was an important long-range program. Our particular emphasis was on small weapons, but that is a program which has no standing in the Commission's program now, and I think we should go ahead with it." He said, "You go back and read that report and you will find that that essentially says that the hydrogen-bomb program is interfering with the small-weapons program, and it has caused me no end of trouble at Los Alamos. It is being used against our program. It is slowing it down and it could easily kill it." I have recently reread that report in the last day, and I am also shocked as was Dr. Teller. I can only say in my defense that I have not spent much time on policy reports, staff papers, and things of that sort, and I am not attuned to them and I didn't catch this implication. I should have done so, and I didn't.

Q. Who wrote it?

A. Dr. Oppenheimer wrote it. I think that probably Dr. Lauritsen and Dr. Bacher and I made minor changes in it, but certainly the main draft was written by Dr. Oppenheimer.

Q. Dr. Alvarez, how well do you know Dr. Edward Teller?

A. I think I know him quite well.

Q. Have you worked with him for many years?

A. I worked with him at Los Alamos, not as an intimate worker. He was in the field of theoretical physics, whereas I was in the experimental program. But he and I often discussed matters of physics and bomb technology. He was my introduction to Los Alamos technology. He and I rode from Chicago to Los Alamos in the same drawing room when I first went there, and he spent the whole time briefing me on the program.

Q. Are you familiar with the work he is now conducting at Livermore?

A. Yes, I am; in some detail.

Q. Do you know other people out at Livermore who also know Dr. Teller and work with him?

A. I do.

Q. Many people?

A. I probably know 100.

Q. There has been a suggestion here by some people that Dr. Teller is a hard man to get along with, a hard man to work with. Have you found that to be true?

A. I can hardly think of a statement that is further from the truth. I am sure that Dr. Teller would be a hard man to work with if the man above him were trying to stop his program and to put obstacles in his path. Then I am sure he would be a very hard man to work with because he would fight strongly for what he thought was right. But in any friendly climate, Dr. Teller is a perfect colleague, scientifically and personally. I can't think of a finer man in almost every respect than Dr. Teller.

Q. Would you say that is his reputation and standing among the people who work with him at Livermore?

A. I can say that is the uniform opinion of everyone at the Livermore Laboratory and at the Radiation Laboratory in Berkeley. I don't think if I searched the laboratory with a fine tooth comb that I could find anyone who had a bad word to say for Edward Teller.

Mr. ROBB. That is all I care to ask, Mr. Chairman.

Mr. GRAY. It is now 25 minutes to 6. I assume you will have some questions to ask?

Mr. SILVERMAN. I think so.

Mr. ROBB. Mr. Chairman, I am sure it could be an accommodation to the chairman if it would be brief, if we could do it now.

Mr. SILVERMAN. I hate to incommode the witness but I really think it will be much shorter if we resume tomorrow morning, sir.

Mr. GRAY. I think we will recess until 9:30 tomorrow.

(Thereupon, at 5:35 p. m., a recess was taken until Friday, April 30, 1954, at 9:30 a. m.)

UNITED STATES ATOMIC ENERGY COMMISSION

PERSONNEL SECURITY BOARD

IN THE MATTER OF J. ROBERT OPPENHEIMER

ATOMIC ENERGY COMMISSION,
BUILDING T-3, ROOM 2022,
Washington, D. C., Friday, April 30, 1954.

The above entitled matter came on for hearing, pursuant to recess, before the board, at 9 : 30 a. m.

Personnel Security Board: Mr. Gordon Gray, chairman; Dr. Ward T. Evans, member; and Mr. Thomas A. Morgan, member.

Present: Roger Robb, and C. A. Rolander, Jr., counsel for the board; J. Robert Oppenheimer, Lloyd K. Garrison, Samuel J. Silverman, and Allan B. Ecker, counsel for J. Robert Oppenheimer; Herbert S. Marks, cocounsel for J. Robert Oppenheimer.

PROCEEDINGS

Mr. GRAY. We will resume.
Mr. ROBB. Mr. Chairman, I have two questions I would like to ask.
Mr. GRAY. I suggest you proceed.
Whereupon,, Luis Walter Alvarez, the witness on the stand at the time of taking the recess, resumed the stand and testified further as follows:

DIRECT EXAMINATION

By Mr. ROBB:

Q. Dr. Alvarez, your dairy showed, and you testified that you talked to various individuals about your plan and the plans of others for the development of the thermonuclear weapon in early October 1949; is that right?
A. Yes, sir.
Q. At that time these individuals were enthusiastic for going ahead with it; is that right?
A. That was my very strong impression.
Q. To your knowledge, were those conversations in advance of any talks that these people had with Dr. Oppenheimer?
A. I think that is so, sir. I am sure it is so in the case of Dr. Serber. I am quite sure in the case of Drs. DuBridge and Bacher, and also in the case of Dr. Rabi.
Q. Subsequently these people changed their views; is that right?
A. Quite drastically; yes.
Q. Did you learn at that time whether in the interim they had talked to Dr. Oppenheimer?
A. I am sure that in the interim they talked with Dr. Oppenheimer, because the interim extends until now.
Mr. ROBB. That is all I care to ask on direct, Mr. Chairman.
Mr. GRAY. Mr. Silverman.

CROSS-EXAMINATION

By Mr. SILVERMAN:

Q. Self evidently these people have talked to a lot of other people?
A. That is absolutely right.
Q. Dr. Alvarez, when you came east with Dr. Lawrence in the trip of which you kept a diary, am I correct in my understanding that the specific thing you were trying to promote for want of a better word, or push, was a reactor pile that would produce excess neutrons?
A. That is right, sir.
Q. Did the Commission thereafter build or cause to be built a reactor to produce excess neutrons?
A. There are some reactors of that general class now under construction at Savannah River; yes.
Q. Is that Savannah River reactor not in operation at all?
A. I don't know. I have no knowledge of this except what I read in the paper. I believe, however, they have not been turned on. That is my impression. This can be checked easily.
Q. Do you know whether the reactor at Savannah River was based largely on Mr. Zinn's design?
A. I haven't really any idea, sir. I would assume that his advice was taken, but I believe that the reactors were designed by the engineers of the du Pont Co., and the only consultant that I know of personally employed by them was Dr. John Wheeler, who was their consultant on the Hanford pile designed during the war. I think it was pretty much of a company design job, rather than an AEC design.
Q. Do you know designs for reactors to produce excess neutrons were fairly well along in October of 1949?
A. The files of the AEC were bulging with designs for reactors; this is just the point that I made. There were designs by the gallon, but no piles.

Q. Do you know whether the Savannah River pile more nearly followed the designs that Mr. Zinn had participated in making, and he was enthusiastic about, than the Chalk River pile or something based on it?
A. My impression is that Mr. Zinn believed strongly that the piles of the future to give excess neutrons should be enriched uranium piles of the type now in operation at Arco, Idaho. Zinn has believed strongly in the small enriched piles as against the Savannah River design, which is along the broad general lines of the Canadian pile.
Q. You think the Savannah River pile is along the broad general lines of the Canadian pile?
A. I have never seen its design, but it is a heavy water moderated pile, using natural uranium, which is certainly what the Canadian pile is, and very definitely different from the many designs which Dr. Zinn had to do with and eventually has constructed.
Q. Are you aware that the GAC did in fact recommend going ahead with the Savannah River project?
A. Oh, yes, I am quite aware of that. I would be interested in the date when that project was supported.
Q. You don't know the date?
A. I don't know the date. I know, however, it was after the Presidential directive, of course.
Q. Do you know that the GAC had been recommending a production facility that would produce excess neutrons for well over a year before the President's directive?
A. I knew that everyone was in favor of piles but nonetheless no piles got built.
Q. The GAC was an advisory committee?
A. Yes, it was.
Q. And it advised that such piles be built?
A. I have never seen their recommendation, sir, so I don't know, but being in favor of piles is like being against sin. I think everyone is for piles, but nonetheless none got built.
Q. But it was not the GAC's job to build them?
A. That is true, yes.
Q. Do you know what the Savannah River pile cost?
A. I would guess it was in the neighborhood of $1½ billion, just from what I see in the newspapers.
Q. And who built the pile?
A. The du Pont Co.
Q. Do you think that the Atomic Energy Commission was perhaps justified in entrusting the building of a billion and a half dollar project to the du Pont Co. rather than to your group?
A. Oh, absolutely. They had tremendous competence in the field, and we had no competence whatsoever in pile design. The only thing we had to offer to the Commission was the ability to build things rapidly in the scientific field. This was a demonstrated capacity of the Radiation Laboratory.
Q. The du Pont Co. had that capacity, too?
A. Yes, to an even greater extent than we did, obviously.
Q. And the du Pont Co. had experience in building piles?
A. Yes, sir.
Q. And you didn't.
A. That is right. The right decision was certainly made there. The du Pont Co. was certainly better equipped to build piles than we were. There is no question about that.
Q. Are you sure that the development of the Savannah River project was not carried at Argonne under Zinn?
A. I have no knowledge of this, but looking at the pile in the broad sense, I would say it doesn't look like a Zinn pile, and the way that an architect would look at a building and say this was not designed by such and such an architect.
Q. It would surprise you to learn that that development was carried out at Argonne under Zinn?
A. It would not surprise me particularly. I would guess that it was not a development of Zinn, but rather of du Pont. This is purely a guess.
Q. That would be purely a guess.
I would like now to turn to the discussions in the panel—I think perhaps you called it the panel on long-range planning, something like that.
A. I believe that was the official name.
Q. I believe you called it that. I am not sure. It may have been referred to at other times as the Military Objectives Committee?

A. Perhaps it was.

Q. In December 1950, you referred to a statement by Dr. Oppenheimer somewhat to the effect that we all agree that the hydrogen-bomb program should be stopped. If we did this and recommended it, it would cause too much disruption at Los Alamos?

A. That is right.

Q. And let it go on and the project would die when the Greenhouse test failed, as Dr. Oppenheimer expected them to. Is that substantially correct?

A. That is substantially the way I remember it, yes.

Q. I would like you to turn to the first part of that statement that we all agree that the hydrogen-bomb program should be stopped. I want to ask you whether it is possible that what Dr. Oppenheimer said was that "We all agree that the hydrogen-bomb program does not look very hopeful now."

A. No, I am quite sure I remember it the other way. It was such a startling statement to me that it is indelibly in my mind. I don't think I could be mistaken on that.

Q. You of course were a representative of the other view?

A. That is right.

Q. And when Dr. Oppenheimer said that "We all agree that the hydrogen-bomb program should be stopped," did you as a member of the panel say, "We don't all agree; I don't".

A. I didn't interrupt him until he finished his statement at the end of which time, as he pointed out, he said he was not going to stop it, and I pointed out since he said he was not going to stop it, there seemed to be no point in arguing about it.

Q. But you did not correct him and say "We do not all agree."

A. No. I am sure from what I have said in this hearing you would know that I did not agree.

Q. It is sometimes necessary on cross examination to emhpasize points.

A. Very well, sir. Had he stopped his statement with that first sentence, I am sure that I would have dissented vigorously.

Q. Was it the fact that everybody there agreed that at that time the hydrogen bomb program did not look very hopeful?

A. I don't know whether everyone did agree on that.

Q. Did you think at that time that the hydrogen bomb program did not look very hopeful?

A. I thought it looked exceedingly hopeful. Again I can only see it through the eyes of people like Edward Teller, who have the technical competence, who know the details of the program. I am not a theoretical physicist. All I can do is base my judgment on people in whom I have great scientific trust.

Q. Wasn't everybody pretty depressed in December 1950?

A. No. I certainly didn't sense that at all, but I was not at Los Alamos. I did not know that things were going very badly. Perhaps they were, I don't know. I was not aware of the fact that people were depressed.

Q. And you had not heard from other people working on the project in December of 1950 that things didn't look so good?

A. I had heard that the requirements for tritium had temporarily taken a turn toward larger quantities being required. But I had seen the requirements go up and down and up and down on many occasions, and this did not disturb me at all.

Q. You had not heard at the time that this was a temporary turn, that it turned out to be temporary?

A. I really couldn't say positively one way or the other.

Q. Did the others at the meeting agree that the hydrogen bomb program did not look hopeful?

A. I can't recall. I do know that Dr. Lauritsen apparently had strong reasons, probably some of a moral nature for not wanting the hydrogen bomb. I do know that Dr. Lauritsen's closest associate, Dr. William Fowler, had been giving lectures on the radio against the hydrogen bomb. I was in Pasadena staying with Dr. Bacher one night when I was giving a lecture at Cal. Tech., and at a dinner party that night all I heard was stories about why you should not have hydrogen bombs, and the fact that the members of the staff at Cal. Tech. were giving public lectures and talking on the radio against the hydrogen bomb. I thought Dr. Lauritsen wanted no part of the hydrogen bomb.

Mr. GRAY. In what period of time was this?

The WITNESS. This was at the time of the panel at the end of 1950.

Br. Mr. SILVERMAN:
Q. Before or after the panel, would you say?
A. During that general period. I could not pinpoint the date precisely.
Q. Did Dr. Lauritsen express any views at this panel meeting as to either whether the hydrogen bomb program should be stopped or as to its feasibility?
A. I discussed the program with him on a number of occasions and I always got the impression that he thought that the small weapons program and the hydrogen bomb were mutually exclusive. The country could not do both of them at the same time; since he had strong reasons for desiring a small weapons program, he felt that the hydrogen bomb program should not go ahead.
Q. Did the report the panel filed say that the small weapons program and the hydrogen bomb program were mutually exclusive?
A. Not in exactly those words, but it certainly pointed out that the hydrogen bomb program was taking manpower and effort of the Los Alamos Laboratory away from the small-weapons program and the panel recommended that it not do so in the future.
Q. And you signed that report?
A. I signed the report, and as I have said, I am sorry I signed it.
Q. Do you recall whether Dr. Lauritsen at the panel said anything about the outlook for feasibility of the hydrogen bomb?
A. Whether Dr. Lauritsen said that it was feasible or not feasible would have made no impression on me, because Dr. Lauritsen like myself was not entitled to scientific opinion. Neither he nor I have enough knowledge in this field to form an opinion ourselves.
Q. And you do not recall whether he said anything about it?
A. No; but had he said so, it would have made no impression on me.
Q. Did Dr. Bacher say anything about what the outlook was at the panel for the feasibility of the hydrogen bomb program?
A. Again, I can't recall for the same reason. Dr. Bacher was not entitled to an opinion, nor am I.
Q. Did Admiral Parsons express a view on that subject?
A. I think Admiral Parsons stayed very neutral throughout the whole thing. He was a good naval officer, and I don't think that he was trying to inject his own personality into this thing.
Q. Did General McCormack express a view?
A. I don't believe so. I don't know.
Q. So that when Dr. Oppenheimer said, "We all agree," they all just sat?
A. Yes. No one commented on this at all. That is to the best of my recollection.
Q. I understand you are testifying from your recollection, sir.
I think you said that Dr. Oppenheimer indicated that he thought that the Greenhouse tests would fail.
A. Yes.
Q. Just what does that mean?
A. That no thermonuclear reaction would take place in the Greenhouse test explosive device. In order for a thermonuclear reaction to take place, very high temperatures must be reached, as you know. I think that Dr. Oppenheimer felt that those high temperatures would not be reached, if you can permit me to read his mind.
Q I would rather you tell us what he said.
A. I have already told you what he said.
Mr. ROBB. Mr. Chairman, everybody else is reading Dr. Oppenheimer's mind.
Mr. GRAY. The Chair will say that there has been a parade of witnesses here who testified on their intimate knowledge of Dr. Oppenheimer, and that they would know exactly what his reaction would be in any particular situation. I do not think this witness should be denied an opportunity to make his own guess about what Dr. Oppenheimer might think.
Mr. SILVERMAN. I do not wish to cut a witness off. I would point out between opinion evidence testimony as to a man's character and evidence as to what a man was thinking about a scientific project.
Mr. GRAY. I will ask you, Mr. Silverman, if you have not asked witnesses in this proceeding what did Dr. Oppenheimer think about so-and-so.
Mr. SILVERMAN. I would certainly not be prepared to say——
Mr. GRAY. Would it surprise you to learn that you have asked such a question?
The WITNESS. Could I be allowed to say what I was going to say in a different way? I testified that Dr. Oppenheimer made a certain statement, that he thought the thing would fail. There are only two possibilities that the thing should fail, as far as I can see. One is that the device misfired. When the

button was pressed, nothing happened. Certainly the atomic bomb primer of the device would work. We have great experience in this line. After that fired, then the temperature of the reactants would rise. If they rose high enough, I doubt if you could find a scientist in the world who would nòt agree that the thermonuclear reaction would take place. It is taking place in the sun all the time. Therefore, when Dr. Oppenheimer said that the thing would fail, it could mean to me only one thing, namely, that he thought the temperature would not rise high enough. That is why I said I thought I could read his mind.

By Mr. SILVERMAN:

Q. Let me suggest this to you, and see whether it does not refresh your recollection as to what Dr. Oppenheimer did say, if he said it; that he thought that the Greenhouse tests wouldn't fail, but fail or not, they would not be particularly relevant to deciding the question of the feasibility of the Super?

A. I am quite sure that he didn't say that.

Q. In fact, the Greenhouse test did not fail, is that not right?

A That is right.

Q. In fact, did they demonstrate the feasibility of the Super?

A. You are asking me a question in a field in which I have no sufficient competence to answer. All I can say is that everyone connected with the Greenhouse tests was elated at the outcome of the Greenhouse tests. I believe that the success of the Greenhouse tests led to the successful tests at Ivy.

Q. Did Dr. Oppenheimer say that he thought the Greenhouse tests were not directly relevant to the determination of the feasibility of the classical Super, but that it was far along and people at Los Alamos had their hearts so much in it that it ought to be allowed to continue; otherwise it would disrupt things too much and discourage them?

A. I testified what I remember Dr. Oppenheimer to say, and I don't see much point in the question, sir.

Q. You ultimately signed the report.

A. Yes.

Q. And there is a part of it that you have regretted signing?

A The thing that I regret is that the report was used to slow down the hydrogen bomb program. The statements having to do with the hydrogen bomb come in the last three paragraphs, save for one rather trivial one.

Q. Did Dr. Oppenheimer use the report to slow down the hydrogen bomb program?

A. I don't know who used the report. I have had Edward Teller tell me, as I said yesterday, that the report was used to slow down the program.

Q. This being a matter where Dr. Oppenheimer personally is very seriously concerned, it becomes a matter of considerable importance as to whether Dr. Oppenheimer used it.

A. Dr. Oppenheimer wrote the report, I am sure. Dr. Oppenheimer ordered the statements presumably in the order of the importance he attached to them, and the super was more or less damned by faint praise.

Q. Did everybody go over the report?

A. On the last day of the meeting with everybody with an airplane ticket in his pocket, one goes over a report and if there is not something that is obviously terribly wrong, one signs it.

Q. There were changes made in the report?

A. Of a rather trivial nature.

Q. You suggested some?

A. I can't recall whether I did. It is possible that I did.

Q. It was a pretty serious matter, this report, obviously.

A. You see, this was the point that I was not sure of. I did not know that this report was anything more than a document to go into the files to be looked at in 2 or 3 years, so that one could see in what direction the program should be then oriented. * * * It was a so-called long-range objective panel. I thought of it as something that would be pulled out of a file in a couple of years, someone would look at it and say, "Well, perhaps we ought to get into some of these things that are in this long-range panel report."

Q. Wasn't this report prepared in the light of a possibility of our being involved in all-out war in the near future?

A. I understood the panel to be calle_ to review the long-range objectives of the military weapons program as it had been reviewed in the past. I believe this was the second or perhaps the third meeting of such a panel.

Q. This panel was meeting just after the Chinese intervention in Korea, wasn't it?

A. When you state that I am sure that you have checked the dates. It would take me some time to be sure of that. Certainly the Korean war was on at the time.

Q. Did you consider the small-weapons program a long-range thing?

A. No. I thought the small-weapons program was a rather simple program to develop compared to the program of developing the implosion weapon in the first place, or developing the hydrogen bomb. The principles of making small weapons were well known. It seemed to me mainly what we call a hardware program. One takes designs which are theoratically good and one builds the small weapons. No fundamental research so far as I know had to be done to implement this program. This is one of the reasons why I thought it should not interfere with the hydrogen-bomb program. It took a different type of man to do the work.

Q. The small-weapons program was one of the major things discussed in this report.

A. That is right.

Q. And since it was mainly a hardware problem, it was not very much of a long-range thing, was it?

A. It was in the zero to 2-year period, which was one of the 2 periods which the program was concerned with.

Q. Would you consider zero to 2 years long range?

A. I believe that our directives were to consider long-range programs in 3 stages, zero to 2 years, which was called the short-range program; 2 to 5 years, called the intermediate program, and beyond that, the long-range program.

Q. As to the zero to 2 years part, that was not a mattter that was going to be long range looked at after some years?

A. No, but I was not setting the agenda of this meeting. That was in Dr. Oppenheimer's hands, and he spent most of the time or a good part of the time talking on this phase. That was not my doing, sir.

Q. I think you said Dr. Oppenheimer invited you as the representative of the opposite view.

A. He said as much. As I said, I admired him for doing that.

Q. And you considered yourself the representative of the opposite view?

A. I think that is true, yes.

Q. And that was the opposite view on the hydrogen bomb?

A. That is right.

Q. As the representative of the opposite view on the hydrogen bomb, weren't you perhaps more interested in what was said in the report about the hydrogen bomb than anything else?

A. I was only interested in seeing that the hydrogen bomb program was not stopped. The hydrogen-bomb program was at that time on the rails. The Greenhouse device was being fabricated; people were working hard to build the instrumentation to tell whether the thermonuclear reaction took place. I thought the hydrogen bomb program at that time was in very good shape. The only thing that could have happened was that it be stopped. It could not have been speeded up tremendously at that point.

Q. Surely that was not the only thing that you were interested in the report in relation to the hydrogen bomb—that it should not be stopped?

A. After I heard Dr. Oppenheimer's statement that was my main interest, yes.

Q. Didn't you read with particular care the portions of the report that referred to the hydrogen bomb?

A. I thought I did, but as I pointed out, one who is not trained in the legal ways of reading documents would not have found this thing to be a document which would slow down the hydrogen bomb program. It turned out to be that.

Q. Was Dr. Oppenheimer a man trained in the legal ways of reading documents?

A. I would certainly say that Dr. Oppenheimer is one of the most skilled document writers that I have ever run across.

Q. That is slightly different from being trained in the legal way of reading and writing documents.

A. If he is trained or not, I say he has the skill. I don't say this in a derogatory sense.

Dr. ROBB. Mr. Chairman, these questions are getting a bit frivolous.

Mr. SILVERMAN. There is nothing frivolous about them. Here is a man that signed the report and didn't know what was in it, although he was the representative of the opposite camp on that precise point.

Mr. GRAY. Mr. Silverman will proceed.

By Mr. SILVERMAN:

Q. As the representative of the opposite camp, did you not read—I withdraw that.
Was it lawyers who were reading the document and misreading it?
A. I really don't know. Someone in the Atomic Energy Commission read the document and apparently tried to reorient the program at Los Alamos to the detriment of the hydrogen bomb program. This I have been told by Edward Teller. That is my only source of information on this point.
Q. Dr. Alvarez, would it be fair to say that the document that you signed was a document which fairly represented the views of the Committee, that afterwards you were informed that it was misused, and that you thereafter regretted that you had signed it?
Mr. ROBB. Could I have that question read back?
(Question read by the reporter.)
The WITNESS. I would say this, sir, that the main emphasis of the document was on the small weapons, and this represented the opinions of most of the members of the Committee. As I said, I was essentially neutral on this point. I had no strong feelings one way or the other. I appreciated the fact that small weapons were useful things.

By Mr. SILVERMAN:

Q. You have not regretted the part about the small weapons?
A. I have certainly not regretted the part about building small weapons. I have regretted the part that recommendations apparently were interpreted to mean that the small weapons had a higher priority than the hydrogen bomb, and therefore were to be allowed to interfere with the hydrogen bomb. That is my objection to the report.
Q. When you read the report at the time, did it seem to reflect the views of the panel, including yourself?
A. As I said, I didn't appreciate this fine point in the emphasis. I signed the thing and therefore I agreed. My name is signed to the thing.
Q. Are you sure that your present disagreement with the report isn't the result of a change of mind on your part?
A. I am completely convinced of that. I have reread the report and knowing now what happened at Los Alamos, I can see why it happened, and I can see that I was not careful enough to guard against this possibility.
Q. That is what I am suggesting to you, that it is what happened afterward that made you regret signing the report; that when you read the report, it did seem to you to reflect the views of the panel.
A. It is quite clear to me that my regrets come from the fact that the report was used this way, and it was used this way because of the lack of vigilance on my part to see that the report did not act adversely to the hydrogen bomb I thought in view of Dr. Oppenheimer's statements that things were under control.
Q. You feel you fell down on the job as the representative of the opposite camp?
A. That is right, and I am reminded of a recent case that has been much in the papers——
Mr. SILVERMAN. We have been stopped——
Mr. ROBB. Wait a minute. I think he has a right to explain the answer.
Mr. SILVERMAN. Mr. McCloy was stopped.
Mr. GRAY. He later testified on the point that I stopped it on.
Mr. SILVERMAN. He never gave the example.
Mr. GRAY. Yes, he did.
Mr. SILVERMAN. All right.
The WITNESS. I said we have a recent example of a man more skilled than I in the political field who thought after having a meeting with another gentleman that he had his points across, he felt very happy about it, he signed the document and went out of the room saying, "I have won my point," and he took a terrible beating in the press. I find that I was in the same position. I thought I had gotten my points across. I signed the document which I thought fairly reflected the views which I heard expressed in the meeting. I found out later that I had been had, if you don't mind my using that expression.

By Mr. SILVERMAN:

Q. Now, you testified to a statement by Dr. Bush. I think you said it was 2 or 3 months after the GAC meeting—a couple of months or so.
A. No; I didn't testify in that way. I testified that it was at the week that Dr. Bush's article was reprinted in Life magazine.

Q. Yes; I remember you said that. You said that would give you the date within a day or two.
A. That is right.
Q. And I thought you said you thought it was a couple of months——
Mr. ROBB. That was my question. I thought that is what it was. I was trying to bring him down to the date.

By Mr. SILVERMAN:
Q. Have you since checked the date of that?
A. No, I haven't. As a matter of fact, I have never even asked whether such a meeting took place. I have never checked with the Atomic Energy Commission or anyone else to find out that such a meeting took place.
Q. What meeting?
A. The meeting to evaluate the effects of the bomb. I am going completely on my memory there.
Q. Did I understand you said that Dr. Bush said that the reason he was Chairman—the reason the President had named him as Chairman was that the President didn't trust Dr. Oppenheimer?
A. That was the reason that he said he thought he had been named Chairman. I rather doubt that the President told him that he didn't trust Dr. Oppenheimer. I think this was Dr. Bush's construction.
Q. Have you heard since that that panel was not named by the President but by the Air Force?
A. I have never heard a single word about this panel, sir. As I said, I refreshed my memory on the long range objective panel. I reread the report. I have never checked at all anything to do with this. As a matter of fact, I had forgotten this thing until recently. I did not mention it to the gentlemen who questioned me in Berkeley some months ago.
Q. You mean Mr. Robb?
A. Yes.
Q. How long ago were you questioned at Berkeley?
A. It was probably in February or March.
Q. You gathered, you said, that Dr. Bush—I withdraw that.
Dr. Bush said that he understood the reason that he, Dr. Bush, had been named Chairman and not Dr. Oppenheimer was that he, Dr. Bush, thought that the President didn't trust Dr. Oppenheimer?
Mr. ROBB. Wait a minute. I don't think the witness so testified.
Mr. GRAY. I think that is correct. That was not the witness' testimony.
Mr. SILVERMAN. I thought the witness just said that.
Mr. ROBB. No, he said he thought that.
Mr. SILVERMAN. I thought that is what I said.
Mr. GRAY. No, you said that he understood. The witness testified that he did not have any reason to believe the President had told Dr. Bush that, that he thought that Dr. Bush said that because he, Dr. Bush, thought it. Is that correct?
The WITNESS. I pointed out the fact that Dr. Bush was trying to justify to himself his chairmanship of this Committee. He pointed out his own limitations and said essentially, "Why have I been chosen? Why wasn't it Dr. Oppenheimer? He is the logical man."

By Mr. SILVERMAN:
Q. This was a pretty important Committee.
A. I think it was a very important Committee.
Q. The President was about to make a momentous announcement.
A. That is right.
Q. And he wanted to be sure he was advised by people he trusted.
A. That is right.
Q. Didn't you say to Dr. Bush, "Look, if the President doesn't trust Dr. Oppenheimer, why does he name him to the Committee at all"?
A. As I pointed out to you, this was the first time I had ever heard Dr. Oppenheimer's trustworthiness challenged. Until that time I had always thought that Dr. Oppenheimer was the most loyal person, the most wonderful man. He is one of my scientific heroes. I had never had any reason to believe that Dr. Oppenheimer would not do anything that was not right.
Q. In any event you did not say to Dr. Bush why did the President appoint him at all if he didn't trust him, and Dr. Bush didn't say why.
A. No, this question didn't come up.
Mr. SILVERMAN. That is all. Thank you.

Mr. GRAY. Dr. Alvarez, for the purposes of the record, references have been made in the direct and cross examination to the panel on which you served, and there has been considerable discussion. I would like to get clear on this point. Would the correct title of this committee have been, as you recall it, Panel on Military Objectives in the Field of Atomic Energy? I am not trying to confuse you.

The WITNESS. That is possible. I believe it is always referred to as the long range objective panel. The precise title I am not clear on, sir.

(Discussion off the record.)

Mr. SILVERMAN. Mr. Chairman, perhaps if it is helpful, may I point this out: There was a panel on long range objectives in 1948 of which Dr. Alvarez was not a member, and which I assume is perhaps what you are looking at. There is a panel in 1950, Research and Development Board, Committee on Atomic Energy, ad hoc panel on military objectives in the field of atomic energy, from November 21, 1950, to January 30, 1951, of which Dr. Oppenheimer was chairman, and of which Dr. Alvarez was a member, and which is the panel I assume Dr. Alvarez was testifying about.

The WITNESS. I believe this is the reason the panel I served on was referred to as the long-range objectives panel, because we considered it to be a continuation of the first panel. At least during that discussion, Dr. Oppenheimer read to us the report of the first panel, and led us to believe that we were the second such panel to be installed.

Mr. GRAY. Thank you. I think that identifies for me and I hope for the record which panel we are talking about.

Mr. SILVERMAN. Mr. Chairman, while we are on this subject of panels, and the biography, I find a slight correction that has to be made in the biography with respect to one of the panels that has been testified about here.

Mr. GRAY. Has it been testified about by this witness?

Mr. SILVERMAN. Yes, sir. It was the panel on the Soviet explosions in 1949, of which Dr. Bush was chairman. In this biography Admiral Parsons is mentioned as chairman. You recall that Dr. Oppenheimer testified that this was gotten up by his secretary, and the biography names Admiral Parsons as a chairman. That is an error. It was Dr. Bush who was chairman. It is the Department of the Air Force AFOAT-1 advisory panel to Gen. Hoyt Vandenberg, review panel on the Soviet explosions, September 1949.

Mr. GRAY. Now, Dr. Alvarez, is it quite clear to you that you signed this report rather than subscribed to it?

The WITNESS. I certainly signed it; yes, sir.

Mr. GRAY. I believe we have had testimony from one member of that panel who was not quite clear as to whether the report was signed by the membership or not, but you are clear on that point?

The WITNESS. I can't remember the physical act of signing it.

Mr. GRAY. You have seen the document recently?

The WITNESS. I have seen the document.

Mr. GRAY. And your name is on it?

The WITNESS. I have seen my name typewritten on the document. I believe that I signed it, and I certainly should have signed it. Whether I went through the physical act or not, I don't recall. I mean I would have signed it. The only reason for not signing it would have been that I had to catch an airplane before the final draft was in or something of that sort.

Mr. GRAY. Has this report or any portions of it been in the record?

Mr. ROBB. No, sir; I don't think so. Has it?

Mr. SILVERMAN. It is a classified report, or am I wrong?

Mr. ROLANDER. To clarify the signature, Dr. Alvarez saw a copy of the report which is in the possession of the AEC. He did not see the original which would have had signatures. In fact, signatures did appear. The record that Dr. Alvarez saw was an official copy.

Mr. GRAY. I understand that, and I think he cannot remember whether he signed so we still don't know whether it was a signed document on the basis of testimony before this board.

The WITNESS. I would certainly not try to get out of my responsibility by saying that I perhaps had not signed it.

Mr. GRAY. This is not my purpose, Dr. Alvarez. I am trying to get it clear in my mind whether this was the kind of a report that each of the members signed, or whether the members more or less left it to the chairman to write the report saying that they subscribed to his summary of it.

The WITNESS. Excuse me. I believe that I do recall now how the signatures took place. I believe the final document was typed up after I had left Washington, and that it was brought to me to sign by a courier of the Atomic Energy Commission. I have this remembrance of it on one occasion having signed a report of a committee in this fashion. Perhaps this happened this time.

Mr. SILVERMAN. Mr. Chairman, my recollecion is that Mr. Robb examined Dr. Kelly, I think, about the same document and perhaps if Mr. Robb has a copy, he can tell it. I don't know. This was done in a classified session.

Mr. ROBB. I don't have a photostat, Mr. Chairman. If that is the report I examined Dr. Kelly about, I had only an excerpt.

The WITNESS. Sir, my memory is now complete. I do remember how I signed this report. It was brought by courier to Pasadena, and I went down and signed it in the office of Dr. Lauritsen together with Dr. Bacher. The three of us signed it in Pasadena.

Mr. GRAY. I think that answers the question.

Dr. Alvarez, what was the period of your service at Los Alamos?

The WITNESS. I believe, sir, that I arrived there in April of 1944, and left in approximately November of 1945.

Mr. GRAY. My next question is one which has not been the subject of testimony at all, by you here this morning, so it is something new to you. First of all, do you remember when the news about the Fuchs treachery took place, or I mean came to you?

The WITNESS. Yes; I do.

Mr. GRAY. Do you remember approximately when that was?

The WITNESS. I think it would take me some little while to find in my memory exactly when that took place.

Mr. GRAY. Let me see if I can help you on that.

Mr. ROLANDER. It was approximately February 1950, when the first news came to the AEC.

Mr. GRAY. When the news first came to the AEC?

Mr. ROLANDER. Yes, from the investigative channels.

Mr. GRAY. Can you remember under what circumstances you first heard about it?

The WITNESS. I read it in the paper, sir.

Mr. GRAY. You never heard any intimation before that about this?

The WITNESS. Absolutely none.

Dr. EVANS. Did you know Fuchs?

The WITNESS. I nodded to him in the halls when we passed in Los Alamos. I had no scientific business with him. He was a very retiring person. He didn't want to make friends for fairly obvious reasons. I understand that when there were parties at Los Alamos, he would take care of the children of the people who went to the parties so he had an excuse not to go. He was not a particularly social person. I had no reason to know him scientifically, and I certainly never got to know him socially. I recognized him and nodded to him in the halls. That is my only recollection of him, sir.

Mr. GRAY. Have you ever heard it intimated that these facts about Fuchs were known to anybody in the scientific community in this country before the public announcement and the events immediately leading up to the public announcement?

The WITNESS. I had never heard any such allegation.

Mr. GRAY. Do you have any questions, Dr. Evans?

Dr. EVANS. I have some questions; yes.

Dr. Alvarez, you have been asked a good many questions and been sitting on that chair quite a time, and that main thing that we have gotten out of you is that you have tried to show that Dr. Oppenheimer was opposed to the development of the super weapon; is that true?

The WITNESS. I believe that has been known for a long time, and I think I just have given some corroborative testimony in this regard.

Dr. EVANS. What does this mean in your mind—anything?

The WITNESS. By itself it means absolutely nothing because I have many other friends in the scientific world who feel precisely this way. The point I was trying to bring out was that every time I have found a person who felt this way, I have seen Dr. Oppenheimer's influence on that person's mind. I don't think there is anything wrong with this. I would certainly try to persuade people of my point of view, and Dr. Oppenheimer is quite free and should try to persuade people of his convictions. I just point out the facts as I see them, that this reaction has always taken place in the people that I know who have been opposed to the bomb.

Dr. EVANS. It doesn't mean that he was disloyal?
The WITNESS. Absolutely not, sir.
Dr. EVANS. Might it mean that he had moral scruples about the development of the atomic bomb?
The WITNESS. I have heard that he has. He has never expressed them to me. I told you the one occasion on which Dr. Oppenheimer expressed to me his reasons for not wanting to build the hydrogen bomb, and it had nothing to do with morals, in the usual sense.
Dr. EVANS. You think it might have been peculiar for him to have moral scruples after he had been so active in developing the atomic bomb?
The WITNESS. I have never had any moral scruples about having worked on the atomic bomb, because I felt that the atomic bomb saved countless lives, both Japanese and American. Had the war gone on for another week, I am sure that the fire raids on the Japanese cities would have killed more people than were killed in the atomic bombs. I am also quite convinced that the atomic bomb stopped the invasion of Japan, and therefore saved well over 100,000 American lives. I believe there are estimates of up to a half million.
Dr. EVANS. Don't we always have moral scruples when a new weapon is produced?
The WITNESS. That is a question I can't answer, sir.
Dr. EVANS. After the battle of Hastings, a little before my time——
Mr. SILVERMAN. Would you give the time, sir?
Dr. EVANS. I cannot give the time, but it was before I was born.
Mr. SILVERMAN. That is 1066, sir.
Dr. EVANS. There was great talk about ostracizing the long bow, because it was so strong that it could fire an arrow with such force, it occasionally pierced armor and killed a man. They felt they ought to outlaw it.
When the Kentucky rifle came in, it was so deadly that they talked of getting rid of it. When we had poison gas, I made a lot of lectures about it, that it was terrible. So we have had that after every new weapon that has been developed.
The WITNESS. Yes, I recognize that.
Dr. EVANS. This opposition that Dr. Oppenheimer had, might he have been jealous that someone else was becoming prominent in this field, rather than himself?
The WITNESS. I don't think so; no.
Dr. EVANS. You don't think so?
The WITNESS. No.
Dr. EVANS. Do you think that Dr. Oppenheimer had considerable power with men like Conant, Bush, and Groves?
The WITNESS. I don't think power is the right word. Dr. Oppenheimer is certainly one of the most persuasive men that has ever lived, and he certainly had influence. They respected his opinions and listened to him.
Dr. EVANS. Looking by hindsight, do you think he showed good judgment in the fact that he opposed this bomb in the light of present conditions?
The WITNESS. I think he showed exceedingly poor judgment. I told him so the first time he told me he was opposed to it. I have continued to think so. The thing which I thought at that time was the overpowering reason for building the hydrogen bomb was that if we did not do it, some day we might wake up and read headlines and see pictures of an explosion such as we saw a month or so ago, only this would be done off the coast of Siberia. I felt sure that this would be one of the most disastrous things that could possibly happen to this country. I thought we must not let this happen.
Dr. EVANS, His opposition to it, might it mean that he feared the spending of a large sum of money and the using of time on a project that would not work and might thus endanger the security of our country by not going ahead with a project that we knew would work?
The WITNESS. I think he has expressed an opinion somewhat as you just stated it.
Dr. EVANS. You see, Dr. Alvarez, as a member of this board, I am trying to get something about what is in your mind and what is in Dr. Oppenheimer's mind.
The WITNESS. Yes, sir.
Dr. EVANS. We have a recommendation to make and we have to do the best we can. You understand that?
The WITNESS. I do, sir.
Dr. EVANS. You mentioned Professor Serber. That is the same Professor Serber that had these leftwing tendencies, or do you know anything about that?
The WITNESS. I know nothing of that personally. I have no personal knowl-

edge of it. I have read and I have been told by other people that this might be so.

Dr. EVANS. Were there a number of other men in the country that could have built the A-bomb?

The WITNESS. I am sure that there are. I don't want in any way to minimize Dr. Oppenheimer's contribution, because to my way of thinking he did a truly outstanding job at Los Alamos. I think he was one of the greatest directors of a military program that this country has ever seen. I stand in awe of the job he did at Los Alamos.

Dr. EVANS. You spoke of Dr. Bush?

The WITNESS. Yes.

Dr. EVANS. Possibly having made a statement—I forget what your statement was—but this is the question I want to ask you. Did Dr. Bush sometimes make statements that are not quite accurate? Do you know anything about that?

The WITNESS. I really could not say. I have great admiration for Dr. Bush as a scientist and as a scientific administrator, and I like him as a man.

Dr. EVANS. That is all I have.

Mr. GRAY. Mr. Robb.

REDIRECT EXAMINATION

By Mr. ROBB:

Q. Dr. Alvarez, Mr. Silverman asked you some questions about the relative competency of you and your group and the du Pont Co. to build reactors. I would like to ask you, sir, were you intending to suggest in any way that you were to be compared with the du Pont Co.?

A. No; that is ridiculous.

Q. Would you care who built the reactors, as long as they were built?

A. Of course not. As a matter of fact, I didn't want to build reactors. I disliked the idea of building reactors. I suggested that we build reactors only because I felt the country needed them and we could be of help.

Q. And if the Government employed the du Pont Co. to come and build them out near San Francisco, you would have been very happy?

A. It would have made no difference where the du Pont Co. built them. I am sure the du Pont Co. would not have asked me for any advice, because I have no special competence in that field.

Q. Your point was that we ought to get going on the hydrogen bomb?

A. That is right.

Q. Whoever did it?

A. That is right.

Q. You testified as others did that Dr. Oppenheimer did a splendid job at Los Alamos. Did it strike you as peculiar that one who had done such a splendid job at Los Alamos could entertain opinions which you considered so wrong in respect of the hydrogen bomb?

A. I was very surprised when I found that he had these opinions, since he had used the super as the primary incentive to get me to join the Manhattan District in the first place. He had spent almost a solid afternoon telling me about the exciting possibilities of the super, and asked me to join and help with the building of such a device. So I was therefore very surprised when I found he had these objections. You will note in my diary that I had no hint of this until essentially the last entry.

Q. To use a homely simile, did it strike you as peculiar that such a wonderful batter as Dr. Oppenheimer should suddenly begin striking out the way he did?

A. It certainly struck me as peculiar.

Q. One further question, Doctor. Have you had any hesitation in answering questions here or in any way restricted your testimony in answer to any question put to you because of the presence here of Dr. Oppenheimer and his counsel?

A. No. I must confess that it is a little hard for personal reasons to say some of the things that I have said, but I have said them anyway.

Mr. ROBB. Thank you.

Mr. GRAY. Mr. Silverman.

RE-CROSS-EXAMINATION

By Mr. SILVERMAN:

Q. Did it strike you as peculiar that Dr. Bacher had these views about the hydrogen bomb?

A. It did, as a matter of fact; yes.

Q. Dr. Lauritsen?

A. Yes.

Q. Dr. Conant?

A. No; not in the case of Dr. Conant for a reason which I will mention now.
Q. If you think it will be helpful.
A. I think it will; yes. I can remember an occasion a few months before the Russian explosion when Dr. Lawrence, Dr. Conant, and I were driving from Berkeley to San Francisco.
Q. Which explosion was this?
A. The first one. The one that led to the hydrogen bomb controversy, in 1949.
Q. You mean the Soviet.
A. The first Soviet operation, Joe. Dr. Lawrence was trying to get a reaction from Dr. Conant on the possibility of radiological warfare and Dr. Conant said he wasn't interested. He didn't want to be bothered with it. I have the strong recollection that Dr. Conanat said something to the effect that he was getting too old and too tired to be an adviser on affairs of this sort. He said, "I did my job during the war" and intimated that he was burned out, and he could not get any enthuhiasm for new projects. So when Dr. Conant disapproved of the hydrogen bomb, I interpreted it in the light of that conversation.

Dr. EVANS. Dr. Conant was not an authority in that field at all. He is an organic chemist; isn't that true?

The WITNESS. Dr. Conant showed to me a remarkable degree of knowledge about the details of nuclear physics and the construction of bombs on the two occasions I talked with him at Los Alamos. I was almost overwhelmed by the detailed knowledge he had on all fields. So although he was trained as an organic chemist, he certainly got to know a lot of weapon technology.

Dr. EVANS. He had been briefed up very well.
The WITNESS. Yes.

By Mr. SILVERMAN:

Q. You say Dr. DuBridge worked on the atom bomb, had he not?
A. No; he had not.
Q. Dr. Fermi had, of course.
A. Yes.
Q. Were you surprised that he was against going ahead with the hydrogen bomb and did that strike you as peculiar?
A. I never knew that Dr. Fermi was. I knew Dr. Fermi worked quite hard at Los Alamos for two summers since the Presidential anouncement.
Q. Didn't you know that he was one of the members of the General Advisory Committee?
A. I knew he was, and I heard that he was 1 of 2 men who signed an appendix to the report expressing views somewhat different from those of the majority group led by Dr. Oppenheimer.
Q. Did you know whether the extent to which there was that difference that perhaps they were even more opposed to the hydrogen bomb than the others?
A. I had not read the report, and I was led to believe that Dr. Fermi did not have such strong objections. I may be wrong on this. That was my impression.
Q. How did Dr. Rabi feel? Had he worked on the atom bomb?
A. He was a consultant to Los Alamos. He would come out occasionally from his job as assistant director of the radar laboratory and talk with people about problems.
Q. Did it strike you as peculiar that he was opposed to going ahead with the hydrogen bomb?
A. As I stated earlier, I was surprised that he changed his mind so drastically after talking with Dr. Oppenheimer. I was not at all surprised by his initial reaction, which was one of enthusiasm.
Q. And you have no way of knowing who else these people consulted?
A. No.

Mr. SILVERMAN. Thank you.
Mr. ROBB. That is all.
Mr. GRAY. Thank you very much, Doctor.
(Witness excused.)

Mr. GRAY. Let me say for the record that in recognition of the fact that Mr. Mitchell has been in and out of the hearing and I want him to be present when we discuss the request for documents which has been earlier referred to, inasmuch as he was involved, and in view of the fact that we at this point are between witnesses, I would like to return to the discussion which was had— whatever day it was—and allow Mr. Garrison to make his request at this time.

Mr. GARRISON. Thank you, Mr. Chairman. If I might just recapitulate for a moment to explain the nature of the request, I previously referred to the fact that back in the middle of February, I asked for the minutes and documents relating to the question of the clearance of Dr. Oppenheimer by the AEC in 1947, and that I was thereafter informed in General Nichols' letter of February 19, 1954, and in a conversation with Mr. Mitchell over the telephone—Mr. Mitchell's letter of February 19, 1954, and in conversation with him over the telephone—that the only information that could be supplied to us would be in the form of a stipulation which has already been read into the record, and which in substance contained the first half, but not the last half of the sentence in the minutes which finally were supplied to us the other day in General Nichols' memorandum to you.

I also would note that in the course of Mr. Lilienthal's cross-examination relevant documents to this whole matter were declassified by the Government on the spot and put into evidence. I think there were four that were put in in that fashion, and then two more at our request that followed that.

The testimony was left in a somewhat uncertain state, I think, and I don't want now to argue its significance, except to say that in my own view the second half of the sentence from the Commission's minutes would indicate to me quite clearly that the Commission, as such, examined further reports, and had taken them into account, and had reached the view that they contained no information which would warrant reconsideration of a clearance which apparently took place in February 1947, which apparently had been reopened as a result of the Hoover letter of March.

Mr. GRAY. I am going to interrupt, Mr. Garrison. I do not wish at this time to discuss the import of minutes. I have repeatedly indicated you will be given an opportunity to address yourself to that. I would like now for you to confine yourself to the request.

Mr. GARRISON. Yes. Mr. Chairman, to put it in nontechnical terms, what I would like to ask the board to request of the Commission that we have a statement in as much detail as classification will permit of the items of derogatory information which were contained in the files that went to the members of the Commission. Those files are referred to in Mr. Jones' memorandum to Mr. Bellesly. I think it is there stated that every member of the Commission received these particular files or reports except two memoranda which were summaries—I am doing this from memory—which Mr. Jones referred to in his memorandum. I should think it would not be a difficult matter for the Commission to look at those reports that we know from the record did go to the board members—I mean to the Commission members—and to ask in as much detail as can properly be given here a description of what the derogatory items consisted of so that we may more clearly determine what was before the board—I mean before the Commission.

I don't want to make a great thing out of this. I am not going to argue to this board that the action which the Commission took in 1947 was in any way conclusive or binding upon this board at all. I don't want to make such an argument. I do say it is quite relevant to consider what those five men who knew Dr. Oppenheimer and went through the report thought and believed at that time.

I think, Mr. Chairman, you raised the question when I started to make this request before as to whether we ought not to make the request directly of the Commission. I should do it any way that you wish, but I do think from the reading of the rules, it seems to me, sir, that it is appropriate and indicated that the Board itself should ask for relevant information. I would refer to section 4.15 (e) which says that the board will ask the individual AEC representatives and other witnesses any questions calculated to obtain the fullest possible disclosure of relevant and material facts.

Then there is another one, (g), the board will admit in evidence this and that and so forth, and then it says, "Every reasonable effort will be made to obtain the best evidence reasonably available."

"(j) The board shall endeavor to obtain all the facts that are reasonably available in order to arrive at its recommendations."

I think those are the principal sections. Perhaps I should refer also to (n). "The board may request the manager to arrange for additional investigation on any points which are material to the deliberations of the board which the board believes need extension or clarification."

It seems to me that the proper procedure is for me to ask the board for this information, and then for the board to try to obtain it.

Mr. GRAY. With respect to Mr. Garrison's request, as I understood it, as we discussed previously, you made particular reference to a conversation which was had with Mr. Mitchell.

Mr. GARRISON. And General Nichols and Mr. Marks.

Mr. GRAY. I had forgotten who else was there. General Nichols and Mr. Marks, with respect to a number of items, and it is my recollection you said seven the other day.

Mr. GARRISON. I think I had five written down on the yellow piece of paper which I showed to Mr. Mitchell the other day, and Mr. Marks had a number on a typewritten memorandum.

Mr. GRAY. I would like to state the impression of the chairman of the board, and be corrected if I am wrong.

Among those items were pertinent GAC reports and/or minutes.

Mr. GARRISON. Yes.

Mr. GRAY. It is my information that Dr. Oppenheimer had been notified officially by the Commission that he could have access to these reports and/or minutes, and that he has not availed himself of this opportunity.

Among the items, as I understand it, requested was the minute which has been read into the record. One of the items involved was Dr. Oppenheimer's testimony before the Joint Committee on Atomic Energy, if that is the proper title of the congressional committee concerned. One of the items was the contents of Mr. Hoover's letter.

It is my impression that with respect to these items, whether 5 or 7 or whatever the number, the request that they be made available in one way or another has been met with respect to all but two, the two being the congressional hearing record, which this board is not at liberty to make available, and the other is the FBI letter, which under the regulations we are not at liberty to make available.

With respect to the congressional testimony, I assume that it is not inappropriate for Dr. Oppenheimer to request of the committee the privilege of seeing those portions of the hearings which contain his own testimony, but this board does not have the power to produce such a document.

I think I have referred to the regulations which specifically cover information from the Federal Bureau of Investigation. So the earlier discussion centered around these requests which were made in a conversation between Atomic Energy Commission officials and Dr. Oppenheimer's representatives, and I think those requests have been met insofar as it is possible for this board to have any influence in meeting them, or any power in meeting them.

Now, with respect to the current request which, if I understand it correctly, is a list of all items of so-called derogatory information about Dr. Oppenheimer in the hands of this board, again I would have to respond that information which is contained in FBI reports cannot be made available.

I think I shall have to stop my observation at that point. It may be that my interpretation of the procedures under which we operate is faulty, and I would ask counsel for the board if he has anything to add to what I said.

Mr. ROBB. I certainly agree that your interpretation is entirely correct, Mr. Chairman. I would add only one observation, which is that so far as we are able to bring it together, all the information and reports which were before the Commission in 1947 are now before this board for its consideration and its evaluation.

Of course, as the chairman has said, the FBI reports under the rules of these hearings may not be made available to counsel for Dr. Oppenheimer or Dr. Oppenheimer.

Mr. GRAY. Let me make one other observation. I suppose it would be reasonable for counsel to assume that the board in its effort to get at the truth with respect to any matter of very material consequence has sought to have light thrown on such a matter of material consequence. This, of course, involves, I am sure, the question of anybody's reliance on the good faith of this board. What I am trying to say is that I do not think you are materially disadvantaged by not having the detailed list of information which you have requested.

Mr. GARRISON. I would like to make just one observation. I want to make it clear, Mr. Chairman, that so far as the fairness of the members of this board and their desire to do the right thing, I have no doubt whatever. My problem is one of knowing what seems to us to be relevant so that we may comment upon it as one should in presenting Dr. Oppenheimer's case, as well as we can. In a process of this kind I should suppose that the adversary process which we seem to be engaged in should be carried out to the fullest extent that it can be done within the limits of the governmental regulations with respect to the preservation of whatever has to be confidential, that this process will aid rather than to the contrary in the deliberations of the board.

I would like to make one or two things clear in the February discussions and correspondence. The Commission did, indeed, say to Dr. Oppenheimer that he might inspect minutes and reports of the GAC meetings in which he participated, and could also see any documents which he himself signed. What I am talking about here is the action of the Commission in 1947. I am not asking that the FBI reports be disclosed. I appreciate the rule that the reports of the Federal Bureau of Investigation shall not be disclosed to the individual or to his representative. I regretfully have to accept that rule. It does seem to me, however, that, since in the very letter of General Nichols with which we are concerned a very lengthy account is given of numerous derogatory items in the file and disclosure has been made of that, I cannot see how it would violate this rule to have us informed as to the derogatory items which were before the board in 1947. I am not asking for a transcript of the reports or a copy of the reports, but simply for a description of what the board acted on—I mean the Commission acted on.

Mr. ROBB. Mr. Chairman, as I interpret Mr. Garrison's last remark, he does not want a copy of the reports or the transcripts of the reports; he merely wants to know their contents, which seems to me to fly right in the fact of the rule. I am sorry.

Mr. GARRISON. Let me ask this final question: Would it fly in the face of the rule if we were limited merely to being told which of the items now before the board were before the Commission in 1947?

Mr. ROBB. I think it would, Mr. Chairman.

Mr. GARRISON. I just don't understand that, Mr. Chairman, as to why we can't be told of these items that such-and-such were before the board and such-and-such were not. What disclosure of FBI reports is that any more than this letter itself is a disclosure of FBI reports?

Mr. GRAY. I believe that what was before the Commission in 1947, and certainly from the testimony here, cannot be certain, because the recollection of the four former Commissioners who have testified here is uniformly hazy as to what happened. I hope that is not an incorrect statement about their testimony. With respect, in any event, to what was before them at that time we are not certain. I believe what was before them at that time was FBI reports. It seems to me that comes into the rule.

I would make this further observation: That if counsel wishes at some subsequent point in these proceedings to argue the import of the actions of the Commission insofar as they can be reconstructed in 1947, whether February, March, or August, that opportunity will be given. As far as this board is concerned, we must be concerned with everything before us; and what the Commission did in 1947 is, of course, important, but, as you say, not conclusive.

I would like to suggest a recess at this point.

(Brief recess.)

Mr. GRAY. Colonel Pash, do you care to testify under oath? You are not required to.

Colonel PASH. Yes, sir.

Mr. GRAY. Would you give me your full name?

Colonel PASH. Boris T. Pash.

Mr. GRAY. Will you raise your right hand? Boris T. Pash, do you swear that the testimony you are to give the board shall be the truth, the whole truth, and nothing but the truth, so help you God?

Colonel PASH. I do.

Whereupon Boris T. Pash was called as a witness, and, having been first duly sworn, was examined and testified as follows:

Mr. GRAY. Will you be seated, please, sir.

It is my duty, Colonel Pash, to remind you of the existence of the so-called perjury statutes. May I assume you are familiar with them and they need not be reviewed?

The WITNESS. Yes, sir.

Mr. GRAY. You understand, I suppose, or you should know, in any event, that there are persons in this room who may not have clearance for certain classified material. I would ask, therefore, in the course of your testimony, if you are getting into classified areas, you seek to notify me in advance so we may take the necessary steps.

The WITNESS. Yes, sir.

Mr. GRAY. Finally, Colonel, I should say to you that we consider this proceeding a confidential matter between the Atomic Energy Commission officials and witnesses, on the one hand, and Dr. Oppenheimer and his representatives, on the other. The Commission is making no releases with respect to these proceedings.

I express the hope on behalf of the local board that witnesses will take the same view.

The WITNESS. I am, sir.

Mr. GRAY. Mr. Robb, will you proceed?

DIRECT EXAMINATION

By Mr. ROBB:

Q. Colonel Pash, will you give us for the record your present station?
A. My present station is Presidio of San Francisco, Calif.
Q. You are an officer in the United States Army?
A. I am.
Q. And have been for how long, sir?
A. I am a Reserve officer on active duty, and I have been on active duty for about 14 years.
Q. What is your present assignment?
A. Presently I am Chief of the Counterintelligence Division in the office of G-2, Headquarters, Sixth Army.
Q. What are your duties in that capacity?
A. In that capacity I review and pass on the activities of my branch offices which are concerned with counteraction against espionage, sabotage, the conduct of personnel security investigations, and industrial security investigations.
Q. Without going into detail for a moment, Colonel, how long have you been engaged in that general sort of work for the Army?
A. About 12 years of the 14.
Q. Let me, if you will, get a little of your personal history. You came on this present tour of duty when?
A. About the 1st of June of 1940.
Q. What was your assignment?
A. I was then for a short time the counterintelligence officer of the Ninth Corps Area.
Q. Where is that?
A. In Presidio of San Francisco. In about March of 1941—I am not sure of that date—I became the Chief of Counterintelligence Branch of the Ninth Corps Area, and later of the Western Defense Command and the Fourth Army.
Q. Will you tell us whether you took any course at about that time in connection with your work?
A. In January of 1941, I took the officers investigate course conducted by the Federal Bureau of Investigation.
Q. Where?
A. In Washington at the Justice Department Building. I think that is between 10th and 9th on Pennnsylvania.
Q. That had to do with espionage and sabotage?
A. Espionage, sabotage, interrogation, writing of reports, securing evidence, the overall investigative course.
Q. What was your next assignment?
A. In November 1943, I left the Fourth Army Western Defense Command and proceeded to Europe where I organized and commanded the scientific intelligence mission of the G-2, War Department, known under the code name of the Alsos mission.
Q. What was that mission, Colonel?
A. The primary mission was to determine the extent of German atomic developments and to find out whether they would or would not use the bomb in World War II, and if possible secure the scientists and documents and any equipment that they may have.
Q. How long did that mission last?
A. The commission was deactivated in December of 1945.
Q. In connection with that work, were you required to interrogate scientists and other personnel?
A. Yes; we did interrogate scientists.
Q. Following the completion of that mission, what did you do?
A. In March of 1946, I went to Japan, where I was assigned as the Chief of the Foreign Liaison Section in G-2, Headquarters, Far East Command. In connection with those activities my primary responsibility as designated by the Chief of Staff was to deal with the Soviet mission. Since I was a colonel, the Chief of Staff felt I could deal with the Commission directly because I speak Russian.
Q. You speak Russian?
A. Yes; I speak Russian fluently.
Q. Were you born in Russia or in this country?

A. I was born in San Francisco.
Q. Your father was a Russian bishop?
A. He arrived in the States in 1894, and in the later years he was known as the Metropolitan, which is the top or senior bishop of the American Orthodox Church, which officially, I believe, the name is the Russian Orthodox Greek Catholic Church of North America.
Q. Was your mother a Russian or American?
A. No; she was born in San Francisco.
Q. In all events, you learned to speak Russian from your father?
A. No; I studied it and had experience, of course.
Q. You say you were in Japan for how long?
A. Two years.
Q. Dealing with the Russians?
A. Primarily. I dealt with all the foreign missions there.
Q. Who was the commanding officer in Japan then?
A. General MacArthur was then commanding.
Q. At the completion of that duty, what did you do?
A. At the completion of that duty I was assigned to G-2, Department of the Army, in the Eurasian Branch.
Q. Will you tell us what your work was there?
A. Study of the Soviet Union and the Soviet Army.
Q. G-2 is Intelligence?
A. Yes, sir.
Q. At the end of the year, where did you go?
A. At the end of that year I was detailed to the Central Intelligence Agency.
Q. How long did you stay there?
A. I served with the Central Intelligence Agency for 3 years.
Q. Are you able within the rules of security regulations to tell us anything about your work there?
A. No, sir; I am not.
Q. You were there for 3 years.
A. Yes, sir.
Q. Until when?
A. Until I believe the 7th of January 1952.
Q. Then where did you go?
A. Then I was assigned to Austria, Headquarters, United States Forces in Austria, was stationed in Saltzburg.
Q. What was your duty there?
A. There I was in G-3, which is the Planning Section.
Q. How long were you there?
A. I returned from Austria in August 1953.
Q. And then you went to your present duty?
A. Went to my present station, reporting to the Presidio in September 1953.
Q. Now, going back to 1943, in what month was it, Colonel, that you reported for duty at San Francisco in 1943?
A. I was in San Francisco at the time in 1943.
Q. What month did you begin your duty as what was it, now?
A. Chief of the Counterintelligence Branch.
Q. Yes, sir.
A. That was in 1941.
Q. And you stayed there until when?
A. I would like to make a correction. I am not sure whether it is early 1941 or late. I mentioned the early part of 1941. I am not sure of that date and I didn't check it.
Q. Coming to May 1943, Colonel, I will ask you whether or not at or about that time you began an investigation into certain reported espionage taking place or which had taken place at the Radiation Laboratory in Berkeley?
A. Yes, sir; we did.
Q. Would you tell us something of how that investigation began and what you did?
A. Yes, sir.
Q. Just tell us in your own way, and I will try not to interrupt you.
A. I believe it was in May of 1943 an officer from the Department of the Army reported to General DeWitt, who was commanding general of the Western Defense Command, requesting that an officer be designated to conduct a special investigation connected with War Department activities.
General DeWitt designated me to take charge of that investigation.

Together with the initiation of this investigation, I received a report from the Department of the Army. I think it was the War Department then, indicating that there had been an attempt to secure information from the radiation laboratory and that the personnel involved were Steve Nelson, of the Communist Party, a prominent Communist Party member in California at the time, and a man by the name of Joe.

We further knew that Joe had furnished some information, including information of a technical nature, which I don't recall clearly, and I would not dare to try to explain anything of the technical nature, and that he had furnished Steve Nelson with a timetable pertaining to activities in which we were to become interested—the technical activities.

We had very little information. The only thing we had definite was that the man's name was Joe, and the fact that he had sisters living in New York, and that he had come from New York.

We started the investigation. We immediately started procuring files of personnel working at the laboratory in order to try to analyze and determine who this man may be. I will not go into the technical details of our surveillance or operational methods except to say that we did conduct an investigation.

We first thought this man may be a man by the name of Lomanitz.

Q. Would you tell us why you thought that?

A. Because of Lomanitz's past history. We were able to procure that. Lomanitz was affiliated with some Communist-front organizations, and actually was reported to be a Communist Party member.

In our operational work, we were able to procure a photograph of 4 men, and I had 1 of our men working on that photograph to determine the background of the personnel in the photograph.

In the meantime we also found out that at some meetings sponsored by either— I forget the organization sponsoring it—it was on Van Ness Avenue, we observed, I believe it was either Bohm or Lomanitz going in with an unidentified man, a man unidentified by us.

Q. Which Bohm was that?

A. His first name slips me. He was closely affiliated socially, and I suppose in the school, with Lomanitz and Weinberg.

Q. By the way, did you ascertain what he was doing at the time, and where he was employed?

A. Yes; we knew that he was employed at the Radiation Laboratory.

Q. How about Bohm?

A. Bohm also.

Q. Go ahead, sir.

A. We had an unidentified man and we had this photograph. As a result of our study we determined and were sure that Joe was Joseph Weinberg.

Q. Where was he employed?

A. He was employed at the Radiation Laboratory.

Q. Were you able to ascertain whether Lomanitz, Weinberg, and Bohm were associates or intimates?

A. Yes; they were. The photograph consisted of Weinberg, Lomanitz, a man by the name of Max Friedman, and I think Bohm.

With that in mind, we started our operational procedures and at the same time a review of the file itself. I reported the identification of Joe to the War Department at the time. This must have been some time in the early part or the first half of June 1943.

Q. What did your investigation disclose with respect to the Communist activities of this group—Weinberg, Lomanitz, Bohm, and Friedman?

A. We determined in the first place that these four men I mentioned were very frequently together. I don't mean constantly with no interruption, but very frequently they were together.

Through our operational procedures, we found out that Lomanitz was a member of the Communist Party. From the conversations we also determined that we had sufficient information to determine that both Weinberg and Bohm were members of the party.

Q. By the way, Colonel, I might ask you whether, under security regulations, you are permitted to disclose investigative techniques or operational procedures?

A. No, sir; I would be glad to present them to the board.

Q. I might ask you, just for the record, Colonel, I assume you are here under orders?

A. Yes, sir; I have been ordered here by the Department of the Army.

Q. But the testimony you are giving is your own testimony, and not what someone told you to say?

A. No, sir; I think I better correct that. The testimony is my own.
Q. Did there come a time when certain steps were taken with respect to the draft status of this man Lomanitz?
A. Yes, sir; when we determined and felt sure that Lomanitz was a member of the Communist Party, we recommended that his draft deferment not be renewed. I made that recommendation to General Groves' office.
Q. What happened then?
A. We received information from General Groves' office that the deferment will be canceled, and we were to keep General Groves advised of the status of the situation.

When Lomanitz heard the fact that his deferment was being canceled, he started contacting a number of people. He contacted members of the union, the FAECT union, which was interested in the Radiation Laboratory. He contacted his friends. He discussed with his friends the situation. He also called and, if I am not mistaken, wrote to Dr. Oppenheimer about it.
Q. What?
A. Called Dr. Oppenheimer about it.
Q. About when was that, if you remember?
A. That was in the early part of August, I think. I don't know the date.
Q. Do you recall whether or not Dr. Oppenheimer manifested any interest in this matter of Lomanitz' defendant?
A. Yes; according to my recollection, Dr. Oppenheimer took some steps to request that deferment be granted to Lomanitz.
Q. Did the activities of Dr. Oppenheimer in that connection strike you as usual or unusual?
A. Not having sufficient knowledge of the technical phase of this particular situation, I am not expressing an opinion which is based on reactions other than any technical reactions.
Q. Yes, sir.
A. Since we were interested in this investigation, we certainly followed very closely the activities as they were proceeding, and we felt at the time that pressure was being put on to keep Lomanitz on the project.
Q. Pressure by whom?
A. By Dr. Oppenheimer, by his associates, Max Friedman, Weinberg, and Bohm.

Mr. SILVERMAN. May I interrupt for one moment? Who do you mean by "his"?

The WITNESS. Lomanitz'.

By Mr. ROBB:

Q. Did it strike you that the pressure put on by Dr. Oppenheimer was ordinary or was out of the ordinary, as it struck you at that time?
A. It was my feeling that there was pressure beyond that which would be normal.
Q. You mentioned the FAECT, the union; did your investigation include any study of the Communist tendencies or influences in that union?
A. We based our evaluation of the FAECT on reports received. We did not investigate the union as such. However, we also received considerable information from discussions among those people who are within our investigative field, and whom we were investigating.
Q. Let me ask you at that point, Colonel, did you have any jurisdiction to investigate or interview anybody who was not either in the Army or connected with the project?
A. Yes, sir; the project was given to us as our responsibility when the officer came out from the Department of the Army.
Q. I don't think you quite caught my question. Did your investigative jurisdiction go beyond that? In other words, could you investigate people who were civilians not connected with the project?
A. No.
Q. All right, sir. Now would you come back to the matter of the union and what you found out about the union?
A. Again based on the information available to us and from reports available to us, we felt that there was a strong Communist influence among a group of people—at least a group of people—in the union, and that the union was attempting to place people in the radiation laboratory.
Q. Do you recall who any of the group of people of the union were that you had in mind?

A. There was a man by the name of Adelson. There was also a woman belonging to the union called Rose—it starts with an "S."
Q. Would it be Segure?
A. Segure; yes.
Q. How about this group that you told us about, the Weinberg-Bohm-Friedman-Lomanitz group; were they in the union?
A. Yes; they were members of the union. To the best of my knowledge they were members of the union.
Q. Was Lomanitz finally drafted?
A. Yes; Lomanitz' deferment was canceled and he was drafted.
Q. At or about that time did you receive certain information from Lt. Lyall Johnson concerning statements made to him by Dr. Oppenheimer?
A. I did. Lyall Johnson reported to me toward the end of August that Dr. Oppenheimer came to him and made some statements which he felt I should know about. My reaction was to request an immediate interview with Dr. Oppenheimer on this matter.
Q. Who was Johnson?
A. Johnson was the intelligence officer for the radiation laboratory.
Q. Do you recall whether or not Johnson gave you any details of that conversation?
A. Johnson told me it concerned a possible espionage effort in connection with the radiation laboratory.
Q. Did you thereafter interview Dr. Oppenheimer?
A. Yes, I interviewed Dr. Oppenheimer on the 26th or 27th of August 1943.
Q. Where did the interview take place, Colonel, and what were the circumstances under which it took place?
A. The interview was conducted on the University of California campus. There was a building in which Lieutenant Johnson had his office. Captain Fidler was a member of the staff. I don't recall his exact capacity at the time. He was in the Army. We used Lieutenant Johnson's office to conduct this interview.
Q. Did you make any arrangements to have it recorded?
A. Yes. We felt that this information was of considerable importance, and we did not want to rely later on on what we may remember, so I made arrangements for an officer in charge of my investigative unit to set up a recording for us.
Q. So far as you know, was that with the knowledge of Dr. Oppenheimer, or was he unaware that it was being recorded?
A. As far as I know, he was unaware.
Q. Subsequent to the interview, were the recordings transcribed?
A. Yes; after hearing what Dr. Oppenheimer had to tell me, I immediately had the recordings transcribed so I could forward them to General Groves' office. I recall we made the first draft off the recordings and we tried to check that as much as we could. Subsequent to that I wanted to hurry this to General Groves, so I recall we started doing a second typing of it, and I stopped the typist and forwarded it by airmail immediately to General Groves' office.
Q. So far as you were able to tell at that time did the draft that you forwarded substantially state or reflect your conversation with Dr. Oppenheimer?
A. It did; yes.
Q. Would you say that every word was right?
A. No; there were a few words missing. I personally made some corrections in the draft.
Q. After you forwarded it?
A. Before I forwarded it. That is before I forwarded this first draft.
Q. I have before me a copy of a memorandum dated August 28, 1943, indicating that on that date you forwarded to Colonel Lansdale the transcript of your interview with Dr. Oppenheimer. Would that enable you to tell us when you did forward it to General Groves?
A. This was forwarded either on the 28th of August or it may be forwarded the day after.
Mr. GRAY. May I ask was this covering memorandum in the record, too?
Mr. ROBB. I am not sure whether it was or not. Do you want me to read it in?

s is:

"HEADQUARTERS WESTERN DEFENSE COMMAND AND FOURTH ARMY,
"OFFICE OF THE ASSISTANT CHIEF OF STAFF G-2,
"*Presidio of San Francisco, Calif.*

"In reply refer to: (CIB).
"August 28, 1943.
"Subject: DSM Project.
"To: Lt. Col. John Lansdale, Jr., Room 2C, 654 Pentagon Building, Washington, D. C.

"1. Transmitted herewith in a transcript of an interview with Dr. J. R. Oppenheimer, held in the office of Captain Fidler, University of California.

"2. No distribution of this was made other than to furnish one copy to Mr. King of the San Francisco field division of the Federal Bureau of Investigation. General Groves will be shown a copy of this transcription when he arrives on the 1st of September 1943.

"3. No comments or conclusions are made until a thorough study is completed. Any such recommendations or conclusions reached will be reported to you.
"For the A. C. of S., G-2:

"(S) BORIS T. PASH,
"*Lt. Col., M. I., Chief, Counter Intelligence Branch.*

"1 Incl: As indicated (dup)."

By Mr. ROBB:

Q. Have you recently refreshed your recollection about this interview by looking over a copy of that transcript?
A. I have.
Q. Do you recall, Colonel, whether or not in that interview Dr. Oppenheimer said anything to you about somebody in the office of the Russian Consul?
A. Of the Soviet Consul, yes.
Q. Is there any question in your mind that was mentioned?
A. No, sir; that was mentioned.
Q. In what connection?
A. Dr. Oppenheimer told me that a man contacted him with the suggestion that technical information can be made available through proper channels to the Soviet Consulate and that there was a man available who was proficient in microfilming, and that there were channels established for the transmission of available information.
Q. Is there any question that Dr. Oppenheimer made that reference to the use of microfilm?
A. No, sir; not in my mind.
Q. Do you recall whether or not Dr. Oppenheimer mentioned to you whether this man who had made the approach had made more than one approach to people on the project?
A. Yes. He indicated three definite approaches that were made.
Q. Is there any question about that in your mind?
A. No, sir.
Q. Did you ask Dr. Oppenheimer who the man was who had made these approaches?
A. Yes, I did. I asked him for the name of the man.
Q. Did he give it to you?
A. No, he did not.
Q. Did he say why he would not give it to you?
A. He stated that this man was a friend of his, he felt that no information was leaking out, and he felt that he did not want to give the man's name under the circumstances since he felt that it wasn't successful in accomplishing his mission.
Q. Were you anxious to know the name?
A. We were. As a matter of fact, I insisted several times and I told Dr. Oppenheimer that without the knowledge of that name our activities were going to be made much more difficult. Since he knew the name of the man, I felt he should furnish it to me. I think we broached that subject through the conversation on several occasions.
Q. Why were you so anxious to know the name?
A. Without the knowledge of the man, our job was extremely difficult. We knew definitely that there were espionage activities conducted in favor of the Soviets in that area. We knew now that there was a new or at least an additional effort being exerted through this man. Our investigative unit was lim-

ited in itself, and if we had to start digging to find out who this man is, it would put a tremendous burden on us.

I also felt, if I may say, that Dr. Oppenheimer knew the name of the man, and it was his duty to report it to me.

Q. Did you thereafter send to General Groves a memorandum on the subject of the importance of obtaining the name of the contact?

A. I did.

Q. I show you a copy of a memorandum dated September 2, 1943, and ask you if that is the memorandum to which you refer?

A. Yes; this is it.

Mr. ROBB. I will read this in this record, if I may, Mr. Chairman.

"SEPTEMBER 2, 1943.

"Memorandum for: General Groves
"Re: DSM project (J. R. Oppenheimer)

"1. It is essential that name of professor be made available in order that investigation can continue properly.

"2. If disposed to talk, also request names of individuals contacted by professor in order to eliminate unnecessary investigation and following of leads which may come to the attention of this office. If names of these people are known, this office will not have to conduct investigation into their activities if such names come to our attention through our own channels.

"3. It is desirable to have names of any people whom it is felt could be contacted by the professor, particularly CP members or sympathizers.

"4. Has anyone approached JRO at any time while he was connected with the project? If so, was it the professor, Eltenton, or some other party?

"B. T. P."

By Mr. ROBB:

Q. B. T. P. was what?

A. My initials.

Q. I call your attention to the use of the word "professor." To whom did you refer by that?

A. The unidentified person. I was told by Dr. Oppenheimer that the man was a member of the staff, or had been a member of the staff of the University of California.

Q. "If disposed to talk"; what did you mean by that? Who was supposed to talk?

A. If when General Groves would ask Dr. Oppenheimer for information, and if Dr. Oppenheimer felt he would give additional information, to get additional information we requested.

O. "Also request names of individuals contacted by the professor"; the individuals were the three contacts?

A. Those three contacts; yes.

Q. Dr. Oppenheimer did not give you those names?

A. No; he did not. He told me at the time that two of the men were down at "Y" that we called it, that was Los Alamos, and that one man had either already gone or was to go to site X, which I believe was Oak Ridge.

Q. Did you conduct any investigation as a result of that lead?

A. Yes; we did. That was another tedious project we had. We had to go through files, try to find out who was going to go to site X. We determined, and I took measures to stop—at least I asked General Groves to stop the man's movement to that area.

Q. What man?

A. The third man. I can't recall the name at this time. I am not sure of the name.

Q. But you felt that you had identified somebody who was about to be moved to the site?

A. Yes. As a matter of fact, we did. But at this point I don't remember the man's name.

Q. And you took steps to stop that transfer?

A. Yes.

Q. Thinking that he was the man referred to?

A. That is right. He was the only one who at the time was scheduled to go.

Q. Referring to the third paragraph of your memorandum, "CP members," that means what?

A. Communist Party members.

Q. Fourth paragraph, "Has anyone approached JRO at any time while he was connected with the project?" Calling your attention to that, Colonel, did you have any suggestion from your interview with Dr. Oppenheimer that he himself had been approached?
A. Yes.
Q. Beg pardon?
A. Yes.
Q. What was that?
A. He told me that this unidentified professor contacted him.
Q. Yes; but aside from that.
A. We felt that this was a vulnerable situation and if he was contacted by one, he may be contacted by others.
Q. Will you tell us whether or not, Colonel, you believed there was any connection between this episode of Dr. Oppenheimer's statement to you and the situation which had recently arisen involving Lomanitz?
A. Definitely.
Q. Would you explain that to us?
A. When we first met in the room, I asked Dr. Oppenheimer or said that I was interested in a certain incident. He immediately started telling me about the Lomanitz situation. I told him then it was not the Lomanitz situation that I was interested in, but other contracts that had been made. If I am in order, as a result of the study of the interview, it was my definite feeling at the time that the interview Dr. Oppenheimer had with me was the result of Lomanitz's situation. I felt definitely at the time that Dr. Oppenheimer knew or had reason to know that we were investigating or making an investigation which was more thorough than a normal background investigation. It was my opinion that Dr. Oppenheimer wanted to present this information to us for the purpose of relieving any pressure that may be brought on him for further investigation of his personal situation.
Q. In that connection, did you prepare a memorandum for General Groves?
A. Yes; I did.
Q. I will show you a copy of a paper dated September 2, 1943, with initials "B. T. P." and ask you if that is the memorandum?
A. Yes.
•Mr. ROBB. I will read this in the record, Mr. Chairman.
"SEPTEMBER 2, 1943.
"Memorandum for: General Groves.
"Re: DSM project (J. R. Oppenheimer).
"1. This office is preparing a memorandum in which it is pointed out that O.'s contact with Colonel Pash, through Lieutenant Johnson, was the result of the following circumstances:
"(a) Lomanitz was denied deferment.
"(b) Lomanitz told O. of this and also told him that he felt he was being investigated for subversive activities.
"(c) O. could conclude that this office is conducting some investigation and would probably determine that contacts have been made.
"(d) O. felt that it was safer to come out with the information at the present time in order to clear himself of any future investigation.
"(e) In this way he would retain the confidence of the Army personnel responsible for this project.
"2. Above, briefly, is a thesis of a memorandum which will be presented to you through Colonel Lansdale in a more detailed form. This office is of the opinion that O. had an ulterior motive in furnishing this information at such a late date and the above explanation seems reasonable. It is not believed that he should be taken fully into the confidence of the Army in the matters pertaining to subversive investigations."

By Mr. ROBB:
Q. "O." in that memorandum refers to whom?
A. Dr. Oppenheimer.
Q. You mentioned a late date. What did you mean by that?
A. When I had the interview with Dr. Oppenheimer, he told me that the incident which he was reporting to me had happened a few months prior to this interview.
Mr. GRAY. Excuse me. Was this memorandum signed or identified?
Mr. ROBB. This is a copy I have here. I assume it was signed.
Mr. GRAY. You didn't read any initials.
Mr. ROBB. Signed "B. T. P." That was you, Colonel?

The WITNESS. Yes.
Mr. ROBB. I previously identified it.
Mr. GRAY. I am sorry.

By Mr. ROBB:
Q. Colonel, had you had this information about the approach to Dr. Oppenheimer immediately after it had taken place, would that have made a difference to you in your investigation?
A. It certainly would.
Q. What difference would it have made?
A. Not having the name, I felt at the time, and I think I still feel impeded seriously our investigation.
Q. Why?
A. We had to start an investigation of a factor which was unknown to us. We knew that there was a man, a professor. There were many professors at the University of California. The only thing I knew was that he was not connected with the radiation laboratory, which put it into the University of California, and the staff was tremendous there.
Q. Did Captain De Silva subsequent or at about that time prepare the analysis to which you referred in your memorandum of September 2?
A. He had; yes.
Q. I will show you a photostat of a document dated September 2, 1943, "Memorandum for Lt. Col. B. T. Pash. Subject: 'J. R. Oppenheimer'," signed by "P. DeS." Is that the analysis prepared by Captain De Silva?
A. Those are his initials. Yes; this is the memorandum that he prepared.
Q. Did you transmit that to General Groves through Colonel Lansdale?
A. I did.
Q. I will show you a memorandum dated September 6, 1943, signed "Boris T. Pash," and ask you if that is your letter of transmittal of Captain De Silva's memorandum?
A. Yes; it is.
Mr. ROBB. These two documents have already been read into the record, Mr. Chairman.
Mr. SILVERMAN. Yes; pages 877 and following, if they are the documents you are talking about.

By Mr. ROBB:
Q. When did you finally learn the name of the unknown professor?
A. The name of the unknown professor was furnished to me by General Groves' office. I can't recall the exact time. I presume it was either the end of September some time——
Q. End of when?
A. September or maybe October. I am not sure of the time.
Q. Let me see if I can refresh your recollection. I will show you a photostat of a teletype addressed to the Area Engineer, University of California, Berkeley, Calif., attention Lt. Lyall Johnson, signed "Nichols," and asked if looking at that you are able now to refresh your recollection about it?
A. Yes; this is the way we received the information.
Q. When was the date?
A. December 13. I must say that I had—there was another somewhat previous—this never reached me.
Q. That never did?
A. No.
Q. How did you get the information?
A. I never got the information—I was gone.
Q. Do you recall that you did receive the information before you went or not?
A. I think I was only informally informed of certain suspicions but I had never received that information.
Q. When did you leave there?
A. About the 26th or 25th of November. It was the end of November.
Q. By the way, was there a Lieutenant Murray in your organization?
A. Yes; Lieutenant Murray was in charge of my investigative unit.
Q. I will show you a photostat of a memorandum dated San Francisco, Calif., November 22, 1943, bearing the signature of James S. Murray. Is that your Lieutenant Murray?
A. That is the same Lieutenant Murray.
Q. I notice that the title of this memorandum is, "Memo for the Officer in Charge. Subject DSM Project. Re Possible Identity of the Unnamed Professor Referred to by Dr. J. Robert Oppenheimer."

Do you recall having seen that memorandum?

A. Yes. Lieutenant Murray's memoranda to me were addressed in this form, and I recall this memorandum.

Q. That would indicate at that time at least you had not received the name of the unidentified professor?

A. No, sir.

Q. Had not, would it not, Colonel?

A. Yes, sir.

Mr. ROBB. I will ask Mr. Rolander if he might read this memorandum.

Mr. ROLANDER. "San Francisco, Calif. November 22, 1943."

Mr. SILVERMAN. Mr. Robb, do you think if we saw the memorandum it might be unnecessary to read it? I don't know what is in it.

Mr. ROBB. No; I think we better have it in the record just for completeness, if the Chairman doesn't mind.

Mr. ROLANDER. "Memorandum for the Officer in Charge.

"Subject: DSM project.

"Re Possible Identity of the Unnamed Professor Referred to by Dr. J. Robert Oppenheimer."

The date is November 22, 1943.

Mr. SILVERMAN. Have you an extra copy?

Mr. ROLANDER. I am sorry; I do not.

Mr. SILVERMAN. Can I look over your shoulder, Mr. Rolander?

Mr. GRAY. I will follow the reading of it. It this a copy that counsel can follow?

Mr. ROLANDER. Mr. Chairman, I didn't get a chance to glance at it again. It may refer to the FBI, and I would have to note that we could not make any mention of the FBI. This may not be the case, but I would have to read it through to be sure.

Mr. GRAY. I will ask you to look at it, and see if there is anything you will have to omit or not.

I can tell you there is some material you will want to leave out. Page 3.

Mr. ROBB. We had not gotten to that yet, sir. That seems to be the only sentence or paragraph.

Mr. GRAY. Can you give pages 1 and 2?

Mr. ROBB. While we are at it, Mr. Chairman, I see attached to that memorandum is a covering memorandum dated November 27, 1943, signed Boris T. Pash. I will show that to the colonel and ask him if he sent that memorandum.

The WITNESS. No; I did not. This was sent by a then Lieutenant or Captain Maharg.

By Mr. ROBB:

Q. He signed your name?

A. Yes. In this investigation he was acting for me.

Q. Do his initials appear?

A. His initials appear below.

Q. Is there any question that this memorandum was sent on the date indicated enclosing the memorandum prepared by Lieutenant Murray?

A. From this record it appears that this was sent. I would have no personal knowledge of the fact.

Mr. ROBB. I think we might read them both, Mr. Chairman, while we are about it.

Mr. GARRISON. May we have a chance to read this before it is read on the record, Mr. Chairman?

Mr. GRAY. Yes.

Mr. ROBB. Mr. Chairman, while my friends are reading that, I might say the purpose of offering this is to show for the board the attempts that were being made to identify this contact and what the knowledge was at that time.

Mr. SILVERMAN. Mr. Robb, I represent only Dr. Oppenheimer, but there are a lot of names of people here I never heard of. I wonder whether in fairness to these people it might not be better when you read the memorandum to say there are then given the names of 10, 11, or whatever number of people there are, of whom Professor Chevalier is one, or is not one.

Mr. ROBB. He is not. I don't care about that. I don't know who these people are, either.

Mr. GRAY. Let me suggest that the first 2 paragraphs be read, which I take it do not involve persons who may not be concerned in this proceeding, that then counsel indicate that there is paragraph 1, name of an individual with 7 or

8 lines of information about him, paragraph 2, and so on. I think counsel's point that Colonel Pash's office or the office of the intelligence people was involved in very extensive investigation to ascertain the name of the unidentified professor is a well taken point. I see no reason——

Mr. ROBB. I have no desire to read them in. I take it the paragraph about Dr. Weinberg might be read.

Mr. GRAY. I think there is no reason why you should not indicate when you came to his name.

Mr. ROBB. Yes, sir.

Mr. ROLANDER. May I proceed?

Mr. GRAY. Please.

Mr. ROLANDER. "San Francisco, Calif. November 22, 1943"——

Mr. GRAY. Did you read the covering memorandum?

Mr. ROLANDER. I beg your pardon. The covering memorandum or letter:
"Army Service Forces, Headquarters, Ninth Service Command, Office of the Director, Intelligence Division, Forward Echelon, Presidio of San Francisco."
The initials "SPRIC: FE."
The date, "November 27, 1943."
The written initials of "CLC" in the right-hand corner. There is also some written comment on the left-hand corner which, since it mentions a name, I will omit.

Mr. GRAY. May I call your attention also to what would appear to be the initials "YL" next to the initials "CLC."

Mr. ROLANDER (reading):

"NOVEMBER 27, 1943.

"Subject: DSM Project. Possible identity of unnamed professor referred to by Dr. J. R. Oppenheimer.

"To: Lt. Col. John R. Lansdale, Jr., 2C654 Pentagon Building, Washington, D. C.

"Enclosed for your information and files find memorandum for the Office in Charge, dated November 22, 1943, subject as above, for the Director, Intelligence Division."

Signature, "Boris T. Pash." Typed, "Boris T. Pash, Lt. Col. M. I.," and then an initial beneath there which was referred to by Colonel Pash, "Chief, Counter Intelligence Branch." One enclosure: "Duplicate, memo as indicated. cc Captain Maharg with enclosure."

The memorandum itself:

"SAN FRANCISCO, CALIF., *November 22, 1943*.

"Memorandum for the Office in Charge.

"Subject: DSM Project.

"Re: Possible identity of unnamed professor referred to by Dr. J. Robert Oppenheimer.

"Reference is made to various conversations and interviews between Dr. J. R. Oppenheimer, head of DSM project at site Y, and Lt. Col. Boris T. Pash, Chief, CIB, Forward Echelon, Ninth Service Command. Reference is also made to conversations and interviews between Dr. J. R. Oppenheimer and Lt. Col. John R. Lansdale, Jr., Chief, Investigations Branch, CIG, MIS. During the above-named interviews, Dr. Oppenheimer has frequently made reference to a professor located at the University of California campus who acted as a go-between for George Eltenton, and 3 unnamed persons working on the DSM project in an endeavor to gain information for Eltenton to transmit to the Soviet Government. On all of the above-named occasions, Dr. Oppenheimer has refused to name the professor or the 3 persons who were contacted. Dr. Oppenheimer stated that the 3 persons did not disclose any information, and therefore they are not pertinent to any investigation promulgated by Military Intelligence Services. Efforts of this office during the past month have been directed in an attempt to ascertain the identity of the professor contact. A record check of all professors and associates in both the physics and chemistry departments at the University of California was made with the Federal Bureau of Investigation and the results thereof contained in a progress report from this office dated October 20, 1943. A continued survey and check has been made and it is believed that it is entirely possible that the professor might be one of the following."

"1." and then a name and 7 lines of discussion.

"2." a name and 7 lines of discussion.

"3." a name and 6 lines of discussion.

"4." The name appears "Joseph W. Weinberg." It states further: "Weinberg has been known to commit at least one espionage act, and on June 28, 1943, he was awarded a Ph. D. degree by the University of California, and assumed an associate professorship there."

"5." A name and 5 lines of discussion.
"6." A name and 7 lines of discussion.
"7." A name and 6 lines of discussion.
"8." A name and 8 lines of discussion.
"9." A name and 5 lines of discussion.

Mr. ROBB. May it be agreed, Mr. Chairman, that none of the names was the name of Haakon Chevalier?

Mr. SILVERMAN. Certainly not on these two pages.

Mr. GRAY. That name does not appear in this memorandum.

Mr. ROBB. That is right, it does not appear in the memorandum.

Mr. GRAY. I would suggest that actually the remainder of this memorandum is not pertinent to the question being put to the witness.

Mr. ROBB. I think not, Mr. Chairman. There is no point of cluttering up the record.

Mr. GRAY. I see there is no point of cluttering up the record.

Mr. SILVERMAN. It has nothing to do with Dr. Oppenheimer.

Mr. GRAY. No, with other individuals. Let me say it does mention some familiar names, Lomanitz, Friedman, Weinberg, Bohm, but really not connected with what we are talking about.

Mr. SILVERMAN. Could I take a look at that part of it to see whether something occurs to me about it, which perhaps may not.

Mr. GRAY. I think you will have to accept my assurance that it would not help you to see the remainder. It is not really related.

Mr. ROBB. Mr. Chairman, would you like to break for lunch?

Mr. GRAY. One of the members of the board has an engagement. Am I right in assuming that you are not at this point finished with your direct examination?

Mr. ROBB. That is correct.

Mr. GRAY. Therefore I think we should recess for lunch at this time, and we shall return at 2 o'clock.

(Thereupon at 12:35 p.m., a recess was taken until 2 p.m., the same day.)

AFTERNOON SESSION

Mr. GRAY. Let the record show that Mr. Garrison is not present at the beginning of the hearing.

Will you proceed, Mr. Robb?

Whereupon, Boris T. Pash, the witness on the stand at the time of taking the recess resumed the stand and testified further as follows:

DIRECT EXAMINATION

By Mr. ROBB:

Q. Colonel, I think I asked you before the noon recess when you first learned the name of Haaken Chevalier, and I believe you said some time in September.
A. Early October or September.
Q. In what connection did that name come to your attention?
A. We were receiving reports of other investigative agencies relating to Communist activities in the area. I don't recall exactly who delivered those reports to us, but they probably came from Washington, from General Grove's office.
Q. What was the purpose of the report about Dr. Chevalier? I don't mean for you to give details.
A. It concerned Communist activities in the area. It concerned contacts with people who were either known or suspected Communists.
Q. I don't want to lead you but I am quite sure you are not very easily led anyway. Was the burden of the report that Dr. Chevalier was in some way connected with Communist activities?
A. That is right.
Q. The identification of Dr. Chevalier as the unknown professor came later?
A. That is right. It didn't come to me then.
Q. It did not come to you?
A. No.
Q. Would you say it came after you left Berkeley?
A. When I returned from a short tour in Europe, after being in the Mediterranean Theater, I was brought up to date on certain things that transpired in my absence.
Q. Is that when you first learned the identity of the unnamed professor?
A. Yes, sir, I believe so.

Q. When did you first begin giving attention and consideration to Dr. Oppenheimer in connection with your investigation of espionage and Communist activities in Berkeley?
A. At the early part of the investigation. It was either late in May or some time early in June.
Mr. GRAY. What year?
The WITNESS. 1943, sir. Excuse me.

By Mr. ROBB:
Q. I will show you a copy of a report with the typewritten signature, "Boris T. Pash" dated June 29, 1943, and ask you whether you recall preparing that report?
A. Yes.
Mr. ROBB. Do you have a copy of this for our friend across the way?
Mr. ROLANDER. I don't believe this can be read in its entirety.
Mr. ROBB. I am sorry; this report has some references to FBI materials.

By Mr. ROBB:
Q. At all events, Colonel, the subject of this report is "Julius Robert Oppenheimer," is that correct?
A. Yes.
Q. Without going into details about it, it concerns investigative information in respect of Dr. Oppenheimer, is that right?
A. That is right.
Q. During the time that you were conducting this investigation, Colonel——
Mr. SILVERMAN. Do you suppose you could read the portions that relate to Dr. Oppenheimer?
Mr. ROBB. The whole thing relates to Dr. Oppenheimer.
Mr. SILVERMAN. Is there some way we could see it without the FBI part?
Mr. GRAY. At this moment, I think this is true. The witness has had his recollection refreshed with respect to a memorandum which he wrote.
Mr. ROBB. Yes, sir.
Mr. GRAY. I don't know what you propose to do.
Mr. ROBB. Nothing further.
Mr. GRAY. Can you do this in a way which will not make it necessary to read it into the record?
Mr. ROBB. My purpose for referring to it was to have some specific date in the record to show that by at least June 29, 1943, Dr. Oppenheimer was under investigation by Colonel Pash's organization in respect of espionage, that is all.
Mr. SILVERMAN. In respect of suspected espionage by Dr. Oppenheimer.
Mr. ROBB. In the context of the espionage investigation that was going on. Is that correct, Colonel?
The WITNESS. Yes.
Mr. SILVERMAN. I really think that in fairness it would be well to read as much of that memorandum into the record as can be read by skipping the references of the FBI. We are somewhat at a disadvantage. Our friends on the other side have the memorandum before them. Doubtless the members of the board have it before them.
Mr. GRAY. I don't know whether the members of the board have or not.
Mr. ROBB. Will you take the best I can do on it, Mr. Silverman?
Mr. SILVERMAN. Yes, sir.
Mr. ROBB. I will do the best I can, and I think it will be all right.
Memorandum June 29, 1943:

"Subject: Julius Robert Oppenheimer.
"To: Lieutenant Colonel Lansdale, Jr., Room 2C 654, Pentagon Building, Washington, D. C.
"1. Information available to this office indicates that subject may still be connected with the Communist Party."

Then I omit the next sentence.
"This is based on the following specific information.
"(a) Bernadette Doyle, organizer of the Communist Party in Alameda County, Calif., has referred to subject and his brother, Frank, as being regularly registered within the party.
"(b) It is known that the Alameda County branch of the party was concerned over the Communist affiliation of subject and his brother, as it was not considered prudent for this connection to be known in view of the highly secret work on which both are engaged.

"2. Results of surveillances conducted on subject, upon arrival in San Francisco on June 12, 1943, indicate further possible Communist Party connections.

"(a) Subject met and is alleged to have spent considerable time with one Jean Tatlock, the record of whom is attached.

"(b) He attempted to contact by phone and was later thought to have visited a David Hawkins, 242 32d Avenue, San Francisco, a party member who has contacts with both Bernadette Doyle and Steve Nelson. A preliminary report on Hawkins is attached.

"3. Further investigations of the possible connections of subject with the Communist Party are being carried out by this office."

I omit the next sentence.

"4. In view of the fact that this office believes that subject still is or may be connected with the Communist Party, and because of the known interest of the Communist Party in this project, together with the interest of the U. S. S. R. in it, the following possibilities are submitted for your consideration:

"(a) All indications on the part of Communist Party members who have expressed themselves with regard to subject lead this office to believe that the Communist Party is making a definite effort to officially divorce subject's affiliation with the party and subject, himself, is not indicating in any way interest in the party. However, if subject's affiliation with the party is definite and he is a member of that party, there is a possibility of his developing a scientific work to a certain extent then turning it over to the party without submitting any phase of it to the United States Government. It is the understanding of this office that subject is the only person who knows the exact progress and results of this research work, and, as a result, is difficult to check.

"(b) In view of the above there exists another possibility that while subject may not be furnishing information to the Communist Party direct he may be making that information available to his other contacts, who, in turn, may be furnishing or will furnish such information, as it is made available to them by subject, to the Communist Party for transmission to the U. S. S. R.

"5. On the basis of the present status of this case and with the limited knowledge available to this office on the organization and administration of the project, the following possible plans of action are recommended:

"(a) That every effort be made to find a suitable replacement for subject and that as soon as such replacement is trained that subject be removed completely from the project and dismissed from employment by the United States Government.

"(b) That subject be told that in view of the importance of the project and the posibility of an accident which may incapacitate or eliminate him, that a second in command be assigned to subject who will share in the knowledge of all developments and processes of interest in the project.

"(c) That subject be called to Washington for purposes of being interviewed by Chief, MIS, and General Groves; that subject first be told of the Espionage Act and its ramifications; of the knowledge MIS has of Communist affiliations and that this Government will not tolerate any leakage of information, either by subject or any of his associates to the Communist Party, whether this be for the purpose of transmitting information as such or of informing the Communist Party of the progress made by its members and, further, that this Government intends to maintain rigid control of the development of the project.

"6. It is the opinion of this office that subject's personal inclinations would be to protect his own future and reputation and the high degree of honor which would be his if his present work is successful, and, consequently, it is felt that he would lend every effort to cooperating with the Government in any plan which would leave him in charge. It is, therefore, recommended that the plan outlined in paragraph 5-c be adopted upon completion of a thorough investigation of subject presently being conducted by this office. This investigation is being made to secure all possible information on subject's background, particularly his past and present affiliations with the Communist Party.

"It is further recommended that regardless of the plan adopted, or whether any of the above-proposed plans are adopted, that subject be told that there exists a possibility of violence on the part of Axis agents who may wish to interfere with this project and, therefore, the War Department deems it advisable to assign to subject two bodyguards. These bodyguards will be selected from specially trained Counter Intelligence Corps agents who will not only serve as bodyguards for subject but also as undercover agents for this office.

"For the A C of S, G-2.

"Boris T. Pash. Lt. Col., M. I., Chief, Counter Intelligence Branch.

"2 Incls:
"#1—Memo, 6–29–43, re Jean Tatlock (dup)
"#2—Memo, 6–29–43, re David Hawkins (dup)
"cc: Capt. H. K. Calvert."

By Mr. ROBB:
Q. Colonel, do you know whether or not the two bodyguards were assigned?
A. No, I don't.
Q. Let me ask you as an expert——
A. I don't think so.
Q. Let me ask you, Colonel, as an expert in these matters, how effective can a surveillance be to prevent the transmission of information?
A. In my opinion, it is impossible to maintain a 100-percent surveillance or maintain a surveillance which would assure 100-percent success.
Q. Why?
A. There are so many different ways in which information can be transmitted and in this particular instance we did not have any qualified men who knew the technical field sufficiently to be able to determine even in an open conversation if any information is being transmitted.
Q. You mean would not understand it?
A. That is right.
Q. You mention in here a thorough investigation of subject. Subject being Dr. Oppenheimer?
A. Yes, sir.
Q. Was that conducted?
A. That was insofar as I was in charge there; that was discontinued on instructions from Washington.
Q. When?
A. I believe some time in the middle of August.
Q. Was any reason given for that?
A. Not to me.
Q. Did all the reports concerning communistic activities at Berkeley concerning Dr. Oppenheimer come across your desk while you were there?
A. I believe so. During this period I had made some short trips. In that case either DeSilva or Maharg would act for me. Normally they tried to bring me up to date when I returned.
Q. You kept yourself thoroughly familiar with the investigation going on?
A. I tried to, yes.
Q. On the basis of the information which you had concerning Dr. Oppenheimer, did you consider him to be a security risk?
A. Yes, I would.
Q. Did you then?
A. Yes, I did.
Q. Do you now?
A. Yes, I think I do. I do, yes.
Q. Going back for a moment to your interview with Dr. Oppenheimer, you mentioned that he had spoken to you or told you that this unnamed professor had mentioned someone in the Russian consulate, microfilm, the three contacts, two of them having gone to Los Alamos and one being about to leave for Oak Ridge; did you have any opinion as to whether or not Dr. Oppenheimer in those respects was truthfully reporting to you what the unnamed professor had said to him?
A. Yes, I was sure of that.
Q. You were sure of that?
A. Yes.
Q. Why?
A. In the first place, Dr. Oppenheimer spoke to Lyall Johnson, telling him that he had something, as Johnson told me, something important to convey concerning espionage. When I arranged for the interview and Dr. Oppenheimer came in, when I told him that I wanted to discuss the incident, he immediately started discussing Lomanitz with me. When I told him it was the other incident where other parties may be interested in this, he immediately started then relating the information he gave me. I don't think there was any break or adjustment at the time. I felt he was giving something he already had or he knew. Furthermore, as I believe I stated before, and reviewing the situation after a while, I felt that he had this information and he felt that he wanted to give it to us because of the fact that he found out we may be making a rather thorough investigation of the whole project and the activities. Finally, the

information given there was rather serious and to a certain extent detailed. It referred to a plan. It included a plan that was supposed to be in existence. It included some details such as the contract, about the availability of contact with the Soviet consulate and the reference to a technical device for purposes of recording what information may be available.

Q. What conclusion did you draw from the fact that the information was in some circumstantial detail? What did that indicate to you?

A. That indicated that it was information already available to a man, and in a field which probably was more operational, and therefore I felt, and feel, that it was transmitted to him rather than made up by him.

Q. Do you still feel that way?

A. Yes, I do.

Q. You had a great deal of experience, have you not, in interrogating witnesses?

A. I have had some experience, yes.

Q. You have been doing it for years, haven't you?

A. For a few years.

Q. You have had a great deal of experience in evaluating statements made by witnesses, have you not, sir?

A. Yes, I have.

Q. Was there then and is there anything now to suggest to you that his statements to you about these details Dr. Oppenheimer was not giving you an accurate report of what he had been told by the unnamed professor?

A. No. I had no reason not to believe they were truthful.

Q. Do you have any now?

A. No, I only know this from newspaper information.

Q. Yes.

A. And whether it is correct, I don't know. But I read in Dr. Oppenheimer's reply to General Nichols he relates this incident. I feel that the information which Dr. Oppenheimer gave me in 1943 was far more damaging to him and to any of his friends than the information as related in the newspaper. If Dr. Oppenheimer was not telling the truth at that time, he was making up a story which would be more damaging to him than it appears the situation was according to the newspaper item. I don't think that that is a normal human reaction. I feel that the story as told then—the story as related in the newspaper probably is in favor of Dr. Oppenheimer. In evaluating that, I felt that the inconsistentcy there in my mind would favor the truth in the preliminary interview, the interview of 1943.

Q. Would you care to elaborate upon your statement that you now consider Dr. Oppenheimer a security risk?

A. As far as I know, Dr. Oppenheimer was affiliated with Communist front activities. I have reason to feel that he was a member of the Communist Party. I have seen no indication which indicates any change from that. I feel that his supposed dropping of the Communist party activities in the early part of the war need not necessarily express his sincere opinions, since that was done by most all members of the Communist party. As a result of that, I feel that the opinion I had back in 1943 probably would stand.

Q. You say was done by most all members of the party. Just what do you mean by that?

A. Members of the party who came into the service, members who continued in Government work, disclaimed any affiliation with the party.

Q. Colonel, did any incident or episode occur shortly after your interview with Dr. Oppenheimer which tended to confirm your doubts about Dr. Oppenheimer?

A. There was an incident which caused me to stop and think. The evaluation was difficult, but the timing and coincidence was an important factor. Joseph Weinberg wrote a note to a man, a Flanigan, also a known Communist, stating— in the letter, it was a card, he did not it, but it was in the letter which he mailed, stating, "Dear A. Please don't contact me," or something to that effect. I can't recall. "Please don't make any contact with me, and pass this message to S and B, only don't mention any names. I will take a walk with you when this matter is all cleared up." That was dated the 6th of September. Of course, we were very concerned over the entire situation and since Weinberg had close contact and association with Dr. Oppenheimer I felt at the time that it was the result of the situation which culminated in my interview with Dr. Oppenheimer.

Q. How did you interpret the expression "take a walk with you"?

A. The Communist people at the time were trying to avoid any discussions. They tried to carry on their discussions either outside or in an automobile or out on the street.

Q. Why?

A. In order to avoid detection. They avoided fixed positions.

Q. Colonel, I will ask you what information you can give us in brief about certain people whose names I will give you. William Schneiderman?

A. William Schneiderman was one of the top Communist functionaries in California. His name appeared quite a bit in the process of our investigation, and it was always Communist connected. I believe he has been tried and convicted for advocating the overthrow of the Government by force and violence, and has been convicted and if I am not mistaken, is now out on appeal.

Q. Rudy Lambert.

A. Rudy Lambert was also in the same class with Schneiderman, same type of individual. He is now also under conviction for the same offense.

Q. Steve Nelson.

A. Steve Nelson, of course, was directly connected with the espionage efforts at the Radiation Laboratory. He was convicted in the East for the offense of advocating the overthrow of the Government by force and violence. I think he was convicted and may be serving a jail term now.

Q. Isaac Folkoff?

A. Isaac Folkoff is a Communist in the bay area, I think in San Francisco—I am not sure—and he was in a business, I believe, and served as an intermediary.

Q. Intermediary for what purpose?

A. For contact between Communists.

Q. Louise Bransten.

A. Louise Bransten is a Communist Party member who has a record of contacts with Soviet officials. She, according to reports I have read, I think, is independently wealthy and has served the Communist cause. She is, I think, in the East now.

Q. Contact with Soviet officials in what connection?

A. I presume that the contact with Soviet officials for the purpose of passing information. She was in contact for instance with a man, Kheifits, who was a Soviet official in San Francisco. I think he took the place of the initial contact of the Soviet official who contacted Nelson.

Q. What was his name?

A. Ivanov.

Q. Joseph Weinberg you have already told us about.

A. Yes.

Q. Dr. Thomas Addis.

A. I don't know much about Dr. Thomas Addis. He was a professor at Sanford University, I think. As far as I can recollect there were allegations that he was a Communist Party member.

(Mr. Garrison returned to the hearing room.)

By Mr. ROBB:

Q. David Jenkins.

A. David Jenkins was a member of the California Labor School. If I am not mistaken, he was the head of it at one time in the early forties.

Q. Do you remember his wife's name?

A. No, I don't.

Q. Did you know of someone named Edith Arnstein?

A. No, I don't.

Q. John Pitman?

A. John Pitman, if I am not mistaken was on the staff of the Peoples World

Q. What was the Peoples World?

A. Peoples World was a Communist Party publication.

Q. Where?

A. In San Francisco.

Q. Hannah Peters.

A. The name Peters is familiar.

Q. And her husband, Bernard Peters.

A. Bernard Peters I know was a scientist, I think, but I don't know enough about him.

Q. David Adelson.

A. David Adelson was very active in the FAECT, the union, Federation of Architects, Engineers, Chemists, and Technicians.

Q. Do you have any information with respect to his Communist connections?
A. There were reports of his Communist connections. He was very active in trying to penetrate the radiation laboratory with members of the union. As a matter of fact, I think he was one of the men who were contacted by Lomanitz and Weinberg, and so forth, when Lomanitz was inducted.
Q. Kenneth May.
A. I remember the name of Kenneth May as being connected with the Communist Party. I don't know any particulars about him.
Mr. ROBB. That is all I care to ask, Mr. Chairman.
Mr. GRAY. Mr. Silverman.

CROSS-EXAMINATION

By Mr. SILVERMAN:
Q. Colonel Pash, how often have you met Dr. Oppenheimer?
A. Once, for this interview.
Q. That was that meeting of August 26, 1943?
A. Yes.
Q. And as far as you can recall until today that is the only time you have ever seen him in your life?
A. Physically, yes.
Q. I think you gave some testimony about four people, Messrs. Lomanitz, Bohm, Friedman.
A. And Weinberg.
Q. And Weinberg, yes. Those people were employed at the radiation laboratory?
A. That is right.
Q. In Berkeley.
A. In Berkeley.
Q. They were not employed at Los Alamos?
A. Not to my knowledge.
Q. So far as you know, did Dr. Oppenheimer have any responsibility for their employment at Berkeley?
A. I don't know enough about personnel administration there. I recall in reviewing the documents available to me at the time that I think he made some comments with reference to Lomanitz.
Q. He didn't hire these people?
A. I don't know who hired them.
Q. He was not the director of the radiation laboratory the way he was at Los Alamos?
A. Not to my knowledge.
Q. You said he made some comments about Lomanitz. I think you said he made some, I don't remember the word now, protest, pressure, or something about it, when Lomanitz' draft deferment was terminated?
A. When it was about to be terminated.
Q. Did Lomanitz' superiors on his job complain about it?
A. Weinberg and Bohm, to my knowledge.
Q. Did Lomanitz's superiors on his job complain about it?
A. I think that Dr. Lawrence may have.
Q. Did anyone else of his superiors?
A. That I don't know.
Q. You have recently had occasion to refresh your recollection as to what Dr. Oppenheimer did about this matter; have you not?
A. Yes.
Q. You have not had occasion to refresh your recollection as to whether—before I finish this question, I want to be perfectly clear I am not and do not intend to make any accusations about any people I am naming here, because I consider all their actions perfectly innocent—you have not had occasion to refresh your recollection recently as to what Dr. Lawrence did about protesting or objecting to Mr. Lomanitz' deferment?
A. The only way that I knew that Dr. Lawrence may have taken part is because Lomanitz mentioned in discussing the matter that Dr. Lawrence was going to state that he was needed or something to that effect.
Q. You knew that Dr. Lawrence was very anxious to see that the work of his laboratory went well?
A. Yes, I realize that.
Q. And Dr. Oppenheimer was very anxious to see that the work of his laboratory went well?
A. I realize that.

Q. And neither one of them would be very happy to lose a good technical man?
A. I presume so.
Q. And were you told that Dr. Oppenheimer said that if Lomantiz is drafted, Dr. Lawrence will want to take somebody from Dr. Oppenheimer's staff?
A. Yes.
Q. And Dr. Oppenheimer didn't like that.
A. That is right.
Q. And he so wrote you?
A. I know he stated that. I don't know whether he wrote it.
Q. I think that is in the record. By the way, in the course of refreshing your recollection, have you also listened to the recording of your conversation?
A. Yes.
Q. When did you do that?
A. I think about 2 days ago.
Q. And you played it over once?
A. Yes.
Q. I would like to come to the incident of September 6 in which Joseph Weinberg wrote a note to Flanigan omewhat to the effect, "Dr. A. Please don't contact me, and pass this message to S and B, and I will take a walk with you" and so on.
A. Yes.
Q. As far as you know, was A, S, or B, Dr. Oppenheimer?
A. No.
Q. You connected this with your talk with Dr. Oppenheimer?
A. Yes, with the situation around that time, which culminated in Dr. Oppenheimer's interview.
Q. One reason for that was the timing?
A. That is right.
Q. Dr. Oppenheimer's interview with you was on August 26th?
A. That is right.
Q. And this letter was 11 days later, September 6?
A. Yes.
Q. Obviously you don't know what other problems Mr. Weinberg was worried about in that period, or what else may have happened in that 11 days to stir him up.
A. That is right.
Q. I think you said that another reason you connected was because of Dr. Weinberg's close contact and association with Dr. Oppenheimer. Would you tell us so far as you know what Dr. Weinberg's association with Dr. Oppenheimer was?
A. Yes. He was a student of Dr. Oppenheimer's at the university. In two, I think, instances when problems arose for him on one instance he went with Bohm to see Dr. Oppenheimer. That was on the 2d of September, in connection with the Lomanitz situation. And from the conversations that were had in the group, my impression was that he discussed Dr. Oppenheimer as sort of a man they could advise with. I recall that was not the 2d of September. It may have been during Dr. Oppenheimer's trip to San Francisco that Bohm and Weinberg saw him on which they said they also feel that the draft may reach them, too.
Q. They also felt, too, what—the draft?
A. The draft may reach them because of their activities.
Q. Would you try to identify the approximate time of this? You say you think it was not September?
A. No, if I am not mistaken it was during the trip of Dr. Oppenheimer to San Francisco.
Q. When was that?
A. It was in those dates of 26th or 27th of August.
Q. So that too was about 10 days before?
A. Yes.
Q. Did Lieutenant Johnson go to Dr. Oppenheimer and question him about Eltenton or did Dr. Oppenheimer come to Lieutenant Johnson?
A. As Lieutenant Johnson related it to me, I don't know, he said Dr. Oppenheimer told me. I don't know the details of where they met or what the circumstances surrounding that was.
Q. Did Lieutenant Johnson tell you that Dr. Oppenheimer at that very first interview mentioned Mr. Eltenton's name?
A. No. I don't know whether it was the first interview he had with Johnson.
Q. Wasn't your interview with Dr. Oppenheimer the day after?

A. Excuse me. I thought you meant Johnson's first interview with Dr. Oppenheimer. It is first because it preceded mine is what you mean, is that right?
Q. Yes.
A. I understand.
Q. And according to Lieutenant Johnson's report, Dr. Oppenheimer came to Lieutenant Johnson and mentioned Eltenton's name?
A. Yes. I don't recall that. He mentioned the espionage activities.
Q. You do not now recall whether Dr. Oppenheimer mentioned to Lieutenant Johnson Eltenton's name on the day before?
A. No, I am sorry, I don't.
Q. In your one interview with Dr. Oppenheimer, Dr. Oppenheimer did mention the name?
A. Yes.
Q. He volunteered the name?
A. Yes.
Q. At that time——
Mr. ROBB. Mr. Chairman, I don't mean to interfere but I think the question whether he volunteered the name is a conclusion. I don't wish to concede——
Mr. SILVERMAN. There have been a fair number of conclusions suggested by you, Mr. Robb.
Mr. ROBB. There certainly have.
Mr. GRAY. Proceed, Mr. Silverman.
Mr. SILVERMAN. Thank you, sir.

By Mr. SILVERMAN:
Q. At the time that Dr. Oppenheimer gave you Mr. Eltenton's name, was Mr. Eltenton already under suveillance by you?
A. We had no connection with Mr. Eltenton. We had his name, but he was not under our surveillance. He was not connected with the radiation laboratory as far as I know.
Q. So that when Dr. Oppenheimer gave you this name, this was an important piece of information for you?
A. No, we had his name, but not in connection with our investigation.
Q. Did you have his name as someone who might be mixed up in an espionage attempt?
A. Yes, as a Communist Party member. We would not have those details as to his activities, because we were not conducting the investigation.
Q. You were conducting an investigation about espionage.
A. Yes, by the limitation agreement we did not investigate people who were not connected with the military or specifically with the radiation laboratory.
Q. So far as you know was there any information—I withdraw that.
You did not have any information that connected Mr. Eltenton with an espionage attempt or approach?
A. We had information which connected him with the contacts of the Soviet contacts, but I personally in my office did not have the details of those contacts.
Q. And did Dr. Oppenheimer say to you that the reason he was not giving you the name of the professor was that he thought the man was innocent?
A. He thought that this was not serious and that he had not achieved anything.
Q. And of course Dr. Oppenheimer was very wrong not to give you that name.
A. Yes.
Q. And I think we would all agree with that. Do you have any information of any leakage of restricted data through Dr. Oppenheimer to any unauthorized person?
Mr. ROBB. May I have that read back?
(Question read by the reporter.)
The WITNESS. No.

By Mr. SILVERMAN:
Q. And Dr. Oppenheimer did tell you that on the one instance when the professor approached him, he refused to have anything to do with it?
A. Yes; he told me that.
Q. And some time in 1943, he did give the professor's name?
A. Yes.
Q. We all agree that Dr. Oppenheimer exercised poor judgment, indeed, and was very wrong not to give you the name of Professor Chevalier. Against that agreement by everyone here, I would like to ask you these questions.

Mr. GRAY. Wait a minute. I take it that everyone here includes the members of this board. The hearing is being conducted for the information of the members of this board in the discharge of its functions. I as chairman have been extremely lenient, perhaps unduly so, in allowing counsel to express an opinion. This is not the first time that you have said, Mr. Silverman, that everyone here agrees on something.

I should like to ask you please to refrain from expressions of opinions, and not to try to give a witness an indication that you speak for anybody but yourself, if you are expressing an opinion.

Mr. SILVERMAN. Very well, sir. I am sorry.
Mr. GRAY. It is all right; proceed.

By Mr. SILVERMAN:

Q. You have had a good deal of experience with security and intelligence matters in the last 12 or 13 years.
A. I have had some experience; yes.
Q. You were pretty new at security matters in 1943?
A. No; I don't think so.
Q. You had a couple of years of experience?
A. I have had past experience, too.
Q. I assume it is fair to say that in the last 12 or 13 years you have learned a good deal about security and intelligence work?
A. Yes, I have.
Q. And perhaps your own opinions have to some extent changed or crystallized over that period?
A. Opinions as to operational procedures?
Q. Yes; and the right things for people to do with respect to security and so on.
A. No; I don't think they have changed much as to the right things to do.
Q. Do you believe it possible that Dr. Oppenheimer's opinions have changed over that period?
A. I don't think I can speak for Dr. Oppenheimer.
Q. You have only seen him once in your life.
A. That is right.
Q. Do you believe that his record since 1943 should properly be weighed against his admitted mistake and failure to make a prompt, frank and full report in 1943, to determine whether he is now a security risk?
A. I don't know which record you are referring to.
Q. Whatever his activities have been since 1943 with which of course you are not familiar.
A. Yes. I again think that is the position of the board——
Q. Exactly.
A. To answer, not mine.
Mr. ROBB. I am sorry. I didn't get that.
The WITNESS. I said that is not my position to answer that.
Mr. SILVERMAN. That is all. Thank you.
Mr. GRAY. Colonel Pash, I would make reference now to your interview with Dr. Oppenheimer. I don't have the date fixed in my mind, but the only interview you had with him.
The WITNESS. Yes, sir.
Mr. GRAY. In your earlier testimony, I believe you indicated that with respect to this interview, Lieutenant Johnson reported to you that he had received some information and you then decided you wished to talk personally to Dr. Oppenheimer.
The WITNESS. Yes. Lieutenant Johnson said he received it from Dr. Oppenheimer.
Mr. GRAY. In the beginning of your interview, it seemed to you that Dr. Oppenheimer thought you wanted to talk to him about Lomanitz?
The WITNESS. Yes, sir.
Mr. GRAY. But that the substance of the interview concerned the so-called Chevalier episode.
The WITNESS. I may not quite understand you.
Mr. GRAY. I am afraid it was not a good question. In the beginning of your interview with Dr. Oppenheimer, there was some mention of Lomanitz, and then you had to make it plain to Dr. Oppenheimer you wanted to talk about the Chevalier incident.
The WITNESS. About the incident which eventually involved Chevalier.

Mr. GRAY. Yes. And you testified also, I think, that it may have occurred to you at the time that the reason Dr. Oppenheimer volunteered to Lieutenant Johnson what he did about the episode was that he may have known there was an investigation going on, and that this might have been found out about in some other way, and therefore he thought he better get the information to the security officers himself.

The WITNESS. Yes, sir; I apologize. Did you mention Johnson's name in connection with that? I may have thought I heard it. May I ask that it be reread? I am sorry.

Mr. GRAY. Yes; you may ask. I am going to be embarrassed when I read it in print. I might as well be embarrassed when I hear it read back now. It was a poor question. Can you read it back?

The WITNESS. I didn't mean to imply that.

Mr. GRAY. There is no need to. I know it.

(Question read by the reporter.)

The WITNESS. Yes.

Mr. GRAY. Is that the substance of what you said?

The WITNESS. Yes, sir; and the question I think was clearly put to me.

Mr. GRAY. Now, if there had been only one person involved in the Chevalier contact, that is, to Dr. Oppenheimer himself, it is unlikely, I suppose, that you would have found out about it, except from Dr. Oppenheimer.

The WITNESS. That is right.

Mr. GRAY. In other words, unless he had volunteered this information to Lieutenant Johnson in the first place, and repeated it to you in the second place, this may never have been a matter of discussion in a possible future hearing?

The WITNESS. Of course, we cannot exclude the possibility if the investigation took some other tangent and that may have come out, but that is just a supposition.

Mr. GRAY. But if the contact had been just between Professor Chevalier and Dr. Oppenheimer in Dr. Oppenheimer's home, it is pretty unlikely that you would have known about it except from Dr. Oppenheimer?

The WITNESS. It is unlikely that we would know about; yes, sir.

Mr. GRAY. Did it occur to you, and if it didn't, I wish you would say so, that the fact that Dr. Oppenheimer in some detail mentioned two other people than the individual who later turned out to be himself—I am not sure it was two other people.

The WITNESS. It was three other people, sir.

Dr. EVANS. Three other people.

The WITNESS. Yes, sir.

Mr. GRAY. Let me rephrase my question. Did it occur to you at the time that the fact that Dr. Oppenheimer mentioned both to Lieutenant Johnson and to you contacts with 3 people for information, 2 of whom were supposed to be at Los Alamos and 1 of whom was supposed soon to go to Oak Ridge, that he was giving you this information thinking that you possibly could find out about these other 3 people? I am afraid that is not a clear question. I am trying to ask you whether it occurred to you at the time that he was giving you the story of the contact because he felt that it might be otherwise discovered, or that he was giving you the story in this kind of detail because he felt these details might be discovered?

The WITNESS. Yes, sir; my impression was that he felt that we would discover in our investigation the fact that there were these contacts, and the extent of them.

Mr. GRAY. Is it true that he said he thought these were innocent contacts, and therefore weren't worth pursuing in his judgment? Is that correct?

The WITNESS. He said that; yes, sir. But the reason—well, excuse me.

Mr. GRAY. You did, indeed, try to find out at least who the individual was who was scheduled to go to Oak Ridge?

The WITNESS. Yes, sir; as I recall we did.

Mr. GRAY. So that at the time you did believe that people other than Dr. Oppenheimer himself were involved in this.

The WITNESS. We didn't believe, sir?

Mr. GRAY. You did believe.

The WITNESS. We did believe. As a matter of fact, we didn't know how many more contacts were made.

Mr. GRAY. But in fact, you never established that there were any other contacts?

The WITNESS. No, sir.

Mr. GRAY. And the man whose orders you held up, who had been scheduled for movement to Oak Ridge, turned out in fact not to have been involved?

The WITNESS. I hate to bring up a name at a sort of very slim recollection, but to emphasize the point, I believe, and in this instance I hope if I am mistaken it is excusable, because I feel it was a man whom we had under suspicion as one of the men who was a Communist Party member or associate, and on whom an investigation was being run. We had never established his contact with Chevalier.

Mr. GRAY. Just for the clarification of the record, Colonel Pash, am I correct in thinking that after receiving Eltenton's name from Dr. Oppenheimer your jurisdictional limitations would have prevented your investigating Mr. Eltenton, whatever your inclinations might have been? Is that correct?

The WITNESS. Yes, sir.

Mr. GRAY. And that this then became a function of some other agency of Government?

The WITNESS. Yes, sir.

Mr. GRAY. Did you communicate with the other agency of Government what you learned?

The WITNESS. I did, sir, yes.

Mr. GRAY. Dr. Evans.

Dr. EVANS. Colonel Pash, did you know Fuchs?

The WITNESS. No, sir.

Mr. EVANS. Did you know Greenglass?

The WITNESS. No, sir; that happened in my absence.

Dr. EVANS. Having been connected with a couple of institutions of learning myself, not radiations laboratories, of course, and not the high powered nuclear physics that was going on here, I am surprised—maybe I should not be—at the number of communists and fellow travelers gathered together at one point in this radiation laboratory. Did that surprise you or is that just normal?

The WITNESS. No, sir; that was a surprise. We did not expect it.

Dr. EVANS. It is a surprise to me. I am still concerned, and I don't understand these three men that Dr. Oppenheimer mentioned, three contacts, is that correct?

The WITNESS. Yes, sir.

Dr. EVANS. Did he mean there were three men besides Chevalier who had approached him, or these other men were approaching somebody else besides Dr. Oppenheimer?

The WITNESS. No, sir; this unknown professor contacted these three men, which proved to be Chevalier later.

Dr. EVANS. He contacted Dr. Oppenheimer, and then he contacted three other men?

The WITNESS. Yes, sir.

Dr. EVANS. Possibly to get information from them.

The WITNESS. Yes, sir.

Dr. EVANS. I just wanted that clear for the record. Maybe everybody understood it, but I didn't. You tried to find out those other three men, didn't you?

The WITNESS. Yes, sir; both from Dr. Oppenheimer and through investigative procedures.

Dr. EVANS. That is all.

Mr. GRAY. Mr. Silverman.

Mr. SILVERMAN. May we take just a moment, sir?

Mr. GRAY. Yes.

Mr. SILVERMAN. I have no further questions.

Mr. ROBB. That is all. Thank you very much.

Mr. GRAY. Thank you, Colonel Pash.

(Witness excused.)

Mr. ROBB. For the record, I think counsel have finally agreed upon the final definitive text of the Pash interview.

Mr. SILVERMAN. Yes, I understand they have.

Mr. ROBB. I think that should be read into the record when we get it typed up, and also I should like to have the Lansdale interview read into the record.

Mr. GARRISON. Mr. Chairman, we had previously requested that it be read aloud. We waived that in the interest of getting along.

Mr. SILVERMAN. Why doesn't the stenographer just copy it?

Mr GRAY. Yes, it will just be copied into the record.

(Brief recess.)

Mr. GRAY. Do you wish to testify under oath?
Mr. BORDEN. I would prefer to testify under oath.
Mr. GRAY. Would you stand and raise your right hand. Give your full name.
Mr. BORDEN. My name is William Liscum Borden.
Mr. GRAY. William Liscum Borden, do you swear that the testimony you are to give the board shall be the truth, the whole truth, and nothing but the truth, so help you God?
Mr. BORDEN. I do.
Whereupon, William Liscum Borden was called as a witness, and having been first duly sworn, was examined and testified as follows:
Mr. GRAY. Will you be seated, please.
It is my duty, Mr. Borden, to remind you of the existence of the so-called perjury statutes. May I assume you are familiar in general with them?
The WITNESS. Yes.
Mr. GRAY. Also I should like to request that if in the course of your testimony it becomes necessary for you to refer to or to disclose restricted data or classified material you notify me in advance so that we may take necessary security measures.
Finally, Mr. Borden, I should say that we treat these proceedings as a confidential matter between the Atomic Energy Commission, its officials and witnesses on the one hand, and Dr. Oppenheimer and his representatives on the other. The Commission is making no releases with respect to these proceedings and on behalf of the board, I express the hope that witnesses will take the same view of the situation.
The WITNESS. You may count on me to observe that suggestion.
Mr. GRAY. Mr. Robb.

DIRECT EXAMINATION

By Mr. ROBB:
Q. Where do you live at present?
A. 711 St. James Street, Pittsburgh, Pa.
Q. What is your present occupation?
A. I work for the Westinghouse Electric Corp. in its atomic power division.
Q. What is your position?
A. My title is assistant to the manager of the Westinghouse atomic power division.
Q. How long have you held that position?
A. Since July 1, 1953.
Q. What are your duties?
A. I assist the manager of the division in planning and coordinating matters, serve as his alter ego as to certain designated matters which he stipulates.
Q. Prior to your assuming that position, what was your position?
A. I was executive director of the Joint Committee on Atomic Energy.
Q. Of the United States Congress?
A. That is correct.
Q. How long did you hold that position?
A. From the last days of January 1949 until about June 1, 1953.
Q. Did you have a staff serving with you?
A. I did.
Q. How many people?
A. Approximately 19 or 20.
Q. In general what was the scope of your work?
A. It was the duty of the staff to collect facts concerning the atomic energy program, and to make recommendations to the chairman and members of the committee.
Q. Prior to assuming those duties, what did you do?
A. I was legislative secretary to Senator Brien McMahon for about 6 months. I believe it was in the middle part of 1948 that I went to work for him.
Q. What is your educational background?
A. I hold an A. B. and LL. B. degree from Yale.
Q. What dates?
A. I got my A. B. in the spring of 1942, and my LL. B. in September 1947.
Q. Where were you in the interim?
A. I was a pilot in the Army Air Force for 3 years during the war.
Q. Where?
A. I served with the 8th United States Air Force based in England.
Q. After you graduated from law school, you went where, with Senator McMahon?

A. No, I went to the Office of Alien Property of the Justice Department.
Q. As an attorney?
A. As an attorney, and I stayed there as I recall from January of 1948 until mid-1948, when I went to work for Senator McMahon.
Q. In your capacity as executive director of the staff of the joint committee, did you give consideration to the matter of Dr. J. Robert Oppenheimer?
A. I did, yes.
Q. Would you say you gave much or little consideration to Dr. Oppenheimer?
A. I would say I gave increasing consideration over a period of years, Mr. Robb.
Q. By the way, I might ask you, Mr. Borden, you are appearing today in response to a subpena?
A. Thank you for giving me an opportunity of emphasizing that a subpena commanding me to appear here has been served on me, and I testify under official compulsion.
Q. As a result of your study of the matter of Dr. Oppenheimer, did you reach certain conclusions in your mind with respect to him?
A. I did, yes.
Q. Did there come a time when you expressed those conclusions in a letter to Mr. J. Edgar Hoover of the Federal Bureau of Investigation?
A. That is correct.
Q. When was that?
A. The letter was dated November 7, 1953.
Q. Was that subsequent to the termination of your connection with the joint committee?
A. That was, yes.
Q. Prior to writing that letter, did you discuss the writing of it with anybody connected with the Atomic Energy Commission?
A. I did not.
Q. Did you in that letter express your conclusions with respect to Mr. J. Robert Oppenheimer?
A. I did.
Q. Were those conclusions your own conclusions?
A. They are.
Q. Were they your honest conclusions arrived at after great thought?
A. That is correct.
Q. Are they still your conclusions?
A. They are.
Q. Do you have a copy of your letter with you?
A. I have one in front of me.
Q. Would you be good enough to read it?
A. This letter is dated November 7, 1953.
Q. While our friends are looking at that, I might ask you whether you know Dr. Oppenheimer personally?
A. I have met him on a few occasions.
Mr. ROBB. May we proceed, Mr. Chairman?
Mr. SILVERMAN. One moment, please.
Mr. GRAY. I would like to ask the counsel what the purpose of delay is. He is simply going to read this.
Mr. SILVERMAN. Mr. Chairman, I can hardly conceive that a letter, with due respect to Mr. Borden, by a gentleman stating what he adds the evidence up to can be enormously helpful to the board which has itself heard the evidence. There are statements in this letter, at least one that I see, which I don't think anybody would be very happy to have go into this record, and under those circumstances, I would like to look at it a minute longer. There may be serious question whether anybody will be helped by having this letter in the record.
Mr. GRAY. I think you are now raising a question that counsel cannot determine, Mr. Silverman.
Mr. SILVERMAN. Of course not, sir.
Mr. GRAY. If you have any argument about it, I shall be glad to have it. If you wish to protest the reading of the letter into the record, you are certainly at liberty to do so. I take it, however, that it is evident that Mr. Borden is before he committee, he states that this letter is his own letter, he wrote it without consultation with the Commission, that it represents the views he held in November 1953, it represents the views he holds today, he is the individual concerned, he is being confronted by Dr. Oppenheimer and Oppenheimer's counsel and will be available for cross-examination. In view of the fact that being

here as he is under subpena, which has been made clear, presumably this being his opinion, this is what he would testify to. I simply don't see the objection to reading the letter. If I am wrong about that, I should be glad to hear it.

Mr. SILVERMAN. Mr. Chairman, much of the material in this letter, or some of the material in this letter, at least, is matter that has already been before the board.

Mr. GRAY. Mr. Silverman, you are not suggesting that we should not hear from any witnesses who will testify to the same matters previous witnesses have testified to?

Mr. SILVERMAN. Let me say it this way. The thing that struck my eye at once is subdivision (e) on page 2. That troubles me going into the record. If you think it will advance things to have it in, all right.

Mr GRAY. I would like to take a moment to consider that objection.

Mr. GARRISON. Mr. Chairman, the third paragraph on page 4, and some comparable material brings in accusations here that have not before been made in this record or even indicted in the Commission's letter.

Mr. GRAY. You are referring to what?

Mr. GARRISOON. To the third paragraph on page 4, and to the first clause on page 4, and also the last clause on page 3.

Mr. GRAY. Mr. Garrison, is there any question in your mind that if this is the view of the witness, he would not so testify?

Mr. GARRISON. I have no question about that.

Mr GRAY. I am puzzled by the objection to his reading the letter he wrote in November 1953, which he states now represents his present views as distinguished from giving his present views at this time. I am just honestly not clear as to what the objection is.

Mr. GARRISON. It is simply my feeling, Mr. Chairman, that if these represent his present views, and the Commission's counsel has brought him here to testify to this board about accusations which are not in the Commission's letter and are not even suggested in them, and have never before been suggested in these proceedings, we now have a new case which it seems to me either does not belong here or should be included in the Commission's letter, either in the first instance or by amendment.

Mr. GRAY. I think now you are making a point that the board should examine, and specifically in that case you refer to material on page 4, is that correct?

Mr. GARRISON. That is correct, and on the bottom of page 3, and the first sentence at the top of page 2.

Mr. GRAY. I repeat you are making a point which you are entitled to have considered by the board; that was certainly not clear to me from anything Mr. Silverman said earlier.

I would therefore ask everyone to retire from the room except the board and counsel for the board.

(All persons with the exception of the board and counsel for the Board left the hearing room, and after a brief time reentered the room.)

Mr. GRAY. In response to the objection raised by counsel for Dr. Oppenheimer, I would have this to say on behalf of the board:

No. 1, the material which the witness was about to read constitutes testimony by the witness, and does not become a part of the letter of notification from the General Manager of the Commission to Dr. Oppenheimer. I would remind counsel that under the regulations pursuant to which this proceeding is conducted the requirements are that this Board makes specific findings with respect to the items in the letter of notification.

I should also remind counsel that much of the testimony here given has not necessarily reflected either items in the letter of the General Manager of the Commission to Dr. Oppenheimer, or Dr. Oppenheimer's reply to that letter. With the exception of the personal items referred to on page 2, and I will have something to say to the witness about that, the material as I understand it specifically referred to by Mr. Garrison is stated as a conclusion of the author of the letter. Again I take it that the witness would be permitted to present his conclusion about matters which are before this board. Witnesses have done so with constancy throughout this proceeding. Therefore, after consultation with the members of the board, the witness will be allowed to read this letter, and all concerned will understand that this is a part of his testimony which is not necessarily accepted by the Commission, does not become a part of the Commission's letter of notification, nor are the conclusions drawn in the testimony necessarily to be considered accepted by the board. It is the conclusion of the witness, one of many whom we have had before the board, with respect to matters concerned in this proceeding.

Mr. GARRISON. May I ask the Chairman a question?

Mr. GRAY. Yes; you certainly may.

Mr. GARRISON. Is it the opinion of the board that the matters which I identified by paragraph and page numbers——

Dr. EVANS. What page is that?

Mr. GARRISON. The passages to which I previously directed your attention. Is it the opinion of the board that those are matters into which inquiry should be directed?

Mr. GRAY. These are conclusions drawn by a witness with respect to material I think all of which in one way or another has been touched upon in testimony before the board.

Mr. GARRISON. The conclusions that are here stated, Mr. Chairman, that I referred to, refer to entirely new topics so far as this proceeding and the letter is concerned about which there has not been one breath in this record. I take it that the rule from which I am reading, paragraph (j) of section 4.15, is for the protection of the individual. Being for the protection of the individual, it is not likely to be disregarded, because the purpose of this is to give full notice of the individual. If we are to be tried here upon the subject matter of these conclusions, this is something that belongs in the criminal courts and not here. But if it must be heard here, then there should be notice of it.

Mr. GRAY. I would say to counsel that it is not my understanding from conversations with the board that testimony of this witness is in any way going to broaden the inquiry of the board.

Mr. GARRISON. How can it avoid it, sir? Supposing you should believe the witness? Here is a witness produced by counsel engaged by the Commission and delegated with the responsibility by this board of calling such witnesses he wishes, and he brings a witness in to make this kind of an accusation not dreamed of in this proceeding up to this point, and not mentioned in the letter. I think if anything could be more of a surprise and more calling for time, if this is to be the subject matter of the inquiry, I don't know what it is.

Mr. GRAY. I should like to ask, Mr. Garrison, whether you knew of the existence of this letter?

Mr. GARRISON. I had heard rumors that Mr. Borden had written a letter; yes, sir. I had no notion that this kind of material was in it.

Mr. GRAY. This is a conclusion of a witness that you are speaking to now.

Mr. GARRISON. Yes; but I take it you are going to permit the witness to adduce his evidence upon these topics. Otherwise, there is no point of his reading the letter unless he is going to testify about it.

I would suggest, Mr. Chairman—I don't want to delay the proceeding——

Mr. GRAY. The board is very much concerned with protecting the interests of the individual concerned, the Government and the general public. So that I do not consider this discussion a matter of delay.

Mr. ROBB. Mr. Chairman, might I suggest one thing? I assume that in the event the witness should be asked whether or not upon the basis of the evidence he has considered that he considers Dr. Oppenheimer a security risks, and he should say that he did, and should then be asked to give his reasons, he might very well give the reasons that he set forth in this letter under conclusions. I can't see much difference. I think it would not be contended the scope of the inquiry is thereby broadened or would be thereby broadened.

Mr. GARRISON. Mr. Robb is making a point of form and not of substance, Mr. Chairman. We are here put on notice in advance—this is the only way in which it happens to come up—that this witness proposes to make accusations of a new character not touched upon in the letter, and not suggested before in these proceedings by anybody, even by the most vigorous critics of Dr. Oppenheimer.

Mr. ROBB. Mr. Chairman, might I say one thing for the record? The witness wrote this letter on his own initiative and his own responsibility, setting out certain matters of evidence, I think all of which, if not all, certainly most all of which, are mentioned in the letter from General Nichols to Dr. Oppenheimer. This letter was to Mr. Hoover. The letter is a part of the files before the board. It is, I think, an important letter. It seemed to the Commission, it seemed to us, that under those circumstances it was only fair to Dr. Oppenheimer and his counsel that this witness, should be presented here, confronted by Dr. Oppenheimer, and his counsel, subjected to cross-examination on the matters set out in this letter.

The conclusions drawn by this witness in his letter are not allegations in the letter from General Nichols to Dr. Oppenheimer. They will not be allegations in any possible amendment of that letter. The conclusions are the conclusions of the witness alone. They are conclusions which he has drawn from the evidence just as other witnesses on behalf of Dr. Oppenheimer have drawn the conclusions that Dr. Oppenheimer is not a security risk, but on the contrary is a man of great honesty, integrity, and patriotism.

I assume that if the witness having written this letter had concluded from the evidence set out by him that Dr. Oppenheimer was not a security risk, that he was a splendid American, a man of honor, that Mr. Garrison would have no objection to reading those conclusions. It seems to me it cuts both ways, Mr. Chairman.

Mr. GARRISON. May I ask how long the Commission has had this letter in its file?

Mr. ROBB. I don't know, Mr. Garrison. Some time, of course.

Mr. GARRISON. Did it have it prior to the letter of December 23, 1953?

Mr. ROBB. Mr. Garrison, I don't think I should be subjected to cross-examination by you, but I can say to you that I am sure Mr. Hoover did not wait 8 months to send it over to the Commission.

Mr. GARRISON. Mr. Chairman, at the bottom of page 3, it says, "From such evidence considered in detail the following conclusions are justified." You can call them conclusions or allegations; it is all the same thing.

Mr. GRAY. This is simply the testimony of a witness.

Mr. GARRISON. This is the testimony of a witness produced by the Commission's counsel to whom this task has been delegated, on his own responsibility bringing in here to make accusations of the kind that I don't think belong here.

Mr. GRAY. I will state to counsel for Dr. Oppenheimer that copies of this letter have been in the possession of the board along with all other material and have been read by members of this board. Mr. Borden's conclusions are, therefore, known to the members of this board. The board has certainly made no suggestion to the Commission and the general manager of the Commission has not otherwise taken the initiative to broaden the inquiry to include these stated conclusions of the witness. If you prefer not to have Dr. Oppenheimer confronted by a witness and cross-examined by his counsel with respect to material which you know is in the possession of the board, of course that would be your decision in what you consider to be the best protection of the interests of Dr. Oppenheimer.

I gather that is what you are saying, because you have been informed by the chairman that a copy of this letter is in the possession of the members of the board. That, again, if I need to repeat this, does not in any way indicate that it is anything more than one part of material consisting of a record which is to be thousands of pages long, and various other data voluminous in nature which are before this board. You may not assume that any of the conclusions of any of the witnesses may necessarily be those of the board. As far as this board is concerned—I hope I may speak for my colleagues—I do not think we will insist on either direct or cross-examination of this witness. The conclusion which we had reached in the period during which you were excused from the room was that we would proceed. However, I shall be glad to consult further with the members of the board to determine whether we shall proceed with the introduction of this letter.

I take it that counsel would not object to direct examination of this witness? You are not objecting to the winess?

Mr. GARRISON. No.

Mr. GRAY. Mr. Morgan has just observed to me that he felt that it was the fairest thing to Dr. Oppenheimer to give him and his counsel the opportunity to examine the witness with respect to this letter which was in the possession of the board. He doesn't insist that we proceed. I have not yet consulted Dr. Evans.

Dr. EVANS. That is all right with me.

Mr. GARRISON. Mr. Chairman, it is needless to say that we would much rather have an opportunity to cross-examine if the board considers that this topic is properly a part of the case. If the board considers that it is, then let us proceed with it. I trust that in view of the circumstances if it be your decision to proceed, that to the extent that we need time here to prepare on this new kind of an allegation, that we may have it.

Mr. GRAY. Yes.

(Discussion off the record.)

Mr. GARRISON. Mr. Chairman, with respect to the objection previously raised by Mr. Silverman, we withdraw that objection and prefer that the letter in its entirety be read, if we are to go ahead with it.

Mr. GRAY. All right, sir.

The WITNESS. This letter is dated November 7, 1943. A copy went to the Joint Committee on Atomic Energy. The original went to Mr. J. Edgar Hoover, Director, Federal Bureau of Investigation, Washington, D. C.:

"DEAR MR. HOOVER: This letter concerns J. Robert Oppenheimer.

"As you know, he has for some years enjoyed access to various critical activities of the National Security Council, the Department of State, the Department of Defense, the Army, Navy, and Air Force, the Research and Development Board, the Atomic Energy Commission, the Central Intelligence Agency, the National Security Resources Board, and the National Science Foundation. His access covers most new weapons being developed by the Armed Forces, war plans at least in comprehensive outline, complete details as to atomic and hydrogen weapons and stockpile data, the evidence on which some of the principal CIA intelligence estimates is based, United States participation in the United Nations and NATO and many other areas of high security sensitivity.

"Because the scope of his access may well be unique, because he has had custody of an immense collection of classified papers——"

Dr. EVANS. Documents. You said papers.

The WITNESS. That is right. Perhaps I should state that the copy I have before me is one that I typed myself, and it is possible that it does not conform.

"Because the scope of his access may well be unique, because he has had custody of an immense collection of classified papers covering military, intelligence, and diplomatic as well as atomic-energy matters, and because he also possesses a scientific background enabling him to grasp the significance of classified data of a technical nature, it seems reasonable to estimate that he is and for some years has been in a position to compromise more vital and detailed information affecting the national defense and security than any other individual in the United States.

"While J. Robert Oppenheimer has not made major contributions to the advancement of science, he holds a respected professional standing among the second rank of American physicists. In terms of his mastery of Government affairs, his close liaison with ranking officials, and his ability to influence high-level thinking, he surely stands in the first rank, not merely among scientists but among all those who have shaped postwar decisions in the military, atomic energy, intelligence, and diplomatic fields. As chairman or as an official or unofficial member of more than 35 important Government committees, panels, study groups, and projects, he has oriented or dominated key policies involving every principal United States security department and agency except the FBI.

"The purpose of this letter is to state my own exhaustively considered opinion, based upon years of study, of the available classified evidence, that more probably than not J. Robert Oppenheimer is an agent of the Soviet Union.

"This opinion considers the following factors, among others:

"(a) He was contributing substantial monthly sums to the Communist Party;

"(b) His ties with communism had survived the Nazi-Soviet Pact and the Soviet attack upon Finland;

"(c) His wife and younger brother were Communists;

"(d) He had no close friends except Communists;

"(e) He had at least one Communist mistress;

"(f) He belonged only to Communist organizations, apart from professional affiliations;

"(g) The people whom he recruited into the early wartime Berkeley atomic project were exclusively Communists;

"(h) He had been instrumental in securing recruits for the Communist Party; and

"(i) He was in frequent contact with Soviet espionage agents.

"2. The evidence indicating that—

"(a) In May 1942, he either stopped contributing funds to the Communist Party or else made his contributions through a new channel not yet discovered;

"(b) In April 1942 his name was formally submitted for security clearance;

"(c) He himself was aware at the time that his name had been so submitted; and

"(d) He thereafter repeatedly gave false information to General Groves, the Manhattan District, and the FBI concerning the 1939–April 1942 period.

"3. The evidence indicating that—

'(a) He was responsible for employing a number of Communists, some of them nontechnical, at wartime Los Alamos;

"(b) He selected one such individual to write the official Los Alamos history;

"(c) He was a vigorous supporter of the H-bomb program until August 6, 1945

(Hiroshima), on which day he personally urged each senior individual working in this field to desist; and

"(d) He was an enthusiastic sponsor of the A-bomb program until the war ended, when he immediately and outspokenly advocated that the Los Alamos Laboratory be disbanded.

"4. The evidence indicating that:

"(a) He was remarkably instrumental in influencing the military authorities and the Atomic Energy Commission essentially to suspend H-bomb development from mid-1946 through January 31, 1950.

"(b) He has worked tirelessly, from January 31, 1950, onward, to retard the United States H-bomb program;

"(c) He has used his potent influence against every postwar effort to expand capacity for producing A-bomb material;

"(d) He has used his potent influence against every postwar effort directed at obtaining larger supplies of uranium raw material; and

"(e) He has used his potent influence against every major postwar effort toward atomic power development, including the nuclear-powered submarine and aircraft programs as well as industrial power projects."

From such evidence, considered in detail, the following conclusions are justified:

"1. Between 1929 and mid-1942, more probably than not, J. Robert Oppenheimer was a sufficiently hardened Communist that he either volunteered espionage information to the Soviets or complied with a request for such information. (This includes the possibility that when he singled out the weapons aspect of atomic development as his personal specialty, he was acting under Soviet instructions.)

"2. More probably than not, he has since been functioning as an espionage agent; and

"3. More probably than not, he has since acted under a Soviet directive in influencing United States military, atomic energy, intelligence, and diplomatic policy.

"It is to be noted that these conclusions correlate with information furnished by Klaus Fuchs, indicating that the Soviets had acquired an agent in Berkeley who informed them about electromagnetic separation research, during 1942 or earlier.

"Needless to say, I appreciate the probabilities identifiable from existing evidence might, with review of future acquired evidence, be reduced to possibilities; or they might also be increased to certainties. The central problem is not whether J. Robert Oppenheimer was ever a Communist; for the existing evidence makes abundantly clear that he was. Even an Atomic Energy Commission analysis prepared in early 1947 reflects this conclusion, although some of the most significant derogatory data had yet to become available. The central problem is assessing the degree of likelihood that he in fact did what a Communist in his circumstances, at Berkeley, would logically have done during the crucial 1939–42 period—that is, whether he became an actual espionage and policy instrument of the Soviets. Thus, as to this central problem, my opinion is that, more probably than not, the worst is in fact the truth.

"I am profoundly aware of the grave nature of these comments. The matter is detestable to me. Having lived with the Oppenheimer case for years, having studied and restudied all data concerning him that your agency made available to the Atomic Energy Commission through May 1953, having endeavored to factor in a mass of additional data assembled from numerous other sources, and looking back upon the case from a perspective in private life, I feel a duty simply to state to the responsible head of the security agency most concerned the conclusions which I have painfully crystalized and which I believe any fairminded man thoroughly familiar with the evidence must also be driven to accept.

"The writing of this letter, to me a solemn step, is exclusively on my own personal initiative and responsibility.

"Very truly yours,

"(Signed) William L. Borden,
"(Typed) WILLIAM L. BORDEN."

Mr. ROLANDER. Mr. Chairman, I had copies of this letter made, and Mr. Borden read from the copies, and I think there is one error in the copy that he read. That begins where the letter says, "This opinion considers the following factors among others: (1) The evidence indicating that as of April of 1942" and then it proceeds.

Mr. SILVERMAN. Indicating that as of what date?

Mr. ROLANDER. "This opinion considers the following factors, among others:
"1. The evidence indicating that as of April 1942 (a)."

Mr. GRAY. Now, I should like to make a statement with respect to this letter which I am authorized to make by the two other members of the board which I think may ease Mr. Garrison's problem as he has seen it in this discussion.

I would say to you that the board has no evidence before it that Dr. Oppenheimer volunteered espionage information to the Soviets or complied with a request for such information; that he has been functioning as an espionage agent or that he has since acted under Soviet directive, with one qualification as to that latter point, which I am sure will not surprise you. That is, there has been testimony by various witnesses as to whether members of the Communist Party, as a matter of policy at the time of the war years or entering into Government or military service, complied with policy or policy directions in that regard. With respect to that qualification, which I believe appears already in the record, and which is certainly no surprise to Dr. Oppenheimer and his counsel, I repeat that the members of the board feel that they have no evidence before them with respect to these matters which I have just recited.

I repeat, therefore, that there are now before the board in the nature of conclusions of the witness, stated to be his own conclusions on the basis of other material which is set forth in some detail, and I believe practically all of which has been referred to without making a judgment whether it has been established or not.

Mr. ROBB. May I proceed?
Mr. GRAY. Yes.

By Mr. ROBB:

Q. Mr. Borden, may I ask you, sir, why you waited until you left the joint committee to write that letter?

A. Mr. Robb, this case has concerned me over a period of years. My concern has increased as time passed. Several actions were taken with respect to it while I was working for the joint committee. It has consisted in the preparation of 400 questions raised on the case. This was the final work that I performed before leaving the committee. I felt at that time that I had not previously fully measured up to my duty on this matter. As of the time I left, the preparation of those questions constituted for me the discharge of the duty. However, no position was taken in the formulation of those questions, or at least if there was a position, it was implicit only.

After I left, I took a month off and this matter pressed on my mind. The feeling grew upon me that I had not fully discharged what was required of me in view of the fact that I had not taken a position.

Accordingly, by approximately mid-October, I had crystalized my thinking to the point where I felt that this step was necessary. There is a letter which I have written to the joint committee on this subject, if you wish me to refer to it, or to read it to you.

Q. Is there anything, Mr. Borden, that you can now add to what you have set out in this letter as your conclusions?

A. I have no desire to add anything.

Q. I am not ask you that, sir. Is there anything that you feel that is appropriate for you to tell this board in addition to what you have set out in that letter?

A. I feel, Mr. Robb, that it is my obvious duty to answer any questions that are asked me. If I were to volunteer information, I think it is obvious that I could talk over a long period of time.

Q. I am not asking you to volunteer, but what I want to know is, Does that letter fully state your conclusions?

A. This letter reflects my conclusions as of now.

Q. Does it fully reflect your conclusions?

A. Yes.

Q. So there is nothing that you feel you should add to it?

A. That is correct. Perhaps I misunderstood you.

Q. Let me see whether or not you feel any hesitation about answering any questions that either have been or may be put to you here, because of the presence of Dr. Oppenheimer and his counsel.

A. I do not.

Q. The answer is no?

A. The answer is no.

Mr. ROBB. I think that is all I care to ask. You may cross-examine.

Mr. GRAY. We will now take a recess until Monday at 2 o'clock for many reasons. One is commitments identified with this enterprise as to schedule. Second, I think it is useful if time is required for Mr. Garrison. I would hope that my statement that I made to the board takes care of most of the difficulties that we discussed.

Mr. GARRISON. Is it to be understood that the witness will be back here on Monday?

Mr. GRAY. The witness is under subpena, and he is not happy to be here in the first place. It is understood that he will be.

We are now in recess until Monday at 2 o'clock.

(Thereupon, at 4:30 p. m., a recess was taken until Monday, May 3, 1954, at 2 p. m.)

UNITED STATES ATOMIC ENERGY COMMISSION

PERSONNEL SECURITY BOARD

In the Matter of J. Robert Oppenheimer

Atomic Energy Commission,
Building T-3, Room 2022,
Washington, D. C., Monday, May 3, 1954.

The above-entitled matter came on for hearing pursuant to recess before the board, at 2 : 30 p. m.

Personnel Security Board: Mr. Gordon Gray, chairman; Dr. Ward T. Evans, member; and Mr. Thomas A. Morgan, member.

Present: Roger Robb and C. A. Rolander, Jr., counsel for board; J. Robert Oppenheimer, Lloyd K. Garrison, Samuel J. Silverman, and Allen B. Ecker, counsel for J. Robert Oppenheimer; Herbert S. Marks, cocounsel for J. Robert Oppenheimer.

PROCEEDINGS

Mr. GRAY. Mr. Garrison.

Mr. GARRISON. Mr. Chairman, I would like to make a short statement, sir.

Over the weekend we have examined Mr. Borden's letter to Mr. Hoover of November 7, 1953, which he read into the record at the last session. Mr. Borden in his brief testimony stated that the letter constituted his conclusions, and that he had nothing to add. It is quite clear that the letter consists not of evidence, but of Mr. Borden's opinions arrived at from studying FBI reports and other unspecified data. These opinions relate essentially to the items contained in General Nichols' letter to Dr. Oppenheimer of December 23, 1953, which have been canvassed in the testimony, and the documents before this board. It is apparent that except for Mr. Borden's conclusions about espionage, for which there is no evidence, and as to which the chairman has assured us there is no evidence before the board, Mr. Borden's opinions represent his interpretation of evidentiary matters which this board has been hearing about for the past 3 weeks from persons who actually participated in the particular events which have been the subject matter of this investigation.

In view of these considerations, it has seemed to us that if we were now to ask Mr. Borden to develop further his opinions and conclusions, we would merely be inviting argument about the interpretation of evidence.

While the board has been lenient in permitting argument by witnesses, it hardly seems to us that we would be justified in provoking or inviting opinions and argument which could run the gamut of all the evidence before the board.

For these reasons it has seemed to us appropriate to respond to Mr. Borden's letter in our rebuttal and summation as we expect to do. Consequently, we shall dispense with cross-examination unless the board should wish to ask Mr. Borden questions, in which event we would like to reserve the right to do ours when the board is through.

Mr. GRAY. Of course, it is the right of Dr. Oppenheimer and counsel to decline to cross-examine any witness before this board. Obviously there is nothing in our procedure which requires cross-examination.

Mr. Garrison has stated that this letter constitutes conclusions of the witness which, I think he has stated, was the case on direct examination. I think, however, it appropriate that the record reflect the fact which would be very obvious to anyone who reads it, that there has been a great deal of testimony here of conclusions with respect to these matters which were contained in General Nichols' letter to Dr. Oppenheimer, and witnesses called by Dr. Oppenheimer, and his counsel, have repeatedly stated that they had certain conclusions with respect to these matters which related to Dr. Oppenheimer's loyalty, character, and associations.

I think the present witness has not sought to state anything other than these are his own conclusions.

Mr. GARRISON. That is right, Mr. Chairman. I did not mean to suggest that other witnesses have not stated their conclusions and opinions. Of course, they have. I meant merely to say that those conclusions were derived from testimony of their own with respect to matters in which they had participated either with Dr. Oppenheimer or in other connections from which they derived their conclusions.

Mr. ROBB. Mr. Chairman, just one perhaps minor remark. I am sure Mr. Garrison did not mean it that way. He stated that there was no evidence of espionage. I think what the chairman said, and I have the transcript before me, is, "I would say to you that the board has no evidence before it that Dr. Oppenheimer volunteered espionage information to the Soviets or complied with a request for such information, that he has been functioning as an espionage agent."

I assume that is what Mr. Garrison referred to.

Mr. GARRISON. That is to which I had reference.

Mr. ROBB. Yes.

Mr. GRAY. From the circumstances, I would say the witness is excused. Thank you very much, Mr. Borden. I offer the apologies of the board for having kept you here through the weekend or having you required to return for this purpose. Thank you very much.

Mr. ROBB. That is all the testimony we have to offer, Mr. Chairman. I would like to talk to Mr. Garrison and his colleagues about the so-called Pash ad Lansdale transcripts. I think we finally worked out the final version of the Pash transcript. I would like to have them appear in the record at this point, if there is no objection.

Mr. GARRISON. I have not seen whatever Mr. Rolander and Mr. Ecker worked out.

Mr. ROBB. Neither have I.

Mr. GARRISON. Subject to looking it over and possible conference that might be necessitated by that between us, I certainly have no objection to its going in. Before it is considered to be finally in, we should have an opportunity to look whatever they have agreed to over.

Mr. ROBB. Surely.

Mr. GRAY. The chairman would like to make a comment on this. It certainly is to be hoped that counsel can agree. Perhaps I had misunderstood. I thought each of you had delegated.

Mr. ROBB. I did, Mr. Chairman.

Mr. GRAY. I had the understanding that whatever Mr. Rolander and Mr. Ecker agreed was to be. If in your examination of it you think there is some material matter, of course, you will not be denied the opportunity to bring it to the attention of the board.

Mr. ROBB. So far as I am concerned, Mr. Chairman, Mr. Rolander's decision is it.

Mr. GRAY. If Mr. Garrison feels compelled to raise questions, then I think it would be well for you to become "it."

Mr. ROBB. Very well.

Mr. GARRISON. I would just like an opportunity to read it over.

Mr. ROLANDER. It is being typed now. There is a possibility for some typographical errors.

(Discussion off the record.)

Mr. ROBB. Mr. Chairman, in the interest of completeness of the record, we feel that the original typewritten transcript as prepared in the office of Colonel Pash in 1943, as he has testified here, should also be set out in the record following the appended table, and I ask that be done.

Mr. GRAY. Very well.

Mr. ROBB. Mr. Chariman, in respect of the so-called Lansdale transcript, which is also being set out in the record, counsel for Dr. Oppenheimer wish the record to reflect that in agreeing to the printing of the Lansdale transcript of September 12, 1943, they do not concede its correctness, since the original recording or tape from which the transcript was made is not available to counsel for purposes of comparison.

Mr. GRAY. I take it that there is agreement among counsel with respect to these matters which you have read, and the record will reflect that.

Mr. ROBB. Yes, Mr. Chairman.

Mr. GARRISON. Yes, Mr. Chairman.

STIPULATION

Counsel for Dr. Oppenheimer and counsel for the Atomic Energy Commission have compared the recording of the so-called Pash interview of August 26, 1943, with the typewritten transcript, portions of which were used in the cross-examination of Dr. Oppenheimer. The following is, as nearly as they can understand the recording, a correct transcription. Where portions did not appear to counsel for Dr. Oppenheimer and for the Commission to be reasonably decipherable, this has been indicated by three asterisks. The appended table reflects the changes from the typewritten transcript.

"UNCLASSIFIED, 4/19/54 CAR,
"San Francisco, Calif., August 27, 1943.

"MEMORANDUM FOR THE OFFICER IN CHARGE

"Subject: D. S. M. Project.
"Re: Transcript of Conversation between Dr. J. R. Oppenheimer, Lt. Col. Boris T. Pash, and Lt. Lyall Johnson.

"Transmitted herewith is the transcript of conversation between Dr. J. R. Oppenheimer, Lt. Colo. Boris T. Pash, and Lt. Lyall Johnson, held in Lt. Johnson's office in the New Class Room Building, University of California, Berkeley, Calif., on August 26, 1943. It is to be noted that in some places the conversation was very indistinct and that the running commentary may be indecisive in these places, but the substance of the material discussed is herewith presented:

"P. This is a pleasure, because I am interested to a certain extent in activities and I feel I have a certain responsibility in a child which I don't know anything about. General Groves has, more or less, I feel, placed a certain responsibility in me and it's like having a child, that you can't see, by remote control. I don't mean to take much of your time——

"O. That's perfectly all right. Whatever time you choose.

"P. Mr. Johnson told me about the little incident, or conversation, taking place yesterday in which I am very much interested and it had me worried all day yesterday since he called me.

"O. I was rather uncertain as to whether I should or should not talk to him [Rossi] when I was here. I was unwilling to do it without authorization. What I wanted to tell this fellow was that he had been indiscreet. I know that that's right that he had revealed information. I know that saying that much might in some cases embarrass him. It doesn't seem to have been capable of embarrassing him—to put it bluntly.

"P. Well, that is not the particular interest I have. It is something a little more, in my opinion, more serious. Mr. Johnson said there was a possibility that there may be some other groups interested.

"O. I think that is true, but I have no first-hand knowledge that would be, for that reason, useful, but I think it is true that a man, whose name I never heard, who was attached to the Soviet consul, has indicated indirectly through intermediary people concerned in this project that he was in a position to transmit, without any danger of a leak, or scandal, or anything of that kind, information, which they might supply. I would take it that it is to be assumed that a man attached to the Soviet consulate might be doing it but since I know it to be a fact, I have been particularly concerned about any indiscretions which took place in circles close enough to be in contact with it. To put it quite frankly—I would feel friendly to the idea of the Commander in Chief informing the Russians that we were working on this problem. At least, I can see that there might be some arguments for doing that, but I do not feel friendly to the idea of having it moved out the back door. I think that it might not hurt to be on the lookout for it.

"P. Could you give me a little more specific information as to exactly what information you have? You can readily realize that phase would be, to me, as interesting, pretty near, as the whole project is to you.

"O. Well, I might say that the approaches were always to other people, who were troubled by them, and sometimes came and discussed them with me; and that the approaches were always quite indirect so I feel that to give more, perhaps, than one name, would be to implicate people whose attitude was one of bewilderment rather than one of cooperation. I know of no case, and I am fairly sure that in all cases where I have heard of these contacts, would not have yielded a single thing. That's as far as I can go on that. Now there is a man, whose name was mentioned to me a couple of times—I don't know of my own knowledge that he was involved as an intermediary. It seems, however, not impossible and if you wanted to watch him it might be the appropriate thing to do. He spent quite a number of years in the Soviet Union. He's an English * * * I think he's a chemical engineer. He was—he may not be here now—at the time I was with him here, employed by the Shell development. His name is Eltenton. I would think that there was a small chance—well, let me put it this way: He has probably been asked to do what he can to provide information. Whether he is successful or not, I do not know, but he talked to a friend of his who is also an acquaintance of one of the men on the project, and that was one of the channels by which this thing went. Now I think that to go beyond that would be to put a lot of names down, of people who are no only innocent but whose attitude was 100-percent cooperative.

"P. Now here's a point. You can readily realize that if we get information like that we have to work in an absolutely discreet manner. In other words we can't afford to even indicate——

"O. That you are concerned.

"P. That we are concerned or through whom we get information. However anything that we may get which would eliminate a lot of research work on our part would necessarily bring to a closer conclusion anything that we are doing.

"O. Well, I'm giving you the one name that I think is, or isn't—I mean I don't know the name of the man attached to the consulate—I think I may have been told or I may not have been told and I have, at least not purposely, but actually forgotten. He is—and he may not be here now. These incidents occurred of the order of about 5, 6, 7, months ago.

"J. I was wondering, Dr. Oppenheimer, if there was a particular person—maybe a person on the project that they were trying to pump information from—that if we knew who those were, would at least know where to look for a leak, not from the standpoint of * * *, but looking at a certain picture

"P. Here's the point that I would feel.

"O. I would feel that the people that they tried to get information from were more or less an accident [interpolation] and I believe I would be making some harm by saying that——

"P. Yes. Here's the thing—we, of course, assume that the people who bring this information to you are 100 percent with you, and therefore, there is no question about their intentions. However, if——

"O. Well, I'll tell you one thing—I have known of 2 or 3 cases, and I think two of the men were with me at Los Alamos—they are men who are very closely associated with me.

"P. Have they told you that either they thought they were contacted for that purpose or they were actually contacted for that purpose?

"O. They told me they were contacted for that purpose.

"P. For that purpose.

"O. That is, let me give you the background. The background was—well, you know how difficult it is with the relations between these two allies, and there are a lot of people who don't feel very friendly to Russia, so that the information—a lot of our secret information, our radar and so on, doesn't get to them, and they are battling for their lives and they would like to have an idea of what is going on and this is just to make up, in other words, for the defects of our official communication. That is the form in which it was presented.

"P. Oh, I see.

"O. Of course, the actual fact is that since it is not a communication which ought to be taking place, it is treasonable. But it wasn't presented in that method. [Garble.] It is a method carrying out a policy which was more or less a policy of the Government and the form in which it came was that an interview be arranged with this man Eltenton who had very good contacts with a man from the embassy attached to the consulate who was a very reliable guy (that's his story) and who had a lot of experience in microfilm work, or whatever the hell.

"P. Well, now I may be getting back to a little systematic picture. * * * These people whom you mentioned, who (two?) are down with you now * * * were they contacted by Eltenton direct?

"O. No.

"P. Through another party?

"O. Yes.

"P. Well, now, could we know through whom that contact was made?

"O. I think it would be a mistake, that is, I think I have told you where the initiative came from and that the other things were almost purely accident and that it would involve people who ought not be involved in this.

"P. This will not involve the people but it would indicate to us Eltenton's channel. We would have to, now that this is definite on Eltenton. We, of course——

"O. It is not definite in the sense that I have seen him do the thing. He may have been misquoted. I don't believe so. Now Eltenton is a member of the FAECT or not?

"P. That's the union.

"O. That's the CIO union. He's a man whose sympathies are certainly very far "left," whatever his affiliations, or he may or may not have regular contacts with a political group.

"P. Well, here's how I feel——

"O. I doubt it. In any case, it is a safe thing to say that the channels that would be followed in this case are those involving people who have been generally sympathetic to the Soviet and somehow connected peripherally with the Communist movements in this country. That's obvious. I don't need to tell you that.

"P. Well, yes, the fact is, this second contact—the contact that Eltonton had to make with these other people—is that person also a member of the project?

"O. No.

"P. That also is an outsider?

"O. It's a member of the faculty, but not on the project.

"P. A member of the faculty here? Eltenton made it through a member of the faculty to the project.

"O. As far as I know—these approaches were—there may have been more than one person involved. I don't know.

"P. Here's how I feel about this leftist inclination. I think that whether a man has "left" or "right" inclinations, if his character which is back of it—if he's willing to do this, it doesn't make any difference what his inclinations are. It's based on his character primarily and not——

"O. A thing like this going on, let us say, with the Nazis would have a somewhat different color. I don't mean to say that it would be any more deserving of attention, or any more dangerous, but it would involve rather different motives.

"P. Oh, yes, sure.

"O. I'm pretty sure that none of the guys here, with the possible exception of the Russian, who is doing probably his duty by his country—but the other guys really were just feeling they didn't do anything but they were considering the step, which they would have regarded as thoroughly in line with the policy of this Government, just making up for the fact that there were a couple of guys in the State Department who might block such communications. You may or may not know that in many projects we share information with the British and some we do not, and there is a great deal of feeling about that, and I don't think that the issues involved here seem to the people very different, except that of course, the people on the project realize the importance and that this is a little bigger and the whole procedure gets away. [Garble.]

"P. Now. Do you feel that would affect—and there could be continued attempts now to establish this type of contract?

"O. I haven't any idea.

"P. You haven't any idea?

"O. As I say, if the guy that was here may by now be in some other town and all that I would have in mind is this—I understood that this man to whom I feel a sense of responsibility, Lomanitz, and I feel it for two reasons. One, he is doing work which he started and which he ought to continue, and, second, since I more or less made a stir about it when the question of his induction came up, that this man may have been indiscreet in circles which would lead to trouble. That is the only thing that I have to say. Because I don't have any doubt that people often approach him, with whom he has contact, I mean whom he sees, might feel it their duty if they got word of something, to let it go further and that is the reason I feel quite strongly that association with the Communist movement is not compatible with the job on a secret war project, it is just that the two loyalties cannot go.

"P. Yes—well——

"O. That is an expression of political opinion, I think that a lot of very brilliant and thoughtful people have seen something in the Communist movement, and that they maybe belong there, maybe it is a good thing for the country. I hope it doesn't belong on the war project——

"P. I get your point. I don't want to seem to you insistent. I want to again sort of explore the possibility of getting the name of the person of the faculty—I'll tell you for what reason. Not for the purpose of taking him to task in any way whether its nonofficially, officially, or openly or not but to try to see Eltenton's method of approach. You may not agree with me, but I can assure you that that is one of the most important steps.

"O. I understand that, but I have to take the following points of view: I think in mentioning Eltenton's name I essentially said about the man that I think that he may be acting in a way which is dangerous to this country, and which should be watched. I'm not going to mention the name of anyone in the same breath, even if you that you will make a distinction. I just can't do that, because in the other cases, I am convinced from the way in which they handled the thing that they themselves thought it was a bad business.

"P. These other people, yes; I realize—but if—here is the point—if that man is trying to make other contacts for Eltenton, it would take us some time to try to——

"O. My honest opinion is that he probably isn't—that he ran into him at a party and they saw each other or something and Eltenton said, "Do you suppose you could help me? This a very serious thing because we know that important work is going on here, and we think this ought to be made available to our allies, and would you see if any of those guys are willing to help us with it—and then it wouldn't have to be much." You see, that is the kind of thing. [Remaining statement unintelligible.]

"P. Were these two people you mentioned—were they contacted at the same time?

"O. They were contacted within a week of each other.

"P. They were contacted at two different times?

"O. Yes; but not in each other's presence.

"P. That's right. And then from what you first heard, there is someone else who probably still remains here who was contacted as well?

"O. I think that is true.

"P. What I am driving at is that means that there was a plan, at least for some length of time, to make these contacts—and we may not have known all the contacts.

"O. That is certainly true. That is why I mentioned it. If I knew all about it, then I would say forget it. I thought it would be appropriate to call to your attention the fact that these channels at one time existed.

"P. Yes.

"O. I really think that I am drawing [garbled].

"P. You see, you understand that I am sort of—you picture me as a bloodhound on the trail, and that I am trying to get out of you everything I possibly can.

"O. That's your duty to a certain extent.

"P. You see what I mean.

"O. It is also my duty not to implicate these people, who are acquaintances, or colleagues and so on of whose position I am absolutely certain—myself and my duty is to protect them.

"P. Oh, yes.

"O. If I thought that—I won't say it—it might be slightly off.

"P. Well, then here's another point, Doctor, if we find that in making these various contacts, that we get some information which would lead us to believe that certain of these men may have either considered it or are still considering it (mind you I do not even know these men, so it can't be personal)——

"O. Well, none of them that I had anything to do with considered it. They were just upset about it. * * * [Garbled.] They have a feeling toward this country and have signed the Espionage Act; they feel this way about it for I think that the intermediary between Eltenton and the project, thought it was the wrong idea, but said that this was the situation. I don' think he supported it. In fact I know it.

"P. He made about at least three contacts that we knew of.

"O. Well, I think that's right, yes.

"P. And two of these contacts are down there. That means we can assume at least there is one of these men contacted still on the project here.

"O. Yes, I believe that this man has gone, or is scheduled to go to Site X.

"P. This third man?

"O. I think so.

"P. Well that is, as I say, if I can't get across that line, I even certainly appreciate this much, because it——

"O. I think it's a thing you ought to know.

"P. Oh, no doubt.

"O. I think it's probably one of those sporadic things and I do not think—I have no way of thinking it was systematic but I got from the way in which it was handled, which was rather loosely, and frankly if I were an agent I would not put much confidence in people who are loose-mouthed or casual. I would not think that this was a very highly organized or very well put-together plan but I don't know and I was very much afraid when I heard of Lomanitz' indiscretion that it might very well be serious. I hope that isn't the case.

"P. You mentioned that this man is a member of this FAECT. Do you think that, as a representative of the organization, he would sort of represent their attitude or do you think he is doing this individually?

"O. Oh, the FAECT is quite a big union and has got all sorts of people in it. I'm pretty sure and I don't think it is conceivable that he could be representing the attitude of the union, but it is——

"P. Well, I don't know enough about it to——

"O. I think that at one time—well, I don't know—they had a strong branch up at the Shell Development Research Laboratories, the FAECT—and I believe it is the union which has got organized on the hill.

"J. Yes, it has been around for some time.

"P. This man Eltenton * * * is a scientist.

"O. I don't know, I would guess he was a sort of a chemical engineer.

"P. Would he be in a position to understand the information furnished him?

"O. I don't know that either. It would depend on how well it was furnished. I mean, he has some scientific training and certainly if you sat down with him and took a little time. My view about this whole damn thing, of course, is that the information that we are working on is probably known to all the governments that care to find out. The information about what we are doing is probably of no use because it is so damn complicated. I don't agree that the security problem on this project is a bitter one, because if one means by the security problem preventing information of technical use to another country from escaping. But I do think that the intensity of our effort and our concern with national investment involved—that is information which might alter the course of the other governments and don't think it would have any effect on Russia * * * it might have a very big effect on Germany, and I am convinced about that and that is as everyone else is.

"P. Oh.

"O. To give it roughly what we're after and I think they don't need to know the technical details because if they were going to do it they would do it in a different way. They wouldn't take our methods—they couldn't because of certain geographical differences so I think the kind of thing that would do the greatest damage if it got out, would just be the magnitude of the problem and of the time schedules which we think we have and that kind of thing. To answer your question—Eltenton if you were picking a man which would be an intermediary he wouldn't be a bad choice, I would mention he had some kind of chemical engineering job in Russia. He was trained in England, was in Russia 4 or 5 years and things like that and here——

"P. Does he speak Russian, do you know?

"O. I don't know—I don't know. He speaks with a slight English accent.

"P. If it is necessary would you mind and would it interfere with your work much if I would have to come down and discuss this with you further. Counter assurance—I mean this is—ah——

"O. This is important?

"P. Oh yes, I not only——

"O. If I may express my own opinion as well as my conviction this is not common knowledge.

"P. No, it isn't.

"J. You see a lot of people have reported it to us * * *.

"P. That's why Mr. Johnson called me up yesterday it sort of——

"O. Yes. I mentioned this to Colonel Lansdale.

"P. You did.

"O. Yes.

P. Aha, well of course right now I say—ah—it is all new and—it has come to me——

"O. Right now it means absolutely nothing but what you now find out at this——

"P. If—but——

"O. I would like to say that if I think that * * * certain affiliations that were incompatible to the best interests of this country and this business would die * * *.

"P. It may be necessary for us to—to take certain steps in trying to trace this down and so forth—if anything would develop where we would have to or would be interested in either your place down at Los Alamos or other places, you feel it would be all right for me to contact you on it so that——

"O. Oh, certainly * * * certain precautions——

"P. Oh, yeah, yes—what I mean is instead of going out on certain steps which may——

"O. Yeah——

"P. Come to your attention and be a little bit disturbing to you, I would rather discuss those with you first so that you will be aware of it. I think that, that—well that——

"O. Well, I hope that won't * * *. If I had reason to believe * * *. I will if anything ever comes up that I am convinced—I can always say that everything I know is absolutely 100 percent negative.

"P. If we should find any information which would lead us to believe that there still may be some of that going on, and if it would be important for us to then know a little more in detail who the contacts were and everything and we could show you and that is important to us, I hope you will then find it possible to——

"O. I am only trying to define our future and I will try to act reasonably.

"P. Fine.

"O. As I say I am trying to draw the line here between people who took some responsibility and the people who were purely pushed around and since nothing occurred and the responses seem to have been 100 percent negative, I think I am perhaps justified in—in——

"P. I am not persistent (ha ha) but——

"O. You are persistent and it is your duty.

"P. That is, there is one point in there, that you say that the responses were 100 percent negative. Do you feel that you know everyone whom this intermediary contacted?

"O. Well, no, but I think it is practical to say that it is not inconceivable that the people whom he contacted would be—would have come to my attention but I am not sure.

"P. Well, I would like to say——

"O. Well, I think it would be [one word missing] to say that I just don't know.

"P. I would like to leave this thought with you, Dr. Oppenheimer, if you at some time find it possible, we certainly would give a lot of thanks and appreciation for the name of that intermediary because it's going to—I tell you—the only reason why I would want it, is not for his sake but to see who his contacts are——

"O. Yes, I see——

"P. I can see that we are going to have to spend a lot of time and effort which we ordinarily would not in trying to——

"O. Well——

"P. In trying to run him down before we even can get on to these others——

"O. You'd better check up on the consulate because that's the only one that Eltenton contacted and without that contact there wouldn' be anyhing * * *

"P. You say his man is not employed in the consulate?

"O. Eltenton?

"P. No, no, I mean this man——

"O. I have never been introduced to him * * * or heard his name or anything but I have been given to understand that he is attached to the consulate.

*   *   *   *   *

"O. Maybe this guy is a military attaché—I don't know.

"P. You don't know anything about him?

"O. I don't know anything about him and never have. I may have been told the name, but it made no impression.

"P. Is this member of the faculty in any way—does he in any way come in contact with your project? Why would he be contacted? Is it because he has contacted these people?

"O. I think that Eltenton must have said to him * * * I don't know—that would be my impression of the thing * * *

"P. Well I think that——

"O. Well, I am sorry. I realize that you would like more information but I have been under a little bit of difficulty. The fact that I did not raise this [one word omitted] for a long time——

"P. That's right.

"O. I have difficulty in * * * serious * * * what to do * * * I think my general point of view is that there are some things there which would bear watching.

"P. That's right.

"O. It is doubtful to me if there is anything there which can't be uncovered.

"P. Well, that—I can see where * * *. We will be hot under the collar until we find out what is going on there. I mean—that's the point of view we have to take——

"O. Well, I don't know. * * * Well I would think * * * that it's conceivable—that it wouldn't hurt to have a man in the local of this union FAECT—to see what may happen and what he can pick up.

"P. You feel there could be something—not in the organization itself but some——

"O. Within it.

"P. Within it.

"O. I don't know, I am sure that if they had 20 members, 19 of them might not be involved in it. But I am not sure of the 20th, you see.

"P. Yes.

"O. Forty members correspondingly and—let me put it this way—the bonds that hold them together are very strong you see, and they talk over their problems with their sisters and brothers and it is rather difficult to maintain a complete security in an outfit like that.

"P. Does this union that is up on the hill, do they have members which are not connected with the hill at all?

"O. Oh, yes; they have an international union and has represpresentatives all over this country.

"P. And the same group then, the same mixture would be of people off and on the project would be in the same——

"O. Oh, I imagine so—I don't know; I don't know about that.

"P. Well, we can——

"O. Ordinarily I think that they would have their own local.

"P. Which would be up there.

"O. Maybe not. Maybe it is all one big local. I'm not sure, but that varies with the union.

"P. Well, that is certainly interesting a—you are going to be here for some time?

"O. Oh, no; I am leaving tonight.

"P. Oh, you are; are you flying?

"O. No; I am not. I have orders not to fly.

"P. At least you get some relaxation in between your project. Well, I think that it may——

"O. I will be very glad to see you there. I have a feeling though, a fellow can be fooled you see. I feel responsible for every detail of this sort of thing down at our place and I will be willing to go quite far in saying that everything is 100 percent in order. That doesn't go for this place up here.

"P. No.

"O. I think that's the truth. If everything weren't being done and if everything weren't proper, I think that I would be perfectly willing to be shot if I had done anything wrong.

"P. Well, ah——

"O. I don't say that about this place. It's a very different situation, a very much harder situation. I don't know the people but it's a hard situation; in particular was put together in a casual way and I think that the problem of being sure that there were no leaks * * * and that pressure can be brought with discretion.

"P. I am then, as I say, I may have the pleasure of visiting your place because it may——

"O. My motto is God bless you.

"P. Well, as I say, if this becomes serious, that is to say, I don't know anything about it, but if it becomes——

"O. My guess is that it wouldn't but if I weren't first absolutely sure that it wouldn't—that it were not serious, I wouldn't——

"P. That's right. Well, if it does become serious I may come down with some of my persistency—I mean I would hate to—I have a responsibility of running things down myself.

"O. I also think the particular way this was—that if there is anything going on it would be very easy to find out. I am not worried about that—we can take care of that ourselves.

"P. No; you wouldn't——

"O. Well, I can handle in a way * * *

"P. But it is a situation which would have to be handled very delicately. That's what makes it so difficult. If it is something that's easy to handle and you don't have to worry about it, why you just sort of bull your way through, but these things one has to be very careful.

"O. That's always the case—wanting to be very careful.

"P. I am not the judge to tell whether they should or should not get the information. My business is to stop it going through illegally.
"O. Well, I think——
"I don't actually know whether, if you were in Washington—asking advice on the question how far should cooperation go. I don't know wherein the right answer lies. I have heard of cases with very strong arguments on both sides.
"P. Yes.
"O. * * * we don't have to worry about * * *
"P. Yes; that's right.
"O. Well, I wish good luck——
"P. We could work a hundred years (I mean) and never get this information. That's where we start you see—I mean—we get this information and we have something to start on—we have something to run down. I certainly appreciate this opportunity to visit you.
"O. I hope it's not a waste of time——
"P. Well, I know it's not a waste of my time and ah——
"O. That's all I meant—perhaps as far as the project is concerned * * * a fair starting point——
"P. Could have——
"O. Why not take an about face?
"P. Do you——
"O. And one could do anything about the attache—that would be the natural thing to watch.
"P. Do you know anyone—and because we like to eliminate unnecessary work if we have to—do you know anyone who is on the project who is connected with the FAECT, Dr. Oppenheimer?
"O. Who would be willing to——
"P. That's right.
"O. I don't know who is in the union at all. I have heard that a boy called Fox is president of it.
"J. David Fox.
"O. David Fox, but I would feel * * * I hope that the trade union isn't tied up in this—and they would not act like this because I think it would give them a very black eye and it is no love of mine from the start, and it might have consequences beyond the reasonable. I doubt whether anyone mixed with the union in good faith would be very sympathetic.
"P. Yes.
"O. This isn't a suggestion that there is anything wrong. I have no reason at all to believe that there is, except that it is inevitable that any left wingers still interested in left-wing activity would join such a union. I think I can be quite sure of it. And I don't think that is due to unions who are seeking a selected group of people——
"P. Yes.
"O. You might get some—of course, this is just my opinion that there is no harm in discussing it—well, I just don't know.
"P. May I just ask then, Doctor, if you would please not discuss this with anyone—so that they would not be aware of this fact that——
"O. No; I would not have raised the question if it had not seemed to me that it deserved looking into.
"P. Yes.
"O. And if I seem uncooperative I think that you can understand that it is because of my insistence in not getting people into trouble——
"P. I can assure you that if something comes to your attention out there——
"O. Let me dispose of that statement which came over the long-distance phone. Frankly, I got—from that boy a promise to stop all this sort of thing when he came on the job. * * * to that promise [garbled]. * * * I do not know what he was doing it for but I thought there was a possibility. He said he understood that * * * I talked to him yesterday. He said he had no connection * * *.
"P. Well, what I mean, if anything does come to your attention in connection with this phase, if you can——
"O. It won't be really necessary——
"P. If, in the first place you will let me know, I will be glad to come down and discuss the matter with you——
"O. Well, I am very glad of that, and we may have other problems which we would like to discuss.
"P. Yes.
"O. But I do not think that there will be any of this nature because really we have * * *.

"P. Well, something may come to your attention relating to this place up here. You may get it down there and I would really prefer to——

"O. There is almost no contact. I have official technical letters but really no personal letters from here. I don't know what's going on, and I think the chance of my being useful in that way is very slight. But you ought to be able to find people here who could have their eyes and ears open and who know what's going on. That would be, I would be, I would be fairly sure that there are quite a few here who would be willing to give you—who would realize the importance of it and—I can't advise you any further.

"P. No; O. K.; as a matter of fact I am not formulating any plans, I am just going to have to digest the whole thing.

"P. Well, we appreciate it and the best of luck.

"O. Thank you very much."

STIPULATED TABLE OF CORRECTIONS OF TRANSCRIPT OF RECORDING OF CONVERSATION BETWEEN DR. J. R. OPPENHEIMER, LT. COL. BORIS T. PASH, AND LT. LYALL JOHNSON OF AUGUST 26, 1953

Original transcript, page 1, line 12 of dialogue:
"O. I was rather uncertain as to whether I should or shouldn't talk to him (Rossi) when I was here."
Should read:
"O. I was rather uncertain as to whether I should or should talk to him [Rossi] when I was here."

* * * * * * *

Original transcript, page 1, line 21:
"O. I think that is the case, but I have no first hand knowledge that would be, for that reason, useful. But I think it is useful, for a man, whose name I never heard, who was attached to the Soviet consul, has indicated indirectly through intermediate people concerned in this project that he was in a position to transmit, without any danger of a leak, or scandal, or anything of that kind, information, which they might supply. Since I know it to be a fact, I have been particularly concerned about any indiscretion which took place in aides close enough to be in contact with it. To put it quite frankly—I would feel friendly to the idea of the Commander in Chief informing the Russians who are working on this problem. At least, I can see that there might be some arguments for doing that, but I don't like the idea of having it moved out the back door. I think that it might not heard to be on the look-out for it."
Should read:
"O. I think that is true, but I have no first hand knowledge that would be, for that reason, useful. But I think it is true that a man, whose name I never heard, who was attached to the Soviet consul, has indicated indirectly through intermediate people concerned in this project that he was in a position to transmit, without any danger of a leak, or scandal, or anything of that kind, information, which they might supply. I would take it that it is to be assumed that a man attached to the Soviet consulate might be doing i, but since I know it to be a fact, I have been particularly concerned about any indiscretions which took place in circles close enough to come in contact with it. To put it quite frankly— I would feel friendly to the idea of the Commander in Chief informing the Russians that we were working on this problem. At least, I can see that there might be some arguments for doing that, but I do not feel friendly to the idea of having it moved out the back door. I think that it might not hurt to be on the look-out for it."

* * * * *

Original transcript, page 2, line 13:
"O. Well, I might say that the approaches were always through other people, who were troubled by them, and sometimes came and discussed them with me; and that the approaches were quite indirect so I feel that to give more, perhaps, than one name, would be to implicate people whose attitudes was one of bewilderment rather than one of cooperation. I know of no case, and I am fairly sure that in all cases where I have heard of these contacts, would not have yielded a single thing. That's as far as I can go on that. Now there is a man, whose name was mentioned to me a couple of times—I don't know of my own knowledge that he was involved as an intermediary. It seems, however, not impossible and if you wanted to watch him it might be the appropriate thing to do. He spent a number of years in the Soviet Union. I think he's a chemical engineer. He was—he may not be here now—at the time I was with him here,

employed by the Shell Development. His name is Eltenton. I would think that there is a small chance—well, let me put it this way—he has probably been asked to do what he can to provide information. Whether he is successful or not, I do not know, but he talked to a friend of his who is also an acquaintance of one of the men on the project, and that was one of the channels by which this thing went. Now I think that to go beyond that would be to put a lot of named down, of people who are not only innocent but whose attitude was 100 percent cooperative."

Should read:

"O. Well, I might say that the approaches were always to other people, who were troubled by them, and sometimes came and discussed them with me; and that the approaches were always quite indirect so I feel that to give more, perhaps, than one name, would be to implicate people whose attitude was one of bewilderment rather than one of cooperation. I know of no case, and I am fairly sure that in all cases where I have heard of these contacts, would not have yielded a single thing. That's as far as I can go on that. Now there is a man, whose name was mentioned to me a couple of times—I don't know of my own knowledge that he was involved as an intermediary. It seems, however, not impossible and if you wanted to watch him it might be the appropriate thing to do. He spent quite a number of years in the Soviet Union. He's an English * * * I think he's a chemical engineer. He was—he may not be here now—at the time I was with him here, employed by the Shell Development. His name is Eltenton. I would think that there was a small chance—well, let me put it this way—he has probably been asked to do what he can to provide information. Whether he is successful or not, I do not know, but he talked to a friend of his who is also an acquaintance of one of the men on the project, and that was one of the channels by which this thing went. Now I think that to go beyond that would be to put a lot of names down, of people who are not only innocent but whose attitude was 100 percent cooperative."

* * * * *

Original transcript, page 2, line 41:

"O. Well, I am giving you the one name that I think is, or isn't—I mean I don't know the name of the man attached to the consulate—I think I may have been told and I may not have been told, and I have, at least not purposely, forgotten. He is—and he might not be there now. These incidents occurred in the order of about 5, 6 or 7 months.

"J. I was wondering, Dr. Oppenheimer, if there is a particular person—maybe a person on the project that you were trying to pump information from—that if we knew who those were, would at least know where to look for a leak, not from the standpoint of fellow hate, but looking at a certain picture.

"P. Here's the point that I would feel——

"O. I would feel that the people that tried to get information from were more or less an accident and I would be making some harm by saying that.

"P. Here's the thing—we of course assume that the people that bring this information to you are 100 percent with you, and therefore, there is no question about their intentions. However, if——

"O. Well, I'll tell you 1 thing—I have known of 2 or 3 cases, and I think 2 of them are with me at Los Alamos—they are men who are closely associated with me."

Should read:

"O. Well, I'm giving you the one name that I think is, or isn't—I mean I don't know the name of the man attached to the consulate—I think I may have been told or I may not have been told and I have, at least not purposely, but actually forgotten. He is—and he may not be here now. These incidents occurred of the order of about 5, 6, 7 months ago.

"J. I was wondering, Dr. Oppenheimer, if there was a particular person—maybe a person on the project that they were trying to pump information from—that if we knew who those were, would at least know where to look for a leak, not from the standpoint of * * * but looking at a certain picture.

"P. Here's the point that I would feel——

"O. I would feel that the people that they tried to get information from were more or less an accident [interpolation] and I believe I would be making some harm by saying that.

"P. Yes. Here's the thing—we of course assume that the people who bring this information to you are 100 percent with you, and therefore, there is no question about their intentions. However, if——

"O. Well, I'll tell you 1 thing—I have known of 2 or 3 cases, and I think of 2 of the men were with me at Los Alamos—they are men who are very closely associated with me."

* * * * *

Original transcript, page 3, line 23:
"O. That is, let me give you the background. The background was—well, you know how difficult it is with the relations between these two allies, and there are a lot of people that don't feel very friendly toward the Russians so that the information—a lot of our secret information, our radar, and so on, doesn't get to them, and they are battling for their lives and they would like to have an idea of what is going on and this is just to make up in other words for the defects of our official communication. That is the form in which it was presented.
"P. Oh, I see.
"O. Of course, the actual fact is that since it is not a communication that ought to be taking place, it is treasonable. But it wasn't presented in that method. It is a method that carrying out a policy which was more or less a policy of the Government and the form in which it came was that could an interview be arranged with this man Eltenton who had very good contacts with a man from the Embassy attached to the consulate who was a very reliable guy and who had a lot of experience with microfilm, that's the story.
"P. Well, now I may be getting back to a little systematic picture * * * these people whom you mentioned, that are down there with you now * * * were they contacted by Eltenton direct?
"O. No.
"P. Through another party.
"O. Yes."
Should read:
"O. That is, let me give you the background. The background was—well, you know how difficult it is with the relations between these two allies, and there are a lot of people who don't feel very friendly to Russia, so that the information—a lot of our secret information, our radar and so on, doesn't get to them, and they are battling for their lives and they would like to have an idea of what is going on and this is just to make up in other words for the defects of our official communication. That is the form in which it was presented.
"P. Oh, I see.
"O. Of course, the actual fact is that since it is not a communication which ought to be taking place, it is treasonable. But it wasn't presented in that method. [Garbled.] It is a method of carrying out a policy which was more or less a policy of the Government and the form in which it came was that an interview be arranged with this man Eltenton who had very good contacts with a man from the embassy attached to the consulate who was a very reliable guy (that's his story) and who had a lot of experience in microfilm work, or whatever the hell.
"P. Well, now I may be getting back to a little systematic picture * * * these people whom you mentioned who [two?] are down there with you now * * * were they contacted by Eltenton direct?
"O. No.
"P. Through another party?
"O. Yes."

* * * * *

Original transcript, page 4, line 16:
"O. It is not definite in the sense that I have seen him do the thing. He may have been misquoted. I don't believe so. Now Eltenton is the member of the FAECT.
"P. That's the union.
"O. That's the CIO union. He's a man whose sympathies are certainly very far left, whatever his affiliations, and he may or may not have regular contacts with a political group. I doubt it. In any case, it is a safe thing to say that the channels to be followed in this case are those involving people who have been generally sympathetic to the Soviet and somehow connected peripherally with Communist movements in this country. That's obvious. I don't need to tell you that."
Should read:
"O. It is not definite in the sense that I have seen him do the thing. He may have been misquoted. I don't believe so. Now, Eltonton is a member of the FAECT or not?

"P. That's the union.

"O. That's the CIO union. He's a man whose sympathies are certainly very far 'left' whatever his affiliations, and he may or may not have regular contacts with a political group.

"P. Well, here's how I feel.

"O. I doubt it. In any case, it is a safe thing to say that the channels that would be followed in this case are those involving people who have been generally sympathetic to the Soviet and somehow connected peripherally with Communist movements in this country. That's obvious. I don't need to tell you that."

* * * * * * *

Original transcript, page 5, line 13:

"O. I am pretty sure that none of the guys here, with possible exception of the Russian, who is probably doing his duty by his country—but the other guys really were just feeling they didn't do anything, but they were considering the step, which they would have regarded as thoroughly in line with the policy of this Government, just making up for the fact that there were a couple of guys in the State Department who would block such communications. You may or may not know that in many projects we share information with the British and some we do not, and there is a great deal of feeling about that and I don't think the issues involved here seemed to the people very different, except that of course the people on the project realize the importance and that this is a little bigger and the whole procedure gets away from them."

Should read:

"O. I'm pretty sure that none of the guys here, with the possible exception of the Russian, who is doing probably his duty by his country—but the other guys, really were just feeling they didn't do anything but they were considering the step, which they would have regarded as thoroughly in line with the policy of this Government, just making up for the fact that there were a couple of guys in the State Department who might block such communications. You may or may not know that in many projects we share information with the British and some we do not, and there is a great deal of feeling about that, and I don't think that the isues involved here seem to the people very different, except that of course the people on the project realize the importance and that is a little bigger and the whole procedure gets away [garbled]."

* * * * * * *

Original transcript, page 6, line 7.

"P. I get your point. I don't want to seem to you insistent. I want to again sort of explore the possibility of getting the name of the person of the faculty—I'll tell you for what reason. Not for the purpose of taking him to task in any way whether its nonofficially, officially, or openly or what but to try to see Eltenton's method of approach. You may not agree with me, but I can assure you that this is one of the more important steps.

"O. I have to take the following points of view: I think in mentioning Eltenton's name I subsequently said about the man that I think that he may be acting in a way which is dangerous to this country and which should be watched. I am not going to mention the name of anyone in the same breath, even if you say that you will make a distinction. I just can't do that, because in the other cases, I am convinced from the way in which they handled the thing that they themselves thought it was a bad business."

Should read:

"P. I get your point. I don't want to seem to you insistent. I want to again sort of explore the possibility of getting the name of the person of the faculty—I'll tell you for what reason. Not for the purpose of taking him to task in any way whether it's nonofficially, officially, or openly or not but to try to see Eltenton's method of approach. You may not agree with me, but I can assure you that is one of the most important steps.

"O. I understand that, but I have to take the following points of view: I think in mentioning Eltenton's name I essentially said about the man that I think he may be acting in a way which is dangerous to this country, and which should be watched. I'm not going to mention the name of anyone in the same breath, even if you say that you will make a distinction. I just can't do that, because in the other cases, I am convinced from the way in which they handled the thing that they themselves thought it was a bad business."

* * * * * * *

Original transcript, page 6, line 33:
"P. Were these two people you mentioned—were they contacted at the same time?
"O. Oh, no. They were contacted within a week of each other."
Should read:
"P. Were these two people you mentioned—were they contacted at the same time?
"O. They were contacted within a week of each other."

*   *   *   *   *   *   *

Original transcript, page 7, line 12:
"O. I really think that I am drawing a line in the right place."
Should read:
"O. I really think that I am drawing [garbled]."

*   *   *   *   *   *   *

Original transcript, page 7, line 18:
"O. It is also my duty not to implicate these people, acquaintances, or colleagues of whose position I am absolutely certain—myself and my duty is to protect them.
"P. O yes.
"O. If I thought that—I won't say it—it might be slightly off.
"P. Well then, here's another point, Doctor, if we find that in making these various contacts, that we get some information which would lead us to believe that certain of these men may have either considered it or are still considering it (mind you, I do not even know these men, so it can't be personal)——
"O. Well, none of them that I had anything to do with even considered it. They were upset about it. They have a feeling toward this country and have signed the Espionage Act; they feel this way about it for I think that the intermediary between Eltenton and the project, thought it was the wrong idea, but said that this was the situation. I don't think he supported it. In fact I know it."
Should read:
"O. It is also my duty not to implicate these people, who are acquaintances, or colleagues, and so on of whose position I am absolutely certain—myself and my duty is to protect them.
"P. Oh, yes.
"O. If I thought that—I won't say it—it might be slightly off.
"P. Well then, here's another point, Doctor, if we find that in making these various contacts, that we get some information which would lead us to believe that certain of these men may have either considered it or are still considering it (mind you, I do not even know these men, so it can't be personal)——
"O. Well, none of them that I had anything to do with considered it. They were just upset about it * * * [garbled]. They have a feeling toward this country and have signed the Espionage Act; they feel this way about it for I think that the intermediary between Eltenton and the project, thought it was the wrong idea, but said that this was the situation. I don't think he supported it. I fact, I know it."

*   *   *   *   *   *   *

Original transcript, page 8, line 1:
"P. And two of these contacts are down there. That means we can assume at least there is one of these men contacted still on the project here.
"O. Yes, I believe that this man has gone, or is scheduled to go to site X.
"P. This third man?
"O. That is right."
Should read:
"P. And two of these contacts are down there. That means we can assume at least there is one of these men contacted still on the project?
"O. Yes, I believe this man has gone, or is scheduled to go to site X.
"P. This third man?
"O. I think so.

*   *   *   *   *   *   *

Oroginal transcript, page 8, line 29:
"P. This man Eltenton * * * is a scientist?
"O. I don't know, I would guess he is some sort of a chemical engineer.
"P. Would he be in a position to understand the information furnished him?
"O. I don't know that either. It would depend on how well it was furnished. I mean, he has some scientific training and certainly if you sat down with him

and took a little time. My view about this whole damn thing, of course, is that the information that we are working on is probably known to all the governments that care to find out. The information about what we are doing is probably of no use because it is so damn complicated. I don't agree that the security problem on this project is a bitter one, because. if one means by the security problem preventing information of technical use to another country from escaping. But I do think that the intensity of our effort and our concern of the international investment involved—that is information which might alter the course of the other governments and don't think it would have any effect on Russia * * * it might have a very big effect on Germany, and I am convinced about that and that is as everyone else is.

"P. Oh.

"O. To give it roughly, what we're after and I think they don't need to know the technical details because if they were going to do it they would do it in a different way. They wouldn't take our methods—they couldn't because of certain geographical differences so I think the kind of thing that would do the greatest damage if it got out, would just be the magnitude of the problem and of the time schedules which we think we have of that kind.

"P. To answer your question—Eltenton if you were picking a man which would be an intermediary wouldn't be a bad choice, I would mention he had some kind of chemical engineering job in Russia. He was trained in England, also in Russia 4 or 5 years and things like that and here."

Should read:

"P. This man Eltenton * * * is a scientist?

"O. I don't know, I would guess he was a sort of a chemical engineer.

"P. Would he be in a position to understand the information furnished him?

"O. I don't know that either. It would depend on how well it was furnished. I mean, he has some scientific training and certainly if you sat down with him and took a little time. My view about this whole damn thing, of course, is that the information that we are working on is probbably known to all the governments that care to find out. The information about what we are doing is probably of no use because it is so damn complicated. I don't agree that the security problem of this project is a bitter one, because if one means by the security problem preventing information of technical use to another country from escaping. But I do think that the intensity of our effort and our concern with national investment involved—that is information which might alter the course of the other governments and don't think it would have any effect on Russia * * * it might have a very big effect on Germany, and I am convinced about that and that it as everyone else is.

"P. Oh.

"O. To give it roughly what we're after and I think they don't need to know the technical details because if they were going to do it they would do it in a different way. They wouldn't take our methods—they couldn't because of certain geographical differences so I think the kind of thing that would do the greatest damage if it got out, would just be the magnitude of the problem and of the time schedules which we think we have and that kind of thing. To answer your question—Eltenton, if you were picking a man which would be an intermediary, he wouldn't be a bad choice, I would mention he had some kind of chemical engineering job in Russia. He was trained in England, was in Russia 4 or 5 years, and things like that and here."

*   *   *   *   *   *   *

Original transcript, page 9, line 35:
"J. You see a lot of people have put it to us."
Should read:
"J. You see a lot of people have reported it to us."

*   *   *   *   *   *   *

Page 10, omission before line 1:
"P. You did.
"O. Yes."

*   *   *   *   *   *   *

Original transcript, page 10, line 6:
"O. I would like to say that if I think that if there are certain affiliations that were incompatible to the best interests of the country and this business would retard."

Should read:

"O. I would like to say that if I think that certain affiliations that are incompatible to the interests of this country and this business would die * * * "

* * * * * * *

Original transcript, page 10, line 14:
"O. Oh, certainly, it is perfectly obvious that certain precautions"——
Should read:
"O. Oh, certainly * * * certain precautions"——

* * * * * * *

Original transcript, page 10, line 21:
"O. Well, that won't. I most fervently hope that they are not in any way and if I had reason to believe that some technical men were involved I would certainly tell you and I will if anything comes up that I am convinced I can always say that I know everything is absolutely 100 percent negative."
Should read:
"O. Well, I hope that won't * * * if I had reason to believe * * * I will if anything ever comes up that I am convinced—I can always say that everything I know is absolutely 100 percent negative."

* * * * * * *

Original transcript, page 10, line 26:
"P. If we should find any information which would lead you to believe that there may be some of that going on, and that it would be important for us to know a little more in detail who the contacts were and everything and we could show you that that is important for us, I hope you then find it possible to"——
Should read:
"P. If we should find any information which would lead us to believe that there may still be some of that going on, and if it would be important for us to then know a little more in detail who the contacts were and everything and we could show you that that is important to us, I hope you will then find it possible to"——

* * * * * * *

Original transcript, page 10, line 33:
"O. As I say I am trying to draw the line here between people who took some responsibility and the people who were purely pushed around and since nothing occurred and the responses seemed to have been 100 percent negative, I think I am perhaps justified in—in"——
Should read:
"O. As I say I am trying to draw the line here between people who took some responsibility and the people who were purely pushed around and since nothing occurred and the responses seem to have been 100 percent negative, I think I am perhaps justified in—in"——

* * * * * * *

Original transcript, page 11, line 3:
"P. That is, there is one in there, that you say that the responses were 100 percent negative. Do you feel that you know everyone whom this intermediary contacted?"
Should read:
"P. That is, there is one point in there, that you say that the responses were 100 percent negative. Do you feel that you know everyone whom this intermediary contacted?"

* * * * * * *

Original transcript, page 11, line 10:
"O. Well, I think it would be creditable to say that I just don't know."
Should read:
"O. Well, I think it would be [one word missing] to say that I just don't know."

* * * * * * *

Original transcript, page 11, line 11:
"P. I would like to leave this thought with you, Dr. Oppenheimer, if you at some time find it possible I certainly would give a lot of thanks and appreciation for the name of that intermediary and I am going to explain to you—I tell you—if it is going to—the only reason I would want it, is not for his sake but to see who his contacts are"——
Should read:
"P. I would like to leave this thought with you, Dr. Oppenheimer, if you at some time find it possible we certainly would give a lot of thanks and appreciation for the name of that intermediary because it's going to—I tell you—the

only reason why I would want it, is not for his sake but to see who his contacts are"——

* * * * * * *

Original transcript, page 11, line 20:
"P. In trying to run him down before we even go on this"——
Should read:
"P. In trying to run him down before we even can get on to these others"——

* * * * * * *

Original transcript, page 11, line 21:
"O. You'd better check up on the consulate because that's the only one that Eltenton contacted and without that contact he would be inefficient and that would be my"——
Should read:
"O. You'd better check up on the consulate because that is the only one that Eltenton contacted and without that contact there wouldn't be anything. * * *"

* * * * *

Original transcript, page 11, line 27:
"O. I have never been introduced to him.
"P. Have you ever heard his name mentioned?
"O. I have never heard his name mentioned, but I have been given to understand that he is attached to the consulate. * * *
"O. Maybe this guy is a military attache—I don't know."
Should read:
"O. I have never been introduced to him * * * or heard his name or anything, but I have been given to understand that he is attached to the consulate. * * *
"O. Maybe this guy is a military attache—I don't know."

* * * * * *

Original transcript, page 12, line 8:
"O. I think that Eltenton must—I said to him 'what can you do about it?' I don't know—that would be my impression of the thing."
Should read:
"O. I think that Eltenton must have said to him * * * I don't know—that would be my impression of the thing. * * *"

* * * * *

Original transcript, page 12, line 11:
"O. Well, I am sorry, I realize you would like more information but I am under a little bit of difficulty deciding what to do about it. The fact that I did not raise this question for a long time——
"P. That's right.
"O. I have been in difficulty about what to do realizing how serious it is. I think my general point of view is that there are some things there which would bear watching.
"P. That's right.
"O. It is doubtful to me if there is anything there which can't be uncovered.
"P. Well, that—I can see where it would be highly difficult to find out what's going on. We will be hot under the collar until we find out what is going on there. I mean—that's the point we have to take——
"O. Well I don't know what a job like this—well I would think it's conceivable—that it wouldn't hurt to have a man in the local of this union, FAECT—to see what may happen and what he can pick up.
"P. You feel there could be something in the organization itself?
"O. Within it."
Should read:
"Well, I am sorry, I realize that you would like more information but I have been under a little bit of difficulty. The fact that I did not raise this [one word omitted] for a long time——
"P. That's right.
"O. I have difficulty in * * * serious * * * what to do * * * I think my general point of view is that there are some things there would would bear watching.
"P. That's right.
"O. It is doubtful to me if there is anything there which can't be uncovered.
"P. Well that—I can see where * * * we will be hot under the collar until we find out what is going on there. I mean—that is the point of views we have to take.

"O. Well, I don't know * * * well I would think * * * that it's conceivable—that it wouldn't hurt to have a man in the local of this Union, FAECT—to see what may happen and what he can pick up.

"P. You feel there could be something not in the organization itself but some——

"O. Within it."

* * * * * * *

Original transcript, page 13, line 5:
"P. Does this union that is up on the hill, is it not connected with the hill at all?

"O. Oh, yes. It is an international union and has representatives all over this country."

Should read:
"P. Does this union that is up on the hill, do they have members which are not connected with the hill at all?

"O. Oh, yes, they have an international union and has representatives all over this country."

* * * * * * *

Original transcript, page 13, line 11:
"O. Oh, I imagine so. I don't know, I don't know about that."

Should read:
"J. Oh, I imagine so.
"O. I don't know. I don't know about that."

* * * * * * *

Original transcript, page 13, line 33:
"P. Well, ah——
"O. It's a very different situation, a very much harder situation. I don't know the people but it is a hard situation and in particular to put together in a casual way and I think that the problem of being sure that there are no leaks there is a real problem and that pressure can be brought with discretion."

Should read:
"P. Well, ah——
"O. I don't say that about this place. It's a very different situation, a very much harder situation. I don't know the people, but it is a hard situation; in particular was put together in a casual way and I think that the problem of being sure that there are no leaks * * * and that pressure can be brought with discretion."

* * * * * * *

Original transcript, page 14, line 12:
"P. That's right. Well, if it does become serious I may come down with some of my persistency—I mean I would hate to—I have a responsibility of running things down there.

"O. I also think the particular way this way—that if there is anything going on it would be very easy to find out. I am not worried about that—we can take care of that ourselves.

"P. No you wouldn't.

"O. Well, I meant in a way which you think best.

"P. But it is a situation which would have to be handled very delicately. That's what makes it so difficult. If it is something that's easy to handle and you don't have to worry about it, why you just sort of bull your way through, but these things have to be, one has to be careful."

Should read:
"P. That's right. Well, if it does become serious I may come down with some of my persistency—I mean I would hate to—I have a responsibility of running things down myself.

"O. I also think the particular way this was—that if there is anything going on it would be very easy to find out. I am not worried about that—we can take care of that ourselves.

"P. No you wouldn't.

"O. Well, I can handle it in a way——

"P. But it is a situation which would have to be handled very delicately. That's what makes it so difficult. If it is something that's easy to handle and you don't have to worry about it, why you just sort of bull your way through, but these things, one has to be very careful."

* * * * * * *

Original transcript, page 14, line 29:

"O. I don't actually know whether, if you were in Washington—after advice on the question how far should cooperation go. I don't know wherein the right answer lies. I have heard of cases with very strong arguments on both sides.
"P. Yes.
"O. That's a particular we don't have to worry about, but in Washington there are more ticklish situations"——
Should read:
"O. I don't actually know whether, if you were in Washington—asking advice on the question how far should cooperation go. I don't know wherein the right answer lies. I have heard of cases with very strong arguments on both sides.
"P. Yes.
"O. * * * we don't have to worry about * * *"

* * * * * * *

Original transcript, page 14:
Omission after last line (after words "more ticklish situations").
Should read:
Insert (after words; "to worry about * * *"):
"P. Yes, that's right.
"O. Well, I wish good luck.
"P. We could work a hundred years (I mean) and never get this information. That's where we start you see—I mean we get this information and we have something to start on—we have something to run down. I certainly appreciate this opportunity to visit you.
"O. I hope it's not a waste of time.
"P. Well, I know it's not a waste of my time and ah——
"O. That's all—I meant—perhaps as far as the project is concerned * * * a fair starting point——
"P. Could have——
"O. Why not take an about face?
"P. Do you——
"O. And one could do anything about the attache—that would be the natural thing to watch."

* * * * * * *

Original transcript, page 15, line 1:
"P. Do you know anyone—and because we like to eliminate unnecessary work if we have to—do you know anyone on the project who is connected with the FAECT. Dr. Oppenheimer?
"O. Who would be willing to——
"P. That's right.
"O. I don't know who is in the union at all. I have heard that a boy called Fox is president of it.
"J. David Fox.
"O. David Fox, but I would feel that that boy could do the trick. I hope that the trade union isn't tied up in this—and they would not act like this because I think it would give them a very black eye and it is no love of mine from the start, and it might have consequences beyond the reasonable. I doubt whether anyone mixed with the union in good faith would be very sympathetic."
Should read:
"P. Do you know anyone—and because we like to eliminate unnecessary work if we have to—do you know anyone on the project who is connected with the FAECT, Dr. Oppenheimer?
"O. Who would be willing to——
"P. That's right.
"O. I don't know who is in the union at all. I have heard that a boy called Fox is president of it.
"J. David Fox.
"O. David Fox, but I would feel—I hope that the trade union isn't tied up in this—and they would not act like this because I think it would give them a very black eye and it is no love of mine from the start, and it might have consequences beyond the reasonable. I doubt whether anyone mixed with the union in good faith would be very sympathetic."

* * * * * * *

Original transcript, page 15, line 27:
"O. No, I would not have raised the question if it didn't seem to me that it deserved looking into——
"P. Yes.

"O. And if I seem uncooperative I think that you can understand that it is because of my insistence in not getting people into trouble——
"P. I can assure you that if anything comes to the attention out there——
"O. Now, wait a minute, let me dispose of that statement which came over the long-distance phone. When I first talked with this boy I extracted from him a promise to stop all this kind of thing when he came on the job. Of course, I can't hold him to that promise * * * [unintelligible].
"P. Well, what I mean, if anything does come to your attention in connection with this phase if you can"——
Should read:
"O. No; I would not have raised the question if it had not seemed to me that it deserved looking into.
"P. Yes.
"O. And if I seem uncooperative I think that you can understand that it is because of my insistence on not getting people into trouble.
"P. I can assure you that if something comes to your attention out there——
"O. Let me dispose of that statement which came over the long-distance phone. Frankly, I got—I extracted from that boy a promise to stop all this sort of thing when he came on the job * * * to that promise * * * I did not know what he was doing it for but I thought that there was a possibility * * * he said he understood that * * * I talked to him yesterday. He said he had no connection * * *.
"P. Well, what I mean, if anything does come to your attention in connection with this phase if you can"——

*   *   *   *   *   *   *

Original transcript, page 16, line 10:
"O. But I do not think there will be any of this nature because really we have very little incentive"——
Should read:
"O. But I do not think there will be any of this nature because really we have * * *

*   *   *   *   *   *   *

(The original typewritten transcript as prepared in the office of Colonel Pash in 1943 is as follows:)

"SAN FRANCISCO, CALIF., *August 27, 1943.*

"MEMORANDUM FOR THE OFFICER IN CHARGE

"Subject: D. S. M. Project.
"Re: Transcript of conversation between Dr. J. R. Oppenheimer, Lt. Col. Boris T. Pash, and Lt. Lyall Johnson.
"Transmitted herewith is the transcript of conversation between Dr. J. R. Oppenheimer, Lt. Col. Boris T. Pash, and Lt. Lyall Johnson held in Lieutenant Johnson's office in the New Class Room Building, University of California, Berkeley, Calif., on August 26, 1943. It is to be noted that in some places the conversation was very indistinct and that the running commentary may be indecisive in these places, but the substance of the material discussed is herewith presented:
"P. This is a pleasure, because I am interested to a certain extent in activities and I feel I have a certain responsibility in a child which I don't know anything about. General Groves has, more or less, I feel, placed a certain responsibility in me and it's like having a child, that you can't see, by remote control. I don't mean to take much of your time——
"O. That's perfectly all right. Whatever time you choose.
"P. Mr. Johnson told me about the little incident, or conversation, taking place yesterday in which I am very much interested and it had me worried all day yesterday since he called me.
"O. I was rather uncertain as to whether I should or should not talk to him [Rossi] when I was here. I was unwilling to do it without authorization. What I wanted to tell this fellow was that he had been indiscreet. I know that that's right that he had revealed information. I know that saying that much might in some cases embarrass him. It doesn't seem to have been capable of embarrassing him—to put it bluntly.
"P. Well, that is not the particular interest I have. It is something a little more, in my opinion, more serious. Mr. Johnson said there was a possibility that there may be some other groups interested.

"O. I think that is the case, but I have no first-hand knowledge that would be, for that reason, useful, but I think it is true that a man, whose name I never heard, who was attached to the Soviet Consul, has indicated indirectly through intermediate people concerned in this project that he was in a position to transmit, without any danger of a leak, or scandal, or anything of that kind, information, which they might supply. Since I know it to be a fact, I have been particularly concerned about any indiscretions which took place in aides close enough to be in contact with it. To put it quite frankly—I would feel friendly to the idea of the Commander in Chief informing the Russians who are working on this problem. At least, I can see that there might be some arguments for doing that, but I don't like the idea of having it moved out the back door. I think that it might not hurt to be on the lookout for it.

"P. Could you give me a little more specific information as to exactly what information you have. You can readily realize that phase would be, to me, as interesting, pretty near, as the whole project is to you.

"O. Well, I might say that the approaches were always through other people, who were troubled by them, and sometimes came and discussed them with me; and that the approaches were quite indirect so I feel that to give more, perhaps, than one name, would be to implicate people whose attitude was one of bewilderment rather than one of cooperation. I know of no case, and I am fairly sure that in all cases where I have heard of these contacts, would not have yielded a single thing. That's as far as I can go on that. Now there is a man, whose name was mentioned to me a couple of times—I don't know of my own knowledge that he was involved as an intermediary. It seems, however, not impossible, and if you wanted to watch him it might be the appropriate thing to do. He spent a number of years in the Soviet Union. I think he's a chemical engineer. He was—he may not be here now—at the time I was with him here, employed by the Shell Development. His name is Eltenton. I would think that there is a small chance—well, let me put it this way—he has probably been asked to do what he can to provide information. Whether he is successful or not, I do not know, but he talked to a friend of his who is also an acquaintance of one of the men on the project, and that was one of the channels by which this thing went. Now I think that to go beyond that would be to put a lot of names down, of people who are not only innocent but whose attitude was 100 percent cooperative.

"P. Now here's a point. You can readily realize that if we get information like that we have to work in an absolutely discreet manner. In other words we can't afford to even indicate——

"O. That you are concerned.

"P. That we are concerned or through whom we get information. However, anything that we may get which would eliminate a lot of research work on our part would necessarily bring to a closer conclusion anything that we are doing.

"O. Well, I'm giving you the one name that I think is, or isn't—I mean I don't know the name of the man attached to the Consulate—I think I may have been told and I may not have been told and I have, at least not purposely, but actually forgotten. He is—and he may not be here now. These incidents occurred in the order of about 5, 6, or 7 months.

"J. I was wondering, Dr. Oppenheimer, if there is a particular person—maybe a person on the project that you were trying to pump information from—that if we knew who those were, would at least know where to look for a lead, not from the standpoint of fellow hate, but looking at a certain picture.

"P. Here's the point that I would feel——

"O. I would feel that the people they tried to get information from were more or less an accident and I would be making some harm by saying that.

"P. Here's the thing—we of course assume that the people who bring this information to you are 100 percent with you, and therefore, there is no question about their intentions. However, if——

"O. Well, I'll tell you one thing—I have known of two or three cases, and I think two of them are with me at Los Alamos—they are men who are closely associated with me.

"P. Have they told you that either they thought that they were contacted for that purpose or they were actually contacted for that purpose?

"O. They told me they were contacted for that purpose.

"P. For that purpose.

"O. That is, let me give you the background. The background was—well you know how difficult it is with the relations between these two allies, and there are a lot of people that don't feel very friendly toward the Russians, so that the information—a lot of our secret information, our radar and so on, doesn't get to them, and they are battling for their lives and they would like to have an idea of what is going on and this is just to make up in other words for the defects of our official communication. That is the form in which it was presented.

"P. Oh, I see.

"O. Of course, the actual fact is that since it is not a communication that ought to be taking place, it is treasonable. But it wasn't presented in that method. It is a method that carrying out a policy which was more or less a policy of the Government and the form in which it came was that could an interview be arranged with this man Eltenton who had very good contacts with a man from the Embassy attached to the consulate who was avery reliable guy and who had a lot of experience with microfilm, that's the story.

"P. Well, now I may be getting back to a little systematic picture * * * These people whom you mentioned, two are down there with you now. Were they contacted by Eltenton direct?

"O. No.

"P. Through another party?

"O. Yes.

"P. Well now, could we know through whom that contact was made?

"O. I think it would be a mistake, that is, I think I have told you where the initiative came from and that the other things were almost purely accident and that it would involve people who ought not be involved in this.

"P. This would not involve the people but it would indicate to us Eltenton's channel. We would have to, now that this is definite on Eltenton.

"O. It is not definite in the sense that I have seen him do the thing. He may have been misquoted. I don't believe so. Now Eltenton is the member of the FAECT.

"P. That's the union.

"O. That's the CIO union. He's a man whose sympathies are certainly very far left, whatever his affiliations, and he may or may not have regular contacts with a political group. I doubt it. In any case, it is a safe thing to say that the channels to be followed in this case are those involving people who have been generally sympathetic to the Soviet and somehow connected peripherally with Communist movements in this country. That's obvious. I don't need to tell you that.

"P. Well, yes. The fact is, this second contact—the contact that Eltenton had to make with these other people—is that person also a member of the project?

"O. No.

"P. That also is an outsider.

"O. It's a member of the faculty, but not on the project.

"P. A member of the faculty here? Eltenton made it through a member of the faculty to the project.

"O. As far as I know—these approaches were—there may have been more than one person involved. I don't know.

"P. Here's how I feel about this leftist inclination. I think that whether a man has left or right inclinations, if his character which is back of it—if he's willing to do this, it doesn't make any difference what his inclinations are. It's based on his character primarily and not——

"O. A thing like this going on, let us say, with the Nazis would have a somewhat different color. I don't mean to say that it would be any more deserving of attention, or any more dangerous, but it would involve rather different motives.

"P. Oh, yes; sure.

"O. I'm pretty sure that none of the guys here, with possible exception of the Russian, who is doing probably his duty by his country—but the other guys really were just feeling they didn't do anything but they were considering the step, which they would have regarded as thoroughly in line with the policy of this Government, just making up for the fact that there were a couple of guys in the State Department who would block such communications. You may or may not know that in many projects we share information with the British and some we do not, and there is a great deal of feeling about that, and I don't think that the issues involved here seem to the people very different, except that, of course, the people on the project realize the importance and that this is a little bigger and the whole procedure gets away from them.

"P. Now, do you feel that would affect—and there could be continued attempts now to establish this type of contact?

"O. I haven't any idea.

"P. You haven't any idea?

"O. As I say, if the guy that was here may by now be in some other town and all that I would have in mind is this—I understood that this man to whom I feel a sense of responsibility, Lomanitz, and I feel it for two reasons. One, he is doing work which he started and which he ought to continue, and second, since I more or less made a stir about it when the question of his induction came up, that this man may have been indiscreet in circles which would lead to trouble. That is the only thing that I have to say. Because I don't have any doubt that people often approached him, with whom he has contacted, I mean whom he sees, might feel it their duty if they got word of something, to let it go further and that is the reason I feel quite strongly that association with the Communist movement is not compatible with the job on a secret war project, it is just that the two loyalties cannot go.

"P. Yes—well——

"O. That is not an expression of political opinion. I think that a lot of very brilliant and thoughtful people have seen something in the Communist movement, and that they may belong there, maybe it is a good thing for the country. They hope that it doesn't belong on the war project.

"P. I get your point. I don't want to seem to you insistent. I want to again sort of explore the possibility of getting the name of the person of the faculty— I'll tell you for what reason. Not for the purpose of taking him to task in any way, whether it's nonofficially, officially, or openly, or what, but to try to see Eltenton's method of approach. You may not agree with me, but I can assure you that that is one of the more important steps.

"O. I have to take the following points of view: I think in mentioning Eltenton's name I subsequently said about the man that I think that he may be acting in a way which is dangerous to this country, and which should be watched. I'm not going to mention the name of anyone in the same breath, even if you say that you will make a distinction. I just can't do that, because in the other cases I am convinced from the way in which they handled the thing that they themselves thought it was a bad business.

"P. These other people, yes, I realize—but if—here's the point—if that man is trying to make other contacts for Eltonton, it would take us some time to try to——

"O. My honest opinion is that he probably isn't—that he ran into him at a party and they saw each other or something and Eltenton said, 'Do you suppose you could help me? This is a very serious thing because we know that important work is going on here, and we think this ought to be made available to our allies, and would you see if any of those guys are willing to help us with it—and then it wouldn't have to be much.' You see, that is the kind of thing [remaining statement unintelligible].

"P. Were these two people you mentioned—were they contacted at the same time?

"O. Oh, no. They were contacted within a week of each other.

"P. They were contacted at two different times?

"O. Yes; but not in each other's presence.

"P. That's right.

"And then from what you first heard, there is someone else who probably still remains here who was contacted as well.

"O. I think that is true.

"P. What I am driving at is that there was a plan, at least for some length of time, to make these contacts—and we may not have known all the contacts.

"O. That is certainly true. That is why I mentioned it. If we knew all about it, then I would say forget it. I thought it would be appropriate to call to your attention the fact that these channels at one time existed.

"P. Yes.

"O. I really think that I am drawing a line in the right place.

"P. You see, you understand that I am sort of—you picture me as a bloodhound on the trail, and that I am trying to get out of you everything I possibly can.

"O. That's your duty to a certain extent.

"P. You see what I mean.

"O. It is also my duty not to implicate these people, acquaintances, or colleagues of whose position I am absolutely certain—myself and my duty is to protect them.

"P. Oh, yes.

"O. If I thought that—I won't say it—it might be slightly off.

"P. Well then, here's another point, Doctor, if we find that in making these various contacts, that we get some information which would lead us to believe that certain of these men may have either considered it or are still considering it (mind you, I do not even know these men, so it can't be personal)——

"O. Well, none of them that I had anything to do with even considered it. They were upset about it. They have a feeling toward this country and have signed the espionage act; they feel this way about it for I think that the intermediary between Eltenton and the project, thought it was the wrong idea, but said that this was the situation. I don't think he supported it. In fact I know it.

"P. He made about at least three contacts that we know of.

"O. Well, I think that's right, yes.

"P. And two of these contacts are down there. That means we can assume at least there is one of these men contacted still on the project here.

"O. Yes, I believe that this man has gone, or is scheduled to go to site X.

"P. This third man?

"O. That is right.

"P. Well that is, as I say, if I can't get across that line, I even certainly appreciate this much, because it——

"O. I think it's a thing you ought to know——

"P. Oh, no doubt——

"O. I think it's probably one of those sporadic things and I do not think—I have no way of thinking it was systematic, but I got from the way it was handled, which was rather loosely, and frankly if I were an agent I would not put much confidence in people who are loose-mouthed or casual. I would not think that this was a very highly organized or very well put-together plan but I don't know and I was very much afraid when I heard of Lomanitz's indiscretion that it might very well be serious. I hope that isn't the case.

"P. You mentioned that this man may be of this FAECT. Do you think that, as a representative of this organization, he would sort of represent their attitude or do you think he is doing this individually?

"O. Oh, the FAECT is quite a big union and has all sorts of people in it. I'm pretty sure and I don't think it is conceivable that he could be representing the attitude of the union, but it is——

"P. Well, I don't know enough about it to——

"O. I think that at one time they had a strong branch up at the Shell Development Research Laboratories, the FAECT—and I believe it is the union which has got organized on the hill.

"J. Yes; it has been around for some time.

"P. This man Eltenton * * * is a scientist?

"O. I don't know, I would guess he is some sort of the chemical engineer.

"P. Would he be in a position to understand the information furnished him?

"O. I don't know that either. It would depend on how well it was furnished. I mean he has some scientific training and certainly if you sat down with him and took a little time. My view about this whole damn thing, of course, is that the information that we are working on is probably known to all the governments that care to find out. The information about what we are doing is probably of no use because it is so damn complicated. I don't agree that the security problem on the project is a bitter one, because if one means by the security problem preventing information of technical use to another country from escaping. But I do think that the intensity of our effort and our concern of the international investment involved—that is information which might alter the course of the other governments and don't think it would have any effect on Russia * * * it might have a very big effect on Germany, and I am convinced about that and that is as everyone else is.

"P. Oh.

"O. To give it roughly what we're after and I think they don't need to know the technical details because if they were going to do it they would do it in a different way. They wouldn't take our methods—they couldn't because of certain geographical differences so that I think the kind of thing that would do the greatest damage if it got out, would just be the magnitude of the problem and of the time schedules which we think we have of that kind.

"P. To answer your question—Eltenton, if you were picking a man which would be an intermediary he wouldn't be a bad choice, I would mention he had some kind of chemical engineering job in Russia. He was trained in England, also in Russia 4 or 5 years and things like that and here——

"P. Does he speak Russian, do you know?

"O. I don't know—I don't know. Speaks with a slight English accent.

"P. If it is necessary would you mind and would it interfere with your work much if I would have to come down and discuss this with you further. Counter-assurance—I mean this is—ah——

"O. This is important?

"P. Oh, yes; I not only——

"O. If I can express my own opinion as well as my conviction this is not common knowledge.

"P. No; it isn't.

"O. You see a lot of people have put it to us.

"P. That's why when Mr. Johnson called me up yesterday it sort of——

"O. Yes. I mentioned this to Colonel Lansdale.

"P. Aha; well, of course, right now I say—ah—it is all new and—it has come to me.

"O. Right now it means absolutely nothing but what you now find out at this——

"P. If—but——

"O. I would like to say that if I think that if there are certain affiliations that were incompatible to the best interests of this country and this business would retard.

"P. It may be necessary for us to—to take certain steps in trying to trace this down and so forth—if anything would develop where we would have to or be interested in either your place down at Los Alamos or other places, you feel it would be all right for me to contact you on it so that——

"O. Oh, certainly, it is perfectly obvious that certain precautions——

"P. Oh, yeah, yes—what I mean is instead of going on certain steps which may——

"O. Yeah——

"P. Come to your attention and be a little bit disturbing to you, I would rather discuss those with you first so that you will be aware of it. I think that, that—well that——

"O. Well, that won't. I most fervently hope they are not in any way and if I had reason to believe that some technical men were involved I would certainly tell you and I will if anything comes up that I am convinced I can always say that I know everything is absolutely 100 percent negative.

"P. If we should find any information which would lead you to believe that there still may be some of that going on, and that it would be important for us to know a little more in detail who the contacts were and everything and we could show you that that is important for us, I hope you will then find it possible to——

"O. I am only trying to define our future and I will try to act reasonably.

"P. Fine.

"O. As I say I am trying to draw the line here between people who took some responsibility and the people who were purely pushed around and since nothing occurred and the responses seemed to have been 100-percent negative, I think I am perhaps justified in—in——

"P. I am not persistent (ha ha) but——

"O. You are persistent and it is your duty.

"P. That is, there is one in there, that you say that the responses were 100-percent negative. Do you feel that you know everyone whom this intermediary contacted?

"O. Well, no, but I think it is practical to say that it is inconceivable that the people whom he contacts would be—would have come to my attention but I am not sure.

"P. Well, I would like to say.

"O. Well, I think it would be creditable to say that I just don't know.

"P. I would like to leave this thought with you, Dr. Oppenheimer, if you at some time find it possible, we certainly would give a lot of thanks and appreciation for the name of that intermediary and I'm going to explain to you—I tell you—if it is going to—The only reason I would want it, is not for his sake but to see who his contacts are——

"O. Yes, I see.

"P. I can see that we are going to have to spend a lot of time and effort which we ordinarily would not in try to——

"O. Well——

"P. In trying to run him down before we even go on this——

"O. You'd better check up on the consulate because that's the only one that Eltenton contacted and without that contact he would be inefficient and that would be my——

"P. You say this man is not employed in the consulate?

"O. Eltenton?

"P. No, no, I mean this man——
"O. I have never been introduced to him.
"P. Have you ever heard his name mentioned?
"O. I have never heard his name mentioned, but I have been given to understand that he is attached to the consulate.

* * * * * * *

"O. Maybe this guy is a military attaché—I don't know.
"P. You don't know anything about him?
"O. I don't know anything about him and never have. I may have been told the name, but it made no impression.
"P. Is this member of the faculty in any way—does he in any way come in contact with your project? Why would he be contacted? Is it because he has contacted these people?
"O. I think that Eltenton must—I said to him 'What can you do about it?' I don't know—that would be my impression of the thing.
"P. Well, I think that——
"O. Well, I am sorry, I realize you would like more information but I am under a little bit of difficulty deciding what to do about it. The fact that I did not raise this question for a long time——
"P. That's right.
"O. I have been in difficulty about what to do, realizing how serious it is. I think my general point of view is that there are some things there which would bear watching.
"P. That's right.
"O. It is doubtful to me if there is anything there which can't be uncovered.
"P. Well that—I can see where it would be highly difficult to find out what's going on. We will be hot under the collar until we find out what is going on there. I mean—that's the point we have to take——
"O. Well, I don't know what a job like this—well I would think that it's conceivable—that it wouldn't hurt to have a man in the local of this union, FAECT—to see what may happen and what he can pick up.
"P. You feel there could be something in the organization itself?
"O. Within it.
"P. Within it.
"O. I don't know, I am sure that if they had 20 members, 19 of them might not be involved in it. But I am not sure of the 20th, you see.
"P. Yes.
"O. Forty members correspondingly and—let me put it this way—the bonds that hold them together are very strong you see, and they talk over their problems with their sisters and brothers and it is rather difficult to maintain a complete security in an outfit like that.
"P. Does this union that is up on the hill, is it not connected with the hill at all?
"O. Oh yes, it is an international union and has representatives all over this country.
"P. And the same group then, the same mixture would be of people off and on the project would be in the same——
"O. Oh. I imagine so—I don't know, I don't know about that.
"P. Well, we can——
"O. Ordinarily I think that they would have their own local.
"P. Which would be up there.
"O. Maybe not. Maybe it is all one big local. I'm not sure, but that varies with the union.
"P. Well, that is certainly interesting and—you are going to be here for some time?
"O. Oh no, I am leaving tonight.
"P. Oh, you are, are you flying?
"O. No I am not. I have orders not to fly.
"P. At least you get some relaxation in between your project. Well, I think that it may——
"O. I will be very glad to see you. I have a feeling though, a fellow can be fooled you see. I feel responsible for every detail of this rot of thing down at our place and I will be willing to go quite far in saying that everything is 100 percent in order. That doesn't go for this place up here.
"P. No.
"O. I think that's the truth. If everything weren't being done and if everything weren't proper, I think that I would be perfectly willing to be shot if I had done anything wrong.

"P. Well, ah——

"O. It's a very different situation, a very much harder situation. I don't know the people but it's a hard situation and in particular to put together in a casual way and I think that the problem of being sure that there are no leaks there is a real problem and that pressure can be brought with discretion.

"P. I am then, as I say, I may have the pleasure of visiting your place because it may——

"O. My motto is God Bless You.

"P. Well, as I say, if this becomes serious, that is to say, I don't known anything about it, but if it becomes——

"O. My guess is that it wouldn't but if I weren't first absolutely sure that it wouldn't, that it were not serious, I wouldn't——

"P. That's right. Well, if it does become serious I may come down with some of my persistency—I mean I would hate to—I have a responsibility of running things down there.

"O. I also think the particular way this way—that if there is anything going on it would be very easy to find out. I am not worried about that—we can take care of that ourselves.

"P. No, you wouldn't——

"O. Well, I meant in a way which you think best.

"P. But it is a situation which would have to be handled very delicately. That's what makes it so difficult. If it is something that's easy to handle and you don't have to worry about it, why you just sort of bull your way through, but these things have to be, one has to be careful.

"O. That's always the case—wanting to be very careful.

"P. I am not the judge to tell whether they should or should not get the information. My business is to stop it going through illegally.

"O. Well, I think—I don't actually know whether, if you were in Washington—after advice on the question how far should cooperation go. I don't know wherein the right answer lies. I have heard of cases with very strong arguments on both sides.

"P. Yes.

"O. That's a particular we don't have to worry about, but in Washington there are more ticklish situations——

"P. Do you know anyone—and because we like to eliminate unnecessary work if we have to—do you know anyone who is on the project who is connected with the FAECT, Dr. Oppenheimer?

"O. Who would be willing to——

"P. That's right.

"O. I don't know who is in the union at all. I have heard that a boy called Fox is president of it.

"P. David Fox.

"O. David Fox, but I would feel that that boy could do the trick. I hope that the trade union isn't tied up in this—and they would not act like this because I think it would give them a very black eye and it is no love of mine from the start, and it might have consequences beyond the reasonable. I doubt whether anyone mixed with the union in good faith would be very sympathetic.

"P. Yes.

"O. This isn't a suggestion that there is anything wrong. I have no reason at all to believe that there is, except that it is inevitable that anyone—that any leftwingers still interested in leftwing activity would join such a union. I think I can be quite sure of it. And I don't think that it is due to unions who are seeking a selected group of people——

"P. Yes.

"O. You might get some—of course, this is just my opinion that there is no harm in discussing it—well, I just don't know.

"P. May I just ask then, Doctor, if you would please not discuss this with anyone—so that they would not be aware of this fact that——

"O. No, I would not have raised the question if it didn't seem to me that it deserved looking into.

"P. Yes.

"O. And if I seem uncooperative I think that you can understand that it is because of my insistence in not getting people into trouble.

"P. I can assure you that if anything comes to the attention out there——

"O. Now, wait a minute, let me dispose of that statement which came over the long-distance phone. When I first talked with this boy I extracted from him a promise to stop all this kind of thing when he came on the job. Of course, I can't hold him to that promise * * * [unintelligible].

"P. Well, what I mean, if anything does come to your attention in connection with this phase, if you can——

"O. It won't be really necessary.

"P. If, in the first place, you will let me know, I will be glad to come down and discuss the matter with you.

"O. Well, I am very glad of that, and we may have other problems which we would like to discuss.

"P. Yes.

"O. But I do not think that there will be any of this nature because really we have very little incentive.

"P. Well, something may come to your attention relating to this place up here. You may get it down there and I would really prefer to——

"O. There is almost no contact. I have official technical letters but really no personal letters from here. I don't know what's going on, and I think the chance of my being useful in that way is very slight. But you ought to be able to find people here who could have their eyes and ears open and who know what's going on. That would be, I would be, I would be fairly sure that there are quite a few here who would be willing to give you—who would realize the importance of it and—I can't advise you any further.

"P. No, okeh, as a matter of fact I am not formulating any plans, I am just going to have to digest the whole thing.

"P. Well, we appreciate it and the best of luck.

"O. Thank you very much."

*   *   *   *   *   *   *

(The transcripts of interviews referred to are as follows:)

"TRANSCRIPT OF INTERVIEW WITH DR. OPPENHEIMER BY LT. COL. LANSDALE, SEPTEMBER 12, 1943

"L. Well the thing I've been thinking very considerably about my earlier conversation with you, then Colonel Pash's memorandum to me of his conversation with you.

"O. Well, the history of that—I spoke to Johnson briefly and I heard quite a little bit about the chain * * * about the nature of the fuss that Lomanitz was making, and I thought it might be a good idea if I talked to him. I thought I might be able to talk him out of some of this foolishness so I asked Johnson for permission to do that. I had a rather long discussion with Lomanitz which I should describe as pretty unsuccessful, or at least only partially successful. And, of course, Johnson had expressed the opinion that he was dangerous and why, and that Pash ought to be brought in on it. So I told Pash some of the reasons why I thought it was dangerous and I suppose that is probably what you mean.

"L. Well, now I want to say this—and without intent of flattery or complimenting or anything else, that you're probably the most intelligent man I ever met, and I'm not sold on myself that I kid you sometimes, see? And I'll admit freely that at the time we had our discussion at Los Alamos I was not perfectly frank with you. My reasons for not being are immaterial now. Since your discussion with Colonel Pash I think that the only sensible thing is to be as frank with you as I can. I'm not going to mention certain names, but I think that you can give us an enormous amount of help, and as I talk you will realize, I think, some of the difficulties that have beset us.

"O. There are some I think I know already.

"L. That's right. Now, I will say this, that we have not been, I might say, asleep at the switch, to a dangerous extent. We did miss some things, but we have known since February that several people were transmitting information about this project to the Soviet Government.

"O. I might say that I have not known that. I knew of this one attempt to obtain information which was earlier, or I don't, I can't remember the date, though I've tried.

"L. Now, we have taken no action yet except with respect to Lomanitz.

"O. Are they people who would be in a position to transmit substantial information?

"L. Yes, I'm so informed, I don't know personally, of course.

"O. Well, Lomanitz by virtue of being a theoretical physicist would probably have a rather broad knowledge of the things he is working on.

"L. I get the impression that Lomanitz has a broad knowledge of the theory of what you're trying to accomplish probably, but apparently a rather limited knowledge of the practical manner in which it is being done.

"O. That's right.

"L. Now, which is the most important to transmit?

"O. There are two things which seem important to me. One is the extent of the interest of this country and the nature of the commitment and the probable time scale. Second, it wold be important to transmit when the situation is suited to the Russian industrial machines, which I think maybe we don't have at all.

"L. All right; now I'll tell you this: They know, we know they know, about Tennessee, about Los Alamos, and Chicago.

"O. And the connection of all that?

"L. And the connection. We know that they know that the method, I may state it wrong, that the spectrographic method, is being used at Berkeley. They know, of course, the method involved. They know that you would be in a position to start practical production in about 6 months from, say, February, and that perhaps 6 months thereafter you would be in a position to go into mass production. Now, you and I know, of course, how accurate those figures are.

"O. All I know is——

"L. When they were reported to Pash, they may have been shaded one way or the other. Now, that is the substance of what they know. Now, we, of course, have acted. The people who are responsible for this thing have been willing to take some risks in the hope of some return. It is essential that we know the channels of communication. We never had any way of knowing whether we have—whether the ones we know about are——

"O. Are the main ones.

"L. Are the main ones, or whether this market will change them from time to time, and so on.

"O. This information which has been transmitted has not been transmitted to the consulate or——

"L. Well, all we know is that it's gone through several hands to the Government, some through consular channels. And, of course, they have many means of transmitting information, perhaps, you know. The fact that it goes to the consulate today doesn't mean that it's going to the consulate tomorrow. The fact that it goes through Joe Doaks today doesn't mean it's going through him tomorrow. Of course, that's our problem.

"O. No; the only thing that it does mean is that an effort is being made to get it.

"L. I can assure you that there's no question of the effort being made. We know enough to know that. It's not simply the Communist Party, U. S. A., off on a frolic of their own. Now, that, of course, presents—I want to get into more specific details later—but that, of course, presents several problems. We know, for instance, that it is the policy of the Communist Party at this time that when a man goes into the Army his official connections with the party are thereupon ipso facto severed.

"O. Well, I was told—I was told by a man who came from my * * * a very prominent man who was a member of the Communist Party in the Middle West, that it was the policy of the party there that when a man entered confidential war kork he was not supposed to remain a member of the party.

"L. That is correct. That was just the next point I was coming to. We know that they do that with the Army, and we have strong suspicions that they do the same with any confidential war work. That severance is not a severance in fact. It's merely to enable the person to state without lying, without perjuring himself, that he is not a member: 'Oh, yes; I was a member, but I'm all over that now. I'm not a member; I don't have any connection with it.'

"O. Well, there are some cases that I know about.

"L. In some cases it may be true.

"O. That I'm quite clear about—not to pull any punches, my brother has made a severance, in fact.

"L. Well, we know that he has been a member.

"O. Yes.

"L. We also know that there has been in recent years no indication that he is still a member.

"O. It's not only that he's not a member. I think he has no contact.

"I know I overwhelmingly urged about 18 months ago, when we started, that she should drop social ones which I regard as dangerous. Whether they have, in fact, done that I don't know.

"L. Well, I'm quite confident that your brother Frank has no connection with the Communists. I'm not so sure about his wife.

"O. I'm not sure, either, but I think it likely some of its importance has left her. And, also, I believe it to be true thta they do not have any—I don't know this for a fact—but if they had, I didn't know it, any well-established contacts in Berkeley. You see, they came from Palo Alto, and they had such contacts there. Then my brother was unemployed for 3 very salutory months, which changed his ideas quite a lot; and when they started in Berkeley, it was for this war job, and I do not know, but think it quite probable that his wife, Jackie, had never had a unit or a group to which she was attached in any way. The thing that worried me as that their friends were very left wing, and I think it is not always necessary to call a unit meeting for it to be a pretty good contact.

"L. Now, I don't want you to feel that any of these questions that I'm going to ask you—I'm going to ask you some pretty pertinent and direct ones—are made for any purpose of embarrassing you in any way. It's only that I feel it my duty to.

"O. I'll answer them as well as I can.

"L. I tried to explain to you my problem, which as you can see is due to the nature of the kind of espionage we're up against, is extremely difficult.

"O. Because it's so ramified.

"L. It's so ramified, and, after all, we're dealing with an allied nation.

"O. And who are the people in the project at Berkeley, are they my former students?

"L. I'm not ready yet to tell you. As I say, I'm going to try to——

"O. I'm concerned if they are people for whom I have some kind of responsibility.

"L. Well, I'm not going to try to fence with you or mislead you at all as I did without success, I feel, to some degree, at Los Alamos.

"O. Well, I felt there was a lot in your mind, and we were talking around, the conversation was quite clear.

"L. It was perfectly obvious that you did read into it more than was stated. Well, however, to refer again to this business concerning the party, those reasons make it clear that the fact that a person says they have severed connection with the party, the fact that they have at present no apparent interest or contact in it does not show where they have unquestionably formerly been members that they are not dangerous to us.

"O. I agree with that.

"L. That again poses a terrific problem because so many of the people * * * you know as well as I do how difficult it is to prove communism. I'm going to discuss yourself with you in a few minutes, and that will serve to illustrate as well as anything some of the difficulties involved. We've got to weigh, we feel, I believe that the first and primary thing to do is to get the job done. That is, the project completed. Now, if that involves taking some risks, why, of course, we'll take the risks. After all, you are risking your lives and everything else to do this, and everything has to be done with a risk so that we don't want to protect the thing to death. But, therefore, all persons who are essential to the project in any marked degree, really unless they cease to make themselves useful there's no use talking about severing their connections while they're there whether we believe they're Communists, pro-Nazi, pro-Fascist, or what.

"O. I won't agree with that, I mean, I think one has to——

"L. You have to weigh, you never know. I mean if you know it's comparatively easy. The only question that remains is whether we shall wait awhile until we take somebody else or how we shall go about it. Shall we try to prosecute him for espionage or shall we just forget about it and weed him out. I mean, you see that once you've made the decision why then the problem is practical. The difficulty is making the decision. Now, I want to know. In the first place I think we know now who the man that you referred to as approaching the other college project was. I wonder if you feel that you're in a position to tell me.

"O. I think it would be wrong.

"L. I'd like to discuss with you your attitude on that for a minute.

"O. It is primarily this, that this came to me in confidence and the actions taken were negative, the actions of this intermediary were reported as essentially negative, and although it would have been really negative not to have touched it, I feel that I would implicate, so to speak, one fellow about whom, who has initiative, would be persecuted.

"L. You mean Eltenton?

"O. Yes; this is the way it came to me straight.

"L. Well now, you see what you stated that he contacted, I believe it was three persons on the project, and they told him to go to hell in substance.

"O. Although probably more politely.

"L. And how do you know that he hasn't contacted others?

'O. I don't. I can't know that. It would seem obvious that he would have.

"L. If you heard about them they unquestionably were not successful.

"O. Yes.

"L. If you didn't hear about them they might be successful or they might at least be thinking about it, don't you see? Now you can, therefore, see from our point of view the importance of knowing what their channel is.

"O. Yes.

"L. And I was wondering, is this man a friend of yours by any chance?

"O. He's an acquaintance of mine, I've known over many years.

"L. Well do you—I mean there are acquaintances and there are friends. In other words, do you hesitate for fear of implicating a friend?

"O. I hesitate to mention any more names because of the fact that the other names I have do not seem to be people who were guilty of anything or people who I would like to get mixed up in it, and in my own views I know that this is a view which you are in a position to doubt. They are not people who are going to get tied up in it in any other way. That is, I have a feeling that this is an extremely erratic and unsystematic thing.

"L. Here is, I want you to in no derogatory way understand my position again.

"O. Well * * * there is a very strong feeling. Putting my finger on it I did it because of a sense of duty. I feel justified * * *.

"L. Now, here is an instance in which there is an actual attempt of espionage against probably the most important thing we're doing. You tell us about it 3 months later.

"O. More than that, I think.

"L. More than that. When the trail is cold it's stopped, when you have no reason not to suppose that these cases which you hear about are unsuccessful, that another attempt was made in which you didn't hear about because it was successful.

"O. Possibly. I am very, very inclined to doubt that it would have gone through this channel.

"L. Why?

"O. Because I had the feeling that this was a cocktail party channel. A couple of guys who saw each other more or less by accident.

"L. Well, people don't usually do things like that at cocktail parties, I know. All the stuff that we've picked up has certainly not been at cocktail parties.

"O. Well, that's where * * * I don't know, there may be many, many other channels besides Eltenton, and I would assume that there would be, but I have the feeling that Eltenton's suggestion to this fellow was whether he was willing to do this, was really a potential suggestion and not a systematic one.

"L. Well, I don't want to draw this out unduly, but I want to examine that proposition for a few minutes. Why would Eltenton working for the Shell Development Co. be interested on a frolic of his own, as it were, in trying to find out for the Soviet Government what's going on?

"O. I don't think it was a frolic of his own, but my answer is that he worked in Russia for 5 years * * * and had some contacts.

"L. That's right, so what I'm getting at is this—he unquestionably was asked or directed.

"O. Depending upon the point of view.

"L. To see what he could find out?

"O. I would think so.

"L. Which would mitigate against any conclusion that this attempt here was a mere casual thing?

"O. No; I don't think, I mean let me put it this way. The reason I mentioned Eltenton's name was because I thought it was likely that Eltention would persevere in this. But the reason I mention no other names is that I have not felt that those people would. That they were all just accidental.

"L. Now, I don't want these names of the people who were contacted or the person who contacted them. Let's stick to the persons that were contacts. To do anything to them because it's perfectly evident to me that they sure as hell would never come tell you about it if they were going to do it.

"O. Yes, that's right.

"L. Now, while I would like to have those names very much it's not as essential as that we know the contact. Because I think, there's one channel, of course, there's other channels, we know of. We don't know that one. Now we've got no way of knowing whether the ones that we've picked up or the names that I know of are identical with this man. Now, that's a simple reason why I want that name, and I want to ask you pointblank if you'll give it to me. If you won't, well O. K., no hard feelings.

"O. No; I've thought about it a good deal because Pash and Groves both asked me for the name, and I feel that I should not give it. I don't mean that I don't hope that if he's still operating that you will find it. I devoutly do. But I would just bet dollars to doughnuts that he isn't still operating.

"L. I don't see how you can have any hesitancy in disclosing the name of the man who has actually been engaged in an attempt at espionage to a foreign power in time of war. I mean, my mind just doesn't run along those channels, and——

"O. I know, it's a tough problem, and I'm worried about it a lot.

"L. I can understand personal loyalty, yet you say he's not a close friend of yours. May I ask, do you know him as a Communist?

"O. I know him as a fellow traveler.

"L. You know him as a fellow traveler. Course in our book, membership in the party is not material, it's whether they follow the party line which is a test.

"O. Well, I don't know whether the fellow has or has not in all detail, but he certainy has so far as I know about it in a general way.

"L. He is now at the university?

"O. I don't know that. That is, I think, I don't know the date on this precisely, but I think it was some time maybe before Christmas of last year that this matter was brought to my attention. I don't know how long it is. There was some talk of his trying to get a job elsewhere * * *.

"L. Well, of course, that's the question. Do you now feel you can tell me who it is?

"O. I do not now feel that I ought to tell you.

"L. In what event would you feel that you should?

"O. If I had any evidence or anything came to my attention which was indicative that something was transmitted * * *.

"L. Well I'm telling you it is. Right today, I can't tell you the last time information was passed, but I think it was about a week ago.

"O. I mean something that there is a reasonable chance is the man whose name I don't want to give to you.

"L. Well, of course, I——

"O. There's a very strong feeling on my part that I ought not to.

"L. I have no way, of course, of knowing.

"O. What I want to say is this—I'm not kidding you and I'm not trying to weasel out. It's my overwhelming judgment that this guy isn't involved. That isn't judgment which is based on hope but his character. If I am wrong, then I am making a very serious mistake, but I think that the chances are very, very small.

"L. Let me ask you a personal question—you don't have to answer it if you don't want to. Is part of your feeling based on the fact, don't be insulted please, that you don't consider that it would be such a catastrophe (*sic*) anyway for us if they did find it out?

"O. That is not my feeling. I think it would be a catastrophe (*sic*) and I made this clear when I talked Pash. If Russia found out except through official channels. I do not know whether what we are now doing with the British is the right thing to do. I do not know whether it would be right to include Russia and China in that. I think we are now reopening negotiations with the British, in fact that is why I'm here. I don't know whether that is right. That is a very hard and tough question. I am sure that it is wrong for the Russians to find out about those things in any way except through official channels.

"L. Well, if you won't do it, you won't do it, but don't think I won't ask you again. Now, I want to ask you this. And again, for the same reason which implies you're here, you may not answer. Who do you know on the project in Berkeley who are now, that's probably a hypothetical question, or have been members of the Communist Party?

"O. I will try to answer that question. The answer will, however, be incomplete. I know for a fact, I know, I learned on my last visit to Berkeley that both Lomanitz and Weinberg were members. I suspected that before, but was not sure. I never had any way of knowing. I will think a minute, there were other people. There was a, I don't know whether she is still employed or was at one time a secretary, who was a member.

"L. Do you recall her name?

"O. Yes; her name was Jane Muir. I am, of course, not sure she was a member, but I think she was. In the case of my brother it is obvious that I know. In the case of the others, it's just things that pile up, that I look at that way. I'm not saying that I couldn't think of other people, it's a hell of a big project. You can raise some names.

"L. Did Lomanitz—was it Weinberg or Bohm?

"O. Weinberg. 1 do not know now, and did not know that Bohm was a member.

"L. Well, did you met both Weinberg and Bohm?

"O. They came over to Lawrence's office.

"L. Yes; I remember that. I think it was General Groves told me about that. Well, anyway, did they tell you at this recent meeting that they were members?

"O. No; what they told me was the following: That they were afraid that Lomanitz was being forced out because he was active in the union and that their history was also somewhat red.

"L. By their you mean the union or Weinberg and Lomanitz?

"O. Weinberg and Lomantiz. That they felt that they, as they put it, would also be framed and they asked my advice as to whether they should leave the project. That is what they came to discuss. I said in my opinion Lomanitz was not being framed, that if they were fulfilling three conditions I thought that they should stay on the project. The conditions were first, that they abided in all strictness all of the security regulations; second, that they had no political activity or contacts of any kind; and third, that they——

"L. Now why isn't that—can you tell me the names of anyone at Los Alamos that have been or are now party members?

"O. I can't tell you the numbers of any who now are, but I know that at least Mrs. Serber was a member. She comes from the Leof family in——

"L. The Leof family in Philadelphia.

"O. And I know that my wife was a member.

"L. That was a long time ago.

"O. Yes—you haven't found out a lot about my wife.

"L. Well we might have missed some points. We were fairly confident that she's not a member now, although she was years ago for a very brief time.

"O. Yes, she was a member for a very brief time. She was married to a fellow who was working in Youngstown and was killed in Spain.

"L. Was that your wife's first husband?

"O. I believe she had an early marriage which was annulled, a very nasty fellow—she has told me very little about it, but I think he was quite talented a musician.

"L. Well I'm really not concerned much with that. She's a very attractive girl I think.

"O. I feel I shouldn't hesitate to say these things.

"L. Now, do you know, was Mr. Serber a member of the party?

"O. I think it possible, but I don't know.

"L. How about Mrs. Woodward?

"O. I don't know. I don't know them very well.

"L. Do you know of anyone who came from Berkeley down there with you, by with you I don't mean the same time, of course, who were members of the party?

"L. I'm afraid I can't give you any names.

"O. No, there was a whole group of people of whom I would be rather astonished if any of them were. Allison, Frankel, Miss Roper.

"L. How about Dave Hawkins?

"O. I don't think he was, I would not say so.

"L. Now, have you yourself ever been a member of the Communist Party?

"O. No.

"L. You've probably belonged to every front organization on the coast.

"O. Just about.

"L. Would you in fact have considered yourself at one time a fellow traveler?

"O. I think so. My association with these things was very brief and very intense.

"L. I should imagine the latter anyway.

"O. It was historically quite brief and quite intense, and I should say I was——

"L. Now I have reason to believe that you yourself were felt out, I don't say asked, but felt out to ascertain how you felt about it, passing a little information, to the party.

"O. You have reason?

"L. I say I have reason to believe, that's as near as I can come to stating it. Am I right or wrong?

"O. If it was, it was so gentle I did not know it.

"L. You don't know. Do you have any one who is close to you, no that's the wrong word, who is an acquaintance of yours, who may have perhaps been a guest in your house, whom you perhaps knew through friends or relatives who is a member of the Communist Party. By that I mean——

"O. Well, my brother, obviously.

"L. Well, no, I don't mean him.

"O. I think probably, you mean someone who just visited for a few hours.

"L. Yes.

"O. Yes; certainly, the answer to that is certainly, yes.

"L. Well, would you care to give me any of their names?

"O. There is a girl called Eldred Nelson.

"L. Suppose I've got a bunch of names here, some of them are right and some of them are wrong, you don't mind treating it that way do you?

"O. No.

"L. Did you know William Schneiderman?

"O. I know who he is. He's the secretary of the Communist Party. I've met him at cocktail parties.

"L. You have no real personal acquaintance with him?

"O. No.

"L. Do you know a fellow named Rudy Lambert?

"O. I'm not sure, do you know what he looks like?

"L. No, I've never seen him. He's a member of the party. Do you know a Dr. Hannah L. Peters?

"O. Yes; I know her quite well.

"L. Do you know that she's a Communist?

"O. I certainly knew that she was very close. I did not know she was a member.

"L. You don't know what her position in the party is?

"O. No; I didn't even know she was a member.

"L. Do you have any more than just an acquaintance with her?

"O. Yes, I know her quite well. Her husband is on the project.

"L. How about a fellow by the name of Isaac Folkoff?

"O. I don't know. I knew a Richard Folkoff who was a member of considerable importance.

"L. How about a man by the name of Steve Nelson?

"O. He is a professional party member; he's an organizer.

"L. Did you know him well at all—under what circumstances did you know him?

"O. He was a friend of my wife's former husband who was killed in Spain. I have a thoroughly unprofessional acquaintance with him.

"L. How about Haakon Chevalier?

"O. Is he a member of the party?

"L. I don't know.

"O. He is a member of the faculty and I know him well. I wouldn't be surprised if he were a member, he is quite a Red.

"L. Do you know Alexander S. Kaun?

"O. I know him, in fact I once rented a house from him about 7 or 8 years ago, but I never had any more relations with him.

"L. Do you know whether he is a member?

"O. No; I don't. I know he's a member of the American Soviet Council.

"L. How about a girl named Jean Tatlock?

"O. She is a close friend of mine, and I'm certain at one time she was a member of the party.

"L. Whether or not she is now or not?

"O. I would rather doubt it. I know she dropped out at one time and I rather think she probably still is.

"L. How about a man by the name of A. Flaniger?

"O. I know who he is, I've never met him but I've heard stories about him.

"L. Do you know who he is?

"O. No.

"L. Is he a professor?

"O. I don't know. I know he was a graduate student at Berkeley at one time. General Groves asked me about him.

"L. Oh, he did.

"O. I don't know anything about him.

"L. Now, have any of these people that I've mentioned ever said anything to you about your work? Snyder, Nelson, Peters, or Folkoff?

"O. Well, I think—let's see, I don't know what words to use. Obviously, Hannah Peters because there was some question of their going to Los Alamos, and I am really rather surprised that she is a member of the party. They have only very recently gotten their citizenship.

"L. What was her original nationality?

"O. German.

"L. Cigarette?

"O. No, I'll smoke my pipe.

"L. Did you say there was some question of them going to Los Alamos?

"O. Yes.

"L. About when was that?

"O. I would think that November would be a good guess on that.

"L. Have they ever been employed on the project?

"O. Peters is on it now.

"L. Oh, he is on it now?

"O. As a matter of fact the reason he didn't come was that Lawrence agreed to his release and then at the last minute changed his mind. He's quite a good mathematician.

"L. Now, I want to ask you to go back to Lomanitz. You told me when I was down there that when you broke the subject to, what do you call him, Rossi?

"O. Rossi.

"L. Rossi. When you first broke the subject to him about going on the place you stated that he was uncertain, he came up to your house and did what you characterized as a good deal of soul searching. I would like to know whether that soul searching or discussion of his own feelings had any relation to his work in the party.

"O. None whatever, I did not know he was a member of the party.

"L. Until just recently.

"O. Yes, and I knew he was extremely Red, but frankly I thought he was a member of the Trotskyite faction.

"L. Which would ipso facto prevent him from——

"O. Being a member of the party. That's what I thought at that time. What he said he wanted at that time was to be a soldier and be one of the American people in that way and help to mold their feelings by being a soldier, and wasn't that more worthwhile than working on this project. I told him he obviously had a lot of talent; he had training that he was throwing right away and that if he could make up his mind and it was a clear-cut decision to use himself as a scientist and nothng else, that then that was the right thing to do.

"L. Now, what led you to exact from him a promise, or to make the condition of giving up political activities?

"O. Because he had distributed leaflets and because it was just generally obvious that he was a member of the union and radical societies.

"L. Now, you have stated to me and also I think to General Groves that in your opinion membership in the party was incompatible with work on the project from a loyalty standpoint.

"O, yes.

"L. Now, do you also go so far as to believe that persons who are not actually members but still retain their loyalty to the party or their adherence to the party line are in the same category?

"O. Let me put it this way. Loyalty to the party, yes; adherence to the party line, maybe no, is that it need not necessarily, although it often is, be the sign of subservience. At the present time I don't know what the party line is in too much detail, but I've heard from Mrs. Tolman, Tolman's wife, that the party line at present is not to discuss postwar affairs. And I would be willing to say that anyone who, well, let me put it this way, whose loyalty is above all else to the party or to Russia obviously is incompatible with loyalty to the United States. This is, I think, the heart of it. The party has its own disciples.

"L. Now, I was coming to that. I would like to hear from you your reasons as to why you believe—let's stick to membership in the party—is incompatible to complete loyalty to the project. When, to state something a little bit foolishly, membership in the Democratic Party certainly wouldn't be.

"O. It's an entirely different party. For one thing * * * I think I'd put it this way. The Democratic Party is the framework of the social customs * * * of this country, and I do not think that is true of the Communist Party. At least, I think that there are certainly many Communists who are above all decent guys, but there are also some who are above all Communists. It's primarily that question of personal honor that I think is involved. I don't know whether that answers the question but my idea is that being a Democrat doesn't guarantee that you're not a floor-flusher [sic] and also it has no suggestion just by virtue of your being a Democrat that you would think it would be all right to cheat other people for a purpose, and I'm not too sure about this with respect to the Communist Party.

"L. Let me ask you this—how in your opinion would the Communists engaged in espionage on this project transmit their information. I want to ask it by a question. Would it be necessary for them to pass it in writing?

"O. To be effective. It depends, I mean gossip could be effective but it could only be effective on the first sort of thing we talked about, namely, the extent and purpose and dates of the project and how many people were involved, where they were involved, and if it were hopeful or not and stuff something like that. But if it were going to be anything of a technical nature well, I won't say it would be impossible but it would be very difficult to find a method of transmission which would preserve the technical details without having some of it written down.

"L. Do you have any real knowledge of the methods used in the party for the transmission of information?

"O. No; I certainly don't.

"L. Such as their ordinary industrial espionage.

"O. I didn't know there was any. I'll put it this way. I would assume that it existed because of their policies, but I couldn't know it existed because I'm not in the party. Why do you look so worried?

"L. Because I'm not getting anywhere.

"O. Well, you're getting, except on that one point, I think that you're getting everywhere than I can get you.

"L. Let we ask.

"O. I do not know, and let me just make it simple. I have never been involved in and I do not know anything about this Communist business. If there is such an effort, and I assume there might be, I would assume that it might be very different in different parts of the country, I don't know.

"L. Do you feel that anything like this would be run by party headquarters or by the Soviets themselves?

"O. I don't know, my general feeling is that the Soviets are too hard headed to trust it to an organization which is as fly by night as the party. But I don't know. That is, you understand Eltenton is also known to me only as a fellow traveler and I see him around a lot of places where I was, and I can't from that assume that his * * * were pretty leftwing and I know he was in Russia. I do not know whether the initiative for what he was doing came from himself or of it was something that he was told to do or something that he was ordered to do.

"L. Well, the fact that he would do it would indicate that he was much more than a fellow traveler.

"O. Well, I'm saying that now. But my initial information.

"L. Well, try to put yourself in our position.

"O. All right, Lansdale.

"L. You're confronted with this situation. You've got a few men that you know are actually doing it. You've got a few other men who are associated with them closely. You can assume that if they're engaged in an organized attempt, they're around propagandizing the party. That any two or more of several people who are engaged jointly in this enterprise are not going to be seen together. Now, with the idea also you know what you have been able to pick up that you haven't got everybody. Now, the only thing we've got to go on are peoples' associations and prior activities. All right, now to use an illustration which is personal but not pointed, you get my distinction, we've got the case of Dr. J. R. Oppenheimer, whose wife was at one time a member of the party anyway, who himself knows many prominent Communists, associates with them who belongs to a large number of so-called front organizations and may perhaps have contributed financially to the party himself, who becomes aware of an espionage attempt by the party 6 months ago and doesn't mention it, and who still won't make a complete disclosure. Now, I'm giving yourself, because by doing that I'm not giving you information about anybody I don't want you to know anything about, and I may say that I've made up my mind that you yourself are O. K. or otherwise I wouldn't be talking to you like this, see? Now.

"O. I'd better be—that's all I've got to say.

"L. Well, that's my idea. Now, what are we to do in a case like that? There are a good many people on this project who are somewhat in the same position, who we have every reason to believe have been party members, who are certainly not now whether for technical reasons or whether actually, who certainly retain their, shall I say their feeling for the common man which probably led them to the Communist fold in the first place. Their feeling that all is not right with the world, and what Wilkie calls the palace on the hill surrounded by the mud huts is all wrong. Now, what attitude are we to take toward these people? What is our position? Here we are; we know that information is streaming out from this place every day. We know about some of it. How much of it is there that we don't know about?

"O. Places other than the west coast?

"L. Sure, we know that definite efforts are being made to find out. They wouldn't be going to those efforts unless they really wanted it. Now, what shall we do? Shall we sit back and say well, my God, maybe the guy recanted, maybe he isn't at all?

"O. Hard for me to say because of my own personal trends, and as I say I know that the Serbers afford a good illustration of this I would hesitate to say to a stranger * * * about another closeup * * * person whose history was the same as that of Mrs. Serber's, sure she's all right but I know the Serbers and I am confident of them. Now, I have worked on rather a personal basis. I don't know the Woodwards are members, I did not know that until General Groves mentioned it the other day that there was some question of it. I feel that in the case of the Serbers I could understand that very well. But I just don't know in a general case; it's impossible to say. I don't know any of these people in Berkeley, I don't know Weinberg or Lomanitz well enough to swear * * *

"L. Why is he moving heaven and earth to keep out of the Army?

"O. He told me that he thought he was being framed, and I said I think that's nonsense why would you be framed, and he said, "Well, part of the general scheme * * * maybe they're after bigger game than the party."

"L. Did you ask him what the bigger game was?

"O. He said he thought you were after the union.

"L. We're not.

"O. Well, I suggest you keep your eyes open.

"O. * * * I persuaded him, I think, that he should not try to stay on the project there.

"L. He's gotten notice of his induction on the 20th of September. Well, suppose we tell you, and I'm not telling you now, of course, for the sake of illustration which is way off the beam, intentionally so, say we told you that Backus (Bacher?) was unquestionably a member of the party, at least up until a short time before he came on the project and he's one guy we don't have anything on, would you concur in his removal?

"O. Just on the basis of his having been a member?

"L. Just on the basis of his having been a member of the party.

"O. That's pretty hard to say. I would try, myself, to get some information about him from people in whom we could have confidence. I wouldn't take it on the face of that until I saw what it looked like. Because Backus (Bacher?) is a pretty valuable man and assuming the fellow is * * *

"L. Yes, that's the reason I used him as an illustration. How close are you to members of the party? Are you close enough to get any information from them?

"O. Well, in a way at Berkeley I could. I don't think I could get information about this business. But I think I could get information about who is doing what.

"L. Could you get information about who is and who isn't a member of the party?

"O. I don't know whether I could now. At one time I could have. I never tried to.

"L. Would you be willing to?

"O. Not in writing, I think that would make a very bad impression.

"L. No; not in writing.

"O. I don't know anyone at Los Alamos who could give information of that kind. I could get partial information.

"L. Do you think that you you'd be in any position to be of assistance in uncovering the ramifications of this case?

"O. It would obviously depend a hell of a lot of where it was.

"L. Well, I was coming to that on the next question.

"O. I think it's like this. I think that my being at Los Alamos very seriously interferes with their actions there, but I wouldn't want to give a general answer.

I wouldn't want to say "no" in a general way; and I can't very well say "yes" in a general way.

"L. My next question which you have already partially answered is wicked, and you've answered at least that you probably would if you were persuaded, but you wouldn't want to.

"O. That's it. That isn't my business at all.

"L. It's not your business and I don't think you ought to be asked unless * * *

"O. Unless it would be a desperate attempt.

"L. Well, we have of course my job operatively is to try to prevent the escape of information, and of course since that is my job, although this project is only part of it, it probably looms larger in my daily problems of course than it does in yours. You have other things to worry about and you ought not to have to worry about this. And the only reason you are being worried about it is because you do have some information. Now I say this that we have been fairly sure for a long time that you knew something you weren't telling us.

"O. How did you know about it because I wouldn't have known. How did you know that?

"L. Well, you don't mind if I don't tell you. It wasn't anything you did or said. And don't you think, I'm coming up on the other beam now, don't you think that you'd be a whole lot happier and have a whole lot less worries if you were in the position of having told us everything you know about it, that could possibly help us, and then forget about the whole damn thing, and not be bothered any more.

"O. I would be bothered by thinking it was something I shouldn't have done. I mean I've told you technically everything except this guy's name.

"L. You haven't told me his name. Now I don't, if your description of your relation with him is so, I'm in an extreme quandary as to whether I know him or not and here we've gone to great risks to try and button this whole thing up, and it's perfectly obvious that here's a big hole in our net that we haven't stopped.

"O. I don't know how much you've got, but it seems to me that you want me to give you more information about many others who are not involved. You may have it, I don't know.

"L. For instance, I don't follow you.

"O. Well, it seems to me that the essential position you have—Eltenton for instance.

"L. Oh, sure. You know the way these things operate. You have one guy here, we'll call him the master spy, he has a lot of people * * * he may have more people working on the same thing all unknown to each other, and they may each have others. That spreads out. They may have missed on these three men that you mentioned. Let me ask you this question. How did this intermediate contact happen to go to these three particular people?

"O. I would suppose it was because the way it was told to me, he was told to. It was well the relations between Russia and the United States are not what they ought to be and we are not giving them half of all of our technical information that is important for an alliance. They know about as much as they can, they're working at great odds and so on. Well, I suppose he picked people who might be susceptible to that approach.

"L. In other words people who were apt to be sympathetic to Russia?

"O. That's right.

"L. Who were apt to feel that the State Department, or whoever was responsible was rapidly leading this country into the position of making a break with the Soviet Union, when it was to the enormous interests of this country and if not to the U. S. S. R. that we maintain the most cordial relations.

"O. I would think it might be.

"L. Now, are these three people to your knowledge members of the party or have been?

"O. No—no.

"L. Are they in the category of what you would call fellow travelers?

"O. I actually don't know except in one case where I would say he was a fellow traveler.

"L. Now, why did they come to you?

"O. I suppose for 2 reasons; 1, because I was more or less responsible for the work, and 2, because they thought I wouldn't hit the roof over it. I might say I did.

"L. I know, of course, that you probably have administratively the best running outfit, in the project, and that it's largely because of the intense personal loyalty which you seem to be able to inculcate in the people that work for you.

"O. I have my troubles.

"L. Yes, I can imagine you do. I can see one of the reasons for it—they stick by you and you stick by them, which after all is the secret of obtaining people's loyalty.

"O. I do have my troubles.

"L. Let me skip to a slightly different subject. Do you know anything about any difficulties at the metallurgical laboratory? Administrative difficulties?

"O. Yes, I do.

"L. What do you know?

"O. I don't know very much, I don't know nearly enough about them to be of any use.

"L. Do you know anything about the incipient, I guess more than incipient, intramural organization, you might say?

"O. I know about it. Allison told me about it last time I was there. I do not know of a single member of it. I know nothing about it except what Allison told me which was that there were too darned many Jews on it.

"L. How do you feel about any intramural organization like that?

"O. Personally, it doesn't bother me so much although * * *.

"O. It is not dangerous in the sense in which the enemy is * * * how it came into all this because of the fact that the FAECT is organized on this project, that I know that one of its most zealous members is Eltenton, and I was frankly afraid that I was being used or might be used for the provocation of leaks, that is the reason why I spoke of these things.

"L. Do you know how close the heads of that union are to the party?

"O. Which union?

"L. The FAECT.

"O. No, I don't. I don't even know who the heads at Berkeley are. I met once the international vice-president, or whatever he was called, Scheres, but I do not know whether he was a party member or not.

"L. And you just don't know anything about any party activities in connection with that?

"O. No, I would somewhat doubt whether there were any, but I don't know.

"L. Of course, you now know that Weinberg and Lomanitz are both members of the party and members of the union.

"O. I didn't know Weinberg was a member.

"L. Well, as a matter of fact, I don't either.

"O. I had a feeling of surprise * * *

"L. He's probably mixed up. He's close to Lomanitz who unquestionably is a member.

"O. Well, that's certain. Lomanitz said to me that he had been very active.

"L. And, of course, nobody can be very active in the union and do the union any good or offer anything unless they do know considerable about the work that's going on.

"O. I feel that any place like this project should not be unionized. * * * I know nothing about the Chicago thing except that it was being a great headache to my good friend Allison.

"L. Do you know of any action to prevent the unionization?

"O. In Chicago or at Berkeley?

"L. No, at Berkeley. I'm talking about Berkeley. These intramural things don't bother me, I mean I've got nothing against unions as such, and as long as its an intramural affair why then it can't have the effects. I mean I don't care whether they demand higher wages.

"O. The union at Berkeley I don't think is getting very far.

"L. I don't seem to make much progress, but I certainly try hard. Well, they've got to know all of the people who are working on the damn thing.

"O. It would seem to me that this may strike you as being a very foolish thing to do, and I don't want to do it if that's the case. If I were in your shoes I would ask to speak to the * * * or counsel or executive committee or whatever they have in that including the officers if possible. Also, who was instrumental in getting it started and I would make a rather clear and helpful expose of why it was an undesirable thing. I'm sure that it would have a very great effect. You might not be able to do it, and if you can't Fidler can do it for you, and maybe some civilian could do it for you, I don't know, but I mean, let me put it this way—if the word gets started at Los Alamos I shall go to the boys and talk to them, I shall tell them some of the reasons which * * *. Whether that would be helpful at Berkeley or not I don't know.

"L. I don't think we can get Lawrence to do it.

"O. No, maybe you can't. And then of course there are some advantages that are of interest to * * *.

"L. There are some advantages to what?
"O. To having a Red history like mine.
"L. So they know that you're not antiunion.
"O. So they know that I'm not a banker or——
"L. Well, of course you're right, you're perfectly right, they know from your past history that whatever anyone might say about you, you're certainly not antilabor.
"O. Well, I was only trying to say that if I were at Berkeley now, as a matter of fact it wasn't my business, but I did try to discourage the FAECT. It was getting started before I left Berkeley, but that was in the form of talking to other people, my brother, and one or two others, and saying this looks like a bad thing. I did nothing at all deliberate about it.
"L. Can you tell me any more, did Weinberg, it was Weinberg and Bohm who came to you, wasn't it?
"O. Yes, they came to me in Lawrence's office.
"L. Yes. Did Weinberg and Bohm say anything? What did they say about the party?
"O. They didn't say anything about the party.
"L. They didn't? Did they talk about the union?
"O. They talked, well they didn't even talk about the union. They talked about, I think I've given you a fairly good, I don't know what they might have said if we had met in the woods some place, but we met after all where there were two secretaries in the room.
"L. Oh, they were there.
"O. I don't know whether the door was closed or not, but it was extremely open interview. I saw Lomanitz more or less, well I saw him first at one of the offices of a man and we walked out to telegraph * * * but his discussion was a little bit more uninhibited than the others. These 2 fellows were concerned with only 1 thing—they said they had worked closely with Rossi, they thought he was a good guy and that they thought he was being framed for his activities in the union and his political sympathies, and they thought that because of this they were also in danger of such a nature that they should get out of the project into some other useful work or were they likely to be treated in the same way.
"L. Now let me ask you this. From what you stated to them, if they were in fact not fulfilling the conditions which you mentioned to them, which you said to them would have been tantamount to telling, then if you are doing that you'd better get out.
"O. Yes.
"L. That is correct, isn't it?
"O. Yes, that is if they were violating any of the three rules which meant active in union, maintaining any contact with Reds, not maintaining discretion, they were useless in the project.
"L. Now, you still don't want to tell me that name?
"O. Not if I can help it.
"L. Well, is there anything else that you believe you can tell me that could give us any assistance?
"O. Let me walk around the room and think.
"L. Sure, it's getting warm isn't it?
"O. I have been thinking about this. I can tell you that I doubt very seriously whether (I don't know Bohm very well), I doubt very seriously whether Weinberg would do anything along the lines of what we were talking about. I * * *.
"L. Well, do you base that on his character or what?
"O. Yes. I should have told you before, but I have told you since, no I haven't, but I will tell you now, you said that Mrs. Peters was a member of the party. I do not know whether her husband is or not, but I know that he was in Germany, and that he was actually in prison there, and I also know that he has always expressed a very great interest in the Communists, and I think whether he is a member or not would perhaps partly depend on whether he was a citizen or whether he was working on a war job. That is, it may just have happened that the has had no period where he could be and that he would otherwise be.
"L. Is his wife also German?
"O. Yes.
"L. How recently did they come over here?
"O. I don't know; they were in New York for some time, I met them first, oh, it must be 4 or 5 years that I met them first, they had been in California some time before that. I believe that they came over very early because immediately

after Mr. Peters was imprisoned in Dukon (sic, Dachau) then he escaped, at least that's the story that I have heard from more than one person and that she was * * * and they tried * * *.

"L. By the way, what was your wife's husband's name that was killed in Spain?

"O. Dallet; I never knew him.

"L. You never knew him.

"O. He was a big shot in the party * * *. I have heard that * * *.

"L. Great for sentiment those boys.

"O. Oh, they were. Those that went over there were I suppose * * * I've met 2 other people, 1 was a young man called Thompson who was working in San Francisco * * *.

"L. Let me ask you a question, I don't want to draw this out unduly, I know you're busy.

"O. Well, you're having trouble.

"L. But you see, that I'm in, you might say mental difficulties over this thing. I feel I have a heavy responsibility and I'm trying to find out everything I can that will help.

"O. I would warn you, you see that the fact that there are some people attached to the Communist machinery, who may, or may be not attached, who may be guilty of passing but its a hell of a bad reason for suspecting everyone.

"L. Of course, of course. For the same reason that it's a hell of, if you're just out for who are Communists and who isn't it's a hell of a bad thing to say well everybody who is a labor union man or everybody who hollers for a second front or some of the other things which so happen to be same things that the Commies are interested in, to say they're Commies. Of course, many people, we have to guard against them all the time who want to say just that and of course it's just dam foolishness. At the same time we're presented with a thing of such terrific importance that * * *.

"O. I think it is in a sense of business and loyalty a terribly serious thing. I have not felt that this information in the hands of the Russians was likely to be dynamite in the way of action because of the fact that that * * * because they're so tied up with the Nazis they wouldn't know what to do with it.

"L. There's something I want to ask you about. Do you believe the Russians having this information would have any greater effect than perhaps, I might say, guiding their foreign relations regarding how far they would be willing to go to hold the friendship of the United States, let us say. Of course * * *.

"O. I think it's like this, I think that once the pressure on Russia is reduced to such that they can turn to * * * that they like any other great nation would probably turn to working on this. I think that at a time like that that any information they had gotten about what we are doing would certainly have an influence, but I don't know, I somehow don't see how in the present war.

"L. In other words, you question seriously the whether any serious effects could come of this. We're not interested in the answer to that question. For instance, I don't give a damn whether Hawkins or Peters or anybody are working on this thing because they think it's a chance of a lifetime to learn physics or whether they are supremely ambitious to see this thing completed above all else, or whether they're doing it because they want to give it to their country, or because they want to win the war because they are against Germany and for the United States. I mean, the question really, and practically is not material.

"O. Well, information going to Russia is a very serious thing even if that information is not used in Russia because we have no control over what happens to it.

"L. And they may make a separate peace.

"O. I don't think for this reason that it's a minor point by any means. I don't personally have the feeling that they're working on it.

"L. There's this thing to be considered, dealing as we must with possibilities—they may next month or 2 months from now make a separate peace with Germany.

"O. I don't think that is utterly impossible. I've thought about it a great deal.

"L. And put themselves in precisely the same position they were in when they precipitated the European war by making an alliance in 1939. When they did that they did everything in their power in a neutral way to help Germany. You might also go so far as to say that they followed a state of nonbelligerency of Allied nonbelligerency where they did everything in their power to further strikes and sabotage in this country and propagandize to prevent this country from getting into a position to defend itself.

"O. I may say that I never understood that policy. Even from the point of view which I think is right; namely, that the intent of Russia is to make a strong Russia. I just don't see the point * * *.

"L. Now, can I ask you one more theoretical question. Do you have any, did you ever arrive at any conclusion as to what peculiar psychological trait, or what advantage held out makes native-born Americans of three generations stock go wild, be members of the Communist Party and act in the sole interest of Russia as distinguished from the interests of the United States?

"O. It could be a lot of things, but there have been examples of this strange loyalty to instructors (?) which you might call treachery. * * * It was certainly true in religious wars and I think also during a period in which your emotions or a rather similar situation * * * where people * * * There was tremendous opposition to this revolution. Remembers the Jacobins? And I think that (sic)——

"L. You mean allegiance to the church accomplished the same thing as allegiance to Russia?

"O. It's not quite the same.

"L. It's more or less of a religious aspect.

"O. I think that the one thing that I was trying to say was that a lot of these people join the party and would seem to have no reason to do so, do so out of a very deep sense of right and wrong which does not express itself * * * it's a combination of religious temperament and actual relations * * * that probably comes closest to it.

"L. Have you ever read any of Ruth McKenny's stuff?

"O. I've read one novel.

"L. Did you ever read Jake Home?

"O. No.

"L. She continually speaks in that novel of the central character feeling the call, much as you might refer——

"O. To an evangelist.

"L. To an evangelist or a minister.

"O. There is something like that involved, and some of the people who are on that list that you read me or who I know well have a very deep fervor.

"L. But I can't understand; here's the particular thing about it. They are not adhering to any constant ideals.

"O. No; I more or less feel.

"L. They may be adhering to Marxism, but they follow the twistings and turnings of a line designed to assist the foreign policy of another country.

"O. This conviction that makes it not only hysterical but * * * I think absolutely unthinkable. My membership in the Communist Party. At the period in which I was involved there were so many positions in which I did fervently believe, in correctments (sic) and aims of the party * * * at that time * * *

"L. Can I ask you what period that was?

"O. That was at the time of the Spanish War, up to the pact.

"L. Up to the pact. That is the time you broke, you might say?

"O. I never broke; I never had anything to break. I gradually disappeared from one after another of the organizations. I didn't like the way some came out and wrote letters to the Republic saying they had seen the light, and I had some personal loyalty involved * * * and, too, some of these organizations which I may say all pretty well cracked up.

"L. Yes; they certainly did let them down during that period. I think I've about talked myself down.

"O. I wish, Colonel, that I could do what you want. I'm thinking about this specific point you want this information. I can't deny that I could give you that information. I wish I could do it.

"L. And don't think it's the last time I'm going to ask you, 'cause it isn't.

"O. I think I believe in what you say that you'd stop asking that question if I answered it.

"L. Well, I want to say that personally I like you very much and I wish you'd stop being so formal and calling me Colonel, 'cause I haven't had it long enough to get used to it.

"O. I remember at first you were a captain, I think.

"L. And it hasn't been so long since I was a first lieutenant, and I wish I could get out of the Army and back to practicing law, where I don't have these troubles.

"O. You've got a very mean job and——

"L. I want you to know that I like you personally, and believe me it's so. I have no suspicions whatsoever, and I don't want you to feel that I have, and——

"O. Well, I know where I stand on these things. At least I'm not worried about that. It is, however, as you have asked me, a question of some past loyalties * * * I would regard it as a low trick to involve someone where I would be dollars to doughnuts he wasn't involved.

"L. O. K., sir."

* * * * * * *

Mr. GRAY. You have concluded witnesses?

Mr. ROBB. Yes, sir.

Mr. GRAY. Mr. Garrison, I feel it my duty to raise a point with respect to these proceedings upon the conclusion of testimony by witnesses called by Mr. Robb. This has to do with the scope of the Commission letter to Dr. Oppenheimer and the testimony which has been adduced before this board. You will recall that in your direct examination of Dr. Oppenheimer and in the examination of witnesses called by him, there developed much discussion about matters in which Dr. Oppenheimer participated in the postwar years, which are not referred to in the Commission's letter, and I have in mind specifically as examples the Vista report, the long-range detection problem, various attitudes and statements about strategic offensive air power, some testimony about relative emphasis in the use of this new kind of energy for military purposes as between sea vessels and aircraft, and so forth. These various items are the ones I refer to. I think that I should say to you that these have necessarily and under the circumstances become material to the matters under consideration by this board.

I think I should say to you that I am quite sure that the board will not disregard the testimony and other material before it with respect to these matters. I am concerned that there be no element of surprise to Dr. Oppenheimer or to his attorneys, and I wish to therefore notify you at the conclusion of the evidence which Mr. Robb has presented that these matters are considered material by the board.

Mr. GARRISON. I think the chairman is quite right and we certainly raise no question of the broadening of the Commission's letter in order to avoid surprise. We make no contention of that sort, Mr. Chairman.

Mr. GRAY. I want to be sure I understand it. I believe it was your statement quite early in these proceedings that with respect to these matters there would be no request that the Commission's letter be broadened. I am not sure we were talking about these precise things, but I do remember your making an observation of that nature. I should want to make it perfectly clear that if the Commission's findings ultimately do concern themselves with these matters, among other things, that no one be surprised that that has happened. I want it to be particularly clear.

Mr. GARRISON. It is clear.

Mr. ROBB. I believe you said the "Commission's findings"; I believe you mean the board's findings.

Mr. GRAY. I am sorry, I certainly did mean the board's findings.

Mr. ROBB. Excuse me for interrupting.

Mr. GRAY. No, I am pleased that you did.

May I ask whether you have anything to say with respect to this?

Mr. ROBB. Not at all. We have no objection whatever to the issues being taken in that light. The matter was thought by Mr. Garrison to be material. I am sure it is. Dr. Oppenheimer testified about it, and a number of his witnesses did. I think those are matters which should have been brought to light. I think it entirely appropriate that these issues should be taken to include those matters.

Mr. GRAY. Now, Mr. Garrison, I indicated to you the other day that I thought it was likely that the board would like to put a few questions both to Dr. Oppenheimer and Mrs. Oppenheimer.

I should like now to ask you what you have in mind with respect to anything else you want to offer as attorney for Dr. Oppenheimer.

Mr. GARRISON. Yes, Mr. Chairman. We were a little at a loss to know quite how to plan, because in the first place, we didn't know whether the board itself might want to go into matters further with Mr. Borden. We also understood from Mr. Robb that he might have another witness, and that he would not be able to tell us until late this morning wether he would have another witness or not. So we were a little uncertain whether we would run over this afternoon or not. We tried to reach Mr. Robb at lunch time and were unable to do so.

Mr. ROBB. I am sorry. I could not have told you then.

Mr. GARRISON. We want to introduce a limited amount of rebuttal testimony. We have in mind calling Dr. Oppenheimer, naturally. We have some very short testimony to put into the record from Dr. Bush, Dr. Zacharias, and Mr.

Hill. Mrs. Oppenheimer will be, of course, available whenever you would like to question her. If it would be acceptable to the board it would be helpful to us if we could adjourn and commence tomorrow morning. I think we could probably be done by certainly the early afternoon with our rebuttal. Then if you would like to have Mrs. Oppenheimer present for futher questioning after that we could do it very easily.

If you would like to put what questions you have to Dr. Oppenheimer we can, of course, do that. The only point is that it would be unlikely to begin rebuttal tomorrow because of the uncertainties of the afternoon and the relative lateness of the hour. Dr. Oppenheimer is available, and he would be very happy to submit himself to your questions if you would like to put some yourself to him.

Mr. GRAY. I think we would like to put such questions as we have to him this afternoon. Would there be any chance of getting Mrs. Oppenheimer this afternoon also, because I don't think these would be long appearances, and perhaps we could finish up with that, and then let your start your rebuttal in the morning.

Mr. GARRISON. May I speak to Dr. Oppenheimer about it?

Mr. GRAY. Yes. We will take a short recess.

(Brief recess.)

Mr. GARRISON. Mr. Chairman, I find Mrs. Oppenheimer is not available this afternoon, but will be in the morning. I am now putting in a telephone call to Dr. Bush, so when you are through asking questions to Dr. Oppenheimer, I think we can get him down here on 5 minutes notice. His testimony will not be long.

Mr. GRAY. All right, sir.

Whereupon, J. Robert Oppenheimer, a witness having been previously duly sworn, was recalled to the stand and testified further as follows:

EXAMINATION

By Mr. GRAY:

Q. Dr. Oppenheimer, I think it is probably my duty to remind you that you are still under oath in this proceeding.

A. Thank you.

Q. I have some questions I would like to ask you, and possibly some other members of the board will.

I want now to go back to the so-called Chevalier incident.

A. Right.

Q. I should like to give you something of a summary of what I believe to have been your testimony before the board. If it is not an accurate summary in your opinion, or your counsel thinks it is not an accurate summary, I would like to know about it. But on the basis of a summary, then, of your testimony, I should like to ask some questions.

The summary would be this: You said that Chevalier was your friend in whom you had confidence, and that you were convinced that his remarks about passing information to the Russians were innocent. For these reasons, you testified, it did not occur to you for a long time that you should report this incident to the security officers, and when you did tell them about it, you declined to name Chevalier, because you were convinced that he was innocent, and in effect wanted to protect him from the harrassment of an investigation because of your belief in his innocence.

You testified on the other hand that the story of the Chevalier incident which you told to Colonel Pash in August 1947, and reaffirmed to Colonel Lansdale in September 1943, was false in certain material respects. Let me repeat, you testified here that that story was false in material respects. I believe you testified that this story was a cock and bull story, and that the whole thing was a pure fabrication except for the name Eltenton, and that this fabrication was in some very considerable circumstanital detail, and your testimony here as to your explanation for this fabrication was that you were an idiot, and that you were reluctant to mention Chevalier and no doubt somewhat reluctant to mention yourself.

However, I believe that your testimony indicated that you agreed that if the story you told Pash had been true, it showed that Chevalier was deeply involved, that it was not just a casual conversation, that it would not under those circumstances just have been an innocent and meaningless contact, and that it was a criminal conspiracy.

In short, with respect to that portion of your testimony I believe you led the board to believe that you thought that if your story to Colonel Pash had been true it looked like a very unsavory situation, to say the very best about it.

Now, here is my question: If Chevalier was your friend and you believed him to be innocent and wanted to protect him, then why did you tell a complicated false story that on the face of it would show that the individual was not innocent, but on the contrary, was rather deeply involved with several people in what might have been a criminal espionage conspiracy?

Or to put the question in another way, I ask you whether it is not a fair inference from your testimony that your story to Pash and Lansdale as far as it went was a true story, and that the fabrication may have been with respect to the current version.

A. Let me take the second part of your question first.

Q. Yes.

A. The story I told to Pash was not a true story. There were not three or more people involved on the project. There was one person involved. That was me. I was at Los Alamos. There was no one else at Los Alamos involved. There was no one in Berkeley involved. When I heard the microfilm or what the hell, it didn't sound to me as to this were reporting anything that Chevalier had said, or at that time the unknown professor had said. I am certain that was not mentioned. I testified that the Soviet consulate had not been mentioned by Chevalier. That is the very best of my recollection. It is conceivable that I knew of Eltenton's connection with the consulate, but I believe I can do no more than say the story told in circumstantial detail, and which was elicited from me in greater and greater detail during this was a false story. It is not easy to say that.

Now, when you ask for a more persuasive argument as to why I did this than that I was an idiot, I am going to have more trouble being understandable. I think I was impelled by 2 or 3 concerns at that time. One was the feeling that I must get across the fact that if there was, as Lansdale indicated, trouble at the Radiation Laboratory, Eltenton was the guy that might very well be involved and it was serious. Whether I embroidered the story in order to underline that seriousness or whether I embroidered it to make it more tolerable that I would not tell the simple facts, namely, Chevalier had talked to me about it, I don't know. There were no other people involved, the conversation with Chevalier was brief, it was in the nature of things not utterly casual, but I think the tone of it and his own sense of not wishing to have anything to do with it, I have correctly communicated.

I think I need to say that it was essential that I tell this story, that I should have told it at once and I should have told it completely accurately, but that it was a matter of conflict for me and I found myself, I believe, trying to give a tip to the intelligence people without realizing that when you give a tip you must tell the whole story. When I was asked to elaborate, I started off on a false pattern.

I may add 1 or 2 things. Chevalier was a friend of mine.

Dr. EVANS. Did you say is a friend.

The WITNESS. He was a friend of mine.

Dr. EVANS. Today?

The WITNESS. He was then. We may talk later of our present relations. He was then a friend of mine. As far as I know he had no close relations with anyone else on the project. The notion that he would go to a number of project people to talk to them instead of coming to me and talking it over as we did would have made no sense whatever. He was an unlikely and absurd intermediary for such a task. I think there are circumstances which indicate that there was no—that there would not have been such a conspiracy—but I am in any case solemnly testifying that there was no such conspiracy in what I knew, and what I know of this matter. I wish I could explain to you better why I falsified and fabricated.

By Mr. GRAY:

Q. Of course, the point I am trying to make with you, and that is the reason for the question I asked, is the inference to be drawn from your motive at the time, as I think you have testified, was the protection of an innocent person, because the story you told was certainly not calculated to lead to the conclusion of innocence on Chevalier's part. These inferences necessarily present themselves.

Let me ask this: First, you heard Colonel Pash testify that as a result of the interview with him in which you indicated that there were three other people involved, he and his associates actually held up orders with respect to an individual who was to transfer to Oak Ridge, I think. Were you aware of that at the time?

A. I was not, not until Friday.

Q. I think a few moments ago, you questioned whether you had discussed microfilm in this interview with Colonel Pash.

A. Then I didn't make myself clear. I asserted that I had not discussed it with Chevalier or Chevalier with me. When I mentioned to Colonel Pash, it came in the form of microfilm or whatever the hell, that was the phrase, which is not very precise. May I add a point, Mr. Chairman?

Mr. GARRISON. Just a minute. You are clear he means the phrase in the recording as it was played?

Mr. GRAY. Yes, I understand.

The WITNESS. May I add a point. When I did identify Chevalier, which was to General Groves, I told him of course that there were no three people, that this had occurred in our house, that this was me. So that when I made this damaging story, it was clearly with the intention of not revealing who was the intermediary.

By Mr. GRAY:

Q. Again with respect to Chevalier, can you recall any efforts you have ever made in his behalf with respect to passport difficulties or problems that he may have had? I think you testified about one.

A. Yes. I remember that at the time when his wife had divorced him and he was determined to go to France, I recommended counsel to him to obtain an American passport. He had also a French passport. Without discussing it with me, nor I believe with anyone else, while the negotiations or the effort to secure an American passport were in process, he did leave on his French passport.

Q. Is it clear to you that in your visit in the late fall of 1953 to Paris, you did not in any way get involved in Dr. Chevalier's passport problems as of the present time?

A. I don't believe I became involved in them. I am not even sure he discussed them with me.

Q. You say he did discuss them with you?

A. I am not even sure he discussed them with me. I am sure he discussed one point with me at length, which was his continued employment at UNESCO.

Q. You don't remember discussing with him the best possible way to get information on his part about a passport, or the way to obtain a passport?

A. That could well have happened and I would have referred him to the embassy.

Q. Did you in fact do so?

A. If I were sure I would tell you.

Q. I am putting some of the same questions to you now, Dr. Oppenheimer, that Mr. Robb put earlier.

A. Right.

Q. You had luncheon I believe with Mr. Wymans of the embassy?

A. That is right.

Q. I believe you testified on the question of Mr. Robb you did not discuss Mr. Chevalier's passport problem with Mr. Wymans?

A. No, I saw Mr. Wymans long before I saw Mr. Chevalier; not long before, but well a week before.

Q. Have you been in communication with Chevalier since the time you had luncheon with Mr. Wymans?

A. Yes; I saw Chevalier after my lunch with Wymans, but not the other way around.

Q. Have you been in communication with Chevalier since the evening you spent with him?

A. The next day we drove out to visit Malraux.

Q. Yes; you testified about that. Have you been in communication with him since that time?

A. No. Well, we had a card from him, just for my birthday.

Dr. EVANS. When did you get that card?

The WITNESS. Around my birthday, which was during these hearings. I don't recall this. I could have advised Chevalier to consult Wymans with regard to his passport.

By Mr. GRAY:

Q. I am sure that you could have, because I believe it to be true that he did, and specifically stated that it was at your suggestion that he do so. I want again to ask you whether you had conversations with anybody else other than Chevalier about his passport problem while you were in Paris in the late fall? I think I am asking you, is it clear to you that you did not?

A. It is quite clear to me. If—I believe I saw no one at the Embassy after seeing Chevalier or no one connected with the Embassy.
Q. Do you have any guess or knowledge as to whether Chevalier today is active in Communist Party affairs?
A. I have a strong, strong guess that he is not. I have no knowledge. His new wife is an extremely sensible, wholly un-Communist girl. The other person we saw together was a man who has become a violent anti-Communist and is now apolitical. I don't have knowledge.
Q. The record shows, I believe, Dr. Oppenheimer, that you continued probably until sometime in 1942 to make financial contributions which went to Communist causes, with money passing to different people, but among others, Folkoff was one who was known to you to be a Communist Party officer.
A. That is right.
Q. Did you discuss these contributions with Mrs. Oppenheimer? Was she aware that you were making these contributions?
A. I would assume that we discussed everything in our life at that time.
Q. Did she make any contributions on her own account?
A. I have no knowledge of that.
Q. I see.
A. I am sure that everything was quite open between us. She has told me that she may have given Steve Nelson some money. She remembers that not as a contribution for a cause, but as something she was giving Nelson for his own use. But I have no recollection of it.
Q. If you made contributions as late as 1942, and this fact were known to Mrs. Oppenheimer, it was certainly clear to her at that time, or should have been clear to her at that time, that these funds were going to Communist Party causes because of her previous membership and presumably full awareness of the methods of operation.
A. I hate to say so, but I think as to this you will have to ask her. My recollection of her Communist Party experience was a very limited one—very hard work with the steel union and mimeographing and things like that—and I doubt whether she was at any time what you would call an expert on how Communists dealt with things.
Q. I don't think I have heard suggested at any time that Mrs. Oppenheimer was politically naive. I don't believe that you have made that suggestion, although there has not been much testimony about her, I might say that anything I have read or heard in or about these proceedings would indicate nothing other than a pretty full knowledge of what she was about. I agree with you, however, that some of these questions should be put to her.
I want to go back now, Dr. Oppenheimer, to a portion of your testimony which related to this matter of ceasing political activity by those who came into the active service on the project. I believe you testified that as to some of these individuals, whose names I don't recall at the moment, you told them that they would have to cease their political activities, and you testified that by that you meant making speeches, et cetera.
Now, do you today take the view that ceasing political activity, whatever is encompassed in that phrase, is an adequate safeguard even though you think you know the individual and trust his innocence and loyalty completely?
A. Today? No. Well, I think there is nothing better to go on than the judgment of a man, but I am not suggesting that it should not be supplemented by whatever evidence is available as to what the man is up to.
Q. Let me put a hypothetical question to you.
A. Perhaps I did not understand you.
Q. Suppose you today had a friend in whom you had the highest degree of confidence as to his loyalty to this country and his discretion and his character; assume further that you could make the judgment to your own satisfaction that this man would never yield in the matter of protecting the security interests of this country? Incidentally, you happen to have reason that he was a member of the Communist Party. I am asking whether again you would say, well, it depends on the individual.
A. Let me first point out an implausibility in the hypothetical question. I would not today suppose that a man who is a member of the Communist Party, was now or recently a member of the Communist Party, whatever his other merits, could put the interests of the United States above those of a foreign power. But if we can relax it a little bit and say that I know a man who once was a member or who I had reason to think was once a member of the Communist Party, and whom I knew well and trusted, and of whom there was question of his employment on serious secret work, I would think it would not

be up to me to determine whether his disengagement from the Communist Party was genuine. I would think that at this time investigation would be called for. But I could have a very strong conviction as to whether that disengagement had in fact occurred and was real or whether the man was fooling me.

I would like that conviction to be supported by other evidence. It should be. In other words, I would not act today as I did in 1943 for a whole lot of reasons.

Q. What would you consider to be adequate in the way of an act of disengagement? What kind of thing?

A. A man's acts, his speech, his values, the way he thinks, the way he talks, and the fact of his disengagement. The fact that there are no longer any threads binding him to an organization or connecting him with an organization. These would be some of the things. And no doubt his candor.

Q. Would you expand on this candor point a little bit because I am wondering whether you are saying that his own statement about disengagement is to be made a primary factor in a determination.

A. That certainly does depend on the man. His statement that there was something to disengage from is something I should think would be relevant.

Q. Let me turn now to the so-called Vista report about which there has been very considerable testimony and not altogether consistent. Did you in fact prepare a draft of an introduction to chapter 5 of the Vista report?

A. Yes; I did. It was not a solitary labor. When I got there, I found a mass of drafts, papers, and notes. People who had written these were Christie, Bacher, Lauritsen, possibly others. But those were the principal ones. Christie had spent quite a lot of time at Los Alamos quite recently. We went over what they wanted to say and sometimes discussed it from the point of view, did they really want to say it, and were they sure that this was what they wanted to say. I think my contribution to the writing of this was that I—well, let me back off.

The principal thing they wanted to say was that atomic weapons would be useful in the defense of Europe, in the antiair campaign, and many other ways that you will know as much about as I do, and that for this to happen, developments of hardware, of tactics, of command structure, of habits of behavior, of exercises needed to be gone into, which would give to our tactical readiness at least a small part of the training and precision which the Strategic Air Force already had. I believe my contribution apart from incidentals to the writing of this report was a notion that occurred very early and I believe has remained in all drafts, and that is still basic to my own views, and that is that this is not a very fully known subject—what atomic weapons will do, either tactically or strategically, that as you go into battle, you will learn a great deal, and the primary preparation must be of two kinds. First that you have capabilities which allow you a lot of options, which give you choices that you can make at the time, and second, that you be so set up that if your guesses have been wrong, your technical preparations are such that you can change quickly in the course of the battle. If you are wrong about the effect of a bomb on an airfield, if you are not getting away with it, that you can make the proper reassignment of fissionable material and hardware and aircraft to do what is effective. These were the two guiding ideas that I believe I brought into the organization of the report.

I then with the help of the others drafted a chapter—either chapter 5 or its introduction, I don't remember which it was called. It was a matter of some 20 pages, I believe, and had some twenty-odd recommendations.

Q. Was there in this draft at any stage the suggestion that the United States, this country, should state that it would not use atomic weapons strategically against the Soviet Union until after such weapons had been used against American cities?

A. Let me say the best of what I recollect was in there. It is related to the question you asked but it is not identical with it. We said that we were in a coalition with the Europeans and that one of the things which we must be alert to is how the Europeans would view the destruction of their own cities by the enemy. Therefore, we needed to envisage the situation that would occur if we used our strategic air as a deterrent to the destruction of Europe's cities, as well as our own, and in that circumstance there was still a great deal that could and should be done with atomic weapons, and that we should be prepared for that contingency. We did not recommend a proclamation.

Q. Was there in the language of the draft at any time a recommendation——

A. I believe this is pretty close to the language of the draft what I have told you. It was not a recommendation that this be the course of history. It was the contemplation of a possible course of history.

Q. Did what you recall and what you have just testified to appear in the final document?

*   *   *   *   *   *   *

I may say, Mr. Chairman, that I think in the papers that were sent down to Washington, there is not only a copy of the final Vista report, but there is a chapter or draft of chapter 5 as we took it to Europe. I have not seen it for several years, but if these are material points, I think that they can be found by you, if not by me.

Q. I have stated earlier, I think, that in the course of this proceeding the board has come to the conclusion that they have a material bearing. I will indicate to you, Dr. Oppenheimer, another respect they might be material.

First of all, it is true that the statement of the strategic use of atomic weapons was very important to the Air Force, particularly, and to the Defense Department. So therefore it was a material thing in this report.

A. Yes.

Q. We have had testimony from witnesses called by you of people who should have known everything that went on from beginning to end of this Vista report, who testified clearly and unequivocally that there was no important change, that it was only a minor language change and a minor change in emphasis. I as a member of the board am confused by this testimony.

A. May I say a few words more?

Q. Yes.

A. We took this chapter 5—I will not say as I had drafted it, but as it had been drafted, and perhaps amended and fiddled with a little bit, but with at least some of the provisions in it which were disturbing and the language which was disturbing. We took it to Europe. We showed it, I think, to General Schuyler, General Gruenther, and General Eisenhower. While we had been on our way to Europe General Norstad had been called home for consultation. It was several days later that we showed it to General Norstad. He expressed an objection. This objection was in rather formal terms. * * * There was so much in the report that was good that he hoped we would take out the things that were not acceptable. He said possibly in a couple of years, a few years, the kind of thing you are thinking about will be realistic, but this is too early. It just won't work.

We said to him, after recovery, we think that this may be a matter of substance, and it may be a matter of language. Let us rewrite this in order to remove from it those phrases and those arrangements of ideas which appear to be bothering you and see if then this statement of the case is one which is satisfactory to you.

We did so. We showed it to him a day or so later and he said, "If I am asked, I will tell the Chief of Staff and the Secretary that I think this is a fine report and very valuable."

Now, it is clear that our critics thought these were substantial changes. It is clear that the authors of the report didn't think so, or they would not have made them.

Q. Was it clear to the authors that the critics thought they were substantial?

A. It was clear to me, sir.

Q. You would never describe these changes as mere changes in language or minor shifts in emphasis; would you? Excuse me for putting the question that way. Would you so describe them?

A. Since the principal purpose of the report was to point out the many ways in which atomic weapons—or at least relatively new ways in which atomic weapons could play a part in the battle of Europe, I think that the heart of the report was wholly unchanged. I still think if we are ever called to fight the battle of Europe, we will have to face up to the questions of how deep, how massive, and of what quality will our atomic source be. I do not think it was necessary to raise this question in that chapter of Vista because our arguments were solid without it. But they were even more solid. The reason we did was that at the time we didn't have much armament. Europe was not easy to defend, and the point that we wished to make was that there was more than one way in which the atom could be used in what might be a very critical campaign.

Have I lost your question?

Q. Yes; it is all right, but I want to get back to it. You did not suggest—I think you are testifying that you did not suggest—in any draft that we make a statement that we would not use these weapons strategically, that is, with respect to the U. S. S. R., unless and until they had first been used against our cities, and industrial centers.

A. We did not recommend such a statement.

Q. Did you by implication recommend such a position?
A. My memory, and I probably should be less categorical than I am, my memory is that we contemplated a situation in which we would in fact not do this. In fact, we would not use an all-out strategic attack, but consider our Strategic Air Force as a deterrent to Soviet attack upon the cities of our allies and our own.

Q. As far as you know, about the final version of the Vista report, did that notion appear?
A. Yes; there was still a remark that the deterrent effect of our Strategic Air Command with regard to the protection of allied capitals would be an important factor to take into consideration along with many others. This is from memory, but they were things I was interested in and I believe I am telling the true story.

Q. As long as your memory serves, did you at the time think we should have a policy, whether publicly announced or not, which would lead us to suffer atomic attack upon our cities before we would make a similar attack upon Soviet cities?
A. I think the question of our own cities, Mr. Gray, never came into this report, or at least was not the prominent thing. The prominent problem——

Q. I didn't ask about the report, then. I asked in your best recollection was this a view you entertained.
A. That we would welcome an attack on our own cities?

Q. No; I don't think that is an accurate restatement of my question. I said that we would suffer an attack upon our cities with the use of atomic weapons before we would ever make a strategic strike against the U. S. S. R.
A. Oh, lord, no. I mean the very first thing we would do against the U. S. S. R. is to go after the strategic air bases and to the extent you can the atomic bases of the U. S. S. R. You would do everything to reduce their power to impose an effective strategic attack upon us.

Q. Which might include attacks on cities and industrial concentrations.
A. It might, although clearly they are not the forward component of the Strategic Air Command.

Q. Perhaps we are tangled up with the question of strategic?
A. I have always been clear that the thing that you do without fail and with certainty is to attack every air base that has planes on it or may have planes on it the first thing. I believe our report said that.

Q. I will try again. Did you have at that time the view that we should not use the atomic weapons against any militarily promising target which might includes cities in the U. S. S. R. until after such weapons had been used against such targets in this country?
A. I think I have never been entirely clear on that. This seemed to me one of the most difficult questions before us. I am sure that I have always felt that it should be a question that we were capable of answering affirmatively and capable of thinking about at the time.

Q. This is not clear in your mind as to what our position should be, you say. Have you ever thought about it in terms of a public announcement as to policy in that regard?
A. This has always struck me as very dangerous.

Q. Then you did not advocate a public announcement?
A. You mean have I publicly advocated it?

Q. No. I mean did you feel that the United States should make a public announcement about its policy, whatever it might be, with regard to the use of atomic weapons against the Soviet Union against whatever targets might present themselves?
A. In the 9 years we have been talking about these things, I have said almost everything on almost every side of every question. I take it you are asking whether in some official document 1 unequivocally recommended that we make a public pronouncement of our policy with regard to this, and to that my best and fairly certain answer is "No."

Q. I really asked you what your own personal view was.
A. I think that we had better not make public announcements about what we are going to do, if and when. But I do think we need to know more about it and think more about it than we had some years ago.

Q. You don't think the import of the original draft of the introduction to chapter 5 was to this effect?
A. No. It was to call very prominently to the attention of the services that there might be considerations against the then present air plan, and that nevertheless there were very important things to do with the atom.

Now, I would feel a little more comfortable if I had a draft of chapter 5 of Vista that we are talking about before me.

Q. I have not seen it my self, Dr. Oppenheimer.

I have asked yo a lot of questions about how the crash program, as the issue, came before the General Advisory Committee in the meeting in October 1949. Perhaps I asked you some questions about that.

A. I think you did.

Q. But in any event, has the testimony, all that you have heard in the last weeks, made it clearer to you how this came as the alternative, crash program or not?

A. I am a little clearer. I think the greatest clarification came from Dr. Alvarez' testimony. It is clearer to me now than it has been before that in the meeting with the Commission, the Commission probably through its chairman— told us what was on their minds. It is clear to me that the Commission was being beseiged by requests to authorize this, to proceed with that, all on the groud that these were the proper ways to expedite the thermonuclear program, and all on the ground that the thermonuclear program was the thing to do. It is clear to me that the Commission asked for our views on this.

Q. Looking back on it, do you feel that the GAC in consistency and with technical integrity could have recommended something short of the crash program, but something at the same time that was more active and productive than the alternate program?

A. Indeed I do. Indeed I do. We could have very well written the report to the following effect, that the present state of the program is such and such as we see it. This we did do. That in order to get on with it, this and this and this and this would need to be done. This we did do. We could have said that the present state of fog about this is such that we don't really know just what the problem is that is to be decided. Let us get to work and remove as much of this fog as fast as possible.

We could further have said the decision as to whether this is the important, the most important, an important, an undesirable or disastrous course involves lots of considerations of which we are dimly aware in the military and political sphere, and we hope that these will be taken into account when the decision is made. We could have written such a report.

I think apart from what personal things, feelings, still of the people involved, the best explanation of why we wrote the kind of report we did was that we said what we thought, rather than pointing out that there were other people who could be asked to evaluate (a) because we thought, and (b) because the pressure, the threat of public discussion, and the feel of the time was such that we thought our stating our own case, which was a negative case, was a good way, and perhaps the only way to insure mature deliberation on the basic problem, should we or shouldn't we.

Q. And your position as reflected in the report under no circumstances should we?

A. I think that is not quite right. I think the report itself limits itself to saying that we are reluctant, we don't think we should make a crash program, we are agreed on that, and that the statement in the majority annex that it would be better if these weapons were never brought into being was a wish, but it was not a statement that there were no circumstances under which we would also have to bring them into being.

Q. Wouldn't you say that the impression that the majority annex was calculated to give was that those who signed it were opposed to anything that would lead to the development of the hydrogen bomb?

A. That is right, under the then existing circumstances.

Q. So that really the majority in effect would not have been sympathetic with any acceleration of the program which would lead to the development of the bomb?

A. Of course. That does not mean that we would not have been sympathetic to studies and clarification. This was a question of whether you were going to set out to make it, test it, and have it.

May I make one other comment? This was not advice to Los Alamos as to what it should or should not study. This was not advice to the Commission as to what it should or should not build. Some such advice we gave in that report. This was an earnest, if not very profound, statement of what the men on that committee thought about the desirability of making a superbomb.

Q. And they felt that it was undesirable?

A. We did.

Q. If the Commission had taken their advice, or if the Government ultimately had taken the advice of the General Advisory Committee, we would not now have it.

A. I am not certain of that, but it is possible.

Q. Your advice, it seems to me, has said, and as I interpreted it, the majority annex was that we should never have it. I would guess if that advice had been taken literally the Commission would have——

A. The majority annex 1 still think never said that we should not have it. I think it said that it would be better if such weapons never existed.

Q. I think this is an important point, and I would like to hold on that.

A. All right. But could we have the context which I also have forgotten?

Q. Yes. I will try not to take it out of context.

Mr. ROBB. Here is the majority annex.

Mr. GARRISON. Mr. Chairman, would it not be helpful if Dr. Oppenheimer could look at the report which he has not seen for some time?

The WITNESS. I saw it the other day.

Mr. GRAY. I will show it to him again. I want to pick out the portions that I think are pertinent here, and let him make any observations about context. The security officer cautioned me that I am really getting on difficult ground. May I interline this?

Mr. ROLANDER. Yes.

The WITNESS. Does the majority annex contain information which should not be on this record?

Mr. GRAY. I do not know.

(Mr. Rolander handed copy of report to Dr. Oppenheimer.)

The WITNESS. I would like to quote the entire paragraph, if that is permissible. I see something—well, I don't know.

Mr. GRAY. I see no reason why the whole paragraph should not be quoted.

Mr. ROLANDER. It is all right.

The WITNESS. This is the fourth paragraph of a six paragraph annex:

"We believe a super bomb should never be produced. Mankind would be far better off not to have a demonstration of the feasibility of such a weapon until the present climate of world opinion changes."

That is that paragraph in its entirety.

By Mr. GRAY:

Q. That language is pretty clear, isn't it, that "We believe a super bomb should never be produced"?

A. Sure it is.

Q. So that there was not any question that the six people of the majority were saying that we should not take steps to develop and produce.

A. Let me indicate to you——

Mr. GARRISON. Mr. Chairman, could he read it once more, because it is the first we heard it.

The WITNESS. This is one paragraph. The document is full of the word "mankind" and this paragraph reads:

"We believe a super bomb should never be produced. Mankind would be far better off not to have a demonstration of the feasibility of such a weapon until the present climate of world opinion changes."

Let me indicate——

By Mr. GRAY:

Q. The question I would ask which would be related to this paragraph is—I am not attacking the motivation of those who held that belief, I am simply saying that the belief is clearly stated there, that the super bomb should never be produced.

A. That a super bomb should never be produced. But look at what that means. If we had had indication that we could not prevent the enemy from doing it, then it was clear that a super bomb would be produced. Then our arguments would be clearly of no avail. This was an exhortation—I will not comment on its wisdom or its folly—to the Government of the United States to seek to prevent the production of super bombs by anyone.

Q. Again, without reference to its wisdom or its folly, is it unreasonable to think that the Commission, reading this report or hearing it made, whichever form it took, would believe that the majority of the General Advisory Committee recommended that the Government not proceed with steps which would lead to the production of a super bomb?

A. That is completely reasonable. We did discuss this point with the Commission on two subsequent occasions. On one occasion we made it clear that nothing in what we had said was meant to obtain, should it be clear or should it be reasonably probable that the enemy was on this trail.

In another, we made it clear that there was a sharp distinction between theoretical study and experiment and invention and production and development on the other hand. So that the Commission, I think, had a little more than this very bald statement to go on.

Q. Dr. Oppenheimer, I am looking at——
A. May I see that, too?
Q. Yes; you may. I am going to show it to you.
A. Do I have it?
Mr. ROLANDER. Yes. Part 1.
The WITNESS. Right. I have it before me.

By Mr. GRAY:

Q. May I ask whether all of Dr. Alvarez' or if none of Dr. Alvarez' testimony was treated as restricted?
Mr. ROBB. No, sir; it was not.

By Mr. GRAY:

Q. Dr. Oppenheimer, in part 1, paragraph 3——
Mr. GARRISON. What document?
The WITNESS. This is a top secret report of the General Advisory Committee dated October 30, 1949. This is the report as such, as distinct from the annexes.
Mr. GRAY. May I read this sentence?
Mr. ROLANDER. The difficulty is that I have not obtained from the Commission the approval to quote directly the minutes of this meeting. It seems to me that if general statements are made with reference to either Dr. Oppenheimer's recollection or general questions are raised, it would be proper.
Mr. GRAY. All right.
The WITNESS. Is it the last sentence?

By Mr. GRAY:

Q. That is right.
A. Fine.
Q. Again, that is pretty clear, isn't it?
A. Indeed it is. I think this has been read into the record by Mr. Robb.
Mr. ROBB. I don't know. I was under the same handicap that Mr. Gray is laboring under. I don't know whether I read it to you or paraphrased it, but you and I knew what we were talking about.
The WITNESS. Yes; it is in the record.
Mr. ROBB. It may have been that it was in the classified portion of the testimony.
The WITNESS. We recommended a certain reactor program, we had a lot of reasons for it, and we said that one of the reasons might be that this would be useful for the super and that reason we did not agree with it, and it was understood that building this reactor was not a step in making the super. That seems to be a paraphrase.

By Mr. GRAY:

Q. If you will look at page 4 of that document, the first sentence in the last paragraph that begins on that page.
A. Right.
Q. Reference to the majority of the committee there makes it clear——
A. Wait now. I am not with you. The second paragraph, page 4?
Q. No; the first sentence in the last.
A. I have only two paragraphs on my page 4.
Q. There is a sentence that begins, "We are somewhat divided"——
A. Right, I have that.
Q. That sentence, and the following sentence.
A. Right.
Q. From that it would appear that the majority of the members of the GAC at that time felt unqualifiedly that they opposed not only the production, but the development.
A. Right.
Q. So that my question to you is, in this proceeding there has been a lot of testimony that the GAC was opposed to a particular crash program. Isn't it clear that it was not only the crash program that the majority of the GAC

found themselves in opposition to, but they were just opposed to a program at all which had to do with thermonuclear weapons?

A. I think it is very clear. May I qualify this?

Q. Yes, you may.

A. I think many things could have qualified our unqualified view. I have mentioned two of them. I will repeat them. One is indications of what the enemy was up to. One of them is a program technically very different from the one that we had before us. One of them a serious and persuasive conclusion that the political effort to which we referred to in our annexes could not be successful.

Q. Now, following the Government's decision in January 1950, would it be unfair to describe your attitude toward the program as one of passive resistance?

A. Yes.

Q. That would be unfair?

A. I think so.

Mr. GARRISON. Unfair, Mr. Chairman?

Mr. GRAY. He said unfair to so describe it.

By Mr. GRAY:

Q. Would it be unfair to describe it as active support?

A. Active could mean a great many things. I was not active as I was during the war. I think it would be fairer to describe it as active support as an adviser to the Commission, active support in my job on the General Advisory Committee. Not active support in the sense that I rolled up my sleeves and went to work and not active support in the sense that I assumed or could assume the job of attracting to the work the people who would have come to a job in response to a man's saying, "I am going to do this; will you help me."

Q. You testified that you did not seek to dissuade anyone from working on the project.

A. Right.

Q. There have been a good many others who have given similar testimony. It also, however, has been testified there there would have been those who would have worked on the project had you encouraged them to do so.

A. There has been testimony that there were people who believed this.

Q. Yes. Do you believe that?

A. I think it possible. Let me illustrate. In the summer of 1952, there was this Lincoln summer study which had to do with continental defense. On a few limited aspects of that I know something. On most I am an ignoramus. I think it was Zacharias that testified that the reason they wanted me associated with it was that that would draw people into it. The fact that I was interested in it would encourage others. In that sense I think that if I had gone out to Los Alamos even if I had done nothing but twiddled my thumbs, if it had been known that I had gone out to promote the super, it might have had an affirmative effect on other people's actions. I don't believe that you can well inspire enthusiasm and recruit people unless you are doing something about it yourself.

Q. Furthermore, it was fairly well known in the community—that is, the community of physicists and people who would work on this—that you had not been in favor of this program prior to the Government's decision. That probably was a factor?

A. I would think inevitably so.

Q. Do you think that it is possible that some of those individuals who were at Princeton whose names were suggested for the project might have gone had they thought you were enthusiastic for the program?

A. I don't believe this was the issue. For one thing, I know that I said to all of them that it was a very interesting program and that they should find out about it. For another—I am talking about a group of people that has been testified to, but as to whom I don't know who they were, I don't know what these names are—but the issue has usually been, should a man give up his basic research in science in favor of applied work, and I believe it was on that ground and on the personality ground as to whether they did or did not want to work with Dr. Teller, and whether they did or did not want to go to Los Alamos, the decisions would have been made. I don't think my lack of enthusiasm—I don't believe I would have manifested any, nor do I believe it would have been either persuasive or decisive. This is in that period after we were going ahead.

Q. Do you remember at approximately what date it was that you offered to resign as chairman of the General Advisory Committee?

A. Yes, approximately. It was when Mr. Dean had taken office, the first time I saw him. That would have been perhaps late summer of 1950. I believe I testified that at the time of the President's decision Dr. Conant told me he had recently talked with the Secretary of State, that the Secretary of State felt that it would be contrary to the national interest if either he or I at that time resigned from the General Advisory Committee; that this would promote a debate on a matter which was settled. The question was how soon after that could this be done.

I talked to Mr. Dean, not primarily about quitting the Advisory Committee, but about quitting the chairmanship about which by then I felt not too comfortable. That would have been August, September of 1950.

Mr. GRAY. I think I have no more questions. Dr. Evans.

Dr. EVANS. Dr. Oppenheimer, you said you had received a birthday card from Chevalier?

The WITNESS. Yes.

Dr. EVANS. He is now in France, is that it?

The WITNESS. Yes.

Dr. EVANS. Is he teaching or writing?

The WITNESS. I remember very much what he is doing because he discussed this with us. He is translating, and part of his job is translating for UNESCO, or was. I don't know that it still is.

Dr. EVANS. May I ask you this question. Have you received any cards or letters from any of these other men like Peters, Hawkins, Weinberg, or Serber?

The WITNESS. We had a birthday card from Mr. and Mrs. Serber, not from the others.

Dr. EVANS. Where is Mr. Serber now?

The WITNESS. He is a professor at Columbia and a consultant to the Atomic Energy Commission establishment at Brookhaven.

Dr. EVANS. And you say you didn't hear from the others?

The WITNESS. No.

Dr. EVANS. This has not much to do with this case. Did you see a little squib in the Washington Post this morning saying if the English had made a superbomb——

The WITNESS. I didn't see it.

Dr. EVANS. I was interested in it. I didn't put much confidence in it, but I was interested.

The WITNESS. I didn't see it.

Mr. EVANS. That is all.

Mr. GRAY. Mr. Robb.

By Mr. ROBB:

Q. Doctor, what was the address on that card from Dr. Chevalier? Was it addressed to you at Princeton or here?

A. I think it was addressed to Princeton and forwarded here. I don't know.

Q. Was there any note with the card?

A. I think there was.

Q. Do you recall what it said?

A. No; I can find this. It is back——

Q. Do you receive a card from him every year at your birthday?

A. No; this was my 50th birthday.

Q. Do you know how he knew that?

A. No.

Q. Do you recall what the note said?

A. Not very much.

Q. Any?

A. It didn't say very much, and I don't recall it. It was written by his wife and it said greetings from our Butte.

Q. Our what?

A. Our Butte. They live on a hill.

Q. Doctor, you testified you didn't feel too comfortable as chairman of GAC in 1950; is that right?

A. Yes.

Q. Why not?

A. Because on a very major point of policy I had expressed myself, had become identified with a view which was not now national policy. I thought that there could be strong arguments for having as chairman of that committee someone who had from the beginning been enthusiastic and affirmative.

Q. Did you feel that others of the scientific community might well feel that you still were not enthusiastic?

A. This is not a consideration that crossed my mind at that time. I think I had more in mind that when on an important thing a man is overruled, his word is not as useful as it was before.

Q. Do you now feel that others in the scientific community might then have believed that you still were not very enthusiastic about the thermonuclear?

A. I know that now.

Q. Do you now feel that your lack of enthusiasm which might have been communicated to other scientists might have discouraged them from throwing themselves into the program?

A. I think this point has been discussed a great deal. I don't have substantive knowledge about it. I think that the critical, technical views which the General Advisory Committee expressed from time to time had a needling effect on the progress at Los Alamos which probably had something to do with the emergence of the brilliant inventions.

Q. To get back to the question, Doctor, would you mind answering that question?

A. Could you say it again?

Mr. ROBB. Would you read it?

(Question read by the reporter.)

The WITNESS. I suppose so.

By Mr. ROBB:

Q. Doctor, you mentioned the brilliant invention. That was Dr. Teller's?

A. It was indeed.

* * * * * * *

Q. Who were the principal, to use the newspaper phrase, architects of the thermonuclear?

A. Teller.

Q. Teller. You would not say you were?

A. No. There is a part of all these things that I did invent. As I testified, it is extremely useful, but it is not very bright.

Q. Is that the one you got the patent on?

A. This is mentioned in the patent, but it is only a part of what we got the patent on. Most of what we got the patent on was wrong.

Q. Doctor, exploring for a bit your work in recent years on the thermonuclear, I believe you testified previously some days ago that you had been thinking about it and trying to learn about the program; is that right?

A. It would be a reasonable thing for me to have said.

Q. When you did do any work for the Atomic Energy Commission, you were on the basis of a per diem consultant, were you?

A. You mean since I left the GAC?

Q. Yes.

A. Yes, sir.

Q. And when you did any work for other agencies, you were on the basis of a per diem consultant?

A. I think with the GAC, with the RDB, with most of these we were paid for days at work and in travel.

Q. Yes.

A. In the case of the Science Advisory Committee, there was no pay. In the case of the State Department panel there was no pay. But there was some kind of subsistence allowance.

Q. Any work that you may have done on the thermonuclear program would have been done for the Atomic Energy Commission, wouldn't it?

A. Any traveling around or anything like that. If I thought about things at home, that would not be charged to the Atomic Energy Commission.

Q. No. I have before me a record showing that in 1953 your total compensation received from the Atomic Energy Commission was $250. Would that accord with your recollection?

A. It would be consistent with it. I would have no recollection.

Q. That would amount to some——

A. Two and a half days. This would certainly correspond to a visit to Los Alamos or Sandia.

Q. Did you in 1953, go to Los Alamos or Sandia in connection with the thermonuclear program?

A. I did in 1953.

Q. But that amount of work would fall within the scope of your statement that you did not take your coat off on this program, wouldn't it?

A. I was thinking of the earlier days when I was a member of the GAC.

Q. Yes.
A. I still didn't take my coat off.
Q. Doctor, I would like to return briefly to Vista. That was a project which was carried out in Pasadena?
A. The headquarters were in Pasadena, and all the activities I know of where in Pasadena. No, no. There were things that I didn't participate in, field trips, inspections.
Q. But your connection with it had to do with Pasadena.
A. It did.
Q. Did you go to Pasadena in November 1951?
A. I went out in the fall. I don't remember the date.
Q. How long were you out there?
A. Not less than a week nor more than two is my best guess. Perhaps only 6 days.
Q. Was that toward the end of the project?
A. It was toward the end of the writing of the report.
Q. Did you complete your answer?
A. Yes, I answered the question.
Q. While you were there on that occasion did you prepare a draft of an introduction to chapter 5 of the report?
A. I prepared what I believe to be a draft or had helped to prepare a draft of chapter 5, not the introduction.
Q. Was that presented to the people who were there by Dr. DuBridge?
A. As to that I have heard only his testimony or your questioning. I was not there.
Q. You were not there?
A. No.
Q. Let me ask you, Doctor, in order that you may have a chance to comment on it on the record, and that the record will be plain, in that draft that you prepared was there anything about dividing the stockpile of atomic weapons into three parts?
A. There was indeed. I think again the phrasing was not quite that. This was something that I found in the working papers when I got there. It had been worked over with great elaborateness. I believe that the phrasing was, we may consider, or we may think of, our stockpile should be thought of as divided roughly into three equal parts. I think that is the way it went.
Q. One part to be held in reserve, one part assigned to the Strategic Air Command, and the third part assigned to the tactical defense of Europe, is that right?
A. To tactical air.
Q. That was in that draft.
A. I believe so. It was certainly in the talk, in the papers that I found there. I am not even sure that it was missing from the final Vista report.
Q. That was my next question. First, was it in the draft of chapter 5 which you testified you prepared after you got there?
A. I believe so, yes.
Q. Was that in the final report?
A. As to that, I don't remember.
Q. The best evidence of that would be the final report.
A. That is right.
Q. Did you inform yourself as to what the final report was?
A. I read it. I had an awful time getting it. Everybody had an awful time getting it. I read it long after it was submitted.
Q. That suggestion as to the division of atomic stockpile was a pretty important matter, wasn't it?
A. We thought of it as rather important because we thought it diverged from the existing policy, and would almost certainly not be accepted in full, but that the direction in which it went was a healthy direction.
Q. It represented in effect some restriction on the freedom of action of the Air Force, didn't it?
A. Very little, because the main emphasis was that whatever you thought, you should be able to convert from one to the other at a minute's notice.
Q. But if the Air Force could use its atomic weapons in any way it chose, it was a restriction to say that you ought to divide it up * * * and assign each part to a particular function, wasn't it?
A. I think this is quite a misrepresentation. We were not given an Air Force which could use its atomic weapons in any way it chose.

* * * * * * *

Q. Given an Air Force which had no such restrictions, this certainly represented a change in policy.
A. If the Air Force had no restrictions, any restriction would be a change of policy.
Q. Was there in the draft of the report which you prepared or your visit to Pasadena in the fall of 1951 any suggestion that the United States should announce that no strategic air attack would be directed against Russia unless such an attack were first started by Russia, either against the European Zone of Interior or against our cities or against our European allies?
A. I have testified on this as fully as I could in response to the chairman's questions.
Q. I want to have it specific, if I may, Doctor—a specific response to that particular question.
Mr. GARRISON. Mr. Chairman, is Mr. Robb reading from the record?
Mr. ROBB. No, sir, I am not. I don't have it. This is a draft, and we can't find this draft.
The WITNESS. I can tell you where you can find it.

By Mr. ROBB:
Q. Before you do that, would you mind answering the question?
A. I would mind answering it, because I have been over this ground as carefuly as I know how. When you say "suggest," I don't know whether you mean recommendation or consideration.
Q. Was there any language in the report to that effect?
A. To what effect; that this might be the state of affairs?
Q. That this might be a good idea.
Mr. GARRISON. What might be a good idea? I am lost.

By Mr. ROBB:
Q. Was there any language in the draft to the effect that it would be a good idea if the United States should announce that no atomic attack would be directed against Russia unless such an attack was first started by Russia either against our Zone of Interior or against our European allies?
A. To the very best of my recollection, we said we may be faced with a situation in which this occurs.
Q. We may be faced with a situation in which that was desirable; is that right?
A. Yes; in which it is wise, or in which it is done.
Q. Was there any language in the final draft or the final report which said that?
A. In the final draft of the final report it said that in the consideration of the use of our strategic airpower, one of the factors should be the deterrent value— I have not got the words—the deterrent value of this strategic air in the protection of European cities.
Q. Do you consider that to be different from the language we have talked about before?
A. It is manifestly different language.
Q. Yes. And don't you think the difference is important?
A. It was very important to our readers.
Q. Was there any language in the draft of the Vista report when you were out there to the effect that at the present state of the art the value of the thermonuclear weapons could not be assessed, and therefore they were not included in your study?
A. This is something which I found written when it was out there. It is not something that I myself wrote, and I don't know whether it was in my draft or not.
Q. Did you agree with it?
A. As far as tactical things, quite definitely. I was not present during the discussions to which Griggs referred at which Teller had talked about it. I don't know whether the value of thermonuclear weapons as tactical weapons has been or can be assessed.
Q. You restrict it to tactical weapons. Suppose you take that restriction off. Was there anything in the report that the value of the thermonuclear weapon could not be assessed?
A. As to that, I don't remember.
Q. Doctor, you testified that Mrs. Oppenheimer has told you that she may have given some money to Steve Nelson; is that correct?
A. Yes.

Q. Did she tell you how much?
A. No.
Q. Did you ask her?
A. Yes.
Q. What did she say?
A. She said she didn't remember. Not that she had told me that she had given, but that she may have given.
Q. Did you ever give Nelson any money?
A. I don't believe so.
Q. Mr. Gray asked you some questions about your contributions that you made from time to time that you told us about before. Let me ask you, did you ever receive any receipt for those contributions?
A. I don't believe so.
Q. Did you ever sign any pledge to make contributions?
A. Oh, no.
Q. Did you ever make any moral agreement with respect to the amount of your contributions?
A. No, I don't think so.
Q. Were these contributions made at any regular interval?
A. There may have been some sometimes when they were more or less regular, but over the time they were not regular.
Q. You say they may have been more or less regular. You mean monthly?
A. I have no reason to think that.
Q. You say you have no reason to think it?
A. Right.
Q. What was the basis for your suggestion that might have been the case?
A. Because I don't remember the timing of it.
Q. It could have been, maybe, or maybe it wasn't; is that your answer?
A. It could not have been monthly over years. It might have been monthly over a few months.
Q. There are 1 or 2 things in the record I would like to clear up a little bit. Has Paul Crouch ever been in your house?
A. I think not.
Q. You mentioned having seen Miss Tatlock on various occasions. Were any of those occasions meetings of Communist groups?
A. No.
Q. Or left-wing groups?
A. If you are willing to include Spanish bazaars. I never saw her at a political meeting.
Q. Did you ever see her at a meeting where a Communist talk was given?
A. I certainly don't remember.
Mr. GARRISON. What kind of a talk?
Mr. ROBB. Communist.
Mr. GARRISON. A Communist talk?
Mr. ROBB. Yes.
The WITNESS. We went together to some CIO affair, but I don't remember who talked.
Mr. GRAY. Could this have been the FAECT?
The WITNESS. No, it wasn't. It was in San Francisco. I don't know what it was.

By Mr. ROBB:

Q. Did you ever go with her to any meeting of any kind at which literature was passed out?
A. The only meeting at which literature was passed out that I recollect is the one at my brother's house, which I described.
Q. Was Miss Tatlock there?
A. No.
Q. What kind of literature was that, Communist literature that was passed out?
A. I think so; yes.
Q. At that meeting were any pledges of contributions made by any of the people present?
A. I am not certain. My impression is that it was some kind of a dues gathering.
Q. I believe you testified to that.
A. I am not certain.

Q. By the way, you mentioned the meeting you went to at the home of Miss Louise Bransten. Do you recall that?
A. Yes.
Q. Who invited you to go to that meeting?
A. I don't remember. I can presume that it was the hostess.
Q. Do you recall how you happened to hear the meeting was going to be held that particular time?
A. We were invited, whether by phone or by personal invitation, by letter, I don't know.
Q. You knew Miss Bransten fairly well?
A. Not very well, not——
Q. Beg pardon?
A. Not well enough to know the things you said about her.
Q. Doctor, did you ever notice a man named Albert Lang Lewis?
A. I don't remember. Can you tell me how or where I might have known him? The name means nothing as you read it.
Q. Who lived in, I think, Los Angeles.
A. It means nothing to me so far.
Q. Did you know a man named Allen Lane?
A. It also means nothing to me.
Q. Did you ever know a man named Melvin Gross?
A. The name doesn't sound as unfamiliar as the others but it rings no bell.
Q. You mentioned the other day a man named Straus.
A. Yes.
Q. I believe you mentioned him as perhaps having been present at one or more of these meetings you attended. Do you remember that?
A. That is right.
Q. Was he a businessman in San Francisco?
A. Or an attorney, I don't know. He was not a college person.
Q. Did you see him around rather frequently?
A. No, I believe I once had dinner at his home, maybe my wife and I had dinner with them once. I think that is the only time.
Q. Do you recall when that was?
A. No.
Q. Why did you think perhaps he might have been present at one of these meetings that you went to?
A. My recollection is that he said something very foolish, but if you press me to try to remember who was at these meetings——
Q. I was curious because you searched your recollection as to who might have been present, and he was one of the men that came up and I wanted to ask you how you happened to remember him.
A. I think either he was involved in an argument or he and my wife were involved in an argument, or he said something that made an impression.
Q. Do you recall what the foolish thing he said was?
A. No, I certainly can't.
Q. Was it before or after that meeting that you had dinner at the house?
A. I don't remember.
Q. Did you ever hear of a man named Bernard Libby?
A. I don't think so.
Q. Doctor, is it your testimony that you told a false story to Colonel Pash so as to stimulate him to investigate Eltenton?
A. That appears not to have been necessary.
Q. Was that your testimony?
A. No, it is not. I testified that I had great difficulty explaining why I told him a false story, but that I believed that I had two things in mind. One was to make it clear that there was something serious, or rather I thought there might be something serious, and the other was not to tell the truth.
Q. Did you have any reason to believe that Colonel Pash would not be active in investigating the story you told?

Mr. GARRISON. Mr. Chairman, isn't this covering ground that has already been gone over this afternoon with you, and already over again in cross-examination? I mean do we have to go on and on with this?

Mr. GRAY. I think that clearly this is one of the important things in the Commission's letter. I think I will ask Mr. Robb to proceed unless he feels he is simply covering ground that has already been covered.

Mr. GARRISON. I think he ought to try as much as possible not to put words in the witness' mouth.

Mr. ROBB. I am cross-examining him.

By Mr. ROBB:
Q. I asked you whether you had any reason to believe that Colonel Pash would not be active in investigating your story?
A. I had no reason to believe anything. I had never met Colonel Pash before.
Q. Are you really serious, as you stated to the Chair, that you told Colonel Pash for the purpose of stimulating him?
A. I have been very serious in all my testimony and certainly not less in this very bizarre incident.
Q. You would agree that testimony is somewhat bizarre, wouldn't you?
A. That is not what I said.
Mr. GARRISON. Mr. Chairman, he is arguing with the witness.
Mr. ROBB. No; I am asking.
Mr. GARRISON. You are asking, wouldn't you agree, and this and that, which seems to me to be argument. I let it go if the chairman thinks not. But it seems to me to be an attempt to make him say what does not come from him in his own natural way.
Mr. ROBB. The word "bizarre" was his, not mine.
The WITNESS. I said the incident was bizarre.

By Mr. ROBB:
Q. Dr. Oppenheimer, you testified in response to a question by Mr. Gray that you told General Groves that there were not three men; is that right?
A. That is right.
Q. To whom did you make the first disclosure of the identity of the unknown professor?
A. I believe General Groves.
Q. What were the circumstances?
A. I think that it was at Los Alamos.
Q. If you told him that there were not 3 men, would you give us your thoughts, Doctor, on why it was that the telegrams that went out announcing the name of Haakon Chevalier all referred to 3 men?
A. I found this quite comprehensible when you read them.
Q. Have you seen Dr. E. U. Condon since 1951?
A. Oh, surely.
Q. Frequently?
A. No.
Q. Did you see him in 1952?
A. I would assume so. He is a member of the visiting committee to the physics department at Harvard of which I am chairman. We see each other at meetings. I would assume I saw him in 1952, but I don't recall.
Q. 1953?
A. As to that I am much less sure.
Q. Have you received any other letters from him other than the letters he wrote you about Peters and the one he wrote you about Lomanitz?
A. Yes; I have had other letters from him.
Q. When?
A. He has recently been having his clearance reviewed.
Q. His what?
A. His clearance reviewed, and he wrote me a letter about it.
Mr. GARRISON. Mr. Chairman, I wonder why we have to go into his relations with Dr. Condon. Are they a part of this case?
Mr. GRAY. Dr. Condon——
Mr. GARRISON. I don't know what this is about.
Mr. GRAY. Dr. Oppenheimer testified earlier——
The WITNESS. I have no reason not to answer these questions.
Mr. GARRISON. I withdraw my objection.
Mr. GRAY. I would like to complete my sentence that it was probably due to Dr. Condon's frantic—I am not sure about the language—at least Dr. Condon's disturbance about Lomanitz that he made the representations on behalf of Lomanitz. I believe that was your testimony.
Mr. GARRISON. I think the testimony was that was Bethe——
The WITNESS. No, that was a different matter.

By Mr. ROBB:
Q. I think you said that Dr. Condon wrote you about his clearance.
A. Right.
Q. I was about to ask you whether he asked you to testify in his behalf.
A. He did.

Q. How long ago was that?
A. It was shortly after my own case was opened.
Q. I assume you wrote him back you had troubles of your own, is that right?
A. No.
Q. What did you write him?
A. I think he asked me not to testify, but to write him a statement. I wrote him a letter outlining a statement that I could put in the form of an affidavit. In the meantime it seemed only fair for him to know about my situation, or at least for his attorneys to know about. I tried to keep this as quiet as I could. Therefore, my counsel got in touch with Dr. Condon's counsel. I believe that they explained the situation to Dr. Condon's counsel. This is——
Q. Your statement that you submitted to him, I suppose, was favorable to him, was it?
A. I am sure it was.
Q. By the way, speaking of counsel, Doctor, there has been some mention here of a Mr. Volpe in connection with the review of your matter in 1947. Has Mr. Volpe represented you since that time?
A. Yes.
Q. Is he now representing you?
A. No.
Q. When did he represent you?
A. He represented me along with Mr. Marks in connection with the Government's action against Weinberg.
Mr. GRAY. Against whom?
The WITNESS. Weinberg, where it seemed possible I might be called as a witness.

By Mr. ROBB:
Q. That was in 1951 or 1952?
A. 1952 and 1953.
Q. Did Mr. Volpe conduct the investigation in New Mexico to determine your whereabouts during the month of July——
A. Mr. Volpe and Mr. Marks had joint responsibility for finding out where I was.
Q. Just a couple of more questions. I am not sure if the record discloses this. If it does, I am sorry.
When you saw Dr. Chevalier, in Paris, as you testified, in November or December of 1953, how did you get in touch with him?
A. I had a letter from him before we left home saying that Professor Bohr——
Q. I believe you did testify.
A. Had told him we were coming to Europe and urging that if we were in Paris we try to have an evening with them. My wife called Mrs. Chevalier, found out that he was away, but that he probably could arrange to return before we left. We then did have dinner with them.
Mr. ROBB. I think that is all.
Mr. GRAY. I have one question. Back to Vista, Dr. Oppenheimer. Is it possible that some of these witnesses who felt there were no material changes in this draft were in effect saying that the draft really was not changed, and the military only thought it was being changed, or that was the essential notion?
The WITNESS. I will simply quote what either DuBridge or I said to General Norstad. We said we were much disturbed by what you said yesterday. We don't know whether there is a difference between us as to real things, or whether there is a difference between us as to the words that are used. We have therefore sought to put our views in a form which will be as little irritating to you as possible and still keep them our views. We don't know whether you will like what we have now written down or not. This is not a literal quotation. I should think that was as good an expression of what we thought we were doing in that change as we could give.
Mr. GRAY. Do you have some questions, Mr. Garrison, because if you do, I want to have a short break.
Mr. GARRISON. I think, Mr. Chairman, we would like to have some rebuttal testimony, but it is now 5 o'clock, and I wonder if we might not do that tomorrow morning.
Mr. GRAY. You have already indicated you would probably call Dr. Oppenheimer tomorrow morning for rebuttal testimony, and that is quite all right.
Mr. ROBB. May I ask one question?
Mr. GRAY. Yes.

By Mr. ROBB:

Q. Was Dr. Condon's counsel Clifford Durr?
A. In this recent undertaking?
Q. Yes.
A. No.
Q. I thought it was.
A. It was not at least the counsel my counsel saw. My counsel saw Hayes. I think I should not testify——

Mr. MARKS. I should state for the record it was I who saw Dr. Condon's counsel, and his counsel was Mr. Henry Fowler and Mr. Alexander Haas.

Mr. TRAY. We will recess now until 9:30 tomorrow morning.

(Thereupon at 5:05 p. m., a recess was taken until Tuesday, May 4, 1954, at 9:30 a. m.)

UNITED STATES ATOMIC ENERGY COMMISSION

PERSONNEL SECURITY BOARD

In the Matter of J. Robert Oppenheimer

Atomic Energy Commission,
Building T-3, Room 2022,
Washington, D. C., Tuesday, May 4, 1954.

The above entitled matter came on for hearing before the board, pursuant to recess, at 9 : 30 a. m.

Personnel Security Board: Dr. Gordon Gray, chairman; Dr. Ward T. Evans, member; and Mr. Thomas A. Morgan, member.

Present: Roger Robb, and C. A. Rolander, Jr., counsel for the board; J. Robert Oppenheimer, Lloyd K. Garrison, Samuel J. Silverman, and Allan B. Ecker, counsel for J. Robert Oppenheimer; Herbert S. Marks, cocounsel for J. Robert Oppenheimer.

PROCEEDINGS

Mr. GRAY. The proceeding will begin.

Mr. ROBB. Mr. Rolander has a brief statement about a matter, Mr. Chairman.

Mr. ROLANDER. During the course of this hearing it has been stated that the transcript of this hearing is being reviewed for declassification purposes by the AEC and other agencies. For the purposes of clarification of the record, in regard to AEC declassification of the transcript of this hearing, it should be stated that the AEC is taking full responsibility for such declassification. When classified information inadvertently enters the record, and when such information is of primary concern to other Government agencies and when the AEC feels that advice is necessary to a proper decision, we are asking the advice of the interested agency as to whether such information should be deleted.

Representatives of these other agencies review only those portions of the record in which the AEC thinks they may have a prime interest. These reviews are being made in the AEC offices, and in the presence of an AEC declassification expert.

Mr. GARRISON. Does Mr. Rolander know when the remaining volumes will be made available to us to take from the building?

Mr. ROLANDER. I understand that they are working on it, Mr. Garrison. I will have to check with the classification official. Perhaps I can do that at recess.

Mr. GRAY. Will you proceed, Mr. Garrison.

Whereupon, Vannevar Bush, a witness having been previously duly sworn, was called in rebuttal, examined and testified as follows:

Mr. GRAY. It is my duty to remind the witness that he continues to be under oath.

The WITNESS. Quite right, sir.

DIRECT EXAMINATION

By Mr. GARRISON:

Q. Dr. Bush, I want to read you from the testimony of Dr. Luis Alvarez before this board a short passage which mentions yourself, and I want to ask you to comment on it. I am reading from the direct testimony at page 2697, and it may run over to 2698. Perhaps the shortest way is to read it to you as it actually is. Recalling a conversation with you, he says: "I can give you some information that will"——

"A. I think I ought to have the time of that and the circumstances.

Q. He says it was perhaps a couple of months after Dr. Alvarez' return from Washington in October 1949. Then he goes on to talk about the date a little more precisely.

Mr. ROBB. That is what I said, Mr. Garrison.

Mr. GARRISON. Mr. Robb said that. Then he goes on to give his own fixation of the date. I think it will become clear when I read this to you.

By Mr. GARRISON:

Q. Dr. Alvarez says: "I can give you some information that will place this conversation to within a day, because Dr. Bush was in California to inspect one of the Carnegie Institution facilities at Stanford University. As you know, Dr. Bush is director of the institution. I remember that when I arrived home after our conversation with Dr. Bush, I found in the mailbox a copy of Life magazine which had a condensation of the book 'Modern Arms and Free Men.' So that places the date within a day."

If I can pause a moment, that date would be approximately when, Dr. Bush, do you recall?

A. I suppose that is along in October 1949.

Q. I don't think it is of any particular moment here.

A. I think that is the date of that article.

Q. Going on quoting: "What Dr. Bush said to Professor Lawrence and me was that he had been appointed by the President to head an ad hoc committee to assess the evidence for the Russian explosion. The Atomic Energy Commission and the Armed Forces, particularly the Air Force, had collected a good deal of information, all of which tended to indicate that the Russians had exploded a bomb, but before announcing that to the public the President wanted to make sure that the evidence was conclusive. If I remember Dr. Bush correctly, he said that he was made chairman of that. If I can paraphrase Dr. Bush's statements and give them in the first person, they went something like this. He said, "You know, it is a funny thing that I should be made head of such a committee, because I really don't know the technical facts in this field. I am not an atomic physicist, and I am not the one to assess these matters." But, he said, "I think the reason the President chose me is that he does not trust Dr. Oppenheimer and he wants to have someone in whom he has trust as head of this committee."

I will stop at that point, because I want to ask you about that. I should say on cross-examination—I will read the passage at pages 2731 or 2730, I guess it begins. This is the question put to Dr. Alvarez:

"Did I understand you said that Dr. Bush said that the reason he was chairman, the reason the President had named his as chairman, was that the President didn't trust Dr. Oppenheimer?

"A. That is the reason he said he thought he had been made chairman. I rather doubt that the President told him that he didn't trust Dr. Oppenheimer. I think this was Dr. Bush's construction

"Q. Have you heard since that panel was not named by the President, but by the Air Force?

"A. I have never heard a single word of this panel, sir. As I said, I refreshed my memory on the long-range objective panel. I read the report. I have never checked at all anything to do with this. As a matter of fact, I had forgotten this thing until recently. I did not mention it to the gentlemen who questioned me in Berkeley some months ago."

I think that is all I need to read, unless Mr. Robb or the chairman thinks there is more.

Mr. ROBB. No.

By Mr. GARRISON:

Q. Dr. Bush, who appointed you to the chairmanship of the committee that is here under discussion?

A. General Vandenberg.

Q. And not the President?

A. No. I had no contact with the President in connection with that matter, either before or after the panel's action.

Q. Did President Truman ever indicate to you any distrust of Dr. Oppenheimer?

A. He did not.

Q. Any doubt about him of any sort?

A. Not at any time.

Q. Did you ever gather from anyone else that President Truman had any doubt about Dr. Oppenheimer?

A. No.

Q. Do you recall having made any statement of this general kind to Dr. Alvarez?

A. I don't remember that conversation in detail, of course. I go to the coast about twice a year, once or twice a year, to visit Carnegie installations. I suppose 3 times out of 4 I see Dr. Lawrence. He is one of my trustees, and I have been a friend of his for many years. Occasionally I see others, including Alvarez, from his group. I don't remember in detail that particular conversation. I am quite sure I didn't say to him that the President had doubts about Dr. Oppenheimer simply because it was not true.

Q. Did anybody in the Air Force at the time of your appointment say that you were being made chairman because of doubts about Dr. Oppenheimer's loyalty?

A. No; they did not. The only thing that occurred there—I think it was General Nelson who visited me in this connection—when he told me of the makeup of the committee, I remember saying to him, "But wouldn't it be more reasonable for Dr. Oppenheimer to be chairman, since he is chairman of the General Advisory Committee," and he said to me something to the effect that they would prefer it the way it was. That is all there was.

Q. Going back now to Dr. Alvarez's direct testimony at page 2697 or 2698:

"Dr. Bush then said that the meetings of the committee were very interesting. In fact, he found them humorous in one respect, because he said, 'I was

ostensibly the chairman of the committee. I called it to order, and as soon as it was called to order, Dr. Oppenheimer took charge as chairman and did most of the questioning,' and I believe Dr. Bush said that Dr. Oppenheimer wrote the report. This was the first time I had ever heard anyone in my life say that Dr. Oppenheimer was not to be trusted."
That is referring back to the alleged statement of President Truman.
Do you recall saying anything of the sort that I have just quoted to you?
A. On the contrary, I am sure I did not make that statement for the same reason as before; the statement is not true.
Q. In what sense is it not true?
A. No part of it is true. The procedure of that panel was one exactly of what one would expect of a panel of that sort. I acted as chairman. I have acted as chairman of a great many meetings. I can't recall any instance where any member of the committee has taken over my functions as chairman while I was chairman. Certainly nothing of the sort occurred at that time. We all questioned witnesses. I think that probably Dr. Bacher, Admiral Parsons, and Dr. Oppenheimer did more questioning than I did, becauae there is just one thing that is correct in there, and that is namely, that I am not a nuclear physicist. Hence they conducted most of the detailed questioning. But I acted as chairman.
When we came to the report, we wrote that report around the table. It was a very brief report. I remember writing a paragraph of it myself. I don't remember who contributed what parts of it today. It was the sort of job that a committee of four would do around the table. Dr. Oppenheimer contributed throughout in a normal and perfectly proper manner.
Mr. GARRISON. That is all, Mr. Chairman.

CROSS-EXAMINATION

By Mr. ROBB:
Q. Dr. Bush, did you ever discuss Dr. Oppenheimer with General Vandenberg?
A. No, sir.
Q. It did strike you as unusual that you were chairman of that committee, instead of Dr. Oppenheimer?
A. Merely because I had no official connection at that time with the United States Government. He was chairman of the General Advisory Committee of the Atomic Energy Commission, and it seemed to me that it would have been more normal for him to have been the chairman of this panel reviewing the evidence. Hence I raised the question. I think it was General Nelson of the Air Force that was talking to me—I can't be sure I have the right general—when he said that the Air Force would prefer the panel the way it stood, we went no further.
Q. Did you ask him why?
A. No.
Q. Do you recall the occasion when you and Dr. Alvarez and Dr. Lawrence were driving in a car after inspecting one of your places?
A. I don't remember in detail. There have been dozens of such occasions and I can't separate that out and recall it in any detail.
Q. You would not question that Dr. Alvarez was correct about that?
A. No; I wouldn't question that he was correct, that he picked me up at Palo Alto and we drove somewhere. Whether it was a hotel—I think you said something about a hotel—I don't remember going to a hotel. But several times—well, quite frequently—Dr. Lawrence would join me at Palo Alto and we would drive over to his laboratory at Berkeley.
Q. And your suggestion is that nothing like that at all happened, and there was no reason for Dr. Alvarez to even have that impression of his conversation with you, is that right?
A. I made it very clear the parts of that statement which I say did not occur. There are two parts and I say those did not occur because neither of them was true. I don't make false statements. Hence I know I didn't make that one.
Q. What I am getting at is do you think you said anything from which Dr. Alvarez might have gotten that impression?
A. No; I certainly do not remember anything of the sort.
Q. You made no remark which was in your opinion susceptible of any such construction?
A. I am sure that I made no remark that would reflect upon Dr. Oppenheimer's loyalty or integrity or judgment in which I have had great confidence for many years.

Q. Did you make any remark, Dr. Bush, which in your opinion was susceptible of the construction which Dr. Alvarez placed upon it in his testimony?
A. I have no recollection of any remark from which he could get any such impressions.
Q. Would you say you didn't make any such remark?
A. I say I don't remember the conversation in detail.
Q. I see. If I might, Dr. Bush, clear up something in the record having to do with your testimony when you came here before. Do you recall you were rather critical of the letter written to Dr. Oppenheimer by Mr. Nichols?
A. Quite right.
Q. And in particular you were critical of the paragraphing?
A. No, sir. I don't remember I was critical of the paragraphing. I was critical of one particular statement in there because I said that it could be interpreted readily by the public, and in my opinion was being thus interpreted, as putting a man on trial for his opinions.
Q. Don't you remember that you made some particular reference to the paragraphing?
A. I don't remember. Can you give it to me?
Q. I will read it to you at page 1984. This was in answer to a question by Mr. Morgan:
"Doctor, on what ground would you ask for a bill of particulars if you didn't know the record?"
And you answered:
"I think that bill of particulars was obviously poorly drawn on the face of it, because it was most certainly open to the interpretation that this man is being tried because he expressed strong opinions."
A. Right.
Q. (Reading.) "The fact that he expressed strong opinions stands in a single paragraph by itself. It is not directly connected. It does not have in that paragraph, through improper motivations he expressed these opinions. It merely says he stated opinion, and I think that is defective drafting and should have been corrected."
Do you recall that?
A. Yes; I remember that.
Q. You had read that particular paragraph in the New York Times, I take it?
A. Yes; I believe I said so.
Q. Yes; I think you did. I want to show you the New York Times for Tuesday, April 13, 1954, page 16, carrying the text of the letter to Dr. Oppenheimer, and ask you if you will show us the paragraph you were talking about. I think you will find it here some place.
A. Yes, sure; this is it through here.
Q. Which is the one paragraph you had in mind?
A. This is the paragraph I referred to, I think, isn't it?
Q. I don't know, Doctor.
A. Yes.
Q. Would you read us the paragraph you had in mind?
A. Let me be sure I have the right one. "It was further reported"—no, wait a minute. Yes. "It was reported that in 1945 you expressed the view that there was a reasonable possibility—" wait a minute. This is the one. "It was further reported that in the autumn of 1949 and subsequently you strongly opposed the development of the hydrogen bomb on moral grounds, by claiming that it was not feasible, by claiming that there were insufficient facilities and scientific personnel to carry on the development, and four, that it was not politically desirable."
Q. That is the paragraph you had in mind?
A. That is the one I referred to.
Q. And you felt that putting that sentence in a separate paragraph was improper and damaging; is that correct?
A. The fact that it was in a separate paragraph was secondary. I feel that statement as a whole is fully open to the interpretation that a man is being tried for his opinions. That any reasonable man, particularly not a man with legal training, reading that entire statement, would feel that this man is being tried because he had strong opinions and expressed them, which I think is an entirely un-American procedure.
Q. But the fact of the matter is, Doctor, that you felt that the paragraphing was of sufficient importance that you made a point of it.
A. I think the paragraphing as I read it emphasized the point, but is not necessary to the point that I am making, which is that the statement as a whole, the letter as a whole, was open to that interpretation.

Q. I am directing your attention to your testimony about the paragraph and you concede, Doctor, you gave that testimony, didn't you?
A. I gave the testimony and I referred to that particular paragraph.
Q. And you were not giving testimony before this board about a matter which you thought was trivial?
A. I was giving testimony about a very important matter, indeed.
Q. Yes, sir. Now, Doctor, you took that paragraphing from the New York Times' didn't you?
A. So I said.
Q. Yes, sir. Now, I am going to show you the letter, the actual text of the letter sent to Dr. Oppenheimer, and ask you if you don't see from that that that paragraph which you read was not a separate paragraph in the letter at all, but was part of a much longer paragraph beginning, "It was reported that in 1945 you expressed the view that there was a reasonable possibility" and so forth, and ending "of which you are the most experienced, most powerful and most effective member, had definitely slowed down its development."
In other words, Doctor——
A. But the wording is the same——
Q. May I finish my question, and then you can finish your answer.
In other words, Doctor, the New York Times in its story broke up the paragraph of General Nichols' letter, into four paragraphs.
A. Without changing the wording.
Q. That is right.
A. I don't need to read that, if you tell me that.
(Document handed to witness.)
The WITNESS. This is a separate paragraph [indicating].

By Mr. ROBB:
Q. Where?
A. Here [indicating].
Q. It starts up here, "It was reported in 1945."
A. Oh, yes. Right.
Q. So you agree, Doctor, that the Times no doubt for greater clarity to its readers or for reasons of newspaper technique broke the paragraph in the Nichols letter into four separate paragraphs.
A. I would have expressed exactly the same opinion had I read the thing you later showed to me, namely, that is fully open to the interpretation that a man is being tried for his opinions.
Q. But if you read the original letter, you would not have made your point about the separate paragraphs.
A. No.
Q. Because it was not based on fact, was it?
A. It was based on what facts I cited.
Q. Yes, sir. Wouldn't you conclude from that, Doctor, that before making such statements it is well to know all the facts?
A. Yes; I think you sitting here, if you find me operating on a basis of a published statement, which is not exact, should have called it to my attention at that time.
Q. That is exactly what I am doing now, Doctor. It was not until after you testified that I realized you had been in error. Thank you.
Mr. GRAY. Dr. Bush, I think I should say to you that this board was confused about some of your testimony, especially on this particular point. I think that no member of the board was aware that this paragraphing change had been made at the time you were here, so this is not an unimportant matter because we have had another distinguished witness before this board, a man of international distinction, who in milder terms, but in somewhat the same spirit, was critical of the general manager's letter. I don't think he went as far as you did in saying that the board should have refused to serve at the call of the country until——
The WITNESS. Mr. Chairman, may I interrupt? I don't think I said that.
Mr. GRAY. You have interrupted me.
The WITNESS. Excuse me, sir.
Mr. GRAY. You said until the letter had been rewritten.
The WITNESS. Excuse me.
Mr. GRAY. I was in the middle of that sentence.
The WITNESS. Excuse me.
Mr. GRAY. But the other witness to whom I refer made a particular point about the construction of the letter. There was no uncertainty in his views whatso-

ever, and the thing that concerns me, also, about all of this is public misapprehension of which I am sure there is a great deal. So that if witnesses before this board have testified in such strong terms about the construction of this letter, before the board, they no doubt are testifying in equally strong terms among their associates, perhaps in the scientific community. This is another case of misapprehension or misunderstanding.

I want to make it clear that this discussion which I am conducting with you is for the purpose of emphasizing the seriousness of some of these misapprehensions, and not in defense of or attack upon the letter which was written by the general manager with which this board was not concerned.

I would like to ask you another question which relates now to the Alvarez testimony.

The WITNESS. I think I might clarify a point if you will let me.

Mr. GRAY. You certainly may.

The WITNESS. I have not discussed the procedure of this board with anyone, of course, while it is going on—scientists or otherwise. I have not given any statement to the press. I have talked over that particular matter which I raised here and which I think is so important with several men, not scientists, as it happens—there was one scientist among them—but men that I have great confidence in, in order to attempt to clarify my own thinking. One of those was a Justice of the Supreme Court. One or two others were men whose names you would recognize.

I realize what an important thing it is that I am calling attention to there. I realize how serious a thing it is in this country if the public gets the impression that a man is being tried for his opinions. Hence, before appearing before you, I talked to a number of men for the purpose of clarifying my own thinking. But otherwise, I have not discussed this matter with scientists, and I certainly have not done so generally in public.

Mr. GRAY. All right, sir.

Mr. GARRISON. Mr. Chairman, would it be appropriate for me to make a statement about this Times paragraphing which I would like to do, but I don't want to interrupt the course of your questioning.

Mr. ROBB. I was about to say that if Mr. Garrison was going to say that he was not responsible for the Times paragraphing, I think that is a fact.

Mr. GARRISON. I would like to say this. It was brought to our attention for the first time yesterday, Mr. Chairman, that this passage in the Times had been broken up into four paragraphs. We checked with Mr. Reston, who verified the fact that the copy which we had given him was a Chinese copy, in the journalistic phrase, of General Nichols' letter, that is, with every page the same and every paragraph the same identically as it appeared. He sent it up to New York to be set up, and without any instructions from him or any knowledge on his part it was broken up into these paragraphs, presumably, he said, because it was so very long.

I should also say that this having come to my attention in this fashion, I showed it to Dr. Bush before the session began to ask him if he wanted to modify his testimony about the effect which the reading of the passage in question made upon him, and he told me he could not.

Needless to say, Mr. Chairman, I regret very much indeed that the matter was broken up in the manner described.

Mr. GRAY. I think the Board understands that the newspaper reconstruction of this thing is frequently done in the press.

Dr. Bush, I would like to go back to the Alvarez testimony about which there was some discussion.

The WITNESS. Yes, sir.

Mr. GRAY. Let me ask this question: If you substitute the name General Vandenberg for President Truman—this is a hypothetical type of question—suppose Dr. Alvarez's testimony had been to the effect that General Vandenberg appointed this committee—I am substituting Vandenberg for Truman—and that your guess was that General Vandenberg appointed you chairman rather than Dr. Oppenheimer because he, Vandenberg, probably did not trust Dr. Oppenheimer, assume for the purpose of the question that Alvarez testimony had been to that effect, is it possible, then, that a conversation with him might have left with him the impression that he testified to?

A. That certainly also was not true, sir, so I know I did not make any such statement to him.

Q. This means, then, I take it, that you have no question in your mind about General Vandenberg's attitude?

A. I have no question in my mind. There was no statement to the contrary. He appointed Oppenheimer as a member of this panel. There was no point at any time questioning Oppenheimer's qualifications or his loyalty or anything else.

Mr. GRAY. I think you are very clear on that in your recollection.
Are there any more questions?
Mr. ROBB. Nothing further.
Mr. GRAY. Thank you, Doctor.
(Witness excused.)
Mr. GARRISON. May we have a couple of minutes, Mr. Chairman?
Mr. GRAY. Yes.
(Short recess.)

Whereupon, Katherine Puening Oppenheimer, a witness, having been previously duly sworn, was recalled to the stand and testified further as follows:

Mr. GRAY. Mrs. Oppenheimer, it is my duty to remind you that you are still under oath in this proceeding.
The WITNESS. Right.

EXAMINATION

By Mr. GRAY:
Q. We have asked you to come before the board again for some further questions.
Do you remember a man named Jack Straus?
A. I have heard him mentioned in the last few days. I could not have said that I remembered him; no.
Q. So you don't recall then, getting into an argument or discussion with him at a meeting, or one of the meetings that Dr. Oppenheimer testified about.
Are you familiar with the fact that he testified that to the best of his recollection, Mr. Straus attended one or two meetings, was it, Mr. Robb, do you remember?
Mr. ROBB. I think the meeting at Miss Bransten's house.
Mr. SILVERMAN. There is also testimony of a meeting at Mr. Chevalier's house earlier.
Mr. ROBB. One or the other.
Mr. SILVERMAN. Perhaps both.

By Mr. GRAY:
Q. In any event, Dr. Oppenheimer testified that he recalled Mr. Straus was there. You say you do not remember Mr. Straus at all?
A. I don't remember Mr. Straus.
Q. Were you personally acquainted with an individual named David Adelson?
A. I think I have met him, but I am not sure.
Q. You don't have any clear recollection?
A. No; I don't.
Q. Do you recall a man named John Steuben?
A. Yes.
Q. Who was he?
A. He was the section organizer of the Communist Party when I was a member of the party in Youngstown, Ohio.
Q. And that was back in the thirties sometime?
A. 1945—1934–35.
Q. When you knew him in Youngstown, did you ever have any association with him following the years when you were in Youngstown?
A. Yes.
Q. Could you tell us about that?
A. I saw him when I returned from Europe in 1937 to go back to school. I saw him in New York.
Q. And didn't see him after 1937?
A. I don't think so.
Q. Did you ever have any telephone conversations with him after 1937 that you recall?
A. No; not that I recall. I am quite sure I didn't.
Q. Did you know a man named Paul Pinsky?
A. As I recall, he also comes up in this letter from General Nichols, and I think I may have met him, too.
Q. Did you ever have any discussion with anybody about Dr. J. Robert Oppenheimer running for Congress from the seventh district, or whatever the appropriate district is?
A. No.

Q. So you would not have received the suggestion from David Adelson and Paul Pinsky to this effect, or you don't recall?
A. I am sorry it makes me giggle, but it does. I have never heard of such a thing.
Q. Do you know someone named Barney Young?
A. Barney?
Q. B-a-r-n-e-y.
A. No.
Q. I want to refer now to the contributions that Dr. Oppenheimer was making through Isaac Folkoff and possibly others as late as sometime in 1942. Were you familiar with the fact that these contributions were being made at the time?
A. I knew that Robert from time to time gave money; yes.
Q. Do you remember whether he gave money on any regular or periodic basis?
A. Do you mean regular, or do you mean periodic?
Q. I really mean regular.
A. I think he did not.
Q. Were you aware that this money was going into Communist Party channels?
A. Through Communist Party channels?
Q. Yes.
A. Yes.
Q. You had yourself broken with the Communist Party as early as 1937, I believe?
A. 1936 I stopped having anything to do with the Communist Party.
Q. Would it be fair to say that Dr. Oppenheimer's contributions in the years as late as possibly 1942 meant that he had not stopped having anything to do with the Communist Party? I don't insist that you answer that yes or no. You can answer that any way you wish.
A. I know that. Thank you. I don't think that the question is properly phrased.
Q. Do you understand what I am trying to get at?
A. Yes; I do.
Q. Why don't you answer it that way?
A. The reason I didn't like the phrase "stopped having anything to do with the Communist Party" because I don't think that Robert ever did——
Dr. EVANS. What was that?
The WITNESS. It is because I don't think Robert ever had anything to do with the Communist Party as such. I know he gave money for Spanish refugees; I know he gave it through the Communist Party.

By Mr. GRAY:
Q. When he gave money to Isaac Folkoff, for example, this was not necessarily for Spanish refugees, was it?
A. I think so.
Q. As late as 1942?
A. I don't think it was that late. I know that is some place in the record.
Q. I may be in error. My recollection is that Dr. Oppenheimer testified that these contributions were as late as 1942. Am I wrong about that?
A. Mr. Gray, Robert and I don't agree about everything. He sometimes remembers something different than the way I remember it.
Q. What you are saying is that you don't recall that the contributions were as late as 1942?
A. That is right.
Q. Are you prepared to say here now that they were not as late as 1942?
A. I am prepared to say that I do not think that they were that late.
Q. But you do think it is possible that they could have been?
A. I think it is possible.
Q. I mean, it is possible, if you don't have a very clear recollection——
Mr. SILVERMAN. Would it be helpful for me to state my recollection of the evidence on this point, or would you rather not, sir?
Mr. GRAY. No, I would prefer to proceed. What I am trying to get at, Mrs. Oppenheimer, is at what point would you say Dr. Oppenheimer's associations or relationships with people in the Communist Party ceased?
The WITNESS. I do not know, Mr. Gray. I know that we still have a friend of whom it has been said that he is a Communist.
Mr. ROBB. I beg your pardon?
The WITNESS. I said I know we still have a friend of whom it has been said that he is a Communist.

Mr. GRAY. You refer to Dr. Chevalier?
The WITNESS. Yes.
Mr. GRAY. I really was not attempting to bring him into the discussion at this point. I believe the import of the testimony you gave the other day was that at one time you felt that the Communist Party in this country was of an indigenous character and was not controlled or directed by international communism.
The WITNESS. That is right.
Mr. GRAY. I think also that you testified that knowing today what you do, you would think it would be a mistake to be identified——
The WITNESS. That is right.
Mr. GRAY. Now, I am trying to get at the point of by what mechanics one who has been associated becomes clearly disassociated.
The WITNESS. I think that varies from person to person, Mr. Gray. Some people do the bump, like that, and even write an article about it. Other people do it quite slowly. I left the Communist Party. I did not leave my past, the friendships, just like that. Some continued for a while. I saw Communists after I left the Communist Party. I think that I did not achieve complete clarity about it until quite a lot later.
Mr. GRAY. About when would that be, do you suppose?
The WITNESS. I find that very hard to say, but I have been thinking about it. I would roughly date a lot of it around Pearl Harbor.
Mr. ROBB. Around what, Mrs. Oppenheimer?
The WITNESS. Pearl Harbor. I mean as sort of an end point. There were other things that happened much earlier that made me feel that the Communist Party was being quite wrong.
Mr. GRAY. Would you attempt to date Dr. Oppenheimer's conclusion to that effect?
The WITNESS. Yes.
Mr. GRAY. About when would that be?
The WITNESS. I thought you said to that effect, meaning Pearl Harbor.
Mr. GRAY. No. I mean by that the conclusion that the Communist Party was quite wrong. At what time would you guess that he came to the same conclusion with clarity?
The WITNESS. I think earlier than I.
Mr. GRAY. Earlier than you?
The WITNESS. Yes.
Mr. GRAY. Which would have been earlier than December 1941?
The WITNESS. Yes.
Mr. GRAY. Mrs. Oppenheimer, a witness testified here as to an opinion he held, which was this: That he felt that you had decided that the most important thing in the world was your husband and his career. That is not an unreasonable assumption. And that he felt that you were determined to help him not make mistakes. Let me say that this is certainly not a verbatim recital of what he said, but I am sure it is the import.
If you had thought that Dr. Oppenheimer's contribution to Folkoff and others would adversely affect his career, would you have attempted to dissuade him from making such contributions?
The WITNESS. If I thought that?
Mr. GRAY. Yes.
The WITNESS. Yes.
Mr. GRAY. Did you ever discuss with him the necessity for avoiding associations with people who were identified with the Communist Party, to your knowledge, or whom you might have suspected that were identified with the Communist Party?
The WITNESS. I do not remember thinking of anybody as being identified with the Communist Party in those days, except people whom I knew were out-and-out Communists.
Mr. GRAY. Yes. And did you ever discuss with him the desirability of not continuing an association with those people?
The WITNESS. I did not think of anybody as being a Communist Party member except certain party functionaries. We have to have that straight.
Mr. GRAY. Let us hold it to the party functionaries. Let us mention the name Folkoff.
The WITNESS. I did not think that Robert's contacts with Folkoff as an association.
Mr. GRAY. You did not consider the contributions to Folkoff as an association?
The WITNESS. No.

Mr. GRAY. What would constitute an association in your judgment?

The WITNESS. Let us take a man like William Schneiderman, who is definitely a Communist in San Francisco. I think if one were friends with him, that would be association with the Communists.

Mr. GRAY. If one gave money to him, would that be an association?

The WITNESS. It would depend for what reason one gave him some money.

Mr. GRAY. If one knew that the money was going into Communist Party channels, would it make any difference for what reason the party membership said the money was going to be used?

The WITNESS. I think so.

Mr. GRAY. You do?

The WITNESS. I do not think so now, but I did then.

Mr. GRAY. Today you would say you would not think so?

The WITNESS. Indeed not.

Mr. GRAY. And you think then that the conclusion you hold now was one that if you had to date it might have come around Pearl Harbor?

The WITNESS. Or later.

Mr. Gray, let me make quite clear that my prorgess of thought has not been a clear chain about these things. I have been quite fuzzy about a lot of things. I have always to differentiate between what I thought at a certain time and what I think now. It is not easy.

Mr. GRAY. I am going back now to John Steuben.

The WITNESS. Stueben?

Mr. GRAY. Steuben. You are quite sure that you do not recall any kind of communication with or from him as late as 1944, 1945, or 1946?

The WITNESS. Yes.

Mr. GRAY. Dr. Evans?

Dr. EVANS. Mrs. Oppenheimer, there has been a lot of talk here about the Communists and fellow travelers. Could you tell me so that you and I can understand the difference between a Communist and a fellow traveler?

The WITNESS. To me, a Communist is a member of the Communist Party who does more or less precisely what he is told.

Dr. EVANS. He does what?

The WITNESS. Rather precisely what he is told to do by the Communist Party. I think a fellow traveler could be described as someone to whom some of the aims of the Communist Party were sympathetic and in this way he knew Communists. For instance, let us take the classic example that is bandied about all the time nowadays; that is, the Spanish War. Many people were on the side of the Republicans during the Spanish War. So were the Communists. I think the people who were not Communists and were on the side are now always known as fellow travelers.

Dr. EVANS. Did you ever try to get your husband to join the party?

The WITNESS. No.

Dr. EVANS. You never did?

The WITNESS. I was not a Communist then.

Dr. EVANS. How is that?

The WITNESS. I was not a Communist then. I would not have dreamed of trying to get anybody to be a member of the Communist Party.

Dr. EVANS. Do you think you have been completely disillusioned now or are you still fuzzy?

The WITNESS. No, I have been disillusioned for a long time.

Dr. EVANS. Did you ever talk to your husband about some of the men that worked at the Radiation Laboratory and the possibility of their being Communists, men like Lomanitz, Peters, Hawkins, and those?

The WITNESS. As being members of the Communist Party?

Dr. EVANS. Yes.

The WITNESS. No.

Dr. EVANS. I am not quite sure, but I thought there was some evidence here that some witness said that Mrs. Oppenheimer tried to talk to her husband about some of these people. Do you remember that? That she tried to get him to stop his association with them. Was there such a thing as that in the record?

Mr. GRAY. I think, Dr. Evans, you probably have reference to Mr. Lansdale's testimony. This is the thing I was referring to.

Dr. EVANS. I just wanted to know. It was Lansdale's testimony. You have ansewered the question. I have no more questions.

Mr. GRAY. I have one more. In early 1944, where would you have been?

The WITNESS. Los Alamos.

Mr. GRAY. Did you stay there pretty constantly and regularly?
The WITNESS. I went away once when my mother had pneumonia, but I forget what year that was.
Mr. GRAY. Where did you go, then?
The WITNESS. To Bethlehem, Pa.
Mr. GRAY. You only left Los Alamos once in the year that you lived there during the war?
The WITNESS. I went to Santa Fe sometimes.
Mr. GRAY. Did you go to Berkeley?
The WITNESS. I do not think so. I would say, "no."
Dr. EVANS. Does your mother still live in Bethlehem?
The WITNESS. She has until—she has come and gone quite a bit the last few years—but until 1 or 2 years ago, they were in Bethlehem.
Mr. ROBB. May I ask a couple of questions?
Mr. GRAY. Yes.

By Mr. ROBB:

Q. Mrs. Oppenheimer, did you used to read the People's Daily World?
A. I have seen it, yes.
Q. That is the west coast Communist newspaper?
A. That is right.
Q. Did you see it around your house in Berkeley?
A. I think it got delivered to our house on Shasta Road.
Q. On where?
A. On Shasta Road.
Q. Who subscribed to it, you or Dr. Oppenheimer?
A. I do not know. I did not subscribe to it. Robert says he did. I sort of doubt it. The reason I have for that is that I know we often sent the Daily Worker to people that we tried to get interested in the Communist Party without their having subscribed to it. So I do not know whether or not Robert subscribed to it. I know it was delivered to the house.
Q. You say "we"; do you mean the Communists? Do you mean when you were a Communist?
A. Yes, that is what I mean.
Q. Tell me, Mrs. Oppenheimer, you said you knew this man Adelson.
A. I think I have met him, yes.
Q. Do you recall how you happen to know those men?
A. Mr. Robb, I have read the letter from General Nichols quite a lot of times and I have naturally thought about a lot of things. The names Adelson and Pinsky were not unfamiliar to me. I do not know how I met them. I think I did.
Q. Do you remember when there was some discussion about Frank Oppenheimer running for Congress?
A. I have heard that since. I do not remember it as of then, no.
Q. You took no part in it?
A. No.
Q. Did you know some people named Bartlett?
A. Bartlett?
Q. Perhaps I can help you. Did they occupy the garage apartment at Frank Oppenheimer's place?
A. I know some people did. I did not know that.
Q. You did not know them?
A. In that connection the name Bartlett does not mean anything. I met the people who occupied that apartment but I do not remember them.
Q. Did you ever discuss Adelson with Dr. Oppenheimer?
A. I do not know.
Q. By the way, did you ever hear of Steve Nelson given the nickname Stephen Decatur?
A. No.
Q. You mentioned that you still had a friend who people say was a Communist. Was that Dr. Chevalier you had in mind?
A. Yes.
Q. You heard it said that he still is a Communist?
A. No, I have heard it said he was.
Q. Did you know anything about his activities in Communist causes?
A. I think he went to Spanish relief parties. I know he had this party at his house at which Schneiderman spoke.
Q. Had you finished your answer?

A. I am trying to think if I knew anything else about him. I think I know no other facts in that direction.
Q. Did you ever see his name in the Daily Worker or the Daily People's World as having endorsed the so-called purge trials in Russia?
A. No.
Q. You saw Dr. Chevalier in France last fall?
A. That is right, in December.
Q. In Paris?
A. In Paris.
Q. How long were you in Paris on that occasion?
A. Well, let's see. We went over—I think we spent 2 days and then went up to Copenhagen and came back, and I think we spent something like a week again. It may have been 5 days or it may have been a little longer than a week; I do not remember.
Q. Was it on the first 2 days that you saw Dr. Chevalier?
A. No.
Q. You mean after you came back from Copenhagen you saw him?
A. I think so, yes.
Q. Do you recall how you happened to get in touch with him?
A. Yes, I do.
Q. Would you tell us that?
A. I called his wife and said we would like to see them. She said that Haakon was in Italy, but she thought he would be back and she would let us know.
Q. Do you remember how you happened to have her telephone number?
A. It was in the book. I think it was in the book. I think I looked it up. On the other hand, I may have had a note from Haakon in my purse with the telephone number on it, which I would have taken along because if we went to Paris we wanted to see them.
Q. Do you recall how you happened to know they were in Paris at all?
A. Yes. I think Haakon wrote us.
Q. How long before you went there?
A. I think he has written us probably 3 or 4 times in the last few years.
Q. I suppose he expressed a hope that if you came there you would look him up?
A. Certainly.
Q. Do you know how he happened to know you might come to Paris?
A. I remember his wife saying to me that they had read in the paper that Robert was giving lectures in England.
Q. This was the occasion of these Reith lectures?
A. R-e-i-t-h.
Q. Do you recall whom else you saw in Paris on that occasion?
A. Yes. Oh, my, now wait. We saw LePrince-Ringuet and we saw a number of physicists. I do not know whether both Auger or Perrin or whether it was just one of them. We went to the apartment of another physicist whose name I can't remember. I will have to ask Robert.
Dr. OPPENHEIMER. May I answer? Goldschmidt.

By Mr. ROBB:
Q. You saw a number of physicists. I don't care about the names.
A. We saw Francois and Yvonne de Rose.
Q. I believe you had lunch with the Chevaliers or dinner.
A. Dinner. We had dinner at their house.
Q. And then did you take them to lunch or something?
A. Oh, yes, no.
Q. Did they take you to lunch?
A. No. Haakon called for us and we went out to see Malraux.
Q. Do you remember any discussion about Dr. Chevalier's passport difficulties?
A. I do not remember it but it has been recalled to me since.
Q. How was it recalled to you?
A. I think Robert mentioned it to me.
Q. Would you tell us what he had to say about it?
A. He said that he had been asked whether Haakon had spoken to him about it and he did not remember it.
Q. Did Dr. Oppenheimer tell you pretty generally what he had been asked about matters of which you had knowledge?
A. Yes.
Q. Did you meet a Mr. Wymans when you were in Paris on that occasion?
A. Yes, I did.
Q. How did you happen to meet him?

A. He is a—I don't know—a classmate or something of Harvard. He was at the embassies. We had lunch with him.
Q. Mrs. Oppenheimer, do you know or have you ever seen Paul Crouch?
A. I do not think so, Mr. Robb. I have seen his picture in the paper a few years ago and I saw his picture in Time recently. He doesn't look to me like anybody I have ever seen.
Q. Do you know this so-called 10 Kenilworth Court episode about which there has been some controversy?
A. Yes.
Q. Do you recall such a meeting having taken place?
A. No.
Q. Would you say it did not?
A. I would say it did not.
Q. So far as you know, Paul Crouch has never been in your house?
A. That is right.
Q. You could not be mistaken about that?
A. I could be mistaken about almost anything, but I do not think I am.
Q. I understand that.
Mr. GRAY. Let me ask a question while he is looking at his paper. When was it that you lived at this address that you gave to which the People's World came?
The WITNESS. When I first got married to Robert.
Mr. GRAY. This was in 1940?
The WITNESS. Yes.
Mr. GRAY. Do you remember seeing People's World in the house as late as 1941?
The WITNESS. I do not know. I think the paper came to the house at 10 Kenilworth Court, too, but how long it came there, I do not know.
Mr. GRAY. You lived at Kenilworth Court after this—I have forgotten the address that you mentioned.
Mr. SILVERMAN. Shasta.
Mr. GRAY. You lived at Kenilworth Court after you lived at Shasta Road?
The WITNESS. Yes.
Mr. GRAY. And you think the People's World came to Kenilworth Court?
The WITNESS. I think so.
Mr. ROBB. That is all I care to ask.
Mr. SILVERMAN. I think I have one or two questions to ask Mrs. Oppenheimer.

By Mr. SILVERMAN:

Q. Mrs. Oppenheimer, Mr. Gray asked you about your leaving Los Alamos, and you referred to a visit to Bethlehem, Pa., when your mother had pneumonia. I think you gave a date in 1944.
A. I gave no date because I do not remember when it was.
Q. I thought that she adopted a date that had been given. It had been suggested that it was May 1945. Would you recall one way or the other?
A. I am afraid I wouldn't.
Q. Did you, in fact, attempt to dissuade your husband from making contributions or having associations with Communist Party people?
A. I think not.
Mr. SILVERMAN. That is all.
Mr. GRAY. Are you familiar with a Thornwall Telephone Co.?
The WITNESS. Cornwall—I think that is a Berkeley exchange.
Mr. GRAY. Thornwall 6236; does that mean anything to you?
The WITNESS. No.
Mr. GRAY. That never was your telephone number?
The WITNESS. I do not know. It does not mean anything to me, Mr. Gray I do not remember our Berkeley telephone number.
Mr. GRAY. Could it have been Dr. Frank Oppenheimer's number?
The WITNESS. All I can say is that I do not know.
Mr. GRAY. Thank you very much, Mrs. Oppenheimer.
(Witness excused.)
Mr. GARRISON. Could we have a short recess?
Mr. GRAY. Yes; we will recess for a few minutes.
(A short recess was taken.)
Whereupon Jerrold R. Zacharias, a witness, having been previously duly sworn, was called in rebuttal, examined, and testified as follows:

Mr. GRAY. I think the record should show that Dr. Zacharias is here, as I take it Dr. Bush was, in the capacity of what we have informally referred to as rebuttal witnesses.

Mr. GARRISON. Yes, Mr. Chairman.

Mr. GRAY. It is my duty, Dr. Zacharias, to remind you that you continue under oath in the proceeding.

The WITNESS. I do.

Mr. GRAY. Mr. Marks.

DIRECT EXAMINATION

By Mr. MARKS:

Q. Dr. Zacharias, I wish you would state for the record whether or not I asked you to read testimony which has been given in these proceedings by Mr. Griggs when you arrived this morning?

A. Yes; you did. You gave me that, and I read a part of the Griggs testimony that had to do with the summer study and the so-called ZORC.

Q. Testimony given before this board by Mr. Griggs described a meeting in the fall of 1952 in Cambridge, a meeting of the Scientific Advisory Board. Were you present at that meeting?

A. I was present for a panel discussion that had no bearing on the subject at issue, a small panel discussion, and present to give a report of the summer study findings to the full Science Advisory Board of the Air Force.

Q. I don't understand what you mean when you say you were present with respect——

A. I am not a member of the Science Advisory Board. There was a 3-day meeting. I was present for a subcommittee meeting which has no bearing on the present discussion and present at a report made by the Lincoln Laboratory to the Science Advisory Board. It is that full discussion of the full committee that I think comes into question here.

Q. Did you make any presentation to the Scientific Advisory Board on that occasion?

A. I did.

Q. Will you tell us whether or not in the course of any of the meetings of the Scientific Advisory Board at that occasion you had occasion to say anything about or do anything about a term that has been used—ZORC?

A. I testified under oath the last time I was here, and I will repeat the testimony, that I had never heard of any such organization or name of organization or anything resembling it until I read it in an article in Fortune magazine.

Q. And when would that have been?

A. When that magazine article came out, in May of 1953, a year after the beginning of the summer study, about.

Mr. ROBB. Pardon me. I suppose that is an answer to the question. Maybe the witness would like to have the question read back. I am not sure that is a direct answer to the question.

Mr. MARKS. Let us read it back, Mr. Reporter.

(Question read by the reporter.)

Mr. MARKS. I would like to ask another question.

The WITNESS. Do you want me——

Mr. ROBB. I just want to make sure the witness understands the question.

Mr. MARKS. May we proceed, Mr. Robb?

Mr. ROBB. Certainly. I just want to be fair to the witness, that is all.

Mr. GRAY. You may proceed, Mr. Marks.

Mr. MARKS. Thank you.

By Mr. MARKS:

Q. Did you or did you not, Dr. Zacharias, on the occasion of the 1952 Scientific Advisory Committee meeting, in the fall of that year in Cambridge, write on the blackboard in the course of that meeting the term "ZORC" and explain it?

A. To the best of my knowledge and belief, I did not write on the board the letters "ZORC." May I state this a little more fully?

Being a school teacher, I naturally emphasize things by writing on the board. This is one of our chief methods for emphasis. I don't remember seeing any reason now why I should have wanted to emphasize my own name. I had been properly introduced and Lauritsen was a member of the Science Advisory Advisory Board, and was present in the audience. So all I can say is to the best of my knowledge and belief, I did not write any such thing on the blackboard. I have even gone so far as to check the memory of a few other people on this very point, and none who has been questioned remembers any such thing.

Q. Are you or are you not clear as to when you first heard the term "ZORC"?

A. I am very clear that I first heard the term "ZORC" when I read it in Fortune magazine of May 1953, 9 months after the meeting of the Science Advisory Board in question.

Mr. ROBB. I am awfully sorry. Could I have that answer read back?

(Question and answer read by the reporter.)

By Mr. MARKS:

Q. Dr. Zacharias, I should be, but I am not, clear in my memory as to whether when you previously appeared in these proceedings you testified concerning your participation in and the circumstances under which the so-called Lincoln Summer Study originated. I would like to ask you now to describe the circumstances or such of them as you know about under which the Lincoln Summer Study originated and the specific purposes, if you know them, of that summer study.

A. I was from the beginning of the Lincoln Laboratory until I resigned shortly after the end of the summer study, associate director of the Lincoln Laboratory.

Q. When did the laboratory originate?

A. It is hard to know exactly. It was in June of 1951, roughly. You can't nail it down too tight. So that for approximately a year and a half I was associate director of the Lincoln Laboratory. In roughly March of 1952, I visited Pasadena-Los Angeles area—in fact, as I remember it, I paid a visit to the Hughes Aircraft Co. One evening Dr. Lauritsen, and I had a discussion about air defense, and the participation of Lincoln and how it would be possible to make an air defense in the face of a growing threat * * *. Dr. Lauritsen and I thought it would be a good idea to set up a study group to investigate the question of defense of the North American continent.

I got in touch with Dr. Hill, then the director.

Q. The director of what?

A. Of the Lincoln Laboratory. More specifically, he was deputy director, but indeed running the laboratory. We decided that it would be a good thing to do, that it would help air defense if we did it, and it would also likely help the Lincoln Laboratory's growth.

We had a discussion about this with Dr. Lauritsen, Dr. Oppenheimer, and Dr. Rabi. I remember that it was in a room in the Hotel Statler. Five of us, as I remember it, certainly Dr. Hill was there.

Q. When would this have been?

A. In early April or the end of March of 1952. We discussed the possibility of going ahead with the study, and one of us, namely either I or Dr. Hill, made the suggestion that the prestige of Drs. Oppenheimer, Rabi, and Lauritsen, would help to bring in some of the bright people who would otherwise find other things to do.

They agreed to help with the study and did, not on a full-time basis. We proceeded to try to recruit people for the study, some from within the Lincoln Laboratory—a few within the Lincoln Laboratory, so as not to deplete the Lincoln Laboratory force—and several or many from the outside.

The summer study got going about the first of July 1952, and continued for 2 months thereafter, with Drs. Oppenheimer, Rabi, and Lauritsen participating on a part-time basis in the initial discussions and in the terminal discussions.

By Mr. MARKS:

Q. What were the specific purposes of the summer study as they were conceived by you in its inception?

A. The purpose of the summer study was simply this. We knew that the Russian threat might grow in a variety of ways. The types of aircraft, the types of delivery means, including ballistic missiles and so on would increase, and we wanted to see whether the kind of air defense planning that was going on and the air defense work going on within Lincoln was appropriate to the growing threat. There is no sense in trying to make an air defense against yesterday's airplanes. The defense that one develops has to be against the airplanes that will be in being and threatening when the air defense is in being. Remember, that technical discussion and technical work has to precede use by a number of years.

Q. Dr. Zacharias, was it ever suggested to you or intimated to you by Dr. Oppenheimer that the summer study should have other purposes?

A. Not that I can possibly remember.

Q. Was it ever a contemplated purpose of the summer study to bring about a reduction in the power of the Strategic Air Command?

A. Certainly not. In fact, it is clear to anyone who tries to think of defense of the continent—let me be a little specific about this—that there are essentially what you might call four possible rings of defense. One is an innermost last-ditch affair, largely from the ground with the aid of missiles or antiaircraft guns; a second ring, which can be provided by interceptor aircraft of short range and moderately close to home; a third ring which is further out away from our shores, and away from our borders; and a fourth which is the destruction of enemy bases by means of long-range bombardment aircraft. All of these elements for defense of the continent are terribly important, regarded as very important by all members of the study group, and the Strategic Air Command is included in the last 1 of the 4. Not last in order of priority, but only last because if you start from the inside out, you get to Russia last.

Q. Was there any purpose in the summer study to effect a reduction in the budget of the Strategic Air Command?

A. There certainly was not.

Q. Was there ever any purpose in the summer study, or was any such purpose ever suggested to you, of studying or considering submarine warfare?

A. * * * Several of us had participated in the project on antisubmarine warfare 2 years prior to this. We saw no reason to examine the situation again. Maybe I have not answered the question quite. You said was it ever suggested. It is very difficult to remember who suggested what. I certainly remember no emphasis at all on the antisubmarine problem.

* * * * * * *

Q. When you speak of argument at the summer session, who do you have in mind as involved in that argument, if anybody?

A. The summer-study group that was full time worked on that. I don't have the names directly at my fingertips. Those of us who were involved full time besides myself were Lloyd Berkner, Brockway McMillan from the Bell Telephone Laboratories, Julian West from the Bell Telephone Laboratories, Wippanie, M. M. Hubbard, of the Lincoln Laboratory. I would rather find a list than to try to cite one here.

Let me say that the detailed discussion of relegating the problem of countering missiles launched from submarines, relegating that to the countersubmarine force, was largely done by the full-time members of the group.

Q. Mr. Griggs has testified that "we;" that is, I take it, he and his associates, whoever might have been, were concerned with the fear that the summer study might get into things which he and his associates regarded as inappropriate for Lincoln, and as of questionable value to the Air Force. He referred specifically to the strategic air arm and allocation of budget between the Strategic Air Command and Air Defense Command.

You have already commented on these matters. I think at this point in his testimony, he went on to say that we also were very much concerned in the early days of the formulation of the Lincoln summer study because it was being done in such a way that had it been allowed to go in the direction in which it was initially going, every indication was that it would have wrecked the effectiveness of the Lincoln Laboratory.

This, Mr. Griggs said, was because of the way the thing was, the summer study was handled administratively.

Mr. Robb. What page are you reading from, Mr. Marks?

Mr. Marks. This is from pages 2617 and 2618 of volume 14.

By Mr. Marks:

Q. He went on to say, "So far as I know, it was not because of any direct action on the part of Dr. Oppenheimer. On the other hand, I felt at the time that Dr. Oppenheimer should have been well enough informed and alert enough to see that this would be disastrous to the Lincoln summer study."

Mr. Robb. Mr. Chairman, I think it fair to say in the interest of accuracy that Mr. Marks was not reading a verbatim portion of the record when he did that. I think you left out several things and paraphrased in other cases, did you not, sir?

Mr. Marks. May I show the witness the transcript?

Mr. Robb. I think the record ought to reflect whether or not you read from the record verbatim or whether or not you paraphrased or omitted certain portions of what you have been reading.

Mr. Marks. This is a rather pointless discussion.

Mr. Robb. It is not pointless to me.

Mr. MARKS. I did change some "we's" to "they." Let me, if I may, show Dr. Zacharias the portion of the transcript from which I was reading.

Mr. GRAY. I think you should read the portion, whatever it was, Mr. Marks, and then put your question to the witness.

By Mr. MARKS:

Q. I would like to read to you, Dr. Zacharias, a portion of the transcript, namely, pages 2617 and 2618, relating to testimony of Mr. Griggs, and ask you whether you have any comment to make on it. Starting at page 2617:

"Q. Was that the main object of the Lincoln summer study, to find ways to improve our air defense?

"A. Yes, sir.

"Q. And did the Lincoln study ever recommend the giving up of any part of our strategic air power?

"A. No, not to my knowledge.

"Q. I think you have already said so far as your knowledge goes, Dr. Oppenheimer did not recommend that?

"A. That is right. I would like to amplify my answer on that for the benefit of the board, since this is the first mention of the summer study in this much detail.

"We were concerned by the thing I have already mentioned, that is, the fear that the summer study might get into these things which we regarded as inappropriate for Lincoln, and as of questionable value to the Air Force—I refer to the giving up of our strategic air arm, and the allocation of budget between the Strategic Air Command and the Air Defense Command—but we were also very much concerned in the early days of the formation of the Lincoln summer study, because it was being done in such a way that had it been allowed to go in the direction in which it was initially going, every indication was that it would have wrecked the effectiveness of the Lincoln Laboratory. This was because of the way the thing was, the summer study was being handled administratively.

"So far as I know, it was not because of any direct action on the part of Dr. Oppenheimer. On the other hand, I felt at the time that Dr. Oppenheimer should have been well enough informed and alert enough to see that this would be disastrous to the Lincoln summer study."

Now, unless Mr. Robb would like me to read more, which I would be glad to do, I would like to ask Dr. Zacharias the question, if he has any comment to make on the passage that I have read.

Mr. ROBB. No, Mr. Chairman, it is not my satisfaction. It is a question that I merely want the record to be accurate.

Mr. GRAY. The witness will proceed with any comment he has to make.

The WITNESS. Those of us who were trying to start the summer study felt— let me say specifically I felt—that we were trying to help air defense and also the Lincoln Laboratory. That the Lincoln Laboratory is an important part of our air defense development system and strengthening the Lincoln Laboratory would strengthen air defense.

Correspondingly we also wanted to see whether the technical means that we were trying to employ were adequate. Remember that this was at a time when the early warning for the Air Force against incoming raids was pitifully short in time. Substantially no warning until enemy bombers might be directly on us. We therefore wanted to look at the early warning, the air battles, and possibilities of defense against new types, new mechanisms of delivery. This was our objective. This is something of interest to the Air Force and specifically of interest to the three services. Remember, the Lincoln Laboratory is an Army, Navy, Air Force laboratory, despite the fact that the Air Force contributes the major share. So we felt that we were helping the Air Force, or that we would help the Air Force by our efforts.

I would like to make the comment that Dr. Griggs, the witness there in question, was then I think called the chief scientist for the Air Force, and as we saw it, or as I saw it—"we" is indefinite, that is why I use "I"—as I saw it, he was doing everything he could to prevent our starting this summer study. He tried to influence people not to join it. He tried to influence President Killian and Provost Stratton to prevent the initiation of the summer study. By his own admission—Dr. Griggs' own admission—the summer study turned out to be a good thing. This is what we thought it would be. You can never promise in advance, before you start a study, what the study will end up with. You can't be sure that it will turn out to be fruitful, whereas this one, in my opinion, did turn out to be fruitful.

Dr. Griggs' efforts—let me use a strong word—to sabotage the summer study from a position of power as chief scientist for the Air Force I regarded as unwise, but not subversive. I would not want to bring up Dr. Griggs on charges of being disloyal in his effort to sabotage an effort in which I was the major promoter. However, let me say rather informally that it is a bit of a pity that dueling has gone out of style. This is a very definite method of settling differences of opinion between people than to try to bring out all the detail in a hearing.

About the administrative part of that question, to my memory there were no administrative changes involved in the initiation of the summer study. We had planned to hold it in the Lincoln Laboratory somewhere, that I was going to direct it, as the director of the Lincoln Laboratory I would thereby report to Dr. Hill on this.

Griggs' efforts to stop the summer study did result in a delay of several weeks, critical weeks, as a matter of fact, in trying to gather the people to form a summer study. Remember a summer starts at a fairly definite time for university people, and a delay of 3 weeks in my opinion then and in my opinion now resulted in our not having as large a group or even as capable a group as we might have had if it were not for obstructive tactics used by Griggs in this matter.

But the administrative detail of the running of the summer study was carried out the way it was initially conceived.

Mr. MARKS. I have no further questions.

Mr. GRAY. Mr. Robb.

CROSS-EXAMINATION

By Mr. ROBB:

Q. Doctor, you either knew or assumed, did you not, that in his position with respect to the Lincoln summer study Dr. Griggs was following out the policies of his superiors in the Air Force?

A. Is that a loaded question, sir? I had no way of knowing whether he was carrying out orders or acting on his own initiative. When I say "knowing" I use the word very carefully. I believe, however, that he was acting on his own initiative.

Q. You think he was just carrying out a personal vendetta?

A. I think not. I think again that he was not doing this because of any personal animosity toward me or to some of the other members of the group. I would not want to go on record to say that he was doing it because of a personal animosity toward Dr. Oppenheimer. I am sure that Dr. Oppenheimer's presence on the group colored Griggs' actions and thoughts considerably.

Q. Why?

A. The question is why do I think so or why were they colored?

Q. Why do you think that?

A. Because Griggs spoke to some people in a very derogatory way regarding Dr. Oppenheimer.

Q. What reason do you have for believing or feeling, whichever it is, that Dr. Griggs' attitude toward the Lincoln summer study was not in accord with the wishes of his superiors in the Air Force?

A. I didn't say that. I said that I thought he was acting on his own initiative.

Q. All right.

A. That the stimulus for doing what he was doing came from him. It is perfectly possible for a man to convince his superiors to do something, or to order him to do something that he wants to do. I have talked with Mr. Finletter a little about the early history here, and his mind was rather vague on the subject, because I wanted to be sure that it was not Mr. Finletter who was directing these delays.

Q. At the time that Dr. Griggs made his position on the summer study known to you, did you communicate with any of Dr. Griggs' superiors in the Air Force to see whether or not Dr. Griggs was carrying out their wishes?

A. No, I don't remember. I remember some discussion with Mr. Norton, but I don't remember the substance of it very much. It certainly would not have been in this form.

Q. Can you tell the Board, Dr. Zacharias, any single specific fact or circumstance which indicated to you that Dr. Griggs' attitude in respect to the Lincoln summer study was not acting in conformity with the wishes or orders of his superiors?

A. Mr. Robb, I would have had to be there to answer that question. When a man is acting or doing something, if he is in military uniform, I think he can always rely—he can always depend on being able to say that he is acting under orders. Civilians in the military don't always do that.

Q. Dr. Zacharias, you have testified, have you not, that in your opinion Dr. Griggs—strike that.

You have testified, have you not, that Dr. Griggs' attitude in your opinion was his personal attitude, and did not necessarily reflect the attitude of his superiors? Is that a fair statement?

A. Yes, I have no proof of that, however.

Q. Yes.

A. I have no proof of that. I have tried to keep that part of the record clear.

Q. Yes. So is it not a fair question, Doctor, to ask you what your proof, if any, is?

Mr. MARKS. The witness has already said he had no proof.

The WITNESS. That is a very telling kind of question in the sense——

By Mr. ROBB:

Q. Thank you.

A. In order to get to the answer I would have to pull into my memory all of the details of the back and forth talk on this and in particular on what Griggs said to me and said to others. In order to get this thing clear, I think it would take a fair time of the committee.

Q. We have lots of time, Doctor. Your answer is that you can't do it as of now; isn't that right?

A. Yes.

Q. Doctor, so that there may be no misunderstanding, may I ask you, sir, is it your testimony that the first time you ever heard this name or expression "ZORC" was when you read it in the Fortune article in May 1953?

A. Yes, sir.

Q. Just so we can be sure we are talking about the same thing, I have before me that article or a photostat of it, and I will read you a few sentences from it to make sure that is what you are talking about. The byline is "ZORC Takes Up the Fight," "A test of Teller's thermonuclear device was scheduled for late 1952 at Eniwetok. Oppenheimer tried to stop the test. In April 1952, Secretary Acheson appointed him to the State Department Disarmament Committee of which he became chairman. Here was generated a proposal that the President should announce that the United States had decided on humanitarian ground not to bring the weapon to final test and that it would regard the detonation of a similar device by any other power as an act of war. Mr. Truman was not persuaded. That project cost Oppenheimer his place on the General Advisory Committee. When his term expired that summer he was not reappointed. Neither were DuBridge nor Conant who supported him throughout. Now came a shift in tactics. At a meeting of scientists——"

Mr. GRAY. Let me ask, are you going to ask a question about the substance of the article, or is this for the purpose of identification?

Mr. ROBB. It is just for the purpose of identification. The first paragraphs I read merely to get the time fixed and I don't intend to question the witness with those.

"Now came a shift in tactics. At a meeting of scientists in Washington that spring there formed around Oppenheimer a group calling themselves ZORC, Z for Zacharias, an MIT physicist, O for Oppenheimer, R for Rabi, and C for Charles Lauritsen."

By Mr. ROBB:

Q. Is that the piece to which you referred?

A. Yes, sir; it is.

Q. When you read that reference to ZORC, were you surprised by that name?

A. Yes.

Q. You never heard it before?

A. I had never heard it before.

Q. You are sure about that?

A. I am sure about that.

Q. You could not reasonably be mistaken about it?

A. I could not be reasonably mistaken about that.

Q. Did that reference rather anger you?

A. Very much so.

Q. Why?

A. Because it implied that there was a cabal group of people who were trying to do things or to influence policy one way or another by existing in a group. To the best of my memory, which certainly is not adequate here, I can't think of any time when those four people sat together alone in a room to discuss anything. In other words, there was so little to their being a group that if there was a time—there may have been—when those four people, including myself, were together alone in a room, it would surprise me very much.

Q. In other words, you thought it was quite a material point whether there had been such a group calling itself ZORC, or not?

A. I felt that it was a journalistic trick to bring into focus the kind of scurrilous charges that were being made in the article.

Q. You thought it was an important point?

A. I thought it was an important journalistic trick. This is very different from its being—yes, I agree, I thought it was—if it were true, if it had been true—it would have been a point. Therefore, maybe to get to what you are thinking I believe it is germane to these hearings.

Q. Yes. In other words, if it were true as you have testified, it tended to show that there was a cabal.

A. Yes.

Q. Was Dr. Griggs present at that meeting of the Scientific Advisory Board in Boston or Cambridge in the fall of 1952?

A. I don't know.

Q. How many people were present there?

A. I don't know exactly. There was rather a roomfull, a room that might hold 50 to 100. A number I think given in Griggs' testimony.

Q. You did address the meeting, I suppose?

A. I did.

Q. And never having heard the expression or dreamt of it, you could not have written it on the blackboard. Is that your testimony?

A. No, sir. Never having heard of something, you could still—letters might go together. Remember, this is a rather technical point here. Let me say I never heard of it and certainly did not write it on the blackboard.

Q. Doctor, if you were surprised and angered when you saw that expression "ZORC" in the Fortune article in May 1953, you could not very well have written it on the blackboard in the fall of 1952, could you?

A. That is my feeling, sir.

Q. Aren't you sure about that?

A. I am as sure of that as I am sure of anything in my memory for which I don't have written documentation. Let me say this, if someone presented me with a photograph of the blackboard at that time with me in front of it, I would say sure, that must be it. But my memory aids in this are simple. I see now no reason why I should have put those initials there for any point of emphasis that I might have wanted to make. Remember that I was at that meeting trying to impress the Science Advisory Board with the sum of the results of the summer study, and that there were tangible results. Some of the people in the group were impressed by those results. I had no need for recourse to prestige. The results stood on their own feet as they still do.

Q. Just to draw the issue plainly, Dr. Griggs has testified here that you wrote those letters on the board and explained that Z was Zacharias, O was Oppenheimer, R was Rabi, and C was Charlie Lauritsen. Did that take place?

A. To the best of my memory, it did not take place.

Q. Could you reasonably be mistaken about it?

A. I am afraid I am a scientist, sir, and I could be mistaken about anything that is not written down in my notebook.

Q. Aren't scientists usually pretty accurate?

A. No more accurate on things of this sort than anyone else. I think if you wanted to establish this point very carefully you might have to call a fair number of the witnesses of the people at that meeting.

Q. Do you recall at that meeting in the fall of 1952 that you were anxious to impress people that Dr. Oppenheimer was participating or had participated in this study in some way?

A. No, sir. I had in my mind two most important things. One was to get going on an early warning system, and the second to get going on a remote intercept system. I wanted those understood in a technical way.

Q. Is there any other meeting that you can think of that that incident described by Dr. Griggs might have occurred?

A. I can think of no other meetings where Dr. Griggs was present, and like this meeting, I can think of no reason for having written such things on the board anywhere.

Q. So far as you know you have never written such letters on the board?

A. As far as I know I have never written any such things on the blackboard. I might do it now because it is a short word and is, as I say, a neat journalistic trick. Whether all journalistic tricks are dirty, I don't know. I rather feel this one was.

Q. So far as you know, you never used that word "ZORC" prior to seeing it in the Fortune article?

A. Yes, sir. I did not use that word prior to seeing it in the Fortune article.

Mr. ROBB. That is all I care to ask.

Mr. GRAY. Dr. Zacharias, if you were today shown a photograph of the blackboard and the letters "ZORC" on the blackboard and you standing beside it with a piece of chalk in your hand, you would say then "I was mistaken in my testimony"?

The WITNESS. What with the present trend in doctoring photographs, I might want to question the photographer.

Mr. GRAY. That was my next question. Would your reaction be, "I did actually do this" or would your first reaction be that must be a doctored photograph?

The WITNESS. My first reaction would be one of considerable surprise to the extent that I would doubt the veracity of the photograph and would want to question the photographer.

Mr. GRAY. Earlier when you were before the board, you testified that you had no knowledge of the origin of the nomenclature. This refers to "ZORC," now. Then you said, "I do know one friend of mine went around to a meeting of the Physical Society and hunted for people that had heard of it, found one, and I would rather not mention his name, because it has nothing to do with this thing. He may have heard it, or it may have been the invention of the man who wrote the article."

You were not asked, Dr. Zacharias, who this man was. I would like now to ask you——

The WITNESS. You would?

Mr. GRAY. I would like to, yes. Who is the man who had heard of it?

The WITNESS. This is a second-hand report. The man who said he had heard of it was Alvarez. My memory of the man who told me of this is James B. Fisk.

Mr. GRAY. I asked you this question because Mr. Griggs testified very clearly that he saw you perform this act of writing the letters on the blackboard, and you have testified pretty strongly that you think it hardly possible that this happened.

Dr. Zacharias, in a rather long response to a question from Mr. Marks, inviting comment on some testimony of Mr. Griggs which was read, you made some observation about dueling having gone out of style. Do you mind telling me— I didn't stop you when you were giving your answer, because I have tried very hard not to restrict witnesses in their answers—what was your reference to dueling?

The WITNESS. I meant that where there are personal differences that are very strong, that in the old days some of these were settled by dueling. Let us take the McCarthy-Stevens difference. * * * It might well have been settled that way rather than at such extensive length. Apropos of this, having read some of Dr. Griggs' testimony, my blood begins to boil a bit. I feel no great liking for Dr. Griggs at this particular point.

Mr. GRAY. Is this only since you have read his testimony that you have no liking for him?

The WITNESS. I would say that my respect for Dr. Griggs has been declining rapidly over the past 2 or 3 years, and it hits a rather low point with this sworn testimony of his.

Mr. GRAY. Did you have this feeling about him at the time of the summer study?

The WITNESS. It certainly was not as strong then as it is now.

Mr. GRAY. If dueling had not gone out of style at the time of the summer study, would you have felt strongly enough to challenge him at the time of the summer study?

The WITNESS. Perhaps.

Mr. GRAY. So you did feel pretty strongly?

The WITNESS. I felt pretty strongly then.

Mr. GRAY. And it is not just his testimony before this board?

The WITNESS. Certainly the testimony has added to it.

Mr. GRAY. You stated that you felt that Dr. Griggs attempted to sabotage this project, I believe.

The WITNESS. I said that I wanted to use a strong word. He tried in every way he could to stop it, to prevent its happening.

Mr. GRAY. Do you wish to withdraw your characterization of it as sabotage?

The WITNESS. I don't know the full implication of withdrawing this.

Mr. GRAY. This is not involved——

The WITNESS. I said it was a strong word with color to it. I think it is more appropriate than not. Let me say it this way. The word sabotage has many implications. One is that it was being done without the knowledge of many others. Griggs was quite open in his opposition to this summer study. In that sense I would only say that he was doing his best to stop or to prevent the project.

Mr. GRAY. I asked you whether you thought that was a very serious matter at the time.

The WITNESS. Yes; I did.

Mr. GRAY. And you felt, I believe you said, that it was resulting in appreciable delay?

The WITNESS. It did, sir.

Mr. GRAY. Appreciable delay was resulting from Mr. Griggs'——

The WITNESS. An appreciable delay did result from it.

Mr. GRAY. Did you discuss these problems of getting underway with the summer study with anybody other than Mr. Griggs who was identified with the Department of Defense?

The WITNESS. I don't remember exactly who we discussed this with.

Mr. GRAY. My purpose in asking the question is not to confuse the situation at all. I am simply asking what you did, if anything, to overcome the obstacles which you felt Mr. Griggs was putting in the way of something that you also felt was extremely important to the security of the country.

The WITNESS. Remember that Dr. Griggs was working on my superiors, namely, Dr. Killian and Dr. Stratton, and I talked with them about it. I would have to look at the record to see whether I talked with General Craigie. I very likely did, but I can't be certain.

Mr. GRAY. So if you had a protest or complaint to make, you would have made it normally through Dr. Killian and Dr. Stratton, and not the Air Force people?

The WITNESS. Griggs' major attempt to stop the project was his trying to influence them, at least from my point of view at that time. I didn't know of all the things that he was doing.

Mr. GRAY. You said you talked with Mr. Finletter about this. When was that?

The WITNESS. I talked with Finletter——

Mr. GRAY. Was it within the last year?

The WITNESS. Within the last 8 months, I believe. He was just vague on the subject, and I didn't press it.

Mr. GRAY. That was not in connection with your appearance before this board?

The WITNESS. It was not; no, sir. It was something like last June.

Mr. GRAY. Have you ever known of a study under contract with the Armed Services, say at MIT, as an example, in which there was official complaint by the services that the reasonable bounds of the study had been exceeded?

The WITNESS. I know of none.

Mr. GRAY. You don't have any?

The WITNESS. I know of no official complaint, not even in this case.

Mr. GRAY. You know of no study, for example, which might have concerned itself with electronic problems which came up with recommendations with respect to foreign policy?

The WITNESS. I know of a study that was concerned with electronics problems and also discussed, questions of foreign policy. I was not a member of that study.

Mr. GRAY. But you have heard of it?

The WITNESS. Yes, sir.

Mr. GRAY. And you never heard that there was any complaint from the Defense Department about the study having exceeded its reasonable bounds?

The WITNESS. I was not a member of that study. I did truly not ever hear of this complaint.

Mr. GRAY. If you were directing a study which had to do with electronics, a pretty clearly denfied field, and it started to come up with recommendations with respect to foreign policy, would you feel that an official of the Defense Department who urged that you stick to electronics was acting with impropriety?

The WITNESS. I think I would not direct a project that was as restrictive as that, sir, as to be restricted only to electronics.

Mr. GRAY. I am not going to press you further, because I don't think it is getting us anywhere.

The WITNESS. No.
Mr. GRAY. The question was related to the somewhat conflicting testimony here about whether the summer study was tending to get into budget matters, for example, as distinguished from what was to have been the main purpose of the study.
The WITNESS. The study did not get into budget matters.
Mr. GRAY. And it did not tend to at any time?
The WITNESS. And did not intend to at any time. One must not confuse the word "budget" with what things might possibly cost. In other words, you can't make a technical evaluation of anything without trying to decide whether it could be afforded, whether it is possible to have that much money available to make what you want. But that is not a budgetary question. That is a technical question.
Mr. GRAY. Do you think that the writer of the Fortune magazine article is the originator of the four letter word we have been discussing?
The WITNESS. I have no idea.
Mr. GRAY. You are saying you don't know where is originated?
The WITNESS. I don't know.
Mr. GRAY. Dr. Evans, do you have any questions?
Dr. EVANS. Dr. Zacharias, did you ever know a Robert M. Zacharias?
The WITNESS. No.
Dr. EVANS. He was a classmate of mine. I just wondered if he might be a relative of yours.
The WITNESS. No, sir. I come from Florida.
Dr. EVANS. I suppose I ought to know this, but I don't. Do you know why Griggs was so opposed to this study?
The WITNESS. I don't know. I think he makes it pretty clear in his testimony. He was opposed to this for one thing because of Dr. Oppenheimer's possible participation, and he was opposed to it because he said he thought it might alter the course of the Lincoln laboratory, an air defense laboratory. This is his own testimony. I only paraphrase it. It is better given there.
Dr. EVANS. That is all.
Mr. GRAY. Mr. Marks.

REDIRECT EXAMINATION

By Mr. MARKS:

Q. Dr. Zacharias, I am not sure I caught one of your answers to question Mr. Robb put, but I think you said something to the effect that you had never been alone in a room with Rabi, Lauritsen, Oppenheimer.
A. I said I don't remember any such circumstance, only to lend weight to the fact that I know of no such organization. It is certainly possible to have any four people in a room, especially physicists who know each other well. I didn't make the point that they had never been together. The point is that the only time I remember we were together there were other people present.
Mr. MARKS. I have no further questions.

RE-CROSS-EXAMINATION

By Mr. ROBB:

Q. Dr. Zacharias, did you undertake to find out who wrote that Fortune article?
A. I didn't undertake to find out. It was found out pretty quickly.
Q. Did you ever talk with that gentleman?
A. No, sir; I never have. I understand that he has recanted considerably.
Q. What was his name?
A. The name is Charles Murphy, as I understand it.
Q. Did you make any protest or representations either to him or to Fortune about the article?
A. No, sir.
Q. You didn't write to the editor or anything like that?
A. No.
Q. You read the article pretty carefully.
A. I read it once or twice.
Mr. ROBB. That is all.
Mr. MARKS. I do have one other question, if I may.
Mr. GRAY. All right.
The WITNESS. Could I interpose?
Mr. GRAY. Yes.

The WITNESS. Mr. Robb's question about my writing to the editor of Fortune, or so on, might be used as a gage of my anger on reading it. I think it is not such a gage. There are many of us who try to work with the military. The more we can do to keep our names and ideas out of the public, away from the public, the better can we get along with the military and work with them. I would not write a letter to the editor in protest or do anything of that sort because of straining relations with people who, like all the rest of us, are people, too, and like to get credit for what is going on. You see, there is a simple theory that you can either get something done and get credit for doing it, and not both. The scientific people who try to work with the military try as much as possible to get credit for what gets done allocated to the military. In this sense, in this kind of context, I would not write a thing of this sort, and therefore my answer to the question could not be used to indicate that my blood pressure didn't hit the top when I read the article.

Mr. ROBB. I was not intending to indicate that. Your point is the fact that you didn't write doesn't show you were not all wrought up about it.

The WITNESS. That is right.

REDIRECT EXAMINATION

By Mr. MARKS:

Q. In response to a question by the Chairman, Dr. Zacharias, I think you indicated that the first use of the term "ZORC" by a scientist that had come to your knowledge was attributed to Alvarez. Did I understand——

Mr. ROBB. I don't think that is what he said.

The WITNESS. No; I didn't say that.

Mr. MARKS. I am sorry. I didn't understand that testimony, Mr. Chairman, and I would like to see if I couldn't understand it.

Mr. GRAY. I think Dr. Zacharias testified that after he first heard about the "ZORC" phrase and tried to find out if anybody else had heard about it, he found one scientist who indicated that he had.

The WITNESS. I found out second hand.

Mr. GRAY. He found out second hand that there was a scientist who had heard of it and that scientist was Alvarez.

Mr. MARKS. Thank you, that clears it up.

Mr. GRAY. I am sorry to hold you, Dr. Zacharias, but that leads me to another question. Credibility of witnesses is now involved. What are your personal relationships with Dr. Alvarez, as you see them? Are you on friendly terms?

The WITNESS. I would say moderately friendly. I would say he and I have never been, that I can remember it, fond of each other.

Mr. GRAY. Have you ever felt strongly enough about it to wish that dueling had not gone of style as far as Dr. Alvarez is concerned?

The WITNESS. Oh, I respect Dr. Alvarez very much. He is a very intelligent man. In his own way, I think, he tries to be reasonable. But he has very strong opinions, and I think it is his arrogance——

Mr. ROBB. His what?

The WITNESS. His arrogance—that bothers me most.

Mr. GRAY. Do you question his veracity?

The WITNESS. I would not question his veracity in the real sense. I believe that if he says something he believes it.

Mr. GRAY. I guess that is a pretty good definition of veracity, isn't it?

The WITNESS. Yes.

Mr. GRAY. Do you question Mr. Griggs' veracity?

The WITNESS. Yes; I would.

Mr. GRAY. You do question his veracity. Are any of the differences you may have with Dr. Alvarez in any way related to Dr. Oppenheimer?

The WITNESS. No.

Mr. GRAY. That would not be involved at all. Let me explain to you why I ask the question. You have testified that you thought that Mr. Griggs felt strongly about the summer study because of the possibility of Dr. Oppenheimer's association with it. I believe therefore that your testimony brings the Griggs-Oppenheimer relationship squarely into this proceeding, or at least Griggs' attitude toward Dr. Oppenheimer. I am trying to find out whether, since Dr. Alvarez has come into this, whether that is at all involved in your difficulties with Dr. Alvarez.

The WITNESS. I have no direct knowledge of what Dr. Alvarez thinks about things specifically germane to the hearing of Dr. Oppenheimer. I think the difference between me and Alvarez are matters of taste and subtle things of that sort. In some cases matters of substance. Dr. Alvarez participated in the Hartwell project the * * * study study that I directed. He picked on a particular part of antisubmarine warfare that he thought should be pushed very hard. Very few of the other members of the Hartwell group agreed with him. I did not agree with him, but this was not anything but a difference of opinion on a technical matter.

Mr. GRAY. That is not related in any way to this hearing.
The WITNESS. It is not related to this at all.
Mr. GRAY. Dr. Evans.
Dr. EVANS. You know Dr. Teller quite well?
The WITNESS. I don't know him very well; I know him.
Dr. EVANS. Do you rather like him or don't you, or can't you answer that?
The WITNESS. That is hard to answer. I don't know how to answer that question, sir. I would think hard to try to do it, if you press me.
Dr. EVANS. I won't press you.
Mr. GRAY. Perhaps this might not be difficult. Do you consider Dr. Teller a difficult man to work with? Have you ever worked with him?
The WITNESS. I have never worked with Dr. Teller.
Mr. MARKS. No further questions.
Mr. ROBB. Nothing further.
Mr. GRAY. Thank you very much.
(Witness excused.)
Mr. GRAY. I want to get on the record a couple of things. I think we have had so much discussion about the Fortune magazine article that that should go in as an exhibit because parts of it have been read into the record and it has been referred to a good deal. I assume nobody objects to that?
Mr. GARRISON. My problem about that is, Mr. Chairman, that if that goes in it seems to me we ought to have a chance to answer it. I just think it is going to prolong the record. I am perfectly content with what was read into the record out of it. I don't ask that the rest of it be put in. If it does, it contains various veiled allegations that I just think ought not to stand in the record without some answer to them. I have not myself read it. I have only got a sense of what it is like.
Mr. GRAY. I think I would say, Mr. Garrison, that I don't think it is in any way prejudicial to Dr. Oppenheimer to have this as an exhibit. I am a little uncomfortable about having so many references to the article.
Mr. GARRISON. All right.
Mr. GRAY. I say to you I don't think you need to make any more answer than you have made or are making.
Mr. GARRISON. It may very well not be worthy of any answer. I haven't read it.
(The document was received as exhibit No. 2.)
Mr. GRAY. Yesterday you asked me about further procedure, particularly with respect to what the board would like to have in the way of proposed findings of fact and briefs. I have read the procedure under which we operate, and they are silent with respect to that matter, as far as the board is concerned. There is some reference to briefs in the event of an appeal to the standing board of the Commission, the Personnel Security Review Board. So I take it there are no requirements in this matter under the procedures. If you wish to present to the board proposed findings of fact, of course, we would certainly consider them. If you wish to present briefs, of course, we would consider them. In that event, if you do wish to file documents of this sort, the board requests that they be filed with the board no later than May 17, which is 2 weeks from yesterday. I am not sure whether that answers the question that you raised yesterday or not.
Mr. GARRISON. What day of the week is that?
Mr. GRAY. That is Monday.
Mr. GARRISON. Mr. Chairman, we will do our best. It is a very tight time schedule, but we will do our best. If there is any possible give on that at all, it would be helpful.
Mr. GRAY. I am authorized to say that this matter has been discussed with the other board members, and the board feels that this is a date we will request you to observe.
Mr. GARRISON. All right. Among our problems is that of transcripts which is a perennial one with us. We can't take them out of the building here except a certain number that have been released. It is fearfully difficult for us to work here out of our offices. I suppose in due course we will get them, but there are these problems.

Mr. GRAY. In recognition of this difficulty, I can only ask Mr. Rolander and his associates to do the best, with all their problems they have, that they can.

Mr. ROBB. Mr. Chairman, I might say just for the record that I think it should be said that we have made available to Mr. Garrison and his associates a room here with a table in it which they have been using as their office in this building.

Mr. GARRISON. I appreciate that.

Mr. ROBB. We have done our best for them.

Mr. GARRISON. I am not raising any question of the courtesy that has been provided, but of the problem of working away from one's headquarters.

Mr. Chairman, did you have any further thought about a hearing of argument and summation by counsel?

Mr. GRAY. I have assumed that you would wish to present a summation to the board. I would assume that it would contain some argument.

Mr. GARRISON. As far as I can put into it.

Mr. GRAY. Yes. I want to have that before we adjourn or recess this series of daily meetings, as it were. We are ready for that when you have finished with your rebuttal witnesses.

Mr. GARRISON. You mean this afternoon?

Mr. GRAY. I would hope we could get started this afternoon.

Mr. GARRISON. I just can't, Mr. Chairman. It is just physically not possible to do it.

Mr. GRAY. May I ask how many more witnesses you will put on rebuttal?

Mr. GARRISON. Mr. Hill and Dr. Oppenheimer. I imagine it will take the afternoon, the way we go.

Mr. GRAY. We then will ask you to start your summation and argument with the morning session tomorrow.

Mr. GARRISON. Would it be possible to do it at the afternoon session, and have the morning free to do a little work? Mr. Chairman, I don't—well, I won't sketch to you our problems, but it has been a matter of night work every night for the last 3½ weeks, apart from the transcripts. I have been with my client, my colleagues and the witnesses, the transcripts have been down here, and I have not even quite finished reading a summary of them prepared by Mr. Ecker, let alone reading the transcripts themselves. I am just so hard pressed to try to gather anything together that would be of use to the board, if I could at least have a half day clear in which to do a little work; it would be a great help. I think in the end to the board also.

Mr. GRAY. I will discuss this with the board during the noon recess.

Mr. GARRISON. I would prefer a whole day if it could be had, but I would greatly prefer to do it on Thursday if it could be done.

Dr. EVANS. May I just say——

Mr. GARRISON. If you are going to be here.

Dr. EVANS. I know just how you are pressed for time, Mr. Garrison, but you must remember that some of us——

Mr. GARRISON. I know that, Dr. Evans; indeed I do.

Dr. EVANS. We are in pretty bad shape, too.

Mr. GARRISON. I know you are. I should say one thing, Mr. Chairman, if you don't mind. At Mr. Baruch's request, Dr. Oppenheimer saw him on Sunday—Mr. Bernard Baruch—and as a result of that conversation, Mr. Baruch said that he would be glad to have me get in touch with him with respect to testifying here. I did as soon as I could reach him. He said—this was last night that I reached him—that the earliest that he can come would be Thursday morning. He could come down on the 10:15 plane and testify. I told him I didn't know whether this would be possible, because of the probable close of testimony today, but I would mention the matter to the board, which I do now.

Mr. GRAY. We should be glad to receive a written statement from Mr. Baruch.

Mr. GARRISON. Thank you, Mr. Chairman.

Mr. GRAY. Can we start at 2:15?

(Thereupon at 12:40 p. m., a recess was taken until 2:15 p. m., the same day.)

AFTERNOON SESSION

Mr. GRAY. Would you be good enough to stand. Give your full name.

Dr. HILL. Albert Gordon Hill.

Mr. GRAY. Albert Gordon Hill, do you swear the testimony you are to give the board shall be the truth, the whole truth, and nothing but the truth, so help you God?

Dr. HILL. I do.

935

Whereupon Albert Gordon Hill was called as a witness, and having been first duly sworn, was examined and testified as follows:

Mr. GRAY. Would you be seated, please.

It is my duty, Dr. Hill, to remind you of the existence of the so-called perjury statutes. I should be glad to review their general provisions with you if it is necessary. I won't do so if you are familiar with them.

The WITNESS. I think I know them generally well.

Mr. GRAY. It is not clear to me, Mr. Hill, whether it is likely that you might get into a discussion of restricted data, but in any event, I should like to request that if in the course of your testimony you find it necessary to disclose classified material, that you notify me in advance so that we may take the necessary steps in the interest of security.

Finally, I should like to say to you that we consider this proceeding a confidential matter between the Atomic Energy Commission and its officials on the one hand, and Dr. Oppenheimer, his representatives and witnesses on the other. The Commission is making no releases about these proceedings, and on behalf of the board I express the hope that witnesses will take the same view.

The WITNESS. Yes, sir.

Mr. GRAY. Mr. Marks.

DIRECT EXAMINATION

By Mr. MARKS:

Q. Dr. Hill, what is your present occupation and position?
A. I am a professor of physics at MIT, and also director of the Lincoln Laboratory.
Q. How long have you been connected with the faculty at MIT?
A. I think 17 years, except for a brief 6 months period before the war.
Q. Without going into detail, what war work did you do?
A. I was at the radiation laboratory during the entire war.
Q. The radiation laboratory where?
A. MIT.
Q. Turning now to more recent days, how long have you had a connection with the Lincoln Laboratory?
A. Since its inception. It was preceded by a Project Charles which began January of 1951. This terminated in the summer * * *. The exact genesis and birthday of Lincoln would be hard to give. Somewhere before September 1, 1951.
Q. What positions have you held in connection with the Lincoln Laboratory?
A. I was assistant director, I guess, when it started, became deputy director in the spring of 1952, and director on July 1, 1952.
Q. Became director when?
A. July 1, 1952.
Q. In your capacity, as you have described it, in connection with the Lincoln Laboratory, did you have anything to do with the inception of the so-called summer study?
A. Yes, quite a bit. I should say the inception took place likely before I became director. It began in the last week in June. I should say that the former director, Dr. Loomis, of the University of Illinois, resigned effective July 1. This was done on March 1, and I was then appointed deputy director and director-elect, if you like. Loomis continued to run the laboratory, but we had a rather firm agreement that things that were going to extend beyond July 1 I would take responsibility for them. So although the inception of the summer study took place while I was not director, I was completely responsible for it as the senior Lincoln person.
Q. What can you tell us about the circumstances of the origination of the summer study?
A. There are probably threefold. A number of us have always worried a bit about how to improve continental defense and the like. I should perhaps parenthetically say that the Lincoln Laboratory is devoted primarily to continental defense and air defense in general.

During the late winter and spring of 1952, Lloyd Berkner, who was then director—I am sorry—who was president of Associated Universities and very active in the East River project, which they ran, this was a study on civil defense, early concluded that civil defense would be very difficult, if not impossible, without some measure of early warning. Lloyd used to come periodically to see a number of us at MIT, at Lincoln, talking about the possibility of early warning. We invented various things on the cuff, found most of them wanting, and it was my feeling that a rather serious study of early warning, whether it was possible or not, should take place.

That was one genesis. Another genesis came from Zacharias talking. I believe first with Charlie Lauritsen on the broad question of whether air defense is possible. Zacharias and I talked over the summer study one night at his house. There may or may not have been others present. I don't know. We agreed it was a good idea. I said I would only go along with it if he would be the head of the summer study which he agreed to. I also insisted that early warning be looked at. He was quite in favor of that.

Q. I meant to ask you to state at the outset, Dr. Hill, whether I asked you when you arrived this morning to look at the transcript of testimony in these proceedings given by Mr. David Griggs.
A. I did look at it, not all of it. I looked mostly at the part that pertained to Lincoln or the summer study.
Q. You have spoken of your interest in the problem of early warning. Did the summer study have any other specific purposes?
A. Oh, yes.

* * * * * * *

I should like to add one thing. Before coming down I thought I was going to be asked to testify only as to the origin of this word "ZORC," and I did refresh my memory on that point. I have not refreshed my memory by referring to files or anything on these general questions about the summer study. I may have to hesitate at points and say I would like to refresh my memory, if that is all right with the committee.

Q. Since you have mentioned "ZORC," what is your memory about that?
A. All the soul and memory searching I can do, I first saw it in an issue of Fortune that came out just about a year ago. I think it was the May 1953 issue of Fortune.
Q. Do you remember a meeting of the Scientific Advisory Committee in Boston in the fall of 1952?
A. If I may correct you, Scientific Advisory Board of the Air Force. Yes, I did.
Q. Did you attend?
A. The session, as I recall, was 3 days. I was not a member of the board, but we were asked to make a presentation from the Lincoln Laboratory.
Q. When you say "we," who do you mean?
A. Well, I was. The presentation occupied about half of one morning's session. I attended certainly all of the Lincoln presentation and most of what came before. I cannot swear I was there all the time before we went on. But I rather chairmaned our presentation which was made by 5 or 6 people.
Q. Did Dr. Zacharias have anything to do with that presentation?
A. Yes. He had the final presentation on the results of the summer study. I believe he spoke for 30 or 40 minutes.
Q. Do you recall any incident occurring during the occasion that you just described of the meeting of the Scientific Advisory Board in which the word "ZORC" or anything like that figured?
A. I cannot recall any such thing. The statement was made in Griggs' testimony that Zacharias wrote this on the blackboard. I cannot believe that, because it would have been a cute trick in a very public and formal meeting, and I know Zacharias well enough to know that I would have been quite angry with him had he done it. I am convinced he did not do it. To the best of my knowledge, as I say, I never saw or heard the word before the Fortune article of last May.
Q. Returning to the inception of the so-called summer study, do you have any recollection of any part that Dr. Oppenheimer played in that?
A. I believe that Zacharias and I approached Charlie Lauritsen, Robert Oppenheimer, and I. I. Rabi, and talked to them about it to get their opinion.
Q. When would that have been?
A. That would have been around the time of the Physical Society meeting in 1952. I think it was that period. That is the first week in May and the last week in April. We discussed it at some length with Robert then.
Q. Discussed at some length with whom?
A. With Dr. Oppenheimer.
Mr. ROBB. He said Robert.
Mr. MARKS. I just didn't understand him, Mr. Robb.

By Mr. MARKS:

Q. Do you remember anything of the views that were expressed at that time about the purposes that should or might be served by the summer study that you were then——

A. I think in general this group agreed with Zacharias and, I think a study would be a worthwhile thing. We talked some about the problems that might be looked at, * * * and I think it was the general opinion of the three gentlemen that Zacharias and I approached that they would support this by joining to the extent that their time permitted, and would help us in any way on call.

Q. Was there any discussion then or at any other time about the relation between the summer study and the problems of the Strategic Air Command?

A. I don't specifically recall in that period that there was such discussion. I can recall other discussions with this group and others, like Dr. Piore of the Navy, and Dr. Haworth of Brookhaven, and Berkner, whom I have already mentioned, of general discussion of offense and defense, and so on. In all these discussions I believe the only positive statement made about the Strategic Air Command was that it should be strengthened.

Having seen Griggs' testimony, I should add that there is some inference somewhere in it that increasing defense might weaken Strategic Air Command, and hence increasing defense is bad, or that some scientists definitely were against the Strategic Air Command, and thought it should be cut or abolished. I have never heard any such statement in my discussion with scientists cleared for military work. As I say, the only thing I can recall in this sense is that in general we thought it should be strengthened.

We also thought air defense should be strengthened.

Dr. Hill, I would like to read you a portion of the testimony given by Mr. Griggs, and I will then ask you a question about it. I am reading from page 2617 of the transcript, and the passage that I intend to read runs from page 2617 to page 2620.

"And did the Lincoln study" (I am reading just a little after the middle of page 2617)——

Mr. ROBB. This is a question by who, Mr. Marks?

Mr. MARKS. This is a question on cross-examination of Mr. Griggs. I believe Mr. Silverman conducted it.

"And did the Lincoln study ever recommend the giving up of any part of our strategic air power?

"A. No, not to my knowledge.

"Q. I think you have already said so far as your knowledge goes, Dr. Oppenheimer did not recommend that.

"A. That is right. I would like to amplify my answer on that for the benefit of the board, since this is the first mention of the summer study in this much detail.

"We were concerned by the thing I have already mentioned, that is, the fear that the summer study might get into these things which we regarded as inappropriate for Lincoln, and as of questionable value to the Air Force—I refer to the giving up of our strategic air arm, and the allocation of budget between the Strategic Air Command and the Air Defense Command—but we were also very much concerned in the early days of the formation of the Lincoln summer study, because it was being done in such a way that had it been allowed to go in the direction in which it was initially going, every indication was that it would have wrecked the effectiveness of the Lincoln Laboratory. This was because of the way the thing was, the summer study was being handled administratively.

"So far as I know, it was not because of any direct action on the part of Dr. Oppenheimer. On the other hand, I felt at the time that Dr. Oppenheimer should have been well enough informed and alert enough to see that this would be disastrous to the Lincoln summer study.

"After having reported this to the Secretary of the Air Force, Mr. Finletter, who had been actively concerned with the summer study, and had been very much—excuse me, made a mistake—I said Mr. Finletter had been actively concerned with the summer study. I meant to say he had been concerned with project Lincoln. He had been in touch with President Killian, and Provost Stratton of MIT on the prosecution of project Lincoln. So I reported this to Mr. Finletter, and he essentially charged me with trying to find out if the summer study was going to be conducted in such a way as to result in a net gain to the effectiveness of Lincoln or a net loss.

"If it looked to me as though it were going to be a net loss, I was asked to inform him so that steps could be taken to correct this condition, or to cancel the summer study if that were necessary.

"I got in touch with Provost Stratton at MIT. I found that he hardly knew about the existence of the plan for the summer study. He undertook to look into it. I told him the things that worried me and worried Mr. Finletter about it. He did look into it. Some corrective action was taken in terms of discussions with people most involved and in terms of changing the organizational structure by which the summer study was to be introduced into the Lincoln project, and at a slightly later date Mr. Killian of MIT called me and told me that he was satisfied partly as a result of the recent activities that he and Dr. Stratton had been engaged in, which I have already mentioned, that the Lincoln summer study would operate to the benefit both of Lincoln and the interests of the Air Force.

"He further said, since I had mentioned that one of the things we were afraid of was that the Lincoln summer-study results might get out of hand, from our standpoint, in the sense that they might be reported directly to higher authority, such as the National Security Council, President Killian reassured me that he had taken steps so that he was sure that the summer study would be—I think his words were 'kept in bounds.'"

Mr. ROBB. I think you ought to read next the paragraph.

Mr. MARKS. I would be glad to (reading):

"On the basis of this assurance we had no further—that is, Mr. Finletter, myself, and General Yates and the other Air Force people—had no further immediate worries about the summer study and we encouraged it."

By Mr. MARKS:

I would like to ask you generally, Dr. Hill, whether you have any comment to make in respect to the passages that I have read to you?

A. In the first place, I should just like to comment on Griggs' ideas of what he thought the summer study was going to be. He evidently was concerned that the purpose would come out with some supermaster plan—I mean the purpose was to come out with a master plan—of how to divide money between Strategic Air Command and Air Defense Command. Such was farthest from our thoughts. We at no time, to the best of my knowledge, considered worrying about the problems of Strategic Air Command any way except insofar as they relate to defense and defense relates to them.

I don't know where Griggs got this idea, and I don't doubt that he had because I know for a while he was quite concerned about this summer study, and about allowing it to be set up. I know this only by hearsay. He never came to me with his qualms. He did talk to a lot of other people. He discouraged some people from participating, so I have been told, and he evidently talked to my superiors at MIT.

The inference is made—I can't quarrel with what Griggs thought—the inference is made that he somehow by this maneuvering changed our purpose. This I deny.

Q. Did you talk to your superiors at MIT about this project?

A. Yes. In setting up this we first talked to our superiors at MIT and very briefly with the Air Force and there seemed to be good support for it. Then I know that this occurred during the physical society meetings. Several people came to me and said they were quite concerned about setting us up. One, that it might wreck the program already going on in a growing laboratory, and, secondly, they were concerned about Dr. Oppenheimer's participation in it for security reasons.

I said it was my practice to leave security matters entirely to those people charged with them; that we would put Dr. Oppenheimer's name in for clearance just as we would anyone else. This created enough of a stir so that Zacharias and I went back to Killian and Stratton, our own superiors, told them about it, and it was then agreed rather than going ahead immediately—I had already prepared letters to send out to people whom we hoped would participate—instead of that, to make sure at the highest levels that we should talk to in the Air Force, Army, and Navy, that we make certain this was all right.

Zach and I spent several weeks seeing all the proper people, and I know the persons I talked to, the senior ones: Admiral Bolster and his associates in the Navy; General Maris in the Army; and first General Putt in the Air Force; and later General Craigie, all of whom expressed a certain amount of concern and a certain amount of enthusiasm, and the net result was that they all agreed we should go ahead.

So instead of starting our recruiting procedure, shall I say, May 1, we started about May 20. So that there was a delay while we reexamined these fears that Griggs and others had raised.

Had I answered the question? It was rather lengthy if you consider the background, and I may have left something out there.

Q. I would like to direct your attention specifically to one matter that was referred to in the passage that I read. In the passage I read to you there occurred at one point the following: This was in one of Mr. Griggs' answers, and I am starting in the middle of the answer on page 2618:

"* * * we were also very much concerned in the early days of the formation of the Lincoln summer study, because it was being done in such a way that had it been allowed to go in the direction in which it was initially going, every indication was that it would have wrecked the effectiveness of the Lincoln Laboratory. This was because of the way the thing was, the summer study was being handled administratively.

"So far as I know, it was not because of any direct action on the part of Dr. Oppenheimer. On the other hand, I felt at the time that Dr. Oppenheimer should have been well enough informed and alert enough to see that this would be disastrous to the Lincoln summer study."

Have you anything to add to what you have already testified that would explain the reference in the passage that I have just read about how things were being handled administratively?

A. I make no claim to knowing all about administrative procedures. It seems to me there are two ways to wreck a laboratory. One is to ruin the morale on the inside, and the other is to ruin the confidence of those on the outside who must support it.

With regard to the former, although Griggs doesn't say so specifically, I think this has to be brought in. Some people were concerned that bringing in a group of some rather high-powered physicists and others, and putting them down in the middle of an organization might be so glamorous that people would neglect their work and so on.

It was my feeling that the ability to bring in outside people of stature in this field was very valuable and rather than hurt morale, would rather help it. I think events have proved me right on this. The amount of time that any members of the laboratory took off or neglected their work because of the presence of this group was completely negligible.

As far as destroying confidence on the outside is concerned, first of all, of course, I must have the confidence of my superiors, and this was carefully cleared with them before any move was made to solicit any help. We had talked to other people and we received advice, but to solicit any help from the outside, not a thing was turned until Zacharias and I felt we had the complete confidence of Killian and Stratton. They in turn said you must get the Air Force, who is the contractor, and the primary support behind you, too.

It was my feeling—here I will have to refresh my memory from the files, gentlemen,—that I would naturally have gone to General Putt in this instance, since he is the chairman of our Military Advisory Committee. I know I went to someone, I believe it was Putt. We discussed the pattern, thought it a good idea.

So this business about administrative procedures, I don't understand. I point out that Griggs was not around the laboratory at any time. He could know nothing of these directly. He never consulted me or asked what we were doing. So I can only tell you what we did. I must leave to your judgment whether it was good administrative policy or not.

After the fuss was made my Griggs around the first of May, then things got in an uproar, and I was called in by Killian and Stratton as you might expect, and we went through it again. Then we had this other go-around which I explained earlier, seeing all the services and seeing them in detail. They bought it.

Q. They bought it?
A. That is what I said. Maybe I better put it in good English. They agreed that what we were planning to do was quite all right, and probably a good thing, and if we wanted to do it, we should be supported.

At no time, I reiterate, did we change what we had started out to accomplish.
Mr. MARKS. I have no further questions.
Mr. GRAY. Mr. Robb.

CROSS-EXAMINATION

By Mr. ROBB:

Q. Doctor, you told us about going to see the various representatives of the services, General Putt, and others, and you said they expressed a certain amount of concern, and a certain amount of enthusiasm. Could you explain that a little bit to us?

A. Let me point out, I believe that Vista was just reporting then. This had created a certatin amount of stir in the military. They were afraid, as I recall, that Vista would carry too much weight with higher authorities that did not understand their problems, and would hurt their program. They were afraid, and they expressed some concern, if we started a program of this sort, to take a general look at a broad military problem, that this in turn might give them a headache rather than do good. I think events have proved that this concern was all right, but there was no undue problem that resulted from it.

Q. You felt, of course, that the Air Force being the contractor who was going to pay for this had a perfect right to be concerned about it?

A. Oh, yes, sir.

Mr. ROBB. That is all. Thank you.

Mr. GRAY. Dr. Hill, have you read Dr. Zacharias' testimony here before this board?

The WITNESS. The testimony that was given last week I skimmed through quickly; yes, sir. I did not discuss with him what he talked about this morning.

Mr. GRAY. How much other testimony have you read besides Griggs' and Zacharias'?

The WITNESS. I confess to glimpsing at some of the others while I was sitting out in the room, the others that were in those two volumes. I can't say that I read any of the testimony so as to remember it. I sort of skimmed a page and read a paragraph.

Mr. GRAY. With respect to "ZORC," you said you were confident that Dr. Zacharias would not use this phrase or go through this procedure which Mr. Griggs testified about. Am I correct in my recollection?

The WITNESS. That is correct. I am also confident, if I may add, that had he done it, I would have been quite annoyed, and would have let him know it. Had I seen him do this, I am sure I would have remembered. That is the point I was trying to make.

Mr. GRAY. If he did it today, would you be annoyed with him?

The WITNESS. I think that would depend entirely on the circumstances, sir.

Mr. GRAY. Is this a matter of time? I think we ought to tell you that he testified today that he might do it today.

The WITNESS. I don't know. If he were in a group with friends, and they were talking about things like that Fortune article, and he happened to write "ZORC" on the blackboard, I would not be surprised. If he did it at a formal meeting, I would be quite surprised.

Mr. GRAY. I see. You said in your direct testimony that you never heard any scientist who was cleared for military work argue for the dissolution of the strategic arm. Have you heard any scientist argue for the dissolution of the strategic arm?

The WITNESS. No, sir.

Mr. GRAY. So the "cleared for military work" had no significance?

The WITNESS. No, sir.

Mr. GRAY. You said that Dr. Griggs had discouraged people from working on the summer study, so you had been told. Who told you that?

The WITNESS. Dr. Getting—I tried to say that this was inferred, and also second hand—Dr. Getting, for instance, had seemed quite enthusiastic about working on this summer study. I know he talked to Griggs at length and after that he cooled off completely.

Mr. GRAY. And you would draw the conclusion from that that it was as a result of talking with Griggs——

The WITNESS. I infer that, yes, sir.

Mr. GRAY. Who told you that he had kept people from working on the project?

The WITNESS. Well, this was the inference of other people, too. I cannot testify that this actually happened. It was inferred by other people. I think Zacharias would say this.

Mr. GRAY. Did Zacharias tell you this? Was he the source of your information?

The WITNESS. I think he did.

Mr. GRAY. Did anybody else tell you that Griggs had been instrumental in persuading people not to work on this project?

The WITNESS. No; but I know one of my colleagues was very bitter about it, and very much set against starting it.

Dr. EVANS. Set against what?

The WITNESS. Set against starting the project. I also know this was shortly after a talk with Griggs. Again this is only inference.

Mr. GRAY. Do you think it is fair to draw such an inference and to conclude from that that he was responsible for people not working on the project?

The WITNESS. No, sir; it is not.

Mr. GRAY. You have me confused now. Do you wish to have it appear that you testified here that Griggs was responsible for people not working on this project?

The WITNESS. If I can state it now, I would like to.

Mr. GRAY. I wish you would. I am not trying to trap you, Dr. Hill.

The WITNESS. I understand that perfectly. In trying to talk around Griggs' testimony which was rather general in spots, I had to give some flavor of my feeling of his activities at that time, too. I tried to make clear that Griggs never talked to me about his concern, and that I never talked to him about my concern about his activities. Therefore, I think it only fair that I drew certain inferences just as he did. I think it would be strictly unfair on my part to accuse him of having dissuaded people from taking part in the study.

Mr. GRAY. If you had come here without knowing what Dr. Zacharias had testified to this board, would you have stated that Griggs had been instrumental, so you had been told, in keeping people from working on this project?

The WITNESS. Sir, about Zacharias' testimony, I skimmed through his earlier testimony of a week ago, and I don't recall from it any mention of Griggs. My statements about Griggs have been drawn from Griggs' testimony and my own memory. As I say, I did not talk to Zacharias about his testimony this morning.

Mr. GRAY. Did you talk with Dr. Oppenheimer's attorneys about Dr. Zacharias' testimony this morning?

The WITNESS. No, sir. I have not talked to anybody about it. Both Zacharias and I talked with Mr. Marks very briefly this morning about the flavor of what might go on.

Mr. GRAY. Was there any mention made of persuading people not to work on this project?

The WITNESS. I can't recall.

Mr. GRAY. This conversation took place this morning and you can't recall?

The WITNESS. Yes, sir. You will recall you asked me a question about Zacharias' testimony and inferences I drew from it. There may have been discussion this morning about Griggs, but if there was, I brought it up from having read Griggs' testimony.

Mr. GRAY. It is not a question of who brought it up. I am asking you whether in your preparation for this appearance there was any discussion of Griggs having been instrumental in persuading people not to work on this project.

The WITNESS. Yes, there was.

Mr. GRAY. And so, therefore, you are unable to tell me who told you other than that conversation in preparation for this that Griggs had been instrumental in persuading people not to work on this project?

The WITNESS. No. Dr. Getting gave me this impression and Dr. Zacharias gave it to me 2 years ago at the time we were setting it up. There we had long conversations on the subject.

Mr. GRAY. Dr. Hill, you testified on direct examination that at a meeting several persons came to you and expressed concern about the project, at least partially on the score of Dr. Oppenheimer's security status. Do you remember who some of these people were?

The WITNESS. If I said several, I was wrong; only one, and that was Dr. Getting.

Mr. GRAY. You did say several.

The WITNESS. I am sorry, then. That was a slip of the tongue.

Mr. GRAY. What was Dr. Getting's official position?

The WITNESS. He is now vice president of the Raytheon Manufacturing Co.

Mr. GRAY. But he was then in the Navy?

The WITNESS. No; he was then at that job, but the year previous he had worked on a staff job in the Air Force.

Mr. GRAY. And he was the only one that mentioned concern about Dr. Oppenheimer's security?

The WITNESS. Yes, sir. Quite a few others mentioned that Griggs was talking about it and had talked to them.

Mr. GRAY. Do you remember who they were?

The WITNESS. I know of one. Dr. Fisk of Bell Laboratories.

Mr. GRAY. Was he concerned?

The WITNESS. He was not concerned about Dr. Oppenheimer. He was very much concerned about Griggs making this sort of statement.

Mr. GRAY. He rejected the notion that there was any question?

The WITNESS. Yes, sir.

Mr. GRAY. I think I should tell you, Dr. Hill, that I am very much concerned, as are my colleagues on the board, about the fact that there is testimony before this board which indicates very clearly that some one or more witnesses have not told the truth to this board. There has now developed in this proceeding a real question in some cases of veracity.

I have another question which is not related to the remark which I made in any way.

The WITNESS. May I ask you a question, sir?

Mr. GRAY. Yes, sir.

The WITNESS. Were you referring to the "ZORC" incident?

Mr. GRAY. Among others; yes. Do you have anything to add about the "ZORC" incident?

The WITNESS. No, sir.

Mr. GRAY. Was Griggs the only person who was responsible for the delay in the beginning of the summer study?

The WITNESS. That is a very difficult question to answer.

Mr. GRAY. Let me remind you that you testified on direct examination that there was a delay of several weeks as a result of the activities—this may not be your words—but as a result of the activities of Griggs and others. If you are uncomfortable about my statement of your testimony, I will be glad to have it read back to you.

The WITNESS. No; I would be very happy to clear this up.

Mr. GRAY. Yes.

The WITNESS. I said that a stir took place around the 1st of May which resulted in a delay.

Mr. GRAY. Yes.

The WITNESS. I later, I think, said I thought that Griggs was in part responsible for that stir. I don't know of others.

Mr. GRAY. So you think Griggs was probably the one responsible.

The WITNESS. As much as anyone. The one I know anything about. I have no way of knowing that there were others.

Mr. GRAY. Would you characterize Griggs' activities in this episode as sabotage?

The WITNESS. No, sir.

Mr. GRAY. What would you call it?

The WITNESS. I would call it difference of opinion.

Mr. GRAY. Would you call it honest difference of opinion?

The WITNESS. I would think so. I would think also, however, that a good deal of misinformation about what we were trying to do, if this present testimony reflects what he thought then.

Mr. GRAY. You didn't question his right as a senior scientist of the Air Force to have an opinion about the shape and form of the study?

The WITNESS. Not at all; no, sir.

Mr. GRAY. Now, after Dr. Killian and Dr. Stratton called you, and perhaps others, in, following the "stir," and you were authorized to go ahead, I believe you said, was there any change whatsoever in the plan of the study?

The WITNESS. No, sir.

Mr. GRAY. Not the slightest change?

The WITNESS. No, sir.

Mr. GRAY. So therefore your interpretation of the situation was that there had simply been delay of several weeks without consequence otherwise?

The WITNESS. That is correct.

Mr. GRAY. Dr. Evans.

Dr. EVANS. Dr. Hill, would you tell us something about your education; where you were educated?

The WITNESS. Yes, sir. I attended Washington University in St. Louis from 1926 to 1930, receiving a bachelor's degree in mechanical engineering. After 3 years working, I came back and took a master's degree in physics, and then went to the University of Rochester and finished a Ph. D., in 1937, in physics.

Dr. EVANS. From what I heard here—I am just trying to get my thinking cleared up—there seems to have been two schools of thought engaged in this work, and there doesn't seem to be much love lost between them; is that true?

The WITNESS. I have heard this, sir. I don't consider myself a member of any school of thought. I have heard that there is quite a difference of opinion among certain groups of physicists.

Dr. EVANS. You would say that if there were two schools of thought, you would say you belong to Dr. Oppenheimer's school; is that it?

The WITNESS. I think I would have to have the definition of the school of thought. If you mean about the H bomb——

Dr. EVANS. No; I mean about this laboratory we are talking about—this summer course, I beg your pardon.
The WITNESS. Summer study.
Dr. EVANS. Yes; the summer study.
The WITNESS. There I certainly can identify myself with a school, and that was that it was a very good thing and needed doing. If Dr. Oppenheimer belongs to that school, then we are joint members.
Dr. EVANS. Would you care to name some of the men besides Griggs that belonged to the other school?
The WITNESS. If you mean now, people who questioned the wisdom of the summer study in the scientific field?
Dr. EVANS. Yes.
The WITNESS. The three I can think of most quickly are Griggs, Getting, and Valley.
Dr. EVANS. Where did Alvarez fit in this?
The WITNESS. I don't recall ever talking to him about it.
Dr. EVANS. You don't know anything about Teller?
The WITNESS. No, sir. He would not—in general, the people we would have talked to about this would have been those more closely associated with electronics than with nuclear weapons. There are some exceptions. So Teller never entered into our discussion, to the best of our knowledge.
Dr. EVANS. I have no other questions.
Mr. GRAY. Mr. Marks.

REDIRECT EXAMINATION

By Mr. MARKS:

Q. Dr. Hill, when you came from the train this morning to Mr. Garrison's office and met me, did I ask you any questions about whether Griggs had discouraged people from working on the summer study?
A. I don't recall that you did.
Mr. MARKS. I have no other questions.
Mr. ROBB. I have no questions.
Mr. GRAY. Is it your testimony, then, Dr. Hill, that you did not discuss with the attorneys this morning this question of discouraging people from working on the study?
The WITNESS. Sir, I have already given you an answer to it.
Mr. GRAY. I don't believe you have given me a clear answer.
The WITNESS. I am trying to clear it up.
Mr. GRAY. I would like to have you clear it up. That is my entire purpose.
The WITNESS. I should really go back to make this completely clear 2 years, to this time when Zacharias and I were trying to set up this summer study. At that time we felt rather clear that Griggs was quite opposed to it and doing what he could to put it in the best light for Griggs to see that it was in its proper perspective. In talking to each other we may have used other words. That brings us up to this morning, and I honestly felt I was here only to testify as to the "ZORC" incident. So I had not reviewed my memory at all, and I have been trying to all day, which is why I hesitate just a little about when who said what to whom. I know after reading Griggs' testimony I made a statement that it looks to me like Dave was really in there pitching and trying his best to keep people from joining. I can't recall that Mr. Marks asked me a question. His question to me was did he ask the question, and I said no.
Now, then, Zacharias and I sat out in the waiting room together and we discussed it some more about Griggs and 2 years ago, you see. So my discussion on that subject with Zacharias and with Marks, I think, mostly my talking. I don't recall what Zacharias said except as a sort of nod agreement. Does that clarify my testimony on this point?
Mr. GRAY. Let me answer your question this way—I am trying to clarify it: On the direct question by Mr. Marks you made the statement that Mr. Griggs had discouraged people from working on the project, so I have been told. I would be glad to have this read back to you if you wish.
Mr. MARKS. Mr. Chairman, I don't think I asked him a question about that. I think that was a reference to the general question which I asked.
Mr. GRAY. In his direct testimony he made this statement. I will ask the reporter to read the statement that has me concerned.
(The reporter thereupon read the record, as follows:)

"By Mr. MARKS:

"Q. I would like to ask you generally, Dr. Hill, whether you have any comment to make in respect to the passages that I have read to you.

"A. In the first place, I should just like to comment on Griggs' ideas of what he thought the summer study was going to be. He evidently was concerned that the purpose would come out with some supermaster plan—I mean the purpose was to come out with a master plan—of how to divide money between Strategic Air Command and Air Defense Command. Such was farthest from our thoughts. We at no time, to the best of my knowledge, considered worrying about the problems of Strategic Air Command any way except insofar as they relate to defense and defense relates to them.

"I don't know where Griggs got this idea, and I don't doubt that he had because I know for a while he was quite concerned about this summer study, and about allowing it to be set up. I know this only by hearsay. He never came to me with his qualms. He did talk to a lot of other people. He discouraged some people from participating, so I have been told, and he evidently talked to my superiors at MIT."

Mr. GRAY. "He discouraged some people from participating, so I have been told."

The WITNESS. Yes, sir. If I could retract my words, I would say that this way. Inferences have been made by me and others that he discouraged other people from working on it. When I say I have been told, I meant as of 2 years ago, and the thing I was trying to bring out was that this was completely inference on my part. Does that clear it up, sir?

Mr. GRAY. I think so.

I should like to say for the record that if in my questioning of this witness I have seemed to impute to Dr. Oppenheimer's attorneys any impropriety, I have no such intention.

Mr. MARKS. Thank you.

Mr. GARRISON. Thank you.

Mr. GRAY. I should say further that I understand that the witness did not discuss with Mr. Marks the question of discouragement of employment at the summer study although it is my understanding that this matter did come up in conversation with the witness with Dr. Zacharias.

The WITNESS. That is right.

Mr. GRAY. Is that a correct statement?

The WITNESS. That is correct.

Mr. MARKS. I think I need to add to that, sir, that I believe some remarks to that effect about discouraging people on the summer study was made in my presence when I first met with Dr. Zacharias and Dr. Hill this morning. I don't recall which of them made it. I didn't pay any attention to it.

Mr. GRAY. Do you have any further questions?

Mr. MARKS. No, sir.

Mr. ROBB. No, sir.

Mr. GRAY. Thank you very much, Dr. Hill.

The WITNESS. Thank you.

(Witness excused.)

Mr. GARRISON. Mr. Chairman, could we just talk about procedure for a minute?

Mr. GRAY. Yes. I have talked with the members of the board at the noon recess, and I may say I am authorized to say we will allow you to start your summation and argument tomorrow afternoon, rather than tomorrow morning, which I believe was your request.

Mr. GARRISON. I appreciate that very much. May I then say that in the lunch hour which we did not spend with Dr. Zacharias and Dr. Hill, I reached the conclusion in my conscience as a lawyer that I just must finish the reading of the summary which I can do in a few hours before reaching a final decision as to whether to ask Dr. Oppenheimer to make a rebuttal or not. I am just not quite clear at this point whether it is going to be necessary. If I could have some means of communicating with you and with the other board members either late this afternoon or very early this evening as to whether or not I would like to put him on or ask him to resume testifying in the morning or not, I would like to leave it in that manner, if it is feasible for you. I realize the inconvenience that this may mean, but I should certainly arrive at the decision early enough this evening—I should think by dinner time.

Mr. GRAY. We will proceed tomorrow afternoon with your summary.

Mr. GARRISON. Yes.

Mr. GRAY. I think I can say on behalf of the board that we will not insist that you tell us now that you will or will not call Dr. Oppenheimer back as a witness tomorrow morning. I would like to know as early as possible about that so that we may make our own plans.

Mr. GARRISON. Thank you, Mr. Chairman.

(Discussion off the record.)

Mr. GRAY. You have presented your witnesses except for possibly Dr. Oppenheimer?

Mr. GARRISON. Yes.

Mr. GRAY. I take it, then, gentlemen, we are in recess until 9:30 tomorrow morning. If you decide in the meantime you will not call Dr. Oppenheimer to the stand, we will meet at 2 o'clock tomorrow afternoon.

Mr. GARRISON. Yes, sir.

(Thereupon at 3:35 p. m., a recess was taken until Wednesday, May 5, 1954, at 9:30 a. m.)

UNITED STATES ATOMIC ENERGY COMMISSION

PERSONNEL SECURITY BOARD

In the Matter of J. Robert Oppenheimer

Atomic Energy Commission,
Building T-3, Room 2022,
Washington, D. C., Wednesday, May 5, 1954.

The above-entitled matter came on for hearing, pursuant to recess, before the board, at 9:30 a. m.

Personnel Security Board: Mr. Gordon Gray, chairman; Dr. Ward T. Evans, member; and Mr. Thomas A. Morgan, member.

Present: Roger Robb and C. A. Rolander, Jr., counsel for the board; J. Robert Oppenheimer; Lloyd K. Garrison, Samuel J. Silverman, and Allan B. Ecker, counsel for J. Robert Oppenheimer; Herbert S. Marks, cocounsel for J. Robert Oppenheimer.

PROCEEDINGS

Mr. GRAY. You may proceed, Mr. Silverman.
Whereupon J. Robert Oppenheimer, a witness, having been previously duly sworn, was called in rebuttal, examined, and testified as follows:

DIRECT EXAMINATION

By Mr. SILVERMAN:

Q. Dr. Oppenheimer, Dr. Alvarez testified that when he came to Los Alamos there was a hydrogen liquefaction plant there. Will you tell us what that was used for?

A. Yes. It was actually one of the first structures erected at Los Alamos, and reflected the opinion, which turned out to be erroneous, that going from the fission weapon to the fusion weapon would not be too tough a step.

Its initial purpose was to make studies of the thermodynamics, and steresis phenomena in the liquefaction of hydrogen isotopes. This work was also conducted by a subcontractor at the University of Ohio.

About halfway through the war, a number of points arose which changed the program. One I think Dr. Teller referred to. He discovered in the work we had earlier done we had left out something very important and very serious, which proved that the ideas we had had about how to make this machine would not work in the form we then had. The pressure on the whole laboratory to get the fission job done and the difficulties of that job both increased. The cryogenic facility actually played a small part in our researches for the fission job but I do not propose to describe it. I think it is classified.

The head of that group, Earl Long, now of the University of Chicago, left the cryogenic job and became director of the shop. I believe that very little was done with the cryogenic facility in the last year before the war ended.

I may, if this is still responsive to your question, describe what else was going on at Los Alamos during the war related to the thermonuclear program.

Q. I wish you would, yes.

A. As nearly as I can recollect, there were two groups in addition to the cryogenic group concerned. One was Dr. Teller's group which toward the end of the war was in the part of the laboratory that Fermi as associate director ran. It was called the advanced development division, and several young people under Teller were figuring and calculating on aspects of the thermonuclear program. There was another group in which there were three members of the British mission, and a number of Americans who were measuring the reactivity of the materials which seemed to us relevant to a hydrogen bomb, and who actually completed some measurements on this before the war was over. I think this is about the whole story.

Q. As a matter of characterization, would you say that at Los Alamos during the war years the laboratory was actively working on the development of the thermonuclear bomb?

A. We planned to be, but we were in fact not.

Q. And why not?

A. I have outlined the two major reasons. First, we didn't know how to do it, and second, we were busy with other things.

Q. At the end of the war, was there any expression to you of Government policy with respect to going ahead with the thermonuclear weapon?

A. I think I have already testified, but I am willing to repeat. After the Trinity test, the Alamagordo test, but before Hiroshima, I went to Chicago to consult General Groves largely about the major mechanics of the overseas mission, and how we would meet our time schedules. In the course of that, I put up to General Groves—I think I had already put in writing an account of the problem—the fact that we had not moved forward, and perhaps had moved somewhat backward on the thermonuclear program, and was this something that he wanted the laboratory to take hold of. This was while the war was still on. He was fairly clear in saying no. I believe—I will not speculate as to his reasons for that, but it was clear to me.

The only other communication to me of a view on the matter was incidental. In August, Dr. Bacher and I had come on to report to General Groves, and it was at that time that I told him that I thought I should not continue as director of the laboratory, and that we began discussing the problem of who was to run it. Just before I flew west, I had a message to consult General Groves. I did so. He told me two things. He had had a conversation with Mr. Byrnes, who was then the President's representative on the Secretary of War's Interim Committee.

Mr. ROBB. Could we have the date on this?
The WITNESS. This would have been after the 15th of August, but not much.
Mr. ROBB. What year?
The WITNESS. 1945. This is all in the period immediately around the surrender.
Mr. MORGAN. Was that General Byrnes?
The WITNESS. No; this was James Byrnes who was very shortly thereafter to be Secretary of State. It was then Justice Byrnes.

Groves said that in the present state of the world, the work on weapons must continue, but that this did not include, he thought, the super. That was about all. These were not formal expressions of opinion; they were from my boss to me in a most informal way at a time when I was preparing not to retain active responsibility.

By Mr. SILVERMAN:

Q. Dr. Teller testified about a board of four people at the end of the war, or near the end of the war, who he understood decided that the thermonuclear program should not be pushed. Can you cast some light on that?
A. I think I can. I think I know what Dr. Teller was talking about.
There was a panel of four people. Their names were Arthur Compton, Ernest Lawrence, Enrico Fermi and me, Robert Oppenheimer. We had been asked to advise on the use of the bombs, on the general nature of the future atomic energy program, but we were asked specifically through Mr. Harrison, on behalf of the Secretary of War, to prepare as detailed an account as we could of everything we knew that could be done or needed doing in the field of atomic energy.
This was not just military things. It involved the use of isotopes and the power problem and the military problems. As a part of this report, we discussed improvements in atomic weapons and in the carrier problem. As a part of this report, we discussed the thermonuclear bomb, the super, as it was called. That was all we had in mind then. I believe that section was written by Fermi. I believe that Dr. Teller correctly testified that his own view on what the problem was, was attached as a slightly dissenting or even strongly dissenting view to our account.
We wrote an account which was not a recommendation of policy at all, as I remember, but was an analysis of where we thought the matter stood. I think General Nichols' letter to me quotes from it, and says this program did not appear on theoretical grounds as certain then as the fission-weapon program had at some earlier stage. This was a rather long and circumstantial account of what we knew about it. It was not intended and was not a statement of what should be done. It was an assessment of the technical state of the problem.
This board had no authority to decide, it was not called on to recommend a decision, it did not decide nor recommend a decision. It described. I think Dr. Teller was a little mistaken about what our function was.

* * * * * * *

By Mr. SILVERMAN:

Q. Between January 1947 and January 1950, which is the first 3 years of your chairmanship of the GAC, how many new reactors were started by the Atomic Energy Commission?
A. This would be better found by reading the Commission's reports, and I have not done so. This work was very slow to get started, but if you include all kinds of reactors, for development, for research, and for production, perhaps around eight.
Q. And did the GAC express its views to the Commission about the slowness of getting started?
A. The GAC wrote reams on the subject of getting the reactor program off the dime. The reams may not have been very sensible, but they were clearly addressed to this problem.
Q. Dr. Libby and Dr. von Neumann are now members of the GAC, are they not?
A. Yes; they are.

Q. And they are both enthusiastic proponents of the hydrogen bomb?
A. Yes; they are. I believe today everybody is an enthusiastic proponent.
Q. But were they when they were appointed?
A. Yes; they were.
Q. Did you have anything to do with their appointment?
A. I don't know. The appointments were presidential. I did, however, include the names of von Neumann and Libby on the list, I believe, of five names that I submitted to Mr. Dean in the summer of 1950.

I should for completeness say that the other people on that list, as I recollect, though very competent, were not identified with enthusiasm for the hydrogen bomb. Bacher, Fermi, and Bethe were also on the list. Libby was appointed in the summer of 1950. Von Neuman was not, but he was appointed as soon as a vacancy appeared through the resignation of Dr. Cyril Smith. Both men served on the GAC for a while while I was chairman.

Q. General Wilson testified, I believe, that at some stage you did not support the installation of 2 of the 3 methods of long-range detection. Did you ultimately support those 2 methods?
A. Yes.
Q. And was your decision about supporting the installation of those 2 or 3 methods made on the basis—on what basis was it made?
A. This is not recollection.
Mr. GRAY. This is not what?
The WITNESS. This is not a recollection. The only ground for holding up the installation of something is doubt as to whether its development had reached the right stage for it to be effective. That is the best answer I can give to you.

By Mr. SILVERMAN:

Q. As to the third method, the one you did support, do you recall the circumstances of the initiation of that method?
A. Yes; I do. This was just after Hiroshima, and we developed at Los Alamos—I believe that the man directly in charge was Kenneth Bainbridge—what we hoped might be an effective long-range detection device. I directed that we try this out with the cooperation of the Air Force, and we did succeed in identifying and describing the Hiroshima explosion by flights over the continental United States.

Later, when I was on the General Advisory Committee, I believe the committee wrote something to this effect, that the problem of detection of foreign explosions was of unparalleled importance. That since this was not clearly a Commission problem, we did not insist on being informed of the progress of the work, but we wished to record our view that progress was urgent and important. It was in the Defense Department that I had a more direct connection with the development of this method. It was completely successful in detecting and describing the first Soviet explosion, at least the first one we know about.

Dr. EVANS. That was radiation detector; was it not?
Mr. ROLANDER. I don't think we should discuss that.
The WITNESS. I am sorry.
Dr. EVANS. Excuse me.

By Mr. SILVERMAN:

Q. Of the three methods, was that first method the one that has furnished the most significant and important information, as far as you know, or is that classified?
A. Let me say simply that it has furnished an enormous amount of information which is technically very valuable. For some purposes the other methods are quite useful in giving supplementary data. I think I can't go further.
Q. Dr. Alvarez testified that at a meeting of the Military Objectives Panel in about December 1950, you said something to the effect that "We all agree that the hydrogen bomb program should be stopped, but to do so will disrupt the people at Los Alamos and other laboratories, so let us wait for the Greenhouse tests, and when those fail that will be the time to stop the program. Can you cast any light on that?
A. I am clear as to what my views were, and therefore fairly clear as to what I would have said, which resembles to some extent what Dr. Alvarez recounted. I did not think the Greenhouse test would fail. It was well conceived technically, and there was no ground such an opinion. * * * I could not have said that I expected it to fail, because I didn't think it would, and I could not have said that I expected it to fail, because this sort of statement about a test is some-

thing none of us ever made. The reason for making the test was that we wanted to find out.

What I did believe, and for the wisdom of this view I am not making an argument, was that the real difficulties with the Super program, as it then appeared, were not going to be tested by this Greenhouse test; that the test was not relevant to the principal question of feasibility. I am fairly sure that in the course of discussions at the panel, we would have commented on this.

On the question of where the Super program stood, on the relevance of that to the Greenhouse test, of the doubts that I felt as to whether this part of the Greenhouse test was a sensible thing technically to do, I would have said that to stop this part of the Greenhouse test, even though it made no technical sense, would be disruptive and destructive of all parts of the Los Alamos program.

I think that is the true story of what I would have said at this panel meeting and Dr. Alvarez' recollection is in some respects mistaken.

Q. What were your views as to the feasibility of the Super at that time?

Mr. ROBB. What time are we talking about?

Mr. SILVERMAN. This is December 1950, at the time of the military objectives panel.

The WITNESS. On the basis of then existing ideas it was highly improbable that this could be made; that we needed new ideas if there was to be real hope of success.

May I add one comment? In actual fact this component of the Greenhouse test had a beneficial effect on the program. This was in part because the confirmation of rather elaborate theoretical prediction encouraged everybody to feel that they understood and when they then made very ambitious inventions, the fact that they had been right in the past gave confidence to their being right in the future.

It may also to a smaller extent have provided technical information that was useful. Certainly its psychological effect was all positive. It would have been a great mistake to stop that test.

By Mr. SILVERMAN:

Q. And you thought so at the time and said so?

A. But not for the right reasons.

Q. There have been discussions on your views on continental defense and tactical and strategic use of weapons and so on. Perhaps if we could do this very briefly, could you give very briefly your views on continental defense?

A. As of when?

Q. As of now, if you like. As of the last year or two.

A. If the board is not saturated with this, I will say a couple of sentences.

Q. As of the time of the Lincoln study.

A. The immediate view after the war was that defense against atomic weapons was going to be a very tough thing. The attrition rates of the Second World War, though high, were wholly inadequate to this new offensive power.

Q. By the attrition rates, you mean the number of attacking airplanes you could shoot down and kill?

A. Precisely. In the spring of 1952, the official views of what we could do were extremely depressing, * * * and there were methods of attack which appeared to be quite open to the enemy where it was doubtful that we would either detect or intercept any substantial fraction of the aircraft at all.

I knew that on some aspects of the defense problem, valuable work was in progress at Lincoln and elsewhere. I knew something of the Charles study. * * * My view is that this is by no means a happy situation, and I know of no reason to think that it ever will be a happy situation, but that the steps that are now being taken and others that will come along as technology develops are immensely worth taking if they only save some American lives, if they only preserve some American cities, and if they only create in the planning of the enemy some doubt as to the effectiveness of their strikes. I don't know whether this answers the question.

Q. I think that answers the question.

A. I have never gone along with the 90 to 95 percent school. I hope they are right, but I have never believed them.

Q. The 90 to 95 percent school is the school——

A. That thinks you can eliminate practically all of the enemy attack.

Q. What did you conceive to be the relation between continental defense and strategic airpower?

A. First, strategic airpower is one of the most important ingredients of continental defense. Both with the battle of Europe and with the intercontinental clearly the best place to destroy aircraft is on the ground on enemy fields, and that is a job for strategic airpower.

Second, at least the warning elements and many of the defensive elements of continental defense are obviously needed to protect the bases, the aircraft, which take part in the strategic air campaign. This is the two-way relation which I think has been testified to by others. This has always been my understanding.

Q. It has been suggested that perhaps you had more interest in the tactical than the strategic use of atomic weapons. Could you comment on that?

A. It has been talked about a great deal. When the war ended, the United States had a weapon which revolutionized strategic air warfare. It got improved a little. The Air Force went hard to work to make best possible use of it. * * * Even during World War II we had a request through General Groves from the Army as to whether we could develop something that would be useful in the event of an invasion of Japan to help the troops that would be faced with an entrenched and determined enemy. The bomb that was developed and embellished in the years 1945 to 1948, and the aircraft that go with it, the whole weapons system, can of course be used on any target, but it is a very inappropriate one for a combat theater. Therefore, there was a problem of developing the weapon, the weapon system, the tactics to give a new capability which would be as appropriate as possible under fire, and in the combat theater. This is not because it is more important. Nothing could be more important than the armament that we had, and which is now to be extended, perhaps to some extent superseded, by thermonuclear weapons. It was simply another job which needed doing, and which is not competitive, ought not to be competitive any more than continental defense is, which is another part of the defense of the country and of the free world. That job was slow in accomplishment. It is accomplished now, or largely accomplished now.

Mr. SILVERMAN. I have no further questions of Dr. Oppenheimer.
Mr. GRAY. I wonder if you have any, Mr. Robb?
Mr. ROBB. I have a few; yes, sir.

CROSS-EXAMINATION

By Mr. ROBB:

Q. Doctor, I want to show you a carbon copy of a letter dated September 20, 1944, addressed to Dr. R. C. Tolman, 2101 Constitution Avenue, Washington, D. C., bearing the typewritten signature, "J. R. Oppenheimer." and ask you if you wrote that.

Mr. SILVERMAN. May I look at it?
Mr. ROBB. I am sorry, it is declassified with certain deletions which have just been circled here.
The WITNESS. I am sure I wrote it. Would you give me the courtesy of letting me read it?

By Mr. ROBB:

Q. You mean read it aloud?
A. No.
Q. Sure, that is why I showed it to you.
A. I remember the circumstances.
Q. Have you read it now?
A. Yes.
Q. Including the portions that were circled?
A. Right; which I think they are relevant to the sense of the whole letter.
Q. Doctor, do you think if we read this into the record that you can paraphrase those portions in some innocuous way?
A. Let us see how it goes.
Q. It doesn't seem to be very much, and we did that once before.
Mr. ROBB. Mr. Chairman, might I ask to have this read by Mr. Rolander? When you get to the portions that are delteted——
Mr. SILVERMAN. I really find this a very disturbing procedure.
Mr. GRAY. All right, you can state your concern.
Mr. SILVERMAN. My concern is that here on what I hope is the last day of the hearing we are suddenly faced with a letter which I have not seen, which I know nothing about, and which is going to be read into the record, and I haven't the vaguest idea of what it is about.
The WITNESS. It is from my file.
Mr. SILVERMAN. There are lots of things in the file.

Mr. ROBB. Mr. Chairman, Dr. Oppenheimer testified, as I understand his testimony, to certain opinions which were expressed to him, and I think by him in the period 1944–45, about the thermonuclear.

The WITNESS. No.

Mr. ROBB. I think there were certain discussions he had with Groves and others.

The WITNESS. In 1945?

Mr. ROBB. In 1945; yes.

I think the letter pertains to that general subject. I think the board ought to have the letters before the board.

Mr. GRAY. There seems to be no question about this is a letter written by Dr. Oppenheimer. I believe he has identified it.

I repeat, Mr. Silverman, what I have said many times, and what I hope has been demonstrated by the conduct of this proceeding, that if you are taken by surprise by anything that happens in this procedure, we will give you an opportunity to meet a difficulty arising.

Mr. SILVERMAN. At this moment I haven't any idea that whether I am going to be taken by surprise. I do think it would have been a very easy matter to give us a paraphrased copy of this letter in advance.

Mr. ROBB. Mr. Chairman, until Dr. Oppenheimer testified about this this morning, we had no idea that this letter would become relevant at this particular time. If Mr. Silverman does not want Dr. Oppenheimer to have a chance to comment on the letter, that is all right with me.

Mr. SILVERMAN. I really think that is not the question at all. The real question that I suggest is that it would have been a very easy thing to let us have some intimation of what this is about, instead of having it just flounder here— I don't know whether we are caught by surprise or not. I don't know what we are talking about.

Mr. ROBB. You know, Mr. Chairman, it seems to me that Mr. Silverman is most anxious to be outraged. I don't know why.

Mr. SILVERMAN. Mr. Chairman, is that remark to remain on the record?

Mr. GRAY. I know we have had frequent exchanges between counsel which are on the record.

Mr. SILVERMAN. The suggestion that I am anxious to be outraged suggests that I am putting on some kind of an act——

Mr. ROBB. Mr. Chairman, there is some suggestion that I have done something improper in anticipating what Dr. Oppenheimer is going to testify.

Mr. SILVERMAN. I frankly am about documents being produced that we have not seen and being produced at the last minute. This is an inquiry and not a trial, and it would not happen at a trial. I still don't know what is in this document. For all I know it is a very helpful document.

Mr. GRAY. It may well be. The Chairman of the board makes this statement, that while this is an inquiry and not a trial, there are involved in this proceeding counsel who have not always agreed. I think I can speak for my colleagues on the board when I say that this board takes cognizance of this fact, and the fact that observations of counsel appear on the record do not in any way indicate agreement or disagreement on the part of this board with observations by counsel. As far as producing the testimony here has been concerned, there has been the greatest amount of latitude afforded both to Dr. Oppenheimer and his counsel and to Mr. Robb throughout. I must say that I don't think frankly that the observations of counsel on either side are matters which will be of too much interest and concern to this board. I suggest that you proceed, Mr. Robb.

Mr. ROBB. Would you go ahead and read it?

Mr. ROLANDER. I will hand Dr. Oppenheimer a copy of this letter.

The WITNESS. Is this an unexpurgated copy?

Mr. ROLANDER. It has the portions that are classified circled. The letter is dated September 20, 1944, addressed to Dr. R. C. Tolman, 2101 Constitution Ave., Washington, D. C.:

"DEAR RICHARD. The accompanying letter makes some suggestions about procedure in the matter of site Y recommendations for postwar work. As you will recognize, the problem of making sensible recommendations is complicated by the fact that we do not know how far this project will get during its present life. It seems a reasonable assumption that we will succeed in making some rather crude forms of the gadget per se, but that the whole complex of problems associated with the super will probably not be pushed by us beyond rather elementary scientific considerations.

"I should like, therefore, to put in writing at an early date the recommendation that the subject of initiating violent thermonuclear reactions be pursued with

vigor and diligence, and promptly. In this connection I should like to point out that gadgets of reasonable efficiency and suitable design can almost certainly induct significance thermonuclear reactions in deuterium even under conditions where these reactions are not self-sustaining"——
Then there is a portion that has been deleted.

By Mr. ROBB:

Q. Can you paraphrase that for us, doctor?
A. Yes. It is a part of the program of site Y to explore this possibility
Mr. ROLANDER. Continuing, "It is not at all clear whether we shall actually make this development during the present project, but it is of great importance that such"—and then there is a blank.
The WITNESS. I think that can just be left out.
Mr. ROLANDER. —"such blank gadgets form an experimentally possible transition from a simple gadget to the super and thus open the possibility of a not purely theoretical approach to the latter.
"In this connection also I should like to remind you of Rabi's proposal for initiating thermonuclear reactions"— and then blanks.

* * * * * * *

Mr. ROLANDER. "At the present time site Y does not contemplate undertaking this, but I believe that with a somewhat longer time scale than our present one, this line of investigation might prove profitable.
"In general, not only for the scientific but for the political evaluation of the possibilities of our project, the critical, prompt, and effective exploration of the extent to which energy can be released by thermonuclear reactions is clearly of profound importance. Several members of this laboratory, notably Teller, Bethe, von Neumann, Rabi, and Fermi have expressed great interest in the problems outlined above and I believe that it would be profitable to have a rather detailed discussion of the present technical status—which I know to be confused—which should be made available to the committee before it draws up its final recommendations.
"Sincerely yours,

"J. R. OPPENHEIMER."

By Mr. ROBB:

Q. Doctor, before we go into any discussion, I will show you a carbon copy of another letter dated October 4, 1944, addressed to Dr. R. C. Tolman, 2101 Constitution Avenue, Washington, D. C., bearing the typewritten signature, "J. R. Oppenheimer," and ask you if you will read that and tell us if you wrote it.
Mr. SILVERMAN. Is this a continuation of the same correspondence, Mr. Robb?
Mr. ROBB. Yes; I think so. I am trying to get this unclassified so I can hand you a copy of it, Mr. Silverman.
Mr. MARKS. When was this document unclassified that you are about to hand to us?
Mr. SILVERMAN. It is being declassified now.
Mr. MARKS. I think we are entitled to an answer to that question.
Mr. ROBB. How is that again?
Mr. MARKS. The question is when was this document unclassified?
Mr. ROBB. I haven't any idea. Do you know, Mr. Rolander?
Mr. ROLANDER. It may appear on the face of the document.
Mr. ROBB. There is a note on there. I don't know when it says.
Mr. ROLANDER. Just a minute.
The WITNESS. I have read the letter.
Mr. ROBB. Does it say on there when it was unclassified?
The WITNESS. April 13, 1954.
Mr. ROBB. Have you a copy of that for Mr. Silverman?
The WITNESS. I will recognize the letter as one that I wrote.
Mr. ROBB. We are handing you a copy of that last letter, Mr. Silverman.

By Mr. ROBB:

Q. You testified that is a letter you wrote, Doctor, or rather a copy of a letter you wrote.
A. I have no reason to doubt it whatever.
Mr. ROBB. Did the Chairman wish me to wait until counsel have had a chance to look at this before it is read or could they follow it as it is read?
Mr. SILVERMAN. I will request that.
Mr. GRAY. All right. We will wait until they get a chance to look at it.
Mr. SILVERMAN. We are ready.
Mr. ROBB. Would you read it, Mr. Rolander?

Mr. ROLANDER. The letter is dated October 4, 1944, addressed to Dr. R. C. Tolman, 2101 Constitution Avenue, Washington, D. C.

"DEAR DR. TOLMAN: In transmitting to you the recommendations of workers at project Y on the technical and scientific developments which should be supported in the postwar period, it would seem unnecessary, in view of the essential unanimity in detail and in emphasis, to provide a summary of our opinions. I should like, however, to emphasize a general point of view which I believe is shared by most of the responsible members of the project"——

Dr. EVANS. Of this project.

Mr. ROLANDER. "Of this project, but which deserves repeated and clear statement.

"It may be difficult for those not directly associated with the efforts of project Y to appreciate how provisional, rudimentary, and crude they have been. I regard this not primarily as criticism of the project, but as an inevitable consequence of our attempt to meet a directive with the greatest possible speed. This has for instance made it impossible for us to embark on methods of assembly and use which require long experience with the active materials. It has furthermore discouraged us from entering into a program of more than the minimum complexity. I believe that these limitations have all been appropriate for this wartime project. What is essential is that they should not be forgotten in evaluating future prospects.

"To make these points somewhat more concrete, it is extremely unlikely that project Y, even if completely successful in its present program, will produce weapons whose explosive effect is equivalent to more than about 10,000 tons of high explosive. It would seem unlikely that we will manage to design weapons in which the efficiency of the reaction is as much as 10 percent. It is almost certain that we shall not in a practical way explore the possibilities of releasing the vastly greater energies available in self-sustaining thermonuclear reactions which should afford energy release some ten thousand times greater than those from presently contemplated designs. Finally, the methods of assembly actually being pursued by this laboratory are complicated, crude and bulky, and we shall probably not develop methods which by incorporating autocatalytic features in assembly may completely alter the nature and difficulty of the problems of delivery.

"The above are specific indications of directions which we now know to be worthy of further research. No one can have witnessed the rapid development of ideas in this project, and the extreme liability of fundamental design, without appreciating that the work of this project constitutes a beginning in a field of great complexity and great novelty. Only when investigations can be pursued in a more leisurely and scientifically sound manner than is possible in war, and only when actual experience with the active materials can be used to supplement theoretical ideas of their behavior, will it be possible to foresee the boundaries of this new field.

"The above considerations are all intended to focus attention at one point. Such technical hegemony as this country may now possess in the scientific and technical aspects of the problem of using nuclear reactors for explosive weapons is the result of a few years of intensive but inevitably poorly planned work. This hegemony can presumably be maintained only by continued development both on the technical and on the fundamental scientific aspects of the problem, for which the availability of the active materials and the participation of qualified scientists and engineers are equally indispensable. No government can adequately fulfill its responsibilities as custodian if it rests upon the wartime achievements of this project, however great they may temporarily seem, to insure future mastery in this field. I believe that this point is one which will readily be appreciated by the members of your committee, but that it is my duty as the director of the project directly concerned with these developments, to insist on it in the clearest possible terms.

"Sincerely yours,

"J. R. OPPENHEIMER."

Mr. SILVERMAN. Just one second. Do we now have the complete correspondence between Dr. Oppenheimer or Dr. Tolman on this matter, or are there more letters?

Mr. ROBB. I haven't the slightest idea whether there were more letters written or not. These are the ones that are available to me now. I may say I never read these letters until this morning myself.

Mr. SILVERMAN. Thank you.

By Mr. ROBB:

Q. Doctor, who was Dr. Tolman?
A. He was a very close and dear friend of mine. He had been Vice Chairman of the National Defense Research Committee. When I assumed the responsibility for Los Alamos I introduced him or saw that he was introduced to General Groves. General Groves asked him to be one of his two scientific consultants. He was a member, possibly secretary, of the Committee of Review, which visited Los Alamos in the spring of 1943, and pointed out some things that we needed to do if we were to be a successful laboratory. He was a frequent and helpful visitor to Los Alamos throughout the war. He was at one time, and I would assume at the time these letters were addressed to him, a member of a committee, possibly chairman of a committee appointed by General Groves which was a precursor to the scientific panel to the interim committee in trying to sketch out for the benefit of the Government what the postwar problems in atomic energy might be. These included military and nonmilitary problems.

I think that these letters were addressed to him in that capacity.
Q. And site Y was what?
A. Los Alamos.
Q. Doctor, have you any comment you wish to make on these letters, and if so, will you please do it?
A. I have a couple of comments. Let us take the first letter, the one of September 20. In the second paragraph, the second sentence—do you have a copy of this?
Mr. SILVERMAN. No.
The WITNESS. I will read it: "In this connection I should like to point out that gadgets of reasonable efficiency and suitable design can almost certainly induce significant deuterium reactions even under conditions where these reactions are not self-sustaining."

That turned out not to be true, and I think it was known by the end of the war.

In the third paragraph it says, "in this connection also I should like to remind you of Rabi's proposal for initiating thermonuclear reactions." * * *

"At the present time site Y does not contemplate undertaking this, but I believe that with a somewhat longer time scale than our present one this line of investigation might prove profitable."

This has been under investigation at Los Alamos both immediately after the war and very recently.

On the general character of the recommendations or views, especially on the second letter, this is the point I made in the testimony before the Stimson committee, that we were at the very beginning. The comments on how successful a wartime effort would be were too conservative. We did substantially better than was here indicated, but the warning that however it looked, it was not right to rest on it was one that I repeated then. I think that we went over all the points that are mentioned in these letters in the report of the scientific panel to the Secretary of War's interim committee. I would think that we went over them in the most careful and complete way that we could. These were some comments.

Mr. SILVERMAN. What was the date of Secretary Stimson's interim committee, approximately?
The WITNESS. Which dates do you want?
Mr. SILVERMAN. The date they started.
The WITNESS. I don't know when they started, but the date that the panel appeared with them was the 1st of June, 1945; the date of filing on this long report to which reference has already been made was perhaps October 1945.

By Mr. ROBB:
Q. Had you completed your comment on these letters?
A. I may need to come back to them, but that is what comes to mind at the moment.
Q. At the time you wrote these letters, you were in favor of going ahead with a program for the development of a thermonuclear weapon, weren't you?
A. The letters speak for themselves. I believe they speak exactly what I meant.
Q. Did you mean that?
A. I meant these letters.
Q. Did you mean that you were in favor of going ahead with the thermonuclear?
A. I would like to read the phrases.

Q. What I am getting at, Doctor, laying aside the technical language, wasn't that the ordinary meaning of that you said, that you though you ought to get busy on the thermonuclear?
A. Among other things.
Q. Yes.
A. With the exploration of the thermonuclear.
Q. Did there come a time when you changed that view in subsequent years?
A. Manifestly by October 29, 1949, I was saying very different things.
Q. Yes. Doctor, something was said about the liquid hydrogen plant at Los Alamos. That was constructed for the purpose of working on a fusion weapon, wasn't it, or hydrogen weapon?
A. For preliminary research on ingredients that we thought would be essential in a hydrogen weapon.
Q. Yes. In the matter of reactors, there are various kinds of reactors, aren't there?
A. Indeed there are.
Q. Those built for commercial purposes, those built for research purposes, and those built for production of weapons purposes, isn't that right?
A. I have yet to see one built for commercial purposes but I hope I some day will.
Q. I am asking for information.
A. There are, as I testified, reactors for the development of reactors, reactors for production, reactors for research, and reactors that serve more than one pupose.
Q. You were asked about how many reactors were built during your tenure as chairman of the GAC and I think you said nine, was it?
A. No. I think you asked me during the entire period how many were started, and I think I said about a dozen and a half. Mr. Silverman asked me up to the first of 1950 how many were started, and I said perhaps eight.
Q. Were those eight built for research or production?
A. This is better found in the reports of the Commission. I believe that 3 or 4 were reactor development reactors, namely, to improve the art of reactor development. A couple, 2 or 3 were for supplementary production, and 2 or 3 were for research.
Q. Was any of them a so-called heavy-water reactor?
A. No. I am not quite sure there was not a research reactor at the Argonne, but there was no production reactor involving heavy water.
Q. You spoke of the long range detection matter and the three methods which we speak of rather cryptically. Is it true, Doctor, that it was the opinion of certain qualified people that the one method which you supported might not detect a Russian explosion if it occurred under certain circumstances?
A. We argued about that, and I advocated that opinion.
Q. That it might not?
A. That the Russians might hide an explosion, that this was unlikely, but that they might do it if we relied only on this one method.
Q. In other words, the other methods were necessary to make sure that you could detect the explosion?
A. That's right. May I add that I know of no instance in which the method I advocated has not detected the explosion and in which the others have.
Q. Do you recall who it was recommended Dr. Libby for appointment to the GAC?
A. I wrote a note to Mr. Dean recommending him. Are you asking how the idea came to me?
Q. I am asking if you recall who it was, if anyone, who brought his name to your attention?
A. Yes, it was Fermi.
Q. Did Dr. Pitzer have anything to do with it?
A. No.
Q. So far as you know.
A. I don't know that he had to do with his being appointed, but he didn't discuss it with me.
Q. Doctor, you have spoken somewhat of strategic and tactical airpower and strategic and tactical uses of weapons and all that; you of course don't conceive yourself to be an expert in war, do you, or military matters?
A. Of course not. I pray that there are experts in war.
Q. Have you from time to time, however, expressed rather strong views one way or the other in the field of military strategy and tactics?

A. I am sure that I have. I don't know what specific views or instances you are refrring to, but I am sure the answer to your question is "Yes."

Q. I am not referring to any for the moment.

A. I am sure the answer to your question is "Yes."

Q. Doctor, I am a little curious and I wish you would tell us why you felt it was your function as a scientist to express views on military strategy and tactics.

A. I felt, perhaps quite strongly, that having played an active part in promoting a revolution in warfare, I needed to be as responsible as I could with regard to what came of this revolution.

Q. To draw a parallel, Doctor, of course you recall that Ericsson designed the first ironclad warship.

A. I don't. I am reminded of it.

Q. Beg pardon?

A. I am reminded of it.

Q. Do you think that would qualify him to plan naval strategy merely because he built the *Monitor*?

Mr. SILVERMAN. Aren't we really getting into argument?

The WITNESS. I don't think that I ever planned military——

Mr. GRAY. Wait just a minute. Are you objecting?

Mr. SILVERMAN. Yes, I think this is argument.

Mr. GRAY. Argument?

Mr. SILVERMAN. Yes, of course.

Mr. GRAY. It seems to me that this board has listened for weeks to witnesses who have probed into Dr. Oppenheimer's mind, have said what he would do under circumstances, have stated with certainty what he would, what his opinions are, witnesses who disagreed on this, and I think that counsel has not failed to ask almost any question of any witness that has appeared here. I can't think of questions that could be remotely related to Dr. Oppenheimer that have not been asked.

My ruling is that Mr. Robb will proceed with his question.

The WITNESS. Now I have forgotten the question.

Mr. ROBB. Perhaps we better have it read back.

(Question read by the reporter.)

The WITNESS. Merely because he built the *Monitor* would not qualify him to plan naval strategy.

By Mr. ROBB:

Q. Doctor, do you think now that perhaps you went beyond the scope of your proper function as a scientist in undertaking to counsel in matters of military strategy and tactics?

A. I am quite prepared to believe that I did, but when we are talking about my counseling on military strategy and tactics, I really think I need to know whom I was counseling and in what terms. I am sure that there will be instances in which I did go beyond, but I do not wish to give the impression that I was making war plans or trying to set up military planning, nor that this practice was a very general one.

Mr. GRAY. I think the witness is entitled to know whether Mr. Robb has in mind committees, panels, and other bodies on which Dr. Oppenheimer served or something else.

Mr. ROBB. I was merely trying to explore in general Dr. Oppenheimer's philosophy in respect of this matter. That is what I had in mind. I was not pinpointing on any particular thing, Doctor, and I wanted to get your views on it as to proper function.

The WITNESS. I served on a great many mixed bodies. This controversial Vista project was not a civilian project. There were a great many military consultants. I learned a great deal from them. The formulation of the views of Vista depend to a very large extent on discussions, day-to-day discussions with working soldiers and staff officers. The committees in the Pentagon on which I sat were usually predominantly committees of military men. I also sat on some bodies where there were no military men. I would have thought that in an undertaking like Vista the joint intelligence, in which I played an extremely small part, of a lot of bright technical and academic people—not all scientists—and of a lot of excellent staff officers and military officers was precisely what gave value to the project.

By Mr. ROBB:

Q. Doctor, you stated in response to a question by Mr. Silverman that among other things the job of the strategic airpower was to destroy enemy aircraft on the fields. Do you recall that?
A. Yes.
Q. Do you confine the job of strategic airpower to that, or would you also include the destruction of enemy cities and centers of manufacture?
A. The Strategic Air Command has not only very secret but extremely secret war plans which define its job.
Q. I am asking you for your views on its job.
A. You mean what it should do?
Q. Yes, sir.
A. I think that it should be prepared to do a great variety of things, and that we should maintain at all times full freedom to decide whether in the actual crisis we are involved in, this or that should be done. It must obviously be capable of destroying everything on enemy territory.
Q. Do you think that it should do that in the event of an attack on this country by Russia?
A. I do.
Mr. ROBB. That is all. Thank you.
Mr. GRAY. I think that the only question I have, Dr. Oppenheimer, really relates to a matter that was discussed briefly at an earlier appearance before the board and not anything that has been asked this morning, but I take it that counsel would not object to my question?
Mr. SILVERMAN. Anything that will enlighten the board we are all for.
Mr. GRAY. I think I know the answer to this, but there was some discussion about Mr. Volpe, the other day.
The WITNESS. Yes. I have not read the transcript of that.
Mr. GRAY. I don't think this will be involved. Is the board correct in thinking that this is the same Mr. Volpe that made a speech the other day to the Physical Society?
The WITNESS. As far as I know, sir. I have not been in communication with Mr. Volpe, but I read it in the newspapers.
Mr. GRAY. The board has discussed this. I think counsel is entitled to know it. The board has assumed that this was the same man.
The WITNESS. It obviously is.
Mr. SILVERMAN. I should say self-evidently the speech was made without our knowledge or consent or instigation.
Mr. GRAY. I think I am willing to state for the record that the Chairman believes that this is the case.
Mr. MARKS. I wanted to add to what Mr. Silverman said, not only without our knowledge or consent, but to our embarrassment.
Mr. GRAY. I think the board recognizes that and my question, I would like to have it clearly understood, was not in any suggestion that you as counsel had anything to do with it. My own belief is that you didn't.
Mr. SILVERMAN. It is a fact, sir.
Mr. GRAY. Dr. Evans?
Dr. EVANS. No questions.
Mr. GRAY. Mr. Silverman?
Mr. SILVERMAN. I have just 1 or 2 questions really.

REDIRECT EXAMINATION

By Mr. SILVERMAN:

Q. Do you think that a scientist can properly do his job of advising the military on the potential of newly developed weapons without having some idea of the use that they are to be put to, and some idea of the tactical and strategic use?
A. It depends. I believe we developed the atomic bomb without any idea at all of military problems. The people who developed radar needed to know precisely, or to have a very good idea of what the actual military campaign and needs were. Certainly you do a much better job if you have a feeling for what the military are up against. In peacetime it is not always clear, even to the military, what they will be up against.
Q. You were shown two letters by Mr. Robb, one dated September 20, 1944, I think, and the other October 4, 1944. Do those letters in any way modify the testimony you gave on direct examination as to the scale and intensity of the thermonuclear effort at Los Alamos?
A. Oh, no.

Mr. SILVERMAN. That is all.
Mr. GRAY. May I have that read back?
(Question and answer read by the reporter.)
The WITNESS. May I amplify? I testified what I could recollect, and I think it is complete, of what was going on at Los Alamos during my period there in the thermonuclear program. I was asked whether these letters caused me to have a different view of what was going on there and I said they did not.
Mr. GRAY. I understand, thank you.
Mr. Robb, do you have any questions?
Mr. ROBB. I have nothing further.
The WITNESS. May I make a comment. I don't care whether it is on the record or off.
Mr. GRAY. Yes.
The WITNESS. I am grateful to, and I hope properly appreciative of the patience and consideration that the board has shown me during this part of the proceedings.
Mr. GRAY. Thank you very much, Dr. Oppenheimer.
Do you have anything else?
Mr. SILVERMAN. There are two or three documents I would like to have go in. I have no further questions of Dr. Oppenheimer.
Mr. GRAY. All right.
(Witness excused.)
Mr. SILVERMAN. Unfortunately I don't have copies of it here.
Mr. ROBB. I don't care.
Mr. SILVERMAN. A letter from Maj. Peer de Silva to Dr. Oppenheimer dated April 11, 1945. I will read it into the record. Do you want to see it first [handing].
Mr. ROBB. Sure.
Mr. SILVERMAN (reading).

"ARMY SERVICE FORCES,
"UNITED STATES ENGINEER OFFICE,
"*P. O. Box 1539, Santa Fe, N. Mex., April 11, 1945.*

"Dr. J. R. OPPENHEIMER,
"*Project Director.*

"DEAR OPPIE: Upon my transfer from duty at the project, I want you to know of my sincere appreciation of the support and encouragement which you have personally given me during my services here. In spite of your many more urgent problems and duties, your consideration and help on matters I have brought to you have been gratifying and have, in fact, contributed much to whatever success my office has had in performing its mission.
"I am sure you know that my interests and thoughts will concern themselves in large measure with the continued progress and ultimate success of the work which you are directing. My service at the project and my association with you and your assistants and fellow workers, are matters which I shall remember with pride.
"I want to wish you and your staff every possible success in your work, upon which so much depends.
"Sincerely,

"(S) PEER, PEER DE SILVA,
"*Major, Corps of Engineers.*

"cc—Maj. Gen. L. R. Groves."

During Dr. Oppenheimer's cross-examination, Mr. Robb questioned Dr. Oppenheimer about certain public statements that Dr. Oppenheimer had made in which there was reference to the hydrogen bomb.
Dr. Oppenheimer referred to appearing on a radio panel with Mrs. Roosevelt and also to a speech which he made before the Science Talent Search, Westinghouse, I think. We have here the precise thing that was said on those two occasions. I thought I would read them into the record insofar as they relate to the hydrogen bomb, so that you would know exactly what it is he said.
Mr. ROBB. May I inquire as to the source of the text?
Mr. SILVERMAN. Yes. I have the text of the radio broadcast in two things. One is the bulletin of the Atomic Scientist, and one appears to be the script of the radio thing. The other, the Science Talent Search thing, is a draft of a talk on the encouragement of science, which comes from Dr. Oppenheimer's files. I understand this was also published in the bulletin of the Atomic Scientist.
Dr. OPPENHEIMER. It was published in Science.

Mr. SILVERMAN. I will read what Dr. Oppenheimer said. Other people have said stuff which I don't know is too important. I will read what Dr. Oppenheimer said on the radio thing with Mrs. Roosevelt which appears to have been on February 12, 1950.

"Dr. Oppenheimer: Of course, we personally agree with you about the fostering of science and basic knowledge of nature and man which is one of the few creative elements of our times. It is very essential to the idea of progress to sustain the rest of the world throughout the last centuries. The growth of science is a condition, a precondition, to the health of our civilization. It is manifestly not a job for the AEC alone. It is manifestly not a primary job of the AEC or the primary reason for interest in atomic energy. These reasons lie a lot deeper.

"The decision to seek or not to seek international control of atomic energy, the decision to try to make or not to make the hydrogen bomb, these are complex technical things, but they touch the very basis of our morality. It is a grave danger for us that these decisions are taken on the basis of facts held secret. This is not because those who contributed to the decisions or make them are lacking in wisdom; it is because wisdom itself cannot flourish and even the truth not be established, without the give and take of debate and criticism. The facts, the relevant facts, are of little use to an enemy, yet they are fundamental to an understanding of the issues of policy. If we are guided by fear alone, we will fail in this time of crisis. The answer to fear can't always lie in the dissipation of its cause; sometimes it lies in courage."

That is the end of what Dr. Oppenheimer said on that occasion.

Mr. ROBB. Mr. Chairman, might I interpose here for a moment. I have before me what I believe to be what is called in the language of the trade "the off-the-air" transcript of that statement. I think it is what Mr. Silverman read substantially, but I do find in this "off-the-air transcript" this sentence at the end of the first paragraph Mr. Silverman read: "It is manifestly not the primary job for the AEC or the primary reason" and then a series of dots and in parentheses "voice drops." Apparently there was something unintelligible that the off-the-air reporter didn't get.

Mr. SILVERMAN. I read that. I didn't say that the voice dropped. "It is manifestly not a job for the AEC alone. It is manifestly not a primary job for the AEC or the primary reason for atomic energy. These reasons lie a lot deeper."

Mr. ROBB. All right.

Mr. SILVERMAN. I will not frighten the board by reading them 6 pages of single-space material. The only reference to the hydrogen bomb in this speech which was given on March 6, 1950, to the Science Talent Search Awards banquet, Washington, D. C.—that is these high school boys, I think—is the second paragraph which I will read into the record.

"I do not propose to talk to you of such topics of the day as the hydrogen bomb and the statutory provisions of the National Science Foundation. If these matters are not in a very different state when you shall have to come to assume the full responsibilities of citizenship, you will have reason to reproach your elders for your inheritance."

That is all. Perhaps we might have the whole speech go into the record, but I won't read it now.

Mr. GRAY. There certainly would be no objection to having the speech appear as an exhibit.

(The document was received as exhibit No. 3.)

Mr. SILVERMAN. I will have some copies made.

Mr. ROBB. I don't think we will need some.

Mr. SILVERMAN. That is all, sir.

Mr. GRAY. I thought you had three documents you referred to.

Mr. SILVERMAN. Didn't I give you three; deSilva's letter——

Mr. GRAY. Oh, I beg your pardon. Does this complete what you have?

Mr. SILVERMAN. Yes.

Mr. GRAY. We will recess now until 2 o'clock, but I want to alert Mr. Garrison that I will at that time wish to raise again the question of any necessity for broadening the Commission's letter not with respect to the points we discussed in an earlier session, but with respect to other points which have been very clearly in this testimony. I don't think there is any surprise, but I want to make sure that we have no misunderstanding about it. I will wait to raise this question at 2 o'clock.

Mr. GARRISON. I wonder if it would not be better if you would raise them now, Mr. Chairman, so I might reflect on it a little.

Mr. GRAY. I would be very glad to.

Mr. GARRISON. I don't mean not to have the recess.
Mr. GRAY. We will recess briefly.
(Brief recess.)
Mr. GRAY. The points I would like to discuss are these. The letter of notification from General Nichols to Dr. Oppenheimer of December 23, 1953, contain some detail about the so-called Chevalier incident. The letter, however, does not, I believe, refer to a matter about which we have had a good deal of testimony, and that is the fabrication in the Pash and Lansdale interviews. I think Dr. Oppenheimer's counsel ought to know that the board considers that an important item, and certainly is one of the innumerable things that will be taken into consideration, I am sure, when we begin our deliberations.

I therefore want to avoid any misunderstanding about the question of whether the letter should be broadened to contain a point about that aspect of the episode. That is the first point I have.

Do you care to comment on that?

Mr. GARRISON. I thought perhaps you would proceed, and let me comment at the end.

Mr. GRAY. All right. The other which you may wish in your summation to address yourself to, Mr. Garrison, is the matter, as well as we have been able to ascertain, of what really happened at the time, the 1947 clearance of Dr. Oppenheimer by the Commission.

Mr. GARRISON. This is for summation, Mr. Chairman?

Mr. GRAY. I am saying you may wish to be aware of the fact—you must be aware of the fact—that the Chairman up to this point has stated that he has been a little confused about the attendant circumstances.

Mr. GARRISON. Yes.

Mr. GRAY. So you may want to bear that in mind in preparation of your summation. There is related to the events in 1947 involving Dr. Oppenheimer's clearance by the Commission the General Groves letter to the Commission at that time, and his testimony before this board. I must confess I am not clear just how this might be involved in a broadening of the letter of specifications and yet at least as of this time we consider these things material without in any way being able to say now how material, but at least material.

Mr. GARRISON. Sir, the letter contains derogatory items and I don't quite understand what in the 1947 clearance might be regarded as derogatory.

Mr. GRAY. I think that is a very good question and is a different kind of thing than the matter I referred to in the Chevalier episode. I suppose, Mr. Garrison, what the board is doing at this time is taking cognizance of statements made to the press, and perhaps otherwise, which have been to the effect that the full picture was known to the Commission in 1947, and it acted on the full picture, therefore leaving at least the impression that if the Nichols letter is taken in connection with these statements, then the only thing considered under those circumstances would be the so-called derogatory information with respect to the hydrogen bomb development.

What I am trying to say is that it is not clear to the board yet that the full file was before the Commission in 1947, and at least the circumstances of the clearance at that time are to me still somewhat hazy.

I think in moving more directly to an answer to the question that you put to me, I suppose this is not a matter of broadening the Commission's letter, and perhaps therefore I am talking at this time only about the Chevalier incident.

Mr. GARRISON. I think I know what I would like to say about that, but if it is completely agreeable to you, Mr. Chairman, I would make my comment when we reconvene.

Mr. GRAY. That is quite all right.

Mr. GARRISON. I have to do a little more work than I anticipated on the 1947 thing. I wonder if it would be agreeable if we could resume at 2:30.

Mr. GRAY. Yes.

Mr. GARRISON. I hope you won't take me amiss if I just ask this for information. If the board is going to be here in any event tomorrow—I don't want to make this as a formal request, because I fully accepted your conclusion that I should sum up this afternoon—I just would like to ask once more if you are going to be here tomorrow, would it be just as convenient to have me sum up tomorrow morning as this afternoon. Please don't misunderstand me. I am not pressing this, and I am not making an argument of it.

Mr. GRAY. I think my answer without having consulted the board as of this moment is that the board would prefer to proceed this afternoon.

(The room was cleared while the board conferred.)

(The persons previously present, with the exception of Messrs. Robb and Rolander, returned to the room.)

Mr. GRAY. The board has had a discussion of this matter of time and procedure, and in the interest, Mr. Garrison, of not pressing you and and not thereby perhaps affecting Dr. Oppenheimer's interest, I think the board is willing to put over until tomorrow, frankly at some considerable inconvenience to the board, your summing up. However, in a sense perhaps I am suggesting a bargain with you, and that is, if we put it over until tomorrow morning, do you think we can be through by 1 o'clock?

Mr. GARRISON. Mr. Chairman, I give you my word on that, and I appreciate very, very much your consideration.

Mr. GRAY. You wish to wait until tomorrow to discuss these points I raised with you?

Mr. GARRISON. I think so. It will only take me a minute.

Mr. GRAY. All right. Then we will be in recess until 9:30 tomorrow morning.

Mr. GARRISON. I might say the longer I have in preparation, the shorter my argument will be.

(Thereupon at 11:45 a. m., a recess was taken until Thursday, April 6, 1954, at 2 p. m.)

UNITED STATES ATOMIC ENERGY COMMISSION

PERSONNEL SECURITY BOARD

In the Matter of J. Robert Oppenheimer

Atomic Energy Commission,
Building T-3, Room 2022,
Washington, D. C., Thursday, May 6, 1954.

The above-entitled matter came on for hearing, pursuant to recess, before the board, at 9:30 a. m.

Personnel Security Board: Mr. Gordon Gray, chairman; Dr. Ward T. Evans, member; and Mr. Thomas A. Morgan, member.

Present: Roger Robb and C. A. Rolander, Jr., counsel for the board; J. Robert Oppenheimer, Lloyd K. Garrison, Samuel J. Silverman, and Allan B. Ecker, counsel for J. Robert Oppenheimer; Herbert S. Marks, cocounsel for J. Robert Oppenheimer.

PROCEEDINGS

Mr. GRAY. Before Mr. Garrison's summation, there are a couple of things I would like to take care of which I do not think will take very long. One concerns the 1947 AEC activities with respect to Dr. Oppenheimer's clearance, and I believe Mr. Marks has a statement that he would make with respect to that matter, inasmuch as he was at that time the General Counsel of the Commission, as I understand it.

Mr. MARKS. That is correct.

As the chairman will recall, I mentioned to him in the course of these proceedings some time ago, during one of the recesses, I believe, that in view of the questions that the board was asking about the 1947 clearance, I thought it might wish me to state, either on the record or otherwise, what recollection I had of the events connected with that matter. I mentioned this subject again this morning informally to the board, and ascertained that they would be interested in my stating what my memory was, and I am glad to do this because, while I think that what I have to report will not add much, if anything, to what the board already has heard, I would prefer for them to judge it, rather than me.

Soon after the Hoover letter to the Commission about the Oppenheimer case, I learned about that letter. This would have been, as the proceedings here have brought out, in March of 1947. Whether I was told about the letter by Mr. Volpe or by the then chairman of the Commission, or at a Commission meeting, I do not recall. I believe that at about that time Mr. Volpe told me of the derogatory information concerning Dr. Oppenheimer as transmitted to the Commission with Mr. Hoover's letter.

I believe also that it was I who then first suggested that consideration be given to establishing a board to review the case. In that proposal, I suggested that such a board might include distinguished jurists. I would not have recommended that members of the Supreme Court be included. Whether I made this suggestion to Mr. Volpe with the expectation that he would communicate it to the Commission, whether I made it to the general manager or to the chairman of the Commission or at a Commission meeting, I do not recall. I certainly made it under circumstances where I expected it to be considered by the Commission.

As general counsel for the Commission, I was naturally concerned with questions of procedure in personnel security cases. At the same time, I believe I am correct in my memory that in this matter I had a quite minor role. This was partly because Mr. Volpe, who was deputy general counsel, and as such my first assistant, was handling the matter to the extent that the office of general counsel was concerned, but perhaps more importantly, because Mr. Volpe, as a result of his experience with the Manhattan District, was in those early days of the Commission organization looked to by the Commission for assistance in security matters, and aspects of security matters outside of the sphere of the office of general counsel.

I have no independent recollection, but there certainly must have come a time when I was aware that the idea of a board had not been adopted, and there must have been a time also when I was aware that the Oppenheimer case had in some way been disposed of by the Commission. I have no independent recollection of the Commission meeting of August 6, 1947, or of the other documents concerning this matter that have come into these proceedings, except that I have a vague memory that I knew that Mr. Lilienthal, and I believe Mr. Volpe, had visited Mr. Hoover about the matter, and I also have a memory that there was consultation or correspondence with Dr. Conant, Dr. Bush, Mr. Patterson, and General Groves about the matter.

I should say also that when I was in Washington during the year I was general counsel in 1947 either Mr. Volpe or I, or both of us, attended regular Commission meetings. If the meeting of August 6 was of that character, it is quite possible that one or both of us attended. Seldom, if ever, did I attend executive sessions of the Commission. I think it quite possible that on one or more occasions this case might have been the subject of conversation between the Chairman of the Commission and me, although I have no memory of it. I rather doubt that there were any extensive discussions either between Mr. Lilienthal and me, or the Commission, because I was surprised to find in one of the documents that came into this proceeding that the idea of a board of review included the notion of having Supreme Court Justices be members. I would certainly have opposed any such idea, simply because I have long felt that the Supreme Court Justices should not take assignments off the Court.

If the board has any questions, I would be glad to try to answer them.

Mr. GRAY. Thank you very much, Mr. Marks. I should say that I recall very clearly that you mentioned this matter to me several days ago, and also, of course, you came informally to us this morning and we discussed it again. I think it appropriate that your statement be made.

I would like to ask a couple of questions.

Do you recall whether you were asked to review the file in the case at that time? Do you have a recollection of whether the material which, I guess, came to the Commission from Mr. Hoover was submitted to you for study and comment?

Mr. MARKS. As to the material that came to the Commission initially from Mr. Hoover, I was certainly told the nature of the derogatory information by someone. I seem to remember that on one occasion Mr. Volpe had that Hoover letter with attachments when he was talking to me. I think he showed me the Hoover letter, and that I may have flipped through the pages of the attachment, but I have no recollection of studying the information in the sense in which I think you inquired, and I doubt very much that I did.

Mr. GRAY. I asked the question because, as I recall the testimony here, the recollection of former Commissioners as to whether they saw the file or what kind of a file they saw was very hazy. I think it is of interest to this board to know how extensively this file really was reviewed by members of the Commission and their principal advisers at the time.

Mr. MARKS. My memory, Mr. Chairman, is that what I saw would have been more or less contemporaneously with the communication from Mr. Hoover, and whether I am now going on my memory or my memory is refreshed by questions that have been asked by Mr. Robb, certainly the impression that I have of the bulk of that particular document is consistent with the questions which Mr. Robb has asked. That is to say, that it was certainly not a document of 100 pages; it was a document of a half inch or quarter thick, speaking now of the Hoover letter, and what was attached to it.

Mr. GRAY. I have asked you this question informally, but I should like to ask you again, you are sure that you did not prepare this unidentified memorandum about which we had very considerable discussion earlier in these proceedings? You know the one to which I have reference which I characterized as not being signed or initialed in any way.

Mr. MARKS. I am quite sure that I did not prepare that. I doubt very much that I ever saw it. It is hard for me to say without not now seeing the document whether I ever saw it, but the description of it here——

Mr. GRAY. It would not have been your practice to prepare a memorandum for the file and put it in the file without in some way indicating that you had seen it or authored it?

Mr. MARKS. Certainly not. I think I was quite meticulous about such matters.

Mr. GRAY. In this connection, I think that Dr. Oppenheimer and counsel ought to know that an effort has been made to learn the authorship of this document that we discussed, and the people who are concerned now in the Commission I think just don't know who prepared it. It was not prepared by Mr. Jones, whose name has come into these hearings, or by Mr. Menke, or by Mr. Uanna. Also, Mr. Belcher did not write it, he says, and nobody can furnish any information that is of any real value apparently as to the identity of the person who wrote the summary or memorandum. The best guess of the people connected with it is that it was probably written by Mr. Volpe, but that is pure guess and speculation. I suppose as far as this proceeding is concerned, the author of the memorandum will remain unidentified. We have done all that is reasonable to do to find out.

Mr. MARKS. I think I ought to say that I would have expected that if Mr. Volpe had prepared a memorandum of the kind that was described here that he would have mentioned it to me. I have no recollection of his ever having done so, or ever having prepared a memorandum of that kind.

Mr. GRAY. Thank you very much, Mr. Marks.

Do you want to proceed, Mr. Robb?

Mr. ROBB. Yes, Mr. Chairman.

After the board adjourned yesterday, we received three documents which I think should be made a part of the record. The first and second of these documents respectively are photostats of a letter from Haakon Chevalier to Mr. Jeffries Wyman, dated February 23, 1954, and the response to that letter from Mr. Wyman to Mr. Chevalier dated March 1, 1954. I will ask to have these read into the record by Mr. Rolander, if you please, sir. I am sorry we haven't copies of these. These just came in this morning.

Mr. ROLANDER. The address is "19, rue du Mont-Cenis, Paris, 18e."

"FEBRUARY 23, 1954.

"Mr. JEFFRIES WYMAN,
"7, Cité Martignac, Paris, 7e.

"DEAR MR. WYMAN: My friend—and yours—Robert Oppenheimer, gave me your name when he was up for dinner here in our apartment early last December, and urged me to get in touch with you if a personal problem of mine which I discussed with him became pressing. He gave me to understand that I could speak to you with the same frankness and fullness as I have with him, and he with me, during the 15 years of our friendship.

"I should not have presumed to follow-up such a suggestion if it had come from anyone else. But as you know, Opje never tosses off such a suggestion lightly.

"If you are in Paris, or will be in the near future, I should, then, like to see you informally and discuss the problem.

"On rereading what I have written, I have a feeling that I have made the thing sound more formidable than it really is. It's just a decision that I have to make, which is fairly important to me, and which Opje in his grandfatherly way suggested that I shouldn't make before consulting you.

"Very sincerely,

"HAAKON CHEVALIER."

There is a signature and then typed name.
The second letter:

"AMERICAN EMBASSY, PARIS, *March 1, 1954.*

"MR. HAAKON CHEVALIER,
"*19, rue du Mont-Cenis, Paris (18e).*

"DEAR MR. CHEVALIER: I have just received your letter of February 23. I shall be delighted to see you and talk over your problem with you. Would you care to have lunch with me at my house on Thursday, the 4th of March, at 1 o'clock? The address is 17, rue Casimir Pèrier, Paris (7e), third story. (The telephone is Invalides 00–10.)

"Time being rather short, will you let me know your answer by telephone either at my house or preferably here at the embassy (Anjou 74–60, extension 249). If the time I suggest is not convenient we will arrange for another.

"You will notice that my address is not that given you by Bob Oppenheimer. I have moved since he was here.

"Yours sincerely,

"JEFFRIES WYMAN, *Science Attaché.*"

Mr. ROBB. Mr. Chairman, the third document is an affidavit dated May 4, 1954, signed and sworn to by Ernest O. Lawrence. Would you read that, please?

Mr. ROLANDER (reading):

"MAY 4, 1954.

"I remember driving up to San Francisco from Palo Alto with L. W. Alvarez and Dr. Vannevar Bush when we discussed Oppenheimer's activities in the nuclear weapons program. At that time we could not understand or make any sense out of the arguments Oppenheimer was using in opposition to the thermo-nuclear program and indeed we felt he was much too lukewarm in pushing the overall AEC program. I recall Dr. Bush being concerned about the matter and in the course of the conversation he mentioned that Gen. Hoyt Vandenberg had insisted that Dr. Bush serve as chairman of a committee to evaluate the evidence for the first Russian atomic explosion, as General Vandenberg did not trust Dr. Oppenheimer. I beleive it was on the basis of the findings of this committee that the President made the announcement that the Soviets had set off their first atomic bomb."

Signed "Ernest O. Lawrence", typed "Ernest O. Lawrence." His signature appears twice signed.

In the bottom left hand corner, "Subscribed to and sworn before me this 4th day of May, 1954," the signature of Elizabeth Odle, the name, and then typed, "Notary public in and for the County of Alameda, State of California. My commission expires Aug. 26, 1956."

The seal appears thereon.

Mr. ROBB. That is all, Mr. Chairman.
Mr. GRAY. Mr. Garrison, do you want to have a recess for a conference?
Mr. GARRISON. It may be a minute of two of discussion.
Mr. GRAY. By all means, take it. We will take a short recess.
(Brief recess.)
Mr. GARRISON. I think Dr. Oppenheimer would like to make a very short statement.
Mr. GRAY. Before he does, I would like to say something about this affidavit which was offered by Mr. Robb. It will be recalled that when Dr. Bush came back before this board as a rebuttal witness, the chairman of the board asked him the question whether if you substitute the name Vandenberg for Truman whether his recollection would be the same, and Dr. Bush said emphatically that his recollection would be the same. I wish it known that there is no way that Dr. Lawrence could have known of my question to Dr. Bush. I wish it also known that I had no knowledge of Dr. Lawrence's affidavit, or that there was to be an affidavit at the time I put the question.
Mr. GARRISON. Mr. Chairman, I think it is correct that Dr. Bush testified on May 4, I believe this affidavit is dated May 4. I assume Mr. Robb, you communicated with Dr. Lawrence about it?
Mr. ROBB. I asked Mr. Rolander to communicate with Mr. Lawrence, yes.
Mr. GARRISON. Did you tell him Dr. Bush's testimony?
Mr. ROLANDER. J communicated with Dr. Lawrence through Dr. Alvarez, during which I asked Dr. Alvarez to check with Lawrence and ask Lawrence to prepare a statement as to his recollection of the conversation that took place in this automobile trip from Palto Alto.
Mr. GARRISON. Did you tell Mr. Alvarez about the nature of the discussion here before the board?
Mr. ROLANDER. I am quite sure that I told him there was some question as to what did take place, but I am also quite sure I did not mention the name "Vandenberg."
Mr. GRAY. Mr. Garrison, are you prepared now to proceed with your summation?
Mr. GARRISON. I would like to clear up just one procedural matter, and then I think Dr. Oppenheimer has a very brief comment to make on the matter of his dinner with Mr. Chevalier.
Mr. GRAY. He will be given that opportunity.
Mr. GARRISON. At the session yesterday, Mr. Chairman, you said to me that the jeneral Nichols letter of December 23 contained some detail about the so-called Chevalier incident. The letter did not, however, refer to a matter about which the board has had a good deal of testimony, and that is the fabrication of the Pash and Lansdale interviews. You informed me that we should know that the board considers this an important item, that it would be one of the innumerable things that would be taken into consideration when you begin your deliberations. You wanted to avoid any misunderstanding about the question whether the letter should be broadened to contain the point about that aspect of the episode, and you asked me if I had a comment to make on that.
My comment is, Mr. Chairman, that in Dr. Oppenheimer's letter of response to General Nichols in which he refers to Eltenton's approaching people on the project through intermediaries and then recounts his own conversation with Chevalier, it is quite clear that he was indicating that he had fabricated the story which he had told, and, therefore, Mr. Chairman, we do not suggest or request that the letter of General Nichols be broadened to contain this point.
Mr. GRAY. I see.
Mr. GARRISON. It is at the bottom of page 22. He has previously in the preceding paragraph described his conversation with Chevalier in which it is clear that he did not believe that Chevalier was seeking information.
Mr. GRAY. Yes. I think you have answered the question which I asked you.
Mr. SILVERMAN. Mr. Chairman, while Mr. Garrison has been making his statement, we have been checking the transcript to see what the testimony is on this business of Chevalier's discussion with Dr. Oppenheimer, and with Mr. Wyman.
Mr. GRAY. Yes.
Mr. SILVERMAN. As Mr. Garrison said, Dr. Oppenheimer was going to take the stand again for a minute to tell what he knows about it, but we find in looking at the transcript that he has already said what he has to say. I would simply call your attention to page 2990 of the transcript. I will wait a moment for you, Mr. Robb.
Mr. ROBB. I have it.

Mr. SILVERMAN. In which, Mr. Chairman, you were questioning Dr. Oppenheimer. I am reading only a part of the questioning on this point, but it is the part I think is material.

"Is it clear to you in your visit in the late fall of 1953 to Paris you did not in any way get involved in Dr. Chevalier's passport problems as of the present time?

"The WITNESS. I don't believe I became involved in them. I am not even sure we discussed them.

"Mr. GRAY. You say he did discuss them with you?

"The WITNESS. I am not even sure he discussed them with me. I am sure he discussed one point with me at length which was his continued employment at UNESCO."

Mr. GRAY. If Dr. Oppenheimer wishes to add to that, we should be glad to hear it.

Mr. GARRISON. I think he would just for a moment. Would you care to comment on this?

Whereupon, J. Robert Oppenheimer, a witness, having been previously duly sworn, resumed the stand and testified further as follows:

The WITNESS. I understand that I am under oath.

The problem that most of the evening with Chevalier was spent in quite scattered talk; there was one thing that was bothering him and his wife. Either a large part or a substantial part of his present employment is as a translator for UNESCO. He understood that if he continued this work as an American citizen, he would be investigated, he would have to be cleared for it, and he was doubtful as to whether he would be cleared for this. He did not wish to renounce his American citizenship. He did wish to keep his job, and he was in a conflict over that. This occupied some of the discussion. This is the only problem that I knew about at that time. I don't know what the problem is that he did consult Wyman about. I believe I should also say that the sense that the Chevalier letter to Wyman gives, that Wyman should act as a personal confidant assistant to him and not as an officer of the Government, could not have been anything that I communicated. It was precisely because Wyman was an officer of the Government that it would have appeared appropriate to me for Chevalier to consult Wyman, precisely because anything that was said would be reported to the Government and would be quite open. That is about all I can remember.

Mr. SILVERMAN. May I add one thing. I note at page 462, when Mr. Robb was originally questioning Dr. Oppenheimer about this matter, let me read the question and answer that I refer to:

"Q. Did you thereafter go to the American Embassy to assist Dr. Chevalier getting a passport to come back to this country?

"A. No."

That is the context of this matter.

The WITNESS. Thank you.

Mr. GRAY. Thank you, Dr. Oppenheimer.

(Witness excused.)

Mr. GRAY. Will you proceed, Mr. Garrison.

SUMMATION

Mr. GARRISON. Mr. Chairman and members of the board, I would like to thank you again for waiting over until this morning to give me a little more time to prepare what I might say to you. I want to thank each of you also for your great patience and courtesy and consideration which you have extended us all through these weeks that we have been together.

I think I should take judicial notice of the fact that unless Dr. Evans has some possible question, that I understand that you did not seek the positions which you are here occupying, and I appreciate the fact that you are rendering a great public service in a difficult and arduous undertaking.

As we approach the end of this period in which we have been together, my mind goes back to a time before the hearings began when the Commission told me that you were going to meet together in Washington for a week before the hearings began here to study the FBI files with the aid of such staff as might be provided. I remember a kind of sinking feeling that I had at that point—the thought of a week's immersion in FBI files which we would never have the privilege of seeing, and of coming to the hearings with that intense background of study of the derogatory information.

I suggested two things to the Commission. One, that I might be permitted to meet with you and participate with you during the week in discussions of the case without, as I knew would have to be the case, actual access to the FBI documents themselves, but at least informally participating with you in discussions about what the files contained.

This the Commission said was quite impractical because of the confidential nature of the material, and I then suggested that I meet with you at your very first session in Washington to give you very informally a little picture of the case as we saw it, so that you might at least have that picture as you went about your task, and also that we might have a chance to explore together the procedures which would be followed in the hearings. That request likewise was not found acceptable.

It was explained to me that the practice in these proceedings was that the board would conduct the inquiry itself and would determine itself whether or not to call witnesses and so forth, and it was therefore necessary for the board to have a thorough mastery of the file ahead of time.

We came together then as strangers at the start of the formal hearings and we found ourselves rather unexpectedly in a proceeding which seemed to us to be adversary in nature. I have previously made some comments upon this procedure. I don't want to repeat them here. I do want to say in all sincerity that I recognize and appreciate very much the fairness which the members of the board have displayed in the conduct of these hearings, and the sincere and intense effort which I know you have been making and will make to come to a just understanding of the issues.

I would like now to discuss very briefly the legal framework in which it seems to me you will be operating. You have two basic documents, I suppose, the Atomic Energy Act of 1946 and Executive Order 10450. The essential provisions of these two enactments are contained in summary form in General Nichols' letter of December 23 in the second paragraph, in which the question before the board is put, I think, in this way. General Nichols in the second paragraph of his letter of December 23 says that, "As a result of the investigation and the review of your personnel security file in the light of the requirements of the Atomic Energy Act and the requirements of Executive Order 10450, there has developed considerable question whether your continued employment on Atomic Energy Commission work will endanger the common defense and security"—that is the language of the act—"and whether such continued employment is clearly consistent with the interests of the national security." That is the language of the Executive order. So that they are both together in that sentence.

Now, I think that the basic question—the question which you have to decide—can be boiled down to a very short form. Dr. Oppenheimer's position is that of a consultant. He is to give advice when his advice is sought. This is up to the Atomic Energy Commission as to when and where and under what circumstances they shall seek his advice. That, of course, is not a question that this board is concerned with. The basic question is whether in the handling of restricted data he is to be trusted. That, it seems to me, is what confronts this board, that bare, blunt question.

In trying to reach your determination, you have some guides, some things that you are to take into consideration. The statute speaks of character, associations, and loyalty. Certainly loyalty is the paramount consideration. If a man is loyal, if in his heart he loves his country and would not knowingly or willingly do anything to injure its security, then associations and character become relatively unimportant, it would seem to me.

I suppose one can imagine a case of a loyal citizen whose associations were so intensely concentrated in Communist Party circles—it is hard for me to suppose this of a loyal citizen, but I suppose one might reach a case where the associations were so intense and so pervasive—that it would create some risk of a chance word or something doing some harm, a slip, and so forth.

In the case of character, I suppose that a loyal citizen could still endanger the national security in the handling of restricted data if he were addicted to drunkenness or to the use of drugs, if he were a pervert. These conditions, we of course don't have here.

I would like to skim through with you, because it seems to me to illuminate the nature of the task before you, the Commission's memorandum of decision regarding Dr. Frank Graham, because this was a case which involved a consideration of loyalty and associations. I have the memorandum of the decision here, which was one, I think, of only two that the Commission has thought it desirable to publish. This is dated December 18, 1948. If the board would like copies of it, I would be glad to pass them up to you. I don't propose to read it all, Mr. Chairman, but to point out what seems to me significant in it.

I would direct your attention to paragraph 4, which follows the brief recital of Dr. Graham's character and it cites the sentence from the Atomic Energy Act with which we are familiar, and refers to the FBI report on character, associations and loyalty. Then it goes on to describe their examination of the security file:

"The five members * * * are fully satisfied that Dr. Graham is a man of upright character and thoroughgoing loyalty to the United States. His career as a leading educator and prominent public figure in the South has, it appears, been marked by controversy, engendered in part by his role in championing freedom of speech and other basic civil or economic rights.

"6. In the course of his vigorous advocacy of the principles in which he believes, Dr. Graham has allied himself, by sponsorship or participation, with large numbers of people and organizations all over the country. In this way he has been associated at times with individuals or organizations influenced by motives or views of Communist derivation. These associations, which in substance are described in various published material, are all referred to in the security file.

"7. 'Associations,' of course, have a probative value in determining whether an individual is a good or bad security risk. But it must be recognized that it is the man himself the Commission is actually concerned with, that the associations are only evidentiary, and that commonsense must be exercised in judging their significance. It does not appear that Dr. Graham ever associated with any such individuals or organizations for improper purposes; on the contrary, the specific purposes for which he had these associations were in keeping with American traditions and principles. Moreover, from the entire record it is clear in Dr. Graham's case that such associations have neither impaired his integrity nor arouse in him the slightest sympathy for Communist or other antidemocratic or subversive doctrines. His record on controversial issues has made this abundantly clear, and his course of conduct during the past two decades leaves no doubt as to his opposition to communism and his attachment to the principles of the Constitution.

"8. All five members of the Commission agree with the conclusion of the General Manager that, in the words of the Atomic Energy Act of 1946, it 'will not endanger the common defense or security' for Dr. Graham to be given security clearance, and that it is very much to the advantage of the country that Dr. Graham continue his participation in the atomic-energy program. Our long-range success in the field of atomic energy depends in large part on our ability to attract into the program men of character and vision with a wide variety of talents and viewpoints."

So I say to you, Mr. Chairman and members of the board, that in the Commission's own view of the matter it is the man himself that is to be considered, commonsense to be exercised in judging the evidence, and that it is appropriate to consider in the final reckoning the fact that our long-range success in the field of atomic energy depends in large part on our ability to attract into the program men of character and vision with a wide variety of talents and viewpoints.

The factors of character, associations, and loyalty are not the only ones that are set forth in the catalog of things that you are to consider. Section 4.16 (a) of the Atomic Energy Commission Rules and Regulations contains two paragraphs about the recommendations of the board, and the very first sentence says that the board shall consider all material before it, including the reports of the Federal Bureau of Investigation, the testimony of all witnesses, the evidence presented by the individual, and the standards set forth in "AEC Personnel Security Clearance Criteria for Determining Eligibility" (14 F. R. 42).

That, it seems to me, means that the standards set forth in this document entitled "AEC Personnel Security Clearance Criteria for Determining Eligibility" are all to be considered. It is, as Mr. Robb pointed out, true that this document in many places refers to the General Manager and what the General Manager shall take into account. I think that it is both sensible and logical and clearly intended by section 4.16 (a) that you, in making your recommendations to the General Manager, would take into account the things which he has to take into account in arriving at the decision.

Mr. GRAY. May I interrupt?
Mr. GARRISON. Yes.
Mr. GRAY. I am very much interested in this point, Mr. Garrison. You earlier, I believe, suggested that the usefulness of a man to the program of the Commission was something that the general manager had to consider. Does this most recent observation you made mean that this board must take into account that kind of thing also, because if you say that this board takes into account every-

thing the general manager takes into account, then it seems to me that is inconsistent with an earlier portion of your argument.

Please don't misunderstand me. I am not arguing with you but I want to have your views clearly on this point because it may be an important one.

Mr. GARRISON. I think, as I said earlier, that in the case of a consultant where it is up to the Commission to decide what advice to seek from him, and when that a commonsense reading of this document would leave that question of the appraisal of his usefulness as an adviser necessarily to the Commission. I should think that would be true. I would not want to make a rigid argument that every sentence in this document must be literally applied in arriving at your opinion. Indeed, what I am going to end up in a moment is, having eliminated all of the things that appear in here, when you add to those the words that appear in the statute, you have really in the end no way of arriving at a judgment except by a commonsense overall judgment, which is what is emphasized in the personnel security clearance document and in the regulations.

If I might just pursue that for a moment, the personnel security clearance criteria include references to the past association of the person with the atomic energy program and the nature of the job he is expected to perform. It is there, I think, that the fact that this is a consultant position does come into the consideration. It goes on to say that the judgment of responsible persons as to the integrity of the individuals should be considered. A little later it talks about the mature viewpoint and responsible judgment of Commission staff members, and then it goes on to list these categories (a) and (b) with numerous subheadings.

I don't think there should be any mystery about these categories. Category (a) does not differ from category (b) except to the extent that items that are established under category (a) create a presumption of security risk, and a presumption, of course, is something which is rebuttable by other evidence. If there is any doubt on that point, I hope the board will let me know.

It would be, I think, a complete misreading of this document to say that if you should find an item established under category (a), let us say, that disposes of the case, because everything in the document and in section 4.16 to which I shall return in the rules and regulations, emphasize that everything in the record is to be considered.

For example, this document entitled, "The Criteria," says that the decision as to security clearance is an overall commonsense judgment made after consideration of all the relevant information as to whether or not there is risk that the granting of security clearance would endanger the common defense or security.

The next paragraph says that cases must be carefully weighed in the light of all the information and a determination must be reached which gives due recognition to the favorable as well as unfavorable information.

Then 4.16 (a) provides that the members of the board as practical men of affairs should be guided by the same consideration that would guide them in making a sound decision in the administration of their objectives. It goes on to instruct the board to consider the manner in which witnesses have testified, their credibility, and so forth. Then that if after considering all the factors that they are of the opnion that it will not endanger the common defense and security to grant security clearance, they should so recommend.

So I think we come down in the end, Mr. Chairman, to the basic acid question before the board, whether in the overall judgment of you three men, after considering and weighing all the evidence, that Dr. Oppenheimer's continued right of access to restricted data in connection with his employment as a consultant would endanger the natonal security and the common defense, or be clearly inconsistent with the national security.

It would seem to me that in approaching that acid question the most impelling single fact that has been established here is that for more than a decade Dr. Oppenheimer has created and has shared secrets of the atomic energy program and has held them inviolable. Not a suggestion of any improper use by him of the restricted data which has been his in the performance of his distinguished and very remarkable public service.

Now, at this moment of time, after more than a decade of service of this character, to question his safety in the possession of restricted data seems to me a rather appalling matter.

I would like to tell you what this case seems to me to look like in short compass. I wish we could dispose of it out of hand on the basis of the fact that I have just mentioned to you, that for more than a decade Dr. Oppenheimer has been trusted, and that he has not failed that trust. That in my judgment is the most persuasive evidence that you could possibly have. But I know that you will have to go into the testimony and the evidence, the matters in the file before you, and I would like to sum up, if I may, that it looks like to me to be like.

Here is a man, beginning in 1943—beginning in 1942, actually—taken suddenly out of the academic world in which up to that time he had lived, and suddenly in 1943 put in charge by General Groves of the vast and complex undertaking of the establishment and operation of the laboratory at Los Alamos, a man who suddenly finds himself in administrative charge of the scientific direction of some 4,000 people in a self-contained community in a desert. He performs by common consent an extraordinary service for his country, both administratively and militarily. After the war he hopes to go back to his academic work, back to physics, but the Government keeps calling upon him almost continuously for service. Secretary Stimson puts him on his Interim Committee on Atomic Energy, the Secretary of State puts him on the consultant group in connection with the program for the control of atomic energy before the U. N., he write a memorandum to Mr. Lilienthal within a month of his appointment which contains the essence of the plan which the United States is to adopt, a plan which would have called for the breaking down of the Iron Curtain, and which was to prove extremely distasteful to the Russians. He serves Mr. Baruch at the United Nations and after Mr. Baruch retires, he served General Osborne, and General Osborne has told us here of his firmness and his realism and his grasp of the problems of the conflict and the difficulties of dealing with the Russians.

He makes speeches and he writes articles setting forth the American program and the essence of it, and supporting it. Some of those you have heard before you.

The President appoints him to the General Advisory Committee in January of 1947, and then he is elected chairman by his fellow members, and he serves on that for 6 years. He helps to put Los Alamos back on its feet. He has earlier supported the May-Johnson bill as a means of insuring that this work at Los Alamos or the work on atomic weapons wherever it be conducted can go forward.

He backs in his official work every move calculated to expand the facilities of the Commission, to enlarge raw material sources, to develop the atomic weapons for long-range detection, so that we may find out what the Russians are doing, if and when they achieve the atomic bomb.

After Korea when we are in the midst of an actual shooting war with a military establishment then found to be very depleted, he interests himself in the development of atomic weapons for the battlefield in connection not merely with our problems of intervention in situations like Korea, but more importantly for the defense of Europe against totalitarian aggression.

Finally, he interests himself in continental defense as a means of helping to preserve the home base from which both strategically and tactically any war must be fought. In these and in other ways through half a dozen other committees he gives something like half his time to the United States Government as a private citizen.

Now he is here in this room and the Government is asking the question, is he fit to be trusted.

How does this case come about? Why is Dr. Oppenheimer subjected to this kind of a scrutiny by the Government he has served so long and so brilliantly? Two main things stand out. His opposition to the H-bomb development in 1949 in the report in which he joined with the other members of the GAC, and his left-wing associations and related incidents through 1943. I emphasize that period because it is there that the real searching questions have been put. These are the two main things, and I am going to concentrate in the remarks that I have to make chiefly on these two main facts of the case.

I would digress for a moment to make a short comment on Mr. Borden's letter. I will say this merely. It appears that this letter was before the Atomic Energy Commission at the time that General Nichols wrote his letter to Dr. Oppenheimer; that to the extent that the items in Mr. Borden's letter are covered in General Nichols' letter, there is adequate testimony before the board in our judgment to shed light on all of them. To the extent that there are items in Mr. Borden's letter not covered by the Nichols letter, I just assume that they were not worthy of credence by the Atomic Energy Commission, and are not worthy of credence here.

Finally, I would point out that the matters contained in his letter are matters of opinion and conclusions without evidentiary testimony or facts.

976

Now, returning to the two central elements in this case, of the H-bomb opposition and the leftwing associations and the related incidents through 1943, I would say this in the shortest possible compass about the H-bomb opposition in 1949—that on the whole record here it represented simply an honest difference of opinion. I don't see how it is possible to arrive at any other conclusion than that; that there are on this record no acts of opposition to this program once the President decided to go ahead with it, and that finally there is evidence of affirmative support for the program, particularly after new inventions had established the practical possibility or the near possibility of the creation of the bomb for the first time.

In respect to the leftwing associations and their related incidents through 1943, I would say in all basic essentials they were known to General Groves, and they were known to Colonel Lansdale, and these two men trusted Dr. Oppenheimer. I propose to show in a moment that in all basic essentials they were known to the Atomic Energy Commission in 1947, and that the Commission cleared him, as I shall argue, and as I believe to be the case from the records.

This perhaps might be enough, and surely should be enough, but in addition, we have the testimony of a long series of witnesses here who have worked with Dr. Oppenheimer and have known him for many years and who have arrived at the kind of judgment of the whole man which is the real task before us.

I would like, if I might, now to develop these very shortly stated observations about first the H-bomb and then the leftwing associations. I hope the board will interrupt me at any point at which you would like to put questions. I hope you will interrupt me at any point when you feel you are getting tired listening to me, and you would like a recess or a few minutes of relaxation.

Mr. GRAY. I would just put a question to you now, Mr. Garrison. Did I understand you to say that you feel that the clearance in 1947, which you are prepared to argue, is clearly established, is sufficient?

Mr. GARRISON. No; I didn't mean to suggest in any way that it forecloses the judgment of this board, or that you are not under a responsibility to consider the whole record. If I conveyed any other impression, I didn't intend to. That is your task. I would have thought as an original proposition that this proceeding ought never to have been instituted in the light of this history and in the light of the clearances and of the whole record. But it has been and it is before you, and it is your responsibility and it is your task. When I said this should have been enough, I meant it should have been enough and this proceeding should never have been brought.

Let me return to the topic of the H-bomb. You have had an enormous quantity of evidence, some of it quite technical and some of it quite complicated, about the pros and cons of proceeding with an intensified H-bomb program in 1949, and I am not going to dream of attempting at this time to recapitulate that evidence. I just want to pick out a few salient points and enlarge on them a little bit.

I want to stress at the outset what I am sure this board must feel, and that is that the members of the General Advisory Committee who appeared here and testified before the board were men deeply convinced of the rightness as of 1949 of the judgments which they then made. Certainly that those judgments were honest judgments, that they were arrived at by each individual, each in his own way. No two men put the case to you in quite the same fashion as to what was in their minds. I am sure you must credit each of them with sincerity, with honesty, and with having made a genuine effort in 1949 to say, and to recommend what each believed to be in the interests of America. Surely that was true of Dr. Conant, who expressed his own views, while Dr. Oppenheimer was still not quite certain of his before the meeting of the GAC, and I think Dr. Alvarez or somebody testified to that effect, who was as strong in his opposition as a man can be, who drafted the majority annex with Dr. DuBridge, and whose rugged and independent character is well known to the country and must be apparent to all of us here.

Dr. Fermi, who spoke of the soul searching for all of us which they went through at that time, and to whom Dr. Conant looked for technical appraisals, who surely must have given this board of the sense of the struggle that they went through at the time to do what they believed to be the right thing.

Dr. Rabi, now chairman of the General Advisory Committee; Mr. Oliver Buckley, who made that very sincerely felt and separately stated statement on September 3 to make sure that the very most precise sense of what he believed was on the record. And of Mr. Hartley Rowe, who told you among other things of his experience with Communists and communism in the Latin American countries, and who certainly felt deeply what he was up to in 1949. And then Dr. Oppenheimer, who by the account of all of the members, did not attempt in any way to impose his own views, to dominate the sessions. On the contrary, there is evidence quite to the contrary of the extent to which he welcomed and stimulated discussion of the most protracted character from all concerned, who unquestionably had the influence which goes with great mastery of the subject and of a character that carries weight and meaning and significance in itself.

But the picture that some would paint of a Svengali or a mastermind manipulating men to do his will just falls apart when one actually hears and sees and talks with the members who served with him on the General Advisory Committee. Honest judgments honestly arrived at by Dr. Oppenheimer and all the others.

I would like to stress now the thoroughgoing nature of the consideration which they brought to this subject. This was not a snap decision. Before the meeting, the record now shows that Dr. Oppenheimer had discussions with all kinds of people, including Dr. Teller, who was of course very much for the program, Dr. Bethe; Dr. Serber came to see him; Dr. Alvarez. Not only that, but all around in the Government, this thing was being discussed and considered. General Wilson has described to us the meeting on October 14 of the Joint Chiefs with the Joint Committee on Atomic Energy, with General Vandenberg for the Joint Chiefs urging the development of the H-bomb. This is 2 weeks before the GAC meeting. General Wilson has described how, on the same day, the chairman of the Military Liaison Committee informed that committee of his visit with General McCormack and Dr. Manley to Dr. Oppenheimer at Princeton where they had discussed the super and other problems to be taken up by the General Advisory Committee.

I quote that verbatim from General Wilson's testimony at page 2354. The chairman of the Military Liaison Committee goes with General McCormack, and with Dr. Manley to see Dr. Oppenheimer at Princeton where they discuss the super and other problems to be taken up by the General Advisory Committee.

Then on October 17, the Joint Congressional Committee writes a letter to the Atomic Energy Commission requesting further information on the super. A copy of this goes to the Military Liaison Committee. Then we have Dr. Alvarez talking with all the members of the GAC, and with most of the AEC Commissioners a couple of days before the meeting, and also a couple of days before the meeting, we have a joint meeting of the Atomic Energy Commission and the Military Liaison Committee, and in General Wilson's testimony, the Atomic Energy Commission—and I am now quoting verbatim—"announced that it had asked the General Advisory Committee to consider the superweapon in the light of recent developments."

Then we have the meeting itself, beginning on October 29, and running for 3 days, beginning with a joint session with the Atomic Energy Commission. There was, for a little while, some doubt in the record which puzzled the chairman particularly, as to how the question of the super arose in the Commission. It was the recollection of Dr. Oppenheimer and of Mr. Rowe, and Mr. Lilienthal, Mr. Dean, none of them perhaps very sharp, that at this joint meeting the chairman of the Atomic Energy Commission, for the Commission, raised the question. Mr. Lilienthal testified about Admiral Strauss' memorandum of October 5 or 6, which asked that this be considered by the General Advisory Committee. But I think General Wilson's testimony, it is quite apparent that informally no doubt this matter was actually at the top of the agenda for the General Advisory Committee.

Then you have this 3 days of discussion, consultation with the State Department, with Intelligence, and the Military Liaison Committee, and after all this is over, these gentlemen of the General Advisory Committee sit down and draft their report, and the annexes expressing their individual points of view. Not a snap decision; a decision arrived at after the most intense kind of discussion with people representing the whole gamut of points of view about it.

And then, not content with that, at this December meeting of the General Advisory Committee, the matter is reviewed once more in the light of all the discussion and reactions that have taken place since October.

We have to take into account in measuring or appraising whether Dr. Oppenheimer, which is the only question you have here, whether his own advice, unlike that of every other member of the GAC, was motivated by a sinister purpose to injure the United States of America, and to help our enemy—the mere utterance of that proposition is somehow shocking to me. But it is the question that has been posed and because it is a shocking question, we have to deal with it in direct and blunt terms.

Not one scrap of evidence to indicate that he differed in his purposes from the other honorable Americans who served on the committee and who went into this matter at such length.

There were other leading men in the country who formed the same kind of judgments. This was not an isolated piece of advice that the General Advisory Committee gave. This was a very, very close, difficult and warmly debated subject, debated by all kinds of men. You heard Dr. Kennan, the author of our containment policy, former Ambassador to Russia, describe his own thinking for the State Department Policy Planning Committee on the subject. You have heard Mr. Winne, of the General Electric, giving in retrospect his views, and Dr. Burke giving in retrospect his, and Hans Bethe and Dr. Lauritsen and Dr. Bacher, Mr. Pike, of the AEC, Mr, Lilienthal, men of the most varied outlooks, experiences, and backgrounds themselves troubled by the whole business of going forward to make this super weapon.

Then you heard also from other men who, while they favored going forward with the H-bomb program, were not in the slightest critical of those like Dr. Oppenheimer, who favored the other course. On the contrary, they expressed themselves of the extraordinarily difficult nature of the problem. Gordon Dean, who favored going ahead with the H-bomb program, joining with Admiral Strauss on the Atomic Energy Commission in that, gave us his view of the difficulty of the decision that confronted everybody. Norris Bradbury, who likewise favored moving forward with it, giving similar testimony. And Dr. von Neumann, in the same vein, Professor Ramsey, who was then with the Science Advisory Committee of the Air Force, describing the closeness of the 55–45 in his own mind.

Now, let us come down to Dr. Oppenheimer himself and the honesty of his own judgment, which seems to me impossible to doubt. Even the most active pro-H-bomb advocates, the strongest critics of the position which Dr. Oppenheimer took in 1949, have not questioned his loyalty, although they have, some of them, in strong terms questioned the wisdom of his judgment. Dr. Teller, Dr. Alvarez, Dr. Pitzer, Professor, Latimer, General McCormack, General Wilson. If you will read the record, you will find that all of those men, critics as they were and strong critics of the position taken, did not doubt Professor Oppenheimer's loyalty in the advice that he gave with his fellows on the GAC.

It seems to me that in the face of all of the long catalog of efforts of Dr. Oppenheimer since 1945, let alone at Los Alamos, but since 1945, to strengthen our defenses, to build up Los Alamos, to expand the weapons program, to make us strong in atomic energy, and strong in weapons and strong in defense, it is fantastic to suppose that in the face of all those efforts he should be harboring a motive to destroy his own country in favor of Russia. Just the mere proposition is unthinkable on its face.

Then, in spite of his strong feelings on the subject, when the President has made the decision to go ahead, the record shows whatever might be the situation in his heart about this matter, difficult for a man to change what is in his mind and his convictions, but no opposition in this record to the carrying forward of the program. On the contrary, affirmative evidence that all members of the GAC including Dr. Oppenheimer went along with it, and when it became by process of unexpected inventions something that could really be talked about in terms of production, Dr. Oppenheimer chairs the meeting and presides over the meeting at Princeton which is called together to really put the stuffing in this program. Dr. Teller himself paid tribute to Dr. Oppenheimer's attitude and efforts that he made at that meeting to get the program going.

What can be made of this H-bomb argument? The only thing that has been suggested has been an alleged pattern of opposition which somehow is intended to imply a sinister and un-American attitude toward the whole safety of the military program of the country. This alleged pattern of opposition comes down to the Lincoln summer session, to the Vista project, to the second laboratory. Those are the three main things that one witness here at least suggested constituted a pattern of opposition which troubled him about Dr. Oppenheimer.

Now, we have looked at these. We have looked at the Lincoln summer session. We have seen that the suspicions that that was somehow going to do something that would impair the Strategic Air Force was unfounded. There was no change in the program at all. It was a matter of suspicion that was simply shown to be completely groundless. Over and above that, the affirmative contributions that the thinking and the planning that went on at that session made to the Lincoln project, which is warmly supported by the Air Force as has been brought out.

Now, in Vista, the business of the atomic weapons for the battlefront. Such minor differences as may have existed between the thinking of the group in Dr. Oppenheimer took a certain but not a leading part were adjusted, the report came out to the satisfaction of all concerned, and the testimony of those who criticized what may have been some suggestions in some portions of the report, although the record is very unclear about the whole business, the testimony was that this chapter 5, the whole business of developing these atomic weapons for the battlefront was a great contribution to the country. Actually the work that was done in Lincoln and Vista has become the official policy of the Military Establishment of the country.

Dr. Oppenheimer, if anything could be said about him, could be said that he was a little ahead of his time.

The second laboratory controversy comes down likewise to a difference of opinion about the building of a new Los Alamos in the desert. Dr. Oppenheimer's position in the matter, as Chairman of the GAC, was no different from that of Dr. Bradbury at Los Alamos, whose respect Dr. Teller testified so warmly about. Dr. Oppenheimer supported the Livermore Laboratory when that was found to be the solution to the whole matter, and in the end the bomb that we have been exploding was produced at Los Alamos.

So this alleged pattern of opposition really falls apart upon examination, and it is the only shred of a suggestion of evidence that Dr. Oppenheimer was pursuing an unpatriotic course.

Now, the alleged opposition by Dr. Oppenheimer after President Truman's go-ahead has also vanished under the miscroscope of the testimony, that he caused to be distributed the GAC report to top personnel to discourage them from working on the H-bomb. That I take by common consent has been dropped out of this because its origin in an unfounded suspicion by Dr. Teller has been made quite apparent. Dr. Manley and Dr. Bradbury have explained precisely how those reports came to be distributed by order of the General Manager of the Atomic Energy Commission.

We have gone over the evidence about recruitment and the suggestion in the letter that Dr. Oppenheimer was instrumental in persuading people not to work on the project has no foundation of fact, and on the contrary, the evidence shows that he took affirmative steps to help in that direction, the difficulties of Dr. Teller as an administrator being recognized as one of the problems that made recruitment difficult, until the Livermore Laboratory was set up, and the administration was handled under Dr. Lawrence's direction.

The Princeton meeting I have already referred to and I shall not mention it again, but as an evidence of the affirmative help to the H-bomb program, I might just mention a little item of Dr. Bradbury's testimony, that the GAC and Dr. Oppenheimer were willing to go further in pushing the new invention than the laboratory itself was at the time. You will find that at page 1582 of the record.

You have also testimony by Gordon Dean and by Dr. Bradbury of the help to the staff at Los Alamos that Dr. Oppenheimer and his colleagues gave. The GAC went to Los Alamos in the summer of 1950 when the H-bomb project was at its lowest point, when there was grave doubt whether the thing could ever be built at all, and went out there to help Dr. Bacher and see what they could do.

In general you have testimony from numerous people—Hartley Rowe, General McCormack, and others—that there was no holding back when the President's decision was made.

Now, just a word about the myth of delay. I trust that Dr. Bradbury's testimony will be studied with particular care by this board, because of all the men who testified here he is the one who knows the most about the actual work at Los Alamos, about the problems of producing the H-bomb at the place where it actually has been produced, and I think that his testimony completely destroys the myth of delay. I shall say no more about that because in any event, it has really nothing to do with the question of Dr. Oppenheimer's clearance. Indeed, none of this has to do with it at all. This whole H-bomb controversy, all of the rest of these things, Vista, Lincoln, and all the rest of them, that we have been talking about, except as an indicating and affirmative attitude, as I believe, toward the strengthening of the United States, have nothing to do with the question of Dr. Oppenheimer's clearance unless you are willing to believe to me the unthinkable thought, and I am sure to you, that in spite of everything he had done to help this country from 1945 on, he suddenly somehow becomes a sinister agent of a foreign power. It is unthinkable.

I think, Mr. Chairman, that you would like a recess.

Mr. GRAY. I was about to ask if we may recess for a few moments.
(Brief recess.)
Mr. GRAY. You may resume, Mr. Garrison.
Mr. GARRISON. Mr. Chairman, I would like to turn now to the topic of leftwing associations and related incidents through 1943. In my previous summary of this topic, I said that the basic facts about Dr. Oppenheimer's background and his actions in relation to persons themselves of leftwing background had been known to General Groves and Lansdale, and that they trusted him knowing these basic facts.

The basic facts I have listed as follows:

1. That Dr. Oppenheimer's wife and brother and sister-in-law had been Communists.
2. That Dr. Oppenheimer had a number of leftwing associations and friends.
3. That Dr. Oppenheimer had brought certain persons with former leftwing associations to Los Alamos.
4. That Dr. Oppenheimer had assigned Hawkins to write the history, with General Groves' consent.
5. That Dr. Oppenheimer had protested Lomanitz' draft deferment, with a notation as I go along, that Dr. Oppenheimer's knowledge of Lomanitz' indiscretions, which is the word used throughout the Lansdale and the Pash interviews by them themselves, whatever these indiscretions may have been, that his knowledge of them came from the security officers as is apparent from those interviews, and that in asking deferment for Lomanitz he took notice of the existence of the objections. He said he understood the objections, but Lomanitz' value as a physicist was so-and-so.

Parenthetically I will observe here that Colonel Lansdale brought out quite forcibly the acute manpower problem in the scientific world that existed in those days, and he testified how persons whom the security officers regarded as dangerous were in particular instances deliberately employed because they had to be. They had this great necessity for manpower, and they were then surrounded with extra special surveillance.

You have also in the record Dr. Ernest Lawrence's great urgencies about manpower for the Berkeley Laboratory. This is all part of the setting of the times which we must not lose sight of.

6. That Dr. Oppenheimer had visited Jean Tatlow during the existence of the period of his work at Los Alamos.
7. That he may have made contributions to or through the Communist Party. This is in the Lansdale interview and appears from Lansdale's own statement.
8. That he had delayed in reporting Eltenton, but had delayed still longer in naming Chevalier, and had not told a frank story. I will come back to this in a moment.

At least the foregoing items and no doubt others were known to Groves and Lansdale. I don't think it would serve any purpose to refine this matter into any greater detail, but Groves and Lansdale certainly had before them these basic facts with which we are now concerned here once again after 11 years. They knew all about them and they trusted Dr. Oppenheimer.

I am going to discuss the Chevalier case in a little detail particularly because the Chairman had raised the question of the possibility that the board intends to consider that the story which Dr. Oppenheimer told Pash and Landsdale was true and that his account to this board of his Chevalier incident was not true.

I want to make the point to begin with that the Chevalier fabrication, if I may use the word, was the statement that there were three persons whom Chevalier had contacted, or "X" as the course of the examination went. The question of the microfilm seems now to have been quite inconsequential.

In Dr. Oppenheimer's cross examination before the Pash transcript had been revealed, he was asked if Chevalier had talked about microfilm with him, and put in that way, creating an image of Chevalier coming about microfilm. He answered no, and he answered honestly. It rang no bell in his recollection. When we get to the actual Pash recording, what do we find, this not even in the typewritten transcript that Dr. Oppenheimer was confronted with—not until we get to the recording do we find him saying to Colonel Pash that he understood that this fellow at the Consulate had some means, microfilm "or whatever the hell" of getting the information to Russia.

That is the most casual kind of remark—microfilm or whatever the hell—and might simply be regarded as another means of saying that this fellow has means of getting secret information to Russia. To blow that up into a lie to this board I think it utterly unfair and not warranted by the course of the proceedings here.

The reference to the Russian consulate, it seems to me, is likewise an inconsequential matter. If Eltenton was a spy, if he was seeking information, it would be perfectly natural that he should have a contact at the consulate whether he did or not. I would like to point out that neither this reference to the consulate nor the reference to the microfilm or whatever the hell appears in the Lansdale interview. It just is of no account.

Dr. Oppenheimer's final testimony to this board, going over this matter again with you, was that it was the very best of his recollection that Chevalier did not mention the consulate, but it was conceivable that he know that Eltenton had some connection with the consulate, although he doesn't remember it. Both of these things seemed to me to be of no significance. The way in which these separate items of the story were broken down and converted into separate lies, and the phrase in cross-examination put into Dr. Oppenheimer's mouth that he told a tissue of lies, I think is a most false characterization of what happened. I think his own characterization is the right one, that the story he told was a fabrication, but it was one story, and it was not a separate series of lies each of them to be held up and looked at with the way one looks at that sort of thing.

Now, as to the story about the 3 contacts which I think this really all boils down to, the record indicates that Chevalier did contact only 1 person, as Dr. Oppenheimer stated to this board. Lansdale testified that in the end the number of contacts by Chevalier definitely came down to only one. The only doubt left in the recollections of himself and General Groves is whether that one was Frank Oppenheimer or Robert Oppenheimer. Lansdale testified that there was only one. He believes, according to his testimony, that it was Frank. But this he had from General Groves. And he conceded that General Groves may have told him not that Robert Oppenheimer had named Frank to General Groves, but only that General Groves thought that when Robert Oppenheimer named himself, he was really protecting his brother Frank who was the one, a suspicion in Groves' mind. But again it is one person.

General Groves testified that his own recollection of what Dr. Oppenheimer told him is in a complete state of confusion.

When we leave out Colonel Pash's speculations about which is the truth and which is the false story, his investigations again bear out or support Dr. Oppenheimer's testimony that the story he told to this board is the truth and what he told Colonel Pash was the invention, because when he was asked if they had ever established that there were any other contacts, Colonel Pash testified, "No, sir."

I submit to you, Mr. Chairman, that upon this close examination of the evidence, looking upon it as reasonable men searching for the truth of the matter, as I know you will, you will reach only the conclusion that Dr. Oppenheimer told you here the truth, and that he did in fact in his anxiety to protect Chevalier invent, embroider a story, fabricate a story, to Colonel Pash and Lansdale.

Now, this whole Chevalier incident has, I am convinced, assumed undue importance, and must be judged in perspective. It has been so extensively analyzed here in cross-examination, in the reading of transcripts of interviews of 11 years ago, the hearing of a recording, Colonel Pash's presence here, it is almost as if this whole Chevalier case brought into this room here at 16th and Constitution Avenue in 1954 had happened yesterday in the setting of today, and that we are judging a man for something that has happened almost in our presence.

I get that illusion of a foreshortening of time here which to me is a grisly matter and very, very misleading. This happened in 1943. It happened in a wholly different atmosphere from that of today. Russia was our so-called gallant ally. The whole attitude toward Russia, toward persons who were sympathetic with Russia, everything was different from what obtains today. I think you must beware above everything of judging by today's standards things that happened in a different time and era.

The next perspective about this story is that Dr. Oppenheimer has surely learned from this experience. People who have known him intimately over the years, who have worked with him as closely as anybody could work with people, have heard of this account with some pain, they have taken it in their stride, they have given their own judgment to you that Dr. Oppenheimer would not today do what he did 11 years ago, and that like all good men and intelligent men, he can learn by the bitter fruits of experience. Surely you must have felt, as you listened to the cross-examination here, the sense of guilt which he bore within himself about this incident, something that he does not like to think about back in his past, that God knows he has outlived in his service to this country, and in the way in which he has deported himself as a servant of the United States.

Getting back again to the judgment of this thing in its perspective, General Groves certainly did not regard the matter as a very urgent one. He testified about the schoolboy attitude of Dr. Oppenheimer. That was the way he characterized this thing, this schoolboy attitude of not telling on one's friends which warped his whole judgment and led him into this unfortunate spinning of a story. He didn't seem to be pressed for time, General Groves. He testified that after the first interview with Dr. Oppenheimer—now I am quoting the testimony—about 2 months later, or some time later—actually I think the record will show that it was 3 months—after much discussion in trying to lead him into it and having then got the situation more or less adjusted, "I told him if you don't tell me, I am going to have to order you to do it, then I got what to me was the final story."

This is at page 542.

The final point of perspective is Groves' and Lansdale's own testimony as to their conviction of Dr. Oppenheimer's loyalty. General Groves was asked the question, "Based on your total acquaintance with him and your experience with him and your knowledge of him, would you say that in your opinion he would ever commit a disloyal act?" Answer, "I would be amazed if he did." That is at page 533.

Now, I know that this incident of 1943 has posed in the minds of some of you, perhaps all of you, this question: Did he put loyalty to a friend above loyalty to his country? He has given the straight answer that he did not in his own mind, which is what we are here analyzing, put loyalty to his friend above loyalty to his country. In his own mind, his friend was innocent and the investigation would be in no way benefited y knowing that it was Haakon Chevalier.

That his fault consisted in, and what he has freely confessed to this board, was his arrogance, if I may use my own word, in putting his judgment as to what the interests of the country required at that point about the judgment of the security officers, but that he thought he was injuring the United States of America, that did not occur to him.

Now, it is true that Colonel Pash was put to some labor and wasted efforts. That was not known to Dr. Oppenheimer. Perhaps he should have known of it. I am not apologizing for this incident. I am not condoning it. I am not saying it is something irrelevant and not to be taken into account. Of course it has to be. I am urging you to make the intellectual effort which, gentlemen, will require effort, to put this whole thing into the perspective where it ought to be and not judge it in the light of today's standards and to take into full account the testimony of General Groves and Lansdale about it.

I think at this point I might just remind you of General Groves' letter to Dr. Oppenheimer of May 18, 1950, just after the Paul Crouch testimony. I am not going to read it to you because you have heard it read, but I want to remind you that this letter was volunteered by General Groves and sent on his own initiative out of feelings about Dr. Oppenheimer that were in his system when this incident occurred in California. Why did he do it if he didn't believe Dr. Oppenheimer to be a loyal American citizen? He authorizes him to make a public statement, and the public statement he authorizes him to make is that General Groves has informed me, Dr. Oppenheimer, that shortly after he took over the responsibility for the development of the atomic bomb he reviewed personally the entire file and all known information concerning me, and immediately ordered that I be cleared for all atomic information in order that I might participate in the development of the atomic bomb. General Groves has also informed me that he personally went over all information concerning me which came to light during the operations of the atomic project—and that includes the whole Chevalier business—and that at no time did he regret his decision.

Colonel Lansdale's conviction about Dr. Oppenheimer's loyalty and basic integrity is to the same effect.

Their judgment about this whole matter should not lightly be disregarded by this Board. It should indeed be taken to heart, because their judgment was made in the context of the times and their judgment took into account all that Dr. Oppenheimer was then doing and then thinking, his life, his surroundings, everything about him, viewed from a more intimate standpoint than any that can now be reconstructed. We cannot here reconstruct Robert Oppenheimer's life and activities in the sense of the time and the pressures under which he was working and laboring and all the rest of it. That is gone forever. No one can reconstruct that, but Groves and Lansdale have that in their minds, and in their memories, and they lived with it, and they have testified about it, and they have given you their solemn sworn testimony about the way they viewed that incident.

Dr. Oppenheimer comes out of the war, he embarks on this continuous career of service to the Government. Like the jobs which Dr. Evans, you, Mr. Gray, and Mr. Morgan now fill, he did not seek these positions. The Government called him into service as it has called you into service, and he goes forward.

He becomes chairman of the GAC and the Atomic Energy Commission has then occasion to consider his clearance under the Atomic Energy Act, which we are here bound by. You asked me to pay particular attention to that, and I therefore am going to discuss it in rather meticulous detail. I am going to begin with the entry in the minutes.

The first sentence, which was the basis of the stipulation which the Commission entered into with us and which we put on the record at the start of these proceedings, and which has been found to have been half of the action that was taken and not all of it. Mr. Bellsley called the Commission's attention to the fact that the Commission's decision to authorize the clearance of J. R. Oppenheimer, chairman of the General Advisory Committee, made in February 1947, had not previously been recorded.

I want to say a word about February 1947. There has been a suggestion and at first I myself thought it was the correct suggestion, that before we had the whole story from the documents which were doled piecemeal during the cross-examination and which were subsequently given to us insofar as they are available at our own request afterwards. But before all that, I had credited the suggestion that the Commission took formal action to clear Dr. Oppenheimer in March and that they had not then recorded it, and woke up to the fact in August that they had not and made a minute to that effect, and that the reference to February was a clerical error.

Now, upon a closer examination of the documents in the case, it seems to me that the rational explanation of this overwhelming probability is that February 1947 was correct. Mr. Pike made the suggestion, or offered the guess that in February 1947, the Commission which was then just getting going, acted upon Dr. Oppenheimer's name and cleared him as a matter of course. They knew him, they knew a great deal about him, he had been appointed by the President, they had no occasion to raise any question, and they cleared him.

Then what happened was that in March, Mr. J. Edgar Hoover raised the question in his letter to Lilienthal, and sent over material about him and so forth, and that precipitated an inquiry into Dr. Oppenheimer's associations, background and so forth, and they in effect opened up the whole question and then disposed of it at the August 6 meeting which I shall come to a little later, and said in substance we have examined all this material from the FBI, we have talked with Dr. Bush and Conant and Groves, and so forth, we have thought about this, we see no reason to alter our original action of February in clearing him, which is, I think, an affirmative act of judgment.

Mr. GRAY. You think that the March memorandum of Mr. Wilson, who was then the general manager, as I recall it, from which it was indicated that the Commission was concerned with this matter for 2 days, one meeting and then a subsequent meeting; that the August statement which you refer to as the second half of the action referred all the way back to the March——

Mr. GARRISON. To February.

Mr. GRAY. I am talking about March now.

Mr. GARRISON. No, I say it did not. I originally thought it did. I originally thought from Mr. Lilienthal's testimony which he had told me about before I called him as a witness and reconstructed this from his diary as best he could, I thought from his statement of the affair that there had been clearance in March. I assumed that this February thing was therefore an error, and that the first time it came up was in March. But then under cross examination of Mr. Lilienthan when these documents began to come out, and when we obtained further documents later on, it now seems to me to have been, as Mr. Sumner Pike suggested, and not as Mr. Lilienthal suggested—and I would like to trace through those documents with you.

Mr. GRAY. I would like to get back to your statement that the August 5 minutes in effect say in the second sentence that we have examined the FBI documents——

Mr. GARRISON. I was attempting to say what I thought the Commissioners had done.

Mr. GRAY. I am not quarreling with your interpretation. I am asking you for my own clarification whether you mean by that, that in August they made a minute referring to action which they had actually taken in March?

Mr. GARRISON. No; I don't think they took action in March, except to study the FBI files and to discuss the matter. They took some action in March.

Mr. GRAY. Not action, but the study took place in March, and they waited until August to weigh——

Mr. GARRISON. No; I think the study as again will be shown probably stretched over quite a period of time because the staff went to work, as these documents show, they got the whole file from Mr. Hoover, and the staff got to work on that. There is a memorandum here that everything in the file, all the reports were seen with the exception of two memoranda that I will come to in a moment. So there was study going on. Nobody knows whether it was in June or July or when it was. But I think it certainly shows that it stretched well beyond March.

Mr. GRAY. Is there anything that reflects any action or activity of the Commission between March and August?

Mr. GARRISON. I would like to come to that, if I may.

Mr. GRAY. All right.

Mr. GARRISON. To answer your question; yes.

Mr. GRAY. I am trying to get the straight of it.

Mr. GARRISON. I really don't think it is so complicated, although it has to take a sort of steppingstone approach.

I am proceeding on the assumption that in February 1947 there was what might be called a sort of an off-the-cuff clearance of Dr. Oppenheimer simply based on the knowledge of him, the fact that the President had appointed him.

Then comes a letter from Mr. Hoover to Mr. Lilienthal dated March 8, 1947, which sends over and draws to his attention the attached copies of summaries of information about Dr. Oppenheimer and his brother Frank. That then comes before the board.

Dr. EVANS. You mean the Commission.

Mr. GARRISON. Comes before the Commission. Thank you, Dr. Evans.

In Mr. Wilson's memorandum of March 10 it shows that the Commission met. The actual FBI file says that the file was delivered to Mr. Jones by the FBI on Saturday morning, March 8. But I don't want to make any point now of what was in the particular documents, and I will limit myself to the summaries of information which, for the moment, Mr. Hoover sent over on March 8. The Wilson memorandum says each of the Commissioners read the rather voluminous summary after they met. You know what happened. They called in Dr. Bush and they called in Dr. Conant. They had rather a long discussion of the matter. They tried to reach General Groves. That ultimately was accomplished by Secretary Patterson. There is set forth here the views of Drs. Bush and Conant, not based apparently on an examination of the summary—at least they don't recall it—they were testifying merely from their knowledge of Dr. Oppenheimer as to his loyalty and the serious consequences that failure of clearance would have and so forth.

Then on March 11, the Commission meets again. They have 2 days of meetings. They arrive at the conclusion on March 11 that Dr. Oppenheimer's loyalty was prima facie clear despite the FBI, and that there was no immediate hazard or any issue requiring immediate action, but that a full and reliable evaluation should be made of the case so that it can be disposed of. It is quite clear that at this meeting they are not trying to dispose of it. They say evaluation should be made. Then they decide to seek written views from Drs. Bush and Conant and General Groves, and they instruct the Chairman to confer with Dr. Bush and Mr. Clifford concerning the establishment of an evaluation board. They go to the White House on that mission, and we know all about that.

Mr. GRAY. Do we know the outcome of that?

Mr. GARRISON. No. I am going into that. I mean we know about the proposal for the board, the discussion with Clifford, and their coming back to the meeting that same afternoon and reporting the results of their discussion with Mr. Clifford.

Then we have this entry. At that meeting, that is 5 o'clock in the afternoon of March 11, the general manager reported that a detailed analysis of the FBI summary was in process of preparation by the Commission's security staff as an aid to evaluation. So they have put their staff to work on the FBI summary to make an evaluation of it.

The next thing that happens is Mr. Lilienthal's minute about his telephone conversation with Clark Clifford about the proposal that they had made. It appears from this that Clifford reported the matter to Truman, that Truman wanted to think about it, that he was busy with the Mediterranean crisis, that Clifford said that the Commission had done all that they were under any reasonable obligation to do, and presented the matter and he would take it up with the President, but if Mr. Lilienthal did not hear from him, he should call and remind him about it.

The next document that throws light on this subject is the memorandum from Mr. Jones, the security officer, to the file, dated March 27. I might say perhaps at this point that as we know, there is no more in the record about what happened to this proposal at the White House. Either the President considered it and thought it quite unnecessary to have a board to evaluate Dr. Oppenheimer's qualifications as a loyal citizen of the United tSates, and that this was reported to the Commission in some way or other, or that in the press of his affairs the President never got around to doing anything about it, and either Mr. Lilienthal didn't call up Mr. Clifford in the end to check or find out, or he may have called him up and Mr. Clifford said, "Well, we are not going to take any action on it." Nobody can remember what happened, and there is no documentary evidence to show.

Now, I want to resume the story of what the Commission and its staff were doing. This next thing is this Jones memorandum of March 27, which talks about Mr. Lilienthal going to see Mr. J. Edgar Hoover on March 25 with representatives of the AEC and the FBI. This meeting was attended both by Mr. Lilienthal and Mr. Hoover, and there was a discussion of the case.

I now want to read to you what seems to me particularly in the light of the discussion of the Chevalier incident to be quite a significant passage in this memorandum which I think has escaped our attention until just now. It says, and this is the third paragraph of the memorandum, and the page in the transcript that this appears is 1231, I think: "In the case of J. Robert, those present all seem keenly alive to the unique contributions he has made and may be expected to continue to make. Further there seems general agreement on his subversive record * * * that while he may at one time have bordered upon the communistic"—this is all language of the security officer—"indications are that for some time he has decidedly moved away from such a position. Mr. Hoover himself appeared to agree on this stand with the one reservation, which he stated with some emphasis, that he could not feel completely satisfied in view of J. Robert's failure to report promptly and accurately what must have seemed to him an attempt at espionage in Berkeley."

Now, we know from the record that the files of the Manhattan District went to the FBI. We know from the record that the transcript of the Pash and Lansdale interviews went to the FBI. So that all of this must be presumed to have been known to Mr. Hoover when he participated in this conference, and he says that Dr. Oppenheimer's failure to report promptly and accurately what took place has given him pause, and that is the only thing apparently in the record that troubled him.

Mr. GRAY. Where does he say this?

Mr. GARRISON. This is as reported by Mr. Jones, the security officer of the AEC in his memorandum of March 27, 1949, from which I have been reading, which is in the record. It is not a verbatim quote from Mr. Hoover. It is obviously Mr. Jones' recollection of the conversation that took place there. Mr. Jones was the security officer of the AEC and he says Mr. Hoover was troubled about Dr. Oppenheimer's failure to report promptly and accurately. This is one more piece of evidence, Mr. Chairman, that Dr. Oppenheimer's story about the Chevalier incident contained the elements of fabrication that we have talked about and that this was known to General Groves and Lansdale as it was known to J. Edgar Hoover.

The next thing that happens—this is March 27, now—is a memorandum again from the security officer, Mr. Jones, and this is at page 1409 of the transcript, a memorandum from Mr. Jones to Mr. Bellsley dated July 18. We are now in the middle of July. This memorandum to Mr. Bellsley, the secretary of the AEC, says, "Herewith a complete investigative file on J. Robert Oppenheimer upon which it is believed the Commission may not have formalized their decision. If the Commission meeting minutes contain indication of Commission action, would you kindly so advise. If they do not, I presume you will wish to docket this case for early consideration."

Now comes the sentence I want to stress:

"Each Commissioner and the General Manager have seen every report in this file with the exception of the summary of July 17, and my memorandum for the file dated July 14, 1947."

That memorandum for the file of July 14 is in the record. It is an account of a discussion with Lansdale in which Lansdale vouches for Robert Oppenheimer's loyalty as an American citizen. So whether they saw that or not does not affect the matter, because it was favorable to Dr. Oppenheimer and not derogatory.

What this summary of July 17 contained, which they may or may not have seen, Mr. Volpe in his sort of return memorandum here, suggests that it be circulated among the Commissioners if Mr. Jones thinks it ought to be. We don't know whether they saw it or not. We don't know what is in it because when we asked that it be produced here, we were told that it was confidential and could not be. The record shows here that each Commissioner and the General Manager had seen every report in this file with the exception of this summary of July 17, and the Lansdale transcript saying Dr. Oppenheimer was loyal. This cannot have amounted to anything very important, because Mr. Volpe, whose job then was security matters as well as Deputy General Counsel, left it to the security officer whether it was important enough to send to the members of the board. So presumably it was not much of a document. And the thing that stands out starkly here is that every report in this file except for this probably not important document had gone to each Commissioner and the General Manager, and that they had seen them. They have seen every report in this file, not just that they received them.

It is this memorandum which leads me to suppose that that after the two meetings in the middle of March, the staff which was at work, as we know, had sent the reports in the investigative file to the members of the Commission. I think this many account, sir, for the testimony here which had a ring of veracity to it, by Dr. Bacher, by Mr. Lilienthal, by Mr. Pike, that what they remembered going through was a thick document—a thick document—it stuck in their memories that this thing was thick.

I think in giving credit to that testimony, as one should that presumably that thick stuff that went through was all the reports in this file that the staff had sent around in the course of time. Again whether this was April, May, June, or when, that these things were sent around and reviewed, I don't know. The record does not show. But that there was more than they had before them, the 12-page summary that Mr. Robb identified here, at the March 10 and 11 meeting, seems to me pretty clear on the face of the record.

Dr. Bacher testifies explicitly that what they saw "was first a summary of information from the FBI and later a quite voluminous file, the file being a fairly thick document," at page 2126. That seems to me that had happened here. They testify, these gentlemen, that they treated this matter seriously. Mr. Pike said they all treated it as a serious thing. I am sure we all did. They would indeed have been derelict in their duty if they had not.

Here they were, operating under the Atomic Energy Act, a new thing, laying duties upon them, conscientious men, J. Edgar Hoover's putting them on notice, his explicit reserve about the Chevalier incident, the staff at work on this, the reports in the file, voluminous, going to them—how can we conclude anything but that they took this seriously as they said they did and acted upon it.

Now I come back to the minutes of that August 6 meeting and read the last sentence of the minutes; this, you will remember, follows the memorandum of July 18, in which Mr. Jones, the security officer, asks that a check be made to see what the Commission has done about this in a formal way, and evidently they did make this check and they saw that no formal action had been taken with respect to the matters that had come from Mr. Hoover.

The Commission then on this meeting of August 6, which follows in due course after this July 18 memorandum, Mr. Bellsley calls their attention to the fact that the decision made in February, which I think we must take as the right date, had not previously been recorded. The Commission directed the Secretary to record the Commission's approval of security clearance in this case, and now here are the key words that were not in the stipulation from the Commission when we asked for information about all this, "and to note that further reports"—that means further FBI reports which we are talking about here—"concerning Dr. Oppenheimer since that date (since February) had contained no information which would warrant reconsideration of the Commission's decision."

If that is not action by the Commission, I will eat my hat. They are saying that they got reports after this business in February, they got FBI reports, that they contained no information which would warrant them to go back and re-do what they had done in February. That surely means, as nearly as words can, that this was considered by the Commissioners, as all the documents here indicate, and that they took a serious action upon the matter, saying, "We have gone all through this stuff, we have looked at it all, we have considered this whole thing, and we will let the February action stand." It is exactly the same thing as saying, "We have looked at it all and we hereby reaffirm what we did in February." There is no difference in it. It is just the form of verbiage.

I don't want to make too much of this action, but I think that this board should not lightly pass over it. I want to tell you why.

It seems to me that you should give great weight to the judgment of these five men who bore the responsibility of the United States Government under the Atomic Energy Act in the administration of the program, the judgment that they formed in 1947. This is not a light matter.

Considering one other factor about this whole business of security clearance, when a man is cleared it seems, as we see in this case, and as we have seen in other cases, that the matter can be brought up again and again and again. I think that is most unfortunate. If a man is solemnly and seriously and deliberately cleared by responsible men, that ought to have a kind of sticking quality—I don't say conclusive for all time at all, I say it can be reexamined in the light of what happens later on—but where, as in this case, it seems to me that nothing has happened since 1947 of import, and I want to argue that in a minute, that the sticking quality of an action of this character should be taken seriously to heart and respected. I say this because this business of haling men before security boards is one of the most terrible ordeals that we can subject fellow citizens to. We all know that. It is not good for the country. It is not good for the whole operation of the country. Once a man has been cleared, unless there are serious things that have happened since, it ought to stick. That I urge upon you to take most seriously.

Needless to say in these proceedings, if a man's clearance is taken away from him, that action probably is final for all time. As a practical matter, when a man is branded as disloyal to his country or is not fit to be trusted with classified data by a board of distinction and character and integrity, like this board, and like the Commission in this case, if that happens to a man, that is the end of that fellow for the rest of his life. It is the end of the country's chance to use him, too. That can't be redone. There are therefore hazards to the country and to the man in dragging him up again and again for these clearance ordeals. I urge upon you that consideration as an additional reason for giving the greatest weight to this action of the Commission in 1947.

Now, what did the Commission have before it? I know that question comes up, and it is a question I can't answer, because the files are not available to us, and I can't argue it. I do want to say that I think this aspect of the case, like all others, needs to be judged in the large and not to hang upon some detail. Supposing that in these reports that went to these Commissioners from this investigative file supposing there was some document or other that gets into the file later that may not have been there, or some document at the time that was not in there, what are we dealing with her basically? We are dealing here with big facts about Dr. Oppenheimer. These basic facts, his wife had been a Communist, his borther had been a Communist, his sister-in-law had been a Communist, all these things that have happened that we are talking about here, can anybody suppose that those things were not in the FBI files that went to the members of the Commission? That is the main thing. These big things were in there, the Chevalier incident, the whole thing, and they acted upon it. That seems to me is what we should go by. Just because we haven't a precise and meticulous enumeration of every document in the file that we can compare with the Nichols letter, I think that should not be regarded as of any moment. I will come to that later.

What has happened since 1947 that this board has before it? There is the whole record of Dr. Oppenheimer's public service since 1947, his service on the GAC, on these various other boards and committees which we have talked about at the greatest length. There has been the controversy over the 1949 report on the H-bomb. I think it was Dr. Conant who testified here, if I am not mistaken, that if the case in 1947 for clearance was strong, the case since 1947 is all the stronger in the light of the record of what Dr. Oppenheimer has done for the whole Defense Establishment, and the inference that he has made as a loyal American to help his country.

The Commission did not have Paul Crouch's testimony before it. I cannot suppose that that would be regarded as a change in the condition of substance though it has to be looked at, of course. I am not going to discuss that incident except to say that I am sure that if this board had any substantial doubt on the validity and the accuracy of Dr. Oppenheimer's sworn reply that Mr. Crouch would have been produced here. I venture the assertion that if he had, Dr. Oppenheimer's case would have become even stronger.

Now, what is left? Some associations, but awfully little, I want to bring this to a close soon, and I am going to say just a little word about Dr. Oppenheimer's associations. The point is really what are these associations now? There is no use going back into the days that now have been cut asunder, the whole Berkeley period, Los Alamos period is over with. What is the situation about these associations?

There have been so many names brought into this record in the form of questions, did you know X, no, did you know Y, no, did you know Z, no, questions put to witness after witness that I have gotten a little bit dizzy listening to all the catalogs of names whose significance I have absolutely no way of judging. But so far as Dr. Oppenheimer is concerned, and that is what we are talking about, his present contacts of a kind that this board should consider are for the most part nearly all of the merely casual contacts inevitable to a man of Dr. Oppenheimer's prominence and professional standing—he goes to a meeting of the physicists once a year, some scientific meeting, and he bumps into a physicist there who may have had some past record of association with Communist causes. This is inevitable in the life any scientist who goes to meetings, that he will meet at these meetings some scientist here or there who at one time had some past associations with the Communist Party. But to say that because of that a man like Dr. Oppenheimer is not fit to be trusted with restricted data just seems to me to reduce the whole business to absurdity.

With respect to only two of the names can it really be said that his present association with them is more than a casual one? One of these is Dr. Chevalier, whom Dr. Oppenheimer believes not to be a Communist, and whom he has seen twice in the last few years. He has described him as a friend. I think he has honored himself in describing him as a friend, and in not trying to say that it is just a casual matter. He has his loyalties, Mr. Chairman.

The other one is Dr. and Mrs. Serber. There has been quite a lot of talk about the Serbers. Dr. Serber, as we know from the record, is a distinguished scientist, professor of physics at Columbia University, consultant to the Atomic Energy Commission at Brookhaven Laboratory, and cleared by the Atomic Energy Commission as a result of a review by a board under the chairmanship of Admiral Nimitz, with John Francis Neyland on it. I have forgotten the third man. You know Mr. Neyland as the protagonist of the teachers' oath and the great controversy at the University of California, and counsel for William Randolph Hearst, and surely not a man soft on leftwingers. He and Admiral Nimitz, and the third man, General Joyce, went over the Serber case back in the late forties for the Commission, and they said he is O. K. This man is a loyal citizen, and give him his Q clearance. They have to take into account Mrs. Serber. If he is fit to associate with Mrs. Serber, I don't know what her background, but if Admiral Nimitz and Neyland and Joyce say that Dr. Serber is fit to associate with his wife and have a Q clearance and work for the Atomic Energy Commission, then why should there be any question about Dr. Oppenheimer once in a while seeing Dr. and Mrs. Serber as he does, maybe once or twice a year?

I am going to wind up, sir, in just a very few minutes. I want to mention and not make anything conclusive of it, but direct seriously to your attention the testimony of Dr. Walter Whitman, who in July 1953, as special assistant to the Secretary of Defense for Research and Development, had to review Dr. Oppenheimer's file under this Executive order that we are operating under, requiring a review of cases with derogatory information in it. He testified here that he went through the file, that it had maybe 50 or 60 pages in it. He read it and reread it, he said, until he had the full significance of it. He examined very carefully General Nichols' letter. He said to the best of his recollection everything in it, except this controversy about the H-bomb, was in this file. He reaches the mature conclusion that Dr. Oppenheimer's clearance should be continued. He makes this recommendation to a review board consisting of Dr. Carnes, Dr. Thompson, and General Hines, and to the best of his information, this board agreed with his recommendation. Certainly the clearance was continued until this unfortunate episode in which we are engaged. I think that, too, is entitled to weight.

Now, I am going to make the briefest kind of mention of the men who have appeared here in Dr. Oppenheimer's behalf. We have had a whole lot of fellows here who have talked about Dr. Oppenheimer for 3½ weeks. Dr. Oppenheimer has sat here day after day and listened to the minute analysis of his character, mind, his background and his past. How he survived it all I don't know. I am not going to elaborate about these people. I want to say this, that they differ from the ordinary character witnesses that we are used to in judicial proceedings, where a man comes in and is asked, "Do you know the reputation in the community of the defendant for whatever it may be," and he says "Yes" and they say, "What is that reputation," and he says, "It is good," or whatever he says about it. This has not been that kind of testimony. I can't emphasize that too much. Every one of these men who has appeared here have been men who have worked with Dr. Oppenheimer, who have seen him on the job and off the job, who have formed judgments about character which is the way human beings do judge one another. How do we learn to trust one another except by knowing each other. How can we define the elements of that trust except to say I know that man, I have worked with that man? That is what it comes down to. How else can you express it? These men have known him and have worked with him, and have lived with him.

I am just going to mention 1 or 2 or 3 that I want to especially comment on. I would like to mention Gordon Dean for one, because among other things, he saw him not only in his relationship as an Atomic Energy Commissioner to Dr. Oppenheimer as the GAC chairman, but he also went through this famous FBI file in 1950 and later. He made it his business to follow that file. He testified that if anything came along, whatever came along, he looked into it, and he took it very, very seriously, as to the responsibility that he bore. He came in here without a shadow of a reservation about Dr. Oppenheimer as a security risk and as a loyal American citizen. He considered the Chevalier incident, and he put it in its place, and looked at it as so many of these men of the highest probity and honor have looked at it and said, "Yes, that is there and we don't like it, but we know Dr. Oppenheimer and we trust him, and we trust him for the United States of America."

Here is Dr. Rabi, present chairman of GAC. He too read this file, 40 pages, he said it was, in January of this year which Admiral Strauss gave him to read. He went all through it. He testified, as you know, of his complete and unwavering faith in Dr. Oppenheimer.

Here is Norris Bradbury, surely a man that this board can tie up to and lean upon, a man of obvious deep probity, good judgment, sound fellow, who has lived at Los Alamos for about the whole shooting match than any other man you have seen here, including Dr. Teller, because he has had the whole thing in his hands, and everything to do with it that Dr. Oppenheimer has had he knows. If anybody was in a position to say this fellow impeded our progress or interfered with us, or was somehow sinister, it would be Bradbury. Exactly the reverse is the case.

I could go on and I think I won't. You will read the record, and I know that you will take these judgments deeply seriously. You had 3½ weeks now with the gentleman on the sofa. You have learned a lot about him. There is a lot about him, too, that you haven't learned, that you don't know. You have not lived any life with him. You have not worked with him. You have not formed those intangible judgments that men form of one another through intimate association, and you can't. It is impossible for you to do so. And I think that you should take most earnestly to heart the judgment of those who have.

Here he is now with his life in one sense in your hands, and you are asked to say whether if he continues to have access to restricted data he may injure the United States of America, and make improper use of that. For over a decade that he has had this position of sharing in the atomic energy information, never a suggestion of an improper use of data. His life has been an open book. General Wilson, one of his critics, on the H-bomb end of things, testified—I have forgotten the exact words, but we probably have it around here—that if anybody had demonstrated his loyalty by affirmative action, it is Dr. Oppenheimer, and this affirmative action runs all through his record.

You have a tough job of applying these rather complicated standards, criteria, and so forth. I know that. I beg of you, as I wind up now my conclusion, to take the straightforward commensense judgment that the Commission took in the case of Dr. Graham, and look at the whole man, and you consider the case, "It must be recognized that it is the man himself that the Commission is actually concerned with. Associations are only evidentiary, and commonsense must be exercised in judging their significance." There is the whole thing in a nutshell.

Now, the concluding sentence, indeed that whole memorandum of decision, breathes a kind of air of largeness of reality of practicality in dealing with this problem. The thing that I would most urge you not to do, in addition to not bringing 1943 into 1954, is to get chopped up into little compartments of categories that will give to this case a perfectly artificial flavor of judgment, that you will treat it in the round and the large with the most careful consideration of the evidence, and then treat it as men would treat a problem of human nature, which can't be cut up into little pieces.

There is more than Dr. Oppenheimer on trial in this room. I use the word "trial" advisedly. The Government of the United States is here on trial also. Our whole security process is on trial here, and is in your keeping as is his life— the two things together. There is an anxiety abroad in the country, and I think I am at liberty to say this to you, because after all, we are all Americans, we are all citizens, and we are all interested here in doing what is in the public interest, and what is best for our country. There is an anxiety abroad that these security procedures will be applied artificially, rigidly, like some monolithic kind of a machine that will result in the destruction of men of great gifts and of great usefulness to the country by the application of rigid and mechanical tests. America must not devour her own children, Mr. Chairman and members of this board. If we are to be strong, powerful, electric, and vital, we must not devour the best and the most gifted of our citizens in some mechanical application of security procedures and mechanisms.

You have in Dr. Oppenheimer an extraordinary individual, a very complicated man, a man that takes a great deal of knowing, a gifted man beyond what nature can ordinarily do more than once in a very great while. Like all gifted men, unique, sole, not conventional, not quite like anybody else that ever was or ever will be. Does this mean that you should apply different standards to him than you would to somebody like me or somebody else that is just ordinary? No, I say not. I say that there must not be favoritism in this business. You must hew to the line and do your duty without favor, without discrimination, if you want to use those words.

But this is the point that if you are to judge the whole man as the Commission itself in its regulations and its decisions really lays upon you the task of doing, you have then a difficult, complicated man, a gifted man to deal with and in judging him, you have to exercise the greatest effort of comprehension. Some men are awfully simple and their acts are simple. That doesn't mean that the standards are any different for them. The standards should be the same. But this man bears the closest kind of examination of what he really is, and what he stands for, and what he means to the country. It is that effort of comprehension of him that I urge upon you.

I am confident, as I said, that when you have done all this, you will answer the blunt and ugly question whether he is fit to be trusted with restricted data, in the affirmative. I believe, members of the board, that in doing so you will most deeply serve the interests of the United States of America, which all of us love and want to protect and further. That I am sure of, and I am sure that is where the upshot of this case must be.

Thank you very much.

Mr. GRAY. Thank you, Mr. Garrison.

I would like to make a couple of observations. I think I should say that at some points in your sum-up, I believe you stated that you were assuming that the board reached some conclusion, and therefore something didn't happen. I have in mind particularly your observation about the Crouch episode. I would have to say to you in the interest of the record that at those points my failure to interrupt and question you did not indicate acquiescence nor disagreement.

On 1 or 2 legal points, it was my recollection that in your reference to the Executive order—were you reading from notes on that point?

Mr. GARRISON. I have the Executive order here, Mr. Chairman.

Mr. SILVERMAN. Mr. Chairman, if you are all thinking about the same thing, I think it was a slip of the tongue by Mr. Garrison.

Mr. GRAY. I am trying to clear it up. I would like to know. In any event, it was a distinction between what the department head should do with respect to clearing an individual or not clearing an individual, and it is my impression you said—I am sorry. I think I would like to check and get the exact reference.

Mr. GARRISON. I think I have the phrase here, Mr. Chairman.

Mr. GRAY. All right. Where is that?

Mr. GARRISON (reading). "The head of the agency has to find that his reinstatement, restoration, or reemployment is clearly consistent with the interests of the national security." If I misquoted that, I beg your pardon.

Mr. GRAY. I believe you stated it in the negative. I just wanted to clear that up.

Mr. GARRISON. Thank you, Mr. Chairman.

Mr. GRAY. Finally, on the legal point involved, you made some argument in that respect. I think that you should know that the board, as to these legal points involved, has asked the opinions of attorneys for the Commission. This reflects some difference which emerged in the questioning of the witnesses. With respect to those persons who have been assisting the board in the course of these proceedings, and particularly in response to a question which you have asked about possible proposed findings of fact which might be submitted by Mr. Robb, Mr. Robb will not submit proposed findings of fact, and I would advert to implications which might be in the question.

The regulations under which this board has operated or these proceedings have been conducted state that no person who has assisted the board shall express an opinion as to the merits of the case, among certain other things stated in that regulation. This board is to be governed by the procedures under which it operates, and we shall have to be the guardians of these duties and obligations put upon us.

I think I am required to make a statement to Dr. Oppenheimer at this point. As I think you know, you will have a copy of the transcript of this proceeding with certain exceptions which relate to classified material in the proceeding, and to certain deletions, I suppose they might be called, of testimony which have to do with security problems. Of course this board will make its deliberations on the entire record of your case, and will submit its recommendations to Mr. K. D. Nichols, General Manager of the Atomic Energy Commission.

In the event of an adverse recommendation, you will be notified of that fact by letter from Mr. Nichols. In such event, you will have an opportunity to review the record made during your appearance before this board, and to request a review of such adverse recommendation by the Atomic Energy Commission Personnel Security Review Board prior to final decision by the general manager.

Under those circumstances, you must notify Mr. Nichols by letter within 5 days from the receipt of notice of an adverse recommendation of your desire for a review of your case by the Atomic Energy Commission Personnel Security Review Board.

In the absence of such a communication by you to Mr. Nichols under such circumstances, it would be assumed that you do not desire further review.

You are further advised that in the event this board or the General Manager of the Atomic Energy Commission desires any further information to be presented to the Board, you will be notified of the time and place of the hearing and of course will be given an opportunity to be present.

Dr. OPPENHEIMER. Thank you, Mr. Chairman.

Mr. GRAY. I believe that this completes the proceedings as of now.

Mr. GARRISON. I have just a couple of details.

Mr. GRAY. All right.

Mr. GARRISON. There are in this transcript quite a number—this is without criticism of our very able and efficient reporter—inevitable garbles and mistakes, some of them quite unimportant, but I assume, Mr. Chairman, that if we should get up a list of them and take it up with Mr. Robb or Mr. Rolander, if he wants to arrange it so, and if we should reach an agreement that a memorandum of errata corrections might be incorporated in the record.

Mr. ROBB. When you are doing it, would you cover the whole record and not just the questions you asked?

Mr. GARRISON. Yes, I will try to.

Mr. GRAY. I assume there is no objection?

Mr. ROBB. No, I think that is a good idea. If I had time, I would have done it, too, because that is inevitable in any long proceeding, no matter how good the reporter is.

Mr. GARRISON. I have been meaning to give to the Board, and through inadvertence I haven't, a collection of excerpts from the speeches and writings of Dr. Oppenheimer, but they were handed in at different times. I have just bound them together, and I would be very glad to leave copies of these with you. It is a convenient way of getting at them. I have compared them carefully. I don't think there is anything that is not in the record except the top page, which is just my own.

Mr. GRAY. We acknowledge receipt of the document you refer to.

Mr. GARRISON. Mr. Chairman, may I thank you again for having borne so patiently with me and for the great consideration you have shown to us throughout the proceedings.

Mr. GRAY. Thank you.

Mr. GARRISON. Mr. Morgan and Dr. Evans, the same.

Dr. EVANS. Thank you.

Mr. MORGAN. Thank you.

Mr. GRAY. We now conclude this phase of the proceedings. I think that I have already indicated to Dr. Oppenheimer that if we require anything further, he will be notified.

We are now in recess.

(Thereupon at 1 : 30 p. m., the hearing was concluded.)

LIST OF WITNESSES

	Page
Luis Walter Alvarez	770–805
Robert Fox Bacher	608–630
Hans Bethe	323–340
William Liscum Borden	832–844
Norris Bradbury	477–494
Oliver E. Buckley	603–608
Vannevar Bush	560–568, 909–915
James B. Conant	384–394
K. T. Compton	256–258
Gordon Dean	300–323
Lee Alvin DuBridge	514–534
Enrico Fermi	394–398
James B. Fisk	340–342
T. Keith Glennan	253–256
David Tressel Griggs	742–770
Leslie R. Groves	160–180
Albert Gordon Hill	935–944
Mervin J. Kelly	57–65
George Frost Kennan	352–372
John Lansdale, Jr	258–281
Wendell Mitchell Latimer	656–672
Charles Christian Lauritsen	577–596
David E. Lilienthal	372–382, 398–425
John J. McCloy	731–742
James McCormack, Jr	633–643
J. Robert Oppenheimer	26–57, 65–160, 180–253, 887–906, 949–961
Katherine Oppenheimer	571–577, 915–921
Frederick Osborn	342–346
Boris T. Pash	808–831
Sumner T. Pike	429–438
Kenneth Sanborn Pitzer	697–709
Isadore Isaac Rabi	451–473
Norman Foster Ramsey, Jr	440–451
Hartley Rowe	507–514
Edward Teller	709–727
John Von Neumann	643–656
Walter G. Whitman	495–507
Roscoe Charles Wilson	679–697
Harry Alonso Winne	542–557
Jerrold R. Zacharias	596–602, 922–933

Certain sections of this transcript pertain to playback of recordings, to stipulations entered by counsel, and to procedural discussions and summation. These items appear as follows:

	Page
Oral transcription plus written corrections plus stipulations on a recorded interview among Lt. Col. Boris T. Pash, Lt. Lyall Johnson, and Dr. J. Robert Oppenheimer	285–300, 844–871
Oral transcription of interview between Lt. Col. John Lansdale, Jr., and Dr. J. Robert Oppenheimer	871–886
Procedural discussions and summation	967–992

UNITED STATES ATOMIC ENERGY COMMISSION

In the Matter of

J. ROBERT OPPENHEIMER

TEXTS OF PRINCIPAL DOCUMENTS
AND LETTERS

of

PERSONNEL SECURITY BOARD
GENERAL MANAGER
COMMISSIONERS

WASHINGTON, D. C.
May 27, 1954, through June 29, 1954

FINDINGS AND RECOMMENDATION
OF THE
PERSONNEL SECURITY BOARD
IN THE MATTER OF
DR. J. ROBERT OPPENHEIMER

UNITED STATES ATOMIC ENERGY COMMISSION,
Washington, D. C., May 27, 1954.

Subject: Findings and recommendation of the Personnel Security Board in the case of Dr. J. Robert Oppenheimer.

Mr. K. D. NICHOLS,
  General Manager, U. S. Atomic Energy Commission,
  1901 Constitution Avenue NW., Washington 25, D. C.

DEAR MR. NICHOLS: On December 23, 1953, Dr. J. Robert Oppenheimer was notified by letter that his security clearance had been suspended. He was furnished a list of items of derogatory information and was advised of his rights to a hearing under AEC procedures. On March 4, 1954, Dr. Oppenheimer requested that he be afforded a hearing. A hearing has been conducted by the Board appointed by you for this purpose, and we submit our findings and recommendation.

Dr. Ward V. Evans dissents from the recommendation of the majority of the Board, and his minority report is attached. He specifically subscribes to the "Findings" of the majority of the Board, and to a portion of the material entitled "Significance of the Findings."

INTRODUCTION

It must be understood that in our world in which the survival of free institutions and of individual rights is at stake, every person must in his own way be a guardian of the national security. It also must be clear that, in the exercise of this stewardship, individuals and institutions must protect, preserve, and defend those human values for which we exist as a nation, as a government, and as a way of life.

The hard requirements of security, and the assertion of freedoms, together thrust upon us a dilemma, not easily resolved. In the present international situation, our security measures exist, in the ultimate analysis, to protect our free institutions and traditions against repressive totalitarianism and its inevitable denial of human values. Thoughtful Americans find themselves uneasy, however, about those policies which must be adopted and those actions which must be taken in the interests of national security, and which at the same time pose a threat to our ideals. This Board has been conscious of these conflicts, presenting as they do some of the grave problems of our times, and has sought to consider them in an atmosphere of decency and safety.

We share the hope that some day we may return to happier times when our free institutions are not threatened and a peaceful and just world order is not such a compelling principal preoccupation. Then security will cease to be a central issue; man's conduct as a citizen will be measured only in the terms of the requirements of our national society; there will be no undue restraints upon freedom of mind and action; and loyalty and security as concepts will cease to have restrictive implications.

This state of affairs seems not to be a matter of early hope. As we meet the present peril, and seek to overcome it, we must realize that at no time can the interests of the protection of all our people be less than paramount to all other considerations. Indeed, action which in some cases may seem to be a denial of the freedoms which our security barriers are erected to protect, may rather be a fulfillment of these freedoms. For, if in our zeal to protect our institutions against our measures to secure them, we lay them open to destruction, we will have lost them all, and will have gained only the empty satisfaction of a meaningless exercise.

We are acutely aware that in a very real sense this case puts the security system of the United States on trial, both as to procedures and as to substance. This notion has been strongly urged upon us by those who recommended clearance for Dr. J. Robert Oppenheimer, and no doubt a similar view is taken by those who feel he should not be cleared.

If we understand the two points of view, they may be stated as follows: There are those who apprehend that our program for security at this point in history consists of an uneasy mixture of fear, prejudice, and arbitrary judgments. They feel that reason and fairness and justice have abdicated and their places have been taken by hysteria and repression. They, thus, believe that security procedures are necessarily without probity and that national sanity and balance can be served only by a finding in favor of the individual concerned. On the other hand, there is a strong belief that in recent times our government has been less than unyielding toward the problem of communism, and that loose and pliable attitudes regarding loyalty and security have prevailed to the danger of our society and its institutions. Thus, they feel that this proceeding presents the unrelinquishable opportunity for a demonstration against communism, almost regardless of the facts developed about the conduct and sympathies of Dr. Oppenheimer.

We find ourselves in agreement with much that underlies both points of view. We believe that the people of our country can be reassured by this proceeding that it is possible to conduct an investigation in calmness, in fairness, in disregard of public clamor and private pressures, and with dignity. We believe that it has been demonstrated that the Government can search its own soul and the soul of an individual whose relationship to his Government is in question with full protection of the rights and interests of both. We believe that loyalty and security can be examined within the frameworks of the traditional and inviolable principles of American justice.

The Board approached its task in the spirit of inquiry, not that of a trial. The Board worked long and arduously. It has heard 40 witnesses including Dr. J. Robert Oppenheimer and compiled over 3,000 pages of testimony in addition to having read the same amount of file material.

Dr. Oppenheimer has been represented by counsel, usually four in number, at all times in the course of the proceedings. He has confronted every witness appearing before the Board, with the privilege of cross-examination. He is familiar with the contents of every relevant document, which was made available to the Board, except those which under governmental necessity cannot be disclosed, such as reports of the Federal Bureau of Investigation. He has, in his own words, received patient and courteous consideration at the hands of the Board. The Board has, in the words of his chief counsel, displayed fairness in the conduct of the hearings. And, finally, perhaps it should be said that the investigation has been conducted under the auspices of the responsible agency which has the obligation of decision.

As it considered substance, the Board has allowed sympathetic consideration for the individual to go hand in hand with an understanding of the necessities for a clear, realistic, and rugged attitude toward subversion, possible subversion, or indeed broader implications of security.

It was with all these considerations in mind that we approached our task.

PROCEDURES GOVERNING THE HEARINGS

This proceeding is based upon the Atomic Energy Act of 1946; upon the Atomic Energy Commission's published Security Clearance Procedures, dated September 12, 1950; and Personnel Security Clearance Criteria for Determining Eligibility, dated November 17, 1950; and upon Executive Order No. 10450, dated April 27, 1953.

Subparagraphs (ii) and (iv) of section 10 (b) (5) (B) of the Atomic Energy Act provide that, except as authorized by the Commission in case of emergency, no individual shall be employed by the Commission until the Civil Service Commission or (in certain instances) the Federal Bureau of Investigation shall have made an investigation and report to the Commission on the "character, associations, and loyalty" of such individual.

The AEC published Procedures provide, among other things, for written notice to the individual (1) listing the items of derogatory information and (2) explaining his rights (*a*) to reply in writing to the information set forth in the Commission's letter, (*b*) to request a hearing before a personnel security board, (*c*) to challenge the appointment of the members of the Board for cause, (*d*) to be present for the duration of the hearing, (*e*) to be represented by counsel of his own choosing, and (*f*) to present evidence in his own behalf through witnesses, or by documents, or by both. The Commission's Procedures further provide that in the event of a recommendation for a denial of security clearance, the individual shall be immediately notified of that fact and of his right to request a review of his case by the AEC Personnel Security Review Board, with the right to submit a brief to that Board before the case goes to the general manager for final determination.

The AEC published Criteria establish the uniform standards to be applied in determining eligibility for clearance. These Criteria, which, of course, are binding on this Board, provide that it is the responsibility of the Atomic Energy Commission to determine whether the common defense or security will be endangered by granting security clearance.

The Executive order requires the head of each department and agency of the Government to establish and maintain within his department or agency an effective program to insure that "the employment and retention in employment of any civilian officer or employee within the department or agency is clearly consistent with the interests of the national security." The Executive order further provides that information on this issue shall relate, but shall not be limited, to certain categories of information set forth in the order.

FINDINGS

In compliance with section 4.16 (c) of the Commission's Security Clearance Procedures, the Board makes the following specific findings as to the allegations contained in Mr. K. D. Nichols' letter of December 23, 1953 to Dr. J. Robert Oppenheimer:

1. It was reported that in 1940 you were listed as a sponsor of the Friends of the Chinese People, an organization which was characterized in 1944 by the House Committee on Un-American Activities as a Communist-front organization.

The Board concludes that this allegation is true.

Dr. Oppenheimer in his answer replied that he had no recollection of the Friends of the Chinese People, or of what, if any, his connection with this organization was.

The Board had before it a four-page pamphlet (undated) entitled, "American Friends of the Chinese People." The fourth page contains a list of sponsors which includes "Prof. J. R. Oppenheimer."

2. It was further reported that in 1940 your name was included on a letterhead of the American Committee for Democracy and Intellectual Freedom as a member of its National Executive Committee. The American Committee for Democracy and Intellectual Freedom was characterized in 1942 by the House Committee on Un-American Activities as a Communist-front which defended Communist teachers, and in 1943 it was characterized as subversive and Un-American by a Special Subcommittee of the House Committee on Appropriations.

The Board concludes that this allegation is true.

Dr. Oppenheimer testified before the Board to having joined the American Committee for Democracy and Intellectual Freedom in 1937. He said that it then stood as a protest against what had happened to intellectuals and professionals in Germany. The Board had before it a letterhead of the "American Committee for Democracy and Intellectual Freedom." The letterhead contains a printed list of the National Executive Committee, which includes "Prof. J. R. Oppenheimer." Dr. Oppenheimer testified that he supposed he accepted membership on this Executive Committee although he did not meet with it.

Dr. Oppenheimer stated in his personnel security questionnaire, which he executed on April 28, 1942, for the purpose of obtaining a clearance for work on the atomic program, that he had joined the "American Committee for Democratic Intellectual Freedom" in 1937 and was still a member on the date the PSQ was executed. He testified that he did not know how long after that he continued to be a member; that, in any event, he was not active thereafter.

3. It was further reported that in 1938 you were a member of the Western Council of the Consumers Union. The Consumers Union was cited in 1944 by the House Committee on Un-American Activities as a Communist-front headed by the Communist Arthur Kallet.

The Board concludes that this allegation is true.

Dr. Oppenheimer in his answer stated that for perhaps a year he had been a member of the Western Council of the Consumers Union. In his personnel security questionnaire, which he executed on April 28, 1942, Dr. Oppenheimer stated that he had been a member of the Consumers Union (Western) in 1938–39.

The Board had before it a photostat of a four-page pamphlet (undated) entitled, "Western Consumers Union," containing a list of Western sponsors, which included the name "Dr. Robert J. Oppenheimer—Internationally-known Physicist at the University of California."

4. It was further reported that you stated in 1943 that you were not a Communist, but had probably belonged to every Communist-front organization on the west coast and had signed many petitions in which Communists were interested.

The Board concludes that this statement was made by Dr. Oppenheimer, and the Board had before it considerable evidence indicating Dr. Oppenheimer's membership in, and association with, Communist-front organizations and activities on the west coast. However, Dr. Oppenheimer, in his answer, claimed that the quotation was not true and that if he had said anything along the lines quoted, it was a half-jocular overstatement.

The Board had before it a memorandum, dated September 14, 1943, prepared by Lt. Col. John Lansdale, Jr., who was then head of Security and Intelligence for the Manhattan District, which reported "Oppenheimer categorically stated (to General Groves) that he himself was not a Communist and never had been, but stated that he had probably belonged to every Communist-front organization on the west coast and signed many petitions concerning matters in which Communists were interested.

The Board also had before it a transcript of an interview between Colonel Lansdale and Dr. Oppenheimer on September 12, 1943, which reflected that Colonel Lansdale had asked Dr. Oppenheimer, "You've probably belonged to every front organization on the coast," to which Dr. Oppenheimer replied, "Just about." The transcript further records that Dr. Oppenheimer also stated that he thought we would have been considered at one time a fellow-traveler and that "my association with these things was very brief and very intense."

Dr. Oppenheimer in his testimony defined "fellow-traveler" as "someone who accepted part of the public program of the Communist Party, who was willing to work with and associate with Communists, but who was not a member of the party." He testified to having been a fellow-traveler from late 1936 or early 1937, with his interest beginning to taper off after 1939, and with very little interest after 1942. He further stated that within the framework of his definition of a fellow-traveler, he would not have considered himself as such after 1942.

He further stated that with respect to things that the Communists were doing, in which he still had an interest, it was not until 1946 that it was clear to him that he would not collaborate with Communists no matter how much he sympathized with what they pretended to represent.

5. It was reported that in 1943 and previously you were intimately associated with Dr. Jean Tatlock, a member of the Communist Party in San Francisco, and that Dr. Tatlock was partially responsible for your association with Communist-front groups.

The Board concludes that this allegation is true.

Dr. Oppenheimer in his testimony before this Board admitted having associated with Jean Tatlock from 1936 until 1943. He stated that he saw her only rarely between 1939 and 1943, but admitted that the association was intimate. He admitted having seen Jean Tatlock under most intimate circumstances in June or July of 1943, during the time when he was Director of the Los Alamos Laboratory, and admitted that he knew she had been a Communist and that there was not any reason for him to believe that she was not at that time still a Communist. He named several Communists, Communist functionaries or Communist sympathizers whom he had met through Jean Tatlock, or as a result of his association with her.

6. It was reported that your wife, Katherine Puening Oppenheimer, was formerly the wife of Joseph Dallet, a member of the Communist Party, who was killed in Spain in 1937 fighting for the Spanish Republican Army.

The Board concludes that this allegation is true.

Mrs. Oppenheimer testified that she was married to Joseph Dallet from 1934 until he was killed in Spain, fighting for the Spanish Republican Army in 1937.

Mrs. Oppenheimer admitted knowing that Dallet was a member of the Communist Party and was actively engaging in Communist Party activities.

7. It was further reported that during the period of her association with Joseph Dallet, your wife became a member of the Communist Party. The Communist Party has been designated by the Attorney General as a subversive organization which seeks to alter the form of government of the United States by unconstitutional means, within the purview of Executive Order 9835 and Executive Order 10450.

The Board concludes that this allegation is true.

Mrs. Oppenheimer testified to having been a member of the Communist Party from about 1934 to June 1936 and having engaged in Communist Party activities in the Youngstown, Ohio, area.

8. It was reported that your brother Frank Friedman Oppenheimer became a member of the Communist Party in 1936 and has served as a party organizer and as educational director of the professional section of the Communist Party in Los Angeles County.

The Board concludes that this allegation is true.

Dr. Frank Friedman Oppenheimer admitted in testimony before the Committee on Un-American Activities of the House of Representatives, on June 14, 1949, that he had been a member of the Communist Party from about 1937 until the early spring of 1941. He testified that he joined under the name of "Frank Folsom."

From information before it, the Board concludes that Dr. Frank Oppenheimer had served as a party organizer and as educational director of the professional section of the Communist Party in Los Angeles County.

9. It was further reported that your brother's wife, Jackie Oppenheimer, was a member of the Communist Party in 1938.

The Board concludes that this allegation is true.

Mrs. Jacquenette Oppenheimer in testimony before the Committee on Un-American Activities, House of Representatives, on June 14, 1949, admitted having been a member of the Communist Party from 1937 until the spring of 1941.

10. and that in August, 1944, Jackie Oppenheimer assisted in the organization of the East Bay branch of the California Labor School.

On the basis of information before it, the Board concludes that this allegation is true.

11. It was further reported that in 1945 Frank and Jackie Oppenheimer were invited to an informal reception at the Russian Consulate, that this invitation was extended by the American-Russian Institute of San Francisco and was for the purpose of introducing famous American scientists to Russian scientists who were delegates to the United Nations Conference on International Organization being held at San Francisco at that time, and that Frank Oppenheimer accepted this invitation.

On the basis of information before it, the Board concludes that this allegation is true.

12. It was further reported that Frank Oppenheimer agreed to give a 6-week course on The Social Implications of Modern Scientific Development at the California Labor School, beginning May 9, 1946, The American-Russian Institute of San Francisco and the California Labor School have been cited by the Attorney General as Communist organizations within the purview of Executive Order 9835 and Executive Order 10450.

On the basis of information before it, the Board concludes that this allegation is true.

13. It was reported that you have associated with members and officials of the Communist Party, including Isaac Folkoff, Steve Nelson, Rudy Lambert, Kenneth May, Jack Manley, and Thomas Addis.

The Board concludes that this allegation is substantially true.

Dr. Oppenheimer in his answer and in his testimony admitted having associated with Isaac Folkoff, Steve Nelson, Rudy Lambert, Kenneth May, and Thomas Addis. He testified that he knew at the time of his association with them that Folkoff, Nelson, Lambert, and May were Communist Party functionaries, and that Addis was either a Communist or close to one. He admitted that his associations with these persons continued until 1942. There was no evidence before the Board with respect to an association with Jack Manley.

Dr. Oppenheimer testified that he made contributions to the Spanish War and Spanish Relief through Isaac Folkoff and Thomas Addis. He testified that he had seen Lambert on half a dozen occasions, and that he discussed such contributions once or twice at luncheon with Lambert and Folkoff.

Dr. Oppenheimer testified that Steve Nelson and his family visited his home on several occasions, the last being probably in 1942; that such visits lasted "a few hours;" that he had met Steve Nelson through his (Oppenheimer's) wife since Nelson had befriended her in Paris at the time of Dallet's death; that he had nothing in common with Nelson "except an affection for my wife."

14. It was reported that you were a subscriber to the Daily People's World, a west coast Communist newspaper, in 1941 and 1942.

The Board concludes that this allegation is true.

Dr. Oppenheimer testified that he had subscribed to the People's World "for several years." He could not recall when the subscription expired and stated that he did not believe he had canceled the subscription. He testified that he knew the Daily People's World was the west coast Communist newspaper.

15. It was reported in 1950 that you stated to an agent of the Federal Bureau of Investigation that you had in the past made contributions to Communist-front organizations, although at the time you did not know of Communist Party control or extent of infiltration of these groups. You further stated to an agent of the Federal Bureau of Investigation that some of these contributions were made through Isaac Folkoff, whom you knew to be a leading Communist Party functionary, because you had been told that this was the most effective and direct way of helping these groups.

The Board finds that Dr. Oppenheimer made the statements attributed to him by the Federal Bureau of Investigation.

The Board concludes that Dr. Oppenheimer in the past made contributions to Communist-front organizations and that some of these contributions were made through Isaac Folkoff, a leading Communist Party functionary.

Dr. Oppenheimer testified that he contributed to Spanish causes through Communist Party channels from the winter of 1937–38 until early in 1942. He said that he had contributed more than $500 and less than $1,000 each year during this period. He testified that he had made the contributions in cash and, in explaining how these contributions came to an end, he said (in referring to Pearl Harbor) that he "didn't like to continue a clandestine operation of any kind at a time when I saw myself with the possibility or prospect of getting more deeply involved in the war."

Dr. Oppenheimer in his answer admitted making the contributions through Thomas Addis and Isaac Folkoff. He testified that he knew Addis was a Communist or very close to a Communist. He knew that Folkoff was connected with the Communist Party. In addition, Dr. Oppenheimer admitted having contributed about $100 in cash to the Strike fund of one of the major strikes of "Bridges' Union" about 1937 or 1938.

16. It was reported that you attended a house-warming party at the home of Kenneth and Ruth May on September 20, 1941, for which there was an admission charge for the benefit of The Peoples World, and that at this party you were in the company of Joseph W. Weinberg and Clarence Hiskey, who were alleged to be members of the Communist Party and to have engaged in espionage on behalf of the Soviet Union. It was further reported that you informed officials of the United States Department of Justice in 1942 that you had no recollection that you had attended such a party, but that since it would have been in character for you to have attended such a party, you would not deny that you were there.

The Board concludes on the basis of information before it, that it was probable that Dr. Oppenheimer attended "the house-warming party" at the home of Kenneth and Ruth May. The Board concludes that Dr. Oppenheimer made the statements to the United States Department of Justice officials attributed to him.

Dr. Oppenheimer did not deny having attended such a party and testified that he knew Kenneth May. He denied knowing Hiskey but testified that he, Oppenheimer, was at parties at which Weinberg was present.

17. It was reported that you attended a closed meeting of the professional section of the Communist Party of Alameda County, Calif., which was held in the latter part of July or early August 1941, at your residence, 19 Kenilworth Court, Berkeley, Calif., for the purpose of hearing an explanation of a change in Communist Party policy. It was further reported that you denied that you attended such a meeting and that such a meeting was held in your home.

The Board is of the opinion that the evidence with respect to this meeting is inconclusive. The Board finds that Dr. Oppenheimer did deny that he attended such a meeting and that such a meeting was held in his home.

18. It was reported that you stated to an agent of the Federal Bureau of Investigation in 1950 that you attended a meeting in 1940 or 1941, which may have taken place at the home of Haakon Chevalier, which was addressed by William Schneiderman, whom you knew to be a leading functionary of the Communist Party. In testimony in 1950 before the California State Senate Committee on Un-American Activities, Haakon Chevalier was identified as a member of the Communist Party in the San Francisco area in the early 1940's.

The Board finds that Dr. Oppenheimer made the statements attributed to him by the Federal Bureau of Investigation.

Dr. Oppenheimer testified that on December 1, 1940, he attended an evening meeting at the home of Haakon Chevalier at which perhaps 20 people were present and at which William Schneiderman, secretary of the Communist Party in California, gave a talk about the Communist Party line. He testified that he thought that possibly Isaac Folkoff, Dr. Addis, and Rudy Lambert were there.

He also testified that "after the end of 1940" he attended a similar meeting at the home of Louise Bransten, "a Communist sympathizer," at which some of the same people were present and at which Schneiderman also spoke and expounded the Communist Party line.

Dr. Oppenheimer testified that sometime between 1937 and 1939 as a guest he attended a Communist Party meeting at the home of his brother, Frank.

19. It was reported that you have consistently denied that you have ever been a member of the Communist Party. It was further reported that you stated to a representative of the Federal Bureau of Investigation in 1946 that you had a change of mind regarding the policies and politics of the Soviet Union about the time of the signing of the Soviet-German Pact in 1939. It was further reported that during 1950 you stated to a representative of the Federal Bureau of Investigation that you had never attended a closed meeting of the Communist Party; and that at the time of the Russo-Finnish War and the subsequent break between Germany and Russia in 1941, you realized the Communist Party infiltration tactics into the alleged anti-Fascist groups and became fed up with the whole thing and lost what little interest you had.

Dr. Oppenheimer testified that he had never been a member of the Communist Party. The Board finds that Dr. Oppenheimer made the statements attributed to him by the Federal Bureau of Investigation.

It was further reported, however, that:

19. (a) Prior to April 1942 you had contributed $150 per month to the Communist Party in the San Francisco area, and that the last such payment was apparently made in April 1942 immediately before your entry into the atomic bomb project.

The Board concludes on the basis of testimony and other information before it that Dr. Oppenheimer made periodic contributions through Communist Party functionaries to the Communist Party in the San Francisco area in amounts aggregating not less than $500 nor more than $1,000 a year during a period of approximately 4 years ending in April 1942. As of April 1942, Dr. Oppenheimer had been for several months participating in Government atomic energy research activities. He executed a questionnaire for Government clearance on April 28, 1942, and subsequently assumed full-time duties with the atomic energy project.

19. (b) During the period 1942–45 various officials of the Communist Party, including Dr. Hannah Peters, organizer of the Professional Section of the Communist Party, Alameda County, Calif., Bernadette Doyle, secretary of the Alameda County Communist Party, Steve Nelson, David Adelson,

Paul Pinsky, Jack Manley, and Katrina Sandow, are reported to have made statements indicating that you were then a member of the Communist Party; that you could not be active in the party at that time; that your name should be removed from the Party mailing list and not mentioned in any way; that you had talked the atomic bomb question over with Party members during this period; and that several years prior to 1945 you had told Steve Nelson that the Army was working on an atomic bomb.

The Board finds that during the period 1942–45, Dr. Hannah Peters, Bernadette Doyle, Steve Nelson, Jack Manley, and Katrina Sandow made statements indicating that Dr. Oppenheimer was then a member of the Communist Party; and that the other statements attributed to officials of the Communist Party in this allegation were made by one or more of them. The Board does not find on the basis of information available to it that such statements were made by David Adelson and Paul Pinsky.

19. (c) You stated in August of 1943 that you did not want anybody working for you on the project who was a member of the Communist Party, since "one always had a question of divided loyalty" and the discipline of the Communist Party was very severe and not compatible with complete loyalty to the project. You further stated at that time that you were referring only to present membership in the Communist Party and not to people who had been members of the party. You stated further that you knew several individuals then at Los Alamos who had been members of the Communist Party. You did not, however, identify such former members of the Communist Party to the appropriate authorities. It was also reported that during the period 1942–45 you were responsible for the employment on the atomic bomb project of individuals who were members of the Communist Party or closely associated with activities of the Communist Party, including Giovanni Rossi Lomanitz, Joseph W. Weinberg, David Bohm, Max Bernard Friedman, and David Hawkins. In the case of Giovanni Rossi Lomanitz, you urged him to work on the project, although you stated that you knew he had been very much of a "Red" when he first came to the University of California and that you emphasized to him that he must forego all political activity if he came on to the project. In August 1943 you protested against the termination of his deferment and requested that he be returned to the project after his entry into the military service.

The Board concludes that Dr. Oppenheimer did state in 1943 that he did not want anybody working for him on the project who was a member of the Communist Party, since "one always had a question of divided loyalty" and the discipline of the Communist Party was very severe and not compatible with complete loyalty to the project. He further stated at that time he was referring only to present membership in the Communist Party and not to people who had been members of the party. He stated further that he knew several individuals then at Los Alamos who had been members of the Communist Party. He did not, however, identify such former members of the Communist Party to the appropriate authorities.

The Board concludes that Dr. Oppenheimer was responsible for the employment on the atom bomb project of Giovanni Rossi Lomanitz at Berkeley and David Hawkins at Los Alamos.

The Board concludes that Dr. Oppenheimer asked for the transfer of David Bohm to Los Alamos, although Bohm was closely associated with the Communist Party. In his answer, Dr. Oppenheimer admitted that while at Berkeley he had assigned David Bohm to a problem of basic science having a bearing on atomic research.

Dr. Oppenheimer testified that he understood that Hawkins had left-wing associations; and that Hawkins "talked about philosophy in a way that indicated an interest and understanding and limited approval anyway of Engels."

The Board does not conclude that Dr. Oppenheimer was responsible for the employment of Friedmann or Weinberg on the atomic energy program.

Dr. Oppenheimer testified that Joseph W. Weinberg was a graduate student of his; that he had heard that Weinberg had been a member of the Young Communist League before coming to Berkeley and the Board had before it a transcript of a conversation with Dr. Oppenheimer indicating that at least by August 1943, he knew Weinberg to be a member of the Communist Party and that he "suspected that before but was not sure." Weinberg gave Oppenheimer as a reference at the time he (Weinberg) obtained employment at the Radiation Laboratory on April 22, 1943.

Dr. Oppenheimer testified that he asked General Groves for the transfer of David Bohm to Los Alamos in 1943, but was told by General Groves that he could not be transferred since he had relatives in Nazi Germany. In March 1944 after a conversation with Bohm at Berkeley (a surveillance report indicated that the talk took place at a sidewalk meeting), he checked with the security officer at Los Alamos to see whether the objections to Bohm still obtained.

Dr. Oppenheimer testified that he thought that in 1946 or 1947 he helped Bohm get a job as Assistant Professor of Physics at Princeton. He testified that he happened to meet Bohm and Lomanitz on the street in Princeton in 1949 just prior to their testifying before the House Committee on Un-American Activities; that he said that "they should tell the truth"; that he later saw Bohm at Princeton and attended a farewell party for him in Princeton; that he would, if asked, have written a letter of recommendation for Bohm as a competent physicist in connection with a job in Brazil, although he knew and was worried about Bohm having pleaded the Fifth Amendment when he testified.

The Board finds that Dr. Oppenheimer did urge Lomanitz to work on the project although he knew he had been very much a "Red" when he first came to the University of California and, in fact, during his attendance at the University, and that Dr. Oppenheimer later stated to a Manhattan District official that he had warned Lomanitz that he must forego all political activity if he came to the project. The Board finds further that in August 1943, Dr. Oppenheimer protested against the termination of Lomanitz' deferment and urgently requested that he be returned to the project after his entry into the military service. It appears from the testimony that Dr. Oppenheimer first learned of the impending induction of Lomanitz in a letter from Dr. E. U. Condon who wrote to him "About it in a great sense of outrage."

20. It was reported that you stated to representatives of the Federal Bureau of Investigation on September 5, 1946, that you had attended a meeting in the East Bay and a meeting in San Francisco at which there were present persons definitely identified with the Communist Party. When asked the purpose of the East Bay meeting and the identity of those in attendance, you declined to answer on the ground that this had no bearing on the matter of interest being discussed.

The Board concludes that this allegation is true. The Board finds that Dr. Oppenheimer did attend a meeting in the East Bay and a meeting in San Francisco (see item 18 above) at which there were present persons definitely identified with the Communist Party and that when he was asked about this meeting by representatives of the Federal Bureau of Investigation on September 5, 1946, he declined to answer on the ground that this had no bearing on the matter of interest being discussed.

The Board finds that Dr. Oppenheimer advised representatives of the FBI of this meeting in a subsequent interview in 1950.

21. It was reported that you attended a meeting at the home of Frank Oppenheimer on January 1, 1946, with David Adelson and Paul Pinsky, both of whom were members of the Communist Party. It was further reported that you analyzed some material which Pinsky hoped to take up with the Legislative Convention in Sacramento, Calif.

The Board concludes that this allegation is true.

22. It was reported in 1946 that you were listed as vice chairman on the letterhead of the Independent Citizens Committee of the Arts, Sciences, and Professions, Inc., which has been cited as a Communist-front by the House Committee on Un-American Activities.

The Board concludes that this allegation is true, although the Board finds that Dr. Oppenheimer advised the organization in a letter on October 11, 1946, that he was not in accord with its policy and wished to resign. He wrote again on December 2, 1946, insisting upon resignation. The resignation was accepted on December 10, 1946.

23. It was reported that prior to March 1, 1943, possibly 3 months prior, Peter Ivanov, secretary at the Soviet Consulate, San Francisco, approached George Charles Eltenton for the purpose of obtaining information regarding

work being done at the Radiation Laboratory for the use of Soviet scientists; that George Charles Eltenton subsequently requested Haakon Chevalier to approach you concerning this matter; that Haakon Chevalier thereupon approached you, either directly or through your brother, Frank Friedman Oppenheimer, in connection with this matter; and that Haakon Chevalier finally advised George Charles Eltenton that there was no chance whatsoever of obtaining the information. It was further reported that you did not report this episode to the appropriate authorities until several months after its occurrence; that when you initially discussed this matter with the appropriate authorities on August 26, 1943, you did not identify yourself as the person who had been approached, and you refused to identify Haakon Chevalier as the individual who had made the approach on behalf of George Charles Eltenton; and that it was not until several months later, when you were ordered by a superior to do so, that you so identified Haakon Chevalier. It was further reported that upon your return to Berkeley following your separation from the Los Alamos project, you were visited by the Chevaliers on several occasions; and that your wife was in contact with Haakon and Barbara Chevalier in 1946 and 1947.

The Board concludes that this allegation is substantially true.

The Board had before it a recording of a conversation between Dr. Oppenheimer and Lt. Col. Boris T. Pash, War Department intelligence officer, who had the responsibility for investigating subversive activities at the Radiation Laboratory, University of California at Berkeley. This conversation took place on August 26, 1943, at the Radiation Laboratory.

It was on this occasion that Dr. Oppenheimer reported the incident to Government authorities. He named Eltenton but refused to identify Chevalier. He also stated that the unnamed contact (Chevalier) had approached three persons on the atomic project and in the course of the interview mentioned other factors, such as the use of microfilm or other means and the involvement of the Russian Consulate.

The Board also had before it a transcript of a conversation between Dr. Oppenheimer and Lieutenant Colonel Lansdale which records that on September 12, 1943, Dr. Oppenheimer again refused to name Chevalier but reported the involvement of three others.

It was not until December 1943, that Dr. Oppenheimer, after being told by General Groves that he would be ordered to divulge the identity of the contact, reported the name of Chevalier. However, the record shows that having been told of the identity of Chevalier by Dr. Oppenheimer, the Manhattan District officials were still of the opinion that Chevalier had contacted three employees on the atomic project.

Dr. Oppenheimer, in his answer, stated that his friend, Haakon Chevalier, with his wife, visited him at his home on Eagle Hill probably in early 1943. He stated further that during the visit Chevalier came into the kitchen and told him that George Eltenton had spoken to him of the possibility of transmitting technical information to Soviet scientists. Dr. Oppenheimer said that he made some strong remark to the effect that this sounded terribly wrong to him, and the discussion ended there.

Dr. Oppenheimer's answer further states that nothing in his long-standing friendship would have led him to believe that Chevalier was actually seeking information, and he was certain that Chevalier had no idea of the work on which Dr. Oppenheimer was engaged.

Dr. Oppenheimer testified that the detailed story of the Chevalier incident which he told to Colonel Pash on August 26, 1943, and affirmed to Colonel Lansdale on September 12, 1943, was false in certain material respects. Dr. Oppenheimer testified that this story was "a cock-and-bull story"; that "the whole thing was a pure fabrication except for the one name, Eltenton." He said that his only explanation for lying was that he "was an idiot" and he "was reluctant to mention Chevalier" and "no doubt somewhat reluctant to mention myself." He admitted on cross examination, however, that if the story he told Colonel Pash had been true, it would have shown that Chevalier "was deeply involved"; that it was not just a casual conversation; that Chevalier was not an innocent contact, and that it was a criminal conspiracy.

Dr. Oppenheimer admitted that if this story to Colonel Pash had been true, it made things look very bad for both Chevalier and himself. He acknowledged that he thought the request for information by Eltenton was "treasonable." He admitted that he knew when he talked to Colonel Pash that his falsification impeded Colonel Pash's investigation.

Dr. Oppenheimer testified that in June or July of 1946 shortly after Chevalier was interviewed by the FBI about the Eltenton-Chevalier Incident, Chevalier came to Oppenheimer's home in Berkeley and told Oppenheimer about the interview; that Chevalier said the FBI had pressed him about whether he talked to anyone besides Oppenheimer; that quite awhile later Dr. Oppenheimer was interviewed by the FBI about the same matter, and at this time he knew from Chevalier substantially what Chevalier had said to the FBI about the incident.

Dr. Oppenheimer testified that he recalled getting a letter from Chevalier in 1950 asking him about Dr. Oppenheimer's testimony before the House Un-American Activities Committee concerning the Chevalier-Eltenton incident. He responded, giving Chevalier a summary of what he, Dr. Oppenheimer, had testified. This letter was later used by Chevalier in support of his application for a passport. Dr. Oppenheimer further testified that at about that time, Chevalier came to Princeton and spent 2 days with Dr. Oppenheimer, discussing Chevalier's personal affairs and that he also then mentioned the matter of his passport. Dr. Oppenheimer said that on this occasion he recommended to Chevalier a lawyer named Joseph Fanelli, who, cross examination disclosed, was the attorney who represented Joseph Weinberg at his trial for perjury. Dr. Oppenheimer testified that he did not know Mr. Fanelli at this time but he had represented Frank Oppenheimer at his appearance before the House Committee on Un-American Activities.

Dr. Oppenheimer testified further that in December of 1953, when he and Mrs. Oppenheimer were in Paris, they had dinner with Dr. and Mrs. Chevalier and, on the following day, went with the Chevaliers to visit a Dr. Malraux. According to Dr. Oppenheimer, Dr. Malraux had given a speech at a "Spanish Relief" meeting in California at which Chevalier presided in about 1938. Dr. Oppenheimer said that since that time, Malraux had undergone "rather major political changes"; that "Malraux became a violent supporter of deGaulle and his great brainman and deserted politics and went into purely philosophic and literary work." It appears also that subsequent to his meeting with Dr. Oppenheimer in Paris in December 1953, Chevalier wrote a letter to an official of the United States Embassy in Paris, reading as follows:

"My friend—and yours—Robert Oppenheimer, gave me your name when he was up for dinner here in our apartment early last December, and urged me to get in touch with you if a personal problem of mine which I discussed with him became pressing. He gave me to understand that I could speak to you with the same frankness and fullness as I have with him, and he with me, during the 15 years of our friendship.

"I should not have presumed to follow up such a suggestion if it had come from anyone else. But, as you know, Opje never tosses off such a suggestion lightly.

"If you are in Paris, or will be in the near future, I should, then, like to see you informally and discuss the problem.

"On rereading what I have written, I have a feeling that I have made the thing sound more formidable than it really is. It's just a decision that I have to make, which is fairly important to me, and which Opje in his grandfatherly way suggested that I shouldn't make before consulting you.

"Very sincerely,

"HAAKON CHEVALIER."

Dr. Oppenheimer testified that the problem which was bothering Chevalier and his wife was that Chevalier was employed as a translator for UNESCO, and he understood that if he continued this work as an American citizen, he would have to be cleared after investigation, and he was doubtful as to whether he would be cleared. He did not wish to renounce his American citizenship but did wish to keep his job, and he was in a conflict about it. Dr. Oppenheimer in his testimony denied going to the American Embassy to assist Dr. Chevalier in getting a passport to return to the United States although he admitted having had lunch with the official in question.

Dr. Oppenheimer also denied discussing with the official in question or anyone else the matter of Chevalier's passport.

Dr. Oppenheimer in his testimony has stated that his association with Chevalier has continued and that he still considers him to be his friend.

24. It was reported that in 1945 you expressed the view that "there is a reasonable possibility that it (the hydrogen bomb) can be made," but that

the feasibility of the hydrogen bomb did not appear, on theoretical grounds, as certain as the fission bomb appeared certain, on theoretical grounds, when the Los Alamos Laboratory was started; and that in the autumn of 1949 the General Advisory Committee expressed the view that "an imaginative and concerted attack on the problem has a better than even chance of producing the weapon within 5 years." It was further reported that in the autumn of 1949, and subsequently, you strongly opposed the development of the hydrogen bomb: (1) on moral grounds, (2) by claiming that it was not feasible, (3) by claiming that there were insufficient facilities and scientific personnel to carry on the development, and (4) that it was not politically desirable. It was further reported that even after it was determined, as a matter of national policy, to proceed with development of a hydrogen bomb, you continued to oppose the project and declined to cooperate fully in the project. It was further reported you departed from your proper role as an adviser to the Commission by causing the distribution, separately and in private, to top personnel at Los Alamos of the majority and minority reports of the General Advisory Committee on development of the hydrogen bomb for the purpose of trying to turn such top personnel against the development of the hydrogen bomb. It was further reported that you were instrumental in persuading other outstanding scientists not to work on the hydrogen bomb project, and that the opposition to the hydrogen bomb, of which you are most experienced, most powerful, and most effective member, has definitely slowed down its development.

In order to assess the influence of Dr. Oppenheimer on the thermonuclear program, it has been necessary for the Board not only to consider the testimony but also to examine many documents and records, most of which are classified. Without disclosing the contents of classified documents, the Board makes the following findings, which it believes to be a sufficient reference to this allegation.

The Board confirms that in 1945 Mr. Oppenheimer expressed the view that " 'there is reasonable possibility that it (the hydrogen bomb) can be made,' but that the feasibility of the hydrogen bomb did not appear, on theoretical grounds, as certain as the fission bomb appeared certain, on theoretical grounds, when the Los Alamos Laboratory was started; and that in August of 1949, the General Advisory Committee expressed the view that 'an imaginative and concerted attack on the problem has a better than even chance of producing the weapon within 5 years.' "

With respect to Dr. Oppenheimer's attitude and activities in relation to the hydrogen bomb in World War II, the evidence shows that Dr. Oppenheimer during this period had no misgivings about a program looking to thermonuclear development and, indeed, during the latter part of the war, he recorded his support of prompt and vigorous action in this connection. When asked under cross examination whether he would have opposed dropping an H-bomb on Hiroshima, he replied that "It would make no sense," and when asked "Why?" replied, "The target is too small." He testified further under cross examination that he believed he would have opposed the dropping of an H-bomb on Japan because of moral scruples although he did not oppose the dropping of an A-bomb on the same grounds. During the postwar period, Dr. Oppenheimer favored, and in fact urged, continued research in the thermonuclear field and seemed to express considerable interest in results that were from time to time discussed with him. However, he was aware that the efforts being put forth in this endeavor were relatively meager and he knew that if research were continued at the same pace, there would be little likelihood of success for many years. Testimony in this connection indicated that there was a feeling on his part that it was more important to go forward with a program for the production of a wider range of atomic bombs.

The Board finds further that in the autumn of 1949, and subsequently, Dr. Oppenheimer strongly opposed the development of the hydrogen bomb on moral grounds; on grounds that it was not politically desirable; he expressed the view that there were insufficient facilities and scientific personnel to carry on the development without seriously interfering with the orderly development of the program for fission bombs; and until the late spring of 1951, he questioned the feasibility of the hydrogen bomb efforts then in progress.

Dr. Oppenheimer testified that what he was opposing in the fall of 1949 was only a "crash program" in the development and production of thermonuclear weapons. In this connection, Dr. Oppenheimer contended that the main question relating to thermonuclear weapons presented to the GAC at its meeting of October 29, 1949, was whether or not the United States should undertake such a crash program. The Board does not believe that Dr. Oppenheimer was entirely candid with the Board in attempting to establish this impression. The record reflects that Dr. Oppenheimer expressed the opinion in writing that the "super bomb should never be produced," and that the commitment to this effect should be unqualified. Moreover, the alternatives available to the GAC were not a choice between an "all-out effort" and no effort at all; there was a middle course which might have been considered.

The Board further concludes that after it was determined, as a matter of national policy (January 31, 1950) to proceed with development of a hydrogen bomb, Dr. Oppenheimer did not oppose the project in a positive or open manner, nor did he decline to cooperate in the project. However, Dr. Oppenheimer is recognized in scientific circles as one of the foremost leaders in the atomic energy field and he has considerable influence on the "policy direction" of the atomic program. The Board finds that his views in opposition to the development of the H-bomb as expressed in 1949 became widely known among scientists, and since he did not make it known that he had abandoned these views, his attitude undoubtedly had an adverse effect on recruitment of scientists and the progress of the scientific effort in this field. In other words, the Board finds, that if Dr. Oppenheimer had enthusiastically supported the thermonuclear program either before or after the determination of national policy, the H-bomb project would have been pursued with considerably more vigor, thus increasing the possibility of earlier success in this field.

The Board finds that Dr. Oppenheimer was not responsible for the distribution, separately and in private, to top personnel at Los Alamos of the majority and minority reports of the General Advisory Committee on development of the hydrogen bomb, but that such distribution was made on the direction of the then general manager of the Atomic Energy Commission, Carroll L. Wilson, apparently in order to prepare the personnel at Los Alamos to discuss the matter with the chairman of the Joint Committee on Atomic Energy of the Congress.

The Board does not find that Dr. Oppenheimer urged other scientists not to work on the program. However, enthusiastic support on his part would perhaps have encouraged other leading scientists to work on the program.

Because of technical questions involved, the Board is unable to make a categorical finding as to whether the opposition of the hydrogen bomb "has definitely slowed down its development." The Board concludes that the opposition to the H-bomb by many persons connected with the atomic energy program, of which Dr. Oppenheimer was the "most experienced, most powerful, and most effective member" did delay the initiation of concerted effort which led to the development of a thermonuclear weapon.

GENERAL CONSIDERATIONS

We do not believe that our findings with respect to the letter of notification provide a full and automatic answer to the categorical question posed to us in these proceedings. Only the dimensions of the problem have perhaps been defined. On the one hand, we find no evidence of disloyalty. Indeed, we have before us much responsible and positive evidence of the loyalty and love of country of the individual concerned. On the other hand, we do not believe that it has been demonstrated that Dr. Oppenheimer has been blameless in the matter of conduct, character, and association.

We could in good conscience, we believe, conclude our difficult undertaking by a brief, clear, and conclusive recommendation to the general manager of the Commission in the following terms: There can be no tampering with the national security, which in times of peril must be absolute, and without concessions for reasons of admiration, gratitude, reward, sympathy, or charity. Any doubts whatsoever must be resolved in favor of the national security. The material and evidence presented to this Board leave reasonable doubts with respect to the individual concerned. We, therefore, do not recommend reinstatement of clearance.

It seemed to us that an alternative recommendation would be possible, if we were allowed to exercise mature practical judgment without the rigid circumscription of regulations and criteria established for us.

In good sense, it could be recommended that Dr. Oppenheimer simply not be used as a consultant, and that therefore there exists no need for a categorical answer to the difficult question posed by the regulations, since there would be no need for access to classified material.

The Board would prefer to report a finding of this nature. We have had a desire to reconcile the hard requirements of security with the compelling urge to avoid harm to a talented citizen.

The Board questioned why the Commission chose to revoke Dr. Oppenheimer's clearance and did not follow the alternative course of declining to make use of his services, assuming it had serious questions in the area of security. To many, this would seem the preferable line of action. We think that the answer of the Commission to this question is pertinent to this recital. It seemed clear that other agencies of Government were extending clearance to Dr. Oppenheimer on the strength of AEC clearance, which in many quarters is supposed to be an approval of the highest order. Furthermore, it was explained that without the positive act of withdrawal of access, he would continue to receive classified reports on Atomic Energy activities as a consultant, even though his services were not specifically and currently engaged. Finally it is said that were his clearance continued, his services would be available to, and probably would be used by, AEC contractors. It is noted that most AEC work is carried on by contractors. Withdrawal of clearance and Dr. Oppenheimer's request for a hearing precipitated this proceeding.

In view of the fact that we must address ourselves to security, we feel constrained to examine some of the great issues and problems brought into focus by the case. Many of these are perhaps more important than the outcome of this inquiry. We believe their examination is a necessary precondition to its disposition on security grounds.

What, within the framework of this case, is meant by loyalty?

Because of widespread confusions and misapprehensions about the security system of the United States, the Board feels that it must state some considerations with respect to loyalty. If a person is considered a security risk in terms of loyalty, the fact or possibility of active disloyalty is assumed, which would involve conduct giving some sort of aid and comfort to a foreign power. The Communist Party is an international conspiracy organized in support of the Soviet Union. It should then be clear that (1) a member of the Communist Party is automatically barred from a position of trust with the United States Government; (2) a fellow-traveler must be declared ineligible for such a position of trust—such a person being described as one who perhaps may not be subject to party discipline, but who is sufficiently close to the party, or sympathetic with its aims, purposes and methods that danger inheres in the situation; (3) any person whose absolute loyalty to the United States is in question, aside from present or former Communist affiliations or associations, should be rejected for Government service; (4) a person whose former status would be encompassed in (1), (2), or (3) above, has the burden of proof of change in position and attitude which must be so clearly borne by him as to leave no reasonable doubt in the minds of those who are called upon to make a governmental decision in the case. If he fails in this demonstration, he must be considered a security risk and denied access to classified information.

One of the important issues presented in cases of this sort is that of rehabilitation

Stated in the context of this proceeding, must we accept the principle that once a Communist, always a Communist, once a fellow-traveler, always a fellow-traveler? Can an individual who has been a member of the Communist Party, or closely enough associated with it to make the difference unimportant at a later time, so comport himself personally, so clearly have demonstrated a renunciation of interest and sympathy, so unequivocally have displayed a zeal for his country and its security as to overcome the necessary presumptions of security risk? We, as a Board, firmly believe that this can be the case, and, if we may be permitted something in the nature of a dictum, we believe that this principle should be a part of the security policy of the United States Government. The necessary but harsh requirements of security should not deny a man the right to have made a mistake, if its recurrence is so remote a possibility as to permit a comfortable prediction as to the sanity and correctness of future conduct.

This Board has been conscious of the atmosphere of the time in which Dr. Oppenheimer's clear-cut Communist affiliations occurred. We have considered his activities against the background of the pervasive disillusionment among many of our people arising out of the effects of the great depression and the perhaps normal tendency of a humanitarian to turn to an organization which seemed to him to be espousing primarily humanitarian causes. We recognize what may have seemed to be at the time a beckoning towards a better social order. We know that many academic people and other intellectuals, honest and moral though they were, misinterpreted the talk, aims and purposes of the Communist Party and its affiliated organizations. We are aware that the fact that the Soviet Union was an ally during some of those years cannot be overlooked. This intellectual exercise has, we think, not been inappropriate because we recognize that 1943 conduct cannot be judged solely in the light of 1954 conditions. At the same time, it must be remembered that standards and procedures of 1943 should not be controlling today.

Another vital question is, can an individual be loyal to the United States and, nevertheless, be considered a security risk?

Because the security interests of this country may be endangered by involuntary act, as well as by positive conduct of a disloyal nature, personal weaknesses of an individual may constitute him a security risk. These would include inordinate use of alcohol or drugs, personal indiscretion (in the sense of careless talk), homosexuality, emotional instability, tendency to yield to pressures of others, unusual attachment for foreign systems. The presence of any of these items would support a finding of security risk, even though in every case accompanied by a deep love of country.

There remains also an aspect of the security system which perhaps has had insufficient public attention. This is the protection and support of the entire system itself. It must include an understanding and an acceptance of security measures adopted by responsible Government agencies. It must include an active cooperation with all agencies of Government properly and reasonably concerned with the security of our country. It must involve a subordination of personal judgment as to the security status of an individual as against a professional judgment in the light of standards and procedures when they have been clearly established by appropriate process. It must entail a wholehearted commitment to the preservation of the security system and the avoidance of conduct tending to confuse or obstruct.

The Board would assert the right of any citizen to be in disagreement with security measures and any other expressed policies of Government. This is all a part of the right of dissent which must be preserved for our people. But the question arises whether an individual who does not accept and abide by the security system should be a part of it.

In this connection, we should acknowledge that in the early war years very few people were aware of the full implications of security or security measures which needed to be undertaken. Even many of those in the military services found themselves for a time in a new field. This was a new concept under strange and alien pressures. We believe that no person should now be held accountable for lack of full knowledgeability in the early years of the war. However, those who have been associated with it during the war years and subsequently and who have been exposed repeatedly to security measures, should not fail to understand the need for their full support of the system.

Another major question posed by these proceedings is whether we should take Calculated risks where the national security is involved

It has been urged upon us that where there is lingering doubt about the security status of an individual in the absence of a finding of disloyalty or a tendency towards indiscretion, we should take a calculated risk in granting clearance to such an individual if he is a man of great attainments and capacity and has rendered outstanding services.

Within the framework of our national philosophy which rests in large part upon the declaration that all men are equal before the bar of justice, can we apply one test to an individual, however brilliant his capacities and however magnificent his contributions, and another test to an individual with more mundane capabilities and lesser contributions? In other words, can a different test for security purposes be justified in the case of the brilliant technical consultant than in the case of the stenographer or clerk? It seems to us that such a distinction can be justified only on the ground of critical national need and that otherwise there can be but one standard for all.

We acknowledge that the national necessity may at times require the taking of a calculated risk. Such a calculated risk was taken in the employment and retention of Dr. Oppenheimer as Director of the Los Alamos Laboratory during the war years, on the ground of the overriding need for his services. The officer-in-charge has said that, had he found the risk becoming a danger, he would have felt impelled to open up the whole project and throw security to the winds rather than lose the talents of the individual. Again, wartime exigencies demanded the use of Nazi scientists before the issues with Germany were settled.

What we have learned in this inquiry makes the present application of this principle inappropriate in the instant case. Notwithstanding the undoubted and unparalleled contributions of Dr. Oppenheimer to the atomic energy program, it appears that his services as a consultant were used by the Atomic Energy Commission during the entire year of 1953 for a period approximating only 2½ days' time. We conclude, therefore, that our recommendation should not be based upon such principle, overriding all other considerations.

Another major issue which has been highlighted by this inquiry is whether a moral principle akin to double jeopardy in the traditional legal sense should have a place in the jurisprudence of security

We properly ask ourselves the question: How many times may the same circumstances of a man's life be examined with a view toward determination of his security status? Once a responsible agency of the Government has made an evaluation, should this not be a bar to later and similar consideration by the same or another agency in the absence of newly discovered evidence or developments. This is an important consideration and the Board has undertaken to examine it with care.

It must be made clear to the public by the Government that its employees and consultants are not to be subject to repeated and capricious reviews of their loyalty or security status. In general, this Board believes that responsible prior clearance should be given great weight and should be virtually considered a settled matter in cases where there is manifestly no new material or developments of consequence. We would not urge this as an absolute principle, however, for the reason that the criminal law concept referred to is for the protection of the individual whereas security measures are for the protection of the country, whose interests should never be foreclosed.

There seems to be a widespread view that such a principle should apply in the case at hand. It has been suggested that the clearance by the Manhattan Engineering District and the subsequent action of the AEC in 1947 should be controlling. We believe this not to be sound.

In the first place, we must acknowledge the important difference between an administrative review of files not involving the personal appearance of the employee and of which he is probably not aware, and a hearing before the Board at which the employee appears and at which testimony is taken. This is the first occasion of review of this case by a Personnel Security Board. Indeed, this is the only time that all of the available evidence regarding Dr. Oppenheimer has been correlated and presented in a package. This latter fact suggests the second reason why Dr. Oppenheimer is not being placed in double jeopardy in a moral sense by this proceeding. It was necessary to the national security that material information not considered in previous clearances be studied.

Third, new developments have occurred since the granting of previous clearances. Among these are changed national and international circumstances and new security standards and criteria which have been published in the interim. We refer specifically to the AEC criteria published since 1947 and the Executive order of the President of April 27, 1953.

It must be recalled that the Manhattan District criteria were primarily loyalty and discretion. Such records as are available with respect to the AEC clearance in 1947 indicate that in general it was based in large part upon the earlier clearance by the Manhattan Engineering District, upon a finding of loyal service to the country, and the risk to the program in the loss of services of the individual.

Fourth, viewed against the background of earlier history, the conduct of the individual subsequently to 1947 has been such as to raise questions of security risk.

Another major issue prompted by these proceedings concerns itself with the extent of the right of a citizen to continued employment by his Government because of loyal and distinguished accomplishment in Government service

There are those who seem deeply convinced that Dr. Oppenheimer has a right to continued employment, in view of his previous contributions and in the light of his brilliant capabilities. Citizens of this country have many inalienable rights, but it is clear that Government service is a privilege and not a right. This principle was simply, but effectively, stated by Oliver Wendell Holmes:

"* * * The petitioner may have a constitutional right to talk politics, but he has no constitutional right to be a policeman * * *"

We deem it, therefore, to be within the power of Government in the absence of Civil Service requirement or contractual relationships to terminate employment of a consultant at any time.

A major question which has repeatedly emerged in our deliberations is whether in determining the security status of an individual who is a scientist, the Government must take into account the reactions of, and the possible impact upon, all other scientists

The Board takes cognizance of the serious alarm expressed to it by witnesses and frequently adverted to in the public press that denial of clearance to Dr. Oppenheimer would do serious harm in the scientific community. This is a matter of vital concern to the Government and the people.

We should express our considered view that, because the loyalty or security risk status of a scientist or any other intellectual may be brought into question, scientists and intellectuals are ill-advised to assert that a reasonable and sane inquiry constitutes an attack upon scientists and intellectuals generally. This Board would deplore deeply any notion that scientists are under attack in this country and that prudent study of any individual's conduct and character within the necessary demands of the national security could be either in fact or in appearance a reflection of anti-intellectualism.

The Board has taken note of the fact that in some cases of this sort groups of scientists have tended toward an almost professional opposition to any inquiry about a member of the group. They thus, by moving in a body to the defense of one of their number, give currency, credence, and support to a notion that they as a group are under attack. A decision of a Board of this sort, whether favorable or unfavorable to the individual whose case is before it, should be considered neither as an exoneration of all scientists from imputations of security risk nor a determination that all scientists are suspect.

We know that scientists, with their unusual talents, are loyal citizens, and, for every pertinent purpose, normal human beings. We must believe that they, the young and the old and all between, will understand that a responsible Government must make responsible decisions. If scientists should believe that such a decision in Goverment, however distasteful with respect to an individual, must be applicable to his whole profession, they misapprehended their own duties and obligations as citizens.

In this connection, the Board has been impressed, and in many ways heartened by the manner in which many scientists have sprung to the defense of one whom many felt was under unfair attack. This is important and encouraging when one is concerned with the vitality of our society. However, the Board feels constrained to express its concern that in this solidarity there have been attitudes so uncompromising in support of science in general, and Dr. Oppenheimer in particular, that some witnesses have, in our judgment, allowed their convictions to supersede what might reasonably have been their recollections.

One important consideration brought into focus by this case is the role of scientists as advisers in the formulation of Government policy.

We must adress ourselves to the natural constraints and the particular difficulties inherent in the AEC program itself. As a Nation we find it necessary to delegate temporary authority with respect to the conduct of the program and the policies to be followed to duly elected representatives and appointive officials as provided for by our Constitution and laws. For the most part, these representatives and officials are not capable of passing judgment on technical matters and, therefore, appropriately look to specialists for advice. We must take notice of the current and inevitable amplification of influence which attaches to those giving advice under these circumstances. These specialists have an exponential amplification of influence which is vastly greater than that of the individual citizen.

It must be understood that such specialists did not, as scientists, deliberately create this condition. For example, Dr. Oppenheimer served his Government because it sought him. The impact of his influence was felt immediately and increased progressively as his services were used. The Nation owes these scientists, we believe, a great debt of gratitude for loyal and magnificent service. This is particularly true with respect to Dr. Oppenheimer.

A question can properly be raised about advice of specialists relating to moral, military and political issues, under circumstances which lend such advice an undue and in some cases decisive weight. Caution must be expressed with respect to judgments which go beyond areas of special and particular competence.

Any man, whether specialist or layman, of course, must have the right to express his deep moral convictions; must have the privilege of voicing his deepest doubts. We can understand the emotional involvement of any scientist who contributed to the development of atomic energy and thus helped to unleash upon the world a force which could be destructive of civilization. Perhaps no American can be entirely guilt-free, and, yet, these weapons did not bring peace nor lessen the threats to the survival of our free institutions. Emotional involvement in the current crisis, like all other things, must yield to the security of the nation.

Dr. Oppenheimer himself testified, "I felt perhaps quite wrongly that having played an active part in promoting a revolution in warfare I needed to be as responsible as I could with regard to what came of this revolution."

We have no doubt that other distinguished and devoted scientists have found themselves beset by a similar conflict.

It is vitally important that Government and scientists alike understand the need for and value of the advice of competent technicians. This need is a present and a continuing one. Yet, those officials in Government who are responsible for the security of the country must be certain that the advice which they seriously seek appropriately reflects special competence on the one hand, and soundly based conviction on the other, uncolored and uninfluenced by considerations of an emotional character.

In evaluating advice from a specialist which departs from the area of his specialty, Government officials charged with the military posture of our country must also be certain that underlying any advice is a genuine conviction that this country cannot in the interest of security have less than the strongest possible offensive capabilities in a time of national danger.

Significance of the findings of the Board

The facts referred to in General Nichols' letter fall clearly into two major areas of concern. The first of these, which is represented by items 1 through 23, involves primarily Dr. Oppenheimer's Communist connections in the earlier years and continued associations arising out of those connections.

The second major area of concern is related to Dr. Oppenheimer's attitudes and activities with respect to the development of the hydrogen bomb.

The Board has found the allegations in the first part of the Commission letter to be substantially true, and attaches the following significance to the findings: There remains little doubt that, from late 1936 or early 1937 to probably April 1942, Dr. Oppenheimer was deeply involved with many people who were active Communists. The record would suggest that the involvement was something more than an intellectual and sympathetic interest in the professed aims of the Communist Party. Although Communist functionaries during this period considered Dr. Oppenheimer to be a Communist, there is no evidence that he was a member of the party in the strict sense of the word.

Using Dr. Oppenheimer's own characterization of his status during that period, he seems to have been an active fellow-traveler. According to him, his sympathies with the Communists seem to have begun to taper off somewhat after 1939, and very much more so after 1942. However, it is not unreasonable to conclude from material presented to this Board that Dr. Oppenheimer's activities ceased as of about the time he executed his Personnel Security Questionnaire in April 1942. He seems to have had the view at that time and subsequently that current involvement with Communist activities was incompatible with service to the Government. However, it also would appear that he felt that former Communist Party membership was of little consequence if the individual concerned was personally trustworthy.

Dr. Oppenheimer's sympathetic interests seemed to have continued beyond 1942 in a diluted and diminishing state until 1946, at which time we find the first affirmative action on his part which would indicate complete rejection. In October 1946, he tendered his resignation from the Independent Citizens Committee of the Arts, Sciences, and Professions, Inc., and he now says it was at this time that he finally realized that he could not collaborate with the Communists, whatever their aims and professed interests. We would prefer to have found an affirmative action at an earlier date.

The Board takes a most serious view of these earlier involvements. Had they occurred in very recent years, we would have found them to be controlling and, in any event, they must be taken into account in evaluating subsequent conduct and attitudes.

The facts before us establish a pattern of conduct falling within the following Personnel Security Clearance criteria: Category A, including instances in which there are grounds sufficient to establish a reasonable belief that an individual or his spouse has (1) Committed or attempted to commit or aided or abetted another who committed or attempted to commit any act of sabotage, espionage, treason, or sedition. (2) Establish an association with espionage agents of a foreign nation * * * (3) Held membership or joined any organization which had been declared by the Attorney General to be * * * Communist, subversive * * * These criteria under the AEC procedures establish a presumption of security risk.

The Board believes, however, that there is no indication of disloyalty on the part of Dr. Oppenheimer by reason of any present Communist affiliation, despite Dr. Oppenheimer's poor judgment in continuing some of his past associations into the present. Furthermore, the Board had before it eloquent and convincing testimony of Dr. Oppenheimer's deep devotion to his country in recent years and a multitude of evidence with respect to active service in all sorts of governmental undertakings to which he was repeatedly called as a participant and as a consultant.

We feel that Dr. Oppenheimer is convinced that the earlier involvements were serious errors and today would consider them an indication of disloyalty. The conclusion of this Board is that Dr. Oppenheimer is a loyal citizen.

With respect to the second portion of General Nichols' letter, the Board believes that Dr. Oppenheimer's opposition to the hydrogen bomb and his related conduct in the postwar period until April 1951, involved no lack of loyalty to the United States or attachment to the Soviet Union. The Board was impressed by the fact that even those who were critical of Dr. Oppenheimer's judgment and activities or lack of activities, without exception, testified to their belief in his loyalty.

The Board concludes that any possible implications to the contrary which might have been read into the second part of General Nichols' letter are not supported by any material which the Board has seen.

The Board wishes to make clear that in attempting to arrive at its findings and their significance with respect to the hydrogen bomb, it has in no way sought to appraise the technical judgments of those who were concerned with the program.*

We cannot dismiss the matter of Dr. Oppenheimer's relationship to the development of the hydrogen bomb simply with the finding that his conduct was not motivated by disloyalty, because it is our conclusion that, whatever the motivation, the security interests of the United States were affected.

We believe that, had Dr. Oppenheimer given his enthusiastic support to the program, a concerted effort would have been initiated at an earlier date.

Following the President's decision, he did not show the enthusiastic support for the program which might have been expected of the chief atomic adviser to the Government under the circumstances. Indeed, a failure to communicate an abandonment of his earlier position undoubtedly had an effect upon other scientists. It is our feeling that Dr. Oppenheimer's influence in the atomic scientific circles with respect to the hydrogen bomb was far greater than he would have led this Board to believe in his testimony before the Board. The Board has reluctantly concluded that Dr. Oppenheimer's candor left much to be desired in his discussions with the Board of his attitude and position in the entire chronology of the hydrogen-bomb problem.

We must make it clear that we do not question Dr. Oppenheimer's right to the opinions he held with respect to the development of this weapon. They were shared by other competent and devoted individuals, both in and out of Government. We are willing to assume that they were motivated by deep moral conviction. We are concerned, however, that he may have departed his role as scientific adviser to exercise highly persuasive influence in matters in which his convictions were not necessarily a reflection of technical judgment, and

---

*This is the end of p. 32 referred to in the minority report of Dr. Ward V. Evans.

also not necessarily related to the protection of the strongest offensive military interests of the country.

In the course of the proceedings, there developed other facts which raised questions of such serious import as to give us concern about whether the retention of Dr. Oppenheimer's services would be clearly consistent with the security interests of the United States.

It must be said that Dr. Oppenheimer seems to have had a high degree of discretion reflecting an unusual ability to keep to himself vital secrets. However, we do find suggestions of a tendency to be coerced, or at least influenced in conduct over a period of years.

By his own testimony, Dr. Oppenheimer was led to protest the induction into military service of Giovanni Rossi Lomanitz in 1943 by the outraged intercession of Dr. Condon. It is to be remembered that, at this time Dr. Oppenheimer knew of Lomanitz's connections and of his indiscretions. In 1949, Dr. Oppenheimer appeared in executive session before the House Un-American Activities Committee, and at that time was asked about his friend, Dr. Bernard Peters. Dr. Oppenheimer confirmed the substance of an interview with the security officer which took place during the war years and in which he had characterized Dr. Peters as a dangerous Red and former Communist. This testimony soon appeared in the Rochester, N. Y., newspapers. At this time, Dr. Peters was on the staff of the University of Rochester. Dr. Oppenheimer, as a result of protestations by Dr. Condon, by Dr. Peters himself, and by other scientists, then wrote a letter for publication to the Rochester newspaper, which, in effect, repudiated his testimony given in secret session. His testimony before this Board indicated that he failed to appreciate the great impropriety of making statements of one character in a secret session and of a different character for publication, and that he believed that the important thing was to protect Dr. Peters' professional status. In that episode, Dr. Condon's letter, which has appeared in the press, contained a severe attack on Dr. Oppenheimer. Nevertheless, he now testifies that he is prepared to support Dr. Condon in the loyalty investigation of the latter.

Executive Order 10450 in listing criteria to be taken into account in cases of this sort indicates in part the following:

"Section 8 (a) (1) (i) any behavior, activities, or associations which tend to show that the individual is not reliable or trustworthy.

(v) Any facts which furnish reason to believe that the individual may be subjected to coercion, influence, or pressure which may cause him to act contrary to the best interest of the national security."

Whether the incidents referred to clearly indicate a susceptibility to influence or coercion within the meaning of the criteria or whether they simply reflect very bad judgment, they clearly raise the question of Dr. Oppenheimer's understanding, acceptance, and enthusiastic support of the security system. Beginning with the Chevalier incident, he has repeatedly exercised an arrogance of his own judgment with respect to the loyalty and reliability of other citizens to an extent which has frustrated and at times impeded the workings of the system. In an interview with agents of the FBI in 1946, which in good part concerned itself with questions about Chevalier, when asked about a meeting which Dr. Oppenheimer had attended, at which Communists and Communist sympathizers were in attendance, he declined to discuss it on the ground that it was irrelevant, although the meeting itself was held in Chevalier's home. In a subsequent interview, he declined to discuss people he had known to be Communists.

Indeed, in the course of this proceeding, Dr. Oppenheimer recalled pertinent details with respect to Communist meetings and with respect to individuals with Communist connections, which he had never previously disclosed in the many interviews with Government authorities, in spite of the fact that he had been interviewed regarding such matters.

In 1946 or 1947, he assisted David Bohm in getting a position at Princeton and, at least on a casual basis, continued his associations with Bohm after he had reason to know of Bohm's security status. He testified that today he would give Bohm a letter of recommendation as a physicist, and, although not asked whether he would also raise questions about Bohm's security status, he in no way indicated that this was a matter of serious import to him.

While his meeting with Lomanitz and Bohm immediately prior to their appearance before the House Un-American Activities Committee in 1949, at which time both pleaded the fifth amendment, may have been a casual one as he testified, he nevertheless discussed with them their testimony before that committee.

Moreover, his current associations with Dr. Chevalier, as discussed in detail in item No. 23, are, we believe, of a high degree of significance. It is not important to determine that Dr. Oppenheimer discussed with Chevalier matters of concern to the security of the United States. What is important is that Chevalier's Communist background and activities were known to Dr. Oppenheimer. While he says he believes Chevalier is not now a Communist, his association with him, on what could not be considered a casual basis, is not the kind of thing that our security system permits on the part of one who customarily has access to information of the highest classification.

Loyalty to one's friends is one of the noblest of qualities. Being loyal to one's friends above reasonable obligations to the country and to the security system, however, is not clearly consistent with the interests of security.

We are aware that in these instances Dr. Oppenheimer may have been sincere in his interpretation that the security interests of the country were not disserved; we must, however, take a most serious view of this kind of continuing judgment.

We are constrained to make a final comment about General Nichols' letter. Unfortunately, in the press accounts in which the letter was printed in full, item No. 24, which consisted of 1 paragraph, was broken down into 4 paragraphs. Many thoughtful people, as a result, felt that the implication of one or more of these paragraphs as they appeared in the press standing alone was that the letter sought to initiate proceedings which would impugn a man on the ground of his holding and forcefully expressing strong opinions. It is regrettable that the language of the letter or the way in which it publicly appeared might have given any credence to such an interpretation. In any event, the Board wishes strongly to record its profound and positive view that no man should be tried for the expression of his opinions.

RECOMMENDATION

In arriving at our recommendation we have sought to address ourselves to the whole question before us and not to consider the problem as a fragmented one either in terms of specific criteria or in terms of any period in Dr. Oppenheimer's life, or to consider loyalty, character, and associations separately.

However, of course, the most serious finding which this Board could make as a result of these proceedings would be that of disloyalty on the part of Dr. Oppenheimer to his country. For that reason, we have given particular attention to the question of his loyalty, and we have come to a clear conclusion, which should be reassuring to the people of this country, that he is a loyal citizen. If this were the only consideration, therefore, we would recommend that the reinstatement of his clearance would not be a danger to the common defense and security.

We have, however, been unable to arrive at the conclusion that it would be clearly consistent with the security interests of the United States to reinstate Dr. Oppenheimer's clearance and, therefore, do not so recommend.

The following considerations have been controlling in leading us to our conclusion:

1. We find that Dr. Oppenheimer's continuing conduct and associations have reflected a serious disregard for the requirements of the security system.

2. We have found a susceptibility to influence which could have serious implications for the security interests of the country.

3. We find his conduct in the hydrogen-bomb program sufficiently disturbing as to raise a doubt as to whether his future participation, if characterized by the same attitudes in a Government program relating to the national defense, would be clearly consistent with the best interests of security.

4. We have regretfully concluded that Dr. Oppenheimer has been less than candid in several instances in his testimony before this Board.

Respectfully submitted.

GORDON GRAY, *Chairman.*
THOMAS A. MORGAN.

MINORITY REPORT OF DR. WARD V. EVANS

I have reached the conclusion that Dr. J. Robert Oppenheimer's clearance should be reinstated and am submitting a minority report in accordance with AEC procedure.

The Board, appointed by the Commission, has worked long and arduously on the Oppenheimer case. We have heard 40 witnesses and have taken some 3,000 pages of testimony in addition to having read a similar number of pages of file material. We have examined carefully the notification letter to Dr. Oppenheimer from Mr. Nichols of December 23, 1953, and all other relevant material.

I am in perfect agreement with the majority report of its "findings" with respect to the allegations in Mr. Nichols' letter and I am in agreement with the statement of the Board concerning the significance of its "findings" to the end of page 32.* I also agree with the last paragraph of this section in which the Board makes a final comment on Mr. Nichols' letter. I do not, however, think it necessary to go into any philosophical discussion to prove points not found in Mr. Nichols' letter.

The derogatory information in this letter consisting of 24 items has all been substantiated except for one item. This refers to a Communist meeting held in Dr. Oppenheimer's home, which he is supposed to have attended.

On the basis of this finding, the Board would have to say that Dr. Oppenheimer should not be cleared.

But this is not all.

Most of this derogatory information was in the hands of the Commission when Dr. Oppenheimer was cleared in 1947. They apparently were aware of his associations and his left-wing policies; yet they cleared him. They took a chance on him because of his special talents and he continued to do a good job. Now when the job is done, we are asked to investigate him for practically the same derogatory information. He did his job in a thorough and painstaking manner. There is not the slightest vestige of information before this Board that would indicate that Dr. Oppenheimer is not a loyal citizen of his country. He hates Russia. He had communistic friends, it is true. He still has some. However, the evidence indicates that he has fewer of them than he had in 1947. He is not as naive as he was then. He has more judgment; no one on the Board doubts his loyalty—even the witnesses adverse to him admit that—and he is certainly less of a security risk than he was in 1947, when he was cleared. To deny him clearance now for what he was cleared for in 1947, when we must know he is less of a security risk now than he was then, seems to be hardly the procedure to be adopted in a free country.

We don't have to go out of our way and invent something to prove that the principle of "double jeopardy" does not apply here. This is not our function, and it is not our function to rewrite any clearance rules. The fact remains he is being investigated twice for the same things. Furthermore, we don't have to dig deeply to find other ways that he may be a security risk outside of loyalty, character, and association. He is loyal, we agree on that. There is, in my estimation, nothing wrong with his character. During the early years of his life, Dr. Oppenheimer devoted himself to study and did not vote or become interested in political matters until he was almost 30. Then, in his ignorance, he embraced many subversive organizations.

His judgment was bad in some cases, and most excellent in others but, in my estimation, it is better now than it was in 1947 and to damn him now and ruin his career and his service, I cannot do it.

His statements in cross examination show him to be still naive, but extremely honest and such statements work to his benefit in my estimation. All people are somewhat of a security risk. I don't think we have to go out of our way to point out how this man might be a security risk.

Dr. Oppenheimer in one place in his testimony said that he had told "a tissue of lies." What he had said was not a tissue of lies; there was one lie. He said on one occasion that he had not heard from Dr. Seaborg, when in fact he had a letter from Dr. Seaborg. In my opinion he had forgotten about the letter or he would never have made this statement for he would have known that the Government had the letter. I do not consider that he lied in this case. He stated that he would have recommended David Bohm as a physicist to Brazil, if asked. I think I would have recommended Bohm as a physicist. Dr. Oppenheimer was not asked if he would have added that Bohm was a Communist. In recent years he went to see Chevalier in Paris. I don't like this, but I cannot

---

*The reference is to p. 32 of the typewritten document. In this reproduction the material referred to is to the end of the seventh full paragraph on p. 19.

condemn him on this ground. I don't like his about face in the matter of Dr. Peters, but I don't think it subversive or disloyal.

He did not hinder the development of the H-bomb and there is absolutely nothing in the testimony to show that he did.

First he was in favor of it in 1944. There is no indication that this opinion changed until 1945. After 1945 he did not favor it for some years perhaps on moral, political or technical grounds. Only time will prove whether he was wrong on the moral and political grounds. After the Presidential directive of January 31, 1950, he worked on this project. If his opposition to the H-bomb caused any people not to work on it, it was because of his intellectual prominence and influence over scientific people and not because of any subversive tendencies.

I personally think that our failure to clear Dr. Oppenheimer will be a black mark on the escutcheon of our country. His witnesses are a considerable segment of the scientific backbone of our Nation and they endorse him. I am worried about the effect an improper decision may have on the scientific development in our country. Nuclear physics is new in our country. Most of our authorities in this field came from overseas. They are with us now. Dr. Oppenheimer got most of his education abroad. We have taken hold of this new development in a very great way. There is no predicting where and how far it may go and what its future potentialities may be. I would very much regret any action to retard or hinder this new scientific development.

I would like to add that this opinion was written before the Bulletin of the Atomic Scientists came out with its statement concerning the Oppenheimer case.

This is my opinion as a citizen of a free country.

I suggest that Dr. Oppenheimer's clearance be restored.

WARD V. EVANS.

LETTER FROM GENERAL MANAGER K. D. NICHOLS, UNITED STATES ATOMIC ENERGY COMMISSION, TO DR. J. ROBERT OPPENHEIMER—FORWARDING FINDINGS AND RECOMMENDATION OF THE PERSONNEL SECURITY BOARD

UNITED STATES ATOMIC ENERGY COMMISSION,
May 28, 1954.

Dr. J. ROBERT OPPENHEIMER,
  The Institute for Advanced Study, Princeton, N. J.

DEAR DR. OPPENHEIMER: I am enclosing herewith a copy of the Findings and Recommendation of the Personnel Security Board which has been considering your case. A majority of the Board recommends that your clearance not be reinstated.

You have the right, under section 4.18 of the Atomic Energy Commission's Security Clearance Procedures, to request review of your case by the Personnel Security Review Board and to submit a brief in support of your contentions. If you wish such review, it is necessary that you submit your request to me within 5 days of your receipt of this letter or June 7, 1954, whichever is later, and that your brief be filed with me not later than 20 days after your receipt of this letter.

If you do not request review of your case by the Personnel Security Review Board within the prescribed time, a final determination will be made on the basis of the existing record.

Upon full consideration of the entire record in the case, including the recommendation of the Personnel Security Review Board in the event you request review by that Board, I shall submit to the Commission by recommendation as to whether or not your clearance should be reinstated. The final determination will be made by the Commission.

I am sending a copy of this letter to Mr. Garrison, along with additional copies of the Personnel Security Board's Findings and Recommendation.

Sincerely yours,

K. D. NICHOLS, *General Manager.*

Cc: Lloyd K. Garrison, Esq., Paul, Weis, Rifkind, Wharton & Garrison, 575 Madison Avenue, New York 22, N. Y. (with 5 copies enclosure).

LETTER FROM DR. OPPENHEIMER'S ATTORNEYS TO GENERAL MANAGER K. D. NICHOLS, UNITED STATES ATOMIC ENERGY COMMISSION, REPLYING TO THE LATTER'S LETTER TO DR. OPPENHEIMER TRANSMITTING THE FINDINGS AND RECOMMENDATION OF THE PERSONNEL SECURITY BOARD

JUNE 1, 1954.

GENERAL K. D. NICHOLS,
  General Manager,
    United States Atomic Energy Commission,
      Washington 25, D. C.

DEAR GENERAL NICHOLS: Dr. Oppenheimer has received your letter of May 28, 1954, in which you enclosed a copy of the "Findings and Recommendation of the Personnel Security Board" dated May 27. In this document the Board unanimously found that Dr. Oppenheimer was a loyal citizen, but by a 2 to 1 vote, Dr. Ward V. Evans dissenting, recommended that Dr. Oppenheimer's clearance should not be reinstated. Dr. Oppenheimer has asked me to send you this reply on his behalf.

You informed Dr. Oppenheimer that he might have until June 7 to notify the Commission whether he would request a review of the case by the Commission's Personnel Security Review Board. You also inform d him that, after considering the record (including the recommendation of the Review Board if a review were taken), you would make your recommendation to the Commission and the Commission would finally determine the matter.

Since the Commission is in any event to decide the case, it seems to us that no useful purpose would be served by our requesting the Review Board to go over the matter afresh and to make a recommendation that would have no more finality to it than that which the Personnel Security Board has already made.

Such a review would entail further delay, which Dr. Oppenheimer is naturally anxious to avoid; moreover, Dr. Oppenheimer's annual contract as a Consultant to the Atomic Energy Commission expires on June 30, the end of the fiscal year, and with it his clearance (now suspended) would automatically expire, so that if this case is not finally determined by June 30, the possibility exists that the question of reinstating Dr. Oppenheimer's clearance might be regarded as moot and might be left in a state of confusion and uncertainty. We do not believe that such an outcome would be in the public interest. Accordingly, Dr. Oppenheimer waives his privilege of review by the Personnel Security Review Board and requests immediate consideration of the case by the Atomic Energy Commission.

In order to assist the Commission in its deliberations, and because of the great public importance of some of the issues raised by the majority and minority opinions of the Personnel Security Board, we request the Commission's permission to file a brief and to make oral argument. A brief is already in the course of preparation and can be delivered to the Commission by June 7 (the latest date specified in your letter to request review). Argument of counsel can, we are sure, be arranged to meet the Commission's convenience.

Meanwhile we think it fitting to identify for the Commission what we conceive to be certain issues of basic importance which are presented by the majority and minority opinions. We believe it essential, however, that the Commission have the benefit of critical analysis and illumination of these issues, which can only be supplied adequately by the brief and oral argument herein requested.

To begin with, the majority's conclusion not to recommend the reinstatement of Dr. Oppenheimer's clearance stands in such stark contrast with the Board's findings regarding Dr. Oppenheimer's loyalty and discretion as to raise doubts about the process of reasoning by which the conclusion was arrived at. All members of the Board agreed:

(1) That the Nation owed scientists "a great debt of gratitude for loyal and magnificent service" and that "This is particularly true with respect to Dr. Oppenheimer" (p. 27).

(2) That "we have before us much responsible and positive evidence of the loyalty and love of country of the individual concerned" (p. 21), and "eloquent and convincing testimony of Dr. Oppenheimer's deep devotion to his country in recent years and a multitude of evidence with respect to active service in all

sorts of governmental undertakings to which he was repeatedly called as a participant and as a consultant" (p. 29).

(3) That "even those who were critical of Dr. Oppenheimer's judgment and activities or lack of activities, without exception, testified to their belief in his loyalty" (p. 30).

(4) That "we have given particular attention to the question of his loyalty, and we have come to a clear conclusion, which should be reassuring to the people of this country, that he is a loyal citizen. If this were the only consideration, therefore, we would recommend that the reinstatement of his clearance would not be a danger to the common defense and security" (p. 33).

(5) That "It must be said that Dr. Oppenheimer seems to have had a high degree of discretion reflecting an unusual ability to keep to himself vital secrets" (p. 31).

In spite of these findings of loyalty and of discretion in the handling of classified data, the majority of the Board reached the conclusion that Dr. Oppenheimer's clearance should not be reinstated. How can this be? The majority advanced four considerations as controlling in leading them to their conclusion (p. 33).

The first two—an alleged "serious disregard for the requirements of the security system," and an alleged "susceptibility to influence"—rest upon an appraisal of the evidence which we do not think is justified by the record. Taking sharp issue, as we do, with the majority's treatment of the incidents cited in support of these two considerations, we cannot undertake here to review the detailed evidence, but propose to do so in the brief.

We would like, however, to draw attention to two of the incidents referred to by the majority in support of these considerations, merely to indicate the care with which we think the record needs to be reviewed by the Commission. The majority held it against Dr. Oppenheimer, apparently as an example of his supposed susceptibility to influence, that despite a severe attack on him by Dr. Edward Condon in 1949, in a letter which appeared in the press, Dr. Oppenheimer is now prepared to support Dr. Condon in the latter's pending loyalty investigation (p. 31). It seems to us strange that a man should be criticized for refusing to let his personal feelings stand in the way of his giving evidence on behalf of a man he believes to be loyal. The majority further criticized Dr. Oppenheimer for his continuing associations and supposed disregard of security requirements in that "In 1946 or 1947 he assisted David Bohm [a former student] in getting a position at Princeton and, at least on a casual basis, continued his associations with Bohm after he had reason to know of Bohm's security status. He testified that today he would give Bohm a letter of recommendation as a physicist, and, although not asked whether he would also raise questions about Bohm's security status, he in no way indicated that this was a matter of serious import to him" (p. 32). Dr. Evans' comment on this incident was: "I think I would have recommended Bohm as a physicist. Dr. Oppenheimer was not asked if he would have added that Bohm was a Communist" (p. 35).

We propose to analyze in detail, in brief and argument, these and other incidents referred to by the majority as bearing on Dr. Oppenheimer's supposed "disregard" for "the security system" and "susceptibility to influence."

The third and fourth considerations advanced by the majority for concluding that Dr. Oppenheimer was a "security risk" warrant more extended comment here.

The third item—Dr. Oppenheimer's "conduct in the hydrogen bomb program," characterized as "disturbing"—and the fourth—alleged "lack of candor" in several instances in his testimony—require discussion, because they involve questions of policy and procedure which we wish particularly to draw to the Commission's attention in a preliminary way.

In the case of the third consideration—Dr. Oppenheimer's so-called disturbing conduct in the hydrogen bomb program—the Board's unanimous findings of fact again stand in stark contrast with the conclusion of the majority. Thus the Board unanimously found:

(1) That Dr. Oppenheimer's opposition to the H-bomb program "involved no lack of loyalty to the United States or attachment to the Soviet Union" (p. 30).

(2) That his opinions regarding the development of the H-bomb "were shared by other competent and devoted individuals, both in and out of Government" (p. 30).

(3) That it could be assumed that these opinions "were motivated by deep moral conviction" (p. 30).

(4) That after the national policy to proceed with the development of the H-bomb had been determined in January 1950, he "did not oppose the project in a positive or open manner, nor did he decline to cooperate in the project" (p. 20).

(5) That the allegations that he urged other scientists not to work on the hydrogen bomb program were unfounded (p. 20).

(6) That he did not, as alleged, distribute copies of the General Advisory Report to key personnel with a view to turning them against the project, but that on the contrary this distribution was made at the Commission's own direction (p. 20).

In short, all the basic allegations set forth in General Nichols' letter to Dr. Oppenheimer on December 23, 1953, regarding any improper action by him in the H-bomb problem were disproved.

In the face of these unanimous findings, the majority then conclude that "the security interests of the United States were affected" by Dr. Oppenheimer's attitude toward the hydrogen bomb program (p. 30). Why? Because, according to the majority of the Board:

"We believe that, had Dr. Oppenheimer given his enthusiastic support to the program, a concerted effort would have been initiated at an earlier date.

"Following the President's decision, he did not show the enthusiastic support for the program which might have been expected of the chief atomic adviser to the Government under the circumstances. Indeed, a failure to communicate an abandonment of his earlier position undoubtedly had an effect upon other scientists" (p. 30).

Without taking into account the factual evidence, which in our opinion should have led the Board to an opposite conclusion, we submit that the injection into a security case of a scientist's alleged lack of enthusiasm for a particular program is fraught with grave consequences to this country. How can a scientist risk advising the Government if he is told that at some later day a security board may weigh in the balance the degree of his enthusiasm for some official program? Or that he may be held accountable for a failure to communicate to the scientific community his full acceptance of such a program?

In addition to Dr. Oppenheimer's alleged lack of "enthusiasm," there are indications that the majority of the Board may also have been influenced in recommending against the reinstatement of Dr. Oppenheimer's clearance by judgments they had formed as to the nature and quality of the advice he gave to the AEC. While the majority of the Board stated—with sincerity, we are sure—that "no man should be tried for the expression of his opinions" (p. 33), it seems to us that portions of the majority opinion do just that.

For example, the opinion says that while the Board can understand "the emotional involvement of any scientist who contributed to the development of atomic energy and thus helped to unleash upon the world a force which could be destructive of civilization," nevertheless "emotional involvement" of this sort in the current crisis "must yield to the security of the Nation"; and Government officials "who are responsible for the security of the country must be certain that the advice which they seriously seek appropriately reflects special competence on the one hand, and soundly based conviction on the other, uncolored and uninfluenced by considerations of an emotional character" (p. 28). Does this mean that a loyal scientist called to advise his Government does so at his peril unless, contrary to all experience, he can guarantee that his views are unaffected by his heart and his spirit?

The opinion further stated that defense officials "must also be certain that underlying any advice is a genuine conviction that this country cannot in the interest of security have less than the strongest possible offensive capabilities in a time of national danger" (p. 28). Does this mean that a loyal scientist called to advise his Government does so at his peril if he happens to believe in the wisdom of maintaining a proper balance between offensive and defensive weapons?

It would appear from the following passage about Dr. Oppenheimer's advice that the majority of the Board assumed affirmative answers to both of the foregoing questions:

"We are concerned, however, that he may have departed his role as scientific adviser to exercise highly persuasive influence in matters in which his convictions were not necessarily a reflection of technical judgment, and also not necessarily related to the protection of the strongest offensive military interests of the country" (p. 30).

This poses a serious issue. If a scientist whose loyalty is unquestioned may nevertheless be considered a security risk because in the judgment of a board he may have given advice which did not necessarily reflect a bare technical judgment, or which did not accord with strategical considerations of a particular kind, then he is being condemned for his opinions. Surely our security requires that expert views, so long as they are honest, be weighed and debated and not that they be barred.

We quite agree with the Board's view that, "because the loyalty or security risk status of a scientist or any other intellectual may be brought into question, scientists and intellectuals are ill-advised to assert that a reasonable and sane inquiry constitutes an attack upon scientists and intellectuals generally" (p. 26). This statement, however, begs the fundamental question as to what are the appropriate limits of a security inquiry under existing statutes and regulations, and under a government of laws and not of men—a question of concern not merely to scientists and intellectuals but to all our people.

*   *   *   *   *   *   *

As to the majority's comments about Dr. Oppenheimer's alleged lack of "candor" in "several instances in his testimony," we shall ask the Commission to take special note of the observations in Dr. Evans' minority opinion that while Dr. Oppenheimer's "statements in cross-examination show him to be still naive," they also show him to be "extremely honest and such statements work to his benefit in my estimation" and that while "his judgment was bad in some cases" it was "most excellent in others but it is better now than it was in 1947," when the Atomic Energy Commission unanimously cleared him. We shall also direct the Commission's attention to the fact that the text of the Board's report contains only three specific references to alleged lack of candor, all having to do with the hydrogen bomb program (pp. 20 and 30). [We should point out that as to two of these references (p. 30) Dr. Evans specifically disassociated himself from the majority (p. 34) and as to the other (p. 20) he did so by clear implication (p. 35).] As to all these matters there was extensive testimony not only by Dr. Oppenheimer but also by others who served with him on the General Advisory Committee, including Dr. James B. Conant, Dr. I. I. Rabi (now Chairman of the GAC), Dr. Enrico Fermi, and Mr. Hartley Rowe of the United Fruit Co.; and also by Dr. Norris Bradbury (Dr. Oppenheimer's successor as director of Los Alamos); by Mr. Gordon Dean (former Chairman of the Atomic Energy Commission); by Dr. Hans Bethe; by Dr. Robert Bacher (a former member of the AEC); and by a number of other distinguished men.

In brief and argument we expect to analyze for the Commission's assistance the evidence they gave, which in our judgment bears out the truth and sincerity of Dr. Oppenheimer's account of the H-bomb controversy.

*   *   *   *   *   *   *

We wish to make two more observations of a general character.

First, we trust that the Commission in weighing the evidence, including the instances of alleged lack of candor, will take into account certain procedural difficulties which beset the presentation of Dr. Oppenheimer's case. Weeks before the hearing commenced we asked you and the Commission's general counsel for much information which we thought relevant to our case but which was denied us—documents and minutes concerning Dr. Oppenheimer's 1947 clearance and a variety of other material. Much of this information did come out in the hearings but usually only in the course of cross-examination when calculated to cause the maximum surprise and confusion and too late to assist us in the orderly presentation of our case. Some of the information which was denied to us before the hearing was declassified at the moment of cross-examination or shortly before and was made available to us only during cross-examination or after.

It is true that Dr. Oppenheimer was accorded the privilege of reexamining, prior to the hearings, reports and other material in the preparation of which he had participated. But he was not given access to the broad range of material actually used and disclosed for the first time at the hearings by the Commission's special counsel who had been retained for the case. And of course Dr. Oppenheimer was not given access to the various documents which, according to the Board's report "under governmental necessity cannot be disclosed, such as reports of the Federal Bureau of Investigation" (p. 31).

The voluminous nature of this undisclosed material appears from the Board's report. It notes that in our hearings the Board heard 40 witnesses and compiled over 3,000 pages of testimony; and we then learn from the report that "in addition" the Board has "read the same amount of file material" (p. 2). We can only speculate as to the contents of this "file material." We cannot avoid the further speculation as to how much of this material might have been disclosed to Dr. Oppenheimer in the interests of justice without any real injury to the security interests of the Government if established rules of exclusion, which the Board felt bound to apply and we to accept, had not stood in the way.

Having in mind the difficulties and handicaps which have been recounted above, we urge upon the Commission as strongly as possible the following:

(1) That in weighing the testimony, and particularly those portions where documents were produced on cross-examination in the manner described above, the Commission should constantly bear in mind how, under such circumstances, the natural fallibility of memory may easily be mistaken for disingenuousness;

(2) That in the consideration of documentary material not disclosed to Dr. Oppenheimer, the Commission should be ever conscious of the unreliability of ex parte reports which have never been seen by Dr. Oppenheimer or his counsel or tested by cross-examination; and

(3) That if in the course of the Commission's deliberations the Commission should conclude that any hitherto undisclosed documents upon which it intends to rely may be disclosed to us without injury to what may be thought to be overriding interests of the National Government, they should be so disclosed before any final decision is made.

* * * * * * *

Our final observation has to do with the general structure of the Board's report, and with what has been omitted from it which we feel the Commission should put in the forefront of its consideration if it is to view this case in anything like the true perspective of history—a history through which Dr. Oppenheimer has lived and which in part he has helped to create.

The Board's opinion, as required by the AEC Procedures, makes specific findings on each allegation of "derogatory information" contained in your letter of December 23, 1953. These findings, which are placed at the beginning of the report, are not thereafter, except in Dr. Evans' dissenting opinion, considered in the context of Dr. Oppenheimer's life as a whole. Dr. Oppenheimer's letter to you of March 4, 1954, in answer to yours of December 23, 1953, stated at the outset that "the items of so-called 'derogatory information' set forth in your letter cannot fairly be understood except in the context of my life and work."

In his letter Dr. Oppenheimer tried to describe the derogatory information about him in that context. There is, in fact, little in the Board's findings that did not appear from what Dr. Oppenheimer volunteered about himself in his original letter to you. Over and above that, he gave a picture of his life and times without which the items of derogatory information cannot fairly be understood—a picture to which many witnesses added who had known him intimately and had worked side by side with him in the positions of high responsibility which the Government, first in war and then in peace and then in the cold war, successively devolved upon him. This picture, which is glimpsed in Dr. Evans' vivid opinion, does not appear at all in the main body of the report, nor is any mention made of the many witnesses who testified at his behest. Some of these we have already named above. The others were Mervin J. Kelly, president of Bell Telephone Laboratories; Gen. Leslie R. Groves; T. Keith Glennan, president of Case Institute of Technology; Karl T. Compton, retired president of Massachusetts Institute of Technology; Col. John Lansdale, Jr., wartime senior security officer for the atomic-bomb project; James B. Fisk, vice president of Bell Telephone Laboratories; Prof. Jerrold B. Zacharias, of MIT; Oliver E. Buckley, of the GAC, retired chairman of Bell Telephone Laboratories; Gen. Frederick H. Osborn; Ambassador George F. Kennan; Prof Walter G. Whitman, of MIT; Harry A. Winne, former vice president of General Electric, chairman of the Defense Department's panel on atomic energy; Dr. Vannevar Bush, president of the Carnegie Institution; Sumner T. Pike, former Atomic Energy Commissioner; David E. Lilienthal, former Chairman of the AEC; Lee A. DuBridge, president of California Institute of Technology; James R. Killian, Jr., president of MIT; Prof. Norman F. Ramsey, Jr., of Harvard; Maj. Gen. James McCormack, Jr., vice commander of the Air Research and Development Command; John J. McCloy, chairman of the Chase National Bank; Prof. John von Neumann of the Institute for Advanced Study; Prof. John H. Manley of the University of Washington, former secretary of the GAC; and Prof. Charles C. Lauritsen of the California Institute of Technology. The witnesses included 10 former and present members of the General Advisory Committee, and 5 former Atomic Energy Commissioners.

Since all these witnesses testified to Dr. Oppenheimer's loyalty and since the Board unanimously found him to be loyal, the omission of their names from the report was understandable; but we mention them here and direct the Commission's attention particularly to their testimony, to which we hope to refer in brief and argument, because they did much more than vouch for Dr. Oppenheimer's loyalty. These, and the other men previously mentioned, were not ordinary character witnesses who tell about a man's reputation in the community. Every one of them had served with Dr. Oppenheimer, either at Los Alamos or on the many governmental boards and committees to which he was later appointed. They saw him on the job and off the job, and in their varied testimony about their contacts with him over many years they helped to fill in the picture of the "man himself" which the Atomic Energy Commission, in its 1948 opinion in Dr. Frank Graham's case, said should be considered in determining whether an individual is a good or bad security risk.

Because we believe that the "man himself" can only be understood, and therefore fairly judged, by the closest attention to the testimony of those who have known him and worked intimately with him, as well as to his own testimony, we are particularly hopeful that the Commission will permit us to file a brief and to be heard.

In closing this letter we wish to record our appreciation of the patience and consideration accorded to Dr. Oppenheimer and his counsel by Mr. Gray, Mr. Morgan and Dr. Evans throughout the nearly 4 weeks of hearings, and our recognition of the sacrifices which they made in the public interest in assuming the long and arduous task assigned to them.

Mr. John W. Davis has authorized me to say that he joins in this letter and will join in the brief.

Very truly yours,

LLOYD K. GARRISON.

cc Adm. Lewis L. Strauss, Chairman.
   Mr. Joseph Campbell.
   Mr. Thomas E. Murray.
   Dr. Henry D. Smyth.
   Mr. Eugene M. Zuckert.

(NOTE.—No deletions were made in reproducing this letter. Ellipses appeared in original document.)

LETTER FROM GENERAL MANAGER K. D. NICHOLS, UNITED STATES ATOMIC ENERGY COMMISSION, TO DR. OPPENHEIMER'S ATTORNEYS CONCERNING PROCEDURES IN THE MATTER OF DR. J. ROBERT OPPENHEIMER

UNITED STATES ATOMIC ENERGY COMMISSION,
Washington 25, D. C., June 3, 1954.

LLOYD K. GARRISON, ESQ.,
  Paul, Weiss, Rifkind, Wharton & Garrison,
  575 Madison Avenue, New York 22, N. Y.

DEAR MR. GARRISON: This will acknowledge receipt of your letter of June 1, 1954, in which you refer to my letter of May 28, 1954, to Dr. Oppenheimer.

In your letter you refer to the fact that review by a Personnel Security Review Board would "entail further delay." As you are fully aware, my letter to Dr. Oppenheimer of December 23, 1953, stated that Dr. Oppenheimer would have 30 days in which to submit a written answer. In your letter to me of January 20, 1954, you asked for an extension of this time to and including February 23. In my letter to you of January 27 this extension was granted. In your letter to me of February 19 you confirmed your telephone request for an extension of time from February 23 to March 1. In my letter to you of February 25 this request was granted. In my letter to you of March 3 I confirmed a telephone conversation with you of the previous day in which the time for your answer was extended to March 5, 1954. I know of no delays other than those I have referred to above.

The Atomic Energy Commission's published Security Clearance Procedures, a copy of which was furnished to Dr. Oppenheimer with my letter of December 23, 1953, provide that if the individual requests a review of his case by the Personnel Security Review Board, he shall have 20 days within which to submit a brief in support of his contentions. The procedures further prescribe that oral argument before the Personnel Security Review Board may be had only in the discretion of that Board. The procedures make no provision for submission of a brief, or for oral argument, when the case then comes to the General Manager for final determination.

Since Dr. Oppenheimer has waived his right to review by the Personnel Security Review Board, our procedures do not contemplate any further presentation by Dr. Oppenheimer, either oral or written. The written brief which you mention in your letter, and which we understand will be received by the Commission on June 7, will be given very careful consideration. The Commission does not feel that it can accede to your suggestion that there be oral argument as well.

Your letter states that Dr. Oppenheimer was not given access to material "actually used and disclosed for the first time at the hearings." I should like to remind you that in my letter to you of February 12, 1954, I stated:

"We have also indicated to you our willingness to make available to you, insofar as our facilities permit us to do so, documents which you reasonably believe are relevant to the matters in issue. You will appreciate, however, that the Commission must in fulfillment of its responsibilities for the maintenance of the common defense and security reserve the right to decide whether particular documents to which you request access are relevant and whether your access to such documents or parts thereof would be consistent with the national interest."

In a letter dated February 19, 1954, to you from William Mitchell, our General Counsel, it was stated:

"Furthermore, Dr. Oppenheimer will be given an opportunity to read the minutes of the GAC meeting of October 1949, by coming to the Commission's offices at his convenience. Arrangements for this purpose may be made with Mr. Nichols."

Dr. Oppenheimer did not avail himself of this opportunity.

Furthermore, you were given an opportunity by the Commission, prior to the hearings, to request security clearance for yourself. You will recall that the question of your clearance was discussed with you on January 18, and that on January 27 I wrote you stating that the Commission was prepared to process your clearance as expeditiously as possible upon receipt of your personnel security questionnaire. On February 3, 1954, you wrote me stating that you had decided not to request clearance. In your letter to me of March 26, 1954, you stated that you had finally decided that one of Dr. Oppenheimer's counsel should be cleared. The Board first convened on April 5. At the time your March 26 letter was received, it was not possible to complete the necessary background investigation, which is a prerequisite to clearance, until after the hearings had been concluded and the Board's report had been submitted. At the hearings themselves, whenever any document was introduced which still bore a security classification, Dr. Oppenheimer himself was permitted to read the document. Since his counsel had not been cleared, they were not, and could not be, given access to such classified documents. I know of no other material, considered by the Gray Board, which could be made available to Dr. Oppenheimer at the present time.

Sincerely yours,

K. D. NICHOLS,
General Manager.

RECOMMENDATIONS OF THE GENERAL MANAGER
TO THE
UNITED STATES ATOMIC ENERGY COMMISSION
IN THE
MATTER OF DR. J. ROBERT OPPENHEIMER

UNITED STATES ATOMIC ENERGY COMMISSION,
Washington 25, D. C., June 12, 1954

Memorandum for: Mr. Strauss, Dr. Smyth, Mr. Murray, Mr. Zuckert, Mr. Campbell.

Subject: Dr. J. Robert Oppenheimer.

GENERAL

On December 23, 1953, Dr. J. Robert Oppenheimer was notified that his security clearance had been suspended, and informed of his right to a hearing under AEC procedures. By telegram dated January 29, 1954, Dr. Oppenheimer requested that he be afforded a hearing and on March 4, 1954, after requesting and receiving three extensions of time, he submitted his answer to my letter of December 23, 1953.

Mr. Gordon Gray, Mr. Thomas A. Morgan and Dr. Ward V. Evans agreed to serve as members of the Personnel Security Board to hear Dr. Oppenheimer's case. The Board submitted its findings and recommendation to me on May 27, 1954. A majority of the Board recommended against reinstatement of clearance, Dr. Evans dissenting.

On May 28, 1954, I notified Dr. Oppenheimer of the recommendation of the Personnel Security Board and forwarded to him a copy of the Board's findings and recommendation. I informed Dr. Oppenheimer of his right to request review of his case by the Personnel Security Review Board and informed him that upon full consideration of the entire record in the case, including the recommendation of the Personnel Security Review Board in the event he requested review by that Board, I would submit to the Commission my recommendation as to whether or not his clearance should be reinstated. I also informed him that the final determination would be made by the Commission.

By letter of June 1, 1954, Dr. Oppenheimer waived his right to a review of his case by the Personnel Security Review Board and requested an immediate consideration of his case by the Commission.

FACTORS CONSIDERED

In making my findings and determination I have considered the question whether a security risk is involved in continued clearance of Dr. Oppenheimer I have taken into account his contributions to the United States atomic energy program and in addition I have, in accordance with AEC procedures, considered the effect which denial of security clearance would have upon the program.

DR. OPPENHEIMER'S WORLD WAR II CONTRIBUTION

Dr. Oppenheimer has been intimately associated with the atomic energy program virtually from its inception. He participated in early weapons research and was selected as the wartime Director of the Los Alamos Laboratory. As district engineer of the wartime Manhattan Engineer District, I was keenly aware of the contribution he made to the initial development of the atomic bomb. His leadership and direction of the Los Alamos weapons program were outstanding; his contributions leading to a successful atomic weapon have properly received worldwide acknowledgment and acclaim.

WORLD WAR II CLEARANCE 1943

As deputy district engineer of the Manhattan District, I was also aware of the circumstances, which have been brought out in the record, surrounding Dr. Oppenheimer's appointment as head of the Los Alamos Laboratory, and his subsequent clearance. He was selected in spite of the fact that he was considered a "calculated risk." He would not have been chosen had he not been considered virtually indispensable to the atomic bomb program. After he was chosen, as

General Groves testified before the Board, Dr. Oppenheimer probably would not have been cleared had he not already been thoroughly steeped in knowledge of weapons research and had he not been considered absolutely essential.

Security officers opposed the clearance of Dr. Oppenheimer and it was not until July of 1943, after he had participated in the program for many months, that the decision to clear him was made by General Groves. I personally signed the directive advising the commanding officer at Los Alamos that there was no objection to Dr. Oppenheimer's employment.

The Manhattan District had one mandate—to build an atomic bomb as quickly as possible. Fears that Germany would build an atomic weapon first and possibly win the war thereby spurred the Manhattan District in what was felt to be a race against the German effort. Communist Russia was also fighting Germany at that time.

General Groves testified before the Board that he did not regret having made the decision to clear Dr. Oppenheimer in consideration of all of the circumstances which confronted him in 1943 but that under the present requirements of the Atomic Energy Act, as he interprets them, he would not clear Dr. Oppenheimer today.

BASIS FOR CLEARANCE UNDER ATOMIC ENERGY ACT AND EXECUTIVE ORDER 10450

In this case as well as in all personnel security cases, the AEC in granting or reinstating a clearance must determine that the common defense and security will not be endangered. Under the Atomic Energy Act, such determination must be made on the basis of the character, associations, and loyalty of the individual concerned. Thus, a finding of loyalty in any given case does not suffice; substantial deficiency in any one of the three factors—character, associations, or loyalty—may prevent the determination that permitting such person to have access to restricted data will not endanger the common defense or security.

In addition, the criteria set up by Executive Order 10450 must be considered. This order requires that a program be established to insure that the retention in employment of any employee is clearly consistent with the interests of national security.

SECURITY FINDINGS

I have reviewed the entire record of the case, including the files, the transcript of the hearing, the findings and recommendation of the Personnel Security Board, and the briefs filed by Dr. Oppenheimer's attorneys on May 17, 1954, and June 7, 1954, and have reached the conclusion that to reinstate the security clearance of Dr. Oppenheimer would not be clearly consistent with the interests of national security and would endanger the common defense and security.

I concur with the findings and recommendation of the majority of the Personnel Security Board and submit them in support of this memorandum. In addition, I refer in particular to the following considerations:

1. *Dr. Oppenheimer's Communist activities.*—The record contains no direct evidence that Dr. Oppenheimer gave secrets to a foreign nation or that he is disloyal to the United States. However, the record does contain substantial evidence of Dr. Oppenheimer's association with Communists, Communist functionaries, and Communists who did engage in espionage. He was not a mere "parlor pink" or student of communism as a result of immaturity and intellectual curiosity, but was deeply and consciously involved with hardened and militant Communists at a time when he was a man of mature judgment.

His relations with these hardened Communists were such that they considered him to be one of their number. He admits that he was a fellow traveler, and that he made substantial cash contributions direct to the Communist Party over a period of 4 years ending in 1942. The record indicates that Dr. Oppenheimer was a Communist in every respect except for the fact that he did not carry a party card.

These facts raise serious questions as to Dr. Oppenheimer's eligibility for clearance reinstatement.

It is suggested that Dr. Oppenheimer has admitted many of the facts concerning his past association with Communists and the Communist Party. Whether this be true or not, it appears to me that Dr. Oppenheimer's admissions in too many cases have followed, rather than preceded, investigation which developed the facts. It appears that he is not inclined to disclose the facts spontaneously, but merely to confirm those already known. I find no great virtue in such a plea of guilt; certainly it does not cause me to dismiss Dr. Oppenheimer's past associations as matters of no consequence simply on the ground that he has admitted them.

2. *The Chevalier incident.*—Dr. Oppenheimer's involvement in the Chevalier incident, and his subsequent conduct with respect to it, raise grave questions of security import.

If in 1943, as he now claims to have done, he knowingly and willfully made false statements to Colonel Pash, a Federal officer, Dr. Oppenheimer violated what was then section 80, title 18, of the United States Code;[1] in other words if his present story is true then he admits he committed a felony in 1943. On the other hand, as Dr. Oppenheimer admitted on cross-examination, if the story Dr. Oppenheimer told Colonel Pash was true, it not only showed that Chevalier was involved in a criminal espionage conspiracy, but also reflected seriously on Dr. Oppenheimer himself.

After reviewing both the 16-page transcript (as accepted by the Board) of the interview between Dr. Oppenheimer and Colonel Pash on August 26, 1943, and recent testimony before the Board, it is difficult to conclude that the detailed and circumstantial account given by Dr. Oppenheimer to Colonel Pash was false and that the story now told by Dr. Oppenheimer is an honest one. Dr. Oppenheimer's story in 1943 was most damaging to Chevalier. If Chevalier was Dr. Oppenheimer's friend and Dr. Oppenheimer, as he now says, believed Chevalier to be innocent and wanted to protect him, why then would he tell such a complicated false story to Colonel Pash? This story showed that Chevalier was not innocent, but on the contrary was deeply involved in an espionage conspiracy. By the same token, why would Dr. Oppenheimer tell a false story to Colonel Pash which showed that he himself was not blameless? Is it reasonable to believe a man will deliberately tell a lie that seriously reflects upon himself and his friend, when he knows that the truth will show them both to be innocent?

It is important to remember also that Dr. Oppenheimer did not give his present version of the story until 1946, shortly after he had learned from Chevalier what Chevalier himself had told the FBI about the incident in question. After learning of this from Chevalier, Dr. Oppenheimer changed his story to conform to that given to the FBI by Chevalier.

From all of these facts and circumstances, it is a fair inference that Dr. Oppenheimer's story to Colonel Pash and other Manhattan District officials was substantially true and that his later statement on the subject to the FBI, and his recent testimony before the Personnel Security Board, were false.

Executive Order 10450 provides:

"SECTION 8 (a). The investigations conducted pursuant to this order shall be designed to develop information as to whether the employment or retention in employment in the Federal service of the person being investigated is clearly consistent with the interests of the national security. Such information shall relate, but shall not be limited, to the following:

(1) Depending on the relation of the Government employment to the national security:

    (i) Any behavior, activities, or associations which tend to show that the individual is not reliable or trustworthy.

    (ii) Any deliberate misrepresentations, falsifications, or omission of material facts.

    (iii) Any criminal, infamous, dishonest, immoral, or notoriously disgraceful conduct, habitual use of intoxicants to excess, drug addiction, or sexual perversion."

In my opinion, Dr. Oppenheimer's behavior in connection with the Chevalier incident shows that he is not reliable or trustworthy; his own testimony shows that he was guilty of deliberate misrepresentations and falsifications either in his interview with Colonel Pash or in his testimony before the Board; and such misrepresentations and falsifications constituted criminal, * * * dishonest * * * conduct.

Further, the significance of the Chevalier incident combined with Dr. Oppenheimer's conflicting testimony from 1943 to 1954 in regard to it were not, of course, available in whole to General Groves in 1943, nor was the complete record on the Chevalier incident considered by the Atomic Energy Commission

---

[1] 18 U. S. Code, sec. 80, provides in pertinent part: "Whoever * * * shall knowingly or willfully falsify or conceal or cover up by any trick, scheme, or device a material fact, or make or cause to be made any false or fraudulent statement or representations * * * in any matter within the jurisdiction or agency of the United States * * * shall be fined not more than $10,000 or imprisoned not more than 10 years, or both."

in 1947. Consideration of the complete record plus a cross-examination of Dr. Oppenheimer under oath were not accomplished by anyone prior to the personnel Security Board hearing in 1954.

3. *Dr. Oppenheimer's veracity.*—A review of the record reveals other instances which raise a question as to the credibility of Dr. Oppenheimer in his appearance before the Personnel Security Board and as to his character and veracity in general.

(*a*) The record suggests a lack of frankness on the part of Dr. Oppenheimer in his interviews with the FBI. It appears that during this hearing he recollected details concerning Communist meetings in the San Francisco area which he did not report in previous interviews with the FBI.

(*b*) Dr. Oppenheimer told the FBI in 1950 that he did not know know that Joseph Weinberg was a Communist until it became a matter of public knowledge. When confronted with the transcript of his interview with Colonel Lansdale on September 12, 1943, he admitted that he had learned prior to that date that Weinberg was a Communist.

(*c*) It is clear from the record that Dr. Oppenheimer was a great deal more active in urging the deferment of Rossi Lomanitz and his retention on the atom bomb project than he said he was in his answer to my letter of December 23, 1953. Furthermore, Dr. Oppenheimer testified that if he had known that Lomanitz was a Communist he would not have written the letter to Colonel Lansdale of the Manhattan District on October 19, 1943, supporting Lomanitz' services for the project. However, the record reflects that Dr. Oppenheimer told Colonel Lansdale of the Manhattan District on September 12, 1943, that he had learned that Lomanitz was a Communist.

(*d*) Dr. Oppenheimer admitted in his testimony before the Board that in 1949 he wrote a letter to a newspaper which might have misled the public concerning his testimony before the House Un-American Activities Committee on Dr. Bernard Peters. He testified that an earlier article in the newspaper which summarized his testimony was accurate, yet the effect of his published letter was to repudiate the earlier article.

(*e*) Dr. Oppenheimer in his answer to my letter of December 23, 1953, and in his testimony before the Board with respect to the H-bomb program undertook to give the impression that in 1949 he and the GAC merely opposed a so-called "crash" program. It is quite clear from the record, however, that the position of the majority of the GAC, including Dr. Oppenheimer, was that a thermonuclear weapon should never be produced, and that the United States should make an unqualified commitment to this effect. In discussing the building of neutron-producing reactors, a majority of the GAC, including Dr. Oppenheimer, expressed the opinion that "the super program itself should not be undertaken and that the Commission and its contractors understand that construction of neutron-producing reactors is not intended as a step in the super program." The testimony of Dr. Oppenheimer viewed in light of the actual record certainly furnished adequate basis for the majority of the Board not believing that Dr. Oppenheimer was entirely candid with them on this point.

(*f*) Dr. Oppenheimer testified before the Board that the GAC was unanimous in its basic position on the H-bomb. He specifically said that Dr. Seaborg had not expressed his views and that there was no communication with him. It should be noted that the statement that "there was no communication with him" was volunteered by Dr. Oppenheimer in his testimony on cross-examination before the Board. However, Dr. Oppenheimer received a letter from Dr. Seaborg, expressing his views, prior to the October 29, 1949, GAC meeting.

4. *Dr. Oppenheimer's continued associations after World War II.*—Dr. Oppenheimer has continued associations which raise a serious question as to his eligibility for clearance. He has associated with Chevalier on a rather intimate basis as recently as December 1953, and at that time lent his name to Chevalier's dealings with the United States Embassy in Paris on a problem which, according to Dr. Oppenheimer, involved Chevalier's clearance. Since the end of World War II he has been in touch with Bernard Peters, Rossi Lomanitz, and David Bohm under circumstances which, to say the least, are disturbing.

5. *Obstruction and disregard of security.*—Dr. Oppenheimer's actions have shown a consistent disregard of a reasonable security system. In addition to the Chevalier incident, he has refused to answer questions put to him by security officers concerning his relationships and knowledge of particular individuals whom he knew to be Communists; and he has repeatedly exercised an arrogance of his own judgment with respect to the loyalty and reliability of his associates and his own conduct which is wholly inconsistent with the obligations necessarily

imposed by an adequate security system on those who occupy high positions of trust and responsibility in the Government.

FINDING OF SECURITY RISK IS NOT BASED ON DR. OPPENHEIMER'S OPINIONS

Upon the foregoing considerations relating to the character and associations of Dr. Oppenheimer, I find that he is a security risk. In making this finding I wish to comment on the item of derogatory information contained in my letter of December 23, 1953, which relates to the hydrogen bomb and, in particular, which alleged that:

"* * * It was further reported that even after it was determined, as a matter of national policy, to proceed with development of a hydrogen bomb, you continued to oppose the project and declined to cooperate fully in the project. It was further reported you departed from your proper role as an adviser to the Commission by causing the distribution, separately and in private, to top personnel at Los Alamos of the majority and minority reports of the General Advisory Committtee on development of the hydrogen bomb for the purpose of trying to turn such top personnel against the development of the hydrogen bomb. It was further reported that you were instrumental in persuading other outstanding scientists not to work on the hydrogen bomb project, and that the opposition to the hydrogen bomb, of which you are the most experienced, most powerful and most effective member, has definitely slowed down its development."

It should be emphasized that at no time has there been any intention on my part or the Board's to draw in question any honest opinion expressed by Dr. Oppenheimer. Technical opinions have no security implications unless they are reflections of sinister motives. However, in view of Dr. Oppenheimer's record coupled with the preceding allegation concerning him, it was necessary to submit this matter for the consideration of the Personnel Security Board in order that the good faith of his technnical opinions might be determined. The Board found that, following the President's decision, Dr. Oppenheimer did not show the enthusiastic support for the program which might have been expected of the chief atomic adviser to the Government under the circumstances; that, had he given his enthusiastic support to the program, a concerted effort would have been initiated at an earlier date, and that, whatever the motivation, the security interests of the United States were affected. In reviewing the record I find that the evidence establishes no sinister motives on the part of Dr. Oppenheimer in his attitude on the hydrogen bomb, either before or after the President's decision. I have considered the testimony and the record on this subject only as evidence bearing upon Dr. Oppenheimer's veracity. In this context I find that such evidence is disturbing.

DR. OPPENHEIMER'S VALUE TO ATOMIC ENERGY OR RELATED PROGRAMS

In addition to determining whether or not Dr. Oppenheimer is a security risk, the General Manager should determine the effect which denial of security clearance would have upon the atomic energy or related programs. In regard to Dr. Oppenheimer's net worth to atomic energy projects, I believe, first, that through World War II he was of tremendous value and absolutely essential. Secondly, I believe that since World War II his value to the Atomic Energy Commission as a scientist or as a consultant has declined because of the rise in competence and skill of other scientists and because of his loss of scientific objectivity probably resulting from the diversion of his efforts to political fields and matters not purely scientific in nature. Further, it should be pointed out that in the past 2 years since he has ceased to be a member of the General Advisory Committee, his services have been utilized by the Atomic Energy Commission on the following occasions only:

October 16 and 17, 1952.
September 1 and 2, 1953.
September 21 and 22, 1953.

I doubt that the Atomic Energy Commission, even if the question of his security clearance had not arisen, would have utilized his services to a markedly greater extent during the next few years. I find, however, that another agency, the Science Advisory Committee, Office of Defense Mobilization, has stated in a letter dated June 4, 1954, signed by Dr. L. A. DuBridge, Chairman, and addressed to Chairman Strauss, that:

"* * * It is, therefore, of great importance to us that Dr. Oppenheimer's 'Q' clearance be restored. This is especially true since our Committee is planning to undertake during the coming months an intensive study of important items related to national security on which Dr. Oppenheimer's knowledge and counsel will be of very critical importance."

Dr. DuBridge further stated that:

"* * * His value is, it seems to me, so enormous as to completely overbalance and override the relatively trivial risks* which the Personnel Security Board reports. In other words, the net benefits to national security will be far greater if Dr. Oppenheimer's clearance is restored than if it is terminated. Even though he served the Government in no other capacity than as a member of the Science Advisory Committee of the Office of Defense Mobilization, the above statements will, I am confident, be true beyond question."

Other Government agencies may also desire to use Dr. Oppenheimer's services if he were to be cleared. In addition, contractors and study groups involved in atomic power activities undoubtedly will from time to time, as one or two have already indicated, desire to clear Dr. Oppenheimer for consulting work.

Dr. Oppenheimer could of course make contributions in all these fields, but he is far from being indispensable.

CONCLUSION

I have conscientiously weighed the record of Dr. Oppenheimer's whole life, his past contributions, and his potential future contributions to the Nation against the security risk that is involved in his continued clearance. In addition, I have given consideration to the nature of the cold war in which we are engaged with communism and Communist Russia and the horrible prospects of hydrogen bomb warfare if all-out war should be forced upon us. From these things a need results to eliminate from classified work any individuals who might endanger the common defense or security or whose retention is not clearly consistent with the interests of national security.

Dr. Oppenheimer's clearance should not be reinstated.

K. D. NICHOLS,
General Manager.

---

*It should be noted that Dr. DuBridge to my knowledge has never had access to the complete file or the transcript of the hearing.

DECISION AND OPINIONS
OF THE
UNITED STATES ATOMIC ENERGY COMMISSION
IN THE
MATTER OF DR. J. ROBERT OPPENHEIMER

STATEMENT BY THE ATOMIC ENERGY COMMISSION

[For immediate release June 29, 1954]

UNITED STATES ATOMIC ENERGY COMMISSION,
Washington 25, D. C.

The Atomic Energy Commission announced today that it had reached a decision in the matter of Dr. J. Robert Oppenheimer.

The Commission by a vote of 4 to 1 decided that Dr. Oppenheimer should be denied access to restricted data. Commissioners Strauss, Murray, Zuckert, and Campbell voted to deny clearance for access to restricted data, and Commissioner Smyth voted to reinstate clearance for access to restricted data. Messrs. Strauss, Zuckert, and Campbell signed the majority opinion; Mr. Murray concurred with the majority decision in a separate opinion. Dr. Smyth supported his conclusion in a minority opinion.

Certain members of the Commission issued additional statements in support of their conclusions. These opinions and statements are attached.

UNITED STATES ATOMIC ENERGY COMMISSION,
Washington 25, D. C., June 29, 1954.

The issue before the Commission is whether the security of the United States warrants Dr. J. Robert Oppenheimer's continued access to restricted data of the Atomic Energy Commission. The data to which Dr. Oppenheimer has had until recently full access include some of the most vital secrets in the possession of the United States.

Having carefully studied the pertinent documents—the transcript of the hearings before the Personnel Security Board (Gray Board), the findings and recommendation of the Board, the briefs of Dr. Oppenheimer's counsel, and the findings and recommendation of the General Manager—we have concluded that Dr. Oppenheimer's clearance for access to restricted data should not be reinstated.

The Atomic Energy Act of 1946 lays upon the Commissioners the duty to reach a determination as to "the character, associations, and loyalty" of the individuals engaged in the work of the Commission. Thus, disloyalty would be one basis for disqualification, but it is only one. Substantial defects of character and imprudent and dangerous associations, particularly with known subversives who place the interests of foreign powers above those of the United States, are also reasons for disqualification.

On the basis of the record before the Commission, comprising the transcript of the hearing before the Gray Board as well as reports of Military Intelligence and the Federal Bureau of Investigation, we find Dr. Oppenheimer is not entitled to the continued confidence of the Government and of this Commission because of the proof of fundamental defects in his "character."

In respect to the criterion of "associations," we find that his associations with persons known to him to be Communists have extended far beyond the tolerable limits of prudence and self-restraint which are to be expected of one holding the high positions that the Government has continuously entrusted to him since 1942. These associations have lasted too long to be justified as merely the intermittent and accidental revival of earlier friendships.

Neither in the deliberations by the full Commission nor in the review of the Gray Board was importance attached to the opinions of Dr. Oppenheimer as they bore upon the 1949 debate within the Government on the question of whether the United States should proceed with the thermonuclear weapon program. In this debate, Dr. Oppenheimer was, of course, entitled to his opinion.

The fundamental issues here are apart from and beyond this episode. The history of their development is as follows:

On December 23, 1953, Dr. Oppenheimer was notified that his security clearance had been suspended, and he was provided with the allegations which had brought his trustworthiness into question. He was also furnished with a copy of the Atomic Energy Commission's security clearance procedures, and was informed of his right to a hearing under those procedures. By telegram dated January 29, 1954, Dr. Oppenheimer requested a hearing. On March 4, 1954, after requesting and receiving three extensions of time, he submitted his answer to the letter of December 23, 1953. On March 15, 1954, Dr. Oppenheimer was informed that Mr. Gordon Gray, Mr. Thomas A. Morgan, and Dr. Ward V. Evans would conduct the hearing.

The hearing before the Gray Board commenced on April 12, 1954, and continued through May 6, 1954. Dr. Oppenheimer was represented by four lawyers. He was present to confront all witnesses; he had the opportunity to cross-examine all witnesses; his counsel made both oral and written argument to the Board.

The Board submitted its findings and recommendation to the General Manager of the Commission on May 27, 1954. A majority of the Board recommended against reinstatement of clearance, Dr. Evans dissenting.

Dr. Oppenheimer had full advantage of the security procedures of the Commission. In our opinion he had a just hearing.

On May 28, 1954, the General Manager notified Dr. Oppenheimer of the adverse recommendation of the Personnel Security Board and forwarded to him a copy of the Board's findings and recommendation. The General Manager informed Dr. Oppenheimer of his right to request review of his case by the Personnel Security Review Board. Dr. Oppenheimer was also informed that upon consideration of the record in the case—including the recommendation of the Personnel Security Review Board in the event review by that Board was requested—the General Manager would submit to the Commission his own recommendation as to whether or not clearance should be reinstated and that the Commission would thereafter make the final determination.

By letter of June 1, 1954, Dr. Oppenheimer waived his right to a review of his case by the Personnel Security Review Board. He requested immediate consideration of his case by the Commission. On June 7, 1954, his counsel submitted a written brief to the Commission. The General Manager reviewed the testimony and the findings and recommendation of the Gray Board and the briefs; his conclusion that Dr. Oppenheimer's clearance should not be reinstated was submitted to the Commission on June 12, 1954.

Prior to these proceedings, the derogatory information in Government files concerning Dr. Oppenheimer had never been weighed by any board on the basis of sworn testimony.

The important result of these hearings was to bring out significant information bearing upon Dr. Oppenheimer's character and associations hitherto unknown to the Commission and presumably unknown also to those who testified as character witnesses on his behalf. These hearings additionally established as fact many matters which previously had been only allegations.

In weighing the matter at issue, we have taken into account Dr. Oppenheimer's past contributions to the atomic energy program. At the same time, we have been mindful of the fact that the positions of high trust and responsibility which Dr. Oppenheimer has occupied carried with them a commensurately high obligation of unequivocal character and conduct on his part. A Government official having access to the most sensitive areas of restricted data and to the innermost details of national war plans and weapons must measure up to exemplary standards of reliability, self-discipline, and trustworthiness. Dr. Oppenheimer has fallen far short of acceptable standards.

The record shows that Dr. Oppenheimer has consistently placed himself outside the rules which govern others. He has falsified in matters wherein he was charged with grave responsibilities in the national interest. In his associations he has repeatedly exhibited a willful disregard of the normal and proper obligations of security.

As to "character"

(1) Dr. Oppenheimer has now admitted under oath that while in charge of the Los Alamos Laboratory and working on the most secret weapon development for the Government, he told Colonel Pash a fabrication of lies. Colonel Pash was an officer of Military Intelligence charged with the duty of protecting the atomic-weapons project against spies. Dr. Oppenheimer told Colonel Pash in circumstantial detail of an attempt by a Soviet agent to obtain from him information about the work on the atom bomb. This was the Haakon Chevalier incident. In the hearings recently concluded, Dr. Oppenheimer under oath swears that the story he told Colonel Pash was a "whole fabrication and tissue of lies" (Tr., p. 149).

It is not clear today whether the account Dr. Oppenheimer gave to Colonel Pash in 1943 concerning the Chevalier incident or the story he told the Gray Board last month is the true version.

If Dr. Oppenheimer lied in 1943, as he now says he did, he committed the crime of knowingly making false and material statements to a Federal officer. If he lied to the Board, he committed perjury in 1954.

(2) Dr. Oppenheimer testified to the Gray Board that if he had known Giovanni Rossi Lomanitz was an active Communist or that Lomanitz had disclosed information about the atomic project to an unauthorized person, he would not have written to Colonel Lansdale of the Manhattan District the letter of October 19, 1943, in which Dr. Oppenheimer supported the desire of Lomanitz to return to the atomic project.

The record shows, however, that on August 26, 1943, Dr. Oppenheimer told Colonel Pash that he (Oppenheimer) knew that Lomanitz had revealed information about the project. Furthermore, on September 12, 1943, Dr. Oppenheimer told Colonel Lansdale that he (Oppenheimer) had previously learned for a fact that Lomanitz was a Communist Party member (Tr. pp. 118, 119, 128, 129, 143, 875).

(3) In 1943, Dr. Oppenheimer indicated to Colonel Lansdale that he did not know Rudy Lambert, a Communist Party functionary. In fact, Dr. Oppenheimer asked Colonel Lansdale what Lambert looked like. Now, however, Dr. Oppenheimer under oath has admitted that he knew and had seen Lambert at least half a dozen times prior to 1943; he supplied a detailed description of Lambert; he said that once or twice he had lunch with Lambert and Isaac Folkoff, another Communist Party functionary, to discuss his (Oppenheimer's) contributions to the Communist Party; and that he knew at the time that Lambert was an official in the Communist Party (Tr. pp. 139, 140, 877).

(4) In 1949 Dr. Oppenheimer testified before a closed session of the House Un-American Activities Committee about the Communist Party membership and activities of Dr. Bernard Peters. A summary of Dr. Oppenheimer's testimony subsequently appeared in a newspaper, the Rochester Times Union. Dr. Oppenheimer then wrote a letter to that newspaper. The effect of that letter was to contradict the testimony he had given a congressional committee (Tr. pp. 210–215).

(5) In connection with the meeting of the General Advisory Committee on October 29, 1949, at which the thermonuclear weapon program was considered, Dr. Oppenheimer testified before the Gray Board that the General Advisory Committee was "surprisingly unanimous" in its recommendation that the United States ought not to take the initiative at that time in a thermonuclear program. Now, however, under cross-examination, Dr. Oppenheimer testifies that he did not know how Dr. Seaborg (1 of the 9 members of Dr. Oppenheimer's committee) then felt about the program because Dr. Seaborg "was in Sweden, and there was no communication with him." On being confronted with a letter from Dr. Seaborg to him dated October 14, 1949—a letter which had been in Dr. Oppenheimer's files—Dr. Oppenheimer admitted having received the letter prior to the General Advisory Committee meeting in 1949. In that letter Dr. Seaborg said: "Although I deplore the prospects of our country putting a tremendous effort into this, I must confess that I have been unable to come to the conclusion that we should not." Yet Dr. Seaborg's view was not mentioned in Dr. Oppenheimer's report for the General Advisory Committee to the Commission in October 1949. In fact the existence of this letter remained unknown to the Commission until it was disclosed during the hearings (Tr. pp. 233, 237–241).

(6) In 1950, Dr. Oppenheimer told an agent of the Federal Bureau of Investigation that he had not known Joseph Weinberg to be a member of the Communist Party until that fact become public knowledge. Yet on September 12, 1943, Dr. Oppenheimer told Colonel Lansdale that Weinberg was a Communist Party member (Tr., p. 875).

The catalog does not end with these six examples. The work of Military Intelligence, the Federal Bureau of Investigation, and the Atomic Energy Commission—all, at one time or another have felt the effect of his falsehoods, evasions, and misrepresentations.

Dr. Oppenheimer's persistent and willful disregard for the obligations of security is evidenced by his obstruction of inquiries by security officials. In the Chevalier incident, Dr. Oppenheimer was questioned in 1943 by Colonel Pash, Colonel Lansdale, and General Groves about the attempt to obtain information from him on the atomic bomb project in the interest of the Soviet Government. He had waited 8 months before mentioning the occurrence to the proper authorities. Thereafter for almost 4 months Dr. Oppenheimer refused to name the individual who had approached him. Under oath he now admits that his refusal to name the individual impeded the Government's investigation of espionage. The record shows other instances where Dr. Oppenheimer has refused to answer inquiries of Federal officials on security matters or has been deliberately misleading.

As to "associations"

"Associations" is a factor which, under the law, must be considered by the Commission. Dr. Oppenheimer's close association with Communists is another part of the pattern of his disregard of the obligations of security.

Dr. Oppenheimer, under oath, admitted to the Gray Board that from 1937 to at least 1942 he made regular and substantial contributions in cash to the Communist Party. He has admitted that he was a "fellow traveler" at least until 1942. He admits that he attended small evening meetings at private homes at which most, if not all, of the others present were Communist Party members. He was in contact with officials of the Communist Party, some of whom had been engaged in espionage. His activities were of such a nature that these Communists looked upon him as one of their number.

However, Dr. Oppenheimer's early Communist associations are not in themselves a controlling reason for our decision.

They take on importance in the context of his persistent and continuing association with Communists, including his admitted meetings with Haakon Chevalier in Paris as recently as last December—the same individual who had been intermediary for the Soviet Consulate in 1943.

On February 25, 1950, Dr. Oppenheimer wrote a letter to Chevalier attempting "to clear the record with regard to your alleged involvement in the atom business." Chevalier used this letter in connection with his application to the State Department for a United States passport. Later that year Chevalier came and stayed with Dr. Oppenheimer for several days at the latter's home. In December 1953, Dr. Oppenheimer visited with Chevalier privately on two occasions in Paris, and lent his name to Chevalier's dealings with the United States Embassy in Paris on a problem which, according to Dr. Oppenheimer, involved Chevalier's clearance. Dr. Oppenheimer admitted that today he has only a "strong guess" that Chevalier is not active in Communist Party affairs.

These episodes separately and together present a serious picture. It is clear that for one who has had access for so long to the most vital defense secrets of the Government and who would retain such access if his clearance were continued, Dr. Oppenheimer has defaulted not once but many times upon the obligations that should and must be willingly borne by citizens in the national service.

Concern for the defense and security of the United States requires that Dr. Oppenheimer's clearance should not be reinstated.

Dr. J. Robert Oppenheimer is hereby denied access to restricted data.

       Lewis L. Strauss, *Chairman.*
       Eugene M. Zuckert,* *Commissioner.*
       Joseph Campbell,* *Commissioner.*

Statement by Commissioner Zuckert

1. BASIS OF AGREEING TO DENY ACCESS

In subscribing to the majority decision and the substance of the Commission opinion, I have considered the evidence as a whole and no single factor as decisive. For example, Dr. Oppenheimer's early Communist associations by themselves would not have led me to my conclusion. The more recent connections, such as those with Lomanitz and Bohm, would not have been decisive. The serious 1943 incident involving Chevalier would not have been conclusive, although most disturbing and certainly aggravated by the continuation of the relationship between Chevalier and Dr. Oppenheimer. Individual instances of lack of veracity, conscious disregard of security considerations, and obstruction of proper security inquiries would not have been decisive.

But when I see such a combination of seriously disturbing actions and events as are present in this case, then I believe the risk to security passes acceptable bounds. All these actions and events and the relation between them make no

---

*See additional statements by Commissioners Zuckert and Compbell and separate opinion, concurring with the decision, by Commissioner Murray.

other conclusion possible, in my opinion, than to deny clearance to Dr. Oppenheimer.

There follow some additional observations of my own which I believe are pertinent in the consideration of this case and the problems underlying it.

It is a source of real sadness to me that my last act as a public official should be participation in the determination of this matter, involving as it does, an individual who has made a substantial contribution to the United States. This matter certainly reflects the difficult times in which we live.

2. "SECURITY" IN 1954

The fact is that this country is faced with a real menace to our national security which manifests itself in a great variety of ways. We are under the necessity of defending ourselves against a competent and ruthless force possessed of the great advantage that accompanies the initiative. There is no opportunity which this force would not exploit to weaken our courage and confuse our strength.

The degree of attention which Dr. Oppenheimer's status has evoked is indication of the extent to which this force has imposed upon us a new degree of intensity of concern with security. There has always been a recognition of the need for security precautions when war threatened or was actually in progress. It is new and disquieting that security must concern us so much in times that have so many of the outward indications of peace. Security must indeed become a daily concern in our lives as far as we can see ahead.

In this Nation, I believe we have really commenced to understand this only within the past 10 years. It would be unrealistic to imagine that in that brief period of time we could have acquired a well-rounded understanding, much less an acceptance, of the implications of such a change in our way of life. It will not prove easy to harmonize the requirements of security with such basic concepts as personal freedom. It will be a long and difficult process to construct a thoroughly articulated security system that will be effective in protecting strength and yet maintain the basic fabric of our liberties.

It is clear that one essential requirement of the struggle in which this Nation is engaged is that we be decisive and yet maintain a difficult balance in our actions. For example, we must maintain a positive armed strength, yet in such a manner that we do not impair our ability to support that strength. We must be vigilant to the dangers and deceits of militant communism without the hysteria that breeds witch hunts. We must strive to maintain that measure of discipline required by real and present-day danger without destroying such freedoms as the freedom of honest thought. Our Nation's problem is more difficult because of a fundamental characteristic of a democratic system: We seek to be a positive force without a dominated uniformity in thought and action dictated by a small group in power.

The decision in this particular matter before us must be made not in 1920 or 1930 or 1940. It has to be made in the year 1954 in the light of the necessities of today and, inevitably, with whatever limitations of viewpoint 1954 creates. One fact that gives me reassurance is that this decision was reached only after the most intensive and concerned study following a course of procedure which gave the most scrupulous attention to our ideas of justice and fair treatment.

The problem before this Commission is whether Dr. Oppenheimer's status as a consultant to the Atomic Energy Commission constitutes a security risk.

3. THE CONCEPT OF "SECURITY RISK"

One of the difficulties in the development of a healthy security system is the achievement of public understanding of the phrase "security risk." It has unfortunately acquired in many minds the connotation of active disloyalty. As a result, it is not realized that the determination of "security risk" must be applied to individuals where the circumstances may be considerably less derogatory than disloyalty. In the case of Dr. Oppenheimer, the evidence which convinced me that his employment was not warranted on security grounds did not justify an accusation of disloyalty.

The "security risk" concept has evolved in recent years as a part of our search for a security system which will add to the protection of the country. In that quest, certain limited guidelines have emerged. With respect to eligibility of people for sensitive positions in our Government we have said, in effect, that there must be a convincing showing that their employment in such positions will not constitute a risk to our security. Except in the clearest of cases, such as present Communist membership, for example, the determination may not be an easy one. In many cases, like the one before us, a complex qualitative determination is required. One inherent difficulty is that every human being is to some degree a security risk. So long as there are normal human feelings like pain, or emotions like love of family, everyone is to some degree vulnerable to influence, and thus a potential risk in some degree to our security.

Under our security system it is our duty to determine how much of a risk is involved in respect to any particular individual and then to determine whether that risk is worth taking in view of what is at stake and the job to be done. It is not possible, except in obvious cases, to determine in what precise manner our security might be endangered. The determination is rather an evaluation of the factors which tend to increase the chance that security might be endangered. Our experience has convinced us that certain types of association and defects of character can materially increase the risk to security.

Those factors—many of which are set forth in the majority opinion—are present in Dr. Oppenheimer's case to such an extent that I agree he is a security risk.

4. POSSIBILITY OF AN ALTERNATIVE ACTION

There have been suggestions that there may be a possible alternative short of finding Dr. Oppenheimer a security risk. One possibility suggested was that the Commission might merely allow Dr. Oppenheimer's consultant's contract to lapse when it expires on June 30, 1954, and thereafter not use his services. I have given the most serious consideration to this possibility and have concluded that it is not practical.

The unique place that Dr. Oppenheimer has built for himself in the scientific world and as a top Government adviser make it necessary that there be a clear-cut determination whether he is to be given access to the security information within the jurisdiction of the Commission.

As a scientist, Dr. Oppenheimer's greatest usefulness has been as a scientific administrator and a scientific critic. He has been looked to for scientific judgment by people within the profession. He is a personality in whom students place particular reliance for leadership and inspiration. These qualities, coupled with a nature that enables him to keep in active touch with great numbers of people in the scientific professions, have given him a unique place in the scientific community.

The Commission's clearance has permitted Dr. Oppenheimer to carry out his role as an active consultant of scientists. For example, Los Alamos Laboratory reports on the most intimate details of the progress of the thermonuclear and fission programs have continued to flow to him. I would gather that these reports were sent to Dr. Oppenheimer because his leadership and scientific judgment were recognized, and it was felt that he should be kept intensively abreast of the development of the weapon art.

I think the Commission is clearly obligated to determine whether Dr. Oppenheimer may continue to carry out this function and whether scientists may continue to call upon him as they have in the past in regard to highly classified material.

In addition, the scope of Dr. Oppenheimer's activities as a top adviser to various agencies of Government on national security policies make imperative a determination of his security status.

After the development of the atomic bomb and the end of World War II, Dr. Oppenheimer was quite suddenly projected into a far more important capacity than he had held as a scientist and laboratory director at Los Alamos. He was given responsibilities for the formulation of international controls of atomic energy. His post as chairman of the General Advisory Committee and a host of other committees in the Defense Establishment made him an adviser on national security problems at the top level of Government. His advice was sought on many matters in which science or technical aspects of atomic energy were important, but important as incidentals and background. With his unique experience, his intellect, his breadth of interests and his articulateness it was almost inevitable that he was consulted on a growing number of national security policy matters. As a result, his degree of access to the detailed essentials of our most secret information was, in my opinion, among the greatest of any individuals in our Government. I doubt that there have been contemporaneously more than a handful of people at the highest levels who have possessed the amount of sensitive information which was given to Dr. Oppenheimer.

Since Dr. Oppenheimer's retirement from the General Advisory Committee he has been employed as a consultant to the Commission. It is true that since 1952 the Commission has used him very little. Commission clearance has, however, been a basis for other agencies using him in connection with delicate problems of national security. It is logical to expect that would continue. For example, the Commission has recently received a letter from Dr. DuBridge, Chairman of the Science Advisory Committee, Office of Defense Mobilization which says:

"Our Committee is planning to undertake during the coming months an intensive study of important matters related to national security on which Dr. Oppenheimer's knowledge and counsel will be of critical importance."

I believe that the outlined facts concerning Dr. Oppenheimer's activities in the scientific profession and employment by the Government demonstrate that the Commission could not decide the matter on any other basis than to grant or deny clearance. Any other action would merely postpone the problem. His activities cannot be compartmented to some particular area of scientific effort. It is only reasonable to expect that he would be used in connection with broad assignments such as he has had in the past. Inevitably the question would arise whether he should be given access to the most sensitive restricted data which is under the Commission's jurisdiction.

Therefore, there must be a determination as to his security status with respect to this data.

All of the facts concerning Dr. Oppenheimer's activities, scientific and governmental, and the consequent access to vital information emphasize the degree of his security responsibility.

For the reasons outlined in the first paragraphs of these comments, I conclude that he falls substantially below the standard required by that responsibility. There seems to me no possible alternative to denying Dr. Oppenheimer clearance.

5. THERMONUCLEAR CONTROVERSY DISREGARDED

There is one final comment which I should add. My decision in this matter was influenced neither by the actions nor by the attitudes of Dr. Oppenheimer concerning the development of thermonuclear weapons. Nor did I consider material any advice given by Dr. Oppenheimer in his capacity as a top level consultant on national security affairs.

In my judgment, it was proper to include Dr. Oppenheimer's activities regarding the thermonuclear program as part of the derogatory allegations that initiated these proceedings. Allegations had been made that Dr. Oppenheimer was improperly motivated.

The Gray Board, although doubting the complete veractiy of Dr. Oppenheimer's explanations, found that these most serious allegations were not substantiated. I have carefully reviewed the evidence and concur in the finding.

Concurring Opinion of Commissioner Campbell

On November 7, 1953, Mr. William L. Borden, legislative secretary to the late Senator Brien McMahon in 1948 and later executive director of the Joint Committee on Atomic Energy from 1949 to June 1953, addressed a letter to the Director of the Federal Bureau of Investigation relative to Dr. J. Robert Oppenheimer.

In this letter Mr. Borden, who had previously had access to the Atomic Energy Commission files and FBI reports concerning Dr. Oppenheimer, made very grave accusations, allegations, and charges pertaining to the character, loyalty, and associations of Dr. Oppenheimer. Upon receipt of this letter, the FBI prepared a summary report on Dr. Oppenheimer and November 30, 1953, distributed that report and the Borden letter to interested agencies of the Government, including the Office of the President.

On December 10, 1953, the Commission unanimously voted to institute the regular procedures of the Commission to determine the veracity or falsity of the charges. At the direction of, and with the unanimous approval of the Commission, the General Manager on December 23, 1953, informed Dr. Oppenheimer of the substance of the information which raised the question concerning his eligibility for employment on Atomic Energy Commission work and notified him of the steps which he could take to assist in the resolution of the question.

At the request of counsel for Dr. Oppenheimer, an extension of time was granted Dr. Oppenheimer for the preparation of his case. Other extensions were subsequently granted. On March 15, 1954, Dr. Oppenheimer was notified that Mr. Thomas A. Morgan, Mr. Gordon Gray, and Dr. Ward V. Evans had been selected for the Personnel Security Board. On March 17, 1954, Dr. Oppenheimer, by letter, advised the Commission that he had received the notification of the membership of the Board and that he knew of no reason why he should challenge any member of that Board, as it was his right to do under the Personnel Security Procedures of the Atomic Energy Commission.

As early as January 18, 1954, Dr. Oppenheimer's counsel discussed the possibility of securing Q clearance with the Chairman and the General Manager of the Commission, and he was notified that clearance would be expedited as rapidly as possible if he would submit the required papers. These papers were not submitted until March 26, 1954—over 60 days later.

During the week of April 5 through 9, the Personnel Security Board met and familiarized themselves with the pertinent files relative to Dr. Oppenheimer. On April 12 the hearings began and were continued until May 6. After a 10-day recess the Board convened again on May 17.

On May 17, 1954, counsel for Dr. Oppenheimer submitted a brief to the Personnel Security Board which was included in the record.

On May 18, 1954, the Commission moved that at each step the case of Dr. Oppenheimer be brought to the Commission for a vote. This motion was carried 3 to 2. I voted against this motion since I felt that this was a very definite change in the official procedures. In my opinion, it was not desirable to change the rules in the midst of the proceedings. At this same Commission meeting on May 18, I moved that the procedures, as published in the Federal Register, be revised to indicate that, after determination had been made by the General Manager, the Commission would make the final determination in this matter. This motion did not carry by a vote of 3 to 2.

A recommendation was submitted by the Personnel Security Board to the General Manager on May 27, 1954. In essence the recommendation of the Personnel Security Board, by a 2 to 1 majority, was that: "We have, however, been unable to arrive at the conclusion that it would be clearly consistent with the security interests of the United States to reinstate Dr. Oppenheimer's clearance and, therefore, do not so recommend."

Upon receipt of the recommendation of the Board, the General Manager notified counsel for Dr. Oppenheimer on May 28 of the majority and minority recommendations of the Board and furnished a copy of the Personnel Security Board report. At the same time notification was given that Dr. Oppenheimer was entitled to make an appeal to the Personnel Security Review Board. The General Manager further stated that following such an appeal he would make a recommendation and the Commission would then make a final determination in the case.

By letter of June 1, counsel for Dr. Oppenheimer responded that they would waive the right of appeal to the Personnel Security Review Board and instead wished to present oral arguments and a written brief directly to the Commission for a final determination.

On June 3, 1954, the Commission denied the counsel for Dr. Oppenheimer the privilege of oral argument before the Commission but granted permission to file a written brief with the provision that the brief be presented on or before June 7. It was my personal opinion that this permission constituted another departure from the procedures, but my view was not sustained by my colleagues.

Counsel for Dr. Oppenheimer filed a brief with the Commission on June 7, 1954.

On June 12, the General Manager submitted his findings to the Commission in which he reaffirmed the recommendation of the Gray Board. The General Manager's letter stated:

"I have reviewed the entire record of the case, including the files, the transcript of the hearing, the findings and recommendation of the Personnel Security Board and the briefs filed by Dr. Oppenheimer's attorneys on May 17, 1954, and June 7, 1954, and have reached the conclusion that to reinstate the security clearance of Dr. Oppenheimer would not be clearly consistent with the interests of national security and would endanger the common defense and security."

In addition, Mr. Nichols stated:

"In regard to Dr. Oppenheimer's net worth to atomic energy projects, I believe, first, that through World War II he was of tremendous value and absolutely essential. Secondly, I believe that since World War II his value to the

Atomic Energy Commission as a scientist or as a consultant has declined because of the rise in competence and skill of other scientists and because of his loss of scientific objectivity probably resulting from the diversion of his efforts to political fields and matters not purely scientific in nature. Further, it should be pointed out that in the past 2 years since he has ceased to be a member of the General Advisory Committee, his services have been utilized by the Atomic Energy Commission on the following occasions only:

"October 16 and 17, 1952.
"September 1 and 2, 1953.
"September 21 and 22, 1953.

"I doubt that the Atomic Energy Commission, even if the question of his security clearance had not arisen, would have utilized his services to a markedly greater extent during the next few years. * * * Dr. Oppenheimer * * * is far from being indispensable. * * * Dr. Oppenheimer's clearance should not be reinstated."

On June 28, 1954, the question of the clearance of Dr. Oppenheimer was presented to the Commission and by a vote of 4 to 1 it was decided that clearance should be denied him.

My vote was to sustain the recommendations of the Gray Board and the General Manager for the following reasons:

1. I have had no personal association with Dr. Oppenheimer and no personal knowledge as to his contributions to the atomic energy program. Neither do I have any personal knowledge as to his character, loyalty, and associations. The responsibility of a Commissioner of the Atomic Energy Commission in a proceeding of this type is, in my view, an appellate responsibility.

2. Having examined the transcript of the hearings, it is established that Dr. Oppenheimer had an opportunity prior to the hearings to challenge the members of the Board and did not choose to do so. At all times, Dr. Oppenheimer was represented by four attorneys. At no time during the course of the hearings has the integrity, honesty, and impartiality of any of the Board members been subject to challenge by any parties to the proceedings. Dr. Oppenheimer, through his counsel, has had the opportunity to produce any witnesses he desired to call on his behalf. Through his counsel he had opportunity to cross-examine any persons who testified on items which he might have considered to be of a derogatory nature. Ample opportunity was given to Dr. Oppenheimer's counsel to present their case. In fact, extensions and delays were granted, which by some might be considered unreasonable, so that there can be no possibility that there was any pressure of time in the presentation of the information which Dr. Oppenheimer desired to place before the Board.

3. From an examination of the transcript and from the report, both majority and minority of the Board, it is evident that the members of the Board were fully aware of the criteria which had been established by the Atomic Energy Commission and by the various executive orders and public laws relative to the clearance of individuals for classified work. At no time was any question raised by any party to the proceedings as to the competence of the Board insofar as its knowledge of the criteria and procedures under which the hearing was being conducted.

4. I have carefully studied the recommendations of the General Manager and have concluded that from the presentation of the testimony before the Personnel Security Board and the information made available to the parties in the proceedings from the investigative files, the General Manager has arrived at the only possible conclusion available to a reasonable and prudent man. The finding, by the General Manager, that the services of Dr. Oppenheimer are not indispensable to the atomic energy program, is compelling.

5. I have read the brief submitted by counsel for Dr. Oppenheimer to the Atomic Energy Commission and though this brief is argumentative and perhaps persuasive to some, it contains no new evidence and it does not directly or indirectly charge that Dr. Oppenheimer has been unfairly treated or deprived of a full and complete opportunity to make the best possible presentation available in his defense.

(I neither concur nor dissent from the findings of the Personnel Security Board and the General Manager relating to the allegation that Dr. Oppenheimer initially opposed and later declined to cooperate in the program for the development of thermonuclear weapons. It is my view that the opinions and judgments of Dr. Oppenheimer on this subject were not relevant to the inquiry. I, therefore, have made my determination as to Dr. Oppenheimer's fitness for continued employment upon other evidence and testimony presented which bears on his loyalty, character, and associations.)

CONCLUSION

I conclude, therefore, that serious charges were brought against Dr. Oppenheimer; that he was afforded every opportunity to refute them; that a board was appointed, composed of men of the highest honor and integrity, and that in their majority opinion Dr. Oppenheimer did not refute the serious charges which faced him; that the record was reviewed by the General Manager, keenly aware of his serious responsibility in this matter, and that he concurred, and even strengthened the findings of the Personnel Security Board.

If the security system of the United States Government is to be successfully operated, the recommendations of personnel security boards must be honored in the absence of compelling circumstances. If the General Manager of the Atomic Energy Commission is to function properly, his decisions must be upheld unless there can be shown new evidence, violations of procedures, or other substantial reasons why they should be reversed.

Therefore, I voted to reaffirm the majority recommendation of the Personnel Security Board and to uphold the decision of the General Manager. Clearance should be denied to Dr. Oppenheimer.

CONCURRING OPINION OF COMMISSIONER THOMAS E. MURRAY

I concur in the conclusion of the majority of the Commission that Dr. J. Robert Oppenheimer's access to restricted data should be denied. However, I have reached this conclusion by my own reasoning which does not coincide with the majority of the Commission. Therefore, I submit my separate opinion.

In my opinion the Personnel Security Board report and the recommendations of the General Manager as well as the majority opinion do not correctly interpret the evidence in the case. They do not make sharply enough certain necessary distinctions. They do not do justice to certain important principles. What is more important they do not meet squarely the primary issue which the case raises.

The primary issue is the meaning of loyalty. I shall define this concept concretely within the conditions created by the present crisis of national and international security. When loyalty is thus concretely defined and when all the evidence is carefully considered in the light of this definition, it will be evident that Dr. Oppenheimer was disloyal.

There is a preliminary question. It concerns Dr. Oppenheimer's opposition to the hydrogen bomb program and his influence on the development of the program. On this count I do not find evidence that would warrant the denial to Dr. Oppenheimer of a security clearance.

I find that the record clearly proves that Dr. Oppenheimer's judgment was in error in several respects. It may well be that the security interests of the United States were adversely affected in consequence of his judgment. But it would be unwise, unjust, and dangerous to admit, as a principle, that errors of judgment, especially in complicated situations, can furnish valid grounds for later indictments of a man's loyalty, character, or status as a security risk. It has happened before in the long history of the United States that the national interests were damaged by errors of judgment committed by Americans in positions of responsibility. But these men did not for this reason cease to merit the trust of their country.

Dr. Oppenheimer advanced technical and political reasons for his attitude to the hydrogen-bomb program. In both respects he has been proved wrong; nothing further need be said.

He also advanced moral reasons. Here two comments are necessary. First, in deciding matters of national policy, it is imperative that the views of experts should always be carefully weighed and never barred from discussion or treated lightly. However, Dr. Oppenheimer's opinions in the field of morality possess no special authority. Second, even though Dr. Oppenheimer is not an expert in morality, he was quite right in advancing moral reasons for his attitude to the hydrogen bomb program. The scientist is a man before he is a technician. Like every man, he ought to be alert to the moral issues that arise in the course of his work. This alertness is part of his general human and civic responsibilities, which go beyond his responsibilities as a scientist. When he has moral doubts, he has a right to voice them. Furthermore, it must be firmly maintained, as a principle both of justice and of religious freedom, that opposition to governmental policies, based on sincerely held moral opinions, need not make a man a security risk.

The issue of Dr. Oppenheimer's lack of enthusiasm for the hydrogen bomb program has been raised; so, too, has the issue of his failure to communicate to other scientists his abandonment of his earlier opposition to the program. Here an important distinction is in order. Government may command a citizen's service in the national interest. But Government cannot command a citizen's enthusiasm for any particular program or policy projected in the national interests. The citizen remains free to be enthusiastic or not at the impulse of his own inner convictions. These convictions remain always immune from governmental judgment or control. Lack of enthusiasm is not a justiciable matter.

The point that I shall later make in another connection is pertinent here. The crisis in which we live, and the security regulations which it has rendered necessary in the interests of the common good, have made it difficult to insure that justice is done to the individual. In this situation it is more than ever necessary to protect at every point the distinction between the external forum of action and omission, and the internal forum of thought and belief. A man's service to his country may come under judgment; it lies in the external forum. A man's enthusiasm for service, or his lack of it, do not come under judgment; they are related to the internal forum of belief, and are therefore remote from all the agencies of law.

The citizen's duty remains always that of reasonable service, just as the citizen's right remains always that of free opinion. There is no requirement, inherent in the idea of civic duty, that would oblige a man to show enthusiasm for particular governmental policies, or to use his influence in their favor, against his own convictions; just as there is no permission, inherent in the idea of intellectual freedom, that would allow a man to block established governmental policies, against the considered judgment of their responsible authors.

The conclusion is that the evidence with regard to Dr. Oppenheimer's attitude toward the hydrogen bomb program, when it is rightly interpreted in the light of sound democratic principles, does not warrant the denial to Dr. Oppenheimer of a security clearance.

The primary question concerns Dr. Oppenheimer's loyalty. This idea must be carefully defined, first, in general, and second, in concrete and contemporary terms.

The idea of loyalty has emotional connotations; it is related to the idea of love, a man's love of his country. However, the substance of loyalty does not reside solely in feeling or sentiment. It cannot be defined solely in terms of love.

The English word "loyal" comes to us from the Latin adjective "legalis," which means "according to the law." In its substance the idea of loyalty is related to the idea of law. To be loyal, in Webster's definition, is to be "faithful to the lawful government or to the sovereign to whom one is subject." This faithfulness is a matter of obligation; it is a duty owed. The root of the obligation and duty is the lawfulness of the government, rationally recognized and freely accepted by the citizens.

The American citizen recognizes that his Government, for all its imperfections, is a government under law, of law, by law; therefore he is loyal to it. Furthermore, he recognizes that his Government, because it is lawful, has the right and the responsibility to protect itself against the action of those who would subvert it. The cooperative effort of the citizen with the rightful action of American Government in its discharge of this primary responsibility also belongs to the very substance of American loyalty. This is the crucial principle in the present case.

This general definition of loyalty assumes a sharper meaning within the special conditions of the present crisis. The premise of the concrete, contemporary definition of loyalty is the fact of the Communist conspiracy. Revolutionary communism has emerged as a world power seeking domination of all mankind. It attacks the whole idea of a social order based upon freedom and justice in the sense in which the liberal tradition of the West has understood these ideas. Moreover, it operates with a new technique of aggression; it has elaborated a new formula for power. It uses all the methods proper to conspiracy, the methods of infiltration and intrigue, of deceit and duplicity, of falsehood and connivance. These are the chosen methods whereby it steadily seeks to undermine, from within, the lawful governments and communities of the free world.

The fact of the Communist conspiracy has put to American Government and to the American people a special problem. It is the problem of protecting the national security, internal and external, against the insidious attack of its Communist enemy. On the domestic front this problem has been met by the erection of a system of laws and Executive orders designed to protect the lawful Government of the United States against the hidden machinery of subversion.

The American citizen in private life, the man who is not engaged in governmental service, is not bound by the requirements of the security system. However, those American citizens who have the privilege of participating in the operations of Government, especially in sensitive agencies, are necessarily subject to this special system of law. Consequently, their faithfulness to the lawful Government of the United States, that is to say their loyalty, must be judged by the standard of their obedience to security regulations. Dr. Oppenheimer was subject to the security system which applies to those engaged in the atomic energy program. The measure of his obedience to the requirements of this system is the decisive measure of his loyalty to his lawful Government. No lesser test will settle the question of his loyalty.

In order to clarify this issue of the meaning of loyalty, the following considerations are necessary. First, the atomic energy program is absolutely vital to the survival of the Nation. Therefore the security regulations which surround it are intentionally severe. No violations can be countenanced. Moreover, the necessity for exact fidelity to these regulations increases as an individual operoperates in more and more sensitive and secret areas of the program. Where responsibility is highest, fidelity should be most perfect.

Second, this security system is not perfect in its structure or in its mode of operation. Perfection would be impossible. We are still relatively unskilled in the methods whereby we may effectively block the conspiratorial efforts of the Communist enemy without damage to our own principles. Moreover, the operation of the system is in the hands of fallible men. It is therefore right and necessary that the system should be under constant scrutiny. Those who are affected by the system have a particular right to criticize it. But they have no right to defy or disregard it.

Third, the premise of the security system is not a dogma but a fact, the fact of the Communist conspiracy. The system itself is only a structure of law, not a set of truths. Therefore this system of law is not, and must not be allowed to become, a form of thought control. It restricts the freedom of association of the governmental employee who is subject to it. It restricts his movements and activities. It restricts his freedom of utterance in matters of security import, not in other matters. It restricts his freedom of personal and family life. It makes special demands on his character, moral virtue, and spirit of sacrifice. But no part of the security system imposes any restrictions on his mind. No law or Executive order inhibits the freedom of the mind to search for the truth in all the great issues that today confront the political and moral intelligence of America. In particular, no security regulations set any limits to the free-ranging scientific intelligence in its search for the truths of nature and for the techniques of power over nature. If they were to do so, the result would be disastrous; for the freedom of science is more than ever essential to the freedom of the American people.

Fourth, the preservation of the ordered freedom of American life requires the cooperation of all American citizens with their Government. The indispensable condition of this cooperation is a spirit of mutual trust and confidence. This trust and confidence must in a special sense obtain between governmental officials and scientists, for their partnership in the atomic-energy program and in other programs is absolutely essential to the security interests of the United States. It would be lamentable if conscientious enforcement of security regulations were to become a danger to the atmosphere of trust and confidence which alone can sustain this partnership. In order to avert this danger, there must be on the part of Government a constant concern for justice to the individual, together with a concern for the high interests of the national community. On the part of scientists there should be a generous disposition to endure with patient understanding the distasteful restrictions which the security system imposes on them.

Finally, it is essential that in the operation of the security system every effort should be made to safeguard the principle that no American citizen is to be penalized for anything except action or omission contrary to the well-defined interests of the United States. However stringent the need for a security system, the system cannot be allowed to introduce into American jurisprudence that hateful concept, the "crime of opinion." The very security of America importantly lies in the steady guaranty, even in a time of crisis, of the citizen's right to freedom of opinion and of honest and responsible utterance. The present time of crisis intensifies the civic duty of obedience to the lawful government in the crucial area of security regulations. But it does not justify abridgment of the civic right of dissent. Government may penalize disobedience in action or omission. It may not penalize dissent in thought and utterance.

When all these distinctions and qualifications have been made, the fact remains that the existence of the security regulations which surround the atomic-energy program puts to those who participate in the program a stern test of loyalty.

Dr. Oppenheimer failed the test. The record of his actions reveals a frequent and deliberate disregard of those security regulations which restrict a man's associations. He was engaged in a highly delicate area of security; within this area he occupied a most sensitive position. The requirement that a man in this position should relinquish the right to the complete freedom of association that would be his in other circumstances is altogether a reasonable and necessary requirement. The exact observance of this requirement is in all cases essential to the integrity of the security system. It was particularly essential in the case of Dr. Oppenheimer.

It will not do to plead that Dr. Oppenheimer revealed no secrets to the Communists and fellow travelers with whom he chose to associate. What is incompatible with obedience to the laws of security is the associations themselves, however innocent in fact. Dr. Oppenheimer was not faithful to the restrictions on the associations of those who come under the security regulations.

There is a further consideration, not unrelated to the foregoing. Those who stand within the security system are not free to refuse their cooperation with the workings of the system, much less to confuse or obstruct them, especially by falsifications and fabrications. It is their duty, at times an unpleasant duty, to cooperate with the governmental officials who are charged with the enforcement of security regulations. This cooperation should be active and honest. If this manner of cooperation is not forthcoming, the security system itself, and therefore the interests of the United States which it protects, inevitably suffer. The record proves Dr. Oppenheimer to have been seriously deficient in his cooperation with the workings of the security system. This defect too is a defect of loyalty to the lawful government in its reasonable efforts to preserve itself in its constitutional existence. No matter how high a man stands in the service of his country he still stands under the law. To permit a man in a position of the highest trust to set himself above any of the laws of security would be to invite the destruction of the whole security system.

In conclusion, the principle that has already been stated must be recalled for the sake of emphasis. In proportion as a man is charged with more and more critical responsibilities, the more urgent becomes the need for that full and exact fidelity to the special demands of security laws which in this overshadowed day goes by the name of loyalty. So too does the need for cooperation with responsible security officers.

Dr. Oppenheimer occupied a position of paramount importance; his relation to the security interests of the United States was the most intimate possible one. It was reasonable to expect that he would manifest the measure of cooperation appropriate to his responsibilities. He did not do so. It was reasonable to expect that he would be particularly scrupulous in his fidelity to security regulations. These regulations are the special test of the loyalty of the American citizen who serves his Government in the sensitive area of the Atomic Energy program. Dr. Oppenheimer did not meet this decisive test. He was disloyal.

I conclude that Dr. Oppenheimer's access to restricted data should be denied.

THOMAS E. MURRAY,
Commissioner.

JUNE 29, 1954.

DISSENTING OPINION OF HENRY DE WOLF SMYTH

I dissent from the action of the Atomic Energy Commission in the matter of Dr. J. Robert Oppenheimer. I agree with the "clear conclusion" of the Gray Board that he is completely loyal and I do not believe he is a security risk. It is my opinion that his clearance for access to restricted data should be restored.

In a case such as this, the Commission is required to look into the future. It must determine whether Dr. Oppenheimer's continued employment by the

Government of the United States is in the interests of the people of the United States. This prediction must balance his potential contribution to the positive strength of the country against the possible danger that he may weaken the country by allowing important secrets to reach our enemies.

Since Dr. Oppenheimer is one of the most knowledgable and lucid physicists we have, his services could be of great value to the country in the future. Therefore, the only question being determined by the Atomic Energy Commission is whether there is a possibility that Dr. Oppenheimer will intentionally or unintentionally reveal secret information to persons who should not have it. To me, this is what is meant within our security system by the term security risk. Character and associations are important only insofar as they bear on the possibility that secret information will be improperly revealed.

In my opinion the most important evidence in this regard is the fact that there is no indication in the entire record that Dr. Oppenheimer has ever divulged any secret information. The past 15 years of his life have been investigated and reinvestigated. For much of the last 11 years he has been under actual surveillance, his movements watched, his conversations noted, his mail and telephone calls checked. This professional review of his actions has been supplemented by enthusiastic amateur help from powerful personal enemies.

After reviewing the massive dossier and after hearing some forty witnesses, the Gray Board reported on May 27, 1954, that Dr. Oppenheimer "seems to have had a high degree of discretion reflecting an unusual ability to keep to himself vital secrets." My own careful reading of the complete dossier and of the testimony leads me to agree with the Gray Board on this point. I am confident that Dr. Oppenheimer will continue to keep to himself all the secrets with which he is entrusted.

The most important allegations of the General Manager's letter of December 23 related to Dr. Oppenheimer's conduct in the so-called H-bomb program. I am not surprised to find that the evidence does not support these allegations in any way. The history of Dr. Oppenheimer's contributions to the development of nuclear weapons stands untarnished.

It is clear that Dr. Oppenheimer's past associations and activites are not newly discovered in any substantial sense. They have been known for years to responsible authorities who have never been persuaded that they rendered Dr. Oppenheimer unfit for public service. Many of the country's outstanding men have expressed their faith in his integrity.

In spite of all this, the majority of the Commission now concludes that Dr. Oppenheimer is a security risk. I cannot accept this conclusion or the fear behind it. In my opinion the conclusion cannot be supported by a fair evaluation of the evidence.

Those who do not accept this view cull from the record of Dr. Oppenheimer's active life over the past 15 years incidents which they construe as "proof of fundamental defects in his character" and as alarming associations. I shall summarize the evidence on these incidents in order that their proper significance may be seen.

Chevalier incident.—The most disturbing incidents of his past are those connected with Haakon Chevalier. In late 1942 or early 1943, Chevalier was asked by George Eltenton to approach Dr. Oppenheimer to see whether he would be willing to make technical information available for the Soviet Union. When Chevalier spoke to Dr. Oppenheimer he was answered by a flat refusal. The incident came to light when Dr. Oppenheimer, of his own accord, reported it to Colonel Pash in August 1943. He did not at that time give Chevalier's name and said that there had been 3 approaches rather than 1. Shortly thereafter, in early September, Dr. Oppenheimer told General Groves that, if ordered, he would reveal the name. Not until December 1943, did General Groves direct him to give the name. It is his testimony that he then told General Groves that the earlier story concerning three approaches had been a "cock and bull story." Not until 1946 were Eltenton, Chevalier, and Dr. Oppenheimer himself interviewed by security officers in this matter. When interviewed by the FBI in 1946, Dr. Oppenheimer recounted the same story of the incident which he has consistently maintained ever since. He stated explicitly in 1946 that the story told to Colonel Pash in 1943 had been a fabrication. In the present hearings before the Gray Board he testified, before the recording of the Pash interview was produced, that the story told to Colonel Pash was a fabrication to protect his friend Chevalier. The letter which he wrote Chevalier in February 1950, concerning Chevalier's role in the 1943 incident, stated only what Dr. Oppenheimer has consistently maintained to the FBI and to the Gray board concerning Chevalier's lack of awareness of the significance of what he was doing.

The Chevalier incident involved temporary concealment of an espionage attempt and admitted lying, and is inexcusable. But that was 11 years ago; there is no subsequent act even faintly similar; Dr. Oppenheimer has repeatedly expressed his shame and regret and has stated flatly that he would never again so act. My conclusion is that of Mr. Hartley Rowe, who testified, "I think a man of Dr. Oppenheimer's character is not going to make the same mistake twice."

Dr. Oppenheimer states that he still considers Chevalier his friend, although he sees him rarely. In 1950 just before Chevalier left this country to take up residence in France, he visited Dr. Oppenheimer for 2 days in Princeton; in December 1953, Dr. Oppenheimer visited with the Chevaliers in Paris at their invitation. These isolated visits may have been unwise, but there is no evidence that they had any security significance. Chevalier was not sought out by Dr. Oppenheimer in Paris but, rather, the meeting was proposed by the Chevaliers in a letter to Mrs. Oppenheimer. The contact consisted of a dinner and, on the following day, driving with Chevalier to meet Andre Malraux, the famous French literary figure for whom Chevalier was a translator. Malraux in the later years of his political life has been an active anti-Communist adviser to General deGaulle. These short visits were followed 2 months later by Chevalier's use of Dr. Oppenheimer's name in connection with clearance for employment by UNESCO. Dr. Oppenheimer's action in this matter seems quite correct. When Chevalier mentioned the problem, Dr. Oppenheimer suggested that the proper place for advice was the American Embassy and that Dr. Geoffrey Wyman, the scientific attaché, might be in a position to give the advice. Before seeing Chevalier, Dr. Oppenheimer had lunched at the Embassy with Dr. Wyman, a former classmate, but it is clear from Dr. Wyman's affidavit in the record that Dr. Oppenheimer did not at that time or later mention or endorse Chevalier.

Associations.—It is stated that a persistent and continuing association with Communists and fellow travelers is part of a pattern in Dr. Oppenheimer's actions which indicates a disregard of the obligations of securtiy. On examination, the record shows that, since the war, beyond the two visits with the Chevaliers, Dr. Oppenheimer's associations with such persons have been limited and infrequent. He sees his brother, Frank Oppenheimer (an admitted former Communist who left the party in 1941) not "much more than once a year" and then only for "an evening together." By chance, while returning from the barber, he ran into Lomanitz and Bohm on the streets of Princeton in May 1949. Dr. Peters called on him once to discuss testimony given by Dr. Oppenheimer before the House Committee on Un-American Activities. He has seen Bohm and 1 or 2 other former students at meetings of professional groups. I find nothing in the foregoing to substantiate the charge that Dr. Oppenheimer has had a "persistent and continuing" association with subversive individuals. These are nothing more than occasional incidents in a complex life, and they were not sought by Dr. Oppenheimer.

Significance has been read into these occasional encounters in the light of Dr. Oppenheimer's activities prior to 1943.

The Gray Board found that he was an active fellow traveler, but that there was no evidence that he was a member of the party in the strict sense of the word. Dr. Oppenheimer's consistent testimony, and the burden of the evidence, shows that his financial contributions in the 1930's and early 1940's were directed to specific causes such as the Spanish Loyalists, even though they may have gone through individual Communists.

The Communists with whom he was deeply involved were all related to him by personal ties: his brother and sister-in-law, his wife (who had left the party before their marriage), and his former fiance, Jean Tatlock. Finally, while there are self-serving claims by Communists on record as to Dr. Oppenheimer's adherence to the party, none of these is attributed to Communists who actually knew him, and Steve Nelson (who did know him) described him in a statement to another Communist as not a Marxist. The evidence supports Dr. Oppenheimer's consistent denial that he was ever a Communist.

Dr. Oppenheimer has been repeatedly interrogated from 1943 on concerning his associations and activities. Beyond the one admitted falsehood told in the Chevalier incident, the voluminous record shows a few contradictions between statements purportedly made in 1943 and subsequent recollections during interrogations in 1950 and 1954. The charges of falsehood concerning Weinberg and Lambert relate to such contradictions, and are dependent on a garbled transcript. In my opinion, these contradictions have been given undue significance.

Peter's letter.—I find it difficult to conclude that the letter written by Dr. Oppenheimer in 1949 following his testimony about Dr. Bernard Peters before a congressional committee is evidence of any fault in character. This carefully composed letter, a copy of which was sent to the congressional committee, was not an attempt to repudiate the testimony relating to Dr. Peters' background but, rather, was a manifestation of a belief that political views should not disqualify a scientist from a teaching job. He was led to this action by the protests of Dr. Bethe, Dr. Weisskopf, and Dr. Peters himself, and of Dr. Condon, and by the "overwhelming belief of the community in which I lived that a man like that ought not to be fired either for his past or for his views, unless the past is criminal or the views led him to wicked actions." One might disagree with this belief without taking it as evidence of untrustworthiness.

Lomanitz deferment.—It is clear that in cross-examination in 1954, Dr. Oppenheimer was led into contradictions concerning the induction into the Army of Rossi Lomanitz in 1943. These contradictions, understandable as errors of memory, are serious only if Dr. Oppenheimer's behavior at that time was improper. Actually, Dr. Oppenheimer's letter to Colonel Lansdale in 1943 says: "Since I am not in possession of the facts which led to Mr. Lomanitz's induction, I am, of course, not able to endorse this request in an absolute way. I can, however, say that Mr. Lomanitz's competence and his past experience on the work in Berkeley should make him a man of real value whose technical service we should make every effort to secure for the project." The letter was sent to Colonel Lansdale, the man to whom Dr. Oppenheimer had given information on Lomanitz' Communist affiliation and the man who had told Dr. Oppenheimer that Lomanitz had been indiscreet with information.

Obstruction of security officers.—The majority opinion cites the Chevalier incident as an instance of obstruction of security officers and states without specification that there are other instances. I have sought to identify these other instances. The only instance I have found is a refusal by Dr. Oppenheimer in 1950 to answer FBI questions about Dr. Thomas Addis and Dr. Jean Tatlock on the ground that they were dead and could not defend themselves. This reticence to discuss the activities of a friend and of a former fiance years after their deaths may have been an error. But in the circumstances, it seems understandable hesitation, and does not indicate a persistent "willful disregard" of security.

Seaborg letter.—Before the October 1949 meeting of the General Advisory Committee at which the H-bomb program was discussed, Dr. Seaborg, a member of the General Advisory Committee who was unable to be present, sent Dr. Oppenheimer a letter on the topics to be discussed. In Dr. Oppenheimer's letter to the Commission reporting the unanimous view of the eight members present at the General Advisory Committee meeting, there is no mention of Dr. Seaborg's views. It is hard to see how Dr. Oppenheimer could have forgotten the letter, but it is still harder to see what purpose he could have hoped to achieve by intentionally suppressing it—and then turning it over to the Commission in his files. At the next meeting of the General Advisory Committee in December 1949, the action of the October meeting was reviewed, and the minutes show that Dr. Seaborg raised no objection. It seems likely that Dr. Seaborg himself did not consider that he had expressed any formal conclusions. His letter of October 14, 1949, opens as follows:

"I will try to give you my thoughts for what they may be worth regarding the next GAC meeting, but I am afraid that there may be more questions than answers—it seems to me that conclusions will be reached, if at all, only after a large amount of give and take discussion at the GAC meeting" (Tr., p. 238).

*   *   *   *   *   *   *

The instances that I have described constitute the whole of the evidence extracted from a lengthy record to support the severe conclusions of the majority that Dr. Oppenheimer has "given proof of fundamental defects in his character" and of "persistent continuing associations." Any implication that these are illustrations only and that further substantial evidence exists in the investigative files to support these charges is unfounded.

With the single exception of the Chevalier incident, the evidence relied upon is thin, whether individual instances are considered separately or in combination. All added together, with the Chevalier incident included, the evidence is singularly unimpressive when viewed in the perspective of the 15 years of active life from which it is drawn. Few men could survive such a period of investigation and interrogation without having many of their actions misinterpreted or misunderstood.

To be effective a security system must be realistic. In the words of the Atomic Energy Commission security criteria:

"The facts of each case must be carefully weighed and determination made in the light of all the information presented, whether favorable or unfavorable. The judgment of responsible persons as to the integrity of the individuals should be considered. The decision as to security clearance is an overall, common-sense judgment, made after consideration of all the relevant information as to whether or not there is risk that the granting of security clearance would endanger the common defense or security."

Application of this standard of overall commonsense judgment to the whole record destroys any pattern of suspicious conduct or catalog of falsehoods and evasions, and leaves a picture of Dr. Oppenheimer as an able, imaginative human being with normal human weaknesses and failings. In my opinion the conclusion drawn by the majority from the evidence is so extreme as to endanger the security system.

If one starts with the assumption that Dr. Oppenheimer is disloyal, the incidents which I have recounted may arouse suspicion. However, if the entire record is read objectively, Dr. Oppenheimer's loyalty and trustworthiness emerge clearly and the various disturbing incidents are shown in their proper light as understandable and unimportant.

The "Chevalier incident" remains reprehensile; but in fairness and on all of the evidence, this one admitted and regretted mistake made many years ago does not predominate in my overall judgment of Dr. Oppenheimer's character and reliability. Unless one confuses a manner of expression with candor, or errors in recollection with lack of veracity, Dr. Oppenheimer's testimony before the Gray Board has the ring of honesty. I urge thoughtful citizens to examine this testimony for themselves, and not be content with summaries or with extracts quoted out of context.

With respect to the alleged disregard of the security system, I would suggest that the system itself is nothing to worship. It is a necessary means to an end. Its sole purpose, apart from the prevention of sabotage, is to protect secrets. If a man protects the secrets he has in his hands and his head, he has shown essential regard for the security system.

In addition, cooperation with security officials in their legitimate activities is to be expected of private citizens and Government employees. The security system has, however, neither the responsibility nor the right to dictate every detail of a man's life. I frankly do not understand the charge made by the majority that Dr. Oppenheimer has shown a persistent and willful disregard for the obligations of security, and that therefore he should be declared a security risk. No gymnastics of rationalization allow me to accept this argument. If in any recent instances, Dr. Oppenheimer has misunderstood his obligation to security, the error is occasion for reproof but not for a finding that he should be debarred from serving his country. Such a finding extends the concept of "security risk" beyond its legitimate justification and constitutes a dangerous precedent.

In these times, failure to employ a man of great talents may impair the strength and power of this country. Yet I would accept this loss if I doubted the loyalty of Dr. Oppenheimer or his ability to hold his tongue. I have no such doubts.

I conclude that Dr. Oppenheimer's employment "will not endanger the common defense and security" and will be "clearly consistent with the interests of the national security." I prefer the positive statement that Dr. Oppenheimer's further employment will continue to strengthen the United States.

I therefore have voted to reinstate Dr. Oppenheimer's clearance.

HENRY D. SMYTH, *Commissioner.*

JUNE 29, 1954.

NAME INDEX

Acheson, Dean, 35, 37, 41, 244, 343–345, 404, 543, 544, 552, 731
Addis, Thomas, 4, 9, 102, 135, 139, 155, 183, 185, 189, 191, 588, 825, 1003–1005, 1064
Addison, Joseph, 372
Adelson, David, 5, 10, 11, 131, 135, 187, 188, 588, 813, 825, 915, 919, 1005, 1007
Alderson, Harold B., 412
Alexander, Archibald, 143
Allison, Samuel, 34, 711, 876, 882
Alsop, Joseph, 54, 55, 57
Alsop, Stewart, 54, 55, 57
Altschul, Frank, 738
Alvarez, Luis, 46, 58, 61, 63–66, 226, 231, 245, 247, 317, 452, 616, 659, 684, 687, 699, 929, 932, 933, 969, 977, 978
  testimony of, 770–805, 909, 910
Amdur, I. I., 672
Amory, Robert, Jr., 738
Anderson, John, 561
Angell, Homer, 176
Armstrong, Hamilton F., 354, 738
Arnold, Henry, 680
Arnstein, Edith, 155, 190, 825
Asbridge, Neil, 198
Atlee, Clement, 35, 561
Austin, Warren, 344
Aydelott, Frank, 26

Bacher, Robert F., 20, 30, 33, 40, 46, 48, 58, 63–66, 75, 83, 84, 89, 227, 231, 246, 304, 345, 352, 374, 379, 437, 495, 521, 527, 545, 557, 584, 683, 684, 687, 747–749, 759, 762, 781, 787, 789, 793, 795, 796, 802, 804, 880, 891, 950, 978, 986, 1032
  testimony of, 608–630, 633
Bainbridge, Kenneth, 102, 440, 443, 951
Baker, Al, 544
Baker, Nicholas, 140. *See also* Bohr, Neils
Baker, Selma, 571
Barlow, Samuel L. M., 107
Barnard, Chester I., 37, 39, 373, 552, 732
Barnett, Henry, 267, 588
Barnett, Shirley, 446, 588
Baruch, Bernard, 16, 39, 40, 41, 44, 105, 326, 343, 374, 377, 385, 419, 975
Beckerley, James, 254, 400, 401, 402, 405, 410, 422, 429, 434, 451, 457
Beckler, David, 687
Befschetz, S., 183
Bellesly [Billsley], Lyle, 424, 434, 610, 611, 806, 983, 985
Benet, William R., 107

Berkner, Lloyd, 94, 765, 924, 935
Bernstein, Leonard, 107
Bernstein, Walter, 107
Bethe, Hans, 11, 20, 27, 30, 76, 84, 85, 87, 89, 212, 215, 229, 231, 234, 242, 245, 253, 299, 300, 305, 323, 456, 462, 497, 520, 545, 645, 706, 712, 714, 715, 719, 725, 782, 783, 977, 978, 1032, 1064
  testimony of, 323–340
Bhabha, Honi, 215
Billings, Henry, 107
Bloch, Felix, 332
Bohm, David, 5, 13, 119, 120, 126, 149–152, 157, 172, 180, 181, 201, 208, 209, 268, 271, 277, 588, 811–813, 820, 826, 827, 876, 883, 1006, 1007, 1018, 1020, 1052, 1063
Bohr, Neils, 140, 166, 280, 735, 740
Borden, William L., letter to FBI of, 833, 835–838, 843, 975
  testimony of, 832–844
Bowers, K. V., 3
Bowie, Robert R., 734, 738
Bowles, Edward L., 743
Boyer, Charles, 107
Bradbury, Norris, 15, 20, 33, 56, 61, 68, 82, 84, 85, 89, 90, 91, 109, 242, 246, 247, 305, 323, 352, 422, 520, 559, 645, 682, 703, 712, 713, 724, 725, 784, 785, 978, 979, 989, 1032
  testimony of, 477–494
Bradley, Omar, 77, 404, 407, 462, 464, 683
Bradley, Robert, 158
Bransten, Louise, 191, 192, 197, 216, 217, 588, 825, 903, 1005
Breit, G., 27
Brereton, Lewis H., 379
Bridges, Calvin, 102, 157
Brobeck, William, 778, 782–784
Buckley, Oliver E., 17, 19, 80, 81, 93, 299, 596, 976, 1033
  testimony of, 603–608
Buckmaster, Henrietta, 107
Bullitt, William C., 361, 362
Bundy [Brundy], McGeorge, 95, 738
Burden, William, 504, 525, 747, 751, 758
Burke, Kenneth, 978
Bush, Vannevar, 11, 28, 29, 31, 36, 45, 46, 56, 75, 76, 93, 94, 96, 135, 180, 247, 257, 307, 352, 373, 375–377, 379, 381, 390, 413, 414, 418–420, 508, 557, 577–579, 611, 613, 615, 733, 760, 786, 878, 799–801, 803, 804, 969, 983, 984, 1037
  testimony of, 560–568, 909–914
Byrnes, James F., 33, 35, 227, 544, 561

1067

Cadogan, Alexander, 344
Cairns, Robert W., 507
Calvert, H. K., 150, 153, 260, 270, 823
Campbell, Joseph, 1034, 1049
  opinion of, 1055–1058
Cantor, Eddie, 107
Carlson, Evans F., 107
Carpenter, Donald, 543
Carroll, H. H., 132
Chadwick, J., 31, 65
Chevalier, Barbara, 6, 10, 14, 189, 627, 1008
Chevalier, Haakon, 4–6, 10, 14, 102, 129–131, 136, 137, 140–143, 145, 146, 149, 152, 153, 155, 168, 189, 191, 193, 201, 206, 209, 217, 220, 251, 252, 264, 265, 279, 309, 380, 414, 466, 471, 472, 522, 548, 549, 588, 627, 628, 637–639, 647, 820, 828, 831, 877, 887–889, 898, 904, 905, 919, 980, 981, 988, 1005, 1009, 1020, 1063
  letter of, 969
Chiang Kai-Shek. *See* Kai-Shek, Chiang
Christie, Robert, 48, 521, 584, 617, 779, 891
Churchill, Winston, 385, 389
Clifford, Clark, 376, 379, 414–416, 419, 420, 422, 623, 984, 985
Cockroft, John, 65, 452, 453
Cohen, Benjamin, 95
Compton, Arthur, 11, 14, 27, 28, 34, 40, 88, 135, 164, 189, 258, 271, 280, 615, 771, 950
Compton, Karl, 257–258, 271, 280, 508, 543, 744, 1033
  testimony of, 257–258
Conant, James B., 12, 17, 19, 28, 36, 41, 45, 56, 68, 76, 80, 81, 86, 180, 231, 242–246, 259, 280, 303, 307, 321, 328, 345, 351, 373, 375–378, 381, 383, 397, 402, 406, 413, 414, 418–420, 470, 508, 509, 517–519, 560, 561, 611, 613, 663, 664, 666, 671, 699, 715, 761, 783, 803–805, 898, 976, 983, 984, 987
  testimony of, 383–394
Condon, Edward, 127, 134, 166, 167, 173, 174, 210, 212–215, 252, 253, 904–906, 1007, 1018, 1064
Condon, Hugh, 215
Connally, Tom, 36
Cooke, Morris L., 107
Cooksey, Don, 778, 788
Corson, Samuel A., 107
Corwin, Norman, 107
Cromwell, John, 107
Crouch, Paul, 16, 216–218, 305, 541, 542, 902, 987, 988, 991
Crouch, Mrs. Paul, 16, 541, 542
Crowther, Bosley, 107
Curie, J. F. Joliot-, 732
Cutler, Robert, 94, 95

Dallet, Joe, 4, 10, 188, 206, 266, 571–574, 884, 1002
Dallet, Kitty, 575. *See also* Oppenheimer, Katherine

Damon, Mrs., 218, 219
Davies, Joseph E., 107
Davis, John W., 1034
Davis, Robert R., 445, 588
Dean, Arthur, 738
Dean, Gordon, 56, 83, 96, 97, 299, 300, 337, 338, 347, 403, 405, 410, 429, 431, 659, 701, 704, 898, 977, 978, 979, 989, 1032
  testimony of, 300–323
de Gaulle, Charles A. J., 141, 1009, 1063
Dennis, William, 197
deSilva [DeSilva, DeSylva], Peer, 119, 120, 121, 149–151, 153, 211, 271–273, 449, 823
  Oppenheimer letter of, 273–275, 961
DeWitt, John L., 810
Dickey, John, 95, 562
Diebold, William, 738
Dirac [Dirae], P. A. M., 183
Dodd, Bella, 602
Doolittle, James, 749
Dorner, Hannah, 105, 106
Dow, David, 198
Doyle, Bernadette, 5, 11, 821, 822, 1005, 1006
DuBridge, Lee A., 17, 19, 48, 66, 80, 93, 94, 231, 246, 307, 352, 496, 504, 505, 584, 590, 602, 616, 747, 748, 751, 759, 761, 762, 771, 774, 781, 793, 905, 976, 1033, 1045, 1046, 1055
  testimony of, 514–534
Duffield, Priscilla, 446
Dulles, Allen, 95, 562, 734
Durr, Clifford, 151, 906

Early, Stephen, 376
Eberstadt, Ferdinand, 40, 734, 738
Ecker, Allen B., 1ff, 51ff, 99ff, 161ff, 221ff, 283ff, 349ff, 427ff, 475ff, 523ff, 569ff, 631ff, 675ff, 729ff, 791ff, 841ff, 907ff, 947ff, 965ff
Eddy, Mildred, 158
Einstein, Albert, 257
Eisenhower, Dwight D., 48, 94, 307, 333, 496, 498, 503, 504, 508, 521, 522, 892
Ellington, Duke, 107
Ellis, Margaret, 156
Elsey, George M., 375, 376, 379
Eltenton, George C., 5, 6, 14, 130, 131, 134–138, 141, 145, 147–150, 152, 167, 251, 264, 274, 288–298, 308, 309, 392, 431, 471, 472, 522, 549, 627, 628, 648, 815, 819, 827, 828, 831, 845–850, 854–860, 865–869, 873, 874, 879, 881, 882, 887, 888, 980, 981, 1007, 1008
Engels, Friedrich, 116
Evans, Ward V., 1ff, 51ff, 99ff, 161ff, 221ff, 283ff, 349ff, 427ff, 475ff, 535ff, 569ff, 631ff, 675ff, 729ff, 791ff, 841ff, 907ff, 947ff, 965ff, 999, 1019, 1032–1034, 1056
Everest, Frank, 682

Fainsod, Merle, 734, 738
Fanelli, Joseph, 142, 143, 1009
Faraday, Michael, 556
Farrell, Thomas, 32, 33, 41, 226, 345
Fast, Howard, 107
Fermi, Enrico, 14, 17, 20, 30, 34, 68, 75, 77, 79, 80, 84, 85, 89, 258, 271, 280, 305, 351, 383, 387, 402, 406, 454, 456, 464, 497, 518, 519, 527, 529, 665, 711, 714, 716–718, 722, 725, 755, 782, 805, 950, 976, 1032
  testimony of, 394–398
Ferrer, Jose, 107
Fidler, Harold, 123, 778
Fine, Paul, 776
Finkelstein, Marina S., 738
Finletter, Tom, 504, 748, 749, 752–754, 758, 926
Fisk, James B., 61, 299, 340, 342, 929, 941
  testimony of, 340–342, 1033
Flanders, Donald, 450, 655
Flannigan, A., 277, 824, 827, 877
Flemming, Arthur S., 94
Folkoff, Isaac, 4, 9, 115, 116, 139, 140, 185, 205, 588, 825, 877, 878, 890, 916, 917, 1003–1005, 1051
Folkoff, Richard, 158, 205, 877
Folsom, Frank. *See* Oppenheimer, Frank F.
Fontaine, Joan, 107
Fontenrose, Joe, 156
Forrestal, James, 344, 354
Fowler, William, 584, 747, 762, 795
Fox, David, 852, 863, 870
Fox, William T. R., 738
Frankel, Stanley, 876
Franklin, George S., 738
Freelon, Allan R., 107
Friedman, Max B., 5, 13, 157, 172, 588, 811–813, 820, 826
Froman, Darol, 85, 91, 109
Frothingham, Channing, 107
Fuchs, Klaus, 174, 175, 177, 220, 233, 278, 331, 444, 450, 486, 490, 492, 494, 593, 626, 651, 652, 671, 708, 802, 831, 838
Furry, Wendell, 442–444, 446, 449, 450

Ganz, Rudolph, 107
Gardner, Trevor, 527, 528
Garrison, Lloyd K., 1ff, 51ff, 99ff, 161ff, 221ff, 283ff, 349ff, 427ff, 475ff, 535ff, 569ff, 631ff, 675ff, 729ff, 791ff, 841ff, 907ff, 947ff, 965ff, 1034
  security clearance for, 1037
  summation of, 977–990
Gasdor, Albert J., 78
Getting, Ivan A., 746, 940, 941
Gladstone, William E., 556
Glennan, Keith, 1033
Glennon, Thomas K., 189, 253–256
Golden, William, 93
Gordon, Lincoln, 345, 778, 783
Graham, Frank, 972, 973
Grauer, Ben, 107
Graves, Alvin, 91

Gray, Gordon, 1ff, 51ff, 99ff, 161ff, 221ff, 283ff, 349ff, 427ff, 475ff, 535ff, 569ff, 631ff, 675ff, 729ff, 791ff, 841ff, 907ff, 947ff, 965ff, 1034, 1041, 1056
Green, Jerry, 215, 539, 540
Greenglass, David, 278, 280, 448, 492, 831
Greenwalt, Crawford, 45
Griggs, David T., 338, 339, 525, 526, 527, 532, 758, 779, 780, 925–927, 929, 932, 942, 943
  Oppenheimer memorandum of, 757
  testimony of, 742–770, 925, 937, 938, 841
Gromyko, Andrei A., 40, 43
Groner, D. Lawrence, 376
Gross, Melvin, 903
Grossman, Aubrey, 155
Groves, Leslie R., 12, 14, 15, 28–33, 36, 41, 45, 76, 119, 120, 129, 137, 151, 153, 159, 160, 172, 196, 198, 207, 208, 212, 220, 227, 259, 260, 261, 263–265, 269, 271, 272, 277, 308, 309, 345, 347, 373, 375–377, 379, 381, 384, 390, 392, 414, 416, 418, 420, 431, 471, 482, 509, 523, 544, 557, 560, 561, 563, 564, 579, 611, 628, 637, 643, 648, 662, 663, 680, 681, 725, 731, 732, 803, 815, 817, 822, 837, 875, 877, 878, 904, 950, 957, 1007, 1008, 1033, 1042, 1051, 1062
  clearance of Oppenheimer and, 1042
  office of, 812, 813
  testimony of, 160–180
Groves, Ralph, 425
Gruenther, Alfred, 48, 496, 498, 892
Gustavson, Reuben G., 107

Hafstad, Larry, 75, 577, 578, 779, 780, 782
Hancock, John M., 40
Hand, George, 189, 190
Hargrove, Marion, 107
Harkins, William, 671
Harmon, John H., 198
Harriman, Averill, 734, 738
Harris, Louis, 107
Harrison, George, 35, 257
Harrison, Richard S., 10, 574
Hart, Moss, 107
Hawkins, David, 5, 13, 116, 196–198, 204, 252, 268, 445, 450, 468, 493, 494, 588, 625, 651, 708, 822, 876, 884, 980, 1006
Hawkins, Frances, 445, 588
Haworth [Hayworth], Leland, 617
Heisenberg, Werner, 467
Hellman, Lillian, 107
Hersey, John, 107, 172
Herskovits, Melville J., 107
Hickerson, J. Allen, 107
Hildebrand, Joel, 131, 135, 190, 280
Hilgard, Ernest, 131, 135
Hill, Albert G., 598, 683, 763, 923
  testimony of, 934–944

Hillenkoetter, Roscoe H., 76
Hines, John, 507
Hinshaw, Carl, 777
Hiskey, Clarence, 4, 10, 172, 194, 1004
Hogness, Thorfin R., 107
Holloway, Marshall, 109
Holmes, Oliver, 1015
Hoover, Herbert, 680
Hoover, J. Edgar, 25, 27, 180, 264, 374, 375, 379, 380, 381, 411–417, 423, 424, 610–612, 983, 984, 985, 986
  March 1947 letter of, 806, 807, 967, 968
Hosford, William, 545
Hubbard, M. M., 924
Hull, John E., 77
Huston, Walter, 107

Ivanov, Peter, 5, 825

Jenkins, David, 189, 190, 780, 825
Jenkins, Francis, 189
Johnson, Crockett, 107
Johnson, Howard G., 738
Johnson, Joseph, 95, 562
Johnson, Louis, 404
Johnson, Lyle, 128, 136, 137, 144, 152, 153, 244, 264, 274, 277, 285, 813, 817, 823, 827–830, 849, 871
  Oppenheimer interview of, 285, 829, 831, 863–871
Johnson, Ralph, 776, 777
Jones, Tom O., 379, 380, 411, 413, 416, 417, 424, 425, 610–612, 985, 986
  memorandum of, 806
Josephs, Devereux, 734, 738

Kai-Shek, Chiang, 105
Kallet, A., 3, 9
Kaneman, Leon, 450
Kaneman, Martin, 450
Karplus, Robert, 709, 724
Katz, Milton, 738
Kaun, Alexander S., 588, 877
Keech, Richmond B., 376
Keith, Dobie, 544
Kellogg, J. M. B., 90
Kelly, Gene, 107
Kelly, Mervin J., 46, 56–66, 86, 342, 686, 687, 738, 787, 802, 1033
Kennan, George F., 26, 41, 77, 299, 351, 352, 978, 1033
Kennedy, Joseph W., 198, 658, 671, 672
Kenney, George, 744
Killian, James R., 352, 602, 672, 673, 765, 925, 1033
Kimball, Robert, 91
King, MacKenzie, 35
Kinsey, Bernard, 777, 778
Kistiakowsky, George, 772, 786
Klein, A. C., 544
Knowland, William F., 242, 243, 245, 782, 783
Knudsen, William, 376
Kolthoff, Isaac M., 107
Konopinski, Emil, 11

Kuntz, C. S., 351
Kyes, Roger M., 77

LaGuardia, Fiorello H., 107
Lambert, Rudy, 4, 9, 139, 140, 155, 183, 205, 589, 825, 877, 1003–1005, 1051, 1063
Lane, Allen, 903
Langer, William L., 738
Langmuir, Irving, 173, 337
Lansdale, John, 14, 31, 117–119, 123, 128, 132, 136, 137, 141, 143, 152, 153, 159, 189, 199, 200, 202–210, 224, 227, 276–280, 308, 425, 471, 549, 627, 637, 648, 813, 816, 819, 821, 849, 976, 980, 981, 983, 1008, 1033, 1044, 1051
  letter of, 276, 278, 1002
  Oppenheimer interview of, 203–208, 281, 347, 523, 831, 844, 871–888, 985, 986
  testimony of, 258–281, 982
Latimer, Wendell M., 190, 671, 699, 774–776, 978
  testimony of, 656–671
Lattes, Caesar, 152
Lattimore, Owen, 371
Laukhuff, Perry, 738
Lauritsen, Charles, 10, 48, 58, 64–66, 231, 352, 496, 521, 526, 598, 600, 616, 617, 659, 687, 688, 747–750, 759, 762, 787–789, 795, 796, 802, 804, 891, 922, 923, 936, 978
  testimony of, 577–596
Lauterbach, Richard, 107
Lawrence, Ernest O., 14, 27, 34, 36, 46, 48, 61, 90, 108, 112, 117, 118, 123, 164, 166, 172, 173, 187, 189, 194, 208, 231, 238, 242–246, 258, 268, 271, 272, 277, 311, 314, 315, 317, 452, 460, 461, 546, 547, 659, 679, 682, 699, 707, 755, 771, 774–780, 782, 783, 785, 787, 793, 826, 876, 878, 882, 950, 979, 980
  affidavit of, 969
LeBaron [Lebarron], Robert, 57, 58, 62, 77, 242, 758, 776
Lee, Robert, 259
Lehmann, Lloyd, 589
LeMay, Curtis, 681, 682
Lemnitzer, Lyman L., 734, 738
Lewis, Albert L., 903
Lewis, Roger, 190
Libby, Bernard, 903
Libby, Willard F., 17, 83, 85, 658, 659, 701, 702, 705
Lilienthal, David E., 27, 37, 39, 86, 109, 168, 169, 180, 229, 238, 244, 251, 303, 320, 344, 345, 351, 387, 390, 429–433, 435, 436, 438, 544, 552, 563, 564, 610, 614, 677, 708, 777, 806, 975, 977, 978, 983, 985, 1033.
  See also Acheson-Lilienthal Report
  letter to FBI of, 611
  testimony of, 372–382, 398–425
List, Eugene, 107

Lomanitz, Giovanni R., 5, 13, 117–119,
    122–129, 132–134, 136, 143, 208,
    209, 212, 268, 271, 272, 274, 276,
    277, 294, 450, 493, 589, 811–813,
    816, 820, 823, 826, 827, 829, 845,
    847, 848, 853, 866, 867, 871, 875,
    876, 878, 880, 882, 883, 980, 1006,
    1007, 1018, 1044, 1051, 1052, 1063
Longmire, Conrad, 83, 481, 485, 719,
    720, 722
Loomis, Alfred, 776
Loomis, F. W., 597
Loomis, Mannette, 776
Lovett, Robert, 59, 344, 495, 496, 505,
    506, 521, 584, 753, 758
Lyon, Peter, 107

MacArthur, Douglas, 33, 509, 744, 810
Mahan, Alfred Thayer, 680
Malina, Frank K., 587
Mallory, Walter H., 738
Malraux, André, 139, 140, 141, 889, 920,
    1009, 1063
Manley, Jack, 4, 5, 11, 81, 88, 1003, 1004,
    1006
Manley, John H., 28, 68, 88, 92, 93, 198,
    242, 246, 396, 402, 480, 481, 683,
    716–718, 725, 775, 977, 979, 1033
    affidavit of, 673–674
    statement of, 88–92
March, Florence E., 107
Mark, Carson, 91
Marks, Herbert S., 55ff, 99ff, 161ff,
    221ff, 283ff, 349ff, 427ff, 475ff,
    535ff, 569ff, 631ff, 675ff, 729ff,
    791ff, 841ff, 907ff, 947ff, 965ff
Marshall, George, 32, 34, 35, 342, 343,
    345, 353, 354, 495, 561
Marshall, James C., 28, 123, 129, 132
Martin, W. T., 672
May, Alan N., 492
May, Andrew, 35
May, Kenneth, 4, 10, 189, 190, 193, 218,
    826, 1003, 1004
May, Mrs. Kenneth, 10, 189, 1004
Maynor, Dorothy, 107
McCarthy, Joseph, 54
McCloy, John J., 36, 37, 373, 738, 1033
    testimony of, 731–742
McCone, John, 521, 758, 759
McCormack, James, 46, 58, 59, 65, 66,
    82, 242, 352, 505, 683, 685, 687,
    703, 706, 776, 796, 977–979, 1033
    testimony of, 633–643
McKenny, Ruth, 885
McMahon, Brien, 15, 90, 91, 242–245,
    257, 717, 718, 776, 777, 832, 833,
    1055
McManus, John T., 107
McMillan, Brockway, 924
McMillan, Edwin, 189, 778
McMorris, Charles H., 46, 48
McNarey [McNary], Joseph T., 270
McNaughton, Andrew, 344, 346
Milliken, R. A., 60, 61, 747
Mitchell, William, 7, 25, 381, 405, 613,
    633, 677, 678, 690, 805–807

Morgan, Thomas A., 1ff, 51ff, 99ff, 161ff,
    221ff, 283ff, 349ff, 427ff, 475ff,
    535ff, 569ff, 631ff, 675ff, 729ff,
    791ff, 841ff, 907ff, 947ff, 965ff
Morrison, Philip [Phillip], 196, 197,
    225–227, 267, 425, 445, 446, 494,
    548, 549, 551, 557, 589, 594, 620,
    621, 625, 626, 651
Mosely, Philip E., 738
Moses, 502
Moss, Stanley, 107
Muir, Jane, 200–202, 876
Murphree, Eger, 17, 83
Murphy, Charles, 931. *See also Fortune*
    magazine
Murray, James S., 817, 818
Murray, Thomas, 84, 311, 312, 1034, 1049
    opinion of, 1058–1061

Nedelsky, Winona, 200
Nelson, Eldred, 204, 205, 589, 877
Nelson, Steve, 4, 5, 10, 13, 75, 116, 180,
    181, 190, 194–196, 205, 206, 268,
    573, 575, 576, 811, 822, 825, 877,
    878, 901, 1003–1006
Neyland, John, 613, 623, 779, 780, 782,
    988
Nichols, Kenneth D., 20, 22, 28, 41, 53,
    58, 66, 77, 107, 117, 122, 123,
    164, 165, 171, 227, 231, 250, 252,
    260, 264, 270, 357, 405, 416, 432,
    463, 505, 506, 616, 677, 678, 681,
    682, 683, 687, 688, 706, 722, 787,
    788, 807, 991. *See also* Nichols
    letter
    memorandum of, 677
    recommendations of, 1040–1046. *See
    also* General Advisory Committee,
    General Manager of
    security findings of, 1042
Niemeyer, Gerhart, 738
Nimitz, Chester W., 613, 623, 988
Nixon, Richard M., 178
Nordheim, Lothar, 255, 305, 338, 484,
    485, 645
Norstad, Lauris, 41, 48, 496, 498, 684,
    685, 892, 905
Norton, Garrison, 504, 525, 747, 751,
    758

Ofstie, Ralph A., 690
O'Leary, Mrs., 270
Oppenheimer, Frank F., 4, 5, 8, 9, 13,
    101, 111, 112, 117, 119, 142, 150,
    159, 167, 168, 186–188, 196, 199,
    204, 212, 215, 216, 219, 220,
    263–265, 277, 380, 381, 412, 417,
    446, 468, 549, 586, 587, 592, 594,
    627, 821, 837, 872, 873, 877, 919,
    980, 981, 984, 987
Oppenheimer, Jackie, 4, 9, 101, 186, 199,
    220, 446, 872, 873, 980, 987
Oppenheimer, J. Robert
    American Committee for Democracy
        and Intellectual Freedom and,1001
    Atomic Energy Commission and, 1045,
        1053, 1056

Oppenheimer, J. Robert (*continued*)
  attorneys of, 1034
    attorneys letter of, 1029–1034
  biographical data on, 24
  David Bohm and, 1030, 1063
  "calculated risk" of, 1014, 1041
  Chevalier incident and, 1043, 1044, 1050, 1051, 1062, 1063, 1064
  clearance for, 999, 1019, 1021
  clearance of counsel, 1037, 1038, 1056
  Communist activities and, 1002, 1005, 1006, 1042, 1051, 1052
  Communist Party and, 101, 885, 902, 916, 917
  Edward Condon and, 1030
  false statements of, 1043, 1051
  as fellow traveler, 876, 1002, 1052, 1063
  felony of, 1051
  Friends of the Chinese People and, 1001
  General Advisory Committee (GAC) and, 520, 526, 533, 660, 663, 664, 670, 687, 748, 879, 951, 958, 1044, 1045, 1051, 1054. *See also* General Advisory Committee
  government service of, 23, 837, 975
  House Un-American Activities Committee, testimony before, 1009, 1018, 1044, 1051, 1063
  Independent Citizens Committee of the Arts, Sciences, and Professions, Inc. and, 1007, 1016
  influence of in thermonuclear program, 1010, 1016, 1017
  internationalizing of atomic energy and, 684, 691
  lectures of, 97, 226, 227, 673
  left-wing associations of, 975, 976, 980
  letters of, 5, 104, 106, 182, 213, 214, 242, 243, 954, 955, 956
  life picture of, 1033
  Lincoln Summer Study and, 749, 750, 764, 936, 938, 941
  Lomanitz and, 1051, 1063, 1064
  at Los Alamos, 692, 712, 732, 768, 949, 950, 957, 960, 961, 1050, 1054. *See also* Los Alamos
  loyalty of, 430, 431, 459, 465, 489, 498–499, 512, 517, 522, 523, 533, 545, 546, 548, 549, 550, 551, 555, 594, 595, 599, 606, 617, 628, 637, 694, 710, 737, 749, 754, 800, 843, 1017, 1020, 1029, 1030, 1059, 1061
  moral opposition of, 1058
  1947 security clearance of, 806, 976
  opposition to thermonuclear program of, 647, 659, 660, 684, 685, 692, 693, 700, 701, 804, 1010, 1016, 1017, 1044, 1058, 1059
  perjury of, 1051
  personnel security questionaire of, 1001, 1016
  resignation of, 33
  restoration of 'Q' clearance, 1046
  secret information and, 1062
  security officers, opposition to, 1042
  security risk, 1045, 1046, 1053, 1054
  Soviet-American relations and, 732, 733, 738
  testimony of, 21–49, 65–88, 92–160, 180–287, 887–906, 949–961, 971
  thermonuclear weapons program and, 702, 958
  value of, 1057
  veracity of, 1044, 1045, 1055
  Vista report and, 747, 748, 759, 762, 891, 959. *See also* Vista, project
  Western Council of the Consumers Union and, 1001, 1002
  witnesses for, 1033
Oppenheimer, Katherine P., 3, 4, 10, 13, 15, 21, 25, 118, 187, 188, 195, 206, 219, 265, 266, 277, 446, 451, 555, 837, 876, 879, 890, 901, 980, 987, 1002
  testimony of, 571–577, 914–920
Osborn, Frederick H., 41, 43, 299, 342, 385, 419, 975, 1033
  testimony of, 342–346, 539
Ott, Henry, 782

Papich, A., 3
Parodi, Alexandre, 344
Parsons, William S., 30, 31, 45, 46, 58, 65, 66, 75, 77, 441, 505, 615, 616, 680, 687, 796, 801
Pascal, Ernest, 107
Pash, Boris T., 128, 137, 138, 141, 143–148, 153, 210, 264, 270, 271, 273, 274, 277, 285, 308, 637–639, 648, 872, 875, 904, 980–982, 1008, 1009, 1043, 1050, 1051
  memoranda of, 814–819, 821, 822, 871
  Oppenheimer interview of, 281, 285, 347, 829, 831, 844, 845–871, 887, 888, 985
  testimony of, 808–831
Patterson, Robert P., 15, 35, 107, 180, 344, 345, 375, 377, 381, 418, 611
Pauling, Linus, 107
Payne, Virginia, 107
Pearson, Drew, 558, 780
Pegram, George, 467
Peierls, Rudolph, 31
Pepper, Stephen, 190
Peters, Bernard, 116, 120–122, 129, 150, 190, 210–215, 252, 253, 450, 468, 589, 594, 708, 818, 825, 883, 884, 1018, 1021, 1044, 1051, 1063, 1064
Peters, Hannah, 5, 120, 121, 190, 205, 210, 251, 589, 825, 877, 878, 883, 1005, 1006
Peters, John P., 107
Pike, Sumner T., 143, 251, 320, 551, 374, 401–403, 423, 677, 978, 983, 1033
  testimony of, 429–438
Pinsky, Paul, 5, 10, 11, 187, 188, 589, 915, 1006, 1007
Pitman, John, 825
Pitzer, Kenneth S., 316, 317, 321, 659, 665, 666, 776, 780, 782, 978
  testimony of, 698–709
Placzek, George, 10, 324, 328

1073

Quarles, Donald, 499, 500, 507, 550
Quesada, Elwood R., 524, 525, 584

Rabi, Isadore I., 17, 19, 33, 34, 68, 76, 79, 80, 81, 94, 231, 314, 320, 352, 395, 396, 517, 518, 519, 525, 526, 527, 529, 589, 598, 600, 665, 740, 750, 751, 778, 779, 793, 805, 936, 976, 989, 1032
  testimony of, 451–473
Radford, Arthur W., 47
Ramsey, Norman F., 31, 351, 597, 680, 978, 1033
  testimony of, 440–451
Rautenstrauch, Walter, 107
Redding, A. David, 738
Reed, Mrs. John, 362
Reston, James, 53, 54, 55, 57
Reynolds, Quentin, 107, 679, 783, 784
Richman, C., 13
Richtmyer, F. K., 337, 484
Ridenour, Louis N., 746
Rider, A., 8
Ridgeway, Matthew B., 38
Riley, William E., 405
Robb, Roger, 1ff, 51ff, 99ff, 161ff, 221ff, 283ff, 349ff, 427ff, 475ff, 535ff, 569ff, 631ff, 675ff, 729ff, 791ff, 841ff, 907ff, 947ff, 965ff
Roberts, Henry L., 738
Roberts, Owen J., 376
Robeson, Paul, 107
Robinson, Gerold T., 738
Rockefeller, Nelson, 94
Rolander, C. A., 1ff, 51ff, 99ff, 161ff, 221ff, 283ff, 349ff, 427ff, 475ff, 535ff, 569ff, 631ff, 675ff, 729ff, 791ff, 841ff, 907ff, 947ff, 965ff
Roosevelt, Eleanor, 229, 270, 961, 962
Roosevelt, Franklin D., 104, 105, 163, 252, 255, 361
  letter from, 29–30
Root, Elihu, 562
Roper, Miss, 876
Rosenbluth, Marshall N., 337
Rowe, Hartley, 17, 19, 31, 80, 352, 518, 976, 977, 979, 1032
  testimony of, 507–514
Rushmore, Howard, 541
Rusk, Dean, 738
Russell, Bertrand, 97
Russell, Katharine, 108

Saint Francis, 368
Saint Paul, 502
Sandow, Katrina, 5, 11, 1006
Schein, Marcel, 10
Schneiderman, William, 4, 10, 139, 140, 191, 192, 205, 216, 217, 219, 590, 825, 877, 918, 919, 1005
Schoenberg, Mario, 181, 182, 183
Schuyler, Cortland, 892
Scott, Hazel, 107
Seaborg, Glenn, 17, 19, 68, 81, 164, 233, 234, 237, 238, 239, 240, 241–243, 402, 462, 657, 658, 659, 664, 665, 666, 778, 1020, 1044, 1051, 1064
  letter of, 1064
Segré, Emilio, 712
  letter to Oppenheimer of, 238, 241
Segure, Rose, 813
Seitz, Frederick, 314
Sen, Sun Yat, 105
Serber, Charlotte, 11, 119, 202, 203, 208, 224, 267, 277, 445, 590, 594, 613, 614, 620, 621–624, 633, 876, 880, 898, 988
Serber, Robert, 203, 204, 208, 225, 226, 231, 245, 246, 247, 267, 277, 549, 590, 594, 613, 614, 623, 624, 633, 778, 780, 783, 784, 785, 793, 803, 876, 898, 977, 988
Shapley, Harlow, 107
Shaver, Dorothy, 257
Silverman, Samuel J., 1ff, 51ff, 99ff, 161ff, 221ff, 283ff, 349ff, 427ff, 475ff, 535ff, 569ff, 631ff, 675ff, 729ff, 791ff, 841ff, 907ff, 947ff, 965ff
Sinatra, Frank, 107
Smith, Cyril, 17, 19, 31, 81, 198, 670
Smyth, Henry D., 83, 84, 314, 403, 404, 408, 430, 431, 432, 740, 1034, 1049
  opinion of, 1061–1065
Snapp, Roy B., 320, 405, 410, 411, 422, 435
Snyder, Hartland, 878
Souers [Sowers], Sidney, 683
Spectorsky, A. C., 107
Spofford, Charles M., 738
Sproul [Spraulle], Robert G., 744
Steuben, John, 915, 918
Stevenson, Adlai, 335
Stimson, Henry L., 14, 31, 34, 70, 171, 342, 377, 561, 564, 731, 732, 733, 975
Stone, Shepard, 738
Stratton, Julius, 764, 925
Straus, Jack, 139, 192, 903, 915
Strauss, Lewis L., 22, 27, 257, 374, 402, 403, 404, 431, 433, 437, 463, 468, 469, 666, 677, 977, 978, 989, 1034, 1045, 1049
Summerfield, Martin, 588
Szilard, Leo, 398

Tamm, Frank, 264, 265
Tatlock, Jean, 4, 8, 116, 121, 122, 153, 154, 155, 156, 190, 264, 267, 822, 877, 902, 980, 1002, 1063, 1064
Teller, Edward, 11, 20, 27, 61, 63, 64, 76, 77, 80, 82, 83, 84, 85, 86, 91, 231, 232, 234, 242, 243, 255, 304, 305, 310, 311, 312, 314, 315, 316, 322, 323, 325, 328, 329, 330–334, 336–339, 437, 460, 461, 481, 484, 488, 497, 520, 645, 662, 669, 670, 682, 699, 701, 703, 704, 706, 707, 753, 758, 774, 775, 776, 783, 786, 789, 795, 797, 799, 899, 977, 979
  discovery of, 331, 334, 338, 462
  at Livermore, 698
  second laboratory and, 755

Teller, Edward (*continued*)
  testimony of, 709–727
  Vista project and, 748, 762, 763
Thomas, Charles A., 34, 39, 40, 373, 544, 552, 732
Thomas, Norman, 532
Thompson, L. T. E., 507, 884, 988
Thoreau, Henry D., 473
Thorez, Maurice, 572
Thorndike [Thorndyke], Alan, 617
Thornton, Robert, 778
Tolman, Ed, 189, 345
Tolman, Richard, 10, 15, 36, 40, 175, 178, 197, 207, 582, 954, 955, 956, 957
Tolman, Mrs. Richard, 207, 878
Truman, Harry S., 34, 35, 41, 60, 96, 257, 330, 331, 433, 437, 438, 441, 516, 520, 532, 672, 910, 911, 985
  hydrogen-bomb decision of, 241, 366, 431, 456, 462, 493, 496, 506, 511, 605, 615, 643, 645, 717, 718, 719, 720, 786, 794, 978, 979
  letter from, 96–97
  Russian atomic explosion announcement of, 399, 562
Tyler, C. L., 90

Ulam, Stanislaw, 305, 669, 775
Unna [Uanna], William, 425, 612
Urey, Harold C., 36, 175, 659
Urquidi, Donald, 738

Valley, George, 769, 770
Vandenberg, Hoyt, 36, 39, 75, 76, 682, 748, 749, 754, 801, 910, 969, 977
Van Doren, Carl, 107
Van Vleck [Van Fleck], John H., 11
Viner, Jacob, 738
Vinograd, Jerome, 131, 135
Volpe, Joseph, 211, 219, 306, 379, 380, 381, 415, 416, 423, 424, 438, 610, 611, 612, 905, 986, 1033
von Neumann, John, 17, 20, 26, 61, 68, 80, 82, 84, 87, 231, 242, 246, 305, 314, 352, 712, 775, 978
  testimony of, 643–656

Walkowicz, T. F., 747
Wallace, Henry A., 16, 104, 105
Waters, John A., 321
Waymack, Wesley W., 374
Webb, Beatrice, 10
Webb, Sidney, 10
Webster, William, 41, 57, 543
Wedemyer, Albert, 584
Weil, George, 782
Weinberg, Alvin, 782

Weinberg, Joseph W., 4, 5, 10, 13, 15, 117, 120, 121, 126, 128, 142, 143, 150, 157, 172, 176, 193, 194, 200, 201, 208, 209, 210, 216, 268, 271, 277, 391, 450, 494, 590, 811, 812, 813, 819, 820, 821, 825, 826, 827, 875, 876, 880, 882, 883, 905, 1004–1006, 1044, 1063
Weisskopf, Victor, 10, 212, 215, 253, 324, 328
Welles [Wells], Orson, 107
West, Julian, 924
Weyermack, W. W., 677
Wheeler, John A., 20, 82, 84, 183, 305, 793
Whitman, Walter G., 17, 48, 58, 66, 80, 83, 151, 352, 477, 521, 522, 687, 787, 988, 1033
  statement of, 672
  testimony of, 495–507
Whitson, Lish, 264, 265
Wigner, Eugene P., 87, 151, 152, 183, 430
Williams, Walter, 544
Wilson, Carroll, 90, 387, 389, 417, 552, 553, 700, 983, 1011
Wilson, Charles E., 93, 495, 500, 501
  press conference of, 501–503
Wilson, Roscoe C., 46, 58, 59, 616, 977, 978
  testimony of, 679–697, 977
Winne, Harry A., 37, 39, 352, 373, 374, 542, 732, 978, 1033
  testimony of, 542–557
Withington, Frederic, 545
Wood, C., 43
Woodward, Mr. and Mrs., 208, 446, 876, 880
Worthington, Hood, 17
Wright, Jerauld K., 545
Wriston, Henry M., 734, 738
Wyman [Wymans], Jeffries [Jeoffrey], 141, 889, 920, 921, 1063
  letter of, 969

York, Herbert, 86, 314, 457, 645
Young, Barney, 188, 916
Young, Gale, 783, 784

Zacharias, Jerrold R., 31, 351, 440, 458, 528, 750, 760, 769, 936, 941, 943
  testimony of, 596–602, 933
Zingrosser, Carl, 107
Zinn, Walter, 90, 780, 782, 783, 784, 793, 794
Zinner, Paul E., 738
Zuckert, Eugene M., 1034, 1049, 1052
  statement of, 1052–1055

SUBJECT INDEX

A-bomb. *See* Bomb, atomic
Abraham Lincoln Brigade, 266, 270
Acheson-Lilienthal report, 16, 38, 39, 40, 87, 95, 345, 384, 385, 691, 721, 722
Advanced Study, Institute for. *See* Institute for Advanced Study
AFOAT-1 advisory panel, 801
Aircraft, nuclear powered, 75, 684, 692, 693, 695
Air Defense Command, 763, 765, 938. *See also* Strategic Air Command
Air Force, 338, 340, 440, 441, 582, 634, 787, 837, 910, 911
   atomic detection and, 695, 951
   Atomic Energy Commission and, 696
   atomic weapon delivery and, 696
   Chief of Staff of, 685, 744
   continental defense and, 598. *See also* Defense, continental
   fighter tactics of, 641
   intelligence of, 694
   Lincoln Summer Study and, 764, 925–927
   at Massachusetts Institute of Technology, 94. *See also* Charles project
   radar and, 743
   rockets and, 578
   Scientific Advisory Board of, 440, 441, 443, 446, 515, 583, 644, 750, 769, 922, 928, 936, 978
   second laboratory and, 746, 754, 755
   strategic, 585, 586. *See also* Weapons, strategic
   thermonuclear bomb and, 447, 682, 744, 751
   Vista report and, 747, 748, 749, 758, 759. *See also* Vista project
Air War College, 680
Alamogordo, testing at, 11, 681, 711, 949. *See also* Testing
Alsop column, 760, 761
American Committee for Democracy and Intellectual Freedom, 3, 9, 158, 159
American Friends of the Chinese People, 1001
American Institute of Physics, 698
American League against War and Fascism, 600
American Mathematical Society, 644
American Philosophical Society, 57, 63
American Physical Society, 193, 324, 327, 339, 440, 443, 515, 566, 587, 589, 600, 604, 614, 723

American-Russian Institute, 4
American Soviet Council, 877
Anticommunism, 386
Appropriations, House Committee on, 3
Architects, Engineers, Chemists and Technicians, Federation of (FAECT). *See* Federation of Architects, Engineers, Chemists and Technicians
Arco, uranium piles at, 794
Argonne Laboratory
   in Chicago, 722, 785
   at Oak Ridge, 67, 73, 75, 87, 91, 241, 341, 777, 780, 794. *See also* Oak Ridge
Armaments, regulation of, 41, 43, 330. *See also* Control
Armed Forces Special Weapons Project, 681, 682, 744
Armed Services, House Committee on, 690
Army, 582, 837, 872
   advisory panel of, 515
   Air Force support of, 680
   Department of, 813
   Information and Educational Division of, 342
   Red, 115, 116
   rockets and, 578
Army Ordnance Corps, 588
Army Signal Corps, 505, 603
Associated Universities, 451
Association of Scientific Workers, 131
Atlee Conference, 561
Atomic Development Authority (ADA), 42
Atomic energy, 97, 178, 305, 356
   British program of, 578
   inspection of work on, 37, 373
   international control of, 36, 40, 44, 227, 231, 303, 344, 345, 355, 356, 359, 360, 373, 544, 694, 695. *See also* Control
   Joint Congressional Committee on, 18, 19, 71, 76, 81, 227, 234, 239, 302, 313, 319, 382, 435, 682, 686, 687, 832, 977
   Panel on Military Objectives in Field of, 801
   power use of, 34, 61, 69
   rehabilitation of, 398
   Secretary of State's committee on, 16, 37, 39. *See also* Acheson-Lilienthal report
   Senate Committee on, 416
   Special committee on, 15
   for weapons, 18. *See also* Weapons

Atomic Energy Act, 3, 65, 178, 179, 334,
  364, 369, 379, 391, 405, 465, 493,
  511, 516, 522, 553, 558, 566, 610,
  739, 972, 973, 983, 986, 987, 1000,
  1042, 1049
 FBI investigation and, 416
 grandfather clause in, 374
Atomic Energy Commission (AEC), 3,
  6, 7, 18, 19, 20, 69, 74, 90, 91, 92,
  96, 211, 213, 231, 232, 254, 300,
  301, 302, 307, 313, 317, 322, 323,
  324, 333, 337, 339, 340, 343, 355,
  533, 683, 693, 983
 Deputy General Counsel of, 438
 Division of Research of, 340, 341,
   698, 699
 fellowships of, 230, 517, 668, 669,
   671, 699
 General Advisory Committee of. *See*
   General Advisory Committee
 General Manager of, 20, 58, 321, 340,
   973, 979
 membership of, 518, 520, 1033
 Personnel Security Clearance Criteria
   for Determining Eligibility, 1000,
   1001, 1014, 1017
 Personnel Security Procedures of, 1056
 rules and regulations of, 22, 973
 security clearance procedures of, 21,
   321, 741, 742, 1000, 1001, 1050
 security criteria of, 1065
Atomic explosions, long-range detection
  of, 692, 694, 695. *See also* Testing
Atomic Scientist, 961
Atomic Weapons Program, 90, 517. *See
  also* Thermonuclear program;
  Weapons, atomic
Attorney General, of United States, 4,
  16, 417, 490
Attrition, policy of, 47

Bank of America, 737
Baruch Plan, 303, 343, 344, 345, 561,
  564, 691
Bell Telephone Laboratories, 17, 57,
  242, 340, 400, 602, 738, 924
Berkeley, 8, 13, 14, 26, 72, 85, 173,
  588, 698, 704, 705. *See also*
  California, University of
 accelerator at, 74
 espionage at, 381, 985
 radical circles at, 101
Berlin University, 353
Bill of Rights, 566
Bikini [Binkini], atomic testing at, 177,
  257, 713
Bolsheviks, 369
Bomb
 atomic, 70, 75, 77, 87, 233, 257, 302, 303,
   304, 305, 315, 319, 324, 325, 436.
   See also Super; Weapons
  definition of, 30
  delivery of, 561
  development of, 378, 510, 579. *See
    also* Thermonuclear program
  German, 261
  military and, 390
  opposition to, 81, 228, 229, 247,
    326
  Soviet, 237, 400, 401. *See also*
    Soviet Union
 buzz, 509
 design for, 73, 163
 electromagnetic, 164, 176
 fission, 77, 400, 422, 483, 1010. *See also*
   Fissionable materials
 fusion, 483. *See also* Fusion process
 gaseous diffusion, 175, 543
 hydrogen, 75, 82, 83, 84, 86, 175,
   227, 229, 231, 233, 249, 250, 302,
   303, 305, 319, 329, 330, 363, 1010,
   1011, 1044
  development of, 23, 25, 55, 85, 87,
    230, 241, 301, 302, 304, 307, 320,
    358–360, 367, 395, 402, 404, 431,
    437, 464, 469, 478, 479, 512, 547,
    564, 616, 725, 785, 795, 798
  feasibility of, 590, 796
  opposition to, 18, 19, 79, 80, 328,
    333, 336, 368, 384, 448, 481, 510,
    546, 547, 563, 591, 788, 797, 799,
    975, 976
  Russian, 370. *See also* Russia;
    Soviet Union
  test of, 562. *See also* Testing
 implosion, 177
 materials for, 75, 176, 356
 plutonium, 303. *See also* Plutonium
 Russian, 67, 144, 145, 174, 175, 176,
   249, 250, 316, 319, 330, 362, 363,
   658, 659. *See also* Russia;
   Soviet Union
 strategic, 697
 super, 67, 774. *See also* Super
 tactical, 697
 thermal diffusion, 164
 thermonuclear, 62, 63, 229, 235,
   246, 247, 248, 714. *See also*
   Thermonuclear program;
   Weapons, thermonuclear
Brazil, University of, 181
British Constitution, 513
British mission, at Los Alamos, 31, 732
British representative, to United
  Nations, 344, 346
Brookhaven, laboratory at, 68, 73, 341,
  445, 451, 614, 617, 988
Brown University, 734, 738
Bruceton Laboratory, 786
Bulletin of Atomic Scientists, 329, 336

Cabinet, members of, 500
California, University of, 5, 7, 17, 26,
  73, 91, 102, 174, 233, 276, 277,
  285, 315, 478, 490, 563, 574, 588,
  613, 657, 698, 710, 724, 744, 779,
  817, 863, 988. *See also* Berkeley
California Labor School, 4, 1003
California Institute of Technology, 7,
  8, 9, 10, 15, 17, 26, 35, 102, 496,
  515, 577, 578, 581, 587, 588, 608,
  698, 708, 779, 781
 Vista project at, 521, 582, 746, 747.
   See also Vista project

Cambridge, University of, 7
Canada
  atomic energy and, 18, 175
  cooperation with, 355, 356, 777
  international control and, 41
  United Nations representative of, 344, 346
Carbide and Carbon Chemicals Co., 257, 544
Carnegie Institute, 560
Case Institute of Technology, 254
Censorship, mail, 261, 262. *See also* Security clearance
Central Intelligence Agency (CIA), 734, 810, 837
Chalk River
  atomic energy plant at, 776, 777, 778, 779, 780, 784, 794
  international meeting at, 238
Charles project, 94, 582, 583, 585, 597, 935
Chase National Bank, 731, 737
Chemical Society, 698
Chevalier incident, 110, 147, 167, 263, 280, 308, 381, 392, 431, 465, 468, 470, 471, 522, 532, 533, 549, 550, 556, 557, 625, 627, 647, 648, 829, 830, 887, 981, 985, 987, 989, 1008, 1009, 1018, 1044, 1050, 1051, 1052, 1064. *See also* Chevalier, Haakon
Chicago, University of, 17, 233, 310, 397, 621, 701, 710, 713
Chicago laboratory, 658, 755
China
  intervention of, 67
  United States troops in, 105
CIO. *See* Union, CIO
Clearance, basis for, 1042
Cold war, 359
Columbia University, 17, 397, 398, 450, 467, 601, 614, 710, 738
Committee on Present Danger, 16
Communism
  dogma of, 361
  movement of, 294
  Soviet, 576
Communist-front organizations, 3, 4, 16, 104, 159, 186, 277, 811, 1001, 1007
Communist Party, 4, 5, 8, 9, 10, 13, 15, 20, 102, 110–113, 184, 211, 212, 217, 225, 252, 276, 277, 306, 362, 389, 466, 592, 601, 650, 1003, 1013
  American, 212, 576, 601, 917
  in California, 4, 5, 10, 101, 102, 139, 191, 192, 195, 821, 1005
  candidate of, 531, 532
  closed meetings of, 216, 217, 218, 219, 318, 319
  espionage of, 111, 152, 176
  functionary of, 1002, 1004, 1005, 1051, 1052
  German National, 211, 212, 213, 214
  Hitler and, 531
  Lomanitz and, 5, 127. *See also* Lomanitz, Giovanni R.
  Los Alamos Project and, 1006

  members of, 119, 139, 153, 155, 159, 181, 183, 187–190, 193–195, 197, 200, 201, 204, 210, 216, 219, 220, 278, 361, 442, 443–445, 522, 549, 551, 555, 651, 652, 671, 672, 815, 882, 883, 915, 916, 1002–1005, 1007, 1012, 1052, 1061
  policy of, 199, 216
  public program of, 1002
  security and, 112, 147, 154
  sympathizers with, 156, 188, 189–192, 217, 252, 279, 292, 293, 385, 1002, 1005
Communists
  association with, 117, 365, 588, 589
  in Central America, 511, 976
  Chinese, 15
  definition of, 266
  employment of, 89, 110, 116
  hysteria regarding, 270
  at Los Alamos, 468, 622, 831, 876
  recognition of, 115
  Soviet, 98, 1012. *See also* Soviet Union
  war projects and, 274, 591, 601
Condon letters, 539, 540. *See also* Condon, Edward
Containment, policy of, 978
Control
  international, 38, 42, 43, 378, 385, 561, 591, 722. *See also* Atomic energy; Inspection
  Committee on, 557, 731
  over proceedings, 55
Cornell University, 324
"Coordinator of Rapid Rupture," 27
Cosmotron, at Brookhaven, 614
Council of Foreign Affairs, 59, 354, 732, 734, 738
Council of National Academy of Sciences, 16
Court-martial, possibility of, 640
Crash program for thermonuclear development, 437, 510, 518, 529, 605, 715, 716. *See also* General Advisory Committee
Crossroads Operation, 493
Crouch incident, 23, 25, 306, 318, 319, 321, 322. *See also* Crouch, Paul
Cyclogenesis, 26
Cyclotrons, 164, 657

Dahlgren, proving ground at, 493
Daily People's World, 1004
Daily Worker, 920
Defense
  continental, 59, 94, 452, 458, 585, 586, 598, 599, 602, 635, 636, 775
  Department of. *See* Department of Defense
  strategic, 458, 463, 602, 635
Delivery, problem of, 34, 400, 561
Denver University, 98
Department of Defense, 94, 315, 316, 355, 422, 439, 440, 464, 498, 499, 501, 506, 525, 550, 563, 616, 731, 837
  Reorganization Plan No. 6 of, 502

Department of Defense (*continued*)
  Weapons System Evaluation Group of, 582
Deuterium, measuring of, 484, 490
Diffusion, gaseous, 301
Disarmament, international, 556. *See also* Atomic energy; Control, international
Division of Military Applications, 84, 634, 635, 703
DSM project, 273, 285, 817, 818, 863
Duke University, 485
duPont Company, 17, 238, 241, 257, 398, 793, 794, 804

Eastman Kodak, 257
East River Project, 451, 935
Eltenton-Chevalier incident, 129, 141, 143, 209, 263, 1008. *See also* Chevalier incident; Eltenton, George C.
Eniwetok operation, 11, 70, 305, 480, 720
Engels, F., 1006
England, 298, 299
  atomic energy and, 18
  bomb project of, 174, 175, 177
  cooperation with, 292, 355, 356, 652, 653, 949
  international control and, 41
Espionage Act, 295, 296, 822, 848, 857
Espionage, 136, 269, 278, 809, 843. *See also* Information
  implication of, 638, 639
  investigation of, 821
  Operation Joe, 242, 811
  of Soviet Union, 274, 838
Ethical Culture School, 7
Ethics of weapon development, 80, 81
Europe, defense of, 386, 387, 498, 504, 591
Executive Order 9835, 1003
Executive Order 10450, 3, 6, 7, 499, 500, 503, 1000, 1001, 1003, 1018, 1042, 1045
Executive Order of the President of April 27, 1953, 1014
Exhibit No. 1, 24

FAECT. *See* Federation of Architects, Engineers, Chemists and Technicians
Fascism, 580
Federal Bureau of Investigation (FBI), 4, 5, 10, 16, 113, 137, 155, 169, 181, 191, 192, 194, 209, 210, 217, 218, 220, 254, 260, 306, 317, 318, 379, 415, 419, 437, 463, 558, 559, 593, 671, 837, 1000, 1004, 1005, 1007, 1032, 1049, 1051, 1055. *See also* Hoover, J. Edgar
  Atomic Energy Commission and, 380, 612
  communist expert of, 264, 265
  officers investigate course of, 809
  Oppenheimer files of, 564, 610, 618, 807, 808, 971, 986
  personnel files of, 412
  record checks by, 819
  Seattle branch of, 673
Federation of Architects, Engineers, Chemists and Technicians (FAECT), 14, 135, 136, 188, 206, 263, 292, 293, 549, 627, 812, 813, 825, 846, 848, 849, 851, 852, 855, 861, 862, 867, 870, 882, 883, 902
Fellow travelers, 113, 114, 140, 155, 182, 192, 203, 204, 217, 219, 490, 506, 556, 593, 624, 831, 875, 879, 881, 918, 1012, 1061
Fission, nuclear, 478, 483, 492, 530. *See also* Bomb, fission
Fissionable materials, 70, 301, 497, 507, 617, 635, 788
Ford Foundation, 738
Foreign Affairs, 45, 95, 96, 356, 734
Foreign Service, 370, 371
Fortune magazine, 922, 923, 927, 928, 931, 936
France
  espionage of, 732
  international control and, 41
  United Nations representative of, 344, 346
Frankfort University, 335
Friends of the Chinese People, 3, 9, 158
Fusion process, 319, 486, 492

General Advisory Committee (GAC), 6, 16, 17, 18, 19, 45, 46, 48, 60, 66, 67, 71, 76, 77, 79–91, 96, 97, 228, 231–233, 243, 245, 247–251, 300–304, 307, 311, 313, 314, 315, 317, 320, 321, 324, 327, 328, 331, 334, 336, 337, 340, 341, 356, 359, 377, 398, 400, 402, 403, 410, 432, 433, 437, 499, 516, 683, 686, 799, 899
  annexes of, 236, 469, 518, 519, 529, 752, 753, 805, 976
  appointment of J. R. Oppenheimer to, 66, 300, 384, 385, 414, 419, 516, 975
  chairmanship of, 386, 396, 461, 482, 489, 490, 517, 518, 520, 606, 717, 983
  crash program and, 240, 423, 432, 433, 435, 436, 454, 510, 518, 637, 644, 725, 786, 896
  December 1949 meeting of, 977
  Fermi and, 395, 396, 397
  first meeting of, 430, 495
  function of, 399, 563, 705
  General Manager of, 82, 84, 553, 628
  letter of, 1025, 1037, 1056, 1057
  recommendations of, 1040–1046, 1049, 1056, 1058
  Joint Chiefs of Staff and, 455, 462
  membership of, 379, 515, 516, 517, 519, 520, 603, 605, 610, 644, 665, 670, 677, 687, 688, 701, 951, 1032, 1033
  October 1949 meeting of, 509, 518, 607, 615, 640, 644, 709, 716, 725, 894
  opposition to thermonuclear program of, 659, 664, 670, 894, 895

General Advisory Committee (GAC) (*continued*)
  reports of, 518, 529, 664, 665, 699, 702, 703, 894, 895, 896
  second laboratory and, 755
  security clearance for, 434
  Soviet policy of, 388
  statements of, 387, 390, 404, 406, 407, 408, 409
  subcommittees of, 68, 82
  thermonuclear development and, 234, 479, 526, 751, 752, 899, 950, 1010, 1011, 1044
  weapons committee of, 517
General Electric Co., 33, 75, 173, 542, 543, 547, 548, 557
General Rubber and Tire Co., 528
German Physical Society, 324
Germany, 298, 299, 849, 858, 867
  atomic development of, 261, 809
G.I. Bill, 73
Goettingen, University of, 7, 257
Government Printing Office, 133
Gray Board, 1049–1052, 1055, 1057, 1061–1063, 1065. *See also* Personnel Security Board
Greenglass case, 261, 262. *See also* Greenglass, David
Greenhouse tests, 524, 525, 796, 797, 798, 951, 952. *See also* Testing
G-2, 259, 260, 273, 810, 822

Hanford, laboratory at, 174, 176, 301, 317, 398, 454, 517, 543, 546, 547, 548, 670, 779
Harper's magazine, 8
Hartwell project, 458, 582, 597, 933
Harvard Law School, 259, 734
Harvard University, 7, 16, 17, 95, 242, 243, 341, 384, 440, 442, 443, 444, 734, 738
H-bomb. *See* Bomb, hydrogen; Thermonuclear program
Hiroshima, bombing of, 14, 15, 196, 197, 226, 235, 333, 400, 422, 505, 667, 733, 773, 838
Hungarian Revolution, 654
Hunter College, 601

Illinois, University of, 88
Implosion, 278
Independent Citizens Committee of the Arts, Sciences, and Professions, Inc., 5, 16, 23, 25, 104, 105–107, 114, 133, 252, 1007, 1016
Industrial Advisory Committee, 603, 604
Information
  classified, 109, 846. *See also* Security clearance
  espionage and, 274, 858
  Microfilm transmission of, 130. *See also* Microfilm
  passing of, 637, 823, 825, 871
Inspection, of atomic weapons, 373, 374, 722
Institute for Advanced Study, at Princeton University, 82, 83, 84, 87, 142, 255, 304, 311, 353, 359, 516, 644, 653, 709, 719, 720, 724, 738, 775
Intelligence, atomic, 109. *See also* Information
Interim Committee, on Atomic Energy, 14, 34, 35, 377, 564, 975
International Atomic Energy Commission, 40
International Agreement on Atomic Weapons, 326, 327
International Brigade, 186
International Control of Atomic Energy, report on. *See* Acheson-Lilienthal report
Interstate Commerce Commission, 437
Inyokern, 582
Iron Curtain, 38, 45, 344, 374, 975
Isotopes
  distribution of, 74
  of hydrogen, 484
  separation of, 779
  use of, 69
Ivy shot, 591, 595, 797. *See also* Testing
Iwo Jima, 578

Jackson committee, 96
Japan, 14
  invasion of, 34
  bombing of, 33, 236
  targets in, 236. *See also* Hiroshima; Kyoto; Nagasaki
JATO (jet-assisted takeoff), 587
Jews, German, 8
Joe, agent, 811
Joe I, 401. *See also* Operation Joe; Soviet Union, atomic explosion of
Joint Academy of Sciences, 63
Joint Army and Navy Committee on Welfare and Recreation, 342
Joint Chiefs of Staff, 257, 453, 454, 462, 682, 977
  New Weapons Committee of, 560
Joint Congressional Committee, on Atomic Energy, 18, 19, 71, 76, 81, 227, 234, 239, 302, 313, 319, 382, 435, 682, 686, 687, 832, 977
Joint Research and Development Board (RDB), 57, 86, 257, 373, 376, 377, 418, 495, 498, 499, 502, 505, 521, 543, 560, 583, 787, 837
  Atomic Energy Committee of, 16, 258, 307, 543–545, 550, 562, 564, 609, 616, 681, 682, 801
Kansas, University of, 657
Kellex Corp., 544
Kenilworth incident, 216
Kilgore-Magnuson Bill, 18
Kilgore-Magnuson Committee, 35
Knolls Laboratory, 543
Korean War, 46, 65, 66, 73, 83, 93, 94, 329, 386, 404, 512, 582, 583, 584, 688, 798, 975
  rocketry in, 578
Kyoto, 33, 236

Laboratory
  bomb, 28. *See also* Los Alamos;
    Manhattan project
  second, 311, 312, 313, 314, 315, 338,
    339, 755, 756
Laboratory Security Committee, at Los
  Alamos, 198
Lederle Laboratories, 467
Leftists, 187, 191, 194, 196, 197, 198,
  199, 203, 224, 225, 335, 975
Lehman Brothers, 300
Leiden, University of, 7
Lexington project, at Massachusetts
  Institute of Technology, 495,
  673, 692, 693
Life magazine, 176, 799, 909
Lilienthal panel, 36, 37, 303, 326
Lilienthal report. *See* Acheson-
  Lilienthal report
Linac, 781
Lincoln project, 94, 315, 443, 447, 451,
  582, 583, 585, 597, 598, 673, 750,
  925, 935, 979
Lincoln Summer Study, 59, 452, 458,
  600, 749, 750, 751, 760, 763, 765,
  897, 923, 924, 925, 926, 935, 943,
  978
Listener, The, 98
Livermore, laboratory at, 85, 86, 311,
  314, 315, 339, 442, 698, 699, 723,
  724, 726, 756, 789, 979
Los Alamos, laboratory at, 5, 6, 12,
  13-15, 17, 18, 20, 27, 29, 69, 70,
  72, 74, 77, 82-85, 87-92, 97,
  109, 163, 171, 197, 202, 203,
  224-226, 290, 296, 297, 304, 305,
  310, 311, 313-315, 319, 322-
  331, 333, 334, 400, 418, 468, 516,
  597, 703, 772, 773, 795. *See also*
    Manhattan project
  administration of, 262, 478, 482
  Alvarez at, 802, 949
  Bomb Physics Division of, 609
  British mission at, 31, 732
  cryogenics building at, 234
  delivery group at, 441
  director of, 703, 804, 975
  Division of Experimental Physics of,
    609
  General Advisory Committee (GAC)
    and, 479, 480, 490. *See also*
    General Advisory Committee
  Germans at, 31, 324
  implosion weapon and, 772
  Italians at, 31
  Kelly at, 60
  military at, 641
  official history of, 837, 980
  pacifists at, 660, 666, 667
  personnel of, 520, 772
  plutonium at, 657, 658
  project compartmentalization at, 165,
    166, 167, 173, 175, 220, 261, 441,
    448, 449, 692
  recruitment at, 268, 621, 979
  reorientation at, 799, 838, 975
  second site and, 86, 248, 249, 255,
    457, 458, 724, 978. *See also*
    Livermore, laboratory at
  security at, 29, 118, 124, 128, 129,
    165, 167, 177, 198, 204, 214, 219,
    220, 260-262, 270, 278, 297, 325,
    449, 522, 732
  small-weapons programs at, 796
  special engineer detachment (SED)
    at, 278, 493
  spectrographic method used at, 872
  Teller at, 710, 711, 714, 774, 780,
    781
  Theoretical Division at, 331, 337, 712
  thermonuclear work at, 323, 328, 331,
    333, 334, 336, 337, 339, 484, 662,
    700, 717, 718, 719, 726, 899, 949,
    960, 961
  water boiler at, 779
Los Alamos, town of, 30
Lovett briefing, 757, 758. *See also*
  Lovett, Robert
Loyalists, in Spain, 8, 9, 157, 184, 224,
  277
Loyalty
  consideration of, 972
  meaning of, 1058-1061

Maginot Line, 315, 316, 387, 447, 740,
  751
Magnetic separation, process of, 543
Manhattan District, mandate of, 1042
Manhattan project, 9, 12, 13, 14, 15, 29,
  74, 129, 163, 168, 169, 170, 177,
  179, 238, 326, 341, 375, 377, 398,
  411, 418, 425, 437, 488, 517, 544,
  557, 560, 561, 622, 658, 681, 771,
  804, 837
  records of, 123, 124, 416, 417, 985
  security of, 110, 117, 259, 260, 318,
    374, 438, 549, 610, 637
Maniac, 775
Marines, in intelligence work, 355, 361
Massachusetts Institute of Technology
  (MIT), 17, 226, 257, 490, 596,
  672, 673
  Lincoln Laboratory and, 764. *See
    also* Lincoln project
May case, 175
May-Johnson Bill, 15, 35, 36, 326, 332
McCarthy hearings, 444
McGraw Award, 543
McMahon Act, 18, 36, 398
McMahon's committee, 35, 39, 228
Medal of Merit, 33, 396, 377, 379, 420,
  611
Mesons, 74
Metallurgical Laboratory, in Chicago,
  11, 88, 174, 271, 771
Michigan, University of, 608
Microfilm, use of, 130, 138, 146, 149,
  290, 291, 309, 522, 628, 823, 865,
  888, 980, 1008
Military
  atomic policy and, 389, 640, 641, 642,
    682, 683
  Intelligence Service of, 274, 475

Military (*continued*)
  Joint Research and Development
    Board of, 45, 46. *See also* Joint
    Research and Development
    Board
  laboratory organization and, 28
  Manhattan project and, 31
  strategy of, 390
Military Affairs, House Committee on, 15
Military Application, Division of, 399,
  400
Military establishment, 12, 16, 18, 69,
  72, 73, 77, 197, 243, 399, 404, 405,
  408, 506, 750, 979
Military Intelligence Services (MIS),
  819, 822, 1049, 1050, 1051
Military Liaison Committee, to Atomic
  Energy Commission, 163, 379,
  455, 456, 462, 543, 544, 583, 636,
  641, 681, 683, 776, 977
Military Policy Committee, 560, 561
Monsanto Chemical Co., 544
Moscow, mission to, 36. *See also* Soviet
  Union
Munich University, 335

Nagasaki, 14, 15, 226, 333, 400, 738,
  779
National Academy of Science, 27, 324,
  566, 644, 657, 726, 751
  Review Committee on Atomic Energy
    of, 603
  special committee of, 11
National Advisory Committee for
  Aeronautics, 560
National Defense Research Committee
  (NDRC), 257, 440, 508, 560,
  577, 578, 957
  Communications Division of, 603
  Massachusetts Institute of Technology
    and, 771
National Laboratories, 238
National Science Foundation, 35, 85,
  231, 706, 837
National security
  calculated risks and, 1013, 1014
  interests of, 1042
National Security Council, 93, 96, 307,
  403, 404, 407, 435, 683, 837
National Security Resources Board, 837
National War College, 42
Naval Research Advisory Committee,
  515
Naval Research Laboratory, 164
Navy, 582, 837
  rockets and, 578
Nazis, 292, 324, 501, 847, 865. *See also*
  Germany
Nazi-Soviet Pact, 10, 114, 115, 186, 188,
  189, 191, 449, 837
NDRC. *See* National Defense Research
  Committee
Neutrons, 70, 242, 301, 302, 785, 793
New York Bar Association, 44
New York Daily News, 215
New York Journal American, 541

New York Life Insurance Company,
  257, 734, 738
New York Times, 40, 53, 54, 165, 336,
  388, 422, 423, 430, 538, 546, 725,
  912, 913
Nichols letter, 3-7, 20, 125, 324, 389,
  459, 465, 513, 522, 533, 537, 546,
  548, 553, 564, 565, 571, 647, 685,
  706, 725, 806, 835, 843, 913, 915,
  972, 975, 988, 1001, 1016, 1017,
  1031, 1062. *See also* Nichols,
  Kenneth D.
  attorneys of J. R. Oppenheimer, 1037
  forwarding findings and recommenda-
    tions, 1025. *See also* General
    Advisory Committee, General
    Manager of
  Oppenheimer reply to, 7-20, 107,
    117, 252, 824
North American Committee, 157, 183
North Atlantic Treaty Organization
  (NATO), 360, 521, 837
Nuclear Development Associates, 784
Nuclear Disarmament Program, 250

Oak Ridge, 11, 12, 13, 60, 68, 73, 75, 87,
  102, 129, 171, 301, 399, 543, 544,
  621, 722, 723, 779, 780, 785
  security at, 270
  as site X, 815
Office of Defense Mobilization, 20, 93
Office of Naval Research, 69, 74
Office of Scientific Research and
  Development, 93, 507, 560, 563
Operation Candor, 307
Operation Joe, 242, 805
Operation Lincoln, 307
Operation Vermont, 401
OSRD. *See* Office of Scientific
  Research and Development

Paducah, laboratory at, 257, 435
Palmer Physical Laboratory, 182
Panel, AFOAT-1 advisory, 801
  of Committee on Atomic Energy,
    686, 687
  Interim Committee Advisory, 377
  Lilienthal, 36, 37, 303, 326
  on Military Objectives in the Field
    of Atomic Energy, 801
Party
  Communist. *See* Communist Party
  Democratic, 207
  Socialist, 224, 225
Pasadena. *See* California Institute of
  Technology
Patent
  application for, 234, 235
  for thermonuclear devices, 336
Pearl Harbor, 9, 112, 186, 353
Pease Laboratory, 467
Pennsylvania, University of, 574
Pentagon, 495, 499, 787
  target planners in, 641
People's World, 4, 8, 155, 157, 825,
  919, 920, 921
Perjury statutes, 253, 256, 742

Personnel Security Board
  majority, findings of, 999–1019
  majority, recommendations of, 1019, 1042, 1046, 1049, 1051, 1056, 1057
  members of, 1041, 1056
  minority, findings and recommendations of, 1019–1021
Personnel Security Questionnaire (PSQ), 444
Pisa, University of, 397
Plutonium, 175, 315, 517, 546, 658, 722
  production of, 255, 301, 657, 658
  purification of, 163, 164
Poland, German invasion of, 581
Pomona College, 490
Potsdam Conference, 31–32, 731, 732, 733
Prentice [Prentiss] Hall Publishing Co., 657
Presidential Directive, 84, 229, 231, 232, 700, 701, 745, 767
Princeton University, 26, 304, 311, 353, 421, 485, 738. *See also* Institute for Advanced Study
  fellowship program at, 706
  June 1951 meeting at, 304, 305, 319, 320, 497, 520, 645, 705, 720
Project Charles. *See* Charles project
Project Y, at Los Alamos, 149, 150, 276, 277, 956, 957
Propulsion, military, 71, 75

Race, arms, 80, 95, 670
Radar, development of, 509
Radiation Laboratory
  at Berkeley, 5, 9, 11, 13, 27, 63, 112, 122, 153, 186, 187, 264, 268, 311, 412, 515, 516, 657, 658, 660, 679, 698, 724, 744, 771, 781, 789, 794, 810, 811, 812, 826, 888, 1007, 1009
  espionage at, 825
  at Massachusetts Institute of Technology, 440, 441, 458, 467, 468, 515, 516, 596, 609, 743, 781
Rand Corporation, 163, 738, 744
Rand project, 744, 753, 780
RDB. *See* Joint Research and Development Board
Reactor Division, of Atomic Energy Commission, 577, 782
Reactors
  development of, 71, 87, 241, 242, 245, 301, 670, 793
  duPont Company and, 804
  at Hanford, 301, 317, 784, 793
  heavy water, 238, 241, 242, 243, 775, 776
  neutron producing, 462
  for plutonium, 27, 301
  at Savannah River, 301, 793
  submarine, 75
  use of, 74, 317
Reactor Safeguard Committee, 706
Regulation of Armaments (RAC), President's Executive Committee on, 344
Reith lectures, 920

Research and Development Board. *See* Joint Research and Development Board
Rochester Institute of International Affairs, 44
Rochester Times Union, 210, 1051
Rochester University, 83, 213, 214
Rockefeller Foundation, 738
Rocketry, 578
Rosenberg Affair, 156, 278
Roumanian-English Oil Company, 655
Russia, 10, 298, 299, 366, 378. *See also* Soviet Union
  as ally, 652, 653
  arms race and, 80, 86, 242
  atomic capability of, 95, 257, 258, 409, 485, 486, 585, 605
  atomic explosion of, 75, 77, 90, 242, 302, 319, 328, 346, 386, 387, 436, 452, 460, 467, 478, 485, 486, 490, 509, 512, 533, 564, 615, 618, 659, 686, 699, 713, 714, 720, 773, 778, 787, 805
  attack on, 503, 504
  Baruch Plan and, 343, 344
  Communist Party in, 110
  consulate of, 138, 144, 146, 147, 148, 177, 191, 250, 309, 471, 522, 1003, 1008
  espionage of, 31, 174, 175, 308, 309, 355, 358
  fission explosion of, 642
  gaseous diffusion plant in, 374
  international arms control and, 38, 39, 41, 303, 326, 327, 329, 344, 345, 385
  menace of, 385, 442, 545
  relations with, 34, 98, 248, 286, 287, 288, 289, 303, 356, 394, 526, 562, 881
  revolution in, 362
  "super weapon" and, 658
  stockpile of, 545
  thermonuclear development of, 370, 453
  totalitarian regime of, 388
  war in, 115
Russian-Nazi Pact, 441. *See also* Nazi-Soviet Pact
Russian Orthodox Church, 810
Russo-Finnish War, 5, 1005
Russo-German Pact, 601. *See also* Soviet-German Pact; Nazi-Soviet Pact

SAM Laboratories, 621
Sandia production facility, 17, 31, 60, 64, 73, 85, 400, 442
Sandstone, atomic testing at, 713
Saturday Evening Post, 751
Savannah River project, 19, 81, 93, 241, 257, 302, 454, 456, 457, 462, 793, 794
Science Advisory Committee, of Office of Defense Mobilization, 20, 95, 341, 451, 496, 499, 515, 516, 528, 602, 604, 672, 1045, 1055
  member of, 1046

Scientific American, 329, 336
Scientists
  association with Nazis of, 501
  as advisors, peril of, 1031
  characteristics of, 491
  duty of, 1059
  enthusiasm of, 1031
  in government, 23, 704
  loyalty of, 1031
Scientist X. *See* Weinberg, Joseph W.
Secretary of Defense, 304, 306, 521, 565, 683
Secretary of State, 20, 86, 95, 247, 248 304, 306, 343, 353, 358, 359, 360, 734, 898
  committee of, 16, 561
  panel of, 562
Secretary of Navy, 641
Secretary of War, 14, 32, 33, 440, 445, 641, 731, 732, 733, 743, 950
Securities and Exchange Commission, 437
Security, system of, 1065
Security clearance
  AEC criteria for, 6, 7, 260, 366, 558, 559, 628, 973
  for Joint Research and Development Board, 507
  as preventive measure, 501
  Q type, 425, 511, 583, 612, 614, 646, 988
Security inquiry, limits of, 1032
Selective Service, President's Advisory Committee on, 342
SHAEF (Supreme Headquarters Allied Expeditionary Forces), 508, 509, 521
SHAPE (Supreme Headquarters Allied Powers, Europe), 496, 498, 503, 505
Shell Development Company, 135, 145, 167, 168, 188, 288, 849, 854, 867
Ships, nuclear powered, 692, 693
Soviet-German Pact, 5, 1005. *See also* Nazi-Soviet Pact
Site X, 297
Site Y, 954, 957
Smyth report, 175. *See also* Smyth, Henry D.
Soviet Union, 5, 10, 15, 20, 288, 289, 354, 356, 357, 1013, 1051. *See also* Russia
  arms control and, 44–45
  atomic explosion of, 15, 18, 76, 237, 399, 432
  atomic threat of, 47, 594, 801
  Communist Party in, 111, 115
  consulate of, 130, 286, 291, 549, 815, 845, 860, 864, 865, 872, 981, 1007, 1052
  espionage of, 4, 233, 274, 277, 278, 354, 1004, 1050
  Germany and, 278
  land power of, 684
  relations with, 358, 359, 732
  scientists of, 1008
  sympathies with, 390

  thermonuclear weapons of, 334
Spanish Civil War, 266, 491, 574, 580
Spanish relief, 9, 139, 140, 158, 159, 183, 184, 186, 191, 203, 224, 225, 885, 1063
Spanish Republican Army, 4, 1002
Squire, Sanders and Dempsey, 259
Standard Oil Development Company, 17
Stanford University, 490, 787, 825
State Department, 75, 95, 353, 354, 358, 361, 418, 419, 422, 439, 453, 455, 526, 738, 837, 856, 977
  Board of Consultants of, 557
  bulletins of, 26
  Policy Planning Committee of, 353, 355, 358, 978
  Statutory Advisory Committee, 701
Stimson's Committee on Atomic Energy, 257, 957
Stockpiling, thermonuclear, 479, 503, 532, 584, 635, 774, 779. *See also* Control
Stone and Webster Company, 544
Strategic Air Command (SAC), 94, 497, 498, 503, 599, 616, 747, 749, 750, 763, 765, 893, 924, 937, 938, 960, 978
Submarine, atomic, 69, 75
Suisun Laboratory, 781, 782
Super, 65, 78, 82, 83, 86, 234, 237, 242, 243, 409, 423, 432, 433, 435, 452, 454, 463, 642. *See also* Weapons, thermonuclear
  opposition to, 519, 797. *See also* Bomb, hydrogen, opposition to
Supreme Court, 419
Surveillance reports, 150

Tata Institute, 215
Television, color demonstration of, 777, 778
Testing, atomic, 562, 564, 591, 722, 778
  at Alamogordo, 11, 711
  at Bikini, 177, 257, 713
  continental sites for, 71, 305
  at Eniwetok, 11, 70, 305, 480, 720
  Greenhouse, 788, 796, 797. *See also* Greenhouse
  at Inyokern, 579
  at Los Alamos, 32
  Pacific sites for, 71, 305, 315, 339
  resumption of, 713
  at Tinian, 33, 441
  at Trinity, 32, 949
Thermonuclear program, 72, 76, 77, 85, 87, 90, 91, 227, 233, 247, 254, 304, 305, 307, 312, 313, 314, 315, 323, 328, 329, 330, 334, 337, 338, 395, 408, 452, 460, 479, 480, 488, 530, 704, 726, 786, 1049. *See also* Bomb, hydrogen; Weapons, thermonuclear
  computer development and, 654, 655
  delays in, 642
  military and, 486, 640–641
Thorium, 175
Time magazine, 8, 457, 723

Treason, 130, 131
Tritium, properties of, 28–29, 32, 301, 778
Trotskyites, 206, 225, 878
Truman Doctrine, 41. See also Truman, Harry S
Twentieth Century Limited, 28

Un-American Activities
  California State Senate Committee on, 4, 16, 217, 1005
  House Committee on, 3, 5, 16, 102, 104, 151, 210, 213, 214, 219, 1001, 1003, 1007, 1018
  Senate Committee on, 197
Union
  Bridges' [International Longshoremen's and Warehousemen's Union], 8, 156
  CIO, 292, 293, 846, 855, 856, 865, 902
  consumer's, 3, 9, 158, 159, 205
  teachers, 139, 156, 188, 208, 277, 602
United Front, of Communist Party, 8
United Fruit Company, 17, 508, 509, 511
United Nations
  atomic energy and, 36, 359, 385
  Atomic Energy Commission of, 16, 44, 326, 342, 343
  atomic inspection and, 373, 374
  conference on international organization of, 4
  discussions on disarmament of, 95
  Educational Scientific and Cultural Organization (UNESCO) of, 76, 140, 589
  international control and, 561. See also Control
  representatives to, 344, 345
United States
  attack on, 46, 47
  Moscow embassy of, 353, 355, 361, 362
  security system of, 1012
United States Atomic Energy Commission, decision and opinions of, 1048–1065
United States Department of Justice, 1004
United States Embassy, Paris, 1009, 1052
Universal Military Service, 386, 387
Uranium
  fission of, 398
  production of, 670
  purification of, 164, 165, 301, 315
  scramble for, 39
U.S. News & World Report, 685

Virginia Military Institute, 259
Vista, project, 48, 65, 94, 447, 496, 503, 504, 505, 520, 524, 582, 583, 585, 590, 591, 594, 597, 616, 617, 746, 748, 760, 900, 905, 978, 979
Vista report, 521, 527, 746, 747, 762, 891, 892, 893, 901
Voice of America, 26

War
  Secretary of. See Secretary of War
  threat of, 37
War College, 362
War Department, 14, 211, 228, 259, 418, 561, 642, 731, 732, 733, 743, 811
Warfare
  antisubmarine, 788
  atomic, defense against, 94
  radiological, 47, 69
Warheads, atomic, 73
War Production Board, 508, 553
Washington, University of, at St. Louis, 671
Washington Post, 178, 538, 539, 542, 898
Washington Star, 453
Washington University, at Seattle, 88
Weapons
  ABC, 360;
  atomic, 18, 69, 73, 83, 89, 339, 340, 362, 366, 458, 787. See also Bomb; Thermonuclear program
    custody of, 249
    development of, 58, 87, 366, 367
    strategic, 585
  biological, 360, 363
  chemical, 360, 363
  fission, 91, 229, 328, 333, 335, 519, 520, 615, 616, 641, 643, 712, 714, 715
  fusion, 62, 229, 255, 303, 430, 432
  GAC subcommittee on, 82, 84, 301, 305, 464
  small-scale, 788, 796, 798, 799
  strategic, 635, 636. See also Defense
  thermonuclear, 64, 77, 78, 91, 92, 236, 244, 250, 251, 302, 303, 305, 333, 335, 340, 457, 481, 486, 504, 636, 778. See also Thermonuclear program
    development of, 234, 235, 241, 250, 303, 366, 453, 487, 488, 696, 711, 793
    limitation of, 591. See also Control, international
    testing of, 85. See also Testing
Weinberg case, 263. See also Weinberg, Joseph W.
Western Electric Company, 545
Western Electric Laboratories, 61, 400, 603. See also Bell Telephone Laboratories
Westinghouse Electric Corp., 127, 174, 212, 257, 832
West Point, 440, 680
White House, 416, 419, 432, 434, 469, 985
Whitman report, 692
Wisconsin, University of, 571
World Bank, 737

Yalta Conference, 163
Young Communist League, 194, 225, 1006

ZORC, 600, 750, 769, 922, 923, 927, 928, 929, 932, 936

ST. MARY'S COLLEGE OF MARYLAND
ST. MARY'S CITY, MARYLAND